GW01218117

Terrestrial Photosynthesis in a Changing Environment

A Molecular, Physiological and Ecological Approach

Understanding how photosynthesis responds to the environment is crucial for improving plant production and for maintaining biodiversity in the context of global change. Covering all aspects of photosynthesis, from basic concepts to methodologies and from the organelle to whole ecosystem levels, this is an integrated guide to photosynthesis in an environmentally dynamic context. Focusing on the ecophysiology of photosynthesis – how photosynthesis varies in time and space, responds and adapts to environmental conditions, and differs among species within an evolutionary context – the book features contributions from leaders in the field. The approach is interdisciplinary, and the topics covered have applications for ecology, environmental sciences, agronomy, forestry, and meteorology. It also addresses applied fields such as climate change, biomass and biofuel production, and genetic engineering, making a valuable contribution to our understanding of the impacts of climate change on the primary productivity of the globe, and on ecosystem stability.

JAUME FLEXAS is Associate Professor of Plant Physiology and Vice-Dean of the Faculty of Sciences at the Universitat de les Illes Balears, Palma de Mallorca, Spain. His research focuses on plant ecophysiology and photosynthesis. He received the Young Scientist Award (plant physiology) from the Federation of European Societies of Plant Biology in 2006.

FRANCESCO LORETO is Research Director at the Consiglio Nazionale delle Ricerche, Istituto per la Protezione delle Piante, Firenze, Italy. He studies the interaction between biosphere and atmosphere with an emphasis on biosynthesis and emissions of biogenic volatile organic compounds (VOC), and on primary and secondary metabolism of plants under environmental constraints.

HIPÓLITO MEDRANO is Professor of Plant Physiology and Director of the Biology Department at the Univesitat de les Illes Balears, Palma de Mallorca, Spain. Over the past 10 years he and his research group, "Plant Ecophysiology under Mediterranean Conditions", have focused on achieving continued progress in the fields of photosynthesis, carbon and water flow, and water-use efficiency (WUE).

Terrestrial Photosynthesis in a Changing Environment

A Molecular, Physiological and Ecological Approach

Edited by
Jaume Flexas
Universitat de les Illes Balears, Palma de Mallorca, Spain
Francesco Loreto
Consiglio Nazionale delle Ricerche, Istituto per la Protezione delle Piante, Firenze, Italy
Hipólito Medrano
Universitat de les Illes Balears, Palma de Mallorca, Spain

CAMBRIDGE UNIVERSITY PRESS
Cambridge, New York, Melbourne, Madrid, Cape Town,
Singapore, São Paulo, Delhi, Mexico City

Cambridge University Press
The Edinburgh Building, Cambridge CB2 8RU, UK

Published in the United States of America by Cambridge University Press, New York

www.cambridge.org
Information on this title: www.cambridge.org/9780521899413

First published 2012

Printed in the United Kingdom at the University Press, Cambridge

A catalogue record for this publication is available from the British Library

Library of Congress Cataloguing in Publication data
Terrestrial photosynthesis in a changing environment : a molecular, physiological, and ecological approach / [edited by]
Jaume Flexas, Francesco Loreto, Hipólito Medrano.
p. cm.
Includes bibliographical references and index.
ISBN 978-0-521-89941-3 (hardback)
1. Ecophysiology. 2. Photosynthesis–Environmental aspects. 3. Space biology. I. Flexas, Jaume.
II. Loreto, Francesco. III. Medrano, Hipólito.
QH541.15.E26T47 2012
571.2–dc23
2012007358

ISBN 978-0-521-89941-3 Hardback

Contents

Contributors

ANDRALOJC, P. JOHN
Department of Biological Sciences IENS, Lancaster University, Lancaster, UK

BARÓN, MATILDE
Department of Biochemistry, Cell and Molecular Biology of Plants, Estación Experimental del Zaidín, Spanish Council for Scientific Research (CSIC), Granada, Spain

BERNACCHI CARL J.
Illinois State Water Survey, Department of Plant Biology, USA

BERNINGER, FRANK
Université du Québec à Montréal, Faculté des Sciences, Département des Sciences Biologiques, Montréal (Québec), Canada

BRUGNOLI, ENRICO
National Research Council (CNR) Institute of Agro-Environmental and Forest Biology, Porano, (TR) Italy

CALFAPIETRA, CARLO
National Research Council (CNR), Institute of Agro-Environmental and Forest Biology, Roma, Italy

CENTRITTO, MAURO
National Research Council (CNR), Institute of Agro-Environmental and Forest Biology, Roma, Italy

CHAVES, MANUELA, M.
Universidade Técnica de Lisboa (UTL), Dep. Botânica e Engenharia Biológica, Instituto Superior de Agronomia (ISA), Tapada da Ajuda, Lisbon, Portugal

CHEESEMAN, JOHN
Department of Plant Biology, Director, IB Honors Biology, University of Illinois, Urbana, Illinois, USA

COLLATZ, G. JAMES
Biospheric Sciences Branch, NASA Goddard Space Flight Center, Greenbelt, Maryland, USA

DELUCIA, EVAN H.
Department of Plant Biology, University of Illinois at Urbana-Champaign, Urbana, Illinois, USA

DÍAZ-ESPEJO, ANTONIO
Instituto de Recursos Naturales y Agrobiología, CSIC, Sevilla, Spain

DUCRUET, JEAN MARC
NRA/CEA, Section de Bioénergétique/DBCM, CE Saclay, France

EARL, HUGH
Department of Plant Agriculture, University of Guelph, Guelph, Ontario, Canada

ENSMINGER, INGO
Department of Biology, Graduate Program in Cell and Systems Biology, University of Toronto, Mississauga Campus, Mississauga, Ontario, Canada

FLEXAS, JAUME
Grup de Recerca en Biologia de les Plantes en Condicions Mediterrànies
Departament de Biologia, Universitat de les Illes Balears
Ctra. Palma de Mallorca, Spain

FOYER, CHRISTINE H.
School of Agriculture, Food and Rural Development, The University of Newcastle upon Tyne, Newcastle upon Tyne, UK

GALMÉS, JERONI
Grup de Recerca en Biologia de les Plantes en Condicions Mediterrànies, Departament de Biologia, Universitat de les Illes, alears, Ctra. -Palma de Mallorca, Spain

GARCÍA-PLAZAOLA, JOSÉ I.
Departamento de Biología Vegetal y Ecología. Universidad del País Vasco UPV/EHU. Bilbao, Spain

GIBSON, ARTHUR C.
Department of Ecology and Evolutionary Biology, University of California, Los Angeles, California, USA

GULÍAS, JAVIER
Grup de Recerca en Biologia de les Plantes en Condicions Mediterrànies, Departament de Biologia, Universitat de les Illes Balears, Ctra. Palma de Mallorca, Spain

HARBINSON, JEREMY
Wageningen University, Department of Plant Sciences, Horticultural Supply Chains Group, Wageningen, The Netherlands

KEELEY, J.E.
U.S. Geological Survey, Western Ecological Research Center, Sequoia-Kings Canyon Field Station, Three Rivers, California, USA and University of California, Department of Ecology and Evolutionary Biology, Los Angeles, California, USA

LORETO, FRANCESCO
Consiglio Nazionale delle Ricerche (CNR), Istituto per la Protezione delle Piante (IPP), Area della Ricerca del CNR di FirenzeVia Madonna del Piano 1050019 Sesto Fiorentino, Firenze, Italy

LÜTTGE, ULRICH
Institute of Botany, Technical University of Darmstadt, Darmstadt, Germany

MANCA, GIOVANNI
National Research Council (CNR), Institute of Atmospheric Pollution Research Rende (CS), Italy

MATTEUCCI, GIORGIO
National Research Council (CNR), Institute for Agricultural and Forest Systems in the Mediterranean, Rende (CS), Italy

MEDRANO, HIPÓLITO
Grup de Recerca en Biologia de les Plantes en Condicions Mediterrànies, Departament de Biologia, Universitat de les Illes Balears, Ctra. Palma de Mallorca, Spain

MONSON, RUSSELL K.
Department of Ecology and Evolutionary Biology and Cooperative Institute for Research in Environmental Sciences, University of Colorado, Boulder, Colorado, USA

MONTGOMERY, REBECCA
Department of Ecology and Evolution, University of Connecticut, Storrs, Connecticut, USA

MORALES, FERMÍN
Estación Experimental Aula Dei, CSIC, Zaragoza, Spain

MOYA, ISMAEL
Laboratoire pour l'Utilisation du Rayonnement Electromagnétique, CNRS, Orsay, France

NIINEMETS, ÜLO
Estonian University of Life Sciences, Institute of Agricultural and Environmental Sciences, Tartu, Estonia

PARRY, MARTIN A.J.
Centre for Crop Genetics, Department of Plant Science, Rothamsted Research, Harpenden, Herts, UK

RIBAS-CARBÓ, MIQUEL
Grup de Recerca en Biologia de les Plantes en Condicions Mediterrànies, Departament de Biologia, Universitat de les Illes Balears, Ctra. Palma de Mallorca, Spain

RUNDEL, PHILIP W.
Department of Ecology and Evolutionary Biology, University of California, Los Angeles, California, USA

SHARKEY, TOM D.
Department of Biochemistry and Molecular Biology, Michigan State University, East Lansing, Michigan, USA

STREB, PETER
Laboratoire d'Ecophysiologie Végétale, Bâtiment 362, UFR Scientifique d'Orsay, Université Paris XI, Orsay, France

TCHERKEZ, GUILLAUME G.B.
Plateforme Métabolisme-Métabolome, Institut de Biotechnologie des Plantes, Université Paris-Sud XI, Orsay, France

TOSENS, TIINA
Institute of Agricultural and Environmental Sciences, Estonian University of Life Sciences, Tartu, Estonia

VALLADARES, FERNANDO
Centro de Ciencias Medioambientales C.S.I.C., Madrid, Spain

WARREN, CHARLES R.
University of Sydney, NSW, Australia

Preface

Photosynthesis is the physiological process that overwhelmingly supports the Earth's primary production. For this reason, photosynthesis research has attracted interest worldwide since the earliest foundations of modern science several centuries ago. Renewed interest in photosynthesis research has been spurred in recent decades. In part this is owing to the impressive technological advancements in analysing the molecular and physiological bases of the photosynthetic processes. But interest in photosynthesis also derives from the need to know how global climate change will impact on the primary productivity of the globe, and on ecosystem stability. Being a highly dynamic process, the precise understanding of how photosynthesis responds to environmental changes is crucial to predict how single plants and entire agro- or ecosystems will be affected in a scenario of rising CO_2, rising temperature, and large disturbances in water and nutrient cycles. Finally, photosynthesis studies are the cornerstone for the development of new crops better suited for novel purposes, primarily high yield for biomass and biofuel use.

Whereas several books have proficiently addressed the photosynthetic process, the recent important advances in photosynthesis are still in need of comprehensive coverage. To mention just a few: new techniques have been tested for remote sensing of photosynthesis; important limitations of traditional gas-exchange analysis have been highlighted and solutions proposed; advances have been made in the measurements of diffusive resistance of CO_2 inside leaves; the mechanistic knowledge of leaf mesophyll conductance to CO_2 has been identified as a decisive but often neglected aspect of leaf photosynthetic; an increasing interest has arisen regarding photosynthetic responses to biotic stresses, a field that has been under-explored or totally unrecognised; photosynthetic responses under leaf development and ontogeny have been assessed; evolutionary trends have been described for several photosynthesis-related processes, such as leaf morphology or stomatal responses to environmental cues; and of course, large-scale studies have been performed regarding photosynthesis at ecosystem level, and the response of ecosystems to climate change.

The present volume reviews the progress made in photosynthesis research, and particularly focuses on understanding how photosynthesis responds, acclimates, and adapts to a rapidly changing environment. Although this is a multi-authored book, it has been designed to cover the whole spectrum of subjects related to photosynthesis responses to the environment as in a textbook or monograph, with the idea of making the book useful not only for professors, graduate students, and researchers, but also for undergraduate students and laymen. The contents of *Terrestrial Photosynthesis in a Changing Environment* have thus been comprehensively structured in six parts comprising 34 chapters, in increasing order of biological complexity. Hence, the introductory chapter provides a brief summary and overview of the historical aspects concerning this subject, and of the structure and contents of the book. Part 1 'Photosynthesis: the process' (Chapters 2 to 8) offers a succinct review of the basics of photosynthesis, with a particular focus on models of leaf photosynthesis and photosynthetic limitation analysis, both important aspects for ecophysiology. Part 2 'Measuring photosynthesis' (Chapters 9 to 15) addresses how to measure photosynthesis and related aspects in ecophysiological studies, providing a detailed presentation of the most useful techniques, their principles, and procedures. Part 3 'Photosynthetic response to single environmental factors' (Chapters 16 to 22) deals with mechanisms underlying the impact of single environmental factors on the photosynthetic process, reviewing current state-of-the-art techniques, especially when using ecophysiological approaches. The following parts, 4 'Photosynthesis in time' (Chapters 23 to 25) and 5 'Photosynthesis in space' (Chapters 26 to 32), show more complex and interacting aspects of photosynthesis, related to photosynthetic variations in time, from ontogeny to evolutionary aspects (Part 4), and in space, covering photosynthesis in crops, as well as in the most important biomes of the world (Part 5). Finally, by bringing the complexity to the global scale, part 6, 'Photosynthesis in a global context' (Chapters 33 and 34), covers worldwide aspects of photosynthesis such as its implications for water-use efficiency and its responses in a climate-change context.

As photosynthesis is the cornerstone of life on Earth, preservation and optimisation of the photosynthetic process will be a challenge and a duty. We hope that this text will fulfil the need of a reference book and will be guidance for new generations of agronomists, foresters and plant scientists, as well as for students who approach photosynthesis studies from a more distant intellectual background.

Jaume Flexas
Francesco Loreto
Hipólito Medrano

Acknowledgements

First of all, we want to acknowledge Cambridge University Press Editor Jacqueline Garget, who first approached us to write a book for the Editorial, and who from the very beginning believed that the subject we chose was an exciting one that deserved priority publication. Thanks to all the other people involved at Cambridge University Press, especially Dominic Lewis and Megan Waddington.

We could not have afforded the sometimes tedious task of formatting all the book sections without the invaluable help of Belén Escutia, Antonia Morro and Violeta Velikova. Many thanks for this! Belén also designed the beautiful cover of the book, for which we are in debt.

And of course, we acknowledge most of all, the contributors. From the very first call, they were all willing not only to contribute one or more chapters, but also to help with multiple revisions to achieve the needed coherence and uniformity for the book. Many thanks to you all!

Last, but not least, a number of colleagues – some of them having authored other chapters too – helped with detailed peer-reviewing of the different chapters before we released them to the Editorial. Many thanks to all: Giovanni Agati (Istituto di Fisica Applicata 'Nello Carrara', Firenze, Italy); Elizabeth Ainsworth (Agricultural Research Service and Photosynthesis Research Unit, Urbana, USA); Owen Atkin (Research School of Biological Sciences, Canberra, Australia); Alberto Battistelli (Istituto di Biologia Agroambientale e Forestale, Porano, Italy); Joe Berry (Carnegie Institution of Washington; Stanford, USA); León Bravo (Universidad de La Frontera, Temuco, Chile); Carlo Calfapietra (Institute of Agro-Environmental and Forest Biology, Monterotondo Scalo, Italy); Mauro Centritto (Institute of Agro-Environmental and Forest Biology, Monterotondo Scalo, Italy); Manuela Chaves (Universidade Técnica de Lisboa, Lisboa, Portugal); Antonio Díaz-Espejo (Instituto de Recursos Naturales y Agrobiología, Sevilla, Spain); Erwin Dreyer (INRA Ecologie et Ecophysiologie Forestières, Vandoeuvre les Nancy, France); Jim Ehleringer (University of Utah, Utah, USA); John Evans (Research School of Biological Sciences, Canberra, Australia); Christine Foyer (University of Leeds, Leeds, UK); Alexander Gallé (Universitat de les Illes Balears, Palma de Mallorca, Spain); Bernard Genty (CEA Cadarache, Saint Paul lez Durance, France); John Grace (School of GeoSciences, The University of Edinburgh, Edinburgh, UK); Howard Griffiths (University of Cambridge, Cambridge, UK); Michel Havaux (CEA Cadarache, Saint Paul lez Durance, France); Brent Helliker (University of Pennsylvania, Philadelphia, USA); Jeffrey Herrick (West Virginia University, Morgantown, USA); Norman Huner (The University of Western Ontario, Ontario, Canada); Christian Körner (University of Basel, Basel, Switzerland); Heinrich Krause (Heinrich-Heine University, Düsseldorf, Germany); Agu Laisk (Institute for Molecular and Cell Biology, Tartu, Estonia); Hans Lambers (The University of Western Australia, Crawley, Australia); Richard Leegood (University of Sheffield, Sheffield, UK); Ulrich Lüttge (Technical University of Darmstadt, Darmstadt, Germany); Stephen Marek (Oklahoma State University, Stillwater, USA), Angelo Massacci (Institute of Agro-environment and Forest Biology, Monterotondo Scalo, Italy); Laurent Misson (CNRS-CEFE, Montpellier, France); Russell Monson (University of Colourado, Boulder, USA); Fermín Morales (Universidad de Navarra, Pamplona, Spain); Barry Osmond (Australian National University, Canberra, Australia); Thijs Pons (Utrecht University, Utrecht, The Netherlands); Francisco Pugnaire (Estación Experimental de Zonas Áridas, Almería, Spain); John Raven (University of Dundee, Dundee, UK); Philip Rundel (University of California, Los Angeles, USA); Rowan Sage (University of Toronto, Toronto, Canada); Eva Sárvári (Eötvös University, Budapest, Hungary); Leonid Savitch (Agriculture et Agroalimentaire Canada, Ontario, Canada); Luca Sabastiani (Scuola Superiore Sant'Anna, Pisa, Italy); Tom Sharkey (Michigan State University, East Lansing, USA); Guillaume Tcherkez (Institut de Biotechnologie des Plantes, Orsay, France); Ichiro Terashima (The University of Tokyo, Tokyo, Japan); Violeta Velikova (Institute of Plant Physiology, Sofia, Bulgaria); Ian Woodward (University of Sheffield, Sheffield, UK); Stan Wullschleger (Oak Ridge National Laboratory, Oak Ridge, USA); and Pablo Zarco-Tejada (Instituto de Agricultura Sostenible, Córdoba, Spain).

Abbreviations

Abbreviation	Concept
A	Antheraxanthin
A	Gross photosynthesis
a_b	Carbon-isotope fractionation owing to diffusion through the boundary layer
ABA	Abscisic acid
A_C	Crown projected area
A_c	Photosynthetic rate limited by carboxylation
a_b	Carbon-isotope fractionation owing to diffusion through the boundary layer
a_d	Carbon-isotope fractionation owing to diffusion through the stomata
ADP	Adenosine diphosphate
AG	Afterglow emission
A	Photosynthetic rate limited by the rate of RuBP regeneration
a_l	Carbon-isotope fractionation in cell water
A_{leaf}	Unit leaf area
$A_{max,M}$	Photosynthetic capacity per dry mass
A_{mes}	Surface area of mesophyll cell walls
AMP	Adenosine monophosphate
A_N	Net photosynthesis
A_{Nmax}	Photosynthetic capacity
AOX	Alternative oxidase
A_p	Photosynthetic rate limited by the rate of triose phosphate utilisation
APX	Ascorbate peroxidase
Asp	Aspartate
ATP	Adenosine triphosphate
b	Net-isotope fractionation by Rubisco and PEP carboxylase
BGF	Blue-green fluorescence
BSA	Bovine serum albumin
BSC	Bundle sheath cells
BWB	Ball-Woodrow-Berry model
BWYV	Beet western yellow virus
BYDV	Barley yellow dwarf virus
B2L	Biosphere 2 laboratory
b_3	Net-isotope fractionation by Rubisco
b_4	Net-isotope fractionation by PEP carboxylase
C_a	Ambient CO_2 concentration
CA	Carbonic anhydrase
cab	Chlorophyll *a/b* binding genes
CABP	Carboxyarabinitol bisphosphate

CA_{leaf}	CA-catalysed hydration of CO_2
CAM	Crassulacean acid metabolism
CaMV	Cauliflower mosaic virus
CAT	Catalase
C_{bs}	CO_2 concentration in the bundle sheath
C_c	Chloroplast CO_2 concentration
CCA1	Circadian clock-associated gene
C_{cs}	CO_2 concentration at chloroplast surface
CEFI	Cyclic electron flow around PSI
CEFII	Cyclic electron flow around PSII
CFBase	Cytosolic fructose-1,6-bisphosphatase
chl.	Chlorophyll
$chl._M$	Chlorophyll content on a leaf dry mass basis
^{3}Chl	Triplet chlorophyll
Chl*	Excited singlet chlorophyll
Chl-F	Chlorophyll fluorescence
CI	Cylindrical inclusion protein
C_i	Leaf internal CO_2 concentration (sub-stomatal cavity)
C_i^*	Apparent (intercellular) CO_2 photocompensation point
C_{in}	CO_2 concentration of air entering the gas-exchange cuvette
CMV	Cucumber mosaic virus
CMV-Y	Cucumber mosaic virus strain Y
C_m	CO_2 mole fractions in the mesophyll
C_o	CO_2 concentration of air leaving the gas-exchange cuvette
COR	Cold-induced gene
CP	Viral coat protein
CP47	Core protein 47 complex
C_s	CO_2 concentration at the leaf surface
Cu-SOD	Superoxide dismutase co-factored with copper
CVC	Citrus variegated chlorosis
C_w	Molar density of water
cyt	Cytochrome
cyt b_6f	Cytochrome b_6f
C_3	C3 plants
C_3	C3 plants
C_4	C4 plants
D	Fraction of energy absorbed by PSII allocated to heat dissipation in the antenna
D_{airCO2}	Diffusion coefficient of CO_2 in air
DCMU	3-(3,4-dichlorophenyl)-1,1-dimethylurea
2DE	Two-dimensional electrophoresis
DER	Double fluorescence ratio
DHAP	Dihydroxyacetone phosphate
DHAR	Dehydroascorbate reductase
DIAL	Differential Absorption Lidar
D_{hw}	Diffusivity of $H_2{}^{18}O$ in water
DIC	Dissolved inorganic carbon
DIRK	Dark interval relaxation kinetics
2D-DIGE	Two-dimensional-difference gel electrophoresis

D_{jCO2}	Diffusion coefficient of CO_2 in a medium j
D_m	Molecular diffusion
$D_{o/c}$	Diffusivity of O_2 relative to CO_2 between the bundle sheath and mesophyll
D_{ox}	Diffusion coefficient of oxygen
D_r	Oxygen-isotope fractionation by respiration
D_s	Leaf-to-air vapour-pressure deficit
D1	PSII reaction-centre protein D1
D2	PSII reaction-centre protein D2
e	Carbon-isotope fractionation owing to mitochondrial respiration
E	Transpiration
e_a	Water-vapour pressure in the surrounding air of the leaf
E_c	Cuticular transpiration
EC	Eddy covariance
ECS	Electrochromic shift
e_i	Water-vapour pressure inside the leaf
E_k	Irradiance at which the maximum photosynthetic quantum yield balances photosynthetic capacity
EL	Electrolyte leakage
e_s	Carbon-isotope fractionation occurring when CO_2 enters in solution
ESA	European Space Agency
esca	Fungal grapevine trunk diseases
ESR	Electron spin resonance
EWT	Equivalent water thickness
E_l	Transpiration from a leaf
f	Fraction of oxygen remaining in a closed cuvette after respiration
f	Carbon-isotope fractionation owing to photorespiration
FACE	Free-air CO_2 enrichment
Fd	Ferredoxin
Fe-SOD	Superoxide dismutase co-factored with iron
FGVI	Fluorescence global vegetation index
FI	Fluorescence imaging
f_{ias}	Fraction of the mesophyll volume that is actually intercellular air spaces
F_{in}	Gross CO_2 influx rate during photosynthesis
f_L	Fraction of plant biomass in foliage
FLD	Fraunhofer line discrimination
FLEX	Fluorescence explorer
F_m	Maximum chlorophyll fluorescence
F'_m	Maximum chlorophyll fluorescence under illumination
FQR	Ferredoxin-plastoquinone-reductase
FR	Far red
F_s	Steady state chlorophyll fluorescence
F_v	Variable chlorophyll fluorescence
F_v/F_m	Maximum PSII efficiency
F_0	Basal chlorophyll fluorescence
g_a	Aerodynamic conductance (canopy boundary layer conductance)
g_b	Boundary layer conductance
g_b	Conductance for the boundary layer
$g_{b,H}$	Conductance for boundary layer to heat transfer
GC	Gas chromatography

GC-TOF-MS	Gas Chromatography Time-of-flight Mass Spectrometry
g_{chl}	Chloroplast conductance to CO_2
$G/\cos(\theta)$	Extinction coefficient dependency on angular distribution of foliage
GDC	Glycine decarboxylase
g^{j}_{CO2}	CO_2 conductance for any given cellular compartment *j*
g_{ias}	CO_2 conductance of intercellular air spaces
g_{leaf}	Leaf conductance (stomatal plus boundary layer)
g_{liq}	Liquid-phase conductance to CO_2
Gln	Glutamine
Glu	Glutamate
Gly	Glycine
g_m	Mesophyll conductance to CO_2 (leaf internal diffusion conductance)
$g_{m,helox}$	Mesophyll conductance to CO_2 in helox
GPP	Gross primary production
GR	Glutathione reductase
g_s	Stomatal conductance
GSH	Reduced glutathione
GSSG	Oxidised glutathione
g_w	Cell-wall conductance to CO_2
G3P	Glyceraldehyde-3-phosphate
H^+-ATPase	Proton-ATPase
HDIAL	Heterodyne differential absorption lidar technique
HPLC	High-performance liquid chromatography
H^+-PPase	Proton-pyrophosphatase
HSFs	Heat shock factors
HSPs	Heat-shock proteins
HIC	*High carbon dioxide gene*
H_2O_2	Hydrogen peroxide
I	Irradiance
ias	Intercellular air spaces
IPCC	International panel on climate change
IR	Infrared
IRGAs	Infrared gas analysers
IRMS	Isotope ratio mass spectrometer
J	Electron transport rate
J_a	Rate of linear electron transport
JA	Jasmonic acid
J_f	Electron transport rate (chlorophyll fluorescence)
J_{max}	Maximum rate of electron transport used in regeneration of RuBP
k	Initial slope of A_N response to C_c
k	Rate constant of carbonic anhydrase
K_c	Michaelis-Menten constant for Rubisco carboxylation
K_{jCO2}	Partition coefficient for CO_2 diffusion in a medium *j*
K_m	Michaelis-Menten constant
K_o	Michaelis-Menten constant for Rubisco oxygenation
K_p	Apparent Michaelis constant of PEPC for CO_2
$k\tau$	Number of hydration reactions achieved per CO_2 molecule
l	Mesophyll thickness

L	Lutein
L_D	Leaf density
LAI	Leaf-area index
LC-MS	Liquid chromatography mass spectrometry
l_{gm}	Limitation of photosynthesis by mesophyll conductance
l_{gs}	Limitation of photosynthesis by stomatal conductance
LHC	Light-harvesting complex
Lhcb1	LHCII type I chlorophyll a/b binding protein
Lhcb2	LHCII type II chlorophyll a/b binding protein
LHCRS	Chlamydomonas light-harvesting complex protein involved in VAZ-related thermal dissipation
LHCI	Light-harvesting complex of PSI
LHCII	Light-harvesting complex of PSII
LHY	Late hypocotyls elongating gene
LIF	Laser-induced fluorescence
L_m	Effective length in the mesophyll
LMA	Leaf dry mass per unit leaf area
LT50	Letal temperature
Lx	Lutein-5-epoxide
MDA	Malondialdehyde
MDH	Malate dehydrogenase
MDHA	Monodehydroascorbate
MDHAR	Monodehydroascorbate reductase
mlo	Broad-spectrum resistance pathway
Mn-SOD	Superoxide dismutase co-factored with manganese
MS	Mass spectrometry
m_T	Total plant mass
NAD^+	Nicotinamide adenine dinucleotide
NADH	Nicotinamide adenine dinucleotide reduced
$NADP^+$	Nicotinamide adenine dinucleotide phosphate
NADPH	Nicotinamide adenine dinucleotide phosphate reduced
NADPme	NADP-malic enzyme
NDVI	Normalised difference vegetation index
NEE	Net ecosystem exchange
NEP	Net ecosystem productivity for CO_2
NMR	Nuclear magnetic resonance
NPP	Net primary productivity
NPQ	Non-photochemical quenching of chlorophyll fluorescence (Stern-Volmer)
npq5	NPQ mutant 5
NtAQP1	Aquaporin 1 from tobacco
NUE	Nitrogen-use efficiency
O	Oxygen concentration
$^3O_2$	Triplet oxygen
$^1O_2$	Singlet oxygen
O_2^-	Superoxide
OAA	Oxaloacetate
OEC	Oxygen-evolving complex
•OH	Hydroxil radical
OJIP	Fluorescence induction

P	Fraction of energy absorbed by PSII allocated to photochemistry
P_{atm}	Atmospheric pressure
P_{turgor}	Turgor pressure
$P(\theta)$	Beam penetration in the canopy for solar zenith angle θ
PAM	Pulse amplitude modulated
PAR	Photosynthetically active radiation
PC	Plastocyanin
PCA	Photosynthetic carbon assimilation
PCR	Polymerase chain reaction
PCRC	Photosynthetic carbon-reduction cycle
PD	Pierce's disease bacterium
PDB	Pee Dee Belemnite
PED	Potato early dying disease
PEP	Phosphoenolpyruvate
PEPC	Phosphoenolpyruvate carboxylase
PEPCk	Phosphoenolpyruvate carboxykinase
PET	Photosynthetic electron transport
3PGA	Phosphoglycerate
PG	Phosphatidylglycerol
P_i	Inorganic phosphate
pI	Isoelectric pH
P^j_{CO2}	Permeability coefficient of CO_2 in a medium *j*
pmf	Proton motive force
PMMoV	Unknown tobamovirus
PMMoV-I	Italian strain of PMMoV
PMMoV-S	Spanish strain of PMMoV
PNRSV	Prunus necrotic ringspot ilarvirus
PPFD	Photosynthetic photon-flux density
Pph	*P. syringae* pv. *phaseolicola*
PPV	Plum pox virus
PQ	Plastoquinone
PQH_2	Plastohydroquinone
PRI	Physiological (or Photochemical) reflectance index
psaA	Anthena protein gene
psaB	Anthena protein gene
psaD	Gene encoding the PsaD subunit of PSI
PsbO	33 kD-extrinsic protein
PsbP	OEC proteins
PsbQ	OEC proteins
PsbS	OEC proteins
PSBS	Antenna protein involved in vaz-associated termal dissipation
PSI	Photosystem I
PSII	Photosystem II
Pto	*P. syringae* pv. *tomato*
PTOX	Plastid terminal oxidase
PVY	Potato virus Y
Q_A	PSII primary quinone acceptor
Q_B	PSII secondary quinone acceptor

qE	Energy dependent quenching of chlorophyll fluorescence
Q_{int}	Daily integrated quantum-flux density
qN	Non-photochemical quenching of chlorophyll fluorescence
qP	Photochemical quenching of chlorophyll fluorescence
QTL	Quantitative trait loci
R_a	Molar ratio of $^{13}CO_2/^{12}CO_2$ in the air
R_A	Autotrophic respiratory processes
R_{air}	Isotope ratios of the atmospheric air
RbcS	Rubisco small subunit genes
RbcL	Rubisco large subunit protein
RC	Reaction centre
R_d	Respiration in the light
R_E	Total ecosystem respiration
RET	Respiratory electron transport
R_g	Effective hydraulic resistance from the soil to the guard cells
RGR	Relative growth rate
R_H	Heterotrophic respiratory processes
RH	Relative humidity
RH_s	Relative humidity at the leaf surface
R_L	Leaf mitochondrial respiration
R_n	Respiration in the dark
R_o	Initial $^{18}O/^{16}O$ ratio
ROS	Reactive oxygen species
R_p	Molar ratios of $^{13}CO_2/^{12}CO_2$ in the photosynthetic product
PR	Photorespiration rate
RP-HPLC-ESI-MS	Reverse phase high-performance liquid chromatography electrospray spectrometry
R_{plant}	Isotope ratios of the plant biomass
R_{pr}	Isotope ratios of the product
$R_{product}$	Molar ratio of $^{13}CO_2/^{12}CO_2$ in the photosynthetic product
R_{so}	Isotope ratios of the source
R_{sm}	Isotope ratios of the sample
R_{st}	Isotope ratios of the standard
R_t	$^{18}O/^{16}O$ ratio at time after respiration in a closed sistem
RTV	Rice tungro virus
Rubisco	Ribulose-1,5-bisphosphate carboxylase/oxygenase
RuBP	Ribulose-1,5-bisphosphate
RUE	Radiation-use efficiency
RWC	Relative water content
s	Surface area of leaf
SA	Salicylic acid
S_c	Surface area of chloroplasts exposed to intercellular air spaces per unit leaf area
S_c	Term representing the formation/destruction processes of the scalar molecules operated by biological sources and sinks present in the air volume (for eddy covariance measurements)
$S_{c/o}$	Rubisco specificity factor
s_l	Carbon-isotope fractionation owing to leakage from the bundle-sheath cells
SLA	Specific leaf area
SLAP	Standard light Anthartic precipitation
S_m	Surface area of mesophyll cell walls exposed to intercellular air spaces per unit leaf area

SMOW	Standard mean ocean water
SOD	Superoxide dismutase
SO_2	Sulphur dioxide
SPS	Sucrose phosphate synthase
STOF	Single turnover flaxes
STOMAGEN	Gene involved in regulating stomatal density during leaf development
Suc	Sucrose
SVIs	Spectral vegetation indices
t	Lag time between the light pulse and photoacoustic detection of the oxygen wave
T	Leaf thickness
TBARS	Thiobarbituric acid-reactive-substances
TCAC	Tricarboxylic acid cycle
T_d	Tissue density
TDLAS	Tunable diode laser absorption spectrometers
TILLING	Targeted induced local lesions
T_l	Leaf temperature
TL	Thermoluminescence
TMV	Tobacco mosaic virus
TOC1	Time of chlorophyll a/b binding gene
ToRSV	Tomato ringspot nepovirus
TP	Triose phosphates
TPU	Triose-phosphate utilisation
TPT	Triose phosphate/phosphate translocator
TSP9	Thylakoid-signaling protein 9
TuMV	Turnip mosaic virus
TYMV	Turnip yellow mosaic tymovirus
TyrZ	Tyrosine Z
u	Air flow into the chamber
U_o	Oxygen uptake
UV	Ultraviolet radiation
UV-B	Ultraviolet radiation B
V	Violaxanthin
VAZ	Xanthophyll pool
V_c	Velocity of carboxylation by Rubisco
VDE	Violaxanthin de-epocidase
$V_{c,max}$	Maximum velocity of carboxylation
VOC	Volatile organic compounds
$V_{o.max}$	Maximum rate of oxygenation
$v_o V_o$	Velocity of oxygenation by Rubisco
V_p	Carboxylation rate of PEPC
VPD	Vapour-pressure deficit
V-PDB	Vienna-PDB
$V_{p.max}$	Maximum velocity of the PEPC reaction at saturating levels of c_m
V-SMOW	Vienna-SMOW
WI	Water index
wt	Wild-type
wv	Water vapour
WUE	Water-use efficiency

w′	Vertical wind speed
x	Distance between the center of the chloroplasts and the intercellular air space
X	Fraction of energy absorbed by PSII allocated to processes other than heat dissipation and photochemistry
Z	Zeaxanthin
Zn-SOD	Superoxide dimutase co-factored with Zinc
α	Canopy assimilation quantum efficiency
α_k	Isotope effects
α_L	Leaf absorptance
α_{TPU}	Proportion of glycerate that is not returned to chloroplast during TPU limitations
β	Constant proportionally related to the tissue density
β_{PEP}	Fraction of CO_2 fixed by PEP carboxylations in C3 plants
ε	Apparent carboxylation efficiency
ε_k	Kinetic fractionation during diffusion through leaf boundary layer and stomata
ε_{PSII}	Fraction of quanta absorbed by PSII
$\varepsilon+$	Equilibrium fractionation associated with the differences in vapor pressure between light ($H_2{}^{16}O$) and heavy ($H_2{}^{18}O$) molecules
δ_a	$\delta^{18}O$ of CO_2 in the overlying air
δ_{chl}	$\delta^{18}O$ of CO_2 in full isotopic equilibrium with water in the chloroplast
δ_e	Isotopic composition of water at evaporative sites
δ_{in}	Isotopic composition of air entering the gas-exchange cuvette
δ_o	Isotopic composition of air leaving the gas-exchange cuvette
δ_p	Isotope composition of plant biomass
δ_R	Isotopic ratio of ecosystem respiration
δ_t	Isotopic composition of transpired water
$\delta^{13}C$	Carbon isotope fractionation
Δ	Isotope discrimination
Δ_a	Oxygen-isotope fractionation by the alternative respiratory pathway
Δ_{bio}	Integrated net biochemical carbon isotope fractionation in CAM plants
Δ_c	Oxygen-isotope fractionation by the cytochrome respiratory patway
Δ_{canopy}	Canopy photosynthetic discrimination
Δ_{ea}	A factor relating the $\delta^{18}O$ of CO_2 in the overlying air with that in full isotopic equilibrium with water in the chloroplast
Δ_i	$^{12}C/^{13}C$ carbon isotope fractionation associated with photosynthesis
Δ_n	Oxygen-isotope fractionation in the absence of respiratory inhibitors
Δ_V	Isotopic enrichment of atmospheric water vapour relative to source water
Δ_x	Thickness of the barrier during diffusion of CO_2
$\Delta^{13}C$	Carbon-isotope fractionation
$\Delta^{13}C_{sug}$	Carbon-isotope discrimination in soluble sugars
$\Delta^{13}{}_{obs}$	Measured instantaneous carbon isotope discrimination
$\Delta^{18}O$	Oxygen-isotope discrimination
$\Delta^{18}O_{es}$	Isotopic enrichment of leaf water at the site of evaporation
$\Delta^{18}O_{LS}$	Average isotope water enrichment in leaf lamina under steady state
$\Delta^{18}O_{obs}$	Observed oxygen-isotope discrimination
$\Delta^{18}O_{pred}$	Predicted oxygen-isotope discrimination
$\Delta^{18}O_{suc}$	Average isotope water enrichment of sucrose in leaves
ϕ	Fraction of CO_2 that leaks out of the bundle sheath cells
$\phi_{CO2/O2}$	Ratio of oxygenation to carboxylation events
Φ_c	Proportion of carbon respired at night by whole plant or during the day by non-photosynthetic organs

Φ_n	Net radiant flux density
Φ_w	Proportion of water transpired at night by stomata or cuticular transpiration
ϕ_{CO2}	Quantum efficiency of CO_2 assimilation
ϕ_{PSII}	Actual PSII efficiency
ϕ_V	Proportion of leaf water associated with small veins
ϕ_x	Proportion of leaf water associated with xylem and surrounding tissues
ϕ_L	Proportion of leaf water associated with lamina mesophyll
χ	Turgor-to-conductance scaling factor
γ	Constant proportionally related to the tissue density
π	Osmotic pressure
π_a	Apoplastic osmotic potential
π_e	Epidermal osmotic potential
π_g	Guard-cell osmotic potential
ρ	Molar density of air
μ	R_d/R_n ratio
τ	Lifetime of excited chlorophyll
τ_a	Partitioning of electrons though the alternative respiratory pathway
τ_c	Chamber response half-time
τ_{CO2}	The residence time of CO_2 in the leaf
Θ	Convexity
$\in$	A factor relating the CO_2 concentration in air with that at the chloroplast surface
ξ	A factor relating the CO_2 concentration of air entering and leaving the gas-exchange cuvette
Ψs	Soil-water potential
Ψw	Leaf-water potential
$\Delta\Psi$	Transmembrane electric field
ΔpH	Transmembrane pH gradient
Γ	CO_2-compensation point
Γ^*	CO_2 photocompensation point
Ω	Spatial clumping index
θ_{eq}	Isotopic equilibrium in the CO_2-H_2O system
σ_{PSII}	Effective absorption cross section of PSII
$\wp$	Péclet effect associated with the lamina mesophyll
$\wp_r$	Péclet effect associated with radial flow
$\wp_{rw}$	Péclet effect associated with veinlets
$\bar{a}$	Weighted-mean diffusional fractionation through the boundary layer, stomata and aqueous leaf media

1 • Terrestrial photosynthesis in a changing environment

J. FLEXAS, F. LORETO AND H. MEDRANO

1.1. HISTORICAL PERSPECTIVE

The study of plant physiological responses to the environment (i.e., 'ecophysiology') has attracted researchers since early times, starting from Hales' proposal that plants took their nourishment from the surrounding air (Hales, 1727), through Charles Darwin's observations on leaf and chloroplast movements in response to external conditions and experimental manipulations (Darwin, 1881a,b; 1882), to Darwin's son Francis' early works on the relationship between transpiration and stomatal aperture (Darwin and Pertz, 1911; Darwin, 1916). From its very early re-foundations in modern times, plant ecophysiology has focused on photosynthesis – and transpiration, its undissociable process in terrestrial environments – as the most central physiological characteristic of plants changing in response to the environment. In parallel, considerable progress in the structural and biochemical basis of photosynthesis enabled improvement in the understanding of photosynthetic processes (Calvin and Benson, 1948), the integration of photosynthesis with transpiration and respiration and the view of photosynthesis as the basis for quantitative models of plant growth and crop production. Monsi and Saeki (1953) developed a theoretical frame to describe the distribution of light within plant communities, as Gaastra (1959) worked on photosynthesis in terms of gas exchange along a series of resistances in and out of leaves. Studies on leaf energy balance were initiated (Raschke, 1956; Gates, 1962) and later developed (Montieth, 1973), when portable infrared (IR) gas analysers for measuring photosynthesis became available (Bosian, 1960), allowing field campaigns of measuring photosynthesis in natural environments around the globe (Tranquilini, 1957; Lange *et al.*, 1969; Björkman *et al.*, 1972; Billings, 1973). In the late sixties and early seventies, the C_4 pathway was discovered (Hatch and Slack, 1966) as well as the oxygenase activity of ribulose-1,5-bisphosphate carboxylase/oxygenase (Rubisco) (Ogren and Bowes, 1971), revealing a greater diversity of photosynthetic pathways than thought. Meanwhile, photosynthesis-based plant growth models were developed (Brouwer and de Wit, 1969; Penning de Vries *et al.* 1974). In the late seventies, an optimisation-theory model to explain stomatal behaviour linking photosynthesis to transpiration was developed by Cowan and Farquhar (1977), and the most widely used leaf photosynthesis model was presented (Farquhar *et al.*, 1980a). After those decades, the importance of photosynthesis and related processes (such as transpiration, respiration and growth) was such that, according to several authors, plant ecophysiology in the eighties was fully focused on photosynthesis-related subjects, such as energy and mass exchange (Mooney *et al.*, 1987; DeLucia *et al.*, 2001; see Fig. 1.1).

Over the last thirty years, the boost of 'molecular biology' studies, in particular addressing the plant's genome and functional genetics development, offered new tools to explore photosynthesis ecophysiology and evolution. The studies on plant ecophysiology in general, and photosynthetic response to the environment in particular, moved from an organismal focus to scales below and above the organism (DeLucia *et al.*, 2001; see Fig. 1.1). The new scenarios of climatic change challenged our capacity to predict plants, communities and global atmospheric changes in the near future. Studies on the ecophysiology of photosynthesis are now frequently backed by phylogenetics, molecular genetics and biochemical analysis, and serve as the basis to describe population dynamics and ecosystem processes (Fig. 1.1). Recent new directions in photosynthesis research include expanding current bi-dimensional gas-exchange models to tri-dimensional views (Verboven *et al.*, 2008; Kaiser, 2009; Tholen and Zhu, 2011), using systems biology approaches for the analysis of potential metabolic responses to the changing

Terrestrial Photosynthesis in a Changing Environment: A Molecular, Physiological and Ecological Approach, ed. J. Flexas, F. Loreto and H. Medrano. Published by Cambridge University Press.

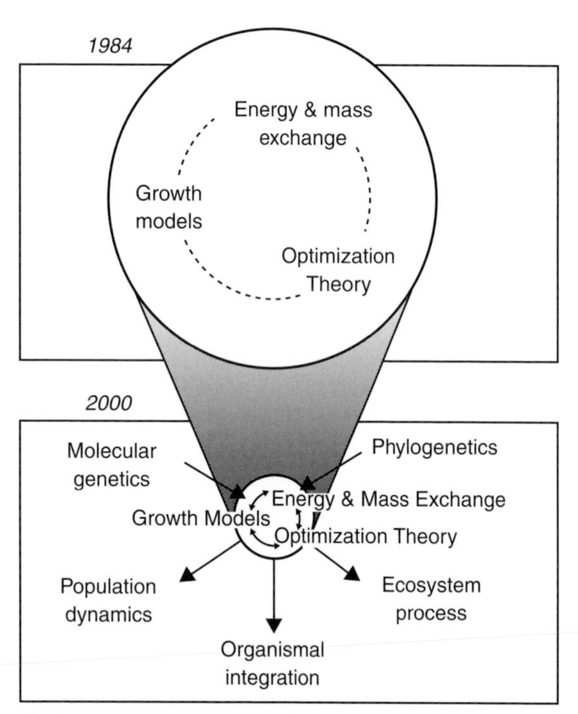

Fig. 1.1. Evolution of the focus of studies on plant ecophysiology from the 1980s to the current century, highlighting the importance of photosynthesis-related aspects, such as energy and mass exchange and growth (after DeLucia *et al.*, 2001).

environment (Luo *et al.*, 2009), and designing and developing both genetically modified plants with improved photosynthesis (Zhu *et al.*, 2010) and synthetic photosynthesis (Bar-Even *et al.*, 2010; Larom *et al.*, 2010). Despite many recent advances essential basic questions remain unresolved, such as which is the natural variation in plant photosynthesis (Flood *et al.*, 2011) or why is the photosynthetic process less energy efficient than human-built photovoltaic devices (Blankenship *et al.*, 2011). For these reasons, knowledge of photosynthesis is of general interest to researchers in plant biology, environmental sciences, agronomy or forestry, among others. Owing to this expansion, the response of photosynthesis to the environment is a very active area of research, producing more than a thousand publications per year. Over the last few years, up to four of the highest impact plant journals published Special Issues on this subject (Annals of Botany Vol. 103, Issue 4; Journal of Experimental Botany Vol. 60, Issue 8; Plant and Cell Physiology Vol. 50, Issue 4; Plant Physiology Vol. 155, Issue 1).

1.2. IMPORTANCE OF TERRESTRIAL PHOTOSYNTHESIS

The reasons why photosynthesis in terrestrial plants deserves so much attention are multiple. Perhaps the most obvious is the fact that photosynthesis has been the cornerstone of evolution of current life on earth, and is nowadays the process supporting most of the Earth's primary production. The world's human population has raised exponentially, overpassing 7 billion people, whereas changing patterns of land use have resulted in a reduced crop area. During the so-called 'Green Revolution', starting in the sixties, global crop productivity increased mostly because of better knowledge of plant nutrition, plant defence against pathogens and genetic improvement of plant structure and harvest index (i.e., the proportion of total plant biomass that is harvestable), rather than by improvements of photosynthesis and total plant biomass (Zhu *et al.*, 2010). However, an increased use of chemical fertilisers and pesticides are now known to have negative effects on the environment and ecological resources (Tilman *et al.*, 2002). The world population has now doubled since the 'Green Revolution' started, largely increasing the global demands for food. Most predictions suggest that such demand will double again by 2030 (Edgerton, 2009; Murchie *et al.*, 2009). Moreover, the demand for plant materials besides food has expanded for many years owing to the increasing importance of, for instance, plant-based biofuels (Edgerton, 2009). On the other side, the strategies implemented during the 'Green Revolution', which had led to steady increases in global cereal production from 1960 to 1990, appear now almost saturated, leading to only minor additional increases in global production (Tilman *et al.*, 2002). Overall, to meet the global demands for plant resources in the immediate future, only two options appear to be available (Edgerton, 2009): (1) expanding the area under production; and (2) improving productivity on existing farmland. Although these two options are not mutually exclusive, the second is particularly challenging as genetic improvement of plant structure and harvest index seems to have already achieved an optimum. Hence, improving productivity would require improving leaf photosynthesis and photosynthetic efficiency in the use of carbon, water and nutrients (Long *et al.*, 2006a; Peterhansel *et al.*, 2008; Murchie *et al.*, 2009; Zhu *et al.*, 2010).

Besides feeding humans, photosynthesis is of broad importance for many other reasons. For instance, the evolution of photosynthesis-related plant traits in response to the

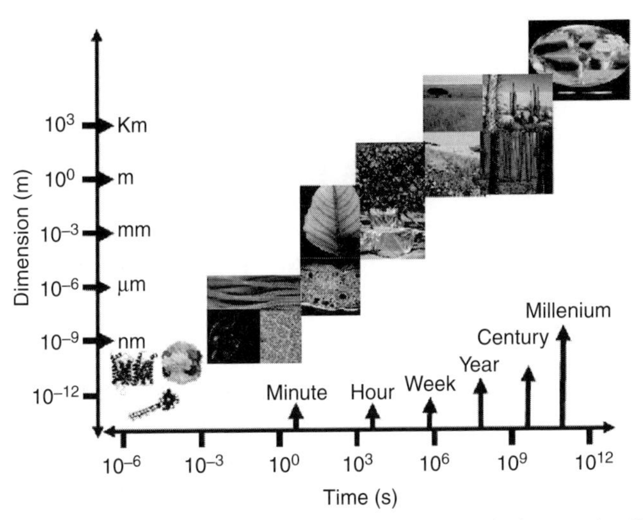

Fig. 1.2. Different time and dimension scales for photosynthetic events (modified from Osmond *et al.*, 2004).

changing conditions during the Earth's history is recognised to have started the oxygenic environment that allowed life as we know it now, and have contributed to shape the current physiognomy of vegetation and its diversity (Beerling *et al.*, 2001; Brodribb *et al.*, 2007; Franks and Farquhar, 2007; Sack *et al.*, 2008; Hohmann-Marriott and Blankenship, 2011). Photosynthesis is therefore an essential component of plant survival and species fitness (Athanasiou *et al.*, 2010; Donovan *et al.*, 2011; McDowell, 2011). Alternatively, terrestrial photosynthesis constitutes about half of the total Earth carbon sinks excluding the atmosphere (Canadell *et al.*, 2007; Beer *et al.*, 2010). Moreover, whereas the ocean sink strength is very controversial, estimations of the terrestrial sink, contributed to by photosynthesis, is rather solid (IPCC, 2007). Therefore, terrestrial photosynthesis is the main natural tool to counteract climate change and to contribute, together with transpiration (Hetherington and Woodward, 2003) and ecosystem carbon dynamics (Heimann and Reichstein, 2008), to plant-driven climate feedbacks (Bonan, 2008; Cao *et al.*, 2010; Rotenberg and Yakir, 2010). This is the reason why current arguments by the COP 15 at the Copenhagen meeting (http://en.cop15.dk/) have addressed issues other than forestation, which is still regarded as the main option to reduce CO_2 increase in the atmosphere. Moreover, because of the tight relationship between photosynthesis and transpiration, which results from the central role of stomata in both processes, photosynthesis is also pivotal for determining WUE (i.e., the amount of biomass obtained per unit water used). Owing to increasing scarcity of fresh water on a global scale (IPCC, 2007), improving water use in crop production through improving photosynthetic efficiency is a requisite for a sustainable agriculture (Rockström *et al.*, 2007; Morison *et al.*, 2008). But on the other hand, and against all these needs, there is ample evidence that human activities are acting in detriment of photosynthetic carbon fixation, delaying for example the carbon cycle of forests (Magnani *et al.*, 2007), or competing for the use of fertile soils by agro-forestry. Clearly, the development of a social consciousness will be required, in addition to research, to achieve the goal of a sustainable but sufficiently productive agriculture coupled with plant-biodiversity preservation.

1.3. RECENT ADVANCES IN STUDIES ON TERRESTRIAL PHOTOSYNTHESIS AND ITS RESPONSES TO THE ENVIRONMENT

In recent decades, important advances have been made in the knowledge of photosynthetic responses to the environment. Studies in this area are currently performed using multidisciplinary approaches, and range from the smallest time-and-space scales to the largest (Fig. 1.2).

At the smallest scales, photosynthetic processes occurring in fractions of seconds can now be properly tracked, even under field conditions, using available spectroscopic techniques (see Chapter 10). The recent and rapid development of molecular analytical tools has also allowed genomic and proteomic analysis that strengthen the links between genotype and phenotype when measuring the photosynthetic capacity and the resistance of the photosynthetic apparatus to stress conditions (e.g., Bogeat-Triboulot *et al.*, 2007; see Chapter 9). At the molecular level too, important advances are being made in understanding the existing variability in the structure and kinetics of photosynthetic enzymes (Galmés *et al.*, 2005) and in photoprotective mechanisms (Bode *et al.*, 2009). Perhaps the most significant advances in recent years have been made thanks to the rapid development of genetically engineered plants, which considerably increases the precision with which single photosynthetic components and their effects on the functionality of photosynthesis can be studied (Kozaki and Takeba, 1996; Horváth *et al.*, 2000; Kebeish *et al.*, 2007; Merlot *et al.*, 2007; Rivero *et al.*, 2007, 2009; etc.).

At the organelle and leaf levels, advances in the last few decades include improved knowledge of stomatal diversity and its functional significance (Franks and Farquhar, 2007; Brodribb *et al.*, 2009; Brodribb and McAdam, 2011); understanding the inter-relations between leaf structure, hydraulic architecture and photosynthetic function (Nikopoulos *et al.*, 2002; Brodribb *et al.*, 2007; Sack *et al.*, 2008); highlighting the crucial importance of mesophyll conductance to CO_2 as a significant and variable limiting factor for photosynthesis (Loreto *et al.*, 1992; Flexas *et al.*, 2008); highlighting the importance of the interactions between photosynthesis and respiration (Krömer *et al.*, 1993; Gallé *et al.*, 2010; Nunes-Nesi *et al.*, 2011); improving basic knowledge about the genetics and physiology of different photosynthetic pathways (Tanz *et al.*, 2009); and further developing leaf photosynthesis models (Yin *et al.*, 2009). Of particular interest, because of the generality of its implications, has been the description of a 'worldwide leaf economics spectrum', consisting of retrieval of tight relationships among leaf traits (including photosynthesis) that are largely independent of the world's regions and climates (Fig. 1.3). These relationships represent trade-offs for which any existing plant can present at a definite time high growth capacity and high resistance to stressful environments. The existence of such a spectrum sets the boundaries of leaf functioning, and helps predict plant distribution and vegetation boundaries under changing land use and climate. It also shows how limited the possibilities are of improving a given leaf trait such as photosynthesis without co-improving or impairing other leaf traits. For example it proves very difficult to improve photosynthesis without increasing stomatal conductance and hence water losses, or to improve photosynthesis of herbivore-resistant

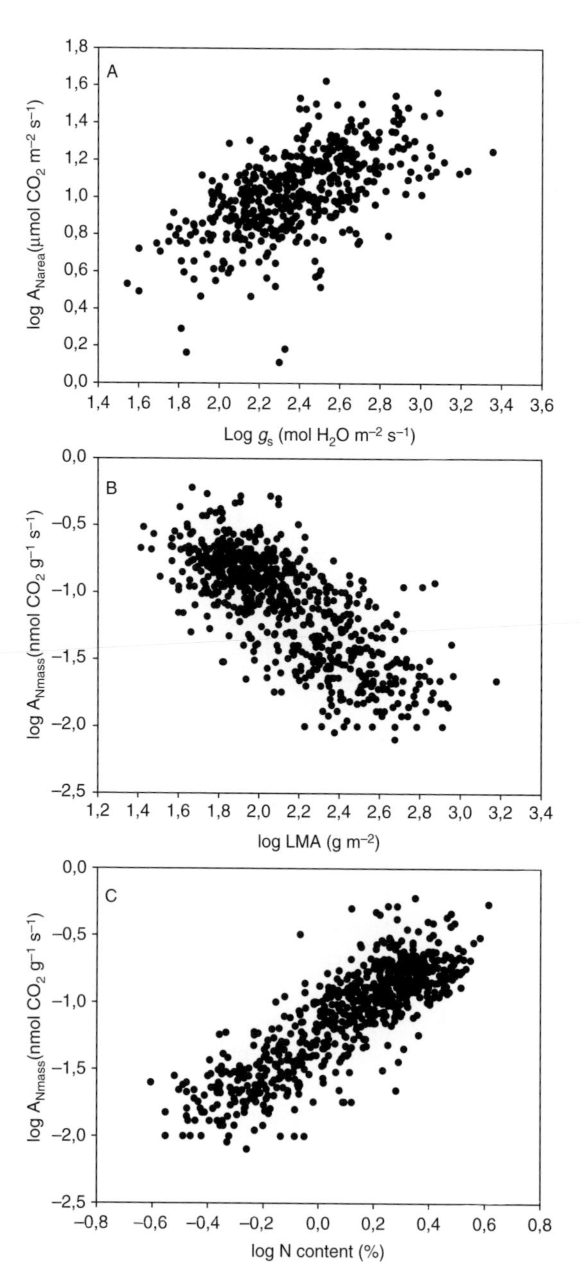

Fig. 1.3. General (worldwide) relationships between photosynthesis and other leaf traits across a large number of species belonging to different functional groups and inhabiting different biomes. **(A)** Log-scale relationship between net photosynthesis on an area basis and stomatal conductance. **(B)** Log-scale relationship between net photosynthesis on a mass basis and leaf mass per area. **(C)** Log-scale relationship between net photosynthesis on a mass basis and leaf nitrogen content. LMA, Leaf dry mass per unit leaf area. Data from Wright *et al.* (2004).

leaves that are characterised by a high leaf mass per area. Nevertheless, there is some scattering in these relationships, suggesting that some variability can be explored in the ratios between any pair of these leaf traits, further improving the photosynthetic efficiency in limiting conditions (see Chapters 16–22).

At the organism and population level, photosynthesis has been recognised as a crucial determinant of genotype fitness in segregating populations. In this sense, photosynthetic differences between male and female plants (Nicotra *et al.*, 2003; Letts *et al.*, 2008) or the impacts of polyploidy (Vyas *et al.*, 2007; Li *et al.*, 2009) on photosynthesis are just a few examples of novel described features of photosynthetic responses. Alternatively, numerous studies have increased our understanding of photosynthesis in organs other than leaves (Chapter 7), and highlighted photosynthetic adaptations of particular plants, increasing the known range of diversity in photosynthetic strategies to cope with environmental constraints (Chapter 7). It is also worth mentioning that important advances have been made relating to systemic signalling systems, including reactive oxygen species (ROS) (Miller *et al.*, 2010) and electrical impulses (Grams *et al.*, 2007) that connect the photosynthetic responses of distant plant parts. Many molecules that have previously been seen as indicators of photosynthetic damage, such as ROS, are indeed now emerging as important signalling molecules activating hypersensitive responses and other reactions of plants to stressful agents (Miller *et al.*, 2010).

Finally, in recent years photosynthesis studies have extended to the largest scales, the canopy and ecosystem levels. At these scales, the introduction of flux-level gas-exchange measurements has been crucial (e.g., Misson *et al.*, 2007; Baldocchi, 2008; see Chapter 14), as well as the development of suitable remote sensing techniques (e.g., Grace *et al.*, 2007; Damm *et al.*, 2010; see Chapter 15). Currently, efforts are being made to map CO_2 fluxes, photosynthetic light and water-use efficiencies at regional and global scales, aided by remote sensing tools, especially to allow assessment of photosynthesis from the space (Gamon *et al.*, 2004; Fuentes *et al.*, 2006; Drolet *et al.*, 2008).

1.4. SCOPE AND STRUCTURE OF THE PRESENT BOOK

The contents of the present book have been designed to try to exhaustively cover the entire span of photosynthesis studies, from the molecular to the ecosystem level, including the basics of terrestrial photosynthesis, its responses to the environment and the most recent advances in the field. The reader may feel lost in the wealth of information of this book, but we have adopted a modular strategy to drive the reader to the information that is specifically needed and at a well-defined level of investigation. Thus, the book has been comprehensively structured in parts comprising 34 chapters, their contents being in increasing order of biological complexity. Hence, the present introductory chapter provides a brief summary and overview of the structure and contents of the book. Part 1 'Photosynthesis: the process' (Chapters 2 to 8) offers a succinct review of the photosynthetic basic of, with a particular focus on models of leaf photosynthesis and photosynthetic limitation analysis, both important aspects for ecophysiology. Part 2 'Measuring photosynthesis' (Chapters 9 to 15) addresses how to measure photosynthesis and related aspects in ecophysiological studies, providing a detailed presentation of the most useful techniques, their principles and procedures. Part 3 'Photosynthesis response to single environmental factors' (Chapters 16 to 22) reviews current state-of-the-art techniques while taking the simplest approach in ecophysiology of photosynthesis. The following Parts, 4 'Photosynthetic in time' (Chapters 23 to 25) and 5 'Photosynthesis in space' (Chapters 26 to 32), show more complex and interacting aspects of photosynthesis related to photosynthesis variations in time, from ontogeny to evolutionary aspects (Part 4), and photosynthetic variations in space, covering photosynthetic in crops, as well as in the most important biomes of the world (Part 5). Finally, increasing the complexity to the global scale, Part 6 'Photosynthesis in a global context' (Chapters 33 and 34) covers worldwide aspects of photosynthesis, such as its implications for WUE and its responses in a climate-change context.

The book is designed to offer a basic, yet informative and updated, view of the primary process driving life on earth to students of all university courses, and to scholars of all biological fields. By transferring current knowledge of photosynthesis to new generations we hope to seed the perception of the importance of this process for the environment, and for the future of agriculture and forestry, the paramount activities sustaining human life on Earth and the mitigation of global change. As for photosynthesis, time and space are crucial parameters that will indicate whether our intents will be met!

Part I
Photosynthesis: the process

2 • Biochemistry and photochemistry of terrestrial photosynthesis: a synopsis

T.D. SHARKEY, J.-M. DUCRUET AND M.A.J. PARRY

2.1. INTRODUCTION

Photosynthesis is typically understood as the light-dependent production of sugar from carbon dioxide (CO_2). The endosymbiotic chloroplast is the cellular location for most of this metabolism in plants, but some additional metabolism occurs in the cytosol to make the sugars that will be transported around the plant, mainly sucrose and also sugar alcohols, such as sorbitol and manitol. There are many processes that can properly be called photosynthesis, but a core set of processes underlie most of the considerations in this book. This chapter will provide an overview of those processes, and many topics covered in this chapter are the subject of more in-depth chapters later on. This chapter begins by describing the initial capture and temporary storage of light energy as highly reactive molecules (nicotinamide adenine dinucleotide phosphate (NADPH) and adenosine triphosphate (ATP)) on carbon. By reducing (i.e., by adding electrons to) carbon from its most oxidised state (CO_2) to the status of sugars $(CH_2O)_n$, the energy initially stored as NADPH and ATP can be stored on the carbon. Additional energy can be stored on each carbon atom by reducing it fully, as happens in the synthesis of oils ($R-CH_2-R$), but this is generally not considered when describing photosynthesis. Finally, issues surrounding uptake of the CO_2 will be addressed.

2.2. PHOTOCHEMISTRY SYNOPSIS

Photochemistry, the capture of light energy and its conversion to chemical energy suitable for reducing CO_2 to sugar, is the source of nearly all energy available to living things. Energy captured by absorbing molecules is stored as the high-energy intermediates NADPH (reducing power) and ATP (sometimes called the energy currency of the cell).

2.2.1. The pigment antennae and photochemical centres I and II

In the first step of photosynthetic electron transport (PET), light is absorbed by chlorophylls (chl.) and auxiliary pigments organised in pigment-protein complexes that constitute the antennae of two photosystems (photosystem I and II (PSI and PSII)) (Figure 2.1). Most chl.s (>99%) are antennae chl.s and do not participate directly in photochemistry. These antenna chl.s are found attached to 'light-harvesting' complexes (LHC), and to core antenna subunits. In cyanobacteria the antennas are found outside of the photosynthetic membrane as phycobilisomes, whereas in algae and plants the antennas are integral membrane protein complexes. Higher-plant LHCIIs (light-harvesting complexes of PSII) are trimers, with each trimer having three membrane-spanning α-helices. Each monomer has a lutein molecule that can absorb light, but its main role is likely protection of the chl. molecules from photodestruction. LHCIs (light-harvesting complexes of PSI) are associated with PSI; LHCII is normally are associated with PSII, but can move within the thylakoid membrane from PSII to PSI. Chl. *b* absorbs shorter red wavelengths and longer blue wavelengths than chlorophyll *a* (Figure 2.1) and is located in the LHC, mainly LHCII, and in the minor chl.-protein complexes (CP) of PSII. LHCI in the peripheral antenna of PSI contains CP complexes with a red absorption shifted towards longer wavelengths. Normally, light excitation is balanced between PSII and PSI, but spectral differences between the two photosystems depending on the spectrum of incoming light or excessive rates of cyclic electron flow around PSI (CEFI) may result in an imbalance. Phosphorylation of some LHCII, which causes it to migrate from PSII to PSI, results in a transfer of light energy from PSII to PSI. This is called a state 1 to state 2

Terrestrial Photosynthesis in a Changing Environment: A Molecular, Physiological and Ecological Approach, ed. J. Flexas, F. Loreto and H. Medrano. Published by Cambridge University Press.

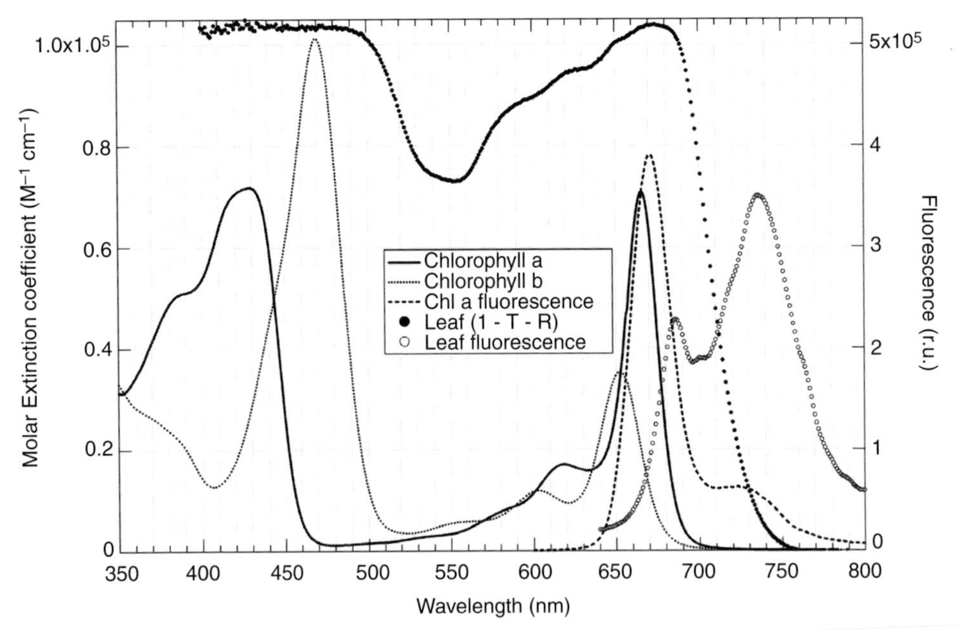

Fig. 2.1. Absorption and fluorescence emission spectra of pure chlorophylls in methanol (lines) and of a pea leaf (circle). Light absorbed by leaf is the fraction of incoming light normalised to one that is not transmitted (T) nor reflected (R). In leaf fluorescence, the 685-nm peak is more strongly decreased by chlorophyll re-absorption than the 740-nm peak.

transition (Allen and Forsberg, 2001; Jensen *et al.*, 2007) and requires a specific subunit of PSI (Lunde *et al.*, 2000).

Light quanta are absorbed, and then migrate randomly in antennae by resonance transfer between chl.s, first to antenna chl.s in the core photosystem complexes, and then to the photochemical traps, called centres, containing a special chl. *a* dimer (except in some bacteria). At the reaction centre (RC), an electron is transferred from the excited singlet Chl* to a primary acceptor, transiently producing a chlorophyll$^+$ cation, which is re-reduced by a primary electron donor. This separation of +/– charge pairs stabilised on electron carriers constitutes the primary step of photosynthesis. The Chl* ↔ chl.$^+$ + e$^-$ charge separation step corresponds to an absorption change of centre chl. dimers, at 680 nm for PSII (P680) and 700 nm for PSI (P700). The P700$^+$/P700 ratio can be measured *in vivo* in the region of 810 to 830 nm in order to monitor PSI activity, and this underlies an optical method for assessing PSI function (Chapter 10). Alternatively, P680$^+$ appears only transiently in undamaged PSII, and is rapidly re-reduced in the submicrosecond range by the primary donor TyrZ connected to the oxygen evolving complex (OEC). PSII is, however, endowed with the unique property of emitting both a variable fluorescence and a luminescence, which reflect its functioning. The variations of the fluorescence yield results from those of the light-induced high fluorescence P680Q_A^- state of PSII centres (Krause and Weis, 1991; Govindjee, 1995; Baker, 2008). A weak delayed fluorescence, generally called luminescence, originates from two Chl* that is created in darkness by the +/– charge pairs separated by a previous illumination and stabilised on PSII electron carriers. Fluorescence reflects the fate of the light quanta in the antenna, whereas luminescence is a probe of stabilised charge pairs produced by a previous illumination.

PSII acts as a photo-water-oxidase-plastoquinone-reductase. It extracts four electrons from two water molecules and raises them successively to a more reducing (negative) redox potential, allowing them to reduce plastoquinone (PQ) to plastohydroquinone (PQH_2), i.e., two electrons per PQ. This requires two buffer systems to store the + and – charges after a photochemical event (Figure 2.2).

Positive charges are stored on the OEC. A tyrosine (called Z) transfers an electron to the oxidised P680$^+$, and then accepts an electron from the OEC. The OEC stores positive charges on a Mn_4O_4Ca cluster endowed with four oxidation steps, the S states S_0 to $S_{4(+)}$ (Rutherford and Boussac, 2004). The transient last step performs water photolysis:

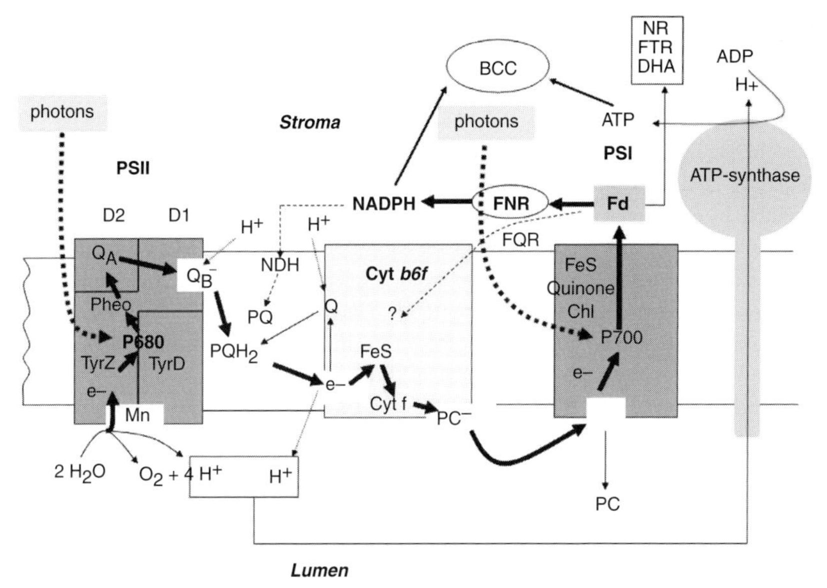

Fig. 2.2. The photosynthetic electron-transfer chain. Electrons extracted to water are raised to a more reducing potential by the two photosystems, PSII and PSI, functioning in series, to finally reduce ferredoxin, then NADPH. This electron transfer is coupled to a proton pumping from stroma to lumen. The proton flow through the ATP synthase drives ATP synthesis. Inducible cyclic pathways are FQR (ferredoxin-plastoquinone-reductase) and NDH (NADPH dehydrogenase). Chl, chlorophyll; Cyt, cytochrome; DHA, mono-dehydro-ascorbate; FNR, ferredoxin; NADP Reductase; FTR, ferredoxin thioredoxin reductase; NR, nitrite reductase; Pheo, pheophytin; PQ, plastoquinone; PQH_2, plastohydroquinone; Q, Q cycle (that involves PQ/PQH_2 and amplifies proton pumping by the cyt b_6f); Tyr, tyrosine.

$$Z^+ + S_3 \rightarrow Z + S_4 + 2H_20 \rightarrow S_0 + O_2 + 4H^+ \qquad \text{[Eqn. 2.1]}$$

The OEC is exposed to the lumen, in which protons are released fractionally during the S_0 to S_4 oxidation steps.

The electron charge is stabilised on the secondary acceptor Q_B formed by a PQ loosely bound to a protein pocket, which is also the binding site for diuron- or atrazine-like herbicides, until a second arrives on the primary acceptor Q_A:

$$Q_A^- + Q_B^- \rightarrow Q_A + Q_B^{2-} + 2H+ \rightarrow Q_A + Q_B\text{–}empty + PQH_2 \qquad \text{[Eqn. 2.2]}$$

The quinonic acceptors are facing the stroma, from which the protons are taken up (Figure 2.2).

In addition to the variable chl. fluorescence and luminescence that reflects its functioning (Krause and Weis, 1991; Govindjee, 1995; Baker, 2008), PSII has the other noteworthy property of being highly sensitive to various climatic or chemical stresses (Aro *et al.*, 1993). The reasons for the latter are many: (1) an elaborate but fragile OEC complex; (2) a fairly long (several nanoseconds) excitonic lifetime in the antenna, increased when the light conversion is slowed by downstream inhibition of electron flow, which produces singlet oxygen; (3) a high local concentration of O_2 in grana where PSII is located; (4) a loose Q_B pocket interacting with the highly unsaturated, therefore fluid and oxidisable, lipids of the thylakoid membrane. This can explain the basic architecture of the core PSII, formed by two 32-kDa polypeptides, reaction-centre proteins D1 and D2, encoded by the chloroplastic genes *PsbA* and *PsbD* respectively. D1 is the high-turnover protein in the whole plant cell, the only one that is rapidly marked by ^{35}S-methionine in a leaf submitted to a strong illumination (Mattoo *et al.*, 1981); it acts as a fuse that bears the centre chlorophyll pair, the OEC and the stable Q_B acceptor, and that it is frequently disassembled and replaced when a high-energy pressure produces damaging reactive species. Photoinhibition occurs when the repairing process cannot cope with damage to D1 caused by high light intensities; it is favoured by stress factors such as cold (see Chapter 18) that both slows down energy draining towards the Calvin cycle and increases the membrane rigidity, which impairs the replacement of D1 (Kanervo *et al.*, 1997).

The core of PSI consists of a large pigment-protein complex of 15 integral proteins, including a central heterodimer of sub-units encoded by genes *PSaA* and *PsaB*

(Jensen *et al.*, 2007). Owing to spectral properties of PSI antenna, FR light (>700 nm) can be used to excite PSI preferentially to PSII. An electron is transferred from the excited dimer chl. centre P700* to a primary acceptor, another chl. molecule (A_0), then to a phylloquinone (A_1) and to iron-sulfur centers (F_X, F_A, F_B). The final electron acceptor of PSI is ferredoxin (Fd), soluble in the stroma, which reduces $NADP^+$ (nicotinamide adenine dinucleotide phosphate) to NADPH via the enzyme ferredoxin-NADP-reductase (FNR; Fig. 2.2). The PSI centre P700 is located close to the lumen, and a soluble plastocyanin molecule has to dock at the interface to re-reduce the $P700^+$ resulting from a charge separation (Bottin and Mathis, 1985). PSI contrasts with PSII by a slower re-reduction of its oxidised dimer $P700^+$ by the mobile plastocyanin and by a faster reoxidation of its primary chl. acceptor A_0 by the secondary phylloquinone acceptor A_1, which leads to an increase of the light-induced $P700^+A_0$ state; PSI has no variable fluorescence. The absorption change owing to P700 ↔ $P700^+$, generally measured on a secondary band around 820 nm, provides a tool to monitor PSI activity (Haveman and Mathis, 1976; Harbinson and Woodward, 1987; Schreiber *et al.*, 1989). Ferredoxin is also an electron donor to other metabolic pathways: nitrite reductase, the ascorbate cycle by reduction of the monodehydroascorbate (MDHA) radical and the thioredoxin redox regulatory system. PSI is less sensitive to heat than PSII and generally considered more tolerant to stresses, although a photoinhibition of PSI also occurs in some species in cold conditions (Havaux and Davaud, 1994; Terashima *et al.*, 1998; Scheller and Haldrup, 2005).

The transmembrane cyt b_6f complex mediates the electron transport from PSII to PSI, receiving electrons from reduced PQ and transferring them to plastocyanin, with separate pathways around PSI for the two electrons and the involvement of a Q cycle (Section 2.1.3). Cyt b_6f appears to also be the entry point of electrons from reduced Fd to the intersystem chain, constituting the specific step of the cyclic electron-transfer pathway around PSI (CEFI). The CEFI was discovered by Arnon (1959) in isolated thylakoids but its role *in vivo* was questioned until optical methods (Chapter 10) allowed it to be monitored in leaves (Bukhov and Carpentier, 2004). It corresponds to the putative ferredoxin-plastoquinone-reductase or FQR (Bendall and Manasse, 1995), an electron-transfer activity to which no protein support could be ascribed up to now, although related mutations have been recently characterised (Shikanai, 2007). A second pathway, the NAD(P)H-dehydrogenase (NDH) that reduces the PQ pool, is supported by a well-characterised membraneous protein complex; it is also active in the dark and constitutes a first step of the chlororespiratory pathway, of which a second step is the oxidation of PQH_2 via a plastid terminal oxidase (PTOX) (Peltier and Cournac, 2002). CEFI performs proton pumping without generation of reducing power, which has two consequences: (1) an increase of the lumen acidity that enhances the protective non-photochemical quenching (NPQ); and (2) an increase of ATP synthesis when needed, for example, for protein synthesis and repair. Indeed, CEFI is triggered in various stress situations (Rumeau *et al.*, 2007).

2.2.2. Provision of reducing power

The photons captured by antennas and transferred to *RC*s provide the energy that allows electrons to flow from water to $NADP^+$ in a process called linear electron flow (LEF). In LEF, the rates of PSII and PSI activity must be matched and there are a number of regulatory mechanisms to achieve this balance. LEF provides reducing power that is used primarily for carbon reduction, but also for other important reduction reactions such as nitrite reduction. Up to 20% of reducing power can be used in nitrogen metabolism (Bloom *et al.*, 2002). The reducing power stored on carbon as CO_2 (fully oxidised) is changed to the redox level of a sugar (partially reduced carbon), and represents the bulk of the energy stored as a result of photosynthesis. Fully reduced carbon (e.g., lipids) is formed in anabolic reactions not strictly associated with photosynthesis.

2.2.3. Provision of adenosine triphosphate

The LEF from water to the reducing side of PSI generates reduced Fd, then NADPH, the source of electrons for the Calvin cycle. However, the Calvin cycle also requires ATP to assimilate CO_2. The photosynthetic electron transfer proceeds along vectorially organised transporters, so that several steps of electron transfer are coupled to proton pumping from the stroma into the lumen, creating a pH gradient (ΔpH) and a transmembrane electric field ΔΨ. Together, these constitute a chemiosmotic potential, or proton motive force (*pmf*), that drives synthesis of ATP as described by Peter Mitchell. In chloroplasts this is called photophosphorylation to distinguish it from the similar process in mitochondria, termed oxidative phosphorylation. The relationship between the ΔpH and ΔΨ is described by the Nernst equation:

$$\Delta\mu = 2.3 \cdot R \cdot T \cdot \log\left(\frac{C_i}{C_o}\right) + z \cdot F \cdot \Delta\Psi \quad \text{[Eqn. 2.3]}$$

where $\Delta\mu$ is the chemical potential difference across the membrane, R is the universal gas constant (=8.314 joules·deg^{-1}·mol^{-1}), T is the absolute temperature (Kelvin), C_i is the concentration inside the cell, C_o is the concentration outside the cell, z is the charge (+1 for H^+), F is the Faraday constant (=96,490 joules mol^{-1} V^{-1}), and $\Delta\Psi$ is the difference in electric potential across the membrane.

At equilibrium $\Delta\mu$=0 and so the following is true:

$$\log\left(\frac{C_i}{C_o}\right) = \frac{z \cdot F \cdot \Delta\Psi}{2.3 \cdot R \cdot T} \quad \text{[Eqn. 2.4]}$$

and a rough guide is that each decade of concentration difference equates to 60 mV. In addition to LEF, other mechanisms, such as the Q cycle around the cyt b_6f complex and CEFI, also contributes to proton pumping and participates in the fine tuning of the ATP/NADPH ratio according to physiological demands.

During LEF, there are two mechanisms by which protons are unequally distributed across the thylakoid membrane. The first occurs because water is split in the thylakoid lumen-releasing protons, whereas protons are taken up from the stroma upon formation of NADPH from $NADP^+$. About twice as many protons are translocated because of the transport of electrons on PQ. PQ is reduced at the Q_B site of PSII and this leads to the uptake of two protons from the stroma. The resulting plastoquinol migrates through the thylakoid membrane and delivers the electron to the cytochrome (cyt) complex while simultaneously depositing the proton to the lumen side of the thylakoid membrane. Thus, electron transport is coupled to proton translocation from the stroma to the thylakoid lumen. For each pair of electrons transported to the cyt complex, one continues to PSI by way of plastocyanin, whereas the other cycles back to reduce more PQ, forming a cycle called the Q cycle. This second electron causes PQ to, once again, take up a proton from the stroma and eventually deposit it on the lumen side of the thylakoid membrane. The net result is that for every electron that moves from PSII to PSI, two protons are translocated to the thylakoid lumen by PQ/plastoquinol shuttling and one proton is liberated from water inside the thylakoid lumen.

The *pmf* is converted to a physical rotation as protons flow through a large protein complex in the thylakoid membrane called the coupling factor, which acts as a chloroplast ATPase. The coupling-factor complex uses rotation to bring adenosine diphosphate (ADP) and phosphate into close proximity, resulting in ATP synthesis. In this way, a physical force (rotation) is converted into a chemical force (ATP). Because the proton translocation needed for ATP synthesis is coupled to electron transport needed for $NADP^+$ reduction, each photon contributes to both reducing power and ATP synthesis.

2.2.4. Stoichiometries

During LEF, three protons are translocated for each electron. Each $NADP^+$ requires two electrons, and each CO_2 reduced requires two NADPH. This means eight photons are required to satisfy the reducing requirements for carbon fixation. This will simultaneously cause 12 protons to appear in the thylakoid lumen, presuming that the Q cycle described above occurs for each PQ used in electron transport. Depending on the stoichiometry of the subunits that make up the coupling factor, the number of protons per ATP can vary; if there are 12 subunits, then 12 protons will be required for one complete turn, which will result in three ATP. Thus, for LEF, production of two NADPH will be accompanied by production of three ATP, which is conveniently the number of ATP theoretically required for carbon fixation. However, the number of subunits may be as high as 14, the Q cycle may not be obligatory and some protons may leak out of the thylakoid lumen without making ATP. Also, ATP is required for other processes. It is generally considered that PET has a deficiency of ATP synthesis, which can be made up by several mechanisms.

The two mechanisms most often invoked for accounting for the ATP deficit are CEFI and the water-water cycle, or Asada cycle (Asada, 1999), involving the Mehler reaction and the donation of electrons to O_2 eventually reforming water, the original source of the electrons. The production of ATP that can accompany the water-water cycle has been called pseudocyclic electron flow or photophosphorylation as, like CEF, there is no net O_2 change, but unlike CEF involves both PSI and PSII. Generally, it is considered that only a small fraction of energy is devoted to the water-water cycle except in some cases, such as during light induction (Makino *et al.*, 2002). Different views about H^+/ATP stoichiometries and their implications for photosynthesis regulation are described in Chapter 3.

2.3. THE PATH OF CARBON IN PHOTOSYNTHESIS

The title of this section was the title of a book published soon after the major steps of photosynthetic carbon metabolism were worked out (Bassham and Calvin, 1957). The reactions of what became known as the 'Calvin cycle' have not changed in the intervening 50 years. Some writers acknowledge the contributions of J.A. Bassham and A. Benson, by calling the cycle the Calvin-Benson cycle, the Calvin-Benson-Basham cycle or the Benson-Calvin cycle. Still others avoid such concerns about credit by referring to it as the photosynthetic carbon reduction cycle (PCRC). All of these names refer to the same set of reactions outlined below.

CO_2 enters the Calvin cycle in just one location, but it can leave in many different forms. The ways in which carbon leaves the Calvin cycle will be discussed in section 2.3.2. Finally, the path of carbon that occurs when O_2 substitutes for CO_2 in the initial reaction of the Calvin cycle, called photorespiration, will be covered in section 2.3.3.

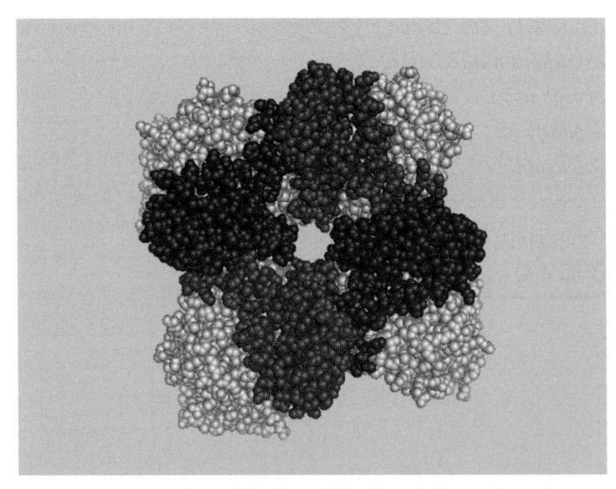

Fig. 2.3. End-on view of Rubisco. The four darker-coloured chains are the small subunits sitting on top of the lighter large subunit dimers. Another four small subunits would be found at the other end of the molecule. Rubisco is therefore sometimes referred to as an (L2)4 (S4)2 enzyme. Figure created in MacPyMOL using the 1IR1.pdb Rubisco crystal structure.

2.3.1. The classic Calvin cycle

The Calvin cycle consists of 13 reactions that reduce CO_2 to the level of a sugar. Evolutionarily, the Calvin cycle is associated with bacteria. Terrestrial plants have the Calvin cycle because of the bacterially derived endosymbiotic chloroplast. All of the steps of the Calvin cycle are found inside chloroplasts. The other ancient lineage in the tree of life, the archea, have many of the steps of the Calvin cycle and some appear to have complete Calvin cycles, though there is reason to believe this may have come about by horizontal gene transfer from bacteria to archea, and that the Calvin cycle evolved after the split between bacteria and archea (Delwiche and Palmer, 1996; Boucher *et al.*, 2003). A number of archea have alternative pathways to the Calvin cycle, but these often require anaerobic or microaerobic conditions and are not important to terrestrial photosynthesis (except in a few hot springs). Some archea have an incomplete Calvin cycle that they use by substituting other reactions (Finn and Tabita, 2004).

The central reaction of the Calvin cycle is the carboxylation of ribulose -1,5-bisphosphate (RuBP). Nearly all photosynthetic organisms use this reaction as the starting point for carbon reduction reactions. There are some exceptions among bacteria, one of the more notable is a reverse Krebs cycle. The carboxylation of RuBP is carried out by the enzyme Rubisco. The minimal active unit is a homodimer, found in some bacteria. In plants, algae and many bacteria, the enzyme exists as a hexadecamer of eight large subunits and eight small subunits. The large subunits occur as four dimers, and the small subunits lie in spaces between the dimers as shown in Fig. 2.3. In some archea Rubisco is made up of five dimers.

Rubisco is considered a slow enzyme, with each active site catalysing about three reactions per second. As a result, plants typically invest significant amounts of nitrogen into this one enzyme. It is not unusual for this one enzyme to comprise 20% of total leaf protein; no other leaf protein comes near to this concentration. Thus, carbon fixation by Rubisco is responsible for the large requirement for nitrogen and why nitrogen availability can determine ecosystem productivity.

Rubisco has a moderately low affinity for its substrate, CO_2, especially when the effect of O_2 is considered. As a result, plants that rely on diffusion for their CO_2 supply (i.e., C_3 plants) must allow significant amounts of gas exchange, inevitably leading to significant water loss. Thus, the properties of Rubisco also are responsible for the very large water usage by most plants.

The other reactions of the Calvin cycle are mostly the same as the nearly universal pentose-phosphate pathway (Table 2.1). This pathway is broken into two branches, the reductive branch, in which glucose6-phosphate (G6P)

Table 2.1. *The Calvin cycle. Reactions of the Calvin cycle with trivial names of enzymes indicated. Reactions are used as little as just one-third the rate of carbon dioxide uptake and as much as twice the rate. DHAP, dihydroxyacetone phosphate; E4P, erythrose 4-phosphate; FBP, fructose 1,6-bisphosphate; F6P, fructose 6-phosphate; GAP, glyceraldehyde 3-phosphate; 3-PGA, 3-phosphoglyceric acid; Ru5P, ribulose 5-phosphate; RuBP, ribulose 1,5-bisphosphate; SBP, sedoheptulose 1,7-bisphosphate; S7P, sedoheptulose 7-phosphate; Xu5P, xylulose 5-phosphate.*

Cycle step (Fig. 2.4)	Reactions per carboxylation	Name	Reaction
1.	1	Rubisco	CO_2 + RuBP → 2 (3-PGA)
2.	2	PGA kinase	PGA + ATP ↔ 1,3-bisPGA
3.	2	GAP dehydrogenase	1,3-bisPGA + NADPH ↔ GAP
4.	1	Triose Phosphate Isomerase	GAP ↔ DHAP
5.	1/3	Aldolase	GAP + DHAP ↔ FBP
6.	1/3	FBPase	FBP → F6P + P_i
7.	1/3	Transketolase	F6P + GAP ↔ Xu5P + E4P
8.	1/3	Aldolase	E4P + DHAP ↔ SBP
9.	1/3	SBPase	SBP → S7P + P_i
10.	1/3	Transketolase	S7P + GAP ↔ Xu5P + R5P
11.	2/3	Ribulose phosphate-3-epimerase	Xu5P ↔ Ru5P
12.	1/3	Ribose-5-phosphate isomerase	R5P ↔ Ru5P
13.	1	Phosphribulokinase	Ru5P + ATP → RuBP

is reduced and decarboxylated to ribulose-5-phosphate (Ru5P), and the non-reductive branch, in which Ru5P is changed into a number of other compounds important in metabolism (e.g., erythrose 4-phosphate) and/or fructose 6-phosphate (F6P), which can be used to regenerate G6P. The Calvin cycle is similar to the non-reductive branch of the pentose-phosphate pathway with one important difference, the Calvin cycle does not use transaldolase. The reactions of the Calvin cycle are needed in different amounts for each CO_2 fixed (Table 2.1) and it is difficult to draw out the reactions as a simple cycle (Figure 2.4).

2.3.2. The products of the Calvin cycle

Any intermediate of the Calvin cycle is a potential product. Several products play an important role in plants. Two products, triose-phosphate (TP) and F6P, are major products supplying sugar to the plant, whereas other products are important as sources for metabolic pathways inside the chloroplast.

CARBON EXPORT FOR SUGAR PRODUCTION

In terms of amount, by far the most significant product of the Calvin cycle is dihydroxyacetone phosphate (DHAP). This is a ketose form of TP and is in equilibrium with an aldose form, glyceraldehyde 3-phosphate (G3P). However, the ketose form is more stable, so there is over 20 times more DHAP than G3P at equilibrium. The TP transporter (TPT) (Flügge, 1999) in the inner membrane of the chloroplast envelope controls the passive exchange of DHAP for phosphate. The TPT will allow DHAP, G3P, phosphate and phosphoglyceric acid (in the –2 ionic form only) to freely exchange across the envelope in a strict one-for-one exchange. In this way, the TPT keeps the total amount of phosphate inside the chloroplast constant. The metabolite exchanged and the net direction is determined by the metabolism on either side of the membrane. Inside the chloroplast DHAP is normally produced in the Calvin cycle during the day. Outside the chloroplast DHAP is metabolised with the release of phosphate. Therefore, the net flux across the TPT is DHAP out, in exchange for phosphate in (Flügge and Heldt, 1991) (Fig. 2.5).

The second most important point in terms of amount, at which carbon leaves the Calvin cycle, is F6P. This is the starting point for starch synthesis. Phospho-gluco-isomerase catalyses the conversion of F6P to G6P, and G6P is outside the Calvin cycle. Although isomerases are often in equilibrium, the chloroplastic phospho-gluco-isomerase catalyses a

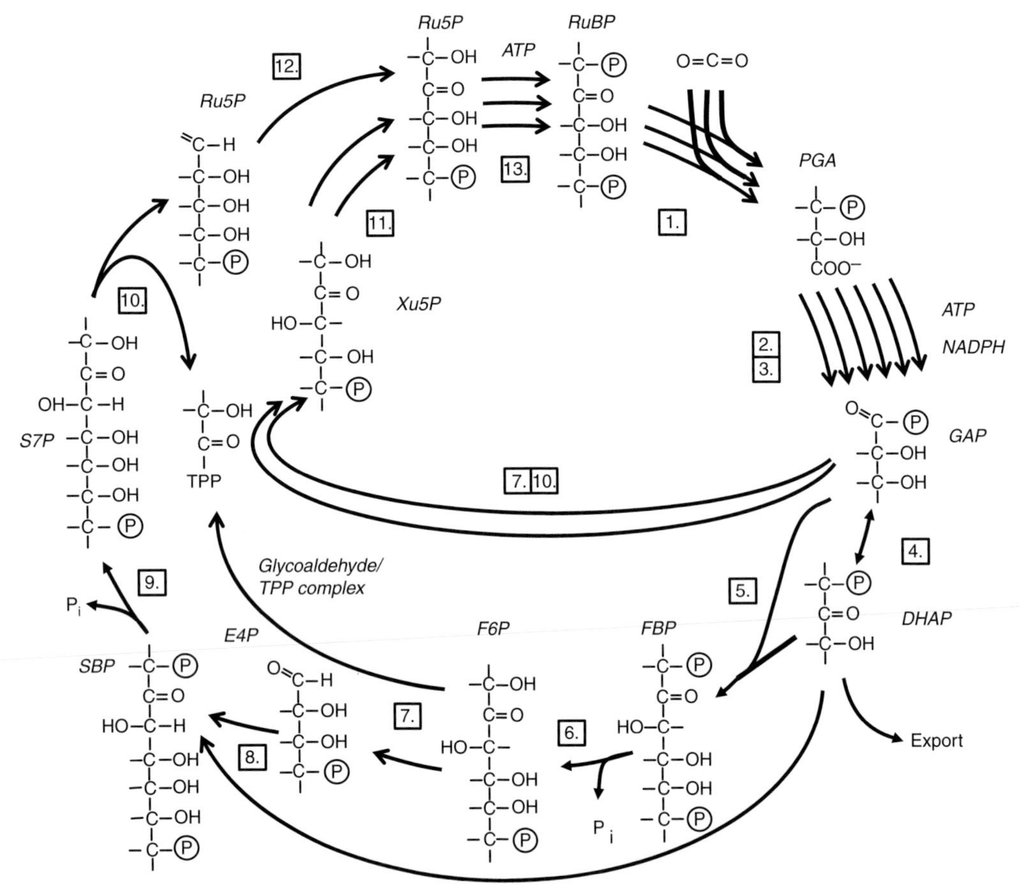

Fig. 2.4. The Calvin cycle. The numbers in boxes refer to the enzymes listed in Table 2.1. The number of arrows show how many reactions occur for fixation of three molecules of CO_2, the smallest number that does not require fractions of reactions. DHAP, dihydroxyacetone phosphate; E4P, erythrose 4-phosphate; FBP, fructose 1,6-bisphosphate; F6P, fructose 6-phosphate; GAP, glyceraldehyde 3-phosphate; 3-PGA, 3-phosphoglyceric acid; Ru5P, ribulose 5-phosphate; RuBP, ribulose 1,5-bisphosphate; SBP, sedoheptulose 1,7-bisphosphate; S7P, sedoheptulose 7-phosphate; Xu5P, xylulose 5-phosphate.

branch point and is highly regulated, especially by inhibition by PGA (Dietz, 1985; Kruckeberg *et al.*, 1989; Schleucher *et al.*, 1999). Starch buildup is typically very constant through the day giving a linear increase in the amount of starch in leaves from morning to night. Overnight the rate of starch degradation is also very constant and regulated so that the amount of transitory starch in leaves is nearly zero at the end of the night (Smith *et al.*, 2005). When measuring starch, it is crucially important to take the samples at the same time each day and at the end of the light period to provide the most information. Breakdown of starch at night can be hydrolytic (making maltose and glucose) for export or phosphorolytic (making glucose 1-phosphate) for metabolism inside the chloroplast (Weise *et al.*, 2006). The maltose is exported from the chloroplast through a specialised carrier (Nittylä *et al.*, 2004) (Fig. 2.5).

Sugars in the cytosol are exported from the leaf through the phloem (Fig. 2.6). Most of the carbon is converted to sucrose in the cytosol, although a number of plants make sugar alcohols such as mannitol and sorbitol. These sugars travel symplastically (through plasmodesmata) up to the veins. There is significant variation in the mechanism of loading sugars and sugar alcohols into veins. Most but not all plants use energy to concentrate sugars in the phloem, which leads to water uptake, and movement through the phloem to sink regions of the plant, where the sugars unload and the water leaves the phloem (Turgeon and Wolf, 2009). Many plants use sucrose-proton co-transport to

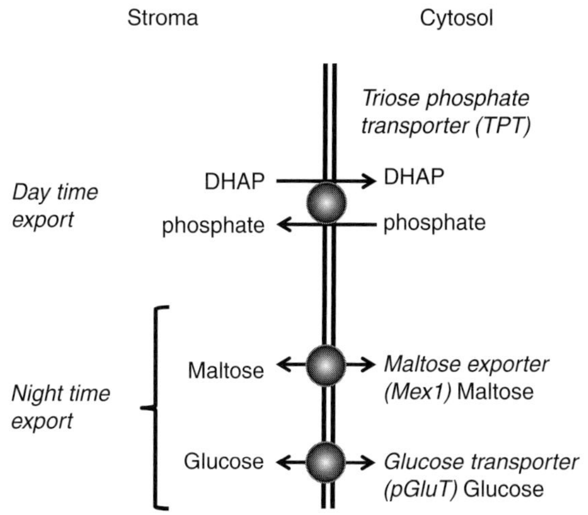

Fig. 2.5. Primary carbon export pathways from chloroplasts. During the day, dihydroxyacetone phosphate (DHAP) is the primary export product while phosphate is the primary import because of the metabolism in the stroma and cytosol. The triose phosphate transporter can transport DHAP, glyceraldehyde 3-phosphate, phosphoglyceric acid and phosphate. At night, DHAP is not available but maltose and, to a lesser degree, glucose are made from starch and exported.

actively accumulate sucrose, whereas a few use a polymer-trap method in which galactose is added to sucrose to make raffinose, stachyose and verbascose. If energy is not used, the concentration of sugar in the cells of the source regions has to be high enough to cause phloem transport. It is generally more efficient to actively accumulate sugars. High sugar levels in photosynthesising cells can result in signals that reduce expression of photosynthetic genes, but there is no known mechanism for high sucrose levels to feedback on photosynthetic rates over very short periods.

CARBON EXPORT FOR OTHER METABOLIC PATHWAYS

In addition to the Calvin cycle's role of producing sugars for the plant, chloroplasts and other plastids, are the sites of synthesis of other families of chemicals important to plant responses to the environment. Perhaps best known is the production of carotenoids, which occurs only in plastids. Plastids use GAP from the Calvin cycle and pyruvate (mostly imported from the cytosol as phosphoenolpyruvate (PEP)) to begin the methyl-erythritol 4-phosphate (MEP) pathway leading to isopentenyl diphosphate and dimethylallyl diphoshate (Lichtenthaler *et al.*, 1997). These can also be produced in the cytosol of plants using the mevalonic-acid pathway, but the bulk of carbon in carotenoids comes from the chloroplastic pathway. Subsequent reactions leading to carotenoid synthesis occurs only in plastids. The MEP pathway is also the source of carbon for diterpenes (C20), monoterpenes (C10) and isoprene (C5). Alternatively, sesquiterpenes (C15) are made in the cytosol using carbon from the mevalonic-acid pathway and so are not as closely related to the Calvin cycle.

The Calvin cycle is also one source of carbon for the synthesis of phenylpropanoids by the shikimic-acid pathway (Herrmann and Weaver, 1999). As with the MEP pathway, the shikimic-acid pathway is a cooperative venture drawing on carbon from the Calvin cycle (erythrose-4-phosphate) and the cytosol (PEP). Products of the shikimic-acid pathway include the essential amino acids phenylalanine, tyrosine and tryptophan. The shikimic-acid pathway is also the source of carbon for lignin, flower colours and fragrances, and benzoic acid and related compounds. Many of these play important roles in ecophysiology.

2.3.3. Photorespiration carbon pathway

Rubisco is a bifunctional enzyme. The 'co' of its name comes from carboxylase/oxygenase (Ogren, 2003). When Rubisco oxygenates RuBP, a series of reactions is started that consumes energy and releases CO_2. This series of reactions is called photorespiration, though it is not respiration in the sense of generating energy for the organism (Ogren, 1984). Photorespiration reduces photosynthesis in three different ways (Sharkey, 1988). First, when Rubisco is oxygenating RuBP, it is not carboxylating and so Rubisco is less efficient as a carboxylating enzyme when oxygenation occurs. Second, energy is consumed in the photorespiratory reactions, reducing the energy available to regenerate RuBP. Third, carbon previously converted into sugars is released as CO_2 in photorespiration, essentially the reverse reaction of photosynthesis.

Photorespiration goes faster at high temperature and is linearly related to the partial pressures of O_2 and CO_2. It is exceedingly difficult to measure photorespiration, but there is plenty of information from which to model rates of photorespiration. Although these models are only as good as the assumptions made, they are still generally much better than any method for measurement in whole leaves. In practice, for ecophysiology, photorespiration is modelled not measured. This makes models of photosynthesis (Chapter 8) very important to ecophysiology.

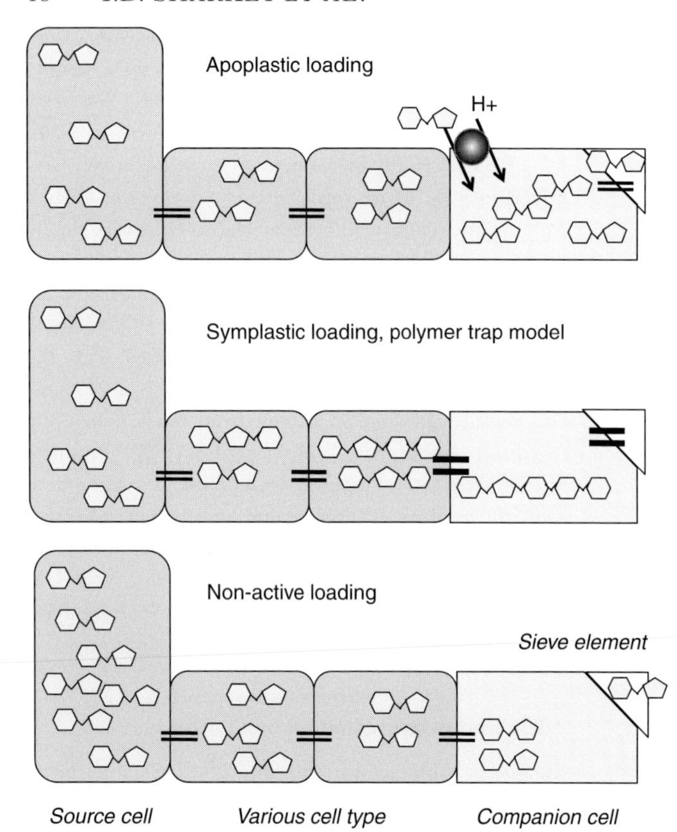

Fig. 2.6. Three types of phloem loading. The pentagons represents fructose, while the hexagons can represent glucose (on the left of fructose) or galactose (on the right of fructose). In apoplastic-loading species at least one cell interface has very few or no plasmodesmata so all sucrose must leave the cells. Sucrose is actively accumulated by a sucrose-proton co-transport mechanism, increasing its concentration for movement into the sieve element cell responsible for transport. In symplastic-loading species galactose is added to sucrose and the larger raffinose molecule cannot diffuse backward through the plasmodesmata just big enough to pass sucrose. Additional galactose molecules can be added to make stachyose and then verbascose. In non-active-loading species the sucrose concentration in the source cells must be high enough to cause phloem transport.

2.4. CARBON ACQUISITION SYSTEMS

It is likely that carbon dioxide was plentiful when photosynthesis evolved (see Chapter 24), but the success of photosynthetic organisms has resulted in long-term storage of a significant amount of carbon as limestone, oil and coal. As a result, the carbon dioxide availability is now a significant issue for most photosynthetic organisms. Most plants rely on diffusion of CO_2 from the air to sites within the cell where it is consumed, and plants have developed mechanisms to minimise resistance to this diffusion. However, other plants have developed mechanisms, such as biochemical and biophysical pumps for using energy to actively accumulate CO_2 against its concentration gradient.

2.4.1. Diffusion (C_3)

Most plants rely on diffusion of CO_2 for uptake from the environment. To balance the cost in terms of water loss that occurs when wet leaf surfaces are exposed to air to allow CO_2 diffusion with the need for CO_2, successful lineages of plants have stomata, adjustable pores in their leaves and other plant parts that open when CO_2 is needed and close when it is not, or when the water cost is too high, for example during drought. Leaves are designed to have lots of cell surface in contact with air, and so the cells of many leaves are very loosely packed allowing for extensive air passageways inside the leaf.

Because water vapour diffuses out of a leaf along the same path that CO_2 diffuses in, reliance on diffusion for carbon dioxide acquisition causes the plant to require a great deal of water. The gradient for water loss is typically 100-times greater than the gradient for CO_2 diffusion, so each CO_2 molecule that enters photosynthesis requires 100 times or more molecules of water. The amount of CO_2 fixed per water-vapour molecule (A_N/E) is often called the instantaneous photosynthetic water-use efficiency (WUE). A related parameter can be calculated by taking the amount of CO_2 fixed divided by the stomatal conductance. This is called the intrinsic WUE and can be used to compare plants at different humidities. The amount of carbon assimilated divided by the amount of water used by a crop can be calculated and this integrated measure of the water cost of carbon is WUE of productivity. An important consequence of reliance diffusion for CO_2 uptake is significant water usage by plants. The regulation of stomata opening is described in Chapter 3, the role of stomata in limiting photosynthesis during drought is described in Chapter 20, and WUE in photosynthesis is the subject of Chapter 33.

As CO_2 diffuses from the intercellular airspaces of the leaf through the mesophyll to Rubisco, it encounters a series of resistances such as the cell wall, the cell membrane, the chloroplast envelope membrane and the stroma, which also limit CO_2 diffusion and uptake. There is an increasing view that the restrictions to photosynthesis imposed by finite CO_2 diffusion inside the mesophyll are of similar importance to those imposed by finite stomatal conductance. The regulation of the so-called mesophyll conductance (g_m) is covered in Chapter 3, and methods for estimating g_m in ecophysiological studies are profusely described in Chapter 12.

2.4.2. Biochemical pump (C_4)

Declining CO_2 in the atmosphere became a critical problem for plants about 30 million years ago. This provided strong evolutionary pressure to overcome the limitations of relying on diffusion for CO_2 uptake. A number of plants developed metabolism that allows them to use energy to concentrate CO_2 in the same compartment as Rubisco. These systems generally have in common the carboxylation of PEP to make oxaloacetate (OAA) and then either malate or aspartate (Asp). Because these are both four-carbon acids the pathways were collectively called the C_4 pathway. By carboxylating PEP in one compartment and releasing that CO_2 by decarboxylation in a different cellular compartment, the concentration in the second compartment can be 10 times higher than in air. The C_4 plants can be subdivided according to which decarboxylation pathway they use. C_4 metabolism is covered in Chapter 5.

2.4.3. Storage (crassulacean acid metabolism)

Another way to actively accumulate CO_2 is to use two separate carboxylations, as in C_4 metabolism, but separate them in terms of time rather than space, as they are in C_4. This approach is called crassulacean acid metabolism (CAM). CAM requires storage of carboxylic acids during the night and so many CAM plants have fleshy leaves. CAM is covered in Chapter 6.

2.4.4. Biophysical pumps of algae and bacteria

CO_2 acquisition is a particular problem for organisms that grow in water. Once again, a range of solutions have evolved in cyanobacteria and algae that allow them to use chemical energy to actively accumulate CO_2 at the location it will be used by Rubisco (Badger *et al.*, 2000a). Once a CO_2-accumulating mechanism evolves, high resistance to CO_2 diffusion becomes advantageous. Thus, CO_2-accumulating metabolism versus reliance on diffusion for CO_2 uptake require opposite traits, and organisms almost never do both. To date, no land plants are known that have a biophysical inorganic carbon-accumulating mechanism. The solution for land plants appears to be biochemical, e.g., C_4 and CAM pathways.

3 • Photosynthetic regulation

C.H. FOYER AND J. HARBINSON

3.1. INTRODUCTION

The evolution of oxygenic photosynthesis played an important role in the oxygenation of the atmosphere of the Earth. This rise in O_2 also had an impact on the subsequent evolution of the photosynthetic organisms themselves, enabling them to develop more efficient bioenergetic systems. Thus, the reduction/oxidation (redox) reactions of the PET chain of green algae and higher plants and their regulation have become adapted to the O_2-rich atmosphere of the Earth. Oxygen is, however, not only a product of photosynthesis but it is also a regulator of PET-chain activity and photosynthetic metabolism. Molecular O_2 ($^3O_2$), is highly reactive and thus inherently toxic, but aerobic cells have evolved the ability to harness the energy potential of aerobic metabolism while minimising potentially harmful effects. Partly this is achieved by using the ROS, such as superoxide (O_2^-), hydrogen peroxide (H_2O_2) and singlet oxygen ($^1O_2$), formed as by-products of photosynthesis as important metabolic signals. The complex interactions of molecular O_2 with the cellular electron-transport and metabolic systems of the cell have become an intrinsic feature of plant redox regulation and homeostasis.

Coordination between energy producing and energy utilising processes is at the heart of the processes that regulate photosynthesis and ensure efficient functioning over a wide range of environmental conditions. Respiration works alongside photosynthesis to secure efficient biological energy production in plant cells. However, unlike the regulation of respiration, which is driven by metabolic substrates that are protected from depletion by effective control mechanisms, the driving force for photosynthesis is the free energy of light, a substrate that cannot be conserved except through light harvesting, efficient charge separation and electron transport. The efficiency of the conversion of the free energy of light into chemical free energy by photosynthesis has been optimised during evolution. A complex network of defence systems protects photosynthesis against the potentially harmful effects of excess light, i.e., light capture that is in excess of the amount that can be used to drive photosynthesis. Efficient dissipation mechanisms are available to protect the photosynthetic membranes and their protein and pigment components by releasing the energy absorbed from light, largely as heat. Photosynthesis is thus able to operate in a highly flexible manner, harvesting energy efficiently at low irradiances and dissipating excess energy at high irradiance. The regulation of the photosynthetic activity of a leaf is the result of a hierarchy of processes, ranging from short-term regulatory adjustments, through mid-term acclimatory responses, to long-term evolutionary and epigenetic changes. A key factor in these interactions is the genome-environment relationship, where many environmental factors, such as irradiance, temperature, CO_2 and water supply, influence the short-term regulation acclimation and ultimately the evolution of photosynthesis.

The PET chain not only produces a strong oxidant (molecular O_2, which is in the triplet ground state), but also strong reductants in the forms of reduced Fd and NADPH, which drive biosynthetic reactions in the chloroplast. The major consumer of NADPH in photosynthesis is the reductive pentose-phosphate pathway or Calvin cycle, in which CO_2 from the atmosphere is assimilated and reduced to the level of sugar phosphates that form the substrates or skeletons for the enormous range of compounds produced in plants. The various sugar phosphates produced by the Calvin cycle are principally used to make carbohydrates such as starch and sucrose, and other quantitatively minor products, such as amino acids. The thermodynamic irreversibility of photosynthesis necessitates precise regulation of output from the Calvin cycle, principally through regulation of the pathways of sucrose and starch synthesis. As in

Terrestrial Photosynthesis in a Changing Environment: A Molecular, Physiological and Ecological Approach, ed. J. Flexas, F. Loreto and H. Medrano. Published by Cambridge University Press.

all irreversible pathways, however, much of the overall control of photosynthesis resides at the point of input, which lies at PSI and PSII.

Photosynthetic carbon assimilation (PCA) drives plant growth and biomass accumulation. An appropriate continuous supply of sucrose from the source organs is essential to support the growth of sink organs, which are net importers of photoassimilate. The diurnal turnover of starch in the leaves is important in matching carbon supply to carbon demand in the whole plant. Control of flux through the Calvin cycle is intimately linked to regulated PET activity and to stomatal functioning (Fig. 3.1). Flux through the Calvin cycle is not only determined by the availability of the products of the PET chain and the availability of CO_2 in the atmosphere but also the ability of the sink tissues to import and use photoassimilate (Fig. 3.1). In the following discussion, we present the current concepts of photosynthetic regulation, focusing on redox systems and the complex interactions with molecular O_2 that are crucial to both short- and long-term control. We also consider the regulation of CO_2 diffusion and carboxylation, particularly in relation to leaf structure and acclimatory control.

The water-splitting system of PSII undertakes the concerted four-electron oxidation of water, a process that is reversed in aerobic respiration, where $^3O_2$ is used as the terminal acceptor and reduced to water. However, many processes in plants catalyse the partial reduction of O_2 and so produce superoxide, H_2O_2 and hydroxyl radicals, all of which are more reactive than ground-state triplet O_2 (Foyer and Noctor, 2009). Although less reactive than the hydroxyl radical (•OH) singlet state ($^1O_2$) is more reactive than O_2^- or H_2O_2, and is largely responsible for chloroplast lipid peroxidation (Triantaphylidès *et al.*, 2008). It has long been considered that $^1O_2$ had only a very limited diffusion. However, recent evidence suggests that $^1O_2$ can apparently diffuse significant distances from the site of production (Fischer *et al.*, 2007).

Oxygen is also a substrate for the oxygenase activity of Rubisco. Oxygenation of RuBP by Rubisco produces 2-phosphoglycolate in addition to 3-phosphoglycerate. Phosphoglycolate is metabolised to 3-phosphoglycerate through a sequence of reactions that are together called 'photorespiration'. This pathway includes H_2O_2 production by glycolate oxidase, so photorespiration is a high-flux pathway of H_2O_2 generation. The relationship of photorespiration to photosynthesis and cellular redox signalling are discussed in Chapter 4 (see also Foyer *et al.*, 2009).

Plant cells contain large amounts of low-molecular-weight antioxidants that work in conjunction with antioxidant enzyme systems to prevent ROS accumulation (Foyer and Noctor, 2009). Low-molecular-weight antioxidants can be defined as molecules that are able to reduce oxidants without themselves having significant pro-oxidant action. Numerous classes of plant cell compounds can act as antioxidants. Of these, ascorbate, glutathione, tocopherols, carotenoids, polyamines, flavonoids and related phenylpropanoid derivatives are all have known to serve antioxidant functions. In addition to peroxiredoxins, the antioxidant functions of ascorbate, glutathione and tocopherol are particularly important to the regulation of photosynthesis (Foyer and Noctor, 2009; Foyer *et al.*, 2009). As well as being a substrate for the enzyme ascorbate peroxidise (APX), ascorbate is a cofactor in the xanthophyll cycle, and is important in the production of zeaxanthin, which participates in the thermal-energy dissipation NPQ processes as discussed below. Light quantity and quality are important determinants of leaf ascorbate contents. The PET chain influences ascorbate synthesis and accumulation (Yabuta *et al.*, 2007). Similarly, mitochondrial electron transport is also important in the control of ascorbate synthesis (Millar *et al.*, 2003).

3.2. REGULATION IN THE PHOTOSYNTHETIC ELECTRON TRANSPORT SYSTEM

Sustained flux through the redox-active carriers of the PET chain demands that the species be present in both oxidised (acceptor) and reduced (donor) forms, and this effect is known as redox poising. Both over-reduction and over-oxidation of PET carriers prevents the smooth running of photosynthesis, necessitating homeostatic mechanisms that regulate cellular redox potential. Moreover, electron transport and metabolic processes have to be balanced with respect to the supply and demand for ATP and reductant. In natural environments there are many situations where this balance will be perturbed. For example, a limitation in the supply of CO_2 will result in a reduction in PCA, which will result in a decreased demand for ATP and reductant. Similarly, restrictions in sucrose synthesis or transport caused by stresses, such as low temperatures or drought, can result in decreased photosynthetic carbon-assimilation activity. In response to such imbalances between supply and demand, regulatory mechanisms function either to increase demand or to dissipate the excess energy available to the system. These regulatory mechanisms usually act by adjusting the rate constants (i.e., the kinetic constraints) of specific processes. In the absence of regulation to accommodate a

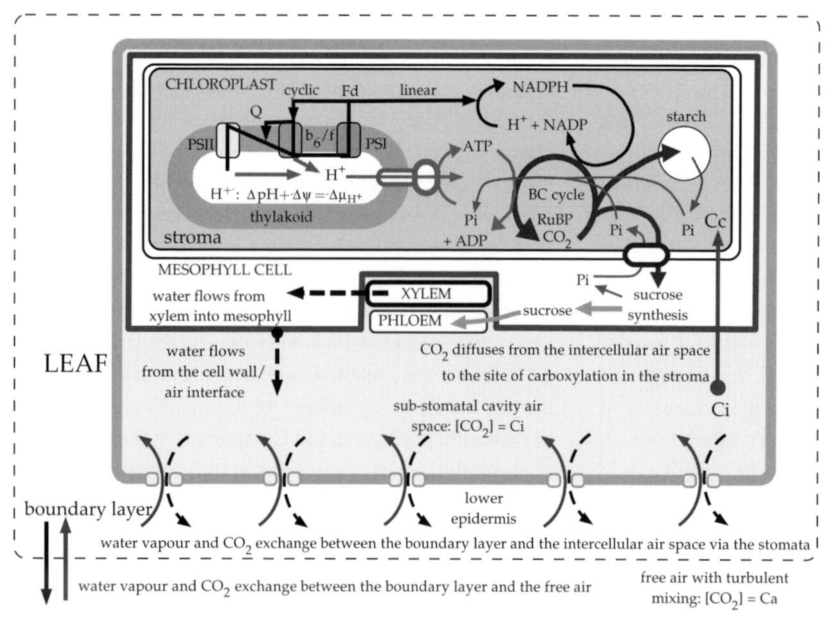

Fig. 3.1. A simple description of the major components involved in the regulation of photosynthesis in C_3 leaves. Light-driven electron transport through photosystem II (PSII), the cytochrome b_6/f complex (b_6/f) and photosystem I (PSI) leads to the reduction of ferredoxin (Fd), which is followed by the reduction of NADP. Proton transport, enhanced by the function of a Q cycle (Q), into the lumen is coupled to this electron transport activity. In addition to the linear path, some electrons can follow a cyclic path around PSI that does not result in the net formation of reductant but that does translocate protons into the thylakoid lumen. Various cyclic electron pathways have been proposed, of which only one (ferredoxin reducing the cytochrome b_6/f complex) is shown. The pumping of protons into the thylakoid lumen generates a transthylakoid membrane proton potential difference ($\Delta\mu_H{}^+$) that is comprised of two components: a pH difference (ΔpH) and a voltage difference (Δψ). The pH difference is significant as the intrathylakoid pH is involved in various regulatory responses of the thylakoid membrane. Proton efflux from the lumen in response to the $\Delta\mu_H{}^+$ drives the formation of ATP by the thylakoid ATPase. The ATP and NADPH drive the continued fixation of CO_2 by the Calvin cycle (C cycle), a process that also regenerates ADP and NADP, that are substrates for the light-driven processes of the thylakoid. The products of Calvin cycle are sugar phosphates (the phosphate groups of which are derived from ATP), and inorganic phosphate (P_i) – a substrate for ATP formation. These are used to either to make starch in the chloroplast or are exported from the chloroplast to make sucrose; both these processes release P_i. In the case of sucrose formation, the P_i is exchanged across the chloroplast membrane for the exported sugar phosphates. However, if sucrose export rates are slow then the release of P_i from the sucrose synthesis pathway can be decreased and the export of carbon from the chloroplast is decreased in turn. In addition to the regeneration of the substrates for ATP and NADPH synthesis, sustained photosynthesis also requires the replacement of CO_2 fixed by ribulose-1,5-bisphosphate by CO_2 from the free, mixed air surrounding the leaf. Carbon dioxide must diffuse from the free air, through the boundary layer, the stomata, the intercellular air spaces and finally through the mesophyll cell to the site of carboxylation in the chloroplast. This process requires a concentration gradient of CO_2 such that $C_a > C_i > C_c$. The diffusion path for CO_2 into the leaf creates a path for the diffusion of water vapour out of the leaf. This lost water needs to be replaced to prevent the leaf from desiccating: this replenishment occurs via the xylem.

change in demand, two things would happen to flux through the PET chain. Initially electron transport and proton translocation will continue until either $\Delta\mu_H{}^+$ or the redox potential of a critical electron transport component is increased to the point where the rates of back-reactions are sufficient to bring the net rate of forward reactions into balance with metabolic demand. In addition, the rate of side-reactions could increase, permitting the forward reactions to continue but coupled to different sinks. These back and side reactions can produce ROS that are potentially toxic, but which also act as powerful signalling molecules. An important side reaction in the electron transport chain is a process called pseudocyclic electron flow or the Mehler reaction. In the Mehler reaction, electron acceptors such as Fd that have a sufficiently low mid-point potential, transfer electrons directly to O_2, producing O_2^-, which is further metabolised to H_2O_2 by the action of SOD. When this reaction is coupled to the ascorbate-glutathione cycle the reaction sequence is termed the 'water-water cycle', or the 'Mehler-peroxidase' reaction.

3.2.1. Over-reduction and oxidation and the regulation of photosynthesis

The concept of reductive stress involves either: (1) over-reduction of redox-active carriers or metabolites that then transfer electrons to O_2 favouring ROS production; or (2) modifications of protein function through a drop in the redox potential of pyridine nucleotides, thioredoxins or glutathione. As an acceptor for electrons, O_2 is a key player in preventing bottlenecks to electron flow in the PET and respiratory electron transport (RET) chains. Over-reduction of electron transport components favours production of ROS, such as $^1O_2$, O_2^- and H_2O_2. Of these, $^1O_2$ produced by PSII is the most damaging, as it is responsible for over 80% of the non-enzymatic lipid peroxidation in leaves (Triantaphylidès *et al.*, 2008) and is believed to be a major cause of damage to the *RC* of PSII (Telfer, 2005).

Superoxide production at the reducing side of PSI plays a crucial role in the poising of carriers of the PET as electron transport to oxygen electron acts as an overflow system. Moreover, in the context of cell-survival signalling, O_2^- production on the reducing side of PSI may serve to induce defence-gene expression and so offsets the influence of $^1O_2$-induced cell-death signals. Singlet oxygen is produced within PSII (Foyer and Harbinson, 1994; Telfer, 2005; Foyer *et al.*, 2006a,b), whereas O_2^- is largely formed at the high potential end of PSI in the PET chain. It is therefore not surprising that they tend to have an inverse action on gene expression (Apel and Hirt, 2004).

3.2.2 The phenomenology of short-term regulation *in vivo*

Components that are central to the integrated regulation of the PET chain and carbon assimilation through the Calvin cycle are illustrated in Figure 3.2. The modulation of electron- and proton-transport processes by back reactions or side reactions is a relatively unsophisticated strategy for regulation. Although this type of limitation would work in theory, it does not occur to any extent under steady state conditions *in vivo*, except perhaps in stress situations. The regulatory network that controls electron- and proton-transport fluxes and brings the photosynthetic formation of reductant and ATP into balance with the demands of metabolism *in vivo* is called 'photosynthetic control' (Foyer *et al.*, 1990). In non-stress conditions the lack of effective photosynthetic control is observed *in vivo* for only brief periods during the early stages of photosynthetic induction (i.e., the start-up of photosynthesis in the light from the dark-adapted state). Fluxes through the PET and respiratory electron transport (RET) systems are subject to adjustments that also affect the parallel process of proton transport (Avenson *et al.*, 2005a,b). This adjustment forms the basis of the photosynthetic control mechanisms. These are short-term (response time of seconds) adjustments that act by controlling the rate constant of the reaction between plastoquinol (reduced PQ) and the cyt b_6f complex and the rate of proton efflux from the lumen via the ATPase. Longer-term regulatory adjustments are also employed and involve changes in the pool sizes of electron transport components.

Chlorophyll fluorescence (Chl-F) is often used to measure the budget of light-use efficiency of PSII (Hendrickson *et al.*, 2004a; Kramer *et al.*, 2004a; Baker *et al.*, 2007; see Chapter 10) and electron transport by PSII. In the context of regulation, the most relevant parameters are qP, also known as qQ or $F_q'/'$ (Baker *et al.*, 2007), and the steady state F_v'/F_m'. It is important to note that the Q_A redox state is non-linearly related to qP, which is strictly an estimate of the probability that a Chl* in PSII will be quenched by photochemistry, i.e., that it will encounter and produce a charge separation in an open *RC*. The term qP is sometimes treated as a direct measure of the Q_A redox state, and though this may be qualitatively acceptable, it is inappropriate for a quantitative analysis. The redox state of Q_A ($Q_A(ox)/Q_A(total)$) can, however, be calculated from qP (Kramer *et al.*, 2004a).

Another important fluorescence parameter is F_v'/F_m'. This parameter is often used as a measure of the quantum yield for charge separation in open PSII traps (Baker *et al.*, 2007). Along with other analogous parameters, such as qN and qE, the F_v'/F_m' parameter is an important measure of the development of NPQ (see Section 3.2.3) of the Chl* in PSII. The efficiency of PSII charge separation (commonly abbreviated as Φ_{PSII}, F_q'/F_m', or $\Delta F/F_m$) is the product of qP and F_v'/F_m' (Baker *et al.*, 2007). Note that no process comparable with NPQ exists in PSI. In the following discussion, we will concentrate on the mechanistic details of the NPQ and regulation of electron transport by the cyt b_6f complex. We will not consider other possible PET limitations, such as the diffusion of PQ and plastocyanin or the possible relevance of membrane crowding by proteins (percolation) for these processes.

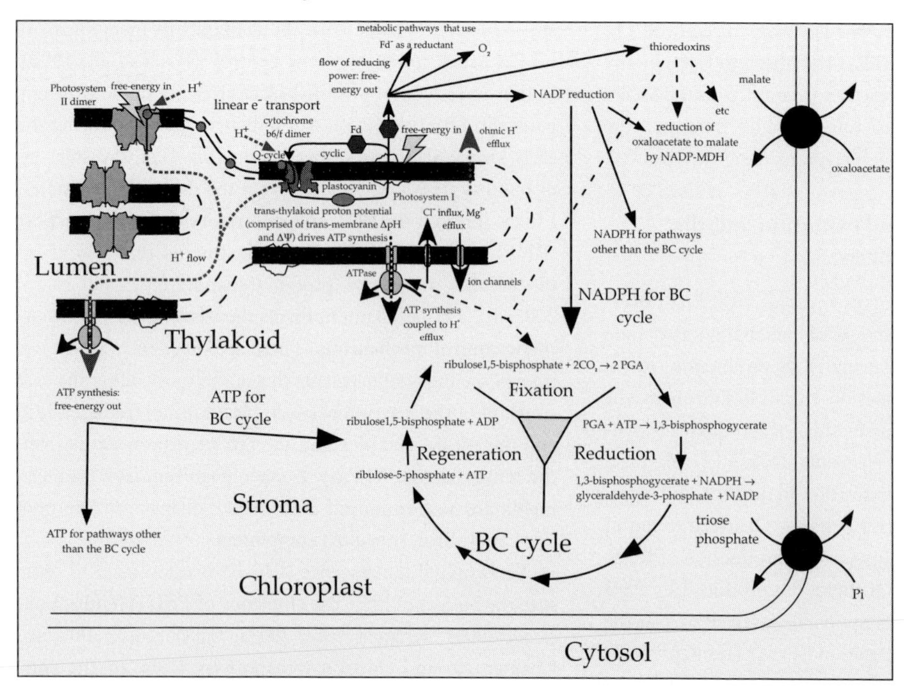

Fig. 3.2. Producer-consumer relationships between the thylakoid photosynthetic electron transport system and carbon metabolism in the chloroplasts. Reduced ferredoxin, produced by the electron transport chain, can be re-oxidised by several mechanisms; it can reduce O_2 (one possible mechanism for pseudocyclic electron transport), it can reduce thioredoxin, it can be used directly as a reductant by metabolic pathways (such as nitrite reduction) and it can reduce NADP (normally this is the predominant use for reduced ferredoxin). The thioredoxins are an important class of proteins that reductively activate and inactivate many chloroplast enzymes. Of these the reductive activation of the ATPase, enzymes of the Calvin cycle (C cycle), and NADP-MDH is noteworthy. Though NADPH is used predominantly as a reductant for the C cycle, it is also used as a reductant by other metabolic pathways in the stroma, such as the reduction of oxaloacetate to malate by NADP-MDH. The malate formed in this reaction can be exported out of the chloroplast in exchange for oxaloacetate, which itself is the product of malate oxidation in the cytosol – by this means, reductant can be exported from the chloroplast. ATP formed from ADP and inorganic phosphate (P_i) is, like reduced ferredoxin, used largely by the C cycle, though some is used by other pathways. The major metabolic process of the chloroplast is the C cycle. This complex pathway is shown in an abbreviated form; it functions to fix CO_2, reduce the carboxylic acid products of that reaction to ketones/aldehydes (the oxidation state of sugars), and then regenerate the ribulose-1,5-bisphosphate substrate for the carboxylation reaction. These processes require ATP and NADPH. The products of the C cycle are NADP, ADP and P_i, which are substrates for ferredoxin oxidation and the ATPase, and triose phosphate and fructose-6-phosphate. The triose phosphate is exported from the chloroplast to feed the process of sucrose synthesis in the cytosol, whereas the fructose-6-phosphate is used for starch synthesis in the chloroplast; both these process releases P_i. The P_i released during sucrose synthesis is exchanged for triose phosphate by a transporter in the inner chloroplast membrane. The release of P_i by metabolism of the C cycle, and sucrose and starch synthesis provides the P_i necessary for ATP synthesis from ADP and P_i.

The redox state of electron donors/acceptors (including reaction-centre components such as Q_A) in the PET chain can be measured by means of light-induced absorbance changes. The principle of this class of techniques is simple: reduction or oxidation of an electron acceptor (or donor) in the PET chain is accompanied by changes in its absorbance spectrum (for details see Chapter 10). These changes are often sufficiently specific and intense to offer a useful means for measuring the oxidation state, or changes in the oxidation state, of that acceptor under a given treatment, such as steady state illumination or a flash. In the case of PSI, whose light-use efficiency cannot be probed by means of fluorescence, light-induced absorbance changes in the near-infrared (NIR) offer a powerful means for measuring the

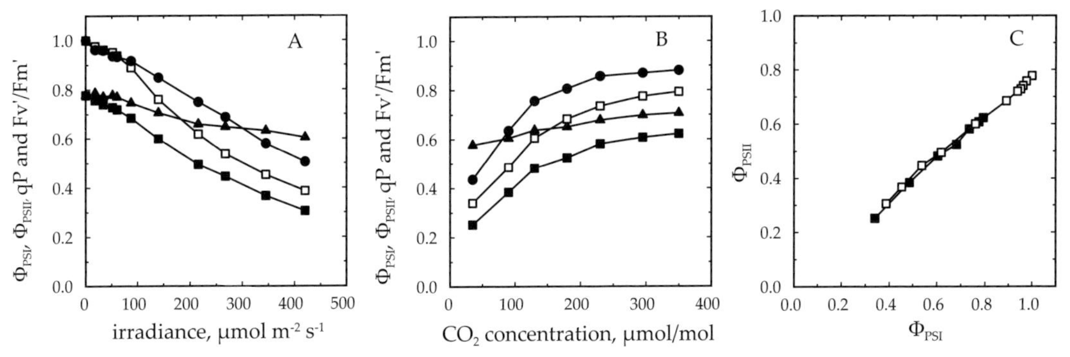

Fig. 3.3. Light (A) and CO_2 (B) response curves for photosynthesis and relationships between the efficiencies of the photosystems (C) in an attached leaf of the epiphytic shrub *Juanulloa mexicana*. The response of the quantum yield for charge separation by photosystem I (Φ_{PSI}, □) and photosystem II (Φ_{PSII}, ■), and q_P (●) and F_v'/F_m' (▲) were measured in relation to an increasing irradiance in an atmosphere of 360 µmol/mol CO_2, 2% O_2 and 98% N_2 (A) or an increasing CO_2 concentration (B) in an atmosphere of 2% O_2 and 98% N_2 and an irradiance of 140 µmol m^{-2} s^{-1}. F_v'/F_m' (the quantum yield for charge separation by open PSII reaction centres) is proportional to the quenching of PSII by non-photochemical quenching and it is used here rather than NPQ because the product of F_v'/F_m' and q_P gives Φ_{PSII}, which makes it easier to understand changes in Φ_{PSII} in terms of changes in q_P and and non-photochemical quenching. Note that with either increasing irradiance or decreasing CO_2 concentration the quantum yields of both photosystems decrease, as do the values of q_P and F_v'/F_m'. C. Whether the decrease in the photochemical efficiencies of either photosystem is produced by an increasing irradiance (□) or a decreasing CO_2 concentration (■) the relationship between the efficiencies of the photosystems is constant and linear, implying a predominant role for linear electron transport in this leaf under conditions of increasing irradiance or during downregulation produced by decreasing concentrations of CO_2 in largely non-photorespiratory conditions.

oxidation state of P700 and therefore the light-use efficiency of PSI (Baker *et al.*, 2007).

A consistent pattern of changes has been observed in the absorbance and fluorescence patterns of photosynthetic tissues, such as leaves, algae or chloroplasts, that points to a consistent mode of regulation. During irradiance the thylakoid lumen becomes more acid, owing to proton transport coupled to PET activity. The transport of electrons also results in a varying degree of reduction of the Q_A and PQ pools, whereas the cyt f, plastocyanin and P700 pools tend to become oxidised, relative to the equilibrium dark-adapted state. These changes in the redox state underlie the changes in certain photosynthetic parameters produced by increasing irradiance. Decreases in qP are caused by increases in the reduced form of Q_A (Fig. 3.3A; Baker *et al.*, 2007) and this accounts for part of the decrease in Φ_{PSII} observed with increasing irradiance, the rest being a result of the increase in NPQ (Fig. 3.3A) (Baker *et al.*, 2007 and see Section 3.2.3). The increase in P700 oxidation results quantitatively in the decrease in Φ_{PSI} observed in response to increasing irradiance (Fig. 3.3). Reduction of the acceptor side of PSI (i.e., the PSI reaction components that transfer electrons to Fd) appears not to limit charge separation in the PSI RC except under certain circumstances (e.g., photosynthetic induction, low CO_2 and O_2 concentrations, and in certain genetically modified organisms). Evidence from absorbance and fluorescence measurements implies that electron transport is subject to a major limitation at the PQ oxidation step, which has been shown from in-vitro measurements to be the rate-limiting step of the PET chain (Haehnel, 1984; Heber *et al.*, 1988).

The organisational coordination of PET and carbon metabolism is perhaps best demonstrated in the A_N/C_i curves of photosynthesis, which typically show two zones or phases of limitation (Farquhar *et al.*, 1980a). The first is observed at low C_i values and corresponds to a limitation by Rubisco. The second phase is observed at high C_i values and corresponds to a limitation of photosynthesis by RuBP regeneration. In many situations, the C_i value at the transition point between these zones occurs at the same C_i value as that present in the leaves under the growth environment conditions. Similarly maximal photosynthesis rates are generally proportional to the leaf Rubisco and cyt b_6f contents (Onoda *et al.*, 2005; Yamori *et al.*, 2005). The presence of correlations between the rates of CO_2 fixation in the leaves measured *in vivo* and the rates of electron transport measured *in vitro* in thylakoids isolated from these leaves implies a complexity of regulation that is based at the level of chloroplast structure and composition (Evans and Terashima, 1988).

Evidence of physiological regulation comes from demonstrations of adjustments in the kinetic parameters that determine the fluxes through the various proton and electron pathways of the chloroplast, for example in response to alterations in the light availability or in the metabolic demand for reductant and ATP. An important predicted consequence of this regulation is that rate constants should be variable. The measured rate constants for PET processes in leaves subjected to increasing irradiance, for example that for $P700^+$ reduction by electrons from the PQH_2 pool (Fig. 3.2), often do not change (Harbinson and Hedley, 1989). The observed changes in the reduction of the Q_A (and PQH_2 pool) and oxidation of the high potential end of the electron transport chain (i.e., from cytf to P700) are therefore the consequence of an increasing electron flux through the PET chain with constant kinetic limitations. However, changes in the short-term regulation of the photosynthetic energy transduction system can be easily observed in certain situations, for example when carbon metabolism limits photosynthesis. This can be demonstrated experimentally in a 2% O_2 atmosphere where photorespiration is largely suppressed. Under these conditions, the same pattern of PET redox changes is observed as that found in leaves illuminated in air; the Q_A pool becomes more reduced, resulting in decreases in qP, whereas the PSI end of the electron transport chain becomes more oxidised, resulting in decreases in Φ_{PSI} (Fig. 3.3B). In this case, however, redox changes are parallelled by changes in the rate constant for electron transport between PQH_2 and the cyt b_6f complex (commonly measured as the rate constant for electron transport between PQH_2 and the $P700^+$ pool (see Genty and Harbinson, 1996)). Electron transport is therefore subject to short-term regulation in response to a decrease in metabolic activity, and this regulation functions at the level of cyt b_6f complex. The redox poise of the PQ pool and the cyt b_6f complex is not only important in the short-term regulatory mechanisms, but is also instrumental in initiating signal cascades that regulate gene expression (Khandelwal *et al.*, 2008; Pfannschmidt *et al.*, 2009).

The key to short-term regulation of electron transport is the pH sensitivity of the reaction between plastoquinol and the Rieske FeS of the cyt b_6f complex. The mechanism of this pH sensitivity, which is found not only in cyt b_6f but in cyt bc_1 complexes in general, has been explained by the need of PQH_2 to form an H-bond with a specific histidine molecule on the Rieske FeS protein adjacent to the binding pocket for the PQH_2 molecule. The formation of an H-bond between a hydroxyl-group proton of the quinol head group of PQH_2 and the lone pair on a nitrogen in the imidazole ring of histidine, is the first step in the oxidation of the bound PQH_2 (Crofts *et al.*, 1999). If the lone pair is protonated this H-bond cannot form. The measured pH optima for the reduction of cyt f lie in the range of pH 6.5–7.25 (Tikhonov *et al.*, 1984; Nishio and Whitmarsh, 1993; Hope *et al.*, 1994; Finazzi, 2002), the highest value being from the alga *Chlamydomonas*. The pH sensitivity of the PQH_2/cyt b_6f reaction coupled with decreases in lumen pH offers a model for the regulation of linear (otherwise called non-cyclic) and cyclic electron transport, which explains the observed changes in the oxidation or reduction of components of the chain during regulation. This model, however, does not explain the mechanism that brings about the changes in lumen pH required for adjustment of the kinetics of the PQH_2 /cyt b_6f.

3.2.3. The relevance of non-photochemical quenching of PSII chlorophyll fluorescence

The chl. *a* fluorescence parameters, F_0 and F_m, decrease relative to the dark-adapted state when leaves are illuminated. This decrease is owing to NPQ processes, of which there are three major components: a rapidly reversible component called qE, a slowly reversible component termed qI and a component with intermediate kinetics denoted as qT (Quick and Stitt, 1989; Walters and Horton, 1991). The high rate constants for quenching of Chl* by qE and qI allows them to efficiently compete with photochemistry and fluorescence (Baker *et al.*, 2007; Baker, 2008); the action of qT is different.

The dominant NPQ component observed in the absence of stress as irradiance is increased or photosynthesis is down-regulated by, for example, low temperatures or diminished CO_2 supply, is qE. This 'energy dependent quenching' process depends on the lumen acidification produced by electron transport and also upon the presence of the PsbS protein of PSII. Protonation of the PsbS protein occurs in response to lumen acidification and this is considered to trigger qE (Li *et al.*, 2004; Horton *et al.*, 2005). In addition, the formation of the xanthophyll pigment, zeaxanthin, that occurs by the de-epoxidation of violaxanthin by violaxanthin de-epoxidase, is also induced by lumen acidification (Kramer *et al.*, 2003). The binding of a fraction of the zeaxanthin pool to an as yet unknown protein facilitates the quenching process induced by lumen acidification (Ruban *et al.*, 2002), resulting in enhanced quenching at higher (though still acidic) lumen pH values. Despite these findings, much uncertainty remains regarding the qE quenching mechanism and several models have been proposed. One model suggests that the formation of a protonated PsbS protein with bound zeaxanthin causes a structural change in the light-harvesting components of PSII that produces a quenching centre based upon the formation of a lutein-chl. *a* pair (Horton *et al.*, 2005; Ruban *et al.*, 2007). In the alternative model zeaxanthin plays a more direct role in the quenching process (Holt *et al.*, 2005; Bode *et al.*, 2009). In this model, bound zeaxanthin is oxidised by an adjacent, excited chl. *a* molecule to form a cation-anion pair that undergoes a recombination process, whereby the energy of the chl* is quenched and lost as heat via zeaxanthin, and chl. *a* returns to the ground state.

The slowly reversible component qI was originally equated with photoinhibition, and thus the slow relaxation of qI was attributed to PSII repair processes (Krause, 1988; Quick and Stitt, 1989; Walters and Horton, 1991; Aro *et al.*, 1993). However, the definition of qI has since been extended to cover other slowly relaxing quenching mechanisms (Demmig-Adams and Adams, 2006). Some plants are able to form a slowly reversible downregulated PSII state that requires zeaxanthin but not PsbS, even though PsbS-like proteins may still be involved. This slowly reversible form of NPQ may also be associated with the degradation of the PSII *RC* in stress situations (Demmig-Adams and Adams, 2006). The model of qE developed by Horton and co-workers incorporates a slowly reversible qI state, where zeaxanthin is bound but there is no conformational change produced by PsbS (Horton *et al.*, 2005). Much remains to be resolved concerning the precise mechanism(s) responsible for the qI and qE states and how these states interact *in vivo*. However, literature data suggest that qE and qI are diverse processes both mechanistically and physiologically.

The form of quenching with intermediate kinetics called qT is generally attributed to state transitions. This mechanism of energy transfer between the photosystems involves transfer of a portion of the LHCII that is associated with PSII in the dark-adapted state, to PSI in the light-adapted state (Allen, 1992). As a consequence, the excitation-energy input into PSII is decreased, and F_m or F_m' is lowered. Unlike qE and qI however, qT does not reduce the efficiency of charge separation by open PSII centres. Rather, it decreases F_m' by channelling energy to PSI instead of PSII, diminishing the F_0 or F_0' fluorescence signal from PSII. The effects of state-transitions on Chl F observed in isolated thylakoids also depend upon the Mg^{2+} concentration of the medium.

The redistribution of LHCII from PSII to PSI is related to the phosphorylation state of the LHCII proteins. In the absence of phosphorylation the LHCII protein binds predominantly to PSII, whereas phosphorylated LHCII proteins are less likely to associate with PSII but can bind reversibly to PSI, increasing the mean PSI antenna size (Allen and Forsberg, 2001; Wollman, 2001). LHCII phosphorylation is facilitated by the action of one or more protein kinases, particularly the STN7 and STN8 kinases (Rochaix, 2007). These protein kinases not only serve to adjust photosystem stoichiometries in the short term so that they are consistent with the light environment, but the kinase activities also trigger signalling cascades that orchestrate re-adjustments in the expression of genes encoding thylakoid proteins and so achieve an appropriate overall composition of LHC in the longer term (Rochaix, 2007).

The thylakoid protein kinases are activated when the PQ pool is reduced. The cyt b_6f is able to detect the redox state of the PQ pool via the binding of plastoquinol to the lumenal Q_o site, where plastoquinol is oxidised (Vener *et al.*, 1997). The resultant conformational changes in the cyt b_6f complex activate the protein kinases that are located on the stromal face of the thylakoid membrane (Finazzi *et al.*, 2001). In this way, a signal is transferred across the thylakoid membrane by the cyt b_6f complex. The LHCII phosphorylation state is determined by the balance between protein phosphorylation owing to protein-kinase activation and de-phosphorylation catalysed by chloroplast protein phosphatases. An excess of PSII excitation relative to that of PSI would result in an increased reduction in the PQ pool, enhancing protein-kinase activity and leading to a higher LHCII phosphorylation state and migration of some LHCII to PSI. This would balance excitation-energy distribution as described previously (Allen, 1992; Tikkanen *et al.*, 2008).

Although state transitions are undoubtedly important in the regulation of photosynthetic electron flow in many types of algae, the role of state-transitions in higher plants remains uncertain. High irradiance would increase the amount of plastoquinol and hence the extent of the state transition. However, in the leaves of higher plants the influence of qT is restricted to low irradiances (Walters and Horton, 1991; Rintamaki *et al.*, 1997). This is related to the regulation of LHCII protein kinase that is subject to thioredoxin-mediated redox regulation; increases in the reduction of the thioredoxin pool occurring in response to increased irradiance (for example) result in inactivation of the kinase (Rintamaki *et al.*, 1997). A large number of chloroplast enzymes are subject to this type of redox regulation, as discussed in more detail below. Whereas there is no doubt that: (1) state transitions in higher plants decrease the antenna size of PSII (Deng and Melis, 1986; Tikkanen *et al.*, 2011); (2) that *Arabidopsis* plants that lack the PSI subunits PsaH, PsaL or PsaO are less able to perform state transitions (Lunde *et al.*, 2000; Jensen *et al.*, 2004); and (3) that state transitions increase the yield of the 77-K fluorescence from PSI (which is often used to demonstrate an increase in PSI cross-section), direct measurements of P700 photo-oxidation kinetics do not always reveal an increase in the antenna size of PSI after a state transition (Haworth and Melis, 1983; Deng and Melis, 1986; but see also Telfer *et al.*, 1984). Similarly state transitions produce no change in the quantum yield for CO_2 fixation, despite evidence of changes to PSII fluorescence (Andrews *et al.*, 1993). Such observations are at odds with the model of excitation-energy redistribution by state transitions that suggests that it should increase the light-use efficiency for electron transport under conditions of excess PSII excitation. Hence, while light-induced phosphorylation of LHCII and other thylakoid membrane proteins occurs and this has an effect on the quenching of F_m, it is still not clear that this process results in any increase in PSI excitation or an increase in light-use efficiency by photosynthesis.

3.2.4. The role of non-photochemical quenching in electron transport regulation

Plastoquinol oxidation by the cyt b_6f complex is generally believed to exercise the dominant limitation on the PET chain. The limitations imposed by NPQ are not considered to be important in the regulation of linear electron transport because qP decreases with increasing irradiance, implying an increase in Q_A reduction, even though NPQ increases. Nonetheless, NPQ will decrease the formation of Q_A^-, which in turn will affect the redox state of the plastoquinol pool. It has been suggested that this could restrict electron transport under light-limiting conditions and could also partially restrict electron transport at higher irradiances as the overall limitation is transferred to the plastquinol/cyt b_6f step (Heber *et al.*, 1988).

PSII activity is often in excess of that of PSI at limiting irradiance because there is a (small) loss of qP, even when CO_2 fixation is light-limited. This loss of qP is a result of a proportionately larger increase in Q_A^- (Kramer *et al.*, 2004a). Moreover, NPQ does not usually develop under limiting light levels but is generally observed when the overall limitation of electron transport shifts to the plastquinol/cyt b_6f reaction. Under light-limiting conditions, therefore, NPQ normally exercises no restriction on light-use efficiency. Under conditions of stress, however, a slowly reversible form of NPQ that relaxes only slowly in the dark could limit light-use efficiency under light-limited conditions (Demmig-Adams and Adams, 2006). It is difficult to determine the extent to which NPQ contributes to the restriction of electron transport under steady state, non-light-limiting conditions. This will depend upon the effect NPQ has on the redox state of the plastoquinol pool and how this affects the plastoquinol/cyt b_6f reaction. Whereas the overall regulation of the plastoquinol pool is poorly understood, it does display a lower than predicted apparent equilibrium constant with respect to the Q_A pool (Joliot *et al.*, 1992; Cleland, 1998). This means that any decrease in the amount of Q_A^- would have a significant effect on the amount of available plastoquinol and limit electron flux. However, limitations on the rate of electron transport caused by plastoquinol deficiency could be compensated by an increased abundance of the cyt b_6f complexes. It has been proposed that even the relatively rapid relaxation of qE could limit light-use efficiency following the sudden decreases in irradiance that occur in natural environments (Long *et al.*, 2006a). However, metabolic limitations might be more significant than any qE effects in these situations (Prinsley *et al.*, 1986).

3.3. INTEGRATION OF PHOTOSYNTHETIC ELECTRON TRANSPORT PROCESSES AND METABOLISM

In addition to the demands of the Calvin cycle, ATP and reducing power are used to drive many other processes including primary nitrogen assimilation and non-sugar metabolism, such as the shikimate, nucleotide, fatty acid and isoprenoid-synthesis pathways (Noctor and Foyer, 1998a; Lewis *et al.*, 2000; Foyer *et al.*, 2006a,b). Processes such as primary nitrogen assimilation integrate metabolic pathways involving carbon and nitrogen and they also serve to balance overall ATP/NADPH ratios. In the cytosol, sucrose synthesis provides the largest sink for triose phosphate (TP) produced by the Calvin cycle, but glycolysis, the anaplerotic pathway, and respiration are also major sinks for assimilated carbon. ATP and reductant production by the chloroplasts are integrated at a cellular level with NADH utilisation and ATP production in the mitochondria, as discussed further in Chapter 4. The PET and RET chains interact directly with metabolism with regard to the production and utilisation of reductant and ATP. Regulation of energy metabolism in the chloroplasts has many parallels with that of the mitochondria in relation to the rest of the cell. Similarly, separate but interfacing pathways of chloroplast to nucleus (Ankele *et al.*, 2007; Koussevitzky *et al.*, 2007) and mitochondria to nucleus retrograde signalling (Rhoads and Subbaiah, 2007) coordinate gene expression and protein-complex formation appropriate to bioenergetic and metabolic requirements of the whole cell.

3.3.1. Fine tuning of metabolism – thioredoxin and the stromal electron transport chain, and modulation of enzyme activity

A number of chloroplast enzymes are regulated by dynamic changes in stromal redox potential. This process generally involves thioredoxin-mediated regulation of enzyme activity in response to changes in light availability (Buchanan and Balmer, 2005; Scheibe *et al.*, 2005). Plastidial thioredoxins also play a role in peroxiredoxin-linked H_2O_2 and lipid peroxide metabolism (Vieira Dos Santos and Rey, 2006). Whereas enzymes of the Calvin cycle were the first enzymes whose activity was shown to be regulated via the PET chain and thioredoxin, there is now an extensive list of thioredoxin targets in the chloroplasts and other cellular compartments (Schürmann and Buchanan, 2008). Thiol-disulfide-exchange reactions modulate the activities of a wide range of chloroplast enzymes, including those of starch synthesis and degradation. Regulation by different members of the thioredoxin family of proteins occurs in the cytosol and mitochondria, as well as chloroplasts (Lemaire *et al.*, 2007). For example, cytosolic thioredoxin has implicated in plant responses to pathogens (Laloi *et al.*, 2004).

Although some thioredoxin-regulated chloroplast enzymes, such as glucose-6-phosphate dehydrogenase, are activated by thiol oxidation, most are activated by disulfide reduction. This involves an increase in the reduction state of a pre-existing pool of thioredoxins mediated by light-driven reduction of Fd (Setterdahl *et al.*, 2003). The thioredoxin-regulated activity of chloroplastic NADP-dependent malate dehydrogenase (MDH) not only participates in the regulation of stromal metabolic pathways, but it is also instrumental in the transfer of reducing power from the stroma to the rest of the cell (Fig. 3.2) via the malate shuttle system (Scheibe *et al.*, 2005; Holtgrefe *et al.*, 2007). Pyridine nucleotides do not cross the inner-chloroplast envelope membrane at rates that are comparable with chloroplast NADP redox cycling. However, the malate shuttle system allows high-flux exchange of reducing equivalents across the membrane, linking changes in stromal and extra-chloroplastic NADP redox states. Such shuttles serve not only to relieve electron pressure in the chloroplast but they also signal a state of high reduction in the chloroplast to the cytosol and nucleus. Their operation also influences a wide range of different processes and cellular functions. For example, the light dependence for certain stress responses might be explained by a dependence on light-driven reductant export from the chloroplast, as discussed by Foyer and Noctor (2009).

Redox gradients in the nicotinamide adenine dinucleotide (NAD(P)H / NAD(P)) status exist between chloroplasts, mitochondria and cytosol (Igamberdiev and Gardeström, 2003). However, the redox states of compartments are not fixed and are likely to change with environmental fluctuations, in particular the balance between light availability and the capacity of metabolism to use light energy. As discussed below mechanisms exist to decrease light-capture efficacy and constrain electron transport when light is in excess of metabolic capacity. Stromal ATP utilisation is linked to light-capture efficiency by changes in the redox state of the PQ pool. Moreover, it has recently been proposed that NADP redox status could also feedback to regulate PET activity (Hald *et al.*, 2007). Although the primary impact of changes in light availability is in the chloroplast, effects on NAD(P) status in extra-chloroplastic compartments are also possible (Igamberdiev and Gardeström, 2003). This could be important in relaying light signals to the cytosol, and could also affect mitochondrial redox state and the regulation of mitochondrial metabolism.

3.3.2. Ribulose-1,5-bisphosphate carboxylase/oxygenase regulation

The fixation of atmospheric CO_2 into sugar phosphate in the chloroplasts is catalysed by the enzyme Rubisco. In this carboxylation reaction, CO_2 reacts with RuBP to form two molecules of phosphoglycerate (3PGA: Tcherkez *et al.*, 2006; see Chapter 2). The activity of the enzyme is however, limited by CO_2 availability under current atmospheric conditions in C_3 plants. This is crucial as Rubisco can also catalyse the incorporation of molecular O_2 into RuBP. This oxygenation reaction produces one molecule of phosphoglycolate and one molecule of 3PGA. Phosphoglycolate is the substrate for photorespiration, which provides an alternate pathway for the production of 3PGA, but unfortunately this pathway uses additional NADPH and ATP and leads to the release of CO_2 and ammonia (Foyer *et al.*, 2009). Thus, the higher plant Rubisco enzymes are notorious for their inefficiency. The slow turnover rate of these enzymes at limiting CO_2 partial pressures and the tendency of the enzyme to confuse the CO_2 substrate with the more abundant molecular O_2 has necessitated first a requirement for the energy intensive photorespiratory pathway to recycle oxygenated products, and second the need to make a large investment of leaf protein in Rubisco. There also appears to be, at least in broad terms, a decrease in turnover number as the specificity of the enzyme for CO_2, rather than O_2, is increased. There have been many attempts to engineer a better Rubisco, but the complexity of the enzyme has proved to be a major impediment. The enzyme requires the coordinated expression and assembly of the eight plastid-expressed and assembled large (L) subunits and the eight nuclear-expressed small (S) subunits, to form a hexadecameric (L_8S_8) enzyme. The L subunits house the catalytic site, but the S subunits, whose precise functions remain poorly understood, are essential for catalytic viability. Despite the problems, the goal of engineering Rubisco for enhanced efficiency remains possible although difficult, and indeed is still being attempted (Whitney and Andrews, 2001; Sharwood *et al.*, 2008). The goal is encouraged by the fact that some evolution towards improved Rubisco efficiency seems to have occurred in C_3 plants in response to environmental selection pressure (Galmés *et al.*, 2005).

Rubisco activity is modulated by the carbamylation of an essential lysine residue at the catalytic site and stabilisation of the resultant carbamate by magnesium ions, forming a catalytically active ternary complex (Parry *et al.*, 2008). It is also regulated by tight-binding endogenous inhibitors that resemble the transition-state intermediates of catalysis. The binding of such inhibitors before or after carbamylation blocks the active site of the enzyme. The best characterised of these is 2-carboxy D-arabinitol-1-phosphate, which is only found in chloroplasts. However, another inhibitor, pentadiulose-1,5-bisphosphate, is important as it is produced via the oxygenation reaction and it accumulates progressively at the active site of the enzyme. Any situation that favours oxygenase activity leads to pentadiulose-1,5-bisphosphate accumulation. Such inhibitors can only be removed from the Rubisco protein by Rubisco activase, which is essential for the activation and maintenance of the catalytic activity of Rubisco. Several forms of Rubisco activase have been described and they play a role in plant responses to high temperature stress (see Chapter 19).

Rubisco activase removes any tightly bound inhibitors and sugar phosphates from the active site in a process that involves ATP hydrolysis. Rubisco is light activated by changes in the stromal redox state and ATP/ADP ratio via Rubisco activase (Portis *et al.*, 2008). The ATPase activity of this enzyme is very sensitive to the stromal ATP/ADP ratio and is modulated by the large alpha isoform of Rubisco activase, which has a C-terminal extension bearing two cysteines that are regulated by the thioredoxin-mediated thiol–disulphide exchange systems of the chloroplast stroma. Other effectors, such as sugar phosphates and some inorganic ions, either modulate Rubisco directly by influencing the extent of carbamylation at sub-saturating CO_2 and/or magnesium-ion concentrations. Sugar phosphates can also influence the activity of the activase. Rubisco synthesis and turnover also subject to multilevel redox regulation (Moreno *et al.*, 2008). The stress-induced oxidative modification of the specific Rubisco protein residues are considered to be particularly important in targeting the protein for degradation (Moreno *et al.*, 2008; Prins *et al.*, 2008).

3.3.3. Flexibility in the production of adenosine triphosphate and reductant

Whereas the Calvin cycle is the major sink for the ATP and reductant produced by the PET chain, photosynthesis is also the driving force for other important pathways such as primary nitrogen assimilation. The Calvin cycle, photorespiration and primary nitrogen assimilation require different amounts of ATP and reductant. The PET chain therefore has to be sufficiently flexible to meet the requirements of the changing demands of these pathways. Similarly, decreased demand for ATP and reductant arises in situations where metabolism is downregulated by, for example, stomata closure, limiting CO_2 supply, or if the capacity to synthesise end-products, particularly sucrose and starch, is decreased. The regulation of electron transport has to be flexible in order to accommodate changes in total demand and ratio of ATP and reductant.

Variations in the relative production of ATP and reductant by the PET chain are possible because of the flexibility of the electron- and proton-transport systems. The linear, cyclic and pseudocyclic pathways of electron transport within the PET chain generate different proton-transport/reductant ratios (see Chapter 2 for details). For example, if a Q cycle is obligate, as is widely accepted (Rich, 1988), linear electron flow to Fd pumps three protons per electron, whereas pseudocyclic electron flow results in no net generation of reductant but is coupled with the transport of three protons. The flux through the pseudocyclic electron transport pathway has been a matter of much debate but it is generally accepted that in most conditions it is only about 10% of the linear flux (Robinson, 1988; Genty and Harbinson, 1996; Badger *et al.*, 2000b; Ort and Baker, 2002). In stress situations, however, much higher pseudocyclic flux rates have been reported (Cheeseman *et al.*, 1997; Fryer *et al.*, 1998; Farage *et al.*, 2006).

Like pseudocyclic electron flow, cyclic electron flow does not generate net reductant, but it translocates protons into the thylakoid lumen. Four possible cyclic pathways have been proposed (Bendall and Manasse, 1995). The H^+/e^- for cyclic electron transport *in vivo* in C_3 leaves is not known but a H^+/e^- value of two is widely assumed. Cyclic electron transport can be quantitatively important, for example in the bundle-sheath chloroplasts of C_4 NADP-malic enzyme species and in algae. In C_3 leaves the functional operation of cyclic electron transport pathway(s) remains controversial but it is believed that it is probably not more than 10% of the linear flux (Baker and Ort, 1992; Avenson *et al.*, 2005a,b). In some circumstances, however, such as photosynthetic induction (Joët *et al.*, 2002) or in the absence of CO_2 (Harbinson and Foyer, 1991), cyclic electron transport rates can be much higher, at least on a relative basis.

Reductant in excess of that required to drive the many biosynthetic pathways located within the chloroplast can be exported to the cytosol via the malate shuttle (Scheibe *et al.*, 2005). In this way, the reductant generated by the PET chain can be used by assimilatory processes outside the chloroplast that increases the effective ATP/reductant ratio available in the chloroplast. The export of reductant depends upon the reduction of OAA to malate by NADP-MDH, a chloroplast enzyme that is subject to reductive activation by the chloroplast thioredoxin system. In this way, the export of reductant from the chloroplast is dependent on the redox state of the stroma. NADP-malate-dehydrogenase activity responds to short-term changes in irradiance and also other factors (Foyer *et al.*, 1992). Estimates of the amount of reductant exported from the chloroplast by the malate shuttle suggest values of about 5% of linear electron transport rates (Fridlyand *et al.*, 1998).

The presence of several pathways through the PET chain results in a range of possible ratios for proton transport to the lumen versus reductant formation in the stroma. Protons accumulated within the thylakoid lumen are used to drive the synthesis of ATP, so the ATP/H^+ is of crucial importance in quantitatively understanding the reductant/ATP ratio generated by the combined action of electron transport, proton transport and ATP synthesis. Studies on the structure and operation of the thylakoid ATPase (Seelert *et al.*, 2000; Müller *et al.*, 2001) yield an estimate of about 4.7 H^+ per ATP (Allen, 2003). Although the structure-based estimate for the H^+/ATP ratio is currently widely accepted, some recent measurements of the H^+/ATP stoichiometry made on the ATPases from *E. coli* and spinach have cast some doubt on the very plausible 4.7 H^+ per ATP derived from structural studies (Steigmiller *et al.*, 2008). The ATPases from *E. coli* and spinach have different stoichiometries for the F0 subunits (10 for *E. coli* and 14 for spinach), but in both cases the H^+/ATP ratio is four (Steigmiller *et al.*, 2008), a value that is consistent with earlier experimental determinations of the H^+/ATP ratio. In addition to the ideal ATP/H^+ of the ATPase, slips and leaks in the thylakoid membrane and ATPase will decrease the ATP/H^+ stoichiometry (Nelson *et al.*, 2002). Slippage and leakage are predicted to increase as the *pmf* increases, as will happen when the electron transport chain is subject to downregulation, though the impact of such processes on overall regulation remains to be quantified.

A simple example will show the consequences of the stoichiometry of the linear electron transport chain for the formation of ATP and reductant. The fixation of one mole of CO_2 requires two moles of NADPH and three of ATP. As linear electron transport translocates three H^+ into the lumen per electron, the formation of two NADPH would be coupled to the synthesis of only 2.6 ATP (assuming the 4.7 H^+ per ATP is correct). This shortfall of 0.4 ATP would need to be met by the pumping of more protons into the lumen by an electron transport pathway that does not generate reductant, such as cyclic electron transport, pseudocyclic electron transport or the export of reductant to the cytosol. Photosynthetic metabolism involves more than just the fixation of CO_2; this is normally accompanied by photorespiration, nitrite reduction and many other minor pathways. Obtaining an estimate of the ratio of metabolic demand for ATP and reductant requires the integration of all photosynthetic pathways and requires the use of models (e.g., Yin *et al.*, 2006) of metabolism. These models can also be used to estimate, based on parameters such as the $H+/e^-$ and ATP/H^+ ratios, the extent to which linear electron transport alone could meet the demands of metabolism.

The presence of different possible electron transport pathways that can adjust ATP and reductant supply is well established, but it is less clear how the relative fluxes through each pathway are regulated. The Mehler reaction, cyclic electron transport and export via the malate shuttle will all tend to increase as the stroma becomes more reduced (Hosler and Yocum, 1987; Joët *et al.*, 2002). Any shortfall in ATP production will result in an increase in the reduction state of the stroma, and this in turn should increase the mechanisms that enhance the rate of ATP synthesis. The hierarchy of regulatory mechanisms that adjust ATP/reductant ratios has been investigated (Backhausen *et al.*, 2000), but it is not clear whether the relative activities of the various electron transport pathways are subject to control (i.e., a change in rate constants) or if they vary as a result of changes in the concentrations of the donors and acceptors.

The mechanisms that could act to adjust the ATP/reductant ratio could also act to downregulate electron transport by decreasing intrathylakoid pH. If the metabolic demand for reductant is diminished then the consequent increase in the reduction state of the stromal electron transport system would increase the rates of the Mehler reaction and export via the Malate shuttle and possibly of cyclic electron transport. This would decrease the intrathylakoid pH, slowing electron transport and resulting in an increase in ATP/ADP. An alternative mechanism suggests that intrathykoid pH can be lowered as a result of increasing the kinetic limitation on the ATPase (Kramer *et al.*, 2003, 2004b). This mechanism would require a greater pH gradient, and thus a lower intrathylakoid pH, for the same rate of ATP synthesis. A decrease in intrathylakoid pH will bring about downregulation of the reaction between plastoquinol and the cyt b_6f complex, thus slowing electron transport of all kinds. As with the balance of ATP/reductant, the hierarchy of mechanisms that act *in vivo* to restrict linear electron transport in response to decreases in metabolic demand needs to be demonstrated. In short, we know what mechanism could work, but not whether they operate *in vivo*.

3.4. DIFFUSIVE LIMITATIONS ON THE CO_2 SUPPLY AND THE REGULATION OF STOMATAL AND MESOPHYLL CONDUCTANCES

Leaf architecture plays an important role in regulation of photosynthesis, providing a structural intercellular framework for the diffusion of gases and optimisation of photosynthetic activity (Evans and Vogelmann, 2006). Atmospheric CO_2 has to move from the atmosphere surrounding the leaf across a boundary layer in the air above the foliage surface, to the sub-stomatal internal cavities through stomata through the leaf mesophyll to the site of carboxylation. The adaxial and abaxial surfaces of leaves can often show marked differences in the regulation of photosynthesis, particularly with respect to light orientation and CO_2 availability. The number and density of stomata can vary on each surface of leaves from monocotyledonous and dicotyledonous species. Moreover, the stomata on each surface may respond differently to environmental triggers such as light intensity, allowing surface-specific regulation of stomatal opening and gas exchange (Wang *et al.*, 1998). Leaf surface-specific variations in photosynthetic regulation in dicotyledonous leaves are related to intercellular gradients in the profile of light absorption (Ögren and Evans, 1993) and/or the diffusion of CO_2 within the leaf. Differences in the surface-specific regulation of photosynthesis are also observed in C_4 monocotyledonous leaves (Driscoll *et al.*, 2006; Soares *et al.*, 2008), which often have a physical restriction to airspace continuity between the upper-adaxial and lower-abaxial surfaces (Long *et al.*, 1989).

Thus far, we have considered the factors that regulate the efficiency with which the free-energy of photosynthetically active radiation (PAR) is used to make energy rich products (ATP and reductant). However, the supply of CO_2 to the chloroplasts and the restrictions to that supply caused by leaf structure are also crucial to the overall regulation of photosynthesis, particularly in C_3 leaves. The competition that exists between O_2 and CO_2 as substrates for Rubisco has important consequences for the quantitative coupling between PET and CO_2 fixation, depending on the ratio of the CO_2 concentration at the site of carboxylation (C_c) and the O_2 concentration at the site of carboxylation (O_c: C_c/O_c), which determines the relative velocities of oxygenation and carboxylation. The mechanisms of O_2 and CO_2 incorporation and competition between the carboxylation and oxygenation reactions have been elegantly encapsulated by the model of C_3 photosynthesis developed by Farquhar *et al.* (1980a). Sustained net CO_2 fixation requires that the irradiance is higher than the light-compensation point, and that the CO_2 consumed by photosynthesis is replaced by atmospheric CO_2. As C_c decreases as CO_2 is consumed by the reaction with RuBP, then oxygenation is enhanced and electron transport will be used increasingly to drive photorespiration. The replacement of CO_2 in the chloroplasts of C_3 plants is dependent upon diffusion and thus upon the difference between the atmospheric CO_2 concentration (C_a) and C_c, and the resistance (or its reciprocal, conductance) for CO_2 diffusion within the leaf. As the diffusive resistance is greater than zero, any flux of CO_2 driven by diffusion will require a finite difference in the CO_2 concentrations of the

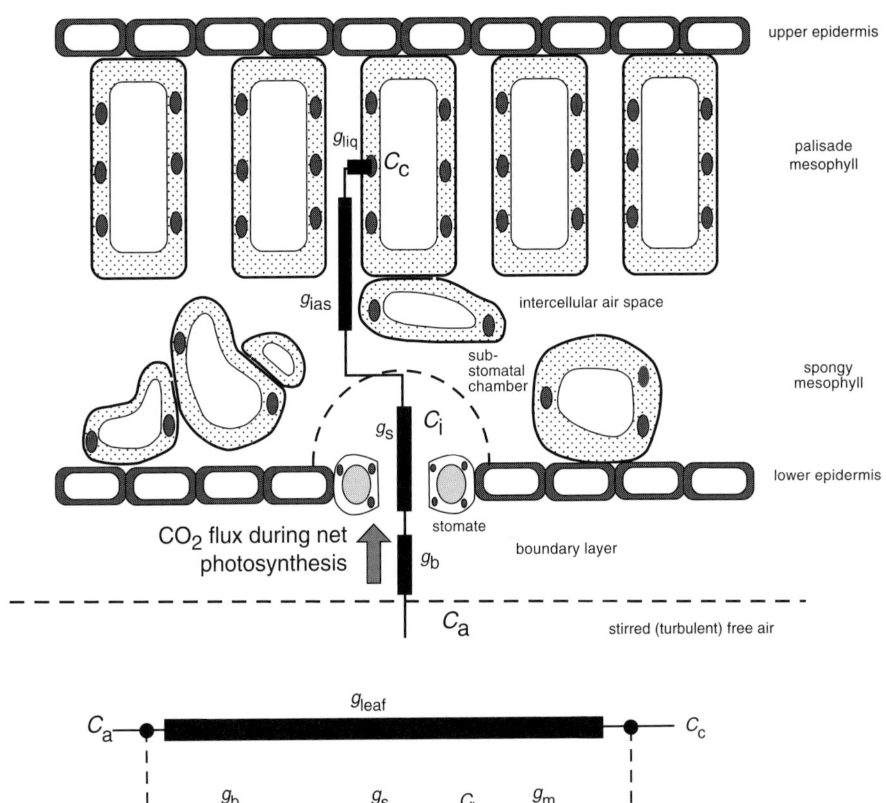

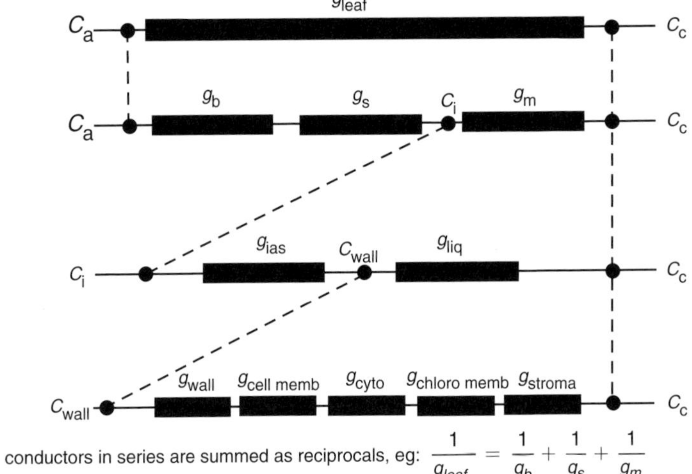

Fig. 3.4. Pathways for CO_2 diffusion in the leaves. **(A)** A simplified cross-section of a leaf showing the principal conductive pathways for CO_2 diffusion within the leaf and the surrounding air. The interface between the boundary layer of air immediately surrounding the leaf and the stirred air of the atmosphere is not discrete (as drawn) but it is dynamic. The boundary between C_i and C_c is also, in reality, not discrete, but a continuum, reflecting the fact that water evaporates from multiple sites close to the stomatal aperture. The sizes of the symbols used indicate that the overall conductance is not proportional to the typical relative sizes of the individual conductances. The linear series of conductance values illustrated here is a symbolic representation of what in reality is a complex, three-dimensional network of pathways. **(B)** The hierarchy of conductance values that together form the total leaf conductance for CO_2 diffusion from the stirred free air lying beyond the boundary layer (C_a) and the CO_2 concentration at the site of carboxylation (C_c). The subordinate conductances must be summed as reciprocals, and the relationship between A_N, the CO_2 concentration difference and conductance is a form of Ohm's law (V=IR), where A_N is the flux (I), the concentration difference is the driving force (V) and the inverse of conductance is the resistance (R).

stroma and the air surrounding the leaf (i.e., $C_c < C_i < C_a$). The carboxylation/oxygenation ratio and the relationship between CO_2 fixation and electron transport depend substantially upon C_c. Differences in O_2 concentration will be approximately the same as CO_2 concentration differences, but because the base O_2 concentration is high, these differences will be trivial and can be ignored. Thus, in any appraisal of the regulation of photosynthesis, it is crucial to know C_c. The difference between C_c and C_a is determined by the flux of CO_2 and the total conductance of the diffusive pathway (Fig. 3.4) as described in detail in Chapter 12. Of the many components that contribute to conductance, only the conductance of stomata (g_s) and that occurring through the apoplast and symplast of the leaf mesophyll cells, involving the cell wall, the cytosol, plastids and the cell membranes (g_m) are subject to short-term regulation. Other components that contribute to leaf conductance such as the boundary layer and intercellular air space are only subject to adjustment during the development of the leaf, but depend on factors external to the plant, such as wind speed in the case of boundary layer conductance. They are thus essentially fixed conductances and they will not be considered in any detail here.

Stomatal opening is subject to a complex network of regulation (Hetherington and Woodward, 2003). This is perhaps not surprising given the pivotal role of stomata in regulating gaseous diffusion across the cuticle, allowing CO_2 to enter and H_2O to leave the leaf. The stomata are thus crucial to leaf and whole-plant water economy, with stomatal regulation playing a major role in the protection of the plant from desiccation when the availability of soil water is restricted. There is a hierarchy in regulation, particularly in situations of water-supply deficits or where excessive transpiration (E) could jeopardise leaf function. In this case, the regulation of stomata is more directed to the needs of whole-plant water balance than the immediate requirements of photosynthesis. The water consumed in PET is approximately one per CO_2 fixed (i.e., ignoring photorespiration), whereas the water lost from the leaf via the stomata is between 100 and 1000 per CO_2 assimilated. In the absence of water stress, the stomata of C_3 leaves are regulated in relation to photosynthesis such that g_s has a linear or near-linear relationship with CO_2 fixation rate (Wong *et al.*, 1985; Hetherington and Woodward, 2003; Baroli *et al.*, 2008; Flexas *et al.*, 2008). This implies that C_i should be relatively constant with changing CO_2 fixation rates and irradiance levels (Wong *et al.*, 1985). In a survey of data obtained from 44 diverse species, Warren (2008a) found that at a C_a value of 360 μL L^{-1} the mean C_i was about 123 μL L^{-1} lower, though leaves of different types had different mean C_i values. Whereas the mechanisms that regulate the stomatal apertures are far from understood, stomatal opening (and thus g_s) in leaves is regulated by the intensity and by the spectrum of the irradiance. The stability of C_i with changing CO_2-fixation rate changes implies that a form of C_i sensing and signalling exists in leaves (Warren, 2008a). The sensing of C_i, rather than C_a or C_i/C_a, by leaves has also been demonstrated previously (Mott, 1988).

Stomata open when leaves perceive light within the range that drives photosynthesis (PAR), but blue light and red light affect stomata differently. Exposure to red light alone induces stomatal opening (Shimazaki *et al.*, 2007). Exposure to weak blue light facilitates the opening of stomata but only after they have been induced by red light at intensities sufficient to drive CO_2 fixation. Blue-light perception appears to be localised in the guard cells (Shimazaki *et al.*, 2007). In contrast, the red light and C_i responses demonstrated by Mott (1988) do not seem to be localised in the stomata alone, but only occur when the epidermis is in contact with the mesophyll (Mott *et al.*, 2008). Such studies imply that the regulation of the stomata is not wholly autonomous, but depends to some extent on the adjacent mesophyll and its photosynthesis rates. An unknown diffusible signal from the mesophyll is believed to mediate the effects of C_i and red light on stomatal opening. The relationship between stomatal behaviour and the mesophyll is, however, not simply related to the photosynthetic activity of mesophyll cells. Tobacco leaves with altered photosynthetic capacities had similar g_s, and therefore a higher (though still constant) C_i, compared with wild-type leaves (von Caemmerer *et al.*, 2004; Baroli *et al.*, 2008). This suggests that whatever the nature of the signal from the mesophyll that influences stomatal opening, it is not primarily photosynthetic in origin. Recently, carbonic anhydrases (CA) have been shown to act as upstream regulators of CO_2-induced stomatal responses, but these CAs do appear to operate in guard cells, and hence are unlikely to be involved with signalling from the mesophyll (Hu *et al.*, 2010). Alternatively, stomata respond directly (i.e., without intervention of the leaf mesophyll) to hydraulic signals, such as those occurring during the so-called Iwanoff effect (a transient stomatal opening owing to decreased turgor in epidermal cells) after excising leaves (Kaiser and Gramms, 2006), or upon changes in relative humidity (RH) in the air surrounding non-excised leaves (Farquhar *et al.*, 1980b; Shope *et al.*, 2008). Indeed, Pieruschka *et al.* (2010) have recently suggested that pure radiation-induced transpiration has a regulatory role on stomatal conductance through a hydraulic effect caused by evaporation of epidermal-cell water, and that differences in stomatal opening in response to blue or red light of similar intensity are simply owing to different associated energies causing different evapration rates. Moreover, stomata opening appears to also respond to electrical signals within the leaf (Kaiser and Gramms, 2006; Grams *et al.*, 2007). Recently, *SLAC1* (*SLOW ANION CHANNEL-ASSOCIATED1*) has been described as a central gene involved in stomatal responses to light, CO_2, abscisic acid (ABA) and H_2O_2, among other elicitors (Vahisalu *et al.*, 2008; Kim *et al.*, 2010), but upstream of *SLAC1* a complex network of redox, chemical and hormonal signals regulates stomatal closure under conditions of water stress (reviewed in Chapter 20) and others (see Section 3.5.1).

The factors that contribute to g_m have been the subject of much debate (Flexas *et al.*, 2008; Warren, 2008a; Evans *et al.*, 2009). The consensus view is that g_m includes all diffusive pathways, including those that might be facilitated in some way, within the mesophyll but excluding g_s. The term g_m is now used to describe only those impediments to CO_2 fixation within the mesophyll that are diffusive in nature. As g_s is measured by means of water vapour fluxes, the path corresponding to g_s only extends as far into the mesophyll as the sites from where water evaporates. These are predominantly in or on the inner non-cuticularised guard-cell walls and possibly the walls of adjacent cells (Meidner, 1975). As C_i is calculated using g_s, C_i would therefore be the CO_2 concentration in the mesophyll air space close to the stomata and the rest of the mesophyll air space, which is the bulk of the mesophyll air space, is part of g_m and specifically the g_{ias} part of g_m. The diffusive resistance owing to the intercellular air space is fixed by leaf development as it depends upon the architecture of mesophyll. Though g_{ias} cannot be adjusted, it does make a finite contribution to g_m and although this is normally considered to be minor (Parkhurst and Mott, 1990; Genty *et al.*, 1998) in sun-adapted walnut leaves it has been shown to be close to 50% of the total g_m (Piel *et al.*, 2002). Most often, however, g_m is not analysed into the g_{ias} and g_{liq} components, but it is presented as a sum value of conductance and it is assumed that the dominant component is g_{liq}.

The leaf components suggested to regulate g_{liq} may be partly structural, such as the cell walls and surface of chloroplasts exposed to intercellular air spaces (Evans *et al.*, 2009), and partly biochemical, such as the plasma membrane and chloroplast-membrane aquaporins (Terashima and Ono, 2002; Uehlein *et al.*, 2003; Hanba *et al.*, 2004; Flexas *et al.*, 2006a; Miyazawa *et al.*, 2008; Uehlein *et al.*, 2008) or CAs (Price *et al.*, 1994; Williams *et al.*, 1996; Gillon and Yakir, 2000a,b), or mixed, such as light-induced chloroplast movements (Tholen *et al.*, 2008). The relative contribution of each of these components is still controversial, and more work is needed to fully understand the regulation of g_m (Flexas *et al.*, 2008; Evans *et al.*, 2009).

As mentioned above, g_m, like g_s is also subject to short-term regulation over the seconds-to-minutes timescale (Flexas *et al.*, 2007a) and it appears to track g_s (Flexas *et al.*, 2008; Warren, 2008b), although sometimes it anticipates g_s, such as in response to varying CO_2 (Flexas *et al.*, 2007a). The g_m parameter largely changes with plant and leaf age (see Chapter 23), and responds to changes in environmental parameters as water availability (see Chapter 20), temperature (Bernacchi *et al.*, 2002; Warren and Dreyer, 2006; Yamori *et al.*, 2006), light (Flexas *et al.*, 2007a; Hassiotou *et al.*, 2009a; Monti *et al.*, 2009) and CO_2 (Flexas *et al.*, 2007a; Hassiotou *et al.*, 2009a; Velikova *et al.*, 2009; Vrábl *et al.*, 2009; Yin *et al.*, 2009). Nevertheless, the responses to light (Loreto *et al.*, 2009; Tazoe *et al.*, 2009) and CO_2 (Bunce, 2009; Tazoe *et al.*, 2009, 2011) are still controversial. g_m is broadly similar in size to g_s (Flexas *et al.*, 2008; Warren, 2008a) so the relative limitations imposed on photosynthesis by g_m and g_s are comparable (Warren, 2007).

It is convenient to assume that C_c and C_i are identical (an infinite g_m), as g_s is relatively easy to measure and g_m is difficult to measure, and in the absence of a measure of g_m its value is usually assumed to be infinite. However, a consequence of the finite value of g_m is that during net photosynthesis C_c will be lower than C_i, and at light saturation the difference is generally 85 ppm (Warren, 2008a). This limitation on CO_2 supply has consequences for the overall operation of photosynthesis. When C_a is at current atmospheric values (currently at a global average of about 382 ppm), the lower C_c produced by the finite g_m means that the velocity of carboxylation will be decreased and that of oxygenation increased compared with the idealised situation, where $C_i = C_c$. Thus, the efficiency with which electron transport drives CO_2 fixation (i.e., the light-use efficiency of CO_2 fixation) is strongly affected by g_m, just as it is by g_s. A quantitative analysis of the light-use efficiency of CO_2 fixation must therefore include consideration of g_m, as must any model of photosynthesis that attempts to translate electron transport rates into a rate of CO_2 fixation (Yin *et al.*, 2006). The fact that g_m (like g_s) is subject to short-term control in response to many common environmental factors

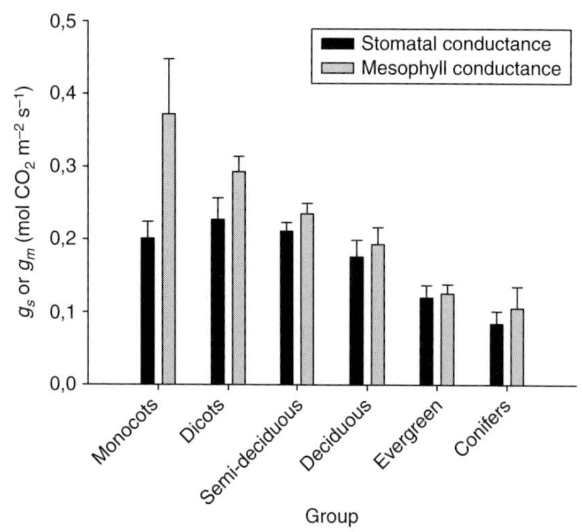

Fig. 3.5. A survey of average stomatal and mesophyll conductances to CO_2 in different plant groups. Examples of average±standard error values for maximum (i.e., under optimum conditions without stress) stomatal (g_s, black bars) and mesophyll (g_m, gray bars) conductances to CO_2 are given in different plant groups, including species for which simultaneous measurements of both conductances are available in the databases, as used by Flexas *et al.* (2008) and Niinemets *et al.* (2009a). These data illustrate the variability of g_s and g_m in nature, and the general differences among plant groups, but not to set limits to the maximum possible conductances. Higher g_s and g_m (>1 mol CO_2 m^{-2} s^{-1}) than those shown here have been eventually described, particularly in some fast-growing herbaceous species.

(Flexas *et al.*, 2007a; 2008) further complicates the complete analysis and the modelling of photosynthetic responses to environmental factors, such as C_a or irradiance. The interaction between g_m and the light-use efficiency for CO_2 fixation has consequences for the photosynthetic efficiency of different leaves where these leaves differ systematically in the ratio of C_a and C_c. Average maximum values of g_s and g_m in different plant groups are shown in Figure 3.5. It is clear that both conductance terms are of comparable magnitude, hence exerting a similar limitation to photosynthesis. Moreover, important differences emerge when comparing different plant groups (e.g., herbaceous monocots versus woody conifers), which may have important ecophysiological consequences. Interestingly, non-stomatal land plants (i.e., no g_s), such as liverworts and hornworts, also present much lower g_m values than any stomatal plant group shown in Figure 3.5 (Meyer *et al.*, 2008). The ecophysiological consequences of g_m are described in detail in the reviews of Warren (2008) and Flexas *et al.* (2008). Specific responses of both g_s and g_m to several environmental variables are covered in Chapters 16 to 22.

3.5. REDOX HOMEOSTASIS AND SIGNALLING

It is important that the enzymes of metabolism operate in an environment where the redox state is well buffered and not subjected to extremes of oxidation or reduction. The redox state of aqueous compartments of the mesophyll cells are protected against oxidation by rapid ROS removal and specialised metabolic cycles involving redox-active components that act as antioxidants, particularly ascorbate, glutathione and peroxiredoxins. Ascorbate, glutathione and pyridine nucleotides are the major redox couples of the aqueous phase of the cell, and the concept of cellular redox homeostasis applies largely to this phase, where the redox potential is maintained between –250 and –300 mV. The turnover of the ascorbate, glutathione and pyridine nucleotides pools occurs within seconds to minutes, and the reactions are maintained close to thermodynamic equilibrium. However, unlike the pyridine nucleotide, peroxiredoxin and thioredoxin pools, the ascorbate and glutathione pools are maintained in a highly reduced state in the light and dark, and their overall redox state is not significantly changed as a result of the immediate impact of light on PET.

The reductive activation of the Calvin cycle by thioredoxins that are reduced in a light-dependent fashion by the PET chain is essentially a reversible process in which O_2 acts as the terminal electron acceptor. Thus, oxidation of thiol-modulated enzymes by oxidised thioredoxin or H_2O_2 with O_2 as the terminal electron acceptor is presumed to reverse the thioredoxin-mediated activation process. However, certain key oxidative enzymes such as thiol oxidases operate in the chloroplasts in the light despite the highly reduced state of the stroma. Virtually nothing is known about how such enzymes form disulphide bonds in proteins in the light in organelles such as chloroplasts.

Chloroplasts produce a wide variety of different signals including ROS. The intermediates of chl. biosynthesis have traditionally been regarded as important signalling molecules involved in the repression of nuclear genes that encode chloroplast proteins (Fernandez and Strand, 2008). They were thus considered to be important in chloroplast-to-nucleus retrograde signalling. However, no correlation has been found between the accumulation of intermediates of chl. biosynthesis and the expression of nuclear genes (Mochizuki *et al.*, 2008; Moulin *et al.*, 2008). It appears that

ROS or redox signalling might modulate gene expression in response to defects in chl. biosynthesis (Moulin *et al.*, 2008).

3.5.1. Roles for redox signals in the regulation of stomatal movements

Stomata are epidermal structures that occur in most aerial organs of terrestrial higher plants. They consist of two guard cells that form a pore, and play an essential role in regulating gas exchange between the plant and the atmosphere. Pore opening depends on increases in the turgor of the guard cells, which in turn is controlled by the flow of water and ions between the guard cells and the neighbouring epidermal cells. Loss of guard-cell turgor triggers stomatal closure. Guard-cell turgor is regulated by many environmental triggers, particularly light, water deficits and CO_2 availability, and endogenous regulators such as ABA. Plants respond to water loss and dehydration primarily by closing the stomatal pores on the leaf surface and thus reducing *E* from the leaves. Light responses in plants are mediated by at least three types of photoreceptors: phytochrome, cryptochrome and phototropin. Of these, phototropin appears to control light-induced movement responses, such as stomatal opening.

Changes in the volume of stomatal guard cells are rapidly reversible and essential for stomatal functioning (Shope *et al.*, 2008). During each cycle of stomatal opening and closing, the guard-cell volume changes by a factor of two or more, which exceeds the elasticity or flexibility of the membrane and requires that membrane vesicles be added to or taken from the cell and vacuolar membranes.

ABA regulates ion channels in the tonoplast and plasma membrane, which allow water movement in and out of the cell, enabling the necessary volume changes for opening and closing of stomatal pores (Fan *et al.*, 2004). This process is regulated by a complex signal transduction network that involves protein kinases, phosphatases and multiple secondary messengers, such as inositol 1,4,5-trisphosphate, heterotrimeric G proteins, calcium, H_2O_2, nitric oxide (NO) (Pei *et al.*, 2000; Fan *et al.*, 2004), a syntaxin protein encoded by *NtSyr1* (Leyman *et al.*, 1999) and SLAC1 (see Section 3.4). The vesicle trafficking process involves integral membrane proteins called 'vesicle associated membrane proteins (VAMP)', which form the major component of the SNARE (soluble N-ethylmaleimide-sensitive factor attachment protein receptor) complexes that function in facilitating vesicle fusion with target membranes (Leshem *et al.*, 2006). The Nt-Syr1 is a t-SNARE that resides in the plasma membrane and participates in the induction of ABA-dependent ion currents.

ABA-dependent production of H_2O_2 is also important for stomatal closure (Pei *et al.*, 2000; Wang and Song, 2008). The application of ABA leads to H_2O_2 accumulation in the guard cells, with the calcium/calmodulin system acting both upstream and downstream of the H_2O_2 signal (Hu *et al.*, 2007; Wang and Song, 2008). The H_2O_2 produced in this is transported within the cell inside the vesicles (Leshem *et al.*, 2006). Trafficking of membrane-bound solutes like H_2O_2 may be essential for appropriate signalling. Moreover, ABA-induced H_2O_2 production leads to NO generation in leaves and stomata (Bright *et al.*, 2006). Therefore, besides hormonal signalling, redox signals have a central role in regulating stomata opening.

3.5.2. Long-term acclimation of photosynthesis to environmental triggers

In addition to the local responses to high light intensity, chloroplasts in leaves that are exposed to high light can also initiate systemic signals that spread through the vascular system and induce stress resistance in other leaves that have not been exposed to high light (Karpinski *et al.*, 1999; Rossel *et al.*, 2007). The systemic light acclimation responses triggered by PET activity in high-light-exposed leaves are mediated to distal tissues via salicylic acid (SA), jasmonic acid (JA) and ethylene (ET)-dependent pathways (Rossel *et al.*, 2007; Muhlenbock *et al.*, 2008). For example, *Arabidopsis* mutants deficient in SA or JA signalling show compromised responses to systemic acquired high-light acclimation (Rossel *et al.*, 2007). The ZAT10 zinc-finger transcription factor is also important in this signalling cascade and in other plant defence responses (Rossel *et al.*, 2007). ZAT10 accumulates in the vascular tissues within minutes of the onset of exposure to high light and is correlated with the enhanced expression of antioxidant genes. The chloroplast protein kinases, STN7 and STN8, that are important in LHCII phosphorylation have also been implicated in the PET-signalling network that coordinates nuclear and chloroplast gene expression, as has a histidine sensor kinase (Chloroplast Sensor Kinase: Puthiyaveetil *et al.*, 2008).

Superoxide, H_2O_2 and $^1O_2$ can act as local and/or systemic signals that can influence and interact with other component systems to coordinate effective protection of chloroplasts in locations remote from those experiencing excess excitation energy. Perhaps by virtue of their intrinsic roles in photosynthetic redox metabolism, ROS and antioxidants are important signals conveying information on organellar redox state (Foyer and Noctor, 2009). For example, H_2O_2 has been implicated in long-distance inter-organ signals in systemic acclimation to excess light.

A possible candidate for such cell-type-specific signalling is the bundle sheath, as systemic light-stress signalling appears to involve H_2O_2 accumulation in the chloroplasts of bundle sheath cells (BSC) (Karpinski *et al.*, 1999; Fryer *et al.*, 2003). Thus, it is possible that the bundle sheath chloroplasts play a role in systemic light-signalling cascades leading to the development of stress tolerance in distal leaves. However, it is unlikely that H_2O_2 and other ROS are transported long distances in plants. A more likely situation is that ROS are continuously generated along the signalling pathways or that there is another mobile signal that is induced by H_2O_2 at the point of departure and that generates H_2O_2 at the point of arrival. An attractive candidate for the long-distance mobile signal in the case of excess light acclimation is ABA, which is an important inducer of H_2O_2 production, for example during stomatal closure. The involvement of the SA-, JA- and ET-dependent signalling systems in systemic light signalling provides also a direct link between light signalling and plant biotic stress responses (Rossel *et al.*, 2007; Muhlenbock *et al.*, 2008).

Photorespiratory H_2O_2 production is important in cellular redox control and signalling (Foyer *et al.*, 2009). The analysis of mutants that are deficient in photorespiratory catalase (CAT) have shown that the photorespiratory H_2O_2 signal interacts with another signal arising from day length (Queval *et al.*, 2007). When CAT-deficient mutants (*cat2*) were grown under short days, there was a strong upregulation of glutathione and defence genes. However when the mutants were grown under long days, H_2O_2 accumulation resulted in programmed cell death (Queval *et al.*, 2007). There is thus significant interplay between photorespiratory H_2O_2, SA and the photoperiod-signalling cascades in determining light- and stress-signalling responses. It will be interesting to see whether such interactions also occur in C_4 plants that have repressed photorespiratory H_2O_2 production because CO_2 is first assimilated into organic acids, and CO_2 is only released from them following transport to BSC. The BSC of some C_3 plants have a C_4-like pathway that serves to diminish photorespiration and hence associated H_2O_2-signalling capacity (Hibberd and Quick, 2002). It is possible that the bundle sheath tissues act as an integrator of peroxide signals arising directly from metabolism or hormone action. Moreover, the accumulation of $^1O_2$ is lower in the BSC than the surrounding mesophyll (Fryer *et al.*, 2002).

3.5.3. Singlet-oxygen signalling

Singlet oxygen is also an important signal originating in chloroplasts that modulates nuclear gene expression. Much of our current understanding of how $^1O_2$ induces programmed cell death has come from studies on the conditional *flu Arabidopsis* mutant, which accumulates photosensitising chl. precursors in the dark and therefore generates $^1O_2$ upon subsequent illumination (Apel and Hirt, 2004). Whereas a relatively large number of genes have been identified as $^1O_2$-inducible, the mechanism of signal transduction is unknown.

Singlet oxygen reacts with many molecules in the chloroplasts that could act as 'redox' sensors, such as of polyunsaturated fatty acids (PUFAs), chl., carotenoids and tocopherol. The oxidation of PUFAs leads to a wide range of metabolites called 'oxylipins', which have signalling functions. Exposure of the conditional *flu* mutant to conditions that allow $^1O_2$ formation causes leaf bleaching and accumulation of oxylipins. Both responses are abrogated by introducing the *executor1* mutation into the *flu* background. However, genetically blocking oxylipin synthesis is not sufficient to prevent the phenotypic response, though it does modify the expression of subsets of $^1O_2$-induced genes. In etiolated *flu* seedlings, in which the $^1O_2$ sensitiser accumulates to very high levels, transfer to light causes tissue collapse that cannot be prevented by the *executor1* mutation (Apel and Hirt, 2004). Singlet oxygen-mediated gene expression accounts for the largest fraction of ROS-inducible genes observed under various abiotic stresses (Gadjev *et al.*, 2006). EXECUTER1 (EX1) and EXECUTER2 (EX2) are chloroplast-localised signalling proteins involved in $^1O_2$-mediated regulation of gene expression in the nucleus related to the initiation of cell-death responses (Lee *et al.*, 2007). Similarly, CRYPTOCHROME1 (CRY1), a blue-light receptor that is important in the modulation of the high-light response (Kleine *et al.*, 2007), is also important in this chloroplast to nucleus signalling cascade (Danon *et al.*, 2006).

3.6. REGULATION OF PHOTOSYNTHESIS BY END-PRODUCT METABOLISM

The activity of the PET chain and the Calvin cycle are carefully regulated by direct feedback or feed-forward mechanisms in order to maintain an appropriate balance between assimilate production and the capacities of sucrose and starch synthesis (Lewis *et al.*, 2000; Foyer *et al.*, 2006a,b). Starch and sugar metabolism have important roles in the regulation of photosynthesis in source leaves. End-product inhibition by high concentrations of leaf sugars and starch has long been a key concept in the regulation of photosynthesis (Paul and Foyer, 2001). Regulatory and feedback mechanisms such as phosphate recycling serve to coordinate carbon metabolism in the chloroplast and cytosol (Paul and Foyer, 2001). There is also integration of respiration and photosynthesis to ensure the assimilation of nitrate and ammonia in primary nitrogen assimilation and also to recycle the carbon metabolites passing through the photorespiration pathways (Noctor and Foyer, 1998b; Foyer *et al.*, 2009). The accumulation of the photosynthetic end products in leaves resulting from decreased demand by more distant sinks leads to the repression of photosynthesis and it also changes photosynthetic gene-expression patterns. However, the physiological significance of direct feedback controls related to end-product accumulation remains debatable, and questions still remain concerning the precise nature of the mechanisms that regulate photosynthesis in relation to sink demand.

The sugar-signalling network that regulates photosynthetic gene expression is highly complex. It involves components such as trehalose-6-phosphate, which responds to light and in relation to carbon status of the cytosol (Lunn *et al.*, 2006) and stimulates starch synthesis via redox activation of the enzyme ADP glucose pyrophosphorylase (Kolbe *et al.*, 2005). Trehalose-6-phosphate also modulates other important components that are involved in sugar signalling, such as hexokinases and the SNF1-related protein kinase 1 (SnRK1), which play fundamental roles in transcriptional, metabolic and developmental regulation in response to energy limitations and carbon starvation (Hardie, 2007). SnRK1 activates the expression of photosynthetic genes and inhibits those involved in RET, the TCAC and in other biosynthetic process, and so regulates cell growth and metabolism in response to carbon availability. In young leaves trehalose-6-phosphate is a SnRK1 inhibitor and so represses the expression of photosynthetic genes, but this repression is lost in mature source leaves possibly owing to the absence of an intermediary factor whose expression is altered during leaf development (Zhang *et al.*, 2009a).

The storage of assimilates as starch in the chloroplast plays a crucial role in balancing supply and demand for carbohydrate in the light, as well as enabling a constant supply of assimilate in darkness to sustain the growth of distant sink organs. In the chloroplasts, starch synthesis and accumulation are regulated during the day, as is starch degradation at night. In this way, the provision of sugars required for growth is maintained at a constant level throughout the diurnal period. The triose phosphate/phosphate translocator (TPT) is the major metabolite transporter of the chloroplast inner-envelope membrane and is the main exporter of assimilate to the cytosol in the light. However, the complete absence of the TPT in mutants and transformed plants can be compensated by an increased turnover of transitory starch and consequently the export of maltose and glucose in the light. However, adg1–1/tpt-1 double mutants remain viable (if small) despite the defects in the ability to synthesise starch or export TP.

3.7. CONCLUSIONS AND PERSPECTIVES

The success of oxygenic photosynthesis is a result of the efficient design of the photosynthetic apparatus, which incorporates extreme flexibility and adaptability both in short-term regulation and longer-term changes in the molecular structure and composition of the PET chain and associated metabolic pathways to changing environmental conditions. Our understanding of the regulation of leaf photosynthesis, particularly in terms of environmental and metabolic controls, has greatly increased over the last twenty years (cf. Foyer *et al.*, 1990). Although there is overwhelming evidence that the photosynthetic energy transduction and metabolic systems are organised so that they are in balance, our current understanding of the systems that achieve this balance in both the short and long term is far from complete.

Considerable evidence now supports the view that limitations at the plastoquinol oxidation step of electron transport have a central role in the intrinsic regulation of the PET chain by photosynthetic control processes. NPQ processes and protein kinase activation, triggered by sensors around the plastoquinol/cyt b_6f steps are key features of the regulatory network that control electron- and proton-transport fluxes short term, and brings the photosynthetic formation of reductant and ATP into balance with the demands of metabolism in the short term. The plastoquinol/cyt b_6f step also plays a central role in initiating the protein kinase signalling cascades that modulate gene expression, in order to achieve longer-term changes in PET processes and metabolic pathways. Similarly, the transfer of energy and/or electrons to O_2 in side reactions associated with the PET chain contributes to the overall protection of the PET system from over-reduction, and pseudocyclic electron flow contributes to overall ATP synthesis. Although the regulation of photosynthesis is geared to avoiding excessive ROS formation through the photosynthetic control of electron transport and thermal-energy dissipation, it is now accepted that the ROS generated in this way are important signalling molecules that participate in the orchestration of gene expression and long-term acclimation processes.

ROS have long been regarded as toxic to the chloroplast and there is a long history of the interpretation of the roles of ROS in chloroplasts simply in terms of 'oxidative or photo-oxidative damage' with non-enzymatic lipid peroxidation as a primary event in photo-induced oxidative stress in chloroplasts. In this concept ROS are considered to exert effects primarily through indiscriminate oxidation and inactivation of cellular processes. However, such a concept is not compatible with an accumulating body of evidence showing that ROS fulfil essential signalling functions in plants, or that specific enzyme systems are induced in response to environmental triggers to enhance ROS generation. Singlet oxygen and H_2O_2 modulate gene expression through signalling components such as kinase cascades (Pitzschke and Hirt, 2009). These ROS-activated kinase-dependent pathways play key roles in hormonal signalling and development. A striking conceptual shift has recently concerned our understanding of $^1O_2$, which until recently was only considered to be a very reactive and toxic molecule. The demonstration that the cell-death responses observed in plants exposed to $^1O_2$ is largely under genetic control (Apel and Hirt, 2004) has led to a shift in the photooxidative stress paradigm. ROS can no longer be considered simply as damaging molecules that cause programmed cell death by indiscriminate oxidation. However, despite considerable evidence to the contrary, the effects of ROS are still often discussed in terms of oxidative damage, rather than signalling processes.

4 • Interactions between photosynthesis and day respiration

G.G.B. TCHERKEZ AND M. RIBAS-CARBÓ

4.1. INTRODUCTION

Although the general metabolic schemes of photosynthetic and respiratory pathways are well known when considered as separate entities, their interactions are one of the conundrums of plant photosynthetic biology. However, such interactions are the cornerstone for nitrogen assimilation by leaves, simply because carbon assimilation produces organic materials (carbohydrates) that are converted to nitrogen acceptors by respiration. Unsurprisingly then, intense efforts are currently devoted to elucidate the metabolic basis of reciprocal influences of photosynthesis and respiration, with the optimisation of nitrogen assimilation for a better yield of crop plants as an ultimate goal (Lawlor, 2002). From a basic science point of view, the interactions between respiration and photosynthesis have been a matter of interest for more than a century. The very first experiments were reported by Pizon (1902): 'The respiration rate appears to be decreased by light. In fact, the quantity of CO_2 evolved by button mushrooms (*Agaricus bisporus*) in the light is always smaller than that evolved in darkness. It is generally the same for green plants; however, difficulties remain to precisely assess the difference between respiratory rates in the light and in the dark, because chl.-dependent assimilation occurs in the light and one does not know how to accurately separate both phenomena.'

As emphasised by Pizon (1902), one major difficulty stems from the limited technical possibilities to measure the rate of day respiration (CO_2 evolution in the light produced by respiration, as opposed to night respiration, i.e., the CO_2 evolution in darkness or R_n), and importantly, from the impossibility to isolate and purify such respired CO_2 molecules for example, ^{14}C or ^{13}C isotopic analyses. Several techniques (gas-exchange or $^{12}C/^{13}C$ techniques) have been developed over nearly 30 years to overcome this difficulty, and they are summarised below. As such, it is currently believed that photosynthesis is accompanied by lower respiration rates (inhibition of day respiration). Molecular techniques have also provided the evidence that the respiratory metabolism changes somewhat during photosynthesis in illuminated leaves, and this correlates with the operation of nitrogen and redox metabolism. At the whole-plant level, respiration is responsible for major carbon losses, but there is now a large body of evidence showing that plant growth and development is also sustained by respiration. Therefore, respiration is often viewed as beneficial for plant carbon gain at both leaf and plant levels. In the following, we intend to describe the main advances that have converged towards such a picture, and the reader may find more detailed information in the quoted references.

4.2. RESPIRATION AS AN ESSENTIAL FACTOR OF PHOTOSYNTHESIS AND PLANT CARBON MASS-BALANCE: GENERAL FEATURES

Two levels at which respiration interacts with plant photosynthesis and primary production should be recognised. First, both R_d and R_n produce CO_2, as opposed to photosynthesis that assimilates CO_2. When heterotrophic organs such as roots and stems are taken into account, it is believed that nearly 40 to 60% of photosynthetically derived CO_2 is lost by respiration in cultivated plants (Lambers and Poorter, 1992; Evans, 1993). Such a proportion nevertheless varies among functional groups and depends upon the prevalence of heterotrophic versus autotrophic organs within plants (Mooney, 1972). At the leaf level, up to 20% of assimilated CO_2 is lost through respiration. From such large figures, however, it should not be concluded that reducing respiration may necessarily optimise plant growth and net carbon gain. At the whole-plant level, respiration is essential

Terrestrial Photosynthesis in a Changing Environment: A Molecular, Physiological and Ecological Approach, ed. J. Flexas, F. Loreto and H. Medrano. Published by Cambridge University Press.

for growth and CO_2 assimilation simply because of energy requirements (ATP synthesis), primary N metabolism and amino-acid synthesis, which is in turn the cornerstone for photosynthesis (the major photosynthetic components contain N, like Rubisco and chl.s). In fact, whereas the relationship between respiratory metabolism and body size follows similar rules in animals and plants, there is indeed a strong relationship between total respiration and total plant N content (or shoot respiration and shoot N content) across a wide range of species (Reich *et al.*, 2006). Such a relationship may, however, depend upon functional groups (e.g., forbs, broad-leafed trees, shrubs, etc.), but as a general feature the leaf R_n rate correlates well with a combination of assimilation-related traits such as life-span, N content or specific leaf area (SLA) (leaf dry matter per surface area); in other words, mass-based R_n rates rise with increasing N content or SLA, which in turn correlate with the maximum carboxylation velocity ($V_{c,max}$) (Reich *et al.*, 1998a,b,c).

Plant respiration is ordinarily split into two terms (the so-called McCree model), namely, maintenance and growth respiration that are associated with plant-cell regeneration (turnover) and the production of new constituents for growth, respectively. Maintenance processes in plant cells (protein turnover, ion gradients, membrane synthesis, etc.) consume a significant amount of plant productivity (near 10%), equalling 30 to 50 mg of glucose per gram of protein per day (for an extensive review, see Amthor, 2000). In photosynthesising leaves, day respiration may also contribute to these maintenance processes, in addition to supporting ATP for sucrose synthesis in the cytoplasm (see below). The energy sustaining role of R_d may also be emphasised when the energy need is high in the cell. This is the case under environmental conditions that induce a stress, because plant-cell repair of stress-induced damage (of, for example, photosystems, membranes and pigments) necessitates an enhanced ATP synthesis (Chaumont *et al.*, 1995). Such a positive view disagrees, however, with the absence of biomass reduction in several respiratory mutants affected in, for example, phosphofructokinase (Hajirezaei *et al.*, 1994) or aconitase (Carrari *et al.*, 2003). Nevertheless, some compensatory processes are likely to occur in mutants (metabolic bypasses) and as such, no firm conclusions may be drawn from their phenotypes or their lack thereof.

Second, respiration interacts with photosynthesis as an integrated component of the photosynthetic metabolism in illuminated leaves (Krömer *et al.*, 1993; Noguchi and Yoshida, 2008; Nunes-Nesi *et al.*, 2008). Such an interaction manisfests itself through the inhibitory effect of light on leaf R_d (see Introduction and below). In fact, it is now widely believed that R_d exhibits at least three cellular interaction levels with photosynthesis: first, it provides carbon skeletons (such as 2-oxoglutarate) for N assimilation, the products of which are in turn needed by photorespiration (level 1 in Fig. 4.1); second, mitochondria may oxidise NADH molecules of photorespiratory origin and export excess reductive power through a malate shuttle (level 2 in Fig. 4.1); and third, it sustains sucrose synthesis by providing ATP molecules in the cytoplasm (level 3 in Fig. 4.1). Based on these principles, it is now accepted (and we will develop below) that in C_3 leaves, photosynthesis inhibits respiratory CO_2 evolution in the light because of a complex regulation set based on redox-poise and effectors between plant-cell compartments. Such an inhibition nevertheless allows a small respiratory metabolic flow to operate in the light, and it is thus assumed that the recycling of stored molecules sustains N assimilation. In other words, R_d is partly uncoupled from current CO_2 assimilation in illuminated leaves, and so is N assimilation. However, the interaction between photosynthesis and respiration may be more complex. On one hand, photosynthesis mutants show alterations in mitochondrial electron transport, particularly under high-light conditions (Yoshida *et al.*, 2007, 2008). On the other hand, respiratory mutants present alterations in their photosynthetic rates. These include complex mitochondrial rearrangements (Juszczuk *et al.*, 2008), complex-I mutations (Priault *et al.*, 2006; Gallé *et al.*, 2010), alternative oxidase (AOX) antisense plants (Giraud *et al.*, 2008; Chai *et al.*, 2010), mitochondrial uncoupling protein (UCP) knockouts (Sweetlove *et al.*, 2006) and plants with altered levels of different tricarboxylic-acid-cycle (TCAC) components (Carrari *et al.*, 2003; Nunes-Nesi *et al.*, 2005, 2007; Studart-Guimerães *et al.*, 2007; Sienkiewicz-Porzucek *et al.*, 2008, 2010). Although impairments of most of these mitochondrial functions result in decreased photosynthesis, at least reduced expression of aconitase and mitochondrial MDH results in enhanced photosynthesis (Carrari *et al.*, 2003; Nunes-Nesi *et al.*, 2005). Alternatively, the mechanisms for decreased photosynthesis seem to differ in each case. For instance, impaired AOX and UCP affect photosynthesis mostly through impairment of thylakoid reactions (Sweetlove *et al.*, 2006; Giraud *et al.*, 2008; Chai *et al.*, 2010; Dinakar *et al.*, 2010; Florez-Sarasa *et al.*, 2011), whereas impaired complex I results in decreased stomatal and mesophyll conductance to CO_2 (Priault *et al.*, 2006; Juszczuk *et al.*, 2008; Gallé *et al.*, 2010). Among the TCAC

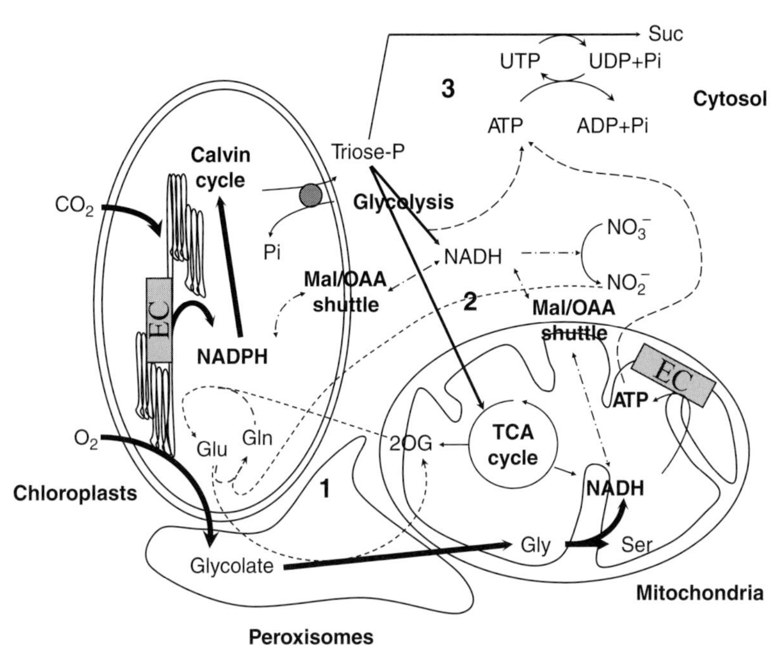

Figure 4.1. Simplified scheme showing the main interaction levels between day respiration and photosynthesis or photorespiration. First, day respiration provides 2OG molecules through the TCA cycle activity, which are required to assimilate -NH_2 groups to Glu. The latter is in turn consumed by the photorespiratory pathway to generate Gly from glycolate. Second, reductive-power excess (NAD(P)H) of the mitochondria may be exchanged with the cytosol or, maybe, peroxisomes, via a malate/oxaloacetate shuttle. Third, day respiratory production of ATP may be used for Suc synthesis via UDPG production. ADP/ATP, adenosine di/triphosphate; EC, electron chain; Mal, malate; NADPH, nicotinamide adenine dinucleotide phosphate; OAA, oxaloacetate; 2OG, 2-oxoglutarate; triose-P, triose phosphates; UDP/UTP, uranyl di/triphosphate.

component reductions, those decreasing photosynthesis apparently result from decreasing stomatal (and perhaps mesophyll) conductance, whereas those increasing photosynthesis involve increments of photosynthetic capacity (Carrari *et al.*, 2003; Nunes-Nesi *et al.*, 2005, 2007; Studart-Guimerães *et al.*, 2007; Sienkiewicz-Porzucek *et al.*, 2008, 2010). Clearly, intensive research is required in this area to unravel the precise levels of interaction between photosynthetic and respiratory metabolism.

4.3. GAS-EXCHANGE AND ISOTOPIC DATA, AND PHOTOSYNTHETIC MODELS INTEGRATING DAY RESPIRATION

4.3.1. How do photosynthesis models integrate day respiration?

The well-known photosynthesis model derived from the works of Farquhar, von Caemmerer and Berry (for a review of such models, see von Caemmerer, 2000) is as follows, where A_N is net CO_2 assimilation:

$$A_N = V_c\left(1-\frac{\Gamma^*}{C_c}\right) - R_d \qquad \text{[Eqn. 4.1]}$$

where V_c is the RuBP-carboxylation rate, C_c the internal chloroplast CO_2 mole fraction (at the carboxylation sites), R_d is day respiration and Γ^* is the CO_2 compensation point in the absence of day respiration. The latter depends on the specificity factor (denoted as $S_{c/o}$) of Rubisco as:

$$\Gamma^* = \frac{[O_2]}{2S_{c/o}} \qquad \text{[Eqn. 4.2]}$$

where $[O_2]$ is the O_2 mole fraction at the site of carboxylation. Equation 4.1 may be converted to a C_i-based equation, with a C_i-based Γ^* value (denoted as C^* in the following). It might be argued that R_d in equation 4.1 appears as a simple additional term and should be increased by a mathematical factor that accounts for possible refixation of day respired CO_2 (for a discussion, see Pinelli and Loreto, 2003). However, that would be incorrect. R_d is the true day respiration rate, and refixation is already taken into account in equation 4.1 simply because in the steady state, C_c integrates day respiratory CO_2 production, photorespiration and gross photosynthetic CO_2 fixation (von Caemmerer, 2000).

In equation 4.1, day respiration is an additional term that decreases A_N. Such an effect is nevertheless small, as the order of magnitude of R_d ordinarily lies below 1 μmol m^{-2} s^{-1} (see also below). However, the ratio R_d/A_N becomes much more important under stress conditions (e.g., drought) when photosynthesis can decrease by nearly 90%, whereas respiration might hardly be affected or even increased (for a review, see Atkin and Macherel, 2008). Nevertheless, it should be recognised that photosynthesis under drought may also be accompanied by an increase of refixation of respired CO_2 owing to stomatal closure (Haupt-Herting *et al.*, 2001), and so uncertainty remains about whether the relative importance of respiration is indeed larger under these circumstances. In fact, R_d is commonly considered as a constant parameter that does not depend upon C_c nor V_c. Whereas several techniques show that this seems to be the case (e.g., a single convergence point is found with Laisk's method), other recent studies indicate that such an assumption may not be valid in usual conditions at the metabolic scale (e.g., day respiratory metabolism changes as the CO_2/O_2 ratio varies, see below).

Using a kinetic model describing V_c as a function of C_c, the maximal carboxylation velocity ($V_{c,max}$), the apparent Michaelis constant for CO_2, it is possible to fit the experimental values from an A_N/C_c curve and get an estimate of R_d. However, such an estimate is not always reliable, as the steep slope at low C_c values may cause large uncertainties for the *y* intercept and the estimate of R_d is quite sensitive to the linear part of the A_N/C_c curve. Thus, other particular techniques are currently used to measure R_d, as described below.

4.3.2. Methods to measure day respiration: a short overview

The five main methods currently used are recalled here (for a newly proposed method using combined gas exchange and Chl-F, see Yin *et al.*, 2011). With Kok's method (Kok, 1948), a light-response curve is generated with very low light levels and ordinarily it can be seen that the A_N shows an abrupt change of slope. The extrapolation to zero light from the low slope part indicates the rate R_d, whereas the steeper part reaches the *y* axis (zero light) at the night respiration rate (R_n) (see Fig. 4.2A for an example).

With Laisk's method (Laisk, 1977), one uses A_N/C_c or A_N/C_i curves at different light levels, that converge to a single point that corresponds to Γ^* (CO_2 compensation point in the absence of day respiration) and $-R_d$ (see Fig. 4.2B for an example). It can indeed be seen from equation 4.2 that Γ^* does not depend on light or CO_2 for a given O_2 concentration in the air. Thus, for $C_c = \Gamma^*$, we have $A = -R_d$ whatever the light level is. However, this technique is difficult to perform, mainly because very slight errors on C_i (or C_c) values lead to large errors in the convergence area, with quite often a triangular convergence area instead of a single point. Moreover, much care should be taken at low CO_2 concentrations because leaks and respiration of surrounding tissue may adulterate measurements with many of the commercially available gas-exchange systems (Hurry *et al.*, 2005, see also Chapter 9). This technique has been improved from a statistical point of view by the use of the CO_2 compensation point, Γ, and a regression analysis (Peisker and Apel, 2001). If equation 4.1 is C_i-based, it may be re-written as:

$$A_N = g_m(C_i - C^*) - R_d \qquad \text{[Eqn. 4.3]}$$

where C^* is the C_i-based Γ^* (compensation point in the absence of day respiration) and g_m is mesophyll conductance, defined here as the slope of the A_N/C_i relationship in the linear area (low C_i values). By definition, A_N=0 at C_i=Γ and so we have:

$$\Gamma = \frac{\mu}{g_m} \times R_n + C^* \qquad \text{[Eqn. 4.4]}$$

where R_n is the rate of night respiration and μ is the R_d/R_n ratio. A dataset made of R_n, g_m and Γ values obtained with different A_N/C_i curves allows one to get an estimate of the slope μ with a regression analysis (Fig. 4.2C).

Following Cornic's method (Cornic, 1973), an illuminated leaf is placed in CO_2-free air in either N_2 (0% O_2) or 21% O_2 and then darkened. The CO_2 production rate in the light is denoted as L_O (in 21% O_2) or L_N (in N_2). As soon as the leaf is darkened, refixation of (photo)respired CO_2 vanishes and one can see a peak of CO_2 production (denoted as *p*). Under several simplifying assumptions, it can be shown that R_d=L_O–L_N–p+R_n (see Fig. 4.3A for an example).

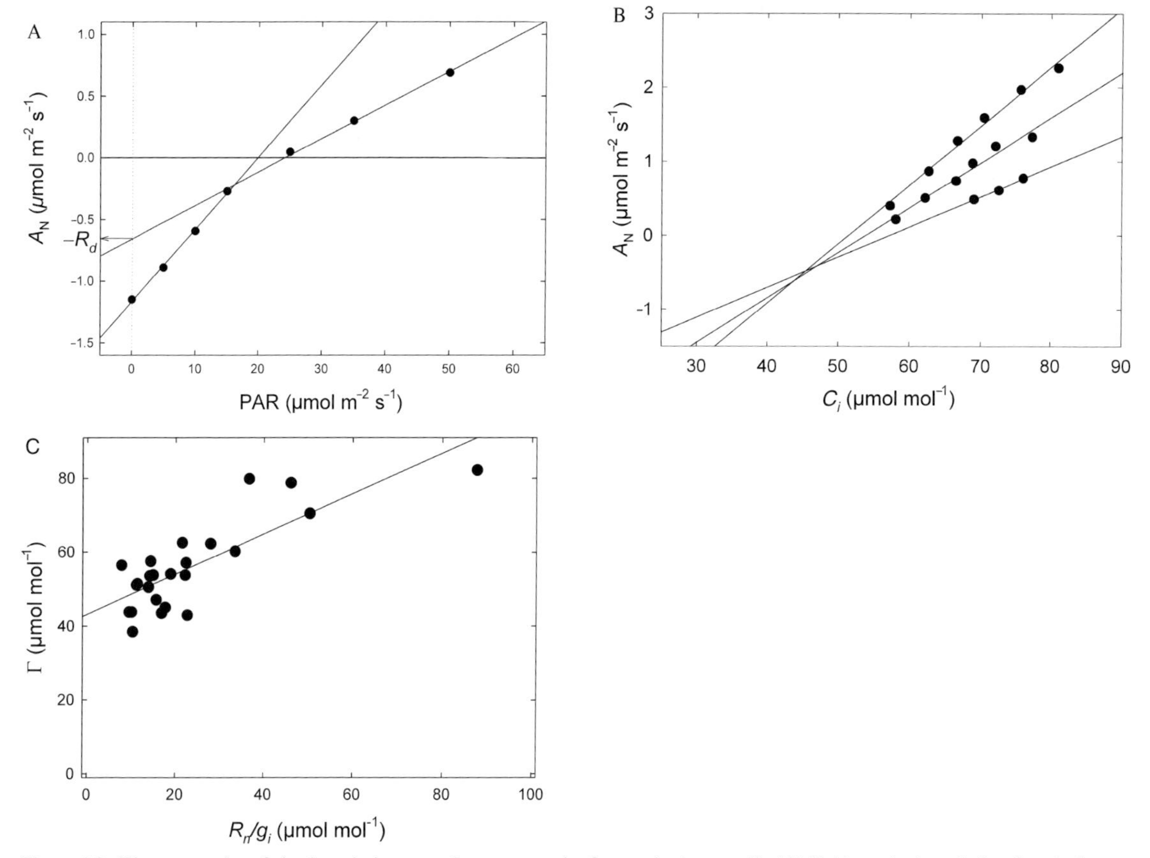

Figure 4.2. Three examples of classic techniques used to measure the day respiration rate R_d. (**A**) Kok's method applied to detached *Phaseolus vulgaris* leaves. The dataset was obtained on *Xanthium strumarium* leaves at 21°C in 21% O_2 (typical graph redrawn from the dataset of Tcherkez *et al.*, 2008). (**B**) Laisk's method. A_N/C_i curves at 200, 400 and 600 μmol m^{-2} s^{-1} (PAR) at 21°C in 21% O_2 in french bean (*Phaseolus vulgaris*). In this example, the day respiration rate R_d is 0.51 μmol m^{-2} s^{-1} and the C_i-based compensation point in the absence of day respiration, C^*, is 47 μmol mol^{-1}. (**C**) The method of Peisker and Apel (2001) applied to detached *Phaseolus vulgaris* leaves. The dataset was obtained from A/C_i curves at 21°C in 21% O_2. The regression line gives a slope μ (=R_d/R_n) of 0.54±0.08 and a C^* value of 43±2 μmol mol^{-1} (r^2=0.81). Data are redrawn from Tcherkez *et al.* (2005).

Loreto's method (Loreto *et al.*, 2001a; Pinelli and Loreto, 2003 and see also Haupt-Herting *et al.*, 2001 for a similar method) takes advantage of the slower turnover of day respiratory substrates (half time in the order of minutes or more) compared with photorespiratory substrates (half time in the order of seconds). Then the production of $^{12}CO_2$ in a $^{13}CO_2$ atmosphere by plants grown in a natural $^{12}CO_2$ atmosphere indicates the R_d rate (with some corrections needed to be taken into account for refixation of respired $^{12}CO_2$).

These methods have contrasting advantages and difficulties: for example, there is a large measurement variability owing to small assimilation rates in both Kok's and Laisk's methods. In addition, Laisk's method requires several light levels and it remains plausible that R_d varies with light. In that sense, the use of equation 4.4 is a clear improvement because it does not necessarily requires different light conditions.

4.3.3. $^{12}C/^{13}C$ isotopic fractionation, photorespiration and day respiration

Carbon isotopes have proven to be useful tools to elucidate plant carbon metabolism, and the use of $^{12}C/^{13}C$

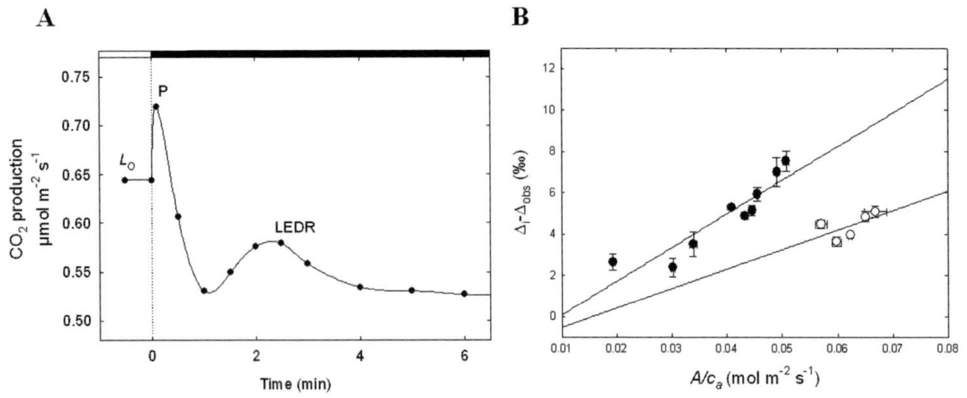

Figure 4.3. (A) Cornic's method to measure the day respiratory rate R_d. CO_2 evolved by white mustard (*Sinapis alba*) leaves in the light in CO_2-free air at 21% O_2 and 25°C (L_O) and thereafter in darkness (note the photorespiratory peak p and then the light-enhanced dark respiration peak LEDR). Assuming that: (1) the peak p represents photorespiratory decarboxylation (Φ) plus night respiration (that is, $\Phi + R_n$); and (2) L_O–L_N comprises photorespiration plus day respiration ($\Phi+R_d$), we have indeed $R_d=L_O-L_N-p+R_n$. Redrawn, from Cornic 1973. (**B**) An example of the relationship between the deviation (Δ_i–Δ_{obs}) of the photosynthetic $^{12}C/^{13}C$ fractionation from the theoretical value (Eqn. 4.7) and the assimilation-to-CO_2 ratio (A/C_a) in *Xanthium strumarium* leaves at 21°C and 400 µmol mol^{-1} CO_2 in either 21% O_2 (closed symbols) or 2% O_2 (open symbols). Plants were cultivated in normal air ($\delta^{13}C$ of CO_2 at –9.5‰), while inlet CO_2 used for the gas-exchange experiments was industrial (at –50.2‰). The y intercept represents the (photo)respiratory fractionation component, that turns out to be slightly negative in 2% O_2. This simply indicates that day respiration oxidises stored material which is enriched in ^{13}C compared with inlet CO_2. In other words, photosynthates produced in the greenhouse (and subsequently decarboxylated by day respiration) had a $\Delta^{13}C$ value of around at –28‰, that is, ^{13}C-enriched by nearly 40‰ compared with current assimilates produced during the experiment. Redrawn, from Tcherkez and Farquhar (2008).

fractionations brought out information on photorespiration and respiration with respect to their relationships with assimilation. The relevant theory summarised in the following has been described elsewhere (Farquhar *et al.*, 1982 and see Chapter 11) and only the main equations are presented here. The 'theoretical' form of the $^{12}C/^{13}C$ carbon-isotope fractionation associated with photosynthesis (denoted as Δ_i), which neglects (photo)respiratory fractionation and the boundary layer resistance, is:

$$\Delta_i = a + (\bar{b} - a)\frac{C_i}{C_a} \quad \text{[Eqn. 4.5]}$$

where the diffusional fractionation a is the fractionation associated with diffusion and $\bar{b}$ is the fractionation associated with carboxylation (the Rubisco-catalysed $^{12}C/^{13}C$ fractionation is 29‰); C_a and C_i are the atmospheric (outlet) and internal mole fractions of CO_2. In equation 4.5, the $\bar{b}$ value bintegrates internal conductance effects (and so is actually slightly less than 29‰). The more complete expression of the isotope discrimination, usually assumed to explain the 'observed' photosynthetic fractionation ($\Delta^{13}{}_{obs}$), is given by:

$$\Delta^{13}{}_{obs} = a_b\frac{C_a - C_b}{C_a} + a\frac{C_b - C_i}{C_a} + (a_l + e_s)\frac{C_i - C_c}{C_a} + b\frac{C_c}{C_a} - \frac{eR_d/k + f\Gamma^*}{C_a} \quad \text{[Eqn. 4.6]}$$

where the subscripts a, b, i and c refer to atmospheric, leaf surface, intercellular and carboxylation site CO_2 mole fractions; a_b is the gaseous diffusional fractionation associated with the boundary layer (2.9‰) and a_l (1.1‰) and e_s (0.7‰) are the fractionations associated with diffusion in the liquid phase and with CO_2 dissolution, respectively; e and f are the fractionations associated with R_d rate and photorespiration, respectively; k is the carboxylation efficiency and Γ^* the CO_2 compensation point in the absence of day respiration; b is the carbon-isotope fractionation associated with Rubisco-catalysed carboxylation (in contrast to equation 4.5), the b value does not integrate internal conductance effects and so is purely related to carboxylation by Rubisco). Under the assumption that $b=\bar{b}$ =29‰, we have:

$$\Delta_i - \Delta_{obs} = \frac{b - e_s - a_l}{g_m} \times \frac{A_N}{C_a} + d \qquad \text{[Eqn. 4.7]}$$

where A_N is net assimilation and $d = (eR_d/k + f\Gamma^*)/C_a$. A linear regression can be used to determine d if C_a is maintained constant during the experiment (otherwise d varies along the A_N/C_a range investigated). Under non-photorespiratory conditions (2% O_2), d comprises only the R_d term eR_d/kc_a. With a known value of e, R_d may be determined or, reciprocally with known values of R_d, a range of e values is obtained. Such values allow one to know the natural ^{13}C abundance of day respired CO_2, that is, the carbon source oxidised by day respiration and/or the $^{12}C/^{13}C$-fractionating metabolic steps involved in day respiration. An example that uses the relationship given in equation 4.7 is shown in Fig. 4.3B, where eR_d/kc_a=–0.5‰ (y intercept in 2% O_2). Under the assumption that the ^{13}C enrichment of day respired CO_2 is around 40‰ (e=–40‰) compared with current fixed CO_2, and supposing a k value of around 0.1 mol m^{-2} s^{-1}, R_d is then around 0.5×0.1×400/40=0.5 μmol m^{-2} s^{-1}.

4.4. DAY RESPIRATION IN ILLUMINATED LEAVES, AS CONTROLLED BY PHOTOSYNTHESIS AND PHOTORESPIRATION

4.4.1. The inhibition of leaf respiration in the light

With the use of the techniques described above, it has been repeatedly shown that the rate R_d is smaller than the rate R_n, the R_d/R_n ratio (denoted as μ) being around 0.5 with a range of values between 0.1 and 0.7 (for a review, see Atkin *et al.*, 2000a). For example, the R_d value obtained in Figure 4.2 is 0.51–0.54 μmol m^{-2} s^{-1}, and that in Figures 4.3A and 4.3B is near 0.4 and 0.5 μmol m^{-2} s^{-1}, respectively. In addition, $^{12}CO_2$ production in a $^{13}CO_2$ atmophere has shown that R_d is less than R_n by 30–70% in typical conditions and additionally, the ratio μ decreases at high CO_2 concentration to nearly 0.1 (Pinelli and Loreto, 2003). Similar are the results obtained with ^{14}C techniques (Pärnik and Keerberg, 2006). Thus light 'inhibits' leaf respiration. This view is consistent with the observation that the respiratory rate usually shows a wide peak within 2–5 minutes after darkening. Such a peak positively depends on illumination and the photosynthesis rate and is thus called the 'light-enhanced dark respiration' (denoted as LEDR; see Fig. 4.3A and Atkin *et al.*, 1998). Pioneering gas-exchange measurements on white mustard (*Sinapis alba*) suggested that some enzymatic activities are inhibited in the light so that substrates accumulate (Cornic, 1973), explaining such a respiratory burst. Recently, rapid ^{13}C analysis techniques (based on tunable diode laser systems) allowed the ^{13}C abundance of CO_2 evolved just after darkening to be measured. It has thus been shown that the CO_2 associated with LEDR is strongly enriched with ^{13}C (Barbour *et al.*, 2007). The ^{13}C-enriched source material is thought to be organic acids (e.g., malate) that accumulate in the illuminated leaf, presumably because of the inhibition of the TCAC (Barbour *et al.*, 2007) and in fact the ^{13}C enrichment of post-darkening CO_2 correlates closely to the decrease of the malate content (Gessler *et al.*, 2009a,b). In other words, the inhibition of R_d is thought to cause the LEDR. It has been consistently found that malic enzyme (which converts malate into pyruvate + CO_2) is inhibited in the light (Hill and Bryce, 1992). By contrast, no clear picture emerges from studies on O_2 consumption associated with day respiration (for a review, see Atkin *et al.*, 2000a). In *Chlamydomonas reinhardtii* cells under saturating CO_2 concentration (non-photorespiratory conditions), respiratory O_2 consumption was the same in the light and in the dark (Peltier and Thibault, 1985). With ^{18}O techniques, Canvin *et al.* (1980) showed that the O_2 consumption rate in the light was virtually zero in detached *Hirschfeldia incana* leaves in low photorespiratory conditions (1000 μmol mol^{-1} CO_2), whereas that in darkness reached 2 μmol m^{-2} s^{-1}. However, it has been argued that the mitochondrial electron transport is likely to change in the light as its role shifts from ATP production (darkness) to redox regulation (light) (Gardeström *et al.*, 2002) and in fact the increase of the NAD(P)H dehydrogenases activity is divorced from ATP synthesis in the light (Hurry *et al.*, 2005). In addition, the activation of the cyanide-resistant AOX by light has been observed in soybean cotyledons (Ribas-Carbo *et al.*, 2000), and this appears to be related to phytochrome-mediated signalling (Ribas-Carbo *et al.*, 2008).

Much uncertainty remains and if respiratory O_2 consumption happens to be affected by light, one may expect a large variability owing to the rate of photorespiration: the mitochondrial electron transport chain can re-oxidise NADH molecules derived from glycine (Gly) oxidation, which in turn depends on the photorespiration rate. Future experimental studies are needed to address this issue.

4.4.2. The biochemical origin of the inhibition of day respiration

Whereas there is very little evidence of a regulation at the genetic transcription level (Rasmusson and Escobar, 2007), the fundamental reasons invoked to explain the inhibition of R_d by light are enzymatic (post-transductional or biochemical). It has indeed been shown in the unicellular alga *Selenastrum minutum* that pyruvate kinase is inhibited in the light by the high level of cytoplasmic DHAP (Lin *et al.*, 1989). In addition, the mitochondrial pyruvate-dehydrogenase complex is partly inactivated by (reversible) phosphorylation in extracts from illuminated leaves (Budde and Randall, 1990; Tovar-Mendez *et al.*, 2003). Photorespiration is probably also involved in the inhibition of pyruvate dehydrogenase as it has been shown that this enzyme is downregulated by NH_3, which is a byproduct of the photorespiratory Gly decarboxylation (Krömer, 1995). Enzymes of the TCAC are also assumed to be inhibited in the light because of the high mitochondrial $NADH/NAD^+$ ratio owing to photorespiratory Gly decarboxylation (Gardeström and Wigge, 1988). Additionally, it has been shown that the mitochondrial isocitrate dehydrogenase is inhibited by the high $NADPH/NADP^+$ ratios that occur in the light (Igamberdiev and Gardeström, 2003). Accordingly, Gly decarboxylase (GDC) antisense lines of potato (*Solanum tuberosum*) have larger decarboxylation rates in the light (as revealed by ^{14}C-labelling experiments), and the ATP/ADP as well as the $NADH/NAD^+$ ratios are both smaller than in the wild type (Bykova *et al.*, 2005). In addition, the predominance of NADH production by Gly oxidation over that by the TCAC has been shown using isolated mitochondria under ADP-limiting conditions (Day *et al.*, 1985). This might reduce the NAD^+ available for the mitochondrial dehydrogenase steps of the TCAC. Such a scenario is consistent with the larger in-vivo decarboxylation rates of ^{13}C-enriched TCAC-substrates under very low, non-physiological O_2 conditions (400 μmol mol^{-1} CO_2, 2% O_2) as compared with high O_2 conditions (1000 μmol mol^{-1} CO_2, 21% O_2) (Tcherkez *et al.*, 2008).

Other physiological experiments have further shown that the inhibition of R_d in the light is associated with a lower TCAC activity. First, O_2 consumption measurements with mitochondria extracted from illuminated spinach leaves (*Spinacia oleracea*) and supplied with either exogenous malate, succinate or citrate show that citrate gives the lowest respiration rate. In addition, when malate is supplied, it is mainly converted to citrate and pyruvate, with less than 1% of isocitrate or fumarate (Hanning and Heldt, 1993). These observations suggest that TCAC reactions are very slow, whereas phosphoenolpyruvate carboxylase (PEPC), MDH or malic enzyme are still active in the light. Consistently, when detached illuminated leaves of *Phaseolus vulgaris* are supplied with [^{13}C-1]-pyruvate, $^{13}CO_2$ is produced in the light, showing the in-vivo activity of pyruvate dehydrogenase. When supplied with [^{13}C-3]-pyruvate, leaves hardly produce $^{13}CO_2$, showing the very weak activity of the TCAC (Tcherkez *et al.*, 2005). Such a ^{13}C-enrichment method has clearly demonstrated that *in vivo*, the TCAC activity is inhibited by 80–95% in the light, whereas pyruvate dehydrogenation is inhibited by light to a much lower extent, by 20–30% (Tcherkez *et al.*, 2005, 2008).

It should be emphasised that CO_2 production by R_d also includes chloroplastic decarboxylases, namely, the chloroplastic pyruvate dehydrogenase and the 6-phosphogluconate-dehydrogenase (that belongs to the oxidative pentose-phosphate cycle and produces ribulose-5-phosphate + CO_2). The latter enzyme is not functional in the light because the first reaction of the oxidative pentose-phosphate cycle is inhibited, preventing 6-phosphogluconate from being produced in the illuminated chloroplast; Singh *et al.* (1993a,b) have indeed shown with [^{14}C]-hexoses that glucose-6-phosphate-dehydrogenase is inhibited by light to nearly zero activity from light levels as low as 5 μmol m^{-2} s^{-1}, and this correlates to the chloroplastic $NADPH/NADP^+$ ratio. The chloroplastic pyruvate dehydrogenase is not controlled by phosphorylation and hence is not downregulated in the light (Tovar-Mendez *et al.*, 2003), and this probably explains the residual 20–30% PDH activity in the data of Tcherkez *et al.* (2005).

4.4.3. Does the inhibition of day respiration vary with CO_2 conditions?

As pointed out in Section 4.3.1, it is often assumed that R_d is a constant parameter in gas-exchange equations (see Eqn. 4.1) and throughout A_N/C_i curves as well. However, R_d is likely to depend upon assimilation and photorespiration, that is, on CO_2 conditions. With several A_N and C_i datasets collected at different light levels, Peisker and Apel (2001) have indeed shown in *Nicotiana tabacum* leaves that μ ($=R_d/R_n$) decreases very slightly from 0.7 to 0.6 as light decreased from 950 to 60 μmol m^{-2} s^{-1}. Pinelli and Loreto (2003) showed in several species that R_d was decreased at a high CO_2 mole fraction, and so it has been supposed that the inhibition of R_d metabolism positively depends on the photosynthetic assimilation rate. Consistently, the phosphorylation (and thus the inhibition) of pyruvate dehydrogenase is lower when the photosynthetic inhibitor DCMU is supplied (Budde and Randall, 1990). Further, with ^{13}C techniques, Tcherkez *et al.* (2008) have shown that the in-vivo metabolic activity of the TCAC is enhanced at a low CO_2 mole fraction (140 μmol mol^{-1} in 21% O_2), and this is associated with a larger commitment to glycolysis and glutamate (Glu) synthesis in *Xanthium strumarium* leaves. In addition, the TCA decarboxylation activity correlates linearly with the dihydroxyacetone phosphate to glucose-6-phosphate ratio. In other words, the larger the CO_2 assimilation, the larger the inhibition of day respiration when CO_2 conditions are to vary.

Nevertheless, as outlined above, photorespiration rates have been assumed to inhibit R_d because of the influence of the NADH/NAD$^+$ ratio on enzyme activities. It is thus likely that the R_d rate (and particularly the activity of the TCAC) is the result of two opposing forces: (1) the stimulation of the TCAC by the glycolytic input and, maybe, allosteric effectors; and (2) the inhibition (of the PDH and TCAC enzymes) caused by enzymatic control through phosphorylation and/or the effect of the redox poise. Because of the very large Gly decarboxylation rate (and thus of NAD$^+$ reduction to NADH in the mitochondrial matrix), the stimulation of the TCAC, as revealed by isotopic-labelling experiments, may appear somewhat surprising under photorespiratory conditions. However, such a stimulation probably reflects an increased need for Glu to feed photorespiratory N recycling when conditions shift to low CO_2 mole fractions, simply because the production of Gly from glycolate requires a higher Glu flow. In this framework, it is unsurprising that the production of the glu precursor 2-oxoglutarate by the TCAC is enhanced at a low CO_2 mole fraction, as revealed by ^{14}C- and ^{13}C-labeling (see Section 4.5.1 below). In addition, mitochondria are currently thought to export the excess of reductive power, thereby maintaining NADH to a rather constant level *in vivo* (see below, Section 4.6).

4.4.4. The effect of temperature on interactions between (day) respiration and photosynthesis

It is believed that temperature has a major impact on photosynthesis/respiration ratios. In fact, increasing leaf temperature above 30 or 35°C leads to altered photosynthesis in many plant species, primarily because both the solubility of CO_2 and the $S_{c/o}$ decrease as temperature increases (Ghashghaie and Cornic, 1994). In addition, leaf R_n is larger at high temperature, with a Q_{10} (scaled increase by 10°C steps) of nearly two (for an extensive review, see Atkin and Tjoelker, 2003). Nevertheless, the extent to which such a relationship remains valid with R_d is poorly known, because measuring R_d at different temperatures is an experimental tour de force. For example, the use of the Laisk method is complicated by the change in Γ^*, as the O_2 mole fraction and $S_{c/o}$ vary with temperature (see Eqn. 4.2, Section 4.3.1). That said, careful experiments (Atkin *et al.*, 2000b; Warren and Dreyer, 2006) have shown that the R_d rate increases slowly with temperature, but remains nearly constant between 10 and 20°C. Thus high temperature (usually, the explored range falls within the 35°C vicinity) accelerates both R_n and R_d. Such an effect does not last, however, as respiratory acclimation eventually occurs (Atkin and Tjoelker, 2003; Atkin *et al.*, 2006).

Quite differently, exposure to low temperature near 10°C, i.e., chilling injury, has a significant impact on primary metabolism, altering the morphologies and ultrastructure of chloroplasts and mitochondria and the photosynthetic apparatus (Lyons, 1973). Photosynthesis and respiration are decreased and less ATP is therefore available for biosynthesis and/or cellular maintenance. Consequently then, whole-plant growth is interrupted when leaf respiration is inhibited by chilling. A strong correlation has been found between the recovery of growth at low temperatures and the potential for mitochondrial ATP synthesis (Graham and Patterson, 1982). Although chilling severely inhibits growth in cold-sensitive species, cold-hardy species such as *Arabidopsis* are able to re-establish growth over extended periods of exposure to low temperatures. Whereas the mitochondria of pre-existing leaves cannot acclimate to low temperatures, growth can be re-established once new leaves that have developed in the cold are formed. Leaves that have developed in the cold have enhanced rates of leaf respiration. The recovery of photosynthesis and growth at chilling temperatures is thus closely coupled to recovery of mitochondrial ATP synthesis, cold-acclimated tissues showing enhanced respiration, a greater mitochondrial volume and the abundance of proteins associated with respiratory ATP synthesis (Kurimoto *et al.*, 2004).

4.5. DAY RESPIRATION INTERACTS WITH PRIMARY NITROGEN METABOLISM AND CELL REDOX BALANCE DURING PHOTOSYNTHESIS

4.5.1. Day respiration, carbon sources and primary nitrogen metabolism

Respiration is often viewed as a key process for N metabolism because the TCAC may provide the 2-oxoglutarate molecules needed for Glu synthesis in the light. The increase of leaf R_n rate when leaves are incubated with NO_3^- – a fact known for half a century – then appears as unsurprising. However, whereas nitrate reduction and assimilation mainly occur in the light, leaf day respiratory decarboxylations involved in the TCAC are much lower than R_n (see Section 4.4 above). Among these decarboxylations is the NAD-dependent isocitrate dehydrogenase (IDH) reaction, which is responsible for 2-oxoglutarate (the key acceptor molecule for Glu synthesis) production by the cycle. These contrasting facts might be explained by the involvement of the cytoplasmic NADP-dependent isocitrate dehydrogenase (ICDH). It may be indeed plausible that the latter enzyme compensates for the TCAC inhibition by producing 2-oxoglutarate in the cytoplasm for N assimilation during the light period (Chen and Gadal, 1990). However, mutants affected at the cytosolic NADP-dependent isocitrate-dehydrogenase level, with a residual enzymatic activity of only up to 10%, do not show any apparent phenotype (Kruse *et al.*, 1998), and so is the case of mutants deficient in the NAD-dependent isocitrate dehydrogenase (Lemaître *et al.*, 2007). Unless there are other prevailing imperatives, this suggests that a rather small metabolic flux associated with I(C)DH is probably enough to sustain N assimilation (Hodges, 2002). In fact, it is striking that the actual flux through the TCAC of illuminated leaves is near 0.05 μmol m^{-2} s^{-1} (Tcherkez *et al.*, 2005, Fig. 4.4) while that of average N assimilation over the lifespan of a leaf is in the same order of magnitude (~4.5 mol N m^{-2} divided by 15 days of 16 h light needed for the development of one leaf; this gives ~ 0.05 μmol m^{-2} s^{-1}) (Tcherkez and Hodges, 2008).

The use of carbon sketelons for N assimilation raises the question of the origin of (iso)citrate molecules simply because the cyclic nature of the TCA 'cycle' does not then hold. It has been shown that some respiratory intermediates such as citrate are accumulated during the night in tobacco leaves and decrease during the day (Scheible *et al.*, 2000; Urbanczyk-Wochniak *et al.*, 2005). Furthermore, it has thus been suggested that the reserve of (vacuolar)

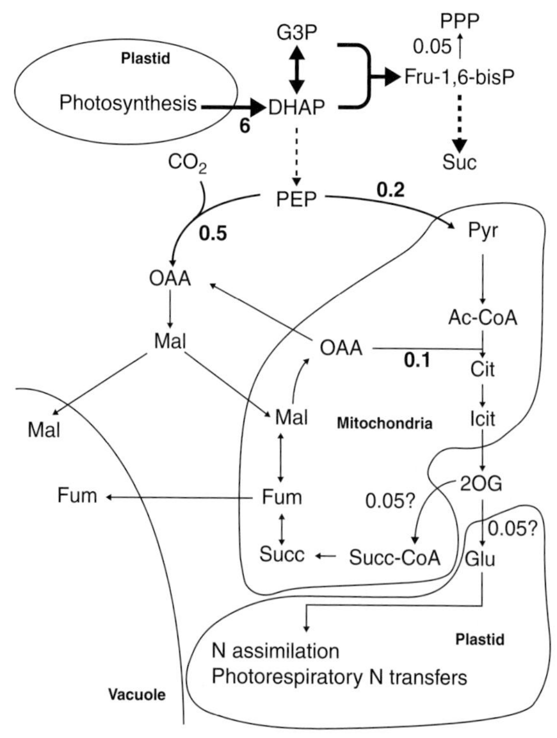

Figure 4.4. Tentative summary of the metabolic fluxes (in μmol m^{-2} s^{-1}) through the TCAC in illuminated leaves, assuming a net photosynthetic flux of 10 μmol m^{-2} s^{-1} of which 40% are directed to starch and 60% to dihydroxyacetone-phosphate export from the chloroplast. Note the uncertainties (question marks) regarding the cellular compartment in which isocitrate dehydrogenation takes place and the partition of 2-oxoglutarate molecules between succinyl-CoA generation and Glu production. Figures are summarised from those found in *Xanthium strumarium* (Tcherkez *et al.*, 2005, 2008). Ac, acetyl; Cit, citrate; DHAP, dihydroxyacetone phosphate; Fum, fumarate; G3P, 3-phosphoglyceraldehyde; Icit, isocitrate; Mal, malate; OAA, oxaloacetate; PEP, phosphoenolpyruvate; PPP, pentose phosphate pathway; Pyr, pyruvate; Succ, succinate.

citrate produced during the night is used during the subsequent light period for Glu synthesis. However, in tobacco (*Nicotiana tabacum*) leaves, the apparent quantity of remobilised citrate (estimated using the drop of citrate concentration) may be insufficient to feed all the Glu synthesis (Stitt *et al.*, 2002). That said, using isotopic methods, decarboxylations associated with R_d have been shown to be partly divorced from current photosynthetic assimilation: (1) using $^{14}CO_2$ labelling, Pärnik *et al.* (2002) and Pärnik and Keerberg (2006) suggested that stored carbon contibutes to respiratory CO_2 evolution in the light up to 25–41% of the rate R_d; (2) similarly, with $^{13}CO_2$ techniques, it has been shown that day respiratory CO_2 inherits partly the $^{12}C/^{13}C$ signature of stored photosynthetic carbon (see Fig. 4.3B for an example). The production of TCAC intermediates is also supplemented by PEPC activity in the light (often assumed to be 5% of the net assimilation rate in C_3 plants, that is, near 0.5 μmol m^{-2} s^{-1}). This enzyme compensates for the consumption of organic acids such as 2-oxoglutarate by Glu synthesis, by providing OAA (or indirectly malate) to feed the TCAC (the so-called anapleurotic function of PEPC). In addition, some OAA molecules may be directly aminated to Asp so that PEPC activity may be viewed as a primary producer of NH_2-acceptors (Huppe and Turpin, 1994). Such a relationship between PEPC and Asp metabolism has now been evidenced by a consistent body of experimental data, reviewed in Tcherkez and Hodges (2008). It is thus plausible that the TCAC does not keep its cyclic nature in illuminated leaves, a substantial part of 2-oxoglutarate molecules being consumed for N assimilation to Glu, while PEPC activity maintains levels of Asp as well as malate and fumarate through the backward reactions of the reversible enzymes MDH and fumarase (see Fig. 4.4 for a summary).

Photorespiration also interacts with N assimilation and day respiration. As quoted above, although R_d is assumed to be constant, several lines of evidence emerge in favour of a promoting effect of photorespiration for N assimilation (on a short-term basis). In fact, labelling experiments with ^{13}C-pyruvate and $^{14}CO_2$ have shown that Glu synthesis is promoted in photorespiratory conditions (Lawyer *et al.*, 1981; Tcherkez *et al.*, 2008). For example, Glu and glutamine (Gln) represent a larger ^{14}C amount and Gln has a higher ^{14}C specific activity after $^{14}CO_2$ labelling in ordinary conditions as compared with non-photorespiratory conditions (Fig. 4.5). Such an effect is probably related to the need of 2-oxoglutarate molecules to run the photorespiratory N cycle when environmental conditions shift to low CO_2. The argument that photorespiration is beneficial for Glu synthesis also agrees with the observed positive correlation between photorespiration and leaf nitrate reduction (Bloom *et al.*, 2002; Rachmilevitch *et al.*, 2004). This scenario is also consistent with the results obtained in the cytoplasmic male sterile CMSII mutant of *Nicotiana sylvestris* (affected in the mitochondrial respiratory complex I) in which the R_d rate is similar or even higher than that of the wild type (Priault *et al.*, 2006), whereas both photorespiration and N metabolism (amino-acid synthesis) are enhanced (Dutilleul *et al.*, 2005). Therefore, while the large reducing power in the mitochondria generated by photorespiration (Gly decarboxylation) and high ATP/ADP and DHAP levels provide a general framework in which leaf glycolysis and respiration are inhibited in the light, leaf day respiration appears to be additionally (and finely) regulated by the CO_2 mole fraction, possibly in order to adjust to N metabolism. It is nevertheless likely that such an effect of CO_2 on R_d disappears on a long-term basis because of acclimation of leaf respiration (Thomas *et al.*, 1993; Shapiro *et al.*, 2004) and, in fact, when measured with ^{13}C techniques the day respiration rate R_d seems to decline slightly under progressive drought in tomato (*Lycopersicon esculentum*) (Haupt-Herting *et al.*, 2001).

4.5.2. Is day respiration involved in redox and phosphate homeostasis in illuminated leaves?

Leaf day respiration is thought to be controlled by several factors (see above), such as phosphorylated metabolites and the redox poise in the mitochondrial matrix (NADH/NAD^+ and NADPH/$NADP^+$). However, uncertainties remain because NAD(P) over-reduction in the matrix may be compensated for by the export of reducing power through the malate-OAA shuttle (Krömer and Heldt, 1991; Noctor *et al.*, 2007). The corresponding reductive power would contribute to convert photorespiratory derived hydroxypyuvate to glycerate in the peroxisome (for a recent review see Reumann and Weber, 2006). Consistently, mitochondria possess organic acid translocators adapted to TCAC intermediates: OAA/malate, malate/citrate and malate/P_i translocators (Hoefnagel *et al.*, 1998). Therefore, although it seems clear that the mitochondrial ATP/ADP ratios are high in the light (Gardeström and Wigge, 1988; Bykova *et al.*, 2005), the hypothesis of a large NADH/NAD^+ ratio as a negative regulator of R_d is unsure. For example, it has been shown on extracted mitochondria of *Solanum tuberosum* that the free NADH amount is relatively constant within the mitochondrial matrix under a range of metabolic conditions (Kasimova *et al.*, 2006). In the respiratory mutant CMS II (*nad1*) of *Nicotiana sylvestris*, the supplemental extrinsic NAD(P) dehydrogenases and the AOXs are thought to partly compensate for the absence of complex I by oxidising more NADH than in the wildtype (Dutilleul *et al.*, 2003). However, the Michaelis-Menten constant (K_m) of such dehydrogenases is higher than that of complex I and so, higher NADH/NAD^+ ratios are expected in the mutant in the photosynthetic steady state, which in turn may impact negatively on the TCAC operation. This would agree with the larger Gln/Glu ratio in the mutant that might reflect

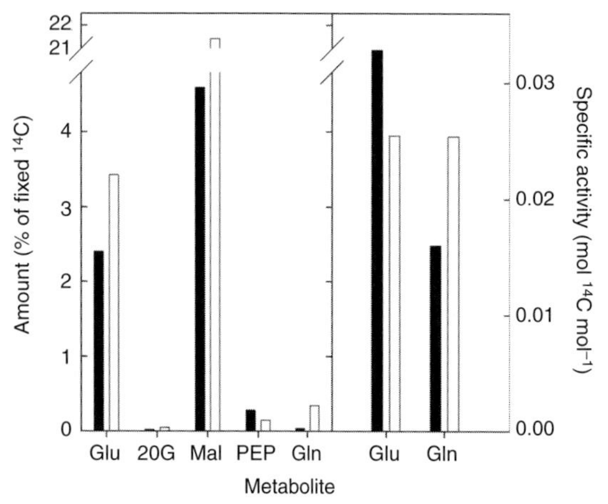

Figure 4.5. ^{14}C labelling of amino acids, 2OG and PEP, after $^{14}CO_2$ feeding to spinach (*Spinacia oleracea*) leaves in non-photorespiratory (high CO_2) conditions (black bars) and ordinary conditions (white bars). Results are presented in % of total fixed ^{14}C (left) or as compound-specific ^{14}C activity (right). Note that ^{14}C in malate (mal) represented less in non-photorespiratory conditions, because the total amount of ^{14}C fixed was much larger and the malate ^{14}C amount hardly doubled. Of note is that both the ^{14}C amount and specific ^{14}C activity of glutamine (Gln) are larger in ordinary conditions, suggesting the stimulating role of photorespiration on Gln synthesis. Data are recalculated and plotted from Lawyer *et al.* (1980). Glu, glutamate; 2OG, 2-oxoglutarate; PEP, phophoenolpyruvate.

the 2-oxoglutarate restriction for the GOGAT reaction (Dutilleul *et al.*, 2005). Nevertheless, as mentioned before, the inhibition of respiration by light is not significantly larger and presumably, the change of amino-acid ratios may be related to the enhancement of photorespiration in the mutant (Priault *et al.*, 2006).

Rather than NADH/NAD$^+$ ratios, it remains plausible that the R_d rate is adjusted to inorganic phosphate (P_i) homeostasis (Raghavendra and Padmasree, 2003), which depends in turn upon the A_N/R_d ratio. This may be illustrated here with rough calculations. On one hand, on average, each mole of TP converted to Suc produces 1.25 moles of P_i. On the other hand, 1 mole of TP committed to the respiratory pathway consumes, on average, 16 moles of P_i within the mitochondria (converted to ATP). The mass-balance on P_i is then reached if the mitochondrial metabolism is reduced to nearly 0.08 (i.e., 1.25/16). As: (1) each respired TP molecule produces three molecules of CO_2; and (2) Suc (made of 12 carbon atoms) represents nearly 60% of the net assimilation flux, we may expect a R_d/A_N ratio of (3×0.08)/(12/0.6), that is, ~3–4%. Whereas such calculations are over-simplified (for example, they do not take into account CO_2 production by the chloroplastic pyruvate dehydrogenase nor any photorespiratory effect), the order of magnitude is correct, pointing to the possibility that P_i homeostasis might be involved in the inhibition of R_d. In such a framework, the involvement of P_i and phosphorylated compounds as regulators of glycolysis comes as no surprise. For example, at high TP/P_i ratios (large assimilation and cytoplasmic P_i restriction), the level of Fru-2,6-bisphosphate decreases and this strongly stimulates the dephosphorylation of Fru-1,6-bisphosphate to Fru-6-bisphosphate, thereby inhibiting glycolysis (see Plaxton, 1996 for a review). In addition to the above-mentioned ATP/P_i balance, it should be emphasised that respiratory metabolism is probably linked to cytoplasmic pH regulation. In heterotrophic cells of *Acer pseudoplatanus*, ATP hydrolysis has been shown to contribute significantly to the cytoplasmic pH decrease (Gout *et al.*, 2001) and such an effect is exacerbated in ATP-limited conditions that restrict the H$^+$-extruding ATPases of the plasma membrane. In addition, the stimulating effect of low intracellular pH on respiratory O_2 consumption has indeed been observed (Gout *et al.*, 1992). In other words, mitochondrial ATP synthesis probably follows a compromise between the above-quoted P_i homeostasis (the adjustment to the ATP demand for Suc synthesis would act against mitochondrial ATP production, see calculations above) and pH regulation (excessive Suc production would decrease cytoplasmic pH and so favour mitochondrial conversion of ADP into ATP).

4.5.3. Interactions between the mitochondrial electron transport and photosynthesis

Although the argument that mitochondria experience large reduction levels within the matrix in illuminated leaves has often been articulated (see above, Section 4.2), there are now views that mitochodria play the role of a photosynthetic safety valve when NADPH/NADP$^+$ ratios happen to be large and ROS may be formed in the chloroplast. During steady state photosynthesis, excess electrons are believed to be exported through a malate/OAA shuttle to the mitochondria where NADH is in turn oxidised (Niyogi, 2000). In this regard, the AOX, which directly catalyses O_2 reduction by ubiquinone, is likely to be crucial as it increases NADH oxidation capacity and therefore impedes the production of ROS within the mitochondrial matrix (Maxwell *et al.*, 1999). So is the case of the additional, rotenone-insensitive NAD(P)H dehydrogenases of the mitochondria, that are thought to accelerate the oxidation of NAD(P)H of photosynthetic origin (Niyogi, 2000). Several respiratory mutants have provided some support to such a scenario. Both *nad4* and *nad7* mutants of *Nicotiana sylvestris* that are affected at the mitochondrial complex-I level (this complex belongs to the mitochondrial electron transport chain and is responsible for internal NADH oxidation), show an enhanced AOX capacity, an impaired Gly oxidation capacity and a decreased photosynthesis rate (Sabar *et al.*, 2000). This suggests that mitochondrial electron transport helps to sustain the photosynthetic activity, although other factors certainly contribute to such an effect on photosynthesis, such as the lower CO_2 conductance. As a matter of fact, *nad1* (CMSII) mutants (similarly affected at the complex-I level) have lower CO_2 assimilation rates and this is fully explained by a lower g_m for CO_2 diffusion, thereby causing larger photorespiration rates (Priault *et al.*, 2006).

Perhaps most convincing are the results obtained with the *ucp1* mutant that lacks the uncoupling protein UCP1. Uncoupling proteins occur in the inner mitochondrial membrane and dissipate the proton gradient across this membrane that is normally used for ATP synthesis. The absence of UCP1 causes the restriction of photorespiration with a decrease in the oxidation rate of photorespiratory Gly in the mitochondria; in addition, there is a reduced photosynthetic assimilation rate (Sweetlove *et al.*, 2006). Because both g_s and C_i are reported to be similar to that in the wild type, these results suggest that photosynthesis and photorespiration require a functional mitochondrial electron transport. Whereas the mechanism of the involvement of UCP1 remains somewhat vague, one may hypothesise that UCP activity leads to a partial dissipation of the proton gradient, which will in turn reduce, but not completely remove, the thermodynamic constraint on the mitochondrial electron flux. It is apparent however that though attractive, the involvement of AOX, external NAD(P)H dehydrogenases and UCPs in impeding photosynthetic overreduction still awaits compelling evidence.

4.6. PERSPECTIVES

Respiratory homeostasis in illuminated C_3 leaves appears to be the result of a photosynthetic compromise between two opposing forces: (1) an inhibition of respiration and glycolysis owing to possible high mitochondrial NADH levels generated by photorespiration (Gly decarboxylation) and elevated ATP/ADP and DHAP levels generated by photosynthetic activity; and (2) a stimulation of the TCAC in order to adjust 2-oxoglutarate production to primary N metabolism and photorespiratory Glu demand. Such a compromise should be very dynamic, adjusting to changes in environmental conditions that modify stomatal closure, thereby altering leaf internal CO_2/O_2 balance. For example, water deficit, that leads to a low C_i, presumably promotes R_d and photorespiration. However, owing to the difficulty to study R_d, only a few studies have focused on the influence of such environmental conditions on respiratory metabolism of illuminated leaves, and similarly, the effect of leaf temperature on R_d is not well known (see Section 4.4.4).

Tcherkez and Hodges (2008) have suggested that isotopic ^{15}N-labelling under different CO_2/O_2 conditions would help to clarify the relationships between day respiration and C/N interactions. Although several techniques have been explored, such as hierarchical metabolomic analyses (for a review, see Stitt and Fernie, 2003), pair-wise correlations between metabolite variations obtained through GC-TOF-MS techniques (Morgenthal *et al.*, 2006) or steady state ^{13}C distribution in metabolites as revealed by NMR detection (see, e.g., Roscher *et al.*, 1998 and Allen *et al.*, 2007 for a review), the control of metabolic fluxes and the control analysis of leaf respiratory metabolism and N assimilation (Glu production) have not been explored so far; more generally, the control of leaf metabolism has not been studied extensively (Rios-Estepa and Lange, 2007). We think it is high time to do so and for such a purpose, presumably, a combination of isotopic and metabolomic methods will provide the missing links in the near future.

5 • The ecophysiology and global biology of C_4 photosynthesis

R.K. MONSON AND G.J. COLLATZ

5.1. INTRODUCTION

The appearance of the C_4 photosynthetic pathway in the Earth's flora represents one of the most impressive and curious examples of evolutionary diversification and biogeographic expansion in the history of life (Ehleringer and Monson, 1983). This complex pathway, involving novel patterns of biochemical compartmentation and anatomical design, has evolved with independent but convergent patterns approximately fifty times during the relatively short geological span of 12–15 million years (Kellogg, 1999; Monson, 1999; Sage, 2004; Christin *et al.*, 2007). The appearance of C_4 photosynthesis has changed the nature of photosynthetic productivity and ecosystem structure on Earth, both regionally and globally. Grassland ecosystems emerged in southwestern Asia, Africa and North America during the mid- to late-Miocene (5–10 Ma) and continued through the Pliocene, (~3Ma), with many of these systems dominated by C_4 species (Cerling, 1999; Beerling and Osborne, 2006). During the appearance of C_4 grasslands, the trophic structures of grazed ecosystems were completely revised, resulting in the emergence of novel mammalian lineages (Cerling *et al.*, 1993; Wang *et al.*, 1994; MacFadden and Cerling, 1996; Ehleringer *et al.*, 1997). Arguably, there is not a better example in the history of life to illustrate the tightly integrated nature of evolutionary novelty and ecological impact, as that shown in C_4 photosynthesis. Clearly, C_4 photosynthesis, though present in only 8,000 of the estimated 250,000 higher plant species, deserves a significant role in the discussion of plant biology.

In this chapter, we focus on a few topics in the C_4 story that are relevant to adaptation, biogeographic expansion and climate change. Our principal aim is to establish conceptual connections among the topics of metabolic design, adaptation and competitive success; an aim that leads us admittedly into the potential pitfalls of teleological interpretation (*sensu* Gould and Lewontin, 1979). With such pitfalls in mind, we will try to develop connections among these topics in as objective a manner as is possible. More specifically stated, our intent is to introduce the advanced botanical student to the principal structural and biochemical features of C_4 photosynthesis, C_4 photosynthesis as a physiological adaptation, the role of C_4 photosynthesis in the historic development of grassland ecosystems and, finally, its role in the global carbon and water cycles in the face of future climate change. We recognise that we cannot do justice to the full slate of C_4 topics, and refer the reader to the book published in 1999 entitled C_4 Plant Biology (Sage and Monson, 1999), for a more complete discussion of C_4 topics (acknowledging that even this book covers discoveries more than a decade old).

5.2. GENERAL OVERVIEW OF C_4 PHOTOSYNTHESIS

The C_4 photosynthetic syndrome refers to the combined influences of anatomical and biochemical compartmentation on the photosynthetic assimilation of CO_2 in a way that overcomes the kinetic limitations associated with C_3 photosynthesis. Given historical fluctuations in the CO_2 concentration of the Earth's atmosphere, and the role of CO_2 as the principal substrate for autotrophic production, the CO_2 assimilation process of plants has been modified significantly during past eons. The enzyme involved in photosynthetic carboxylation, Rubisco, probably arose in photosynthetic autotrophs at least 2.9 billion years ago (Nisbet *et al.*, 2007). In that ancient CO_2-rich atmosphere, the catalytic mechanism that evolved at the active site of Rubisco, was unaffected by the presence of atmospheric O_2 (Sage, 1999; Igamberdiev and Lea, 2006). During subsequent millennia, as atmospheric O_2 increased in abundance (owing to oxygenic photosynthesis and saturation of geochemical O_2 uptake owing to a buildup of oxidised iron) and atmospheric

Terrestrial Photosynthesis in a Changing Environment: A Molecular, Physiological and Ecological Approach, ed. J. Flexas, F. Loreto and H. Medrano. Published by Cambridge University Press.

CO_2 decreased in abundance (owing to geochemical weathering), the reaction catalysed by Rubisco became increasingly susceptible to oxygenation, at the expense of carboxylation. The decreased availability of CO_2 and increased availability of O_2 caused Rubisco to operate at progressively reduced carboxylation potentials, presumably limiting the potential for growth in the Earth's marine and terrestrial autotrophs. In the face of these metabolic limitations, novel anatomical and biochemical mechanisms evolved to enhance the carboxylation potential of photosynthesis, including the emergence of CO_2-concentrating mechanisms in aquatic algae and bacteria (Giordano *et al.*, 2005), and C_4 photosynthesis and CAM in terrestrial plants (Chapter 6).

The principal steps required of a C_4 CO_2-concentrating mechanism are shown in Figure 5.1. Atmospheric CO_2 is fixed to a C_3 acceptor molecule, PEP, in a reaction catalysed in the cytosol of leaf mesophyll cells by the enzyme PEPC. By convention, we will refer to the C_4 tissue (typically mesophyll) in which atmospheric CO_2 is assimilated as the PCA tissue (Hattersley *et al.*, 1977). (Species with complete C_4 photosynthesis limited to a single cell have been described, although we do not consider them here as we focus on broader aspects of C_4 ecology and physiology; see Voznesenskaya *et al.*, 2001; Freitag and Stichler, 2002; Akhani *et al.*, 2005; Voznesenskaya *et al.*, 2005a,b.) PEPC utilises HCO_3^- as its inorganic carbon substrate and, with a $K_m(HCO_3^-)$ of approximately 0.02 mM (Uedan and Sugiyama, 1976), it is able to operate approximately 100 times faster at modern atmospheric CO_2 concentrations than the rate of its K_m value. Catalytic carboxylation using atmospheric CO_2 by PEPC is advantageous, compared with that by Rubisco, which exhibits a $K_m(CO_2)$ (near 25°C) of approximately 15 μM (Yeoh *et al.*, 1980) [1], and operates at rates approximately 0.5 times below those of its K_m value at current atmospheric CO_2 concentrations. This illustrates the catalytic constraint of assimilating CO_2 from the modern atmosphere using Rubisco instead of PEPC. In the atmosphere of ancient times, when Rubisco first appeared in photosynthetic bacteria, CO_2 concentrations were considerably higher, perhaps up to 800 times higher (Berner and Canfield, 1989); under these conditions Rubisco would have been able to catalyse the assimilation of atmospheric CO_2 at rates far above those that occur at its K_m (Sage, 1999).

Following the initial carboxylation by PEPC, C_4 acids (as either malate or Asp; see Meister *et al.*, 1996) are transported through plasmodesmatal connections, from the PCA tissue to specialised cells, where enzymatic decarboxylation occurs and the freed CO_2 is re-assimilated by Rubisco. Again, by convention, we will refer to the C_4 tissue (typically bundle sheath) in which Rubisco carboxylation occurs, as the PCRC tissue (Hattersley *et al.*, 1977). The higher velocity of PEP carboxylation in the PCA cells, compared with that for Rubisco in the PCRC cells, and the fact that more than one PCA cell typically transports C_4 acids to a PCRC cell, results in relatively high CO_2 concentrations in the PCRC cells (~70 μM CO_2 in PCRC cells, compared with ~7 μM CO_2 in PCA cells or about 2000 ppm and 200 ppm respectively at 25°C) (Jenkins *et al.*, 1989). The higher CO_2 concentration in PCRC cells allows Rubisco to operate at rates approximately five times above, rather than 0.5 times below, the rate of its K_m. Thus, C_4 photosynthesis provides an effective CO_2-concentrating mechanism, which in turn provides two benefits to the carboxylation function of Rubisco: (1) it can operate at higher catalytic velocities; and (2) it can operate with reduced competition from O_2. The products from decarboxylation in the PCRC cell include the three-carbon compound, pyruvate, which is returned to the PCA cells (in some species, pyruvate is converted to the amino acid alanine for the return to PCA cells. Once in the PCA cells, alanine is converted back to pyruvate).

The CO_2-concentrating mechanism of C_4 photosynthesis incurs an energetic cost. The pyruvate that is returned to the PCA cells exists at a free energy level approximately 60 kJ mol^{-1} below that of PEP; thus, the equivalent of two ATP molecules (each releasing 30 kJ mol^{-1}) is required to re-establish the initial substrate for C_4 carboxylation (the actual additional ATP cost per CO_2 fixed in C_4 photosynthesis is likely higher than two, owing to leakage of CO_2 from the PCRC cells and subsequent re-fixation; see Kanai and Edwards, 1999).

Efficient C_4 photosynthesis is dependent on the capacity to retain CO_2 at high concentration in the PCRC tissue.

[1] The abundance of CO_2 is often expressed in units of partial pressure (p in units of Pa), molar concentration ($[CO_2]$ in units of moles/volume) and parts per million (moles CO_2/moles of air = partial pressure of CO_2/atmospheric pressure = volume of CO_2/volume of air). Henry's Law states that the concentration of a gas dissolved in water is proportional to its partial pressure in the gas phase in equilibrium with the solution: $P = k_T[CO_2]$, where k_T is the temperature dependent Henry's Constant that for CO_2 is equal to about 30 × 10^5 L Pa/mol at 25°C. Twelve micromolar CO_2 in solution is equivalent to 35 Pa (partial pressure) and at standard atmospheric pressure (1 × 10^5 Pa) is 350 ppm (expressed as a fraction of volume, pressure or moles of atmosphere).

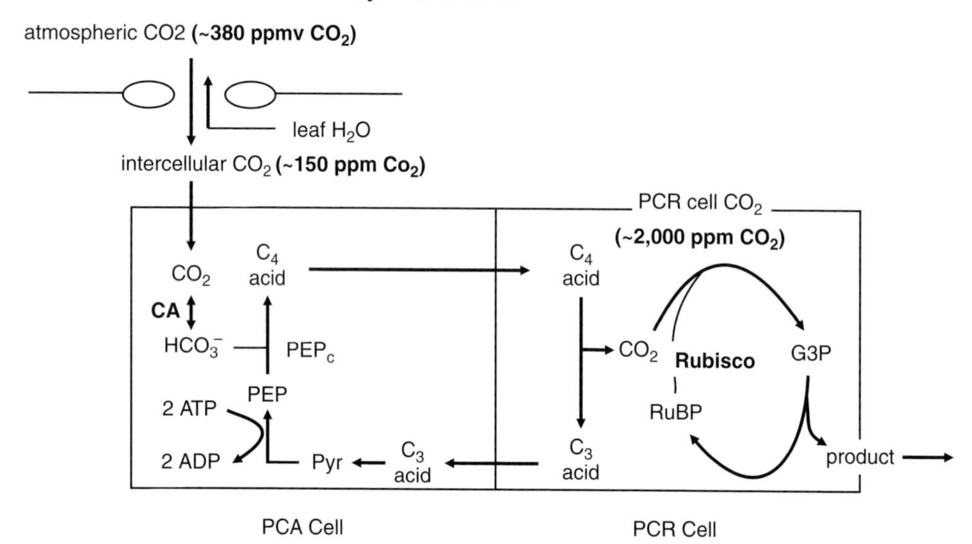

Fig. 5.1. Schematic diagram of the principle steps in the physiology of C_4 photosynthesis. During photosynthesis CO_2 diffuses into the intercellular spaces of the leaf through the stomata (also allowing H_2O vapour to escape in the process of transpiration). The CO_2 dissolves into solution in the mesophyll cells where it is rapidly converted to HCO_3^- by carbonic anhydrase and assimilated into the C_4 photosynthetic process via PEPC. The resulting C_4 acid is transported into relatively gas-tight cells containing Rubisco. The C_4 acid is decarboxylated to form CO_2 that enters the photosynthetic carbon-reduction cycle through the reaction catalysed by Rubisco, ultimately producing photosynthate that is delivered to the rest of the plant. The gas-tight characteristics of the PCR cell allows CO_2 to reach levels ~10x greater than the intercellular spaces in the mesophyll, thus suppressing Rubisco-catalysed oxygenation and subsequent photorespiration. There is an extra light-energy requirement beyond those of the PCRC to regenerate PEP, the substrate for the production of the C_4 acid catalysed by PEPC. ADP/ATP, adenosine di/triphosphate; CA, carbonic anhydrase; G3P, glyceraldehyde-3-phosphate; PCA, photosynthetic carbon assimilation; PCR, photosynthetic carbon reduction; PEP, phosphor-enol-pyruvate; Pyr, pyruvate; Rubisco, ribulose-1,5-bisphosphate carboxylase/oxygenase; RuBP, ribulose-1,5-bisphosphate.

Given that each CO_2 molecule 'pumped' into the PCRC tissue requires energy expenditure in the PCA tissue, leakage of CO_2 from the PCRC tissue will reduce the energetic efficiency of C_4 photosynthesis, a constraint that may be particularly important in habitats where the availability of light is limited (see Ehleringer, 1978; von Caemmerer and Furbank, 2003). In general, the diffusive resistance encountered when CO_2 diffuses out of the PCRC cells is high (Furbank *et al.*, 1989; Jenkins *et al.*, 1989), although it can decrease depending on environmental stress (Bowman *et al.*, 1989; Buchmann *et al.*, 1996; Meinzer and Zhu, 1998; Fravolini *et al.*, 2002). Some studies have provided evidence that the diffusive resistance in the PCRC cells varies among different types of C_4 species (Hattersley, 1982; Oshugi *et al.*, 1988), and this may explain why some C_4 types occur more frequently in wetter habitats and others more frequently in drier habitats (Vogel *et al.*, 1978; Ellis *et al.*, 1980; Prendergast, 1989; Schulze *et al.*, 1996; Taub, 2000); the functional connection between CO_2 leakage from PCRC cells, however, and differential adaptation to moisture gradients has not been firmly established.

5.2.1. C_4 subtypes

Within the general C_4 phenotype, several variant subtypes have been described (Fig. 5.2). The subtypes appear to have evolved in distinct form because of the use of different enzymes to carry out the C_4-acid decarboxylation step. It is worth noting that the enzymes/genes involved in the C_4 syndrome did not evolve *de novo* but were co-opted from existing metabolic pathways located in leaf cytoplasm, mitochondria and chloroplasts (Monson, 2003). In the NADP-malic enzyme (NADPme) subtype, an enzyme that was used in C_3 leaves for the cytosolic production of malate during ammonia assimilation (Chopra *et al.*, 2002) was co-opted for use in C_4 leaves as a decarboxylase; the use of this enzyme created a new source for NADPH as malate was oxidised to OAA. An increase in NADPH production from

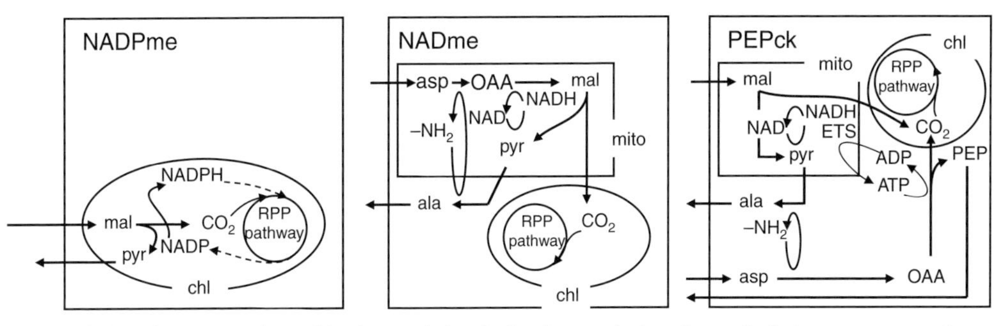

Fig. 5.2. Schematic representations of the three variations in the photosynthetic carbon-assimilation steps among subtypes of C_4 photosynthesis. RPP refers to the reductive pentose-phosphate pathway, otherwise known as the Calvin cycle.

C_4-acid decarboxylation, in turn, relaxed the requirement for photo-generated NADPH, and resulted in the evolutionary reduction of granal densities from PCRC chloroplasts in this subtype (Kanai and Edwards, 1999).

In the NAD malic-enzyme subtype, a mitochondrial enzyme with a role in generating pyruvate from malate was co-opted for use in the photosynthetic decarboxylation of malate in C_4 PCRC cells. Additionally, the evolution of C_4 metabolism in this subtype resulted in the use of Asp, rather than malate, for the transport of carbon (and in this case N) from the PCA cells to the PCRC cells; this, in turn, created the requirement for the amination of pyruvate to form alanine in PCRC cells, and the return flux of alanine (and N) to PCA cells.

In C_4 eudicots, the NADPme and NADme subtypes are the only decarboxylation types described to date. In C_4 monocots, a third subtype has been identified, that relies on both PEP carboxykinase (PEPCk) and NADme as decarboxylating enzymes. In C_3 plants, PEPCk is an enzyme with a crucial role in gluconeogenesis; in C_4 plants, it has been co-opted to catalyse the decarboxylation of OAA to form PEP and CO_2. The use of ATP from the mitochondria of PCRC cells in PEPC monocots, creates a demand for additional mitochondrial NADH in order to maintain ATP production during mitochondrial electron transport; this in turn has resulted in the increased reliance on NADme as a means of generating NADH from the mitochondrial decarboxylation of malate. It is clear that in all three C_4 subtypes, the evolutionary co-option of one piece of ancestral C_3 plant metabolism for use in a novel C_4 role creates various requirements for co-option of other enzymes to maintain metabolite and energetic balance. Thus, one of the principal challenges in studying the evolution of the C_4 form and function has been to reconcile the constraint of interwoven dependencies in the adaptive remodelling of photosynthetic metabolism.

5.3. THE EMERGENCE AND BIOGEOGRAPHIC EXPANSION OF C_4 PHOTOSYNTHESIS DURING THE NEOGENE PERIOD

The oldest reported C_4 fossil is from 12.5 million years ago (Nambudiri *et al.*, 1978). An analysis using molecular-clock techniques and genetic variation within groups in the Poaceae shows that C_4 grasses may have arisen much earlier, perhaps 25–32 million years ago (Gaut and Doebley, 1997; Kellogg, 2001; Christin *et al.*, 2008; Vicentini *et al.*, 2008). Evidence from an examination of plant fossils and $^{13}C/^{12}C$ ratios in aged soil carbonates, fossil tooth enamel of ancient grazers and fossil egg shells from herbivorous birds show that C_4 plants did not appear as major floristic components until 6–8 million years ago (Cerling, 1999; Behrensmeyer *et al.*, 2007). On the basis of the carbon-isotope ratios in the fossilised remains of mammals, C_4 grasses did not make up a significant component of grazed landscapes prior to 8 million years ago (Cerling *et al.*, 1998). Analysis of fossils and soil carbonates that date as far back as 6 million years ago, however, show evidence of C_4 dominance in the grazed ecosystems of Asia (Cerling *et al.*, 1993, 1997), Africa (Morgan *et al.*, 1994), North America (MacFadden and Cerling, 1996) and South America (Latorre *et al.*, 1997). Thus, there was a major global-scale turnover in the composition of temperate and tropical grasslands with regard to the C_3 and C_4 photosynthetic types between 6 and 8 million years ago. Shifts in the photosynthetic composition of grazed ecosystems were accompanied by global turnover in major faunal groups (Vrba *et al.*, 1995; Janis *et al.*, 2000). Arguments have been made that the cause of this global shift in C_3/C_4 dominance was a result of 'CO_2 starvation' when global atmospheric CO_2 concentrations dropped below a critical value for the maintenance of C_3 biomass, and allowed competitive expansion of C_4 plants (Cerling *et al.*, 1998). More recently, an alternative hypothesis has been offered in which the late-Miocene decrease in global CO_2 concentrations was accompanied by increases in global aridity and increased fire frequency; the increased fire activity may have fostered the replacement of post-Miocene woodlands with C_4 grasslands (Keeley and Rundel, 2005; Beerling and Osborne, 2006).

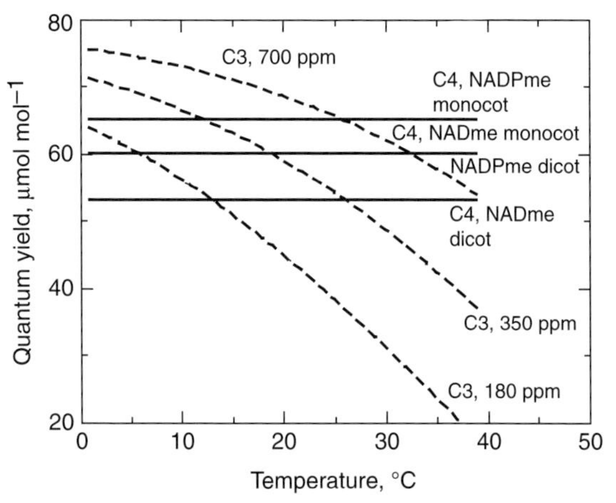

Fig. 5.3. Response of light-limited leaf photosynthesis to temperature. The responses of various subtypes of C_4 photosynthesis (solid lines) are unaffected by atmospheric CO_2 mole fraction. The response of light-limited C_3 photosynthesis at atmospheric current CO_2 levels (350 ppm in 1990), at levels that occurred during the last ice age (~180 ppm, ~18 ka BP) as well as the response predicted for a doubling of current levels (700 ppm) are shown.

5.4. DETERMINANTS OF COMPETITIVE RELATIONS BETWEEN C_3 AND C_4 SPECIES

5.4.1. Photosynthetic responses to CO_2 and temperature

In early work conducted just after the recognition that C_4 physiology provides advantages in terms of reducing photorespiration, Ehleringer and Björkman (1977) reported that the quantum yield for net CO_2 uptake (mol CO_2 assimilated per mol photons absorbed) is relatively independent of temperature in C_4 species, but highly dependent on temperature in C_3 species (Fig. 5.3). These contrasting patterns produce an advantage for C_4 species at higher temperatures and lower CO_2 concentrations, but an advantage for C_3 species at lower temperatures and higher CO_2 concentrations. At higher temperatures and lower CO_2 concentrations, photorespiration rates are predicted to be high in C_3 species, reducing the potential for net CO_2 assimilation (Sage, 1999); this is largely because for Rubisco the K_m for CO_2 increases more with increasing temperature than the K_m for O_2 (Brooks and Farquhar, 1985). The CO_2-concentrating mechanism of C_4 plants minimises photorespiration, allowing them to maintain relatively high rates of CO_2 assimilation at higher temperatures; the higher ATP cost for CO_2 assimilation, however, in C_4 plants causes them to be at a disadvantage at lower temperatures and higher CO_2 concentrations, whereby C_3 plants have relatively low photorespiration rates and lower ATP costs for CO_2 assimilation.

Ehleringer (1978) used these physiological tradeoffs to correctly predict the latitudinal transition from C_3-dominated grasslands to C_4-dominated grasslands in the Great Plains of the U.S.A. (also see Ehleringer *et al.*, 1997); a crossover from 'C_3 advantage' to 'C_4 advantage' was predicted to occur between 43 and 45° N latitude, which was later supported by measurements of C_3 versus C_4 standing biomass (Tieszen *et al.*, 1997). Ehleringer and Pearcy (1983) noted that the quantum yields for NADPme grasses were considerably higher than those for C_4 dicots, as well as NADme C_4 monocots; the differences in quantum yield may be owing to differences in CO_2 leakage from the bundle sheath cells (Hattersley, 1982), or differences in veinal spacing (see Fig. 5.3, Ehleringer *et al.*, 1997); whatever, the cause of the differences, they can be used to predict geographic distribution patterns in C_3 versus C_4 species, as well as the possible expansion of C_3 and C_4 groups during past geologic time in response to changes in the atmospheric CO_2 concentration and temperature (Ehleringer *et al.*, 1997; Collatz *et al.*, 1998). One prediction is that the C_4 pathway in dicots would not have become advantageous with regard to CO_2 assimilation rate, compared with the C_3 pathway in dicots, until atmospheric CO_2 concentrations dropped below approximately 180 ppm during glacial periods of the past 3 million years (Ehleringer *et al.*, 1997). In fact, C_4-dominated ecosystems in general may have been more prevalent during

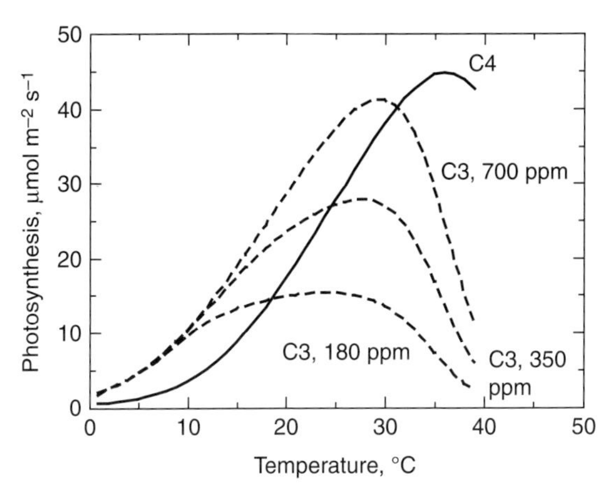

Fig. 5.4. Response of leaf photosynthesis to temperature at full sunlight. The response of C_4 photosynthesis (solid line) is unaffected by atmospheric CO_2 mole fraction. The response of C_3 photosynthesis at CO_2 levels for 1990 (350 ppm), for the last ice age (180 ppm) and for a doubling of 1990 levels (700 ppm) are shown.

the last glacial maximum, when global CO_2 concentrations reached extremely low values, compared with those of today (Cole and Monger, 1994). It is likely that ecosystems at tropical or sub-tropical latitudes would have been the sources of progenitor C_4 taxa during these glacial periods, as northern ecosystems were mostly covered with ice.

The case for quantum yield as a photosynthetic trait that influences the competitive success of C_3 versus C_4 plants is dependent on the assumption that light-limited photosynthesis occurs frequently enough, or at critical periods of growth and development in a plant to limit its overall fitness. In canopies of grasses, this may indeed be the case, as light intensities have been shown to decrease significantly with height (Ryel *et al.*, 1994; Jurik and Kliebenstein, 2000) and where the growth habit of C_3 and C_4 grasses is similar (we note that even at photosynthetic light saturation, however, the temperature-dependent tradeoffs between C_3 and C_4 photosynthesis and higher photorespiration rates in C_3 plants remain; see Fig. 5.4). However, the quantum yield model has been criticised as not adequate to explain the displacement of C_3 woodlands by C_4 grasslands, where growth habit is not similar, and which may indeed have been more significant in the global C_4 expansion following the Miocene (Keeley and Rundel, 2005). In the latter case, increases in global aridity and fire frequency are suggested as having been more important in driving the expansion of C_4 grasslands, than C_3 versus C_4 competition based on differential photosynthesis rates.

5.4.2. Photosynthetic nitrogen-use and water-use efficiencies

Since the earliest days of C_4 discovery, the advantages that are provided by the pathway in terms of improved efficiencies of water and nitrogen use, especially at high temperatures and high light intensities, have been recognised (Björkman, 1971; Downton, 1971; Brown, 1978). Theoretical analyses have indicated that the CO_2-concentrating function of C_4 photosynthesis should allow C_4 leaves to achieve the same photosynthetic rate as C_3 leaves with 20–50% of the Rubisco concentration, depending on temperature (Long, 1991); thus, C_4 leaves have the potential to assimilate atmospheric CO_2 with lower investments of nitrogen (N) in the Rubisco enzyme. Comparative observations on pairs of related C_3 and C_4 species have shown that these theoretical advantages of C_4 photosynthesis are not always supported. At high soil N availability, some C_4 species have been shown to exhibit higher rates of photosynthesis per unit of leaf N, compared with C_3 species, but not at low soil N availability (Sage and Pearcy, 1987a,b; Wong and Osmond, 1991). In considering nitrogen-use efficiency (NUE) in terms of biomass production per unit of applied N, C_4 crops tend to produce more biomass at high rates of N application compared with C_3 crops (Long, 1999), though the case for improved production of C_4 crops at lower fertilisation rates is not as clear. When taken as a whole, the evidence suggests that the high photosynthetic NUE of C_4 plants may not be supported in conditions of low N availability, possibly owing to the production of a faulty C_4 metabolic system (Long, 1999). This latter conclusion was drawn from studies of C_4 crops. In comparisons of native C_3 and C_4 grasses, there is evidence for competitive advantages of C_4 species compared with C_3 species, in regimes of low N availability (Tilman and Wedin, 1991; Wedin and Tilman, 1993). In N-fertilised grasslands, C_3 grasses will often out-compete C_4 grasses, apparently able to overcome any inherent disadvantage owing to photosynthetic NUE (Wedin and Tilman, 1996). The C_3 advantage in the presence of high N availability is likely facilitated by the potential for C_3 grasses to grow at cooler temperatures earlier in the season, at least in temperate grasslands, and thus establish a competitive advantage over later-growing C_4 species (Sage *et al.*, 1999).

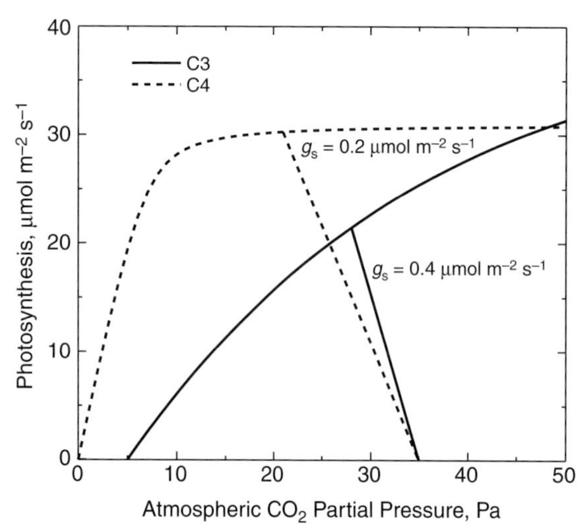

Fig. 5.5. The response of photosynthesis and intercellular CO_2 partial pressure to atmospheric CO_2 and stomatal conductance (g_s). The curved lines represent the response of leaf photosynthesis to atmospheric CO_2 partial pressure under full sunlight, 25°C leaf temperature and no water stress. The straight lines indicate the trajectory of intercellular CO_2 partial pressure in response to photosynthesis, that is, as photosynthesis decreases CO_2 in the intercellular spaces approaches that outside the leaf (here specified as 35 Pa). The slope of the straight lines is defined as $-1\times g_s$. The intersection of the curved and straight lines indicates the operating photosynthetic rate and intercellular CO_2 partial pressure under these conditions. C_4 photosynthesis generally is higher, operates at lower intercellular CO_2 and has lower g_s than C_3 photosynthesis.

The CO_2-concentrating function of C_4 photosynthesis also allows C_4 leaves to achieve higher CO_2 assimilation rates at lower stomatal conductances (g_s). Theoretically, this photosynthetic advantage permits C_4 leaves to lose less water through transpiration (E), compared with C_3 leaves, and on average C_4 leaves exhibit lower g_s compared with C_3 leaves (Körner *et al.*, 1979; Long, 1999). The fundamental difference between C_3 and C_4 leaves in this regard is reflected in the ratio of net photosynthesis to g_s for a given set of environmental conditions; for C_4 leaves, this ratio is generally more than twice that of C_3 leaves (Collatz *et al.*, 1992). These phenomena are illustrated in Fig. 5.5. The response of photosynthesis (A_N) to intercellular CO_2 partial pressure (C_i) is an asymptotic function that can be represented simply as:

$$A_N = f(C_i) \qquad \text{[Eqn. 5.1]}$$

The exact forms of this equation are discussed in Chapter 8. C_4 photosynthesis saturates at lower CO_2 partial pressures, whereas C_3 photosynthesis saturates at CO_2 partial pressures much higher than current atmospheric partial pressures (P_{atm}). The diffusion of CO_2 into the leaf to support photosynthesis is controlled by g_s and the relationship can be expressed as:

$$A_N = \frac{(C_a - C_i)}{P_{atm}} g_s \qquad \text{[Eqn. 5.2]}$$

where C_a is the CO_2 partial pressure at the leaf external surface and P_{atm} is atmospheric pressure. For a constant g_s, equation 5.2 predicts that as $A_N \rightarrow 0$, $C_i \rightarrow C_a$. Fig. 5.5 shows this as the straight lines with negative slopes ($-g_s$) that span between the lines representing $A_N = f(C_i)$ functions and C_a. The CO_2 partial pressure at the intersection of the A_N functions represents the non-stressed light saturated operating C_i. Three important conclusions can be drawn from this figure: (1) under normal (non-stressed) operating conditions photosynthesis is higher in C_4 plants; (2) C_4 plants operate at lower C_i levels; and (3) C_4 plants generally operate at lower g_s and thus exhibit lower potential E rates. The higher WUE of C_4 photosynthesis results from the combination of higher photosynthetic rates at ambient (current) CO_2 partial pressures and lower stomatal conductances. The advantages of lower g_s in C_4 species, may extend to non-steady state environments; it has been shown that C_4 grasses can potentially exhibit smaller variation in g_s in response to variable, non-steady state levels of PPFD, allowing for improved WUE (Knapp, 1993; Nippert *et al.*, 2007).

From the early part of the twentieth century, quantitative observations have shown that some crops and forage grasses, now known to be C_4, produce more biomass for a given amount of water transpired than other crops and forage grasses, now known to be C_3. The most complete set of studies came from careful tracking of the mass of water lost and biomass produced in large pots of plants on the Great Plains of Colorado (Briggs and Shantz, 1914; Shantz and Piemeisel, 1927); potted C_4 plants produced approximately twice the biomass per gram of water lost, compared with potted C_3 plants (also see Downes, 1969). Since these early studies, leaf-level gas-exchange measurements have confirmed the higher rate of photosynthesis per unit of E in C_4 leaves using numerous C_3/C_4 comparisons (Long, 1999).

Field observations of whole-plant production as a function of plant water loss are too sparse to provide generalised conclusions. There are general observations indicating that in grassland ecosystems, C_4 grasses dominate in seasons or regions that receive major precipitation during warm weather, whereas C_3 grasses dominate in seasons or regions that receive major precipitation during cool weather (Williams, 1974; Kemp and Williams, 1980; Monson *et al.*, 1983, Sage *et al.*, 1999). There is no conclusive evidence that C_4 species are better adapted to extremes of drought (see Kalapos *et al.*, 1996; Ripley *et al.*, 2007). In theory, the adaptive advantage to the higher WUE in C_4 species is that they can achieve higher productivity rates during warm but wet weather, and extend their periods of productivity following periodic, but limited rain events during warm weather (Kalapos *et al.*, 1996; White *et al.*, 2000; Knapp *et al.*, 2001). Alteration of the local precipitation regime, toward greater spacing among summer precipitation events, has been shown to cause reductions in net productivity of the dominant C_4 grass, *Andropogon gerardii*, in the North American Great Plains, with accompanied expansion of rarer C_3 forbs (Knapp *et al.*, 2002). Thus, community shifts in this C_4-dominated ecosystem may be more driven by performance related to water stress tolerance of the dominant species, and not traits associated with the photosynthetic pathway *per se*. Dynamics in community composition in this grassland ecosystem may be determined by species traits involving biomass allocation and stress tolerance, rather than competitive interactions that involve differential C_3 and C_4 resource-use efficiency (Smith *et al.*, 2004).

C_4 species tend to be well represented in saline habitats. In some cases this tolerance to salinity allows C_4 plants to dominate in areas that would normally be predicted to be too cool for C_4 success. An example includes the salt-marsh grass *Spartina*, which thrives in high-latitude (>60° N) salt marshes and the prevalence of C_4 shrub species in the cold, arid Great Basin region of the U.S.A. (Sage *et al.*, 1999). This tolerance to salinity leading to dominance in saline environments is generally attributed to the intrinsically higher WUE of C_4 plants (Long, 1999).

All of the studies discussed above have focused on the WUE and NUE of fully evolved C_3 and C_4 species. We also have access, however, to groups of C_3–C_4 intermediate species that represent the initial phenotypes of C_4 evolution, and can thus provide insight into whether improved WUE and NUE represented the earliest adaptive driving forces in C_4 evolution (Monson, 1999; Monson and Rawsthorne, 2000; Sage, 2001). In C_3–C_4 intermediates in the grass genus, *Panicum*, improved photosynthetic WUEs were only observed when C_i was relatively low and there was no advantage in terms of photosynthetic NUE compared with C_3 *Panicum* species (Brown and Simmons, 1979). Similar results were found for C_3–C_4 intermediate species in the genus, *Flaveria* (Monson, 1989a); this species exhibited a slight improvement in photosynthetic WUE (generally 10–15%, but up to 55% in one species), compared with C_3 species, when the C_i was below 150 ppm. Using a model of the C_3–C_4 intermediate photorespiratory recycling system, Schuster and Monson (1990) demonstrated that these improvements in both WUE and NUE can increase substantially at higher leaf temperatures. At a leaf temperature of 35°C, the photosynthetic NUE of a C_3–C_4 intermediate species can be as much as 30% higher. Thus, in environments with high temperatures and limited water (and thus low C_i), both of which promote photorespiration and drive high rates of photorespiratory CO_2 recycling in C_3–C_4 intermediate species, enhanced photosynthetic WUE and NUE can be significant and may represent adaptive advantages.

5.4.3. Differential herbivory between C_3 and C_4 species

Stable isotope ratios from the fossilised teeth of mammalian grazers show that a major shift from C_3- to C_4-dominated diet occurred approximately 8 million years ago, at the same time that C_4 grasslands expanded in several regions of the Earth (MacFadden, 2000). The relationship between these events is typically interpreted in terms of climate-, CO_2- or fire-driven causes to C_4-grassland expansion (Cerling *et al.*, 1998; Keeley and Rundel, 2005), followed by adaptation to new grazing habits within the mammalian fauna (MacFadden and Cerling, 1996). Although we do not debate this form of causality as being likely, we want to note that there are scenarios that could foster herbivory as a direct selective agent, possibly even driving the evolution of C_4 photosynthesis forward. There is evidence even in components of the fossil record that mammalian preferences tend toward C_3 species when given a choice (e.g., Scott and Vogel, 2000). Theory has been developed to predict avoidance by herbivores of C_4 plants relative to C_3 plants (Caswell *et al.*, 1973), and there is some support for the theory based on the relative indigestibility of C_4 PCRC cells (Caswell and Reed, 1975, 1976). The theory put forward by Caswell and co-workers might be called the 'nutritional-quality hypothesis' (see Scheirs *et al.*, 2001), as it postulates that either because of inaccessibility or simply lower concentrations, nutrients are less available to herbivores in C_4 species. Several studies have been conducted to test this hypothesis, with results being mixed in terms of support (Heidorn and Jorn, 1984; Pinder and Kroh, 1987; Pinder and Jackson, 1988; Barbehenn *et al.*, 2004). An alternative hypothesis, termed the 'physical-constraint hypothesis', was presented by Scheirs *et al.* (2001), in which the evolution of a novel leaf anatomy with more closely spaced veins in C_4 grasses, causes herbivores with mouth parts that fit the interveinal distances of C_3 grasses to be excluded from feeding on the C_4 grasses; this hypothesis was supported with observations on herbivory by grass leaf miners. The latter hypothesis has tremendous potential to explain the rapid expansion of C_4 grasses, driven by differential herbivory in the face of devastating outbreaks of insects that might move across a landscape, defoliating C_3 grasses while avoiding C_4 grasses. Whereas such scenarios are tempting at the level of speculation, there is no evidence for such historic influences on C_4-grassland expansion during past epochs.

5.5. THE DISTRIBUTION OF C_4 PHOTOSYNTHESIS: COMPETITION BETWEEN C_3 AND C_4 PLANTS – WHO WINS?

In the above sections we discussed the differences between C_3 and C_4 plants in terms of the general controls by atmospheric CO_2 concentration and mean temperature on leaf photosynthesis, WUE, NUE and salinity tolerance. Another important difference that will be discussed below is woodiness and the ability to compete for sunlight. To determine how these factors ultimately influence the geographic distributions of C_3 and C_4 plants, we assume that conditions favouring the A_N of one type results in greater productivity and fitness at the expense of the other type. Using this assumption, we can infer patterns in the vegetation distribution that can be explained in terms of climate, atmospheric CO_2 concentration and disturbance. We do note, however, that there is need for a more nuanced approach to interpreting interactions between some C_3 and C_4 species. As noted above, in some cases, traits other than resource-use efficiency, including stress tolerance, biomass allocation and canopy morphology, may be more important in determining C_3 and C_4 plant community dynamics. Although we acknowledge these cases, we also make the argument that when viewed in a global context, there are clear influences on the photosynthetic pathway and differential resource-use efficiency that emerge as causal factors determining historical C_3 versus C_4 plant distribution patterns. We will focus on these global patterns.

5.5.1. Climate effects

At the coarsest scale, the current mean global atmospheric CO_2 concentration mediates the climatological constraint on the distribution of C_4 vegetation through effects of temperature on the photosynthetic efficiency of C_3 plants (Figs 5.4 and 5.5). At higher atmospheric CO_2 concentrations, such as will exist in the future, the temperature optimum of C_3 photosynthesis shifts to higher temperatures, causing C_3 photosynthesis to be more efficient in warm climates. An additional constraint on C_3 versus C_4 growth is available soil moisture. Seasonality in precipitation and temperature determines whether soil moisture will be available when temperatures are favourable for photosynthesis. Thus, using the combination of temperature and precipitation, we can infer general patterns in the past, current and future distributions of C_4 plants (Collatz *et al.*, 1998).

The combined black and grey shaded regions shown in Fig. 5.6 experience surface mean monthly air temperatures above 22°C for one or more months in a year; this is the temperature above which the growth of C_4 plants is favoured over C_3 plants. However, warm temperatures alone without sufficient moisture cannot support vegetation growth (e.g., the case of the Saharan Desert), so including the constraint that at least 25 mm of precipitation falls during months with temperatures above 22°C gives a reasonable approximation of climates where C_4 plants have a photosynthetic advantage, at least some of the time during the growing season (Fig. 5.7, grey regions). These regions include the U.S. Great Plains and the tropical and subtropical savannas and grasslands of the world. Climates in which the warm season is dry tend to be dominated by C_3 vegetation, as in those places with Mediterranean climates (Mediterranean, South Africa, Western Australia, California). These maps do not show regions where both C_3 and C_4 plants may dominate at different times of the year. In herbaceous ecosystems that occur in climates that exhibit both cool and warm growing seasons, C_3–C_4 dominance patterns can change with the seasons (Kemp and Williams, 1980; Monson *et al.*, 1983); cool moist parts of

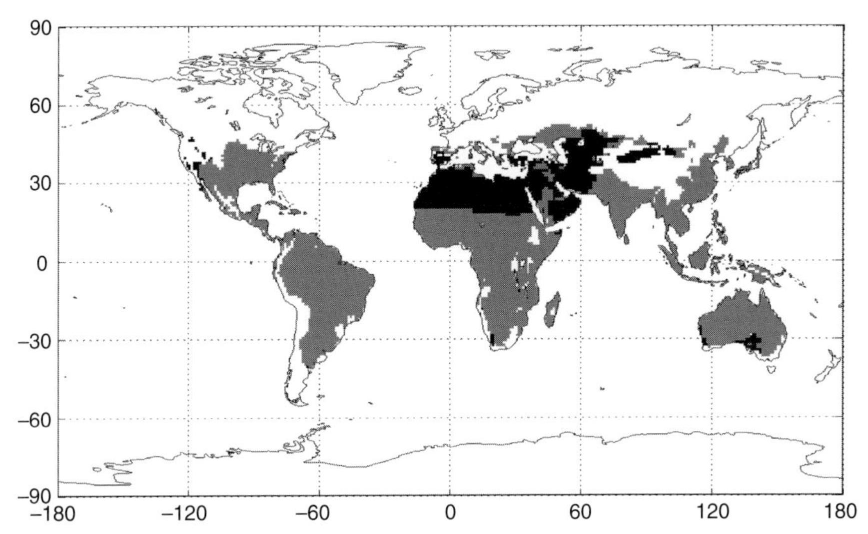

Fig. 5.6. Map of land surfaces with climate conditions that favour C_4 photosynthesis. The combined black and grey regions have a mean air temperature >22°C for at least one month of the year. The grey areas indicate regions that also have a mean precipitation rate of >25 mm for months when air temperature is >22°C. The black regions are too dry during warm months to support significant C_4 vegetation.

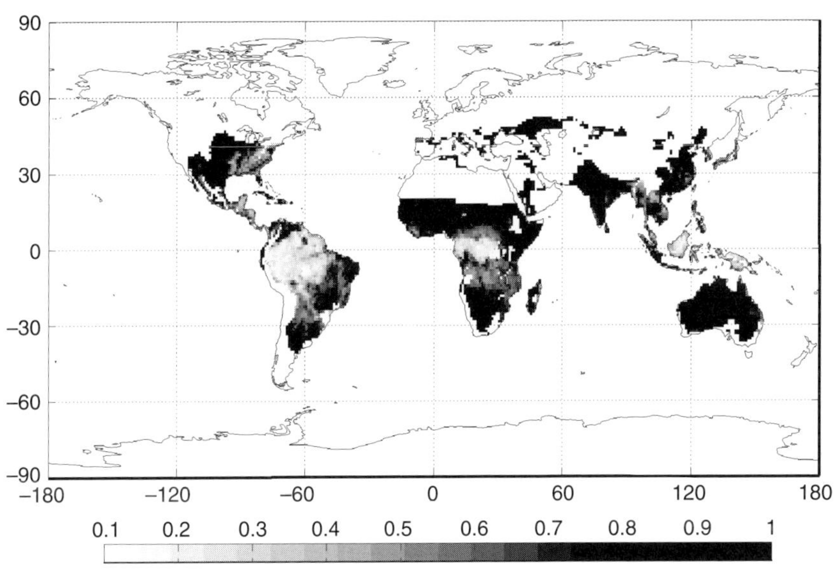

Fig. 5.7. Map of the fractional coverage (shown as grey tones) of C_4 vegetation. The map is derived from the map of favourable climate regions show in Figure 5.6 overlain with a fractional woody vegetation map. Woody vegetation (C_3 photosynthetic type) excludes C_4 herbaceous vegetation in climates that are otherwise favourable for C_4 vegetation.

the year promote the growth of herbaceous C_3 plants, then as temperatures warm, and yet moisture is still sufficient, C_4 plant growth is promoted. These seasonal changes in the differential growth of photosynthetic types occur in the North American Great Plains and temperate grasslands of South America and Asia (Collatz *et al.*, 1998). Similarly, the Sonoran Desert of the southwestern U.S.A. experiences a winter precipitation season associated with C_3 herbaceous plants and late summer precipitation associated with C_4 herbaceous plants (Mulroy and Rundel, 1977).

This emphasis on the temperature/precipitation interactions provides a coarse filter by which to identify sites for potential C_4 dominance. However, consideration of temperature and precipitation alone omits one very important component of C_3-versus-C_4 competition; the influence of plant-growth form on competition for light. In general, with sufficient moisture plants compete for light (absorption of solar radiation for photosynthesis) by means of vertical growth, allowing them to overtop neighbouring plants (Collatz *et al.*, 1998; Sage and Kubien, 2003). Plants with a woody growth form have an advantage over those with herbaceous growth forms, in allowing them to position photosynthetic tissues at taller heights. Whereas woody shrubs are well represented among C_4 species, especially in the Chenopodiaceae, tree-like C_4 plants are not (an exception being some species in the Euphorbiaceae; see Sage *et al.*, 1999). Thus, in terms of competition for light, C_3 trees have a clear advantage over C_4 shrubs and herbs. This has led to the suggestion that C_4 species can dominate over C_3 species in those sites where temperatures are warm, and moisture is available during the warm periods, but only if the amount of moisture is also low enough to prevent establishment of C_3 trees (Sage and Kubien, 2003), or if fire intercedes to prevent C_3 woody plant establishment (Danner and Knapp, 2001).

In Fig. 5.7 a map of woody vegetation fraction (DeFries *et al.*, 1999) is superimposed on the climatology of Fig 5.6 to show where C_3 woody vegetation excludes C_4 plants on the basis of moisture availability and temperature. For example, the tropical rain forests of the globe occur in climates that would favour C_4 plants were it not for the high-precipitation conditions allowing forests to grow, which reduces light availability. The woody vegetation map used here is based on satellite observations from 1992–1993 and as such includes reduced woody vegetation resulting from disturbance. As we discuss below, permanent removal of forests in these C_4-type climates (e.g., through clear-cut logging and/or burning) can potentially lead to replacement of forest by C_4-dominated savannas (see Maas, 1995).

5.5.2. Disturbance effects

Disturbance generally favours the re-establishment of a site with C_4 versus C_3 vegetation in regions where climate favours C_4 photosynthesis (i.e., a moist, warm growing season); disturbances that most likely include herbivory, fire and deforestation. Examples of the importance of disturbance in determining the outcome of competition between C_3 and C_4 plants has been presented in analyses using both current observations of vegetation distribution and future changes as projected by a dynamic vegetation model. In the former case, Hanan *et al.* (2005) provided evidence that in Africa the potential for woody vegetation cover is currently defined by mean annual precipitation; however, relatively large regions persist as C_4-dominated savannas as a result of fire and herbivory. In the latter case, Bond *et al.* (2005) simulated woody vegetation distributions globally with and without wild fires assuming current climate conditions, and predicted where forests or grasslands would occur in the absence of wild fires based solely on climate and competitive responses. The result of their simulation showed a clear expansion of woody vegetation at the expense of the savannas in South America, Africa and South East Asia in the absence of fire. These results can be used to imply that in those regions where water is sufficient to support woody vegetation, but fire cycles intervene, the competitive advantage of taller woody C_3 species is mitigated, and C_4 grasses can potentially express their competitive photosynthetic advantage at warmer temperatures. The authors of this study point out that their conclusions are supported by fire exclusion experiments, the success of plantation forestry and the occurrence of invasive woody vegetation in landscapes currently dominated by herbaceous C_4 vegetation.

Interactions among fire, competition for light and C_4-grass establishment has been described in relation to the ecology of an invasive C_4 species in the Hawaiian Islands (D'Antonio *et al.*, 2001). An introduced C_4 grass, *Schizachyrium condensatum*, is able to establish in relatively open, but nonetheless seasonally shaded, dry woodland habitats on the island of Hawaii. Once in these sites, dense swards of *S. condensatum* effectively exclude a second C_4-introduced grass, *Melinis minutiflora*. Owing to its high rate of productivity in these warm sites, *S. condensatum* promotes an accelerated fire cycle. Burning, in turn, resets the competitive relationship between these two C_4 grasses, allowing *M. minutiflora* to replace *S. condensatum* as the dominant understory grass. Together, the effects of both of these grasses on the local fire regime suppress the dominant woodland C_3 tree species, *Metrosideros polymorpha*, from persisting, and over the period of 20–30 years woodland sites have undergone a transition to grassland sites.

Although not typically considered in ecological treatments of C_4 photosynthesis, agricultural land-use change is another type of disturbance that can affect C_3 and C_4 plant coverage. Major agriculture crops include both physiological types (some C_4 crops are corn, sorghum and sugar cane, whereas C_3 crops include soybean, wheat and rice). Decisions as to which crops are grown in which agricultural areas are largely dependent on social, technological and economic factors, along with climatic constraint. Whereas the replacement of C_3-dominated grasslands with C_4 crops has the potential to change the overall global distributions of C_3 and C_4 species, it has an even more important role in efforts aimed at discerning the relative effects of C_3 versus C_4 ecosystems on the atmospheric carbon-isotope composition (see below Section 5.7.1). It has been estimated that C_4 crops account for approximately 12% of the total area coverage of global C_4 ecosystems (Still *et al.*, 2003).

At finer geographic scales, such as that of the landscape or local habitat, factors other than climate and disturbance can play important roles in the relative distributions of C_3 and C_4 plants. This is particularly the case in cooler climates where C_3 photosynthesis is more efficient, but extreme habitat conditions (e.g., flooding or extreme climate events) reduce or eliminate its dominance. In a few limited places, C_3 vegetation is unable to survive at all; in such places, the surface would be completely barren were it not for C_4 tolerance to extreme conditions (Sage, 2004).

5.6. IMPACTS OF C_4 DISTRIBUTIONS ON THE PHYSICAL CLIMATE

Intrinsic differences in the A_N/g_s ratio between C_3 and C_4 plants (as discussed above in Section 5.4.2) have been shown to cause a feedback on climate in the face of vegetation changes both from observations and modelling experiments. Sellers *et al.* (1992) showed that the spring-to-summer shift from C_3 to C_4 vegetation, in parallel with increasing temperatures, was associated with reductions in surface g_s and, concomitantly, reductions in surface latent heat flux relative to absorbed solar energy in the U.S. Great Plains. In a study by Cowling *et al.* (2007), a coupled global climate and dynamic vegetation model was used to explore the effect of current C_3- and C_4- grassland distributions on global climate dynamics. In general, these simulations showed increased temperatures, decreased latent heat fluxes and decreased precipitation in tropical and subtropical regions owing to the existence of C_4 grasslands. In arid subtropical regions, elimination of C_4 plants did not result in re-colonisation by C_3 plants, leaving significant bare soil regions in places like Sub-Sahelian Africa and northern Australia. These results may have been produced by a combination of physiological intolerance of C_3 photosynthesis to high temperatures and climate feedbacks that exacerbate warming in these regions. This latter effect was also demonstrated in a simulation study by Bounoua *et al.* (2002) in which the effects of historical land-cover changes on climate were investigated. In tropical and subtropical regions, where C_3 forests have been converted to C_4 croplands and pasture, simulated land-surface temperatures increased in part because of the lower g_s, and therefore lower latent heat fluxes from C_4-dominated landscapes. These results have important implications for understanding the potential effects of deforestation and subsequent establishment of C_4 savannas on the global climate.

5.7. THE CONTRIBUTION OF C_4 PHOTOSYNTHESIS TO GLOBAL GROSS PRIMARY PRODUCTION AND IMPACTS ON THE ISOTOPIC COMPOSITION OF THE ATMOSPHERE

Understanding the global coverage and distribution of C_4 ecosystems has emerged as an important aspect of recent modelling exercises aimed at understanding the magnitude and distribution of terrestrial and marine CO_2 sinks and their influence on the global CO_2 budget. Tans *et al.* (1990) published a seminal paper describing the long-term net uptake of CO_2 from the atmosphere by terrestrial ecosystems; they estimated a terrestrial 'carbon sink' of about one-third the rate of annual fossil-fuel emission (which was ~5.3 PgC/yr in the late 1980s, and as of 2007 was ~7.5 PgC/yr). In that study, the increased storage of carbon on land was implied from estimates of ocean CO_2 exchange, inter-hemisphere atmospheric CO_2 gradients, inter-hemisphere differences in the proportion of land to ocean and a reputed source of CO_2 owing to tropical deforestation. Since the publication of this paper, many studies have corroborated their results, although much uncertainty remains as to the underlying mechanisms and the exact locations of the carbon sinks.

One corroborating approach has used concurrent measurements of the global trends of atmospheric CO_2 concentration and the $^{13}C/^{12}C$ ratio of atmospheric CO_2 to discern the magnitude and distribution of CO_2 sinks between land and ocean. Simultaneous solution of the atmospheric budget for carbon (C_a) as CO_2 and the $^{13}C/^{12}C$ provides a means to partition sources and sinks between ocean and land. Two equations representing the global net flux of CO_2 and $^{13}C/^{12}C$ expressed as δ^{13} (see Chapter 11) are given as: (Tans *et al.*, 1993; Still *et al.*, 2003). Column I represents the decadal trends in atmospheric carbon (Eqn. 5.3) and ^{13}C (Eqn. 5.4). Terms in column II represent the fossil-fuel

$$\begin{array}{cccccc} \text{I} & \text{II} & \text{III} & \text{IV} & \text{V} & \text{VI} \end{array}$$

$$\frac{dC_a}{dt} = F_{ff} \cdots\cdots\cdots\cdots + F_{def} \cdots\cdots\cdots - F_{ao} \ldots - F_{al} \qquad \text{[Eqn. 5.3]}$$

$$C_a \frac{d\delta_a^{13}}{dt} \approx F_{ff}\,(\delta_{ff}^{13} - \delta_a^{13}) + F_{def}\,(\delta_{def}^{13} - \delta_a^{13}) - F_{ao}\,\varepsilon_{ao} - F_{al}\,\varepsilon_{al} + G_{diseq} \qquad \text{[Eqn. 5.4]}$$

flux into the atmosphere, in column III the deforestation flux, in column IV the net uptake by the oceans, and in column V the net uptake by land vegetation; $\delta_{xx}^{13} - \delta_a^{13}$ represents the difference between the source and atmospheric δ^{13}. The term, G_{diseq}, referred to as the 'disequilibrium flux' represents the flux of older carbon from the ocean and soils into the atmosphere. This old carbon was taken up in the past when the atmosphere contained less CO_2 derived from fossil-fuel combustion and therefore had a higher $^{13}C/^{12}C$ ratio than the current atmosphere. The disequilbrium flux is a positive net flux of ^{13}C into the atmosphere, but zero net flux of total carbon ($^{12}C+^{13}C$) and as such does not appear as a term in equation 5.3. Terms I, II, III, ε_{ao}, ε_{al} and G_{diseq} are specified in the model *a priori*; terms IV and V can be solved because photosynthesis discriminates against ^{13}C relative to ^{12}C more strongly than diffusion between the atmosphere and ocean (ε_{ao}=–2 ‰ and ε_{la}=–18 ‰).

The solution for the land and ocean carbon sinks *(F_{al}* and F_{ao}, Eqn. 5.3) in this double deconvolution can be expressed graphically as shown in Fig. 5.8 (see Still *et al.*, 2003). Fossil fuels are largely derived from ancient C_3 photosynthesis and therefore carry a ^{13}C-depleted signature (see Section 5.2 above). Thus, as atmospheric CO_2 concentration has increased owing to fossil-fuel burning, the $^{13}C/^{12}C$ ratio of atmospheric CO_2 has decreased; the rate of decrease should have been proportional to the rate of fossil fuel burned if no other processes are in operation (F_{ff} in Fig. 5.8, Terms II in Eqn. 5.3 and 5.4). Additionally, deforestation releases CO_2 (estimated to be ~1.5 PgC/yr), and that released CO_2 carries a $^{13}C/^{12}C$ ratio in similar proportion to fossil-fuel burning, as trees carry the C_3 $^{13}C/^{12}C$ signature (F_{def} in Fig. 5.8, Terms III in equations 5.3 and 5.4). When we combine these effects, however, they are not reconciled with actual measurements of atmospheric CO_2 concentration, which show that the annual increase of CO_2 into the atmosphere is lower (2.6 PgC/yr, $\frac{dC_a}{dt}$ in Fig. 5.8) than our estimates of the rates fossil-fuel burning and deforestation (total of ~7.9 PgC/yr added to the atmosphere, the sum of F_{ff} and F_{def} in Fig. 5.8); this implies the existence of a significant net CO_2 sink. We define sinks as processes occurring at the ocean and land surfaces that cause a net removal of CO_2 from the atmosphere. With respect to the double deconvolution analysis of Fig. 5.8, we need to reconcile the gap between emissions into the atmosphere and the observed changes in the atmosphere (dashed arrow in Fig. 5.8). The gap can be filled by three processes represented in equations 5.3 and 5.4 as terms IV, V and VI. Each of these terms represents a line slope characterised by their respective discrimination against ^{13}C as shown in the inset in Fig. 5.8. The length of the lines in the horizontal direction represents the magnitude of the net flux of carbon between the surface and the atmosphere. Oceans were probably close to equilibrium with the atmosphere prior to the industrial era, but have been forced out of equilibrium owing to the rapid increase in atmospheric CO_2 from fossil-fuel emissions over the last century. In addition to lower-than-expected rates of atmospheric CO_2 increase ($\frac{dC_a}{dt}$ in Fig. 5.8), there is a lower-than-expected decrease in the $^{13}C/^{12}C$ ratio in atmospheric CO_2 since pre-industrial times. Old carbon in isotopic disequilibrium with the current atmosphere increases the $^{13}C/^{12}C$ in the atmosphere with no net change in the CO_2 flux (G_{diseq} in equation 5.4, and Fig. 5.8). The oceans do not exhibit strong selectivity in the absorption of ^{13}C relative to ^{12}C (ε_{ao} in Fig. 5.8, terms IV in equations 5.3 and 5.4), but photosynthesis of terrestrial ecosystems discriminates strongly against ^{13}C (ε_{la}). The lower-than-expected decrease in the atmospheric $^{13}C/^{12}C$ ratio can be explained by a C_3 carbon sink on land (F_{la} in Fig. 5.8, terms V in equations 5.3 and 5.4). This forms the fundamental logic of tracing CO_2 concentration and isotopic dynamics to specific sources and sinks.

C_4 photosynthesis can potential modify two terms in equation 5.4; the ε_{la} in term V and term VI, G_{diseq}. If the land sink includes appreciable C_4 photosynthesis then ε_{la} will be less negative (Fig. 5.8 inset) because C_4 photosynthesis does not discriminate as strongly against ^{13}C compared with C_3 photosynthesis (as discussed in Section

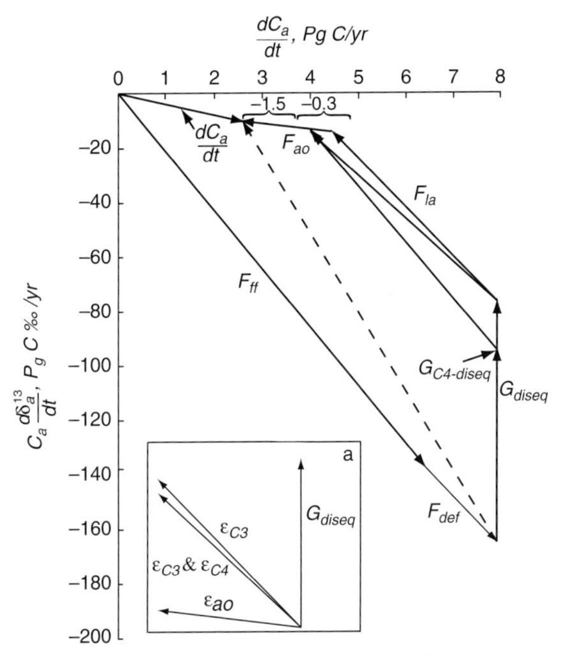

Fig. 5.8. A graphic solution to the global CO_2 and ^{13}C budget of the atmosphere for the 1990s. The abscissa is the flux of carbon as CO_2 into the atmosphere. The ordinate represents the flux of ^{13}C into the atmosphere. The horizontal brackets with numbers at the top of the graph indicate the effect of C_4 photosynthesis on the derivation of the ocean carbon sink. The dashed arrow shows the difference between carbon emissions into the atmosphere and observed changes in atmospheric composition. The inset shows the discrimination by various processes that combine to fill the gap represented by the dashed arrow. The slope of these lines represents the isotope fractionation by the various processes.

5.2 above). If we assume that the contribution of C_4 photosynthesis to net land flux (sink) is proportional to the fraction of global GPP from C_4 photosynthesis (approximately one-third, see Still *et al.*, 2003) then the isotopic discrimination of the CO_2 net flux is determined by the discrimination of C_3 and C_4 weighted by their respective contributions to global GPP (ε_{C3} *and* ε_{C4} in Fig. 5.8 inset). The net result of considering the effect of global C_4 photosynthesis is to reduce the slope (ε_{la} from ε_{C3} to ε_{C3} and ε_{C4} as shown in Fig. 5.8 inset) of the land-sink vector, resulting in a re-apportionment of about 0.3 PgC/yr of ocean sink to the terrestrial sink (Fig. 5.8).

The second way that C_4 photosynthesis can influence the interpretation of the double deconvolution analysis arises from the potential effect of C_4 photosynthesis on G_{diseq} that can arise when tropical deforestation leads to the replacement of the C_3 forests with C_4 pasture grasses and crops (Townsend *et al.*, 2002; Scholze *et al.*, 2008). For decades after such events respiration of soil carbon (into the atmosphere) will carry the C_3 ^{13}C signature. This will cause a decrease in the G_{diseq} term (VI in equation 5.4, to $G_{C4\text{-}diseq}$ in Fig. 5.8) and increase the inferred land carbon sink. The analysis above reflects conditions during the 1990s.

5.7.1. Estimating global gross primary production

Differences in the interaction of C_3 and C_4 photosynthesis with atmospheric CO_2 also has an important effect on research efforts aimed at directly measuring global GPP. As discussed above (in Section 5.7), the relative contributions of C_3 and C_4 photosynthesis to terrestrial carbon fluxes, and hence to the $^{13}C/^{12}C$ signature of the atmosphere, has a significant effect on how we partition atmospheric CO_2 sinks between land and ocean. Similarly, the degree to which we attribute C_4 photosynthesis to global GPP is affected by C_4 influences on the isotopic composition of O_2 in atmospheric CO_2, which in turn reflects the relative contributions of C_3 and C_4 photosynthesis to leaf intercellular CO_2 partial pressure (C_i, see Section 5.4.2 above). As we discussed previously the A_N/g_s ratios are intrinsically different between these two photosynthetic types. As g_s tends to vary proportionally with photosynthesis (but at a different proportionality for C_4 compared with C_3 plants; see Wong *et al.*, 1979; Ball *et al.*, 1987; Collatz *et al.*, 1992), estimation of g_s or E provides an indirect measure of photosynthesis or GPP. One method that exploits these relationships to estimate global GPP relies on the process by which leaf water is enriched in $H_2^{18}O$ during E owing to preferential evaporation of $H_2^{16}O$. CA in leaves catalyses the rapid exchange of O from leaf water with that in CO_2 molecules. Soil water is generally less enriched in ^{18}O because soil evaporation tends to occur at slower rates than E under conditions of significant vegetation cover. Respired CO_2 from stems, from leaves (at night) and from soils should reflect the ^{18}O signature of soil water. Using measurements of atmospheric $^{13}CO^{18}O$, the gross one-way flux of CO_2 into the vegetation can be estimated independent of respiration (Francey and Tans, 1987; Farquhar *et al.*, 1993). Using this method Francey and Tans (1987) estimated global land GPP to be around 200 Pg C/yr, though other modelling studies put it closer to 150 Pg C/yr (see Farquhar *et al.*, 1993; Still *et al.*, 2003).

The exchange of O between leaf water and intercellular CO_2 differs between C_3 and C_4 leaves because of the differential role of CA (Gillon and Yakir, 2000b, but see Edwards *et al.*, 2007). In general, C_4 leaves have lower amounts of CA activity than C_3 leaves; this results in reduced potential for discrimination against ^{18}O during photosynthesis. When this effect is taken into account in the global distribution of $CO^{18}O$, the contribution of C_4 grasslands to global GPP can be estimated as 33% (Gillon and Yakir, 2001). Taking the analysis a bit further, Gillon and Yakir (2001) estimated that the global replacement of C_3 tropical and subtropical forests by C_4-dominated grasslands may be detectable as a small signal (~0.02‰ yr^{-1}) in the atmospheric $CO^{18}O$ content. Leaf structural differences between C_3 and C_4 have been shown to result in greater enrichment in C_4 grasses potentially off-setting some part of the CA effect (Helliker and Ehleringer, 2000).

Another potential method for estimating global GPP that exploits the conservative nature of A_N/g_s proportionality involves measurements of carbonyl sulfide (COS) in conjunction with CO_2 in the atmosphere (Kettle *et al.*, 2002; Montzka *et al.*, 2007; Stimler *et al.*, 2010). COS is generated from a number of sources including ocean plankton, fires and fossil-fuel burning. By far, the major sink for atmospheric COS is leaves. COS diffuses into leaves where it is rapidly converted to CO_2 and H_2S by CA. This sink for COS produces a concentration gradient between the atmosphere and the intercellular spaces of leaves. The absorption of COS by leaves is benign to plants, but depends on leaf stomatal conductance; once again, using the assumption that the leaf photosynthesis rate and stomatal conductance tend to scale proportionally, COS uptake can be used as an indirect measure of photosynthesis and GPP. However, because the ratio of photosynthesis to stomatal conductance is higher in C_4 compared with C_3 leaves, accurate prediction of GPP from atmospheric COS measurements requires knowledge of the relative contributions of C_4 and C_3 to the global flux of CO_2 and COS.

5.8. C_4 PHOTOSYNTHESIS IN RELATION TO FUTURE CHANGES IN CLIMATE AND CO_2 CONCENTRATION

Global atmospheric CO_2 mole fraction has increased from 280 ppm pre-industrial to 350 ppm in 1990 to 380 ppm in 2007. Over the next century, the Earth's climate is expected to warm as atmospheric CO_2 concentration continues to increase. However, this warming will likely be expressed through complicated and poorly understood interactions among various components of the Earth's climate system and will likely lead to regionally variable patterns in precipitation amount and timing, as well as altering the frequencies of extreme climate events. Future climate and atmospheric composition changes combined with human responses to climate change and changing socio-political factors are likely to cause changes in the global distribution and abundance of C_4 plants in ways that are difficult to predict. In this section we will focus on the potential effects of climate change and a brief consideration of disturbance and human influences. Future human activities could be as important as climate change in influencing the distributions of C_4 plants, but are even more speculative and difficult to predict than climate change itself (Sage and Kubien, 2003; Sage, 2004 for a general discussion on C_3 versus C_4 interactions in the face of past, present and future climate change).

5.8.1. Increasing CO_2

The marked increase in photosynthesis of C_3 leaves when exposed to CO_2 concentrations higher than current ambient levels implies that the threshold controlling the competitive advantage of C_3 over C_4 photosynthesis may be exceeded in the near future. The stimulation of C_3 photosynthesis by increased atmospheric CO_2 partial pressure is often called the 'CO_2 fertilisation' effect. Many studies have been made of the responses of vegetation to prolonged exposure to elevated atmospheric CO_2 concentrations. The most credible are the Free-air CO_2 enrichment (FACE) studies in which CO_2 is injected into the canopy air space in open stands of vegetation to estimate responses of entire ecosystems to elevated CO_2 (McLeod and Long, 1999). Though photosynthesis in C_3 plants is often initially stimulated by higher CO_2 levels, in some cases this response is reduced over time as a result of a 'downregulation' of photosynthetic capacity of leaves, perhaps because other physiological processes are unable to 'upregulate' to accommodate the higher photosynthetic rates. Results obtained from CO_2 enrichment studies vary depending on experimental technique, species and exposure duration, among other things, but a recent summary analysis supports the hypothesis that higher CO_2 levels will, in general, increase photosynthesis and growth more in C_3 than in C_4 plants (Long *et al.*, 2004). Both photosynthetic types exhibit reduced g_s when exposed to high CO_2, and this has been shown both theoretically and experimentally to reduce soil water stress in water-limited systems (Owensby *et al.*, 1997, 1999; Morgan *et al.*, 2004); lower E rates caused by reduced g_s increase WUE allowing both photosynthetic types to gain more carbon in water-limited conditions. In regions that would otherwise be moist enough to support woody vegetation, the increase in WUE together with increased productivity caused by CO_2 fertilisation could lead to a greater abundance of C_3 woody vegetation at the expense of C_4 herbaceous vegetation.

5.8.2. Climate change, temperature and precipitation

While it is clear that the world will get warmer in the future as atmospheric CO_2 concentration increases, the patterns of precipitation, especially in the tropics and subtropics are highly variable among the different models used to predict future climate; this has lead to different predictions of C_3 and C_4 distributions. More precipitation in the Sahara could extend C_4 grasslands further into areas that are currently relatively barren. Increased precipitation in regions that currently experience moderate moisture limitations in tropical/subtropical climates could lead to increased woodiness and less coverage by C_4 biomass. Highly variable precipitation patterns could lead to periodic water stress and/or catastrophic fires, thus maintaining herbaceous C_4 dominance. Increases in precipitation during the cooler periods of the growing season or warm seasons are expected to favour C_3 or C_4 biomass, respectively (Winslow *et al.*, 2003). Temperature increases will tend to magnify the advantage of C_3 physiology at high CO_2 (Figs 5.3 and 5.4). Clearly, the interactions ultimately controlling the responses of C_3 and C_4 species to future climate are complex and difficult to predict without the caveat of broad uncertainty.

Many other processes responding to elevated CO_2 and climate change could further complicate the responses. The potential interactions between productivity, WUE and NUE on nutrient availability are complex (Kirschbaum, 2004). In general, plants grown at elevated CO_2 produce tissues with higher C/N ratios, that are less easily decomposed by soil heterotrophs, making nutrients such as N less available to sustain higher rates of photosynthesis and growth. Alternatively, increased productivity and warmer, wetter conditions caused by climate change in some regions could lead to greater rates of decomposition and nutrient availability. Over long periods of time total productivity may increase despite higher detrital C/N ratios because ecosystems tend to sequester nutrients (e.g., Schimel *et al.*, 1996; Rastetter *et al.*, 1997; Luo *et al.*, 2006), shifting the limitation on productivity to factors such as water availability. If water is not limiting, then C_3 plants may dominate in the long run because higher productivity and its associated accumulation of nutrients may shift the most limiting resource to light availability. However, these nutrient-accumulation responses are likely to take centuries while atmospheric CO_2 is increasing at the timescale of decades.

A number of climate model simulations have been carried out that predict changes in the distributions of vegetation, and also inferring changes in the relative distributions of C_3 and C_4 plants (e.g., Lucht *et al.*, 2006; Alo and Wang, 2008). Generally, these models have projected the expansion of drought deciduous trees (C_3) at the expense of evergreen forests (C_3) and grasslands/savannas (C_4). The uncertainties in these predictions are a result not only of the highly variable climate predictions themselves, but also of the assumptions built into the dynamic vegetation models; these include the response of C_3 plants to increased atmospheric CO_2 and the growth of woody vegetation in response to changes in precipitation.

5.8.3. Future patterns of disturbance

Future disturbance regimes in the face of climate change and human pressures will likely result in altered distribution patterns of C_3 woody vegetation, C_4 herbaceous vegetation and crops (both C_3 and C_4). Altered patterns in temperature, precipitation and extreme events will influence wild-fire frequency in a manner that favours C_3 (less fire) or C_4 (more fire) in tropical, subtropical and wet warm-season temperate climates. It has been argued that an increase in the atmospheric CO_2 concentration will allow C_3 woody plants to resist fire by increasing their vigour, and causing a feedback that in itself will reduce fire frequency (Bond *et al.*, 2005). This type of change may allow woodland C_3 trees to be more competitive with C_4 grasses, and through their advantages in low-light, understory environments, allow woodlands to replace C_4-dominated savannas. Dynamic vegetation models driven by future climate scenarios generally predict increased wild-fire emissions owing to higher fuel loads (more C_3 woody vegetation growth in response to CO_2 fertilisation effects) and drier conditions; in some of these projections, fire frequency and severity will increase in sufficient magnitude to block the re-establishment of woody vegetation (Lucht *et al.*, 2006; Alo and Wang, 2008).

Human-driven deforestation in the tropics is likely to continue into the future caused by the demands of increasing population and economic pressures. It is unclear whether conversion of these regions to agricultural use will favour C_4 plants. Clearing of forests for pastures will lead to increases in the coverage of C_4 pasture grasses, but in the case of expanding crop lands, forests may be replaced by C_3 crops (e.g., soybean, palm oil plantations) or C_4 crops (e.g., sugar cane, corn). In any case economic market forces will likely determine what grows on deforested tropical lands of the future. In temperate climates, grazing and fire suppression favour expansion of woody C_3 vegetation. Expanding populations, especially in non-urban areas, will likely increase the political pressure to suppress wild fires to avoid property damage and will lead to increases in the coverage by C_3 woody species at the expense of C_4 herbaceous plants. It is important to point out again that future human activities could be at least as important as climate change in influencing the distributions of C_4 plants through deforestation, aforestation, management of grazing land and crop selection among other influences.

6 • Ecophysiology of CAM photosynthesis

U. LÜTTGE

6.1. CRASSULACEAN-ACID METABOLISM IS A SPECIFIC MODE OF PHOTOSYNTHESIS

Essentially a loop of CO_2 flow via organic acids, mainly malate, is switched before assimilation of CO_2 in the Calvin cycle of C_3 photosynthesis (see Chapter 2). This allows fixation of CO_2 during darkness in the night. The nocturnally fixed CO_2 is transiently stored in the form of organic acids in the cell-sap vacuole. It is remobilised again during the day and assimilated in the light (Fig. 6.1). In the evolution of vascular plants CAM has arisen polyphyletically, i.e., independently many times. In all the branches of the evolutionary tree of vascular plants, beginning with the pteridophyta, we find taxa performing CAM (Fig. 6.2). Polyphyletic evolution of CAM not only occurred between higher taxa such as subclasses and families (Fig. 6.2), but also within families and subfamilies and even within genera, e.g., in the bromeliads (Smith, 1989; Crayn *et al.*, 2000, 2004), the Clusiaceae (Holtum *et al.*, 2004; Gustafsson *et al.*, 2007) and the orchids (Silvera *et al.*, 2009), and there were also evolutionary reversions from CAM back to C_3 photosynthesis. CAM emerges as a good example of Darwinian evolutionary adaptive radiation. It must have been a rather simple step to evolve performance of CAM. In fact, there are basically no new metabolic requirements for CAM as compared with the phylogenetically older general metabolism and C_3 photosynthesis.

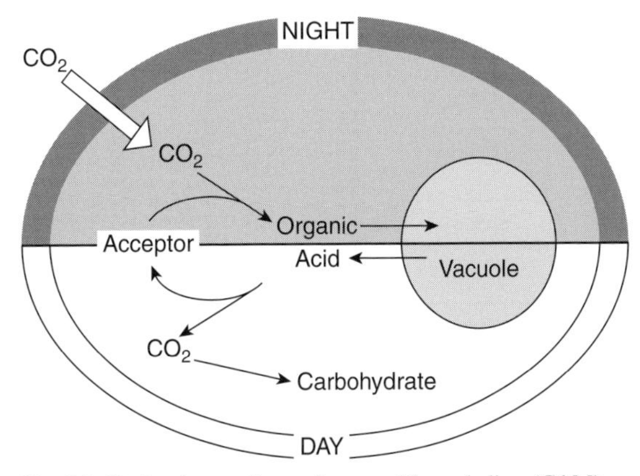

Fig. 6.1. Basic scheme of crassulacean acid metabolism (CAM) with primary fixation of CO_2 in the night, transient nocturnal storage in the form of organic acids and remobilisation and assimilation in the day.

Thus, it can be argued that CAM is nothing more than a different management of available metabolic functions or the particular use of possible metabolic switches. A minimal model of metabolism incorporating only the most essential pools of metabolites and the flows between them shows that the switches are localised at the pools of the metabolites phosphoenol-pyruvate (PEP)/pyruvate, malate and CO_2 (Fig. 6.3). The key element is phosphoenol-pyruvate carboxylase (PEPC) that fixes CO_2 in the dark, i.e., independent of light energy, and thereby serves the generation of malate. Among living organisms PEPC is a ubiquitous enzyme. Other functions of general metabolism that are necessary for the performance of CAM are glycolysis, gluconeogenesis and the fixation of CO_2 by ribulose-1,5-bisphosphate carboxylase/oxygenase (Rubisco) and assimilation to carbohydrate in the Calvin cycle (see Chapter 2). Malate is a key metabolite in general metabolism (Lüttge, 1987a). Among many other functions its synthesis via PEPC is required for anapleurotic reactions feeding into the TCAC, when metabolic pathways branching off from the cycle withdraw metabolites, e.g., α-oxo-acids for the synthesis of amino acids. The switch for performing CAM then is that malate formed in the dark period is stored in the vacuoles, remobilised in the light period and the CO_2 regained is used for assimilatory synthesis of

Terrestrial Photosynthesis in a Changing Environment: A Molecular, Physiological and Ecological Approach, ed. J. Flexas, F. Loreto and H. Medrano. Published by Cambridge University Press.

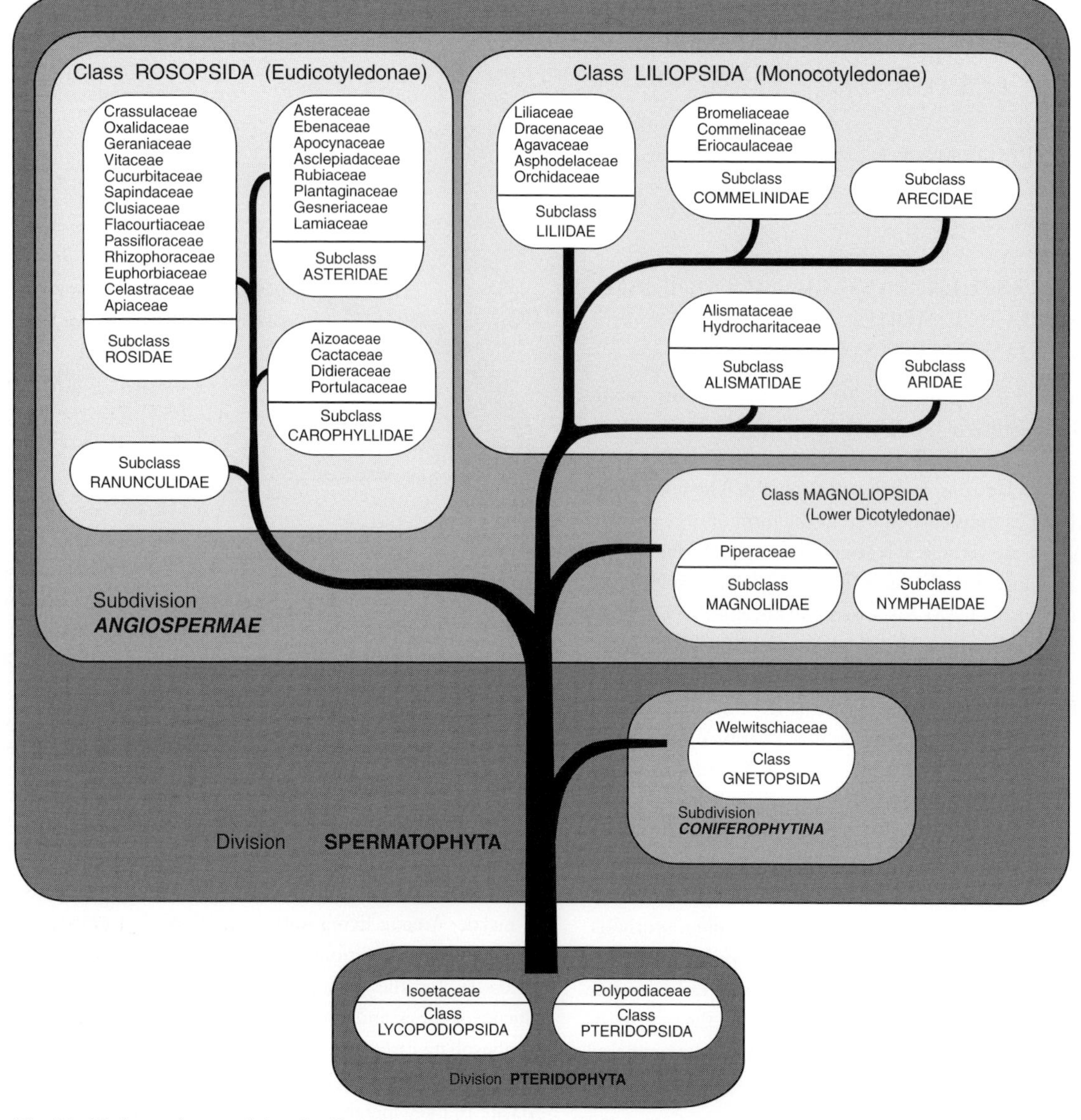

Fig. 6.2. Phylogenetic tree of plant families with crassulacean acid metabolism. (From Lüttge, 2008a, with kind permission of Springer-Verlag, Heidelberg.)

carbohydrates. Vacuolar accumulation of malic acid in the dark period requires energy dependent transport at the tonoplast membrane powered by H^+-transporting pumps, namely the vacuolar H^+-ATPase and H^+-pyrophosphatase (H^+-PPase) and mediated by an inward rectifying malate anion channel (Hafke *et al.*, 2003). However, this is also not unique to CAM because vacuolar organic acid accumulation has manifold functions in plants including osmotic functions of the vacuoles, e.g., in movements such as those of stomatal guard cells and of pulvini.

Thus, we can conclude here that indeed an evolution of particular CAM enzymes was not required for the origin of CAM. We must note at this stage, however, that evolutionary shaping is nevertheless noticeable because many of the enzymes involved in the metabolic cycle of CAM are specific isoenzymes.

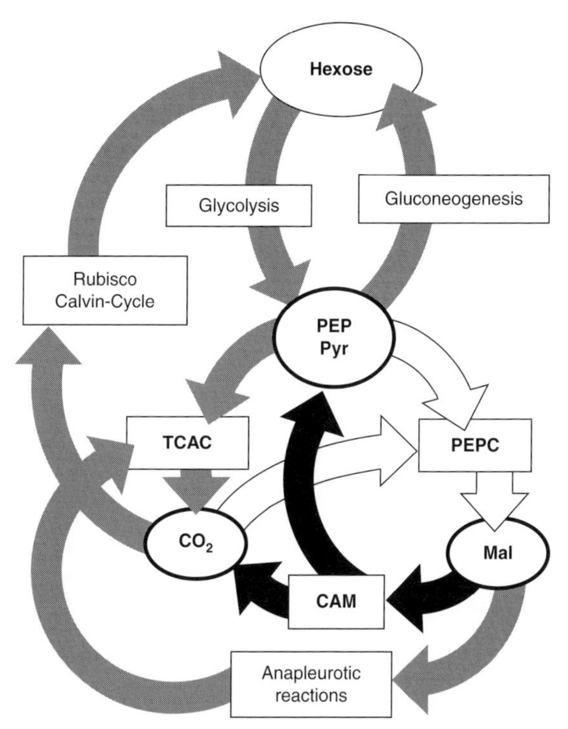

Fig. 6.3. Minimal model of metabolism showing the switch of general metabolism ↔ crassulacean acid metabolism (CAM) by the selective incorporation of only the very essential pools of metabolites (in ovals) and the flows between them (arrows) as mediated by basic functional modules (in rectangles). White arrows, the key element phosphorenolpyruvate carboxylase (PEPC); grey arrows, functions of general metabolism; black arrows, the switch to CAM. Mal, malate; Pyr, pyruvate; PEP, phosphorenolpyruvate; TCAC, tricarbonic acid cycle.

6.2. SWITCHES OF CRASSULACEAN ACID METABOLISM PHASES AND MODES

For understanding CAM phases and modes we need a more detailed view of the metabolism as given in Figures 6.1 and 6.3 (see Fig. 6.4). The CO_2 acceptor, PEP, for nocturnal CO_2 fixation by PEPC is formed via glycolysis. PEPC first generates OAA, which is reduced to malate by cytosolic NAD-dependent MDH. Thereby a new carboxyl group is formed that is dissociated at the slightly alkaline pH of the cytosol. With the extensive formation of the malic acid this would unduly acidify the cytosol. Moreover, malate exerts a feedback inhibition on PEPC. For these two reasons, the malic acid must be removed from the cytosol that is mediated by vacuolar accumulation. In some CAM plants, in addition to malic acid, citric acid also accumulates. Total nocturnal acid accumulation in CAM can be quite substantial. The highest value recorded is just above 1.4 molar titratable protons (Borland *et al.*, 1992). The vacuole is acidified during nocturnal acid accumulation and the organic acids stored get largely protonated. Thus, in the subsequent light period the efflux can occur in the form of the non-dissociated acids. Decarboxylation is brought about by different enzymes (for simplicity Fig. 6.4 only shows cytosolic NAD-dependent malic enzyme). The CO_2 regenerated is fixed via Rubisco. The precursors for the formation of PEP in the next dark period are synthesised by gluconeogenesis (from pyruvate) and by photosynthesis (from CO_2).

This covers the two major phases of CAM (Osmond, 1979). Phase I is the nocturnal uptake of atmospheric CO_2 when stomata are open, with primary fixation via PEPC and vacuolar storage of organic acid. Phase III is the day time regeneration of the CO_2 when stomata are closed with assimilation of the inorganic carbon. There are two additional phases in between, namely phase II in the early morning and phase IV in the afternoon, which are phases of transition (Fischer and Kluge, 1984; Littlejohn and Ku, 1984; Smith and Lüttge, 1985; Lüttge, 1987b; Borland *et al.*, 1993; Roberts *et al.*, 1997). During phase II, PEPC and possibly also the vacuolar H^+-pumps are downregulated and Rubisco is upregulated. PEPC is phosphorylated in the night by a PEPC kinase and dephosphorylated in the day by a phosphatase. It only has high CO_2 affinity and low sensitivity to the feedback inhibitor malate in the phosphorylated stage. Phase IV occurs in the afternoon when the nocturnally stored organic acid is already consumed. Stomata then may open and atmospheric CO_2 is taken up and assimilated directly via Rubisco, just as in C_3 photosynthesis.

The expression of the four CAM phases is modulated by environmental conditions, particularly the availability of water (Fig. 6.4, Smith and Lüttge, 1985). Basically CAM is a water-saving and CO_2-concentrating mechanism. Nocturnal opening of stomata for fixation of atmospheric CO_2 (phase I) and day time stomatal closure (phase III) greatly reduces transpiratory loss of water. The generation of CO_2 from organic acid behind closed stomata in phase III can lead to intercellular CO_2 concentrations in the CAM organs of 2–60 times the external atmospheric CO_2 concentration (Lüttge, 2002a). When water availability becomes limiting the performance at phase IV is suppressed because with open stomata in this phase of the light period evaporative demand for transpiration (E) is high. When drought is more severe the onset of phase I at night is delayed and the amplitude is strongly reduced. Metabolic switches in the

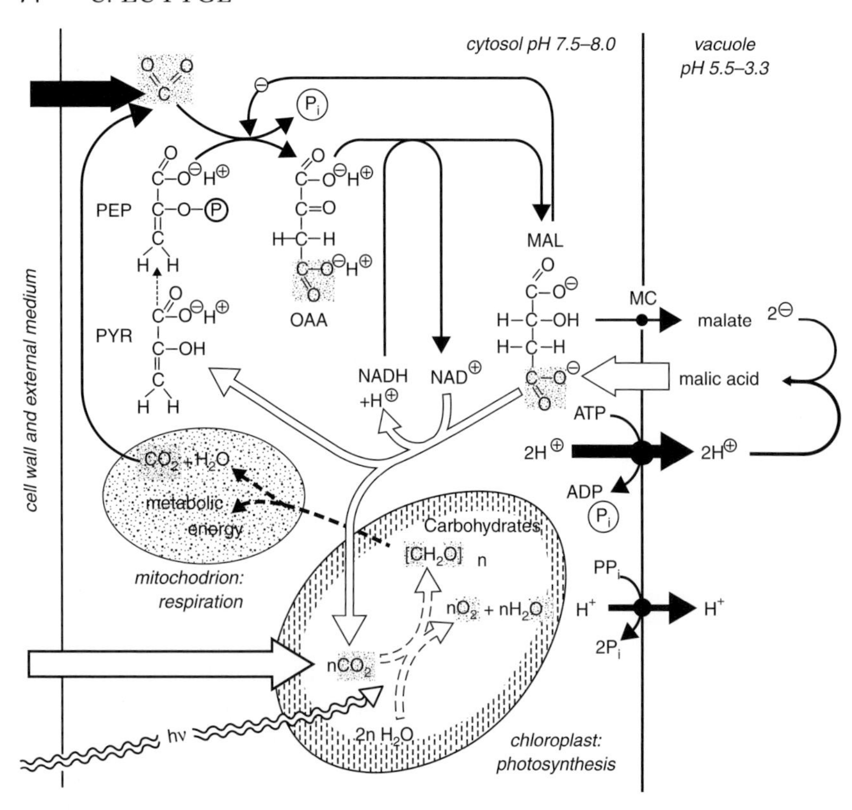

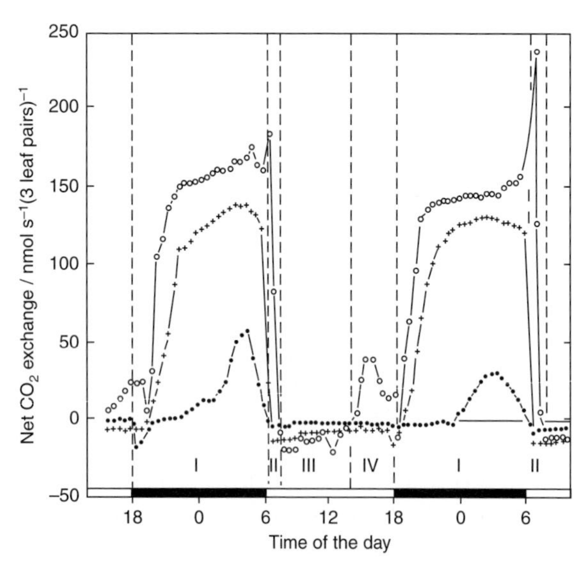

Fig. 6.4. Metabolic scheme of crassulacean acid metabolism (CAM) (above) and expression of the four CAM phases (below) as indicated by net CO_2 exchange of a plant of *Kalanchoë daigremontiana* under increasing drought stress. Open circles, well-watered; crosses, low drought stress; closed circles, high drought stress; dark arrows and white arrows indicate processes of the dark and light period, respectively. MAL, malate; MC, malate-anion carrier; OAA, oxaloacetate; PEP, phosphorenolpyruvate; PYR, pyruvate. (From Lüttge, 2008a; with kind permission of Springer-Verlag, Heidelberg).

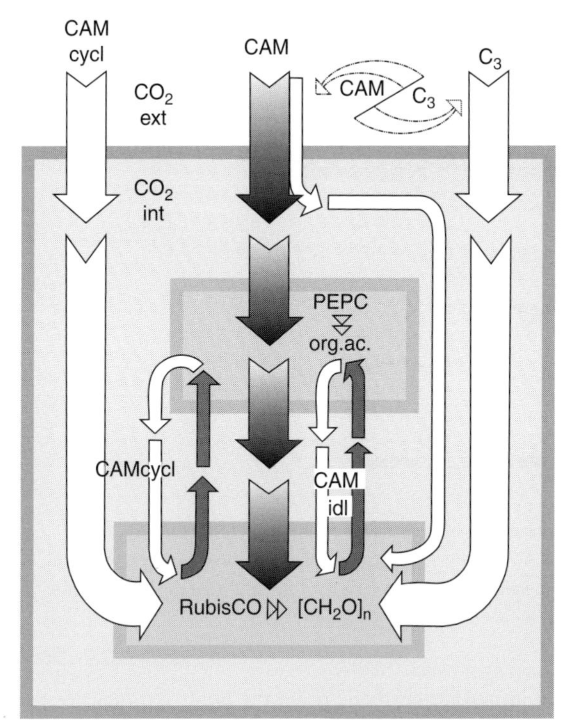

Fig. 6.5. Comparative scheme showing the four crassulacean acid metabolism (CAM) modes from left to right: CAM cycling, obligate CAM with CAM idling and C_3/CAM-intermediacy. The major modules involved are nocturnal synthesis of organic acid and daytime carbon assimilation to carbohydrate ($[CH_2O]_n$). Black arrows and white arrows are processes in the dark and in the light respectively. cycl, cycling; ext, external; idl, idling; int, internal; org. ac., organic acid; PEPC, phosphorenolpyruvate carboxylase; Rubisco, ribulose-1,5-bisphosphate carboxylase/oxygenase. From Lüttge, 2006; with kind permission of the New Phytologist.

expression of CAM phases allow plasticity in the responses of the CAM plants.

When the environmental or developmental modification of the expression of CAM phases is more extreme, we may speak of CAM modes (Fig. 6.5, reviewed in Lüttge, 2002a, 2004, 2006):

- obligate CAM;
- CAM idling;
- CAM cycling;
- C_3/CAM intermediacy.

As we have seen above in experiments with *Kalanchoë daigremontiana* (Fig. 6.4) under drought stress, obligate CAM plants can suppress the expression of CAM phases including an increased reduction of phase I. This is also observed among CAM plants in the field. For example, in a coastal alluvial sand plain of Venezuela in the dry season the CAM bromeliad *Bromelia humilis* (see Section 6.5.4), when growing in the shade, had pronounced phases I and II, and when growing under full sun exposure it had a much reduced phase I and only a very faint indication of phase II. Stomata can become increasingly closed in phase I. Under these conditions the CAM plants can internally recycle respiratory CO_2 via organic acid in the night and re-assimilation in the day. In the shaded *B. humilis*, 56% of the nocturnally stored malate was owing to fixation of internally recycled CO_2, and in the exposed plants this was 87% (Lee *et al.*, 1989). In the extreme stomata may remain completely closed continuously in the dark as well as in the light and plants only recycle respiratory CO_2. This is the mode of so called CAM idling. It is driven by solar radiation energy and can continue for many days, weeks and even months. In this way the plants do not gain carbon but greatly reduce their loss of water. Thus, they can overcome periods of drought in a seasonal precipitation climate. Some plants use the possible flexibility in a different way. They close stomata in the dark period and recapture respiratory CO_2 via PEPC and malate accumulation. They have stomata open in the light period and perform normal C_3 photosynthesis, but in addition the remobilisation of the nocturnally stored organic acid provides CO_2 for inorganic carbon assimilation. This is the mode of so-called CAM cycling. Quite a number of plants, e.g., the bromeliad *Guzmania monostachia*, Crassulaceae of the genera *Sedum* and *Kalanchoë*, *Portulaca*, many species of the genus *Clusia* (Lüttge, 2006, 2007) and some halophytes (Section 6.5.4 below), are capable of genuine and often reversible switches between pure C_3 photosynthesis and pure CAM (Lüttge, 2004). This is the mode of so called C_3/CAM intermediacy.

6.3. ENDOGENOUS CIRCADIAN RHYTHMICITY AND RESPONSES TO ENVIRONMENTAL CUES

The cycle of CAM is not only running under normal ambient night–day changes, but also oscillating as a free-running endogenous rhythm under constant environmental conditions, remarkably under continuous light and constant temperature (Lüttge and Beck, 1992; Wilkins, 1992). It is a circadian rhythm as the endogenous cycle has a period length of close to ('circa') 24 hours ('diem'=day). In the obligate CAM plant, *Kalanchoë daigremontiana*, the free-running CAM rhythm can continue for many endogenous periods

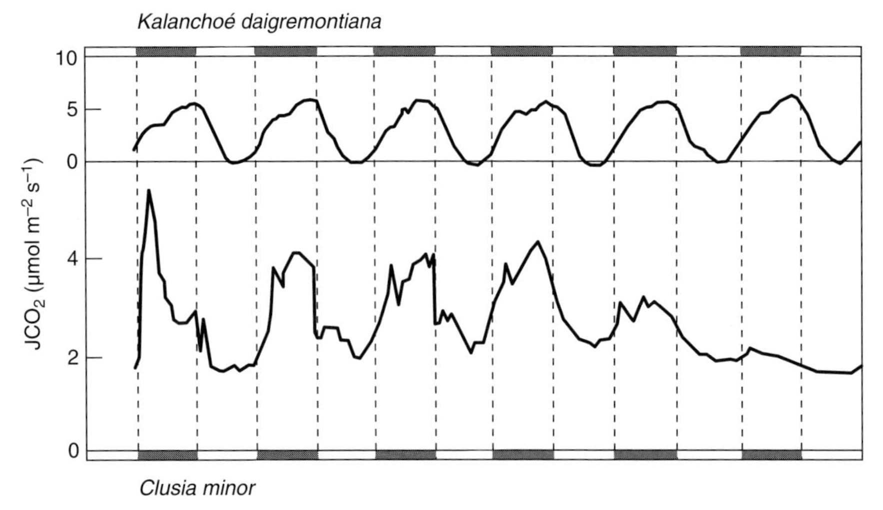

Fig. 6.6. Long-lasting endogenous rhythm of net CO_2 exchange in the obligate crassulacean acid metabolism (CAM) plant *Kalanchoë daigremontiana* (upper curve), and strongly dampened rhythm in the C_3/CAM intermediate plant *Clusia minor* (lower curve). Dark and white bars are each 12 h and indicate subjective dark periods and subjective light periods, respectively, of the endogenous rhythm. (After data of Lüttge and Beck, 1992, and Duarte and Lüttge, 2007.)

over several weeks, whereas in the C_3/CAM-intermediate species, *Clusia minor*, it is strongly dampened after only a few periods (Fig. 6.6).

The circadian rhythm is highly sensitive to the external control parameters irradiance and temperature. It only works below a certain threshold of photosynthetic photon-flux density (PPFD) (irradiance), above which the behaviour of *K. daigremontiana* becomes arrhythmic (Lüttge and Beck, 1992). For temperature there is a window (Fig. 6.7). *K. daigremontiana* is rhythmic below an upper and above a lower threshold temperature (Grams *et al.*, 1997). Above the upper temperature threshold the plants are arrhythmic with a low vacuolar malic acid concentration. Fluidity and hence permeability for malic acid of the tonoplast membrane is increased at higher temperature (Kluge *et al.*, 1991; Kliemchen *et al.*, 1993; Schomburg and Kluge, 1994) so that efflux empties the vacuoles with respect to malic acid, when plants become arrhythmic. This is reversible, and when temperature is lowered again into the rhythmic realm the plants recommence the rhythm by increasing CO_2 uptake to synthesise malate. Conversely, below the lower temperature threshold the plants are arrhythmic with a high vacuolar malic acid concentration. Fluidity, and hence permeability for malic acid of the tonoplast membrane is reduced at lower temperature (Kluge *et al.*, 1991; Kliemchen *et al.*, 1993; Schomburg and Kluge, 1994), so that the vacuoles remain filled with respect to malic acid when plants get arrhythmic. This is also reversible, and when temperature is increased again into the rhythmic realm the plants recommence the rhythm by decreasing CO_2 uptake as CO_2 is produced internally by remobilisation and decarboxylation of vacuolar malic acid (Fig. 6.7).

The plants of *K. daigremontiana* have several circadian oscillators. First, each leaf cell has a metabolic oscillator (Rascher *et al.*, 2001). The essential elements are: (1) internal CO_2 concentration; (2) cytosolic malate levels determined by malate synthesis and a feedback loop of product inhibition by malate on its own production (Section 6.2); and (3) vaculoar malic acid levels, where a temperature-dependent permeability switch at the tonoplast membrane feeds back on cytoplasmic and vacuolar malate levels (Blasius *et al.*, 1999; Lüttge, 2000). The control parameters irradiance and temperature act at the levels of photosynthetic CO_2 reduction and tonoplast permeability, respectively. Second, there is a molecular oscillator with circadian changes of the expression of the central rhythm genes of plants (Boxall *et al.*, 2005). Expression of the *TOC1* gene (*time of chlorophyll a/b binding*) is high in the evening, and expression of the couple of the *CCA1* and *LHY* genes (*circadian clock-associated, late hypocotyl-elongating*) is high in the morning. Both have feedback on each other and downstream regulate other evening and morning genes, respectively. The metabolic and the molecular/genetic oscillator are most likely not operating in a hierarchical order but are interlinked heterarchically in a regulatory network.

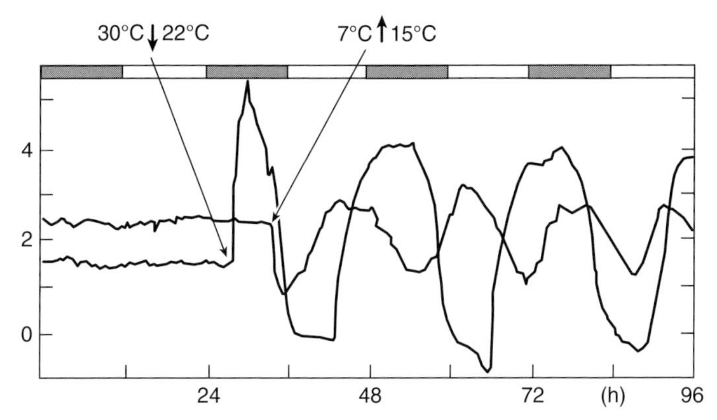

Fig. 6.7. Reactivation of endogenous rhythmicity when plants of *Kalanchoë daigremontiana* were returned into the temperature realm of rhythmic behaviour by decreasing ambient temperature from too high 30↓22°C or increasing temperature from too low 7↑15°C, respectively (data from Grams *et al.*, 1997).

It is generally assumed in evaluations of the ecophysiological importance of circadian rhythmicity, that the endogenous clocks provide preparedness for diurnally changing conditions, and hence fitness when plants are tuned to be ready to respond to requirements of light and day and dark and night, respectively. The control parameters irradiance and temperature have been mainly considered in relation to the mechanisms of the endogenous clocks. However, it will also be important to view this with respect to ecological fitness. Moreover, it is intriguing that by contrast to the obligate CAM plant *K. daigremontiana* in the C_3/CAM-intermediate species *C. minor*, the endogenous rhythm is highly dampened, and thus not so strongly expressed. The very advantage of *C. minor* in ecological niche occupation appears to be its high flexibility (Section 6.6), and it may well be that a strong endogenous pacemaker would hinder this and prevent flexibility.

6.4. TRENDS OF CRASSULACEAN ACID METABOLISM EVOLUTION

6.4.1. Low CO_2 driving crassulacean acid metabolism evolution

We have already seen in Section 6.2 that phase III of CAM is a CO_2-concentrating mechanism. It is widely assumed that the environmental driving force for the very earliest evolution of CAM was low ambient atmospheric CO_2. In geological times a period of atmospheric CO_2 as low as we have it at present occurred about 300–250 million years ago (Fig. 6.8), when CAM may have begun to evolve in some Pteridophyta. This possibly started in submerged aquatic life forms (Sections 6.2 and 6.5.1, Fig. 6.2) that were among the most basic extant tracheophytes performing CAM. In the class of the Lycopodiopsida of the Pteridophyta there are freshwater species of *Isoëtes*. The problem of low CO_2 is amplified in aqueous media owing to constraints on diffusion.

However, why would low atmospheric CO_2 have become an effective driving force for the evolution of CO_2-concentrating mechanisms? Rubisco, the key enzyme of inorganic carbon assimilation in plants, has a very low affinity to its substrate CO_2, and all plants with C_3 photosynthesis that are not having a CO_2-concentrating mechanism operate much below substrate saturation. Rubisco evolved about 3.5×10^9 years ago, when the first litho-autotrophic bacteria began to assimilate CO_2. At that time the earth had a very different atmosphere with a CO_2 partial pressure of 5 to 10 bar, as compared with a present value of about 3.75×10^{-4} bar and a total pressure of the atmosphere of only 1 bar. About 500 million years ago when the large groups of algae evolved, atmospheric CO_2 concentration was up to 20 times the present concentration and the concentration occurring 300–250 million years ago. Thus, up to 350 million years ago atmospheric CO_2 concentration was well above the concentration currently observed to saturate C_3 photosynthesis, but then it dropped below that (Fig. 6.8) and CO_2-concentrating mechanisms became important. Photosynthesising prokaryotic cyanobacteria and eukaryotic algae evolved CO_2-concentrating mechanisms based on transport functions and CA catalysing the $CO_2 \leftrightarrow HCO_3^-$ equilibrium, some aspects of which are also retained in chloroplasts.

However, the most effective CO_2-concentrating mechanisms that evolved in parallel among the vascular plants are

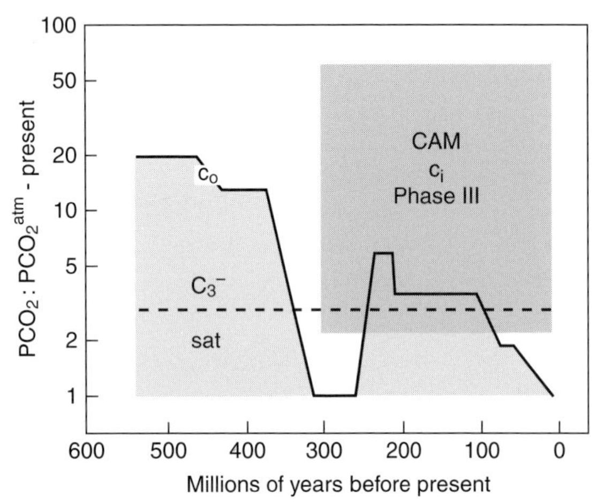

Fig. 6.8. Average partial pressures of CO_2 (PCO_2) related to the current partial pressure of atmospheric CO_2 (PCO_2^{atm}-present) (logarithmic ordinate) of the Earth's atmosphere during geological times (c_o: solid line), of saturation of C_3-photosynthes (C_3-sat: dashed line) and the range observed in leaves of crassulacean-acid metabolism plants in phase III (shaded box).

CAM and C_4-photosynthesis (Chapter 5) based on particular metabolic pathways. With the CO_2 concentrating in phase III of CAM, plants are close to and in most cases well above CO_2 concentrations saturating inorganic carbon assimilation by Rubisco (Fig. 6.8). Succulent plants generally have low intercellular airspace, so internal transport of CO_2 may be largely confined to much slower diffusion in cell-wall water (Rascher *et al.*, 2001; Nelson and Sage, 2008). For the extant CAM water plants certainly CO_2 acquisition is still the decisive advantage of CAM. However, for terrestrial CAM plants the water saving related to the CO_2 concentrating of CAM appears to be the much more important facet. Moreover, as CAM evolved polyphyletically (Section 6.2) it is certainly not sufficient to consider just one selective driving force.

6.4.2. Requirement of economic water use driving crassulacean acid metabolism evolution

In terrestrial plants the phenomenon of CAM cycling is likely a remnant of past epochs with low CO_2, in that it appears to be a means of retrieving precious respiratory CO_2 during the night. This allows closing of stomata during the night, which however, is significantly not as important for water relations as closing during the day. Thus, it was thought that CO_2-saving CAM cycling is a kind of prelude to real CAM, and that it might have been a starting point for CAM evolution (Guralnick *et al.*, 1986; Monson, 1989b; Guralnick and Jackson, 2001). However, in most terrestrial CAM plants the major evolutionary driving force for CAM must have been the requirement of economic water use that is so well achieved by nocturnal CO_2 fixation with night-time opening and daytime closure of stomata because the evaporative demand for *E* is so much lower during the night than during the day. This allows CAM plants to operate with a very high water-use efficiency, especially during phase-I CO_2 acquisition in the dark period (Table 6.1).

Morphological/anatomical traits of CAM plants supporting water relations are the following:

(1) non-green water storage tissues or hydrenchymas with large vacuoles of their cells, where present in the stems or as particular cell layers of leaves of succulent CAM species serve long-term water storage and support CAM idling for survival of dry periods;
(2) tanks formed by overlapping leaf bases of rosettes especially in many bromeliads, catch precipitation and serve medium-term water storage;
(3) large central cell sap vacuoles, involved in short-term water storage during the diurnal CAM cycle;
(4) hydraulic architecture.

The latter two require special comments. With respect to the third point we note that there are diurnal cycles of osmotic relations. The large vacuolar concentrations of nocturnally accumulated organic acids are osmotically active. The increased osmotic pressure (π) drives water into the cells, which is associated with increased turgor pressure (P_{turgor}). This allows acquisition of water, particularly towards the end of the dark period when vacuolar organic acid levels reach their peak values that in moist tropical forests often coincides in time with the formation of dew, which then can be used effectively by epiphytic CAM plants. During acid remobilisation in phase III of CAM, P_{turgor} and π decline again but the water gained is available to the plants (Eller and Ruess, 1986; Lüttge, 1986; Ruess *et al.*, 1988; Eller *et al.*, 1992; Murphy and Smith, 1998).

Stem hydraulic architecture is the morphological trait determining water supply to canopies. Thus, with respect to the last point, observations comparing hemi-epiphytic C_3 species of the genus *Ficus* and the hemi- epiphytic C_3/CAM intermediate species *Clusia uvitana* are quite interesting (Patiño *et al.*, 1995). Making use of the water-saving CAM option in the scarcity of water-prone epiphytic habitat, *C. uvitana* affords much lower specific stem conductivity (Ks, kg s^{-1} m^{-1} Mpa^{-1}) and conductive stem per unit of leaf area (Kl, kg s^{-1} m^{-1} Mpa^{-1}) than *Ficus* species. The ratios of *Ficus*/*C. uvitana* for the two parameters were between 6 and 30 and 4 and 15, respectively.

Table 6.1. *Productivity and water-use efficiency (WUE) of C_3, C_4 and CAM plants (Black, 1973; Nobel, 1996).*

Type of photosynthesis	Daily productivity (g dry matter per $m^2 day^{-1}$)	WUE (mol CO_2/mol $H_2O \times 10^3$)	Crops (Mg $ha^{-1} year^{-1}$)
C_3-photosynthesis	50–200	0.6–1.3	35–45
C_4-photosynthesis	400–500	1.7–2.4	50–90
Crassulacean acid metabolism	1.5–1.8	darkness: 6–30 light: 1–4	40–50

6.4.3. Selection for plasticity?

The relations between plasticity, speciation and the development of biodiversity are currently much debated also with particular respect to CAM (Lüttge, 2005, 2006, 2007). There is the question of whether or not phenotypic plasticity is a major mechanism for evolution (De Jong, 2005). Alternatively, plasticity itself is a trait subject to evolution. An ecologically important question in this context is whether plasticity itself is adaptive or not. For each particular case cost/benefit evaluations need to be made. Considering CAM it is noteworthy that plasticity, as it is given by the options of flexible expression of CAM phases and modes, is one of its most conspicuous characteristics. This may make CAM plants so successful for example in the epiphytic environments of tropical forests. In these habitats it is not a single stress factor that is dominating and adaptation to which will support fitness. It is rather an array of factors, especially water, light, temperature, nutrients, that also underlies complex spatiotemporal variations. Plastic and flexible responses are necessary under these conditions of multiple and variable stress as it is particularly seen in the genus *Clusia*, where species exist that can switch back and forth between C_3 photosynthesis and CAM (Borland *et al.*, 1992; Lüttge, 2006, 2007, 2008a). This obviously increases ecological niche occupation width (Section 6.6).

6.4.4 Responses to increasing atmospheric CO_2 concentrations

We live at a time of a dramatic increase of CO_2 concentration in the atmosphere of the Earth, although on the background of the last 600 million years we are at a comparatively very low level (Fig. 6.8). Nevertheless, we should consider the consequences of a further increase of atmospheric CO_2 concentrations for CAM. If internal CO_2 concentrating of CAM is a benefit at low atmospheric CO_2 concentrations, it might be expected that current manmade increases in atmospheric CO_2 concentrations attenuates this advantage of CAM. Such possible direct consequences are evident from Fig. 6.8. If the current level is doubled this already brings us to the lower end of the range of CO_2 concentrating in phase III of CAM. If it is tripled the level of substrate saturation of C_3 photosyntheis without CO_2 concentrating is reached.

How will this effect CAM performance? An experimental doubling of CO_2 concentration from 0.037 to 0.075 % had no significant effects on the CAM plant *Agave vilmoriniana* (Szarek *et al.*, 1987). However, non intuitively, other studies indicate a growth stimulation of CAM plants by elevated CO_2 concentrations (Drennan and Nobel, 2000). Doubling ambient CO_2 concentration increased the productivity of the CAM cactus *Opuntia ficus-indica* by 35% on average, mostly as a result of increases in both night-time (phase-I) and daytime (phase-IV) CO_2 uptake, whereas in the CAM Crassulaceae *Kalanchoë daigremontiana* increased CO_2 concentration mainly augmented phase-IV CO_2 uptake (Winter *et al.*, 1997). The responses of CAM plants to increased CO_2 concentration are similar to those predicted for C_3 plants, but much greater than those of C_4 plants. There appears to be no expectation of downward acclimation (Wang and Nobel, 1996).

However, we should not only consider CO_2 *per se* and its possible direct effects, but also other environmental factors that may also be affected by globally increasing CO_2 concentration, so that there are many indirect effects of CO_2, particularly via water availability and temperature.

The importance of the temperature factor originally was mainly assessed by looking at temperature optima of key enzymes in the CAM cycle. Brandon (1967) suggested long ago that for optimal performance of CAM the plants require relatively low night temperatures and high day temperatures, because in in-vitro studies the temperature optima of the key enzymes were such that PEPC, the enzyme of nocturnal CO_2 fixation, reached the temperature optimum at 35°C, and the decarboxylating enzymes operative during the daytime reached optimum at 53°C. From overall performance of the counteracting enzymes of carboxylation and decarboxylation, where lower temperatures favour the former and higher temperatures the latter, Buchanan-Bollig *et al.* (1984) concluded that rather cool night temperatures, somewhat below

20°C, would be most favourable for dark fixation in CAM. The temperature optima of both PEPC and decarboxylating enzymes measured by Brandon (1967) are both relatively high. However, in-vitro PEPC enzymology also shows that the active phosphorylated form of the enzyme is stabilised at low temperature (3°C or less), whereas higher temperature promotes dephosphorylation and downregulation of activity (Carter *et al.*, 1995), so that inhibition of the enzyme by allosteric effectors is also lower at low temperatures and higher at high temperatures (Buchanan-Bollig and Kluge, 1981; Carter *et al.*, 1995).

However, owing to the complexity of temperature interactions and the frequent temperature acclimation, considering temperature optima of enzymes *in vitro* is a simplification. This becomes immediately evident when one considers habitats where CAM plants actually occur. CAM plants do not only thrive in deserts with cool nights (Section 6.5.3), but also in moist tropical forests (Section 6.5.7) where night temperatures may remain between 25 and 30°C. Comparing three growth temperature regimes, Israel and Nobel (1995) found that PEPC and Rubisco had maximal activities at 45/35°C day/night. In contrast, total daily CO_2 uptake was greater at 30/20°C and 15/5°C. This shows that in addition to temperature dependence of the carboxylating enzymes, other responses of the overall machinery of daily CO_2 acquisition to temperature were also involved.

Increased atmospheric CO_2 concentrations may contribute to temperature increases in the habitats of CAM plants, but it was observed that increased CO_2 concentration compensated for the inhibitory effect of increased temperature on nocturnal CO_2 fixation (Zhu *et al.*, 1999).

Temperature effects on air humidity may largely determine relationships between temperature and stomatal opening. For the CAM plants of *Opuntia* sp., Conde and Kramer (1975) and Osmond *et al.* (1979) showed that lower tissue-to-atmosphere water VPDs stimulated dark CO_2 fixation. Stomata of the CAM plant *Kalanchoë pinnata* are highly sensitive to air humidity (Medina, 1982). Maximum rates of dark CO_2 fixation were similar at all temperatures between 12 and 25°C within a given range of leaf/air water-vapour-pressure differences. However, the onset of nocturnal net dark CO_2 fixation and the time to reach a peak rate were delayed as temperature increased, and hence total CO_2 uptake and malate accumulation were reduced with increasing temperature during the dark period.

The timescales of all these observations are much too short to allow any predictions on future changes affecting habitat occupation of CAM plants. Overall, on a percentage basis, effects are very small and in detail there are many uncertainties (Poorter and Navas, 2003; Osmond *et al.*, 2008). Currently, with the timescales of ecosystem responses given, ecologically noticeable changes of global distribution of CAM plants may not be expected.

6.5. ECOSYSTEMS WITH CRASSULACEAN ACID METABOLISM PLANTS

The discussion of the flexible expression of CAM phases and modes has shown us above (Section 6.2) that CAM offers ways for plants to respond to two major environmental factors, namely CO_2 and H_2O. The recapturing of respiratory CO_2 in CAM cycling and the internal concentrating of CO_2 in pure CAM deal with the limitations given by CO_2 supply owing to the low affinity of Rubisco to CO_2, which at ambient CO_2 concentration operates far below its maximum activity (Chapter 2). The modulation of CAM phases in pure CAM and CAM idling effectively deals with the water factor. Adaptation to scarcity of CO_2 rather than conservation of water must have been an early predominant driving force for the evolution of CAM (see Section 6.4.1) because, as described below, CAM is found in submerged aquatic plants. However for the vast majority of extant CAM tracheophytes, water is the most determinant environmental cue. Together with other stress factors or stressors, such as temperature, irradiance and nutrient supply, it determines performance of CAM plants at the community level. CAM plants occur in many different plant communities, some of which can be characterised by the dominating CAM plants (Lüttge, 2004). A survey of ecosystems with CAM plants, the prevailing stressors, the dominant CAM life forms and morphotypes, and the major CAM taxa are given in Table 6.2.

6.5.1. Submerged aquatic sites

Freshwater sites with submerged CAM plants are seasonal pools, oligotrophic lakes, freshwater tidal creeks, irrigation channels and the like found all over the world, including high latitudes (Keeley, 1996). The life form of CAM plants from the genus *Isoëtes* and various families of the angiosperms is an isoëtid rosette morphotype. The major stressor here is availability of CO_2. Carbon dioxide gas dissolves in water in the form of bicarbonate, HCO_3^-. However, the concentration remains low, and resistance to diffusion restricts the supply to photosynthesising water plants. CO_2 diffusion in water is about 8,640 times slower than in air. Performance of CAM helps in two ways. First, it allows acquisition of inorganic carbon in the dark period when more is available owing to the respiration of microbes, animals and non-CAM plants

Table 6.2. *Ecosystems with Crassulacean acid metabolism (CAM) plants.*

Ecosystems	Stressors	Life forms	Major CAM taxa
Submerged aquatic sites	CO_2 supply	Isoëtid rosette growth form	*Isoëtes*, Alismataceae, Hydrocharitaceae, Crassulaceae, Apiaceae, Plantaginaceae
Temperate environments	Seasonal temperature, irradiance, H_2O supply	Leaf and stem succulents	Crassulaceae (*Sedum*, *Sempervivum*), Cactaceae
Deserts	H_2O supply	Leaf and stem succulents, 'CAM-trees'	Agavaceae, Cactaceae Euphorbiaceae, Didieraceae, Asteraceae, Apocynaceae, Asclepiadaceae
Salinas	Salinity	Annuals, leaf and stem succulents, epiphytes	Aizoaceae, Cactaceae, Bromeliaceae, Orchidaceae
Restingas	H_2O supply, irradiance	Leaf and stem succulents, trees, epiphytes	Clusiaceae, Bromeliaceae, Cactaceae
Savannas and cerrados	H_2O supply, fire	Dicot trees epiphytes	Clusiaceae, Bromeliaceae
Dry and moist forests	H_2O and nutrient supply, irradiance	Leaf and stem succulents, 'CAM trees', dicot trees, epiphytes	Cactaceae, Euphorbiaceae, Bromeliaceae Orchidaceae, Crassulaceae, Piperaceae, Polypodiaceae
Inselbergs	Irradiance, temperature, H_2O-supply	Leaf and stem succulents, dicot trees	Cactaceae, Crassulaceae, Euphorbiaceae, Agavaceae, Bromeliaceae, Orchidaceae, Clusiaceae
Páramos	Temperature	Leaf and stem succulents	Cactaceae, Crassulaceae

in the biotope, and at the same time there is no competition for it by water plants performing C_3 photosynthesis. Second, the concentrating of CO_2 by CAM overcomes the problem of low ambient levels. Root uptake of inorganic carbon and transfer as CO_2 to the shoots also occurs in most isoëtids.

6.5.2. Temperate environments

CAM plants are mainly concentrated in the subtropics and tropics. However, some CAM plants also intrude into certain habitats in the temperate zones. There they are mainly found in northern deserts and in prairies on the American continent (Cactaceae, Crassulaceae), where CAM may be a useful adaptive trait under limited supply of water, and in mountains in the old world (Crassulaceae: *Sedum, Sempervivum*). An important stressor is temperature. Seasonal cold hardening with acclimation to sub-freezing temperatures in the winter is important for those cacti that reach northern latitudes in southern Canada and the eastern U.S.A. (Nobel and Smith, 1983). The lowest temperatures tolerated by cacti are −10°C by *Opuntia ficus-indica* and *Opuntia streptacantha* and −24°C by *Opuntia humifusa* (Goldstein and Nobel, 1994). Low temperatures inhibit daytime remobilisation of vacuolar organic acid because they reduce membrane fluidity and hence permeability of the tonoplast (Wagner and Larcher, 1981). At high elevations in the mountains irradiance is an additional stressor and phenolics may be formed as UV protectants (Bachereau *et al.*, 1998).

6.5.3. Deserts

The physiognomy of some deserts and, particularly semi-deserts, on the American continent is strongly characterised by CAM-performing large Agavaceae and arborescent cacti, in Africa by stem-succulent Euphorbiaceae and in Madagascar by the Didieraceae, where the cliché view has developed that CAM plants and desert plants are almost synonymous. This view is completely wrong. Ellenberg (1981) has investigated the reasons for the phytogeographic distribution of stem succulents in the arid zones of the world. The large succulents are not truly drought resistant. They require regular seasonal precipitation regimes. They grow best with an average annual precipitation between 75 and 500 mm. They can withstand low precipitation, but precipitation must be regular. CAM succulents can overcome many weeks and even months by CAM idling (Szarek and Ting, 1975). However, even with permanently closed stomata they loose some water by cuticular E, and they must be able to refill the water stores of their hydrenchyma. Cacti die when they loose more than 54 % of their tissue water (Holthe and Szarek, 1985). Thus in very extreme deserts there are few or no CAM succulents. In the Negev desert in Israel, a few species of *Caralluma*, including *Caralluma negvensis*, are the only stem-succulent CAM species, and they grow protected in the shade of rocks and benefit from the formation of dew after cold nights in these niches (Lange *et al.*, 1975). Specific morphological/anatomical adaptations and hydraulic architectures of C_3 plants are more suitable for survival in extreme deserts, with high variability of precipitation over many years than the water requiring succulent morphotype of CAM plants. Overall, desert succulents Cactaceae and Agavaceae together comprise about 1,800 species, almost all of which are CAM species, but this total number is only less than a fifth of the number of CAM species occurring in most tropical forests (Section 6.5.7).

An additional stressor is temperature, especially the large diurnal temperature amplitude given in deserts with often frosty nights and very hot days. Cacti can tolerate a remarkable range of temperatures. The highest soil temperature observed to be tolerated by cacti was 74°C (Nobel *et al.*, 1986) and the lowest temperatures may be below –20°C (Section 6.5.2).

6.5.4. Salinas

With their succulence and other traits allowing adaptation to limited water supply and osmotic stress one would expect that CAM plants would be highly suited to inhabit saline habitats. However, observations do not support this expectation as, in general, halophytes are not CAM plants and CAM plants are not halophytes. Most CAM plants, including desert succulents, are highly salt sensitive (Nobel, 1983; Nobel and Berry, 1985). A single conspicuous exception is the annual facultative halophyte *Mesembryanthemun crystallinum* (Lüttge, 1993; Adams *et al.*, 1998; Cushman and Bohnert, 2002; Lüttge, 2002b). It is a C_3/CAM-intermediate species. It can grow well in the absence of NaCl but has its growth optimum at several hundred mM of NaCl in the medium, and can complete its lifecycle at 500 mM NaCl (Winter, 1973; Lüttge 2002b). In saline arid sites it starts its lifecycle with C_3 photosynthesis when germinated after sufficient winter rains and later, when the dry period proceeds, it switches to CAM and can survive much longer than similar annuals with C_3 photosynthesis bringing a large number of seeds to maturity. This response may be shared by a few other halophytes, such as *Calandrinia* spp., *Carpobrotus* sp. and *Disphyma,* engaged in CAM in response to salinity and water stress, as found in a survey of Australian succulent native halophytes (Winter *et al.*, 1981).

Alternatively one can observe that CAM plants may determine the physiognomy of tropical salinas inland or near the coast, e.g., large columnar cacti of *Pilosocereus ottonis* and often a dense groundcover by tank-forming bromeliads in vegetation islands on saline sand plains (Medina *et al.*, 1989; Lüttge, 2008a). The columnar cacti *Stetsonia coryne* and *Cereus validus* are characteristic elements of the vegetation at the rim of the Salinas Grandes in Argentina (Ellenberg, 1981; Yensen *et al.* 1981). However, all of these CAM plants are not halophytes but strict salinity stress avoiders. *P. ottonis* is a salt excluder at the root level. It even sacrifices absorptive fine roots during periods of strong salinity in the dry season, survives by CAM idling and rapidly grows new roots in the wet season (Lüttge *et al.*, 1989). The rosettes of the terrestrial CAM bromeliad *Bromelia humilis* just lie on the ground and do not produce absorptive roots, but form tanks and have tank roots fed by rain (Lee *et al.*, 1989). The CAM bromeliad *Tillandsia flexuosa* and the CAM orchid *Schomburgkia humboldtiana* in coastal salinas of Venezuela are epiphytes, and thus not at all affected by soil salinity. It should be noted, however, that the strategy of salt-stress avoidance in these CAM plants of salinas is supported by CAM, especially by the option of CAM idling.

6.5.5. Restingas

Restingas are marine sandy deposits and dunes of quaternary origin on the Brazilian coast. They have a high species diversity of CAM plants, which is overwhelming at some sites that contain terrestrial and epiphytic bromeliads and cacti, and among the cacti several epiphytic species, orchids,

Piperaceae and several species of *Clusia* (Lüttge, 2007). Although C_3 vegetation is also abundant, restingas are one of the most unique CAM domains known (Reinert *et al.*, 1997). The major stressors are water supply and high irradiance, and it is a great problem for plants to get established on the bare sand that is often very dry even in the wetter season. The water-saving CAM plants, especially the woody shrubs and trees of *Clusia*, serve as nurse plants, which protect the establishment of other plants. The flexibility of CAM and its associated niche width (Section 6.6) may have made CAM plants particularly suited to invade these sandy dry habitats (Lüttge, 2007). On the sand dune ridges of restingas, the establishment of vegetation starts in the form of a mosaic of vegetation islands, where frequently *Clusia* species are the pioneers and nurse plants under which other vegetation can develop. *Clusia hilariana* and *Clusia fluminensis* are CAM species that are dominant nurse shrubs and small trees in the restingas (Zaluar and Scarano, 2000; Liebig *et al.*, 2001).

6.5.6. Savannas and cerrados

CAM plants are found in tropical savannas and the Brazilian savanna-like cerrados. However, these are not typical CAM habitats. CAM succulents have problems becoming established as they are often overgrown by grasses with C_3 and typically often C_4 photosynthesis dominating the savannas, and by C_3 trees and shrubs (Ellenberg, 1981). CAM species in savannas and cerrados can be the dicotyledonous CAM trees of *Clusia* (Lüttge, 2007).The large flexibility offered by the CAM option confers a large niche width on the C_3/CAM-intermediate species *Clusia minor* that can intrude from dry forest into savannas (Section 6.6). In addition to strong seasonal variation in precipitation, frequent fires are a major stress factor and this may also be the real reason why the CAM tree *Clusia criuva* was not found to be able to penetrate much into the cerrado from an adjacent gallery forest where it occured (Herzog *et al.*, 1999a).

6.5.7. Forests

CAM plants can be dominant in various types of tropical forests, which include dry forests, gallery forests, moist or rain forests and even specific 'CAM forests'. The latter is an intriguing facet because it requires that there are CAM trees. The only bona fide dicotyledonous CAM-performing trees with a typical secondary growth are the many CAM species of the genus *Clusia* (Lüttge, 2007). In addition, the CAM-performing Joshua tree of the genus *Yucca* has a special monocotyledonous type of secondary thickening. By contrast, all the stem-succulent CAM-performing dicotyledons (Lüttge, 2008b) have no secondary growth. However, they are arborescent life forms and can reach a spectacular size, for example columnar cacti, candelabrum euphorbias and some Didieraceae. They have been called 'tree succulents' (Ellenberg, 1981). Spectacular ecosystems dominated by tall columnar and candelabrous cacti, e.g., in Mexico and Venezuela, have been named 'cactus forests' (Vareschi, 1980). However, the sensation of being in a forest with a closed canopy when walking in such habitats does not so much originate from the columnar cacti, but rather from the acacias and other woody plants scattered among them. Thus, much depends on the definition of forest one may wish to use (Vareschi, 1980).

The role of non-arborescent CAM species in the vegetation of various forests is significant. Semi-deciduous, deciduous and scrub forests may have a very dense groundcover of CAM plants, e.g., *Bromelia humilis,* together with abundant epiphytic CAM bromeliads in a tropical dry forest in Venezuela (Lüttge, 2008a). Terrestrial CAM plants are also found in moist tropical forests and rain forests. There, however, the epiphytes constitute the most important species diversity and biomass of CAM plants. The dominating families are Orchidaceae and Bromeliaceae. About half of all species of these families are CAM species. It was estimated that about 10% of all vascular plants are epiphytes, i.e., approximately 23,500 species (Lüttge, 1989). Of these vascular epiphytes, 57% are CAM plants, i.e., 13,400 species (Lüttge, 2003). In some wet tropical forests, 50% of all leaf biomass may be owing to epiphytes or close to 30% to CAM plants, depending on the abundance of CAM among epiphytes. The major stressors are all those that are typical for the epiphytic habitats, such as supply of water and nutrients, and the specific light climate with either limiting irradiance in shaded parts of the canopy of excess irradiance in sun-exposed parts. In very wet cloud forests CAM taxa are scarce, which suggests that it is a recent radiation of this photosynthetic pathway, and occupation of new niches where the plasticity inherent in CAM is pre-adapting more xerophylic physiotypes to wetter habitats.

6.5.8. Inselbergs

Inselbergs are rock outcrops, several tens to several hundred metres high, emerging from savannas or rain forests (Porembski and Barthlott, 2000; Lüttge, 2008a). They are strongly fragmented into a variety of smaller ecological units, such as humus accumulating in cracks and hollows, vegetation islands, shrubberies and even small forests on their tops. The major stress factors are high irradiance and temperatures above the bare rocks of the outcrops and scarcity of water owing to runoff even in the wet seasons. This makes CAM an appropriate way of coping. Many CAM families, i.e., Cactaceae, Crassulaceae, Euphorbiaceae, Agavaceae, Bromeliaceae and Orchidaceae, are frequently found on inselbergs in the paleotropics and the neotropics (Kluge and Brulfert, 2000). CAM species of *Clusia* often prevail in the formation of shrubbery and vegetation islands on inselbergs in the neotropics (Lüttge, 2007).

6.5.9. Páramos

The occurrence of CAM plants in the tropical mountains at very high altitudes, especially in the páramos of South America is intriguing. There are the cacti in the Andes (Keeley and Keeley, 1989) and the Crassulaceae *Echeveria columbiana* (Medina and Delgado, 1976). These high altitudes at 3,500 m above sea level and above are characterised by a regular diurnal climate change with 'summer every day and winter every night' (Hedberg, 1964). The obligate CAM plants in these habitats must perform the nocturnal part of the CAM cycle, with glycolysis and dark fixation of CO_2 by PEPC and subsequent active accumulation of malate in the vacuole every night at sub-freezing temperatures, whereas the daytime part of the cycle runs at temperatures between 20 and 30°C. There is no information about how this functions and is regulated diurnally as it cannot be achieved by long-term acclimation, as in winter in the temperate regions (Section 6.5.2). Biochemical reactions of CAM must be maintained in a state of super cooling at temperatures below 0°C, as nocturnally measured in the tissue of cacti in the high Andes (Keeley and Keeley, 1989).

6.6. CONCLUSIONS: PRODUCTIVITY, PLASTICITY AND NICHE OCCUPATION

In principle, in the field the advantage of CAM is flexible adaptation to stress and ability to occupy ecological niches determined by stressful conditions, and not a means for considerable production of biomass. Productivity of CAM plants is generally much lower than that of C_3 and C_4 plants (Table 6.1). There are, however, some cultivated crops, mainly cacti and agaves, that may reach productivities comparable with those of C_3 plants (Table 6.1). However, this occurs under technically advanced agricultural management and relies mainly on the C_3-type phase IV of CO_2 acquisition (Nobel, 1996).

The conclusion that the real ecophysiological impact of CAM is its very flexibility is best supported by finally describing a comparative study of the obligate C_3 *Clusia muliflora* and the C_3/CAM-intermediate *C. minor* (Herzog *et al.*, 1999b). When both are grown in a glass house at low irradiance and then transferred to a phytotron growth chamber at high irradiance they show a quite contrasting performance. Photosynthesis of *C. multiflora* is strongly reduced and the leaves eventually get necrotic and die. *C. minor* switches from C_3 photosynthesis to CAM, and the leaves do not show any damage. In the field in a secondary savanna in Venezuela, *C. multiflora* was observed to only occupy the open sun-exposed sites, whereas – counterintuitively, when we consider CAM as an adaptation to stress – *C. minor* was mainly found inside semi-deciduous dry forest. However, *C. minor* could intrude into the open habitat of *C. multiflora* and could form a *Clusia* shrubbery together with the obligate C_3 species. Conversely, *C. multiflora* was not found in the more shaded dry forest. Evidently, *C. multiflora* performs best in the open exposed habitat, but it needs to grow its leaves under the conditions of exposure that are then constructed as sun leaves adapted to the high irradiance. Indeed, in the experiment (Herzog *et al.*, 1999b) where the leaves had become necrotic after transfer to high irradiance, the plants could grow new leaves from dormant buds that were then performing appropriately. The story tells that quite evidently the C_3/CAM-intermediate species has a wider niche occupation than the obligate C_3 species. It appears that in the field C_3/CAM-intermediate species, e.g., *Clusia cylindrica* and *Clusia pratensis*, for most of the time perform C_3 photosynthesis, and for shorter periods perform strong CAM as environmental conditions require (Winter *et al.*, 2009).

7 • Special photosynthetic adaptations

J.I. GARCÍA-PLAZAOLA AND J. FLEXAS

7.1. INTRODUCTION

Photosynthesis studies and models are usually restricted to green leaves. However, a closer observation of nature reveals that most plants also photosynthesise through non-foliar structures, and photosynthetic tissues are not always green. Non-foliar photosynthesis is performed by a variety of organs such as stems, bark, roots, petioles, fruits or flowers. These structures contribute positively to the carbon balance, and in some leafless species these are the only photosynthetic organs. In general, two main groups of photosynthetically active organs can be discerned (Aschan and Pfanz, 2003): those optimised for photosynthetic performance that achieve a net positive carbon gain (such as green petals, leaves or stems); and those involved in the internal recycling of CO_2 released by respiration (such as roots, fruits or chlorophyll-containing bark) (Table 7.1). With a few exceptions, chlorophyll is the only green pigment in plants, implying that all green tissues are chlorophyllous, and all chlorophyll-containing organs are, to some extent, photosynthetic. Nevertheless, photosynthetic organs may not appear green externally because of chlorophyll being masked by outer layers of pigmented cells, waxy cuticles or bark. In addition to plants with whole leaves or parts of leaves (variegated leaves) presenting special photosynthetic characteristics, unique photosynthetic adaptations are also shown by parasitic and submerged plants. Also, stomatal cells from typical leaves of higher plants display specific photosynthetic features, different from those of mesophyll cells.

7.2. NON-FOLIAR PHOTOSYNTHESIS

Non-foliar photosynthesis is an ancient characteristic developed by the earliest terrestrial plants that lacked specialised photosynthetic organs. As a probable consequence of the continuous fall in atmospheric CO_2, the laminate leaf structure evolved in the Devonian, around 360 million years ago (Osborne *et al.*, 2004a,b). However, the capacity to perform photosynthesis in non-foliar structures has been retained, and in fact most, if not all, plants show some degree of photosynthetic activity in their stems, flowers or fruits. As will be discussed in the next sections, several estimations have shown that non-foliar photosynthesis, as net CO_2 assimilation or internal refixation of endogenously produced CO_2, contributes significantly to plant-carbon assimilation (Table 7.1).

7.2.1. Structural organs

STEM PHOTOSYNTHESIS

Many tropical CAM species are completely devoid of leaves, and photosynthesis, is exclusively performed in stems by a peripheral, photosynthetically active chlorenchyma (Chapter 6). Apart from succulent species, many others also present green stems, and even some angiosperms (in particular legumes from arid ecosystems) have evolved to be completely leafless, relying only on stems (Nilsen *et al.*, 1989) or phyllodes (Montagu and Woo, 1993) for photosynthesis. Stems and phyllodes of these species contain abundant stomata and a chlorenchyma structurally similar to that of leaves. These characteristics allow them to perform a photosynthetic activity equivalent in regulation and responses to 'normal' leaf photosynthesis. In current-year stems of the Mediterranean shrub *Retama sphaerocarpa*, maximum photosynthesis is 20–26 μmol CO_2 m^{-2} s^{-1}, and even during the driest season it reached 10 μmol CO_2 m^{-2} s^{-1} (Haase *et al.*, 1999a). In phyllodes of *Acacia auriculiformis* photosynthesis, rates declined from 24 μmol CO_2 m^{-2} s^{-1} in the rain season and 5 μmol CO_2 m^{-2} s^{-1} during the dry season (Montagu and Woo, 1993).

The simultaneous occurrence of photosynthesis in leaves and stems is common in nature, with the rates of

Terrestrial Photosynthesis in a Changing Environment: A Molecular, Physiological and Ecological Approach, ed. J. Flexas, F. Loreto and H. Medrano. Published by Cambridge University Press.

Table 7.1. *Photosynthetic characteristics of non-foliar structures.*

Structure	Photosynthesising structure(s)	CO_2 source	Maximum A_N ($\mu mol\ CO_2\ m^{-2}\ s^{-1}$)	Respiratory CO_2 refixation (%)
Typical leaves	Mesophyll cells	Atmospheric	5–50	—
Shade leaves	Mesophyll cells	Atmospheric	0.5–4	—
Structural organs/ stems	Twigs, petioles, green stems	Atmospheric	6–12	—
Phyllodes	Mesophyll cells	Atmospheric	5–24	—
Bark	Chlorenchymatous cells located beneath the periderm	Respiratory	0–4	30–90
Wood	Ray cells, ring-shaped halos around/inside the pits	Respiratory	—	5–10
Roots	Aerial roots, pneumatophores, stilt roots of pals	Respiratory	—	50–70
Reproductive organs Flowers	Mostly green petals, but also sepals, anthers, tepals, buds, calyx lobes, bracts, corolla, capsules and flower stalks	Atmospheric	0–5	—
Fruits	Mostly exocarp (peel)	Respiratory (with few exceptions)	—	1–50
Seeds	Testa, embryo supposedly low	Respiratory	—	—

stem photosynthesis usually being lower than that of leaves (Nilsen, 1992), though stem contribution to growth and biomass production is still significant (Nilsen *et al.*, 1996). Apart from the contribution of stem photosynthesis to the overall carbon gain, the main advantage of possessing photosynthetic stems is to meet different needs from leaf photosynthesis. For example, whereas drought stress promotes leaf abscission in some desert species (Nilsen *et al.*, 1996), their stems are less sensitive to water stress (Nilsen, 1992), accumulating osmoprotective solutes and remaining photosynthetically active. Stem photosynthesis also reduces the impact of foliar loss (Eyles *et al.*, 2009). During these leafless periods, stems are the only sources of carbon gain. The occurrence of photosynthetic phyllodes is more uncommon, but their photosynthetic rates are comparable with those of leaves. As occur in the stems of desert shrubs, phyllodes are less sensitive to drought than leaves (Brodribb and Hill, 1993).

BARK PHOTOSYNTHESIS

The presence of chlorophyll is universal in developing stems of woody plants. The so-called corticular or bark photosynthesis is performed by a layer of chlorenchymatous cells located beneath the periderm. This photosynthetic activity is limited by periderm resistance to gas diffusion and by light penetration.

Light transmission through peridermal layers mostly depends on structural features, such as bark thickness and colour, cork peeling, cracks and openings in the bark (Aschan and Pfanz, 2003), pubescence or branch age (Filippou *et al.*, 2007). Other external factors also influence transmission, e.g., a lichen cover that reduces light penetration up to 10% (Solhaug *et al.*, 1998) or twig wetting that may increase light transmission by 45–50% (Manetas, 2004a,b; Filippou *et al.*, 2007). As a result of the interaction between such factors, periderm transmittance ranges between 1% and 60%, with most observations in the range 10–20% (Pfanz *et al.*, 2002). In general, longer wavelengths penetrate deeper into the stems, whereas shorter wavelengths are absorbed in the outermost layers (Aschan and Pfanz, 2003). In a recent study, Manetas and Pfanz (2005) hypothesised that lenticels may act as gates for light penetration but, in contrast to their prediction, light transmittance through lenticels was much lower than in the surrounding periderm areas, and they concluded that their main role is to facilitate gas penetration through peridermal tissues. Low light penetration allows a shade-adapted bark photosynthesis characterised by a low chlorophyll *a/b* ratio and a low light-compensation point (Filippou *et al.*, 2007). Chl. concentration, expressed on an area basis, was 50–70% lower than that of leaves on the same plants. However, there is also some degree of sun/shade acclimation, with internal PPFD one order of magnitude lower in the shade- than in the sun-exposed sites. In a study on *Eucalyptus nitens* stems, Tausz *et al.* (2005a) determined that the sun-exposed side of the trunk experiences significant photoinhibition in the field, and the xanthophyll cycle is fully operative in this tissue.

Several experiments have demonstrated that low permeability to gas exchange (including CO_2 diffusion) of periderm makes photosynthesis very unlikely, and the main function of bark photosynthesis is probably to re-assimilate internal CO_2 produced by branch respiration or CO_2 diffusion from the ascending transpiration stream (Pfanz *et al.*, 2002; Wittmann *et al.*, 2006). Low gas exchange allows internal CO_2 concentrations to reach extremely high partial pressures (higher than 20%) (Teskey *et al.*, 2008) that limit photorespiration, but potentially inhibit carbon fixation and the development of photoprotective processes. With these limitations, gross photosynthesis (A) in bark reaches values comparable with those of shade leaves (0.7–3.7 μmol CO_2 m^{-2} s^{-1}) (Berveiller *et al.*, 2007a). Thus, photosynthesis can contribute significantly to compensate for stem-growth respiration and plays a major role in the annual carbon balance. For example, in a beech stand it represents approximately one third of the ecosystem carbon loss from respiration (Damesin *et al.*, 2002) and averages 55% of dark respiration in *Pinus monticola* (Cernusak and Marshall, 2000). Oxygen evolution from stem photosynthesis using internally sourced CO_2 could also be of importance for the avoidance of stem internal anaerobiosis (Pfanz *et al.*, 2002; Teskey *et al.*, 2008) and to provide a mechanism of aeration in submerged roots (Armstrong and Armstrong, 2005). Recent studies using light exclusion in trunks of several woody species suggest that stem photosynthesis significantly contributes to the total carbon income of these plants, so that impairing stem photosynthesis results in reduced growth (Saveyn *et al.*, 2010; Cernusak and Hutley, 2011).

Bark photosynthesis is also affected by seasonal changes in environmental conditions, being strongly reduced by suboptimal (<5°C) and supraoptimal (>30°C) temperatures (Wittmann and Pfanz, 2007). Inhibitory high temperatures may be achieved in sun-exposed sites because of the lack of evaporative cooling, but low irradiance may also limit carbon fixation (Cerasoli *et al.*, 2009). Alternatively, winter reductions of photochemical efficiency have been described for the bark of the temperate deciduous species *Betula pendula* (Wittmann and Pfanz, 2007), *Fagus sylvatica* (Berveiller *et al.*, 2007b) and *Populus tremula* (Solhaug and Haugen, 1998). Changes in light conditions related with leaf abscission could exacerbate light stress in winter. In spring, resumption of bark photosynthesis precedes that of leaves, as has been shown in *P. tremula* (Solhaug and Haugen, 1998), whose stems recovered one month before bud-break.

WOOD AND ROOT PHOTOSYNTHESIS

Apart from the bark, chlorophyll is also present in the rays and in parenchymatic cells of the wood of some species. Light is transmitted to deep layers in the trunk through vascular tissues that behave as fibre-optic cables, allowing photosynthesis to proceed (Sun *et al.*, 2003). Avoidance of anaerobiosis and fermentation is probably the main function of this metabolic process.

Chlorophyll is also found in roots, whenever light has access to these tissues. Thus, well-developed photosynthetic capacity is also found in aerial roots, such as those of epiphytes, and in pneumatophores of mangroves, with chlorophyll concentrations ranging from 9 to 55% of that in the leaves of the same plants (Aschan and Pfanz, 2003). This is also the case in underground roots in which light is transmitted from the surface by vessel elements (Sun *et al.*, 2003). Low gas exchange from aerial roots suggests that recycling of respired CO_2 is probably the main function of root photosynthesis, but there are some leafless epiphytes, such as *Aranda* orchid, in which CO_2 fixation is performed by aerial roots (Hew *et al.*, 1991).

7.2.2. Reproductive organs

PHOTOSYNTHESIS IN FLOWERS

Although the primary function of flowers is reproduction, some plants have green flowers and other photosynthetic structures associated with the inflorescence, and these structures may supply a significant fraction of the total carbon and energy required for reproduction (Bazzaz *et al.*, 1979; Marcelis and Hofman-Eijer, 1995). Contrary to wood, roots and most fruits, flowers show net CO_2 assimilation (A_N), as observed for many species in sterile as well as fertile parts of the inflorescence (Aschan and Pfanz, 2003), including calyx lobes, bracts, sepals, petals (in early stages of flower development), anthers, corolla, capsules and flower stalks (Table 7.1). In the sacred lotus (*Nelumbo nucifera)* floral receptacles are initially heterotrophic but become photosynthetic after anthesis (Miller *et al.*, 2009). All these examples of floral photosynthesis are achieved by the presence of functional chloroplasts in flower structures (although only at 10–40% of the concentration in leaves, Aschan and Pfanz, 2003). Stomata are also present at a density in some flowers similar to, or just slightly lower than, that of leaves (Venmos and Goldwin, 1993; Antlfinger and Wendel, 1997), but that in others is substantially lower (Hew *et al.*, 1980; Aschan *et al.*, 2005). In contrast to leaf stomata, flower stomata fail to respond to light intensity, VPD, CO_2 and ABA, and hence are suggested to be permanently open (Hew *et al.*, 1980). Water loss during flowering is an important part of seasonal water loss in succulents with CAM (Nobel, 1988).

Despite some similarities with leaves, photosynthetic rates of flowers are usually lower than those of leaves. For instance, although the electron transport rate in the chloroplasts (J) in sepals of some tomato cultivars achieved high values similar to those of leaves (Smillie *et al.*, 1999), the rates of photosynthesis in calyx and sepals of other cultivars were only 60% of those of leaves (Hetherington *et al.*, 1998). In *Helleborus viridis* (Fig. 7.1), J of sepals was up to 80% of that of leaves (Aschan *et al.*, 2005), whereas A_N ranged from 0 to 5 μmol CO_2 m^{-2} s^{-1} in flower structures, i.e., about an order of magnitude lower than the range found in typical leaves (Table 7.1). This was explained by a 70–80 % lower stomatal density of sepals in comparison with leaves, suggesting that, even when positive A_N is achieved by many flowers, substantial refixation of internally produced CO_2 also occurs (see Chapters 4 and 12).

In summary, flower photosynthesis provides only a small portion of the carbon required for reproduction, and hence flowers of most plants are heterotrophic and require imported carbohydrates for their development, which are derived mostly from adjacent leaves. Nevertheless, the proportion of total carbon required for production of mature seeds that is provided by flower photosynthesis can be significant, ranging between 2 and 65%, depending on the species (Bazzaz *et al.*, 1979; Watson and Caspar, 1984). Alternatively, in some green-flowered species, like *Helleborus viridis*, as the basal leaves emerge late during fruit development, the photosynthetically active sepals are a major source of assimilates, contributing more than 60% of whole-plant CO_2 gain in early spring (Aschan *et al.*, 2005).

FRUIT PHOTOSYNTHESIS

Young fruits are usually green, owing to the presence of functional chloroplasts. During ripening, chlorophyll is degraded and fruit chloroplasts develop into chromoplasts that accumulate carotenoids. Anthocyanin synthesis also contributes to the colour shift and enhances photoprotection (Steyn *et al.*, 2009). However, mature fruits of many species may retain some chlorophyll and perform photosynthesis (Bean *et al.*, 1963; Blanke and Lenz, 1989; Aschan and Pfanz, 2003).

In most fleshy fruits both light transmission and CO_2 diffusion are very limited. For instance, peel transmission ranges from only 1 to 47% of incident PPFD, and that of the mesocarp is less than 5% (Aschan and Pfanz, 2003). Gas exchange is greatly limited by an outer epidermis coated with a thin cuticle impregnated with waxy or greasy layers. Stomata are present, but their frequency is 10- to 100-fold less than in the abaxial epidermis of leaves of the same species (Blanke and Lenz, 1989). In addition, both light transmission (Aschan and Pfanz, 2003) and stomatal density (Blanke, 2002) decrease during fruit development, owing to the accumulation of pigments and surface expansion, respectively. Therefore, fruit photosynthesis occurs mainly in the exocarp, and the main function is regarded as internal recycling of respiratory produced CO_2, which is regarded as an energy saving mechanism as it provides a substantial part of the carbohydrates required for fruit growth, hence allowing more leaf-produced carbohydrates to be used for growth and maintenance of other plant tissues (Blanke and Lenz, 1989; Aschan and Pfanz, 2003). This is demonstrated by comparing the fluorescence-estimated electron transport rate (J_f), which may achieve values up to 70–100% of that of leaves, with the absence of A_N (Carrara *et al.*, 2001; Aschan and Pfanz, 2003). In a few species fruits may reach positive A_N, such as in some tropical green-ripe fruits (Cipollini and Levey, 1991), young tomato fruits (Xu *et al.*, 1997), developing peach fruits (Pavel and Dejong, 1993) and young, green blueberries (Birkhold *et al.*, 1992). Contrary to fleshy fruits, dry and dehiscent fruits, such as pea pods, show A_N during most of their development (Flinn *et al.*, 1977) that is probably related to their higher light transmission (at saturating light, up to 23% of incident light reaches seeds in pea pods, Atkins *et al.*, 1977).

Photosynthesis in fruits exhibits some characteristics of C_4 photosynthesis (see Chapter 5), such as β-carboxylation, and a CO_2-concentrating mechanism owing to enhanced activities of C_4 enzymes (Blanke and Lenz, 1989), particularly an efficient phosphoenolpyruvate carboxylase (PEPC), but also other enzymes such as malic dehydrogenase and

Fig. 7.1. Special photosynthetic adaptations: **(A)** Photosynthetic flowers of *Helleborus viridis*. **(B)**, **(C)** and **(D)** Examples of understory plants with variegated leaves. In the case of *Erythronium dens-canis* (B) variegation consists of red patches. In the other species it is owing to the presence of whitish sections. **(E)** and **(F)** Examples of a holoparasitic stem (*Cuscuta planiflora*) and a hemiparasitic stem (*Viscum album*).

phosphoenolpyruvate carboxykinase (PEPCk). Fruits also exhibit some CAM-like characteristics, such as malic-acid storage in vacuoles, but the kinetic properties of fruit PEPC are those typical of C_3 plants. Hence, Blanke and Lenz (1989) proposed a distinctive type of photosynthesis for fruits, intermediate between the common types in leaves. The green fruits of cacti show normal CAM (Nobel, 1988).

Several estimates of photosynthetic efficiency in fruits have been summarised in Aschan and Pfanz (2003). In large fruits, such as avocados, mass-based photosynthetic rates are only 0.4–2.5% of those of avocado leaves (Whiley *et al.*, 1992). Other fruits, such as those of cotton, cucumber and orange, attain up to 30–50% of leaf photosynthetic efficiency (Aschan and Pfanz, 2003). By contrast, small-seeded rape fruits produce nearly all of their carbon through their own photosynthesis (Singh, 1993). Fruit photosynthesis can provide a significant proportion of the carbon requirement of reproduction, and may positively contribute to the whole-plant carbon budget. Refixation rates of CO_2 internally produced by respiration typically reach 50–80% in some fruits (Araus *et al.*, 1993; Bort *et al.*, 1996; Proietti *et al.*, 1999; Zotz *et al.*, 2003), and the contribution of fruit photosynthesis to the total-fruit carbon requirement can be as high as 20–70% (Aschan and Pfanz, 2003). The proportional contribution of fruit over leaf photosynthesis to total carbon balance can be even higher in desert and semi-desert plants under field conditions. For instance, there are interesting examples of hollow fruits with an elevated-CO_2 internal environment (500 to 4000 ppm) that favours CO_2 fixation (Goldstein *et al.*, 1991). There are few studies of fruit photosynthesis responses to irradiance, temperature and ontogeny (Pavel and Dejong, 1993; Aschan and Pfanz, 2003), whereas others have focused on the heritability of fruit photosynthetic characteristics (Zangerl *et al.*, 2003). Fruit photosynthesis estimated by Chl-F has emerged as a practical tool to follow fruit maturation and to assess the presence of infections in fruits prior to the appearance of other visual symptoms (Nedbal *et al.*, 2000a).

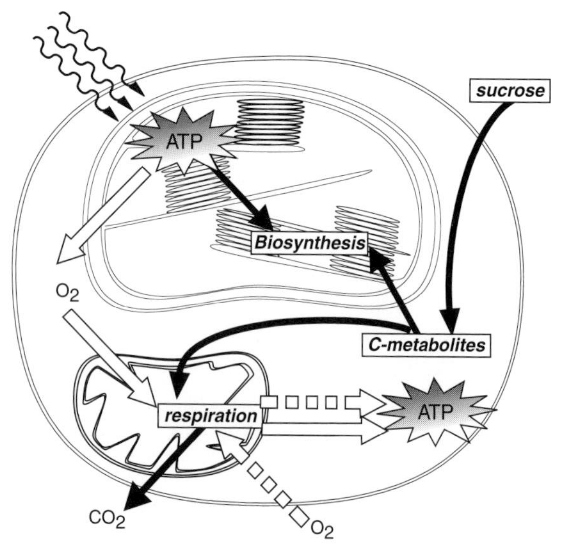

Fig. 7.2. Photoheterotrophic metabolism in embryos: photosynthetic metabolism in legume embryos may function to increase oxygen supply for adenosine triphosphate (ATP) synthesis necessary for fatty acid synthesis and storage activity.

PHOTOSYNTHESIS IN SEEDS

Seeds of many plant species are green during embryogenesis, but they tend to lose chlorophyll during maturation. Gradients in chlorophyll, PET and O_2 release are observed within seeds. In cereal grains, for instance, photosynthetic activity is restricted to distinct cell layers within the pericarp, the developing embryo and endosperm (main storage organ). Indeed, Chl-F of the testa of cabbage seeds was found to be inversely correlated with germination capacity (Jalink *et al.*, 1998), and a method was patented to determine seed maturity and seed quality based on Chl-F measurements. Light has a clear stimulatory effect on storage of carbohydrates in seeds (Fader and Koller, 1985; Willms *et al.*, 1999). Using inhibitors, Fuhrman *et al.* (1994) showed that this effect is directly linked with photosynthetic activity within seeds. However, owing to the low light supply to the seeds (owing to the low light transmittance of fruits, see previous section) it has been questioned whether photosynthesis can play this role *in vivo* (Eastmond and Rawsthorne, 1998).

One important question is the contribution and/or regulatory function of seed photosynthesis, and several hypotheses have been postulated. As for bark and fruit photosynthesis, it has been proposed that photosynthesis within seeds predominantly refixes carbon derived from respiration and metabolism of the seed itself (King *et al.*, 1998; Schwender *et al.*, 2004), thus improving carbon balance. Also, as plastidic synthesis of fatty acids requires high amounts of ATP and reducing equivalents, photosynthesis was proposed as an energy supply (Asokanthan *et al.*, 1997; Ruuska *et al.*, 2004; see Fig. 7.2). Alternatively, it has been suggested that redox signals mediated via the ferredoxin/thioredoxin system could modulate the activity of biosynthetic enzymes (Ruuska *et al.*, 2004). Finally, it has been proposed that photosynthetic oxygen release can promote viability of seeds in storage by mitigating internal hypoxia (Rolletschek *et al.*, 2003).

Although there have been few estimates, it seems unlikely that seed photosynthesis has a significant function in carbon refixation. Rubisco has been shown to be highly activated in both light and dark, possibly owing to high internal CO_2 concentrations in seeds, and it has been argued that the observed activities of Rubisco and phosphoribulokinase are sufficient to account for significant refixation of CO_2 produced during *Brassica napus* oil biosynthesis (Ruuska *et al.*, 2004). In addition, a low chlorophyll *a*/*b* ratio and the presence of substantial amounts of carotenoid pigments in seeds can provide additional capture of the green light that filters through siliques (Ruuska *et al.*, 2004). Nevertheless, even with excess light, the rates of J_f determined by Chl-F in seeds are extremely low (Borisjuk *et al.*, 2002). Low rates of CO_2 refixation may not contribute significantly to seed carbon balance, but perhaps could alleviate excess CO_2 accumulation, which otherwise might induce cellular acidification (Ruuska *et al.*, 2004).

A role of seed photosynthesis in fatty acid synthesis and storage is possible, and Ruuska *et al.* (2004) estimated that the rate of ATP and NADPH production by seed photosynthesis matches the requirements for the synthesis of fatty acids. Based on measurements of light transmittance and activation of enzymes, these authors concluded that even the low level of light reaching seeds plays a substantial role in activating light-regulated enzymes, thus also contributing to increase fatty acid synthesis. Alternatively, it has been shown that photosynthetic capacity temporally and spatially corresponds to the ATP distribution, and the highest biosynthetic fluxes are observed when the embryos turn green and contain high concentrations of ATP (Borisjuk *et al.*, 2002). Nevertheless, an inverse relationship has been observed between photosynthesis and lipid biosynthesis/deposition (Borisjuk *et al.*, 2002), which argues against a direct metabolic involvement of photosynthesis in lipid biosynthesis via redox modulation or energy supply. Alternatively, seed photosynthesis can promote storage via O_2 release (Rolletschek *et al.*, 2005). Developing legume seeds contain very low internal O_2 levels (Rolletschek *et al.*, 2003) that can fall below 2.5 μM (<0.1% of atmospheric saturation). Despite the fact that J is low and saturates at low light intensities, photosynthetic O_2 release is sufficient for significant increase in internal O_2 levels, thereby preventing hypoxia (Rolletschek *et al.*, 2005). Seed photosynthesis leads to supra-ambient O_2 levels (up to 250%) within pericarp, and supplies the interiorly located endosperm with O_2. This allows the endosperm to overcome internal hypoxia and to generate high ATP levels required during storage (Rolletschek *et al.*, 2004).

7.3. PHOTOSYNTHESIS IN NON-GREEN TISSUES

7.3.1. Masked chlorophyll: anthocyanin-containing and waxy leaves

In contrast to algae, in which diverse forms of chlorophyll and other pigments confer common names of brown, red, yellow or blue-green, terrestrial photosynthesis is predominantly restricted to green-coloured structures. It has been proposed that green leaves are particularly well adapted to use the highly dynamic light environment of land habitats (Nishio, 2000). This hypothesis is based on the fact that, under saturating light conditions, most of the absorbed blue and red light (these wavelengths are preferentially absorbed by chl.) is dissipated as heat in the upper mesophyll, and photosynthesis relies mainly on green light reaching the bottom portion of the leaf. The importance of green light for photosynthesis is frequently underestimated, and in fact as much as 80–90% of green photons are absorbed by leaves. Contrasting with aquatic photosynthesis, in which alternative photosynthetic pigments absorb all wavelengths, in terrestrial photosynthesis this 'green window' allows light to reach deep layers in the leaf or canopy (Terashima *et al.*, 2009).

Nevertheless, many plant species possess, either temporarily or permanently, photosynthetic organs that appear to be non-green owing to pigments in outer layers (epidermis, upper mesophyll or trichomes) that mask chlorophyll and other photosynthetic pigments. Two main types of chlorophyll masking exist: light reflection owing to epicuticular waxes, salt crystals, trichomes or scales; and reddish colouration owing to internal factors such as the vacuolar (and sometimes chromoplastidic) accumulation of red pigments (Fig. 7.3). Very rarely, the simultaneous presence of extremely high concentrations of red anthocyanins and green chlorophyll generate black foliage, as is the case of *Ophiopogon planiscapus* (Hatier and Gould, 2007), and leaves of some plants show a transition from early internal anthocyanin colouration to high external reflectivity owing to epidermal waxes later in development (Barker *et al.*, 1997).

RED PIGMENTS

Apart from photosynthetic pigments (chlorophyll and carotenoids), leaves frequently contain red pigments. These are mainly anthocyanins, but also betacyanins that accumulate in vacuoles (Lee and Collins, 2001), or some carotenoids that are present in chromoplastic plastoglobuli (Ida *et al.*, 1995). Red cells are typically observed in adaxial epidermes, palisade parenchyma and more rarely in spongy mesophyll and vascular bundles (Lee and Collins, 2001). In general, red pigments appear transiently in leaves, typically during leaf expansion in tropical plants, as well as during bud break and senescence in temperate deciduous species. Reddening also occurs transiently during photoinhibitory periods, such as drought, disease or winter stresses and as a consequence of nutrient deficiencies, with light exposure as a prerequisite for the activation of red pigment biosynthesis. After stress periods, red pigments disappear with the resumption of photosynthesis or organ abscission. In the case of developing leaves, redness disappears with the expansion and the metabolic transition from sink to source, whereas in senescing leaves it is usually retained until leaf abscission. The permanent presence of anthocyanins is restricted to organs in which carbon assimilation is not the main function, such as

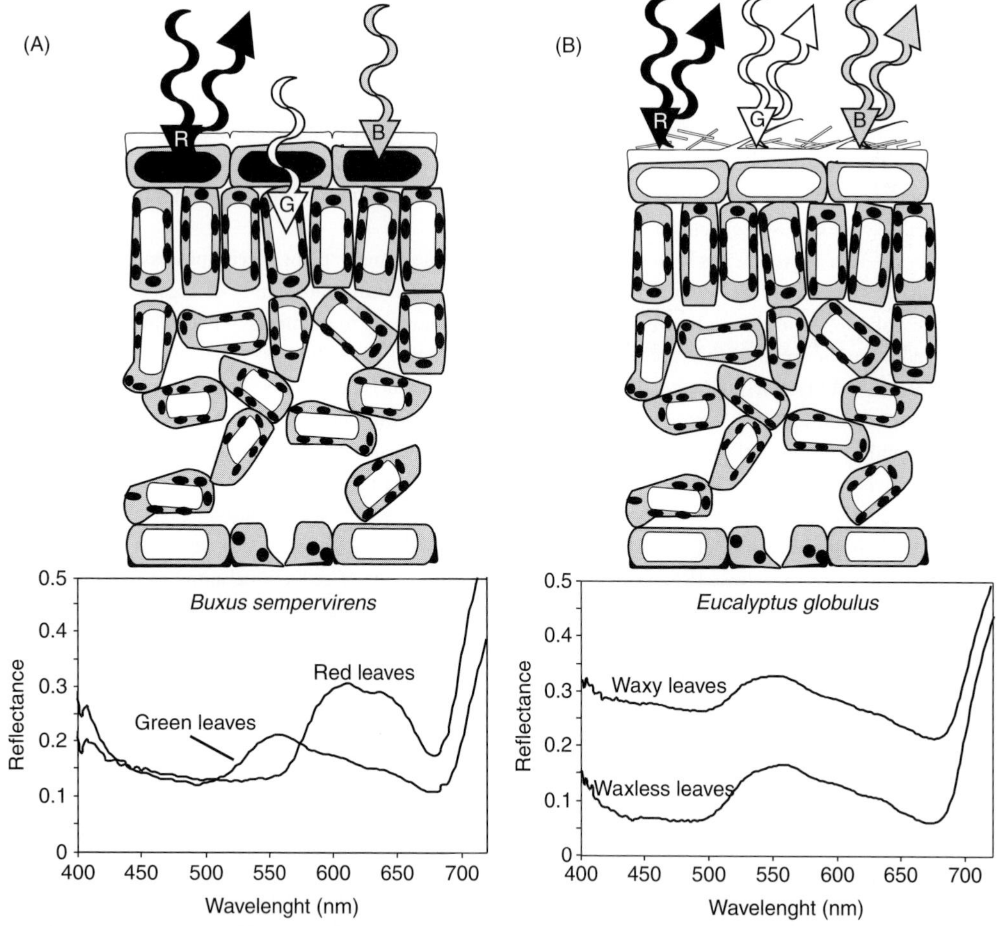

Fig. 7.3. Optical properties of anthocyanin-containing (**A**) and glaucous (**B**) leaves. Leaf sections illustrate how and where red, blue and green photons are absorbed and/or reflected. Lower panels illustrate an example of reflectance spectra of red and green leaves of the same species (**A**) and of glaucous leaves before and after wax removal (**B**).

petioles, veins, stems or roots. Apart from its presence in the upper surface of leaves, some understory plants also present an abaxial layer of red cells, as is the case in *Saxifraga hirsuta* (Fig. 7.4). In this case, it has been suggested that anthocyanins are able to increase light capture through backscattering of red photons (Lee *et al.*, 1979). However, this is unlikely because in shade environments only green light will reach the abaxial layers, and would be absorbed by red pigments. Experimental evidences also suggest a different role for abaxial anthocyanin accumlation (Hughes *et al.*, 2008). An alternative point of view suggests that this would represent a mechanism to reduce leaf transmittance and competition with other individuals or to increase leaf temperature (Pietrini and Massacci, 1998).

The role of chlorophyll masking is not clear, and different hypotheses have been proposed and tested. In the case of red pigments, there is a trend to accumulate red pigments linked to the occurrence of periods of high excitation pressure and increased risk of photooxidative damage (Steyn *et al.*, 2002), suggesting a protective function. Three main hypotheses have been suggested to explain such accumulation (Chalker-Scott, 1999; Steyn *et al.*, 2002; Archetti, 2009): first, these pigments act as passive filters that reduce photosynthetic light reaching the inner mesophyll; second, these compounds act as efficient scavengers of free radicals and ROS (Yamasaki, 1997; Gould, 2004); and third, these compounds are visualised by herbivores allowing some degree of communication between plants and insects (Schaefer and

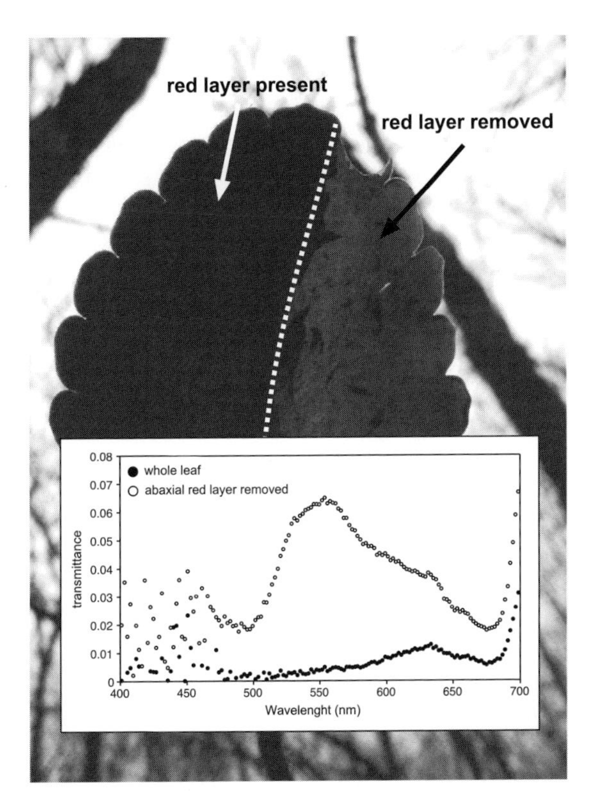

Fig. 7.4. The 'green window' of land plants. *Saxifraga hirsuta* presents an abaxial layer of red cells. When this layer is superimposed on to a green mesophyll, green photons are absorbed by red pigments and visible light cannot pass through the leaf. When the red layer is removed (right) a substantial amount of green light is transmitted. Transmittance spectra were kindly performed by Marta Pintó (University of Barcelona).

Wilkinson, 2004). Anthocyanins also could provide protection against harmful UV-B radiation. Although all the above functions have been tested, it is unlikely that any one of them alone will be able to individually explain leaf reddening. In fact, there is no need for an antioxidant or a UV-B absorber to be red (Steyn *et al.*, 2002), and the best chlorophyll protection would be achieved by a green pigment, not by a red one (Hormaetxe *et al.*, 2005a).

Red pigments probably confer an array of protective functions. For instance, chlorophyll masking by anthocyanins in red senescing leaves reduces the risk of photo-oxidative damage when photosynthetic apparatus is being dismantled and temperatures starting to decrease, allowing efficient nutrient retrieval from senescing leaves (Feild *et al.*, 2001). However some reports indicate a higher susceptibility of red leaves to photoinhibition (Esteban *et al.*, 2008; Zeliou *et al.*, 2009). Irrespective of the protective functions, there is no doubt that red pigments, mostly located in the outer cell layers of the leaf, reduce the amount of light that can be used by chlorophyll molecules located deeper in the leaf, by increasing the reflectance of red light (Woodall *et al.*, 1998) and the absorption of blue wavelengths (Feild *et al.*, 2001). Theoretically, this must imply a photosynthetic cost. In fact, reductions in photosynthetic rates of red leaves have been described in several species (Chalker-Scott, 1999).

GLAUCOUS AND PUBESCENT LEAVES

Photosynthetic organs of some species appear pale green or even whitish. This trait, is generated by the supra-epidermal presence epicuticular waxes (glaucousness), hairs (pubescence), scales or salt crystals. Among their functions, the main roles seem to be the limitation of water evaporation, defence from pathogens and herbivores, mechanical resistance, thermal regulation, water repellency and photoprotection. Quality and quantity of wax deposition is environmentally regulated by temperature, irradiance and water stress, but there is also a seasonal and ontogenetic trend with maximum levels during leaf expansion (Shepherd and Griffiths, 2006). Damage by weathering or acid rain influences the thickness of wax deposition.

The main effect of glaucousness and pubescence in photosynthetic performance is to modify leaf reflectance and consequently spectral properties and quantity of light transmitted to the inner mesophyll (Holmes and Keiller, 2002). Presence of hairs or scales also enhances boundary layer thickness between the leaf and the atmosphere (Febrero *et al.*, 1998). Both effects also modify indirectly an essential regulatory factor of photosynthesis: leaf temperature. On one hand, enhanced reflectance lowers leaf temperature and reduces the risk of overheating, but on the other hand it reduces vapour-pressure difference between air and tissue reducing transpiration (E) and consequently evaporative cooling (Shepherd and Griffiths, 2006). The cuticular component of E is strongly influenced by the presence of waxes, and this effect becomes significant under water stress when stomatal conductance is very low.

Glaucousness or pubescence increase reflectance typically by 10–20% (Johnson *et al.*, 1983; Holmes and Keiller, 2002), but some species from arid or high- altitude environments may exceed 50%, as is the case of *Encelia farinosa* (Ehleringer and Björkman, 1978a,b; Björkman and Demmig-Adams, 1994; Barker *et al.*, 1997). In general, glaucous leaves are more efficient reflecting UV, whereas hairy leaves are more reflective of photosynthetic radiation (Holmes and Keiller, 2002). Light reflection might also contribute to illuminate leaves located in shaded positions within the canopy (Shepherd and Griffiths, 2006). In most species glaucousness is insufficient to affect photosynthesis at saturating light (Johnson *et al.*, 1983), but it might limit carbon assimilation at lower irradiances. Photoprotective effects of decreased light absorption by pubescence are also relevant as has been probed by wax or trichome removal of intact leaves (Robinson and Osmond, 1994; Morales *et al.*, 2002).

VARIEGATED LEAVES

Photosynthetic surfaces are usually uniformly coloured, but some species show an irregular pattern of white, yellow or red patches, the so-called variegated leaves. Red patches are formed when a layer of anthocyanin-containing cells is present, whereas white and yellow portions occur in the absence of chlorophyll or in some species when palisade cells are loosely packed and air spaces are generated increasing light reflection (bubble effect) (Tsukaya *et al.*, 2004). These sections are generated as a result of genotypic characteristics, somatic mutations or virus infections. Variegated leaves occur rarely in nature but are relatively frequent among shade plants (Fig. 7.1); this characteristic has been selected for in ornamentals. The benefits, if any, of such foliar characteristics are not clear, but its frequency in understory leaves suggests a photoprotective role. Thus it has been speculated that this could be a mechanism to maintain in the same leaves highly efficient green sections and highly protected variegated portions, that would be more relevant for carbon uptake during steady state shade periods and during sunflecks, respectively. However experimental evidence seems to be inconsistent with such a hypothesis (Esteban *et al.*, 2008).

7.4. PHOTOSYNTHESIS IN HOLO- AND HEMIPARASITIC PLANTS

Over 4,500 angiosperm species distributed among 19 families are considered to be parasitic. Plant parasitism has evolved several times; parasitic plants are either holo- or hemiparasites (Stewart and Press, 1990), with the first group being phloem-dependent and obligate parasites with low or no capacity for photosynthetic assimilation and entirely dependent on their host for reduced carbon. The second group comprises chlorophyllous obligate or facultative parasites that are xylem-dependent and able in some cases to photosynthesise at rates comparable with their hosts (Fig. 7.1), but often rates are considerably less, with significant carbon import via the xylem from the host plant. Among parasitic plants there is a high degree of variation in photosynthetic capacity. Carbon dependence from the host ranges from facultative hemiparasites, such as *Rhinanthus minor* that possess a complete and functional photosynthetic apparatus and can grow without a host providing reduced carbon (Seel *et al.*, 1993), to completely achlorophyllous plants unable to photosynthesise.

Hemiparasites are by far the commonest group, comprising more than 70% of all parasitic species mostly mistletoes and Orobanchaceae (witchweeds and broomrapes). Mistletoes belong to two families (Loranthaceae with 900 species and Viscaceae with 550 species) representing the major group among parasitic plants. Hemiparasites feed on the E stream of their host, extracting nutrients from the host xylem via a specialised organ, the haustorium. Obligate hemiparasites rely on their hosts only for water and mineral nutrition, whereas carbon uptake depends on the photosynthetic activity of the parasitic plant and the host. It has been estimated that obtain between 24 and 62% of carbon from the host, mainly in the form of organic nitrogen compounds that are transported in the xylem but also by the presence of phloem connections (Lüttge, 1997). Rates of photosynthesis are in general lower (0.5–5 μmol CO_2 m^{-2} s^{-1}), compared with their hosts, and are coupled with high rates of E, leading to extremely low WUE. To compete for transpirational flux and obtain water from the host, mistletoes commonly have lower water potential than their hosts (Strong *et al.*, 2000) owing to the high content of compatible solutes, such as cyclitols that do not interfere with cell metabolism (Richter and Popp, 1992). Lower rates of CO_2 assimilation are frequently associated with low numbers of plastids, lower chlorophyll content and a shade acclimation syndrome characterised by lower light-saturation intensities and chlorophyll *a*/*b* ratios (Strong *et al.*, 2000). Mistletoes from temperate regions are evergreen but they frequently parasitise deciduous trees, and the leaf phenology of the host implies dramatic changes in the light environment to which mistletoes respond by the seasonal activation of different photoprotection mechanisms, including the operation of the xanthophyll cycle and sustained downregulation of PSII (Matsubara *et al.*, 2002).

Most holoparasites do not posses any chlorophyll or functional chloroplasts, but in some species the presence of Rubisco and/or chlorophyll has been described (Stewart and Press, 1990). This is the case of *Cuscuta reflexa* Roxb. (dodder), in which photosynthesis, based on the re-assimilation of carbon released by the respiration of its own cells, is performed by a band of cells adjacent to vascular bundles that contain functional chloroplasts (Hibberd *et al.*, 1998a). Fixation of external CO_2 is unlikely because

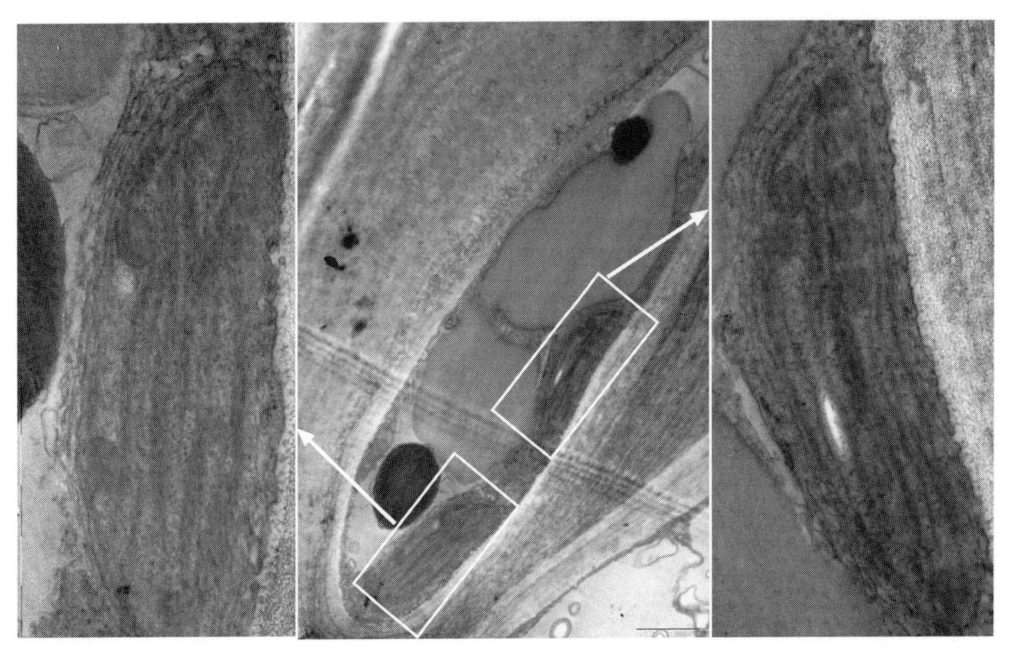

Fig. 7.5. Guard cell of *Buxus sempervirens* leaves showing two chloroplasts. Original micrographs obtained by Unai Artetxe (UPV/EHU).

of the low rates of CO_2 diffusion from the atmosphere and lack of intercellular spaces, and the parasite obtains carbon almost exclusively from the host (99% in *C. reflexa*). This photosynthetic activity also demands photoprotection, which is fulfilled by the operation of the xanthophyll cycle (Bungard *et al.*, 1999). In addition, this species possesses a unique carotenoid composition with a complete absence of neoxanthin (which replaced by 9-*cis*-violaxanthin in the binding sites of the major LHC proteins) and the presence of lutein 5-epoxide (Lx) (Snyder *et al.*, 2005). The latter carotenoid is involved in an alternative xanthophyll cycle, the so-called Lx cycle, that involves light-driven conversion of lutein epoxide into lutein (Bungard *et al.*, 1999). This cycle has been identified in other holo- and hemiparasitic angiosperms (Matsubara *et al.*, 2001).

In general, parasitism negatively affects host photosynthesis (see Chapter 22) that cannot compensate water and carbon losses to the parasite. However, in some species parasitism may stimulate host photosynthesis (Jeschke and Hilpert, 1997).

7.5. PHOTOSYNTHESIS IN STOMATAL CELLS

Unlike other epidermal cells, stomatal cells contain chlorophyll and well-developed chloroplasts and are capable of photosynthetic CO_2 assimilation (Shaw and MacLachan, 1954; Yemm and Willis, 1954). Also, Rubisco and other Calvin-cycle enzymes are found in guard cells. However, there are less than 20 chloroplasts per guard cell, compared with 30–70 in a mesophyll cell. In addition, guard-cell chloroplasts are generally smaller, and with less granal stacking than those in mesophyll cells (Fig. 7.5). Photosynthetic pigment composition is similar in both types of cells (Willmer and Fricker, 1996). Because of these differences and the specific function of guard cells, the role of stomatal photosynthesis has long been controversial (Zeiger *et al.*, 2002). It has been hypothesised that guard-cell chloroplasts may contribute to the control of stomatal opening (Zeiger *et al.*, 2002; Messinger *et al.*, 2006) via at least four possible mechanisms: (1) photosynthetic production of ATP and reductants (Tominaga *et al.*, 2001; Buckley *et al.*, 2003); (2) production of osmotically active sugars by PCA (Talbott and Zeiger, 1998); (3) zeaxanthin functioning as the low-intensity blue-light photoreceptor (Zhu *et al.*, 1998; Zeiger *et al.*, 2002); and (4) starch accumulation becomes available upon hydrolysis to synthesise malate as a counter ion to K^+ (Hedrich *et al.*, 1994).

Whereas these mechanisms and functions are still under debate, data are becoming available concerning the photosynthetic activity of guard cells. For instance, the rates of photophosphorylation on a per chlorophyll basis can be as high as 70–90% of that of mesophyll chloroplasts. The ATP produced by guard-cell chloroplast photophosphorylation is used by the plasma membrane H^+-ATPase and is therefore important for the ion transport that determines aperture (Tominaga *et al.*, 2001). Additionally, chloroplast NADPH synthesis can also contribute to mitochondrial plasma membrane H^+-ATPase activity via export through an OAA/malate shuttle (see Chapter 4) to the mitochondria (Gautier *et al.*, 1991).

Using Chl-F imaging techniques (see Chapter 10) it has been shown in leaves of many different species that the guard-cell photosynthetic efficiency is approximately 80% of that of the mesophyll cells, irrespective of PPFD (Goh *et al.*, 1999; Lawson *et al.*, 2002, 2003). In addition, guard-cell photosynthesis responds to changes in light, CO_2, O_2, air humidity and water stress, with a pattern that is essentially similar to that of mesophyll chloroplasts (Lawson *et al.*, 2002, 2003). Moreover, low O_2 concentration (2%) increases photosynthetic efficiency in both guard cells and mesophyll cells, but when CO_2 supply is high then changing the O_2 concentration has little effect, strongly suggesting that photorespiration occurs and the Calvin cycle is a major sink for the ATP and NADPH produced through J in the guard-cell chloroplasts, just as in the mesophyll chloroplasts (Lawson *et al.*, 2002, 2003).

7.6. SPECIAL CO_2-ACQUISITION SYSTEMS

Throughout this chapter, we have highlighted how plant structures possessing stomata generally capture atmospheric CO_2 and support net photosynthesis, whereas structures containing chlorophyll but lacking stomata (or presenting very low stomatal densities) mainly re-fix metabolically produced CO_2. However, land plants may also exhibit CO_2 acquisition without stomata. Some examples will be briefly described here: (1) CO_2 acquisition from the sediment/roots in partially submerged species, as well as in totally emerged species; (2) CO_2 uptake from the sediment and/or atmosphere in land plants lacking stomata; and (3) adaptive plant morphologies that may allow concentrating CO_2-surrounding stomata above atmospheric levels.

In temporary and partially submerged macrophytes (i.e., roots and part of the stem submerged, aerial parts emerged) CO_2 released from the sediment or root respiration can represent a substantial source, sometimes more important than atmospheric CO_2. In some of these plants, such as water lilies that have an internal ventilation system, leaves assimilate CO_2 that arrives from the roots. For most emergent wetland species such as *Scirpus lacustris* and *Cyperus papyrus* (Farmer, 1996), fixation of CO_2 from the sediment is minimal (0.25% of the total CO_2 uptake in photosynthesis). In contrast, for some species characterised by a high root-to-shoot ratio, such as *Stratiotes aloides*, *Lilaeopsis macloviana* or *Ludwigia repens*, the sediment is a major source of CO_2, although it is only taken after diffusion from sediment into the water column (Prins and de Guia, 1986; Winkel and Borum, 2009). Recent estimates by Winkel and Borum (2009) using ^{14}C labelling showed that for species of the genus *Lobelia*, *Lilaeopsis* and *Vallisneria*, CO_2 uptake from the sediment covered >75% of their total CO_2 requirements. A different behaviour has been shown in *Rumex palustris*, whose submerged leaves have about double the stomatal density of emerged leaves (Mommer *et al.*, 2005). Some amphibious species such as *Ranunculus flabellaris* (Bruni *et al.*, 1996) show chloroplast re-orientation towards the epidermial layers directly facing the water column. Taken together, these results suggest that reducing the CO_2 diffusion pathway, thus decreasing photorespiration, allows submerged leaves of partially submerged plants to take most of their CO_2 directly from the water column rather than from the sediment (Mommer *et al.*, 2005; Winkel and Borum, 2009). Other species present gas films on submerged leaves, that provide an interface for CO_2 and O_2 exchange, enhancing submerged photosynthesis up to sixfold (Colmer and Pedersen, 2008). Alternatively, even totally terrestrial C_3 plants, such as tobacco, show photosynthetic CO_2 assimilation in cells adjacent to the vascular system. Indeed, a C_4-like pathway using either malate or CO_2 transported from roots seems involved, the latter probably being supplied by either root respiration or entering the roots from the substrate (Hibberd and Quick, 2002).

In contrast to many amphibious plants, submerged macrophytes of the isoetid life form (quillworts) receive a very large portion (60–100%) of their carbon for photosynthesis directly from the sediment (Keeley *et al.*, 1984). CO_2 uptake via their roots seems not to be significant (Farmer, 1996), but CO_2 diffuses from the sediment via the lacunal air system to the submerged leaves, which, in contrast to those of *Rumex palustris* (Mommer *et al.*, 2005), lack functional stomata, possess thick cuticles and contain large air spaces inside facilitating internal diffusion and hampering gas exchange with the atmosphere. The air spaces in the leaves are connected with those in stems and roots, thus facilitating the transport of CO_2 from the sediment to the leaves. At night, only part of the CO_2 coming up from the sediment via the roots through the lacunal system is fixed, the rest being lost to the atmosphere. Some of these aquatics have a CAM photosynthetic metabolism (see Chapter 6), they accumulate malic acid during the night, and have rates of CO_2 fixation during the night that are similar in magnitude as those during the day, when the CO_2 supply from the water is very low (Keeley, 1990; Keeley and Busch, 1984).

Interestingly, it has been demonstrated that *Stylites andicola*, a land plant totally lacking stomata (also belonging to Isoetaceae), derives nearly all of its photosynthetic carbon through its roots, also exhibiting CAM-like metabolism (Keeley *et al.*, 1984). This species forms dense colonies over substrates of highly decomposed peat, and the plants are so much embedded in the peat that the bulk of the plant is underground, the only exposed surfaces being the upper-third chlorophyllous portion of the leaves. Other land ferns lacking stomata, such as those filmy ferns of the genus *Hymenoglossum*, do present some (although low) positive A_N rates (Parra *et al.*, 2009), and hence they are capable of atmospheric CO_2 uptake, but it is unknown whether they also have some additional CO_2 capture from the sediment.

Even in stomata-containing emerged leaves, current atmospheric CO_2 concentration limits photosynthesis (see Chapters 2 and 8). It has been suggested (Rossa and von Willert, 1999) that growth forms resulting in leaves prostrate on the soil surface may result in increased CO_2 concentration reaching stomata, particularly under high temperature and humidity (favouring soil microbial activity and CO_2 release), thus supposing a benefit for plant CO_2 acquisition (Fig. 7.6). Although this hypothesis is attractive, experimental results based on carbon-isotope ratios in the prostrate-leaved geophyte *Brunsvigia orientalis* have shown that only 7% of the leaf carbon comes from soil-derived CO_2. This is consistent with the high proportion (75%) of adaxial stomata in this species, hence favouring atmosphere-leaf over soil-leaf CO_2 exchange (Cramer *et al.*, 2007). Nevertheless, this 7% carbon uptake occurs with minimal *E*, therefore increasing plant WUE (Cramer *et al.*, 2007).

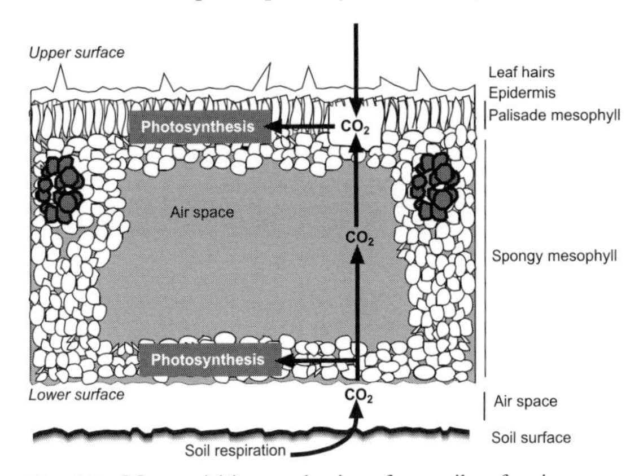

Fig. 7.6. CO_2 acquisition mechanisms from soil surface in prostrate-leaved plants.

7.7. SUMMARY: ECOPHYSIOLOGICAL IMPORTANCE OF THE FUNCTIONAL DIVERSITY OF PHOTOSYNTHETIC ORGANS

Throughout this chapter several examples of the functional biodiversity of photosynthesis have been shown. From an ecological perspective these examples illustrate how plants maximise carbon gain by developing photosynthetic tissues in the majority of the light-exposed structures. With internal or atmospheric CO_2 as a carbon source, non-foliar photosynthesis contributes significantly to the carbon balance of the whole plant. The importance of such organs, mainly in adverse environments, is great not only because they increase photosynthetic surface but also because they provide a wider range of stress responses. In many plant species this photosynthetic complexity is enhanced by the presence of unusual cells and leaves that improve gas exchange or modify optical properties. In fact, most if not all higher plants present some manifestations of functional biodiversity of photosynthesis. Consideration that photosynthesis is not restricted to 'typical' green leaves helps better understand plant responses to the environment.

8 • Models of photosynthesis

A. DIAZ-ESPEJO, C.J. BERNACCHI, G.J. COLLATZ AND T.D. SHARKEY

8.1. INTRODUCTION: WHY USE PHOTOSYNTHESIS MODELS?

One of the goals of science is prediction, and quantitative predictions are based on models. Models vary in the degree to which they represent underlying processes from purely empirical regression models to those that represent physical, chemical and biological processes (mechanistic or process models). Mechanistic models tend to be more complex and are empirical at the level of the processes they include (parameters). Mechanistic models can also test whether our current knowledge is sufficient to explain experimental data and generate and test new hypotheses that lead to progress in understanding. Prediction of carbon fixation of vegetation by photosynthesis in a changing environment is important to assess ecological matters like global climate change or agronomic issues such as yield improvements. Models of photosynthesis play a key role in predicting primary production of vegetation and crop yield in a variable climate. They have also been implemented in weather-forecast models, improving predictions of humidity and temperature anywhere from 5 to 10% (Goudarzi, 2006). The level of detail of these models is largely dependent upon the objectives of the studies for which the models were designed. Questions concerning effects of environmental conditions on the process of photosynthesis itself can be addressed using single-leaf photosynthesis models. Vegetation canopies are composed of leaves and of support structures (stems, trunks) that represent a greater challenge for predicting photosynthesis. The simplest canopy models invoke the so-called 'big-leaf' assumption, and may include various simple approximations for structural and physiological complexity (e.g., Sellers *et al.*, 1992). More complicated models include multiple canopy layers each with different properties (e.g., Medlyn *et al.*, 2005). Layered models require detailed knowledge of the variation in parameters throughout the canopy depending on position and age, as well as knowledge of the distribution of foliage and environmental variables in the canopy. The objective of modelling is double. On one hand, models can be used in a bottom-up approach to predict photosynthesis variations in response to environmental variables. To do so, it is needed to first parameterise the photosynthesis model itself, obtaining representative values of the maximum carboxylation efficiency ($V_{c,max}$), the maximum rate of electron transport (J_{max}) and the capacity for TPU (sections 8.3–8.6). Once these are obtained, the model allows to estimate net photosynthesis (A_N) at any given C_i. As the latter is controlled not only by photosynthetic activity but also by stomatal conductance (g_s), the photosynthesis model needs to be coupled to one of the existing models of g_s responses to ambient conditions (Section 8.7). On the other hand, a top-down approach can be followed with the aim of using model parameterisation to assess what is limiting photosynthesis under a given condition. That is, rather than estimating A_N, it is experimentally determined. If A_N decreases as compared with a reference value, the model can then be used to unravel which factor(s) (e.g., impaired $V_{c,max}$, J_{max}, g_m or g_s) are provoking such decreases. To do so, the basic model is coupled to a photosynthesis limitation model (section 8.8).

This chapter is not intended to be a comprehensive review of all published models of photosynthesis so far, but rather a guide for scientists who are making their first contact with models.

8.2. PHOTOSYNTHESIS MODELS: ONE MODEL FOR EACH APPLICATION?

At the leaf level, we are generally interested in understanding the mechanisms involved in the response of photosynthesis to environmental variables, diseases or other physiological variables, such as phenology. In this case, the

Terrestrial Photosynthesis in a Changing Environment: A Molecular, Physiological and Ecological Approach, ed. J. Flexas, F. Loreto and H. Medrano. Published by Cambridge University Press.

use of mechanistic models is best as they are based on prior knowledge about the response of the photosynthetic processes to any of the above mentioned variables. However, when questions concern changes in canopy structure, competition for light or the effects of culture practices on plant growth or yield, the use of empirical photosynthesis models can be sufficient.

Among the empirical models, the model proposed by Thornley and Johnson (1990) is a popular one that has given satisfactory fits to a wide range of A_N data.

$$A_N = \frac{\phi_{CO2}\, PPFD\, C_i \varepsilon}{\phi_{CO2}\, PPFD + C_i \varepsilon} - R_d \qquad \text{[Eqn. 8.1]}$$

In equation 8.1 of this model: φ_{CO2} is the apparent quantum yield; PPFD is the photosynthetic photon-flux density; C_i the intercellular CO_2 concentration; ε is the apparent carboxylation efficiency; and R_n is the mitochondrial respiration.

Another often-used model of photosynthesis is that of Goudriaan *et al.* (1985) that is a less-detailed version of the model of Farquhar *et al.* (1980a). However, the model satisfactorily describes most of the essential response characteristics of photosynthesis. The modelling approach of Goudriaan *et al.* (1985) is based on the distinction between two essentially different conditions: (1) light is the limiting factor (at relatively high CO_2 concentrations); and (2) CO_2 is the limiting factor (at relatively high light intensity). It has been used in models like ForGro, which is a physiologically based carbon-balance model of forest growth (Vandervoet and Mohren, 1994) and because the Goudriaan model directly relies on conductance to describe the diffusion of CO_2 between the air and chloroplasts, it could be thought that it is more closely linked with meteorological research than biochemical models. Most current SVATs (Soil Vegetation Atmosphere Transfer) however use the biochemical photosynthesis model to control canopy conductance (e.g., Sellers *et al.*, 1996, 1997 (SiB2 model), Foley *et al.*, 1996 (IBIS model), Bonan *et al.*, 1997 (NCAR model) or Sitch *et al.*, 2003 (LPJ model)).

8.3. C_3-PHOTOSYNTHESIS MODEL (FARQUHAR *ET AL.*)

The most widely used mechanistic model for a leaf C_3 photosynthesis is that of Farquhar *et al.* (1980a). In the words of the authors, the model has had an impact and seen application that far exceeded their expectations (Farquhar *et al.*, 2001). The success of the model is not only owing to its solid theoretical basis, but also to its simplicity, which is a necessary prerequisite to make any model useful (von Caemmerer, 2000). As originally developed, the model is based on the assumption that photosynthesis rate is limited either by the Rubisco activity (A_c) or RuBP regeneration (A_j). In the first case the rate of carboxylation is determined by the RuBP-saturated kinetic properties of Rubisco and the concentrations of CO_2 and O_2. The rate of RuBP regeneration is assumed to be dependent on the rate of regeneration of ATP, which is usually limited by the electron transport capacity, and therefore related to incident irradiance. However, further investigations revealed that RuBP regeneration can be limited by Calvin-cycle reactions, as was assumed in the modelling of Zhu *et al.* (2007). The rate at which triose phosphate (TP) is utilised (A_p) was identified as a third biochemical limitation to photosynthesis (Badger *et al.*, 1984) and was subsequently incorporated into the C_3-photosynthesis model (Sharkey, 1985; Harley and Sharkey, 1991). This limitation comes about when RuBP regeneration is limited by the availability of inorganic phosphate (P_i) for photophosphorylation, which (in the short term) is determined by the relative rates of the production of TPs and P_i release, as TPs are used in starch and sucrose production (Sharkey, 1985).

In an A_N/C_c response curve (e.g., Fig. 8.1), A_N can be determined by the minimum of these three potential rates:

$$A_N = \min\{A_c, A_j, A_p\} \qquad \text{[Eqn. 8.2]}$$

where:

$$A_c = V_{c.max}\left[\frac{C_c - \Gamma^*}{C_c + K_c\left(1 + \frac{O}{K_o}\right)}\right] - R_d \qquad \text{[Eqn. 8.3]}$$

$V_{c,max}$ is the maximum velocity of Rubisco for carboxylation; C_c is the CO_2 partial pressure at Rubisco; Γ^* is the CO_2 concentration at which the photorespiratory efflux of CO_2 equals the rate of photosynthetic CO_2 uptake; K_c is the Michaelis constant of Rubisco for CO_2; O is the partial pressure of O_2 at Rubisco; K_o is the Michaelis constant of Rubisco for O_2; and R_d is the mitochondrial respiration under light conditions. This equation is based on a modified Michaelis-Menten response curve of a substrate (CO_2) with a competitive inhibitor (O_2).

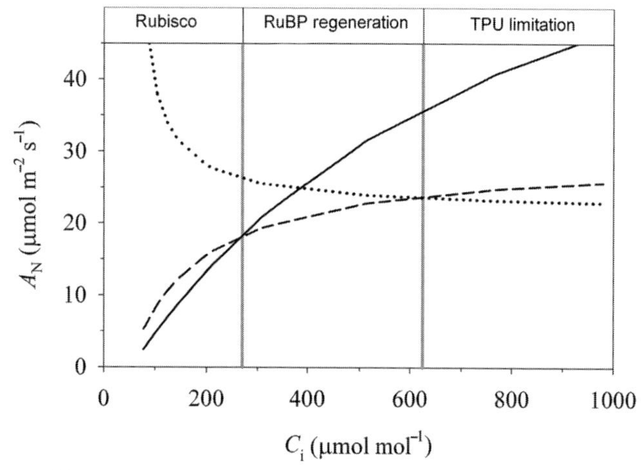

Fig. 8.1. Theoretical A_N/C_i curve on which the regions where Rubisco (A_c), regeneration of RuBP (A_j) and triose phosphate utilisation (A_p) are limiting A_N have been identified. Rubisco, ribulose-1,5-bisphosphate carboxylase/oxygenase; RuBP, ribulose-1,5-bisphosphate; TPU, triose phosphate utilisation.

The equation representing RuBP-regeneration-limited photosynthesis is given as:

$$A_j = J\frac{C_c - \Gamma^*}{4C_c + 8\Gamma^*} - R_d \quad \text{[Eqn. 8.4]}$$

where J is the rate of electron transport. The numbers four and eight in the denominator of equation 8.4 represent the stoichiometry of the number of electrons needed for the regeneration of ATP and NADPH. It is important to consider that other stoichiometries are possible and that the exact values to use are still uncertain (e.g., von Caemmerer, 2000). At light saturation, J approaches the theoretical maximum achievable rate of electron transport, J_{max}, but at lower light levels, J may be estimated for PPFD from a non-rectangular hyperbola (Long and Bernacchi, 2003, see Fig. 8.2):

$$J = \frac{PPFD_2 + J_{max} - \sqrt{(PPFD_2 + J_{max})^2 - 4\phi_{PSII} PPFD_2 J_{max}}}{2\phi_{PSII}} \quad \text{[Eqn. 8.5]}$$

where ϕ_{PSII} represents a curvature factor and $PPFD_2$ represents the incident quanta utilised in electron transport through PSII:

$$PPFD_2 = PPFD\alpha_L \phi_{PSII,max} \varepsilon_{PSII} \quad \text{[Eqn. 8.6]}$$

where α_L is leaf absorptance; ϕ_{PSII} is the quantum efficiency of PSII; and ε_{PSII} is the fraction of absorbed quanta that go to PSII. The model describing the limitation imposed by A_j is more empirically based than for A_c as it relies on a quadratic equation to describe the relationship between J and PPFD. Nevertheless, this model has been widely validated over a range of conditions and, for the purposes of most modelling exercises, is sufficient to make reasonable predictions. Photosynthesis rate defined by A_p is mathematically expressed as:

$$A_p = 3TPU - R_d \quad \text{[Eqn. 8.7]}$$

where TPU is the rate of use of TP. This equation fits the final portion of A_N/C_c curves where an increase in C_c does not produce any further increase in A_N. However, a decline in A_N at high CO_2 concentrations has also been reported, which cannot be explained solely by the phosphate limitation proposed above. In these cases, it is likely that a decrease in photorespiratory activity is limiting the amount of P_i available. Typically, the photorespiratory pathway is involved in recycling P_i to the stroma almost immediately. One of the initial stages in the photorespiratory pathway is to remove a P_i group from the phosphoglycolate formed as a result of the oxygenation event (see Chapter 2). As this occurs in the stroma this immediate recycling helps to minimise the occurrence of TPU-limited photosynthesis, but as oxygenation events decrease as a result of higher CO_2 concentrations this immediate recycling of P_i also decreases, resulting in a higher likelihood that TPU-limited photosynthesis might occur (Harley and Sharkey, 1991). Thus, the above equation is altered to account for the proportion of glycerate that is not returned to chloroplast (α_{TPU}); von Caemmerer, 2000).

$$A_p = \frac{3TPU \times C_c}{(C_c - (1+3\alpha_{TPU}/2)\Gamma^*} - R_d \quad \text{[Eqn. 8.8]}$$

A second way that reverse CO_2 sensitivity can occur is by PGA inhibition of chloroplast phosphoglucomutase (Sharkey and Vassey, 1989).

Another version of the Farquhar *et al.* (1980a) model that has been widely used in soil-vegetation-atmosphere-transfer models coupled to atmospheric and climate models (e.g., Sellers *et al.*, 1996; Boucher *et al.*, 2009) is that of Collatz *et al.* (1991). This version of the model assumes the J_{max} is never limiting (J of equation 8.5 equal to $PPFD_2$ of equation 8.6) and approximates A_p as a proportional to $V_{c,max}$.

8.4. MESOPHYLL CONDUCTANCE: THE LAST KEY VARIABLE IN PHOTOSYNTHESIS MODELS

In the former equations, A_N has been expressed as a function of C_c. The CO_2 concentration at the site of carboxylation is less than in the intercellular spaces (C_i). In their initial publication, Farquhar *et al.* (1980a) discussed the degree to which C_c might be less than C_i, but there was little information available. Some estimates had C_c as low as zero, others at the compensation point, but given what was known about the kinetics of Rubisco and photorespiration, C_c had to be much closer to C_i than those estimates. The authors of the original model devised an online fractionation of the carbon-isotopes method, which is still one of the most used methods of determining g_m (Evans *et al.*, 1986). For some plants, C_c may be close enough to C_i that it could be used in place of C_c. One example is *Phaseolus vulgaris*, which is the species that was used in early work testing the model (von Caemmerer and Farquhar, 1981). Many subsequent modelling studies assumed that C_c and C_i were identical; this is a simplification that does not work for many, possibly most species (Ethier and Livingston, 2004).

Inside the leaf, CO_2 has to diffuse from the intercellular spaces surrounding the mesophyll cells to the sites of carboxylation in the chloroplasts. Each of these steps can impose a resistance to CO_2 diffusion (Nobel, 1991; Flexas *et al.*, 2008). Because the resistances encountered by CO_2 are in series, they can be summed to a total resistance. This resistance is called the mesophyll resistance and its inverse is g_m. This topic is covered in Chapter 12, but the effect on modelling will be described here. Mesophyll conductance can be used to determine C_c as a function of C_i and A_N:

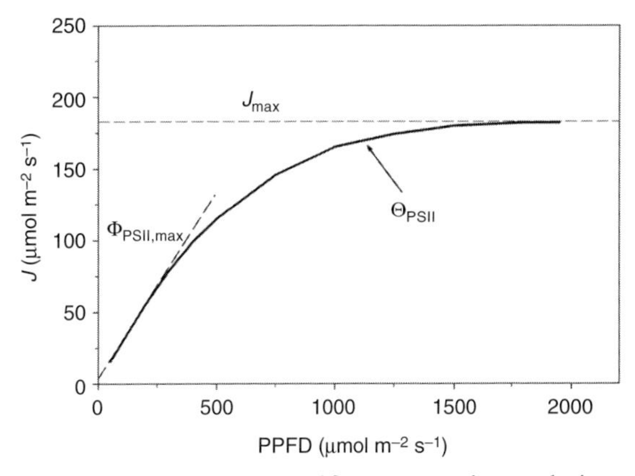

Fig. 8.2. Electron transport rate (J) response to photosynthetic photon-flux density (PPFD). Maximum electron transport rate (J_{max}), curvature factor (Θ_{PSII}) and maximum quantum yield of photosystem II ($\Phi_{PSII,max}$) are indicated. PSII, photosystem II.

$$C_c = C_i - A_N / g_m \quad \text{[Eqn. 8.9]}$$

Mesophyll conductance generally limits A_N by as much as 20% (Bernacchi *et al.*, 2002; Grassi and Magnani, 2005; Diaz-Espejo *et al.*, 2007), although this value has been reported to be higher than 50% under severe drought conditions (Galmés *et al.*, 2007a). It also has a large effect on in-vivo estimates of $V_{c,max}$ and of the Michaelis-Menten constants, K_c and K_o (Bernacchi *et al.*, 2002; Ethier and Livingston, 2004). Manter and Kerrigan (2004) showed for 19 conifer and broadleaf species that $V_{c,max}$ estimated in A_N/C_c increased around 25% compared with A_N/C_i, and that the differences in $V_{c,max}$ between the two groups of species was dramatically reduced when considering g_m, as conifers tend to have lower g_m than broadleaves. Flexas *et al.* (2008) showed how $V_{c,max}$ can be underestimated if g_m is ignored under drought conditions. Similarly, leaf aging can decrease g_m (Loreto *et al.*, 1994) and produce erroneous conclusions about the evolution of photosynthetic capacity of leaves when $V_{c,max}$ is calculated from A_N/C_i curves instead of A_N/C_c (Flexas *et al.*, 2007a). Bongi and Loreto (1989) also showed that the increase in the photosynthesis limitation observed in olive saplings under salt stress was a consequence of a simultaneous reduction in g_s and g_m, and Centritto *et al.* (2003) demonstrated that when $V_{c,max}$ and J_{max} were analysed in an A_N/C_c curve they remained unaffected in salt-stressed olive plants, thus confirming that all limitations of photosynthesis were owing to diffusive restrictions to CO_2 entry. In summary, it is important to analyse the response of A_N to CO_2 on a C_c basis to reach valid conclusions related to the photosynthesis performance to leaves.

But, despite the importance of g_m is widely accepted, it has not yet been satisfactorily included in photosynthesis models. That is, we still do not know how to model g_m as a function of a changing environment. However, there are some examples where g_m plays a role when modelling A_N. Williams *et al.* (1996), while modelling the soil-plant-atmosphere continuum of a mixed forest, included a fixed value of g_m. Ohsumi *et al.* (2007) linked g_m to morphological aspects of the leaves, and made g_m function of nitrogen content. Cai *et al.* (2008) took a step forward and interpreted the reported dynamic behaviour of g_m as a function of g_s. More recently, Keenan *et al.* (2010a,b) have included the important role played by g_m during drought.

8.5. MODEL PARAMETERISATION

Since the photosynthesis model presented by Farquhar *et al.* (1980) has become so widely used at scales ranging from the chloroplast to the globe, proper parameterisation is becoming increasingly more relevant. The model as a whole consists of various parameters, some of which are relevant for specific limitations (e.g., $V_{c,max}$ for Rubisco-limited photosynthesis), some relevant for more than one limitation (e.g., Γ^* for both Rubisco- and RuBP regeneration-limited photosynthesis) and some for all limiting processes (e.g., R_d). The model is only as accurate as the parameterisation employed, and thus, prior to any modelling exercise, a parameterisation needs to be selected and justified based on the objectives of the modelling run. In the following sections, the parameters that are relatively conserved among C_3 species will be discussed first, followed by the parameters that are highly variable among species, growth environment and leaves on a plant.

8.5.1. K_c, K_o and Γ^*

The terms K_c and K_o represent the Michaelis constants (K_m) for carboxylation and oxygenation, respectively. The photosynthetic CO_2 compensation point, Γ^*, is derived from these terms and from the maximum rates of carboxylation ($V_{c,max}$) and oxygenation ($V_{o.max}$) as:

$$\Gamma^* = \frac{0.5 O K_c V_{o,max}}{V_{c,max} K_o} \quad \text{[Eqn. 8.10]}$$

These kinetic parameters were derived initially from in-vitro data (Farquhar *et al.*, 1980), but these conditions are quite dissimilar from conditions experienced *in vivo*. More recently gas-exchange systems have become accurate enough to elucidate the kinetics from in-vivo measurements of photosynthetic CO_2 responses. The initial portion of the photosynthetic CO_2 response is usually Rubisco-limited. This portion of the curve represents a range of CO_2 concentrations that fall well below the values of K_c; thus, determining K_c from this portion of the curve is statistically weak and small measurement errors can result in large errors in the derived kinetic parameters (Long and Bernacchi, 2003). To account for this, von Caemmerer *et al.* (1994) determined Rubisco kinetic parameters for tobacco using plants transformed to have reduced Rubisco, ensuring that Rubisco was limiting over a wider range of CO_2, that included the CO_2 concentrations represented by K_c. Similar plants were then used to determine the temperature responses of Rubisco kinetics over a wide range of temperatures (e.g., Bernacchi *et al.*, 2001).

The parameters that represent the kinetic properties of Rubisco (e.g., K_c, K_o and Γ^*) are generally consistent enough among species that literature values can be used (Bernacchi *et al.*, 2001). This eliminates the need for detailed and time-consuming experiments specific to each species investigated. The basis for this assumption is that C_3-plant species are closely related and that the Rubisco did not deviate substantially (von Caemmerer, 2000). However, this assumption was shown not to hold for all C_3 species and for all growth conditions (e.g., Jordan and Ogren, 1981; Spreitzer and Salvucci, 2002; Galmés *et al.*, 2005). Whether the slight differences among C_3 species should be considered for any particular modelling exercise needs to be specifically addressed by the modeller, ideally through a sensitivity analysis or by other means of assessing whether the introduced error is large enough to alter the model output.

8.5.2. $V_{c,max}$, J_{max}, triose-phosphate utilisation and R_d

Although K_c, K_o and Γ^* are all properties of Rubisco and more or less conserved among C_3 species, $V_{c,max}$, J_{max} and R_d are physiological properties related to amounts of different proteins present in the leaf and various regulatory mechanisms. These parameters vary considerably among species, individuals within a species and leaves within a plant. Variation in $V_{c,max}$ is quite common over both temporal and spatial scales and it is generally driven by changing concentrations of Rubisco in the chloroplasts, but changes in the activation state of Rubisco will also influence $V_{c,max}$. Although $V_{c,max}$ is highly variable, this variability has no influence on Γ^* (e.g., Eqn. 8.10) as any changes in $V_{c,max}$ will have a similar relative influence on $V_{o,max}$. As a result, $V_{o,max}$ is generally expressed as a percentage of $V_{c,max}$ (Badger and Collatz, 1977; Bernacchi *et al.*, 2001).

A significant difficulty in estimating the plant-specific parameters, $V_{c,max}$, J_{max}, TPU, R_d and g_m from gas-exchange data is to know which factor is controlling the response of photosynthesis to CO_2 at any given point. As we have indicated above and according to the Farquhar *et al.* (1980) model, carboxylation rates are limited by A_c, A_j or A_p. It has been suggested as a basis to set the transition point between A_c and A_j at 200–250 μmol mol^{-1} (von Caemmerer and Farquhar, 1981), but this is not reliable. The designation *a priori* of a C_c threshold at which the A_N/C_c curve switches between the A_c region to the A_j has been reported to be a source of error in the calculation of $V_{c,max}$ and J_{max} by Manter and Kerrigan (2004). Therefore, these authors suggested not using a constant C_c value but a variable one. But still, this arbitrary subsetting of the data is less than optimal for two reasons. First, it does not proceed from the data themselves or from the fit of the model to them. Second, each segment is estimated using only partial data. Dubois *et al.* (2007) suggested that the entire data set should be used to estimate the three parameters of the model ($V_{c,max}$, J_{max} and R_d) without the predetermination of transitions by the investigator. Although Dubois *et al.* (2007) did not consider g_m in their analysis, g_m could be included in the routine of parameters fitting, as the authors mention in their work. Alternatively, the approach taken by Sharkey *et al.* (2007) is to allow the researcher to designate which process is controlling at each point. This approach requires judgements to be made but it avoids solutions that may not fit with other known leaf variables. With five parameters to vary, $V_{c,max}$, J_{max}, triose phosphate utilisation (TPU), R_d and g_m, solutions with infinite g_m and low $V_{c,max}$ are sometimes given when imposing different control factors, as some points give a more realistic g_m and essentially the same judgement entirely out of the procedure. The development and refinement of fitting methods are still very active, and improvements are being proposed in the recent years. Miao *et al.* (2009) compared the six most-employed fitting methods described in the literature to estimate Farquhar's model parameters. They concluded that using the entire range of A_N/C_c data points simultaneously and increasing the number of data points at the low C_i end will significantly improve the fitting and estimation of $V_{c,max}$. Yin *et al.* (2009) suggested a new approach. Previous methods estimated Farquhar's model parameters using gas-exchange data only, whereas these authors present an integrated approach combining gas-exchange and Chl-F measurements. The new method allows for some alternative electron transport. Equation 8.4 assumes that RuBP regeneration is limited because of insufficient NADPH and implies 100% linear electron transport: the non-cyclic e-flux used for carbon reduction and photorespiration. However, Yin *et al.* (2009) method assumes that ATP is insufficient: its two forms result from different assumptions about the operation of the Q cycle and the number of protons (H^+) required for synthesising an ATP (von Caemmerer, 2000). Although the new methodology requires some time for testing and generalisation to all conditions and species, the method shows much promise, and it can be used to estimate some other features of the photosynthetic apparatus, such as the occurrence of alternative electron transport. Finally, Chen *et al.* (2008) warned that leaf-level photosynthetic heterogeneity within the mesophyll could lead to underestimation of $V_{c,max}$, J_{max} or TPU calculated by fitting the Farquhar model to the A_N/C_i response of photosynthesis. However, it was concluded that implications for both crop-production and Earth-system models would only take place if the $V_{c,max}/J_{max}$ ratio or TPU itself was heterogeneous within the leaf. It is expected that variation in $V_{c,max}$ and J_{max} would normally be well coupled, therefore the impact in models will be very small.

8.5.3. Effect of temperature on photosynthetic parameters

Photosynthesis is a temperature-dependent process with optimum at around 25°C in most species, although acclimation processes to growing conditions can modify this optimum (Kattge and Knorr, 2007). The effect of temperature on photosynthesis models is often poorly taken into account because, although it is well known that photosynthetic temperature responses have an important role in the response to environmental conditions and differ among species, the parameterisation of the temperature response is not straightforward.

The most used in-vivo temperature dependence of the Michaelis-Menten coefficients of Rubisco, K_c and K_o is the one measured in tobacco (Bernacchi *et al.*, 2001, 2002). The first one in 2001 was determined on a C_i basis that used transgenic plants that expressed only 10% of wild type levels of Rubisco, which ensured that photosynthesis would be Rubisco-limited over a wide range of environmental conditions (Bernacchi *et al.*, 2001). The latter of the two papers was based on C_c, which utilised the temperature functions from Bernacchi *et al.* (2001) and incorporated the measured temperature response of g_m to calculate the kinetics from C_c. Therefore, depending on how we have calculated photosynthetic parameters, A_N/C_i or A_N/C_c, we will choose one or the other. However, the C_i-based kinetic constant (Bernacchi *et al.*, 2001) is not universally applicable (Warren and Dreyer, 2006). For C_c-based estimation of $V_{c,max}$ and J_{max}, the appropriate kinetic constants are a function of the relationship between g_m and photosynthetic capacity ($V_{c,max}$ and J_{max}), and therefore are only appropriate for plants that have the same relationship between g_m and photosynthetic capacity ($V_{c,max}$, J_{max}) and the same temperature response of g_m.

The temperature dependence of $V_{c,max}$ and J_{max} has been described by an Arrhenius equation (Medlyn *et al.*, 2002):

$$V_{c,\max} = V_{c,\max}(25)\exp\left[\frac{E_a(T_l - 25)}{298RT_l}\right] \qquad \text{[Eqn. 8.11]}$$

where $V_{c,max}$ (25) is the value of $V_{c,max}$ at 25°C; E_a is the activation energy of $V_{c,max}$; R is the gas constant (8.314 J mol^{-1} K^{-1}); and T_l is the leaf temperature.

Many times, especially in the case of J_{max}, decreasing photosynthetic capacity has been reported above an optimum temperature (Walcroft *et al.*, 1997; Leuning, 2002; Diaz-Espejo *et al.*, 2006). This has been described using a modified Arrhenius equation (Walcroft *et al.*, 1997):

$$V_{c,\max} = V_{c,\max 0}\frac{\exp[(H_a / RT_0)(1 - T_0 / T_l)]}{1 + \exp[(S_v T_l - H_d)/(RT_l)]} \qquad \text{[Eqn. 8.12]}$$

where $V_{c,max0}$ is the value of $V_{c,max}$ at the reference temperature (usually 25°C); H_a is the enthalpy of activation (conceptually similar to energy of activation); H_d is the enthalpy of deactivation; and S_v is an entropy term. Although this equation is widely used, June *et al.* (2004) suggested a much simpler equation for J_{max}:

$$J_{\max} = J_{\max}(T_0)e^{-\left(\frac{T_l - T_0}{\Omega_T}\right)^2} \qquad \text{[Eqn. 8.13]}$$

where J_{max} (T_0) is the rate of electron transport at the optimum temperature T_0, and Ω_T is the difference in temperature from T_0 to the temperature at which J_{max} equals to e^{-1} of its value at T_0.

8.6. C_4-PHOTOSYNTHESIS MODEL

8.6.1. Overview of C_4 models

C_4-photosynthetic physiology represents a complex combination of leaf biochemistry, compartmentation and structure that differs from C_3 physiology enough to make direct use of C_3 models inappropriate for modelling C_4 photosynthesis. The unique features of C_4 physiology are described in more detail in Chapter 5. Briefly, C_4 leaves differ from C_3 in that the complete photosynthetic cycle is partitioned between two basic cell types that are separated biochemically, spatially and diffusively from one another. CO_2 is fixed initially in one cell type (the mesophyll cells) to form a four-carbon acid that is transported to another cell type (the bundle sheath cells, BSC) where decarboxylation and refixation into the Calvin cycle occurs thus completing the photosynthetic cycle.

This process is, in effect, a light-driven (energy driven) pump for CO_2 that minimises photorespiration and competitive inhibition of Rubisco carboxylation by oxygen. Further discussion of the evolutionary development is considered in Chapter 24.

Intercellular CO_2 dissolves into the mesophyll cell surfaces where it is rapidly converted to HCO_3^- by CA. C_4 mesophyll cells contain high PEPC activities whose substrates are HCO_3^- and PEP. The product of this reaction is the four-carbon compound OAA that is normally converted to malate. Malate is transported into the BSC of the leaf where it undergoes decarboxylation, producing CO_2. The BSC are pneumatically isolated from the mesophyll, and levels of CO_2 can build up much higher than in the intercellular spaces. These high CO_2 levels effectively saturate the carboxylation reaction of Rubisco, inhibit oxygenation and thereby nearly eliminate photorespiration. This physiology results in higher photosynthetic rates for a given amount of Rubisco, but also has an additional light-energy requirement for the recycling of the bundle sheath decarboxyation product pyruvate back into PEP. The ecological conditions that favour C_4 over C_3 photosynthesis are discussed in Chapter 25.

The bundle sheath chloroplasts are fully capable of C_3 photosynthesis and as such are effectively represented by the equations 8.2–8.6 in the previous section. However, leaf-level responses to CO_2, light and temperature are significantly different from C_3 as a result of the operation of the light-driven C_4 pump.

For simplicity, in the following sections we will ignore mitochondrial respiration that occurs in the bundle sheath and mesophyll, though it contributes to the O_2 and CO_2 levels in these tissues and becomes important when photosynthetic rates are low (low light, low CO_2). Typically leaf respiration rates are in the order of <10% of the light saturated leaf photosynthesis in C_4 leaves (Byrd *et al.*, 1992).

A comprehensive description of C_4-photosynthesis modelling is given by von Caemmerer and Furbank (1999), and below we summarise some of the salient aspects.

8.6.2 Rubisco activity in the bundle sheath

At steady state the rate of C_3 photosynthesis (A_N) in the BSC is

$$A = V_c - 0.5V_o \qquad \text{[Eqn. 8.14]}$$

where V_c is the Rubisco carboxylation rate and V_o is the rate at which CO_2 is generated by photorespiratory recycling of the products of oxygenation. The CO_2-limited rate expresses both the activity of Rubisco and photorespiratory recycling of the products of oxygenation (Farquhar *et al.*, 1980):

$$A = \frac{V_{c,max}(C_{bs} - \frac{0.5O_{bs}}{S_{c/o}})}{C_{bs} + K_c(1 + \frac{O_{bs}}{K_o})} \qquad \text{[Eqn. 8.15]}$$

and is dependent on the CO_2 and O_2 concentration in the bundle sheath (C_{bs} and O_{bs} respectively; see equation 8.10 and note that $\Gamma^* = 0.5O_{bs}/S_{c/o}$, where $S_{c/o}$ is the specificity of Rubisco for CO_2 relative to O_2 is equal to $\frac{V_{c,max}K_o}{K_c V_{o,max}}$).

The light-limited Rubisco rate is determined by the light-limited rate of regeneration of RuBP, which is dependent on the J. Electron transport rate is often represented by a non-rectangular hyperbolic function (see Eqn. 8.5), and as co-limitation between light absorption and J_{max} capacity becomes small, the following approximation holds:

$$J \approx \min\begin{cases} J_{max} \\ a_L\, 0.5\text{PPFD} \end{cases} \qquad \text{[Eqn. 8.16]}$$

where a_L is leaf absorptance. The light-limited Rubsico activity in the bundle sheath, taking into account the effects of oxygenase activity, and expressed in terms of the ATP required to generate RuBP is:

$$A = \frac{f\,J(C_s - \frac{0.5O_s}{S_{c/o}})}{4.5C_s + 10.5\frac{0.5O_s}{S_{c/o}}} \qquad \text{[Eqn. 8.17]}$$

or equation 8.4 above, assuming NADPH limitation on RuBP generation rather than ATP. In this case f is the fraction of ATP generation rate that is used to regenerate RuBP while the rest is used to regenerate PEP (see below).

8.6.3 Phosphoenolpyruvate carboxylase activity in the mesophyll

Carbon fixation can also be defined in terms of the initial carboxylation rate of PEPC (V_p) minus the leakage of CO_2 out of the bundle sheaths back into the mesophyll (L_{bs}):

$$A = V_p - L_{bs} \qquad \text{[Eqn. 8.18]}$$

where:

$$L_{bs} = (C_s - C_m)g_{sCO_2}, \qquad \text{[Eqn. 8.19]}$$

and g_{sco_2} is the conductance (1/resistance) to the diffusion of CO_2 between the bundle sheath and the mesophyll. The CO_2-limited PEPC reaction can be expressed as:

$$V_p = \frac{V_{p,max}C_m}{C_m + K_p} \qquad \text{[Eqn. 8.20]}$$

where $V_{p,max}$ is the maximum velocity of the PEPC reaction at saturating levels of C_m and K_p is the apparent Michaelis constant of PEPC for CO_2. The kinetic properties of PEPC are such that it is able to fix HCO_3^- originally derived from CO_2 at 100 times the rate of Rubisco at current atmospheric levels of CO_2. The conversion of CO_2 to HCO_3^- by CA is so rapid (>1000× faster than potential leaf photosynthesis) that it is not considered limiting and therefore is not included in models.

The light-limited PEPC reaction based on the quantum requirement for the production of PEP may be expressed as:

$$V_p = (1 - f)\frac{J}{2} \qquad \text{[Eqn. 8.21]}$$

The formation of PEP from pyruvate requires approximately two ATP.

Note that equations 8.14, 8.15, 8.17 and 8.18 are equal during steady state photosynthesis in C_4 plants.

Another important relationship to express is the partial pressure of O_2 in the bundle sheath, as it can influence the rate of photorespiration.

$$\lambda A = (O_{bs} - O_i) g_{sCO_2} D_{o/c} \qquad \text{[Eqn. 8.22]}$$

where λ is the fraction of the total net O_2 production that occurs in the bundle sheath and can vary between 0 and 0.5 (von Caemmerer and Furbank, 1999) and $D_{o/c}$ is the diffusivity of O_2 relative to CO_2 between the bundle sheath and mesophyll.

If we assume that at low light the response of J (Eqn. 8.16) is linear and at low CO_2 the response of PEPC to C_m (Eqn. 8.20) is linear then the equations 8.14, 8.15, 8.17–8.22, 8.9 and 8.11–8.17 can be solved analytically as a pair of quadratic equations, one representing the CO_2-limited conditions and other light-limited conditions (Collatz *et. al.*, 1992).

Theoretical considerations and experimental observations have shown that generally the CO_2 partial pressure in the bundle sheath approaches saturation with respect to Rubisco carboxylation, the CO_2 leak rate out of the bundle sheaths is low and the carboxylation capacity of PEPC far exceeds that of Rubisco. Given these conclusions we make the following assumptions.

(1) At rate-limiting light intensities the rate of leaf photosynthesis is determined by the intrinsic quantum requirement for CO_2 fixation and leaf PPFD absorptance.
(2) At low C_m, the rate of leaf photosynthesis is limited by the activity of PEPC.
(3) At high light intensities and ambient levels of CO_2, leaf photosynthesis is limited by the maximum Rubisco capacity, which can be expressed as three potential limitations on leaf photosynthesis:

$$A_N \approx \min \begin{cases} a\,\alpha_{c4}\,\text{PPFD} \\ k_p C_m \\ V_{c,max} \end{cases} - R_d \qquad \text{[Eqn. 8.23]}$$

$a\alpha_{c4}$PPFD is a simplification of equations 8.12 and 8.16 from the first assumption, where α_{c4} represents the total quantum requirement for the regeneration of RuBP and PEP.

The second limitation in equation 8.23 is derived from equation 8.20 and the second assumption where $k_p \approx \frac{V_{p,max}}{K_p}$.

The third limitation is derived from equation 8.15 and the third assumption.

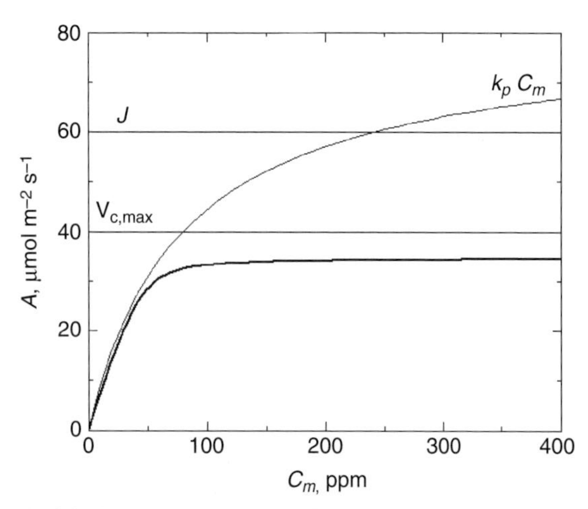

Fig. 8.3. Characteristics of a typical response of C_4 photosynthesis to mesophyll CO_2 partial pressure (C_m) at high levels of photosynthetic photon-flux density (PPFD). The curved line represents the resultant overall response given the interactions between three major limitations (shown as straight lines): electron transport (J), Rubisco capacity ($V_{c,max}$) and phosphoenolpyruvate carboxylase (PEPC) activity ($k_p c_m$).

Transitions between these three limitations are a function of how much the processes are co-limiting. Empirical evidence from leaf measurements indicate that photosynthesis is linear with respect to CO_2 at limiting partial pressures and saturates relatively abruptly, indicating low co-limitation between PEPC activity and either $V_{c,max}$ or light limitation (Fig. 8.3). Light-response measurements, however, do show significant curvature for transitions between light limitation and PEPC- or $V_{c,max}$- limitation (Fig. 8.4). These transitions can be accounted for empirically through curvature parameters when the limitations of equations 8.28 are combined as quadratic functions (Collatz *et al.* 1990, 1991, 1992).

The behaviour of this simplified leaf model is shown in Figs 8.3 and 8.4. Representative values of the three limitations and their overall effect on photosynthesis is shown (see Collatz *et al.*, 1992; von Caemmerer and Furbank, 1999; Massad *et al.*, 2007 for values of model parameters). At low CO_2 and low light, the respective limitations dominate the overall photosynthetic rate. However, at high light and ambient CO_2 levels, co-limitation by the main three processes depicted in equation 8.23 prevent photosynthesis from fully approaching the maximum capacity specified by the remaining limitations.

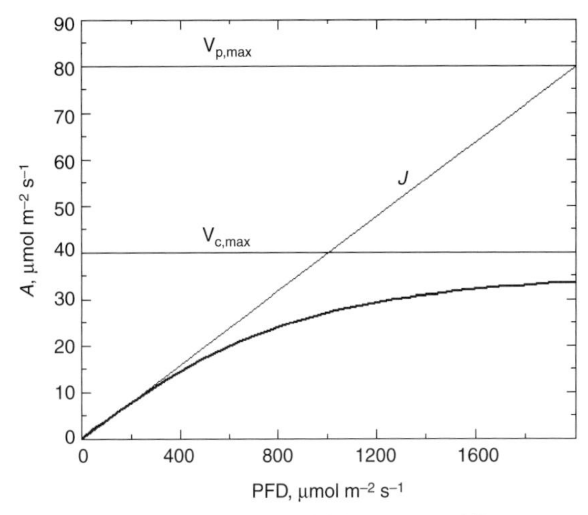

Fig. 8.4. Characteristics of a typical light response of C_4 photosynthesis. The curved line represents the resulting overall response given the interactions between the three major limitations (shown as straight lines): maximum capacities of phosphoenolpyruvate carboxylase (PEPC) ($V_{p,max}$), Rubisco ($V_{c,max}$) and electron transport (J).

It is important to note that although we have expressed gas levels in terms of mole fraction in this chapter, in actuality enzymatic reactions are a function of the activity of the reactants that is represented by their partial pressures rather than mole fractions (Badger and Collatz, 1977; Sharkey *et al.*, 2007). Mole fraction (C) is a function of a gas partial pressure (p): $C = \frac{p}{P_{atm}}$ where P_{atm} is atmospheric pressure (see footnote in Chapter 5). Therefore all variables and parameters expressed in units of mole fraction are based on 1 atmosphere of pressure (P_{atm}=1 atm, 1013 mbar, 101.3 kPa)

8.6.4 Temperature responses

Over a range of non-damaging temperatures (reversible temperature responses), C_4-photosynthetic response is largely governed by the temperature responses of the maximum capacities of two of the three limiting processes, $V_{c,max}$ and $V_{p,max}$ (Eqns 8.15 and 8.20). Estimates of the responses of J_{max} and J to temperature indicate that they do not limit leaf photosynthesis over normal operating ranges (Massad *et al.*, 2007). Also, under typical atmospheric levels of CO_2, PEPC does not limit C_4 photosynthesis, leaving most of the temperature response caused by $V_{c,max}$.

8.6.5 Parameterisation of models

Parameterising equation 8.23 generally involves measuring the response of leaves to varying CO_2, light and temperature, but in-vitro assays of the activities or contents of Rubisco, PEPC electron transport capacity have also been used to parameterise C_4 models (see von Caemmerer and Furbank, 1999). Here we will briefly discuss empirical approaches to fitting measured intact leaf responses to the limitation-based model (Eqn. 8.23).

The linear portion of the initial slope of the light response curve (Fig. 8.3) is a function of leaf absorptance and the intrinsic quantum efficiency of photosynthesis ($\frac{dA_N}{dPPFD} = \alpha\alpha_{c4}$). The response tends to be linear up to about one-quarter full sunlight (~500 μmol photons m^{-2} s^{-1}); however, at <0.1 full sunlight, quantum efficiencies may tend to decrease. In general the intrinsic quantum efficiency of C_4 plants varies around three distinct values associated with the specific C_4 type and dicot/monocot distinction as show in Fig. 8.3. In general $a\alpha_{c4}$ is independent of temperature over non-damaging ranges (Ehleringer and Pearcy, 1983).

The linear portion of the initial slope of the CO_2 response curve (Fig. 8.4) is a function of the capacity of PEPC to fix carbon (Eqns 8.20 and 8.23). However, as discussed for C_3-leaf photosynthesis earlier in this chapter, CO_2 encounters a finite resistance to diffusion as it moves from the intercellular spaces to the cytoplasm of mesophyll cells. This will cause C_m to be lower than C_i during active photosynthesis. Considering this added limitation, the initial slope becomes:

$$\frac{dA_N}{dC_i} = \frac{K_p g_m}{K_p + g_m} \quad \text{[Eqn. 8.24]}$$

Where K_p is Michaelis–Menten constant of PEPC for CO_2.

Pfeffer and Peisker (1998) estimate g_m to be in the same order as found in C_3 plants, but C_4 plants exhibit larger slopes because $\frac{V_{p,max}}{K_p} \gg \frac{V_{c,max}}{K_c(1+O_2/K_o)+\Gamma^*}$, so the impact of g_m on the slope will be relatively larger for C_4. Numerical curve-fitting routines can be used to estimate the combination of parameters that best fit leaf measurements (e.g., Massad *et al.*, 2007)

Finally, predicting leaf CO_2 fluxes (A_N) requires consideration of leaf respiration (A_N=A–R_d). Leaf respiration can be estimated from extrapolation of light response curves to zero light and from measurements in the dark. Typically respiration rates are higher immediately after exposure to light with high photosynthetic rates, but then decrease with time in the dark (hours). Leaf respiration is strongly temperature dependent and generally doubles for every 10°C increase over non-damaging temperature ranges ($Q_{10} \approx 2$).

8.7. COUPLING PHOTOSYNTHESIS MODELS TO STOMATAL CONDUCTANCE MODELS

Photosynthesis and transpiration are coordinated processes as the stomatal aperture exerts a control on both. If we want to run a photosynthesis model based exclusively on environmental variables, a stomatal conductance model must be coupled with it in order to calculate C_i. Therefore, three variables (A_N, g_s and C_i) are unknown, but an analytical solution exists as we have three equations: equations 8.2 or 8.3 for A_N, equation 8.8 for C_i and one from the following section for g_s. In the following subsections the three main available options will be briefly reviewed. Some of them (like Jarvis and BWB approaches) have been widely used and there is much information in the literature about their performance. The hydromechanical model of g_s proposed recently has not been so rigorously tested to date, but this approach undoubtedly represents a future option in leaf gas-exchange models.

8.7.1. Jarvis approach

Before Jarvis proposed his model (Jarvis, 1976), it was well known that illuminated stomata responded to environmental stimuli among which were light, CO_2 concentration, air VPD, leaf temperature, soil water status (Ψ_s) and leaf water potential (Ψ_w). The shape of the functions describing the response of g_s to each of these functions is well established, and stomata generally show a hyperbolic response to PPFD, an exponential decay response to VPD, a sigmoidal response to soil-water status and a peak-function or bell-shape-function response to T_l. Jarvis (1976) assumed that each of the environmental variables affected g_s independently (that is, without any synergistic interactions).

$$g_s\,(\text{PPFD}, \text{VPD}, T_l, \Psi_s) = g_s\,(\text{PPFD})\; g_s\,(\text{VPD})\; g_s\,(\text{T}_l)\; g_s\,(C_a)\; g_s\,(\Psi_s) \qquad \text{[Eqn. 8.25]}$$

This simple approach has been used very much since its publication (Stewart, 1988; Van Wijk, 2000; Moriana *et al.*, 2002), and despite of its empiricism it has provided satisfactory results. Currently, it is still widely used in different applications, such us for modelling ozone pollution (Fuhrer, 2009; Mészáros *et al.*, 2009), modelling transpiration in forests (Kumagai *et al.*, 2008; Matsumoto *et al.*, 2008; Zhou *et al.*, 2008), grassland ecosystems H_2O fluxes (Zhang *et al.*, 2009b), hydrological studies (Winsemius *et al.*, 2008) and agriculture (van der Velde *et al.*, 2006; Maruyama and Kuwagata, 2008; Massonnet *et al.*, 2008; Zhang *et al.*, 2008c). Owing to its formulation based on the response of g_s to environmental variables, it has had a large success among micrometeorologists and hydrologists.

The two main approaches to obtain the parameters of the Jarvis model are: (1) boundary layer analysis of the response of g_s to each environmental variable; and (2) measurements of g_s response curves to each environmental variable while the other variables are kept constant. Jarvis (1976) performed a boundary layer analysis in his original model. When measured, g_s under field conditions is plotted against any environmental variable, for example PPFD, a scatter diagram results owing to the effect of any of the other variables that have a major influence in g_s; for example, at a certain time of day VPD is limiting g_s whereas PPFD is not. This can be seen in Fig. 8.5a, where the response of g_s to PPFD is plotted. Provided that enough measurements have been made adequately to cover the variable space, the upper limit of the scatter diagram indicates the response to the particular independent variable when the other variables are not limiting. However, they are not always available for field data for the entire range, and the relationship of g_s with any environmental variable is best determined in controlled environments (Fig. 8.6b). Once the form of the relationship is known, the parameters can be easily determined by using non-linear least squares. Usually, the Jarvis model is expressed as a product of a reference g_s, which could be the maximum ($g_{s,max}$), multiplied by a series of functions to each variable describing the relative response to it or how that reference g_s is increased or decreased as that variable changes (Le Roux *et al.*, 1999; Diaz-Espejo *et al.*, 2006). One of the main difficulties and shortcomings reported about this model is the calculation of $g_{s,max}$. Schulze *et al.* (1994) reported a positive relationship between leaf nitrogen content and $g_{s,max}$. A strong relationship with leaf chl. content was found in *Quercus serrata* (Matsumoto *et al.*, 2005), which was interpreted by the authors as the dependence of g_s not only on environmental variables, but also on physiological properties. A similar conclusion was reached by Maruyama and Kuwagata (2008) in rice, who proposed that the phenological stage of the plant, which affected g_s, can be approximated by the plant chl. content.

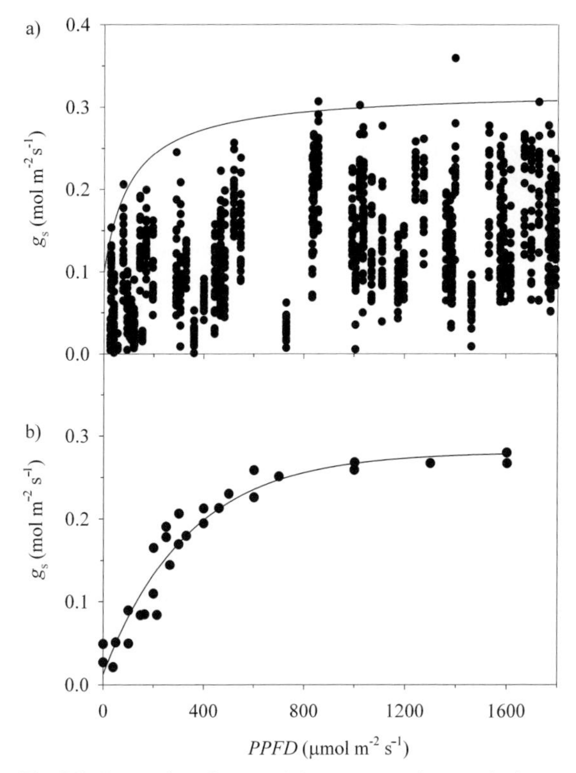

Fig. 8.5. Stomatal conductance (g_s) response to photosynthetic photon-flux density (PPFD) in olive leaves. **(a)** Data from survey measurements in plants under different levels of water stress throughout the day. **(b)** Data from g_s/PPFD response curve under controlled conditions of three control plants.

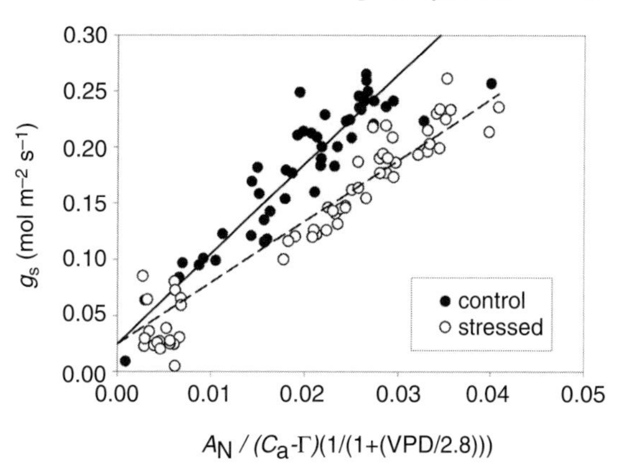

Fig. 8.6. Stomatal conductance (g_s) versus Ball-Woodrow-Berry (BWB) index, modified by Leuning, in plants under control and stress conditions. Data from survey measurements throughout the day. Note that the slope is reduced in the stressed plants.

It has also been criticised as to whether it is appropriate to couple a version of the Jarvis model where there is not explicitly a function of response to C_i with a photosynthesis model (Zhou *et al.*, 2008) as there is no feedback between g_s and C_i. Most of the works do not even consider C_a, which is obviously relatively constant under natural conditions. However, this point highlights again that most of the applications of this model are solely based on environmental variables and do not consider that the phenological and physiological characteristics of the plant can modulate the response to each of these environmental variables.

8.7.2. Ball-Woodrow-Berry and Leuning approaches

The observed correlation between leaf conductance and leaf photosynthetic activity, and the coordinated response of stomata to light, humidity and CO_2 led to the development of an empirical g_s model (Ball *et al.*, 1987). The Ball-Woodrow-Berry (BWB) model is expressed as:

$$g_s = g_0 + a\, A_N\, \mathrm{RH}_s / C_a \qquad \text{[Eqn. 8.26]}$$

where a is an empirical coefficient; RH_s is RH at leaf surface; and g_0 is g_s when $A_N = 0$ at the light compensation point. This model has acquired a good reputation and it is the preferred option of modellers when the main interest is to model A_N. As the model includes a relationship with A_N, it has been called physiological or semi-empirical (Leuning, 1995). However, Aphalo and Jarvis (1993) stated conclusively that the BWB model cannot be used for defining causal relationships.

One of the concerns caused by the application of the BWB model comes because it is widely accepted that stomata respond to VPD rather than RH_s (Aphalo and Jarvis, 1991). Furthermore, Mott and Parkhurst (1991) demonstrated that stomata respond to atmospheric humidity through evaporation from the leaf, and this is why there is a closer link with VPD, although it is not a direct response to VPD.

Leuning (1995), echoing this fact, substituted RH_s with VPD in the BWB equation. Previously, the inclusion of the CO_2 compensation point improved the behaviour of the model at low values of C_a (Leuning, 1990). Therefore, g_s is described by the Leuning model by:

$$g_s = g_0 + a_1 A_N f(VPD)/(C_a - \Gamma) \quad \text{[Eqn. 8.27]}$$

where f(VPD) is a function of response of g_s to air humidity. There are several options proposed for this function (see Leuning, 1995), but the hyperbolic form suggested by Lohammer *et al.* (1980) appears as the best option.

$$f(VPD) = (1 + VPD/VPD_0)^{-1} \quad \text{[Eqn. 8.28]}$$

where VPD_0 is an empirically determined coefficient. This function leads to a linear relationship between evaporation rate and g_s, and it is consistent with current understanding of the interaction between these two variables (Mott and Parkhurst, 1991).

Among physiologists, the BWB and Leuning models are the most common options chosen. Despite the response of g_s to RH_s, the BWB approach is still widely used (Wilson *et al.*, 2001; Misson *et al.*, 2004; Ye and Yu, 2008). However, it is the version proposed by Leuning that is used most often. Applying the Leuning model coupled with the Farquhar *et al.* (1980) photosynthesis model it is as simple as plotting g_s versus $A_N/(C_a\text{-}\Gamma)(1+VPD/VPD_0)$. The slope of this relationship is a and the y intercept is g_0 (Fig. 8.6). As said before, we need to couple g_s and A_N models. There are three unknowns, g_s A_N and C_i, so we need equation 8.9 to have a set of three equations. The analytical solution can be seen in Leuning (1990), substituting RH_s with f(VPD).

The main shortcoming of the BWB and Leuning models is the absence of an explicit dependence of g_s on soil-water potential. The slope, a_1, is usually around ten in well-watered plants of diverse C_3 species of various biomes (Misson *et al.*, 2004), but is drastically reduced under drought conditions (Sala and Tenhunen, 1996; van Wijk *et al.*, 2000; Gutschick and Simonneau, 2002). This parameter has been modelled as a function of leaf (Ψ_w) or soil (Ψ_s) water potential (Sala and Tenhunen, 1996), or, as proposed by Gutschick and Simonneau (2002), as a function of [ABA]. It is important to note, however, that the decrease of g_s under water-limiting situations might be foreseen to occur as a result of a reduction of either the photosynthetic capacity and/or as a response to plant endogenous factors. The first response is more likely to happen under long-term imposition of water stress, when acclimation processes take place, or during the ontogenesis of new foliage under less favourable conditions, as the response photosynthetic capacity of leaves in the short-term has been reported to occur only under severe water stress (Flexas *et al.*, 2009). Therefore, when the water stress is imposed in the short-term or to leaves acclimated to non-stressful conditions, a decrease in the slope, a, is expected. Sala and Tenhunen (1996) found a seasonal decrease in a_1 that was linearly related to pre-dawn Ψ_w. It has been reported that stomata sense a decrease in Ψ_s and root water potential by synthesising ABA in roots (Davies and Zhang, 1991; Tardieu and Simonneau, 1998). Gutschick and Simonneau (2002) made a good effort to include the effect of the water stress in BWB-Leuning models by relating the factor a_1 explicitly to [ABA]. Simultaneously, Dewar (2002) proposed a combination of the BWB-Leuning model with Tardieu-Davies model considering [ABA], extending his analysis of the g_s model in terms of guard-cell function (Dewar, 1995). Dewar's model (2002) presented some novel features not included in previous approaches, such as stomatal sensing of C_i, explanation of the three regimes of the stomatal response to air humidity, incorporation of xylem embolism and maintenance of hydraulic homeostasis by combined hydraulic and chemical signalling in leaves. Although Dewar coupled his model with the Thornley and Johnson model (1990) for simplicity, it can be combined with Farquhar *et al.* (1980). The model of Dewar (2002) can be considered as a good representation of the new generation of models called to get into stage in the near future, which are introduced in the next subsection.

8.7.3. Hydromechanical model of stomatal conductance

Although at the beginning of this century there was not sufficient information for a mechanistic model of stomatal functioning (Farquhar *et al.*, 2001), in the last years some physiologically based models have been proposed (Dewar, 2002; Gao *et al.*, 2002; Buckley *et al.*, 2003). These models attempt to simulate g_s in a more mechanistically explicit fashion. The model developed by Buckley *et al.* (2003) captures all current knowledge about stomatal hydromechanics, and therefore, it is able to predict both steady state and transient responses of stomata to environmental perturbations (Buckley, 2005). This model predicts g_s from the balance of opposing hydromechanical and biochemical influences in and around guard cells.

All three models mentioned above assume that g_s is proportional to guard cell turgor pressure, which in turn depends on conductance through the effects of transpiration on water potential. These assumptions, taken alone, lead to the following basic equation

$$g = \chi \frac{\psi_s + \pi_g}{1 + \chi R_g D_s} \quad \text{[Eqn. 8.29]}$$

where χ is a turgor-to-conductance scaling factor; Ψ_s is the soil-water potential; π_g is guard cell osmotic potential; R_g is the effective hydraulic resistance from the soil to the guard cells; and D_s is the leaf to air VPD. This equation is consistent with the negative stomatal response to soil drought conditions (low Ψ_s), to reduced hydraulic conductivity (high R_g), to high air evaporative demand (high D_s) and to reduced light intensity, which is accompanied by a decrease in guard-cell osmotic pressure (low π_g).

However, the model in the form presented in equation 8.29 is not able to explain by itself all the 'wrong-way' stomatal responses initially observed when Ψ_s is decreased and R_g or D_s increased. Under these conditions, g_s first increases and then slowly decreases following the well-known steady state response. These observations require consideration of epidermal backpressure. A new equation proposed by Buckley *et al.* (2003) and subsequently modified by Barbour and Buckley (2007) includes all of these features:

$$g_s = \chi \frac{(B-M)(\psi_s + \pi_e) - \pi_e + \pi_a}{1 + \chi R D_s (B - M + \rho)} \quad \text{[Eqn. 8.30]}$$

where B is a variable describing the sensitivity of π_g to epidermal turgor pressure and its responses to light and CO_2; M is residual epidermal mechanical advantage; R is the hydraulic resistance from the soil to the epidermis; π_e is the epidermal osmotic potential; π_a is the apoplastic osmotic potential; and ρ is the guard cell resistive advantage (epidermis to guard-cell resistance as a fraction of R). Although the parameters involved have a physiological meaning, both models can be criticised for being more difficult to apply to ecological or agronomical studies than models described in Sections 8.7.1 and 8.7.2. This is essentially right as parameters involved are difficult to measure directly. However, this should not be an excuse to ignore the application of these models, because there are ways that the parameters can be estimated by gas exchange, similar to the Farquhar *et al.* (1980) model of photosynthesis. It is possible to fit the parameters using nonlinear regression or any other fitting procedure.

Gao *et al.* (2002) show how this is achieved in their version of the model based on equation 8.29. Three parameters are obtained by nonlinear regression analysis for eleven species that belong to three functional vegetation types. Parameters were suggested to reflect the elastic property of guard cell structure, the sensitivity of solute concentration and osmotic potential of guard cells to photosynthetic activities and the efficiency of transportation from soil to leaf. Although confirmation of the meaning of these parameters remains hypothetical, undoubtedly this is a step forward in the use of g_s models. However, it must be said that the Buckley *et al.* model (2003) described by equation 8.30 can be used in a similar fashion, for which some simplifications could be assumed for applications that are not concerned with resolving questions about stomatal behaviour (Buckley, personal communication). For example, the parameters M, ρ and π_a can be set to zero with little effect on the steady state behaviour of the model. When these parameters are fixed, the model has only two empirical parameters: χ and B. The parameter B can be simulated in relation to mesophyll ATP as shown by Buckley *et al.* (2003), or it can be related empirically to irradiance and CO_2. The other terms are physiological or environmental (Ψ_s, π_e, R_g). In short, the complete model is meant to be mechanistically rigorous, but it can still do a very good job when some simplifying assumptions are made to reduce the number of parameters.

8.8. ASSESSMENT OF LIMITATIONS TO PHOTOSYNTHESIS

Modelling the different limitations imposed on leaf photosynthesis can help assess the impact of stresses (Farquhar and Sharkey, 1982; Martin and Ruiz-Torres, 1992) and to accurately predict net primary production (Grassi and Magnani, 2005; Diaz-Espejo *et al.*, 2006). Different approaches have been applied. The most widely used was proposed by Farquhar and Sharkey (1982). In their model, stomatal (l_{gs}) and Rubisco (l_{ce}) limitations can be estimated by relating the actual values of A_{Nmax} to the theoretical value that would result when assuming an infinite g_s or an infinite $V_{c,max}$, respectively (Jones, 1985).

$$l_{gs}(l_{ce}) = \frac{A_{Nmax} - A_N}{A_{Nmax}} \quad \text{[Eqn. 8.31]}$$

where A_{Nmax} is the assimilation rate that would occur with an infinite g_s, or $V_{c,max}$. Not only Rubisco-related limitations can be assessed, also RuBP limitations have been estimated (Martin and Ruiz-Torres, 1992), g_m (Bernacchi *et al.*, 2002) and all four limitations (Sampol *et al.*, 2003). This approach has been criticised for two aspects: (1) it relies on unrealistic conditions, i.e., g_s is never infinite; and (2) stomatal and non-stomatal limitations calculated with this approach are not additive (i.e., their total could be higher than 100%). By this approach, the maximum possible increase that could occur with, for example, increased conductance is determined but the relative effect of small increases in conductance are not determined.

Jones (1992) proposed an alternative option based on the theory of control analysis that could avoid the unrealistic conditions of assuming infinite g_s. The limitation to photosynthesis imposed by g_s would be defined as the relative sensitivity of A_N to a small change in g_s. In terms of resistances, this can be expressed as

$$l_{gs} = \frac{r_g}{r_g - r^*} \times 100 \qquad \text{[Eqn. 8.32]}$$

where r_g is the gas-phase CO_2 transfer resistance ($1/g_{sCO2}$), which defines the CO_2 supply function and r^* is the inverse slope of the A_N/C_i response $(dA_N/dC_i)^{-1}$ at the operating point, where the supply function intercepts the A_N/C_i response function.

Grassi and Magnani (2005) suggested a new approach whose maximum innovation was the quantification of each of the main components affecting photosynthesis limitation. Relative changes in light-saturated assimilation (dA_N/A_N) can be expressed as relative contributions of stomatal (S_L), mesophyll conductance (MC_L) and biochemical (B_L)

$$\frac{dA_N}{A_N} = S_L + MC_L + B_L = l_{gs}\frac{dg_{sc}}{g_{sc}} + l_{mc}\frac{dg_m}{g_m} + l_b\frac{dV_{c\max}}{V_{c\max}} \qquad \text{[Eqn. 8.33]}$$

Each of these limitations was defined, following Jones (1985), combining a proper parameterisation of the Farquhar *et al.* (1980) model of photosynthesis with estimates of stomatal and mesophyll conductance, and were described as

$$l_{gs} = \frac{g_{tot} / g_{sCO_2} \cdot \partial A_N / \partial C_c}{g_{tot} + \partial A_N / \partial C_c} \qquad \text{[Eqn. 8.34]}$$

$$l_{gm} = \frac{g_{tot} / g_m \cdot \partial A_N / \partial C_c}{g_{tot} + \partial A_N / \partial C_c} \qquad \text{[Eqn. 8.35]}$$

$$l_b = \frac{g_{tot}}{g_{tot} + \partial A_N / \partial C_c} \qquad \text{[Eqn. 8.36]}$$

where g_{tot} is total conductance to CO_2 between the leaf surface and carboxylation sites ($1/g_{tot} = 1/g_{sCO2} + 1/g_m$); and l_{gs}, l_{gm} and l_b are the stomatal, mesophyll conductance and biochemical limitations, respectively, with values between zero and one. This approach has been used in ecophysiology for seasonal acclimation (Grassi and Magnani, 2005; Diaz-Espejo *et al.*, 2007), water stress and recovery (Galmés *et al.*, 2007a; Flexas *et al.*, 2009; Gallé *et al.*, 2009) and leaf age (Flexas *et al.*, 2007c). The only handicap of this method is that a control treatment is needed to estimate the highest values of g_s, g_m and $V_{c,max}$. Previous approaches (Eqns 8.31 and 8.32) have the advantage that only one treatment is necessary and it is semi-quantitative, but they do not give information about the relative contribution of each component of limitation involved.

8.9. CONCLUDING REMARKS

Over the past 30 years there has been significant work on models of many photosynthetic processes. This has resulted in an interchange between models on the one hand and experimentation and observation on the other, significantly enriching our understanding of photosynthesis and improving our ability to predict it. Some of the same models have been useful for studies at the scale of individual chloroplasts through to global photosynthesis. Significant effort has been made to base the models on known physiological mechanisms though most models still have some empirical relationships built in. There is sufficient detail in models now to allow sophisticated predictions and it is likely that photosynthesis models will continue to enrich research into mechanisms and control of photosynthesis.

Part II
Measuring photosynthesis

9 • Gas-exchange analysis: basics and problems

C. BERNACCHI, A. DIAZ-ESPEJO AND J. FLEXAS

9.1. THE DEVELOPMENT OF GAS-EXCHANGE SYSTEMS AND THEIR PRACTICAL APPLICATIONS

Gas-exchange systems have come a long way since the technology for real-time measures of gas concentrations in air became available. The method of choice among most plant biologists involves the use of IR gas analysers (IRGAs) integrated into gas-exchange systems to measure concentrations of CO_2 taken up by photosynthesis and water released via transpiration (E) over a range of conditions that can be manipulated by the researcher. Typically, gas-exchange systems rely on steady state conditions surrounding photosynthetic tissues, however, unique systems have been developed to measure non-steady state gas exchange for rapid responses (e.g., seconds and faster) to changing conditions (e.g., Laisk and Oja, 1998). Less common, but never-the-less available and quite useful, are methods that rely on oxygen analysers to measure photosynthetic oxygen release. Furthermore, there are methods for determining rates of photosynthesis that rely on Chl-F, reflective indices (e.g., photochemical reflectance index, PRI) and isotopic analysis, each of which are discussed in other chapters. The objective of this chapter is to present and discuss the current methods specific to gas exchange for measuring photosynthesis. This objective includes discussions on the current state of technology, a summary of the equations and theories behind the measurements, potential sources of error and the variables of interest that can be extracted from the most common measurements. Although the focus of this chapter is to provide an 'entry point' into the common gas-exchange techniques, it is not intended to provide a complete description of all gas-exchange systems and all physiologically meaningful data that can be collected. Rather, the reader is recommended towards a wide variety of excellent texts that detail many aspects of gas exchange (e.g., Long and Hällgren, 1993; Laisk and Oja, 1998; Long and Bernacchi, 2003).

The development of gas-exchange systems has allowed researchers to understand the mechanisms of photosynthesis and to investigate environmental effects on photosynthesis. Given the proven technology employed for measuring photosynthetic gas exchange and the ability to vary widely the conditions surrounding photosynthetic tissue, there are numerous physiologically meaningful photosynthetic parameters that can be determined *in vivo*. Thus, gas-exchange measurements are useful for model parameterisation, in-situ measurements, biochemical analysis, understanding photosynthetic responses to changes in the environment and/or resources and a range of other studies. Combining gas-exchange measurements with other techniques allows for an in-depth understanding of numerous processes related to photosynthesis beyond what gas exchange alone can provide.

Commercially available gas-exchange systems are based on measuring either the photosynthetic uptake of CO_2 or evolution of O_2. Despite the direct link of photosynthesis with both of these gases, the purpose of a CO_2- versus O_2-based system varies, and whether one is used over another is determined by the specific research question. Both systems allow for meaningful data to be collected, but each system also has its own errors and assumptions. Given the popularity of the CO_2-based systems, the majority of this chapter will focus on these. At the end of this chapter, we will briefly describe O_2-based systems, the current state of these systems and the types of measurements that can be obtained. The objective of this chapter is to provide a description of the current technology of gas-exchange systems, the physiological basis of the measurements, techniques of gas exchange and common errors associated with measurements.

Terrestrial Photosynthesis in a Changing Environment: A Molecular, Physiological and Ecological Approach, ed. J. Flexas, F. Loreto and H. Medrano. Published by Cambridge University Press.

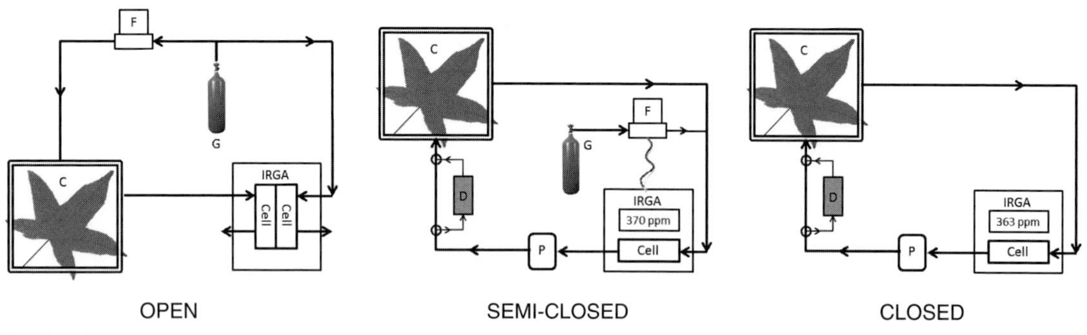

Fig. 9.1. Schematics showing three types of gas-exchange system. The open gas-exchange system consists of an air inlet, often with capabilities to alter gas concentrations (G), a mass flow controller (F), a leaf cuvette (C) and an infrared gas analyser with two separate cells, one for the reference and one for the sample air flow. The semi-closed system has the same components with a few exceptions: the cylinder (G) only provides CO_2, the mass flow controller (F) changes based on CO_2 assimilation of the leaf, the IRGA has only one cell and a pressure sensor (P) determines changes in the pressure of the system. The closed system is the simplest system, containing an IRGA similar to the semi-closed system and a pressure sensor (P). As shown in these diagrams, the air in the semi-closed and closed systems is recycled through the system, often using dessicant (D) to avoid excessive humidity, whereas it is released to the atmosphere for the open system. Figure redrawn from Long and Hällgren (1993).

9.2. GAS-EXCHANGE SYSTEMS

Gas-exchange measurements are designed to measure the fluxes of CO_2 and water vapour as they relate to photosynthesis and E for photosynthetic material. Most systems are designed to make these measurements over a varied set of conditions surrounding the leaf that are established by the researcher. Commercially available gas-exchange systems are able to measure photosynthesis and E, as well as other processes associated with photosynthesis.

9.2.1. Closed, open and semi-closed gas-exchange systems

There are three main types of gas-exchange systems: closed, open and semi-closed systems (Fig. 9.1). Closed systems estimate rates of photosynthesis, respiration, or E based on the rate of CO_2 draw-down, CO_2 accumulation or water-vapour accumulation, respectively. In these systems, air is continually recycled throughout a system containing a cuvette in which a portion of or a whole plant is placed. Cuvettes are usually constructed using metal with high thermal conductivity and often painted white to reflect much of the solar radiation, although they can range from small leaf-sized cuvettes to large chambers in which many plants can be grown (Tholl *et al.*, 2006). Although a thorough investigation of the properties of cuvettes has been discussed previously (Tholl *et al.*, 2006), some key characteristics are necessary to allow for proper gas-exchange measurements. Most notably, cuvettes must allow for uniform conditions across the leaf including light and temperature, allow for turbulent mixing of the air to minimise the leaf boundary layer, have a small volume to prevent stagnant air and have little opportunity for air to diffuse in or out.

With the closed system, after air flows through the cuvette, it is analysed using an IRGA and is repeatedly recycled through the system. This system uses an IRGA to measure CO_2 and water vapour and is relatively simple to use. The rates of carbon assimilation (A_N) and E are calculated as the change in CO_2 and water vapour, respectively, in the system over a period of time. However, changes in temperature and pressure can occur over the measurement's timescale and need to be accounted for (Long and Hällgren, 1993). The equation, including the influences of changes in temperature and pressure, to solve for A_N using a closed gas-exchange system is:

$$A_N = \left[\frac{C_{o,1}\left(P_{ATM,1}/T_1\right) - C_{o,2}(P_{ATM,2}/T_2)V \cdot T_V)}{(t_2 - t_1)s \cdot P_V} \right] \quad \text{[Eqn. 9.1]}$$

where C_o is the CO_2 concentration (μmol CO_2 mol^{-1} air) in the air leaving the cuvette; P_{atm} is pressure (kPa); T is air temperature (°C); s is the surface area of the leaf (m^2); and V is volume (mol) of the system at a given temperature (T_v) and pressure (P_v) (Long and Hällgren, 1993). The subscript values 1 and 2 represent the values at the first (t_1)

and second (t_2) measurement timepoints. In principle, the difference between the two measurement points should be large enough that the differences in $C_{o,1}$ and $C_{o,2}$ are greater than the variability in the measurements, but not so great that substantial differences in the environmental conditions between the timepoints occur. As this is a closed system, the leaf will continue to draw down [CO_2] until the CO_2 compensation point, Γ, is reached. Therefore, a criticism of closed systems is that A_N for these leaves may never achieve steady state (Long and Hällgren, 1993). Another problem with closed systems is that a leak, regardless of how minor it may be, can have a tremendous impact on rates of photosynthesis given the fact that the air is continuously recirculated and that the system is quite sensitive to fluctuations in pressure and CO_2.

The simplicity of the closed systems made it useful in many laboratories that conducted in-situ photosynthesis measurements. Advances in IRGA technology and computational power led to the development of commercially available open gas-exchange systems, which have largely replaced the closed systems. These systems employ analysis of the gas concentrations in two different chambers: a reference chamber, which has an air stream that is not modified by the presence of a leaf, and a sample chamber, which contains a leaf. The difference between the [CO_2] in the reference (C_e) and sample (C_o) chambers are used to calculate A_N (μmol m^{-2} s^{-1}) as:

$$A_N = \frac{u(C_e - C_o)}{s} \quad \text{[Eqn. 9.2]}$$

where u (mol s^{-1}) is the mass flow of air into the chamber.

Because there is invariably an increase in water-vapour partial pressure in the sample cuvette as a result of leaf E, it is important to consider its impact on gas-exchange measurements. For a closed system, the increase in water vapour will dilute the concentration of CO_2 in the air stream. The release of water vapour will also increase the pressure of the system at a rate that, for all practical purposes, equals the release of water vapour. Therefore, this can be safely ignored in a closed system provided that the system does not contain any leaks. As pressure in the system builds, any leaks will result in an overestimation of A_N, as the increase in pressure will drive air out of the system resulting in no change in pressure, but an increase in water vapour will dilute the CO_2 concentration. In an open system the increase in water vapour will dilute the CO_2 in the system without a pressure increase and if not accounted for it will cause measurement error. Therefore, a correction that accounts for this dilution effect can be incorporated as:

$$A_N = \frac{u}{s}\left(C_e - C_o \frac{(1-w_e)}{(1-w_o)}\right) \quad \text{[Eqn. 9.3]}$$

where w_e and w_o are the water-vapour concentrations (mmol mol^{-1}) in the reference and sample chambers, respectively. Similar to the closed system, leaks can result in incorrect estimates of A_N in open systems, although depending on the location in the gas-exchange system, leaks can drive responses that are not as predictable as with closed systems (i.e., they do not always overestimate A_N). This topic will be addressed in more detail later in this chapter.

The final category of gas-exchange system is the semi-closed system. This type of system works in much the same way as a closed system, only the CO_2 concentration surrounding the leaf is kept constant. This requires a source of CO_2 that is highly accurate and a feedback loop that provides only as much CO_2 as is assimilated by A_N, which is calculated as:

$$A_N = \frac{u_{CO2}}{s} \quad \text{[Eqn. 9.4]}$$

where u_{CO2} is the flow of CO_2 into the air volume of the system. The semi-closed system, often referred to as a compensating system, has distinct advantages over closed systems in that minor leaks will not accumulate over time and that steady state measurements can be made over a wide range of CO_2 values. For example, the desired CO_2 concentration can be set and the system can equilibrate to this level. Given the feedback between the leaf and the gas-exchange system, these changes are not always immediate, and slight changes in the environment surrounding the leaf requires patience on behalf of the researcher (Long and Hällgren, 1993). Further, as with closed systems, an accumulation of water vapour in the system will occur and often the leaves may respond to this increase.

9.2.2. Basics of infrared gas analysis

Modern gas-exchange systems employ a wide range of techniques to measure the CO_2 and water-vapour fluxes associated with A_N and E, with each system having specific strengths and weaknesses. The variety and ease of use of modern gas-exchange systems makes these quite useful. However, as is the case with any research equipment,

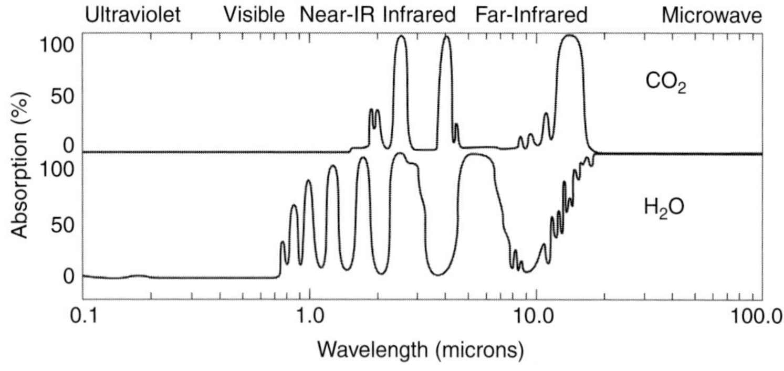

Fig. 9.2. Absorption of CO_2 and H_2O over a range of wavelengths.

simplicity often results in greater opportunity for error. It is difficult to verify the validity of measurements obtained from commercial gas-exchange systems without fully understanding the possible sources of errors. Specific attributes of a commercial gas-exchange system should therefore be studied in detail prior to its utilisation. A complete comparison of commercial gas-exchange systems and their individual strengths and weaknesses is outside the scope of this chapter; however, some basic principles of how IRGAs work are provided below.

Both CO_2 and H_2O absorb radiation at specific IR wavebands that is determined from the absorption spectrum (Fig. 9.2). The general design of an IRGA consists of an IR source, a cell through which the gas being measured flows and a detector. The loss of IR radiation over the path-length of the IRGA is a function of the amount of gas that is absorbing in the wavelength being measured (e.g., Sestak *et al.*, 1971). The output of IR from the source must remain constant over time in order for the accuracy of the system to remain stable, and the accuracy of the detector will vary with temperature. Therefore, most gas-exchange systems will either keep the temperature of the IRGA constant or they incorporate a correction for temperature. Modern IRGAs will provide simultaneous partial pressures of CO_2 and H_2O by using a number of different methods. In some instances, a spinning chopping wheel with filtered windows of a specific wavelength will pulse the beam of IR radiation at wavelengths specific to CO_2 and H_2O. These chopper wheels spin at very specific speeds and the timing is linked to the detector. At any given time, all of the light minus the waveband associated with either CO_2 or H_2O is filtered out. Other methods employ splitting a beam of IR light to two different detectors that have filters that remove all but the waveband specific to either CO_2 or H_2O. These systems will often employ a third filter that allows for a waveband near those for CO_2 and H_2O, but at which neither CO_2 nor H_2O will absorb. This 'null' waveband is then used to correct for pollen, dust or any other particle that might interfere with the IR beam. The benefit of the filtered chopper wheel or the beam-split protocols is that they allow for a single beam of light for analysis of both CO_2 and H_2O. Other systems will employ a separate analysis cell each for CO_2 and H_2O with filters on either the source or detector end of the IRGA appropriate for each gas. These systems require the use of four separate IRGAs incorporated into the system.

9.3. RELATIONSHIPS BETWEEN INFRARED GAS-ANALYSER MEASUREMENTS AND GAS EXCHANGE

The number of physiologically meaningful measurements that can be elucidated from a gas-exchange system is quite large, as will be discussed in later sections. These measurements are all rooted in the two fluxes measured with gas-exchange systems, A_N and E. As explained above, the type of gas-exchange system used will determine which calculations are needed to determine A_N. Even within a single type of gas-exchange system, for example a closed system, schematics and equations employed for calculating fluxes will vary by manufacturer. Fortunately, manufacturers of the most popular gas-exchange systems provide the equations specific to their system. The basic equations that are used to determine fluxes of A_N and E, as well as other physiological meaningful variables, have been developed, compiled and published elsewhere (von Caemmerer and Farquhar, 1981; Long and Hällgren, 1993) and are described below.

The rate of E from a leaf is a physiologically meaningful parameter in terms of determining absolute water use and WUE (A_N/E). As stated previously, the release of water vapour into the cuvette increases the total pressure in the cuvette. For a closed system this results in an increase in the system pressure. For an open system, this results in an increased flow rate through the sample cuvette. The E for a surface area (s) is calculated as:

$$sE = u_o w_o - u_e w_e, \quad \text{[Eqn. 9.5]}$$

where the subscripts $_o$ and $_e$ represents the air as it exits and enters the cuvette. As the flow rate leaving the cuvette, u_o, is higher than u_e by the amount released via transpiration, sE:

$$u_o = u_e + sE \quad \text{[Eqn. 9.6]}$$

Combing equations 9.5 and 9.6 yields:

$$sE = (u_e + sE)w_o - u_e w_e \quad \text{[Eqn. 9.7]}$$

that rearranges to:

$$E = \frac{u_e(w_o - w_e)}{s(1 - w_o)} \quad \text{[Eqn. 9.8]}$$

This equation simplifies measurements in that only u_e is measured, minimising errors associated with using two separate flowmeters.

Comparisons of photosynthetic rates among differing treatments will often require that the intercellular, or even chloroplastic, concentrations of CO_2 are known. This is particularly important when two leaves demonstrate different A_N and the researcher is interested in whether this difference is driven by changes in photosynthetic processes or by the supply of CO_2 into the leaf or into the chloroplast. For example, differing rates of photosynthesis under ambient conditions might be driven by different concentrations or activities of photosynthetic enzymes or by differences in g_s of the leaves. Using a resistance analogy, E is expressed using the following equation:

$$E = (w_i - w_a) / r_{l,wv} \quad \text{[Eqn. 9.9]}$$

where w_i and w_a are the water-vapour concentrations inside and outside the leaf, respectively and $r_{l,wv}$ is the resistance of water-vapour transport from the intercellular airspaces to the air surrounding the leaf. Values for E and w_a are measured directly from gas-exchange analysis, and w_i is assumed to be saturating based on leaf temperature (T_l). The evaporation of water in the intercellular airspaces increases the pressure inside the leaf and drives a net outward flow of water vapour resulting in greater E than predicted from equation 9.9. This error will generally be less than about 3% (von Caemmerer and Farquhar, 1981) and can be corrected for by using the following equation:

$$E = (w_i - w_a) / r_{l,wv} + \bar{w}E \quad \text{[Eqn. 9.10]}$$

where $\bar{w} = (w_i + w_a)/2$. The total resistance of water-vapour transport from inside the leaf to the atmosphere can be calculated by rearranging equation 9.10 to:

$$r_{l,wv} = \frac{(w_i - w_a)}{E(1 - \bar{w})} = \frac{1}{g_{l,wv}} \quad \text{[Eqn. 9.11]}$$

The term $r_{l,wv}$ represents the sum of the stomatal and leaf boundary layer resistances, and the conductance term is presented at the reciprocal to the resistance as $1/g_{l,wv}$. As the dimensions and circulation of air of a gas-exchange cuvette are well known, boundary layer resistance is usually provided by the manufacturer of the gas-exchange system. The stomatal conductance to water vapour is then determined as:

$$g_{l,wv} = g_{l,wv} \cdot g_{b,wv} / (g_{b,wv} - g_{l,wv}) \quad \text{[Eqn. 9.12]}$$

With the water-vapour conductances for the boundary layer and stomatal conductances known, the conductances for CO_2 through the same pathways are determined based on the ratio of the binary gas diffusivities of water vapour to air and CO_2 to air, which works out to be 1.61 for the stomatal component and 1.37 for the boundary layer component. Thus, $g_{s,CO2} = 1.61 g_{s,wv}$ and $g_{b,co2} = 1.37 g_{b,wv}$. The latter value differs from the former in that the boundary layer consists of both transfer associated with diffusion and turbulent transfer (von Caemmerer and Farquhar, 1981; Long *et al.*, 1996). Analogous to equation 9.10, the rate of PCA, A_N (μmol m^{-2} s^{-1}), is described by the equation:

$$A_N = g_{l,CO2}(C_a - C_i) - \bar{C}E \quad \text{[Eqn. 9.13]}$$

that rearranges to solve for the intercellular CO_2 concentration, C_i, as (von Caemmerer and Farquhar, 1981):

$$C_i = \frac{(g_{l,CO2} - E/2)C_a - A_N}{(g_{l,CO2} + E/2)} \quad \text{[Eqn. 9.14]}$$

The importance of C_i in analysing certain photosynthetic measurements will be discussed in later sections of this chapter.

9.4. TECHNIQUES FOR GAS-EXCHANGE MEASUREMENTS

System calibration and verifying the absence of leaks or other problems prior to measuring gas exchange is crucial for accurate measurements. This is then followed by configuring the gas-exchange system according to the experiment purposes, i.e., type of cuvette employed, light intensity, air flow into the cuvette, leaf temperature, air cuvette humidity, etc. These conditions will vary based on the leaf, growth conditions, experimental questions and types of measurements desired, including but not limited to light, CO_2, temperature response curves or in-situ measurements comparing treatments or species performance. In the next sections we will describe briefly many important features and tips for successful gas-exchange measurements.

9.4.1. Calibration and system set-up

Modern IRGAs do not need calibrating as frequently as older instruments that were based on gas-filled detectors; however, regular calibration is recommended and frequent verification of calibrations are strongly recommended to ensure accuracy. Modern gas-exchange systems typically have established protocols to be executed before measurements take place. These protocols provide calibration and verification of IRGAs, flowmeter, leaf temperature thermocouple and light source. Manufacturers recommend checking these components regularly, which, in addition to verifying calibrations, are very useful as diagnostic tools. A complete IRGA calibration includes zeroing with a source of H_2O- and CO_2-free air and verifying the span with a source of air containing a precisely known mole fraction of H_2O and CO_2. A proper calibration generally requires sending the unit to the manufacturer for a full calibration, but verifying whether calibration for a system is necessary can be accomplished using gas of a known CO_2 concentration contained in cylinders or an air stream with a known water vapour. Gas cylinders with certified mixtures of CO_2 in air should also be verified independently as some certifications are often only established at 5% or 10% of a target CO_2 concentration, which is beyond the range of error that should be used for calibrating IRGAs. Calibrating an IRGA for H_2O requires the use of a dew-point generator to provide a stream of air with precise water vapour. Minor changes in the span can be done by the user for some commercial systems, but significant deviation in the IRGA reading and span gases indicate that a factory calibration should be performed. In practice, IRGAs are zeroed routinely when measurements are made either daily or throughout the day when conditions surrounding the gas-exchange system are variable (i.e., diurnal measurements of A_N and g_s). Zeroing the IRGAs in a gas-exchange system is commonly performed using magnesium perchlorate, drierite or silica gel to remove H_2O and soda lime to remove CO_2. It is important that the chemicals are fresh to ensure that CO_2 and H_2O removal is complete. It is also important to consider that the chemicals that remove H_2O will often release CO_2 and the chemicals that remove CO_2 will release H_2O. Therefore, when zeroing, it is best to bypass the soda lime when zeroing the IRGAs for H_2O and to bypass the desiccant when zeroing for CO_2. In practice, removing the CO_2 and H_2O concurrently saves time provided that the CO_2 in the airstream is scrubbed first, followed by the desiccation of the air as the CO_2 released from the desiccant is substantially less than the H_2O from the soda lime. After the IRGA is zeroed specifically for H_2O, the desiccant can be bypassed to allow for the source of CO_2 to be removed and then the IRGAs can be zeroed for CO_2. Another alternative is to use a gas cylinder to zero the IRGA. Once calibration is completed, the desired conditions in the system should be set. Light intensity in the cuvette, leaf temperature, air VPD, air flow into the cuvette and CO_2 concentration are the most commonly controlled variables. In the next sections, we will detail the most important points based on the goal of the measurements.

9.4.2. In-situ photosynthesis measurements

Measurements of instantaneous A_N and g_s are useful to characterise a community, to compare plants under different treatments or to monitor changes over time and/or space. These measurements often require sampling many leaves over a short time period and spending a minimum amount of time on each leaf to maximise the sample size. When using natural light, particular attention should be paid to ensure that leaf orientation in the chamber is similar to natural conditions, including irradiance and leaf orientation. Shading by the cuvette should be avoided as it creates a heterogeneous light environment across the leaf. Photosynthesis responds rapidly to variations in light conditions, but stomata take longer to respond. Assuming that leaves are near steady state conditions in the field, any significant change in A_N without a change in g_s might have implications in calculations of C_i. Leaf to air VPD and leaf temperature are unavoidably modified once the leaf has

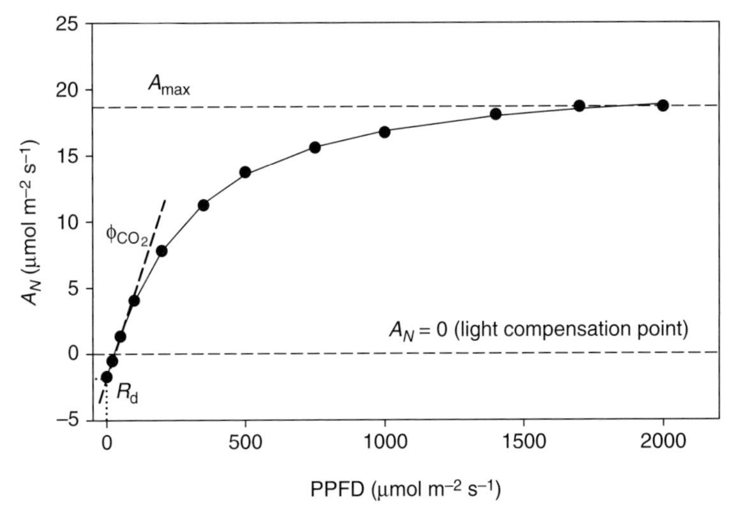

Fig. 9.3. Light response curve with representative parameters indicated: maximum photosynthetic rate (A_{max}); dark respiration rate (R_n); light compensation point; quantum yield (initial slope, ϕ); and the convexity term (θ). PPFD, photosynthesis photon-flux density.

been enclosed in the cuvette. In order to minimise this, high air-flow rates help to keep VPD close to ambient, and the use of a cuvette or leaf temperature control in systems should be employed when available.

In the next two subsections, we will focus on the most widely reported photosynthetic response curves, photosynthetic light (A_N/PPFD) and photosynthetic CO_2 (A_N/C) response curves. However, in-situ measurements at standard field conditions, photosynthetic temperature response curves, photosynthetic VPD response curves or a range of other measurements can be collected depending on the hypotheses generated by the researcher.

9.4.3. Photosynthetic light response curves

Photosynthetic light response (A_N/PPFD) curves provide useful information about characteristic parameters of a leaf related to many factors influencing the development of a leaf, including but not limited to position in a canopy, temperature in which the leaf developed and nutrient availability. There are several empirically based models for describing the A_N/PPFD response curve (Gomes *et al.*, 2006). The non-rectangular hyperbola is the most used for research purposes as it contains an important qualitative parameter, the convexity term (Ögren, 1993). From an A_N/PPFD curve, leaf mitochondrial respiration, (R_d), Γ, quantum yield (ϕ_{CO2}), convexity (Θ) and the maximum photosynthetic rate (A_{SAT}) can all be determined (Fig. 9.3) by fitting a curve to the equation:

$$A_N = \frac{PPFD \cdot \phi_{CO2} + A_{SAT} - [(\phi_{CO2} \cdot PPFD + A_{SAT})^2 - PPFD \cdot 4\phi_{CO2} \cdot A_{SAT} \cdot \Theta]^{0.5}}{2\Theta} - R_d \quad \text{[Eqn. 9.15]}$$

Initially the leaf should be allowed to reach steady state conditions at saturating PPFD. It is important to determine *a priori* the saturating PPFD, as too much light can photoinhibit the leaves and too little light can drive the deactivation of photosynthetic enzymes and cause stomatal closure. Steady state conditions at each light level are compulsory to achieve accurate calculations of the parameters associated with a light response curve (Rascher *et al.*, 2000). But in practice, this is difficult to achieve in the field owing to changing environmental conditions and because it is time consuming; therefore fast response curves are typically performed, which are usually sufficient for comparison purposes.

Measurement of ϕ_{CO2} in a conventional portable system has three potential sources of error (Long *et al.*, 1996). The first involves the incorrect measurement of PPFD at leaf surface. Quantum sensors measuring PPFD are not usually positioned at the same level as the leaf surface, and therefore the measured radiation intensity can be overestimated. The second involves the variability that can be introduced from A_N being measured close to the light compensation point to determine ϕ_{CO2}. Under these conditions, the CO_2 differentials between the sample and reference IRGAs in open systems or the rate of change in CO_2 with closed systems are minimal, and the inherent error associated with gas-exchange systems are maximal. Third, ϕ_{CO2} determined by a conventional portable gas-exchange system will be the apparent maximum ϕ_{CO2}, as it is based on incident and not absorbed light (Long *et al.*, 1996). Actual ϕ_{CO2} of a leaf may be determined in an integrating sphere leaf chamber or by measuring absorptance separately in an integrating sphere, where reflected and transmitted light can be determined. The use of an integrating sphere is particularly necessary when measuring light response curves in needle-shaped leaves such as conifers, to eliminate shading among needles resulting from direct illumination (Ethier *et al.*, 2006). However, there are other aspects that usually yield a substantial underestimation of ϕ_{CO2}, as reported by Singsaas *et al.* (2001). Including both data points at low PPFD (usually <20 μmol m^{-2} s^{-1}) that are influenced by the Kok effect (see Chapter 4) and data that extends beyond the linear region of the A_N/PPFD response curve produces errors in the calculation of ϕ_{CO2}.

The convexity term, which is unitless and constrained between zero and one, describes the inflection between light-limited and light-saturated photosynthesis. A value of Θ approaching one results in photosynthesis transitioning from light-limited rates to light-saturated rates more rapidly. Alternatively, leaves that have Θ values closer to zero result in a slower transition to light-saturated photosynthesis, with some leaves not reaching light saturation at full sunlight. The value of Θ associated with a leaf is related to the intensity and direction of light during growth and development (Ögren, 1993). There are many well-known relationships between Θ and environmental conditions, but a full mechanistic understanding of this term is lacking. Therefore, it is generally treated as an empirical term (Cannell and Thornley, 1998). A leaf that expresses a high value of Θ approaching one is generally assumed to have a sharp inflection between Rubisco- and RuBP-limited photosynthesis (Ögren, 1993). At lower PPFD, photosynthesis is limited by electron transport rate, and at higher PPFD it is limited by Rubisco. As the PPFD transitions from low to high, an inflection on an A_N/PPFD curve should theoretically occur. In reality there exist gradients of light and CO_2 within a leaf, so the whole transect of the leaf will not experience this transition at the same point (Ögren, 1993) and will cause Θ to be less than one. Further, some leaves do not experience Rubisco limitation under ambient [CO_2] and saturating PPFD. In these leaves, Θ will be lower as the transition from low to high light will be driven by biophysical properties of the electron transport chain. For example, when leaves are grown in elevated [CO_2] they experience RuBP-limited photosynthesis over the entire range of PPFD, which results in a lower Θ than plants grown under non-elevated [CO_2] (Ögren, 1993).

9.4.4. Photosynthetic CO_2 response curves (A_N/C_i)

The A_N/C_i response curve represents the demand function of photosynthesis for CO_2. The model proposed by Farquhar *et al.* (1980a) provides a mechanistic explanation of the observed data points at various stages along this curve. Key biochemical kinetic variables determining photosynthetic rate in leaves *in vivo*, $V_{c,max}$, J_{max}, TPU and R_d, can all be obtained from an A_N/C_i response curve (Sharkey *et al.*, 2007). The first three parameters, usually used as proxies of photosynthetic capacity of leaves, are deduced from three different regions of the A_N/C_i response curve where these processes are limiting (see Chapter 8 and Fig. 9.4). Each curve in its entirety should be made at constant leaf temperature, saturating irradiance and constant and reasonably low VPDs. High VPD values are known to induce stomatal closure, and although calculations of C_i take the influence of g_s into account, the reductions of A_N can cause problems of accuracy at low [CO_2]. Leaf temperature should be constant during the measurement of the whole curve because photosynthesis is quite sensitive to temperature. Kinetic parameters of photosynthetic enzymes used in A_N/C_i curves analysis, e.g., Michaelis-Menten constant coefficients of carboxylation (K_c) and oxygenation (K_o) or the photosynthetic CO_2 compensation point (Γ^*), are highly dependent on temperature. These parameters represent kinetic properties of Rubisco, and for this reason they are commonly considered to be relatively similar among terrestrial C_3 species. Saturating irradiance, determined from A_N/PPFD curves, is necessary to direcly measure J_{max}, i.e., maximum RuBP regeneration limited by potential whole-chain electron transport rate. A_N/C_i response curves are usually made starting at ambient CO_2 concentration and decreasing to values as low as 50 μmol mol^{-1}, the concentration representing a value similar to Γ. The measurement at each CO_2 step must be completed quickly (within 2 min) in order to ensure a steady state activation of Rubisco (Long and Bernacchi, 2003). Then CO_2 is increased back to values of ambient CO_2 and increased stepwise again until around 1200 μmol mol^{-1}. The second measurement at ambient CO_2 is used to confirm that no deactivation of Rubisco has occurred at low CO_2

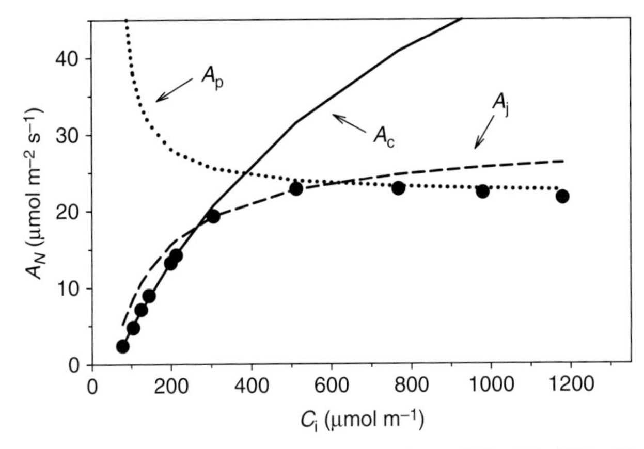

Fig. 9.4. Rate of CO_2 assimilation (A_N) versus intercellular CO_2 (C_i) in *Olea europaea* measured at photosynthetic photon-flux density of 1600 μmol m^{-2} s^{-1}. The measured values are represented by black circles, and lines represent each of the rates of photosynthesis that would be achieved depending on whether Rubisco (A_c), RuBP (A_j) or TPU (A_p) are limiting.

concentrations. The rapid measurements associated with an A_N/C_i curve do not allow for the stomata to reach steady state. However, most gas-exchange systems are able to make accurate measures given the slow rate of change of stomata for most plants (Long and Bernacchi, 2003).

When interpreting A_N/C_i response curves, the goal is generally to estimate the biochemical and biophysical parameters that are predicted to change based on treatments, development or any other experimental factor. The basis of an A_N/C_i curve is to determine these parameters based on CO_2 concentrations in the intercellular air spaces. However, observed differences in $V_{c,max}$ between two treatments can be driven by changes in the amount or activation state of Rubisco, as well as the changes in mesophyll conductance (g_m; Singsaas *et al.*, 2001; Flexas *et al.*, 2007a,b,c). As a result, it is important to consider whether A_N/C_i curves are sufficient when comparing two treatments or whether measurements based on chloroplastic CO_2 concentrations, as will be discussed in Chapter 12, need to be made.

9.4.5. Photosynthetic CO_2 compensation point (Γ^*), dark and light respiration

Respiration in a general sense can occur over a couple of different pathways and occurs under dark and light conditions. For the sake of clarity, we define the different respiratory components as follows: night respiration (R_n) is mitochondrial-driven CO_2 efflux in the absence of light; leaf respiration (R_d) is mitochondrial-driven CO_2 efflux in the presence of light; and photorespiration (R_p) is the release of CO_2 from the photorespiratory pathway that occurs after an oxygenation of Rubisco takes place.

The Γ^* represents the CO_2 concentration at which photosynthesis and photorespiration offset each other and is a function of the kinetic properties of Rubisco related to the specificity for CO_2 versus O_2, or the Rubisco Specificity Factor ($S_{c/o}$) (Farquhar, 1980). It is important to note that the Γ^* is not the same as the CO_2 compensation point, Γ. The difference between these two terms is represented by the fact that Γ includes the amount of photosynthetic CO_2 uptake necessary to offset the combined release of CO_2 from photorespiration and mitochondrial respiration. Its link with $S_{c/o}$, Γ^* is important for analysis of many photosynthetic and photosynthesis-related measurements including but not limited to A_N/C_i curve analysis and determination of g_m. It has long been known that $S_{c/o}$ is variable among species (Jordan and Ogren, 1981a,b; Spreitzer and Salvucci, 2002). It has recently been demonstrated that conditions to which a species is adapted, e.g., drier habitats, are correlated to some degree with Γ^* (Galmés *et al.*, 2005). As Γ^* is a kinetic property of Rubisco, it is likely that it does not change within a species, but adaptation of populations within a species should be considered when assessing whether to rely on previously published values of Γ^* for a species or to make direct measurements. Another consideration involves the degree of accuracy needed for Γ^*. Although some protocols require highly accurate values of Γ^* for a species, the fact that the range of values for C_3 species is relatively limited might result in relative insensitivity with errors

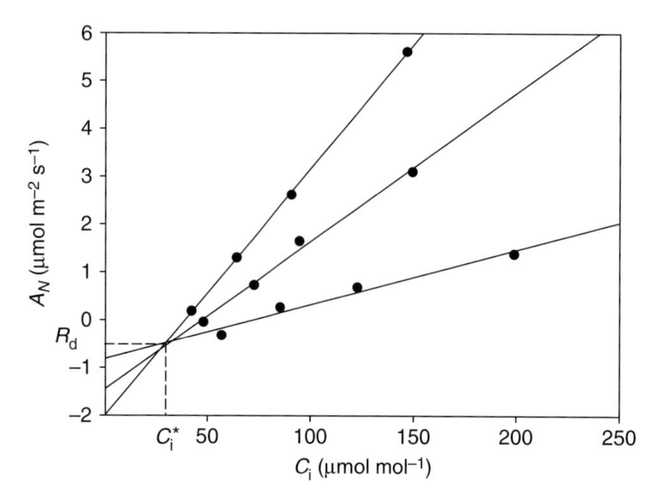

Fig. 9.5. Estimate of intercellular photocompensation point (C_i*) and light respiration (R_d) following the Laisk method. Three A_N/C_i curves are completed at [CO_2] lower than ambient and at three different photosynthetic photon-flux densities: 500, 150, 50 μmol m^{-2} s^{-1}. The intersection of the initial linear portion of the three A_N/C_i response curves generated determines both parameters.

in Γ*. Whether a general published value of Γ* should be used should be assessed according to what the experiment requires.

The most widely used method to calculate Γ* is based on a gas-exchange technique developed by Laisk (1977) and Brooks and Farquhar (1985). The method consists of three partial A_N/C_i response curves (C_a values ranging from 300 to 40 μmol CO_2 mol^{-1}) made at three different PPFDs (e.g., 500, 200, 50 μmol m^{-2} s^{-1}; Fig. 9.5). The intersection point of the three A_N/C_i curves is used to determine C_i* (x-axis) and R_d (y axis). C_i* differs from Γ* proportionally to R_d and g_m ($\Gamma^* = C_i^* - R_d/g_m$). As these measurements require limiting light and low rates of photosynthesis, it is important to consider the impact that small errors will have on measurements.

Respiration measurements in the absence of light (R_n) can be measured simply by placing a leaf in a gas-exchange cuvette. As the transition from light-acclimated to dark-acclimated mitochondrial respiration can be lengthy, it is suggested that a previously light-acclimated leaf be kept in the dark for a sufficient period of time to ensure steady state has been achieved. In some cases, this period can extend well beyond 30 min. Some authors have used R_n values as proxies of R_d (Niinemets *et al.*, 2006a). However, as discussed thoroughly in Chapter 4, the temperature responses of R_d and R_n may not be similar (Tjoelker *et al.*, 2001; Zaragoza-Castells *et al.*, 2008), and it has been shown that R_n and R_d do not acclimate in the same proportion to changes in temperature or growth irradiances (Zaragoza-Castells *et al.*, 2007). For more specific concerns with respiration measurements, see Chapter 4.

9.4.6. Non-steady state measurements

Although this section has focused on steady state measurements of photosynthesis, there are a wide range of research questions that can be addressed through measuring dynamic changes in photosynthesis to a wide range of conditions. Most if not all gas-exchange systems are developed to measure gas exchange when the photosynthetic tissue has achieved steady state values or when the changes in the fluxes of either CO_2 or H_2O are changing at a rate that is lower than the variation that the system can detect. The design of larger leaf cuvettes imposes a chamber response time that makes steady state slower to achieve (Weiss *et al.*, 2009). These systems perform poorly when rapid changes in the leaf fluxes occur as a result of the large volume of air that exists in the cuvettes and/or low flow rates through the cuvettes. Unique gas-exchange-system designs have been presented (e.g., Laisk and Oja, 1998) that allow for fluxes to be measured at the second or faster time resolution.

9.5. COMMON PROBLEMS ENCOUNTERED IN GAS-EXCHANGE MEASUREMENTS

Gas-exchange measurements are subjected to a number of technical concerns and problems that can be divided into three main groups: those associated with the fundamentals of the technique, including calibration issues; those associated with the specific design of the leaf cuvettes used in most commercially available systems; and those associated with leaf physiology, which are independent of calibration and cuvette designs. These will be discussed in more detail in the following sections.

9.5.1. Problems related to calibration and leaf-temperature determinations

Leaf temperature is an important environmental variable affecting leaf gas exchange on several aspects. Leaf temperature has a direct effect on photosynthetic rates through the enhancement of carboxylation rate of Rubisco. This enhancement, however, peaks at a relatively low maximum temperature, from which a decline is usually observed. Plants do acclimate at different growing conditions and can often modify the optimal temperature for photosynthesis (Bernacchi *et al.*, 2003a; Yamori *et al.*, 2006). Measured leaf temperature also determines leaf-to-air VPD, which has a direct effect on g_s.

Measurements of VPD are based on the leaf temperature as it is required to determine the saturation leaf vapour pressure in the leaf. Stomatal conductance is 17-fold more sensitive to an error in leaf temperature than to an error in air temperature (McDermitt, 1990). This is because of the highly temperature-dependent phase change from liquid water to water vapour that occurs in the leaf. In most portable gas-exchange systems, leaf temperature is measured with a thermocouple, which is gently pressed to the lower surface of the leaf clamped in the cuvette. This thermocouple usually uses an air-temperature thermistor or similar device as a reference junction. The thermocouple and the reference thermistor must be matched. Therefore, these should be checked according to the instructions specific to each system. Even under optimal conditions, measured leaf temperature is shown to deviate from actual leaf temperature as a result of the microenvironment of the leaf cuvette and of erroneous measurement with a porometer thermocouple (Tyree and Wilmot, 1990). Tyree and Wilmot (1990) show that placing the leaf in the porometer modified the leaf temperature, decreasing or increasing it by 2 to 3°C, depending on whether previous measurements were made in the shade or in the sun. Further studies suggest that temperature differences between the porometer head and leaf surface as low as 0.5°C, can cause a error in g_s of 25–50% depending on the range of g_s being measured (Verhoef, 1997). In general, it is recommended to minimise the effect of cuvette temperature on leaf temperature by keeping the cuvette open in the shade with the fans running between measurements (McDermitt, 1990).

Under full sunlight conditions, it is important to consider the impact that solar loading will have on the gas-exchange system, most notably the cuvette. Although some gas-exchange systems allow for the temperature of the cuvette to be controlled, there will nevertheless be circumstances where the level of control is lacking. It is crucial that leaf temperature be recorded and that the researcher is aware of the conditions that the leaf is experiencing during measurements. Even the best designed gas-exchange systems will result in large-scale alterations of the leaf owing to the cuvette, and as such it is important to consider these impacts on measurement. If measurements are made in rapid sequence then it is possible to obtain leaf gas-exchange measurements representative of field conditions as changes in stomatal conductance are generally slow. This allows for confidence in the measures of A_N and C_i, but care must be taken in interpreting values obtained for E as the microclimate in the gas-exchange cuvette is not representative of those in the field.

9.5.2. Leaf-cuvette design-associated problems

In contrast to earlier custom-built gas-exchange systems, the modern commercially available systems enclose small leaf areas typically less than 10 cm^2, often as small as 2 cm^2 (Long *et al.*, 1996; Long and Bernacchi, 2003). This implies a large ratio of cuvette-edge surface to leaf surface enclosed in the cuvette (Fig. 9.6). A main assumption of gas-exchange measurements and calculations is that gas exchange occurs only in the illuminated area *inside* the cuvette and between the leaf and the surrounding air over that same area (i.e., vertically). However, if some CO_2 diffuses laterally, either through the gaskets, through the leaves or between the leaf and the gaskets (Fig. 9.6), this will have a significant effect on the calculation of photosynthetic parameters, contrary to what would occur in large cuvettes with a small edge-to-inner-surface ratio. There is evidence that different types of such problems occur in commercially available instruments (Jahnke, 2001; Jahnke and Krewitt, 2002; Pons and

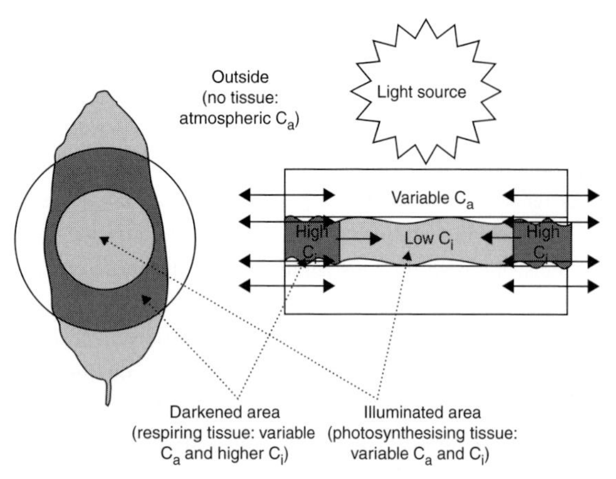

Fig. 9.6. Diagram showing how a leaf is positioned inside a small gas-exchange cuvette (left shows top view and right shows lateral view). White areas illustrate the leaf-cuvette gaskets. Dark-grey areas show darkened leaf tissue below the gaskets, while pale-grey areas show illuminated leaf tissue inside the cuvette. Notice that the darkened area is similar to or even higher than the illuminated area in such small gas-exchange cuvettes. Thick arrows in the right diagram illustrate potential CO_2 (and, to a lesser extent, H_2O) fluxes between inside and outside the cuvette not associated with photosynthesis and hence inducing potential errors in its measurement.

Welschen, 2002; Long and Bernacchi, 2003; Pieruschka *et al.*, 2005; Galmés *et al.*, 2006; Jahnke and Pieruschka, 2006; Flexas *et al.*, 2007b).

LEAF-CUVETTE LEAKAGE

Leaf cuvettes are not completely sealed from the air outside, and both CO_2 and H_2O will diffuse to some degree between the inside and outside. The magnitude of leakage is particularly important when CO_2 gradients exist between the inside and outside of the leaf chamber, as occurs when performing A_N/C_i curves. When using the smallest leaf chambers, the 'apparent' A_N of an empty chamber can be as high as 3–5 μmol CO_2 m^{-2} s^{-1} at high CO_2 mole fractions entering the chamber, e.g., at 2000 μmol CO_2 mol^{-1} air (Flexas *et al.*, 2007b; Rodeghiero *et al.*, 2007). For a leaf displaying photosynthetic rates of 30–50 μmol CO_2 m^{-2} s^{-1} at these CO_2 concentrations, A_N may be overestimated by up to 10% as a result. For leaves with lower A_N, the effect will be even greater. To account for this problem, two solutions have been proposed: to avoid leakage by minimising the gradient of CO_2 concentrations inside and outside the chamber; or to determine the magnitude of leaks and use the information to correct the rates (Long and Bernacchi, 2003).

Enclosing the leaf chamber in a bag or container filled with the same gas mixture that is being introduced into the chamber should eliminate CO_2 gradients and, hence, leakage. This can be accomplished by allowing the gas purged from an open system to remain surrounding the cuvette during measurements. However, this method can be complicated when working under field conditions or when working with plants where the leaves do not extend far beyond the stem. It is difficult to enclose the gas-exchange chamber in another chamber when the whole plant and the pot are in close proximity to the cuvette. Even with an ideal plant, only partial reductions in leakage occur given the need to work around a plant (Flexas *et al.*, 2007b; Rodeghiero *et al.*, 2007).

For most purposes, estimating leaks and correcting the rates accordingly will be sufficient to account for leaks. Including a leaf between the gaskets in the leaf chamber may increase leakage, particularly leaves with prominent veins where the formation of small air channels between the gaskets and the sides of the veins can occur (Long and Bernacchi, 2003). It was suggested that a dead leaf formed by rapidly drying a live specimen could be used to assess the true magnitude of leaks (Long and Bernacchi, 2003; Rodeghiero *et al.*, 2007). However drying the leaf may unavoidably result in reduced leaf thickness and result in a different gasket-leaf interaction as compared with a living leaf (Flexas *et al.*, 2007b). Thermally killed leaves obtained by immersion into boiling water until no variable Chl-F is detectable has either no effect (Rodeghiero *et al.*, 2007) or reduces (Flexas *et al.*, 2007b) leakage relative to an empty chamber. This suggests that leaves increase the resistance of diffusion through the gaskets. In summary, it is important to understand the sources of leaks that can occur with gas-exchange systems, and to understand the implications that physical properties of leaves may have on leakage. Visual inspection of the leaf-gasket interface should provide confidence that no large air gaps exist, which can be aided through the use of an artificial light source integrated into the cuvette.

EDGE EFFECTS AND LATERAL FLUXES THROUGH LEAVES

Besides leakage of gases into or out of the gaskets, lateral diffusion of CO_2 from respiring leaf tissue underneath the gaskets occurs in what is commonly referred to as 'edge effects' (Pons and Welschen, 2002). In addition, lateral diffusion of CO_2 inside the leaves can also occur (Jahnke and Krewitt,

2002; Pieruschka *et al.*, 2005; Morison *et al.*, 2005, 2007; Morison and Lawson, 2007; Pieruschka *et al.*, 2008). These effects are particularly important given that illuminated leaf areas inside the chamber are usually surrounded by areas of leaf between the gaskets that are darkened and thus respiring and results in erroneous estimates of A_N and C_i.

In the small chambers of most commercial gas-exchange systems, the area of leaf between the upper and lower gaskets can be similar to or greater than the area enclosed in the chamber (Pons and Welschen, 2002). Hence, respired CO_2 will decrease the measured net flux and lead to underestimated A_N and erroneous C_i. Pons and Welschen (2002) showed that the effect was important at low light (about 25% underestimation of A_N) but negligible at high light. This effect also causes an important overestimation (up to 30–40%) of R_n, as the actual leaf area contributing to the observed CO_2 flow is much larger than that assumed in the calculations (Pons and Welschen, 2002).

Lateral diffusion through air space in the mesophyll has been shown to occur, particularly in homobaric species (Jahnke, 2001; Jahnke and Krewitt, 2002; Pieruschka *et al.*, 2005; Morison and Lawson, 2007; Morison *et al.*, 2007; Pieruschka *et al.*, 2008). This occurs owing to the fact that illuminated mesophyll patches inside the leaf chamber have a lower C_i than the non-photosynthesising mesophyll patches darkened under the leaf-chamber gaskets. This gradient causes a lateral flux of CO_2 from the darkened air spaces to the illuminated air spaces, leading to an underestimation of photosynthesis. Lateral diffusion also causes an important underestimation of dark respiration when measured at high CO_2 concentrations. Previously stated decreases of mitochondrial respiration in response to high CO_2 (Amthor, 1991) were almost entirely owing to an artifact of the measuring system (Jahnke, 2001; Jahnke and Krewitt, 2002).

PRESSURE-RELATED EFFECTS

Air pressure in leaf chambers appears to affect the above mentioned problems, i.e., both leakage and lateral diffusion. Altering the pressure between inside and outside the leaf chamber for a clamp-on leaf chamber influenced both A_N and E (Jahnke and Pieruschka, 2006). When g_s is high, a pressure difference is predicted to act as if the effective surface area inside the chamber is reduced, leading to underestimations of A_N that can exceed 10% with a ΔP_{atm} as small as 0.14 kPa (Jahnke and Pieruschka, 2006). At low g_s, lateral gas flow driven by pressure differences between leaf chamber and ambient air is only minor because of the high flow resistance of the stomata. In systems where the IRGA is integrated into the leaf chamber, the gas lines are very short and ΔP_{atm} can be considered to be zero. In contrast, in systems in which the IRGA cells are connected to the leaf chamber by relatively long tubing, ΔP_{atm} can be considerably larger than zero depending on the flow rate and the flow resistance in the outgoing gas line.

9.5.3. Problems associated with leaf physiology

Several errors in gas-exchange measurements that are associated with leaf physiology have been described that lead to impaired C_i calculation, including heterogeneous stomatal closure (patchiness), cuticular conductance and leaf shrinkage (Laisk, 1983; Cheeseman, 1991; Terashima, 1992; Mott, 1995; Boyer *et al.*, 1997; Buckley *et al.*, 1997; Tang and Boyer, 2007). All these problems are especially prominent under some stress conditions, but some issues including cuticular conductance could be problematic under any condition in which g_s is low.

The density of stomatal apertures over leaf area is not homogeneous (Ishihara *et al.*, 1971; Laisk *et al.*, 1980). Moreover, it has been shown that neighboring stomata react to environmental stresses to a different extent (Omasa *et al.*, 1985; Kappen *et al.*, 1987). The possibility that variability in the distribution of stomatal conductance could influence the measurement of A_N and the calculation of C_i was first suggested by Laisk (1983), and extended by Downton *et al.* (1988a,b) and Terashima *et al.* (1988). These researchers noted that A_N often declined with stomatal closure, whereas C_i appeared to be non-limiting. This was especially evident during rapid leaf dehydration or when feeding leaves with ABA. An example of deviation in the A_N/C_i relationship induced by patchiness of stomatal opening upon cutting a leaf through the petiole in air is shown in Fig. 9.7. In the experiment represented in Fig. 9.7, a leaf was dehydrated rapidly and this was not shown to induce any change in Rubisco activity (Flexas *et al.*, 2006b). Therefore, the decreased A_N to C_i ratio was a result of erroneous estimations of C_i from stomatal patchiness, driven by the curvilinear relationship between A_N and g_s. To better illustrate this, Table 9.1 shows modelled output for this leaf assuming various different scenarios for stomatal patchiness. Assuming that g_s (0.15 mol CO_2 m^{-2} s^{-1}) is homogeneous over the leaf blade (i.e., no patchiness), then A_N is 9.7 μmol CO_2 m^{-2} s^{-1} and C_i is 335.3 μmol CO_2 mol^{-1} air. However, these numbers change based

Table 9.1. *Modelled A_N and C_i based on varying degrees of stomatal patchiness, where patchiness suggests that variability in the degree of stomatal opening exists.*
Average g_s: 0.15 mol CO_2 m^{-2} s^{-1} (C_a=400 μmol CO_2 mol^{-1} air)

Distribution of g_s (patchiness) (mol CO_2 m^{-2} s^{-1})	Calculated A_N (μmol CO_2 m^{-2} s^{-1})	Calculated C_i (μmol CO_2 mol^{-1} air)
100% 0.15 (no patchiness)	9.7	335.3
50% 0.10+50% 0.20	9.4	336.0
50% 0.05 +50% 0.25	8.5	339.4
25% 0.05+50% 0.10+25% 0.35	8.1	335.7
62.5% 0.01+12.5% 0.35+25% 0.40	6.0	378.1

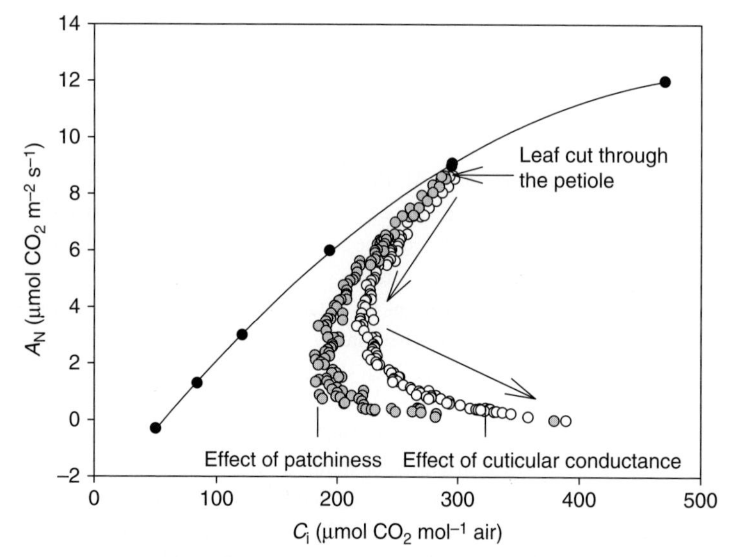

Fig. 9.7. The effects of stomatal patchiness and cuticular conductance on the apparent A_N/C_i response of a bean (*Phaseolus vulgaris*) leaf (Flexas, unpublished results). The A_N/C_i response was determined in an attached leaf (black circles, line), then the leaf was cut through the petiole in air, inducing patchy stomatal closure during dehydration (arrows indicate the progression of gas-exchange measurements after cutting). The apparent A_N/C_i response was strongly deviated from the original owing to patchiness and cuticular effects (white circles). Even after determining cuticular conductance and correcting values accordingly, patchiness alone induced a strong bias in the apparent A_N/C_i response (grey circles).

on the heterogeneity of g_s. For example, when half of the stomata have a g_s of 0.2 mol CO_2 m^{-2} s^{-1} and the others a g_s of 0.1 mol CO_2 m^{-2} s^{-1}, then the measured g_s would still be 0.15 mol CO_2 m^{-2} s^{-1}, but A_N would be lower (9.4 μmol CO_2 m^{-2} s^{-1}) and the calculated C_i would be slightly higher (336 μmol CO_2 mol^{-1} air). As shown with the other scenarios, the degree and skew of stomatal patchiness can have dramatic effects on calculations of A_N and C_i despite the mean g_s across the leaf averaging 0.15 mol CO_2 m^{-2} s^{-1}.

Whether stomatal patchiness actually ever occurs under field conditions has yet to be determined (Gunasekera and Berkowitz, 1992; Kaiser and Kappen, 2000), but the potential effects of patchiness on gas-exchange calculations suggest that the researcher needs to be aware of the phenomena and how to identify its occurrence. The presence of patchiness can be revealed by images of starch staining or $^{14}CO_2$ incorporation over the leaf blade (Downton *et al.*, 1988a,b; Terashima *et al.*, 1988) by Chl-F imaging

(Mott, 1995; Buckley *et al.*, 1997) or by sampling Chl-F variations along the leaf blade with a non-imaging system (Flexas *et al.*, 2002a). A convenient test consists of measuring the response of A_N to C_i at different VPDs (Grassi and Magnani, 2005). The rationale is that changing leaf-to-air VPD will not influence Rubisco directly, at least in the short term, and hence if differences in the A_N to C_i ratio appear under these conditions, these may be owing to erroneous estimations of C_i induced by patchiness. A simpler test is to make A_N/C_i measurements with numerous data points near the inflection between Rubisco- and RuBP-limited photosynthesis. If the inflection point is very precise, as in Fig. 9.4, then it is likely that stomatal patchiness is not occurring. Alternatively, if the inflection is a very gradual shift from one limitation to another, it is possible that the C_i in some parts of the leaf is low enough that those areas are Rubisco-limited and that C_i in other parts are high enough that they are RuBP-limited.

Another cause of overestimation of C_i results in a higher conductance of the leaf cuticle to H_2O than to CO_2 (Boyer *et al.*, 1997). For example C_i values determined with and without correcting for cuticular conductance results in different values (Fig 9.7). Calculations assume that all gas exchange occurs through the stomata and not through the cuticle. Through the stomata, the ratio of diffusivities between H_2O and CO_2 is 1.6, as described above. Through the cuticle, however, this ratio can be as high as 15–40 (Boyer *et al.*, 1997), assuming a ratio of 1.6 can lead to overestimation of C_i. Such an overestimation is particularly important when cuticular conductance is the main determinant of leaf CO_2 uptake and E (e.g., the stomata are closed). Thus, it is likely that such effects are important under severe water stress. To account for this problem, cuticular conductance must be determined and C_i data re-calculated according to the formula given by Boyer *et al.* (1997):

$$C_i = \frac{C_a - 1.6A_N}{[(w_i - w_a)/(E_l - 2E_c)]} \quad \text{[Eqn. 9.16]}$$

where E_l is the leaf transpiration and E_c is the cuticular transpiration. The factor 2 accounts for the fact that cuticular E_c occurs in both leaf sides. Estimates of E_c can be obtained by several procedures. In hypostomatous leaves, a suitable method consists of sealing the lower epidermis with silicone vacuum grease and a polyethylene stuck, and then gas exchange is measured in the upper, non-stomatal-containing leaf side (Boyer *et al.*, 1997). This method assumes identical cuticular conductance in both leaf sides, which may not necessarily be true (Šantrůček *et al.*, 2004) and cannot be used in amphistomatous leaves. Another method consists of allowing a leaf to desiccate until full turgor loss, which causes water loss to decrease linearly with time and until all stomata are closed (Burghardt and Riederer, 2003). Although this method generally yields reliable estimates, for some species bias from leaf shrinkage reducing porosity and transpiration (Boyer *et al.*, 1997) or by significant residual stomatal transpiration (Burghardt and Riederer, 2003) can occur. More accurate methods require isolation of cuticular membranes (Burghardt and Riederer, 2003; Frost-Christensen and Floto, 2007) or the use of high-resolution fluorescence imaging (FI) (Meyer and Genty, 1998).

9.5.4. Oxygen measurement systems

The oxygen electrode is a polarographic sensor that uses electric properties to elucidate a chemical reaction. Specifically, the electrical current that flows between a platinum or zirconium cathode and a silver anode is proportional to the amount of oxygen that is reduced by the cathode. The rate in which this reduction occurs will be linearly related to the amount of oxygen surrounding the cathode, i.e., the oxygen in the volume of the leaf chamber. Thus, photosynthetic tissue placed in the oxygen-electrode chamber alters the oxygen concentration in the chamber in a manner that can be used to determine the fluxes of oxygen. There exist two main types of oxygen electrodes, one which is designed to support photosynthetic tissue in aqueous solution and one designed for leaf discs. The principle behind the measurements of these two systems is similar although the design differs only to allow for measurement of submerged or dry tissues in turn allowing for measurement of aquatic-based organisms (e.g., algae) versus leaf cutting of terrestrial plants.

The design and calibration of these systems is provided in detail elsewhere (Walker, 1993). Instead, we will focus on a brief summary of the physiological basis of these measurements. Photosynthesis, at its most basic, follows that CO_2 will be assimilated into carbohydrate, while H_2O will be cleaved to ultimately yield O_2 to the air surrounding the leaf as represented in the general formula:

$$CO_2 + H_2O \rightarrow CH_2O + O_2 \quad \text{[Eqn. 9.17]}$$

Despite the simplicity of this equation, the reduction of CO_2 into carbohydrate occurs in the photosynthetic carbon reduction cycle (PCRC), while the cleavage of H_2O occurs in the light reactions. Although these two metabolic pathways are intricately linked, there are conditions where the ratio of O_2 release and CO_2 assimilation will not be constant with changing conditions (Walker, 1993). For example, with changes in the PFD, there is a lag between oxygen evolution and carbon uptake (Kirschbaum and Pearcy, 1988). Thus, the oxygen electrode is useful in assessing the rate at which the evolution of oxygen occurs under given circumstances, although the practical applications may be relatively limited. Current oxygen-electrode systems allow for control of light and temperature, but other environmental variables such as [CO_2] and humidity are relatively uncontrolled, and more importantly the traditional oxygen electrode does not work with intact plant material.

More current oxygen-based measurement systems provide the opportunity for differential oxygen measurements similar to an open IRGA system. These systems allow for a typical cuvette to be attached to an analyser that allows the researcher to measure the generation of oxygen during photosynthesis or the consumption of oxygen during respiration. Although there are few studies that have utilised this system in photosynthesis research (but see Cen *et al.*, 2001), the fact that strong gradients between oxygen concentration inside and outside of a cuvette do not exist even with high respiration rates, this system has been used frequently for measuring plant respiration in the dark (e.g., Davey *et al.*, 2004).

9.6. SUMMARY

Gas-exchange measurements can provide a wealth of knowledge about the physiology of photosynthesis and plant water use. Measurements can be made at scales ranging from the leaf to the canopy, provided instrumentation is available to enclose these scales. Commercial gas-exchange systems are useful research tools provided that the errors associated with measurements are known and that the research can identify these. Most problems associated with gas exchange are of minimal importance when gas-exchange rates are high. Alternatively, when measurements of stressed plants are made the number of measurement errors increase. Although these errors are not insurmountable, it is crucial that the variance or bias imposed by these errors is considered.

10 • Optical methods for investigation of leaf photosynthesis

J.M. DUCRUET, M. BARON, E.H. DELUCIA, F. MORALES AND T.D. SHARKEY

10.1. OPTICAL METHODS

Light absorbed by chl. antenna is converted almost instantaneously into charge pairs in photochemical centres, which makes possible excitation by short flashes or modulated light thus generating various absorption, fluorescence or luminescence responses. The possibility of manipulating light excitation at will, together with the abundance of chromophores in the photosynthetic machinery, has favoured the development of optical monitoring of photosynthesis *in vivo*. The three complementary dimensions of optical methods, spectral, kinetic and imaging, provide unique tools to investigate the photosynthetic energy metabolism and, beyond, the bioenergetic status of the whole cell.

There is not a one-to-one correspondence between the measuring opportunities offered by various chromophores embedded in the thylakoid membranes and the aspects of the photosynthetic process they allow monitoring (Table 10.1). Starting from the mechanisms of light interactions with pigments, we will introduce the optical methods that have proved useful for studying leaf photosynthesis, using 'intensive' parameters (ratios such as F_v/F_m, kinetic amplitudes and time constants) that automatically compensate for variations of signal intensity between leaf samples. Other non-photosynthetic optical methods that can be used simultaneously (blue-green fluorescence (BGF), IR reflectance) will be briefly mentioned. Applications of leaf reflectance and fluorescence to remote sensing will be addressed in more detail in Chapter 15 and light absorption by leaves and canopies in Chapter 16.

10.2. ABSORPTION AND REFLECTION OF LEAVES

10.2.1. Leaf optics

Leaves have evolved to maximise light capture under most conditions, and a number of properties of leaves are related to this. Perhaps most striking is the common bifacial leaf in which the 'top' (usually the adaxial surface) has palisade cells that form light guides between them, whereas the 'bottom' (usually abaxial) surface has spongy mesophyll cells that maximise light scattering (Fig. 10.1A). This arrangement enhances light penetration from the top through the palisade layer, but enhances absorption of light that gets into the spongy layer. This effect works for collimated light like that experienced by leaves in direct sunlight. However, it does not work for diffuse radiation found in understories. Not surprisingly leaves exposed to full sun often have very distinct palisade layers while leaves in shade environments may have little or no palisade layer, even when they are from the same plant.

As a result of the internal structures of leaves, light being reflected from the top of a leaf may have travelled through the palisade layer, been scattered in the spongy layer and travelled through the palisade layer again. Light reflected from the bottom of a leaf may be reflected from much shallower regions as it encounters the scattering spongy layer as soon as it gets through the epidermis. Many leaves look much lighter on the bottom than the top because of the enhanced scattering in the spongy mesophyll layer (Vogelmann, 1993). The lighter appearance indicates greater reflectance, for example, a leaf of *Xanthium strumarium* was found to have a reflectance of 6% of 681 nm light from the adaxial surface, but 10% from the abaxial surface. Transmissivity was not affected by whether the light was adaxial or abaxial, and so the difference in reflectivity indicated a difference in absorptivity, 88.5% when illuminated on the adaxial surface but just 84.5% when illuminated on the abaxial surface (Sharkey, 1979). This demonstrates the utility of the internal structure of leaves for increasing light absorption.

The light scattering in the mesophyll layer, and to a lesser extent scattering caused by structures within cells,

Terrestrial Photosynthesis in a Changing Environment: A Molecular, Physiological and Ecological Approach, ed. J. Flexas, F. Loreto and H. Medrano. Published by Cambridge University Press.

Table 10.1 *Optical methods and the photosynthetic mechanisms they allow to monitor in leaves. X: relevant. (X): indirect indication.*

The corresponding sub-chapters are indicated in the first column. **PSII, PSI**: *intrinsic properties of the photosystems and alterations by stresses.* **PSI, PSI ET**: *electron transfer through illuminated photosystems. Particular excitation wavelengths can be used to favour PSII (blue 480 nm) or PSI (far-red > 700nm) activity.* **PMF**: *Proton Motive Force. The two components* **ΔpH** *(thylakoid transmembrane proton gradient) and ΔΨ (thylakoid transmembrane electric field) can be selectively monitored. Under illumination, the pH can reach 7.8 in the stroma and 4.5 in the lumen. A residual ΔpH can remain in the dark, maintained by a chlororespiratory electron flow.* **AP**: *Assimilatory potential NADPH + ATP in equilibrium with the pool of triose-phosphates (Gerst et al., 1994).* **ST**: *State transitions, state 1 to state 2 corresponds to the migration of the peripheral antenna of PSII towards PSI. State 2 favours CEFI.* **CEFI**: *electron transfer around PSI under light, generates ATP only.* **Ind CEFI, CR**: *induction of CEFI pathways, chlororespiration. The cyclic pathway(s) around PSI may be open although there is no electron flow in absence of light. This induction can be monitored in the dark by a weak electron transfer, closely related to chlororespiration (from stroma to PQ pool).* **Thylakoid membrane**: *these lipids are highly unsaturated, which contributes to the membrane leakiness to protons (measured by green light absorption), to the heat sensitivity of PSII (curves F_0/T)and to lipid peroxidation (High temperature ThermoLuminescence).* **Commercial instruments**: *commercially available optical instruments dedicated to photosynthesis research.*

		PSII	PSII ET	PSI	PSI ET	PMF: ΔpH	PMF: ΔΨ	AP	ST	CEFI	Ind. CEF1 CR	Thylakoid membrane	Commercial instruments
Absorption and related													
1.2	P700 (A820)			X	X			(X) (light)		X (light)	X (dark)		H W B
1.3	Green 505–518–535					X	X			(X)		X Leak	W B
1.4	PRI		(X)		(X)	(X)	(X)						P S
1.5	NADPH (F500)							X					W
1.6	Photoacoustic Spectroscopy	X	X	X	X				X	X	X		
Chlorophyll fluorescence													
2.1	spectra	X		X					X				B
2.2	lifetime	X	X	X									
2.3	0→P	X											B H O P QW
2.4	Modulated	X	X			X (qE)			X (qT)				H L O P W QT
2.5	F_0	X									X F_0/t	X F_0/T	→ id
Chlorophyll luminescence													
3.1	DL	X				X (dark)							
3.2	TL (A,Q,B,C)	X				(X) (dark)							F P Q
3.3	Afterglow (AG)							X		(X)	X		→ id
3.4	Modulated (DLE)	X				X (light)	X (light)						H
3.5	HTL bands											X Peroxid.	F

Commercial instruments for photosynthesis. B: www.bio-logic.info, F: www.fan-gmbh.de, H: www.hansatech-instruments.com, L: www.licor.com O: www.optisci.com, P: www.psi.cz, Q: www.qubitsystems.com, S: www.skyeinstruments.com, T: www.technologica.co.uk, W: www.walz.com

Fig. 10.1. Leaf optics. **(A)** Light paths through leaves. Light rays that pass through the epidermis and encounter the space between palisade cells will pass through the upper layer of the leaf. The irregular shapes of the spongy cells will maximise the chance that light will enter a cell. The spaces between the palisade cells will act as light guides because the large difference in index of refraction will cause light to reflect if it hits at a shallow angle (as depicted in the space between the left and centre palisade cell). Light that reflects off one spongy mesophyll cell is likely to hit another cell at a steep angle and so enter the cell. Chloroplast position can also affect light absorption. If all chloroplasts are along the anticlinal walls, the centre of a palisade cell can act as a light guide, just like the spaces between the palisade cells (shown in centre palisade cell). In addition to the light-guide effect, the fact that essentially all of the absorbing material (e.g., chlorophyll) is located in relatively small spaces (chloroplasts) gives rise to the sieve effect. **(B)** Epidermal lensing. Parallel light rays can be concentrated in some regions of the mesophyll cells because the shape of epidermal cells can cause them to act as lenses. The advantage of increasing the non-uniformity of light distribution through the leaf is not clear (Brodersen and Vogelmann, 2007).

can increase the effective pathlength of light as it travels through a leaf. The effect is strongest for leaves with low concentrations of chl. and for wavelengths that are not strongly absorbed. Thus, for blue or red light the effective pathlength may by roughly the same as the depth of the leaf, whereas for green light the effective pathlength may be five times greater than the depth of the leaf (Vogelmann, 1993). One result of this arrangement is that photosynthesis at different depths in the leaf may be driven by different light quality, with green light extending deeper into the leaf than red or blue (Terashima *et al.*, 2009).

Leaf epidermal cells are often curved, concentrating light within leaves (Fig. 10.1B). However, the advantage of concentrating light is not clear as high light levels in some regions would be balanced by low light elsewhere in the leaf. Another possibility is that the lens-shaped epidermal cells might enhance penetration of light coming from low angles into the leaf, but experimental evidence does not support this view (Brodersen and Vogelmann, 2007). These effects and others will affect the optical methods used to probe photosynthesis. For example, fluorescence from leaves will come from superficial layers of the leaf, whereas CO_2 fixation will be an average for all cell layers (Peguero-Pina *et al.*, 2009, and references therein; see also Chapter 16).

Absorption (Section 10.2) is classically measured as transmittance through the leaf, absorptance being defined as $\log_{10}$ (1/transmittance), whereas fluorescence (Section 10.3) and luminescence (Section 10.4) are measured from the illuminated side. However, taking advantage of light diffusion by leaf tissues, absorption can be detected as reflectance, i.e., light diffused backwards for experimental convenience (e.g., using multifurcated light guides). Because light is strongly absorbed and re-absorbed in the blue and red absorption bands of chl. and diffused, even more transmitted light can be detected in leaves mainly in the near IR (P700 measured at 820 nm) and green (505 nm, 518 nm, 535 nm, (PRI)) spectral regions. Photoacoustic spectroscopy also relies on light absorption, but detection by sound emission is quite different.

10.2.2. Absorption of P700 (820 nm): redox kinetics of the photosystem I centre

Photochemical charge separation, produced by a quantum of light energy absorbed in the pigment antenna and migrating randomly to a photochemical centre, consists of the transfer of an electron from the excited reaction centre chl. molecule to a primary acceptor:

$$Chl^* + A \rightarrow Chl^+ + A^- \quad \text{[Eqn. 10.1]}$$

The transient Chl^+ cation (more exactly a positive charge delocalised over a chl. dimer) is rapidly re-reduced by a primary electron donor. These redox changes result in an absorption change at 680 nm for PSII (P680) and 700 nm for PSI (P700), and increased absorption band around 820 nm that offers the experimental advantage of being outside the domain of strong chl. absorption.

The transiently oxidised primary donor $P680^+$ in PSII is re-reduced to P680 in the nanosecond range by the closely bound primary donor tyrosine Z, producing the high fluorescence state $P680Q_A^-$, the origin of variable fluorescence (Chl-F: see Section 10.3 and Chapter 2). In contrast, in PSI, $P700^+$ being re-reduced by a mobile plastocyanin molecule in the lumen has a longer lifetime, in the microsecond range, whereas the primary acceptor A_0^- is rapidly reoxidised by the secondary acceptor A_1, resulting in $P700^+A_0$. No variable fluorescence arises from PSI. The electron produced by the charge separation is transferred through protein-bound electron carriers to a ferredoxin (Fd) molecule soluble in the stroma. This gives rise to the common practice of using fluorescence to study PSII, but absorbance at 820 nm to study PSI.

Absorbance changes at 820 nm, outside bulk chl. absorption, primarily reflect P700 $\leftrightarrow$ $P700^+$ oxydoreduction, but this 820 nm secondary band of P700 overlaps with the broad absorption band of plastocyanin: reduced noise and improved specificity for P700 can be attained using a dual-wavelength differential detection system. The ratio of $P700^+/P700$ reflects the photochemical activity of PSI under steady state illumination (Schreiber *et al.*, 1989). In the dark, P700 is fully reduced. It can be fully oxidised into $P700^+$ by a light-saturating pulse when the donation of electrons by plastocyanin to $P700^+$ is the limiting step (donor-side limitation). However, the turnover of PSI can be also acceptor-side limited, for example, after inactivation of the Calvin cycle by a period of darkness, and the total amount of P700 has then to be assessed by a combination of far-red (FR) illumination and saturating pulse (Klughammer and Schreiber, 1994). Three relative values of P700 absorbance measured by an unactinic 820 nm light, first without, then with actinic light and finally during a saturating pulse, are needed to determine the $P700^+/P700$ ratio under a steady state illumination, in order to calculate the electron-transfer activity of PSI.

Cyclic electron flow around PSI was discovered *in vitro* by Arnon (1959), but its role *in vivo* has been questioned until optical methods (mainly P700 absorption and photoacoustic spectroscopy (Section 10.2.6), more indirectly basal Chl-F F_0 and luminescence increase in darkness) made its monitoring in leaves possible. This has provided evidence for the roles of CEFI in various stress situations: (1) enhancing the protective NPQ of fluorescence by increased proton pumping into the lumen (Section 10.3.4); and (2) provision of additional ATP for protein synthesis and repair (Rumeau *et al.*, 2007). The molecular basis of CEFI are still partly unknown (Chapters 2 and 3, review by Shikanai, 2007).

An important tool for detecting an induction of CEFI is provided by kinetics of $P700^+$ re-reduction in the dark immediately after an illumination. After white light, dark reduction of $P700^+$ is multiphasic, with fast phases down to the sub-ms range corresponding to electron flow from PSII (Govindachary *et al.*, 2007; Baker *et al.*, 2007). As FR light (>700 nm) excites PSI at a greater extent than PSII, it can be used to favour PSI activity. Kinetics of $P700^+$ reduction after FR light are slower than after white light, and consist of two phases with $t_{1/2}$ around ~2 s and ~12 s and a 2:3–1:3 ratio of amplitudes respectively (Cornic *et al.*, 2000; Fig. 10.2A,B). A faster phase (~100 ms) emerges after a time of dark adaptation corresponding to the inactivation of the Calvin cycle (Chow and Hope, 2004); it can be ascribed to a cyclic "overflow" of electrons from the over-reduced PSI acceptor pool to P700+ (Breyton *et al.*, 2006). An acceleration up to fivefold of $P700^+$ biphasic (2s and 12s) reduction occurs in conditions that trigger the pathway(s) for CEFI (pre-illumination by high-intensity white light, anoxia, warming, etc): electrons coming from reductants stored in the stroma are enabled to flow to the intersystem chain, then reduce $P700^+$ (Fig. 10.2). A faster $P700^+$ reduction in the dark indicates that cyclic pathway(s) are open, even though the actual electron flow is very weak in the absence of light. $P700^+$ reduction rate can also be recorded under continous illumination during 100 ms dark intervals (Golding and Johnson, 2003).

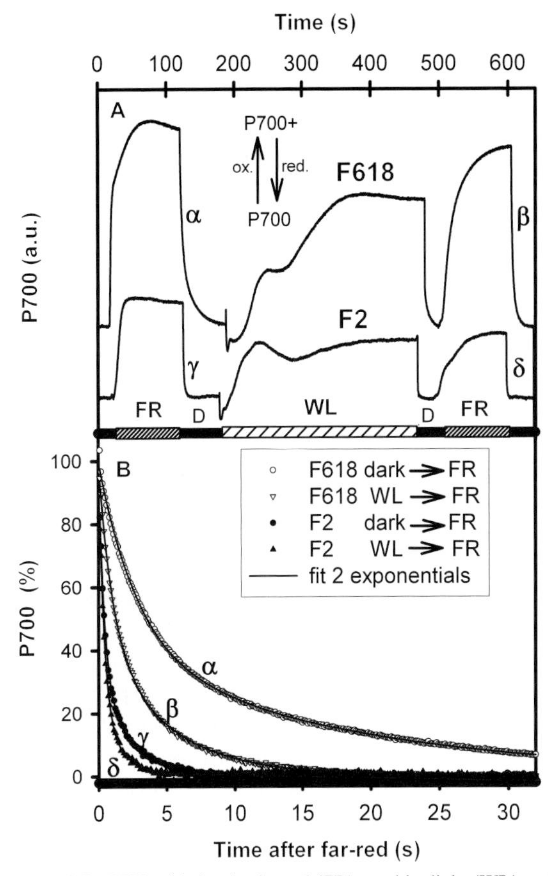

Fig. 10.2. P700 oxidation by far-red (FR) or white light (WL) and re-reduction in the dark (D), monitored by absorption at 820 nm. F618 and F2 maize inbred lines, sensitive and tolerant to chilling temperatures, resp. Plants transferred from illuminated growth chamber (17°C, 290 μmol m^{-2} s^{-1}) were maintained in darkness for 1 h. Measurements on attached leaves at an approximate 21°C room temperature. **(A)** P700 redox kinetics during a light sequence: dark → 100s FR → dark → 5min WL → dark →100s FR → dark. Kinetics α – δ are enlarged in B. **(B)** Reduction kinetics following 100 s FR, before and after 5 min white light, fitted by two exponentials. From Ducruet *et al.* (2005) with permission.

The pathways that support an ATP-generating CEFI in the light (review by Bukhov and Carpentier, 2004) may be shared by a chlororespiratory electron flow in the dark from stroma reductants to the PQ pool, subsequently reoxidised by oxygen through a plastidial terminal oxidase (PTOX) (review by Peltier and Cournac, 2002). This stroma-to-PQ part of the chlororespiratory pathway can be investigated by the same methods as CEFI.

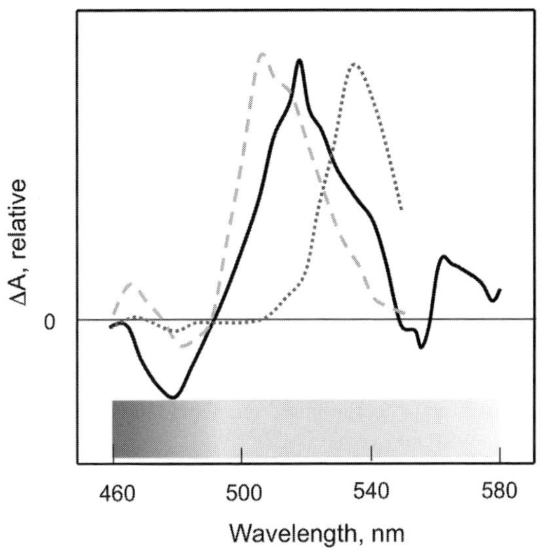

Fig. 10.3. Absorbance phenomena observable in green wavelengths. The solid line is the absorption spectrum for the electrochromic shift, the dashed line is the absorbance change caused by zeaxanthin formation and the dotted line is the absorption change caused by chloroplast swelling. Data are taken from Wise and Ort (1989), Bilger *et al.* (1989).

Activity of CEFI (and also CEFII) can also be deduced from the difference between PSI and PSII electron transfer, monitored by P700 absorption and Chl-F (Section 10.3.4), respectively, on the same leaf spot.

10.2.3. Green-light absorption methods.

Changes in optical density (absorbance [A] or extinction [E]) in the green region of the spectrum can give information about several different photosynthetic processes. Three signals that can be distinguished are the electrochromic shift (ECS) reflecting the electrical potential of the thylakoid membrane, light scattering, reflecting chloroplast swelling and chloroplast movement and the formation of zeaxanthin.

The ECS was discovered by Junge and Witt (1968). They found that absorption changes centred at 518 nm (ΔA518) were linearly related to the transthylakoid electrical potential (ΔΨ). These changes come about because the electric field changes the absorption spectrum of carotenoids. Although the effect is small, at the steeply declining slope of the absorption spectrum, around 518 nm, the change in optical density is easily measured (Fig. 10.3).

The ΔA518 measurement can be made with dark-adapted leaves to obtain information on the efficiency of PSII and the integrity of the thylakoid membrane. In dark-adapted leaves the ATP synthase is fully deactivated and the ΔΨ resulting from a very brief flash of light is dissipated through leakiness of the membrane. Wise and Ort (1989) used ΔA518 (referenced to 540 nm) to show that chilling reduced the efficiency of PSII without affecting membrane integrity. Ortiz-Lopez *et al.* (1991) used it to show that photophosphorylation is not impaired by water stress in field-grown sunflowers. Schrader *et al.* (2004) used ΔA518 measurements with heat-stressed leaves to show the opposite, that heating compromised membrane integrity before reducing PSII efficiency.

In light-adapted leaves the ΔA518 is measured during short dark intervals, and the technique has been called dark interval relaxation kinetics (DIRK) (Sacksteder and Kramer, 2000). The $DIRK_{ECS}$ reflects the steady state *pmf* and the rate of proton flux through the ATP synthase. When actinic light is turned off the *pmf* continues to drive protons through the ATP synthase until the entire *pmf* has been dissipated (strictly speaking, until the *pmf* equals the ΔG of ATP formation). The ΔΨ plus the ΔpH make up the *pmf*. Ion movements that occur in light-adapted leaves can convert some of the *pmf* from ΔΨ to ΔpH. After darkening a leaf, the *pmf* is dissipated in approximately 30 ms, while the ΔpH component relaxes over several seconds reversing the ECS signal (Baker *et al.*, 2007). Therefore, the rapid decline in the ECS signal following darkening reflects the total *pmf*, whereas the subsequent reverse ECS reflects the ΔpH. The difference is the ΔΨ (Baker *et al.*, 2007). The ECS signal is very rapid, and the measure of *pmf* can be made simply by measuring the absorbance at 518 nm. This technique has now been used to investigate both water stress and heat stress on the proton fluxes in PET (Kohzuma, *et al.*, 2009; Zhang and Sharkey, 2009; Zhang *et al.*, 2009c).

However, other changes in the green absorption spectrum are much larger (>0.1 OD[1] are common) than the ECS signal (commonly <0.01 OD), and so even though much slower, measuring the partitioning of *pmf* between ΔΨ and ΔpH requires measuring at at least two wavelengths, and it is even better to measure at three wavelengths (505 nm for zeaxanthin changes, 518 nm for ECS, and 535 nm for scattering changes; Fig. 10.3).

In intact leaves large absorption changes in the green (and other wavelength ranges) can be caused by chloroplast movement (Wada *et al.*, 2003). *In vivo* (in chloroplasts and leaves), the light-minus-dark-difference absorption spectra have three peaks in green light, at 505 nm (with a shoulder at 535 nm), 468 nm and 437 nm (Yamamoto *et al.*, 1972; Morales *et al.*, 1990). These three peaks reflect the conversion of violaxanthin to zeaxanthin (Yamamoto *et al.*, 1972). The broad shoulder at 535 nm is likely caused by changes in light scattering related to ΔpH-mediated chloroplast changes. Light scattering caused by chloroplast swelling and/or thylakoid stacking, may be related to the ΔpH component of the *pmf* (Crofts et *al.*, 1967; Deamer *et al.*, 1967; Heber, 1969). Changes in the aggregation state of antenna pigment-protein complexes mediated by an accumulation of de-epoxidised forms of the xanthophyll-cycle molecules (Gamon *et al.*, 1990; Morales *et al.*, 1990; Ruban *et al.*, 1993) can also influence the light-scattering effect seen at 535 nm. The details of the causes of chloroplast swelling and shrinking remains under debate (Ruban *et al.*, 2002).

These effects (plus other minor absorption changes in the green) vary in extent and speed of change. The ECS change generally can be measured over hundreds of milliseconds but, over seconds, the much larger changes of chloroplast movement and chloroplast swelling and shrinking can overwhelm the ECS signal. A two-wavelength system, with an additional third reference wavelength, is now commercially available (emitter unit DUAL-EP515 and the detector unit DUAL-DP515, Heinz Walz GmbH, Effeltrich, Germany), but three measuring wavelengths are needed to accurately separate the different effects that give rise to changes in green-light absorption of leaves.

10.2.4. Photosynthetic reflectance indexes

Chlorophyll fluorescence (see Section 10.3 and Chapter 16) is an intrinsic probe of photosynthetic efficiency. However, crop biomass is not only determined by leaf photosynthetic activity but by other factors as well, such as PAR absorbed and leaf area index (LAI). In many cases, in response to environmental factors, leaf growth stops before leaf photosynthesis is affected. Therefore, a convincing global indicator of the physiological state of plants should be able to monitor simultaneously both photosynthetic efficiency and

[1] OD (Optical density) or sometimes *E* (extinction) is the measure of absorptivity from the Beer-Lambert equation, $OD = \log_{10} (I_0/I_1)$, where I_0 is the light intensity coming into the system (e.g., the leaf) and I_1 is the light intensity coming out of the system. If a leaf absorbs 90% of the incoming light the OD is one, for 99% absorption the OD is two.

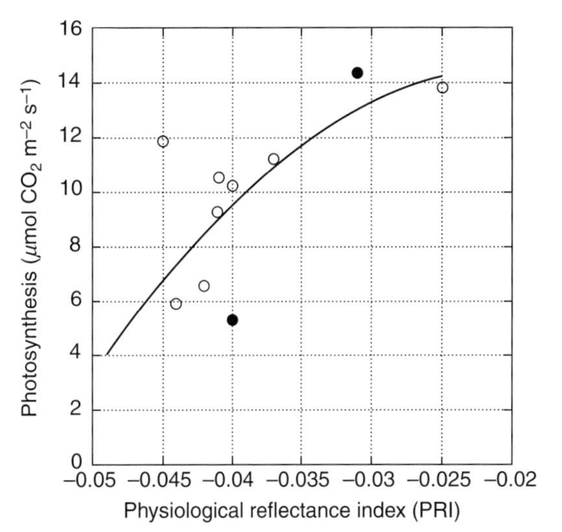

Fig. 10.4. Relationship between the physiological reflectance index (PRI) and photosynthesis. Kermes oak seedlings experiencing a progressive drought stress (open circles) and during two different days after re-irrigation (solid circles). From Peguero-Pina *et al.* (2008).

leaf growth. This can be achieved measuring hyper or multi-spectral reflectance in the visible area, where both normalised difference vegetation index (NDVI) and PRI can be obtained by using hyper-spectral or, at least, multi-spectral reflectance.

With the development of hyper-spectral reflectance in the 1990s, it has become possible to quantify simultaneously different leaf properties, including those related to photosynthesis (greenness, ie., chl. concentration, photosynthetic efficiency, etc.) (Gamon *et al.*, 1990; Peguero-Pina *et al.*, 2008). Several vegetation-monitoring techniques have been developed to quantify green vegetation and to map its spatial distribution, with the aim of estimating canopy photosynthesis and/or net primary productivity (NPP) (Grace *et al.*, 2007). Many vegetation studies were focused to detect only canopy greenness (e.g., NDVI=(R750−R705)/(R750+R705), where R750 and R705 represent reflectance at 750 and 705 nm, respectively). NDVI is not necessarily a good indicator of crop physiology because cessation of leaf growth and impaired photosynthetic rates often occur before chl. net destruction (see Chapter 15).

The so-called PRI=(R531−R570)/(R531+R570), where R531 and R570 are reflectance signals at 531 and 570 nm respectively, may be used to monitor photosynthetic dynamic changes (Fig. 10.4), with reflectance sensors designed to measure at certain distances (Evain *et al.*, 2004). Thus, PRI provides a quick and non-destructive assessment of leaf and canopy physiological properties (Evain *et al.*, 2004; Louis *et al.*, 2005) for a wide range of species (Gamon *et al.*, 1997). The physiological bases for PRI changes began to be investigated in the 1990s (see Section 10.2.3). Rapid vegetation reflectance changes around 531 nm (PRI changes) owing to sudden changes in incident light or upon imposition of a given stress condition could be sensed remotely and passively using a portable radiometer (Peguero-Pina *et al.*, 2008). PRI changes have been shown to correlate well with other photosynthetic activity parameters such as NPQ, the stationary fluorescence level F_s (Section 10.3.4) and de-epoxidation state of the xanthophyll cycle (Peguero-Pina *et al.*, 2008), and proved to be a good indicator of the increasing CO_2 assimilation of a pine forest during spring (Louis *et al.*, 2005) or of water stress level (Peguero-Pina *et al.*, 2008).

In contrast to the three green absorption signals described above (Section 10.2.4) that vary in the sub-second/second range and return specific information, the PRI is a global indicator of photosynthetic activity, varying within seconds/minutes. PRI measurements can be done by clipping the leaf or remotely.

Other simple reflectance indices have been proposed to estimate photosynthetic activity and to track plant stress. Dobrowski *et al.* (2005) reported that R690/R600 and R740/R800 correlated well with F_s and CO_2 assimilation rate in a water-stress experiment. Obviously, reflectance at 690 and 740 nm contains Chl-F, and therefore correlation with other photosynthetic parameters is not surprising. R690/R600 and R740/R800 changed only 4–13% with fourfold decreases in photosynthetic rates and twofold decreases in F_s induced by the water stress imposed (Dobrowski *et al.*, 2005). Tools for investigating dynamic photosynthetic changes from remote changes using reflectance and Chl-F are discussed in detail in Chapter 15.

Leaf clipping chl. meters that estimate chl. content rely on leaf transmittances at two red and near infrared wavelengths (Steele *et al.*, 2008).

10.2.5. NADPH/NADP: near-ultraviolet absorption and blue fluorescence

Leaves emit BGF when illuminated with UV light. There are two types of information to be obtained from BGF. The first relates to plant phenolic and flavonoid compounds bound to the cell walls or located in vacuoles of epidermal cells (Cerovic *et al.*, 1999). The second concerns the variation of the concentration and redox state of pyridine nucleotides (NADH and NADPH) and requires very sensitive fluorimeters (Cerovic *et al.*, 1993; Schreiber *et al.*, 1993; review by Cerovic *et al.*, 1999). These cofactors are good BGF fluorophores but, owing to their localisation in mesophyll cells (especially for NADPH in the chloroplasts), they are less accessible to UV excitation and their BGF is re-absorbed by photosynthetic pigments. These facts limit pyridine-nucleotide contribution to BGF in intact leaves to less than 10% in most cases (Cerovic *et al.*, 1993). NADPH-dependent BGF can be changed by applying actinic light to drive photosynthesis (Cerovic *et al.*, 1998).

10.2.6. Photoacoustic spectroscopy

Photoacoustic spectroscopy consists in detecting a sound emitted from a light-absorbing material illuminated with a modulated light, usually with a microphone. Light energy at absorbed wavelengths is converted to heat, so that thermal dilatation causes a small volume change that increases and decreases at the frequency of the modulated light oscillations, transmitted to air, thus producing a sound with an intensity proportional to light absorption.

This is called the photothermal effect, which is commonly detected in plant leaves at high modulation frequencies (typically above 100 Hz, reviewed by Malkin and Canaani, 1994). The signal is insensitive to light scattering and depends on absorbed photons, not on transmitted or reflected photons. The fraction of absorbed light energy that is photochemically converted does not produce heat and can be thereby calculated by subtracting the energy converted into heat from the total energy absorbed by antenna pigments. This latter quantity can be measured by adding a strong non-modulated light to the modulated measuring light that will saturate photochemistry and hence will induce maximal energy dissipation in the form of heat. Discrimination between the contributions of the two photosystems can be achieved by using a monochromatic wavelength for excitation. FR light excites mainly PSI, so that changes in the FR-induced photoacoustic signal reflect mainly variations in the activity of PSI. This is particularly useful to follow induction of CEFI (Herbert *et al.*, 1990; Ravenel *et al.*, 1994; Joët *et al.*, 2001) or transitions from state 1 to state 2 corresponding to migration of the peripheral antenna from PSII towards PSI (Veeranjaneyulu *et al.*, 1991; Section 10.3.1).

Photosynthetic material evolves oxygen under light, and the changes of gas volume associated with photosynthetic oxygen formation also produce a sound under modulated light at frequencies below 100 Hz, slow enough to separate the bursts of oxygen (the half-time of oxygen evolution after a flash is ~5 ms). This low-frequency photoacoustic signal specific to oxygenic photosynthesis is called a photobaric signal and reflects gross photosynthesis. The photobaric signal is particularly useful to measure the quantum yield of oxygen evolution or the so-called Emerson enhancement, which gives information on the PSI/PSII activity ratio (Canaani and Malkin, 1984).

10.3. CHLOROPHYLL FLUORESCENCE

Chlorophylls in solution emit red/FR fluorescence when excited by photosynthetically active radiations (see Chapter 2, Fig. 2.1). However, the unique property of Chl-F *in vivo* is to be variable, as its yield[2] depends on the functioning of the PSII centre and antenna from which it originates. At the onset of an illumination, the fluorescence yield is minimum (F_0) owing to 'photochemical quenching' (qP): energy quanta migrating through the antenna chl.s are trapped to separate a charge pair +/- by 'open' PSII centres, which become temporarily 'closed', i.e., unable to perform another charge separation. The proportion of closed centres increases during illumination and reaches 100% when light is strong enough, corresponding to the maximum fluorescence (F_m). The lifetime of excited chl.s, hence their probability to deactivate as fluorescence (Section 10.3.2), is inversely proportional to qP, which varies from one (F_0) to zero (F_m). The decrease of fluorescence yield following the maximum is ascribable not only to the reoxidation of the primary acceptor Q_A^- by downstream electron transfer, but also to NPQ, a downregulation mechanism that prevents damage as a result of an excess of excited chl.s in the antenna (Section 10.3.4).

The emergence of the modulated fluorescence technique (Schreiber *et al.*, 1986), which allowed the qP and NPQ[3] to be separated, has prompted a great deal of applications in plant physiological and ecophysiological research, as well

[2] Fluorescence yield = fluorescence photons / absorbed photons

[3] Historically, the non-photochemical quenching has been investigated in the early 1980s, using 'light-doubling' intensity instead of saturating pulses.

as in agricultural and environmental monitoring. Many reviews are available (e.g., Govindjee, 1995; book edited by Papageorgiou and Govindjee, 2004; Baker, 2008), including aspects of molecular mechanisms of variable Chl-F emission (Krause and Weis, 1991; Dau, 1994).

10.3.1. Fluorescence excitation and emission spectra

Chl. molecules of peripheral and core antennae are bound to proteins, forming complexes with characteristic absorption and fluorescence spectra (see Fig. 2.1). Absorption of one photon, from blue to red, by a chl. molecule produces one singlet excited Chl* storing a quantum of energy corresponding to that of a red photon; this quantum migrates by resonance as an exciton through the antenna, until it is trapped by a photochemical centre or converted into heat or into a red fluorescence photon. Low-temperature (liquid nitrogen: 77K) fluorescence emission spectra of dilute thylakoid suspensions exhibit a sharp band at 685 nm with a shoulder at 695 nm, both corresponding to PSII emission, and a broader band at 730/740 nm, almost absent at ambient temperatures, corresponding mainly to PSI. The two photosystems do not have the same absorption spectra (specifically: more chl. *b* for PSII, chl. *a*–protein complexes absorbing at long wavelengths for PSI), which may result in an imbalance of their photochemical activities, depending on incoming light spectrum. State transition is a corrective mechanism, by which the LHCII connected to the PSII antenna (state 1) becomes phosphorylated when the PQ pool is reduced and then migrates towards the PSI antenna (state 2); this leads to an increase of the 730 nm/685 nm ratio, i.e., PSI/PSII 77K fluorescence emission. Therefore low-temperature fluorescence spectroscopy with photoacoustic spectroscopy (Section 10.2.6) provides a suitable tool to study state transitions. This technique cannot be applied as such to whole leaves, because of the strong reabsorption of the 685/695 nm emission by chl.s, but it works well on a diluted leaf powder obtained by grinding a few mm^2 leaf spot with water and quartz particles in liquid nitrogen (Weis, 1985a).

At ambient temperature, the variable Chl-F (F_v) in leaves at 685 and 735 nm emanates from PSII (Krause and Weis, 1991), with a 30 to 50% PSI contribution to the F_0 fluorescence level in the FR region (Genty *et al.*, 1990; Pfündel, 1998; Agati *et al.*, 2000). PSI has no F_v; this is why all the parameters derived from F_v reflect PSII activity. The 685 nm fluorescence is much more re-absorbed by chl.s than 735 nm fluorescence, resulting in a negative curvilinear relation between their ratio and the leaf chl. content. Therefore, F690/F730 ratio is a remote-sensing indicator of leaf chl. content (Lichtenthaler *et al.*, 1990).

10.3.2. Lifetime and quantum yield

Picosecond and nanosecond events are the time domains in which chl. emits fluorescence when bound to proteins in the light-harvesting pigment-protein complexes (LHC) of the photosynthetic apparatus. At present, a major limitation for using Chl-F lifetime (τ) for investigation of the photosynthetic process is the complex and expensive electronic system needed for measurements.

A major interest of lifetime measurements in ecophysiology lies in the τ/ϕ_{PSII} relationship. The average lifetime, τ, of excited chl. in the antenna increases when photochemistry slows down, e.g., is inversely proportional to ϕ_{PSII}, the quantum yield of PSII (Section 10.3.4).

Most of the time-resolved Chl-F measurements reported to date have been carried out in plant materials in which fluorescence has reached a steady state level, including dark-adapted (F_0) and illuminated (F_m and F_s, in presence and absence of DCMU, respectively) samples (Holzwarth *et al.*, 1985; Moya *et al.*, 1986). Such reports have shown the existence of different (in most cases four) Chl-F components with lifetimes ranging from picoseconds to nanoseconds. The fastest lifetime component (τ1, 60 ps) is associated with PSI antenna fluorescence (Hodges and Moya, 1986; Agati *et al.*, 2000). τ2 (270 ps) and τ3 (550 ps) are components modified by PSII photochemistry and usually associated with PSII (Hodges and Moya, 1986). τ4 (1.30–3.30 ns) is a minor component associated with free chl. not connected to the functional photosynthetic apparatus.

Chl-F lifetime studies on leaves have been performed mainly under laboratory or greenhouse conditions (Cerovic *et al.*, 1999). τ-LIDAR devices use very short laser pulses (35 ps) and rapid detectors and electronics (streak camera or photomultiplier tube). This permits comparison of the kinetics of emitted Chl-F with the excitation pulse, and allows the average lifetime τ to be deduced. Also, special fluorimeters have been designed and mounted using time-correlated single-photon-counting detection to resolve the different leaf Chl-F lifetimes (Briantais *et al.*, 1996). One alternative to time-correlated single-photon-counting or direct-decay

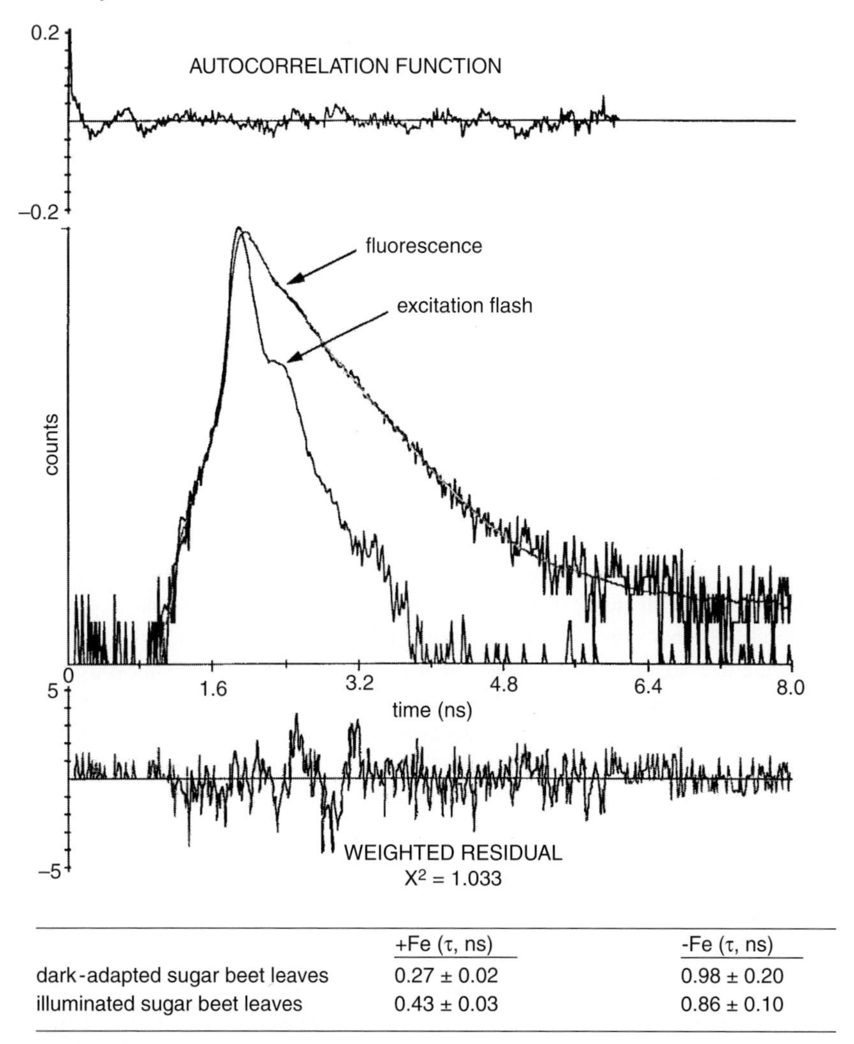

	+Fe (τ, ns)	-Fe (τ, ns)
dark-adapted sugar beet leaves	0.27 ± 0.02	0.98 ± 0.20
illuminated sugar beet leaves	0.43 ± 0.03	0.86 ± 0.10

Fig. 10.5. Typical chlorophyll-fluorescence (Chl-F) decay and excitation flash from which mean fluorescence lifetime (τ, ns) and their corresponding lifetime components are resolved, using a time-correlated, single-photon counting system (see Morales *et al.*, 2001 for details). Excitation and detection wavelengths were 635 nm and the whole red band of the spectrum, respectively. The figure also shows the mean Chl-F lifetime (τ, ns) in dark-adapted and illuminated sugar beet leaves in plants affected by iron deficiency (control; +Fe, healthy plants; –Fe, Fe-deficient plants). Data are mean±SD. From Morales *et al.* (1999, 2001).

analysis is multifrequency phase modulation fluorimetry, an emerging technique not yet applied in ecophysiology (Moise and Moya, 2004a,b).

There is not much information in the scientific literature on stress-induced changes in Chl-F lifetimes in intact leaves. Increases in Chl-F lifetimes in spruce and pine needles after exposition to high O_3 doses (Schneckenburger and Frenz, 1986), heat-stressed barley (Briantais *et al.*, 1996) and in Fe-deficient sugar beet (Morales *et al.*, 1999; Fig. 10.5) and decreases in illuminated, water-stressed leaves (Cerovic *et al.*, 1999) are reported in the literature.

10.3.3. The chlorophyll-*a* fluorescence induction (Kautsky effect)

Upon blue illumination of a dark-adapted leaf, Kautsky and Hirsch (1931) observed a red fluorescence emission of variable intensity, which first rose to a maximum then decayed more slowly. This constitutes the fluorescence induction (now called OJIP): at the onset of a constant illumination, the fluorescence yield starts increasing from a minimum level O (F_0: constant fluorescence) to reach, within hundreds of milliseconds, a maximum level P (F_p). P grows with excitation intensity until it saturates between 300 and 1000 μmol m^{-2} s^{-1} depending on light spectrum and plant material, to an absolute maximum level F_m (Fig. 10.6A). The fluorescence rise reflects the reduction of the primary acceptor Q_A (open centre) to Q_A^- (closed centre). The increase of the fluorescence yield under light corresponds to a build-up of the high fluorescence P680Q_A^- state of PSII centres, resulting from the fast re-reduction of $P680^+$ by the primary donor Tyr-Z (see Section 10.2.2, also Chapter 2). Closed PSII centre are unable to trap another incoming light quantum, which stays longer in the antenna with a higher probability to deactivate by emitting a fluorescence photon (Section 10.3.2). The difference $F_v=F_m-F_0$ is the variable fluorescence. A ratio $F_v/F_m=0.84$ has been measured in several angiosperm species (Björkman and Demmig, 1987). In practice, a value of ~0.8, obtained under a saturating light pulse after dark adaptation in leaves, where emission around 730 nm is prominent (cf Section 10.3.2), indicates a fully efficient PSII. The shape of the fluorescence rise from F_0 to F_m depends on the excitation intensity (Fig. 10.6A). It appears biphasic at moderate light intensities (<1000 μmol m^{-2} s^{-1}), and triphasic at high light intensities in intact organisms and leaves (>1000 μmol m^{-2} s^{-1}), with two intermediate levels originally called I_1 and I_2 (Neubauer and Schreiber, 1987; Schreiber and Neubauer, 1987), and subsequently J and I (Strasser *et al.*, 1995). The fluorescence intensity decreases after P, exhibiting several characteristic points S, M and T that can be observed especially at moderate light intensities (Papageorgiou *et al.*, 2007).

Freezing the sample in liquid nitrogen or infiltrating it with DCMU to block the electron transfer from Q_A to Q_B converts the bi- or triphasic kinetics into a single photochemical rise reflecting the reduction of the primary quinone acceptor Q_A (see Chapter 2). The complementary area (area of rectangle with F_0 and F_m opposite corners *minus* area under the rise curve) is a measure of the pool of electron acceptors of PSII: this area without DCMU is 10 to 15-fold that with DCMU, consistent with the PQ/Q_A ratio, i.e., the number of PQ molecules per PSII centre.

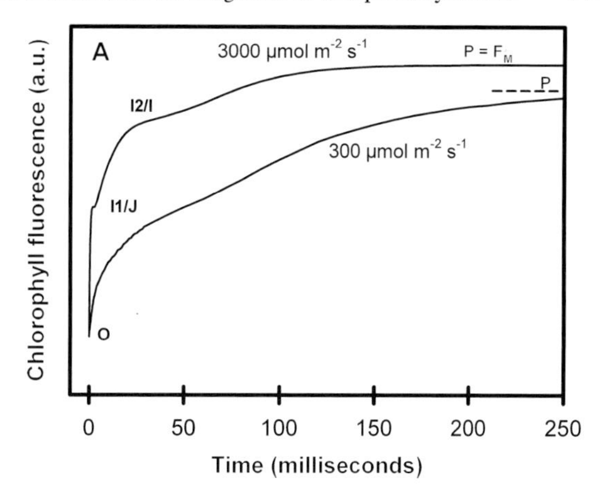

Fig. 10.6. Fluorescence induction and variation of its determining parameters. **(A)** Chlorophyll-fluorescence (Chl-F) induction transients recorded during 1 second on pea leaves, normalised to F_0. Excitation by 300 and 3000 μmoles m^{-2} s^{-1} continuous red light (650 nm peak wavelength). Fluorescence emission measured above 700 nm. The first reliably measured point of the fluorescence transient is at 20 μs, taken as F_0. Handy PEA instrument (Hansatech Instruments Ltd, UK). Maximum P for lower curve is at 800 ms. Credit S.Z. Toth. **(B)** Fluorescence induction kinetics. **(C to E)**: Schematic representations of key factors causing the fluorescence variation shown in panel **(B)**. The fluorescence trace in **(B)** was recorded with a dark-acclimated leaf. First, in the dark, minimum fluorescence (F_0) was established (conditions: PSII reaction centres closed, low trans-thylakoid pH gradient (ΔpH), and low zeaxanthin content), see **(C to E)**. Also in the dark-acclimated state, a strong but relatively short (0.6 s) saturating pulse (SP) was applied to close PSII reaction centres transiently and, thus, to increase fluorescence to the maximum level, F_m. Thereafter, 180 s of actinic light was given (*cf.* 'ACTINIC LIGHT' on top of figure): during illumination variations in the closure state of PSII reaction centres **(C)**, ΔpH **(D)** and zeaxanthin **(E)** occurred, which produced a typical Kautsky type fluorescence induction kinetics. At the end of illumination, the maximum fluorescence level (F_m') was elicited by an SP. Immediately thereafter, actinic light was switched off and the minimal fluorescence in the light-acclimated state (F_0') was assessed during post-pulse exposure to far red light: the FR predominantly drives PSI with the effect that electrons are withdrawn from the inter system electron chain and that PSII reaction centres become fully open while the factors involved in non-photochemical fluorescence quenching (ΔpH and zeaxanthin) are still active. Source of panels B to E: E. Pfündel, JUNIOR-PAM Handbook of operation, Walz; with permission.

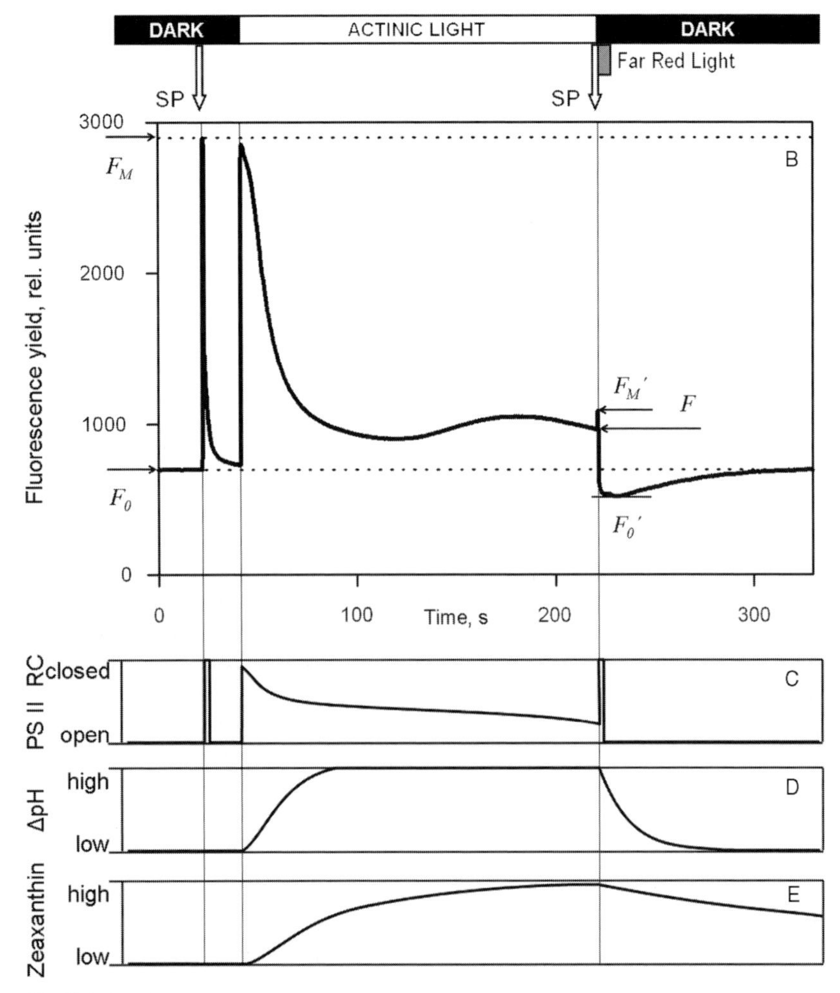

Fig 10.6. (*cont.*)

The I_1(J) level (also called step) unambiguously corresponds to that previously called I[4] (Delosme, 1967) and reflects the reoxidation of Q_A^- by the PQ pool. The half-time of the photochemical O → I_1 phase progressively shifts from tens of milliseconds to hundreds of microseconds when excitation intensity is increased. Consistently, I_1(J) is specifically increased by low concentrations of DCMU that binds to the Q_B site or by the triazine-resistance mutation Ser264Gly in the Q_B site (Hiraki *et al.*, 2003) that slows down the Q_A^- → Q_B transfer.

The I_2(I) level at about 30 ms appears under light intensities above 1000 μmol m^{-2} s^{-1} in dark-adapted leaves. The I_2(I) → P phase corresponds to the reduction of the pool of electron acceptors of PSI (Schansker *et al.*, 2005). I_2(I) vanishes in pre-illuminated leaves, because the Calvin cycle activated by light reoxidises the PSI acceptors and in disrupted chloroplasts as a result of the loss of Fd.

A new step K exhibiting a maximum at 300–400 μs (Guissé *et al.*, 1995; Tóth *et al.*, 2007) substitutes the I_1(J) step when the oxygen-evolving complex (OEC) is damaged, for example by heat treatment.

A proper tuning of excitation light intensity is important to get appropriate responses from OJIP kinetics. Intensities between 200 and 1000 μmol m^{-2} s^{-1} raise the P level close

[4] It should be noticed that the *I* level in the classical Chl-F literature corresponds to *J* not *I* in the newer OJIP nomenclature.

to F_m while keeping the I_1(J) level low; these conditions are optimal to detect an inhibition at the acceptor side of PSII and to monitor the penetration, migration and detoxication of DCMU-like herbicides in plants (Ducruet *et al.*, 1984; Habash *et al.*, 1985). Higher light intensities reveal the two I_1(J) and I_2(I) levels, with a shape depending on plant species (Tyystjärvi *et al.*, 1999).

The Q_A^- to Q_B and Q_A^- to Q_B^- electron-transfer steps can be directly investigated after a few microseconds saturating flash that forms Q_A^-, corresponding to a high fluorescence-state yield. The decay of fluorescence yield detected by a weak analytical excitation reflects the reoxidation of Q_A^-, with a main phase in the hundreds of microseconds range (Bowes and Croft, 1980).

10.3.4. Modulated fluorescence and quenching analysis

The technique of pulse amplitude modulation (PAM) emerged in the mid-1980s (Schreiber *et al.*, 1986) and made it possible to measure fluorescence yield by a modulated analytical excitation during a continuous actinic illumination. LED pulses of variable frequency[5] constitute the measuring (analytical) light: the small fluorescence surges corresponding to light pulses appear superimposed on background emission and provide a measure for the fluorescence yield. Saturating light pulses lasting a few hundreds of milliseconds, also called flashes, reveal two types of fluorescence quenching, qP and NPQ (Schreiber *et al.*, 1986). This technique allowed the actual PSII efficiency (ϕ_{PSII}) of illuminated leaves to be estimated (Genty *et al.*, 1989). Measurements are done with a portable or laboratory PAM-type fluorimeter and involve two successive measurements on the same leaf (Fig. 10.6B). First, fluorescence is measured after dark adaptation to obtain the F_0 and F_m fluorescence levels (corresponding to open Q_A and closed Q_A^- centres, respectively, as defined above). Later on, modulated Chl-F is measured under continuous illumination until the stationary state (F_s) is attained. During continuous illumination, saturating light pulses (400 to 1200 ms) are applied to establish F_m', lower than the dark-measured F_m owing to NPQ. After turning out the actinic light, a short interval of FR light can be used to obtain the F_0' fluorescence level. ϕ_{PSII} is calculated as the quotient $(F_m'-F)/F_m'$. ϕ_{PSII} depends on two factors: (1) the proportion of open, oxidised PSII *RC*s (estimated by the so-called photochemical quenching, qP, calculated as quotient $(F_m'-F)/F_v'$, where F_v' is $F_m'-F_0'$); and (2) the intrinsic PSII efficiency that is estimated by the F'_v/F'_m ratio. ϕ_{PSII} is therefore equal to qP x F'_v/F'_m (Genty *et al.*, 1989). Under high light intensities, such as those encountered in field conditions, the light pulse may not be strong enough to close all PSII *RC*s in order to reveal the real F_m' level. This problem can be solved by using successive non-saturating pulses (Markgraf and Berry, 1990; Earl and Ennahli, 2004) or a multiphase single pulse (Loriaux *et al.*, 2006) with decreasing intensities, then extrapolating these apparent F_m' values to infinite light intensity to determine F_m' exactly.

The access to ϕ_{PSII} allows the electron-transfer rate through PSII ($\mathcal{J}_f$) to be calculated, a major application of modulated Chl-F (Genty *et al.*, 1989). The $\mathcal{J}_f$ is defined as $\varphi_{PSII} \times PPFD \times \alpha_L \times \varepsilon_{PSII}$ where PPFD is the PAR photon-flux density; α_L is the fraction of incoming PAR that is absorbed (not transmitted nor reflected); and ε_{PSII} is the fraction of these quanta absorbed by PSII, $1-\varepsilon_{PSII}$ by PSI. In a first approximation, it can be considered that $\alpha_L=0.85$ and $\varepsilon_{PSII}=0.5$. The product $\alpha_L \times \varepsilon_{PSII}$ can be more accurately determined (Valentini *et al.*, 1995a) from the relationship between ϕ_{PSII} and ϕ_{CO2} obtained by varying light intensity in an atmosphere containing 1% O_2 in order to eliminate photorespiration. The possibility to measure $\mathcal{J}$ through PSII by fluorescence allowed, for example, to demonstrate that $\mathcal{J}_f$ was much less reduced by moderate drought than the assimilation of CO_2 inhibited by stomatal closure, the electrons being routed towards photorespiration (Cornic and Briantais, 1991).

NPQ, the rate constant of thermal deactivation, can also be estimated from modulated Chl-F data $((F_m/F_m')-1)$. A major component of NPQ is the electrochemical quenching qE that depends both on an acidic lumen (ΔpH) and of conversion of violaxanthin to zeaxanthin (Fig. 10.6B); it relaxes within a few minutes in the dark. qT as a result of state transitions (Section 10.3.1) and qI as a result of photoinhibition have longer dark relaxation times. NPQ is reduced by uncoupling or stress-induced membrane leakiness; it is increased when a strong proton gradient builds up upon triggering of CEFI (Makino

[5] Analytical or unactinic excitation = measuring light with an intensity low enough not to influence the observed phenomenon. An LED pulsed at a low frequency (e.g., 9 kHz), not actinic in darkness owing to the very low average light intensity, is used to measure F_0, but it can be set at a higher frequency (e.g., 100 kHz) under illumination to improve the signal-to-noise ratio.

et al., 2002). Applications of modulated fluorescence to leaf photosynthesis have led to thousands of published works, especially concerning stress physiology, which cannot be briefly summarised here (see book edited by Papageorgiou and Govindjee, 2004; reviews by Baker, 2008; and those cited in Chapters 16 and 21). Improvements and possible pitfalls in particular conditions are still being discussed (e.g., see Buckley and Farquhar, 2004; Peguero-Pina *et al.*, 2009).

10.3.5. Thermofluorescence: F_0, F_m during warming or freezing

Under a very weak exciting light, the rate of closure of PSII traps by incoming photons is slower than their reopening by downstream electron transfer or charge recombination, so that this fluorescence level stays minimum (F_0)[6]. At temperatures above 40°C however, F_0 progressively increases. This can be optimally resolved during a slow warming (e.g., 1°C/min) in the form of F_0 versus temperature (F_0/T) curves (Schreiber and Berry, 1977; Berry and Björkmann, 1980; Briantais *et al.*, 1996) that exhibit a maximum emission at about 55–60°C followed by a shoulder, or sometimes a second maximum near 70°C. The intersection of the tangent to the fluorescence rise with the horizontal initial F_0 level defines a characteristic temperature Tc, between 40 and 50°C, correlated with the heat sensitivity of leaves measured afterwards by leaf necrosis (Bilger *et al.*, 1984). The fluorescence F_m level starts decreasing with temperature above 30°C, so that the F_m/T curve meets the F_0/T curve at its maximum near 60°C.

F_0/T curves are a tool to characterise the thermal sensitivity of PSII in leaves, which depends on the thermal adaptation of plant species and on previous heat acclimation (Bilger *et al.*, 1984; Terzaghi *et al.*, 1989). They are shifted towards higher temperature by illumination or drought (Havaux, 1992; Rekika *et al.*, 1997; review by Ducruet *et al.*, 2007). An acidic pH in the lumen or non-electrolytes solutes such as sugars protect PSII from heat damage (Williams *et al.*, 1992).

A temperature-dependent increase of F_0 and decrease of F_m is also observed in leaves during slow freezing (Sundbom *et al.*, 1982; Pospisil *et al.*, 1998; Neuner and Pramsohler, 2006), corresponding to the nucleation temperature at which water in a supercooled state almost instantly crystallises with a temperature burst detectable by IR thermography.

10.4. CHLOROPHYLL LUMINESCENCE (DELAYED FLUORESCENCE)

In leaves previously illuminated, a faint glow of red light decaying within minutes can be recorded in darkness with sensitive detectors. Its emission spectrum is similar to that of chl. 'prompt' fluorescence emitted during light excitation. This dark emission is indeed a 'delayed' fluorescence, called luminescence for clarity, and also originates from PSII. Luminescence results from the recreation, with a low yield, of an excited singlet chl. by the recombination of a +/- charge pair previously separated by the photochemical conversion of a light quantum. The positive charges are stabilised on the S states of the OEC ($S_{2(+)}$, $S_{3(+)}$, storing two or three positive charges on a manganese complex) of PSII while an electron is usually stored on the secondary quinone acceptor Q_B^- (Chapter 2). Each type of charge pair produces a characteristic luminescence decay phase or a thermoluminescence (TL) band. While prompt fluorescence depends on the state of PSII antenna and centre, delayed fluorescence or luminescence depends on the charge distribution on PSII electron carriers beyond the photochemical tarp. This distribution can be controlled by using 'single turnover' flashes (STOF) powerful and short enough to induce only one charge separation in every centre[7].

10.4.1. Luminescence-decay kinetics

Luminescence-decay kinetics can be recorded easily after illumination of a leaf piece maintained at a constant temperature in a dark room or box, simply by opening a shutter that protects the detector (usually a photomultiplier) during illumination. Temperature control is essential because

---

[6] F_0 can be measured either by a modulated fluorimeter operated at low frequency or by coupling an ultra-low blue exciting light to a sensitive fluorescence detection (photomutiplier) behind a red filter.

[7] This can be achieved using xenon bulbs that deliver a few microsecond flashes. They emit, however, a weak long-lasting glow or tail that interfers with low-level luminescence measurements in the millisecond range, making a flash shutter necessary. LEDs have become powerful enough to produce saturating square light pulses shorter than 10 μs without tail. The STOF flashes are different from the saturating flashes used for modulated fluorescence, that last for hundreds of milliseconds. For controversial reasons, F_m cannot be reached upon one STOF.

Table 10.2. *Origin of thermoluminescence bands. T_M values correspond to a 0.5°C/s thermoluminescence heating rate.*

Name	T_m range	Origin	PSII	Comments
A_T	–10 to –20°C	TyrZ + Q_A^-	+	Damage to Mn oxygen-evolving complex (TyrZ is the functional donor to PSII center)
A	~ –15°C	$S_3Q_A^-$?	+	—
Q	+2 to 10°C	$S_2Q_A^-$	+	Damage to secondary Q_B quinonic acceptor or inhibition by DCMU-like herbicides
B	30 to 38°C	$S_{2/3}Q_B^-$	+	Lumen pH neutral
B2	28 to 32°C	$S_2Q_B^-$	+	Lumen pH slightly acidic
B1	20 to 30°C	$S_3Q_B^-$	+	" "
AG	+45°C (→+35°C)	$S_2/S_3Q_B + e^-$	(+)	e^- from stroma, in intact chloroplasts or cells
C	+52/55°C	TyrD + Q_A^-	+	Minor band, increased by DCMU or damage (TyrD is the non-functional donor to PSII centre)
HTL1	60 to 85°C	?	—	Different bands of unknown origin
HTL2	120 to 40°C	Lipid peroxides	—	Thermolysis: -C-O-O- → *C=O + Chl → *Chl

of the high-temperature dependence of luminescence emission. Recording decay phases faster than 0.1 s requires special technical precautions (Section 10.4.3).

Luminescence decays can be theoretically decomposed into exponentials, corresponding to TL bands (Section 10.4.2). Practically this proves unreliable in leaves owing to the occurrence of non-exponential decay phases. Therefore TL is often preferred for its better resolving capacity, even though DL as a non-destructive technique may be needed for example when periodical measurements have to be done on the same leaf or for non-contact sensing in darkness.

10.4.2. Thermoluminescence

The +/– charge pairs produced by illumination are stabilised in energetic wells, with 'activation-energy' barriers limiting the back reaction, which makes photosynthesis possible. Increasing temperature, i.e., vibrational energy, enhances the recombination over these barriers according to the Arrhenius equation. TL is a technique that consists of illuminating the sample at a temperature cold enough to practically block the recombination of the investigated charge pairs and to reveal them by a progressive warming (0.1 to 0.6°C/s) as successive TL bands (see Table 10.2) peaking at increasing temperatures (reviews by Vass and Govindjee, 1996; Ducruet, 2003; Gilbert *et al.*, 2004; Tyystjärvi and Vass, 2004). These bands often overlap and can be separated by fitting elementary components governed by three parameters (activation energy E_A; pre-exponential factor; area, from which the peak temperature Tm can be derived) calculated by analytical or numerical methods (Vass *et al.*, 1981; Ducruet, 2003).

In a healthy photosynthetic organism, PSII charge pairs +/- are stabilised as $S_2Q_B^-$ and $S_3Q_B^-$ states able to recombine with luminescence emission, corresponding to one or two B bands observed in the 20 to 38°C temperature range (Rutherford *et al.*, 1982, 1984; see Chapter 2). The intensity of the B band oscillates with a period 4 according to the number of flashes, reflecting the four-step turnover of S state of the OEC. The B band often splits into a B_1 band (~20–25°C) corresponding to $S_3Q_B^-$ and a B_2 band (~30°C) to $S_2Q_B^-$. This split can be explained by a greater proton uptake (Lavergne and Junge, 1992) in $S_3Q_B^-$ than in $S_2Q_B^-$ recombination steps. The temperature downshift of the $S_3Q_B^-$ band induced by two or three flashes is a direct probe of the dark-stable pH of lumen in unfrozen leaves.

Destruction of PSII under stress conditions results in a decrease of the B-band(s) intensity along with the F_v/F_m fluorescence ratio, and may be accompanied by the enhancement of minor bands. The Q band at ~5°C, produced by the inhibitor DCMU that blocks the Q_B pocket, corresponds to $S_2Q_A^-$ and indicates damage to the PSII acceptor side. Bands A/At (–20°C) reflects damage to PSII OEC.

A major requirement for using (thermo)luminescence for *in vivo* photosynthesis studies is to prevent freezing below the nucleation temperature (~–5°C), under which ice crystals break thylakoid membranes. This dissipates the proton and ionic gradients and in some species grossly distorts the signal (Homann, 1999; Janda *et al.*, 2004).

An 'afterglow' emission (AG) can be observed in unfrozen leaves, algal cells or intact chloroplasts, appearing as a delayed luminescence burst superimposed to luminescence decays (Bertsch and Azzi, 1965), or as a sharp TL band at about 45°C (Fig. 10.7). It is generally induced by FR light, or sometimes by flashes or white light pulses depending on plant species and growth conditions. It originates from the PSII 'silent'centres S_2Q_B and S_3Q_B, unable to emit luminescence; these centres become enabled to undergo +/- charge recombination and hence to emit luminescence when an electron is back-transferred to Q_B (Sundblad *et al.*, 1988). Electrons follow the cyclic/chlororespiratory pathway(s) connecting the pool of reductive compounds stored in the stroma to electron carriers of the intersystem chain. Temperatures above 30°C stimulate this pathway (Weis, 1985b; Havaux, 1996). Consistently the afterglow emission (AG) is enhanced by warming and can be optimally resolved as a sharp TL band peaking at 45°C with a 0.5°C/s warming rate (Miranda and Ducruet, 1995). Cyclic/chlororespiratory pathways can be induced at ambient temperatures by stress treatments (e.g., high light, anoxia, cold, drought). For example, anoxia induces CEFI (Joët *et al.*, 2002) and thereby allows the electron transfer from stroma to Q_B to occur at lower temperatures, resulting in a temperature downshift of the AG band that ultimately fuses with the B band (Fig. 10.7A). This downshift of AG is correlated to a faster re-reduction of $P700^+$ after a FR-light excitation (Ducruet *et al.*, 2005; Apostol *et al.*, 2006) that also characterises an opening of cyclic pathways in the dark (Section 10.2.2).

The variations of the NADPH $^+$ ATP assimilatory potential, in equilibrium with the pool of ATP in the stroma (Gerst *et al.*, 1994), determine those of AG intensity (Krieger *et al.*, 1998). Its increase reflects a slower CO_2 fixation in case of, for example, low CO_2 in algae (Palmqvist *et al.*, 1986), induction of CAM metabolism (Krieger *et al.*, 1998), viral infection (Sajnani *et al.*, 2007) and drought (Fig. 10.7B).

Chl-F and (thermo)luminescence both originate from PSII, but provide complementary information on a wider range of the photosynthetic metabolism.

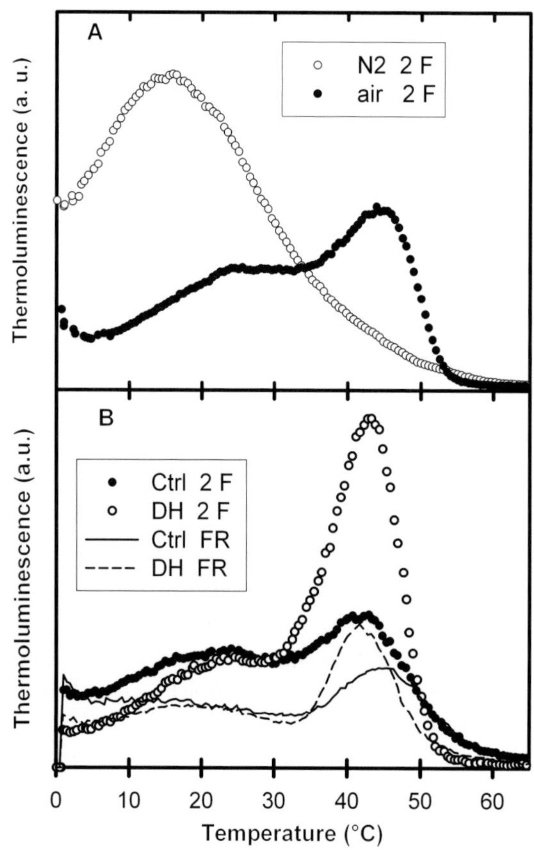

Fig. 10.7. Induction of cyclic pathway(s) by anoxia and increase of NADPH+ATP potential by drought, monitored by the afterglow thermoluminescence band. **(A)** Effect of anoxia (2 min nitrogen gas) on pea. Paired leaflets, 1-hour dark adapted. Control in air shows a B band as a shoulder at 25°C and an AG band at 46°C. Anoxia causes the AG to fuse with the B band, with a downshift of B caused by a lower lumen pH. **(B)** Effect of drought on barley leaves. Plants watered (ctrl) or not (drought, DH) after 6 days (8-hours dark adaptation). The AG band at ~40°C induced by two flashes (2F) is increased by drought, reflecting a higher NADPH+ATP potential that is not dissipated by the Calvin cycle owing to stomatal closure. The AG band induced both by two flashes or 10 s FR is downshifted by ~5°C, indicating a slight induction of cyclic pathway(s) (corresponding to faster re-reduction of P700 after FR, not shown). The S_2/S_3 luminescence emitting states of OEC are 100% of PSII centres after two flashes, 50% after FR. (V. Peeva *et al.*, unpublished).

10.4.3. Fast luminescence decay and modulated luminescence (DLE)

Fast luminescence-decay phases exist in the millisecond and submillisecond time domain. Historically, a so-called delayed light emission was measured under a steady state periodical excitation using a rotating-disc phosphoroscope. The availability of light emitting diodes (LED) brought more flexibility and allowed fluorescence OJIP rise and luminescence decay to be recorded on an LED pulse. Several luminescence-decay phases in the millisecond and microsecond time range (Zankel, 1973) offer insight into the early steps of oxygen evolution. A phase recorded in the 0.5- to 5-ms time interval (Boussac *et al.*, 1985) has proved useful to follow the building up of the ΔpH under light. It occurs after three single turnover flashes that produce the putative transient state S_4, more exactly Z + S_3, leading to O_2 evolution within ~5 ms after the last flash:

$$S_3 + h\nu \leftrightarrow S_3 + Z^+ + n\,H^+ \rightarrow S_4 \rightarrow S_0 + O_2 \quad \text{[Eqn. 10.2]}$$

with n=1.5 to 2 protons taken up during the reverse luminescence-emitting step (Lavergne and Junge, 1992). This makes the 5-ms luminescence decay from S_3Z^+ a good sensor of lumen pH. It is possible to record stroboscopically both a fluorescence rise during the illumination periods and a luminescence rise during the dark periods (Malkin *et al.*, 1994), the latter reflecting the light-driven 'energisation' of thylakoids (e.g., building up of the *pmf*, see Section 10.2.3). Steady state modulated luminescence (DLE) recorded at increasing temperatures shows a maximum near 15°C in chilling-sensitive species only, and another one around 40°C: both indicate an increased membrane leakiness as a result of fluid ↔ gel phase transitions of saturated membrane lipids and temperature-induced membrane disorder, respectively (Havaux and Lannoye, 1983; Fork *et al.*, 1985; Fork and Murata, 1990; review by Ducruet *et al.*, 2007). DLE sensing of the transmembrane electrical field ΔΨ and proton gradient ΔpH is related to that performed by green-light-absorption methods (Section 10.2.3) and to fluorescence NPQ (Section 10.3.4).

10.4.4. High-temperature thermoluminescence and oxidative stress

Heat damage to PSII starts at ~40°C, with full destruction above 60°C (Section 10.3.5). Nevertheless, TL bands can be observed at higher temperature without prior illumination as a heat-induced chemiluminescence (Vavilin *et al.*, 1991; Vavilin and Ducruet, 1998; Havaux and Niyogi, 1999; review by Havaux, 2003). This emission results from a radiative thermolysis of peroxide-containing -O-O-C- bonds that generate excited carbonyls, which subsequently excites red Chl-F in leaves or blue or green direct-light emission (Abeles, 1986). This high-temperature thermoluminescence (HTL) exhibits a main band close to 130°C (HTL2), its area being correlated with the amount of peroxides measured by chemical techniques (reaction with thio-barbituric acid, ethane evolution). The leaf sample should be allowed to dry during warming to prevent non-radiative hydrolysis that increases with temperature and competes with the luminescence-emitting radiolysis. This chemiluminescence stimulated by warming also occurs at ambient temperature as an ultra-weak luminescence, detectable by highly sensitive, albeit costly, photomultipliers or cameras (Abeles, 1984; Flor-Henry *et al.*, 2004; Havaux *et al.*, 2006). The HTL and ultra-weak luminescence can be used for imaging oxidative stress in leaves (Section 10.6).

10.5. EXPERIMENTAL ASPECTS OF OPTICAL MEASUREMENTS ON LEAVES

10.5.1. Summary of other optical methods not related to photosynthesis

Chl. can be used as an internal probe to estimate the epidermal transmittance from the ratio of Chl-F yields induced by UV and blue-, green- or red-excitation lights (Bilger *et al.*, 1997; Ounis *et al.*, 2001a; Pfündel *et al.*, 2007) either in leaves or fruits (Cerovic *et al.*, 2002; Kolb *et al.*, 2003; Agati *et al.*, 2007).

Leaves fluoresce not only red light from chl., but also blue-green light from polyphenolic compounds (Lichtenthaler and Miehé, 1997; Cerovic *et al.*, 1999) when illuminated with UV light or blue light (Sections 10.2.5 and 10.6).

Assessing optically the content of various polyphenols in epidermis is in growing use in agriculture, e.g., for optimising nitrogen fertilisation of crops, early detection of pathogen infection (see Chapter 22) and detection of fruit ripeness.

Infrared thermometry allows to measure leaf temperature remotely and monitor its increase by closure of stomata and decreased transpiration. It can be combined with photosynthetic optical measurements, such as Chl-F (Chaerle *et al.*, 2006, Section 10.6) or visible and NIR spectroscopy, especially in remote sensing.

10.5.2. Instruments: optronics for optical measurements

Applications of optical methods outside biophysical laboratories have been hindered by the lack of user-friendly, (trans)portable and affordable instruments. Tremendous advances in optronics during the last 25 years and emergence of a wider market for agricultural and environmental monitoring have both favoured the development of commercial instruments.

LED and laser diodes that can be easily modulated now cover the whole PAR spectrum, from IR to UV, and have become powerful enough to replace lamps and xenon flashes. In addition to now widely used solid-state silicon photodiodes (PIN) more sensitive avalanche photodiodes (APD) and pocket-size photomultiplier modules with integrated power supplies and amplifiers are available for weak signals (luminescence, phytoplankton fluorescence, remote sensing). The temperature control of samples has been greatly simplified and made flexible by solid-state 'Peltier' thermoelectric elements. Computer interfaces allowing both signal recording and instrument driving with real-time signal analysis have played a major role in the diffusion of optical techniques outside laboratories.

In a similar way to medical indicators (stethoscope, thermometer, etc), several optical indicators are necessary to diagnose correctly the physiological status of vegetation. Complementary measurements on the same leaf spot, using mutifurcated fiber optics or inserting various LEDs and detectors in a compact measuring head, rules out the variability between leaf samples especially in wild species.

Beyond ecophysiological research, optical methods are expanding in environmental survey, precision farming and genetic phenotyping: for these applications proximal non-contact sensing is often preferred to more time-consuming and somewhat tricky leaf clipping.

10.6. IMAGING TECHNIQUES FOR PLANT-STRESS DETECTION

Measurement of Chl-F from leaves reveals basic information about the component processes in the light reactions of photosynthesis, and non-imaging chl. *a* fluorometry has been recognised as a non-invasive, highly sensitive probe of the responses of photosynthetic processes to stress (see the book *Chl a Fluorescence: A Signature of Photosynthesis*, Papageorgiou and Govindjee (eds), 2004). However, single-point measurements are not representative for a whole sample, and only imaging provides the information about photosynthetic and metabolic gradients (induced by stress or plant development) of different fluorescence signals over a sample (Lichtenthaler *et al.*, 2005). Visualisation of the light signals emitted by a plant can track the movement of herbicides (Fig. 10.8) or the spreading of a pathogen through its host (Chaerle *et al.*, 2006; Pérez-Bueno *et al.*, 2006; Pineda *et al.*, 2008a,b). The limitations of conventional spectrofluorometers, recording Chl-F and BGF emission spectra or gas-exchange devices measuring stomatal conductance and CO_2 assimilation in very small areas, could be overcome by multispectral fluorescence or thermal imaging. In the last two decades, new *in vivo* imaging approaches (Jones and Morison, 2007) from the subcellular to the ecophysiological level have been developed to study different plant processes and their modulation by the environment. The future is a multisensor approach where a combination of at least reflectance, fluorescence and thermal imaging will facilitate basic research, field screening and stress diagnosis in the frame of a precision agriculture with a lower environmental impact.

10.6.1. Chlorophyll-fluorescence imaging

Techniques allowing the visualisation of the spatial distribution of fluorescence have been implemented both in laboratories as well as in field experiments. Fluorescence can be captured from individual chloroplast, single cells, microalgae, plant organs, whole plants or canopies, and can be applied from the microscopic to the remote-sensing scale (Daley, 1995; Nilsson, 1995; Baker *et al.*, 2001; Chaerle and Van Der Straeten, 2001; Nedbal and Whitmarsh, 2004; Aldea *et al.*, 2006a; Baker, 2008). FI is becoming a promising tool to be used in screening programmes for crop-yield management (Baker and Rosenqvist, 2004). It is applied for mutant screening, analysis of fruit quality and post-harvest damage, study of secondary metabolism, visualisation of source-sink relationships and monitoring of abiotic stress factors including herbicides, wounding, oxidative stress, ozone, nutrient deficiencies and others. The Chl-FI tool is especially valuable in the case of biotic stress where non-uniform alterations of photosynthetic activity and symptom development is expected (Section 22.8 of this book is devoted to imaging methods for biotic stress detection).

The technology and methodology to image Chl-F and obtain the spatial pattern of a range of fluorescence parameters are reviewed in Nedbal and Whitmarsh (2004) and Oxborough (2004). Commercial Chl-F instruments are available (Table 10.1). A number of Chl-FI instruments have been developed by different research groups for research at the microscopic level (Oxborough and Baker, 1997; Osmond *et al.*, 1999; Holub *et al.*, 2000; Küpper *et al.*, 2000; Rolfe and Scholes, 2002; reviewed in Oxborough, 2004) or at low resolution (Omasa *et al.*, 1987; Daley *et al.*, 1989; Fenton and Crofts, 1990; Genty and Meyer, 1995; Siebke and Weiss, 1995; Scholes and Rolfe, 1996; Osmond *et al.*, 1998; Nedbal *et al.*, 2000b; Chaerle and van der Straeten, 2001; Zangerl *et al.*, 2002; Valcke, 2003; Oxborough, 2004).

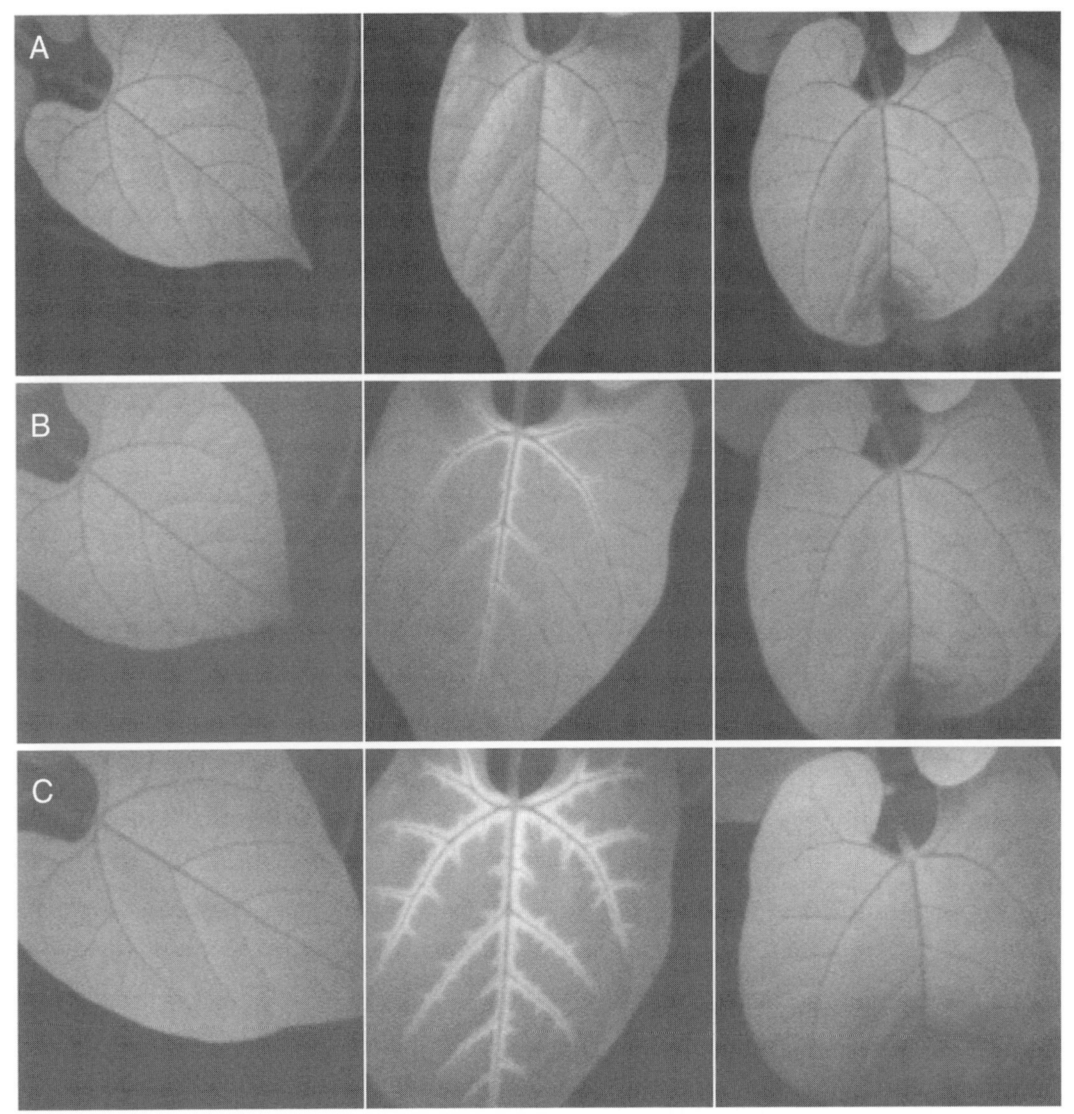

Fig. 10.8. Visualisation of linuron degradation by the bacterial strain *Variovorax paradoxus* WDL1 using multispectral imaging. Images of chlorophyll fluorescence (Chl-F) under excitation light with a photosynthetic photon-flux density of 250 μmol m^{-2} s^{-1} from primary leaves of bean plants. First column, control; second column, leaf treated with the DCMU-like herbicide linuron; last column, linuron+*V. paradoxus*-treated leaf. Leaf sizes increase gradually with time owing to growth expansion. (A) Primary leaves 10 days after planting; (B) images captured after 10 h; (C) images captured 16 h after (A). Modified with permission from Chaerle *et al.* (2003).

In the field of underwater imagery, submersible equipment has investigated the spatial distribution of phytoplankton as well as coral bleaching after recording multispectral images of induced fluorescence (Zawada, 2003).

Chl-FI instruments usually are designed to obtain the two-dimensional maps of standard fluorescence values: F_0, F_0', F_m, F_m' and F_s. To estimate rates and efficiency of the photosynthetic process, F_v/F_m, Rfd, ϕ_{PSII} and NPQ are calculated for each pixel of the image and presented in a black and white or a false-colour scale (Nedbal and Whitmarsh, 2004; Oxborough, 2004). Nedbal and Březina (2002) have also imaged the harmonically forced oscillations in fluorescence emission to map regulation in light capture. The choice of the adequate fluorescence parameter to follow a physiological process or to evaluate stress-induced damage depends on the experimental and growth conditions and the stage of plant development, as well as the stress factor analysed (Nedbal and Whitmarsh, 2004). Experimental algorithms that identify the combination of fluorescence parameters provide the highest contrast between different plant samples or segments (Soukupová *et al.*, 2003; Berger *et al.*, 2007; Pineda *et al.*, 2008a,b; see Section 22.8). Some basic statistic tools, such as fluorescence profiles in a given sample and histograms (plotting the number of pixels with selected fluorescence values; Lichtenthaler and Babani, 2000), could also be used to present image data.

10.6.2. Multicolour fluorescence imaging

Fluorescence images taken at different spectral bands (multispectral or multicolour fluorescence imaging, MCFI) increase the number of plant processes than can be simultaneously analysed, and provide a promising way to detect specific *signatures* for a particular stress (Lenk *et al.*, 2007). The excitation of leaves with long-wavelength UV radiation (320–400 nm) results in four characteristic fluorescence bands with peaks near to 440 nm (blue; F440), 520 nm (green; F520), 690 nm (red; F690) and 740 nm (FR; F740). BGF is not evenly distributed over the leaf area, but primarily emanates from the main and side veins. In contrast, the red and FR Chl-F is predominantly emitted from the vein-free leaf regions (Lichtenthaler and Miehé, 1997). The use of the fluorescence ratios F440/F520, F440/F690, F440/F740 and F690/F740 is widespread, as BGF and Chl-F have distinct origins and can change independently in response to stress factors and during plant development. F440/F690 and F440/F740 are very early stress indicators and F440/F520 could change after long stress exposure. The F690/F740 ratio presents an inverse relationship with the leaf chl. content (Cerovic *et al.*, 1999; Buschmann *et al.*, 2000). MCFI can provide information about gradients and distribution of the different fluorescence signals related to photosynthetic (Chl-F) and secondary (BGF) metabolism. During the stress response, different phenolic compounds from the phenylpropanoid pathway can accumulate and generally emit BGF after UV excitation (Lichtenthaler and Miehé, 1997; Cerovic *et al.*, 1999). Microscopic studies of UV-induced fluorescence have revealed some of the main BGF emitters and their location in the plant cell as ferulic acid (Morales *et al.*, 1996) or chlorogenic acid (Morales *et al.*, 2005).

Changes in fluorescence ratios and values were analysed in plants suffering under different plant developmental stages (Buschmann *et al.*, 2000; Meyer *et al.*, 2003) and abiotic stress conditions: light (Middleton *et al.*, 2005; Lenk and Buschmann, 2006), drought and temperature (Lichtenthaler and Miehé, 1997), herbicides (Lichtenthaler *et al.*, 2005), as well as mineral deficiencies and excesses (Lichtenthaler *et al.*, 2005).

Multispectral fluorescence and reflectance imaging have been successfully used for classification or sorting of agricultural products (Kim *et al.*, 2003; Ariana *et al.*, 2006), evaluation of UV-screening and activation of the secondary metabolism (Lichtenthaler and Miehé, 1997; Agati *et al.*, 2002).

10.6.3. Other related imaging techniques applied to measurements of plant physiology

The ability to collect spatially resolved data for a wide range of molecular, physiological and biophysical processes is increasing dramatically (Chaerle and Van Der Straeten, 2000). Water limitation or other localised changes in leaf chemistry affect stomatal conductance, and thermal imaging offers a powerful tool for mapping changes in temperature associated with variation in latent heat flux across leaf surfaces (Omasa and Takayama, 2003). With proper calibration, thermal maps can be converted directly into maps of stomatal conductance (Jones, 2004a).

A wealth of information about photosynthesis and other aspects of leaf physiology could be obtained by applying complementary imaging methods in the same leaf. Combining different images with different resolution is, however, challenging. One approach is to construct simple regressions between the values in aggregate pixels in one image with aggregate pixels in another image. West *et al.* (2005) applied this approach to an examination of the effect of stomatal patchiness (thermal image) on photosynthesis (fluorescence image). Deeper insight can be gained by applying methods of geographical image analysis to physiological data (Omasa and Takayama, 2003; Leinonen and Jones 2004; Aldea *et al.*, 2006a). By registering and re-sampling images taken with different instruments, different images can be aligned precisely and expressed at a common resolution.

The spatial pattern of other components of the photosynthetic machinery, including chl. content and engagement of the xanthophyll cycle (Gamon *et al.*, 1997), are readily mapped with hyperspectral imaging (Chaerle and Van Der Straeten

2000; Schuerger *et al.*, 2003; Blackburn, 2007), though this has not yet been applied to variation within single leaves.

Highly sensitive cameras now make it possible to image the autoluminescence or ultra-weak spontaneous luminescence owing to oxidative stress and lipid peroxidation in plants (Flor-Henry *et al.*, 2004; Havaux *et al.*, 2006). This new imaging approach enables the screening of mutants with variable tolerance to oxidative stress, as well as following the generation of excited states during stress-induced oxidative reactions (Ohya *et al.*, 2002; Reverberi *et al.*, 2005). Various tracers and dyes also allow mapping ROS (Miller *et al.*, 2005; Flors *et al.*, 2006; Hideg *et al.*, 2006).

Examples of different physiological processes (water and metabolite movement, root–soil–water interactions, gene expression, presence of defence and signalling compounds) that can be monitored remotely *in vivo* with a range of imaging approaches (nuclear magnetic resonance (NMR) imaging, geophysical techniques, use of intrinsically fluorescent proteins, LIDAR imaging) are presented in Chaerle and Van der Straeten (2001), Aldea *et al.* (2006a), as well as in the special issue of Journal of Experimental Botany devoted to *Imaging Techniques for Understanding Plant Responses to Stresses* edited by Jones and Morison (2007).

11 • Stable isotopic compositions related to photosynthesis, photorespiration and respiration

E. BRUGNOLI, F. LORETO AND M. RIBAS-CARBÓ

11.1. HISTORICAL OVERVIEW OF STABLE ISOTOPIC STUDIES IN PLANT SCIENCE

Isotopes are atoms of the same element having the same numbers of protons and electrons but a different number of neutrons. Stable isotopes are those that do not decay to other isotopes on the geological timescales, but can themselves be originated from the decay of radioactive isotopes.

The stable isotopes of carbon and oxygen have major relevance for photosynthesis, respiration and photorespiration, and link the carbon and water cycles. Hence, this chapter will attempt to summarise the current knowledge of processes affecting the stable carbon and oxygen isotopic composition of plants and the exchange between plants (and ecosystems) and the atmosphere.

In addition, hydrogen, nitrogen and sulphur isotopes also have major relevance for plant biology and physiological ecology, but these will not be covered here as these subjects are out of the scope of this book.

Early studies and measurements of stable isotopes in plants were initiated by scientists outside biology, with the initial interest developed in physical science and subsequently extended to geology. Studies were focused on deuterium and its variation in nature, as it was the first isotope to be discovered and was the easiest to measure with the equipment then available, mainly spectrographs. It was not until the late 1930s/early 1940s, with the development of a modern sector field-isotope mass spectrometer by Alfred Nier (1940), that precise measurements of other isotopes such as carbon and oxygen became possible.

11.1.1. Carbon isotopes

Nier and Gulbransen (1939) were the first to observe that ^{13}C was depleted in plants compared with inorganic carbonaceous material. The earliest studies in plants were conducted by geochemists, who were mostly interested in exploring the variation in isotopic composition in natural inorganic and biological material. Over the last three decades, the analysis of natural abundances of stable isotopes began to be widespread among botanists, plant biologists and physiological ecologists. Early systematic isotope-abundance measurements in plants were reported by Wickman (1952) and Craig (1953), who showed systematic ^{13}C depletion in plants compared with inorganic substances, and wide variation between different species and among plants collected in different regions of the world (e.g., boreal versus tropics). Following Craig's work, Park and Epstein (1960) proposed a three-step model to explain isotope fractionation during photosynthesis, with diffusive and biochemical fractionations playing major roles, with fractionation during secondary metabolism accounting for only a small fraction of total fractionation. They also pointed to a fractionation during translocation of carbohydrates and argued that CO_2 respired by roots may be enriched in ^{13}C though they were unable to confirm the latter hypothesis. This question remained open for nearly 40 years, when Duranceau *et al.* (1999) showed the apparent fractionation associated with respiration, leading to a variable enrichment in ^{13}C in respiratory CO_2.

The C_4-photosynthetic pathway had not yet been discovered when several reports showed 'unusual $\delta^{13}C$ values' for several plant species (Wickman, 1952; Craig, 1953). It was only later, after the concomitant discovery of the C_4-photosynthetic pathway by Hatch and Slack (1966) and Kortschak (1965), that those early observations were explained. Shortly after, Bender (1968, 1971) and Smith and Epstein (1971) reported the early systematic surveys of carbon-isotope compositions in C_3 and C_4 plants, with C_4 species found in 13 different families, therefore suggesting parallel evolution rather than a common origin. Other reports showed that some succulent plants also

Terrestrial Photosynthesis in a Changing Environment: A Molecular, Physiological and Ecological Approach, ed. J. Flexas, F. Loreto and H. Medrano. Published by Cambridge University Press.

presented anomalous ^{13}C contents (Vogel and Lerman, 1969; Bender, 1971) and these were related to the presence of CAM (Bender *et al.*, 1973). Following those early studies, carbon-isotope composition became a classical method to distinguish plants on the basis of different photosynthetic pathways, and to study their taxonomic, geographical and ecological distributions. At the same time, it also became evident that substantial variation in isotopic composition occurred within plants and species with C_3 photosynthesis, whereas much less variation occurred in plants with the C_4 pathway. During the following years several studies contributed to fill gaps in the knowledge on the kinetic isotopic fractionation during photosynthesis and models were developed to describe the processes involved. In 1982, Graham Farquhar, Marion O'Leary and Joe Berry published a work showing that $\delta^{13}C$ variation in C_3 plants correlated with the ratio of intercellular and atmospheric partial pressures of CO_2, and proposed the well-known model that stimulated a real explosion of interest on stable isotopic studies in plant physiology. Based on this model, a large number of studies focused on the physiological and environmental effects on stable-isotope composition and on the relationship between carbon-isotope discrimination and plant and canopy WUE (Farquhar *et al.*, 1982; Farquhar and Richards, 1984).

Starting around 1990, many efforts concentrated on the study of atmosphere-biosphere gas exchange, coupling micrometeorological and isotopic methods, to separate different sources of CO_2 and to study the ecosystem C budget. Nowadays, the isotopic fluxes are major components of the global C-cycle modelling (e.g., Ciais *et al.*, 1995a,b; Diefendorf *et al.*, 2010; Werner and Máguas, 2010).

11.1.2. Oxygen and hydrogen isotopes

Early studies of hydrogen and oxygen isotopes in water were started around 1950. Urey (1947) first published the concept of using oxygen isotopes to reconstruct temperature in paleoclimate studies. Subsequently, Epstein and Mayeda (1953) working on $^{18}O/^{16}O$ and Friedman (1953) on D/H, measured the isotopic composition in different water samples. They were able to show that freshwater samples generally contained less of the heavy isotopes compared with seawater samples. Subsequently, the systematic studies by Dansgaard (1954) led to understanding the geographic isotopic variation in meteoric waters.

Dole *et al.* (1954) measured the isotopic composition of diatomic oxygen and found that it was enriched by approximately 23‰ compared with ocean water (the Dole effect). It was soon evident that oxygen originated from photosynthesis derived from leaf water. In 1965, Craig and Gordon published their landmark paper about the evaporative isotopic enrichment. They also studied the causes of variation in isotopic composition of surface water (e.g., precipitation, evaporation, mixing and freezing) and characterised the isotopic composition of water vapour and the underlying isotopic effects during evaporation and condensation. Gonfiantini *et al.* (1965) were the first to demonstrate an ^{18}O enrichment in leaf water relative to soil water and since then many authors started to analyse the isotopic composition of leaf water to measure the enrichment during transpiration. Subsequently the Craig-Gordon model has been extensively applied to transpiring leaves and its theoretical basis have been verified and proven to be robust over a wide range of environmental conditions (Flanagan *et al.*, 1991a,b; Roden and Ehleringer, 1999; Barbour and Farquhar, 2000). Several studies showed that there can be strong spatial variation in the stable-isotope composition (both $^{18}O/^{16}O$ and D/H) of water along a single leaf both in monocots and dicots (Yakir *et al.*, 1989; Bariac *et al.*, 1994; Wang and Yakir, 1995; Helliker and Ehleringer, 2000; Gan *et al.*, 2002; Farquhar and Gan, 2003; Affek *et al.*, 2006; Ferrio *et al.*, 2009) and that this spatial pattern can be reflected in variation in the isotopic composition of leaf cellulose. Based on that knowledge, Farquhar *et al.* (1993) showed that the equilibration of ^{18}O between CO_2 and leaf water could be used to partition the gross fluxes of CO_2 from vegetation on a global basis and recently many laboratories use ^{18}O in conjunction with ^{13}C to study carbon and hydrological cycles, and it has become evident that oxygen isotopes represent the major link between the two cycles. Recently, Helliker and Richter (2008) have developed a method to estimate whole-canopy leaf temperature in trees based on ^{18}O in cellulose.

11.2. STABLE ISOTOPES IN NATURE

Stable isotopes are differently distributed in nature and variations occur in the different pools owing to physical, chemical and biological transformations. For example, during the evaporation and condensation of water, the concentration of molecules containing different oxygen and hydrogen isotopes changes, because molecules with lighter isotopes concentrate in the vapour phase leaving the heavier ones in the liquid phase. Similarly, wide variations in the carbon-isotope composition occur in the organic and inorganic carbon pools. The most abundant carbon reservoir is represented by the Earth's crust, with a bulk $\delta^{13}C$ of about –5‰, whereas carbonate sediments have an isotopic composition close to 0‰ (Schidlowski, 1988). Atmospheric CO_2 is fixed by photosynthetic organisms, which provide energy directly or indirectly to the entire biosphere. Hence, atmospheric CO_2 represents the major link between the inorganic and the organic pools and between marine and terrestrial ecosystems (Siegenthaler and Sarmiento, 1993). It is well known that photosynthetic fractionation is responsible for the overall isotopic fractionation in terrestrial carbon with the organic matter produced by autotrophic organisms being around –25‰. Global isotopic mass balance requires that partitioning between inorganic and organic components of the sedimentary C should be approximately 20 and 80% respectively (0.2×(–25‰) + 0.8×(0‰)=–5‰) (Schidlowski, 1988, 1995; Yakir, 2003).

Apart from crustal carbon, dissolved inorganic carbon (DIC) is the largest pool in the global carbon cycle and it is represented by bicarbonate, carbonate and dissolved CO_2. The isotopic composition of both DIC and marine carbonate sediments is close to 0‰. Because of photosynthetic fractionation, organic carbon in marine environment is strongly depleted in ^{13}C compared with inorganic carbon.

The exchange of CO_2 between the atmosphere and the ocean surface involves an equilibrium fractionation so that atmospheric CO_2 would be depleted by approximately 7‰ compared with total DIC (Mook *et al.*, 1974; Mook, 1986; Peterson and Fry, 1987). Because the oceanic inorganic carbon is a much larger carbon pool than atmospheric CO_2, then the isotopic composition of the latter is largely determined by ocean-atmosphere exchange, and especially by the exchange between atmosphere and the terrestrial biosphere.

Initially, measurements on atmospheric CO_2 were focused on ^{13}C, but subsequently it became evident that the ^{18}O content of atmospheric CO_2 showed altitudinal and latitudinal variations governed by the exchange between atmosphere and the land biosphere (Francey and Tans, 1987; Friedli *et al.*, 1987). It is now well documented that the ^{18}O composition of atmospheric CO_2 reflects the equilibration of CO_2 and water in leaves. After dissolution CO_2 exchanges virtually all oxygen atoms with water and, as water in leaves is highly enriched in ^{18}O owing to evaporative fractionation (Gonfiantini *et al.*, 1965; Dongmann, 1974), the CO_2 released in the atmosphere has a contrasting ^{18}O signature depending on the source (leaves, trunk, root and soil respiration).

The isotopic composition of CO_2 varies seasonally as a consequence of variation in photosynthesis, respiration and evaporation-transpiration, especially of the terrestrial biosphere (Keeling *et al.*, 1996). The seasonal change in ^{13}C signal is antiparallel with fluctuation in CO_2 concentration (troughs in concentration correspond to maximum photosynthetic activity by terrestrial biosphere, which in turn determine peaks in ^{13}C enrichment in atmospheric CO_2 owing to photosynthetic fractionation), whereas the ^{18}O signal is parallel to the CO_2 concentration with the troughs coinciding with maximum evaporative enrichment in spring–summer. Oscillations in the three signals (concentration, ^{13}C and ^{18}O) in the Southern hemisphere are six months out of phase and much less pronounced compared with the Northern hemisphere because there is less land in the Southern hemisphere and most of the terrestrial photosynthetic activity is concentrated in the tropics, where seasonality is much less marked.

The isotopic composition of atmospheric CO_2 is also subjected to long-term trends with a clear decline in $\delta^{13}C$ associated with the well-known increase in CO_2 concentration, mostly owing to anthropogenic fossil-fuel combustion and land-use change, mainly deforestation (Keeling, 1958; Keeling *et al.*, 1979; Mook *et al.*, 1983; Francey, 1985; Friedli *et al.*, 1987). Fossil fuels are variably depleted in ^{13}C and, therefore, anthropogenic combustions are diluting the ^{13}C content of atmospheric CO_2, and this is also known as the 'Suess effect' (Suess, 1955). Oxygen isotopes are also subjected to a long-term trend but its relationship with increasing CO_2 is less clear.

After photosynthetic fractionation, invariably leading to a ^{13}C depletion in organic carbon, CO_2 is cycled through heterotrophic organisms and then mineralised back by various saprophytic organisms. The food web and trophic cycling roughly maintain the isotopic signature without major fractionation processes for carbon and, therefore, it is approximately valid to say that 'you are what you eat' with a slight trend to an enrichment with trophic levels (e.g., Peterson and Fry, 1987; Eggers and Jones, 2000).

11.3. PRINCIPLE AND APPLICATIONS OF STABLE-ISOTOPE RATIO MASS SPECTROMETRY

Isotopic compositions are generally measured using special mass spectrometers (Isotope Ratio Mass Spectrometers, IRMS) that determine the isotope abundance ratio of the sample. Modern IRMS are based on the field-sector mass

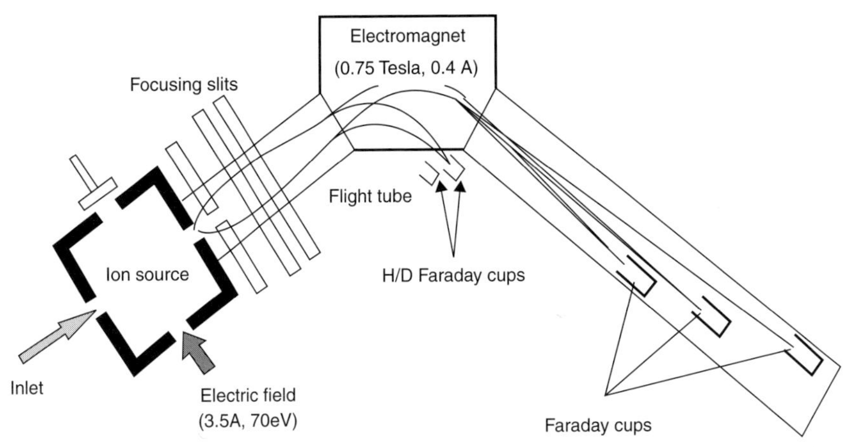

Fig. 11.1. Diagram showing the principle of stable-isotope ratio mass spectrometry (IRMS), with the source, the flight tube, the electromagnet and the Faraday collectors. Figure is a modification of the drawing provided by Elementar, Isoprime Ltd., Cheadle, UK.

spectrometers developed by Nier in the 1940s and modified by McKinney *et al.* (1950). Since then, enormous technological advances in electronics, vacuum and ion optics have improved precision by more than tenfold. Advances in electronics have been especially relevant as they have allowed the full automation of mass spectrometers.

Samples are normally analysed as a gas and injected into the IRMS inlet. Under high vacuum, the gas is ionised and accelerated by an electric field along an ion beam, then deflected by a magnetic or electromagnetic field and collected by a proper detection system (Fig. 11.1). These processes take place in different compartments of the IRMS. The ionisation, beam formation and acceleration processes occur in the source, whereas magnetic deflection occurs in the flight tube. Ionisation is commonly obtained by passing a beam of electrons through the gas sample. Normally, positive ions formed are accelerated and aligned as a beam by a positive potential through a series of focusing slits. The ion beam is accelerated and then reaches the flight tube where a magnetic field separates ions into different beams, each with a different radius proportional to its mass. Finally, the separated beams are analysed in the detector by resolving slits and detected by Faraday cups differing for position and dimension (Fig. 11.1). The ion current generated in each cup is proportional to the number of incident ions and, therefore, to the partial pressure of the correspondent isotopic molecular species in the gas injected. In modern IRMS, data are acquired and processed automatically by computers.

Although IRMS are the most commonly used equipment for isotopic ratio measurements, recently tunable diode laser absorption spectrometers (TDLAS) are becoming more and more common (see Chapter 10) for measuring isotopic ratios of gaseous samples and, especially, for taking measurements at high frequency, such as measuring isotopic fluxes in ecosystems (Bowling *et al.*, 2003). Using IR absorption, the TDLAS system determines the mole fractions of individual isotopic species (e.g., $^{12}CO_2$ and $^{13}CO_2$, $H_2{}^{16}O$ and $H_2{}^{18}O$) independently, rather than their ratio as in IRMS.

Generally, with IRMS the isotopic ratio of carbon in plant samples is determined on CO_2 after oxidation of the sample, whereas oxygen isotope ratios are determined either on CO_2, CO or O_2. Oxygen from water is normally exchanged after equilibration with CO_2, with all the oxygen of CO_2 being replaced by oxygen of water. The isotopic analysis of CO_2 provides a good measure of the $\delta^{18}O$ of water. The analysis of $\delta^{18}O$ in organic material is normally achieved by pyrolysis, in specially adapted equipments where organic material is pyrolysed in the absence of any external oxygen to carbon monoxide, so that all the oxygen of CO derives from that contained in the sample molecules. Similarly, pyrolysis is applied to water samples and the oxygen isotopic composition is analysed as CO at masses 28, 29 and 30.

All the above peripherals (elemental analysers, pyrolysers, trace-gas interfaces etc.) are usually coupled to IRMS in continuous-flow (CF), with a flow of carrier gas (helium) transporting the sample after its preparation and conditioning to the IRMS. In special applications, when the required precision is higher than that of continuous flow, dual inlet is used. In such cases the gaseous sample is analysed statically and compared with the isotopic ratio of a standard gas.

11.4. DEFINITIONS OF ISOTOPIC COMPOSITIONS AND ISOTOPE EFFECTS

Mass spectrometers provide high-precision measurements of the isotope-abundance ratio R, defined as the ratio of the less-abundant isotopic molecular species over the most abundant one (e.g., $^{13}CO_2/^{12}CO_2$, D/H, $C^{18}O^{16}O/C^{16}O^{16}O$ etc.) relative to a standard. In practice, the IRMS determines the isotopic composition as the deviation of the sample isotopic ratio from that of a standard, with the isotopic composition defined as:

$$\delta = R_{sm}/R_{st} - 1 \qquad \text{[Eqn. 11.1]}$$

where R_{sm} is the isotopic ratio of the sample and R_{st} is that of the standard. With carbon ($\delta^{13}C$), the isotopic ratio $^{13}CO_2/^{12}CO_2$ is measured in the sample relative to the standard, whereas in the case of oxygen ($\delta^{18}O$), the ratio is $^{18}O/^{16}O$. The isotopic standard for carbon used to be a cretaceous belemnite fossile called Pee Dee Belemnite (PDB), whereas for oxygen both PDB and standard mean ocean water (SMOW) are used. However, as the supply of these standards has been exhausted, the International Union of Pure and Applied Chemistry recommended the use of Vienna-PDB (VPDB) and Vienna-SMOW (VSMOW) (Coplen, 1995). These materials are produced and certified by the International Atomic Energy Agency in Vienna. VPDB is defined taking a value of +1.95‰ for NBS-19 carbonate standard relative to VPDB, whereas VSMOW is defined using a normalised $\delta^{18}O$ value of –55.5‰ for SLAP (Standard Light Antarctic Precipitation) that is depleted in the heavy isotopes (Coplen, 1995).

Fractionation during physical, chemical and biological transformations causes variation in the isotopic abundances in nature. For example, $^{12}CO_2$ diffuses faster than $^{13}CO_2$ and the same is true for oxygen isotopes. These fractionation processes are described as isotope effects (α_k), the ratio of the rate constants for reactions of different isotopic species. In the case of kinetic isotopic effects and when the source is a reservoir large enough to not be significantly affected by the formation of product (or at the very beginning of the reaction when time tends to zero), then the isotope effect α_k is:

$$\alpha_k = k^n/k^{n*} = R_{so}/R_{pr} \qquad \text{[Eqn. 11.2]}$$

where, k^n and k^{n*} represent the rate constants for the substrate containing the lighter and the heavier isotopes, respectively (e.g., k^{12} and k^{13} for reaction with molecules containing ^{12}C and ^{13}C, respectively), and it is equal to the ratio of isotope ratios of the source (R_{so}) and of the product (R_{pr}). Similarly, the equilibrium isotope effects are defined as the ratio of the equilibrium constants for the molecules containing the different isotopes. Because equilibrium isotope effects are the resultants of two opposing kinetic effects, equilibrium effects are generally smaller than kinetic ones. Detailed discussion of isotope effects can be found in Farquhar *et al.* (1989) and O'Leary (1993).

The use of isotope effect is not very convenient in plant physiology, and Farquhar and Richards (1984) proposed the use of carbon-isotope discrimination (Δ), defined as:

$$\Delta = \alpha_k - 1 \qquad \text{[Eqn. 11.3]}$$

which is more practical as normally in plant studies α_k exceeds unity by a small number (e.g., if α_k=1.027, then Δ=0.027 or 27‰). In the study of photosynthesis, the source is atmospheric CO_2 with isotopic ratio R_{air}, and the product is plant biomass (e.g., photosynthetic carbohydrates) with isotopic ratio R_{plant}; then from equation 11.2:

$$\Delta = R_{air}/R_{plant} - 1 \qquad \text{[Eqn. 11.4]}$$

In practice, the value of Δ is calculated using the isotopic composition of atmospheric CO_2 (δ_a) and that of the plant sample (δ_p) measured by IRMS. Combining equation 11.4 with equation 11.1 for the isotopic composition of air CO_2 and of plant biomass, it derives:

$$\Delta = \frac{\delta_a - \delta_p}{1 + \delta_p} \qquad \text{[Eqn. 11.5]}$$

Both, the isotope composition δ and isotope discrimination Δ are dimensionless, and results are expressed for convenience as the value multiplied by 10^{-3} or ‰ (per mil). For example, a δ value of –0.030 is normally presented as –30‰ or -30×10^{-3}, but these notations are not strictly speaking units, as often erroneously reported in the literature. It is also worth noting that in plants, while δ is normally negative, Δ has generally a positive value, because photosynthesis discriminates against ^{13}C. However, it should be noted that while the value of δ is dependent on the isotopic composition of the source (e.g., atmospheric CO_2 for photosynthesis in the field or thank CO_2 for growth-chamber experiments), the value of Δ expressing the deviation between source and product is independent of the source and, therefore, Δ values, if properly calculated, can be compared across different experiments even with very different sources of CO_2.

11.5. CARBON ISOTOPES IN THE STUDY OF PHOTOSYNTHESIS, PHOTORESPIRATION AND RESPIRATION

There are two natural stable isotopes of carbon, ^{12}C, representing about 98.9% of total C in nature, and ^{13}C, which is present in a minor fraction of about 1.1%. As discussed above, plants are invariably depleted in ^{13}C compared with atmospheric CO_2 because of photosynthetic fractionation. Fractionation occurs during diffusion, with $^{13}CO_2$ having lower diffusivity compared with $^{12}CO_2$ (the same is true for oxygen, with lighter isotopes diffusing faster than heavier ones), as the diffusivity is inversely proportional to the square root of the reduced masses of CO_2 and air (Mason and Marrero, 1970). After diffusion fractionations (in air but also in the liquid phase) there are further substantial isotope effects owing to enzymatic carboxylations.

11.5.1. Carbon isotopes and photosynthesis

The isotope effect owing to carboxylation reactions is widely variable depending on the enzyme involved, on concentrating mechanisms and, therefore, on the photosynthetic pathway. As stated in previous paragraphs, there is a wide variation of approximately 14–15‰ between plants with the C_3- and the C_4-photosynthetic pathway (Brugnoli and Farquhar, 2000). This is mainly attributable to different kinetic isotope effects shown by Rubisco compared with PEPC, the primary carboxylating enzymes in C_3 and C_4 plants, respectively. In addition, the substrate for Rubisco is gaseous CO_2, whereas that for PEPC is bicarbonate (HCO_3^-) with a different $\delta^{13}C$ because of fractionation during hydration of CO_2 to HCO_3^-.

C_3 PHOTOSYNTHESIS

Carbon-isotope discrimination in C_3 plants is determined by several fractionation processes occurring during diffusion, from the free atmosphere to the carboxylation sites, and then to carboxylation itself. Mechanistic understanding of this process is robust, and Farquhar *et al.* (1982) developed a model accounting for different fractionations, divided into diffusive and carboxylation components, such that carbon-isotope discrimination is given by (see also Chapter 12):

$$\Delta = a_b \frac{C_a - C_s}{C_a} + a \frac{C_s - C_i}{C_a} + (e_s + a_l) \frac{C_i - C_c}{C_a} + b \frac{C_c}{C_a} - \frac{\frac{eR_d}{k} + f\Gamma_*}{C_a} \quad \text{[Eqn. 11.6]}$$

where C_a, C_s, C_i and C_c are the CO_2 mole fractions in the free atmosphere, at the leaf surface within the boundary layer, in the intercellular air spaces before it enters in solution and at the sites of carboxylation, in that order; a_b is the discrimination occurring during diffusion in the boundary layer (2.9‰, Farquhar, 1982); a_d is the fractionation occurring during diffusion in still air (4.4‰, Craig, 1954); e_s is the fractionation occurring when CO_2 enters in solution (1.1‰, at 25°C, Vogel, 1980); a_l is the fractionation occurring during diffusion in the liquid phase (0.7‰, O'Leary, 1984); b is the net discrimination occurring during carboxylations in C_3 plants; e and f are the fractionations occurring during dark respiration (R_n) and photorespiration (R_p), respectively; κ is the carboxylation efficiency; and Γ^* is the CO_2 compensation point in the absence of dark respiration (Brooks and Farquhar, 1985). Therefore, the model predicts a linear relationship between Δ and the ratio of intercellular and atmospheric CO_2 concentrations.

There are some uncertainties with some terms of equation 11.6, for example, the value of b is still debated. This value is not simply the discrimination by Rubisco, because in C_3 plants a variable amount of carbon is fixed by PEPC (Nalborczyk, 1978; Farquhar and Richards, 1984). In C_3 plants, this enzyme operates in parallel with Rubisco, affecting the isotopic composition of fixed carbon (Brugnoli and Farquhar, 2000). Anaplerotic CO_2 fixation by PEPC is not the only contribution other than Rubisco in C_3 plants, but other carboxylases may contribute for a small proportion of fixed carbon (Raven and Farquhar, 1990). Provided that β_{PEP} is the proportion of PEP carboxylations over the total carbon fixed in C_3 plants, then, according to Farquhar and Richards (1984), b is given by:

$$b = (1-\beta_{PEP})\, b_3 + \beta_{PEP}\, b_4 = b_3 - \beta_{PEP}\,(b_3 - b_4) \quad \text{[Eqn. 11.7]}$$

where b_3 and b_4 are the isotopic fractionations associated with Rubisco and PEPC, respectively. Equation 11.7 shows that in C_3 plants the total fractionation during carbon fixation is lower than that by Rubisco by a quantity that depends on the proportion of PEP carboxylations and by the difference in fractionation between Rubisco and PEPC (b_3–$b_4 \approx 36$‰), provided that subsequent decarboxylations have no effect on the isotopic composition of the products of PEPC. If, for example, organic acids produced are subsequently decarboxylated removing the C atom added by β-carboxylation, then the remaining molecule would have a C_3-type isotopic composition.

The value of β_{PEP} varies in C_3 plants depending on intrinsic and environmental factors, such as stress effects and the source of nitrogen nutrition. A value of β_{PEP} close to 0.05 is often used for most C_3 plants (Brugnoli and Farquhar, 2000). The value of b_3 shows some uncertainties, and variations in its value have been reported in the literature (see review by Brugnoli and Farquhar, 2000) and confirmed recently, with C_3 and C_4 Rubisco showing slightly different isotope effects (Tcherkez *et al.*, 2006; McNevin *et al.*, 2007). However, in higher plants the value of discrimination by Rubisco is likely very close to 30‰ with respect to gaseous CO_2 (Brugnoli *et al.*, 1988; Guy *et al.*, 1993; Brugnoli and Farquhar, 2000).

The value of b_4 varies with temperature because it depends on the isotope effect associated with the equilibrium between gaseous CO_2 and HCO_3^-, which is strongly temperature dependent (Farquhar, 1983). PEPC discriminates against ^{13}C by approximately 2.2‰, whereas the isotopic equilibrium favours ^{13}C enrichment in HCO_3^- by 7.9‰ at 25°C. Therefore, the net isotope effect during PEP carboxylation taking CO_2 as the source is b_4=–5.7‰, which actually favours ^{13}C. Consequently, using the above values of β_{PEP} (=0.05), b_3 (=30‰) and b_4 (=–5.7‰) a value of b=28.2‰ is obtained. However, if there is some decarboxylation of carbon atoms fixed by PEPC then the value of b would fall between 28.2 and 30‰, and this can partly explain different values reported in the literature (Brugnoli and Farquhar, 2000).

The fractionation factors associated with respiration (*e*) and photorespiration (*f*) in equation 11.6 have been considered very small and negligible for a long time, based on indirect or direct observations (von Caemmerer and Evans, 1991; Lin and Ehleringer, 1997). However, recent extensive studies have shown that R_n (at night) produces CO_2 enriched in ^{13}C by a variable extent (e.g., *e*=–6‰) depending on species and environmental conditions, affecting the $\delta^{13}C$ of the remaining material (Duranceau *et al.*, 1999; Brugnoli and Farquhar, 2000; Ghashghaie *et al.*, 2001, 2003; Gessler *et al.*, 2009a and citations therein). However, the actual possible fractionation during respiration in the light remains unknown, and it has been argued that even a fractionation against ^{13}C may occur (Tcherkez *et al.*, 2004).

Similarly, it has been shown by several authors (see Gillon and Griffiths, 1997; Igamberdiev *et al.*, 2004; Tcherkez, 2006; Lanigan *et al.*, 2008; for reviews see Brugnoli and Farquahar, 2000 and Ghashghaie *et al.*, 2003) that the fractionation factor associated with photorespiration can be high (*f*=10‰).

Equation 11.6 represents the complete Farquhar's model, taking into account all the fractionation processes

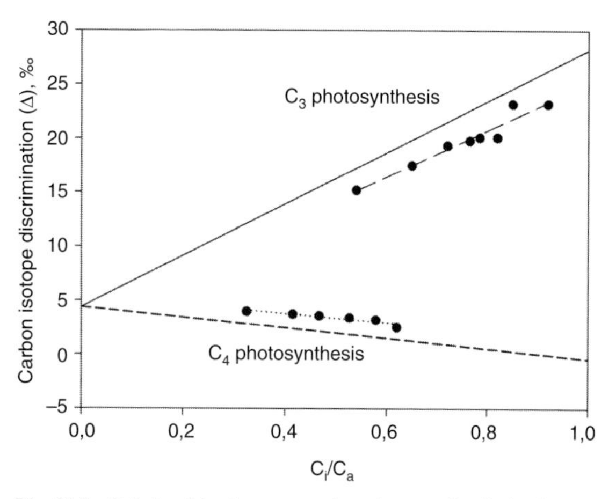

Fig. 11.2. Relationships between carbon-isotope discrimination (Δ) and the ratio of intercellular and atmospheric concentrations of CO_2 in C_3 (beech, *Fagus sylvatica* L.) and C_4 plants (corn, *Zea mays*). The solid line represents the expected $\Delta = a + (b - a)C_i/C_a$, with b=28.2‰; the dotted line is the expected Δ from the model $\Delta = a + (b_4 + \phi(b_3 - s) - a)C_i/C_a$, with ϕ=0.2 and b_4=–5.7‰. The dashed lines represent the regression fitted to actual data.

occurring during movement of CO_2 from the atmosphere to the chloroplast and during subsequent carboxylation, respiration and photorespiration of various substrates. However, in many studies, several fractionation factors of equation 11.6 are considered negligible, and then a simplified and a very useful version of the model is given by:

$$\Delta = a + (b - a)\frac{C_i}{C_a} \qquad \text{[Eqn. 11.8]}$$

This equation also assumes an infinite internal conductance to CO_2. A large amount of reports have shown that, although equation 11.8 is a simplified version and omits factors not always negligible, it works in most C_3 species and conditions as a reasonable approximation. This model shows (Fig. 11.2) that when the ratio of intercellular and atmospheric concentrations of CO_2 is small, Δ is dominated by diffusive fractionation (Δ tends to 4.4‰ when C_i/C_a is near zero), whereas at high C_i/C_a (near unity) the fractionation by carboxylation becomes prevailing (Δ≈ *b*).

C_4 PHOTOSYNTHESIS

In C_4 plants, the primary carboxylating enzyme located in mesophyll cells, PEPC, uses bicarbonate as a substrate. As previously documented, the net isotopic effect of this carboxylation with respect to gaseous CO_2 (i.e., starting from CO_2, then its dissolution and conversion to bicarbonate and then carboxylation by the enzyme) is –5.7‰ at 25°C (Farquhar, 1983), therefore favouring ^{13}C. When the C_4 skeleton enters into the BSC it is decarboxylated and the CO_2 is carboxylated by Rubisco. In the absence of CO_2 leakage from the BSC, there would be no further fractionation by Rubisco, because in a perfectly closed system all CO_2 molecules are fixed irrespective of their carbon isotopes. However, as there is always some CO_2 leaking out of the BSC then Rubisco can discriminate against ^{13}C. Finally, also during leakage from BSC there will be some fractionation during back diffusion of CO_2. A model for C_4 plants is given by:

$$\Delta = a_b \frac{C_a - C_s}{C_a} + a \frac{C_s - C_i}{C_a} + (e_s + a_l) \frac{C_i - C_m}{C_a} + (b_4 + b_3\phi - s_l\,\phi) \frac{C_m}{C_a} \quad \text{[Eqn. 11.9]}$$

where C_a, C_s, C_i and C_m are the CO_2 mole fractions in the free atmosphere, at the leaf surface, in the intercellular air spaces before it enters in solution and in the mesophyll cytoplasm, in that order; b_4 and b_3 are the fractionations associated with PEP and RuBP carboxylations, respectively; ϕ is the fraction of CO_2 that leaks out of the BSC; and s_l is the fractionation during leakage (likely to be very close to the sum $e_s + a_l$, Henderson *et al.*, 1992). Also in the case of C_4 photosynthesis, a simplified model ignores the effects owing to diffusion in the boundary layer and diffusion from the intercellular air spaces and mesophyll cytoplasm. In that case, Δ would be:

$$\Delta = a + (b_4 + \phi\,(b_3 - s_l) - a) \frac{C_i}{C_a} \quad \text{[Eqn. 11.10]}$$

It is noteworthy that the slope of the relationship between Δ and C_i/C_a (Fig. 11.2) is dependent on the fraction of CO_2 that actually leaks out of BSC and is normally slightly negative. Often, the term s_l in equation 11.10 is neglected. Several values of ϕ have been estimated by several authors in various species. Although reports using online discrimination (see Chapter 12) found ϕ falling in the range from 0.2 to 0.3 in most species (Henderson *et al.*, 1992), estimations using incorporation of $^{14}CO_2$ (Hatch *et al.*, 1995) resulted in much lower values in the range between 0.08 and 0.14. Several explanation can be found for such variation (Brugnoli and Farquhar, 2000), however, it seems clear that ϕ is smaller than reported using isotopic measurements in dry matter. For a detailed review about variation in ϕ see Brugnoli and Farquhar (2000).

CRASSULACEAN-ACID METABOLISM PLANTS

Calculations of the isotope fractionation in plants exhibiting the CAM is complicated by the presence of C_3 and C_4 cycles in the same individual but temporally separated (see Chapter 6). Δ of fixed carbon can vary from the typical C_4-type values during dark CO_2 fixation, to typical C_3 values when stomata can re-open and some CO_2 can be fixed directly by Rubisco. Obviously all the intermediate values can occur during transition from phase I to phase IV of CAM.

Discrimination in CAM plants has been studied by several authors (Farquhar *et al.*, 1989; Griffiths *et al.*, 1990; Griffiths, 1992; Borland and Griffiths, 1997; Maxwell *et al.*, 1997).

More recently, a model defining carbon-isotope discrimination in CAM plants has been developed and experimentally validated in *Kalanchoë daigremontiana* (Griffiths *et al.*, 2007). The model proposed for CAM discrimination is given by:

$$\Delta = a' + (e_s + a_l - a') \frac{C_i}{C_a} + (\Delta_{bio} - e_s - a_l) \frac{C_m}{C_a} \quad \text{[Eqn. 11.11]}$$

where Δ_{bio} is the integrated net biochemical discrimination, depending on the biochemistry of CO_2 fixation with the expression differing for the four CAM phases as described in Griffiths *et al.* (2007); a' is given by the expression

$$a' = \frac{(1 + a_b)(C_a - C_s) + (1 + a)(C_s - C_i)}{C_a - C_i} \quad \text{[Eqn. 11.12]}$$

11.5.2. Measurements of Δ in plants and validation of the Farquhar's model

In the foregoing, most of the isotopic studies were focused on measurements of Δ in plant biomass (e.g., mostly leaf or whole-plant biomass) to be compared with gas-exchange measurements. Several studies validated the relationships between Δ and C_i/C_a in C_3 and C_4 species under a wide range of environmental conditions and environmental stresses (reviewed by Brugnoli and Farquhar, 2000). In these studies, the relationships between Δ and C_i/C_a were reasonably in agreement with equation 11.8, especially considering that the isotopic signature of plant dry matter provides an assimilation-weighted integration of the photosynthetic behaviour and C_i/C_a over the entire period during which the carbon being measured had been fixed. This time period can range from weeks to the entire lifespan of the plant or the organ, whereas usually gas-exchange measurements are taken during short time intervals (minutes to hours).

Furthermore, the model has been validated more precisely using simultaneous measurements of leaf gas exchange and carbon-isotope discrimination, called 'online discrimination' (Fig. 11.2; see also Chapter 12). This method gives an almost instantaneous measurement of Δ and of photosynthetic physiology, in contrast with bulk biomass. In early studies, this method consisted of collecting CO_2 samples simultaneously with gas-exchange measurements, using a series of cryogenic traps: to separate water vapour a series of alcohol dry ice traps were used, and then the CO_2 was frozen in a second series of traps kept at liquid-nitrogen temperature and evacuated under high vacuum while frozen (Evans *et al.*, 1986). Subsequently, the CO_2 collected was injected in a mass spectrometer to determine $\delta^{13}C$. Because of isotopic discrimination during photosynthesis, $\delta^{13}C$ will be different in the air before and after it passes over a leaf in a well-stirred gas-exchange chamber: the air leaving the chamber will be enriched in ^{13}C compared with that entering the chamber. Measuring this difference and measuring the CO_2 concentrations in the air entering (C_{in}) and leaving (C_o) the chamber, by, for example, IR gas analysis, makes it possible to estimate the online discrimination as:

$$\Delta = \frac{\xi(\delta_o - \delta_{in})}{1 + \delta_o - \xi(\delta_o - \delta_{in})} \quad \text{[Eqn. 11.13]}$$

with $\xi = \frac{C_{in}}{C_{in} - C_o}$, and δ_{in} and δ_o being the isotopic composition of CO_2 in the air entering and outgoing the leaf chamber, respectively. A different expression of online Δ to account for refixation of CO_2 respired and photorespired and to correct possible overestimation of measured Δ under special conditions has been developed by Gillon and Griffiths (1997). This is important when the isotopic composition of source CO_2 from cylinders differs significantly from that of atmospheric CO_2, and the CO_2 respired is derived from previously fixed carbon (Tazoe *et al.*, 2009). However, the correction is normally minor and it is only relevant under conditions of low stomatal conductance.

The use of continuous flow IRMS coupled with gas-chromatographs allows the direct injection of air CO_2 from gas-exchange systems allowing for almost real-time simultaneous measurements of online discrimination and gas exchange (e.g., the system described by Cousins *et al.*, 2006). Nowadays, also TDLAS can be used to detect continuous measurements of online Δ (e.g., Bowling *et al.*, 2003; Flexas *et al.*, 2006a; Barbour *et al.*, 2007; Schaeffer *et al.*, 2008).

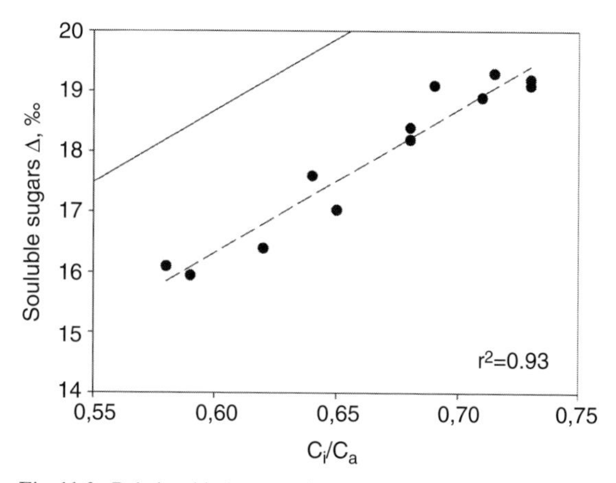

Fig. 11.3. Relationship between Δ measured in leaf sugars and C_i/C_a in leaves of European beech (*Fagus sylvatica*). The solid line represents the expected Δ on the basis of the simple Farquahar's model, the dashed line represents the regression equation fitted to the data (r^2=0.93). The deviation between expected and observed Δ is attributable to parameters unaccounted for, such as mesophyll conductance that is actually relatively low in trees.

Other studies demonstrated that carbon-isotope discrimination measured in leaf soluble sugars and starch correlated with an assimilation-weighted average of C_i/C_a over a period of 1–2 days (Brugnoli *et al.*, 1988, Fig. 11.3). Therefore, Δ in leaf soluble sugars and starch provides an assimilation-weighted average of C_i/C_a intermediate between that of bulk biomass and that measured online. To measure isotopic discrimination in leaf sugars and starch, different methodologies are available. The $\delta^{13}C$ of soluble sugars can be measured after preparing a low molecular-weight neutral fraction; the leaf extract is passed through successive cation- and anion-exchange chromatography columns to remove the ionic fractions (e.g., organic acids and amino acids), as described by Brugnoli *et al.* (1988). Other methods use compound-specific carbon-isotope analysis of sucrose, glucose and fructose after either HPLC or GC separation (LC-IRMS or GC-IRMS). These two compound-specific methods are rather different because GC-IRMS requires derivatisation of sugars, whereas LC-IRMS does not. For this reason the newly available oxidation systems used to interface HPLC with IRMS offer a very promising perspective in the study of $\delta^{13}C$ of plant metabolites. Compound-specific analysis should also be preferred because it produces sugars of higher purity and will not include sugar-like substances, such as polyols (e.g., myo-inositol), and oligosaccharides such as fructans, as is the case with the preparation of bulk-sugar fraction by water extraction. Detailed comparisons of different methods and of their respective advantages and disadvantages have been recently published by Richter *et al.* (2009).

There are also several methods to extract and purify starch from plant material and for determining its isotopic composition. For starch extraction, an acid or an enzymatic hydrolysis can be used. Each of these two methods, although different in the procedure, share the common feature to be based on a three-step procedure (Brugnoli *et al.*, 1988; Wanek *et al.*, 2001): (1) removal of soluble sugars by excessive washing with water and/or hydrophilic organic solvents; (2) acids or enzymatic hydrolysis of starch; and (3) purification by precipitation of hydrolysed starch (acid hydrolysis) or by removal of the enzyme by dialysis or ultra-filtration (enzymatic hydrolysis). Both methods are easy to use but can also have some potential problems, such as contamination or incomplete hydrolysis that can affect the isotopic analysis (Richter *et al.*, 2009). These methodological problems may also explain why the analysis of leaf sugars normally gives a better correlation with C_i/C_a compared with starch (Brugnoli *et al.*, 1988).

Irrespective of the problems mentioned above, compound-specific analysis of starch and, especially, leaf sugars has proven to be a very useful method to assess a relatively short-term measure of photosynthetic discrimination in ecophysiological and ecological applications (Fig. 11.3). It can be easily applied to a large number of samples collected directly in the field and extracted later in the laboratory. It can be a useful substitute of gas-exchange measurements where it is necessary to assess C_i/C_a (Fig. 11.3).

11.5.3. Carbon isotopes and water-use efficiency

The relationship between Δ and C_i/C_a in C_3 species also implies a negative relationship between Δ and photosynthetic WUE, also called transpiration efficiency (for a detailed discussion of this subject see Chapter 33). WUE is defined as the ratio of carbon gain to water loss and is given approximately by:

$$WUE = \frac{A_N}{E} = \frac{C_a(1-\frac{C_i}{C_a})}{1.6(e_i - e_a)} \quad \text{[Eqn. 11.14]}$$

where A_N and E are the rates of CO_2 assimilation and of transpiration, respectively; e_i and e_a are the water-vapour pressures inside the leaf and in the surrounding air, respectively; and the factor 1.6 is the ratio of binary diffusivities of water vapour and CO_2 in air, respectively, and therefore, takes into account the ratio of the conductance to water vapour to that to CO_2. Equation 11.14 shows that WUE is negatively correlated with the C_i/C_a ratio and, therefore, a negative correlation with Δ should be also expected (Farquhar *et al.*, 1982). Equation 11.14 expresses the instantaneous ratio of carbon gain to water loss, whereas at the plant level the proportion (Φ_c) of carbon respired at night by whole plants or during the day by non-photosynthetic organs and the proportion (Φ_w) of water transpired at night by stomata or cuticular transpiration, independent of photosynthetic CO_2 fixation, must be taken into consideration (Brugnoli and Farquhar, 2000). Hence, plant WUE is given by:

$$WUE = \frac{C_a(\frac{b-\Delta}{b-a})(1-\Phi_c)}{1.6(e_i - e_a)(1+\Phi_w)} \quad \text{[Eqn. 11.15]}$$

and it is obtained by combining equation 11.14 with 11.8, taking into account Φ_c and Φ_w. In C_4 species, given the negative correlation between Δ and C_i/C_a, the correlation between Δ and WUE should be positive (Brugnoli and Farquhar, 2000).

The formulation of the discrimination model, Δ, has been proposed as an indicator of plant and canopy WUE, and several studies have explored this possibility in many plant species, showing that Δ may be used as a surrogate for WUE (Brugnoli and Farquhar, 2000). The final objective of this application would be the selection for improved yield in water-limited environments. Phenotypic correlation between Δ and grain yield in cereals, for example, is generally high and positive or negative, depending on the tissue analysed and the yield potential of the environment under consideration (for a review see Condon *et al.*, 2004; Seibt *et al.*, 2008). Alternatively, significant variations of mesophyll conductance (g_m) along the day and in response to environmental conditions (see Chapters 3 and 12) lead to non-negligible discrepancies between the actual Δ and that predicted using the simple model neglecting g_m (Bickford *et al.*, 2010). Notwithstanding complications in the correlations between Δ, WUE and yield, much progress has been achieved in this field and some genotypes selected for improved WUE based on isotopic discrimination have been registered and are now commercially available.

11.5.4. Stable carbon isotopes in labelling studies

Analysis of natural abundance of stable isotopes delivers useful information about photosynthesis, as documented in the other sections of this chapter. However, a wealth of information can also be inferred by enrichment of heavier stable isotopes that in nature are less abundant than lighter ones. Retrieval of the enriched isotope in compounds other than that supplied unambiguously reveals intermediates of the biochemical pathway under investigation. In the case of photosynthesis, research has focused on carbon isotopes of its primary substrate, CO_2. It is worth noting that in nature, heavier isotopes ($^{13}CO_2$ or $^{14}CO_2$) represent only a minimal fraction of CO_2. Early experiments have used CO_2 labelled by the radioactive isotope ^{14}C. Seminal studies of timecourse and amount of ^{14}C incorporation into photosynthetic metabolites have allowed determining the sequence of appearance of these metabolites, which has in turn unravelled the early steps of photosynthetic carbon fixation (see Chapter 13). Recent advances in mass spectrometry have made possible the use of the non-radioactive ^{13}C isotope to perform the same labelling studies, avoiding contamination with radioactive material. Measurements of ^{12}C or ^{13}C require mass-spectrometric measurements, but some IRGA are also able to discriminate $^{12}CO_2$ and $^{13}CO_2$, as they are virtually blind to ^{13}C (the wavelengths normally used by IRGAs are not absorbed by $^{13}CO_2$).

The time course of $^{14}CO_2$ or $^{13}CO_2$ fixation has also allowed to partition fluxes of CO_2 between photosynthesis, photorespiration and respiration. This is a very difficult task when leaves are photosynthesising, as the CO_2 uptake by photosynthesis overwhelmingly exceeds the CO_2 release by photorespiration and respiration. As a result, the net exchange of CO_2 between leaf and air is normally measured, and is defined as net photosynthesis. As photorespiration and respiration are labelled more slowly than photosynthesis (the actual delay depends on the flux supplied and on the residence time of the labelled compound in the enclosure used to monitor gas exchanges between leaf and air), quantification of the release of unlabelled CO_2 ($^{12}CO_2$) at different times after labelling of photosynthesis by $^{13}CO_2$ or $^{14}CO_2$ is complete, allows measuring the contribution of the other two processes to the net exchange of CO_2 by photosynthesising leaves (Ludwig and Canvin 1971; Loreto *et al.*, 1999, 2001a; Haupt-Herting *et al.*, 2001).

An interesting implication is that these direct measurements of photorespiratory and respiratory CO_2 emission ($^{12}CO_2$) after supplying enriched CO_2, are often lower than the rates of photorespiration and respiration indirectly estimated by other methods (Loreto *et al.*, 1999, Chapter 4). This is interpreted as evidence that the photorespiratory and respiratory CO_2 is partially recycled by photosynthesis. Accordingly, Loreto *et al.* (2001a) have shown that the amount of $^{12}CO_2$ emitted by respiration is inversely related to the photosynthetic rates of healthy leaves. In stressed leaves, in which photosynthesis is heavily impaired, refixation does not account for the well-recognised low rates of day respiration in the light, and the respiratory CO_2 is entirely emitted (see Chapter 4).

11.6. OXYGEN ISOTOPES IN CO_2, H_2O AND O_2

Oxygen has three natural stable isotopic forms ^{16}O, ^{17}O and ^{18}O, with average concentrations of approx. 99.759%, 0.037% and 0.204%, respectively. Only a small proportion of water molecules contain the heavy isotopes of oxygen (and of hydrogen).

The ^{18}O composition of local precipitation at any given site is approximately dependent on the temperature of droplet formation for rain or snow falling at that site. Water vapour evaporated from oceanic water and advected toward continents undergoes a series of condensations that preferentially removes heavy isotopes (e.g., ^{18}O and deuterium). It is well known that precipitation becomes more depleted in ^{18}O as temperature decreases with altitude and latitude. Hence, in a cloud front heavy water will fall first leaving progressively more depleted precipitation as air masses move across continents (Dansgaard, 1964). There is also an amount effect (Augusti and Schleucher, 2007), with sites characterised by high amounts of precipitation, such as in the tropics, showing more ^{18}O-depleted rainfall than expected at local temperatures. Therefore, there is a significant correlation between $\delta^{18}O$ of precipitation and air temperature below 15°C (Ciais and Meijer, 1998), whereas above 15°C in the tropics there is a negative correlation between $\delta^{18}O$ and the amount of rainfall.

Massive amounts of data are available on isotopic composition of water from global sampling networks (e.g., IAEA, Global network of isotopes in precipitation; http://www-naweb.iaea.org/napc/ih/IHS_resources_gnip.html). Despite the variation depicted above, the mean $\delta^{18}O$ in a given site is rather conservative and dependent on geographic location (Craig, 1961).

In the soil, additional fractionation associated with evaporation from the soil surface may determine an isotopic enrichment and variation in $\delta^{18}O$ along the soil profile. Indeed, soil evaporation results in an isotopic gradient in the top soil layer for approximately 1 m. This variation is useful to assess different water sources used by plants, because plants using water from different depths will show differences in $\delta^{18}O$ and this may be applied to ecophysiological and ecological studies (see reviews by Dawson, 1993; Ehleringer *et al.*, 2000; Dawson *et al.*, 2002). The evaporative enrichment can be very strong near the soil surface but usually evaporation also causes the soil moisture to drop to very low levels (below the wilting point) not useful for plant uptake. Therefore, most of this strong isotopic enrichment is just not 'sensed' by plants and cannot be found in plant xylem water.

It is assumed that there is no substantial isotopic fractionation during water uptake and its transport in the xylem, so that the isotopic composition of water collected from the xylem or any non-transpiring tissue represents a weighted average of the isotopic compositions of different water sources available. Hence, from measurements of $\delta^{18}O$ in xylem water, the source of water mostly used by the plant under those specific conditions can be assessed. Most studies showed no isotopic modifications during water uptake and transport, although several exceptions have been reported owing to evaporation from stems or to exchange with enriched water flowing down the phloem from leaves (see Yakir, 1998 and references herein).

In leaves, water is transpired and an isotope effect occurs, causing a strong enrichment in the heavy isotope ^{18}O in leaf water (Gonfiantini *et al.*, 1965; Farris and Strain, 1978; Barbour, 2007; Farquhar *et al.*, 2007). This is relevant for ecological and global studies because in leaves carbon, water and oxygen cycles are linked and the isotopic enrichment of leaf water is transferred to CO_2 during hydration catalysed by CA, determining the oxygen isotopic exchange between CO_2 and water (De Niro and Epstein, 1979). Therefore, the organic matter produced by photosynthesis will carry the isotopic enrichment of leaf water. Only part of the CO_2 that diffuses into leaves is fixed by photosynthesis, some of it will escape back to the atmosphere, carrying the ^{18}O signature of leaf water. This process of 'retrodiffusion' is very useful to assess the exchange between the atmosphere and the biosphere (e.g., Farquhar *et al.*, 1993; Ciais *et al.*, 1997a,b; Francey and Tans, 1987). Leaf water in the chloroplast is also the substrate for the water-splitting reaction in photosynthesis and determines the ^{18}O composition of photosynthetically produced O_2 (Guy *et al.*, 1987, 1993). This represents the contribution by land vegetation to the Dole effect on a global scale (see Section 11.6.5), i.e., the ^{18}O enrichment in the $\delta^{18}O$ of atmospheric O_2 by 23.5‰ with respect to that of mean ocean water. Hence, this signal can be used to assess changes in the balance between oceanic and terrestrial productivity as terrestrial vegetation will produce O_2 enriched in ^{18}O, whereas oxygen-evolving oceanic organisms produce O_2 roughly of the same isotopic composition of oceanic water (e.g., Ciais *et al.*, 1997a,b).

11.6.1. Isotopic enrichment in leaf water

When water evaporates from the leaf, heavier molecules containing ^{18}O or deuterium (D) tend to be left behind because the vapour pressure of heavy water is less than that of the water molecules containing ^{16}O and ^{1}H. This process continues until the leaf water becomes enriched enough so that the evaporation of heavy and light water molecules matches the isotopic composition of source water provided through the xylem to leaves. In this case leaves have reached the isotopic steady state. The isotopic enrichment of leaf water relative to source water was first shown by Gonfiantini *et al.* (1965), and the theoretical basis is derived from the model of evaporative enrichment on a free-water surface developed by Craig and Gordon (1965), with several modifications by Farquhar *et al.* (1993), Dongmann (1974) and Flanagan *et al.* (1991a). For a more extensive treatise of this subject there is some recent work and excellent reviews also taking into account non-steady state conditions (e.g., Farquhar and Cernusak, 2005; Barbour, 2007; Farquhar *et al.*, 2007; Ogee *et al.*, 2007; Ferrio *et al.*, 2009). The isotopic enrichment at the sites of evaporation can be approximated by:

$$\Delta^{18}O_{es} = \varepsilon^{+} + \varepsilon_k + (\Delta_V - \varepsilon_k)\frac{e_a}{e_i} \qquad \text{[Eqn. 11.16]}$$

where $\Delta^{18}O_{es}$ denotes the Craig–Gordon estimate of the isotopic enrichment of leaf water above source water; Δ_V is the enrichment of atmospheric water vapour relative to source water; ε_k is the kinetic fractionation during diffusion through stomata and leaf boundary layer; $\varepsilon+$ is the equilibrium fractionation associated with the differences in vapour pressure between light ($H_2{}^{16}O$) and heavy molecules ($H_2{}^{18}O$); and e_a and e_i are the water-vapour pressures in the atmosphere and in the intercellular spaces, respectively. Equation 11.16 shows that under constant air-vapour pressure (e_a), leaf isotopic enrichment is negatively related to stomatal conductance, because, for example, increasing the stomatal aperture will cause a decrease in leaf temperature and in e_i owing to increased transpiration. The model predicts general trends in isotopic enrichment in leaf water rather well, but in several experiments substantial deviations from the expected enrichment have been observed (reviewed by Barbour, 2007). In some cases, the observed enrichment was lower than that calculated on the basis of equation 11.16, whereas in others the enrichment was higher than expected. Moreover, strong spatial variations in the isotopic composition (both $^{18}O/^{16}O$ and D/H) along a single leaf have been observed, with a progressive isotopic enrichment from the base to the tip and from the midrib to the edge of the leaf (see review by Barbour, 2007). Several explanations have been invoked, such as the existence of different pools of water within a leaf, with unenriched water in veins and the convection of unenriched water to the sites of evaporation, opposed by backward diffusion of enriched water from the evaporation sites (Péclet effect) resulting in lower enrichment of bulk leaf water. The Péclet effect describes an increased discrepancy between the enrichment predicted by equation 11.16 and that measured in leaf lamina with increasing transpiration rate. It is a dimensionless number, expressed by the ratio of convection to diffusion ($\wp$):

$$\wp = \frac{L_m E}{C_w D_{hw}} \qquad \text{[Eqn. 11.17]}$$

where E is the transpiration rate (mol m^{-2} s^{-1}); L_m is the effective length in the mesophyll (m) over which the effect is evident; C_w is the molar density of water (55.5×10^3 mol m^{-3}); and D_{hw} is the diffusivity of $H_2{}^{18}O$ in water (2.66×10^{-9} m^2 s^{-1}).

The average enrichment in leaf lamina ($\Delta^{18}O_{LS}$) under steady state conditions can be derived as:

$$\Delta^{18}O_{LS} = \frac{\Delta^{18}O_{es}(1-e^{-\wp})}{\wp} \qquad \text{[Eqn. 11.18]}$$

A model including longitudinal and radial Péclet effects was developed by Farquhar and Gan (2003), showing that bulk leaf-water enrichment ($\Delta^{18}O_B$) is given by:

$$\Delta^{18}O_B = \Delta^{18}O_{es}[\phi_x e^{-\wp_r} + \phi_V \frac{e^{\wp_{rv}}-1}{\wp_{rv} e^{\wp_r}} + \phi_L \frac{1-e^{-\wp}}{\wp}] \qquad \text{[Eqn. 11.19]}$$

where ϕ is the proportion of leaf water associated with different tissues and the subscripts x, V and L refer to xylem and surrounding tissue, small veins and lamina mesophyll in that order; $\wp_r$ is the Péclet effect associated with radial flow (i.e., from the veins to the sites of evaporation), gradients in the enrichment in the veinlets ($\wp_{rv}$) and the lamina mesophyll ($\wp$) such that:

$$\wp_r = \wp + \wp_{rv} \qquad \text{[Eqn. 11.20]}$$

However, since the steady state water model did not provide satisfactory predictions, especially under field conditions, a non-steady state model has been developed (Farquhar and Cernusak, 2005) and it gave good agreement with measured leaf-water enrichments.

Recently, Ogee *et al.* (2007) presented a two-dimensional model of isotopic leaf-water enrichment incorporating radial diffusion in the xylem, longitudinal diffusion in the mesophyll, non-uniform gas-exchange parameters and non-steady state effects. Essentially, this model also combined the Gan *et al.* (2002) approach with the Farquhar and Cernusak (2005) non-steady state model. This model accurately predicts most of the published measurements of leaf $\delta^{18}O$ in monocots, except in the leaf tip. Ogee *et al.* (2007) also suggested that differences in enrichment between C_3 and C_4 plants reflected differences in mesophyll tortuosity, which is the most sensitive parameter, rather than in length or in interveinal distance.

11.6.2. Oxygen isotopes in CO_2

The initial interest in the study of isotopic composition of CO_2 was focused on ^{13}C, and only more recently the study of $C^{18}O^{16}O$ has become of increasing interest, following the observation that the ^{18}O signature of atmospheric CO_2 was strongly influenced by the exchange between the atmosphere and the land biosphere (Francey and Tans, 1987; Friedli *et al.*, 1987). It is now well established that the ^{18}O signal in atmospheric CO_2 reflects the coupling between global carbon and hydrological cycles, occurring in the terrestrial vegetation and in soil. As water in leaves is strongly enriched in ^{18}O compared with soil water, it is possible to distinguish, for example, the CO_2 flux released to the atmosphere from the soil and the leaves by analysing its ^{18}O composition (Farquhar *et al.*, 1993; Yakir and Wang, 1996; Ciais *et al.*, 1997a; Yakir, 2003).

During photosynthesis the CO_2 diffusing into leaves readily dissolves into water allowing the rapid exchange of its oxygen with H_2O in the cytoplasm and in the chloroplast, a process strongly dependent on the catalysis by the enzyme CA. In fact, the short residence time of CO_2 in leaves requires the catalysis by CA to fully equilibrate CO_2 on water. Hence, CO_2 will bring a signal strongly enriched owing to the effect of evaporation on leaf water, as previously discussed. Only about one-third of CO_2 diffusing into leaves is fixed by photosynthesis, whereas the other two-thirds diffuse back into the atmosphere after equilibration with water, and, therefore, carrying a signal which is strongly enriched compared with soil CO_2 (Yakir, 2003). This signal would be dependent on the isotopic composition of leaf water, the CO_2 concentration at the equilibration sites and the extent of isotopic equilibrium between CO_2 and H_2O. It is nevertheless believed that equilibrium is reached owing to the very large enzymatic efficiency of CA.

CO_2 in the soil originates from both root respiration and decomposition, and its $\delta^{18}O$ is strongly influenced by the isotopic composition of soil water around it, which is in turn influenced by the isotopic composition of precipitation. Water in a drying soil may turn out to be highly enriched in ^{18}O near the surface, causing a sharp isotopic gradient in soil water (about 10–15‰ enrichments are usual; Yakir, 2003). Nevertheless, it has been shown (Miller *et al.*, 1999) that the ^{18}O enrichment in the top soil layer cannot transfer all of its isotopic signature to the CO_2 released from the soil, because the diffusion of CO_2 out of the soil is faster than the uncatalysed hydration of CO_2.

11.6.3. ^{18}O in organic molecules

The oxygen isotopic composition of plant material is dependent on several factors, not only the ^{18}O composition of the source water taken up by the plant. Indeed, the $\delta^{18}O$ composition of plant material is also controlled by the evaporative enrichment in ^{18}O of leaf water, which is passed to CO_2 through the activity of CA. Carbohydrates formed during photosynthesis should be in equilibrium with the water in which they are formed (see Barbour, 2007), that is, should reflect the equilibrium fractionation of about +27‰ (Sternberg and De Niro, 1983; Sternberg *et al.*, 1986). Subsequently, a fraction of the oxygen atoms in plant carbohydrates are exchanged with those of local water during subsequent metabolism (Barbour *et al.*, 2005). The isotopic exchange occurs between carbonyl oxygen and water (Sternberg *et al.*, 1986), whereas the oxygen atoms of other functional groups, such as carboxyl, hydroxyl and phosphate groups, are not normally exchangeable at cellular conditions of temperature and pH.

The exchange of oxygen between carbonyl groups and water occurs via the formation of short-lived gem-diol intermediates. In triose-phosphates one of the three oxygen atoms are on carbonyl groups and the half-time of equilibration is fast (Sternberg *et al.*, 1986; Farquhar *et al.*, 1998), and it has been demonstrated that exported sucrose is nearly at isotopic equilibrium with average leaf water (Barbour *et al.*, 2000a; Cernusak *et al.*, 2003) with an enrichment of sucrose above source water given by:

$$\Delta^{18}O_{suc}=1.027\ \Delta^{18}O_{LS}+27‰ \qquad \text{[Eqn. 11.21]}$$

While sucrose directly reflects leaf-water enrichment, cellulose is not at full isotopic equilibrium with leaf water, because during sucrose breakdown some oxygen atoms became exchangeable with local unenriched water and can be replaced (Barbour and Farquhar, 2000; Barbour, 2007). This re-exchange of oxygen atoms in sink cells forming new cellulose will dampen the leaf-water signal in plant tissue. This effect has been modelled by Barbour and Farquhar (2000) and extensively studied (e.g., Cernusak *et al.*, 2002, 2005) and it has recently been reviewed (Barbour, 2007).

As the $\delta^{18}O$ of organic matter roughly reflects that of leaf water, which in turn depends on stomata, source water and atmospheric water (Eqn. 11.16), the oxygen-isotope composition in plant material (leaf and phloem sugars, cellulose, bulk biomass) can provide records of variation in stomatal conductance and environmental variables (air humidity, temperature, but see Helliker and Richter, 2008). In particular, the combined study of $\delta^{13}C$ and $\delta^{18}O$ can allow separation of stomatal and photosynthetic effects and limitations. The ^{18}O composition of leaf sugars and of phloem sap is very promising in ecophysiological and ecological applications to study the influence of environmental variables on plant and ecosystem function.

11.6.4. ^{18}O in O_2 in the study of respiration

The last reaction of respiration is the reduction of oxygen to water. In plants and fungi, there are two terminal oxidases, the cyanide-sensitive cytochrome oxidase (COX) and the cyanide-resistant AOX. These reactions show an isotopic fractionation, based on the differential energy required to break the O=O bond. Thus, because the bond $^{18}O{=}^{16}O$ is harder to break than $^{16}O{=}^{16}O$, in a closed system where respiration is taking place, the ratio $^{18}O/^{16}O$ of the remaining O_2 will increase as oxygen is being consumed. The application of the oxygen-isotope fractionation technique to plant respiration was first introduced by Guy *et al.* (1989), observing that the two oxidases present different isotope fractionation against ^{18}O, being larger for the AOX (25‰ for non-green and 30‰ for green tissues) than for the cytochrome oxidase (18–20‰). This differential fractionation permits to measure the activity of each oxidase during respiration.

Presently, this is the only reliable technique to measure the relative participation of the two respiratory pathways (Day *et al.*, 1996; Ribas-Carbo *et al.*, 2005a).

Unlike most of the stable-isotope techniques where the isotopic composition of both substrate and product can be easily measured, the product of the respiratory oxidases is water and hence difficult to obtain. Consequently, oxygen-isotope studies of respiration are performed in a closed system where changes in the substrate are measured (Guy *et al.*, 1989; Ribas-Carbo *et al.*, 2005a).

The respiratory isotope fractionation can be obtained by measuring the oxygen-isotope ratio (R) and the fraction of molecular O_2 remaining at different times during the course of the reaction.

As described by Guy *et al.* (1989), the oxygen-isotope fractionation during respiration will be calculated as:

$$D_r = \ln (R_t/R_o) / -\ln f \qquad \text{Eqn. [11.22]}$$

where R_o and R_t are the $^{18}O/^{16}O$ ratios at the beginning of the experiment and at time t, respectively; f is the fraction of the remaining oxygen at time t: $f = [O_2]/[O_2]_o$; and D_r can be determined by the slope of the linear regression of a plot of ln R_t/R_o versus –ln f, without forcing this line through the origin (Henry *et al.*, 1999). Accurate determinations of D_r can be achieved with experiments consisting of six measurements, provided the r^2 of the linear regression is 0.995 or higher (Ribas-Carbo *et al.*, 1995; Henry *et al.*, 1999).

As it is common practice in the botanical literature to express isotope fractionation as 'Δ' notation, the fractionation factors, D_r, are converted to Δ as described by Guy *et al.* (1993):

$$\Delta = \frac{D_r}{1 - \frac{D_r}{1000}} \qquad \text{[Eqn. 11.23]}$$

The partitioning between the cyt and the alternative respiratory pathways (τ_a) is obtained as described by Ribas-Carbo *et al.* (1997):

$$\tau_a = \frac{\Delta_n - \Delta_c}{\Delta_a - \Delta_c} \qquad \text{[Eqn. 11.24]}$$

where Δ_n is the oxygen-isotope fractionation measured in the absence of inhibitors; and Δ_c and Δ_a are the fractionation by the cyt and alternative pathway, respectively.

Oxygen-isotope fractionation during respiration is thought to play a role in the isotopic composition of oxygen in the atmosphere.

11.6.5. The Dole effect

Oxygen is the second most abundant gas in the atmosphere after nitrogen. Whereas N_2 is almost inert, O_2 is under continuous exchange. On a global scale, O_2 is produced biologically by oxidation of water by oxygenic photosynthesis while being consumed during respiration (Guy *et al.*, 1993; Falkowski and Isozaki, 2008). Although the concentration of O_2 has changed through thousands or millions of years, in the short term, it can be considered that the O_2 concentration in the atmosphere reflects a global 'compensation point' (Guy *et al.*, 1993).

The isotopic composition of O_2 in the atmosphere ($^{18}O/^{16}O$) is higher than that of average seawater by 23.5‰ (Kroopnick and Craig, 1972) and this is called the Dole effect (Dole, 1935; Morita, 1935). No isotopic discrimination has been observed during the production of O_2 during photosynthesis (Stevens *et al.*, 1975; Guy *et al.*, 1993; Tcherkez and Farquhar, 2007). Therefore, it is thought to be mainly attributable to: (1) isotopic discrimination during respiration (Guy *et al.*, 1989, 1993), which preferentially removes ^{16}O from the air discriminating against ^{18}O (Lane and Dole, 1956); (2) the ^{18}O leaf-water enrichment during evaporation or transpiration (Dongmanm, 1974; Förstel, 1978); (3) the isotopic fractionation associated with the equilibrium between dissolved and gaseous O_2 (Bender *et al.*, 1994); and (4) photorespiration and the Mehler reaction that both discriminate against ^{18}O (Canvin *et al.*, 1980; Guy *et al.*, 1993). Although the origin of the Dole effect appears to be complex, in the short-term the isotopic composition of O_2 is quite constant with a variability as low as 0.25‰ (Dole *et al.*, 1954; Kroopnick and Craig, 1972), therefore suggesting that atmospheric O_2 is nearly at a steady state.

In the long-term scale of millions of years, the presence of O_2 in the atmosphere is owing to a larger O_2 production by photosynthesis than oxygen consumption by respiration or other oxidations. These imbalances can be explained by tectonic movement in which organic matter deposited in the sediments is buried in the continents (Falkowski and Isozaki, 2008).

11.7. STABLE ISOTOPES IN ECOSYSTEM STUDIES

The study of carbon and oxygen isotopes in plants, soil and atmospheric CO_2 can help gain insight into the mechanistic processes, both physical and biological, responsible for the exchange of CO_2 and water between the atmosphere and terrestrial vegetation. Stable-isotope analysis has been used to identify global carbon sources and sinks and to separate different component fluxes (Ciais *et al.*, 1995b; Lloyd *et al.* 1996; Fung *et al.*, 1997; Ciais *et al.*, 1997a,b). Distinguishing photosynthetic and respiratory fluxes is relevant to obtain insights in the processes governing ecosystem responses to climatic conditions and to global change. Combining $\delta^{13}C$ and $\delta^{18}O$ analyses and concentration measurements makes it possible to separate the net ecosystem exchange (NEE) into gross photosynthetic and respiratory flux components, and to identify the autotrophic and heterotrophic components affecting these fluxes (Yakir and Wang, 1996; Bowling *et al.*, 2001; Yakir, 2003).

At the global level, the mass-balance approach can be used to separate the relative magnitude of different fluxes (ocean exchange, land biosphere, biomass and fossil-fuel combustion, stratospheric reactions). At the ecosystem level, measurements of NEE are obtained routinely by the eddy covariance (EC) approach as part of the FLUXNET network (http://www.fluxnet.ornl.gov). However, partitioning NEE into photosynthetic and respiratory fluxes is a relevant and difficult challenge, and stable isotopes provide an independent and useful methodology for this purpose. The combination of isotopic and micrometeorological measurements (Yakir and Wang, 1996; Bowling *et al.*, 2001) can allow partitioning of NEE into component fluxes. For this purpose, it is essential to obtain reliable assessment of: (1) canopy photosynthetic discrimination (Δ_{canopy}); (2) the isoflux of ^{13}C, an approximation of the vertical isotopic flux; and (3) the isotopic ratio of ecosystem respiration (δ_R).

Estimates of Δ_{canopy} can be obtained directly by measuring online Δ at the canopy level (Lloyd *et al.*, 1996), or indirectly by estimating the integrated canopy stomatal conductance through the Penman-Monteith equation (Bowling *et al.*, 2001). Alternatively, one could measure the $\Delta^{13}C$ of ecosystem components (e.g., leaves, shoots, roots, soil organic matter, litter, etc.) and pools and scale-up to the ecosystem level. Such measurements would need to integrate autotrophic and heterotrophic ecosystem components, and many studies have traditionally been focused on total bulk carbon of various components. More recently, the study of recently fixed carbon has been shown to be useful to estimate short-term photosynthetic Δ, because it gives insights into the prevailing environmental conditions at the time of sampling or immediately before (i.e., one day before). Several studies have measured $\delta^{13}C$ and $\delta^{18}O$ in leaf sugars and phloem sucrose to estimate photosynthetic Δ and to separate changes in stomatal limitations and in photosynthetic rates (Pate and

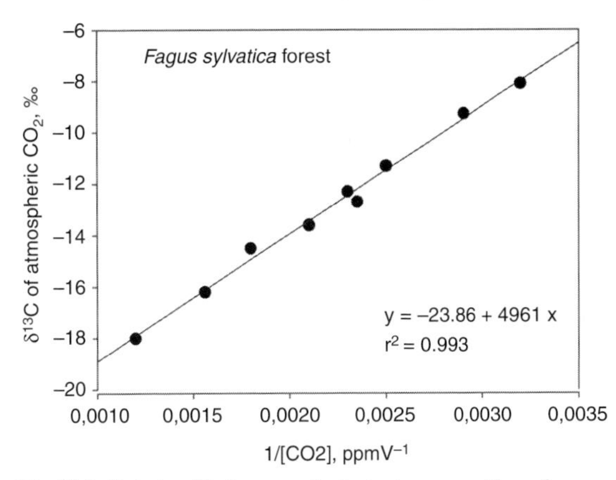

Fig. 11.4. Relationship between the isotopic composition of atmospheric CO_2 and $1/[CO_2]$ (Keeling plot). The intercept with y axis provides an estimate of the isotopic composition of respired CO_2 (δ_R).

Arthur, 1998; Gessler *et al.*, 2001; Cernusak *et al.*, 2003; Keitel *et al.*, 2003; Scartazza *et al.*, 2004; Keitel *et al.*, 2006; Gessler *et al.*, 2008). In particular, these reports have shown that phloem carbon integrates over the whole canopy in the short term and provides a reliable estimate of canopy C_i/C_a and/or canopy stomatal conductance. As in the short term phloem carbon would provide substrates for autotrophic respiration, phloem $\delta^{13}C$ can be compared with δ_R to assess the autotrophic and heterotrophic contribution to ecosystem respiration, and to estimate the time lag between carbon fixation and its subsequent respiration by various plant organs (Scartazza *et al.*, 2004; Gessler *et al.*, 2009b).

Estimates of isoflux traditionally rely on flask sampling and the isoflux is given by:

$$Isoflux = \overline{\rho w'(\delta_a C_a)'} \qquad \text{[Eqn. 11.25]}$$

where ρ is the molar air density; w is the vertical wind speed; and δ_a and C_a are the isotopic composition and concentration of ambient CO_2, respectively. The primes indicate fluctuations from the mean values and the overbar indicates time averaging.

The isotopic ratio of ecosystem respiration can be estimated using the Keeling plot method, as the y intercept of the mixing relationship between $\delta^{13}C$ and the inverse of concentration of atmospheric CO_2 (Fig. 11.4; Keeling, 1958).

Past approaches to quantify the isotopic ecosystem–atmosphere exchange, such as relaxed eddy accumulation, the flask-based isoflux method and the flux-gradient technique relied on flask sampling and subsequent laboratory analysis. More recently, the development of TDLAS allows continuous measurements of stable-isotope ratios at high temporal resolution (Bowling *et al.*, 2003; Zhang *et al.*, 2006; Griffis *et al.*, 2008). This relatively new technology is very promising for obtaining continuous long-term measurements of isotopic CO_2 and water-vapour exchange (Wen *et al.*, 2008) between ecosystems and the atmosphere, and to study photosynthesis and respiration at the ecosystem level.

11.8. CONCLUDING REMARKS AND FUTURE PERSPECTIVES

The study of stable isotopes, together with other information obtained from physiological, biochemical, molecular and modelling data can provide relevant insights into photosynthetic and respiratory processes at very different levels of complexity, from the organelle (chloroplasts and mitochondria) to leaves and whole plants, and scaling-up to ecosystem, regional and global-level studies. In particular, significant advancements in these fields are expected from the quantitative understanding and association of carbon and oxygen isotopes in plants and in photosynthesis/respiration. Further promising perspectives are expected from modelling approaches and from an increasing collaboration between experimentalists and modellers.

New technological developments, such as compound-specific isotope analysis (GC-IRMS, GC-C-IRMS, LC-IRMS), are expected to allow significant advancements in our understanding of photosynthetic and respiratory isotope fractionations and into the subsequent isotopic effects during plant metabolism. The recent development and increasingly wide availability of TDLAS techniques allows the continuous measurements of isotopic mixing ratios of CO_2 and water vapour in ecophysiological and ecological studies. These will provide further insight into the partitioning of photosynthetic and respiratory fluxes and ecosystem discrimination, allowing the quantitative approach to ecosystem carbon budget relevant for global-change studies.

12 • Mesophyll conductance to CO_2

J. FLEXAS, E. BRUGNOLI AND C.R. WARREN

12.1. HISTORICAL VIEW: THE CONCEPT OF MESOPHYLL CONDUCTANCE TO CO_2 AND ITS PERCEPTION IN PLANT SCIENCE

During photosynthesis, CO_2 has to move from the atmosphere surrounding the leaf, across a boundary layer in the air above the foliage surface, to the sub-stomatal internal cavities through the stomata (Fig. 12.1A) and from there to the site of carboxylation inside the chloroplast stroma through the leaf mesophyll (Fig. 12.1B). From Fick's first law of diffusion, the net photosynthetic flux at steady state (A_N) can be expressed as:

$$A_N = g_s (C_a - C_i) = g_m (C_i - C_c) \quad \text{[Eqn. 12.1]}$$

where g_s and g_m are the stomatal (g_s) and mesophyll (g_m) conductance to CO_2 diffusion; and C_a, C_i and C_c are the CO_2 concentrations in the atmosphere, in the sub-stomatal internal cavity and in the chloroplast stroma, respectively (Long and Bernacchi, 2003).

In Gaastra's (1959) pioneering work on leaf photosynthesis, g_m (and its inverse, mesophyll resistance) was defined essentially as a diffusion component of the photosynthesis pathway, and it was considered to be an important factor in determining leaf photosynthesis. Later, Troughton and Slatyer (1969) extended the use of the term to a 'mixed' diffusion-biochemical component, i.e., to refer to the initial slope of A_N versus C_i relationship. Following Gaastra's assumption that CO_2 concentration inside the chloroplast was near zero or close to the CO_2 compensation concentration, Jones and Slatyer (1972) proposed a method to separate the transport and carboxylation components of what continued to be called mesophyll or intracellular conductance. Using this method, earlier conclusions by Gaastra that mesophyll-transport resistance was a limiting factor for photosynthesis were confirmed (Jones and Slatyer 1971, 1972; Samsuddin and Impens, 1979), and it was suggested for the first time that the leaf internal resistance to CO_2 transfer could be variable, and respond for instance to water stress (Jones, 1973). The nomenclature was further complicated by Samsuddin and Impens (1979), who termed internal resistance a combined-diffusion biochemical term, while restoring the use of the term mesophyll resistance to refer to the transfer component alone. Such a semantic confusion continues nowadays, and several more-recent papers still call 'mesophyll conductance' or 'internal conductance' the initial slope of A_N versus C_i relationship. We propose, as did Farquhar and Sharkey (1982) and Parkhurst (1994), to restrict the term 'mesophyll conductance' to the diffusion of CO_2 through leaf mesophyll, including intercellular air spaces (g_{ias}), cell wall (g_w) and the intracellular liquid pathway (g_{liq}). That is, g_m should be viewed as synonymous of 'leaf internal-diffusion conductance', whereas terms such as 'apparent carboxylation efficiency' or 'apparent carboxylation conductance' more adequately refer to the initial slope of the A_N versus C_i relationship.

With the introduction of the most commonly used leaf photosynthesis model by Farquhar *et al.* (1980a), the early assumption that CO_2 concentration in the chloroplasts was close to zero or to the compensation point was rejected, as was later confirmed by direct measurements of C_i (Sharkey *et al.*, 1982; Lauer and Boyer, 1992). Thereafter, most gas-exchange studies have usually assumed that g_m was large and constant, i.e., that C_i=C_c. However, different studies in the late eighties and early nineties already suggested that C_c is significantly less than C_i, although not close to zero. Evans (1983) and Evans and Terashima (1988) reached this conclusion by comparing the initial slope of the A_N versus C_i curve in wheat and spinach leaves with the activity of Rubisco as determined *in vitro*. Evans *et al.* (1986) reached the same conclusion by comparing online carbon-isotope

Terrestrial Photosynthesis in a Changing Environment: A Molecular, Physiological and Ecological Approach, ed. J. Flexas, F. Loreto and H. Medrano. Published by Cambridge University Press.

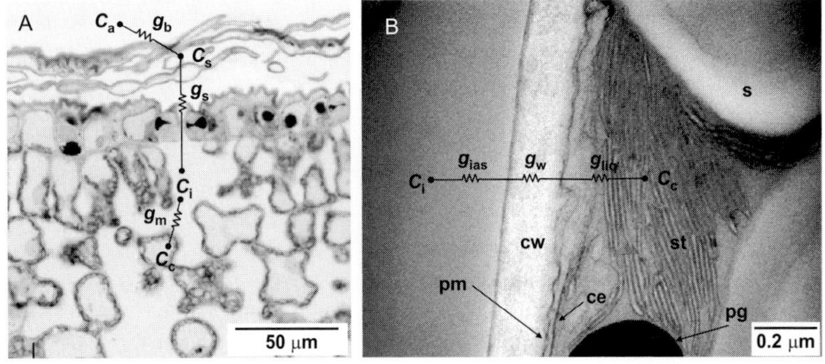

Fig. 12.1. **(A)** Micrograph of the abaxial surface of an olive leaf (bottom side up), where stomata can be seen, as well as the pathway of CO_2 from ambient (C_a) through leaf surface (C_s) and intercellular air spaces (C_i) to chloroplast (C_c). Boundary layer conductance (g_b), stomatal conductance (g_s) and mesophyll conductance (g_m) are indicated. **(B)** Electron micrograph of a grapevine leaf where cell wall (cw), plasma membrane (pm), the chloroplast envelope (ce) and stroma thylakoid (st) can be observed. The pathway of CO_2 from C_i to chloroplastic CO_2 (C_c) is characterised by intercellular air-space conductance to CO_2 (g_{ias}), through the cell wall (g_w) and through the liquid phase inside the cell (g_{liq}). A grain of starch (s) and a plastoglobule (pg) can be also observed in the picture (Photos by A. Diaz-Espejo; from Flexas *et al.* 2008).

discrimination of photosynthesising leaves with the theoretical discrimination that would be expected if C_i=C_c. With the advent of pulse-amplitude fluorometers, the comparison of Ch-F with gas-exchange measurements also indicated that C_c was lower than C_i, and that their difference increased under conditions of water stress or salinity (Bongi and Loreto, 1989; Cornic *et al.*, 1989; Di Marco *et al.*, 1990). The water stress-induced decrease of C_c below C_i was independently demonstrated by measuring leaf^{18}O uptake (Renou *et al.*, 1990).

12.2. CURRENT STATE-OF-THE-ART OF STUDIES ON MESOPHYLL CONDUCTANCE OF CO_2

After the founding works mentioned in the previous section, much evidence during the last decade has accumulated showing that g_m is sufficiently small as to significantly decrease C_c relative to C_i, therefore limiting photosynthesis (Evans and Loreto, 2000). Moreover, g_m is not constant as it has been shown to acclimate during leaf development (Miyazawa and Terashima, 2001; Marchi *et al.*, 2008) and senescence (Loreto *et al.*, 1994; Flexas *et al.*, 2007c; Zhang *et al.*, 2008b; Egea *et al.*, 2011; Whitehead *et al.*, 2011), as well as to light (Piel *et al.*, 2002; Li *et al.*, 2008; Monti *et al.*, 2009), temperature (Yamori *et al.*, 2006; Bunce, 2008) and CO_2 (Singsaas *et al.*, 2004; Velikova *et al.*, 2009) conditions during growth. Within large trees, g_m also varies depending on the tree size, as well as on the leaf height within a tree (Mullin *et al.*, 2009; Woodruff *et al.*, 2009; Han, 2011; Whitehead *et al.*, 2011). There is also evidence for rapid variation of g_m in response to drought (Brugnoli *et al.*, 1998; Flexas *et al.*, 2002a; Warren *et al.*, 2004; Fleck *et al.*, 2010), salinity (Bongi and Loreto, 1989; Loreto *et al.*, 2003), leaf temperature (Bernacchi *et al.*, 2002; Warren and Dreyer, 2006), CO_2 concentration (Centritto *et al.*, 2003; Flexas *et al.*, 2007a; Vrábl *et al.*, 2009; Yin *et al.*, 2009; Bunce, 2010; Schäufele *et al.*, 2011; Tholen and Zhou, 2011 – but see Tazoe *et al.*, 2009, 2011) and light intensity and quality (Flexas *et al.*, 2007a; Hassiotou *et al.*, 2009a; Loreto *et al.*, 2009). Mesophyll conductance also varies when plants grow with different nitrogen (Yin *et al.*, 2009; Yamori *et al.*, 2011), phosphorous (Bown *et al.*, 2009) and zinc concentrations (Sagardoy *et al.*, 2010). As a consequence of its responses to many environmental factors, g_m varies diurnally (Araujo *et al.*, 2008; Bickford *et al.*, 2009; Grassi *et al.*, 2009) and seasonally (Montpied *et al.*, 2009). Moreover, the response of g_m has been shown to be as rapid and reversible as that of g_s (Centritto *et al.*, 2003), or even faster (Flexas *et al.*, 2007a). Owing to its variability and importance as a limiting factor for photosynthesis, efforts have commenced to adapt photosynthesis and isotope-discrimination models to accommodate for a variable g_m (Bickford *et al.*, 2009; Niinemets *et al.*, 2009a; Yin *et al.*, 2009; Keenan *et al.*, 2010a,b).

Despite substantial evidence for large variability of g_m, the mechanistic basis of these variations remains unclear. Early literature highlights that leaf structural properties correlate with variations in g_m (Lloyd *et al.*, 1992; von

Caemmerer and Evans, 1991). In fact, g_m has been shown to correlate with some leaf structural properties, such as phytochrome-related mutation-induced chloroplast re-arrangements (Sharkey *et al.*, 1991), the chloroplast surface directly exposed to intercellular air spaces (Evans and Loreto, 2000), leaf dry mass per area (LMA) (Flexas *et al.*, 2008; Hassiotou *et al.*, 2009a; Niinemets *et al.*, 2009b) and cell-wall thickness (Evans *et al.*, 2009; Scafaro *et al.*, 2011), as well as with some dynamic structural changes such as chloroplast movements (Tholen *et al.*, 2008). In view of all these evidences, leaf functional anatomy has recently attained high attention to explain variations of g_m (Terashima *et al.*, 2011), and, indeed, a three-dimensional model considering the basic structure of a cell and its components, together with simple diffusion coefficients/permeabilities of CO_2 in the different cell media (intercellular air, lipid membranes, cytosol and chloroplast stroma water) reasonably estimates g_m (Tholen and Zhu, 2011). However, although structural properties could certainly be involved in adaptive and acclimation responses, they could not account for the rapid variations observed in response to varying environmental conditions. Most likely, a metabolic process is involved in g_m changes in these cases. Based on a temperature-response coefficient (Q_{10}) of approximately 2.2 for g_m in tobacco leaves, Bernacchi *et al.* (2002) speculated that enzymatic or protein-facilitated diffusion of CO_2 controls g_m. The most likely candidates for the most dynamic g_m changes would be CA and aquaporins.

Some authors have suggested that CA activity is closely associated with g_m in C_3 plants (Makino *et al.*, 1992; Sasaki *et al.*, 1996). However, modification of CA activity revealed little or no change in g_m and photosynthesis (Price *et al.*, 1994; Williams *et al.*, 1996). Lately, Gillon and Yakir (2000a) showed that the relative contribution of CA to the overall g_m is species dependent. They hypothesised that CA-mediated CO_2 diffusion may be more important when g_m is low owing to structural properties of the leaves, as seemed to be the case for woody species, where cell-wall conductance was suggested to be much lower than chloroplast conductance.

Concerning a possible role for aquaporins, the first indirect evidence was provided by Terashima and Ono (2002), who impaired g_m to CO_2 by $HgCl_2$ (a non-specific inhibitor of some aquaporins). Uehlein *et al.* (2003) demonstrated that tobacco aquaporin NtAQP1 facilitates transmembrane CO_2 transport by expression in *Xenopus* oocytes. More recently, substantial evidence has been compiled clearly demonstrating a role of some specific aquaporins in the regulation of g_m (Hanba *et al.*, 2004; Flexas *et al.*, 2006a; Uehlein *et al.*, 2008; Heckwolf *et al.*, 2011), which can be specifically important under drought (Miyazawa *et al.*, 2008). Rapid aquaporin-mediated changes of g_m could be explained by aquaporin gating mediated by light, ROS, phosphorylation and pH (Tournaire-Roux *et al.*, 2003; Törnroth-Horsefield *et al.*, 2006; Maurel *et al.*, 2008; Kim and Steudle, 2009), although there is still some controversy about these mechanisms (Fischer and Kaldenhoff, 2008).

In summary, during the last decade it has been evidenced that g_m is sufficiently small as to significantly limit photosynthesis, and moreover it is not constant so that its contribution as a limiting factor depends on environmental factors. The mechanistic basis of these variations remains unclear. Although structural properties could be involved in adaptive and acclimation responses, the most likely candidates for the most dynamic g_m changes would be CA and aquaporins. For more detailed reviews on the current state-of-the-art on g_m to CO_2 see Flexas *et al.* (2008); Warren (2008); Evans *et al.* (2009); Niinemets *et al.* (2009c); and Pons *et al.* (2009).

12.3. EARLY METHODS TO ESTIMATE MESOPHYLL CONDUCTANCE TO CO_2

12.3.1. Anatomical methods

The very earliest estimates of g_m were based on the assumption that only anatomical properties of leaves influenced g_m, i.e., g_m could not be regulated in the short term (Raven and Glidewell, 1981; Nobel, 2005). Although there is now ample evidence that g_m is actually determined by a complex interaction of biochemical and anatomical factors and not simply by leaf anatomy (Evans *et al.*, 2009), we will briefly describe these early methods here. Notice that all the parameters involved are either physical constants or anatomical characteristics that can be determined in leaf sections with the help of a microscope (see Evans *et al.*, 1994; Syvertsen *et al.*, 1995 for details on such measurements).

Nobel *et al.* (1975) showed that the CO_2 assimilation rate could be correlated with the surface area of mesophyll cell walls (A_{mes}) exposed to intercellular air spaces per unit leaf area (A_{leaf}), i.e., $S_m = A_{mes}/A_{leaf}$. Based on this finding, and the fact that the resistance to diffusion of CO_2 across a given barrier, *j* (e.g., the mesophyll, the chloroplast, etc.), equals the reciprocal of its permeability coefficient (P^j_{CO2}), Nobel (1991b) described a simple method to estimate the CO_2 conductance for any given cellular compartment, j (g^j_{CO2}), expressed on a leaf-area basis, as follows:

$$g^{j}{}_{CO2} = \frac{S_m}{P^{j}{}_{CO2}}$$ [Eqn. 12.2]

following the definition of the permeability coefficient ($P^{j}{}_{CO2}$) as:

$$P^{j}{}_{CO2} = \frac{D_{jCO2}K_{jCO2}}{\Delta_x}$$ [Eqn. 12.3]

where Δ_x is the thickness of the barrier x; D_{jCO2} is the diffusion coefficient of CO_2 in it; and K_{jCO2} is a suitably defined partition coefficient.

Using a similar approach but increasing complexity, Syvertsen *et al.* (1995) defined equations to assess g_{ias} and g_{liq} separately based on anatomical measurements. Concerning g_{ias}, it was defined as follows:

$$g_{ias} = \frac{(f_{ias})^{1.55}}{\alpha pl}$$ [Eqn. 12.4]

where *l* represents the mesophyll thickness; f_{ias} is the fraction of the mesophyll volume that is actually intercellular air spaces; the 1.55 power accounts for a modelled tortuosity in the diffusion path through small pores, and the parameter αp is a fitted constant equivalent to:

$$\alpha p = \frac{P_{atm}}{\rho D_{airCO2}}$$ [Eqn. 12.5]

where P_{atm} is the atmospheric pressure; ρ is the molar density of air; and D_{airCO2} is the diffusion coefficient of CO_2 in air.

Concerning g_{liq}, it was defined either as a function of S_m as by Nobel (1991), or as a function of the surface area of chloroplasts exposed to ias per A_{leaf} (S_c), following the conclusions by von Caemmerer and Evans (1991) that this parameter was a better estimator of the effective surface area available for CO_2 diffusion than S_m. The resulting equation was:

$$g_{liq} = \frac{S_m}{\beta} \text{ or } g_{liq} = \frac{S_c}{\gamma}$$ [Eqn. 12.6]

where β and ϒ are constants proportionally related to the tissue density T_d.

12.3.2. Measurements using normal air and helox

In the presence of 21% O_2, the diffusivity of CO_2 is about 2.33 times greater in a mixture of helium and O_2 than in nitrogen/O_2. Based on this effect, Parkhurst and Mott (1990) developed a method to assess the relative limitation imposed to photosynthesis by restricted CO_2 diffusion in ias (hence, strictly speaking, this was not a method to assess g_m or g_{ias}, but to assess g_{ias}-related photosynthesis limitation). Essentially, the method consisted of performing gas-exchange measurements in leaves under two different gas compositions: normal air and helox (an air mixture where N_2 is replaced by He). In the latter medium, A_N was usually slightly increased, which means that finite g_{ias} limits photosynthesis to some extent. Because the diffusivity of CO_2 in helox is increased, but is not infinite, g_{ias} limitation is not totally eliminated. Actually, the intercellular diffusion would be completely non-limiting when $1/D_{CO2}=0$. Assuming that A_N is approximately a linear function of the reciprocal of CO_2 diffusivity, a rough estimate of the total intercellular diffusion limitation can be obtained by plotting A_N versus $1/D_{CO2}$ using data from air and helox, and extrapolating A_N when $1/D_{CO2}=0$.

12.3.3. The initial-slope method

Evans (1983) and Evans and Terashima (1988) proposed a method to estimate g_m based on the comparison of the initial slope of A_N response to C_i at low C_i (ε), and the theoretical initial slope of A_N response to C_c (*k*), the latter derived from in-vitro measurements of Rubisco activity (Table 12.1). *k* can be described using Michaelis-Menten kinetics (Farquhar *et al.*, 1980a) as:

$$k = \frac{V_{c,\max}}{\Gamma^* + K_c(1 + O/K_o)}$$ [Eqn. 12.7]

$V_{c,max}$ is determined either from measured Rubisco content and assumed specific activity ($V_{c,max}$ = Rubisco content × specific activity), or in-vitro measurements of Rubisco activity (see methods in Chapter 13).

Therefore, *k* reflects $V_{c,max}$, i.e., Rubisco activity. By contrast, the initial slope of A_N to C_i (ε) may not directly reflect $V_{c,max}$ because finite g_m reduces ε compared with *k*. Assuming that the overall discrepancy between ε and *k* originates from g_m, the latter can be calculated by comparing ε and *k* as follows:

$$g_m = \frac{k\varepsilon}{k - \varepsilon}$$ [Eqn. 12.8]

This procedure only works properly if there is a sufficiently large range in ε and *k* (typically greater than twofold), which many times would imply comparing different leaves (e.g.,

Table 12.1. *Characteristics of the methods available for the estimation of mesophyll conductance.*

	Photocomp. point method	Curve-fitting method	Initial-slope method	Single-point instantaneous Δ^{13}	Slope-based instantaneous Δ^{13}	Δ^{13} in soluble sugars	^{18}O consumption	Constant J method	Variable J method
Methods required	Gas exchange	Gas exchange	Gas exchange Rubisco activity / amount	Gas exchange Isotope discrimination	Gas exchange Isotope discrimination	Gas exchange Sugar extraction Isotope discrimination	Gas exchange Atm. with $^{18}O_2$ Isotope consumption	Gas exchange Chl-F	Gas exchange Chl-F
Assumptions concerning the responses of g_m itself	g_m does not respond to CO_2 concentration	g_m does not respond to CO_2 concentration	g_m does not respond to CO_2 concentration		g_m does not respond to light intensity and / or CO_2 concentration	g_m does not change fast with time	g_m does not change in an atmosphere with $^{18}O_2$ only	g_m does not respond to CO_2 concentration	
Assumed facts and values (not measured)	Different kind of leaves (e.g., old/young, sun/shade) are comparable	K_c, K_o and their temperature-dependencies	Different kind of leaves (e.g. old/young, sun/shade) are comparable Quantitative estimates of Rubisco activity are possible *in vitro* K_c, K_o and their temperature-dependencies	Fractionations in: -boundary layer -diffusion in air -dissolution and diffusion of CO_2 in water -Rubisco + PEP carboxylation -mitochondrial respiration (negligible) -photorespiration (negligible)	Fractionations in: -boundary layer -diffusion in air -dissolution and diffusion of CO_2 in water - Rubisco + PEP carboxylation	Fractionations in: -boundary layer -diffusion in air -dissolution and diffusion of CO_2 in water -Rubisco + PEP carboxylation -mitochondrial respiration -photorespiration	All products from J are used in carboxylation or oxygenation (i.e., alternative electron sinks are negligible)	All products from J are used in carboxylation or oxygenation (i.e., alternative electron sinks are negligible)	All products from J are used in carboxylation or oxygenation (i.e., alternative electron sinks are negligible) 4–6 electrons are required per each carboxylation Chl-F and gas exchange are comparable The contribution of PSI to Chl.-F is negligible
Required parameters (measured or estimated)	A_N C_i R_D C_i^* and Γ^*	A_N C_i R_D Γ^* or $S_{c/o}$	A_N C_i R_D Γ^* or $S_{c/o}$ Rubisco activity	A_N C_i Δ_{obs}	A_N C_i Δ_{obs}	A_N C_i Δ_{obs}	A_N C_i R_D Γ^* U_o	A_N C_i R_D Γ^* or $S_{c/o}$	A_N C_i R_D Γ^* or $S_{c/o}$ J (ϕ_{PSII}, PPFD, α, β)

young versus old, sun versus shade). In addition to this difficulty, the use of this method is further complicated by the fact that substantial amounts of Rubisco are lost during common extraction procedures, leading to underestimations of the true in-vivo Rubisco activity (Rogers *et al.*, 2001; Chapter 13).

12.4. ISOTOPIC METHODS

12.4.1. The online ^{13}C-discrimination method

Estimation of internal conductance from carbon-isotope discrimination is based on simultaneous measurements of leaf gas exchange and carbon-isotope discrimination, for which a mass spectrometer (see Chapter 11) or a tunable laser diode (Chapter 10) must be used. Discrimination occurs owing to different diffusion and carboxylation rates of $^{12}CO_2$ and $^{13}CO_2$. $^{13}CO_2$ diffuses more slowly through the boundary layer and stomata and through the liquid phase, and is carboxylated much more slowly than $^{12}CO_2$ by Rubisco and PEPC (Farquhar *et al.*, 1982; Evans *et al.*, 1986).

In mathematical terms, carbon-isotope discrimination (Δ) is (see also Chapter 11):

$$\Delta = \frac{a_b(C_a - C_s)}{C_a} + \frac{a(C_s - C_i)}{C_a} + \frac{a_i(C_i - C_c)}{C_a} + \frac{bC_c}{C_a} - \frac{(eR_d / k + f\Gamma^*)}{C_a} \quad \text{[Eqn. 12.9]}$$

where $\Delta = R_{air}/R_{product} - 1$ and R_{air} and $R_{product}$ are the molar ratios of $^{13}CO_2/^{12}CO_2$ in the air and the photosynthetic product, respectively. In this model, discrimination is a function of (Table 12.1) the concentrations of CO_2 in air (C_a), at the leaf surface (C_s), in the intercellular air spaces (C_i), in the chloroplast (C_c); and fractionations owing to diffusion through the boundary layer (a_b, 2.9‰), diffusion through stomata (a, 4.4‰), diffusion and dissolution of CO_2 into water (a_i, 1.8‰), net fractionation by Rubisco and PEPC (b, 27–30‰), fractionation owing to mitochondrial respiration (e), and fractionation owing to photorespiration (f). If one ignores fractionation owing to the boundary layer, this simplifies to:

$$\Delta = \frac{a(C_a - C_i)}{C_a} + \frac{a_i(C_i - C_c)}{C_a} + \frac{bC_c}{C_a} - \frac{(eR_d / k + f\Gamma^*)}{C_a} \quad \text{[Eqn. 12.10]}$$

Therefore, the measured instantaneous carbon-isotope discrimination ($\Delta^{13}{}_{obs}$) must be proportional to the concentration of CO_2 in chloroplasts (C_c), whereas standard gas-exchange measurements estimate the concentration of CO_2 in the intercellular spaces (C_i) and net photosynthesis (A_N), allowing calculation of g_m re-arranging equation 12.10 using equation 12.1:

$$\Delta_i - \Delta_{obs} = \frac{\frac{A_N(b - a_i)}{C_a}}{g_m} + \frac{(eR_d / k + f\Gamma^*)}{C_a} \quad \text{[Eqn. 12.11]}$$

where Δ_i is the expected carbon-isotope discrimination when g_m is infinite (i.e., when $C_i = C_c$):

$$\Delta_i = \frac{a(C_a - C_i)}{C_a} + \frac{bC_i}{C_a} \quad \text{[Eqn. 12.12]}$$

Equation 12.12 is akin to a null hypothesis. If $\Delta^{13}{}_{obs}$ were the same as Δ_i, we would conclude that g_m is infinite. In reality, however, $\Delta^{13}{}_{obs}$ is lower than Δ_i (Fig. 12.2), and it is this difference that allows us to calculate g_m (Eqn. 12.11). g_m may be calculated directly from equation 12.11 by ignoring the terms involving respiration and photorespiration ('single-point' method). Alternatively, from different measurements (e.g., at different light intensities or CO_2 concentrations), one can plot ($\Delta_i - \Delta^{13}{}_{obs}$) as a function of A_N/C_a, with the slope yielding an estimate for $1/g_m$, and the A_N/C_a intercept an estimate of the combined respiratory terms ('slope-based' method). In principle, ignoring the respiratory terms can lead to biased estimations (Gillon and Griffiths, 1997), although Flexas *et al.* (2007b) have shown that using the 'single-point' method in an O_2-free atmosphere (to decrease the respiration and the photorespiration components of discrimination) results in estimations of g_m similar to those in normal air, suggesting that ignoring the respiratory terms is not affecting g_m estimates significantly. Alternatively, using the 'slope-based' method requires measuring the same leaves under different conditions, such as different light intensities or CO_2 concentrations, under the assumption that g_m does not respond to these environmental variables. However, it has been recently shown that these variables strongly affect g_m, which questions the use of the 'slope-based' method. Nevertheless, the scarce studies in which the 'single-point' and the 'slope-based' method have been compared over the same plants have revealed similar results using either method (e.g., Flexas *et al.*, 2006a) but not in every case, so it cannot be simply ignored (Tazoe *et al.*, 2009).

Regardless of whether the 'single-point' or the 'slope-based' method is used, to determine g_m from discrimination,

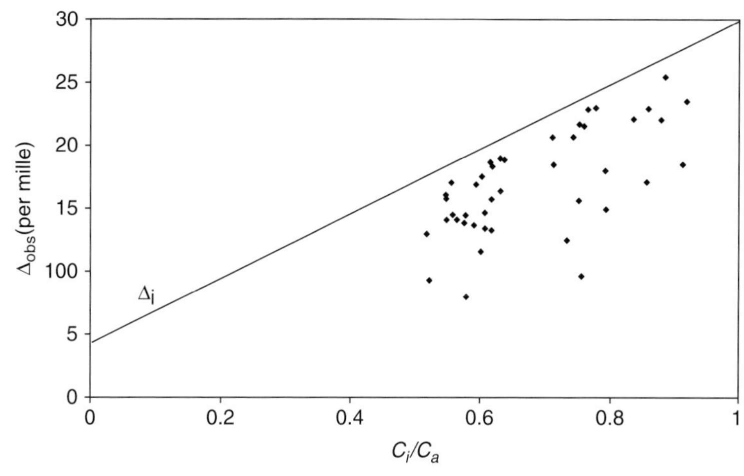

Fig. 12.2. Relationship between the ratio of intercellular to ambient CO_2 concentrations (C_i/C_a) and measured discrimination against $^{13}CO_2$ during photosynthesis (Δ_{obs}) for a single Douglas fir tree. The straight line indicates the discrimination expected if internal conductance were infinite (i.e., Δ_i=4.4+(30–4.4) C_i/C_a). Note that Δ_{obs} is less than Δ_i; this is because internal conductance lowers CO_2 concentration, thereby reducing discrimination. Each datapoint is a single simultaneous measurement of carbon-isotope discrimination and gas exchange. Modified from Warren *et al.* (2003).

one requires simultaneous measurements of gas exchange and instantaneous Δ. Δ and gas exchange may be determined simulateously by connecting a gas-exchange system to an IRMS (e.g., Cousins *et al.*, 2006), or by collecting gas samples while gas exchange is measured and then analysing gas samples elsewhere at a later date (Warren *et al.*, 2003a; Flexas *et al.*, 2007a,b). The non-simultaneous ('off-line') systems are more flexible because one may work outside the laboratory, but there is greater risk of contamination and diffusion leaks. As an alternative to mass spectrometry, it is possible to use tuneable diode laser spectroscopy (TDLS) and other optical methods, such as cavity ring-down spectroscopy to measure g_m in the field, but it is not so straightforward because TDLS requires frequent calibration and a very constant temperature, and is not as precise as IRMS (Pons *et al.*, 2009). A third alternative, that constitutes a somewhat different method, consists in integrating g_m over longer time periods, namely measuring Δ^{13} in recently synthesised sugars (see next subsection). The difficulties shared by these approaches include: (1) the need to maximise a discrimination that is usually small; and (2) uncertainties in the assumed values of discrimination along the CO_2 path.

The amount of discrimination owing to photosynthesis is typically rather small (a few ‰) and must be maximised to ensure good estimates of g_m. The only way to do this is by having a large draw-down in CO_2, which is possible if flow rate through the chamber is slow, although this comes at the cost of increased diffusion leaks (see Chapter 9), and thus one must strike a compromise between flow rate and diffusion leaks. Greater discrimination can also be achieved by using a large leaf area, which in many cases may mean using a custom-built chamber capable of accommodating a large leaf area (e.g., >10 cm^2). Finally, increased precision to determine a small discrimination can be attained by using a 'dual-inlet' instead of a 'continuous-flow' mass spectrometry system (Flexas *et al.*, 2007a,c).

Alternatively, estimates of g_m are affected by a number of assumed fractionations along the CO_2 path. There is a large consensus concerning fractionations owing to diffusion through the boundary layer (a_b, 2.9‰), diffusion through stomata (a, 4.4‰) and diffusion and dissolution of CO_2 into water (a_i, 1.8‰), and values of fractionation owing to mitochondrial respiration (e), and fractionation owing to photorespiration (f) can be either ignored ('single-point' method) or estimated ('slope-based' method). On the contrary, there is no consensus concerning net fractionation by Rubisco and PEPC (b), with estimates ranging from 27 to 30‰. As it is difficult to measure b most authors have to assign it *a priori*, but for these a value has to be chosen from 27 to 30‰. Such variability seems to be really present in nature, reflecting different structures of Rubisco (Tcherkez *et al.*, 2006), and not owing to artifacts in estimations. However, as measuring b is impracticable, the only way to characterise this uncertainty is to perform a sensitivity analysis in which g_m is calculated using b=27 or b=30.

Recently, the ^{13}C discrimination method has been extended to estimate canopy scale g_m either by determining ^{13}C discrimination of whole plants enclosed in closed cabinets forming part of the discrimination system (Schäufele *et al.*, 2011) or by determining phloem ^{13}C contents (Ubierna and Marshall, 2011).

12.4.2. ^{13}C content on recently assimilated sugars

On one hand, online carbon-isotope discrimination analyses are not available in many labs, and they are certainly difficult to apply when working under field conditions (see Warren *et al.*, 2003a). On the other hand, measuring carbon-isotope discrimination in leaf dry matter is easy, as the samples can be transported easily and stored for long periods of time, and the service is available in many labs.

In addition, online discrimination analysis provides an instantaneous assessment of fractionation during photosynthesis. At the other extreme, the isotopic signal in leaf dry matter integrates plant behaviour over very long times, up to the entire lifespan of leaves or plants (see Chapter 11), and hence, cannot be used to describe rapidly changing parameters such as g_m. Brugnoli *et al.* (1988) proposed an intermediate approach, namely measuring carbon-isotope discrimination in leaf soluble carbohydrates and starch. These are accumulated during photosynthesising periods (i.e., during the day). It is well known that most of the carbon fixed by photosynthesis, up to 80–90%, is found in leaf starch and sucrose fractions, leaving only minor amounts in ionised fraction (e.g., organic acids and amino acids). Hence, the isotopic composition of these photosynthetic products reflects the isotopic fractionation processes occurring mainly during gaseous- and liquid-phase diffusion, enzymatic carboxylation, photorespiration and respiration. The isotopic signature of photosynthetic carbohydrates brings a signal integrated over a much shorter time (approximately the length of the photoperiod) compared with that obtained by the analysis of bulk dry matter. Therefore, their isotopic signal reflects the leaf behaviour during a few hours, and the method has the same advantages of sampling, storing and analysing as leaf dry matter.

Among the other abovementioned parameters, the isotopic composition of sucrose and starch also reflects mesophyll CO_2-transfer conductance, and, hence, it is possible to estimate g_m using the 'single-point' method illustrated above. This approach was first used by Brugnoli *et al.* (1994) to estimate g_m and to compare several herbaceous and woody species. Subsequently, the method has been applied to several species like *Castanea sativa* (Lauteri *et al.*, 1997), cotton, chestnut, sunflower (Brugnoli *et al.*, 1998) and rice cultivars (Scartazza *et al.*, 1998), subjected to various environmental stresses. More recently, the same methodology has been used to estimate g_m in sugar beet subjected to transient and continuous drought stress (Monti *et al.*, 2006).

They used equation 12.11, as described in the previous section, except that in this case Δo_{bs} is that of soluble sugars ($\Delta^{13}C_{sug}$) and not of air collected during photosynthesis measurements. Soluble sugars (mainly sucrose plus small amounts of glucose and fructose) may be extracted and purified according to different methods, either based on bulk-sugar extraction and subsequent purification on chromatography columns, or compound-specific HPLC-IRMS analysis (Brugnoli *et al.*, 1988; Duranceau *et al.*, 1999; Wanek *et al.*, 2001).

Irrespective of the extraction procedure, this method yields g_m values very similar to those obtained by the common isotope-discrimination method (Brugnoli *et al.*, 1998; Scartazza *et al.*, 1998) and some quite good agreement with Ch-F methods (Centritto *et al.*, 2009). However, the main difference compared with the online method is that the values obtained using carbohydrates $\delta^{13}C$ may be considered as average values over the photosynthesising period (i.e., the method does not allow to follow rapid – minutes to hours – variations of g_m). Nevertheless, this method is very useful in ecophysiological applications and in field studies to compare different species or genotypes and to obtain average estimates of g_m, especially where many measurements are needed. In fact, it is very simple to collect a large number of leaves after gas-exchange measurements and store them appropriately, for extraction and purification at a later stage in the laboratory. With this method, no electricity or complex equipment is needed in the field, but only ice or dry ice to store leaves and limit respiratory losses.

12.4.3. ^{18}O uptake in combination with gas exchange

Oxygen uptake (U_o) occurs in leaves owing to (at least) three processes: photorespiration, O_2 photoreduction (also known as the Mehler reaction, see Chapters 2 and 3) and mitochondrial respiration. Renou *et al.* (1990) defined a method based on measuring U_o mass spectrometrically using $^{18}O_2$, to determine C_c and, hence, estimating g_m from equation

12.1. Using a gas-exchange cuvette, with an atmosphere containing only $^{18}O_2$, coupled to a mass spectrometer, total U_o can be determined from the rate of $^{18}O_2$ consumption in the cuvette, which could not be done using normal air because O_2 (mostly $^{16}O_2$) is released simultaneously with uptake owing to photosynthetic activity. Such measurements should be done coupled with normal gas-exchange measurements to determine A_N and C_i. An assumption of this method is that O_2 photoreduction is negligible, and independent estimates of R_d and Γ^* are required (Table 12.1), which can be obtained using some of the methods outlined in Chapters 9 and 11. If O_2 photoreduction is negligible, then U_o is given by the sum of velocity of oxygenation by Rubisco (V_o), oxygen consumed in the oxidation of glycolate (0.5 V_o) and O_2 uptake driven by mitochondrial respiration in the light (R_d):

$$U_o = V_o + 0.5V_o + R_d \quad \text{[Eqn. 12.13]}$$

Alternatively, A_N can be expressed as:

$$A_N = V_c - 0.5V_o - R_d \quad \text{[Eqn. 12.14]}$$

Solving equation 12.13 for V_o and substituting in equation 12.14, solving this in turn for V_c gives:

$$V_c = A_N + \left(\frac{U_o}{3}\right) + \left(\frac{2R_d}{3}\right) \quad \text{[Eqn. 12.15]}$$

By definition, however (Farquhar and von Caemmerer, 1982):

$$C_c = \frac{2\Gamma^*}{V_o V_c} \quad \text{[Eqn. 12.16]}$$

Finally, combining equations 12.15 and 12.16 gives:

$$C_c = \frac{\Gamma^*(3A_N + U_o + 2R_D)}{(U_o - R_D)} \quad \text{[Eqn. 12.17]}$$

12.5. CHLOROPHYLL-FLUORESCENCE METHODS

12.5.1. The variable J method

The variable J method is based on Chl-F measurements of the rate of electron transport (J), gas-exchange measurements and the known kinetic properties of Rubisco (Table 12.1). The variable J method requires that photorespiration is present, and it works better with increasing proportions of photorespiration to photosynthesis. Laisk *et al.* (2002), for example, argued that the optimal choice is the measurement of g_m near the CO_2 photocompensation concentration (Γ^*) at 21% O_2. The electron transport rate is estimated from Chl-F, usually considering that a proportion of this rate is going to alternative electron acceptors, and the majority is used in photosynthesis and photorespiration (i.e., alternative electron sinks including the Mehler reaction and nitrate reduction are considered negligible). The relative proportions of photosynthesis and photorespiration are a function of the substrate concentrations (i.e., CO_2 and O_2 in the chloroplast) and the relative specificity of Rubisco for CO_2 and O_2 ($S_{c/o}$). Photorespiration can be estimated from the (measured) rates of electron transport and photosynthesis. Then, C_c can be calculated from the rate of photorespiration and specificity of Rubisco, which then allows calculation of g_m.

In mathematical terms, the variable J method rests upon the assumption that the rate of linear electron transport (J_a) is a function of gross photosynthesis, C_c and the 'known' CO_2/O_2 specificity of Rubisco that is normally described by Γ^* ($S_{c/o}$=0.5O/Γ^*) (Di Marco *et al.*, 1990; Harley *et al.*, 1992a):

$$J_a = (A_N + R_d)\frac{4(C_c + 2\Gamma^*)}{C_c - \Gamma^*} \quad \text{[Eqn. 12.18]}$$

Substituting C_c with $C_i - A_N/g_m$, equation 12.18 becomes:

$$J_a = (A_N + R_d)\frac{4((C_i - A_N/g_m) + 2\Gamma^*)}{(C_i - A_N/g_m) - \Gamma^*} \quad \text{[Eqn. 12.19]}$$

Rearranging equation 12.19 allows g_m to be calculated directly:

$$g_m = \frac{A_N}{C_i - \dfrac{\Gamma^*(J_a + 8(A_N + R_d))}{J_a - 4(A_N + R_d)}} \quad \text{[Eqn. 12.20]}$$

Directly measuring J_a under normal respiratory conditions (21% O_2) is difficult, and thus Chl-F is used to provide an indirect measurement, as described in Chapter 10.

An alternative formulation, intrinsically similar but perhaps more intuitive, was proposed by Epron *et al.* (1995) and has been described in Chapter 9.

The variable J method has the problem that there are uncertainties in the relationship of J_f with J_a (Fig. 12.3). The crucial factors are that the use of a Chl-F-derived electron transport rate (J_f) assumes that alternative electron sinks are negligible, that Chl-F arises from the same cells contributing to the measured gas exchange (which is not true, as all leaf cells contribute to gas exchange, whereas only some cells contribute to fluorescence owing to intra-leaf reabsorption) and that there is no contribution of PSI to

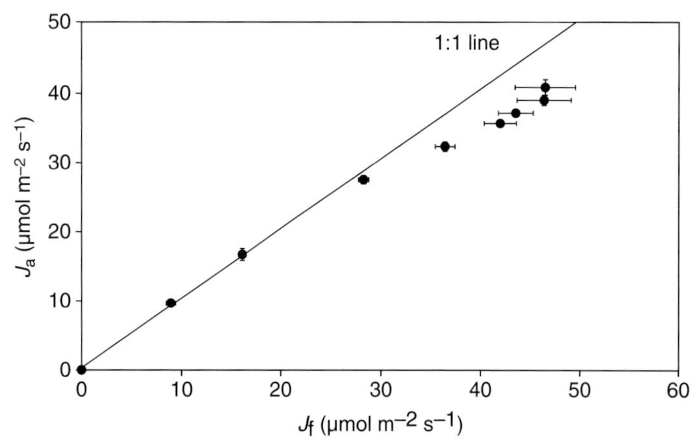

Fig. 12.3. The relationship between rates of electron transport estimated from chlorophyll fluorescence (J_f) and gross CO_2 assimilation (J_a) in seedlings of Douglas fir. Measurements were made with an open gas-exchange system (LI-6400) and integrated fluorescence chamber head (LI-6400–40) at 25°C, leaf-to-air vapour pressure deficit at 1.5–1.9 kPa and C_a of 425–435 µmol mol^{-1}. J_f=φPSII α 0.5 PAR, where $\varphi_{PSII}=(F_m'-F)/F_m'$ and α is total leaf absorptance (measured with a spectroradiometer). A_N was measured under non-photorespiratory conditions (1% O_2) and J_a was estimated as $4(A_N+R_d)$, where R_d is mitochondrial respiration in the light determined by the Laisk method. The solid line in the figure is the 1:1 line. Note that the relationship of J_f with J_a is not strictly 1:1. Curvilinearity of the relationship, especially at high PAR (i.e., high J_f and J_a), is owing to alternative electron sinks. Modified from Warren *et al.* (2004).

chl. fluorescence at room temperature (but see Agati *et al.*, 2000; Franck *et al.*, 2002). In addition, it requires accurate estimates of α_L and of the distribution of light between the two photosystems. Hence, much of this uncertainty can be eliminated by measuring α_L. Uncertainties in distribution of light between the two photosystems can also be eliminated by measuring the PSII optical cross-section (Laisk and Loreto, 1996; Eichelmann and Laisk, 2000), but this is comparatively rare. However, even when these precautions are taken there may not be a 1:1 relationship between J_f and J_a, because the described assumptions may not be totally true. Hence, J_f is best regarded as an indicator of *relative* rates of J_a. A partial solution to these problems is to use a 'calibration curve' of the relationship between J_f and J_a under non-photorespiratory conditions (1–2% O_2, Fig. 12.3). Under non-photorespiratory conditions J_a is associated with Rubisco carboxylation and alternative electron sinks. The rate of J_a owing to Rubisco carboxylation is given by $4(A_N+R_d)$. Curvilinearity in the relationship of J_f with J_a (e.g., Fig. 12.3) can be attributed to alternative electron sinks; whereas a deviation from the 1:1 line might be owing to inaccuracies in the optical cross-section or α_L but also to the non-representativeness of fluorescence. The latter is owing to the fact that fluorescence does not arise equally from all cells in the leaf, but actually from a very limited cell layer, and indeed φ_{PSII} calculated from the fluorescence signals of the irradiated surface is often very much smaller than that for the other side of the leaf (Tsuyama *et al.*, 2003; see Pons *et al.*, 2009 for further details).

Estimates of g_m are sensitive to Γ* (Harley *et al.*, 1992a). Most studies take literature values of Γ* as it is difficult to measure and it is argued to be an invariant property of all C_3 species. Measurements of Γ* for the species in question would be preferable, but this has been attempted by few. In general we have few estimates of Γ*, and temperature responses are especially scarce (Jordan and Ogren, 1984; Bernacchi *et al.*, 2002). A common approach is to measure the intercellular photocompensation point (C_i*) with the Laisk method and then use this as a surrogate for Γ*, but this approach is logically flawed because $\Gamma^*=C_i^*+R_d/g_m$. Hence, others substitute $C_i^*+R_d/g_m$ for Γ* and then solve for Γ* and g_m simultaneously. The potential error introduced by errors in Γ* can be estimated by calculating g_m with alternative values of Γ* (i.e., a sensitivity analysis). Finally, another alternative is to estimate Γ* from published values of $S_{c/o}$ ($\Gamma^*=0.5O/S_{c/o}$), which are available for an increasing amount of species, although mostly at 25°C only (Galmés *et al.*, 2005).

Estimates of g_m are also sensitive to R_d (Harley *et al.*, 1992a; Flexas *et al.*, 2004a), and thus it is imperative that R_d is estimated or measured accurately. R_d may be directly measured

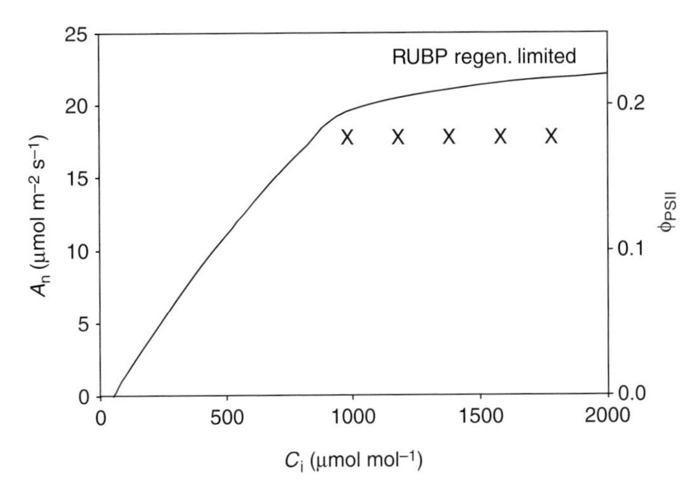

Fig. 12.4. Typical response of net photosynthesis (A_N, solid line) and quantum yield of PSII (φ_{PSII}, crosses) to intercellular CO_2 concentration (C_i). The constant J method estimates g_m from the ribulose biphosphate (RuBP)-limited portion of the response curve, where J is constant. θ_{PSII} is used as an indicator of J because it is directly proportional to linear electron transport (J=φ_{PSII} α_L 0.5 PAR).

with one of the isotopic methods (Loreto *et al.*, 1999, 2001a; Haupt-Herting *et al.*, 2001; see Chapter 11). Unfortunately, R_d is rarely *directly measured* with most authors choosing to *estimate* R_d. The Laisk method is the most common means of estimating R_d. Indirect estimates of R_d based on measured R_n or a percentage of measured R_n (e.g., R_d=50% of R_n) are common but poorly defensible. Once again, a sensitivity analysis to estimate the likely error in g_m that results from alternative values of R_d would be recommendable.

12.5.2. The constant J method

The constant J method shares many common assumptions with the variable J method. Both methods require accurate and precise measurements of A_N, C_i, R_d and Γ^* (Table 12.1), and both methods are sensitive to errors in their estimation. However, in the constant J method Chl-F is only used to establish the range over which J_a is constant (this method is only valid when J_a is constant and photosynthesis is limited solely by RuBP regeneration) and not to obtain quantitative estimates of J_a, assumptions about Chl-F are less strict than for the variable J method.

At high CO_2, when A_N is limited by RuBP regeneration and J_a is constant (Fig. 12.4), the response of A_N to CO_2 depends on C_c and the CO_2/O_2 specificity of Rubisco (Bongi and Loreto, 1989). g_m is determined based on equation 12.20 and several measurements of A_N and C_i at three or more (high) C_i. A 'dummy' value of g_m (between zero and one) is substituted into equation 12.6, and the variance in J_a calculated:

$$\sum_{i=1}^{n}(J_{\bar{x}} - J_i)^2 / (n-1) \qquad \text{[Eqn. 12.21]}$$

where $J_{\bar{x}}$ is the mean value of J_a for the three or more C_i values; and J_i is the value of J_a for each C_i. g_m is determined as the value that gives the minimum variance. In practice this can be done graphically by plotting variance as a function of g_m (Harley *et al.*, 1992a), or else the minimum can be found iteratively by using a commercial software package (e.g., the solver add-in of Microsoft Excel), the latter method being quickest. Recent statistical variants of this method have been proposed that increase its accuracy (Yin and Struik, 2009).

12.6. GAS-EXCHANGE METHODS

12.6.1. The photocompensation-point method

One can estimate g_m from the difference between the Γ^* and C_i^*. This method requires only a gas-exchange system capable of accurate and precise measurements, especially at low CO_2 concentrations (Table 12.1). The C_i^* is most commonly estimated via the Laisk method (Laisk, 1977; see Chapter 9). The original formulation (Laisk, 1977) assumes infinite g_m. If one considers finite g_m, then the Laisk method measures the C_i^* that is related to Γ^* by R_d and g_m (von Caemmerer *et al.*, 1994; Peisker and Apel, 2001):

$$C_i^* = \Gamma^* - \frac{R_d}{g_m} \qquad \text{[Eqn. 12.22]}$$

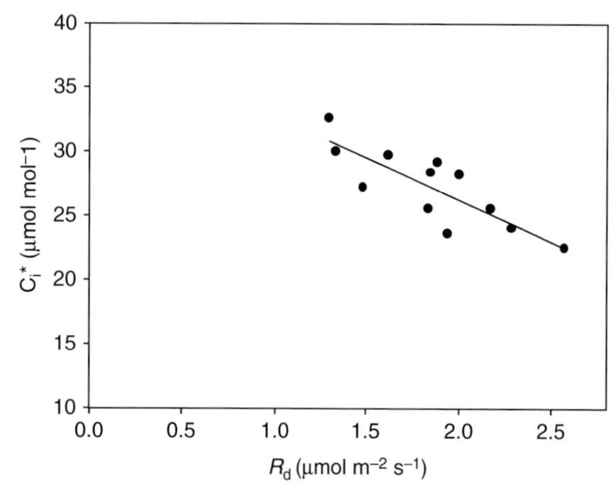

Fig. 12.5. Relationship between the intercellular photocompensation point (C_i^*) and respiration in the light (R_d) in Douglas fir. The straight line is a linear regression fitted to the Douglas fir data: $C_i^*=39.4+6.5\ R_d$, $r^2=0.67$. Mean g_m is given by 1/slope (i.e., $1/6.5=0.14$ mol m^{-2} s^{-1}). Modified from Warren *et al.* (2003).

Following equation 12.22, a mean Γ^* can be determined from the *y* intercept of a C_i^* versus R_d plot, whereas the slope yields an estimate of the *mean* $1/\ g_m$ (Fig. 12.5). This procedure only works if there is a sufficiently large range in R_d (e.g., greater than twofold). To obtain a large range in R_d one must work with different leaves (e.g., young and old, or sun and shade). Where the range in R_d is large, the slope will give a single g_m estimate: the mean. This is problematic if the aim is to examine differences in g_m (e.g., between young and old, or sun and shade). An alternative approach (Peisker and Apel, 2001; Warren *et al.*, 2003a) is to determine g_m of individual plants or leaves based on measured C_i^*, measured R_d and the mean Γ^* (determined as the *y* intercept, Eqn. 12.22). This method assumes that Γ^* does not vary among plants or leaves, but that g_m does. This is somewhat tricky because the very same regression is used to estimate a single constant mean Γ^* but many different individual g_m. The primary assumption of the photocompensation-point method is the accuracy of C_i^* and R_d estimates. De-activation of Rubisco at low CO_2 concentrations (Sage *et al.*, 2002) is potentially a problem for the Laisk method, and stomatal closure leads to inaccurate estimations (Galmés *et al.*, 2006; see Chapter 9).

12.6.2. The curve-fitting method

Ethier and Livingston (2004) introduced quadratic equations for estimating g_m based on a modification of the Farquhar *et al.* (1980a) model of C_3 photosynthesis (see Chapter 8). Recently Sharkey and co-workers (Sharkey *et al.*, 2007) introduced a fundamentally similar method that relies on non-linear curve fits rather than quadratic equations. The fundamental premise of the two curve-fitting methods is that g_m reduces the curvature of an A_N/C_i response curve (Fig. 12.6). Like the photocompensation-point method, the curve-fitting methods require no specialised equipment besides an accurate and precise gas-exchange system (Table 12.1). The Farquhar *et al.* (1980a) model of C_3 photosynthesis states that photosynthesis is the minimum of rates limited by Rubisco carboxylation (A_c) or the rate of RuBP regeneration (A_j):

$$A_N = \min\{A_c, A_j\} \qquad \text{[Eqn. 12.23]}$$

The rates of A_c and A_j are conventionally given by non-linear equations (e.g., Sharkey *et al.*, 2007), but can also be formulated as a quadratic (Ethier and Livingston, 2004):

$$A_c = \frac{-b+\sqrt{b^2-4ac}}{2a} \qquad \text{[Eqn. 12.24]}$$

where $a\ =-1/g_m$; $b=\frac{V_{c,\max}-R_d}{g_m}+C_i+K_c(1+O/K_o$; and

$c=R_d[C_i+K_c(1+O/K_o)]-V_{c,\max}(C_i-\Gamma^*)$

$$A_j = \frac{-b+\sqrt{b^2-4ac}}{2a} \qquad \text{[Eqn. 12.25]}$$

where $a\ -1/g_m$; $b=(J/4-R_d)/g_m+C_i+2\Gamma^*$; and $c=R_d(C_i+2\Gamma^*)-J/4(C_i-\Gamma^*)$ and where $V_{c,\max}$ is the maximum rate of carboxylation; K_c is the Michaelis-Menten constant for carboxylation; K_o is the Michaelis-Menten constant for oxygenation; J is the electron transport rate under RuBP-limited conditions; and O is oxygen concentration.

Equation 12.24 is fitted to the Rubisco-limited portion of an A_N/C_i response, whreas equation 12.25 is fitted to the RuBP regeneration-limited portion. As CO_2 alters g_m, estimates performed at the two different parts of the curve are different and may be considered only as mean g_m for a given range of C_i (Flexas *et al.*, 2007a). To use this method, Rubisco kinetic constants (K_c, K_o and Γ^*) and R_d must be known (Table 12.1). Choice of kinetic constants is problematic as these have only been estimated in a few species and there is quite some range in estimates (von Caemmerer, 2000; see Chapter 8).

The primary assumptions of the curve-fitting methods relate to the validity of the Farquhar *et al.* (1980a)

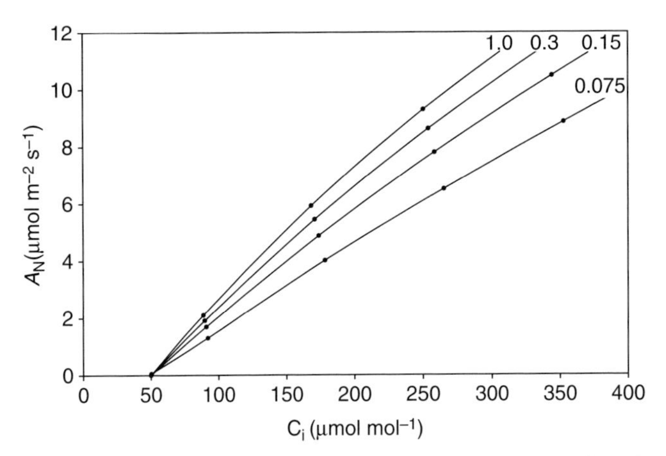

Fig. 12.6. The effect of mesophyll conductance (g_m) on the response of net photosynthesis (A_N) to intercellular CO_2 concentration (C_i). Values of g_m (in mol m^{-2} s^{-1}) are given in the upper right corner of the figure. Note how mesophyll conductance affects not only the initial slope, but also the curvature of the response curve. $V_{c,max}$=50 μmol m^{-2} s^{-1}; J_{max}=100 μmol m^{-2} s^{-1}; R_d=0.6 μmol m^{-2} s^{-1}.

model of C_3 photosynthesis. For example, deactivation of Rubisco at low and high CO_2 concentrations will affect estimates of g_m. It would be preferable to omit affected parts of the CO_2-response curve, but unequivocal identification is difficult without direct measurements of, for example, activation state. In a similar fashion it is necessary to avoid those portions of a CO_2-response curve affected by TPU limitation, but TPU limitation is notoriously difficult to identify unequivocally. The attraction of the curve-fitting methods is their simplicity, but it is actually quite technically demanding. Poor precision of A_N and C_i estimates will have a large effect because the aim is to define the curvature of an A_N/C_i response. The curve-fitting methods are so sensitive that instrument noise alone can affect estimates by ±10% of the true value (C. Warren unpublished data).

12.6.3. Photosynthetic response to oxygen

Recently, a new method has been proposed based on the idea that the sensitivity of CO_2-limited photosynthesis to a change in the concentration of O_2 provides information about C_c (Bunce, 2009). In short, the method consists of estimating $V_{c,max}$ and J_{max} at 21% O_2 using A_N response to C_i, i.e., assuming infinite g_m (see Chapters 8 and 9). The obtained $V_{c,max}$ is used then to estimate the expected A_N at 2% O_2. If the predicted value of A_N at low O_2 exceeds the observed value, an arbitrary estimated value of g_m is chosen and used to calculate C_c at 21% O_2 and find the new $V_{c,max}$ (or J_{max}) value that fits A at 21% O_2 at that C_c. The new model value of A_N at 2% O_2 is then compared with the observed value at the C_c at 2% O_2. If the modelled value of A_N at 2% O_2 is less than the observed value, then the estimate of g_m is too low, and vice versa. The procedure is repeated until a g_m value is found that yields a $V_{c,max}$ (or J_{max}) value that estimates A_N at 2% so that it fits the measured value.

12.7. SEPARATING GAS-PHASE AND LIQUID-PHASE COMPONENTS OF g_m

In addition of determining an 'average' g_m using any of the methods described, methods are needed to separate gas-phase and liquid-phase components of g_m, i.e., to separately estimate g_{ias} and g_{liq}. This is a task that only few have attempted for the moment. Still, some techniques have been proposed for this purpose that are briefly explained in the following sections. All of them are based on the assumption of a rather simplistic model in one dimension, defining g_m as:

$$g_m^{-1} = g_{ias}^{-1} + g_{liq}^{-1} \text{ (or } g_w^{-1}\text{)} \quad \text{[Eqn. 12.26]}$$

that is, no method has been described up to now to separate g_m into g_{ias}, g_w and g_{liq}, but the latter two components are considered in combination. Initial estimates of the relative contributions of the two components were done by Syvertsen *et al.* (1995) using anatomical measurements only. However, their method was based on the assumption that both g_{ias} and q_{liq} were strictly dependent on leaf anatomical factors, and hence not regulated in the short term. Although this may probably be true for g_{ias}, it is certainly not for g_{liq}. Hence, we will focus here in describing only those methods relying on the assumption that at least g_{liq} can be variable independently of leaf anatomy.

12.7.1. Combining measurements in helox with other methods

Genty *et al.* (1998) proposed a method based on combining gas-exchange and Chl-F measurements in air and helox. Based on the assumption that changing the leaves from normal air to helox (see Section 12.3.2) changes only g_{ias} but not g_{liq}, one can make estimates of g_m using any of the methods explained (Genty *et al.*, 1998) used the 'variable J' method as described by Harley *et al.* (1992a) under two different conditions: normal air and helox. From the estimated g_m in air and helox (g_m and $g_{m,helox}$, respectively), and from the reported ratio of binary diffusion ratios of CO_2 for helox/air, g_{ias} is calculated as:

$$g_{ias} = \frac{\left(1-K^{-1}\right)}{\left(g_m^{-1} - g_{m,helox}^{-1}\right)} \quad \text{[Eqn. 12.27]}$$

where K is the binary diffusion ratio for CO_2 (2.33 for helox/air). Finally, g_{liq} is calculated from g_m and g_{ias} by rearranging equation 12.26.

12.7.2. Combining ^{18}O discrimination with other methods

Gillon and Yakir (2000a) proposed a method based on simultaneous ^{13}C and ^{18}O discrimination combined with gas exchange. In their original work, Gillon and Yakir (2000a) state that their method serves to separate g_w and the chloroplast conductance, g_{chl}. However, this is because they considered g_{ias} negligible (and hence their term g_w actually includes g_{ias}), and they replaced the term g_{liq} with g_{chl} after the assumption that all CO_2 transfer in the liquid phase occurs within chloroplasts, because these are tightly coupled to cell surfaces directly facing intercellular air spaces. Therefore, essentially this is also a method that separates a combined g_{ias} and g_w from g_{liq}.

First, an estimate of C_c must be obtained using ^{13}C discrimination, as described. Simultaneous to ^{13}C discrimination, $\Delta^{18}O$ of CO_2 is also measured online in a method equivalent to the one for ^{13}C (Evans *et al.*, 1986):

$$\Delta^{18}O_{obs} = \left(\frac{\xi(\delta_o - \delta_{in})}{\left(1000 + \delta_o - \xi(\delta_o - \delta_{in})\right)}\right)1000 \quad \text{[Eqn. 12.28]}$$

where $\xi = C_{in}/(C_o - C_{in})$ and C_{in}, C_o and δ_o, δ_{in} refer to the CO_2 concentration and isotopic composition of air entering and leaving the gas-exchange cuvette, respectively. Alternatively, $\Delta^{18}O$ can also be predicted according to Farquhar and Lloyd (1993) as:

$$\Delta^{18}O_{pred} = \frac{\left(\bar{a} + \varepsilon\Delta_{ea}\right)}{\left(1 - \left(\varepsilon\Delta_{ea}/1000\right)\right)} \quad \text{[Eqn. 12.29]}$$

where Δ_{ea}=1000 [(δ_{chl} / 1000 + 1) / (δ_a / 1000 + 1) -1]; $\epsilon = C_{cs}/(C_a - C_{cs})$ and δ_a and δ_{chl} represent the $\delta^{18}O$ of CO_2 in the overlying air and in full isotopic equilibrium with water in the chloroplast, respectively, and C_a and C_{cs} the air and chloroplast surface CO_2 concentrations; and $\bar{a}$ is the weighted-mean diffusional fractionation through the boundary layer (5.8‰), stomata (8.8‰) and aqueous leaf media (0.8‰). Therefore, $\Delta^{18}O$ is a function of C_{cs}, and the measured and predicted values should be equal using $\Delta^{13}C$-derived C_c whenever the latter is equal to the CO_2 concentration at the site of oxygen exchange, i.e., the chloroplast surface (C_{cs}). Similarly to the principle of the ^{13}C-discrimination method, this is akin to a null hypothesis, and the present method consists in estimating the difference between C_c and C_{cs} from the usually observed discrepancy between $\Delta^{18}O_{pred}$ and $\Delta^{18}O_{obs}$.

However, application of this method is not easy because, as seen in equation 12.29, one must know the $\delta^{18}O$ of CO_2 in full isotopic equilibrium with water in the chloroplast. Because of the proximity of chloroplasts to the liquid-air interface of leaves, the isotopic composition of water at evaporative sites (δ_e) may be a good proxy for the isotopic composition of water in the chloroplasts (δ_{chl}). δ_e may be estimated from the Craig and Gordon (1965) model of evaporative enrichment (see Chapter 11 for the equations), which requires a number of physical constants and determining the isotopic composition of transpired water (δ_t). Therefore, in addition to measuring gas exchange and discrimination of ^{13}C and ^{18}O of CO_2, the method requires measuring the isotopic composition of transpired water (δ_t). To achieve this purpose, the air exiting the gas-exchange cuvette may be derived using a vent to a trapping loop for

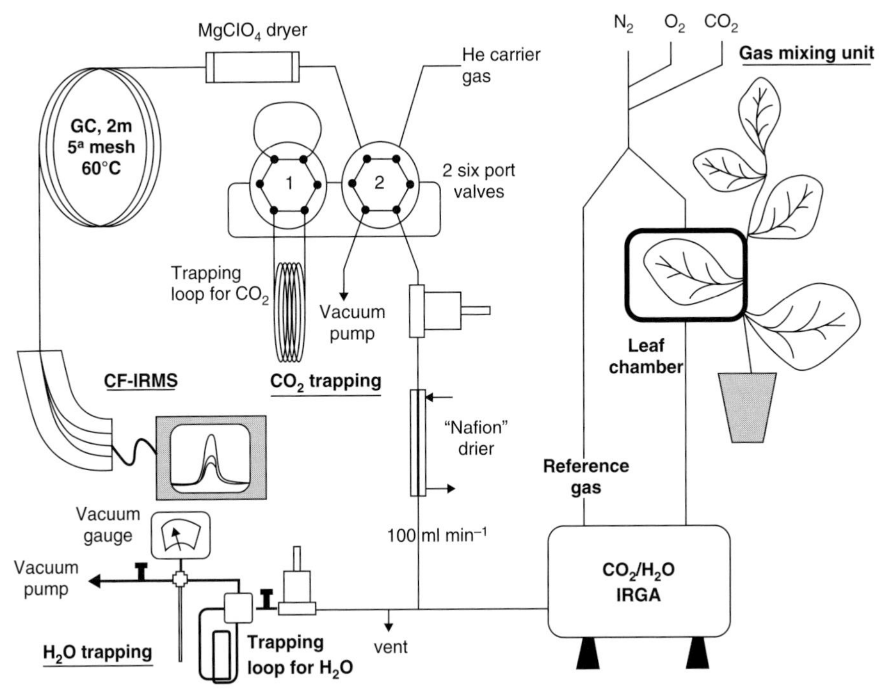

Fig. 12.7. Arrangement of online CO_2-trapping and off-line H_2O-trapping apparatus for continuous-flow CO_2 isotopic analysis, in conjunction with leaf chamber and gas-exchange system. IRGA, infrared gas analyser; IRMS, infrared mass spectrometer. Modified after Gillon and Yakir (2000).

water for off-line determination of δ_t, while at the same time a proportion may be derived after passing through a desiccant to a CO_2 trapper for online measurement of oxygen discrimination. The overall experimental design required is shown in Fig. 12.7.

It is unclear, however, whether oxygen-isotope exchange between CO_2–H_2O in the chloroplast is complete, i.e., if the condition of full isotopic equilibrium with water in the chloroplast required by equation 12.29 is fulfilled. Although some reports have suggested full isotopic equilibrium >95% (Farquhar and Lloyd, 1993), given the potential uncertainties in $\delta^{18}O$ of water and CO_2 concentration in the chloroplast when interpreting $\Delta^{18}O$, Gillon and Yakir (2000a) proposed using an independent method to test this assumption. The isotopic equilibrium in the CO_2–H_2O system can be derived from Mills and Urey (1940) as:

$$\theta_{eq}=1-e^{-k\tau/3} \quad \text{[Eqn. 12.30]}$$

which describes the fractional approach to full equilibrium ($\theta_{eq}=1$) as a function of the number of hydration reactions achieved per CO_2 molecule ($k\tau$). This coefficient of hydration may be calculated for a leaf by calculating the rate constant k from biochemical measurements of CA activity (see Chapter 13), and estimating the residence time of CO_2 in the leaf (τ_{CO2}) from gas-exchange measurements of the CO_2 flux. During photosynthesis, the gross CO_2 influx rate (F_{in}) regulates the residence time of CO_2 in the aqueous medium, whereas the CA-catalysed hydration of CO_2 (CA_{leaf}) determines the efficiency of oxygen exchange. The rate constant for CA (k) is equivalent to CA_{leaf} / C_{cs}, and the τ_{CO2} is given by C_{cs} / F_{in}. Thus, the product $kt=CA_{leaf}$. Therefore, the method also requires collecting leaf samples after gas-exchange measurements to analyse CA_{leaf} using biochemical methods (see Chapter 13). F_{in} is calculated from gas-exchange measurements from the product of external CO_2 concentration (C_a) and total conductance to the site of H_2O equilibrium (g_l), where g_l is the combination of the boundary (g_b) and stomatal (g_s) conductances determined by gas exchange, and the mesophyll conductance (g_m) determined by ^{13}C discrimination.

Once θ_{eq} is determined, it is possible to incorporate it in equation 12.29, resulting in:

$$\Delta^{18}O=1000\frac{\bar{a}+\varepsilon\left(\theta_{eq}\Delta_{ea}-(1-\theta_{eq})\frac{\bar{a}}{\varepsilon+1}\right)}{1000-\varepsilon\left(\theta_{eq}\Delta_{ea}-(1-\theta_{eq})\frac{\bar{a}}{\varepsilon+1}\right)} \quad \text{[Eqn. 12.31]}$$

Now, from the measured values of $\Delta^{18}O$, Δ_{ea} and θ_{eq}, € can be derived, and hence C_{cs}, the effective CO_2 concentration at the site of CO_2–H_2O equilibrium. Finally, g_{ias} and g_{liq} are calculated as follows:

$$g_{ias} = A_N (C_i - C_{cs}) \quad \text{[Eqn. 12.32]}$$

$$g_{liq} = A_N (C_{cs} - C_c) \quad \text{[Eqn. 12.33]}$$

12.7.3. Combining photoacoustic measurements with other methods

Gorton *et al.* (2003) have proposed a photoacoustic method to estimate g_{liq}. Combining this method with any of the methods described to determine g_m allows estimating g_{ias} by subtraction. Photoacoustic methods (see Chapter 10) analyse pressure waves of photosynthetic origin. These are generated by the conversion of absorbed light to heat and by the evolution of oxygen, both phenomena occurring at PSII. The thermal photoacoustic signal arises from conversion of absorbed light into heat at PSII, which is followed by heat diffusion to the internal gas space of the leaf, thermal expansion of the gas and propagation of the signal as a pressure wave to the microphone detector. The oxygen signal arises from oxygen evolution at PSII, diffusion of O_2 to the internal gas space of the leaf and propagation of the signal as a pressure wave. Pulsed photoacoustics use a short pulse of light to generate a single complex pressure wave to which both heat and oxygen contribute. A second measurement using a background saturating light suppresses the oxygen signal because the measuring pulse generates no further oxygen. Then, subtracting the signal obtained in the presence of saturating background light from that obtained in its absence yields a pure oxygen signal. What is measured is the lag time (t) between the light pulse and photoacoustic detection of the oxygen wave. The lag time for diffusion can also be modelled using a one-dimensional model treating the diffusion of a planar front of a substance (here, O_2) along one axis, which seems appropriate to represent diffusion between chloroplasts and the neighbouring air space. According to a standard equation derived from Fick's second law of diffusion (Nobel, 2005):

$$t = x^2 / (4\, D_{ox}) \quad \text{[Eqn. 12.34]}$$

where x is the distance between the center of the chloroplasts and the intercellular air space (determined by microscopy); and D_{ox} is the diffusion coefficient of oxygen.

The method assumes that CO_2 and O_2 follow the same diffusion pathway but in opposite directions (which is actually unknown). If so, then based on theoretical considerations, the overall permeability to CO_2 is 21.1 times the permeability for oxygen in the liquid phase of leaf tissue (i.e., $D_{CO2}=21.1\, D_{ox}$). Finally, one can convert this diffusion coefficient and the known diffusion distance to conductance:

$$g_{liq} = D_{CO2} / x \quad \text{[Eqn. 12.35]}$$

12.8. A COMPARISON OF METHODS. ADVANTAGES AND INCONVENIENCIES: WHICH METHOD SHOULD BE USED?

The different technologies required for each method, their underlying assumptions and required parameters are summarised in Table 12.1. In principle, they all have many assumptions and technical difficulties. Nevertheless, the fact that comparing several of these methods, which rely on substantially different assumptions, yields similar results and reinforces the idea that they all provide reasonable estimates of g_m in leaves (Loreto *et al.*, 1992; Flexas *et al.*, 2006a, 2007a,b; Warren and Dreyer, 2006). However, in this section we will show that not all these assumptions and difficulties are of the same magnitude, and that different methods may serve different purposes or be preferable depending on the conditions. For instance, if one aims to compare g_m among old and young leaves, the photocompensation-point method may not be useful s it provides an average value pooling different kinds of leaves together. However, if one only has available a gas-exchange system, the only possible options are the photocompensation-point method, the curve-fitting method and perhaps (depending on laboratory facilities) the initial-slope method.

However, some methods may be a-priori preferable over others. For instance, considering the assumptions underlying each method that refer to the responses of g_m itself (Table 12.1), only the 'single-point' ^{13}C-discrimination method and the 'variable J' method have none of these assumptions. All the other methods have some implicit assumptions, many of them now known to be untrue. For instance, up to five of the methods (the photocompensation-point method, the curve-fitting method, the initial-slope method, the 'slope-based' ^{13}C-discrimination method and the 'constant J' method) work on the assumption that g_m does not change with CO_2 concentration and/or light intensity. However, Flexas *et al.* (2007b) have clearly demonstrated that g_m strongly (up to an order of magnitude) responds to CO_2 within the concentrations typically used with these methods, which has been confirmed by several other laboratories (Hassiotou *et al.*, 2009a; Vrábl *et al.*, 2009; Yin *et al.*, 2009) but not in others (Bunce, 2009; Tazoe *et al.*, 2009). This fact, in principle, makes these methods less recommendable, although they may be used to provide an *average* estimate of g_m over a range of CO_2 concentrations.

Still, among these methods some may be preferred over others. For instance as already stated the use of the photocompensation-point method faces other limitations, including those common to any other use of the 'Laisk' method (see Chapter 9), and the fact that one needs to pool measurements in different leaves to obtain an 'average' value of g_m. The latter limitation is shared with the initial-slope method, which in addition has the problem of requiring absolute quantitative estimations of Rubisco activity *in vitro*, which as discussed is very difficult to achieve. Therefore, these two are probably the less recommendable methods for many purposes. Instead, the 'slope-based' discrimination method, the 'constant J' fluorescence method and the curve-fitting method may be more suitable, particularly if the range of CO_2 concentrations/light intensities used is constrained. However, this may not always be achievable when using the 'slope-based' discrimination method, as a large range of $\Delta_i - \Delta^{13}_{obs}$ and A_N/C_a values are required to obtain good correlations, which can only be achieved by large changes in light intensity and/or CO_2 concentrations. Concerning the 'constant J' fluorescence method, the range of CO_2 concentrations used are smaller, but the referred concentrations are high, corresponding to electron transport-limited photosynthesis. Therefore, this method could not be used to estimate g_m at current ambient CO_2 concentrations. Instead, the curve-fitting method works well both at low (i.e., in the carboxylation-limited region) and high CO_2 concentrations (electron transport-limited region), and an estimate can be obtained for each region separately (Flexas *et al.*, 2007a). Therefore, of the five methods assuming that g_m does not respond to CO_2, the curve-fitting method could be the most flexible, allowing reasonable estimations at different ranges of CO_2. Unfortunately, as discussed above, it is one of the most sensitive to instrument noise and estimations of R_d.

Concerning the method of measuring Δ^{13} in soluble sugars, it is now well known that g_m changes fast in response to many environmental variables, changing along a diurnal course. Therefore, although the underlying assumptions of this method are similar to those of the other two ^{13}C-discrimination methods, this one can be used to provide only an 'average' daily integrated g_m value, and is not suitable to study rapid changes of g_m in response to environmental variables. Contrarily, the assumption of the ^{18}O-consumption method that g_m does not change in an atmosphere containing $^{18}O_2$ only is probably accurate, and no discrepancy with it has been published.

In summary, based on their a-priori assumptions, the most suitable methods may be the ^{18}O-consumption method, the 'single-point' ^{13}C-discrimination method and the 'variable J' fluorescence method. The latter two are actually the most commonly used, particularly the variable J method. However, even these methods have some underlying assumptions and require very accurate estimations of several independent parameters, which are not easy to determine. In practice, any estimation should be accompanied by sensitivity analyses considering the most susceptible parameters (i.e., *b*, *e* and *f* for the 'single-point' ^{13}C-discrimination method; α_L and light partitioning between photosystems for the 'variable J' fluorescence method; and R_d and Γ^* for all three methods). In addition, to support relevant conclusions involving variations of g_m, using at least two different independent methods would be preferable.

13 • Biochemical and molecular techniques for the study of photosynthetic processes

M.A.J. PARRY, P.J. ANDRALOJC, C.H. FOYER, J. GALMÉS AND T.D. SHARKEY

13.1. INTRODUCTION

Increasing understanding of the many molecular and biochemical processes that respond in a purposive way to the changing environment has given rise to an appreciation that many, if not all, environmental cues evoke primary responses at a molecular level, and that it is these responses that result in changes in gross plant physiology and morphology. Likewise, changes in the relative proportions of metabolites and ions within intracellular compartments in response to such environmental cues also give rise to multiple changes in gene expression. The interaction between these levels of complexity in response to changes in the external environment is illustrated in Scheme 13.1.

This chapter describes and discusses approaches: (1) for the unbiased analyses of gene, protein and metabolite function facilitated by a variety of high-throughput approaches; and (2) for the focused analyses of specific genes, gene products and metabolites. The former approaches seek to identify hitherto unknown genes and molecular interactions, while the latter are used to probe those elements that we currently consider most important in understanding and interpreting how photosynthetic processes relate to ecophysiological questions. In particular, we discuss aspects of the isolation and assay of the carboxylating enzymes, Rubisco and phosphorenolpyruvate carboxylase (PEPC), owing to their pivotal roles in assimilation and to the continuing interest in their measurement. In general, we have selected methods and approaches that have been applied in our laboratories, but acknowledge that many alternative methods could have been described, which are equally reliable and quantitative.

13.2. HIGH-THROUGHPUT APPROACHES

13.2.1. Metabolomics

Metabolomics aims to identify and quantify the global metabolite profile in plant tissues and can use a wide array of analytical chemistry techniques (e.g., gas chromatography, (GC) high-performance liquid chromatography, (HPLC) nuclear magnetic resonance (NMR); mass spectrometry, (MS). Currently, NMR and MS are the preferred technology platforms that can deliver fast, effective and unbiased analyses for metabolomic studies (Krishnan *et al.*, 2005).

^{1}H-NMR is an effective tool for the identification and quantification of plant metabolites in plant tissue and tissue extracts. The simplest approach is to use the ^{1}H-NMR spectra of total extracts of plant tissues to give fingerprints of the metabolite profile. Use of multivariate analysis tools provide a simple way of comparing global changes in metabolites from many samples arising from different species or genotypes or of assessing the impact of the environment (nutrients, stress etc.) on the plant metabolome (Fig. 13.1). The differences between fingerprints are revealed and then used to identify the parts of the metabolome for further study by higher resolution MS-based techniques. The method is fast and well suited to the characterisation of the abundant metabolites (e.g., sugars and amino acids), but less suitable for metabolites that are found at low concentrations (e.g., signalling molecules) (Ward *et al.*, 2007). Typically, more than 40 metabolites can be immediately identified within an unpurified plant-extract ^{1}H-NMR profile. Where sample preparation usually involves prolonged extraction in warm methanol, the method should not be relied on for the quantification of volatile compounds. Most published NMR metabolomics work utilises polar-solvent extracts, often from freeze-dried tissue. Consequently, lipids and volatiles are not revealed by this technique.

MS is a more appropriate approach where greater sensitivity and more targeted analysis are required. MS separates molecules and fragments according to their mass-to-charge ratio, and is very sensitive and effective over a wide dynamic range. To increase its utility further, MS has been combined with other methods, such as liquid and GC, to generate a

Terrestrial Photosynthesis in a Changing Environment: A Molecular, Physiological and Ecological Approach, ed. J. Flexas, F. Loreto and H. Medrano. Published by Cambridge University Press.

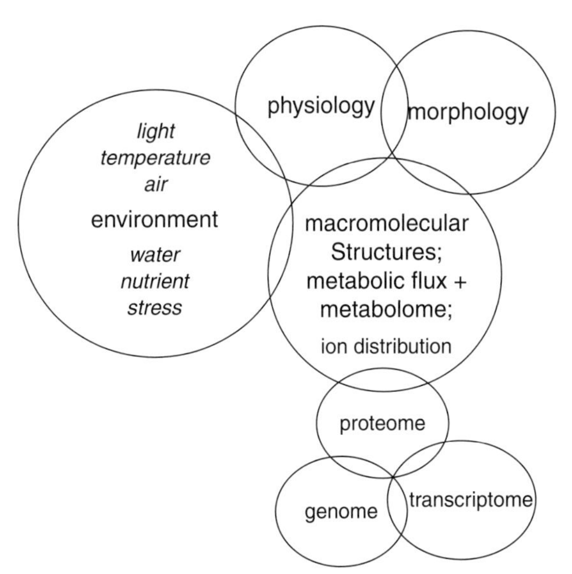

Scheme 13.1. Spheres of influence. Impact of the environment on and crosstalk between supramolecular, molecular and biochemical plant processes.

wider portfolio of methods that are optimised for specific uses (e.g., liquid chromatography mass spectrometry (LC-MS) and GC mass spectrometry (GC-MS)). MS-based techniques give molecular weights of compounds. However, structures can only be confidently assigned if appropriate standards are available for direct comparison by chromatography.

Knowledge of the subcellular localisation of metabolites is important in understanding the impact of metabolite concentrations, and non-aqueous fractionation can be effectively combined with GC-MS (Benkeblia *et al.*, 2007). Metabolomics can be usefully combined with ecophysiologial measurements to assess plant-photosynthesis responses (Berger *et al.*, 2010; Pinheiro and Chaves, 2011).

13.2.2. Molecular approaches and platforms

PROTEOMICS

There is growing recognition that the abundance of mRNA transcripts is not always representative of cognate protein levels and that mechanisms of post-translational regulation must also play an important role. Therefore, proteomic technologies are routinely used as a complement to investigate the genomic, transcriptomic and metabolomic global response of photosynthesis (as any other physiological aspect) under an ecophysiological perspective.

High-throughput protein identification has become possible, together with improved protein extraction and purification protocols and the development of genomic sequence databases, for peptide-mass matches. Recent proteome analysis performed in plants has provided new ways to assess the changes in protein types and phosphorylation status. Thus proteomics aims to not only identify and quantify the global protein profile of plants but also to determine their localisation, post-translational modifications and activities. This provides a simple way of comparing changes between different species or genotypes or to assess the impact of environmental stresses on the plant proteome (Hall *et al.*, 2007). Protein composition can be determined using a range of biochemical and chemical techniques (e.g., chromatography, two-dimensional electrophoresis (2DE), microarrays, MS). To increase its utility, methods are frequently combined (e.g., LC or 2DE coupled with MS).

The HPLC techniques, given their wide versatility, relative ease of use, reproducibility, fast analysis time and high resolution, are valuable tools for characterisation of hydrophobic proteins and where isoelectric pH (pI) is very close. A range of biochemical methods have been used to separate specific protein complexes prior to their investigation by HPLC-MS. As an example, Timperio and co-workers (2007) isolated PSI and PSII antennae proteins from spinach leaves by sucrose-gradient ultracentrifugation to further resolve protein composition by reverse-phase HPLC electrospray spectrometry (RP-HPLC-ESI-MS).

Two-dimensional electrophoresis can be used to successfully separate and visualise up to 10,000 proteins in a single experiment, separating and displaying the components of large protein complexes, which can be identified following careful excision by MS. With the advantages of simplicity, a wide size range (10 kDa to 500 kDa) and both moderately hydrophobic or very acidic or basic proteins can be isolated and visualised. However, the limitations of gel-based techniques include poor resolution of low-abundance proteins, limited pI range and absence of membrane proteins. Moreover, the excision and identification of proteins is time consuming and requires substantial manual editing; it works best for the most abundant proteins (Dowsey *et al.*, 2006). To be reliable, it is best to use pre-cast gels and commercially prepared reagents to ensure consistent and reproducible results. Two-dimensional-difference gel electrophoresis (2D-DIGE) is a recent advance in proteomic technology that addresses several of the problems encountered in the reliability, quantitative analysis and efficiency of proteomic research (see Renaut *et al.*, 2006 for a review).

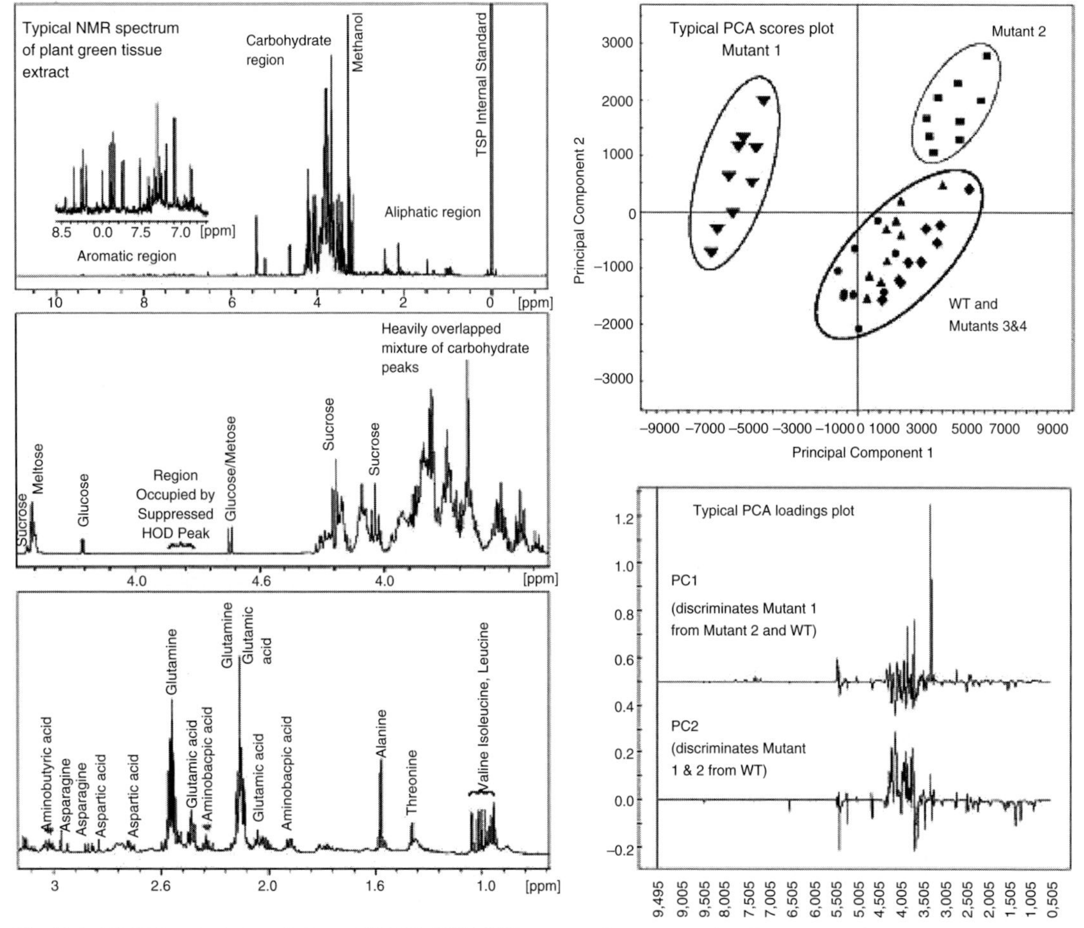

Fig. 13.1. NMR fingerprinting of plants. 600MHz [^{1}H]-NMR spectra of aqueous methanol extracts of plants are very reproducible and contain much information about polar compounds including carbohydrates, amino acids and aromatic compounds, such as flavonoids, phenylpropanoids and nucleosides. The best way to compare many NMR spectra of plant lines is by PCA analysis. The scores plot depicts clusters of similar spectra. The loadings plots describe the differences between clusters in the scores plot. Loadings plots for the principal components have the same scale as the original NMR data and depict spectra of the metabolites that are increased and decreased (supplied by J. Ward and M. Beale, National Centre for Plant and Microbial Metabolomics, UK).

Two-dimensional electrophoresis coupled to MS is well suited to ecophysiological studies, and the number of studies devoted to characterise the proteome is currently increasing steeply. Although most of these ecophysiological approaches are inherently aimed at identifying differential expression patterns of virtually all the proteome, they also serve in detecting photosynthesis-related proteins, and therefore their relative abundance with respect to the remaining proteome can be assessed. For instance, Blödner *et al.* (2007) showed in needles of *Picea abies* subjected to drought, that most of the early responsive proteins were related to photosynthesis. Using 2DE-MS, Castro *et al.* (2005) showed that some photosynthesis-related proteins were among the most responsive proteins in grapevine leaves subjected to the herbicide flumioxazin, together with several enzymatic antioxidant systems, and the abundance of some enzymes of the photorespiration pathway. In a similar approach, Vincent *et al.* (2007), studying the proteome of grapevines subjected to water deficit and salinity, demonstrated the diversity of response of different enzymes involved in the photosynthetic pathway, and that protein expression was highly affected by cultivar. Nevertheless, the lack of sufficient bioinformatics data still compromises the characterisation of protein composition for most natural, non-crop species.

High-density protein microarrays provide a fast and high-throughput alternative method of profiling proteins.

Complex mixtures of proteins in extracts can be profiled using analytical arrays of library antibodies (Bertone and Snyder, 2007). The biggest problem with this approach for ecophysiological studies is that antibodies with the necessary specificity are not available for many of the proteins or species. In addition, false negatives are common because proteins require different conditions for binding. Nevertheless, this is a powerful approach and the technology is being developed to widen the range of capture molecules (e.g., peptides, nucleic acids).

GENE EXPRESSION BY ARRAY HYBRIDISATION AND CDNA SYNTHESIS

An alternative approach to proteomics is to determine patterns of gene transcription in different genotypes, or in response to environmental variables including stress. High-density oligonucleotide arrays, or microarrays, have become synonymous with and are the most widely adopted technology platform to deliver data for global transcript profiling. Commercial microarrays are available for the full or partial genome of *Arabidopsis* and the major crop plants. Unfortunately, microarrays are not generally available for many other species. The production of microarrays is expensive and requires a substantial amount of background information on ESTs or a sequenced genome. Fortunately, a considerable amount of success has been achieved in applying microarrays across different species (Becher *et al.*, 2004); the best results are most likely to be achieved with the most closely related species. Similar to proteomic methodology, although also regarded as a high-throughput approach, array hybridisation of multiple transcripts has allowed new insights in the expression of photosynthesis-related genes. In a comprehensive study, Bogeout-Triboulot *et al.* (2008) integrated characterisation of ecophysiology and molecular responses – at transcriptomic, proteomic and metabolomic level – of *Populus euphratica* subjected to water deprivation and recovery. Such an integrated study allowed specific insights into: (1) the crossing relations between the expression of photosynthesis-related and non-related genes; and (2) the relationship between gene expression and protein abundance for several photosynthesis-related genes.

Most of the array methodology is focused on the analysis of samples from entire plants or even from a whole organ, resulting in data that represent an average of multiple cell types. However, physiological processes are often restricted to specific tissues. To solve this concern, high spatial resolution analyses have emerged for a better understanding of biological processes, such as adaptation to environmental conditions. In-situ hybridisation (Shu *et al.*, 1999) and in-situ RT-PCR (Koltai and McKenzie Bird, 2000) in fixed tissue sections have been used to determine the tissue-specific expression of individual transcripts. In a more refined methodology, cDNA array hybridisation and an amplification strategy using reverse transcriptase PCR are merged with single-cell spatial resolution sampling from undamaged plant tissue. Using this approach in *Arabidopsis* leaves, Brandt *et al.* (1999) differentiated preferentially epidermis- and mesophyll-cell expression of some photosynthesis-related genes.

The emergence of second and third generation sequencers (e.g., array based and massively parallel pyrosequencing or sequencing-by synthesis) developed for example by 454 Life Sciences and Illumina Technology that can provide fast, low-cost sequences of thousands of DNA base pairs, offers the possibility of sequencing entire (cDNA) pools and whole genomes (ultra-broad PCR), as well as the detection of specific gene mutations at extremely low levels (ultra-deep PCR). This technique does not require any prior knowledge of the sequences investigated, obviating the need for generation of any microarrays, is extremely cost-effective and appears to be the ideal tool for many ecophysiologists. As with all emerging technologies, some caution in using this approach will be necessary until accepted standards have been established.

13.3. GENETIC PLATFORMS

The rapid growth in genomic sequence information has revealed numerous genes for which the function is unclear. Revealing the function of these genes requires reverse-genetic approaches, such as gene silencing, overexpression, reporter gene expression or targeting induced local lesions in genomes (TILLING).

ECOTILLING is an adaption of TILLING that can identify natural polymorphisms in different populations or between closely related species (Comai *et al.*, 2004).

13.3.1 Mutagenesis and targeted induced local lesions

Targeted induced local lesions and ECOTILLING are reverse-genetic approaches that can be used to identify allelic variation in target genes to further explore their function or to identify the molecular mechanisms of their action (McCallum *et al.*, 2000a,b). It aims to identify the functional effects of single nucleotide changes to a particular gene. TILLING is a high-throughput low-cost procedure that is more widely applicable than other reverse genetic methods (Haughn and Gilchrist, 2006). TILLING has been successfully applied to many organisms, including all major crop plants. It requires the screening of a mutant population and exploits the power of mutagenesis to generate novel variation. Once created, the population can be used to screen for allelic variation in any target gene. TILLING is based on PCR: a region of approximately 1 kb of the target gene is amplified using pooled genomic DNA from several individuals as a template, after subsequent heating and annealing, any heteroduplexes formed are cut by an endonuclease (CEL1 or ENDO1) at mismatches then cut fragments visualised. Typically fluorescently labelled primers are used for the amplification to aid visualisation (e.g., on Li-Cor gel analysers), but non-labelled primers and fast capillary gel electrophoresis have also been used successfully (Suzuki *et al.*, 2008). The size of the fragment indicates the approximate location of nucleotide polymorphisms that are then confirmed by sequencing. As TILLING does not require other genomic resources, it is well suited to ecophysiological studies. For annual herbs, the creation of mutant populations is normally simple, but this approach is not appropriate for species that require many years to become fertile and produce seeds. Either chemicals or radiation can be used to cause mutations; to minimise the size of the population that needs to be screened to identify mutations in any target gene, the exposure should be great enough to give a high mutation load while retaining the viability of the material. Higher exposures can be used in polyploid species that have genetic redundancy. An adequate knowledge of genomic sequences is essential in designing appropriate PCR primers and in targeting coding regions.

13.3.2. Quantitative-trait-loci analysis

Dissection of quantitative trait loci (QTLs) is a powerful tool to enable the identification of genomic regions that contain genes that make a significant contribution to the expression of traits (Xu, 2002). This approach attempts to describe which genomic regions or genes segregate together with the physiological traits of interest. Typically this involves the selection of genetically divergent parent lines and the development of an F_2, backcross or recombinant inbred mapping population for QTL analysis. The approach is generally more effective for out-crossing species (e.g., maize) with large numbers of polymorphisms, than for inbreeding species with smaller numbers of polymorphisms (e.g., wheat). As the F_2 and backcross populations are not eternal, the populations of recombinant inbred lines are most useful. A large population is necessary to ensure sufficient recombination events. Quantitative trait analyses are performed by comparing the linkage of molecular markers and traits to reveal regions of genomic DNA influencing specific traits. For instance, genetic analysis of QTL markers has been used in conjunction with stable-isotope analysis to evaluate the physiological basis of variation in photosynthetic WUE in a large number of crop species, such as tomato (Martin *et al.*, 1989) or *Brassica oleracea* (Hall *et al.*, 2005).

The greater the density of the markers and the higher the recombination events, the greater the precision of the QTLs that can be identified. Although the QTLs are often determined for complex traits (e.g., drought tolerance), specific genes contributing to the trait can be identified by parallel gene-expression profiling using DNA microarrays to yield expression QTLs (e-QTLs). Similarly, metabolite QTLs (mQTLs) can be obtained through the metabolite profiling of mapping populations, and associate the occurrence of specific metabolites with QTLs. With all these approaches, it is important to test how robust the QTLs are in different physical environments and in different genetic backgrounds, and to establish their heritability. Ultimately, constituent genes underlying robust QTLs can be used to isolate the specific genes determining the trait in question.

13.4. FOCUSED ANALYSES

One should be aware from the outset that the variability of plant material – determined by environmental conditions and species differences – means that no single extraction or analytical protocol exists that will work satisfactorily for all samples. However, acceptable results can often be obtained after the introduction of minor modifications to accommodate such differences. For efficient use of time and resource, any analytical approach should be critically assessed for quantitative and qualitative feasibility at the outset. The approach to tissue extraction will depend on the analyses being undertaken, as indicated in Fig. 13.2 and explained in more detail below. In general, the study of whole-leaf processes in conjunction with the underpinning biochemical processes – such as photosynthetic CO_2 assimilation in relation to metabolite abundance, enzyme activity or gene expression – should be designed to enable the acquisition of as much complementary information from a single sample as possible, thus increasing the credibility of consequent correlations (Fig. 13.2).

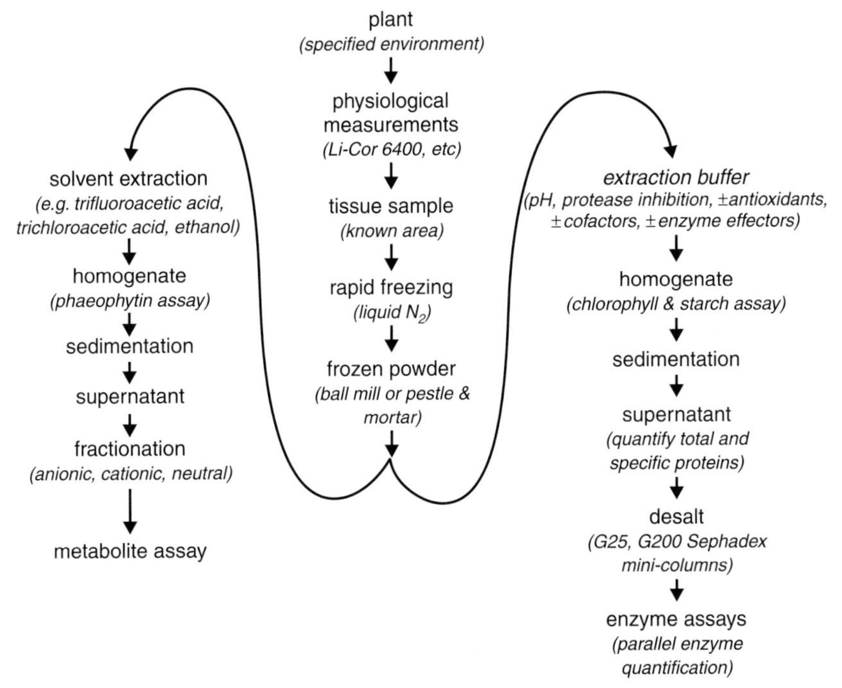

Fig. 13.2. General approaches to plant-tissue extraction for the study of metabolite, pigment, protein and enzyme activity.

13.4.1. Metabolites

Starch and sucrose are the major end products of photosynthetic carbon fixation, amounts of which change relatively slowly. In certain plant families considerable quantities of other carbohydrates may accumulate – such as fructans in Geramineae and polyols such as mannitol and sorbitol in a diverse range of plant species. This latter group may also contribute towards amelioration to osmotic stress. Where previous analyses indicate the occurrence of other significant pools of assimilated carbon, these should be included in the overall assessment of photosynthetic metabolism. Measurements of leaf starch and sucrose give an integrated estimate of recent photosynthetic activity less respiration, metabolism and translocation. In photosynthesising leaves sucrose represents the balance between synthesis and export, while starch represents the daily integral of starch synthesis. Comparisons of starch and sucrose synthesis rates require a kinetic method, especially an assessment of the amount of starch and sucrose made in response to a short pulse of $^{14}CO_2$. The amounts of Calvin-cycle intermediates change very rapidly and therefore provide specific information about photosynthetic function at any moment in time; for example monitoring RuBP concentrations can indicate whether the Calvin cycle is limited by RuBP utilisation or regeneration.

Intra- and extracellular processes operate all the time to maintain a state of redox homeostasis. These are a series of coordinated processes whose function is to ameliorate the effects of oxidative side reactions that are an unavoidable consequence of oxygenic photosynthesis in an oxygen-rich environment. The relative amounts of metabolite components of these processes are diagnostic of the stress status of plants, and their measurement and interaction are considered in the final section. Thus, metabolite analyses can be very useful for ecophysiological studies and are increasingly used (Hummel *et al.*, 2010; Pinheiro and Chaves, 2011; Pinheiro *et al.*, 2011; Warren *et al.*, 2011).

As the amounts of some metabolites can change rapidly it is essential that samples are collected, stored and extracted in such a way as to eliminate artifacts. We therefore recommend freeze clamping, storage at –80°C and acid extraction at low temperature. To enable a valid comparison between metabolite, protein or enzyme measurements from disparate leaf samples, data should be expressed relative to leaf area or fresh weight, and so these parameters – far from being trivial – must be carefully recorded at the outset. There is no substitute for numerous biological replicates for estimation of sample mean and variance. When multiple, discrete measurements are required on a single sample, material from a representative selection of leaves can be combined, frozen and powdered (in liquid nitrogen) and portions of this used for each type of measurement. This can be useful when comparing different methodologies, ensuring a uniform starting material.

13.4.2. Hexose and starch determination

The concentration of hexose solutions can most easily be determined using a modification of the anthrone method of Hansen and Moller (1975), as described below. It should be noted that this assay is applicable to any hexose or acid-hydrolysable, hexose-containing polysaccharide, including starch. Therefore, in combination with extract fractionation (e.g., solvent-based separation of simple sugars from starch) it can give quantitative information about two assimilate pools. Add 1 ml of anthrone reagent (0.2% (w/v) anthrone in 13.5 M H_2SO_4) to 0.2 ml of an aqueous extract in a microfuge tube and mix. Transfer the microfuge tubes to a heating block at 100°C for exactly 11 min then cool on ice immediately. The absorbance of the samples is measured at 630 nm and the concentration determined by reference to a standard curve from samples containing 0–200 nmoles of glucose. Ensure that the absorbance of the samples falls within the range of linear proportionality with the standards and, if not, alter the amount of extract until this condition is satisfied. More detailed information regarding the relative proportions of glucose, fructose and sucrose can be obtained by HPLC analyses of leaf extracts. The most straightforward of these employs a resin-based immobile phase (e.g., Bio-Rad Aminex HPX-87H) and a dilute-acid mobile phase (5 mM H_2SO_4) in conjunction with a refractive-index detector, the use of which requires little additional sample preparation. The sequential use of a combination of commercially available enzymes (such as hexokinase, phosphoglucose isomerase, invertase and amyloglucosidase) has been used to convert starch, sucrose, glucose and fructose into glucose 6-phosphate, whose subsequent oxidation can be linked to the reduction of $NADP^+$ to NADPH by the action of G6P dehydrogenase (Jones *et al.*, 1977). These reactions can be easily and quantitatively monitored spectrophotometrically and can easily be adapted for application with ELISA plate readers. It must be emphasised that more detailed information about relative proportions of sucrose, glucose, fructose and other simple sugars are only meaningful if the extraction processes have been chosen to minimise any artefactual changes in the amounts of these compounds. For example, aqueous extraction is unsuitable as enzyme-mediated interconversion could take place. Sucrose is acid labile and so the use of acidic extraction media should be avoided in the measurement of sucrose. Hot ethanolic solutions or chloroform/methanol are suitable for the extraction of simple sugars from complex mixtures.

13.4.3. RuBP and Calvin-cycle intermediates

The amount of RuBP in samples can be quantified using activated Rubisco together with cosubstrate $^{14}CO_2$ and determining the amount of radiolabel that is rendered acid stable (i.e., 3-phosphoglycerate (PGA), formation, Sicher *et al.*, 1979). To be reliable it is essential that oxygen is excluded from the reaction and that the reaction is taken to completion. This is accomplished by ensuring high $^{14}CO_2$ concentration (high $H^{14}CO_3$ in the presence of carbonic anhydrase), the use of a reaction buffer pre-treated by bubbling with pure nitrogen and septum reaction vials pre-gassed with pure nitrogen. As RuBP breaks down at neutral and alkaline pH, it is also essential that it is not kept at these pHs for longer than necessary and that sufficient active Rubisco is used to ensure rapid conversion to PGA during the assay. We also suggest that samples have no more than 100 μM RuBP.

It should be remembered that, apart from RuBP, the intermediates of the Calvin cycle are not unique to the chloroplast and so their measurement is not necessarily diagnostic of Calvin-cycle status. For this, prior non-aqueous fractionation would be required. Nevertheless, quantification of fructose, glucose, sedoheptulose and ribulose mono- and bisphosphates by anion-exchange chromatography (Dionex CarboPac PA1 analytical column) with 10 mM sodium hydroxide and a 100–800 mM sodium-acetate gradient together with peak detection by pulsed amperometry (Dionex) has proven satisfactory, although phosphorylated organic acids including phosphoglycerate and phosphoglycolate escape detection. Alternative anion-exchange chemistry (Dionex IonPac AS11-HC) with 5–100 mM sodium-hydroxide gradients and detection by peak conductivity are suitable. In both cases, additional sample preparation is essential, particularly to remove lipophilic residues (C18 minicolumns) and divalent metals cations (Dowex-50 (H^+) minicolumns) that would otherwise poison the HPLC columns.

The phosphorylated intermediates of the Calvin cycle can also be assayed spectrophotometrically using a series of enzyme-mediated interconversions, which ultimately cause the reduction or oxidation of a nicotinamide adenine dinucleotide, with the concomitant rise or fall in absorbance at 340 nm, respectively. For more detail on such assays, the reader is referred elsewhere (Leegood, 1993).

13.4.4. Isotopic labelling to investigate assimilate metabolism

Soon after becoming available, the 'tracer' isotope ^{14}C was used to investigate carbon metabolism (Calvin *et al.*, 1949). This isotope – in the form of $^{14}CO_2$ – was used by Calvin and co-workers to establish the primary products and the sequence of reactions comprising C_3 photosynthesis, culminating in the elucidation of the photosynthetic carbon-reduction cycle (Calvin, 1962). The metabolism of ^{14}C-containing compounds within leaves can also provide information about the relative rates of formation and degradation of specific intermediates, and about the relative flux through competing pathways. All such experiments require the acquisition of tissue samples at specific intervals after isotope application. Key to these approaches is the manner by which the isotopically enriched compounds gain access to metabolically relevant compartments within the leaf. Vacuum infiltration has been used to effect the rapid entry of isotopically enriched compounds, via the stomata and intercellular air spaces (Andralojc *et al.*, 1996), although entry through the vascular tissue of the cut petiole may be more physiological albeit slower (Moore and Seemann, 1992). The introduction of a pulse of $^{14}CO_2$ into the stream of air passing over leaves during steady state photosynthesis is undoubtedly the most natural means by which ^{14}C may gain access to photosynthetic tissue, although the duration of exposure should be limited to prevent saturation of all primary metabolites with the radioisotope. The value of all such approaches is conditional upon the subsequent identification of the isotopically enriched products, usually by means of thin-layer or HPLC. A combination of these approaches has been used to elucidate a biosynthetic pathway for the naturally occurring Rubisco inhibitor, 2-carboxy-D-arabinitol 1-phosphate (Andralojc *et al.*, 2002). An identical approach can be applied using the non-radioactive carbon isotope, ^{13}C, followed by mass analysis of the products. In this case, however, prior fractionation of the extracted metabolites may be unnecessary, if the relevant parent ions have distinct mass/charge ratios.

13.4.5. Pigments and proteins

PIGMENTS EXTRACTION

For chl. and carotenoid determinations, leaves or leaf discs are typically sampled with a cork borer and then extracted either by grinding the leaf material with an organic solvent or by incubating leaf pieces for a few days in the solvent. Two major points should be considered when making pigment extractions. First, commonly used organic solvents may have traces of acid that can cause sample acidification, and consequent pigment decomposition, e.g., when water is introduced to the ground tissue in acetone. In addition, when plant tissue is ground, acidic compounds from the vacuole are released into the grinding medium and may also significantly lower the extract pH. To avoid acidification, it is desirable to add a buffering agent as a precautionary measure, typically calcium carbonate, sodium carbonate or sodium ascorbate (around 0.1 g g^{-1} plant material). Second, pigments in organic solvents are light sensitive so should be kept away from light during storage (e.g., wrapped in aluminium foil) and subsequent analysis (e.g., under low light intensities).

Chlorophylls and carotenoids are located in a lipidic environment, but thylakoids are surrounded by an aqueous medium. Therefore, chl. and carotenoids are extractable from thylakoids only by organic solvents able to mix with the water contained in the plant tissue. A wide range of solvents have been used for this purpose, but acetone and methanol are the most commonly used. There is some controversy regarding the relative efficiency of pigment extraction using pure or water-diluted solvents, but most evidence suggests that keeping the water content to a minimum is preferable. The presence of an excess of water aggravates the negative effects of low pH and light, and the extraction of the most non-polar pigments (chl. *a* and β-carotene) using solvents diluted with water may be incomplete in some plant materials (Lichtenthaler, 1987). In addition, it must be considered that fresh plant tissue already contains more than 70% water, which will be incorporated into the extraction media immediately after grinding the tissue and could negatively affect the extraction process.

For the extraction of samples, a known amount of leaf tissue is homogenised in a chilled pestle with a few millilitres of bulk cold (0°C) organic solvent and a small amount of buffering agent. The homogenate is then poured into a volumetric flask and acetone is added to attain the desired volume. The mixture is clarified to remove suspension particles (until minimal absorbance is measured at 730 or 750 nm) by either quick centrifugation or filtration through 5 μm filter. In both cases, it must be ensured that thorough pigment extraction has occurred, by checking that the pellet (if centrifuged) or retained solids (if filtered) are colourless.

As well as leaf area and leaf dry weight, leaf biochemical properties are also frequently expressed relative to the accompanying chl. content. Accurate and consistent methodology is therefore important.

PIGMENT DETERMINATION

Chlorophyll and phaeophytin determination by spectrophotometry

A set of equations can be used to calculate the chl. content for any solvent used (Holden, 1976; Lichtenthaler, 1987). Each of these equations is based on the absorbance reading at specific wavelengths. For instance, the total chl. content (the sum of chl. *a* and *b*) can be determined spectrophotometrically at a single wavelength using neutral (pH 7 to 8) extracts either in acetone (by a modification (Bruinsma, 1961) of the method of Arnon (1949)):

$$\text{total chl. } (\mu g/ml) = 1000 \times A_{652} / 36) \quad \text{[Eqn. 13.1]}$$

or in 96% (v/v) ethanol (as reported by Wintermans and De Mots (1965)):

$$\text{total chl. } (\mu g/ml) = 1000 \times A_{654} / 39.8 \quad \text{[Eqn. 13.2]}$$

These two approaches are suited to dilute and concentrated leaf extracts, respectively. These sort of equations present some inherent problems, related to the purity of standards used to solve the equations or to the spectrophotometer resolution. As a consequence, the correct coefficients and reading wavelengths for the equations have been reinvestigated a number of times over the years (Lichtenthaler, 1987).

In acid, the Mg^{2+}ion of chl. porphyrin is replaced by a pair of protons, to yield phaeophytin (types *a* and *b*). In excess acid, the total concentration of phaeophytin is equivalent to the original concentration of chl. and can be determined in acetone by measuring the light absorbance at two wavelengths, according to Vernon (1960):

$$\text{total phaeophytin } (\mu g/ml) = 77.58 \times A_{536} - 0.33 \times A_{666} \quad \text{[Eqn. 13.3]}$$

When samples undergo some breakdown of chl. to phaeophytin because, for example, of acidity in the leaf samples, it can be convenient to convert all chl. to phaeophytin to overcome this problem.

Chlorophyll and carotenoid determination by high-performance liquid chromatography

The reverse-phase HPLC method resolves the major higher plant photosynthetic pigments in leaves, including neoxanthin, violaxanthin, taraxanthin, antheraxanthin, zeaxanthin, lutein and carotenes. Chl. *a* and *b* can also be quantified using this approach, and therefore a complete characterisation of pigment concentration can be performed on the same sample (Fig. 13.3). For the separation of tocopherols, suitable HPLC methods commonly use normal phases, because with reversed phases it is hardly possible to separate *a*- and *b*-tocopherols (Kamal-Eldin *et al.*, 2000). Pigments are then identified by their UV/visible absorption spectra in at least two different solvents, their retention times compared with standards and, where necessary, by MS. Finally, they are quantified using published extinction coefficients (Davies, 1976). Leaf concentrations of violaxanthin (V), antheraxanthin (A) and zeaxanthin (Z), as well as the total pool (VAZ), are usually expressed in relation to total chl. (chl. *a*+*b*) and on a dry weight basis. The de-epoxidation state of the xanthophyll-cycle pigments (DPS) is calculated as the ratio of antheraxanthin and zeaxanthin to the total xanthophyll-cycle pool as described by Adams *et al.* (1995).

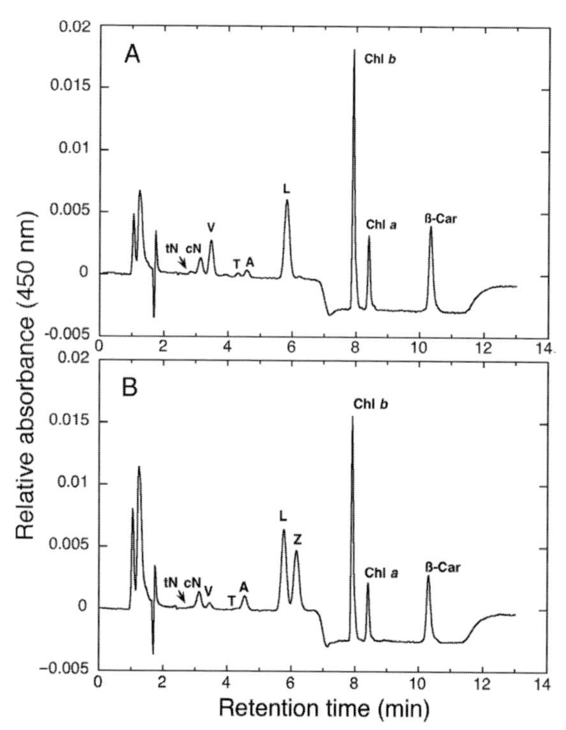

Fig. 13.3. HPLC separation of pigments from dark-adapted leaf samples from *Beta maritima* under well-watered conditions **(A)** and water stress **(B)**. A, antheraxanthin; B-Car, β-carotene; Chl *a*, chlorophyll *a*; Chl *b*, chlorophyll *b*; L, lutein; Lx, lutein-5,6-epoxide; N, neoxanthin; V, violaxanthin; Z, zeaxanthin. Note the conversion of V to Z under water stress by comparing graph A and B.

This approach has been used to assess the effects of a myriad of environmental stresses on leaf-pigment concentrations (e.g., Elvira *et al.*, 1998; García-Plazaola and Becerril, 2001; Galmés *et al.*, 2007b).

More precise determinations, e.g., identification of isomeric forms of carotene and xanthophyll, require the use of chromatographic systems replete with MS (LC-MS).

PROTEIN EXTRACTION

In all cases it is important that the protein to be measured is protected from proteolytic degradation, modification by free radicals or interaction with phenolic compounds prior to assay. Thus extracts must contain appropriate protease inhibitors (proprietary protease-inhibitor cocktails, metal chaelators), reducing agents (dithiothreitol, mercaptoethanol, reduced glutathione (GSH)) and polymeric protectants (soluble or insoluble polyvinylpyrrolidone, polyethylene glycol). Glycerol solutions stabilise most enzymes, e.g., PEPC by 20% (v/v) glycerol. Addition of alternative or sacrificial protein substrates – such as bovine serum albumin (BSA) or milk protein (casein) – during the extraction process, can help to protect endogenous proteins from degradation, but yields extracts whose proteins are less pure in the initial stages of fractionation and renders meaningless any estimation of total soluble protein.

PROTEIN DETERMINATION

Total soluble protein is another frequently used reference parameter that can easily be measured in clarified leaf extracts by the method of Bradford (1976), modified for use with a microplate spectrophotometer. Sample is added to water/extraction buffer in the wells of a microtitre plate, to a total volume of 50 μL. Three hundred microlitres of the Bradford Reagent is added, and after 5 min the absorbance at 595 nm is recorded. Protein concentrations are determined by reference to a standard curve made with samples containing up to 12 μg BSA, prepared in parallel. It is essential that the absorbance of the samples falls within the linear range of the standards and if not that more (or less) extract is used until this condition is satisfied. The protein concentration of unfractionated leaf extracts should be determined before freezing, as protein precipitation is frequently observed when frozen extracts are thawed leading to underestimation. Interference by detergents or reductant is a frequent source of error in such determinations. Such problems can be overcome by prior precipitation of the protein using either acid (e.g., trichloroacetic) or cold organic solvent (e.g., ethanol at –20°C). The sedimented protein can then be redissolved in solvents that do not contain the interfering agent and the protein assay resumed.

Accurate quantification of specific enzymes usually requires the availability of a highly purified standard and of specific antisera; amounts can then be determined accurately using a range of immunological approaches (e.g., Western blotting, immunoprecipitation). Unfortunately, the range of commercially available antisera is limited. Consequently, many authors estimate amounts of enzyme by measuring the corresponding catalytic activity. However, ideally, both activity and amount should be known, particularly if enzyme activity is regulated post-translationally or by the presence of other compounds that inhibit or otherwise affect enzyme activity.

RUBISCO QUANTIFICATION

Because of its unparalleled abundance, Rubisco can be quantified directly by protein gel electrophoresis, comparing the band intensity of the large subunit with that of purified Rubisco (Rintamaki *et al.*, 1988). To be reliable, it is essential that the Rubisco subunits are resolved and that quantitative staining can be demonstrated.

Rubisco content is, however, most accurately determined by incubating the activated enzyme with ^{14}C-labelled carboxyarabinitol bisphosphate (CABP) – a tight binding inhibitor of Rubisco – according to the method of Yokota and Calvin (1985). To be reliable, it is essential that: (1) the CABP is in the open-chain conformation rather than the corresponding lactone – delactonisation requires overnight incubation at pH 8; and (2) it is incubated with CABP long enough for all the active sites to become occupied. CABP is synthesised by alkaline hydrolysis of the cyanohydrin of RuBP and potassium cyanide (the latter with or without ^{14}C) as described by Pierce *et al.* (1980). A working specific radioactivity of 37kBq (1 μCi)/μmol CABP is suitable for most purposes. Crucially, the amount of ^{14}C-CABP present must be in excess of the Rubisco active-site concentration, otherwise the number of active sites will be underestimated. We recommend at least a twofold excess.

13.4.6. Enzyme activity

Before sampling, one must consider the source of material, the time of day, the enzyme activity to be determined and whether, in the case of regulated enzymes, the in-vivo activity or the maximum catalytic capacity (or both) are required. Considerable care must be exercised in determining the composition of extraction buffers. In addition to the protective measures described above, buffer concentration, buffer pH range, buffer chemistry and acidity of the tissue undergoing extraction must be carefully considered. For example, the leaves of CAM plants accumulate C_4 acids during night, requiring high concentrations of buffer to avoid acid denaturation. We have also found that buffers containing primary amines (principally Trizma) may form Schiff's bases with carbonyl compounds, leading to artifacts. Furthermore, it may be necessary to consider the effect of naturally occurring cofactors and enzyme effectors – present in extraction buffers – on the activation state of the enzymes of interest.

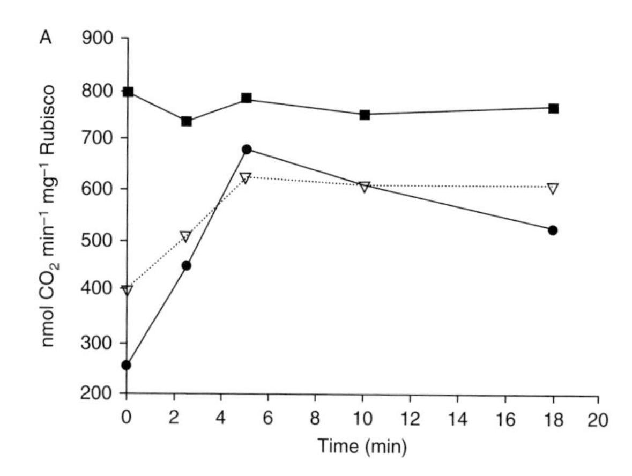

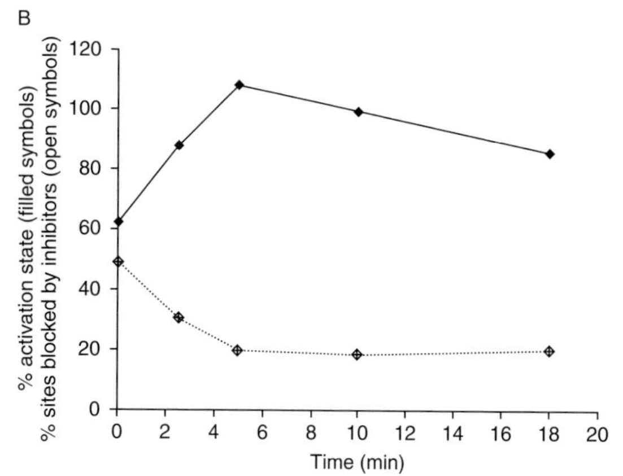

Fig. 13.4. **(A)** Changes in initial (closed circle), total (open triangle) and maximal (closed square) Rubisco activity of tobacco following illumination. **(B)** Changes in the activation state (filled diamonds) and active sites blocked by inhibitors of tobacco Rubisco following illumination.

RUBISCO ACTIVITY

Initial, total and maximum Rubisco activities can be determined by measuring the amount of $^{14}CO_2$ incorporated into acid-stable products according to Parry *et al.* (1997). Initial activity is assayed as soon as possible after tissue extraction and reflects the activity of the enzyme at the time of sampling. Total activity is determined in a similar way, except the extract is briefly incubated with Mg^{2+} ions and bicarbonate before the assay is initiated by addition of RuBP. The 'activation state' is the ratio of initial-to-total activity, and provides an estimate of the amount of enzyme that is carbamylated (Butz and Sharkey, 1989). Determination of the maximum catalytic capacity of extracted Rubisco requires prior removal of tight-binding inhibitors by treatment with high concentrations of SO_4^{2-} ions, thorough desalting (as SO_4^{2-} ions are inhibitory) and finally Rubisco reactivation with Mg^{2+} ions and bicarbonate, prior to the activity assay. The ratio of total-to-maximal activity can be used to estimate the number of catalytic sites that are blocked by tight-binding inhibitors (Fig. 13.4). In all such assays, to avoid 'fallover', i.e., a progressive loss of activity during catalysis (Edmondson *et al.*, 1990), it is essential that assay times are kept short (e.g., 1 min). The extent to which the accumulation of tight-binding inhibitors at the catalytic site of Rubisco occurs, is also determined by the ratio of Rubisco and Rubisco activase. This ratio is diminished at elevated temperature, but also appears to be developmentally regulated (Fig. 13.5).

RUBISCO KINETIC CONSTANTS AND SPECIFICITY FACTOR

By measuring the activity of purified and fully activated Rubisco at various concentrations of CO_2, O_2 and RuBP, the corresponding kinetic constants can be estimated (e.g., Bird *et al.*, 1982; Jordan and Ogren, 1984; Makino *et al.*, 1985). To be reliable, it is essential that the O_2 and CO_2 concentration of each reaction is correctly established. We achieve this by prior buffer equilibration using CO_2-free nitrogen to which O_2 has been added by means of a mass-flow gas-mixing device. Gas of the same composition is also passed through a series of septum reaction vials, replete with miniature magnetic stirring bars in a temperature-controlled water bath with submersible stirrer. After injection of gas-equilibrated buffer, the stream of gas through the vials is stopped and bicarbonate of known specific radioactivity injected through the gas-tight septum. Sufficient time (1 h) is allowed for equilibration between HCO_3^- and CO_2 in the gas and liquid phases, so that a known CO_2 concentration prevails in the liquid phase when the assay is initiated. CA is included in all buffers at the outset to facilitate rapid equilibration. RuBP is added and the assay initiated by injection of activated Rubisco. Reaction times are short ($\leq$1 min) so that no

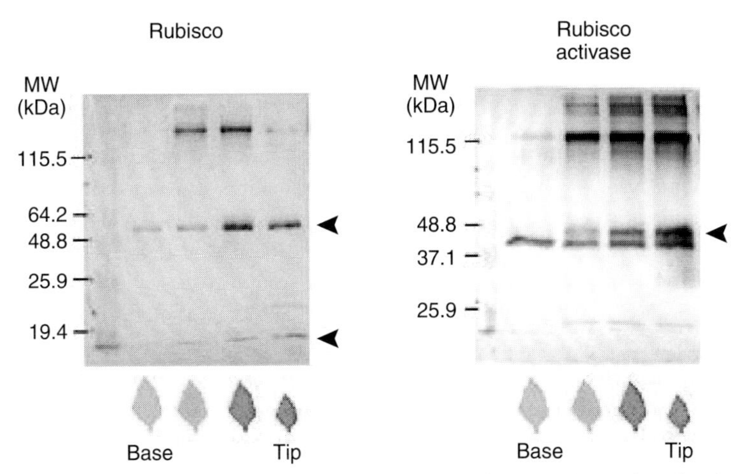

Fig. 13.5. Comparison of the relative abundance of Rubisco and Rubisco activase in tobacco leaves at different positions along the main stem and, hence, different stages of development. The positions of Rubisco large subunit, small subunit and Rubisco activase are indicated beside the corresponding westerns (kindly provided by Anneke Prins).

more than 10% of any substrate is consumed and enzyme decarbamylation is restricted. Reactions are quenched by injection of concentrated formic acid. It is important that the pH (normally between pH 8.00 and 8.20) of all reaction buffers is precisely measured, as the concentration of CO_2 in the liquid phase is highly pH dependent.

For determination of the specificity factor, Rubisco carboxylase and oxygenase activities can be determined simultaneously with an oxygen electrode by measuring the consumption of oxygen accompanying the total consumption of a known amount of RuBP (Parry *et al.*, 1989). This assay can be enhanced by the use of $^{14}CO_2$, and following the incorporation of ^{14}C into acid-stable products (Lorimer *et al.*, 1977). Other approaches include the HPLC resolution of reaction products, phosphoglycolate (PG) and PGA, in conjunction with either a sensitive conductivity detector (Uemura *et al.*, 1996), or [1–^{14}C] RuBP as co-substrate and measuring the radioactivity of the fractionated products (Kane *et al.*, 1994). This approach is particularly useful for enzymes with low activities and can be used to reveal novel reaction products.

It has been found that absolute rates of catalysis by Rubisco may fall during the process of purification. While this is considered not to affect measurement of relative (CO_2/O_2) specificity, it would decrease the rates (k_{cat}) of oxygenase and carboxylase activities of the fully activated enzyme. For this reason, it may be preferable to use rapidly isolated enzyme instead, although the possibility of competing side reactions in these relatively crude extracts must be carefully monitored, corrected and/or eliminated to ensure reliable results. For example, we have detected significant phosphogycolate-phosphatase activity in certain grass extracts, but these were quickly and effectively removed using a small column of G-200 Sephadex.

For many applications, a non-radioactive, rapid, sensitive and convenient microtitre-plate-based assay for Rubisco is available (Sulpice *et al.*, 2007) and should be considered. In its current form, however, it would not be suitable for determination of K_m or V_{max} for either CO_2 or O_2 (owing to the requirement for precise concentration of dissolved CO_2 and O_2) or for maximal Rubisco-activity determinations.

PHOSPHO-ENOL-PYRUVATE CARBOXYLASE ACTIVITY

The time of day that C_4 and CAM leaves are harvested for measurement of PEPC activity is important, as the catalytic activity and response of the enzyme to effectors glucose 6-phosphate (positive) and malate (negative) varies greatly, with the phosphorylation state of a serine residue near the N-terminus, which in turn varies with light intensity (C_4) or the status of a circadian clock (CAM), through the action of an upstream kinase. Furthermore, the same N-terminal region is particularly susceptible to proteolytic removal during isolation, although protease inhibitors – especially chymostatin – provide protection. The enzyme requires the cofactor Mg^{2+} and is inhibited by inorganic phosphate and compounds resembling the substrate, PEP. PEPC activity is assayed in a coupling system linking the production of OAA to NADH oxidation using MDH (Lane *et al.*, 1969). The reaction is monitored: (1) instantaneously, by measuring the change in A_{340} in a thermostatically controlled spectrophotometer/microplate reader; or (2) by the use of sodium ^{14}C-bicarbonate as co-substrate, acid quenching (formic acid), removal of acid-labile ^{14}C by evaporation and finally liquid scintillation counting. While the latter method has greater sensitivity, it is relatively slow and requires care to minimise unintentional changes in specific radioactivity of bicarbonate (owing to atmospheric CO_2), yet still requires the presence of the coupling components (MDH and NADH) as OAA is prone to non-enzymatic decarboxylation. The in-vivo activity of PEPC – relative to the maximum potential activity – is dependent on the extent of protein phosphorylation, the concentration of PEP and HCO_3^- and of effectors, malate, G6P and P_i. Estimates of the $V_{p,max}$ for PEPC (Carmo-Silva *et al.*, 2008a,b) are necessary in the determination of the activation state ($V_{initial}/V_{p,max}$) of PEPC, and are approximated by measuring catalytic activity at pH 8 with saturating substrate (e.g., 10 mM PEP, HCO_3^- and Mg^{2+}) using light-adapted leaf material.

REDOX-REGULATED ENZYMES OF THE CALVIN CYCLE

In the absence of light, it is vital that the energy consuming reactions of the Calvin cycle are inactivated, to prevent the cycle from being driven by the respiration of previously fixed carbon. Conversely, to prevent the loss of recently assimilated carbon (as CO_2) the oxidative pentose-phosphate cycle must not operate while the Calvin cycle is active. The enzymes involved in these processes – which are regulated by reduced thioredoxin – are fructose- and sedoheptulose-bisphosphate phosphatases, phosphoribulokinase, glyceraldehyde phosphate dehydrogenase and G6P-dehydrogenase. For a detailed exposition on these processes, we commend the work of others (Buchanan, 1984; Scheibe, 1990). In general, to understand the degree of activation of any of these enzymes, it is necessary to know their instantaneous activity relative to their maximum and minimum activities (i.e., when in the fully reduced or oxidised state). To achieve these extremes of activation *in vitro*, we have found that pre-incubation at physiological pH with high concentrations (e.g., 50 mM) of DL-dithiothreitol (reductant) or oxidised glutathione (GSSG) (oxidant) for 30–60 minutes prior to assay is often effective.

13.4.7. Reactive oxygen species, antioxidants and oxidative-stress markers

Cellular reduction and oxidation (redox) balance, oxidative signalling and oxidative stress are influenced by the presence of any of a number of ROS, which are metabolised by a battery of antioxidant enzymes and low-molecular-weight antioxidants. These serve to maintain cellular redox homeostasis and counterbalance excessive oxidation or reduction. A portfolio of classical biochemical assays, and in-vivo and in-situ techniques can be used for the detection of ROS and antioxidants (see Shulaev and Oliver, 2006 for a review). Uncontrolled oxidation can lead to damage of many types of molecules including DNA, RNA, proteins and lipids, and these also can be used as *oxidative* markers in the assessment of the extent of cellular oxidation. In addition to the direct measurement of oxidants, antioxidants and marker molecules, measurements of more global molecular and metabolic oxidative-stress markers are becoming increasingly popular. The last 10 years have seen the publication of a large number of papers describing the oxidative-stress transcriptome, proteome and metabolome in plants (for example, Desikan *et al.*, 2001; Romero-Puertas *et al.*, 2002; Mittler *et al.*, 2004; Job *et al.*, 2005; Millar *et al.*, 2005; Förster *et al.*, 2006; Mahalingam *et al.*, 2006; Baxter *et al.*, 2007; Queval *et al.*, 2007). As well as providing useful markers for global changes in the cellular redox state, the use of 'marker transcripts' that are induced specifically by different ROS forms have been used to distinguish between the operations of different ROS-signalling pathways in various stress conditions (Gadjev *et al.*, 2006). In the future, it would be very helpful if such marker-transcript promoters could be used to drive reporter genes that would provide an accurate quantification of ROS, either *in vivo* or in extracts, as ROS estimation remains problematic in plants (Queval *et al.*, 2008). Although in-situ staining techniques for superoxide and H_2O_2 are a very popular tool used to obtain information on the intracellular localisation of ROS, particularly in responses to environmental or metabolic triggers, such data remain qualitative (Garmier *et al.*, 2008)

SPECTROPHOTOMETERIC QUANTIFICATION OF REACTIVE OXYGEN SPECIES

The estimation of ROS is complicated by high reactivity, poor stability and relatively low abundance. Further complexities arise from the presence of a highly active antioxidative system, coupled to poor assay specificity and artifactual interference during both extraction and assay. Although H_2O_2 is the least reactive of the different ROS, the quantification of even this relatively stable ROS is problematic because of a range of potential artifacts (Queval *et al.*, 2008). A list of procedures aimed at minimising, or at least identifying potential, assay problems has been outlined by Queval *et al.* (2008). Despite the questionable specificity information derived from direct determinations of superoxide and H_2O_2, such values are often used as indicators of the degree of 'generalised oxidative stress' (Shulaev and Oliver, 2006). However, this conclusion remains debatable and there is still no consensus of opinion on probable leaf superoxide and H_2O_2 contents, or on the relative concentrations of these ROS in the different intracellular compartments (Queval *et al.*, 2008).

The high sensitivities of some chloroplast enzymes to oxidative inactivation and the relatively high affinities of peroxidases would suggest that intracellular H_2O_2 concentrations are unlikely to exceed 100 μM. Literature evidence supports the view that considerably higher concentrations can persist at other sites in the cell, particularly in the apoplast/cell wall. Values within the range 0.5–5 $\mu mol\ g^{-1}$ fresh weight have been reported for several species (Cheeseman, 2006). However, cellular H_2O_2 contents as high as 1 $mmol\ g^{-1}$ fresh weight have important implications, suggesting that plants are able to tolerate much higher intracellular H_2O_2 concentrations than other organisms. Alternatively, marked concentration gradients in ROS might exist between intracellular and extracellular compartments, with preferential accumulation in the apoplast and/or vascular tissues (Queval *et al.*, 2008). If basal ROS levels really are as high as some reports suggest (Cheeseman, 2006), data concerning extractable-tissue superoxide or H_2O_2 levels may be of limited value as an indicator of oxidative stress or of oxidative signalling.

Other important signalling ROS, such as singlet oxygen, are even more difficult to measure than H_2O_2. Singlet-oxygen measurements can be performed using the compound 2,2,6,6-tetramethyl-piperidine-dansyl (DanePy), which shows good selectivity for singlet oxygen and works best with leaves, and not with isolated membranes such as thylakoids (Hideg *et al.*, 2001). However, DanePy is not commercially available and its synthesis is not trivial. Recent measurements suggest that singlet oxygen may be more stable than initially thought (Fischer *et al.*, 2007). Tissue levels of singlet oxygen, like those of the hydroxyl radical, are thought to be essentially controlled by non-enzymatic detoxification systems. Current methods for detecting hydroxyl radicals, such as electron spin resonance spectroscopy (ESR), which measures the electron paramagnetic resonance spectrum of a spin-adduct derivative after trapping, are relatively insensitive. Moreover, ESR determinations are not quantitative because the hydroxyl-radical adduct is not stable. A more popular method involves aromatic hydroxylation – by reaction with compounds such as phenol, benzoic acid or SA – followed by separation by GC or LC. Although more sensitive, such determinations are complicated by multiple reaction products and secondary generation of superoxide. Although dimethyl sulfoxide (DMSO) has some advantages as a probe for the detection of hydroxyl radicals (Shen *et al.*, 1997), problems associated with effective extraction of hydroxyl radicals still persist because of their very high reactivity and poor stability.

QUANTIFICATION OF LOW-MOLECULAR-WEIGHT WATER-SOLUBLE ANTIOXIDANTS AND PYRIDINE NUCLEOTIDES

Given the problems associated with accurate determination of cellular ROS concentrations, the use of other markers of cellular redox state is often more informative. Commonly used markers of cellular redox state include redox buffers, such as ascorbate and glutathione, specific lipid peroxides or derivatives, the abundance of transcripts known to be induced by ROS and oxidation of protein amino-acid residues (Gadjev *et al.*, 2006; Shulaev and Oliver, 2006). Although the accurate estimation of the oxidised and reduced forms of some key redox pools such as thioredoxin remains notoriously difficult, others such as the pyridine nucleotide pools, ascorbate and glutathione are more easily determined (Foyer *et al.*, 2008). Leaves, such as those in an *Arabidopsis* rosette, maintain substantial pools of the redox buffers, asorbate and glutathione, and these are largely reduced in all except the most-stressful conditions (Fig. 13.6). Rapid and careful extraction procedures of such antioxidants are crucial as failure to prevent oxidation during these stages can lead to an overestimation of the oxidised forms of ascorbate and glutathione. Redox-sensitive green fluorescent proteins (GFPs), which are considered to report the cytosolic glutathione redox potential *in vivo*, are being increasingly used to probe the redox state of various cellular compartments particularly the cytosol (Meyer *et al.*, 2007). These measurements suggest that the cytosolic GSSG concentrations are in the nanomolar range or lower, implying that the true GSH/GSSG is equal to or greater than 10^6. A decrease in this ratio to 10^3 would represent an increase in absolute GSSG concentrations of about 1000-fold, a change extremely difficult to measure in homogenised extracts with commonly used techniques. In contrast to ascorbate and glutathione, the pyridine nucleotide pools are much smaller and they are substantially oxidised (Fig. 13.6), consistent with their roles in numerous redox reactions associated with photosynthesis and leaf metabolism. Other indicators can be used effectively to estimate the redox state of specific intracellular compartments. For example, the activation state of

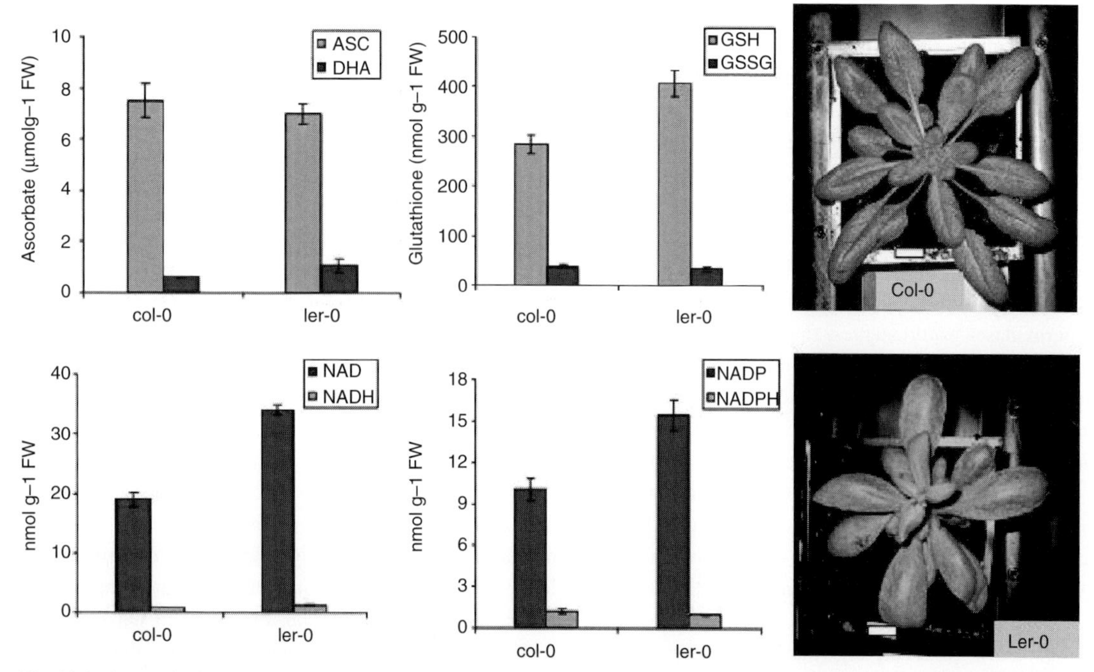

Fig. 13.6. A quantitative comparison of the ascorbate (ASC and DHA), glutathione (GSH and GSSG) and pyridine nucleotide (NAD/NADH and NADP/NADPH) pools in 4-week old plants of the Landsberg erecta (Ler) and Colombia (Col) accessions of *Arabidopsis thaliana*.

the chloroplast NADP-MDHs, which is regulated by thioredoxins and by the redox state of the NADP/NADPH pool (Scheibe *et al.*, 2005) has been used as a marker of the redox state of the chloroplast stroma (Foyer *et al.*, 1992).

MARKER ENZYMES AND PROTEINS

The activity or abundance of proteins or transcripts encoding enzymes that play important roles in H_2O_2 removal, or otherwise participate in antioxidative metabolism and related redox reactions, are often used as estimates of antioxidant capacity or oxidative stress (Noctor and Foyer, 1998a). Ideally, such enzyme markers should act primarily as antioxidative enzymes: (1) using superoxide, H_2O_2 or organic peroxide as substrates; (2) maintaining the redox state by supplying or regenerating reduced forms of cellular antioxidants; or (3) controlling the release of secondary metabolite signals, e.g. conjugases. Within the first category, the most studied plant primary antioxidative enzymes are the superoxide dismutases (SODs) and CATs that catalyse dismutation reactions, and the ascorbate peroxidases (APX), which require reductant in the form of ascorbate. However, many other types of peroxidases exist in plants. Among these, the peroxiredoxins are considered to have an important function alongside APX in H_2O_2 removal, particularly in chloroplasts (Dietz *et al.*, 2006). However, the activities of peroxiredoxins are more difficult to measure than those of the well-characterised APX and less information is available on their enzymatic properties. The second category of antioxidative enzymes includes dehydroascorbate reductase (DHAR), glutathione reductase (GR) and NADPH-generating dehydrogenases, as well as some glutaredoxins and thioredoxins. The third category of potentially useful markers includes enzymes such as glyoxylases, aldo/keto reductase, cyt P450s, conjugase-type GTs and glycosyl transferases. Only enzymes that catalyse ROS removal or associated reductive reactions should strictly be considered as 'antioxidative'. However, other activities that decrease the probability of ROS production, such as the mitochondrial AOX or genes that are induced by oxidising conditions, though more generally involved in 'cell rescue/defence' are often used as redox marker enzymes, particularly in transcriptomics studies.

Antioxidative enzyme markers can be very useful indicators of oxidative signalling *in vivo*. In contrast, the genes encoding the core antioxidant system are generally either not induced or only moderately induced by stress. For example, a recent analysis of transcriptomic responses in a range of different organisms revealed that, while external H_2O_2 induced antioxidative genes in unicellular organisms, comparable responses were less clear or absent in multicellular organisms (Vandenbroucke *et al.*, 2007).

PROTEIN OXIDATION

The degree of oxidative modification to proteins, commonly referred to as *oxidative damage* to proteins, is often used as a measure of oxidative stress. Oxidative damage occurs primarily on side chains of amino acids such as Pro, His, Arg, Lys and Thr, and produces ketone or aldehyde derivatives (protein carbonyls) that are reactive with 2, 4-dinitrophenylhydrazine (DNPH). Other forms of protein oxidative damage can occur by reaction with lipid-peroxidation products, or by conjugation with sugars (glycation) or their oxidation products (glycoxidation). However, although the measurement of the number of DNPH-reacting protein carbonyl groups is relatively simple and straightforward, the interpretation of such results must be regarded with caution as the extent of protein oxidation is controlled by development in species such as *Arabidopsis* (Johansson *et al.*, 2004). Similarly, the oxidative modifications of protein Cys thiol groups can involve formation of disulfides with other protein thiol groups or soluble thiols such as glutathione, as well as production of more oxidised sulfur states (sulfenic, sulfinic and sulfonic groups). Such highly oxidised proteins, which are found in all cellular compartments, are generally assumed to lose their catalytic activity. The formation of the more highly oxidised Cys residues (sulfinic and sulfonic groups) was once classed as irreversible 'damage'. However, it is now known that sulfinic acids are part of the catalytic cycle of certain proteins such as peroxiredoxins.

LIPID PEROXIDATION

Measuring the end products of lipid peroxidation is one of the most widely accepted assays for oxidative damage and has been extensively used in plants. ROS cause peroxidation of polyunsaturated fatty acids, producing aldehydic secondary products, such as 4-hydroxynonenal (4-HNE) and malondialdehyde (MDA), commonly considered useful markers of oxidative stress (Del Río *et al.*, 2005). Several analytical techniques can be used to assay for lipid peroxidation (see Halliwell and Whiteman, 2004 for a review).

A common method for measuring MDA, referred to as the thiobarbituric acid-reactive-substances (TBARS) assay, is to react it with thiobar-bituric acid (TBA) and record the absorbance at 532 nm (Draper and Hadley, 1990). The TBARS assay or its modifications have been widely used to measure lipid peroxidation in plants subjected to different stresses. For instance, cotton seedlings exposed to temperature stress (Mahan and Mauget, 2005), and transgenic tobacco expressing glutathione S-transferase/glutathione peroxidase under stress conditions (Roxas *et al.*, 2000).

Results of the simple TBA assay should be interpreted with caution because it was shown that, in many cases, TBA-reactive substances are not related to lipid peroxidation (Halliwell and Whiteman, 2004). An additional problem of this spectrophotometric method is caused by the fact that many plants contain interfering compounds that also absorb at 532 nm, leading to overestimation of MDA values. Hodges *et al.* (1999) described a method to correct for these interferences by subtracting the absorbance at 532 nm of a solution containing plant extract incubated without TBA from an identical solution containing TBA.

The recent development of sensitive MS methods for measuring lipid peroxidation marked significant progress in the field, as prompted the development of more accurate and sensitive methods for MDA, 4-HNE and other lipid-peroxidation products (Deighton *et al.*, 1997; Liu *et al.*, 1997). The main advantage of MS-based methods is the possibility to identify individual lipid species targeted by ROS and to detect the various oxidative products formed (Byrdwell and Neff, 2002).

ELECTROLYTE LEAKAGE

Electrolyte leakage (EL) is used to assess membrane permeability and stability. A common procedure is described below. Sampled leaf discs are lightly rinsed in distilled water, gently blotted with paper and placed in test tubes (one disc per tube). Then, 20 mL of distilled water is added to each test tube. Samples are then vacuum infiltrated to allow uniform diffusion of electrolytes and shaken on a gyratory shaker at 250 rpm for 4 h at room temperature. After incubation, electrical conductivity of each solution is measured using a conductivity meter. After measuring initial EL ($C1$), samples are heat killed (autoclaved at 121°C, 124 kPa for 15 min) and final EL ($C2$) measured at room temperature. Ion leakage is calculated using the equation:

$$EL\ (\%) = C1/C2 \times 100 \qquad \text{[Eqn. 13.4]}$$

EL has been widely used in ecophysiological studies, together with other photosynthesis-related parameters. As an example, Epron and Dreyer (1992) studied the effects of dehydration on photosynthesis in *Quercus petraea* by measuring concomitantly Chl-F and EL.

Table 13.1. *Recent transgenic approaches used to study the influence of stomatal and mesophyll conductance on the control of photosynthesis and to assess the response of photosynthetic mechanisms to a variety of environmental stresses.*

Targeted process	Species	Transgenic molecular trait	Reference
Stomatal conductance and photosynthesis			
	Nicotiana plumbaginifolia	Zeaxanthin epoxidase	Borel and Simonneau (2002)
	Nicotiana tabacum	NAD(P)H dehydrogenase complex	Horváth *et al.* (2000)
	Solanum lycopersicum	Fumarate hydratase	Nunes-Nesi *et al.* (2007)
Mesophyll conductance and photosynthesis			
	Nicotiana tabacum	Aquaporin NtAQP1	Uehlein *et al.* (2003)
	Nicotiana tabacum	Aquaporin NtAQP1	Flexas *et al.* (2006)
Photosynthesis response to water/salt stress			
	Arabidopsis thaliana	HT1 protein kinase	Hashimoto *et al.* (2006)
	Arabidopsis thaliana	Phytoene synthase	Han *et al.* (2008)
	Citrus sinensis × *Poncirus trifoliata*	*GA20ox* gene	Huerta *et al.* (2008)
	Gossypium hirsutum	Vacuolar sodium/proton antiporter AtNHX1	He *et al.* (2005)
	Gossypium hirsutum	*H1-PPase* gene	Lv *et al.* (2008)
	Lycopersicum esculentum	C repeat/dehydration-responsive element binding factor 1	Hsieh *et al.* (2002)
	Nicotiana tabacum	Aquaporin PIP1b	Aharon *et al.* (2003)
	Nicotiana tabacum	Ectoine biosynthetic genes (*ect. ABC*)	Moghaieb *et al.* (2008)
	Nicotiana tabacum	Choline monooxygenase	Zhang *et al.* (2008a)
	Nicotiana tabacum	Betaine aldehyde dehydrogenase	Yang *et al.* (2008)
	Nicotiana tabacum	Senescence-Associated Receptor Kinase:Isopentenyltransferase	Rivero *et al.* (2009)
	Oryza sativa	Manganese superoxide dismutase	Wang *et al.* (2005)
	Oryza sativa	Stress responsive gene *SNAC1*	Hu *et al.* (2006)
	Solanum lycopersicum	Trehalose-6-phosphate synthase	Stiller *et al.* (2008)
	Vicia fava	Putative aquaporin gene *VfPIP1*	Cui *et al.* (2008)

Photosynthesis response to temperature stress			
	Arabidopsis thaliana	α- and β-isoform of Rubisco activase	Salvucci *et al.* (2006)
	Flaveria bidentis	Rubisco activase	Hendrickson *et al.* (2008)
	Nicotiana tabacum	Rubisco activase	Sharkey *et al.* (2001)
	Nicotiana tabacum	Betaine aldehyde dehydrogenase	Yang *et al.* (2007)
	Nicotiana tabacum	Glycerol-3-phosphate acyltransferase	Yan *et al.* (2008)
	Oryza sativa	Sedoheptulose-1,7-bisphosphatase	Feng *et al.* (2007a)
	Oryza sativa	Rubisco small subunit	Makino and Sage (2007)
	Triticum aestivum	Plastidal protein synthesis elongation factor	Fu *et al.* (2008)
Photosynthesis response to chemical stress			
	Nicotiana tabacum	Glutathione reductase	Ding *et al.* (2008)
Photosynthesis response to CO_2 stress			
	Nicotiana tabacum	Ethylene response gene	Tholen *et al.* (2007)
	Oryza sativa	Rubisco small subunit	Makino *et al.* (2000)
Photosynthesis response to light stress			
	Arabidopsis thaliana	Chlorophyllide a oxygenase	Hirashima *et al.* (2006)
	Arabidopsis thaliana	NADPH thioredoxin reductase	Pβrez-Ruíz *et al.* (2006)
	Lycopersicum esculentum	Selenium-independent glutathione peroxidase	Herbette *et al.* (2007)
	Nicotiana tabacum	Cytochrome b_6f complex and Rubisco small subunits	Baroli *et al.* (2008)
	Nicotiana tabacum	Ferredoxin-NADP(H) reductase	Rodríguez *et al.* (2007)
	Oryza sativa	NADP-malic enzyme, phosphoenol-pyruvate carboxylase and pyruvate orthophosphate dikinase	Jiao *et al.* (2002)
Photosynthesis response to water/salt and temperature stress			
	Arabidopsis thaliana	ABA-responsive-element binding protein 9	Zhang *et al.* (2008b)
	Glycine max	Arabidopsis L-Δ1-pyrroline-5-carboxylate reductase	De Ronde *et al.* (2004)

Table 13.1. (*cont.*)

Targeted process	Species	Transgenic molecular trait	Reference
Photosynthesis response to light and temperature stress			
	Nicotiana tabacum	Chloroplast-localised small heat-shock protein	Guo *et al.* (2007)
Photosynthesis response to water/salt and chemical stress			
	Arabidopsis thaliana	Monodehydroascorbate reductase	Eltayeb *et al.* (2007)
	Brassica campestris ssp. pekinensis	Superoxide dismutase and catalase	Tseng *et al.* (2007)
Photosynthesis response to light and CO_2 stress			
	Nicotiana tabacum	Sedoheptulose-1,7-bisphosphatase	Lawson *et al.* (2008)
Photosynthesis response to multiple stresses			
	Nicotiana tabacum	Violaxanthin deepoxidase	Sun *et al.* (2001)
	Nicotiana tabacum	S-adenosylmethionine decarboxylase	Wi *et al.* (2006)

13.4.8. Transformation and transfection

Gene silencing (antisense, co-suppression, RNAi and transposon tagging) overexpression and reporter-gene studies rely on the stable or transient expression of heterologous genes in plants. This necessitates approaches that enable both the delivery and function of such gene constructs in plant host cells. Stable expression requires the recovery of transgenic plants whose progeny also contain and express the introduced DNA. The correct targeting of expression requires careful choice of appropriate promoters. Several techniques have been developed to enable the stable integration of foreign DNA into the host-cell genome. Although protocols have been developed for the transformation of all the major food crops, protocols do not yet exist for many other species. Developing successful protocols can be a labour-intensive and time-consuming task. The simplest and most frequently employed method for the stable integration of single copies of DNA is that of *Agrobacterium*-mediated transformation. Unfortunately, not all species are transformable by this method, but wherever possible this should be the method of choice. Particle bombardment is effective for a wider range of species. It relies on the physical delivery of DNA to the plant tissue. The transformation efficiency is often lower than with *Agrobacterium*-mediated transformation, and multiple copies of the construct may become integrated. However, the presence of multiple copies can be advantageous, offering a wide range of expression levels in the transformed progeny, which can be exploited in the determination of gene function. In addition, DNA can be introduced into plant cells by viruses. DNA introduced in this way is not normally integrated into the host genome and not therefore heritable.

Table 13.1 highlights recent studies – performed with transformed (genetically modified) plants – on the influence of stomatal and mesophyll conductance on the control of photosynthesis and on the response of photosynthetic mechanisms to a variety of environmental stresses.

13.4.9. Quantitative polymerase chain reaction

Quantitative polymerase chain reaction (qPCR) has replaced northern-blot analysis to quantify individual transcripts. qPCR combines real-time PCR with reverse transcription PCR to enable the abundance of individual transcripts (mRNA) to be quantified by the detection of a fluorescent reporter molecule that accumulates as the PCR reaction proceeds (Higuchi *et al.*, 1992). The reporter molecules can be either non-specifc intercalating dyes that flouresce when bound to DNA (e.g., ethidium bromide and SYBR® Green) or specific oligonucleotide probes that are labelled with both a reporter fluorescent dye and a quencher dye (dual-labelled fluorogenic probes), cleavage of the probe after binding by the polymerase removes the quencher and permits detection (e.g., TaqMan® probes). Although absolute values can be obtained, quantification is normally only relative; the fluorescence is used to quantify the amount of transcript relative to a specific housekeeping gene whose transcript is believed to be of constant abundance. The specific oligonucleotide probes can give more reliable results as the fluorescence of intercalating dyes bound to non-specific products can increase the background and decrease the sensitivity and reliability of this approach. In contrast, specific reporter probes bind only to the specific gene product. The specific reporter probes also enable several qPCR reactions to occur in a single experiment (multiplexing) by using probes with specific fluorescent tags.

13.5. CONCLUDING REMARKS

In general, plant ecophysiology and specifically photosynthesis responses to the environment have been historically investigated either by using organ- or plant-level approaches. In parallel, biochemical approaches have been mainly used to assess plant processes, and photosynthesis has been used for physiological processes at a more fine level. After initial studies, where sufficient knowledge has been gained over the molecular plantomic technology, the next devised steps should utilise single experiments to verify observations at all possible levels. This should allow unravelling of the biochemical and molecular basis of organ, plant and ecosystem responses to the environment. Working in reverse will give the ability to predict organ, plant and ecosystem consequences of given biochemical and molecular processes. Using such multiple-level studies, the photosynthetic processes must be one of the main targeted issues.

14 • Measuring CO_2 exchange at canopy scale: the eddy covariance technique

G. MATTEUCCI AND G. MANCA

14.1. INTRODUCTION

The interest of researchers for a more precise estimation of primary productivity of terrestrial ecosystems and its underlying mechanisms dates back to the 1960s, when the International Biological Programme was launched (Lieth and Whittaker, 1975). At that time, primary productivity was assessed almost exclusively by destructive sampling of plant biomass and measurements of growth, whereas gas-exchange measurements were focused on investigating physiological responses and on model parameterisation. Indeed, estimates of photosynthesis and/or respiration of entire plants or ecosystem was limited by technical problems, and measurements performed on single leaves or plant parts were scaled up using knowledge on plant architecture and ecosystem-structure parameters, such as leaf area index or wood area index (Schulze and Koch, 1971; Tenhunen *et al.*, 1990; Pearcy and Sims, 1994; Matteucci *et al.*, 1995). Nevertheless, knowledge on primary productivity of the terrestrial biosphere increased significantly, and the first attempts of regression modelling were performed (Lieth, 1975; Reichle, 1981).

Since then, the interest for scaling physiological processes in time and space has increased strongly (Ehleringer and Field, 1993; Jarvis, 1995) and a series of new technologies have expanded the spatial scale of observations from leaves to canopy, from ecosystems to globe (e.g., Waring and Running, 1998), and the availability of new measurements, such as canopy fluxes through eddy covariance (EC), remote sensing of absorbed radiation and atmospheric CO_2 concentration measurements by tall towers, aircraft or high elevation stations, have improved the understanding of biosphere processes at larger and longer scales, providing better constraints to the estimation of photosynthetic fluxes and carbon-budget components by biogeochemical models (Running *et al.*, 1999; Griffith and Jarvis, 2005).

Since the early 1990s, the development of micrometeorological theories and techniques boosted the studies of vegetation-atmosphere interactions (Baldocchi *et al.*, 1988; Baldocchi and Valentini, 1996) and the EC technique is today largely used to quantify carbon exchange at the ecosystem scale (Baldocchi *et al.*, 2001). The success of the technique is based on three main factors: (1) it allows the direct, non-destructive measurement of the net carbon exchange between vegetation and atmosphere – the measured flux being the balance between gross photosynthesis and respiration (autotrophic and heterotrophic); (2) the flux is representative of a relatively large area (depending on the height of measurements); and (3) the technique can be operated continuously, giving the possibility to derive daily, seasonal and annual carbon budgets and to assess interannual and long-term variability.

Furthermore, flux measurements at canopy scale have proved to be particularly useful to study ecophysiological responses at ecosystem level, such as those to temperature, water availability and solar radiation (Law *et al.*, 2002; Valentini, 2003; Reichstein *et al.*, 2007), to nitrogen deposition (Magnani *et al.*, 2007) and to management and stand age (Kolari *et al.*, 2004; Kowalski *et al.*, 2004).

This chapter will drive the reader through the measurements of photosynthesis-related processes at canopy scale, providing the fundamentals of the technique allowing this and including hints on instrumentation and system set-up.

14.2. THE EC TECHNIQUE FOR MEASURING CARBON-DIOXIDE FLUXES

14.2.1. Brief history

The theoretical framework was established at the end of the 19th century by Reynolds (Reynolds, 1895). After the Second World War, with the development of fast-response

Terrestrial Photosynthesis in a Changing Environment: A Molecular, Physiological and Ecological Approach, ed. J. Flexas, F. Loreto and H. Medrano. Published by Cambridge University Press.

anemometers and thermometers based on the hot-wire technology, EC measurements were conducted over short vegetation, focusing on turbulence and heat fluxes (Swinbank, 1951). CO_2 exchanges of terrestrial ecosystems were measured using the flux-gradient method until the early 1970s, when the first EC measurements were conducted (Baldocchi, 2003). Key innovations started to become available in the eighties, with the development of sonic anemometers and open-path CO_2 analysers able to rapidly sample CO_2 fluctuations. The first applications, limited to short campaigns, were performed over crops (Anderson and Verma, 1986) and forests (Verma *et al.*, 1986; Valentini *et al.*, 1991).

Measurements of canopy level CO_2 exchange by EC on a continuous basis (24 h a day) and on longer timeframes (seasonal to pluriannual scales) started in the 1990s, generally over forests (Wosfy *et al.*, 1993; several papers in Baldocchi and Valentini, 1996). In the second half of the 1990s, regional networks of flux-measurement sites started to become operative in Europe (Valentini, 2003) and North America (Running *et al.*, 1999). Currently, most of the operating sites are participating in the FLUXNET programme (Baldocchi *et al.*, 2001), a global network of sites that use EC methods to measure the exchanges of CO_2, *water vapour* and energy between terrestrial ecosystem and atmosphere. At present, over 400 tower sites distributed in regional networks are operating on a long-term and continuous basis (http://www.fluxnet.ornl.gov/).

14.2.2. Basic theory

A thorough treatment of the EC theory is beyond the scope of this chapter. Below we report some theoretical basis, addressing the reader to the cited references for details and in-depth considerations (Aubinet *et al.*, 2000; Baldocchi, 2003).

In the atmosphere, trace gases, such as CO_2 (scalars), are transported by turbulent motions of upward and downward moving air. Practically, the EC technique samples this turbulence to determine the net difference of material moving across the canopy atmosphere interface (Baldocchi, 2003). Equation 14.1, representing the mass conservation of a scalar, is at the basis of the EC technique:

$$\underset{\text{I}}{\frac{\partial \overline{C}}{\partial t}} + \underset{\text{II}}{\left(\bar{u}\frac{\partial \overline{C}}{\partial x} + \bar{v}\frac{\partial \overline{C}}{\partial y} + \bar{w}\frac{\partial \overline{C}}{\partial z} \right)} = \underset{\text{III}}{\left(\frac{\partial \overline{u'C'}}{\partial x} + \frac{\partial \overline{v'C'}}{\partial y} + \frac{\partial \overline{w'C'}}{\partial z} \right)} + \underset{\text{IV}}{Dm} + \underset{\text{V}}{S_c} \qquad \text{[Eqn. 14.1]}$$

where C is the mixing ratio of the scalar of interest; u, v and w are the three components of wind speed on the horizontal (x, y) and vertical (z) directions; *Dm* represents molecular diffusion; and S_c includes formation/destruction processes of the scalar molecules operated by biological sources and sinks present in the air volume. In equation 14.1, the temporal variation of concentration of the scalar C (I) and the local advection (II) are balanced by the mean horizontal and vertical divergence (or convergence) of turbulent flux (III) by molecular diffusion (*D*, IV) and by a S_c (V). All the processes are considered within an infinitesimal air volume. The variables are averaged along time (as indicated by horizontal bars). In term III, variables are included as instantaneous fluctuations around means (as denoted by the apexes), and their products result in statistical covariances among the component of wind speed (u, v, w) and the scalar concentration. These covariances are representing the turbulent flux along the space dimensions (x, y, z).

Compared with the other terms of equation 14.1, molecular diffusion and lateral concentration gradients along y are generally negligible (Baldocchi *et al.*, 1988; Kaimal and Finnigan, 1994). Assuming that the horizontal gradients (along x and y) of turbulent flux are zero and integrating equation 14.1 along the vertical, z, equation 14.2 is obtained:

$$\underset{\text{I}}{\int_0^{z_m} S_c dz} = \underset{\text{II}}{\overline{w'C'}} + \underset{\text{III}}{\int_0^{z_m} \frac{\partial \overline{C}}{\partial t} dz} + \underset{\text{IV}}{\int_0^{z_m} \bar{u}\frac{\partial \overline{C}}{\partial x} dz} + \underset{\text{V}}{\int_0^{z_m} \bar{w}\frac{\partial \overline{C}}{\partial z} dz} \qquad \text{[Eqn. 14.2]}$$

The Net Ecosystem Exchange (NEE) of the scalar within the air volume between the ground and z_m (I) is equal to the vertical turbulent flux (II) measured at z_m (the height at which the instrumentation for EC is located) plus a term representing the *storage* of the scalar (III, generally measured by temporal variation of concentration between the ground and z_m). The last two terms of equation 14.2 represent the advective fluxes in the x-horizontal and z-vertical directions. In conditions of horizontal homogeneity (no horizontal concentration gradients) and atmospheric stationarity (the ideal conditions for EC), equation 14.2 can be simplified as follows:

$$\int_0^{z_m} S_c dz = \overline{w'C'} \qquad \text{[Eqn. 14.3]}$$

The term on the right is the covariance that is directly measured by EC systems currently in use, composed of an ultrasonic anemometer, an analyser for the scalar of interest (generally CO_2 and *water vapour*) and devices for data logging. All the instruments must have fast-response times as the turbulent transport in the atmosphere is inherently fast (fractions of a second, 10–20 Hz). In a certain point in space, the turbulent motion of air masses (eddies) causes stochastic fluctuations of scalar concentrations and wind speed (the vertical turbulent flux, the terms in equation 14.3), and the estimated flux will be statistically more correct if a higher number of fluctuations will be measured in a certain time interval.

The major assumptions used to derive equation 14.3 from the complete mass-balance equation (Eqns 14.1 and 14.2) are: (1) horizontal homogeneity of the site; and (2) atmospheric stationarity. To respect the stationarity criteria, the mean value of physical variables describing the state of the atmosphere should be constant over time, a condition that is verified for time windows less than 60 minutes (Kaimal and Finnigan, 1994; Finnigan *et al.*, 2003). For this reason, EC fluxes are generally estimated in half-hourly intervals.

Horizontal homogeneity is related to a uniform spatial distribution of biological sources and sinks of CO_2 and to ecosystem topography and roughness, and ideally EC measurements should be performed on flat sites. However, the interest of the scientific community to investigate ecosystem processes in very diverse landscapes (particularly for forests) has boosted research on methods to correct EC measurements performed on complex sites (Aubinet *et al.*, 2000, 2003; Baldocchi *et al.*, 2000; Massman and Lee, 2002). The main corrections are generally applied to night-time fluxes, when atmospheric turbulent mixing is low or absent causing a build-up of respired CO_2 below the instrument height: in this case, the storage term of equation-14.2 terms (III) cannot be considered negligible with respect to the turbulent flux (term II) and needs to be estimated. Storage of CO_2 within a canopy can be measured in a relatively easy way using a vertical profile of concentration (at least three levels) or with the discrete approach (comparing changes of concentration at z_m) (Finnigan, 2006). Alternatively, when EC measurements are performed over heterogenous and/or non-flat terrains, where horizontal gradients of wind field and/or concentration may develop during a period of weak turbulence, advection (term IV and V in Eqn. 14.2) may occur. Generally, the quantitative estimation of the advective terms need complex experimental set-up (Aubinet *et al.*, 2005), although simpler alternatives have been proposed (van Gorsel *et al.*, 2007). In most correction procedures, EC-flux data measured in non-ideal conditions are replaced with data estimated using an empirical relationship, correlating fluxes measured in good conditions with climatic variables. Non-ideal conditions for the EC technique can be recognised empirically using friction velocity values (obtained by sonic anemometers) that represent a good indicator of turbulence intensity (Goulden *et al.*, 1996). It is worth noting that, also in complex, non-ideal terrains, the higher the intensity of atmospheric turbulence, the lower the difference between the EC flux and the total ecosystem exchange.

14.2.3. Main advantages and limits of the technique

The main features of the EC technique for studying the CO_2 exchange of plant canopies can be summarised as follows.

- It is a canopy scale technique providing a measure of CO_2 and *water-vapour* exchange over relatively extended areas (from hundreds of meters to a km, Schmid, 2002) and it is then able to characterise fluxes at ecosystem level.
- It can be implemented on a continuous basis (24 h a day) and along seasons and years, sampling the variability of climatic and ecophysiological canopy conditions.
- The data flow is practically instantaneous, giving the possibility of studying the response of ecosystems to fast changes in driving variables or climatic extreme events (e.g., Ciais *et al.*, 2005; Del Pierre *et al.*, 2009).
- It is possible to study ecophysiological response at canopy level (e.g., Law *et al.*, 2002; Reichstein *et al.*, 2007).
- Integration of CO_2 exchange data over time provide net carbon budget of ecosystem over seasonal to annual and pluriannual scales.
- Canopy fluxes are measured more correctly over homogeneous surfaces, a limited heterogeneity can be faced by currently available correction schemes.
- When applied in complex topographic conditions, advection or katabatic flows may occur, limiting the reliability of the technique.
- Under stable conditions, particularly at night, the correct evaluation of fluxes is hindered by limited or absent turbulence. In this condition, biological CO_2 can accumulate below the EC instruments and the storage component needs to be measured/estimated (Finnigan, 2006).
- It can also be operated at remote sites, powered by solar panels, batteries and generators. However, the best

continuity of measurements is granted on sites equipped with line power.

- Instrumentations and site set-up are relatively complex, requiring regular, although simple, servicing and calibration.
- Owing to the massive data flow and the possibility of ecophysiological and carbon-budget studies, EC can be considered as 'value for money' when compared with the other techniques necessary to provide a similar level of information.

As all the experimental techniques, EC is also prone to random and systematic errors. These errors are associated with measurements, sampling and theoretical issues connected to the application of the EC technique in non-ideal situations (Baldocchi, 2003). Detailed analysis of errors and uncertainties can be found in several papers and reviews (Goulden *et al.*, 1996; Moncrieff *et al.*, 1996; Baldocchi, 2003; Loescher *et al.*, 2006; Richardson *et al.*, 2006): here we provide a summary of the most relevant error sources.

Calibration errors of regularly serviced gas analysers are in the order of 2–3%. Errors related to the time lag between the sensors for wind speed and trace gases, whose sampling can be placed some centimetres apart from each other, are limited to 2%. Covariance measurements errors are less than 7–12%, whereas the natural variability of turbulence is in the order of 10–20%, limiting the comparability of fluxes measured under similar turbulence conditions (Baldocchi, 2003). Random errors can be limited by system design and method implementation, and longer-term averages of flux measurements may reduce random sampling errors to ±5% (Goulden *et al.*, 1996; Moncrieff *et al.*, 1996; Baldocchi, 2003).

Ecosystem flux measured by EC provides an accurate estimate of net carbon exchange when the theoretical basis of the technique is met. In some conditions, there could be a measurement bias between the total and the turbulent flux (the one measured by EC), particularly at night when atmospheric turbulence is weak or absent (Goulden *et al.*, 1996; Moncrieff *et al.*, 1996; Aubinet *et al.*, 2005; van Gorsel *et al.*, 2008). In ideal sites, the bias can be corrected simply by measuring CO_2 storage within the canopy (term III in Eqn. 14.2) below the EC sampling point, whereas in complex terrains EC fluxes have to be empirically corrected using friction velocity threshold criteria or there is the need to measure additional flux components, such as advection (term IV and V in Eqn. 14.2, Aubinet *et al.*, 2005).

Another bias can be related to the lack of energy balance closure when comparing latent and sensible heat fluxes measured by EC to independent estimates of available energy (Wilson *et al.*, 2002). Differences in energy balance closure can be related also to advection and different footprints sampled by EC and available energy sensors. In this respect, the comparison of EC measurements with independent estimation of evaporative fluxes resulted in reasonable agreement, supporting the accuracy of daytime EC measurements (Baldocchi, 2003).

The reasons for the possible underestimation of nighttime fluxes by EC have been briefly presented in the previous paragraph. Currently, it is a common procedure to use empirical relationships to correct this underestimation (Papale *et al.*, 2006; van Gorsel *et al.*, 2009). Unreliable data can be replaced using temperature-dependent respiration functions calculated on good quality data sets or using ancillary measurements with soil chambers. Recently, standardised procedures have been proposed for estimation of ecosystem respiration, using nighttime data in good turbulence conditions (Reichstein *et al.*, 2005) or the x-intercept of the daytime light curves of canopy fluxes (Lasslop *et al.*, 2010).

14.2.4. Instrumentation and set-up

In order to derive the covariances of the EC equation (Eqn. 14.3), we need to measure the fluctuating components of concentration, wind speed and temperature (for sensible heat flux). These variables must be measured by fast-response sensors in order to sample most of the fluctuating turbulent structure, which is responsible for the fluxes. A system for EC is made of several instruments (anemometer, analyser, data logging device, etc.) and an overall accepted standard is not available. Nevertheless, regional networks have tried to develop common standards, such as, for example, Euroflux (Aubinet *et al.*, 2000; Valentini, 2003), or to compare set-up of various sites against a roving standardised system (Billesbach *et al.*, 2004; Ocheltree and Loescher, 2007).

The three components of wind speed (u, v, w) can be measured using a three-dimensional ultrasonic anemometer. The instrument determines u, v and w through the distortion that wind causes in the ultrasound signal travelling between the three pairs of the anemometer's transducers. Air temperature can be derived by the speed of sound obtained with this type of anemometers. Alternatively, a fast-response fine-wire thermometer can be used to measure temperature fluctuation in the anemometer sampling volume. Ultrasonic anemometers sample the wind-speed signal with a frequency of 20 to 100 times per second (20–100 Hz). There are several companies producing ultrasonic anemometers. The most used are CSAT-3 (Campbell Scientific, Logan, UT, USA) and various models from Gill (R3, HS, Windmaster, Gill Instruments Ltd, Lymington, Hampshire, UK). Also Metek (METEK GmbH, Elmshorn, Germany) and Young (R.M. Young Company, Traverse City, MI, USA) anemometers can be used for EC measurements. A performance comparison of different anemometers can be found in Loescher *et al.* (2005).

Concentrations of CO_2 and *water vapour* in the air are measured using fast-response IRGA. Originally, fast-response instruments were of the open-path type, where concentration is measured directly in air, the sampling volume being the volume between the transmitter and the receiver of the IR source (normally a few cm apart). Open-path analysers can be placed very close to sonic anemometers, limiting possible errors related to the distance between sensors (ideally the two sensors should sample the same air volume). Fluxes measured by open-path analysers need to be corrected for air-density fluctuations (Webb *et al.*, 1980), and data collected under wet conditions (rain, snow, condensating humidity) need to be discarded and gap-filled. Open-path analysers are commercially available from Li-Cor (model LI-7500, LI-COR, Lincoln, Nebraska USA, Campbell Scientific (model EC 150, Campbell Scientific, Logan, Utah, USA) and ADC (model OP-2, ADC, Hoddesdon, UK), and are also custom-made by several research groups.

Since the early 1990s, fast-response closed-path analysers have become available to be used in EC systems (Valentini *et al.*, 1996). This kind of instrument is characterised by sampling cells of small size where the air can be flushed through at several litres per minute to maintain the turbulent characteristics of the sampled air parcel. The air inlet is positioned close to the anemometer sampling volume, and a pump is used to suck air down a sampling tube of appropriate material (not absorbing CO_2 and H_2O, like Teflon, high-density Rilsan or brass) through the analyser. Hence, EC set-up using closed-path analysers need an additional pump to complete the system. This will increase energy consumption when compared with systems using open-path analysers. Analysers are generally located a few meters apart from sonic anemometers, though it is better to maintain a short length of sampling tube. The air flow is made isothermal before entering the analyser cell, limiting the density correction compared to open-path sensors (Ibrom *et al.*, 2007). Finally, closed-path instruments can be calibrated on a daily basis by a tank-valve automatic system, are robust and versatile, are resistant to rain and can be left unattended for long periods (Moncrieff *et al.*, 1996; Aubinet *et al.*, 2000). Closed-path analysers are commercially available by Li-Cor (model Li-7000, LI-COR, Lincoln, Nebraska USA, Campbell Scientific (model EC 155, Campbell Scientific, Logan, Utah, USA) and Picarro (model G1301-f, multi-gas, Picarro Inc., Sunnyvale, CA, USA).

Open- and closed-path analysers have been tested for their comparative performance (Suyker and Verma, 1993; Leuning and Judd, 1996). In order to ensure reliable long-term measurements of eddy fluxes, careful design and maintenance is required for both systems. Open-path analysers provide generally better estimates of latent heat fluxes, whereas closed-path analysers are more reliable (less data gaps) for long-term carbon budget. Recently a new gas analyser that combines the benefits of open and closed-path devices has been put on the market by LI-COR (model Li-7200, LI-COR, Lincoln, Nebraska USA) after a successful comparison with other analysers (Clement *et al.*, 2009).

Anemometers, open-path analysers or the inlet of the sampling tube of a closed path analyser must be placed above a canopy using a 'simple' mast or tripod over crops, grasslands or similar short-vegetation ecosystems, whereas walk-up towers, scaffolds or pylons are needed to deploy EC systems over forests. The height at which sensors should be placed is related to the homogeneous ecosystem area to be sampled (*fetch*). Generally, the higher the sampling height, the larger the measured footprint. However, fluxes have to be measured in the so-called *constant flux layer*, that is the layer where measured flux is independent from the distance between the EC sensors and the ecosystem. Under ideal conditions, the height of this layer is approximately equal to 10% of the fetch radius (Kaimal and Finnigan, 1994). When the fetch is limited (e.g., relatively 'small' fields or forest stands), height and distance of the EC sampling point have to be properly selected to avoid the influence of discontinuities at the ecosystem border. A commonly used rule of thumb indicates that the homogeneity of the sampled ecosystem should extend approximately 100 times the measurement height (Businger, 1986), although more precise and less-stringent assumptions may apply (Monteith and Unsworth, 1990). Furthermore, the instruments must not be placed too close to the ecosystem to avoid influences of canopy roughness on turbulent motions. Depending on the variability of ecosystem roughness, EC systems are placed 2–6 m above crops and grasslands and 8–20 m above forests.

The massive dataflow from the instruments (generally six variables at 20 Hz, a total of 36,200 data lines each half-hour) is collected using data loggers (e.g., Campbell Scientific, Logan, UT, USA), PC or laptops through serial

Table 14.1. *List of core and optional measurements at eddy covariance (EC) sites (modified from Baldocchi et al., 1996).*

Eddy Fluxes

Core measurements: Carbon dioxide, *water vapour* (latent heat), sensible heat, momentum (friction velocity)

Storage fluxes

Core measurements: CO_2 storage within the canopy (concentration profiles); heat and *water vapour* within the canopy (discrete approach); canopy heat in forests (bole temperature)

Optional measurements: *water vapour* and latent heat storage with concentration and temperature profile

Soil fluxes

Core measurements: soil heat flux

Optional measurements: soil CO_2 efflux

Meteorology

Core measurements: global radiation, wind speed and direction, air temperature, relative humidity, precipitation, net radiation, air pressure

Optional measurements: photon-flux densities (direct and diffuse), absorbed photon-flux density, longwave – shortwave radiation components, snow depth and water equivalent, canopy radiative temperature, bole temperature (forest)

Biology

Core measurements: leaf-area index, canopy height, biomass, species composition, foliage nitrogen

Optional measurements: distribution of leaf area along canopy profile, leaf optical properties, phenology, cuvette measurements of photosynthesis, net primary production

Soil physics

Core measurements: profiles of soil temperature and soil-water content

Optional measurements: soil bulk density, soil texture and structure, hydraulic conductivity, soil chemistry, root biomass, litter and soil organic carbon

interface or equipped with appropriate data-input boards. If the anemometer is equipped with additional analogue-input channels, the analogue output of the analyser can be collected through the anemometer that will send the data stream through serial outputs to the computers or loggers. Alternatively, analogue outputs from EC sensors are collected separately. In both cases, the data stream is then elaborated by maximising the covariances of the two signals (w and CO_2), in order to take into account the time delay between the measurements of wind velocity and CO_2 concentration. Calculation of fluxes can be performed online, or raw data can be later elaborated in the lab. In both cases it is recommended to store raw data as it will be possible to re-calculate fluxes in the future when methodological updates become available.

EC systems can be installed for short-term campaigns but, more often, they are installed at permanent long-term research and monitoring sites. At those sites, generally, several ancillary environmental, soil and biological measurements are collected concurrently with canopy fluxes to improve data interpretation, investigate flux drivers and provide variables for model parameterisation and testing (Baldocchi *et al.*, 1996; Baldocchi *et al.*, 2001; Valentini, 2003). A list of those measurements is reported in Table 14.1.

14.2.5. Data treatment

At the beginning of EC-technique application, the focus was on studying turbulence characteristics and ecosystem response on short-term campaigns. Since the mid 1990s, the interest on deriving long-term carbon budgets from EC measurements has increased, and most of the research projects funded since then have addressed the issue of carbon sinks and sources in terrestrial ecosystems over annual to multi-annual scales (Baldocchi *et al.*, 2001; Valentini, 2003; Luyssaert *et al.*, 2007). This has brought up the necessity of summing eddy fluxes over time to derive long-term budgets, posing new challenges to the EC scientific community (Baldocchi, 2003).

High-frequency wind speed and concentration data (10–20 Hz) are elaborated into half-hourly mean fluxes of CO_2, *water vapour* and latent heat fluxes using assumptions and equations developed from the last 20 to 30 years (Aubinet *et al.*, 2000). Data are then summed up to derive daily, monthly, seasonal and annual budgets, and this is particularly true for carbon, that is the quantity of greatest interest in EC studies (Baldocchi, 2003).

Gaps in long-term data records are generally inevitable. Along a year, typical data coverage ranges between 65 and 75% (Falge *et al.*, 2001; Moffat *et al.*, 2007), and data gaps are related to instrument breakdown or servicing (e.g., calibration), data rejection according to quality criteria or data collected during low-turbulence periods. Several methods have been applied to fill data gaps, and regional networks are trying to standardise procedures for deriving annual carbon budgets (Papale *et al.*, 2006; Moffat *et al.*, 2007). If the data population is adequately sampled, different gap-filling procedures produce repeatable annual sums (Falge *et al.*, 2001), whereas several methods working on artificially created gaps from a few hours to 12 days have resulted in a difference in the range of ±25 gC m^{-2} $year^{-1}$ (C, carbon) (Moffat *et al.*, 2007). Overall errors on annual sums range between ±30 and ±180 gC m^{-2} $year^{-1}$, with larger errors for sites in less-ideal conditions (Baldocchi, 2003).

Analysis of data for investigating short-term ecophysiological responses to climatic and/or site variables has to be restricted to directly measured data, properly screened for overall quality. In fact, all gap-filling procedures include relationships with driving variables, and then gap-filled data will not be independent from other measured data meaning cross-correlation may limit the usefulness of gap-filled data series for short-term data analysis.

14.3. ECOLOGICAL QUANTITIES MEASURED AND DERIVED USING EC

The ecological quantity that is measured by EC of CO_2 over plant canopies is the net ecosystem exchange (NEE). Indeed, an analyser located above an ecosystem cannot discriminate between photosynthesis and respiration, but measures the net flow between gross photosynthesis (gross primary production, GPP) and total ecosystem respiration (R_E):

$$NEE = GPP - R_E \qquad \text{[Eqn. 14.4]}$$

Following micrometeorological conventions, NEE is negative when the flow is from the atmosphere to the ecosystem (carbon absorption) and positive when the system emits CO_2. Similarly to leaf-level photosynthesis measurements, half-hourly values of NEE are generally expressed in μmol CO_2 m^{-2} s^{-1}, where the unit area refers to the ground surface over which the EC system is deployed. When values are cumulated at daily or longer scales, NEE is calculated in g CO_2 m^{-2} or gC m^{-2} over the time unit of integration (day, month, year).

Integrated over a year or over a full vegetation cycle of a seasonal canopy (e.g., a crop from seeding to harvesting), the NEE represents the quantity of carbon that is finally stored (or eventually emitted) over that integration period: indeed, the strong interest in EC research in the last 15 years is linked to the estimation of that quantity over forest, cropland and grassland ecosystems (Valentini *et al.*, 2000; Luyssaert *et al.*, 2007; Soussana *et al.*, 2007).

During the night and dormancy periods for deciduous canopies GPP is zero, and EC systems under proper turbulence conditions measure ecosystem respiration (R_E) caused by the sum of autotrophic (R_A) and heterotrophic (R_H) respiratory processes. At site level, R_E is correlated to air or soil temperature (Law *et al.*, 2002). The relationship between nighttime R_E and temperature can then be used to calculate R_E during daytime and to estimate gross photosynthesis at an ecosystem level (Eqn. 14.4).

The interest in deriving ecosystem-component fluxes (GPP, R_E) from EC measurements is related to the fact that physiology drives photosynthesis and respiration, whereas NEE results from a combination of those processes and of the way they are influenced by ecophysiological drivers. Nevertheless, up until now relationships among climatic drivers and EC-component fluxes have proved to be significant for GPP and R_E and not for NEE (Law *et al.*, 2002; Reichstein *et al.*, 2007). Furthermore, researchers have also been interested in providing validation parameters to modellers (Running *et al.*, 1999; Baldocchi, 2003), as ecosystem models provide the net flux by difference of the simulated GPP and respiratory processes (R_A, R_H) estimated separately.

Estimation of GPP and R_E from NEE can be based on R_E-temperature relationships calculated on a restricted time window to ensure a limited impact of phenological changes (Reichstein *et al.*, 2005) or, more recently, on the analysis of NEE-light-response function (Gilmanov *et al.*, 2007; Lasslop *et al.*, 2010). Several methods to separate NEE into assimilation and respiration have been recently compared and resulted in differences of less than 10% in estimates of both GPP and R_E, providing increased confidence in previously published multi-site comparisons and syntheses (Desai *et al.*, 2008).

It is thus possible to measure photosynthesis-related processes at canopy scale and to investigate ecophysiology at ecosystem scale (Law *et al.*, 2002; Reichstein *et al.*, 2007).

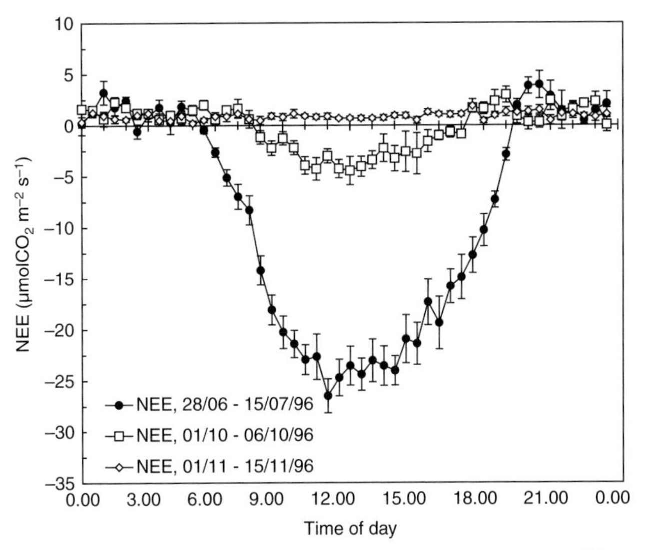

Fig. 14.1. Mean diurnal trends of net ecosystem exchange (NEE) for different periods of the year. Data measured at the Collelongo beech forest in 1996 (Valentini *et al.*, 1996; Matteucci, 1998) and averaged for each half hour and for the period indicated in the inset. Bars represent two standard errors. The sign follows the micrometeorological convention, being negative for CO_2 absorbed by the ecosystem.

14.4. EXAMPLES OF CO_2-EXCHANGE PROCESSES MEASURED BY EC

In this paragraph, we will provide some examples of data and results obtained with the EC technique over forests to show how EC is able to assess photosynthesis-related processes. Since the mid 1990s, the number of papers presenting results on canopy level fluxes has increased steadily. If in 2003 a citation search with the term 'EC' provided nearly 900 results (Baldocchi, 2003), today the number is at least doubled. If earlier work focused on single sites generally for one year (Valentini *et al.*, 1995b; several papers in Baldocchi and Valentini, 1996), later the creation of flux networks and the availability of longer-term data sets stimulated the production of papers addressing comparison of multiple sites, patterns at regional level, interannual variability and drivers at single sites, environmental controls over fluxes of terrestrial vegetation, response to stand age and management (Valentini *et al.*, 2000; Law *et al.*, 2002; Valentini, 2003; Kolari *et al.*, 2004; Kowalski *et al.*, 2004; Ciais *et al.*, 2005; Luyssaert *et al.*, 2007; Magnani *et al.*, 2007; Reichstein *et al.*, 2007; Soussana *et al.*, 2007; Del Pierre *et al.*, 2009).

Typical daily courses of instantaneous CO_2 exchange are presented in Fig. 14.1. Data were measured above a beech forest in Central Italy (Valentini *et al.*, 1996; Matteucci, 1998; Scartazza *et al.*, 2004).

At the growing-season maximum (early July), the ecosystems absorbed CO_2 from 5.30 to 19.00, with a maximum of –23 $\mu molCO_2\ m^{-2}\ s^{-1}$ between 10.30 and 14.00. At night, ecosystem respiration reached on average 2–3 $\mu molCO_2\ m^{-2}\ s^{-1}$, being higher soon after sunset.

Differences in CO_2 exchange owing to seasonality can be appreciated comparing daily trends recorded in contrasting periods of the season. In early October (Fig. 14.1), the growing season was close to the end, temperature and radiation were less favourable and the beech forest absorbed carbon from 7.30 to 16.30, with a maximum of –3.5 $\mu molCO_2\ m^{-2}\ s^{-1}$. Compared with early July, night respiration was lower (1–1.5 $\mu molCO_2\ m^{-2}\ s^{-1}$). In November, after leaf shedding, NEE was relatively constant and positive over the whole day (average 1–1.5 $\mu molCO_2\ m^{-2}\ s^{-1}$, Fig. 14.1). It is interesting to note how the EC technique is able to catch the *dynamic* of NEE in different conditions, going from early July, when the beech forest was a strong sink for CO_2, to early November, when the forest became a net source of CO_2 to the atmosphere.

The relationship between *instantaneous* NEE and PPFD is shown in Fig. 14.2. The data are the same as presented in Fig. 14.1 for early July and were measured generally during bright days, when the forest NEE was particularly high. In that period, mean daily temperature ranged between 11.5 and 18.2°C, maybe causing some of the scatter that can be seen in the graph.

Nevertheless, the relationship between NEE and PPFD is clear, and it is interesting to note the similarity

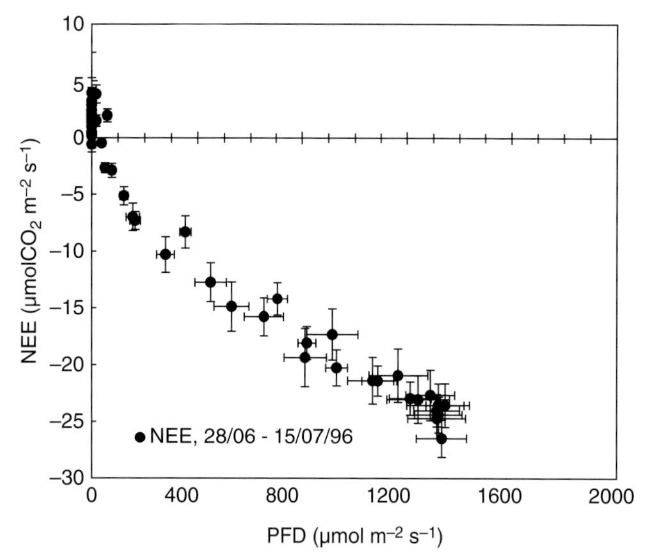

Fig. 14.2. Light response curve for net ecosystem exchange (NEE) at growing-season maximum. Data measured at the Collelongo beech forest in 1996 (Valentini *et al.*, 1996; Matteucci, 1998). Bars represent two standard errors of NEE (y-axis) and PFD (x-axis). The sign follows the micrometeorological convention, being negative for CO_2 absorbed by the ecosystem.

of ecosystem-level response with those reported for single leaves (see Chapter 9). Data were fitted with a rectangular hyperbola:

$$NEE = R_d - \frac{\alpha \cdot PPFD \cdot NEE_{max}}{\alpha \cdot PPFD + NEE_{max}} \quad \text{[Eqn. 14.5]}$$

in which NEE_{max} represents the maximum carbon flux at light saturation; R_d is the ecosystem respiration in the dark; and α represents the canopy quantum efficiency (μmol_{CO2} μmol_{PPFD}^{-1}). For early July, NEE_{max} was –38.8±2.2 μmolCO_2 m^{-2} s^{-1}; R_d 1.9±0.3 μmolCO_2 m^{-2} s^{-1}; the compensation point for light was 40 μmol_{PPFD} m^{-2} s^{-1}; and canopy quantum efficiency 0.052±0.004 μmol_{CO2} μmol_{PPFD}^{-1}. The r^2 of the relationship was 0.97.

When equation 14.5 is applied to the half-hourly values of NEE and PPFD measured in the different months of the year, it is possible to evaluate seasonal variation of the light-response parameters. Figure 14.3 shows the results of this analysis applied to a turkey oak forest in Central Italy during 2002 (Manca, 2003).

NEE_{max} grows up between April (starting of the growing season) and May, whereas the decrease in June and July is related to unfavourable vapour-pressure deficit. In August, NEE_{max} is similar to the values in May and reaches the seasonal maximum in September (_25.0±2.9 μmol CO_2 m^{-2} s^{-1}). The observed summer increase of NEE_{max} should not occur in a typical Mediterranean summer but, in 2002, August and September were the wettest months in the year (360 mm in two months). R_d values ranged between 2.5±0.8 and 4.97±2.1 μmol CO_2 m^{-2} s^{-1} measured respectively in November and May. Comparing trends of NEE_{max} and R_d, it is evident that photosynthetic carbon uptake was the flux component more sensitive to an anomalous wet summer. Excluding the months of strong phenological changes (April and November), canopy quantum efficiency ranged from 0.036±0.006 μmol_{CO2} μmol_{PPFD}^{-1} in September to 0.064±0.003 μmol_{CO2} μmol_{PPFD}^{-1} in June (Fig. 14.3).

As explained in previous paragraphs, half-hourly CO_2-exchange data can be summed up at daily and longer timescales. At daily scale, data are integrated and converted in g C m^{-2}. Missing or discarded data are replaced by gap-filling procedures (Falge *et al.*, 2001; Papale *et al.*, 2006; Moffat *et al.*, 2007) to provide a complete daily budget.

The annual trend of daily integrated carbon fluxes of a beech-forest ecosystem is presented in Fig. 14.4.

In Winter 2001, NEE (respiration) was relatively low and constant at 1–1.5 g C m^{-2} day^{-1} until early April when snow completely melted, air and soil temperature increased and the ecosystem phenological state started to change with onset of fine-root growth. In mid April, NEE presented a clear peak of respiratory flux, and soon after the ecosystem started to absorb carbon (day 120–123). Day 130 (10

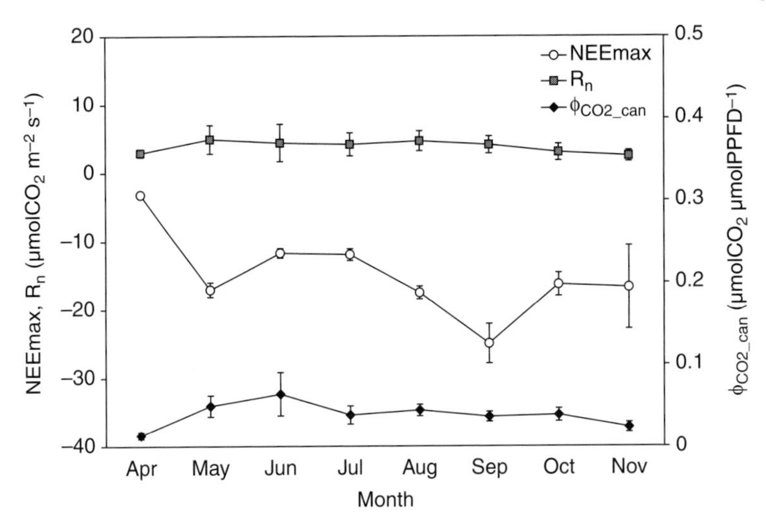

Fig. 14.3. Seasonal course of maximum carbon flux at light saturation (NEEmax), ecosystem respiration in the dark (R_n) and canopy quantum efficiency (α) as calculated with NEE light response curves. Data measured on a closed stand of turkey oak located in Roccarespampani, Central Italy, in 2002.

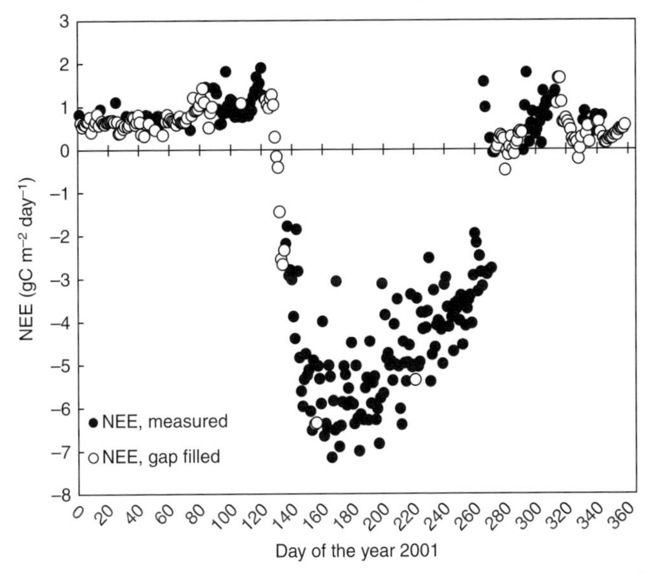

Fig. 14.4. Seasonal trend of daily cumulated net ecosystem exchange (NEE). Data measured at the Collelongo beech forest in 2001 (Scartazza *et al.*, 2004) and gap filled according to standard CarboEurope-IP procedures (Papale *et al.*, 2006). The sign follows the micrometeorological convention, being negative for CO_2 absorbed by the ecosystem. Measured data: 80% or more of the daily half-hours from direct measurements.

May) was the first of the season in which the ecosystem was a carbon sink. During rapid leaf development in May, daily carbon gain increased steadily, reaching –6.5 gC m^{-2} day^{-1} at the end of the month (Fig. 14.4). A seasonal maximum of –6.5 to –7 gC m^{-2} day^{-1} was maintained, with some day to-day variability until the end of July. This confirms previous findings at the same site, that showed a seasonal maximum between mid June and mid July (Valentini *et al.*, 1996; Matteucci, 1998; Granier *et al.*, 2003). In August, NEE decreased in response to higher VPD and limiting

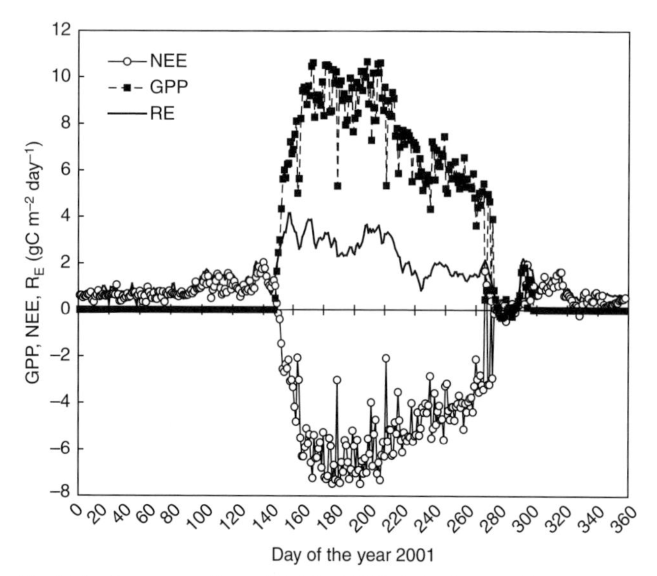

Fig. 14.5. Component fluxes (GPP, R_E, NEE) derived from eddy covariance (EC) measurements at the Collelongo beech forest in 2001 (Scartazza *et al.*, 2004). Separation and gap filling was performed according to standard CarboEurope-IP procedures (Reichstein *et al.*, 2005; Papale *et al.*, 2006). For NEE and R_E, the sign follows the micrometeorological convention, whereas for GPP the sign for reversed to enhance figure readability.

soil-water content (Scartazza *et al.*, 2004). In September, the start of leaf senescence, rainfalls restablishing soil-water content and boosting soil respiration and declining temperatures drove the decrease of NEE to –2.5 g C m^{-2} day^{-1}. On 23–24 September, heavy rainfall (50 mm in 2 days) suddenly turned the ecosystem from carbon sink to carbon source (1.6 gC m^{-2} day^{-1}) for three days. In 2001, the ecosystem ceased completely to be a carbon sink around mid October (Fig. 14.4). In November, availability of fresh litter coupled with mild temperatures favoured respiration processes that, later on, decreased following falling soil and air temperatures. In 2001, annual NEE summed up to approximately –498 g C m^{-2} year^{-1}, indicating that the ecosystem was a significant carbon sink.

NEE can be partitioned into its component fluxes, GPP, e.g., gross photosynthesis, and R_E. Values of nighttime NEE measured in periods of good turbulence mixing can be related to soil and/or air temperature. The relationship can then be used to calculate daytime R_E that, summed to growing season daytime NEE, will give GPP.

At Collelongo in 2001, flux partitioning was calculated with the methodology used within the CarboEurope Integrated Project (Reichstein *et al.*, 2005). Annual trends of flux components are reported in Fig. 14.5.

Trends in NEE have been discussed above. Here we want to signal some interesting features of GPP and R_E. It is worth noting that respiration peaks in May, which is not the period with maximum temperature. Nevertheless, May is the month of most active canopy development. Furthermore, in that month soil-water content was at its maximum for the 2001 growing season and hence non-limiting for soil processes (Scartazza *et al.*, 2004). After the rapid increase from the onset of the growing season, gross photosynthesis peaked at the end of May (10.6 g C m^{-2} day^{-1}, Fig. 14.5). GPP remained around maximum values of 9.5–10.5 gC m^{-2} day^{-1} from 27 May to 19 July. During this period of 55 days, representing one-third of the growing season, the ecosystem put together approximately 50% of total GPP and 70% of total annual net budget. In mid October, low GPP equalled R_E and the system became carbon neutral with some day by day variability. However, in that period partitioning of NEE into GPP and R_E became highly uncertain. In 2001, annual totals of GPP and R_E were 1045 g C m^{-2} and 547 g C m^{-2}, respectively.

Comparison of trends of NEE values cumulated throughout the year has been used to study the dynamics of carbon balance of managed ecosystems at different stages of their lifecycle (Kolari *et al.*, 2004). An example of annual trends of cumulated NEE for two stands within a chronosequence of turkey oak is presented in Fig. 14.6. The chronosequence is within a coppice-with-standards forest managed for fuelwood production, has an extension of 1350 hectares and is

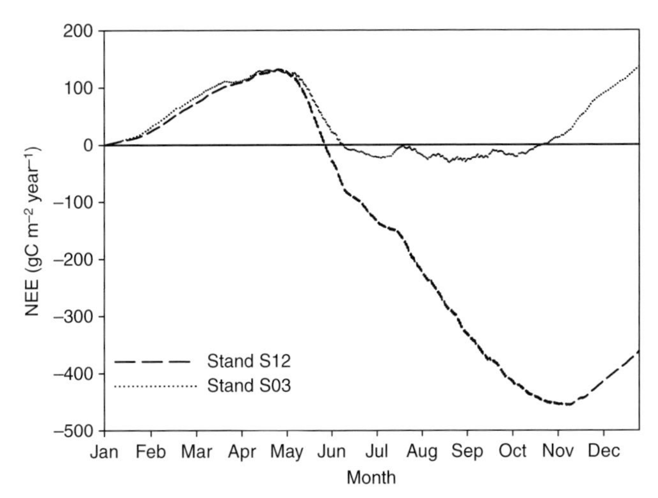

Fig. 14.6. Trends of cumulated net ecosystem exchange (NEE) measured in two stands of a turkey oak chronosequence located in Roccarespampani, Central Italy, in 2002. Stand S12: 12 years after coppicing; Stand S03: 3 years after coppicing. On the same stand, coppicing occurs approximately every 18–20 years and the coppice-with-standards management treatment leaves 50 to 80 20-year-old trees per hectare (the 'standards') that are removed 18–20 years later.

divided into 15 stands of various ages. Two EC systems were installed in two stands of different age: S03, harvested in 1999–2000, and S12, harvested in 1990–91. Data are from year 2002 (Manca, 2003).

The two stands shows the same trend of cumulate NEE between January and April, namely the period preceding the beginning of the growing season (Fig. 14.6).

In the following months of the growing season, the oldest stand (S12) shows a greater net carbon uptake, whereas the net balance of stand S03 remains close to a source/sink equilibrium. At the end of the year, the net balance was –362 g C m^{-2} $year^{-1}$ for S12, whereas the youngest stand (S03) was a carbon source (137 gC m^{-2} $year^{-1}$). Interestingly, when the net flux was partitioned into GPP and R_E, it turned out that the two stands had similar gross photosynthesis, but respiration was 50% higher in the more-recently harvested stand (S03) (Manca, 2003). At the same site, soil respiration and its response to temperature decreased with time after coppicing, whereas the heterotrophic fraction was relatively higher in the more-recently coppiced stand (Tedeschi *et al.*, 2006). Hence, ecosystem respiration was boosted by harvesting-related effects on soil structure and microclimate and on woody debris decomposition. These factors turned the S03 stand into a carbon source. During the forest development, the ecosystem may turn again to be a carbon sink as the carbon balance of S12 demonstrated (Manca, 2003).

As shown, EC is able to estimate gross photosynthesis and other related processes at canopy level. At present, over 400 tower sites distributed in regional networks are operating on a long-term and continuous basis. Forests are the more represented ecosystem. However, fluxes are being measured from several ecosystem types (cropland, grassland, shrubland, wetland). As an example, within the recently completed European Commission Integrated Project CarboEurope-IP, EC towers were operated in 50 forests, 28 grasslands, 22 croplands and 3 wetlands/peatlands (http://www.carboeurope.org). Some of the towers have been active since the 1990s and the number of available flux/years in centralised databases is increasing month after month (http://www.fluxnet.ornl.gov/fluxnet/index.cfm).

Profiting of the large mass of annual flux and ancillary data, several studies have performed multi-site analyses to highlight environmental controls on ecosystem carbon fluxes. In one of the first global analyses, Law and colleagues (2002) found general relationships between monthly R_E and air temperature for deciduous forests (r^2=0.61), whereas annual mean temperature was able to explain 50% of variation in annual GPP for forests and grasslands. Furthermore, variation in GPP was also well correlated (r^2=0.53) with the product of annual mean temperature and site water balance (Law *et al.*, 2002). Reichstein and colleagues (2007), analysing data from flux towers in Europe, found a linear relationship among an index of site-water

Table 14.2. *Gross primary production (GPP), net ecosystem exchange (NEE) and ecosystem respiration (R_E) for different forest biomes. Standard deviation indicates the variability around mean values (data from Luyssaert et al., 2007).*

Ecosystem type	GPP (gC m^{-2} year^{-1})	NEE (gC m^{-2} year^{-1})	R_E (gC m^{-2} year^{-1})
Boreal humid evergreen	973±83	–131±79	824±112
Boreal semiarid evergreen	773±35	–40±30	734±37
Boreal semiarid deciduous	1201±23	–178±NA	1029±NA
Temperate humid evergreen	1762±56	–398±42	1336±57
Temperate humid deciduous	1375±56	–311±38	1048±64
Temperate semiarid evergreen	1228±286	–133±47	1104±260
Mediterranean warm evergreen	1478±136	–380±73	1112±100
Tropical humid evergreen	3551±160	–403±102	3061±162

Values are n±SD. NA, not applicable.

availability and annual GPP (and R_E) for temperate and Mediterranean sites, whereas annual fluxes at boreal sites were highly correlated with mean annual temperature (Reichstein *et al.*, 2007). To date, however, it has not being possible to find a reasonable direct relationship between the NEE and environmental variables. Nevertheless, although gross photosynthesis and respiration are directly related to ecophysiological processes that are controlled by plants (e.g., genetic, molecular) and environmental factors, NEE is resulting from those several processes differently influenced by the environment.

A recent paper put together a comprehensive global database for forest ecosystems, which includes carbon-budget variables (fluxes and stocks), ecosystem traits (e.g., leaf-area index (LAI), age) and ancillary site information from more than 500 sites. The database is publicly available and can be used for several purposes, such as to quantify global, regional or biome-specific carbon budgets (Luyssaert *et al.*, 2007). Some of the carbon-budget data reported on that paper are presented in Table 14.2.

GPP ranges between 770 gC m^{-2} year^{-1} for boreal semi-arid evergreen forests to 3550 gC m^{-2} year^{-1} of tropical humid forests. The two forest types show also the minimum and maximum for ecosystem respiration. Temperate and Mediterranean forests GPP is between 1230 gC m^{-2} year^{-1} and 1760 g C m^{-2} year^{-1}. Interestingly, from the data available so far, all biomes are net carbon sink ranging from 40±30 gC m^{-2} year^{-1} of boreal semi-arid evergreen to 400 gC m^{-2} year^{-1} of temperate and tropical humid evergreen forests (Table 14.2, Luyssaert *et al.*, 2007).

14.5. CONCLUDING REMARKS

In the last 20 years, the EC technique has been used to measure canopy level processes on a growing number of sites located both in close-to-ideal and in more problematic conditions. The data that are continuously gathered at those sites are contributing to a deeper understanding of ecosystem-scale determinants of fluxes. The scientific community is actively working to improve the methodology, particularly for flux correction in low turbulence periods and for advection (Aubinet *et al.*, 2005; Richardson *et al.*, 2006; van Gorsel *et al.*, 2009).

Large databases of flux data are more and more utilised by modellers, and there is the need to integrate various techniques for assessing carbon balance to provide multiple constraints to the measured fluxes (Waring and Running, 1998; Baldocchi, 2001, 2003).

Continuous and long-term research and monitoring at flux sites will allow to quantify the impact of extreme events (heat wave, drought spells, pests) on carbon exchanges as, for example, the recent analysis of the 2003 heatwave on European terrestrial ecosystems (Ciais *et al.*, 2005) or the role of anomalous seasons (e.g., Del Pierre *et al.*, 2009). Furthermore, in the long term, EC-flux sites will provide information on how terrestrial ecosystems will respond to global changes.

15 • Remote sensing of photosynthesis

I. MOYA AND J. FLEXAS

15.1. INTRODUCTION

Remote sensing of photosynthesis consists of directly measuring or indirectly estimating photosynthesis rates or photosynthesis-related parameters using non-contact devices positioned far from the plant or canopy. This definition is a broad one, and includes the gas-exchange measurements at the ecosystem level described in Chapter 14, but the term is more commonly used to describe radiation measurements with spectrometric techniques. Three main groups of spectrometric techniques have emerged to remotely assess photosynthesis or photosynthesis-related parameters: reflectance, fluorescence and thermal imagery (Fig. 15.1). Each of these principles has advantages and inconveniences as compared with the others. These are summarised in Table 15.1, and will be discussed in more detail throughout the chapter.

Remote sensing has been identified as an essential tool to complement gas-exchange and atmospheric-circulation models, in order to monitor accurately the spatial and temporal changes in biosphere primary production (Sellers *et al.*, 1992, 1997; Baldocchi, 2008; Malenovsky *et al.*, 2009). Although there have been important advances in this field during the past thirty years, a comprehensive review of the techniques available and their scope is lacking. Partial reviews on remote sensing of reflectance (Peñuelas and Filella, 1998; Gamon and Qiu, 1999), fluorescence (Cerovic *et al.*, 1999; Moya and Cerovic, 2004) and thermometry (Jones, 1999, 2004a) have been published. However, none of these reviews includes recent advances, and none has treated all these techniques together. An early review by Lichtenthaler *et al.* (1998) considered reflectance and fluorescence, and these two techniques have been recently evaluated together with the focus of their potential for measuring terrestrial photosynthesis from space (Grace *et al.*, 2007). Chaerle *et al.* (2007a) have reviewed the usefulness of combined thermal and Chl-F imaging for monitoring and screening plant populations. All three types of techniques were considered together in a review by Cifre *et al.* (2005), but these authors focused only on their specific use in irrigation scheduling in grapevines. Meroni *et al.* (2009) have reviewed remote sensing of solar-induced Chl-F, whereas several recent reviews have focused on light-use efficiency and PRI (Coops *et al.*, 2010; Hilker *et al.*, 2010; Garbulsky *et al.*, 2011). Only a recent review has considered together the scientific and technical challenges in remote sensing of both plant reflectance and fluorescence (Malenovsky *et al.*, 2009).

In this chapter all these emerging techniques will be examined, considering their potential for application at different integration levels, from leaf to canopy, and from ground measurements to satellite detection. The basic principles of all these techniques have been described in Chapter 10, and it is far from the scope of the chapter to provide comprehensive technical details. However, for readers interested in taking the subject to a deeper or more practical level, we have endeavoured to provide useful references. We have tried to review recent advances and actual limitations with some emphasis on the necessary preparatory steps towards a space mission.

15.2. ASSESSING PHOTOSYNTHETIC ACTIVITY BY PHOTOMETRIC CO_2-UPTAKE MEASUREMENTS: CO_2-CONCENTRATION MEASUREMENTS WITH LIDARS

Measurements of CO_2 exchange and separation of respiratory fluxes from photosynthetic fluxes are often conducted from the ground using in-situ sensors and instrumented towers in the framework of regional or global networks (Conway *et al.*, 1994; Lambert *et al.*, 1995). In addition,

Terrestrial Photosynthesis in a Changing Environment: A Molecular, Physiological and Ecological Approach, ed. J. Flexas, F. Loreto and H. Medrano. Published by Cambridge University Press.

Table 15.1. *A comparison of the three main principles used for remote sensing of photosynthesis. Features indicated as 'high' may be considered advantageous as compared with those indicated as 'low'.*

Variable	Reflectance	Fluorescence	Thermometry
Proximity of the measured parameters to actual photosynthesis	MEDIUM	HIGH	LOW
Availability and economy of adequate instrumentation	MEDIUM	LOW (except contact fluorometers)	HIGH
Range of detection	HIGH	LOW to HIGH	MEDIUM
Simplicity of required software / data analysis	MEDIUM	HIGH	LOW

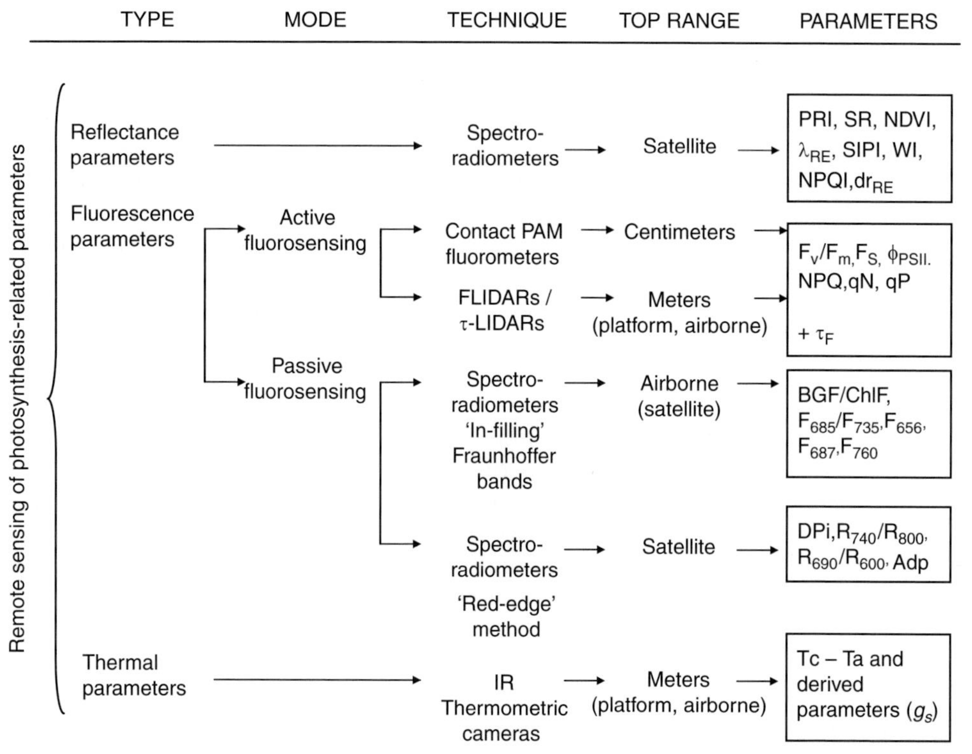

Fig. 15.1. A diagram showing the main three types of techniques for remote sensing of photosynthesis, highlighting the specific techniques within each type, the top range of measurements and the most commonly used parameters (see text for details). BGF, blue-green fluorescence; Chl-F, chlorophyll fluorescence; DPI, double peak index; IR, infrared; NDVI, normalised difference vegetation index; NPQ, non-photochemical quenching; PRI, physiological reflectance index; qN, non-photochemical quenching; qP, photochemical quenching; SR, spectroradiometer; SIPI, structure insensitive pigment index; WI, water index.

airborne measurements also using in-situ sensors complement the current ground-based networks. These airborne measurements are conducted on a regular basis in some locations or during dedicated field campaigns (Matsueda and Inoue 1996; Schmitgen *et al.*, 2004). Lidar measurements have demonstrated great promise, and shown to have the capability of determining the vertical mean CO_2-mixing ratio in the atmospheric boundary layer with a 2% measurement precision (Gibert *et al.*, 2006). The authors used a 2-μm heterodyne differential absorption lidar technique (HDIAL). The HDIAL consists of a single-mode power oscillator in a ring cavity alternatively seeded by a 2-micron continuous-wave laser at 2063.7 nm and 2062.0 nm, corresponding respectively to an online and offline absorption of CO_2. The method yields mean NEE that agrees well with EC measurements (Gibert *et al.*, 2007). Daytime NEE can be used to follow the vegetation growth and the change in the ratio of diffuse/direct radiation. The precision of the HDIAL method can be greatly improved by averaging a large number of measurements thanks to the increase of the excitation frequency from 5 Hz to the KHz domain (Gibert *et al.*, 2006).

Global monitoring, ultimately from space, is foreseen as a means to quantify CO_2 sources and sinks on a regional scale and to better understand the links between the various components of the carbon cycle. A new satellite mission is under evaluation by the European Space Agency (ESA). The A-SCOPE mission (Flamant *et al.*, 2005) aims to observe the total CO_2 column with a nadir-looking pulsed carbon-dioxide differential absorption lidar (DIAL) for a better understanding of the global carbon cycle and regional CO_2 fluxes, as well as for the validation of greenhouse-gas-emission inventories.

15.3. ASSESSING PHOTOSYNTHETIC ACTIVITY USING REFLECTANCE-VEGETATION INDICES

Remote sensing typically involves non-contact measurement of electromagnetic radiation, either in an imaging or non-imaging mode (Gamon and Qiu, 1999). Although not as directly related to photosynthesis as Chl-F, reflectance-vegetation indices have the advantages of a large range of detection and the availability of many commercial instruments capable of their measurement (Table 15.1). Also, satellite platforms are all equipped with sensors of electromagnetic radiation at several wavelengths or 'windows' (Table 15.2). As the most important source of electromagnetic radiation *towards* the vegetation is solar radiation, that which is captured *from* the vegetation is the reflected proportion of solar-radiation incident over such vegetation, i.e., the reflected radiance, which can be converted to reflectance if incident radiance is known.

Many different vegetation-reflectance indices (or SVIs, spectral-vegetation indices) have been used in photosynthesis studies although they are not directly related to photosynthesis detection (reviewed in Peñuelas and Filella, 1998). Among these, the water index (WI=R900/R970) and several other reflectance parameters centred at bands from 970 to 2000 nm have been shown to be related to the vegetation relative water content (RWC) or the equivalent water thickness (EWT), for which they may be envisaged not as estimators of photosynthesis, but useful for water stress and fire-risk assessment (Peñuelas *et al.*, 1993a; Zarco-Tejada *et al.*, 2003a; Ustin *et al.*, 2004; Riaño *et al.*, 2005; Rodríguez-Pérez *et al.*, 2007). The parameters most directly related to photosynthesis are those related to leaf photochemistry, and to some extent also to photosynthetic pigments and/or biomass (Peñuelas and Filella, 1998).

15.3.1. Chlorophyll-concentration indexes

Several vegetation-reflectance parameters have been proposed as indicators of chl. concentration and green biomass. For instance, the wavelength of the red-edge peak (i.e., λ_{RE}, the position wavelength of the maximum slope in the increase of reflectance from red to near IR) has been shown to correlate with chl. content, whereas the amplitude of the same red-edge peak (dr_{RE}) correlates with LAI in plants with different water status and nitrogen availability (Filella and Peñuelas, 1994). Alternatively, the so-called structural independent pigment index (SIPI=(R800-R445)/(R800-R680)) has been shown to correlate with the ratio of carotenoids to chl. a (Peñuelas *et al.*, 1995a), and the normalised phaeophytinisation index (NPQI=(R415-R435)/(R415+R435) reflects chl. degradation (Peñuelas *et al.*, 1995b). Similarly, the greenness index (GI=R554/R677) or the Zarco-Tejada and Miller index (ZM=R750/R710) have been shown to be correlated with chl. contents (Zarco-Tejada *et al.*, 2001, 2005a,b). The brown pigment index (BPI, calculated as the wavelength difference between the wavelengths, where the second derivative of the reflectance in the FR region intercepts zero) has been proposed as an indicator of β-carotene and oxidative stress (Peñuelas *et al.*, 2004).

Besides these, the most used vegetation indices among chl. and biomass estimators are the simple ratio (SR=NIR/R) and the normalised difference vegetation index (NDVI=(NIR-R)/(NIR+R)). These ratios, rather than centring at specific wavelengths, broadly compare the vegetation reflectance in the NIR and red (R) parts of the spectrum. They were first introduced in the late 1960s as indicators of greenness (Birth and McVey, 1968) and LAI (Jordan, 1969), and the measurement of earliest versions of such indices was soon incorporated to NASA's early LANDSAT satellite (reviewed in Cohen and Goward, 2004). These parameters have been repeatedly shown to correlate with stand green biomass and LAI in terrestrial and aquatic vegetation (Tucker *et al.*, 1985; Peñuelas *et al.*,

Table 15.2. *Technical specifications of some selected hyperspectral facilities of potential interest for remote sensing of photosynthesis.*

Platform	Sensor	Approx IFOV	Bands (n)	Wavelength range (μm)	Time repeat (days)	Source	Web page
Airborne	ADAR System-5500 (Airborne Data Acquisition and Registration)	0.5–3.0	4	0.45–1.00	On demand	Positive Systems Inc., USA	http://www.possys.com/
Airborne	AHS (Airborne Hyperspectral Scanner)	2	80	0.43–2.45	On demand	INTA, Spain	http://www.crepad.rcanaria.es/es/npoc/adquisicion_ahs.html
Airborne	AVIRIS (Airborne Visible InfraRed Imaging Spectrometer)	20	224	0.40–2.50	On demand	NASA, USA	http://aviris.jpl.nasa.gov/
Airborne	CASI-2 (Compact Airborne Spectrographic Imager)	Variable	288	0.43–0.87	On demand	ITRES Res., Ltd., Canada	http://www.itres.com/CASI_550
Airborne	AIRFLEX (passive fluorescence sensor)	20	6	0.685–0.770	On demand	ESA (EU)	http://esamultimedia.esa.int/esaLP/SEMEDC0CYTE_LPgmes_0.html
Satellite (Regional)	ASTER (Advanced Spaceborne Thermal Emission and Reflection Radiometer)	15–90	14	0.52–11.65	16	NASA, USA	http://asterweb.jpl.nasa.gov/
Satellite (Regional)	LANDSAT	15–120	8	0.45–12.50	16	NASA, USA	http://landsat.gsfc.nasa.gov/
Satellite (Global)	AVHRR (Advanced Very High Resolution Radiometer)	1100	5	0.58–12.50	4 times per day	NOAA, USA	http://noaasis.noaa.gov/NOAASIS/ml/avhrr.html
Satellite (Global)	MODIS (Moderate Resolution Imaging Spectrometer)	250–2250	36	0.40–14.38	2 times a day	NASA, USA	http://modis.gsfc.nasa.gov/
Satellite (Global)	SPOT 5 (Satellite Pour l'Observation de la Terre)	1150	4	0.43–1.75	26	CNES, France	http://www.cnes.fr/web/1415-spot.php

1993b; Gamon *et al.*, 1995), as well as with canopy stomatal conductance and photosynthesis under non-stress conditions (Sellers *et al.*, 1992; Verma *et al.*, 1993) and under stress conditions where canopy development and photosynthetic activity are in synchrony (Gamon *et al.*, 1995; Peñuelas *et al.*, 1997a; Running *et al.*, 2004).

The mechanistic explanation for the correlation between NDVI and canopy photosynthesis is based on Monteith's (1972) suggestion that the GPP of well-watered and fertilised annual crops is linearly related to the amount of solar energy that the plants absorb over a growing season. Sellers (1987) was the first to show that NDVI was closely related to the fraction of absorbed PPFD, i.e., NDVI ≈ absorbed PPFD/incident PPFD. Gamon *et al.* (1995) showed that, indeed, NDVI specifically reflects the fraction of absorbed PPFD by green tissues (i.e., the fraction that can be potentially converted into biomass through photosynthesis). Hence, if incident PPFD is known (e.g., from ground meteorological stations, from reflectance of a reference panel, etc.), using remotely sensed NDVI as a proxy for the fraction of absorbed PPFD, GPP can be estimated as (Running *et al.*, 2004):

$$GPP = \alpha \times NDVI \times PPFD \qquad \text{[Eqn. 15.1]}$$

where α is a conversion efficiency (empirically determined in the field for each vegetation type), expressed as the grams of carbon photosynthesised per absorbed megajoule of photosynthetic radiation.

As mentioned above, NDVI closely tracks canopy photosynthesis in many situations, and it can therefore be used to produce large-scale estimates of GPP and NPP. At the global scale, a satellite-based Earth NPP monitoring using this system has existed for more than 15 years (Running and Nemani, 1988; Myneni *et al.*, 2002; Running *et al.*, 2004; http://www.ntsg.umt.edu/). This is producing global-scale mean NPP maps at 8-day intervals, using the MODIS system onboard NASA's Terra Satellite Platform (Table 15.2). However, the first satellite-based NDVI regional images (Tucker *et al.*, 1985; Running and Nemani, 1988) were taken using the AVHRR system (Table 15.2), and, lately, the same algorithm described above and used for MODIS was applied to AVHRR images to reconstruct past (1982–1999) trends in global NPP, showing a clear global-change-related increase by 6% (Nemani *et al.*, 2003).

Despite the successful use of NDVI as a proxy for NPP at the global scale, the fraction of incident PPFD that is actually utilised in photosynthesis (i.e., α) strongly varies throughout the year (from >90% to <25%), particularly in evergreen ecosystems (Runyon *et al.*, 1994). Hence it has been shown that NDVI fails to estimate NPP in such cases (Gamon *et al.*, 1995). To track NPP in such ecosystems, it will be necessary to have an indicator of dynamic changes in α, which may be also possible using proper SVIs, as explained in the next section.

15.3.2. A reflectance index related with dissipation of excess absorbed energy (PRI)

The physiological (sometimes also called 'photochemical') RI (PRI=R531-R570/R531+R570) is strongly correlated with the de-epoxidation state of the xanthophyll cycle (Gamon *et al.*, 1990, 1992; Peñuelas *et al.*, 1994; Peñuelas and Filella, 1998; Stylinski *et al.*, 2000, 2002), although it has a xanthophyll-independent fast component, which is possibly related to the generation of trans-thylakoid pH gradients (Evain *et al.*, 2004; Peguero-Pina *et al.*, 2008), and, at large spatial and temporal scales, it is also influenced by seasonal changes in pigment contents and canopy structure (Gamon *et al.*, 2004). Because xanthophyll-related thermal dissipation is often negatively related to leaf photochemistry, PRI has been generally shown to correlate with F_v/F_m (Winkel *et al.*, 2002), Φ_{PSII} (Peñuelas *et al.*, 1995c, 1998; Stylinski *et al.*, 2002; Evain *et al.*, 2004; Nichol *et al.*, 2006), qN or NPQ (Peñuelas *et al.*, 1997a; Evain *et al.*, 2004; Peguero-Pina *et al.*, 2008), F_s (Evain *et al.*, 2004; Dobrowski *et al.*, 2005; Peguero-Pina *et al.*, 2008), photosynthesis and photosynthetic capacity (Gamon *et al.*, 1992; Filella *et al.*, 1996; Stylinski *et al.*, 2000, 2002; Peñuelas *et al.*, 2004; Louis *et al.*, 2005; Sims *et al.*, 2006) and g_s (Peguero-Pina *et al.*, 2008; Suárez *et al.*, 2008).

Therefore, PRI is a potentially promising candidate for remote assessment of photosynthesis, and indeed has been shown to track daily (Peñuelas *et al.*, 1995c; Filella *et al.*, 1996) and seasonal (Stylinski *et al.*, 2002; Louis *et al.*, 2005) variations of photosynthesis, as well as photosynthesis changes owing to water stress (Peñuelas *et al.*, 1997b, 1998; Winkel *et al.*, 2002; Evain *et al.*, 2004; Fuentes *et al.*, 2006; Sims *et al.*, 2006; Peguero-Pina *et al.*, 2008; Suárez *et al.*, 2008), nitrogen deficiency (Gamon *et al.*, 1992, Peñuelas *et al.*, 1994), dithiothreitol infiltration (Evain *et al.*, 2004), leaf development (Winkel *et al.*, 2002) or spring recovery of photosynthesis in boreal forests (Louis *et al.*, 2005). However, the PRI does not unequivocally reflect photosynthesis. First, the absolute PRI values are extremely variable. For instance, the range of PRI values has been described to be between –0.34 and –0.24 (Peñuelas *et al.*, 1994), between –0.20 and 0 (Sims *et al.*, 2006), between –0.03 and +0.02 (Peñuelas *et al.*, 1995c), or between +0.06 and +0.10 (Gamon *et al.*, 1992), and all these ranges corresponded to similar variations in photosynthesis and the de-epoxidation state of the xanthophylls. It is yet unclear what causes this variation, whether it depends on the species, the conditions of measurement, the utilised instrument or other. Second, PRI has been shown to correlate either positively (Stylinski *et al.*, 2002; Evain *et al.*, 2004; Peguero-Pina *et al.*, 2008) or negatively (Peñuelas *et al.*, 1997a, 1998) to photosynthesis. This is simply owing to the fact that the latter authors use an inverse definition for the parameter (i.e., R570–R531/R531+R570 instead of R531–R570/R531+R570), but it adds some confusion when comparing data from different studies. Third, the exact relationship between PRI and both photosynthesis and the de-epoxidation state of the xanthophylls depends on the species analysed (Peñuelas *et al.*, 1995c; Stylinski *et al.*, 2002), the prevailing irradiance (Evain *et al.*, 2004) and the treatments applied (Gamon *et al.*, 1992). Moreover, the PRI index sometimes fails to follow water-stress-induced changes in photosynthesis (Gamon *et al.*, 1992; Peñuelas *et al.*, 1994). This was originally thought to be a wilting-dependent effect, as leaf wilting may change the optical properties of leaves (Gamon *et al.*, 1992; Peñuelas *et al.*, 1994), but it has been recently shown to occur as well in non-wilting leaves of olive trees (Sun *et al.*, 2008). Finally, PRI at canopy level is strongly affected by the geometry of solar illumination and by specific leaf-angle distributions in the canopy, as well as by detector viewing angles. Therefore, PRI at canopy level sometimes fails to track variations in photosynthesis, for which some authors prefer reliance on fluorescence measurements (Rascher and Pieruschka, 2008).

Despite these potential pitfalls, there is a growing interest in PRI as the reflectance parameter that most dynamically tracks photosynthesis. Presently, PRI can be measured at near-contact or short distance using commercially available spectroradiometers (Gamon *et al.*, 1990; Nichol *et al.*, 2006), laboratory built two-channel radiometers (Méthy *et al.*, 1999; Méthy, 2000a) or CCD digital cameras with appropriate filters (Méthy *et al.*, 1999). Evain *et al.* (2004) introduced another instrument, based on commercially available interference filters and photodiode detectors coupled to a small-sighting telescope. The reflectance signal is calculated by dividing the spectral radiance of the leaf or the canopy by the radiance of a reference panel, for which the instrument introduces a flip-flop mirror to alternate the field of view between the reference (calibrated against Spectralon, Labsphere, North Dutton, N.H.) and the sample. With this device, PRI can be measured continuously and at distance. To date, it has been shown to measure properly at a distance of upto 50 m from the canopy (Louis *et al.*, 2005). Attempts have been made to extend the measuring distance for PRI. Nichol *et al.* (2000) mounted a portable spectroradiometer on a helicopter and used it to assess canopy PRI with a ground resolution of 79 m, from an altitude of 300 m over boreal forests. Because of the filter specifications of the device used, their definition of PRI was somewhat different of the generic one (PRI=(R569–R529)/(R569+R529)). Despite this, the data correlated well with stand measurements of photosynthesis. Later, Nichol *et al.* (2002) mounted a similar device on a helicopter to provide PRI measurements closer to typical ones (PRI=(R570–R530)/(R570+R530)). Using a simulation from the helicopter-based results, Nichol *et al.* (2000) suggested that it would be possible to assess PRI from an aircraft using the AVIRIS sensor (Table 15.2). Later, Rahman *et al.* (2001) and Fuentes *et al.* (2006) tested the potential of AVIRIS data for PRI assessment, finding that they track drought-induced changes well in net CO_2 fluxes in a chaparral ecosystem. Equally successful PRI estimates have been obtained from other aircraft-mounted sensors (Table 15.2), such as ADAR (Sims *et al.*, 2006) and AHS (Suárez *et al.*, 2008), from altitudes of 1000 m and spatial resolutions of 2 m. Also based on airborne data, the work of Suárez *et al.* (2009) discussed the effect of canopy structure on modelling PRI by using radiative transfer models. Finally, satellite detection of PRI seems now feasible. Rahman *et al.* (2004) used MODIS (Table 15.2) reflectance bands originally intended for ocean observations and calculated PRI over a temperate deciduous forest. They took MODIS bands #11 (bandwidth 526–536 nm) and #12 (546–556 nm), and elaborated a surrogate of PRI as:

$$\mathrm{MODPRI} = (R_{band}\#_{11} - R_{band}\#_{12}) / (R_{band}\#_{11} + R_{band}\#_{12}) \quad \text{[Eqn. 15.2]}$$

They found that, using MODPRI as a surrogate for α, and NDVI as an estimator of the canopy absorbed PPFD (aPPFD), the product MODPRI×NDVI (i.e., α×aPPFD) strongly correlated with the ecosystem NEP determined with tower eddy fluxes. In this ecosystem, the typical MODIS-derived NPP (i.e., based on NDVI alone) failed to track tower-based NPP measurements, as NDVI did not change over the study period. MODPRI was validated as a remote indicator of photosynthesis by Drolet *et al.* (2005) in a boreal forest. Drolet *et al.* (2008) have recently used the same approach to map the gross light-use efficiency of forested areas in Canada. Goerner *et al.* (2009) have also used MODPRI to successfully track ecosystem light-use efficiency even during severe summer drought in a Mediterranean forest.

15.4. CHL-F MEASUREMENTS

15.4.1. Active Chl-F remote sensing

Active fluorosensing has greatly benefited from recent progress in optoelectronics, and particularly from the development of lasers. In order to measure fluorescence under natural conditions (i.e., in the presence of daylight) most systems have relied on a brief laser excitation pulse (<1 μs), which plays a role similar to that of the modulated light in PAM fluorometers (see Chapter 10). Synchronised detection is generally used to extract the fluorescence signal from the background. It is possible to avoid significant perturbation of the light climate of the sample by keeping the frequency of the additional light excitation sufficiently low. Measurements are thus non-invasive and can be applied during extended periods of time. The excitation wavelength is often chosen below 400 nm (e.g., 355 nm, the third harmonic of Nd:YAG laser). This is because the international safety regulation for the use of lasers has favoured the use of UV lasers, mainly owing to the high eye sensitivity for visible light and despite the higher energy of UV radiation. For instance, the permissible energy flux for pulsed beams is ten-thousand times larger below 400 nm than above this threshold wavelength (ICNIRP, 1996).

Active fluorosensing measures fluorescence intensities and, by consequence, remains influenced by factors including distance, atmospheric transmission, geometry and movements of the plant. These limitations can be overcome by using emission or excitation ratios as fluorescence signatures.

THE EMISSION RATIO RF/FRF

The RF/FRF (or F685/F735) ratio was the first signature introduced by Lichtenthaler *et al.* (1986). It depends on the chl. content of the leaf and on leaf anatomy. Owing to the selective re-absorption of red fluorescence by chl. molecules relative to FR fluorescence (see Chapter 10), there is an inverse curvilinear relationship between this ratio and the leaf chl. content observed in many plant species (Gitelson *et al.*, 1998). The RF/FRF ratio is largely affected by changes in chl. content at values lower than 250 mg m^{-2}, i.e., in pale (chlorotic) leaves. Unfortunately, the sensitivity of the RF/FRF ratio decreases dramatically at chl. contents higher than 300 mg m^{-2}, where practically no differences can be detected by this fluorescence ratio (Gitelson *et al.*, 1998). Nevertheless, the RF/FRF ratio was useful to monitor the degradation of chl., carotenoids and the photosynthetic apparatus in leaves of *Xerophyta scabrida* during slow desiccation (Csintalan *et al.*, 1998). It must be mentioned, however, that most environmental stresses are expected to modify chl. concentration only slowly. The use of the RF/FRF ratio has also been criticised by Rosema and Verhouf (1991), who used a canopy fluorescence model (FLSAIL, Rosema *et al.*, 1991) to show the simple reflectance index (R770–R690)/R690 to be much more sensitive to the chl. content of leaves, particularly in the case of dark-green leaves.

The RF/FRF ratio can also show changes of small amplitude at constant chl. content. Changes in the RF/FRF ratio have been detected during the diurnal cycle (in the range of 0.75 to 1.15) (Valentini *et al.*, 1994; Agati *et al.*, 1995) and under temperature stress (Agati *et al.*, 1995, 2000). The dependence of this signature on irradiance indicates that these changes are owing to a change in the level of NPQ of Chl-F.

TWO-WAVELENGTH EXCITATION OF CHL-F

Most environmental conditions that induce long-term changes of photosynthesis, involve variation in carbon and nitrogen economy. It is known that high irradiance and soil-nitrogen deficiency stimulate production and accumulation of phenolic compounds and reduce mass-based protein content (see Koricheva *et al.*, 1998 for a review).

Phenolic compounds, mainly hydroxycinnamic-acid esters and flavonoid glycosides, presenting mainly in the cell walls and vacuoles, respectively, are good UV absorbers and will therefore efficiently screen UV light before reaching the chl. located in the mesophyll (cf. Cerovic *et al.*, 1999). As a consequence, Chl-F excited below 400 nm is one-to-two orders of magnitude lower than that excited in the visible part of the spectrum (Cerovic *et al.*, 1999). This screening effect can be exploited for a quantitative assessment of phenolic compounds, present mainly in the epidermis, by comparing leaf Chl-F excited in the UV and visible part of the spectrum (Sheahan, 1996; Bilger *et al.*, 1997). New FLIDARs using double- (Ounis *et al.*, 2001a) or multiple-excitation (Samson *et al.*, 2000; Corp *et al.*, 2003) wavelengths were designed and used to validate this approach for fluorosensing (Ounis *et al.*, 2001a), and to follow the accumulation of phenolic compounds in nitrogen-deficient maize plants (Samson *et al.*, 2000; Corp *et al.*, 2003) and other species (Cartelat *et al.*, 2005; Meyer *et al.*, 2006). In the work of Ounis *et al.* (2001a), the DE-FLIDAR (dual-excitation LIDAR) was also used to check a new fluorescence signature that combines the double-excitation and double-emission approach. The double-fluorescence ratio (DER), red (685 nm) to FR (735 nm) fluorescence-emission ratio, excited at 355 and 532 nm, was shown to be linearly dependent on the leaf chl. content, for a large range of 200 to 700 mg m^{-2}. The determination of DER makes use of 532 nm excitation, which is a wavelength of low chl. absorption. As a result, the DER ratio is more robust to saturation than the red/FR fluorescence ratio usually determined under blue excitation (Lichtenthaler *et al.*, 1986).

MEASUREMENT OF THE EMISSION RATIO BGF/RF (OR BGF/FRF)

UV excitation of green leaves induces two distinct, fundamentally different but still complementary types of fluorescence: a BGF in the 400–630 nm range and the chl. *a* fluorescence in its red and FR region (630–800 nm) of the spectrum. As it has been reviewed by Cerovic *et al.* (1999), the relative intensities of these two types of fluorescence are highly sensitive to intrinsic leaf properties and environemental factors. Therefore, fluorescence-emission spectra induced by UV light can be considered as a complex fluorescence signature but can reveal much about the physiological state of plants.

An interesting approach to explain the changes of UV-induced fluorescence in nutrient-deficient plants is to use the frame of the carbon-nutrient balance hypothesis that is widely accepted by chemical ecologists (Bryant *et al.*, 1987; Baas, 1989; Price *et al.*, 1989). According to this hypothesis, excess of fixed carbon relative to the plant's resources (high carbon-nutrient ratio) stimulates the shikimate pathway and therefore the production of plant phenolics (Price *et al.*, 1989). Increases of total-plant phenolics have indeed been observed in nutrient-deficient plants (Bryant *et al.*, 1987; Price *et al.*, 1989). The increases of the BGF/Chl-F ratios observed in N-deficient plants (Chappelle *et al.*, 1984; Corp *et al.*, 1997) are also consistent with the carbon-nutrient balance hypothesis.

Based on these works, a new vegetation index, the fluorescence global vegetation index (FGVI) (Cerovic *et al.*, 1999) was proposed. It uses a normalised fluorescence ratio, (Chl-F–BGF)/(Chl-F+BGF) as a global vegetation index of nutrient shortage and presence of stress, which is now strengthened by the knowledge of the UV-screening effect of the secondary metabolites described above. Low FGVI or a high simple BGF/Chl-F ratio is a global indication of nutrient shortage induced either by an accumulation of fluorescing phenolics, which contribute to BGF increase, or accumulation of non-fluorescent phenolics, which by screening UV-excitation of chl., contributes to Chl-F decrease. The FGVI would be interesting on the regional and global level, like the reflectance signature NDVI, but its practical application will depend on the development of airborne lidars at the regional level, and passive fluorosensing in the Fraunhofer lines (see below) for the global level (Moya *et al.*, 1992).

UTILISATION OF CHL-F LIFETIME

An alternative to the measurement of the Chl-F yield (Φ) is to use the mean fluorescence lifetime, τ (see Chapter 10 for an exhaustive definition of this parameter). The use of τ in Chl-F remote sensing is based on several studies showing that in most physiological situations, τ is roughly proportional to Φ (Moya, 1974; Cerovic *et al.*, 1996). As τ is an intensive physical property of the fluorescence decay, it is independent of the geometry of the leaf, the concentration or the absorption of the fluorophore and of the atmospheric transmission. Importantly, τ has been shown to be almost insensitive to Chl-F reabsorption (Terjung, 1998), which greatly helps quantitative measurements on leaves.

A system for measuring the Chl-F lifetime of plants from a distance (τ–FLIDAR) was successfully developed (Moya *et al.*, 1995; Goulas *et al.*, 1997). It was based on the use of a laser delivering very short pulses (<100 ps) and on the temporal analysis of fluorescence and backscattered signals detected by a fast photomultiplier and a digital transient recorder. The overall time response (0.35 ns) imposes the requirement of deconvolution techniques. To fluorosense complex targets that contain several leaves at different surface planes and even stems in the field of view, a special two-step deconvolution program was developed to retrieve the lifetime parameter. A first deconvolution was applied to the backscattered laser light from the target, which provides the level and the relative area of leaves illuminated by the laser beam. In a second step, this information on the position of the illuminated leaves was used to calculate the mean Chl-F lifetime (Camenen *et al.*, 1996; Goulas *et al.*, 1997).

The τ–FLIDAR was used on individual leaves from a distance of 15 m to track the variation of the mean Chl-F lifetime as a function of the plant-water status (Cerovic *et al.*, 1996). As an important outcome of this work, the parameter τ_s (Chl-F lifetime in the steady state) was shown to be a good indicator of stomata closure and, consequently, it can be used as an indicator of water stress (Cerovic *et al.*, 1996). Another important implication is the possibility to determine the spatial position of leaves within the canopy by analysing the backscattered signal after picosecond excitation. The structure of canopies is of particular interest for the evaluation of the intercepted radiation in radiative-transfer models. Another set-up for measuring the mean Chl-F lifetime at distance has been developed by Sowinska *et al.*, (1996).

Among other LIDAR techniques, remote sensing of the mean fluorescence lifetime (by τ-FLIDAR) appears to be the most robust method, at the cost, however, of a sophisticated and expensive technology necessary to achieve the sub-nanosecond temporal resolution. For a broader application of this method, further technological progress is still necessary.

IMAGING SYSTEMS

Although conventional PAM imaging systems can be used at a certain distance, for instance to detect biotic stresses (see Chapter 22), it is with FLIDAR imaging systems that a true remote sensing of fluorescence images becomes possible. The FLIDAR imaging systems are based on bi-dimensional matrix detectors (Saito *et al.*, 1999; Sowinska *et al.*, 1999) or on 'raster' systems that reconstitute the image pixel by pixel (Johansson *et al.*, 1996). Their spatial resolution allows taking into account the leaf or canopy fluorescence heterogeneity. Imaging FLIDARs use many emission-fluorescence ratios. Of particular interest is the truck-based laser-induced fluorescence (LIF)-imaging system described in the paper by Sowinska *et al.* (1999). A remote-sensing unit devoted to distance measurements records images of fluorescence excited at 355 nm and detected consecutively at four wavelengths (440, 520, 690 and 730 nm) through band-pass filters. In addition, a second detection unit is installed inside the truck for measurements on detached leaves under laboratory like conditions. The system was used to discriminate between wheat plants submitted to different fertilisation rates (Heisel *et al.*, 1997). A similar instrumental approach was developed by Corp *et al.* (2003) and applied in site-specific management of N-fertilisation programs.

MICROLIDARS

The method of modulate-light and saturating-light pulses (PAM, see Chapter 10), introduced more than two decades ago by Schreiber *et al.* (1986), has multiplied the possibilities of measuring Chl-F in the laboratory and in the field. Several commercial instruments allow precise measurements of variations of fluorescence yield under full daylight conditions, from which one can deduce the photosynthetic electron-transfer rate (Genty *et al.*, 1989). Although these instruments are intended to operate at near contact, they can be adapted for remote-sensing purposes. For example, a PAM 101 fluorimeter (Heinz Walz, Effeltrich, Germany) has been modified for continuous monitoring of the steady state Chl-F (F_s) at distances upto several meters (Ounis *et al.*, 2001b). It is based on a laser diode emitting at 638 nm, and a Fresnel lens coupled to the regular detection unit. The detection unit has also been modified to measure simultaneously both the modulated fluorescence and the light reflected by the leaf. Reflected light showed a good estimation of the PAR measured exactly at the same area as the fluorescence. However, in this transformation the possibility to measure Chl-F at distance has been obtained at the expense of the possibility to saturate PSII photochemistry. Several instruments have been specifically developed for remote sensing, which are detailed below.

Frequency-induced pulse amplitude modulation (FIPAM)

Variable Chl-F measurements by the light-doubling method has also been shown to be possible at distances up to 5 metres thanks to a specifically developed fluorimeter, the FIPAM (Cavender-Bares *et al.*, 1999; Flexas *et al.*, 2000; Moya *et al.*, 2004). The FIPAM is a small laser diode-based FLIDAR, able to measure the F_s level or the minimum (F_0) level in the dark at a very low repetition rate (1 Hz). The maximum Chl-F level (F_m or F_m') is obtained by increasing the frequency of the modulated light up to 100 kHz that fully saturates PSII photochemistry. This system was intended as a reference tool for other lidars, or to calibrate passive fluorescence measurements (see next sections).

The FIPAM fluorimeter was used to follow F_s and F_m Chl-F parameters of a potted grapevine continuously during 17 days of water-stress development. Figure 15.2 shows the relationship between Φ_{PSII} and light intensity in the afternoon. Two different days, before and after water-stress development, were compared. It is observed that the stressed plants exhibit a lower capacity to recover after high light exposure. According to Cornic and Briantais (1991), electron transport to O_2 should increase during the desiccation of the leaf. This alternative sink for electrons should be large enough to maintain high rates of electron transport during most of the day. By contrast to what happened under irrigated conditions, such an increase in electron transport

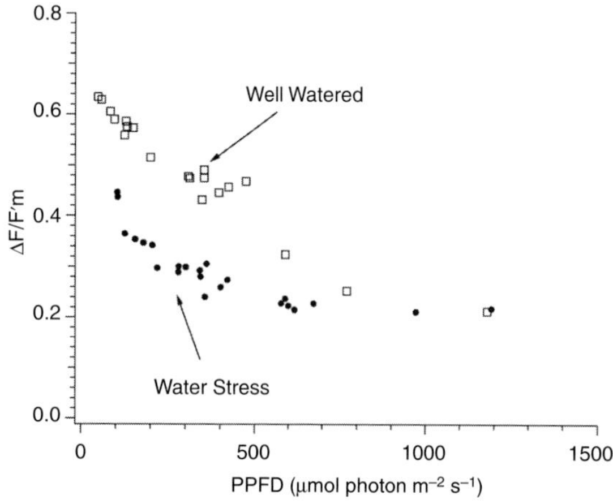

Fig. 15.2. The $\phi_{PSII}=(F'_m-F_s)/F'_m$ parameter plotted against photosynthetic photon-flux density (PPFD), where F_s is the stationary chlorophyll-fluorescence level in light, and F'_m the maximum chlorophyll-fluorescence level during a short saturating flash. Afternoon data of a diurnal cycle of a potted vine plant (*Vitis vinifera* L.) are presented. Open squares, well-watered plant; closed circles, the same plant after water was withheld for several days. After Moya *et al.* (2004).

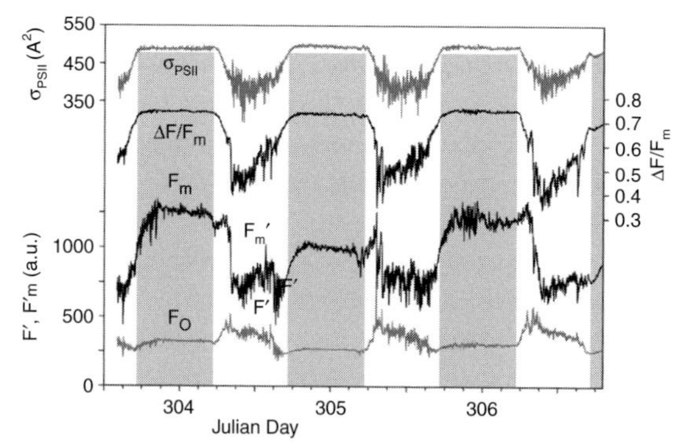

Fig. 15.3. Example of long-term, autonomous LIFT measurement on a free-standing oak tree (*Quercus* sp.) growing outdoors at Biosphere 2 Laboratory at a distance of about 40 m. The shaded areas indicate night periods. F′ and Fm′, and F_0 and F_m, are the initial and the maximal levels of variable fluorescence (in the light and darkness respectively), σ_{PSII} is the functional absorption cross section of PSII and ϕ_{PSII} is the maximum electron-transfer efficiency of PSII. After Kolber *et al.* (2005).

to O_2 was not able to protect leaves from photoinhibition during drought stress, as witnessed by only a partial recovery of the afternoon quenching (Fig. 15.2; Flexas *et al.*, 2000). Although this was a situation of severe water stress, as photoinhibition is uncommon under most water-stress situations (see Chapter 20), it illustrates the possibilities of remote sensing using the FIPAM.

Laser-induced fluorescence transient (LIFT) technique

Kolber *et al.* (2005) developed a laser-induced fluorescence transient (LIFT) instrument to remotely measure photosynthetic properties in terrestrial vegetation at a distance of up to 50 m. The LIFT method was developed primarily for continuous monitoring of the photosynthetic properties in the relatively inaccessible outer canopy of trees over long time periods (Osmond *et al.*, 2004). The LIFT set-up uses a 665-nm laser diode to project a collimated, 100-mm diameter excitation beam onto leaves of the targeted plant. The laser beam is pulse modulated in the μs range (Ananyev *et al.*, 2005), but contrary to the FIPAM, it cannot saturate the PSII photochemistry as the excitation power density has been reduced to satisfy the guidelines regarding eye safety. Fluorescence emission at 690 nm is collected by a 250-mm telescope and processed in real time by numerically fitting the measured fluorescence yields to a theoretical model described previously (Kolber *et al.*, 1998). However, the authors experience some difficulties in estimating ϕ_{PSII} under high irradiances, where LIFT estimates of ϕ_{PSII} are usually noisier than those measured by the PAM method. The main problem seems to be the large noise owing to background fluorescence produced by the ambient light. Interestingly, the LIFT instrument is designed to either be controlled remotely, from any terminal connected to the internet, or to operate automatically to execute a set of preprogrammed measurements. The acquired data are immediately available by logging into the LIFT server. Several versions of the instrument have been declined that differed by the number of laser diodes, including an airborne version operated at 150 m altitude.

Figure 15.3 shows an example of long-term autonomous LIFT measurements on a free-standing oak tree (*Quercus* sp.) growing outdoors at the Biosphere 2 laboratory (B2L), at a distance of about 40 m. The shaded areas indicate night periods. The LIFT set-up was also used to observe the diel courses of fluorescence parameters in leaves of two tropical-forest dominants, *Inga* and *Pterocarpus*, growing in the enclosed B2L. Unexpectedly, mid-canopy leaves of both trees showed an unusual afternoon increase in NPQ in the shade, which was ascribed to reversible inhibition of photosynthesis at high leaf temperatures in the enclosed canopy (Ananyev *et al.*, 2005).

Eye-safety reasons preclude the use of high-power focused laser beams required to saturate PSII photochemistry. This

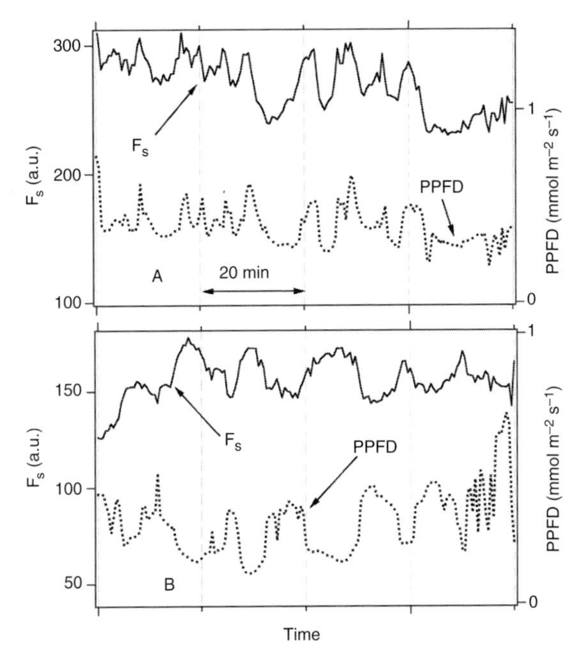

Fig. 15.4. Comparison of the stationary chlorophyll fluorescence (Chl-F) (F_s) (left ordinate, solid line) and photosynthetic photon-flux density (PPFD) (right ordinate, dotted line) in a potted vine plant (*Vitis vinifera* L.) as a function of time. Measurements were done with the FIPAM fluorometer at a distance of 1 m. PPFD variations are owing to clouds. **(A)** Well-watered plant. **(B)** Same plant after several days of withholding water. Note the anti-parallel correlation between F_s and PPFD. a.u, arbitrary units. After Moya *et al.* (2004).

is why most of the remote-sensing applications of fluorescence under field conditions are based on the measurement of F_s. Although lower than that of F_m, the variations of F_s can be up to 100% (Cerovic *et al.*, 1996). Several studies have been devoted to the understanding of F_s changes in relation to the constraints to which plants are subjected (Cerovic *et al.*, 1996; Rosema *et al.*, 1998; Flexas *et al.*, 1999a, 2000, 2002b; Soukupová *et al.*, 2008). Although on a seasonal basis F_s tracks roughly the activation/deactivation of the photosynthetic apparatus in evergreen plants at the beginning and end of the growing season (Soukupová *et al.*, 2008; Malenovsky *et al.*, 2009), more dynamic changes can be observed on a daily basis. Figure 15.4 shows that valuable information on the water status of a vine plant can be obtained by the simple measurement of F_s as a function of natural-light (PPFD) changes. In this experiment the two parameters were measured on exactly the same part of the leaf, which is one of the special features of the FIPAM. The results shown in Fig. 15.4 can be interpreted as follows (Flexas *et al.*, 2002b): when the stomata are open (Fig. 15.4A), photochemical quenching (qP) mostly determines the actual fluorescence level. An increase in the light intensity modifies the equilibrium of the electron transport chain in the direction of a reduction, which is accompanied by an increase of Chl-F (see Chapter 10). In a water-stressed plant (Fig. 15.4B), the stomata are closed. To dissipate the excess of absorbed energy, a mechanism of non-radiative (heat) dissipation takes place at the level of the LHCII antennae (NPQ), inducing a decrease of F_s (note the different scales in Fig. 15.4A and B). When this mechanism is reinforced following an increase of light, F_s decreases further. Reciprocally, F_s increases when the NPQ relaxes after a decrease of light.

Thanks to several commercially available instruments, active fluorosensing has been extensively used at the leaf level in laboratory studies. This possibility is lacking for fluorescence measurement at the canopy level where only a limited number of proprietary instruments with sufficient footprint and ranging are under operation. This situation greatly reduces the availability of the database required for long-term studies of varied cultivated fields or biomes.

15.4.2. Passive Chl-F remote sensing

‘IN-FILLING’ THE FRAUNHOFER LINES

Under natural-sunlight illumination, the amount of Chl-F emitted by vegetation represents a very small fraction (<1%) of the reflected light in the NIR part of the spectrum. However, at certain wavelengths, where the solar spectrum is attenuated (Fraunhofer lines), the fluorescence signal can be quantified. The positions of the main Fraunhofer lines in the solar spectrum are shown in Fig. 15.5A. In the red and NIR part of the solar spectrum three main absorption bands are present: the Hα line at 656 nm, which is owing to absorption by the hydrogen in the solar atmosphere, whereas the two bands at 687 and 760 nm are owing to absorption by the molecular oxygen of the terrestrial atmosphere. These bands largely overlap the Chl-F emission spectrum of leaves (Fig. 15.5B). One way to obtain information on the fluorescence from the whole radiance signal is to use the FLD (Fraunhofer Line Discrimination) method (Fig. 15.6A). In short, this method compares the depth of the band in the solar-irradiance spectrum with the depth of the band in the radiance spectrum of the plant (Plascyk, 1975; Carter *et al.*, 1990, 1996; Moya *et al.*, 1992).

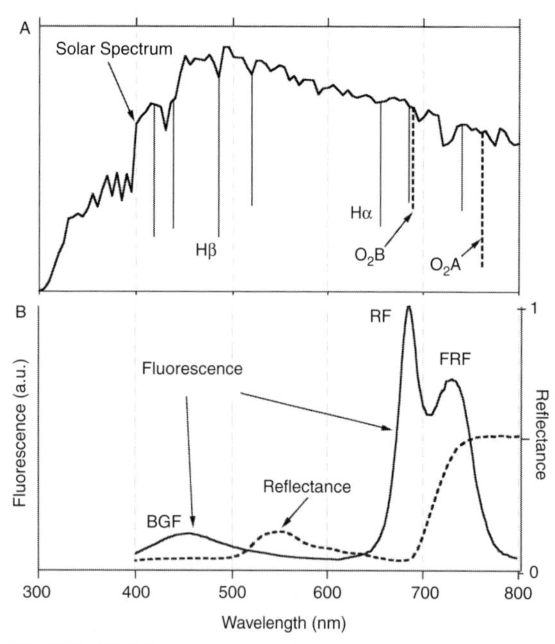

Fig. 15.5. **(A)** Solar spectrum at sea level. Dashed vertical lines, oxygen absorption bands owing to terrestrial atmosphere; continuous vertical lines, absorption lines owing to the solar atmosphere. **(B)** Continuous line, fluorescence-emission spectrum of a grapevine leaf excited at 355 nm; dashed line, reflectance spectrum of the same leaf; BGF, blue-green fluorescence; RF, red fluorescence; FRF, far-red fluorescence. All curves are in relative units.

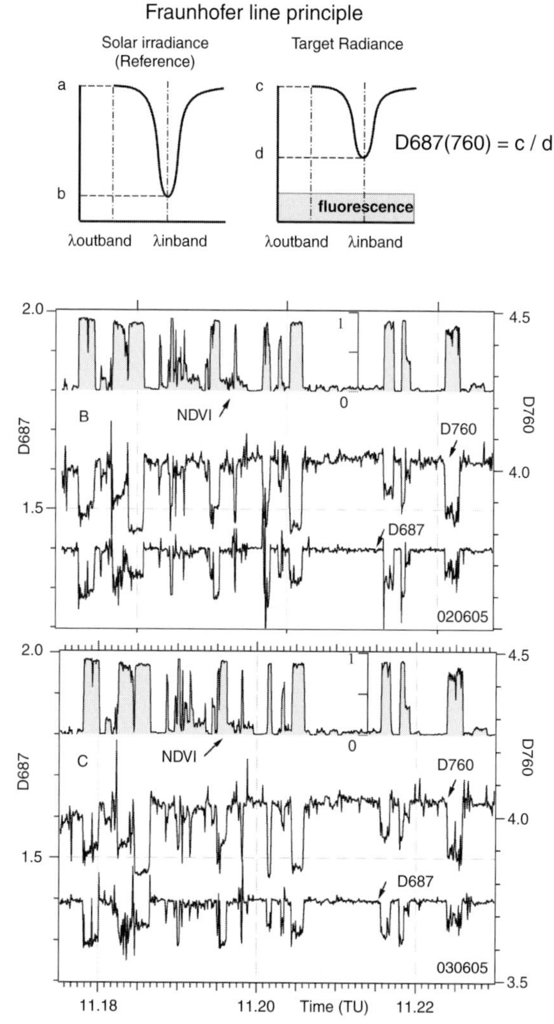

Fig. 15.6. **(A)** Fraunhofer-line principle. The method is based on the partial in-filling of the absorption band by the sun-excited fluorescence of the target (vegetation). **(B)** Airflex flights over Barrax at the altitude of 300 m at almost solar noon. 2 June 2005. Variations along the track of the NDVI index, depth at 687 nm (D687) and at 760 nm (D760). **(C)** 3 June 2005. Similar flight the day after. Note the excellent reproducibility of NDVI and depths. After Moya *et al.* (2006).

Oxygen-absorption bands of the terrestrial atmosphere have been recently used for fluorosensing. Moya *et al.* (1999) were the first to build a new instrument based on interference filters, for the determination of Chl-F using the in-filling of the atmospheric oxygen O_2A band. The width of this band (≈1 nm) and its spectral position (760 nm) makes it a better candidate for Chl-F detection than the Hα line (Moya *et al.*, 2004). Later, a more complete passive multi-wavelength fluorescence detector (PMFD) instrument was built, also based on interference filters. This instrument aimed at measuring fluorescence and reflectance at 760 nm and 687 nm, together with the PRI. It was successfully applied to monitor the photosynthetic-activity recovery of the boreal forest during the SIFLEX measuring campaign organised by ESA during spring 2002 (Louis *et al.*, 2005). The sensor at the top of a 20-m tall tower built in the forest was directed towards a target formed of several trees at a distance of 50 m. In this work the time series of the fluorescence yields ΦF687 stayed constant, whereas ΦF760 fluorescence slightly increased, but much less than the CO_2 assimilation that marked a step increase after a warmer and sunny period at the end of May (Fig. 15.7A). Importantly, a good correlation between the PRI variation and the net CO_2 assimilation was also found (Fig. 15.7B). In parallel a decrease of the carotenoid pool and especially of the lutein pigment was also observed during the spring recovery, in agreement with previous work (Ottander *et al.*, 1995).

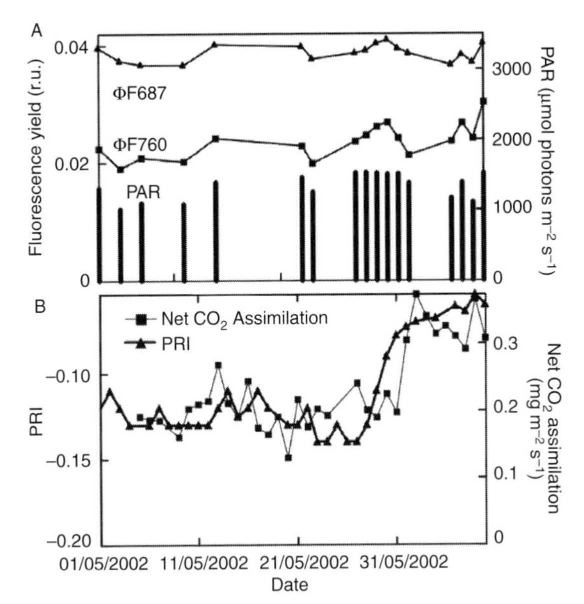

Fig. 15.7. **(A)** Time series of fluorescence yields and PAR integrated over 4 h around solar noon (14:00 local time). Both yields were correlated with PAR (except for the sunny days at the end of May). ΦF760 increased at the end of the campaign, whereas ΦF687 stayed fairly constant. **(B)** Parallel plots of the PRI-index and CO_2-uptake time series during the campaign. Observe the parallel increase at the end of May. After Louis *et al.* (2005). PAR, photosynthetically active radiation; PRI, photochemical reflectance index.

One may suppose PRI is sensitive to a long-term sustained quenching. The steep increase of PRI in June correlates with the relaxation of this NPQ, and thus with the increase of the Scots pine photosynthetic activity.

A further development within the interference filter-based instruments was 'AIRFLEX', an airborne sensor that measures fluorescence at 687 and 760 nm simultaneously (Moya *et al.*, 2006). AIRFLEX is basically a six-channel radiometer aimed to measure the 'in-filling' of the atmospheric O_2 bands. A set of three different interference filters was associated with each absorption band: one filter in-band and another two filters out-band, immediately before and after the O_2-absorption feature. AIRFLEX was operated in the frame of the Sent2flex campaign organised by the ESA during the summer of 2005. The sensor was fixed on the floor of a CESSNA Grand Caravan airplane of the German Aerospace Centre (DLR). AIRFLEX carried out several flights during June and July over cultivated fields in Barrax within La Mancha, Spain. The track flew over a succession of cultivated fields including alfalfa, sugarbeet and wheat. At the altitude of 300 m the target footprint diameter was ≈10 m. Adjacent bare fields were used as references. It is worth noting the good reproducibility of the data (≈2%) when repeating the flights on different days, as judged by the band depths (Fig. 15.6B and C). Band depth contains most of the fluorescence information, and could be used with very little processing as a signature of the vegetation. A deeper analysis consists in retrieving the actual fluorescence that should be measured at ground.

Before calculating fluorescence signals, radiance fluxes should be corrected for several effects including atmospheric transmission, scattering (including aerosol), ground elevation along the track and flight altitude. These corrections were evaluated using the Modtran4 atmospheric model (PcModWin 4.0 v1r1, Ontar Corporation, North Andover, MA – USA). It is out of the scope of this chapter to go deeper into the technical aspects of passive fluorescence retrieval. Additional information can be found in Meroni *et al.* (2006) and Guanter *et al.* (2010).

Figure 15.8 shows the retrieved fluorescence signals. In order to better discriminate green fields from bare fields, both fluorescence fluxes at 687 and 760 nm have been divided by the radiance out-band at 685 nm, L(685). This 'fluorescence fraction', FF687=F687/L(685), represents the amount of fluorescence in the vegetation-radiance continuum at the vicinity of the O_2B band. It can be seen that FF687 accounts for ≈10% of the vegetation-radiance continuum. The fluorescence fraction at 760 nm has been calculated in the same way: FF760=F760/L(685), and can be directly compared with FF687. One may observe the good correlation between FF and NDVI, which is not a trivial result as FF results mainly from the variation of the band depths. The radiometric calibration of AIRFLEX allowed calculating the fluorescence flux of green fields in absolute units at ground level: F687 ≈1–2 W $sr^{-1}m^{-2}\mu m^{-1}$ and F760 ≈3–6 W $sr^{-1}m^{-2}\mu m^{-1}$.

One striking result is the value of the ratio FF760/FF687≈2.5–3 measured over almost all fields, which is at least twice the value measured on the ground, on the same species at the leaf level. This makes fluorescence ratios measured at the canopy level (i.e., over a three-dimensional target) very different to the fluorescence ratios measured on an isolated single leaf. Two effects may contribute to explain this discrepancy: (1) the strong absorption of vegetation at 687 nm makes F687 primarily dependent on actual sunlight absorbed by leaves of the target in direct view of the sensor; and/or (2) the fluorescence at 760 nm is the sum of the emission resulting from direct absorption by

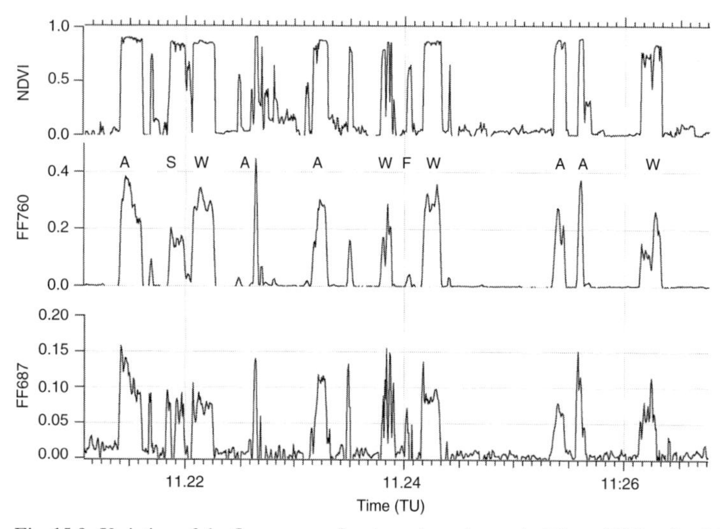

Fig. 15.8. Variation of the fluorescence fractions along the track, 1 June, 2005 at the altitude 300 m. One may observe, around 11:22, a similar NDVI value for sugarbeet and wheat fields, both with the same leaf-level chlorophyll concentration of about 350–400 mg.m^{-2}. FF687 is also similar for both fields. However FF760 is about twice for wheat compared with sugarbeet. This reproducible observation over all the flights is attributed to a different contribution of the canopy structure for the two fields (see text). A, alfalfa; F, fescue; S, sugarbeet; W, wheat.

the target plus the emission generated on other illuminated leaves not in view of the sensor and transmitted to leaves in direct view, as a result of the high reflectivity and high transmission of leaves at 760 nm. Also sunlight reflected by the ground may generate preferential fluorescence at F760 as predicted by canopy fluorescence models (Rosema *et al.*, 1991). In line with the hypothesis is a lower value of the FF760/FF687 ratio measured over a closely cropped grassland field.

Data collected during the AIRFLEX campaign in Barrax were processed by Daumard *et al.* (unpublished), who analysed for the first time the effect of changing the altitude from 300 to 600 m and 1200 and 3000 m on the retrieved fluorescence fluxes. Although the signal-to-noise ratio decreased with altitude, fluorescence was still detectable at 3000 m. At this altitude atmospheric corrections attained were upto 50% of those required for a top-of-the-atmosphere detection. These results are the first direct proof of quantitative Chl-F measurements from an airborne platform. They encourage us to believe in the technological possibility of the in-filling method applied in the atmospheric oxygen-absorption bands to detect vegetation fluorescence from a space-borne platform. However, the strong interference of the three-dimensional canopy structure with the actual fluorescence flux is a major concern for retrieving data, for example, the chl. content from fluorescence-ratio measurements as previously proposed by Lichtenthaler *et al.* (1990).

Finally it is worth noting the possibility to obtain images of Chl-F using a narrow-band multispectral camera as reported by Zarco-Tejada *et al.* (2009). Although the actual radiometric resolution of the camera (10 bits) was too low for an accurate encoding of the depth of the O_2A band, experiments were conducted on olive and peach orchards that demonstrated the usefulness of the approach. The unmanned aerial vehicle (helicopter) used in this work (Berni *et al.*, 2009) should also be mentioned as a cost-effective and efficient alternative to regular airplanes for airborne mezzo-scale vegetation remote sensing.

MEASUREMENTS WITH HIGH-RESOLUTION SPECTROMETERS

An alternative to the development of dedicated instruments based on interference filters consists of using small-footprint high-resolution fiber-optic spectrometers, nowadays available from several companies. The detectability of Chl-F has been considered by Liu *et al.* (2005) using the FLD method and an ASD FieldSpec NIR spectrometer with a spectral resolution of 3 nm. However the insufficient spectral resolution yielded quantitative Chl-F measurements of low quality.

Zarco-Tejada *et al.* (2000a,b) illustrated the Chl-F effects on leaf-level reflectance and transmittance, developing a FRT leaf-level radiative-transfer model. These authors also suggested that a double-peak feature at the 690–710 nm spectral region seen in the derivative of the apparent reflectance was possibly owing to the effect of fluorescence emission, and proposed the use of this parameter for vegetation-stress detection (Zarco-Tejada *et al.*, 2003b; Dobrowski *et al.*, 2005). In the work reported by Pérez-Priego *et al.* (2005), detection of water stress in orchard trees was conducted with a narrow-band spectrometer, enabling subnanometer spectral resolution. Spectral measurements of downwelling irradiance and upwelling crown radiance permitted the evaluation of the fluorescence *in-filling* effects in the O_2A band. Diurnal and seasonal measurements using a PAM fluorimeter showed consistently lower F_s and quantum yield in water-stressed trees. The authors showed that the chlorophyll fluorescence *in-filling* in the O_2A band was sensitive to diurnal variations of fluorescence and water stress, demonstrating the close relationships between F_s and the *in-filling* of the O_2A band at the crown level.

Simultaneous determination of Chl-F at 687 and 760 nm was presented in Meroni *et al.* (2006). These authors used a couple of high-resolution fiber-optic spectrometers featuring a full-width high maximum of 0.06 nm. DCMU addition to control bean plants induced a fourfold increase in both Chl-F at 687 and 760 nm, in agreement with previous measurements (Moya *et al.*, 2004). However these results were difficult to extrapolate to the canopy level because the measurement distance was only 3 cm.

At the canopy level quantitative measurements of F687 and F760 using the in-filling method were performed on developed (≈0.6 m height) *Holcus lanatus L.* natural grassland (Fournier *et al.*, 2012). They installed a set-up also based on a high-resolution fiber-optic spectrometer at the top of a scaffolding, which was measuring in a Nadir-viewing direction on a target of 1.1 m of diameter. A special feature was the use of three wavelengths outband to allow for a parabolic fit of the reflectance spectrum in the vicinity of the O_2B band. On a grassland canopy, a fluorescence ratio (F760/F687) ≈2.9 was obtained by passive measurements at solar noon. This value was compared with the ratio measured at the leaf level on samples issued from the same target with an active fluorimeter using the same light intensity. The F760/F687 at leaf level was ≈1.45, only 50% the value measured at canopy level. These results confirm previous airborne measurements with the AIRFLEX instrument (Moya *et al.*, 2006). This work illustrates that passive and active fluorescence measurements at the canopy level, although observing the same target and the same physical fluorescence phenomenon, may sense very different signals. Another run of simultaneous passive and active fluorescence measurements, confirming a good agreement between the two, has been performed during the CEFLES2 campaign (Rascher *et al.*, 2009). Remote sensing of sun-induced passive fluorescence has been recently shown to aid improving the modelling of diurnal courses of gross primary production (GPP) at the ecosystem level (Damm *et al.*, 2010, Rossini *et al.*, 2012). Overall, the results suggest that passive fluorosensing is a very promising tool to develop intensive and accurate remote monitoring of photosynthesis in the near future.

15.4.3. The need for vegetation-fluorescence models

A study was launched in October 2002 by the ESA to advance the underlying science of a possible future vegetation-fluorescence space mission by addressing the need for an integrated canopy fluorescence model. The objective of this study was to review and advance existing fluorescence models at the leaf level and to integrate these into canopy models in order to simulate the combined spectral reflected radiance and fluorescence-emission signals.

The resulting leaf fluorescence model (FluorMODleaf, Pedrós *et al.*, 2009) is based on a modified version of PROSPECT (Jacquemoud and Baret, 1990), which includes an elementary fluorescence-emission spectrum (Pedrós *et al.*, 2008) and allows a dependence on some physiological variables, e.g., canopy temperature and illumination light levels. The canopy model is based on the turbid medium SAIL model (FluorSAIL), modified to internally compute the illumination levels through the canopy, to couple with FluorMODleaf and through MODTRAN, to provide a means of simulating incoming irradiance at the top of the canopy as well as the up-welling radiance above the atmosphere at a spectral resolution of 1 nm in the 400–1000 nm wavelength range. Since the beginning of the project, a graphical user interface has been developed (FluorMODGUI V.3.0, Zarco-Tejada *et al.*, 2006) as a consistent means of linking leaf and canopy simulations. The current models and documentation can be accessed from a dedicated web page at http://www.ias.csic.es/fluormod, which provides the software and related publications developed under the FluorMOD project. More recently, a sophisticated model of soil-canopy spectral radiances, photosynthesis, fluorescence, temperature and energy balance has been presented by van der Tol *et al.* (2009).

15.4.4. New space-mission projects for photosynthesis remote sensing

Despite the fact of the several missions already dedicated to global terrestrial-vegetation monitoring, the derived information is mostly related to the amount of vegetation or to the *potential* photosynthetic activity. A remaining topic to be covered in global vegetation monitoring is the measurement of the *actual* photosynthesis from space.

A new satellite mission is under evaluation by the ESA. The FLuorescence EXplorer (FLEX) mission proposes to launch a satellite for the global monitoring of F_s in terrestrial vegetation. The main aim of the mission is global remote sensing of vegetation photosynthesis, which is an important component of the global carbon cycle and also closely linked to the hydrological cycle through transpiration.

For a global coverage, the optimal revisit time for a mission like FLEX would be in the range 5–10 days, taking into account the phenological development of vegetation and the variations in environmental conditions. In the particular case of carbon exchanges between surface and atmosphere, the combination of requirements provides an optimum spatial resolution in the range 100–300 m, enough to identify vegetated patches while resolving interactions at scales comparable with typical heights of boundary layer conditions. The orbit planned is sun synchronous, at 10.00, with swath width of 220 km. FLEX encompasses up to five instruments viewing the same area simultaneously.

(1) FIS (Fraunhofer imaging spectrometer) is the primary instrument. It covers the O_2A and O_2B absorption lines at 761 and 687 nm with 0.1 nm resolution to determine the fluorescence signal. A spectral interval of 20 nm will be measured over each window to guarantee improved retrievals, proper spectral calibration and enough interval for radiometric cross-calibration with the other instruments. The main challenge regarding the FIS concept comes from the fact that a quite large field-of-view will be required to guarantee global coverage, whereas high spectral resolution and spectral stability (absolute spectral position) are required all across the field of view (all along the ground swath), so that the spectral requirements become essential to guarantee successful results.
(2) VNIR (visible/NIR) spectrometer has a continuous spectral coverage from 450 to 1000 nm, at a spectral resolution of 5 nm. The main role of VIS is the normalisation of fluorescence measurements by the total amount of light absorbed by vegetation, in order to compute the fluorescence quantum efficiency, but VNIR will also provide key additional information like PRI or NDVI vegetation indices. Although the measurements from FIS are quite innovative, those provided by VNIR are already quite consolidated. Emphasis has also been put on providing the necessary capabilities for appropriate cloud screening and atmospheric correction procedures.
(3) SWIR (short wave IR) separates the bands of non-photosynthetic material of vegetation from the fraction of green-vegetation component that is responsible for vegetation photosynthesis. Six appropriated bands have been chosen from 1360 to 2220 nm for an improved cloud screening, which is essential for accurate fluorescence retrievals.
(4) The TIR (thermal imager) will provide TOA radiances in four bands in the thermal domain (8.5–12.5 μm) in order to allow estimation of foliage and soil temperatures, and to specifically compensate for atmospheric effects and spectral variations in surface emissivity.
(5) The ONI (off-Nadir imager) instrument will be used in conjunction with the VNIR spectrometer to estimate aerosol optical thickness (AOT). It has three spectral bands at 450, 660 and 1665 nm, and looks forward along-track under an observation zenith angle of 50 degrees to increase the sensitivity to aerosols.

The retrieval of vegetation fluorescence from measured spectral radiances at the top of the atmosphere requires a proper understanding of the signal and the different steps that the radiation follow from the incoming solar irradiance at the top of the atmosphere, the illumination levels at the surface, then the surface reflection and the associated fluorescence emission, and finally back through the atmosphere up to the satellite (Guanter *et al.* 2010).

15.5. THERMAL INDICES

All bodies with a temperature above absolute zero emit thermal radiation as a function of their temperature. At the temperatures encountered at the Earth's surface, such radiation is IR, therefore IR radiation can be used for remote sensing of a body's temperature.

The temperature of a leaf or canopy is highly dependent on the rate of transpiration. This is because the energy balance for a leaf is (Jones, 2004a):

$$\Phi_n + M - \lambda E - C = S \qquad \text{[Eqn. 15.3]}$$

where Φ_n is the net radiant-flux density absorbed by the leaf; M is the metabolic heat generated per A_{leaf}; λE is the rate of heat loss through the evaporation of water; C is the rate of heat loss by conduction and convection to the environment; and S is the rate of increase of heat content of the tissue. Alternatively:

$$C = g_{b,H}\, c_p\, (T_l - T_a) \quad \text{[Eqn. 15.4]}$$

$$E = g_{l,water\ vapour}\, (x_l - x_a) \quad \text{[Eqn. 15.5]}$$

where $g_{b,H}$ is the boundary layer conductance to heat transfer; c_p is the molar specific heat of air; T_l and T_a are leaf and air temperature, respectively; E is the rate of transpiration; $g_{l,water\ vapour}$ is the overall conductance to *water-vapour* loss (including stomatal and boundary layer components); and x_l and x_a are the mole fractions of water inside the leaf and in the surrounding air, respectively.

From the above equations, it is clear that leaf temperature, as determined by IR thermometry, can be used to estimate $g_{l,water\ vapour}$ provided that all other terms of the equations can be estimated. Hence, it is potentially an indicator of stomatal opening, which in turn is strongly correlated to photosynthesis. An advantage of this approach is that commercial systems for its measurement are common (Table 15.1). Accordingly, IR thermometry has been developed as a means of estimating stomatal conductance to develop irrigation-scheduling programmes (Jones, 1999; Jones *et al.*, 2002). It is important to stress the needs of knowing many variables in order to apply this model to estimate stomatal conductance. For instance, air-temperature and VPD have to be determined, and the net radiant-flux density absorbed by the leaf as well as the boundary layer estimated. Estimating the absorbed radiation is far from easy, and a method has been proposed to avoid the need of an estimate, based on measuring the temperature of a dry reference assumed to have similar aerodynamic and optical properties to the leaf (Jones, 1999; Leinonen *et al.*, 2006). The requirement for ancillary meteorological data can be further decreased by using the temperature of a wet in addition to a dry surface (Jones, 1999). In this case, the only environmental variables required are the air temperature and the boundary layer conductance.

Taking all this into account, Jones *et al.* (2002) and Grant *et al.* (2007) have shown reasonable estimates of whole-canopy g_s based on thermometric determinations. However, although T_l–T_a provides a straight way to assess relative variations of stomatal conductance, the absolute values estimated this way vary depending on the model used to estimate the required meteorological parameters or to adjust the parameters of the energy balance (Leinonen *et al.*, 2006). Work is in progress to improve g_s estimates from IR thermometry.

Meanwhile, Chaerle *et al.* (2007a) have recently proposed combining thermal and Chl-F imaging as a powerful system to study photosynthesis. Among the applications proposed are diagnosis of pre-symptomatic responses to biotic and abiotic stresses, screening and detection for stomatal or photosynthetic mutants or remotely assessing leaf WUE. Recently, Sepulcre-Cantó *et al.* (2007) have demonstrated the feasibility of measuring T_l–T_a in olive and peach orchards from airborne platforms using the AHS sensor (Table 15.2). Moreover, they have simulated the possibility of extrapolating these methods to the ASTER satellite (Table 15.2), reaching the conclusion that, in principle, it should be possible to use it for global monitoring of open tree canopies.

15.6. CONCLUDING REMARKS

Knowledge on the functioning of vegetation is a prerequisite to the understanding of the role of terrestrial vegetation in sequestering atmospheric carbon. Satellite measurements appear to be the best way to obtain the necessary repetition and the spatial coverage for the estimation of vegetation functioning at the regional or global level. As described in the above sections, satellite determination of reflectance-based indices has already started, particularly in the present decade. Passive fluorosensing is also based on reflectance, and therefore its satellite detection may be also possible. Here we summarise the most important current constraints and highlight future needs to standardise satellite remote sensing of photosynthesis.

Sun-induced vegetation-fluorescence detection using the in-filling method was first proposed by Plascyk, 1975 and discussed by Stoll *et al.* (1999) using the Hα Fraunhofer line. Nowadays, passive chl. fluorescence detection from space is principally envisaged in the atmospheric O_2-absorption bands (Drush *et al.*, 2008). However, a new step has been made thanks to the high resolution (0.022 nm) Fourier transform spectrometer on board the Japanese satellite GOSAT that provides the first observations of global and seasonal terrestrial chl. flourescence from space using the K I Fraunhofer line near 770 nm (Joiner *et al.*, 2011; Frankenberg *et al.*, 2011).

To advance in the knowledge of sun-induced vegetation fluorescence, there is an urgent need for novel dedicated instruments encompassing all the crucial channels to extract not only fluorescence but also the relevant optical information on vegetation (reflectance, PRI, temperature, etc). It is interesting to mention, as an example, the crane-based set-up on the jib of which several sensors can be fitted at 22 m. This crane is installed at the INRA agricultural station in Avignon (France) and can move on a 120 m railway in the middle of cultivated fields (Daumard *et al.*, 2010.). In addition to developing optical methods, the platform has several specific objectives including addressing issues of sampling scale, comparing optical sampling methods, relating optical signals to CO_2 fluxes and exploring spatial and temporal patterns. Extrapolating recent improvements of airborne passive-fluorescence measurements, we may expect in the near future a great development for novel low-altitude unmanned platforms that will extend the possibilities of tower-based instruments. Such development is highly recommendable and should encompass, in addition to the fluorescence parameters, all the other optical indexes related to the vegetation status. But can these measurements be extended at TOA? To answer the question we evaluate the impact of atmospheric effects in the retrieved signal for an instrument having the optical specification of AIRFLEX.

These effects may be grouped into four main types: (1) attenuation of the signal; (2) albedo of the air column; (3) scattering by the surrounding environment; and (4) line filling. The first three phenomena are relatively well known and several models have been developed for their correction, in particular MODTRAN, that accounts for multiple scattering and has sufficient spectral resolution. Line filling has several origins, including Rayleigh-Brillouin scattering, Raman scattering and aerosol fluorescence. These effects are described in the literature as the 'ring effect' (Grainger and Ring, 1962). The strongest contribution is assigned to rotational Raman scattering by nitrogen and oxygen molecules. The *in-filling* of Fraunhofer lines by the above-described effects can be modelled (Vountas *et al.*, 1998; Sioris and Evans, 1999). Still, it is important to consider a proper model, as raw spectral radiance data as measured at TOA differ substantially from radiance measured on the ground (Guanter *et al.*, 2010).

For the example of the AIRFLEX sensor chosen here, fluxes may increase (80%) or decrease (50%) compared with ground radiance, depending on whether the contribution of the air-column radiance dominates (O_2B band) or the absorption dominates (O_2-A band), principally by the atmospheric oxygen. In addition to the effect on the solar radiances, the atmosphere also attenuates the fluorescence flux by a factor of approximately five (O_2A) or two (O_2B). Taking into account the uncertainty of the calculated corrections and the fluorescence attenuation, the fluorescence signal-to-noise ratio at TOA should be decreased by a factor of three to four when compared with ground measurements with the same instrument (Daumard, 2010). In view of the recent progress in airborne passive-fluorescence remote sensing, one can close positively for the time being the question of the feasibility of vegetation-fluorescence detection from TOA, although it should be a real technical challenge.

The question arises to optimise the design of the sensor in orbit in order to mitigate the decreased signal-to-noise ratio at TOA. Among the great number of parameters of the FLEX project still under debate, there is one of crucial importance: the optimum spectral resolution required to optimise the signal-to-noise ratio. This question has been examined by Goulas *et al.* (unpublished) in the case of a detector having a photon-limited noise figure (noise=√(number of photons)), who found an optimum of about 1 nm in case of the O_2A band and about 0.3 nm for the O_2B band.

Current satellite detection systems present a series of constraints that can be summarised as lack of appropriate wavelengths and broadness of wavelength windows. Concerning the appropriate wavelengths, none of the currently used platforms (Table 15.2) presents a band where the classical PRI can be measured. Therefore, only PRI proxies can be measured using MODIS (Rahman *et al.*, 2004; Drolet *et al.*, 2005, 2008). Therefore, newly designed satellite platforms should be implemented to contain 10 nm bands centered at 531 nm and 570 nm, respectively (Grace *et al.*, 2007), and to use the proposed algorithms for net primary production estimates (Rahman *et al.*, 2004; Fuentes *et al.*, 2006; Sims *et al.*, 2006). It is worth noting that FLEX not only addresses the question of vegetation-fluorescence flux retrieval (FIS sensor) but proposes other important sensors devoted to fulfil this goal.

Part III
Photosynthetic response to single environmental factors

16 • Photosynthetic responses to radiation

F. VALLADARES, J.I. GARCÍA-PLAZAOLA, F. MORALES AND Ü. NIINEMETS

16.1. RADIATION AND PLANTS: AN INTRODUCTION TO PHOTOBIOLOGY

16.1.1. Basic characteristics of solar radiation

Among the factors affecting plants, solar radiation is perhaps the most heterogeneous in space and time. Important parts of solar radiation provide energy for photosynthesis and serve as signals in photoregulation of plant growth and development. The sun radiates energy in the spectral range from 280 to 4000 nm, with a maximum in the blue-green (480 nm; Fig. 16.1). Within the PAR, solar radiation peaks at ca. 590 nm (Fig. 16.1). Solar radiation can be segregated into direct solar radiation and diffuse sky radiation, which reaches the ground after multiple scattering on atmospheric particles and clouds, reflection from the ground surface and additional scattering in the atmosphere (Ross, 1981).

The widespread, albeit vague, term light is used for the portion of the electromagnetic spectrum in the vicinity of visible light (Kohen *et al*., 1995). Many past ecological and physiological studies were based on measurements that represent the stimulation of the human eye by radiant energy, a measure called illuminance and expressed in foot-candles (English system) or luxes (metric system). The human eye is most sensitive in the green spectral region, centered around 550 nm, whereas any quanta in the spectral region of 400–700 nm have enough energy to drive photosynthesis, so illuminance is obviously not well suited for plant science.

With further developments in understanding the way solar radiation drives photosynthesis, PAR has been used extensively in plant biology and ecology. Roughly half of the solar radiation is in the region of PAR and the rest in the NIR radiation (Fig. 16.1). PAR in W m^{-2} is measured by radiation sensors (pyranometers) equipped with specific filters to remove ultraviolet (UV) and IR spectral parts (Pearcy, 1989). In most countries, PAR is defined as radiation in the spectral interval 400–700 nm, whereas in the former Soviet Union and socialist countries, PAR was defined as radiation between 380–710 nm (Ross and Sulev, 2000). This discrepancy leads to measurements 5–7% higher when the extended interval is considered instead of the most common one.

However, the amount of quanta that ultimately drives photosynthesis can be different for the same PAR values depending on the spectrum of light, e.g., diffuse solar irradiance enriched by blue quanta with higher energy contains fewer quanta at a given PAR than direct solar irradiance, with more-uniform spectral distribution. At low sun elevations, light is enriched by orange, red and FR photons (end-of-day effect) with lower energy, and accordingly there are more quanta at a given PAR. Different light sources used in plant science also have different quantum/energy conversions. Therefore, photosynthetically active quantum-flux density (PPFD, μmol m^{-2} s^{-1}) is currently the most widely used quantity to characterise plant-light availability. The most reliable but at the same time costly way of estimating PPFD is by measuring the radiation with a spectroradiometer and integrating spectral irradiance over the region 400–700 nm. The most common estimations of PPFD are conducted by quantum sensors based on silicon photodiodes, equipped with specific filters to remove IR- and UV-spectrum parts (Pontailler, 1990).

Solar radiation reaches the top of the Earth atmosphere at a rate of 1396 W m^{-2}, the so-called solar constant. The solar 'constant' varies about 5% depending on the distance between the sun and the Earth, and on the activity of the sun. On cloudless days, only 800 to 1200 W m^{-2} of total radiation reach the Earth's surface owing to radiation absorption by the atmosphere. The light environment in a given site is then a consequence of the latitude of the site, time of the day, neighbouring plants and surrounding objects casting shade, and of a number of properties of the low and variable

Terrestrial Photosynthesis in a Changing Environment: A Molecular, Physiological and Ecological Approach, ed. J. Flexas, F. Loreto and H. Medrano. Published by Cambridge University Press.

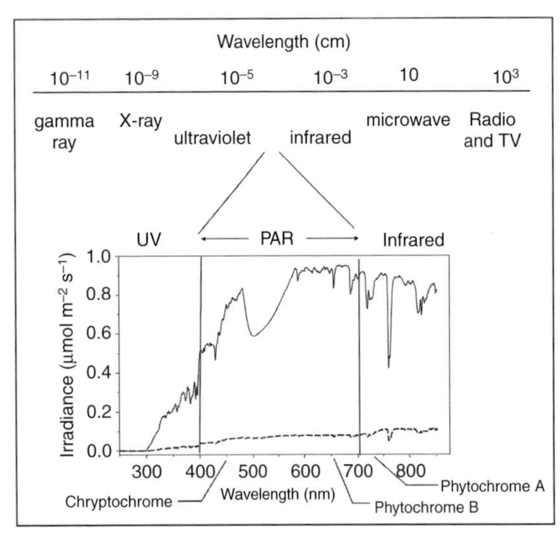

Fig. 16.1. Electromagnetic spectrum of solar irradiance with detailed view of the region relevant for photosynthesis and plant performance (UV, ultraviolet; PAR, photosynthetically active radiation; and infrared). The detailed sunlight spectrum for the 300–800 nm region corresponds to a typical clear summer day at 40 degrees latitude both in the open (continuous line) and in the understorey of an evergreen rainforest in Chile (Puyehue, 40° 39′S, 72° 11′W; dashed line). Plants are particularly sensitive to irradiance of certain wavelengths owing to specific phytochromes and cryptochromes (main photoreceptors indicated in the graph, see text for further details).

atmosphere, mainly cloudiness and transparency. All these factors affect the radiation spectra, the intensity and direction of beam radiation, the distribution of the radiation between beam and diffuse components and the duration of the photoperiod. The amount of PPFD in total solar radiation is a highly variable quantity. For instance, idealised conversion coefficients (PPFD/Total solar radiation, μmol/J) of 1.814 for global, 1.758 for direct, 2.127 for diffuse and 0.462 for reflected solar radiation have been proposed (Ross and Sulev, 2000), but these coefficients differ under different environmental conditions.

Many of the factors altering the amount of total solar radiation that drives leaf energy balance and evaporation, and PPFD that drives photosynthesis have been considered in a number of models to estimate the radiation below and/or reflected by a canopy, but theoretical estimations and actual measurements often do not coincide well (see references below). For instance, the sun disc is completely blocked by an object at a theoretical distance of 108 times the object diameter, but empirical data show that this distance is reduced to 50–70 times the diameter of the object. When direct sunlight is blocked by clouds or by a canopy, light reaching a plant comes from the entire sky hemisphere (diffuse light); in such conditions, the shadow of an object vanishes at a distance equal to its diameter (Horn, 1971; Valladares and Niinemets, 2007). Directionality or the lack of directionality and the so-called penumbral effects are important sources for discrepancies between the theory and the reality of a particular light environment (Oker-Blom, 1984; Smith *et al.*, 1989). In particular, it is very hard to model stochastic events, such as changes in beam irradiance, owing to movement of clouds and movement of canopy elements, e.g., as a result of wind. Several simplifications used in modelling diffuse irradiance, such as uniformly overcast sky conditions or standard overcast sky conditions (the latter accounting for brighter sky towards the sun) are very crude simplifications (Hutchison *et al.*, 1980; Spitters *et al.*, 1986). Nevertheless, long-term light estimates, for instance for the entire growing season, can generally be estimated rather well even with crude assumptions (e.g. Roderick, 1999).

16.1.2. Solar radiation within vegetation

RADIATION QUANTITY

One of the most important factors affecting the light environment is the presence of plant canopy that harvests the light, but also modifies the intensity and spectral quality of the penetrating radiation. Accurate description of the light environment in a forest understorey is more complex than in the open, not only because of its intrinsic spatial and temporal heterogeneity, but also because of forest-type specific structural features that alter the correlations between light environment and aggregated-canopy characteristics, such as canopy height, density and LAI (the amount of leaf area per unit ground area). PAR in the understory ranges from 50–80% of full sunlight under leafless deciduous trees to 10–15% in even-aged pine stands, 2.5% in closed spruce canopies, 0.2–0.4% in dense beech forests and even less than 0.1% in certain tropical rainforests (Barnes *et al.*, 1998). These values vary in dependence on factors such as wind speed and cloudiness. On windy days, more beam irradiance can penetrate into deeper canopy horizons (Tong and Hipps, 1996), whereas on cloudy or overcast days, the relative percent of radiation transmitted by forest canopies is higher than on clear days (Endler, 1993; Federer and Tanner, 1966). It is well known that the percentage of sunlight available in a forest canopy decreases exponentially as the cumulative LAI (LAI_{cum}) increases (e.g., Baldocchi and Collineau, 1994; Cescatti and Niinemets, 2004). Initially, simple exponential relationships following the Lambert-Beer law were proposed

(Monsi and Saeki, 1953; now an English translation of this classical paper is available Monsi and Saeki, 2005) describing the relationship between canopy to-incident radiation (Q_0) and any level inside the canopy, Q, as $Q=Q_0\exp(-k\text{LAI}_{\text{cum}})$, where k is the extinction coefficient depending on solar angle and leaf angular distribution. This equation assumes that foliage is randomly dispersed. However, foliage in natural communities is often clumped, intercepting less light for any given LAI_{cum} than the foliage with random dispersion; the foliage may also be regularly dispersed, intercepting more light than the foliage with random dispersion (Cescatti and Niinemets, 2004). Parameterisation of such differences requires at least one more parameter. In addition, such one-dimensional models often fail in more complex heterogeneous canopies, where three-dimensional models provide a much better but still stochastic description of canopy light environment (Cescatti and Niinemets, 2004). Ray tracing models, describing the location of every foliage element in space and providing the best correspondence between actual and simulated light environment (North, 1996; Pearcy and Yang, 1996) have won popularity, especially for studies of functional significance of plant form (Falster and Westoby, 2003; Valladares and Pearcy, 2000). However, such models require huge (and often impractical) effort to parameterise at stand scale (Niinemets and Anten, 2009), although more-efficient parameterisation routines are currently under development (Casella and Sinoquet, 2003; Cescatti and Niinemets, 2004).

RADIATION QUALITY

Light transmitted through a canopy experiences remarkable spectral alterations owing to enhanced capture of photons within the PAR region, and becomes enriched by radiation in green wavelengths and NIR, where leaves absorb radiation inefficiently. Although overall PAR transmittance in a Costa Rican rainforest at solar elevation near the zenith was as low as 0.5%, transmittance in the FR or NIR (ca. 730 nm) was about 4% (Lee, 1987). Spectral changes in transmitted light depend not only on canopy LAI, but also on spectral quality of incident radiation, for instance on cloudiness. On cloudy days, spectral changes in light transmitted through the canopy are minimised (Federer and Tanner, 1966; Barnes *et al.*, 1998). In addition, various species alter the light gradient to a different degree. For instance, conifers with thick foliage elements that are essentially optically black in PAR and NIR, and that are also strongly aggregated into shoots with a relatively large gap fraction inside the canopy, do not modify the spectrum to the same extent as deciduous species (Federer and Tanner, 1966; Barnes *et al.*, 1998).

According to the colour of light, a total of five basic light environments have been described in terrestrial ecosystems (Endler, 1993; Kiltie, 1993): (1) forest shade characterised by greenish or yellow-greenish light owing to selective absorption of red and blue by vegetation; (2) woodland shade with bluish or bluish-grey light owing to the dominance of the radiation from the sky; (3) small gaps characterised by yellowish-reddish light owing to direct sunlight; (4) large gaps, open or any habitat under cloudy conditions where the light is whitish owing to the combination of sun and sky light, or because of the dominance of the white light radiating from clouds; and (5) any habitat early and late in the day, when the sun is below 10° from the horizon and light is mainly orange, red, FR and, under certain atmospheric conditions, purple.

Although visible light, more specifically PAR, is the most relevant portion of the electromagnetic spectrum for plants, radiation of shorter UV or longer FR wavelengths is important in a variety of signalling responses. Plants have three main photoreceptors: cryptochromes sensitive to blue-light and UV-A (450 nm and <400 nm), phytochrome B (mainly in green tissues), sensitive to red and FR light (660 and 730 nm, respectively) with reversibility, and sensitive to very high irradiance of FR, and phytochrome A (mainly in de-etiolation processes) sensitive to a wide range of wavelengths including FR (very low fluence rate reaction) and red-FR reversible reaction (germination) (Whitelam, 1995; Lin, 2000). Briggs and Olney (2001) have further shown that up to nine photoreceptors (five phytochromes, two cryptochromes, and one phototropin, plus one superchrome specific to ferns) are known in plants. Phototropins are important for chloroplast movements, stomatal opening, leaf unrolling, etc. Studies with mutants have revealed complex interactions between these photoreceptors, interactions that buffer differences in spectral composition and intensity of light providing informational homeostasis (Mazzella and Casal, 2001). Unlike phytochromes, plants have various blue-light-UVA receptors that appear to be derived from more than one evolutionary lineage (Lin, 2000).

16.2. PHOTOSYNTHETIC RESPONSES TO LIGHT

16.2.1. The basics

Light responses of photosynthesis depend on conversion of photons into chemical energy, i.e., formation of reductive and energy equivalents NADPH and ATP, and use of this energy to fix CO_2. Photochemical reactions are typically studied by Chl-F techniques, whereas gas-exchange methods are used to investigate CO_2 fixation. However, caution must be exerted with photosynthetic estimations from the former technique because the fluorescence signal is a mixture of contributions from different depths and layers within the mesophyll of the leaf (Evans, 2009; Oguchi *et al.*, 2011; see below on light gradients within the leaf).

During the photosynthetic process, sunlight is gathered by large arrays of light-harvesting pigment-protein complexes (LHC) and transferred to the reaction centres of PSI and PSII, where charge separation and stabilisation take place (see Chapter 2). Chl-F methods are based on re-emission of light as fluorescence from the red region of the spectrum (from 660 to 800 nm) owing to chl. *a*, mostly from PSII (see Chapter 10). Monitoring the yield of fluorescence makes it possible to estimate the fractions of absorbed light energy used for photosynthesis (photochemistry) and dissipated thermally as heat. There is always a balance between three possible routes for the absorbed light energy: photochemistry, heat dissipation and Chl-F. These three processes must cover 100% of the energy absorbed by the LHC of the photosynthetic apparatus.

In a fully functional photosynthetic system under low light (in dark-adapted state), most of the absorbed light energy is used for photochemistry, and only a minor fraction is thermally dissipated or re-emitted as fluorescence (Papageorgiou, 1975). However, all leaves have a certain finite capacity for photochemistry. As PPFD increases, and increasingly more PSII centres become closed, the fraction of light energy dissipated as heat and re-radiated as fluorescence increases. Under intense sunlight, more than 50% of the absorbed irradiance can be thermally dissipated (see Section 16.4 on excessive light and photoprotection). When the capability of the photosynthetic apparatus to convert the absorbed light energy photochemically is further impaired by any stress factor, the fraction of energy that is thermally dissipated can further increase to more than 90% (Demmig-Adams and Adams, 2006).

Over the past decades, the study of several parameters of the Chl-F emission together with gas-exchange measurements has become a rapid, sensitive and non-destructive method to investigate numerous aspects of the photosynthetic functioning (see Chapter 10), including the estimation of the photosynthetic electron transport rate (Genty *et al.*, 1989; Krall and Edwards, 1992; Evans, 2009). One of the most classical approaches is the measurement of gas exchange (CO_2 fixation or O_2 evolution) and PET as a function of light intensity (Genty *et al.*, 1989; Morales *et al.*, 1991, 1998, 2006).

16.2.2. Electron transport and gas-exchange rates as a function of light

The leaf-photosynthetic rate increases curvilinearly with increasing PPFD exhibiting saturation at intermediate to high light intensities. This response can be most commonly described by a non-rectangular hyperbola (e.g., Leith and Reynolds, 1987) that has three characteristic regions (Fig. 16.2): the initial slope (maximum quantum yield) where photosynthesis increases linearly with light; the curvature region, where photosynthesis starts to saturate with light; and maximum rate (maximum CO_2-fixation or O_2-evolution rate), where photosynthesis reaches an apparent plateau (Fig. 16.2). Measurements are made at a relatively low and narrow PPFD range, when the aim is to determine quantum yield, or increasing PPFD stepwise up to full sunlight, when the aim is to get more complete information on the light response curve of photosynthesis.

The quantum yield of photosynthesis (ϕ_{CO2}) is a measure of the efficiency of the photosynthetic process expressed in moles of photons absorbed within the PAR region per mol of CO_2 fixed or O_2 evolved (Ehleringer and Pearcy, 1983). The maximum quantum yield is measured when photosynthesis is light-limited (Fig. 16.2). The maximum quantum yields are calculated by linear regression from the initial part of the net photosynthesis versus light response curve, typically using a PPFD range from 20–30 up to 100–120 μmol photons m^{-2} s^{-1} (Morales *et al.*, 1991; Fig. 16.2). Quantum yields measured on an incident-light basis (called apparent quantum yields) are further corrected for leaf absorptance. In herbaceous species with thin leaves, leaf absorptance is ca. 80–85% of the incident light (Björkman and Demmig, 1987), but can be more than 90% for thick-leaved sclerophylls and conifers (Mesarch *et al.*, 1999; Niinemets *et al.*, 2005a,b).

The theoretical value of maximum quantum yield under CO_2-saturated conditions (in the absence of photorespiration) is 0.125 mol mol^{-1}, implying that 8 moles of photons are required to reduce 1 mole of CO_2 (Bolton and Hall, 1991). In general, the maximum value is closer to 0.112 owing to cyclic photophosphorylation, and it does not apply to C_4 species owing to the higher energy requirements of their CO_2-concentrating mechanism (Long *et al.*, 1993). Given also leaf absorptances of PPFD of the order of 0.85,

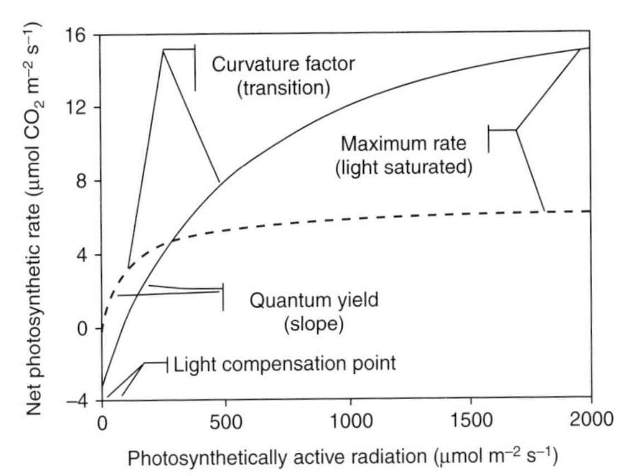

Fig. 16.2. Idealised net photosynthetic responses to photosynthetically active quantum-flux density (PPFD) of plant leaves acclimated to high light (continuous line) and low light (dashed line). Key parts of the hyperbolic relationship of photosynthesis are also shown: initial slope (quantum yield), curvature factor (transition from light-limited to light-saturated photosynthetic rate) and the asymptote (maximum photosynthetic rate). Light compensation point is the PPFD value at which net assimilation rate is zero (gross photosynthesis rate equals the respiration rate). Dark respiration rate is the rate of gas exchange at zero PPFD.

and other losses within the leaf (not all PPFD absorbed is absorbed by photosynthetic pigments), in practice maximum conversion efficiency of incident light is ca. 8–9% (Bolton and Hall, 1991; Ehleringer and Björkman, 1977; Ehleringer and Pearcy, 1983). In production-biology studies, the efficiencies are frequently calculated considering total solar radiation, which leads to efficiencies of ca. 1.8 times less, i.e., to maximum efficiency of ca. 4.5% for total solar radiation.

Quantum yields calculated from O_2 evolution might be higher than those calculated from CO_2 fixation because oxygen is evolved earlier and thus it is affected by fewer processes than CO_2 assimilation. Use of quantum yields calculated from CO_2 and methodological problems, such as including data extending beyond the linear region of the photosynthetic light response in quantum yield calculations, or changing levels of intercellular CO_2 concentrations during the measurements, can lead to underestimations of the maximum energy conversion efficiency. Strong correlations between the quantum yields and maximum potential PSII efficiencies, F_v/F_m ratios obtained by Chl-F in dark-adapted leaves, have often been found (Björkman and Demmig, 1987; Ögren, 1988; Demmig-Adams *et al.*, 1989). Measuring Chl-F to estimate maximum-potential PSII efficiency is less time consuming than measuring quantum yields, as a single shot of strong light of 1–2 s is required in the former, and generally at least six gas-exchange measurements at different light intensities (several minutes each) are required in the latter (Morales *et al.*, 1991). This is why in most ecophysiology laboratories Chl-F has replaced gas exchange as the default method to estimate maximum-potential photosynthetic efficiency (Adams *et al.*, 1990a).

Beyond initial slope, photosynthesis starts to saturate as soon as the photosynthesis in the first cell is light saturated. Owing to the light gradients within the leaves, mesophyll cells in different leaf layers are saturated at different quantum-flux densities (see Section 16.2.3) leading to the curvature of the light response curve of the leaf (Leverenz, 1988; Evans *et al.*, 1993; Ögren and Evans, 1993). In fact, there is often no apparent saturation of photosynthesis up to full sunlight in optically thick leaves, such as high-light-acclimated foliage (Fig. 16.2). For practical purposes, PPFD for 95% of theoretical maximum photosynthetic rate (estimated by fitting an empirical photosynthesis versus light-response-curve model to the data) is used as saturating irradiance in such cases (e.g., Niinemets and Tenhunen, 1997). Plants with aggregated foliage such as conifers constitute a challenge for measurements of photochemical yields and curvature of the light response. Owing to shading between foliage elements, awkwardly low quantum yields and curvature values can be obtained using standard measurement techniques suitable for broad-leaved species, commonly involving unilateral illumination with beam irradiance. For such species, correct estimates of photosynthetic light response can be obtained when the shoots are measured in the diffuse radiation field by placing the shoots into an integrated sphere (Öquist *et al.*, 1978; Leverenz, 1988).

Both gas-exchange and fluorescence methods can be used to study the responses of photosynthesis to light at curvature and saturating parts of the light response curve. Genty *et al.* (1989) found over a broad PPFD range a linear relationship between the PSII efficiency in light-adapted leaves, ϕ_{PSII} and the quantum efficiency of CO_2 assimilation under the same conditions, ϕ_{CO2}. This finding has been confirmed and extended by several authors for different plant species and for a variety of physiological conditions (see references in Morales *et al.*, 1998). However, different factors, such as water stress, SO_2 fumigation, cold stress and iron deficiency (Adams *et al.*, 1989, 1990a; Morales *et al.*, 1998), can decrease more ϕ_{CO2} (or ϕ_{O2}) than ϕ_{PSII}. Commercial portable instruments are available to measure Chl-F and CO_2

fixation simultaneously. They are tools to estimate photosynthetic electron transport rate (Krall and Edwards, 1992; see Evans, 2009 for an analysis of errors on assessing electron transport rate), quantify different electron-consuming processes (photosynthesis, respiration, photorespiration, etc.) (Valentini *et al.*, 1995a; Medrano *et al.*, 2002a) and identify any possible stress-mediated imbalance between electron generation and consumption (see Morales *et al.*, 2006, and references therein).

16.2.3. Light induction of photosynthesis

Upon illumination of a dark-adapted leaf, photosynthesis increases progressively for several minutes until reaching a steady state level, a process called photosynthetic induction. This process is important as photosynthetic responses to light depend on the degree of photosynthetic induction. Fully induced leaves respond immediately to any change in incident quantum-flux density, whereas induction significantly slows down the responses of photosynthesis to light, reducing the potential quantum-use efficiency. This plays a particularly important role in daily carbon gain in the understory, where leaves exposed to overall low diffuse irradiance sustain periods with high PPFD (lightflecks), and thus leaf photosynthetic apparatus undergoes repeated induction and deinduction cycles (Pearcy *et al.*, 1997).

During the induction, photosynthetic apparatus passes through different transitory stages. Electron transport reactions of photosynthesis become induced relatively rapidly. Their induction is reflected in changes in Chl-F yield (see Chapter 10) and lifetime (Morales *et al.*, 1999, 2001; Moise and Moya, 2004a,b). Upon sudden illumination of dark-adapted leaves by actinic light, Chl-F intensity increases up to six times (within a (few) second(s)) and then decreases to a stationary level (within a few min. after illumination). This transient Chl-F induction, known as 'Kautsky effect', reflects the PSII photochemical activity (Krause and Weis, 1991). The rapid kinetics of dark-adapted leaves has been shown to be an excellent tool to estimate the maximum potential PSII efficiency through F_v/F_m ratio (Morales *et al.*, 1991). Moise and Moya (2004a,b) have measured Chl-F yield and lifetime changes during the 'Kautsky effect' using phase fluorometry, and demonstrated that conformational changes in PSII and associated pigment-binding complexes occur during the dark-to-light transition.

The slower kinetics of electron transport, up to a few minutes, are reflected in reduction of Chl-F down to a steady state (F_s) level, owing to: (1) the partial re-oxidation of Q_A^- (photochemical quenching increases); and (2) the development of non-photochemical mechanisms (dissipation of excess excitation energy) (Buschmann, 1999). This is considered an essential regulatory mechanism of the photosynthetic control (Foyer *et al.*, 1990). Engagement of NPQ results in safe dissipation of absorbed light energy not used for photochemistry.

Simultaneously with commencing electron transport activities, Calvin-cycle enzymes, in particular Rubisco, ribulose-1-phosphate kinase etc. become activated (Leegood, 1990; Sassenrath-Cole *et al.*, 1994). Rubisco commonly exhibits the slowest rate of activation, and typically the Calvin cycle reaches full induction state within 5–10 min. after illumination (Leegood, 1990; Sassenrath-Cole *et al.*, 1994). The slowest process affecting photosynthetic light induction is commonly stomatal opening (Tinoco-Ojanguren and Pearcy, 1992, 1993a), where non-stressed plants take 5–10 min., but in water-stressed plants and under low humidity can take more than 30 min. (Tinoco-Ojanguren and Pearcy, 1993a; Aasamaa *et al.*, 2002). PET and carbon fixation are tightly coupled during photosynthetic induction, and low CO_2 concentrations owing to closed stomata reduce both the activity of PET (Ott *et al.*, 1999) (while enhancing energy dissipation) and Rubisco (Sage *et al.*, 2002). With the application of the saturation pulse method using modulated (PAM) fluorimeters (Ögren and Baker, 1985; Schreiber *et al.*, 1986), the relative importance of photochemistry and non-photochemical processes can be routinely determined at any time during the photosynthetic induction (see Chapter 10 for formulae and further explanations).

16.2.4. Gradients of light and photosynthesis within the leaf

Plant leaves have a complex three-dimensional architecture characterised by multiple cell layers consisting of cells with various sizes and shapes, differently absorbing and scattering light within the leaves and collectively leading to complex within-leaf light gradients. This importance of leaf optical properties and features of the internal light microenvironment have often been discussed (Fukshansky, 1981; Vogelmann, 1989; Sun *et al.*, 1996; Nishio, 2000). Leaf internal architecture and distribution of pigments determine integrated leaf optical characteristics such as absorptance, reflectance and transmittance, but they can also importantly modify the responses of photosynthesis to light with differing intensity, spectral composition and directionality. Why leaves are green has been an intriguing question for a long time, and the steep gradients of light within a leaf could provide a possible evolutionary explanation, because green light penetrates further into the leaf than red or blue light. By using chl.s that absorb green light weakly, a green leaf can achieve two conflicting requirements to maximise photosynthesis of the entire leaf: to increase radiation absorptance for photosynthesis and to maximise photosynthesis over all the chloroplasts of the leaf (Terashima *et al.*, 2009).

In leaves lacking specialised anatomical features such as high waxiness, pubescence etc., total pigment content is the main determinant of bulk-leaf optical traits (Evans and Poorter, 2001). Nevertheless, leaf structure and anatomy influence the capture and internal processing of absorbed light (Vogelmann, 1989, 1993; Vogelmann *et al.*, 1996). Depending on leaf optical properties and light spectral quality, the shape of the light gradients within a leaf can be relatively steep or gradual, exponential or linear. Leaf epidermal cells with thickened walls can act as lenses, focusing light into leaf-interior mesophyll cells (Poulson and Vogelmann, 1990). Sclereides and leaf bundle-sheath extensions can also act as optical fibres guiding light into the deeper interior layers (Karabourniotis *et al.*, 1994; Nikolopoulos *et al.*, 2002). In addition, long palisade cells can also canalise light into lower mesophyll layers enhancing photosynthesis (Vogelmann *et al.*, 1996; Smith *et al.*, 1997). Recent work has shown that at a leaf level, direct irradiance may have higher quantum efficiency than diffuse irradiance (Brodersen *et al.*, 2008). This has been explained by canalisation of beam irradiance into the leaves by palisade cells (Brodersen *et al.*, 2008; Brodersen and Vogelmann, 2010).

Thick leaves and those with a large proportion of spongy mesophyll with numerous intercellular air spaces, favour light scattering and promote light trapping. Owing to low absorptance of FR light (centered at 730 nm) by leaf pigments and leaf structures, the gradient in FR light is approximately linear, whereas blue light (centered at 450 nm) decreases exponentially with depth (Seyfried and Fukshansky, 1983; Vogelmann *et al.*, 1989). Vogelmann *et al.* (1989) and Cui *et al.* (1991) have reported that blue light was largely attenuated by the initial 50–100 μm of leaf, consisting of the upper epidermal-cell layer and one-half to one palisade-cell layer. Green light is weakly absorbed, consistent with the absorption spectrum of chloroplasts, and penetrates deeper into the mesophyll than blue light (Evans, 1999). Red light strongly absorbed by leaf pigments creates steep gradients, although data reported to date indicate that red-light gradients are intermediate between those of blue and green (Evans, 1999; Evans and Vogelmann, 2006; Vogelmann and Han, 2000). In leaves irradiated with white artificial light or with sunlight, transmitted light is depleted in red and blue, and light scattered in the forward or backward directions consists predominantly of green and FR light. Thus, green and FR light dominate the light environment within the leaf both in the palisade and spongy mesophyll layers. These results indicate that blue and red light provide energy for photosynthesis in the cell layer(s) near the upper leaf surface, whereas green light is a particularly important light energy source deep within the leaf (Sun *et al.*, 1998; Nishio, 2000). Although the chl. molar-absorption coefficient is small in the green spectral region, leaf absorptance is only moderately smaller (Morales *et al.*, 1991). This is a result of the generally high leaf chl. concentration and to a larger effect of multiple scattering in spectral regions with weak absorption (see Louis *et al.*, 2006, and references therein).

Cell composition and chemistry varies among different leaf layers. In *Spinacia oleracea*, Terashima and Inoue (1985a,b) and Terashima *et al.* (1986) have reported that chloroplast morphology and electron transport components gradually change from typical high-light-acclimated chloroplasts at upper mesophyll cells to low-light-acclimated chloroplasts in lower mesophyll cells (see Fig. 16.3). In paradermal leaf sections, chl. content per unit leaf layer reached a maximum at 75–105 ± 15 μm from the upper leaf surface (Vogelmann and Martin, 1993). Nishio *et al.* (1993) reported in paradermal leaf sections a gradual decrease in the chl. *a*/*b* ratio from the top to the bottom of the leaf. In addition, photosynthetic capacity also decreases from the upper- to the lower-leaf side (Evans *et al.*, 1993; Ögren and Evans, 1993; Han *et al.*, 1999; Evans and Vogelmann, 2003), but the gradients in light absorption and photosynthetic capacity do not necessarily match (Vogelmann and Evans, 2002; Evans and Vogelmann, 2003, 2006), implying that the leaf photosynthetic rate responds differently to light with varying spectral quality, resulting in changes in the curvature in the photosynthetic light response curve (Ögren and Evans, 1993). The mismatch between light absorption and photosynthetic capacity is especially pronounced for blue and red light that are already mostly absorbed by the upper-leaf layers, and is less for green light that is more uniformly distributed throughout the leaf (Evans and Vogelmann, 2003, 2006; Vogelmann and Evans, 2002). Such effects can be further amplified by CO_2 gradients within the hypostomatous leaves with stomata commonly on the lower-leaf surface (see Aalto and Juurola, 2002 for numerical simulation of within-leaf CO_2 gradients).

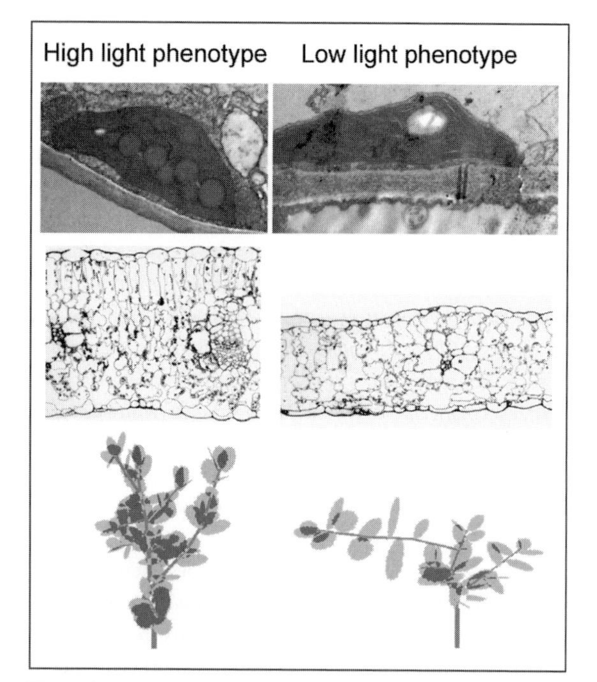

Fig. 16.3. Plant responses to light availability range from modifications at the level of organelles within the cells to modifications at whole-plant level. Phenotypes of a given plant growing in high light exhibit distinctive chloroplasts with highly stacked granal lamellae (upper images of transmission electron micrographs), thick, multilayered leaves (central images of light transmission micrographs) and dense crowns with steep branches and leaves (lower images of three-dimensional computer reconstructions of seedlings). Opposite suites of traits are observed in the same plant species grown under low light (right images).

Gradients in cell chemistry also have important implications for the photoinhibition sensitivity of upper and lower surfaces of bifacial leaves. Comparing the upper- and lower-leaf sides of *Phaseolus vulgaris* leaves, Louis *et al.* (2006) reported differences in chl.-excitation spectra, and ascribed them to a higher pool of carotenoids or to a greater conversion of violaxanthin to zeaxanthin on the upper side. Such gradients in photoprotection capacity along with gradients in photosynthetic capacity likely explain why the upper-leaf surface is more tolerant than the lower-leaf surface to high light exposure (Evans *et al.*, 1993; Sun *et al.*, 1996). Moreover, Oguchi *et al.* (2011) have recently shown that photoinhibiting leaves with lights of different colours results in different degrees of photoinhibition along leaf depth. Hence, although blue light induced the greatest photoinhibition, both near the adaxial surface and in the deeper tissue, red light induced greater photoinhibiton than green light near the adaxial surface but the opposite occurred in the deeper tissue.

16.3. PHOTOSYNTHESIS WHEN LIGHT IS SCARCE

Plant leaves growing in different light environments, ranging from more than 2000 μmol m^{-2} s^{-1} in full sunlight to less than 5 μmol m^{-2} s^{-1} in the forest understory, develop a set of morphological, physiological and biochemical characters improving photosynthesis in such challenging conditions. Under high light, plants optimise the use of light energy in photosynthesis (photosynthetic capacity) and energy dissipation, whereas under low light plants maximise light capture (Fig. 16.3). In this section, acclimation responses enhancing light-capture efficiency will be described with special reference to low light. As light is an extremely variable factor varying during the day, between days and during the season, plants growing in open places frequently encounter situations in which light is limiting for photosynthesis so its absorption must be guaranteed and kept sufficiently high to feed the high photosynthetic capacity. Also, there are always gaps in the canopy such that plants in the deep understory encounter bright lightflecks during the day and also demand photoprotection. A quantitative relationship between photoprotective energy dissipation (via the xanthophyll cycle, see below) and the characteristic growth light environments was shown across a large group of closely related species in the Hawaiian lobeliads, providing evidence for adaptive diversification in photosynthetic physiology (Montgomery *et al.*, 2008).

To obtain a positive carbon balance in low light, acclimated leaves display a set of morphological, biochemical and physiological adaptations (Valladares and Niinemets, 2008). Most of these traits are not easily reversible, and there are some 'obligate shade plants' for which low light is mandatory, often because of interacting stresses accompanying high-light environments. For instance, many bryophytes that lack advanced water-conducting elements and stomata are rapidly dried out under high irradiance (Proctor, 1984) and therefore prefer low-light environments. High-light acclimation of several 'shade plants' has also been shown to be limited by nutrient availability (Osmond, 1983). Nevertheless, most plants growing in the shade are 'shade tolerant' as they are able to develop leaves plastically and structurally, and are physiologically adapted to the whole spectrum of irradiances (Valladares and Niinemets, 2008).

Morphological characteristics of low-light-acclimated leaves include thinner mesophyll with fewer or no layers of palisade parenchyma and fewer chloroplasts per area with larger grana and more stromal thylakoids (Fig. 16.3). Despite a decreased leaf mass per unit area and a low thickness, mesophyll conductance is often low in shade leaves, which contributes to a low photosynthetic capacity (Monti *et al.*, 2009). There are major differences in pigment composition between low-light and high-light-acclimated leaves in several respects: (1) chl. *a*/*b* ratio is commonly lower in low-light-acclimated leaves (e.g., Demmig-Adams, 1998) as a consequence of higher antenna size (LHCII enriched by chl. *b*) relative to reaction centers (PSI and PSII depleted from chl. *b*); (2) the ratios of photoprotective carotenoids to chl. are lower in low-light-acclimated leaves, reflecting lower demand for photoprotection. This is especially the case for the xanthophyll-cycle pigments, whose content relative to chl. can be four times lower than that in high-light leaves of the same species, indicating the increasing emphasis on light collection versus energy dissipation (Demmig-Adams, 1998; Niinemets *et al.*, 2003); and (3) presence in high amounts of some taxonomically restricted carotenoids. This is the case for α-carotene, lactucaxanthin (Demmig-Adams and Adams, 1996a), lutein 5-epoxide (García-Plazaola *et al.*, 2007) or *trans*-neoxanthin. In some species α-carotene may replace β-carotene, and lactucaxanthin may replace lutein in low-light versus high-light-acclimated leaves (Demmig-Adams, 1998). The presence of lutein epoxide increases the efficiency of light-energy use and simultaneously represents a reservoir for photoprotective lutein (Matsubara *et al.*, 2007) (see 16.5). In addition to these differences, some deep-shade tolerant plants also possess leaves with a reddish lower surface owing to the presence of a layer of red cells, characterised by high vacuolar anthocyanin content. This layer may serve to reflect red light, possibly increasing light scattering, penetrating the leaf again thereby favouring its re-absorption. However, the function of this red layer is still a matter of debate, for a deeper discussion see Chapter 7. Although the physiological meaning of all these modifications is still not completely understood, this evidence underscores the profound modifications in light-harvesting and the photoprotective-pigment system in acclimation to low light.

Apart from modification in pigment composition, acclimation to low light usually results in enhanced chl. content on a leaf dry mass basis ($chl._M$) (Niinemets, 2007; Hallik *et al.*, 2009). At the same time, thinner mesophyll is associated with lower LMA (Poorter *et al.*, 2009) allowing the plants to construct more foliar area with a given biomass in leaves. As the result of simultaneous modifications in LMA and $chl._M$, chl. content per leaf area ($LMA*chl._M$) that scales positively with leaf absorptance is generally weakly associated with light availability (Valladares *et al.*, 2002; Hallik *et al.*, 2009). Nevertheless, the light availability of single cells scales with leaf absorptance per unit leaf mass. Given that the absorptance per leaf mass scales with $chl._M$, leaf cells under low light do harvest light more efficiently than the leaves under high light (Evans and Poorter, 2001; Niinemets, 2007).

Biochemical differences between high-light and low-light-acclimated leaves reflect a trade-off in nitrogen allocation among light-harvesting and carbon-assimilating enzymes. Chloroplasts in low-light-acclimated leaves contain a larger proportion of antennae with fewer reaction centres, lower ATPase and lower Rubisco content, resulting in a greater fraction of nitrogen in LHC (Evans and Poorter, 2001; Eichelmann *et al.*, 2005; Niinemets, 2007). In low light, major antenna complexes of LHCII are dephosphorylated and aggregated with PSII core complexes, adjusting the energy partition between photosystems. Differences in nitrogen distribution among pigment-binding complexes and rate-limiting proteins of photosynthetic machinery, along with differences in LMA are responsible for the differences in photosynthetic-light responses between high-light and low-light-acclimated leaves: low-light-acclimated leaves have lower light compensation points, lower rates of dark respiration and photosynthetic capacity at saturating irradiance, but higher quantum yields for an incident PPFD (Björkman, 1981; Björkman and Demmig-Adams, 1994; Osmond *et al.*, 1999; Niinemets, 2007).

16.4. PHOTOSYNTHESIS WHEN LIGHT IS EXCESSIVE: PHOTOPROTECTION AND PHOTOINHIBITION

Photosynthetic tissues absorb light and transfer the excitation energy to the reaction centres under a wide and fluctuating range of PPFDs (Fig. 16.4). Photon absorption results in single-state excitation of chl. ($^1Chl^*$), which may return to its ground state by transferring the energy to reaction centres resulting in charge separation and commencement of PET. Light capture and energy conversion must be efficiently coupled to the use of resulting chemical energy in photosynthesis. Whenever light-energy absorption by chl. surpasses its utilisation, the excess excitation energy can lead to generation of harmful ROS. In particular, high light intensities increase the risk of ROS production that

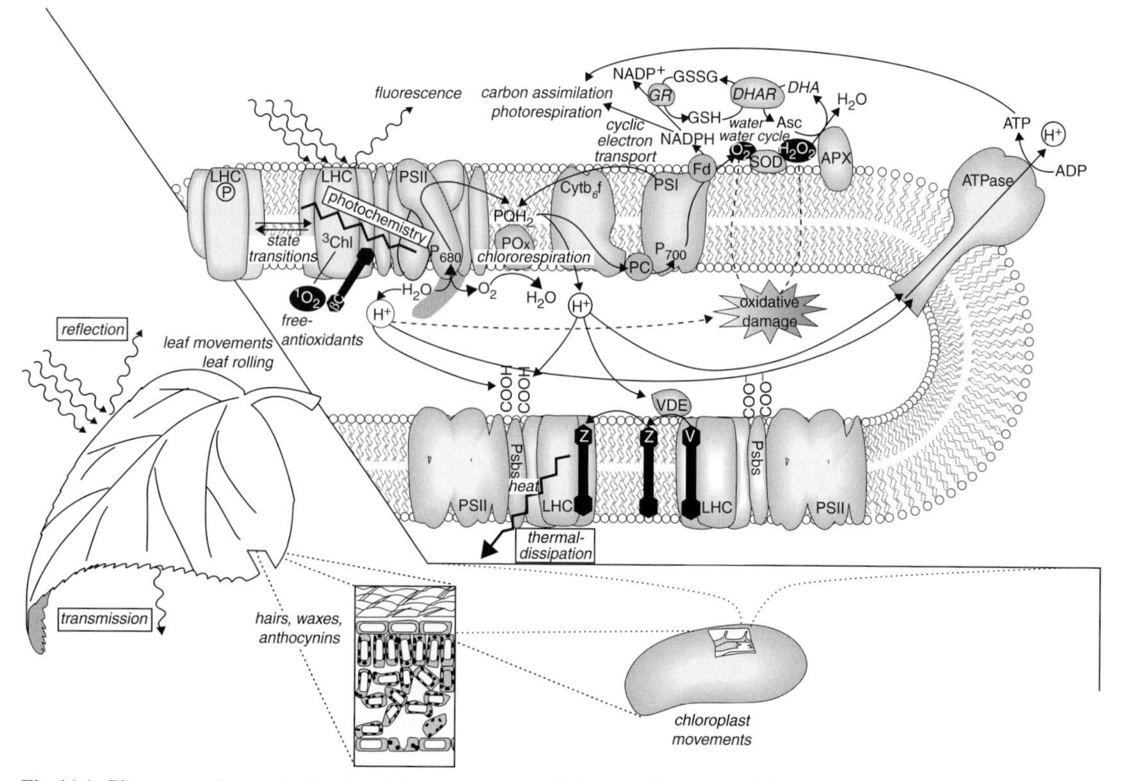

Fig. 16.4. Photoprotection mechanisms in photosynthetic tissues. When the photosynthetically active radiation (PAR) impinges on the leaf, only a fraction of it is reflected or transmitted. Morphological modifications, such as leaf rolling or movements together with chloroplast movements, and accumulation of reflective structures and compounds, such as hairs, waxes or anthocyanins, modulate the amount of light absorbed by chloroplast pigments. Redistribution of energy between both photosystems (state transitions), also contribute to reduce the excess of energy mostly under low-light conditions. Still the absorbed PAR might exceed the intensity that can be used in photosynthesis and a fraction must be dissipated as heat or canalised in alternative pathways to prevent damage to the photosystems. Thermal dissipation is modulated by the transthylakoid pH gradient and the operation of the xanthophyll (VAZ) cycle, whereas alternative energy sinks include the channelling of electrons through the water-water cycle, cyclic electron transport or chlororespiration. The last line of defence is constituted by the antioxidant metabolism and the action of repair mechanisms.

ultimately will result in generalised cellular damage, the so-called photodamage. Stress factors such as drought, high and low temperatures, air pollution, herbicides or mineral deficiencies reduce the rate of carbon assimilation, thereby exacerbating the risk of ROS formation. Photoprotection is achieved by a plethora of mechanisms: (1) structural avoidance of light interception; (2) enhanced metabolic usage of light energy; (3) conversion of absorbed energy into fluorescence or heat that is dissipated in the pigment bed; and (4) its conversion into ROS and subsequent scavenging. All these mechanisms act separately or simultaneously and at different time-scales ranging from seconds, in the case of heat dissipation, to long-term acclimation responses. They represent a trade-off between photosynthetic efficiency and photoprotection. In general, most morphological modifications to reduce light capture are considered as more static, often irreversible acclimation responses, whereas biochemical alterations constitute highly flexible and reversible acclimation to a fluctuating environment.

16.4.1. Avoidance of light capture

Leaves developed under high light absorb much more energy than they are able to use, and are acclimating to these high photon loads both biochemically and structurally. Biochemical avoidance of light capture is mainly associated with smaller antenna size and a higher carotenoid/chl. ratio, overall reducing light harvest at a given incident quantum-flux density (Kitajima and Hogan, 2003; Niinemets, 2007; Takahashi and Badger, 2011).

Morphological avoidance of light interception is frequently the result of plastic and irreversible acclimation mechanisms (Fig. 16.3), serving to reduce the amount of light intercepted in leaves growing in open habitats. In general, high-light-exposed leaves of many species have steeper leaf angles in the top of the canopy to reduce interception of midday irradiances with highest intensity, whereas the leaves deeper in the canopy become more horizontal (Cescatti and Niinemets, 2004; Fleck *et al.*, 2003). Björkman and coworkers reported that such modifications in foliar angle can be reversible in herbaceous species, even at a timescale of hours or minutes (Björkman and Demmig-Adams, 1994). For instance, *Oxalis oregana*, an understory herb species, adjusts leaf angle in seconds by turgor pressure changes in response to sunflecks (Björkman and Demmig-Adams, 1994). Other high-light-adapted species exhibit paraheliotropic (away from sun) diurnal or seasonal variation in leaf angles (Pastenes *et al.*, 2005), thereby significantly reducing light harvesting. Petiole curvature determining leaf movements is controlled by two processes, the induction and the cessation of curvature, and at least three photoreceptor systems are involved in the processes (Fujita *et al.*, 2008). In many species, high-light exposed leaves are more rolled with lower effective area for light interception (Fleck *et al.*, 2003). In herbaceous species, such rolling responses can occur relatively fast as the result of modification of turgor in bulliform cells in leaf epidermis, thereby reducing the light-harvesting surface when the potential excess irradiance load is the highest, for instance, during water stress, and increasing again the surface area under non-stressed conditions (Turgut and Kadioglu, 1998). In woody species with more rigid leaves that lack such specialised anatomical adjustments, leaf rolling is often irreversible.

At the sub-cellular level, chloroplast migration within photosynthetic cells represents the fastest morphological response to fluctuating PPFD. At low PPFDs chloroplasts in mesophyll cells are generally positioned side-by side perpendicular to the penetrating beam irradiance to maximise light interception. Upon exposure to high irradiance, chloroplasts move towards the vertical side walls of mesophyll cells, parallel to light direction. Chloroplast movements in plants are induced by blue light in a process mediated by a group of blue-light photoreceptors, the phototropins (Wada *et al.*, 2003). In thin leaves, chloroplast movements may reduce leaf absorptance by up to 20% in a few minutes (Björkman and Demmig-Adams, 1994), representing a cheap photoprotective strategy in fluctuating light environments. The adaptive value of chloroplast movement has been probed by the use of mutants defective in chloroplast avoidance (Kasahara *et al.*, 2002). However, in plants with optically thick leaves, chloroplast movement in upper mesophyll layers changes light-distribution gradients, but not necessarily much whole-leaf light absorptance.

Over time, many species regulate energy interception by changes in leaf reflectance. This is mainly achieved by the presence of trichomes, scales, air-filled spaces, salt crystals or epicuticular waxes (see Chapter 7). With a few exceptions, all these structures are spectrally neutral generating whitish (glaucous) leaves, which increase reflectance for the entire photosynthetic light spectrum. Through this process, leaves may reflect more than 50% of incident light, as occurs in the Mediterranean shrub *Encelia farinosa* (Björkman and Demmig-Adams, 1994). However, these modifications are slow or not reversible when light becomes limiting, and may become a disadvantage to plants under limiting light conditions.

An alternative mechanism to increase light reflectance is the accumulation of red pigments in the outer cell Reddening usually occurs transiently during leaf expansion or senescence but also under different environmental stresses (mineral deficiencies, drought, high or low temperatures or pathogens). Differently from structural features (e.g., hairiness) and compounds (epicuticular waxes) leading to glaucousness, accumulation of red pigments is easily reversible, and disappears once the stress is over. However, there is still a controversy regarding their photoprotective functions, with two main hypotheses that are mutual but not exclusive: red pigments act as passive light filters or as antioxidants (Steyn *et al.*, 2002; Gould, 2004) (see Chapter 7). These pigments are mainly anthocyanins, but also betacyanins and carotenoids (Steyn *et al.*, 2002). These compounds increase red-light reflectance and absorb blue and green wavelengths, attenuating green light that can penetrate more deeply into the mesophyll.

16.4.2. Energy consumption by metabolism

Enhanced photosynthetic capacity allows for an increased use of photosynthetic energy. The main metabolic contribution to light-energy consumption ('dissipation') is CO_2 assimilation, whose potential capacity is commonly larger in leaves growing at higher light (see Niinemets, 2007 for a review). Photosynthetic capacity also increases in plants acclimated to sub-optimal temperature conditions

(Holladay *et al.*, 1992). However, under several other stresses limiting CO_2 entry into the chloroplasts, such as water stress, CO_2 fixation becomes heavily reduced and alternative metabolic processes gain importance. Other potentially major processes of metabolic energy dissipation are photorespiration (in C_3 plants; Kozaki and Takeba, 1996), nitrate reduction, the Mehler reaction (Osmond *et al.*, 1997), electron transport to ascorbate (Tóth *et al.*, 2009) and cyclic electron flow (Takahashi *et al.*, 2009). All these processes consume energy and reducing power, but their activities do not result in net carbon assimilation. They represent important energy sinks under conditions of restricted carbon assimilation. For instance, under water stress, photorespiration may consume the same amount (or even higher) of energy as that used in non-stressed conditions for carbon assimilation (Björkman and Demmig-Adams, 1994). The importance of photorespiration also increases with increasing temperature in parallel with the increased affinity of Rubisco for O_2. It is widely recognised that photorespiration protects C_3 plants from photooxidation, and new accounts for ecologically relevant situations of this protective feature are continuously reported (e.g., Zhang *et al.*, 2009d).

Other metabolic processes that enhance dissipation by acting as electron sinks are chlororespiration and cyclic electron transport (Bennoun, 2001; Joët *et al.*, 2002). The first represents a shortcut of the electron transport chain, in which a PQ oxidase oxidises PQ, transferring electrons to O_2. The second consists of the CEFI or CEFII, with re-entry of electrons into the PQ pool or reduction of excited PSII ($P680^+$) by cyt b_{559}. Both mechanisms may act as a 'safety valve' to avoid over-reduction of PSII. In fact, *Nicotiana tabacum* mutants lacking chlororespiration were more sensitive to high light intensity (Peltier and Cournac, 2002). The mitochondrial AOX-respiratory pathway has been shown to protect the photosynthetic electron transport chain from the harmful effects of excess light, particularly under drought (Bartoli *et al.*, 2005).

16.4.3. Regulation of the efficiency of energy conversion

The efficiency of light-energy conversion into chemical energy can be finely regulated by chloroplasts (Fig. 16.4). The regulation of light energy conversion can be achieved by reorganisation of photosynthetic machinery by phosphorylation of thylakoid proteins. This process, called state transition, results in separation of LHCII and lateral migration towards PSI, and alterations in energy partition between PSI and PSII (Haldrup *et al.*, 2001). State transitions are important under low light, loosing importance under high light.

When excess light is absorbed, two additional main mechanisms reduce the yield of light-energy conversion: dissipation of energy as heat, which is proportional to the so-called NPQ, and the re-emission of photons as fluorescence. Although in terms of total energy, the part attributable to fluorescence is minor (ca. 3%), it is easily quantified and provides an essential tool for photosynthesis studies. NPQ is able to dissipate up to 99% of energy absorbed by chl. The non-photochemical energy dissipation mechanism is a rapid and flexible means for adjustment to the extremely variable light environment, switching from a highly efficient light-harvesting antenna to a low-efficiency dissipating system. *Arabidopsis* mutants lacking NPQ have demonstrated that this mechanism confers plants a strong fitness advantage as it increases plant tolerance to variations in light intensity (Külheim *et al.*, 2002).

Development of NPQ requires the build up of ΔpH across thylakoid membranes together with the presence of PsbS protein (Li *et al.*, 2000). Consequently, rates of NPQ increase whenever products of electron transport are not consumed in metabolic processes. Dissipation of excess energy as heat in the antenna is stabilised and amplified by the operation of the xanthophyll (or VAZ) cycle. In this cycle zeaxanthin is produced by light-induced de-epoxidation of violaxanthin, through the intermediate antheraxanthin, by the enzyme V de-epoxidase (VDE) (Müller *et al.*, 2001; Demmig-Adams and Adams, 2006). This reaction is controlled by the acidification of the thylakoid lumen (ΔpH) that occurs when chloroplasts are illuminated. Epoxidation back to V occurs in the dark or under decreased light pressure when the proton gradient is dissipated, completing the VAZ cycle and returning the leaf to a non-dissipating condition. Apart from the ubiquitous VAZ cycle, a parallel cycle that involves the light-induced de-epoxidation of lutein epoxide (Lx) to form lutein (L) has been described in some taxa of higher plants (García-Plazaola *et al.*, 2007; Förster *et al.*, 2009). The exact biophysical mode by which NPQ takes place is not completely understood, and several models have been proposed. Among others, the following mechanisms have been proposed: energy transfer from chl. to a low-energy carotenoid's excited state in LHCII (Ruban *et al.*, 2007), the charge transfer from chls to a chl.-Z heterodimer in minor antenna components (Ahn *et al.*, 2008) or the direct dissipation within the PSII core complex (Finazzi *et al.*, 2004). These mechanisms are not mutually exclusive and all of them may operate simultaneously.

In addition to the ubiquitous mechanism of flexible NPQ associated with PsbS protein of PSII and controlled by ΔpH, under unfavourable conditions (specifically under low temperature) some perennial evergreens develop a process of sustained thermal-energy dissipation that does not relax with the darkening of leaves and disappearance of trans-thylakoid ΔpH (Demmig-Adams and Adams, 2006). Differently from ΔpH-dependent NPQ, sustained NPQ is neither flexible nor reversible. This mechanism is associated with PSII core-protein reorganisation and/or degradation and is characterised by a great increase in PsbS level, absence of ΔpH control, partial loss of D1 protein units from PSII core, a decrease in total chl. and the retention of high concentrations of de-epoxidised xanthophylls (A and Z) (Öquist and Huner, 2003). This mechanism is typically found under severe-stress conditions, such as those encountered by temperate evergreens in winter, leading to a sustained decrease of photosynthetic rates, the so-called 'winter photoinhibition' that covers the whole winter period.

Apart from their participation in energy dissipation, other mechanisms relating A and Z with photoprotection have been proposed, such as membrane stabilisation (Havaux, 1998) and antioxidant activity (Havaux and Niyogi, 1999; Havaux *et al.*, 2007). *Arabidopsis* plants overexpressing the gene responsible for Z biosynthesis (β-carotene hydroxylase) were also more tolerant to high light owing to the function of Z in prevention of oxidative damage of membranes (Davison *et al.*, 2002).

16.4.4. Detoxification of ROS

Sustained excess light can overwhelm the capacity for photoprotection, with the subsequent over-excitation of photosynthetic apparatus. This can lead to over-reduction of the electron transport chain and the conversion of singlet excited chl. ($^1Chl^*$) into long-lived triplet chl. (^3Chl; Fig. 16.4). These effects ultimately result in the production of harmful ROS: singlet oxygen ($^1O_2$), the major ROS involved in photo-oxidative damage to plants (Triantaphylidès *et al.*, 2008), superoxide (O_2–), hydroxyl radical (·OH) and hydrogen peroxide (H_2O_2). The production of $^1O_2$ occurs mainly in the antenna by energy transfer from ^3Chl, whereas O_2– is primarily produced by direct oxygen reduction by PSI when the pool of NADP is mostly reduced (Mittler, 2002). The origin of H_2O_2 is photorespiration and the spontaneous or SOD catalysed dismutation of the O_2–. Some free metal ions in excess, such as Fe and Cu, and the Fe-S centre in PSI, may catalyse the production of ·OH by H_2O_2 reduction. A Fenton reaction may also occur between reduced Fe-S centres and H_2O_2. This reaction seems to trigger the photoinhibition of PSI (Sonoike, 1996).

If not chemically quenched, ROS can damage cellular components by direct oxidation of molecules. This is the case of the generation of peroxidation chains of polyunsaturated lipids. Even under normal growing conditions, ROS are produced as an unavoidable result of aerobic conditions, but any condition leading to a reduction in metabolic activity also enhances ROS generation (Osmond *et al.*, 1997). This process may threaten cell integrity, however, it also triggers acclimation responses, because ROS also participate in signalling events implicated in the process of acclimation to high light intensity (Mullineaux and Karpinski, 2002). In addition, it has been suggested that the degree of oxidation of antioxidant pools (e.g., ascorbate/dehydroascorbate) or elements of the electron transport (PQ pool) chain can serve as signals (Mittler, 2002). Thus, the role of antioxidant mechanisms is not only ROS detoxification, but also modulation of such stress signals, with photosynthesis playing a dual role in energy conversion and in the perception of environmental information.

To avoid damage, direct detoxification of ROS is carried out by a set of enzymatic mechanisms (SOD, CAT and ascorbate-glutathione pathway), which constitute the so-called water-water cycle, called this because the overall reaction implies water oxidation by PSII and water generation from ROS reduction (Asada, 1999). Photorespiratory H_2O_2 is deactivated by the enzyme CAT, whereas chloroplastic H_2O_2 is reduced to water by APX through the oxidation of ascorbate. This compound can be directly regenerated by monodehydroascorbate reductase (MDHAR) with NADPH as electron donor, or by the DHAR using GSH as electron donor. GSSG is again reduced with NADPH by GR. Active oxygen species can also be removed by several antioxidant molecules, either lipophilic (tocopherols and β-carotene, which can quench free radicals in membranes) or hydrophilic (ascorbate and glutathione). β-carotene is an efficient quencher of $^1O_2$ present in the core complex of PSII and in thylakoid membranes. Tocopherols also protect thylakoid membranes against ROS, and are able to terminate lipid-peroxidation chains. In high-light exposed leaves, the pools of tocopherol are several-fold higher than in low-light-exposed leaves, suggesting that a higher capacity for membrane protection is needed at higher light (García-Plazaola *et al.*, 2004). Ascorbate is also an effective scavenger for peroxy radicals and $^1O_2$, which is able to act in synergy with tocopherols and is the electron donor for reactions catalysed by APX and VDE. Glutathione participates in direct or enzymatic detoxification of ROS, and acts indirectly as an antioxidant in its involvement in metal sequestration. Both antioxidants and enzymatic defence systems typically increase under stress conditions, but the accumulation of these molecules varied greatly with plant species and environmental conditions. Moreover, compelling evidence indicates that the set of physiological antioxidant molecules is much broader than previously thought, with many non-ubiquitous compounds being relevant in photoprotection, as is the case of hydrophilic flavonoids and lipophilic isoprene, monoterpenes and diterpenes (Peñuelas and Munné-Bosch, 2005).

16.4.5. Repair mechanisms

Repair mechanisms, such as the rapid replacement of photodamaged D1 protein in PSII core by newly synthesised units, represent the last line of plant defence against photo-oxidative damage. This protein is preferably damaged owing to the oxidation–reduction reactions that occur in the D1/D2 complex. The rate of photodamage of D1 protein depends on the redox state of the electron transport chain (Melis, 1999). Rapid D1 turnover in chloroplasts (60 minutes on average) contributes to avoidance of photodamage although D1 is in constant turnover. This process includes not only the synthesis of new D1, but also the complex disassembly of PSII complexes and removal of damaged D1, followed by reinsertion of newly synthesised D1 in the thylakoids and the reconstitution of the multi-protein reaction centers. Photodamage of D1 is a continuous process, but its rate increases linearly with irradiance (Takahashi and Murata, 2008) and with the size of the antenna (Melis, 1999). Retardation of the recovery of photoinactivated PSII can be a result of restricted mobility of PSII in the thylakoid membrane (Oguchi *et al.*, 2008).

Adverse conditions such as low or high temperatures also impair energy balance accentuating photodamage to D1, but recent studies also suggest that these stresses inhibit the repair of PSII by downregulation of the D1 synthesis (Takahashi and Murata, 2008). The balance between the rate of D1 damage relative to the rate of D1 repair finally determines whether sustained photodamage does occur in the leaves (Yokthongwattana and Melis, 2006).

16.5. SHORT-TERM VERSUS LONG-TERM RESPONSES TO RADIATION INTENSITY

Plants exhibit a wide range of phenotypic differences when growing in high versus low light (Fig. 16.3). Although some of these changes are slow and mostly irreversible, such as those involving the development of tissues and organs, others are relatively quick and reversible, such as those involving photosynthetic pigments and enzymes. Although all these changes reflect different aspects of phenotypic plasticity (i.e., the capacity of a given genotype to render different phenotypes under different environments), the former are classically included within the notion of plasticity (typically, developmental plasticity) and the latter are referred to as acclimation responses (Valladares and Niinemets, 2008). Whereas plasticity has been shown to increase light capture and photosynthetic utilisation (e.g., Valladares and Pearcy, 1998), not all plants exhibit the same levels of plasticity because there are many limits to and costs of a highly plastic response to a changing environment (Valladares *et al.*, 2007).

Leaf morphological and physiological adaptations to high and low irradiances received intense attention during the 1970s and 1980s, which led to a thorough description of the so-called sun and shade types of leaf (Björkman, 1981). Of course, light is a continuous variable, and leaf structural and physiological responses to light availability are also continuous (Niinemets, 2007), such that there is a spectrum of foliage structural and physiological variation between high- ('sun') and low- ('shade') light availabilities. This acclimation of foliages to their light environment is an important part of the potential phenotypic plasticity that plants can carry out in response to light availability (Valladares *et al.*, 2000a,b, 2002).

Leaves can acclimate to their light environment at several levels. For instance, they can modulate the leaf area per unit biomass and they can change the relative distribution of nitrogen between photosynthetic components (Evans and Poorter, 2001; Niinemets and Tenhunen, 1997). It was widely accepted that one of the most important differences between high- and low-light-acclimated leaves is the much lower respiration rates of the latter (Björkman, 1981). However, many studies have shown transient transfers to low or high light quickly reverse the patterns (e.g., Pearcy and Sims, 1994), and it seems that it is total daily carbon gain that drives respiration rates more than acclimation. Besides, comparisons can be misleading as respiration occurs in response to growth and maintenance processes, which are restrained by light in the shade (Zaragoza-Castells *et al.*, 2007). The low respiration rates observed in the shade mostly reflect the low resource supply, but whether the capacity for respiration also adjusts is less certain (Pearcy, 2007). Another clearly established difference between high- and low-light-acclimated leaves is the leaf mass per unit area, which is larger in the former than in the latter. Leaves developed in the shade are generally thin, with a loosely organised mesophyll (Fig. 16.3; Niinemets, 2007; Valladares and Pearcy, 1999). The large differences in photosynthetic capacities per unit area of high- and low-light-acclimated leaves are reduced or absent when photosynthetic rates are expressed on a leaf-mass basis (Pearcy, 2007). This is consistent with a rather constant amount of photosynthetic machinery per unit leaf mass, and with a greater leaf mass per unit area of high-light versus low-light leaves. As leaf turnover is faster in high light, leaves developed under different PPFD also differ in their mean longevity; this has been related to the fact that the payback of foliage-construction costs under low light takes longer than under high light (Williams *et al.*, 1989).

Comparisons across many species reveal no consistent differences in leaf chl. concentration per A_{leaf} among leaves in contrasting light environments (Björkman, 1981). Average chl. concentrations per A_{leaf} are sufficient to absorb 80–85% of the incident PAR (Agustí *et al.*, 1994), the remainder being lost by either reflection or transmission. Although leaves in the shade could benefit from an enhanced light capture by an increased chl. concentration per unit mass, the costs are likely to exceed the benefits. A doubling of the chl. concentration is needed to increase leaf absorptance by just 5% from the 80–85% of absorptance of an average leaf (Niinemets, 2007; Pearcy, 2007). Although there are only 4 mol of nitrogen in a mol of chls, the associated proteins in chl.-protein complexes contain 21–79 mmol nitrogen per mol of chls (Evans, 1986). Thus, light harvesting by chl.-protein complexes involves an important fraction of the foliar nitrogen that may even be up to 60% of total leaf nitrogen (Niinemets and Tenhunen, 1997). The lower chl. *a*/*b* ratio observed in low-light-acclimated leaves has a negligible impact per se on light harvesting and quantum yield, but it reflects the cost of light capture per unit nitrogen invested: LHCII, where many chl. *b* molecules are located, contains only 25 mmol N per mol of chl., whereas PSII core complex, where many chl. *a* molecules are located, contains 83 mmol N per mol of chl. (Evans, 1986; Pearcy, 2007). Sun leaves cannot minimise the investment in PSII and save nitrogen in this way because they need the high electron transport capacity of PSII. There are clear pressures for increasing construction costs in both high-light and low-light environments: high-light leaves can be expensive because of their high concentrations of proteins (primarily photosynthetic enzymes), whereas low-light leaves can be expensive because of their high concentrations of anti-herbivore compounds associated with their long lifespan (Chabot and Hicks, 1982). High-light-developed leaves are more costly to construct on a per unit area basis, largely because of their greater mass per unit area, but on a per unit mass basis differences are minor. In broad surveys of leaf-construction costs, specific differences are small and no clear trends among habitats or growth forms have been identified, suggesting that leaf mass per unit area is more important than construction costs in determining differences in the carbon balance (Villar and Merino, 2001). This lack of consistent pattern seems to be owing to the variety of trade-offs between investments in structural, biochemical and protective compounds in different leaves from a given environment.

Plant responses to available light involve a suite of traits scaling from the cell to the whole plant. Photosynthetic capacity is linked not only to maintenance respiration but also to leaf-area ratio and leaf mass per area among the most relevant features involved in light acclimation (Pearcy and Sims, 1994). Relative growth rate (RGR) of plants is differentially affected by these four traits, depending on the light environment to which the plant is acclimated (Fig. 16.5). Photosynthetic capacity has a large effect on growth in high but not in low light, whereas the reverse is true for respiration. By contrast, the light environment has little effect on the influence of leaf-area ratio on growth, but model simulations reveal close links between these key plant features (Pearcy and Sims, 1994).

16.6. SUNFLECKS: PHOTOSYNTHESIS IN HIGHLY DYNAMIC LIGHT ENVIRONMENTS

Changes in the light environment associated with successional changes are relatively slow and predictable, which allow individuals to anticipate and respond; and the same applies to both seasonal and diurnal variations. But other changes can take place more rapidly and in a less predictable

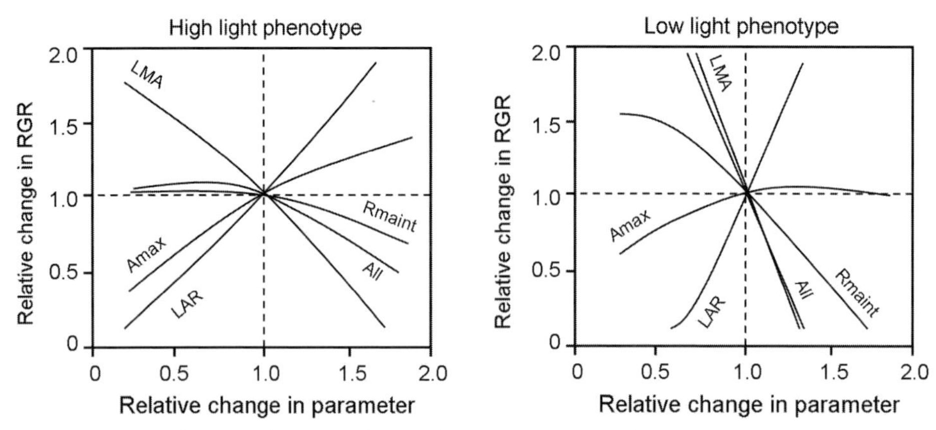

Fig. 16.5. Relative change in relative growth rate (RGR) in plants acclimated to high or low light as a function of a relative change in photosynthetic capacity (A_{max}), maintenance respiration (Rmaint), leaf area ratio (LAR), leaf mass per area (LMA) and all the four parameters according to a concerted acclimation response (all). The graphs represent a sensitivity analysis of RGR for the tropical plant *Alocasia macrorrhiza* modelled after experimental acclimation to contrasting light conditions. Adapted from Pearcy and Sims (1994).

manner (Fig. 16.6). Because the leaf area is heterogeneously distributed in the canopy, the resulting gaps allow penetration of sunflecks of various durations and intensities to lower canopy layers and the understory. Sunflecks cause the most rapid scale of temporal heterogeneity: typically from seconds to minutes. Sunflecks often contribute a substantial fraction of the total light available in the understory (Chazdon, 1988; Pearcy, 1983). The constraints imposed by and the ability to respond to this fine-grain temporal heterogeneity are crucial to sunfleck utilisation and eventually to survival in dark understories (Pearcy *et al.*, 1994; Valladares *et al.*, 1997).

Most studies of the spatial heterogeneity of light focused on the variation of light regimes on large scales, comparing gaps with the understory or gaps of contrasting sizes, but a multilayered forest canopy may result in variations of light regimes on much smaller spatial scales (Canham *et al.*, 1994; Tang *et al.*, 1999). The size of the gap not only affects the duration of the sunfleck but also its intensity, because if it is not large enough (>0.5°) only radiation from a fraction of the solar disc will be received and the understory will be in partial shadow or penumbra. Most sunflecks in temperate and tropical forests are penumbral light, with maximum PAR intensity of 0.1 to 0.5 of full sunlight (Pearcy, 1990). As penumbral light is at a lower intensity than full sunlight, it can be used in photosynthesis more efficiently and can result in proportionally more carbon gain in the understory. The spatial scale of sunflecks typically vary from 0.1 to 1 m, so that often only part of the crown of an understory plant will be influenced by a given sunfleck (Baldocchi and Collineau, 1994).

Most of our knowledge on photosynthesis has been obtained from studies under steady light conditions, but light and photosynthesis are highly dynamic in nature. When light is suddenly increased during a sunfleck, photosynthesis will accelerate, but if the leaf has been in the shade for a long period, this initial increase can be small or almost missing because the stomatal opening is too small and because the enzymes of the photosynthetic carbon-reduction cycle are inactivated. Stomatal opening and activation of these enzymes is a relatively slow process, leading to the well-known induction requirement of photosynthesis in which the assimilation rate requires 10–30 minutes to reach its maximum value (Pearcy, 1988). If a fully induced leaf is shaded for a few minutes and then re-exposed to saturating light, only a few seconds are required for the assimilation rate to recover to its maximum value (Pearcy, 2007). Once the enzymes are activated, photosynthesis can continue in the shade at a high rate although for a very brief period of time in the so-called post-illumination carbon-fixation process. Post-illumination carbon fixation can contribute to as much as a 6% enhancement of carbon gain when very short sunflecks make an important contribution to the available light, as in the case of quaking aspen canopies under windy conditions (Pearcy, 2007). In the understory, induction limitations (stomatal limitations and Rubisco inactivation) predominate and significantly constrain carbon gain during sunflecks. Species differ in

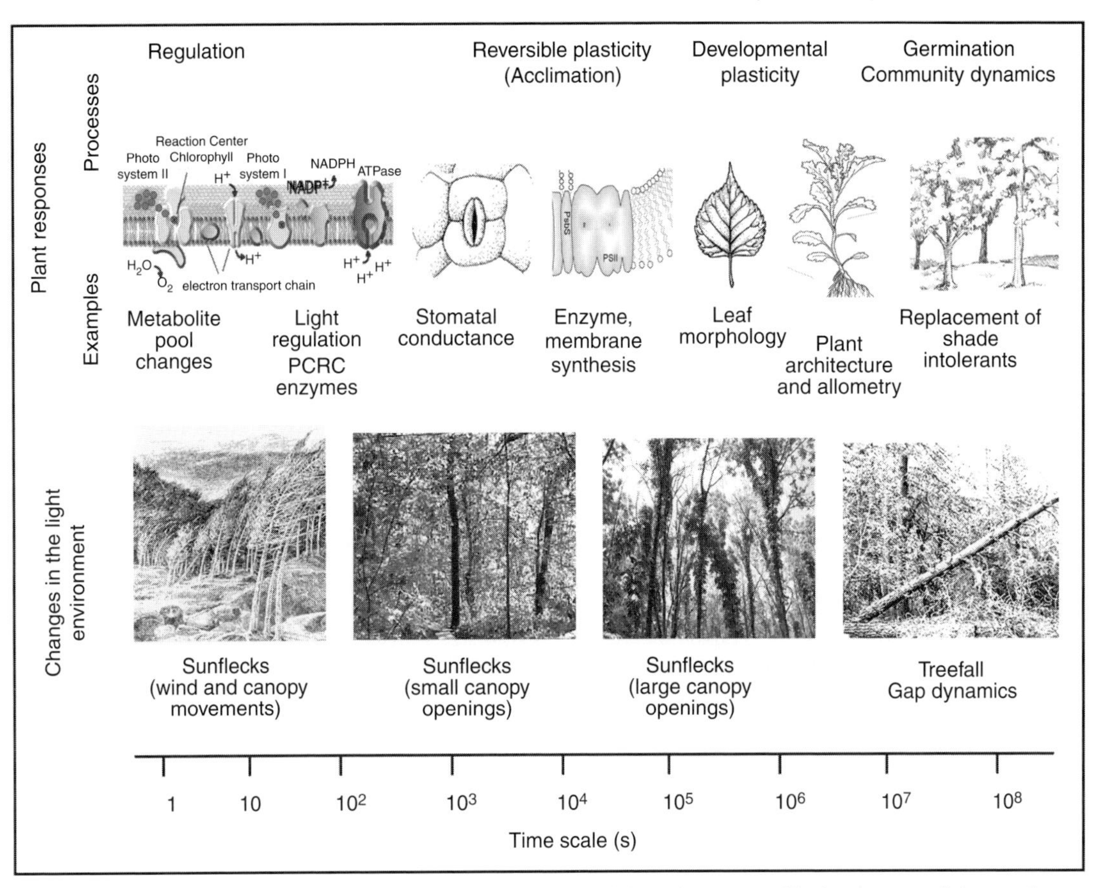

Fig. 16.6. Timescale from fractions of a second to years of changes in the light environment and in the responses of plants to these changes. Representative examples are given to illustrate both the sources of light temporal heterogeneity and the processes involved in plant responses.

their response to dynamic light and it has been shown that low-light species exhibit faster induction dynamics, remain induced for longer and are more efficient in the photosynthetic use of sunflecks than high-light species (Valladares *et al.*, 1997; Pearcy, 2007).

16.7. GLOBAL CHANGE AND RADIATION

Climate change coupled with changes in land use and atmospheric features are inducing remarkable alterations of the light environment experienced by plants all over the world. They are leading either to an increased light availability (e.g., logging, clearing, fires, droughts) in certain areas, or to a decrease of available light in others. The latter is the case for certain temperate zones such as the Mediterranean, where land abandonment, reforestation programmes and strict conservation of habitats is reducing light heterogeneity and mean values of radiation available in the understory (Valladares, 2003). Whether many Mediterranean, typically high-light plants will be able to persist under this combination of global dimming and more extensive shade conditions is uncertain. Solar radiation reaching the Earth's surface has been significantly reduced by 0.51 W m^{-2} per year, equivalent to a reduction of 2.7% per decade (Stanhill and Cohen, 2001). The causes for this global dimming are unclear, but they seem to be primarily attributable to human-made aerosols and changes in cloudiness. And even less clear are the potential consequences of this reduction on ecosystem function in general, and on plant productivity in particular, especially in arid sites. Although productivity at high latitudes could be diminished by this reduction of solar radiation, in high-radiation arid climates productivity could increase owing to reductions of water stress and photoinhibition (Stanhill and Cohen, 2001). The increased fraction of diffuse light associated with this global dimming is expected to increase productivity of complex plant crowns as transmission of diffuse light through the foliage is more efficient than that of direct light, but the issue is contentious and yet poorly understood (Roderick *et al.*, 2001; Gu *et al.*, 2002). Even the global trend on radiation is controversial because certain regions are experiencing a global 'brightening' (Wild *et al.*, 2005). Shade tolerance of the species and their photosynthetic utilisation of available light are inextricably involved in the responses of ecosystems to global change (see discussion in Valladares and Niinemets, 2008).

Many tropical and subtropical forests are experiencing an increased frequency and intensity of clearings and openings in the canopy. This perturbation is promoting the successful establishment of certain invasive plants, which seem to be remarkably plastic in their capacity to exploit the light available (Yamashita *et al.*, 2000; Durand and Goldstein, 2001). And the situation could be taking place in temperate forests as well. In a study of two contrasting populations of *Rhododendron ponticum*, an economically and ecologically relevant invasive species in many moist temperate areas, leaf-level plasticity in response to light was significantly higher than in *Ilex aquifolium*, a co-occurring evergreen shrub. A similar difference was also found between the two populations of *R. ponticum*, one autochthonous and the other one native and invasive: plasticity was linked to invasibility (Niinemets *et al.*, 2003).

Because photosynthetic (chls and carotenoids) and photoprotective (carotenoids and anthocyanins) pigments are positioned to capture solar energy they are well suited for detection from above (see Chapter 15). Using spectral reflectance, not only chls and carotenoids but also flavonoids, water, lignin, nitrogen-containing compounds and cellulose can be estimated by non-contact measurements of electromagnetic radiation (Gamon *et al.*, 1997; Gamon and Surfus, 1999). With the adequate ecological and physiological information for each particular case, remote sensing can be used to detect and quantify relevant biological processes in the canopy (productivity, water content and risk of fire, physiological status, radiation-use efficiency, chl. degradation), which is fostering research on ecosystem functioning at the landscape scale, a key issue to understand global-change impact on terrestrial ecosystems (Gamon *et al.*, 1997; Peñuelas and Filella, 1998).

The study of the biological effects of UV radiation, especially UV-B (280–315 nm), has attracted considerable attention during the last decades because the thinning of the stratospheric ozone layer leads to elevated solar UV-B at ground level (McKenzie *et al.*, 1999). Intense sunlight is associated with elevated UV-B, especially at high latitude or altitude. Field studies have shown that solar UV-B can reduce growth of certain species (Ballare *et al.*, 1996), but it can induce resistance against photoinhibition (Mendez *et al.*, 1999) or it can have few effects in other species (Papadopoulos *et al.*, 1999; Rousseaux *et al.*, 2001). Some of the most dramatic effects of solar UV-B have been observed in the Antarctic Peninsula with the grass *Deschampsia antarctica* (Day *et al.*, 1999).

Radiation is both a crucial resource and a heterogeneous factor, so even though plants depend on it, they more often than not must cope with either too much or too little energy. Global change is pushing the physiological, ecological and evolutionary limits of plants coping with light heterogeneity a bit further.

17 • Photosynthetic responses to increased CO_2 and air pollutants

C. CALFAPIETRA, C.J. BERNACCHI, M. CENTRITTO AND T.D. SHARKEY

17.1. HISTORICAL VIEW: PLANTS IN A CHANGING ENVIRONMENT

After centuries of relatively stable atmospheric composition, the Industrial Revolution has driven a need for plants to cope with a changing atmosphere. However, plants not only passively undergo global climatic changes but are also driving factors that may influence the course of climatic change. This is particularly true for trees and forests as they play a significant role in the global carbon cycle and in the control of CO_2 concentration in the atmosphere. The scientific community has therefore sought to assess and quantify both the response and the contribution of the ecosystems and particularly of the forests to the increase of CO_2 and air pollutants (Dixon *et al.*, 1994; Table 17.1). If the studies on the impact of climate change on forests have environmental understanding as the main objective, the studies on crops have yield and quality of the products as main goals. Moreover studies on herbaceous plants are usually easier than those on trees and can be particularly important when we try to identify the mechanisms of response. In some cases, such as tree crops commonly planted for bioenergy, these objectives can merge and the great challenge becomes maximising future production in a changing climate, while minimising depletion of soil and water resources (Cassman *et al.*, 2003).

Photosynthesis is the first factor driving productivity, thus any influence on the former is likely to influence the latter heavily. However, the ecosystem productivity is also regulated by the amount of photosynthesising leaf area within the canopy, or more generally from the LAI and from the respiration rates at ecosystem level. For instance elevated-[CO_2] experiments proved that the effects on productivity might be much lower than what is hypothesised looking simply at the stimulation on photosynthesis, suggesting that several factors might co-occur (Centritto *et al.*, 1999a; Long *et al.*, 2005, 2006b).

When we mention climatic changes we often refer to one or few factors, whereas the phenomenon involves complex interactions among various factors such as atmospheric CO_2 and tropospheric ozone (O_3) concentrations, air temperature, UV-B radiation, environmental pollutants, extreme atmospheric events, etc. In this chapter we will focus on the main air pollutants intended as the compounds or the elements able to alter physical, chemical or biological properties of the air, leading to injuries or alteration of plant life. In particular we will discuss the effects of elevated [CO_2] and air pollutants on photosynthetic properties, focusing both on the rationale of the response and on the observations presented in the literature to date. Atmospheric [CO_2] has risen from a pre-industrial value of about 280 parts per million (ppm) to a current concentration of 380 ppm, and will likely reach a concentration of about 700 ppm by the end of this century (IPCC, 2007). More recent results suggest that concentrations might potentially exceed the IPCC scenarios as emissions are increasing at rates beyond previous worst-case scenarios (Canadell *et al.*, 2007). Another potent pollutant on which we will focus, O_3, has grown from pre-industrial ground-level concentrations less than 10 ppb (Volz and Kley, 1988) to concentrations of 40 ppb as representative values of the mean daytime ambient background [O_3] during summer months (Fowler *et al.*, 1999a). We will also discuss the effects of NO_x, products of fossil-fuel combustion, SO_2 and particulate matter on photosynthetic properties of the leaves.

Before starting it is useful to clarify three concepts: (1) because the topic is the response of biological organisms, it is likely that genetic diversity plays a crucial role, thus the response is never unique but falls within a range that increases with the increase of biological diversity; (2) the response is likely changing with time because of acclimation processes or simply because of the previous changes

Terrestrial Photosynthesis in a Changing Environment: A Molecular, Physiological and Ecological Approach, ed. J. Flexas, F. Loreto and H. Medrano. Published by Cambridge University Press.

Table 17.1. *A summary of the described effects of several abiotic stresses on photosynthetic parameters. The relative importance of each factor is indicated (–, effect not described; X, effect described occasionally; XX, effect described several times; XXX, main effect). Notice that the fact that a factor is not indicated as important may not necessary mean that it is not, since the literature available is fragmentary and some of the effects have never been analysed under some of the abiotic stresses.*

	Defoliation, leaf destruction / necrosis*	Stomatal and mesophyll conductance limitations	Non-stomatal limitations				
STRESS FACTOR	Importance	Importance	Importance	Structural effects*	Photosynthetic primary reactions	Carboxylation/ Calvin-cycle reactions	Carbohydrate metabolism/ transport
High CO_2	—	XX	XXX	XX	XXX	XXX	XX
High O_3	XXX	XXX	XX	XX	XXX	XXX	XX
NO_x	X	X	X	—	X	—	—
SO_2	XXX	X	XX	XX	XX	X	—
Particulate matter	X	XXX	—	X	—	—	—

* Including leaf shape and size, leaf internal structure and organelle form and distribution.

that lead the plant to differ from the onset of the experiment; and (3) the response changes with the levels of CO_2 or of the pollutant leading to a relationship that can be linear, exponential, logarithmic or parabolic. The combination of length of exposure and levels of the pollutant leads to the concept of dose that influences the uptake and thus the damage. However, in the case of a pollutant, the effect of a chronic exposure (low levels for a long period) can be very different from that of an acute exposure (high levels for a short period).

17.2. OBSERVATIONS ON THE STUDIES OF PHOTOSYNTHESIS UNDER CHANGING ATMOSPHERE: NATURAL OBSERVATIONS VERSUS MANIPULATIVE EXPERIMENTS

Photosynthetic responses to air pollutants have been investigated in the past using chambers where the environmental conditions were manipulated around a small canopy, plant, leaf or a portion thereof (e.g., Long and Drake, 1992; Krupa *et al.*, 1998; Elagoz and Manning, 2005; Bussotti *et al.*, 2007). However, the response of photosynthesis can vary considerably whether we change the conditions around a single leaf or around the entire plant and whether we investigate a short-term or a long-term response (Ainsworth *et al.*, 2002; Morgan *et al.*, 2003).

The duration, volume of the plant fumigated and length of exposure must be described for each study because the feedbacks and the interactions that might occur at plant level and/or over a long-term period significantly influence the response. For this reason most studies moved recently to a larger scale and were prolonged for multiple growing seasons (Fig. 17.1). In some cases it is possible to carry out observations on the effects of global change on plants in natural (not manipulated) environments. This is the case of the natural CO_2 springs where it is possible to test the effects of elevated [CO_2] on photosynthesis and other parameters of the plants grown nearby the CO_2 source. The problems in these studies are the frequent fluctuations of [CO_2] that make it difficult to establish a target concentration, the presence of pollutants emitted together with the CO_2 and, above all, the lack of control plots at ambient [CO_2] with similar characteristics (soil properties, water and nutrient availability, plant traits, etc.) of the ones at elevated [CO_2].

In the case of O_3 pollution, natural observations can be carried out using natural gradients that can be elevational gradients (Winner *et al.*, 1989) or regional gradients (Karnosky *et al.*, 1999; Arbaugh *et al.*, 2003). In these studies, however, as the concentration of O_3 is variable day by day depending on the atmospheric conditions, it is crucial to monitor the O_3 concentration in the air to make sure that the gradient of O_3 concentration is conserved during measurements. The effect of some particular pollutants on photosynthesis can be evaluated using plants grown nearby the source of pollution or in polluted urban environment. The Baltimore ecosystem study is an example of this approach (www.beslter.org). Unfortunately, like in the case of CO_2 springs, the concentration of the pollutant is often fluctuating and the availability of suitable control plants is usually scarce.

For the reasons above, most of the studies on the effects of elevated [CO_2] and air pollutants on photosynthesis derive from manipulative experiments where the concentrations in the air of CO_2 or of pollutants are controlled artificially.

The first studies involved manipulation within cuvettes or small environments, such as phytotrons or branch bags. Later on the diffusion of the open-top chambers (OTC) allowed fumigation of entire plants or even field-grown trees (so-called tree chambers) for long periods. In the 1990s, the FACE (free-air CO_2 enrichment) technique allowed fumigation of large portions of crops or even small forests with elevated concentrations of CO_2 without any alteration of the microclimatic or edaphic conditions (Table 17.2). This technique has been also tested for O_3 (Hendrey *et al.*, 1989).

Among the aspects that distinguish trees from agricultural crops, the size is probably the most evident and this increases the difficulties when manipulative experiments to simulate climate change are planned. This is the reason why most of the studies available in the literature are carried out in laboratories with tree seedlings, even though they often show different responses from mature trees (Ainsworth and Long, 2005). Another limitation is the length of experiments that in many cases last for just a few days or for one growing season, whereas we learnt that some responses, especially the photosynthetic properties, are affected by seasonality or change deeply over multiple years because of acclimation processes. Furthermore, many experiments are biased by changes in the microclimatic conditions, constraints at the root system (typical in plants grown in pots) or limitations in light, water or nutrient availability.

One or more of these problems can deeply compromise the response of trees to climate changes and this is particularly true in the case of fast-growing species or species with indeterminate growth, as they usually respond more prominently to environmental changes than other species.

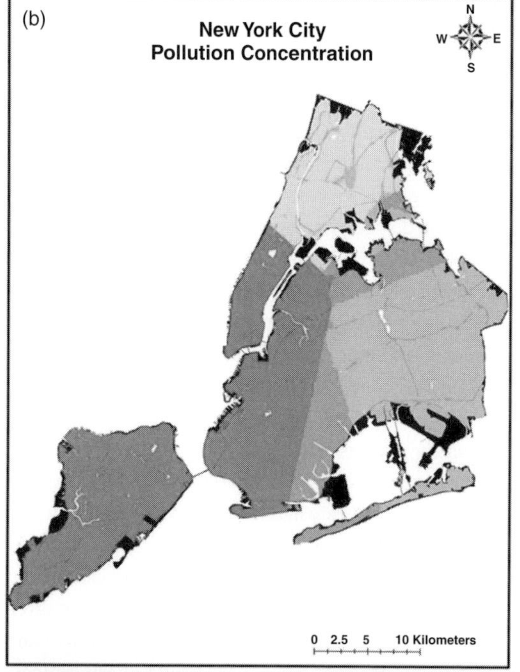

Fig. 17.1. 'Ambient' (a-c) versus 'controlled' (d-f) environments to study the effects of climatic change factors on photosynthesis **(a)** CO_2 spring; **(b)** O_3 gradient; **(c)** polluted site; **(d)** cuvette; **(e)** open-top chamber; **(f)** free-air CO_2 enrichment (FACE).

Table 17.2. *List of the main free-air CO_2 enrichment (FACE) experiments carried out on plants. Additional treatments applied at the experimental site besides CO_2 are reported as: O_3, ozone fumigation; N, fertilisation supply; T, temperature manipulation; W, water manipulation.*

Project	Location	Factors besides CO_2	Ecosystem
AspenFACE	Rhinelander, WI, USA	O_3	*Aspen plantation*
DukeFACE	Durham, NC, USA	N	*Loblolly pine forest*
POP-EUROFACE	Viterbo, Italy	N	*Poplar plantation*
ORNL-FACE	Oakridge, TN, USA	*Sweetgum plantation*	
Basel-WebFACE	Basel, Switzerland	*Broadleaf mixed forest*	
BangFACE	Abergwyngregyn, UK	*Beech, birch and alder plantation*	
SoyFACE	Urbana, IL, USA	O_3	*Soybean and corn*
SwissFACE	Eschikon, Switzerland	N	*Grassland*
AzFACE	Maricopa, AZ, USA	N, W	*Cotton, wheat, sorghum*
RiceFACE	Shizukuishi, Japan	N	*Rice and paddy field*
FALFACE	Braunschweig, Germany	N, W	*Crop rotation*
NewZealandFACE	Bulls, New Zealand	T	*Grazed pasture grassland*
NevadaFACE	Mojave Desert, NV, USA	*Desert ecosystem*	
TasFACE	Pontville, Tasmania, Australia	T	*Native grassland*
GiFACE	Giessen, Germany	W	*Grassland*
ChinaFACE	Wuxi, China	N	*Rice and winter wheat*
OzFACE	Yabulu, Australia	N	*Tropical savanna*
BioCon	Cedar Creek Natural History Area, MN, USA	N	*C_3 and C_4 grasses*
AGFACE	Horsham, Australia	N, W	*Rainfed wheat*

When experiments are carried out in the field only one or two factors are usually applied, whereas in laboratory conditions or in closed environments it is possible to combine multiple factors at the same time. However it is known that the same plant can exhibit very different responses if tested alone or if tested as part of an ecosystem, as the interactions at the ecosystem level are complex, and often exhibit negative or positive feedbacks. We believe that for a broad perspective on the effects of elevated [CO_2] or pollutants on photosynthesis we should focus mainly on the studies from large-scale experiments. Smaller-scale experiments will be also mentioned to refer to particular mechanisms or to show the likely response to multiple factors when tested in combination.

17.3. OBSERVATIONS ON THE EFFECTS OF ELEVATED [CO_2] ON PHOTOSYNTHESIS

The effect of CO_2 fertilisation on primary physiological processes of C_3 species, i.e., photosynthesis and stomata movement, is now firmly established (Drake *et al.*, 1997; Ainsworth and Long, 2005). Photosynthesis is the primary process in which CO_2 enters the biosphere (Sharkey *et al.*, 2004). As C_3 photosynthesis is not yet CO_2 saturated under current atmospheric concentrations, increasing atmospheric [CO_2] are expected to increase photosynthesis above current rates (Fig. 17.2). However several different aspects can interfere with this hypothesis. In the following sections the direct and long-term responses of photosynthesis to elevated [CO_2] will be examined.

17.3.1. Direct effects of CO_2 concentration on CO_2 assimilation

In C_3 species, short-term responses of the photosynthetic rate (A) to changes in intercellular CO_2 concentrations (C_i) are well known (Farquhar *et al.*, 1980a). Carbon dioxide serves as a substrate for photosynthesis and may be directly perceived by the surface of the guard cells of stomata and in the sub-stomatal cavity. Morison (1985), by analysing

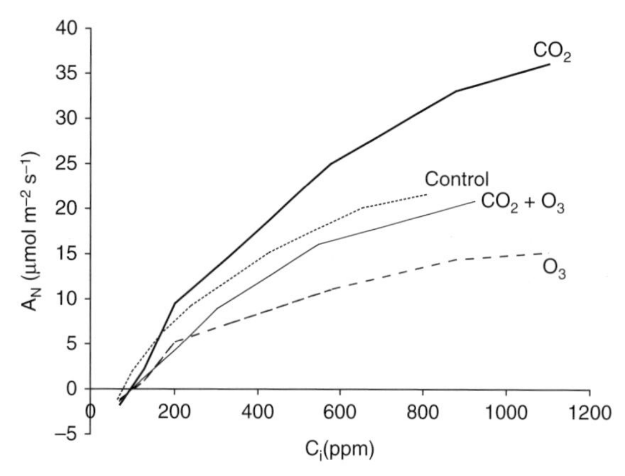

Fig. 17.2. Typical A_N/C_i response curves measured on *Populus tremuloides* trees (mean of different clones) grown for 8 years at the AspenFACE experiment under control (C), elevated CO_2 (CO_2), elevated O_3 (O_3) and elevated CO_2 + elevated O_3 (CO_2+O_3) conditions (adapted from Darbah *et al.*, submitted).

the response of A to PPFD, found that C_i reached a plateau value at low irradiance whereas A was still increasing, and suggested that stomatal conductance (g_s) is controlled in order to maintain a quasi-constant C_i in a changing environment. Mott (1988) showed that stomatal aperture responds to the C_i, such that the ratio of C_a to C_i remains approximately constant. Decreased g_s associated with high C_i is an adaptive response to the increase of C_a, by which diffusional limitations to A are adjusted in response to changes in mesophyll demand for CO_2. These adjustements often result in an increase in instantaneous transpiration efficiency (ITE) (Centritto *et al.*, 1999b,c, 2002; Wullschleger *et al.*, 2002).

However, because of the properties of Rubisco, photosynthesis is limited by substrate supply in ambient [CO_2]. In fact, because the K_c (Michaelis-Menten constant for CO_2) values measured for Rubisco are about 405–460 µmol mol^{-1} at 25°C (Long and Drake, 1992; Bernacchi *et al.*, 2001), Rubisco is not CO_2 saturated in current atmospheric [CO_2]. The response of Rubisco activation to [CO_2] was determined already two decades ago and follows a saturating curve with different slopes according to the presence of light and of RuBP (Portis *et al.*, 1986).

Carbon dioxide is in competition with oxygen for the active sites of Rubisco, thus increases in [CO_2] will shift the balance towards carboxylation and reduce photorespiratory loss (Stitt, 1991). However, the greenhouse effect brought about by anthropogenic emissions of CO_2 includes correlated increases in temperature. Increase in [CO_2] and temperature will have contrasting effects on the ratio of photorespiration to photosynthesis. Rising temperature will increase the solubility of O_2 and especially the specificity of Rubisco for O_2 relative to CO_2, and this will decrease the RuBP-saturated and the RuBP-limited rates of carboxylation, favouring oxygenation and thus increasing the proportion of photosynthesis lost to photorespiration (Jordan and Ogren, 1984).

As the atmospheric [CO_2] increases, carboxylation by Rubisco will be favoured and the depression of the rate of oxygenation relative to carboxylation by elevated [CO_2] will produce an upward shift in the temperature optimum of photosynthesis (Long, 1991). Ehleringer and Björkman (1977) have also shown that the maximum quantum yield (ϕ_{CO2}) of C_3 species decreases with an increase in temperature, as increasing amounts of the NADPH and ATP produced by electron transport are diverted into photorespiration. However, by decreasing photorespiration, elevated [CO_2] will reduce the decline in ϕ_{CO2} at all temperatures (Ehleringer and Björkman, 1977). Consequently, also the compensation PPFD is depressed at all temperatures by elevated [CO_2], and, as for photosynthesis and ϕ_{CO2}, the effect will be larger at higher temperatures (Long, 1991).

Responses observed at the chloroplast or leaf level do not necessarily scale to the whole-plant or ecosystem scale as carbohydrate-feedback inhibition of photosynthesis might occur when plants become sink-limited (Davey *et al.*, 2006). This implies for instance that fumigation experiments with potted plants might be biased by the constraints at root level that could in turn affect photosynthesis, especially over the long term. Not only can pots exacerbate limitations but, for instance in *Populus*, the stimulation of elevated [CO_2] on photosynthesis has been much higher in field experiments than in closed chambers, probably because of the limitations in sink strength in most of the experiments in closed environments (Gielen and Ceulemans, 2001).

17.3.2. Short-term response versus long-term acclimation

In short-term experiments elevated [CO_2] stimulates C_3 photosynthesis and the evidence for this is overwhelming. Long (1991) showed that A_{sat} can increase by 20% at 10°C and by 105% at 35°C, when [CO_2] is doubled from ambient values. Ceulemans and Mousseau (1994) reported that elevated [CO_2] stimulated photosynthesis by about 40% in conifers and 61% in broad-leaved species. In a survey of 60 experiments, Drake *et al.* (1997) found that photosynthesis was increased by about 58% in elevated [CO_2]. However, a recent meta-analysis review, focusing on large-scale FACE studies, showed a much lower stimulation of C_3 photosynthesis in response to elevated [CO_2] (Ainsworth and Long, 2005) and, consequently, highlighted the necessity of a more realistic assessment of the effects of elevated [CO_2] to minimise the experimental artefacts. Ainsworth and Long (2005) showed also that there is a strong difference in the CO_2 effect on PPFD-saturated photosynthesis (A_{sat}) among different functional groups. Although A_{sat} is usually stimulated by elevated [CO_2] because of the increased substrate availability for carboxylation, the stimulation ranged from almost 50% in trees, to a negligible effect in C_4 species, whereas for non-tree C_3 species the effect ranged from 10% to 40%. Because of the many factors that could limit A_{sat}, it is not surprising to find a wide variety of responses to elevated [CO_2].

Elevated [CO_2] was reported to stimulate photosynthesis in many field experiments with trees by about 50% as reported by Gunderson and Wullschleger, (1994) and in a meta-analysis by Medlyn *et al.* (1999), and fast-growing trees are hypothesised to be stimulated even more by elevated [CO_2] (Tjoelker *et al.*, 1999) because of the indeterminate growth and the theoretical unlimited sink capacity. However this idea that the stimulation by elevated [CO_2] usually stimulates photosynthesis and growth of plants with indeterminate growth more than for plants with determinate growth, presumably because of differences in sink strength, was introduced already two decades ago (Oechel and Strain, 1985). This is also probably one of the reasons why meta-analyses show that woody plants exhibit stronger stimulations in both photosynthesis (Fig. 17.3) and growth under elevated [CO_2] compared with the herbaceous plants (Ainsworth and Long, 2005).

Stimulation of photosynthesis by CO_2 is conditioned by many different factors, such as water status and VPD, nutrient status, temperature and presence of pollutants.

Results from the Biosphere 2 experiment show a stimulation by elevated [CO_2] on photosynthesis of *Populus deltoides* across different VPD treatments of 40%, with higher effects when high VPD and low soil-water content or high soil-water content and low VPD were tested in combination (Murthy *et al.*, 2005). A good nutrient status usually increases the stimulation by elevated [CO_2] simply because enhanced photosynthesis and growth might require additional uptake of nutrients (particularly nitrogen) that might become limiting over the long term (Finzi *et al.*, 2007). As C_3 plants invest between 10 and 30% of leaf N in Rubisco (Evans, 1989), a decrease of N availability can impact on the

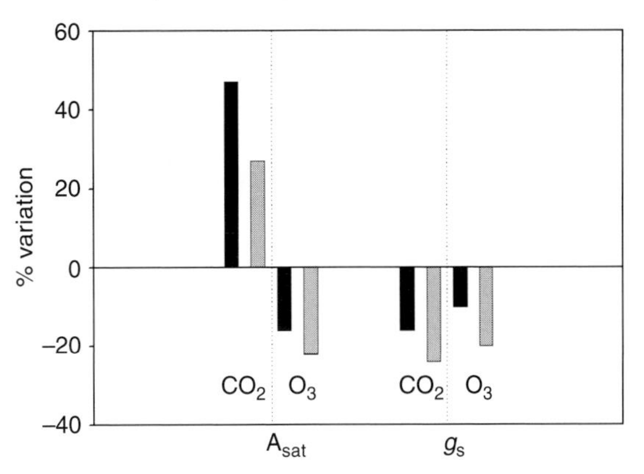

Fig. 17.3. Effects of elevated CO_2 and elevated O_3 concentrations on A_{sat} and g_s in trees (black bars) and herbaceous species (grey bars) according to the recent meta-analyses by Morgan *et al.*, (2003); Ainsworth *et al.*, (2005); Wittig *et al.*, (2007); Ainsworth, (2008).

total amount of Rubisco under elevated [CO_2]. Ainsworth and Long (2005) reported that stimulation in A_{sat} was lower by 23% under low N availability compared with conditions of good N availability.

The effect of O_3 and other pollutants on photosynthesis will be discussed below but it is evident that elevated concentrations of pollutants might suppress the stimulation of photosynthesis by elevated [CO_2]. The CO_2 fumigation under high ozone conditions has been tested in open-field conditions both at the AspenFACE and at the SoyFACE. While in aspen the stimulation by elevated [CO_2] on photosynthesis was basically cancelled by high O_3 levels (Noormets *et al.*, 2001a,b), in soybean the inhibition by O_3 was slight (Bernacchi *et al.*, 2007).

Besides photosynthesis, g_s is the parameter that has a crucial importance both for crops and natural ecosystems, not only because it is strictly linked with the CO_2 assimilation but also because of its relationship with water consumption, which is one of the main concerns in dry environments (Centritto *et al.*, 1999b; Wullschleger *et al.*, 2002). Elevated [CO_2] decreased g_s in open-field experiments with annual crops, grasses, shrub and tree species, although the effect was less than 20% for the last two groups (Ainsworth and Rogers, 2007). In some cases such decline was negligible, such as in three poplar species during the first rotation cycle at the POP-EUROFACE experiment (Bernacchi *et al.*, 2003b). However in the same experiment but in the second rotation cycle *Populus* × *euramericana* exhibited a significant decline in g_s both in the upper and lower canopy and both under fertilised and unfertilised conditions (Calfapietra *et al.*, 2005). It is evident that the response of g_s to elevated [CO_2] is influenced by the environmental conditions (Medlyn *et al.*, 2001; Nowak *et al.*, 2004) and in particular by water status, with smaller effects during dry periods (Centritto *et al.*, 1999b; Tognetti *et al.*, 1999; Centritto *et al.*, 2002; Gunderson *et al.*, 2002; Leakey *et al.*, 2004).

However, long-term growth in elevated [CO_2] may induce loss of photosynthetic capacity in C_3 species in an indirect manner through secondary plant metabolic adjustments that modulate photosynthetic gene expression leading to downward acclimation of photosynthesis (Moore *et al.*, 1999). This process, which is likely related to unbalanced source-sink relations, is seemingly highly variable and may occur when N uptake does not keep pace with carbon uptake resulting in an increase in the sugar-to-amino-acid ratio in leaves which, in turn, likely induces a decreased expression of photosynthetic genes through a hexokinase sensory system (Paul and Driscoll, 1997; Moore *et al.*, 1999; Stitt and Krapp, 1999; Long *et al.*, 2004; Ainsworth and Rogers, 2007). In addition, downregulation of maximal stomatal and mesophyll conductance to CO_2 has been reported in leaves of *Platanus orientalis* emerged under high CO_2 (Velikova *et al.*, 2009). However, this downregulation of photosynthesis rarely makes up completely for the stimulation of A in elevated [CO_2]. Both inadequate potting volume (Arp, 1991) and nutrition (Drake *et al.*, 1997), by altering the source/sink balance, can contribute to acclimation in the photosynthetic apparatus. In plants well-supplied with nutrients, acclimation of photosynthesis has not commonly been found under elevated [CO_2]. When plants growing in elevated [CO_2] are rooted in the ground and adequate sinks are available, downward acclimation of photosynthetic capacity does not generally occur (Farage *et al.*, 1998; Centritto *et al.*, 2004; Ghannoun *et al.*, 2010a,b). Besides photosynthesis, long-term acclimation to high CO_2 also results in a transcriptional reprogramming of metabolism that stimulates sugar metabolism and respiration (Leakey *et al.*, 2009a,b), which is accompanied by a greater number of mitochondria per cell (Griffin *et al.*, 2001).

According to the mechanistic model of CO_2 assimilation proposed by Farquhar *et al.* (1980a), the operating C_i in current atmospheric [CO_2] is usually at the transition between the limitations of photosynthesis caused by Rubisco activity and RuBP-regeneration capacity. Thus, as [CO_2] increases the control of photosynthesis shifts from being limited by the quantity of active Rubisco to being limited by the capacity for RuBP regeneration (Long and Drake, 1992; von Caemmerer and Quick, 2000), as determined by the maximum rate of electron transport and by stroma P_i cycling. Long and Drake (1992) calculated that about 35% of Rubisco content can be lost in elevated [CO_2] before Rubisco will co-limit photosynthesis. Masle *et al.* (1993) showed that growth in elevated [CO_2] of transgenic plants of *Nicotiana tabacum*, transformed to produce 13–18% less of the small subunit of Rubisco, was similar to that of the wildtype. This suggests that acclimation represents an optimisation of the distribution of the resources within the chloroplast to avoid that either Rubisco or the apparatus for the regeneration of RuBP are in excess (Sage *et al.*, 1989). Bernacchi *et al.* (2005a) have recently found that photosynthesis increased but $V_{c,max}$ (the maximum rate of carboxylation of Rubisco per leaf area) decreased in soybean, a nitrogen-fixing species grown under FACE conditions. This raised the operating C_i at which Rubisco-limited and RubP-regeneration-limited photosynthesis occurred. Following reduction in Rubisco, pigments of the light-harvesting complexes are usually decreased by elevated [CO_2]. Moreover, as protein turnover is energetically costly, this downward acclimation in photosynthetic capacity leads to reduced maintenance respiration in elevated [CO_2] through reduced amounts of tissue protein (Ziska and Bunce, 1994). Nitrogen redistribution away from non-limiting components may greatly increase NUE, as more carbon is assimilated per unit of leaf nitrogen, irrespective of the availability of nitrogen in the soil (Drake *et al.*, 1997).

Increasing CO_2 will shift the balance away from photorespiration and toward photosynthesis, increasing the net rate of photosynthesis (A_N), but this effect is typically smaller than the decreased photorespiration. Thus [CO_2] at high levels becomes less important in controlling the photosynthetic rates. Also, it is known that C_4 plants will take little advantage of the increase of CO_2 as they are already CO_2 saturated at current [CO_2] and because they already avoid photorespiration that is partially inhibited under elevated CO_2 for C_3 plants (Long, 1991).

Most commonly, downregulation of photosynthesis involves a reduction of $V_{c,max}$ and/or of the maximum electron transport rate (J_{max}), which often occur owing to restrictive experimental conditions, or by the end of the growing season owing to reduction in the sink strength. Generally, signs of downregulation have been observed in experiments using plants grown in pots or in small chambers (Norby *et al.*, 1999), although in some cases this has been shown also in open-field experiments. This is the case of *Lolium perenne* grown for 10 years in FACE conditions, which exhibited a decline of $V_{c,max}$ by 18% and of J_{max} by 9% under elevated [CO_2]. Nonetheless, even in experiments where downregulation is observed, photosynthesis under elevated [CO_2] is still stimulated relative to controls for long-term experiments (Medlyn *et al.*, 1999; Bernacchi *et al.*, 2005a), even after 7–10 years of treatment (Ainsworth *et al.*, 2003; Crous and Ellsworth, 2004). This occurred when the operating point of photosynthesis in elevated CO_2 was RuBP limited. When rates of electron transport used in the regeneration of RuBP limits photosynthesis in elevated [CO_2], downregulation of Rubisco may not necessarily lead to any loss in photosynthetic capacity. The opposite scenario, no loss in photosynthetic capacity with downregulation of the mechanisms controlling RuBP-limited photosynthesis, is not as likely under the hypothesis that future [CO_2] will drive photosynthesis to be RuBP-regeneration limited, especially when light is not saturating photosynthesis.

Downregulating responses under elevated [CO_2] in fast-growing species grown in open-field experiments have been observed just a few times, such as on *Populus tremuloides* (Ellsworth *et al.*, 2004) and on *P.* × *euramericana*, but only at the end of the growing season and only in the shaded leaves (Calfapietra *et al.*, 2005). However, in the POP/EUROFACE experiment, the level of soluble carbohydrates of both young and fully expanded leaves did not increase, but in contrast large increases in starch levels were observed (Davey *et al.*, 2006). In addition, no changes in levels of Calvin-cycle proteins or in the starch synthetic enzyme, ADP-glucose pyrophosphorylase, (AGPase) were observed (Davey *et al.*, 2006), which explained the sustained stimulation of photosynthesis after six years of experimentation (Liberloo *et al.*, 2007). Davey *et al.* (2006) suggested that the reductions in photosynthetic potential observed occasionally in the studies of Bernacchi *et al.* (2003b) and Calfapietra *et al.* (2005), were caused by short-term feedback responses occurring at the level of enzyme activity, rather than a change at the level of photosynthetic-protein content.

Generally we might assess that photosynthetic downregulation appears to be both growth-form and environment specific (Nowak *et al.*, 2004). In the shorter term the effect of CO_2 is direct and related to its availability or concentration. In the longer term the effects of elevated [CO_2] become more and more indirect and are mediated by source-sink interactions within plants, resources (nutrients, water), temperature, microbes, herbivores and land-use management practices. This becomes even more predominant when we upscale the carbon uptake from the leaf to the canopy level.

17.3.3. Upscaling the measurements and elevated [CO_2] response observed at leaf level

When we discuss the effect of elevated [CO_2] on photosynthesis we usually refer to leaf level, that is leaf gas-exchange measurements carried out with small cuvettes usually enclosing a leaf or a portion of it. When only the cuvette is fumigated, which usually means for short periods, measurements focus on the direct effects of elevated [CO_2] on the mechanisms of assimilation in the leaf, but do not take into account the feedbacks at plant level that might have strong repercussions on photosynthesis at leaf level. For instance, the effect of elevated [CO_2] on the water status of the plant could condition the stomatal opening, or the rate of translocation of carbohydrates outside the leaf might favour or suppress downregulation.

When the CO_2 enrichment is carried out at the level of the entire plant or entire ecosystem the interactions and feedbacks on photosynthesis owing to the effects of elevated [CO_2] outside the leaves are intrinsically included in the measurements. The problem, however, rises when we want to upscale the results from the leaf level to the entire plant or to the entire ecosystem. For instance, a parameter like WUE (carbon gain per unit water loss) at leaf level is often increased under elevated [CO_2] owing to concurrent increase in assimilation and decrease in transpiration (Centritto *et al.*, 1999b,c; Centritto *et al.*, 2002; Wullschleger *et al.*, 2002). However, at the stand level, elevated [CO_2] acts to produce a smaller effect on transpiration than what may be predicted from leaf-level measurements of g_s. Indeed, Wullschleger and Norby (2001), reported that canopy transpiration was significantly reduced in 12-year-old sweetgum trees exposed to FACE for a 7-month growing season only at the beginning of the growing season. Moreover, at ecosystem level the stimulation of LAI under elevated [CO_2] can be higher than (Bernacchi *et al.*, 2007), or potentially counterbalance (Kimball and Bernacchi, 2006) the decrease in g_s, increasing the overall transpiration of the canopy.

Estimations based on the measurements from a single, isolated leaf or plant are not good predictors of how well those same plants perform under intra- and inter-specific competition (Poorter and Navas, 2003). Moreover, plant canopy can be affected at a different extent by elevated [CO_2], which makes the extrapolation of results collected at leaf level more difficult. Data from the AspenFACE experiment for instance report a stimulation for the upper canopy leaves (26%) but not for the lower canopy leaves (Takeuchi *et al.*, 2001). Thus, even if we know that one of the major responses of the ecosystems under elevated [CO_2] is the increase in LAI, which might increase the assimilation of CO_2 by the whole canopy, we should distinguish the effect of elevated [CO_2] on each canopy layer.

Besides upscaling the CO_2 effects on photosynthesis on a ground base, it is also useful to upscale on a time base as leaf gas-exchange measurements are usually expressed per unit of time. In the first rotation cycle, at the POP-EUROFACE experiment, the stimulation of A_{max} by elevated [CO_2] in a multispecies short-rotation forestry (SRF) plantation was by 38% on average for the three species of poplars, but daily integrated rates of in-situ photosynthesis were even increased by 40 to almost 90% (Bernacchi *et al.*, 2003b).

The combination of spatial and temporal upscaling provides a more exhaustive estimation of the effect of elevated [CO_2] on photosynthesis at ecosystem level during one or multiple years, which is defined as GPP. This exercise is carried out through the use of models that usually take into account, besides gas-exchange measurements, LAI values and meteorological conditions. Given the combination of different factors, the effect of elevated [CO_2] on GPP can be different from the effect on leaf-level photosynthesis. It is likely that the stimulative effect of elevated [CO_2] will be very strong at the onset of the experiment when LAI of the ecosystem is largely expanding, whereas it decreases when stimulation of LAI or of photosynthesis decreases because of limitations by competition (Noormets *et al.*, 2001a,b; Wittig *et al.*, 2007). In many cases however, even when LAI is not stimulated, GPP can be stimulated as a result of increased light-use efficiency under elevated [CO_2] (Norby *et al.*, 2003, 2005).

17.4. OBSERVATIONS ON THE EFFECTS OF O_3 ON PHOTOSYNTHESIS

17.4.1. The effects of O_3 on the photosynthetic apparatus

Ozone is becoming a crucial factor for the vulnerability and productivity of natural and managed ecosystems for all areas of the planet (IPCC, 2007). The impact of O_3 on productivity had been largely overlooked for a long time because the focus was on changing [CO_2] as the main driver for climate change. A number of experiments are now providing evidence that O_3 can counteract the positive CO_2 effects on vegetation (Long *et al.*, 2005). Moreover it is becoming clear that O_3 pollution is potentially more of an issue in rural areas than in urban areas, and that local concentrations can result from transport of the precursors to O_3 formation (Gregg *et al.*, 2003).

Ozone diffuses into the leaf by entering through the stomatal aperture where it elicits a damage response. Before entering the leaf, O_3 induces little damage as it is only slightly damaging to cuticles. How the leaf responds to elevated [O_3] will vary based on numerous factors, including dose, species, location in the canopy and a range of other factors.

The first impact of O_3 is usually on the stomata and in particular on guard cells. Contrary to what is usually expected, the reduction of g_s that is often observed under high [O_3] is the result and not the cause of reduced assimilation, as the stomata close in response to increased C_i that occurs because of the reduced photosynthetic activity caused by O_3. Moreover, the effect of O_3 on stomatal opening might also be owing to the formation of H_2O_2 in guard-cell-activating plasma-membrane Ca-permeable channels (McAinsh *et al.*, 2002).

Once O_3 enters the leaf through the stomata from the atmosphere, it dissolves in water surrounding the cells before entering the cells inducing direct cellular damage. The reactions of O_3 with cell walls and the plasma-membrane component produce a wide range of compounds including hydrogen peroxide and hydroxyl radicals that can reach the cells if not detoxified and rapidly damage cell components such as lipids, proteins and nucleic acids (Pell *et al.*, 1997; Loreto *et al.*, 2004a). Damage to the photosynthetic machinery also occurs as a result of the generation of the highly reactive compounds and this in turn is shown to feedback on other aspects of leaf physiology. These feedbacks include a reduction in g_s as a result of high C_i (Long and Naidu, 2002). The damage to the photosynthetic apparatus induced by O_3 involves a decrease in Rubisco activity owing to a loss of the large and small subunits of Rubisco, which is often considered as a symptom of a premature senescence (Heath, 1996). Premature senescence is the most recognised effect of O_3 stress on leaves, and usually involves not only the decrease in Rubisco activity but also loss of chl. and soluble proteins.

Photosynthetic electron transport is also strongly inhibited by O_3 stress, as evidenced both by the PSII functionality and by the xanthophyll cycle. Many studies found a decrease in F_v/F_m, which is linearly related to the maximum light-capture efficiencies of PSII (Fiscus *et al.*, 2005), but this response might be transient owing to relatively fast reversible downregulation or rapidly repaired damage (Osmond, 1994). In particular a thermal dissipation by the xanthophyll cycle would lower F_0 (the basal fluorescence) whereas damage to PSII would increase F_0 (Guidi *et al.*, 2002). Decreased efficiency of PSII by O_3 damage is often accompanied by visible injuries, especially in the sensitive species and at high O_3 concentrations for all species. However the damage of photosynthetic apparatus by O_3 can often occur without any visible symptoms on the leaves, inducing decreases in A_N and yields.

Ozone has been found to generally reduce photosynthesis in many plant species, both trees and herbs, although the extent of this reduction varies based on sensitivity and the exposure conditions (Betzelberger *et al.*, 2010). In a recent meta-analysis Wittig *et al.* (2007) reported the effect of O_3 on leaf gas exchange from different experiments and listed the genus *Populus* as one of the most sensitive among the tree species studied, with a mean decrease of A_{sat} of 27% when comparing plants grown under ambient O_3 versus charcoal-filtered air. Reductions of 21% were reported when comparing plants exposed to elevated versus ambient $[O_3]$. Significant differences were also reported among different genotypes within a species. For example, *P. tremuloides* clones demonstrated a wide range of responses in OTC experiments, and at the AspenFACE experiment with sensitive genotypes exhibiting a strong decrease in A_{sat} and tolerant genotypes showing slight differences under high O_3 relative to the controls (Karnosky *et al.*, 2003; Calfapietra *et al.*, 2008). These differences have been related to different defence mechanisms that regulate the sensitivity to O_3 and that will be discussed in detail in the following sections.

The direct and indirect influences of O_3 damage on plants have been shown to reduce growth by 3% for conifers, 13% for hardwoods and as much as 30% for crops (Reich and Amudson, 1985). The large effects observed for many crops compared with tree species are probably related to the higher g_s and to the lower defence mechanisms, such as the production of antioxidant compounds, as crops have been selected generally to maximise yields regardless of the exposure to abiotic stresses (Ainsworth, 2008). Recently, large manipulative experimental facilities have been constructed to investigate the effects of O_3 on some important crops such as soybean, rice, wheat and corn. These experiments are designed to assess the yield loss predicted for O_3 concentrations predicted for 2050. Decreases in yields ranged from 4 to 12% depending on the crop as a result of decreased photosynthesis (Long *et al.*, 2005). Two meta-analyses, one with rice and one with soybean, showed similar decreases in g_s and A for these two crops under elevated $[O_3]$ (Morgan *et al.*, 2003; Ainsworth, 2008).

17.4.2. The response of photosynthesis to O_3 is not only related to O_3 exposure

Assessing the impact of O_3 on photosynthesis is very difficult because $[O_3]$ can be highly variable, and other factors such as drought, nutrient status or age of the plant and even of the leaf might heavily condition the response. Apart from the variability that is potentially induced by these factors and the species investigated, experiments studying O_3 effects on photosynthesis might differ considerably based on the O_3 treatment, which makes it extremely difficult to interpret the findings. For example, some experiments test ambient $[O_3]$ versus filtered air (= no ozone), whereas others test elevated $[O_3]$ versus ambient $[O_3]$ taken as control. Additionally, most experiments fumigate with differing $[O_3]$ and for varying time periods, including duration and time of day.

The impact of O_3 on A varies drastically whether acute or chronic exposure is applied. Typically the first is characteristic of laboratory experiments, whereas the latter is more common in field experiments. Some O_3-sensitive species show declines in A with treatments that would be considered mild for other species. For example, soybean shows significant declines in A and consequently yield for chronic elevation of O_3 (40–60 ppb) (Ashmore, 2002), which is likely to have little or no impact on corn.

Although decreases in g_s, A and Rubisco activity are among the earliest effects of O_3 exposure, decreases in leaf chl. content will mainly occur during prolonged exposures to O_3 (Loreto *et al.*, 2004a). Decreases in A and Rubisco activity/content are early symptoms of O_3 exposure, which may be followed by accelerated senescence, decreased leaf area and imbalances in carbon partitioning that might exacerbate the negative effect of O_3 over the long-term period (Long and Naidu, 2002). Simultaneously, enzymes associated with the scavenging of the ROS in response to plant stress are shown to increase with O_3 exposure (Sandermann, 1996).

In some cases the reduction in A increases linearly with concentrations of background O_3, consistent with the exposure – response relationship described by Mills *et al.* (2000). For some tree species, such as yellow poplar, loblolly pine and white pine, there is a strong correlation between $[O_3]$, visible injury and reductions in A and growth, although this correlation is not apparent for all species (Felzer *et al.*, 2007), possibly as a result of plants adopting a series of defence mechanisms. There is now evidence from genetic studies that response to O_3 is controlled by a complex genetic regulation, and that O_3 sensitivity can be conferred by deficiencies in a number of different metabolic pathways (Overmyer *et al.*, 2008). This concept is crucial because until now most of the risk assessments have been based on the relationship between external concentration of the pollutant and plant response, which is evidently an inadequate index (Ashmore, 2005).

Contrary to other pollutants, O_3 is highly reactive, which implies that it can be scavenged immediately after entering the substomatal cavities. Thus, the effective O_3 reaching the cells, which in most cases is considerably lower than the O_3 outside the leaf, is to be considered. When g_s is very high, a saturation of the scavenging capacity of the pollutant inside leaves may be reached, and O_3 starts to accumulate inside the mesophyll (Loreto and Fares, 2007). This recent finding disproves the established notion that O_3 immediately reacts after entering stomata (Laisk *et al.*, 1989), and opens new perspectives about the estimation of an internal [O_3], similarly to what was done for CO_2 (Fares *et al.*, 2006).

Moreover, in the presence of antioxidants, the rate of O_3 removal should be higher, despite the lower damage (Fig. 17.4), thus increasing the O_3 flux and decreasing the saturation inside the intercellular spaces, at least under maximal measurable levels of g_s. This has been demonstrated for example with isoprene and monoterpenes, two classes of volatile compounds (Loreto and Fares, 2007). It may be concluded that the higher O_3 flux is not generally reflected by a higher level of O_3 damage, therefore disproving the firmly established notion that O_3 flux is directly associated with O_3 damage (Emberson *et al.*, 2000).

When A is integrated over a whole plant or even an entire ecosystem, O_3 effects are usually exacerbated compared with the single-leaf measurements, because of the negative effects on LAI and generally on imbalances in carbon partitioning (Karnosky *et al.*, 2003).

Finally, different climatic and atmospheric conditions may lead to variable impacts of O_3 on A. Given the increase of atmospheric CO_2 concentration that usually induces a decrease in g_s, it is likely that the impact of O_3 on plants will be less severe in the future. This assumption implies that [O_3] remains constant, which appears unlikely as the industrialising nations have yet to adopt clean-air standards and long-distance transportation can create problems in parts of the world that may emit fewer ozone precursors. Another factor that might reduce the g_s and thus the potential impact of O_3 on A in future conditions is the increase of drought episodes in several areas across the planet.

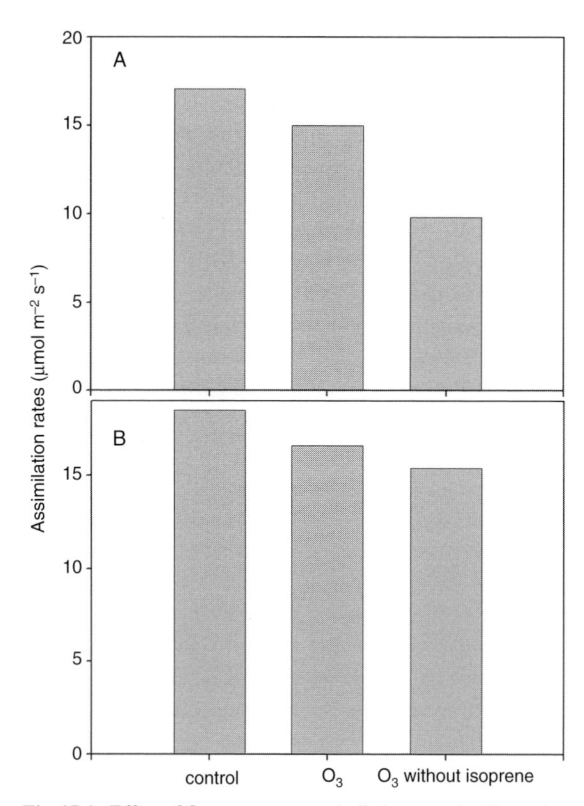

Fig. 17.4. Effect of O_3 exposure on assimilation rates in *Phragmites australis* under control, under high O_3 levels and under high O_3 levels inhibiting isoprene synthesis with the use of fosmidomycin. **(a)** 300 ppb O_3/3h; **(b)** 100 ppb O_3/8h. Adapted from Loreto and Velikova, (2001).

17.5. THE EFFECTS OF OTHER AIR POLLUTANTS ON PHOTOSYNTHESIS

17.5.1. The compounds of nitrogen

The main nitrogen compounds considered as air pollutants include nitrogen oxides, ammonia and all the products deriving from their reactions, such as acids and peroxyacyl nitrates (PAN). Nitrogen oxides include various oxygenated nitrogen compounds, and are common products of combustion (e.g., car exhausts). One subset, nitrogen dioxide (NO_2) and nitric oxide (NO), are collectively called NO_x. During the day, NO_2 is converted to NO and the extra oxygen radical that is released can join with O_2 to make O_3. When there is a significant concentration of NO_x, hydrocarbons in the atmosphere will react with this pollutant, increasing the formation of O_3. Thus, there is a strong interaction between NO_x, isoprene and anthropogenic hydrocarbons and O_3 pollution. In addition, wet deposition of NO_x can occur with precipitation that fertilises ecosystems and may directly affect them through photosynthetic changes.

It has been shown that NO_x can stimulate A at very low doses especially when there is nitrogen deficiency in the soil, but become toxic as the dose increases (Mansfield, 1999). In other cases NO_2 may cause disorder in some important regulatory mechanisms, resulting in decreased A (Mansfield, 2002). Saxe and Christensen (1985) and Saxe (1986) carried out experiments trying to distinguish the impact of NO_2 and NO, and they found that 1 ppm of NO inhibitied A on average four times more than NO_2. In an experiment with lettuce, Caporn *et al.* (1994) showed that an addition of 2 ppm of NO caused a rapid but reversible decrease of A_N, especially at lower temperatures, despite the lower uptake of the pollutant compared with control. Kume *et al.* (2000) observed that maximum net photosynthesis of *Pinus densiflora* in several areas declined by about 30% as compared with other areas, which was explained by the decrease in maximum stomatal conductance owing to NO_x-related substances rather than O_3. Sakugawa and Cape (2007) showed that the main effects of nitrous-acid (HONO) fumigation was a decrease of the ratio of chl. *a* to chl. *b*, and PSII (F_v/F_m).

Ammonia (NH_3) has long been known for its toxicity to plants and it was suggested that it might impact on protein breakdown (Levitt, 1972). However a mild exposure, such as to 1000 ppb of NH_3, resulted in an increased A of plants in *Triticum aestivum* (Clement *et al.*, 1995). This stimulative effect was confirmed in a review by Krupa (2003), which presents a number of other studies where an increase of both g_s and A_N has been shown. Ammonia is efficiently taken up from the atmosphere and can be incorporated into amino acids through the normal nitrogen-uptake metabolism (Farquhar *et al.*, 1980c).

Peroxyacyl nitrate (PAN) has received growing attention both for its potentially toxic effect on human health and plants, but despite the field evidence for damaging effects on vegetation (Cape, 2003) there have been few long-term experimental studies. Previous studies revealed that the main effect of PAN was an inhibition of cyclic photophosphorylation in isolated chloroplasts, and that stronger effects were noticed on younger expanding tissues (Levitt, 1972). Decreased A has been reported in studies with lichens both with chronic and acute exposures (Eversman and Sigal, 1984), whereas very brief (approximately 5–10 min) exposure to PAN during uptake experiments did not affect photosynthesis, transpiration or conductance to *water vapour* in different herbaceous species (Teklemariam and Sparks, 2004).

It is evident that there is still too little information on the spatial and temporal variation in PAN concentrations worldwide and on the real effects on plants to assess the long-term and short-term risks to vegetation from exposure to PAN and related photooxidants (Cape, 2003).

17.5.2. The compounds of sulphur

Sulphur dioxide (SO_2), which is released primarily from burning fuels that contain sulfur (e.g., coal, oil and diesel fuel), is considered as the most important phytotoxic molecule (Legge *et al.*, 1998). When SO_2 is present at low concentrations in the atmosphere, plants that are unable to obtain adequate S nutrition from the soil can fulfil their requirements through stomatal uptake (DeKok *et al.*, 1998). However higher levels of SO_2 can produce serious damage to the photosynthetic apparatus, even inducing visible injuries on the leaves often occurring when leaf surfaces are wet and SO_2 is converted in sulphuric acid (H_2SO_4). Both acute and chronic exposure to SO_2 induce a decrease in A (Winner *et al.*, 1985), but the effects on g_s can be adverse. Mansfield and Pearson (1996) observed that short-term exposures often cause an increase in stomatal opening, whereas long-term exposure tends to decrease stomatal opening.

In a chamber experiment with *Medicago sativa*, A decreased when SO_2 concentrations were higher than 60 ppb. This was the result of a strong, negative effect on g_s rather than a limitation in the photosynthetic machinery (Nali *et al.*, 1995). Generally the effect of SO_2 fumigation does not compromise the photosynthetic apparatus through drastic modifications in the chloroplast or in the chl. content, although a typical symptom often observed was the occurrence of dilated thylakoids (Müller *et al.*, 1997). It is also unlikely that the inhibitory effect of SO_2 on A can be attributed to biochemical impairments, for instance to the inactivation of ribulose bisphosphate carboxylase, or to lower levels of the substrate RuBP (Darrall, 1989). In a number of herbaceous species, such as *Triticum aestivum* and *Brassica napus*, A was only inhibited at SO_2 concentrations of 500 ppb or above, but in other species A was inhibited by SO_2

treatment by 2–3 times compared with control (Darrall, 1989).

Decreases in g_s and A were also registered in experiments on trees such as *Pinus halepensis* and *Picea abies*, although in this study chl. content usually remained unaltered (Barrantes *et al.*, 1997). Also changes in PSII activity and leaf-reflectance features of several subtropical woody plants have been observed under simulated SO_2 treatment (Liu *et al.*, 2006). A long-term case study of SO_2 and elemental S deposition effects on a boral pine forest in Canada showed a decrease in A and in the needle ATP content, with a gradient of response exhibiting maximum ATP concentrations closer to the sources of S gas and S dust (Legge and Krupa, 2002).

17.5.3. Particulate matter

Particulate matter (PM) is a mixture of particles that can adversely effect human health, damage materials and also compromise the normal leaf gas exchange in plants. Though PM is a major pollutant, there is little indication that it may be harmful to vegetation. PM is distinguished according to the size of particles in micrometres ($PM_{2.5}$, PM_{10}, etc.) and, while the smallest particles could be damaging by entering the stomata and reacting in the substomatal cavities, large particles can simply occlude leaf-stomata aperture (Farmer, 2002).

Besides stomatal effects, reduced A in leaves exposed to particles can be the result of simple shading or a decline in metabolic function owing to cell structural damage or toxicity. This depends also on the chemical composition of particles. Particles might include toxic compounds that directly damage leaves, such as caustic dust or coal dust containing high levels of sulphur and fluoride (Rao, 1971). Reduced A is reported in a number of studies, including both herbaceous and tree species, as a result of exposure to a series of particulates, such as cement kiln, inert dust and inert silica gel (Farmer *et al.*, 2002).

Cement dust liberates calcium hydroxide on hydration, raising leaf-surface alkalinity, in some cases to pH 12. This level of alkalinity can hydrolyse lipid and wax components, penetrate the cuticle and denature proteins, ultimately plasmolysing the leaf (Guderian, 1986). In one experiment short-term application (2–3 days) of cement kiln dust (Darley, 1966) yielded dose-specific response curves between A_N inhibition or foliar damage and dust application rate, whereas in another experiment the photosynthesis system and the cell membranes showed no changes as a result of the different treatments of particulates in Rosa and Petunia (Lobel *et al.*, 2001).

The effects of PM on trees in urban environments have also been extensively reported in a review by Farmer (1993). The main responses observed have been limited to reduction in A, diffusive resistance and an increase in leaf temperature, the latter two effects making the tree more likely to be susceptible to drought.

17.5.4. Mixtures of gases: synergistic and antagonistic effects

The effects of the main air pollutants on the photosynthetic apparatus and on assimilation rates have been presented above, but a mixture of the gases is likely to impact A in a manner that is quite difficult to determine. Polluted environments are difficult to simulate in large-scale manipulated experiments, and observations are usually limited to plants grown in urban areas. This raises difficulties in finding plants grown in a clean atmosphere to act as control.

Most experiments focus on one or two factors at one time, especially in the case of large-scale facilities. Long-term exposure to a combination of elevated $[CO_2]$ and $[O_3]$ in the field resulted in soybean having a slightly smaller increase in average assimilation than when CO_2 alone was fumigated, and was significantly greater than the control on 67% of days (Bernacchi *et al.*, 2006). This positive response on A by combination of elevated CO_2 and O_3 was confirmed in aspen. However, A_{max} only increased in young aspen leaves exposed to the combination of CO_2 and O_3, but not in older leaves, and the effect on A at whole-canopy level was not significant (Karnosky *et al.*, 2003). It has been also demonstrated that variations in climatic conditions exacerbate the positive effect of CO_2 relative to O_3 or vice versa (Kubiske *et al.*, 2006).

Most of the literature on mixture of pollutant gases has focused in the last decades on the combination of SO_2, NO_2 and O_3 in particular because of the widespread occurrence of these gases. However, a simultaneous occurrence of air pollutants like SO_2, NO, NO_2, O_3 or NH_3 at phytotoxic levels is unusual (Fangmeier *et al.*, 2002). Many gases such as NO_x, NH_3 and SO_2 induce acidification of subcellular compartments, whereas others such as O_3 and PAN result in oxidative stress (Barnes *et al.*, 2007). Both mechanisms can induce strong inhibition in the photosynthetic processes, but it has been recognised that NO_2 and SO_2 can have synergistic or additive interactions, whereas O_3 has shown either synergistic or antagonistic effects with respect to these two gases (Marshall, 2002).

It is known that the effects of pollutants on the photosynthetic activity are often caused by changes in the expression of defence-related genes and post-translational modification of enzymes such as phosphorylation/dephosphorylation (Kangasjärvi *et al.*, 1994). However, some pollutants (e.g., SO_2 and NO_2) are much less effective than others (O_3) in eliciting changes in antioxidant gene expression (Schrauder *et al.*, 1997). In conclusion, the understanding of the response of A to exposure to combined gases would probably gain from the modelling approach provided that new models include enough parameters to take into account most of the mechanisms affected by the pollutants.

17.6. PHOTOSYNTHESIS AND ATMOSPHERIC COMPOSITION: A DYNAMIC INTERACTION IN A CHANGING ATMOSPHERE

In this chapter it has been shown that photosynthetic adaptation to change of atmospheric composition mainly occurs at the metabolic and biochemical level or through morphological modifications. It was also highlighted that, when moving from the small to the large scale (i.e., from cell to community) the overall response of photosynthesis to changes in the atmospheric composition is driven mainly by indirect effects rather than by direct effects. However, the impact that carbon assimilated by vegetation can have on atmospheric composition itself at a global scale is often disregarded. Carbon uptake through photosynthesis is a crucial component of the global carbon cycle, although the stimulation in photosynthesis with rising CO_2 is not able to offset the anthropogenic emissions. Without CO_2 assimilation by plants however, CO_2 in the atmosphere would be much higher than current concentrations (Schimel, 2007).

In addition, although rising [CO_2] might have positive repercussions on both leaf- and community-level photosynthesis, it can alter competition among different species in natural ecosystems, e.g., enhancing proliferation of some weedy species. Finally, in the long term, rising [CO_2] might lead to a progressive nitrogen limitation owing to increased growth and consequent additional nitrogen uptake, which might negatively impact on photosynthesis itself (Luo *et al.*, 2004).

To limit future increment of [CO_2], large plantations are being established in many areas of the planet, especially using fast-growing species such as poplars, willows and eucalyptuses, which are able to assimilate large amounts of carbon in a relatively short time (FAO, 2005). Unfortunately, most of these plantations are known to emit large amounts of volatile organic compounds that can increase the potential of O_3 formation and other photochemical pollutants, especially in areas with a high presence of NO_x. Thus, the use of large-scale plantations might limit the increment of [CO_2] in the atmosphere but might exacerbate the pollution by photochemical compounds, with possible negative consequences on the capacity of plants for CO_2 assimilation over the long-term period. Moreover, the development of biofuel feedstocks utilising tree and perennial crops are anticipated to decrease the CO_2 emissions from fossil fuels. Owing to the uncertainties and the difficulties in estimating such effects and feedbacks, it is important that these processes and interactions be included in climate modelling efforts to appropriately improve terrestrial-ecosystem carbon-sequestration capacities over time.

18 • Response of photosynthesis to low temperature

I. ENSMINGER, F. BERNINGER AND P. STREB

18.1. LOW-TEMPERATURE EFFECTS ON PHOTOSYNTHESIS: AN INTRODUCTION

18.1.1. Chilling and freezing temperatures

Low temperatures represent a major abiotic constraint to the distribution, development and productivity of many plant species. Plants have evolved adaptations to cope with chilling or freezing and to be able to acclimate to low temperature (Allen and Ort 2001; Wisniewski *et al.*, 2003; Slot *et al.*, 2005; Ruelland *et al.*, 2009). These include dormancy, rapid acclimation and maintenance of cold hardiness during prolonged low-temperature or freezing periods (Howe *et al.*, 2003; Wisniewski *et al.*, 2003). The responses to chilling differ from the responses to freezing temperatures. In temperate regions, chilling refers to non-freezing temperatures (0–12°C) during the growing season with the lowest temperatures typically occurring during the night (Allen and Ort, 2001). In contrast, freezing or frost requires temperatures below 0°C. Adaptation to freezing includes mechanisms to prevent freezing injuries from the formation of ice inside the plant cell, which would result in deleterious damage or death of the cell (for recent reviews on acclimation to freezing temperatures see Kalberer *et al.*, 2006; Thomashow, 1999; Xin and Browse, 2000). Other mechanisms involved in this process include e.g., changes in lipid composition or the reactive-oxygen-scavenging-system. Together this suite of acclimation responses involving various metabolic, physiological and developmental aspects is called cold hardening or cold acclimation (Xin and Browse, 2000).

The ability of a species to acclimate to low temperatures via the cold hardening process allows one to distinguish cold-hardy from non-hardy plants, e.g., plants that have the genetic capacity to acclimate to chilling or freezing versus plants lacking this genetic information (Allen and Ort, 2001). These cold-hardy plants are typically from cold or temperate environments. Importantly, even cold-tolerant plants are at risk of being killed by low temperatures. For example, unhardened wheat plants, which are genetically adapted to sustain freezing temperatures, will not be able to survive temperatures of –5°C if they are not acclimated to low temperatures through an adequate cold-hardening period (Thomashow, 1999).

Exposure to low temperature and light primarily requires adjustments in the rates of photochemical processes to a decreased metabolic-sink capacity, e.g., the consumption of photoassimilates. Mechanisms to achieve this adjustment during short-term and long-term exposure to low temperature include changes in energy absorption and photochemical transformation through energy partitioning and concomitant changes in chloroplastic carbon metabolism, allocation and partitioning. Thus, the mechanisms involved in photosynthetic acclimation to low temperature range from modifications within the thylakoid membrane system affecting PET (Huner *et al.*, 1998, 2003), to post-transcriptional activation and increased expression of enzymes for sucrose synthesis, altered expression of the enzymes of the Calvin cycle, changes in leaf protein content as well as the signals that trigger these processes (Stitt and Hurry, 2002). As revealed by large-scale genetic analysis, all these mechanisms are orchestrated by signalling pathways involved in the process of low-temperature acclimation of photosynthesis and the network that actually controls photosynthesis. As we will discuss later, there is a spatial integration of signals derived from the redox state of the chloroplast and those originating from the sugar status of the cell. In contrast to cold-tolerant species, most plants from tropical or subtropical regions lack this ability to acclimate to low or freezing temperatures and can be injured by temperatures below 10°C (Xin and Browse, 2000).

Terrestrial Photosynthesis in a Changing Environment: A Molecular, Physiological and Ecological Approach, ed. J. Flexas, F. Loreto and H. Medrano. Published by Cambridge University Press.

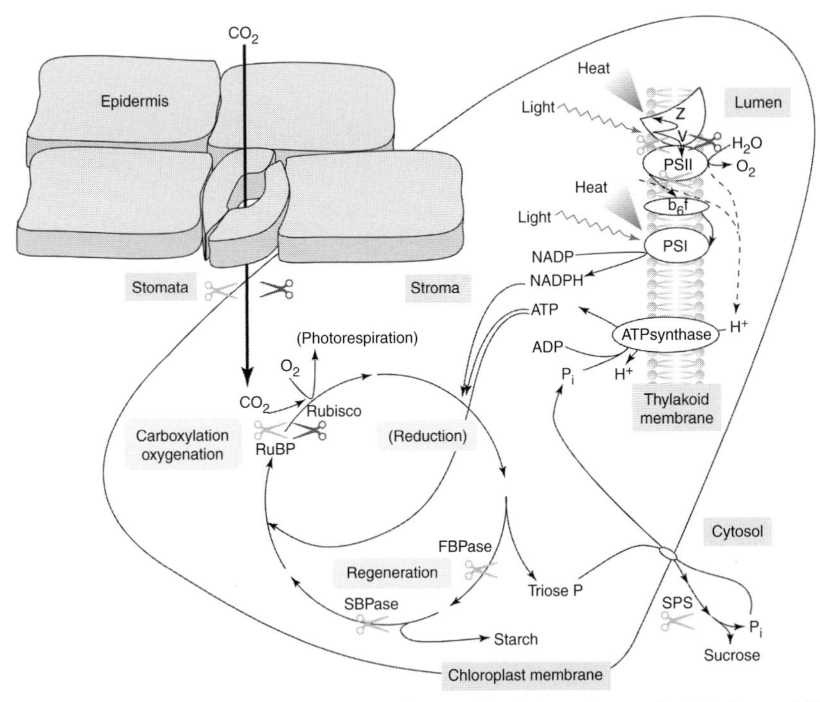

Fig. 18.1. Primary effects of a short chill in the light and the dark on photosynthesis in thermophilic plants. Chilling effects are apparent within the processes of photophosphorylation in the thylakoid membrane, the carbon reduction cycle in the stroma, carbohydrate use in the cytosol and the CO_2 supply to the chloroplast through the stomata. Abbreviations: ATPsynthase, chloroplast ATP synthase; b_6f, cytochrome b_6f complex; FBPase, chloroplast fructose 1,6-bisphosphatase; P_i, inorganic phosphate; PSI, photosystem-I complex; PSII, photosystem-II complex; RuBP, ribulose 1,5-bisphosphate; SBPase, sedoheptulose 1,7-bisphosphatase; V, violaxanthin; Z, zeaxanthin and antheraxanthin; light-grey scissors represent the primary impact of a light chill; dark-grey scissors represent the primary impact of a dark chill (from Allen and Ort, 2001).

Before we discuss the various cold-acclimation mechanisms and strategies in detail, we will briefly focus on the components of photosynthesis that are directly affected by chilling. Many of the studies that focus on chilling effects have been carried out in plants adapted to warm climates and not in cold-tolerant temperate plants, because in tropical and subtropical species the effects of low temperatures are not masked by a multitude of protective mechanisms (Allen and Ort, 2001). The components of photosynthesis that are directly affected by low temperature include PET, the Calvin cycle and stomatal and mesophyll conductance (Fig. 18.1). We will also briefly adress the effect of chilling during illumination versus chilling during the night in each section and demonstrate that night chilling can effect subsequent day photosynthesis. Although there is a wealth of literature covering the response of C_4 photosynthesis to low temperature (e.g. Kubien and Sage, 2004; Sage and Kubien, 2007; Wang *et al.*, 2008), here we will focus on C_3 photosynthesis only. General aspects of C_4 versus C_3 photosynthesis are covered in Chapter 5.

18.1.2. Low-temperature effects on photosynthetic electron transport

Photosynthesis generally responds to low temperature with decreased rates of carbon fixation. In grapevine, maximum photosynthetic carbon fixation decreases by almost 80% in chilled plants (5°C) compared with control plants (25°C) (Fig. 18.2A). A similar response is apparent in the rates of PET (Fig. 18.2D). Nonetheless, in grapevine this effect usually reflects dynamic photoinhibition as photosynthesis after chilling is able to recover within days (Flexas *et al.*, 1999b). Apparently, there are remarkable differences in the degree and the mechanisms involved in the plants response to chilling in the light and chilling in the dark (Allen and Ort, 2001). For example, in grapevine a sequence of chilling

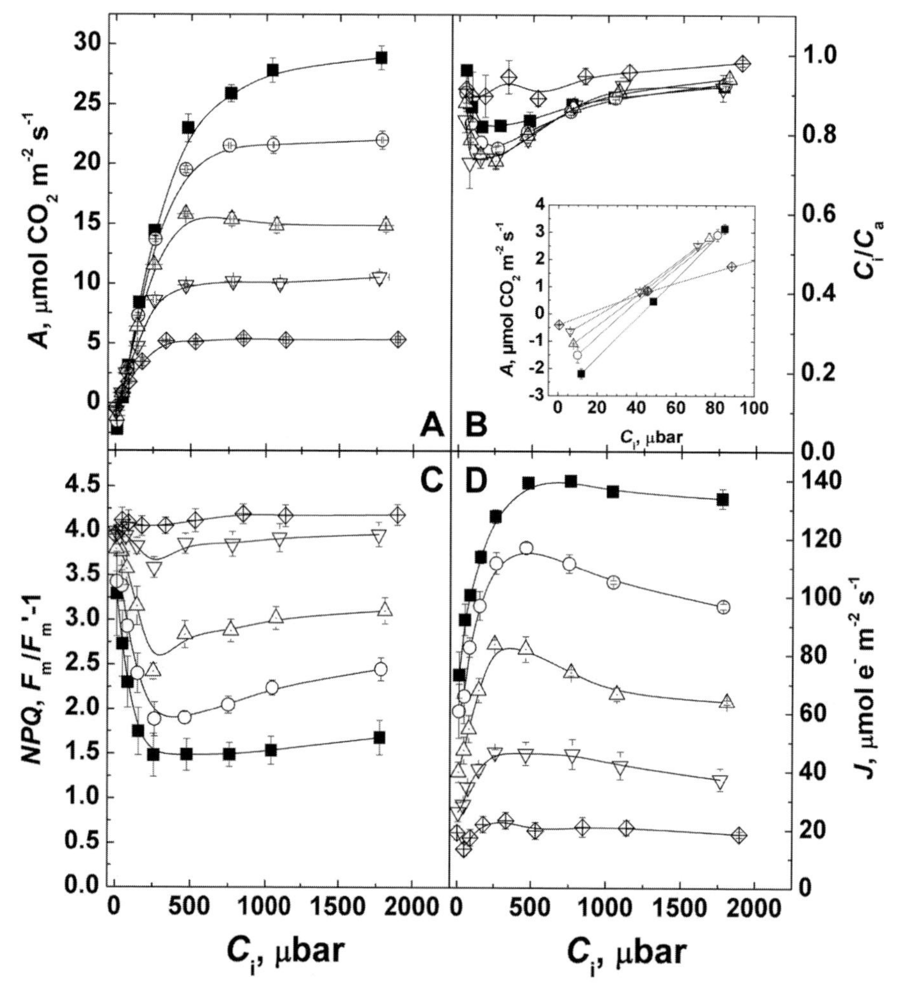

Fig. 18.2. CO_2 and temperature dependence of photosynthetic carbon fixation, *A* (**A**), intercellular to atmospheric CO_2 partial pressure ratio, C_i/C_a (**B**), non-photochemical quenching, NPQ (**C**) and linear electron transport rate, J_f (**D**) for mature grapevine leaves. Leaves were illuminated at 800 μmol quanta m^{-2} s^{-1} and measured at 25 (■), 20 (○), 15 (Δ), 10 (∇) and 5°C (◊). Average intrinsic quantum use efficiency, F_v/F_m, of the leaves was 0.81 ± 0.01. The inset in (**B**) is an expansion of the linear part of the CO_2 response curve in (**A**). Each point is the mean±SE of three leaves (from Hendrickson *et al.*, 2004).

nights normally does not result in any immediate photodamage but a transient and reversible decrease of light-saturated photosynthesis (Flexas *et al.*, 1999b). The inhibitory effects of chilling were greater after a period of illumination. In contrast, an inhibition of PSII photochemistry that is not rapidly reversible is observed when non-hardy plants such as tomato or mango are exposed to the combination of chilling temperatures and high light (Sassenrath *et al.*, 1990; Allen *et al.*, 2000). Inhibition that is not rapidly reversible is also termed chronic photoinhibition of PSII and is tightly coupled to the damage and repair of the PSII reaction-centre protein D1 (Aro *et al.*, 1993).

Why is there a difference in the response to chilling in the dark versus chilling in the light? In general, a decrease in temperature decreases the enzymatic reaction rates and hence CO_2 fixation and photorespiration as the main sinks for the absorbed excitation energy when the plant is exposed to light (Huner *et al.*, 1998). Any decrease in the rates of CO_2 fixation or photorespiration reduces the consumption of electrons generated through temperature-insensitive photochemical reactions and PET. Optimum plant performance requires a balance in the rates of source-versus-sink processes, but as the electrons cannot be removed fast enough by metabolic sinks, the risk of photooxidative damage to PSII (e.g., by direct photoreduction of O_2) increases. In this process the D1 reaction-centre protein of PSII represents the primary target for photodamage. Loss of D1 is a severe impairment of PSII, severely affecting the photochemical quenching (Aro *et al.*, 1993; Long *et al.*, 1994). In addition, low temperature tends to inhibit electron transport via increased membrane viscosity through alterations in the biophysical properties of thylakoid lipids. This further decreases the rates of the enzymatically catalysed biochemical reactions involved in C, N and S reduction more strongly than it inhibits photophysical processes involved in light absorption, energy transfer and its photochemical transformation (Huner *et al.*, 1998; Hendrickson *et al.*, 2006).

In contrast, overnight chill induces a combination of inhibition of photosynthetic metabolism (see Section 18.1.3) and stomatal closure (Section 18.1.4), but not in electron transport capacity. For example, in isolated chloroplasts of tomato that were exposed to chilling in the dark, electron transport capacity always exceeded that required to support light- and CO_2-saturated photosynthesis (Kee *et al.*, 1986).

18.1.3. Inhibition of the Calvin cycle and carbohydrate metabolism

Carbohydrate metabolism is highly sensitive to sudden low-temperature exposure. Sassenrath *et al.* (1990) demonstrated in tomato the particular involvement of the two stromal bisphosphatases, SBPase and FBPase, in the low-temperature downregulation of photosynthesis. They observed a decrease in the activity of SBPase and FBPase caused by a decreased light-dependent activation of these enzymes by the ferredoxin-thioredoxin system, which directly resulted from low-temperature exposure at high light (Sassenrath *et al.*, 1990). Similarly, in maize low-temperature exposure and high light for 24 h caused a more than 50 percent decrease in maximum FBPase activity after rewarming, which was attributed to a net decrease in the enzyme (Kingston-Smith, 1997). From decreased rates of Rubisco carboxylation at all temperatures below 25°C and particularly at 5°C in grapevine leaves, Hendrickson *et al.* (2004b) also concluded an impaired Rubisco activation state by low-temperature-induced end-product limitiation.

In chilling-sensitive species, chilling during the night can also cause severe impairment of the diurnal regulation of the carbohydrate metabolism. In tomato, overnight chilling impairs circadian rhythms in enzyme activities of sucrose phosphate synthase (SPS) and nitrate reductase (Jones *et al.*, 1998), as well as in mRNA expression of Rubisco activase (Martino-Catt and Ort, 1992).

18.1.4. Stomatal and mesophyll diffusion limitations

In many species sensitive to chilling, such as grapevine, maize and rice, the direct effect of chilling is a decrease of stomatal conductance (g_s) that is considered to alleviate chilling-induced water stress (Capell and Dörffling, 1993; Hendrickson *et al.*, 2006). This chill-induced stomatal closure can result from two different causes (Allen and Ort, 2001). On one hand, there can be a direct inhibition of mesophyll photosynthesis (e.g. as described by Sassenrath *et al.*, 1990 and Hutchison *et al.*, 2000), which helps to increase intracellular CO_2 concentration (C_i). On the other hand, stomata themselves could act as the potential target of the chill by, for example, altered apoplastic calcium uptake by guard cells as observed by Wilkinson *et al.* (2001) in cold-tolerant *Commelina communis* leaves. Stomatal closure then results in a decrease of C_i, thereby limiting photosynthesis in the mesophyll. This stomatal limitation is actually indicated by the fact that net photosynthesis (A_N) is more reduced than thylakoid electron transport rate, suggesting that decreased C_i results in increased Mehler reaction in C_4 plants (Fryer *et al.*, 1998) and increased photorespiration and/or Mehler reaction in C_3 plants (Flexas *et al.*, 1999b). However, in practice it is difficult to assess which mechanism actually causes the stomatal limitation of photosynthesis that is therefore frequently attributed to a combination of stomatal and non-stomatal effects (Allen and Ort, 2001). The decreased mesophyll conductance to CO_2 (g_m) in response to short-term decreases of leaf temperature (Bernacchi *et al.*, 2002; Warren and Dreyer, 2006), as well as its further, species-dependent decrease during acclimation to low temperatures (Yamori *et al.*, 2006; Flexas *et al.*, 2008; Warren, 2008a), makes it even more difficult to distinguish between stomatal and non-stomatal effects, or between diffusional and photochemical/biochemical effects.

Interestingly, chill-induced stomatal closure does not only occur upon exposure to the light. It can also result from an overnight chill and thus effect subsequent day photosynthesis (Flexas *et al.*, 1999b; Allen *et al.* 2000). In mango (*Mangifera indica* L.) an inhibition of photosynthesis does not occur immediately following the dark chill. Instead, photosynthesis decreased only by midday and coincided with an increase in stomatal limitation of A_N resulting from altered guard-cell sensitivity to CO_2 following the overnight chill (Allen *et al.*, 2000).

A combination of stomatal and non-stomatal effects can also be observed when plants are exposed to low soil temperature only. This occurs under field conditions, e.g., in boreal conifers during the spring warming, when air temperature increases much faster than it takes for the soil to warm up underneath massive packs of insulating snow (Lopushinsky and Kaufmann, 1984; Ensminger *et al.*, 2008). Ensminger *et al.* (2008) shifted winter-acclimated Jack pine seedlings from freezing temperature (−3°C) to 15°C air temperature while cooling the soil temperature to +1 or −2°C. They observed stomatal limitations indicated by low g_s in seedlings exposed to +1 and −2°C soil temperature compared with seedlings exposed to warmer soil temperature. However, lower rates of the maximum carboxylation by Rubisco indicated additionally non-stomatal limitations of photosynthesis caused by low soil temperature.

18.2. DIVERSITY OF RESPONSES OF PLANTS TO LOW TEMPERATURE

18.2.1. Differences between plant functional types

Acclimation is the result of adjustments to achieve a homeostatic state. It represents the temporal integration of short-term and long-term fluctuations in an environmental signal, such as changes in temperature. Plants have evolved diverse strategies to cope with the consequences of stress and acclimation to low temperatures (Öquist and Huner, 2003) and different plant functional types deploy diverse strategies. In herbaceous winter annuals exploitation of the autumn can be observed, which allows plants to establish site occupancy. This strategy relies on the capacity to enhance photosynthetic carbon metabolism together with coordinated changes in respiratory metabolism (Strand *et al.*, 1997; see also Table 18.1). In contrast, in evergreen conifers photosynthesis is downregulated via the inactivation of PSII *RC*s and a reorganisation of the LHC, from pigment-protein complexes efficient in light harvesting and energy transfer to complexes primed for energy quenching during autumn and winter (Ottander *et al.*, 1995; Savitch *et al.*, 2002; Ensminger *et al.*, 2004; see also Table 18.1).

Such changes in photosynthesis can be directly characterised by CO_2/O_2 gas-exchange and/or by Chl-F measurements. Under controlled laboratory conditions, decreased rates in light-saturated CO_2 uptake with minimal changes in apparent quantum yield for CO_2 uptake are typically observed in cold-tolerant annual plants, such as wheat, rye and *Arabidopsis thaliana*, in response to a sudden shift from warm to cold temperatures. Once leaves are acclimated, photosynthesis recovers under low temperature (Öquist and Huner, 1993; Hurry *et al.*, 2000). In contrast, transfer of evergreen conifers such as *Pinus sylvestris* to low temperature results in considerably decreased rates of photosynthesis without any recovery under the low-temperature conditions, even when plants are fully acclimated (Savitch *et al.*, 2002; Ensminger *et al.*, 2005; Sveshnikov *et al.*, 2005).

18.2.2. Low-temperature responses under field conditions

Experiments under field conditions reflect the much more complex and dynamic environment experienced by naturally grown plants. Farage and Long (1991) assessed photosynthetic-CO_2 uptake in field-grown *Brassica napus*. The quantum yield of CO_2 uptake did not substantially change from October until November, when maximum temperatures decreased from about 20°C to about 2°C. In their experiments a decrease in CO_2 uptake only occurred during mid winter when plants experienced frequent freezing and PPFD received by the canopy caused photoinhibition. Nonetheless, when air temperatures intermittently increased above 2°C, CO_2 uptake always recovered within days.

Conifer photosynthesis under field conditions can suddenly decrease upon exposure to low temperatures. For example, field-grown *Picea abies* showed a decline in maximum photochemical efficiency (F_v/F_m) during cold autumn conditions when air temperature was below 5°C (Lundmark *et al.*, 1998). However, F_v/F_m values did not decrease under much warmer air temperature during the subsequent autumn, and values remained close to summer levels until the end of October (Lundmark *et al.*, 1998). Apparently, temperature alone triggers the timing of seasonal changes in photochemical efficiency, and photoperiod has minimal effects on the downregulation of functional PSII. Studying the effects of photoperiod and temperature on the autumn transition in *Pinus banksiana* seedlings under controlled experimental conditions, low temperature was found to be the important trigger for the autumn downregulation of photochemical efficiency and the degradation of PSII reaction-centre protein D1 as compared with photoperiod (Busch *et al.*, 2007, 2008). Whole-canopy gas-exchange data from EC measurements support this view. Under boreal winter

Table 18.1. *Summary of the suite of ecophysiological mechanisms of photosynthesis to adjust the flow of energy in response to low temperature. In this simplified overview, short-term responses and responses of plants grown and developed at warm temperatures are generally considered as the same in various plant categories. In contrast, long-term responses of plants grown and developed at low temperature differ considerably in various plant categories and are therefore seperated in herbaceous plants (herbs), crop plants (crops) and evergreen conifers (conifers). +, positive response or increase in capacity; –, negative response or decreased capacity; (+/-), different responses between species; ?, unknown (Multiple symbols indicate the intensity of the response);* n.r., *indicates that a process is physiologically not relevant on the respective time-scale.*

		Low temperature response of photosynthesis		
Mechanism	**Short-term** (min to hours)	**Long-term** (days to weeks) Growth and development at warm temperature	**Long-term** (several weeks) Growth and development at low temperature	**Ref.**
State transitions	++	+	?	Kagul and Barber (2008); Chapter 3 (this book)
Cyclic electron transport	+	+	Herbs +/– Crops + Conifers +	Ivanov *et al.* (1998); Savitch *et al.* (2001); Sveshnikov *et al.* (2005); Govindachary *et al.* (2007)
Photorespiration	+	+	Herbs ++ Crops - Conifers ?	Streb *et al.* (1998); Flexas *et al.* (1999); Savitch *et al.* (2000b); Chapter 5 (this book)
Scavenging ROS a) Enzymatic antioxidant capacity b) Non-enzymatic antioxidant capacity)	a) + b) +	a) + b) +	a) Herbs ? Crops + Conifers ? b) Herbs ? Crops + Conifers ?	Guy *et al.* (1984); Schöner and Krause (1990); Doulis *et al.* (1993); Zhao and Blumwald (1998); Streb *et al.* (1999); Leipner *et al.* (2000); Janda *et al.* (2003)

Table 18.1. (*cont.*)

Mechanism	Short-term (min to hours)	Low temperature response of photosynthesis **Long-term** (days to weeks) Growth and development at warm temperature	**Long-term** (several weeks) Growth and development at low temperature	Ref.
Non-photochemical quenching a) dynamic photoinhibition b) sustained thermal dissipation c) reaction center quenching d) chronic photoinhibition	a) +++ b) *n.r.* c) *n.r.* d) +	a) + b) + c) + d) +	a) Herbs *n.r.* Crops + Conifers *n.r.* b) Herbs + Crops ++ Conifers +++ c) Herbs ? Crops + Conifers + d) Herbs *n.r.* Crops *n.r.* Conifers *n.r.*	Aro *et al.* (1993); Demmig-Adams and Adams (1996); Flexas *et al.* (1999); Gilmore and Ball (2000); Ivanov *et al.* (2002); Sveshnikov *et al.* (2005); Savitch *et al.* (2009)
Carbon metabolism and partitioning	—	+++/ –	Herbs ++ Crops +++ Conifers –	Savitch *et al.* (1997); Strand *et al.* (1997); Savitch *et al.* (2002); Stitt and Hurry (2002); Ensminger *et al.* (2004); Gray and Heath (2005)
Rubisco	—	++/ –	Herbs ++ Crops ++ Conifers -	Huner and Öquist (2003); Hendrickson *et al.* (2004); Gray and Heath (2005)

Calvin cycle	—	+++/ –	Herbs ++ Crops + Conifers –	Kingston-Smith (1997); Jones *et al.* (1998); Hurry *et al.* (2000); Ensminger *et al.* (2008)
Stomatal limitations	+++	++	Herbs ++ Crops ++ Conifers ++	Flexas *et al.* (1995); Allen and Ort (2001); Wilkinson *et al.* (2001); Hendrickson *et al.* (2006); Ensminger *et al.* (2008)
Non-stomatal limitations	++	—	Herbs ++ Crops ++ Conifers ++	Jones *et al.* (1998); Allen and Ort (2001); Hendrickson *et al.* (2004); Ensminger *et al.* (2008)
Mesophyll conductance	+++	?	Herbs ? Crops ? Conifers ?	Warren and Dreyer (2006); Yamori *et al.* (2006); Bernacchi *et al.* (2008); Warren (2008)

conditions, A_N in Scots pine (*Pinus sylvestris*) is completely suppressed during subfreezing temperatures from October until about mid May (Lloyd *et al.*, 2002; Ensminger *et al.*, 2004; Monson *et al.*, 2005). In a maritime ecotype of a Scots pine forest in the Netherlands, whole-canopy gas-exchange measurements also indicated the autumn downregulation of photosynthesis (Dolman *et al.*, 2002). In contrast to the boreal pine forests, Dolman *et al.* (2002) also observed a substantial recovery of A_N of this Scots pine forest during warm winter days in January. However, the magnitude of this carbon uptake during wintertime under temperate climate conditions is much smaller than the uptake rates observed under summer conditions (Dolman *et al.*, 2002).

Bauer *et al.* (1994) compared the two distinct strategies of woody and herbaceous plants to cope with low temperatures. They proposed that perennial woody plants tend to invest in efficient frost resistance and upregulation of photoprotective systems instead of efficiency of photosynthesis to survive the winter. In contrast, annual herbaceous plants retain a higher photosynthetic activity and adjust photosynthesis to maximise carbon gain at the risk of frost damage. In this respect it is important to consider that conifers retain their needles for several years, whereas the leaves of cold-tolerant, overwintering plants, such as wheat and rye, die in the subsequent spring (Öquist and Huner, 2003). This gives rise to new leaves that are acclimated to spring growth conditions under both controlled environment and field conditions (Huner *et al.*, 1993; Savitch *et al.*, 2000a). Carbohydrates stored in the crown tissues of these cereals provide the initial energy for this early spring growth. Thus, the persistence of conifer needles is essential to their winter survival, whereas in cold-tolerant cereals, the persistence of the crown as a carbon storage organ and not the leaves is essential for winter survival.

18.3. HOW PHOTOSYNTHESIS SENSES LOW TEMPERATURE

Photosynthesis has to maintain an energy balance between light captured in PSII and PSI, its transformation into NADPH and ATP and its subsequent utilisation in metabolism and growth (Fig. 18.3). This energetic balance between photophysical and photochemical processes transforming light and the energy consuming metabolic sinks is called photostasis (Öquist and Huner, 2003; Ensminger *et al.*, 2006). The photochemical reactions are temperature-independent and extremely rapid, whereas the temperature-dependent biochemical reactions are relatively slow. As the latter reactions typically limit the rate of energy flow, low temperature potentially causes an energy imbalance by differentially affecting energy consumption by metabolism (Fig. 18.3).

How can imbalances in the energy flow be re-balanced? The PQ pool within the photosynthetic electron transport chain acts as a redox sensor in phototrophic cells (Escoubas *et al.*, 1995; Maxwell *et al.*, 1995; Allen and Nilsson, 1997). When light energy is converted to redox potential energy, the rate-limiting step is the oxidation of Q_A^- by the PQ pool and intersystem electron transport. The reduction state of Q_A reflects the redox state of the free PQ pool. The so-called excitation pressure (or the excessive excitation energy) is a relative measure of the reduction state of Q_A and reflects the redox state of the PQ pool (Huner *et al.*, 1998). This can be measured by pulse-amplitude modulated fluorescence as 1-qP. Therefore the excitation pressure reflects the imbalance in energy flow. Figure 18.3 provides a schematic diagram to illustrate the concept of the balance of energy flow as defined by the equation $I > E_k$, and thus, $\sigma_{PSII} \cdot I > \tau^{-1}$.

The PQ pool as the primary energy sensor transduces the electron transport signal into biochemical signals, and any environmental stress causing an imbalance in the flow of energy will affect PQ redox state and thus signalling (e.g., Escoubas *et al.*, 1995; Maxwell *et al.*, 1995; Beck, 2005). The energetic imbalance caused by low-temperature stress will therefore translate into a PQ pool mediated by redox signals, which then regulates transcription of genes in the chloroplast and the nucleus involved in cold acclimation (e.g., Soitamo *et al.*, 2008). In addition to the PQ redox state other signals are also involved in the regulation of acclimation processes in the chloroplast, such as H_2O_2, gluthathione and fatty acid signalling resulting from lipid peroxidation (reviewed e.g. in Beck, 2005; Ledford and Niyogi, 2005). To maintain an energy balance, regulation of photosynthetic genes can either affect adjustment of the source and hence primary photosynthetic reactions and redox components of the photosynthetic electron transport chain, or the regulation of the sink capacity and hence enzymes involved in chloroplastic and cytosolic carbon metabolism. In fact, photoautotrophs exploit any one or a combination of these mechanisms to balance the energy flow under low-temperature conditions (Huner *et al.*, 1998, 2003; Ensminger *et al.*, 2006).

In addition to the redox signal from the PQ pool, which regulates chloroplast and nuclear photosynthetic and non-photosynthetic genes involved in cold acclimation (Escoubas *et al.*, 1995; Maxwell *et al.*, 1995; Allen and Nilsson, 1997; Huner *et al.*, 1998; Fey *et al.*, 2005a), the thylakoid ΔpH can

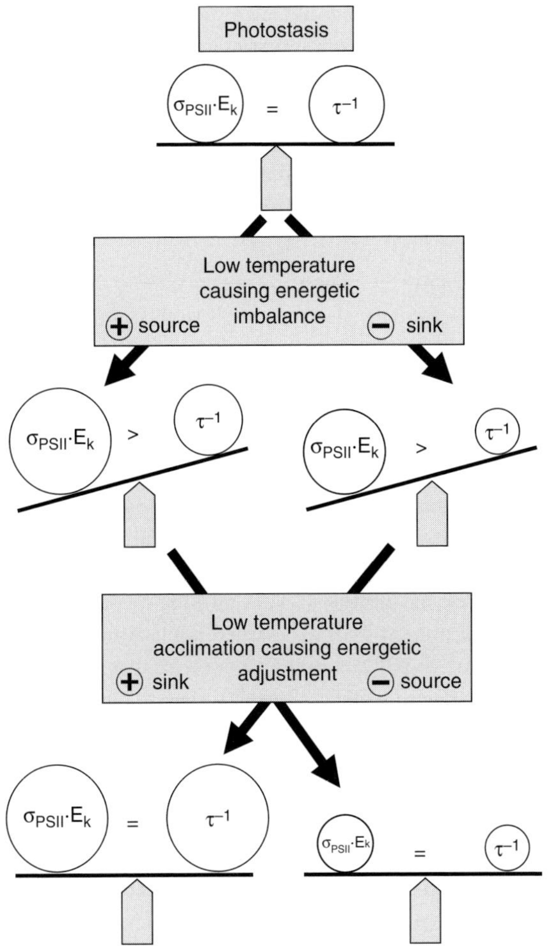

Fig. 18.3. Schematic illustration of the concept of photostasis as defined by the equation σ_{PSII} x $E_k = \tau^{-1}$ (Falkowski and Chen, 2003). Photostasis is achieved through the balance of energy flow from sources to sinks where σ_{PSII} is the effective absorption cross section of PSII, E_K is the irradiance (I) at which the maximum photosynthetic quantum yield balances photosynthetic capacity (estimated from a photosynthetic light response curve) and τ^{-1} is the rate at which photosynthetic electrons are consumed by a terminal electron acceptor such as CO_2 under light saturated conditions. An imbalance between energy absorbed versus energy utilised will occur whenever the rate at which the energy absorbed through PSII and the rate at which electrons are injected into photosynthetic electron transport exceeds temperature-dependent metabolic electron sink capacity (whenever σ_{PSII} x $E_k > \tau^{-1}$). Such an imbalance can be created by increasing the growth irradiance to exceed E_k at a given σ_{PSII} or by lowering the growth temperature at a constant irradiance whereby σ_{PSII} x $E_k > \tau^{-1}$ because of a temperature dependent decrease in τ^{-1}. Adjustments of photosynthesis to balance the flow of energy and to obtain photostasis can either occur via an increase in the rate of sink processes, a decrease in the rate of energy provided through the source processes or a combination of both (adapted from Ensminger *et al.*, 2006).

also act as an important chloroplastic signal. As has been shown previously, the transthylakoid ΔpH is a key regulator of NPQ through the xanthophyll cycle (Krause and Weis, 1991; Horton *et al.*, 1999).

Other mechanisms for sensing temperature are independent from photosynthesis and the PQ pool. These temperature sensors are sensitive to low-temperature-induced changes in membrane viscosity (Murata and Los, 1997) and might involve HIK33/HIK19, a histidine kinase that is responsive to membrane viscosity that has not been detected in plants yet, but has been observed in the outer membrane of cyanobacteria, (e.g., Murata and Los, 2006). Alternative mechanisms suggest action of plasmamembrane Ca^{2+} channels acting as temperature sensors (Monroy and Dhindsa, 1995).

18.4. ACCLIMATION OF PHOTOSYNTHESIS TO LOW TEMPERATURE

High light or low temperature can impose a comparable high excitation pressure resulting in photoinhibition. Rapidly reversible photoinhibition occurs through the process of thermal dissipation of energy (or NPQ). Increased NPQ leads to a decrease in the effective absorption cross section of PSII (σ_{PSII}) and downregulation of PSII activity (Öquist and Huner, 1993), and involves the photoprotective xanthophyll cycle (Demmig-Adams and Adams, 1996b). Whenever the absorbed energy exceeds the capacity of photochemistry as well as for NPQ, this results in photodamage owing to the greater rate of destruction versus repair of the D1 PSII reaction-centre protein (Aro *et al.*, 1993; Melis, 1999). Apparently, acclimation to low temperature mimics acclimation to high light, which also attempts to balance the energy absorbed versus the energy either utilised through metabolism or dissipated via NPQ to avoid photodamage (Huner *et al.*, 1998).

18.4.1. Antioxidants in cold acclimation

Low-temperature stress in particular in combination with light, increases the production of ROS and the need for antioxidative protection (Prasad, 1996; Wise, 1995). ROS can cause lipid peroxidation, DNA damage and protein denaturation. Under chilling conditions, acclimated plants are able to keep ROS at a steady state level through increased pools of antioxidant metabolites (Prasad *et al.*, 1996). Accordingly, increases in contents of antioxidants, like ascorbate, glutathione, α-tocopherol and β-carotene, as well as antioxidative enzymes such as SOD, CAT, GR and ascorbate peroxidise, are observed in plant cells exposed to low temperature (Schöner and Krause, 1990; Doulis *et al.*, 1993; Noctor *et al.*, 1998; Streb *et al.*, 1999; Kocsy *et al.*, 2001). However, there is no unequivocal evidence for a direct role of the antioxidants in protecting the cell from chilling-induced ROS. For example, in plants with artificially increased ascorbate and glutathione contents and in transgenic lines with increased SOD activity, a higher tolerance to paraquat-induced oxidative stress was observed, but almost no improvement of chilling tolerance in the light (van Camp *et al.*, 1996; Payton, 1997; Streb and Feierabend, 1999; Leipner *et al.*, 2000). In contrast, an improved protection against chilling stress was observed in cotton plants overexpressing chloroplastic APX and rice leaves overexpressing the CAT enzyme (Matsumura *et al.*, 2002; Kornyeyev *et al.*, 2003). As low temperature in combination with light induces redox changes of ascorbate and glutathione that might be involved in signal transduction, and since non-scavenged ROS might be directly involved in cellular signalling (Ruelland *et al.*, 2009), the content and redox status of the antioxidants of the leaf might be more important in mediating acclimation to cold stress than in directly preventing ROS damage (Apel and Hirt, 2004; Foyer and Noctor, 2005).

18.4.2. Dissipation of excess energy by modulation of the functional absorption cross section of PSII

Adjustments in the efficiency of energy transfer from LHCs to PSII RCs without changing the physical size of the LHCs are an important mechanism to acclimate photosynthesis at the level of the energy source. A rapid mechanism to balance energy flow is through dissipation of excess energy via NPQ within the LHC. This photoprotective mechanism is facilitated by the light-dependent conversion of violaxanthin to antheraxanthin and the photoprotective zeaxanthin to decrease the energy transfer to chl. *a* in the PSII RC (Horton *et al.*, 1999; Niyogi *et al.*, 2005). Thus, the xanthophyll cycle modulates excitation pressure by affecting the σ_{PSII} and protects PSII. The actual mechanism of energy quenching via NPQ remains a controversial issue. Holt *et al.* (2005) developed a model where direct energy quenching via NPQ depends on a carotenoid radical cation that only forms during excitation of chl. under high steady state NPQ conditions. In contrast, an alternative model discussed by Horton *et al.* (2008) suggests a dynamic structure comprising several components of the light-harvesting antenna and their dynamic organisation in response to thylakoid lumen protonation and controlled by the xanthophyll-cycle carotenoids.

In contrast to this dynamic modulation of excitation pressure, many overwintering plants develop sustained, xanthophyll-dependent energy dissipation (Demmig-Adams and Adams, 1996a,b). Mesophytic plants such as *Malva*, *Arabidopsis* and winter cereals reveal a 'cold-sustained' NPQ that is rapidly reversible upon warming. Sclerophytic evergreens additionally exhibit a sustained zeaxanthin-dependent NPQ during winter that is not rapidly reversible upon warming. The sustained persistent quenching is associated with the reorganisation of the LHCII into xanthophyll-containing aggregates (Gilmore and Ball, 2000; Öquist and Huner, 2003).

18.4.3. Short-term and long-term modulation of functional absorption cross section of PSII

Within minutes, photosynthesis can adjust to sudden changes in light and/or temperature that may result in high excitation pressure. This short-term modulation can be achieved by reducing energy transfer efficiency to PSII via state transitions to divert energy from PSII to PSI. In the field, this mechanism serves as a rapid response preceding a photoprotective adjustment by NPQ during exposure to excess light (Miloslavina *et al.*, 2007). In experiments, state transitions can be induced by sudden changes in light and/or temperature and by using light with a wavelength composition that excites preferentially PSII relative to PSI in order to over-reduce the redox sensor PQ. In higher plants overreduction of the redox sensor PQ activates a thylkoid-bound protein kinase STN7, which in turn phosphorylates LHCII and the minor light-harvesting protein CP29. Phospho-LHCII together with phosphorylated CP29 then migrate and dock onto PSI (Lunde *et al.*, 2000; Kargul and Barber, 2008). This redistributes the light-harvesting chl. to PSI at the expense of PSII (and thus decreases the functional absorption cross section of PSII) resulting in a balanced excitation of PSII and PSI to ensure optimal quantum efficiency for PET.

On a longer timescale, σ_{PSII} is modulated by affecting the physical size of LHCII per PSII RC. This involves the regulation of transcription and translation of the *Lhcb* family of nuclear genes encoding LHCII polypeptides. Downregulation of these genes results in a smaller LHC that protects PSII from excessive excitation by decreasing the light absorption. This is reflected in increases in the chl. *a*/*b* ratio and a decrease in the quantum yields for CO_2 fixation and O_2 evolution, and can be observed in the low-temperature and high-light response of conifers and certain herbaceous plants (Sveshnikov *et al.*, 2005) as well as in green algae (Escoubas *et al.*, 1995; Maxwell *et al.*, 1995).

Zeaxanthin-independent RC quenching is another quenching mechanism complementing the antenna quenching via the xanthophyll cycle and PsbS (Ivanov *et al.*, 2001; Lee *et al.*, 2001a; Ivanov *et al.*, 2002; Sane *et al.*, 2003; Huner *et al.*, 2006). Originally proposed by Briantais *et al.* (1979) and Krause and Weis (1991), RC quenching in field-grown pine is at a maximum in cold-acclimated needles during the winter months and decreases to a minimum upon spring recovery in late spring when pine needles exhibit maximum photosynthetic activity (Ivanov *et al.*, 2002; Sveshnikov *et al.*, 2006).

18.4.4. Aggregation of components of the photosynthetic apparatus under low temperature

Cold acclimation of conifers and cessation of growth (Mellerowicz *et al.*, 1992) decrease sink demand for photoassimilates. This reduces metabolic energy consumption and eventually causes accumulation of phosphorylated photosynthates in leaves, leading to feedback inhibition of photosynthesis (Savitch *et al.*, 2002; Hjelm and Ögren, 2003). Conifers undergo long-term changes in the organisation of the photosynthetic apparatus to attain photostasis under these conditions. These changes involve decreases in the number of PSII RCs (Fig. 18.4), loss of light-harvesting chl.s and aggregation of thylakoid peptides involving LHCII, PSII and PSI (Ottander *et al.*, 1995; Savitch *et al.*, 2002; Ensminger *et al.*, 2004; Ebbert *et al.*, 2005). This aggregation is fully reversible and associated with increased levels of the thylakoid membrane protein PsbS, and the formation of a sustained NPQ via increased levels of zeaxanthin (Ottander *et al.*, 1995; Öquist and Huner, 2003). In naturally grown conifers, initially high PsbS levels can decrease upon exposure to severe subfreezing temperatures. This is associated with concomitant loss of functional PSII once trees have attained the fully acclimated, photoinhibited state (Fig. 18.4) (Ensminger *et al.*, 2004; Ebbert *et al.*, 2005). The reorganisation of the photosynthetic apparatus leads to constitutive thermal dissipation of excess energy to protect conifer needles (Öquist and Huner, 2003).

A contrasting strategy is realised in winter cereals, such as rye, which continue to grow and develop during the cold-acclimation period and attain maximum freezing tolerance. Thereby winter cereals maintain a high demand for photoassimilates during low-temperature acclimation and maximum efficiency and capacity for light absorption (σ_{PSII}). This is indicated by increased chl. per leaf area, minimal changes in chl. *a*/*b* ratios and in the relative contents of Lhcb1, Lhcb2, D1, Cyt_f, PC, PsaA/PsaB heterodimer and the β-subunit of the ATPase complex on a chl. basis (Gray *et al.*, 1998).

18.5. CHANGES IN ELECTRON-SINK CAPACITY DURING ACCLIMATION OF PHOTOSYNTHESIS TO LOW TEMPERATURE

In the previous section we described the different strategies of acclimation to low temperature during autumn and winter in conifers and cold-hardy herbaceous plants. Such different strategies underscore the importance of both source and sink as control points to regulate photosynthesis under low temperature to maintain photostasis (Huner *et al.*, 1998; Paul and Foyer, 2001). The focus of the following section will be on low-temperature effects on sink capacity during cold acclimation, and address temperature-dependent alterations in carbon metabolism and the subsequent alterations in allocation and partitioning of carbohydrates. Other

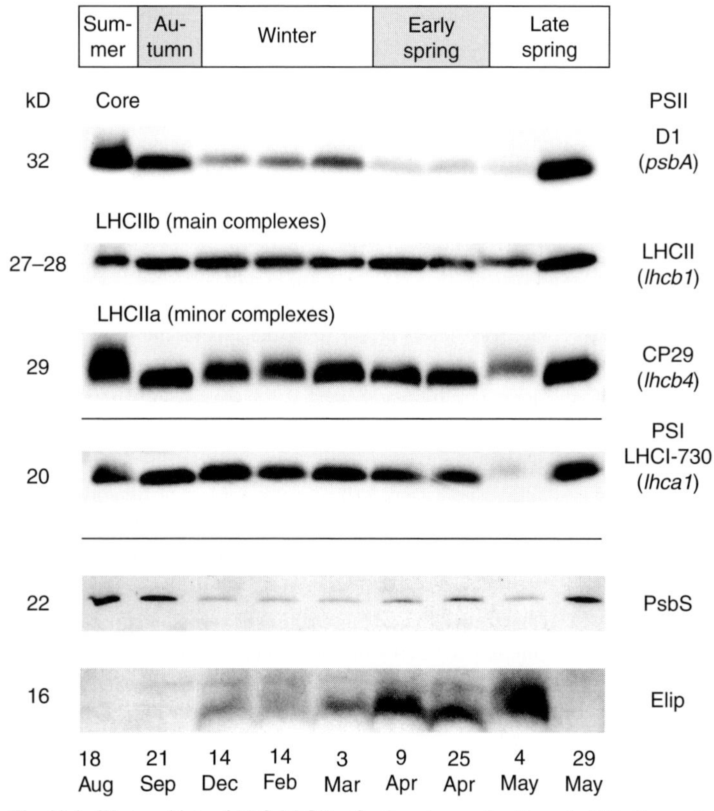

Fig. 18.4. Western blots of SDS-PAGE reflecting changes in chlorophyll binding and reaction-centre proteins during the seasons. Samples were probed with antibodies raised against the reaction-centre D1 protein of photosystem II (PSII), the light-harvesting chlorophyll (chl.)-binding proteins LHCII, CP29, LHCI-730 and the PsbS and early light-induced protein (Elip) protein (corresponding gene in brackets). Samples were collected on indicated dates at solar noon. Lanes were loaded on equal protein basis. LHC, light-harvesting complex. From Ensminger *et al.*, 2004.

mechanisms also play a significant role during low-temperature acclimation by providing alternative sinks for electrons that cannot be used in photochemical processes and carbon allocation and partitioning. These mechanisms include photorespiration, the Mehler reaction and cyclic electron transport around PSI (see Table 18.1 for a summary of these mechanisms and their response to low temperature). However, these mechanisms are beyond the scope of this chapter, but are discussed in detail in Chapters 2, 3 and 4.

18.5.1. Carbohydrate metabolism and the allocation and partitioning of carbon under low temperature

Photosynthesis requires strict regulation of carbon allocation between chloroplastic and cytosolic carbon metabolism. Therefore the rates of carbon fixation in the chloroplast and cytosolic sucrose synthesis are carefully balanced to provide sufficient metabolic sinks in the photosynthetic energy transformation process (Fig. 18.5). A stimulation of cytosolic sucrose synthesis from hexose-phosphate liberates P_i, which can be reimported into the chloroplast where it can then be used to regenerate ATP and RuBP. Conversely, inadequate sucrose synthesis leads to accumulation of phosphorylated intermediates in the cytosol and depletion of P_i in the chloroplast, resulting in inhibition of ATP-synthesis, accumulation of glycerate-3-P and inactivation of Rubisco (Fig. 18.5).

Interestingly, in cold-hardy herbaceous plants, low-temperature acclimation of leaves depends on whether they have been developed under warm-temperature conditions and were shifted subsequently to cold or whether they have actually developed under cold-temperature conditions (Gray and Heath, 2005). An inhibition of sucrose synthesis

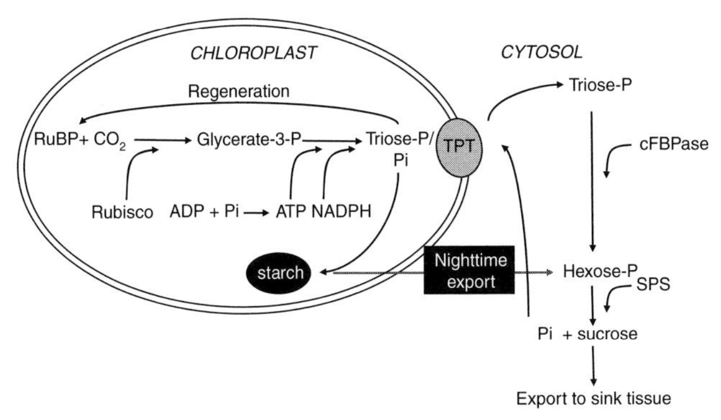

Fig. 18.5. Sugars act as metabolic sinks in the photosynthetic energy transformation process. Photosynthesis generates glycerate-3-phosphate (glycerate-3-P), which is reduced to produce triose-phosphate (triose-P) using NADPH and ATP. Most of the triose-P remains in the chloroplast to regenerate ribulose 1,5 bisphosphate (RuBP). The surplus is converted to end products, thereby releasing inorganic phosphate (P_i), which is used to regenerate ATP. Export of triose-P via the triose-phosphate transporter (TPT) to the cytosol is the dominant end-product pathway. Triose-P is then converted to sucrose via cytosolic fructose-1,6-bisphosphatase (cFBPase) and sucrose phosphate synthase (SPS). When the rate of photosynthesis exceeds triose-P-export, transient starch accumulates in the chloroplast and can be degraded during the night to maltose and glucose, which are then exported to the cytosol (Zeeman *et al.*, 2004). Cytosolic sucrose levels are under the control of (amongst other factors) the efficiency of assimilate export. Limited sink demand will cease sucrose export and thereby its biosynthesis, resulting in decreased triose-P export from the chloroplast, accumulation of glycerate-3-phosphate and starch. The rate of starch biosynthesis can be an important safety valve to maintain high photosynthetic capacity when sink demand is limited owing to, e.g., cold temperatures and low sugar export from source leaves, which favours the accumulation of soluble sugars in the cytosol. Adapted from Ensminger *et al.*, 2006.

in the cytosol was observed in *Arabidopsis* shifted to low-temperature stress causing decreased P_i cycling between the cytosol and the chloroplast (Hurry *et al.*, 2000). As a result, the chloroplast becomes P_i-limited impeding ATP-synthesis, which in turn is required to regenerate RuBP. This decreases the rate of electron transport and causes feedback inhibition of photosynthesis (Hurry *et al.*, 2000). When leaves had grown and developed at low temperature, expression and subsequent activity of Calvin-cycle enzymes was increased (Hurry *et al.*, 2000). In addition, cold-tolerant herbaceous plants grown at cold temperatures exhibit an increase in P_i availability in the chloroplast (Stitt and Hurry, 2002), in adenylates, in phosphorylated intermediates and in the capacity for the regeneration of RuBP (Hurry *et al.*, 1994). Cold-grown plants also exhibit an increase in sucrose-phosphate-synthase (SPS) activity, SPS activation state as well as an increase in the cytosolic hexose-P pool leading to sucrose biosynthesis (Stitt and Hurry, 2002). In cereals, cold acclimation also leads to the stimulation of fructan biosynthesis. Thus, both a stimulation of sucrose and fructan biosynthesis support the recovery of CO_2-assimilation rates by increasing the flux of newly fixed carbon through these biosynthetic pathways (Savitch *et al.*, 2000a).

Concomitant with the upregulation of carbon metabolism, cold-acclimated winter wheat also exhibits a stimulation of carbon export from the leaf compared with cold-stressed non-acclimated plants (Leonardos *et al.*, 2003). This is reflected in an increased capacity to assimilate CO_2 and an increase in plant biomass with minimal changes in photosynthetic efficiency measured as the apparent quantum yield of either CO_2 fixation or O_2 evolution (Öquist and Huner, 1993; Gray *et al.*, 1997; Savitch *et al.*, 2000a). The increased biomass production and carbon export capacity finally translates into significant changes in plant morphology, leaf anatomy and stomatal distribution in *Arabidopsis* (Strand *et al.*, 1999) and winter cereals (Gray *et al.*, 1997).

18.5.2. Alleviation of photoinhibition under low temperature through upregulation of carbon metabolism

Cold acclimation has been shown to enhance resistance to dehydration as well as freezing stress (Thomashow, 1999). The alterations mentioned in the previous section also result in increased WUE in cold-acclimated winter wheat plants (Leonardos *et al.*, 2003). The extent to which cereals are able to increase their photosynthetic capacity during cold acclimation is directly correlated to their freezing tolerance (Öquist and Huner, 1993). Thus, cold-tolerant winter cereals maintain photostasis by upregulating the sink capacity and carbon export, while keeping σ_{PSII} relatively constant compared with cold-stressed plants (Leonardos *et al.*, 2003).

From a series of cold-acclimation experiments in *Arabidopsis thaliana,* Strand *et al.* (2003) concluded that an increased capacity for sucrose biosynthesis at low temperature reduces the inhibition of photosynthesis when coupled to the mobilisation of carbohydrates from source leaves to sinks and increases the rate at which freezing tolerance develops in *Arabidopsis.* The orchestrated response of these plants to low temperature is further supported by the concomitant increased expression of a plastidic triose-phosphate transporter (AtTPT) and a Suc transporter (AtSUC1) (Lundmark *et al.*, 2006).

Concomitant with the upregulation of carbon metabolism, Savitch *et al.* (2000b) observed the suppression of photorespiration in cold-acclimated winter wheat. Interestingly, the suppression of photorespiration at low temperature did not occur in plants grown at high light and comparable excitation pressure. This indicates that the suppression of photorespiration exhibited in cold-acclimated winter wheat is not a response to excitation pressure but rather a specific response to low temperature (Savitch *et al.*, 2000b).

In non-hardy plants, such as tobacco and cold-girdled spinach, sink strength regulates photosynthetic capacity and sugars act as a signal to repress photosynthetic gene expression (Stitt and Hurry, 2002). Long-term growth of *A. thaliana* at low temperature increases soluble sugars in leaves without a concomitant inhibition of photosynthesis or repression of photosynthetic genes (Strand *et al.*, 1997). Low temperature might act here as a signal to release the suppression of photosynthesis and gene expression in developing leaves. Hence, under low temperature photosynthesis can recover because of increased activity and higher levels of photosynthetic and carbon metabolism enzymes. The recovery is further facilitated by increased pools of phosphorylated intermediates. This provides high rates of metabolic carbon flux even at low temperatures (Strand *et al.*, 2003).

In addition, cold-acclimated wheat exhibits decreased accumulation of glucose polymers (starch) in the chloroplast, but increased leaf vacuolar carbon storage through the polymerisation of sucrose to fructans together with an increased fructan accumulation in the crown (Savitch *et al.*, 2000a; Leonardos *et al.*, 2003). This is reflected in an increase in the volume of cytoplasm relative to the vacuole, an increase in specific leaf weight, an increase in cryoprotective soluble sugars and a decrease in leaf-water content (Xin and Browse, 2000; Stitt and Hurry, 2002; Leonados *et al.*, 2003). This demonstrates altered partitioning of carbon within the source leaf as well as to alternative sinks and reflects a mechanism that allows release from the expected suppression of photosynthesis in overwintering herbaceous plants owing to the accumulation of high levels of sucrose.

Woody species show a different strategy of carbon partitioning during cold acclimation and in response to low temperature. Although low temperature creates a surplus in carbohydrate in both trees and grasses, trees convert this surplus to storage polysaccharides, whereas grasses tend to accumulate more soluble carbohydrates (Fig. 18.6a) (Hjelm and Ögren, 2003). Furthermore, grasses show only limited capacity to adjust the photosynthetic apparatus of leaves developed at warm temperatures and subsequently exposed to low temperatures (Fig. 18.6b). The leaves senesce over time at low temperature and are replaced by new, cold-acclimated leaves developed at low temperature. These different strategies provide contrasting benefits: grasses increase their freezing tolerance from the accumulation of soluble carbohydrates. In contrast, trees benefit from the rapid conversion of carbohydrate surplus into storage polysaccharides such as starch by allowing photosynthesis to be regulated independently of growth as growth rates are modulated by temperature (Hjelm and Ögren, 2003).

18.6. INTERACTION BETWEEN PHOTOSYNTHETIC REDOX AND COLD-ACCLIMATION SIGNALLING PATHWAYS

Although the signalling of redox regulation is still not clear, there is consensus that the redox state of the PQ pool regulates short- and long-term photosynthetic acclimation responses (Escoubas *et al.*, 1995; Maxwell *et al.*, 1995; Huner *et al.*, 1998; Falkowski and Chen, 2003; Ruelland *et al.*, 2009). Recently, Carlberg *et al.* (2003) characterised a 9-kDA protein TSP9, which putatively is the retrograde redox signal from the thylakoid membrane that is transported to the RNA polymerase. There is also evidence for several redox signals derived from the electron transport chain via, e.g., ROS, the redox state of PQ as well as the accumulation of Mg-protoporphyrin IX regulating the accumulation of LHC polypeptides at low temperatures (Wilson *et al.*, 2003; Beck, 2005; Fey *et al.*, 2005b).

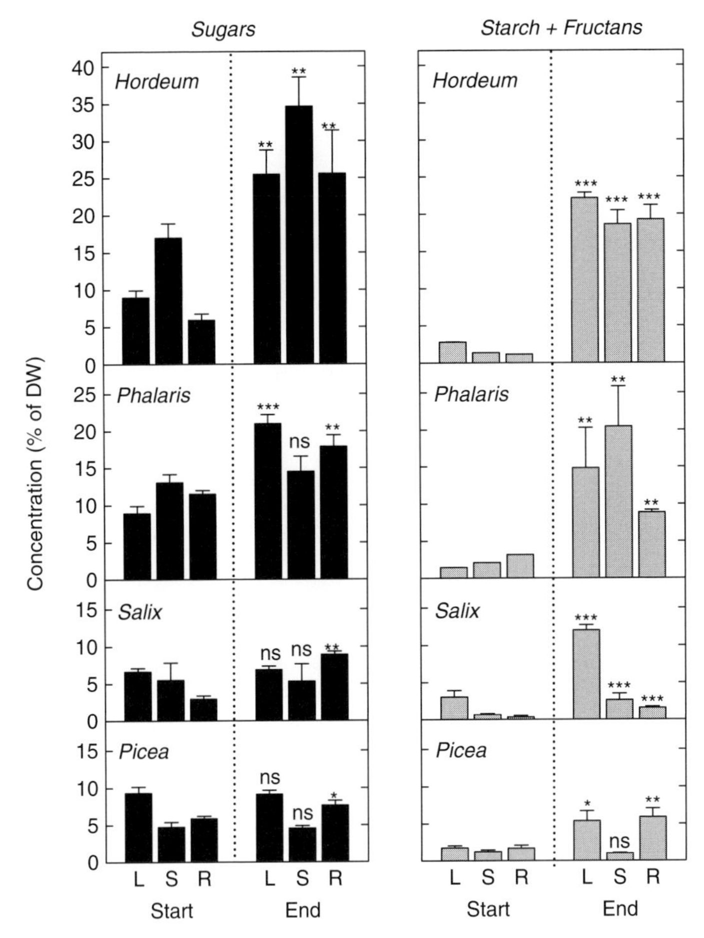

Fig. 18.6. (a) Concentrations of major sugars and storage carbohydrates of leaves (L), stems (S) and roots (R) assessed at the start and end of low-temperature treatments (10/3 degrees C day/night) lasting 28 days. Parallel plants kept at control conditions (20/10 degrees C day/night) showed unchanged values over a 14-day period. Mean±SE values are given for nine replicate plants. Significant differences between end and start values are indicated by asterisks: *, $P<0.05$; **, $P<0.01$; ***, $P<0.001$; ns, $P>0.05$ (from Hjelm and Oegren, 2003).

In addition, there might be an interaction between the so-called *COR* (cold-induced) gene expression originating from a cold-induced signalling pathway and the redox state of the chloroplast. *COR14* mRNA accumulates in barley grown under cold in the dark, but the COR14 protein accumulates in the chloroplast stroma only after brief exposure to light (Crosatti *et al.*, 1995). A recent study showed that the redox state of PQ actually promotes the accumulation of COR14B protein in the light, whereas the steady state level of COR14 mRNA is independent of the redox state (Dal Bosco *et al.*, 2003). This reflects post-transcriptional modification of the transcript of the *COR* gene, which in turn is triggered through the DREB/CBF transcription factors (Kim, 2007; Welling and Palva, 2008).

A characteristic of cold acclimation in herbaceous plants is the accumulation of soluble sugars. Thus, sugar-signalling pathways are likely to be integral to both photosynthetic acclimation as well as cold acclimation. In *Arabidopsis*, the expression of nuclear-encoded photosynthetic genes (e.g., *CAB2*, *rbcS*) is inversely correlated to intercellular soluble-sugar levels (Oswald *et al.*, 2001). This pattern is specific for photosynthesis genes, while nitrate reductase (NR) is positively correlated with sugar levels. The inhibition of PET by DCMU in sucrose-starved cells abolishes starvation-induced increases in chl. *a*/*b*-binding protein (CAB) and ribulose-1,5-bisphosphate carboxylase/oxygenase (rbcS) transcript levels, but not NR transcript abundance. This suggests that sucrose-starvation-induced increases in transcript

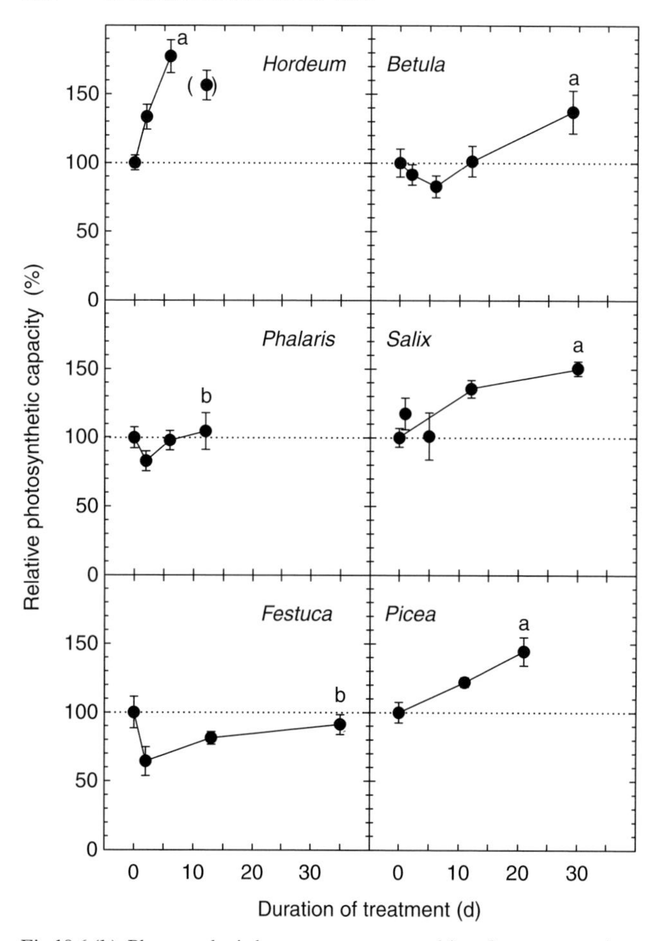

Fig 18.6 **(b)** Photosynthetic low-temperature capacities of warm-grown leaves plotted as a function of duration of low-temperature treatment and expressed relative to the start of treatment. Parentheses denote *Hordeum* leaves that had started to senesce. Mean±SE values are given for 4–6 replicate plants. Significant interspecific differences in low-temperature responses are indicated by different letters ($P<0.05$) (from Hjelm and Ögren, 2003).

levels of nuclear-encoded genes requires PET and is modulated through a plastid-derived signal (Oswald *et al.*, 2001). Thus, plastid redox state regulates nuclear gene expression and actually overrides the sugar signal. When soluble sugar levels rise, e.g., because of an impairment of phloem loading by either genetic or environmentally determined growth cessation as under low temperatures, synthesis of new photosynthetic apparatus will be curtailed until the demand of carbon increases. Therefore, Oswald *et al.* (2001) suggest an interaction between sugar-, photosynthetic electron transport- and ABA-derived regulatory mechanisms to optimise photosynthetic gene expression in response to demand.

Clearly, there is complex crosstalk between the photosynthetic redox, cold acclimation and sugar-signalling pathways to regulate overall plant acclimation to low temperature. In addition, there are overlapping relationships between cold acclimation and responses to dehydration and salt stress in terms of the biochemical changes and in the transcriptional regulation of the responses (Browse and Xin, 2001; Kacperska, 2004). Although the perception of low temperatures and transduction of the signals that initiate the suite of responses are still incomplete, there is evidence for multiple signalling pathways and probably multiple temperature sensors being used in higher plants to initiate and control cold acclimation. Thus, the process of low-temperature acclimation includes the increased expression of many genes, decrease or cessation of growth, increases in ABA concentrations and changes in lipid-membrane composition, accumulation of osmolytes including soluble sugars and increased levels of antioxidants (Browse and Xin, 2001; Kacperska, 2004). Apparently, photosynthetic acclimation to low temperature integrates a complex suite of sensors and interacting signalling pathways.

18.7. CONCLUSIONS

Low-temperature-sensitive plants represent an excellent system to unravel the effects of low temperature on the components of photosynthesis. These include carbohydrate metabolism, inhibition of Rubisco activity and stomatal and mesophyll diffusion behaviour, as well as increases in thermal dissipation of excess energy in the light-harvesting antenna. Cold-tolerant species are able to acclimate to low temperature and to balance energy absorbed by the source and energy used by the metabolic sinks. Importantly, the magnitude and the duration of the low-temperature stress exert different mechanisms in plants to handle any energetic imbalance. Within the timescale of seconds to minutes, sudden decreases in temperature may increase the reduction of the PQ pool or other components of the photosynthetic electron transport chain. On the timescale of hours to days, these short-term photoprotective processes are insufficient to keep the PQ pool oxidised and will be supported or replaced by longer-term adjustments involving, e.g., down-regulation of *Lhcb* expression to decrease the antenna size of PSII.

The regulation of low-temperature acclimation of photosynthesis is controlled by a complex cross-talk of the photosynthetic redox state, cold acclimation and sugar-signalling pathways. Work by Browse and Xin (2001) and Kacperska (2004) on the biochemical and transcriptional changes further suggest overlapping relationships between cold acclimation and water as well as salt stress, all this further stressing the complex integration of a suite of sensing and signalling pathways that respond in a time-dependent manner. Transcript profiling has been used to assess differential gene expression for various model plants exposed to drought and cold stress (Seki *et al.*, 2002; Shinozaki *et al.*, 2003), photosynthetic redox regulation (Singh *et al.*, 2004) or the interaction of photosynthesis and cold-acclimation (Savitch *et al.*, 2005). Nonetheless, owing to the integration of various stress-signalling pathways the understanding of the molecular mechanisms underlying low-temperature acclimation of photosynthesis can be achieved only by considering the ecophysiological responses discussed here.

19 • Photosynthetic responses to high temperature

T.D. SHARKEY AND C.J. BERNACCHI

19.1. PREDICTING PHOTOSYNTHESIS AT HIGH TEMPERATURE

High temperature is known to have a detrimental effect on photosynthesis. When temperatures exceed the optimum for photosynthesis, lower rates of photosynthesis invariably occur (Fig. 19.1). The rapid increase in mean global temperatures and the increased frequencies of temperature extremes and droughts as a result of anthropogenic activities (IPCC, 2007) make understanding supraoptimal temperature effects on photosynthesis crucial because temperature-induced loss of carbon assimilation is likely to become more frequent.

In understanding photosynthetic responses to temperature it is often necessary to mathematically describe the response of photosynthesis or some component process to temperature. Temperature responses are typically logarithmic, that is, processes increase by a set proportion for a given increase in temperature. Descriptions of temperature responses have taken many mathematical forms. One of the simplest was suggested by van't Hoff (discussed in Lloyd and Taylor, 1994) by considering the effect of a 10°C increase in temperature, a parameter called the Q_{10} is calculated as:

$$Q_{10} = (R_2 / R_1)^{\left(\frac{10}{T_2 - T_1}\right)} \qquad \text{[Eqn. 19.1]}$$

where R_2 and R_1 are the rates at the higher and lower temperature (T_2 and T_1). The Q_{10} often underestimates the effect of temperature on a reaction.

van't Hoff later proposed the equation

$$\frac{d(\ln(k))}{dT} = \frac{E}{RT^2} + c \qquad \text{[Eqn. 19.2]}$$

which Arrhenius developed into his equation (Lloyd and Taylor, 1994):

$$k = Ae^{-\frac{E_a}{RT}} \qquad \text{[Eqn. 19.3]}$$

Where k is the rate of the reaction or process; A is a constant that accounts for effects that do not scale exponentially; E_a is the activation energy; R is the gas constant; and T is temperature in Kelvin. The theory developed to explain the Arrhenius equation is that reactants must have a certain amount of energy before they are able to overcome activation barriers between two steady states. Statistical mechanics indicates that, assuming a Maxwell-Boltzman distribution of energies among molecules, the fraction of molecules with a sufficiently high energy is proportional to $e^{-Ea/RT}$.

A later formulation that preserves the exponential response comes from transition-state theory developed by Eyring, Evans and Polyani (Laidler and King, 1983). An equation based on this theory originally published by Johnson *et al.* (1942) in modified form was applied to photosynthesis parameters (Farquhar *et al.*, 1980a; Harley *et al.*, 1985):

$$k = \frac{e^{c - \frac{\Delta Ha}{RT}}}{1 + e^{\frac{\Delta S}{R} - \frac{\Delta Hd}{RT}}} \qquad \text{[Eqn. 19.4]}$$

where c includes a number of theoretical variables but is treated as an arbitrary scaling constant for these purposes; ΔH_a is the enthalpy of activation; ΔS is an entropy term; and ΔH_d is the enthalpy of deactivation. This equation can model both the increase in rate at moderate temperature and the decline in rate at supraoptimal temperature. If a high-temperature-rate decline is not observed or not needed for modelling, the denominator takes on the value of one and the equation is simplified, and is essentially identical to the Arrhenius equation. This approach has been used to model temperature sensitivities of many of the parameters that are needed in many photosynthesis models (Bernacchi *et al.*,

Terrestrial Photosynthesis in a Changing Environment: A Molecular, Physiological and Ecological Approach, ed. J. Flexas, F. Loreto and H. Medrano. Published by Cambridge University Press.

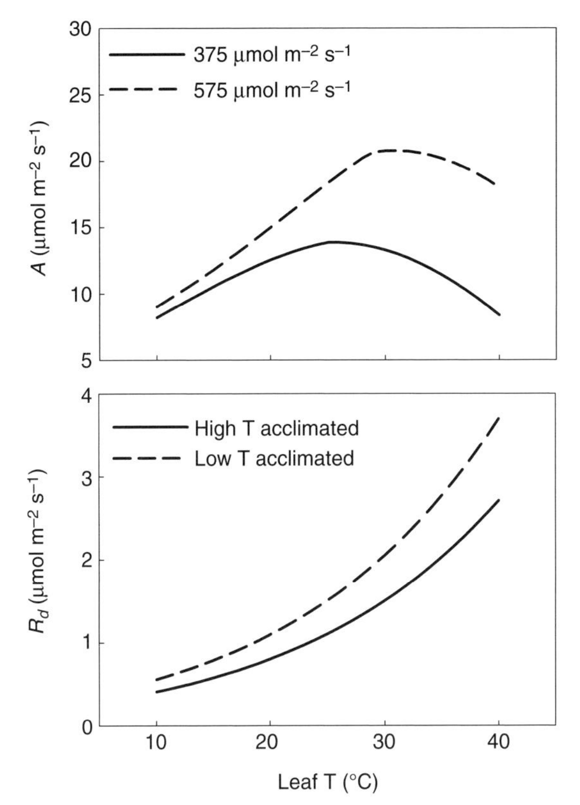

Fig. 19.1. The modelled temperature response of C_3 photosynthesis. The rise in A_N to the maximum is driven by an increase in the activity of photosynthetic enzymes with temperature. At temperatures above the thermal optimum of approximately 27°C, the decline in A_N is driven by the kinetics of Rubisco-favouring oxygenation over carboxylation resulting in increasing rates of photorespiration. This modelling scenario assumes a constant supply of CO_2, i.e., the effect of temperature on stomatal conductance is not considered.

2001, 2002, 2003a) and in models of terpene emissions from plants (Guenther *et al.*, 1993).

Equation 19.4 is relatively complicated (Medlyn *et al.*, 2002a,b) and a simpler equation has been proposed to model photosynthetic electron transport response to temperature (June *et al.*, 2004):

$$\mathcal{J}(T_l) = \mathcal{J}(T_0)e^{-\left(\frac{(T_l - T_0)}{\Omega}\right)^2} \qquad \text{[Eqn. 19.5]}$$

Where T_l is the temperature of the leaf; T_o is the temperature optimum; and Ω is the difference in temperature required to cause the rate to fall to 0.37 (e^{-1}) times its value at T_o.

The approach of June *et al.* (2004) is in line with others who simply fit data with a polynomial, typically second order (Brooks and Farquhar, 1985; Cen and Sage, 2005). Given the complex processes underlying some of the parameters to be modelled, this is a reasonable alternative to the theoretical treatment that may be more appropriate for single chemical reactions rather than complex model parameters.

Photosynthesis of C_3 plants often shows a broad optimum when measured in today's level of CO_2 rather than the significant increase with temperature that is common for biological processes. The reasons for this relatively modest temperature response are complex. At low temperature, the ability of leaves to make starch and sucrose from the products of the Calvin-cycle can often limit photosynthesis causing O_2- and CO_2-insensitive photosynthesis. At the high end of the temperature range (~35°C), Rubisco deactivates and electron transport reactions change significantly. In between, photosynthesis is often relatively insensitive to temperature. Rubisco activity *per se* increases with temperature, like most other enzymes, but photorespiration as a proportion of photosynthesis increases with temperature, and so the response of net CO_2 assimilation (A_N) to temperature is complex. The light-saturated potential for photosynthetic electron transport (PET) increases with temperature up to a point and then generally decreases after that. These interacting factors make it difficult to predict photosynthetic rates as a function of temperature.

In this section, we will focus on the impact of higher temperature on photosynthesis, ranging from the thermal optimum to the temperatures that result in irreversible damage. Later sections will focus on the impact of high temperature on Rubisco activity and on irreversible damage to the photosynthetic machinery (Table 19.1).

19.1.1. The impact of photorespiration on carbon assimilation at high temperatures

The initial decrease in photosynthesis as temperature exceeds the optimum is often driven by the kinetics of Rubisco for two competing reactions, carboxylation and oxygenation. At the active site of Rubisco, a molecule of ribulose 1,5-bisphosphate (RuBP) becomes enolised to form a 2,3-enediol during the first stage of the reaction. At this point, CO_2 will bind to the 2,3-enediol resulting in carboxylation producing two phosphglycerate (PGA) molecules. However, Rubisco will also allow O_2 to bind to the 2,3-enediol intermediate resulting in oxygenation producing one PGA and one phosphoglycolate molecule. The phosphoglycolate molecule enters the photorespiratory cycle. Whereas photosynthesis leads directly to productivity, photorespiration is a wasteful process with

Table 19.1. *Heat effects on photosynthetic parameters.*

Parameter	Effect	Range	Consequence
Carbon metabolism			
Rubisco specificity	Declines	Entire temperature range	Increased photorespiration
Rubisco activation	Declines	35°C and above	Reduced photosynthesis
Other carbon metabolism	Few documented effects	—	—
Electron transport			
PSII	Declines	45°C and above	Irreversible damage
PSI	Increases	35°C and above	Protection
Thylakoids	Become leaky	35°C and above	Counteracted by cyclic electron flow
Stomatal conductance	Variable	—	–
Mesophyll conductance	Decreases	35°C and above	Reduced photosynthesis

little or no benefit to the plant under most circumstances. Despite the presence of O_2 for a significant part of the evolutionary history of photosynthetic organisms, no known form of Rubisco exists that prevents oxygenation from occurring, although, as mentioned in Chapters 5 and 6, there are mechanisms to minimise or eliminate oxygenation. For C_3 plants, however, oxygenation is unavoidable and increasingly favoured as temperature increases.

As both CO_2 and O_2 will enter the same active site and have the ability to bind to the same substrate, they are competitive inhibitors of one another. The oxygenation and carboxylation reactions each have their own kinetic properties and the manner in which these properties change with temperature influences how photosynthesis responds to changes in temperature. At standard temperature, the K_m for CO_2 is three orders of magnitude less (K_c ~272 μmol mol^{-1}) than for O_2 (K_o ~166 mmol mol^{-1}; Fig. 19.2a). The lower Michaelis constant for CO_2 results in carboxylation being highly favoured under conditions where CO_2 and O_2 are in equal concentrations. Under atmospheric conditions, however, the O_2 concentration is three orders of magnitude greater than CO_2 and this attenuates the differences in the Michaelis constants that favour carboxylation. The maximum reaction velocity of oxygenation ($V_{o,max}$) is only about 28% of that for carboxylation ($V_{c,max}$; Fig. 19.2b), which favours carboxylation at the photosynthetic temperature optimum. As temperatures increase, however, the kinetics of Rubisco for CO_2 and O_2 change in a predictable manner that is not uniform for the carboxylation and oxygenation reactions. (When kinetic parameters are expressed in molar terms, a difference in temperature effect on the solubility of O_2 and CO_2 must also be considered, when the terms are expressed in partial pressure this does not apply.) When both K_c and K_o are normalised to one at 25°C, the degree to which K_c increases relative to K_o becomes apparent (Fig. 19.2). As the affinity of an active site is negatively correlated with the Michaelis constant, the kinetics increasingly favour oxygenation at higher temperatures. The increases in V_{max} with temperature for carboxylation and oxygenation are relatively consistent (Fig. 19.2), which demonstrates that affinity for oxygenation versus carboxylation dominates the observed responses with rising temperature.

The different kinetics for carboxylation and oxygenation are best expressed using a term representing the specificity of Rubisco for carboxylation versus oxygenation ($S_{c/o}$). This term accounts for the maximum velocities and Michaelis constants for both reactions as:

$$S_{c/o} = \frac{K_o}{V_{o,\max}} \frac{V_{c,\max}}{K_c} \qquad \text{[Eqn. 19.6]}$$

In general, it is assumed that the kinetics that make up $S_{c/o}$ are conserved among C_3 species; although this is not strictly true (Jordan and Ogren, 1981a,b; Spreitzer and Salvucci, 2002; Galmés *et al.*, 2005). Despite the small differences among C_3 species, the impact of temperature on $S_{c/o}$ will consistently decrease as temperatures rise (Fig. 19.3). For any given set of conditions, the ratio of oxygenation to carboxylation ($\phi_{CO2/O2}$) is calculated using the concentrations of CO_2 and O_2, and the specificity factor as:

$$\phi_{CO2/O2} = \left(\frac{1}{S_{c/o}}\right)\frac{O}{C} = \frac{2\Gamma^*}{C} \qquad \text{[Eqn. 19.7]}$$

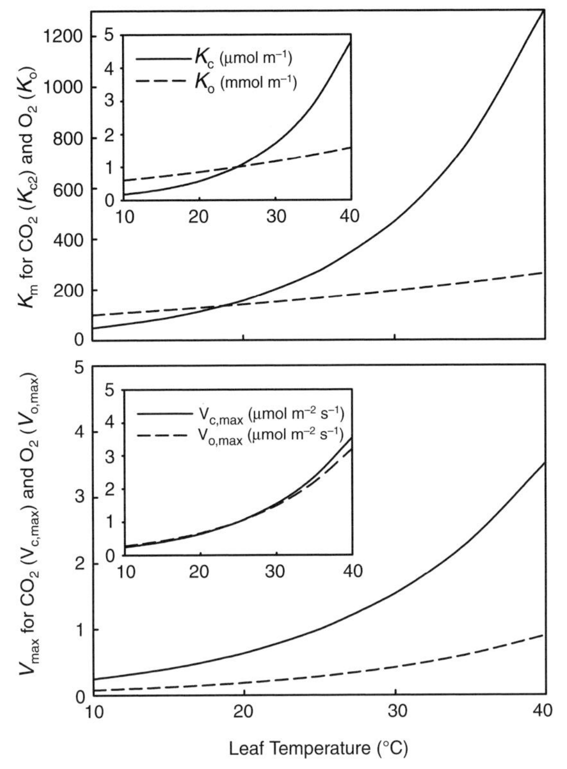

Fig. 19.2. The temperature responses of the Michaelis constants (K_m) for carboxylation (K_c) and oxygenation (K_o; top panel) and of the maximum reaction velocities (V_{max}) for carboxylation ($V_{c,max}$) and oxygenation ($V_{o,max}$; bottom panel). For both graphs, the inset figures represent the relative temperature responses for the larger graphs normalised to unity at 25°C.

where O and C are the partial pressures of O_2 and CO_2, respectively (Farquhar *et al.*, 1980a; von Caemmerer, 2000). Whereas $\phi_{CO2/O2}$ represents the ratio of oxygenation to carboxylation events, two oxygenation events are required to release one CO_2. Thus, the amount of CO_2 that is released from $\phi_{CO2/O2}$ oxygenations will be $\phi_{CO2/O2}$ /2, which is equal to Γ^*/C (von Caemmerer, 2000). The term Γ^* represents the CO_2 concentration where photosynthetic carbon assimilation (PCA) and photorespiratory carbon release offset each other and is important for modelling photosynthesis as (Farquhar *et al.*, 1980a):

$$A_N = \left(1 - \frac{\Gamma^*}{C}\right)V_c - R_d \qquad \text{[Eqn. 19.8]}$$

The term '$(1-\Gamma^*/C)$' represents the relative A_N after accounting for CO_2 loss in photorespiration. These terms are all important when considering the impact of

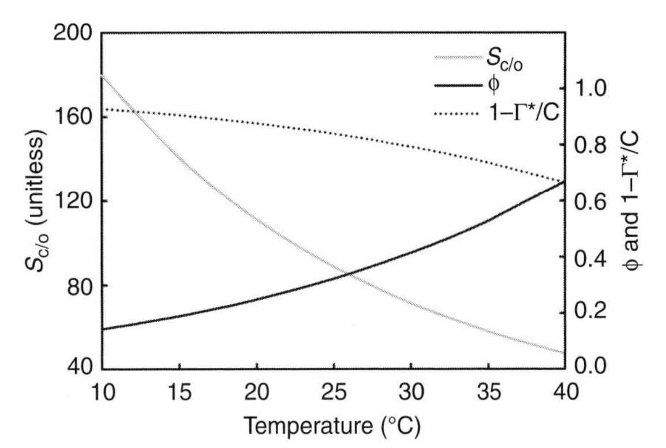

Fig. 19.3. The temperature responses for Rubisco specificity ($S_{c/o}$), the ratio of oxygenation to carboxylation ($\phi_{CO2/O2}$) and the proportion of CO_2 that is initially fixed by Rubisco that is released via photorespiration ($1-\Gamma^*/C$).

rising temperatures on photosynthesis as they represent the increased rate of oxygenation with temperature (Fig. 19.3). When temperatures increase from 25 to 40°C there is a two-fold decrease in $S_{c/o}$ which results in a 50% increase in the amount of CO_2 released via photorespiration (Fig. 19.3).

That the term '$1-\Gamma^*/C$' decreases with temperature signifies a preference for oxygenation at higher temperatures, but increases in temperature also influence the rate of carboxylation. Although the proportion of assimilated carbon that remains assimilated, $(1-\Gamma^*/C)$, decreases with temperature, the amount of carbon that is initially assimilated also decreases. This is demonstrated by the modelled representations of V_c (Farquhar *et al.* 1980a):

$$V_c = \left\{ \frac{C \cdot V_{c,\max}}{C + K_c(1 + (O/K_o)}, \frac{C \cdot J}{4C + 8\Gamma^*} \right\} \qquad \text{[Eqn. 19.9]}$$

where the first part of the equation in the brackets represents the Rubisco-limited and the second part the RuBP-limited photosynthesis model (Farquhar *et al.*, 1980a). For the first part, the value K_c is increased by a term representing competitive inhibition of oxygenation. This ultimately increases the Michaelis constant and means Rubisco has a lower affinity for CO_2. Where K_c increases with temperature, K_o increases much less, so that the effective K_m for CO_2 increases with temperature. The RuBP-limited equation also yields a lower rate as temperatures rise because Γ^* is located in the denominator of this part of the equation.

Photosynthetic electron transport can be independent of temperature at limiting light, but at saturating light electron transport exhibits a peak in capacity at a relatively moderate temperature. The response of $V_{c,max}$ (Bernacchi *et al.*, 2002)

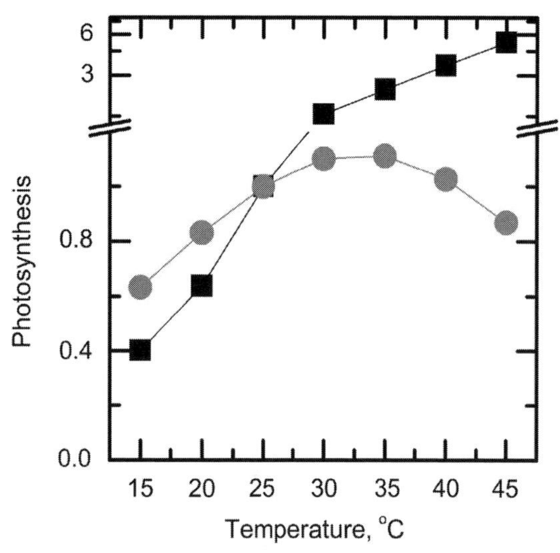

Fig. 19.4. Effect of temperature on $V_{c,max}$ and J_{max} on photosynthesis (relative units). Data are modelled rates normalised to one at 25°C with $V_{c,max}$ (squares) taken from Bernacchi *et al.* (2002) and J_{max} (circles) taken from June *et al.* (2004).

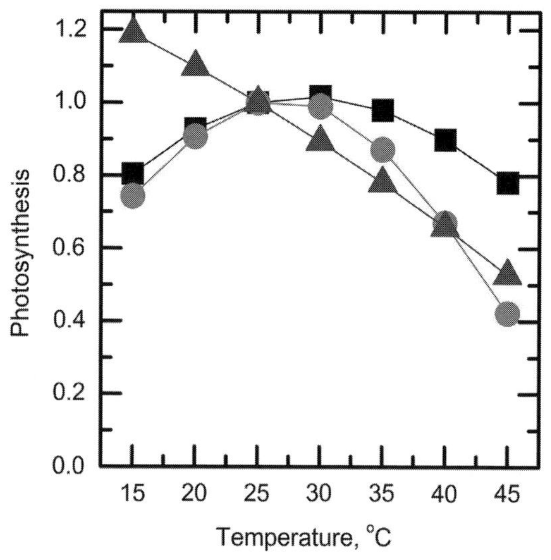

Fig. 19.5. Modelled effect of temperature on photosynthesis with photorespiration. The data were modelled using equations used for Figure 19.4 and in Sharkey *et al.* (2007) and normalised to one at 25°C. The triangles show predicted response of photorespiration with constant RuBP use (condition 2.b.ii, see text). The squares show the predicted response under condition 1.b. The increasing $V_{c,max}$ with temperature becomes overwhelmed by the decreasing efficiency resulting from photorespiration above 30°C. The circles show the predicted response for condition 2.a.i. In this case the decline of J_{max} accentuates the decline resulting from photorespiration causing the steepest decline of overall photosynthesis of any of the conditions.

and J_{max} (June *et al.*, 2004) to temperature, normalised to one at 25°C, is shown in Fig. 19.4.

The net effect of temperature will depend upon which process is controlling the behaviour of photosynthesis at any given condition. Therefore, we must define many conditions that could have different temperature responses:

1. Rubisco controlled:
   a. saturating CO_2 ($V_{c,max}$) (Fig. 19.4 squares)
   b. limiting CO_2 (with K_m effects and photorespiration) (Fig. 19.5 squares);
2. RuBP-regeneration controlled:
   a. saturating light
      i. thylakoid reactions (J_{max})
         - saturating CO_2 (no photorespiration) (Fig. 19.4 circles)
         - limiting CO_2 (with photorespiration) (Fig. 19.5 circles)
      ii. Calvin-cycle reactions (other than Rubisco)
         - saturating CO_2 (no photorespiration)
         - limiting CO_2 (with photorespiration)
   b. limiting light (J):
      i. saturating CO_2 (no photorespiration)
      ii. limiting CO_2 (with photorespiration) (Fig. 19.5 triangles);
3. end-product-synthesis controlled.

Conditions 1.a and 2.a.ii are shown in Fig. 19.4. Conditions 2.b.ii, 1.b and 2.a.i are shown in Fig. 19.5. Condition 2.b.i is insensitive to temperature, where there is very little information on the temperature response for condition 3, but it is strongly temperature dependent (Harley *et al.*, 1992b; Cen and Sage, 2005). Each of these scenarios is a correct depiction of predicted photosynthetic response to temperature without invoking any damage caused by temperature.

Cen and Sage (2005) have examined which processes control the response to heat in sweet potato. They found that at today's CO_2 level, Rubisco limitations dominated the temperature response curve except under moderate stress (Fig. 19.6). At lower CO_2, presumed to represent the CO_2 concentration that has shaped current plant physiology through evolution, Rubisco limitations described the entire temperature response curve. At elevated CO_2, which might represent the future for plant physiological responses, the limiting factor switched abruptly from starch-plus-sucrose synthesis limitation to RuBP-regeneration limitation

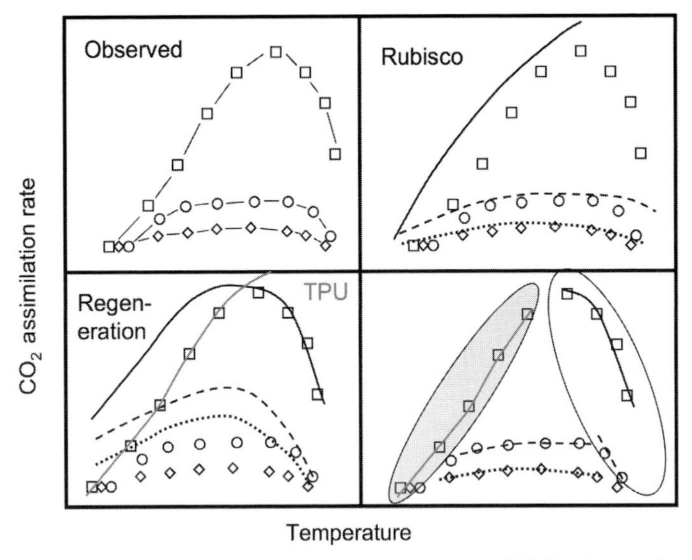

Fig. 19.6. Sweet potato gas-exchange temperature response. Relative photosynthesis rates at intercellular CO_2 concentrations of 140 ppm (triangles), 250 ppm (circles) and 500 ppm (squares) as temperature increases from left to right (based on data in Cen and Sage, 2005). These correspond to ambient CO_2 concentrations that represent when plants evolved, today's CO_2 and saturating CO_2 respectively. The upper-right panel shows that Rubisco kinetics described much of the data at the lower two CO_2 concentrations but not the upper CO_2 concentration. RuBP regeneration (the black line in the lower left panel) describes the decline in photosynthesis at saturating CO_2 and likely also the highest two points of the middle CO_2 concentration. TPU limitation (primarily starch and sucrose synthesis) describes the increase in photosynthesis at lower temperature and saturating CO_2. In the lower right panel the RuBP regeneration (electron transport) limitation space is circled and TPU limitation space is circled and grey.

(electron transport) without ever exhibiting a Rubisco limitation (Fig. 19.6). All of the potential ways photosynthesis can be controlled must be considered for a complete understanding of photosynthetic responses to temperature.

19.2. PHOTOSYNTHETIC STRESS RESPONSES

With the background of how photosynthesis should respond to temperature based on mechanistic models, the observed responses can be examined (see Table 19.1). It is found that photosynthesis can decline more at moderately high temperature (typically between 35 and 40°C, depending on species and growth conditions) than is predicted by the considerations in Section 19.1 (Mohanty *et al.*, 2002; Salvucci and Crafts-Brandner, 2004a,b; Schrader *et al.*, 2004). Empirical studies of how this occurs will be described in this section.

19.2.1. Photosynthetic electron transport responses to heat

Photosystem II is considered a highly stress-sensitive component of PET. If moderate temperature stress is given in the dark (Weis, 1982; Havaux *et al.*, 1991; Wise *et al.*, 2004) or the stress temperature in excess of 45°C, damage to PSII will occur (Santarius, 1975; Santarius and Müller, 1979; Berry and Björkman, 1980; Enami *et al.*, 1994) potentially by causing part of the oxygen-evolving complex (OEC) to become detached from the rest of PSII (Enami *et al.*, 1994). However, PSII damage in light-adapted systems tends to occur only at very high temperature (Terzaghi *et al.*, 1989; Thompson *et al.*, 1989; Gombos *et al.*, 1994; Yamane *et al.*, 1998). Thus, the following situation occurs:

1. The main damage of temperature in excess of 45°C to PET is damage to PSII.
2. Little or no damage to PSII is detected at 40°C.
3. Photosynthesis is reduced beyond what can be predicted based on mechanistic models at 40°C.

These facts led to the conclusion that Rubisco activase may be the primary mechanism of damage to photosynthesis by heat at 40°C. However, there are many easily observed changes in thylakoid reactions that occur at 40°C that need to be considered. First, the thylakoid membranes undergo changes in structure (Armond *et al.*, 1980; Gounaris *et al.*, 1984; Quinn *et al.*, 1985; Sundby and Andersson, 1985).

In addition, there are indications that thylakoid membranes become leaky (Bukhov *et al.*, 1999; Schrader *et al.*, 2004; Zhang and Sharkey, 2009; Zhang *et al.*, 2009c; Hüve *et al.*, 2011) and that the pH component of the *pmf* is lost (Zhang *et al.*, 2009c). The increased leakiness appears to be compensated by increased cyclic electron flow around PSI (CEFI) (Havaux, 1996; Pastenes and Horton, 1996; Bukhov *et al.*, 1999). This cyclic electron flow is supported by increased flow of electrons from the stroma back to the electron transport chain (Havaux, 1996), which results in the stroma becoming more oxidised (Pastenes and Horton, 1999; Ergova *et al.*, 2003; Schrader *et al.*, 2004). The balance of light energy is also changed to favour PSI by a state 1 to state 2 transition (Ovaska *et al.*, 1990; Mohanty *et al.*, 2002; Schrader *et al.*, 2004; Haldimann *et al.*, 2008). The increased efficiency of PSI requires changes in membrane structure and can be prevented by hydrogenation of thylakoid membrane lipids that reduces the fluidity of the membranes (Vígh *et al.*, 1989). The increased CEFI can compensate for thylakoid leakiness, but at the cost of energy diverted from PSII, thus making PET less efficient at high temperature. The reduction of photosynthesis caused by a state transition is not included in any of the modelling of temperature responses above.

19.2.2. Rubisco-activase responses to heat

Rubisco activase is an important enzyme in that it facilitates the removal of inhibitors from Rubisco catalytic sites, allowing for the enzyme to become active or, when already active, allows for RuBP to bind. Rubisco activase, however, is shown to be highly heat labile (Robinson and Portis, 1989) and it is suggested that this causes a major limitation to photosynthesis at higher temperatures (Crafts-Brandner and Salvucci, 2000). Although there is little doubt that higher temperatures result in a loss of activity for Rubisco activase, the impact on photosynthesis of these higher temperatures are debated (Sage and Kubien, 2007). Some evidence suggests that temperatures as low as 30°C can result in loss of Rubisco activity as a function of Rubisco activase being heat labile (Salvucci and Crafts-Brandner, 2004a,b). However, more recent work shows that any losses in Rubisco-activase activity at moderate temperatures (up to 40°C) are not sufficient to cause limitations to photosynthesis. Although increases in temperature are shown to deactivate Rubisco activase, carefully modelled and measured rates of photosynthesis show that the degree of deactivation is not sufficient to become limiting to photosynthesis (Cen and Sage, 2005), with a possible exception in black spruce (Sage *et al.*, 2008). In the study by Cen and Sage (2005), it was demonstrated that perturbations in the ratio of RuBP generation to consumption alters the activation state of Rubisco. This suggests that Rubisco activation is regulated in response to the supply of RuBP, and that any change in the activity of Rubisco activase is a consequence of the supply of RuBP (Cen and Sage, 2005). The data for black spruce (Sage *et al.*, 2008) could be explained as indicating RuBP-regeneration limitation even at low CO_2, where heat-induced deactivation was assessed.

19.2.3. Heat flecks

When air temperature remains moderate, leaves can still experience heat stress because of heat flecks. Photosynthesis requires sunlight, and a large leaf area can improve interception of sunlight for the plant. If the large leaves are also thin, this economises resources but such leaves have very little heat capacity and so are subject to rapid heating when exposed to sunlight. These have been termed heat flecks, analogous to (and sometimes coincident with) sunflecks. Sunflecks are highly likely to cause heat flecks, but in addition, heat flecks can occur without changes in light because of changes in the heat balance of leaves, especially when episodic breezes occur (Sharkey and Schrader, 2006). Because heat flecks can exceed 10°C and be very frequent, they can be a significant component of thermal stress that leaves must cope with. Two mechanisms that seem especially well suited to protecting against heat flecks are production of zeaxanthin during the first exposure to heat stress that protects against subsequent heat stress (Havaux, 1993; Havaux *et al.*, 1996) and isoprene emission that can stabilise membranes (Siwko *et al.*, 2007; Sharkey *et al.*, 2008).

19.2.4. Sustained high-temperature stress

When air temperature is high, plants will experience sustained high-temperature stress, and longer-term responses will become important. This difference is especially clear when considering isoprene emission. Isoprene emission is very common in trees of temperate regions likely to experience heat flecks, but less likely to experience sustained heat stress, such as oaks and aspen. However, isoprene emission is uncommon in hot desert plants, plants more likely to experience sustained temperature stress (Sharkey and Yeh, 2001). Plants, like most organisms, have an extensive array of heat shock proteins (HSPs) that are induced by temperature stress (Lindquist and Craig, 1988; Vierling, 1991). These are often induced as the result of a complex cascade of signals involving heat shock factors (HSFs) (Lee and Schöffl, 1996; Prändl *et al.*, 1998; Zhang *et al.*, 2003; Port *et al.*, 2004).

The heat shock proteins (HSPs) are several families of proteins made exclusively or predominantly during heat stress. The largest proteins belong to the HSP 100 family and there is evidence for members of this family conferring heat tolerance in plants (Queitsch *et al.*, 2000). HSP 90 often works together with other HSPs and could be important for evolution (Queitsch *et al.*, 2002). HSP 70 is known to be involved in protein import into organelles, whereas HSP 60 is essential for the correct folding of Rubisco and other proteins. Plants are unusual among eukaryotes in having a very large family of low-molecular-weight HSPs. These are targeted to many different compartments within plants. Of interest here is the chloroplast-targeted low-molecular-weight HSP. It is distinguished by having a methionine-rich region and an alpha-crystalin domain, a section that resembles the lens protein in eyes (Waters *et al.*, 1996). There is evidence that this HSP can improve thermotolerance of photosynthetic processes (Heckathorn *et al.*, 1998; Heckathorn *et al.*, 2002; Wang and Luthe, 2003).

The function most associated with HSPs is protein folding and renaturation (Vierling, 1991; Boston *et al.*, 1996; Glover and Lindquist, 1998). However, it is becoming increasingly clear that HSPs also protect membrane function (Tsvetkova *et al.*, 2002; Hong *et al.*, 2003; Török *et al.*, 2003; Basha *et al.*, 2004). Thus, the widespread occurrence of HSPs is not evidence for protein damage over membrane integrity as the weak link in photosynthetic function during heat stress.

19.3. SUPPLY SIDE OF PHOTOSYNTHESIS AT HIGH TEMPERATURE

Any discussion that focuses on the impact of rising temperatures on photosynthesis would be incomplete without discussing the supply side. There are numerous barriers to the diffusion of CO_2 from the atmosphere to the site of carboxylation (Nobel, 2005). The two conductances that are controlled physiologically, at least to some extent, include stomatal and mesophyll conductances. Stomatal conductance (g_s) represents the conductance of CO_2 as it diffuses from the leaf surface into the intercellular airspace. Mesophyll conductance (g_m) represents the pathway starting at the intercellular airspace and ending in the chloroplast where assimilation occurs. Although both g_s and g_m are shown to change with temperature, these changes for g_s are likely to be indirectly influenced by temperature and for g_m it is unclear whether temperature is directly or indirectly influencing changes.

A well-known and highly validated model for g_s has been widely used since first being presented (Ball *et al.*, 1987). This model is empirically based and predicts the change in g_s as:

$$g_s = g_0 + m\frac{A_N RH_s}{C_s} \qquad \text{[Eqn. 19.10]}$$

A_N is net photosynthesis; RH_s is humidity at the leaf surface; C_s is the CO_2 concentration at the leaf surface; g_0 is the y axis intercept; and m is the slope. Although temperature is not included in this model, both A_N and RH_s will change with temperature and ultimately have an influence on g_s. A direct effect of temperature on g_s has only been shown if leaf-to-air VPD is held constant, and even then the results are highly variable (Sage and Sharkey, 1987; Aphalo and Jarvis, 1991; Šantrůček and Sage, 1996; Bunce, 2000). However, most findings suggest that g_s will change with A_N such that if increasing temperatures increase A_N it will increase g_s, and *vice versa* for decreasing A_N with temperature (Wong *et al.*, 1985; Šantrůček and Sage, 1996). Under most conditions, however, an increase in temperature coincides with an increase in VPD, which will generally dominate over temperature (Farquhar and Sharkey, 1982; Farquhar and Wong, 1984; Aphalo and Jarvis, 1991; Jones, 1998). No model for g_m exists as it does for g_s. However, numerous studies have demonstrated the impact of changing conditions surrounding the leaf on g_m (Chapter 12), with temperature being one (Bernacchi *et al.*, 2002; Warren, 2006a, 2008a). Increasing temperature is shown to increase g_m until about 35°C, after which it begins to decrease (Bernacchi *et al.*, 2002) or remain constant (Warren, 2008a). Despite the increases in g_m with temperature, the limitation imposed on photosynthesis continues to increase as temperatures rise (Bernacchi *et al.*, 2002; Warren, 2008a), which, added to the increasing limitation of decreasing g_s with temperature (Long, 1991), results in a potentially substantial limitation to photosynthesis (Fig. 19.7).

19.4. HEAT DAMAGE VERSUS THERMOTOLERANCE MECHANISMS (SYNTHESIS)

Photosynthetic responses to high temperature are very complex. There are several concerns. Proteins and membranes are held together by forces that are inherently temperature

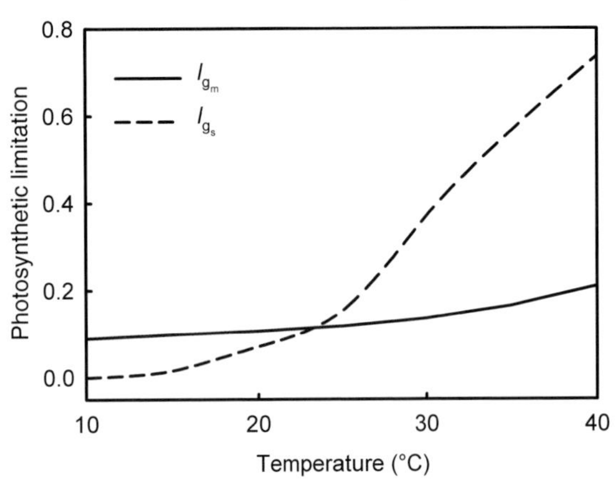

Fig. 19.7. Effect of temperature on the limitation imposed on photosynthesis by mesophyll conductance l_{gm} or stomata l_{gs}.

sensitive (Quinn, 1988; Pace *et al.*, 1996) and so both can denature at high temperature. Membranes are subject to momentary aqueous pores that can damage metabolism (Weaver *et al.*, 1984). High temperature can lead to oxidation stress (Mishra and Singhal, 1993) though when leaves are heated in darkness, the heat is much more damaging and less reversible than when they are heated in the light (Havaux *et al.*, 1991). In addition, very high rates of oxygenation of RuBP can burden the photorespiratory reactions and lead to toxic levels of some metabolites. But by far the two best-known effects of high temperature are destruction of PSII and loss of Rubisco activation. It is not clear how the various observations of temperature effects should be viewed. It is possible that all of the reversible mechanisms that have been discovered that reduce photosynthesis at moderate heat stress below what is predicted by mechanistically based models are adaptive by ensuring that any decline is rapidly recovered when the heat stress is relieved. Thus, as temperature rises we might expect to see adaptive responses, which could be unrelated to the damage responses that occur once the capacity to adapt is exceeded. The distinction between adaptive responses and damage may come down to speed of recovery. There is much less information available on recovery of photosynthesis from different heat stresses and there is much need for work in this area.

19.5. CONCLUDING REMARKS

The response of photosynthesis to high temperature will be a significant component of overall plant response to high temperature, though other plant processes also need to be considered. Significant effects of high temperature can be seen at relatively modest temperature. The first responses are consequences of Rubisco kinetics and electron transport capacity. These are easily predicted and fully reversible. For most plants, once the temperature rises much above 35°C, a significant deactivation of Rubisco becomes easily measured, the stroma becomes more oxidised, a state 1 to state 2 transition occurs and PSI becomes more reduced. Some of these changes appear to be adaptive, and photosynthetic decline is generally reversible if the stress is not too severe. At higher temperatures, irreversible damage occurs. Understanding all of these responses is necessary in order to make reliable predictions about how photosynthesis will be affected by increasing temperature.

20 • Photosynthesis under water deficits, flooding and salinity

M.M. CHAVES, J. FLEXAS, J. GULÍAS, F. LORETO AND H. MEDRANO

20.1. INTRODUCTION

Drought, salinity and flooding are among the most important abiotic stresses affecting plant growth and survival in large areas of the globe. They affect natural ecosystems, plantations and croplands, therefore posing large problems to nature conservation and to farmers and foresters. Drought is a major concern not only in the arid and semiarid zones, but is also increasingly affecting temperate regions that are now subjected to occasional severe drought spells (IPCC2007 www.ipcc.ch). Alternatively, the same IPCC report indicates an increased occurrence of heavy precipitation and tropical cyclone activity, leading to more frequent flooding events. Salinity and flooding are often secondary stresses following years of incorrect and unsustainable irrigation practices, leading to increased soil salinity and the rise of phreatic soil-water. Flooding can have catastrophic impacts on the productivity of arable farmland, as most crops are intolerant to excess water (Voesenek *et al.*, 2006). Moreover, drought and salt stresses are commonly accompanied by high temperatures and high irradiances that exacerbate the negative impact of each stress acting in isolation (Mittler, 2006).

The primary effects of drought and salinity are similar, as salinity in the soil reduces osmotic potential, making it harder for roots to extract water (Munns, 2002; Munns and Tester, 2008). On the contrary, under flooding water availability is high but O_2 availability for root respiration is restricted owing to its slow diffusion in water, limiting root growth (Blom and Voesenek, 1996; Bailey Serres and Voesenek, 2008).

Together with growth, photosynthesis is affected by water and salt stress, as well as by flooding. Photosynthesis limitations have been shown to arise predominantly from stomatal closure (Chaves, 1991; Cornic, 2000; Munns, 2002; Chaves *et al.*, 2009) or diffusion limitations in the mesophyll (Delfine *et al.*, 1999; Flexas *et al.*, 2008), both leading to decreased CO_2 availability to the chloroplasts. However, metabolic impairment (Lawlor and Cornic, 2002; Lawlor and Tezara, 2009) often becomes prominent as stress progresses. In the case of salinity, in addition to dehydration, photosynthesis may also be impaired by hyper-ionic and hyper-osmotic stress (Munns *et al.*, 2006). Under flooding the production of ATP and regeneration of NAD^+ through anaerobic respiration leads to generalised metabolic changes in plants (Bailey Serres and Voesenek, 2008). Limitations to photosynthesis in response to water and salt stresses can also arise via secondary effects, namely decreased hydraulic conductance or oxidative stress, particularly when plants are suffering from multiple environmental constraints.

Plant resistance to water and salt stresses is species specific depending on the ability to escape, avoid or tolerate the stress (Levitt, 1980). The avoidance strategy generally enables the plant to maintain a high water status by stomatal closure and/or reduction of leaf area, but only the first mechanism allows quick recovery of plant functions following stress relief. Plant resistance to water and salt stress also depends on stress duration, intensity and rate of imposition and intermittence, the latter being the factor that most influences acclimation capacity (Chaves *et al.*, 2002, 2003). The phase of plant development when stress strikes exerts a large influence on the degree of injuries; normally it shows more severe effects when interrupting reproductive development.

In this chapter we will highlight similarities and differences of the effects of drought, salinity and flooding on photosynthesis, focussing in particular on primary and secondary limitations and on the underlying signalling events. Concerning flooding we will restrict our analysis to short-term flooding of the root system and will not address shoot submersion, whose peculiarities are outside the scope of

Terrestrial Photosynthesis in a Changing Environment: A Molecular, Physiological and Ecological Approach, ed. J. Flexas, F. Loreto and H. Medrano. Published by Cambridge University Press.

Table 20.1. *A summary of the described signalling factors inducing changes in growth, stomatal closure and decreased photosynthesis in response to drought/salinity and flooding. The relative importance of each factor is indicated (–, effect not described; X, effect described occasionally; XX, effect described several times; XXX, main effect). Notice that if a factor is not indicated as important may not necessary mean that it is not, since the literature available is fragmentary and some of the factors have been poorly studied under some of these stresses.*

	Drought/salinity	Flooding
ABA – root-to-shoot	XXX	—
ABA – shoot tissues	XXX	XX
Ethylene	X	XX
Sap pH	XX	X
Hydraulic signals	XXX	XXX
Electrical signals	X	—
Decreased nutrient availability	X	XXX

this chapter (see reviews by Voesenek *et al.*, 2006 and Bailey Serres and Voesenek, 2008).

20.2. REGULATION AND SIGNALLING EVENTS IN THE RESPONSE TO DROUGHT, FLOODING AND SALINITY

Drought, flooding and salinity stress occur all in the soil where plants are rooted. Therefore, stomatal and photosynthetic responses in the leaves require root-to-shoot signalling. Several molecules have been suggested as possible mediators of root-to-shoot signalling, depending on the type of stress (Table 20.1).

Plant hormones and other compounds moving along the xylem stream are among the most commonly described signalling molecules in response to abiotic stresses. Among these, ABA, ethylene, cytokinins, JA, linolenic acid, amino acids, organic acids, anions and cations and xylem pH have been suggested to play a role in drought-, salt- and/or flooding-induced stomatal closure, photosynthesis reduction and growth decline (Schurr *et al.*, 1992; Schurr and Schulze, 1996; Herde *et al.*, 1997; Jackson, 2002; Blanke and Cooke, 2004; Voesenek *et al.*, 2004). In particular, ABA has been shown to be involved in drought and salinity responses (Wilkinson and Davies, 2002).

The mechanism by which ABA triggers stomatal closure is clearly complex and not fully understood (Schroeder *et al.*, 2001a,b; Hetherington and Woodward, 2003). It has been suggested that ABA synthesis follows the release of signalling molecules that typically accumulate under stress, such as H_2O_2 and nitric oxide (Desikan *et al.*, 2004; Besson-Bard *et al.*, 2008; Neill *et al.*, 2008), and it also interacts with ozone levels (Wilkinson and Davies, 2010). Very recently, the understanding of the molecular mechanism of ABA-induced stomatal closure has been largely clarified with the identification of ABA receptors, which are proteins of the PYR/PYL/RCAR family (Fujii *et al.*, 2009; Santiago *et al.*, 2009). In the absence of ABA, PYR/PYL/RCAR presents a configuration that cannot bind protein phosphatases of type 2C (PP2C), so that they can dephosphorylate SnRK2 kinases, which are downstream-positive elicitors of ABA-mediated responses, blocking their function. Instead, when ABA is present in the PYR/PYL/RCAR receptor, the configuration of the latter changes so that it binds PP2Cs, liberating SnRK2 kinases (Sheard and Zheng, 2009; Cutler *et al.*, 2010), which then induce further responses including Ca^{2+}-dependent and independent pathways, ending up with the induction of SLAC1 S-type-anion channel-mediated stomatal closure (Vahisalu *et al.*, 2008; Ache *et al.*, 2010; Kim *et al.*, 2010; Klingler *et al.*, 2010). According to this regulation, PP2C-defective plants present hypersensitivity to ABA (Sáez *et al.*, 2004, 2006; Rubio *et al.*, 2009), SnRK2 defective plants (such as *OST* – Open Stomata – mutants) are insensitive to ABA (Mustilli *et al.*, 2002; Merlot *et al.*, 2007) and *slac*-1 mutants are insensitive to both ABA and other stimuli, like CO_2, H_2O_2 or light (Vahisalu *et al.*, 2008; Ache *et al.*, 2010).

Irrespective of the mechanism and pathways for ABA-induced stomatal closure, all evidences available in the literature make clear that ABA plays a crucial role in stress-induced stomatal closure. However, it is also clear that ABA alone may not explain the wide range of stomatal movements in leaves exposed to water and salt stress. Nutrients such as potassium can regulate the relationship between ABA concentration and stomatal closure (Wilson *et al.*, 1978). Additional signals are likely involved in the observed responses, possibly of hydraulic (Tyree and Sperry, 1989; Hubbard *et al.*, 2001; Rodrigues *et al.*, 2008; Brodribb and Cochard, 2009; Parent *et al.*, 2009; Resco *et al.*, 2009) or electrical origin (Fromm and Lautner, 2007).

Dehydrating roots in drying or saline soil synthesise ABA more rapidly than fully turgid tissue. The consequent increase in the ABA concentration of xylem sap flowing towards the still-turgid shoot constitutes a chemical signal to the leaves, inducing stomatal closure and decreasing photosynthesis (Davies and Zhang, 1991). Further evidence for the role of ABA has been collected: exogenous application of ABA to either the soil or nutrient solutions in well-watered plants or to detached leaves through the petiole reproduces the observed relationships between endogenous xylem ABA content and stomatal conductance (g_s) (Borel *et al.*, 2001; Flexas *et al.*, 2006b). Furthermore, in *Nicotiana plumbaginifolia*, exogenous application of ABA to both wildtype (wt) and several mutants deficient in endogenous ABA synthesis demonstrated that manipulating the capacity to synthesise ABA does not change stomatal sensitivity to ABA (Borel *et al.*, 2001). However, there is also evidence that leaf-bulk ABA and not xylem ABA mostly controls stomatal closure. For instance, Trejo and Davies (1991) observed that stomatal closure preceded both decreased water potential and increased xylem ABA in *Phaseolus vulgaris* L. Perks *et al.* (2002), measuring sap flow and ABA concentration in tall trees such as Scots pine, calculated that under normal conditions (well-watered plants) a signal produced in the roots would take as much as 12 days to reach the leaves, this time increasing to as much as 6 weeks under drought conditions. It was concluded that the root-ABA signalling model for short-term stomatal response to soil drying was not applicable in mature conifer trees because the signal transmission was too slow.

Alternatively, Holbrook *et al.* (2002) constructed grafted tomato plants from ABA-deficient tomato mutants and wild-type parents. Wildtype shoots grafted on mutant roots and mutant shoots grafted on wild-type roots were assayed. In each of a series of experiments, stomata closed in accordance to the genotype of the shoot not to that of the root. Moreover gradients in g_s, xylem sap ABA and bulk leaf ABA have been shown to exist in grapevine shoots. Using a combination of physiological and molecular approaches, it was demonstrated that gradients in g_s along the cane were correlated with ABA concentration gradients in mature leaves and xylem sap, and that these ABA gradients were, at least partly, owing to differences in local synthesis (Soar *et al.*, 2004). Christmann *et al.* (2005) used a noninvasive system to monitor the generation and distribution of physiologically active pools of ABA, testing wild-type plants responding to ABA (abi1–1) and ABA-deficient mutants (aba2–1) of *Arabidopsis thaliana* in the presence and absence of water stress. Exposure of seedlings to exogenous ABA resulted in a uniform pattern of reporter expression in all tissues. In marked contrast, reporter expression in response to drought stress was predominantly confined to the vasculature and stomata. Surprisingly, water stress applied to the root system resulted in the generation of ABA pools in the shoot but not in the root.

Recently, it has been shown that two pools of ABA may co-exist in leaves. Foliar ABA results from the cleavage of xanthoxin, a xanthophyll formed through the chloroplastic pathway of isoprenoid biosynthesis. Inhibition of isoprene biosynthesis in leaves led to a reduction of about 50% of total ABA content in leaves (Barta and Loreto, 2006). Modulation of isoprene synthesis also reproduces ABA levels under water stress and during midday depression of photosynthesis that occurs under low air humidity (Raschke and Resemann, 1986). This suggests that the isoprenoid-dependent foliar pool of ABA is involved in mechanisms inducing stomatal closure under rapid environmental changes. However, stomatal closure was independent of changes of ABA induced by isoprene inhibition when leaves were darkened or exposed to elevated CO_2, indicating that the labile ABA pool is not the unique mechanism regulating stomatal opening.

Whereas root-to-shoot ABA and/or shoot or leaf-synthesised ABA can be important in inducing stomatal closure under drought and salinity stresses, only shoot-synthesised ABA seems to be involved in flooding-induced stomatal closure (Table 20.1). Indeed, flooding induces large decreases in ABA delivery from roots to shoots (Else *et al.*, 1994; Jackson, 2002; Blanke and Cooke, 2004). However, flooding induces a large increase of the concentration of foliar ABA. Although the mechanisms of signal transduction from roots to shoots is unknown, ABA seems to be involved in stomatal responses to flooding as stomatal closure is accompanied by increased concentrations of foliar ABA. Also, ABA-deficient mutants of pea have shown impaired stomatal responses to flooding (Jackson and Hall, 1987).

Because flooding increases the release of ethylene, this molecule has also been suggested to be involved in stomatal closure (Blanke and Cooke, 2004). However, under drought ethylene has been shown to be implicated in growth reduction but not in stomatal responses. Indeed, increased root ABA keeps ethylene levels low, allowing root growth to continue under drought, whereas ethylene accumulation in shoots inhibits their growth (Sharp, 2002). Moreover, it has been recently shown that ethylene, whose synthesis can be induced by increased ozone, is an inhibitor of the ABA-induced and soil-drying-induced stomatal closure (Wilkinson and Davies, 2009).

Another signal possibly involved in ABA-induced responses and stomatal closure is xylem-sap pH. Early studies demonstrated that xylem-sap pH increased by 0.5–1 units during drought stress (Gollan *et al.*, 1992). Potentially, such a change in pH could influence ABA function, changing the distribution of endogenous ABA among different cell compartments, displacing the equilibrium from ABA lipophilic un-dissociated form (ABAH) towards its lipophobic form (ABA-), and modifying synthesis and/or catabolism of ABA. Conductance of transpiring leaves is lower in pH 7.0 than in pH 6.0 buffers, despite the fact that bulk-leaf ABA and shoot-water status are unaffected by pH (Wilkinson and Davies, 1997), implying that the pH control of g_s is independent of ABA. However, it has been shown recently that spraying leaves with alkaline buffers reduces g_s, but not in the *flacca* mutants of tomato that lack endogenous ABA. This demonstrates that ABA plays an active role in pH-dependent regulation of stomatal closure (Wilkinson and Davies, 2008). Finally, sugars travelling in the xylem of droughted plants, or sugars that might increase in the apoplast of guard cells under high light are also likely to exert an important influence on stomatal sensitivity to ABA (Wilkinson and Davies, 2002).

Concerning hydraulic signals, it was initially proposed that xylem cavitation could underlie stomatal regulation in response to water stress and flooding (Tyree and Sperry, 1989). A few studies have specifically demonstrated that decreasing leaf hydraulic conductivity (Kh) results in stomatal closure (Hubbard *et al.*, 2001). Salleo *et al.* (2000) followed g_s and leaf-water potential (Ψw) upon illumination of dark-adapted leaves, either in detached branches or in intact plants. Stomata increasingly opened after illumination, whereas Ψw progressively decreased owing to leaf transpiration. Only once leaves reached the Ψw at which stem cavitation was triggered, was g_s decreased. This was taken as evidence that early cavitation events trigger stomatal regulation so that Ψw is returned to values just above the cavitation threshold. Similarly, Hubbard *et al.* (2001) showed that inducing xylem cavitation by means of air injection into the xylem vessels resulted in almost immediate stomatal closure. Brodribb and Holbrook (2003) also suggested that drought-induced stomatal closure is mostly dependent on leaf-vein cavitation. Cochard *et al.* (2002) specifically showed that g_s responded to plant hydraulics in such a way that the water pressure in the leaf rachis xylem, which is the most vulnerable organ to cavitation, was maintained above its cavitation threshold. Cochard *et al.* (2007) showed that aquaporins, proteins that specifically vehicle water across cellular compartments, play a dominant role in the regulation of these dynamic changes in hydraulic conductance of leaves. Moreover, Shatil-Cohen *et al.* (2011) have recently shown that, under drought stress, xylem-sap ABA induces specific downregulation of bundle-sheath – but not mesophyll – aquaporins, leading to decreased hydraulic conductivity and stomatal closure. Christmann *et al.* (2007) described another drought-induced hydraulic signal, which is related to decreased shoot pressure and is involved in root-to-shoot water-stress signalling, and perhaps also in flooded plants.

There are evidences that both water stress and salinity, as well as low soil O_2 occurring during flooding, induces fast decreases in root and plant Kh (Jackson, 2002; Parent *et al.*, 2009). Under flooding, the negative hydraulic message could be avoided and stomatal closure delayed by applying pneumatic pressure to flooded roots of *Ricinus communis*. However, this was not the case when using tomato. Thus, at least in some plant species, a hydraulic signal can be also involved in stomatal closure (Else *et al.*, 2001).

Finally, electrical signals have also been suggested to be involved in root-to-shoot signalling and stomatal closure, particularly in leaves undergoing water stress and recovering from stress. Fromm and Fei (1998) found that water stress induced a decreased electric potential difference along a maize leaf, which was restored upon re-watering and correlated with g_s variations. More recently, Grams *et al.* (2007) have specifically demonstrated that electrical signals may play the main role in regulating stomatal re-opening after a water-stress period in maize. Gil *et al.* (2008) showed that, in avocado, a sudden change in soil-water content induced by root drying and re-watering was accompanied by a slow, significant change in the recorded electric potential difference between roots and shoot, which inversely correlated with the g_s difference measured before and after each soil-drying treatment. Plants that were girdled to interrupt the phloem continuity

Table 20.2. *A summary of the described limitations of drought, salt and flooding stresses to photosynthesis. The relative importance of each factor is indicated (–, effect not described; X, effect described occasionally; XX, effect described several times; XXX, main effect). Notice that if a factor is not indicated as important may not necessary mean that it is not, since the literature available is fragmentary and some of the limitations have been poorly studied under some of these stresses.*

	Drought		Salinity		Flooding	
	Mild	Severe	Low [salt]	High [salt]	Short term	Long term
Stomatal limitations	XXX	X	XXX	X	XXX	X
Non-stomatal limitations	X	XXX	X	XXX	XX	XXX
Mesophyll conductance	XX	XX	XX	XX	?	X
Leaf photochemistry	—	X	—	X	X	XXX
Rubisco	—	X	X	XXX	—	X
Calvin cycle	—	X	X	XXX	X	XXX
Feedback inhibition	—	X	X	X	—	X

and then irrigated, tended to have lower electric potential differences over time than non-girdled irrigated plants, suggesting that the electrical signal was transmitted in the phloem.

20.3. PRIMARY AND SECONDARY EFFECTS ON PHOTOSYNTHESIS: COMMON RESPONSES AND SPECIFICITIES OF EACH STRESS

There has been some controversy regarding the main physiological targets responsible for photosynthetic impairment under drought and/or salinity (Chaves, 1991; Flexas and Medrano, 2002a; Lawlor and Cornic, 2002). However, CO_2 diffusion from the atmosphere to the site of carboxylation is always reduced, and is often the main cause for decreased photosynthesis under water stress (Chaves and Oliveira, 2004; Flexas *et al.*, 2004a, 2009; Grassi and Magnani, 2005; Aganchich *et al.*, 2009; Chaves *et al.*, 2009; Resco *et al.*, 2009) and salt stress (Delfine *et al.*, 1999; Stepien and Johnson, 2009) in C_3 plants. A similar pattern of response has been described for C_4 plants (Hura *et al.*, 2006; Ghannoun, 2009), except that leaf photosynthetic metabolism of C_4 plants – particularly Rubisco in BSC – seems somewhat more sensitive to water stress than it is in C_3 plants (Carmo-Silva *et al.*, 2008a,b), whereas stomatal closure is sometimes described as more sensitive in C_3 than C_4 plants (Taylor *et al.*, 2011). CAM plants respond differently, owing to their particular physiological adaptation (see Chapter 6). Reduced g_s is also the main factor leading to decreased photosynthesis during early stages of flooding (Bradford, 1983; Kozlowski, 1984; Vu and Yelenosky, 1991). Metabolic impairment of photosynthesis seems to appear earlier under long-term flooding than long-term drought (Blanke and Cooke, 2004). The factors inducing photosynthesis limitations under mild-to-moderate drought, salinity and short-term flooding are listed in Table 20.2.

Reduced leaf diffusive capacity is owing to at least two components that are often coregulated: stomatal closure and reduced mesophyll conductance (g_m). Stomatal closure has been known for a long time to be one of the first responses of plants to soil-water shortage, and the mechanisms possibly inducing stomatal closure have been revised above. In an early work, Jones (1973) suggested that g_m was depressed under water-stress conditions, but this factor has only recently been recognised as an equally important cause for reduced CO_2 diffusion under drought (Flexas *et al.*, 2002a; Warren, 2008b; Cai *et al.*, 2010), salinity (Bongi and Loreto, 1989; Delfine *et al.*, 1999; Centritto *et al.*, 2003) and long-term flooding (Black *et al.*, 2005). The determinants of g_m are unclear (see Chapters 4 and 12). Reduction of g_m under water-stress conditions may be linked to physical alterations in the structure of the intercellular spaces owing to leaf shrinkage (Lawlor and Cornic, 2002), changes in cell-wall porosity and thickness

(Niinemets *et al.*, 2009b,c) or cell and/or organelle reorganisation (Geissler *et al.*, 2009; Zellnig *et al.*, 2010). However, fine and rapid regulation of g_m to CO_2 in response to varying environmental conditions could also be related to the expression and/or regulation of plasma-membrane aquaporins, which also regulate water diffusion (Uehlein *et al.*, 2003; Flexas *et al.*, 2006a). Although g_s is always reduced under water stresses, g_m is often but not always reduced. For instance, g_s was reduced but g_m was not in olive trees subjected to partial root drying (Aganchich *et al.*, 2009) or in the woody legume *Prosopis velutina* under severe summer stress in Arizona (Resco *et al.*, 2009). Moreover, water-stress effects on g_m are attenuated or even totally cancelled under conditions of low VPD (Perez-Martin *et al.*, 2009), non-saturating light (Gallé *et al.*, 2009) or both (Flexas *et al.*, 2009).

As a consequence of reduced CO_2 concentration and unaltered metabolism, apparently futile cycles, like photorespiration (Osmond *et al.*, 1980a; Rivero *et al.*, 2009) and/or other electron-consuming processes, such as the Mehler reaction (Miyake and Yokota, 2000; Haupt-Herting *et al.*, 2001 but see Driever and Baker, 2011), chlororespiration (Peltier and Cournac, 2002; Stepien and Johnson, 2009), CEFI (Golding and Johnson, 2003; Jia *et al.*, 2008; Kohzuma *et al.*, 2009) or CEFII (Canaani and Havaux, 1990) and increased electron partitioning towards AOX in the mitochondria (Ribas-Carbo *et al.*, 2005b; Atkin and Macherel, 2009), may increase in water- and salt-stressed plants. An exception would be photorespiration in C_4 plants, which remains very low even under severe water stress (Carmo-Silva *et al.*, 2008a).

Restricted CO_2 diffusion across leaves is likely to be the primary, earlier and most usual cause for decreased photosynthesis rates in stressed leaves, but metabolic impairment may also occur, particularly under acute or severe stress (Tezara *et al.*, 1999; Lawlor and Cornic, 2002; Chaves *et al.*, 2009; Lawlor and Tezara, 2009; Zheng *et al.*, 2009; Dias and Brüggemann, 2010; Hu *et al.*, 2010), prolonged drought (Damour *et al.*, 2009), long-term flooding (Blanke and Cooke, 2004; Herrera *et al.*, 2008; Arbona *et al.*, 2009) or under prolonged salt stress in which plants respond, in addition to dehydration and to hyper-ionic and hyper-osmotic stress (Tattini *et al.*, 1997; Munns *et al.*, 2006). Well-adapted plants, however, do not seem to develop metabolic limitations to photosynthesis after successive prolonged cycles of drought and rewatering (Liu *et al.*, 2010) or even after several years of acclimation to dry climate (Grant *et al.*, 2010; Limousin *et al.*, 2010).

Under water and salt stress it has been postulated that metabolic impairment of photosynthesis only occurs when maximum daily g_s is very low (in many C_3 species and regardless of the water status of leaf tissues when g_s drops below 0.05–0.10 mol H_2O m^{-2} s^{-1}, as shown by Bota *et al.*, 2004 and Flexas *et al.*, 2004a,b). In some studies with white poplar it was even shown that photosynthesis was limited by biochemical factors only when there was virtually no soil water available to support plant transpiration (Brilli *et al.*, 2007). It should also be said that several Mediterranean plant species and other plants that are adapted to arid environments may escape this generalisation, with g_s also being drastically restricted in unstressed plants namely under midday conditions, when high light and temperature are superimposed. It is remarkable that regardless of the species analysed, both photochemical (chl. content, photochemical efficiency) and biochemical (contents of ATP, RuBP and total soluble protein; activities of Rubisco, SPS, fructose bisphosphatase, nitrate reductase) components of photosynthesis are impaired at the same g_s threshold under water stress (Flexas *et al.*, 2004a,b). Using a photosynthesis-limitation analysis, Grassi and Magnani (2005) confirmed the same patterns of photosynthesis response to drought during three consecutive years in field-grown ash and oak trees. They showed that during summer stress development, whenever g_s was higher than 0.1 mol H_2O m^{-2} s^{-1}, biochemical limitations to photosynthesis were not detectable, and the sum of limitations imposed by stomatal closure and decreased g_m accounted for the entire decrease in photosynthesis. Stomatal limitations accounted for about two-thirds of the observed decline, whereas decreased g_m accounted for the other one-third decrease. At more severe stress conditions, when g_s was below 0.1 mol H_2O m^{-2} s^{-1}, biochemical limitations were detectable, although they rarely accounted for more than 15–20% of total photosynthetic limitations. Very similar results using the same approach were obtained by Galmés *et al.* (2007a) in water-stressed plants of ten Mediterranean species grown in pots (Fig. 20.1), and by Misson *et al.* (2010) in *Quercus ilex* stands subject to different rain exclusion treatments. The effect of severe water stress on Rubisco initial activity can be reproduced by inducing stomatal closure to levels below the indicated threshold by addition of ABA to the nutrient solution of unstressed plants, or by cutting the petiole in air to rapidly dehydrate leaves (Flexas *et al.*, 2006b) so that it appears that low chloroplast CO_2 concentration during drought and not tissue dehydration is inducing decreased Rubisco initial activity (Flexas *et al.*, 2006b; Galmes *et al.*, 2010). Recently, Zhou *et al.* (2007a) demonstrated that the extent of photosynthetic-metabolism (i.e., the activity of Rubisco) impairment under drought is totally dependent on the prevailing light regime, i.e., the higher the PPFD, the higher the metabolic inhibition. These authors showed that decreased Rubisco activity of water-stressed leaves is highly correlated with the increase of ROS, namely H_2O_2. It is however unclear whether ROS accumulate before Rubisco de-activation. In this case, ROS accumulation may again be a result of reduced CO_2 availability in the mesophyll, which makes the biochemistry of photosynthesis (Rubisco) unable to cope with the large photon-flux photochemically absorbed (see also Section 20.5). Some authors consider that reduced chloroplast ATP_{ase} activity – rather than decreased Rubisco – is the main cause for biochemical impairment of photosynthesis under drought (Tezara *et al.*, 1999; Lawlor and Tezara, 2009), as well as under salt stress, waterlogging or a combination of both (Zheng *et al.*, 2009). These stresses increase proton fluxes into the lumen mostly owing to enhanced CEFI (Jia *et al.*, 2008), particularly in stress-sensitive plants (Stepien and Johnson, 2009). This, coupled with impaired ATP_{ase} that results in decreased thylakoid membrane conductance to protons, leads to a high proton concentration in the lumen triggering increased qE (Kohzuma *et al.*, 2009). Similar to Rubisco (see above), increased CEFI and impaired ATP_{ase} have been shown only in plants with g_s lower than 0.1 mol H_2O m^{-2} s^{-1} (Tezara *et al.*, 1999; Kohzuma *et al.*, 2009; Stepien and Johnson, 2009; Zheng *et al.*, 2009) and tightly coupled with ROS accumulation (Zheng *et al.*, 2009). This has led to the hypothesis that oxidative damage to chloroplast ATP_{ase} produced by ROS under conditions of low CO_2 and excess light causes a water-stress-induced decrease of photosynthetic capacity (Lawlor and Tezara, 2009).

The response of photosynthesis to water stress in C_4 plants may be to some extent different to that of C_3 plants. Although generally stomatal limitations are also present, and may be dominant at the very early stages of water-stress imposition (Marques da Silva and Arrabaça, 2004), non-stomatal limitations often appear also at early stages of drought, immediately after stomatal limitations (Du *et al.*, 1996; Lal and Edwards, 1996; Carmo-Silva *et al.*, 2008a,b). In particular, Rubisco seems very sensitive to water stress, declining linearly with stress intensity in many C_4 species (Du *et al.*, 1996) but not in others (Lal and Edwards, 1996). By contrast, PEPC and C_4 acid decarboxylases are more resistant to water stress, declining only eventually under severe stress conditions (Carmo-Silva *et al.*, 2008b). Although g_m is thought to be not limiting in C_4 plants because of their

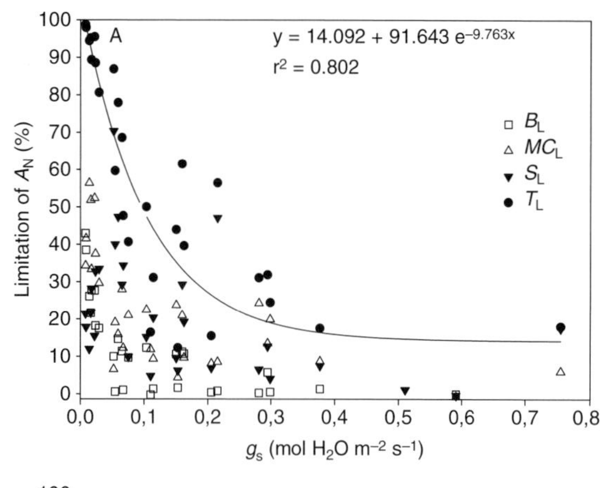

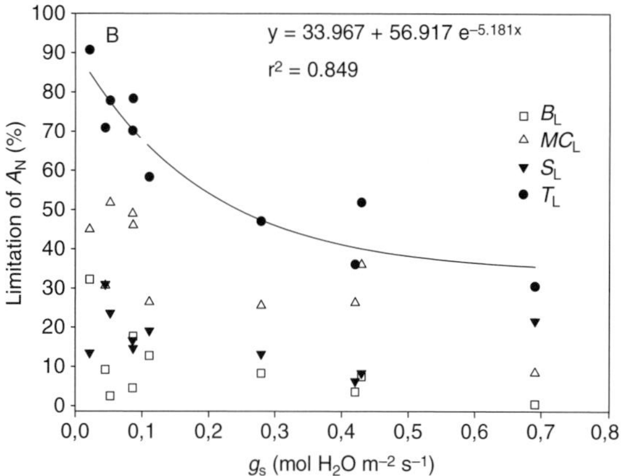

Fig. 20.1. Relationship between limitations of A_N (%) and stomatal conductance (g_s) considering all ten species. **(A)** Values obtained from mild, moderate and severe water-stress treatments. **(B)** Values obtained from rewatering treatment. The regression coefficient and significance of the relationships between total limitations and stomatal conductance are shown. B_L, biochemical limitation; MC_L, mesophyll limitation; S_L, stomatal limitation; T_L, total limitations. After Galmés *et al.* (2007).

CO_2-concentrating mechanisms, it has been suggested that bundle-sheath CO_2 leakiness could increase under water stress, leading to reduced CO_2 availability and contributing to a less-efficient fixation under water stress (Carmo-Silva *et al.*, 2008a). Potential differences in the response to stress among different C_4 subtypes remain elusive (Carmo-Silva *et al.*, 2008a,b).

Although the combination of reduced g_s and g_m is the main determinant of photosynthesis reduction also under salt (NaCl) stress, when plants are chronically or recurrently exposed to salinity for longer periods, stress-specific responses appear. At similar concentrations, alkaline salts

($NaHCO_3$, $NaCO_3$) exert much more severe reductions over photosynthesis than neutral salts (NaCl, Na_2SO_4) in barley (Yang *et al.*, 2009). Alternatively, some plants are able to prevent salt entry by salt exclusion at the whole-plant or the cellular level, or minimise salt concentration in the cytoplasm by compartmentalising it in the vacuoles. These plants are able to withstand very high salt concentrations avoiding toxic effects on photosynthesis and other key metabolic processes. Some plants concentrate toxic ions in leaves that are then shed. Often salt accumulates in old leaves and does not reach the apex, which also allows unrestricted growth in saline environments. Even cultivated plants such as olive adopt this mechanism for survival (Bongi and Loreto, 1989). However, in plants that are not able to operate any of the above mechanisms, Na^+ at a concentration above 100 mM severely inhibits the photosynthetic enzymes (Munns *et al.*, 2006). In particular, the enzymes that require K_+ as a cofactor are sensitive to high concentrations of Na^+ or high ratios of Na^+/K^+. Recently, Pérez-Pérez *et al.* (2007) have shown in *Citrus* sp. that combining drought and salinity stresses result in a much larger photosynthesis depression than under any of the two stresses separately, and Zheng *et al.* (2009) have shown the same for combined salt stress and waterlogging. These larger photosynthesis depressions when the two stresses act in combination are not associated with further stomatal closure, suggesting that they are a result of direct effects of stress on the photosynthetic metabolism.

Among the stresses under consideration in this chapter, flooding is often characterised by photosynthesis limitations that are not caused by diffusion of CO_2. As such, decreased photosynthesis is often accompanied by increased C_i even at relatively high g_s, indicating non-stomatal limitations to photosynthesis (Blanke and Cooke, 2004; Herrera *et al.*, 2008; Arbona *et al.*, 2009). In this case, leaf photochemistry and the Calvin cycle are often more strongly impaired than carboxylation (Table 20.2; but see Vu and Yelenosky, 1991). As it was also the case for the combination of water and salt stress, there are evidences that salt stress and flooding act in a cumulative way, as saltwater flooding results in stronger inhibition of photosynthesis than freshwater flooding (Naumann *et al.*, 2008). A particular case of photosynthesis inhibition caused by flooding is complete submergence of leaves, which has been described in Chapter 7 (Section 7.5).

To further check the importance of metabolic regulation under stress, it is important to combine the traditional, more-descriptive physiological approaches with the techniques of functional genomics, namely the high-throughput methods for transcriptomic, proteomic, metabolomic and ionomic analysis (Chaves *et al.*, 2003, 2009). Using this integrated analysis under drought and salinity stress (no data are available for flooding), it has been shown that genes or proteins associated with photosynthetic pathways are in general not altered by the stress (Kilian *et al.*, 2007). For example, in *Thellungiella* (a stress tolerant relative of *Arabidopsis*), only 15% of all genes downregulated are involved in photosynthesis (Wong *et al.*, 2006). In rice, alterations in photosynthesis-related genes are mostly associated with stress recovery but not with stress imposition (Zhou *et al.*, 2007b). Recently we have reviewed the available information on gene expression under drought and salinity (Chaves *et al.*, 2009). The number of stress-responsive genes or proteins depends on the stress intensity. As a general trend, both drought and salt stresses led to gene downregulation, most of the changes being small (ratio threshold lower than one, Fig. 20.2) possibly reflecting that only mild stress was generally imposed in these stress experiments. This is certainly applicable to agricultural crops that do not undergo heavy stress, whereas may not reflect a stress response by natural ecosystems of harsh environments. Alternatively, a recent proteomics study in peanut cultivars has suggested that the response can also differ in drought-sensitive and drought-tolerant genotypes (Kottapalli *et al.*, 2009). For instance, while Rubisco large and small subunits were decreased in both types of cultivars, some PSII proteins were decreased and ATP synthase increased only in the tolerant genotype (Kottapalli *et al.*, 2009). When compared with drought, salinity was found to affect a higher number of genes and more intensely, possibly reflecting the two components of salt stress (dehydration and osmotic stress). Still, precaution has to be taken when analysing gene-expression data, since they may not necessarily have a reflection on protein contents and physiological properties. For instance, Bogeat-Triboulot *et al.* (2007) combined the analysis of gene expression, protein profiles and ecophysiological performance of *Populus euphratica* subjected to gradual soil-water depletion. They observed that acclimation to water deficits involved the regulation of different networks of genes linked to protection and function maintenance of roots and shoots. Drought successively induced shoot-growth cessation, stomatal closure, moderate increases in ROS, decreases in photosynthesis and in root growth. These alterations were parallelled by transcriptional changes in 1.2% of the genes on the array that were fully reversible upon re-watering. However, no correlation was observed between the abundance of transcripts and

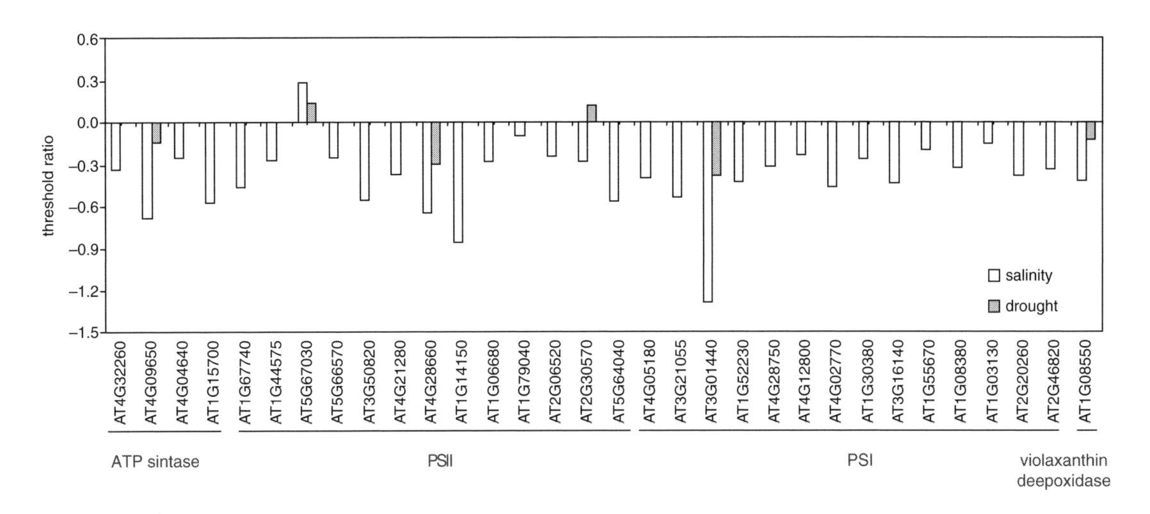

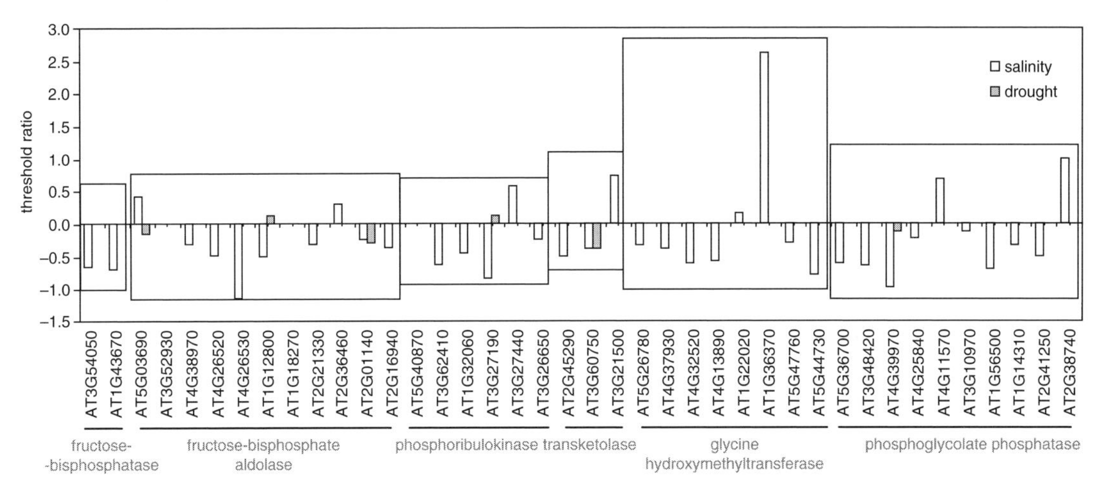

Fig. 20.2. AtGenExpress: salt and drought effects on **(A)** several primary photosynthesis rections related genes (ATP synthesis coupled to proton transport, light reaction and xanthophyll cycle) and **(B)** Calvin-cycle and photorespiration related genes. Data are based on Killian *et al.* (2007) and workout with www.genevestigator.ethz.ch. From Chaves *et al.* (2009).

proteins in this experiment. Alternatively, a study on the expression of a few photosynthetic genes – *rbcL*, *OEC* and *Fd* – has shown their strong reduction under both continuous and periodic flooding (Li *et al.*, 2007).

Overall, the available evidence suggests that, under many situations of drought stress, salinity and flooding, restricted CO_2 diffusion to the carboxylating sites is the main limiting factor for photosynthesis. Reverse genetics experiments also suggest that the maintenance of a high photosynthesis rate under water stress depends mostly on keeping a high CO_2 supply to chloroplasts. Hence, most transgenic plants showing more stable photosynthesis under water stress than wildtype relatives present a higher g_s and instantaneous WUE, with little or no differences in $V_{c,max}$. Examples of this include plant-nuclear-factor-Y (NF-Y) B *Arabidopsis* mutants (Nelson *et al.*, 2007), isopentenyltransferase tobacco mutants with delayed drought-induced senescence (Rivero *et al.*, 2007, 2009), transgenic *Arabidopsis thaliana* plants constitutively expressing a bZIP transcription factor, *ABP9* (Zhang *et al.*, 2008c), or transgenic soybean plants overexpressing the endoplasmic-reticulum-resident binding protein, BiP (Valente *et al.*,

Table 20.3. *Characteristics of photosynthesis recovery after drought, salt and flooding (literature survey).*

	Drought	Salinity	Flooding
Average g_s before recovery (mol H_2O m^{-2} s^{-1})	0.063	0.120	0.042
% of studies in which 75% recovery of A_N was achieved	64	100	57
% of studies in which 75% recovery of g_s was achieved	50	86	83
% of studies in which 100% recovery of A_N was achieved	45	100	50
% of studies in which 100% recovery of g_s was achieved	27	86	86
Average time to reach 75% recovery of A_N (days)	4.0	15.5	8.7
Average time to reach 75% recovery of g_s (days)	6.5	13.7	15
Average time to reach 100% recovery of A_N (days)	9.2	19.2	21.5
Average time to reach 100% recovery of g_s (days)	7.8	15.3	17.5

Studies included: (**Drought**) Angelopulos *et al.* (1996); Souza *et al.* (2004); Cai *et al.* (2005, 2007); Miyashita *et al.* (2005); Guida dos Santos *et al.* (2006); Grzesiak *et al.* (2006); Hura *et al.* (2006); Bogeat-Triboulot *et al.* (2007); Flexas *et al.* (2009); Gallé *et al.* (2007a,b, 2009); Galmés *et al.* (2007a); Gómez-del-Campo *et al.* (2007); Montanaro *et al.* (2007); Pérez-Pérez *et al.* (2007); Gomes *et al.* (2008); Pou *et al.* (2008); Xu *et al.* (2009); (**Salinity**) Tattini *et al.* (1997); Delfine *et al.* (1999); Hernández and Almansa (2002); Pérez-Pérez *et al.* (2007); (**Flooding**) Crane and Davies (1988); Domingo *et al.* (2002); Nicolás *et al.* (2005); Arbona *et al.*, (2009).

2009). Metabolic regulation/impairment occurs eventually – depending on the species and/or environmental conditions – when stress is more severe or when two or more of these stresses are combined. Nevertheless, this happens earlier and more intensely under salt and flooding stress than under drought, possibly reflecting an ion-toxicity mediated inhibition mechanism under salt stress and an unknown mechanism under flooding. In this sense, overexpression of sedoheptulose-1,7-bisphosphatase enhances photosynthesis under salt stress by increasing Rubisco activation and carboxylation efficiency (Feng *et al.*, 2007a). Under drought conditions, metabolic impairment corresponds invariably to, and seems caused by, low g_s and occurs mostly under high irradiance likely as a consequence of oxidative stress.

20.4. PHOTOSYNTHESIS LIMITATIONS DURING RECOVERY AFTER DROUGHT, FLOODING AND SALINITY

The carbon balance of a plant following a period of water, salt or flooding stress may depend as much on the velocity and degree of the recovery of photosynthesis after stress relief, as it depends on the degree and velocity of photosynthesis decline during stress occurrence (Flexas *et al.*, 2006c). Surprisingly, since early studies by Kirschbaum (1987, 1988), studies analysing the capacity of recovery of photosynthesis from different stresses are scant.

Stress intensity and/or duration are crucial factors affecting both the velocity and the extent of recovery after stress relief, although some differences between the three types of stress may occur (Table 20.3). In general, plants subjected to severe water stress recover only 40–60% of the maximum photosynthesis rate during the day after re-watering, and recovery continues during the next days, but maximum photosynthesis rates are sometimes not recovered (Kirschbaum, 1987, 1988; Sofo *et al.*, 2004; Grzesiak *et al.*, 2006; Bogeat-Triboulot *et al.*, 2007; Gallé *et al.*, 2007). This suggests that water stress induces some irreversible damage to the photosynthetic apparatus. However, as incomplete recovery of photosynthesis is more frequent in fast growing (e.g., herbaceous) species, it is possible that water stress (and in general all stresses) accelerates leaf senescence, with the consequent irreversible loss of photosynthetic performances. When averaging all available data, the time to recover 75% of the pre-stress rates of photosynthesis is 4 days in plants subjected to drought (Table 20.3), and 100% recovery (when occurs) takes up to 10 days. These periods are similar to the recovery of g_s.

The time period required for photosynthesis recovery after salt stress is much higher (up to 15–20 days), even when the average g_s before the onset of recovery was threefold that observed in drought and flooding studies (Table 20.3). This likely reflects the metabolic nature of a large proportion of photosynthesis limitations under salinity. Surprisingly and despite the slower recovery a 100% recovery occurred in all studies analysed, whereas it occurred in only about half the studies on drought.

Recovery after flooding has been scarcely studied (Domingo *et al.*, 2002; Nicolás *et al.*, 2005; Arbona *et al.*, 2009). The few data available suggest that recovery capacity may be intermediate between water deficit and salt stress, recovery of photosynthesis being similar to that after drought, but recovery of g_s similar to that after salinity (Table 20.3).

In the case of drought, for which most of the data are available, high asymptotic correlations can be found between the value of A_N or g_s at the onset of re-watering, and A_N after 5 days of recovery (Fig. 20.3). Whenever A_N during water stress remains higher than 3 μmol CO_2 m^{-2} s^{-1} or g_s higher than 0.05 mol H_2O m^{-2} s^{-1} (i.e., when metabolic limitations are unlikely to have occurred, see previous section), an almost complete photosynthesis recovery occurs within this short time period. When the stress has been more severe, however, the recovery is incomplete and linearly correlated to the strength of the stress. The influence of previous water-stress severity in the velocity and extent of photosynthesis recovery was illustrated by Miyashita *et al.* (2005) and Grzesiak *et al.* (2006). The interaction of salt and water stress strongly reduces a plant's capacity to recover photosynthesis after stress alleviation as compared with plants subjected to a single stress (Pérez-Pérez *et al.*, 2007). However, in these studies, the physiological mechanisms limiting recovery were not assessed. Kirschbaum (1987, 1988) showed that recovery after a severe dehydration was a two-stage process: the first stage occurs during the first days upon re-watering, and is associated to recovery of water status and stomata re-opening; the second stage lasts several days and likely requires de-novo synthesis of photosynthetic proteins.

Concerning the first of these two phases, however, in some species a sustained downregulation of g_s after re-watering imposes substantial limitations to photosynthesis while increasing intrinsic WUE (Bogeat-Triboulot *et al.*, 2007; Gallé and Feller, 2007; Gallé *et al.*, 2007, 2009; Galmés *et al.*, 2007a,c; Flexas *et al.*, 2009; Xu *et al.*, 2009). Orange trees that have endured severe water stress do not ever fully recover g_s after two months of re-watering (Fereres *et al.*, 1979). In some of these cases, limited recovery of leaf-specific hydraulic conductivity is the likely cause for the long-term downregulation of g_s after re-watering (Galmés *et al.*, 2007c; Pou *et al.*, 2008), and aquaporins may play a dominant role in this regulation (Cochard *et al.*, 2007; Galmés

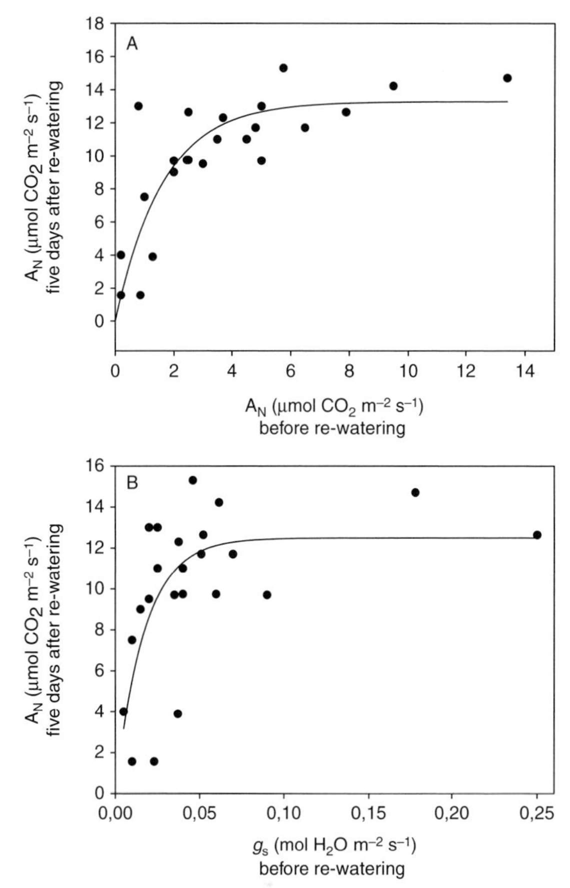

Fig. 20.3. The effects of **(A)** net photosynthesis (A_N) and **(B)** stomatal conductance (g_s) values achieved during water stress on subsequent net photosynthesis values 5 days after rewatering. Data from different species (literature survey as in Table 20.3).

et al., 2007d; Parent *et al.*, 2009). Resco *et al.* (2009) have demonstrated specifically that the number of days necessary to reach maximum photosynthesis after a pulse of rain (i.e., the inverse of velocity of recovery) depends on antecedent, drought-induced percentage loss of hydraulic conductance, and Brodribb and Cochard (2009) reached a similar conclusion regarding conifers. Alternatively, in beech stomatal occlusions formed during water stress restrain recovery of g_s after rewatering (Gallé and Feller, 2007). In other species, notably in many Mediterranean species belonging to different growth forms and functional groups, also g_m to CO_2 is slowly reversible, and may constitute an additional, important factor limiting photosynthesis recovery after severe water stress (Galmés *et al.*, 2007a).

As for the second phase, Bogeat-Triboulot *et al.* (2007) showed that recovery after water stress, determined 10 days after re-watering, was accompanied by increases in some photosynthetic proteins, particularly Rubisco activase and proteins of the water-splitting complex, although increased proteins transcripts were not detected. In the cases where photosynthesis recovery is slow and/or incomplete, photoinhibition and/or oxidative stress have been suggested as possible causes (Sofo *et al.*, 2004; Gallé *et al.*, 2007; Galmés *et al.*, 2007a).

20.5. OXIDATIVE STRESS: A COMMON FEATURE UNDER MULTI- AND SEVERE STRESS CONDITIONS

When CO_2 availability to the chloroplasts is drastically reduced under drought, salinity or flooding the photosynthetic sink for electrons decreases. Excess electrons can be diverted to other processes, such as photorespiration or thermal dissipation, the latter being considered a major process in plant photoprotection under stress conditions (representing up to 80% of total absorbed light). However, when these processes saturate light harvesting, electron transport components become over-energised, resulting in electrons being transferred to oxygen at PSI or via the Mehler reaction. This generates ROS, such as superoxide ($O_2^{\bullet -}$), hydrogen peroxide (H_2O_2) and the hydroxyl radical (•OH), that can cause oxidative damage to the photosynthetic apparatus if the plant is not efficient in scavenging these molecules. When each of these stresses is combined either with high light or high temperature, synergistic effects are likely to occur (Mittler, 2006). The methods to detect ROS are described in Chapter 13.

It was proposed that ROS accumulation and the consequent oxidative stress may function as a genetic modulator of plant response to stress, inducing for example genes that actively protect leaves during stress (e.g., heat-shock factors), and increasing the chances of producing better-adapted plants to the particular stress. Therefore, they will act as a major driving force in plant evolution (Levine, 1999).

Depending on the duration and severity of the stress, the plant species, the specific tissue/organ and the phase of development, ROS can elicit antioxidant responses, typically observed under mild stress, or can lead to accelerated senescence (Munné-Bosch and Alegre, 2002), programmed cell death or necrosis (Levine, 1999). Acclimation to stress is generally associated with enhanced activity of the antioxidant molecules, which are able to remove oxygen radicals or repair the damage, thus keeping ROS concentration relatively low (Smirnoff, 1998; Dat *et al.*, 2000), although seasonal patterns of leaf H_2O_2 contents reveal variations of up to sevenfold (Cheeseman, 2009). Antioxidant activity includes enzymatic and non-enzymatic mechanisms, such as the SODs that catalyse the dismutation of $O_2^{\bullet -}$ to H_2O_2, CATs responsible for removal of H_2O_2 and the enzymes and metabolites of the ascorbate-glutathione cycle that are also involved in the removal of H_2O_2 (Foyer and Noctor, 2003).

ROS can also serve as secondary messengers in the signalling for the activation of defence responses (Dat *et al.*, 2000). This dual function of ROS, first described in responses to pathogens and later demonstrated in several abiotic stresses, presumably plays an important role on the acclimation processes (Dat *et al.*, 2000). Redox signals are early warnings exerting control over the energy balance of a leaf. Alterations in the redox state of redox-active compounds regulate the expression of several genes linked to photosynthesis (both in the chloroplast and in the nucleus), thus providing the basis for the feedback response of photosynthesis to the environment, or in other words, the adjustment of energy production to consumption (Chaves and Oliveira, 2004). It must be pointed out that the data on the redox regulation of photosynthesis genes is still contradictory, suggesting a highly complex signalling network (see the review by Pfannschmidt, 2003). Redox-signalling molecules include some key electron carriers, such as the PQ pool or electron acceptors (e.g., ferredoxin/thioredoxin system), as well as ROS (e.g., H_2O_2).

The intracellular concentrations of the ROS are controlled by the plant detoxifying system, which includes ascorbate and glutathione pools. Increased activities of plant antioxidant systems under stress have been considered characteristic of drought-resistant (e.g., Gao *et al.*, 2009) and waterlogging-resistant species (e.g., Kumutha *et al.*, 2009). Accumulating evidence suggests that these compounds are implicated in the redox signal transduction, acting as secondary messengers in hormone-mediated events (Foyer and Noctor, 2003), namely stomatal movements (Pei *et al.*, 2000). H_2O_2 acts as a local or systemic signal for leaf stomata closure, leaf acclimation to high irradiance and the induction of HSPs (see the review by Pastori and Foyer, 2002). However, when the production of H_2O_2 exceeds a threshold, programmed cell death might follow.

In a recent work by Rivero *et al.* (2007) the expression of isopentenyltransferase (*IPT*), which catalyses the rate-limiting step in cytokinin (CK) synthesis, led to increased concentrations of enzymes associated with the glutathione-ascorbate cycle in transgenic plants and resulted in the suppression of drought-induced leaf senescence and in outstanding drought-tolerance. In transgenic plants with elevated CK production that were subjected to drought, 20% of the upregulated transcripts were related to ROS metabolism. The efficient scavenging of ROS protects the photosynthetic apparatus during drought stress, leading to improved WUE of the transgenic plants during and after stress.

Nitric oxide, a reactive nitrogen species, also acts as a signalling molecule, in particular by mediating the effects of hormones and other primary signalling molecules in response to environmental stimuli. It may act by increasing cell sensitivity to these molecules (Neill *et al.*, 2008). NO was shown to play a role as an intermediate of ABA effects on guard cells (Neill *et al.*, 2008). Although the links between dehydration and NO are not yet fully resolved, it seems that some of signalling components downstream of NO (and H_2O_2) in the ABA-induced stomatal closure are calcium, protein kinases and cyclic GMP (Desikan *et al.*, 2004). NO accumulation has often been detected in response to oxidative stresses induced by abiotic (e.g., ozone) or biotic stresses, but has never been reported in plants undergoing reversible water or salt stress. Thus the importance of this molecule in stress responses driven by these environmental factors remains questionable.

20.6. CONCLUDING REMARKS

Major progress in the understanding of the physiological and molecular limitations and the signalling events underlying photosynthetic responses to drought, salt and flooding has been achieved over the last decade(s). Overall, the available evidence suggests that under many situations of drought stress, salinity and flooding, restricted CO_2 diffusion to the carboxylation sites is the main limiting factor for photosynthesis, especially during the early phases of stress. Metabolic regulation/impairment occurs eventually – depending on the species and/or environmental conditions – when the stress is more severe or when two or more of these stresses are combined. Metabolic impairment of photosynthesis seems to appear earlier under long-term flooding than long-term drought. Moreover, the production of ATP and regeneration of NAD^+ through anaerobic respiration under flooding leads to generalised metabolic changes in plants. Under severe stress ROS may accumulate and function as a genetic modulator of plant response to stress. Our understanding of the factors limiting photosynthesis recovery after water, salt and flooding stresses still needs improvement. In general, a better integration of the molecular/cellular information with the whole-plant responses, both under controlled and field conditions is required. The latter studies are essential to test for cross-resistance responses as this is what generally occurs in nature. Precaution also has to be taken when analysing gene-expression data in plants under stress, as they may not necessarily have a reflection in protein contents and physiological properties.

21 • Photosynthetic responses to nutrient deprivation and toxicities

F. MORALES AND C.R. WARREN

21.1. INTRODUCTION

21.1.1. General aspects of plant responses to nutrients

Plant growth requires the incorporation of elements (nutrients) into plant organs. In non-woody plants, 15–20% of fresh weight is made from such elements, the rest being water. There are two criteria to consider an element as essential. First, an element is essential if a plant cannot complete its lifecycle (till viable seeds) in its absence. Second, an element is essential if necessary to synthesise molecules that cannot be replaced by other element(s) (for example, N in proteins). In natural ecosystems, soil-nutrient availability is rather heterogeneous, and plants may adapt their growth to nutrients taken up by roots exploring a determined soil volume. In agricultural areas, the situation is different. Lack of or excess nutrients are frequent, owing to soil characteristics (which may immobilise nutrients) or to growers' applications, respectively.

An idealised representation of plant growth-rate response to availability of any given nutrient would show three different zones: namely (1) deficient; (2) adequate; and (3) toxic. In the range of low nutrient concentrations (deficient zone), growth and the plant nutrient concentration markedly increase as soil nutrient availability increases. As availability increases further the so-called critical concentration is reached. This corresponds to the lowest concentration of nutrient in plant tissue that gives almost maximal growth. Above this point, increases in soil nutrient availability do not affect growth (adequate zone). In the adequate zone, there is a plentiful supply of nutrients, and the excess nutrients may be taken up and stored in leaf vacuoles, special storage proteins in bark or uptake may be downregulated so as to avoid taking up excess nutrients. This zone is fairly wide for macronutrients, but narrower for micronutrients. If nutrient availability increases more and uptake cannot be controlled, toxicity appears and growth is reduced (toxic zone).

21.1.2. Concepts of deficiency and toxicity

Deficiency and toxicity are the two extremes of the relationship between nutrients and growth. Whereas in natural ecosystems, scarcity of macronutrients often limits plant growth, N, P, and K are the most commonly applied fertilisers in modern agriculture. The intense, periodic addition of such fertilisers means that in many cases growers do use these nutrients in excess, leading to significant environmental damage including emissions of greenhouse gases such as N_2O, decreasing soil and stream pH and increasing nutrient loads in streams, estuaries and oceans. The significant economic and environmental costs associated with excessive fertiliser use and reduced growth and economic return that results from nutrient deficiencies are strong incentives to optimise fertiliser application by striking a balance between too little and too much. The ultimate aim should be to adjust use of fertiliser, while still avoiding deficiencies and their consequences for photosynthesis, growth and crop yield.

Iron, Mn, Cu and Zn are essential plant micronutrients. Iron, Mn and Zn deficiencies are widespread constraints in plants grown in high-pH calcareous soils, causing reductions in crop production (White and Zasoski, 1999). Copper and Mn are commonly deficient in the light soils of Australia (Kriedemann and Anderson, 1988).

Iron toxicity is typical of flooded soils, flooding may increase soluble Fe from 0.1 to 50–100 ppm in a few weeks (Kim and Jung, 1993). Various soil conditions often present in acid and volcanic soils, and areas of persistent acid rain can lead to Mn reduction and create Mn toxicity in natural

Terrestrial Photosynthesis in a Changing Environment: A Molecular, Physiological and Ecological Approach, ed. J. Flexas, F. Loreto and H. Medrano. Published by Cambridge University Press.

and agricultural systems (Foy *et al.*, 1978). Copper excess may be a result of geological anomalies or human activities such as urban, metal mining, smelting operations and manufacturing. Zinc is a major industrial pollutant of the terrestrial and aquatic environments (Barak and Helmke, 1993).

Cadmium accumulation in the environment may come from different sources, including industrial effluent, mining (e.g., Cd is a by product of Zn mining), burning and leakage of urban and industrial wastes (galvanic batteries, dyes), air pollutants and soil applications of phosphate commercial fertilisers, sewage sludge, manure and lime. Lead contamination originates from industrial activities, such as metal mining and smelting and automobile use. In acid soils, the harmful effects on plant growth are closely related to Al toxicity (Bona *et al.*, 1994), which can be also mediated by acid rain. Global atmospheric Hg burdens may have increased up to fivefold since the Industrial Revolution. This is a result of fossil-fuel combustion, ore smelting and waste incineration among other causes (Dunagan *et al.*, 2007).

The photosynthetic response of a plant to a given nutrient increase will depend much on the zone in which the plants are growing. In general, plants under nutrient deficiency or toxicity conditions decrease photosynthesis and develop mechanisms of photoprotection aimed at avoiding photoinhibitory damage. Limitation of growth by N or P does not necessarily mean that photosynthesis is limited too. Growth and leaf production of some evergreen forest species such as *Pinus* spp. is so strongly coordinated with nutrient supply that nutrient-deficient foliage is rarely produced (Warren *et al.*, 2003b). Because such species do not produce foliage that is nutrient deficient (in terms of photosynthesis), relationships of photosynthesis with nutrient concentrations are weak (Schoettle and Smith, 1999). Growth of other species is not so closely coupled with nutrient supply, deficient foliage may be produced and the relationships of photosynthesis with nutrient supply are stronger.

In this chapter, we will review changes observed in photosynthesis, photoprotection and photoinhibitory mechanisms in response to nutrient deficiencies, toxicities and heavy metals. In particular, we have focused it on N, P and K deficiencies, Fe, Mn, Cu and Zn deficiencies and toxicities and effects of Cd, Pb, Al and Hg. Other nutrient deficiencies or toxicities have not been considered, but readers may refer to papers published elsewhere (Marschner, 1995).

21.2. PHOTOSYNTHETIC RESPONSES TO MACRONUTRIENT DEFICIENCY

21.2.1. Nitrogen deficiency

Plants require 2–5% average nitrogen (N) content in dry matter, depending on plant species, phenological stage and organ (Marschner, 1995). Under suboptimal supply, growth is retarded (actively growing leaves are smaller and thicker), N is retranslocated from mature leaves to new growth areas and senescence of older leaves is enhanced (Marschner, 1995). The restricted development of N-deficient plants has been ascribed to low rates of leaf expansion, rather than to declines in photosynthesis (Sage and Pearcy, 1987a). Nitrogen shortage, however, does result in a marked decrease in photosynthesis in some species. This is to be expected because more than half of the total leaf N is allocated to the photosynthetic apparatus (Makino and Osmond, 1991). Photosynthetic capacity and the total amount of leaf N per A_{leaf} are often correlated (Field and Mooney, 1986; Sage and Pearcy, 1987b; Walcroft *et al.*, 1997). The enhanced photosynthetic rates with high N supplies have been recently attributed to a higher chloroplastic CO_2 concentration, related to a higher CO_2 mesophyll conductance (g_m) owing to an increased chloroplast size (Li *et al.*, 2009). The C_i/C_a ratios either increase or did not change under N limitation, indicating the absence of stomatal limitations (Sage and Pearcy, 1987b). Warren (2004) showed that diffusional limitations (stomatal plus mesophyll) increased with N deficiency. There is substantial evidence nowadays for N-deficiency mediated decreases in CO_2 g_m (Urban *et al.*, 2008; Bown *et al.*, 2009; Li *et al.*, 2009; Yin *et al.*, 2009; Yamori *et al.*, 2011).

Evidence for a broad, almost universal relationship between maximum rates of photosynthesis (A_{max}) and N among species has come from several large data sets (Field and Mooney, 1986; Reich *et al.*, 1997). However, A_{max}–N relationships differ among species and functional groups (Evans, 1989; Reich *et al.*, 1998a,b,c), and the 'universal' A_{max}–N relationship is, in fact, made up of a series of nested relationships with increasing slopes as SLA (and usually N concentration) increases (Reich *et al.*, 1998a,b,c). Generally too, the strength of these relationships decreases as the range in N concentration decreases (Reich, 1993), and thus relationships within functional groups may be weak.

Changes in different components of the photosynthetic machinery are coordinated, and both chl. (light harvesting) and Rubisco (energy transduction) decrease with N deficiency. Conventional wisdom suggests that N is 'used' most efficiently when the carboxylation and light-harvesting capabilities are matched (von Caemmerer and Farquhar, 1981). Typically, values of $V_{c,max}$ and J_{max} derived from the CO_2 response of photosynthesis are closely coupled (e.g., Wullschleger, 1993; Walcroft *et al.*, 1997) irrespective of nutrient supply. Such findings lend support to the argument of tight control in the allocation of N between carboxylation capacity (Rubisco and other Calvin-cycle enzymes) and light harvesting (chl. and pigment-protein complexes). Effects of N deficiency on Rubisco and other photosynthetic enzymes, however, are often larger than those on chl. (Ferrar and Osmond, 1986; Evans and Terashima, 1987), and part of the decreased photosynthetic capacity can be ascribed to the diminished amounts of Calvin-cycle enzymes (Terashima and Evans, 1988). Chloroplasts are fewer in number (Al-Abbas *et al.*, 1974), and the amount of thylakoids and thylakoid protein per chloroplast are decreased (Evans and Terashima, 1987; Terashima and Evans, 1988).

Also, there is clear evidence that N deficiency induces sink limitation within the whole plant, owing to decreased growth (Paul and Driscoll, 1997; Logan *et al.*, 1999). This leads, in turn, to feedback downregulation of photosynthesis – a phenomenon that has been known for 40 years (Neals and Incoll, 1968), yet only now is widely recognised.

One factor that is important for how plants perform at any given N supply is their rates of photosynthesis per unit nitrogen (PNUE, photosynthetic NUE). A general observation is that plants from water and/or nutrient deficient habitats have a poorer PNUE than species from nutrient- or water-rich habitats. An obvious question to ask is: why? Unfortunately, there is no single answer. Multiple factors combine to cause low PNUE in stressed plants. Species from xeric and/or oligotrophic habitats may allocate a smaller fraction of N to photosynthetic functions (Field and Mooney, 1986; Hikosaka *et al.*, 1998). Stress tolerance can also be enhanced by leaves that are structurally reinforced (Coley, 1988). However, the structural reinforcement has implications for photosynthetic physiology. Greater allocation of carbon to structural tissues (e.g., sclerenchyma and collenchyma) generally increases leaf density and decreases, proportionally, the amount of photosynthetic mesophyll tissue, the content per unit dry mass of water and N and N allocation to Rubisco. Low PNUE can also be a consequence of low CO_2 sub-stomatal concentrations (C_i) owing to more conservative water use (higher WUE), although this may not be the situation in plants enduring N deficiency (Field *et al.*, 1983).

N-deficient plants with decreased photosynthetic rates may suffer from photoinhibition when they are exposed to high PPFDs. Incoming light can be reflected on the leaf surface (reflectance), transmitted through (transmittance) or absorbed within the leaf (absorptance) (Fig. 21.1). The absorbed light (usually 80–85% in absence of stress) can be used in photochemistry, thermally dissipated or diverted to potentially deleterious mechanisms (see below) (Fig. 21.1).

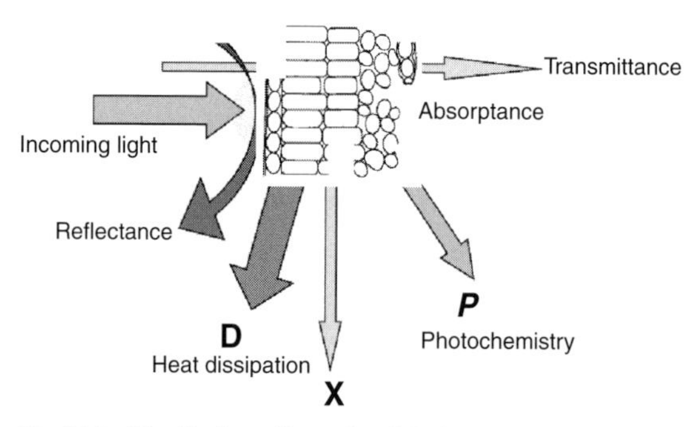

Fig. 21.1. Distribution of incoming light into several fractions. Although most of the light is absorbed by leaves (usually 80–85% in absence of stress; absorptance), minor fractions are reflected (reflectance) or transmitted (transmittance). Absorbed light can be used in photochemistry (*P*), dissipated as heat (D) or not used photochemically nor thermally dissipated (X) (see text for details).

In N-deficient spinach some photoprotection occurs through a decrease in light-harvesting capacity owing to decreased leaf chl. (Bottrill *et al.*, 1970; Verhoeven *et al.*, 1997; Demmig-Adams and Adams, 2003). Light absorptance, however, is less affected than leaf chl. concentrations (Al-Abbas *et al.*, 1974; Verhoeven *et al.*, 1997). This increases the average number of photons absorbed by each chl. molecule (and, of course, modifies the large gradient in light absorption per chl. molecule occurring within the leaf; see Chapter 16) and the excess of absorbed PPFD compared with the amount that can be used in photosynthesis (Verhoeven *et al.*, 1997).

Demmig-Adams *et al.* (1996a) proposed a modulated Chl-F approach to allocate the energy absorbed by PSII in three different fractions: i.e., a fraction that is dissipated thermally by the antenna (D); a fraction used in photochemistry (*P*, including electrons used in RuBP carboxylation and oxygenation, Mehler reaction, chlororespiration and any other electron-consuming processes); and a third fraction that is neither used in photochemistry nor dissipated by the PSII antenna (X). The latter probably determines the rate of PSII photoinactivation (Kato *et al.*, 2002) and may include potentially deleterious mechanisms, such as the formation of triplet chl. excited state (Demmig-Adams *et al.*,

1996b), which may lead to singlet oxygen formation, $^1O_2^*$ (Demmig-Adams *et al.*, 1996a), and in turn to PSII damage by photo-oxidative and degradation processes. The repair process is also very sensitive to these ROS.

In N-deficient leaves, there is a rapidly reversible, xanthophyll-cycle-dependent, thermal energy dissipation within the chl. pigment bed (Verhoeven *et al.*, 1997; Chen *et al.*, 2003; Demmig-Adams and Adams, 2003). In N-deficient leaves, D may account for 35% and 60–73% of the light absorbed by the PSII antenna at limiting and saturating light, respectively, whereas in controls it is 25% and in the range 36–59% (Verhoeven *et al.*, 1997; Lu and Zhang, 2010; data recalculated from Chen *et al.*, 2001). *P* decreases with N deficiency (Verhoeven *et al.*, 1997). The partitioning of electron flow between CO_2 assimilation and photorespiration was not affected by N deficiency in apple leaves (Chen *et al.*, 2001) and respiration decreased in spinach leaves (Bottrill *et al.*, 1970). X in spinach and maize would be approximately 10–15% and 13–19% in high- and low-N leaves, respectively (recalculated from Verhoeven *et al.*, 1997 and Lu and Zhang, 2000).

Antioxidative plant systems are activated in N-deficient leaves. Increases in Cu-Zn superoxide dismutase (Cu-ZnSOD) activity have been reported in response to an enhanced formation rate of superoxide radical ($O_2^{\cdot-}$) in N-deficient coffee plants (Ramalho *et al.*, 1998). The fatty acid composition of the chloroplastic membranes was also modified, which may have made them less susceptible to peroxidation. In line with this, N-limitation in spinach caused decreases in the APX and GR activities when data were expressed on a leaf area or dry weight basis, activities were similar when expressed on a chl. basis, but higher in N-limited plants on a total protein basis (Logan *et al.*, 1999). Nitrogen-deficient, low-chl. leaves also have increases in the carotenoid-to-chl. molar ratios (Khamis *et al.*, 1990; Verhoeven *et al.*, 1997), which may help to prevent $^1O_2^*$ formation.

Plants growing under N limitation have low chl. concentrations and a decreased quantum yield of photosynthesis (Terashima and Evans, 1988; Khamis *et al.*, 1990; Verhoeven *et al.*, 1997; Lu and Zhang, 2000; Chen *et al.*, 2001; Demmig-Adams and Adams, 2003). These features are found in sun-exposed fast-growing leaves but not in shaded ones (Terashima and Evans, 1988), suggesting that photodamage may occur. Alternatively, this observation may reflect a differential N-deficiency mediated feedback downregulation of photosynthesis in sun and shade leaves.

The susceptibility to photoinhibition is larger in plants grown with low N than in those grown with high N (Balachandran and Osmond, 1994). Also, when transferred from low to high PPFDs, plants grown under N limitation are more susceptible to photoinhibition, and furthermore the recovery is slower than in N-sufficient controls (Ferrar and Osmond, 1986; Khamis *et al.*, 1990). Seedlings fertilised with N also show faster recovery from photoinhibition than those that were unfertilised (Close *et al.*, 2003). Photoinhibition was also enhanced in N-deficient, field-grown rice, when incident PPFD was increased by altering the leaf angle (Chen *et al.*, 2003).

No sustained photoinhibition occurs under N deficiency in spinach, as judged by the high F_v/F_m ratios after dark adaptation (Verhoeven *et al.*, 1997). Slightly low F_v/F_m ratios at predawn are likely associated with overnight retention ('lock-in') of zeaxanthin and antheraxanthin and sustained NPQ in N-deficient leaves (Verhoeven *et al.*, 1997). Nitrogen availability, however, did not change D1-turnover rates in leaves of *Chenopodium album* (Kato *et al.*, 2002), and the amounts of D1 were similar in N-replete and N-starved *Anacystis nidulans* (Biswal *et al.*, 1994). It seems therefore that the potential occurrence of photoinhibition in response to N deficiency will depend much on the species and on the previous high PPFDs acclimation, which may confer a better capacity for D1 renewal (Long *et al.*, 1994) and repair processes.

In summary, photosynthesis is affected under N deficiency through several routes (Table 21.1). Chl., Rubisco, the mesophyll conductance to CO_2 and the amount of Calvin-cycle enzymes are decreased, and the reduction in growth induces sink limitation. Photoprotection in N-deficient plants is mediated by decreases in light gathering, increases in xanthophyll-cycle-dependent energy dissipation (competing with photochemistry) and induction of antioxidative systems (Table 21.2). The susceptibility or occurrence of photoinhibition under N deficiency depends on the species and on the previous light acclimation of the plants.

21.2.2. Phosphorous deficiency

The phosphorous (P) requirement for optimal growth is in the range 0.3–0.5% of the plant dry weight (Marschner, 1995). In a plant with an adequate P supply, 85–95% of the total inorganic P (P_i) is located in vacuoles. Release of P_i from vacuoles is usually slow. Under optimal P nutrition, P_i concentration in the cytosol is approximately 6 mM. In the

Table 21.1. *Summary of the effects of different nutritional deficiencies, toxicities and heavy metals on photosynthesis. –, no effect; n.d., effect not described in the literature; *, described occasionally; **, several times; ***, main effect.*

			Non-stomatal limitations			
Nutritional stress	Chlorosis/ necrosis	Stomatal limitations	Ultrastructural effects and mesophyll conductance limitations	Primary reactions	Carboxylation/ Calvin-cycle reactions	Carbohydrate metabolism/ transport
–N	*	n.d.	**	**	***	**
–P	n.d.	**	*	n.d.	**	*
–K	*	**	**	n.d.	*	**
–Fe	**	*	**	***	**	*
Excess Fe	*	n.d.	n.d.	*	*	*
–Mn	**	—	*	**	*	n.d.
Excess Mn	*	*	n.d.	n.d.	n.d.	n.d.
–Cu	*	*	*	*	*	n.d.
Excess Cu	*	*	*	*	*	*
–Zn	**	**	**	*	**	*
Excess Zn	**	**	**	**	**	n.d.
Cd	**	*	**	***	**	*
Pb	*	*	n.d.	—	n.d.	n.d.
Al	**	*	**	*	*	n.d.
Hg	**	*	n.d.	*	n.d.	n.d.

Table 21.2. *Summary of the effects of different nutritional deficiencies, toxicities and heavy metals on the distribution of the incoming light into several fractions: absorptance, reflectance, transmittance, P (absorbed light used in photochemistry), D (dissipated as heat) and X (not used photochemically nor thermally dissipated). –, no effect; n.d., effect not described in the literature; inc, increase; dec, decrease.*

Nutritional stress	Effects on transmittance	Reflectance	Absorptance	P	D	X
–N	inc	inc	dec	dec	inc	inc
–P	inc	inc	dec	n.d.	inc	n.d.
–K	inc	inc	dec	n.d.	inc	n.d.
–Fe	inc	inc	dec	dec	inc	—
Excess Fe	n.d.	n.d.	n.d.	dec	inc	n.d.
–Mn	inc	inc	dec	dec	—	n.d.
Excess Mn	n.d.	inc	n.d.	n.d.	n.d.	n.d.
–Cu	n.d.	n.d.	n.d.	dec	n.d.	n.d.
Excess Cu	n.d.	inc	n.d.	dec	inc	n.d.
–Zn	n.d.	inc	n.d.	dec	n.d.	n.d.
Excess Zn	n.d.	inc	n.d.	dec	—	n.d.
Cd	n.d.	inc	n.d.	dec	inc	—
Pb	n.d.	inc	n.d.	—	—	—
Al	n.d.	n.d.	n.d.	dec	inc	n.d.
Hg	n.d.	inc	n.d.	n.d.	n.d.	n.d.

chloroplast stroma, severe inhibition of starch biosynthesis occurs below approximately 5 mM P_i (Marschner, 1995).

Low P nutrition reduces crop yields. Rates of P supply are seldom continuously sufficient to maintain metabolic processes at maximum rates. Halsted and Lynch (1996) suggested that P limitation manifests first in growth and later in photosynthesis. Hence, inadequate P supply typically reduces growth rates (Bottrill *et al.*, 1970; Plesnicar *et al.*, 1994) owing to inhibition of leaf cell division and expansion (Terry and Ulrich, 1973a). This may or may not be accompanied by slower photosynthesis (Bottrill *et al.*, 1970; Brooks, 1986; Lauer *et al.*, 1989). At least in some species, reductions in A_{max} owing to sub-optimal P supply can be large, for example in *Eucalyptus globulus* it was reduced by 55% (Turnbull *et al.*, 2007a).

Some studies have argued that P affects photosynthesis via effects on stomatal function. This may be because P_i is required for the biosynthesis of ATP, which is involved in stomatal physiology (Agbariah and Roth-Bejerano, 1990). Hence, in many cases P deficiency leads to significant decreases in leaf stomatal conductance (g_s) (Brooks, 1986; Rao and Terry, 1989; Xu *et al.*, 2007). Mesophyll conductance to CO_2 is, however, hardly affected by P deficiency (Bown *et al.*, 2009).

The simplest explanation for positive relationships of P with photosynthesis is that P_i is one of the primary substrates of photosynthesis (Walker and Sivak, 1986). Hence, there are dramatic and rapid effects of sudden changes in internal concentrations of P_i (Rao and Terry, 1995). Phosphorus-deficiency mediated decreases in photosynthesis were accompanied by lower Rubisco content and specific activity and carboxylation efficiency (Brooks, 1986; Lauer *et al.*, 1989), suggesting that P_i limitation has a greater effect on the carbon-reduction reactions than on the light reactions (Lauer *et al.*, 1989). It is known that the activation of Rubisco is accelerated in the presence of P_i (Sawada *et al.*, 1992). Both accumulation of starch (and/or sucrose) in leaves owing to reduced sink demand (Pieters *et al.*, 2001), or insufficient P_i for operation of TL transport (Herold, 1980), can impair photosynthesis. Furthermore, photosynthetic metabolism is particularly sensitive to P_i insofar as RuBP pool size (Brooks, 1986; Fredeen *et al.*, 1989) and regeneration efficiency (Brooks, 1986; Plesnicar *et al.*, 1994) can be reduced in P-limited plants. Nevertheless, the amount and fraction of P involved in the primary processes of photosynthesis are variable and often small (Bieleski, 1973), which makes it less clear why correlations of P with photosynthesis are so strong. In line with this, levels of P_i that reduce photosynthesis in sucrose-accumulating species may hardly influence photosynthesis in starch-accumulating species (Walker and Sivak, 1985). This is because starch biosynthesis liberates P_i from reduced carbon, thereby making P_i available for different metabolic reactions.

In many cases positive relationships between P supply and photosynthesis are explained best by leaf P content, not by active (cytoplasmic located) P pool (Turnbull *et al.*, 2007b). This may be because, at least in some cases, P affects photosynthesis indirectly via effects on N allocation to the photosynthetic machinery. In *Pinus pinaster*, for example, there is a strong positive correlation between P supply and P storage as orthophosphate and relative and absolute amounts of N allocated to Rubisco (Warren and Adams, 2002). This may come about because at longer timescales P deficiency causes accumulation of carbohydrates that downregulate expression of genes coding for the photosynthetic machinery (Krapp and Stitt, 1995).

There is not much data to conclude how P-deficient plants protect themselves from the possible energy excesses. P-deficient soybean leaves exhibit paraheliotropic (leaflet laminae parallel to incident radiation) orientation on bright, sunny days at midday rather than the normal diaheliotropic (solar-tracking) orientation of control leaves and of P-deficient leaves early and late in the day (Lauer *et al.*, 1989). Decreases in absorptance, other than those owing to leaf movements, in P-deficient plants are rather small and paralleled to increases in reflectance and transmittance (Al-Abbas *et al.*, 1974). Short-term P-deficiency induced decreases in intrinsic PSII efficiency and increases in NPQ have been reported (Xu *et al.*, 2007), when compared with control plants, suggesting xanthophyll-cycle-mediated increases in energy excess thermal dissipation. Also, Takizawa *et al.* (2008) have reported that depletion of stromal P_i decreases proton conductivity at ATP synthase level, leading to an acidification of the thylakoid lumen (see also Walker and Sivak, 1985) and subsequent downregulation of photosynthetic light capture, which were accompanied by thermal-dissipation increases. Rao *et al.* (1986) concluded that P-deficient sugar beet plants have altered the ability to dissipate the intrathylakoid proton gradient. The carotenoid-to-chl. ratio increased in P-deficient plants (Xu *et al.*, 2007). However, no studies have so far tackled the functioning of the xanthophyll cycle and its associated energy dissipation in plants affected by P deficiency. An increase in the CO_2 compensation point may suggest elevated respiration and/or photorespiration in low-P plants (Lauer *et al.*, 1989). Dark respiration (and photorespiration, see below) is unlikely to contribute to PSII photoprotection under P deficiency because P_i is itself a substrate for mitochondrial respiration (excluding AOX, which is not regulated by adenylate level). Both dark respiration and photorespiration (Bottrill *et al.*, 1970; Terry and Ulrich, 1973a) have been shown to decrease with respect to the controls under P deficiency (it should be noted, however, that at that time it was not revealed that Rubisco catalyses both carboxylation and oxygenation). Alternatively, the ratio of electron transport rate (J_f, estimated from Chl-F measurements) to photosynthesis (A_N), J_f/A_N, was reported to decrease under P deficiency from approximately 7–9 to 6 (recalculated from Xu *et al.*, 2007), which suggests that a potential risk of ROS generation from excess electrons not consumed in CO_2 assimilation was unlikely in P-deficient plants. However, increases in SOD and APX were shown in response to P deficiency (Xu *et al.*, 2007). The origin of such apparent discrepancy could be that antioxidant enzymes were expressed on protein basis (Xu *et al.*, 2007), and protein decreased in response to P deficiency (Xu *et al.*, 2007). Further research is required to elucidate the role that other electron-consuming processes may play in photoprotection under P deficiency.

Under P deficiency, leaf chl. concentrations usually increase (Bottrill *et al.*, 1970; Marschner, 1995), but not always (Al-Abbas *et al.*, 1974; Brooks, 1986; Lauer *et al.*, 1989; Xu *et al.*, 2007). Chl.*a*/*b* ratio does not change (Brooks, 1986), whereas photosynthesis saturates at lower PPFD values than in the P-sufficient controls (Walker and Sivak, 1985; Fredeen *et al.*, 1989; Lauer *et al.*, 1989; Rao and Terry, 1989). These factors would result in an excess of energy excitation when P-deficient leaves are exposed to relatively high PPFD, leading in turn to a potential photoinhibition. However, the scant literature available suggests that no sustained photoinhibition occurs under P deficiency, as judged by the unaffected quantum efficiency in soybean (Fredeen *et al.*, 1989), and the high F_v/F_m ratios found in rice after dark adaptation (Xu *et al.*, 2007). On the contrary, in P-deficient spinach (Brooks, 1986) and soybean (Lauer *et al.*, 1989) the quantum efficiency decreased, estimated from the initial response of gas exchange, O_2 or CO_2, to light. However, Brooks (1986) also reported that the F_v/F_m ratio (recalculated from Brooks' data) only decreased from 0.79 to 0.76 under P starvation in spinach.

In summary, P deficiency affects plant growth but this is not always accompanied by photosynthesis decreases. This may be largely dependent on (starch- or sucrose-accumulating) species. Stomatal function, Rubisco content and specific activity, RuBP pool size and regeneration and sugar transport have been reported to impair photosynthesis under P limitation (Table 21.1). Leaf movements and xanthophyll-cycle-mediated thermal-energy dissipation (indirect evidences) seem to protect the photosynthetic apparatus under limited P supply (Table 21.2), with no increased risk of oxidative damage and no sustained photoinhibition reported to date.

21.2.3. Potassium deficiency

Potassium (K) may limit growth on some sites, especially those with sandy soil, owing to the high mobility of K in both soil and plant and generally large-plant requirements. Potassium normally constitutes 1–2% of the total plant dry weight (Marschner, 1995). Cytoplasmic K concentrations are maintained in a narrow range of 100–120 mM, whereas in the chloroplasts they are more variable, between 20 and 200 mM (Marschner, 1995). In many species K deficiency not only decreases growth but also photosynthesis (Bottrill *et al.*, 1970; Basile *et al.*, 2003; Kanai *et al.*, 2007; Weng *et al.*, 2007). In some species, the decreased photosynthesis is associated with low chl., damaged chloroplasts and restricted long-distance sugar transport (Marschner, 1995).

Potassium is the most abundant univalent cation in plant cells, and plays a significant role in regulating stomatal function (Macrobbie, 1998). As the light-dependent uptake of K into the guard cells is a crucial step for stomatal opening (Schroeder, 2003), it is possible that K deficiency causes stomatal limitations. In fact, Gates (1970) reported an increase of leaf temperature of ca. 0.5–1.5°C in K-deficient sugarcane when compared with the controls, which suggests stomatal closing. There are reports showing that K deficiency reduces photosynthesis by decreasing g_s to CO_2 (Terry and Ulrich, 1973b).

In other cases, changes in g_s are not responsible for the decline in photosynthesis. In these species, transpiration increases if K is in poor supply (Lösch *et al.*, 1992). There are also reports showing K starvation increases g_s in olive trees (Arquero *et al.*, 2006). The large K requirement for protein synthesis (Leigh and Wyn Jones, 1984) is probably responsible for strong correlations between concentrations of K and proteins such as Rubisco (Flaig and Mohr, 1992).

How do K-deficient plants protect themselves from the possible energy excesses? There is not much information about this. Al-Abbas *et al.* (1974) reported a decreased absorptance, accompanied by increases in reflectance and transmittance, in K-deficient plants. Increases in NPQ and decreases in the intrinsic PSII efficiency under K deficiency in rice suggest xanthophyll-cycle-mediated increases in thermal-energy dissipation as one possible mechanism (Weng *et al.*, 2007), although no work has been done investigating the functioning of the xanthophyll cycle under K deficiency. Activities of two of the key enzymes involved in scavenging ROS, SOD and APX, were shown to increase in response to K deficiency (Weng *et al.*, 2007). Dark respiration and photorespiration are unlikely to contribute to the protection of PSII. Similarly to what happens with P deficiency (see Section 21.2.2), dark respiration and photorespiration rates have been shown to decrease with respect to the controls (Terry and Ulrich, 1973b). However, there is one report showing marked increases in dark respiration in K-deficient plants (Bottrill *et al.*, 1970). Further research is required to elucidate the role that electron-consuming processes may play in photoprotection under K deficiency.

In almond, the development of K deficiency depends on leaf position. Leaves from the upper part of the canopy develop symptoms of deficiency, whereas those of the lower part do not (Basile *et al.*, 2003). This might suggest that the upper part of the canopy more exposed to sunlight could be affected by photoinhibitory damage. A feedback downregulation of photosynthesis, similar to that occurring in N-deficient leaves (see Section 21.2.1), cannot be excluded (Marschner, 1995). In fact, Kanai *et al.* (2007) have reported that K deficiency in tomato decreases fruit growth, a diminished sink activity that led to leaf-sugars accumulation and a concomitant fall in photosynthesis.

In summary, growth and photosynthesis are impaired in K-deficient plants. Damages to chloroplast ultrastructure and chl. have been reported in some cases (Table 21.1). In some species, K deficiency diminishes CO_2 g_s (Table 21.1). In other species, g_s increases in K-deficient leaves, and decreases in photosynthesis are mediated by effects on Rubisco concentration and activity (Table 21.1). Photoprotection mechanisms under K deficiency are related to decreases in absorptance, possibly increases in thermal energy dissipation within the PSII antenna (indirect evidences) and increases of enzymatic activities scavenging ROS (Table 21.2). Some observations that can be interpreted as photoinhibition processes under K deficiency could also be the result of a restricted long-distance sugar transport.

21.3. PHOTOSYNTHETIC RESPONSES TO MICRONUTRIENT DEFICIENCY AND TOXICITY

21.3.1. Iron deficiency

The deficiency level of leaf iron (Fe) is in the range 50–150 $\mu g\ g^{-1}$ dry weight (Marschner, 1995). Approximately 80% of the plant Fe is located in the chloroplast, where it is a constituent of cyts, Fe-S centres and others (Terry and Abadía, 1986). When Fe is deficient, the amount of photosynthetic membranes per chloroplast decreases. This is accompanied by decreases in all membrane components, including the thylakoid pigments, chls and carotenoids (Morales *et al.*, 1990, 1994), proteins associated to them in PSI and PSII (Timperio *et al.*, 2007) and the electron carriers of the photosynthetic electron transport chain (Terry and Abadía, 1986). Iron deficiency also decreases RuBP-carboxylation capacity, both through reduced Rubisco-enzyme activation (Taylor and Terry, 1986) and downregulation of gene expression (Winder and Nishio, 1995). The Fe-deficiency mediated decreases in light harvesting, electron transport and carbon-fixation capacities seem to be well coordinated (Larbi *et al.*, 2006). Sugars (starch, sucrose and glucose) are drastically depleted in Fe-deficient leaves (Arulanantham *et al.*, 1990), which rules out sugar accumulation as a possible cause for the decreased photosynthetic rates under Fe deficiency.

As a consequence of all these changes, Fe-deficient leaves have low photosynthetic rates and have correspondingly small dissipation of energy by means of photosynthesis. The decreases in leaf pigment concentrations occurring with Fe deficiency may provide some protection through decreases in light-harvesting capacity (see below). Pigment decreases, however, may also involve some dangers, as these pigments have a general protective function, acting as 'sunglasses' for the abaxial leaf cell layers (Nishio, 2000). In fact, the number of blue and red photons absorbed per chl. increases with Fe deficiency, because decreases in absorptance at these wavelengths are less marked than those found for photosynthetic pigments (Morales *et al.*, 1991; Abadía *et al.*, 1999). Therefore, Fe-deficient plants are prone to be exposed to an excessive PPFD.

Iron-deficient leaves are first protected to some extent against excess PPFD by decreases in light absorptance, associated with the decreases in chls and carotenoids (Terry, 1980; Morales *et al.*, 1991; Masoni *et al.*, 1996; Abadía *et al.*, 1999). Data from Morales *et al.* (1991) and Abadía *et al.*

(1999) show that 40–80% of the incident PPFD is simply not absorbed by the low-chl. leaves. Also, Fe-deficient leaves show increases in light reflectance and transmittance (Morales *et al.*, 1991; Masoni *et al.*, 1996; Abadía *et al.*, 1999) that deserve further investigation.

A large part of the energy absorbed by PSII is dissipated thermally within the PSII antenna in Fe-deficient plants. With severe Fe deficiency, D may reach up to 75–80% of the light absorbed by PSII at midday. In Fe-sufficient controls, D only accounts for approximately 25 and 54–57% under low and high PPFD, respectively. The Fe-deficiency mediated relative increases in D have been related to the increases in the Z+A-to-chl. molar ratio and to the extent of de-epoxidation of violaxanthin (V) into A+Z, which occur in daily cycles in Fe-deficient plants (Morales *et al.*, 1998, 2000). At low PPFDs, *P* decreases from 66 to 21% when Fe is deficient, whereas under full-sunlight field conditions, *P* decreases from 29–38 to 10%. Both in Fe-deficient and control leaves, X accounts for 4–14% of the PPFD absorbed by PSII. The xanthophyll-cycle-related thermal dissipation found in leaves of Fe-deficient sugar beet grown at low PPFDs is not 'locked in' overnight (Morales *et al.*, 1990, 1998), indicating that it constitutes a photoprotective mechanism. In pear grown in the field at high PPFDs, it is not maintained overnight in moderately Fe-deficient leaves, but appears to be 'locked in' in severely deficient leaves (Morales *et al.*, 1994, 2000).

Evidence for a different mechanism of photoprotection has arisen from studies with very severe Fe-deficient leaves. These leaves show sustained decreases in dark-adapted F_v/F_m ratios, which cannot be relieved by FR pre-illumination. This finding was attributed to the presence of a constant PSII emission, with a lifetime of approximately 3.3 ns, which would account for approximately 15% of the total fluorescence (Morales *et al.*, 2001). When F_v/F_m values were corrected to eliminate such emission, the measured F_v/F_m value of 0.66 becomes very similar to those of the controls, in the range 0.73–0.83. Only in some leaves F_v/F_m are still lower than the controls, possibly owing to the presence of some closed PSII *RC*s in dark-adapted Fe-deficient leaves (Belkhodja *et al.*, 1998). It was hypothesised that the constant PSII-fluorescence emission in severely Fe-deficient sugar beet leaves would come from internal antenna complexes fully disconnected from the PSII reaction centre (Morales *et al.*, 2001).

In spite of the excess PPFD experienced by Fe-deficient leaves, the extent of oxidative damage seems very limited. For instance, Fe-deficient pea leaves did not show accumulation of oxidatively damaged lipids and proteins (Iturbe-Ormaetxe *et al.*, 1995). Of course, the concentrations of some anti-oxidative elements containing Fe, including Fe-SOD (Kurepa *et al.*, 1997) and APX (Ranieri *et al.*, 2001), are reduced in Fe-deficient plants. The decrease in some of these elements is compensated by increases in alternative enzymes, such as Mn- and Cu/Zn-SODs, which may take on the role of Fe-SOD. Fe-deficient leaves could be protected against ROS by relatively high concentrations of antioxidant molecules, such as ascorbate and glutathione (Iturbe-Ormaetxe *et al.*, 1995). The possible photoprotective role of Z as an antioxidant (Havaux and Niyogi, 1999) or as a ROS-signalling-modulating substance (Demmig-Adams and Adams, 2002) has not been investigated in Fe-deficient leaves. The ratio of protective enzymes and antioxidants to chl. could become very large in Fe-deficient plants. Also, the carotenoid/chl. ratios increase markedly, and carotenoids can directly de-excite triplet chl. (Foyer *et al.*, 1994). All these data suggest that Fe-deficient leaves are well protected against oxidative damage, and contribute to explain the observation that chlorotic leaves of Fe-deficient trees can remain stable in the field for a few months. Most of these leaves appear otherwise healthy, and only extremely deficient leaves show necrotic spots.

Some reports indicate that severe Fe deficiency could induce PSII photoinhibition in photosynthetic organisms. For instance, in isolated thylakoids of severely Fe-deficient peach, grapevine and tomato, decreases in the abundance of D1 and other PSII polypeptides have been found (Ferraro *et al.*, 2003). The turnover of D1 was also increased in Fe-deficient maize (Jiang *et al.*, 2001). These studies, however, were carried out with SDS-PAGE of isolated thylakoids, which may not be fully representative of the whole-leaf cell population (see Section 21.5). Studies on Cu excess, which causes a Cu-induced Fe deficiency (see Section 21.3.6), indicate that Fe-deficient, low-chl. plants could be photoinhibited faster than the controls in the presence of lincomycin (Patsikka *et al.*, 2002). In isolated thylakoids from the same plants, however, susceptibility to photoinhibition was not affected (Patsikka *et al.*, 2002). This suggests that Fe deficiency increases the susceptibility of PSII towards photoinhibition mainly through the decrease in leaf chl. This is likely related to the increases in the amount of photons absorbed per chl., which is markedly higher in Fe-deficient leaves.

Moderately Fe-deficient leaves have no signs of significant photoinhibitory damage, as judged by both Chl-F and gas-exchange parameters. Evidence for the maintenance of a good maximum PSII energy conversion efficiency comes from measurements of the F_v/F_m ratio after dark adaptation, both in controlled environments (Morales *et al.*, 2001) and in the field (Morales *et al.*, 2000). Also, high quantum yields of CO_2 fixation (Terry, 1980) and O_2 evolution (Morales *et al.*, 1991) occur in moderately Fe-deficient leaves. Some data in the literature had previously suggested that Fe deficiency could cause general decreases in maximum potential PSII efficiency. For instance, markedly low F_v/F_m ratios were reported in Fe-deficient sugar beet (Morales *et al.*, 1991). These values, however, were likely owing to inaccurately high F_0 values, caused by the reduction of the PQ pool occurring in Fe-deficient organisms during dark adaptation, which is only prevented by FR pre-illumination (Belkhodja *et al.*, 1998). No detailed study has been carried out so far on the effects of different levels of Fe deficiency on specific photoinhibition-related sites as a consequence of excessive PPFD levels.

In Fe-deficient leaves of pear trees grown at high PPFDs in the field, a significant part of the Z pool is maintained ('locked in') overnight (Morales *et al.*, 1994, 2000). This might suggest that Fe-deficient plants may experience processes that could be considered as photoinhibitory. In these cases, however, dark-adapted F_v/F_m ratios still remained high (Morales *et al.*, 2000).

In summary, Fe deficiency affects photosynthesis through well-coordinated decreases of light harvesting, electron transport and carbon fixation (Table 21.1). A large part of the energy absorbed by PSII is dissipated, thermally mediated by the xanthophyll cycle or is not transferred from the internal PSII antenna complexes to the PSII reaction centres (Table 21.2). The antioxidant enzymes and molecules to chl. ratios are markedly increased in Fe-deficient plants. Iron deficiency increases the PSII susceptibility towards photoinhibition, although some signs of photoinhibitory processes may only appear when Fe deficiency is very severe.

21.3.2. Iron toxicity

The crucial Fe-toxicity levels are approximately 300–500 $\mu g\ g^{-1}$ dry weight (Marschner, 1995). Plants grown under excess Fe show a variety of symptoms of Fe toxicity depending on species: leaf bronzing (rice), freckle leaf (sugar cane), leaf black spots (navy bean), leaf brown and necrotic spots (*Nicotiana plumbaginifolia*), purple blight (pea), decreased leaf strength with dark brown to purple leaves (tobacco) and red to brown spots on darkened foliage and badly stunted top growth (mung bean). Excess Fe accumulated in chloroplasts incorporated into the stroma as phytoferritin (Marschner, 1995) and thylakoids as non-heme Fe (Kim and Jung, 1993), later identified as cyt b_6f complex (Suh *et al.*, 2002) containing both haem and non-haem Fe.

Effects of Fe toxicity on photosynthesis depend on species and light intensity (Kim and Jung, 1993; Kampfenkel *et al.*, 1995). Seedlings under excess Fe at low light did not lose photosynthetic activity (Kim and Jung, 1993). Exposure to sunlight resulted in a reduction in their photosynthetic capacity both in PET and key enzymes of the Calvin cycle (Kim and Jung, 1993). The decreased electron transport rate could be related to an increased PSII reduction, and higher thylakoid energisation (Kampfenkel *et al.*, 1995). In cases where photosynthesis decreases, it is not owing to a direct inhibition of starch or sucrose (Kampfenkel *et al.*, 1995).

Fe excess seems to stimulate alternative electron sinks. The oxygenase activity of Rubisco produces 2-phosphoglycolate and 3-phosphoglycerate, the former after hydrolisis to glycolate is oxidised to glyoxylate producing H_2O_2. An indirect suggestion for increased photorespiration under Fe toxicity could be, therefore, the twofold increase in CAT and APX activity (Kampfenkel *et al.*, 1995; Kumar *et al.*, 2008). They are also indicative of an increased requirement to detoxify H_2O_2 in leaves. Respiration rates also increase (Kampfenkel *et al.*, 1995). NPQ was also increased in plants grown with excess Fe (Kampfenkel *et al.*, 1995). Contents of the antioxidants ascorbate and glutathione were, however, reduced (Kampfenkel *et al.*, 1995).

The extent of the abovementioned photoprotection mechanisms should not be enough because Fe excess is believed to generate oxidative stress (Kampfenkel *et al.*, 1995). Photoinactivation was the result of a combination of Fe excess and high light (Kim and Jung, 1993). Thylakoids from Fe-overloaded plants were more susceptible to loss of D1 and several other thylakoid proteins (Suh *et al.*, 2002). Oxidative damage in Fe-poisoned plants was related to the generation of $^1O_2^*$ (Kim and Jung, 1993), which has a substantially long lifetime to diffuse far from its generation site and react with potential target molecules (Suh *et al.*, 2002). Also, it was related to increases of H_2O_2, leading to lipid peroxidation (Kumar *et al.*, 2008). However, chl. (Kampfenkel *et al.*, 1995; Kumar *et al.*, 2008) and carotenoid (Kumar *et al.*, 2008) concentrations were almost unaffected, and the maximum potential PSII efficiency decreased only slightly (F_v/F_m values decreased from 0.80 to 0.76, recalculated from Kampfenkel *et al.*, 1995; or from 0.81 to 0.79, Suh *et al.*, 2002), which questions the extent and physiological significance of PSII photoinhibition-mediated oxidative stress under Fe excess. In Fe-poisoned plants, activities of H_2O_2-scavenging enzymes (peroxidase, CAT, and APX) were increased (Kumar *et al.*, 2008), which have probably provided enough antioxidant protection to the whole photosynthetic machinery.

In summary, the effects of Fe toxicity on photosynthesis are only evident in the presence of high light, decreasing electron transport and activities of key enzymes of the Calvin cycle (Table 21.1). Although it is thought that Fe toxicity induces oxidative damage, data reported to date seem to indicate that photoprotection processes (increased NPQ (Table 21.2), increased rates of respiration and photorespiration and increased activities of ROS-scavenging enzymes) limit the extent and physiological significance of photoinhibition-mediated oxidative stress under Fe excess.

21.3.3. Manganese deficiency

Critical manganese (Mn) deficiency levels are approximately 10–20 $\mu g\ g^{-1}$ dry weight, regardless of plant species or prevailing environmental conditions (Marschner, 1995). Growth is reduced under Mn deficiency (Ohki, 1985; Yu *et al.*, 1998). Manganese deficiency decreases leaf chl. (Bottrill *et al.*, 1970; Ohki *et al.*, 1981; Ohki, 1985; Kriedemann and Anderson, 1988; Pérez *et al.*, 1993; Masoni *et al.*, 1996; Henriques, 2003) and carotenoid (Henriques, 2003) concentrations. Chlorosis usually occurs in the oldest leaves, and only the main veins remain green. Terry and Ulrich (1974) reported chlorosis in young leaves.

Photosynthesis declines under Mn deficiency (Bottrill *et al.*, 1970; Terry and Ulrich, 1974; Ohki *et al.*, 1981; Ohki, 1985; Kriedemann and Anderson, 1988; Henriques, 2003). The causes for such impaired photosynthesis are not fully understood. On one hand, some authors ruled a decreased CO_2 availability at the carboxylation sites out as a possible cause because no reduction in transpiration rates or g_s were observed under Mn deficiency (Terry and Ulrich, 1974; Ohki *et al.*, 1981; Ohki, 1985; Henriques, 2003). On the other hand, CO_2 diffusion rates from intercellular spaces to the chloroplast stroma were decreased (i.e., CO_2 g_m decreased) in Mn-deficient plants (see Terry and Ulrich, 1974), taking into account that at that time mesophyll resistance was not properly assessed and definitions may be different. Both initial slope and CO_2-saturated phases of A versus C_i curves were similarly affected under Mn deficiency (Kriedemann and Anderson, 1988).

The optical properties of Mn-deficient leaves are also different to their respective controls. Manganese deficiency reduced leaf absorptance in the range 2–29% depending on species and increased reflectance (Adams *et al.*, 1993) and transmittance, in line with the observed changes in chl. concentration (Masoni *et al.*, 1996). Bottrill *et al.* (1970) and Kriedemann and co-workers (Kriedemann *et al.*, 1985; Kriedemann and Anderson, 1988) have reported a reduction of the chl. *a*/*b* ratio in spinach and cereals, whereas in soybean (Pérez *et al.*, 1993), peach (Val *et al.*, 1995) and pecan (Henriques, 2003) the ratio remained unchanged or was only slightly affected. It is noteworthy that under Mn deficiency, F_s values are lower than initial F_0 values (Kriedemann and Anderson, 1988), which may reflect either the development of strong NPQ during illumination or the existence of dark processes that largely increase F_0, like those reported for Fe-deficient materials (Belkhodja *et al.*, 1998), or other causes. Concerning the latter hypothesis, the PQ pool size associated with PSII was markedly reduced under Mn deprivation but the F_v/F_m ratios were unaffected (Henriques, 2003). Xanthophyll-cycle-mediated increases in thermal-energy dissipation, and therefore NPQ under Mn deficiency are possibly small. The A+Z/(V+A+Z) ratio increased from 0.06–0.09 (control) to 0.08–0.19 (Mn-deficient soybean), whereas the V+A+Z/chl. and lutein/chl. ratios remained fairly constant (recalculated from Pérez *et al.*, 1993). Also, the total carotenoid/chl. ratio was only slightly increased in Mn-deficient leaves (Henriques, 2003). Photorespiration and respiration are not likely to contribute to photoprotection. The former decreased (Terry and Ulrich, 1974), and the latter decreased (Bottrill *et al.*, 1970) or remained unchanged (Terry and Ulrich, 1974; Ohki *et al.*, 1981) under Mn deficiency.

Photosynthetic electron transport is likely to be affected by Mn deficiency, mostly owing to the participation of Mn in water photolysis. In most species, Mn deficiency greatly increases F_0 and decreases F_v (Kriedemann *et al.*, 1985; Adams *et al.*, 1993; Val *et al.*, 1995), greatly decreasing the maximum potential PSII efficiency. Also, the capacity to evolve O_2 is clearly reduced under Mn deficiency (Kriedemann *et al.*, 1985). These changes were accompanied by a substantial loss of PSII *RC*s (Abadía *et al.*, 1986), possibly associated with the disappearance of the thylakoid grana. In some other species, however, the maximum potential PSII efficiency undergoes only small decreases with increasing Mn deficiency (Henriques, 2003). In these species, the impaired photosynthetic rates were associated to a decreased number of PSII units, owing to a decreased number of chloroplasts per A_{leaf} (Henriques, 2003).

The photosynthetic apparatus of Mn-deficient leaves may become vulnerable under high-incident light intensities. Thus, it has been suggested that Mn deficiency increases peroxidation of thylakoid-membrane lipids through the highly reactive $O_2^{\cdot -}$ and the $\cdot OH$ (Eyster *et al.*, 1958), but experiments in pecan leaves have concluded that this is not the case (Henriques, 2003). In line with the former findings, APX and SOD activities were found to decline drastically in Mn-deficient tobacco plants (Yu *et al.*, 1998). Increased activities of these two antioxidant enzymes in pea and *Picea abies* agree with the latter (discussed by Yu *et al.*, 1998).

In summary, Mn deficiency decreases photosynthesis mainly through decreases of PET, owing to the participation of Mn in water photolysis, and possibly through decreases of CO_2 mesophyll conductance (Table 21.1). Effects on electron transport can be mediated by either decreases of the PSII efficiency or decreases of the number of PSII units (maintaining fairly constant PSII efficiency), depending on species. Decreases of leaf absorptance have been reported in some cases (Table 21.2). Plants under Mn deficiency do not always respond with increased activities of antioxidant enzymes and, hence, occurrence (or not) of oxidative damage, leading to lipid peroxidation, depends on the species experiencing Mn deficiency.

21.3.4. Manganese toxicity

Critical Mn toxicity levels vary widely among plant species and environmental conditions (Marschner, 1995). Critical toxicity levels (in mg g^{-1} dry weight), associated with a 10% reduction in dry matter production, have been reported for several species: maize (200), pigeon pea (300), soybean (600), cotton (750), sweet potato (1380) and sunflower (5300). High external Mn supply induces appearance of Mn toxicity symptoms in leaves, including stunted growth, chlorosis, crinkled leaves and brown lesions or speckles (Ohki, 1985; González *et al.*, 1998). Mn toxicity inhibits photosynthesis and transpiration (Ohki, 1985).

Effects of light intensity on Mn toxicity were first reported in 1935. McCool (1935) reported that low-light-grown plants displayed fewer symptoms of excess Mn than those grown at high light (see also González *et al.*, 1998, and references therein). Chl. reduction ranged from 48% at high light to nil at low light (González *et al.*, 1998). This fact suggests that excess Mn induces photo-oxidative processes.

There are not enough available data on photoprotective processes under Mn toxicity. Excess Mn increases leaf reflectance only at certain wavelengths within the visible range (Schwaller *et al.*, 1983). Excess Mn largely increased the leaf SOD and APX activities (González *et al.*, 1998). The sequential operation of these two antioxidant enzymes should control H_2O_2 accumulation under Mn toxicity.

Some reports suggest a role for Mn toxicity in induction of oxidative stress. Panda *et al.* (1986) reported lipid peroxidation in isolated chloroplasts treated with excess Mn. Lipid peroxidation was, however, not detected in floating leaf discs of leaves exposed to excess Mn. In this latter case, the marked reduction of chl. at high light could be related to decreased ascorbate concentrations and, possibly, to the activation of the cytosolic fraction, not the chloroplastic one, of APX and SOD (González *et al.*, 1998).

In summary, symptoms of excess Mn have only been reported in combination with high PPFDs. Excess Mn reduces photosynthesis and transpiration (Table 21.1). Induction of oxidative stress, with lipid peroxidation and net loss of chl., is not always observed (Table 21.1). In some cases, increases of leaf SOD and APX are enough to control ROS.

21.3.5. Copper deficiency

Critical copper (Cu) deficiency level is generally in the range 3–5 μg g^{-1} dry weight, although it can be larger depending on species, organ, developmental stage and N supply (Marschner, 1995). Cu deficiency reduces plant growth (Yu *et al.*, 1998). Effects of Cu deficiency on leaf chl. were reported to be moderate in cereals (Kriedemann and Anderson, 1988), very important in sugar beet (Henriques, 1989) and with no effects in spinach (Bottrill *et al.*, 1970). Copper deficiency decreases g_s and photosynthetic rates (Bottrill *et al.*, 1970; Kriedemann and Anderson, 1988). Similar decreases were observed on both initial-slope and CO_2-saturated phases of A_N/C_i curves under Cu deficiency (Kriedemann and Anderson, 1988), suggesting a general reduction in Rubisco activity and rates of RuBP regeneration. Also, Cu deficiency interferes with pigment and lipid biosynthesis (Barón *et al.*, 1995) and, consequently, with chloroplast ultrastructure, greatly reducing the number of thylakoids per granum and most stacks showing some swelling (Henriques, 1989).

Chlorophyll *a*/*b* ratios decline under Cu deficiency in cereals (Kriedemann and Anderson, 1988), but not in spinach (Bottrill *et al.*, 1970) or sugar beet (Henriques, 1989). In Cu-deficient plants, fluorescence quenching subsequent to the maximum Chl-F during the Kautsky induction curve (F_p) was slight so that F_s remained close to F_p (Kriedemann and Anderson, 1988). As plastocyanin, a Cu-containing protein, fulfils a key role as a carrier in PSII-to-PSI electron transport, and in CEFI, this behaviour may reflect a slower reoxidation of Q_A maintaining reduced the PSII acceptor side. This is consistent with a higher reduction of PSI activity compared with that of PSII under Cu deficiency (Henriques, 1989).

Few data are available about alternative electron sinks and photoprotection in Cu-deficient plants. Dark respiration was not affected under Cu deficiency (Bottrill *et al.*, 1970). APX and SOD activities were reported to increase in Cu-deficient plants (Yu *et al.*, 1998). Cu-deficient leaves showed minor changes in F_0, lack of the typical biphasic rise to F_p and variable decreases in F_v, although the maximum potential PSII efficiency was only marginally affected (Kriedemann and Anderson, 1988). The latter would suggest the absence of PSII irreversible damages under Cu deficiency.

In summary, effects of Cu deficiency on plant physiology depend largely on species, some of them showing growth reduction and chlorosis and other species showing no changes in chl. (Table 21.1). Cu-deficient plants have decreases in photosynthesis, associated with one of the following processes: thylakoid degradation and swelling, reduced CO_2 stomatal conductance, decreased PSI (more than PSII) activity and/or lower Rubisco activity and RuBP regeneration (Table 21.1). Some antioxidative enzymes are upregulated, and irreversible damages to PSII have not been observed.

21.3.6. Copper toxicity

Critical toxicity levels of Cu in leaves is considered to be approximately 20–30 μg g^{-1} dry weight (Marschner, 1995). Copper toxicity inhibits growth; symptoms include leaf wilting (Kumar *et al.*, 2008), stunted plants, darkening of leaf veins (Horler *et al.*, 1980), chlorosis (Ouzounidou *et al.*, 1994) and leaf necrotic scorching (Kumar *et al.*, 2008). Chlorosis has been associated to Fe deficiency (see Section 21.3.1) induced by excess Cu (Ouzounidou *et al.*, 1994; Patsikka *et al.*, 2002). Amelioration of Cu toxicity by excess Fe was reported in spinach and maize (Ouzounidou *et al.*, 1998; Kumar *et al.*, 2008).

Photosynthesis is highly sensitive to sub-lethal Cu concentrations (Ouzounidou, 1996). Stomatal conductance and transpiration rates were reduced under Cu toxicity (Moustakas *et al.*, 1997). Excess Cu reduces leaf chl. concentration and causes thylakoid degradation (Ouzounidou *et al.*, 1994). Thylakoid electron transport from Cu-overloaded, low-light-grown plants, unlike Fe overload (see 21.3.2), was remarkably affected (Kim and Jung, 1993). Both, PSI and PSII electron transport between pheophytin and Q_A were affected (Yruela *et al.*, 1993; Ouzounidou *et al.*, 1994; Barón *et al.*, 1995; Ouzounidou, 1996). PSI activity was depressed more than PSII, leading to closure of PSII centres under Cu excess (Ouzounidou, 1996). Loss of phosphorylation efficiency has been reported in Cu-treated plants, owing to a direct interaction of Cu with the coupling factor (Uribe and Stark, 1982). Rubisco activity and the Calvin cycle might also be important primary sites of Cu action (discussed by Ouzounidou, 1996 and Moustakas *et al.*, 1997). Starch, soluble sugars and lipids were reduced in plants grown under Cu toxicity (Moustakas *et al.*, 1997).

Leaf reflectance increases in response to Cu excess (Horler *et al.*, 1980; Schwaller *et al.*, 1983). Energy thermally dissipated seems to be increased under Cu excess, as photoacoustic spectroscopy estimated heat emission, whereas qN and D increase in response to Cu toxicity (Ouzounidou, 1996; Ouzounidou *et al.*, 1997; Moustakas *et al.*, 1997). The latter increases from 35–40 to 45–52% (recalculated from Ouzounidou, 1996 and Ouzounidou *et al.*, 1997). Photorespiration is not likely contributing to photoprotection, as CAT activity was depressed in Cu-overloaded leaves (Kumar *et al.*, 2008). APX and SOD activities increased in plants growing with Cu excess (Kumar *et al.*, 2008).

Signs of membrane damage have been reported under Cu excess. Sandmann and Boger (1980) reported lipid peroxidation of photosynthetic membranes when isolated spinach chloroplasts were treated with Cu, very likely mediated by H_2O_2 (Kumar *et al.*, 2008). Excess Cu decreased the chl. *a*/*b* ratio (Horler *et al.*, 1980; Moustakas *et al.*, 1997) with exceptions (Kumar *et al.*, 2008), which could indicate a larger sensitivity of chl. *a* than chl. *b* to Cu toxicity. Carotenoid concentration was also diminished under excess Cu (Kumar *et al.*, 2008). The maximum potential PSII efficiency (F_v/F_m ratio) was, however, unaffected or slightly affected by Cu excess in intact leaves (Ouzounidou, 1996; Ouzounidou *et al.*, 1997). This is similar to what happens with Fe deficiency. Recent studies have shown that Cu excess causes Fe deficiency (see Section 21.3.1).

In summary, symptoms exhibited by plants grown under excess Cu resemble those of Fe-deficient plants. These include reductions in light harvesting, PET and carboxylation (Table 21.1). Also, a large part of the absorbed light is thermally dissipated (Table 21.2) and the antioxidative system is activated. However, signs of membrane damage are reported, which was not observed in Fe-deficient plants.

21.3.7. Zinc deficiency

The critical Zinc (Zn) deficiency levels are approximately 15–20 μg g^{-1} leaf dry weight (Marschner, 1995). Zn deficiency reduces plant growth and inhibits photosynthesis in a wide variety of plants (Bottrill *et al.*, 1970; Yu *et al.*, 1998; Schuerger *et al.*, 2003; Chen *et al.*, 2008). Increases in light intensity rapidly induced chlorosis and necrosis in Zn-deficient plants (Marschner and Cakmak, 1989). This is accompanied by thylakoid disorganisation and degradation (Henriques, 2001). Zinc-deficiency mediated inhibition of photosynthesis was associated with a decreased g_s, and sub-stomatal CO_2 concentration decreased in cauliflower (Sharma *et al.*, 1995) but increased in rice (Chen *et al.*, 2008). In Zn-deficient rice, more excitation energy was distributed to PSII than to PSI (Chen *et al.*, 2008). Possibly related to this unbalance, Zn deficiency has been reported to diminish PSII photochemistry (Schuerger *et al.*, 2003; Chen *et al.*, 2008). Effects of Zn deficiency on CA (Ohki, 1976) and Rubisco (Marschner, 1995) enzymes point to carboxylation reactions as bases for the observed decreased photosynthesis. Finally, a feedback downregulation of photosynthesis under Zn deficiency cannot be excluded as an accumulation of sugars in leaves has been related to decreases in photosynthetic CO_2 fixation (Marschner, 1995).

Leaves from Zn-deficient plants have increased visible reflectance (cited by Masoni *et al.*, 1996). Plant respiration decreases under Zn deficiency (Bottrill *et al.*, 1970). CAT activity increases in Zn-deficient plants (Chen *et al.*, 2008), which may suggest increased photorespiratory rates. The J_f/A_N ratios increased from 5 to 11 in the C_4 *Paspalum notatum* Flugge, affected by Zn deficiency (recalculated from Schuerger *et al.*, 2003). From a physiological point of view, this increase should not suggest an increased risk of oxidative damage.

However, signs of oxidative damage have been reported in Zn-deficient plants. Zn deficiency increases the production of O_2^- (Marschner and Cakmak, 1989; Wang and Jin, 2005) and the concentration of H_2O_2 (Chen *et al.*, 2008). Lipid peroxidation and damage to cell and chloroplast membranes were observed under Zn deficiency (Henriques, 2001; Wang and Jin, 2005; Chen *et al.*, 2008). In line with this, Zn deficiency decreases APX (Yu *et al.*, 1998) and SOD (Yu *et al.*, 1998; Wang and Jin, 2005; Chen *et al.*, 2008) activities. Zinc application to Zn-deficient plants enhanced SOD activities and reduced O_2^- content (Wang and Jin, 2005). APX and SOD activities, however, increased in Zn-deficient rice (Chen *et al.*, 2008).

Other signs could be indicative of oxidative damage under Zn deficiency. The maximum potential PSII efficiency, F_v/F_m, was slightly or moderately diminished in plants affected by Zn deficiency (Schuerger *et al.*, 2003; Wang and Jin, 2005; Chen *et al.*, 2008). Also, leaf chl. (Bottrill *et al.*, 1970; Schuerger *et al.*, 2003; Wang and Jin, 2005; Chen *et al.*, 2008) and carotenoid (Schuerger *et al.*, 2003) concentrations were negatively affected by Zn deficiency. As the chl. *a*/*b* ratio was decreased (Bottrill *et al.*, 1970; Schuerger *et al.*, 2003; Chen *et al.*, 2008), Zn deficiency affects more chl. *a* than chl. *b*. In Zn-deficient maize, however, this ratio increases (Wang and Jin, 2005).

In summary, Zn deficiency may decrease photosynthesis in several ways. Stomatal conductance, PSII photochemistry, CA and Rubisco were affected in Zn-deficient plants (Table 21.1). Also, an accumulation of leaf sugars may induce photosynthesis feedback downregulation under Zn deficiency (Table 21.1). The photoprotection mechanisms of Zn-deficient plants have not been elucidated (Table 21.2), and some data available in the literature are contradictory. In any case, Zn-deficient plants show signs of oxidative damage: production or accumulation of ROS, lipid peroxidation, damages to chloroplast and cell membranes, and decreases of the maximum potential PSII efficiency, which were accompanied by chlorosis (both chl. and carotenoids decreasing) and degradation of chloroplast thylakoids (Table 21.1).

21.3.8. Zinc toxicity

Critical toxicity levels of leaf Zn in crop plants are more than 400–500 μg g^{-1} dry weight (Marschner, 1995). In pea, excess Zn reduces growth, leading to stunted plants and chlorotic leaves (Horler *et al.*, 1980). Chlorosis in young leaves was suggested to result from a Zn-induced Fe or Mg deficiency, based on the fact that the three metals have similar ion *radii* (Marschner, 1995). Chlorosis can be the result of the reduction of the number or size of chloroplasts (Prasad and Strzalka, 1999) or of direct effects on chl. biosynthesis pathways (van Assche and Clijsters, 1990; Prasad and Strzalka, 1999). In soybean, however, Zn-treated plants were dark green, suggesting that inhibition of leaf expansion predominated over chl. breakdown or lack of biosynthesis (Horler *et al.*, 1980).

Zinc toxicity inhibits photosynthesis at various steps and through different mechanisms. Stomatal conductance and transpiration were impaired (Schuerger *et al.*, 2003; Vaillant *et al.*, 2005; Sagardoy *et al.*, 2009, 2010; Disante *et al.*, 2011), and mesophyll conductance was impaired to a lesser extent (Sagardoy *et al.*, 2010). Thylakoid electron transport from Zn-overloaded, low-light-grown plants was remarkably affected (Kim and Jung, 1993). Specific effects on PSII photochemistry were related to competitive substitution of Mn by Zn on the water photolysis site, inhibiting electron transport and oxygen evolution (van Assche and Clijsters, 1986a). In line with this, Zn excess decreased ϕ_{PSII} (Schuerger *et al.*, 2003; Sagardoy *et al.*, 2009). Non-cyclic photophosphorylation was significantly inhibited as a result of Zn treatment (van Assche and Clijsters, 1986a). Specific effects on the carboxylase capacity of Rubisco (van Assche and Clijsters, 1986b) and Calvin cycle (Chaney, 1993) have been reported.

Leaf reflectance increases under Zn toxicity (Horler *et al.*, 1980). The small effects on the V cycle suggest that the formation of the thylakoid pH gradient may be strongly impaired by excess Zn (Sagardoy *et al.*, 2009). Monnet *et al.* (2001) reported that excess Zn reduces Rubisco carboxylase activity to nearly nil, possibly owing to a decreased affinity of Rubisco for CO_2, whereas its oxygenase activity was unaffected or even increased. They concluded that the oxygenase activity of Rubisco appears to be an essential mechanism for the protection of the photosynthetic apparatus under Zn excess. However, CAT activity was decreased in Zn-poisoned plants (Chaoui *et al.*, 1997), which may suggest a decreased photorespiratory activity. The J_f/A_N ratios increased under Zn excess, from approximately 9 in the controls to 31–77 (recalculated from Monnet *et al.*, 2001), which may suggest an increased risk of suffering oxidative damage. It should be noted that, in cases where stress-induced gradients within the leaf do occur (see Chapter 16), J_f will be severely underestimated if the surface chloroplasts are chronically photodamaged. Chaoui *et al.* (1997) reported the induction of Mn-SOD and APX in plants exposed to high levels of Zn.

Excess Zn decreased the maximum potential PSII efficiency to different extents, and greatly reduced leaf chl. and carotenoid concentrations (Monnet *et al.*, 2001; Schuerger *et al.*, 2003; Vaillant *et al.*, 2005; Sagardoy *et al.*, 2009), which can be interpreted as signs of photodamage. The chl. *a*/*b* ratio decreased from 2.8–3.4 to 2.1–2.4 (Monnet *et al.*, 2001; Schuerger *et al.*, 2003; Vaillant *et al.*, 2005), indicating that chl. *a* is more sensitive to Zn exposure than chl. *b*. Contrary to this, the quantum yield of PSII-related electron transport at low PPFD was unaffected (Van Assche and Clijsters, 1986a). This latter report is in line with that of Schuerger *et al.* (2003) and Sagardoy *et al.* (2009), who showed decreases of the F_v/F_m ratio from 0.82 to 0.70 under excess Zn.

Excess Zn generates ROS and/or displaces metals from active sites. Lipid peroxidation was enhanced in plants treated with excess Zn (Chaoui *et al.*, 1997). It has also been suggested that chl. biosynthesis can be affected through Fe depletion or substitution of the central Mg ion with Zn (Prasad and Strzalka, 1999).

In summary, excess Zn may induce chlorosis (both chl. and carotenoids decreasing) or plants may become dark green, depending on how and to what extent the number/size of chloroplasts, chl. biosynthesis and leaf-expansion processes are affected (Table 21.1). Photosynthetic decreases under Zn excess have been related to lower CO_2 stomatal and mesophyll conductances, PSII photochemistry, non-cyclic photophosphorylation, Rubisco carboxylation capacity and Calvin-cycle enzymatic activities (Table 21.1). Data reported to date seem to indicate an increased risk of oxidative damage, with lipid peroxidation and, not always, decreases of the maximum potential PSII efficiency. In some cases, antioxidant enzymatic activities are increased under excess Zn.

21.4. PHOTOSYNTHETIC RESPONSES TO TOXICITIES OF NON-ESSENTIAL ELEMENTS

21.4.1. Cadmium toxicity

Sugar beet plants grown in the presence of cadmium (Cd) show toxic effects at leaf Cd concentrations of approximately 400 $\mu g\ g^{-1}$ dry weight (Larbi *et al.*, 2002). Physiological effects of Cd toxicity in plants include major reductions in growth (Huang *et al.*, 1974; Horler *et al.*, 1980; Larbi *et al.*, 2002). Leaf yellowing has also been reported (Carlson *et al.*, 1975; Horler *et al.*, 1980; Smeets *et al.*, 2005) with 200 μg Cd g^{-1} dry weight, attributed to a Cd-induced Fe deficiency (Larbi *et al.*, 2002). Chlorosis was reflected in major decreases in photosynthetic pigments (Larbi *et al.*, 2002) and disappearance of granal stacks (Baszynski *et al.*, 1980; Geiken *et al.*, 1998). Growth reductions could also be a consequence of Cd interference with processes such as photosynthesis and translocation of photosynthetic products (discussed by He *et al.*, 2008).

Cadmium treatments markedly decrease photosynthesis (Huang *et al.*, 1974; Greger and Ögren, 1991; Larbi *et al.*, 2002), g_s (Schlegel *et al.*, 1987) and transpiration (Carlson *et al.*, 1975; Larbi *et al.*, 2002). Stomatal opening, measured on epidermal strips, decreases with Cd (Carlson *et al.*, 1975). However, there are works showing increased rates of transpiration under Cd toxicity (Greger and Johansson, 1992) or no changes in g_s (Di Cagno *et al.*, 2001), despite an increase of defective and undeveloped stomata, which were ascribed to an increased cuticular transpiration (Greger and Johansson, 1992). Other reports state that Cd affects photosynthesis by decreasing the content of CA (Lee *et al.*, 1976). Research has concentrated efforts on the inhibitory effect of Cd on the level of some Calvin-cycle enzymes, the greatest inhibition being observed in Rubisco (discussed by Krupa *et al.*, 1993; Di Cagno *et al.*, 2001). Sigfridsson *et al.* (2004) localised different Cd-binding sites in PSII. Chl-F measurements revealed Cd-mediated decreases in ϕ_{PSII} (Krupa *et al.*, 1993; Di Cagno *et al.*, 2001; Larbi *et al.*, 2002), owing to decreases of intrinsic PSII efficiency (Krupa *et al.*, 1993; Larbi *et al.*, 2002) and qP (Krupa *et al.*, 1993; Di Cagno *et al.*, 2001). As a consequence, the electron transport rate decreased with Cd toxicity (recalculated from Larbi *et al.*, 2002, using absorptance values reported by Morales *et al.* (1991) for comparable chl. contents). The J_f/A_N ratio increased from approximately 11–15 to 11–21 (recalculated from Larbi *et al.*, 2002 and He *et al.*, 2008), which suggests that the potential risk of oxidative damage has not increased much from a physiological point of view. In fact, ascorbate concentration (Di Cagno *et al.*, 2001), CAT (Chaoui *et al.*, 1997; León *et al.*, 2002) and SOD (León *et al.*, 2002) activities were shown to decrease under Cd toxicity. APX, however, increased in sunflower (Di Cagno *et al.*, 2001) and bean (Smeets *et al.*, 2005) affected by Cd.

Cadmium toxicity increases leaf reflectance (Horler *et al.*, 1980). Increases in ratios of protective carotenoids to chl. and de-epoxidation of xanthophyll-cycle pigments under Cd toxicity (Larbi *et al.*, 2002) were similar to that reported for Fe-deficient plants with comparable chl. concentrations (see Section 21.3.1). A higher amount of de-epoxidated xanthophylls may be attributed to the diminished zeaxanthin-epoxidase activity induced by Cd (Latowski *et al.*, 2005). These changes were accompanied by increases in D from 22–36 to 33–65% (recalculated from Krupa *et al.*, 1993, Di Cagno *et al.*, 2001 and Larbi *et al.*, 2002). *P* decreased from 65 to 35%, whereas X remained at 5% in sugar beet plants grown in presence of Cd (recalculated from Larbi *et al.*, 2002). NPQ also increased (Krupa *et al.*, 1993; Di Cagno *et al.*, 2001; Larbi *et al.*, 2002). Dark respiration increased in protoplasts isolated from Cd-treated sugar beet plants (Greger and Ögren, 1991). Photorespiration is unlikely to contribute to photoprotection, because CAT activity decreases in plants grown with Cd (Chaoui *et al.*, 1997; León *et al.*, 2002). In pepper plants, the activity of two key photorespiratory enzymes, glycolate oxidase and hydroxypyruvate reductase, was either unchanged or decreased in response to Cd (León *et al.*, 2002). In pea, the photorespiration rate was not affected in Cd-poisoned plants (McCarthy *et al.*, 2001). However, Weigel (1985) used isolated chloroplast to postulate a shift of Rubisco more towards its oxygenase function under Cd toxicity.

Cadmium is thought to cause oxidative damage. Cadmium directly inhibits chl. biosynthesis in barley (Stobart *et al.*, 1985) in short-term experiments with leaf segments. Chl. and carotenoids were observed to decrease in Cd-treated plants (Krupa *et al.*, 1993; Di Cagno *et al.*, 2001; He *et al.*, 2008). The chl. *a*/*b* ratio decreased under conditions of Cd stress in pea and rice (Horler *et al.*, 1980; He *et al.*, 2008), suggesting that Cd is more toxic for chl. *a* than for chl. *b*. However, it increased under Cd toxicity in sunflower (Di Cagno *et al.*, 2001) and sugar beet (Larbi *et al.*, 2002).

Lipid peroxidation and protein-oxidation products (Chaoui *et al.*, 1997; Di Cagno *et al.*, 2001; Smeets *et al.*, 2005) accumulated after Cd exposure, possibly owing to the action of O_2^- and H_2O_2 (Romero-Puertas *et al.*, 2004). In this sense, APX doubled its activity under Cd stress (Chaoui *et al.*, 1997), but this is not always the case (Di Cagno *et al.*, 2001). Greger and Ögren (1991) reported a decreased maximum quantum yield of CO_2 assimilation in sugar beet, whereas F_v/F_m ratios remained unaltered or were only slightly affected in several species (Greger and Ögren, 1991; Krupa *et al.*, 1993; Di Cagno *et al.*, 2001; Larbi *et al.*, 2002). When Cd-mediated F_v/F_m reductions were much larger, they were accompanied by damage to the OEC and changes in D1 turnover (Geiken *et al.*, 1998). However, it should be noted that these experiments were made by cutting petioles at their base and keeping petioles in solutions containing Cd. The low effects of Cd in F_v/F_m ratios of intact plants could be related to the subcellular localisation of the generated O_2^- and H_2O_2. Accumulation of O_2^- and H_2O_2 was observed mainly in the plasma membrane and tonoplast of different plant cell types but not in chloroplasts (Romero-Puertas *et al.*, 2004). It can be a result of the chloroplast capacity to remove ROS using an efficient antioxidant system (Romero-Puertas *et al.*, 2004).

In summary, Cd effects are concentration dependent, with low concentrations inducing Fe deficiency and high concentrations being toxic. Cd may affect photosynthesis through its effects on CO_2 stomatal conductance, with exceptions PSII photochemistry, the content of CA, Rubisco and, possibly, through inhibition of translocation of photosynthetic products (Table 21.1). A large part of the energy absorbed by PSII is safely dissipated as heat mediated by zeaxanthin (Table 21.2); the activity of the zeaxanthin epoxidase enzyme is inhibited in Cd-treated plants. Dark respiration is also increased. These photoprotection mechanisms do not seem to be enough, because lipid peroxidation and protein-oxidation products accumulate.

21.4.2. Lead toxicity

Root-growth inhibition and plant mortality have been reported for some species under excess lead (Pb) (Horler *et al.*, 1980). Pb has been reported to decrease shoot Fe, Mn, Cu and Zn (Cseh *et al.*, 2000; Larbi *et al.*, 2002). Also, Pb affects the water status of plants (Larbi *et al.*, 2002), possibly through effects on water channels (Cseh *et al.*, 2000).

In sugar beet, Pb has little effect on plant growth up to concentrations as high as 1–2 mM in the growing media and 500 μg Pb g^{-1} shoot dry weight (Larbi *et al.*, 2002), causing only moderate changes in pigment composition and photosynthesis-related parameters (Carlson *et al.*, 1975; Horler *et al.*, 1980; Larbi *et al.*, 2002). *P*, D and X were unaffected (recalculated from Larbi *et al.*, 2002). However, under other circumstances, Pb inhibits growth (Horler *et al.*, 1980) and photosynthesis (Huang *et al.*, 1974), in line with effects on stomatal opening of epidermal strips (Carlson *et al.*, 1975), isolated chloroplasts (Miles *et al.*, 1972) and mitochondria (Koeppe and Miller, 1970).

Excess Pb increases leaf reflectance (Horler *et al.*, 1980). Xanthophyll cycle and thermal dissipation were unaffected in Pb-stressed plants (Larbi *et al.*, 2002). The possible role of photorespiration as an alternative electron sink under Pb stress is uncertain and indirect, because CAT activities increased or decreased depending on ecotype (Liu *et al.*, 2008). SOD activities increased under excess Pb (Liu *et al.*, 2008).

Altered chl. biosynthesis has been reported in response to Pb (Prasad and Prasad, 1987). In other works, changes in chl. biosynthesis depend on ecotype (Liu *et al.*, 2008). Lipid peroxidation and damage to cell membranes, chloroplasts and mitochondria have been reported under excess Pb (Liu *et al.*, 2008). However, in sugar beet plants grown in the presence of Pb, the F_v/F_m ratio remained unchanged (Larbi *et al.*, 2002), which evidences absence of irreversible or permanent damages to PSII under excess Pb.

In summary, effects of Pb on photosynthesis depend on the experimental system used. Most of the effects reported to date have been observed with epidermal strips, isolated chloroplasts and mitochondria (Table 21.1). When intact plants are used photosynthesis is hardly affected, the consequence being that mechanisms of photoprotection are not elicited and PSII permanent damages are not observed (Table 21.2).

21.4.3. Aluminium toxicity

Cultivated species respond to aluminium (Al) with growth reductions (Moustakas *et al.*, 1995). Although thylakoid degradation (Haug, 1984) and concomitant intrathylakoid acidification (Moustakas *et al.*, 1995) were postulated as mediating mechanisms for Al photosynthesis inhibition, a detailed investigation of the mechanisms of Al inhibition of photosynthesis is still lacking. Al toxicity in plants is often characterised by symptoms resembling those of P (see Section 21.2.2) or Ca deficiency (Foy *et al.*, 1978). In wheat, growth, chl., photosynthesis and transpiration were negatively correlated with leaf Al concentrations (Ohki, 1986). Contrary to that, Al-mediated photosynthesis and PSI electron transport decreases in maize were observed without Al translocation to leaves (Lidon *et al.*, 1999), concluding that toxic effects were mostly indirect and owing to the concurrent inhibition of N, P and Fe translocation mechanisms. Recently, Ali *et al.* (2008) have reported decreases of the activity of CA under Al stress.

The maximum potential PSII deficiency showed either non-significant changes or slight changes under Al toxicity (Moustakas *et al.*, 1995; Lidon *et al.*, 1999). These data indicate that photoinhibitory damage to PSII is very limited. This could be possibly owing to some type of photoprotection mechanism plants develop. In that respect, Moustakas and Ouzonidou (1994) reported an increased NPQ in leaves of Al-stressed wheat. Also, the activities of CAT and SOD have been reported to increase under excess Al (Ali *et al.*, 2008). Recently, Jiang *et al.* (2008) have reported increased C_i values under Al stress that, together with the CAT increases (Ali *et al.*, 2008), would indicate a relatively high photorespiration (and/or respiration) in these plants.

The existence of oxidative damage in Al-stressed plants is unclear. In isolated chloroplasts and intact leaves, Lidon *et al.* (1999) reported an increased lipid peroxidation in

Al-stressed maize, although some antioxidant enzymes were also increased under the same experimental conditions. Chl., one of the main and first targets of oxidative damage, was reported to decrease in wheat (Ohki, 1986) and *Citrus grandis* (Jiang *et al.*, 2008), but not in cover crops (Vieira *et al.*, 2008) under Al toxicity.

In summary, little is known about the effects of Al on photosynthesis. In some cases, growth, chl., photosynthesis and transpiration decreases are correlated with leaf Al concentrations, whereas symptoms of P or Ca deficiency are observed in other species in presence of Al without translocation from roots to shoots (Table 21.1). Other Al effects reported in the literature include thylakoid degradation, intrathylakoid acidification and decreases of CA activity (Table 21.1). On the photoprotection side, increases of NPQ, and SOD and CAT activities have been reported (Table 21.2). On the photoinhibition side, it has been reported that Al mediated increases of lipid peroxidation, although effects on the F_v/F_m ratios are minor.

21.4.4. Mercury toxicity

Mercury (Hg) is a very toxic heavy metal even at low concentrations (Marschner, 1995). Documented critical levels in plant tissues range from 0.5–3 μg Hg g^{-1} dry weight (Dunagan *et al.*, 2007). Symptoms of Hg toxicity include chlorosis, leaf elongation, curling leaf edges, necrotic leaf tips and stunted growth (Dunagan *et al.*, 2007). Hg suppressed growth of both shoots and roots (Cho and Park, 2000). Turgid root tips are necessary to keep stomata open (Davies *et al.*, 1986). This is why some authors have ascribed Hg toxicity to indirect effects on roots (Godbold and Hüttermann, 1988), although direct effects on chl. (Prasad and Prasad, 1987) and protein biosynthesis (Beauford *et al.*, 1977) have been reported. Godbold and Hüttermann (1988) related photosynthesis and transpiration declines in spruce seedlings, associated to stomatal closure, to root damage leading to an impaired water and nutrients supply to the needles. It should be also mentioned that Hg ($HgCl_2$) acts as an aquaporin blocker, which may alter water relations in the whole plant and CO_2 transfer across plasma membranes (Terashima and Ono, 2002). Other authors localise the site of Hg action on the donor side of PSII and on P700, the chl. dimer in the PSI core (Sersen *et al.*, 1998).

Only some clues are known nowadays on the photoprotection responses of plants to Hg. Leaf reflectance at 550 nm increased when grown in the presence of Hg (Dunagan *et al.*, 2007). Leaf respiration remained unaffected or decreased in response to Hg (Godbold and Hüttermann, 1988). Substantial increases of SOD, CAT and peroxidase activities were observed in Hg-stressed plants (Cho and Park, 2000).

Photoprotection responses to Hg do not seem to be enough for avoiding oxidative damage. Chl. concentration was decreased in spruce needles (Godbold and Hüttermann, 1988) and tomato seedlings (Cho and Park, 2000). Chl. decreases may be the result of an enhanced production of ROS (particularly H_2O_2) and subsequent lipid peroxidation (Cho and Park, 2000), or direct chl.-biosynthesis inhibition (Prasad and Prasad, 1987).

In summary, Hg decreases photosynthesis and transpiration associated to stomatal closure and related to root damage (Table 21.1). ROS production is enhanced under Hg toxicity and, despite increases of antioxidant enzyme activities, chlorosis and lipid peroxidation have been reported in Hg-treated plants (Table 21.1).

21.5. CONCLUSIONS AND FUTURE RESEARCH

When plants are under nutrient stress, photosynthesis decreases for a variety of reasons (Table 21.1). At the same time, in most cases they cannot avoid continuing to gather sunlight. In some species, low-P leaves respond by actively avoiding direct radiation during the period of maximum solar intensities, placing leaves parallel to sunlight (Lauer *et al.*, 1989). This is an aspect poorly investigated for other nutrient stresses. In most cases, when performing measurements, researchers fix leaves perpendicularly to sunlight, not taking into account any possible paraheliotropic plant response. In other common nutrient stresses, such as N or Fe deficiency, the amount of chl. decreases, in turn decreasing leaf absorptance (Table 21.2). As a consequence of the imbalance between light absorption and energy utilisation, plants experience what the research community has called, in a wide sense, photoinhibition. Photoinhibition may reflect either photoprotection mechanisms or photodamage. Data reported to date seem to indicate that photoprotection mechanisms are far more important (Table 21.2). Some signs of photodamage are often found, but its relative importance depends on the nutrient or the heavy metal considered and the plant species experiencing stress.

Plants under stress generally dissipate a large part of the light absorbed by PSII as heat, a process mediated by the xanthophyll-cycle pigments and ΔpH. Increases in thermal-energy dissipation under nutrient or heavy metal stress are summarised in Table 21.2. This has been confirmed for N and Fe deficiency and for Cd and Cu excess. The mechanism remains largely unexplored for some of the stresses reported here.

One aspect that should be taken into account when investigating effects of nutrients on photosynthesis is the possible interactions between them. For instance, low Cu or excess Zn is often accompanied by increases in Mn, exceeding even that of the controls (Kriedemann and Anderson, 1988; Monnet *et al.*, 2001), but in other cases excess Zn induced Mn deficiency (Van Assche and Clijsters, 1986a). When planning experiments with low or excess nutrients, it should be taken into account the presence of other nutrients, which may mimic or counteract the possible response. Cd toxicity has been reported to reduce Fe concentration in several species (Larbi *et al.*, 2002) causing Fe deficiency. The inhibition of the water-splitting reaction disappears after Mn supply to Cd-treated plants (Baszynski *et al.*, 1980). Also, an excess supply of Fe to plants grown under excess Cu recovered almost completely the physiological (Ouzounidou *et al.*, 1998) and biochemical characteristics (Kumar *et al.*, 2008) of control plants. Recently, it has been reported that excess Fe supply is a promising method to recover the Cd-mediated damage in developing leaves of poplar plants (Solti *et al.*, 2008). An Indian saying appears to be true: 'poison kills poison and iron cuts iron' (Kumar *et al.*, 2008).

The spatial patterning of deficiency symptoms across leaves complicates biochemical analyses. In leaves with different pigmentation zones, such as those with interveinal chlorosis, it may be difficult to obtain a sample of thylakoids or chloroplasts representative of the leaf owing to the heterogeneous chloroplast populations. Green zones may account for the majority of the final pellet after several centrifugations, and these will have similar chloroplasts to those of control leaves. Conclusions obtained from isolated thylakoids or chloroplasts may differ substantially from those obtained with intact plants. For example, in whole plants toxic heavy metals must cross many biological membranes and cope with different antioxidant strategies in very different tissues, from the rhizosphere to the chloroplast, where it is expected to be toxic for different photosynthetic processes.

Changes in photosynthetic metabolism occurring when plant status gradually declines over days or weeks may well be different from those seen by incubating otherwise uniform chloroplasts in solutions containing different nutrient concentrations for some minutes (discussed for P by Brooks, 1986). Similar conclusions can be obtained for other macro- or micronutrients and for heavy metals.

22 • Photosynthetic responses to biotic stress

M. BARÓN, J. FLEXAS AND E.H. DELUCIA

22.1. INTRODUCTION TO PLANT BIOTIC STRESS

Agricultural and native plants are subject to a myriad of biotic stresses inflicted by other living organisms, from viruses to mammals, and many of these damaging agents affect photosynthesis, either by altering its underlying metabolism (primary photochemistry, electron transport, Calvin cycle) or gas diffusion, or by reducing photosynthetic leaf area. Pathogens (fungal, bacteria or viral agents) and animal pests causes on average, a 15% and 18% reduction in crop yield, respectively (Oerke and Dehne, 2004). Biotic stresses on plants have had enormous repercussions for humanity. For example, the potato blight (*Phytophthora infestans*) caused widespread famine in England, Ireland and Belgium, or the introduction of grape phylloxera (*Daktulosphaira vitifoliae* or *Phylloxera vastatrix*) from America in the mid 19th century nearly put an end to the French wine industry. In addition, it is predicted that plant-pathogen interactions will favour pathogens under the high CO_2 conditions expected for the next decades (Lake and Wade, 2009).

This chapter reviews the effects of parasitic plants, pathogens (virus, bacteria and fungi) and arthropods on photosynthesis. While competition among plants clearly is a 'biotic interaction', we have not included it in this discussion as competition often manifests itself through reduced availability of growth-limiting resources and does not affect photosynthesis *per se*. The underlying mechanisms by which biotic agents affect photosynthesis vary widely and it may be useful to classify these agents into different 'damage guilds' (Table 22.1), although it has been claimed that biotic stresses generally downregulate photosynthesis genes (Bilgin *et al.*, 2010). Chewing insects, for example, reduce carbon gain primarily by reducing leaf area, where virus infections rarely reduce leaf area but instead depress the rate of photosynthesis per unit leaf area. Among fungi, leaf-disease fungi (rust, mildew, etc.) behave like herbivores in their ability to reduce photosynthetic leaf area, where vascular-wilt fungi compromise plant water transport and reduce photosynthesis by inducing stomatal closure. In addition to exploring how different biotic agents affect photosynthesis, this chapter concludes with a brief discussion of different methods for detecting the effects of biotic stress on photosynthesis, sometimes in advance of the development of visual symptoms.

22.2. EFFECTS OF PARASITIC PLANTS ON THEIR HOSTS' PHOTOSYNTHESIS

Parasitic plants depend on their hosts for carbohydrates and other resources, and while the mechanisms by which they affect photosynthesis in their hosts are not well understood, they have enormous potential to manipulate source-sink balance. This taxonomically diverse group represents about 1% of all flowering plants (~4500 species); they invade host tissues, above- or belowground, and remove resources via a specialised structure known as the haustorium (Watling and Press, 2001). The degree of dependence on the host varies, from holoparasitic species that have no capacity for independent photosynthesis, to hemiparasitic species that retain some capacity to fix carbon. There is some evidence that hemiparasitic and homoparasitic species affect photosynthesis in their hosts differently.

Holoparasitic species lack chl. and are totally dependent on the host for photosynthate (Watling and Press, 2001; Bungard, 2004). The loss of photosynthetic capacity seems to depend on several evolutionary steps towards holoparasitism. For instance, some gene losses (such as those related to chlororespiration) occurred in the early stages in the evolution of parasitism in *Cuscuta reflexa*, below the loss of photosynthetic capacity (Bungard, 2004). Other species, like *Lathraea clandestina*, have lost photosynthetic function but

Terrestrial Photosynthesis in a Changing Environment: A Molecular, Physiological and Ecological Approach, ed. J. Flexas, F. Loreto and H. Medrano. Published by Cambridge University Press.

Table 22.1. *A summary of the described effects of several biotic stresses on photosynthetic parameters. The relative importance of each factor is indicated (–, effect not described; X, effect described occasionally; XX, effect described several times; XXX, main effect). Notice that a factor not indicated as important may not necessarily mean that it is not, as the literature available is fragmentary and some of the effects have never been analysed under some of the biotic stresses. In the case of endophytic fungi, both positive (i.e., symbiotic) and negative stomatal and non-stomatal effects have been described, depending on the fungus-host combination and/or environmental conditions.*

Biotic stress	Defoliation, leaf destruction/ necrosis*	Stomatal limitations		Non-stomatal limitations			
	Importance	Importance	Importance	Structural effects*	Photosynthetic primary reactions	Carboxylation/ Calvin-cycle reactions	Carbohydrate metabolism/ transport
Virus	—	—	XXX	XXX	XXX	XX	XX
Bacteria	—	XX	XXX	X	XXX	X	X
Fungi – endophytic	—	X (+/–effects)	X (+/–effects)	—	X	—	X
Fungi – vascular wilt	X	XXX	X	X	X	X	—
Fungi – leaf disease	XXX	X	XXX	X	X	XX	XX
Nematodes	X	X	—	—	—	—	—

* Including leaf shape and size, leaf internal structure, and organelle form and distribution

produce functional Rubisco large subunits (RbcL), while *Epifagus virginiana* have lost all trace of the gene encoding the RbcL protein (*rbc*L gene) (Bungard, 2004). Because they depend on host plants for carbohydrates, it is likely that holoparasitic plants interact with host photosynthesis mostly through source-sink interactions.

The few studies of holoparasitic plants do not point to a common mechanism for reducing photosynthesis in their hosts. Biomass, particularly leaf mass of tomato plants parasitised by *Orobanche aegyptiaca*, is severely reduced (Barker *et al.*, 1996). In contrast, tobacco infected with *Orobanche cernua* maintained the same leaf area as uninfected plants and increased carbon allocation to roots by 77%, of which 73% was removed by the parasite (Hibberd *et al.*, 1998b). In addition, leaf senescence was delayed, which resulted in a 20% increase in canopy photosynthesis compared with uninfected plants (Hibberd *et al.*, 1998b). Sink stimulation of host photosynthesis was also observed in *Lupinus alba* infected with the stem holoparasite *Cuscuta reflexa* (Jeschke *et al.*, 1997). In this case and despite a reduction of host nitrogen, the effect was induced through increased leaf-area-based photosynthesis in the host, which was associated with increased chl. contents. Unlike other holoparastic plants, infection by the stem holoparasite *Pilostyles ingae* had no effect on photosynthesis of its host, *Mimosa naguirei* (Fernandes *et al.*, 1998).

Even thought hemiparasitic species contain chl. and can fix carbon autotrophically (Strong *et al.*, 2000), they are obligate parasites either because they pass through a holoparasitic stage during development or because they depend on the host for resources other than carbon, such as water and mineral nutrients (Watling and Press, 2001). Although some proportion of their carbon can be obtained from host plants, it is likely that the effect of hemiparasitic species on host photosynthesis does not occur through source-sink interactions. In fact, they often reduce rather than stimulate host photosynthesis.

Striga hermonthica, the most extensively studied hemiparasitic species, causes severe losses of yield in cereal crops such as maize and sorghum in the semi-arid tropics. The removal of resources (i.e., carbon and others) by the parasitic plant is only responsible for 20% of the observed decrease in host growth, while the other 80% is attributed to the impact of the parasite on host photosynthesis (Graves *et al.*, 1989), reducing its leaf area and leaf photosynthesis (Seel and Press, 1996; Watling and Press, 2001). The reduction in photosynthesis is achieved by parasite-induced stomatal closure, associated with increased ABA synthesis in the host (Taylor *et al.*, 1996; Frost *et al.*, 1997). That photosynthesis is inhibited by stomatal closure is reinforced by the observations that infected and non-infected plants have the same A_N/C_i responses (Frost *et al.*, 1997; but see Cechin and Press, 1993); in addition, increasing nitrogen availability, which typically stimulates photosynthetic capacity, had no effect on photosynthesis in infected plants (Gurney *et al.*, 1995). There is, however, one example of non-stomatal reduction in photosynthesis by a hemiparasite. The reduction in photosynthesis of the grass *Phleum bertolinii* by the facultative hemiparasitic plant *Rhinanthus minor* was associated with substantial reduction in chl. and Rubisco content (Cameron *et al.*, 2007).

Summarising, it is evident that many parasitic plants manipulate the photosynthetic performance of their hosts, the magnitude of this effect, as well as the underlying mechanisms, being highly variable. Increasing the span of parasitic – host systems studied, both holo- and hemiparasitic, would be necessary for a better understanding of the mechanisms leading to photosynthesis regulation in these interactive plant systems.

22.3. EFFECT OF VIRAL INFECTION ON A HOST'S PHOTOSYNTHESIS

Goodman *et al.* (1986) summarised changes in the photosynthetic behaviour of the host plants induced by virus, bacteria and fungi, showing many points of similarity. Early works demonstrated that leaf photosynthesis rates are reduced in host plants infected by a number of viral families (Table 22.2).

Viruses cause only small reductions of A_N in host plants at early stages of infection (He *et al.*, 2004; Rowland *et al.*, 2005), but at later stages A_N is often depressed by as much as 50–85% compared with non-infected plants (Smith and Neales, 1977; Balachandran *et al.*, 1997; Sampol *et al.*, 2003; Bertamini *et al.*, 2004; Rowland *et al.*, 2005). Gas-exchange and Chl-F analysis have revealed that virus-induced depressions of photosynthesis are caused almost exclusively by non-stomatal limitations (Table 22.1). Further evidence for the non-stomatal inhibition of photosynthesis by viruses is provided by the strong interaction between virus infection and plant nitrogen availability: the effects of viruses are pronounced when superimposed with nitrogen deficiency (Balachandran *et al.*, 1997; Sampol *et al.*, 2003). The nature of these non-stomatal limitations can be variable, possibly depending on factors such as the virus sp. or strain, the resistance of the host plant and the environmental variations.

Reductions in photosynthesis by viral infection are typically accompanied by chlorosis and structural changes in photosynthetic organelles (Goodman *et al.*, 1986), decreased mesophyll conductance to CO_2 (Sampol *et al.*, 2003), altered photochemistry and primary photosynthetic reactions (Balachandran *et al.*, 1997; Rahoutei *et al.*, 1999, 2000; Bertamini *et al.*, 2004; Pérez Bueno *et al.*, 2006; Sajnani *et al.*, 2007), inhibition of Rubisco and other photosynthetic enzymes (Balachandran *et al.*, 1997; Sampol *et al.*, 2003; Bertamini *et al.*, 2004) as well as inhibition of carbohydrate export (Shalitin and Wolf, 2000). Indeed, gene-expression studies show that virus infection induces downregulation of numerous genes involved in photon capture and thylakoid processes, CO_2 uptake and photosynthesis metabolism (Golem and Culver, 2003; Whitham *et al.*, 2003; Kokkinos *et al.*, 2006; Espinoza *et al.*, 2007).

22.3.1. Viral-induced changes on chloroplast ultrastructure and starch accumulation

Viral pathogens can affect chloroplast number, size, morphology and content, as well as the size and number of chloroplast inclusions (plastoglobuli, starch grains, etc). Almási *et al.* (2001) summarised chloroplast aberrations in virus-infected plants; they differ both qualitatively and quantitatively among host-pathogen systems. Alterations in both starch accumulation and metabolism appear as a common feature of pathogen infection.

Goodman *et al.* (1986) cited a number of early studies that reported starch accumulation during viral pathogenesis (Carroll, 1970; Conti *et al.*, 1972; Favali *et al.*, 1975; Tomlinson and Webb, 1978). Starch lesions or rings spots on inoculated leaves could appear before the visible disease symptoms (Cohen and Loebenstein, 1975). In marrow inoculated with the *Cucumber mosaic virus* (CMV), starch-accumulating cells display increased photosynthetic capacity relative to uninfected cells, or cells in which virus replication is actively occurring (Técsi *et al.*, 1996). During infection of a starch-depleted mutant line of *Arabidopsis thaliana* with *Turnip vein-clearing virus* (TVCV), CMV or *Cauliflower mosaic virus* (CaMV), Handford and Carr (2007) demonstrated that starch accumulation during infection is not required for successful viral infection; however, carbohydrate metabolism does influence symptom development.

Table 22.2. *Summary of the most representative studies on the effect of different viral families on the photosynthetic process.*

Family/Virus	Host plant	Effects on photosynthesis-related parameters	Reference(s)
Tobamovirus			
Tobacco mosaic virus (TMV)	Tobacco	Chloroplast malformations Loss of Chl-protein complexes OEC alterations PSII alterations Decreased PSII quantum yield Increased NPQ Altered Calvin cycle Starch accumulation BGF increase during HR Leaf temperature increase	Zaitlin and Jagendorf, 1960; Esau, 1968; Carroll, 1970; Carroll and Kosuge, 1989; Hodgson *et al.*, 1989; Koiwa *et al.*, 1989; Montalbini and Lupattelli, 1989; Chaerle *et al.*, 1999; van Kooten *et al.*, 1990; Balachandran *et al.*, 1997; Abbink *et al.*, 2002; Hlaváckoá *et al.*, 2002; Lehto *et al.*, 2003; Wilhelmová *et al.*, 2005; Chaerle *et al.*, 2007
	Cucumber	Swollen and deformed plastids Accumulation of osmiophilic plastoglobuli Starch lesions or ringspots Large and irregular starch grains	Cohen and Loebenstein, 1975
	Tomato	Loss of Chl-protein complexes	Koiwa *et al.*, 1992
	Arabidopsis thaliana	Repression of plastid-associated genes	Golem and Culver, 2003
Turnip vein-clearing virus (TVCV)	*Arabidopsis thaliana*	Altered starch accumulation	Handford and Carr, 2007
Pepper mild mottle virus (PMMoV)	*Nicotiana benthamiana*	Swollen and deformed plastids Accumulation of osmiophilic plastoglobuli Large and irregular starch grains OEC alterations PSII alterations Decreased PSII quantum yield BGF increase NPQ increase Leaf temperature increase Changes of the assimilatory potential of infected leaves	Rahoutei *et al.*, 1999, 2000; Pérez-Bueno *et al.*, 2004, 2006; Chaerle *et al.*, 2006; Pineda *et al.*, 2008 a,b; Sajnani *et al.*, 2007.
Calimovirus			
Cauliflower mosaic virus (CaMV)	Cabbage	Enlargement and accumulation of starch grains	Conti *et al.*, 1972
	Arabidopsis thaliana	Altered starch accumulation	Handford and Carr, 2007

Luteovirus			
Beet western yellow virus (BWYV)	Lettuce	Starch accumulation	Tomlinson and Webb, 1978
Barley yellow dwarf virus (BYDV)	Barley	Starch accumulation	Jensen, 1972
Cucumovirus			
Cucumber mosaic virus (CMV)	Cucumber	Reduced carbohydrate export from leaves	Shalitin and Wolf, 2000
	Marrow	Inhibition of lamellar development	Técsi *et al.*, 1996
		Starch lesions or ringspots	
	Tobacco	OEC alterations	Ehara and Misawa, 1975; Takahashi and Ehara, 1992; Roberts and Wood, 1982
		Inhibition of lamellar development	
	Arabidopsis thaliana	Repression of plastid-associated genes	Whitham *et al.*, 2003
		Altered starch accumulation	Handford and Carr, 2007
Hordeovirus			
Barley stripe mosaic virus (BSMV)	Barley	Cytoplasmatic invaginations	Carroll, 1970; MacMullen *et al.*, 1978
Tymovirus			
Turnip yellow mosaic tymovirus (TYMV)	Chinese cabbage	Cytoplasmatic invaginations	Goffeau and Bové, 1965; Matthews and Sarkar, 1976
		Inhibition of Hill reactions	
		Changes in the ratios Chl-F/BGF	
Potyvirus			
Potato virus Y (PVY)	Potato	Abundance of plastoglobuli	Schnablová *et al.*, 2005
Peanut green mosaic virus (PGMV)	Peanut	Inhibition of PSII electron transport	Naidu *et al.*, 1986
		OEC alterations	
Plum pox virus (PPV)	*Nicotiana benthamiana*	Repression of plastid-associated genes	Jiménez *et al.*, 2006; Dardick, 2007
		PSII alterations	
Sweet potato feathery mottle virus (SPFMV)	Sweet potato	Repression of plastid-associated genes	Kokkinos *et al.*, 2006
Turnip mosaic (TuMV)	*Arabidopsis thaliana*	Repression of plastid-associated genes	Whitham *et al.*, 2003; Yang *et al.*, 2007
Tobacco etch virus (TEV)	Tobacco	Inhibition of photosynthesis	Owen, 1957
Maize dwarf mosaic virus (MDMV)	Corn	Inhibition of photosynthesis	Tu and Ford, 1968

Table 22.2 (*cont.*)

Family/Virus	Host plant	Effects on photosynthesis-related parameters	Reference(s)
Geminivirus			
Tobacco leaf curl virus (TLCV)	*Eupatorium makinoi*	Loss of Chl-protein complexes	Funayama *et al.*, 1997a,b
Abutilon Mosaic Virus (AbMV)	*Abutilon striatum*	Impaired NPQ reflecting the state of symptom development	Osmond *et al.*, 1998; Lohaus *et al.*, 2000
Closterovirus			
Grapevine leaf roll associated virus (GLRaV)	Grapevine	Decreased mesophyll conductance to CO_2 Decreased chlorophyll content OEC alterations Decreased PSII quantum yield Decreased activity of Rubisco and nitrate reductase Repression of plastid-associated genes	Sampol *et al.*, 2003; Bertamini *et al.*, 2004; Espinoza *et al.*, 2007
Nepovirus			
Tomato ringspot nepovirus (ToRSV)	Tobacco	Inhibition of photosynthesis	Roberts and Corbett, 1965
Tomato ringspot nepovirus (ToRSV)	*Nicotiana benthamiana*	Repression of plastid-associated genes	Dardick, 2007
Grape fanleaf yellow-mosaic virus (GFYM)	Grapevine	Inhibition of CO_2 fixation	Pozsar *et al.*, 1969
Carmovirus			
Turnip crinkle- and saguaro cactus carmovirus	*Nicotiana benthamiana, Brassica pekinensis, Chenopodium amaranticolour*	Abundance of plastoglobuli	Russo and Martelli, 1982
Crinivirus			
Sweet potato chlorotic stunt virus (SPCSV)	Sweet potato	Repression of plastid-associated genes	Kokkinos *et al.*, 2006

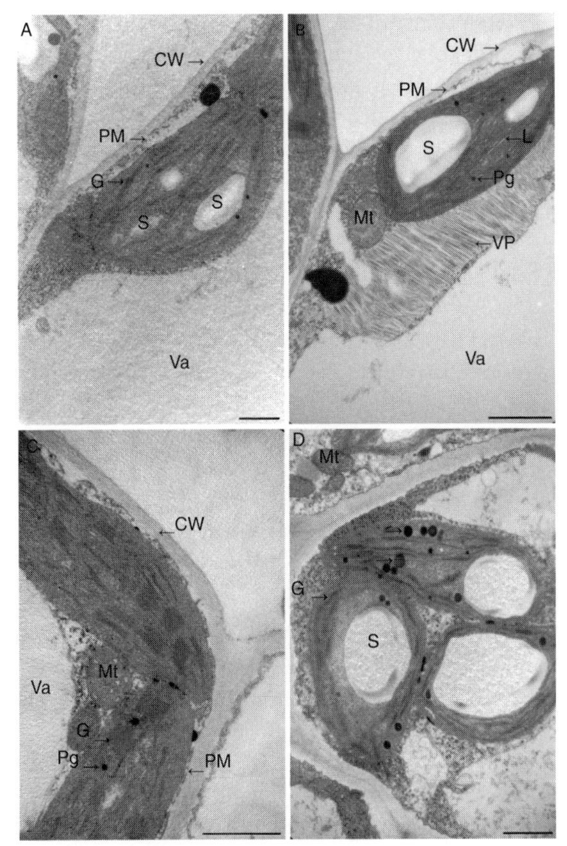

Fig. 22.1. Transmission electron micrographs of *N. benthamiana* chloroplasts. Control young leaves (**A**) and the corresponding symptomatic leaves of PMMoV-I infected plants at 7 dpi (**B**). Control old leaves (**C**) and the equivalent asymptomatic leaves of PMMoV-I infected plants at 17 dpi (**D**). C, chloroplast; CW, cell wall; L, stroma *lamellae* ; G, *grana*; M, mitochondrion; Pg, *plastoglobuli*; Pm, plasmatic membrane; S, starch grains; VP, virus particles; Va, vacuole. Scale, 1 μm. (Pictures by M.L. Pérez Bueno and N. Steffanie. R.Valcke´s lab. Hasselt University).

Starch accumulation also occurs in infected tissue in the absence of symptoms. Changes in *Pepper mild mottle virus* (PMMoV)-infected *N. benthamiana* cells from asymptomatic leaves (Pérez-Bueno *et al.*, 2006; Fig. 22.1), included the occurrence of large and irregular starch grains in swollen and deformed plastids, accumulation of osmiophilic plastoglobuli and disorganisation of chloroplast lamellar structures; these changes in chloroplast structure are similar to those detected in TMV-infected cucumber cotyledons (Cohen and Loebenstein, 1974), where the infection is restricted to the inoculated leaves forming starch lesions (Lindner *et al.*, 1959). Different metabolic processes, such as alterations in the Benson-Calvin cycle, modifications in the permeability of chloroplast membranes and disturbances of source-sink relationships were found to be correlated with starch accumulation (Wright *et al.*, 1995; Hull, 2002), but the precise responsible mechanisms remain unknown.

Chloroplasts of infected cells also show a wide range of alterations in their shape and inner structure, from swelling to complete rearrangement of the thylakoids (Esau, 1968; Koiwa *et al.*, 1992; Lehto *et al.*, 2003). The inhibition of lamellar development depends on when during development the leaf became infected (Ehara and Misawa, 1975; Roberts and Woods, 1982). The presence of vesicles within or attached to plastids was reported by various authors cited by Goodman *et al.* (1986) and Almási *et al.* (2001). In barley infected with *Barley stripe mosaic virus* (BSMV) (Carroll, 1970; McMullen *et al.*, 1978), chloroplasts became swollen, aggregated with parts of the cytoplasm trapped between adjacent organelles and contained cytoplasmatic invaginations. These changes resembled those observed in Chinese cabbage infected with *Turnip yellow mosaic tymovirus* (TYMV) (Matthews and Sarkar, 1976). In some cells, the vesicles were so large that they filled almost the entire space of the chloroplast, the membrane structure could become pushed to one side, producing a 'sickling effect'. It was proposed that these vesicles serve as sites of viral replication, but this is contested (Almási *et al.*, 2001).

Abundant plastoglobuli also occur in the swollen chloroplasts of infected leaves (Russo and Martelli, 1982; Schnablová *et al.*, 2005; Pérez-Bueno *et al.*, 2006). The formation of plastoglobuli is thought to be linked to the breakdown of thylakoids that accompanies senescence (del Río *et al.*, 1998). Thylakoid-membrane degradation by lipid peroxidation has been shown to take place at the final infection steps by PMMoV (Rahoutei *et al.*, 1999). Alternatively, the abundance of plastoglobules and disorganisation of chloroplast lamellar structures resemble those of chloroplasts from senescent leaves (del Río *et al.*, 1998; Almási *et al.*, 2001 and references therein). Recently, Bréhélin *et al.* (2007) suggested that plastoglobules participate in diverse secondary metabolism pathways and stress responses, and are not merely a 'passive storage' compartment. Lehto *et al.* (2003) assigned the chloroplast malformations in tobacco plants infected with the flavum strain of *Tobacco mosaic virus* (TMV) to deficient synthesis and assembly of thylakoid proteins; Koiwa *et al.* (1992) and Funayama *et al.* (1997b) suggested that the impact of infection on the chloroplast with tobamovirus or geminivirus, respectively, could be related to the loss of chl.-protein complexes, mainly from the LHC. Disturbances of other photosynthetic complexes, such as the OEC of PSII, during early infection stages in PMMoV-infected *N. benthamiana* plants have also been demonstrated (Rahoutei *et al.*, 2000; Pérez-Bueno *et al.*, 2004).

22.3.2. Photosynthetic electron transport during viral pathogenesis

Studies of several plant–virus systems have demonstrated lower photosynthetic electron transport rates in infected plants, essentially at the PSII level (Table 22.2). The involvement of specific viral gene products in the inhibition of the photosynthetic J has been proposed frequently (Hodgson *et al.*, 1989; Reinero and Beachy, 1989).

That the decreased rate of photosynthesis in viral-infected plants is not simply caused by reduced chl. content was already evident in early studies (Spikes and Stout, 1955). Photosynthetic phosphorylation and the Hill reaction were decreased in chloroplasts isolated from TMV-infected plants (Zaitlin and Jagendorf, 1960; Montalbini and Lupattelli, 1989), as well as in Chinese cabbage leaves infected with TYMV (Goffeau and Bové, 1965).

Studies of *peanut green mosaic virus* (PGMV)-infected peanut revealed that reduced electron transport rates was caused primarily by direct inhibition of PSII (Naidu *et al.*, 1986 and references therein). For some authors (Takahashi and Ehara, 1992) disturbances of the OEC protein pattern are associated to the primary molecular processes of symptom expression during infection. The inactivation of PSII associated to a decrease in PsbO (a 33-kDa extrinsic protein) was also described in grapevine plants infected with the *grapevine leafroll virus* (Bertamini *et al.*, 2004).

Analysing the chloroplast proteome and the transcript profile of chloroplast proteins in *Nicotiana benthamiana* infected with two strains of PMMoV (Spanish and Italian strain, PMMoV-S and I, respectively), it was established that the OEC is the main target of the tobamovirus. PMMMoV-infected plants displayed a decrease of OEC proteins (PsbO, PsbP and PsbQ), whose synthesis could be transcriptionally regulated during pathogenesis. The OEC proteins were found to be products of three multigene families in *N. benthamiana*. The PsbP family was differentially regulated during the infection process (Rahoutei *et al.*, 2000; Pérez-Bueno *et al.*, 2004). Changes on the TL characteristics of chloroplasts isolated from PMMoV-infected plants were analysed, concluding that the formation of the higher S states of the OEC was inhibited during infection with both strains of the virus. The simultaneous appearance of high-temperature TL bands is indicative of lipid peroxidation in the photosynthetic membranes (Rahoutei *et al.*, 1999).

In addition to viral-induced changes affecting OEC, measurements of Chl-F revealed that viral infection caused a decrease in PSII photochemical efficiency and an increase in NPQ (van Kooten *et al.*, 1990; Balachandran *et al.*, 1997; Rahoutei *et al.*, 2000; Hlaváckovà *et al.*, 2002; Pérez-Bueno *et al.*, 2004; Wilhelmová *et al.*, 2005). PSII quantum yield and NPQ are affected to a different extent depending upon the virus and the developmental and growing condition of the plant, as well as leaf age. PSI electron transport was only slightly affected during pathogenesis.

Funayama *et al.* (1997a,b) found higher chl. *a*/*b* ratios and a preferential loss of LHCII in a *geminivirus*-infected *Eupatorium makinoi*. They suggested that decreased PSII quantum yield was mainly owing to the decreased energy allocation to PSII, altering the energy distribution balance between PSII and PSI. Subsequent growth experiments revealed that the performance of infected *E. makinoi* plants varied largely with growth light environment (Funayama *et al.*, 1997b).

Recently, Dardick (2007) has studied the gene expression profiles of *N. benthamiana* leaves systemically infected with three different fruit-tree viruses (*Plum pox potyvirus*, PPV, *Tomatoringspot nepovirus*, ToRSV and *Prunus necrotic ringspot ilarvirus*, PNRSV). Consistent with the severity of the symptoms, repression of plastid-associated genes (mainly nuclear-encoded proteins involved in electron transport, light harvesting and Benson-Calvin cycle) was observed for both PPV and ToRSV, but not for PNRSV. Yang *et al.* (2007) carried out a spatial analysis of *Arabidopsis thaliana* gene expression in response to *Turnip mosaic virus* (TuMV) infection tagged with the GFP. Downregulated genes are those associated with chloroplast functions, sulphate utilisation or cell-wall expansion. The extent to which TuMV-responsive genes were up- or downregulated primarily correlated with the amount of virus accumulation regardless of gene function.

Summarising, virus-induced depressions of photosynthesis are caused mainly by non-stomatal limitations of different nature. Structural changes in photosynthetic organelles, altered photochemistry, inhibition of the activity from Rubisco and other photosynthetic enzymes, as well as inhibition of carbohydrate export and changes in source-sink relationships are evident in plants infected with different viral families.

22.4. IMPACT OF FUNGAL INFECTION ON A HOST'S PHOTOSYNTHESIS

In contrast to viruses, the effects of fungi on host photosynthesis are considerably more varied. To characterise common patterns and modes of action, we have divided fungi into three damage guilds (i.e., such division does not have any taxonomic or systematic basis, but rather it is based on the effects on host plants): endophytic fungi, vascular-wilt fungi and leaf-disease fungi (Table 22.3). Their effects on photosynthesis-related parameters are summarised in Table 22.3.

Endophytic fungi are often considered symbiotic as infected plants perform better than non-infected plants. Infection induces increases in host plant A_N, although this is not always the case. Vascular-wilt fungi include a number of fungi that grow mostly on xylem tissues of host plants, causing a decrease in hydraulic conductance that leads to severe wilting of the foliage. Reductions of A_N induced by wilt-disease fungi can be as large as 30–90% and typically caused by reduction in stomatal conductance (g_s). As in the case of limited soil-water availability (see Chapter 20), non-stomatal limitations to photosynthesis (Table 22.1) appear only secondarily, when the infection is severe and long-lasting. Leaf-disease fungi live in mesophyll cells and their effects on photosynthesis are well characterised. Leaf diseases largely reduce CO_2 assimilation of host plants mostly by reducing the photosynthesising leaf area (Table 22.1), but they also induce a decrease in A_N in the remaining green-leaf areas that in some cases can be as high as 30–50% (see references for the three types of fungi in Table 22.3).

22.4.1 Fungal-induced changes on chloroplast ultrastructure

Chloroplast degeneration is also associated with decreased photosynthetic rates in fungal-infected plants. Goodman *et al.* (1986) describes a general damage or ageing of the photosynthetic machinery in fungal-infected plants. However, infected leaves are heterogeneous, consisting of regions of cells directly invaded by the pathogen and regions remote from the fungal colony; consequently, alterations in photosynthesis are often spatially and temporally complex and depend upon the particular host/pathogen interaction (Scholes and Rolfe, 1996; Chou *et al.*, 2000; Swarbrick *et al.*, 2006). Rust and powdery mildew induce the formation of *green islands* at the infection site, which are photosynthetically active with a higher chl. concentration and starch accumulation than in the uninfected cells, while the rest of the leaf remains chlorotic. Chloroplasts in the green island of rust-infected bean leaves are functional but contain abundant peripheral reticula (Sziráki *et al.*, 1984). In rust diseases starch content varies along the infection, increasing during the sporulation time in the chloroplast of the host cells adjacent to the fungal hyphae. Donald and Strobel (1970) found that the activity of the ADP-glucose pyrophosphorylase was similar to the pattern of starch accumulation, and was almost the inverse of the variation observed in inorganic phosphate in diseased leaves during the infection process.

22.4.2. Fungal-induced alterations on photochemistry and related processes

Fungal infections have been shown to induce alterations of leaf photochemistry. For instance, the decline in the rate of photosynthesis and the corresponding increase in NPQ during infection of *Arabidopsis thaliana* leaves with *Albugo candida* (Chou *et al.*, 2000) was restricted closely to invaded regions of the leaf. Changes in NPQ were interpreted as evidence of a greater reduction in the activity of the Calvin cycle relative to other components of the photosynthetic apparatus, and were associated with a lower amount of mRNA encoding the small Rubisco subunit (RbcS). The expression of *cab* (chl. *a/b*-binding protein) genes was also repressed during fungal pathogenesis. Soluble carbohydrates accumulated in the infected region, whereas the amount of starch declined. The reverse was seen in uninfected regions of the infected leaf. Aldea *et al.* (2006b) has summarised the effect on actual PSII efficiency (ϕ_{PSII}) of different biotic agents, including fungal infection (*Phyllosticta, Cercospora, Gymnosporangium*), on leaf tissues adjacent to the site of direct damage.

Phaeomoniella chlamydospora, *Phaeoacremonium aleophilum* and *Fomitiporia mediterranea* cause *esca* disease, a grapevine trunk disease (Petit *et al.*, 2006). Christen *et al.* (2007) compared the photosynthetic responses to *esca* and to drought stress; although both stresses modified the PSII performance, there was a differentiated functional pattern of PSII for the two stress types. Furthermore, infection of *Solanum tuberosum* with *Phytophora infestans* was shown to induce the reduction of the standard Chl-F parameters (F_v/F_m, F_v, F_m: maximum PSII efficiency, variable chl. fluorescence, maximum chl. fluorescence), as well as increase of qP (photochemical quenching of Chl-F) at an early stage of

Table 22.3. *Summary of the most representative studies on the effect of fungi on host photosynthesis, considering different damage guilds.*

Fungi	Host plant	Effects on photosynthesis-related parameters	References(s)
Leaf-disease fungi			
Uncinula necator, Erysiphe graminis, Plasmopara viticola	Grapevine	Reduction of CO_2 assimilation	Shtienberg, 1992; Scholes *et al.*, 1994; Moriondo *et al.*, 2005; Petit *et al.*, 2006; Agati *et al.*, 2008
Puccinia triticina	Wheat		Shtienberg, 1992; Robert *et al.*, 2005
Alternaria spp., *Blumeriella* spp., *Septoria* spp.	Several species		Niederleitner and Knoppik, 1997; Robert *et al.*, 2005; Roloff *et al.*, 2004
Phaeocryptopus gaeumannii	Douglas fir		Manter and Kavanagh, 2003
M. nubilosa, M. cryptica	Eucalyptus		Pinkard and Mohammed, 2006
Puccinia coronata	Oat	Alteration in several photosynthetic parameters in both infected and non-infected tissues	Scholes and Rolfe, 1996
Albugo candida	*Arabidopsis thaliana*	Alterations in phochemistry (NPQ), down-regulation of cab and *Calvin*-cycle genes	Chou *et al.*, 2000
Phytophthora ramorum	Tobacco SR1, *Rhododendron macrophyllum, Lithocarpus densiflorus, Umbellularia californica*	Disturbances on PSII function	Manter *et al.*, 2007
Phytophora infestans	*Solanum tuberosum*		Koch *et al.*, 1994
Blumeria graminis (during HR)	Barley	Alteration in source-sink relationships and carbon utilisation, changes on invertase activity, down-regulation of Calvin-cycle genes	Swarbrick *et al.*, 2006
Phytophtora nicotianae (during HR)	Tobacco	Stomatal closure, PSI inhibition, inactivation of Calvin cycle	Scharte *et al.*, 2005
Endophytic fungi			
Acremonium coenophialum,	*Festuca arundinacea*	*Changes* on A_N (light-saturated net photosynthesis)	Marks and Clay, 1996

Table 22.3. (*cont.*)

Fungi	Host plant	Effects on photosynthesis-related parameters	References(s)
Vascular-wilt fungi			
Fusarium oxysporum f. Sp. lycopersici	Tomato	A_N reduction by decreased stomatal conductance	Duniway and Slatyer, 1971; Lorenzini *et al.*, 1997; Nogués *et al.*, 2002
Verticillium dahliae	Potato		Bowden *et al.*, 1990; Haverkort *et al.*, 1990; Bowden and Rouse, 1991; Gent *et al.*, 1995, 1999; Saeed *et al.*, 1999; Goicoechea *et al.*, 2001; Rotenberg *et al.*, 2004
Phytophtora capsici	Pepper		Aguirreolea *et al.*, 1995
Phaeomoniella chlamydospora	Grapevine		Edwards *et al.*, 2007a,b

the disease, indicating a disturbance of PSII function (Koch *et al.*, 1994).

Comparatively little is known about the consequences for photosynthetic metabolism of activating resistance responses in plants challenged with pathogenic fungi. Swarbrick *et al.* (2006) studied barley leaves infected with *Blumeria graminis*. During resistance, photosynthesis was most severely inhibited in cells directly associated with attempted penetration of the fungus but also in surrounding cells as a result of cell death but also to an alteration in source-sink relationships and carbon utilisation. Invertase activity increased more rapidly and to a much greater extent than in infected susceptible leaves and was accompanied by an accumulation of hexoses and downregulation in the expression of *rbcS* and *cab* genes (to a lesser extent than in a compatible interaction).

Scharte *et al.* (2005) found in source leaves of *Nicotiana tabacum* infected with *Phytophthora nicotianae* that hypersensitive cell death (HR) did not appear until photosynthesis completely declined. The decline in assimilation occurs in two steps: first by stomatal closure and later by inhibition of electron donation to PSI, which prevents H_2O_2 release at PSI (Mehler reaction), and kept the stroma in an oxidised state inactivating the Calvin cycle. Moderate photoinhibition of PSII at the infection site could be a consequence, rather than a primary cause for restricted electron transport.

Beech seedlings infected with the root-rot pathogen *Phytophthora citricola* showed a decrease in the rate of A_N in the very early infection steps, indicating the involvement of a mobile signal from the root; later in infection, PSII electron quantum yield, leaf-water potential and total water consumption were slightly impaired and wilt symptoms occurred (Fleischmann *et al.*, 2005).

Giving the main points of the varied effects of fungi on host photosynthesis, it is evident that leaf-disease fungi (rust, mildew, etc.) reduce the photosynthetic leaf area and consequently the CO_2 assimilation. Although these fungi induce photosynthetic alterations spatially and temporally and depending upon the particular host/pathogen interaction, repression of photosynthetic genes, changes in chloroplast structure and starch metabolism as well as altered J were common to different host plants. In contrast, wilt fungi compromise plant water transport and reduce photosynthesis by inducing stomatal closure, and endophytic fungi, often considered symbiotic, could increase the photosynthetic performance of the host plant during infection.

22.5. IMPACT OF BACTERIAL CHALLENGE ON PHOTOSYNTHESIS

22.5.1. The effect of pathogenic bacteria on gas exchange and CO_2 assimilation

The nature of photosynthetic limitations imposed by bacteria are variable (Table 22.1, Table 22.4), although probably non-stomatal effects, mostly associated with leaf chlorosis,

Table 22.4. *Summary of the most representative studies on the effect of different pathogenic bacteria on the photosynthetic process.*

Pathogenic bacteria	Host plant	Effects on photosynthesis-related parameters	References(s)
Xanthomonas			
Xanthomonas campestris pv. *phaseoli*	Bean	Decreased in chl. content	Berova *et al.*, 2007
Xanthomonas campestris pv. *pelargonii*	Geranium	Decreased in CO_2 export from source leaves	Jiao *et al.*, 1999
Xanthomonas axonopodis pv. *manihotis*	Cassava	Downregulation of photosynthesis genes	López *et al.*, 2005
Pseudomonas			
Pseudomonas syringae pv. *tomato*	Tomato	Downregulation of photosynthetic genes	Berger *et al.*, 2004
Pseudomonas syringae pv. *glycinea*	Soybean		Zou *et al.*, 2005
Pseudomonas syringae pv. *tomato*	*Arabidopsis thaliana*	Decreased F_v/F_m, ϕ_{PSII} and NPQ	Bonfig *et al.*, 2006; Berger *et al.*, 2007
Pseudomonas syringae pv. *phaseolicola* and pv. tomato	Bean	Alterations in NPQ pattern	Rodriguez-Moreno *et al.*, 2008
Pseudomonas syringae pv. *syringae*	Tobacco	Change of ferredoxin levels	Huang *et al.*, 2007
Xylella			
Xylella fastidiosa	Sweet orange	Lower leaf-water potential, carboxylation efficiency and stomatal conductance, impairment of the biochemical reactions of photosynthesis, downregulation of genes encoding LHCII, RuBisco activase and PsaO	Habermann *et al.*, 2003a,b; Ribeiro *et al.*, 2003a,b; de Souza *et al.*, 2007
Xylella fastidiosa + temperature stress	*Parthenocissus quinquefolia*	Lower CO_2 assimilation rates and stomatal conductance	Hopkins, 1989; Ribeiro *et al.*, 2004
Xylella fastidiosa + water stress		Reduced photosynthesis by stomatal and not stomatal limitations (depending on the disease and drought severity)	McElrone *et al.*, 2003; McElrone and Forseth, 2004
Erwinia			
Erwinia carotovora ssp. *carotovora*	Potato	Decreased mRNA levels of PsaD, accumulation of hydrogen peroxide in chloroplasts	Montesano *et al.*, 2004
	Tobacco	Change of ferredoxin levels	Huang *et al.*, 2007

predominate (Goodman *et al.*, 1986). Different strains of *Xanthomonas campestris* produce almost exclusively non-stomatal limitations on photosynthesis. Nevertheless, the specific limitations may be different: *X. campestris pv. phaseoli* caused strong reductions in chl. content in beans, whereas *X. campestris pv. pelargonii* did not affect chl. concentration, but largely decreased carbon export from leaves. *X. axonopodis pv. manihotis* was shown to induce downregulation of photosynthetic genes in cassava, while genes against oxidative stress were upregulated (López *et al.*, 2005).

In the case of *Xylella fastidiosa*, causing Pierce's disease of grape and leaf scorch of different plants, little differences in hydraulic conductivity and g_s were observed between healthy and infected plants in the absence of additional stress (Table 22.4).

Another disease induced by *Xylella fastidiosa* is the citrus variegated chlorosis (CVC). Diseased sweet orange (*Citrus sinensis* cv. Pera) plants showed stomatal disfunction as well as lower carboxylation efficiency (Habermann *et al.*, 2003b), related with impairment of the biochemical reactions of photosynthesis (Ribeiro *et al.*, 2004). The pathogenicity of *X. fastidiosa* was aggravated by the occurrence of additional stresses (Hopkins, 1989; Ribeiro *et al.*, 2004).

Bacterial infection has been shown to reduce plant CO_2 assimilation. The magnitude of the effect depends on the severity and timing of infection, but also on the particular type of bacteria and on genotype-associated resistance of the host plant (Jiao *et al.*, 1999; McElrone and Forseth, 2004; Berova *et al.*, 2007).

In addition, a potentially synergistic effect of CO_2 with bacterial infection was investigated in geranium (*Pelargonium × domesticum, 'ScarletOrbit Improved'*) plants infected with *Xanthomonas campestris pv. pelargonii.* High CO_2 lowered the bacterial number in infected leaves, the reductions in photosynthesis and export of photoassimilates from 'source leaves' being greater than at ambient CO_2 (Jiao *et al.*, 1999).

Despite the lack of proper gas-exchange analysis in plants infected with *Pseudomonas syringae*, it can be deduced that it affects photosynthesis by both stomatal and non-stomatal limitations. Stomatal closure is part of a plant innate immune response to restrict bacterial invasion. To circumvent this innate immune response, plant pathogenic bacteria have evolved specific virulence factors to induce stomatal reopening (Melotto *et al.*, 2006). In view of the interaction between bacterial infection and high CO_2, as well as the importance of stomatal regulation on the plant innate immune response to the pathogen, it is likely that in a future secenario of climatic change the impact of bacterial infection on plant photosynthesis will be lower. Nevertheless, this interaction needs further studies.

22.5.2. Bacterial-induced alterations of the PET in the host plant

Bacterial infection strongly affects the photochemical steps of photosynthesis. In fact, downregulation of genes encoding photosynthetic functions (Tao *et al.*, 2003; Zou *et al.*, 2005; Truman *et al.*, 2006) as well as changes in PSII proteins (Jones *et al.*, 2006) have been reported in *P. syringae*-infected plants. Bacterial effects on fluorescence parameters have been associated with a decrease in ϕ_{PSII}, as well as an increase in NPQ in leaf areas infiltrated with an avirulent *P. syringae pv. glycinea* inducing the HR, although little effect was observed during a compatible interaction (Zou *et al.*, 2005). In contrast, Bonfig and collaborators (2006) found that maximum F_v/F_m, ϕ_{PSII} and NPQ decreased in *Arabidopsis* plants infected with either virulent or avirulent *P. syringae pv. tomato.* A novel combination of Chl-F imaging (Chl-FI) and statistical analysis (Matouš *et al.*, 2006) used to study *P. syringae* infection of *Arabidopsis* leaves resolved very early and late phases of the plant response to infection, allowing for pathogen detection before the appearance of visual symptoms (Berger *et al.*, 2007). Rodríguez-Moreno *et al.* (2008) could differentiate a compatible from an incompatible plant-bacteria interaction in asymptomatic leaf tissues from *Phaseolus vulgaris* plants inoculated with either *P. syringae* pv. *phaseolicola* or *P. syringae* pv. *tomato* by Chl-FI analysis. A decrease in NPQ, evident in both infiltrated and non-infiltrated leaf areas, was observed in Pph-infected plants as compared with corresponding values from controls and Pto-infected plants.

De Souza *et al.* (2007) found that transcripts of LHCII, RuBisco activase (Rca) and PsaO were downregulated in sweet orange plants exhibiting CVC symptoms; in contrast, the phosphoribulokinase of the Calvin cycle was upregulated. Queiroz-Voltan and Paradela-Filho (1999) found chloroplasts totally damaged in chlorotic regions present in CVC-symptomatic leaves. Therefore the downregulation of these photosynthesis-associated genes is perhaps a consequence of disorders that occur in the photosynthetic apparatus.

A downregulation of PSI has been demonstrated in *Solanum tuberosum* treated with *Erwinia carotovora* ssp. *carotovora*. The expression of the *psaD*, a nuclear gene encoding the PsaD subunit of PSI, was downregulated and this correlated with an accumulation of H_2O_2 in chloroplasts (Montesano *et al.*, 2004). The levels of another PSI protein, ferredoxin (Fd), seem to play an important role in plant defence against bacterial infection (Huang *et al.*, 2007).

Summarising, while the changes caused by infection with biotrophic fungi and viruses on the photosynthetic machinery are the best understood, more research is needed to elucidate the interaction with virulent as well as avirulent bacterial strains. The nature of photosynthetic limitations imposed by bacteria is variable, although probably non-stomatal effects predominate. Studies on the differential gene expression during HR versus susceptible interaction show that bacterial challenge has a strong impact on photosynthesis through the downregulation of genes encoding photosynthetic functions. Bacterial effects on Chl-F parameters of the host plant have been associated with an inhibition of the photosynthetic $\mathcal{J}$ and changes on the non-photochemical processes of energy dissipation. Regarding bacterial-induced stomatal limitations of the photosynthetic process, the importance of stomatal regulation on the plant innate immune response to a pathogen is a promising research field for the future.

22.6. THE INFLUENCE OF NEMATODE INFESTATION ON THE HOST PHOTOSYNTHETIC METABOLISM

Infection with parasitic nematodes causes decreased transpiration and photosynthesis (Fatemy *et al.*, 1985; Postuka *et al.*, 1986; Melakeberhan *et al.*, 1990; Schans and Arntzen, 1991; Asmus and Ferraz, 2002), and associated decreases in leaf chl. content (Siddiqui and Mahmood, 1999) and plant growth (Decker, 1969; Melakeberhan *et al.*, 1990; Schans and Arntzen, 1991). The effect of the nematodes on physiology and morphology of the host increases with the duration and level of infection.

Soybean plants inoculated with the soybean cyst nematode *Heterodera glycines* showed a marked reduction in photosynthetic rate and chl. content, evident as leaf yellowing (Asmus and Ferraz, 2002). The reduced photosynthetic activity was primarily related to a lesser amount of nutrients, particularly nitrogen, either absorbed or translocated by the infected roots (Koenning and Barker, 1995). These results agree with those found for other plant-nematode interactions, such as *Meloidogyne incognita* Chitwood on vine (*Vitis vinifera* L.) varieties (Melakeberhan *et al.*, 1990) and *Globodera pallida* on potato (*Solanum tuberosum* L.) varieties (Schans and Arntzen, 1991). Alternatively, Bird (1974) hypothesised that the reduced photosynthetic rate could be related to partial closure of the stomata caused by water stress owing to the nematode-damaged roots.

Schmitz *et al.* (2006) examined the effect of infection of sugar beet leaves by *Heterodera schachtii* using laser-induced and pulse-amplitude-modulated (PAM) Chl-F. Sugar beet plants initially responded to *H. schachtii* infestation with a decrease in photosynthetic rate and later with a reduction in nitrogen uptake and chl. concentration.

Mazzafera *et al.* (2004) studied CO_2 fixation and photoassimilate partition in coffee (*Coffea arabica*) seedlings infested with the lesion nematode *Pratylenchus coffeae,* exposing the plants to $^{14}CO_2$. At the highest level of infestation, the carbon fixation in the leaves and partitioning to the roots were decreased.

Potato early dying (PED) is a vascular-wilt disease caused primarily by the fungus *Verticillium dahlia* (Rotenberg *et al.*, 2004). The lesion nematode *Pratylenchus penetrans* interacts synergistically with the fungus to enhance the development of visual PED symptoms and reduce photosynthesis and CO_2 exchange rates in plants co-infected with both pathogens. At early stages of infection, only a small decrease in photosynthetic rate was observed in diseased leaves, with no evidence of non-stomatal limitation to photosynthesis. Later, the joint infection seems to affect the Rubisco activity (Saeed *et al.*, 1999).

In contrast to the number of studies on pathogens and herbivores, there is a lack of studies on the impact of nematodes on photosynthesis. This makes it difficult to define their general mechanism of action on the chloroplast of the host plant. The effects of lesion nematodes on the carbon assimilation and partition are quite distinct from those with root-knot nematodes. In contrast to these nematodes, where the feeding sites are regarded as strong metabolic sinks, even leading to an increase of photosynthesis in the beginning of infestation, the root-lesion nematodes caused a rapid detrimental effect on carbon fixation and photoassimilate distribution in the plant owing to direct damage of the roots.

22.7. EFFECTS OF ARTHROPOD HERBIVORY ON PHOTOSYNTHESIS

Any herbivorous animal will obviously affect plant photosynthesis by leaf removal. Insect herbivory, in addition to inducing a general defoliation or feeding on specific tissues (e.g., phloem or xylem), triggers a complex and interacting array of molecular and physiological responses in plants (Fig. 22.2). This particular aspect, how insect herbivory affects plant photosynthesis in the remaing leaf, will be the scope of this subchapter. These responses potentially reduce the photosynthetic capacity in remaining leaf tissues to a greater extent than the direct removal of photosynthetic surface area. For example, the removal of only 5% of the area of an individual wild parsnip leaf by caterpillars reduced photosynthesis by 20% in the remaining foliage (Zangerl *et al.*, 2002; Nabity *et al.*, 2009). The mechanisms reducing photosynthesis in remaining leaf tissues are multifaceted, ranging from disruptions in fluid or nutrient transport to self-inflicted reductions in metabolic processes.

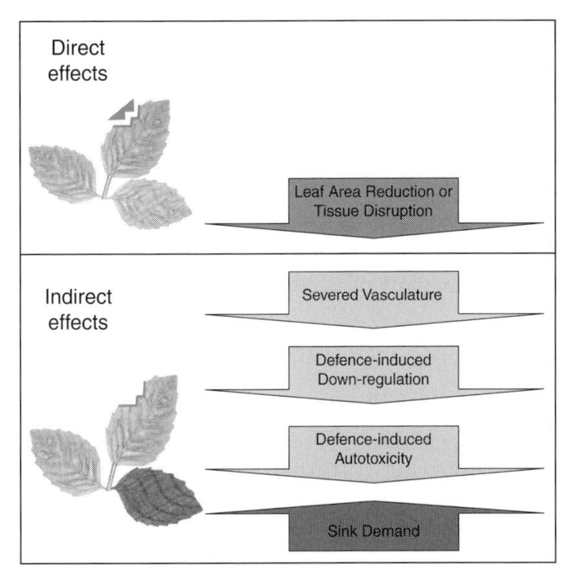

Fig. 22.2. Conceptual model of the direct effect of herbivory (removal of leaf area) and the indirect effects of herbivore damage to foliage on photosynthesis in the remaining leaf tissues.

Plant responses to arthropod herbivory traditionally have been assessed from the guild perspective, where different insect guilds are defined by their feeding mechanisms (Welter, 1989; Peterson, 2001). These guilds (e.g., chewing damage, piercing damage, etc.) were established in an effort to recognise 'homogeneity in physiological response' between different attacking agents (arthropods) that alter plant physiological processes in a similar manner. Welter (1989) examined an extensive body of literature across multiple guilds and found over 50% of all plant–insect interactions resulted in a loss of photosynthetic capacity. Defoliation generally increases photosynthesis of the remaining leaves, whereas specialised cell-content feeding decreases photosynthesis. Since then, several studies have examined plant responses to different insect-feeding guilds and even to different insects within guilds in an effort to develop models for predicting plant response to different feeding mechanisms. A brief review of the recent literature is not entirely consistent with the conclusions stated by Welter (1989). Feeding on specialised tissues typically reduces photosynthesis, regardless of whether the attacked component is the phloem or xylem (Heng-Moss *et al.*, 2006), the stem (Macedo *et al.*, 2007) or general leaf fluids (Haile and Higley, 2003). In contrast, defoliation injury often does not alter photosynthetic capacity, within plant families (e.g., legumes) or between hardwoods and crops (Peterson *et al.*, 2004); however, there are examples where defoliation decreased (Delaney and Higley, 2006) or increased photosynthesis.

The removal of leaf tissue by herbivores represents a 'direct' reduction of photosynthetic capacity. We define the suppression of photosynthesis in remaining leaf tissue by any one of a number of processes, including damage to the vasculature supplying that tissue, as an 'indirect' effect of herbivory. Arthropods damage xylem or phloem (Welter, 1989), which may alter water transport, stomatal aperture and sucrose transport and loading, thereby reducing photosynthesis in the remaining leaf tissue. Tissue disruption by severing vasculature alters leaf hydraulics and, subsequently, nutrient or osmotica transport (Sack and Holbrook, 2006). If insect feeding is subtle enough to avoid outright cell rupture, modulation of nutrients sequestered by feeding will alter plant osmotica or sink-source relationships (Dorchin *et al.*, 2006). These effects also may be mediated by the plant's response. Insect attack, or even the perception of attack, can induce a myriad of defence-related responses, while concomitantly reducing the expression of photosynthesis-related genes (Kessler and Baldwin, 2002). In instances where plant defences are constituently expressed, the release of biocidal compounds against attackers may damage photosynthetic or homeostatic mechanisms vital for plant function (e.g., Zangerl *et al.*, 2002). We assigned indirect effects of herbivory into four classes: severed vasculature; altered sink demand; defence-related autotoxicity; and defence-induced downregulation of photosynthesis (Fig. 22.2).

22.7.1. Insect herbivores can affect photosynthesis by severing leaf vasculature

Damage to leaf venation alters leaf hydraulic conductance, reducing g_s and photosynthesis. In the absence of alternative pathways for water transport, the consequences of damage to venation can persist for weeks after the initial injury and lead to desiccation (Sack and Holbrook, 2006). Defoliation injury that severs venation indiscriminately or by feeding on specific tissues, may physically obstruct fluid flow with insect mouthparts (stylets) or cell fragments and alter photosynthesis and water balance in the remaining leaf tissue (Delaney and Higley, 2006). In *Glycine max* (soybean) a form of defoliation (skeletonisation) that removes patches of tissue reduced photosynthesis in the remaining tissue on damaged leaves and on adjacent undamaged leaflets (Peterson *et al.*, 1998). Interestingly, soybean increased carbon-uptake rates and transpiration in the remaining leaf tissue when one or two leaflets were completely lost (Suwignyo *et al.*, 1995), but when leaf-area removal (no patches) occurred to only part of a leaflet, CO_2 uptake did not decrease in the remaining leaflet tissue (Peterson *et al.*, 2004).

Aldea *et al.* (2005) confirmed that skeletonising soybean leaves by Japanese beetles substantially increased water loss from the cut edges. Damaging the interveinal tissue had no effect on CO_2 exchange, but increased transpiration by 150% for up to four days post injury. In contrast to the response to interveinal damage, severed vasculature caused a decrease in CO_2 exchange and an increase in ϕ_{PSII}, suggesting that insect damage transiently decoupled photosynthetic J from carbon assimilation (Aldea *et al.*, 2005). Damage to *Arabidopsis* by cabbage looper (*Trichoplusia ni*) larvae also increased water loss from the cut edges and caused a localised reduction in ϕ_{PSII} (Tang *et al.*, 2006). That the reduction in ϕ_{PSII} could be reversed by exposing the leaf to higher concentrations of CO_2, suggests that profligate water loss near cut edges reduced ϕ_{PSII} and increased NPQ by causing localised stomatal closure in the remaining undamaged leaf tissue.

22.7.2. Changes in sink demand by insect herbivory can affect photosynthesis

In instances where plants respond to herbivory with increased CO_2 uptake, the mechanism typically is linked to compensation or an increase in the sink demand within the leaf (Trumble *et al.*, 1993). For some gall-forming insects, gall tissue itself increases photosynthesis relative to uninjured tissue. In *Ilex aquifolium* (holly), increased ϕ_{PSII} and J enhanced carbon assimilation (Retuerto *et al.*, 2004), whereas a reduction in respiration in *Acacia pycnantha* galls contributed to an increase in A_N (Dorchin *et al.*, 2006).

In other galls of hardwoods, feeding damage reduced photosynthesis and altered water balance. Gall formation in red maple, pignut hickory and black oak reduced ϕ_{PSII} and increased NPQ, indicating a downregulation of the PSII reaction centres in the area around galls (Aldea *et al.*, 2006b). A sharp reduction in leaf temperature near galls suggests that transpiration was greater and fluid and nutrient transport increased near the point of damage. In contrast to gall-forming insects, a leaf-mining moth that lives enclosed within leaf tissue of apple trees, reduced carbon assimilation rates by decreasing transpiration (Pincebourd *et al.*, 2006).

Defoliation of plants may also increase photosynthesis by altering sink demand, but concerns over how remaining tissues were measured have been noted (Welter, 1989). By enclosing severed edges within gas-exchange cuvettes or measuring treatment effects on leaves where adjacent leaves were removed (within-plant controls), the data may not accurately describe plant responses specific to the herbivory treatment. Despite these challenges and potential limitations, data suggest defoliation may improve photosynthesis in the remaining leaf tissue (Thomson *et al.*, 2003) as a result of increased carboxylation efficiency and RuBP regeneration (Turnbull *et al.*, 2007a).

22.7.3. Autotoxicity following herbivory may reduce photosynthesis

Plants invest heavily in chemical defences (Berenbaum and Zangerl, 2008) and they run the risk of autotoxicity because of the biocidal properties of many secondary compounds employed in defence. Although in-vivo studies of autotoxicity are limited, photosynthesis may be severely reduced for some species. For example, wild parsnip (*Pastinaca sativa*) contains an arsenal of defence compounds including furanocoumarins, which are photoactivated and biocidal against a variety of organisms. Furanocoumarins are contained in oil tubes under positive pressure and bleed profusely from the wounding site (Gog *et al.*, 2005). When herbivores sever these tubes, the release of furanocoumarins reduces ϕ_{PSII} and gas exchange at considerable distances from the actual point of insect damage (Zangerl *et al.*, 2002; Gog *et al.*, 2005).

The autotoxic effect of defensive compounds on photosynthesis is highly species specific. Essential oils derived from parsley (*Petroselinum crispum*), wild parsnip and rough lemon (*Citrus jambhiri*) reduce ϕ_{PSII} when applied to leaves of conspecifics; however, oils from parsley affected a two-fold greater area than the other species (Gog *et al.*, 2005). Baldwin and Callahan (1993) fed nicotine to two species of tobacco (*Nicotiana sylvestris, N. glauca*) that naturally synthesised this alkaloid as a defence, and to two other solanaceous species lacking nicotine (*Datura stramonium, Lycopersicon esculentum*). Photosynthetic rates declined in both species that synthesised nicotine but only in one that did not (*L. esculentum*). Reduced photosynthesis contributed to reduced growth and fitness.

22.7.4. Herbivory and the defence response cause downregulation of photosynthesis-related genes

Jasmonates play a central role in regulating plant-defence responses to herbivores. The mechanism by which herbivore-induced jasmonate synthesis promotes global reprogramming of defence gene expression, as well as the regulation of this response, have been reviewed recently (Howe and Jander, 2008). Although jasmonates induce defences, they also inhibit growth and photosynthesis (Giri *et al.*, 2006).

Transcription analysis of plant–herbivore interactions revealed that photosynthesis-related genes are downregulated after attack; however, few studies have demonstrated the effects of herbivore attack on photosynthesis at the proteome and physiological levels. Attack by herbivores reduces Rubisco transcription (Hui *et al.*, 2003). Using two-dimensional electrophoresis, Giri *et al.* (2006) observed that herbivory reduced the abundance of the gene coding for Rubisco activase (*rca*) in *N. attenuata*.

Partial defoliation of individual leaves by herbivores largely increases evapotranspiration via enhanced water loss from cut edges and produces leaf dehydration (Aldea *et al.*, 2005), which not only reduces photosynthesis by causing stomata to close, but also by initiating senescence signalling (Lim *et al.*, 2007). A number of genes are induced by endogenous ABA in response to dehydration through the synthesis of the regulating transcription factors MYC and MYB (Yamaguchi-Shinozaki and Shinozaki, 2006). Both MYC and MYB function as *cis*-acting elements that regulate transcription of dehydration-related genes. Transgenic plants overproducing MYC and MYB had higher osmotic stress tolerance, and microarray analysis indicated the presence of ABA- and JA-inducible genes (Abe *et al.*, 2003). It has been suggested that cross talk occurs on AtMYC2 between ABA- and JA-responsive gene expression at the MYC recognition sites in the promoters, and that AtMYC2 is a common transcription factor of ABA and JA pathways in *Arabidopsis* (Yamaguchi-Shinozaki and Shinozaki, 2006).

The lipoxygenase pathway leading to the production of JA is differentially induced depending on the attacking agent (Kempema *et al.*, 2007), and the initiation of jasmonate signalling reduces photosynthesis and vegetative growth.

Plants treated with methyl jasmonate develop shorter petioles than control plants (Cipollini, 2005), and *Arabidopsis* mutants that accumulate higher JA concentrations have shorter petioles than wildtype (Bonaventure *et al.*, 2007); these effects of JA on plant growth are modulated by the gene JASMONATE-ASSOCIATED1 (*JAS1*; Yan *et al.*, 2007). It has been suggested that the slower growth and downregulation of photosynthetic-related genes by herbivore elicitation may be required to free up resources for defence-related processes (Baldwin, 2001). It is not clear whether the change in carbon allocation affects photosynthetic rate *per se*, but growth reduction would affect leaf expansion and total plant photosynthesis.

22.8. IMAGING METHODS FOR BIOTIC-STRESS DETECTION

22.8.1. Chl-F imaging

Pathogens induce in their hosts a wide range of foliar visual symptoms with a heterogeneous distribution (mottle, mosaic, chlorosis, necrosis, etc.) and different Chl-FI prototypes have provided valuable tools to follow the infection and investigate the spatial and temporal heterogeneity of the foliar photosynthetic efficiency during pathogenesis (see reviews Nilsson, 1995; Nedbal and Whitmarsh, 2004). Measurements have been carried out in both inoculated and systemically infected leaves during pathogen challenge, showing photosynthesis impairment in symptomatic and asymptomatic areas during pathogenesis (Esfeld *et al.*, 1995; Scholes and Rolfe, 1996; Osmond *et al.*, 1998; Chou *et al.*, 2000; Lohaus *et al.*, 2000). In some cases, the rate of photosynthetic inhibition is associated to the severity of the symptoms; however, changes in some fluorescence parameters could precede the symptom development and allow a presymptomatic diagnosis of the disease (Chaerle *et al.*, 2007b; Fig. 22.3). A correlation between the pattern of Chl-F quenchings and virus distribution in leaves was found during the infection of *Nicotiana benthamiana* with PMMoV (Fig. 22.4; Pérez-Bueno *et al.*, 2006; Pineda *et al.*, 2008b), as well as in *Abutilon* mosaic virus-infected *Abutilon striatum* leaves (Osmond *et al.*, 1998; Lohaus *et al.*, 2000).

Alterations in photosynthesis of fungi-infected plants mapped by Chl-FI are also spatially and temporally complex. Infected leaves consist of regions of cells directly invaded by the pathogen and regions remote from the fungal colony (e.g., Scholes and Rolfe 1996; Osmond *et al.*, 1998; Chou *et al.*, 2000; Meyer *et al.*, 2001).

Imaging analysis of the changes on photosynthesis parameters during bacterial challenge in compatible and incompatible interactions has also deserved special attention (Zou *et al.*, 2005; Bonfig *et al.*, 2006; Rodriguez-Moreno *et al.*, 2008).

The fluorescent parameter best suited for either evaluating damage or carrying out presymptomatic diagnosis depends on the type of infection. Invasion of bean leaves by rust fungi (*Uromyces appendiculatus*) was revealed by changes in the fluorescence-induction kinetics (Peterson and Aylor, 1995). Cedar needles (*Torreya taxifolia*) infected by the fungus *Pestaliopsis* spp. were identified by an empirical estimate

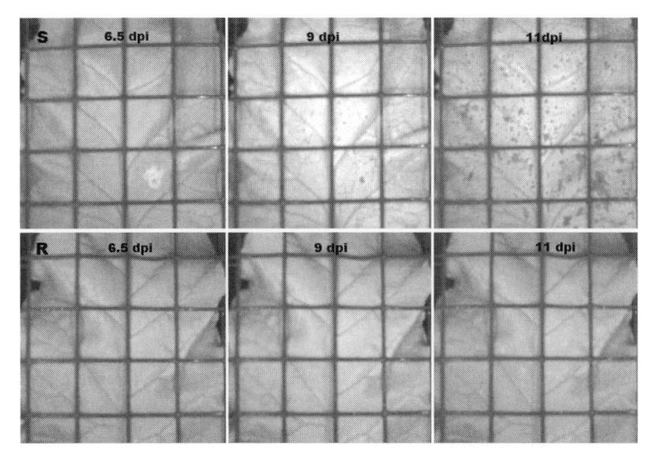

Fig. 22.3. Presymptomatic chlorophyll-fluorescence (Chl-F) increase upon *Cercospora* infection of attached sugar beet leaves. Chl-F images of infected leaves are captured at different days post-infection (dpi) in susceptible (S) and resistant (R) plants. The first Chl-F symptoms appear at 6 dpi on the S leaf. At 8 dpi, a general increase of Chl-F intensity was apparent. At 11 dpi widespread cell death (as indicated by low intensity spots) was visualised in the S leaf. Modified with permission from Chaerle *et al.* (2007a).

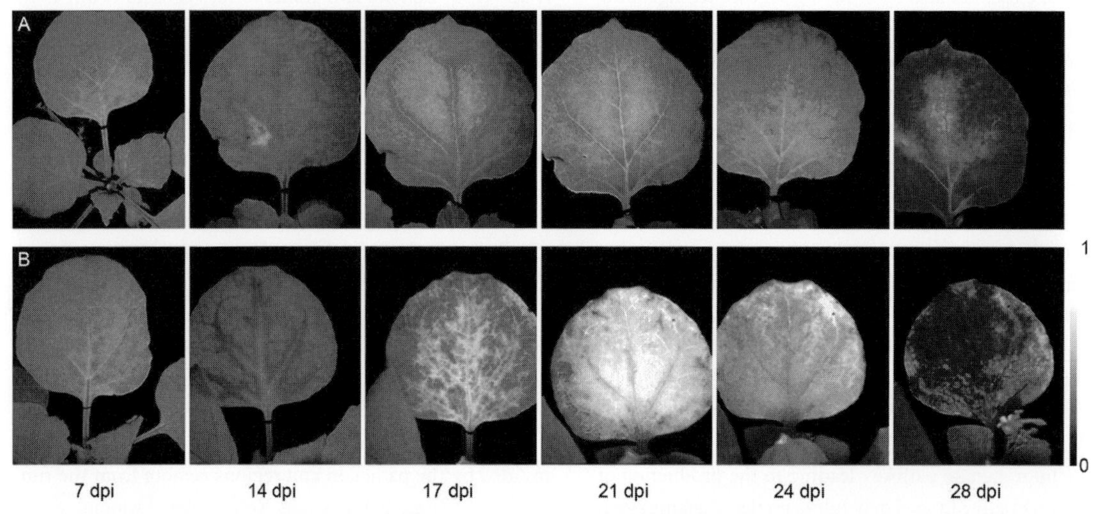

Fig. 22.4. Images of NPQ_{300} (at 300 s during the fluorescence-induction kinetics) from healthy (**A**) and asymptomatic (**B**) leaves of PMMoV-I infected *Nicotiana benthamiana* plants during pathogenesis. Modified, with permission from Pérez –Bueno *et al.* (2006).

of quantum yield (Ning *et al.*, 1995), which also was used to visualise the impact of fungal phytotoxins in hibiscus leaves (*Hibiscus sabdariffa*) (Bowyer *et al.*, 1998). Soukupová *et al.* (2003) proposed an experimental algorithm to identify the combination of fluorescence parameters providing the highest contrast between affected and unaffected plants in canola (*Brassica napus*) and white mustard (*Sinapis alba*) leaves exposed to phytotoxins of *Alternaria brassicae.* Imaging of ϕ_{PSII} of chickpea leaves was used to assess the impact of a fungal pathogen from *Ascochyta rabiei* that altered source-sink distribution (Esfeld *et al.*, 1995; Weis *et al.*, 1998). Changes of photosynthetic parameters were also visualised during host resistance (Repka, 2002; Swarbrick *et al.*, 2006).

22.8.2. Multicolour fluorescence imaging

There is a lack of multicolour fluorescence (MCF) studies, including the analysis of BGF signals in pathogen-infected

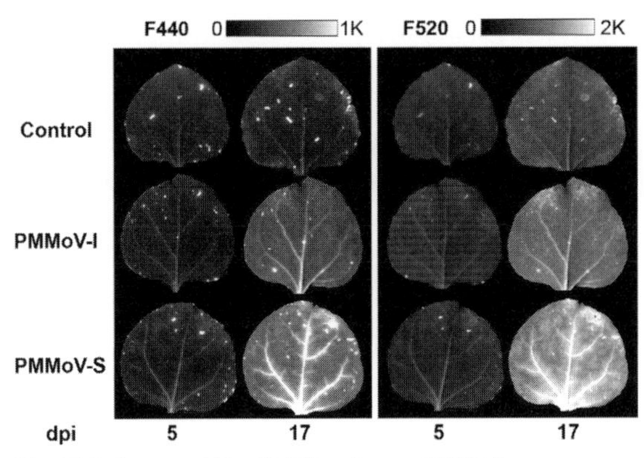

Fig. 22.5. Images of blue (F440) and green (F520) fluorescence emission from the abaxial surface of asymptomatic leaves from *Nicotiana benthamiana* control and PMMoV-infected plants. Images obtained before (5 dpi) and after (17 dpi) fluorescence changes occurred are displayed. Modified, with permission from Pineda *et al.* (2008a).

plants. Fungal infection has been reported to increase F440 either by the fungus autofluorescence (Lüdeker *et al.*, 1996) or by the production of plant phytoalexins (Niemann *et al.*, 1991). Buschmann and Lichtenthaler (1998) reported an early visualisation by MCFI of small punctures made by tobacco flies on leaves and mite attack on bean. Pineda *et al.* (2008a) monitored a systemic viral infection in PMMoV-infected *Nicotiana benthamiana* plants using a compact flash-lamp MCF system (Fig. 22.5). BGF increase linked to the accumulation of different phenolic compounds was monitored during the HR in tobacco plants challenged with TMV (Chaerle *et al.*, 2007c), *Phytophtora megasperma*, (Dorey *et al.*, 1997) and *Phytophthora nicotianae* (Scharte *et al.*, 2005).

22.8.3. Other imaging techniques for visualisation of the plant-pathogen interaction and herbivore attack

Each representative of the wide range of plant-pathogen and plant-insect interactions likely affects different plant physiological processes to a varying extent. Using several imaging techniques in parallel could reveal *disease signatures*, allowing diagnosis in the absence of symptoms in the visible spectrum (Nedbal and Whitmarsh, 2004; Aldea *et al.*, 2006a; Chaerle *et al.*, 2007c). Parallel measurements of thermal and Chl-F imaging permit to study the spatial and temporal heterogeneity of leaf transpiration and photosynthesis under biotic stress. Downy mildew infection in cucumber was visualised at an early stage by thermal imaging (Lindenthal *et al.*, 2005; Oerke *et al.*, 2006). In PMMoV-infected plants, virus immunolocalisation on tissue prints matched well with the concomitant pattern of thermal and Chl-F increase (Chaerle *et al.*, 2006). Infection with the fungus *Phyllosticta* of two conifer species induces a temperature increase in areas surrounding the inoculation point (Aldea *et al.*, 2006b); the corresponding stomata closure is associated to a decrease of ϕ_{PSII} and NPQ increase.

Development of cell death was visualised by Chl-F and thermal imaging during the HR of potato to *Phytophthora* (Scharte *et al.*, 2005), as well as in tobacco challenged with TMV (Chaerle *et al.*, 1999) and *Nicotiana sylvestris* inoculated with *Erwinia amylovora* (Boccara *et al.*, 2001). Chaerle *et al.* (2004) observed opposite effects on leaf temperature in a necrotrophic fungal infection (*Cercospora*-bean) versus a viral-induced HR (resistant tobacco-TMV).

A number of new imaging techniques have been used for following the interaction of the host plant with either virulent of avirulent pathogens: autoluminescence or biophoton imaging, associated with oxidative stress reactions (Mansfield, 2005; Havaux *et al.*, 2006 and references therein; Kobayashi *et al.*, 2007); NIR imaging for field applications (Zandonadi *et al.*, 2005; Pethybridge *et al.*, 2008); and hyperspectral imagery (Franke and Menz, 2007; Huang *et al.*, 2007). Obtaining an overview of spatial in-field variability would thus be a pre-requisite for site-specific disease management. This will be important in the context of precision agriculture, where different imaging techniques could be combined into a multispectral visualisation approach.

23.9. CONCLUSIONS

The effects of pathogens and insects on photosynthesis are as varied as their life forms and feeding behaviours, but some intriguing commonalities are emerging. Although virtually all forms os biotic damage applied directly to foliage cause some form of direct reduction of photosynthesis by removing or killing portions of the leaf, the reduction of photosynthesis in the remain tissue represents a 'hidden' and potentially considerable consequence of biotic stress. Although there are exceptions, many forms of biotic damage downregulate genes coding for the component processes of photosynthesis. The effects of biotic damage on photosynthesis are notoriously heterogeneous. Imaging the patterns of Chl-F and leaf temperature across damaged leaves has greatly increased our understanding of the plant responses to biotic attack. Understanding the underlying molecular and biochemical mechanisms governing the response of photosynthesis has not kept pace because of a more limited ability to measure the spatial patterns of these processes. Future advances in the ability to map transcriptional and proteomic responses will shed new light on how photosynthesis is regulated following biotic damage.

Part IV
Photosynthesis in time

23 • Photosynthesis during leaf development and ageing

Ü. NIINEMETS, J. I. GARCÍA-PLAZAOLA AND T. TOSENS

23.1. INTRODUCTION

Periods of leaf development and senescence comprise a significant fraction of leaf lifespan. Therefore, leaf lifetime carbon gain is importantly modified by the overall duration and time kinetics of these processes (Wilson *et al.*, 2001; Morecroft *et al.*, 2003; Grassi and Magnani, 2005). In addition, significant time-dependent changes occur in leaf function in mature non-senescent leaves owing to continuous accumulation of cell walls and concomitant reductions in mesophyll-diffusion conductance, as well as owing to re-acclimation of foliage to dynamically changing environmental conditions. Such modifications are of particular importance in evergreen species supporting foliage for several growing seasons, but foliage structure and physiological potentials also change in mature non-senescent leaves in deciduous species (Flexas *et al.*, 2001; Wilson *et al.*, 2001; Niinemets *et al.*, 2004a).

A large body of information of fine-scale regulation of leaf development and senescence has become available (Dengler and Kang, 2001; Kessler and Sinha, 2004; Fleming, 2005; Lim *et al.*, 2007). Although these studies cover in depth the regulatory sequences and signalling pathways during leaf development and senescence, the last comprehensive series of reviews on leaf photosynthetic modifications in developing leaves was published in 1985 (Shesták, 1985). Furthermore, the available treatises of leaf ontogenetic effects on photosynthesis have focused on herbaceous plants or on fast-growing deciduous trees. However, the rate of developmental and ageing processes largely differs among species with varying leaf longevity and structure (Miyazawa *et al.*, 2003). Consideration of these functional-type specific variation patterns in leaf development is of major importance for prediction of plant photosynthetic productivity of highly structured natural plant communities consisting of species with varying leaf longevity and architectural constitution.

We review the current knowledge of the alterations in foliage photosynthetic function during leaf development, age-dependent changes in fully developed non-senescent leaves and decline of photosynthetic function during foliar senescence. Special attention is paid to enormous variation in the time-kinetics and in the sensitivity of development and ageing to environmental conditions in plant species with differing leaf lifespan and structure.

23.2. DEVELOPMENT OF LEAF PHOTOSYNTHETIC CAPACITY IN GROWING LEAVES

A plethora of modifications in foliage structural and physiological characteristics occurs in developing leaves. As several leaf characteristics, such as cell size, tissue composition and leaf thickness, are determined during this stage of leaf ontogeny (Yamashita *et al.*, 2002; Fleming, 2005), leaf developmental processes play a key role in shaping leaf photosynthetic capacity of mature leaves. During leaf development, synthesis of chlorophyll, pigment-binding complexes and photosynthetic enzymes occur simultaneously with multiplication of chloroplasts, expansion of leaf surface area and formation of leaf internal structure. While all these processes are highly coordinated throughout the leaf development, they proceed with varying time kinetics and alter distinct partial processes of leaf photosynthetic apparatus, e.g., light reactions versus dark reactions of photosynthesis. In the following, the key processes during leaf development are reviewed and the time kinetics of these processes in different plant species are analysed, starting from qualitative biochemical changes (leaf greening and accumulation of rate-limiting enzymes), followed by structural modifications (expansion of leaf area and formation of internal architecture) and then analyzing the

Terrestrial Photosynthesis in a Changing Environment: A Molecular, Physiological and Ecological Approach, ed. J. Flexas, F. Loreto and H. Medrano. Published by Cambridge University Press.

combined effects of biochemical and structural alterations on development of leaf photosynthetic capacity and the implications of structural modifications on leaf mesophyll diffusion conductance (g_m). We conclude with the examination of the variation in developmental processes among plants with differing foliage longevity.

23.2.1. Chloroplast biogenesis

Chloroplasts, the photosynthetic organelles in plant cells, originate from proplastids. Proplastids are small colourless organelles that lack thylakoids and only possess a few internal membranes formed by invaginations of the inner envelope. The transformation of proplastids into chloroplasts is initiated in meristematic cells that typically contain 10–20 proplastids (Biswal *et al.*, 2003). Differentiation of meristematic cells to photosynthetically active mesophyll cells is associated with active division of chloroplasts in each cell to reach a value between 100–200 chloroplasts in mature leaves (Osteryoung and Nunnari, 2003). In parallel with chloroplast multiplication, thylakoid lipids are synthesised in the cytosol and further transferred to chloroplasts where they are desaturated by specific desaturases and embedded in thylakoids (e.g., Xu *et al.*, 2003; Heilmann *et al.*, 2004). In addition to multiplication, chloroplasts also increase in size in developing leaves (Ellis and Leech, 1985; Pyke, 1999). Changes in the size of the chloroplasts track lamina expansion rate (Ellis and Leech, 1985).

During chloroplast formation, pigments and pigment-binding proteins are synthesised and integrated into thylakoids. As a consequence of the massive gene transfer from the endosymbiont to the nucleus, the number of nuclear genes encoding chloroplast proteins ranges from 2500 to 4000 (Stern *et al.*, 2004). The remnant plastid genome generally encodes approximately 80–100 proteins, among which there are many essential components of the photosynthetic apparatus, such as most subunits of PSII (Palmer, 1985), but highly reduced chloroplastic genomes lacking most of the photosynthetic genes have been reported for some parasitic plants (Wolfe *et al.*, 1992). The coordinated expression of nuclear and plastid genes, and the regulatory signals that interconnect plastids and nucleus lead to a sequential organisation of chloroplast structure and function (Marín-Navarro *et al.*, 2007). Studies with mutants have demonstrated that highly coordinated expression of both genomes is necessary for development of functional chloroplasts (Shirano *et al.*, 2000).

23.2.2. Leaf greening and assembly of pigment-binding complexes

Leaf greening owing to the biosynthesis of chlorophyll and carotenoids in chloroplasts is the most remarkable visual aspect of chloroplast formation, development and multiplication (Fig. 23.1). During chlorophyll synthesis, protochlorophyllide oxidoreductases (POR) that catalyse the reduction of protochlorophyllide to chlorophyllide resulting in leaf greening, play a key role in the regulation of chlorophyll synthesis and thylakoid formation (Fig. 23.2) (Schoefs and Franck, 2003). In angiosperms, only light-dependent POR is present, and protochlorophyllide accumulates in the dark, forming a supramolecular-aggregated complex with POR. During light, this complex captures light that is used for protochlorophyllide photoreduction to chlorophyllide in the presence of NADPH, but it also dissipates excess light energy (Reinbothe *et al.*, 1999; Schoefs and Franck, 2003). In many other photosynthetic organisms, including gymnosperms, a second light-independent POR is present in addition to the light-dependent enzyme and thus, greening of photosynthetic tissues can also occur in the darkness (Schoefs and Franck, 2003).

Chlorophyll biosynthetic intermediates are highly photodynamic and may generate ROS if their synthesis exceeds the accumulation of chlorophyll-binding apoproteins. Therefore, chlorophyll formation is efficiently coordinated with the synthesis of pigment-binding proteins, with chlorophyll molecules regulating the synthesis, accumulation, import and stability of these proteins (Tanaka and Tanaka, 2007). As carotenoids also form the integral part of pigment-binding complexes, and both carotenoids and phytol residues of chlorophyll are synthesised in the chloroplast by the same isoprenoid synthesis pathway (2-methylerythritol 4-phosphate (MEP) pathway, Lichtenthaler, 1999), accumulation of carotenoids and chls occurs simultaneously during leaf development (Welsch *et al.*, 2000; Hirschberg, 2001). Throughout the pigment synthesis, chlorophyll and carotenoids bind to different apoproteins to form primary complexes that in latter stages assemble with others to form the final light-harvesting and electron transport complexes in thylakoids.

In addition to formation of functionally active pigment-binding complexes, photo-oxidative risk in young leaves is remedied by enhanced accumulation of antioxidative flavonoids (Neill *et al.*, 2002), energy dissipative carotenoids, antheraxanthin and zeaxanthin (Yoo *et al.*, 2003), and lipid- (tocopherol) (Hormaetxe *et al.*, 2005b) and water-soluble (glutathione, ascorbate) antioxidants (Fig. 23.1).

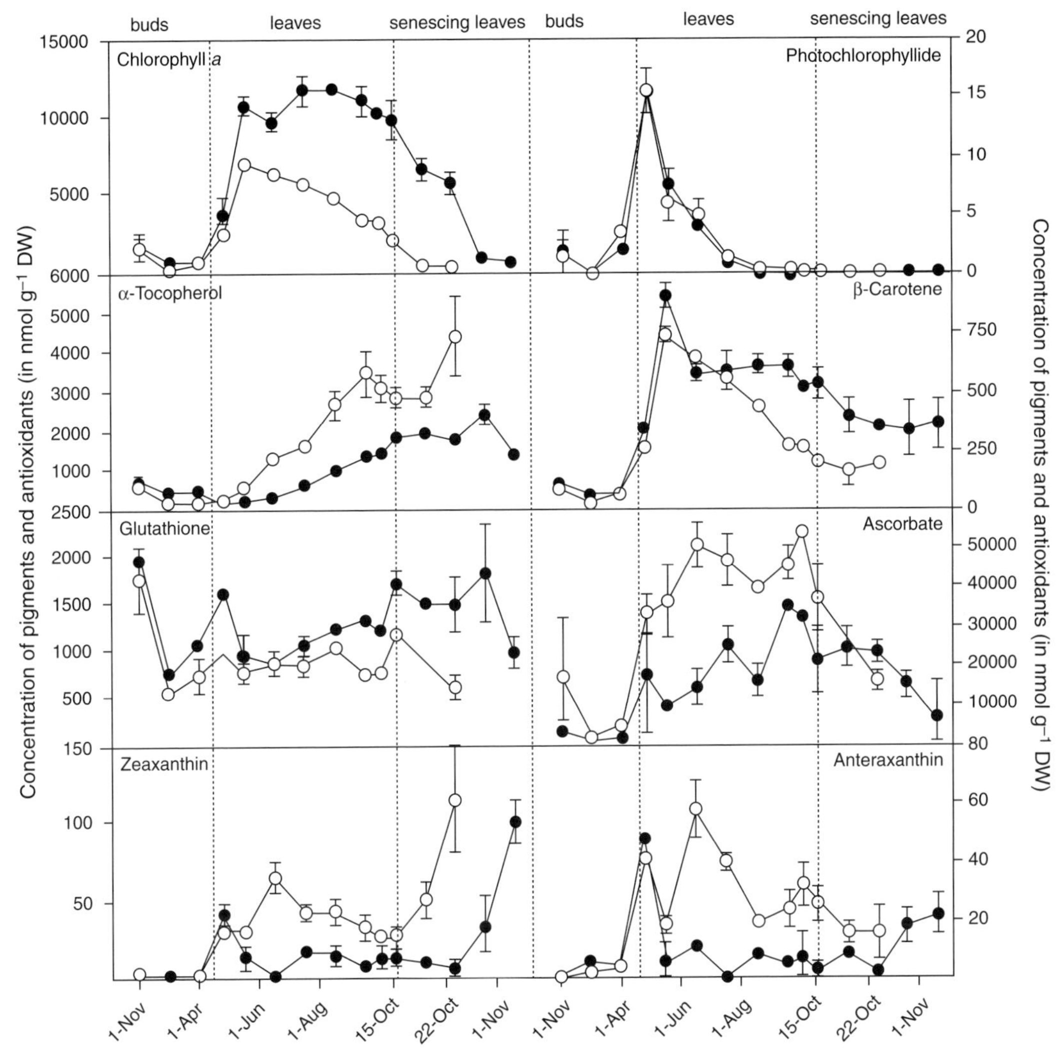

Fig. 23.1. Age-dependent changes in pigments (in nmol g^{-1} dry weight) in high-light (open symbols) and low-light (filled symbols) exposed leaves of temperate deciduous tree *Fagus sylvatica*. Dotted lines represent bud-break in April and the onset of leaf senescence in October. Both periods are characterised by a transient increase of some photoprotective compounds. Modified from Garcia-Plazaola and Becerril (2001).

In particular, there is an initial peak of contents of photoprotective carotenoids and some other antioxidants, such as glutathione and ascorbate, in an early leaf developmental phase, matching the period of highest protochlorophyllide accumulation when the light-harvesting complexes have not yet fully developed (Fig. 23.1), followed by more gradual changes throughout the rest of the leaf developmental period (Fig. 23.1).

23.2.3. Synthesis of rate-limiting photosynthetic proteins

Chlorophyll accumulation and thylakoid formation are accompanied by synthesis of other key components of photosynthetic machinery. The concentrations of rate-limiting proteins of thylakoid electron transport machinery, cyt *f* and ATPase (Jiang *et al.*, 1997) and the carboxylating enzyme,

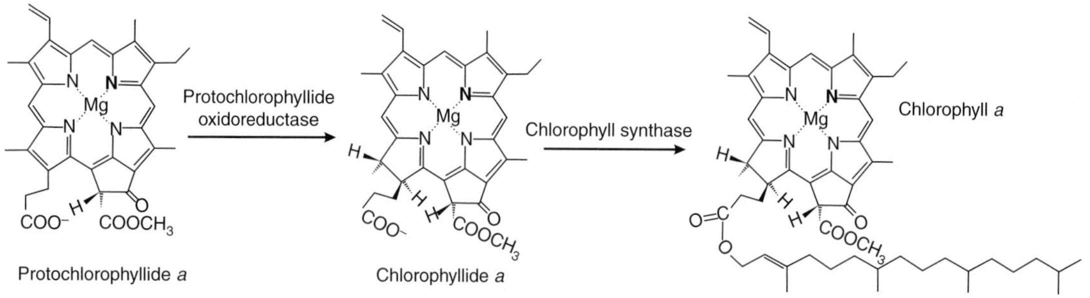

Protochlorophyllide *a* — Protochlorophyllide oxidoreductase → Chlorophyllide *a* — Chlorophyll synthase → Chlorophyll *a*

Fig. 23.2. Terminal steps of chlorophyll *a* biosynthesis in plants. Protochlorophyllide *a* is converted to chlorophyllide *a* by protochlorophyllide oxidoreductases. Angiosperms have only the light-dependent NADPH:protochlorophyllide oxidoreductase, and protochlorophyllide *a* accumulates in the dark. Gymnosperms and several other photosynthetic organisms have both light-dependent and light-independent protochlorophyllide oxidoreductases. Chlorophyllide *a* is further converted to chlorophyll *a* by chlorophyll synthase that catalyses esterification with the phytol residue (modified from Schoefs and Franck, 2003; Rüdiger *et al.*, 2005).

Rubisco, increase with similar time kinetics as leaf chlorophyll concentrations (Jiang *et al.*, 1997; Valjakka *et al.*, 1999; Eichelmann *et al.*, 2004). The increase in Rubisco content in developing leaves is further matched with the increase in Rubisco activase (Zielinski *et al.*, 1989). This coordinated expression of photosynthetic genes is achieved by multiple signals, including nuclear-encoded protein factors that regulate transcriptional and post-transcriptional steps of chloroplast gene expression, and light-dependent changes in the redox status of key components of photosynthetic electron transport chain, generation of proton gradient, ATP/ADP ratio and assembly related autoinhibition, i.e., the excess amount of a certain subunit inhibits its own synthesis (Marín-Navarro *et al.*, 2007). Nevertheless, as studies in mutants with genetically altered components of photosynthetic machinery demonstrate, syntheses of key parts of photosynthetic machinery are not strictly interdependent (Jiang *et al.*, 1997). Furthermore, environmental factors such as light intensity and temperature can importantly alter the gene expression and resulting stoichiometry of photosynthetic machinery (Sugita and Sugiura, 1996, Section 23.3).

23.2.4. Leaf expansion and formation of internal architecture

Leaf growth is a key process providing the highly organised scaffolding for leaf photosynthetic machinery. Leaf primordia are composed of tightly compressed cells with a weakly developed internal structure. Rapid leaf-expansion growth after bud-burst leads to formation of characteristic leaf shape and size in a highly coordinated process, still including many unknowns in the complex signal-transduction network (Fleming, 2005; Micol, 2009). Leaf growth consists of cell division and expansion growth. Although both processes contributing to leaf expansion occur simultaneously and are closely synchronised (Li *et al.*, 2005), cell division occurs with a faster rate in initial stages of leaf development and stops earlier than cell expansion (Granier *et al.*, 2000; Schurr *et al.*, 2000). Leaf-expansion growth is characterised by pronounced daily rhythm, whereas the timing of the maximum growth activity during the day widely varies among species owing to reasons not yet fully understood (Walter and Schurr, 2005).

Leaf-expansion growth reflects periclinal cell division and lateral expansion. However, during leaf expansion cells also divide anticlinally and extend dorsiventrally. As a result, the number of cell layers and overall leaf thickness increase throughout leaf development (Fig. 23.3). The increase in the thickness is reflected in increases in leaf dry mass per unit leaf area (LMA) in developing leaves (Niinemets *et al.*, 2004a; Katahata *et al.*, 2007). Commonly, leaf-expansion growth is completed earlier than the growth in leaf thickness, especially in species with thick leaves (Dengler, 1980; Eichelmann *et al.*, 2004). In initial stages of leaf-expansion growth, the internal architecture still remains weakly developed (Fig. 23.3). With increasing leaf thickness, spongy and palisade mesophyll layers are differentiated (Psaras and Rhizopoulou, 1995), and intercellular air space is gradually formed in the leaves as the result of schizogeny (cell separation) (Kozela and Regan, 2003). In some species, there is also evidence of lysogenic activity (programmed cell death, Gunawardena, 2008) in intercellular air-space formation (de Chalain and Berjak, 1979). At full maturation, internal air space forms on average 25% of total leaf volume and 30% of mesophyll volume across a variety of vascular-plant species covering all plant major functional types (Niinemets, 1999 for a meta-analysis). Given this high

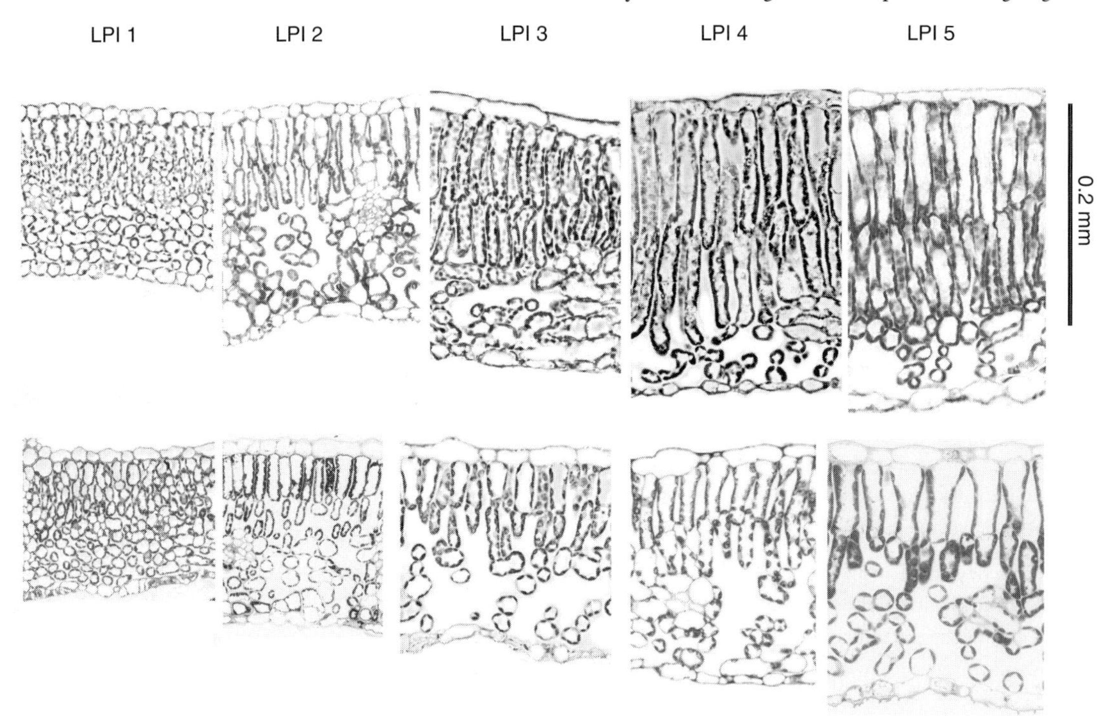

Fig. 23.3. Changes in internal leaf architecture during leaf development in temperate deciduous tree *Populus tremula* grown under high light (46.8 mol m^{-2} d^{-1}, upper cross-sections) and low light (9.36 mol m^{-2} d^{-1}, lower cross-sections). Leaf age is quantified by leaf plastochron index (LPI, Maksymowych, 1973) that provides an estimate of leaf biological age. LPI for a given leaf is approximately equal to n_1-n_0, where n_1 is the number of leaves produced after the formation of the given leaf, and n_0 is the number of these leaves equal to or less than the arbitrarily chosen threshold leaf size. For *P. tremula*, the threshold for leaf width was set to 20 mm, i.e., a leaf of 20 mm width has a LPI of zero. Unpublished data of Tiina Tosens, Vivian Vislap and Ülo Niinemets.

percentage, surprisingly little information is available about the mechanisms of formation of intercellular air space in plants (Kozela and Regan, 2003).

Significant changes in mesophyll cell-wall architecture and chemistry occur during foliage growth. During rapid leaf-expansion phase, cell walls stay relatively thin, in the order of 0.05–0.1 μm. These thin cell walls are compliant to cell-wall loosening agents, such as expansins, facilitating rapid regulation of growth (Cosgrove, 2000, 2005). With cessation of leaf-expansion growth, cell walls rigidify and become cross-linked (Cosgrove, 2005). Cell-wall thickening and deposition of pectic intrusions occurs well beyond the cessation of leaf-expansion growth (Miyazawa and Terashima, 2001; Miyazawa *et al.*, 2003). The thickness of mesophyll walls in fully mature leaves can reach 0.1–0.4 μm or more depending on species (Hoffmann-Benning *et al.*, 1997; Moghaddam and Wilman, 1998; Paoletti, 1998; Miyazawa and Terashima, 2001; Miyazawa *et al.*, 2003). Although ultrastructural characteristics, such as porosity and cell-wall chemical composition, are difficult to measure in whole leaves, cell-suspension studies demonstrate that cell-wall porosity is decreasing with cell age, likely reflecting cross-linking of cell-wall polysaccharides at the end of the growth cycle (Titel *et al.*, 1997). In addition, the concentration of phenolic lignin-like polymers in mesophyll cell walls also increases in some species with increasing leaf age (Gordon *et al.*, 1985), possibly further reducing porosity.

Most effort in leaf-growth studies has been devoted to understanding the mechanisms of and environmental controls on leaf-expansion growth (Schurr, 1997; Walter and Schurr, 2000). Kinetics of foliage expansion and final size of leaves are important as leaf size alters leaf temperature and transpiration rate through modifications in leaf boundary layer conductance for heat and gaseous exchange (Stokes *et al.*, 2006), and at whole-plant level the total light-harvesting surface area (Niinemets and Anten, 2009). However, the variation in foliage photosynthetic potentials in leaves of given size mainly depends on leaf thickness and leaf internal architecture (packing of cells per A_{leaf}, and the cell-wall thickness of mesophyll cells). Thicker leaves commonly contain more photosynthetic mesophyll per A_{leaf}, and have greater rates of photosynthesis per unit area, whereas dense leaves contain more structural and less photosynthetic biomass and have lower rates of photosynthesis per dry mass (Niinemets, 1999 for review). Currently, the factors controlling leaf thickness and density are much less understood than the factors altering leaf-expansion growth. Although the expansion of leaf surface area, dorsiventral expansion and formation of internal architecture are parts of the same multifaceted leaf developmental process, the timing of and the environmental effects on these processes can be largely different as analysed below (Section 23.3).

23.2.5. General framework to study simultaneous biochemical and structural modifications in photosynthetic machinery in developing leaves

To analyse the development of leaf photosynthetic activity, it is useful to distinguish between the changes in the capacities for the light reactions and dark reactions of photosynthesis. According to the widely used process-based leaf photosynthesis model (Farquhar *et al.*, 1980a), the capacity for the light reactions is determined by the maximum photosynthetic electron transport rate (J_{max}), while the capacity for the dark reactions is determined by the maximum Rubisco carboxylase activity. Both J_{max} and $V_{c,max}$ determine leaf photosynthetic capacity per ($A_{max,A}$), while leaf light-harvesting efficiency for an incident light (initial quantum yield of photosynthesis) is mainly driven by leaf chlorophyll content per area. As J_{max} and $V_{c,max}$ per unit area ($J_{max,A}$ and $V_{c,max,A}$) are the products of the rate per unit dry mass ($J_{max,M}$ and $V_{c,max,M}$) and LMA, both biochemical and structural changes alter the development of $J_{max,A}$ and $V_{c,max,A}$. The capacities per unit dry mass are mainly driven by the concentration of rate-limiting proteins in average leaf cells, while LMA characterises the developmental modifications in leaf density and thickness. Analogously, chlorophyll content per area (Chl/area) is the product of mass-based content and LMA. As the photosynthetic proteins and chlorophyll contain a large fraction of nitrogen (N) that must be transported to the growing leaves from the rest of the plant, changes in mass-based concentrations of proteins and chlorophyll can be a result of developmental changes in leaf N content per dry mass (N_M) and/or owing to modifications in the fraction of N in photosynthetic proteins. To separate between these controls imposed by structure, N and biochemical investments, J_{max}/area and $V_{c,max}$/area can be expressed as the products of LMA, N_M, the fractions of N in rate-limiting proteins and protein-specific activities (Niinemets and Tenhunen, 1997):

$$J_{max} / \text{area} = 8.06 J_{mc} \cdot \text{LMA} \cdot F_B \cdot N_M \quad \text{[Eqn. 23.1]}$$

$$V_{c.max} / \text{area} = 6.25 V_{cr} \cdot \text{LMA} \cdot F_R \cdot N_M \quad \text{[Eqn. 23.2]}$$

where F_B is the fraction of leaf N in rate-limiting proteins of PET and F_R the fraction of leaf N in Rubisco; J_{mc} is the capacity for PET per unit cyt *f*; and V_{cr} is the maximum rate of ribulose-1,5-bisphosphate carboxylation per unit Rubisco protein (specific activity of Rubisco). The coefficients 8.06 and 6.25 are determined on the basis of N content of proteins and the stoichiometry of rate-limiting proteins (Niinemets and Tenhunen, 1997). These equations demonstrate that leaf photosynthetic capacity in developing leaves may vary because of simultaneous changes in N_M, LMA and the fraction of leaf N in photosynthetic machinery, F_B and F_R.

Analogously, Chl/area can be expressed as:

$$\text{Chl/area} = \text{LMA} \cdot F_L \cdot C_B \cdot N_M \quad \text{[Eqn. 23.3]}$$

where F_L is the fraction of leaf N in light harvesting; and C_B (mmol chlorophyll (g N)$^{-1}$), chlorophyll binding, is the amount of chlorophyll equivalent to the amount of N invested in light harvesting. C_B is determined by the nitrogen cost of chlorophyll and chlorophyll-binding proteins, and varies in dependence on the stoichiometry of light-harvesting pigment-binding protein complexes that each have a different N cost. C_B is commonly between 2.1–2.5 mmol g^{-1} (Niinemets and Tenhunen, 1997; Niinemets *et al.*, 1998).

23.2.6. Combined effects of biochemical and structural modifications on the development of leaf photosynthetic competence

The equations (23.1–23.3) provide a quantitative method to characterise variations in photosynthetic potentials and leaf pigmentation owing to N allocation, N partitioning and leaf structural development. Young leaves immediately after bud-burst have very high N contents per dry mass, even up to 4–5% of total dry mass (Reich *et al.*, 1991; Miyazawa *et al.*, 1998; Niinemets *et al.*, 2004a; Marchi *et al.*, 2008), but J_{max}/area, $V_{c,max}$/area and Chl/area are low owing to small LMA and low fractions of N in the components of photosynthetic machinery (Fig. 23.4, Niinemets *et al.*, 2004a; Katahata *et al.*, 2007). Leaf net assimilation rates are even relatively lower than can be expected on the basis of J_{max} and $V_{c,max}$, because the rate of dark respiration is very high in growing leaves actively synthesising proteins, lipids and cell-wall

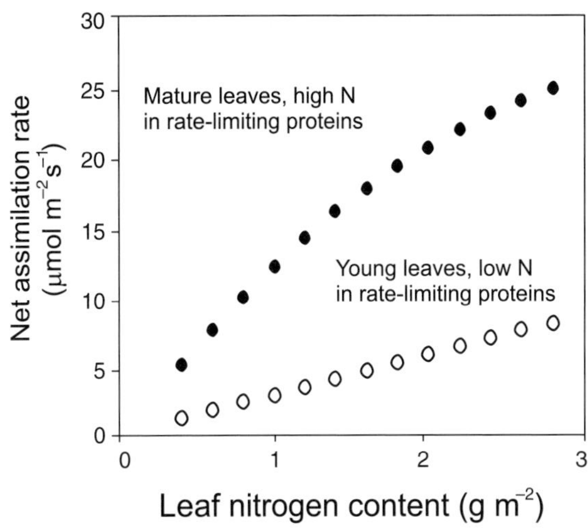

Fig. 23.4. Simulated leaf photosynthetic capacity (net assimilation rate at saturating light) in dependence on leaf nitrogen content per area for two leaves with differing fractional allocation of leaf nitrogen in the rate-limiting proteins of photosynthetic electron transport (F_B, Eqn. 23.1) and in ribulose-1,5-bisphophate carboxylase/oxygenase (Rubisco, F_R, Eqn. 23.2). For the young leaf, F_B was set to 0.02, and F_R to 0.05, whereas for the mature leaf, F_B was set to 0.07 and F_R to 0.25 and the capacity for photosynthetic electron transport (J_{max}) and maximum carboxylase activity of Rubisco ($V_{c,max}$), were calculated by equations 23.1 and 23.2. Net assimilation rate was simulated according to Farquhar *et al.* (1980) photosynthesis model for chloroplastic CO_2 concentration of 250 µmol mol^{-1}, leaf temperature of 25°C and incident quantum flux density of 1000 µmol m^{-2} s^{-1}. The quantum yield for photosynthetic electron transport was set to 0.245 mol mol^{-1} and the dark respiration rate continuing in light was taken as 0.0143 $V_{c,max}$ (Niinemets *et al.*, 1998).

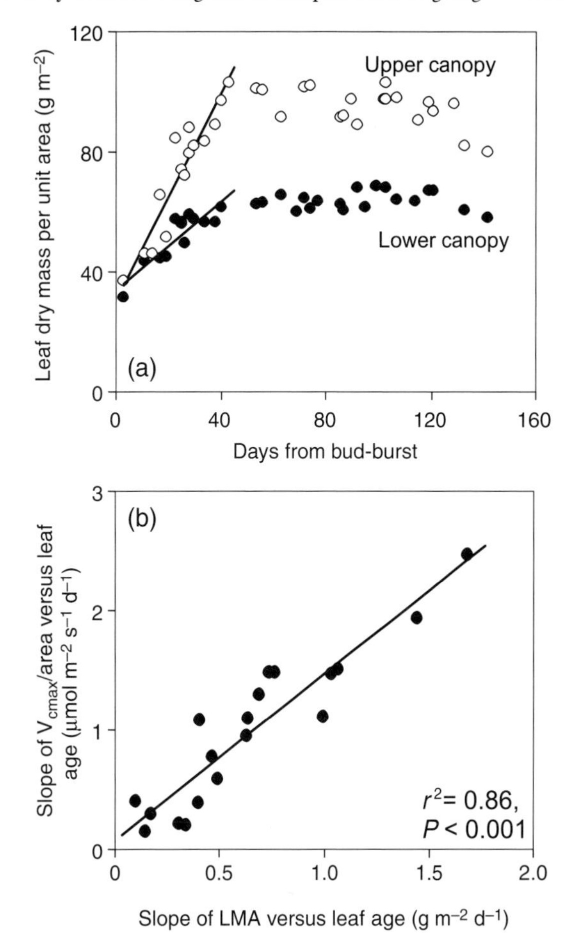

Fig. 23.5. Representative seasonal time-courses of leaf dry mass per unit area (LMA) in the upper (height 23.5–27 m, ca. 83% of above-canopy light) and lower (14–20 m, ca. 28% of above-canopy light) canopy of temperate deciduous tree *Populus tremula* **(a)** and a meta-analysis highlighting the coordinated developmental modifications in LMA and maximum carboxylase activity of Rubisco per unit leaf area ($V_{c,max}$/area) **(b)**. The rate of change in LMA is the initial slope of the relationship between LMA and leaf age as highlighted in (a). The rate of change in $V_{c,max}$/area is calculated analogously. The meta-analysis is based on 13 temperate deciduous tree species (Niinemets *et al.*, 2004a for original references). Modified from Niinemets *et al.* (2004a).

polysaccharides (Dickmann, 1971; Niinemets *et al.*, 2004a; Katahata *et al.*, 2007). N_M values further decrease to a stable value as carbon is accumulating in growing leaves and LMA is increasing (Fig. 23.5a for sample increases in LMA with leaf age), and this is also associated with decreases in dark-respiration rates (Niinemets *et al.*, 2004a; Katahata *et al.*, 2007). Despite the reductions in N_M, N content per area increases, demonstrating that leaves still remain a N sink for a large part of leaf developmental period (Reich *et al.*, 1991; Niinemets *et al.*, 2004a; Marchi *et al.*, 2008). Parallel to the increase of N content per area, foliage photosynthetic potentials and Chl/area increase, reflecting overall accumulation of photosynthetic proteins per leaf area as well as a larger fraction of leaf N in photosynthetic machinery in older leaves (Fig. 23.4, Terashima, 2001; Miyazawa *et al.*, 2003; Kitaoka and Koike, 2004; Miyazawa and Niinemets *et al.*, 2004a; Katahata *et al.*, 2007; Marchi *et al.*, 2008).

Increases in leaf photosynthetic potentials and Chl/area in developing leaves parallel changes in leaf internal architecture. In particular, the rate of increase in photosynthetic potentials is essentially proportional to the rate of increase in leaf thickness and LMA in different plant species and in leaves grown under different environmental conditions (Fig. 23.5b, Niinemets *et al.*, 2004a), demonstrating that the accumulation of photosynthetic proteins is matched with concomitant alterations in leaf internal structure.

23.2.7. Modifications in leaf mesophyll-diffusion conductance

Apart from changes in foliage photosynthetic potentials, J_{max} and $V_{c,max}$, realised maximum net assimilation rates, A_{max}, in developing leaves also depend on changes in CO_2 concentration in chloroplasts (C_c):

$$A_{max} = g_m(C_i - C_c) \quad \text{[Eqn. 23.4]}$$

where C_i is the CO_2 concentration in the sub-stomatal cavities; and g_m is the mesophyll-diffusion conductance from sub-stomatal cavities to chloroplasts. The CO_2 drawdown from the sub-stomatal cavities to chloroplasts, $C_i - C_c = A_{max}/g_m$, is the measure of the limitation of photosynthesis owing to mesophyll diffusion (Niinemets *et al.*, 2009b). An important implication of changes in leaf internal architecture and cell-wall characteristics is conspicuous changes in g_m in growing leaves. Although mesophyll cells of young, developing leaves have thin cell walls likely exerting low resistance to CO_2 diffusion, mesophyll cells are tightly packed with small air-space fraction, and also with low exposed chloroplast to total surface area. Therefore, the g_m initially increases with increasing leaf age (Miyazawa and Terashima, 2001; Miyazawa *et al.*, 2003; Eichelmann *et al.*, 2004; Parida *et al.*, 2004; Marchi *et al.*, 2008), reflecting increases in air-space gas-phase volume and exposed area during leaf development (Miyazawa and Terashima, 2001; Miyazawa *et al.*, 2003).

However, among the studies, there is a certain discrepancy in how much photosynthetic limitations owing to g_m change in young leaves. Miyazawa and Terashima (2001) for broad-leaved evergreen species *Castanopsis sieboldii* and Marchi *et al.* (2008) for broad-leaved evergreen *Olea europaea* estimated that CO_2 drawdown, $C_i - C_c$, was lower in younger leaves owing to lower A_{max}, implying that g_m limited photosynthesis less in younger leaves. In contrast, Marchi *et al.* (2008) for winter-deciduous species *Prunus persica* and Eichelmann *et al.* (2004) for winter-deciduous species *Betula pendula* have observed a larger CO_2 drawdown in very young compared with older developing leaves, suggesting that photosynthesis was more strongly limited by g_m in young leaves. These differences possibly reflect longer leaf developmental period (Section 23.2.8), and more extended and more extensive accumulation of cell-wall material in evergreens, as this results in more severe g_m limitations of photosynthesis in older leaf ages (Section 23.3.2).

After maximum g_m has been reached in fully expanded leaves, g_m decreases with further increases in leaf age owing to enhanced cell-wall thickening (Miyazawa and Terashima, 2001) and possibly also because of reduction of cell-wall porosity (Section 23.2.4). The changes in CO_2 drawdown in young fully developed leaves are initially small (Hanba *et al.*, 2001; Miyazawa and Terashima, 2001), but the CO_2 drawdowns and reductions in g_m increase with increasing leaf age (Loreto *et al.*, 1994). Overall, these data demonstrate that g_m can be a significant photosynthetic constraint in developing leaves. Clearly, more work investigating the dynamics of g_m in relation to development of foliage photosynthetic potentials and leaf internal architecture is needed.

23.2.8. Timing of leaf development in plants with varying leaf longevity

Plant species widely differ in the rate of development of photosynthetic processes. Under non-stressed conditions, the period for full leaf expansion typically ranges from 10 days to 30–40 days (Tardieu *et al.*, 1999, Fig. 23.6a), and the rate of leaf expansion does not vary among different plant functional types (Fig. 23.6a). However, as the development of leaf internal structure, chloroplast multiplication and accumulation of pigments and photosynthetic proteins can proceed with time kinetics different from the expansion growth, the time for photosynthetic maturity not necessarily matches the time for leaf full expansion. In herbaceous annuals and perennials, full photosynthetic maturity is observed 10–25 days after leaf unfolding, sometimes even before full leaf expansion, while this time period can be 30–40 days in deciduous woody species and 30–80 days in evergreen conifers and broad-leaved evergreen woody species (Fig. 23.6a). Because full photosynthetic competence is reached later than full leaf expansion, the photosynthetic development in these species has been denoted as 'delayed greening' (Kursar and Coley, 1992a,b, Fig. 23.6c; Miyazawa *et al.*, 1998; Miyazawa and Terashima, 2001). Detailed anatomical studies demonstrate that the delay in reaching full photosynthetic potential is associated with belated development of internal leaf architecture. In fact, in addition to the positive correlation with leaf longevity (Fig. 23.6c,d), the delay in foliage development is positively related to LMA (Miyazawa *et al.*, 1998; Miyazawa and Terashima, 2001), demonstrating that larger thickness and more robust structure of leaves with larger lifespan likely explains the longer time period for leaf development in species with greater longevity. Although the

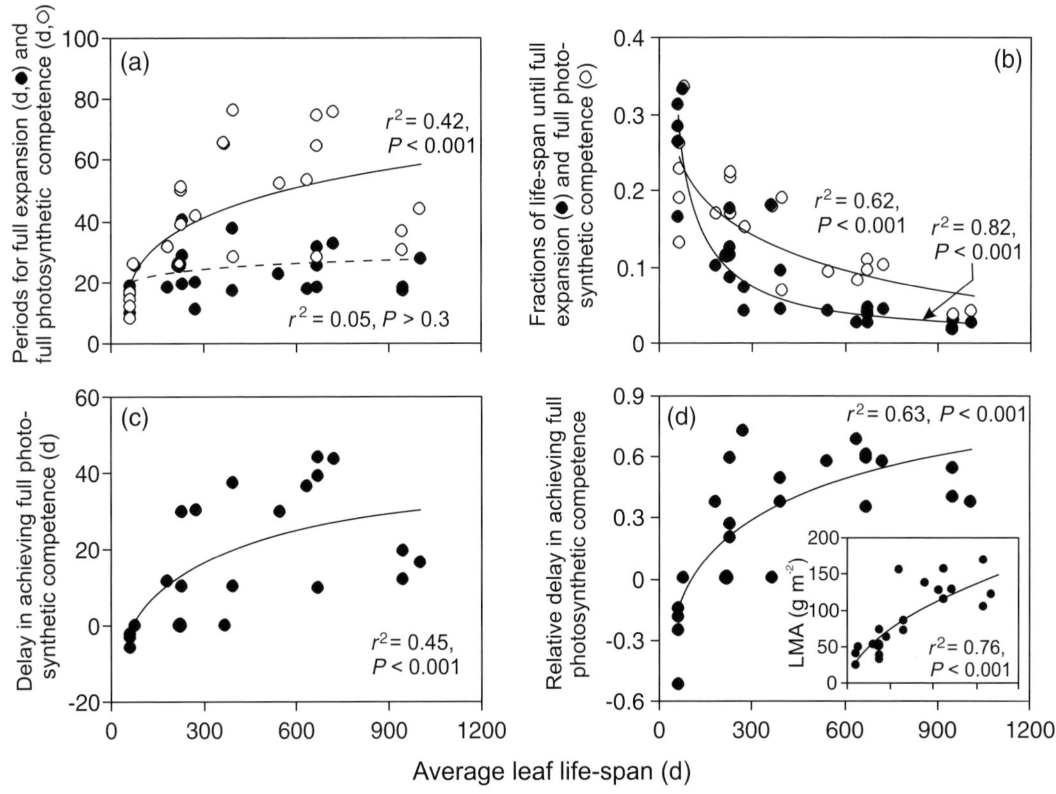

Fig. 23.6. Meta-analysis of the relationships between average leaf longevity (LL) and **(a)** the time for full leaf expansion (τ_E, filled symbols) and full photosynthetic competence (τ_P, open symbols); **(b)** the fraction of leaf lifespan until full expansion (τ_E/LL, filled symbols) and full photosynthetic competence (τ_P/LL, open symbols); **(c)** the delay period in photosynthetic maturation (the difference between the times for full photosynthetic competence and full expansion, $\tau_D=\tau_P-\tau_E$); and **(d)** the relative delay period (fraction of total photosynthetic maturation, τ_D/τ_P) in warm-temperate and tropical broad-leaved species. The inset in (d) demonstrates the correlation between LL and leaf dry mass per unit area. The dataset of leaf developmental kinetics (n=25) is mainly based on Miyazawa *et al.* (1998) who provided information for leaf growth dynamics and photosynthetic development for six original species and reviewed information for 14 additional species. All species were included from Miyazawa *et al.* (1998), except for the deciduous temperate vine *Actinidia deliciosa*, for which other literature data provided contrasting information (Greer, 1996; Greer *et al.*, 2004). In addition, information from three additional studies was included: Cromer *et al.* (1993) – *Gmelina arborea*; Miyazawa and Terashima (2001) – *Castanopsis sieboldii*, *Phaseolus vulgaris*; Miyazawa *et al.* (2003) – *Quercus glauca*, *P. vulgaris*. Data on average leaf life-span were obtained from Kitajima *et al.* (1997); Miyaji *et al.* (1997); Nitta and Ohsawa, (1997); Wright *et al.* (2004); Yukawa and Tsuda (1986) and on the basis of authors' knowledge of species biology. The data were fitted by asymptotic non-linear regressions in the forms of $y=a+b\mathrm{Log}(x)$ and $y=ax^b$. The non-significant regression in (a) is denoted by a dashed line.

achievement of full photosynthetic competence is more time consuming in an absolute time scale in such species, relative to total leaf lifespan, structurally more robust long-living leaves spend a shorter period of their lifetime for foliage construction (Fig. 23.6b). Thus, from the perspective of leaf longevity, 'delayed' foliage development is not necessarily a disadvantage as it allows the plants to form robust foliage that 'catches up' during the rest of its lifetime.

As shown in the Section 23.3, many environmental factors can either delay or speed up the rate of foliage development. Thus, in seeking general correlations between the rate of foliage development and leaf longevity and structure of mature leaves, it is important to compare species from similar climates and similar environmental conditions. Ideally, for testing the fundamental relationships, the measurements should be conducted under controlled conditions, as leaf area dynamics may be very complex under realistic field conditions (Greer *et al.*, 2004). The longevity/developmental-rate analysis depicted in Fig. 23.6 is based on species from warm temperate to tropical climates. To gain further insight into the variations in the rate of development of leaf photosynthetic potentials in dependence on leaf structure and longevity, we suggest that analogous relationships should be developed for other climatic zones.

23.3. LEAF DEVELOPMENT IN RELATION TO LEAF GROWTH ENVIRONMENT

The interaction of endogenous factors and environmental signals plays a key role in the development and differentiation of chloroplasts and development of photosynthetic capacity. Leaf growth and development belong to the most sensitive plant processes affected by virtually all environmental factors. Environmental drivers affect both leaf-expansion growth and development of internal structure and accumulation of photosynthetic proteins (Tardieu *et al.*, 1999; Van Volkenburgh, 1999; Walter and Schurr, 2005). Owing to the strong dependency of the rate of leaf-expansion growth on the environment, leaf developmental status is often not characterised on the basis of chronological age but biological age, or more correctly 'morphological' age, using so-called leaf plastochron index (LPI, Maksymowych, 1973). LPI for a given leaf is approximately equal to the number of leaves with defined size (typically 10–20 mm) formed after the formation of the given leaf. While the LPI is a useful concept for analysing leaf-expansion growth, it does not consider the environmental effects on leaf internal architecture that may follow kinetics different from leaf expansion (e.g., Sims *et al.*, 1998a). Light, water and nutrient availability, ambient CO_2 concentration and temperature belong to the key factors affecting leaf-area expansion and internal architecture (Dale, 1982; Woodward *et al.*, 1986; Cromer *et al.*, 1993; Sims *et al.*, 1998a,b; Van Volkenburgh, 1999; Walter and Schurr, 1999, 2005; Schurr *et al.*, 2000; Poorter *et al.*, 2009). Here we analyse the effects of varying light to exemplify the importance of environment on the development of photosynthetic apparatus. Apart from the relevance of considering single-factor responses, environmental drivers can interactively affect leaf development (Sims *et al.*, 1998a,b; Walter *et al.*, 2007), and understanding such interactions is crucial to predict the development of leaf photosynthetic competence in field conditions.

In addition to the direct environmental effects on leaf development during an active leaf-growth period, leaf development can also depend on previous environmental conditions prevailing during the formation of leaf primordia (Hansen, 1959; Uemura *et al.*, 2000). The degree of predetermination of developmental processes largely varies among species and can even vary among the leaves of a given plant. Thus, for understanding the response of leaf photosynthetic development to the immediate leaf-growth environment it is often important to consider previous environmental conditions. Furthermore, during the growing season in the field, environmental conditions often change, and leaves acclimated to certain conditions during their formation become non-acclimated to the changed environment. The capacity for re-acclimation to modified conditions is strongly driven by leaf characteristics determined by the previous leaf environment, implying a significant interaction of past and present environmental conditions on leaf photosynthetic plasticity.

23.3.1. Responses of photosynthetic-machinery formation to immediate growth light environment

Large gradients in light occur along gap to understory continua and within plant canopies. Leaf-expansion growth is only moderately affected by medium to high growth irradiances or even somewhat inhibited in high light (Taylor and Davies, 1986; Walter *et al.*, 2007), while very low-light availabilities can inhibit leaf expansion in shade-intolerant species (Granier and Tardieu, 1999). At the same time, leaf thickness growth and biomass accumulation per area occur with a faster rate in high than under low light (Fig. 23.3 and Fig 23.5a, Niinemets *et al.*, 2004a; Grassi *et al.*, 2005). In tree canopies producing leaves at different canopy position, almost simultaneously, such as in temperate deciduous trees or in temperate conifers, the within-canopy variation in leaf thickness and LMA is generally small immediately after bud-burst, but within-canopy differences in leaf structure increase with increasing leaf age, resulting in several-fold variation in leaf structural traits between canopy top and bottom (Fig. 23.5a, Niinemets *et al.*, 2004a). As detailed anatomical studies demonstrate, greater biomass accumulation in higher light is associated with enhanced development of palisade parenchyma in broad-leaved species (Dengler, 1980), and with overall extension of mesophyll cells and increase of cell number in leaf cross-section in needle-leaved species with less differentiated mesophyll layers (Niinemets *et al.*, 2005c). Chloroplast multiplication and accumulation of photosynthetic proteins matches the increase in leaf thickness and LMA, and as a result, the rate of development in leaf photosynthetic capacity scales with the rate of change in LMA and the amount of light received by the given leaf (Fig 23.5a, Niinemets *et al.*, 2004a).

In addition to quantitative modifications owing to accumulation of photosynthetic biomass per unit area, expression of genes coding photosynthetic proteins is controlled by the redox status of the photosynthetic apparatus (Marín-Navarro *et al.*, 2007). This results in changes in the stoichiometry of the main components of the photosynthetic apparatus. In particular, the fraction of N allocated to the proteins limiting foliage photosynthetic potentials, capacity for J_{max} and $V_{c,max}$ (Eqns 23.1–23.2) is higher at high light. In contrast, the fraction of N in chlorophyll and light-harvesting pigment-binding complexes (Eqn. 23.3) is higher in low light (Makino *et al.*, 1997; Niinemets *et al.*, 2004a; Katahata *et al.*, 2007).

As leaves developing at higher light have greater excitation pressures and greater probability for photodamage, development of photoprotective capacity also varies in dependence on light availability. In very young developing leaves undergoing greening, carotenoid and glutathione concentrations are already high with small differences at different light availabilities. However, the concentrations of xanthophyll-cycle carotenoids, zeaxanthin and antheraxanthin, and ascorbate increase more strongly with further increases in leaf age in high-light exposed leaves than in low-light leaves (Fig. 23.1, Garcia-Plazaola and Becerril, 2001). Alternatively, under low light, light-harvesting β,ε-carotenoids lutein and its precursor α-carotene can accumulate (Garcia-Plazaola and Becerril, 2001; Matsubara *et al.*, 2009). Such a modification in carotenoid composition is associated with upregulation of the β-cyclases with respect to the ε-cyclases under high light (Hirschberg, 2001; Cunningham, 2002), leading to an enhanced proportion of photoprotective β,β-carotenoids zeaxanthin and antheraxanthin relative to light-harvesting β,ε-carotenoids.

23.3.2. Predetermination of foliage characteristics by former environmental conditions and foliage re-acclimation

In several species, the structure of mature leaves is already partly determined by the environment during bud formation (Kimura *et al.*, 1998; Uemura *et al.*, 2000), and the environmental conditions during leaf development are mainly expected to modify the length of the leaf developmental period. In other species, and in cases where leaves develop without an intervening period of dormancy, leaf structure and development of leaf photosynthetic apparatus is mainly driven by the environmental conditions during foliage development. In continuously expanding canopies, there may even be mixed responses with leaf internal structure being determined by the light environment experienced by older leaves and chloroplast structure by immediate leaf-growth environment (Yano and Terashima, 2001), a response possibly mediated by sugars transported from older leaves to developing foliage (Yano and Terashima, 2004). Nevertheless, even in species in which the structure of leaf primordia is determined by a previous environment, the conditions during bud-burst and leaf development can still alter the structure of mature leaves (Labrecque *et al.*, 1989).

In the field, environmental factors fluctuate during a day and between days, weeks and seasons. Although leaf photosynthetic apparatus does not track very rapid environmental alterations, leaves can adjust to longer-term environmental signals. In particular, gaps are often formed in plant canopies exposing shade-adapted leaves to high irradiance. Alternatively, in rapidly expanding canopies and in evergreen canopies, older foliage becomes gradually shaded by newly developed foliage, emphasising the importance of re-acclimation of older leaves to new environmental conditions. In the following, we analyse foliage re-acclimation to altered irradiance.

In young developing leaves, foliage photosynthetic capacities may essentially fully adjust to modified irradiance through alterations in leaf thickness and LMA (Sims and Pearcy, 1992; Yamashita *et al.*, 2002; Oguchi *et al.*, 2005). However, the capacity for such structural re-acclimation depends on leaf age at the time of modification of light availability (Section 23.3.3; Sims and Pearcy, 1992; Yamashita *et al.*, 2002). In mature leaves, structural re-acclimation of photosynthetic capacity is inherently limited owing to crosslinking and lignification of cell walls (Sims and Pearcy, 1992; Yamashita *et al.*, 2002; Oguchi *et al.*, 2005), and photosynthetic capacity can mainly be modified by changes in the number of chloroplasts per leaf area and by alteration of the fraction of N in Rubisco and limiting components of photosynthetic electron transport chain (Fig. 23.4 for changes in photosynthetic capacity as the result of altered N partitioning). Given that chloroplasts should be adhered close to mesophyll cell walls to maximise liquid-phase diffusion conductance for CO_2 (Terashima *et al.*, 2005), acclimation of low-light-adapted leaves to high irradiance is dependent on the availability of the space unfilled by chloroplasts along the cell wall, i.e., on the ratio of chloroplast exposed surface area to inner mesophyll-cell-wall area (S_c/S_m) (Oguchi *et al.*, 2003; Oguchi *et al.*, 2005). Because the outer mesophyll-cell surface can be often fully covered by chloroplasts even in shade-adapted leaves ($S_c/S_m \approx 1$) (Oguchi *et al.*, 2005), anatomical constraints can limit the re-acclimation of photosynthetic capacity of such leaves to high irradiance. Alternatively, no such anatomical limits exist for re-acclimation of foliage photoprotective capacity. Foliage antioxidant and xanthophyll-cycle carotenoid pools can essentially fully re-acclimate to altered irradiance, although the acclimation kinetics may depend on previous leaf light environment (García-Plazaola *et al.*, 2004; García-Plazaola *et al.*, 2008a).

Acclimation to low light is mainly associated with redistribution of N from Rubisco and other rate-limiting proteins to light-harvesting chlorophyll-binding pigment-protein complexes (Pons and Pearcy, 1994), which is reflected in increases in foliage chlorophyll content and chlorophyll to N ratio (Brooks *et al.*, 1994, 1996). However, light pathlength is necessarily long in thicker leaves developed in higher light. This entails that high internal shading within the leaves may limit shade acclimation of high-light-developed older foliage. These examples collectively demonstrate that determination of foliage characteristics during leaf developmental period can significantly constrain foliage photosynthetic plasticity for the rest of the growing season.

23.4. AGE-DEPENDENT ALTERATIONS IN LEAF PHOTOSYNTHETIC ACTIVITY IN NON-SENESCENT LEAVES

After rigidification of cell walls, profound modifications in leaf structure become impossible, but still a series of physiological and structural changes occur in fully developed mature leaves, reflecting foliage ageing. Typically, foliage photosynthetic potentials of non-senescent mature leaves continuously decrease at a low rate until the onset of leaf senescence (Teskey *et al.*, 1984; Bungard *et al.*, 2002; Niinemets *et al.*, 2004a). These changes in foliage photosynthetic potentials occur at much slower rate compared with rapid reductions in the rate of photosynthesis after induction of senescence. Nevertheless, such changes can potentially importantly alter whole-season carbon gain and need consideration in larger-scale models (Wilson *et al.*, 2000a,b, 2001; Morecroft *et al.*, 2003; Grassi and Magnani, 2005). Especially in evergreen species, where older foliage can constitute the bulk of whole-canopy leaf biomass and the reductions of foliage photosynthetic rates occur over multiple growing seasons, such age-dependent modifications can have a large effect on whole-canopy carbon acquisition.

23.4.1. Changes in foliage function in non-evergreen leaves

In broad-leaved deciduous woody species, leaf N and chlorophyll contents per dry mass continuously decrease after leaf maturation (Wilson *et al.*, 2000b; Kitajima *et al.*, 2002; Niinemets *et al.*, 2004a; Grassi *et al.*, 2005). This reduction is associated with decreases in $V_{c,max}$ and J_{max} and foliage photosynthetic capacity (Wilson *et al.*, 2000b, 2001; Kitajima *et al.*, 2002; Morecroft *et al.*, 2003; Niinemets *et al.*, 2004a; Grassi and Magnani, 2005). Although such declines in non-senescent leaves during the growing season have been consistently observed, these reductions are generally in the order of only 10–30%, which is relatively small compared with reductions of 50–90% during leaf senescence (Section 23.5).

In herbaceous species with shorter leaf lifespan and continuous development of new leaves, reductions in foliage photosynthetic potentials can start even before full leaf expansion (Miyazawa and Terashima, 2001; Miyazawa *et al.*, 2003; negative 'delay' periods in Fig. 23.6c,d). Such rapid decreases in photosynthetic potentials are compatible with the overall high turnover of leaves and reallocation of N to growing leaves (Hikosaka, 2003).

Apart from the reductions in the content of chlorophyll and proteins and in photosynthetic capacity, there is evidence of a greater degree of foliage lignification (larger foliage carbon contents, presumably reflecting accumulation of carbon-rich lignin especially in high-light exposed leaves, Jayasekera and Schleser, 1991) and higher LMA and greater density (mass per unit volume) (Wilson *et al.*, 2000a; Kitajima *et al.*, 2002; Niinemets *et al.*, 2004a; Saito *et al.*, 2006) of older leaves. In addition, content of non-mobile nutrients such as calcium and boron also continuously accumulate throughout the leaf lifespan (Tew, 1970; Oleksyn *et al.*, 2000). Calcium and boron are mainly accumulated in the cell wall and their accumulation reflects crosslinking of cell-wall polysaccharides (Demarty *et al.*, 1984; Fleischer *et al.*, 1998). In addition, older leaves that have sustained environmental stress in the past owing to, for instance, exposure to high ozone concentrations (Günthardt-Goerg *et al.*, 1997; Ljubešic and Britvec, 2006) can have higher cell-wall thickness. Thicker and more strongly lignified and crosslinked cell walls with a lower pore fraction likely provide the explanation for reduced mesophyll diffusion conductance from leaf internal air space to the carboxylation sites in the chloroplasts in older non-senescent leaves (Grassi and Magnani, 2005). Lower g_m further constrains foliage photosynthesis rates at given biochemical capacities of photosynthetic apparatus. Qualitatively similar

g_m reduction patterns have been observed independently of whether leaf ageing occurs over months as in deciduous trees (Grassi and Magnani, 2005), over weeks as in the annual herb with medium longevity *Spinacia oleracea* (Delfine *et al.*, 1999) or over days as in the annual short-living herb *Arabidopsis thaliana* (Flexas *et al.*, 2007c). Biochemical and structural limitations together can result in up to 50% lower photosynthetic rates immediately before the onset of leaf senescence than the rates observed in young fully developed leaves (Grassi and Magnani, 2005).

23.4.2. Modifications in foliage structure and chemistry in evergreen species

Although the age-dependent modifications in foliage photosynthetic potentials occur qualitatively similarly in species with short and long lifespan, evergreen foliage sustains a considerably larger range and more cycles of environmental stress than deciduous foliage. In addition, age-dependent modifications occur over a longer time period. As a result, the changes in foliage function observed before leaf senescence are more profound in evergreen species. For instance, foliage N contents per dry mass are approximately 1.5-fold lower in 6-year-old needles of conifer *Pinus heldreichii* compared with current-year needles, and these changes are accompanied by an almost fourfold reduction in foliage photosynthetic capacity (Oleksyn *et al.*, 1997). Analogously, foliage photosynthetic capacity of 9-year-old needles is reduced by more than an order of magnitude relative to current-year needles in conifer *Abies amabilis* (Teskey *et al.*, 1984). Reductions in photosynthetic capacity are also accompanied by reductions in light-harvesting chlorophyll and carotenoid contents (Oleksyn *et al.*, 1997), while there is almost liner accumulation of α-tocopherol content with leaf age (Hormaetxe *et al.*, 2005b). Accumulation of α-tocopherol may serve as a molecular marker of leaf ageing (Hormaetxe *et al.*, 2005b). These changes in pigment system and accumulation of certain lipophilic antioxidants are reflected in chloroplast ultrastructural changes. The chloroplasts in older leaves typically have more plastoglobuli and more densely stacked thylakoid lamellae (Fig. 23.7).

The age-dependent chemical and photosynthetic modifications are also associated with larger LMA and leaf density (Niinemets *et al.*, 2004b, Fig. 23.8) and greater leaf dry to fresh mass ratio (Teskey *et al.*, 1984) in older leaves. In addition, cell-wall thickness is larger in older non-senescent leaves of evergreens (Saito *et al.*, 2006, Fig. 23.7). Apart from qualitative similarity of these structural alterations

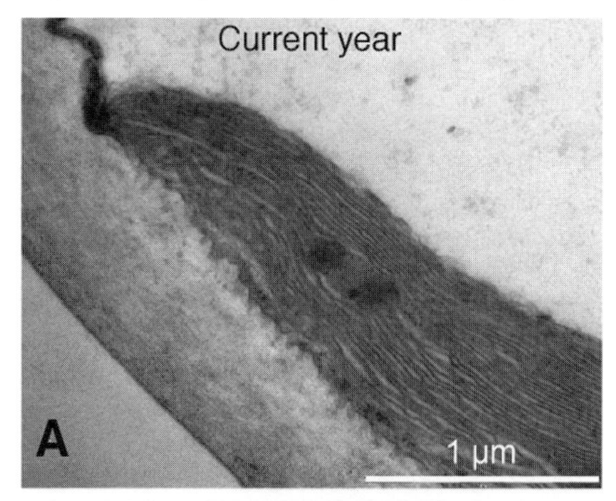

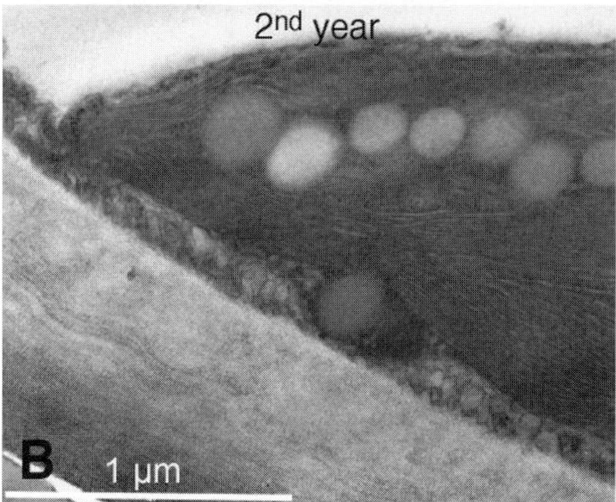

Fig. 23.7. Chloroplasts from current (A) and 2-year-old (B) leaves of *Buxus sempervirens*. Leaf ageing is accompanied by extensive ultrastructural changes in chloroplasts, among which the most conspicuous is the accumulation of plastoglobuli. Increases in cell-wall thickness in older leaves are also visible. Micrographs by Unai Artetxe (UPV/EHU). Hormaetxe *et al.* (2005b) provides the description of the methods.

in evergreen and deciduous leaves, evergreen conifers have unique age-dependent modifications in foliage structure. Needle thickness and width continuously increase with increasing needle age (e.g., Niinemets, 1997a) as a result of secondary growth in the needle central cylinder (Ewers, 1982; Gilmore *et al.*, 1995). Although the needles do expand as the result of secondary growth, the endodermal belt of the central cylinder is strongly lignified, curbing needle secondary expansion. Secondary cambium activity within the limited space results in crushing of needle xylem and phloem (Ewers, 1982; Gilmore *et al.*, 1995), potentially leading to reductions in needle hydraulic conductance and

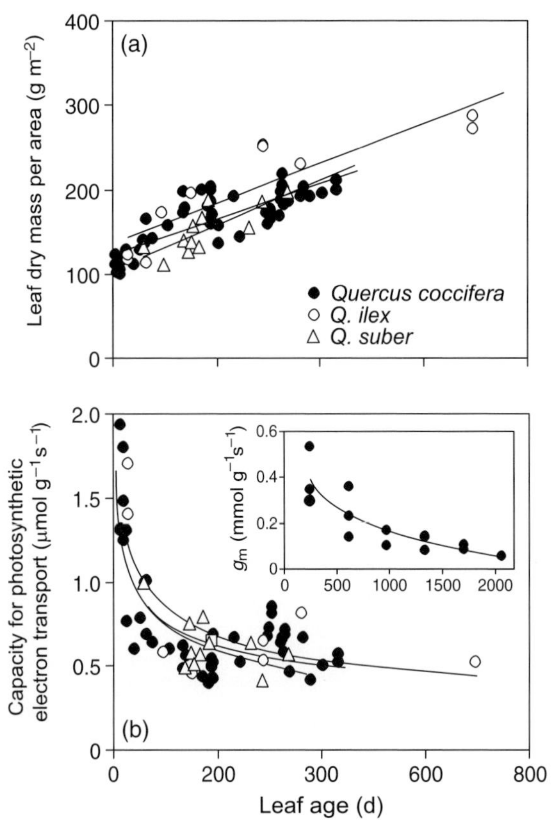

Fig. 23.8. Age-dependent modifications in leaf dry mass per unit area **(a)** and concomitant changes in the capacity for photosynthetic electron transport per unit dry mass (J_{max}/mass) **(b)** and mesophyll-diffusion conductance from sub-stomatal cavities to chloroplasts (g_m, inset in b) in non-senescent leaves of three Mediterranean evergreen broad-leaved sclerophyll oak species. Modified from Niinemets *et al.* (2004b, data in main panels) and Niinemets *et al.* (2005a, data in the inset).

in the capacity for assimilate translocation. Limited hydraulic conductance can explain the age-dependent reductions in stomatal conductance in older foliage (Ethier *et al.*, 2006) and reductions of photosynthesis with tree ageing (Drake *et al.*, 2010), whereas limited phloem transport capacity can explain greater non-structural carbohydrate concentrations in older needles (Niinemets, 1997a).

Owing to the outlined structural modifications, g_m is strongly curbed in older foliage of conifers (Ethier *et al.*, 2006) and broad-leaved evergreens (Niinemets *et al.*, 2005a, Fig. 23.8). Some studies suggest that age-dependent modifications in photosynthetic capacity and g_m occur in parallel, such that the CO_2 drawdown from sub-stomatal cavities to chloroplasts, A_{max}/g_m (Eqn. 23.4), is weakly affected by decreases in g_m in evergreens (Ethier *et al.*, 2006). However, other studies have observed that the reduced g_m does limit photosynthesis more strongly in older leaves (Niinemets *et al.*, 2005a; Flexas *et al.*, 2007c).

This evidence collectively demonstrates that age-dependent changes in non-senescent leaves can importantly modify foliage photosynthesis. Differences in photosynthetic potentials of different-aged evergreen leaves can be considered in some large-scale photosynthesis models (Medlyn, 2004), but none of the current models considers age-dependent alterations in g_m. Also, continuous changes in photosynthetic capacity and g_m in non-senescent deciduous leaves are not considered in large-scale simulation analyses. We suggest that it is inevitable to take the dynamics of photosynthetic potentials in non-senescent leaves into account in future carbon-gain model developments.

23.5. PHOTOSYNTHESIS DURING LEAF SENESCENCE

Senescence that follows the photosynthetically active period of plants is a pre-programmed process that can be triggered by internal or external factors. It is a specific type of programmed cell death (Yoshida, 2003; Lim *et al.*, 2007), but abiotic or biotic stress may also induce apoptosis-like cell death that is associated with early disruption of vacuole and very rapid leaf death (Fukuda, 2000; Gunawardena, 2008). In practice, the apoptosis-like cell death and normal senescence are sometimes difficult to separate (Günthardt-Goerg and Vollenweider, 2007), although programmed cell death has often localised appearance characterised by necrotic spots or chlorotic flecks (Günthardt-Goerg and Vollenweider, 2007), whereas senescence involves the entire leaf.

Rapid reduction of photosynthetic rates and chlorophyll are the characteristic modifications in foliage photosynthetic apparatus in senescing leaves (Valjakka *et al.*, 1999; Herrick and Thomas, 2003; Katahata *et al.*, 2007). Massive retranslocation of N, phosphorus and other mobile nutrients occurs during the senescence, culminating with the death of the organ or even the entire individual (Munné-Bosch, 2007). Senescence is an active deteriorative process consisting of a regulated sequence of events involving activation of specific metabolic pathways. In fact, senescence is preceded by a peak in transcriptional activity (Andersson *et al.*, 2004; Lim *et al.*, 2007) and involves a major shift in gene expression, with the activation of more than 2,000 genes in *A. thaliana* (Gepstein,

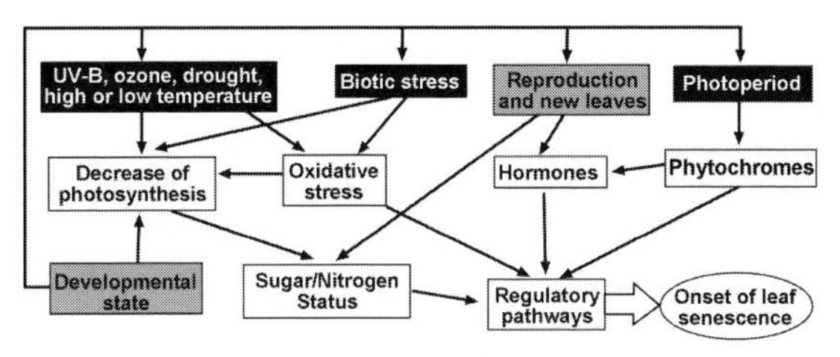

Fig. 23.9. A model for the induction of leaf senescence. The onset of leaf senescence is a complex process that results from the interaction between internal (grey boxes) and environmental (black boxes) factors. As a consequence of these interactions, leaf senescence can be accelerated or delayed. Based on the information in Lim *et al.* (2007) and Munné-Bosch (2008).

2004). These genes belong to five major groups: macromolecule degradation, nutrient recycling, defences, transcriptional regulation and signal transduction (Guo *et al.*, 2004).

Senescence can be observed at the organ level (flowers, leaves, fruits), occurring repeatedly during plant lifespan (polycarpic senescence), or affecting the whole plant and resulting in plant death (monocarpic senescence). Organ-level senescence may end with the abscission of the organs (e.g., leaves) or with maturation (e.g., fruits). Senescence of reproductive organs is in general irreversible, but leaf senescence may be fully reversible, especially during the initial stages of senescence (Thomas *et al.*, 2003). In the following, we will concentrate on mono- and polycarpic senescence of photosynthetic organs (leaves). In polycarpic senescence, nutrients are either reallocated to developing leaves during leaf-growth period or in storage organs to fuel the growth in the subsequent season. In monocarpic senescence, nutrients are translocated to growing fruits. All these types of senescence have many mechanisms and processes in common.

23.5.1. Determination of the onset of leaf senescence: modified environment versus seasonality

The timing of leaf senescence is generally controlled by developmental age, but it is still unknown how the process is initiated (Lim *et al.*, 2007). Apart from the developmental age, stress factors such as drought, high or low temperatures, pathogens, ozone or mineral deficiencies may initiate early senescence (Munné-Bosch *et al.*, 2001; Munné-Bosch and Alegre, 2004). Enhanced oxidative stress during unfavourable periods can be involved in the induction of early senescence (Fig. 23.9, Munné-Bosch *et al.*, 2001). This has been indirectly shown by the delay in senescence in mutants with reduced metabolic activity and consequently, with lower capacity for generation of ROS (Lim *et al.*, 2007).

In actively growing plants, the initiation of senescence also strongly depends on endogenous factors. In particular, on source-sink relationships and plant C-N balance (Wingler *et al.*, 2004, Fig. 23.9). Formation of new leaves and reproductive organs typically induces nutrient remobilisation from older leaves leading to senescence. Removal of young leaves or reproductive structures typically delays or reverses the senescence (Weikert *et al.*, 1989; Gwathmey *et al.*, 1992). Sugar levels, either high or low, have also been hypothesised to be involved in initiation of senescence (van Doorn, 2008; Wingler *et al.*, 2009). Thus, ageing or stress contributing to the loss of photosynthesis and reduced sugar levels (sugar 'starvation') might initiate senescence (Quirino *et al.*, 2000). From this perspective, when photosynthesis declines below a threshold value, senescence is initiated. For example, leaf shading in rapidly expanding canopies results in initiation of senescence of shaded leaves (Frantz *et al.*, 2000; Weaver and Amasino, 2001). Alternatively, sugar accumulation, for instance in high-light exposed leaves or in response to environmental stress such as low nutrient availability inhibiting growth more than photosynthesis, may also induce senescence (Miller, 2000; van Doorn, 2008; Wingler *et al.*, 2009, Fig. 23.9). This can occur through feedback inhibition of photosynthesis, but induction of senescence in leaves accumulating sugars may also reflect altered sink-source relationships (Miller, 2000; van Doorn, 2008, Wingler *et al.*, 2009).

In perennials, leaf senescence can be triggered by stress factors and sink-source relationships similarly to monocarpic plants (Munné-Bosch and Alegre, 2004; Munné-Bosch, 2008). However, autumn senescence in deciduous trees constitutes a special case with a stronger interaction of endogenous and environmental factors (Fig. 23.9). The onset of autumn senescence is controlled by temperature and day length in an interactive manner (Rosenthal and Camm, 1996). Both higher temperature and extended day length can delay the onset of autumn senescence (Rosenthal and Camm, 1996). Phytochromes play a key role in the perception of photoperiod and the initiation of the process (Briggs, 2005, Fig. 23.10 for the visualisation of enhanced leaf retention owing to artificial lighting; Olsen and Junttila, 2002). Furthermore, the sensitivity to temperature and photoperiod of senescence induction is under strong genetic control. When grown in a common garden, northern provenances of Northern-hemisphere temperate trees senesce earlier than the southern provenances of given species (Ovaska *et al.*, 2005; Fracheboud *et al.*, 2009). Such within-species

Fig. 23.10. A change in day length is an important environmental signal triggering leaf senescence in temperate deciduous trees. Illustration of delayed autumnal leaf senescence in temperate deciduous tree *Liquidambar styraciflua* as the result of artificial lighting. Leaves exposed to artificial light are retained for several weeks longer than the leaves perceiving normal photoperiod, a response likely mediated by phytochromes (Briggs, 2005).

variations in leaf phenology can play an important role in species adaptation to globally changing environmental conditions and also have a large practical value as selection criteria for more productive provenances under global change.

23.5.2. Photosynthetic apparatus during leaf senescence: from chloroplasts to gerontoplasts

As the extensive overlap in gene-expression patterns during senescence induced in response to several stresses demonstrates, once induced, the development of senescence is independent of the initiating events (Lim *et al.*, 2003). Even before any visible signs of senescence, transcripts from photosynthetic genes start to decrease (Valjakka *et al.*, 1999; Miller *et al.*, 2000) while the transcripts of degrading enzymes and stress hormones, such as ethylene and senescence-associate genes (SAG), many with still unknown function, are increasing (Lee *et al.*, 2001b; Andersson *et al.*, 2004; Lim *et al.*, 2007). Proteolysis in plastids starts early, but mitochondrial respiration is maintained or increased until the final stages of leaf senescence to supply energy for the degradation and export of recycled chemical constituents (Collier and Thibodeau, 1995; Yoshida, 2003, Fig. 23.11). The nucleus also remains functional to allow the expression of senescence-related genes. Enhanced proteolysis is accompanied by accelerated nutrient recycling as evidenced by greater phloemsap amino acid concentrations (Sandström, 2000), as well as by higher concentrations of mobile mineral nutrients in phloem cells (Eschrich *et al.*, 1988). The loss of functional phloem connections, necessary to export nutrients outside the leaf, determines the end of the process, although in some species phloem blockage can start relatively early (Jongebloed *et al.*, 2004).

The senescing plastids in photosynthetic tissues are called 'gerontoplasts' (Thomas *et al.*, 2003). During conversion of chloroplasts to gerontoplasts, a successive sequence of morphological changes denotes dramatic alterations in photosynthetic apparatus. The contact between the grana is lost and thylakoids disappear progressively forming plastoglobuli, which increase in number and volume (Park *et al.*, 2007). Plastoglobuli are not only passive accumulation structures, but also play an active metabolic role in synthesis and degradation of carotenoids and other isoprenoid-derived molecules, such as tocochromanol (vitamin E) and lipids (Ytterberg *et al.*, 2006). Remaining thylakoids are less structured and starch granules tend to disappear. In final stages, plastoglobuli extrude in blobs from the gerontoplasts (Park *et al.*, 2007). Most of these modifications are reversible and gerontoplasts retain the capacity for re-conversion to chloroplasts in re-greening tissues (Biswal *et al.*, 2003).

23.5.3. Dismantling of photosynthetic enzymes and chlorophyll degradation in senescing tissues

Up to 75% of leaf N is located in chloroplast proteins with Rubisco and pigment-binding proteins as the main fractions (Hörtensteiner and Feller, 2002). Stromal enzymes are degraded first leading to interruption of photosynthetic activity. It is postulated that Rubisco degradation is initiated by ROS (Ishida *et al.*, 1999; Nakano *et al.*, 2006) and completed by the action of endo- and ecto-peptidases that generate remobilisable free amino acids (Minamikawa *et al.*, 2001). Besides proteolytic activities, protein susceptibility to proteolysis affected by the presence of protein stabilisers or enzyme inhibitors also plays an important role in protein degradation (Hörtensteiner and Feller, 2002).

Breakdown of thylakoid proteins proceeds later than stromal proteins and starts with the loss of cyt b_6f complex and ATPase (Humbeck *et al.*, 1996). Chlorophyll-binding proteins are stabilised by chlorophyll and carotenoids, and their disassembly requires pigment release. Thus, the content of pigment-binding proteins decreases synchronously with chlorophyll degradation (Park *et al.*, 2007). Chlorophyll *a*/*b* ratio decreases during leaf senescence, indicating that the antenna proteins degrade at a slower rate than the reaction centres (Keskitalo *et al.*, 2005).

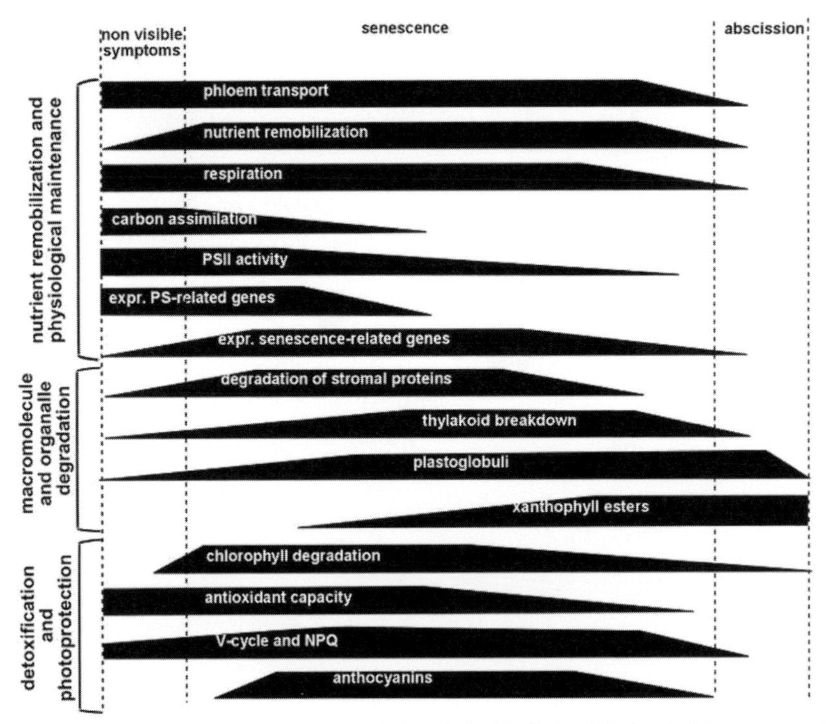

Fig. 23.11. Simplified representation of the timing of physiological and biochemical changes occurring during leaf senescence. Progressive decreases of photosynthetic capacity in senescing leaves are accompanied by remobilisation of nitrogen from chloroplasts and detoxification of chlorophyll catabolites. Based on the information presented by Keskitalo *et al.* (2005) and Munné-Bosch (2008).

Chlorophyll degradation and the resulting unmasking of carotenoids that leads to leaf yellowing is the most conspicuous marker of leaf senescence. This process occurs in several sequential highly regulated enzymatic steps, yielding progressively less photodynamic tetrapyrrole derivatives (Fig. 23.12). The immediate intermediates of chlorophyll degradation never reach significant concentrations in leaves, whereas non-fluorescent catabolites accumulate in vacuoles. The four N atoms from the chlorophyll tetrapyrrolic structure that represent 1–2% of total leaf N are not recovered from the final tetrapyrrole or monopyrrole pool in vacuoles, while remobilisation of N from chlorophyll-binding proteins can contribute 20% or more of cellular N recycled (Hörtensteiner, 2006). Therefore, the main function of chlorophyll degradation is the detoxification of photodynamic chlorophyll and its catabolites that could otherwise hamper the whole process of N remobilisation from chloroplast proteins. In fact, mutants defective in enzymes of chlorophyll catabolism develop lesions owing to the unusual accumulation of phototoxic catabolites (Hörtensteiner, 2006; Hörtensteiner and Feller, 2002). Alternatively reduction of chlorophyll content strongly decreases the efficiency of light harvesting (leaf absorptance) and is also correlated with reductions in the initial quantum yield of photosynthesis for an incident light as well as with the quantum efficiency of PSII (Adams *et al.*, 1990b; Greer, 1996; Keskitalo *et al.*, 2005).

The concentration of carotenoids released from thylakoids also decreases in senescing leaves, but with a slower rate than chlorophyll (Lee *et al.*, 2003, Fig. 23.1). Part of the carotenoid pool, in particular neoxanthin and violaxanthin, can be esterified with fatty acids and accumulate in plastoglobuli, while photoprotective carotenoids (antheraxanthin and zeaxanthin) remain unaltered or increase (Garcia-Plazaola *et al.*, 2003a; Keskitalo *et al.*, 2005, Fig. 23.1).

23.5.4. Photoprotective demand during senescence

Different components of the photosynthetic apparatus are degraded with varying rates and their degradation is induced at different times (Fig. 23.11). This can result in imbalances between the capacities for light capture and the rate of carbon assimilation, in turn increasing the fraction of excess excitation energy and leaf sensitivity to photodamage. This situation is exacerbated by the presence of

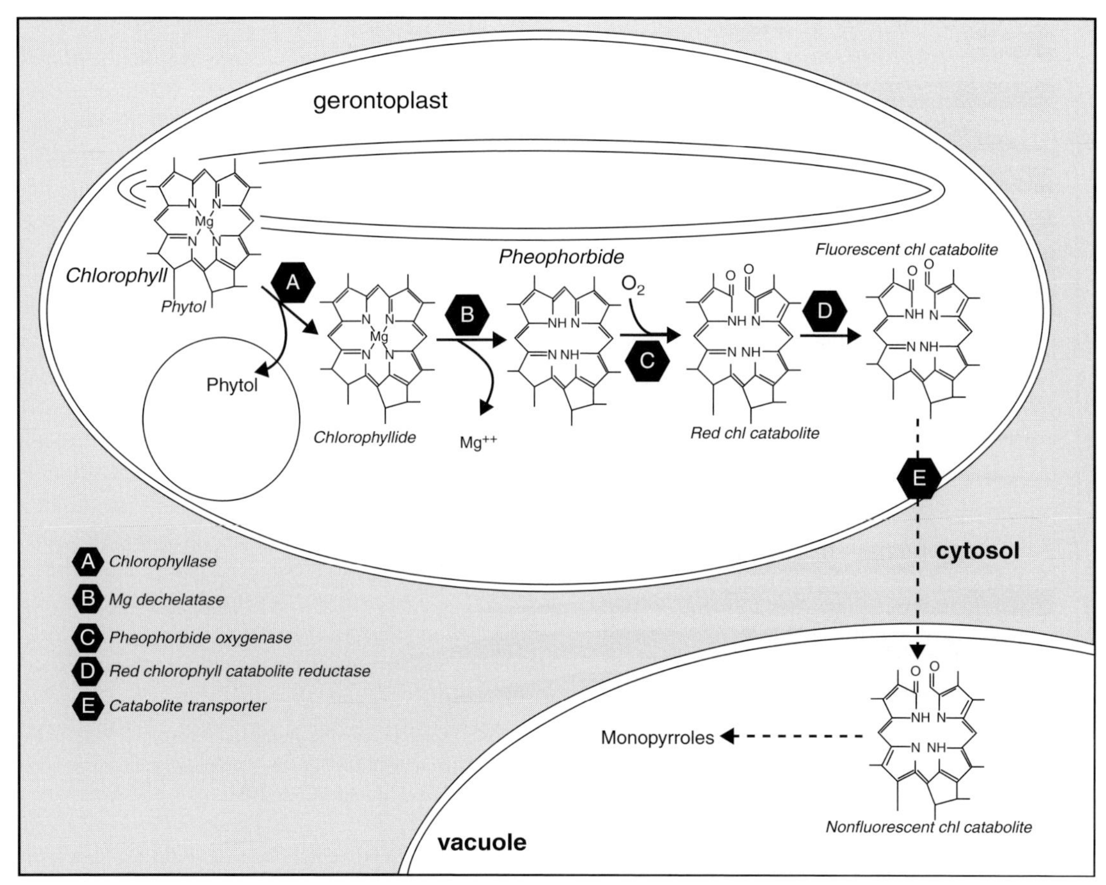

Fig. 23.12. Chlorophyll-breakdown pathways in higher plants. Chlorophyll degradation occurs in senescing chloroplasts called gerontoplasts, where chlorophyll catabolism generates phototoxic intermediates that are progressively detoxified to maintain low oxidative status in the leaves during senescence and allow for nitrogen remobilisation from the chloroplasts. The first step of chlorophyll degradation is the reduction of chlorophyll *b* to form chlorophyll *a*. Phytol chain is then removed by the action of chlorophyllase yielding a much more polar molecule chlorophyllide, whose central Mg atom is thereafter removed by a specific dechelatase. Subsequently, a series of colourless and progressively less photodynamic catabolites are generated after the opening of the tetrapyrrolic ring by an oxygenase and actively transported to the vacuole, where fluorescent intermediates are converted to non-fluorescent catabolites by several non-enzymatic reactions under acidic pH. The colourless non-phototoxic catabolites accumulate in the vacuole until leaf abscission. In some cases, the presence of monopyrrolic degradation products has also been detected (see Hörtensteiner, 2006 for more details; Matile *et al.*, 1999). An alternative extra-plastidial chlorophyll oxidative route catalysed by peroxidases has been also proposed, but it is probably restricted to cell autolysis (Takamiya *et al.*, 2000). The scheme is based on the information in Hörstensteiner and Feller (2002) and Hörtensteiner (2006).

photodynamic chlorophyll catabolites that can generate singlet oxygen ($^1O_2$) (Matile *et al.*, 1999). To maintain efficient nutrient resorption, chloroplast breakdown must be finely coordinated and protected against photodamage.

Senescing leaves are able to increase photoprotection by different means. The most conspicuous alteration is the accumulation of red-coloured anthocyanins, which shade the epidermal and upper palisade layers from blue and red light (Hoch *et al.*, 2001; Lee *et al.*, 2003; Keskitalo *et al.*, 2005). These flavonoids reduce the intensity of light reaching photoactive molecules such as chlorophyll and its catabolites, and also have a high antioxidative potential. Analogously, the increase of carotenoids relative to chlorophyll in senescing leaves reduces the amount of potentially harmful high-energy blue light reaching the chls (Merzlyak and Gitelson, 1995). Carotenoids are also involved in the observed increment in the xanthophyll cycle mediated non-photochemical excitation energy dissipation, reflecting a high content of zeaxanthin in the last stages of senescence (Lu *et al.*, 2001; Garcia-Plazaola *et al.*, 2003a; Fig. 23.1). In fact, NPQ increases progressively and remains high until the last stages of senescence (Wingler *et al.*, 2004). In the case of antioxidants, water-soluble compounds may decrease (ascorbate) or remain relatively invariable (glutathione), whereas lipid-soluble α-tocopherol contents increase before leaf fall in several species (Hormaetxe *et al.*, 2005b, Fig. 23.1). Despite all these chemical modifications in senescing leaves owing to low photosynthetic capacity, the fraction of excess light not used for photosynthesis is still larger in senescing than in fully mature leaves. As a result, senescing leaves are more sensitive to photoinhibition (Greer, 1996).

23.5.5. Alterations in mesophyll-diffusion conductance in senescing leaves

Dismantling of foliage photosynthetic apparatus and reduction of chloroplast number is expected to result in reduced chloroplast to total leaf surface-area ratio S_c/S and accordingly in reductions of mesophyll diffusion conductance from sub-stomatal cavities to chloroplasts, g_m. Furthermore, local thickening of cell walls owing to pectic intrusions has been observed in senescing leaves (Günthardt-Goerg and Vollenweider, 2007), especially in leaves where senescence has been accelerated by environmental stress, such as high ambient ozone concentrations (Günthardt-Goerg *et al.*, 1997; Ljubešic and Britvec, 2006). Such modifications can further reduce the liquid-phase diffusion conductance. Large reductions of g_m in senescent leaves have been observed in several studies (Loreto *et al.*, 1994; Evans and Vellen, 1996; Delfine *et al.*, 1999; Flexas *et al.*, 2007c; Zhang *et al.*, 2008b).

However, the reduction in g_m are parallelled by decreases in foliage photosynthetic rate, A, owing to protein breakdown. The key question is whether the CO_2 drawdown from sub-stomatal cavities to chloroplasts (C_i–C_c=A/g_m, Eqn. 23.4) varies during senescence, i.e., whether photosynthesis in senescent leaves is more strongly limited by g_m. Available evidence suggests that the drawdown generally increases with increasing leaf age (Loreto *et al.*, 1994; Delfine *et al.*, 1999; Flexas *et al.*, 2007c; Zhang *et al.*, 2008b), but not always (Evans and Vellen, 1996). The distribution of photosynthetic limitations between chemical and structural traits obviously depends on the extent to which senescence alters the content of photosynthetic enzymes, cell-wall properties, chloroplast number and size and the location of chloroplasts along the cell walls. So far, the existing data suggest that reduction in g_m is an additional important limitation of photosynthesis in senescing leaves.

23.5.6. Variations in the speed of senescence among and within species

Senescence commonly takes several weeks to complete, but there is a large variation in the rate of senescence and the 'completeness' of senescence indicated by the degree of removal of essential mineral nutrients before leaf abscission (Herrick and Thomas, 2003; Killingbeck, 1996). In temperate deciduous species, timing of autumnal senescence reflects a trade-off between the efficient N remobilisation at the expense of reduced photosynthesis in early senescent species and a higher carbon fixation at the expense of an enhanced risk of premature leaf death and loss of a significant fraction of plant N in late-senescent species (Keskitalo *et al.*, 2005; Niinemets and Tamm, 2005). In some species, very low rates of N resorption occur, and leaves containing 25–50% of pre-senescent chlorophyll and up to 75% of pre-senescent N contents are abscised. For instance, in N-fixing trees, such as *Alnus* spp., leaves are abscised during a short time period and only a moderate fraction of N is recovered before leaf abscission (Garcia-Plazaola *et al.*, 2003a; Niinemets and Tamm, 2005). Such a behaviour is also observed in 'stay green' phenotypes that retain antenna proteins and chlorophyll by blocking chlorophyll catabolism, while other proteins such as Rubisco are degraded (Bachmann *et al.*, 1994). In other species that retain leaves for longer time periods, as much as 80% leaf N can be remobilised and stored for the next growing season (del Arco *et al.*, 1991; Niinemets and Tamm, 2005) contributing significantly to plant fitness (May and Killingbeck, 1992). This is also the case with stress-induced leaf senescence, where nutrient removal from older leaves undergoing senescence fuels the growth of the new leaf flush expanding after the stress (Munné-Bosch and Alegre, 2004).

Nitrogen remobilisation can also vary with growth conditions, but is generally lower in shade positions in the canopy and in plants growing under low light (Weinbaum *et al.*, 1994; Yasumura *et al.*, 2007), probably because of the lower energy supply that limits catabolic activities as well as greater proportions of N associated with light-harvesting pigment-binding complexes that are degraded with a slower rate (Section 23.5.3). In addition, within-canopy variation in senescence can depend on species successional position, with the autumnal senescence starting from the outer part of the crown in shade-tolerant species and from the canopy interior in shade-intolerant species (Koike, 1990; Koike *et al.*, 2007). Although such inter- and intraspecific patterns in the rate and timing of leaf senescence importantly alter the amount of essential nutrients recovered from senescing leaves, mechanistic explanations for such species' differences in performance are largely lacking.

23.6. SUMMARY: SIGNIFICANCE OF LEAF DEVELOPMENTAL PROCESSES AND AGEING ON PLANT CARBON GAIN

Two key periods of leaf ontogeny, development and senescence, are characterised by a series of highly dynamic physiological and structural photosynthetic limitations. In the case of physiological factors, formation and destruction of photosynthetic machinery represent a basic compromise between the efficient protection from photodamage and maximisation of carbon assimilation. The presence of photodynamic chlorophyll molecules and their biosynthetic intermediates and catabolites exposes leaves to a continuous risk of photo-oxidative damage, especially when photosynthetic activity is impaired as can be the case for young and old leaves. Early bud-burst and formation of photosynthetic tissues in developing leaves can extend the growing period, but also entails the risk of damage and deterioration of photosynthetic structures owing to the incidence of early season frost. It further seems that the construction of long-living structurally robust leaves is simply time consuming. A long developmental period is the cost these leaves pay for their overall longer life-span. Alternatively, longer retention of senescing leaves can also extend the photosynthetic period. However, this can enhance the probability for photo-oxidative damage and occurrence of late-season frosts that can reduce the efficiency of N resorption and accordingly the plant success in the subsequent growing season. Transient increases in the protective capacity of chloroplasts during leaf expansion and senescence can compensate for this higher demand of photoprotection, fine-tuning the balance between environmental signals and timing of plant photosynthetic responses.

Alterations in leaf internal architecture, chloroplast multiplication and chloroplast and cell-wall ultrastructure play a further key role in the distribution of photosynthetic limitations between physiological (photosynthetic potentials of single leaf cells) and structural (stacking of photosynthetic tissues, mesophyll diffusion limitations) factors. The leaf internal scaffolding formed during the leaf developmental period affects leaf photosynthetic potentials and leaf capacity to readjust to altered environmental conditions. There is evidence that mesophyll diffusion limits photosynthesis more in very young and senescent leaves than in fully mature leaves, adding g_m to the list of key photosynthetic limitations during the most dynamic periods of leaf development. Overall, leaf development and senescence comprise significant periods of leaf lifetime, and photosynthetic modifications during these dynamic periods of leaf ontogeny need to be included in any large-scale carbon gain model.

24 • Evolution of photosynthesis I: basic leaf morphological traits and diffusion and photosynthetic structures

J. FLEXAS AND J.E. KEELEY

24.1. INTRODUCTION

The photochemical and biochemical processes that utilise solar energy for the synthesis of complex organic molecules have been on Earth for more than 3.5 billion years (Blankenship, 1992). The original photosynthetic mechanisms are thought to have been similar to those of contemporary cyanobacteria, with an oxygen-evolving photosystem that was responsible for the oxygenation of our early atmosphere, although physical processes have also been suggested (Kump, 2008). Evidence of the ancient origins of photosynthesis is seen in the fossil records of stromatolites of Western Australia (Awramik, 1992). Molecular studies show that it is increasingly more likely that photosynthesis evolved after chemolithotrophy (Xiong and Bauer, 2002). The evolutionary path of type-I and type-II reaction-centre systems remains unresolved and there is some debate on whether the earliest O_2-producing cyanobacteria used water or bicarbonate as the terminal reductant that led to an aerobic atmosphere (Dismukes *et al.*, 2001). There is evidence that the reaction giving rise to molecular oxygen, and responsible for our contemporary aerobic atmosphere, arose only once and the structural characteristics of its catalytic centre and its mechanism have been conserved ever since (Barber, 2008a,b). The origin of this oxygenic photosynthesis is under debate, with some authors suggesting an origin as early as 3.8 GA (Buick, 2008).

The Calvin-Benson or C_3 cycle of carbon reduction evolved early in the history of life and this is reflected in its universal presence in photosynthetic plants. Evolution of the carbon-reducing steps of the Calvin cycle is thought to have occurred when the Earth's atmosphere was hypoxic and rich in CO_2, which may account for the extreme sensitivity of Rubisco, the enzyme responsible for CO_2 entry into the Calvin cycle, to contemporary oxygen levels (Ogren, 1984). Presumably, a long period of evolution in a hypoxic atmosphere resulted in a complex pathway of tightly coupled reactions that have not been amenable to evolutionary modification in a way that reduces O_2 inhibition of photosynthesis while retaining the original carbon-fixation function. Under current atmospheric conditions, the O_2 inhibition of photosynthesis occurs through oxygenation of RuBP and subsequent loss of CO_2 through the reactions of photorespiration.

Although the basic structures and mechanisms associated with oxygenic photosynthesis evolved in aquatic plants far before the appearance of land plants, the colonisation of terrestrial habitats would have required adaptations to new stressors. These included increased risk of desiccation owing to the severe water deficit of the atmosphere and increased risk of photoinhibition and photooxidation. All of these constraints are intrinsically linked to the ecophysiology of photosynthesis and are likely to have played a key role in the development and evolution of structures directly involved in photosynthesis (stomata, leaves), photosynthetic WUE (stomata, cuticles, leaf structure) and photo-protection (leaf structure, modifications of chloroplasts, pigments and antioxidants). The earliest plants on land appeared during the Silurian approximately 470 MA (Fig. 24.1), and consisted of filamentous terrestrial algae, lichens and thalloid liverworts (Marchantiopsida), which were followed by hornworts (Anthocerotopsida) and mosses (Bryopsida) (Kenrich and Crane, 1997; Yoshinaga and Kugita, 2004). The origins of vascular plants (Tracheophytes) date to around the mid-Silurian (approximately 428 MA, Fig. 24.1). The ancestral vascular plants are often classified within the Rhyniophytina or Rhyniopsids, and include the extinct genus *Cooksonia* (often considered the earliest vascular plant), *Rhynia* and *Stockmansella* (Kenrich and Crane, 1997; Edwards *et al.*, 1998; Niklas and Kutschera, 2009a,b). The first vascular plants with leaves, microphyll

Terrestrial Photosynthesis in a Changing Environment: A Molecular, Physiological and Ecological Approach, ed. J. Flexas, F. Loreto and H. Medrano. Published by Cambridge University Press.

Era	Period	Epoch	Began (MA)	Event
Cenozoic	Quaternary	Holocene	0.01	
		Pleistocene	1.8	
		Pliocene	5.3	Expansion of C_4plants (8-5)
		Miocene	23	
		Oligocene	34	Origin of C_4plants (30)
		Eocene	54	Cryptostomata (55)
	Tertiary	Paleocene	65	Poaceae and 'dumb-bell-shaped 'stomata (70)
Mesozoic	Cretaceous		145	Probable origin of terrestrial CAM plants
	Jurassic		200	First angiosperms (200)
	Triassic		251	
Paleozoic	Permian		299	First conifers (290)
	Carboniferous		359	Megaphylls (360)
	Devonian		416	Microphylls and 'kidney-shape'stomata (410)
	Silurian		443	Early vascular plants (420)
	Ordovician		488	Early land plants (470)
	Cambrian		542	

Fig. 24.1. Variations in atmospheric CO_2 (solid line) and O_2 (dashed line) over the past 400 million years predicted by the geochemical mass-balance model (from Beerling *et al.*, 1998 as reproduced in Keeley and Rundel, 2005).

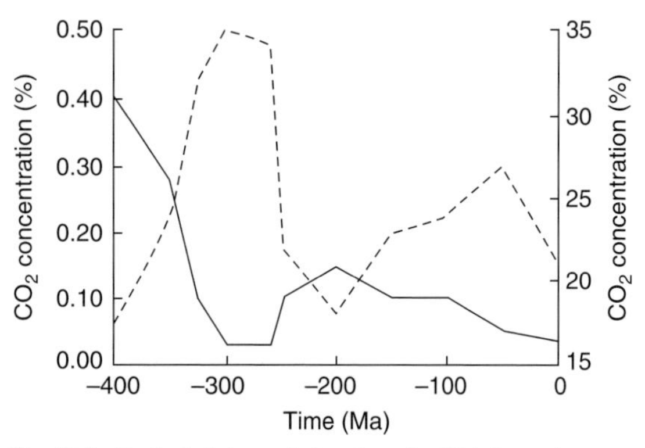

Fig. 24.2. Geological timescale based on the 2004 timescale endorsed by the International Commission on Stratigraphy. The approximate time of occurrence of the main evolutionary events related with photosynthesis is indicated.

plants or lycophytes date from the Devonian (approximately 410 MA, Fig. 24.1). Megaphyll plants or euphyllophytes did not appear until the Carboniferous (approximately 360 MA, Fig. 24.1). Euphyllophytes include Psilophyta (whisk ferns), Equisetophyta (horsetails), Ophioglossophyta (adder's tongues and grape ferns), Pteridophyta (true ferns) and seed plants, including gymnosperms (Gnetophyta, Cycadophyta, Ginkgophyta and Coniferophyta) and angiosperms (Pryer *et al.*, 2001). The earliest fossil records of horsetails, ferns and Cycadophyta date from the Carboniferous (approximately 360 MA, Fig. 24.1) and of Ginkgophyta and Coniferophyta from the Permian (approximately 290 MA, Fig. 24.1). The origin of Gnetophyta and its phylogenetic relations with other groups is under debate. They may have originated from Coniferophyta (Chaw *et al.*, 2000; Nickrent *et al.*, 2000) or be a sister group of them (Soltis *et al.*, 1999; Soltis and Soltis, 2003; Crane *et al.*, 2004) and sometimes are considered the closest to angiosperms among gymnosperms (Soltis *et al.*, 2002). Angiosperms did not appear until the boundary between Triassic and Jurassic (approximately 200 MA, Fig. 24.1), and their phylogenetic origin still remains debated (Friis *et al.*, 2005). *Amborella trichopoda* (Amborellaceae), a shrub with unisexual flowers and vesselless xylem endemic to New Caledonia is considered to be the closest extant relative to the phylogenetic root of angiosperms, closely followed by water lilies or Nymphaeales (Matthews and Donoghue, 1999).

All of these early land plants were undoubtedly C_3 plants. Changes in the composition of the atmosphere may have triggered the evolution of megaphylls first (Beerling, 2005) and later on CO_2-concentrating mechanisms. Indeed, a significant part of photosynthetic evolution in vascular plants has been the development of mechanisms for reducing photorespiration by concentrating CO_2 around Rubisco, thus returning this enzyme to an atmospheric condition that resembles the primitive earth. These CO_2-concentrating mechanisms are known as C_4 photosynthesis and CAM (Chapters 5 and 6) and their evolution is covered in the next chapter. Other CO_2-concentrating mechanisms exist in some algae and cyanobacteria (Kaplan and Reinhold, 1999; Raven *et al.*, 2008), but these are mostly important for aquatic habitats and hence are out of the scope of this book. An exception would be the presence of pyrenoids in hornworts, which are a unique feature of early land plants (Vaughn *et al.*, 1992).

The present chapter focuses on the evolution of land-plant traits directly or indirectly related to photosynthesis, based on paleontological and ecophysiological evidence. The latter is based on the study of extant plants belonging to the different evolutionary groups of land plants (with the exception of Rhyniopsids, which are all extinct). Therefore, ecophysiological aspects must be viewed with some care as current plants within a given group may have suffered evolutionary changes as compared with their ancestors. Even within a given species, examples of ecotypic variation have shown that both morphological and ecological evolution can occur rapidly in decades or even less. Therefore, it is possible, although not demonstrable, that the ecophysiological responses of the ancestors of present plants were different from contemporary representatives. The leaf traits to be covered here are: (1) chloroplast biochemistry, including Rubisco, photosystems and photoprotective mechanisms; (2) plant cuticles and stomata; and (3) mesophyll structure and leaf form.

24.2. CHLOROPLAST BIOCHEMISTRY: RUBISCO, PHOTOSYSTEMS AND PHOTOPROTECTIVE MECHANISMS

Despite some debate as to the evolutionary path of type-I and type-II reaction-centre systems (Blankenship, 1992) and as to the age of Rubisco (Nisbet *et al.*, 2007), it is clear that the basic molecular components of chloroplasts evolved early in the history of life, probably approximately 3 billion years ago. This was long before chloroplasts originated from a cyanobacterium through endosymbiosis (Raven and Allen, 2003) and therefore before plants colonised the land. Following the emergence of land plants, continued evolution of these components is evident.

24.2.1. Rubisco

Concerning Rubisco, although the basic structure of form I in higher plants is similar to that of cyanobacteria (Nisbet *et al.*, 2007), substantial divergence has occurred including variations in the organelle location of the genes and their arrangement, mechanism of Rubisco synthesis, polypeptide sequence and efficacy of the substrates' binding and inhibitor action (Newman and Cattolico, 1990). For instance, in eucaryotes, contrary to ancestral cyanobacteria, the large subunit of Rubisco is encoded by multiple identical copies of *rbc*L in the chloroplast genome (Eilenberg *et al.*, 1998), whereas the small subunit *rbc*S gene family, which has between 2 and 12 nuclear genes, encodes small subunit peptides that are synthesised as precursor polypeptides in the cytosolic ribosomes and imported into the chloroplast in an ATP-dependent reaction (Spreitzer and Salvucci, 2002). Moreover, RuBP has no effect on the in-vitro activation of Rubisco by CO_2 and Mg^{2+} in cyanobacteria, but it induces a large decrease in the rate of activation in land plants (Newman and Cattolico, 1990).

Most relevant to the ecophysiology of photosynthesis is the evolutionary divergence in the functional properties of Rubisco, i.e., its specificity factor for CO_2 over O_2 ($S_{c/o}$) and catalytic constants (Raven, 2000). Nature has been unable to avoid the lack of specificity of Rubisco for CO_2. The reason is probably a result of the low oxygen concentration present in the atmosphere during the early evolution of Rubisco.

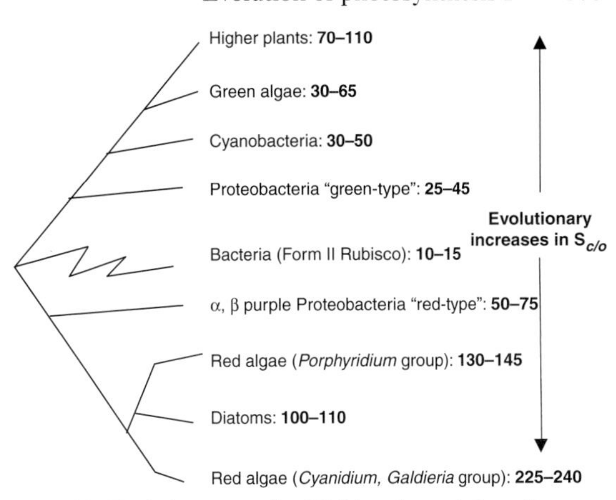

Fig. 24.3. Evolutionary trends of Rubisco $S_{c/o}$ variations. $S_{c/o}$ ranges obtained from various sources.

Comparison of $S_{c/o}$ values (Fig. 24.3), which vary over a 20-fold range from divergent photosynthetic organisms, shows that Rubisco specificity increases from the lower photosynthetic forms to higher plant forms (Jordan and Ogren, 1981a,b, 1983, 1984; Tortell, 2000), although the highest known specificities are for the red algae *Galdieria partita* (Uemura *et al.*, 1997).

Despite the strong phylogeny dependence of Rubisco specificity, it has been hypothesised that the enzyme could have evolved according to the organisms' specific needs for CO_2 assimilation. These would have depended on the environmental conditions in which the organism evolved, which would explain the large specificity in *Galdieria*, an organism inhabiting acidic hot springs rich in sulfur dioxide and low in CO_2 (Horken and Tabita, 1999). This view is also supported by comparing $S_{c/o}$ values along the green-plant lineage. Within this evolutionary line, chlorophyte algae containing CO_2-concentrating mechanisms (CCMs) present significantly lower $S_{c/o}$ than algae lacking CCMs (Raven, 2000). The latter, in turn, present high specificities similar to those of terrestrial C_3 plants including horsetails, gymnosperms and angiosperms (Kent and Tomany, 1995; Raven, 2000). Similarly among higher plants, those having evolved new CCMs in more recent times, such as C_4 metabolism, present lower $S_{c/o}$ values than C_3 species (Kent and Tomany, 1995; Raven, 2000; see Chapters 5, 6 and 25). Even within C_3 plants, Delgado *et al.* (1995) and Kent and Tomany (1995) hypothesised that hot environments together with water stress may impose an increased selection pressure on Rubisco for improved specificity. This was recently supported by Galmés *et al.* (2005) in a survey of mediterranean C_3 species. It was shown that the diversity found in $S_{c/o}$ (from the lowest reported values in C_3 plants around 75 in several herbaceous species to the highest of 110 in the evergreen sclerophyll semi-shrub *Limonium gibertii*) was not related to the phylogeny of the studied plants, but strongly and positively correlated to the aridity of each species natural habitat and to leaf sclerophylly. Both aridity (see Chapters 20 and 29) and increased leaf thickness (see Chapter 12) result in low CO_2 availability in chloroplasts, which in the long term may exert a selection pressure towards more efficient Rubisco, just as it may have occurred when atmospheric CO_2 concentration dropped after land colonisation by plants. Further support for environmentally driven evolution of Rubisco comes from phylogenetic and maximum likelihood analyses of codon substitution models in the *rbc*L gene, which suggest that this gene has evolved under positive Darwinian selection (Kapralov *et al.*, 2006, 2007; Iida *et al.*, 2009). Still, *rbc*L is a slowly evolving gene, allowing it to be used to construct phylogenies and to assess the evolutionary potential of floras in biodiversity hotspots (Forest *et al.*, 2007; Nickrent *et al.*, 2000).

In general, a negative relationship has been described between $S_{c/o}$ and the catalytic rate of carboxylation (Raven, 2000; Zhu *et al.*, 2004; Tcherkez *et al.*, 2006), so that species with a higher affinity for CO_2 present much lower rates of photosynthesis. However, it has been shown recently that both *Galdieria* and *Limonium* simultaneously maintain both high $S_{c/o}$ and catalytic rates, which may open the possibility for genetically engineering plants with improved Rubisco carboxylation (Parry *et al.*, 2007). The variability found in Rubisco functional attributes deserve more attention in future studies as current models of leaf photosynthesis assume constant $S_{c/o}$ values among C_3 plants (Long and Bernacchi, 2003, see Chapter 8).

24.2.2. Photochemistry and photoprotection

Similar to Rubisco, the structure of PSI and II has not significantly changed from cyanobacteria to higher plants (Blankenship, 1992). Similarly, most of the known photoprotective mechanisms of higher plants (see Chapter 16), such as the xanthophyll cycle (VAZ), several tocopherols and all forms of SODs, are already present with little variability in all green algae and many diatoms (Asada, 2000; Baroli and Niyogi, 2000). As an example of evolutionary variations in the operation of VAZ-related energy dissipation processes, higher plants require protein PsbS for its operation (Li *et al.*, 2000). However, green algae, such as *Chlamydomonas*, lack PsbS and require a different protein, LHCRS, for VAZ-related energy dissipation. The moss *Physcomitrella patens* presents LHCRS and PsbS, both being functional and necessary for energy dissipation, suggesting that upon land colonisation photosynthetic organisms evolved a unique mechanism for excess energy dissipation before losing the ancestral one found in algae (Alboresi *et al.*, 2010). Other photoprotective cycles, such as the lutein epoxide cycle, may have evolved more recently as they are exclusive to some families of higher plants (Garcia-Plazaola *et al.*, 2007). Interestingly, lutein epoxide concentrations are highest in basal angiosperms with lower concentrations in gymnosperms and angiosperms, although no clear trend among subgroups of the latter has been found. This irregular distribution of the lutein epoxide cycle suggests that it is not determined phylogenetically but rather by ecological constraints (Esteban *et al.*, 2009). In addition, Esteban *et al.* (2009) have recently shown that although α-tocopherol and the VAZ pigments are present in all green plants, from green algae to angiosperms, there is an evolutionary trend to increase the content of α-tocopherol while decreasing VAZ pigments, which suggests a complementary role of these two photoprotective mechanisms. Hence, VAZ pigments are more abundant in green algae, liverworts and mosses, whereas α-tocopherol is more abundant in ferns, gymnosperms and angiosperms, the highest concentrations corresponding to dicots.

In addition to the evolution of molecular photoprotective mechanisms, it is likely that morphological traits conferring photoprotection (e.g., leaf hairiness, leaf angles, leaf veination) may have suffered different selection pressures depending on the habitat conditions of plants (e.g., Morales *et al.*, 2002; Esteban *et al.*, 2008), but any evolutionary evidence concerning these traits is lacking.

24.3. CUTICLES AND STOMATA

Ever since vascular plants colonised land approximately 400 million years ago, they have been protected by a cuticle, whereas all forms of non-terrestrial plants lack a cuticle (Edwards *et al.*, 1982). On the other hand the earliest well-authenticated stomata in fossil plants date from 410 million years ago, although stomata may have probably appeared earlier (Edwards *et al.*, 1998). Therefore, both cuticles and stomata evolved early after the colonisation of terrestrial environments and far before the appearance of leaves, which has led to the suggestion that the primary function of both structures was to protect plants from desiccation (Kerstiens,

1996, 2006; Raven, 2002a). Indeed, desiccation tolerance is a character apparently being lost during the course of evolution. Although it is common in lichens, bryophytes and lycophytes, less than 1% of all other ferns possess such an ability in their sporophyte stage (Proctor and Pence, 2002) although it is widely exhibited in the gametophyte stage (Watkins Jr. *et al.*, 2007). In higher plants, desiccation tolerance is restricted to seeds, with some exceptions like the genus *Craterostigma*, which shows vegetative desiccation tolerance (Bartels and Salamini, 2001).

24.3.1. Cuticles

Although we understand a great deal about stomatal evolution, much less is known about the evolution of plant cuticles. Kerstiens (2006) pointed out that our understanding of the totality of evolutionary pressures acting on the formation of the cuticle, in past and present environments, is poor. It is likely that cuticles had early roles in defence against parasites and grazers in reflecting UV-B and in water repellency, in addition to preserving plants from desiccation (Raven, 2002a). Regardless of its function, there is evidence that early land plants had cuticles with a similar thickness to present-day plants (Edwards *et al.*, 1998). The water permeability of cuticles in extant plants ranges 1000-fold, from 0.1 to 100 m $s^{-1} \times 10^5$ (Kerstiens, 1996). It is still unclear whether these differences arise from differences in cuticle thickness, composition and/or the number and size of aqueous pores (Kerstiens, 2006).

Such pores occur in many plant cuticles, and they are normally small – 1 to 2 nm in diameter – although they can reach much larger sizes, such as the 'giant' cuticular pores in the Proteaceae *Eidothea zoexylocarya* that can reach diameters of 1 μm (Carpenter *et al.*, 2007). Using the bryophyte gametophyte as the best model of ancestrally astomatous land plants to model photosynthesis, Raven (2002a) suggested that these pores may have permitted higher photosynthetic rates per unit ground area in early land plants under high CO_2 conditions.

Although values of cuticular water permeability in different species are relatively scarce, apparently there are no significant differences in the range of permeability between different evolutionary groups (Kerstiens, 1996, 2006). This suggests that cuticular evolution may have responded mostly to stress conditions, which is supported by comparisons of parental species with different habitat, showing that cuticular waxes (Pearce *et al.*, 2006), hypodermis and cuticular pores (Carpenter *et al.*, 2007) are more evenly distributed in leaf abaxial and adaxial sides in species inhabiting drier habitats. Some well-preserved fossil records from the Jurassic (Wang *et al.*, 2005) and the Cretaceous (Upchurch, 1984; Yang *et al.*, 2009) demonstrate that cuticles of some Ginkgophyta, Coniferophyta and extinct angiosperms already possessed secretory cells comparable with modern-plant oil cells, waxes, fibrils, trichomes and stomatal types (see below), i.e., similar characteristics to those found in extant basal angiosperms (Carpenter, 2006).

24.3.2. Stomata

Contrary to cuticles, much is known about the evolution of stomata. Functional considerations suggest that stomata evolved from pores in the epidermis, and cladistic analyses are consistent with a unique origin of stomata (Edwards *et al.*, 1998; Raven, 2002a). Stomata are absent in liverworts and in the gametophyte phase of hornworts and mosses, but present in the sporophyte phase of the latter two groups and in all known groups of vascular plants, including the early extinct Rhyniopsids. This implies later loss of stomata in the evolution of some lycopsids such as *Stylites*, and ferns such as *Hymenophyllum* (Keeley *et al.*, 1984; Kessler *et al.*, 2007; see Chapter 6), as well as in many submerged aquatic vascular plants (Raven, 2002a; see Chapter 6). The proposed functional roles of stomata include maintenance of hydration, restricting the occurrence of xylem embolism, optimising carbon fixation per unit water lost, cooling leaves and transporting nutrients from roots to shoots (Raven, 2002a). Since their early appearance in land plants, stomata have suffered substantial evolution concerning their numbers and density, morphology, position and function (Hetherington and Woodward, 2003). Stomatal densities range from between 5 and 1000 mm^{-2} of epidermis, depending on species and environmental conditions (Hetherington and Woodward, 2003). It is well known that stomatal density decreases with increasing atmospheric CO_2 and vice-versa (Woodward, 1987a; Woodward and Kelly, 1995; Beerling and Woodward, 1996; Franks and Beerling, 2009); a response already present in the so-called 'living fossil' *Ginkgo biloba* (Beerling *et al.*, 1998) and other gymnosperms (Woodward and Kelly, 1995). This stomatal tracking of CO_2 has been utilised as a proxy signal for paleo-atmospheric changes (Royer *et al.* 2001).

CO_2 perception and stomatal development are linked by the *HIC* (high carbon dioxide) gene-signalling pathway (Gray *et al.*, 2000), although other CO_2-independent genes, such as *STOMAGEN*, are involved in regulating stomatal density during leaf development (Sugano *et al.*, 2010). Similarly, stomatal densities in a particular species may increase with elevation and aridity (Quarrie and Jones, 1977; Franks and Farquhar, 2001; Kessler *et al.*, 2007; Xu and Zhou, 2008). These changes have important implications for photosynthesis, as higher stomatal densities are often associated with higher maximum stomatal conductance, net photosynthesis and WUE, as well as with better stomatal regulation in response to water stress (Pearce *et al.*, 2006; Galmés *et al.*, 2007c; Xu and Zhou, 2008). Moreover, decreasing CO_2 in the atmosphere in the late Devonian may have allowed the development of large stomatal densities, triggering the evolution of megaphylls (Beerling, 2005, see Section 24.4.2.1) and small stomatal sizes, allowing increasing gas-exchange capacity (Franks and Beerling, 2009).

Morphologically, stomata differ in size (from about 10 to 80 μm in length), shape and cellular composition (Hetherington and Woodward, 2003). There is a trade-off between guard-cell length and stomatal density, and hence differences in size are mostly owing to acclimation responses, as described for stomatal density. In contrast, stomatal morphologies have clear evolutionary trends (Hetherington and Woodward, 2003; Carpenter, 2005; Franks and Farquhar, 2007). Stomata in early land plants (Silurian and Devonian fossils) presented lower densities, they were 'kidney shaped' (more or less circular or elongate) and positioned superficially over leaves (Phanerostomata), and they were anomocytic, i.e., they presented no anatomically distinct subsidiary cells (Edwards *et al.*, 1998; Hetherington and Woodward, 2003). The cuticle was generally thinner over the outer periclinal walls of guard cells, but thicker elsewhere. Notably, these plants lacked an extensive substomatal cavity, but instead the stomatal pore led into a narrow canal formed by hypodermal cells with thickened walls, opening into a deep chamber occupied by parenchyma (Edwards *et al.*, 1998). Despite the fact that early land plants are often considered to have inhabited shady and humid environments, because of rapid drainage and little organic matter in soils these habitats were often water-stressed. Consistently, this unusual tissue playing the role of a substomatal cavity has been interpreted as an adaptation to increase WUE, by creating a large internal surface area per unit biomass that would increase efficiency of CO_2 transport to carboxylation sites (Edwards *et al.*, 1998). However, this explanation is not clear, as it has been suggested that large kidney shaped stomata with low densities, such as those encountered in extant ferns from deep-shade environments, may be an important feature of plants of humid and deep shade conditions, but unfavourable under dry conditions (Hetherington and Woodward, 2003).

Anomocytic stomata, such as those described for extinct early land plants, are present in extant lycophytes and ferns, as well as in some basal angiosperms (Carpenter, 2005; Franks and Farquhar, 2007). In Lycopods (Fig. 24.4A), anomocytic stomata are characterised by big guard cells with a small stomatal pore, and present minimal lateral movement or swelling during their aperture. In ferns (Fig. 24.4B) stomata are also anomocytic, but guard cells are smaller and the stomatal pore wider. Some basal angiosperms also have anomocytic stomata, but most have stephanocytic (a more or less well-defined rosette of four or more weakly specialised subsidiary cells), laterocytic (three or more subsidiary cells) or paracytic (one or two lateral subsidiary cells oriented parallel to guard cells) stomata (Carpenter, 2005). Paracytic stomata seem to represent the most evolved form (Carpenter, 2005). Therefore, most angiosperms have 'kidney shaped' stomata but with subsidiary cells clearly defined (Fig. 24.4C).

Two important evolutionary events concerning stomata occurred in the Cenozoic era (~70–35 MA), when there were profound changes in the global climate and terrestrial flora and fauna (Hetherington and Woodward, 2003). These consisted in the evolution of encrypted stomata in the Proteaceae and the evolution of 'dumbell-shaped' stomata in the Poaceae. For species of *Banksia* and *Dryandra* in the Proteaceae of Australia, two clades of species are currently recognised by differences in stomatal distributions: Phanerostomata and Cryptostomata. In the latter, stomata occur in shallow pits or in crypts, whereas the Phanerostomata have a more superficial distribution. Species of the clade Cryptostomata occur in much drier climates and probably diverged from the Phanerostomata clade 55–35 MA ago, at a time when the climate was becoming more arid (Mast and Givnish, 2002). A recent study with a large number of Proteaceae species (Jordan *et al.*, 2008) has shown that deep encryption of stomata evolved at least 11 times, always in very dry environments, whereas other forms of stomatal protection (sunken but not closely encrypted stomata, papillae and layers of hairs covering the stomata) also evolved repeatedly, but had no systematic association with dry climates. It has been recently shown that stomatal encryption rather than creating a favourable humid microenvironment around stomata, facilitates diffusion of CO_2 to adaxial mesophyll cells in thick sclerophylls, i.e., it improves photosynthesis and photosynthetic WUE by means of increasing mesophyll conductance to CO_2 (Hassiotou *et al.*,

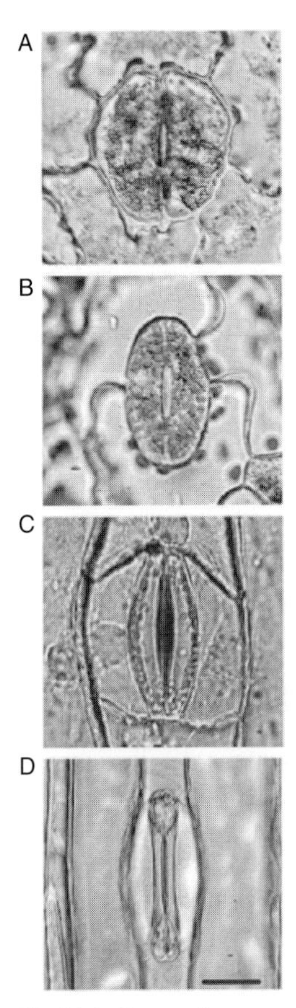

Fig. 24.4. Stomata from extant plants representative of stomatal evolution. **(A)** Anomocytic 'kidney shaped' stomata of the lycopod *Huperzia prolifera* with big guard cells and small stomatal pore. **(B)** Anomocytic 'kidney shaped' stomata of the fern *Nephlorepis exaltata* with smaller guard cells and larger pore. **(C)** Paracytic 'kidney shaped' stomata of the herbaceous angiosperm *Tradescantia virginiana*. **(D)** Paracytic 'dumbell-shaped' stomata of the grass *Triticum aeastivum*. Fig. taken from Franks and Farquhar (2007).

2009a,b, 2010; Roth-Nebelsick *et al.*, 2009). The environmental correlates of the differences in stomatal distribution seen for the Proteaceae are also supported by observations in *Cistus incanus*, for which similar differences in stomatal distribution occur, but between the summer and winter in the mediterranean climate (Aronne and De Micco, 2001). Leaves produced in winter are large and flat with abundant stomata on the adaxial leaf surface, whereas leaves developed in the hot and dry summer are crimped and partially rolled, forming a crypt on the abaxial surface where stomata are located.

The Poaceae or grasses comprise a group of about 10,000 species that originated between about 55 and 70 MA, leading to lineages that were understory plants of tropical forests (Kellogg, 2001). The 'dumbell-shaped' stomata of grasses (Fig. 24.4D) are generally believed to represent a more evolutionary advanced form than the 'kidney shaped' stomata, and indeed during development, grass guard cells adopt a transient 'kidney shaped' phase before assuming their mature 'dumbell-shaped' aspect.

Regarding stomatal functioning, it has been demonstrated that functional improvement parallels morphological evolution from anomocytic 'kidney shaped' (Fig. 24.4A) to paracytic 'dumbell-shaped' (Fig. 24.4B) stomata (Franks and Farquhar, 2007). The anomocytic stomata with large guard cells found in Lycopods (Fig. 24.4A) present a small open-stomata pore area and a small ratio of open pore-to-total-stomata area. Pore area and pore-to-stomata ratios are about double in the anomocytic stomata, with smaller guard cells found in ferns (Fig. 24.4B), but these are still much lower than those found in the paracytic 'kidney shaped' stomata of herbs (Fig. 24.4C) and especially in the paracytic 'dumbell-shaped' stomata of grasses (Fig. 24.4D). This means that for a given leaf area, species with paracytic 'kidney shaped' stomata and especially grasses with 'dumbell-shaped' stomata can have larger total pore areas than anomocytic stomata. This leads to higher maximum stomatal conductance and photosynthesis. However, stomatal apertures are virtually unaffected by subsidiary cell pressure in anomocytic stomata, but substantially influenced in paracytic stomata, which is known as the mechanical advantage of epidermal cells over guard cells. This massive mechanical counteraction is a side effect arising from the need for greater lateral displacement of guard cells to create a larger stomatal pore and could potentially eliminate much of the gain from a more mobile guard-cell pair. However, it is suggested that subsidiary cell turgor is much lower during maximum stomatal aperture, which may be achieved by an active osmotic mechanism that seems to operate better in grasses than in any other species (Franks and Farquhar, 2007). Therefore, the 'dumbell' design magnifies small changes in width to cause large openings, and maximises the potential of stomata to track changes in environmental conditions. Smaller changes in guard and subsidiary cell turgor lead to greater increases in stomatal aperture in the 'dumbell-shaped' stomata than in 'kidney shaped' stomata. This efficiency and speed of stomatal opening in grasses enhances photosynthesis and WUE compared with non-grass species (Grantz and Assmann, 1991). A rapid stomatal response to blue light augments photosynthesis in early morning and under intermittent sunlight, in which light has an enhanced blue-light content and that would have characterised the understory environment during the early evolution of grasses (Grantz and Assmann, 1991). In view of the functional advantages provided by 'dumbell-shaped' stomata, with the capability to respond quicker and more efficiently to the enhanced light conditions of newly open habitats, but with the capacity to avoid the increased likelihood of drought, grasses may have enhanced their spread and diversification of grasses during global aridification, 30–45 MA (Hetherington and Woodward, 2003; Franks and Farquhar, 2007).

Besides these differences among stomatal types in response to guard-cell pressure, most known stomata close in response to darkness and increasing VPD, CO_2 or ABA, and open in response to increasing light intensity and blue light. However, there are doubts as to the functionality of stomata in some bryophyte sporophytes, including all hornworts, but the stomata of the moss *Funaria hygrometrica* respond normally to light, dark and ABA (Raven, 2002a). Despite showing some ABA responsiveness, stomata of these plants are not considered to have a strong regulatory function (Hartung, 2010). Some ferns, such as *Adiantum capillus-veneris*, lack the blue-light-specific opening and CO_2 response, but respond normally to red light (Doi and Shimazaki, 2008). Recently, Brodribb *et al.* (2009) have shown that stomatal responsiveness to low CO_2 has increased from ferns and lycopodes to conifers, and from these to angiosperms. Only the stomata of the latter close in response to high CO_2, which lead them to increased WUE under conditions that depress photosynthesis increasing C_i, such as photoinhibition and fluctuating light and temperature. More recently, Brodribb and McAdam (2011) have analysed in depth stomata responses to leaf excision, ABA, increasing VPD and darkness. They found that lycophytes and ferns generally lacked ABA responses and presented a slower stomatal closure upon transfer to darkness than angiosperms and conifers, but plants of all groups responded similarly to leaf excision and increased VPD. All together, these evidences suggest that some functional attributes, including the response to light/dark, VPD and desiccation, appeared early in the evolution of stomata, whereas other responses such as the blue-light-specific opening and the CO_2 response, may have appeared later, although specifically designed experiments are required to obtain conclusive evidence as for the evolution of stomatal functional attributes. The appearance of ABA responses remains more controversial.

24.4. LEAF FORM AND MESOPHYLL COMPONENTS

Since the endosymbiontic event that led to the origin of photosynthetic eukaryotes, the cellular components of mesophyll cells (i.e., cell walls and chloroplasts) have been present in all plants, even before plants colonised the land and far before the appearance of leaves. However, it is likely that some evolution may have occurred since then in the composition, size and distribution of mesophyll elements (Vaughn *et al.*, 1992; Popper and Fry, 2003). Alternatively, leaves have evolved independently in lycophytes and euphyllophytes (Dolan, 2009) and have suffered substantial evolution according to fossil leaf records (Osborne *et al.*, 2004a,b; Feild and Arens, 2007; Royer *et al.*, 2007) and ecophysiological studies in extant plants (Brodribb *et al.*, 2007; Sack *et al.*, 2008).

24.4.1. Cell walls and chloroplasts

Little is known about the evolution of cell walls and chloroplasts. However, recent studies (Popper and Fry, 2003; Popper *et al.*, 2011) showed that major changes in primary cell-wall composition accompanied major evolutionary steps. In this sense, cell walls in charophytes are mostly composed of galacturonic acid and, in smaller amounts, glucuronic acid and mannose, while lacking xyloglucan. Xyloglucan is present in all land-plant lineages, from hornworts to euphyllophytes. Hornworts present the highest content of galacturonic and glucuronic acids among any group, and are the only group containing the unusual glucuronic acid-$\alpha(1\rightarrow3)$-galactose. Mannose is most abundant in liverworts, mosses and lycophytes. Euphyllophytes present the lowest amounts of all these compounds. It is unclear how these differences in cell-wall composition may affect photosynthesis. However, it has been suggested that the resistance to CO_2 diffusion imposed by cell walls (i.e., the inverse of cell-wall conductance to CO_2, see Chapter 12) is directly proportional to mesophyll cell-wall thickness and to the tortuosity of the cell-wall pores, and inversely proportional to the porosity of the wall (Evans *et al.*, 2009; Terashima *et al.*, 2011). Values of cell-wall porosity and the tortuosity of the pores through cell walls are unknown, but likely affected by the composition of the primary cell wall.

Concerning chloroplasts, hornworts, the extant representatives of one of the most ancient groups of land plants, have unique chloroplast features that suggest an advanced evolution of chloroplast numbers and morphologies in later evolutionary groups (Vaughn *et al.*, 1992). Unlike any other land plant, but similar to many algae, most hornworts have pyrenoids, i.e., the site of accumulation of Rubisco. But unlike most algae, the hornwort pyrenoid is composed of distinct subunits, numbering up to several hundred. Another unique feature of the hornwort chloroplasts is the presence of thylakoids that connect adjacent granal stacks at right angles to the long axis of the granum. It is unclear how these features affect photosynthesis in hornworts owing to the lack of sufficient ecophysiological studies.

Also, similar to some algae (Haupt and Scheuerlein, 1990) and contrary to most vascular plants, liverworts, hornworts and some lycophytes, such as some *Selaginella* species, have a single or a few large chloroplasts per cell (Haupt and Scheuerlein, 1990; Vaughn *et al.*, 1992). It has been suggested that the presence of a large population of small chloroplasts allows for more effective chloroplast movement in response to high light than fewer large chloroplasts (Jeong *et al.*, 2002), although such movements are also present in some algae species having a single chloroplast per cell (Haupt and Scheuerlein, 1990). Chloroplast reorientation in response to high light, i.e., moving from a face pattern under low light to a profile pattern under high light, is a very effective photoprotective mechanism (Kasahara *et al.*, 2002) that is present in many algae and almost all land-plant lineages, including mosses, such as *Funaria hygrometrica* (Haupt and Scheuerlein, 1990), ferns (Agustynowicz and Gabrys, 1999), monocots and dicots (Inoue and Shibata, 1974). In hornworts, however, chloroplast photoactivity is rare, although it has been described in the genus *Megaceros*, possibly being the reason why hornworts within this genus have lost pyrenoids (Vaughn *et al.*, 1992). Although ecophysiological studies comparing photosynthesis in ancestral plants with few large chloroplasts with vascular plants that present many small chloroplasts are scarce, studies with the *arc* mutations of *Arabidopsis* (containing only one to three chloroplasts per mesophyll cell 20-fold larger than wild-type chloroplasts) have shown that mutants have normal phenotype and growth (Pyke *et al.*, 1994), whereas photosynthesis rates are similar to wild-type plants (Austin II and Webber, 2005). Indeed, it has been suggested that few large chloroplasts would result in higher mesophyll conductance to CO_2 (see Chapter 12) and thus higher photosynthesis than plants with many small chloroplasts (Sharkey, personal communication). Moreover, in *Arabidopsis* plants it has been demonstrated that chloroplast movements towards the high-light position results in decreased mesophyll conductance (Tholen *et al.*, 2008). Although mesophyll conductance has yet to be determined in lycophytes or ferns, preliminary measurements in liverworts and hornworts show very reduced g_m rather than an increased one (Meyer *et al.*, 2008). Nevertheless, if an evolutionary trend exists from few large chloroplasts – conferring or not large CO_2 diffusion conductance – towards many smaller chloroplasts facilitating chloroplast movements, this would support the idea that excess irradiance has exerted more selective pressure than low CO_2 availability on land plants.

24.4.2. Leaf form

MICROPHYLLS AND MEGAPHYLLS

Two types of leaves are found in vascular plants: microphylls and macrophylls (Tomescu, 2008; Dolan, 2009; Niklas and Kutschera, 2009a,b). Microphylls are defined as leaves of small size with simple venation (one vein), while megaphylls are generally larger in size and with more complex venation (Tomescu, 2008). Generally, microphylls are leaves of lycophytes, whereas megaphylls are leaves of euphyllophytes, although some 'megaphyll-leaved' species, such as *Equisetum* and several extant and fossil gymnosperms have highly reduced leaves supported by one vein, whereas some 'microphyll-leaved' species have large leaves, such as the extinct lepidodendrales (up to 1 m) or extant *Isoetes* (up to 0.5 m), or complex venation patterns, such as those of some *Selaginella* species (Tomescu, 2008). Despite these exceptions, the terms microphyll and megaphyll are still useful to distinguish two groups of leaves that evolved independently, microphylls about 410 MA, i.e., about 50 MA after land colonisation by plants and approximately 10 MA after the appearance of the first vascular plants (Dolan, 2009), and megaphylls about 360 MA, i.e., about 50 MA after the appearance of microphylls (Beerling, 2005).

Megaphyll leaf photosynthesis allows more efficient light interception and higher rates of photosynthesis than microphylls (Micol, 2009). However, both microphylls and megaphylls are produced by determinate growth of small subpopulations of cells on the flanks of indeterminate shoot apical meristems. Indeterminate growth of the shoot apical meristem is controlled by the class-1 knotted-like homebox (*KNOX*) gene family as well as by BELLRINGER class proteins. *Class-1 KNOX* genes and BELLRINGER are expressed in and around the meristem. As these genes control indeterminate growth, their repression is required for leaf development. This repression is mediated by a Myb transcription factor, which originates Myb proteins collectively known as ARPs (Tomescu, 2008; Dolan, 2009; Micol, 2009).

Class-1 KNOX genes are already present in some algae (e.g., *Acetabularia*), mosses, ferns, gymnosperms and angiosperms, and restricted to the development of the diploid phase of the lifecycle. For instance, in the fern *Ceratopteris richardii*, *Class-1 KNOX* genes are expressed in the meristem, leaf primordia, vascular bundles and leaf margins of the sporophyte, but not in the gametophyte. In other words, the pattern of *Class-1 KNOX* gene expression is similar in lycophytes and euphyllophytes. Given that microphylls and megaphylls evolved independently, it might have been expected that different mechanisms of *KNOX* repression would have evolved in each group. Contrary to that expectation, *ARP* genes are similarly expressed in the lateral primordia of both lycophytes and euphyllophytes. *Class-III HD-ZIP* genes control the initiation of lateral appendages, the differentiation of the top from the bottom of the leaf and the development of vasculature. Moreover, these genes are present in both microphyll and megaphyll species, but their patterns of gene expression in the developing leaf and vasculature are very different in lycophytes and euphyllophytes, suggesting that these genes may have different roles in each group, possibly leading to the observed differences between microphylls and megaphylls (Dolan, 2009). Moreover, it has been suggested that differences may also arise from different *KNOW-ARP* interactions in the different groups (Tomescu, 2008).

Despite the fact that all land plants share most of the genes required for developing either microphylls or megaphylls, the latter did not appear until approximately 50 MA after the appearance of microphylls (Osborne *et al.*, 2004a,b). The appearance of megaphylls in the late Devonian (360 MA) coincides with a large drop of atmospheric CO_2 concentration (Fig. 24.2). Beerling *et al.* (2001) proposed that this CO_2 drop was indeed the requisite for megaphyll formation. The theory is based on the observation that stomatal density is inversely proportional to CO_2 concentration (see Section 24.3.2) and a leaf energy balance model. In plants, energy dissipation takes place mostly through convection (that decreases with increasing leaf size as friction across the surface slows the passage of air and the transfer of heat) and transpiration (whose capacity depends on maximum stomatal conductance, which in turn is directly proportional to stomatal density). Even without invoking a higher global temperature (as a consequence of the 'greenhouse effect' owing to high CO_2), it was calculated that the low stomatal densities found prior to the late Devonian as a consequence of high CO_2 would not allow for the high transpiration rates needed to cool a megaphyll below a lethal temperature threshold. This may have restricted the abundance of megaphyll to a few isolated examples, such as the rare early Devonian plant *Eophyllophyton bellum* (Beerling, 2005). The large fall in CO_2 corresponded with a marked rise in stomatal density, from 5–10 mm^{-2} on early vascular plants (Edwards *et al.*, 1998) to 30–40 mm^{-2} on late Devonian (Osborne *et al.*, 2004a,b) and 800–1000 mm^{-2} on late Carboniferous megaphylls (McElwain and Chaloner, 1995). The rise in stomatal density permitted greater evaporative cooling and alleviated the requirement for convective loss, allowing the spread of megaphylls.

Following a similar but reverse argument, it has been proposed that the later fourfold increase in CO_2 concentration around 200 MA in the Triassic-Jurassic boundary (Fig. 24.2) resulted in a 3° to 4°C 'greenhouse' warming that may have induced another evolutionary constraint on megaphylls (McElwain *et al.*, 1999). Based on the leaf energy balance, these environmental conditions are calculated to have raised leaf temperatures above a lethal threshold, possibly contributing to the observed massive extinction and more than 95% species turnover of the Triassic-Jurasic megaflora. For instance, large leaves (>3 to 4 cm) in full sunlight of the upper canopy (e.g., Ginkgo) or plants in open habitats could have reached noon temperatures at least 10°C above air temperature, which in a time of global warming (summer temperatures >30ºC) may have resulted in lethal temperatures. Indeed, fossil records indicate that this transition favoured highly dissected and/or narrow leaves over large entire leaves (McElwain *et al.*, 1999). It was after massive plant extinctions in the Triassic-Jurassic boundary that angiosperms radiated into a riot of unprecedented variation in foliar morphological diversity (Feild and Arens, 2007). Leaf anatomy of basal angiosperms reveals features linked to high performance in wet, shady habitats. For example, a mesophyll dominated by spongy parenchyma tissue (i.e., no palisade layers) is ancestral among angiosperms. Other features of early angiosperms are stephanocytic stomata with hypostomy and low stomatal densities (15–80 mm^{-2}), low photosynthetic capacities and low light-saturation points (around 20% of full sunlight). Feild and Arens (2007) have suggested that the damp, dark and disturbed ancestral habitat of angiosperms allowed the origin and refinement of ecophysiological traits, such as vessels and broad, net-veined leaves, that permitted them expanding in understory habitats. These traits may have later been co-opted or modified as angiosperms and broke out into more demanding environments. For instance, to break into brighter and more evaporative terrestrial habitats, leaves must have evolved palisade mesophyll and amphistomy to increase photosynthetic rate per A_{leaf} (Smith *et al.*, 1997), while evolution of smaller stomata and more precise guard-cell regulation may have increased WUE (Sack *et al.*, 2003). In addition, it has been suggested that coordinated shifts in hydraulic and photosynthetic performances are necessary to make a substantial movement out of wet habitats (Sack *et al.*, 2003; Feild and Arens, 2007).

LEAF SCLEROPHYLLY

In agreement with the above considerations, plants differ largely in LMA, leaf thickness and density (Beerling and Kelly, 1996; Niinemets, 1999, 2001; Wright *et al.*, 2004). LMA is indeed the product of leaf density and leaf thickness (Niinemets, 1999, 2001). Increasing the complexity of leaf mesophyll may ultimately result in sclerophylly (thick, rigid leaves with high LMA and thick cuticles). This has been debated as to whether it does or does not represent an evolutionary character (Salleo *et al.*, 1997; Salleo and Nardini, 2000; Jordan *et al.*, 2005). The earliest fossil records of sclerophyll leaves date from the late Permian (250 MA), i.e., around 100 MA after the appearance of megaphylls. They are found in Antarctica, South Africa and Texas (Retallack, 2005). Other sclerophyll-rich fossil floras are found in the Triassic-Jurassic boundary (McElwain *et al.*, 1999) and in some Eocene beds (Royer *et al.*, 2007). All these floras correspond with dry and warm (Retallack, 2005; Royer *et al.*, 2007) or warm and high-CO_2 (McElwain *et al.*, 1999) environments.

However, studies in extant plants have questioned the drought-adaptive value of sclerophylly. First, sclerophyll leaves are found in all biomes on Earth, not just in dry environments (Reich *et al.*, 2003; Wright *et al.*, 2004). Second, sclerophyll leaves do not present better water-relation traits than non-sclerophyll leaves (Salleo *et al.*, 1997; Salleo and Nardini, 2000; Galmés *et al.*, 2007c), although other authors have found positive correlations between LMA, leaf density and leaf thickness and the bulk modulus of elasticity (Niinemets, 2001; Corcuera *et al.*, 2002), which is often considered as a water-saving strategy. In addition to water scarcity, other possible determinants of leaf sclerophylly include shortage of nutrients (Loveless, 1962; Wright *et al.*, 2004), increased attack by herbivores or fungi (Turner, 1994; Royer *et al.*, 2007), increased heat stress and increased exposure to the sun (Beerling and Kelly, 1996; Smith *et al.*, 1997; Jordan *et al.*, 2005). For instance, Royer *et al.* (2007) found in Eocene floras a negative correlation between LMA and indices of insect damage to leaves. Alternatively, tests of correlated evolution based on molecular phylogenies suggested that scleromorphic leaf anatomies in Proteaceae have evolved many times, always in association with open vegetation (i.e., high irradiance), but not with dry habitats (Jordan *et al.*, 2005). A positive effect of high irradiance on LMA would also explain why LMA tends to increase with elevation (Beerling and Kelly, 1996; Dunbar-Co *et al.*, 2009). Global analysis shows that LMA is negatively correlated with precipitation and positively correlated with both mean temperature and mean solar radiation (Niinemets, 2001). Interestingly, leaf density correlates better with precipitation, whereas thickness correlates better with temperature and radiation, suggesting that the two components of LMA may respond to different selection pressures (Niinemets, 2001).

Regardless of its adaptive value, LMA shows a global positive relationship with leaf lifespan and trade-offs with nutrient concentrations and, notably, photosynthetic capacity (Reich *et al.*, 2003; Wright *et al.*, 2004). Increasing mesophyll thickness increases the pathways of light and CO_2 from their sites of interception by leaves to the average site of their use in photosynthesis by mesophyll cells (Smith *et al.*, 1997). It also decreases the proportion of leaf N invested in Rubisco, but not the proportion invested in cell walls (Harrison *et al.*, 2009), although a positive correlation has been found between LMA and cell-wall nitrogen allocation (Hikosaka and Shigeno, 2009), and a trade-off between N investment in Rubisco and cell walls has been suggested (Feng *et al.*, 2009; but see Harrison *et al.*, 2009 and Hikosaka and Shigeno, 2009). However, increasing leaf thickness also increases the amount of mesophyll cells per A_{leaf}, hence increasing the overall photosynthetic capacity per area. Moreover, the optical properties of mesophyll cells also appear to regulate the internal distribution of sunlight for enhanced photosynthesis (Vogelmann *et al.*, 1996; Smith *et al.*, 1997). Indeed, Niinemets (1999) showed in a global analysis of extant woody plants, that leaf thickness was poorly related to photosynthetic capacity. This can be partly explained by the fact that many sclerophyll leaves are heterobaric, and the bundle-sheath extensions have been proposed to increase light-transferring capacity inside leaves, favouring photosynthesis and partly counteracting the negative effects of decreasing the percentage of photosynthetically active leaf area in these leaves (Nikolopoulos *et al.*, 2002). Instead, leaf density was found to be strongly negatively correlated with photosynthesis, which determines the negative relationship between photosynthesis and LMA (Niinemets, 1999). Recent global (Flexas *et al.*, 2008) or specific surveys in Australian sclerophyll species (Niinemets *et al.*, 2009c) show that this is a result of a reduction of mesophyll conductance to CO_2 (see Chapter 3) as LMA increases.

LEAF FORM AND VENATION

In addition to variations in leaf thickness and density, plants also differ in leaf blade size, form and venation patterns (reviewed in Roth-Nebelsick *et al.*, 2001). The latter includes dendritic open-venation patterns, in which each vein is connected to a single higher-order vein and ramifies into one or more non-interconnected lower-order veins; and anastomosing close patterns with cyclically connected veins (i.e., a single vein can be connected to several different veins of different orders). The dendritic open pattern represents the primitive architecture. In the Devonian and early Carboniferous almost all plants showed this type of venation, whereas anastomosing patterns appeared later in the Upper Carboniferous (Roth-Nebelsick *et al.*, 2001). Both patterns are found in either fossil or extant species of ferns, gymnosperms and angiosperms. In addition, pinnately veined leaves differ from palmate-veined leaves (Sack *et al.*, 2008). The former, which are more common, have a single first-order vein, from which all secondary veins branch. The latter, accounting for up to 30% of regional floras, have multiple first-order veins branching from the petiole. Palmate venation has evolved many times in different lineages. Single-veined leaves or phyllides are more frequent in bryophytes and lycophytes, whereas multiple-veined leaves are more frequent in ferns and angiosperms (Brodribb *et al.*, 2007). Many gymnosperms present single-veined leaves, but with a unique accessory transfusion tissue, consisting of lignified mesophyll cells that greatly increase conductivity of water far from xylem vessels (Brodribb *et al.*, 2007). Different vein-branching patterns, together with many different leaf shapes from lobed to highly dissected (e.g., Nicotra *et al.*, 2008), result in a large variability in total vein density and redundancy in leaf venation.

In a broad study including different evolutionary groups, it has been shown that leaf vein density is positively correlated with leaf maximum photosynthetic rate (Brodribb *et al.*, 2007). This is owing to the fact that increased leaf density increases total leaf hydraulic conductivity by reducing the average hydraulic pathway through the hydraulically inefficient mesophyll tissue (Brodribb *et al.*, 2007). This in turn is intrinsically and genetically linked to stomatal conductance and photosynthesis (Brodribb and Jordan, 2008; Maherali *et al.*, 2008). This will also explain the positive correlations found between a leaf dissection index and photosynthesis in *Pelargonium* species (Nicotra *et al.*, 2008), and the positive correlation between leaf size and leaf hydraulic conductivity in Hawaiian *Plantago* taxa (Dunbar-Co *et al.*, 2009). Moreover, heterogeneity in photosynthesis is observed even within different parts of a single leaf, in coincidence with different vein densities (Nardini *et al.*, 2008).

In addition, it has recently been shown that leaf palmate venation confers tolerance to hydraulic disruption as compared with pinnate venation (Sack *et al.*, 2008). It was shown that severing the midrib resulted in large decreases in hydraulic conductivity, stomatal conductance and photosynthesis in pinnately veined leaves, the effect being smaller in palmately veined leaves. These results suggest that evolution of palmately veined leaves may have been an adaptive trait to withstand situations favourable for cavitation, such as drought or cold environments, as well as insectivory (Sack *et al.*, 2008). Indeed, in seven taxa of the Hawaiian *Plantago* radiation, a negative correlation was found between total vein density and the mean annual rainfall in each species habitat (Dunbar-Co *et al.*, 2009).

Altogether, evidence suggests that redundancy in leaf venation and increased leaf density results in higher photosynthesis rates over time, which may have been a key feature for the evolution of high photosynthetic capacities (Sack *et al.*, 2008).

24.5. OTHER EVOLUTIONARY EVENTS AFFECTING PHOTOSYNTHESIS

In the previous sections, the crucial evolutionary aspects of photosynthesis have been covered. However, other events and evolutionary mechanisms also exist that lead to effects on photosynthesis. For instance, there has long been recognised that there is environmentally driven phenotypic selection of photosynthetic traits, leading to adaptive radiation and genetic intra-specific variation of photosynthesis characteristics and WUE in plant natural populations (Dudley, 1996a; Geber and Dawson, 1997; Lauteri *et al.*, 1997; Ares *et al.*, 2000; Heschel *et al.*, 2002; Donovan *et al.*, 2007, 2009; Montgomery and Givnish, 2008) as well as in crops (Koç *et al.*, 2003; Poormohammad Kiani *et al.*, 2007).

Also, gene and genome duplications leading to plant polyploidy have been recognised to have an ancient origin (Adams and Wendel, 2005; Soltis, 2005) and to be of crucial importance for plant evolution (Masterson, 1994; Adams and Wendel, 2005; Moore and Purugganan, 2005).

Polyploidy has extensive effects on gene expression and gene silencing (Adams and Wendel, 2005) and sometimes results in gene redundancy that may be of adaptive value (Moore and Purugganan, 2005). Polyploidy also has effects on photosynthesis. For instance, there are strong positive relationships between the number of chromosomes (Masterson, 1994) or the genome size (Lomax *et al.*, 2009) and the stomatal size, so that polyploid plants or plants with large genomes may have a greater gas-exchange capacity. It has indeed long been recognised that polyploids often present higher growth and photosynthesis rates than diploids (the so-called 'hybrid vigour'), which is related to increased Rubisco, chl., chloroplast numbers and stomatal and mesophyll conductance to CO_2 (Warner *et al.*, 1987; Warner and Edwards, 1989, 1993; Vyas *et al.*, 2007).

Alternatively more recent evolutionary events have taken place in angiosperms only that lead to reduced photosynthetic capacity. One of these is represented by the evolution of carnivorous plants. Carnivory has evolved independently at least six times in five angiosperms orders (Ellison and Gotelli, 2009). This represents an adaptation to particularly nutrient-poor environments, but it has a cost in terms of plant energetics. Particularly, it has been shown that carnivorous plants have lower photosynthesis capacity and photosynthetic NUE than their non-carnivorous relatives (Farnsworth and Ellison, 2008; Ellison and Gotelli, 2009; Bruzzese *et al.*, 2010), and they fall out of the general relationships between photosynthesis and LMA or nitrogen content described as part of the 'worldwide leaf-economics spectrum' (Wright *et al.*, 2004). Low photosynthesis in carnivorous plants appears associated to replacement of chl.-containing cells with digestive glands, low chl. content, low stomatal density and a compacted mesophyll with a small proportion of intercellular spaces, probably resulting in low mesophyll conductance to CO_2 (Pavlovic *et al.*, 2007). An extreme case of loss of chloroplast and photosynthetic function is that of parasitic plants (Bungard, 2004), which have been covered in Chapter 7.

24.6. CONCLUDING REMARKS

In summary, the evolution of photosynthesis has been crucial for plant colonisation of terrestrial ecosystems. This evolution has implied a multitude of adaptations having led to important changes not only in the function but also in the morphology of vascular plants, and resulted in an extraordinary diversity of forms in the plant kingdom. Hence, in addition to variations in the basic photosynthetic elements already present in unicellular algae and early forms of pluricellular aquatic plants, such as Rubisco or photosynthetic pigments, the evolution of photosynthesis on land has comported the development and diversification of plant-atmosphere interface structures, such as the cuticle and stomata, a complex vascular system and the unique structure known as the leaf, which in turn has evolved and diversified a multitude of traits, structures, etc. Still, perhaps the most specific features concerning the evolution of photosynthesis itself are the appearance of photosynthetic types different to the so-called C_3 metabolism, namely C_4 and CAM, whose evolution is covered in the next chapter.

25 • Evolution of photosynthesis II: evolution and expansion of CAM and C_4 photosynthetic types

J.E. KEELEY, R.K. MONSON AND P.W. RUNDEL

25.1. INTRODUCTION

The evolutionary traits described in the previous chapter are common to all photosynthetic types, and evolved originally in C_3-like species. Under current atmospheric conditions, the O_2 inhibition of photosynthesis occurs through oxygenation of RuBP and subsequent loss of CO_2 through the reactions of photorespiration in C_3 plants. Consequently, a very significant part of photosynthetic evolution in vascular plants has been the development of mechanisms for reducing photorespiration by concentrating CO_2 around Rubisco, thus returning this enzyme to an atmospheric condition that resembles the primitive earth. These CO_2-concentrating mechanisms are known as CAM and C_4 photosynthesis, while all other plants are referred to as C_3. Other CO_2-concentrating mechanisms exist in some algae and cyanobacteria (Kaplan and Reinhold, 1999; Raven *et al.*, 2008), but these will not be discussed here.

The evolutionary origins of CAM and C_4 are presumably tied to changes in paleoclimates and atmospheres, particularly to historic variations in CO_2 and O_2, and locally warm climates (Fig. 24.2 in previous chapter). Under higher CO_2 partial pressure and/or lower temperature the C_3 pathway fixation does not exhibit limitations that would put a premium on coupling it with a CAM or C_4 pathway. The origins of vascular plants date to around the mid-Silurian (~440 MA, Fig. 24.1 in previous chapter) and these plants were likely C_3, although the astomatous CAM plant *Stylites andicola* has been suggested as a possible model of early plant evolution (Keeley *et al.*, 1984). Coupling the C_3 pathway with one of the CO_2-concentrating mechanisms, CAM or C_4, occurred subsequent to the emergence of vascular plants and possibly arose more than once over the past 400 million years.

25.2. EVOLUTION OF CAM

CAM is a pathway specialised for the capture and storage of CO_2 at night. The CO_2 is fixed to a 3-C 'acceptor' molecule by the enzyme phosphoenolpyruvate carboxylase (PEPC) and stored as the 4-C compound malic acid in the large central vacuole of CAM cells. During the subsequent day, the malic acid is decarboxylated, freeing the CO_2 internally, and the cell utilises light energy to produce sugar through traditional C_3 photosynthesis. Two ecological conditions are known to select for CAM. Most well known is that of persistent water limitation, which has led to the most conventional form of CAM by which plants close stomata during the day to conserve water (see Gibson and Rundel, Chapter 28) and open them at night when the atmospheric water deficit is lower. Another condition occurs in aquatic freshwater environments where CO_2 is rapidly depleted during the day owing to high photosynthetic demand by algae and other aquatic plants, and the high diffusive resistance of water hinders the replenishment of CO_2 from the atmosphere. Aquatic CAM plants lack functional stomata and are able to take advantage of the higher night-time availability of CO_2 and lower water temperatures. The distant phylogenetic relationship of terrestrial and aquatic CAM plants suggests that the pathway arose in these taxa through independent but convergent evolution (Keeley and Rundel, 2003).

25.2.1. Terrestrial CAM-plant origins

In the case of terrestrial CAM plants, arid microhabitats selecting for CAM have undoubtedly been present throughout the Earth's history, although changes in paleoatmospheric conditions would have amplified or muted the selective value of CAM at different points in time (Raven and Spicer, 1996). High CO_2 and/or low O_2 partial pressures

Terrestrial Photosynthesis in a Changing Environment: A Molecular, Physiological and Ecological Approach, ed. J. Flexas, F. Loreto and H. Medrano. Published by Cambridge University Press.

will increase net CO_2 fixation for a given stomatal conductance and thus increase carbon gain per unit of water loss. Under these conditions one might expect a less selective value of CAM, and such would have been the case during early vascular plant evolution. The late Carboniferous period (~300 MA) marks the first time vascular plants were exposed to extremes of low CO_2 and high O_2 that might have put a premium on CO_2-concentrating mechanisms such as CAM. However, none of the current CAM taxa are from clades that date to this period. Although atmospheric conditions during the Mesozoic era were not strongly conducive to the advantages of CAM, local selective environments such as those with topographic rain shadows or those caused by unique plant-growth forms, such as occurs with the epiphytic growth habitat, could have favoured the emergence of CAM. Indeed, this seems likely considering experimental results that show photosynthesis of CAM plants comparable with that of C_3 plants under doubled current atmospheric CO_2, (Drennan and Nobel, 2000). However, we lack clear models predicting the relative changes in selective value of CAM under changing atmospheric conditions, temperature and water availability and thus the early stages of CAM evolution, and associated forcing variables are largely a matter of speculation. As these parameters vary we might expect some conditions favouring strict night-time uptake and others a weaker CAM expression.

Since the biochemical steps behind CAM are evident in C_3 stomatal guard cells it is probable that the origins of CAM were founded on altered patterns of gene expression in new cell types. Genomic studies show CAM induction in C_3-CAM facultative plants is largely a matter of transcriptional activation (Cushman, 2001). It has been postulated that the precursor to fully expressed CAM was CAM-like cycling of CO_2 between stored organic acids and freed CO_2 in otherwise C_3 plants (Monson, 1999).

Succulence is an obvious structural trait that typically occurs in terrestrial CAM species from arid climate zones to facilitate storage of organic acids formed with nocturnal carbon fixation. Recent studies have examined functional constraints of CAM leaf anatomy on the level of CAM function as seen in weak and strong modes of nocturnal carbon fixation (Silvera *et al.*, 2005). Plants with strong modes of CAM function have been defined as those with more than 70% of their carbon fixed in nocturnal uptake, whereas weak modes of CAM function, often termed CAM flexibility, are those species where less than one-third of carbon is fixed using nocturnal CAM (Nelson and Sage, 2008). Such facultative CAM species only utilise the CAM pathway when induced by stress, typically water stress (Cushman and Borland, 2002). Two traits often associated with CAM function, cell size and degree of succulence, were not found to be related to the degree of CAM function, whereas reduced intercellular air space in photosynthetic tissues and a reduction in the surface of mesophyll cells exposed to intercellular air space were positively related to CAM function and negatively related to C_3 function (Nelson and Sage, 2005, 2008). Both of these anatomical traits act to reduce internal CO_2 conductance owing to an enhancement of carbon economy, and thus appear to have been a strong selective force in evolving the efficiency of CAM function (Griffiths, 1989; Maxwell *et al.*, 1997).

Perhaps an example of this early evolution of CAM is the unusual *Welwitschia mirabilis*, a species long known to utilise CAM cycling but now established to fix nocturnal carbon to a limited extent in the field (von Willert *et al.*, 2005). This desert plant in the Gnetophyta (sister group to the angiosperms) has origins that date to the Cretaceous (Crane, 1996). Traits in the Gnetophyta are thought to be the result of seasonally extreme environments (Doyle, 1996), further suggesting conditions early in the Mesozoic era may have been conducive to CAM. Some of these early plants may have been herbaceous species with poor fossil records (Crane, 1996), but the distribution of *Welwtschia* suggests evolutionary origins in open environments where plants might have been subjected to higher temperatures and drought stress leading to selection for CAM as a means to scavenge respiratory CO_2. Despite this evolutionary history in primitive vascular plants, modern lineages of CAM species appear to have largely evolved in the past 35 MA.

Since the mid-Cretaceous the partial pressure of CO_2 has dropped from over 1000 ppm to below 600 ppm through the later half of the Miocene. This decline became very marked during the Pleistocene with full glacial episodes having levels below 200 ppm, significantly below pre-industrial levels and representing a time of strong selection for CO_2-concentrating mechanisms such as CAM (Raven and Spicer, 1996).

CAM is extremely widespread both phylogenetically and geographically, occurring in more than 30 diverse plant families and about 20,000 species of vascular plants, indicating that CAM has evolved independently multiple times (Winter and Smith, 1996; Sayed, 2001). Aridity has clearly been one of the major selective factors in this evolution, taking advantage of the relatively high WUE of CAM species. The iconic CAM habitat would be the succulent dominated deserts of the American Southwest or the Succulent Karoo of South Africa, where there are widespread areas with CAM species as ecosystem dominants (Esler and Rundel, 1999). Environmental selection for CAM function, however, clearly has phylogenetic limitations. Extensive areas of arid lands in the world, such as, for example, those of central Australia, are relatively lacking in CAM species. Although it has been hypothesised that nutrient poverty and intense fires have selected against CAM species in central Australia (Orians and Milewski, 2007), there is little to support this idea. Moreover, introduced *Opuntia stricta* thrived over millions of hectares of arid Australia rangeland before being brought under biological control (Parsons and Cuthberson, 2001).

Tropical mesic environments present epiphytic microhabitats where aridity exists and CAM plants are adaptive because of physiological drought associated with their aerial growth habit (Lüttge, 1989; Benzing, 1990; Martin, 1994). Such a condition can be seen in wet and dry Neotropical forests where CAM plants may be abundant. Epiphytes may comprise up to 35–50% of canopy leaf biomass and taxonomically represent more than one-third of the flora in such forests (Gentry and Dodson, 1987), with CAM epiphytes common in Bromeliaceae (1144 epiphytic species and an estimated 50% are CAM), Cactaceae (120 species, 100% CAM), and Orchidaceae (>10,000 epiphyte species, approximately 60% CAM) (Griffiths, 1989). Asian tropical forests have abundant epiphytes but CAM is less common, where it is best represented in epiphytic Asclepiadaceae (135 species, 90% CAM) and a few epiphytic ferns have been found to have CAM (Holtum and Winter, 1999).

The Bromeliaceae represent an excellent example of adaptive radiation into multiple terrestrial and epiphytic habitats, with CAM as one of the key innovations that has allowed diversification. Both the epiphytic growth habit and CAM photosynthesis have evolved at least three times from an ancestral terrestrial C_3 mesophyte, likely in response to increasing aridity and other climate changes in the late Tertiary. The great majority of CAM epiphytes in the Bromeliaceae occur in the subfamilies Tillansioideae and Bromelioideae. In the former, C_3 photosynthesis was the ancestral state and CAM evolved later in extreme habitats, while in the latter, CAM and a terrestrial growth habit predated the evolution of the epiphytic growth habit (Crayn *et al.*, 2004). Subsequent radiation of the Bromelioideae into more mesic habitats led to a reversion to C_3 photosynthesis.

Although both CAM and C_4 metabolic pathways, the latter as described below, have evolved independently in many lineages, it is interesting to note that both pathways have evolved in four plant families. These are the Aizoaceae, Asteraceae, Euphorbiaceae and Portulacaceae. Within families containing many CAM representatives there remain open questions in some groups as to whether CAM has multiple origins or a single origin that has been switched on and off in various lineages (Reinert *et al.*, 2003).

25.2.2. Aquatic CAM-plant origins

Aquatic CAM is restricted to the Tracheophyta and the earliest appearing taxa are in the genus *Isoetes* (Keeley, 1998a), with fossils dated to the Triassic (>230 Ma) (Wang, 1996; Retallack, 1997). These early representatives of *Isoetes* and related taxa in the Lycophyta were amphibious species restricted to shallow seasonal pools that would have undergone sharp daytime CO_2 depletion even if atmospheric CO_2 levels were high (Keeley, 1998a). This is because of the high demand for CO_2 by the pool flora, coupled with the diffusional limitations of water, both of which inhibit rapid equilibrium with atmospheric CO_2 levels. In addition, high Triassic temperatures (Spicer, 1993; McElwain *et al.*, 1999) would have exacerbated the daytime depletion of CO_2 availability in these pools, as CO_2 is substantially less soluble in warm water compared with cool water.

In summary, considering both terrestrial and aquatic CAM species, the CAM pathway likely evolved earliest in aquatic plants and has persisted in particular lineages that show preference for shallow seasonal pools and oligotriphic lakes where carbon in the water column is low (Keeley, 1998a). Certainly, of extant CAM species, the aquatic genus *Isoetes* is representative of the earliest origins of CAM.

Angiosperms that are aquatic CAM plants include monocot species in the Alismataceae (Sagittaria spp.), dicots in the Crassulaceae (annual species of *Crassula*) and Plantaginaceae (*Littorela* spp.) (Keeley, 1998a). Other than the *Crassula* species, aquatic CAM plants share a common morphology known as the isoetid growth form that includes a short rosette of cylindrical leaves. In light of this it would be most interesting to investigate what is considered the most basal angiosperm family, Hydatellaceae, comprising annual aquatic isoetids in Australia and New Zealand (Rudall *et al.*, 2007; Friedman, 2008).

25.3. C_4 EVOLUTION

C_4 photosynthesis has evolved approximately 45 times independently in 19 families and nine orders of the angiosperms (Sage, 2004). With an estimated age of 125 million years for

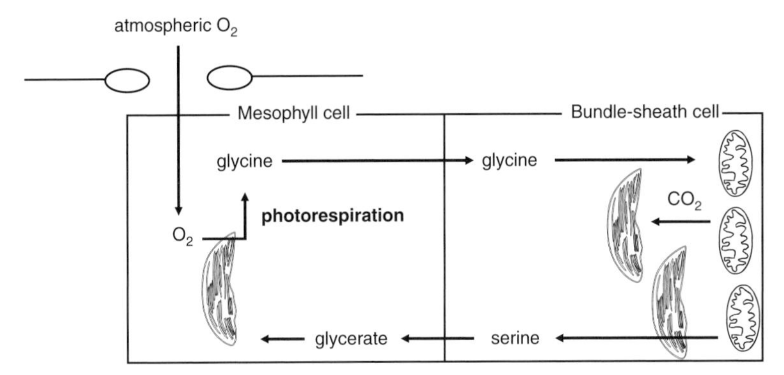

Fig. 25.1. Schematic showing the fundamental steps of the glycine decarboxylase, CO_2-recycling system in C_3-C_4 intermediate species. Glycine decarboxylase activity is isolated in bundle sheath cells. A gradient in the photorespiratory metabolite, glycine, forces CO_2 from photorespiration to be released by mitochondria in the bundle sheath cells. When rates of photorespiration are greater than the rate of glycine decarboxylation and/or in response to diffusive resistance of CO_2 leakage from the bundle sheath cells, the bundle sheath CO_2 concentration can increase; this improves the carboxylation capacity of Rubisco in chloroplasts of the bundle sheath cells.

the angiosperms (Sun *et al.*, 2002), the independent origin of a similar set of traits so many times provides a remarkable example of evolutionary convergence (Christin *et al.*, 2007). This becomes even more significant when one considers the fact that C_4 photosynthesis is a complex, integrated set of morphological, biochemical and physiological traits, probably requiring the structural conversion and/or genomic rewiring of tens to hundreds of genes (e.g., Wyrich *et al.*, 1998). How and why did this set of traits emerge so many times independently in such a short span of geological time? What do the lessons of C_4 tell us about the interaction of terrestrial plant photosynthesis and environmental change? What lessons can we glean from the C_4 story with regard to molecular architecture, evolvability of specific lineages and the nature of selection? These are fundamental questions within the broad topic of plant evolution.

25.3.1. Molecular evolution

C_4 photosynthesis has not arisen *de novo* from newly transcribed evolving genes but rather from the duplication of non-photosynthetic C_3 genes and modification of their expression (Monson, 2003). Presumably the biochemical steps in C_4 were widespread in angiosperms long before this pathway was fully expressed. Some distantly related C_3 species exhibit C_4 carbon assimilation from vascular stem cells and this may be interpreted as one precursor to the evolution of fully expressed C_4 (Raven, 2002b), and some genetic evidence has been interpreted as evidence of such a piecemeal evolution of the pathway (Taylor *et al.*, 1997). Of particular significance here is the fact that some aquatic species exhibit C_4 photosynthesis without the presence of Kranz anatomy, and the CO_2-concentrating mechanism appears to be tied to diffusional constraints characteristic of the aquatic milieu (Keeley, 1998b).

An alternative view is suggested by the study of C_3–C_4 intermediate species that strongly suggest C_4 evolution has involved coordinated changes in morphological, biochemical and genetics (Monson, 1999, 2003). More than 30 species of terrestrial plants in seven different families, both monocot and eudicot, have been reported as having photosynthetic phenotypes intermediate in one way or another to those of fully expressed C_3 or C_4 species (Sage *et al.*, 1999). All of the intermediates described to date share the same glycine decarboxylase CO_2-recycling system illustrated in Fig. 25.1. It is important to note that this recycling system is fundamentally different in many of its biochemical and anatomical features from the fully expressed C_4 pathway, though it appears to share the same advantage of the C_4 pathway in providing elevated CO_2 concentrations at the active site of Rubisco; in the intermediates, however, only a small fraction of the leaf's Rubisco benefits from the elevated CO_2 concentration. This CO_2-concentrating system of C_3–C_4 intermediate species is driven by a gradient in one of the key photorespiratory metabolites (Gly), and is therefore dependent on high rates of photorespiration for its maintenance. This insight presents an interesting piece of irony to the story of C_4 evolution: while it appears that one of the ultimate benefits of the appearance of fully expressed C_4 photosynthesis is a reduced photorespiration rate, high rates of photorespiration were required to sustain the earliest forms of the C_4 pathway.

The biochemical and anatomical features of C_3–C_4 intermediates have been covered in recent reviews (e.g., Monson, 1999; Monson and Rawsthorne, 2000). In this chapter, we take the opportunity to expand the discussion beyond the specific traits associated with C_3–C_4 intermediacy, taking up the broader evolutionary issues of evolvability and predisposition to C_4 evolution. In other words, what are the genotypic, phenotypic, climatic and ecological traits that might allow us to predict whether C_4 photosynthesis will emerge in a specific taxonomic line?

Although the ecologic and climatic contexts for C_4 evolution have been actively debated in recent years, the genetic and life-history traits that might have favoured C_4 evolution have been less discussed. It is clear from the several studies that have been conducted on gene families that code for those enzymes important in C_4 photosynthesis, that C_4-specific genes have most commonly evolved from pre-existing C_3-specific genes through duplication and subsequent neofunctionalisation (Monson, 2003). In some cases, such as that for NADP MDH in the C_4 species, *Flaveria trinervia*, a single C_4 isoform is present without C_3-type isoforms (McGonigle and Nelson, 1995). This may indicate evolutionary modification of a single-copy gene or it may be the result of past duplication and subsequent loss of the C_3 progenitor gene from the C_4 genome. Nonetheless, most C_4 genes belong to larger gene families that include C_3-specific forms. Several studies conducted in Peter Westhoff's laboratory in Dusseldorf have suggested that the genetic modifications required to produce C_4-specific genes from C_3-specific genes are relatively minor for those genes that code for several key enzymes. In the genus *Flaveria* (Asteraceae), the key C_4 enzyme PEPC is coded by three genes, *ppcA*, *ppcB* and *ppcC*, with the latter two sharing common sequences in both C_3 and C_4 species (Bläsing *et al.*, 2002). It is the *ppcA* gene that, while being present in both C_3 and C_4 species, has been modified for specific C_4 function (Westhoff and Gowik, 2004). One of the key events that occurred in the evolution of C_4 *ppcA* was the modification of cis-regulatory elements in the gene promoter providing for enhanced expression in mesophyll cells (Gowik *et al.*, 2004). A functional analysis of the *ppcA* promoter in C_3 and C_4 *Flaveria* species has revealed that a 41 base-pair segment (MEM1) in the C_4 promoter is responsible for mesophyll-specific expression. This segment exhibits three key differences between the C_3 and C_4 forms: (1) the C_3 MEM1 sequence includes a segment of approximately 100 base pairs in the middle of the sequence, which is missing in the C_4 gene; (2) the first nucleotide in the sequence is guanine in the C_4 gene, but adenine in the C_3 gene; and (3) the C_4 gene contains a four-base CACT sequence in one part of the MEM1, which is missing in the C_3 gene. Thus, the totality of genetic modifications possibly required for the conversion of a C_3-type promoter to a C_4-type promoter may be as little as a deletion of a 100-base-pair segment, a single-point mutation and addition of a four-base segment.

Of even more relevance to the early evolution of C_4 photosynthesis is the recent analysis of promoter expression patterns for the gene coding for glycine decarboxylase (*GDC*). As discussed earlier, one of the earliest steps in C_4 evolution appears to be a switch in expression patterns of *GDC* from non-specific expression in leaf cells in the C_3 state to bundle-sheath-specific expression in the C_4 state. Glycine decarboxylase (GDC) is a multi-subunit enzyme, and the differential expression of GDC activity in C_3–C_4 intermediate and C_4 plants is owing to the P-subunit, which carries the decarboxylase active site (Rawsthorne *et al.*, 1988, 1998). Thus, the key evolutionary innovation in this case appears to involve a change in cellular expression pattern (from all cells to just BSC) and potentially addition of an expression enhancer (in BSC). Recently, Englemann *et al.* (2008) conducted a functional analysis of the promoter region of the active P-subunit gene for glycine decarboxylase (*GLDPA*) focusing on expression patterns in the C_4 species *Flaveria bidentis* and the C_3 species *Arabidopsis thaliana*. It was discovered that the C_4 *GLDPA* is expressed in the BSC of *A. thaliana*, indicating that in this C_3 species, the transcription factors are present for bundle-sheath-specific expression of GDC. In other words, it appears that evolutionary changes to the cis-regulatory elements in the gene promoter are all that is needed to accomplish the switch in expression pattern from fully expressed C_3 to C_3–C_4 intermediate and ultimately fully expressed C_4. Using deletion and recombination analysis, these workers were able to dissect the approximately 1,600-base-pair region of the 5′-flanking region of the *GLDPA* promoter, and show that yet-to-be-identified changes in a 433-base-pair segment at the distal end of the promoter and changes in a 212-base-pair segment in the middle of the promoter cause most of the expression enhancement in the bundle sheath and expression repression in the mesophyll.

These studies to date provide evidence that much of the evolutionary change required at the genetic level during the transition from C_3 to C_4 photosynthesis is owing to modifications to the cis-regulatory elements in gene promoter sequences. There are some exceptions to this generalisation, such as the cases for differential cell expression of the small subunit for Rubisco (Patel *et al.*, 2006) and NADP-specific MDH (Nomura *et al.*, 2005). However, most cases of C_4-specific enzyme isoforms that have been examined point to changes in cis-regulatory elements as the primary agent of C_4 evolution (Monson, 1999, 2003). Recent syntheses in the field of evolutionary developmental biology, particularly those dealing with morphological evolution, have concluded that much of the evolutionary novelty that emerges in developmental pathways occurs through changes in the function of cis-regulatory elements (Prud'homme *et al.*, 2007). Processes of gene duplication and neofunctionalisation using pre-existing promoter sequences and transcription factors appear to be the principle processes leading to evolutionary novelty. These generalisations have been criticised (Hoekstra and Coyne, 2007), but at least for the case of C_4 evolution, they explain much of the recent analysis of C_3 versus C_4 trait expression. In the face of these emerging principles we might conclude that the broad and repeating patterns of C_4 evolution over the past several million years are in large partly owing to the pre-existence of a narrow set of biochemical options, the potential for cis-regulatory modification to exert large influences on phenotype and strong selection in the face of global climatic and biogeochemical change.

One aspect of C_4 evolution that has not been considered in much depth in past discussions concerns life-history traits and the potential for certain types of traits to foster the rates of mutation and genetic neofunctionalisation required to force high rates of photosynthetic diversification. For example, in the genus *Flaveria*, a small taxonomic group with 23 closely related species, diversification from C_3 to C_4 has occurred up to three times independently and all three possible occurrences appear to be associated with the evolution from perennial to annual lifecycle and tendency toward self-fertilisation (McKown *et al.*, 2005). Similarly, in the genus *Boerhavia* (Nyctaginaceae), where both C_3 and C_4 species exist, C_4 species are correlated with the annual lifecycle, whereas C_3 photosynthesis is associated with the perennial lifecycle (Douglas and Manos, 2007). Monson (2003) relied on arguments that gene duplication occurs at the highest rates in large populations, and the coupling between neofunctionalised genes and environmental selection would be tightest in small populations of plants with short generation times. Conventional wisdom suggests that self-fertilisation in a species may actually reduce fitness owing to an increased fixation of deleterious alleles (Charlesworth *et al.*, 1993; Nordborg, 2000; Charlesworth and Wright, 2001). However, it is also possible that the evolution of self-fertilisation reduces genome size and in the process simplifies genetic relations that might otherwise be deleterious (Wright *et al.*, 2008); thus, enhancing fitness in the face of neofunctionalised traits. The ultimate reasons for the correlations among changes in life history, mating system and photosynthetic diversification remain to be determined. However, it is clear that this aspect of C_4 evolution has the potential to constrain rates of photosynthetic diversification and may be as important as first-order relations between climate and rates of photosynthetic carbon uptake.

25.3.2. Origins of C_4 photosynthesis

Central to the question of C_4 evolution is the pattern of change in past atmospheric CO_2 concentrations. Conventional theories of C_4 evolution tend to follow the quantum-efficiency model (Ehleringer *et al.*, 1991), which predicts threshold CO_2 concentrations where C_4 photosynthesis has advantages over C_3 photosynthesis in the moles of CO_2 assimilated per mole of solar quanta absorbed. This threshold is constrained by high atmospheric-CO_2 concentrations to favour C_3 photosynthesis throughout much of the Earth's history. However, windows of opportunity for C_4 evolution occurred both early and late in land-plant evolution. Paleoatmospheric models by Beerling and Woodward (2001) predict that significant parts of the globe would have favoured C_4 over C_3 during the Carboniferous (Devonian to Permian Periods). Isotopically heavy carbon deposited during this period may indicate the presence of a C_4-like carbon-concentrating mechanism (Jones, 1994). However, if C_4 taxa evolved under these conditions they are not evident from the fossil record and direct descendants are unknown as none of the contemporary C_4 taxa bear any close relationship to taxa that were extant at that time.

Throughout the Mesozoic era, atmospheric CO_2 values remained relatively high and continued into the Cenozoic. From the middle to late Eocene epoch of the Tertiary period, atmospheric values were between 1000–1500 ppm, but decreased in steps through the Oligocene epoch to modern levels by the mid-Miocene (Pagani *et al.*, 2005). The cyclical pattern of CO_2 drawdown during glaciation events of the Oligocene (Pälike *et al.*, 2006) may have led to conditions favouring C_4 evolution (Ehleringer *et al.*, 1997). Consistent with this scenario is the inferred phylogenic tree of the Poaceae, which suggests the evolution of C_4 photosynthesis occurred in the subfamily Chloridoideae,

32–25 MA in the Oligocene (Christin *et al.*, 2008) and a similar molecular clock date for the origin of the largely C_4 Andropogoneae (Kellogg, 1999). Other evidence in support of an Oligocene origin includes isotopic signatures from soil carbonates (Kleinert and Strecker, 2001) and tooth enamel (Passey *et al.*, 2002).

25.3.3. C_4-grassland expansion

Contemporary tropical and subtropical grasslands are dominated by C_4 grasses. Such ecosystems appear to be relatively recent on a geological timescale, dating in most parts of the world to the late Miocene, 8–5 MA (Ehleringer *et al.*, 1991, 1997; Cerling *et al.*, 1993; Cerling, 1999). Much of our knowledge about the paleo-distributions of C_4 grasslands is based on carbon isotopic ratios measured in fossils of calcareous soils, tooth enamel and egg shells (Tipple and Pagani, 2008). Although there is general agreement on the timing of C_4 grassland dominance, there is far less agreement on the causes.

Worldwide, two-thirds of all C_4 species are grasses and it is therefore little wonder that so much attention has been focused on C_4-grassland expansion during the Miocene period and the associated climate and atmospheric factors that drove this expansion. One thing that is clear is that the timing of C_4-grassland expansion is unrelated to the timing of the origin of the C_4 pathway in grass clades (Cerling, 1999). This mismatch between the timing of the origin of C_4 in grasses and their subsequent expansion argues strongly that environments in the late Miocene changed in ways that provided C_4 grasses a significant ecological advantage over other species. Much of the controversy centres on whether this rise in C_4 grasslands was owing to a shift in conditions favouring C_4 grasses over C_3 grasses, or a shift from woodlands (most of which would have been C_3) to grasslands.

The earliest hypothesis on the emergence of C_4 grasslands was that there was an atmospheric drawdown in CO_2 during the late Miocene. Coupled with higher temperatures in tropical and subtropical latitudes, CO_2 partial pressures reached a critical threshold where C_4 species were at a competitive advantage over C_3 species (Cerling *et al.*, 1997; Ehleringer *et al.*, 1997). This idea was based on the greater quantum efficiency for C_4 photosynthesis under low CO_2 concentrations and high temperature, compared with C_3 photosynthesis. The primary support for this hypothesis was that carbon-isotope data suggested a synchronous rise in C_4 grasslands on three different continents, which would be expected owing to global atmospheric mixing and consequent global climate change.

One limitation to applying the quantum-efficiency model to C_4-grassland expansion is that it considers only one of many competitive interactions and thus is most appropriately applied to interactions between C_3 and C_4 species of similar growth form. For example, under CO_2 partial pressures within the range of those estimated for the much of the Miocene, C_3 trees commonly out-compete C_4 grasses, and once a tree has overtopped the grass canopy light levels and temperature are changed to favour C_3 photosynthesis in the understory (Belsky *et al.*, 1989). These considerations are important because numerous lines of evidence show that the rise in C_4 dominance was coupled with a major vegetation-type conversion from woodland to grassland; this appears to have been the case in Asia, Africa, South America and North America (Morgan *et al.*, 1994; Quade *et al.*, 1995; Hoorn *et al.*, 2000; Janis *et al.*, 2000; Retallack, 2001; Fox and Koch, 2003).

However, Bond and others (Bond and Midgley, 2000; Bond *et al.*, 2003) have suggested the quantum-efficiency model would still be a viable explanation if coupled with fire. They argued that even in a mixed woodland and grassland setting, the impact of fire would be greater on C_3 woodland species, because under declining CO_2 levels C_3 species would be unable to grow quickly enough to shade out their C_4 grass competitors. In addition, Retallack (2001) and Sage (2001) suggested that the combination of reduced CO_2, fire and drought all combined to favour C_4 grasses during the late Miocene.

Despite the important contributions of the quantum-efficiency model (Ehleringer and Bjorkman, 1977), paleoatmospheric CO_2 and temperatures during the late Miocene cast considerable doubt on the model as an explanation for C_4-grassland expansion (Keeley and Rundel, 2003, 2005; Osborne and Beerling, 2006). The primary reason is that recent estimates based on numerous approaches have failed to detect a CO_2 drawdown during the late Miocene (Pagani *et al.*, 1999, 2005; Demicco, 2003). Additionally, average temperatures appear to have declined over that period (Zachos *et al.*, 2001), further weakening the explanatory power of this model.

In place of declining CO_2 levels, drought has been proposed as the major factor in this late Miocene C_4-grassland expansion (Sage, 2001; Edwards and Still, 2008). This derives largely from the greater WUE of C_4 over C_3 species of similar growth form. However, if drought alone were driving this phenomenon, modelling studies of late

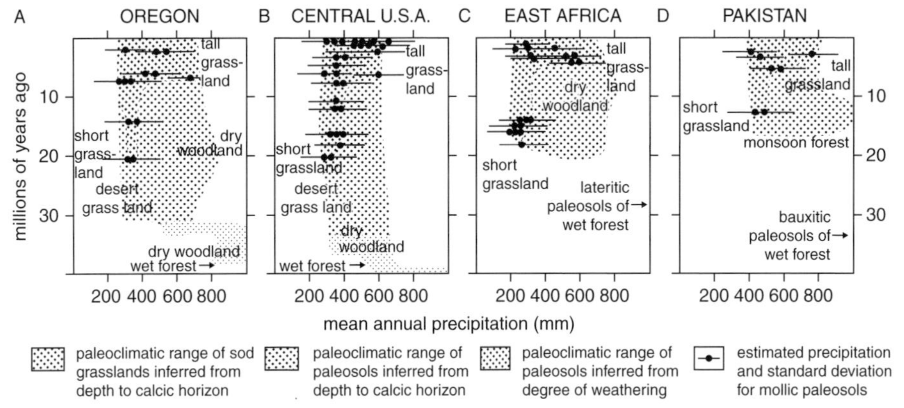

Fig. 25.2. Climatic expansion of sod grasslands through the Miocene in: **(A)** Oregon; **(B)** North American Great Plains; **(C)** East Africa; and **(D)** Pakistan, based on mollic epipedons and depth to calcic horizon of paleosols. A late Miocene advent of tall grasslands is inferred for each area and replaced preexisting dry woodlands (from Retallack, 2001).

Oligocene climates predict that C_4 grasslands should have expanded far earlier than they did (Lunt *et al.*, 2007).

Keeley and Rundel (2003, 2005) acknowledge the importance of drought to the Oligocene origins of the C_4 pathway, but contend that by itself it is inadequate to explain the expansion of grasslands. The primary evidence against drought as a major factor is the fact that C_4-grassland expansion involved more than just an increase in the distribution of C_4 species, but also a rather significant habitat shift. Paleosol data from Asia, Africa and North America all show a temporal change in grassland distribution from semi-arid to more mesic conditions (Retallack, 2001). Under this reconstruction (Fig. 25.2) the late Oligocene to the early Miocene grasslands comprised desert or shortgrass grasslands in environments of 300–400 mm precipitation per year, which is consistent with other data on the origins of C_4 grasses (Cerling, 1999). However, the late Miocene expansion of C_4 grasslands involved a change to tallgrass ecosystems and in regions with 600–800 mm precipitation per year. In other words the C_4-grassland expansion does not appear to be the result of a global increase in the amount of arid habitat, but rather an expansion of C_4 grasses from drier habitats into more mesic habitats.

Both empirical studies and modelling exercises support the conclusion that semi-tropical environments with 600–800 mm precipitation can support dense woodlands (Bond *et al.*, 2005), which implies that the rise of C_4 grasslands involved the invasion of woodland habitats. Fossil records of changes in wood samples during the late Miocene also support this conclusion of a shift from woodlands to grasslands (Quade *et al.*, 1995). In addition, a shift from browser- to grazer-dominated faunas further supports this scenario (Gunnell *et al.*, 1995; Barry *et al.*, 2002).

There are two factors that can drive a shift from woodlands to grasslands; herbivory and fire (Bond and Keeley, 2005; Bond, 2005). In the absence of driving forces from these perturbations, woodlands persist, as illustrated on contemporary landscapes where woodlands readily replace grasslands when all disturbance is removed (Walker and Noy Meir, 1982; Changnon *et al.*, 2002; Furley and Metcalfe, 2007).

A model of how fire and herbivory may have acted in concert to bring about the rise in C_4 grasslands is proposed in Fig. 25.3. Crucial to fire being a keystone process in this phenomenon is development of a markedly seasonal climate. Fire environments have a wet season producing sufficient biomass, followed by a dry season that converts this biomass to available fuel. In continuously arid environments primary production fails to produce fuels sufficient to spread fire, and in a constantly humid environment potential fuels fail to dry sufficiently to become available fuels. One of the characteristics noted with the late Miocene expansion of C_4 grasslands in widely disjunct parts of the globe is increased climatic seasonality. This conclusion is inferred from Paleosol data (Retallack, 2001), geomorphical changes (Barry *et al.*, 2002) and dietary characteristics inferred from carbon and oxygen isotopes of equid teeth (Nelson, 2005). In some regions this seasonality was generated by monsoon climates, but seasonality developed under other climates that underwent a late Miocene rise in C_4 grasslands (Retallack, 2001).

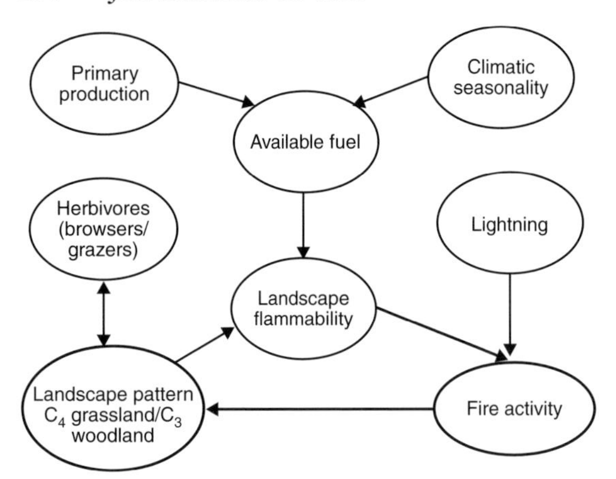

Fig. 25.3. Path model relating factors hypothesised to be the key determinants driving the late Miocene expansion of C_4 grasslands in subtropical and tropical regions.

Increased climatic seasonality from the middle Miocene onward set the stage for a major fire-induced shift in vegetation by creating an annual cycle in which biomass dried sufficiently to become available fuels. However, not all vegetation types produce equally flammable fuels. Grasses generate fine fuels that are capable of igniting and spreading fire much more predictably than coarse woody fuels or evergreen leaves with higher moisture as in woodland species. Therefore the proportion of grass biomass to tree biomass and their spatial pattern would affect flammability of the landscape. If one considers a landscape where patches of highly flammable grasses are embedded within a matrix of less-flammable woodland, then the probability of those grassland patches being ignited by lightning would be affected by the size of the patches. As patches increase their extent they in turn cause an ever-increasing probability of burning. Such an effect has been shown empirically on islands, where fire probability is a function of island size because larger islands receive a greater number of lightning strikes than smaller islands (Wardle *et al.*, 2003). A similar model could apply to Miocene savanna 'islands' within less-flammable woodlands.

Herbivores also would have played a role in opening up woodlands and increasing sites for C_4-grass establishment, in a manner similar to the contemporary impact of mega-herbivores in African woodlands and savannas (Owen-Smith, 1988). This effect, however, would feed back into impacts on the herbivore community and with increasing grass availability cause a shift from browser-dominated to grazer-dominated communities (Barry, 2002). With increasing size of grassland patches and increasing landscape flammability, fire frequency and fire intensity would increase. This would exacerbate the loss of woodland, and thus feed back into an ever increasing expansion of grasslands until ultimately woodland/savannas were converted to grasslands.

Consistent with this model (Fig. 25.3) is the increasing evidence that C_4-grassland expansion was not globally synchronous; rather, grassland expansion was separated by several million years on different continents (Morgan *et al.*, 1994; Jia *et al.*, 2003; Wang and Deng, 2005; Behrensmeyer *et al.*, 2007; Hopley *et al.*, 2007; Tipple and Pagani, 2007). Our model is consistent with this pattern in that it is not dependent on globally synchronous changes in paleoatmospheric conditions, but rather more local effects of climatic seasonality interacting with geomorphology, which would affect fire spread and faunal changes, all of which would be specific to a region. Also consistent with this model is the structure and rate of C_4-grassland expansion within a region. Behrensmeyer *et al.* (2007) have shown that during the late Miocene this expansion was variable at a fine spatial scale, and the complete displacement of C_3 woodlands with C_4 grasslands occurred gradually over a long period of time.

Consistent with this model (Fig. 25.3) is the observation that faunal composition comprised mixtures of browsers and grazers that persisted through much of the Miocene and only gradually shifted towards dominance by grazers at the end of the epoch (Barry *et al.*, 2002). In Kenya, C_4 was a component of herbivore diets as early as 15 MA and shifted to exclusively C_4 between 8.5 and 6.5 MA (Jacobs, 2004). In most regions that underwent a shift to C_4-grassland dominance, faunal communities comprised mixtures of C_3 and C_4 diets in the mid-Miocene indicative of woodland and grassland mixes, which ultimately shifted to C_4-dominated diets indicative of grasslands by the late Miocene (Morgan *et al.*, 1994; Retallack, 2001; Barry *et al.*, 2002).

Fossil evidence of fires synchronous with C_4-grassland expansion is crucial to this hypothesis. There is evidence of an increase in charcoal deposition in late Miocene marine cores in the western Pacific basin (Herring, 1985), but these deposition sites are far removed from sites of grassland expansion and relatively little is known about the area over which this charcoal was drawn. However, in the Niger Delta of west tropical Africa there is a close correlation between charred grass cuticle and grass pollen over the last 8 MA (Germeraad *et al.*, 1967; Morley and Richards, 1993). Also important is the demonstration on the Asian continent of a late Miocene (3 MA) increase in C_4 grasses concomitant with a rise in charred grass from the South China Sea (Jia *et al.*, 2003).

Perhaps the most convincing evidence for the role of fire in C_4-grassland expansion is the evidence that contemporary C_4 grasslands are heavily dependent on fire for their persistence (Hoffman, 1999; Moreira, 2000; Sankaran *et al.*, 2005; Furley, 2006). When fires are present, woodlands convert to grasslands and when fires are excluded, grasslands shift back to woodlands. D'Antonio (2000) has demonstrated this process of fire-induced C_4-grass invasion into less flammable woodlands and ultimately the displacement of woodlands with grasslands in subtropical settings. In these instances grasslands have displaced woodlands on sites that are climatically capable of sustaining woodlands and this has happened in the absence of declining atmospheric CO_2 or increasing aridity. The process is driven entirely by fire opening up forest canopies, which in turn causes changes in fuel availability and further increases the probability of subsequent burning, as is observed at forest edges in tropical regions (Cochrane *et al.*, 1999). These empirical studies suggest that the role of fire, as outlined in Fig. 25.3, would have been sufficient to account for the late Miocene C_4-grassland expansion.

Alternatively, Beerling and Osborne (2006; also Osborne, 2008) have suggested that fire activity would have increased during the late Miocene owing to atmospheric feedbacks generated by fire itself. Their models contend that fire would have promoted climatic characteristics that reduce regional precipitation thus increasing aridity, and this would further favour fire. Their model is a qualitative description that presents an interesting hypothesis, but lacks sufficient information on many parameters for an analytical evaluation. Although atmospheric feedbacks are potentially important in this global expansion of grasslands, their model predicts that grasslands in the late Miocene were expanding owing to an increase in aridity, which is contradicted by the findings of Retallack *et al.* (2001) that late Miocene grasslands were expanding into regions of higher rainfall, not lower rainfall. In addition, they hypothesised that increase in fire activity would lead to atmospheric changes that decreased rainfall, and they interpreted this as a positive feedback on fire, however, in subtropical savannas fire activity is commonly associated with increased rainfall owing to greater grass biomass production (e.g., Duffin, 2008; Harris *et al.*, 2008). These models, however, raise many interesting feedback processes that need further exploration in refining our thinking about the late Miocene C_4-grassland expansion.

25.4. CONCLUSIONS

Photosynthetic carbon reduction evolved under atmospheric levels of O_2 and CO_2 markedly unlike contemporary conditions. This Calvin-Benson cycle is universal in all photosynthetic plants and comprises a complex pathway of tightly coupled reactions that evolved very early in the history of life, and this complexity perhaps has inhibited the ability of plants to alter these reactions as atmospheric conditions became inhibitory during certain periods of the Earth's history. The evolutionary response has been to retain the Calvin-Benson cycle in all photosynthetic cells, but under some conditions certain taxa have evolved various mechanisms for enhancing fitness under atmospheric levels of O_2 and CO_2 inhibitory to carbon reduction. These carbon concentrating mechanisms are diverse, but much of the variation can be circumscribed under the modes of CAM and C_4 photosynthesis, which contrast with the majority of C_3 plants. Opportunities for evolution of these alternative pathways varied through the geological record as atmospheric conditions changed and the timing of their origins has likely varied greatly between aquatic and terrestrial representatives of both pathways.

Part V
Photosynthesis in space

26 • Whole-plant photosynthesis: potentials, limitations and physiological and structural controls

Ü. NIINEMETS

26.1. INTRODUCTION

Whole-plant photosynthesis is a complex process depending on photosynthetic activity of single leaves, plant architecture and plant biomass distribution between support and assimilative tissues. This chapter reviews the importance of whole-plant photosynthesis in ecology and plant science, the possible ways of estimating whole-plant carbon-gain rates and the determinants of whole-tree carbon gain. It further analyses the changes in whole-tree carbon gain in different environments and with plant aging and increasing size. The main message of this chapter is that whole-plant photosynthetic productivity is determined collectively by a series of physiological and structural traits, by within-canopy variation in environmental drivers and by foliage acclimation to the within-plant environmental heterogeneity. Therefore, whole-plant photosynthesis responds differently to the environment than does the sum of single-leaf photosynthetic responses.

26.1.1. Whole-plant photosynthesis: importance for large-scale carbon fluxes

Driven by the need to understand and predict global change, there is strong interest in determinants of vegetation carbon gain at higher scales ranging from whole plants to canopies, landscapes, biomes and globe (e.g., Williams *et al.*, 2004; Ollinger *et al.*, 2008; Duursma *et al.*, 2009). There is a large variation in physiological activity among the leaves of the same plant owing to differences in leaf ontogenetic status, as well as owing to leaf acclimation to within-canopy light, temperature and humidity gradients. Because of this large variation among leaves, the whole-plant performance is difficult to assess from single-leaf measurements (Klingeman *et al.*, 2000), and poor correspondence of single-leaf gas-exchange rates and plant growth has been observed in numerous studies (e.g., Lambers and Poorter, 1992; Lawlor, 1995).

Whole-plant measurements have been frequently conducted in the past, but with the development of portable gas-exchange systems (e.g., Schulze *et al.*, 1982) and with the emergence of the robust process-based photosynthesis model of Farquhar *et al.* (1980a), the focus in gas-exchange studies has shifted from the whole plant towards single-leaf measurements to determine the photosynthetic response curves to light, CO_2 and temperature, and parameterise the Farquhar *et al.* (1980a) photosynthesis model to scale it up to whole trees, canopies, landscapes and biomes (e.g., Wohlfahrt *et al.*, 2000; Ciais *et al.*, 2001; Friend, 2001; Davi *et al.*, 2005). Flux measurements done by eddy covariance (EC) systems are commonly employed to test these scaling-up models (e.g., Baldocchi *et al.*, 1996; Davi *et al.*, 2005). Alternatively, eddy flux data together with scaling-up models are used increasingly to inverse-model average leaf physiological characteristics for larger scale integrations (Raupach *et al.*, 2005; Lasslop *et al.*, 2008; Wu *et al.*, 2009).

This approach considers vegetation as a canopy, either with aggregate properties (big-leaf models, e.g., Jarvis and McNaughton, 1986; Amthor, 1994) or with layered structure consisting of individual leaf layers with specific physiological characteristics (Baldocchi and Harley, 1995; Harley and Baldocchi, 1995; Falge *et al.*, 1997). Such a leaf-to-canopy scaling actually does not need an individual plant scale and so far there have been very few attempts to verify scaling-up models at the whole-plant scale (e.g., Lloyd *et al.*, 1995; Rodriguez *et al.*, 2001). However, individual plants can largely differ in physiological activity. In particular, there is evidence that tree carbon gain decreases with increasing plant age and size, most likely owing to reductions in foliage photosynthetic rates (Bond, 2000; Niinemets, 2002), and similar age-dependent changes in plant physiological

Terrestrial Photosynthesis in a Changing Environment: A Molecular, Physiological and Ecological Approach, ed. J. Flexas, F. Loreto and H. Medrano. Published by Cambridge University Press.

activity are evident in perennial herbs (e.g. Niinemets, 2004). Plants of different size can also have access to different soil-water pools and therefore can respond differently to soil drought (Donovan and Ehleringer, 1991, 1992). In addition, the phenology of juvenile and mature plants can differ (Borchert, 1991; Augspurger and Bartlett, 2003). As natural communities always consist of plants with different age and size, it is important to gain insight into whole-plant level controls on carbon gain to predict long-term carbon fluxes at canopy scale.

26.1.2. Whole-plant photosynthesis assessments for growth and stress studies

Apart from the need to predict large-scale carbon fluxes, a variety of processes are currently studied at the whole-plant level. Traditionally, whole-plant gas-exchange studies have been employed to gain mechanistic insight into the determinants of whole-plant growth rate (McCree, 1986; Dutton *et al.*, 1988; Monje and Bugbee, 1998; Loveys *et al.*, 2001; van Iersel, 2003; Usuda, 2004). Continuous measurements in growing plants allow estimation of the balance between respiration and photosynthesis and determination of whole-plant carbon-use efficiency, i.e., the amount of carbon fixed in photosynthesis that ends up in plant dry mass (McCree, 1988; Monje and Bugbee, 1998; van Iersel, 2003). Further mechanistic insight into the drivers of whole-plant growth and gas-exchange rates has been achieved from simultaneous measurements of above- and belowground plant gas-exchange rates, biomass allocation and construction cost of various plant tissues (Poorter *et al.*, 1990; Lambers and Poorter, 1992; Loveys *et al.*, 2001; Atkin *et al.*, 2007). Whole-plant studies also play a key role in understanding the linkages between above- and belowground plant processes, such as coordinated variation in plant photosynthetic activity and water and nutrient uptake (Ågren and Ingestad, 1987), mycorrhizal colonisation (Eissenstat *et al.*, 1993) and plant photosynthetic activity and N fixation (Gordon and Wheeler, 1978; Dawson and Gordon, 1979). Overall, these studies have demonstrated a good correspondence between whole-plant growth rate and daily integrated plant net carbon gain (Usuda, 2004; Atkin *et al.*, 2007), underscoring the potentials of whole-plant analysis.

In addition to understanding the fundamental principles of plant growth, whole-plant measurements integrating the performance of all leaves in the plant can provide key insight into temporal variations in whole-plant physiological activity as driven by the expansion of new foliage and modifications in physiological activity of leaves (Fig. 26.1, Campbell *et al.*, 1990; Rodriguez *et al.*, 2001; see Chapter 23). As developing plants commonly bear leaves of different developmental stages, disentangling the effects of growth and development and foliage physiological activity is difficult on the basis of leaf-level measurements alone. Experiments under different climates simulating past and future environments can be conducted to investigate the influence of varying developmental and seasonal patterns on whole-plant carbon gain (Royer *et al.*, 2005).

Furthermore, in all cases the environmental or biotic factors alter growth of new organs and biomass allocation, whole-plant measurements provide complimentary insight into the combined effects of growth and photosynthetic activity of single leaves on whole-plant carbon gain. Manipulations that can be conducted at whole-plant level include treatments with different temperatures (Gifford, 1995; Ziska and Bunce, 1995; Ziska and Bunce, 1998; Timlin *et al.*, 2006; Atkin *et al.*, 2007), light (Öquist *et al.*, 1982; Sun *et al.*, 1999; Baltzer and Thomas, 2007a) and UV radiation (Poulson *et al.*, 2006), and mineral salts (Hwang and Morris, 1994) and water (Al-Hazmi *et al.*, 1997; Perez-Peña and Tarara, 2004; Klingeman *et al.*, 2005) availabilities. In addition, whole-plant responses to different atmospheric compositions can be investigated, including experiments under elevated CO_2 concentration (Caporn, 1989; Campbell *et al.*, 1990; den Hertog *et al.*, 1993; Leadley and Drake, 1993; Overdieck, 1993; Garcia *et al.*, 1994; Reid and Strain, 1994; Gifford, 1995; Ziska and Bunce, 1995; Monje and Bugbee, 1998; Ziska and Bunce, 1998; Cheng *et al.*, 2000; Roumet *et al.*, 2000; Rodriguez *et al.*, 2001; Usuda, 2004; Royer *et al.*, 2005; Kim *et al.*, 2006) and studies on the influence of trace gases like volatile hormone ethylene (Woodrow *et al.*, 1989) and atmospheric pollutants O_3 (Beauchamp *et al.*, 2005; Kollist *et al.*, 2007) and NO (Caporn, 1989).

In studies investigating the responses to biotic stress, such as herbivory feeding, whole-plant experiments can be particularly relevant owing to the highly variable degree of damage in different leaves and because biotic attacks often elicit a systemic response in neighbouring leaves not immediately attacked by herbivores (e.g., Arimura *et al.*, 2002; Staudt and Lhoutellier, 2007). Applied studies analysing the effect of insecticides and fungicides to control biotic stress are also important to carry out with whole plants to reduce inherent variation among single leaves (Bednarz and van Iersel, 1999; Klingeman *et al.*, 2000).

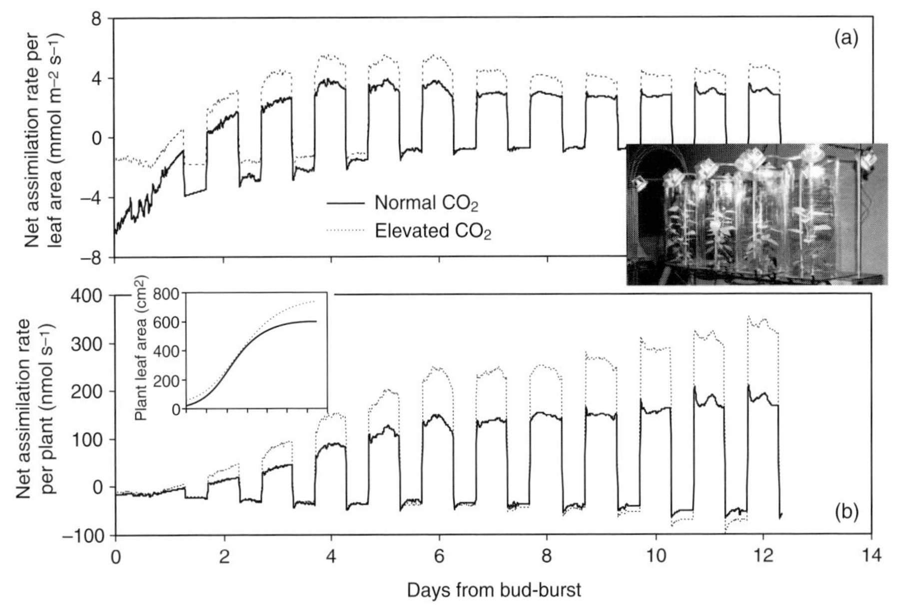

Fig. 26.1. Representative time-dependent changes of foliage photosynthetic rate per unit leaf area **(a)** and per unit whole plant **(b)** in developing canopies of hybrid aspen *Populus tremula* L. × *P. tremuloides* Michx. seedlings under ambient CO_2 concentration of 390 μmol mol^{-1} and elevated CO_2 concentration of 780 μmol mol^{-1} and at an incident quantum flux density of 500 μmol m^{-2} s^{-1} provided during a 14 h photoperiod. One-year-old dormant seedlings were enclosed in four 8 L chambers (two plants at elevated and two at ambient CO_2) as demonstrated in the inset, and the measurements are reported from the time of bud-burst (unpublished data of Hüve, Randjärv, Rasulov, Noe and Niinemets). Plant carbon gain was measured with an open gas-exchange system every 7.5 min and averages of two plants per treatment are shown (Rasulov *et al.*, 2009 for details of the gas-exchange system). Leaf area growth was measured by a technique combining direct measurements of leaf length, regressions between leaf length and area and photographically estimated whole-plant silhouette areas and measured total-plant leaf areas. Leaf-area estimation methods were calibrated in separate plants. The time-dependent changes in leaf-area growth were fitted by a four-parameter sigmoidal regression (Cao *et al.*, 1988). The data demonstrate that enhanced whole plant carbon gain at elevated CO_2 is associated with both greater net assimilation rates per unit leaf area and greater whole-plant leaf area.

26.1.3. Small plants and whole-plant measurements

In plant biology, there has been a new wave of interest in whole-plant measurements powered by the need to investigate the physiological activity of the key model species *Arabidopsis thaliana*. Owing to small size of the leaves and prostrate rosette growth habit, measurements with miniature clip-on cuvettes are difficult and not reliable owing to large diffusion fluxes through the cuvette gaskets (Flexas *et al.*, 2007b; Rodeghiero *et al.*, 2007). Several whole-plant measurement cuvettes have been designed for this species (Donahue *et al.*, 1997; Tocquin and Périlleux, 2004; Kollist *et al.*, 2007; Masclaux-Daubresse *et al.*, 2007; Rossel *et al.*, 2007; Woo *et al.*, 2008). In fact, there is a series of other photosynthetic organisms, such as lichens and mosses (Lange *et al.*, 1997; Balaguer *et al.*, 1999), succulents (Zotz and Winter, 1993a; Eller and Ferrari, 1997), needle-leaved conifers (Leverenz, 1995) etc., in which photosynthetic measurements are commonly conducted on whole plants or on plant parts with multiple leaves.

26.2. ESTIMATION OF WHOLE-PLANT PHOTOSYNTHESIS AND CARBON USE

Plant body mass varies almost twenty orders of magnitude from unicellular algae to large trees (Niklas, 2001). Even among higher plants, the size range between miniature plants from the genus *Wolffia* and massive trees from the genera *Sequoiadendron* and *Sequoia* is more than 14 orders of magnitude. Measurement of whole-plant photosynthesis is a relatively trivial task for plants with a small canopy, such as short-statured herbs, but estimation of photosynthesis of mature trees constitutes a formidable task. Several techniques are available to measure whole-plant photosynthesis, each with specific positive features and pitfalls. Chapter 14 provides a detailed methodological description of approaches to the whole-plant and ecosystem-level measurements. To demonstrate the potentials and limitations of whole-plant measurements and ways of scaling, here different methodologies are summarised with the main emphasis on chamber measurements. Although plant chambers inevitably alter the environmental conditions, whole-plant enclosures provide the only option for direct assessment of whole-plant carbon gain. Indirect methods, including the estimations of whole-plant photosynthesis from chlorophyll fluorescence (Chl-F), sapflow, EC flux and using scaling-up models are also briefly introduced, and benefits and difficulties associated with each specific method are analysed.

26.2.1. Chamber measurements

Enclosure of a plant in a sealed cuvette and monitoring the change of chamber CO_2 concentration owing to plant activity is the direct way to assess whole-plant assimilation rate. The key requirements for reliable whole-plant measurements are the same as for leaf-level studies: good seal, high turbulence of air in the chamber and accurate estimations of gas concentrations and flow rate. In particular, in chambers enclosing multiple leaves, high turbulence is crucial to achieve the homogeneity of air around the foliage and avoid dead air parcels in the chamber. In addition, for the field measurements the chambers should alter the environmental conditions as little as possible.

Whole-plant gas exchange has previously been monitored either using open, closed or semi-closed gas-exchange systems (Mitchell, 1992 for a comparison of various measurement principles). In open systems, the photosynthesis rate is calculated from the CO_2-concentration difference between the reference line and the chamber and the bulk flow rate through the chamber. In closed systems, the photosynthesis rate is determined from the rate of reduction of CO_2 concentration inside the chamber, whereas in semi-closed systems, CO_2 reduction owing to plant activity is compensated by CO_2 injections and the plant photosynthesis is determined from the amount of compensatory CO_2 needed. In the early experiments, closed systems were often employed for whole-plant measurements (Gerbaud and Andre, 1979; Dutton *et al.*, 1988; Berard and Thurtell, 1990; Corey and Wheeler, 1992; Mitchell, 1992; Wheeler *et al.*, 1993). However, the measurements in closed and semi-closed systems are very sensitive to bulk mass flow and diffusion leaks and to non-uniform aerodynamic conditions inside the chamber. In addition, lack of air flow can lead to humidity build-up and increased chamber temperatures. Plant-generated volatile hormones ethylene and NO (Munné-Bosch *et al.*, 2004) and other volatile compounds with partly unknown functions, such as isoprene and monoterpenes emitted in constitutive emitters (Kesselmeier and Staudt, 1999) and mono- and sesquiterpenes induced under stress (Beauchamp *et al.*, 2005), can also accumulate in closed chambers during long-term measurements. Given all these problems, whole-plant photosynthesis measurements are currently commonly conducted with open systems (Table 26.1, Bugbee, 1992; Mitchell, 1992; Long *et al.*, 1996; Takahashi *et al.*, 2008). Nevertheless, owing to the large cost of operation of open systems and in particular when plants are grown and/or measured in atmospheres with different CO_2 concentration, closed or semi-closed environmentally controlled systems have been occasionally used for larger plants and patches of ecosystems (Corey and Wheeler, 1992; Overdieck, 1993; Wheeler *et al.*, 1993; Rosenthal *et al.*, 1999; Zabel *et al.*, 1999; Wallin *et al.*, 2001; Medhurst *et al.*, 2006). In the following, the capacity of various systems and their environmental responsiveness is analysed, mainly focusing on the chambers used in open systems (Table 26.1). As the system responsiveness may interfere with plant activity, understanding the potentials and limitations of the measurement systems is important in interpreting the measured whole-plant environmental responses.

AVAILABLE HARDWARE TO MEASURE DIFFERENT-SIZED PLANTS

As discussed in Section 26.1.3, measurements on individual leaves are either impossible or impractical in photosynthetic organisms with overall small size, such as mosses and lichens and in species having small leaves (e.g., short-needled conifers), short petioles and prostrate growth form (e.g., *Arabidopsis*) or thick succulent foliage, photosynthetic stems and cladodes. In these organisms, photosynthetic measurements have been commonly conducted on shoots or whole plants using self-constructed or commercially available cuvettes with a relatively small volume. Typically, vigorously mixed cuvettes with volumes of 0.02–0.16 L, such as the commercially available lichen/moss/conifer cuvettes from PP systems, Ltd., Walz GmbH and Li-Cor, Inc., and custom-made cuvettes (e.g., Donahue *et al.*, 1997; Tocquin and Périlleux, 2004; Kollist *et al.*, 2007) have been used to measure the gas-exchange rates of mosses and lichens and small herbs such as *Arabidopsis* (Table 26.1).

Table 26.1. *Comparison of various gas-exchange cuvettes and enclosures employed in open gas-exchange systems for measurement of plant gas-exchange rates at various level of biological organisation.*

Measurement scale	Volume (L)*	Maximum flow rate (L min^{-1})	Time for steady state (s) ††	Environmental control†	Reference
Single leaf	$3.75{\cdot}10^{-3}$	$5.00{\cdot}10^{-1}$	$1.25{\cdot}10^{0}$	F	Ciras 2 standard 2.5 cm^2 cuvette, PP Systems, Ltd.
Single leaf	$4.36{\cdot}10^{-3}$	$6.72{\cdot}10^{-1}$	$1.08{\cdot}10^{0}$	F	Laisk and Oja (1998)
Single leaf	$1.20{\cdot}10^{-2}$	$9.41{\cdot}10^{-1}$	$2.12{\cdot}10^{0}$	F	LI-6400 standard 2 × 3 cm cuvette, Li-Cor, Inc.
Single leaf	$4.00{\cdot}10^{-2}$	$9.41{\cdot}10^{-1}$	$7.07{\cdot}10^{0}$	F	GFS-3000 standard cuvette, Walz GmbH
Single leaf	$6.04{\cdot}10^{-1}$	$1.34{\cdot}10^{0}$	$7.48{\cdot}10^{1}$	N	Sangsing *et al.* (2004)
Single leaf	$1.77{\cdot}10^{0}$	$1.08{\cdot}10^{0}$	$2.72{\cdot}10^{2}$	F	Idle and Proctor (1983)
Single leaf	$1.82{\cdot}10^{0}$	$1.34{\cdot}10^{0}$	$2.25{\cdot}10^{2}$	N	Sangsing *et al.* (2004)
Single leaf	$6.00{\cdot}10^{0}$	$7.00{\cdot}10^{0}$	$4.79{\cdot}10^{2}$	F	GWK-8, Walz GmbH, Jurik *et al.* (1984)
Single leaf/shoot	$5.00{\cdot}10^{-1}$	$1.00{\cdot}10^{0}$	$8.32{\cdot}10^{1}$	F	Compact minicuvette, Walz GmbH, Schröder *et al.* (2005)
Shoot	$6.44{\cdot}10^{-2}$	$1.21{\cdot}10^{0}$	$8.85{\cdot}10^{0}$	F	GFS-3000 conifer cuvette, Walz GmbH
Shoot	$1.55{\cdot}10^{-1}$	$9.41{\cdot}10^{-1}$	$2.73{\cdot}10^{1}$	F	LI-6400 conifer cuvette, Li-Cor, Inc.
Shoot	$1.57{\cdot}10^{-1}$	$5.00{\cdot}10^{-1}$	$5.23{\cdot}10^{1}$	F	PLC5 (C) conifer cuvette, PP Systems, Ltd.
Shoot	$2.40{\cdot}10^{0}$	$9.00{\cdot}10^{-1}$	$4.44{\cdot}10^{2}$	N	Weiss *et al.* (2009)
Shoot	$3.50{\cdot}10^{0}$	$1.03{\cdot}10^{0}$	$5.64{\cdot}10^{2}$	N	Hari *et al.* (1999)
Shoot	$3.54{\cdot}10^{0}$	$1.50{\cdot}10^{0}$	$3.93{\cdot}10^{2}$	N	Zha *et al.* (2007)
Shoot	$7.00{\cdot}10^{0}$	$4.17{\cdot}10^{0}$	$2.79{\cdot}10^{2}$	F	Koch *et al.* (1971)
Shoot	$1.20{\cdot}10^{1}$	$4.17{\cdot}10^{0}$	$4.79{\cdot}10^{2}$	F	Koch *et al.* (1971)
Shoot/whole plant	$3.00{\cdot}10^{0}$	$3.00{\cdot}10^{0}$	$1.66{\cdot}10^{2}$	F	GWK-3 M cuvette, Walz GmbH, Zotz and Winter (1993); Winter *et al.* (1997)
Shoot/whole plant	$2.50{\cdot}10^{1}$	$2.00{\cdot}10^{1}$	$2.08{\cdot}10^{2}$	F	GK-8, gas-exchange measuring station, Walz GmbH
Whole plant (moss/lichen)	$2.17{\cdot}10^{-2}$	$1.21{\cdot}10^{0}$	$2.99{\cdot}10^{0}$	F	GFS-3000 moss cuvette, Walz GmbH
Whole plant (moss/lichen)	$4.03{\cdot}10^{-2}$	$5.00{\cdot}10^{-1}$	$1.34{\cdot}10^{1}$	F	PLC moss cuvette, PP Systems, Ltd.
Whole plant (moss/lichen)	$1.90{\cdot}10^{-1}$	$5.00{\cdot}10^{-1}$	$6.32{\cdot}10^{1}$	N	Lange *et al.* (1997)
Whole plant (herb/grass)	$2.00{\cdot}10^{-2}$	$2.50{\cdot}10^{-1}$	$1.33{\cdot}10^{1}$	P	Tocquin and Périlleux (2004)
Whole plant (herb/grass)	$3.00{\cdot}10^{-2}$	$3.50{\cdot}10^{-1}$	$1.43{\cdot}10^{1}$	P	Tocquin and Périlleux (2004)

Table 26.1. (*cont.*)

Measurement scale	Volume (L)*	Maximum flow rate (L min^{-1})	Time for steady state (s) ††	Environmental control†	Reference
Whole plant (herb/grass)	$5.77 \cdot 10^{-2}$	$9.41 \cdot 10^{-1}$	$1.02 \cdot 10^{1}$	F	LI-6400 whole plant *Arabidopsis* cuvette, Li-Cor Inc. (2008)
Whole plant (herb/grass)	$1.20 \cdot 10^{-1}$	$4.50 \cdot 10^{-1}$	$4.44 \cdot 10^{1}$	P	Tocquin and Périlleux (2004)
Whole plant (herb/grass)	$1.67 \cdot 10^{-1}$	$1.10 \cdot 10^{0}$	$2.53 \cdot 10^{1}$	P	Kollist *et al.* (2007)
Whole plant (herb/grass)	$3.00 \cdot 10^{-1}$	$1.00 \cdot 10^{0}$	$4.99 \cdot 10^{1}$	F	Donahue *et al.* (1997); Sun *et al.* (1999)
Whole plant (herb/grass)	$7.15 \cdot 10^{-1}$	$9.41 \cdot 10^{-1}$	$1.26 \cdot 10^{2}$	F	Xiong *et al.* (2000)
Whole plant (herb/grass)	$1.18 \cdot 10^{0}$	$1.20 \cdot 10^{0}$	$1.63 \cdot 10^{2}$	P	Hwang and Morris (1994)
Whole plant (herb/grass)	$5.00 \cdot 10^{0}$	$1.61 \cdot 10^{0}$	$5.16 \cdot 10^{2}$	P	Hüve *et al.* (2007)
Whole plant (herb/grass)	$9.50 \cdot 10^{0}$	$1.67 \cdot 10^{1}$	$9.48 \cdot 10^{1}$	F	Karolin and Moldau (1976)
Whole plant (herb/grass)fi	$1.40 \cdot 10^{1}$	$4.00 \cdot 10^{0}$	$5.82 \cdot 10^{2}$	N	Van Oosten *et al.* (1997)
Whole plant (herb/grass)	$1.60 \cdot 10^{1}$	$1.50 \cdot 10^{0}$	$1.77 \cdot 10^{3}$	F	Cen *et al.* (2001)
Whole plant (herb/grass)fi	$9.60 \cdot 10^{1}$	$3.60 \cdot 10^{1}$	$4.44 \cdot 10^{2}$	P	van Iersel (2003)
Whole plant (herb/grass)	$1.30 \cdot 10^{2}$	$2.00 \cdot 10^{1}$	$1.08 \cdot 10^{3}$	F	McCree (1986)
Whole plant (herb/grass)	$1.50 \cdot 10^{2}$	$1.20 \cdot 10^{3}$	$2.08 \cdot 10^{1}$	N	Alterio *et al.* (2006)
Whole plant (herb/grass)	$1.92 \cdot 10^{2}$	$3.00 \cdot 10^{1}$	$1.06 \cdot 10^{3}$	P	Dutton *et al.* (1988); Woodrow *et al.* (1989)
Whole plant (herb/grass)	$3.00 \cdot 10^{2}$	$8.00 \cdot 10^{1}$	$6.24 \cdot 10^{2}$	N	Usuda (2004)
Whole plant (tree seedling)	$3.00 \cdot 10^{0}$	$3.00 \cdot 10^{0}$	$1.66 \cdot 10^{2}$	P	Rasulov *et al.* (2009)
Whole plant (tree seedling)	$3.23 \cdot 10^{0}$	$7.65 \cdot 10^{-1}$	$7.03 \cdot 10^{2}$	F	Reid and Strain (1994)
Whole plant (tree seedling)	$3.98 \cdot 10^{0}$	$4.0 \cdot 10^{0}$	$1.65 \cdot 10^{2}$	N	Royer *et al.* (2005)
Whole plant (tree seedling)	$8.31 \cdot 10^{0}$	$1.08 \cdot 10^{0}$	$1.29 \cdot 10^{3}$	N	Baltzer and Thomas (2007b)
Whole plant (tree seedling)	$1.20 \cdot 10^{1}$	$2.00 \cdot 10^{0}$	$9.98 \cdot 10^{2}$	P	Gordon and Wheeler (1978)
Whole plant (tree seedling)	$1.10 \cdot 10^{3}$	$4.01 \cdot 10^{1}$	$4.57 \cdot 10^{3}$	P	Hüve *et al.* (2007)
Whole plant (woody vine)	$8.00 \cdot 10^{3}$	$2.70 \cdot 10^{4}$	$4.93 \cdot 10^{1}$	N	Perez-Peña and Tarara (2004)
Whole plant (mature tree)	$8.00 \cdot 10^{2}$	$1.20 \cdot 10^{3}$	$1.11 \cdot 10^{2}$	N	Corelli-Grappadelli and Magnanini (1993)
Whole plant (mature tree)	$1.25 \cdot 10^{3}$	$9.00 \cdot 10^{2}$	$2.31 \cdot 10^{2}$	N	Poni *et al.* (2009)
Whole plant (mature tree)	$4.00 \cdot 10^{3}$	$2.20 \cdot 10^{4}$	$3.02 \cdot 10^{1}$	N	Wünsche and Palmer (1997)
Whole plant (mature tree)	$4.29 \cdot 10^{4}$	$2.00 \cdot 10^{2}$	$3.57 \cdot 10^{4}$	N	Lloyd *et al.* (1995)
Whole plant (mature tree)	$5.63 \cdot 10^{4}$	$9.00 \cdot 10^{2}$	$1.04 \cdot 10^{4}$	F*	Wallin *et al.* (2001); Medhurst *et al.* (2006)
Whole plant (mature tree)	$9.60 \cdot 10^{4}$	$1.38 \cdot 10^{4}$	$1.16 \cdot 10^{3}$	N	Goulden and Field (1994)
Ecosystem (herbaceous)	$3.20 \cdot 10^{1}$	$2.75 \cdot 10^{1}$	$1.94 \cdot 10^{2}$	N	Jiang *et al.* (1997)

Ecosystem (herbaceous)	$6.70 \cdot 10^{2}$	$1.5 \cdot 10^{3}$	$7.43 \cdot 10^{1}$	N	Wedler *et al.* (1996)
Ecosystem (herbaceous)	$8.44 \cdot 10^{2}$	$1.67 \cdot 10^{3}$	$8.43 \cdot 10^{1}$	F*	Müller *et al.* (2009)
Ecosystem (herbaceous)	$1.20 \cdot 10^{3}$	$6.00 \cdot 10^{3}$	$3.33 \cdot 10^{1}$	N	Burkart *et al.* (2007)
Ecosystem (herbaceous)	$6.36 \cdot 10^{4}$	$1.44 \cdot 10^{5}$	$7.35 \cdot 10^{1}$	N	Ham *et al.* (1993)
Ecosystem (herbaceous)	$1.61 \cdot 10^{5}$	$1.12 \cdot 10^{5}$	$2.39 \cdot 10^{2}$	F*	Griffin *et al.* (1996)
Ecosystem (woody)	$1.65 \cdot 10^{4}$	$2.70 \cdot 10^{4}$	$1.02 \cdot 10^{2}$	N	Dore *et al.* (2003)
Ecosystem (multiple)	$2.04 \cdot 10^{8}$	$5.58 \cdot 10^{5}$	$6.08 \cdot 10^{4}$	F*	Rosenthal *et al.* (1999); Zabel *et al.* (1999)

For whole-plant measurements, many different chambers with varying responsiveness and turbulence characteristics have been constructed. All these systems can be used to track plant gas-exchange fluxes and realistic daily time courses of photosynthesis have been observed even with the largest chambers (Fig. 26.3). Nevertheless, in interpreting the chamber measurements, it is important to consider the chamber environmental responsiveness and possible modifications in environmental conditions. Also, accurate leaf-area estimation is crucial for comparison of experiments in various-sized plants and during periods of leaf development and senescence.

ɐ Only the empty leaf-chamber volume. The actual volume affecting the whole-system kinetics also includes other parts of the system, including the volumes of the infrared gas-analyser cells, cooling unit etc. This affects in particular the kinetics of the systems with smaller leaf chambers. For instance, the total system volume of LI-6400 gas-exchange system with the standard 2×3 cm leaf cuvette is $8 \cdot 10^{-2}$ L, and the system reaches a steady state in approximately 15 s (Li-Cor Inc., 2001). During photosynthesis measurements with the plant(s) enclosed in the chamber, the chamber volume is less, but the presence of plant(s) may reduce the uniformity of air mixing in the chamber.

ɔ The time to reach the steady state in the leaf chamber was taken as 4τ, where τ is the half-time of the chamber response (Eqn. 26.B4). This time corresponds to approximately 94% of full response. For all chambers, 4τ was calculated for the maximum flow rate reported, corresponding thus, to the shortest chamber response time.

† Key environmental factors controlled by plant gas-exchange systems are temperature, humidity, gas composition and light intensity. F, full environmental control; F*, full control, except light (naturally lit chambers); P, partial control (some factors controlled); N, no environmental control.

ﬁ multiple plants normally enclosed in the chambers during measurements.

For photosynthetic measurements in larger plants, including tall herbs and grasses, tree seedlings and mature trees, a variety of gas-exchange chambers have been constructed with volumes from a few litres to thousands and tens of thousands of litres (Table 26.1). The largest single-plant chambers enclosing between 8–14 m^2 plant leaf area have been used to measure the photosynthesis of large trees and vines including Norway spruce (*Picea abies*) (Wallin *et al.*, 2001; Medhurst *et al.*, 2006), apple (*Malus domestica*) (Corelli-Grappadelli and Magnanini, 1993; Wünsche and Palmer, 1997), oaks (*Quercus agrifolia* and *Q. durata*) (Goulden and Field, 1994), macadamia (*Macadamia integrifolia*) and lychee (*Litchi chinensis*) (Lloyd *et al.*, 1995) and grapevines (*Vitis vinifera*) (Al-Hazmi *et al.*, 1997; Poni *et al.*, 2003, 2009; Perez-Peña and Tarara, 2004). As an extension of the whole-plant chambers, even entire miniature ecosystems, e.g., patches of grassland and herb crops (e.g., Drake and Leadley, 1991; Ham *et al.*, 1993; Leadley and Drake, 1993; Fredeen *et al.*, 1995; Monje and Bugbee, 1998; Burkart *et al.*, 2000, 2007; Cheng *et al.*, 2000; Kim *et al.*, 2006; Baker *et al.*, 2009; Müller *et al.*, 2009) and young woody stands (Overdieck, 1993; Overdieck and Forstreuter, 1994; Dore *et al.*, 2003) have been measured. At the extreme end of the whole-ecosystem gas-exchange enclosures, large single ecosystem chambers with volumes in the order of 10^5 L have been constructed, such as the NASA's biomass production chamber with $1.13 \cdot 10^5$ L (Corey and Wheeler, 1992; Wheeler, 1992), the EcoCell chamber with $1.6 \cdot 10^5$ L (Griffin *et al.*, 1996; Cheng *et al.*, 2000) and biosphere 2 laboratory (B2L), a facility enclosing multiple artificial ecosystems with a total volume of around $2 \cdot 10^8$ L (Rosenthal *et al.*, 1999; Zabel *et al.*, 1999).

Apart from the size, plant gas-exchange systems differ in handling the above- and belowground plant compartments during measurements. In some cases, entire potted plants are enclosed in the chamber (Caspar *et al.*, 1985; Woodrow *et al.*, 1989). However, the disadvantage of these measurements is that soil and root respiration interferes with the photosynthesis measurements. To best correlate the gas-exchange measurements with plant growth rate, custom-built chambers with two halves allowing separate measurement of gas-exchange of above- and belowground plant compartments have been developed (Bugbee, 1992; Mitchell, 1992; den Hertog *et al.*, 1993; Atkin *et al.*, 2007). Separation of above- and belowground plant parts can sometimes be difficult, in particular, for plants with a rosette leaf form such as *Arabidopsis*. In *Arabidopsis*, the gas exchange of the entire plant with roots and rooting medium has been measured (Donahue *et al.*, 1997; Van Oosten *et al.*, 1997). In other studies, pottery modelling clay or petroleum jelly have been used to isolate soil surface from leaf rosette (Eckardt *et al.*, 1997; Sun *et al.*, 1999; Li-Cor Inc., 2008). For completely non-intrusive enclosure of plants, the leaf rosette grows into the measurement chamber in Kollist *et al.* (2007) system, assuring almost a perfect seal. Growth of plants on hydroponic culture also simplifies enclosure of the leaf rosette without damaging the plant parts (Tocquin and Périlleux, 2004).

Owing to the high cost of construction, especially for large chambers with environmental control, lack of replication can be a problem in whole-plant studies. To cope with this limitation, systems running with several chambers in parallel have been employed (e.g., Fig. 26.1, Dutton *et al.*, 1988; Woodrow *et al.*, 1989; Hwang and Morris, 1994; Perez-Peña and Tarara, 2004; Royer *et al.*, 2005; Kollist *et al.*, 2007). Alternatively, large portable whole-tree chambers have been constructed that can be disassembled and mounted on new plants relatively rapidly (Corelli-Grappadelli and Magnanini, 1993; Wünsche and Palmer, 1997), making it feasible to study multiple plants during the growing season.

PERFORMANCE OF DIFFERENT CHAMBERS IN RELATION TO SIZE

Increasing the chamber size is needed for direct measurements of photosynthesis in larger plants, but the measurements with larger chambers can constitute several difficulties. In particular, the gas-exchange chamber responds to changes in gas concentration and accordingly to modifications in plant physiological activity with a finite time constant. For open gas-exchange systems, the chamber responsiveness depends on the chamber size and the flow rate through the chamber (Box 26.1, Weiss *et al.*, 2009). In theory, fast-responding gas-exchange chambers can be made of any size. In practice, larger chambers typically have higher mean air residence time and take longer to reach a steady state (Box 26.1, Fig. 26.2a). Inherently slow air turnover rate in whole-tree and ecosystem chambers (equivalent to a large storage term) entails that the gas-exchange measurements by big chambers may inadequately track the modifications in environmental conditions, virtually shifting and smoothing out the whole-plant signal in time. This may lead to underestimation of the responsiveness of whole-plant photosynthesis to environmental drivers. The time to reach a steady state varies more than four orders of magnitude for different plant enclosures (Fig. 26.2a, Table 26.1) with the slowest whole-plant systems taking approximately 30 minutes or more to reach the steady state.

Box 26.1. Response time of gas-exchange systems in dependence on chamber size, flow rate and tubing length.

Gas-exchange measurements in an open gas-exchange system are based on measuring the *water-vapour* and CO_2-concentration differences between incoming and outgoing air. These measurements should be conducted in steady-state conditions when the concentration differences are entirely owing to plant responses. The key question in assessing the performance of different plant chambers is how long does it take to reach a steady state?

For an open gas-exchange system, the CO_2 mass balance of the empty chamber with fully turbulent air mixing and without mass flow and diffusion leaks is at any given time t given by:

$$\frac{dC_o V}{dt} = F_1 C_i - F_2 C_o \quad \text{[Eqn. 26.B1]}$$

where C_o is the chamber CO_2 concentration (μL L^{-1}); C_i is the CO_2 concentration of incoming air (μL L^{-1}); V is the chamber volume (L); and F is the volumetric air flow rate (L s^{-1}). Owing to turbulent mixing, the CO_2 concentration at the chamber outlet is equal to the chamber CO_2 concentration. Assuming that there is no change in flow rate as might occur for instance when *water vapour* condenses on the chamber walls (or *water vapour* is added to the gas stream when plants are present in the chamber), $F_1=F_2=F$ and dividing by V, equation 26.B1 becomes:

$$\frac{dC_o}{dt} = \frac{F}{V}(C_i - C_o) \quad \text{[Eqn. 26.B2]}$$

The ratio, F/V is the first order rate constant (κ) for the time-dependent changes in chamber CO_2 concentration. Integrating equation 26.B2 and evaluating the integration constant at t=0 and the initial chamber CO_2 concentration of $C_{o,s}$, yields the following expression for time-dependent changes of chamber CO_2 concentration (Li-Cor Inc., 2001):

$$C_o(t) = C_i - (C_i - C_{o,s})e^{\left(-\frac{F}{V}t\right)} \quad \text{[Eqn. 26.B3]}$$

The chamber response half-time (τ_c) is given as:

$$\tau_c = \mathrm{Ln}(2)/\kappa \quad \text{[Eqn. 26.B4]}$$

Equation 26.B3 suggests that the chamber CO_2 concentration reaches the steady state faster the smaller the chamber volume and the larger the flow rate (Fig. 26.B.1.a). A time of $4\tau_c$ corresponds to around 94% full-system response and can be taken as a satisfactory approximation of reaching steady-state conditions in the chamber.

Equation 26.B3 is valid only for fully turbulent air mixing in the chamber. As Fig. 26.B.1.b demonstrates, this assumption is essentially satisfied for the relatively small 3 L chamber, where the simulation of the chamber time response with actual values of F and V (τ_c=42 s) results in an excellent fit. In contrast, for the larger chamber, the predicted τ_c is 18 s (dotted line in Fig. 26.B1b), but the actual chamber response is slower. The best-fit estimate of τ_c was 67 s (dashed line), but the fit was poor. In fact, the chamber response can be modelled as the sum of several exponentials (data not shown, see also Ham *et al.*, 1993), indicating that aerodynamically, the chamber consists of several virtual compartments with a differing degree of air mixing. The gradients in air mixing in chambers inevitably slow down the overall system response.

Apart from the chamber effects, limited flow rate in the tubes further leads to time lags in the plant and the system response. The average rate of gas flow in the tubing (f, m s^{-1}), is given as:

$$f = \frac{F}{S_T} \quad \text{[Eqn. 26.B5]}$$

where S_T is the tube cross-section area (m^2). For instance, for a volumetric flow rate of 1 L min^{-1} ($1.67 \cdot 10^{-5}$ m^3 s^{-1}), the linear gas flow rate in a tube with 1 cm diameter is 0.21 m s^{-1}. For a 3 m total piping length (including gas source to chamber and chamber to gas analyser), this flow rate leads to an offset of 15 s. Of course, such a consideration of flow in pipes is a simplification because the flow in pipes is rarely fully turbulent, and thus, the linear flow rate can vary in dependence on the distance from the pipe walls, reflecting larger shear stress closer to the walls (Wongwises *et al.*, 1999; Alterio *et al.*, 2006). This non-uniformity of flow in piping also interferes with the initial chamber response during the first few seconds the gas analyser starts to measure the air with altered gas composition. However, for whole-plant chambers with a relatively long response time, flow heterogeneity in pipes is likely to be of minor significance.

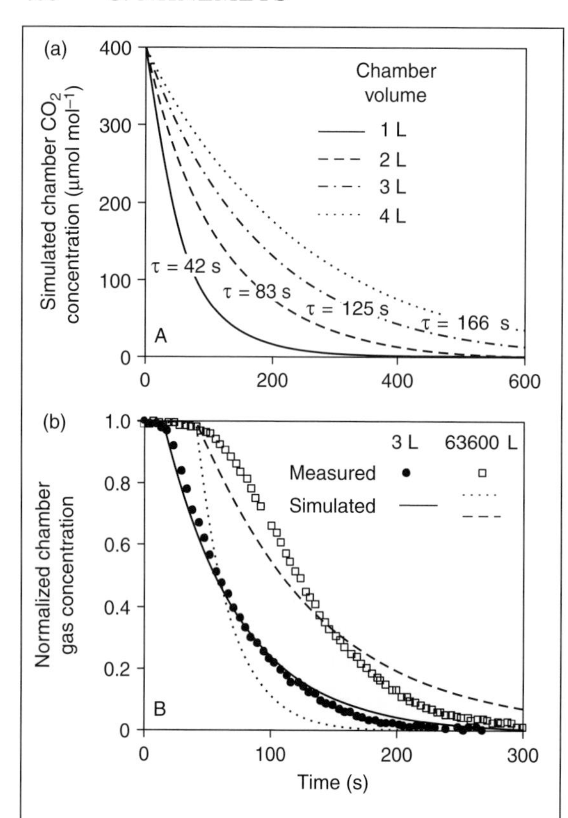

Fig. 26.B.1. Simulated changes in chamber CO_2 concentration after immediate replacement of incoming air with a CO_2 concentration of 400 μmol mol^{-1} by CO_2-free air for leaf chambers with varying volume (V) (a) and measured chamber time responses for two contrasting gas-exchange chambers. In (a), the flow rate (F) was 1 L min^{-1} in all cases and chamber CO_2 concentration was simulated by equation 26.B3 and chamber half-times (τc) were calculated according to equation 26.B4. In (b), the experiments with either a 3 L whole-plant cuvette and flow rate of 3 L min^{-1} (Rasulov et al., 2009) and a large ecosystem open-top chamber with a volume of 63600 L and a flow rate of 144000 L min^{-1} (Ham et al., 1993) were analysed. For the small chamber, the chamber response was assessed by establishing a steady state isoprene flow through the chamber and abruptly stopping the flow at t=0, while for the large chamber, the response was assessed by an analogous experiment with CO_2. In theory, the ratio F/V is the rate constant of that chamber. For the 3 L chamber, the response is simulated by equation 26.B3 using the actual values of F/V (solid line). For the larger chamber, the simulations with both the actual F/V (dotted line) and best-fit value (dashed line) are shown. In these simulations, offsets of 15 s for the 3 L chamber, and 34 s for the large chamber were used to account for the time lags owing to finite linear gas flow rates in the piping.

The time constants reported in Table 26.1 are calculated for completely uniform air fields inside the chambers (Box 26.1), and thus provide the smallest theoretical response time for any given chamber. Any heterogeneity in the chamber gas-concentration field increases the chamber response time for a given bulk flow rate and volume (Box 26.1). For a big whole-ecosystem chamber, the actual time to reach a steady state was more than threefold larger than predicted for wholly uniform gas mixing (Ham *et al.*, 1993, Fig. 26.B1b). Significant profiles of wind velocity and gas concentration have been demonstrated within large chambers (Kellomäki *et al.*, 2000) denoting limited turbulence that inevitably slows down the overall system response.

In Table 26.1, the chamber response time was simulated for an empty chamber. However, the presence of vegetation may reduce the turbulence and the uniformity of the air field inside the chamber. The degree of alteration of turbulence conditions inside the chamber has been shown to depend on the architecture of enclosed plant(s) (Whiting and Lang, 2001). The evidence collectively suggests that non-ideal chamber turbulence fields, especially for chambers including large amount of plant biomass relative to chamber volume, can importantly slow down the chamber responsiveness. Even among the large chambers considered fast (Table 26.1), there is some evidence of chamber effect on whole-tree photosynthesis measurements (Fig. 26.3a).

In addition to the chamber response, offsets owing to limited flow rate in extensive piping with a large cross-section can shift the photosynthetic signal relative to measured environmental drivers (Box 26.1). These offsets need to be corrected when analysing plant response to fluctuations in chamber and ambient environmental conditions, for example, the whole-plant photosynthetic responses to ambient light climate (Fig. 26.3b). As the environmental drivers such as temperature and light can be measured instantaneously, it is highly relevant to correct for finite system response time in seeking for correlations between environmental variables and whole-plant photosynthesis (Müller *et al.*, 2009).

HOW DO CHAMBERS ALTER A PLANT'S ENVIRONMENT?

All plant chambers alter environmental conditions. The reduction of light penetration is 10–60% depending on chamber-wall materials (Leadley and Drake, 1993; Monje and Bugbee, 1998; Kellomaki *et al.*, 2000; Dore *et al.*, 2003; Perez-Peña and Tarara, 2004; Medhurst *et al.*, 2006; Baker *et al.*, 2009) and this may lead to significant underestimation of actual plant productivity.

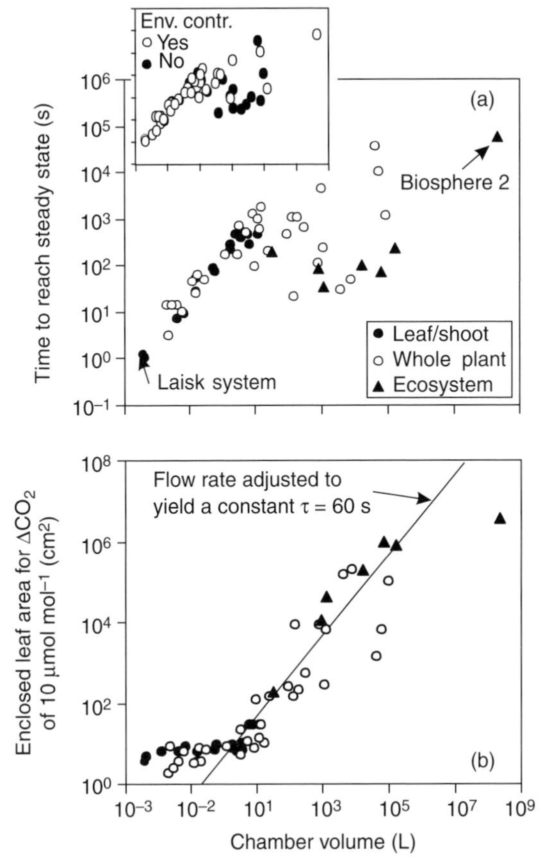

Fig. 26.2. A meta-analysis of the correlations of the time to reach a steady state in the gas-exchange chamber **(a)** and of the amount of enclosed leaf area needed to achieve a steady state CO_2 drawdown (ΔCO_2) of 10 μmol mol^{-1} in the chamber **(b)** with chamber volume (V) across a variety of gas-exchange chambers and plant enclosures used in leaf and shoot (filled circles), whole-plant (open circles) and ecosystem (filled triangles) studies (Table 26.1 for the data sources). The inset in (a) shows the same correlation between the time to reach the steady state and V for chambers with full (temperature, humidity, light and gas composition) and partial (some of the four key environmental drivers) environmental control (open circles) and without environmental control. The time needed to reach a steady state depends both on V and flow rate through the chamber and was calculated by equation 26.B4 (Box. 26.1, system half-time, τ), and was taken as 4τ which yields ≅94% of full system response. The amount of enclosed leaf area needed (S_r) for a certain CO_2 difference between the reference and the chamber depends on the flow rate and net assimilation rate (Eqn. 26.1), and was calculated for a constant leaf-level net assimilation rate of 10 μmol m^{-2} s^{-1}. Laisk and Oja's (1998) chamber is probably the fastest chamber specially designed for rapid photosynthesis measurements at leaf level (4τ=1.1 s), while Biosphere 2 with a total volume of $2.04 \cdot 10^8$ L is a large facility enclosing several ecosystems (Rosenthal *et al.*, 1999; Zabel *et al.*, 1999). All systems included are typically operated in open (flow-through) mode, except for the Biosphere 2 and Flakaliden whole tree chamber (Wallin *et al.*, 2001; Medhurst *et al.*, 2006), where photosynthesis measurements are normally conducted in closed system mode. The estimates of the chamber parameters determined for these systems are for hypothetical measurements in open mode.

The degree of turbulence inside the chamber also affects plant boundary layer conductance for heat transfer and gaseous exchange, and thereby alters leaf temperatures and plant gaseous exchange. Boundary layer conductance can be enhanced by vigorous air mixing by fan(s) installed inside the chamber. In addition, the temperature inside the chamber depends on the air flow rate through the chamber. Owing to limited turbulence and flow rate and low transmittance of chamber material for back-radiating long-waved radiation, i.e., the 'greenhouse effect', the chambers typically warm up during the day with chambers without environmental control being typically 2–5°C higher during day time than the ambient environment (Leadley and Drake, 1993; Dore *et al.*, 2003; Perez-Peña and Tarara, 2004; Alterio *et al.*, 2006; Baker *et al.*, 2009). In addition, warmer temperatures result in greater day time water vapour-pressure deficits (VPDs) inside the chambers (Dore *et al.*, 2003). To cope with these difficulties, chambers with environmental control can be constructed, but such systems become increasingly expensive with increasing size. Only a few large chambers with environmental control have ever been manufactured (Table 26.1; Corey and Wheeler, 1992; Wheeler, 1992; Griffin *et al.*, 1996; Rosenthal *et al.*, 1999; Zabel *et al.*, 1999; Kellomäki *et al.*, 2000; Wallin *et al.*, 2001; Baker *et al.*, 2004; Medhurst *et al.*, 2006). Nevertheless, even in some large chambers with environmental control, temperature deviations in the order of 2–3°C between the chamber and ambient environment can be observed (Kellomaki *et al.*, 2000; Müller *et al.*, 2009). Alternatively, the flow rate through and air mixing in the chamber can be increased, enhancing both boundary layer conductance and convective cooling of the chamber environment (Table 26.1, Corelli-Grappadelli and Magnanini, 1993). This is the choice of most large whole-plant and ecosystem chambers lacking environmental control (Fig. 26.2.a). In addition to improved chamber conditions, systems with high flow-through also respond faster and accordingly are expected to track photosynthetic responses to the environment more reliably.

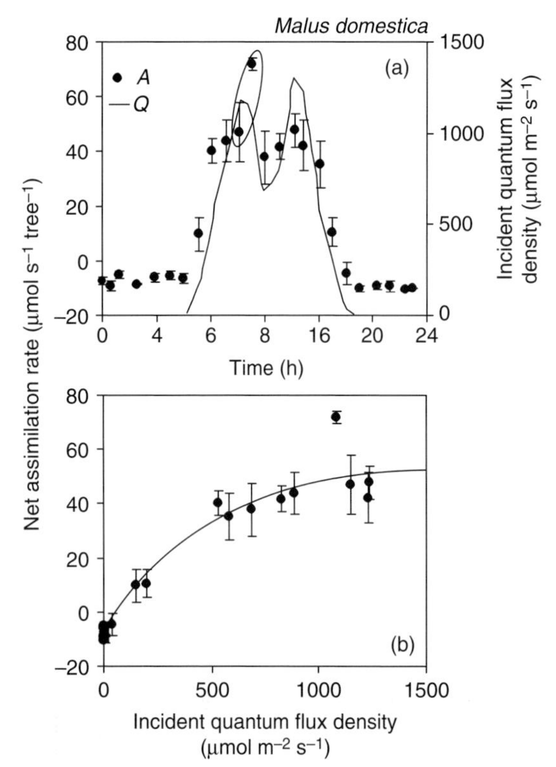

Fig. 26.3. Sample daily time courses of whole-tree net assimilation rate (A, symbols) and incident quantum flux density (Q, line) **(a)** and the whole-tree A versus Q relationship **(b)** measured by a portable tree chamber for mature apple (*Malus domestica*) trees with 9.5–13.2 m^2 leaf area (recalculated from Wünsche and Palmer, 1997, system properties outlined in Table 26.1). Data are averages of measurements in four trees and the error bars show ±SE. Overall, the system tracks photosynthetic responses to light reasonably well, although there is evidence of some time lags between plant response and the measurements at mid-day (denoted by an ellipse). Such time lags were not that visible in the evening when photosynthesis was reduced at given Q owing to decreased stomatal conductance according to whole tree transpiration measurements (Wünsche and Palmer, 1997), masking the shifts between A and Q. Data in (b) were fitted by an empirical hyperbolic relationship (Hanson *et al.*, 1987; Niinemets and Tenhunen, 1997) in the form $A = A_{max}[1-(1+R_n/A_{max})1-\frac{Q}{Q_C}]$, where A_{max} is the light-saturated net assimilation rate (56.3 µmol s^{-1} $tree^{-1}$), R_n is the dark respiration rate (8.4 µmol s^{-1} $tree^{-1}$) and Q_C is the light compensation point (66.4 µmol m^{-2} s^{-1}).

HOW MUCH BIOMASS IS NECESSARY FOR RELIABLE MEASUREMENTS?

Although the system response time, boundary layer conductance for heat and gaseous exchange and overall rate of convective cooling can be enhanced by increasing the flow rate through the system, there is a trade-off between the flow rate and the amount of biomass needed for reliable gas-exchange measurements. In the steady state, the total leaf area needed for a certain CO_2 concentration difference between the reference and chamber air (S_r) depends on the flow rate (F) and net assimilation rate (A_N) as

$$S_r = \frac{F}{A_N}\Delta CO_2 \qquad \text{[Eqn. 26.1]}$$

The modern gas analysers can resolve small differences in CO_2 concentrations in the order of 0.1–0.2 µmol mol^{-1}. Nevertheless, quasi-stable incoming air CO_2 concentrations even with large buffer volumes and other errors, e.g., leaks, temperature gradients etc., imply that the maximum CO_2 concentration difference should be kept relatively large, in particular for field measurements that are conducted through periods with and without environmental stresses that strongly curb photosynthetic activity and reduce the chamber CO_2 drawdown owing to vegetation. Typically, the maximum CO_2 concentration differences in large flow-through chambers observed during highest solar irradiances from non-stressed vegetation are kept between 10–30 µmol mol^{-1} (Corelli-Grappadelli and Magnanini, 1993; Wünsche and Palmer, 1997; Alterio *et al.*, 2006; Takahashi *et al.*, 2008). For a series of available gas-exchange chambers of different size, the amount of leaf area needed for the given CO_2 difference varies over seven orders of magnitude and strongly scales with the chamber volume (Fig. 26.2.b). In particular, larger chambers lacking environmental control and therefore having high air flow-through rates need to include particularly large amounts of biomass in the chamber (Fig. 26.2.b).

BASIS OF EXPRESSION OF WHOLE-PLANT ASSIMILATION RATES IN CHAMBER MEASUREMENTS

With whole-plant enclosures, the rates of gas exchange are straightforward to express per unit whole plant (e.g., Corelli-Grappadelli and Magnanini, 1993; Wünsche and Palmer, 1997) or per unit chamber ground area (e.g., Leadley and Drake, 1993; Fredeen *et al.*, 1995; Burkart *et al.*, 2007; Müller *et al.*, 2009). However, whole-plant photosynthesis can vary not only because of differences in plant physiological activity, but also because of differences in leaf area per plant and because of differences in plant architecture (see Section 26.3.2). To separate the physiological and structural controls on whole-plant gas exchange, the rates of whole-plant gas exchange should also be expressed per leaf area. Leaf-area estimation can introduce large errors in determination of average area-based photosynthetic rates, and the magnitude of errors owing to leaf-area estimation can be comparable with the precision of gas-exchange measurements (Long *et al.*, 1996). Estimation of extensive leaf areas that can be tens of m^2 for whole trees enclosed in large chambers may be particularly problematic, especially for long-term measurements in developing and growing plants constantly changing leaf area. Allometric relationships have been used for leaf-area estimation in large trees in chambers (Wünsche and Palmer, 1997; Wallin *et al.*, 2001), but allometry has large inherent tree-to-tree variability.

Photographic methods can provide an important tool for estimation of plant surface area in photosynthesis studies. In particular, plant silhouette area (S_S) can be estimated from digital images of the enclosed plants, while in separate plants, S_S and one-sided plant leaf area (S_T) can be measured and the S_S/S_T ratio calculated. S_S/S_T values can be further employed to derive S_T estimates for plants enclosed in the gas-exchange system. For simplicity, it has even been argued that in whole-plant studies, S_S can be used as a substitute of S_T (Donahue *et al.*, 1997; Tocquin and Périlleux, 2004; Poulson *et al.*, 2006). However, S_S/S_T ratio varies with the degree of leaf overlap and leaf inclination angle, implying that the rates of plant photosynthesis expressed per unit S_S can differ merely because of differences in plant structural features. Clearly, leaf-area estimation as an integral part of whole-plant measurements needs special attention in whole plant studies.

26.2.2. Estimations from whole-plant Chl-F

Chl-F is a chamber-free method for assessment of foliage physiological activity (Chapters 10 and 16 for detailed description of Chl-F methods). The fluorescence methods are based on measuring the variation in the amount of energy re-radiated as fluorescence owing to changes in the status of foliage photosynthetic machinery. To measure Chl-F under a high background of ambient illumination, pulse amplitude modulated (PAM) fluorimeters are used that employ the saturating pulse method to determine the effective quantum yields of PSII (ϕ_{PSII}) (Schreiber *et al.*, 1994; Maxwell and Johnson, 2000). First, the steady state value of Chl-F of an illuminated sample (F_s), is measured, followed by a high pulse of intensity white light, typically 8000–10000 μmol m^{-2} s^{-1}, for around 1 s to close all PSII centers and obtain the maximum value of light-adapted PSII quantum yield, F_m'. From these measurements, ϕ_{PSII} is determined as $\phi_{PSII}=(F_m'-F_s)/F_m'$ (Genty *et al.*, 1989; Schreiber *et al.*, 1994). The rate of linear PET (J_f), at given incident quantum flux density, Q, is further calculated as:

$$J_f = \Phi_{PSII} \cdot \alpha_L Q \varepsilon_{PSII} \qquad \text{[Eqn. 26.2]}$$

where α_L is leaf absorptance; and ε_{PSII} is the fraction of light absorbed by PSII (typically taken as 0.5). As the bulk of PET is used for carbon assimilation and for photorespiration, strong correlations between net assimilation rate and J_f have been observed in several studies (e.g., Edwards and Baker, 1993; Donahue *et al.*, 1997; Sun *et al.*, 1999). Nevertheless, as the relationship between photosynthesis and J_f is affected by α, ε_{PSII}, the fraction of electrons growing to photorespiration and to alternative electron sinks, application of equation 26.2 to assess foliage photosynthetic rate commonly requires calibrations of the fluorescence signal with actual measurements of photosynthesis. A major difficulty with this method for long-term field measurements is that the scaling between J_f and carbon assimilation can strongly vary. For instance, this scaling differs for leaves with varying stomatal conductance for CO_2 ($g_{s,c}$), because $g_{s,c}$ alters J and gross photosynthesis (A) less than it affects A_N (Edwards and Baker, 1993).

Pulse-amplitude methods are commonly applied at the leaf scale, but imaging PAM fluorimeters with powerful LED arrays to provide uniform actinic light and high-intensity saturating pulses of light have become available (e.g., systems from H. Walz GmbH and Photon Systems Instruments) that can be used to assess photosynthetic activity in small photosynthetic organisms that have flat canopies with all foliage essentially on the same plane, such as lichens (Barták *et al.*, 2005) and *Arabidopsis* (Chaerle *et al.*, 2007a; Masclaux-Daubresse *et al.*, 2007; Rossel *et al.*, 2007; Woo *et al.*, 2008). However, in plants with complex three-dimensional canopies, Chl-F analysis is associated with several difficulties. In particular, the intensity of exciting light and fluorescence from the leaves decreases with the square of the distance from the sample (Gates, 1980). Even the most powerful LED systems used in chlorophyll fluorimeters provide relatively weak light of about 4000 μmol m^{-2} s^{-1} on the top of the canopy. This irradiance can be enough for upper leaves, but the irradiance declines to low values at the bottom of the plant canopy. Thus, leaves at different distances from the light source and fluorescence detector inevitably will show different fluorescence profiles. In addition, self-shading within the plant alters the fluorescence signal much more than it alters the interception of light as quanta re-emitted as fluorescence travel the distance between the source and the sample twice. Using an integrating sphere for plant illumination, fluorescence excitation and measurement can be the way to partly overcome these problems as the plant is essentially uniformly illuminated by the actinic and exciting light, and the fluorescence signal originates from all parts of the canopy (Salvetat *et al.*, 1998). However, problems with a too weak modulated fluorescence signal, even if the detector covers a relatively large area of the sphere surface, may complicate routine application of this method (Salvetat *et al.*, 1998).

In summary, Chl-F is a promising indirect method for rapid assessment of plant physiological status. The current technology is essentially mature for long-term monitoring of Chl-F for small plants with a flat surface. Application of this methodology to larger plants with complex geometry is currently still bound to limitations inherent to optical methods.

26.2.3. Estimations from whole-tree water use

In large plants, whole-plant transpiration rates (sap flow) can be measured by several methods based on heating the stem segments and monitoring the resulting temperature gradient or measuring the amount of heating energy needed to maintain a constant temperature gradient (for reviews see Granier *et al.*, 1996; Köstner *et al.*, 1998). Overall, the chamber and sapflow measurements of whole-tree transpiration rate are strongly correlated (Goulden and Field, 1994). With information on canopy micrometeorology, canopy conductance to *water vapour* (g_C) can be derived from these measurements (e.g., Oren *et al.*, 1998; Ewers *et al.*, 2001). Given that photosynthesis and stomatal conductance are tightly related (Wong *et al.*, 1979; Ball *et al.*, 1987), changes in canopy conductance can be used to diagnose the modifications in whole-tree photosynthetic activity. For instance, the reduction in photosynthetic production in larger and older trees is strongly coupled to reductions in whole-tree transpiration rate (Schäfer *et al.*, 2000; Köstner *et al.*, 2002).

In addition, based on the correlations between photosynthesis and stomatal conductance, the variations in whole-tree photosynthesis in diurnal to seasonal timescales can be potentially derived from sap-flow measurements using modelling approaches (e.g., Falge *et al.*, 1997, 2000; Tenhunen *et al.*, 2001). Sapflow-derived estimates of whole-plant transpiration rates can also be used to verify scaling-up models of whole-plant and stand gas-exchange rates (Falge *et al.*, 2000; Tenhunen *et al.*, 2001; Davi *et al.*, 2005). However, for prediction of carbon gain using sap flow, the proportionality between photosynthesis and stomatal conductance should be known. Leaf-level measurements can be used to derive the proportionality factor (e.g., Sala and Tenhunen, 1996). However, the major disadvantage limiting the use of sap flow for quantitative assessment of photosynthesis is that the proportionality factor between photosynthesis and stomatal conductance may change during the season, for example with advancing drought (Sala and Tenhunen, 1996) or even during a single day (Mencuccini *et al.*, 2000). Thus, sap-flow measurements can provide important insight into the physiological status of large plants and can be used to verify existing models along with other tests, but sap-flow measurements alone cannot immediately be used as a surrogate of whole-tree carbon gain.

26.2.4. Scaling up from the leaves to the whole plant by numerical models

Apart from the direct and indirect methods, whole-tree photosynthetic rates can also be found using modelling approaches, by scaling up from leaf-level measurements. In simple scaling approaches employed for small plants, such as seedlings growing in growth chambers under constant light, a whole-plant assimilation rate has been found as the average assimilation rate measured *in situ* across multiple leaves multiplied by total-plant leaf area (e.g., Walters *et al.*, 1993a,b). In-situ average photosynthetic rate correctly estimates whole-canopy average photosynthetic rate per unit area, but this scaling method is not feasible for large plants growing in the field where continuous variations in environmental conditions simultaneously modify the contribution of different leaves and alter the average canopy photosynthetic rate.

To predict plant photosynthesis in field environments, a leaf-level photosynthesis model describing the responses of photosynthesis to incident quantum flux density, leaf temperature and ambient CO_2 concentration is coupled to the data on canopy micrometeorology or to a model describing variation in canopy microclimatic conditions. Typically, the process-based photosynthesis model of Farquhar *et al.* (1980a) coupled to a stomatal model (e.g., Ball *et al.*, 1987; Baldocchi, 1994) is used in current scaling-up models, and then linked to a one-dimensional layered model that predicts environmental drivers for the leaf layers, or to a three-dimensional model that provides environmental data for three-dimensional canopy elements, voxels (for reviews see Cescatti and Niinemets,

2004; Niinemets and Anten, 2009). As foliage photosynthetic potentials typically decline with light availability in the canopy (Niinemets, 2007 for review), within-canopy variation in foliage photosynthetic characteristics needs to be considered in the outlined scaling-up schemes. Detailed scaling-up models have been shown to realistically simulate whole-canopy photosynthesis with strong correspondence between measured and simulated values (e.g., Lloyd *et al.*, 1995; Wohlfahrt *et al.*, 2000).

As a simplified alternative, big-leaf models have been used that need only the photosynthetic capacity of the upper-canopy leaves and assume that foliage photosynthetic potentials are directly proportional to the average light availability, i.e., that the within-canopy distribution of photosynthetic capacity is optimal (Amthor, 1994; Kull and Jarvis, 1995). These models have been recently modified to consider sunlit and shaded leaf fractions in the canopy, resulting in so-called two-leaf models (de Pury and Farquhar, 1997; Dai *et al.*, 2004). However, as photosynthetic capacity does generally not vary within the canopy in direct proportion to intercepted light (Niinemets and Valladares, 2004 for review), big-leaf models tend to overestimate whole-plant photosynthesis, and fudge factors are needed to get numerical correspondence between the measurements and big-leaf model predictions (Friend, 2001; Niinemets and Anten, 2009).

26.2.5. EC technology to measure large-scale fluxes and test models

The scaling-up models provide a viable solution to the challenging task of estimating photosynthesis in large plants, but the validation of these models can be difficult, in particular owing to logistical problems associated with direct whole-tree measurements, for instance altered environmental conditions and slow responses of larger chambers (see Section 26.2.1, Dore *et al.*, 2003). In addition, plants growing in dense stands shade each other and the crowns can be intermixed, implying that separate measurements on individual plants may simply not be possible. Separation of individual plant contributions can also be complicated in layered models, and thus many scaling-up models are currently applied at the canopy level (Chapter 14). At the canopy scale, the EC technique may be used to validate the scaling-up models (e.g., Amthor *et al.*, 1994; Aber *et al.*, 1996).

The EC technique is a micrometeorological method that estimates the whole-ecosystem CO_2 flux from fast above-canopy measurements of the vertical component of wind speed (m s^{-1}) and CO_2 concentration (μmol mol^{-1}), and from these fast measurements, typically 30–60-minute averages of stand-level carbon gain are calculated (for a detailed review on the topic see e.g., Aubinet *et al.*, 2000). Eddy flux data provide the entire stand carbon balance including plants and soil. To obtain canopy net carbon assimilation rates, soil respiration rates should be estimated by alternative methods, e.g., by soil chamber measurements or by other modelling approaches. Several studies have demonstrated a good correspondence between canopy-level assimilation rates simulated by scaling-up models parameterised either by leaf- or branch-level measurements and between-canopy assimilation rates obtained by EC techniques (Falge *et al.*, 2000; Lebaube *et al.*, 2000; Baldocchi and Amthor, 2001; Sampson *et al.*, 2001; Zha *et al.*, 2007). In addition, large-chamber and eddy flux measurements have shown reasonably good correspondence (Wallin *et al.*, 2001; Dore *et al.*, 2003).

There is a worldwide network of eddy flux measurement sites (Fluxnet) consisting of over 400 measurement tower sites (e.g., Baldocchi *et al.*, 2001; Friend *et al.*, 2007), implying a large potential for the use of eddy flux data for testing canopy models in different climates and ecosystems. Although the eddy flux-measurement protocols have been standardised as much as possible, there is so far no uniform EC methodology (e.g., Baldocchi *et al.*, 2001). Inherent assumptions associated with EC technology and random errors (Luyssaert *et al.*, 2009), result in eddy flux uncertainties of 10–20% under turbulent conditions and 20–100% under non-turbulent conditions (Richardson *et al.*, 2006; Lasslop *et al.*, 2008). In addition, the belowground carbon fluxes need to be assessed separately, and this adds further to the uncertainties of canopy carbon gain that is estimated as the difference between total ecosystem flux and belowground flux (Raupach *et al.*, 2005). This level of uncertainty is tolerable for qualitative testing of the canopy scaling-up routines, but makes the overall confidence intervals for the canopy carbon-gain estimates based on eddy flux data too large to accept or reject a certain canopy model.

26.3. PHYSIOLOGICAL VERSUS STRUCTURAL DETERMINANTS OF WHOLE-PLANT CARBON GAIN

Whole-plant net carbon gain depends on the total amount and functional activity of foliage, but also on the total quantity and respiratory activity of support biomass (roots, branches, stem) and generative organs. In addition, foliage daily carbon gain also depends on the efficiency with which foliage captures light and how foliage responds to

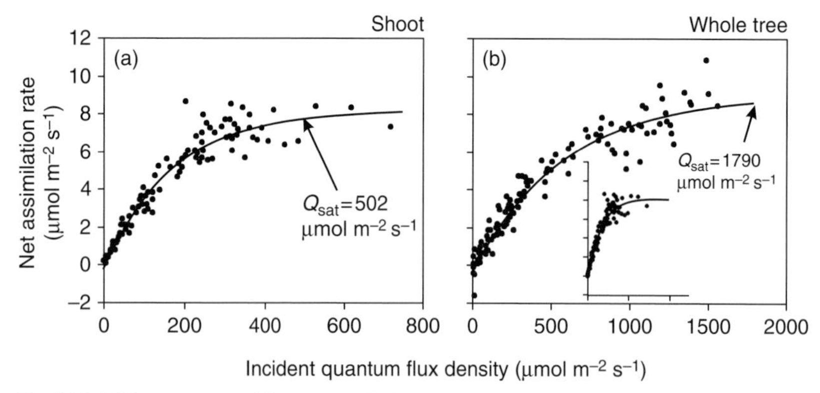

Fig. 26.4. Light responses of the net assimilation rate expressed per unit needle projected area for shoots **(a)** and whole young trees **(b)** of Norway spruce (*Picea abies*) (modified from Wallin *et al.*, 2001). Data were fitted as in Fig. 26.3, and saturating quantum flux density (Q_{sat}) was taken as the light intensity needed to achieve 95% of maximum net assimilation rate. Light responses were obtained as in Fig. 26.3b employing daily variation in A and Q incident to the shoots (a, fifth whorl from the top) and tree (b). The inset in (b) demonstrates the relationship between A and Q for shoots in the same scale as in the tree-level measurements. Total tree projected leaf area was 9.5 m^2 (Wallin *et al.*, 2001 for further details).

environmental stresses in the canopy. The focus of recent plant-science research has mainly been on estimation of the photosynthetic capacity of single leaves. However, the way foliage responds to periods unfavourable to carbon gain (Valladares and Pearcy, 1997; Werner *et al.*, 2001a), canopy architectural patterns (Cescatti and Niinemets, 2004; Valladares and Niinemets, 2007) and whole-plant biomass allocation (Körner, 1991) can be at least as important as the photosynthetic potentials of single leaves.

26.3.1. Environmental responses of whole plants: not just the sum of single leaves

To understand whole-plant photosynthetic responses to temporal variations in environmental conditions, one needs to understand how plants respond to light, temperature and ambient CO_2 concentration. The whole-plant responses to these environmental drivers have not been routinely measured, and there is surprisingly little information of comparisons of single leaf, shoot and whole-plant responses to environmental drivers.

PHOTOSYNTHETIC LIGHT RESPONSES OF WHOLE PLANTS VERSUS SINGLE LEAVES

Photosynthetic light responses of single leaves and canopies differ in several important aspects. First, for single leaves the majority of light compensation points of photosynthesis reported are between 5–40 μmol m^{-2} s^{-1} (Craine and Reich, 2005), while for large whole plants, 65–200 μmol m^{-2} s^{-1} have been observed (calculated from Corelli-Grappadelli and Magnanini, 1993; Wheeler *et al.*, 1993; Garcia *et al.*, 1994; Wünsche and Palmer, 1997). These large values reflect the greater support to assimilative biomass in whole plants relative to single leaves and accordingly greater dark respiration rates of whole plants.

A second important difference is in the curvature and the irradiance required for saturation of photosynthesis rate. Typically, in plants with mesophytic leaves, the rates of leaf photosynthesis become light-saturated at quantum flux densities of 200–600 μmol m^{-2} s^{-1} (e.g., Chapter 16). As chamber measurements in large plants demonstrate, the light saturation of whole-plant photosynthesis occurs at much higher quantum flux densities than the light saturation of single leaves and shoots (Fig. 26.4 for a comparison among light responses for shoots and single leaves). In a whole young tree of Norway spruce (*Picea abies*) with total tree project leaf area of 9.5 m^2, the quantum flux density for 95% saturation of photosynthesis (Q_{sat}) at CO_2 concentrations close to ambient was 1790 μmol m^{-2} s^{-1}, while for apple (*Malus domestica*) trees with leaf areas of 9–13 m^2, Q_{sat} occurred between 1500–1550 μmol m^{-2} s^{-1} (Fig. 26.3.b and Corelli-Grappadelli and Magnanini, 1993), and for grapevine (*Vitis vinifera*) plants with leaf area of about 8 m^2, Q_{sat} occurred at 1830 μmol m^{-2} s^{-1} (recalculated from Perez-Peña and Tarara, 2004). Nevertheless, high saturation of photosynthesis can also be observed in tree saplings and in herbs. For instance in saplings of Eldarica pine (*Pinus eldarica*), Q_{sat} was 1890 μmol m^{-2} s^{-1} (recalculated from Garcia *et al.*, 1994), and values as high as 2100–2700 μmol m^{-2} s^{-1} can be observed in dense herbaceous canopies (recalculated from Campbell *et al.*, 1990; Müller *et al.*, 2009). Analogously, EC measurements also demonstrate high light saturation points of canopy level carbon gain (e.g., Aber *et al.*, 1996; Valentini *et al.*, 1996; Wallin *et al.*, 2001). In fact, productivity of herbaceous stands has also been observed to increase up to 2080 μmol m^{-2} s^{-1} for a 20 h photoperiod, summing up to approximately 150 mol m^{-2} d^{-1} (Bugbee and Salisbury, 1988).

Higher values of Q_{sat} for photosynthesis at any given value of foliage biochemical capacity can result from changes in the initial quantum yield of photosynthesis and the curvature of the photosynthetic light response curve for incident quantum flux density. At the leaf scale, several studies have conclusively demonstrated that the initial quantum yield for photosynthesis on an absorbed light basis is highly conserved among C_3 plants (Ehleringer and Björkman, 1977; Long *et al.*, 1993). However, there are large changes in photosynthetic quantum yield for an incident light owing to variations in leaf absorptance (Long *et al.*, 1996; Singsaas *et al.*, 2001). Measurements in aggregated structures like conifer shoots have further demonstrated that the quantum yield for an incident light also depends on the directionality of light being much lower for direct unilateral irradiance than for diffuse irradiance obtained by placing the soot inside an integrating sphere (e.g., Leverenz, 1995). Lower quantum yields under unilateral irradiance are associated with higher needle self-shading and accordingly reduction in light interception at any given value of incident quantum flux density.

The situation is analogous with whole-plant canopies where light decreases exponentially with cumulative leaf area (e.g., Stenberg *et al.*, 1995; Terashima and Hikosaka, 1995), and leaves positioned lower in the plant crown receive progressively lower irradiances. The effects of shading on the shape of a whole-plant light-response curve can be demonstrated by a simple two-leaf model. In a canopy with part of the leaves exposed to full sunlight and part being shaded, the whole-plant photosynthesis has lower initial quantum yield and higher Q_{sat} (Fig. 26.5.a). This response is also reproduced by more complex models that divide the canopy into multiple leaf layers, each exposed to progressively lower incident irradiances at any given above-canopy light (Fig. 26.5.b). These simulations suggest that self-shading among the leaves within the plant crown is responsible for greater saturated irradiance and lower quantum yields in whole-plant responses, and that single leaf and whole-plant light responses are inherently different. Thus, simulating whole-plant photosynthesis as a single big leaf (e.g., Amthor, 1994; Pachepsky *et al.*, 1994) is inherently biased (see also de Pury and Farquhar, 1997; Friend, 2001).

The importance of within-canopy shading further becomes evident when a whole-plant response to light with differing directionality is analysed. The above analysis was based on direct solar radiation, penetration of which is strongly driven by leaf inclination angles and the degree of foliage aggregation. As the penetration of diffuse radiation that comes from multiple sky directions is less sensitive to canopy structure, light is distributed more uniformly within the canopy in overcast sky conditions, suggesting that whole-canopy photosynthesis may respond more strongly to diffuse light. Higher efficiency of diffuse irradiance in driving photosynthesis has been observed at whole-plant (Lloyd *et al.*, 1995) and ecosystem (Gu *et al.*, 2003) levels. In contrast, single leaves have been demonstrated to use diffuse light less efficiently than direct light (Brodersen *et al.*, 2008). For single leaves in broad-leaved species, the greater efficiency of direct light has been explained by specialised foliage anatomical structure. Specifically, extended palisade cells canalise direct light perpendicular to leaf surface deeper into the leaf interior, thereby making the whole-leaf light environment more uniform (Smith *et al.*, 1997; Brodersen *et al.*, 2008), but such anatomical constitution does not aid to diffuse light penetration.

PHOTOSYNTHETIC CO_2 AND TEMPERATURE RESPONSES OF WHOLE PLANTS: IMPLICATIONS FOR MODEL PARAMETER DERIVATION

As in single leaves, whole-plant CO_2 and temperature response curves have been measured in several studies (Campbell *et al.*, 1990; Corey and Wheeler, 1992; Wheeler *et al.*, 1993; van Iersel and Lindstrom, 1999; Bednarz and van Iersel, 2001; Miller *et al.*, 2001). However, whole-plant CO_2 and temperature responses have not been routinely compared with leaf-level responses. Does the self-shading within the canopy also modify the CO_2 and temperature responses for whole plants? According to the Farquhar *et al.* (1980a) photosynthesis model, RuBP carboxylation in photosynthesis is the smaller of the two potential carboxylation rates limited either by Rubisco activity (A_c) or by RuBP regeneration (A_j). The capacity for A_c is driven by the maximum carboxylase activity of Rubisco ($V_{c,max}$), and the capacity for A_j by the maximum photosynthetic electron transport rate (J_{max}). Typically, for CO_2-response curves measured under high light, the initial part of the CO_2

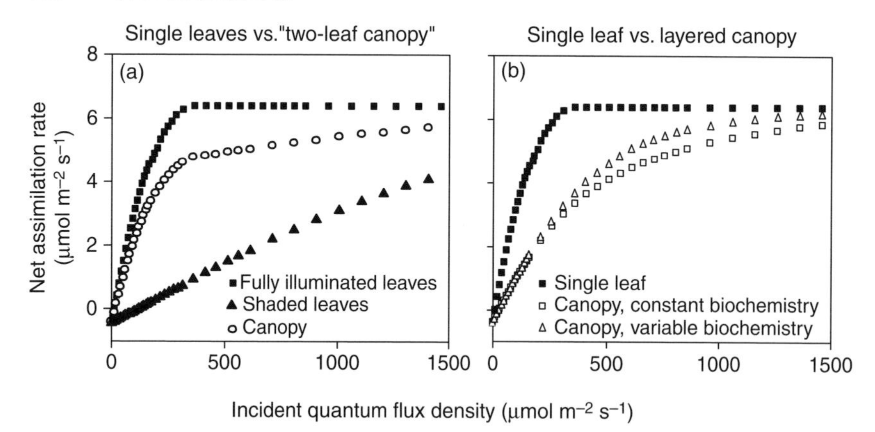

Fig. 26.5. Comparisons of net assimilation rate per unit leaf area versus light response curves between **(a)** fully illuminated leaves, shaded leaves and a hypothetical canopy consisting of 70% of illuminated leaf class and 30% of shaded leaf class, and between **(b)** single leaves and a layered canopy with a total LAI (leaf area per unit ground area) of 5 m^2 m^{-2}, either assuming that foliage photosynthetic potentials are constant for all leaf layers in the canopy and equal to those of the single leaf (constant biochemistry) or that leaf photosynthetic potentials increase linearly with light availability in the canopy (variable biochemistry). Foliage assimilation rates were simulated according to the Farquhar *et al.* (1980) photosynthesis model at a constant leaf temperature of 25°C and CO_2 concentration in chloroplasts (C_c) of 280 µmol mol^{-1} and using the Rubisco constants as revised in Niinemets and Tenhunen (1997). In single-leaf simulations in (a) and (b), and for a constant canopy biochemistry simulation in (b), a value of 0.2482 mol mol^{-1} was used for the initial quantum yield of photosynthetic electron transport on an incident light basis (Niinemets and Tenhunen, 1997), and a value of 20 µmol m^{-2} s^{-1} was used for the maximum carboxylase activity of Rubisco, while the capacity for photosynthetic electron transport was scaled as $2.5V_{c,max}$, and non-photorespiratory respiration rate (R_d) as $0.02V_{c,max}$. In the simulations in (a), foliage photosynthetic potentials were the same for fully illuminated and shaded leaves, but the shaded leaf class received 10% of above-canopy light. In the canopy level simulations, quantum flux density, Q, decreased exponentially with cumulative LAI according to simple Lambert-Beer approximation assuming that foliage is randomly dispersed (no effect of spatial aggregation, equation 26.3: $Q=Q_0e^{-kLAI}$, where Q_0 is the above-canopy quantum flux density and k is the extinction coefficient (k=0.5 in this simulation). Canopy was divided into 50 layers with equal leaf area, and photosynthesis rate was calculated for each canopy layer. In the simulations with variable biochemistry, $V_{c,max}$, J_{max} and R_d varied linearly with light in the canopy, altogether 2.5-fold between the canopy top and bottom, corresponding to the average range in plant communities (Niinemets, 2007), while the average photosynthetic potentials for whole canopy were equal to those used in the single leaf and constant biochemistry simulations.

response curve is limited by A_c, while the saturated-part by A_j. However, light availability at any given CO_2 concentration modifies the cross-over point of A_j to A_c limitation. The lower the light, the lower A_j is and the lower the CO_2 concentration is at which A_c limitation crosses over to A_j limitation. As the simulations with a simple two-leaf canopy model and a layered model demonstrate (Fig. 26.6), shading within the canopy results in overall reductions in A_j and lower net assimilation rates, in particular at higher CO_2 concentrations. Even if the biochemical potentials of all leaves in the canopy were the same, the shape of single-leaf photosynthesis versus CO_2 response curve differs from the whole-canopy CO_2 response (Fig. 26.6).

As discussed in Section 26.1.3, photosynthetic measurements in several plant groups such as plants with rosette prostrate leaf form like *Arabidopsis* are conventionally conducted in whole plants. From these measurements, estimates of foliage photosynthetic potentials, J_{max} and $V_{c,max}$, are derived (Tocquin and Périlleux, 2004). However, different CO_2 responsiveness of whole-plant photosynthesis relative to single leaves has major implications for determination of $V_{c,max}$ and J_{max} from whole-plant measurements. According to the simulations in Fig. 26.6, both $V_{c,max}$ and J_{max} determined from whole-plant measurements are underestimated relative to true leaf biochemical potentials. This underestimation is larger when the light gradients are deeper and self-shading within the canopy is larger. Thus, whole-plant estimates of J_{max} and $V_{c,max}$ depend both on foliage physiological characteristics, and on plant structural features including leaf inclination, spatial aggregation etc., that alter the degree of self-shading. In laboratory studies employing artificial light sources, $V_{c,max}$ and J_{max} can be especially strongly underestimated as light intensity decreases with the square of the distance from the light source (Gates, 1980),

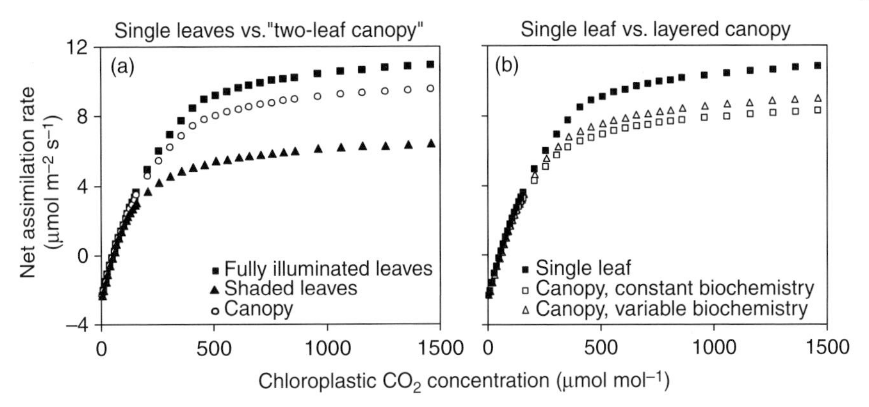

Fig. 26.6. Comparisons of net assimilation rate per unit leaf area versus chloroplastic CO_2 response curves between **(a)** fully illuminated leaves, shaded leaves and a hypothetical canopy consisting of two leaf classes, one (70% of total) fully illuminated, and the other (30% of total) shaded, and between **(b)** single leaves and two hypothetical canopies, one with a constant foliage photosynthetic potential within the canopy (constant biochemistry) and the other with the photosynthetic capacity increasing linearly with increasing light availability in the canopy (variable biochemistry). The CO_2 response curves were simulated by the Farquhar *et al.* (1980) photosynthetic model. Environmental conditions, model parameters and canopy properties are the same as in Figure 26.5. Refitting the canopy level assimilation versus CO_2 response curves in (b) again by Farquhar *et al.* (1980) photosynthesis model demonstrated that the apparent whole-canopy $V_{c,max}$ was underestimated by 7% for the constant canopy, and 3% for the variable canopy, R_d was overestimated by 9% for constant and underestimated by 1% for the variable canopy, while J_{max} was underestimated by 24% for the constant and 17% for the variable canopy.

and accordingly the within-canopy light gradients are particularly large.

In growing plants and in plants acclimating to various environmental conditions, both physiological traits and structure can change (e.g., Mullen *et al.*, 2006 for changes in foliage inclination in response to shading in *Arabidopsis*). Accordingly, developmental and environmental modifications in J_{max} and $V_{c,max}$ measured in whole plants do not necessarily reflect changes in the physiological potentials of leaves. For instance, in whole-plant measurements of *Arabidopsis*, the ratio of $J_{max}/V_{c,max}$ decreased from 3.60 to 1.92 with increasing plant age from 26 d to 40 d (calculated form the data of Tocquin and Périlleux, 2004). Increases in plant age were accompanied by overall increases in plant leaf area and greater self-shading, suggesting that the observed changes in $J_{max}/V_{c,max}$ ratio mainly reflect the underestimation of true foliage biochemical potentials, in particular J_{max}, owing to increased light gradients.

Another major difficulty of derivation of foliage photosynthetic potentials from whole-plant CO_2 response curves is determination of appropriate intercellular CO_2 (C_i) and chloroplastic CO_2 concentrations (C_c). Conventionally, C_i at any value of leaf net assimilation and ambient CO_2 concentration, C_a, is calculated as C_a–$1.56A/g_s$, where g_s is the stomatal conductance for *water vapour* and 1.56 corrects for different diffusivities of CO_2 and *water vapour*. g_s in turn is calculated from measurements of leaf transpiration rate, E, as $g_s=E/(W_L-W_a)$, where W_a is the ambient air *water-vapour* concentration and W_L is the saturated *water-vapour* concentration at the leaf temperature. Saturated *water-vapour* concentration is an exponentially increasing function of temperature. Thus, an appropriate leaf temperature is needed for determination of a reliable value of C_i. Similarly to light, leaf temperatures are also expected to vary from the top to the bottom of the plant canopy owing to variations in the amount of energy absorbed and the degree of foliage coupling to the atmosphere (Niinemets *et al.*, 1999a,b; Baldocchi *et al.*, 2002). In some studies, average temperature is calculated from leaf energy balance (Kollist *et al.*, 2007). However, owing to non-linear scaling of saturated vapour pressure with temperature, the use of an average temperature for the whole canopy does not provide a correct whole-canopy average g_s (Niinemets and Anten, 2009 for integration errors in non-linear relationships). Possibly, with further advancement of thermal-imaging techniques (Jones and Leinonen, 2003; Leinonen and Jones, 2004; Chaerle *et al.*, 2007a; Jones and Schofield, 2008), sensing of temperature can be much improved in whole-plant studies.

After determination of C_i, one further needs to estimate C_c that depends non-linearly on mesophyll-diffusion conductance, g_m, as $C_c=C_i-A/g_m$ (see Chapter 12). For determination of g_m, gas-exchange measurements together with

Chl-F or online carbon discrimination are commonly used (e.g., Warren, 2006b). For whole-plant studies, one would also need spatial resolution that cannot be achieved with current methodology. Clearly, derivation of whole-canopy C_i and C_c values introduces some error in J_{max} and $V_{c,max}$ derivation, and the magnitude of these errors is currently hard to estimate.

As with CO_2 response curves, photosynthesis at different temperatures is limited by different partial processes of photosynthesis. At a leaf level, the cross-over point of A_c- versus A_j-limited photosynthesis varies with temperature owing to different temperature responses of J_{max} and $V_{c,max}$, and owing to varying partitioning of photosynthetic nitrogen between Rubisco and proteins limiting PET (Hikosaka, 1997; Hikosaka *et al.*, 1999). Within-canopy shading that modifies the cross-over point of A_c- and A_j-limited photosynthesis, can thus also alter the shape of whole-canopy temperature response curve relative to leaves, the exact effect depending on the shapes of temperature response curves for J_{max} and $V_{c,max}$ and the overall capacity of Rubisco and electron transport limited processes (Cannell and Thornley, 1998 for detailed analysis). The bottom-line message of this analysis is that the values of J_{max} and $V_{c,max}$ determined for aggregate structures such as shoot, whole plant or canopy are not equivalent to these Farquhar *et al.* (1980a) model parameters originally defined for leaves. Canopy level estimates of $V_{c,max}$ and J_{max} inevitably confound physiological and whole-plant level structural effects, and may vary further owing to the heterogeneity in g_s and g_m within the plant crown.

26.3.2. How canopy structure alters whole-plant photosynthesis

The previous section demonstrated that canopy structure can importantly alter whole-plant environmental responses. How can the structural effects on whole-canopy photosynthesis be quantitatively characterised? Light, temperature, humidity, wind and CO_2 profiles are affected by plant architecture (Baldocchi *et al.*, 2002; Niinemets and Valladares, 2004; Long *et al.*, 2006a), and comprehensive assessment of structural effects on the within-plant microclimate in a plant's natural location in the field requires parameterisation of a full soil-vegetation-atmosphere transfer (SVAT) model that explicitly describes energy and mass transfer within plant canopies (e.g., Baldocchi *et al.*, 2002; Marcolla *et al.*, 2003). For simplicity, only the role of structure in affecting light interception is considered here. Neglecting the radiation scattering and considering the sun as a point light source, the incident quantum flux density at a given cumulative leaf-area index (LAI) from the top of the canopy (leaf area per unit ground area, $m^2\ m^{-2}$) and at a given solar zenith angle θ can be expressed as (Campbell and Norman, 1998):

$$Q(\theta) = Q_0 e^{-k(\theta)\,\Omega\,LAI_c} \quad \text{[Eqn. 26.3]}$$

where Q_0 is the incident quantum flux density; $k(\theta)$ is the canopy extinction coefficient; and Ω is the spatial clumping index (Campbell and Norman, 1998; Nilson, 1971). The extinction coefficient depends on angular distribution of foliage, being larger for more horizontal foliage and smaller for more vertical foliage. Ω characterises the degree of deviation of foliage dispersion from random dispersion. Ω is equal to one for canopies with randomly dispersed foliage, $1>\Omega\geq0$ for aggregated canopies and $\Omega>1$ for regularly dispersed canopies (Nilson, 1971). All else being equal, canopies with regular dispersion intercept more light and canopies with aggregated dispersion intercept less light than the canopies with random dispersion.

Equation 26.3 demonstrates that the variation in light transmission among plants can result from differences in total leaf area, in foliage inclination angle distributions and in the degree of foliage aggregation (clumping). Both inclination angles and clumping are highly variable among plant species (Cescatti and Niinemets, 2004 for a review) and these variations have a major effect on light distribution and whole-plant responses to environmental drivers. The efficiency of light interception of plants with widely varying architecture can be described by simple integrated measures. Shoot or plant silhouette area relative to leaf projected area (S_P) has proved to be particularly useful in characterising the light-harvesting efficiency of objects with complex geometry (Farque *et al.*, 2001; Sonohat *et al.*, 2002; Cescatti and Niinemets, 2004; Sinoquet *et al.*, 2005).

With decreasing S_P, the fraction of foliage exposed to light decreases, implying greater self-shading and accordingly increasingly lower apparent quantum yields and greater saturated irradiances compared with single flat leaves. The importance of structure on the shape of photosynthesis versus light response curve has been nicely demonstrated in nine conifers with varying shoot structure (Leverenz, 1995). In this study, the initial quantum yield and saturated irradiance (lower convexity) for unilateral irradiance increased with shoot silhouette-to-projected-area ratio, but the quantum yield and saturated irradiance were independent of shoot S_P for measurements in the integrated sphere that provides a highly uniform, diffuse-radiation

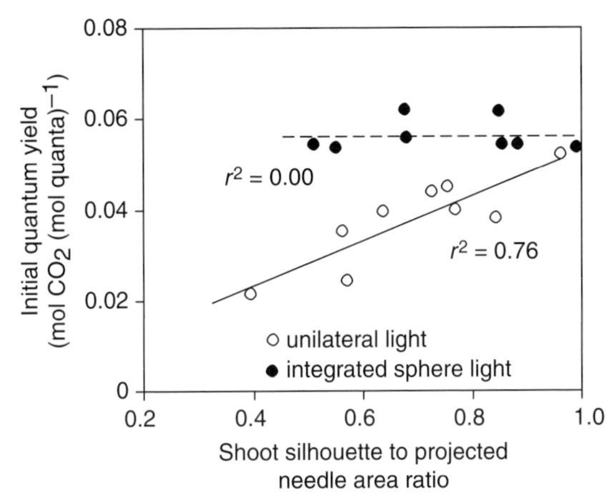

Fig. 26.7. Initial quantum yield of net assimilation for an incident light for shoots measured using either unilateral direct illumination or diffuse illumination in an integrated sphere in relation to shoot silhouette to total needle projected area in nine temperate evergreen conifers with contrasting shoot structure (modified from Leverenz, 1995). Shoot silhouette area is the amount of shade cast by the shoot for a perpendicular light source, while the projected area is the area of the detached needles laid on a flat surface. Shoot silhouette to projected needle area decreases with increasing needle packing and needle aggregation within the shoot and is a measure of within-shoot shading (e.g., Cescatti and Niinemets, 2004; Niinemets *et al.*, 2006b). Within-shoot shading is considerably less in the integrated sphere where the uniform radiation comes from all directions of the sphere. The measurements were conducted at 18°C air temperature and 350 μmol mol^{-1} ambient CO_2 concentration (Leverenz, 1995 for further details). Data were fitted by linear regressions.

environment that reduces the self-shading within the needles in the shoot (Fig. 26.7). The role of whole-plant structure in altering whole-plant photosynthesis has further been directly confirmed by a comparison of whole-plant photosynthesis under a unidirectional and spherical light field in an integrated sphere (Brunes *et al.*, 1980, Fig. 26.8). The initial quantum yields were much smaller and saturated irradiances were much larger for unidirectional incident irradiance than for absorbed light in both the seedlings of the temperate deciduous broad-leaved *Betula pendula* with flat leaves and the temperate evergreen conifer *Pinus sylvestris* having needles with a hemi-elliptical cross-section (Fig. 26.8.a). In addition, the comparison among these contrasting species demonstrated that the species differences in plant photosynthesis rate were larger for the incident than for the absorbed light (cf. Fig. 26.8.a and Fig. 26.8.b). Thus, the higher photosynthetic rate of *Betula pendula* trees under incident light was both because of higher photosynthetic capacity of leaves and because of more efficient light harvesting (Brunes *et al.*, 1980, Fig. 26.8).

Equation 26.3 provides a highly simplified description of light in plant crowns. Real plants often have strongly asymmetric crowns with spatially variable leaf area, leaf inclination angle and leaf aggregation distributions that are difficult to describe precisely by statistical light interception models. A more advanced description of plant architecture can be achieved by novel techniques employing three-dimensional digitisation of canopies of entire plants (Rakocevic *et al.*, 2000; Valladares and Pearcy, 2000; Sonohat *et al.*, 2002; Pearcy *et al.*, 2005; Sinoquet *et al.*, 2005, 2007). With a digitised three-dimensional plant structure, crown light-harvesting efficiencies for different radiation fields can be precisely determined and effects of environmental heterogeneity and environmental stress on whole-plant photosynthesis studied by numerical models (Pearcy and Yang, 1996; Valladares and Pearcy, 1998; Sinoquet *et al.*, 2001). Although these models are tedious to parameterise and the parameterisations are not general, a fundamental insight into the role of crown architecture has been achieved by such models (e.g., Valladares and Pearcy, 2000; Valladares and Niinemets, 2007). Clearly, differences in architecture constitute a major source of variation in whole-plant photosynthesis, and studies on whole-plant photosynthesis need to be combined with whole-plant architectural studies.

26.3.3. Importance of within-canopy profiles of photosynthetic capacity

A whole-plant photosynthetic response reflects the contributions of all leaves in the canopy. As the light conditions vary throughout the canopy, foliage structural and physiological characteristics in the canopy are acclimated to the long-term variation in light. In particular, foliage photosynthetic capacity, A_{max}, increases with increasing light availability in the canopy owing to greater $V_{c,max}$ and J_{max} in higher light (Meir *et al.*, 2002; Niinemets, 2007 for review). This way, leaves with higher A_{max} that require greater irradiances to saturate photosynthesis intercept more light, while leaves with lower A_{max} and lower saturation irradiance intercept less light. Simulation studies have demonstrated that this variation pattern results in larger whole-tree photosynthesis than a constant A_{max} for all leaves in the canopy (Fig. 26.5.b and Fig. 26.6.b for a comparison between average leaf photosynthesis for constant and variable J_{max} and $V_{c,max}$ in

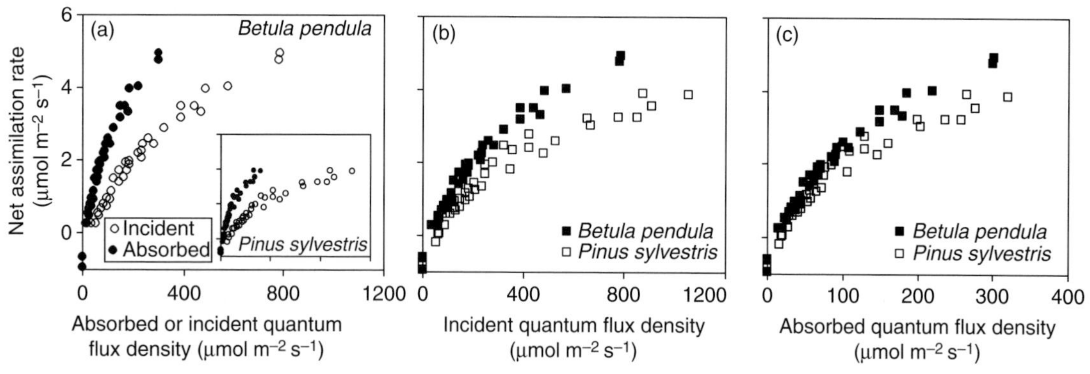

Fig. 26.8. Comparison **(a)** of the responses of net assimilation rate per unit total leaf area to absorbed and incident quantum-flux density in deciduous temperate broad-leaved tree *Betula pendula* and temperate evergreen conifer *Pinus sylvestris* (inset), and the comparisons of the assimilation responses to incident **(b)** and absorbed **(c)** quantum-flux density among *B. pendula* and *P. sylvestris*. Young plants with total leaf (all-sided) area of 0.025–0.0172 m^2 were either measured in situ with unilateral irradiance or in a large integrated sphere providing high intensity diffuse light. Absorbed irradiance was measured from whole plant light absorption measurements (modified from Brunes *et al.*, 1980). Total leaf area rather than projected leaf area was used in the original study to simplify the comparison of the broad-leaved and needle-leaved species.

the canopy) (Baldocchi and Harley, 1995; Gutschick and Wiegel, 1988; Hirose and Werger, 1987).

A_{max} can be increased by investing more nitrogen in photosynthetic machinery or by making leaves thicker with more photosynthetic biomass per leaf area. It has been demonstrated that for maximisation of canopy photosynthesis for a given total plant nitrogen or foliar biomass in the leaves, A_{max} should vary in direct proportion with the daily integrated quantum flux density (Q_{int}) (Farquhar, 1989; Niinemets and Anten, 2009; Sands, 1995). However, numerous experimental studies have demonstrated that the relationships between Q_{int} and A_{max} are curvilinear not proportional (Niinemets and Anten, 2009 for a review). The curvilinearity of A_{max} versus Q_{int} relationships can reflect other interacting environmental factors at high light such as heat and water stresses that limit photosynthetic acclimation (Niinemets and Valladares, 2004 for a review). Owing to non-linearity of A_{max} variations, the theoretical optimal whole-tree photosynthesis is somewhat higher than observed for actual distribution of photosynthetic potentials in the canopy (for reviews see Badeck, 1995; Niinemets and Anten, 2009). Although the plants cannot achieve the theoretical optima, light-driven plasticity in A_{max} is a major adaptive response that significantly enhances whole-plant photosynthesis.

In species with flush-type leaf development, such as most of the deciduous temperate trees where all leaves develop at about the same time, leaf age is similar for all leaves in the canopy. In these species, within-canopy variation in LMA varies commonly two- to four-fold from canopy bottom to canopy top, reflecting enhanced thickness and a more strongly developed palisade layer with tighter mesophyll-cell packing (Niinemets, 2007 for a review). In contrast, leaf nitrogen concentrations per dry mass are considerably less variable along the canopy (Niinemets, 2007 for a review). Thus, photosynthetic capacity per unit dry mass ($A_{max,M}$) is relatively constant in the canopy, and the within-canopy change in photosynthetic capacity per unit leaf area ($A_{max,A}$) is mainly driven by modifications in LMA ($A_{max,A} = A_{max,M}$LMA) (Niinemets and Anten, 2009 for a review).

In several plant functional types, the correlation between light and LMA is not always strong. In flush-type evergreen species, older leaves become shaded by new foliage flushes, and the correlation of LMA with actual light progressively vanishes with increasing leaf age (e.g., Brooks *et al.*, 1996; Niinemets *et al.*, 2006a). Thus, the degree to which $A_{max,A}$ is linked to current light availability depends mainly on the redistribution of nitrogen between the components of photosynthetic machinery and on nitrogen redistribution in the canopy (Field, 1983; Niinemets *et al.*, 2006a). In addition, in most herbs and fast-growing trees and shrubs with continuous leaf development, the gradient in light availability is inevitably accompanied by a gradients in leaf age. In these species, all leaves have a similar LMA, and variation in $A_{max,A}$ mainly depends on the modifications in nitrogen content per dry mass and nitrogen partitioning within the leaves (Niinemets and Anten, 2009 for a review).

Apart from altering the maximum photosynthetic rate, lower LMA in lower light in species with flush-type leaf development also implies greater foliage area, and thus, greater light interception for a given whole-plant biomass investment in leaves. In addition, chlorophyll contents per foliage mass increase with decreasing light availability in all plant functional types, reflecting increases in nitrogen investment in light harvesting (Niinemets *et al.*, 1998; Hallik *et al.*, 2009). These modifications increase light interception per unit foliage mass in low light and further contribute to enhanced whole-plant photosynthesis.

26.4. INFLUENCE OF PLANT AGE AND SIZE ON PHOTOSYNTHESIS

Plant photosynthetic capacity significantly changes during the periods of leaf development when the leaves achieve full photosynthetic potential and during senescence when foliage photosynthetic capacity decreases. Apart from these changes in growing plants, a plethora of structural and functional modifications occur in mature non-senescent plants with increasing plant size and age. There is a general consensus that plant growth rate and productivity decrease with increasing size and age both in herbaceous (Coleman *et al.*, 1993; Cheng *et al.*, 2000; van Iersel, 2003; Usuda, 2004) and in woody species (Ryan *et al.*, 1997 for review). The following sections review age- and size-dependent modifications in plant photosynthetic activity and the mechanisms responsible for the ontogenetic changes in plant photosynthesis.

26.4.1. Modification of biomass partitioning and light interception with plant size

Owing to disproportionate scaling requirements for mechanical stability with increasing size, the fractional investment of plant production in construction of stem, branches and roots increases with increasing plant size (Gerrish, 1990; Miller *et al.*, 1995; Vanninen *et al.*, 1996; Bartelink, 1998; Delagrange *et al.*, 2004; Peichl and Arain, 2007). As the result of these modifications, the fraction of plant biomass in foliage (f_L) decreases with increasing plant size (e.g., Cao and Ohkubo, 1998; Niinemets, 1998; Sterck and Bongers, 1998). In addition, LMA increases with increasing plant size and age (e.g., Niinemets, 2002). Although much of the work on age- and size-dependent modifications in biomass partitioning has been conducted with trees, analogous modifications in biomass allocation and leaf morphology also occur in herbaceous species (Rice and Bazzaz, 1989; Niinemets, 2004).

The alterations in LMA and f_L have important implications for plant LAI (m^2 m^{-2}):

$$\mathrm{LAI} = \frac{f_L \cdot m_T}{\mathrm{LMA} \cdot S_C} \qquad \text{[Eqn. 26.4]}$$

where m_T is total plant mass and S_C is crown projected area. Total plant mass increases with increasing plant size and this leads initially to larger total plant leaf area, despite the increase in LMA and decrease in f_L. However, LAI levels off and decreases with further increases in plant size (Ryan *et al.*, 2004; Funk *et al.*, 2006; Nock *et al.*, 2008). As any change in LAI alters plant light interception, a reduction of LAI in older plant stands has been suggested to provide an explanation for age-dependent reductions of whole-plant photosynthetic capacity (Ryan *et al.*, 2004).

Apart from LAI, there is evidence of stronger shoot-level aggregation of foliage (lower Ω) in canopies of larger trees (Niinemets and Kull, 1995a; Niinemets *et al.*, 2005b). Branch-level aggregation is also expected to increase as the biomass cost of lateral branches becomes increasingly expensive with increasing their length and leaf mass allocated further from the stem (Gerrish, 1989). Thus, enhanced self-shading owing to stronger foliage aggregation can also reduce canopy light interception (Eqn. 26.3) and be partly responsible for the reduced carbon gain in older trees.

26.4.2. Changes in whole-plant respiration with plant age and size

Age- and size-dependent increases in the ratio of respiring support biomass to assimilative biomass have been suggested to result in a larger fraction of photosynthetic production used for maintenance respiration, thereby reducing whole-plant photosynthetic productivity (Ryan *et al.*, 1997 for a review of the hypothesis). The increase in the fraction of carbon lost owing to respiration has been observed in several herbaceous species. In soybean (*Glycine max*), the fraction of daily GPP lost owing to respiration increased from approximately 0.2 to 0.5–0.6 with increasing plant age from 25 d to 50 d (Cheng *et al.*, 2000). An analogous increase in respiration-to-carbon-gain ratio was observed with increasing plant mass in lettuce (*Lactuca sativa*) (van Iersel, 2003). Thus, enhanced maintenance requirements may partly explain reduced productivity of older and larger non-senescent herbaceous plants.

In herbaceous species, a large proportion of support biomass is involved in assimilate translocation and storage and consists of living cells. In contrast, in woody plants a large part of the standing biomass such as heartwood may consist of dead tissues with lower maintenance requirements. Respiration measurements in trees of varying size suggest that respiration constitutes a similar fraction of photosynthetic production in trees of varying size and age (Ryan and Waring, 1992), suggesting that in trees, the changes in the contribution of respiration do not play a significant role in the age- and size-dependent decline of productivity.

26.4.3. Age-dependent changes in plant hydraulics and photosynthesis

To maintain water flow from the soil to the leaves, leaf-water potential must be lower than soil-water potential. Increasing height implies larger water-potential differences between the soil and foliage owing to the gravitational component of water potential and limited hydraulic conductance within the stem, branches and leaves. Gravitational component of water potential increases by 0.1 MPa per 10 m increase in height. Thus, in larger plants, the gravitational component of water potential may significantly constrain water transport to the upper foliage (Bond, 2000; Koch *et al.*, 2004). Furthermore, water-potential gradients as large as 0.1–0.2 MPa m^{-1} have been observed in trees (Hellkvist *et al.*, 1974), demonstrating that limited hydraulic conductance further importantly constrains the water flow in larger plants. The extent to which water potentials decrease owing to limited hydraulic conductance varies among species, and it has been suggested that the maximum height of trees is driven by the species-specific maximum values of hydraulic conductance (Ryan and Yoder, 1997; Bond and Ryan, 2000).

To cope with low water potentials, g_s decreases with increasing tree size to conserve water in larger trees (Yoder *et al.*, 1994; Bauerle *et al.*, 1999; Ryan *et al.*, 2000; Hubbard *et al.*, 2001). The reduction in g_s also results in reductions in leaf assimilation rates (Yoder *et al.*, 1994; Ryan *et al.*, 2000; Hubbard *et al.*, 2001; Mullin *et al.*, 2009; Woodruff *et al.*, 2009). Reduced leaf photosynthesis owing to limited hydraulic conductance has been suggested to be the primary reason for the age-dependent decline in stand productivity (Bond, 2000 for review). However, foliage photosynthetic capacities (Fig. 26.9), J_{max} and $V_{c,max}$, also decrease with increasing tree size and age (Niinemets, 2002; Mullin *et al.*, 2009; Woodruff *et al.*, 2009), indicating that the reductions in g_s

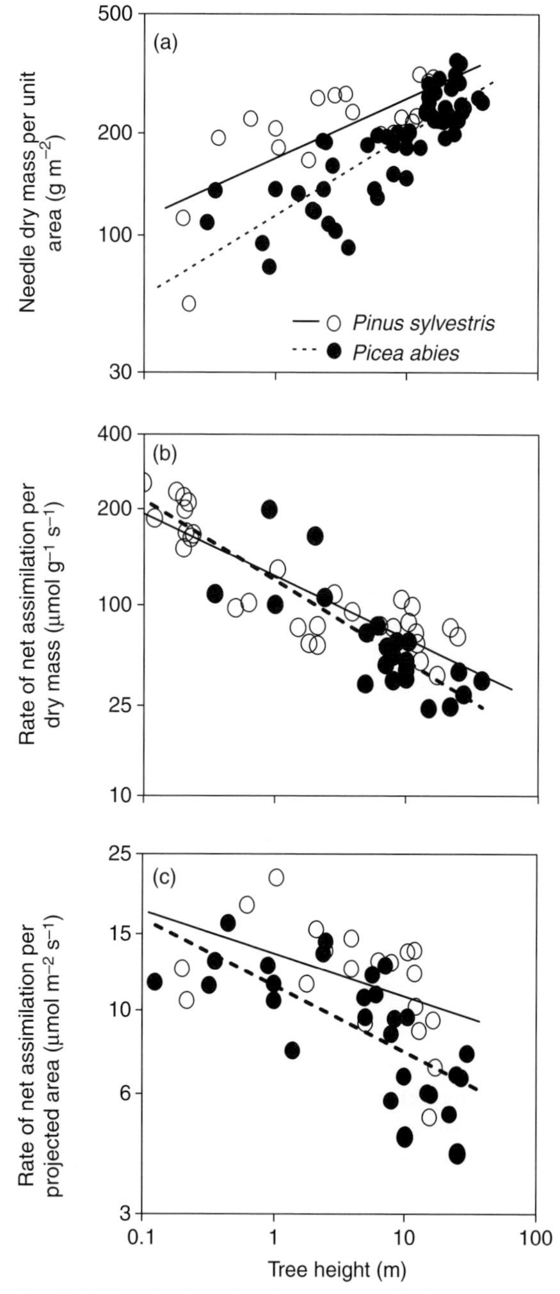

Fig. 26.9. Tree-size dependent increase in needle dry mass per unit projected area (LMA) **(a)** and the decline in foliage photosynthetic capacity (A_{max}) per unit dry mass **(b)** and projected area **(c)** for current year needles of evergreen conifers *Pinus sylvestris* and *Picea abies* according to a meta-analysis on published photosynthesis data in different-sized trees (modified from Niinemets, 2002). A_{max} per unit area is the product of A_{max} per unit dry mass and LMA.

alone are not enough to explain the decrease in photosynthetic rates. In addition, g_m has also been shown to decrease with increasing tree size (Mullin *et al.*, 2009; Woodruff *et al.*, 2009), although this decline seems to be proportional with reductions in A_{max} such that the CO_2 concentration in chloroplasts ($C_c = C_i - A_{max}/g_m$) does not necessarily vary with tree size (Niinemets *et al.*, 2009b).

The reduction of photosynthetic capacity in larger trees is associated with enhanced LMA (Fig. 26.9.a) (Mullin *et al.*, 2009; Woodruff *et al.*, 2009) and leaf lignification (Niinemets, 1997b,c). Separate examination of the components of LMA, leaf density (L_D) and thickness (T, LMA=$L_D T$), indicates that age-dependent changes in LMA primarily result from modifications in L_D (Niinemets, 1997b; Niinemets and Kull, 1995b). This is an important difference as an increase in T is compatible with a larger number of cell layers, while the increase in L_D is usually associated with smaller and more tightly packed cells with a greater cell-wall thickness. All these structural modifications are compatible with enhanced tolerance of low leaf-water potentials, but they also inevitably reduce foliage photosynthetic capacities.

Contrary to the hypothesis that reduced photosynthesis is responsible for the decline in tree productivity, it has been proposed that low water availabilities limit plant growth more than photosynthesis in large trees and that reduction in growth is responsible for the decrease in tree productivity (Sala and Hoch, 2009). This alternative hypothesis is based on the observation of greater non-structural carbohydrate concentrations in taller and older trees (Niinemets, 1997b,c; Hoch *et al.*, 2003; Sala and Hoch, 2009), suggested to be indicative of lack of carbon limiations in large trees (Hoch *et al.*, 2003; Sala and Hoch, 2009). At the same time, the increase in soluble carbohydrates that have low carbon cost for construction (40% carbon) is relatively moderate with increasing tree size compared with the increase in lignin content with high carbon cost (65.5% carbon), and the overall carbon requirement for biomass construction increases with tree size (Niinemets, 1997c) contradicting the suggestion that carbon is in surplus in large trees. In addition, soluble carbohydrates also serve as important osmotica, and their increase in larger trees may reflect acclimation to low plant water potentials. These controversial issues in the causes and effects of whole-tree productivity require further experimental and modelling work combining light interception, hydraulics, photosynthesis, respiration, osmotic adjustment and growth to quantitatively separate the sources of productivity decline in older trees.

26.5. SUMMARY: INTEGRATION OF LEAF AND WHOLE-PLANT STUDIES TO UNDERSTAND VARIATIONS IN WHOLE-PLANT NET CARBON GAIN

This review emphasises the importance of whole-plant measurements to gain insight into large-scale photosynthetic responses of plant stands. Although key insight into physiological controls of photosynthesis can be gained by leaf-level measurements of CO_2, light and temperature responses, these responses alone do not provide information of whole-plant responses to environmental drivers. Discrepancies between single-leaf and whole-plant responses result from inherent light gradients within the plants that alter the shape of whole-plant light, CO_2 and temperature responses. In addition, within-plant gradients in temperature and humidity can also alter whole-plant responses. The magnitude of environmental gradients is driven by plant structure and thus, whole-plant responses to the environment reflect both leaf-level physiology and plant structure.

With development of EC technology and expansion of the coverage of eddy flux measuring stations across earth ecosystems, there is enhanced interest in simple parameterisations of whole-ecosystem photosynthesis models and inversion of these models to gain insight into vegetation physiological activity using eddy flux measurements. As natural plant stands always consist of plants with varying age and size, and there are dramatic shifts in plant functional activity with age and size, it is important to consider the whole-plant scale in large-scale carbon gain models. In addition to leaf-level measurements, whole-plant measurements provide important complimentary information for parameterisation and validation of large-scale photosynthesis models, and have a large potential in development and verification of process-based models of ecosystem function.

27 • Ecophysiology of photosynthesis in the tropics

J. CHEESEMAN AND R. MONTGOMERY

27.1. INTRODUCTION

The 'tropics' are the most diverse and productive ecosystems on earth. They are broadly defined as those ecosystems that lie between 23.4° N and S latitude. The development of tropical ecosystems as we know them occurred shortly after the rise to dominance of angiosperms and coincided with the diversification of many angiosperm lineages. During this period, between 100 and 36 MA, warm paleoclimates supported much larger areas of tropical vegetation globally. Tropical forests were found not only in equatorial regions but also in mid latitudinal bands north and south of subtropical arid belts (Morley, 2000). Tropical climates are broadly characterised as having more diurnal than seasonal variation in temperatures, but often large seasonal variations in rainfall. Within the tropics, ecosystems range from hot, humid-lowland wet-tropical rain forests to cold, dry high-alpine paramos. Figure 27.1, for example, shows 30-year monthly means for temperature and precipitation for four stations within 1°C of the equator. The Kenyan station had the greatest temperature variations, but even then, the range was only 4°C. Izobamba, Ecuador, at 3050 m elevation, was nearly 15°C cooler than the other sites. Rainfall totals varied eightfold between Garissa, Kenya and Padang, Indonesia, and even within Indonesia precipitation varied by 40% between the sites.

Although tropical plants face similar basic ecophysiological challenges to those in non-tropical ecosystems, the tropics are unique in the sheer diversity of species and life forms and the general lack of freezing temperatures (except in high-mountain ecosystems). In this chapter, we will concentrate our discussion on three major types of tropical environments (wet tropical forests, savannas and mangroves). In each ecosystem, we explore major areas of research related to the ecophysiology of photosynthesis.

27.2. TROPICAL LOWLAND EVERGREEN RAIN FOREST

When the public hears the words 'rain forest' the image conjured is usually of the forests characterised as wet tropical or lowland evergreen rain forests. These forests are located at elevations less than 750 m and have an aseasonal pattern of rainfall, with all months receiving at least 100 mm of rain. Temperatures during the coldest months rarely dip below 18°C, and diurnal variation in temperatures are generally less than 5°C (Whitmore, 1990). Because of warm, wet climate conditions these forests are lush, supporting a structurally complex community with high aboveground biomass and productivity. Canopy height ranges from 25 to 45 m, with emergent trees in some forests reaching 60–80 m (Whitmore, 1990). Lowland evergreen rain forests tend to be nutrient poor owing to old, highly weathered soils and high rainfall. They have strong vertical environmental gradients ranging from relatively cool, moist deeply shaded understories to hot, dry forest canopies. Research on the ecophysiology of photosynthesis has focused on mechanisms of shade tolerance, the importance of variable light and comparative photosynthesis of the diverse lifeforms that occupy these forests (e.g., herbs, palms, shrubs, trees, lianas, epiphytes). Eddy covariance (EC) methods have made it possible to estimate photosynthesis and carbon uptake at the whole-ecosystem level.

27.2.1. Photosynthesis in the forest understory

The forest understory of wet tropical forest is characterised by extremely low light levels (Chazdon and Fetcher, 1984; Montgomery, 2004a). The plant community consists of various plant lifeforms (e.g., herbs, shrubs, tree seedlings,

Terrestrial Photosynthesis in a Changing Environment: A Molecular, Physiological and Ecological Approach, ed. J. Flexas, F. Loreto and H. Medrano. Published by Cambridge University Press.

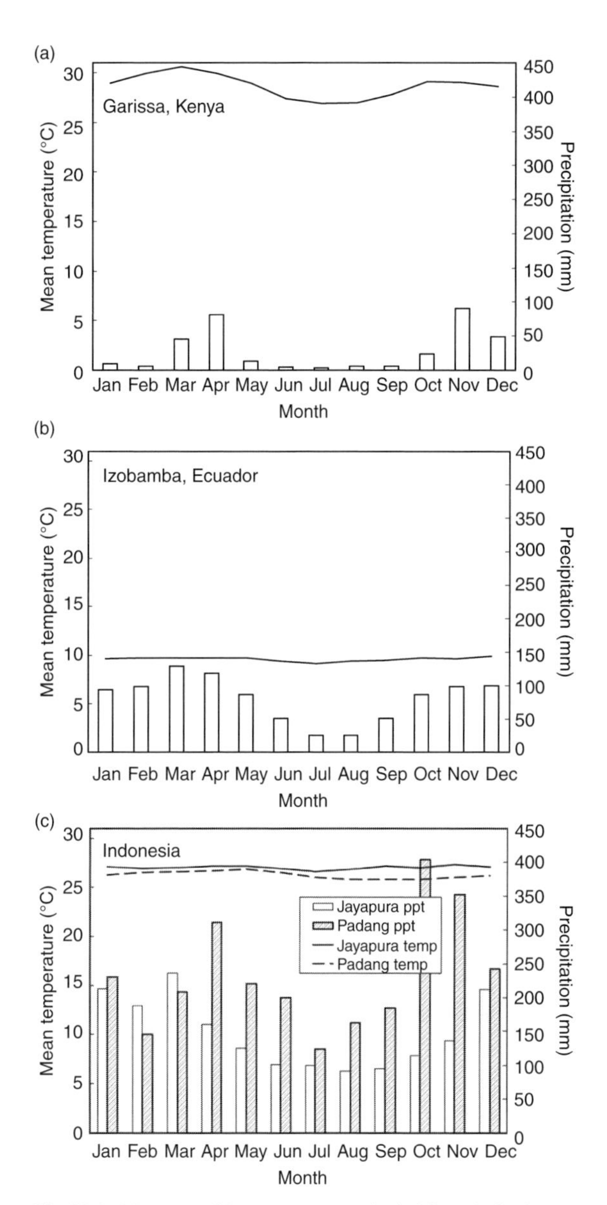

Fig. 27.1. Mean monthly temperature and rainfall totals for four sites within 1° of the equator. Values are 30-year means taken between 1960 and 1990 from WMO records, and were downloaded from the National Climate Data Center (http://cdo.ncdc.noaa.gov/CDO/cdo). WMO Station IDs are: 63723 (Garissa, Kenya) **(a)**; 84088 (Izobamba, Ecuador) **(b)**; 96163 (Padang Tabing, Indonesia) **(c)**; and 97690 (Jayapura, Indonesia) **(c)**.

palms, liana seedlings, bromeliads). Some of these groups complete their entire lifecycle in the shade and others are transient, passing through the understory as they grow towards the canopy. The latter group includes most tree seedlings and has been the focus of much ecophysiological research in tropical forests.

The resource-poor forest understory environment favours plant strategies that reduce metabolic rates, lengthen leaf lifespans and allocate resources to protect long-lived leaves and to survive damage. Maximum photosynthetic rates rarely rise above ten, averaging closer to 3 or 4 μmol^{-2} s^{-1} and low light levels in the understory mean that plants spend large proportions of the day photosynthesising at rates <1 μmol m^{-2} s^{-1}. Leaf lifespans range from less than one to greater than eight years (Reich *et al.*, 2004). Delayed greening of leaves and allocation of defensive compounds protect these long-lived leaves from herbivory (Coley *et al.*, 1985; Kursar and Coley, 1992c). Allocation of non-structural carbohydrates that are then stored in plant tissues provide insurance against future damage (Myers and Kitajima, 2007), and large hypogeal cotyledons also enhance resprout ability of very young seedlings when photosynthetic structures are removed through physical damage or herbivory (Dalling and Harms, 1999; Harms and Dalling, 1997).

In the low light of the understory, light capture is crucial for photosynthesis and carbon gain. Efficient light capture is achieved through leaf display and leaf adaptations that enhance light capture. Patterns of leaf display in the understory that maximise light capture include horizontal orientation of leaves (Chazdon, 1985; Ackerly, 1997; Poorter and Werger, 1999; Takenaka *et al.*, 2001; Valladares *et al.*, 2002; Galvez and Pearcy, 2003; Pearcy *et al.*, 2004, 2005), minimisation of self shading (Chazdon, 1985; Valladares *et al.*, 2002; Galvez and Pearcy, 2003; Pearcy *et al.*, 2004, 2005) and investment in leaves with high SLA (cm g^{-1}) relative to conspecifics growing in high-light microsites (Chazdon, 1992; Montgomery, 2004b). Light capture in shade is further facilitated by increased proportion of spongy mesophyll and low chl. *a*/*b* ratios (Chazdon, 1992). Increased spongy mesophyll causes internal scattering of light and leads to more efficient use of the limited amounts of incoming PAR. Low chl. *a*/*b* ratios are related to high levels of granal stacking in shade leaves and a shift in the ratio of PSI to PSII (Anderson *et al.*, 1988). High levels of α-carotene and lutein epoxidase have also been found in a number of tropical shade-tolerant species and in shade-acclimated leaves. It is hypothesised that they enhance light harvesting (Krause *et al.*, 2001; Matsubara *et al.*, 2008, 2009).

Forest understory environments are characterised not only by low overall PFD levels but also by sunflecks, short temporal bursts of high light that result when beams of light pass through small holes in the forest canopy. Sunflecks in lowland wet tropical forests account for 10–85% of daily total PAR (Pearcy *et al.*, 1987; Chazdon, 1988) and contribute significantly to daily carbon gain (Pearcy *et al.*, 1987). Photosynthesis in plants growing in the forest understory involves rapid induction of photosynthesis (Chazdon and Pearcy, 1986a,b; Poorter and Oberbauer, 1993; Montgomery and Givnish, 2008) and high photosynthetic light-use efficiency in short sunflecks (Chazdon and Pearcy, 1986a,b; Montgomery and Givnish, 2008). Both enhance carbon gain in the light-limited environment of the forest understory. Sunflecks also generate excess excitation energy that could lead to photoinhibition of plants with low inherent rates of photochemistry. This excess energy is dissipated via photoprotective mechanisms such as the xanthophyll cycle. However, sunflecks can lead to short-term reductions in potential efficiency of PSII, manifested in declines of F_v/F_m (Watling *et al.*, 1997). Although dissipation of excess PAR is an important aspect of the observed declines in F_v/F_m, exposure to UV radiation also increases the degree of engagement of photoprotective mechanisms and leads to longer recovery times of F_v/F_m compared with plants shielded from UV (Krause *et al.*, 1999).

At the whole-plant level, positive carbon gain and long-term survival in the shaded understory are supported by low whole-plant compensation points (Baltzer and Thomas, 2007a) and the minimisation of respiration relative to photosynthesis. Whole-plant light compensation points for growth and survival are positively correlated to each other and strongly predicted by leaf-level photosynthetic and respiration rates (Baltzer and Thomas, 2007b). Low photosynthetic and respiration rates and low whole-plant light compensation points are associated with slow growth and high survival in low-light conditions.

Although the role of light in determining photosynthetic rates in the forest understory remains paramount, recent work has demonstrated that drought tolerance, or lack thereof, influences physiology and distribution of species in wet tropical forests (Engelbrecht *et al.*, 2007; Baltzer *et al.*, 2008). On the Thai-Malay peninsula, species restricted to lowland wet forests lack tolerance to desiccation when compared with more widespread species that occur in both aseasonal and seasonal forests (Baltzer *et al.*, 2008). Similarly, nutrient heterogeneity of soils can induce local variation of species composition through impacts on carbon gain and growth. In Borneo, generalist species and specialists on more nutrient-rich soils have significantly higher rates of photosynthesis and growth than specialists on nutrient poor soils (Baltzer *et al.*, 2005).

27.2.2. Photosynthesis and changing light environments

Tropical forests are highly dynamic in space and time. This leads to variation in resource availability within the forest that is important for understanding variation in photosynthesis within and among species. The degree of variability is largely dependent on gap size (Barton *et al.*, 1989). When a gap forms in the canopy, light levels increase drastically. The increase in radiation also increases temperature, reduces humidity and increases VPD. Availability of belowground resources can also increase as there is less vegetation taking up water and nutrients (Vitousek and Denslow, 1986; Uhl *et al.*, 1988). Such microenvironmental changes are short-lived as vegetation quickly fills the gap. The rapidity of gap closure also depends largely on gap size: small gaps may be filled simply by ingrowth of surrounding crowns, whereas large gaps may persist for significantly longer.

There are a small number of species that specialise in gaps for regeneration (pioneers) and have corresponding photosynthetic adaptations (see Chapter 16). Most tropical species fall along a continuum from shade intolerant to shade tolerant and possess varied abilities to survive and grow in shade and to respond to increases in resource availability associated with gap formation. Recruitment in gaps is largely through saplings that are established prior to gap formation (Uhl *et al.*, 1988), the advanced regeneration making plasticity of photosynthetic responses crucial to plant response to gaps.

Plastic responses of photosynthesis to changing light environments have been well documented and involve biochemical and anatomical changes at the leaf and whole-plant level. Plants possess a suite of photosynthetic traits that shift with growth irradiance and theoretically maximise survival, carbon gain or growth at a given irradiance (Boardman, 1977; Givnish, 1988; Strauss-Debenedetti and Bazzaz, 1991; Sims and Pearcy, 1991; Kamaluddin and Grace, 1992; Kitajima, 1994). Briefly, plants grown in high-light environments have high photosynthetic capacity, high respiration rates and require more light to saturate photosynthesis compared with plants grown in low light (Boardman, 1977; Givnish, 1988; Strauss-Debenedetti and Bazzaz, 1991; Sims and Pearcy, 1991; Kamaluddin and Grace, 1992). Within species, leaves decrease SLA as growth irradiance increases, such that in sunny microsites a greater mass is displayed over a smaller area (Jurik and Chabot, 1986). As a result, species often show increases in $A_{max,A}$ with increasing irradiance. From an economic perspective, as irradiance increases, carbon gained per unit of carbon invested (e.g., $A_{max,M}$) should also increase (Givnish, 1988); however, plants differ in their ability to implement this strategy (Chazdon and Field, 1987; Chazdon, 1992; Mulkey *et al.*, 1993; Newell *et al.*, 1993; Ducrey, 1994; Davies, 1998; Sims *et al.*, 1998a,b).

Although generally regarded as positive, leading to increases in carbon gain and growth, shifts to high light from low can also be stressful. A number of studies have documented strong photoprotective responses in tropical plants to abrupt increases in light availability (Langenheim *et al.*, 1984; Lovelock *et al.*, 1994; Krause *et al.*, 2001, 2003a,b). Such photoprotective mechanisms are engaged in response to excess PAR and UV (Krause *et al.*, 2003a,b; 2004; 2007). Early work by Langenheim *et al.* (1984) demonstrated sustained reductions (>24 h) in Chl-F following transfers from low to high irradiance in Amazonian and Australian rainforest tree seedlings. In much longer assessments, some species exposed to high light levels common of gap disturbances did not recover F_v/F_m even after more than 20 days (Lovelock *et al.*, 1994). Species that experienced sustained reductions in F_v/F_m tended to be those classified as the most shade tolerant (Lovelock *et al.*, 1994). Such species may require development of new leaves to cope with new light conditions. Others have found rapid recovery of F_v/F_m and little sustained reductions following transfer to gaps (Krause and Winter, 1996; Thiele *et al.*, 1998; Krause *et al.*, 2001). However, the reductions of F_v/F_m and the time required for recovery appear to be greater in late successional species than pioneers (Krause *et al.*, 2001). There is also evidence that mature shade leaves can acclimate after gap formation (Krause *et al.*, 2004). Rapid recovery and acclimation are associated with differences or changes in pigment pools. Sun leaves and mature shade leaves acclimated to high light have lower levels of α-carotene, higher levels of β-carotene, larger pool sizes of xanthophylls and more rapid turnover of xanthophylls compared with shade-acclimated leaves (Krause *et al.*, 2004). Moreover, plants increase UV-B shielding compounds in response to transfer to high light, reducing the degree of decline of potential efficiency of PSII (Krause *et al.*, 2004, 2007). Sudden exposure to UV-B has also been shown to lower rates of CO_2 uptake independent of declines in potential efficiency of PSII (Krause *et al.*, 2003a).

27.2.3 Photosynthesis in the forest canopy

In addition to the spatially and temporally variable nature of gaps, tropical forests (like all forests) have predictable, strong vertical gradients. Many of the tree seedlings studied in shaded undestories grow up through this gradient, eventually reaching the canopy. In addition, the canopy is also home to lianas, hemiepiphytes, epiphytic bromeliads, orchids, ferns and cacti. The canopy environment is drastically different from the forest understory: high light, high temperature, high VPD and low humidity impose different stresses on plants that occupy the strata. Owing to the difficulty of access, much less research has been done on photosynthesis in canopy trees and other canopy plants.

Shifts in physiology as trees grow from the understory into the canopy involve ontogenetic changes in photosynthesis and acclimation of physiology to different light environments. It is often a challenge to separate these two mechanisms, yet doing so is crucial to understanding mechanisms that influence changes in growth and forest productivity with age. A meta-analysis of studies that compared species responses through ontogeny found that in tropical trees A_{mass} decreased, while g_s and leaf mass per area (LMA; g m^{-2}) increased regardless of whether the comparison was between understory saplings and canopy trees or open-grown saplings and canopy trees (Thomas and Winner, 2002). This analysis suggests a strong role of ontogeny in shaping canopy physiology. The differences in LMA were the most striking and were consistent across tropical evergreen, temperate deciduous and temperate evergreen species. Declines in LMA with ontogeny results from limitations on the cell-expansion rate as increased hydraulic resistance in tall trees leads to a decline in turgor pressure in leaves in the canopy (Thomas and Winner, 2002).

Owing to shifts in LMA, $A_{max,A}$ is generally considerably higher in the canopy: rates >10 μmol^{-2} s^{-1} are not uncommon in canopy trees and lianas. In a study that included diverse tree species in a Panamanian wet forest, maximum photosynthetic rates varied from 9.7–18.3 μmol^{-2} s^{-1} (Santiago *et al.*, 2004). Rates of $A_{max,A}$ were positively related to leaf specific hydraulic conductivity and negatively related to wood density (Santiago *et al.*, 2004), supporting a growing literature on trade-offs in leaf form and function (Reich *et al.*, 1999; Wright *et al.*, 2004; Chave *et al.*, 2009).

Leaves that develop in the forest canopy face multiple stresses including the high irradiance impinging on newly emerging leaves. To dissipate excess energy and protect developing tissues from UV, young leaves have high levels of xanthophyll-cycle pigments and UV-A-shielding compounds (Krause *et al.*, 2003a). High levels of photoprotection are manifested as large reductions in F_v/F_m midday in young compared with mature canopy leaves (Krause *et al.*, 1995). In general, exposure to UV radiation decreases the efficiency of PSII and photoprotection acts to minimise UV damage for canopy leaves (Krause *et al.*, 1999).

In open environments such as the forest canopy, variation in light availability is largely a function of daily and seasonal fluctuations in weather. Recent work on canopy photosynthesis in Panama has shown that during rainy and cloudy periods canopy photosynthesis is light limited (Graham *et al.*, 2003). In that study lamps were used to illuminate the canopy and leaf gas exchange was measured monthly for 2 years. Illuminated leaves had significantly higher rates of photosynthesis, especially in mornings and afternoons. Leaves that developed under illumination also had higher rates of maximum photosynthesis. Higher rates of photosynthesis translated into 31–67% greater extension growth for illuminated compared with control branches (Graham *et al.*, 2003).

27.2.4. Photosynthetic research on a diversity of lifeforms

While the most prominent structural component of tropical wet forests are trees, tropical forests supporting a lush cover of epiphytes and climbing lianas are common. These lifeforms provide unique perspectives on the ecophysiology of photosynthesis in the wet tropics.

A large body of work has developed on photosynthetic physiology in epiphytes and hemiepiphytes of the tropical forest canopy (Winter *et al.*, 1983; Medina, 1987; Zotz and Winter, 1993b, 1994a,b; Andrade and Nobel, 1996; Zotz *et al.*, 1997; Stuntz and Zotz, 2001; Lüttge, 2006; Watkins *et al.*, 2007). Epiphytes and hemiepiphytes by virtue of their growth form tend to be water and nutrient limited. In addition, those that occur near the top of the canopy face high radiation loads. Early explorations of tropical forest-canopy environments revealed that CAM is a common photosynthetic pathway among epiphytes and hemiepiphytes. This water-saving pathway had mainly been associated with deserts and other arid systems. In fact, a large fraction of the world's CAM species are found in tropical rainforest canopies. A survey of 0.4 ha of tropical lowland rainforest found >13,000 individual epiphytes representing 103 species. Of these species, 19.4% showed evidence of CAM metabolism. Many of the species that possess CAM are rare or small in size. Thus, the proportion of individual CAM epiphytes was considerably less (3.6%) and their contribution to total epiphyte biomass was 3.0% (Zotz, 2007).

Lianas compose a relatively small fraction of forest basal area (5.5–14%) (Restom and Nepstad, 2001; Gerwing and Farias, 2000) but owing to climbing habit and ability to shade the upper canopy of their support trees they have the potential to play an important role shaping whole-forest carbon budgets. Owing to difficulty of access there has been relatively little work done on liana photosynthesis. Studies of water transport in lianas show that they tend to have higher sapwood specific hydraulic conductivity and thus are expected to have higher photosynthetic rates compared with the trees on which they grow (Patiño *et al.*, 1995). However, a broad survey of tree and liana physiology using canopy cranes in Panama found that canopy liana leaves have lower $A_{max,A}$ and g_s than canopy trees despite similar SLA (Santiago and Wright, 2007). This may result from hydraulic limitation if lianas support a larger leaf area per unit sapwood compared with trees or if the pathlength for water transport is longer in lianas. Lianas do have larger pool sizes of xanthophyll-cycle pigments compared with co-occurring trees suggesting greater requirements for photoprotection (Matsubara *et al.*, 2009). Further study of physiology of lianas is necessary.

27.2.5. Photosynthesis at the ecosystem scale

The development of EC methods for estimating whole-ecosystem gas exchange has provided a forest ecosystem level view of photosynthesis. There are currently more than 140 eddy flux towers worldwide, more than 20 of which are located in tropical forests. The largest effort has been conducted in the Amazon (the 'Large-scale Biosphere-Atmosphere Experiment in Amazonia' or [LBA]).

The EC technique allows researchers to assess whether tropical forests represent a net source or sink for CO_2 and how environmental change can influence ecosystem uptake or release of CO_2, and thus source-sink status. To date, it appears that tropical forests represent a net sink. Early results from the Amazon measured annual NEE between –1.0 and –5.9 t C ha^{-1} yr^{-1} (Fan *et al.*, 1990; Grace *et al.*, 1995; Malhi *et al.*, 1998), while comparable research in Costa Rica found that the rain forest ranged from being

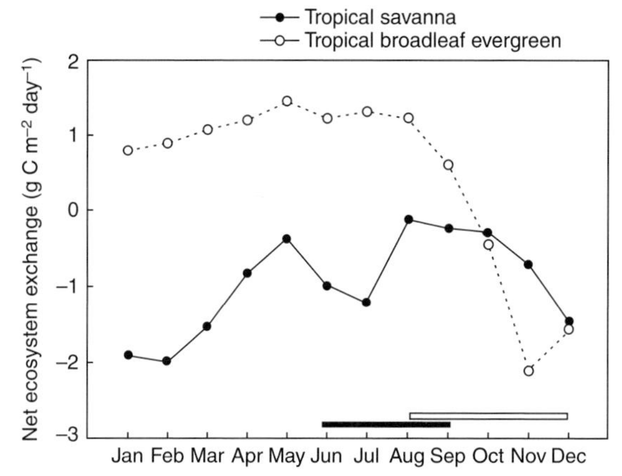

Fig. 27.2. Seasonal variation in net ecosystem exchange (NEE; g C m^{-2} s^{-1}) in two tropical ecosystems. Data were extracted from Beringer *et al.* (2007) for tropical savanna (closed symbols; Howard Springs, Northern Terrritory, Australia) and from the Ameriflux database for tropical evergreen broadleaf forest (open symbols; LBA Tapajos KM67 Mature Forest, Brazil). Data are from a single year (2004). Horizontal bars indicate dry season duration at each site.

a mild source (0.05 t C ha^{-1} yr^{-1}) to a strong sink (–6.91 t C ha^{-1} yr^{-1}) (Loescher *et al.*, 2003).

Interannual variation was hypothesised to result from macro-climate variation associated with an ENSO event, with warm, dry years having less negative NEE (Loescher *et al.*, 2003). A recent review found similar flux rates in Asian evergreen broadleaf forests (MAT=25°C; n=8 sites) with a mean NEE of –1.56±3.79 (Kato and Tang, 2008). Diel patterns of NEE are strongly related to PFD: carbon uptake increases with increasing PFD (Loescher *et al.*, 2003; Goulden *et al.*, 2004; Hutyra *et al.*, 2007; Kosugi *et al.*, 2008). Phenology (Hutyra *et al.*, 2007) and climate anomalies (Loescher *et al.*, 2003; Hutyra *et al.*, 2007) were also important correlates of seasonal patterns of NEE (Fig. 27.2). Water availability appears to influence NEE, mainly through effects on respiration: ecosystem respiration increases with soil moisture (Saleska *et al.*, 2003; Kosugi *et al.*, 2008). Although most agree that tropical forests represent a net sink, daily and annual NEE estimates in tropical forests may be misleading because they underestimate night-time fluxes owing to poor mixing of air between the canopy and the atmosphere. Better estimation of CO_2 losses would likely reduce the sink strength of tropical forests, and some argue that EC methods in tropical forests are best used for understanding hourly controls of NEE (Goulden *et al.*, 2004)

27.3. TROPICAL SAVANNAS

Savannas are grassland ecosystems with scattered trees and shrubs, but with no closed woody canopy. Thus, savanna light environments can be complex and horizontally variable. Savanna community composition and boundaries are determined largely by seasonal moisture patterns, while major disturbances, especially fire and grazing, keep woody plants below a water-defined upper limit (Sankaran *et al.*, 2005). Savanna development and C_4-grassland expansion began 4–6 MA in the Late Miocene, associated with the changing seasonality, increasing aridity and increasing disturbance (especially fire) that eliminated overtopping tree canopies (Osborne, 2008).

Worldwide, savannas cover 16 million km^2, or an eighth of the world's land surface (Scholes and Archer, 1997). In Africa, savannas account for nearly 20% of the total land area, and they are equally vast throughout the tropics – the Venezuelan llanos occupies more than 310,000 km^2, the Cairns to Kimberley savanna in northern Australia is more than 1.5 million km^2 and the Brazilian cerrado is more than 2 million km^2. As a result of global climate change and deforestation, significant replacement of rainforests by savannas in South America has been predicted for the next 30–50 years (Magrin, 2007).

27.3.1. Photosynthesis and C_4 grasses

In tropical savannas, both understory and open ground are dominated by C_4 grasses, reflecting their inherently higher WUE and temperature tolerance (Osborne, 2008; Beerling and Osborne, 2006). These contribute up to nearly 60% of overall net primary productivy (NPP), with the greatest relative contribution in the Neotropics. In Africa, a greater proportion is attributable to woody plants, whereas Australian savannas are intermediate. These differences reflect large scale regional differences in soil fertility and rainfall (Lloyd *et al.*, 2008).

Not all C_4 species or pathways are created equal however, and the distribution of C_4 grasses in the Panicoideae (NAD-malic enzyme type, NAD-ME) and Chloridoidea (NADP-ME) is highly correlated with annual rainfall (negatively and positively, respectively), while phosphoenolpyruvate carboxykinase (PEPCk) species are intermediate (Veenendaal *et al.*, 1993; Osborne, 2008). In mixed stands (i.e., with trees present), NAD-ME species grow in the more fully exposed, drier conditions (Veenendaal *et al.*, 1993). Within the Panicoideae, there is further division: the tribe Andropogoneae dominates areas with >500 mm summer rainfall – areas where disturbance is required for maintenance of the savanna – while the tribe Paniceae dominates areas with lower rainfall. At the leaf level, photosynthetic capacities can be high (approximately 25 μmol CO_2 $m^{-2}s^{-1}$) despite low leaf nitrogen (N) content and the low nutrient status of the humid savanna soils. Seasonal variations in photosynthesis reflected changes in N content (Leroux and Mordelet, 1995), although to a less pronounced degree than in savanna trees (Simioni *et al.*, 2004).

Correlations with rainfall are also reflected in instantaneous WUE, which is enhanced significantly more by drought in NAD-ME (1.20-fold) than NADP-ME (1.11-fold) grasses. These differences are not, however, directly reflected in growth, which is similarly inhibited by drought regardless of C_4 pathway. In contrast, inherent WUE, as reflected in $\delta^{13}C$, is somewhat greater in the NADP-ME group ($\delta^{13}C$=–12.2‰ versus –13.3‰ in NAD-ME species), but responds similarly to drought in the two groups (Ghannoum *et al.*, 2002).

Both photosynthetic and whole-plant NUE are also greater on average in NADP-ME than NAD-ME species when N supplies were adequate. NADP-ME leaves also show both greater assimilation rates per Rubisco catalytic site and faster Rubisco turnover (Ghannoum *et al.*, 2005). Photosynthetic and whole-plant NUE are however both ratios, and these differences reflect both lower N and soluble protein content in NADP-ME leaves, and greater allocation of N and chl. to the mesophyll (Ghannoum *et al.*, 2005).

27.3.2. Woody vegetation

The woody vegetation, both trees and shrubs, is all C_3, and their photosynthetic capacity and WUE vary considerably throughout the year and across rainfall gradients. In the Australian savanna, the large number of *Eucalyptus* species allowed comparison of $\delta^{13}C$ values across a rainfall gradient suggesting potential physiological approaches to explain species distributions and replacements. Broadly, species replacements were marked by a shift in $\delta^{13}C$, but because both stomatal conductance and photosynthetic capacity influence this parameter, it was insufficient as a simple measure of physiological limits on tree distribution patterns (Miller *et al.*, 2001).

A comparison of two species, *Godmania macrocarpa* and *Curatella americana*, in the Venezuelan savanna emphasises the complexity of physiology/performance interactions. With drought, *Godmania* maintains higher g_s and photosynthesises at higher xylem water tensions than *Curatella*. It also allocates more N to leaves, and has consistently higher instantaneous photosynthesis rates in the wet season reflecting a correlation common to savanna trees, but not to grasses (Simioni *et al.*, 2004). WUE (as $\delta^{13}C$), however, is similar in the two species, reflecting the higher photosynthetic rates by *Godmania*, but also lower g_s in *Curatella*. Despite that, *Curatella* is ubiquitously distributed within the grassland whereas *Godmania* is not. Instead, it is restricted by other life-history characteristics to fire-protected areas (Medina and Francisco, 1994).

27.3.3. Environmental factors and savanna vegetation

DROUGHT

Photosynthesis is one important though seldom defining factor in understanding the physiological ecology of savannas, but it provides a useful structure for framing this understanding. Comparing photosynthesis in two dominant grasses and two trees (*Crossopteryx febrifuga* and *Cussonia arborea*) in a West African savanna, for example, Simioni *et al.* (2004) found no systematic differences between the groups for g_s under standard environmental conditions, or stomatal response to light or VPD. Although there were the expected differences in carboxylation efficiency at low C_i, the were no C_3/C_4 differences in A_{max}. *Crossopteryx* seedlings, however, grow in open areas, *Cussonia* only under tree clumps. Thus, other factors contribute deterministically, e.g., shade or high light tolerance, heat tolerance and drought and nutrient availability.

Although drought tolerance, loosely defined, is essential for membership in a savanna community, persistence reflects numerous other life-history traits, such as leaf longevity, seed germination and re-growth capacity, and responses to grazing and fire (e.g., Baruch and Bilbao, 1999). Interspecific relationships involving simultaneous competition for light and stress reduction by exploitation of understory shading highlight the symbiotic nature of savannas. No one strategy is ubiquitously more successful than others. In the *Godmania/Curatella* comparison, for example, the absence of *Godmania* from open grasslands, despite its high potential productivity, reflected unfavourable germination conditions, poor seedling survival and sensitivity to fire (Medina and Francisco, 1994). Moreover, although *Godmania* was more drought tolerant during the wet season, it is drought deciduous, delaying resumption of activity after the dry season.

In general, evergreen savanna woody species tend to have shallow root systems and are more sensitive to the seasonal rainfall than deciduous species whose roots are deeper (Myers *et al.*, 1997; Jackson *et al.*, 1999). Shallow systems may lead to rapid and extreme Ψw declines in the dry season accompanied by sharp and deep declines in A_{max} and carboxylation efficiency (Monteiro and Prado, 2006). Except in extreme drought however F_v/F_m generally remains high, indicating lack of photoinhibition (Sobrado, 1996; Liang and Zhang, 1999). Actual results may vary, dependent on stem hydraulic architecture, on the cost of constructing efficient stem sapwood (Macinnis-Ng *et al.*, 2004) or thick, fire-resistant bark (Hoffmann *et al.*, 2003) and on root phenology patterns (West *et al.*, 2003). Shallow root systems also enable rapid response to rain, both in and after the dry season. For example, in the cerrado *Miconia albicans* experienced nearly 75 and 99% dry season declines in A_{max} and carboxylation efficiency, but with mid-season rain both recovered quickly, leaving the stomates as the principle limitation to photosynthesis (Monteiro and Prado, 2006).

Rapid recovery with minimum costs for infrastructure are important elements of adaptive strategies. For example, photosynthesis by *Acacia auriculiformis* phyllodes in the Australian savanna declined 80% as chl. and soluble protein were lost during the dry season. By its end, chl. and soluble protein had declined 73 and 52% respectively, stomatal limitations became extreme and the C_i/C_a ratio dropped to below 0.5 (Montagu and Woo, 1999). More than half the phyllodes survived, however, and with the first, small rains, stomatal limitations were eliminated after only 1–8 days. Even after five months of drought, photosynthesis rates recovered within four weeks by 70–95%, parallelled by restoration of chl. and soluble-protein contents. This recovery capacity is crucial to trees like *Acacia*, as replacement leaf initiation does not begin for another 2.5 months.

DISTURBANCE

As noted earlier, savannas formed in the Late Miocene owing to changing seasonality, increasing aridity and increasing disturbance, especially fire (Osborne, 2008). Today, savanna 'stability' reflects disturbance-based limitations on tree distribution in rainfall-limited areas or disturbance-dependence where annual precipitation suffices for forest encroachment (Sankaran *et al.*, 2005). The plants themselves play a role in maintaining disturbance. Grasses in the Andropogoneae, for example, depend on fire to maintain their presence, and their low-nutrient, high tannin (and thus, poorly digestible) litter leads to fuel accretion. Without this fuel, woodland encroachment is more likely to occur. In both grasses and trees, fire resistance often reflects maintenance of reserves below ground to support rapid re-sprouting and growth (Alonso and Machado, 2007). In one South African savanna, for example, more than 90% of the *Acacia* trees re-sprout from roots after fires (Meyer *et al.*, 2005). Re-sprouting also supports tree regrowth after other disturbances, as in East African miombo woodlands after clear felling or grazing (Chidumayo, 2004).

THE LIGHT ENVIRONMENT AND PHOTOPROTECTION

Photoprotection, as indicated by sustained, high values of F_v/F_m, is the result of integrated strategies for amelioration and avoidance of excess irradiance. Amelioration depends heavily on xanthophyll-cycle activities. Avoidance, by leaf-angle adjustment for example, reduces the need for those and thus the costs of xanthophyll biosynthesis (Liu *et al.*, 2003). Alternately, because savanna tree shade is relatively light, mixed savanna communities have greater total light interception than pure stands (Tournebize and Sinoquet, 1995). Although shading favours trees and shrubs in competition with grasses and shrubs, it also reduces understory grass leaf temperatures, thus moderating the drought impacts. Shade effects on grass photosynthetic performance involve numerous trade-offs: light limitations under the canopy are an alternative to effects of reduced g_s in the open (Mordelet, 1993). However, shade differentially effects different types of grasses. In Botswana, for example, shade tolerating NADP-ME and PEPCk grasses have larger caryopses (seeds) than sun-tolerating NAD-ME species, ameliorating the effects of lower initial RGRs. Reproduction by tillering also increases with shading, and rapid re-growth after grazing helps exclude encroachment of other species (Veenendaal *et al.*, 1993). Dispersed tree canopies also create another, more complex niche along run-off flow lines. In the same Botswana study, the NADP-ME grass *Digitria eriantha* occupies the drip line and drainage flows leading away from the trees, adapting photosynthetically to sun or shade: in shade, the saturating irradiance is much lower as are maximal CO_2 fixation rates, low-light quantum efficiencies are higher and light compensation points are lower, consistent with true shade adaptation (Veenendaal *et al.*, 1993).

Canopy-shading effects extend more broadly, however. Trees add high-quality litter to the soil surface, a benefit that may outweigh light reduction, especially under thin-canopied trees such as acacias (Veenendaal *et al.*, 1993; Mordelet and Menaut, 1995). Indeed, soil nutritional enhancement by trees, particularly legumes, can speed shrub establishment, even long after the trees themselves (and hence their shade) are eliminated (Barnes and Archer, 1996). Planting tree seedlings under remnant trees in restoration of abandoned agricultural lands is a practical exploitation of this effect (Loik and Holl, 1999).

27.4. MANGROVES

Mangroves are a diverse group of intertidal trees, represented in at least 15 families, including ferns, monocots and dicots. Originating on the Sea of Tethys approximately 65 MA (Plaziat *et al.*, 2001), today they are grouped according to two major geographic areas, the Neotropics and West Africa, and Indo-West Pacific tropics. In any given mangrove forest, species composition, zonation and succession are complex functions rooted in geomorphology (Woodroffe, 1992). Mangrove forest area has been reduced some 35% over the last 20 years by anthropogenic destruction primarily for mariculture (especially shrimp farming) and harvesting wood products (Valiela *et al.*, 2001). Left undisturbed, mangroves play a crucial role in coastal zone protection during catastrophic disturbances; they are intensely rooted and difficult to dislodge, and dramatically increase surface roughness, dissipating energy carried by wind and waves.

Mangrove-ecosystem productivity is usually considered to be 'high', but this varies immensely with site quality (Clough, 1992). In the highly productive Hinchinbrook Channel in Queensland, the coastal system is net autotrophic, and mangroves produce 65% of the 82,000 tonnes of organic carbon exported to near-shore waters each year, making mangroves an important source of organic material for the Great Barrier Reef (Alongi *et al.*, 1998).

27.4.1. Photosynthesis, salinity and temperature

The tropical intertidal zone encompasses extreme and complex environmental constraints, including high and variable salinities, high irradiance (visible and UV), high temperatures, prolonged, sometimes total seawater inundation, low nutrient inputs, anaerobic soils with poor nutrient retention and oxidative stress induced by each of these. Salinity is the most obvious concern, but most species survive even serious hyper-salinity (e.g., twice seawater). In the course of any year, trees along rivers may experience both fresh water and full seawater (Ball, 2002; Krauss *et al.*, 2008).

Mangroves are C_3 plants with photosynthetic capacities comparable with many mesophytic trees (Andrews and Muller, 1985). Mean light-saturated rates range from 2.5 to 27 μmol $m^{-2}s^{-1}$ dependent on species, salinity and VPD (Clough and Sim, 1989). Instantaneous net photosynthesis (A_N) and WUE generally decrease with increasing salinity (e.g., Ewe and Sternberg, 2005; Clough and Sim, 1989), although sun-adapted species (e.g., *Rhizophora mangle*) may show no variation in A_{max} with either salinity (2 to 32‰) or shading (Krauss and Allen, 2003). Increasing light may increase A_{max} for plants grown at low salinity, but decrease it under hypersaline (>150% seawater) conditions (Lopez-Hoffman *et al.*, 2007). Owing in part to increasing respiration with salinity and light, photosynthesis alone is insufficient to predict comparative ecological performance (Krauss and Allen, 2003). The salt sensitivity of photosynthesis is also complicated by factors such as reducing soil conditions and overall nutrient levels (e.g., Lin and Sternberg, 1992a).

High leaf temperature exacerbates salinity effects. Net photosynthetic rates show a broad peak between at about 27 and 35°C, with positive net CO_2 exchange up to nearly 44°C (Cheeseman *et al.*, 1997). Whether this is actually the upper temperature limit depends on stem hydraulic conductivity and stomatal responses; as in other plants, purified Rubisco has a much higher temperature optimum, and mangrove leaf enzymes in general may be stabilised by heat-shock proteins (HSPs) or other chaperones (e.g., Pearse *et al.*, 2005). Mangroves clearly survive long periods at supra-optimal temperatures and photosynthesis recovers quickly after extremes. At the low-temperature end, photosynthesis responses *per se* have not apparently been carefully examined, although low temperatures play a defining role in mangrove distributions (Krauss *et al.*, 2008).

27.4.2. Water-use efficiency

Because of an expectation that water and salt uptakes should be linked, WUE has been a recurring theme in the mangrove literature. The relationship between A_N and g_s tends to be linear and identical for all species within a forest, as are intercellular CO_2 levels, and hence, instantaneous WUE (Clough and Sim, 1989). There may, however, be significant inconsistencies between WUE estimated as A_N/g_s, and WUE estimated from $\delta^{13}C$.

In the Caribbean, for example, stunted interior mangrove forests occur in close proximity to much taller trees fringing shorelines or channels, providing a platform for examining this discrepancy. Reflecting their generally more stressful growth environment, stunted trees of all three neo-tropical species have lower g_s than tall forms, with significantly higher (less negative) $\delta^{13}C$, indicating higher long-term WUE (Lin and Sternberg, 1992b). $\delta^{13}C$, however, reflects the conditions under which leaves develop (Martin and Thorstenson, 1988). Because mangrove leaves may be produced at any time of year, and thus develop under very different temperature and rainfall conditions, $\delta^{13}C$ varies seasonally (Cheeseman and Lovelock, 2004).

Salinity also affects WUE. In a comparison of two neotropical species as an example, *Avicennia germinans* had the most negative $\delta^{13}C$ value at about 20% seawater, which was also broadly the salinity for maximal growth. Alternatively, for *R. mangle* $\delta^{13}C$ was constant and indicative of relatively high stomatal conductance, at 0 and 20% seawater. Only at higher salinity (45% seawater), did ^{13}C discrimination suggest a stomatal limitation (Ish-Shalom-Gordon *et al.*, 1992).

27.4.3. Photoprotection and light avoidance

Mangrove net CO_2-exchange light saturates at relatively low irradiance, usually below 1000 μmol $m^{-2}s^{-1}$. Even fully exposed, unshaded leaves behave more like shade species, saturating at 400–500 μmol $m^{-2}s^{-1}$ (Krauss and Allen, 2003; Cheeseman *et al.*, 1991, 1997; Cheeseman and Lovelock, 2004). Maximal rates often decline by 85% or more by late morning, accompanied by increasing C_i, and cannot be explained by leaf temperature, g_s or water stress alone. Attiwill *et al.* (1980) suggested the decline may reflect photoinhibition. However, in-situ fluorescence measurements using portable fluorometers, have shown that leaves transferred to darkness at midday show full recovery of F_v/F_m within 30 min (Cheeseman *et al.*, 1991) and even top-of-the-canopy leaves monitored *in situ* fully recover within 30 min of sundown (Cheeseman *et al.*, 1997; Cheeseman and Lovelock, 2004).

Fluorescence also suggests that the coupling between photochemistry and biochemistry is less robust, at least in Rhizophoracean mangroves, than in most systems: even when CO_2 fluxes are low, 2% O_2 is used to eliminate photorespiration, g_s is very low and C_i is increasing toward C_a. Electron transport calculated from ϕ_{PSII} ($\mathcal{J}_f$) remains a tight linear function of irradiance up to approximately 500 μmol $m^{-2}s^{-1}$ (Cheeseman *et al.*, 1997). As SOD levels in field-grown mangroves are 10 to 40 times higher than those in other plants, a controlled contribution of the Mehler reaction may be part of the photoprotective strategy (Cheeseman *et al.*, 1997).

Leaf orientation and anatomy also contribute to photoprotection, reducing direct light exposure and leaf heating. Mangrove leaf anatomy is generally more complex than

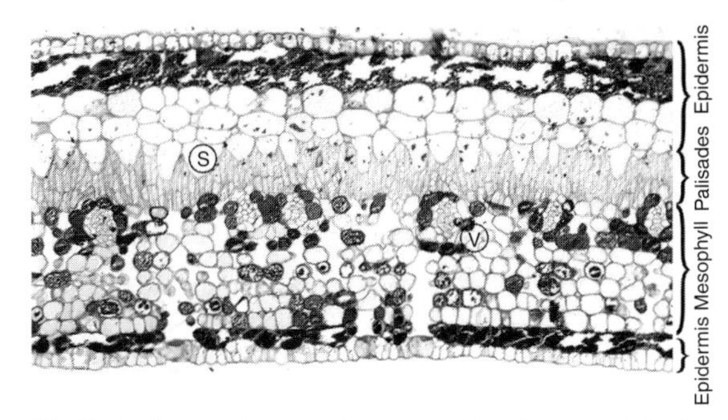

Fig. 27.3. Cross section of an immature *Rhizophora mangle* leaf showing the complexity of the epidermal layers, the small size of the palisades cells and the intercolated pear-shaped 'slime' cells (s) (Roth, 1992). Dark stain is proanthocyanidins (tannins) localised especially to epidermal layers and bundle sheaths (v).

other C_3 species. In *R. mucronata*, for example, back scattering within multiple epidermal layers (Fig. 27.3) created a light gradient that reduced light reaching assimilatory cells by up to 90% (Koizumi *et al.*, 1998). In-situ fluorescence analysis nevertheless showed that the leaves were structured well to capture weak light. Figure 27.3 shows other potentially significant structural features. Tannins, for example, accumulate particularly in the inner layers of the abaxial and adaxial epidermes, and in the cells surrounding the minor veins. These and polyphenolics (which are more ubiquitously distributed) are strong UV absorbers (Lovelock *et al.*, 1992; Pearse *et al.*, 2005).

Both orientation and structure reduce the need for more complex energy dissipation processes such as the xanthophyll cycle (see Section 27.3.3.3). Lovelock *et al.* (1992), for example, found that the sum of the xanthophyll-cycle components per unit chl. correlates with the projection of a mangrove leaf on a horizontal plane, and at steeper angles, a smaller fraction of the xanthophyll pool is de-epoxidised at midday. As the leaves at the top of a *Rhizophora* canopy, for example, may be essentially vertical, this results in a significant reduction in necessary investment. Alternatively, species that maintain more horizontal leaf orientation even in exposed conditions (e.g., *Bruguiera parviflora*), invest more in xanthophyll-cycle pigments, especially in sun leaves.

27.4.4. Shade tolerance and seedling persistence

Mangroves often persist as seedlings in their own understory for years, and regeneration after moderate disturbances is largely from this pool. This release of 'advance recruits' can generate near monocultures (Baldwin, 2001).

Shade persistence reflects photosynthesis at maintenance levels at very low light and utilisation of sunflecks without experiencing photoinhibition (Ball and Critchley, 1982). Pioneer species (e.g., *Rhizophora stylosa* and *Sonneratia alba* in Indo-West Pacific systems) have higher light-saturated J_f than species that persist in the understory (*R. apiculata* and *Bruguiera gymnorrhiza*). These, in turn, have higher J_f than *Xylocarpus granatum*, a canopy species. The recovery of F_v/F_m after high light exposure and the response of qP to incident light show the same ordering. Based on this, Kitao *et al.* (2003) asserted that coordination of leaf absorptance, thermal dissipation and electron transport was central to acclimation of mangroves in exposed habitats.

In Hawaii, Krauss and Allen (2003) compared two invading mangroves, *R. mangle* and *Bruguiera sexangula*. Stomatal control determined photosynthetic capacity in the latter; in *R. mangle*, photosynthesis was regulated largely by light. *B. sexangula* grew slowly regardless of conditions, and photosynthetic parameters (ϕ_{CO2}, A_{max} and light compensation point) responded less to salinity and light conditions. *R. mangle* depended on faster growth under all growth conditions and had greater photosynthetic responses to light and salt. Thus, while both species could perform well at low light, at high salinity, *R. mangle*'s capacity to use light was an additional resource to tolerate the stress.

27.4.5. Photosynthesis, oligotrophy and secondary metabolites

Throughout this chapter, achieving higher WUE has been a repeated theme. Although the associated carbon cost is obvious, whether that is physiologically significant or not depends on other factors limiting organismal performance. The belief that high A_{max} should translate into faster growth is difficult to discard, but over evolutionary time mangrove ecosystems have generally been oligotrophic, i.e., low in available nutrients, especially N and P. This means that their photosynthetic capacity may greatly exceed their ability to make and support new 'productive' tissues, i.e., leaves or roots. The result is accumulation of what is essentially excess carbon. Mangroves have a number of ways to utilise this, especially production of secondary metabolites in all tissues, such as proanthocyanidins (condensed tannins) and flavonoids (Basak *et al.*, 1996; Kandil *et al.*, 2004). These accumulate in leaves to between 10 and 55% of the total dry mass. On the one hand, they serve a defensive function as feeding inhibitors, anti-fungal agents, antioxidants and UV screens. On the other hand, their production consumes a large amount of energy: production of dihydroflavonol, the last common precursor to tannins and flavonoids, requires 121 ATP equivalents. But as Haslam (1986, 1985) argued, shunting metabolic intermediates (e.g., pyruvate, PEP, acetyl-CoA and 3-phosphoglycerate) prevents imbalances between carbon supply and utilisation. However, such flavonoid accumulation also requires from 10 to 25% of all the 'structural' carbon in the leaves to pass through phenylalanine and phenylalanine-ammonia lyase, with the release of NH_3. As with NH_3 released through photorespiration, recycling this is crucial.

27.5. CONCLUDING REMARKS

Two of the major themes of this chapter are that the tropics are very big, and that they are very complex. We have focused on only three biomes within the tropics and on only one aspect of ecophysiology, i.e., photosynthesis. As mentioned at the outset, a broader view of ecosystem physiology is now possible using remote sensing and landscape-level technologies. With those tools, scientists are in a position to model and understand the response of tropical systems to climate change and anthropogenic alteration and the role of tropical ecosystems globally. At this point, there is a significant danger in overestimating what we know and what we can do with such knowledge. The most important major dynamic in the tropics today is ecosystem alteration and destruction. On the one hand, we can hope that regional studies will help convince policy makers to change the way of doing daily business. On the other hand, such studies may serve only to allow the detailed documentation of how we destroyed the planet. Because of the complexity of the tropics, models alone are generally poor tools for restoration efforts.

We must be similarly cautious about complacency at smaller scales. The ecophysiology of photosynthesis is easy to study: leaves are above ground and visible, instrumentation is easily accessible and large datasets can be generated in relatively short times. Photosynthesis is, however, only acquisition of reduced carbon, the plant equivalent to a human diet of bleached, white rice. To fully understand plant organismal biology, we must fill the rest of the gaps and attend to local details. Only then can we apply our knowledge to the crucial problems of ecosystem protection or restoration.

28 • Ecophysiology of photosynthesis in desert ecosystems

A.C. GIBSON AND P.W. RUNDEL

28.1. INTRODUCTION

One of the most interesting aspects of desert-plant ecophysiology is the complex nature of adaptations that allow them not only to survive but also to reproduce and maintain their populations under the extreme conditions of environmental stress present in these ecosystems. Of course, deserts are defined by having limited rainfall (typically <250 mm annually) and thus produce significant climatic drought stress, but the seasonality of rainfall patterns and thus drought varies across desert regions, as do the plant growth forms that dominate the landscape (Fig. 28.1).

Terrestrial vascular plants, in any environment where water is limiting for growth, face a dilemma. Their uptake of CO_2 from the atmosphere can only occur by opening stomata on their exposed organs regulated by guard cells. However, at the same time that stomata are open and CO_2 diffuses into green tissues for photosynthesis, water vapour diffuses out of the plant into the surrounding, drier air. A plant's strategy can be to reduce transpirational loss by keeping its stomata closed and thereby lose no water vapour, but during that time it also fixes no atmospheric CO_2 and thus cannot produce new sugars necessary for growth and respiration. Obviously then, stomata must be opened for photosynthesis to occur but at a significant cost to water levels. The hotter and drier the outside air, the more rapid will be the rate of water-vapour loss from plant tissues.

Despite the temptation with desert plants to focus on defensive strategies that minimise water loss, it is important to draw attention to the converse, i.e., offensive strategies that maximise rates of photosynthesis under favourable environmental conditions. For most desert plants, opportunistic offensive strategies are the key to interpreting many structural and anatomical adaptations of desert plants (Gibson, 1996, 1998).

An energy balance of desert plants may involve two problems related to the conditions of intense solar radiation of a desert region. The first of these problems relates to high tissue temperature. Although some species are more heat tolerant than others, leaf temperatures above 45°C are commonly lethal, unless special HSPs are present (Knight and Ackerly, 2003a). Although only in certain desert regions do ambient temperatures persist above such levels, intense solar radiation can heat leaves and other plant organs well above air temperatures to potentially lethal levels (Gates, 1980; Downton *et al.*, 1984), which begins the shutdown of that shoot. The other problem comes from photoinhibition, where too much irradiance absorbed interferes directly with chloroplast integrity and thereby fundamental processes of photosynthesis (Fig. 28.2). Whereas short-term photoinhibition is reversible, long-term exposure to excess energy of very intense light is permanently damaging (Ehleringer and Cooper, 1992).

28.2. PLANT TYPES AND PHOTOSYNTHETIC CAPACITY

28.2.1. Research on the ecophysiology of desert photosynthesis

Much ecophysiological knowledge of photosynthesis in desert plants comes from research in the Sonoran and Mojave deserts of North America. In recent years, however, there have been an increasing number of studies on desert ecophysiology in southern Africa and South America. Whereas the great majority of world deserts have climatic regimes dominated by monsoonal summer storms or biseasonal rainfall patterns, a disproportionate amount of ecophysiological research has been conducted in deserts exhibiting a cool, winter rainfall regime and hot, dry summers (Fig. 28.3). Environmental challenges arise because winter temperatures are often too low for growth during

Terrestrial Photosynthesis in a Changing Environment: A Molecular, Physiological and Ecological Approach, ed. J. Flexas, F. Loreto and H. Medrano. Published by Cambridge University Press.

Fig. 28.1. Desert landscapes showing widely spaced perennials. **(a)** Desert scrub dominated by evergreen creosote bush (*Larrea tridentata*) on a bajada in the Sonoran Desert of Southern California. **(b)** Desert scrub dominated by drought-deciduous shrubs (*Ambrosia dumosa* and *Atriplex* spp.) on a playa margin in the Mojave Desert of Southern California. **(c)** Interior region of the Sonoran Desert of Baja California with stem-succulent cacti (*Pachycereus pringlei* and *Myrtillocactus cochal*) and boojum (*Fouquieria columnaris*) and leaf succulent species of *Agave*. **(d)** The Succulent Karoo of southwestern Africa dominated by leaf succulent shrubs.

months when rainfall is most likely to occur. For much of the Mojave Desert in the United States, for example, three-quarters of annual rainfall has already occurred before temperatures warm sufficiently in spring to allow new growth (Esler and Rundel, 1999). In contrast, major portions of the Sonoran Desert experience a rainfall regime that is either centred on hot summer months or biseasonal, with distinct peaks of rainfall in both winter and summer and allowing for an autumn growing season. A similar climate regime occurs as well in the Chihuahuan Desert in Mexico, the Monte in Argentina and the Nama Karoo in South Africa.

28.2.2. Maximising photosynthesis

When ecologists first began to study the desert biome a century ago, they naturally first assumed that external form and anatomy of desert plants could be viewed as adaptations to limit water loss – to conserve water under the arid conditions present. Many general biology textbooks and popular accounts of the natural history of desert plants continue to focus on this basic principle. Such texts often emphasise adaptations to water stress by categorising desert species as drought evaders, or drought avoiders, as contrasted with drought endurers. That simplistic conceptual framework has been gradually replaced with a very different view of desert-plant adaptive structure and ecophysiology. Although it is true that desert plants clearly show adaptations to help them conserve and use water efficiently, these adaptations are generally ones that operate through physiological mechanisms rather than gross structural or anatomical traits. Instead, researchers have now shown that many adaptive traits of leaves and photosynthetic stems in desert

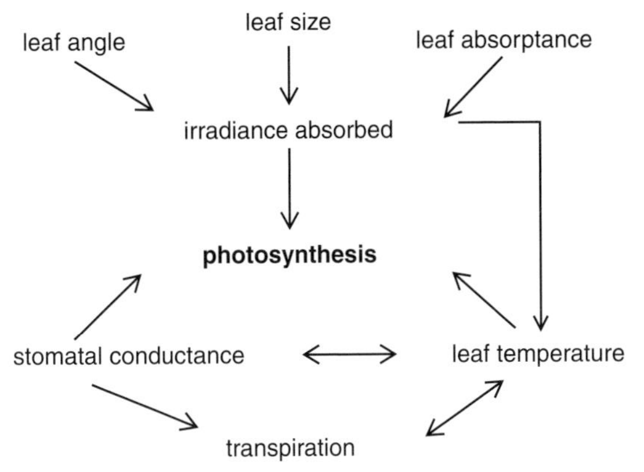

Fig. 28.2. Leaf energy balance as influenced by environmental and ecophysiological factors.

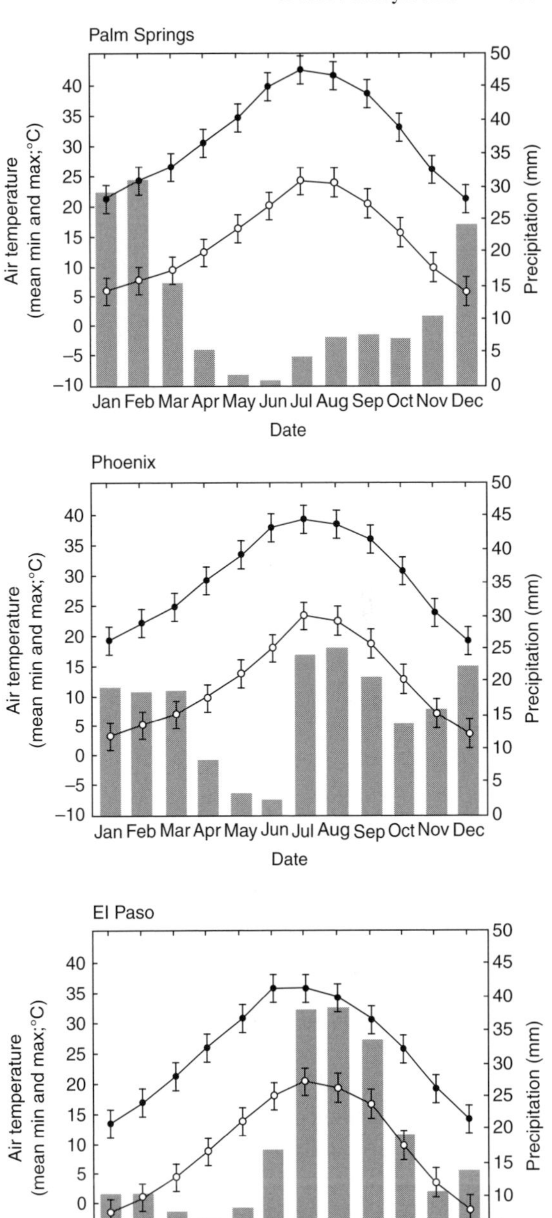

Fig. 28.3. Climographs for mean monthly rainfall and mean monthly maximum and minimum temperatures for Palm Springs (129 m elevation, mean annual precipitation 144 mm) in the winter rainfall area of the Sonoran Desert in California, Phoenix (334 m elevation, mean annual precipitation 207 mm) in the biseasonal area of rainfall in the Sonoran of Arizona and El Paso (1194 m elevation, mean annual precipitation 218 mm) in the summer rainfall Chihuahuan Desert of western Texas.

plants have evolved to regulate the energy balance of leaves and maximise rates of photosynthesis during those periods when soil water is available.

If we look closely at the design of desert leaves we can get many clues about the importance of maximising photosynthesis under conditions when and where water is available. Leaves of most desert species, including all growth forms from large woody perennials to the tiniest annuals, are amphistomatic, having relatively high densities/frequencies of stomata on both upper and lower surfaces (Gibson, 1996). This occurrence of stomata at high densities/frequencies and often in equal numbers on both leaf surfaces seems at first counterintuitive. If minimising loss of water vapour was the principal strategy of a desert leaf, one would first expect to find low densities/frequencies of stomata and a restriction to the lower, shaded leaf surface.

Under conditions where water and light are not limiting, photosynthesis per unit of leaf surface can be substantially increased by having stomata on both surfaces, to allow rapid diffusion of CO_2 (Mott *et al.*, 1982). In contrast, plants often grow in communities where canopies are closed and light does not readily reach lower leaves. This is the case in tropical rainforest, temperate rainforest and oak woodland habitats. Plants in alpine and subalpine fellfield habitats, where canopies are relatively open and growth is limited to a short summer season, share the desert-plant structure of having stomata on both upper and lower leaf surfaces (Rundel *et al.*, 2005; Gibson *et al.*, 2008).

A transection of a desert leaf typically has palisade layers of mesophyll cells adjacent to upper and lower epidermis, i.e., isolateral or isobilateral mesophyll (Fig. 28.4). The multiple layers and efficient packing of these cylindrical cells

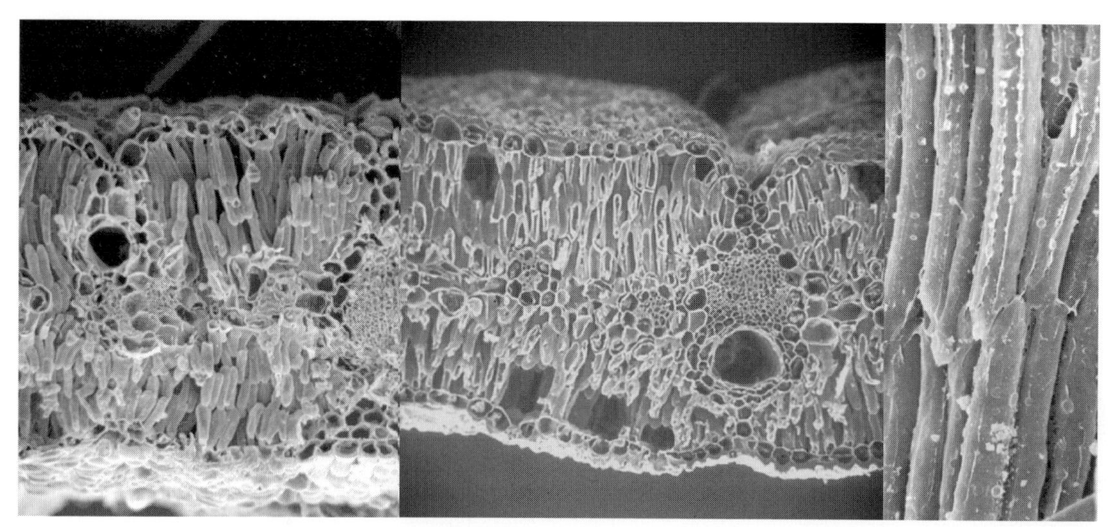

Fig. 28.4. Scanning-electron-micrograph cross-sections of desert leaves. Typical desert leaves have palisade chlorenchyma on both upper and lower sides of mesophyll (isolateral mesophyll), which yield a very high value of internal area of mesophyll plasma membrane for CO_2 diffusion per surface area of leaf (high A_{mes}/A), and permits rapid CO_2 uptake. Left, *Flourensia cernua* (Chihuahuan Desert). Centre, *Ericameria linearifolia* (Sonoran Desert); note large air spaces in mesophyll inside stomata on both surfaces of the leaf blade. Right, *Acacia aneura* (mulga of Australian deserts), elongate palisade mesophyll of a phyllode.

allow a high density of chloroplasts per unit of leaf area. Leaves with a high density of palisade cells are characterised as having a high ratio of mesophyll cell-surface area to leaf area (high A_{mes}/A_{leaf}), and have more chloroplasts per A_{leaf} than species with lower ratios (Slaton and Smith, 2002; Nobel, 2005). There may be as many as 7,000 palisade cells mm^{-2}, with greater than 90% of the cell walls exposed to intercellular air spaces (Gibson, 1996) and A_{mes}/A_{leaf} ratios sometimes exceeding 60 maximising liquid-phase CO_2 diffusion. As long as sunlight is not limiting, leaves with this internal design are capable of very high rates of photosynthesis. Combining what we see with stomatal densities/frequencies and positions and structure of the chl.-rich palisade cell layers, it seems clear that structure of the typical desert leaf has evolved to maximise rates of carbon uptake and thus photosynthesis (Gibson, 1998).

28.2.3. Stems as photosynthetic organs

Whereas leaves are the typical photosynthetic organ, many nonsucculent desert perennials utilise green stems as an important or even the primary photosynthetic organs (Gibson, 1983). These species commonly have very small ephemeral leaves, often scalelike, characteristically present only briefly during the year and hence most days appear leafless. The major component of the annual carbon uptake comes from green stems. At least 36 different dicotyledonous families plus the gymnosperm *Ephedra* have desert species that rely on stem photosynthesis (Gibson, 1996).

Aphylly (the leafless condition) has often been explained as a strategy to reduce transpiration by eliminating A_{leaf}, and thus a morphologic adaptation for water conservation. This hypothesis is clearly not true for many aphyllous species, which have green stems that display an impressive amount of total photosynthetic surface (Adams and Strain, 1968). An alternative hypothesis is that species with photosynthetic stems have evolved to become leafless to prevent leaf shading of the green stem surface (Gibson and Nobel, 1986). Because a cylindrical stem is less efficient than a planar leaf for intercepting direct sunlight, and only half of the stem can be fully illuminated at any particular moment, self-shading is a problem even in the absence of leaves. This explanation, however, begs the question of why photosynthetic stems have evolved.

Although one might assume that a green stem should lose relatively little water because it is essentially leafless, some desert plant stems have moderately high transpiration rates, with stomatal conductance levels similar to that of leaves on the same plant (Comstock and Ehleringer, 1988). Consequently, these shrubs, with their high total

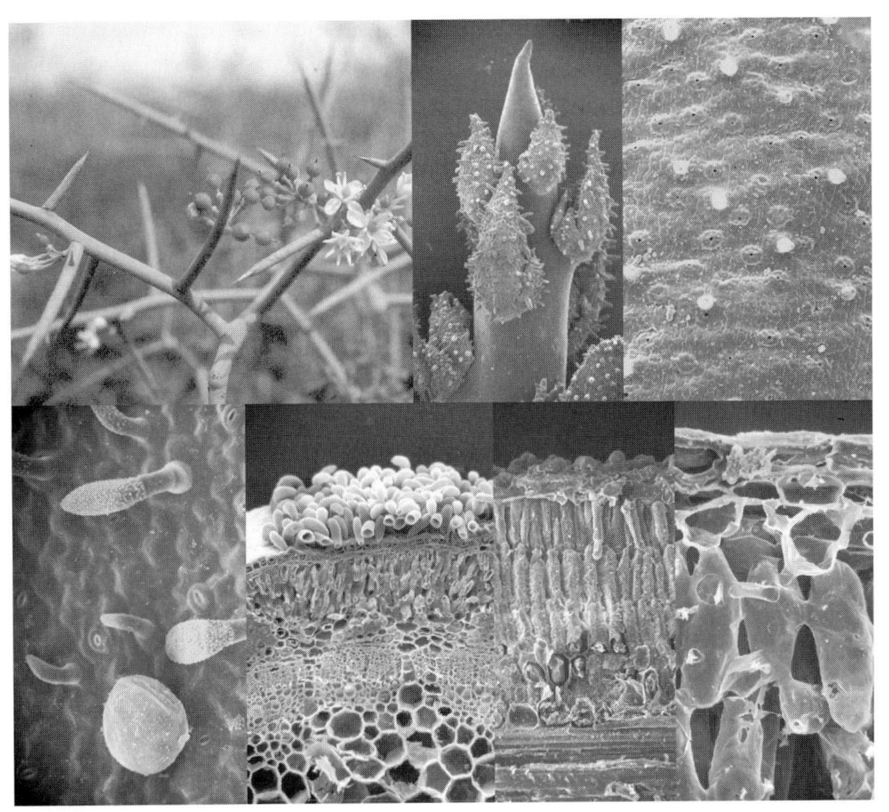

Fig. 28.5. Nonsucculent photosynthetic stems. Top row, *Koeberlinia spinosa* var. *wivaggii* (Chihuahuan Desert); left, typical aphyllous, spinescent, green-stemmed design; centre, SEM of young shoot showing ephemeral leaves before they abscise; right, SEM of stem surface showing stomata and widely scattered, conic hairs. Bottom row, left, *Salazaria mexicana* (Sonoran Desert), SEM of stem surface with stomata and three types of hairs; second, *Senna armata* (Sonoran Desert), SEM cross-section of stem, with palisade-like cortical chlorenchyma beneath a patch of hairs; third, *Ephedra californica* (Mojave Desert), longisection SEM with layers of palisade-like chlorenchyma; right, *Zilla spinosa* (Sahara Desert), SEM cross-section of stem, with large air spaces within palisade-like chlorenchyma.

stem surface area, also have relatively high water requirements, which may explain why the majority of these species either inhabit washes and other water drainage channels or have deep roots. Many green-stemmed shrubs and trees have relatively wide vessels in the xylem to efficiently transport water from wet soil to the transpiring shoots (Gibson, 1996). Stomatal densities/frequencies are typically lower on green stems than on desert leaves, but stomatal pores are slightly larger, permitting CO_2 diffusion into the airy photosynthetic cortex of the stem (Fig. 28.5).

Photosynthetic stems are often covered with an extremely thick cuticle, whose role has been interpreted as reducing transpiration. However, the thick cuticle may also help to protect the stem surface as it increases in diameter and thereby stretches or splits during secondary growth. In many species, surface waxes give a grayish cast to the stem, suggesting that a thick cuticle may reduce absorptance, thereby lowering stem temperature. The cuticle contains UV-absorbing phenolics and may prevent radiation damage to long-lived photosynthetic organs.

Nearly all photosynthetic stems of desert plants have sunken stomata (unlike non-succulent desert leaves, which do not; Gibson, 1996), and scientists have often concluded that sunken stomata or stomata hidden in longitudinal furrows are an adaptation to reduce transpiration. Any adaptation that lowers water-vapour conductance simultaneously lowers CO_2 conductance and net photosynthesis (A_N), a result that is counterproductive for a stem that has become specialised as the chief photosynthetic organ. In addition, if sunken stomata have reduced stomatal conductance (*gs*), then deep pores have the greatest proportional effect when growing conditions are optimal, i.e., when soils are wet. The effects of recessing stomata would, however, be negligible during drought, when the adaptation presumably would have been the most crucial for reducing transpiration. When the probable slowing effect of sunken stomata on water-vapour loss is calculated, most types of sunken stomata yield physiologically insignificant results because the distance of the pathway is substantially increased in only a few extreme examples (Gibson, 1983, 1996). There are, in fact, reasons to hypothesise that sunken stomata are an adaptation to protect the guard cells from direct contact with dry air, thereby preventing them from closing under arid conditions, hence sunken stomata are an adaptation to improve CO_2 uptake by stems. Moreover, furrowing of stems greatly increases the potential photosynthetic area of a cylinder and therefore may be an adaptation to maximise A_N.

A desert shrub that has both drought-deciduous leaves and photosynthetic stems would seem to have an excellent overall strategy for annual carbon gain, and this combination is very common, particularly in semiwoody Asteraceae and Fabaceae. Because of its anatomical design, a leaf is optimally suited to intercept maximum solar radiation, to take up maximum CO_2 with the lowest resistance and to regulate temperature better than a stem. However, thin leaves are more prone to herbivory or damage by wind than are stems, which are mechanically strengthened. Integrated WUE (carbon gain per unit water loss) of green stems is typically higher than that of the leaves, and stems are better at water conservation and therefore being persistent photosynthetic organs during drought months (Comstock and Ehleringer, 1988).

28.2.4. Crassulacean acid metabolism

Crassulacean acid metabolism is widely present in desert succulents growing in arid environments, especially where freezing is absent or very limited, and where midday high temperatures and low RH lead to stomatal closure that restricts CO_2 entry. Lower atmospheric water-vapour deficits at night have selected for reverse stomatal behaviour with the associated increase in WUE, as carbon fixation occurs at lower ambient VPDs and high efficiency of carbon fixation by PEPC. With CAM and night-time opening and daylight closing of stomata, the reciprocal of typical plants, many water-conservation benefits accrue, because during the night ambient temperature is lower, humidity is higher and thus water loss is lowered by a reduced driving force for diffusion (Gibson and Nobel, 1986; Nobel, 1994). During daylight, organic acids formed and stored in darkness are the carbon source for photosynthesis, and thus there is no need for stomata to open. This metabolic system makes CAM succulents exceptionally efficient in their use of water for growth, as well as allowing for good stomatal control at night because of the efficiency of PEPC for CO_2 uptake.

Terrestrial CAM plants are well represented in deserts both as stem succulents and as subshrubs, perennial herbs and small woody species with succulent leaves. Because the primary advantage of this pathway is the ability to continue photosynthetic carbon gain under extreme water stress, summer-dormant herbaceous perennials are seldom CAM, although there are small summer-dormant succulents that utilise CAM (Rundel *et al.*, 1999). Under terrestrial conditions, annuals are only rarely CAM, perhaps because the ephemeral growth selects against slow growth rates. Notable exceptions are annual species of *Mesembryanthemum* (Aizoaceae), which switch developmentally from C_3 to CAM as they experience water stress (Edwards *et al.*, 1996). In Cactaceae, seedlings are typically C_3 before switching permanently to CAM (Altesor *et al.*, 1992); however, it is unclear whether this is owing to stress or a programmed developmental change or both.

Biogeography of CAM desert plants shows interesting patterns of adaptation. Including tropical environments, where CAM often exists in epiphytes, at least 40 plant families have evolved species with CAM. While Cactaceace and *Agave*, classic CAM desert succulents (Nobel, 1994), do not occur in Old World deserts, there has been a remarkable parallel evolution of succulent *Euphorbia* and *Aloe* in semi-arid habitats, which show many of the same morphologic adaptations of a thick cuticle and succulent water-storage tissues as well as CAM.

Vacuolar storage demands of CAM metabolism require a tight evolutionary coupling of this pathway with succulent tissues, a linkage present in the majority of Cactaceae, Crassulaceae, succulent asclepiads and *Euphorbia*, and monocot genera such as *Agave* and *Aloe*. However, this form-function linkage cannot always be assumed, even among desert plants. Some tall-stem succulents with deciduous leaves are typically C_3, including species of *Fouquieria* (Fouquieriaceae), *Pachycormus* (Anacardiaceae; Gibson, 1981) and *Bursera* (Burseraceae) in the deserts and dry forests of Mexico. Hundreds of species of leaf-succulent evergreen shrubs and subshrubs in the Succulent Karoo include species with CAM, CAM-flexible and C_3 forms of metabolism (Rundel *et al.*, 1999).

Whereas it is often assumed that CAM plants have lower daily rates of carbon fixation than those of morphologically similar C_3 plants growing in the same habitat, this may not always be the case. Experimental studies comparing two South African desert succulents, *Cotyledon orbiculata* (CAM) and *Othonna opima* (C_3), found that daily net carbon fixation was similar when water was readily available, and that the two species had similar WUE under hot conditions with low humidity (Eller and Ferrari, 1997). Under non-limited resource conditions, CAM plants may have surprisingly high rates of NPP comparable with that of C_3 crop plants (Nobel, 1991).

28.2.5. C_4 metabolism

C_4 metabolism has evolved in at least 40 different plant families, many of which can be found in desert or semiarid dryland communities. In most respects C_4 metabolism is virtually identical biochemically with CAM metabolism, but with daylight stomatal opening and spatial separation of carbon fixation by PEPC and Rubisco and in different cells. Like C_3 species, desert C_4 species are amphistomatic, although mesophyll structure differs radically because C_4 species exhibit Kranz anatomy with a highly derived bundle sheath within which glucose is manufactured (Johnson and Brown, 1973; Dengler and Wilson, 1999).

C_4 metabolism provides two advantages that can be extremely useful for desert plants. First is that the highly efficient manner of carbon uptake by the initial C_4 enzymes is not inhibited by high temperatures. Plants with more typical C_3 metabolism have an optimal temperature for photosynthesis, but these rates decline at higher temperatures as the Calvin-cycle enzymes become less efficient and release CO_2 through photorespiration. C_4 plants avoid loss of carbon via photorespiration and are thus highly efficient at high temperatures. The desert bunchgrass *Hilaria rigida* has the ability to achieve exceedingly high rates of photosynthesis (Nobel, 1980), although this is probably not typical of most desert species. Another potential advantage of C_4 metabolism comes from the highly efficient fixation of carbon by the C_4 enzymes; with this efficiency, C_4 plants can alternatively open their stomata very little and still achieve the same amount of carbon uptake present in typical plants without this metabolic system. Under these conditions, C_4 plants can exhibit very high rates of WUE to aid their success in desert environments.

C_4 metabolism represents an important ecological strategy in a number of desert shrubs, most notably Chenopodiaceae, e.g., *Atriplex*, which often are important community dominants in areas of alkaline or saline soils. In xerophalophytes, the key adaptation of C_4 metabolism is the ability to maintain growth under high summer temperatures and drought conditions at a time when C_3 shrubs are dormant, thus reducing interspecific competition for water (Caldwell *et al.*, 1977), and they can take advantage of rare summertime rains. Maximal rates of photosynthesis in these desert C_4 shrubs are moderate and typically no higher than that of co-occurring C_3 shrubs, but their WUE is far greater.

There are two other ecologically important groups of desert species with C_4 metabolism. These are summer annuals and summer-active perennial grasses. C_4 annual plants are a conspicuous component of the summer flora of desert regions with summer rainfall regimes, and a smaller set of these species are present in winter rainfall deserts as well, where they germinate in response to irregular summer rains (Mulroy and Rundel, 1977). Desert annual florulas commonly include C_4 species of Amaranthaceae, Asteraceae, Euphorbiaceae, Nyctaginaceae and Poaceae, and they are able to achieve extremely high rates of photosynthesis owing to the absence of photorespiration in C_4 metabolism.

28.3. ECOPHYSIOLOGY AND ENVIRONMENTAL FACTORS

28.3.1. Leaf energy balance

The combined stresses of high summer temperatures and low soil-water availability have led to a variety of important morphologic and physiologic traits of desert leaves, aiding them in maintaining a favourable energy balance under these conditions. Leaf size, angle of orientation and surface reflectance can all affect leaf temperature (Fig. 28.2), as do ecophysiological traits of g_s for water loss through transpiration.

LEAF SIZE

Leaves and leaflets of desert plants are overall smaller than those of plants growing in wetter habitats, importantly in terms of their widths. A survey of blade size in nonsucculent woody desert plants from Southern California showed that 76% have blade widths less than 5 mm, and nearly 90% have widths less than 10 mm (Gibson, 1996). This is not automatically an adaptation to reduce the total A_{leaf} on a plant, because there may be no difference in total A_{leaf} between plants with many small, narrow leaves versus those with fewer broad leaves. The adaptive significance of narrow leaves is that they closely track ambient temperatures by the simple physical process of convective heat transfer to the surrounding air across a very thin boundary layer, a mechanism that requires no plant energy or transpirational water use. Broader leaves in contrast cannot cool efficiently through convection, and without substantial transpiration can quickly reach lethal tissue temperatures under hot summer conditions (Gates and Papian, 1971).

There is a second aspect of adaptive advantage in narrow leaves for desert plants. If narrow leaves can be kept cooler than a broader leaf, they experience a smaller water-vapour gradient from leaf to surrounding air, and thus lose less water when their stomata are open because of the smaller driving force for transpirational water loss (Nobel, 2005). In addition, under high levels of ambient temperature and solar irradiance, desert plants are better able to maintain leaf temperatures closer to optimum thermal conditions, which are about 25–30°C for photosynthetic enzymes (Knight and Ackerly, 2003b).

Can desert plants have broad leaves? The answer is yes, but broad leaves can only exist if the plant can use a reliable source of water so that transpiration can cool leaves and keep them at sublethal temperatures during summer. Such conditions exist for a plant with a root system that taps a permanent supply of groundwater, or for palms at desert oases. There are a number of desert species with broad leaves growing along washes (wadis, arroyos) with subsurface water, able to cool their leaves as much as 5–10°C below ambient temperatures in summer through high rates of transpiration. A broad-leaved species can be successful during hot days or summertime with deep or fleshy belowground organs, as seen with desert cucurbits such as *Citrullus colocynthis* in Saudi Arabia (Althawadi and Grace, 1986) and *Cucurbita*, *Rumex hymenosepalus* and *Datura* in North American deserts. These species make little or no attempt to limit their water use when moisture is available and maintain high rates of photosynthesis with their broad leaves under hot ambient conditions. Closing stomata when water is no longer sufficient results in overheating of leaves and to plant senescence.

Comparative studies of the SLA of desert leaves (blade area per unit of dry mass) have shown that desert leaves are not only typically small but also have a lower SLA than congeners in nondesert environments (Knight and Ackerly, 2003a). This combination of small size and relatively low SLA allows these leaves to better withstand high-temperature stress and recover PET. Relatively low values of SLA are associated with stress-tolerant life-history strategies and may aid in nutrient retention and long-term photosynthetic NUE.

LEAF ORIENTATION

Although most desert plants have leaves in fixed positions, the orientation of leaves can have a profound impact on solar irradiance. Blade orientation in desert plants is only rarely horizontal and may often deploy at steep angles. Whereas one possible interpretation of vertical orientation is to minimise solar irradiance and heating at midday (Gates, 1980), this orientation is better interpreted as a strategy to maximise diurnal photosynthesis (Nobel, 1982; Gibson, 1996, 1998). As described above, vertical orientation can have little impact on the temperatures of narrow leaves that closely track ambient temperatures. Instead, vertical orientation of leaves allows for high levels of solar irradiance on a majority of the plant's leaves during both the morning and afternoon hours, when temperatures are lower and the vapour-pressure gradient from leaf to air is relatively low. Vertical orientation also reduces potential photorespiration losses at high levels of midday solar irradiance (Björkman *et al.*, 1980; Ehleringer and Cooper, 1992).

Certain desert plants exhibit a fixed leaf orientation that has been shown to maximise solar-radiation absorption and thus photosynthesis. One example is *Larrea tridentata* that displays its foliage toward the southeast (or equator) with steeply inclined branches. This orientation minimises self-shading of leaves during morning hours, when leaf-water potentials are highest and maximises total daily photosynthesis by the canopy especially during winter months (Neufeld *et al.*, 1988). Not surprisingly, species of *Larrea* in the Monte Desert of Argentina use the same strategy by orienting photosynthetic surfaces toward the north and the equator (Ezcurra *et al.*, 1991). The pattern of foliage display over seasonal cycles in *Larrea* absorption of solar irradiance is more complex, however, because of leaflet movement associated with drought stress (Ezcurra *et al.*, 1991).

An unusual case of leaf orientation to maximise photosynthetic carbon gain can be seen in *Pachypodium namaquanum*, a peculiar arborescent stem succulent in the Succulent Karoo of South Africa and adjacent Namibia. This species is characterised by a striking curvature toward the north of the terminal 20–60 cm of the trunk, with a single terminal cluster whorl of drought-deciduous leaves angled at a mean inclination of 55° from horizontal (Rundel *et al.*, 1995). Fixed leaf orientation in *P. namaquanum* more than doubles the amount of solar irradiance received during the winter months when leaves are present.

Extending this to photosynthetic stems, a favoured direction of orientation can also be seen in many cactus species, as with the cacti *Ferocactus cylindraceus* in the

Sonoran Desert and *Copiapoa cinerea* in the Atacama Desert (Lewis and Nobel, 1977; Ehleringer *et al.*, 1980). In these cases, the solitary succulent stems of the cacti lean toward the equator, maximising solar radiation on the growing tip and flower buds while minimising heat loads to the sides of the stem.

A number of desert annuals exhibit diaheliotropism, i.e, solar tracking by leaves, with leaf movement that allows them to be oriented perpendicular to direct sunlight at all times throughout the day. This trait maximises interception of solar radiation, and thus total daily photosynthesis (Wainwright, 1977; Mooney and Ehleringer, 1978; Forseth and Ehleringer, 1983). For annuals of desert communities, diaheliotropism becomes more frequent as the length of the growing season decreases. As predicted, leaves perpendicular to the sun's direct rays for tracking can achieve high photosynthetic rates throughout the day, as compared with leaves parallel to the sun's rays, which have reduced leaf temperatures and transpirational water losses. Rarely, some desert annuals use an avoidance variant of this solar tracking termed 'paraheliotropism', orienting perpendicular to the sun in the morning and afternoon hours when solar intensity is lower, and orienting edge-on to the sun under intense midday conditions.

LEAF ABSORPTANCE

Another way by which leaves of desert plants can reduce the amount of solar radiation they receive is through leaf surface hairs, scales or waxes, all which can increase reflectance and therefore reduce absorptance of solar radiation (Ehleringer and Mooney, 1978). Such leaves are grey, silvery or whitish in appearance. The older literature often assumed that thick mats of reflective trichomes on the leaf surface of desert plants were an adaptation to lower transpiration rates by increasing the boundary layer thickness around the blade. Now researchers acknowledge a more important role of these trichomes, at least in larger leaves, is to reduce the heat load on leaves. The instantaneous benefit of lower leaf temperatures resulting from increased IR reflectance would be to reduce the vapour-pressure gradient and thus reduce transpiration (Fig. 28.2), but not by the mechanism originally assumed. The negative effect of such hairs is that they block PFD/PPF irradiance and thus sharply limit rates of photosynthesis (Ehleringer *et al.*, 1976; Ehleringer *et al.*, 1981).

The best example of a species using this reflectance strategy can be seen with *Encelia farinosa*, which forms broad, green winter leaves and smaller, silvery spring leaves approaching summer drought (Ehleringer *et al.*, 1976; Ehleringer, 1977, 1982; Ehleringer and Mooney, 1978). Under conditions of increasing drought stress, the smaller leaves of *E. farinosa* have a denser covering of trichomes. Looking in cross section, the mass of trichomes on both the upper and lower blade surfaces may be thicker than the structure of the leaf tissue itself (Fig. 28.6). These reflective hairs cut out as much as 71% of the solar radiation, that would otherwise be absorbed by the leaf. The result is a cooler leaf with a reduced loss of water through transpiration, but at a cost of a sharp reduction in rates of photosynthesis.

28.3.2. Photosynthetic capacity

Despite their stressful environment, many desert plants achieve surprisingly high maximum rates of net carbon assimilation under optimal field conditions of available soil moisture, bright sun, and moderate air temperatures. Woody species tend to have maximum net rate of photosynthesis (A_{max}) above 20 μmol m^{-2} s^{-1} (Pearcy *et al.*, 1974; Sharifi *et al.*, 1997; Gibson, 1998), which is equivalent to that of many agricultural row crops under irrigation. Likewise, desert riparian trees have relatively high instantaneous rates of photosynthesis (Horton *et al.*, 2001). However, among the woody desert plants, investigators have measured some relatively low rates of leaf photosynthesis. Most notable is creosote bush, *Larrea tridentata*, the dominant evergreen shrub of North American warm deserts, with typical A_{max} of 14–16 μmol m^{-2} s^{-1} under benign conditions (Rundel and Sharifi, 1993), with highest maximum value of 21.8 μmol m^{-2} s^{-1} (Franco *et al.*, 1994). But despite its blade anatomy, i.e., amphistomatic leaves and high A_{mes}/A_{leaf} ratio, which could enable reasonably high A_{max}, *Larrea tridentata* typically photosynthesises at one-quarter of that rate during drought (Sharifi *et al.*, 1997). *Larrea* and probably other long-lived evergreen desert shrubs have a strategy for gas exchange in which high instantaneous rates of net carbon gain have been traded for much lower rates of net carbon gain throughout the year. This slow but steady pattern of gas exchange keeps pace with the cost of maintaining evergreen leaves (Ehleringer, 1985).

The A_{max} of desert annuals and herbaceous perennials are typically higher than rates in woody species and commonly have rates of 30–45 μmol m^{-2} s^{-1}, which is at the high end of the range for any C_3 plant (Mooney *et al.*, 1981; Toft and Pearcy, 1982; Werk *et al.*, 1983; Gibson, 1998). Some of the highest rates of A_N have been measured under optimal field conditions in the Sonoran and Mojave deserts of the United States. *Camissonia claviformis*, a C_3 winter annual growing in Death Valley, had a A_{max} of 59 μmol m^{-2} s^{-1} (Mooney

Fig. 28.6. Desert brittlebush (*Encelia farinosa*) of California. Left, a comparison of a green winter-leaf form (left) and a smaller, highly reflective, silvery spring-leaf form (right) from the same shrub; the silvery leaf can remain several degrees cooler than the green leaf under equivalent conditions of full sun and high ambient temperature. Right, scanning-electron-micrograph cross-section of a silvery leaf showing dense layers of curly hairs, which reflect essentially all infrared radiation, thus reducing the heat load but also reducing PPF by as much as 50% (from Gibson, 1996).

et al., 1976). Laboratory measurements of rates higher than 50 μmol m^{-2} s^{-1} have been measured in several winter annuals (Forseth and Ehleringer, 1983; Werk *et al.*, 1983). Under controlled conditions, the highest photosynthetic rate recorded at ambient CO_2 concentrations was 81 μmol m^{-2} s^{-1} in the C_4 summer desert annual *Amaranthus palmeri* (Ehleringer, 1983).

Maximum photosynthetic rates in stem tissues of woody desert shrubs and trees are lower than those of leaf tissues, but nevertheless respectably high. These rates measured under field conditions range from about 10–24 μmol m^{-2} s^{-1} (Ehleringer *et al.*, 1987; Nilsen *et al.*, 1989, 1996). These rates apply, however, only to young first- and second-year stems; older, strongly lignified green stems in arborescent legumes such as *Psorothamnus spinosus* and *Parkinsonia florida* (=*Cercidium floridum*) are very low.

28.3.3. Seasonal-temperature acclimation

Many desert plants have an adaptation to deal with their dynamic thermal environment using a physiologic strategy involving seasonal changes in the biochemical responses of photosynthetic enzyme systems to temperature (Downton *et al.*, 1984). This seasonal acclimation was first shown in several species from the Negev Desert (Lange *et al.*, 1974) and later in *Larrea tridentata*, where optimal temperatures for photosynthesis changed from 22°C during winter to 32°C during summer (Mooney *et al.*, 1978). Other studies have found seasonal-temperature acclimation in the C_4 halophyte *Atriplex lentiformis*, the perennial C_4 bunchgrass *Pleuraphis rigida* (=*Hilaria rigida*) (Nobel, 1980), the arborescent monocot *Yucca brevifolia* (Smith *et al.*, 1983) and the desert fern *Notholaena parryi* (Nobel, 1978). Researchers in the past have failed to find evidence of such acclimation in drought deciduous and winter deciduous shrub species, suggesting that it might not be associated with these phenological strategies. However, a recent experimental growth study in South Africa with three evergreen leaf-succulent and a drought-deciduous shrub species have found temperature acclimation in the form of increasing optimal temperatures for photosynthesis (Bowie *et al.*, 2000; Wand *et al.*, 2001).

Mooney *et al.* (1978) and Lange *et al.* (1974) suggested that seasonal-temperature acclimation in desert plants is primarily the result of altered photosynthetic capacity and not stomatal activity. A more recent study has suggested that seasonal changes in stomatal behaviour may be more important to photosynthetic temperature acclimation in *Larrea* than previously recognised (Ogle and Reynolds, 2002).

28.3.4. Water-use efficiency

As soils dry and shoots are therefore under greater water stress, photosynthetic rates of desert species are reduced but remain comparatively high compared with species from many other ecosystems. At levels of water stress at which the leaves of most non-desert plants would die or experience little or no net carbon gain, leaves of desert plants still experience substantial rates of CO_2 uptake. This ability reflects one of the most important adaptations of desert plants: through physiological and not structural mechanisms, desert leaves can open stomata at much greater water stress than can typical plants, and they thereby take advantage of leaf architecture designed for high photosynthetic rate.

Water-use efficiency is one measure of water conservation in desert plants. As the shoot experiences greater water stress, the ratio of moles of CO_2 gained to moles of H_2O lost per photosynthetic surface area would be expected to increase if plants use water more efficiently as water stress increases. There are several ways to express WUE, with three of these being point measurements. These are: (1) instantaneous WUE, which is the ratio of assimilation to transpiration; (2) intrinsic WUE, which is the ratio of assimilation to stomatal conductance; and (3) the ratio of internal CO_2 concentration (C_i) to ambient CO_2 (C_a), which reflects the degree of stomatal control. There is also widespread use of integrated WUE, that is, WUE for an entire growing season, which is estimated using carbon-isotope ratios ($\delta^{13}C$), with less negative values of $\delta^{13}C$ indicating higher WUE (Rundel *et al.*, 1988). Floristic surveys of $\delta^{13}C$ have shown that desert plants exhibit a wide range of WUE, often with interspecific and intraspecific relationships to plant functional group, plant longevity, foliage morphology, drought conditions and plant age (Ehleringer and Cooper, 1988; Schuster *et al.*, 1992; Rundel and Sharifi, 1993; Sandquist *et al.*, 1993; Nilsen and Sharifi, 1997; Rundel, 1999).

Relatively high values of instantaneous and intrinsic WUE can be observed among desert ephemerals when photosynthetic rates are very high and the VPD between leaf and air is low (Huxman *et al.*, 2008). When soil water reaches field capacity such as during a rainy season, non-succulent desert leaves use water somewhat more efficiently than do plants of more mesic habitats because of high rates of A_{max} in desert leaves, although they nevertheless have reasonably high transpiration rates. During the course of a day, the WUE of many desert plants characteristically decreases from midmorning when A_N is typically highest, to late afternoon when the photosynthetic rate can be substantially lower but stomata are still open. Therefore, there is no evidence indicating that desert species are necessarily superior to plants from wetter habitats in limiting water loss.

Dating back to the seminal research on xerophytes by Maximov (1931), biologists have been well aware that desert plants can exhibit high transpiration rates when soil water is plentiful, but they are conservative in their water use as the soil dries. For a typical non-succulent desert plant, stomata open daily at levels of leaf hydration that plants from mesic habitats would not tolerate and would result in leaf abscission. Many desert shrubs show substantial uptake of CO_2 even when soils are relatively dry and shoot water potentials drop to –5 MPa or below, well below the maximum leaf water-stress resistance of typical mesophytes. Certain Zygophyllaceae are extreme examples of desert-drought tolerance, with *Larrea* in North America and *Zygophyllum* in Africa able to maintain positive rates of A_N at water potentials well below –7 MPa (Odening *et al.*, 1974; Wand *et al.*, 1999). This capability to open stomata somewhat during extreme water stress is probably what predisposes these plant species to success in arid habitats.

Non-succulent desert leaves do not contain special structural devices for storing or conserving water. In general, such leaves possess no water reservoirs, i.e., no cells with large vacuoles to supply water to leaves during periods of peak transpiration; instead, they have narrow mesophyll cells with small volumes and, therefore, the high surface-to-volume ratios that are optimal for rapid CO_2 uptake. Cuticle coats the epidermis, of course, but this wax layer is not noticeably thicker than that of plants from other ecosystems, although it does appear to be more effective in reducing cuticular transpiration than cuticle on plants from many other habitats (Nobel, 2005).

One form of specialisation for water conservation is leaves coated with resin, as in *Larrea tridentata* (Meinzer *et al.*, 1990). Resin on it leaf surfaces, 2–4 μm in thickness and present within the epidermal cells, not only reduces epicuticular transpiration but also repels herbivores. However, there is a potential negative effect of surface resins as they can seal stomata (Gibson, 1998), reducing water loss but also limiting rates of A_N. This effect may explain why few desert species have resin-covered leaves.

28.4. GLOBAL-CHANGE IMPACTS

It has been suggested that desert plants may be particularly sensitive to global change owing to the strong effect of elevated CO_2 levels on instantaneous WUE, resulting from increased rates of photosynthesis and/or lowered levels of g_s (Strain and Bazzaz, 1983). Despite this suggestion,

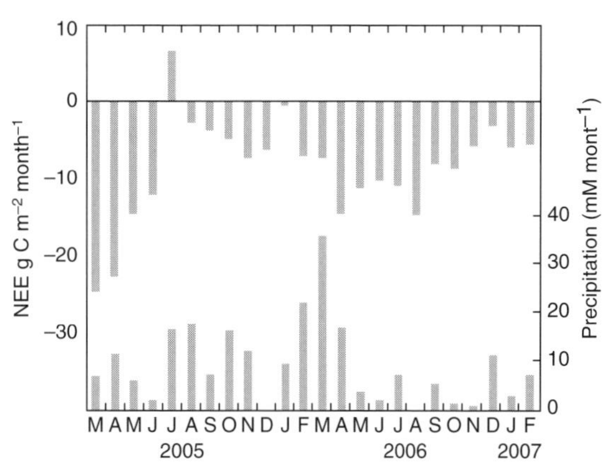

Fig. 28.7. Monthly net ecosystem CO_2 exchange (NEE, upper bars) and mean monthly precipitation in mm (lower bars) during a 2-year study period (2005–07) at the Nevada Test Site in the eastern Mojave Desert. Error bars of the soil-water content refer to the standard deviation around the mean. Redrawn from Wohlfahrt *et al.* (2008).

there have been relatively few studies relating these complex interaction increases in ambient concentrations of CO_2 with associated impacts of high temperatures that may occur with global change for desert plants.

The primary effect of higher temperatures in decreasing photosynthetic rates for C_3 plants comes as the ratio of oxygenation to carboxylation reactions of Rubisco increase as a result of comparatively greater increases in oxygen affinity and solubility at high temperatures. Given this response of Rubisco, one would expect elevated CO_2 to enhance photosynthesis relatively more at high rather than at moderate temperatures (Long, 1991). However, this suggested impact may not be so clear when elevated CO_2 interacts with coincident drought (Huxman *et al.*, 1998, 2008; Hamerlynck *et al.*, 2000). Recent studies with *Larrea tridentata* have suggested that drought can diminish photosynthetic downregulation to elevated CO_2, resulting in seasonally transient patterns of enhanced carbon gain. Comparative studies of the long-term impacts of increased ambient CO_2 on Mojave Desert shrub species suggest that responses will be complex, but increased rates of integrated photosynthesis are greatest under conditions of above-average rainfall (Naumburg *et al.*, 2003, 2004; Barker *et al.*, 2006; Housman *et al.*, 2006). Thus, water status may ultimately control the photosynthetic response of desert ecosystems to rising CO_2 (Huxman *et al.*, 2008).

There have been extensive studies of carbon and biomass cycling in desert ecosystems, generally suggesting relatively low rates of NPP (see Rundel and Gibson, 1996), although high rates of productivity do occur in desert habitats where subsurface water is not limiting (Sharifi *et al.*, 1982). There have been intriguing recent reports of high net carbon uptake by desert ecosystems that have focused media attention on deserts as the location of the long-sought 'missing sink' for atmospheric carbon sequestration. Net uptake of carbon in the Mojave Desert measured in EC studies over a 3-year period (Fig. 28.7) found rates of net carbon uptake ranging from 102 to 127 gC m^{-2} yr^{-1} (Jasoni *et al.*, 2005; Wohlfahrt *et al.*, 2008a). These high values are equivalent to the net ecosystem production of many forest ecosystems with much higher biomass.

High rates of carbon uptake have also been found in a study of saline and alkaline desert soils in western China, with reported carbon uptake values of 62–622 gC m^{-2} yr^{-1}, occurring through abiotic processes at night (Xie *et al.*, 2008). These rates are orders of magnitude greater than net accumulations of soil carbonates of desert ecosystems worldwide. Although it is not clear that there have been methodological errors in measurements from these two studies, the reported values of carbon uptake are incompatible with existing measurements of carbon pools and NPP in desert ecosystems. The comparisons suggest that gas-exchange measurements should be used with caution with appropriate validation if they are expected to quantify the magnitude of carbon sink in these ecosystems (Schlesinger *et al.*, 2009).

28.5. CONCLUSIONS

The photosynthetic strategies of desert plants are, of course, only part of the explanations for how their populations persist under stressful conditions of soil and air dryness, and high ambient and soil-surface temperature. Although much is known about germination patterns in desert annuals, there has been little recent research on germination, establishment strategies and heat resistance of desert perennials. Similarly, little is known about gas exchange and water relations of dominant species from most deserts outside of the western United States to determine whether the ecophysiological patterns presented here apply to every type of desert. Structural analyses suggest that non-succulent desert leaves worldwide present an opportunistic strategy designed to fix as much carbon as possible when soil water is available. As such, most species of desert plants are not, like succulents, water conservers in the broad sense. The ultimate desert synthesis will also impact interpretations of other biomes from which many desert taxa have evolved, which are commonly compared or contrasted with the desert biome. Comparative, ecophysiological and structure-function comparisons between congeners within and outside deserts would provide a valued means to understand the phylogenetic origins of evolutionary adaptations to desert environments. At the ecosystem level, expanded studies are sorely needed to assess the significance of deserts in global carbon cycling.

29 • Ecophysiology of photosynthesis in semi-arid environments

J. GALMÉS, J. FLEXAS, H. MEDRANO, Ü. NIINEMETS AND F. VALLADARES

29.1. CHARACTERISTIC FEATURES OF SEMI-ARID ENVIRONMENTS

29.1.1. Introduction

Arid and semi-arid environments currently cover a third of terrestrial Earth surface. By definition, 'semi-arid' refers to environments where insufficient water is available for vegetation growth. Semi-arid regions are characterised by being intermediates between desert (arid) and humid climates (Fig. 29.1), with an annual precipitation (250–1000 mm year^{-1}) typically lower than the potential evapotranspiration (PET). Furthermore, precipitation is concentrated in specific periods of the year, inducing interruptions of the growing season when water availability reaches the threshold that dramatically limits ecosystem functioning. In addition to pronounced seasonality, a third component is the unpredictability of precipitation, resulting in short drought periods even during the humid season. This unpredictability also refers to high year-to-year variability, which increases with decreasing annual precipitation, often leading to alternation of dry and humid cycles lasting several years. The inter-annual variability is also mirrored in actual evapotranspiration (AET).

The availability of precipitation and the topography of the site are the major factors determining the amount of water available for plants. However, a more detailed division of semi-arid biomes should also consider other components of climate. Temperature is a major climatic element differentiating semi-arid ecosystems. Aside from water, low temperatures become a limiting factor for plant productivity and growth in the coolest semi-arid zones, whereas heat stress can limit plant production in savannas and Mediterranean environments. According to Köppen (1936) classical classification, major biomes in semi-arid climates are savannas (Aw according to Köppen), steppes (BS) and Mediterranean-type ecosystems (Cs). Oceanic and tropical influences prevent low temperatures in Mediterranean regions and especially in savannas. Steppes are characterised by continental influences with wide seasonal and daily ranges in temperature.

In addition to climatic factors, a common feature of semi-arid ecosystems is a reduced nutrient availability in the soil and an important disturbance regime mainly owing to fire and grazing (Section 29.4.5).

29.1.2. Main environmental limits in savannas, steppes and Mediterranean ecosystems

The largest savanna areas spread over sub-tropical regions of South America, Africa and Australia. Savannas are characterised by the absence of freezing temperatures and by a relatively high annual precipitation (Köppen, 1936). In some cases, actual precipitation exceeds evapotranspiration (Baldocchi *et al.*, 2004; Fig. 29.1). These climatic characteristics, along with other features such as deep and well-drained soils, permit the establishment of relatively complex and structured communities.

Steppes have a lower annual precipitation than savannas and are characteristic transition zones between savannas and deserts, with typical examples being the Sahel fringing the Sahara in Africa and similar semi-arid areas around the Thar Desert in the Indian subcontinent. Other boundary steppes represent gradients between Mediterranean-type ecosystems to deserts, such as Tijuana (Baja California, Mexico). The world's largest steppes are in temperate Eurasia (part of Ukraine, southwest Russia and neighbouring countries in Central Asia) and North America (Great Plains). While all steppes are characterised by the virtual absence of tall woody vegetation, this ample geographical distribution involves largely different climates. Steppes are divergent in terms of co-occurrence of high temperatures and seasonal precipitation. For instance,

Terrestrial Photosynthesis in a Changing Environment: A Molecular, Physiological and Ecological Approach, ed. J. Flexas, F. Loreto and H. Medrano. Published by Cambridge University Press.

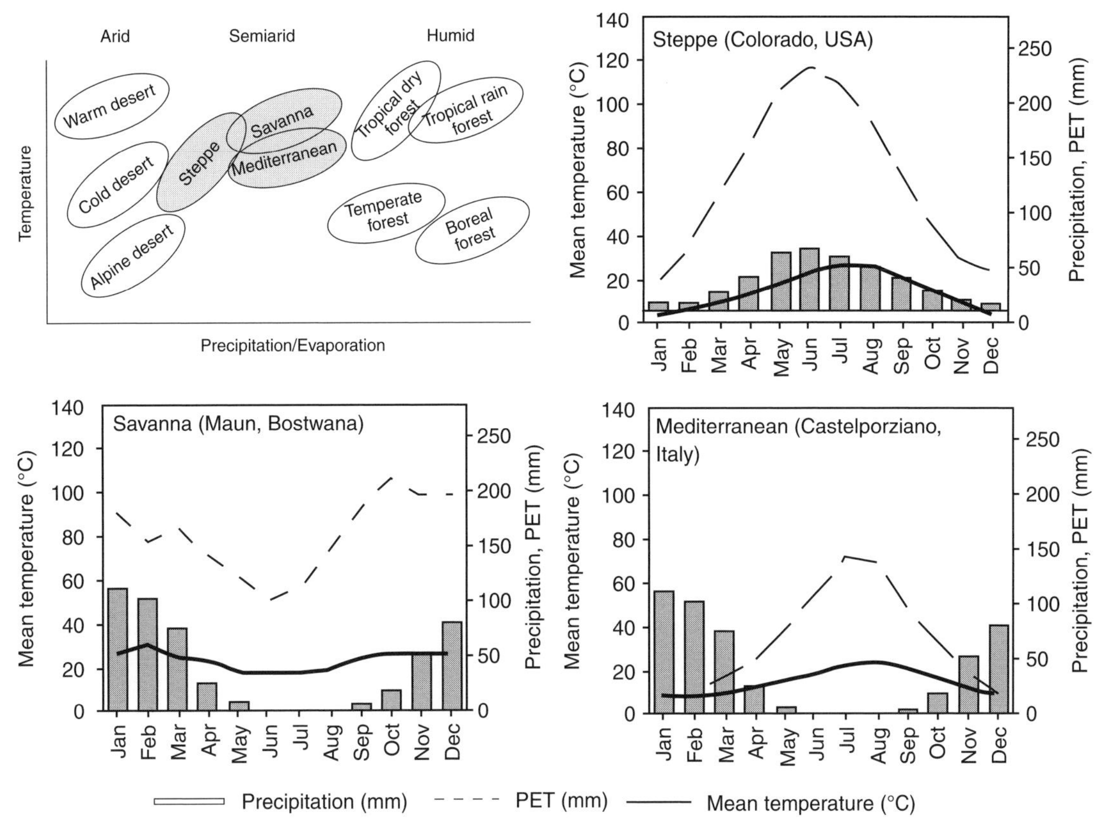

Fig. 29.1. A plot of the ecosystem types that are expected to occur along a climate gradient of temperature and the ratio of the incoming precipitation and potential evapotranspiration. The three semi-arid ecosystems included in the present chapter are shown in grey. Representative ombrothermic diagrams for each of these semi-arid ecosystems are shown below.

in Patagonian steppes the dry season coincides with hot temperatures during summer, whereas in the steppes of Great Plains and in Asian steppes optimal temperature and precipitation can co-exist in some places but not in the others. Freezing winter temperatures can significantly limit vegetation growth in continental steppes in lowlands of temperate zones in Great Plains, Mongolia, Patagonia and Russia.

The Mediterranean climate (or dry summer subtropical climate) is characterised by a hot and dry period in summer and a cool and wet period in winter. One of the most distinctive traits of this climate is the lack of any period of the year when optimal temperature and precipitation co-exist (Fig. 29.1). This type of climate occupies less than 5% of the global land surface and includes specific areas in North and South America, Europe, Australia and Africa, all of them located on the western or south-western coasts of these continents.

29.2. PLANT FUNCTIONAL AND PHOTOSYNTHETIC TYPES INHABITING SEMI-ARID AREAS

29.2.1. Major plant functional types and species co-existence

Vegetation in semi-arid ecosystems has characteristic adaptations to limited water and nutrient availabilities, spanning from physiological (photosynthesis, stomatal regulation) to structural (plant stature, leaf morphology, foliage aggregation and inclination, rooting systems) and phenological (evergreenness, period of physiological activity) features. Ecosystem formations in semi-arid regions differ substantially in species composition, but these ecosystems share some characteristic plant functional types. Although there can be a high diversity of growth forms and leaf habits in semi-arid ecosystems, the dominant growth forms are, however, low-stature woody plants and herbaceous forbs and grasses. The proportions among these major functional types depend on water and nutrient availabilities.

Although the combinations of specific micro and meso-climatic factors can be more favourable to certain plant functional types, several plant functional types with their unique adaptive responses to environmental constraints (e.g., rooting system, photosynthetic pathway type) and phenologies can successfully co-exist in the same habitat. Unique adaptive responses and exploitation of resources in different canopy and root-zone layers and at different times during the season reduce the competition for limiting resources, permitting the coexistence of contrasting plant functional types (Noy Meir, 1973). For instance, grasses and forbs rely on the water in upper soil layers that is highly variable, while shrubs and trees rely on more stable deep-soil-water stores. This conceptual explanation of the coexistence of functional types is supported by evidence from studies in steppe communities (Sala *et al.*, 1988), savannas (Knoop and Walker, 1985) and Mediterranean ecosystems (Gordon *et al.*, 1989). Direct competition for the same resources by different functional types does occur in semi-arid ecosystems, but this is exceptional (McCarron *et al.*, 2001).

29.2.2. Characteristics of major plant functional types

Herbaceous species have a relatively shallow root system, and are unable to tap deep water sources. Main adaptations to seasonally low water availability in herbaceous species are phenological, with plant growth being activated during the periods of high water availability in upper soil layers, and being arrested when water becomes scarcer. Annuals or therophytes allocate a very large proportion of their net biomass production to the growth of new leaves and reproductive organs. Their success is directly related to the capacity to grow quickly during the wettest season, circumventing drought by early completion of lifecycle and releasing a large number of seeds before the onset of the dry period.

Perennial forbs and grasses differ from annuals in their bigger nutrient and carbohydrate reserves. Greater allocation to storage results in lower growth rates, but permits faster and earlier leaf and root development during the wet parts of the year. In semi-arid ecosystems, perennial herbs usually resist the unfavourable season as underground bulbs or tubers (geophytes) or buds near the soil surface (hemicryptophytes).

Among herbaceous species, C_3 is by far the dominant photosynthetic type in semi-arid ecosystems (Flexas *et al.*, 2003), especially in those areas located at higher altitudes, in steppes and under Mediterranean climates. C_4 species tend to be favoured over C_3 plants in warmer and more humid climates and thus dominate savannas (Grace *et al.*, 1998). North- and South-American and Asian steppes also have a relatively low C_3/C_4 ratio (Suyker and Verma, 2001). Among C_4 subtypes, C_4 grasses are characterised by NAD-malic enzyme and PEPCk subtypes in most steppes (Pyankov *et al.*, 2000).

Woody species or phanerophytes are represented by shrubs and trees. Shrubs and trees with extensive and deep-root systems are less limited by seasonal variations in soil-water availability than herbs. Depending on winter minimum temperatures, phanerophytes are either evergreen or winter-deciduous, although depending on water availability phanerophytes can become drought-deciduous or semi-deciduous.

Evergreen species tolerate stress conditions by retaining green leaves throughout the year. Evergreen leaf habit allows retention of nutrients from season to season and has been described as a more conservative water and nutrient-use strategy than the drought-deciduous strategy (Lloyd and Farquhar, 1994). Several studies have demonstrated that evergreen species dominate sites where resources are in short supply, like semi-arid environments (Beadle, 1966). Overall, the share of drought-deciduous or evergreen leaf habits in savannas and Mediterranean ecosystems depends on the availability of groundwater. Shrubs and trees able to tap water stored in deep soil profiles experience less seasonality in moisture and are evergreen, whereas woody species without access to groundwater during the drought period are commonly drought-deciduous.

Drought- or semi-deciduous phanerophytes are best adapted to the driest conditions of semi-arid environments, where they replace the evergreen communities. A special subtype of drought-deciduous xerophytes is the 'leafless non-succulent switch shrubs' (Oppenheimer, 1960). This subgroup consists of perennial species that shed mesophytic leaves during drought, but keep green stems and thereby maintain photosynthetic activity during the dry season (Yiotis *et al.*, 2006). Typically, the drought-deciduous species in semi-arid environments possess relatively mesophytic leaves. However, some semi-deciduous species have seasonal leaf dimorphism, forming sclerophyllous leaves in summer and mesophytic leaves in winter (Aronne and De Micco, 2001). Seasonal dimorphism in semi-deciduous species avoids excessive water loss with

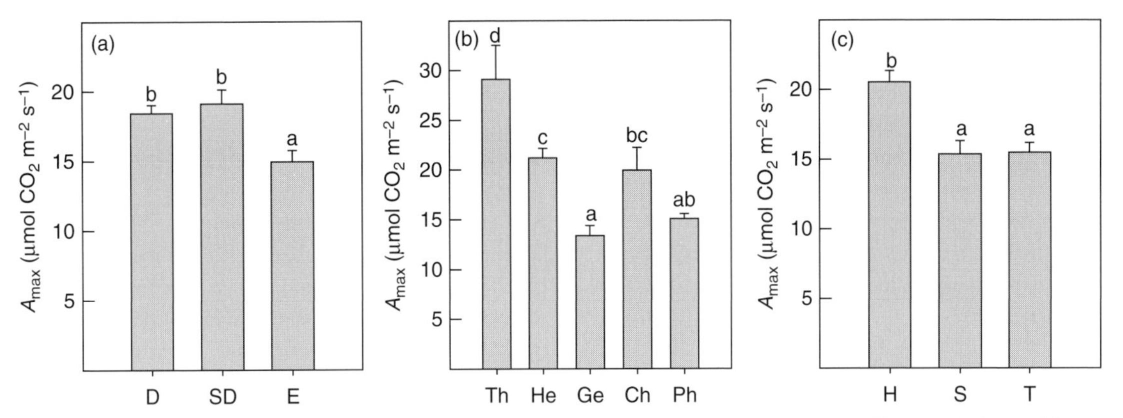

Fig. 29.2. Mean (±SE) photosynthetic rates per unit area of well-watered, non-stressed semi-arid species. Photosynthetic capacities are separately plotted according to **(a)** leaf habit (D, deciduous, n=223; SD, semi-deciduous, n=34; E, evergreen, n=71); **(b)** life form (Th, therophytes; n=6; He, hemicryptophytes, n=123; Ge, geophytes, n=19; Ch, chamaephytes, n=15 for Ch; Ph, phanerophytes, n=165); and **(c)** growth form (H, herbaceous, n=150; S, shrubs, n=115 for S; T, trees, n=63). For the leaf habit, both annuals and perennials with deciduous leaves have been included in group D. Values were taken from previous studies performed in the field and in controlled environments. When different estimates were reported for the same species, the highest value was considered. Varieties or cultivars of the same species were considered as a unique entry. Only the measurements performed at mid-morning were considered. Different letters denote significant differences among the means according to Duncan's test (P<0.05).

a reduction of transpiring surface through partial fall of leaves during the dry period.

Winter deciduousness is a typical feature of cold climates, but clearly disadvantageous in semi-arid regions where precipitation occurs during the winter, such as the Mediterranean climate. In these areas, the combination of winter deciduousness and dry summers would lead to a very short carbon-assimilation period that in some cases is insufficient to pay off the costs of leaf construction and maintenance (Flexas *et al.*, 2003). In fact, woody vegetation is essentially lacking in some environments, such as continental steppes, where winter temperatures are not compatible with wintergreenness, and foliage cannot also be supported during the entire summer owing to excessively low water availabilities.

Most woody species in semi-arid environments possess C_3 metabolism, with only few C_4 shrubs such as *Atriplex* species growing in ecotones between semi-arid environments and deserts (e.g., Akhani *et al.*, 1997).

Succulents with CAM photosynthetic type are relatively abundant in the ecotones between semi-arid ecosystems and deserts, but they are usually characteristic to more continental inner locations. In the Mediterranean Basin, suitable conditions for CAM species are only found in rocky, coastal areas (Flexas *et al.*, 2003). Most of the CAM species found in semi-arid areas are better defined as facultative CAM, with the C_3-photosynthetic pathway dominating during most of the year, and CAM being expressed only under water and salt stress.

29.2.3. Differences in photosynthetic potentials among plant functional types

A revision of maximum photosynthesis rates per unit leaf area (A_{Nmax}) for 328 different species found in semi-arid environments demonstrates that A_{Nmax} is similar among species with annual (including deciduous) and semi-deciduous leaves, and significantly lower in species with evergreen leaves (Fig. 29.2.a). This species ranking is in agreement with previous reports showing lower photosynthetic capacities for evergreens (e.g., Reich *et al.*, 1997; Flexas *et al.*, 2003). The higher leaf lifespan of evergreen leaves is achieved by rigidifying photosynthetic tissues, which results in a lower fraction of photosynthetic biomass within the leaves and reduced internal CO_2-diffusion conductance from substomatal cavities to the site of carboxylation (Niinemets and Sack, 2006; Section 29.2.4).

Among the life form groups, geophytes and phanerophytes had the lowest A_{Nmax}, whereas therophytes and hemicryptophytes had the highest (Fig. 29.2.b). Low A_{Nmax} for semi-arid geophytes has been confirmed in several surveys (Forseth and Ehleringer, 1983; Gulías *et al.*, 2003).

Among growth forms, shrubs and trees had similar values for A_{Nmax} and significantly lower values than the herbaceous species (Fig. 29.2C) in agreement with previous reports (e.g., Jiang *et al.*, 1999). The main differences among the life forms and growth forms are driven by differences in foliage longevity among these species groups (Fig. 29.2.a). Although herbaceous species are characterised by fast turnover of foliage, leaf longevity is significantly larger in woody perennials, especially in woody evergreens (Diemer and Körner, 1996).

29.2.4. Leaf structural adaptations to semi-arid environments that influence photosynthesis

In stressful environments, a variety of selection pressures operate on leaf morphological and anatomical design with important consequences for foliage photosynthetic activity. Plant photosynthetic activity adjusts to semi-arid environments by physiological photoprotection to avoid excess light intensities (Section 29.2.5) and by leaf and crown morphological adaptations to reduce transpiration and/or light interception and improve tolerance of low leaf-water potentials. Structural avoidance of excess radiation absorption is important in protecting against photoinhibition and in avoiding excessively high leaf temperatures and transpiration rates (Valladares and Niinemets, 2007). Thus, structural adjustments permit the plants to achieve an efficient compromise between maximisation of carbon gain and minimisation of exposure to high solar radiation.

Apart from evolutionary adaptations to limiting water availability and to other associated stresses, plants in semi-arid environments have large phenotypic plasticity in leaf and shoot architecture (Valladares *et al.*, 2005). High capacity for foliage adjustment in response to environmental stresses is present in all major functional types dominating semi-arid ecosystems – sclerophyll evergreen shrubs, malacophyll summer-deciduous shrubs and herbaceous species. At the extreme, plant species with dimorphic leaves possess foliage with completely different structure during the dry season in summer (crimped and partially rolled lamina with crypts in the abaxial surface where stomata are located, higher pubescence, smaller tightly packed mesophyll cells forming palisade parenchyma on both sides of the lamina in summer leaves) than during the wet season in winter when the leaves have more mesophytic in appearance (Aronne and De Micco, 2001).

In the following, we examine the key evolutionary and phenotypic modifications in foliage architecture in semi-arid environments.

LEAF MASS PER AREA AND RELATED MORPHOLOGICAL TRAITS

A vast number of studies have shown that LMA is larger in hotter and drier sites (e.g., Nobel, 1977; Wright *et al.*, 2005, Fig. 29.3). Such increases in LMA, both within species and across species, are well documented for key plant functional types in Mediterranean ecosystems (Salleo and Lo Gullo, 1990), steppe communities (Vendramini *et al.*, 2001) and savannas (Read *et al.*, 2006). High LMA has been classically related to sclerophylly, a plant trait in semi-arid ecosystems improving foliage longevity and resistance to low water availabilities (Scholes *et al.*, 2004). Although more robust foliage structure of semi-arid species is often associated with greater foliage longevity in such environments, both evergreen and deciduous species in semi-arid biomes have larger LMA than those in tropical or temperate biomes (Fig. 29.3).

LMA ($g\ m^{-2}$) is a product of leaf density (D, $g\ cm^{-3}$) and thickness (T, μm) that can vary independently (Niinemets, 1999; Fig. 29.3). Separation between these components of LMA is relevant as foliage photosynthetic potentials per area commonly increase with thickness owing to accumulation of mesophyll tissue, while photosynthetic potentials per mass decrease with density owing to accumulation of non-photosynthetic biomass (Niinemets, 1999). LMA increases with site aridity mainly because of modifications in leaf density (Niinemets, 2001, Fig. 29.3). Increases in leaf density in drier sites are commonly associated with thickening of cuticle and epidermis, with increased fractional sclerenchyma investments and smaller and more tightly packed mesophyll cells that also have thicker cell walls and greater pubescence (Niinemets and Sack, 2006).

Biome contrasts suggest that leaves are also commonly thicker in semi-arid environments than either in temperate or in tropical biomes (Fig. 29.3). In particular, a larger LMA of drought-deciduous species in semi-arid biomes than those in drought-deciduous species in tropics or winter-deciduous species in temperate ecosystems is associated with greater foliage thickness not with greater density (Fig. 29.3). Global-scale relationships between foliage structure and climate suggest that greater thickness in such ecosystems is not necessarily linked to water availability, but may result from enhanced radiation loads in semi-arid environments (Niinemets, 2001).

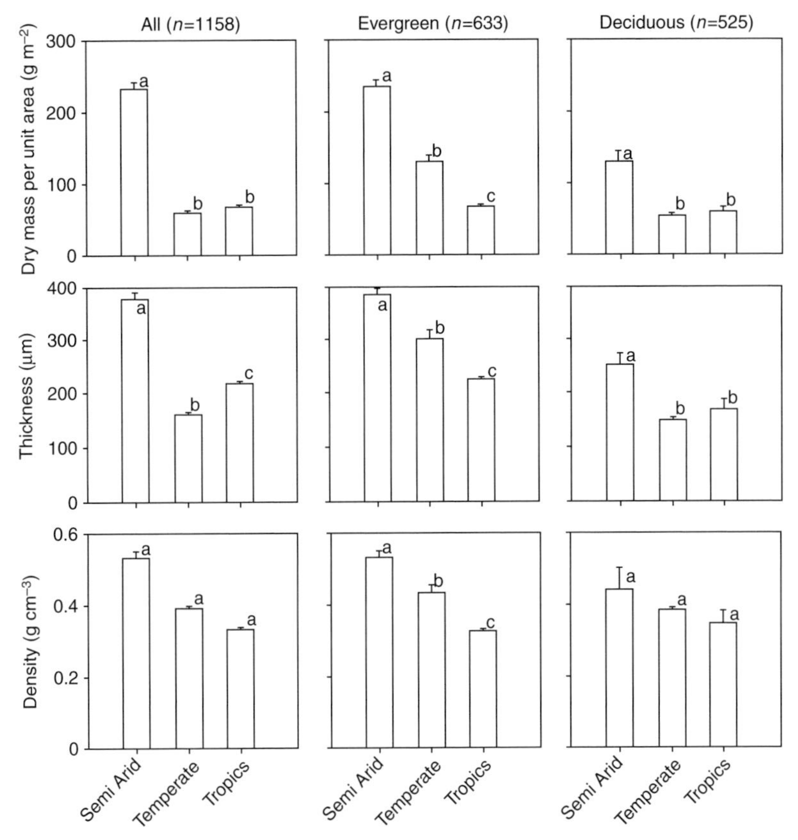

Fig. 29.3. Average (±SE) leaf dry mass per unit area (LMA) and its components, leaf thickness (T) and density (D) (LMA=TD), for a large number of angiosperm species from semi-arid (n=367), temperate (n=523) and tropical (n=268) biomes (database of Niinemets, 1999). The values of foliage morphological traits are separately given for evergreen (n=344 for semi-arid, n=35 for temperate and n=254 for tropical biomes) and deciduous (n=23 for semi-arid, n=488 for temperate and n=14 for tropical biomes) species. In all cases, foliage projected area has been used. Bars with the same letter are not significantly different according to ANOVA (P>0.05).

Benefits of high LMA

The main benefits of high LMA in semi-arid environments are as follows:

1. Larger leaf lifespan (Reich *et al.*, 1997), probably as the result of enhanced tolerance of mechanical injures and herbivory by more robust leaves (Wright and Cannon, 2001).
2. Leaf rigidification that renders the leaves more resistant to pressure-driven changes in leaf volume and water content (Salleo *et al.*, 1997). High LMA reduces tissue compression in dehydrating leaves and favours leaf recovery after drought-induced cavitation (Salleo *et al.*, 1997). Furthermore, less-elastic leaves lose less water for common change in leaf-water potential. Thus, these leaves can extract water from drier soil than more mesophytic leaves (Niinemets, 2001).
3. Greater degree of mesophyll stacking (Smith *et al.*, 1998) resulting in lower surface-to-volume ratio (1/thickness in broad-leaved species). Lower surface-to-volume ratio is associated with reduced transpiration rates and higher photosynthetic WUE as observed in perennial herbaceous steppe species (Voronin *et al.*, 2003), savanna trees (Hoffman *et al.*, 2005) and woody Mediterranean species (Paula and Pausas, 2006).

Negative consequences of high LMA

Although positive effects of high LMA in semi-arid ecosystems are mainly associated with adaptations to limited water availability, the consequence of high LMA is typically low photosynthetic capacity per dry mass (Wright and Cannon, 2001). Negative relationships between LMA and photosynthetic capacity have been postulated to result from the following:

1. Higher resistance to CO_2 internal diffusion (i.e., lower g_m) owing to denser packing of mesophyll cells and thicker cell walls (Niinemets and Sack, 2006).
2. Lower leaf nitrogen (N) concentration and higher investment of N in non-photosynthetic components (Niinemets, 1999).
3. Lower chlorophyll content and lower foliage light capture per mass (Wright and Cannon, 2001).

The last two negative effects of high LMA likely result from greater proportion of support tissue relative to physiologically active biomass.

LEAF SIZE AND SHAPE

Leaf energy balance significantly depends on leaf size and shape. In high radiation loads, large entire leaves have a thicker boundary layer for heat and water exchange, and thus require greater transpiratory cooling to maintain leaf temperatures within the optimal limits for photosynthesis than smaller leaves or leaves with dissected lamina (Givnish, 1979). This may not be possible in semi-arid environments where latent heat loss via transpiration is impeded by drought-driven stomatal closure. Consequently, there is a general trend of decreasing leaf size with increasing site aridity (Scholes *et al.*, 2004), and plants typical of semi-arid habitats commonly possess relatively small leaves (Tenhunen *et al.*, 1987).

Apart from direct influences of leaf size and shape on leaf temperature, thermal optima of photosynthesis may be different in leaves of various habits. In a comparison among eight species of *Pelargonium* from South-African semi-arid ecosystems, Nicotra *et al.* (2008) showed that the species with more dissected leaves had higher thermal optima of photosynthesis and greater rates of carbon gain and water loss. Higher thermal optima may reflect selection pressure to protect photosynthetic machinery against excessive leaf temperatures when stomata close in response to water stress.

LEAF PUBESCENCE

Leaf pubescence has been considered to be an adaptive trait in water-limited environments. A hairy surface acts as a spectrally neutral reflector for the UV, visible and IR parts of the spectrum, reducing the radiant energy absorbed by the leaf (Karabourniotis and Bornman, 1999). In consequence, presence of trichomes on leaf surface reduces heat load, and thus the reliance on transpiratory cooling to avoid high leaf temperatures (Ehleringer and Mooney, 1978). This permits maintenance of photosynthetic activity longer into the drought period. Moreover, reduced absorption of visible and UV-B radiation results in a lower degree of photosynthetic inhibition (Karabourniotis and Bornman, 1999). Further possible benefits associated with pubescence are reduced herbivory (Levin, 1973) and increased probability of water uptake by leaves (Savé *et al.*, 2000).

Although pubescence enhances plant performance in stressful environments, there are trade-offs associated with it. These include the additional costs for construction of pubescent leaves and lower rates of photosynthesis at common light intensity, especially when water availability is higher (Sandquist and Ehleringer, 2003). Overall, positive or negative effects associated with leaf trichomes on productivity and fitness depend on the level of environmental stresses, such as the degree of drought.

STOMATAL CHARACTERISTICS

Stomatal traits, such as density and size of stomata, largely influence stomatal conductance to gaseous transport (g_s). Therefore, modifications in stomatal number and dimensions can importantly alter leaf-water loss and photosynthetic rates. In a range of Mediterranean species belonging to different plant functional types, maximal g_s was highly correlated with stomatal area index (i.e., stomatal density × stomatal size) (Galmés *et al.*, 2007c) and photosynthetic capacity (Galmés *et al.*, 2007a). Small and abundant stomata enhance fine regulation of plant water use (Pearce *et al.*, 2006). Hence, species from semi-arid environments typically have leaves with higher density and smaller stomata than species from humid sites (Sundberg, 1986). However, exceptions to this rule exist, with some species well-adapted to highly stressed conditions having low stomatal densities (Galmés *et al.*, 2007c) and/or stomata with large pore sizes (Rhizopoulou and Psaras, 2003).

Although the plant species dominating semi-arid habitats are generally hypostomatous, possessing stomata commonly on the lower leaf surface only, amphistomatousness becomes more common with further reductions in water availability (James and Bell, 2000). Characteristically, amphistomatous species in xeric habitats have thick foliage elements, and accordingly, amphistomatousness is an adaptive feature shortening the distance of CO_2 diffusion to mesophyll cells in these species (Parkhurst *et al.*, 1988).

Specialised modifications in semi-arid species are sunken stomata and stomatal location in crypts (Fahn and Cutler, 1992). Compared with leaves having stomata on leaf surface, leaves with stomata embedded within mesophyll have vastly reduced boundary layer conductance for water, significantly curbing plant water loss.

CUTICULAR CONDUCTANCE TO WATER

The ability of plants to survive severe drought periods is affected by the ability to restrict residual water loss through the cuticula after stomata have closed. Low cuticular conductances to water have been reported as an adaptive trait in high water- and light-stressed semi-arid environments, and corresponds well with the degree of xeromorphism (Bolhàr-Nordenkampf and Draxler, 1993). In Mediterranean sclerophylls, low cuticular and stomatal conductances commonly co-occur (Levitt, 1980).

Although being beneficial in limiting the non-stomatal transpiration, the disadvantage of impermeable cuticula can be the constrained leaf water-absorption capacity. High capacity to adsorb water from fog, dew and infrequent precipitation has been proposed as an advantageous trait for species from arid and semi-arid ecosystems (Monk, 1966). Munné-Bosch *et al.* (1999) found improvements in water relations of drought-stressed Mediterranean plants after dew formation on the leaf surface. In these plants, the photosynthesis rate increased as the result of water uptake from the leaf surface. In species from semi-arid ecosystems of Venezuela, Díaz and Granadillo (2005) showed that productivity of canopy irrigated trees was even greater than that of soil-irrigated trees. However, foliar water uptake may rely more strongly on uptake through stomata than through cuticula (Peschel *et al.*, 2003). Thus, the capacity of foliar water uptake will more strongly depend on plant ability for rapid stomatal opening in response to humid conditions rather than on cuticular water permeability.

29.2.5. Leaf physiological traits that influence photosynthesis

In addition to structural alterations, species inhabiting semi-arid ecosystems have developed a number of physiological adaptations to cope with co-occurring high-temperature and solar-irradiance stresses. These physiological modifications either serve to minimise the capture and processing of light or directly influence carbon assimilation. In addition, structural adaptation to semi-arid conditions leads to changes in internal CO_2-diffusion conductance thereby further altering leaf-assimilation characteristics.

LEAF PHYSIOLOGICAL TRAITS CONTROLLING THE CAPTURE AND PROCESSING OF LIGHT

Under conditions of high PPFD and limited CO_2 entry through the stomata, the amount of light absorbed by leaves rapidly exceeds the capacity of cells to use the photochemical energy for photosynthesis and photorespiration. Under such conditions, foliage photosynthetic apparatus is prone to photoinhibition. To cope with enhanced risk of photoinhibition, plants in semi-arid ecosystems display a number of physiological adaptations: (1) reduction of light-harvesting pigment-binding complexes; (2) increased alternative electron transport capacity; (3) increased thermal dissipation of absorbed light energy; (4) increased antioxidant defence systems; and (5) enhanced synthesis of volatile compounds.

A reduction of the content of light-harvesting pigment-protein complexes relative to the rest of photosynthetic machinery is an effective way of diminishing the absorption of excess light. Such a reduction of an effective cross-section of photosystems is manifested in decreased chlorophyll and carotenoid contents, frequently observed during drought periods in a series of semi-deciduous (Kyparissis *et al.*, 1995) and evergreen (Munné-Bosch and Alegre, 2000) species in semi-arid environments.

In addition to avoidance of light capture, plants have also evolved other mechanisms to cope with the excess of absorbed light. In drought-adapted species, the photosynthetic electron transport rate is reduced by drought to a lesser extent than CO_2 assimilation (Gulías *et al.*, 2002). Lower sensitivity of electron transport to drought is associated with enhanced electron transport to alternative electron sinks, mostly to photorespiration and the Mehler reaction. Increased engagement of these alternative electron sinks reduces the fraction of excess excitation energy and thus provides photoprotection against excess light (Flexas and Medrano, 2002b).

The rate of thermal dissipation of absorbed light energy is also enhanced in semi-arid species (Galmés *et al.*, 2007b). The capacity for thermal dissipation of excitation energy is commonly explained on the basis of xanthophyll cycle – light-dependent conversion of violaxanthin to zeaxanthin through antheraxanthin – resulting in changes in conformation of pigment-binding proteins (Demmig-Adams and Adams, 2006). Enhanced capacity for thermal excitation-energy quenching is typically linked to higher content of xanthophyll-cycle carotenoids (violaxanthin, antheraxanthin and zeaxanthin) per A_{leaf}, chlorophyll. and total carotenoids and a greater de-epoxidation state of xanthophyll-cycle carotenoids (greater fraction of zeaxanthin) (Demmig-Adams and Adams, 2006). The need for greater capacity of thermal energy dissipation is not only constrained to drought periods in the summer, but can also be relevant on winter days with freezing temperatures (García-Plazaola *et al.*, 2003b). Under such conditions, low temperatures limit CO_2 assimilation and the fraction of excess light can be large.

Increased thermal-energy dissipation and reduced quantum efficiency under excess light generally imply increased foliage photoprotection ('dynamic photoinhibition', *sensu* Osmond, 1994). Under extreme conditions, the capacity for foliage electron transport is also reduced manifesting photodamage ('chronic photoinhibition', *sensu* Osmond, 1994). In semi-arid plants, drought and low temperatures generally result in 'dynamic photoinhibition' and, to a lesser extent – at least in well-acclimated leaves developed under high irradiances – 'in chronic photoinhibition' (Werner *et al.*, 2002).

Other photoprotective mechanisms observed in semi-arid species are increased antioxidant activity (Munné-Bosch and Peñuelas, 2003) and synthesis of volatile compounds (Loreto and Sharkey, 1993). Lipid-soluble antioxidants such as tocopherol quench free radicals and lipid peroxides formed in membranes, while water-soluble antioxidants such as glutathione and ascorbate quench active oxygen species formed in the leaf liquid phase in response to excess energy driven oxidative stress (Havaux and Niyogi, 1999). Several lipid-soluble volatile compounds such as monoterpenes also function as antioxidants in the leaf lipid phase (Loreto *et al.*, 2004a). Enhanced volatile-compound synthesis has been hypothesised to lessen the possible over-reduction of photosynthetic apparatus under stress conditions, functioning thus as a sort of metabolic safety valve (Rosenstiel *et al.*, 2004). Yet only a relatively small fraction of electrons enter volatile-compound synthesis, and thus the contribution of emission of volatile organics to photoprotection is likely to be minor (Niinemets *et al.*, 2002).

LEAF PHYSIOLOGICAL TRAITS INFLUENCING CARBON DIFFUSION AND ASSIMILATION

Two main barriers limit the diffusion of CO_2 from the ambient atmosphere to the site of carboxylation in the chloroplast: stomata and leaf mesophyll. Apart from morphological adaptations of stomata described in the previous section, stomatal closure in response to soil-water shortage is a common response among plants, but the extent and velocity of this response differ among species, resulting in widely varying efficiencies of water use. The species from semi-arid ecosystems typically have higher stomatal responsiveness to water deficit, resulting in greater WUE that allows them to use water longer into the drought period (Galmés *et al.*, 2007c).

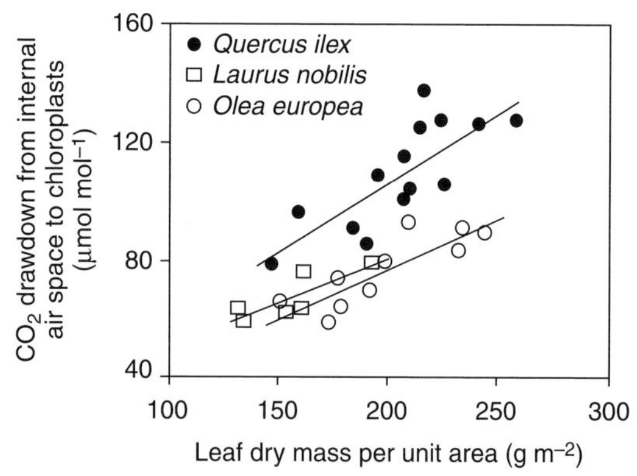

Fig. 29.4. Scaling of the CO_2 drawdown from internal air space to chloroplasts (ΔC) with leaf dry mass per unit area (LMA) in three Mediterranean sclerophyllous species (modified from Niinemets *et al.*, 2005). Within-species variation in LMA was mainly owing to variation in leaf age. ΔC is the difference between CO_2 concentrations in internal air space (C_i) and chloroplasts (C_c). Given that net assimilation, A_N, is given as $A_N=g_m(C_i-C_c)=g_m\Delta C$, ΔC provides a measure of the extent to which photosynthesis is limited by g_m for leaves with a given photosynthetic capacity ($\Delta C=A_N/g_m$). All measurements were conducted at a leaf temperature of 25°C and ambient CO_2 concentration (ca. 350 μmol mol^{-1}).

Once CO_2 has entered into the leaf, it must overcome a series of gas-, liquid- and lipid-phase resistances from the sub-stomatal cavities to the chloroplast stroma. Mesophyll internal diffusion conductance (g_m) accounts for these constraints on the diffusion pathway. The internal diffusion conductance depends on leaf anatomy, with long-lived leaves having lower values of g_m than deciduous species or annuals (Warren *et al.*, 2007; Flexas *et al.*, 2008, see also Section 29.2.4). Stronger limitation of diffusion in leaves with a more robust structure is mainly associated with thicker cell walls and a lower internal gas-phase volume than those in more mesophytic leaves (Syvertsen *et al.*, 1995). The degree to which photosynthesis is limited by g_m (drawdown of CO_2 concentration from sub-stomatal cavities to chloroplasts) is positively associated with LMA in semi-arid species (Fig. 29.4). Lower g_m of these species is the cost these leaves 'pay' for greater longevity, cavitation resistance and capacity to extract water from drying soil.

In addition to leaf anatomy, g_m can also rapidly acclimate to a variety of environmental conditions (see Chapter 4 for a detailed explanation). In semi-arid environments, short-term low water availability and high temperatures can induce large reductions in g_m. As a result, semi-arid species temporarily suffer from noticeably low CO_2 concentrations at the carboxylation sites (Galmés *et al.*, 2007a), strongly limiting CO_2 assimilation rates during water stress. This is especially significant in species with high LMA where already anatomical constraints seriously limit g_m (Galmés *et al.*, 2007a), for which these species can experience a larger advantage than mesic species of increased CO_2 concentration in the atmosphere in terms of both photosynthesis and WUE (Niinemets *et al.*, 2011).

Although low chloroplastic CO_2 concentrations limit carboxylation, low CO_2 stimulates oxygenation (photorespiration), further reducing net carbon fixation. Under these conditions, a higher Rubisco specificity towards CO_2 than to O_2 that competes for the same enzyme active site could importantly increase a species' competitive potential in semi-arid habitats. So far, the evidence of greater Rubisco specificity in ecosystems with high temperature and low water is limited (Galmés *et al.*, 2005), but the number of species screened so far is also clearly insufficient.

Alternatively, CO_2-concentration mechanisms present in several species in semi-arid ecosystems (C_4 and CAM pathway) can importantly increase the ratio of carboxylation to oxygenation and thereby significantly enhance photosynthesis during dry and hot periods (Sage, 2001). C_4 and CAM photosynthesis types are energetically more expensive and require more light quanta per mole fixed CO_2. However, in relatively open semi-arid environments with high solar-energy input, this high energetic cost is vastly outweighed by improved carbon gain during dry periods.

29.3. SEASONAL VARIATIONS AND RESPONSES OF PHOTOSYNTHESIS TO SINGLE ENVIRONMENTAL FACTORS

In addition to differences in the photosynthetic capacity among plant functional types (Section 29.2.3), annual carbon gain also depends on functional type-specific seasonal modifications in photosynthesis. In semi-arid environments (Mediterranean and savanna-type ecosystems) with similar annual amounts and variation in precipitation, there is a vast variation in seasonal timecourses of A_N among annual, drought-deciduous, semi-deciduous, winter-deciduous and evergreen species (Fig. 29.5). Clearly, the annuals have the highest maximum A_N, but they photosynthesise for only a few weeks/months a year. Winter-deciduous woody species have lower maximum rates, but maintain positive net photosynthesis (A_N) for 6–9 months per year (Fig. 29.5). In contrast, evergreen species exhibit even lower maximum photosynthesis rates, but photosynthesise all year round with a significant fraction of annual carbon fixation performed in the winter months (Tenhunen *et al.*, 1987). This indicates that a type of compensatory relationship between maximum photosynthetic capacity, leaf lifespan and seasonal duration of photosynthetic activity is evident.

In semi-arid environments, leaf photosynthesis always exhibits a marked depression during dry months regardless of the functional group and site. However, the relative importance of this depression depends on both the functional type and the climate (Fig. 29.5). The lowest reduction in A_N can be observed in savanna ecosystems with deep soils, and relatively high annual precipitation, allowing species with deep root systems to reach the soil-water table (e.g., *E. tetrodonta* in Fig. 29.5).

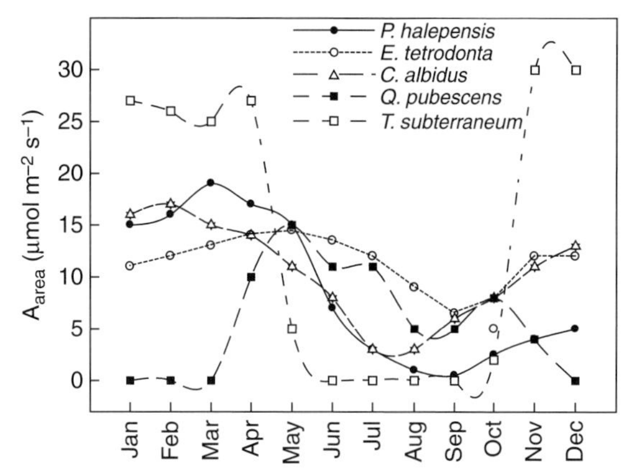

Fig. 29.5. Seasonal time courses of net CO_2 assimilation rate (A_{area}) for semi-arid species belonging to different plant functional types: evergreen conifer *Pinus halepensis* (data of Klein *et al.*, 2005, sampled in Israel, mean annual precipitation 270 mm and mean annual temperature 15°C); broad-leaved evergreen tree *Eucalyptus tetrodonta* (Prior *et al.*, 1997, savanna in Australia, 1300 mm and 26°C); semi-deciduous shrub *Cistus albidus* (Gulías *et al.*, 2009, Mediterranean macchia in Spain, 470 mm and 17°C); winter-deciduous tree *Quercus pubescens* (Damesin and Rambal, 1995, deciduous oak forest in southern France, 1100 mm and 13°C); and summer-deciduous perennial herb *Trifolium subterraneum* (Flexas *et al.*, 2003, Mediterranean grassland in Spain, 500 mm and 17°C). In the semi-deciduous species, the values of A_{area} only correspond to retained leaves.

Although winter depression of photosynthesis is generally considered as a major limitation in Mediterranean ecosystems, the reductions in photosynthetic potentials (A_N) *per se* are relatively moderate (Fig. 29.5). However, reduced light intensities and lower than optimum temperatures still reduce the realised carbon gain (Tenhunen *et al.*, 1987). Nevertheless, in semi-arid ecosystems that can support evergreen vegetation, summer drought is generally a more important limitation of annual carbon gain than winter temperatures (Flexas *et al.*, 2001). In colder semi-arid regions, particularly in steppes with temperate continental climates, low winter temperatures limit photosynthesis more strongly and can curb annual productivity more than summer drought (Méthy, 2000b). In semi-arid environments, chilling and freezing sensitivity of photosynthesis is higher for tropical and sub-tropical flora components than for species typical of temperate climates (Gulías *et al.*, 2009). Sensitivity to photoinhibition is another cause for decreased photosynthesis during winter (García-Plazaola *et al.*, 2003b), although semi-deciduous species seem to be more vulnerable to winter photoinhibition than evergreens (Werner *et al.*, 2002).

Some semi-arid environments such as Mediterranean ecosystems are often associated with complex site topography. To understand the responses of species groups to climate, it is further important to consider that local climatic differences can be highly variable within short distances in complex terrains. Such local climatic differences can affect annual photosynthesis of a single species by the same order of magnitude as the within-site variation in annual photosynthesis among species of different plant functional type (Flexas *et al.*, 2003). These between-site differences likely translate into large variation of the annual carbon balance between local populations of the same species.

29.4. WHOLE-PLANT FEATURES AND ECOSYSTEM GAS EXCHANGE

The consequences of variations in crown architectural traits for light interception and photosynthetic efficiency have been traditionally analysed in terms of total leaf area index (LAI), leaf-inclination-angle distribution and leaf aggregation (spatial clumping) (Valladares and Pearcy, 1998; Cescatti and Niinemets, 2004).

A series of architectural traits at leaf, shoot and whole-canopy scale affect these three key sets of traits. In particular branching pattern, leaf-insertion angle, leaf rolling and phyllotaxis can importantly modify angular distributions and spatial clumping of foliage and thereby alter light-harvesting efficiency. Especially in ecosystems where LAI is small, foliage geometry strongly affects the diurnal and seasonal patterns of light capture and potential carbon gain (Valladares and Pugnaire, 1999).

29.4.1. Leaf-area index

Whole-community LAI is typically low in semi-arid environments owing to low water availability. However, LAI in semi-arid environments is highly non-uniformly distributed in space with areas exhibiting very low plant cover or even bare ground during drought periods intermixed with areas having high LAI (Asner and Wessman, 1997). High spatial heterogeneity in LAI can be found in savannas and Mediterranean shrublands with sparsely distributed evergreen woody species (Asner *et al.*, 2000). Evergreenness is the primary reason for accumulation of large leaf area within the canopy of semi-arid species (Rambal, 2001). Deciduous counterparts have significantly lower LAI in such environments (Damesin *et al.*, 1998).

Typically, whole-community LAI between 1–6 m^2 m^{-2} are found in Mediterranean macchia and forest ecosystems with values of 2–3 m^2 m^{-2} being most common (Rambal, 2001), 0.8–1.7 m^2 m^{-2} in savannas (Scholes *et al.*, 2004) and between 1–4 m^2 m^{-2} for semi-arid steppes and grasslands (Xu and Baldocchi, 2004). However, LAI within any community type increases with increasing precipitation, with the transition zones to humid biomes having the highest LAI values (Damesin *et al.*, 1998).

Apart from spatial heterogeneity, semi-arid environments also exhibit large temporal heterogeneity in LAI. Species from semi-arid ecosystems often partly shed leaves in response to drought and co-occurring high light stress (Werner *et al.*, 1999). Although instantaneous canopy photosynthesis can be linearly related to LAI in semi-arid species until relatively high values of LAI (Goulden, 1996), dropping some but not all leaves allows the plants to diminish the overall transpiratory water loss and continue photosynthesis longer into the dry season. Thus, having lower LAI during drought periods potentially maximising photosynthesis when enough water was available, results in larger annual carbon gain than temporarily invariable LAI (Tenhunen *et al.*, 1990).

Partial leaf loss has been described in all semi-arid ecosystems in a variety of plant functional types such as perennial grasses (Balaguer *et al.*, 2002) and evergreen sclerophylls and conifers (Werner *et al.*, 1999). Many Mediterranean semi-deciduous species have seasonal crown dimorphism, characterised by different types of branches and leaves during different seasons (Aronne and De Micco, 2001), but also reduced leaf area and size during the dry season (Gratani and Bombelli, 2000). Some drought-deciduous species retain green stems (see Chapter 7) after shedding leaves during a dry summer (Gibson, 1983). Reductions of LAI and increased reliance on the use of drought-specialised photosynthetic stems lead to water conservation and persistent photosynthesis during drought in semi-arid species (Comstock and Ehleringer, 1988).

At extremes, drought can result in complete loss of photosynthesising surface in drought-deciduous species (Werner *et al.*, 1999). However, drought-deciduous species from savannas, where multiple severe drought cycles can occur during a year, have been shown to be very opportunistic in their water use by retaining non-dormant buds. This fact results in very fast bud-break and leaf flush after significant rainfall events (Díaz and Granadillo, 2005).

Some authors have argued that all these long-term changes in LAI, and not physiological parameters, are actually the main response of many species well adapated to semi-arid conditions. In this sense, the ecological optimal theory given by Eagleson (1982) has been contrasted by modelling LAI in semi-arid environments, showing that this parameter is finelly tuned depending on the soil-water availability (Hoff and Rambal, 2003).

29.4.2. Leaf-inclination angle distributions

Modification of leaf-inclination-angle distributions is a powerful way to alter the total light interception and temporal variation of the occurrence of peak irradiances on the leaf surface. Horizontally exposed leaves represent an effective strategy for light interception, but horizontal leaves have their highest irradiances on the leaf surface at midday when other co-occurring stresses, such as drought and heat stress, are most severe. Thus, horizontal leaf-inclination angles in semi-arid environments can result in irreversible photodamage and pigment destruction (Valladares and Pugnaire, 1999). In dry and high-light environments, steep leaf-inclination angles provide a viable solution to enhance whole-plant carbon gain and reduce overheating and photoinhibition (Niinemets *et al.*, 2006a,b)

Leaf-inclination angle is generally considered a static trait as it cannot be strongly modified after petiolar and lamina tissues have been rigidified by lignification. Nevertheless, there are seasonal and daily modifications in leaf-inclination angle in several species (Forseth and Ehleringer, 1982). Such modifications have been mainly assigned to drought-deciduous plants (Werner *et al.*, 1999), but can actually occur in evergreen sclerophylls (e.g., Gratani and Ghia, 2002), semi-arid grasses (Ryel and Beyschlag, 1995) and savanna deciduous trees (Scholes *et al.*, 2004). In species with strongly lignified foliage elements, such modifications in inclination-angle distributions can be partly ascribed to preferential shedding of leaves that have a less favourable exposure. In semi-deciduous dimorphic species, summer leaves have steeper leaf angles (Gratani and Bombelli, 2000), and the angles of the photosynthetic surface also become steeper during dry season in drought-deciduous species retaining green stems (Gibson, 1983).

Apart from the dry season, the structural photoprotection by vertical leaf orientation has also been shown to be an efficient photoprotective mechanism during winter when foliage photosynthetic activity is decreased by low temperatures (Oliveira and Peñuelas, 2002).

Although a steep inclination angle is a beneficial feature for high-light exposed foliage, significant within-canopy light gradients also exist in the canopies of semi-arid species. In low light, where most of the light arrives from high solar-inclination angles, vertical foliage orientation is clearly disadvantageous. In fact, there is large phenotypic plasticity in foliage-inclination angles with foliage orientation gradually shifting from vertical in high-light exposed shoots to horizontal in low-light exposed shoots (Valladares and Pearcy, 1998).

29.4.3. Leaf aggregation

Foliage in species dominating semi-arid environments is generally more densely packed and more strongly aggregated in space than in mesic species (Falster and Westoby, 2003). Such modifications result in reduction of mean irradiance on the leaf surface (Cescatti and Niinemets, 2004), and thus in a lower risk of photoinhibition both during the dry and cold seasons. The degree of foliage aggregation (spatial clumping) can be increased by several structural modifications. Leaf rolling or folding is a common response to stresses in semi-arid ecosystems, and has been observed in herbaceous (Haase *et al.*, 1999b) and woody perennials (Kyparissis and Manetas, 1993). Rolling protects leaves from high irradiance levels by decreasing the area exposed to light (Pereira and Chaves, 1993). Rolling also results in concealing stomata in perennial grasses (Haase *et al.*, 1999b) and drought-deciduous species (Aronne and De Micco,

2001), thus minimising water loss at the times photosynthesis is severely restricted. The degree of leaf rolling is proportional to the RWC of leaves (Pugnaire *et al.*, 1996), and therefore this mechanism is a potent way of regulating light interception and water use.

A high degree of foliage aggregation also results from smaller and more tightly packed leaves in semi-arid species (Givnish, 1984). Self-shading (and the degree of aggregation) by shoot axis inevitably increases with decreasing the length of foliage elements (Takenaka, 1994). High frequency of branching, short internodes and spiral phyllotaxis can further enhance the degree of foliage aggregation in the canopies of semi-arid species (Niinemets *et al.*, 2006a,b). Such characteristic architectural features observed among a variety of semi-arid species, such as tussock grass (*Stipa tenacissima*), leafless shrub (*Retama sphaerocarpa*) and evergreen shrub (*Quercus coccifera*), result in high self-shading and significantly reduced mean leaf irradiance (Valladares and Pugnaire, 1999).

Structural photoprotection through enhanced foliage aggregation and steeper inclination has an opportunity cost in terms of reduced carbon gain during periods when the environmental conditions are favourable. Nevertheless, characteristic high light stress in combination with drought and extreme temperatures makes such photoprotective strategies highly adaptive in semi-arid environments (Valladares and Pugnaire, 1999).

29.4.4. Ecosystem productivity and its seasonal and inter-annual variability

Semi-arid ecosystems have intermediate net primary production (NPP) among world biomes (see Grace *et al.*, 2006 and references therein). Only tropical and some temperate forests have higher productivity than semi-arid biomes (Grace *et al.*, 2006). Among semi-arid biomes, savannas exhibit the highest annual NPP with 720 g C m^{-2} $year^{-1}$, Mediterranean-type ecosystems are intermediate with 500 g C m^{-2} $year^{-1}$, and steppes the lowest with 380 g C m^{-2} $year^{-1}$. All semi-arid ecosystems together are responsible for 40% of world annual NPP (Grace *et al.*, 2006).

Net ecosystem productivity for CO_2 (NEP) results from the difference between gross primary productivity (GPP) and ecosystem respiration (R_{eco}). Typically, semi-arid ecosystems exhibit a GPP peak during late spring, when high temperatures co-exist with available water (Fig. 29.6). In summer, water deficits, high light and elevated temperatures lead to reduced photosynthetic rates of evergreen trees and

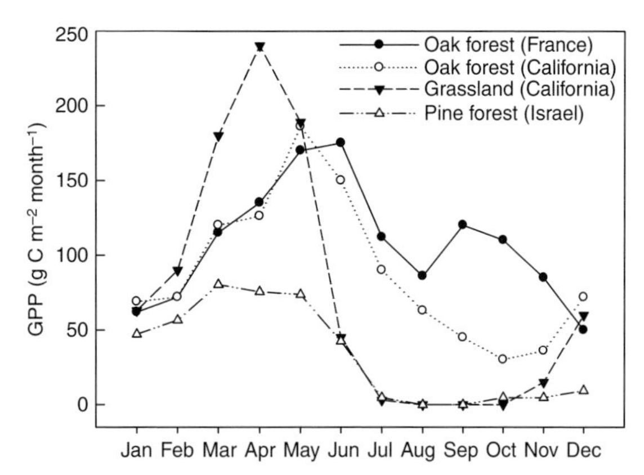

Fig. 29.6. Seasonal variation in gross primary productivity (GPP) in four semi-arid Mediterranean ecosystems. Data are from Ma *et al.* (2007) for Californian oak forest and grassland (with an annual precipitation of 560 mm), Allard *et al.* (2008) for French oak forest (900 mm) and Grünzweig *et al.* (2003) for Israeli pine forest (270 mm).

shrubs and to leaf senescence in shallow-rooted herbaceous annuals and perennials and in drought-deciduous shrubs and trees. This explains a general summer decrease in GPP in various semi-arid ecosystems dominated by different plant functional groups (Fig. 29.6).

Seasonal variability of R_{eco} is more complex than that of GPP, though both parameters are usually highly correleted (Ma *et al.*, 2007). This is partially owing to the limitation imposed by summer conditions on R_{eco}, which nevertheless does not counterbalance for the typically higher decrease in GPP, resulting in semi-arid ecosystems acting as net carbon source (i.e., negative NEP) during summer and early autumn (Pereira *et al.*, 2007). Upon soil re-watering in autumn, vegetation gradually recovers towards winter. For instance, French oak forest exhibited a secondary peak in GPP after the onset of autumn rainfall events (Fig. 29.6). Such a trend for an increased GPP after first rain events in autumn is related to the ability of semi-arid ecosystems to recover from drought effects if the length of drought is not excessive (Luo *et al.*, 2007). This is also related to high rain-use efficiency of several Mediterranean-type ecosystems, especially after the driest periods (Pereira *et al.*, 2007).

The degree of autumn and winter recovery of GPP depends on the amount of rainfall during this season as well as on winter temperatures. In some ecosystems autumn/winter recovery may be lacking, such as in the pine forest in Fig. 29.6, where there was no enhancement of precipitation in the autumn. Autumn recovery was also missing in the grassland ecosystem, where the grasses were dormant after summer drought (Fig. 29.6). Furthermore, there is a clear depression in GPP during late autumn and early winter in the French oakland (Fig. 29.6), reflecting inhibition of photosynthesis by low temperatures.

Similarly to annual GPP, seasonal trends in GPP of semi-arid ecosystems depend on the capacity of plants to reach groundwater reserves, and therefore, on soil texture and plant functional-type-specific rooting systems. Ecosystems dominated by deep-rooted species sustain drought-induced reductions in GPP later than shallow rooted species (e.g., oak forest versus grassland in Fig. 29.6). Inability of steppe and savanna grasses to reach the soil-water table enhances and accelerates summer drought-induced GPP limitations (Suyker and Verma, 2001; Veenendaal *et al.*, 2004).

Inter-annual variability in NEP is extreme for semi-arid ecosystems, especially for semi-arid grasslands (Knapp and Smith, 2001). This is owing to the intrinsic variability of the precipitation patterns in semi-arid environments, and their differential effects on GPP and R_{eco}. For instance, after small pulses of rain, NEP is a source of carbon because R_{eco} is reactivated, but not GPP. Whereas for a big pulse of precipitation, GPP is reactivated and finally leads NEP into being a sink (Knapp and Smith, 2001). Extremely dry years can result in semi-arid ecosystems acting as carbon sources (Xu and Baldocchi, 2004). Such negative effects of unusually severe drought on NEP can even extend into the following years (Luo *et al.*, 2007).

29.4.5. Effects of disturbance on ecosystem productivity

Fire and grazing are the main disturbances in Mediterranean, savanna and steppe ecosystems (e.g., LeCain *et al.*, 2002; Grace *et al.*, 2006). Both fire and grazing exert profound influences on ecosystem structure, in particular on the composition of dominant plant functional types. For instance, grazing and fire reduce the abundance of C_3 woody species in steppe ecosystems (LeCain *et al.*, 2002). Such profound modifications in dominance of plant functional groups can cause dramatic changes in NPP.

The primary consequence of fire and grazing stress is the reduction in aboveground plant biomass. Initially, the decrease in LAI results in a temporary decline in ecosystem photosynthesis, leading to reduced CO_2-sequestration capacity (Grace *et al.*, 2006). Reduction in LAI also results in decreases in ecosystem annual evapotranspiration (AET) (Bremer *et al.*, 2001) and increases in soil-water stores that can partly compensate for the effects of reduced LAI on productivity (Fleck *et al.*, 1998).

As a secondary response, the decrease in LAI stimulates re-growth in tolerant species. Grazing- and fire-tolerant species have a large tillering capacity, allowing them to quickly recover after grazing (McNaughton *et al.*, 1996). Apart from the immediate response after the disturbance, greater plant turnover and more numerous non-dormant buds in grazed ecosystems are associated with earlier 'green-up' after winter and summer stresses. Such an early onset of growth enhances early and late-season photosynthesis (LeCain *et al.*, 2000).

Young resprouts also often have higher A_N (Nowak and Candwell, 1984). Such a compensatory increase in A_N during the first season after fire or grazing is associated with reduced intra- and interspecific competition, resulting in higher light, water (De Souza *et al.*, 1986) and nutrient (Hulbert, 1988) (especially nitrogen (Hastings *et al.*, 1989)) availabilities for resprouts. For instance, improved water availability after fire was associated with higher g_s and g_m, greater A_N and reduced degree of summer photoinhibition in resprouts of Mediterranean evergreen species (Fleck *et al.*, 1998). In addition to immediate increases in photosynthetic capacity by improved nutrition, larger nutrient availability has also been shown to delay leaf senescence (Bremer *et al.*, 2001), thereby further increasing ecosystem carbon gain.

Both enhanced water and nutrient availabilities are necessarily short-lived with the effects becoming gradually less with increasing canopy leaf area and increasing competition. For instance, with increasing LAI the benefits of resprouts are lost as increased evapotranspiration owing to larger canopy conductance of resprouts leading to depletion of soil-water stores (Bremer *et al.*, 2001).

29.5. WATER-USE EFFICIENCY AND CLIMATE-CHANGE PERSPECTIVES

Global climate change associatied with elevated CO_2 concentrations results in increased temperature and altered precipitation patterns, which can profoundly affect the productivity of semi-arid ecosystems. Trends of increased temperature and decreased soil moisture have been predicted and documented globally (Jung *et al.*, 2010) and specifically for semi-arid regions (Osborne *et al.*, 2000). As a consequence of climate change, arid and semi-arid areas are expected to increase in worldwide coverage, but there are still large uncertainties in understanding how semi-arid ecosystems will respond to this global change, although most evidences suggest negative effects. For instance, climate-change-induced increased drought has been reported to decrease the global terrestrial NPP from 2000 through 2009 (Zhao and Running, 2000) and to increase defoliation and mortality in trees from 1990 to 2007, particularly in Southern Europe (Carnicer *et al.*, 2011).

29.5.1. Water-use efficiency

The long standing soil-water shortage and high leaf-to-atmosphere VPD have exerted an important evolutionary pressure for improved carbon assimilation over water losses in species typical of semi-arid environments (Tsialtas *et al.*, 2001). Improved WUE is assimilation rate, which can occur as a result of both adaptation and acclimation processes (Flexas *et al.*, 2003).

With respect to adaptation, as explained in Section 29.2.4, species of semi-arid environments typically possess a number of structural features in the stomatal apparatus that result in a relatively low intrinsic g_s. The negative effect of reduced g_s on conductance to CO_2 is overcountered by the positive effect on leaf-water losses. Hence, when compared with species from other biomes, in-situ measurements of the intrinsic WUE (A_N/g_s) under non-stressing conditions showed that most of the Mediterranean species lay in the region with the highest WUE (Gulías *et al.*, 2003). Additionally, there are many other leaf structural and biochemical features that can indirectly affect WUE by limiting A_N. For instance, Medrano *et al.* (2009) showed that those Mediterranean species with the highest drought-induced increases in LMA presented the lowest increase in A_N/g_s, probably owing to increased internal leaf resistances to CO_2 transfer towards the active sites of carboxylation.

Regarding acclimation processes, stomatal closure is well-known as a common response to drought. In spite of such a general trend, there are important differences among species in their stomatal responsiveness to decreases in soil or plant water status. Historically, these differences have been referred as two different strageties, with drought-avoidant species presenting high stomatal responsiveness to increasing drought, and drought-tolerant species with low stomatal responsiveness to increasing drought. Both strategies are certainly found in semi-arid environments and work in tight coordination with other plant traits, such as xylem hydraulic conductivities, root-system extension or capacity for osmotic adjustments (Galmés *et al.*, 2007c). In some cases, these different strategies are related to growth forms and leaf habits. For instance, for Mediterranean species there is a trend for an increased A_N/g_s from herbaceous species, through shrubs to trees (Medrano *et al.*, 2009). Within shrubs and trees, there is also a trend related to leaf habits, with evergreen species presenting higher A_N/g_s than deciduous ones. The lowest values of A_N/g_s for herbs and semi-deciduous shrubs under well-watered conditions may be attributable to their high g_s. Under drought conditions, all growth forms increase A_N/g_s except herbs, with semi-deciduous shrubs and deciduous trees presenting the highest relative increase (Medrano *et al.*, 2009). This is related to the drought-escape strategy of herbaceous species. However, Medrano *et al.* (2009) showed that herbaceous species typical of Mediterranean coastal environments presented the highest relative increase in A_N/g_s as drought progressed, and suggested that the capacity of withstanding water limitation may be an adaptation of all Mediterranean plants regardless of their growth form and leaf habit.

At the ecosystem scale, an important aspect in semi-arid biomes is ecosystem WUE (EWUE), defined as the ratio of net carbon flux to evapotranspiration. Decrease in EWUE with increasing VPD has been reported in Mediterranean ecosystems (Reichstein *et al.*, 2002) and savannas along a precipitation gradient (Scanlon and Albertson, 2004). Again, differences in a plant functional type's composition among studied ecosystems may be the cause for the observed differences in EWUE along the aridity gradient.

29.5.2. Elevated CO_2 effects on plant water use

Arid and semi-arid ecosystems are believed to be among the most responsive to elevated CO_2, and there is evidence that rising CO_2 has already led to significant increases of NPP in semi-arid biomes during the last century (Osborne *et al.*, 2000). This large CO_2-sensitivity is associated with the circumstance that these strongly water-limited ecosystems potentially benefit the most from CO_2-driven enhancements in plant WUE (Melillo *et al.*, 1993). Numerous studies demonstrate that stomata close under elevated CO_2 resulting in reduced water use, while photosynthesis is stimulated and relatively more carbon can be assimilated with a given water loss, resulting in increased WUE and improved plant water availability (Nelson *et al.*, 2004).

Improved WUE is believed to be the primary factor explaining the positive effects of elevated CO_2 on NPP. As less water is used for assimilation of a given amount of carbon, more water is potentially left in the soil (Nelson *et al.*, 2004). In Mediterranean-type ecosystems, actual evapotranspiration of a community is significantly reduced under elevated CO_2 (Grünzweig and Körner, 2001). This remaining water will permit for an extended period of water extraction into the dry season. The extension of the growing season increases summer canopy photosynthesis and therefore annual NPP and community biomass (Joel *et al.*, 2001).

No acclimation has been found in the stomatal sensitivity to elevated CO_2 in semi-arid species (Scarascia-Mugnozza *et al.*, 1996), suggesting that the reduced stomatal openness is maintained over long term. However, the improved soil-water storage is not always observed. In fact, larger soil-water availability has resulted in increased g_s in some semi-arid grasslands (LeCain *et al.*, 2003) and shrublands (Pataki *et al.*, 2000). Also, the sensitivity to water-vapour pressure was reduced at high CO_2 (Tognetti *et al.*, 2000). It is important to consider that stomatal responses to CO_2 involve both direct effects leading to stomatal closure and indirect responses mediated by long-term changes in soil-water availability. Such long-term effects can result in reversal of the closure responses (Morgan *et al.*, 2004). Although these responses are general for most plants, overall stomatal and mesophyll conductance limitations are greater in evergreen sclerophyll species than in deciduous species. Therefore evergreens have been suggested to improve their photosynthesis and WUE in response to increased CO_2 much more than deciduous, which is supported by meta-analysis data on doubling CO_2 experiments (Niinemets *et al.*, 2011). These differences could lead to shifts in the vegetation boundaries within semi-arid ecosystems, and indeed some evergeen species like *Quercus ilex*, *Ilex aquifolium*, *Hedera helix*, *Rhododendron ponticum* and *Prunus lauroceraus* have experienced northwards-expanded distribution in Europe in the recent decades (Niinemets *et al.*, 2011).

Photosynthetic downregulation or acclimation to high CO_2 and changes in carbon allocation patterns will also reduce potential gains in NPP (Poorter, 1993). Photosynthetic acclimation is often related to decreased leaf N content and photosynthetic enzyme activity (Moore *et al.*, 1999). Decline in foliage N content under CO_2 enrichment is common in semi-arid ecosystems (King *et al.*, 2004). Reduced N can limit the production of new sink tissues when assimilation rates are improved (Lee *et al.*, 2001c). This source/sink imbalance causes accumulation of carbohydrates, leading to feedback-limited photosynthesis (Causin *et al.*, 2004). Available evidence demonstrates that photosynthetic apparatus does acclimate to elevated CO_2 in semi-arid species, but the extent of this acclimation is not entirely known. In Mediterranean species, photosynthetic capacity is commonly lower under elevated CO_2 (Niinemets *et al.*, 1999c; Tognetti *et al.*, 2000), although in steppe species some studies have found (LeCain *et al.*, 2003), but others have not (Anderson *et al.*, 2001) the reduction in the photosynthetic capacity.

29.5.3. Elevated CO_2 influences on LAI and plant functional types

Improved NPP can also be explained by enhanced LAI under elevated CO_2 (Osborne *et al.*, 2000). High whole-season LAI is associated with larger soil-water storage as well as delayed leaf senescence at elevated CO_2 (Zavaleta, 2001). However, larger LAI can offset the reduction in water use per A_{leaf} under elevated CO_2. Over the long term this can lead to reductions in soil-water content under elevated CO_2 (Ham *et al.*, 1995). This in turn will alter LAI until a new equilibrium between soil water and community LAI is achieved.

CO_2 enrichment can importantly alter plant-community structure and composition of key plant functional types in semi-arid environments. For instance, the growth and expansion of woody vegetation is accelerated in steppe ecosystems, possibly as the result of positive effects of increased soil-water stores on species with tap roots (Nelson *et al.*, 2004). Significant changes in plant-functional-type spectra can have major consequences on NPP of semi-arid ecosystems (Lloret *et al.*, 2004).

Among species with different photosynthetic pathways there is abundant evidence that productivity of C_3 species responds more to increased CO_2 than that of C_4 species (e.g., Morgan *et al.*, 2007). Substantial reduction of the area occupied by C_4 grasses is expected to occur under elevated CO_2 (Owensby *et al.*, 1999). Nevertheless, growth of C_4 species is also enhanced by CO_2, with the sensitivity often being similar to C_3 species (Morgan *et al.*, 2001).

In addition, low N concentration in semi-arid ecosystems may limit the photosynthetic advantage of C_3 grasses under elevated CO_2 (LeCain *et al.*, 2003). However, N-fixing C_3 species are expected to become more competitive under enhanced N limitations. CO_2 enrichment in several Mediterranean grasslands has resulted in increased coverage of legumes (Grünzweig and Körner, 2001).

29.5.4. Warming and reduced precipitation

While elevated CO_2 is expected to preferentially affect productivity of semi-arid ecosystems, predicted warming and reductions in precipitation can negatively affect GPP, offsetting the positive effects of increased atmospheric CO_2 (Allard *et al.*, 2008). For instance, in the Mediterranean basin a large decrease of April–September total rainfall is predicted, increasing the duration and severity of drought (Christensen *et al.*, 2007). Abnormally hot years that occurred recently in the Mediterranean basin led to negative NEP (Peñuelas *et al.*, 2007), challenging the hypothesis that future climate change will result in an enhancement of plant growth and carbon sequestration in semi-arid environments. Desertification of semi-arid ecosystems during the past decades is estimated to have contributed to approximately 20% of the global anthropogenic CO_2 effect on the atmosphere over the same period (Rotenberg and Yakir, 2010).

Alternatively, higher temperatures can enhance photosynthetic activity during winter and can promote early spring enhancement in NPP, especially at low latitudes. In semi-arid environments, the main growing season is expected to move towards winter months. This will affect canopy photosynthetic potentials over the seasons and alter the transitions from carbon sink to source, and *vice versa* (Ham and Knapp, 1998).

Higher temperatures in winter/spring may also increase the available thermal budget for growth and reproduction of insect herbivores. The increase of herbivory pressure on the vegetation will impact semi-arid ecosystems through reductions in LAI and whole-canopy carbon assimilation (Allard *et al.*, 2008). Alternatively, higher concentrations of carbon-based protective compounds in plants under elevated CO_2 may reduce herbivory pressure (McDonald *et al.*, 1999).

Increases in temperature may also alter the dominance of plant functional and photosynthetic types with profound influences on NPP of semi-arid environments. For instance, an increase of 2°C in mean annual temperature in Great Plains is expected to reduce the coverage of C_3 grasses by 50% (Epstein *et al.*, 1997).

Overall, global change (i.e., not only climate change but also changes in land use, perturbation regime, habitat degradation etc.) in semi-arid environments includes a wide array of factors influencing NPP at various spatial and temporal scales. Intricate interactions between these drivers complicate the prediction of global-change effects on NPP of semi-arid ecosystems, so more experimental and modelling work is needed to understand the complex interplay between the key climatic and global-change drivers on semi-arid ecosystems.

29.6. CONCLUDING REMARKS

Semi-arid refers to a heterogeneous collection of ecosystems with unique precipitation and temperature regimes. Although the term embraces a wide range of environments, seasonal water limitation and excess irradiance are the main limitations in all these environments, while heat and cold stress may also affect productivity in some of the semi-arid ecosystems. Key structural adaptations to cope with these stressful environments include sclerophylly, deep root systems, pubescence, vertical foliage-inclination angles and stronger foliage aggregation, while key physiological modifications include advanced stomatal regulation of water use, modifications in photosynthesis type and effective photoprotection and antioxidative capacity. As photosynthesis is strongly limited by water in these ecosystems, semi-arid ecosystems are expected to respond particularly strongly to elevated atmospheric CO_2 concentrations. However, the positive effects of elevated CO_2 can be offset by concurrent increases in temperature and reductions in precipitation.

30 • Ecophysiology of photosynthesis in temperate forests

C.R. WARREN, J.I. GARCÍA-PLAZAOLA AND Ü. NIINEMETS

30.1. THE TEMPERATE-FOREST ENVIRONMENT

The temperate zone is characterised by pronounced seasonality with temperatures of the warmest month generally higher than 10°C and temperatures of the coldest month generally between –10 and 10°C (Köppen, 1936; Russell, 1931). Temperature is arguably the most important climatic variable in temperate forests. Temperatures during warm and cold periods are strongly variable within the temperate-forest biome, depending on continentality, latitude and topography (Fig. 30.1). Total precipitation is generally greater than 50–75 cm year^{-1} and is more uniformly distributed over the year than in arid (Chapter 28) and in semi-arid (including Mediterranean ecosystems) (Chapter 29) habitats. The annual input of solar radiation is between 2500–6000 MJ m^{-2}, varying with site latitude, cloudiness and topography (Jarvis and Leverenz, 1983).

Temperate forests are dominated by deciduous trees in oceanic and continental areas of the Northern hemisphere, while evergreens dominate in warmer locations and in the Southern hemisphere. In the edges of temperate biome, mixed forest may appear. Thus, on the cold border the transition to steppes is characterised by open conifer or deciduous forests, while there are mixed conifer-deciduous woodlands in the transition to the boreal biome. In the warm border, the transition is characterised by subtropical evergreen forests in humid locations and by the presence of deciduous Mediterranean oaks in more arid sites.

The productivity and biomass of temperate forests is the second largest of all forested biomes – only being exceeded by tropical forests (Saugier *et al.*, 2001). High average productivity of temperate forests is associated with benign environmental conditions that result in long growing seasons of 140–200 days in temperate deciduous forests and from 200 to more than 250 days in temperate evergreen forests (versus 100 for desert and arctic tundra) (Chapin *et al.*, 2002). Another factor correlated with the fast growth of temperate forests is their high average leaf area index (LAI). The average LAI of temperate forests of 6.0 m^2 m^{-2} is similar to the average LAI of tropical forests (Gower, 2002).

Although these average values characterise the temperate-forest biome as a whole, net primary productivity (NPP), biomass and LAI differ widely within the temperate-forest biome (Reich and Bolstad, 2001). The primary reasons for the heterogeneity among temperate forests are as follows:

1. The temperate-forest biome encompasses a wide range in temperatures and annual precipitation and differs in the extent of seasonality. This leads to differences in the frequency and severity of exposure to stresses and thus large differences in growth rates, standing biomass and LAI. For example, annual NPP tends to be larger in temperate forests with higher mean annual temperature (Reich and Bolstad, 2001), whereas stand LAI and productivity decrease with site aridity.
2. Temperate forests occur on a wide range of soils, owing to variation in soil parent material, age of the soil, climate and topography (Jenny, 1941). As the result of variation in these soil characteristics, the absolute amounts and proportions of different nutrients and water-holding capacity of the soil vary and this has important implications for photosynthesis and growth.
3. The heterogeneity among temperate forests is also driven by historical factors such as the maximum extent of ice during the major ice ages and presence of refugia, isolation and differences in speciation rate (Qian and Ricklefs, 2000; Ricklefs, 2004). Isolation has led to differences in average phylogenetic distances among the forests. Such differences can be large, for example, angiosperm-dominated versus gymnosperm-dominated forests (see Fig. 30.2) or more subtle, such as differences

Terrestrial Photosynthesis in a Changing Environment: A Molecular, Physiological and Ecological Approach, ed. J. Flexas, F. Loreto and H. Medrano. Published by Cambridge University Press.

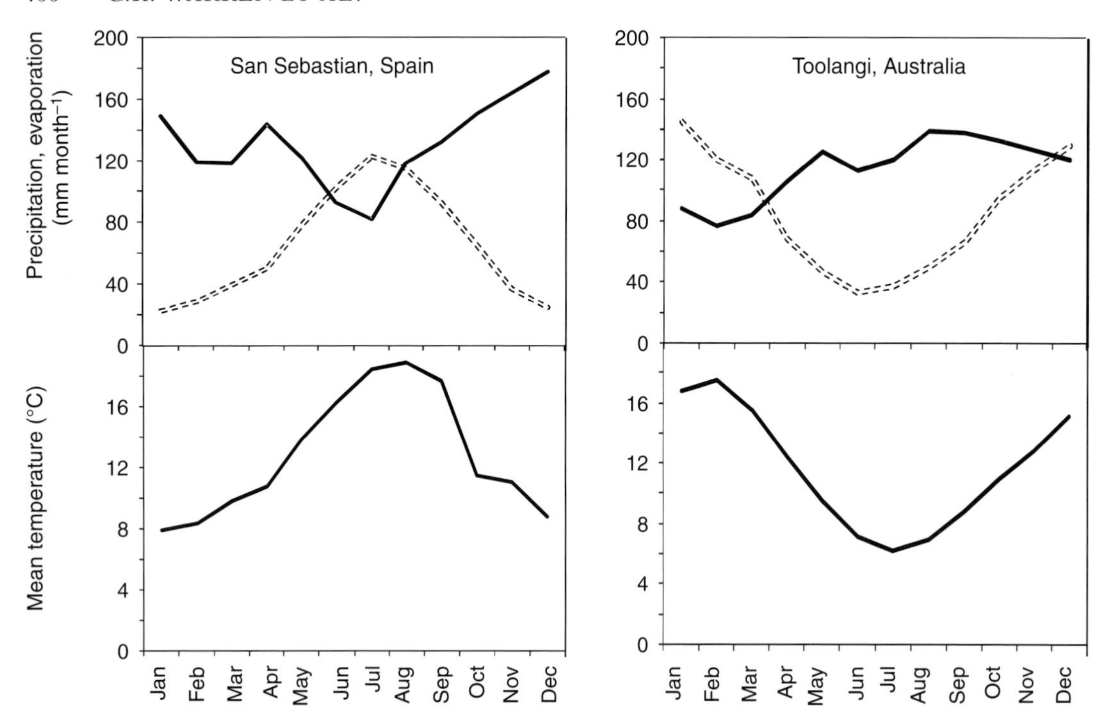

Fig. 30.1. Representative ombrothermic diagrams for Toolangi (south-eastern Australia) and San Sebastian (Spain). Precipitation is given by a solid line and evaporation by a dashed line.

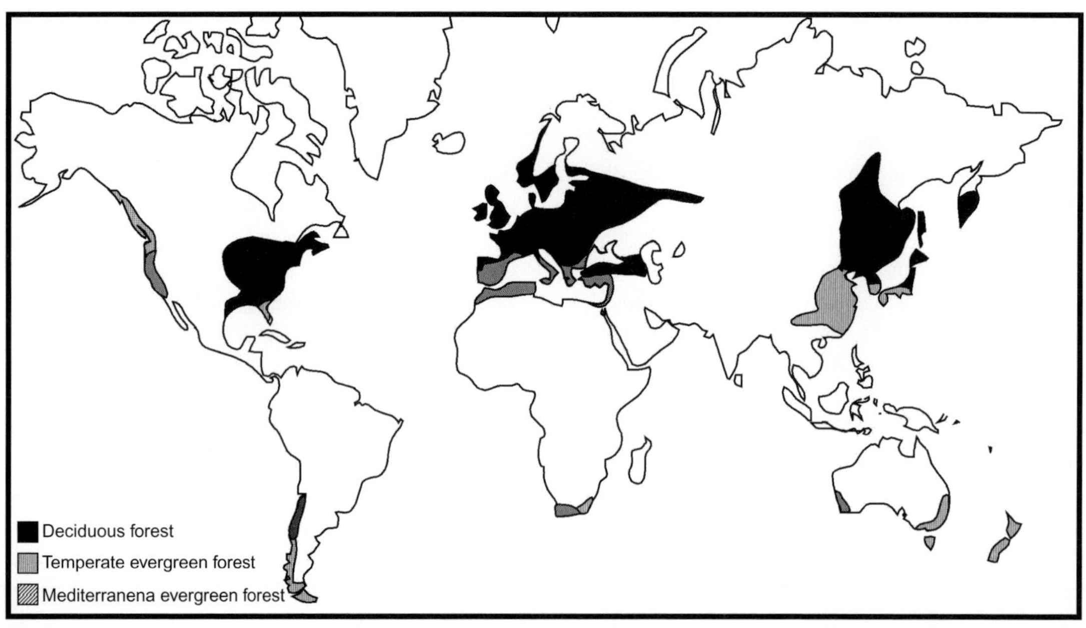

Fig. 30.2. World distribution of temperate forests. Deciduous forests dominate temperate regions, being replaced by conifers in more continental or dry sites, and by broad-leaf (or conifer) evergreen rainforest in warmer locations. Mediterranean vegetation (Chapter 29), that can be also considered 'temperate', occurs where summer precipitation is low.

Table 30.1. *Some examples of the effect of environmental conditions on net photosynthesis at saturating light (A_N) in leaves of beech* (Fagus sylvatica). *Data are expressed as percentage of their respective control values that ranged between 4.3 and 10.8 μmol CO_2 m^{-2} s^{-1}.*

Factor	% of control A_N	Source
UV-B (23% higher biological effective UV-B)	76	Zeuthen *et al.*, 1997
Ozone (71 vs 32 ppb)	79	Zeuthen *et al.*, 1997
Ageing (mid Sept) relative to early summer	64	Zeuthen *et al.*, 1997
Shade (15% full sunlight)	47	Valladares *et al.*, 2002
Change to high light	118	Aranda *et al.*, 2004
Water stress (mesic population)	35	Tognetti *et al.*, 1995
Water stress (xeric population)	50	Tognetti *et al.*, 1995
Drought (late summer relative to late spring)	26	Raftoyannis and Radoglou 2002
High CO_2 (740 μmol mol^{-1}) relative to 400 μmol mol^{-1}	185	Leverenz *et al.*, 1999
Increased temperature (+4.8°C) relative to ambient temperature.	70	Leverenz *et al.*, 1999

in dominance of certain genera or families. Functional characteristics such as evergreenness or deciduousness can also be partly associated with historical factors (Axelrod, 1966).

These data collectively demonstrate that the limitations of photosynthesis vary among temperate forests. In many cases, these limitations can vary just as much within forests owing to spatial and temporal constraints. These spatial and temporal constraints and how they affect gas exchange of forest canopies and ecosystems are dealt with at length in Section 30.4. What follows is a brief review of the key limitation of photosynthesis in temperate-forest species. As an introduction to the section we have summarised the key limitations of photosynthesis in *Fagus sylvatica* (Table 30.1). While *F. sylvatica* may not be representative of all temperate-forest species, it is widespread, is one of the best-studied species and serves as a useful example of the many limitations of photosynthesis.

30.2. DIFFERENCES IN PHOTOSYNTHESIS AMONG AND WITHIN FUNCTIONAL GROUPS

30.2.1. Functional-type spectrum in temperate forests

Tree species give the classic physiognomic appearance to temperate forests, but temperate forests also support a rich understory of lichens, mosses, herbs and graminoids, and woody and dwarf shrubs. At the warm, lower latitude limit of temperate forests vines and mistletoes can also be important forest components. Major functional groups among woody species are evergreen conifers, broad-leaved deciduous and broad-leaved evergreen angiosperms.

The functional-type spectrum of temperate forests is largely a function of temperature and water availability. Minimum temperatures during winter constrain the dispersal of broad-leaved evergreen shrubs, trees and vines at the cooler end of the temperate biome (Box, 1981), whereas maximum temperatures often interact with low water availability at the warmer end of the temperate biome (Buckland *et al.*, 1997; Peñuelas and Boada, 2003; García-Plazaola *et al.*, 2008b) and constrain the dispersal of more mesic species. Warmer winter temperatures are associated with temperate forests that are generally darker (i.e., larger LAI) and taller (Ohsawa and Nitta, 1997; Hiroki and Ichino, 1998; Lusk and Smith, 1998). This comes about because warmer winter temperatures lengthen the growing season, thereby allowing leaves to pay back for their construction cost and achieve a positive carbon balance in lower light. In contrast, when hotter maximum temperatures interact with drought, the forests become more open and the fraction of drought-tolerant shrubby species increases (Specht and Moll, 1983; Jarvis and Sandford, 1986).

Within any given plant functional type there is large variability in species tolerance to key environmental drivers (Fig. 30.3). These species differences in tolerance determine the equilibrium species composition at different sites and also largely govern the responsiveness of these communities to disturbance (succession) (Bellingham *et al.*, 1996; Lusk, 1996; Sykes and Prentice, 1996) and to exceptionally strong stresses such as very severe drought (Buckland *et al.*, 1997) and flooding (Vervuren *et al.*, 2003). As species of varying tolerance

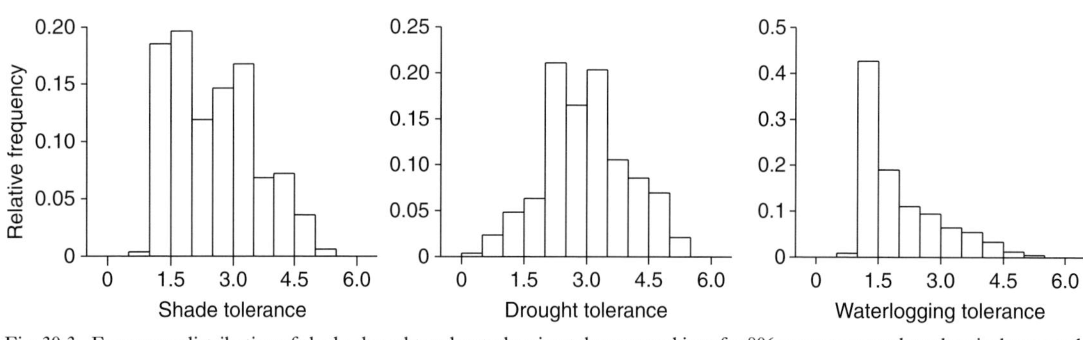

Fig. 30.3. Frequency distribution of shade, drought and waterlogging tolerance rankings for 806 temperate northern hemisphere woody species (data from Niinemets and Valladares, 2006). Tolerance increases with increasing the ranking value (1=very intolerant, 5=very tolerant). Species tolerance is a relative estimate based on species capacity to survive and grow under given environmental limitations relative to other species. Consensus tolerance estimates have been obtained on the basis of a large number of cross-calibrated tolerance rankings.

also often have unique combinations of foliage structural and physiological traits (e.g., Latham, 1992; Niinemets and Kull, 1994), large variation in foliage photosynthetic characteristics is present within any given plant functional type.

30.2.2. Differences in photosynthetic potentials among functional types

Temperate forest species vary widely in maximum net photosynthesis (A_N). There is a general trend for broad-leaved deciduous species to have higher A_N than evergreen broad-leaved species and conifers (Niinemets *et al.*, 2007, Fig. 30.4). Differences in A_N among functional types are consistent with the negative correlation of leaf A_N with leaf longevity (Wright *et al.*, 2004), suggesting that part of the answer may lie in understanding how longevity affects A_N. Long leaf lifespans are associated with lower foliage nitrogen (N) contents and photosynthetic nutrient use efficiency (NUE) (A_N/N). Poor photosynthetic NUE probably reflects fundamental constraints of long leaf lifespans on the N economy of leaves. Nitrogen concentrations are smaller in the long-lived foliage of conifers and evergreen broad-leaved species than the short-lived foliage of broad-leaved deciduous species (Reich *et al.*, 1995; Wright *et al.*, 2004, for the comparison in Fig. 30.5, average±SE A_N/N was 5.77±0.15 μmol g N^{-1} s^{-1} for the deciduous broad-leaved species *Populus tremula* and 3.34±0.05 μmol g N^{-1} s^{-1} for the evergreen conifer *Pinus sylvestris*). This reflects the requirement of long-lived foliage for a larger fraction of support mass (Niinemets *et al.*, 2007). For example, around 50% of the volume of a long-lived conifer needle is support tissue such as the central cylinder, epidermis and hypodermis (Niinemets *et al.*, 2007), while long-lived leaves of broad-leaved evergreens have a large fraction of leaf mass in the cell walls (Takashima *et al.*, 2004). Lower photosynthetic NUE in species with long-living foliage likely results from investing a smaller fraction of total foliage N in compounds with photosynthetic functions (i.e., the photosynthetic machinery) and from a smaller mesophyll-diffusion conductance (g_m). Having a smaller fraction of N in the photosynthetic machinery may be an unavoidable consequence of long-lived foliage having to allocate a larger fraction of N to cell walls (Takashima *et al.*, 2004) and/or N-rich secondary metabolites (Gleadow *et al.*, 1998). Structural constraints of having long-lived foliage may also explain smaller g_m. In the case of conifers, smaller g_m is likely owing to their thicker leaves with more diluted photosynthetic enzymes, whereas long-lived foliage of broad-leaved evergreens has similarly small g_m owing to their tightly packed mesophyll cells with thick walls (Niinemets and Sack, 2006; Niinemets *et al.*, 2009b,c). Smaller g_m causes lower chloroplastic CO_2 concentrations and lower use efficiency of photosynthetic N (Warren and Adams, 2006) in leaves with more robust structures (Niinemets and Sack, 2006; Niinemets *et al.*, 2009b), such as long-lived leaves. This evidence collectively suggests that lower photosynthetic NUE is an inherent constraint of leaves with greater longevity (Warren and Adams, 2004).

30.2.3. Differences in light harvesting among functional types

Evergreen species compensate for their slower photosynthesis by having longer-lived foliage that allows accumulation of more foliage area (Schulze *et al.*, 1977). This is supported by a meta-analysis that demonstrates LAI is greater in species with longer leaf lifespans (Fig. 30.5A, Niinemets, 2009). This relationship is particularly important for conifers with large interspecific variation in longevity (Fig. 30.5A). As a

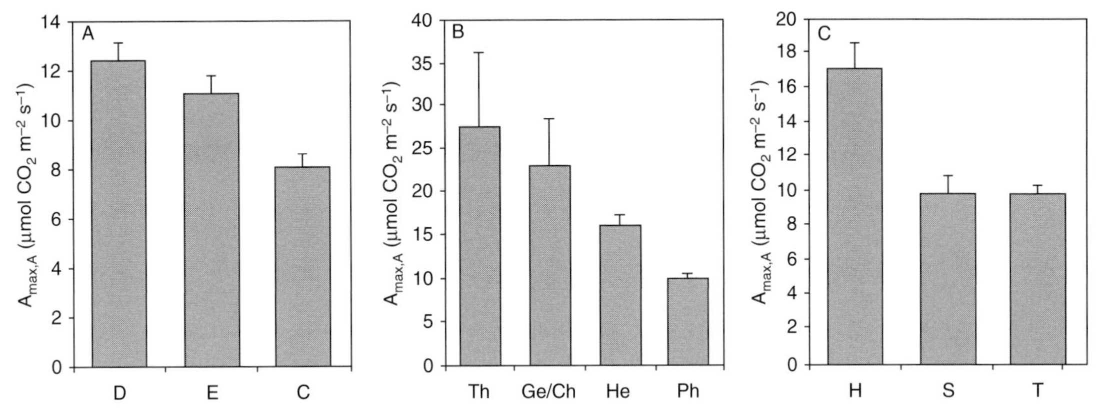

Fig. 30.4. Mean (±SE) photosynthetic rates per unit area of well-watered, non-stressed temperate forest species. Photosynthetic capacities are separately plotted in dependence on **(A)** leaf habit (D, deciduous, n=83; C, conifers, n=27; E, evergreen, n=64); **(B)** life form (Th, therophytes, n=2; He, hemicryptophytes, n=27; Ge/Ch, geophytes+chamaephytes, n=3; Ph, phanerophytes, n=142); and **(C)** growth form (H, herbaceous, n=33; S, shrubs, n=20; T, trees, n=122). Values were taken from previous studies performed in the field and in controlled environments. When different estimates were reported for the same species, the highest value was considered. Varieties or cultivars of the same species were considered as a unique entry. Only the measurements performed at mid-morning were considered. Different letters denote significant differences among the means according to Duncan's test (P>0.05).

trade-off for greater longevity and larger foliage area, foliage A_N tends to be lower in species with greater LAI (Fig. 30.4B, Reich *et al.*, 1992). This compromise between longevity and A_N is evident for temperate conifers and deciduous broad-leaved species, and can partly explain why temperate tree stands dominated by species with varying maximum LAI do not appreciably differ in NPP (Reich and Bolstad, 2001).

Extensive foliage area can also bring about stronger self-shading within the canopy. Most simple light-interception models assume that foliage is randomly dispersed. A canopy with random dispersion of foliage elements that have uniform angular distribution (the extinction coefficient, k in Lambert-Beer equation is equal to 0.5), intercept 95% of light when LAI is 6 m^2 m^{-2} (LAI for 95% light interception, L_{95}=–ln0.05/0.5). Yet, temperate evergreen conifer stands commonly support LAI more than 10 m^2 m^{-2} (Jarvis and Leverenz, 1983). Evergreen trees with small leaves and high foliage area compensate for potentially high self-shading by stronger aggregation of leaves in shoots, branches and crowns (Cescatti, 1998; Kucharik *et al.*, 1999; Cescatti and Niinemets, 2004). Yet, stronger aggregation also means lower light-harvesting efficiency relative to randomly or regularly dispersed foliage. Besides the lower A_N of leaves with greater longevity (lower light-use efficiency), lower light-harvesting efficiency of strongly aggregated canopies can also provide an explanation of similar productivity of evergreen and deciduous stands.

Variations in spatial aggregation and size of foliage elements affect the way photosynthesis responds to variation in the relative amount of direct and diffuse irradiance. Diffuse irradiance penetrates deeper into the canopies, such that canopy scale photosynthesis increases as the ratio of diffuse to direct irradiance increases (Gu *et al.*, 2003; Pielke Sr. *et al.*, 2007). In particular, more-strongly aggregated canopies and canopies with larger LAI are expected to respond more strongly to diffuse irradiance than less-clumped canopies (Gu *et al.*, 2003; Cescatti and Niinemets, 2004). Although it is currently somewhat unclear how diffuse/direct irradiance ratio will be affected by climate change (Stanhill and Cohen, 2001; Pielke Sr. *et al.*, 2007), temporal modifications in the diffuse/direct-irradiance ratio, such as after enhanced volcanic activity (Gu *et al.*, 2003), can favour evergreen species with clumped canopies and high LAI. In addition, contrasting responses of the diffuse/direct-light ratio have important implications for the competitive potentials of plants colonising sites with long-term differences in the diffuse/direct-radiation ratio, e.g., temperate oceanic climates where the fraction of diffuse radiation is high owing to frequent cloudiness relative to more continental forests where clear, cloud-free periods dominate.

However, these predictions of vegetation response to the diffuse/direct-light ratio do not consider penumbra, i.e., the semi-shade between sunlit- and shaded-leaf area fractions owing to the solar disc not being a point source but of finite size. The degree of penumbral versus direct-beam radiation is larger in canopies with a smaller leaf-size-to-canopy-height ratio (Cescatti and Niinemets,

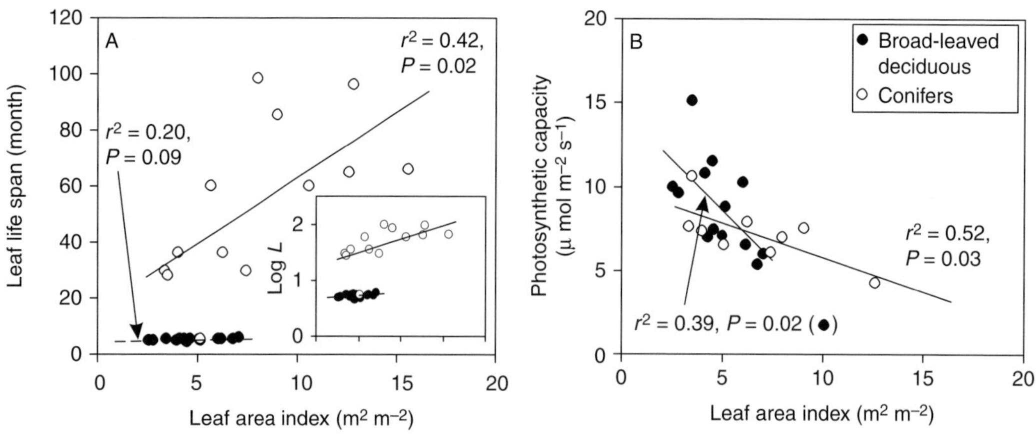

Fig. 30.5. Correlations between leaf longevity (**A**) (inset for correlation with log-transformed longevity, Log*L*) and leaf photosynthetic capacity of current year foliage (**B**) with stand leaf-area index in temperate broad-leaved deciduous trees (filled symbols) and temperate conifers (open symbols). Leaf longevity and photosynthetic capacity data come from the Wright *et al.* (2004, Glopnet database) and Niinemets (1999) databases. All data are expressed per unit projected leaf area. The deciduous broad-leaved species included were *Acer platanoides* (the data of stand leaf-area index from Rauner, 1976), *Acer rubrum* (Jurik, 1986), *Acer saccharum* (Ellsworth and Reich, 1993), *Betula ermanii* (Tadaki, 1977), *Betula pendula* (Rauner, 1976), *Betula platyphylla* (Tadaki, 1977), *Fagus crenata* (Kira, 1975), *Fagus sylvatica* (Bouriaud *et al.*, 2003; Møller, 1946), *Liriodendron tulipifera* (Hutchinson and Matt, 1977), *Populus grandidentata* (Jurik, 1986), *Populus tremula* (Niinemets and Tamm, 2005; Rauner, 1976), *Populus tremuloides* (Jurik, 1986), *Quercus robur* (Rauner, 1976), *Quercus rubra* (Gower *et al.*, 1993; Jurik, 1986). Temperate conifers included were *Abies alba* (Cescatti and Zorer, 2003), *Abies lasiocarpa* (Jack and Long, 1992), *Larix decidua* (Gower *et al.*, 1993), *Picea abies* (Jarvis and Leverenz, 1983; Møller, 1946; Nihlgård, 1972), *Picea sitchensis* (Leverenz and Hinckley, 1990; Norman and Jarvis, 1974), *Pinus banksiana* (Magnussen *et al.*, 1986), *Pinus contorta* (Jack and Long, 1992; Leverenz and Hinckley, 1990), *Pinus ponderosa* (Oren *et al.* 1987), *Pinus resinosa* (Gower *et al.*, 1993), *Pinus strobus* (Gower *et al.*, 1993), *Pinus sylvestris* (Møller, 1946; Jarvis and Leverenz, 1983), *Pseudotsuga menziesii* (Ungs, 1981), *Tsuga heterophylla* (Gholz, 1982). When several estimates of leaf-area index were available, averages were calculated. Data were fitted by linear regressions. The correlations for all data pooled were r^2=0.51, P<0.001 for A and r^2=0.38, P<0.001 for B.

2004). Although the above-canopy light field can be dominated by high-intensity beam radiation, most of the light deeper in the canopy of conifers with small leaf size is penumbral (Cescatti and Niinemets, 2004). This may reduce the differences between conifers and broad-leaved species to varying diffuse/direct-irradiance ratios. Clearly more numerical simulations and experimental work are needed to understand the responses of temperate canopies with various leaf physiognomy, LAI and aggregation to variations in radiation geometry.

30.2.4. Phenological and age-dependent differences

Temperate evergreen forests have low canopy transmittance throughout the year, as the understory is always shaded. In contrast, light availability in the understory of temperate deciduous forests is strongly variable in time: for most of the growing season only 1–5% of light reaches the understory, whereas when the overstory is leafless between 30–50% of light reaches the understory. This temporal variability in light availability is significant in early and late season when air temperatures are high enough to support photosynthesis. Overstory individuals of temperate tree species generally have a conservative phenological strategy, with budburst occurring relatively late in the spring thereby reducing the risk of late-spring frosts that may damage the first leaf flush (Murray *et al.*, 1994). In contrast, herbs and seedlings and saplings of several temperate deciduous species leaf earlier than overstory species, and can thereby take advantage of higher light availabilities at the forest floor (Augspurger and Bartlett, 2003; Augspurger *et al.*, 2005). One study showed that the earlier bud burst and foliage development of deciduous understory plants meant that they intercepted 36–98% of their total annual light interception when the overstory was leafless during spring (Augspurger *et al.*, 2005).

In warm temperate forests, evergreen species can exhibit positive photosynthesis for the majority of the winter period. Winter photosynthesis in the understory can be particularly significant when winter conditions are associated with improved light availabilities. In several temperate forests with moderately cold winters, the canopy is often dominated by evergreen and deciduous species, whereas broad-leaved evergreen shrubs and short trees dominate the understory (Miyazawa and Kikuzawa, 2006). The understory evergreens can fix more than 50% of their annual carbon gain during winter when light availability is higher owing to partial leaf loss in the overstory (Lassoie *et al.*, 1983; Miyazawa and Kikuzawa, 2005a, 2006).

The world-scale general patterns of leaf functioning have been developed for current-year leaves exposed to full sunlight. In addition to the differences in photosynthetic potentials in any specific snapshot of time and canopy location, it is also important to be aware of temporal differences in photosynthetic potentials. So far, little attention has been paid to functional-group differences in age-dependent changes in A_N. In broad-leaved deciduous species, foliage A_N stay constant after leaf expansion during most of the growing season (Wilson *et al.*, 2001; Niinemets *et al.*, 2004a) and then rapidly decline in senescent leaves. In evergreen species, A_N decline slowly with leaf age until leaf senescence shortly before leaf abscission (Brooks *et al.*, 1994; Niinemets *et al.*, 2005a; Warren, 2006a). In broad-leaved evergreens, age-dependent decline is mostly associated with decreases in g_m (Niinemets *et al.*, 2005a). In conifers, this decline is often also associated with reductions in N content per A_N and possibly also with inactivation of Rubisco protein (Brooks *et al.*, 1994; Niinemets *et al.*, 2001; Ethier *et al.*, 2006; Chapter 21). Age-dependent reduction in A_N in older leaves can be an additional factor explaining why the NPP of temperate deciduous stands with lower LAI and evergreen conifer stands with larger LAI is similar in comparable climates (Reich and Bolstad, 2001).

Different species may have similar integrated whole-canopy productivity under any given set of environmental conditions, but these species may respond differently to modifications in the environment if they have differing age spectra of photosynthetic biomass. There is evidence that the re-acclimation potential of fully developed mature leaves to altered environmental conditions is limited (Brooks *et al.*, 1994, 1996; Oguchi *et al.*, 2005; Niinemets *et al.*, 2006a,b), i.e., that mature leaves do not fully adjust to environmental conditions. This is because as cell walls become thicker and more lignified with increasing leaf age, foliage re-acclimation potential decreases (Niinemets *et al.*, 2006a,b). Thus, the whole-canopy acclimation response to altered environmental conditions is expected to decrease with increasing average canopy leaf age. Evergreen species with large numbers of leaf cohorts require several growing seasons to replace all the foliage in their canopy, which highlights that full adjustment to long-term environmental perturbations necessarily requires several growing seasons. Limited stress adjustment can partly explain greater sensitivity of coniferous vegetation to environmental pollution (Bussotti and Ferretti, 1998; Slovik, 1996) and catastrophic stress events.

30.2.5. Differences in photosynthetic potentials within functional types

There is large variation in A_N within each plant functional type. Variation in A_N is related to species-specific capacities to tolerate environmental stresses (Fig. 30.3 for species distributions according to tolerance to major environmental stresses). In particular, shade-intolerant early successional species generally have larger A_N at any given canopy position than shade-tolerant late-successional species (Niinemets *et al.*, 1998, Fig. 30.6; Reich *et al.*, 2003). In mature plants of temperate broad-leaved species, larger A_N of early successional species can partly result from greater LMA and associated larger N contents per area, i.e., stacking of photosynthetic mesophyll (Niinemets, 2006; Valladares and Niinemets, 2008). Typically, larger LMA of mature shade-intolerant temperate species is in marked contrast to the variation patterns in temperate seedlings and in tropics, where LMA of shade-intolerant species is generally lower than that of tolerant species (Lusk and Warton, 2007 for review). In addition, greater photosynthetic NUE of shade-intolerant species can contribute to larger A_N (Niinemets *et al.*, 1998; Reich *et al.*, 2003, Fig. 30.6). Differences in NUE between shade-intolerant and tolerant species reflect the conflicting N requirements for larger A_N, that require large investments of photosynthetic N in Rubisco and other rate-limiting proteins, and for enhanced light-harvesting efficiency that requires large N investments in chlorophyll and pigment-binding proteins (Evans and Seemann, 1989; Hikosaka and Terashima, 1996).

Apart from shade tolerance, less is known of the correlations between A_N and tolerance to other environmental factors. Shade, drought and waterlogging tolerance tend to be reversely correlated across temperate species (Niinemets and Valladares, 2006), but whether this is associated with analogous reverse rankings in A_N is not clear. Commonly, drought-tolerant species are characterised by a

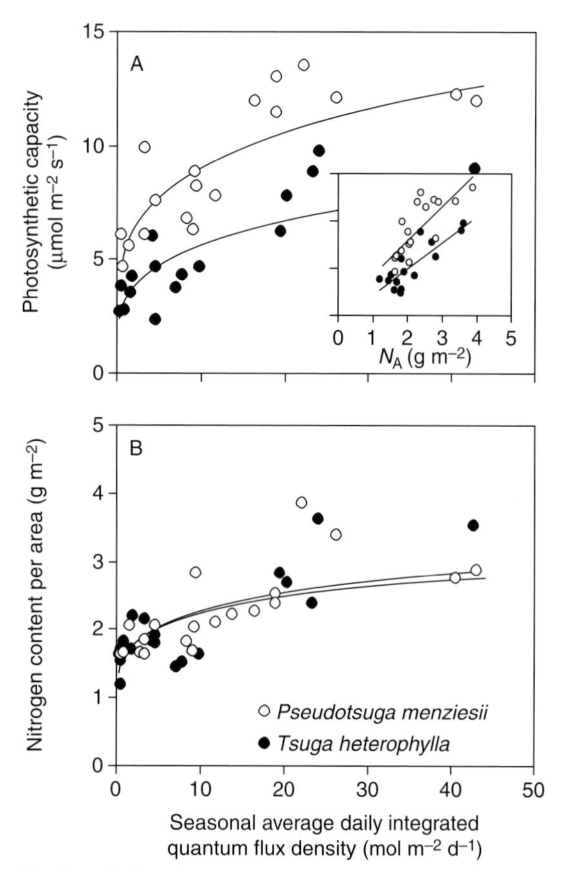

Fig. 30.6. Foliage photosynthetic capacity per unit needle projected area (A_N) (**A**) and needle nitrogen content per unit projected area (N_A) (**B**) in relation to seasonal average daily light availability in relatively shade-intolerant temperate evergreen conifer *Pseudotsuga menziesii* (shade tolerance rank 2.8, Niinemets and Valladares, 2006) and in very tolerant temperate evergreen conifer *Tsuga heterophylla* (tolerance rank 5.0) (data modified from Bond *et al.*, 1999). The inset demonstrates the correlation between A_N and N_A. Data were fitted by non-linear regressions in the main panel and by linear regressions in the inset (all significant at $P<0.001$).

more robust leaf structure that improves tolerance of low leaf-water potentials and improves water extraction from drying soil (e.g., Niinemets, 2001; Lamont *et al.*, 2002). The more robust foliage of drought-tolerant species can result in stronger limitation of A_N by g_m (Niinemets and Sack, 2006). Alternatively, there is evidence of greater investment of nitrogen per A_N and larger A_N in drought-adapted species (Wright *et al.*, 2003; Xu and Baldocchi, 2003). High A_N of drought-tolerant plants allows these plants to assimilate carbon with high rates during short periods of year when soil-water availability is high (Xu and Baldocchi, 2003).

In addition to the variation in A_N of fully mature leaves, important differences can also exist in species phenology. Broad-leaved deciduous early successional species are commonly characterised by a shorter leaf lifespan and greater leaf turnover (Kikuzawa, 1983; Koike, 1987, 1990). Spatial variation in leaf senescence, starting from the outer-canopy surface or from the canopy interior, can also vary among species of different successional positions (Koike, 1990). Such phenological differences can significantly alter the temporal variation in canopy photosynthetic productivity as well as whole-canopy photosynthetic production.

30.3. KEY LIMITATIONS OF PHOTOSYNTHESIS IN TEMPERATE-FOREST SPECIES

30.3.1. Light

Temperate trees have a series of mechanisms that allow them to respond to variation in light in both the short term (e.g., sunflecks) and in the longer term (e.g., spatially within the canopy, seasonally and among years). Sunflecks do not result in an instantaneous increase in photosynthesis; instead it takes some minutes for photosynthesis to become fully inducted. Some temperate-forest species respond to sunflecks very quickly, whereas others do not. For example, the photosynthetic response of the temperate rainforest species *Nothofagus cunninghamii* to simulated sunflecks was rapid, around 90% complete within 3 minutes (Tausz *et al.*, 2005b). Other species, such as *Liquidambar styraciflua*, *Thuja plicata* and *Tsuga heterophylla*, are among the slowest species and take 30 or more minutes for photosynthesis to reach 90% of its maximum after light is rapidly increased (Naumburg and Ellsworth, 2000; Pepin and Livingston, 1997). Some authors have suggested that the rate of response to sunflecks is faster in shade-tolerant than shade-intolerant species; however, a review of published data did not find such a trend (Naumburg and Ellsworth, 2000). The only consistent trend is that gymnosperms tend to be slower to respond to sunflecks than angiosperms.

Sunflecks are potentially harmful to plants because of the time lag between the beginning of a sunfleck and the realisation of maximum rates of photosynthesis. This time lag leads to the quantum of absorbed irradiance being, at least temporarily, in excess of what is used in photochemistry (Watling *et al.*, 1997), and thus may give rise to photoinhibition (Osmond *et al.*, 1999). Few studies have examined photoinhibition in relation to sunflecks. In *Nothofagus cunninghamii*, photoprotective systems responded rapidly to excess excitation pressure during sunflecks, but photoinhibition was still not completely avoided (Tausz *et al.*, 2005b). Sunfleck-induced reduction of photochemical efficiency was retained even after 30 minutes in the dark, and was associated with a high degree of de-epoxidation state of xanthophyll pigments (Tausz *et al.*, 2005b) as is common in photoinhibited leaves (Demmig-Adams and Adams, 2006). Given that even modest photoinhibition may reduce light-limited photosynthesis by 15–20%, sunfleck-induced reductions in quantum efficiency warrant further research (Niinemets and Kull, 2001; Tausz *et al.*, 2005b).

In the longer term, leaves of temperate-forest species acclimate to the within-canopy gradient in light. This may take the form of variation in A_N per unit leaf area and/or variation in leaf light-harvesting efficiency per unit leaf dry mass (chlorophyll content per dry mass). Typically, A_N is positively related to light availability and may be two to four times greater in leaves from the top of the canopy than leaves from the bottom of the canopy (e.g., Niinemets, 2007; Warren *et al.*, 2007). In many species, the height-related increase in A_N may be almost entirely a function of the positive correlation of LMA and leaf thickness with light availability during leaf growth (e.g., Niinemets, 2007; Warren *et al.*, 2007). Trends in A_N are also partially owing to the positive correlation of photosynthetic capacity per dry mass ($A_{max,M}$), with light availability owing to a greater fractional investment of N in rate-limiting proteins of photosynthetic machinery in higher light. Nevertheless, in most species $A_{max,M}$ varies less than leaf mass per unit area (LMA) and thus the strong positive correlation of A_N with light is mainly driven by LMA (Evans and Poorter, 2001; Niinemets, 2007). The large A_N of leaves that develop under high irradiance results in a corresponding larger demand for CO_2. To supply these leaves with more CO_2, g_s and g_m is positively related to light availability (Niinemets *et al.*, 1998; Warren *et al.*, 2003a; Niinemets *et al.*, 2006a). Leaf chlorophyll content per dry mass is negatively related to light availability during leaf growth so as to increase light harvesting per unit leaf mass in low light (Niinemets, 2007; Hallik *et al.*, 2008). As the result of opposing trends in chlorophyll/mass and LMA with light, changes in chlorophyll per area (chl./area=chl./mass×LMA) are generally modest (Niinemets, 2007; Hallik *et al.*, 2008). With increasing light, plants increase the potential for safe dissipation of excess excitation energy by increasing the amount of xanthophyll-cycle carotenoids and tocopherols relative to chlorophyll and total carotenoids (Niinemets *et al.*, 2003; García-Plazaola *et al.*, 2004).

Leaves of temperate-forest species may also acclimate to changes in light availability occurring after full leaf expansion. For example, deciduous and evergreen species may face increases in light owing to gap formation in the overstory, while leaves of evergreen species are perennially faced with the problem of shading as new leaves are added to the expanding canopy (Schoettle and Smith, 1991; Niinemets *et al.*, 2006a,b). The acclimation of leaf structure to light is largely irreversible (Brooks *et al.*, 1996; Niinemets, 1997a; Niinemets *et al.*, 2006b); hence, re-acclimation of A_N and light-harvesting efficiency of mature leaves to modified light primarily involves N partitioning among photosynthetic proteins and alteration of chloroplast number. For example, A_N of shade-developed leaves gradually increases after gap formation owing to the re-allocation of Rubisco and rate-limiting components of photosynthetic electron transport (PET) from chlorophyll and pigment-binding complexes (Naidu and DeLucia, 1997; Oguchi *et al.*, 2005; Oguchi *et al.*, 2006). Such re-acclimation of photosynthesis of fully mature leaves to increased light availability also occurs in wintergreen understory species that make a significant part of their yearly carbon gain during autumn to spring months when the overstory is leafless (Katahata *et al.*, 2005; Miyazawa and Kikuzawa, 2005b). However, full re-acclimation of A_N to increased light intensity may be inherently constrained in fully developed leaves. This is because the leaves acclimated to low light have a small area of mesophyll cells exposed to intercellular air space, implying that extra chloroplasts can be added only to the extent that there are gaps along the exposed mesophyll cell wall (Oguchi *et al.*, 2005; Oguchi *et al.*, 2006).

Re-acclimation of A_N is slow and limited by anatomical constraints, and thus gap formation results in a temporary increase in excitation pressure. Plants combat this increase in excitation pressure by increasing pools of xanthophyll-cycle carotenoids and antioxidants. In one study, the pools of xanthophyll-cycle carotenoids and tocopherols increased within a day of increasing the light intensity, whereas full acclimation took 5–11 days (Niinemets *et al.*, 2003; García-Plazaola *et al.*, 2004). To an understory plant, the summer-to-autumn transition in a deciduous forest results in a rapid increase in light that is analogous to gap formation. The summer-to-autumn transition increases the risk of photoinhibition in understory evergreens not only by increasing light intensity but also by the occurrence of low winter temperatures. In contrast, the winter-to-spring transition decreases the risk of photoinhibition in understory plants via the reduction in irradiance as leaf area accumulates in the overstory and the increase in air temperatures. Hence, the winter-to-spring transition is associated with reductions in photoprotective pigment concentrations in understory plants (García-Plazaola *et al.*, 1999; Kyparissis *et al.*, 2000). These dynamic changes in xanthophylls and tocopherols provide a means for rapid acclimation to excess excitation energy (Demmig-Adams and Adams, 2006).

Canopy expansion causes older foliage to be shaded by younger foliage. Shading of non-senescent foliage causes reallocation of N from Rubisco and components of PET to light-harvesting pigment-binding complexes (Fig. 30.7). These changes are associated with reductions in A_N owing to decreases in maximum Rubisco carboxylase activity and the photosynthetic electron transport capacity (Niinemets *et al.*, 2006a,b), and increases in leaf absorptance as a result of increased chlorophyll content (Brooks *et al.*, 1996). Shading also decreases dark respiration, thereby improving net carbon balance (Brooks *et al.*, 1996). Reductions in light availability can also trigger leaf senescence and lead to retranslocation of N from older, shaded foliage to sunlit foliage (Weaver and Amasino, 2001). This mechanism has been suggested to play a major role in optimisation of plant carbon gain (Field, 1983). Although foliage N concentrations continuously decrease with leaf age, decreases in N contents per area are smaller (e.g., Niinemets *et al.*, 2005a,b,c) because the changes in N concentration are partially a function of dilution owing to age-dependent carbon accumulation in cell walls. Larger fluctuations in N concentrations can occur in evergreen species during bud burst, when N is retranslocated from older foliage to support the growth of new leaves (Weikert *et al.*, 1989). However, the N pools of old leaves are filled during the rest of the growing season (Weikert *et al.*, 1989). Given that both senescence and N retranslocation during new foliage growth are short-term phenomena, we conclude that bulk N reallocation plays a minor role in canopy carbon gain in temperate trees.

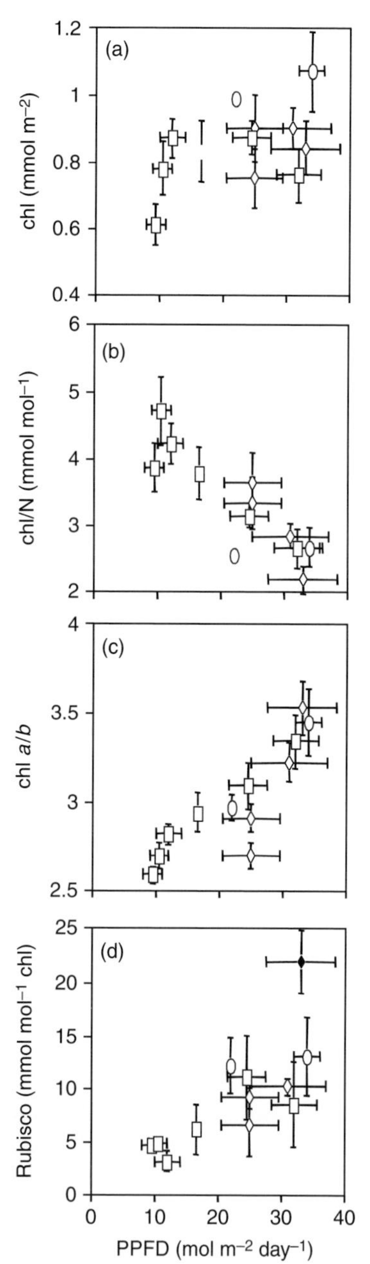

Fig. 30.7. Effects of within-crown variation in light availability on needle properties in a temperate evergreen conifer *Pinus pinaster*. Relationship between average daily PPFD and **(a)** the content of chlorophyll *a* and *b* per unit area (chlorophyll; **(b)** the quotient of total chlorophyll to total nitrogen (chlorophyll/N); **(c)** the quotient of chlorophyll *a* to *b* (chlorophyll *a*/*b*); and **(d)** ratio of Rubisco to chlorophyll. All measurements were made in August 1998. Foliage was collected from the lower canopy (□), middle canopy (◊) and upper canopy (○). Data are means of three replicates, error bars are ±SE. Redrawn from Warren and Adams (2001a).

30.3.2. Temperature

Ambient temperature varies diurnally, between days and seasons, among forests and owing to infrequent, but recurrent, extreme high or low temperatures. In fact, catastrophic episodes of high or low temperatures may restrict and define species-distribution ranges (Hamerlynck and Knapp, 1994). Within forests there are also important horizontal and vertical gradients in temperature. For instance, in a mixed deciduous canopy in Estonia, temperature was 5.5°C higher at

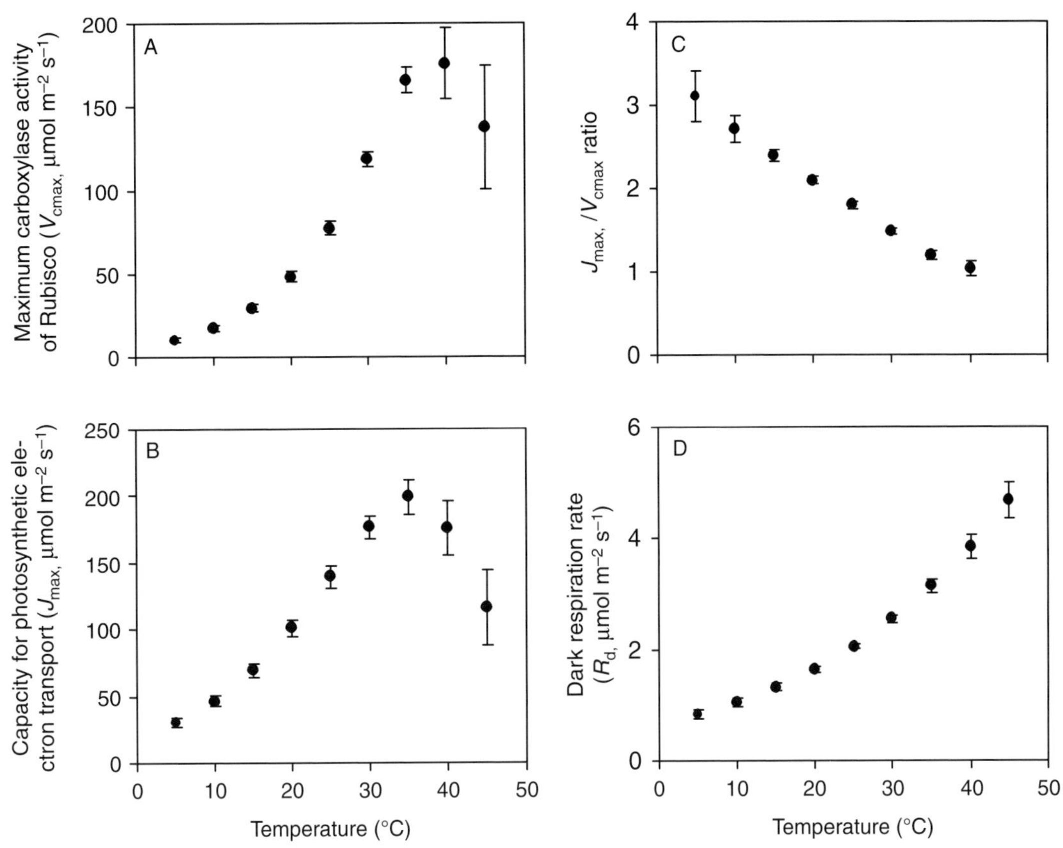

Fig. 30.8. Temperature responses of foliage photosynthetic potentials, maximum carboxylase activity of Rubisco ($V_{c,max}$) **(A)** and the capacity for photosynthetic electron transport (J_{max}) **(B)**, $J_{max}/V_{c,max}$ ratio **(C)** and dark respiration rate (R_d) **(D)** in temperate tree species (Dreyer *et al.*, 2001). Temperature responses of seven temperate tree species (*Acer pseudoplatanus*, *Betula pendula*, *Fagus sylvatica*, *Fraxinus excelsior*, *Juglans regia*, *Quercus petraea* and *Quercus robur*) were calculated at each temperature according to the temperature response functions reported by Dreyer *et al.* (2001) and average (±SE) values were calculated for each temperature.

the top compared with the bottom of the canopy (Niinemets *et al.*, 1999b). This difference may be even larger at the leaf level because of the occurrence of drought-dependent stomatal closure in the upper canopy (Niinemets, 2007). Thus, understanding temperature responses is a key factor to predict the photosynthetic productivity of temperate forests. In addition, tolerance of temperature extremes is an important factor altering species productivity and constraining species-distribution ranges in temperate forests.

GENERAL PATTERNS

Among temperate species, the optimum temperature for the maximum carboxylase activity of Rubisco ($V_{c,max}$), varies between 36 and 45°C, and that for the capacity of PET (J_{max}) between 32 and 43°C (Dreyer *et al.*, 2001, Fig. 30.8). At the same time, temperature curves of light-saturated photosynthetic activity, A_N, at current ambient CO_2 concentrations typically peak at around 25°C in temperate species (Haldimann and Feller, 2004; Warren and Dreyer, 2006, Fig. 30.9). At this temperature, A_N ranges between 3.5 µmol m^{-2} s^{-1} and 20 µmol m^{-2} s^{-1} in deciduous species, and between 3 µmol m^{-2} s^{-1} and 15 µmol m^{-2} s^{-1} in evergreen species.

Stomatal and non-stomatal factors determine the optimum temperature for A_N. The cooler optimum temperature of A_N compared with J_{max} and $V_{c,max}$ is owing to faster increase of photorespiration than photosynthesis with temperature, and the exponential increase of dark respiration with temperature (Fig. 30.8). Internal CO_2 concentration affects the relationship of photosynthesis with photorespiration and thus the temperature response of A_N. Decreasing internal CO_2 concentrations, as occurs owing to water stress, increases photorespiration and reduces photosynthesis

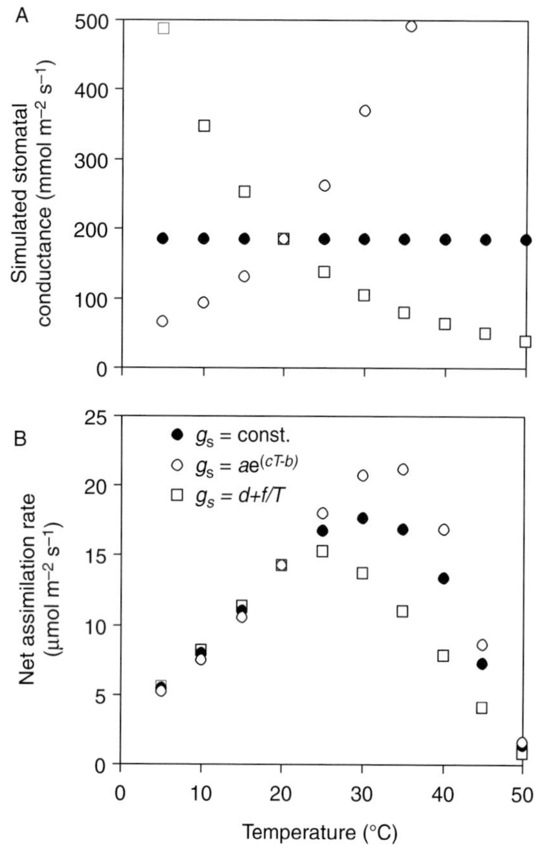

Fig. 30.9. Simulated dependencies of stomatal conductance (g_s) (**A**) and net assimilation rate (A_N) (**B**) on temperature in temperate tree species. Three hypothetical temperature dependencies of stomatal conductance (g_s) – constant with temperature, increasing (Q_{10}=2) and decreasing with increasing temperature – were employed to simulate A_N according to Farquhar *et al.* (1980) photosynthesis model. These three contrasting scenarios of g_s temperature dependence correspond to different patterns of variation of vapour-pressure deficit with leaf temperature. g_s is positively associated with temperature when vapour pressure is constant (e.g., Bunce, 2000; Lloyd, 1991), while g_s decreases with increasing temperature when air becomes drier at higher temperatures. In these simulations, values of J_{max}, $V_{c,max}$ and R_d reported in Figure 30.3 were used and incident quantum flux density was set at 1500 µmol m^{-2} s^{-1}. A_N was calculated according to the analytical expression of photosynthesis of Niinemets *et al.* (2004b) that directly links A_N, g_s and foliage photosynthetic potentials J_{max} and $V_{c,max}$.

thereby shifting the optimum temperature of A_N to cooler temperatures (Fig. 30.9, Nobel *et al.*, 1978). At constant water-vapour concentration, g_s is expected to increase with increasing temperature (Bunce, 2000, 2001; Lloyd, 1991). However, water-vapour deficit is generally larger at higher temperatures as is common in upper-canopy exposed leaves (Niinemets and Valladares, 2004). This results in reductions in g_s with increasing temperature (e.g., Schulze and Hall, 1982), thereby further reducing the optimum temperatures for A_N (Fig. 30.9). The temperature dependence of the J_{max} to $V_{c,max}$ ratio generally decreases with increasing temperature (Fig 30.8). This means that the temperature optimum of A_N can also be modified by alterations in N allocation within the photosynthetic machinery (Hikosaka *et al.*, 1999). Light acclimation of the photosynthetic apparatus can modify the temperature dependence of A_N because of variation in N partitioning within the components of photosynthetic machinery (Section 30.3.1).

The temperature dependence of photosynthesis is affected by growth temperatures such that growth at elevated temperatures increases the optimum temperature (Berry and Björkman, 1980; Gunderson *et al.*, 2000). The strong variation in temperature among and within temperate forests likely results in a spectrum of optimum temperatures for various forest ecosystems and for leaves at different canopy positions (e.g., Niinemets *et al.*, 1999b). Seasonal variation in temperature affects optimum temperatures of photosynthesis (Björkman *et al.*, 1980; Medlyn *et al.*, 2002a,b) with the size of seasonal-temperature acclimation generally proportional to the seasonal variation in temperature (Badger *et al.*, 1982). Overall, the capacity for photosynthetic-temperature acclimation can vary even among closely related genotypes (Slatyer, 1977a,b; Slatyer and Ferrar, 1977; Ferrar *et al.*, 1989), making prediction of temperature-acclimation responses difficult.

Temperature acclimation can also modify A_N itself. Acclimation to low temperatures may increase A_N at all temperatures (Ferrar *et al.*, 1989; Warren *et al.*, 1998), while high temperatures reduce A_N (Osmond *et al.*, 1980b). Such responses may result from anatomical modifications and/or from changes in fluidity of thylakoid membranes and/or alterations of N partitioning within photosynthetic machinery. One reason for higher A_N is that leaves that develop at low temperatures are thicker with more layers of photosynthetic mesophyll (Körner and Pelaez Menendez-Riedl, 1990; Weih and Karlsson, 2001). In addition, over physiological temperature range, greater membrane fluidity of low-temperature acclimated plants (e.g., Raison *et al.*, 1982) can result in higher PET at any given fraction of N in components of electron transport.

RESPONSES TO SUB-OPTIMAL TEMPERATURES

Sub-optimal temperature may limit photosynthesis in temperate ecosystems. Leaves of winter deciduous species avoid the coldest parts of the year, but may still be exposed to temperatures lower than 5°C during spring and early autumn, while leaves of evergreens experience the full effects of cold winter conditions. Rates of net photosynthesis (A_N) are typically zero or negative at temperatures around 0°C, while electron transport can be measured at sub-zero temperatures (Warren *et al.*, 1998). Reduced rates of Calvin-cycle reactions are largely responsible for slower photosynthesis at low temperatures and for increased probability of photoinhibition. A second factor increasing the risk of photoinhibition in autumn and spring is slower engagement of photoprotective zeaxanthin-dependent non-photochemical quenching (NPQ) (Warren *et al.*, 1998) as the result of reduced activity of violaxanthin de-epoxidaze enzyme (Bilger and Björkman, 1991). In addition, late-spring frosts can reduce seasonal carbon gain, competitiveness and species dispersal of winter-deciduous species in temperate forests (Kramer *et al.*, 1996).

In evergreens, lower temperatures affect carbon gain similarly to deciduous species but evergreen foliage must resist considerably lower temperatures. There are essentially two strategies that evergreens use to acclimate to low temperatures. In the first strategy, the onset of winter causes strong reductions in A_N and photochemical efficiency and accumulation of the xanthophyll zeaxanthin (Adams and Demmig-Adams, 1994; Verhoeven *et al.*, 1999). This strategy of winter acclimation occurs in evergreen broad-leaved species (Groom *et al.*, 1991) and evergreen conifers (Huner *et al.*, 1993; Adams and Demmig-Adams, 1994). As the result of sustained downregulation of photosynthetic activity, photosynthesis of such species responds only moderately to warmer periods in winter. In the second strategy of evergreens, A_N increases in winter. This strategy of increasing A_N is common in species from warmer, less continental habitats. Examples of such species are the evergreen trees *Eucalyptus nitens* and *E. pauciflora* (Warren *et al.*, 1998), the evergreen shrubs *Camellia japonica* and *Photinia glabra* (Miyazawa and Kikuzawa, 2005b) and the evergreen herb *Heuchera americana* (Skillman *et al.*, 1996). Enhanced A_N allows these species to achieve a high rate of carbon gain in warmer than average winter days, especially in sites where winter-light availability is higher owing to leafless winter-deciduous overstory (Miyazawa and Kikuzawa, 2005b).

RESPONSES TO SUPRA-OPTIMAL TEMPERATURES

A_N decreases at supra-optimal temperatures and becomes negative at temperatures greater than 40–42°C (Fig. 30.9). PSII is the most thermolabile component of the photosynthetic apparatus, and its denaturation and/or thylakoid destabilisation determine the upper temperature limit of photosynthesis (e.g., Briantais *et al.*, 1996; Yamane *et al.*, 1997). The heat-dependent reversible decrease of the activation state of Rubisco can also partly explain reductions of photosynthesis at high temperatures, for example, in the temperate trees *Quercus pubescens* (Haldimann and Feller, 2004) and *Acer rubrum* (Weston *et al.*, 2007). In the latter the rate of Rubisco inactivation also depended on the geoclimatic origin of each provenance. Supra-optimal temperatures can also cause overproduction of reactive oxygen species (ROS) (Rennenberg *et al.*, 2006) that generate oxidative stress. Species differ in heat tolerance, but the range of variation is quite small: in temperate species the limit of heat-stress tolerance is around 50°C (Knight and Ackerly, 2002). Such high temperatures are rare in temperate ecosystems, but prolonged exposure to high temperature may depress photosynthesis for several weeks.

Leaves acclimate to supra-optimal temperatures by adjusting the temperature response of A_N to warmer temperatures (see Section 30.3.2), increasing their antioxidative defences and accumulating soluble sugars. For example, ascorbate peroxidase (APX), a key enzyme in the metabolism of water-soluble antioxidants, is temperature-regulated in *Fagus sylvatica* (Peltzer and Polle, 2001). The temperature-dependent production of lipid-soluble volatile isoprenoids, isoprene and monoterpenes may also protect trees from high temperatures (Singsaas and Sharkey, 2000; Copolovici *et al.*, 2005) by quenching free radicals and ROS in membranes (Loreto and Velikova, 2001). Other non-volatile lipid-soluble isoprenoid compounds, such as tocopherols, zeaxanthin or diterpenes also increase the heat stability of the photosynthetic apparatus (Peñuelas and Munné-Bosch, 2005). In the longer term, species may prevent ROS formation in heat-stressed leaves by reducing their chlorophyll contents and antenna size (Hikosaka *et al.*, 1999; Peltzer and Polle, 2001). Accumulation of soluble sugars may increase the heat resistance of photosynthesis in temperate trees (Hüve *et al.*, 2006). This is supported by direct experimental manipulation of foliage sugar content (Hüve *et al.*, 2006) and observations that sugar contents are greater in water-stressed foliage, and in the upper than in the lower canopy (e.g., Dietz and Keller, 1997; Niinemets and Kull, 1998). The interactive effects of heat and water stress on sugar accumulation may be especially important given that high temperatures and limited water availability frequently co-occur in temperate forests.

30.3.3. Nutrient availability

Chapter 21 reviews the dependencies of photosynthesis on nutrients. Here we briefly cover limitations of photosynthesis of temperate forests by nitrogen and phosphorus. A focus on N is reasonable as it is required in larger amounts than any other nutrient and commonly limits photosynthesis (Chapin *et al.*, 1987), whereas in many other ecosystems, including much of the southern hemisphere, phosphorus (P) more commonly limits photosynthesis (Attiwill and Adams, 1993). In addition, N and P co-limitations occur in a large number of temperate ecosystems (Güsewell, 2004; Niinemets and Kull, 2005; Portsmuth *et al.*, 2005).

There is a strong positive interspecific correlation between leaf A_N and N concentration (Field and Mooney, 1986). The mechanistic interpretation of this relationship relies on the bulk of leaf N having photosynthetic functions. It is argued that up to 75% of leaf N is present in the chloroplasts, and most of the chloroplastic N is invested in the photosynthetic machinery (Evans, 1989). However, there is wide variation among temperate-forest species in N allocation. For example, in temperate evergreen tree species, the proportion of foliar N in Rubisco varies enormously: 24% for *Eucalyptus camaldulensis*; 21% for *Eucalyptus decipiens*; 9% for *Corymbia calophylla* and *Eucalyptus torquata* (Warren *et al.*, 2000); 7–20% for *Pinus pinaster*; 9–32% for *Pinus radiata* (Warren and Adams, 2000); and 16–17% for *Pseudotsuga menziesii* (Warren *et al.*, 2003a). Variation in N allocation to Rubisco probably occurs as N is required for essential functions other than photosynthesis. For example, stress-tolerant species may allocate more N to cell walls (Onoda *et al.*, 2004) and/or N-rich secondary metabolites (Gleadow *et al.*, 1998) than stress-sensitive species. Following similar logic, species with long-lived foliage allocate more N to cell walls than species with short-lived foliage (Takashima *et al.*, 2004). In all of these cases, investment of N in cell walls or secondary metabolites leaves less N for allocation to Rubisco and other photosynthetic compounds.

Although the correlative relationships between N and A_N have been developed for large numbers of species with contrasting structure, physiology and ecology (e.g., Field and Mooney, 1986; Wright *et al.*, 2004), the relationships between N and A_N can be weak within any narrower range of species (e.g., Reich *et al.*, 1995; Warren *et al.*, 2000; Warren *et al.*, 2003b). In terms of temperate forest species, A_N–N relationships are typically weak in conifers and other sclerophyllous species (e.g., Reich and Schoettle, 1988; Warren and Adams, 2001). Most of these species are from sites with strong limitations by nutrients other than N and include many species from the southern hemisphere owing to its predominant limitation by P (e.g., species from Australia and New Zealand, Warren *et al.*, 2000; Tissue *et al.*, 2005). Part of the reason for weak A_N–N relationships is the comparatively small range in N concentration in these studies (Reich, 1993), but weak relationships are also owing to other features of the evolution of sclerophyllous species. For example, stress tolerance and efficient use of nutrients seem to have been as important in their evolution as the optimisation of photosynthesis. Stress tolerance and efficient nutrient use are manifest as a small and variable investment of N in photosynthetic functions, and small g_m (Warren and Adams, 2004) traits that decouple A_N from N and reduce rates of photosynthesis per unit N.

Rates of P supply are not fast enough to maintain metabolic processes at maximum rates. Phosphorus is required for protein synthesis because it is a component of nucleic acids and for energy transfer and C metabolism owing to its function in energetic equivalents (ATP) and sugar synthesis. Halsted and Lynch (1996) suggested that P limitation manifests first in growth and later in photosynthesis. Hence, inadequate P supply often results in reduced leaf area (Fredeen *et al.*, 1989) and reduced growth rates (Plesnicar *et al.*, 1994), while reduced photosynthesis may (Brooks, 1986; Kirschbaum and Tompkins, 1990) or may not be observed (Chandler and Dale, 1993; Mulligan, 1989). One explanation for the inconsistency of P versus photosynthesis relationships is that in many species growth and leaf production are so tightly linked with P supply that P-deficient foliage is not produced. For example, in a glasshouse experiment with the evergreen tree *Pinus pinaster*, A_N did not differ despite a 16-fold variation in P supply and twofold variation in growth rate (Warren and Adams, 2002).

When P deficiency is severe or when growth is not tightly linked to P supply, photosynthesis is positively related to P. Positive relationships between P and A_N have been observed in many temperate-forest species, these include angiosperms such as the deciduous tree *Alnus rubra* (Brown and Courtin, 2003), evergreens such as *Eucalyptus* spp. (Thomas *et al.*, 2006; Turnbull *et al.*, 2007a) and *Quercus suber* (Niinemets

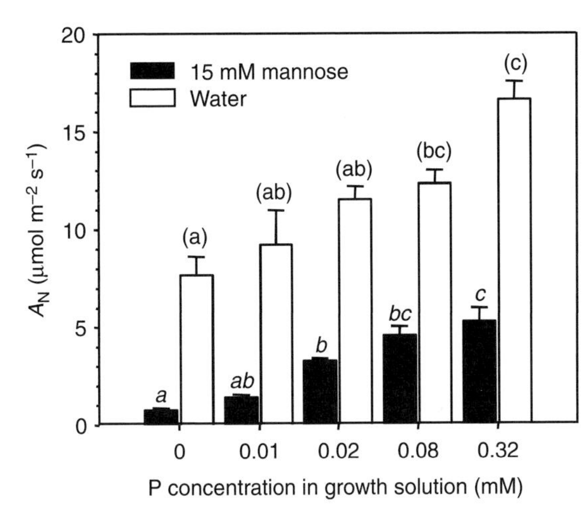

Fig. 30.10. Effect of short- and long-term P deficiency on maximum net photosynthesis (A_N) of 4-month-old temperate evergreen broad-leaved species *Eucalyptus globulus*. Seedlings were grown in P-free acid-washed sand and received one of five different nutrient solutions resulting in long-term differences in P availability. To induce short-term P deficiency, seedlings were fed 15 mM mannose to sequester cellular P_i. Data are means±SE (n=6). Significant effects of P concentration during growth and mannose feeding were determined using a two-way ANOVA. Within-treatment differences in mean values between substrate concentrations are indicated by italicised letters for mannose-fed plants and by letters in parentheses for water-fed plants. Figure re-drawn from Turnbull *et al.* (2007).

et al., 1999c) and many gymnosperms including *Pinus* spp. (Reich and Schoettle, 1988; Conroy *et al.*, 1990; Loustau *et al.*, 1999) and *Dacrydium cupressinum* (Carswell *et al.*, 2005). Reductions in A_N owing to suboptimal P supply can be large, for example in *Eucalyptus globulus* A_N was reduced by 55% by growing seedlings with an inadequate supply of P (Turnbull *et al.*, 2007a). Although reductions in A_N owing to P deficiency may be large, the reason(s) for positive relationships between P and photosynthesis are less clear than for N because only a small amount of P is involved with the primary processes of photosynthesis (Bieleski, 1973). Photosynthesis requires a finely balanced concentration of inorganic phosphate (P_i) in the cytosol (Walker and Robinson, 1978), which is maintained by transport to and from the vacuole and by metabolic processes that change the rate of sucrose synthesis (Foyer and Spencer, 1986). The clear limitation of photosynthesis by cytosolic P_i was demonstrated in *Eucalyptus globulus* by sequestering cytosolic P_i with mannose. When P_i was sequestered with mannose, A_N was reduced by 83% (Turnbull *et al.*, 2007a, Fig. 30.10).

While artificial manipulation of cytosolic P_i clearly demonstrates how photosynthesis is affected by P deficiency in the short term, it is less clear how P affects photosynthesis at longer timescales. Analyses of A/C_i responses indicate that P limitation affects $V_{c,max}$ and triose phosphate utilisation (TPU) (Lewis *et al.*, 1994), $V_{c,max}$ and J_{max} (Loustau *et al.*, 1999) or $V_{c,max}$, J_{max} and TPU (Niinemets *et al.*, 1999c). Additional clues come from glasshouse and field experiments with *Pinus pinaster* that showed there was a strong positive correlation between P supply and the relative and absolute amounts of N allocated to Rubisco (Warren and Adams, 2002b). Hence, it may be the case that P may partially affect photosynthesis in an indirect way by affecting allocation of N to the photosynthetic machinery (e.g., owing to the P requirement of protein synthesis). This may come about because P deficiency decreases sink demand and causes end-product inhibition of photosynthesis (and triose phosphate utilisation (TPU) limitation) in the short term (Pieters *et al.*, 2001). At longer timescales P deficiency causes accumulation of carbohydrates that downregulate expression of genes coding for the photosynthetic machinery (Krapp and Stitt, 1995).

30.3.4. Soil-water availability

Growth and photosynthesis of temperate forests are sometimes limited by water availability. As a general rule, temperate tree species are not as drought tolerant as species from arid and semi-arid ecosystems. Nevertheless temperate species frequently encounter situations in which water limits photosynthesis, and there is a large variability in drought tolerance among temperate-forest species (Section 30.2). This may modify species composition of temperate forests (Abrams, 1996; Buckland *et al.*, 1997). The major effects of water stress on photosynthesis have been analysed in detail in Chapter 20. Here we provide a brief overview of the effects of water limitations in the context of temperate forests.

Stomatal closure in response to reduction of cellular water content is the most common response of plants to water limitation (e.g., Slatyer, 1967). Experiments with small herbaceous plants have observed that root-borne xylem-transmitted signals (e.g., ABA) play a key role in signalling soil-water stress and eliciting stomatal closure (Khalil and Grace, 1993; e.g., Tardieu *et al.*, 1991a,b). Experiments with seedlings and saplings of temperate tree species (e.g., Khalil and Grace, 1993; Croker *et al.*, 1998) and shorter individuals of temperate trees and shrubs (Wartinger *et al.*, 1990;

Tenhunen *et al.*, 1994) have also indicated close correlation between xylem-sap ABA concentration and g_s. However, studies in tall temperate trees have found no evidence for root-borne chemical signals (Fuchs and Livingston, 1996; Fuchs *et al.*, 1999; Niinemets *et al.*, 1999d; Perks *et al.*, 2002). This has been explained by the argument that root-borne signals would be of little use to tall trees because they would take several days to reach leaves. Experiments with other species of tall temperate trees have demonstrated that while xylem-sap ABA concentration is not correlated with g_s, leaf ABA concentrations are correlated with g_s (Aasamaa *et al.*, 2004). In addition to root-borne chemical signals, hydraulic signals have been suggested as candidates for long-distance signalling of water limitations (Comstock, 2002). There is currently some experimental evidence supporting a dominant role of hydraulic signals in controlling stomatal conductance in tall trees (Whitehead *et al.*, 1996; Whitehead, 1998), but there is no consensus on the relative importance of hydraulic versus chemical signals and whether this varies among and within functional groups. One limitation is that many studies comparing chemical and hydraulic signals have not quantified foliage endogenous ABA concentration. This is problematic because hydraulic conductance and foliage endogenous ABA concentration are positively correlated in some temperate trees (Aasamaa *et al.*, 2002, 2004), suggesting that foliage endogenous ABA may play a role in long-term regulation of g_s in water-stressed temperate trees – even if ABA does not serve as a direct root-to-shoot messenger.

Irrespective of the precise response mechanism(s), stomatal closure simultaneously slows the rates of CO_2 diffusion into the leaf and water loss from the leaf. In most cases stomatal closure is the primary cause of reduced photosynthesis under mild water deficits (Gaastra, 1959; Chaves, 1991; Flexas *et al.*, 2006c). A plethora of non-stomatal limitations on photosynthetic metabolism have been reported in moderately water-stressed plants (e.g., Tenhunen *et al.*, 1984). However, these reductions in carboxylation efficiency and photosynthetic electron transport activity were later found to be associated with non-uniform (patchy) stomatal closure (Beyschlag *et al.*, 1992, 1994) that led to overestimation of internal CO_2 concentration and underestimation of foliage biochemical potentials. Other experiments suggest that drought-dependent reductions in g_m also reduce photosynthesis in drought-stressed plants (Warren *et al.*, 2004; Flexas *et al.*, 2006c). Experiments with very high ambient CO_2 concentrations that override diffusion limitations have shown that problems in estimating conductance from ambient air to chloroplasts (either owing to patchiness or overestimation of g_m) are responsible for apparent non-stomatal limitations in drought-stressed plants (Cornic, 2000).

Sustained drought leads to reductions in photosynthetic potentials and light-harvesting antenna size (Flexas and Medrano, 2002b). By reducing the proportion of light utilised in photosynthesis, stomatal closure promotes photorespiration and results in a greater excess of light energy, thereby predisposing plants to photoinhibition. Most studies of photoinhibition in drought-stressed plants have observed only rapidly reversible declines in quantum yields characteristic of non-photochemical dissipation and/or increased photorespiration (Epron *et al.*, 1992). Slowly reversible or irreversible declines in quantum yield and photosynthetic capacity indicative of PSII damage are rare in temperate-forest species in the field, even in severely drought-stressed plants. For example, in deciduous oak *Quercus pubescens*, photosynthetic efficiency was reduced only over a short term and the efficiency always recovered by sunset despite pre-dawn leaf-water potentials as low as –4.5 MPa (Damesin and Rambal, 1995). Similar resilience is found in the evergreen tree species *Arbutus menziesii* and *Quercus agrifolia* (Björkman, 1994). Even under extreme photoinhibitory conditions of long-term drought and high-irradiance, non-radiative energy dissipation, presumably with a high rate of repair, prevented any significant net photodamage to the leaves of the drought-stressed plants (Björkman, 1994). Nevertheless, downregulation of photochemical efficiency in drought-stressed foliage does lead to reduced daily PET and carbon gain (Niinemets and Kull, 2001), and such reductions in photochemical efficiency are important to consider in modelling tree carbon gain (Werner *et al.*, 2001b).

The effects of drought on plant growth and metabolism have been intensively studied (Schulze, 1986), but the recovery of photosynthesis after drought has been studied less. This is perplexing as the recovery of photosynthesis from water deficits can be costly in terms of whole-plant carbon assimilation if recovery is slow. The extent and rate of recovery in photosynthesis depends on the genotype (Fan and Grossnickle, 1998; Ngugi *et al.*, 2004) and the degree of dehydration reached at the end of the drying cycle (Kriedemann and Downton, 1981), as well as the degree of structural modifications at the end of drought cycle. Recent experiments with *Fagus sylvatica* saplings showed that several weeks were needed for A_N to recover from a 5-week water-stress treatment, and that g_s did not completely recover to control values (Gallé and Feller, 2007). Similar experiments with *Quercus pubescens* also found severe limitations of photosynthesis under water stress, but the recovery of photosynthesis was quicker and there were no lasting reductions in g_s (Gallé *et al.*, 2007). These data provide a tempting hypothesis that there is a relationship between drought tolerance and the rate of recovery from drought. More data are needed to test this hypothesis.

It is important to consider that some of the drought-dependent modifications are not fully reversible upon rehydration. For instance, there is a continuous reduction of g_m during the season in temperate deciduous trees that have experienced a series of drought cycles during the growing season (Grassi and Magnani, 2005). In parallel to these modifications, LMA increases mainly owing to carbon accumulation in cell walls, resulting in a more robust leaf structure (Wilson *et al.*, 2000a; Niinemets *et al.*, 2004a; Grassi and Magnani, 2005). Analogous but more profound modifications occur in the foliage of temperate evergreen trees, where foliage is exposed to many more drought cycles during its lifespan (Niinemets *et al.*, 2004c).

30.3.5. Atmospheric [CO_2] and pollutants

Because several reviews have examined responses to elevated CO_2, global change and pollutants (e.g., Saxe *et al.*, 1998; Ceulemans *et al.*, 1999; Medlyn *et al.*, 2001; Ainsworth and Long, 2005; Wittig *et al.*, 2007), here we only briefly highlight the key issues specific to temperate forests.

RESPONSES TO ELEVATED CO_2 AND CLIMATE CHANGE

Average surface temperatures in temperate forests are expected to increase by 1.4°C to 5.8°C by the end of this century, and will be accompanied by changes in the distribution and amount of precipitation (Meehl *et al.*, 2007; Trenberth and Jones, 2007). One of the predicted consequences of climate warming is an increased frequency of heatwaves, similar to the heatwave of 2003 in Western Europe at the southern border of the Northern hemisphere's temperate forests. Such heatwaves have diverse negative effects on photosynthesis, growth and survival of temperate trees (Rennenberg *et al.*, 2006). Temperate deciduous species are less tolerant to combined heat and drought stress than most Mediterranean evergreens, as the latter have mechanisms for acclimation to such combined stresses (García-Plazaola *et al.*, 2008b). There is, for example, a dramatic decrease in growth and increased mortality of *Fagus sylvatica* at the southern limit of its distribution in Spain (Peñuelas and Boada, 2003). Alternatively, winter minimum temperatures are expected to rise and the occurrence of exceptionally cold winters is expected to decrease in the northern border of temperate forests of the Northern hemisphere (Kjellström *et al.*, 2007; Meehl *et al.*, 2007). These changes are expected to shift temperate biomes to higher latitudes and higher altitudes.

CO_2 enrichment has direct effects on photosynthetic carbon metabolism (see Chapter 17). This is especially important in trees that are the functional group most responsive to high CO_2 in terms of photosynthesis and biomass production (Ainsworth and Rogers, 2007) because of their capacity to store photosynthates. In the case of temperate trees, several meta-analytic studies of field-based elevated CO_2 experiments (Medlyn *et al.*, 1999, 2001; Ainsworth and Long, 2005) have shown a general but species-dependent increase in leaf-level light-saturated carbon assimilation (47%) together with a significant decrease (21%) in g_s. This was accompanied by a slight reduction in biochemical capacities of photosynthesis, J_{max} and $V_{c,max}$, perhaps owing to decreased foliage N content (Medlyn *et al.*, 1999, 2001). Although increased photosynthesis under CO_2 enrichment does not necessarily increase biomass (Körner *et al.*, 2005), most studies indicate increased allocation to wood and leaves, with a 21% increase in LAI (Ainsworth and Long, 2005). This effect implies a more closed canopy, which could alter competitive relationships among forest species, but also qualitative and quantitative differences in litterfall. Analyses have shown that responses were stronger in young rather than in old trees and in deciduous species than in conifers (Medlyn *et al.*, 2001), and also differed among species or provenances (Leverenz *et al.*, 1999; Hättenschwiler, 2001). There are further important interactive effects with light, drought, temperature and biotic factors; and this makes it difficult to predict responses to elevated CO_2 (Ceulemans *et al.*, 1999). Among these factors, elevated temperatures and high nutrient supply also increased A_N response (Ainsworth and Long, 2005). Other effects of high CO_2 are that it reduces g_s and this may reduce the negative effects of ozone and reduce water loss by canopies. However, the 'benefits' of the increase in atmospheric CO_2 in the temperate biome may not compensate for the enhanced risk of drought periods and heatwaves.

ATMOSPHERIC POLLUTANTS

Many temperate regions are among the most densely populated and heavily polluted in the world. The most important gaseous pollutants are sulphur dioxide (SO_2), nitrogen oxides (NO_x), ozone (O_3) and some organic compounds.

These emissions are a major problem for human health, but are problematic for plants at much lower concentrations. Recent decades have seen decreased emissions of SO_2, but stable or increasing emissions of NO_x and O_3. Ozone has increased three- to fourfold in the temperate region of the Northern hemisphere during the last century (Volz and Kley, 1988). Tropospheric O_3 is formed by the interaction of VOC, NO_x, solar radiation and oxygen (Chameides *et al.*, 1988, 1992). As both solar radiation and VOC and NO_x concentrations vary spatially and temporally, O_3 concentration varies as well, following seasonal and diurnal cycles with maximum production at high light and temperature. In temperate regions, O_3 concentration is higher in spring and summer, and often a daily cycle is observed with a peak at noon owing to high light. There is also significant geographical variation in O_3 concentration because of the irregular distribution of the sources of VOC and NO_x, distribution of O_3 by atmospheric movements and differences in solar radiation. For instance, production of O_3 is exacerbated at high altitudes because of intense solar radiation. Many important temperate plants such as oaks, poplars, eucalypts and conifers contribute to tropospheric O_3 formation by emitting VOCs (Simpson, 1995; He *et al.*, 2000; Lerdau and Slobodkin, 2002; Peñuelas and Llusià, 2003). Worldwide VOC emission by plants is predicted to exceed human-emitted VOC emission by more than an order of magnitude (Guenther *et al.*, 1995; Arneth *et al.*, 2007), implying that plants play an important role in tropospheric O_3 formation (Chameides *et al.*, 1988, 1992; Simpson, 1995).

Trees, by virtue of their long lifetimes, suffer chronic damage induced by air pollutants. In general, temperate deciduous trees are more resistant to sulphur and nitrogen emissions than conifers, but alternatively deciduous trees are more sensitive to O_3 than coniferous trees (Landlot *et al.*, 2000). Much of the recent decline in the crown condition of European forests (most noticeably in the case of *Fagus sylvatica*, a dominant tree in European forests with a percentage of only 34% undamaged stands) has been attributed to O_3 (Dittmar *et al.*, 2003). Thus, damage symptoms in O_3-treated deciduous trees have been described in *Quercus rubra*, *Betula pendula*, *Acer saccharinum*, *Populus deltoides*, *Fraxinus excelsior*, *Fagus sylvatica* or *Populus nigra*, while conifers such as *Picea abies*, *Pseudotsuga menziesii*, *Abies alba* or *Pinus sylvestris* are slightly or not affected (Bortier *et al.*, 2000; Landolt *et al.*, 2000; Wittig *et al.*, 2007). A recent meta-analysis of 55 studies showed the elevation of O_3 since the industrial revolution is decreasing A_N by an average of 11% and g_s by 13% (Wittig *et al.*, 2007). Furthermore, gymnosperms are less affected by O_3 than angiosperms and young trees are less affected than old trees (Wittig *et al.*, 2007). Ozone sensitivity varies considerably among plants, but in general, fast-growing species are more susceptible and responsive to ozone than slower-growing species (Bortier *et al.*, 2000). In addition to slow growth rates in some of the conifers, higher O_3 tolerance of conifers and several constitutively emitting broad-leaved species can also be associated with their strong emissions of volatile isoprenoids that can react with O_3, such that less O_3 reaches the interior of the leaves (Loreto *et al.*, 2001b, 2004a,b; Loreto and Velikova, 2001).

The amount of damage typically depends on cumulative O_3 uptake. Ozone enters the leaf through the stomata and therefore factors that decrease g_s, such as drought or elevated CO_2, decrease O_3 damage (Wittig *et al.*, 2007). Hence, O_3 damage of *Fagus sylvatica* in Western Europe was reduced during the exceptional summer drought of 2003 (Löw *et al.*, 2006). Once in the leaf, O_3 causes oxidation that leads to chlorophyll bleaching and oxidation of proteins and unsaturated lipids. It affects several photosynthetic processes, such as carboxylation efficiency, electron transport activity and stomatal conductance (Farage, 1996; Löw *et al.*, 2006). This reduces photosynthesis rates, and in the longer-term accelerates leaf senescence and leaf necrosis and reduces growth and primary production (Mikkelsen and Heide-Jųrgensen, 1996; Bortier *et al.*, 2000).

These data collectively emphasise the importance of gaseous pollutants on the performance of temperate forests. An interesting aspect is that plants emitting VOC can themselves affect atmospheric O_3 concentrations in NO_x-polluted air. Because the VOC-emitting plants have higher tolerance to O_3, species capable of emittting VOCs are expected to become more abundant in future forests (Lerdau, 2007).

30.4. SCALING FROM LEAVES TO ECOSYSTEMS

Previous sections have primarily dealt with leaf-level photosynthesis; whereas this section is interested in scaling from leaves to whole plants to canopies, and ultimately forest ecosystems. Modelling and scaling of canopy and ecosystem gas exchange are beyond the scope of this chapter, but it is worthwhile knowing that more complicated models of forest canopies are largely driven by within-canopy gradients in the limitations of photosynthesis. Hence, much of this section is interested in within-canopy spatial and temporal patterns in the key limitations of photosynthesis. This section concludes by considering canopy and ecosystem-level gas exchange.

30.4.1. Spatial patterns

Temperate forests are spatially structured and have strong spatial gradients in the key resources and limitations of photosynthesis. The strongest spatial gradient is the exponential reduction in PPFD from the top to the bottom of forest canopies. Average irradiance on the floor of a temperate forest may be as little as 1–5% of full sunlight, whereas the leaves in the upper canopy receive full sunlight (Niinemets and Valladares, 2004, Fig. 30.11A; Tausz *et al.*, 2005b). The gradients in PPFD from the sunlit exterior of the canopy to the shaded interior vary depending on plant architecture and LAI. The gradients of light are large in stands with high LAI, such as temperate coniferous forests of the Northern Hemisphere growing on nutrient-rich sites (LAI >6 m^2 m^{-2}), while the gradients may be weak in stands with smaller LAI such as temperate *Eucalyptus* forests of Australia (LAI typically ≤3 m^2 m^{-2}).

Temperature and air humidity vary as a function of canopy height and thus co-vary with PPFD (e.g., Baldocchi *et al.*, 2002, Fig. 30.11B,C; Niinemets and Valladares, 2004). Leaves in the upper canopy are exposed to higher temperatures during the day (Fig. 30.11B) and are at greater risk of heat stress (Niinemets *et al.*, 1999b). Leaves exposed to high PPFD may be more than 10°C above the ambient temperature on sunny days (Hamerlynck and Knapp, 1994; Singsaas *et al.*, 1999). Warmer leaf temperatures along with lower humidity cause greater leaf-to-atmosphere VPD and increase the risk of water stress in the upper canopy (Niinemets *et al.*, 2004b). These data collectively emphasise the complex nature of temperate-forest canopies, especially in more continental sites where diurnal variations in within-canopy temperature and humidity can be large.

Whole-canopy factors significantly modify the within-canopy environment. Friction between the forest canopy and the atmosphere results in the canopy having a boundary layer analogous to that of individual leaves. This aerodynamic conductance (g_a) affects exchange of mass and energy between the canopy and the atmosphere. This comes about because there is a certain resistance to CO_2, H_2O and energy exchange between the canopy and the bulk atmosphere, which means that the canopy environment can be different from the bulk atmosphere. For example, in a rapidly

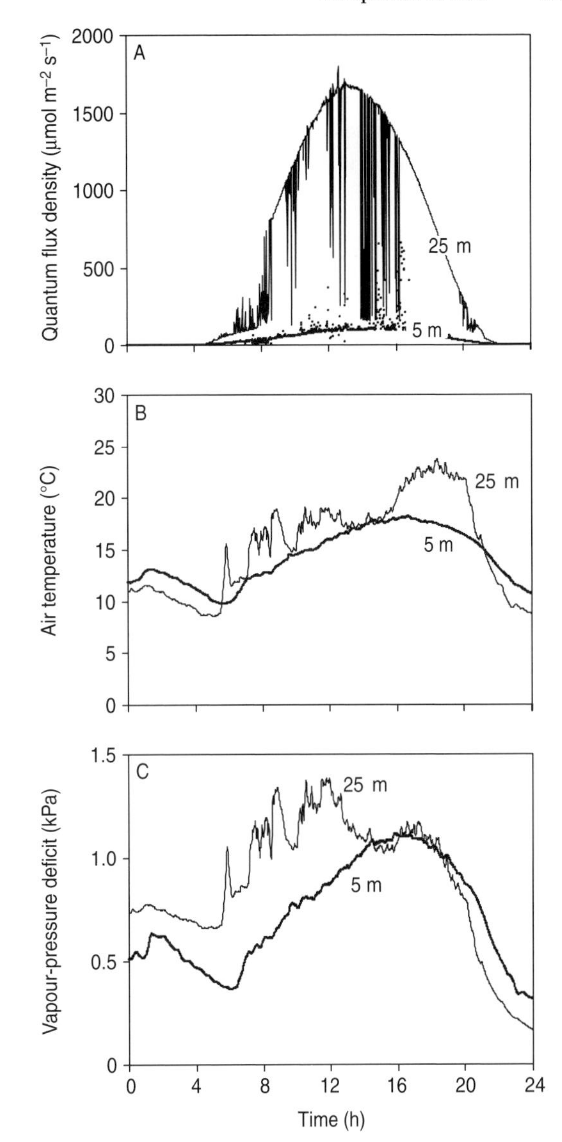

Fig. 30.11. Diurnal variation in photosynthetic photon-flux density (**A**), air temperature (**B**) and air water vapour-pressure deficit (**C**) at the top (25 m) and bottom (5 m) of a temperate deciduous forest dominated by *Populus tremula*, *Betula pendula* and *Tilia cordata* in Järvselja, Estonia (58°16′N, 27°20′E) in July, 15, 1995 (unpublished data of Niinemets and Kull, see Niinemets *et al.*, 1999a for details of the site). Daily integrated quantum flux density at the top of the canopy was 49 mol m^{-2}, and 4.3 mol m^{-2} at the bottom. Daily average air temperature (± SD) was 15.5±4.5°C at the top and 14.3±2.3°C at the bottom, while average vapour pressure deficit was 0.89±0.31 kPa at the top and 0.73±0.23 kPa at the bottom.

photosynthesising and transpiring forest canopy, the concentrations of CO_2 are lower and H_2O higher than in the bulk atmosphere above the canopy. The size of the difference in concentrations among the bulk atmosphere, canopy top and canopy bottom is a function of canopy architecture (aerodynamically 'rough' versus 'smooth'), LAI and wind speed (Jarvis *et al.*, 1976). It is commonly argued that the aerodynamically 'rough' canopies of most conifers are well coupled to the bulk atmosphere, whereas those of deciduous species with large LAI such as *Populus* are poorly coupled to the bulk atmosphere (Whitehead, 1998). How well coupled a canopy is to the atmosphere affects the sensitivity of stand transpiration and photosynthesis to leaf-level changes in g_s (Jarvis and McNaughton, 1986).

30.4.2. Temporal patterns

The temperate-forest environment is highly variable in time. Temporal variations occur at timescales ranging from seconds (e.g., sunflecks) to years (e.g., El Niño/La Niña cycles, global climate change). On the timescale of seconds, plants must cope with light intensity varying by several orders of magnitude (Tausz *et al.*, 2005b; Pearcy, 2007), leaf temperatures varying by as much as ten degrees (Singsaas *et al.*, 1999; Chelle, 2005) and wind speeds varying by several metres per second (C. Warren, unpublished data). Light intensity at the leaf surface is highly variable because of varying cloudiness and diurnal variations in beam pathlength through the atmosphere and canopy. In addition, heterogeneous distribution of gaps in the canopy, differences in leaf height in the canopy and movement of canopy elements in wind provide additional variation in leaf light availability. As a result of these factors, leaves are intermittently shaded or exposed to full sunlight (sunflecks) as the sun tracks across the sky hemisphere and trees move in the wind (Fig. 30.11A). Such sunflecks account for a significant proportion of daily carbon gain, occasionally even more than 40% in understory leaves/species (Tausz *et al.*, 2005b; Pearcy, 2007).

In temperate forests, especially in deciduous forests, there are predictable changes in irradiance owing to seasonal changes in leaf area, day length and solar elevation angle. Budburst in spring and leaf fall in autumn affect the quantity and quality of light reaching the forest floor, and this can alter the carbon gain in understory plants (Koizumi and Oshima, 1985; Katahata *et al.*, 2005). Budburst and leaf fall occur at different times and with differing rates in the understory and the overstory, and this causes important differences in the temporal patterns of photosynthesis (Section 30.2.4).

In addition to light, temperate forests experience consistent seasonal variation in temperature and water availability – these have important implications for photosynthesis and growth. As discussed in Section 30.1.1, 'temperate' is a very broad classification, and thus the size of these seasonal trends varies enormously among temperate forests, mainly as a function of latitude and continentality. In areas with strong maritime influence, temperatures may vary as little as 15°C between winter and summer; whereas in continental areas, temperatures may vary by more than 50°C. Owing to the diversity of climatic conditions, temperate forests vary in the degree to which photosynthesis and growth are limited by adverse conditions, such as hot and cold, drought and waterlogging. Seasonality of growth and photosynthesis are strongest for deciduous forests in continental areas with strong seasonal variations in water availability, temperature and leaf area; whereas seasonality is smaller for evergreen forests in maritime areas.

The means of temperature and rainfall may vary among years. Such differences in yearly means can have important implications for forest growth and C uptake. For example, temperatures a few degrees above and below average have a large effect on net C balance of temperate forests (Morgenstern *et al.*, 2004). In addition, extreme events, such as drought spells, catastrophic flooding, heat waves and extreme frosts, can have devastating effects on temperate forests (Palmer and Räisänen, 2002; Gutschick and BassiriRad, 2003; Vervuren *et al.*, 2003; Kjellström *et al.*, 2007). Certain extreme events can be predictable, such as extremes of rainfall owing to large-scale patterns of climatic circulation and anomalies in the Pacific basin that simultaneously affect large regions of temperate forest (e.g., El Niño/La Niña cycles) (e.g., Armstrong *et al.*, 1988). In other cases, catastrophic events are not very predictable (Gutschick and BassiriRad, 2003; Vavrus *et al.*, 2006; Kjellström *et al.*, 2007) and this complicates long-term predictions of photosynthesis in temperate forests.

30.4.3. Gas exchange of temperate-forest canopies and ecosystems

At very broad scales, the primary determinant of photosynthesis by a forest canopy (i.e., GPP) is the area of photosynthetic material or light intercepted (e.g., as indicated by LAI). This is supported by good correlations among leaf area, amount of light intercepted and growth of various forest species (Landsberg, 1996). There is, however, wide variation in growth per unit radiation intercepted (Whitehead and Beadle, 2004), indicating that LAI is not the sole determinant of canopy photosynthesis. Perhaps the strongest limiter of the amount of canopy photosynthesis is simply the length of the photosynthetic season. What determines the length of the photosynthetic season? Most ecosystems experience periods that are too cold, too hot, too dry, too wet for significant photosynthesis. Environmental controls over GPP during the growing season are the same as those for individual leaves (e.g., diurnal variation in light, seasonal variation in water availability). Soil resources affect GPP via their effects on leaf-level photosynthetic potential (e.g., amounts of N and Rubisco) and leaf area. Hence, on a broad geographic scale, photosynthesis by canopies is largely a function of LAI and climatic factors that slow (or stop) photosynthesis.

The primacy of LAI and climate in determining canopy photosynthesis does not imply that differences in leaf-level photosynthesis are trivial or irrelevant. On the contrary, it is leaf-level photosynthesis that underpins whole-canopy photosynthesis. However, the spatial and temporal scaling involved in going from a leaf to an entire canopy means that leaf-level photosynthesis may not be correlated with canopy photosynthesis. This is largely a reflection of correlations among LAI, A_N, LMA and leaf longevity (see Section 30.2) that mean that plants with very different A_N can have similar canopy photosynthesis, and vice versa.

The proportion of GPP that is lost as autotrophic respiration does not vary significantly among forest trees (Waring and Running, 1998), but is consistently near 50%. What remains is NPP. Because NPP is a (more or less) consistent fraction of GPP, the determinants of NPP are essentially the same as the determinants of GPP ('canopy photosynthesis'): LAI, season length and stresses that reduce photosynthesis. The balance between NPP and heterotrophic respiration determines the net carbon accumulation by an ecosystem (i.e., ecosystem C balance) and its effect on atmospheric CO_2. This is commonly referred to as NEE of CO_2 (Fig. 30.12).

Net ecosystem exchange varies among different temperate forests. In temperate forests with mild winters and minimal water limitations, forests may be net sink for CO_2 all year round (i.e., negative NEE). In *Eucalyptus delegatensis* forests of south-eastern Australia, for example, NEE is negative even in winter when average daily temperatures are

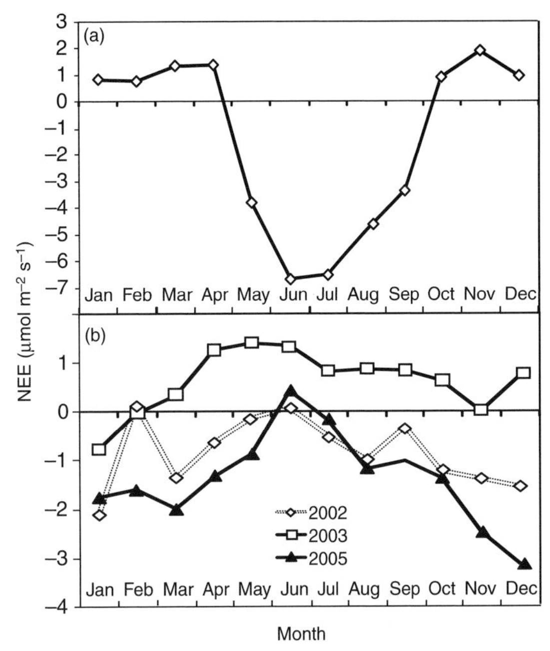

Fig. 30.12. Monthly averages of net ecosystem exchange (NEE) of evergreen broad-leaved *Eucalyptus delegatensis* forest from Tumbarumba, south-eastern Australia **(B)** (Kirschbaum *et al.* 2007; Leuning *et al.* 2005; van Gorsel *et al.* 2008) and deciduous broad-leaved forest **(A)** (Morgan Monroe, Indiana, USA). Data for *Eucalyptus delegatensis* forest are shown for three years (2002, 2003 and 2005) to illustrate large variability between years. 2003 was characterised by severe insect damage and drought, 2002 and 2005 were comparatively normal. *E. delegatensis* data are gap filled, with nocturnal respiration calculated as the CO_2 flux from the maximum of the eddy flux plus change in the storage term in the period after sunset when stable stratification develops (van Gorsel *et al.* 2008).

≤2°C (Leuning *et al.*, 1995). This is a function of substantial GPP (i.e., photosynthesis) at temperatures between 0 and 5°C (perhaps because of strong temperature acclimation of photosynthesis) and the strong limitation of heterotrophic respiration by temperature. In areas with colder winters and/or deciduous vegetation, NEE is positive during the winter months and only negative during the warm, vegetated period (Fig. 30.12).

Annual NEE may vary enormously among years owing to abiotic and biotic factors. In *Eucalyptus delegatensis* forest, NEE over a 5-year period varied between –750 g m^{-2} a^{-1} (strong sink) to +230 g m^{-2} a^{-1} (mild source) owing to drought and/or insect attack (van Gorsel *et al.*, 2007). In

Pseudotsuga menziesii interannual forest differences in NEE over a 4-year period were mainly owing to differences in ecosystem (i.e., autotrophic + heterophic) respiration (Morgenstern *et al.*, 2004). Years with high air temperatures have led to high annual respiration, but only slight increases in annual GPP. In the case of *Pseudotsuga menziesii*, this has led to warm years being the weakest sinks (NEE=–270 g m^{-2} a^{-1} in a warm year versus –390 to –420 g m^{-2} a^{-1} for normal years).

30.5. OUTLOOK: WATER-USE EFFICIENCY, CLIMATE CHANGE AND PHOTOSYNTHETIC CONSTRAINTS IN TEMPERATE FORESTS

There is wide variation in WUE within species (Silim *et al.*, 2001; Brendel *et al.*, 2008), among species (Warren *et al.*, 2006; Gyenge *et al.*, 2008) and among sites (Stewart *et al.*, 1995). This wide variation in WUE has received attention from forest scientists and geneticists interested in selecting high-WUE trees for water-limited sites. Studies in *Populus* have shown that WUE is highly heritable (Monclus *et al.*, 2005), and that WUE is generally unrelated to productivity (Monclus *et al.*, 2006) (i.e., there is no trade-off between WUE and productivity). Thus it is theoretically possible to select *Populus* genotypes with high WUE and productivity. In the future, selection of high-WUE genotypes may be aided by the knowledge of the genetic determination of WUE (Bonhomme *et al.*, 2008; Brendel *et al.*, 2008).

High WUE is frequently proposed as a trait of species from low-rainfall habitats (Sun *et al.*, 1996; Wright *et al.*, 2003), yet many studies have found no relationship of WUE with drought tolerance or aridity (Lauteri *et al.*, 1997; Warren *et al.*, 2001; Monclus *et al.*, 2005). To expect uniformly strong correlations between WUE and drought tolerance is arguably simplistic given that WUE is only one component of drought tolerance. Another explanation is that WUE is affected by all environmental variables that affect A_N and g_s, not just water stress. For example, in a thinning experiment with *Pinus pinaster* and *Pinus radiata* trees it was found that decreased water stress owing to thinning did not decrease WUE as expected because it was counterbalanced by an increase in WUE owing to increased PAR (Warren *et al.*, 2001).

Photosynthetic productivity of temperate forests is mainly limited by temperature and light, but nutrient and water limitations are significant in some temperate forests. These limitations have short-term effects on photosynthesis (Table 30.1), but plants can also acclimate and partially alleviate long-term limitations. For example, plants respond to short-term, rapid increases in light (sunflecks) by opening stomata and increasing rates of photosynthesis, whereas long-term differences in light can cause acclimation by altering leaf structure (as indicated by LMA) and leaf biochemistry (absolute amounts of N, and allocation of that N to compounds with different functions). The effect of temperature on photosynthesis is important because temperate forests must cope with a wide diurnal and seasonal range in temperatures. Photosynthesis is decreased by short-term exposure to temperatures above and below the optimum, whereas plants have some capacity to acclimate longer term (weeks to months) to unfavourable temperatures by shifting the temperature optimum of photosynthesis. Age is another major limitation of photosynthesis that is often overlooked. Many temperate-forest species retain leaves for multiple years and over this time there are consistent decreases in photosynthesis.

Human-managed temperate forests are often monotypic plantations, but natural temperate forests consist of a complex mixture of various plant functional types with the mix of functional types at any site being driven by local environmental conditions. The major plant functional types in temperate forests, broad-leaved deciduous, evergreens and conifers, have contrasting photosynthetic potentials, leaf longevity, leaf aggregation and whole-stand LAI. Although current evidence suggests that different combinations of leaf and stand attributes result in similar stand-level productivity, these differences result in different responsiveness to variation in environmental drivers. Large variation in A_N and foliage structure also occurs within any plant functional type. These variations in foliage photosynthetic characteristics are linked to plant tolerance of environmental conditions, in particular to shade and drought tolerance.

Across the world, temperate forests have been exposed to the strongest human influence over centuries resulting in a series of other stresses, in particular atmospheric pollution, that significantly depress photosynthesis. Human activities also affect photosynthetic performance by the continuous rise in atmospheric CO_2 whose long-term effects are difficult to predict. An indirect effect of CO_2 emission is climate warming. Warming is inducing a dramatic shift in biome distribution, whose effects can be observed in the border of the distribution range of temperate forests. The increased frequency of exceptional events, such as the heatwave of 2003 in western Europe, is probably accelerating this process. However, anthropogenic factors that lower stomatal conductance, such as climate-change-related droughts or elevated CO_2, also diminish damage caused by pollutants leading to an intricate pattern of interactions.

31 • Ecophysiology of photosynthesis in boreal, arctic and alpine ecosystems

F. BERNINGER, P. STREB AND I. ENSMINGER

31.1. PHOTOSYNTHESIS IN THE COLD AND THE SNOW?

In this chapter we discuss photosynthesis and productivity of plants in arctic, alpine and boreal environments. These environments are characterised by cold climate and dominated by plants that thrive in these harsh and often not very productive environments. We concentrate on alpine plants from 'non-tropical' mountains and on higher plants, reflecting mostly our own experience and the larger body of literature, which is available for the non-tropical mountains, as well as the need to focus on general aspects. It is noteworthy to mention that there is a growing number of fascinating studies on tropical and subtropical mountain ecosystems, such as the Andes or the Tibetean plateau. However, here we aim to focus on the general features of alpine environments. We will not be able to address the vast diversity of microclimatic and local differences occuring thoughout the number of temperate and tropical alpine ecosystems. We will also omit the effects of UV-B radiation on plants, as this important abiotic factor is not exclusive to mountain environments.

Climatic conditions of arctic, boreal and alpine environments challenge photosynthesis of plants in several ways. During the winter, plants are either covered by snow, which creates a relatively sheltered environment, or they are exposed to cold temperatures as well as ice and snow particles blown by wind. These particles might cause mechanical damage by abrasion on leave surfaces of evergreen plants. As Thomas Elliot wrote: 'April is the cruelest month. Winter kept us warm, covered by the forgetfulness of snow'. Spring does not necessarily mean less-harsh conditions and smaller challenges for photosynthesis. Temperatures are changing rapidly from sub-zero to high temperatures, and irradiance levels can be extremely high, imposing stress and causing photoinhibitory damage to photosynthetic tisssue. At the same time, plants must prepare themselves rapidly for life during a short summer, during which they need to acquire all the carbon required for reproduction as well as growth and respiration through the next winter.

31.2. CLIMATIC CONDITIONS OF ALPINE AND BOREAL ECOSYSTEMS

31.2.1. High altitudes

Approximately 20% of the Earth's land surface is covered by mountains ranging from the polar to the tropics (Friend and Woodward, 1990; Barry, 2008). As mentioned above, climatic conditions through this wide range of mountains are very different, especially the microclimate for individual plants. For an extensive presentation of mountain climate, the interested reader is referred to Körner (1999) or Barry (2008). Most physiological and meteorological studies on alpine plants were carried out in the European Alps and in North America at altitudes higher than 2000 m. The climatic conditions discussed here therefore mainly characterise these regions and should not be regarded as a general feature of alpine climate in all mountains.

The alpine environment is separated by the timberline from the subalpine and by the snowline from the nival zone (Billings and Mooney, 1968). The altitude of the alpine life zone is highest around the equator and lowest in polar regions (Körner, 1999; Barry, 2008). With increasing altitude, mean air temperature and atmospheric pressure decline, whereas low-temperature extremes, maximum light intensity, UV-radiation and maximum windspeed as well as precipitation increase (Billings and Mooney, 1968; Friend and Woodward, 1990; Körner, 1999). Precipitation is strongly dependent on the regional climate. In most parts of the European alps annual precipitation can be up to 2000 mm (Fig. 31.1). With the exception of lower mean air

Terrestrial Photosynthesis in a Changing Environment: A Molecular, Physiological and Ecological Approach, ed. J. Flexas, F. Loreto and H. Medrano. Published by Cambridge University Press.

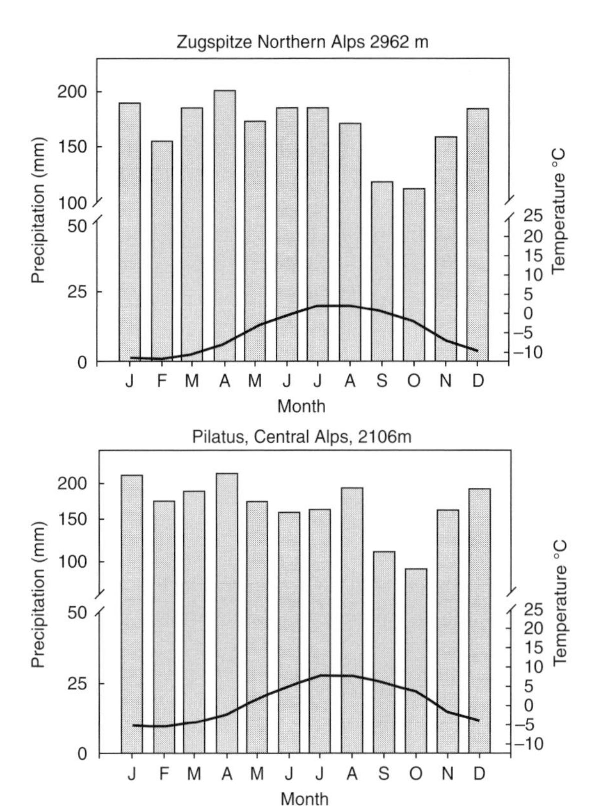

Fig. 31.1. Monthly temperature and precipitations from two alpine locations Zugspitze (northern alps) and Pilatus (central alps).

temperature and lower atmospheric pressure as compared with lower altitude, all other climatic conditions depend strongly on the regional climate, in particular on the cloud cover and the slope of the investigated plants (Barry, 2008). The mean annual air temperature in the alpine life zone is about 0°C or lower (Fig. 31.1).

In contrast to arctic plants, surprisingly high-temperature extremes could be measured in mountain plants on cloudless summer days (Körner, 1999). The leaf temperature of alpine plants may vary from up to 40°C during the day to subzero during the night. Large temperature gradients can even be found within the same plant, from heated leaves to roots resting in a frozen soil, or from one day with a warm temperature to another day with snowfall. Such extreme temperature gradients may co-occur with extreme light intensities being as low as the light compensation point of photosynthesis up to extremely high intensities of 3000 μmol m^{-2} s^{-1} PAR at 2400 m elevation (Fig. 31.2). Moser (1973) measured a remarkably flexible and efficient photosynthetic assimilation rate of *Ranunculus glacialis* leaves in the nival zone of the Austrian alps under widely varying conditions of temperature and light, suggesting that the ability to cope with rapidly changing climate extremes is the most challenging feature for photosynthesis of alpine plant species in particular, as the lifecycle has to be completed within a short vegetation period in non-tropical mountains.

The decrease in atmospheric pressure with the increase in altitude also decreases the partial pressures of CO_2 and H_2O in the ambient air. It has been suggested that high-altitude plants, similar to alpine climbers suffering from a lack of O_2, might be challenged in aquiring sufficient CO_2 under these low atmospheric-pressure conditions. However, the diffusion coefficient of CO_2 increases with decreasing pressure and rates of diffusion of CO_2 into leaves (with identical leaf anatomy and temperature) will change to a much smaller degree (Terashima *et al.*, 1995). In parallel to the decrease in the partial pressure of CO_2, the partial pressure of O_2 also decreases, potentially decreasing the proportion of photorespiration. Altogether, Terashima *et al.* (1995) argued that the decrease in the CO_2 concentration with altitude is therefore not as important for photosynthesis as previously suggested. The same is not true for the diffusion of *water vapour.* The air inside the stomatal opening will remain almost saturated with H_2O, but the saturating *water-vapour* pressure will depend only on temperature (this is the reason why water boils at lower temperatures in very high altitudes). This means that a larger amount of water may be contained in the volume of air at lower pressures, provided that temperature does not change. However, in the field air temperature usually decreases with altitude, this being the reason for the so-called 'adiabatic lapse rate' (e.g., Gale, 1972). Gale (1972) showed that transpiration rates will increase when the temperature decrease is less than the theoretical temperature decrease owing to reductions in air pressure.

31.2.2. Differences of mountain climates and lowland boreal and arctic climates

Northern boreal climates and mountain climates have strong affinities, and floras of both areas have many common species that are missing in the temperate areas. Billings (1974) estimated that half of the 1,000 arctic species also occur in mountain areas further south.

Boreal vegetation is distributed in areas with average summer temperatures between 13 and 20°C. Rainfall

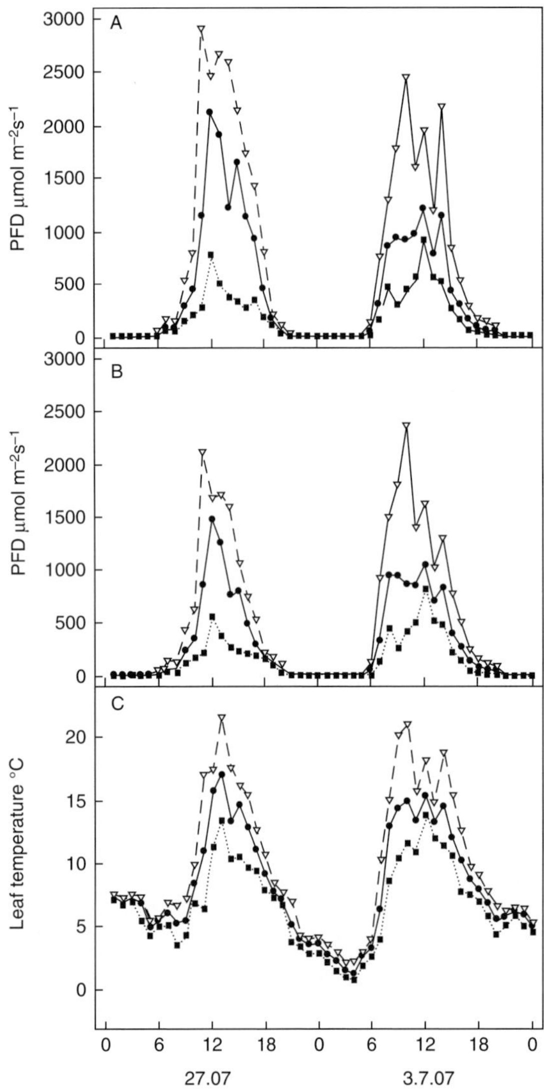

Fig. 31.2. Variation of photon-flux density (PFD) and temperature at 2400 m altitude during two consecutive days in the French alps. Maximum values, ∇; Mean values, •; and minimum values, ■ per hour. The Julian day and hour are indicated. **(A)** PFD at the ground level; **(B)** PFD at the level of a *S. alpina* leaf; and **(C)** the corresponding leaf temperature. At the leaf level PFD varied by more than 1500 µmol $m^{-2}s^{-1}$ per hour and temperature by more than 10°C. Maximum temperature variation between day and night-time was 21°C (Streb, unpublished results).

might vary from 200 to more than 1600 mm a year, with considerable differences in the annual temperature range. Nevertheless, the largest areas of boreal forests are in continental areas with relatively low rainfall (600 mm and less). Although soil drought is not necessarily predominant in the boreal forest, evapotranspiration may exceed precipitation during the summer months in many areas. The spring soil-water reservoir (replenished by snowfall) may, however, greatly reduce the effects of seasonal drought (e.g., Berninger, 1997).

From a plant's perspective, a boreal climate might be less stressful than an alpine climate, as the exposure to high irradiance and to high UV levels is less extreme. Also, daily temperature ranges are more predictable and less important owing to the thicker, isolating atmosphere. Soils are, however, mostly colder in boreal arctic zones than in their analogous alpine zones; and at many locations in the north, trees grow on permafrost.

The shortness of the vegetation period is quite limiting in boreal or arctic climates, where it may last less than three months. However, during this period the day length is very long (as boreal and arctic locations are at high latitudes) and can sustain high daily rates of photosynthetic production (Billings, 1974).

31.3. PHOTOSYNTHESIS OF KEY FUNCTIONAL GROUPS IN THE BOREAL AND ALPINE BIOMES

Boreal and alpine biomes represent a mixture of various different vegetation and plant functional types, but they share floristic similarities. Boreal forests usually have a large component of evergreen coniferous trees (mostly from the genera of *Pinus, Picea, Abies*), but deciduous trees (conifers like *Larix*, and broadleafs like *Betula, Nothofagus* and *Populus*) are also common. All these trees accumulate carbon during a short growing season. In addition, evergreen trees need to protect their leaves from the adversities of winter. While evergreen coniferous trees dominate the boreal and mountain forests in boreal and temperate zones, considerable areas within these biomes are actually covered by deciduous trees. Birch forms the treeline thoughout much of Scandinavia, Iceland, Greenland and Kamshatka (e.g., Krestov, 2003).

The understory of boreal forests is dominated by dwarf shrubs and mosses (or lichen). Dwarf shrubs are mostly from the *Ericaceae* family and might be deciduous as well as evergreen (like *Vaccinium myrtilloides* (deciduous) and *Vaccinium vitis-idae* (evergreen)). Many of the species are circumpolar and also occur in mountain environments. Mosses are mostly feathermoss, and lichen may dominate the understory of northern boreal and very dry forests. These belong mostly to the *Cladina* and *Cladonia* genera.

Soils in the boreal area are mostly podzolic and may be covered by thick organic layers. Herbaceous plants may dominate mostly in young successional ecosystems or fertile sites. Further north in the lower arctic, dwarf shrubs and lichen continue to dominate the vegetation.

About one-third of the boreal-forest area is dominated by peatlands, with peat accumulations that are more than 30 cm deep. The main peat-forming organisms in these ecosystems are usually *Sphagnum* mosses, although sedges (*Carex sp.*) and even dwarf shrubs can be important. Altogether, the alpine and boreal vegetations are functionally diverse. We will explore how these different functional groups respond to the challenges imposed by climate to maintain their photosynthetic production in the rest of the chapter.

The alpine vegetation belt is defined as the zone between the treeline and the nival zone, which is snowcovered all year long. This region has its latitudinal analogue in arctic areas that lie north of the treeline. Compared with boreal and most arctic systems, alpine flora have a higher abundance of herbaceous vegetation. According to Körner (1999) the most important families are *Asteracea* and *Poaceae*, but also *Brassicaceae*, *Caryophyllaceae*, *Rosaceae* and *Ranuncalaceae*. Alpine-shrub vegetation is dominated by *Ericaceae* and *Asteraceae*. Although, there is a high similarity between the alpine and arctic floras, arctic and alpine species are almost always ecotypically different (Billings, 1974). Differences and definitions of arctic and alpine systems are not always clear. For example, Bliss (1962) reported that the eastern North American mountain floras have a high degree of affinity to the arctic floras, whereas mountain floras of the southern Rocky Mountains have certain affinities to the adjacent desert floras.

31.4. PHOTOSYNTHESIS OF ALPINE AND BOREAL HERBACEOUS PLANTS AND DWARF SHRUBS

31.4.1. Adaptation at the leaf and whole-plant level

Most alpine-plant species are comparatively small and grow in cushions, tussocks and rosettes (Körner, 1999). This growth form helps to decrease the amount and the intensity of light energy absorbed and generally ameliorates the microclimate for growth in alpine environments (Beck *et al.*, 1982; Körner, 1999; Germino and Smith, 2001). The small SLA of alpine plants, together with high levels of respiration, is the main cause of the relatively low growth rate in alpine species (Atkin *et al.*, 1996). However, not all species match this typical growth form and are more or less exposed to extreme climatic changes. These plants must have at least developed physiological mechanisms of adaptation. The majority of alpine plants are perennials and show a C_3 type of assimilation, whereas C_4 and CAM metabolism is rare (Körner, 1999).

Leaf structure is one of the main adaptations of alpine plants to their environments. The challenges at high altitudes are the cold and rapidly changing climatic conditions, the high irradiance and the low atmospheric pressures with lower concentrations of CO_2 and, hence, the lower CO_2 supply rates, but also lower O_2 partial pressures. Stomatal density and conductance are known to increase with altitude in most temperate mountains, but not in tropical mountain ranges (Körner *et al.*, 1983; Körner, 1999). Similar changes of stomatal density have been observed in responses to changing CO_2 concentrations in experimental and paleoecological research (e.g., Beerling and Rundgren, 2000). Nevertheless, Körner (2007) argued that this increase is rather an acclimation to higher irradiances at high altitude in temperate mountains, as tropical alpine plants do not show these trends. Stenström *et al.* (2002) have analysed the leaf anatomy of arctic *Carex* species and shown that stomatal density decreased with decreasing temperature in arctic populations of different closely related *Carex* species. Stomatal size tended to increase with increasing temperature. They explain these trends as an adaptation to higher water stress from transpiration in the warmer sites. Au (1969) compared stomatal densities of arctic and alpine populations of *Oxyria digyna* in a common garden experiment and observed an increase in stomatal density from north towards south.

For plants of the same species, SLA decreases with altitude, in other words, leaves with the same area get heavier (and usually thicker) at higher altitudes. Woodward (1979) showed that the decrease in SLA was, at least partially, caused by changes in temperature: leaf specific areas of *Poa bertheloni* and *Poa alpinum* increased by about 2 $dm^2\ g^{-1}$ from 2.6 and 4.7 $dm^2\ g^{-1}$, respectively, when air temperature in the growth chamber increased from 10°C to 20°C. This kind of trend seems to hold in large scale analyses of leaf anatomic characteristics (Wright *et al.*, 2004). Thicker leaf sizes may, however, decrease the mesophyll conductance to CO_2 diffusion (g_m) in alpine species (Sakata, 2002) as shown for *Polygonum cuspidatum* (Kogami *et al.*, 2001; Terashima *et al.*, 2005). Furthermore, the CO_2 partial pressure at the cell surface and at the sites of CO_2 fixation is lower in alpine compared with lowland species (Körner, 1999; Sakata *et al.*, 2002). In contrast, in *Buddleja davidii* the g_m increased with altitude and compensated for a lower stomatal conductance (g_s) at high altitude in comparison to low altitude (Shi *et al.*, 2006). In addition to leaf anatomical structures, aquaporins facilitate the CO_2 diffusion into the cell (Flexas *et al.*, 2006a, 2008). Low temperatures at high altitude and in the boreal zone might therefore affect the regulation of aquaporins and hence CO_2 diffusion (Terashima *et al.*, 2005).

In addition, plants from higher altitudes have usually higher nitrogen (N) contents per leaf area than plants of the same species at lower altitudes. The increase of N concentrations with altitude is probably caused to a large degree by the alpine plant strategies for resource acquisition and storage (Mooney and Billings, 1961), since N mineralisation will normally decrease with altitude (e.g., Haselwandter *et al.*, 1983; see Morecroft and Woodward, 1996 for a diverging opinion). Nitrogen contents of alpine and arctic herbaceous plants are usually higher than 2.5 % (Körner, 1999). These high N contents go along with high rates of photosynthesis (Körner and Diemer, 1987) and with a high activity of the Rubisco enzyme (Westbeek *et al.*, 1999; Sakata *et al.*, 2006). It also seems that the ratio of J_{max} to $V_{c,max}$ decreases with altitude for different *Poa* species (Westbeek *et al.*, 1999). It is noteworthy that high rates of photosynthesis do not mean high RGRs. For example, Westbeek *et al.* (1999) observed a negative correlation between photosynthetic capacity (A_{max}) and RGR. High Rubisco activity (and high acticities of other enzymes of the dark cycle) may, to some extent, help plants to maintain photosynthesis at low temperatures when the rate of the dark reactions efficiency decreases more than that of the light reactions (e.g., Chapter 18). The high N contents may be responsible for the comparatively high rates of respiration of many alpine plants (Körner, 1999).

^{13}C discrimination generally increases at high altitudes, indicating that the ratio of ambient to mesophyll CO_2 decreases at high altitudes. This means that g_s and g_m are more limiting for photosynthesis at high altitudes than at low altitudes. According to Körner (1999), this increase in ^{13}C discrimination can be quite large, with a gradient of 0.78 ppm per km of altitude. The carbon-isotopic discrimination measures the ratio of ambient to mesophyll CO_2 concentrations and the observed gradient corresponds to a decrease of the mesophyll CO_2 concentration of 10.7 ppm per km of altitude (in addition there is a decrease of the atmospheric pressure that reduces the amount of CO_2 in the air). As both the absolute CO_2 concentrations in the air and the ratio of substomatal-to-ambient-CO_2 concentration decrease with altitude, plants are left with a low supply of carbon. Altogether, it seems that the effects of high Rubisco activities and thick leaves with potentially high mesophyll resistances (Sakata *et al.*, 2002) lead to a decrease of the ratio of mesophyll to ambient CO_2. It is also important to note that the low CO_2 concentrations in the mesophyll of many alpine plants aggravate the stress owing to low temperatures and high light. Low CO_2 concentrations in the mesophyll reduce the rate of dark reactions that represent a biochemical sink for the energy produced by the light reactions. Therefore plants have evolved various mechanisms to balance the flow of energy and to avoid energetic imbalances and photoinhibitory stress imposed by low temperatures (see Chapter 18 of this book for a detailed discussion of these mechanisms).

31.4.2. Photosynthetic activity

Mean maximum rates of photosynthesis measured under standard conditions are similar in alpine plants compared with lowland plants, but show a high variability (Körner, 1999; Fig. 31.3). In two Antarctic vascular plant species, assimilation rates at optimal temperature were lower than mean values for various alpine plants but interestingly, the Antarctic variety of *Colobanthus quitensis* had higher assimilation rates than an Andean high-mountain variety (Xiong *et al.*, 1999; Bravo *et al.*, 2007). The photosynthetic temperature optimum of alpine and arctic plants is shifted to lower temperatures compared with lowland and tropical plants, which corresponds to the general lower mean air temperature measured in mountains (Körner and Diemer, 1987; Körner, 1999; Xiong *et al.*, 1999). A survey of different alpine species investigated under the same conditions as low elevation plants showed higher carbon-use efficiencies and higher assimilation rates at saturating CO_2 (Mächler and Nösberger, 1977; Körner and Diemer, 1987). However, low temperature, high light and possibly N deficiency reduce the net-carbon gain of alpine plants at high altitude compared with lower altitude (Cabrera *et al.*, 1998). Rawat and Purohit (1991) measured species-dependent differences in CO_2 assimilation at high and low altitude. Similar to carbon assimilation, Rubisco activity may also differ between alpine and low elevation species. Although Rubisco activity increases with altitude in highland populations of the Andean species *Espeletia schultzii* (Castrillo, 1995), no such correlation was observed in grassland species in the Alps (Sage *et al.*, 1997).

Photosynthesis per leaf area increases in alpine species compared with low-altitude species. This might be explained

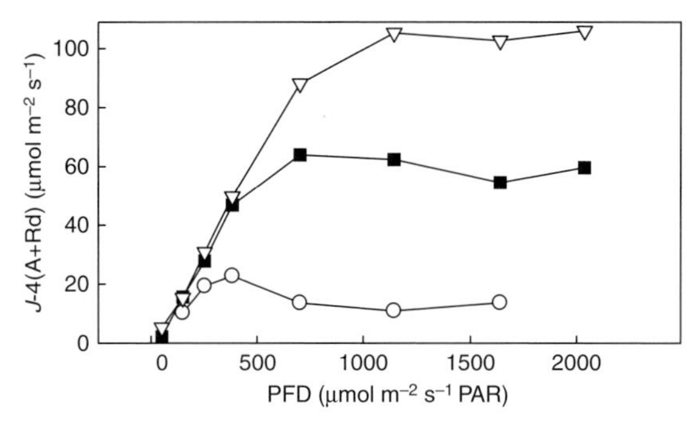

Fig. 31.3. Excess electron transport in *R. glacialis* leaves as a function of PFD at 23°C (▽), 15°C (■) and 10°C (○) leaf temperature. Excess electron transport (*J*) was determined as the difference between whole-chain electron transport and electrons used for assimilation and dark respiration as measured by gas exchange together with chlorophyll fluorescence (from Streb *et al.*, 2005).

by their thicker leaf size, which is usally a result of the cold acclimation process (Körner, 1999). By contrast, a comparison of O_2-evolution capacity in saturating CO_2 on a chl. instead of a leaf-area basis did not show a clear difference between alpine and lowland species (Streb *et al.*, 1997). Light saturation of photosynthesis is often higher in alpine compared with lowland species and can exceed 1200 µmol $m^{-2}s^{-1}$ PAR (Körner and Diemer, 1987). Several alpine plant species, including *R. glacialis*, showed no saturation up to 2000 µmol $m^{-2}s^{-1}$ PAR at normal CO_2 atmospheric pressure (Körner and Diemer, 1987; Körner, 1999; Germino and Smith, 2001; Streb *et al.*, 2005). Furthermore, alpine species in contrast to lowland plant species showed no indication of phosphate limitation during illumination at low temperature. While phosphorylated intermediates accumulate in *Pisum sativum* leaves and the ATP/ADP ratio declines, the pool of nucleotide triphosphate increased in *R. glacialis* and *S. alpina* leaves at low temperature in the light and phosphorylated metabolites did not accumulate (Streb *et al.*, 2003a). Nevertheless, also in *R. glacialis,* PET exceeded electron consumption by carbon metabolism at various light intensities and temperatures (Streb *et al.*, 2005; Fig. 31.3). Similar responses can be expected for other alpine species, at least during periods of varying temperature and photon-flux extremes.

31.4.3. Photoinhibition in alpine plants

Metabolic reactions depend on temperature. At low temperature, biochemical reactions are slowed down, whereas at high temperature the stability of membranes and enzymes are affected when light is still absorbed by the photosystems (Falk *et al.*, 1996; Havaux, 1998). Such conditions may cause an imbalance between absorbed light energy and the ability to use this energy for metabolism especially when PPFD exceeds A_{max} (Wise, 1995; Falk *et al.*, 1996; Huner *et al.*, 1998). As a consequence, the photosynthetic electron transport chain becomes reduced and excess excitation energy leads to the formation of ROS, which in turn can damage membranes, pigments and proteins, as well as blocking repair (Wise, 1995; Asada, 1996; Murata *et al.*, 2007). The alpine climate therefore predisposes plants to compensate for periods of excess excitation energy.

Photoinhibition of PSII and photoinactivation of the enzyme CAT are two early indicators of photooxidative stress (Aro *et al.*, 1993; Anderson and Barber, 1996; Feierabend *et al.*, 1996). Leaves of alpine-plant species are remarkably tolerant to high light intensity at low temperature as measured by photoinactivation of PSII, CAT and of more general photodamage *in vivo* (Heber *et al.*, 1996; Streb *et al.*, 1997; Manuel *et al.*, 1999; Germino and Smith, 2000). This tolerance does not depend on structural differences of the PSII *RC* as PSII activity in isolated chloroplasts or thyalkoids from alpine plants have remained light sensitive (Streb *et al.*, 1997). In contrast, a CAT protein with different primary structure isolated from *Homogyne alpina* leaves was much less light sensitive than CATs from other sources (Feierabend *et al.*, 1996; Streb *et al.*, 1997; Engel *et al.*, 2006), showing that a direct protein modification might, in some cases, contribute to the increased tolerance of high-light stress in alpine plants.

Furthermore, the in-vivo tolerance against photoinactivation does not depend on higher turnover rates of the D1 protein of PSII in alpine plants. In conclusion, alpine plants are better protected from inactivation than lowland plants (Streb *et al.*, 1997; Shang and Feierabend, 1998). In cold-acclimated plants, the desaturation of membrane lipids contributes to maintain membrane fluidity and to protect against photoinhibition, especially in cyanobacteria (Nishida and Murata, 1996). Compared with lowland plant species, however, no different composition of lipids and of their desaturation level was found in several typical alpine species (Dorne *et al.*, 1986), as was generally supposed for cold-acclimated higher-plant species in contrast to cyanobacteria (Falk *et al.*, 1996).

In contrast to light exposure at low temperature, alpine species show severe photoinhibition during illumination at high temperature (Streb *et al.*, 2003b). While irreversible heat damage occurs at temperatures well above 40°C in a great variety of species, including alpine plants, heat-induced photoinhibition is apparent at 38°C in *R. glacialis* and *S. alpina* (Gauslaa, 1984; Larcher *et al.*, 1997; Streb *et al.*, 2003b). Heat sensitivity is higher in *R. glacialis* than in *S. alpina* leaves, and the species differ strongly in the ability to repair photoinactivated PSII. This suggests that occasional high-temperature extremes at low altitude restrain the distribution of *R. glacialis* to high altitudes in the alps (Streb *et al.*, 2003b).

Lundell *et al.* (2008) showed that the A_{max} of lingonberry (*Vaccinium vitis-ideae*) in the understory of a boreal pine forest was pretty high while the plants were covered by snow in the winter. There were, however, signs of photoinhibition after snowmelt. Starr and Oberbauer (2003) showed that F_v/F_m values of evergreen dwarf shurbs in Alaska may reach reasonably high values of 0.4, indicating that PSII function may be impaired but still be good enough to sustain photochemical energy conversion while plants are still covered by snow. They also observed a drawdown of CO_2 under the snowcover and argue that the combination of high CO_2 concentrations, reasonably high irradiance values and temperatures of about 0°C may deduce that an important fraction of the annual photosynthetic production is performed while snow is still on the ground.

31.4.4. Photoprotection in alpine plants

Strategies of photoprotection in alpine plants involve mechanisms to reduce or avoid absorption of excess light and the generation of a favourable microclimate. The morphological structure, the small growth size, the increase in leaf thickness and the organisation of leaves in cushions and rosettes all contribute to avoid adverse photooxidative stress that may result from excess light. Examples of such strategies are, e.g., rosette closure observed in *Saxifraga paniculata*, which is induced by drought stress to protect this plant by shading from heat-induced damage in high light (Neuner *et al.*, 1999). Another strategy is observed in leaves of *Eryophorum angustifolium* that induce chl. synthesis only at warm temperatures thus avoiding light absorption at low temperature (Lütz, 1996).

The harmless dissipation of excess light energy via heat is an important mechanism to avoid over-reduction of the photosynthetic electron transport chain and to prevent specific and general destruction. Energy dissipation via heat is often measured by NPQ and correlates with the generation of a pH gradient, the synthesis of zeaxanthin and the presence of the PsbS protein (Niyogi, 1999; Müller *et al.*, 2001; Adams *et al.*, 2002). The importance of this photoprotective mechansism is highlighted by its actual fraction of the energy quenching processes. According to Niyogi (1999), more than 75% of absorbed photons may be dissipated via non-photochemical dissipation. In some alpine plant species, high contents of xanthophyll-cycle pigments and a high de-epoxidation status correlate well with NPQ under stress conditions, however, the importance of this pathway is not always uniform (Streb *et al.*, 1997, 1998; Williams *et al.*, 2003). Cold-acclimated overwintering plants and high mountain plants retain zeaxanthin contents during the night and show a high NPQ in the morning (Fetene *et al.*, 1997; Germino and Smith, 2000; Heber *et al.*, 2000; Adams *et al.*, 2002; Williams *et al.*, 2003). The alpine-plant species *Geum montanum* uses CEFI under high light intensity to increase the pH gradient and to modulate NPQ (Manuel *et al.*, 1999). The protecting capacity of NPQ decreases, however, often at high PPFD, requiring alternative protection by additional electron sinks (Ort and Baker, 2002; Streb *et al.*, 2005; Baker *et al.*, 2007). A high NPQ and a pronounced xanthophyll-cycle activity were measured in some alpine species such as *Soldanella alpina*, in different Tasmanian alpine shrubs and also in Antarctic *Colobanthus quitensis*, but far less in other alpine species, such as *R. glacialis*, *Ozothamnus ledifolius* and Antarctic *Deschampsia antarctica* (Streb *et al.*, 1998; Williams *et al.*, 2003; Perez-Torres *et al.*, 2007). This clearly shows that xanthophyll-cycle dependent photoprotection is not the only strategy of photoprotection in alpine and arctic plants.

Another important component of protection against damage of leaves by excess photon-flux densities and extreme temperature is the capacity of leaves to scavenge ROS resulting from excess light. ROS are formed in chloroplasts mainly during the Mehler reaction and in photosystems by spin changes generating singlet oxygen (Asada, 1996; Ort and Baker, 2002). Components of the antioxidant system investigated in alpine- and arctic-plant species involve prominent enzymes, such as CAT, SOD, GR and APX, the water-soluble antioxidants ascorbate and glutathione and lipid-soluble antioxidants, such as α-tocopherol and carotenoids including zeaxanthin. Zeaxanthin in this context has been shown to have a higher capacity for antioxidative protection than other carotenoids (Havaux *et al.*, 2007). Contents of ascorbate, glutathione, α-tocopherol and carotenoids including zexanthin are higher in leaves of alpine-plant species collected at high compared with low altitude (Wildi and Lütz, 1996), but their quantity differs greatly in different species (Wildi and Lütz, 1996;

Streb *et al.*, 1997, 1998). Leaves of *S. alpina* and *S. minima* have exceptionally high contents of ascorbate, which represent the second major soluble carbon metabolite in *S. alpina* leaves (Streb *et al.*, 2003c). Also, antioxidative enzyme activities are high in *S. alpina* leaves (Streb *et al.*, 1997), but in leaves of other alpine plant species, such as *R. glacialis*, components of the antioxidative protection system are exceptionally weak (Wildi and Lütz, 1996; Streb *et al.*, 1997, 1998). The antioxidative protection capacity in alpine plants positively correlates with the resistance against the superoxide-generating herbicide paraquat (Streb *et al.*, 1998), demonstrating their importance in protection against ROS. Similarly, the Antarctic plant *Deschampsia antarctica* showed high antioxidative protection when acclimated to low temperature (Pérez-Torres *et al.*, 2004).

Transport of excess electrons to O_2 also contributes to photoprotection during excessive light absorption (Ort and Baker, 2002). In addition to the Mehler reaction, photorespiration is particularly important when the O_2/CO_2 ratio increases in leaves owing to stomatal closure. However, photorespiration is favoured at high temperature and suppressed at low temperature (Leegood and Edwards, 1996), but is nevertheless important in alpine-plant species. Photorespiration is active at low temperature in *R. glacialis* and in alpine *Chenopodium bonus-henricus* at ambient temperature when stomata are closed (Heber *et al.*, 1996; Streb *et al.*, 2005). Furthermore, higher rates of photorespiration were measured in alpine plants compared with lowland ecotypes of *Trifolium repens* (Mächler *et al.*, 1977; Mächler and Nösberger, 1978). The occurrence of proliferations of the chloroplast envelope called protrusions, which are widely distributed in leaves of various alpine-plant species and which are modulated by temperature, were supposed to increase the chloroplast surface in order to facilitate metabolite exchange during photorespiration (Buchner *et al.*, 2007).

Another possibility of electron consumption is the malate shuttle that can transfer excess electrons from chloroplasts to other cell compartments. These electrons may be finally oxidised by the mitochondrial electron transport chain (Scheibe, 2004). As malate contents are very high in *R. glacialis* leaves (Streb *et al.*, 2003a), the malate shuttle might be involved in electron consumption of this species. However, the corresponding activity of NADP-malate deshydrogenase was comparatively low in *R. glacialis* leaves. Furthermore, malate was not labelled or degraded during photosynthesis, suggesting a low turnover and excluding the activity of PEPC for increasing malate contents (Nogués *et al.*, 2006).

Alternatively, the chloroplast-located terminal oxidase (PTOX), homologous to the mitochondrial AOX (Carol and Kuntz, 2001), might be of particular importance in alpine-plant species as a chloroplast-located electron acceptor, transferring electrons to O_2 and avoiding the production of ROS. Several alpine-plant species have much higher PTOX contents than lowland species, in particular *R. glacialis* (Streb *et al.*, 2005). This species transports electrons at low and ambient temperature to O_2 via a yet unknown pathway that may well be PTOX. During deacclimation, PTOX contents decline while the sensitivity to photoinhibition increases (Streb *et al.*, 2003a, 2005). Other species such as *G. montanum* increase their PTOX content with altitude (Streb *et al.*, 2005). In all known non-alpine plant species, PTOX contents are very low and their potential activity was estimated to contribute only up to 0.3% of the electron transport capacity (Ort and Baker, 2002). But in alpine environments at high elevations, in which low temperature and high irradiance may combine with a limited supply of CO_2 alternative electron sinks, such as photorespiration, and chlororespiration may play a substantial photo-protective role.

31.4.5. Synthesis of unusual metabolic products

Metabolic profiling by NMR-spectroscopy (Bligny and Douce, 2001) was applied to identify the major soluble carbon metabolites in different alpine-plant species. Surprisingly, several alpine-plant species accumulate uncommon metabolites with yet unknown functions. Leaves of *G. montanum* synthesise methylglucose in high concentrations (Aubert *et al.*, 2004). The major carbon metabolites in *R. glacialis* leaves, representing more than 50% of total soluble carbon, are ranunculin and malate (Streb *et al.*, 2003a), which are not primary products of photosynthesis (Nogués *et al.*, 2006). The importance of ascorbate relative to other carbon metabolites in *S. alpina* leaves was measured by the same method (Streb *et al.*, 2003a). The sub-antarctic cabbage, *Pringlea antiscorbutica*, accumulates proline (Aubert *et al.*, 1999). As newly assimilated carbon in *R. glacialis* is only weakly respired but transported to sink organs (Nogues *et al.*, 2006), these unusual metabolites might serve as respiratory reserves in leaves or contribute as compatible solutes to freezing tolerance.

31.4.6. Freezing temperatures

Besides low non-freezing chilling temperatures during the photoperiod, temperatures may regularly fall below 0°C at night during the vegetation period and alpine plants have to cope with the risk of freezing damage. A comparison of several alpine-plant species showed a freezing tolerance in leaves between –3°C and –10°C during the summer. This is not much different from plants in other temperate zones (Sakai and Larcher, 1987; Taschler and Neuner, 2004) and suggests the absence of specific mechanisms to decrease the freezing temperature. Interestingly, Mediterranean, high-Andes species tolerated a lower freezing temperature between –8.2°C and –19.5°C, which was explained with the drought effect in arid mountains sharing common responses to cold acclimation (Sierra-Almeida *et al.*, 2009). C_4 plant species of the Mongolian high plateau showed much lower freezing resistances than C_3 species and this might be, in combination with lower leaf temperatures during photosynthesis, a reason why C_4 plants are rare in dry mountain ecosystems (Liu and Osborne, 2008). The freezing tolerance of some hardened alpine species, such as *Saxifraga oppositifolia* and *Silene acaulis*, growing on exposed snowfree sites in winter may, however, be as low as liquid-nitrogen temperature (Sakai and Larcher, 1987; Körner, 1999). Most alpine plants survive the winter period under a thick snowcover and only few species, such ase *S. alpina*, *G. montanum* and *H. alpine*, keep their leaves green during this period. The maximum freezing tolerance of these plants is considerably higher at around –20°C (Körner, 1999). The snowcover protects plants from low-temperature extremes depending on the thickness of the snowcover (Sakai and Larcher, 1987; Körner, 1999).

The high winter-freezing tolerance of leaves is rapidly lost when plants are exposed to elevated growth temperature within a few days (Körner, 1999), suggesting physiological acclimation to freezing. Although freezing tolerance has been determined in a great variety of alpine-plant species, less is known about the mechanisms of this tolerance. Antifreeze proteins may be present in alpine plants as was shown for Antarctic *Deschampsia antarctica* (Bravo and Griffith, 2005). Compatible solutes such as proline and sugars may increase the osmotic potential and decrease the freezing temperature. However, compatible solutes may decrease the freezing temperature only by up to 2°C, which may be sufficiently low for plants during the summer, but is much less than the observed difference in freezing tolerance between summer and winter plants (Körner, 1999). The avoidance of ice formation in tissues, called supercooling, was shown for plants in the tropical-alpine life zone of the Andes; grasses in the same life zone had low supercooling capacity (Körner, 1999; Marquez *et al.*, 2006). Finally, the morphological organisation of leaves not only protects against light absorption during the day but also prevents heat loss during the night and may keep leaf temperature higher than air temperature (Sakai and Larcher, 1987; Körner, 1999).

31.4.7. Long-term adaptations and the annual cycle

Körner (1999) tried to classify the photosynthetic strategies of alpine plants in response to spring conditions (and particularly snowpack). Evergreen plants may remain green and: (1) retain their A_{max} under the snowpack (e.g., *Soldanella*); or (2) deactivate their A_{max} over winter (e.g., *Loiseloiria procumbens* and evergreen trees). Deciduous plants, on the other hand, can: (3) flush after snowmelt to avoid the most stressful period (e.g., most deciduous alpine, arctic and boreal shrubs, *Carex* spp.); or (4) initiate leaf expansion before snowmelt but start greening and photosynthesis immediately (e.g., snowbed *Ranunculus* sp.). Another strategy is: (5) to initiate leaf expansion at or before snowmelt, but activate photosynthesis and greening of leaves later (many herbs), as shown for *Eryophorum angustifolium* (Lütz, 1996). There are also a few annual species that: (6) only germinate after snowmelt.

Lifetime carbon balances for alpine and boreal understory plants have been estimated by Diemer and Körner (1996). They conclude that photosynthetic rates per unit time in high alpine *Ranunculus* species are similar as for *Ranunculus* species from lower altitudes. According to Diemer and Körner (1996), most of the differences in productivity observed between high-alpine and low-altitude species result from the shorter length of the growing season in alpine environments.

31.4.8. Whole-plant gas exchange and ecosystem-flux measurements

There are only few studies reporting measurements of the whole gas exchange of true alpine vegetation. Most of the data available were presented by the recent synthesis paper of Wohlfahrt *et al.* (2008b), although these data originate from subalpine semi-natural grasslands. These grasslands have, however, a similar floristic composition as true alpine vegetation of the lower alpine zone, however LAI and photosynthetic rates are likely to be higher than cited in the paper. Altogether, the Wohlfahrt paper documented very variable rates of photosynthetic production for alpine ecosystems from the European Alps. There are a few other observations from tundra and Tibetan-mountain ecosystems (Harazono *et al.*, 2003; Zamolodchikov *et al.*, 2003;

Kato *et al.*, 2004). Generally speaking the data indicate a large variation in maximum ecosystem photosynthesis rates (with productivities approaching zero when going towards the nival zone). Maximum ecosystem photosynthesis seems to depend on the LAI (or the green area index) of the vegetation and is correlated with respiration (Wohlfahrt *et al.*, 2008b). Drought probably does not affect photosynthesis of alpine herbs and sedges very much, but changes in the water level of wet ecosystems may increase soil respiration and will induce carbon losses from many tundra and wet alpine ecosystems (e.g., Rogiers *et al.*, 2005).

31.4.9 Water-use efficiencies

Water use efficiencies (WUE) of alpine plants have received surprisingly little attention. Rundel *et al.* (2003, 2005) described high WUE for dry mountain areas in the Californian mountains and Chile. However, the authors argued that predawn and midday water potentials of the Californian fellfield plants were relatively high and did not indicate strong water limitations. Körner *et al.* (1991) surveyed carbon-isotope discrimination across the globe. Their worldwide survey indicated that generally carbon-isotope discriminations decrease with altitude, which would mean that WUE should increase in higher altitudes. It is, however, noteworthy that most of the sites in Körner's data were rather humid, and that the effects of changes in site water balance with temperature are therefore not necessarily important. We think that this increase in carbon-isotope discrimination is linked to thicker, more-resistant leaves and increases in mesophyll resistance caused by these thicker leaves.

31.5. PHOTOSYNTHESIS BY EVERGREEN AND DECIDUOUS TREES

31.5.1. Adaptation at the whole-plant level

Kikuzawa (1991) demonstrated that there is a general trade-off between leaf longevity and A_{max}. In other words, plants with short leaf lifespans have a higher A_{max}. He optimised the production per investment of carbon into leaves and predicted, based on this optimisation, that northern latitudes should be dominated by evergreen trees. The trade-offs between different leaf traits have been discussed recently in the paper of Wright *et al.* (2004), who analysed a large database of different leaf traits of higher plants. Wright and co-authors showed that different leaf traits are highly correlated with each other. They confirmed a negative relationship of leaf longevity and A_{max}. Traits were highly correlated and in

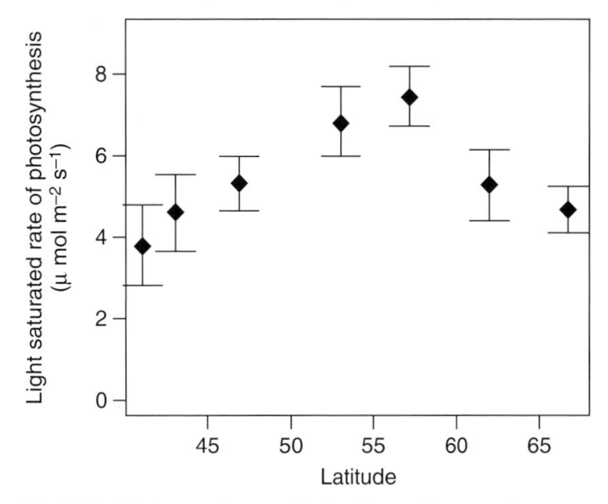

Fig. 31.4. Maximum photosynthesis of Scots pine trees along a geographical gradient from Spain to the treeline in northern Finland. The photosynthesis was measured under light saturating conditions at ambient temperature and is per total leaf area (redrawn from Luoma, 1997).

a principle component analysis, the component explained 77% of the variation of the six leaf traits considered in the analysis (leaf mass per area, leaf longevity, A_{max} per leaf mass, respiration per leaf mass and the gravimetric N and P concentrations). This coordination of leaf traits was largely independent of climate, i.e., plants with any leaf trait can occur in any climate, but leaf traits were well coordinated between each other.

Photosynthesis of Scots pine in temperate regions was highest in the southern boreal zone (Luoma, 1997; Palmroth *et al.*, 1999) and decreased both northwards and southwards (Fig. 31.4). For conifers, deciduous and evergreen shrubs in the Himalaya, Li *et al.* (2004) reported that there seemed to be an optimal zone where trees experienced only little water and cold stress. Changes in A_{max}, N content and carbon-isotope ratios indicate that trees do become more xeromorphic when approaching the treeline. These changes in leaf structure seem to have large genetic components (Luo *et al.*, 2006). For tropical altitudinal gradients, leaf thickness tends to increase when going upwards, as shown for *Metrosideros polymorpha* (Cordell *et al.*, 1998). Luoma (1997) did not find a very clear gradient of A_{max} in Scots pine from temperate and boreal regions, and Palmroth *et al.* (1999) showed that between Scots pine provenances, WUE decreased from northern areas southwards. This trend seems to have certain genetic components, as a common garden and an in-situ gradient behaved in a similar way. The boreal broadleaf Sitka alder (*Alnus sinuata* Rydb.) and paper birch (*Betula papyfera* Marsh.) show a negative correlation between maximal A_{max} in July and frost hardiness two months later (Benowicz *et al.*, 2000). Altogether, it is not clear to what extent the maximal A_{max} of these plants is constrained by resources (such as N) and other processes (for example, frost hardiness). The more xeromorphic behaviour of northern boreal plants could be an adaptation to limited water availability in cold soil or an artifact of cold adaptation: many genes that protect plants from freezing are also linked to better adaptations to drought (e.g., Watkinson *et al.*, 2003). Therefore, the apparent adaptation of northern boreal and subalpine trees to drought might be an artefact owing to the close coupling of the adaptations to drought and cold. As for herbaceous alpine plants we believe that increases in mesophyll resistance caused by thicker more-resistant (and longer-lived) leaves could cause the reduction in A_{max} in northern areas. Resistances to late frosts and lower N availability could be other factors that decrease A_{max} in the north. This seems to be different to the mostly increasing trend of A_{max} with altitude in alpine plants.

Körner (1998, 1999) claimed that alpine or subartic treeline trees are source limited. This means that the photosynthetic production of these trees is higher than their capacity to transport and utilise the products of photosynthesis. This hypothesis remains controversial: Hoch and his co-workers (Hoch *et al.*, 2002; Hoch and Körner, 2003) described that in accordance with the hypothesis, sugar concentrations increase when approaching the treeline. However, Handa *et al.* (2005) found that the responses of *Pinus cembra* L. and *Larix decidua* Mill. provided conflicting evidence for the sink-limitation hypotheses. Investigations into whether there are signs of sink limitation on the A_{max} of trees showed mostly that there are no limitations on photosynthesis (Susiluoto *et al.*, 2007; Basal and Germino, 2008). Altogether, the role of sink limitation in the production of boreal and subalpine ecosystems will remain a subject of discussion for a few years to come.

31.5.2. Photosynthetic capacity

The A_{max} of conifers is usually lower than in broadleafs. Both broadleafs and conifers in boreal and non-tropical sub alpine forests are limited by N concentrations (e.g.,Tan and Hogan, 1995; Vapaavuori *et al.*, 1995; Kull and Niinimets, 1998). Martin *et al.* (2007) also report N limitation for the tropical Hawaiian *Metrosidos polymorpha,* but it is not clear to what extent tropical mountains are limited by N and P. The relation between A_{max} and N concentration of conifers is, however, obscured to some degree by other roles of N in the foliage (Vapaavuori *et al.*, 1995). Therefore, care is required to make direct relationships between A_{max} and N concentration in these plants. Differences in the A_{max} are larger when seen on a mass basis, but they disappear when seen on a per N basis. Lusk *et al.* (2003) compared the photosynthesis and resource utilisation of evergreen angiosperms and evergreen gymnospems in different subalpine forests and concluded that angiosperms have an advantage in resource-rich habitats owing to higher production. In low-resource habitats, the higher nutrient-use efficiency per unit of photosynthesis would be advantageous for conifers. Therefore, evergreen conifers dominate in the boreal forest, but there are a fair amount of deciduous forests in the boreal domain.

31.5.3. Adaptations of photosynthesis to cold stress

Deciduous trees escape the cold winter periods by shedding their leaves and the challenge of maintaining a photosynthetic apparatus during winter. Both boreal conifers and broadleaf trees are able to maintain their A_{max} at a high level and to acclimate to cold spells by actually increasing their A_{max} in the cold (Hjelm and Ögren, 2003). By contrast, evergreen conifers retain their leaves, which are exposed to a potential energetic imbalance between light energy captured by the photosystem and the dark reactions in the spring, when light is high and temperatures are low. This causes photooxydative stress by the production of ROS. Evergreen conifers protect themselves largely by a reduction of their A_{max} and a protection of their photosystem by various mechanisms, e.g., the photoprotective xanthophyll cycle. The regulation of a variety of the involved photoprotective mechanisms differs from their regulation in many herbaceous plants (as e.g., winter wheat and many alpine herbs) that maintain a large A_{max} over the winter (Huner *et al.*, 1998, Chapter 18 of this book).

Ensminger *et al.* (2004) described the downscaling of the photosystem of *Pinus sylvestris* in Siberia in detail. Surprisingly, many of the observed physiological changes occur in the late winter/early spring but not during the coldest period in mid winter. This can be attributed to the higher levels of irradiance during late winter/early spring that are more likely to cause photoinhibitory damage. The reaction-centre protein D1 of PSII was at its lowest level, whereas the ratios of caroteinoids and xanthophylls per chl.

were the highest during spring. Zeaxanthin was the dominant xanthophyll-cycle pigment during spring and winter (Fig. 31.5). This indicates downregulation of PSII together with a large capacity to dissipate excess energy thermally via zeaxanthin. During late spring, when air temperatures were higher than 0°C, there was a gradual recovery of the A_{max} (as seen from fluorescence measurements and photosynthetic O_2 evolution from needles). NPQ decreased during this period, but increased sometimes during days with high irradiance. NPQ increases in light are normal responses. Similar patterns in F_v/F_m and pigment contents were described by Robakowski (2005) for three conifer species in Poland. In addition to the light reactions of photosynthesis, the dark reactions of photosynthesis were also affected by the cold, and carboxlation capacity of Rubisco ($V_{c,max}$) was reduced during the winter (Bigras and Bertrand, 2006; Ensminger *et al.*, 2008). Actually, the recovery of the Rubisco activity appears to delay the recovery of photosynthesis in the spring (Monson *et al.*, 2005). The recovery of the Rubisco activity in boreal evergreens clearly requires more attention and our quantitative understanding of the recovery process is still quite fragmented. Alternatively, early light-induced proteins (ELIPs) are more abundant in conifers during early spring (Zarter *et al.*, 2006). These proteins act in *Aradopsis* as regulators of chl. synthesis and prevent the accumulation of free chl. (Ensminger *et al.*, 2004; Tzvetkova-Chevolleau *et al.*, 2007)

Electron transport capacities of 'visually green leaves' in winter may be very low. The decrease of the electron-transport capacity in conifers during winter reflects the excitation pressure on PSII and results from the combination of light and low temperatures. This was demonstrated in an elegant experiment under field conditions by Porcar-Castell *et al.* (2008). They investigated the recovery of photosynthetic activity during winter and early spring and recorded the variation of fluorescence in experimentally shaded and unshaded control trees in boreal *Pinus sylvestris*. Throughout winter F_v/F_m values were highest in fully shaded trees compared with any other treatment and had fully recovered to their unstressed summer values more than two weeks earlier than e.g., unshaded trees. Unshaded trees also had a higher de-epoxidation status of the xanthophyll cycle and a higher amount of xanthophyll-cycle pigments per chl. during the spring, this indicated the interaction of light and low temperature. The same trends could be observed when comparing (unshaded) upper-canopy foliage with (naturally shaded) lower-canopy foliage (Slot *et al.*, 2005; Porcar-Castell *et al.*, 2008). Similar results were found

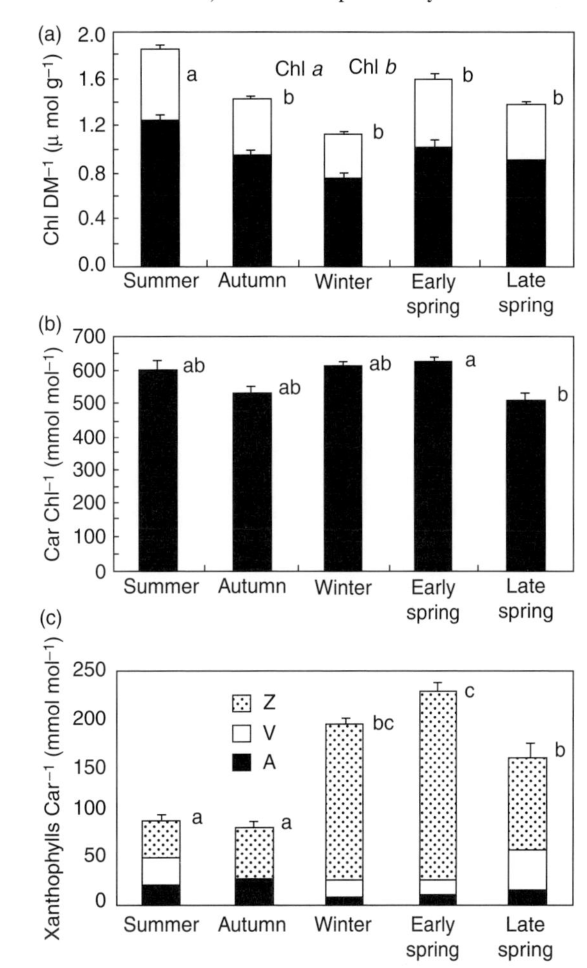

Fig. 31.5. Annual pattern of the composition of photosynthetic pigments of Siberian Scots pine. **(a)** Total chlorophyll per dry mass (chl. *a*+*b* DM^{-1}); **(b)** total carotenoids per total chl. (Car Chl^{-1}); **(c)** xanthophyll-cycle pigments per total carotenoids ((V+A+Z) Car^{-1}). Summer 11 June–06 September (n=11±SE); autumn 5 September–10 December (n=22±SE); winter 11 December–15 March (n=12±SE); early spring 16 March–30 April (n=60±SE); late spring 01 May–01 June (n=53±SE). Letters are used to indicate significantly differing groups ($P<0.05$, Tukey's post hoc test). A, antheraxanthin; V, violaxanthin; Z, zeaxanthin (from Ensminger *et al.*, 2004).

for Norway spruce by Lundmark (1998). The exposure to high light and low temperature, probably in combination with pollutants such as SO_2, can lead to visible foliage damage of trees, especially in the mountains of east Northern America, where large-scale foliage dieback of *Picea rubens* was shown to be related to a combination of light and cold stress (Schaberg *et al.*, 2002). In transgenic *Arabidopsis* plants and in leaves of the boreal Jack pine, enhanced PTOX contents did not contribute to electron consumption and to the maintenance of a low excitation pressure at PSII (Rosso *et al.*, 2006; Busch *et al.*, 2008). However, PTOX activity competes with PSI for electrons and a fine tuning has to be postulated in order to maintain assimilation. Such a fine regulation was suggested for tomato leaves (Shahbazi *et al.*, 2007) and may be absent in transgenic *Arabidopsis* plants, whereas in Jack pine, PTOX might be involved in redox regulation and carotenoid synthesis (Busch *et al.*, 2008).

Ensminger *et al.* (2008) analysed the role of fluctuating temperatures (both cold nights and intermittent frost periods) and soil temperatures on the recovery process of *Pinus sylvestris* in a laboratory setting, and found that cold soils, as well as intermittent cold periods, reduced the recovery rate of both dark and light reactions.

31.5.4. Long-term adaptations and the annual cycle

Evergreen conifers are challenged by the need to maintain green foliage through the harsh conditions of winter. To maintain their green needles thoughout the year, specific adaptations and acclimation processes are required by boreal trees in order to cope with the environmental constraints. As a result, conifers may even survive exposure to liquid nitrogen (e.g., Sakai, 1960). It is important the freezing process occurs slowly to allow for supercooling of the tissue and to prevent nucleation. However, EL (leakage of ions through the cell membrane into the intercellular spaces) shows that some damage to tissues occurs at cold temperatures in the foliage of most plants although this damage is not lethal (Strimbeck *et al.*, 2007).

Winter at subfreezing temperatures makes transport of water to the foliage and leaves difficult or even impossible as most of the water in the stem is frozen. There are sizable reservoirs of liquid water in stem and soil (e.g., Boyce and Lucero, 2002), but we doubt that this water can be transported efficiently. Also, membrane permeability seems to make water uptake from cold soils more difficult (Magnani and Borghetti, 1995). During the spring, when night-time temperatures are clearly below zero and daytime temperatures are quite high, there is the risk of freeze drying of trees. In other words, the soils and parts of the xylem are frozen and do not permit transport of water. The foliage, however, transpires during the daytime. A normal plant response would be to close the stomata, but it seems that peristomatal and cuticular conductances are sometimes large enough to induce freeze drying. These damages could be induced by physical damages to the leaf surface by wind action (e.g., Grace, 1990; Van Gardingen *et al.*, 1991), or as a result of interacting air pollution (e.g., Reinikäinen and Hutunen, 1989). Freezing, and especially recurrent freeze-thaw cycles, lead to the formation of air bubbles in the xylem of trees. This freezing-induced embolism (Tyree and Zimmermann, 2002) should be more severe for broadleaf trees and limit the distribution of ring porous trees in particular in the northern latitudes (Brodribb and Hill, 1999; Taneda and Tateno, 2005). The importance of the freeze-dry embolism is, however, disputed. Mayr *et al.* (2003) found that Norway spruce also suffers from severe embolism, whereas Taneda and Tateno (2005) claim that the effects are rather minor in Japanese conifers. Freeze drying of evergreen conifer leaves and damages to leaves by cold, light or freeze drying are frequent in conifers, and trees in the boreal forest may be drought stressed during the spring. According to our observations frost damage and freeze drying in northern boreal trees are sporadic events that may cause increased damage to trees in some years and be absent during other years.

These acclimation processes are partially active and rely on environmental signals. In animal ecology there is suggestion of hints from the environment. The dominant 'hint' for plants seems to be air temperature, which explains large portions of the recovery of trees from the winter. The reaction to temperature is not direct but involves time lags and reactions to the averages over longer time periods. Suni *et al.* (2003) empirically compared the start of the photosynthetically active period for different spruce and pine forests in Eurasia. The data showed that there was a large intra- and interannual variabiliy in the onset of the photosynthetically active period (up to 60 days for intersite variability and 30 days as intra-site variability). The onset of the photosynthetically active period was best explained by air temperature, and there was no evidence that low soil temperatures could further delay the start of that period. It has been argued that water uptake from frozen soils is difficult and that 'drought stress' in these soils limits the recovery of photosynthesis from the winter stress. To demonstrate the effect of soil temperature on the spring recovery of photosynthesis in Scots pine, Strand *et al.* (2002) manipulated soil temperatures in the field during winter time by altering snowdepth in their treatments. Warming of the soil had little effect on soil temperature, but artificial soil cooling during the spring (through an increase in snow depth) limited the recovery of photosynthesis. This recovery was associated with lower values of the substomatal CO_2 concentration, indicating a possible water stress owing to frozen soils. This experiment was done with relatively large trees in nature. By contrast, experiments in growth chambers using small seedlings indicate decreased rates of photosynthesis and water use in response to low soil temperature (e.g., Repo *et al.*, 2004; Ensminger *et al.*, 2008). The reason for the difference between the results obtained from seedlings compared with larger trees is probably owing to the the properties of the sapwood of large trees. Their sapwood contains a considerable amount of water, while spring transpiration of small seedlings and saplings growing in frozen or low-temperature soil has to be sustained by water supply provided by the roots, which are potentially impaired by low temperature. Under natural conditions early snow melt may also provide a considerable source of liquid water to sustain the recovery of photosynthesis and respiration. Monson *et al.* (2002) reported that the onset of photosynthesis was an indirect effect owing to warmer soil temperatures in a high-elevation subalpine forest in Colourado with harsh winter growth conditions. In this high-elevation forest, reactivation of photosynthesis in spring was tied to soil temperature and the availability of liquid water after the snow melt. Studies across a range of Eurasian boreal forests have shown that once snow melt has commenced, soil temperatures typically rise immediately to close to 0–1°C, remaining almost invariant at this 'zero-curtain' level until snow melt is completed. During this time, there is almost certainly sufficient liquid water available to support any air-temperature-controlled photosynthetic recovery (Suni *et al.*, 2003; Ensminger *et al.*, 2004). It is, however, unknown to what extent roots can take up water under these conditions. When the snow has melted the upper soil layers warm up rapidly, but soil frost may persist lower in the soil profile.

Empirical research shows that air temperature is the factor that explains best the recovery of photosynthesis from winter conditions in the field (Suni *et al.*, 2003). In this respect the recovery of photosynthesis is similar to other phenological processes such as budbreak. However, the recovery of photosynthesis is usually a reversible process, i.e., foliage may fall back to dormancy once temperatures are getting cold again. Although part of this process is a direct and immediate effect of air temperature (through the dependence of photosynthesis on temperature), there are also delayed effects of air temperature owing to a slow recovery of the photosynthetic apparatus. Models of photosynthetic production that do not account for these delayed effects are likely to be severely biased (e.g., Berninger, 1997). Grace *et al.* (2002) demonstrated that differences in air temperature induce important differences in the length of the growing season during which photosynthesis is active. These differences are conserved over the growing season (Fig. 31.6). Mäkelä *et al.* (2004) analysed changes in gas exchange of a Scots pine stand close to the subarctic treeline in Finland. Photosynthesis was determined by air temperature but acclimation to the temperature was lagged. In other words, it took several days after a change in temperature for photosynthesis to reach a new steady state. According to their model it would take more than two weeks before photosynthesis adapted to a step change in temperature. Ensminger *et al.* (2008) stressed in laboratory experiments that short frost events may delay photosynthesis for long periods.

For deciduous trees the development of photosynthesis logically goes hand in hand with the development of leaf area. Budbreak and development of leaf area are temperature-dependent processes. In boreal areas, budbreak is usually adequately described by a temperature-sum approach (e.g., Hänninen, 1989) and the governing processes are quite similar for conifers and broadleaf trees. It should be stressed that there are quite large genetic differences between budbreak of different provenances of trees (e.g., Beuker, 1994). Leaf senescence for deciduous trees seems to be badly understood and could be largely related to day length. Interannual variation of photosynthetic production of broadleaf trees seems to depend on the duration of leaves, and interannual variation in photosynthetic production seems to be higher for deciduous broadleaf trees than for evergreen conifers and more controlled by the length of the growing season (Black *et al.* 2005; Welp *et al.*, 2007).

During autumn many deciduous species develop spectacular autumn colours. These red or yellow colours are caused by anthocyanins and carotenoids, pigments that are known to be involved in the photoprotection of leaves against high light stress. Scientists are still discussing the exact role of these pigments during autumn senecence. Archetti (2009) reviews the different hypotheses that may explain autumn colouration. Some of these hypotheses are related to defence against insects, increase of leaf temperature by higher light absorption or drought resistance. Hypotheses linking anthocyanins to photosynthesis state that anthocyanins increase the absorption of light by non-photosynthetic pigments (although not necessarily in a very efficient way). They also protect the leaves against photoinhibition and facilitate the repair of photoinhibitive damages (Feild *et al.*,

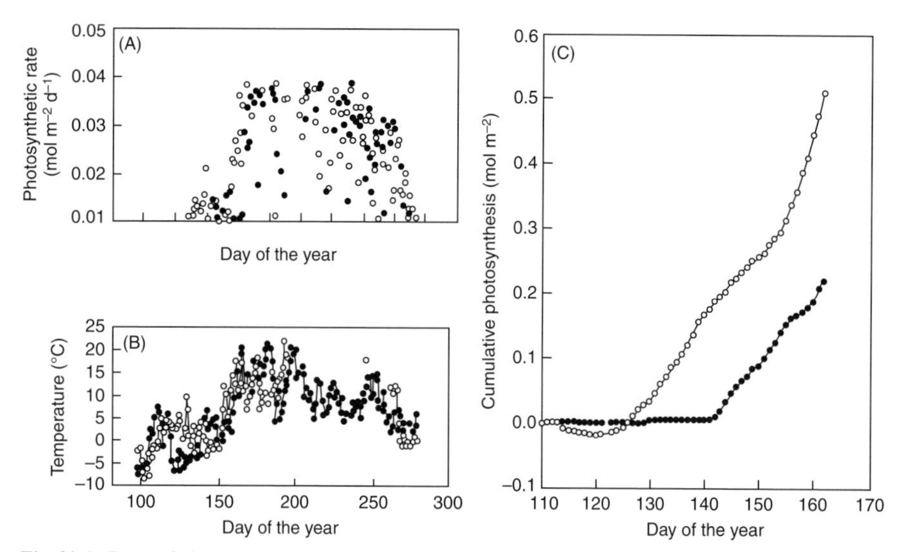

Fig. 31.6. Rates of photosynthesis near to the northern limit of trees at Värrio, northern Finland (67°46′ N, 29°35′ E, 390 m a.s.l.) in contrasting years: 1999 (closed circles) and 2001 (open circles). Graphs show daily totals of photosynthesis, measured with in-situ branch chambers (**A**); annual patterns of temperature (**B**); and cumulative photosynthesis over the first half of the growing season (**C**). From Grace *et al.* (2002).

2001), allowing a more efficient recovery of nutrients from senescing leaves possible (Hoch *et al.*, 2003b).

31.5.5. Whole-plant aspects of photosynthesis and whole-ecosystem gas exchange

Owing to the importance in terms of area, the whole-ecosystem carbon exchange of the boreal forest has received lots of attention and there have been extensive modelling approaches to understand the dependence of the production of boreal forests on the environment and light absorption by the leaves (see Chapter 14).

As shown by, for example, Welp *et al.* (2007), the composition of the stand in terms of broadleaf and coniferous trees stongly affects the timing of the photosynthetic production of the forests (Fig. 31.7). Broadleaf trees seem to be more able to benefit from a longer growing season than conifers.

It has been questioned to what extent the carbon fixation of boreal forests decreases with age. Kolari *et al.* (2004) analysed the carbon balance of Scots pine stands between 0 and 80 years of age and found that there were no signs of a decrease of carbon fixation in old stands within the economic rotation. Alternatively, the carbon balance of wet forests seems to depend largely on the water balance. Interestingly, changes in photosynthesis do not account for this difference but the changes in soil respiration do (Lindroth *et al.*, 1998).

31.5.6. Water-use efficiencies

Boreal forests are usually not water limited and water use does not depend much on soil-water contents. Granier *et al.* (2007) analysed photosynthesis and transpiration of several European forests during a drought year. The GPP of boreal forests during that particular record drought was reduced by less than 20%, whereas the GPP of temperate and Mediterranean ecosystems was reduced by more than 50%. However, boreal forests have relatively large WUE (e.g., Lloyd and Farquhar, 1994). These WUE seem to be higher for conifers than for broadleaf trees (Ponton *et al.*, 2006). The WUE of an adjacent grassland was, however, way lower than that of the trees. Low hydraulic conductivities of the xylem from the soil to the leaves is a possible reason for the high WUE of conifers (e.g., Yoder *et al.*, 1994).

31.6. PHOTOSYNTHESIS BY LICHENS AND MOSSES

A substantial portion of the production in boreal and alpine is made by poikilohydric plants like lichens and

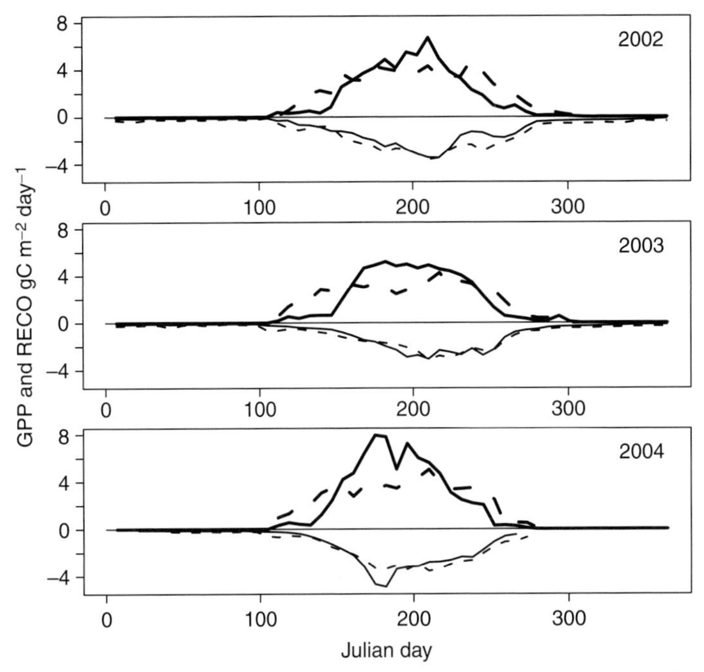

Fig. 31.7. Gross primary production (GPP) of a deciduous (aspen and willow) and an evergreen coniferous (black spruce) forest in Alaska for three contrasting years in Alaska. Solid lines are black spruce gross primary production (thick positive lines) and ecosystem respiration (thiner negative lines). Broken lines are GPP and ecosystem respiration of the deciduous plants respectively. Note that the higher peak production rates of the deciduous plants are partially offset by the earlier start of the black spruce. Spring temperatures increased from 2002 to 2004 and a severe drought occurred in 2004. From Welp *et al.* (2007).

mosses. The peculiarities of photosynthesis in these plants are described in detail by Palmquist (2000). In a study of different larch-dominated ecosystems in Siberia, Vedrova *et al.* (2006) showed that the understory vegetation in four ecosystems in the Yenissey area was responsible for more than half of the NPP. Over longer time periods, the production of the forest floor and that of the trees are not necessarily correlated. Knorre *et al.* (2006) found that mosses had distinct interannual patterns of productivity that were not correlated with the interannual production of trees. Alternatively, Susiluoto *et al.* (2008) showed that the GPP of mountain tundra in northern Finland was dominated by dwarf shrubs whereas lichens that dominated the biomass of the ecosystem had only little impact on the ecosystem carbon exchange.

Mosses and lichens have neither roots nor stomata and their abilites to regulate their water balance are therefore limited. Proctor (2000) stated that: 'bryophytes represent a radically different way in doing things'. Although higher plants are often geared towards higher production rates, bryophytes and lichens try to persist in difficult environments that allow them to survive and grow slowly. Alternatively, lichens are able to photosynthesise under low temperatures. Barták *et al.* (2006) found that the Antarctic lichens *Umbilicaria antarctica* and *Xanthoria elegans* were able to photosynthesise at temperatures as low as –5°C. However, the optimal temperature for lichen photosynthesis is well above 0°C. During the summer season the water balance is a main determinant of the productivity of mosses and lichens, whereas water balance is much less important for the trees in the ecosystem. It has been long known that the carbon balance of mosses and lichens depends strongly on their water content (Skre and Oechel, 1981). Low water content inhibits photosynthesis, but even a very high water content can impair photosynthesis, albeit less radically. Reiter *et al.* (2008) showed that during summer under natural conditions, the high alpine lichens *Xanthoria elegans*, *Brodoa atrofusca* and *Umbilicaria cylindrica* were photosynthesising only during or after rain and fog periods (which accounted for about a quarter of the study time) and were totally inactive during more than half of the summer (Fig. 31.8).

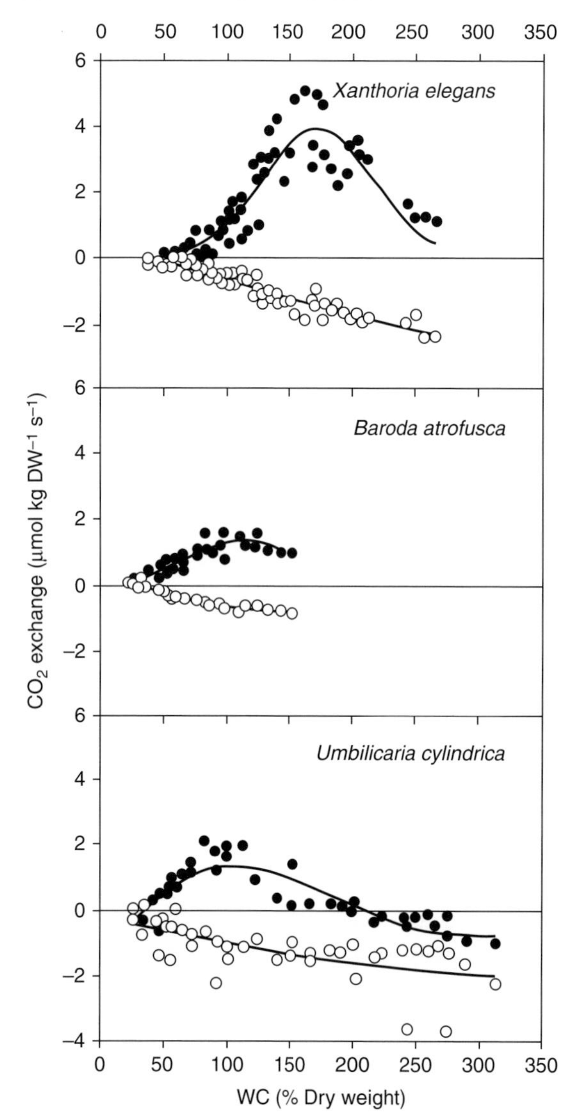

Fig. 31.8. Dependency of net CO_2 exchange on thallus water content (dry weight basis) for the lichens *Xanthoria elegans*, *Brodoa atrofusca* and *Umbilicaria cylindrica*. Measurements were carried out at 10°C and 800 mmol m^{-2} s^{-1} PPFD. For each species three different thalli were measured individually. Filled symbols represent net photosynthesis and empty symbols represent dark respiration (from Reiter *et al.*, 2008).

Although the strategies and physiology of poikilohydric plants seem radically different, the environmental challenges remain much the same. Lovelock *et al.* (1995) demonstrated that freezing and light stresses interacted on photosynthesis of the moss *Grimmia antarctici*. Freezing at higher light levels led to lower levels of fluorescence and slower recovery of F_v/F_m. Xanthophyll-cycle pigments play an important role in the protection of plants under cold and high light conditions (Lovelock *et al.* 1995; MacKenzie *et al.*, 2002; Barták *et al.*, 2006). It should also be noted that, although lichens have low photosynthetic rates, chl. concentrations are high enough to absorb 85–90% of incoming PAR (Valladares *et al.*, 1996).

31.7. COMPARISON OF FUNCTIONAL TYPES AND THE EFFECTS OF CLIMATE CHANGE

The fundamental challenges of plants in boreal and alpine regions come from the need to protect themselves from freezing and photoinhibitory stress. Plants are exposed to stresses of varying degrees, with the ranges of these stresses being larger in mountain ecosystems and perhaps less severe in boreal areas. Although the fundamental stresses are the same, different plant groups have developed very different strategies to cope with these stresses as can be seen from the vastly differing rates of photosynthesis (Fig. 31.9). All plants, including lichens and mosses in high latitudes and altitudes, have evolved mechanisms for the safe dissipation of light energy absorbed in excess of A_{max}. Interestingly, the photoprotective xanthophyll cycle constitutes an important mechanism to facilitate this thermal dissipation of excess light in all plants. An increase in leaf thickness is common among all higher plants in response to low temperature at higher altitudes and latitudes.

The most radically different strategy to cope with stresses is realised by poikilohydric lichen and mosses that depend on frequent precipitation events to remain photosynthetically active or regain A_{max} after drought. This strategy represents persistence, as these poikilohydric plants manage to escape biochemically and metabolically from unfavourable conditions through dormancy.

Herbaceous plants can escape certain periods of cold and high light under the snow. However, they are required to develop and grow in cold temperatures and some may grow even under the snow. Nevertheless, many herbaceous plants reduce their A_{max} and shut down their photosystems during winter and spring. During summer, they combine a relatively high A_{max} with a high level of photoprotection. A relatively high capacity to convert light into carbon fixation seems to be a key element of the strategy of many herbaceous plants in addition to antioxidative protection, dissipation of excess light energy via heat and alternative sinks for excess electrons and

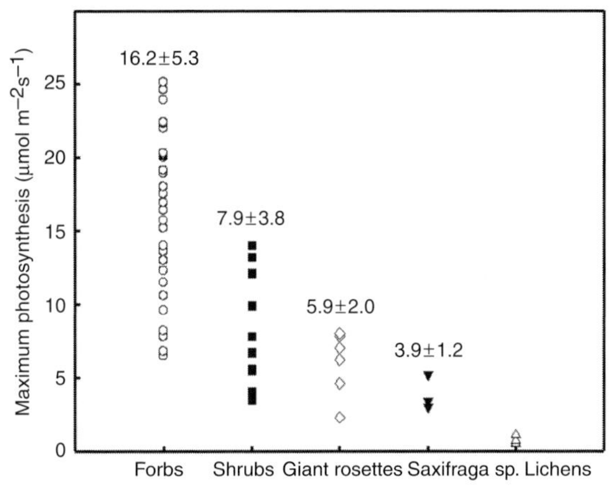

Fig. 31.9. In situ photosynthetic capacity of alpine plants measured under local pressure at altitudes between 2000 and 4300 m. The data were redrawn from Figure 11.1 in Körner (1999) and references therein. Every data point shows a measurement of an individual alpine species. Mean values and standard deviations of Forbs, Shrubs and Giant rosette plants as well as in the genus *Saxifraga* are shown.

ATP, such as PTOX and photorespiration. In this context the effects of climate change on the relatively high photorespiratory activity of alpine plants is difficult to estimate. Although increasing CO_2 concentrations will decrease photorespiratory activity, higher temperatures will increase photorespiration.

Evergreen conifers maintain a longer photosythetically active period than deciduous trees at the price of a need to protect their photosystems during the adverse conditions of winter and spring. A degradation and reconstitution of the photosynthetic machinery during these periods are key elements of the strategy.

Long (1991) showed that the ratio of photorespiration to photosynthesis increases with temperature. Therefore, plants in warm (or dry) areas would be sensitive to increases in the atmospheric CO_2 concentrations, whereas plants in wet and cold areas, as many mountain and boreal areas are, will be less sensitive. The fact that boreal and alpine ecosystems are cold and limited by the length of the growing season makes them potentially sensitive to climate change. However, increases in spring and autumn temperatures will increase the length of the growing season and increase photosynthetic production. Photosynthetic production will increase with an incease in the length of the growing season, however, in the long term climatic changes will also be accompanied by changes in the structure of the vegetation. For example, Grabherr *et al.* (1994) showed that alpine-plant distributions have been shifting upward in the European alps. Eutrophication of mountain and boreal ecosystems seems to be another factor that notably affects the productivity and composition of these nutrient-poor systems (e.g., van Wijk *et al.*, 2004 for a recent meta-analysis). Warming will reduce the importance of lichens and mosses and increase the importance of other functional types. In forests broadleaf tree species may be favoured over conifers. Both changes may increase ecosystem productivity and photosynthetic production.

Photosynthetic production and carbon balance of ecosystems are often considered to go hand in hand. However, both boreal and alpine ecosystems have large carbon pools in the soil. Carbon dioxide is emitted from the soil by respiration of the plants (the so-called autotrophic respiration) and by microorganisms that decompose the organic matter (the so-called heterotrophic respiration). In the short term all respiration is temperature dependent and will increase when temperature is increasing. Autotrophic soil respiration is closely coupled to plant productivity (Bahn *et al.*, 2008 (mountain grasslands); Högberg *et al.*, 2001 (boreal pine forests)). However, heterotrophic respiration is quite important and even large increases in photosynthesis can be easily offset by small decreases in soil organic-matter turnover and the resulting increases in soil respiration. In addition, disturbances like fire, insect outbreaks, windthrow or man-made disturbances, such as forest cuttings, have large effects on the carbon balance of boreal ecosystems.

32 • Crop photosynthesis

H. EARL, C. J. BERNACCHI AND H. MEDRANO

32.1. CROP PHOTOSYNTHESIS

Human societies are dependent upon the productivity of domesticated plant species, including cereals, oilseeds, pulse crops, fruits, vegetables and nuts consumed directly by people; also domesticated animals kept for milk, egg and meat production are reared on feed derived from plants, including grains and leguminous and non-leguminous forage species. Thus, the worldwide availability of calories, protein, dietary fats and other nutrients for human sustenance depends directly on the growth, and therefore the photosynthetic activity, of crops. Accordingly, carbon assimilation of crop species has been extensively studied as a matter of practical importance. Most recently, the expanding use of agricultural crops and crop residues as feedstocks for biofuel production has rejuvenated interest in the potential for increased fixation of atmospheric carbon and capture of solar energy by crop species and agronomic systems (Brown *et al.*, 2000; Heaton *et al.*, 2004). All together, the importance of crop productivity has led in recent years to increased efforts for improving crop photosynthesis by means of mutation discovery and/or transgenic (Parry *et al.*, 2009; Mittler and Blumwald, 2010; see Chapter 13) or epigenetic (Hauben *et al.*, 2009; Mittler and Blumwald, 2010) approaches.

The most relevant spatial scale for the study of crop photosynthesis is often the whole-canopy (plant-community) scale. Communities of crop plants are unique in that they tend to be extremely uniform, usually consisting of a single species (indeed, most often a single genotype), with all members closely synchronised in terms of their phenological development. This unusual uniformity greatly simplifies the study and modelling of crop photosynthesis and growth at the community scale, and has allowed the relevant theory to advance very quickly in comparison with other areas of plant ecology. For example, because crop canopies are spatially quite uniform, they can for some purposes be treated as homogeneous surfaces. This allows photosynthesis of crop canopies to be characterised using theoretical approaches similar to those used for single leaves, quantifying such aspects as light absorption and gas exchange in one dimension (i.e., fluxes per unit area, in this case ground area).

Experimentally, net primary productivity (NPP) of uniform crop canopies can be quantified by making periodic destructive measurements of biomass accumulation over time. The timescale for measuring crop growth in this way is in the order of weeks, and the practical spatial scale is generally several square metres. Although such biomass-based measurements of crop growth have served as indirect measures of canopy photosynthesis for over 80 years (Watson, 1952), practical methods for directly measuring photosynthetic CO_2 assimilation *in situ* have only been widely available since the early 1980s. Such measurements can now be made at the leaf, whole-plant, plant-community or landscape scales, often using timescales as short as a few seconds.

The ability to integrate traditional crop-growth analysis information with measurements of photosynthesis at various levels of organisation and on short timescales has facilitated a rapid advancement in our understanding of the physiological processes determining crop yield, and has led to some surprising conclusions about how single-leaf photosynthesis and whole-crop growth rates are related. Crop physiologists now understand that many properties determining crop growth emerge only at the plant community level of organisation, and cannot be deduced from measurements made at the leaf or, in many cases, even the whole-plant level. The integration of growth analysis data and measurements of crop photosynthesis have also helped us understand the physiological basis of the genetic improvement of crop species realised by plant breeders over the decades, and hopefully will point us towards the most likely avenues for future genetic improvements, be

Terrestrial Photosynthesis in a Changing Environment: A Molecular, Physiological and Ecological Approach, ed. J. Flexas, F. Loreto and H. Medrano. Published by Cambridge University Press.

they achieved through traditional plant-breeding methods or through more targeted genetic transformations.

In this chapter, our objective is to review how easily measured canopy scale processes, such as radiation capture and conversion, are quantitatively related to crop growth and yield, and then to demonstrate how measurements of whole-canopy crop CO_2 exchange can measure these processes on short timescales. We will then briefly review crop-photosynthesis measurements at the leaf scale, and how these are related to whole-canopy carbon fixation and crop growth. Implications for improving crop productivity are also considered.

32.2. CROP DRY MATTER ACCUMULATION AND RADIATION-USE EFFICIENCY

32.2.1. Crop dry matter accumulation as a determinant of yield

A central concept in crop physiology is the simple equation describing how absorption and use of incident PAR determines crop yield:

$$Yield = \int_{t=p}^{t=h} (PAR \cdot ABS_C \cdot RUE)\, \partial t \cdot HI \qquad \text{[Eqn. 32.1]}$$

where ABS_C is the fraction of PAR that is absorbed by the crop canopy; RUE is the radiation-use efficiency (e.g., g of crop dry matter synthesised per MJ radiation absorbed); and HI is the harvest index, the fraction of accumulated crop dry matter that is allocated to the harvested portion of the crop (such as the grain in the case of cereal crops, fruit in the case of tree fruits, vines, etc.). The integration occurs over the time period from planting (p) until harvest (h) in the case of annual crops, or can be calculated seasonally for perennials. The integrated part of the equation represents total crop dry matter accumulation during the period of interest. Figure 32.1 shows how the components of equation 32.1 might vary over the course of a growing season for a typical maize crop.

An important inference of equation 32.1 is that there are two fundamentally different ways to increase crop yield: either by increasing total seasonal dry matter accumulation (i.e., alter the integrated part of the equation), or by increasing the fraction of accumulated crop dry matter that is allocated to the harvested portion of the crop (i.e., increase HI). Genetic improvement in the yield potential of crop varieties has exploited both of these avenues, but to varying degrees for different species. For example, improvement of bread wheat (*Triticum aestivum* L.) has primarily been a consequence of increased HI (Slafer *et al.*, 1994; Araus *et al.*, 2008; Reynolds *et al.*, 2009), whereas yield increases in hybrid maize (*Zea mays* L.) over the last seven decades derive almost entirely from increased dry matter accumulation (Tollenaar *et al.*, 1994). For other species, both avenues of improvement have been found to be important, sometimes depending on the production region. Abeledo *et al.* (2003) reported that increased yield potential of barley (*Hordeum vulgare* L.) varieties in Argentina was associated with increased biomass but not HI, whereas for barley in eastern Canada both biomass and HI were found to have increased (Bulman *et al.*, 1993). In Canadian soybean (*Glycine max* L. Merr.) different studies have found either biomass or HI to be central to genetic improvement (Morrison *et al.*, 1999; Kumudini *et al.*, 2001). For Malaysian oil palm (*Elaeis guineensis* Jacq.) both biomass and HI have contributed (Corley and Lee, 1992). For annual crops, in most cases where total crop biomass has been found to increase as a result of genetic improvement, the differences in crop growth rates between old and new varieties are greatest during the last half of the season, often with no differences in biomass accumulation at all at mid season.

Equation 32.1 also provides a logical framework for considering how environmental stresses can decrease crop yields. Stresses such as drought, freezing, chilling, high temperatures, mechanical damage, insect or disease pressures can affect different parts of the yield equation. For example, drought or high-temperature stresses that occur during crucial reproductive phases in grain crops can reduce seed set and increase barrenness, thus reducing HI. Defoliation events may primarily reduce ABS_C, with little effect on RUE, whereas any stress that reduces photosynthetic CO_2 fixation rates at the leaf level will ultimately reduce whole-canopy RUE. In this context, it is important to understand that ideal growth conditions rarely, if ever, persist for an entire growing season and so resilience of the various components of equation 32.1 in the face of yield-limiting stresses is generally more important than the maximum possible values of those components under non-stress conditions. This is one reason among many why early attempts to increase crop productivity by selecting cultivars for high rates of photosynthesis (usually under ideal conditions) were unsuccessful (see Sections 32.4.2 and 32.5.1).

32.2.2. Crop growth curves

Figure 32.1 shows how PAR, ABS_C and RUE interact quantitatively according to equation 32.1 to produce a typical dry matter-accumulation curve for an annual crop. The slope

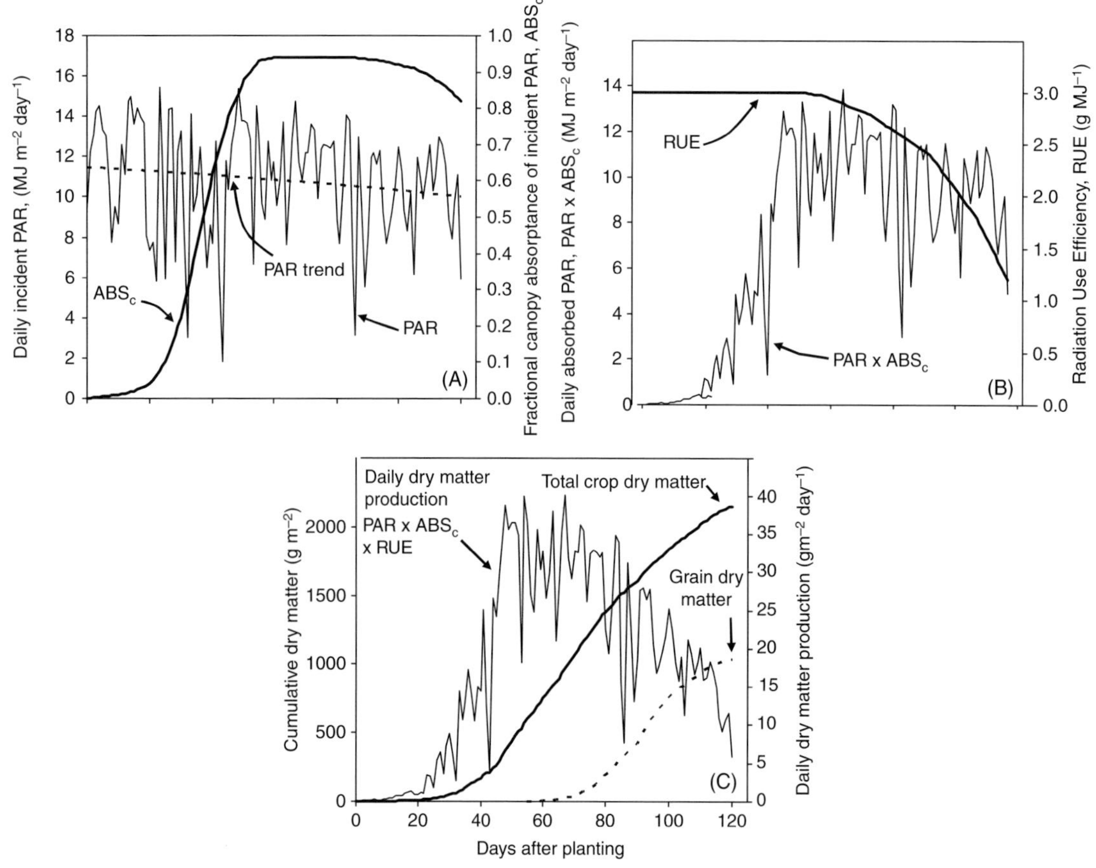

Fig. 32.1. Relationship between daily incident photosynthetically active radiation (PAR), fractional canopy absorption of PAR (ABS_c), radiation-use efficiency (RUE) and crop dry matter accumulation for a typical maize crop. In all panels, area units (m^2) refer to ground area. (A): PAR, which varies on a daily basis owing to time of year and cloud cover, and canopy absorptance (ABS_c), which changes over time as a result of changing leaf area. (B): the daily product of PAR and ABS_c, and the seasonal trend in RUE, which is initially high as new leaves are produced at the top of the canopy, then declines as leaf production ceases and existing leaves begin to senesce. The seasonal trend is graphed for clarity, but in practice RUE is slightly lower on days with higher PAR, and may also be affected by fluctuations in temperature, humidity and soil-water availability. (C): daily dry matter accumulation (or, the crop growth rate, CGR) that is the product of the two quantities in the middle panel, and the crop growth curve (crop dry matter accumulation over time). Dashed line shows the dry matter allocated to the grain. The harvest index (final grain dry matter/final total crop dry matter) is approximately 50%, typical for maize and most other cereals.

of the curve at any point is an estimate of the current crop growth rate (CGR), with units of g of dry matter per m^2 ground area per day. The curve has three distinct phases. (1) During the initial lag phase, canopy photosynthesis (and therefore CGR) is low because canopy leaf area is insufficient to maximise ABS_C. CGR increases during this phase as leaf area develops. (2) Once canopy closure occurs, a linear phase of dry matter accumulation ensues. ABS_C and RUE are at their maximum values, and so CGR is also at its maximum. (3) Eventually, CGR begins to decline, and the senescence phase begins, which continues until physiological maturity, when growth generally stops. Initially, the reduction in CGR is attributable entirely to reduced RUE. RUE declines during this phase because: (1) as leaves age their photosynthetic activity declines; and (2) as the crop grows an increasing fraction of new photosynthate is used to meet the maintenance respiratory requirements of the existing biomass, and is therefore unavailable to support new growth. During the latter parts of the senescence phase, loss of leaf area may also reduce ABS_C and therefore CGR. Finally, for annual crops in temperate regions, reductions in PAR in the latter half of the season can contribute to the declining CGR.

The essentially sigmoidal dry matter-accumulation curve is typical of most determinant crops in field experiments, and it has been described using a number of different mathematical functions (Yin *et al.*, 2003). Of course, dry matter accumulation of a specific crop may deviate from this typical pattern if a stress condition produces a permanent or temporary reduction in the CGR at some point during the season. There is also evidence that some crop species have a period of reduced CGR during the early reproductive phases. In the case of oilseed rape (*Brassica napus*) this may occur because the dense display of highly reflective flowers at the top of the canopy shades the photosynthetic organs lower down (Mendham *et al.*, 1991). Rice (*Oryza sativa* L.) has also been reported to show a temporary reduction in CGR during flowering (Sheehy *et al.*, 2004), although in that case the physiological reason is less clear.

32.2.3. Canopy absorption of photosynthetically active radiation

For annual crops established from seed, the primary limitation to photosynthesis and therefore growth during the early developmental phases is insufficient absorption of the incident solar radiation by the crop canopy (ABS_c in Fig. 32.1), owing to inadequate leaf area. A mechanistic understanding of crop photosynthesis requires some sort of quantitative description of PAR absorption at the plant-community (crop-canopy) level of organisation, as well as the distribution of absorbed PAR among individual leaf elements. Although in general leaf canopies are too structurally complex to describe completely (i.e., in a way that accounts for the vertical and horizontal position, inclination and azimuth and optical properties of every individual photosynthetic surface), the application of some simplifying assumptions about canopy structure and statistical distributions of leaf orientations can yield canopy models that present useful approximations of reality.

An important advance in the conceptualisation of crop canopies as photosynthetic surfaces with their own properties, as opposed merely to collections of individual plants, was the concept of the LAI, introduced by Watson in 1947. The leaf-area index (LAI) is simply the (single-sided) leaf surface area divided by the ground area, and it increases as vegetative development proceeds. The influence of LAI on ABS_c is illustrated by the following simple model:

$$ABS_C = 1 - e^{-k \cdot LAI} \qquad \text{[Eqn. 32.2]}$$

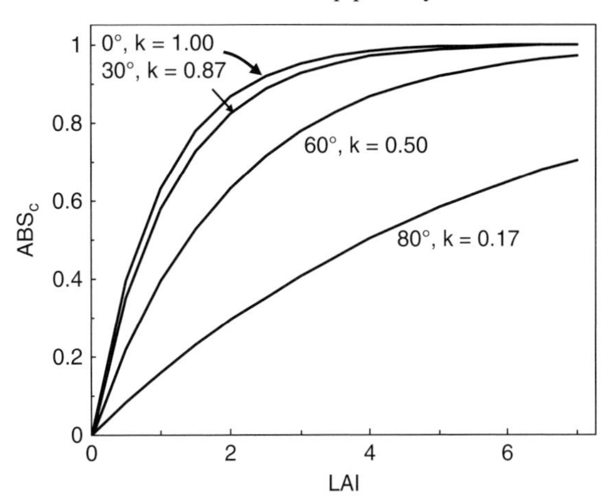

Fig. 32.2. Dependence of canopy fractional absorptance of incident solar radiation (ABS_c) on the leaf-area index (LAI) for various leaf angles. Leaf angles are given as degrees of inclination from horizontal, and the resulting extinction coefficient (k) is simply the cosine of the leaf angle. Model assumes all radiation is received as beam radiation with a beam elevation (zenith) of 90°, and leaves are distributed randomly over the ground area. ABS_c is calculated using equation 32.2.

where k is the extinction coefficient. In this model, k serves as an expression of the efficiency of a unit of leaf area in intercepting incident PAR, which is a function of the leaf orientation relative to the sun. For example, in the simplest case where the sun is directly overhead (beam radiation perpendicular to the ground) and every leaf in the canopy has the same leaf angle (inclination relative to the ground), k is mathematically equal to the cosine of the leaf angle. In other words, k is equal to the ratio of the projected area (the shadow) of the leaf to the leaf's actual surface area. Figure 32.2 shows how ABS_c varies with different values of k and LAI under these conditions. This very basic model can accommodate more realistic canopy structures and radiation environments by altering the value of k accordingly. For example, natural canopies of course tend to have a diversity of leaf orientations, rather than a single uniform leaf angle, and the sun's zenith angle (beam elevation) is rarely 90°. If the distribution of leaf angles and azimuths can be represented by a mathematical function, then the resulting value of k can be calculated for any given zenith angle. One common approach is to mathematically model the crop leaf area as if it were distributed over the surface of an ellipse (Campbell, 1986). This yields an equal probability of any azimuth (N-S-E-W orientation) and a continuous distribution of leaf angles. An elliptical shape with a horizontal

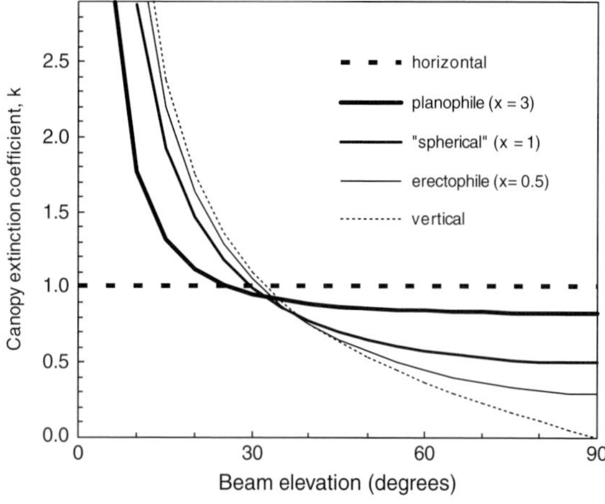

Fig. 32.3. Dependence of the canopy extinction coefficient, k, on the beam elevation (zenith) and the leaf-angle distribution. The modelled value of k is constant at 1.0 for a canopy with completely horizontal leaves (heavy dashed line), and approaches zero for a canopy with completely vertical leaves when the beam elevation is 90° (light dashed line). The solid lines represent distributions of leaf angles, equivalent to the distributions of surface angles on ellipses of different proportions. The spherical leaf-angle distribution is for an ellipse with equal horizontal and vertical axes (x=1). The planophile distribution is modelled after an ellipse with a horizontal axis three times as long as the vertical axis (x=3), and the erectophile distribution is for an ellipse that is half as wide as it is tall (x=0.5).

axis that is long relative to its vertical axis models a more *planophile* canopy structure (i.e., a tendency towards more horizontal leaves), whereas an ellipse with a longer vertical axis models a more *erectophile* canopy (i.e., a tendency towards more upright leaves). The dependence of k on the zenith angle for various ellipsoidal leaf-angle distributions is presented in Fig. 32.3.

The model represented by Fig. 32.3 and equation 32.2 still includes many important simplifications. Most obviously it assumes that the canopy leaf area is distributed uniformly in horizontal space, a condition that is reasonably approximated by crops with dense plant populations at later vegetative stages of development, but that is not accurate for row crops planted in wide rows, especially early in the season when LAI is still low. For these 'hedgerow' canopy structures and also other types of discontinuous canopies, such as orchards, other models have been devised that divide the ground area into different sections – those directly shaded by the crop canopy and those that are directly sunlit. The shaded area alone is then treated in a manner analogous to equation 32.2 (Jackson and Palmer, 1979; Boote and Pickering, 1994). Also, the simple model described above deals only with beam radiation from a single source (the sun), ignoring diffuse radiation from the rest of the sky (significant even on cloudless days, especially at low beam elevations), from scattering of PAR in the canopy owing to reflection from and transmittance through leaves, and penumbral effects. Heliotropisms and other types of leaf movements can also materially alter canopy structure on an hourly basis in some species. Further discussion of these problems and their solutions can be found in Russell *et al.* (1989) and Chapter 2 of Jones (1992).

Although modelling the canopy absorption of PAR based on such variables as the zenith angle, LAI and leaf-angle distribution can be very complex, physical measurement of canopy absorption of incident PAR is in practice quite simple. For example, incoming PAR (PAR_i) can be recorded continuously with a point sensor at the top of the canopy, and PAR transmitted through the canopy (PAR_t) may be measured at ground level with one or more appropriately positioned line sensors (line sensors are used to record the average transmitted PAR over a suitably large area, as the light environment below the canopy is heterogeneous). If sensors are also positioned to record the PAR reflected upwards from the soil (PAR_s) and the PAR reflected upward from the top of the canopy (PAR_c), ABS_c may be calculated as:

$$ABS_C = \frac{PAR_i - PAR_t - PAR_c + PAR_s}{PAR_i} \quad \text{[Eqn. 32.3]}$$

It should be noted that measured PAR_c in equation 32.3 includes not only PAR reflected from the vegetation, but also a small amount of PAR reflected from the soil surface and re-transmitted upward again through the entire canopy. PAR_s is generally difficult to measure, and so is often estimated as PAR_t multiplied by the fractional reflectance of bare soil measured previously in the absence of a plant stand (Gallo and Daughtry, 1986).

Ideally these radiation terms are measured continuously over the entire growth period of interest. However, a useful finding is that for uniform canopies ABS_c measurements made anytime within a 4-hour period centred on solar noon often provide an accurate estimate of daily mean ABS_c (Russell *et al.*, 1989; Daughtry *et al.*, 1992; Earl and Davis, 2003). For discontinuous canopies, solar angle will have a stronger effect on canopy interception, and so physical measurements at different times of day (e.g., Mariscal *et al.*, 2000) or modelling approaches (e.g., Scholberg *et al.*, 2000) may be required to accurately estimate ABS_c.

32.2.4. Radiation-use efficiency

The part of equation 32.1 most directly influenced by photosynthesis at the leaf level is RUE, so estimation of RUE in field experiments can provide some insight into how leaf or whole-canopy photosynthesis influences crop growth at various times during the season, and ultimately crop yield. In field experiments RUE is almost always based only on aboveground (shoot) dry matter, because root biomass is extremely difficult to measure. RUE (in g of dry matter per MJ PAR absorbed) for a particular period can be estimated from destructive measurements of shoot dry matter at the beginning and end of the period, continuous measurements of incident PAR and either continuous or daily measurements of ABS_c:

$$RUE = \frac{DM_y - DM_x}{\sum_{d=x}^{d=y}(PAR_{(d)} \cdot ABS_{C(d)})} \quad \text{[Eqn. 32.4]}$$

where DM_x and DM_y are above ground crop dry matter (g m^{-2}) at the beginning and end of the period, respectively; $PAR_{(d)}$ is the daily incident PAR (MJ m^{-2}); and $ABS_{c(d)}$ is the daily canopy absorptance of incident PAR. ABS_c may be determined daily or, alternatively, by interpolation between, for example, weekly measurements. Frequent ABS_c measurements are more important during the early growth stages when LAI (and therefore ABS_c) is changing rapidly. Field measurements of crop RUE are conceptually very simple, but as many authors have pointed out are also susceptible to systematic measurement errors (especially associated with PAR and ABS_c measurements) that may render comparisons between different studies somewhat tenuous (Gallo *et al.*, 1993; Sinclair and Muchow, 1999; Loomis and Amthor, 1999; Bonhomme, 2000). Comparisons between species, or indeed even between genotypes within a species, may also be complicated by variation in the 'energy content' of the new biomass produced. For example, during the seed-filling phase of oilseed crops much of the new dry matter may be in the form of protein or oil, which require more primary photosynthate (and therefore more photosynthesis) for their manufacture than do starch or cellulose (Penning de Vries *et al.*, 1974). In such cases, the numerator of equation 32.4 should be expressed in energy units or glucose equivalents rather than dry matter to account for any important differences in chemical composition of the new crop biomass.

Although generally more challenging than with annual crops, RUE measurements have also been reported for trees and other perennials. In the case of trees, the numerator of equation 32.4 is sometimes determined using allometric approaches, where plant biomass is estimated from non-destructive measurements of, for example, trunk volume rather than from destructive harvests (e.g., Palmer, 1992; Kiniry, 1998).

Many computer simulations of crop growth rely on RUE estimates for predicting biomass production. Therefore, to support the earliest crop-growth modelling efforts, field studies were undertaken to estimate RUE for most important crop species. The results generally point to higher RUE for C_4 than for C_3 species. For example, in a review of a large number of field studies, Kiniry *et al.* (1989) found typical RUE estimates of 2.2, 2.2 and 2.8 g MJ^{-1} for sunflower, rice and wheat, respectively (all C_3 species), whereas estimates for sorghum and maize (both C_4 species) were 2.8 and 3.5 g MJ^{-1}, respectively. RUE has been studied in maize perhaps more than in any other species, and values exceeding 3.8 g MJ^{-1} have been reported under optimal growing conditions (e.g., Lindquist *et al.*, 2005).

Loomis and Amthor (1999) argue that RUE of C_3 crops should be more variable than for C_4 crops, because the quantum requirement (mol photons required per mol CO_2 fixed) varies greatly with temperature in C_3 species owing to changes in photorespiration, while the quantum requirement is relatively stable in C_4 species (see Chapters 2 and 5). They also point out that RUE should be a strong function of maintenance respiration, which varies with temperature, size (biomass) of the crop and the metabolic activity of the standing biomass (see Chapter 4). Based on such theoretical considerations, they estimated maximum potential RUE for maize of around 4.6 g MJ^{-1}. How closely an actual crop approaches the maximum potential RUE under otherwise ideal conditions, may depend strongly on the prevailing light environment. Although crop-biomass accumulation is sometimes assumed to be a linear function of $ABS_c \times$ PAR, field measurements of whole-canopy CO_2 assimilation reveal a non-linear response (e.g., Rochette *et al.*, 1996; see also Section 32.3.2 below). This indicates that the highest RUE values may occur where either climatic conditions (e.g., cloud cover) or canopy structure (e.g., predominance of upright leaves) result in most leaves being exposed to relatively low PAR levels so that the quantum requirement is minimised.

The most important limitation to RUE as a measure of crop photosynthetic activity is its poor temporal resolution. RUE can only be estimated using equation 32.4 over time intervals long enough to produce measureable differences in total crop biomass; that is, the time period must be long enough that the change in biomass is large relative to the error in its estimation. A typical minimum interval might be 10 days or even longer. This precludes using dry matter-based estimates of RUE for meaningful studies of the effects on canopy photosynthesis of environmental effects that change on timescales of minutes, hours or even days (changes in PAR levels owing to time of day or cloud cover; air-temperature changes; transient water stresses). Such investigations require measurements of canopy level photosynthesis on much shorter timescales, as described in the following section. Alternative methods for frequent assessment of RUE are under development based on remote-sensing techniques (see Chapter 15).

32.3. CANOPY LEVEL CO_2 FLUX AND ITS MEASUREMENT

32.3.1. Methods for measuring canopy level photosynthesis

Canopy photosynthesis measurements for crop species are generally useful in model parameterisation and validation (Leuning *et al.*, 1998; Reddy *et al.*, 2001). At longer timescales (e.g., years), measurements can provide assessments of carbon-sequestration potential of crop ecosystems (Goulden *et al.*, 1996; Baldocchi, 2003; Bernacchi *et al.*, 2005b; Hollinger *et al.* 2005; Verma *et al.*, 2005) as well as indicators of crop health within a growing season by providing estimates of how fluxes deviate from normal with atypical environmental conditions (Hollinger *et al.*, 2005). Measuring crop-canopy photosynthesis is usually simpler than with other ecosystems as the canopies are generally homogeneous with respect to canopy architecture, nutrient availability, treatment of pests and diseases and a range of other factors. These characteristics are typical of highly managed ecosystems and, together with generally short canopy heights, allow for measurements of canopy photosynthesis that are representative of the field or ecosystem as a whole. This is in contrast to ecosystems where species diversity, as well as heterogeneity of growth stages, nutrient availabilities, species distributions, light environments or a range of other factors, would result in high degrees of variation in canopy photosynthetic rates (Baldocchi *et al.*, 1988). As crops are usually annual species, an entire growth cycle can occur in a matter of months rather than years. Crops have also been bred to accommodate a high planting density, allowing for numerous individuals in close proximity and rapid canopy closure. Given these characteristics, many options for measuring crop-canopy photosynthesis exist, and although these methods are sometimes utilised for unmanaged ecosystems or for forest ecosystems, it often requires a great deal more complexity or limiting measurements to the seedling stages.

Greenhouses and growth cabinets are seldom appropriate for measuring canopy fluxes given the associated leakages and artificial growth environments. To deal with these leakages, soil-plant-atmosphere research (SPAR) units (Fig. 32.4A) have been developed for planting vegetation in high densities typical of agriculture but in an enclosed chamber (Phene *et al.*, 1978). These chambers allow for plants to be grown under natural sunlight, however all other facets of growth can be manipulated to create a customised growth environment (Phene *et al.*, 1978; Reddy *et al.*, 2001; Kim *et al.*, 2006). These chambers are constructed to be resistant to leaks, so that in addition to being a growth chamber, the SPAR systems act as gas-exchange cuvettes to measure canopy photosynthesis. SPAR chambers allow for relatively realistic growth conditions given their natural-light environment and the deep rooting depths that are often built into the construction design.

Open-top chambers (Fig. 32.4B) are semi-enclosed chambers that allow for manipulation of the canopy growth environment under field conditions (Drake, 1992). The use of open-top chambers allows for in-situ manipulation of growth conditions by altering, filtering or adding to the gas concentrations in the air surrounding the vegetation. These chambers condition air that is pulled from below the canopy and recycle this altered air to the top of the canopy. These systems are often designed to allow for measurements of canopy gas exchange by covering the chamber with a transparent lid (Leadley and Drake, 1993). This encloses the canopy in a sealed chamber allowing for the open-top chamber to act as a large gas-exchange cuvette. Open-top chambers, however, are not constructed to maintain canopy temperatures; this requires that gas-exchange measurements be made in a timely manner to avoid canopy heating.

Micrometeorological techniques have been employed to measure canopy gas exchange in a manner that does not manipulate the environmental conditions surrounding the vegetation (Baldocchi *et al.*, 1988). Micrometeorological techniques rely on measuring the turbulent transfer of various gases between plant canopies and the atmosphere. As wind moves across a landscape, turbulent eddies move in horizontal and vertical directions. For the example of CO_2, the upward-moving eddies will contain lower than average CO_2 concentrations, and the downward-moving eddies will

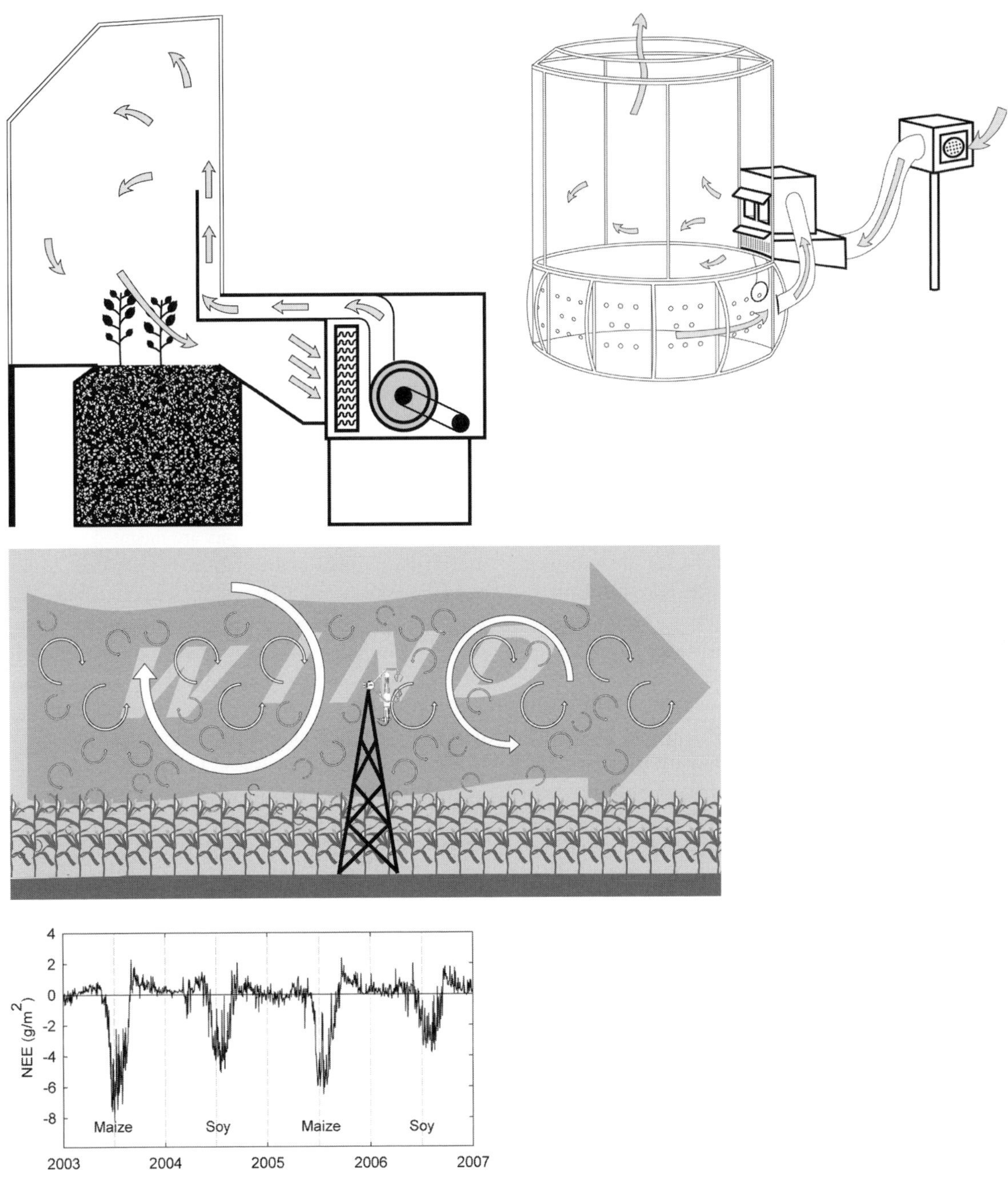

Fig. 32.4. Three methods for assessing canopy photosynthesis. **(A)** A soil-plant-atmosphere research (SPAR) unit that allows for control over many environmental factors and acts as a canopy gas-exchange cuvette for determining fluxes of CO_2, H_2O or other trace gases. **(B)** An open-top chamber that allows for manipulation of the plant-growth environment similar to SPAR chambers, but under field conditions. Enclosing the top portion of the open-top chamber allows for it to act as a gas exchange cuvette as well. **(C)** Eddy covariance measures the speed and trace gas concentrations in the air as it 'rolls' across the landscape. These measurements allow for determination of fluxes of the trace gases, including CO_2. **(D)** Representative net ecosystem exchange (NEE) data collected from an eddy covariance experiment in Bondville, IL, USA. Images redrawn from (A) (Reddy *et al.*, 2001) and (B) (Drake *et al.*, 1989).

be higher than average, and these concentrations will vary with the speed of the wind in the vertical direction. Based on these movements and concentration gradients, different micrometeorological techniques can be employed to assess the rate of exchange of different trace gases between the atmosphere and a plant canopy. While a complete description of the micrometeorological techniques is outside the scope of this chapter, the basis for the most commonly used techniques and the requirements, assumptions and errors will be discussed.

Gradient-flux analysis is based on measuring the trace gas concentrations of interest at two heights above the canopy, coupled with protocols to assess the eddy diffusivity that by analogy to molecular diffusion would correspond to the conductance of gas transport between the two measurement heights. The two measurement heights correspond to the notion that any sink for CO_2 at the canopy level will result in a CO_2 gradient that increases with height above the canopy. Sampling CO_2 concentrations at the two heights provides an estimation of this gradient, but the resistance between the two heights is required to calculate the flux. Eddy diffusivity can be determined using a variety of techniques including sensor-based measurements and modelling approaches (Ottoni *et al.*, 1992; Steduto and Hsiao, 1998). One of the general assumptions associated with eddy diffusivity is that, unlike simple diffusion, all trace gases are predicted to have the same diffusivity as the mode of transport is via the movement of eddies (i.e., diffusion on a molecular scale is negligible relative to the mixing from turbulent eddies). This is shown to be the case except under extreme conditions (Rosenberg *et al.*, 1983). When both temperature and water vapour are measured at two heights, the Bowen ratio can be calculated without knowledge of the eddy diffusivity and once calculated, the Bowen ratio can be used in an energy balance equation to estimate this term (Dugas, 1993). This can then be applied to any other trace gas measurements that are made at the same heights as the sensible and latent heat-flux measurements. In general, the difference in the trace gases between the two heights can be quite small and within the normal range of error associated with typical gas analysers. Therefore, a manifold is usually constructed that allows for the gas from two heights to be sampled in series by the same analyser. Although this does not allow for simultaneous measurements at two heights, it does minimise the influence of sensor error on fluxes.

Eddy covariance (EC) (Fig. 32.4C) is a micrometeorological method that relies on high temporal resolution of measurements of vertical wind speed coupled with trace gas measurements. The general theory behind EC is that the trace gases that are, in the case of CO_2, assimilated by the canopy, will generate a negative covariance with wind speed. This covariance component is a direct measure of the vertical turbulent flux of the trace gases being measured. This method is shown to be more reliable than other methods, however, high-frequency measurements in the order of 10 Hz or greater are required to provide a reasonable statistical result. The convention with EC data is that negative fluxes correspond to a net downward flux of CO_2, i.e., photosynthesis (Fig. 32.4D). Other micrometeorological methods exist and descriptions are provided in detail elsewhere (Baldocchi *et al.*, 1988).

Although many limitations exist when using micrometeorological techniques, they are tremendously useful as they measure large plots of land simultaneously, are continuous and do not alter the canopy microclimate. The area of land that is measured at any timepoint (the footprint) will be a function of how high the sensors are mounted above the ground, the direction of the wind and of turbulence. Under various conditions, the footprint can range from metres to kilometres. Problems associated with micrometeorological conditions include a low boundary layer and advection. If the sensors are mounted above the mixing layer of a canopy as can occur under a range of conditions, then the measurements are completely uncoupled from the vegetation.

Advection is a problem when CO_2 or any other trace gas of interest moves in a horizontal pattern. As CO_2 is heavier than air, advection typically occurs with respiratory CO_2 release from vegetation and from soil at night under calm conditions.

32.3.2. Instantaneous whole-canopy CO_2 flux

A crop canopy at closure is generally quite efficient at intercepting a large fraction of incident irradiance, as discussed in Section 32.2, so ABS_c may be nearly maximised for much of the growing season. However, the specific manner in which radiation is absorbed – that is, how the absorbed radiation is distributed across the total leaf area of the canopy – can have a dramatic impact on crop photosynthesis and therefore the RUE component of equation 32.1. Typically, single-leaf photosynthesis saturates at an irradiance that is well below full sunlight (Fig. 32.5). Thus, if the upper leaves in the canopy are oriented essentially perpendicular to the incoming radiation, much of the light energy they absorb is in excess and will be wasted as heat. Also, the extinction coefficient (k in Eqn. 32.2) of the upper-canopy layers will be high (i.e.,

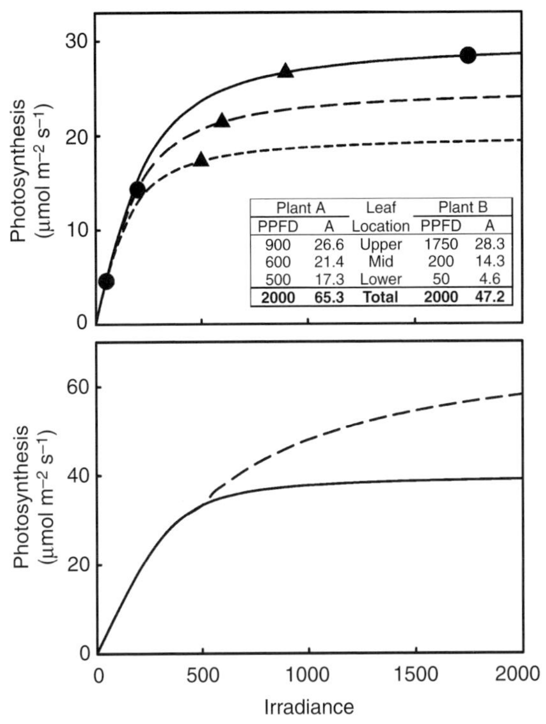

Fig. 32.5. Top panel: leaf-level photosynthetic rates for three different leaves, one each in the upper, mid and lower canopy, as a function of irradiance for two different theoretical plant canopies. Plant A (triangles) is situated in a canopy where upper leaves are angled such that they receive relatively low irradiance, but light attenuates through to the lower leaves; plant B (circles) is situated in a canopy where light is predominately intercepted by the leaves in the upper canopy, and so irradiance received by lower leaves is much reduced. Despite both canopies intercepting all available light, the total photosynthetic rate of the plant in canopy A is much higher than that in B (inset table). The figure is modelled after Long *et al.* (2006). Bottom panel: canopy photosynthetic rates as a function of PPFD for a canopy that absorbs most light in the upper leaf layers (solid line) and one that distributes the absorbed radiation over a greater leaf area (dashed line). Spreading the light over a greater leaf area reduces the irradiance per unit sunlit leaf area, and so canopy photosynthesis saturates at higher incident irradiance; it also increases the light-saturated canopy photosynthetic rate owing to the increase in the sunlit leaf area.

size of the leaf 'shadow' is maximised), so lower leaves will be shaded. If the canopy architecture is altered such that upper leaves intercept less radiation (e.g., by increasing the leaf angle), more radiation is available to leaves lower in the canopy, resulting in an increase in the total sunlit leaf area and therefore the whole-canopy photosynthetic rate (Fig. 32.5). Again, this occurs because the single-leaf photosynthetic response to irradiance is asymptotic; leaf-level radiation-use efficiency (photosynthetic rate per unit irradiance absorbed) is higher when irradiance at the leaf surface is lower. Thus, a canopy structure that distributes the same amount of light over a larger leaf area results in a higher whole-canopy radiation-use efficiency. Canopy architecture can thus have a profound influence on the maximum potential rates of crop photosynthesis. Indeed, some of the greatest breakthroughs in breeding crops to maximise productivity have involved reorganising the canopy architecture to distribute the absorbed radiation over more leaf area (Nobel *et al.*, 1993). This alters the whole-canopy photosynthetic response such that it saturates at much higher levels of irradiance (Fig. 32.5).

32.3.3. Seasonal trends in net CO_2 flux of crop canopies

Crops have been selected for high productivity, and as a result they generally exhibit high photosynthetic rates although substantial variability exists among crops. The amount of CO_2 assimilated by a crop canopy varies throughout the growing season, and these differences have large effects on the RUE of the canopy. For example, total seasonal carbon assimilation for maize is over twice that of soybean, which is driven mostly by maize being a C_4 species and partially by its longer growing seasons (Hollinger *et al.*, 1999, 2005). The higher rates of photosynthesis for maize and the more rapid canopy development results in substantially higher crop growth rates over the entire growing season, with values over three- to fourfold higher during certain time periods (Hollinger *et al.*, 2005).

For many crop-growing regions, it is difficult to assess the impact that crop developmental stages have on canopy photosynthesis and instantaneous RUE as numerous meteorological or climatic changes generally occur throughout a growing season. These factors tend to influence the rates of canopy photosynthesis much more than plant developmental stage, however, the relationship between light availability and canopy photosynthesis is shown to be quite different at three different growth stages for two difference crops, maize and soybean (Gitelson *et al.*, 2006). In this example, it was shown that the relationship of canopy photosynthesis to PAR was linear for two growth stages, the vegetative and senescent stages, but not for the reproductive growth stage (Gitelson *et al.*, 2006). Despite the reproductive growth stage not showing a linear trend, it appeared that the relationship did not differ from the vegetative growth phase, which suggests that meteorological and climatic conditions dictate canopy photosynthetic rates more than developmental stages. During senescence, however, it appears that canopy photosynthetic responses to meteorological conditions begin to deviate from patterns observed during the rest of the season. This is expected given that the onset of senescence is coupled with translocation of nutrients to the reproductive structures (Noodén and Guiamét, 1997). The key biological variable that determines canopy photosynthesis is the amount of photosynthesising leaf tissue present in the canopy. This is observed from numerous studies that have investigated seasonal CO_2 flux for crop ecosystems using micrometerological techniques (e.g., Gao *et al.*, 2003; Hollinger *et al.*, 2005; Verma *et al.*, 2005).

32.4. LEAF-LEVEL MEASUREMENTS OF CROP PHOTOSYNTHESIS

32.4.1. Single-leaf measurements in field studies

Although whole-canopy gas-exchange measurements have the advantage of providing an estimate of photosynthesis that is spatially integrated over the entire canopy and are therefore most predictive of whole-crop dry matter accumulation, such measurements cannot provide much insight into the physiological or biochemical basis of any observed difference in canopy photosynthesis. For that, single-leaf gas exchange and Chl-F measurements are used, as discussed in detail elsewhere in this volume. Field portable instrumentation is available that permits such measurements *in situ*, with automated control of the major environmental parameters within the measurement cuvette – PAR, temperature, humidity and CO_2 concentration. Thus, using field-grown plants *in situ*, it is feasible to construct leaf photosynthesis/PAR curves or photosynthesis/leaf internal CO_2 curves, study stomatal responses to environmental conditions, estimate thylakoid electron transport rates or to quantify the relative magnitudes of gas-phase-diffusive (stomatal), mesophyll-diffusive and biochemical limitations to leaf photosynthesis.

However, to make use of such measurements in the context of crop physiology (as opposed to plant physiology), the leaf-level findings must be somehow associated with what can be measured at higher levels of organisation – crop growth, crop yield or canopy level photosynthesis. This usually proves challenging as the leaf-to-leaf variation for measureable parameters related to photosynthesis is large within a typical canopy, owing to variation in such factors as the light environment and leaf age. This makes it extremely difficult to achieve the spatial and temporal resolution required to meaningfully characterise any aspect of canopy photosynthetic activity based on leaf-level measurements. There is generally a trade-off between the spatial resolution that is achieved with leaf-level measurements and the level of detail that can be pursued in characterising the photosynthetic activity of each leaf sampled. For example, 'survey'-type measurements that seek to quantify only the net CO_2 assimilation rate via gas exchange or the quantum efficiency of PSII via chl. fluorimetry under ambient light conditions may take only a few seconds each (e.g., Earl and Tollenaar, 1999). An attempt to construct a leaf photosynthesis/PAR response curve would take much longer (often ten minutes or more), as the leaf must be allowed sufficient time to reach its new steady state condition at each new PAR level. Also, attempts to apply more advanced techniques such as sensitivity analysis to quantify relative gas-phase and residual limitations (Jones, 1985) or combined fluorescence and gas-exchange measurements to estimate mesophyll-diffusive resistance (Harley *et al.*, 1992a) require even more time per leaf, owing to the requirement for very accurate measurements at true steady state conditions over a range of cuvette environmental conditions. Even instantaneous measurements, if done in all leaves within a plant to get an estimate of whole-canopy photosynthesis, can be time consuming (Escalona *et al.*, 2003; Greer and Sicard, 2009).

Each of the three different types of measurements we have considered – crop dry matter-based RUE, whole-canopy gas exchange and leaf-level measurements of photosynthesis – has its own unique advantages and disadvantages with respect to the types of information that may be gleaned. These are summarised in Table 32.1.

32.4.2. How are RUE, canopy photosynthesis and leaf photosynthesis measurements related?

Although it seems intuitively obvious that crop growth should be a function of photosynthetic activity at the leaf level, early attempts to correlate single-leaf photosynthetic rates, measured using gas-exchange techniques with whole-crop dry matter accumulation or yield were uniformly unsuccessful (Evans, 1975; Elmore, 1980). Indeed, even today there are no examples of single-leaf photosynthesis being used successfully as, for example, a screening method to select potentially high-yielding genotypes

Table 32.1. *Comparison of three categories of measurements used to characterise crop photosynthesis.*

Criterion	Methods		
	Dry matter-based RUE	Whole-canopy gas exchange	Leaf-level photosynthesis
1. Correlation with crop yield:	High	Moderate	Low
2. Temporal resolution/ integration:	Highly integrative – timescale of weeks	Can be highly integrative or provide high temporal resolution, depending on analyses – timescale of minutes to entire season:	Very short timescales (seconds). Integration depends on measurement frequency, but generally low.
3. Spatial resolution/integration:	Low spatial resolution, highly integrative (entire canopy)		Small spatial scale – part of a single leaf. Spatial resolution within canopy depends on sampling strategy – generally low to moderate.
4. Level of mechanistic information:	Very low	Low to moderate	Moderate to high, depending on methods.
5. Applicability to replicated field experiments with multiple treatments:	Most applicable	Least applicable, owing to space and/or cost limitations. SPAR more applicable than micromete-orological methods.	Applicable for low to moderate numbers of treatments/replications, depending on specific measurements.

in a crop-breeding programme. However, Zelitch (1982) pointed out that most such studies were flawed, in that they involved measurements at only one or a few leaf positions, at only one or a few times during the season and often under conditions that were different from those used to assess dry matter production of the crop. In contrast, when whole-canopy photosynthesis has been measured under field conditions at regular intervals during the season, correlations with crop growth rates and yield are generally high (e.g., Puckeridge, 1971; Wells *et al.*, 1982; Ashley and Boerma, 1989). This contrast highlights the difficulty in 'scaling up' from leaf-level measurements to predict relative performance of entire crop canopies.

Another apparent (and related) paradox emerges when one tries to reconcile late-season single-leaf measurements of photosynthesis with dry matter-based estimates of RUE in field experiments. For example, when single-leaf measurements are made on sunlit leaf tissue of maize in field experiments, leaf photosynthetic rates are universally found to decline from mid-season, when production of new leaves ends, right through the grain-filling period, as senescence of the existing leaves proceeds (Earl and Tollenaar, 1999; Ying *et al.*, 2000; Moreno-Sotomayor *et al.*, 2002; Earl and Davis, 2003). In contrast, whole-crop DM-based RUE of maize is occasionally found to be just as high during the grain-filling period as during earlier developmental stages (e.g., Tollenaar and Aguilera, 1992; Muchow and Sinclair, 1994; Lindquist *et al.*, 2005), in one case (Earl and Davis, 2003) in the same experiment where leaf-level measurements indicated a decline in photosynthesis. That is, the seasonal pattern of RUE shown in Fig. 32.1 is not in fact universally observed. Even when RUE is observed to decline during grain filling (e.g., Cirilo and Andrade, 1994; Otegui *et al.*, 1995), the decline is not proportional to the typical measured reduction in leaf-level photosynthesis. How is it then that measured leaf photosynthesis does not appear to be indicative of whole-canopy dry matter accumulation? Again, the answer probably lies in the type of leaf-level measurements that were made. Because leaf photosynthesis is a strong function of PAR, as a practical matter in field studies leaf gas exchange is usually measured at some chosen constant (usually high) PAR level or even saturating PAR. This is generally not representative of the canopy light environment where, owing to leaf orientation and light scattering, much of the total photosynthesis occurs at relatively low PAR levels. Importantly, even though the light-saturated rate of photosynthesis declines strongly as maize leaves senesce, *the photosynthetic efficiency at low PAR hardly changes at all with leaf age* (Stirling *et al.*, 1994; Earl and Tollenaar, 1999; Moreno-Sotomayor *et al.*, 2002). This then is consistent with a relatively stable RUE during grain filling, if indeed the canopy light environment is such that little photosynthesis occurs at high PAR levels. It should also be noted that in temperate regions the daily incident PAR declines during the season (e.g., PAR trend in Fig. 32.1), which would tend to increase RUE with other factors being equal.

32.5. PHOTOSYNTHESIS IN GENETIC IMPROVEMENT OF CROP SPECIES

32.5.1. Selecting for high-photosynthesis phenotypes

As pointed out above, crop yield is a final product of leaf photosynthesis and, shortly following the unprecedented improvements in crop yield during the 'green revolution', an effort was begun to select genotypes for higher photosynthesis. This was seen as a promising means to further increase crop yield (Nasyrov, 1978). Much effort was devoted to this task, but in the end without substantial impact on the improvement of crop production. On the basis of the lack of correlation between crop productivity and photosynthesis in many different crops, Gifford and Evans (1981) questioned the validity of this rationale to improve yield capacity as well as the relationship between photosynthesis and crop yield itself, arguing that genetic improvement of yield potential had been achieved mainly by alterations in the allocation of biomass to harvestable products. As mentioned previously, this lack of correlation was challenged by Zelitch (1982), who argued that measurements of photosynthesis rate collected under field conditions on multiple days and on many leaves during the growth period provided better relationships. These results also illustrate that it is a very large step in scale and complexity from single-leaf CO_2 assimilation to crop yield. Crop production results from the generation of photosynthate as well as its utilisation and final accumulation as dry matter. Therefore, the intervening steps that can modify the benefit of an increased rate of leaf photosynthesis must also be taken into account.

Clearly, the relationship between single-leaf photosynthesis and crop yield is made extremely complex by such factors as described above – effects of canopy architecture and leaf display, variation in photosynthetic response to irradiance both spatially and temporally, etc. However, even the relationship between leaf photosynthesis characteristics and single-plant dry matter production is complicated, and selection for 'improved photosynthesis' can produce surprising results. As an example, haploid plants of tobacco (*Nicotiana tabacum* L.) derived from in-vitro mutagen treated anthers were screened for survival at atmospheric CO_2 concentrations close to the compensation point. The rationale was to increase the genetic variability by the use of mutagenised haploid populations and to then apply selection pressure that was quite specific for an important photosynthesis characteristic, the CO_2 compensation point. The surviving genotypes were diploidised and double-haploid self-pollinated to obtain genotypes that showed significant improvements in dry mass (14 to 36%) and several other growth parameters (Medrano and Primo Millo, 1985). Later measurements under field and greenhouse conditions showed this increase more related to a greater leaf area, a lower respiration rate, a higher rate of photosynthesis along the lifespan and a greater number of mesophyll cells of smaller size in the selected plants. However, it could not be related to reduced photorespiration rates and/or CO_2 compensation point or to Rubisco properties, which the selection method was designed to achieve (Delgado *et al.*, 1993; Medrano *et al.*, 1995).

The recent increase in availability of inexpensive, portable IR CO_2 analysers has provided an opportunity for wider use of leaf photosynthesis rate as a selection criterion for crop improvement (Long *et al.*, 1996). However, the direct selection pressure for higher leaf photosynthetic rates has not directly led to progress in crop yield, leading to a consensus that the complex interactions and integration of canopy architecture, RUE, biochemistry and molecular biology of the photosynthetic process, including assimilate partitioning and dark respiration, render unlikely the achievement of significant progress solely by selection for increased photosynthetic rates (Horton, 2000; Long *et al.*, 2006a; Amthor, 2007).

32.5.2. Feedback (sink) limitations to crop photosynthesis

As discussed in the earlier sections of this chapter, genetic improvement in the yield potential of many crop species has occurred through the increase of HI rather than total biomass accumulation. In these cases, crop yield could be considered sink limited rather than source limited (Egli, 1998). However, source and sink strength can also be interdependent; photosynthetic rates can be limited by the sink capacity to use recently produced photosynthates (Evans, 1993), as demonstrated by removing filling grains (Peet and Kramer, 1980) or fruits (DeJong, 1986; Palmer *et al.*, 1997). A recent analysis of seed dry weight changes in response to different assimilate availability during seed filling concluded that yield was more limited by sink than by source (Borrás *et al.*, 2004). The source/sink relationships and feedback limitations are a matter of current controversy (Amthor, 2007) because, as has been recently reported, even for high-yielding crops there are some unfilled seeds (Sharma-Natu and Ghildiyal, 2005) indicating some source limitation to yield. At the same time, other observations seem to support the predominance of sink limitations (Richards, 2000).

Recent experiments on wheat, soybean and other crops under high CO_2 conditions have provided additional insights into the issue of source/sink limitations of crop yield. A meta-analysis of different experiments in enriched CO_2 atmospheres has revealed parallel increases in photosynthesis rates and yield (Ainsworth *et al.*, 2002; Ainsworth and Long, 2005). These results showed that sustained increases in plant photosynthesis result in concomitant increases in crop dry mass production and also that sink size increases in correspondence with source improvements. Combining genetic manipulation of sink capacity and enriched CO_2 experiments, Ainsworth *et al.* (2004) showed that expected yield increases under high CO_2 conditions were not observed when the sink size was genetically reduced, leading to the conclusion that increases in source capability lead to yield increases unless sink capacity prevents the advantage from being realised. Source and sink capacity should therefore be coordinated to avoid feedback limitations to crop yield as photosynthetic capacity is increased.

32.5.3. Genetic engineering for increased photosynthesis

As recognised by different authors, further increases in yield potential of major crops will depend largely on increasing crop photosynthesis (Long *et al.*, 2006a; Amthor, 2007; Zhu *et al.*, 2010). Recent experiments performed at high CO_2 showed that the positive relationship between photosynthesis and yield is evident when the photosynthesis increase occurs at the whole-canopy level. In this context, biotechnological approaches for specific transformation of some crops are now of interest (Horton, 2000; Sinclair *et al.*, 2004; Ainsworth *et al.*, 2008a; Peterhansel *et al.*, 2008; Flexas *et al.*, 2010). Model plants transformed for single genes are providing increasing evidence that leaf photosynthetic rate can show significant increases with only minor modifications of specific characters and that some of them result in measurable increases in dry matter accumulation (see recent reviews of Raines, 2006; Peterhansel *et al.*; 2008 and Zhu *et al.*, 2010).

Experience with both crop breeding and plant transformation indicates that photosynthesis increases can be achieved in many different ways including changes to light harvesting, mesophyll CO_2 diffusion, carboxylation and use of carbon. Chida *et al.* (2007) reported that overexpressed cyt c_6 in *Arabidopsis* resulted in higher contents of photosynthetic metabolites and improved Chl-F characteristics, while also increasing shoot and root growth.

The conductance to CO_2 diffusion from the substomatal cavities to the carboxylation site (commonly named mesophyll conductance, g_m) is now recognised as a main limitation to photosynthesis (Flexas *et al.*, 2008), and aquaporins (specific plasma-membrane proteins involved in water but also in CO_2 transport) seem to play an important role in CO_2 diffusion and g_m regulation (Terahima and Ono, 2002; Flexas *et al.*, 2006a). Recent experiments with tobacco plants overexpressing specific aquaporins showed substantial increases in g_m both in rice (Handa *et al.*, 2005) and in tobacco plants (Flexas *et al.*, 2006a).

The CO_2/O_2 specificity factor of Rubisco is an important bottleneck for carboxylation efficiency, but shows little variation in higher plants. However, a survey of Mediterranean species adapted to high temperature and drought showed that Rubisco specificity factor was greater for plants from hot, arid environments, and some of the genus *Limonium* were found to have significant higher specificity factor than crop plants (Galmés *et al.*, 2005). Transformation of tobacco or wheat crops with the specific Rubisco genes from *Limonium gibertii* could mean significant increases in photosynthesis rates mainly under water-stress conditions.

Recently Kebeish *et al.* (2007) showed a novel respiratory pathway in *E. coli* that converts glycolate directly to glycerate (Eisenhut, *et al.*, 2006). Expression of this novel pathway in *Arabidopsis* chloroplasts results in a reduction of photorespiratory flow, thus increasing photosynthesis by the release of CO_2 in the vicinity of Rubisco. The transformed plants showed larger leaf area, increased rosette diameter and greater dry mass production (Khan, 2007).

The above examples are far from conclusive but represent the potential of biotechnology to amplify our knowledge on the regulation of photosynthesis and possibly to increase crop photosynthesis and yield. This old challenge is gaining new significance in the context of new crop types for biofuel production and with respect to the potential competition between food and fuel production on the planet's finite area of productive land.

32.6. SUMMARY

It is self evident that photosynthetic carbon fixation underlies crop growth and yield, but decades of research have demonstrated that the mechanistic relationships between chloroplast and leaf-level processes on the one hand and crop productivity on the other are neither simple nor intuitive. Traits that enhance instantaneous photosynthesis of single leaves, or even augment dry matter accumulation of whole plants, may provide little or no benefit when evaluated on the spatial scales (crop canopies) and timescales (entire growth seasons) that are relevant for crop-yield determination. As described in this chapter, many important properties that determine radiation capture, radiation-use efficiency, dry matter partitioning and yield emerge only at the plant-community level of organisation.

The novel methods of biotechnology offer new possibilities for enhancing photosynthesis of crop plants by directly targeting specific biochemical pathways. However, every such achievement, no matter how promising in the laboratory in the context of single plants, remains only a hypothetical advancement until its effects are evaluated in the field. As a corollary to this, continued 'vertical integration' of understanding of the yield-limiting factors that manifest themselves at the crop, whole-plant, leaf and organelle levels of organisation is the key to identifying specific traits and genes to which the methods of genetic transformation can most profitably be applied.

Part VI
Photosynthesis in a global context

33 • Photosynthetic water-use efficiency

H. MEDRANO, J. GULÍAS, M.M. CHAVES, J. GALMÉS AND J. FLEXAS

33.1. INTRODUCTION

Gas exchange is tightly coupled to evaporation in all living organisms (Woods and Smith, 2010). Photosynthesis of terrestrial plants is associated with water loss because the CO_2 needed to be fixed into carbohydrates enters the leaf through stomata with the consequent loss of water from the sub-stomatal cavity to the atmosphere. Leaf-to-air *water-vapour* gradient is about 100 times larger than the CO_2 gradient. Consequently, plants have to tightly regulate stomatal opening in order to avoid leaf dehydration. This causes a wide variation of the ratio between the rate of CO_2 uptake (photosynthesis) and the rate of water-vapour loss (transpiration). This ratio expresses the efficiency of the carbon gain with respect to water loss, i.e., water-use efficiency (WUE).

Plant growth and biomass production are thus largely conditioned by the water resources, which are extremely variable in time and space around the globe, therefore water availability along the growing season is a determinant factor for plant-biome distribution and GPP. In general, it is widely established that ecosystem or crop production is closely dependent on soil-water availability (Beer *et al.*, 2007).

Figure 33.1 shows a list of the main biological determinants, related with plant photosynthesis and transpiration characteristics, as well as the main environmental conditions that determine the specific values of WUE and its wide range of variation.

Variations of WUE are relevant both for plant survival capacity in drought-prone environments and plant (crop) productivity in water-limited lands. In evolutionary terms, it appears that some selection pressure towards improving WUE has occurred (see Chapter 24). For instance, an increased stomatal sensitivity to varying CO_2 from almost no sensitivity in ferns and lycopod through sensitivity to low but not to high CO_2 in conifers, to the greatest sensitivity at both extremes of CO_2 concentration in angiosperms has its major consequences in terms of an increased WUE in the latter ones (Brodribb *et al.*, 2009). The short-term human perspective sees an increasing need of food production caused by the exponential growth of its population. This is unavoidably linked to an increased water consumption for agriculture that, under the increasing water shortage in most agricultural areas, demands a further improvement of WUE as a necessary requisite to secure food supply (Rockström *et al.*, 2007), as well as for other plant uses including urban ecosystem services (McCarthy *et al.*, 2011) despite some claims against this argument (Blum, 2009).

Fresh-water resources are finite, and agriculture is by far the largest current user of available fresh water accounting for 70% of human water consumption (WRI, 2005). A key point to consider is that irrigated lands comprise only 18% of world cultivated areas and produce around 45% of global food (Döll and Siebert, 2002). Therefore, a reduction of the irrigated area would lead to serious problems for global food supply. In addition, there is an increasing demand for water usage by other human activities, intensifying the need for improvement of the efficiency of water use for food production (Araus, 2004). Moreover, the emerging use of agricultural products for biofuel production foresees additional necessities of arable areas and water and nutrient resources, and will therefore increase water use for agricultural activities in the future. Climate change predicts higher temperature and lower humidity in many regions, which will further increase water demand in these cropping areas (IPCC, 2007). The conflict between food (plus biofuel) production and other demands for water will become dramatic especially in semi-arid areas. Consequently, improving WUE in plants is an urgent objective of research, claimed from local politicians to UN–FAO organisations with the slogan

Terrestrial Photosynthesis in a Changing Environment: A Molecular, Physiological and Ecological Approach, ed. J. Flexas, F. Loreto and H. Medrano. Published by Cambridge University Press.

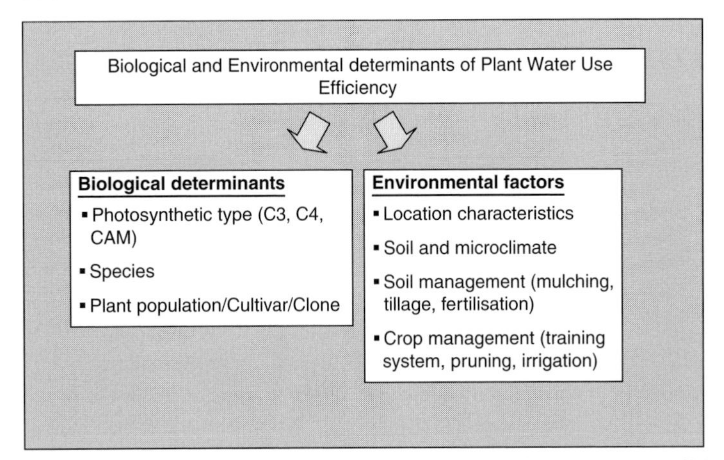

Fig. 33.1. Main biological and environmental determinants affecting water-use efficiency.

of 'more crop per drop' or calling for a blue revolution to follow the green revolution of the sixties (UNIS, 2000).

Optimisation of water used for food (and biofuel) production or 'water footprint' (Gerbens-Leenes *et al.*, 2009; Jongschaap *et al.*, 2009) is a key point requiring multidisciplinary approaches, including institutional, agronomic, physiologic and genetic contributions. Large improvements in water use are coming from educational programs, irrigation-system transformation or land- and soil-management activities focused to enhance water storage and/or to reduce direct soil-water loss (Fig. 33.1). In this sense, the increasing necessity for a more sustainable use of available water for food production is reflected in recent publications showing the wide development of new insights in the improvement of WUE in different crops (Davies and Sharp, 2000; Davies *et al.*, 2007; Hsiao *et al.*, 2007; Morison *et al.*, 2008; Flexas *et al.*, 2010).

Pioneer studies measuring 'water requirement' or 'transpiration coefficient' of crops were reported between 1890 and 1910 at different agricultural research stations, many of which used small lysimeters under both greenhouse or field conditions. Compilation of previous results by Briggs and Shantz (1913) showed important effects of climatic conditions, soil fertility and other agronomic conditions on water requirements. Later, Maximov (1929) showed important variations in WUE among different crops; these variations can nowadays be explained by the different photosynthesis and stomatal conductance (g_s) values of C_4, CAM and C_3 species (reviewed in Jones, 2004a). As pointed out by Tanner and Sinclair (1983) and Jones (2004b), pioneer data have important limitations as they usually only accounted the production of aboveground dry matter or because they were only taken under particular environmental conditions. In those surveys, environmental variation in WUE was primarily related to changes in atmospheric humidity, later named 'evaporative demand of the atmosphere' and quantified as potential evapotranspiration (ETp) by Penman (1948).

The expression 'water-use efficiency' is still matter of controversy (Tanner and Sinclair, 1983; Jones, 2004b; Blum, 2005; Morison *et al.*, 2008). The concept of efficiency of the water use is widely used in the literature, and always reflects the balance between harvested biomass or carbon gain and water lost by plants. However, agronomists and irrigation engineers have a wider consideration as this concept is also applied to the whole water-storage system, conveyance, irrigation facilities and soil management, which can count for a loss of around 60% of total water used (Morison *et al.*, 2008).

This chapter focuses on WUE by plants at different time and space scales (from leaf to plant, from instantaneous measurements to lifecycle), concentrating on the physiological and genetic basis of plant WUE and their relationship with photosynthesis. Finally, potential ways to improve WUE are discussed.

33.2. METHODOLOGY FOR ASSESSING WUE

Assessing WUE requires several considerations concerning the particularity that WUE is a ratio between two different variables that can be formulated at different times and space scales depending on its focus and application (Fig. 33.2). In agronomic, ecological and ecophysiological studies, it is generally assumed that full crop season or lifecycle measurements are desirable. Therefore, assessing WUE

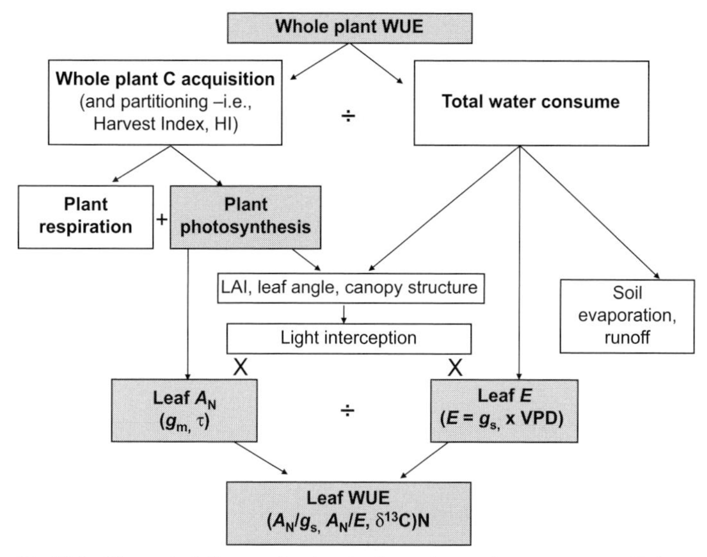

Fig. 33.2. Theoretical diagram showing the dependency of crop water-use efficiency (WUE_C) on different processes and its links with leaf-level WUE (some important parameters related to different processes are indicated in parenthesis). Grey boxes indicated the processes more extensively reviewed in the present chapter. Modified after Flexas *et al.* (2009a).

implies estimating the total biomass produced as well as the real plant-water consumption. Measurements of plant-water consumption under field conditions should take into account the amount of water lost by runoff, percolation and direct soil evaporation (Fig. 33.2) that makes the determination of the actual amount of water consumed by the plants difficult. Soil-water availability is largely dependent on soil characteristics, and the larger and deeper the root system the more difficult it is to measure the total water used by the plant accurately. In addition, determination of biomass production also presents several difficulties, such as root biomass accumulation and the leaf and fruit biomass losses owing to pests and diseases along their living cycle.

The use of lysimeters is perhaps the best approach for reliable measurements of plant (crop)-water consumption, although its use in woody species is limited because of its extensive root system that requires expensive equipment, limiting its utilisation. Conversely, sap flow measurements provide reasonable estimations of daily and seasonal plant-water use for most woody crops and forest applications (Smith and Allen, 1996).

For large temporal and spatial scales, measurements of carbon and water diffusive fluxes between the canopy and its surrounding atmosphere provides reasonable estimations of daily and seasonal WUE, which can be applied to crops and natural communities. Although, the accuracy of these methods depends on environmental conditions, the evidences accumulated support EC techniques as being suitable to derive global carbon and water fluxes at the local scale (Valentini *et al.*, 2000; Beer *et al.*, 2007, 2009). Thus, this is a useful approach for estimations of CO_2 fluxes and GPP, as well as ETp of crops and natural communities, enabling provision of gross estimates of the community WUE (Pruitt *et al.*, 1987; Held *et al.*, 1990; Baldocchi, 1993; Xu *et al.*, 1999; Reichstein *et al.*, 2002). Scaling up of those values to regional and continental levels combined with remote-sensing data may offer interesting data at the watershed, regional or continental scales (Papale and Valentini, 2003; Beer *et al.*, 2007, 2009; Lu and Zhuang, 2010).

At the leaf level, instantaneous values of net carbon assimilation (A_N) and transpiration (E) are easily measured with portable gas-exchange equipments. The ratio between carbon influx and water efflux from leaves, A_N/E, is termed instantaneous leaf WUE or leaf-transpiration efficiency. Because each flux is the product of g_s by the gradient, this value is given by the relationship between CO_2 and *water-vapour* conductances multiplied by the quotient of the respective gradients (see Chapter 9 for details):

$$\frac{A_N}{E} = \frac{g_{s,CO_2}}{g_{s,H_2O}} \times \frac{C_a - C_i}{Cwv_i - Cwv_e} = \frac{1}{1.6} \times \frac{C_a - C_i}{Cwv_i - Cwv_e}$$

[Eqn. 33.1]

Table 33.1. *Variation range of water-use efficiency for different plant groups measured as intrinsic leaf water-use efficiency, (A_N/g_s), instantaneous leaf transpiration efficiency (A_N/E) and leaf discrimination against ^{13}C ($^{13}C\Delta$). Minimum (min), maximum, (max) and average (m) values are also shown. Values taken from experimental data of our group and literature survey.*

		A/g (μmol CO_2 mol H_2O^{-1})			*A*/*E* (μmol CO_2 mmol H_2O^{-1})			∂^{13}C (‰)		
Plants	n	min	max	m±d.s.	min	max	m±d.s.	min	max	m±d.s.
General	1382	2.7	172.7	61.8±28.6	0.1	7.1	2.8±1.5	–29.8	–21.8	–25.2±2.1
Crops	359	2.7	134.1	53.6±28.7	0.1	7.1	2.5±1.5	–29.8	–21.8	–25.2± 2.1
Natural vegetation	1020	6.4	172.7	63.1±28.0	0.3	5.9	3.5± 1.2	—	—	—
Anuals	145	2.7	98.9	33.6±31.3	0.4	7.1	2.7± 1.9	–28.9	–21.8	–24.4±1.7
Deciduous	192	4.2	134.1	54.2±23.1	0.1	6.1	2.4± 1.3	–29.8	–26.8	–28.6± 0.9
Hemideciduous	151	19.7	134.9	70.6±22.8	1.0	5.6	3.5± 1.3	–	—	—
Perennials	617	6.4	172.7	70.2±27.3	0.3	5.9	3.5± 1.1	–	—	—
Herbs	295	2.7	158.6	44.4±29.4	0.3	7.1	2.9± 1.7	–28.9	–21.8	–24.4±1.7
Shrubs	635	6.4	172.7	67.6±24.6	1.0	6.1	3.1± 1.2	—	—	—
Trees	346	4.1	156.9	60.5±30.9	0.1	5.8	1.8± 1.6	–29.8	–26.8	–28.6± 0.9

with A_N being net CO_2 assimilation rate; E, transpiration rate; g_{s,CO_2} and g_{s,H_2O}, stomatal conductances to CO_2 and H_2O, respectively; C_a and C_i, CO_2 concentrations in the surrounding leaf atmosphere and in the air spaces of the substomatal cavity, respectively; and Cwv_i and Cwv_e, the concentrations of *water vapour* in the air spaces of the substomatal cavity and in the surrounding leaf atmosphere, respectively.

The difference in partial pressure is about 100-fold larger for *water vapour* than for CO_2 owing to the low partial pressure of CO_2 in Earth's current atmosphere (around 38 Pa) compared with the *water vapour* (500–4000 Pa, depending on the air humidity and temperature). Consequently, relatively small differences in air humidity (mainly at high temperatures) will lead to important differences in instantaneous WUE (A_N/E) at similar A_N. Alternatively, air humidity and temperature (or VPD) largely affect E, introducing an important interference when comparing instantaneous WUE values from different locations or climatic conditions. This limitation can be overcome by using the ratio A_N/g_s, called 'intrinsic WUE' (Osmond *et al.*, 1980b), because it only depends on the leaf capacity to fix carbon at a certain stomatal aperture. At a given C_a, the ratio A_N/g_s is linearly related with C_i, reflecting the balance between the assimilation capacity and the diffusion flux. Therefore, at similar g_s (and E) the lower C_i is, the higher the carbon flux and the higher the WUE.

For measurements with portable IRGAs, both A_N/E and A_N/g_s are widely used to characterise environmental and genetic effects on WUE as it is straightforward to perform (Condon *et al.*, 2002; Chaves *et al.*, 2004; Flexas *et al.*, 2004b; Galmés *et al.*, 2007c; Morison *et al.*, 2008). Table 33.1 shows a summary of WUE estimated with different methods in several plant groups presenting wide ranges of variation and clear differences among crops, natural vegetation and different functional groups. For Mediterranean vegetation and for grapevines, similar data have already been published (Pou *et al.*, 2008; Medrano *et al.*, 2009).

Discrimination of photosynthesis against the heavy ^{13}C isotope enables assessing WUE by the quantification of the ratio of ^{12}C and ^{13}C isotopes abundance in the atmosphere and the leaf or plant biomass (Farquhar *et al.*, 1982; see Chapter 11). During diffusion from air to inside the leaves, and during carboxylation by Rubisco and/or PEPC, the heavier ^{13}C is discriminated against so that the CO_2 fixed has a lower $^{13}C/^{12}C$ ratio than atmospheric CO_2. This $^{13}C/^{12}C$ ratio (or $\delta^{13}C$) depends on the balance between CO_2 diffusion into the leaf and CO_2 fixation by Rubisco, thus being conceptually similar to the ratio A_N/g_s whenever mesophyll conductance (g_m) is considered to be invariable (and infinite). Although it has been recently stressed that this later assumption is often not the case and that this causes large biases in the relationship between $^{13}C/^{12}C$ and WUE (Flexas *et al.*, 2008; Seibt *et al.*, 2008; Roussel *et al.*, 2009; Soolanayakanahally *et al.*, 2009), the isotope ratio is still currently and frequently used to assess long-term variations in WUE (Rowell *et al.*, 2009) or differences in WUE among sites (Gouveia and Freitas, 2009; Soolanayakanahally *et al.*, 2009), among different plant groups (Golluscio and Oesterheld, 2007; Cernusak *et al.*, 2008) and, especially, among genotypes (Ares *et al.*, 2000; Gibberd *et al.*, 2001; Pieters and Núñez, 2008; Rajabi *et al.*, 2009; Roussel *et al.*, 2009) and phenotypes within a given species (Donovan *et al.*, 2007, 2009). The amount of this discrimination (Δ) refers to the change on the $^{13}C/^{12}C$ ratio during a process and it is defined by:

$$\Delta = \frac{(^{13}C/^{12}C)\ \text{reactants}}{(^{13}C/^{12}C)\ \text{products}} - 1 \qquad \text{[Eqn. 33.2]}$$

The values of Δ usually expressed as parts per thousand (‰), are usually between 13 and 28‰ for C_3 plants, and between –1 and 7‰ for C_4 plants. Discrimination is dependent on the ratio between C_i and C_a (Farquhar *et al.*, 1982), approximately according to:

$$\Delta = 0.0044 + 0.0256\ (C_i/C_a) \qquad \text{[Eqn. 33.3]}$$

The major advantadge of using stable-isotope discrimination is that it provides an integrative measurement of WUE at leaf level combined with an easy and rapid determination in the laboratory, allowing the analyses of large collections of samples (i.e., genotypes under a gradient of environmental conditions). Furthermore, the use of this methodology opens the possibility to study historical changes in WUE from analysis of $\delta^{13}C$ in herbarium collections (Peñuelas and Azcón-Bieto, 1992; Woodward, 1993) and tree-ring samples (Hietz *et al.*, 2005; Cullen *et al.*, 2008; Rowell *et al.*, 2009). Effectively, as demonstrated for different species under a wide range of experimental conditions, $\Delta^{13}C$ values tightly correlate with A_N/g_s (Nainanayake, 2004; see also Fig. 33.3). However, the relationship between $\Delta^{13}C$ and estimations of intrinsic and instantaneous leaf WUE measured by gas exchange is not fully clear (Condon *et al.*, 2004; Moreno *et al.*, 2006). This could be owing to the different time integral of the two methods as gas-exchange methods provide instantaneous measurements, while Δ represents an integrative parameter affected by daily and seasonal variations of CO_2 assimilation rates. Moreover, several reports have shown that

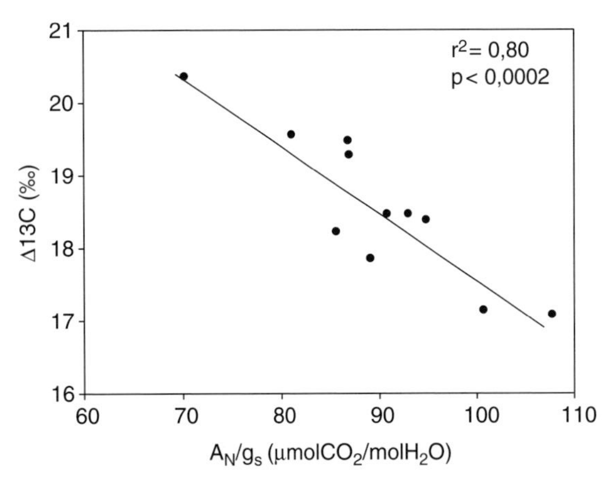

Fig. 33.3. Relationship between leaf ^{13}C discrimination and intrinsic water-use efficiency (A_N/g_s) in several Mediterranean species, including *Pistacia lentiscus*, *Olea europaea* var. *sylvestris* and *Cistus monspeliensis* (from Gulias, J., Flexas, J., Jonasson, S., Medrano, H. unpublished results).

g_m is not constant across environmental conditions and can vary to a great extent among different species and cultivars (Warren and Adams, 2006; Flexas *et al.*, 2008; see Chapter 3). As a consequence, estimation of A_N/g_s from $\delta^{13}C$ must be carefully performed because Δ is affected by the CO_2 concentration at the carboxylation sites (C_c), which is no longer equivalent to C_i (Seibt *et al.*, 2008; Centritto *et al.*, 2009).

$\delta^{13}C$ measurements alone do not provide an answer to whether variations in WUE are caused by changes in the photosynthetic activity or in the g_s and g_m. Parallel measurements of oxygen-isotope discrimination (^{18}O against ^{16}O) can provide supplementary information of *E*. Essentially, the ratio $^{18}O/^{16}O$ of organic matter is related to *E* and leaf-to-air VPD (see Chapter 11 for details), as leaf water tends to become enriched in ^{18}O because ^{16}O evaporates more easily (Farquhar *et al.*, 1989, 2007; Price *et al.*, 2002). Simultaneous measurements of discrimination of ^{13}C and ^{18}O have been used to determine which factor – stomatal regulation or photosynthesis – was more determinant for the differences of WUE (estimated from $\delta^{13}C$) observed among breeding cultivars of wheat (Barbour *et al.*, 2000b) or different tropical species (Cernusak *et al.*, 2008), as well as among years in long series of regional climate-change transitions (Cullen *et al.*, 2008).

Similarly to that observed in the measurement of instantaneous and intrinsic WUE, a key point is to determine to what extent $\delta^{13}C$ is representative of the whole-plant WUE. The relationship among these parameters has been largely discussed by several authors and it is not straightforward (Tanner and Sinclair, 1983; Farquhar *et al.*, 1989). The lack of a simple relationship is partly owing to the fact that the ratio of accumulated biomass/consumed water integrates, to a large extent, processes such as assimilate partitioning, plant carbon balance, changes in the maintenance and growth components of respiration, etc., that are not reflected in the ^{13}C signature.

33.3. CARBON ACQUISITION VERSUS WATER LOSS: HOW TO REGULATE THE COMPROMISE

Fine control of water losses by stomata has enabled plants to colonise terrestrial habitats with fluctuating soil-water availability and air humidity. Regardless of a wide diversity of forms, plants maintain a tight stomatal regulation to optimise carbon and water fluxes to survive in terrestrial ecosystems (Hetherington and Woodward, 2003; Nicotra and Davidson, 2010). Although in opposite directions, CO_2 entering the leaf and *water vapour* exiting it use approximately the same pathway, from the substomatal cavity to free air. Water loss is many times greater than CO_2 entrance for two main reasons. First, the driving force (gradient) is 50–100 times larger for *water vapour*, and second its diffusion constant is 1.6 times larger than for CO_2 owing to its lower molecular weight. The balance reflected by WUE is determined by this gradient and, consequntly, modulation of the stomatal aperture plays a key role in optimising the amount of water lost per carbon gain.

Stomata morphology, density and position seem more conservative than, e.g., leaf shape. Since the appearance of stomata in terrestrial plants around 400 million years ago (Edwards *et al.*, 1998; see Chapter 24), two main broad morphological types have been developed: the dumbell-shape, typical of grasses, and the kidney shape found in other higher plants. It is widely accepted that the linear dumbell-shaped stomata represents a more evolutionarily advanced form, and in general provides more immediate responses to environmental changes than the kidney-shaped morphology. Stomata size ranges from 10 to 80 µm in length and they occur at densities between 5 and 1000 mm^{-2}, usually showing an inverse relationship between density and size, which is considered as compensatory by different authors (Hetherington and Woodward, 2003; Beerling and Woodward, 1997). Whether these morphological characteristics determine leaf WUE is still under discussion, although it is generally assumed that higher stomatal density induces

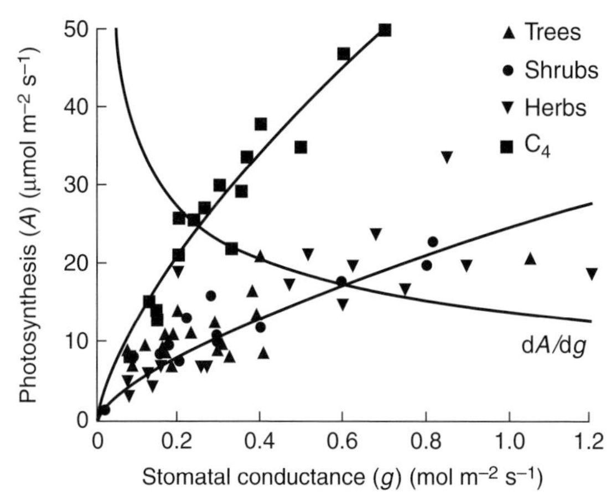

Fig. 33.4. Relationship between maximum photosynthesis and stomatal conductance in C_3 and C_4 species from different life forms (solid lines). The dotted line is the dA_N/dg_s for C_3 species (From Hetherington and Woodward, 2003).

faster responses to environmental changes (Woodward, 1987a,b; Aasamaa *et al.*, 2001; Galmés *et al.*, 2007c).

Stomatal conductance (g_s) is subject to fast and strong regulation (see Chapters 3 and 20) and is widely variable depending on the time of the day, leaf position or transient environmental conditions, like occasional shade or midday dehydration. g_s is usually related to photosynthetic capacity (Wong *et al.*, 1979; Cowan, 1986; Field and Mooney, 1986). When g_s and A_N are compared among different species, the positive curvilinear relationship is clearly dependent on the photosynthetic type (C_3, C_4). Within each group, the correlation between g_s and A_{max} is apparent, regardless of the life form (Fig. 33.4). This is in accordance with the existence of a close matching between the diffusional supply of CO_2 to the chloroplast and the demand from the carboxylation activity in the stroma, leading to a quite constant ratio, C_i/C_a. This ratio falls between 0.65 and 0.8 for non-stressed C_3 species (Farquhar, *et al.*, 1989). Reductions in g_s (under similar A_{max}) or increases in A_{max} for similar g_s lead to lower values of C_i/C_a.

It is widely accepted for different species and environmental conditions that to some extent stomata are able to balance the CO_2 requirement for photosynthesis in the stroma with the need to avoid dehydration in the leaf tissues by excessive water losses, raising the hypothesis that optimisation of water use has been a near-permanent selection pressure and that C_i and the intrinsic WUE would remain constant (Cowan and Farquhar, 1977). It is argued that for a particular leaf type stomatal responses to daily variations in light, humidity and temperature will result in maximal daily CO_2 gain for a certain water use (Farquhar *et al.*, 1980b; Hall and Schulze, 1980). In fact, the range of values of the ratio of C_i/C_a in healthy, unstressed C_3 plants was shown to remain in a narrow band (Leuning *et al.*, 1995). From these arguments, it could be deduced that the capacity for improvement of leaf WUE should be limited. However, following the A_N versus g_s relationship, relatively small changes in C_i/C_a, from 0.7 to 0.6, would lead to a theoretical WUE increase of around 33% even though the expected increases in leaf temperature associated to stomatal closure could lead to E increases, which in turn reduces the expected benefit (Condon *et al.*, 2002). Field experiments comparing different plant species and genotypes support such potential increases in leaf WUE (Dudley, 1996b; Hetherington and Woodward, 2003) even though, the main advantages are only present in drought-prone environments (Condon *et al.*, 2002). Moreover, because of the curvilinear nature of the A_N versus g_s relationship (Fig. 33.5), whenever g_s declines A_N/g_s increases. From this relationship it can be concluded that there are two main ways to improve leaf WUE (Parry *et al.*, 2005): increasing A_N while maintaining or reducing g_s, or reducing g_s while maintaining or increasing A_N (Fig. 33.6). Under well-watered conditions, reducing g_s to nearly 40% of g_{smax}, A_N will be only slightly reduced, therefore improving A_N/g_s. In fact, this is the physiological basis of deficit irrigation (DI) treatments, widely recommended as the easiest and quickest way to improve WUE in some irrigated crops (Dry *et al.*, 2001; Maroco *et al.*, 2002; Cifre *et al.*, 2005; Souza *et al.*, 2005).

Under severe water stress, g_s reductions are followed by large A_N reductions and therefore any increase in A_N/g_s will imply important penalties in terms of plant productivity. Moreover, direct measurements of A_N/g_s at severe water stress reveal that there is a threshold where A_N/g_s is highest: although theoretically a sustained increase could be expected, a dramatic reduction of A_N/g_s occurs (Fig. 33.5B). This reduction has been observed in different plant species and varieties (Medrano *et al.*, 2002b; Gulías *et al.*, 2009) and it takes place when stomata are practically closed and carboxylation suffers from metabolic inhibitions, as reflected by increased C_i (see Chapter 20). In addition, when photosynthesis is largely reduced, respiration rates become relatively much more important. Finally, cuticular water losses can also be exacerbated because of increased leaf temperature.

These environmental conditions causing dramatic reductions of WUE are not unusual. In fact, under Mediterranean summer conditions severe drops in A_N/g_s

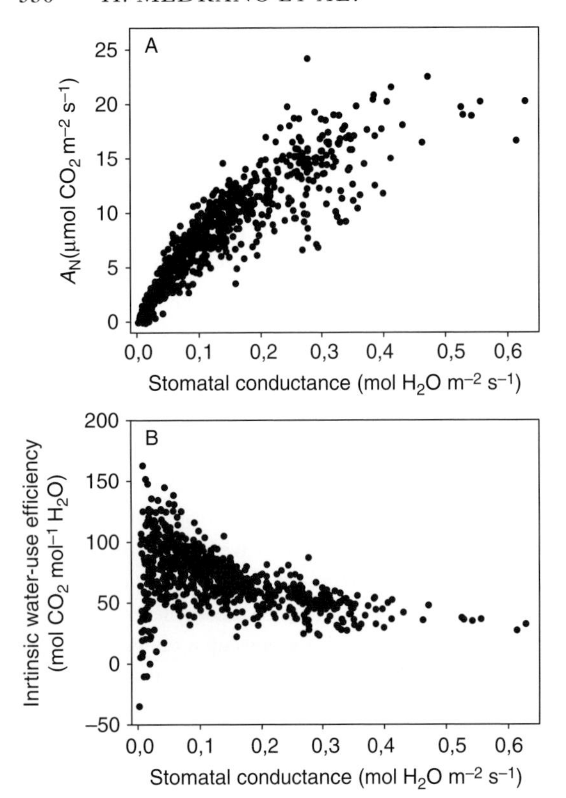

Fig. 33.5. Relationship between **(A)** net photosynthsis (A_N) and stomatal conductance (g_s) and **(B)** intrinsic water-use efficiency (A_N/g_s) and stomatal conductance (g_s). All data is from C_3 species (data from Cifre *et al.*, 2005).

have been observed even in well-watered plants (Lefi *et al.*, 2004). This is owing to a higher water demand from the atmosphere than the supply capacity of the plant vascular system. Moreover, decreases in WUE at the whole-ecosystem level as determined using EC measurements have been shown in Mediterranean forests in summer (Reichstein *et al.*, 2002). Generally, it seems that carbon balance is more severely reduced at the whole-plant level than at the leaf level under water stress, perhaps because dark-respiration expenses of the plant are similar or even increased under drought both in leaves and in whole plants as a result of enhanced root growth (Galmés *et al.*, 2007e). The overall result is a severe reduction of plant carbon balance that results in lower plant WUE. This means that plant water-saving responses to drought cannot avoid dramatic inefficiencies of water use at the plant level under severe water stress. In fact, after Mediterranean summer drought it is not rare to observe native drought-adapted plants to dry up.

33.4. WATER-USE EFFICIENCY IN AN ECOPHYSIOLOGICAL CONTEXT: IMPACTS OF RADIATION, CO_2 AND NITROGEN AVAILABILITY

Plant growth and biomass production is largely controlled by the availability of resources such as radiation, water, mineral nutrients and CO_2 (Lambers *et al.*, 2008). WUE variations with respect to the availability of these resources have been widely studied for water, but less intensively for other resources (Monteith, 1984; Wright *et al.*, 2003). Radiation has an effect on WUE owing to the dependency of photosynthesis to light. Light response curves of photosynthesis show that under suboptimal light levels (below saturation threshold) A_N decreases steeply while stomata remain open. Consequently, photosynthetic WUE is usually lower at light levels below saturation. Inside the canopy, vertical distribution of leaves creates different levels of light availability, usually followed by variations in the leaf N content, which also induce differences in WUE along the canopy (Niinemets *et al.*, 2004c; Woodruff *et al.*, 2009). In field-grown grapevines, for instance, it was observed that leaf WUE was greatly reduced in the most shaded leaves, which were particularly abundant in the canopy and therefore contributed to reducing their whole-plant WUE (Escalona *et al.*, 2003).

Regarding CO_2, high C_a causes A_N increases concomitant with decreases in g_s, resulting in enhanced WUE. This is observed at the leaf level by direct measurements, as well as in FACE approaches (Bernacchi *et al.*, 2005a, 2007). The interest in studying the effects of CO_2 on WUE is related to global-change predictions. In this sense, an increase in C_a would presumably enhance assimilation rates and therefore A_N/g_s. Hietz *et al.* (2005), studying tropical-tree species in a Brazilian rainforest, reported a decrease of $\delta^{13}C$ in wood cellulose during the last decades. In contrast, Sarris *et al.* (2007) reported a decline of *Pinus brutia* growth during the last decades in the Eastern Mediterranean basin, together with a decline in precipitation. Therefore, the effects of the increment in C_a might be counteracted by lower precipitation in this region. Furthermore, Allen *et al.* (2003) showed that the decrease in E induced by higher C_a was counteracted by the effects of rising temperature over ETp, leading to non-significant changes of WUE. Therefore, many physiological aspects of plant response to long-time exposures

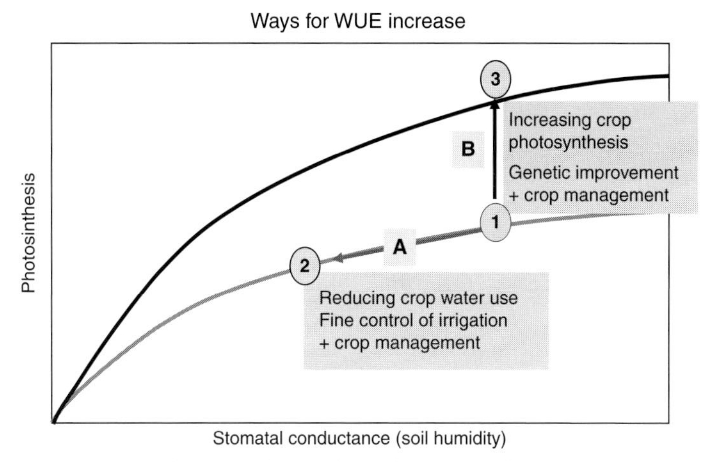

Fig. 33.6. Theoretical ways to improve intrinsic water-use efficiency by genetic improvement (increasing photosynthesis) and agronomical management (reducing stomatal conductance). Modified after Parry *et al.* (2005).

at high C_a are not fully understood. Stomatal and overall photosynthesis-acclimation processes to high CO_2 have been observed to occur differently depending on the species (Chaves *et al.*, 2004). The interaction between high C_a and ambient temperature, or the response of different functional groups to these changes, have to be solved for better predictions of the effects of global change on WUE at the leaf, plant or community levels.

Among mineral nutrients, nitrogen (N) plays a pivotal role. Photosynthetic capacity and leaf N content are positively correlated because the largest bulk of leaf N is found as Rubisco and thylacoid light-harvesting complexes (Lawlor, 2001; Lambers *et al.*, 2008). In this sense, photosynthetic capacities increase with improved N supply combined with a slight or negligible increase in g_s, leading to higher WUE (Ranjith *et al.*, 1995). The relationship between leaf N content and WUE has been widely studied, but strong interactions between the effects of N and other resources such as light and water are found, which makes it difficult to summarise in a simple model (Chaves *et al.*, 2004). In fact, some contradictory results have been reported: a higher N availability may lead to a lower WUE as it usually favours excessive leaf growth, and therefore higher E and water needs (Meinzer and Zhu, 1998). Alternatively, reducing water losses to improve WUE leads to decreased water and therefore nutrient transport, this in turn leads to a reduced N supply to leaves (Cramer *et al.*, 2009).

Alternatively, in those ecosystems where a WUE increase is predicted as a consequence of enhanced assimilation owing to high C_a, it seems that nutrients, mainly N, would limit plant growth as a consequence of the increasing demand by plants growing at faster rates (Luo *et al.*, 2004). Obviously, this limitation will be more severe in ecosystems where soil fertility is already low, such as many tropical lands.

In summary, although the effects of light intensity and to some extent CO_2 concentrations on WUE are well known, the effects of N and other nutrients are yet to be understood. Other factors recently reported to affect plant WUE by mechanisms not fully understood include the infection with endophytic fungi (Swarthout *et al.*, 2009; see also Chapter 22) and phytochrome B (Boccalandro *et al.*, 2009). These and other factors deserve further attention in the future for a proper understanding of the regulation of WUE in plants at the leaf level.

33.5. VARIATION OF PHOTOSYNTHESIS AND WUE IN DIFFERENT FUNCTIONAL PLANT GROUPS

The photosynthetic pathway, i.e., C_3, C_4 or CAM, is one of the most important traits determining the WUE of a species (Chaves *et al.*, 2004; Lambers *et al.*, 2008). However, there are many other plant traits conditioning WUE. Among them leaf habit, life form, symbiotic abilities and canopy status are some of the most relevant.

It is well known that species with different photosynthetic pathways can show large differences in WUE. As a general pattern, C_3 species exhibit lower WUE values than C_4 and CAM species (see, for instance, Lambers *et al.*, 2008 and references therein). CAM species achieve high WUE values by opening stomata at night when atmospheric *water-vapour* pressure is lowest, so E is also lower for a given stomatal aperture (see Chapter 6). There are few reports comparing WUE of CAM species with C_3 or C_4. Osmond *et al.* (1979) and Nobel (1988) reported higher instantaneous WUE in CAM species. Winter *et al.* (2005) also showed lower E (g H_2O g^{-1} dry mass) in constitutive CAM species than in their C_3 and C_4 counterparts growing at the same site. By contrast, C_4 species achieve higher WUE than C_3 plants by increasing partial pressure of CO_2 at the Rubisco sites. Consequently, C_4 species show higher $\delta^{13}C$ values than C_3 species, –10 to –18‰ and –24 to –32‰ for well irrigated C_4 and C_3 plants, respectively. Accordingly, within the same leaf habit or life form and under similar environmental conditions, there are many reports confirming that C_4 species present greater photosynthetic WUE (A_N/E and/or A_N/g_s) than C_3 species (Osmond *et al.*, 1982; Schulze and Hall, 1982; Gulías *et al.*, 2003; Yu *et al.*, 2005; Cunnif *et al.*, 2008). Although C_4 species usually show both high A_N and WUE, the enhanced WUE in CAM species is usually penalised with low photosynthetic and growth capacities (see Chapter 6). Alternatively, C_4 species are mostly restricted to warm areas where photorespiration can achieve high rates, thus conferring an advantage when compared with C_3 species and compensating for the higher energetic cost of C_4 species (see Chapter 5). In this sense, several authors have reported a clear relationship between temperature and C_4-species dominance in a site (Chapter 5). Nevertheless, Edwards and Still (2008) have also pointed out the importance of high WUE of C_4 grasses as a key trait in their widespread expansion in the late Miocene and in their present-day distribution (see also Chapter 25).

Species showing short-lived leaves, mainly herbaceous, woody deciduous and semi-deciduous, usually present higher A_{max} than long-lived leaf species, compensating the fact that they have less time to pay off their 'construction cost' (Brooks *et al.*, 1997; Gulías *et al.*, 2003). This is usually achieved by more efficient photosynthetic machinery and high g_s allowing CO_2 influx to ensure its supply. This usually implies lower A_N/g_s, according to the general A_N-g_s relationship (Figs 33.4, 33.5). In addition, leaf traits related with lifespan are also expected to be related with WUE. For instance, A_N/g_s was positively related to LMA in 73 C_3 Mediterranean species (Fig. 33.7). Moreover, sclerophyll leaves, showing high LMA and long lifespan, have been reported to present higher WUE than malacophyllous *Cistus sp.* along a precipitation gradient under Mediterranean conditions (Gulías *et al.*, 2009). Similarly, Aranda *et al.* (2007)

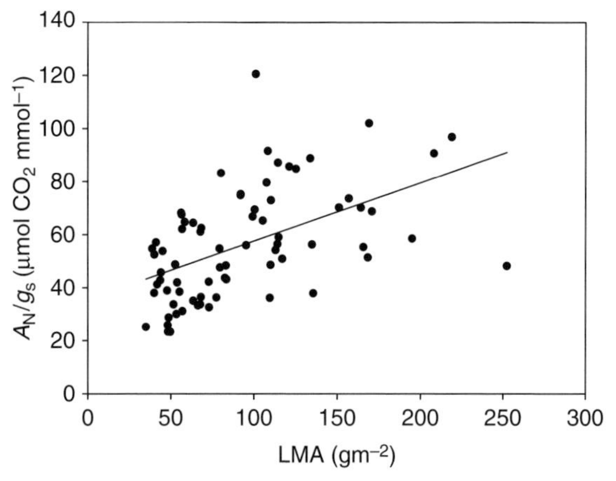

Fig. 33.7. Relationship between intrinsic water-use efficiency (A_N/g_s) and leaf mass area (LMA) in 73 C_3 Mediterranean species (data from Gulías *et al.*, 2003).

reported a positive relationship between $\delta^{13}C$ and LMA in *Quercus suber* plants under different water and light availabilities, implying higher WUE at high LMA. Similar positive relationship between WUE and LMA has been reported in different *Amaranthus* species (Liu and Stützel, 2004; Omami *et al.*, 2006). Other leaf traits, like leaf pubescence, can also modify WUE values. For instance, Galmés *et al.* (2007f), comparing two varieties of *Digitalis minor* differing in the presence of leaf trichomes, showed higher intrinsic WUE in the pubescent rather than in the glabrous variety. Some reports have also shown a significant effect of life and growth forms or leaf habits on WUE (Brooks *et al.*, 1997; Golluscio and Oesterheld, 2007; Medrano *et al.*, 2009).

Symbiotic relations between plants and mycorrhizal fungi are expected to enhance plant A_{max} owing to their higher nutrient supply, as well as to the increased plant ability to uptake soil water when it is scarce (Allen, 1991). Therefore, WUE of mycorrhizal-colonised plants is expected to be higher than that of non-colonised plants. Nevertheless, specific differences to mycorrhisation were shown, and some species have shown to increase their WUE when inoculated, such as *Olea europaea* and *Citrullus lanatus*, whereas other species, such as *Rhamnus lycioides* or *Pistacia lentiscus*, have shown little or no response to mycorrhizal infection (Querejeta *et al.*, 2003, 2007; Kaya *et al.*, 2003). Similarly, Chaves *et al.* (2004) have suggested symbiotic relations with N_2-fixing bacteria to be a factor affecting WUE.

Pioneer species usually present higher A_{max} than late successional species. Their high A_N is often achieved by high g_s and low WUE. Nevertheless, although some reports showed lower $\delta^{13}C$ in pioneer species than in late successional ones

(Huc *et al.*, 1994), the opposite has also been reported (Guehl *et al.*, 1998; Nogueira *et al.*, 2004). This apparent contradiction is related to the specific environmental conditions in each study, particularly for light intensity (see Section 33.4).

33.6. IMPROVING WUE IN CROPS: CROP MANAGEMENT

WUE of crop species is an important trait for the sustainability of agriculture around the world, especially in arid and semi-arid areas. In fact, several efforts have already been performed during the last decade as a consequence of the increasing concern of the scarcity of water resources (Condon *et al.*, 2002; Turner, 2004). In agricultural systems, and specifically in irrigated crops, the final objective is to increase water productivity, i.e., economic yield per unit of consumed water. There are several methods to achieve this goal, from irrigation and agronomic engineering methods improving the efficiency of irrigation systems or reducing soil-water evaporation, to plant breeders' methods finding new cultivars with higher WUE.

Observing the relationships between A_N and g_s (Fig. 33.5), it can be deduced that for a certain g_s (i.e., soil-water availability) there are two main ways to increase A_N/g_s: (1) by reducing g_s; or (2) by increasing A_N, which can be associated to: (1) improvements of agronomic practices; and (2) genetic improvement, respectively (Fig. 33.6).

Most agronomic practices oriented to save water lead to increases in the use of water available in the soil more than to increase on WUE itself, leading to significant reductions of the irrigation necessities. Some specific agronomic practices, such as soil organic-matter increase, soil-porosity management and mulching have a positive impact by increasing the capacity of soil to store water and reducing direct water loss. These practices are important for both rain-fed crops as well as irrigated crops.

Alternatively, the evaporative demand of the atmosphere (VPD) largely affects E and instantaneous WUE (A_N/E), therefore reducing VPD can be an effective method to increase WUE. In Mediterranean areas with dry and hot summers (i.e., high VPD) avoidance of this period as much as possible by early planting can be a useful method to increase WUE (Soriano *et al.*, 2004). This suggests that selecting early season cold-resistant genotypes for these areas could lead to WUE improvements.

Changing environmental conditions by growing crops in greenhouses has been proposed as an effective method to enhance photosynthetic WUE (Hsiao *et al.*, 2007). In fact, for greenhouse crops the higher the control of environmental conditions, i.e., nutrients (hydroponics), radiation, humidity and CO_2, the higher the crop WUE is. Under highly controlled conditions, WUE can be more than double compared with outdoors conditions (Stanghellini, 2005).

WUE can be improved in irrigated crops with several steps. First, selecting an adequate irrigation system (dripping tubes seem to be the best water savers) and tightly controlling unnecessary losses (sometimes more than 30% of water used). Second, defining the most adequate irrigation dosage and frequency to maintain the desired plant-water status. To achieve this, it is necessary to find adequate indicators of plant-water status and thereafter monitor them throughout the cropping season. The final goal is to match the irrigation schedule to the plant physiological status, avoiding over-irrigation that would be lost by percolation. These procedures allow reducing irrigation volumes without any decrease in the crop growth and yield. In this sense, Fig. 33.6A shows that g_s can be reduced by as much as 50% causing only small reductions in A_N (<20%). Following this approach, plants will suffer mild to moderate water stress while achieving important improvements in crop WUE. Actually, this is the basis for deficit irrigation programs and partial root-drying approaches (Costa *et al.*, 2007). Vast experience in this irrigation management supports the use of controlled-deficit irrigation to improve WUE and save water without significant losses in the final harvest. Moreover, in some cases, such as in grapevine, improvements of final fruit quality have been reported (Chaves and Oliveira, 2004; Cifre *et al.*, 2005). Similar results have been observed in other crops, such as potato and sunflower (Karam *et al.*, 2007).

In conclusion, under field conditions improving the capacity to maintain a certain (moderate) level of crop water stress would contribute to large improvements in crop WUE. Defining the most adequate thresholds for a sustained yield with minimal watering is the subject of current and future agronomic research. Automatic devices to detect plant-water status and adjust its water dosage by 'precision agriculture' is a promising field to achieve significant improvements of crop WUE, finally leading to more food per drop.

33.7. OPPORTUNITIES FOR GENETIC IMPROVEMENTS OF WUE

Genetic advances of WUE aim to improve A_N/g_s provided that there is evidence for independence variation in photosynthetic capacity and stomatal conductance (Gilbert *et al.*, 2011). Although improvements of WUE can be achieved by downregulating g_s, more benefits would be expected if this enhancement is mainly owing to improvements of A_N, as reducing g_s might potentially reduce A_N and consequently plant growth and yield.

Based on the strong and general relationship between A_N and both E and g_s, potential improvements of WUE by stomatal regulation are relatively small (Cowan, 1986). Still, photosynthetic WUE has been largely evaluated in many cultivar collections and crop plants, and significant genetic variability has been observed (Hubick *et al.*, 1988; Donatelli *et al.*, 1992; Ismail and Hall, 1993; Bota *et al.*, 2001; Stiller *et al.*, 2005; Massonnet *et al.*, 2007). This variability represents an opportunity to select for more efficient genotypes and improve yield under well-watered and/or water-limited conditions. Nevertheless, further selection efforts have been difficult because the relationship between WUE and yield is yet unclear, reducing the utility of WUE as a selection trait (Stiller *et al.*, 2005). Changes in WUE under drought and salinity are a key point when evaluating different crop genotypes. As stated previously, A_N/g_s increases at moderate water stress when stomatal limitations to A_N are predominant, but decreases when water stress intensifies and non-stomatal limitations dominate plant responses (Flexas *et al.*, 2004a,b). Therefore, leaf intrinsic and instantaneous WUE are alternatively found to increase or decrease under water stress, depending on the level of water shortage. Variable WUE under salt stress depending on the genotype has been observed in *Atriplex halimus* (Martínez *et al.*, 2003) and in *Amaranthus spp.* (Liu and Stützel, 2004). In addition, when WUE is assessed at the plant level (g dry matter kg^{-1} H_2O) many other factors play a role in determining its response to drought/salt stress (Fig. 33.2). Bacelar *et al.* (2007) found that under water stress only one olive-tree cultivar increased whole-plant WUE, although the three cultivars studied presented increased intrinsic WUE. The uncoupling of both parameters limits the use of leaf intrinsic WUE as a selection parameter. Similarly, Centritto *et al.* (1999) showed that instantaneous transpiration-efficiency improvements did not correspond with whole-plant WUE.

The aim of genetic improvement of WUE has been extensively studied and widely presented in the literature (Condon *et al.*, 2002, 2004; Rebetzke *et al.*, 2002; Morison *et al.*, 2008). However, in contrast with this large effort, practical progress is still modest (Richards, 2008). The application of selection criteria based on physiological characteristics directly related with drought resistance, molecular approaches and quantitative-trait-loci (QTL) analysis are increasing the capacity for practical progress, and important advances have been made in relation to the detection of genetic variability, QTL and molecular-markers identification to accelerate selection programmes of different parameters that influence WUE (Serraj *et al.*, 2005; Steele *et al.*, 2006).

Although some authors are sceptical about the possibilities for successful genetic improvement of WUE, there is growing evidence that targeting specific traits in a multidisciplinary programme may lead to elite genotypes, as is the rationale for different wheat-breeding programs (Royo *et al.*, 2002; Araus *et al.*, 2003; Condon *et al.*, 2004). In fact, some success has been attained in discovering useful new varieties applying these approaches, namely using $\Delta^{13}C$ discrimination as selection criteria for higher WUE (Condon *et al.*, 2002; Rebetzke *et al.*, 2002), including the recent new Drysdale variety of wheat that shows great improvement in WUE (Richards, 2008). As expected, the highest advantages of this genotype are displayed in drier environments with yield increases of around 23% with respect to current varieties.

Important efforts have been focused to identify QTL that contain associated genes that could confer particular improvements of WUE, even though only a few genes responsible for such QTL have been identified (Salvi and Tuberosa, 2005). Another strategy chosen by different groups has been to identify, by genetic manipulation in model plants (i.e., *Arabidopsis*), specific genes affecting WUE with potential to be transferred to crops (Schroeder *et al.*, 2001a,b; Sáez *et al.*, 2004, 2006; Rubio *et al.*, 2009). However, the success of manipulation of a single gene on a well-recognised multigene character as WUE is questionable. Some crucial comments to this procedure are related to the fact that molecular biologists identify genes associated with immediate response usually to rapid dehydration, so the relevance for crop productivity under field-drought conditions is questionable (Bray, 2002; Sinclair and Purcell, 2005).

Although many drought-tolerance genes have been found to be unrelated to increased WUE (Blum, 2005), increased drought-tolerance has been associated with higher WUE in P_{SAG12}-*IPT* transgenic tobacco (Rivero *et al.*, 2007), transgenic *Arabidopsis* expressing the bZIP transcription factor *ABP9* (Zhang *et al.*, 2008c) and transgenic *Arabidopsis* and tomato plants expressing tobacco aquaporin NtAQP1 (Sade *et al.*, 2010) or *Eucalyptus* plants expressing several radish aquaporins (Tsuchihira *et al.*, 2010). Moreover, characterisation of *Arabidopsis* with the *ERECTA* mutation showed changes in WUE ($\Delta^{13}C$), reducing g_s and controlling leaf-photosynthetic capacity (Masle *et al.*, 2005). Similarly, in transgenic tobacco plants aimed to delay drought-induced leaf senescence by a drought-inducible promoter coupled to an isopentenyl transferase gene to produce high cytokinin levels, these plants were able to maintain higher water content and photosynthetic activity during drought, with minimal yield losses under severe drought conditions (Rivero *et al.*, 2007). More general genotype attributes, such as reduced night leaf-epidermal conductance (Fish and Earl, 2009) or heterosis (Araus *et al.*, 2010), also seem to improve WUE. Other approaches have consisted in manipulating g_s, by means of modifying different genes implicated in the signal cascade of ABA (Verslues and Bray, 2006; Nilson and Assman, 2007). Among genes involved in ABA signalling and ABA-mediated stomatal closure, protein kinases (like *OST1*) act as positive regulators whereas type-2C protein phosphatases (*PP2C*) and farnesyltransferases (like *ERA1*) act as negative regulators (Schroeder *et al.*, 2001a,b). For instance, Sáez *et al.* (2004, 2006) have shown that *Arabidopsis* single and double mutants affected in *PP2C* present normal growth under well-watered conditions with increased WUE under reduced water availability, while Rubio *et al.* (2009) have shown that triple mutants present a penalty in growth. The benefits of these mutations are increased under water stress, where wild-type plants dehydrate while mutants keep some growth with small water losses. These differences are owed to lower g_s and higher A_N/g_s in the mutants. However, there seems to be a limit in the capacity of improving WUE while maintaining growth with this approach, as triple mutants show constitutive penalties in growth, even under well-watered conditions. These results also reinforce the idea that optimal improvement in WUE may involve increased photosynthesis.

Recently, physiological targets to achieve simultaneous increase of photosynthesis and WUE have been identified. Among them, improving CO_2 diffusion in the mesophyll through either inducing C_4-like photosynthetic metabolism in C_3 plants (Long *et al.*, 2006a) or increasing the g_m from sub-stomatal cavities to chloroplasts (Flexas *et al.*, 2008; Galmés *et al.*, 2011) and/or improving carboxylation efficiency. In turn, increased carboxylation efficiency can be achieved through increased Rubisco catalytic rate (Parry *et al.*, 2007) and/or specificity for CO_2 (Delgado *et al.*, 1995; Galmés *et al.*, 2005; Parry *et al.*, 2007), reducing or bypassing photorespiration (Long *et al.*, 2006a; Kebeish *et al.*, 2007), increasing the capacity for RuBP regeneration (Feng *et al.*, 2007b; Pertehansel *et al.*, 2008) or reducing the photoprotective state upon high- to low-light transitions (Long *et al.*, 2006a; Zhu *et al.*, 2010).

Some elegant biotechnological approaches are seeking to exploit the introduction of genes for all of the steps of C_4 metabolism into single cells within leaves of C_3 plants (Leegood, 2002). Even if all of the steps are introduced and are functional in leaves of C_3 plants, it seems unlikely that WUE will be significantly improved without additional structural changes (Long *et al.*, 2006a; Murchie *et al.*, 2009; Hibberd and Covshoff, 2010). An alternative to improve CO_2 availability in the chloroplasts would be to increase the diffusion of CO_2 from the sub-stomatal cavities to the Rubisco carboxylation site, i.e., g_m (Flexas *et al.*, 2008; see Chapter 3). This component of CO_2 diffusion has been traditionally considered large and invariable in different photosynthesis quantification models. However, it has recently been recognised that g_m is finite and variable, imposing a limitation to photosynthesis similar to that imposed by g_s and responding to factors similar to those to which g_s responds (Flexas *et al.*, 2007a,c, 2008; Niinemets *et al.*, 2009b). A recent study by Levi *et al.* (2009) has shown that indeed in some cases improved WUE in near-isogenic cotton lines obtained in QTL-aided breeding programs is associated with increased g_m, and Galmés *et al.* (2011) have shown that differences in WUE between tomato genotypes strongly correlate with the ratio g_m/g_s. Moreover, g_m decreases as drought progresses, therefore being an even more important limitation to photosynthesis and hence WUE under these conditions (Flexas *et al.*, 2002a, 2009b; Galmés *et al.*, 2007a; Gallé *et al.*, 2009). Although the nature of g_m regulation is still largely unknown, although presumably complex (Evans *et al.*, 2009), an increase in water loss while improving CO_2 diffusion is not expected. Therefore, it would be assumable that higher g_m would lead to WUE increment. Thus, environmental conditions that increase g_m or genetic modifications favouring higher values should lead to leaf WUE improvements. This opens an interesting field of research to elucidate the mechanisms involved in g_m regulation and the responses to different environmental stimuli. Particularly promising gene targets for g_m improvement are aquaporins. Several reports have demonstrated that some but not all of these proteins are involved in the regulation of g_m (Terashima and Ono, 2002; Uehlein *et al.*, 2003, 2008; Hanba *et al.*, 2004; Flexas *et al.*, 2006a), and at least one study suggests that they are particularly involved under drought conditions (Miyazawa *et al.*, 2008). Photosynthetic analysis of aquaporin over-expressing transformants showed that A_N and g_m were higher than in wild types, but unfortunately A_N/g_s was similar owing to compensating changes in g_s in the mutants (Hanba *et al.*, 2004; Flexas *et al.*, 2006a). It remains to be studied whether aquaporin over-expressing plants show benefits in WUE when subject to water stress. However, some statistical evidence has been presented relating natural variations of WUE among genotypes of a given species with variations of g_m (Duan *et al.*, 2009; Soolanayakanahally *et al.*, 2009; Barbour *et al.*, 2010; Flexas *et al.*, 2010; Galmés *et al.*, 2011), which suggests that this may represent a promising way for improving WUE.

For a given stomatal aperture, a higher CO_2-assimilation capacity would result in enhanced A_N/g_s. In this sense, improving the capacity for CO_2 assimilation through engineering of Rubisco kinetics has been argued as a promising way to increase leaf WUE (Long *et al.*, 2006a; Peterhansel *et al.*, 2008). This approach presents the interest that enhancement of A_N/g_s would be more relevant when water is limiting, i.e., at lower stomatal and internal conductances to CO_2, and therefore lower C_c, whenever the engineering of Rubisco is directed toward an increased affinity for the CO_2 substrate. As an example, applying the mathematical model of Farquhar *et al.* (1980a) to a hypothetical replacement of the endogenous Rubisco in *Vitis vinifera* with that from *Limonium gibertii*, the calculations predict a twofold increase in A_N/g_s when g_s is decreased from 0.2 to 0.05 mol m^{-2} s^{-1} (Flexas *et al.*, 2010). This simulation is additionally supported by A_N/g_s data measured in *L. gibertii*, with values significantly higher than the remaining plant functional types from Mediterranean environments (Medrano *et al.*, 2009). Nevertheless, this approach currently presents two major limitations. First, the number of species whose Rubisco catalytic properties have been screened is clearly insufficient, although some of the studies already indicate those habitats where *better* Rubiscos might be found (Galmés *et al.*, 2005). Second, in spite of the important progresses achieved during recent years on the methodology for engineering Rubisco genes – specifically those coding for the large subunits – attempts to increase the specificity for CO_2 have yet to be successful (Peterhansel *et al.*, 2008). Still, while the range of species whose plastome can be transformed continues to expand (Koop *et al.*, 2008), transplantation of foreign or modified Rubisco variants into higher-plant plastids has been exclusively performed in tobacco where transformation efficiency is highest (Whitney and Sharwood, 2008). Better knowledge about the complex processes involved in the folding of Rubisco subunits would certainly increase the chances of success in this direction. A similar approach to increase WUE is to induce C_4 syndrome in C_3 plants, which would increase CO_2 concentration surrounding Rubisco, thus raising the light-saturation of A_N and greatly reduce photorespiration (Leegood, 2008). An additional agricultural benefit of shifting from C_3 to C_4 is that C_4 requires less Rubisco and hence less N.

The recent release of wheat varieties with enhanced WUE and good performance under field conditions, the massive progress in the identification of new genes and QTL enabling unexpected increases of WUE in pot experiments and the finding of new targets for selection and transformation procedures, as is the case of g_m and Rubisco specificity-factor modifications, maintains this topic as a very promising field for research in the near future, as well as brings light to the challenge of the coming difficulties to satisfy water requirements for crops.

34 • Global change and photosynthesis

C.J. BERNACCHI, C. CALFAPIETRA, M. CENTRITTO AND F. VALLADARES

The phrase global change is generally associated with alterations of climate (temperatures, fluctuations in precipitation, etc.) that stem from changes in atmospheric composition. In reality, global change also encompasses more than changes in climate or atmospheric composition; any global-scale change that influences biota directly or indirectly can be considered global change. Global change has influenced the biosphere throughout geological time, with changes occurring over periods that allow for either species to evolve to these changes when they occur over long periods, to adapt or acclimate to the changes or to perish when neither of the previous two responses is effective. Although we are currently in the midst of abrupt global change, it certainly is not the first time that rapid global change has occurred. What differentiates the current changes to our planet from all other global-change events is that these abrupt changes are brought about through anthropogenic influences and that they are occurring more rapidly than in previous occasions.

One of the major challenges of a chapter focused on plants and global change is that the topic is extensive. Many chapters of this book focus on photosynthesis responses to most of the predicted global-change scenarios, and entire volumes can be devoted to these and other global-change scenarios. Further, many reviews have been published that address photosynthetic responses to single environmental-change factors (Ceulemans *et al.*, 1999; Saxe *et al.*, 2001; Ainsworth *et al.*, 2002; Long *et al.*, 2004; Ainsworth and Long, 2005; Hikosaka *et al.*, 2006; Sage and Kubien, 2007; Wittig *et al.*, 2007). Our goals in writing this chapter are: (1) to provide a synopsis of global-change trends; (2) to identify the potential impacts of these trends on photosynthesis; (3) to identify the potential for photosynthesis to mitigate global change; and (4) to present ecosystem and general-circulation models as predictors of the interactions between plants and global change.

34.1. GLOBAL CHANGE: A BRIEF SYNOPSIS OF OBSERVED AND PREDICTED GLOBAL-CHANGE TRENDS

There are numerous changes that are occurring throughout the planet as a result of anthropogenic activities. The combination of large-scale changes in the composition of atmospheric constituents, changes in climate for many regions of the planet and clear evidence of the impacts of these changes on ecosystems, ice fields, sea levels and a number of other events all confirm that global change is occurring.

34.1.1. Atmospheric CO_2 is increasing

Atmospheric concentrations of CO_2 are increasing at a rate unprecedented since the Oligocene-Miocene transition approximately 20–25 million years ago. The rise in CO_2 is a direct consequence of two main activities, the burning of fossil fuels and land-use change. Since the beginning of large-scale industrialisation, concentrations of CO_2 have increased by over 30% (Fig. 34.1). Fossil fuels consist of high-energy organic molecules that are the remnants of ancient biological organisms. Over long time periods, the remains of these organisms were buried under sediment and the carbon in these remains was removed from the global carbon cycle for millions of years. The total amount of organic matter stored in sediment is unknown; however, evidence suggests that the size of this pool is large. For example, during the Cambrian era (~500 MA) atmospheric CO_2 concentrations were as much as 12 times above present (Berner, 1990). It was during this period and the following periods of the Paleozoic era that most fossil-fuel reserves were formed. The component of the rise in CO_2 that is attributed to the burning of fossil fuels is a direct consequence of releasing carbon that was sequestered for millions of years. Although formed under different epochs and attributed to different

Terrestrial Photosynthesis in a Changing Environment: A Molecular, Physiological and Ecological Approach, ed. J. Flexas, F. Loreto and H. Medrano. Published by Cambridge University Press.

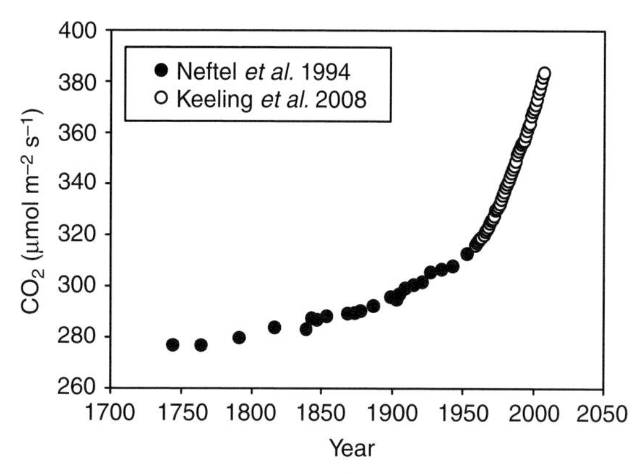

Fig. 34.1. Plots showing the increase in atmospheric concentrations of CO_2 since the Industrial Revolution. Data in the filled symbols represent ice core data from the Siple Station Antarctic ice cores (Neftel *et al.*, 1994) and the open symbols from measurement made atop Mauna Loa in Hawaii (Keeling *et al.*, 2009).

organisms, burning of coal and natural gas are similarly influencing global atmospheric trace gas concentrations.

Changes in land use have impacted atmospheric CO_2 concentrations as a result of numerous activities, including deforestation and the spread of agriculture. Despite the occasional disturbance, most ecosystems prior to large-scale industrialisation and the spread of agriculture were under steady state conditions. When the forests or grasslands were cleared for urbanisation or agriculture, these ecosystems were disturbed from their steady state conditions. The carbon sequestered in plant tissues and soils was released to the atmosphere during this disturbance. For example, simulations suggest that the soil carbon pool in Midwestern U.S. agricultural soils has decreased by 60% since it was converted from pre-settlement prairie (Donigian *et al.*, 1994; Lal, 2004a).

The IPCC has presented numerous model scenarios to predict how CO_2 will continue to increase over the next century (Meehl *et al.*, 2007). Model scenarios for the year 2100 based on the IPCC Fourth Assessment Report predict an increase in atmospheric CO_2 ranging from 700 to over 1000 μmol mol^{-1} (Meehl *et al.*, 2007), which is higher than the range predicted from the IPCC Third Assessment Report (Prentice *et al.*, 2001). Current evidence shows that increasing demand for fossil fuels is resulting in the rates of CO_2 emissions to exceed the 'worst-case' scenarios from the most recent IPCC report (Raupach *et al.*, 2007). This suggests that the increase in atmospheric CO_2 that has been observed since the dawn of the industrial revolution is not showing any signs of stabilisation and that the rise might actually continue at an increasing rate.

The premise that CO_2 will rise at an unprecedented rate throughout this century suggests that vegetation responses to these changes will become more apparent. CO_2 has risen over 30% over the last two centuries and the implications to date are uncertain owing to a lack of preliminary data collected prior to the Industrial Revolution. Therefore, effects of global change and of the green revolution are confounded and make assessments of current responses difficult (e.g., Phillips *et al.*, 2008). Large-scale field campaigns are now providing baseline data in order to assess the implications of changes in CO_2 in the coming decades. In addition to the long-term experiments associated with the steady increases in global CO_2, numerous studies employing a variety of techniques, including FACE, open-top chambers, growth chambers, greenhouses and natural CO_2 sources are allowing researchers to predict how vegetation will respond to substantially higher concentrations at the leaf, whole-plant, ecosystem and global scales.

34.1.2. Tropospheric ozone is increasing

Tropospheric ozone (O_3) is a highly reactive gas that damages any living tissue with which it comes in contact. It has been increasing at a faster rate than CO_2. Prior to large-scale emissions of pollutants, O_3 concentrations were relatively low; however, since the Industrial Revolution concentrations have increased several-fold (Wang and Jacob, 1998). Although O_3 is a naturally occurring constituent of the atmosphere, fossil-fuel combustion has increased the emission of the precursors to ozone formation by more than an order of magnitude from pre-industrial concentrations (Fowler *et al.*, 1998). In order for O_3 to form, both VOC and NO_x must occur in the presence of sunlight. VOC are released to the atmosphere by both anthropogenic activities and natural processes, including fuel consumption, industrial processes, forest fires and natural geologic and biological emissions (Etiope and Ciccioli, 2009). The release of NO_x occurs naturally as a result of biological activity in soils, lightning, forest fires, etc., but the release as a result of anthropogenic activities far exceeds that from natural sources (Fowler *et al.*, 1999b). Ozone formation is relatively complex and is dependent on the relative concentrations of the precursors. For example, maximum O_3 formation will occur with optimal concentrations of NO_x for a given concentration of VOC (Fig. 34.2). Higher or lower concentrations of either of these

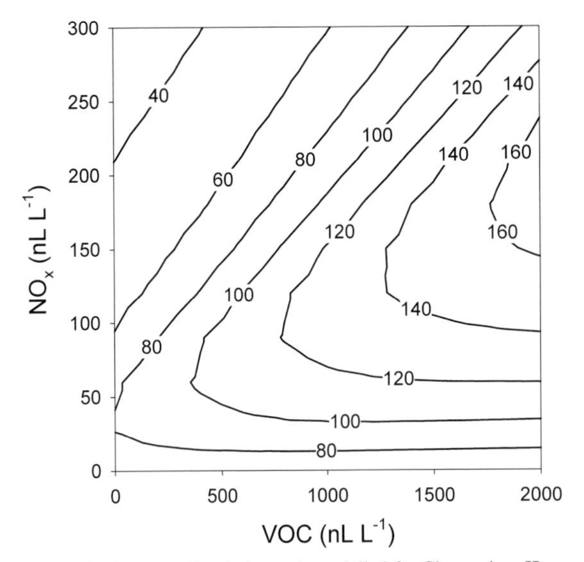

Fig. 34.2. An ozone isopleth graph modelled for Champaign, IL USA on August 15, 2008. The lines on the plot show the predicted ozone concentration (nL L^{-1}) for a given concentration of the two main precursors, nitrogen oxides (NO_x) and volatile oxygenic compounds (VOCs). Note that for a given concentration of VOC, data is modelled using the Ozone Isopleth Plotting Program (OZIP; http://www.shodor.org/ekma/).

substrates reduce the formation of O_3. Ozone concentrations are predicted to continue increasing over the next 50 years with average concentrations reaching 20% over current (Grewe *et al.*, 2001; Hauglustaine and Brasseur, 2001), although changes in O_3 formation are highly variable. Some locations are experiencing rapid increases in O_3 formation while other locations are experiencing decreases (Forster *et al.*, 2007). Thus, unlike rising CO_2, the impacts of rising O_3 will be highly variable and will depend on local and regional factors that drive the abundance of the substrates to O_3 formation, as well as the meteorological conditions that might favour the formation of O_3.

34.1.3. Global warming will continue

As a potent greenhouse gas, CO_2 is the major driver for global warming and combined with emissions of other important greenhouse gases, e.g., N_2O, CH_4, halocarbons and O_3, along with other anthropogenic activities, it will result in a net increase in radiative forcing of over 1.5 W m^{-2} (Forster *et al.*, 2007). Radiative forcing represents a shift in the balance of incoming versus outgoing energy at the ground surface. While 1.5 W m^{-2} is small relative to peak total solar radiation, it is substantially large enough to have already impacted global mean temperatures (Forster *et al.*, 2007). It is highly likely that global temperatures will continue to increase (Meehl *et al.*, 2007). However, not all parts of the planet are expected to experience temperature increases uniformly. For example, warming will be greatest at northern latitudes, and land surfaces are expected to experience temperature increases double that of the oceans (Meehl *et al.*, 2007). The rate of predicted temperature increase over land is variable, with mean increases ranging from about 1.8°C to over 6°C by the year 2100 depending on the region (Christensen *et al.*, 2007). In addition to variability over terrestrial regions, there is predicted seasonality to temperature increases although the most prominent seasonality is predicted in the arctic region, Eastern Canada and Greenland (Christensen *et al.*, 2007).

34.1.4. Extreme events are likely to increase

Although most GCM predictions suggest gradual changes, there is increasing evidence that extreme events are likely to increase in frequency and magnitude in the coming decades. Of the potential extreme events, both high temperature and precipitation are predicted to have the most significant impact on vegetation (Diffenbaugh *et al.*, 2005), although higher probabilities of high O_3 days can induce highly damaging acute responses to vegetation.

34.2. IMPLICATIONS OF GLOBAL CHANGE

Each of the above global-change constituents are likely to have both direct and indirect implications for vegetation. Most of these effects have been described in previous chapters of this book however not always in the context of global change scenarios. Thus, the effects of these changes are described briefly below, followed by a more thorough discussion of interacting factors.

34.2.1. CO_2

As discussed in Chapter 17, rising CO_2 generally increases rates of photosynthesis, which in turn increases productivity. Early studies demonstrated that increases in photosynthesis were quite large and that these increases led to much higher productivity for most C_3 species (e.g., Curtis, 1996; Ainsworth *et al.*, 2002). Recent research in which C_3 plants are grown in elevated CO_2 under field conditions in experiments that utilise the FACE design suggests increases in photosynthesis are common (Kimball *et al.*, 1995; Man and Lieffers, 1997; Noormets *et al.*, 2001a,b; Bernacchi *et al.*, 2003b, 2005a, 2006; Centritto *et al.*, 2004). However, it was observed that photosynthetic rates do not increase as much as enclosure-based studies and that the observed increase in biomass is less than earlier non-FACE studies (Long *et al.*, 2006b, 2007; Ainsworth *et al.*, 2008b). Experiments in which C_4 plants are grown in elevated CO_2 using FACE are much less common; however, the limited studies on this topic suggest little or no increase in photosynthesis (Leakey *et al.*, 2004, 2006; Ainsworth and Long, 2005). Evidence does suggest that under certain conditions, i.e., water stress, elevated CO_2 does increase photosynthetic rates of C_4 plants (Leakey *et al.*, 2004), suggesting that impacts of global change should be considered not only in the context of individual factors but also in the context of interacting factors.

Ecosystem functions and plant productivity have likely responded to the increase in CO_2 that has already occurred, although clear documentation of any change to the present is relatively difficult to assess. This is mostly owing to the lack of baseline measurements prior to the industrial revolution and large-scale changes in land use dominating ecosystem responses. Agricultural productivity has increased by more than an order of magnitude over the last century; however, the portion of this increase that is attributed to rising CO_2 is small relative to the impact that the 'Green Revolution' had on productivity with the use of high-yielding crop varieties, chemical fertilisers and pesticides, irrigation, etc. (Matson *et al.*, 2007). Forest ecosystems are likely to have responded to the increase in CO_2 since pre-industrial times, but separating the climate-change component (i.e., temperature, N deposition, etc.) from the CO_2 increase has been shown to be difficult and uncertain (Jacoby and D'Arrigo, 1997). Some evidence for these changes is apparent (Phillips *et al.*, 2008) and suggests that CO_2-induced changes in plant physiology are already altering various aspects of the environment. Despite the difference in photosynthetic responses for C_3 and C_4 species, there appears to be a consistent response toward lower water use for both types of vegetation in response to growth in elevated CO_2. For example, evidence from modelling exercises suggest that the decrease in g_s associated with elevated CO_2 results in an increase in runoff from major continental regions (Gedney *et al.*, 2006; Betts *et al.*, 2007). This modelling exercise assumes that the decrease in g_s is enough to lower evapotranspiration at the canopy scale, which has been demonstrated by numerous experiments (Hungate *et al.*, 2002; Burkart *et al.*, 2004; Triggs *et al.*, 2004; Yoshimoto *et al.*, 2005; Kimball and Bernacchi, 2006; Bernacchi *et al.*, 2007).

34.2.3. Ozone

Approximately one-quarter of the forested portion of the planet is presently at risk from tropospheric concentrations of O_3 that exceed 60 nL L^{-1} (Fowler *et al.*, 1999b). Despite the continual increases of O_3 in the coming decades, spatial variability is predicted to be high. For example, current trends in tropospheric O_3 demonstrate a divergent response, with some areas of the planet experiencing increases and others decreases or no change (Vingarzan, 2004). The areas that have shown steady or decreasing O_3 concentrations over the last few decades are attributed to changing emissions standards resulting in fewer precursors to O_3 being emitted (Vingarzan, 2004). This suggests that policy decisions can have a large impact on regional concentrations of O_3.

The damage that is imposed on vegetation from exposure to O_3 can be quite severe for many species. Exposure to increased O_3 induces damage and stress responses that reduce photosynthesis, growth, yields and water use (Feng *et al.*, 2003; Morgan *et al.*; 2004; Fiscus *et al.*, 2005; Long *et al.*, 2005; Dermody *et al.*, 2006; Flowers *et al.*, 2007; Low *et al.*, 2007; Wang *et al.*, 2007). However, there are interactions between O_3 and secondary metabolites emitted by plants that in turn may affect photosynthesis. Plants emitting isoprenoids take up more O_3 from the atmosphere, but because of the higher reactivity of these gases O_3 is efficiently scavenged and damage to photosynthesis is lower in isoprenoid-emitting plants (Loreto and Fares, 2007).

34.2.4. Temperature

The impact of increasing temperature on photosynthesis is less certain than the impact of other global-change scenarios. Whereas elevated CO_2 is known to stimulate and elevated O_3 decrease or have no net effect on photosynthesis, the complex interplay between the three major biochemical pathways involved in CO_2 fluxes into and out of the leaf (photosynthesis, photorespiration and mitochondrial respiration) make temperature more complex. Thus, increases in temperature can lead to a wide variety of responses for net photosynthesis.

The impact of rising temperatures on photosynthesis is dependent on whether the increase in temperature results in photosynthesis increasing toward or beyond its optimal temperature. It is also important to consider the extent to which thermal acclimation might occur and the impact that rising temperatures will have on the supply of CO_2 into the chloroplast and on metabolic pathways downstream from photosynthesis. Most metabolic pathways are highly temperature-dependent over short time periods, and over longer periods thermal acclimation will exhibit varying responses for each pathway. For example, a prolonged increase in temperature can result in thermal acclimation of photosynthesis in order to maximise photosynthetic rates for a given temperature. Consequently, optimum temperatures of photosynthesis may be higher in plants acclimated to higher growth temperatures (Kirschbaum, 2004). Respiration, however, can increase and remain high even after acclimation, which results in decreased biomass accumulation. Thus, inferring the response of temperature on photosynthesis may not be sufficient to accurately predict productivity.

34.2.5. Extreme events

Extreme events can represent a wide range of scenarios, including meteorological or climatic conditions or fluctuations in reactive gases that can have large-scale detrimental effects on vegetation. Although plants are well adapted to changes in their environment, events that cause conditions to extend beyond those in which plants can cope can lead to damage or death (McDowell, 2011). Drought is considered a climate extreme that causes large-scale destruction (Dale *et al.*, 2001; Xu and Baldocchi, 2003; Kunkel *et al.*, 2006), but moisture excess can also result in large-scale declines in productivity, particularly for crops (Rosenzweig *et al.*, 2002). In addition to extreme fluctuations in precipitation for a given area, there are implications for temperature extremes on vegetation. For example, increasing global mean temperature is increasing growing-season length over many areas. Unseasonably warm temperatures may result in the initiation of plant growth earlier in the season that makes plants more prone to damage induced by early frosts. An example of this is reported for the Eastern U.S. in 2007 in which warm early season temperatures followed by a late-spring severe frost caused large-scale damage to a number of ecosystems in the region (Gu *et al.*, 2008).

Direct impacts of extreme events on photosynthesis are highly variable. For example, as described earlier maize showed higher photosynthetic rates when grown in elevated CO_2 but only when drought conditions were present (Leakey *et al.*, 2004). Drought conditions result in a loss of transpiratory cooling of leaf surfaces that, combined with warmer temperatures, can induce high-temperature stress more rapidly in future environments. Further, elevated CO_2 is also shown to decrease g_s having a net warming effect on plant canopies (e.g., Bernacchi *et al.*, 2007) that might potentially lead to higher frequencies of lethal leaf temperatures. Despite the range of potential scenarios, few studies have demonstrated the impact of extreme events in the context of combined global-change scenarios.

In addition to the damage imposed on vegetation health, extreme events can impact ecosystem services. For example, it is well documented that numerous ecosystems across the globe are acting as carbon sinks, which help to mitigate some of the CO_2 emissions to the atmosphere every year (Cox *et al.*, 2000; Dufresne *et al.*, 2002). However, a heatwave that affected Europe over the summer of 2003 was associated with the most important productivity crisis based on over a century of modelled global net primary production. It caused a 30% reduction of GPP, becoming a 0.5 Pg C net source of CO_2, which is equivalent to four years of carbon sink by the terrestrial ecosystems of Europe (Fig. 34.3; Ciais *et al.*, 2005). Thus, although extreme events can happen over relatively short time periods, the effects can be far reaching and long lasting.

34.3. THE INTERACTIVE EFFECTS OF GLOBAL-CHANGE SCENARIOS

Although many chapters of this book focus on the responses of photosynthesis to single factors that are relevant to global-change scenarios, we focus here on the interactive effects of these factors on photosynthesis. Ecosystems indeed experience a combination of numerous environmental changes. Studies that address only single responses are useful in determining the underlying physiological responses of vegetation to these global-change scenarios. This information is then used to feed data into models that can incorporate interactive influences of simultaneous environmental changes. However, plant responses are often confounded by multiple environmental factors. For example, the impact of rising O_3 on photosynthesis might differ based on whether there is a concomitant increase in CO_2.

A challenge exists in determining plant responses to interacting factors for many reasons. There is a great deal of uncertainty with regard to the amount of change that is expected for each direct effect. Thus, projecting CO_2 concentrations for 50 years into the future and determining the temperature that will correspond with this CO_2 concentration is highly uncertain. The addition of O_3 or water

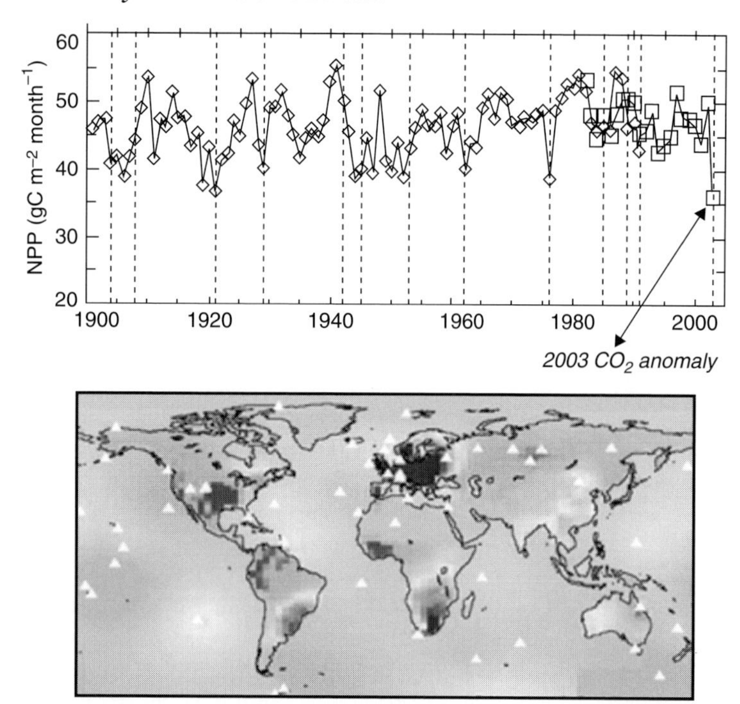

Fig. 34.3. Mean monthly net primary production from 1901 to 2002 and for the drought year, 2003. Data collected from modelled (Mitchell *et al.*, 2004) and measured data (Nemani *et al.*, 2003) for the land surfaces of Italy and France.

availability can further add to uncertainty. Applying numerous interacting treatments to vegetation in a controlled manner is difficult, expensive and often impractical. Various methods for measuring plant responses to their environment have been utilised, including growth chambers, SPAR chambers, open-top chambers, FACE rings, etc., but designing experiments that test two, three or four factors simultaneously result in increasing chambers or plots.

Despite these problems many studies have employed a range of techniques to study interactive effects of global change on photosynthesis. The most commonly studied interactions are presented in more detail below.

34.3.1. The interactive effects of CO_2 and O_3 on photosynthesis

Elevated CO_2 increases the rate of photosynthesis, whereas increases in O_3 result in damage to the photosynthetic apparatus (e.g., Chapter 17). It was originally hypothesised that when these two gases are provided together, a CO_2-induced reduction in g_s would result in less opportunity for O_3 to diffuse into the intercellular air spaces (McKee *et al.*, 1995). This hypothesis led to the prediction that elevated CO_2 would provide some protection against the damaging effects of O_3 on photosynthetic tissues. Numerous experiments have provided support for this hypothesis, showing that in elevated CO_2 lower g_s allows for less diffusion of O_3 into the leaf (McKee *et al.*, 1995, 1997; Fiscus *et al.*, 1997; Morgan *et al.*, 2003). Based on meta-analytic studies of the responses of photosynthesis to combined increases in CO_2 and O_3, the presence of O_3 alone resulted in over a threefold decrease in photosynthetic rates relative to the combination of CO_2 and O_3 for soybean (Morgan *et al.*, 2003). A meta-analytic study focusing on trees also showed an approximate 20% reduction in saturated photosynthesis (A_{SAT}) when exposed to a mean O_3 concentration of 87 ppb, whereas the combined CO_2 and O_3 treatments (mean O_3 concentration of 84 ppb) showed no statistically significant changes in A_{SAT} (Wittig *et al.*, 2007). Although the evidence suggests that CO_2 confers protection from elevated O_3 on photosynthesis, such protective responses can be misleading. While CO_2 protects against O_3, photosynthetic rates in the combined treatment are lower than for elevated CO_2 alone (Ainsworth *et al.*, 2002; Morgan *et al.*, 2003; Ainsworth, 2008).

34.3.2. The interactive effects of CO_2 and temperature on photosynthesis

The implications of increased temperature concomitant with rising CO_2 suggest, based on theoretical modelling, that photosynthetic rates will increase more than can be attributed to CO_2 itself (Long, 1991). When the photosynthetic model of leaf photosynthesis is considered, net carbon assimilation is a function of three separate biochemical pathways: the gross uptake of CO_2 via photosynthesis; the release of CO_2 via photorespiration; and the release of CO_2 via mitochondrial respiration (see Chapter 4). The synergistic influence of increases in CO_2 and temperature is driven by an increase in the amount of CO_2 as a substrate for photosynthesis, resulting in a higher rate of gross photosynthesis and the suppression of photorespiratory CO_2 release at higher temperatures resulting from fewer oxygenation events. Higher concentrations of CO_2 increase the thermal optimum of photosynthesis and, provided that photosynthesis is operating below the thermal optimum, the higher temperatures will result in higher photosynthetic rates that would be experienced if CO_2 and thus the thermal optimum were lower (Fig. 34.4). The basic understanding of this synergism is elucidated from employing the leaf model of photosynthesis (Farquhar *et al.*, 1980a), assuming that the underlying photosynthetic physiology is not influenced by the elevated temperature (Long, 1991). Modelling results also suggest that even a 40% reduction in the amount of Rubisco would yield higher photosynthetic rates at elevated temperature and CO_2 (Long, 1991).

Experimental evidence shows that the CO_2 enhancement of photosynthesis is increasingly favoured at higher temperatures for soybean, but not for rice (Vu *et al.*, 1997). Using natural variability in climatic conditions across three growing seasons for soybean, a clear trend toward higher CO_2-induced stimulation in photosynthesis was observed with increasing daytime maximum temperatures (Bernacchi *et al.*, 2006). Research on trees shows variable responses to combined increases in CO_2 and temperature. For example, pine trees showed variable responses in enhancement of photosynthesis to elevated CO_2 and temperature relative to controls (Teskey, 1997). While species that employ the C_3 photosynthetic pathway show a variety of responses to combined CO_2 and temperature, it is likely that a combination of various factors cause the observed variability in results, including downregulation of photosynthetic machinery, stomatal responses, metabolic activities downstream from photosynthesis, canopy architecture and coupling between air and leaf temperatures and a range of other factors.

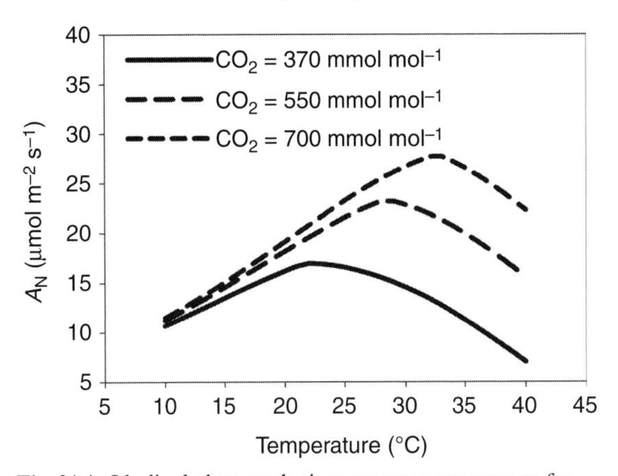

Fig. 34.4. Idealised photosynthetic response to temperature for three CO_2 concentrations using the leaf photosynthesis model of Farquhar *et al.* (1980). For all three scenarios, the parameters included in the model were identical (see Chapter 8) and the temperature responses of the parameters were based on Bernacchi *et al.* (2001, 2003).

34.3.3. Modelling plant responses to interactive effects

The reality of global change is that more than two factors that influence photosynthesis will occur. In the above sections, analyses were limited to two main interacting effects owing to the challenges of manipulating three or more global-change scenarios. Given the challenges, the scientific community must rely heavily on models. The mechanistic model of leaf photosynthesis (Farquhar *et al.*, 1980a) can be and has been used to predict responses to a wide range of environmental conditions (e.g., Long, 1991). However, this model is only useful to assess the responses to global-change scenarios if the changes in underlying photosynthetic physiology are well understood. For example, the maximum velocity for carboxylation ($V_{c,max}$, see Chapter 8) at 25°C is required for parameterisation of the model. There is evidence for some species that $V_{c,max}$ changes with growth in elevated CO_2. Thus, without knowledge of how $V_{c,max}$ responds to specific growth conditions, it is difficult to model photosynthesis for these conditions. Although a large number of studies report values of $V_{c,max}$ under single environmental changes or to two changes simultaneously, little data is available to assess how changes in three conditions, e.g., simultaneous increases CO_2, O_3 and temperature, might influence $V_{c,max}$.

Scaling from the leaf to the plant, canopy or ecosystem results in increasing complexity and modelling at higher scales adds to the uncertainty in modelling. While controlled experiments can be used to assess leaf and whole-plant photosynthetic responses to global change, the observed responses to these treatments at the smaller scales do not often scale to the canopy or ecosystem. Smaller-scale models are also unable to describe the interactions among various different species within an ecosystem. Thus ecosystem models are required to predict the possible influences of interacting global-change effects on plant productivity. For example, four separate ecosystem models were employed to address the impacts on productivity for numerous different ecosystems to varying global-change interactions (Luo *et al.*, 2008). While all four models were fairly consistent in estimation of main effects, e.g., increased CO_2 alone, the authors reported greater variability and inconsistency among the models when interacting effects were modelled. The results of modelling studies that focus on the interactions of numerous global-change scenarios are useful to identify weaknesses in mechanistic understanding of how plants might respond to interacting effects. Models might help investigators to fill in the knowledge gaps, but the model uncertainty is too great for using these tools to predict future ecosystem responses.

34.4. POTENTIAL FEEDBACKS THAT PHOTOSYNTHESIS CAN HAVE ON THE CARBON CYCLE

There is little doubt that if the global-change scenarios presented in the last report of the Intergovernmental Panel on Climate Change (IPCC, 2007) are realised, then further disruptions in ecosystem function will occur in coming decades. Arguably, the increases in CO_2 and temperature are going to have the most profound impact on ecosystem functioning and these two global-change factors are not mutually exclusive. While this book addresses photosynthetic responses to global-change scenarios, there is a growing body of research that addresses the potential for vegetation to mitigate many of the changes occurring (e.g., Canadell and Raupach, 2008). Although it is widely accepted that plants cannot reverse the trend toward increasing CO_2 and warmer environments – only improvements in fuel efficiency and alternatives to fossil fuels can accomplish this – there is strong potential for plants to provide offsets to the rise in CO_2. This section will address three main research areas that use vegetation to: (1) sequester carbon; (2) provide renewable energy sources; and (3) to use vegetation for purposes of geological engineering. It is important to note that these three research areas contain some degree of overlap.

34.4.1. Using photosynthesis to sequester CO_2

In addition to their importance for food and shelter, plants have long been utilised as a source of energy. The combination of these major demands placed upon vegetation has resulted in food, fuel, fibre and wood industries that are not sustainable. Agricultural demand has resulted in an extensive release of CO_2 into the atmosphere. This includes carbon released through deforestation, which accounts for the second largest flux of CO_2 into the atmosphere behind fossil-fuel consumption (IPCC, 2007), and through the release of carbon from agricultural soils, which are predicted to have lost at least 50% of the organic matter relative to their natural state (Lal, 2004b). Given the amount of CO_2 emitted into the atmosphere as a result of ecosystem disturbances, questions remain whether a similar magnitude of CO_2 storage in woody tissues and soils can once again be achieved to mitigate a proportion of the increases in atmospheric CO_2.

The opportunity to sequester carbon in soils is anticipated to be quite large provided proper management strategies and proper species are utilised. Predictions suggest that up to 0.9 Pg of carbon can be sequestered per year through a combination of various management practices that include the use of cover crops, optimal nutrient management and a range of other factors (Lal, 2004c). This would offset a large portion of the net ~3 Pg of carbon that is being emitted and retained by the atmosphere each year. In addition to sequestration of carbon in soils, a large amount of carbon can be sequestered in plant tissues through reforestation, restoration of native ecosystems and the establishment of perennial agricultural species to meet the demands for increasing populations and alternative renewable fuels. Further, current estimates suggest that, owing to combinations in CO_2 fertilisation and recovery from past disturbance events, tropical forests alone are sequestering ca 1.3 Pg C yr^{-1} (Lewis *et al.*, 2009). Regardless of the technique employed to minimise the loss of carbon from soils, the input of carbon is based on photosynthetic uptake by plants. Given that natural ecosystems once contained a large amount of stored carbon in both soils and in plant tissue, the question remains whether we can utilise ecosystems' services to return carbon to a sequestered state.

34.4.2. Photosynthesis as a renewable source of energy

As mentioned earlier in this chapter, the fossil-fuel reserves are quite likely large given that when these reserves were formed, atmospheric concentrations were over 12 times their present value (Berner, 1990). Nevertheless, the ability to efficiently extract them from the ground is diminishing, resulting in more emissions associated with extraction of energy. Even with efficient extraction technologies, continued reliance on fossil fuels will have continued negative impacts on the global environment.

Photosynthesis is increasingly considered a viable alternative to reliance on fossil fuels. Both fossil fuels and biofuels are products of photosynthesis, however the latter is renewable and in theory does not result in increased emissions of greenhouse gases to the atmosphere. In reality, the agriculture industry relies heavily on fossil fuels for items ranging from diesel for tractors, combines and other farm machinery, nutrient and pesticide production, and transport and processing of the product. Thus, while the carbon emitted from combustion of alternative fuels would be recycled via photosynthesis during the next growing season or seasons, various carbon-lifecycle analyses of plant-based fuel sources estimate a wide range of energy efficiencies. Despite some concern that biofuels provide no net benefit or may consume more energy compared with using fossil fuels, most lifecycle analyses are shown to be inconsistent in their assumptions and are not based on known ecological processes (Davis *et al.*, 2009).

With regards to the traits that make a species 'ideal' for biomass production, long canopy duration, C_4 photosynthesis and high radiation-use efficiency were included among important attributes (Heaton *et al.*, 2004). Each of these three traits is directly related to photosynthesis. Long canopy duration maximises the amount of sunlight absorbed by the canopy over the growing season. Annual species are highly productive over a very short period of time but perennials, with longer growing seasons, can yield much higher rates of season-integrated photosynthesis (Fig. 34.5). While C_4 photosynthesis is listed as one of the traits suited optimally for biomass crops, it must also be considered that C_4 species are not well suited for all environments. Finally, high radiation-use efficiency is a trait that allows for the intercepted light to be utilised by photosynthesis with little being penetrated in the ground or reflected away. In a recent paper, it was estimated that maximum efficiency of C_3 photosynthesis is 4.6% and of C_4 species is 6.0% based on calculations from the total sunlight hitting a leaf to carbohydrate formation (Zhu *et al.*, 2008). Considering that total sunlight is seldom captured by a plant canopy and in most productive regions of the planet plants are growing for only a few months per year, the efficiency drops to much lower values. Thus, increasing or selecting species that exploit each of these three components can help to ensure that photosynthesis is maximised to increase growth and productivity.

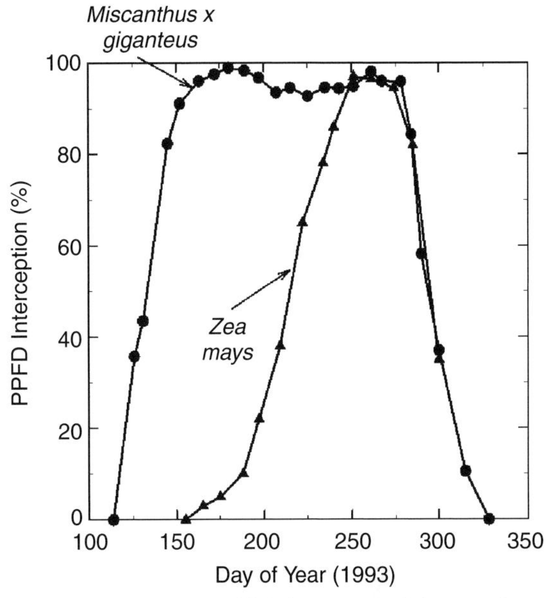

Fig. 34.5. The percentage of light intercepted relative to total available by three different plant canopies over the 1993 growing season. Data is presented for an annual species, *Zea mays*, and a perennial grass species, *Miscanthus* × *giganteus*. With the longer growing season associated with the perennials, more opportunity was available to convert sunlight into biomass via photosynthesis. Figure redrawn from U.S. DOE (2006).

References

Aalto, T. and Juurola, E. (2002). A three-dimensional model of CO_2 transport in airspaces and mesophyll cells of a silver birch leaf. *Plant, Cell and Environment*, **25**, 1399–1409.

Aasamaa, K., Sober, A. and Rahi, M. (2001). Leaf anatomical characteristics associated with hydraulic conductance, stomatal conductance and stomatal sensitivity to changes of leaf water status in temperate deciduous trees. *Australian Journal of Plant Physiology*, **28**, 765–774.

Aasamaa, K., Sõber, A., Hartung, W., *et al.* (2002). Rate of stomatal opening, shoot hydraulic conductance and photosynthesis characteristics in relation to leaf abscisic acid concentration in six temperate deciduous trees. *Tree Physiology*, **22**, 267–276.

Aasamaa, K., Sõber, A., Hartung, W., *et al.* (2004). Drought acclimation of two deciduous tree species of different layers in a temperate forest canopy. *Trees: Structure and Function*, **18**, 93–101.

Abadía, J., Morales, F. and Abadía, A. (1999). Photosystem II efficiency in low chlorophyll, iron-deficient leaves. *Plant and Soil*, **215**, 183–192.

Abadía, J., Nishio, J.N. and Terry, N. (1986). Chlorophyll-protein and polypeptide composition of Mn-deficient sugar beet thylakoids. *Photosynthesis Research*, **7**, 379–381.

Abbink, T.E.M., Peart, J.R., Mos, T.N.M., *et al.* (2002). Silencing of a gene encoding a protein component of the oxygen-evolving complex of Photosystem II enhances virus replication in plants. *Virology*, **295**, 307–319.

Abe, H., Urao, T., Ito, T., *et al.* (2003). Arabidopsis AtMYC2 (bHLH) and AtMYB2 (MYB) function as transcriptional activators in abscisic acid signaling. *Plant Cell*, **15**, 63–78.

Abeledo, J.G., Calderini, D.F. and Slafer, G.A. (2003). Genetic improvement of barley yield potential and its physiological determinants in Argentina (1944–1998). *Euphytica*, **130**, 325–334.

Abeles, F.B. (1986). Plant Chemiluminescence. *Annual Review of Plant Physiology*, **37**, 49–72.

Aber, J.D., Reich, P.B. and Goulden, M.L. (1996). Extrapolating leaf CO_2 exchange to the canopy: a generalized model of forest photosynthesis compared with measurements by eddy correlation. *Oecologia*, **106**, 257–265.

Abrams, M.D. (1996). Distribution, historical development and ecophysiological attributes of oak species in the eastern United States. *Annales des Sciences Forestieres*, **53**, 487–512.

Ache, P., Baucr, H., Kollist, H., *et al.* (2010). Stomatal action directly feeds back on leaf turgor: new insights into the regulation of the plant water status from non-invasive pressure probe measurements. *The Plant Journal*, **62**, 1072–1082.

Ackerly, D. (1997). Allocation, leaf display, and growth in fluctuating light environments. In: *Plant Resource Allocation* (eds Bazzaz, F. A. and Grace, J.). Academic Press, San Diego.

Adams, K.L. and Wendel, J.F. (2005). Polyploidy and genome evolution in plants. *Current Opinion in Plant Biology*, **8**, 135–141.

Adams, M.L., Norvell, W.A., Peverly, J.H., *et al.* (1993). Fluorescence and reflectance characteristics of manganese deficient soybean leaves: effects of leaf age and choice of leaflet. *Plant and Soil*, **155/156**, 235–238.

Adams, M.S. and Strain, B.R. (1968). Photosynthesis in stems and leaves of *Cercidium floridum*: spring and summer diurnal field response and relation to temperature. *Oecologia Plantarum*, **3**, 285–297.

Adams, P., Nelson, D.E., Yamada, S., *et al.* (1998). Growth and development of *Mesembryanthemum crystallinum* (Aizoaceae). *New Phytologist*, **138**, 171–190.

Adams, W.W. III and Demmig-Adams, B. (1994). Carotenoid composition and down regulation of photosystem II in three conifer species during the winter. *Physiologia Plantarum*, **92**, 451–458.

Adams W.W. III, Demmig-Adams, B., Rosenstiel, T.N., *et al.* (2002). Photosynthesis and photoprotection in overwintering plants. *Plant Biology*, **4**, 535–654.

Adams, W.W. III, Demmig-Adams, B., Verhoeven, A.S., *et al.* (1995). 'Photoinhibition' during winter stress: involvement of sustained xanthophyll cycle-dependent energy dissipation. *Australian Journal of Plant Physiology*, **22**, 261–276.

Adams, W.W. III, Demmig-Adams, B., Winter, K., *et al.* (1990a). The ratio of variable to maximum chlorophyll fluorescence from photosystem II, measured at room temperature and 77K, as indicator of the photon yield of photosynthesis. *Planta*, **180**, 166–174.

Adams, W.W. III, Winter, K. and Lanzl, A. (1989). Light and the maintenance of photosynthetic competence in leaves of Populus balsamifera L. during short-term exposures to high concentrations of sulfur dioxide. *Planta*, **177**, 91–97.

Adams, W.W. III, Winter, K., Schreiber, U., *et al.* (1990b). Photosynthesis and chlorophyll fluorescence characteristics in relationship to changes in pigment and element composition of leaves of *Platanus occidentalis* L. during autumnal leaf senescence. *Plant Physiology*, **93**, 1184–1190.

Affek, H.P., Krisch, M.J. and Yakir, D. (2006). Effects of intraleaf variations in carbonic anhydrase activity and gas echange on leaf $C^{18}OO$ isoflux in *Zea mays*. *New Phytologist*, **169**, 321–329.

Aganchich, B., Wahbi, S., Loreto, F., *et al.* (2009). Partial root zone drying: regulation of photosynthetic limitations and antioxidant enzymatic activities in young olive (*Olea europaea*) saplings. *Tree Physiology*, **29**, 685–696.

Agati, G., Cerovic, Z.G. and Moya, I. (2000). The effect of decreasing temperature up to chilling values on the in vivo F685/F735 chlorophyll fluorescence ratio in *Phaseolus vulgaris* and *Pisum sativum*. The role of the photosystem I contribution to the 735 nm fluorescence band. *Photochemistry and Photobiology*, **72**, 75–84.

Agati, G., Cerovic, Z.G., Dalla Marta, A., *et al.* (2008). Optically assessed preformed flavonoids and susceptibility of grapevine to *Plasmopora viticola* under different light regimes. *Functional Plant Biology*, **35**, 77–84.

Agati, G., Galardi, C., Gravano, E., *et al.* (2002). Flavonoid distribution in tissues of *Phillyrea latifolia* L. leaves as estimated by microspectrofluorometry and multispectral fluorescence microimaging. *Photochemistry and Photobiology*, **76**, 350–360.

Agati, G., Mazzinghi, P., Fusi, F., *et al.* (1995). The F685/730 chlorophyll fluorescence ratio as a tool in plant physiology: response to physiological and environmental factors. *Journal of Plant Physiology*, **145**, 228–238.

Agati, G., Meyer, S., Matteini, P., *et al.* (2007). Assessment of anthocyanins in grape (*Vitis vinifera* L.) berries using a noninvasive chlorophyll fluorescence method. *Journal of Agricultural and Food Chemistry*, **55**, 1053–1061.

Agbariah, K.T. and Roth-Bejerano, N. (1990). The effect of blue light on energy levels in epidermal strips. *Physiologia Plantarum*, **78**, 100–104.

Ågren, G.I. and Ingestad, T. (1987). Root:shoot ratio as a balance between nitrogen productivity and photosynthesis. *Plant Cell Environment*, **10**, 579–586.

Aguirreolea, J., Irigoyen, J., Sánchez-Díaz, M., *et al.* (1995). Physiological alterations in pepper during wilt induced by *Phytophthora capsici* and soil water deficit. *Plant Pathology*, **44**, 587–596.

Agustí, S., Enríquez, S., Frost-Christensen, H., *et al.* (1994). Light harvesting among photosynthetic organisms. *Functional Ecology*, **8**, 273–279.

Agustynowicz, J. and Gabrys, H. (1999). Chloroplast movements in fern leaves: correlation of movement dynamics and environmental flexibility of the species. *Plant, Cell and Environment*, **22**, 1239–1248.

Aharon, R., Shahak, Y., Wininger, S., *et al.* (2003). Overexpression of a plasma membrane aquaporin in transgenic tobacco improves plant vigor under favourable growth conditions but not under drought or salt stress. *The Plant Cell*, **15**, 439–447.

Ahn, T.K., Avenson, T.J., Ballottari, M., *et al.* (2008). Architecture of a charge-transfer state regulating light harvesting in a plant antenna protein. *Science*, **320**, 794–797.

Ainsworth, E., Davey, P., Bernacchi, C., *et al.* (2002). A meta-analysis of elevated [CO_2] effects on soybean (*Glycine max*) physiology, growth and yield. *Global Change Biology*, **8**, 695–709.

Ainsworth, E., Leakey, A., Ort, D., *et al.* (2008b). FACE-ing the facts: inconsistencies and interdependence among field, chamber and modeling studies of elevated [CO_2] impacts on crop yield and food supply. *New Phytologist*, **179**, 5–9.

Ainsworth, E.A. (2008). Rice production in a changing climate: a meta-analysis of responses to elevated carbon dioxide and elevated ozone concentration. *Global Change Biology*, **14**, 1642–1650.

Ainsworth, E.A. and Long, S.P. (2005). What have we learned from 15 years of free-air CO_2 enrichment (FACE)? A meta-analytic review of the responses of photosynthesis, canopy properties and plant production to rising CO_2. *New Phytologist*, **165**, 351–372.

Ainsworth, E.A. and Rogers, A. (2007). The response of photosynthesis and stomatal conductance to rising [CO_2]: mechanisms and environmental interactions. *Plant, Cell and Environment*, **30**, 258–270.

Ainsworth, E.A., Davey, P.A., Hymus, G.J., *et al.* (2003). Is stimulation of leaf photosynthesis by elevated carbon dioxide concentration maintained in the long term? A test with *Lolium perenne* grown for 10 years at two nitrogen fertilization levels under free air CO_2 enrichment (FACE). *Plant, Cell and Environment*, **26**, 705–714.

Ainsworth, E.A., Rogers, A. and Leakey, A.D.B. (2008a). Targets for crop biotechnology in a future high CO_2 and HighO_3 world. *Plant Physiology*, **147**, 13–19.

Ainsworth, E.A., Rogers, A. Nelson, R., et al. (2004). Testing the "source-sink" hypothesis of down-regulation of photosynthesis in elevated CO_2 in the field with single gene substitutions in *Glycine max*. *Agricultural and Forest meteorology*, **122**, 85–94.

Akhani, H., Barroca, J., Koteeva, N., *et al.* (2005). *Bienertia sinuspersici* (Chenopodiaceae): A new species from southwest Asia and discovery of a third terrestrial C_4 plant without Kranz anatomy. *Systematic Botany*, **30**, 290–301.

Akhani, H., Trimborn, P. and Ziegler, H. (1997). Photosynthetic pathways in *Chenopodiaceae* from Africa, Asia and Europe with their ecological, phytogeographical and taxonomical importance. *Plant Systematics and Evolution*, **206**, 187–221.

Al-Abbas, A.H., Barr, R., Hall, J.D., *et al.* (1974). Spectra of normal and nutrient deficient maize leaves. *Agronomy Journal*, **66**, 16–20.

Alboresi, A., Gerotto, C., Giacometti, G.M., *et al.* (2010). *Physcomitrella patens* mutants affected on heat dissipation clarify the evolution of photoprotection mechanisms upon land colonization. *Proceedings of the National Academy of Sciences USA*, **107**, 11128–11133.

Aldea, M., Frank, T.D. and DeLucia, E.H. (2006a). A method for quantitative analysis of spatially variable physiological processes across leaf surfaces. *Photosynthesis Research*, **90**, 161–172.

Aldea, M., Hamilton, J.G., Resti, J.P., *et al.* (2005). Indirect effects of insect herbivory on leaf gas exchange in soybean. *Plant, Cell and Environment*, **28**, 402–411.

Aldea, M., Hamilton, J.G., Resti, J.P., *et al.* (2006b). Comparison of photosynthetic damage from arthropod herbivory and pathogen infection in understory hardwood saplings. *Oecologia*, **149**, 221–232.

Al-Hazmi, M., Lakso, A.N. and Denning, S.S. (1997). Whole canopy versus single leaf gas exchange responses to water stress in Cabernet Sauvignon grapevines. In: *Proceedings of the IVth International Symposium on Cool Climate Viticulture and Enology* (eds Edson, C., Wolf, T., Pool, R., Reynolds, A., Henick-Kling, T., Acree, T., Reisch, B. and Harkness, E.), Eastern Section, Am. Soc. Enol. Vitic., Geneva, NY, USA, pp. II.47-II.48.

Ali, B., Hasan, S.A., Hayat, S., *et al.* (2008). A role for brassinosteroids in the amelioration of aluminium stress through antioxidant system in mung bean (*Vigna radiata* L. Wilczek). *Environmental and Experimental Botany*, **62**, 153–159.

Allard, V., Ourcival, J.M., Rambal, S., *et al.* (2008). Seasonal and annual variation of carbon exchange in an evergreen Mediterranean forest in southern France. *Global Change Biology*, **14**, 1–12.

Allen, D.J. and Ort, D.R. (2001). Impacts of chilling temperatures on photosynthesis in warm-climate plants. *Trends Plant Sciences*, **6**, 36–42.

Allen, D.J., Ratner, K., Giller, Y.E., *et al.* (2000). An overnight chill induces a delayed inhibition of photosynthesis at midday in mango (*Mangifera indica* L.). *Journal of Experimental Botany*, **51**, 1893–1902.

Allen, D.K., Shachar-Hill, Y. and Ohlrogge, J.B. (2007). Compartment-specific labeling information in ^{13}C metabolic flux analysis of plants. *Phytochemistry*, **68**, 2197–2210.

Allen, J.F. (1992). How does protein phosphorylation regulate photosynthesis? *Trends in Biochemical Sciences*, **17**, 12–17.

Allen, J.F. (2003). Cyclic, pseudocyclic and noncyclic photophosphorylation: new links in the chain. *Trends Plant Sciences*, **8**, 15–19.

Allen, J.F. and Forsberg, J. (2001). Molecular recognition in thylakoid structure and function. *Trends Plant Sciences*, **6**, 317–326.

Allen, J.F. and Nilsson, A. (1997). Redox signalling and the structural basis of regulation of photosynthesis by protein phosphorylation. *Physiologia Plantarum*, **100**, 863–868.

Allen, L.H., Pan, D., Boote, K.J., *et al.* (2003). Carbon dioxide and temperature effects on evapotranspiration and water use efficiency of soybean. *Agronomy Journal*, **95**, 1071–1081.

Allen, M.F. (1991). *The Ecology of Mycorrhizae*. Cambridge University Press, Cambridge, UK.

Almási, A., Harsányi, A. and Gáborjányi, R. (2001). Photosynthetic alterations on virus infected plants. *Acta Phytopathologica et Entomologica Hungarica*, **36**, 15–29.

Alo, C.A. and Wang, G.L. (2008). Potential future changes of the terrestrial ecosystem based on climate projections by eight general circulation models. *Journal of Geophysical Research – Biogeosciences*, **113**, Article number G01004.

Alongi, D.M., Ayukai, T., Brunskill, G.J., *et al.* (1998). Sources, sinks and export of organic carbon through a tropical, semi-enclosed delta (Hinchinbrook Channel, Australia). *Mangrove and Salt Marshes*, **2**, 237–242.

Alonso, A.A. and Machado, S.R. (2007). Morphological and developmental investigations of the underground system of *Erythroxylum* species from Brazilian cerrado. *Australian Journal of Botany*, **55**, 749–758.

Alterio, G., Giorio, P. and Sorrentino, G. (2006). Open-system chamber for measurements of gas exchanges at plant level. *Environmental Science and Technology*, **40**, 1950–1955.

Altesor, A., Ezcurra, E. and Silva, C. (1992). Changes in the photosynthetic metabolism during early ontogeny of four cactus species. *Acta Oecologica*, **13**, 777–785.

Althawadi, A.M. and Grace, J. (1986). Water use by the desert cucurbit *Citrullus colocynthis* (L.) Schrad. *Oecologia*, **70**, 475–480.

Amthor, J.S. (1991). Respiration in a future, higher-CO_2 world. *Plant Cell and Environment*, **14**, 13–20.

Amthor, J.S. (1994). Scaling CO_2-photosynthesis relationships from the leaf to the canopy. *Photosynthesis Research*, **39**, 321–350.

Amthor, J.S. (2000). The McCree–de Wit–Penning de Vries–Thornley respiration paradigms: 30 years later. *Annals of Botany*, **86**, 1–20.

Amthor, J.S. (2007). Improving photosynthesis and yield potential. In: *Improvement of Crop Plants for Industrial End Uses*, (ed. Ramalli, P.), Springer, NY, USA, pp. 27–58.

Amthor, J.S., Goulden, M.L., Munger, J.W., *et al.* (1994). Testing a mechanistic model of forest-canopy mass and energy exchange using eddy correlation: carbon dioxide and ozone uptake by a mixed oak-maple stand. *Australian Journal of Plant Physiology*, **21**, 623–651.

Ananyev, G.S., Kolber, Z.S., Klimov, D., *et al.* (2005). Remote sensing of heterogeneity in photosynthetic efficiency, electron transport and dissipation of excess light in *Populus deltoides* stands under ambient and elevated CO_2 concentrations, and in a tropical forest canopy, using a new laser-induced fluorescence transient device. *Global Change Biology*, **11**, 1195–1206.

Anderson, D.E. and Verma, S.B. (1986). Carbon dioxide, water vapor and sensible heat exchanges of a grain sorghum canopy. *Boundary Layer Meteorology*, **34**, 317–331.

Anderson, J.M., Chow, W.S., and Goodchild, D.J. (1988). Thylakoid membrane organization in sun shade acclimation. *Australian Journal of Plant Physiology*, **15**, 11–26.

Anderson, L.J., Maherali, H., Johnson, H.B., *et al.* (2001). Gas exchange and photosynthetic acclimation over subambient to elevated CO_2 in a C_3-C_4 grassland. *Global Change Biology*, **7**, 693–707.

Andersson, A., Keskitalo, J., Sjödin, A., *et al.* (2004). A transcriptional timetable of autumn senescence. *Genome Biology*, **5**, R24.

Andersson, B. and Barber, J. (1996). Mechanisms of photodamage and protein degradation during photoinhibition of photosystem II. In: *Advances in Photosynthesis Vol.5: Photosynthesis and the Environment* (ed. Baker, N.R.), Kluwer Academic Publishers, Dordrecht, Netherlands, pp. 101–121.

Andrade, J.L. and Nobel, P.S. (1996). Habitat, CO_2 uptake and growth for the CAM epiphytic cactus *Epiphyllum phyllanthus* in a Panamanian tropical forest. *Journal of Tropical Ecology*, **12**, 291–306.

Andralojc, P.J., Keys, A.J., Kossmann, J., *et al.* (2002). Elucidating the biosynthesis of 2-carboxyarabinitol 1-phosphate through reduced expression of chloroplastic fructose 1,6-bisphosphate phosphatase and radiotracer studies with $^{14}CO_2$. *Proceedings of National Academy of Sciences of USA*, **99**, 4742–4747.

Andralojc, P.J., Keys, A.J., Martindale, W., *et al.* (1996). Conversion of Δ-hamamelose into 2-carboxy D-arabinitol and 2-carboxy D-arabinitol 1-phosphate in leaves of *Phaseolus vulgaris* L. *Journal of Biological Chemistry*, **271**, 26803–26809.

Andrews, J.R., Bredenkamp, G.J. and Baker, N.R. (1993). Evaluation of the role of State transitions in determining the efficiency of light utilisation for CO_2 assimilation in leaves. *Photosynthesis Research*, **38**, 15–26.

Andrews, T.J. and Muller, G.J. (1985). Photosynthetic gas exchange of the mangrove, *Rhizophora stylosa*, in its natural environment. *Oecologia*, **65**, 449–455.

Angelopulos, K., Dichio, B. and Xiloyannis, C. (1996). Inhibition of photosynthesis in olive trees (*Olea europaea* L.) during water stress and rewatering. *Journal of Experimental Botany*, **47**, 1093–1100.

Ankele, E., Kindgren, P., Pesquet, E., *et al.* (2007). In vivo visualisation of Mg-protoporphyrin IX, a coordinator of photosynthetic gene expression in the nucleus and the chloroplast. *The Plant Cell*, **19**, 1964–1979.

Antlfinger, A.E. and Wendel, L.F. (1997). Reproductive effort and floral photosynthesis in *Spiranthes cernua* (Orchidaceae). *American Journal of Botany*, **84**, 769–780.

Apel, K. and Hirt, H. (2004). Reactive oxygen species: Metabolism, oxidative stress and signal transduction. *Annual Reviews of Plant Biology*, **55**, 373–399.

Aphalo, P.J. and Jarvis, P.G. (1991). Do stomata respond to relative humidity. *Plant, Cell and Environment*, **14**, 127–132.

Aphalo, P.J. and Jarvis, P.G. (1993). An analysis of Ball's empirical model of stomatal conductance. *Annals of Botany*, **72**, 321–327.

Apostol, S., Szalai, G., Sujbert, L., *et al.* (2006). Non-invasive monitoring of the light-induced cyclic photosynthetic electron flow during cold hardening in wheat leaves. *Zeitschrift für Naturforschung*, **61c**, 734–740.

Aranda, I., Pardo, F., Gil, L., *et al.* (2004). Anatomical basis of the change in leaf mass per area and nitrogen investment with relative irradiance within the canopy of eight temperate tree species. *Acta Oecologica*, **25**, 187–195.

Aranda, I., Pardos, M., Puértolas, J., *et al.* (2007). Water-use efficiency in cork oak (*Quercus suber*) is modified by the interaction of water and light availability. *Tree Physiology*, **27**, 671–677.

Araujo, W.L., Dias, P.C., Moraes, G.A.B.K., *et al.* (2008). Limitations to photosynthesis in coffee leaves from different canopy positions. *Plant Physiology and Biochemistry*, **46**, 884–890.

Araus, J.L. (2004). The problems of sustainable water use in the Mediterranean and research requirements agriculture. *Annals of Applied Bio*logy, **144**, 259–272.

Araus, J.L., Bort, J., Steduto, P., *et al.* (2003). Breeding cereals for Mediterranean conditions: ecophysiological clues for biotechnology application. *Annals of Applied Biology*, **142**, 129–141.

Araus, J.L., Brown, H.R., Febrero, A., *et al.* (1993). Ear photosynthesis, carbon isotope discrimination and the contribution of respiratory CO_2 to differences in grain mass in Durum Wheat. *Plant, Cell and Environment*, **16**, 383–392.

Araus, J.L., Sánchez, C. and Cabrera-Bosquet, LL. (2010). Is heterosis in maize mediated through better water use? *New Phytologist*, **187**, 392–406.

Araus, J.L., Slafer, G.A., Royo, C., *et al.* (2008). Breeding for yield potential and stress adaptation in cereals. *Critical Reviews in Plant Science*, **27**, 377–412.

Arbaugh, M., Bytnerowicz, A., Grulke, N., *et al.* (2003). Photochemical smog effects in mixed conifer forests along a natural gradient of ozone and nitrogen deposition in the San Bernardino Mountains. *Environment International*, **29**, 401–406.

Arbona, V., López-Climent, M.F., Pérez-Clemente, R.M., *et al.* (2009). Maintenance of a high photosynthetic performance is linked to flooding tolerance in citrus. *Environmental and Experimental Botany*, **66**, 135–142.

Archetti, M. (2009). Classification of hypotheses on the evolution of autumn colours. *Oikos*, **118**, 328–333.

Archetti, M., Doring, T.F., Hagen, S.B., *et al.* (2009). Unravelling the evolution of autumn colours: an interdisciplinary approach. *Trends Ecology Evolution*, **24**, 166–173.

Ares, A., Fownes, J.H. and Sun, W. (2000). Genetic differentiation of intrinsic water-use efficiency in the Hawaiian native *Acacia koa*. *International Journal of Plant Sciences*, **161**, 909–915.

Ariana, D., Guyer, D.E. and Shrestha, B. (2006). Integrating multispectral reflectance and fluorescence imaging for defect detection on apples. *Computers and Electronics in Agriculture*, **50**, 148–161.

Arimura, G.I., Ozawa, R., Nishioka, T., *et al.* (2002). Herbivore-induced volatiles induce the emission of ethylene in neigbouring lima bean plants. *Plant Journal*, **29**, 87–98.

Armond, P.A., Björkman, O. and Staehelin, H.A. (1980) Dissociation of supramolecular complexes in chloroplast membranes. A manifestation of heat damage to the photosynthetic apparatus. *Biochimica et Biophysica Acta*, **601**, 433–442.

Armstrong, J.K., Williams, K., Huenneke, L.F., *et al.* (1988). Topographic position effects on growth depression of California Sierra Nevada pines during the 1982–83 El Niño. *Arctic and Alpine Research*, **20**, 352–357.

Armstrong, W. and Armstrong, J. (2005). Stem photosynthesis not pressurized ventilation is responsible for light-enhanced oxygen supply to submerged roots of alder (*Alnus glutinosa*). *Annals of Botany*, **96**, 591–612.

Arneth, A., Niinemets, Ü. and Pressley, S. (2007). Process-based estimates of terrestrial ecosystem isoprene emissions: incorporating the effects of a direct CO_2-isoprene interaction. *Atmospheric Chemistry and Physics*, **7**, 31–53.

Arnon, D.I. (1949). Copper enzymes in isolated chloroplasts. Polyphenoloxidase in *Beta vulgaris*. *Plant Physiology*, **24**, 1–15.

Arnon, D.I. (1959). Conversion of light into chemical energy in photosynthesis. *Nature*, **184**, 10–21.

Aro, E.M., Virgin, I. and Andersson, B. (1993). Photoinhibition of photosystem II. Inactivation, protein damage and turnover. *Biochicima et Biophysica Acta. (Bioenegetics)*, **1143**, 113–134.

Aronne, G. and De Micco, V. (2001). Seasonal dimorphism in the Mediterranean *Cistus incanus* L. subsp. *incanus*. *Annals of Botany*, **87**, 789–794.

Arp, W.J. (1991). Effects of source-sink relations on photosynthetic acclimation to elevated CO_2. *Plant Cell and Environment*, **14**, 869–875.

Arquero, O., Barranco, D. and Benlloch, M. (2006). Potassium starvation increases stomatal conductance in olive trees. *Hortscience*, **41**, 433–436.

Arulanantham, A., Rao, I. and Terry, N. (1990). Limiting factors in photosynthesis. VI. Regeneration of ribulose 1, 5-bisphosphate limits photosynthesis at low photochemical capacity. *Plant Physiology*, **93**, 1466–1475.

Asada, K. (1996). Radical production and scavenging in the chloroplasts. In: *Advances in Photosynthesis Vol.5: Photosynthesis and the Environment* (ed. Baker, N.R.), Kluwer Academic Publishers, Dordrecht, Netherlands, pp. 123–150.

Asada, K. (1999). The water-water cycle in chloroplasts: scavenging of active oxygen and dissipation of excess photons. *Annual Review of Plant Physiology and Plant Molecular Biology*, **50**, 601–639.

Asada, K. (2000). The water–water cycle as alternative photon and electron sinks. *Philosophical Transactions of the Royal Society of London B*, **355**, 11419–1432.

Aschan, G. and Pfanz, H. (2003). Non-foliar photosynthesis: a strategy of additional carbon acquisition. *Flora*, **198**, 81–97.

Aschan, G., Pfanz, H., Vodnik, D., *et al.* (2005). Photosynthetic performance of vegetative and reproductive structures of green hellebore (*Helleborus viridis* L. agg.). *Photosynthetica*, **43**, 55–64.

Ashley, D.A. and Boerma, H.R. (1989). Canopy photosynthesis and its association with seed yield in advanced generations of a soybean cross. *Crop Science*, **29**, 1042–1045.

Ashmore, M.R. (2002). Effects of oxidants at the whole plant and community level. In: *Air pollution and Plant Life* 2nd edn (eds Bell, J.N.B. and Treshow, M.), John Wiley and Sons, Chichester, USA, pp. 89–118.

Ashmore, M.R. (2005). Assessing the future global impacts of ozone on vegetation. *Plant, Cell and Environment*, **28**, 949–964.

Asmus, G.L. and Ferraz, L.C.C.B. (2002). Effect of population densities of *Heterodera glycines* race 3 on leaf area, photosynthesis and yield of soybean. *Fitopatologia Brasileira*, **27**, 273–278.

Asner, G.P. and Wessman, C.A. (1997). Scaling PAR absorption from the leaf to landscape level in spatially heterogeneous ecosystems. *Ecological Modelling*, **103**, 81–97.

Asner, G.P., Wessman, C.A., Bateson, A.C., *et al.* (2000). Impact of tissue, canopy, and landscape factors on the hyperspectral reflectance variability of arid ecosystems. *Remote Sensing of Environment*, **74**, 69–84.

Asokanthan, P., Johnson, R.W., Griffith, M., *et al.* (1997). The photosynthetic potential of canola embryos. *Physiologia Plantarum*, **101**, 353–360.

Athanasiou, K., Dyson, B.C., Webster, R.E. *et al.* (2010). Dynamic acclimation of photosynthesis increases plant fitness in changing environments. *Plant Physiology*, **152**, 366–373.

Atkin, O.K. and Macherel, D. (2009). The crucial role of plant mitochondria in orchestrating drought tolerance. *Annals of Botany*, **103**, 581–597.

Atkin, O.K. and Tjoelker, M.G. (2003). Thermal acclimation and the dynamic response of plant respiration to temperature. *Trends Plant Science*, **8**, 343–351.

Atkin, O.K., Botman, B. and Lambers, H. (1996). The causes of inherently slow growth in alpine plants: an analysis based on the underlying carbon economies of alpine and lowland *Poa* species. *Functional Ecology*, **10**, 698–707.

Atkin, O.K., Evans, J.R. and Siebke, K. (1998). Relationship between the inhibition of leaf respiration by light and enhancement of leaf dark respiration following light treatment. *Australian Journal of Plant Physiology*, **25**, 437–443.

Atkin, O.K., Scheurwater, I. and Pons, T.L. (2006). High thermal acclimation potential of both photosynthesis and respiration in two lowland *Plantago* species in contrast to an alpine congeneric. *Global Change Biology*, **12**, 500–515.

Atkin, O.K., Scheurwater, I. and Pons, T.L. (2007). Respiration as a percentage of daily photosynthesis in whole plants is homeostatic at moderate, but not high, growth temperatures. *New Phytologist*, **174**, 367–380.

Atkin, O.K., Evans, J.R., Ball, M.C., *et al.* (2000b). Leaf respiration of snow gum in the light and dark. Interactions between temperature and irradiance. *Plant Physiology*, **122**, 915–923.

Atkin, O.K., Millar, A.H., Gärdestrom, P., *et al.* (2000a). Photosynthesis, carbohydrate metabolism and respiration in leaves of higher plants. In: *Photosynthesis, Physiology and Metabolism* (eds Leegood, R. C., Sharkey, T. D. and von Caemmerer, S.) Kluwer Academic Publisher, London.

Atkins, C.A., Kuo, J., Pate, J.S., *et al.* (1977). Photosynthetic pod wall of pea (*Pisum sativum* L.). Distribution of carbon dioxide-fixing enzymes in relation to pod structure. *Plant Physiology*, **60**, 779–786.

Attiwill, P.M. and Adams, M.A. (1993). Nutrient cycling in forests. *New Phytologist*, **124**, 561–582.

Attiwill, P.M. and Clough, B.F. (1980). Carbon dioxide and water vapor exchange in the white mangrove (*Avicennia marina*). *Photosynthetica*, **14**, 40–47.

Au, S.F. (1969). Internal leaf surface and stomatal abundance in arctic and alpine populations of *Oxyria digyna*. *Ecology*, **50**, 131–134.

Aubert, S., Assard, N., Boutin, J.P., *et al.* (1999). Carbon metabolism in the subantarctic Kerguelen cabbage *Pringlea antiscorbutica* R.Br.: environmental controls over carbohydrates and proline contents and relation to phenology. *Plant Cell and Environment*, **22**, 243–254.

Aubert, S., Choler, P., Pratt, J., *et al.* (2004). Methyl-beta-D-glucopyranoside in higher plants: accumulation and intracellular localization in *Geum montanum* L. leaves and in model systems studied by ^{13}C nuclear magnetic resonance. *Journal of Experimental Botany*, **55**, 2179–2189.

Aubinet, M., Heinesch, B. and Yernaux, M. (2003). Horizontal and vertical CO_2 advection in a sloping forest. *Boundary Layer Meteorology*, **108**, 397–417.

Aubinet, M., Berbigier, P., Bernhofer, Ch., *et al.* (2005). Comparing CO_2 storage and advection conditions at night at different Carboeuroflux sites. *Boundary Layer Meteorology*, **116**, 63–94.

Aubinet, M., Grelle, A., Ibrom, A., *et al.* (2000). Estimates of the annual net carbon and water exchange of forests: the EUROFLUX methodology. *Advances in Ecological Research*, **30**, 113–175.

Augspurger, C.K. and Bartlett, E.A. (2003). Differences in leaf phenology between juvenile and adult trees in a temperate deciduous forest. *Tree Physiology*, **23**, 517–525.

Augspurger, C.K., Cheeseman, J.M. and Salk, C.F. (2005). Light gains and physiological capacity of understory woody plants during physiological avoidance of canopy shade. *Functional Ecology*, **19**, 537–546.

Augusti, A. and Schleucher, J. (2007). The ins and outs of stable isotopes in plants. *New Phytologist*, **174**, 473–475.

Austin II, J. and Webber, A.N. (2005). Photosynthesis in *Arabidopsis thaliana* mutants with reduced chloroplast number. *Photosynthesis Research*, **85**, 373–384.

Avenson, T.J., Cruz, J.A., Kanazawa, A., *et al.* (2005a). Regulating the proton budget of higher plant photosynthesis. *Proceedings of the National Academy of Sciences* (USA), **102**, 9709–9713.

Avenson, T.J., Kanazawa, A., Cruz, J.A., *et al.* (2005b). Integrating the proton circuit into photosynthesis: progress and challenges. *Plant, Cell and Environment*, **28**, 97–109.

Awramik, S.M. (1992). The oldest records of photosynthesis. *Photosynthesis Research*, **33**, 75–89.

Axelrod, D.I. (1966). Origin of deciduous and evergreen habits in temperate forests. *Evolution*, **20**, 1–15.

Baas, W.J. (1989). Secondary plant compounds, their ecological significance and consequences for the carbon budget. In: *Causes and Consequences of Variation in Growth Rate and Productivity of Higher Plants* (ed. Lambers, H.), SPB Academic Publishing, The Hague, pp. 313–340.

Bacelar, E.A., Mountibho-Pereira, J.M., Gonçalves, B.C., *et al.* (2007). Changes in growth, gas exchange, xylem hydraulic properties and water use efficiency of three olive cultivars under contrasting water availability regimes. *Environmental and Experimental Botany*, **60**, 183–192.

Bachereau, F., Marigo, G. and Asta, J. (1998). Effect of solar radiation (UV and visible) at high altitude on CAM-cycling and phenolic compound biosynthesis in *Sedum album*. *Physiologia Plantarum*, **104**, 203–210.

Bachmann, A., Fernándes-López, J., Ginsburg, J., *et al.* (1994). Stay green genotypes of *Phaseolus vulgaris* L.: chloroplast proteins and chlorophyll catabolites during foliar senescence. *New Phytologist*, **126**, 593–600.

Backhausen, J., Kitzmann, C., Horton, P., *et al.* (2000). Electron acceptors in isolated intact spinach chloroplasts act hierarchically to prevent over-reduction and competition for electrons. *Photosynthesis Research*, **64**, 1–13.

Badeck, F.W. (1995). Intra-leaf gradient of assimilation rate and optimal allocation of canopy nitrogen: a model on the implications of the use of homogeneous assimilation functions. *Australian Journal of Plant Physiology*, **22**, 425–439.

Badger, M.R. and Collatz, G.J. (1977). Studies on the kinetic mechanism of ribulose-1,5-bisphosphate carboxylase and oxygenase reactions, with particular reference to the effect of temperature on kinetic parameters. *Carnegie Institution of Washington Year Book*, **76**, 355–361.

Badger, M.R., Björkman, O. and Armond, P.A. (1982). An analysis of photosynthetic response and adaptation to temperature in higher plants: temperature acclimation in the desert evergreen *Nerium oleander* L. *Plant, Cell and Environment*, **5**, 85–99.

Badger, M.R., Sharkey, T.D. and von Caemmerer S. (1984). The relationship between steady state gas exchange of bean leaves and the levels of carbon-reduction cycle intermediates. *Planta*, **160**, 305–313.

Badger, M.R., Spalding, M.H., Leegood, R.C., *et al.* (2000a). CO_2 acquisition, concentration and fixation in cyanobacteria and algae. In: *Advances in Photosynthesis, vol.*

9: Photosynthesis, Physiology and Metabolism (ed. Leegood, R.C., Sharkey, T.D. and von Caemmerer, S.), Kluwer Academic, Dordrecht, Netherlands, pp. 369–397.

Badger, M.R., von Caemmerer, S., Ruuska, S., *et al.* (2000b). Electron flow to oxygen in higher plants and algae: rates and control of direct photoreduction (Mehler reaction) and Rubisco oxygenase. *Philosophical Transactions of The Royal Society*, **355**, 1433–1446.

Bahn, M., Rodeghiero, M., Anderson, M., *et al.* (2008). Soil respiration in European grasslands in relation to climate and assimilate supply. *Ecosystems*, **11**, 1352–1367.

Bailey Serres, J. and Voesenek, L.A.C.J. (2008). Flooding stress: acclimations and genetic diversity. *Annual Review of Plant Biology*, **59**, 313–339.

Baker, J.T., Kimb, S.H., Gitz, D.C., *et al.* (2004). A method for estimating carbon dioxide leakage rates in controlled-environment chambers using nitrous oxide. *Environmental Experimental Botany*, **51**, 103–110.

Baker, J.T., Van Pelt, S., Gitz, D.C., *et al.* (2009). Canopy gas exchange measurements of cotton in an open system. *Agronomy Journal*, **101**, 52–59.

Baker, N.R. (2008). Chlorophyll fluorescence: a probe of photosynthesis in vivo. *Annual Reviews of Plant Biology*, **59**, 89–113.

Baker, N.R. and Ort, D.R. (1992). Light and crop photosynthetic performance. In *Topics in Photosynthesis. Crop Photosynthesis: Spatial and Temporal Determinants.* (ed. Baker, N.R. and Thomas, H.), Elsevier, UK, pp. 289–312.

Baker, N.R. and Rosenqvist, E. (2004). Applications of chlorophyll fluorescence can improve crop production strategies: an examination of future possibilities. *Journal of Experimental Botany*, **55**, 1607–1621.

Baker, N.R., Harbinson, J. and Kramer, D.M. (2007). Determining the limitations and regulation of photosynthetic energy transduction in leaves. *Plant, Cell and Environment*, **30**, 1107–1125.

Baker, N.R., Oxborough, K., Lawson, T., *et al.* (2001). High resolution imaging of photosynthetic activities of tissues, cells and chloroplasts in leaves. *Journal of Experimental Botany*, **52**, 615–621.

Balachandran, S. and Osmond, C.B. (1994). Susceptibility of tobacco leaves to photoinhibition following infection with two strains of tobacco mosaic virus under different light and nitrogen nutrition regimes. *Plant Physiology*, **104**, 1051–1057.

Balachandran, S., Hurry, V.M., Kelley, S.E., *et al.* (1997). Concepts of plant biotic stress: some insights into the stress physiology of virus-infected plants, from the perspective of photosynthesis. *Physiologia Plantarum*, **100**, 203–213.

Balaguer, L., Manrique, E., de los Rios, A., *et al.* (1999). Long-term responses of the green-algal lichen *Parmelia caperata* to natural CO_2 enrichment. *Oecologia*, **119**, 166–174.

Balaguer, L., Pugnaire, F. I., Martínez-Ferri, E., *et al.* (2002). Ecophysiological significance of chlorophyll loss and reduced photochemical efficiency under extreme aridity in *Stipa tenacissima* L. *Plant and Soil*, **240**, 343–352.

Baldocchi, D. (1994). An analytical solution for coupled leaf photosynthesis and stomatal conductance models. *Tree Physiology*, **14**, 1069–1079.

Baldocchi, D. (2003). Assessing the eddy covariance technique for evaluating carbon dioxide exchange rates of ecosystems: past, present and future. *Global Change Biology*, **9**, 479–492.

Baldocchi, D. and Collineau, S. (1994). The physical nature of solar radiation in heterogeneous canopies: spatial and temporal attributes. In: *Exploitation of Environmental Heterogeneity by Plants* (eds Caldwell, M.M. and Pearcy, R.W.), Academic Press, San Diego, California, pp. 21–71.

Baldocchi, D.D. (1993). Scaling water vapour and carbon dioxide exchanges from leaves to canopy: rules and tool. In: *Scaling Physiological Processes: Leaf to Global* (eds Ehleringer, J.R. and Field, C.B.), Academic Press, San Diego, USA, pp. 77–114.

Baldocchi, D.D. (2008). 'Breathing' of the terrestrial biosphere: lessons learned from a global network of carbon dioxide flux measurement systems. *Australian Journal of Botany*, **56**, 1–26.

Baldocchi, D.D. and Amthor, J.S. (2001). Canopy photosynthesis: history, measurements, and models. In: *Terrestrial Global Productivity: Past, Present, and Future* (eds Mooney, H.A., Saugier, B. and Roy, J.), Academic Press, Inc, San Diego. pp. 9–31.

Baldocchi, D.D. and Harley, P.C. (1995). Scaling carbon dioxide and water vapour exchange from leaf to canopy in a deciduous forest. II. Model testing and application. *Plant Cell and Environment*, **18**, 1157–1173.

Baldocchi, D.D. and Valentini, R. (associated eds.) (1996). Thematic issue: Strategies for monitoring and modelling CO_2 and water vapour fluxes over terrestrial ecosystems. *Global Change Biology*, **2**, 159–318.

Baldocchi, D.D., Wilson, K.B. and Gu, L. (2002). How the environment, canopy structure and canopy physiological functioning influence carbon, water and energy fluxes of a temperate broad-leaved deciduous forest – an assessment with the biophysical model CANOAK. *Tree Physiology*, **22**, 1065–1077.

Baldocchi, D.D., Xu, L. and Kiang, N.Y. (2004). How plant functional-type, weather, seasonal drought, and soil physical properties alter water and energy fluxes of an oak-grass savanna and an annual grassland. *Agricultural and Forest Meteorology*, **123**, 13–39.

Baldocchi, D.D., Falge, E., Gu, L., *et al.* (2001). FLUXNET: a new tool to study the temporal and spatial variability of ecosystem-scale carbon dioxide, water vapor and energy flux densities. *Bulletin of the American Meteorological Society*, **82**, 2415–2434.

Baldocchi, D.D., Finnigan, J.J., Wilson, K., *et al.* (2000). On measuring net ecosystem carbon exchange over tall vegetation on complex terrain. *Boundary Layer Meteorology*, **96**, 257–291.

Baldocchi, D.D., Hicks, B.B. and Meyers, T.P. (1988). Measuring biosphere-atmosphere exchanges of biologically related gases with micrometeorological methods. *Ecology*, **69**, 1331–1340.

Baldocchi, D.D., Valentini, R., Running, S., *et al.* (1996). Strategies for measuring and modelling carbon dioxide and water vapour fluxes over terrestrial ecosystems. *Global Change Biology*, **2**, 159–168.

Baldwin, A., Egnotovich, M., Ford, M., *et al.* (2001). Regeneration in fringe mangrove forests damaged by Hurricane Andrew. *Plant Ecology*, **157**, 149–162.

Baldwin, I.T. (2001). An ecologically motivated analysis of plant-herbivore interactions in native tobacco. *Plant Physiology*, **127**, 1449–1458.

Baldwin, I.T. and Callahan, P. (1993). Autotoxicity and chemical defense: nicotine accumulation and carbon gain in solanaceous plants. *Oecologia*, **94**, 534–541.

Ball, J.T., Woodrow, I.E. and Berry J.A. (1987). A model predicting stomatal conductance and its contribution to the control of photosynthesis under different environmental conditions. In: *Progress in Photosynthesis Research*, Vol. IV (ed. Biggins, I.), Martinus-Nijhoff Publishers, Dordrecht, Netherlands, pp. 221–224.

Ball, M.C. (2002). Interactive effects of salinity and irradiance on growth: implications for mangrove forest structure along salinity gradients. *Trees*, **16**, 126–139.

Ball, M.C. and Critchley, C. (1982). Photosynthetic responses to irradiance by the grey mangrove, *Avicennia marina*, grown under different light regimes. *Plant Physiology*, **70**, 1101–1106.

Ballare, C.L., Scopel, A.L., Stapleton, A.E., *et al.* (1996). Solar ultraviolet-B radiation affects seedling emergence, DNA integrity, plant morphology, growth rate, and attractiveness to herbivore insects in *Datura ferox*. *Plant Physiology*, **112**, 161–170.

Baltzer, J.L. and Thomas, S.C. (2007a). Determinants of whole-plant light requirements in Bornean rain forest tree saplings. *Journal of Ecology*, **95**, 1208–1221.

Baltzer, J.L. and Thomas, S.C. (2007b). Physiological and morphological correlates of whole-plant light compensation point in temperate deciduous tree seedlings. *Oecologia*, **153**, 209–223.

Baltzer, J.L., Davies, S.J., Bunyavejchewin, S., *et al.* (2008). The role of desiccation tolerance in determining tree species distributions along the Malay Thai Peninsula. *Functional Ecology*, **22**, 221–231.

Baltzer, J.L., Thomas, S.C., Nilus, R., *et al.* (2005). Edaphic specialization in tropical trees: physiological correlates and responses to reciprocal transplantation. *Ecology*, **86**, 3063–3077.

Bansal, S. and Germino, M.J. (2008). Carbon balance of conifer seedlings at timberline: relative changes in uptake, storage, and utilization. *Oecologia*, **158**, 217–227.

Barak, P. and Helmke, P.A. (1993). The chemistry of zinc. In: *Zinc in Soils and Plants, Developments in Plants and Soil Sciences* (ed. Robson, A). Kluwer Academic Press, New York, USA, pp. 1–13.

Barbehenn, R.V., Karowe, D.N. and Zhong, C. (2004). Performance of a generalist grasshopper on a C_3 and a C_4 grass: compensation for the effects of elevated CO_2 on plant nutritional quality. *Oecologia*, **140**, 96–103.

Barber, J. (2008a). Photosynthetic generation of oxygen. *Philosophical Transactions of the Royal Society B*, **363**, 2665–2674.

Barber, J. (2008b). Crystal structure of the oxygen-evolving complex of Photosystem II. *Inorganic Chemistry*, **47**, 1700–1710.

Barbour, M.M. (2007). Stable oxygen isotope composition of plant tissue: a review. *Functional Plant Biology*, **34**, 83–94.

Barbour, M.M. and Buckley, T.N. (2007). The stomatal response to evaporative demand persists at night in *Ricinus communis* plants with high nocturnal conductance. *Plant, Cell and Environment*, **30**, 711–721.

Barbour, M.M. and Farquhar, G.D. (2000). Relative humidity and ABA-induced variation in carbon and oxygen isotope ratios of cotton leaves. *Plant, Cell and Environment*, **23**, 473–485.

Barbour, M.M., Cernusak, L.A. and Farquhar, G.D. (2005). Factors affecting the oxygen isotope ratio of plant organic material. In: *Stable Isotopes and Biosphere-Atmosphere Interactions* (eds Flanagan, L.B., Ehleringer, J.R. and

Pataki, D.E.), Elsevier Academic Press, San Diego, pp. 9–28.

Barbour, M.M., Fischer, R.A., Sayre, K.D., *et al.* (2000b). Oxygen isotope ratio of leaf and grain material correlates with stomatal conductance and grain yield in irrigated wheat. *Australian Journal of Plant Physiology*, **27**, 625–637.

Barbour, M.M., McDowell, N.G., Tcherkez, G., *et al.* (2007). A new measurement technique reveals rapid post-illumination changes in the carbon isotope composition of leaf-respired CO_2. *Plant, Cell and Environment*, **30**, 469–482.

Barbour, M.M., Shurr, U., Henry, B.K., *et al.* (2000a). Variation in the oxygen isotope ratio of phloem sap sucrose from castor bean: evidence in support of the Péclet effect. *Plant Physiology*, **123**, 671–679.

Barbour, M.M., Warren, C.R., Farquhar, G.D., *et al.* (2010). Variability in mesophyll conductance between barley genotypes, and effects on transpiration efficiency and carbon isotope discrimination. *Plant, Cell and Environment*, **33**, 1176–1185.

Bar-Even, A., Noor, E., Lewis, N.E., *et al.* (2010). Design and analysis of synthetic carbon fixation pathways. *Proceedings of the National Academy of Sciences USA*, **107**, 8889–8894.

Bariac, T., Gonzales-Dunia, J., Tardieu, F., *et al.* (1994). Spatial variation of the isotopic composition of water (^{18}O, ^{2}H) in organs of aerophytic plants. 1. Assessment under laboratory conditions. *Chemical Geology*, **115**, 307–315.

Barker, D.A., Seaton, G.G.R. and Robinson, S.A. (1997). Internal and external photoprotection in developing leaves of the CAM plant *Cotyledon orbiculata*. *Plant Cell and Environment*, **20**, 617–624.

Barker, D.H., Vanier, C., Naumburg, E., *et al.* (2006). Enhanced monsoon precipitation and nitrogen deposition affect leaf traits and photosynthesis differently in spring and summer in the desert shrub *Larrea tridentata*. *New Phytologist*, **169**, 799–808.

Barker, E.R., Press, M.C., Scholes, J.D., *et al.* (1996). Interactions between the parasitic angiosperm *Orobanche aegyptiaca* and its tomato host: growth and biomass allocation. *New Phytologist*, **133**, 637–642.

Barnes, B.B., Zak, D.R., Denton, S.R., *et al.* (1998). *Forest Ecology*, 4th edn. John Wiley and Sons Inc., New York, USA.

Barnes, J., Davison, A., Balaguer, L., *et al.* (2007). Resistance to air pollutants: from cell to community. In: *Functional Plant Ecology* (eds Pugnaire, F.I. and Valladares, F.) 2nd edn: CRC Press, USA, pp. 601–626.

Barnes, P.W. and Archer, S. (1996). Influence of an overstorey tree (*Prosopis glandulosa*) on associated shrubs in a savanna parkland: implications for patch dynamics. *Oecologia*, **105**, 493–500.

Baroli, I. and Niyogi, K.K. (2000). Molecular genetics of xanthophyll-dependent photoprotection in green algae and plants. *Philosophical Transactions of the Royal Society B*, **355**, 1385–1394.

Baroli, I., Price, G.D., Badger, M.R., *et al.* (2008). The contribution of photosynthesis to the red light response of stomatal conductance. *Plant Physiology*, **146**, 737–747.

Barón, M., Arellano, J.B. and López Gorgé, J. (1995). Copper and photosystem II: a controversial relationship. *Physiologia Plantarum*, **94**, 174–180.

Barrantes, O., Moliner, E., Plaza, M., *et al.* (1997). Preliminary results of an SO_2 experiment with *Pinus halepensis* Mill. seedlings in open top chambers. In: *Impact of Global Change on Tree Physiology and Forest Ecosystems* (eds Mohren, G.M.J., Kramer, K. and Sabaté, S.), Kluwer Academic Publishers, Dordrecht, Netherlands, pp. 111–118.

Barry, J.C., Morgan, M.E., Flynn, L.J., *et al.* (2002). Faunal and environmental change in the late Miocene Siwaliks of northern Pakistan. *Paleobiology Memoirs*, **28**, 1–55.

Barry, R.G. (2008). *Mountain, weather and climate.* 3rd edn. Cambridge University Press, Cambridge, UK.

Barta, C. and Loreto, F. (2006).The relationship between the methyl-erythritol phosphate pathway leading to emission of volatile isoprenoids and abscisic acid content in leaves. *Plant Physiology*, **141**, 1676–1683.

Barták, M., Gloser, J. and Hájek, J. (2005). Visualized photosynthetic characteristics of the lichen *Xanthoria elegans* related to daily courses of light, temperature and hydration: a field study from Galindez Island, maritime Antarctica. *Lichenologist*, **37**, 433–443.

Barták, M.L., Váczi, P., Hájek, J., *et al.* (2006). Low-temperature limitation of primary photosynthetic processes in Antarctic lichens *Umbilicaria antarctica* and *Xanthoria elegans*. *Polar Biology*, **31**, 47–51.

Bartelink, H.H. (1998). A model of dry matter partitioning in trees. *Tree Physiology*, **18**, 91–101.

Bartels, D. and Salamini, F. (2001). Desiccation tolerance in the resurrection plant *Craterostigma plantagineum*: a contribution to the study of drought tolerance at the molecular level. *Plant Physiology*, **127**, 1346–1353.

Bartoli, C.G., Gómez, F., Gergoff, G., *et al.* (2005). Up-regulation of the mitochondrial alternative oxidase pathway enhances photosynthetic electron transport under drought conditions. *Journal of Experimental Botany*, **56**, 1269–1276.

Barton, A.M., Fetcher, N. and Redhead, S. (1989). The relationship between treefall gap size and light-flux in a Neotropical rain-forest in Costa-Rica. *Journal of Tropical Ecology*, **5**, 437–439.

Baruch, Z. and Bilbao, B. (1999). Effects of fire and defoliation on the life history of native and invader C-4 grasses in a Neotropical savanna. *Oecologia*, **119**, 510–520.

Basak, U.C., Das, A.B. and Das, P. (1996). Chlorophylls, carotenoids, proteins and secondary metabolites in leaves of 14 species of mangrove. *Bulletin of Marine Science*, **58**, 654–659.

Basha, E., Lee, G.J., Breci, L.A., *et al.* (2004). The identity of proteins associated with a small heat shock protein during heat stress in vivo indicates that these chaperones protect a wide range of cellular functions. *Journal of Biological Chemistry*, **279**, 7566–7575.

Basile, B., Reidel, E.J., Weinbaum, S.A., *et al.* (2003). Leaf potassium concentration, CO_2 exchange and light interception in almond trees (*Prunus dulcis* (Mill) D.A. Webb). *Scientia Horticulturae*, **98**, 185–194.

Bassham, J.A. and Calvin, M. (1957). *The Path of Carbon in Photosynthesis*. Prentice-Hall, N.J., USA, p. 104.

Baszynski, T., Wajda, L., Krol, M., *et al.* (1980). Photosynthetic activities of cadmium-treated tomato plants. *Physiologia Plantarum*, **48**, 365–370.

Bauer, H., Nagele, M., Comploj, M., *et al.* (1994). Photosynthesis in cold-acclimated leaves of plants with various degrees of freezing tolerance. *Physiologia Plantarum*, **91**, 403–412.

Bauerle, W.L., Hinckley, T.M., Chermák, J., *et al.* (1999). The canopy water relations of old-growth Douglas-fir trees. *Trees*, **13**, 211–217.

Baxter, C.J., Redestig, H., Schauer, N., *et al.* (2007). The metabolic response of heterotrophic *Arabidopsis* cells to oxidative stress. *Plant Physiology*, **143**, 312–325.

Bazzaz, F.A., Carlson, R.W. and Harper, J.L. (1979). Contribution to reproductive effort by photosynthesis of flowers and fruits. *Nature*, **279**, 554–555.

Beadle, N.C.W. (1966). Soil phosphate and its role in molding segments of the Australian flora and vegetation with special reference to xeromorphy and sclerophylly. *Ecology*, **47**, 992–1007.

Bean, R.C., Porter, G.G. and Barr, K.B. (1963). Photosynthesis and respiration in developing fruits. III. Variations in photosynthesis capacity during colour change. *Plant Physiology*, **38**, 285–290.

Beauchamp, J., Wisthaler, A., Hansel, A., *et al.* (2005). Ozone induced emissions of biogenic VOC from tobacco: relations between ozone uptake and emission of LOX products. *Plant, Cell and Environment*, **28**, 1334–1343.

Beauford, W., Barber, J. and Barringer, A.R. (1977). Uptake and distribution of mercury within higher plants. *Physiologia Plantarum*, **39**, 261–265.

Becher, M., Talke, I.N., Krall, L., *et al.* (2004). Cross-species microarray transcript profiling reveals high constitutive expression of metal homeostasis genes in shoots of the zinc hyperaccumulator *Arabidopsis halleri*. *Plant Journal*, **37**, 251–268.

Beck, C. (2005). Signaling pathways from the chloroplast to the nucleus. *Planta*, **222**, 743–756.

Beck, E., Senser, M., Scheibe, R., *et al.* (1982). Frost avoidance and freezing tolerance in Afroalpine 'giant rosette' plants. *Plant Cell and Environment*, **5**, 215–222.

Bednarz, C.W. and van Iersel, M.W. (1999). Continuous whole plant carbon dioxide exchange rates in cotton treated with pyrithiobac. *The Journal of Cotton Science*, **3**, 53–59.

Bednarz, C.W. and van Iersel, M.W. (2001). Temperature response of whole-plant CO_2 exchange rates of four upland cotton cultivars differing in leaf shape and leaf pubescence. *Communications in Soil Science and Plant Analysis*, **32**, 2485–2501.

Beer, C., Ciais, P., Reichstein, M., *et al.* (2009). Temporal and among-site variability of inherent water use efficiency at the ecosystem level. *Global Biogeochemical Cycles*, **23**, GB2018.

Beer, C., Reichstein, M., Ciais, P., *et al.* (2007). Mean annual GPP of Europe derived from its water balance. *Geophysical Research Letters*, **34**, L05401.

Beer, C., Reichstein, M., Tomelleri, E. *et al.* (2010). Terrestrial gross carbon dioxide uptake: global distribution and covariation with climate. *Science*, **329**, 834–838.

Beerling, D.J. (2005). Leaf evolution: gases, genes and geochemistry. *Annals of Botany*, **96**, 345–352.

Beerling, D.J. and Kelly, C.K. (1996). Evolutionary comparative analyses of the relationship between leaf structure and function. *The New Phytologist*, **134**, 35–51.

Beerling, D.J. and Osborne, C.P. (2006). The origin of the savanna biome. *Global Change Biology*, **12**, 2023–2031.

Beerling, D.J. and Rundgren, M. (2000). Leaf metabolic and morphological responses of dwarf willow (*Salix herbacea*) in the Sub-Arctic to the past 9000 years of global environmental change. *New Phytologist*, **145**, 257–269.

Beerling, D.J. and Woodward, F.I. (1996). Paleo-ecophysiological perspectives on plant responses to global change. *Trends in Ecology and Evolution*, **11**, 20–23.

Beerling, D.J. and Woodward, F.I. (1997). Changes in land plant function over the Phanerozoic: recontructions based on the fossil record. *Botanical Journal of the Linnean Society*, **124**, 137–153.

Beerling, D.J. and Woodward, F.I. (2001). *Vegetation and the Terrestrial Carbon Cycle: Modelling the First 400 Million Years*. Cambridge University Press, Cambridge, UK.

Beerling, D.J., McElwain, J.C. and Osborne, C.P. (1998). Stomatal responses of the 'living fossil' *Ginkgo biloba* L. to changes in atmospheric CO_2 concentrations. *Journal of Experimental Botany*, 49, 1603–1607.

Beerling, D.J., Osborne, C.P., Chaloner, W.G. (2001). Evolution of leaf-form in land plants linked to atmospheric CO_2 decline in the Late Paleozoic era. *Nature*, **410**, 352–354.

Behrensmeyer, A.K., Quade, J., Cerling, T.E., *et al.* (2007). The structure and rate of late Miocene expansion of C_4 plants: evidence from lateral variation in stable isotopes in paleosols of the Siwalik Group, northern Pakistan. *Geological Society of America Bulletin*, **119**, 1486–1505.

Belkhodja, R., Morales, F., Quílez, R., *et al.* (1998). Iron deficiency causes changes in chlorophyll fluorescence due to the reduction in the dark of the photosystem II acceptor side. *Photosynthesis Research*, **56**, 265–276.

Bellingham, P.J., Kohyama, T. and Aiba, S.I. (1996). The effects of a typhoon on Japanese warm temperate rainforests. *Ecological Research*, **11**, 229–247.

Belsky, A.J., Amundson, R.G. Duxbury, J.M. *et al.* (1989). The effects of trees on their physical, chemical, and biological environments in a semi-arid savanna in Kenya. *Journal of Applied Ecology*, **26**, 1005–1024.

Bendall, D.S. and Manasse, R.S. (1995). Cyclic photophosphorylation and electron transport. *Biophysica et Biochimica Acta (Bioenergetics)*, **1229**, 23–38.

Bender, M., Sowers, T. and Labeyrie, L. (1994). The Dole effect and its variations during the last 130,000 years as measured in the Vostok ice core. *Global Biogeochemical Cycles*, 8, 363–376.

Bender, M.M. (1968). Mass spectrometric studies of carbon-13 variations in corn and other grasses. *Radiocarbon*, **10**, 468–472.

Bender, M.M. (1971). Variations in the $^{13}C/^{12}C$ ratios of plants in relations to the pathway of photosynthetic carbon dioxide fixation. *Phytochemistry*, **10**, 1239–1244.

Bender, M.M., Rouhani, I., Vines, H.M., *et al.* (1973). $^{13}C/^{12}C$ ratio changes in crassulacean acid metabolism plants. *Plant Phisiology*, **52**, 427–430.

Benkeblia, N., Shinano, T. and Osaki, M. (2007). Metabolite profiling and assessment of metabolome compartmentation of soybean leaves using non-aqueous fractionation and GC-MS analysis. *Metabolomics*, **2**, 297–305.

Bennoun, P. (2001). Chlororespiration and the process of carotenoid biosynthesis. *Biochimica et Biophysica Acta – Bioenergetics*, **1506**, 133–142.

Benowicz, A., Guy, R.D. and El-Kassaby, Y.A. (2000). Geographic pattern of genetic variation in photosynthetic capacity and growth in two hardwood species from British Columbia. *Oecologia*, **123**, 168–174.

Benzing, D.H. (1990). *Vascular Epiphytes: General Biology and Related Biota*. Cambridge University Press, Cambridge, UK.

Berard, R.G. and Thurtell, G.W. (1990). Respiration measurements of maize plants using a whole-plant enclosure system. *Agronomy Journal*, **82**, 641–643.

Berenbaum, M.R. and Zangerl, A.R. (2008). Facing the future of plant-insect interaction research: le retour a la "raison d'etre". *Plant Physiology*, **146**, 804–811.

Berger, B., Parent, B. and Tester, M. (2010). High-throughput shoot imaging to study drought responses. *Journal of Experimental Botany*, **61**, 3519–3528.

Berger, S., Benediktyová, Z., Matouš, K., *et al.* (2007). Visualization of dynamics of plant-pathogen interaction by novel combination of chlorophyll fluorescence imaging and statistical analysis: differential effects of virulent an avirulent strains of *P. syringae* and of oxylipins on *A. thaliana*. *Journal of Experimental Botany*, **58**, 797–806.

Berger, S., Papadopoulos, M., Schreiber, U., *et al.* (2004). Complex regulation of gene expression, photosynthesis and sugar levels by pathogen infection in tomato. *Physiologia Plantarum*, **122**, 419–428.

Beringer, J., Hutley, L.B., Tapper, N.J., *et al.* (2007). Savanna fires and their impact on net ecosystem productivity in North Australia. *Global Change Biology*, **13**, 990–1004.

Bernacchi, C.J., Calfapietra, C., Davey, P.A., *et al.* (2003b). Photosynthesis and stomatal conductance responses of poplars to free-air CO_2 enrichment (PopFACE) during the first growth cycle and immediately following coppice. *New Phytologist*, **159**, 609–621.

Bernacchi, C.J., Hollinger, S.E. and Meyers, T. (2005b). The conversion of the corn/soybean ecosystem to no-till agriculture may result in a carbon sink. *Global Change Biology*, **11**, 1867–1872.

Bernacchi, C.J., Kimball, B., Quarles, D., *et al.* (2007). Decreases in stomatal conductance of soybean under open-air elevation of [CO_2] are closely coupled with decreases in ecosystem evapotranspiration. *Plant Physiology*, **143**, 134–144.

Bernacchi, C.J., Leakey, A.D.B., Heady, L.E., *et al.* (2006). Hourly and seasonal variation in photosynthesis and stomatal conductance of soybean grown at future CO_2 and ozone concentrations for 3 years under fully open-air field conditions. *Plant, Cell and Environment*, **29**, 2077–2090.

Bernacchi, C.J., Morgan, P.B., Ort, D.R., *et al.* (2005a). The growth of soybean under free air [CO_2] enrichment (FACE) stimulates photosynthesis while decreasing in vivo Rubisco capacity. *Planta*, **220**, 434–446.

Bernacchi, C.J., Pimentel, C. and Long, S.P. (2003a). In vivo temperature response functions of parameters required to model RuBP-limited photosynthesis. *Plant, Cell and Environment*, **26**, 1419–1430.

Bernacchi, C.J., Portis, A.R., Nakano, H., *et al.* (2002). Temperature response of mesophyll conductance: implications for the determination of Rubisco enzyme kinetics and for limitations to photosynthesis *in vivo*. *Plant Physiology*, **130**, 1992–1998.

Bernacchi, C.J., Singsaas, E.L., Pimentel, C., *et al.* (2001). Improved temperature response functions for models of Rubisco-limited photosynthesis. *Plant Cell and Environment*, **24**, 253–259.

Berner, R.A. (1990). Atmospheric carbon dioxide levels over phanerozoic time. *Science*, **249**, 1382–1386.

Berner, R.A. and Canfield, D.E. (1989). A new model for atmospheric oxygen over phaenerozoic time. *American Journal of Science*, **289**, 333–361.

Berni, J.A., Zarco-Tejada, P.J., Suarez, L., *et al.* (2009). Thermal and narrowband multispectral remote sensing for vegetation monitoring from an unmanned aerial vehicle. *IEEE Transactions on Geoscience and Remote Sensing*, **47**, 722–738.

Berninger, F. (1997). Effects of drought and phenology on GPP a simulation study along a geographical gradient. *Functional Ecology*, **11**, 33–42.

Berova, M., Stoeva, N., Zlatev, Z., *et al.* (2007). Physiological changes in bean (*Phaseolus vulgaris* L.) leaves, infected by the most important bean diseases. *Journal of Central European Agriculture*, **8**, 57–62.

Berry, J.A. and Björkman, O. (1980). Photosynthetic response and adaptation to temperature in higher plants. *Annual Review of Plant Physiology*, **31**, 491–543.

Bertamini, M., Muthuchelian, K. and Nedunchezhian, N. (2004). Effect of grapevine leafroll on the photosynthesis of field grown grapevine plants (*Vitis vinifera* L. cv. Lagrein). *Journal of Phytopathology*, **152**, 145–152.

Bertone, P. and Snyder, M. (2007). Prospects and challenges in proteomics. *Plant Physiology*, **138**, 560–562.

Bertsch, W.F. and Azzi, J.R. (1965). A relative maximum in the decay of long-term delayed light emission from the photosynthetic apparatus. *Biochimica et Biophysica Acta*, **94**, 15–26.

Berveiller, D., Kierzkowski, D. and Damesin, C. (2007a). Interspecific variability of stem photosynthesis among tree species. *Tree Physiology*, **27**, 53–61.

Berveiller, D., Vidal, J., Degrouard, J., *et al.* (2007b). Tree stem phosphoenolpyruvate carboxylase (PEPC): lack of biochemical and localization evidence for a C4-like photosynthesis system. *New Phytologist*, **176**, 775–781.

Besson-Bard, A., Pugin, A. and Wendehenne, D. (2008). New insights into nitric oxide signaling in plants. *Annual Review of Plant Biology*, **59**, 21–39.

Betts, R.A., Boucher, O., Collins, M., *et al.* (2007). Projected increase in continental runoff due to plant responses to increasing carbon dioxide. *Nature*, **448**, 1037–1041.

Betzelberger, A.M., Gillespie, K.M., McGrath, J.M., *et al.* (2010). Effects of chronic elevated ozone concentration on antioxidant capacity, photosynthesis and seed yield of 10 soybean cultivars. *Plant, Cell and Environment*, **33**, 1569–1581.

Beuker, E. (1994). Adaptations to climate change of the timing of bud burst of *Pinus sylvestris* L. And *Picea abies* Karst. *Trees*, **14**, 961–970.

Beyschlag, W., Kresse, F., Ryel, R.J., *et al.* (1994). Stomatal patchiness in conifers: experiments with *Picea abies* (L.) Karst. and *Abies alba* Mill. *Trees: Structure and Function*, **8**, 132–138.

Beyschlag, W., Pfanz, H. and Ryel, R.J. (1992). Stomatal patchiness in Mediterranean evergreen sclerophylls. Phenomenology and consequences for the interpretation of the midday depression in photosynthesis and transpiration. *Planta*, **187**, 546–553.

Bickford, C.P., Hanson, D.T. and McDowell, N.G. (2010). Influence of diurnal variation in mesophyll conductance on modeled ^{13}C discrimination: results from a field study. *Journal of Experimental Botany*, **61**, 3223–3233.

Bickford, C.P., McDowell, N.G., Erhardt, E.B., *et al.* (2009). High-frequency field measurements of diurnal carbon isotope discrimination and internal conductance in a semi-arid species, *Juniperus monosperma*. *Plant, Cell and Environment*, **32**, 796–810.

Bieleski, R.L. (1973). Phosphate pools, phosphate transport, and phosphate availability. *Annual Review of Plant Physiology*, **24**, 225–252.

Bigras, F.J. and Bertrand, A. (2006). Responses of *Picea mariana* to elevated CO_2 concentration during growth, cold hardening and dehardening: phenology, cold tolerance, photosynthesis and growth. *Tree Physiology*, **26**, 875–888.

Bilger, H.W., Schreiber, U. and Lange, O.L. (1984). Determination of leaf heat resistance: comparative investigation of chlorophyll fluorescence changes and tissue necrosis methods. *Oecologia*, **63**, 256–262.

Bilger, W. and Björkman, O. (1991). Temperature dependence of violaxanthin de-epoxidation and non-photochemical fluorescence quenching in intact leaves of *Gossypium hirsutum* L. and *Malva parviflora* L. *Planta*, **184**, 226–234.

Bilger, W., Björkmann, O. and Thayer, S.S. (1989). Light-induced spectral absorbance changes in relation to photosynthesis and the epoxidation state of xanthophyll cycle components in cotton leaves. *Plant Physiology*, **91**, 542–551.

Bilger, W., Veit, M., Schreiber, L., *et al.* (1997). Measurement of leaf epidermal transmittance of UV radiation by chlorophyll fluorescence. *Physiologia Plantarum*, **101**, 754–763.

Bilgin, D.D., Zavala, J.A., Zhu, J., *et al.* (2010). Biotic stress globally downregulates photosynthesis genes. *Plant, Cell and Environment*, **33**, 1597–1613.

Billesbach, D.P., Fischer, M.L., Torn, M.S., *et al.* (2004). A portable eddy covariance system for the measurement of ecosystem–atmosphere exchange of CO_2, water vapor, and Energy. *Journal of Atmospheric and Oceanic Technology*, **21**, 639–650.

Billings, W.D. (1973). Arctic and alpine vegetations: similarities, differences and susceptibility to disturbance. *BioScience*, **37**, 58–67.

Billings, W.D. (1974). Adaptations and origins of alpine plants. *Arctic and Alpine Research*, **6**, 129–142.

Billings, W.D. and Mooney, H.A. (1968). The ecology of arctic and alpine plants. *Biological Reviews*, 481–529.

Bird, A.F. (1974). Plant response to root-knot nematode. *Annual Review of Phytopathology*, **12**, 69–85.

Bird, I.F., Cornelius, M.J. and Keys, A.J. (1982). Affinity of RuBP carboxylases for carbon dioxide and inhibition of the enzymes by oxygen. *Journal of Experimental Botany*, **33**, 1004–1013.

Birkhold, K.T., Koch, K.E. and Darnell, R.L. (1992). Carbon and nitrogen economy of developing rabbiteye blueberry fruit. *Journal of the American Society of Horticultural Sciences*, **117**, 139–145.

Birth, G.S. and McVey, G.R. (1968). Measuring the colour of growing turf with a reflectance spectrophotometer. *Agronomy Journal*, **60**, 640–643.

Biswal, B., Smith, A.J. and Rogers, L.J. (1994). Changes in carotenoids but not in D1 protein in response to nitrogen depletion and recovery in a cyanobacterium. FEMS *Microbiology Letters*, **116**, 341–347.

Biswal, U.C., Biswal, B. and Raval, M.K. (2003). *Chloroplast Biogenesis: From Protoplastid to Gerontoplast*, Kluwer Academic Publishers, Dordrecht, Netherlands.

Björkman, O. (1971). Comparative photosynthetic CO_2 exchange in higher plants. In: *Photosynthesis and Photorespiration* (eds Hatch, M.D., Osmond, C.B. and Slatyer, R.O.), Wiley Interscience, New York, USA. pp. 18–32.

Björkman, O. (1981). Responses to different quantum flux densities. In: *Physiological Plant Ecology, vol I, Encyclopedia of Plant Physiology*, 12A (eds Lange, O.L., Nobel, P.S., Osmond, C.B. and Ziegler, H.), Springer-Verlag, Berlin, pp. 57–107.

Björkman, O. (1994). Responses to long-term drought and high-irradiance stress in natural vegetation. *Carnegie Institution of Washington Yearbook*, **94**, 62–63.

Björkman, O. and Demmig, B. (1987). Photon yield of O_2 evolution and chlorophyll fluorescence characteristics at 77K among vascular plants of diverse origins. *Planta*, **170**, 489–504.

Björkman, O. and Demmig-Adams, B. (1994). Regulation of photosynthetic light energy capture, conversion and dissipation in leaves of higher plants. In: *Ecophysiology of Photosynthesis* (eds Schulze, E.-D. and Caldwell, M.M.), Springer-Verlag. Berlin. pp. 17–47.

Björkman, O., Badger, M.R. and Armond, P.A. (1980). Response and adaptation of photosynthesis to high temperature. In: *Adaptations of Plants to Water and High Temperature Stress* (eds Turner, N.C. and Kramer, P.J.), Wiley Interscience, NY, USA, pp. 223–249.

Björkman, O., Pearcy, R.W, Harrison, A.T., *et al.* (1972). Photosynthetic adaptation to high temperatures: a field study in Death Valley, California. *Science*, **175**, 786–789.

Black, C.C. (1973). Photosynthetic carbon fixation in relation to net CO_2 uptake. *Annual Review of Plant Physiology*, **24**, 253–286.

Black, K., Davis, P., McGrath, J., *et al.* (2005). Interactive effects of irradiance and water availability on the photosynthetic performance of *Picea sitchensis* seedlings: implications for seedling establishment under different management practices. *Annals of Forest Science*, **62**, 413–422.

Black, T.A., Gaumont-Guay, D., Jassla, R., *et al.* (2005). Measurement of CO_2 exchange between the boreal forest and the atmosphere. In: *The Carbon Balance of Forest*

Biomes (eds Griffiths, H. and Jarvis, P.G., Taylor & Francis, Oxon.

Blackburn, G.A. (2007). Hyperspectral remote sensing of plant pigments. *Journal of Experimental Botany*, **58**, 855–867.

Blanke, M.M. (2002). Photosynthesis of strawberry fruit. *Acta Horticulturae*, **567**, 373–376.

Blanke, M.M. and Cooke, D.T. (2004). Effects of flooding and drought on stomatal activity, transpiration, photosynthesis, water potential and water channel activity in strawberry stolons and leaves. *Plant Growth Regulation*, **42**, 153–160.

Blanke, M.M. and Lenz, F. (1989). Fruit photosynthesis. *Plant, Cell and Environment*, **12**, 31–46.

Blankenship, R.E. (1992). Origin and early evolution of photosynthesis. *Photosynthesis Research*, **33**, 91–111.

Blankenship, R.E., Tiede, D.M., Barber, J. *et al.* (2011). Comparing photosynthetic and photovoltaic efficiencies and recognizing the potential for improvement. *Science*, **332**, 805–809.

Bläsing, O.E., Ernst, K., Streubel, M., *et al.* (2002). The non-photosynthetic phosphoenolpyruvate carboxylases of the C_4 dicot *Flaveria trinervia*: implications for the evolution of C_4 photosynthesis. *Planta*, **215**, 448–456.

Blasius, B., Neff, R., Beck, F., *et al.* (1999). Oscillatory model of crassulacean acid metabolism with a dynamic hysteresis switch. *Proceedings of the Royal Society of London Series B*, **266**, 93–101.

Bligny, R. and Douce, R. (2001). NMR and plant metabolism. Current Opinion in *Plant Biology*, **4**, 191–196.

Bliss, L.C. (1962). Adaptations of arctic and alpine plants to environmental conditions. *Arctic*, **15**, 117–144.

Blödner, C., Majcherczyk, A., Kües, U., *et al.* (2007). Early drought induced changes to the needle proteome of Norway spruce. *Tree Physiology*, **27**, 1423–1431.

Blom, C.W.P.M. and Voesenek, L.A.C.J. (1996). Flooding: the survival strategies of plants. *Trends in Ecology and Evolution*, **11**, 290–295.

Bloom, A.J., Smart, D.R., Nguyen, D.T., *et al.* (2002). Nitrogen assimilation and growth of wheat under elevated carbon dioxide. *Proceedings of the National Academy of Sciences of the United States of America*, **99**, 1730–1735.

Blum, A. (2005). Drought resistance, water-use efficiency, and yield potential: are they compatible, dissonant, or mutually exclusive? *Australian Journal of Agricultural Research*, **56**, 1159–1168.

Blum, A. (2009). Effective use of water (EUW) and not water-use efficiency (WUE) is the target of crop yield improvement under drought stress. *Field Crops Research*, **112**, 119–123.

Boardman, N.K. (1977). Comparative photosynthesis of sun and shade plants. *Annual Review of Plant Physiology*, **28**, 355–377.

Boccalandro, H.E., Rugnone, M.L., Moreno, J.E., *et al.* (2009). Phytochrome B enhances photosynthesis at the expense of water-use efficiency in Arabidopsis. *Plant Physiology*, **150**, 1083–1092.

Boccara, M., Boue, C., Garmier, M., *et al.* (2001). Infrared thermography revealed a role for mitochondria in pre-symptomatic cooling during harpin-induced hypersensitive response. *Plant Journal*, **28**, 663–670.

Bode, S., Quentmeier, C.C., Liao, P.-N., *et al.* (2009). On the regulation of photosynthesis by excitonic interactions between carotenoids and chlorophylls. *Proceedings of the National Academy of Sciences USA*, **106**, 12311–12316.

Bogeat-Triboulot, M.B, Brosché, M., Renaut, J., *et al.* (2007). Gradual soil water depletion results in reversible changes of gene expression, protein profiles, ecophysiology, and growth performance in *Populus euphratica*, a poplar growing in arid regions. *Plant Physiology*, **143**, 876–892.

Bolhàr-Nordenkampf, H.R. and Draxler, G. (1993). Functional leaf anatomy. In: *Photosynthesis and Production in a Changing Environment* (eds Hall, D.O., Scurlock, J.M.O., Bolhàr-Nordenkampf, H.R., Leegood, R.C. and Long S. P.) Chapman and Hall, London, pp. 91–111.

Bolton, J.R. and Hall, D.O. (1991). The maximum efficiency of photosynthesis. *Photochemistry and Photobiology*, **53**, 545–548.

Bona, L., Carver, B.F., Wright, R.J., *et al.* (1994). Aluminum tolerance of segregating wheat populations in acidic soil and nutrient solutions. *Communications in Soil Science and Plant Analysis*, **25**, 327–339.

Bonan, G.B. (2008). Forests and climate change: forcings, feedbacks, and the climate benefits of forests. *Science*, **320**, 1444–1449.

Bonan, G.B., Davis, K.J., Baldocchi, D., *et al.* (1997). Comparison of the NCAR LSM1 land surface model with BOREAS aspen and jack pine tower fluxes. *Journal of Geophysical Research-Atmospheres*, **102**, 29065–29075.

Bonaventure, G., Gfeller, A., Proebsting, W.M., *et al.* (2007). A gain of function allele of TPC1 activates oxylipin biogenesis after leaf wounding in *Arabidopsis. The Plant Journal*, **49**, 889–898.

Bond, B.J. (2000). Age-related changes in photosynthesis of woody plants. *Trends in Plant Science*, **5**, 349–353.

Bond, B.J. and Ryan, M.G. (2000). Comment on 'Hydraulic limitation of tree height: a critique' by Becker, Meinzer and Wullschleger. *Functional Ecology*, **14**, 137–140.

Bond, B.J., Farnsworth, B.T., Coulombe, R.A., *et al.* (1999). Foliage physiology and biochemistry in response to light gradients in conifers with varying shade tolerance. *Oecologia*, **120**, 183–192.

Bond, W.J. (2005). Large parts of the world are brown or black: A different view on the 'Green World' hypothesis. *Journal of Vegetation Science*, **16**, 261–266.

Bond, W.J. and Keeley, J.E. (2005). Fire as global 'herbivore': the ecology and evolution of flammable ecosystems. *Trends in Ecology and Evolution*, **20**, 387–394.

Bond, W.J. and Midgley, G.F. (2000). A proposed CO_2-controlled mechanism of woody plant invasion in grasslands and savannas. *Global Change Biology*, **6**, 865–870.

Bond, W.J., Midgley, G.F. and Woodward, F.I. (2003). The importance of low atmospheric CO_2 and fire in promoting the spread of grasslands and savannas. *Global Change Biology*, **9**, 973–982.

Bond, W.J., Woodward, F.I. and Midgley, G.F. (2005). The global distribution of ecosystems in a world without fire. *New Phytologist*, **165**, 525–538.

Bonfig, K.B., Schreiber, U., Gabler, A., *et al.* (2006). Infection with virulent and avirulent *P. syringae* strains differentially effects photosynthesis and sink metabolism in *Arabidopsis* leaves. *Planta*, **225**, 1–12.

Bongi, G. and Loreto, F. (1989). Gas-exchange properties of salt-stressed olive (*Olea europaea* L.) leaves. *Plant Physiology*, **90**, 1408–1416.

Bonhomme L., Barbaroux, C. and Monclus, R. (2008) Genetic variation in productivity, leaf traits and carbon isotope discrimination in hybrid poplars cultivated on contrasting sites. *Annals of Forest Science*, **65**, 1–9.

Bonhomme, R. (2000). Beware of comparing RUE values calculated from PAR vs solar radiation or absorbed vs intercepted radiation. *Field Crops Research*, **68**, 247–252.

Boote, K.J. and Pickering, N.B. (1994). Modeling photosynthesis of row crop canopies. *HortScience*, **29**, 1423–1434.

Borchert, R. (1991). Growth periodicity and dormancy. In: *Physiology of Trees* (ed. Raghavendra, A.S.), John Wiley and Sons Inc., New York, USA, pp. 221–245.

Borel, C. and Simonneau, T. (2002). Is the ABA concentration in the sap collected by pressurizing leaves relevant for analysing drought effects on stomata? Evidence from ABA-fed leaves of transgenic plants with modified capacities to synthesize ABA. *Journal of Experimental Botany*, **53**, 287–296.

Borel, C., Frey, A., Marion-Poll, A., *et al.* (2001). Does engineering abscisic acid biosynthesis in *Nicotiana plumbaginifolia* modify stomatal response to drought? *Plant Cell and Environment*, **24**, 477–489.

Borisjuk, L., Walenta, S., Rolletschek, H., *et al.* (2002). Spatial analysis of plant metabolism: sucrose imaging within *Vicia faba* cotyledons reveals specific developmental patterns. *The Plant Journal*, **29**, 521–530.

Borland, A.M. and Griffiths, H. (1997). A comparative study on the regulation of C_3 and C_4 carboxylation processes in the constitutive crassulacean acid metabolism (CAM) plant *Kalanchoë daigremontiana* and the C_3-CAM intermediate *Clusia minor. Planta*, **201**, 368–378.

Borland, A.M., Griffiths, H., Broadmeadow, M.S.J., *et al.* (1993). Short-term changes in carbon-isotope discrimination in the C_3-CAM intermediate *Clusia minor* L. growing in Trinidad. *Oecologia*, **95**, 444–453.

Borland, A.M., Griffiths, H., Maxwell, C., *et al.* (1992). On the ecophysiology of the Clusiaceae in Trinidad: expression of CAM in *Clusia minor* L. during the transition from wet to dry season and characterization of three endemic species. *New Phytologist*, **122**, 349–357.

Borrás, L., Slafer, G.A. and Otegui, M.E. (2004). Seed dry weight response to source-sink manipulation in wheat, maize and soybean: a quantitative reappraisal. *Field Crops Research*, **86**, 131–146.

Bort, J., Brown, R.H. and Araus, J.L. (1996). Refixation of respiratory CO_2 in the ears of C_3-cereals. *Journal of Experimental Botany*, **47**, 1567–1575.

Bortier, K., Ceulemans, R. and De Temmerman, L. (2000). Effects of ozone exposure on growth and photosynthesis of beach seedlings (*Fagus sylvatica*). *The New Phytologist*, **146**, 271–280.

Bosian, G. (1960). Züm kuvettenklimaproblem: beweisführung für die nichtexistenz 2- gipfeliger assimilationskurven bei verwendung von klimatisierten. küvetten. *Flora*, **149**, 167–188.

Boston, R.S., Viitanen, P.V. and Vierling, E. (1996). Molecular chaperones and protein folding in plants. *Plant Molecular Biology*, **32**, 191–222.

Bota, J., Flexas, J. and Medrano, H. (2001). Genetic variability of photosynthesis and water use in Balearic grapevine cultivars. *Annals of Applied Biology*, **138**, 353–365.

Bota, J., Medrano, H. and Flexas, J. (2004). Is photosynthesis limited by decreased Rubisco activity and RuBP content under progressive water stress? *New Phytologist*, **162**, 671–681.

Bottin, H. and Mathis, P. (1985). Interaction of plastocyanin with the photosystem I reaction center: a kinetic study by flash absorption spectroscopy. *Biochemistry*, **24**, 6453–6460.

Bottrill, D.E., Possingham, J.V. and Kriedemann, P.E. (1970). The effect of nutrient deficiencies on photosynthesis and respiration in spinach. *Plant and Soil*, **32**, 424–438.

Boucher, O., Jones, A. and Betts, R.A. (2009). Climate response to the physiological impact of carbon dioxide on plants in the Met Office Unified Model HadCM3. *Climate Dynamics*, **32**, 237–249.

Boucher, Y., Douady, C.J., Papke, R.T., *et al.* (2003). Lateral gene transfer and the origins of prokaryotic groups. *Annual Review of Genetics*, **37**, 283–328.

Bounoua, L., DeFries, R., Collatz, G.J., *et al.* (2002). Effects of land cover conversion on surface climate. *Climatic Change*, **52**, 29–64.

Bouriaud, O., Soudani, K. and Bréda, N. (2003). Leaf-area index from litter collection: impact of specific leaf area variability within a beach stand. *Canadian Journal of Remote Sensing*, **29**, 371–380.

Boussac, A., Maison-Peteri, B., Vernotte, C., *et al.* (1985). The charge accumulation in NaCl - washed and in Ca^{2}+ - reactivated Photosystem-II particles. *Biochimica et Biophysica Acta*, **808**, 225–230.

Bowden, R.L. and Rouse, D.I. (1991). Effects of *Verticillium dahliae* on gas exchange of potato. *Phytopathology*, **81**, 293–301.

Bowden, R.L., Rouse, D.I. and Sharkey, T.D. (1990). Mechanism of photosynthesis decrease by *Verticillium dahliae* in potato. *Plant Physiology*, **94**, 1048–1055.

Bowes, J.M. and Crofts A.R. (1980). Binary oscillations in the rate of reoxidation of the primary acceptor of photosystem II. *Biochimica et Biophysica Acta*, **590**, 373–384.

Bowie, M.R., Wand, S.J.E. and Esler, K.J. (2000). Seasonal gas exchange responses under three different temperature treatments in a leaf-succulent and a drought-deciduous shrub from the Succulent Karoo. *South African Journal of Botany*, **66**, 118–123.

Bowling, D.R., Sargent, S.D., Tanner, B.D., *et al.* (2003). Tunable diode laser absorption spectroscopy for stable isotope studies of ecosystem–atmosphere CO_2 exchange. *Agricultural and Forest Meteorology*, **118**, 1–19.

Bowling, D.R., Tans, P.P. and Monson, R.K. (2001). Partitioning net ecosystem carbon exchange with isotopic fluxes of CO_2. *Global Change Biology*, **7**, 127–145.

Bowman, W.D., Hubick, K.T., von Caemmerer, S., *et al.* (1989). Short-term changes in leaf carbon isotope discrimination in salt- and water-stressed C_4 grasses. *Plant Physiology*, **90**, 162–166.

Bown, H.E., Watt, M.S., Clinton, P.W., *et al.* (2009). The influence of N and P supply and genotype on carbon flux and partitioning in potted *Pinus radiata* plants. *Tree Physiology*, **29**, 1143–1151.

Bowyer, W.J., Ning, L., Daley, L.S., *et al.* (1998). In vivo fluorescence imaging for detection of damage to leaves by fungal phytotoxins. *Spectroscopy*, **13**, 36–44.

Box, E.O. (1981). *Macroclimate and Plant Forms: An Introduction to Predictive Modelling in Phytogeography*, Dr. W. Junk Publishers, The Hague – Boston – London.

Boxall, S.F., Foster, J.M., Bohnert, H.J., *et al.* (2005). Conservation and divergence of circadian clock operation in a stress-inducible crassulacean acid metabolism species reveals clock compensation against stress. *Plant Physiology*, **137**, 969–982.

Boyce, R. and Lucero, S. (2002). Role of roots in winter water relations of Engelmann spruce saplings. *Tree Physiology*, **19**, 893–898.

Boyer, J.S., Wong, S.C. and Farquhar, G.D. (1997). CO_2 and water vapour exchange across leaf cuticle (epidermis) at various water potentials. *Plant Physiology*, **114**, 185–191.

Bradford, K.J. (1983). Effects of flooding on leaf gas exchange of tomato plants. *Plant Physiology*, **73**, 475–479.

Bradford, M.M. (1976). A rapid and sensitive method for the quantitation of microgram quantities of protein utilizing the principle of protein-dye binding. *Analytical Biochemistry*, **72**, 248–254.

Brandon, P.C. (1967). Temperature features of enzymes affecting crassulacean acid metabolism. *Plant Physiology*, **42**, 977–984.

Brandt, S., Kehr, J., Walz, C., *et al.* (1999). A rapid method for detection of plant gene transcripts from single epidermal, mesophyll and companion cells of intact leaves. *Plant Journal*, **20**, 245–250.

Bravo, L.A. and Griffith, M. (2005). Characterization of antifreeze activity in Antarctic plants. *Journal of Experimental Botany*, **56**, 1189–1196.

Bravo, L.A., Saavedra-Mella, F.A., Vera, F., *et al.* (2007). Effect of cold acclimation on the photosynthetic performance of two ecotypes of *Colobanthus quitensis* (Kunth) Bartl. *Journal of Experimental Botany*, **58**, 3581–3590.

Bray, E.A. (2002). Abscisic acid regulation of gene expression during water-deficit stress in the era of the *Arabidopsis* genome. *Plant Cell and the Environment*, **25**, 153–161.

Bréhélin, C., Kessler, F. and van Wijk, K.J. (2007). Plastoglobules: versatile lipoprotein particles in plastids. *Trends in Plant Sciences*, **12**, 260–166.

Bremer, D.J., Auen, L.M., Ham, J.M., *et al.* (2001). Evapotranspiration in a prairie ecosystem: effects of grazing by cattle. *Agronomy Journal*, **93**, 338–348.

Brendel, O., Le Thiec, D., Scotti-Saintagne, C., *et al.* (2008). Quantitative trait loci controlling water use efficiency and related traits in *Quercus robur* L. *Tree Genetics and Genomes*, **4**, 263–278.

Breyton, C., Nandha, B, Johnson, G.N., Joliot, P. and Finazzi, G. (2006). Redox modulation of cyclic electron flow around photosystem I in C3 plants. *Biochemistry*, **45**, 13465–13475.

Briantais, J.M., Dacosta, J., Goulas, Y., *et al.* (1996). Heat stress induces in leaves an increase of the minimum level of chlorophyll fluorescence, F-0: a time-resolved analysis. *Photosynthesis Research*, **48**, 189–196.

Briantais, J.M., Vernotte, C., Picaud, M., *et al.* (1979). Quantitative study of the slow decline of chlorophyll alpha-fluorescence in isolated-chloroplasts. *Biochim Biophys Acta*, **548**, 128–138.

Briggs, L.J. and Shantz, H.J. (1914). Relative water requirements of plants. *Journal Agriculture Research*, **3**, 1–63.

Briggs, L.J. and Shantz, HJL. (1913). The water requeriments of plants: a review of the literature. *USDA Plant Industries Bulletin* 285.

Briggs, W.R. (2005). Physiology of plant responses to artificial lighting. In: *Ecological Consequences of Artificial Night Lighting* (eds Rich, C. and Longcore, T.), Island Press, Washington, DC, USA, pp. 389–412.

Briggs, W.R. and Olney, M.A. (2001). Photoreceptors in plant photomorphogenesis to date: five phytochromes, two cryptochromes, one phototropin, and one superchrome. *Plant Physiology*, **125**, 85–88.

Bright, J., Desikan, R., Hancock, J.T., *et al.* (2006). ABA-induced NO generation and stomatal closure in *Arabidopsis* are dependent on H_2O_2 synthesis. *Plant Journal*, **45**, 113–122.

Brilli, F., Barta, C., Fortunati, A., *et al.* (2007). Response of isoprene emission and carbon metabolism to drought in white poplar (*Populus alba*) saplings. *New Phytologist*, **175**, 244–254.

Brodersen, C.R. and Vogelmann, T.C. (2007). Do epidermal lens cells facilitate the absorptance of diffuse light? *American Journal of Botany*, **94**, 1061–1066.

Brodersen, C.R. and Vogelmann, T.C. (2010). Do changes in light direction affect absorption profiles in leaves? *Functional Plant Biology*, **37**, 403–412.

Brodersen, C.R., Vogelmann, T.C., Williams, W.E., *et al.* (2008). A new paradigm in leaf-level photosynthesis: direct and diffuse lights are not equal. *Plant Cell Environ*, **31**, 159–164.

Brodribb, T. and Hill, R.S. (1993). A physiological comparison of leaves and phyllodes in *Acacia melanoxylon*. *Australian Journal of Botany*, **41**, 293–305.

Brodribb, T. and Hill, R.S. (1999). The importance of xylem constraints in the distribution of coniferous plants. *New Phytologist*, **143**, 365–372.

Brodribb, T.J. and Cochard, H. (2009). Hydraulic failure defines the recovery and point of death in water-stressed conifers. *Plant Physiology*, **149**, 575–584.

Brodribb, T.J. and Holbrook, N.M. (2003). Stomatal closure during leaf dehydration, correlation with other leaf physiological traits. *Plant Physiology*, **132**, 2166–2173.

Brodribb, T.J. and Jordan, G.J. (2008). Internal coordination between hydraulics and stomatal control in leaves. *Plant, Cell and Environment*, **31**, 1557–1564.

Brodribb, T.J. and McAdam, S.A.M. (2011). Passive origins of stomatal control in vascular plants. *Science*, **331**, 582–585.

Brodribb, T.J., Field, T.S. and Jordan, G.J. (2007). Leaf maximum photosynthetic rate and venation are linked by hydraulics. *Plant Physiology*, **144**, 1890–1898.

Brodribb, T.J., McAdam, S.A.M., Jordan, G.J., *et al.* (2009). Evolution of stomatal responsiveness to CO_2 and optimization of water-use efficiency among land plants. *The New Phytologist*, **183**, 839–847.

Brooks, A. (1986). Effects of phosphorus nutrition on ribulose 1,5-bisphosphate carboxylase activation, photosynthetic quantum yield and amounts of some Calvin cycle metabolites in spinach leaves. *Australian Journal of Plant Physiology*, **13**, 221–237.

Brooks, A. and Farquhar, G.D. (1985). Effect of temperature on the CO_2/O_2 specificity of ribulose-1,5-bisphosphate carboxylase/oxygenase and the rate of respiration in the light. *Planta*, **165**, 397–406.

Brooks, J.R., Flanagan, L.B., Buchmann, N., *et al.* (1997). Carbon isotope composition of Boreal plants: functional grouping of life forms. *Oecologia*, **110**, 301–311.

Brooks, J.R., Hinckley, T.M. and Sprugel, D.G. (1994). Acclimation responses of mature *Abies amabilis* sun foliage to shading. *Oecologia*, **100**, 316–324.

Brooks, J.R., Sprugel, D.G. and Hinckley, T.M. (1996). The effects of light acclimation during and after foliage expansion on photosynthesis of *Abies amabilis* foliage within the canopy. *Oecologia*, **107**, 21–32.

Brouwer, R. and de Wit, C.T. (1969). A simulation model of plant growth with special attention to root growth and its consequences. In: *Root Growth* (ed. Whittington, W.J.), Plenum, New York, USA, pp. 224–244.

Brown, K.R. and Courtin, P.J. (2003). Effects of phosphorus fertilization and liming on growth, mineral nutrition, and gas exchange of *Alnus rubra* seedlings grown in soils from mature alluvial *Alnus* stands. *Canadian Journal of Forest Research-Revue Canadienne De Recherche Forestiere*, **33**, 2089–2096.

Brown, R.A., Rosenberg, N.J., Hays, C.J., *et al.* (2000). Potential production and environmental effects of switchgrass and traditional crops under current and greenhouse-altered climate in the central United States: a simulation study. *Agriculture, Ecosystems and Environment*, **78**, 31–47.

Brown, R.H. (1978). A difference in N use efficiency in C_3 and C_4 plants and its implications in adaptation and evolution. *Crop Science*, **18**, 93–98.

Brown, R.H. and Simmons, R.E. (1979). Photosynthesis of grass species differing in CO_2 fixation pathways. I. Water-use efficiency. *Crop Science*, **19**, 375–379.

Browse, J. and Xin, Z. (2001). Temperature sensing and cold acclimation. *Current Opinion in Plant Biology*, **4**, 241–246.

Brugnoli, E. and Farquhar, G.D. (2000). Photosynthetic fractionation of carbon isotopes. In: *Photosynthesis Physiology and Metabolism, Advances in Photosynthesis* (eds Leegood, R.C., Sharkey, T.D. and von Caemmerer, S.), Kluwer Academic Publishers, Dordrecht, Netherlands, pp. 399–434.

Brugnoli, E., Hubick, K.T., von Caemmerer, S., *et al.* (1988). Correlation between the carbon isotope discrimination in leaf starch and sugars of C_3 plants and the ratio of intercellular and atmospheric partial pressures of carbon dioxide. *Plant Physiology*, **88**, 1418–1424.

Brugnoli, E., Lauteri, M. and Guido M.C. (1994). Carbon isotpe discrimination and photosynthesis: response and adaptation to environmental stress. In: *Plant Sciences 1994 Second General Colloquium on Plant Sciences* (eds de Kouchkovsky, Y. and Larher, F.). SFPV Université de Rennes, Rennes, France, pp. 269–272.

Brugnoli, E., Scartazza, A., Lauteri, M., *et al.* (1998). Carbon isotope discrimination in structural and non-structural carbohydrates in relation to productivity and adaptation to unfavourable conditions. In: *Stable Isotopes: Integration of Biological, Ecological and Geochemical Processes* (ed. Griffiths, H.), BIOS Scientific Publishers, Oxford, UK, pp. 133–144.

Bruinsma, J. (1961). A comment on the spectrophotometric determination of chlorophyll. *Biochemistry and Biophysics Acta*, **52**, 576–578.

Brunes, L., Öquist, G. and Eliasson, L. (1980). On the reason for the different photosynthetic rates of seedlings of *Pinus sylvestris* and *Betula verrucosa*. *Plant Physiology*, **66**, 940–944.

Bruni, N.C., Young, J.P. and Dengler, N.C. (1996). Leaf developmental plasticity of *Ranunculus flabellaris* in response to terrestrial and submerged environments. *Canadian Journal of Botany*, **74**, 823–837.

Bruzzese, B.M., Bowler, R., Massicotte, H.B. *et al.* (2010). Photosynthetic light response in three carnivorous plant species: *Drosera rotundifolia*, *D. Capensis* and *Sarracenia leucophylla*. *Photosynthetica*, **48**, 103–109.

Bryant, J.P., Chapin, F.S.I., Reichardt, P.B., *et al.* (1987). Response of winter chemical defense in Alaska paper birch and green alder to manipulation of carbon/nutrient balance. *Oecologia*, **72**, 510–514.

Buchanan, B.B (1984). The ferredoxin/thioredoxin system: a key element in the regulatory function of light in photosynthesis. *BioScience*, **34**, 378–383.

Buchanan, B.B. and Balmer, Y. (2005). Redox regulation: a broadening horizon. *Annual Review Plant Biology*, **56**, 187–220.

Buchanan-Bollig, I.C. and Kluge, M. (1981). Crassulacean acid metabolism (CAM) in *Kalanchoë daigremontiana*: temperature response of phosphoenolpyruvate (PEP)-carboxylase in relation to allosteric effectors. *Planta*, **152**, 181–188.

Buchanan-Bollig, I.C., Kluge, M. and Müller, D. (1984). Kinetic changes with temperature of phosphoenolpyruvate carboxylase from a CAM plant. *Plant, Cell and Environment* **7**, 63–70.

Buchmann, N., Brooks, J.R., Rapp, K.D., *et al.* (1996). Carbon isotope composition of C_4 grasses is influenced by light and water supply. *Plant, Cell and Environment*, **19**, 392–402.

Buchner, O., Holzinger, A. and Lütz, C. (2007). Effects of temperature and light on the formation of chloroplast protrusions in leaf mesophyll cells of high alpine plants. *Plant, Cell and Environment*, **30**, 1347–1356.

Buckland, S.M., Grime, J.P., Hodgson, J.G., *et al.* (1997). A comparison of plant responses to the extreme drought of 1995 in northern England. *Journal of Ecology* **85**, 875–882.

Buckley, T.N. (2005). The control of stomata by water balance. *New Phytologist*, **168**, 275–292.

Buckley, T.N. and Farquhar, G.D. (2004). A new analytical model for whole leaf potential electron transport rate. *Plant Cell and Environment*, **27**, 1487–1502.

Buckley, T.N., Farquhar, G.D. and Mott, K.A. (1997). Qualitative effects of patchy stomatal conductance distribution features on gas-exchange calculations. *Plant, Cell and Environment*, **20**, 867–880.

Buckley, T.N., Mott, K.A. and Farquhar, G.D. (2003). A hydromechanical and biochemical model of stomatal conductance. *Plant, Cell and Environment*, **26**, 1767–1786.

Budde, R.J.A. and Randall, D.D. (1990). Pea leaf mitochondrial PDH complex is inactivated in vivo in a light-dependent manner. *Proceedings of the National Academy of Sciences USA*, **87**, 673–676.

Bugbee, B. (1992). Steady state canopy gas exchange: system design and operation. *HortScience*, **27**, 770–776.

Bugbee, B.G. and Salisbury, F.B. (1988). Exploring the limits of crop productivity. I. Photosynthetic efficiency of wheat in high irradiance environments. *Plant Physiolopgy*, **88**, 869–878.

Buick, R. (2008). When did oxygenic photosynthesis evolve? *Philosophical Transactions of the Royal Society B*, **363**, 2731–2743.

Bukhov, N. and Carpentier, R. (2004). Alternative photosystem-I driven electron transport routes: mechanisms and functions. *Photosynthesis Research*, **82**, 17–33.

Bukhov, N.G., Wiese, C., Neimanis, S., *et al.* (1999). Heat sensitivity of chloroplasts and leaves: leakage of protons from thylakoids and reversible activation of cyclic electron transport. *Photosynthesis Research*, **59**, 81–93.

Bulman, P., Mather, D.E. and Smith, D.L. (1993). Genetic improvement of spring barley cultivars grown in eastern Canada from 1910 to 1988. *Euphytica*, **71**, 35–48.

Bunce, J.A. (2000). Responses of stomatal conductance to light, humidity and temperature in winter wheat and barley grown at three concentrations of carbon dioxide in the field. *Global Change Biology*, **6**, 371–382.

Bunce, J.A. (2001). Seasonal patterns of photosynthetic response and acclimation to elevated carbon dioxide in field-grown strawberry. *Photosynthesis Research*, **68**, 237–245.

Bunce, J.A. (2008). Acclimation of photosynthesis to temperature in *Arabidopsis thaliana* and *Brassica oleracea*. *Photosynthetica*, **46**, 517–524.

Bunce, J.A. (2009). Use of the response of photosynthesis to oxygen to estimate mesophyll conductance to carbon dioxide in water-stressed soybean leaves. *Plant, Cell and Environment*, **32**, 875–881.

Bunce, J.A. (2010). Variable responses of mesophyll conductance to substomatal carbon dioxide concentration in common bean and soybean. *Photosynthetica*, **48**, 507–512.

Bungard, R.A. (2004). Photosynthetic evolution in parasitic plants: insight from the chloroplast genome. *BioEssays*, **26**, 235–247.

Bungard, R.A., Ruban, A.V., Hibberd, J.M., *et al.* (1999). Unusual carotenoid composition and a new type of xanthophyll cycle in plants. *Proceedings of the National Academy of Sciences USA*, **96**, 1135–1139.

Bungard, R.A., Zipperlen, S.A., Press, M.C., *et al.* (2002). The influence of nutrients on growth and photosynthesis of seedlings of two rainforest dipterocarp species. *Functional Plant Biology*, **29**, 501–515.

Burghardt, M. and Riederer, M. (2003). Ecophysiological relevance of cuticular transpiration of deciduous and evergreen plants in relation to stomatal closure and leaf water potential. *Journal of Experimental Botany*, **54**, 1941–1949.

Burkart, S., Manderscheid, R. and Weigel, H.J. (2000). Interacting effects of photosynthetic photon-flux density and temperature on canopy CO_2 exchange rate of spring wheat under different CO_2-concentrations. *Journal of Plant Physiology*, **157**, 31–39.

Burkart, S., Manderscheid, R. and Weigel, H.J. (2004). Interactive effects of elevated atmospheric CO_2 concentrations and plant available soil water content on canopy evapotranspiration and conductance of spring wheat. *European Journal of Agronomy*, **21**, 401–417.

Burkart, S., Manderscheid, R. and Weigel, H.J. (2007). Design and performance of a portable gas exchange chamber system for CO_2- and H_2O-flux measurements in crop canopies. *Environmental Experimental Botany*, **61**, 25–34.

Busch, F., Huner, N.P.A. and Ensminger, I. (2007). Increased air temperature during simulated autumn conditions does not increase photosynthetic carbon gain but affects the dissipation of excess energy in seedlings of the evergreen conifer jack pine. *Plant Physiology*, **143**, 1242–1251.

Busch, F., Hüner, N.P.A. and Ensminger, I. (2008). Increased air temperature during simulated autumn conditions impairs photosynthetic electron transport between photosystem II and photosystem I. *Plant Physiology*, **147**, 402–414.

Buschmann, C. (1999). Thermal dissipation during photosynthetic induction and subsequent dark recovery as measured by photoacoustic signals. *Photosynthetica*, **36**, 149–161.

Buschmann, C. and Lichtenthaler, H.K. (1998). Principles and characteristics of multi-colour fluorescence imaging of plants. *Journal of Plant Physiology*, **152**, 297–314.

Buschmann, C., Langsdorf, G. and Lichtenthaler, H.K. (2000). Imaging of the blue, green, and red fluorescence emission of plants: an overview. *Photosynthetica*, **38**, 483–491.

Businger, J.A. (1986). Evaluation of the accuracy with wich dry deposition can be measured with current micrometeorological techniques. *Journal of Climate and Applied Meteorology*, **25**, 1100–1124.

Bussotti, F. and Ferretti, M. (1998). Air pollution, forest condition and forest decline in Southern Europe: an overview. *Environmental Pollution*, **101**, 49–65.

Bussotti, F., Desotgiu, R., Cascio, C., *et al.* (2007). Photosynthesis responses to ozone in young trees of three species with different sensitivities, in a 2-year open-top chamber experiment (Curno, Italy). *Physiologia Plantarum*, **130**, 122–135.

Butz, N.D. and Sharkey, T.D. (1989). Activity ratios of ribulose-1,5-bisphosphate carboxylase accurately reflect carbamylation ratios. *Plant Physiology*, **89**, 735–739.

Bykova, N.V., Keerberg, O., Pärnik, T., *et al.* (2005). Interaction between photorespiration and respiration in transgenic potato plants with antisense reduction in glycine decarboxylase. *Planta*, **222**, 130–140.

Byrd, G.T., Sage, R.F. and Brown R.H. (1992). A comparison of dark respiration between C_3 and C_4 plants. *Plant Physiology*, **100**, 191–198.

Byrdwell, W.C. and Neff, W.E. (2002). Dual parallel electrospray ionization and atmospheric pressure chemical ionization mass spectrometry (MS), MS/MS and MS/MS/MS for the analysis of triacylglycerols and triacylglycerol oxidation products. *Rapid Communications in Mass Spectrometry*, **16**, 300–319.

Cabrera, H.M., Rada, F. and Cavieres, L. (1998). Effects of temperature on photosynthesis of two morphologically contrasting plant species along an altidudinal gradient in the tropical high Andes. *Oecologia*, **114**, 145–152.

Cai, H., Biswas, D.K., Shang, A.Q., *et al.* (2007). Photosynthetic response to water stress and changes in metabolites in *Jasminum sambac. Photosynthetica*, **45**, 503–509.

Cai, T., Flanagan, L.B., Jassal, R.S., *et al.* (2008). Modelling environmental controls on ecosystem photosynthesis and the carbon isotope composition of ecosystem-respired CO_2 in a coastal Douglas fir forest. *Plant, Cell and Environment*, **31**, 435–453.

Cai, Y.F., Zhang, S.B., Hu, H., *et al.* (2010). Photosynthetic performance and acclimation of *Incarvillea delavayi* to water stress. *Biologia Plantarum*, **54**, 89–96.

Cai, Z.Q., Chen, Y.J., Guo, Y.H., *et al.* (2005). Responses of two field-grown coffee species to drought and tobacco. *Photosynthetica*, **43**, 187–193.

Caldwell, M.M., White, R.S., Moore, R.T., *et al.* (1977). Carbon balance, productivity, and water use of cold-winter desert shrub communities dominated by C_3 and C_4 species. *Oecologia*, **29**, 275–300.

Calfapietra, C., Mugnozza, G.S., Karnosky, D.F., *et al.* (2008). Isoprene emission rates under elevated CO_2 and O_3 in two field-grown aspen clones differing in their sensitivity to O_3. *New Phytologist*, **179**, 55–61.

Calfapietra, C., Tulva, I., Eensalu, E., *et al.* (2005). Canopy profiles of photosynthetic parameters under elevated CO_2 and N fertilization in a poplar plantation. *Environmental Pollution*, **137**, 525–535.

Calvin, M. (1962). The path of carbon in photosynthesis. *Science*, **135**, 879–889.

Calvin, M. and Benson. A. A. (1948). The path of carbon in photosynthesis. *Science*, **107**, 476-480.

Calvin, M., Heidelberger, C., Reid, J.C., *et al.* (1949). Isotopic carbon techniques. In: *Measurement and Chemical Manipulation*, John Wiley and Sons, Inc, New York, USA.

Camenen, L., Goulas, Y., Guyot, G., *et al.* (1996) Estimation of the chlorophyll fluorescence lifetime of plant canopies: validation of a deconvolution method based on the use of a 3-D canopy mockup. *Remote Sensing of the Environment*, **57**, 79–87.

Cameron, D.D., Geniez, J.M., Seel, W.E., *et al.* (2007). Suppression of host photosynthesis by the parasitic plant *Rhinanthus minor. Annals of Botany*, **101**, 573–578.

Campbell, G.S. (1986). Extinction coefficients for radiation in plant canopies calculated using an ellipsoidal inclination angle distribution. *Agricultural and Forest Meteorology*, **36**, 317–321.

Campbell, G.S. and Norman, J.M. (1998). *An Introduction to Environmental Biophysics.* Springer-Verlag, New York, USA.

Campbell, W.J., Allen, L.H. Jr. and Bowes, G. (1990). Response of soybean canopy photosynthesis to CO_2 concentration, light and temperature. *Journal of Experimenal Botany*, **41**, 427–433.

Canaani, O. and Havaux, M. (1990). Evidence for a biological role in photosynthesis for cytochrome b-559, a component of photosystem II reaction center. *Proceedings of the National Academy of Sciences USA*, **87**, 9295–9299.

Canadell, J.G. and Raupach, M.R. (2008). Managing forests for climate change mitigation. *Science*, **320**, 1456–1457.

Canadell, J.P., Le Quéré, C., Raupach, M.R., *et al.* (2007). Contributions to accelerating atmospheric CO_2 growth from economic activity, carbon intensity, and efficiency of natural sinks. *Proceedings of the National Academy of Sciences of the USA*, **104**, 18866–18870.

Canham, C.D., Finzi, A.C., Pacala, S.W., *et al.* (1994). Causes and consequences of resource heterogeneity in forests: interspecific variation in light transmission by canopy trees. *Canadian Journal of Forest Research*, **24**, 337–349.

Cannani, O. and Malkin, S. (1984). Distribution of light excitation in an intact leaf between the two photosystems of photosynthesis: changes in absorption cross-section following state 1 and state 2 transitions. *Biochimica et Biophysica Acta*, **766**, 513–524.

Cannell, M.G.R. and Thornley, J.H.M. (1998). Temperature and CO_2 responses of leaf and canopy photosynthesis: a clarification using the non-rectangular hyperbola model of photosynthesis. *Annals of Botany*, **82**, 883–892.

Canvin, D.T., Berry, J.A., Badger, M.R., *et al.* (1980). Oxygen exchange in leaves in the light. *Plant Physiology*, **66**, 302–307.

Cao, J., Hesketh, J.D., Zur, B., *et al.* (1988). Leaf area development in maize and soybean plants. *Biotronics*, **17**, 9–15.

Cao, K.F. and Ohkubo, T. (1998). Allometry, root/shoot ratio and root architecture in understory saplings of deciduous dicotyledonous trees in central Japan. *Ecological Research*, **13**, 217–227.

Cao, L., Bala, G., Caldeira, K., *et al.* (2010). Importance of carbon dioxide physiological forcing to future climate change. *Proceedings of the National Academy of Sciences USA*, **107**, 9513–9518.

Cape, J.N. (2003). Effects of airborne volatile organic compounds on plants. *Environmental Pollution*, **122**, 145–157.

Capell, B. and Dorffling, K. (1993). Genotype-specific differences in chilling tolerance of maize in relation to chilling-induced changes in water status and abscisic-acid accumulation. *Physiologia Plantarum*, **88**, 638–646.

Caporn, S.J.M. (1989). The effects of oxides of nitrogen and carbon dioxide enrichment on photosynthesis and growth of lettuce (*Lactuca sativa* L.). *New Phytologist*, **111**, 473–481.

Caporn, S.J.M., Risager, M. and Lee, J.A. (1994). Effect of atmospheric nitrogen deposition on frost hardiness in *Calluna vulgaris*. *New Phytologist*, **128**, 461–468.

Carlberg, I., Hansson, M., Kieselbach, T., *et al.* (2003). A novel plant protein undergoing light-induced phosphorylation and release from the photosynthetic thylakoid membranes. *Proceedings of the National Academy of Sciences* USA, **100**, 757–762.

Carlson, R.W., Bazzaz, F.A. and Rolfe, G.L. (1975). The effect of heavy metals on plants. II. Net photosynthesis and transpiration of whole corn and sunflower plants treated with Pb, Cd, Ni, and Tl. *Environmental Research*, **10**, 113–120.

Carmo-Silva, A.E., da Silva, A.B., Keys, A.J., *et al.* (2008b). The activities of PEP carboxylase and the C_4 acid decarboxylases are little changed by drought stress in three C_4 grasses of different subtypes. *Photosynthesis Research*, **97**, 223–233.

Carmo-Silva, A.E., Powers, S.J., Keys, A.J., *et al.* (2008a). Photorespiration in C_4 grasses remains slow under drought conditions. *Plant, Cell and Environment*, **31**, 925–940.

Carnicer, J., Coll, M., Ninyerola, M. *et al.* (2011). Widespread crown condition decline, food web disruption, and amplified tree mortality with increased climate change-type drought. *Proceedings of the National Academy of Sciences USA*, **108**, 1474–1478.

Carol, P. and Kuntz, M. (2001). A plastid terminal oxidase comes to light: implications for carotenoid biosynthesis and chlororespiration. *Trends in Plant Science*, **6**, 31–36.

Carpenter, K.J. (2005). Stomatal architecture and evolution in basal angiosperms. *American Journal of Botany*, **92**, 1595–1615.

Carpenter, K.J. (2006). Specialized structures in the leaf of basal angiosperms: morphology, distribution, and homology. *American Journal of Botany*, **93**, 665–681.

Carpenter, R.J., Jordan, G.J., Leigh, A., *et al.* (2007). Giant cuticle pores in *Eidothea zoexylocarya* (Proteaceae) leaves. *American Journal of Botany*, **94**, 1282–1288.

Carrara, S., Pardosi, A., Soldatini, G.F., *et al.* (2001). Photosynthetic activity of ripening tomato fruit. *Photosynthetica*, **39**, 75–78.

Carrari, F., Nunes-Nesi, A., Gibon, Y., *et al.* (2003). Reduced expression of aconitase results in an enhanced rate of photosynthesis and ed leaves of wild species tomato. *Plant Physiology*, **133**, 1322–1335.

Carroll, T.W. (1970). Relation of barley stripe mosaic virus to plastids. *Virology*, **42**, 1015–1022.

Carroll, T.W. and Kosuge, T. (1969). Changes in structure of chloroplasts accompanying necrosis of tobacco leaves systemically infected with tobacco mosaic virus. *Phytopathology*, **59**, 953–962.

Carswell, F.E., Whitehead, D., Rogers, G.N.D., *et al.* (2005). Plasticity in photosynthetic response to nutrient supply of seedlings from a mixed conifer-angiosperm forest. *Austral Ecology*, **30**, 426–434.

Cartelat, A., Cerovic, Z., Goulas, Y., *et al.* (2005). Optically assessed polyphenolics and chlorophyll content of leaves as indicators of nitrogen deficiency in wheat. *Field Crops Research*, **91**, 35–49.

Carter, G.A., Jones, J.H., Mitchell, R.J., *et al.* (1996). Detection of solar-excited chlorophyll a fluorescence and leaf photosynthetic capacity using a Fraunhofer line radiometer. *Remote Sensing of the Environment*, **55**, 89–92.

Carter, G.A., Theisen, A.F. and Mitchell, R.J. (1990). Chlorophyll fluorescence measured using the Fraunhofer line-depth principle and relationship to photosynthetic rate in the field. *Plant Cell Environment*, **13**, 79–83.

Carter, P.J., Wilkins, M.B., Nimmo, H.G., *et al.* (1995). Effects of temperature on the activity of phosphoenolpyruvate carboxylase and on the control of CO_2 fixation in *Bryophyllum fedtschenkoi*. *Planta*, **196**, 375–380.

Casella, E. and Sinoquet, H. (2003). A method for describing the canopy architecture of coppice poplar with allometric relationships. *Tree Physiology*, **23**, 1153–1170.

Caspar, T., Huber, S.C. and Somerville, C. (1985). Alterations in growth, photosynthesis, and respiration in a starchless mutant of *Arabidopsis thaliana* (L.) deficient in chloroplast phosphoglucomutase activity. *Plant Physiology*, **79**, 11–17.

Cassman, K.G., Dobermann, A., Walters, D.T., *et al.* (2003). Meeting cereal demand while protecting natural resources and improving environmental quality. *Annual Review of Environment and Resources*, **28**, 315–358.

Castrillo, M. (1995). Ribulose-1,5-bisphosphate carboxylase activity in altitudinal populations of *Espeletia schultzii* Wedd. *Oecologia*, **101**, 193–196.

Castro, A.J., Carapito, C., Zorn, N., *et al.* (2005). Proteomic analysis of grapevine (*Vitis vinifera* L.) tissues subjected to herbicide stress. *Journal of Experimental Botany*, **56**, 2783–2795.

Caswell, H. and Reed, F.C. (1975). Indigestibility of C_4 bundle sheath cells by the grasshopper *Melanoplus confusus*. *Annals of the Entomological Society of America*, **68**, 686–688.

Caswell, H. and Reed, F.C. (1976). Plant-herbivore interactions: the indigestibilty of C_4 bundle sheath cells by grasshoppers. *Oecologia*, **26**, 151–156.

Caswell, H., Reed, F., Stephenson, S.N., *et al.* (1973). Photosynthetic pathways and selective herbivory: a hypothesis. *American Naturalist*, **107**, 465–480.

Causin, H.B., Tremmel, D.C., Rufty, T.W., *et al.* (2004). Growth, nitrogen uptake, and metabolism in two semiarid shrubs grown at ambient and elevated atmospheric CO_2 concentrations: effects of nitrogen supply and source. *American Journal of Botany*, **91**, 565–572.

Cavender-Bares, J., Apostol, S., Moya, I., *et al.* (1999). Chilling-induced photoinhibition in two oak species: are evergreen leaves better protected than deciduous leaves? *Photosynthetica*, **36**, 587–596.

Cechin, I. and Press, M.C. (1993). Influence of nitrogen on growth and photosynthesis of a C_3 cereal, *Oryza sativa*, infected with the root hemiparasite *Striga hermonthica*. *Journal of Experimental Botany*, **45**, 925–930.

Cen, Y.P. and Sage, R.F. (2005). The regulation of Rubisco activity in response to variation in temperature and atmospheric CO_2 partial pressure in sweet potato. *Plant Physiology*, **139**, 979–990.

Cen, Y.P., Turpin, D.H. and Layzell D.B. (2001). Whole-plant gas exchange and reductive biosynthesis in white lupin. *Plant Physiology*, **126**, 1555–1565.

Centritto, M., Lauteri, M., Monteverdi, M.C., *et al.* (2009). Leaf gas exchange, carbon isotope discrimination, and grain yield in contrasting rice genotypes subjected to water deficits during the reproductive stress. *Journal of Experimental Botany*, **60**, 2325–2339.

Centritto, M., Lee, H.S.J. and Jarvis, P.G. (1999a). Increased growth in elevated [CO_2]: an early, short-term response? *Global Change Biology*, **5**, 623–633.

Centritto, M., Lee, H.S.J. and Jarvis, P.G. (1999b). Interactive effects of elevated CO_2 and drought on cherry (*Prunus avium*) seedlings: I. Growth, wholeplant water use efficiency and water loss. *New Phytologist*, **141**, 129–140.

Centritto, M., Loreto, F. and Chartzoulakis, K. (2003). The use of low [CO_2] to estimate diffusional and non-diffusional limitations of photosynthetic capacity of salt-stressed olive saplings. *Plant, Cell and Environment*, **26**, 585–594.

Centritto, M., Lucas, M.E. and Jarvis, P.G. (2002). Gas exchange, biomass, whole-plant water-use efficiency and water uptake of peach (*Prunus persica*) seedlings in response to elevated carbon dioxide concentration and water availability. *Tree physiology*, **22**, 699–706.

Centritto, M., Magnani, F., Lee, H.S.J., *et al.* (1999c). Interactive effects of elevated [CO_2] and drought on cherry (*Prunus avium*) seedlings. II. Photosynthetic capacity and water relations. *New Phytologist*, **141**, 141–153.

Centritto, M., Nascetti, P., Petrilli, L., *et al.* (2004). Profiles of isoprene emission and photosynthetic parameters in hybrid poplars exposed to free-air CO_2 enrichment. *Plant, Cell and Environment*, **27**, 403–412.

Cerasoli, S., McGuire, M.A., Faria, J., *et al.* (2009). CO_2 efflux, CO_2 concentration and photosynthetic refixation in stems of *Eucalyptus globulus* (Labill.). *Journal of Experimental Botany*, **60**, 99–105.

Cerling, T.E. (1999). Paleorecords of C_4 plants and ecosystems. In: *C_4 Plant Biology* (eds Sage, R.F. and Monson, R.K.), Academic Press, San Diego, CA, USA, pp. 445–469.

Cerling, T.E., Ehleringer, J.R. and Harris, J.M. (1998). Carbon dioxide starvation, the development of C_4 ecosystems, and mammalian evolution. *Philosophical Transactions of the Royal Society of London, Series B.*, **353**, 159–170.

Cerling, T.E., Harris, J.M. MacFadden, B.J., *et al.* (1997). Global vegetation change through the Miocene/Pliocene boundary. *Nature*, **389**, 153–158.

Cerling, T.E., Wang, Y. and Quade, J. (1993). Expansion of C_4 ecosystems as an indicator of global ecological change in the late Miocene. *Nature*, **361**, 344–345.

Cernusak, L.A. and Hutley, L.B. (2011). Stable isotopes reveal the contribution of corticular photosynthesis to growth in branches of *Eucalyptus miniata*. *Plant Physiology*, **155**, 515–523.

Cernusak, L.A. and Marshall, J.D. (2000). Photosynthetic refixation in branches of western white pine. *Functional Ecology*, **14**, 300–311.

Cernusak, L.A., Arthur, D.J., Pate J.S., *et al.* (2003). Water relations link carbon and oxygen isotope discrimination to phloem sap sugars concentration in *Eucalyptus globulus*. *Plant Physiology*, **134**, 1544–1554.

Cernusak, L.A., Farquhar, G.D. and Pate, J.S. (2005). Environmental and physiological controls over the oxygen and carbon isotope composition of the Tasmanian blue gum, *Eucalyptus globulus*. *Tree Physiology*, **25**, 129–146.

Cernusak, L.A., Pate J.S. and Farquhar, G.D. (2002). Diurnal variation in the stable isotope composition of water and dry matterin fruiting *Lupinus angustifolius* under field conditions. *Plant, Cell and the Environment*, **25**, 893–907.

Cernusak, L.A., Winter, K., Aranda, J., *et al.* (2008). Conifers, angiosperm trees, and lianas: growth, whole-plant water and nitrogen use efficiency, and stable isotope composition ($\delta^{13}C$ and $\delta^{18}O$) of seedlings grown in a tropical environment. *Plant Physiology*, **148**, 642–659.

Cerovic, Z.G., Bergher, M., Goulas, Y., *et al.* (1993). Simultaneous measurement of changes in red and blue fluorescence in illuminated isolated chloroplasts and leaf pieces: the contribution of NADPH to the blue fluorescence signal. *Photosynthesis Research*, **36**, 193–204.

Cerovic, Z.G., Goulas, Y., Gorbunov, M., *et al.* (1996). Fluorosensing of water stress in plants. Diurnal changes of the mean lifetime and yield of chlorophyll fluorescence, measured simultaneously and at distance with a t-LIDAR and a modified PAM-fluorimeter, in maize, sugar beet and Kalanchoë *J. Remote Sensing of Environment*, **58**, 311–321.

Cerovic, Z.G., Langrand, E., Latouche, G., *et al.* (1998). Spectral characterization of NAD(P)H fluorescence in intact isolated chloroplasts and leaves: effect of chlorophyll concentration on reabsorption of blue-green fluorescence. *Photosynthesis Research*, **56**, 291–301.

Cerovic, Z.G., Ounis, A., Cartelat, A., *et al.* (2002). The use of chlorophyll fluorescence excitation spectra for the non-destructive in situ assessment of UV-absorbing compounds in leaves. *Plant Cell Environment*, **25**, 1663–1676.

Cerovic, Z.G., Samson, G., Morales, F., *et al.* (1999). Ultraviolet-induced fluorescence for plant monitoring: present state and prospects. *Agronomie*, **19**, 543–578.

Cescatti, A. (1998). Effects of needle clumping in shoots and crowns on the radiative regime of a Norway spruce canopy. *Annales des Sciences Forestieres*, **55**, 89–102.

Cescatti, A. and Niinemets, Ü. (2004). Sunlight capture. Leaf to landscape. In: *Photosynthetic Adaptation: Chloroplast to Landscape* (eds Smith, W. K., Vogelmann, T. C. and Chritchley, C.), *Ecological Studies*, **178**, 42–85, Springer Verlag, Berlin.

Cescatti, A. and Zorer, R. (2003). Structural acclimation and radiation regime of silver fir (*Abies alba* Mill.) shoots along a light gradient. *Plant, Cell and Environment*, **26**, 429–442.

Ceulemans, R. and Mousseau, M. (1994). Effects of elevated atmospheric CO_2 on woody plants. Tansley Review No. 71. *New Phytologist*, **127**, 425–446.

Ceulemans, R., Janssens, I.A. and Jach, M.E. (1999). Effects of CO_2 enrichment on trees and forests: lessons to be learned in view of future ecosystem studies. *Annals of Botany*, **84**, 577–590.

Chabot, B.F. and Hicks, D.J. (1982). The ecology of leaf lifespans. *Annual Review of Ecology and Systematics*, **13**, 229–259.

Chaerle, L. and Van Der Straeten, D. (2000). Imaging techniques and the early detection of plant stress. *Trends in Plant Science*, **5**, 495–501.

Chaerle, L. and Van Der Straeten, D. (2001). Seeing is believing: imaging techniques to monitor plant health. *Biochimica et Biophysica Acta*, **1519**, 153–166.

Chaerle, L., Hagenbeek, D., De Bruyne, E., *et al.* (2004). Thermal and chlorophyll-fluorescence imaging distinguish plant-pathogen interactions at an early stage. *Plant Cell Physiology*, **45**, 887–896.

Chaerle, L., Hagenbeek, D., De Bruyne, E., *et al.* (2007b). Chlorophyll fluorescence imaging for disease-resistance screening of sugar beet. *Plant Cell Tissue and Organ Culture*, **91**, 97–106.

Chaerle, L., Leinonen, I., Jones, H.G., *et al.* (2007a). Monitoring and screening plant populations with combined thermal and chlorophyll fluorescence imaging. *Journal of Experimental Botany*, **58**, 773–784.

Chaerle, L., Lenk, S., Buschmann, C., *et al.* (2007c). Multicolour fluorescence imaging for early detection of the hypersensitive reaction to tobacco mosaic virus. *Journal of Plant Physiology*, **164**, 253–262.

Chaerle, L., Pineda, M., Romero-Aranda, R., *et al.* (2006). Robotized thermal and chlorophyll fluorescence imaging of *pepper mild mottle virus* infection in *Nicotiana benthamiana*. *Plant Cell Physiology*, **47**, 1323–1336.

Chaerle, L., Van Caeneghem, W., Messens, E., *et al.*(1999). Presymptomatic visualization of plant-virus interactions by thermography. *Nature Biotechnology*, **17**, 813–816.

Chai, T.T., Simmonds, D., Day, D.A., *et al.* (2010). Photosynthetic performance and fertility are repressed in *GmAOX2b* antisense soybean. *Plant Physiology*, **152**, 1638–1649.

Chalker-Scott, L. (1999). Environmental significance of anthocyanins in plant stress responses. *Photochemistry and Photobiology*, **70**, 1–9.

Chameides, W., Lindsay, R., Richardson, J., *et al.* (1988). The role of biogenic hydrocarbons in urban photochemical smog: Atlanta as a case study. *Science*, **241**, 1473–1475.

Chameides, W.L., Fehsenfeld, F., Rodgers, M.O., *et al.* (1992). Ozone precursor relationships in the ambient atmosphere. *Journal of Geophysical Research*, **97**, 6037–6055.

Chandler, J.W. and Dale, J.E. (1993). Photosynthesis and nutrient supply in needles of Sitka spruce [*Picea sitchensis* (Bong.) Carr.]. *The New Phytologist*, **125**, 101–111.

Chaney, R.L. (1993). Zinc phytotoxicity. In: *Zinc in Soil and Plants* (ed. Robson, A.D.), Kluwer Academic Publishers, Dordrecht, Netherlands, pp. 131–150.

Changnon, S.A., Kunkel, K.E. and Winstanley, D. (2002). Climate factors that caused the unique tall grass prairie in the central United States. *Physical Geography*, **23**, 259–280.

Chaoui, A., Mazhoudi, S., Ghorbal, M.H., *et al.* (1997). Cadmium and zinc induction of lipid peroxidation and effects on antioxidant enzyme activities in bean (*Phaseolus vulgaris* L.). *Plant Science*, **127**, 139–147.

Chapin, F.S. III, Bloom, A.J., Field, C.B., *et al.* (1987). Plant responses to multiple environmental factors. *BioScience*, **37**, 49–57.

Chapin, F.S., Matson, P.A. and Mooney, H.A. (2002). *Principles of Terrestrial Ecosystem Ecology*. Springer-Verlag, New York, USA.

Chappelle, E.W., Wood, F.M., McMurtrey, J.E., *et al.* (1984). Laser-induced fluorescence of green plants. 1. A technique for the remote detection of plant stress and species differentiation. *Applied Optics*, **23**, 134–138.

Charlesworth, B., Morgan, M.T. and Charlesworth, D. (1993). The effects of deleterious mutations on neutral molecular variation. *Genetics*, **134**, 1289–1303.

Charlesworth, D. and Wright, S. (2001). Breeding systems and genome evolution. *Current Opinions in Genetics and Development*, **11**, 685–690.

Chaumont, M., Morot-Gaudry, J.F. and Foyer, C.H. (1995). Effects of photoinhibitory treatment on CO_2 assimilation, the quantum yield of CO_2 assimilation, D-1 protein, ascorbate, glutathione and xanthophyll contents and the electron transport rate in vine leaves. *Plant Cell Environment*, **18**, 1358–1366.

Chave, J., Coomes, D., Jansen, S., *et al.* (2009) Towards a worldwide wood economics spectrum. *Ecology Letters*, **12**, 351–366.

Chaves, M.M. (1991). Effects of water deficits on carbon assimilation. *Journal of Experimental Botany*, **42**, 1–16.

Chaves, M.M. and Oliveira, M.M. (2004). Mechanisms underlying plant resilience to water deficits: prospects for water-saving agriculture. *Journal of Experimental Botany*, **55**, 2365–2384.

Chaves, M.M., Flexas, J. and Pinheiro, C. (2009). Photosynthesis under drought and salt stress: regulation mechanisms from whole plant to cell. *Annals of Botany*, **103**, 551–560.

Chaves, M.M., Maroco, J.P. and Pereira, J.S. (2003). Understanding plant responses to drought: from genes to the whole plant. *Functional Plant Biology*, **30**, 239–264.

Chaves, M.M., Osório, J. and Pereira, J.S. (2004). Water use efficiency and photosynthesis. In: *Water Use Efficiency in Plant Biology* (ed. Bacon, M.A.) Blackwell Publishing, Oxford, UK, pp. 42–74.

Chaves, M.M., Pereira, J.S., Maroco, J., *et al.* (2002). How plants cope with water stress in the field. Photosynthesis and growth. *Annals of Botany*, **89**, 1–10.

Chaw, S.M., Parkinson, C.L., Cheng, Y., *et al.* (2000). Seed plant phylogeny inferred from all three plant genomes: monophyly of extant gymnosperms and origin of Gnetales from conifers. *Proceedings of the National Academy of Sciences USA*, **97**, 4086–4091.

Chazdon, R.L. (1985). Leaf display, canopy structure and ligth interception of two understory palm species. *American Journal of Botany*, **72**, 1493–1502.

Chazdon, R.L. (1988). Sunflecks and their importance to forest understory plants. *Advances in Ecological Research*, **18**, 1–63.

Chazdon, R.L. (1992). Photosynthetic plasticity of two rain forest shrubs across natural gap transects. *Oecologia*, **92**, 586–595.

Chazdon, R.L. and Fetcher, N. (1984). Photosynthetic light environments in a lowland tropical rainforest in Costa Rica. *Journal of Ecology*, **72**, 553–564.

Chazdon, R.L. and Field, C.B. (1987). Determinants of photosynthetic capacity in six rainforest *Piper* species. *Oecologia*, **73**, 222–230.

Chazdon, R.L. and Pearcy, R.W. (1986a). Photosynthetic Responses to Light Variation in Rain-Forest Species. I. Induction under constant and fluctuating light conditions. *Oecologia*, **69**, 517–523.

Chazdon, R.L. and Pearcy, R.W. (1986b). Photosynthetic responses to light variation in rain forest species. II. Carbon gain and photosynthetic efficiency during lightflecks. *Oecologia*, **69**, 524–531.

Chazdon, R.L., Williams, K. and Field, C.B. (1988). Interactions between crown structure and light environment in five rain forest *Piper* species. *American Journal of Botany*, **75**, 1459–1471.

Cheeseman, J.M. (1991). PATCHY: simulating and visualizing the effects of stomatal patchiness on photosynthetic CO_2 exchange studies. *Plant, Cell and Environment*, **14**, 593–599.

Cheeseman, J.M. (2006). Hydrogen peroxide concentrations in leaves under natural conditions. *Journal of Experimental Botany*, **57**, 2435–2444.

Cheeseman, J.M. (2009). Seasonal patterns of leaf H_2O_2 content: reflections of leaf phenology, or environmental stress? *Functional Plant Biology*, **36**, 721–731.

Cheeseman, J.M. and Lovelock, C.E. (2004). Photosynthetic characteristics of dwarf and fringe *Rhizophora mangle* L. in a Belizean mangrove. *Plant Cell and Environment*, **27**, 769–780.

Cheeseman, J.M., Clough, B.F., Carter, D.R., *et al.* (1991). The analysis of photosynthetic performance in leaves under field conditions: a case study using Bruguiera mangroves. *Photosynthesis Research*, **29**, 11–22.

Cheeseman, J.M., Herendeen, L.B., Cheeseman, A.T., *et al.* (1997). Photosynthesis and photoprotection in mangroves under field conditions. *Plant, Cell and Environment*, **20**, 579–588.

Chelle, M. (2005). Phylloclimate or the climate perceived by individual plant organs: What is it? How to model it? What for? *New Phytologist*, **166**, 781–790.

Chen, C.P., Zhu, X.G. and Long, S.P. (2008). The effect of leaf-level spatial variability in photosynthetic capacity on biochemical parameter estimates using the Farquhar model: A theoretical analysis. *Plant Physiology*, **148**, 1139–1147.

Chen, L., Fuchigami, L.H. and Breen, P.J. (2001). The relationship between photosystem II efficiency and quantum yield for CO_2 assimilation is not affected by nitrogen content in apple leaves. *Journal of Experimental Botany*, **52**, 1865–1872.

Chen, M., Wang, R., Yang, L., *et al.* (2003). Development of east Asian summer monsoon environments in the late Miocene: radiolarian evidence from Site 1143 of ODP Leg 184. *Marine Geology*, **201**, 169–177.

Chen, R.D. and Gadal, P. (1990). Do mitochondria provide 2-oxoglutarate needed for glutamate synthesis in higher plant chloroplasts? *Plant Physiology and Biochemistry*, **28**, 141–145.

Chen, W.R., Yang, X., He, Z.L., *et al.* (2008). Differential changes in photosynthetic capacity, 77 K chlorophyll fluorescence and chloroplast ultrastructure between Zn-efficient and Zn-inefficient rice genotypes (*Oryza sativa*) under low zinc stress. *Physiologia Plantarum*, **132**, 89–101.

Chen, Y.Z., Murchie, E.H., Hubbart, S., *et al.* (2003). Effects of season-dependent irradiance levels and nitrogen-deficiency on photosynthesis and photoinhibition in field-grown rice (*Oryza sativa*). *Physiologia Plantarum*, **117**, 343–351.

Cheng, W., Sims, D.A., Luo, Y., *et al.* (2000). Photosynthesis, respiration, and net primary production of sunflower stands in ambient and elevated atmospheric CO_2 concentrations: an invariant NPP:GPP ratio? *Global Change Biology*, **6**, 931–941.

Chida, H., Nakazawa, A., Akazak, H., *et al.* (2007). Expression of algal cytochrome c6 in *Arabidopsis* enhances photosynthesis and growth. *Plant Cell Physiology*, **48**, 948–957.

Chidumayo, E.N. (2004). Development of *Brachystegia-Julbernardia* woodland after clear-felling in central Zambia:

Evidence for high resilience. *Applied Vegetation Science*, **7**, 237–242.

Cho, U.H. and Park, J.O. (2000). Mercury induced oxidative stress in tomato seedlings. *Plant Science*, **156**, 1–9.

Chopra, J., Kaur, N. and Gupta, A.K. (2002). A comparative developmental pattern of enzymes of carbon metabolism and pentose phosphate pathway in mung bean and lentil nodules. *Acta Physiologia Plantarum*, **24**, 67–72.

Chou, H.M., Bundock, N., Rolfe, A.S., *et al.* (2000). Infection of *Arabidopsis thaliana* leaves with *Albugo candida* (white blister rust) causes a reprogramming of host metabolism. *Molecular Plant Pathology*, **2**, 99–113.

Chow, W.S. and Hope, A.B. (2004). Electron fluxes through photosystem I in cucumber leaf discs probed by far-red light. *Photosynthesis Research*, **81**, 77–89.

Christen, D., Schonmann, S., Jermini, M., *et al.* (2007). Characterization and early detection of grapevine (*Vitis vinifera*) stress responses to esca disease by in situ chlorophyll fluorescence and comparison with drought stress. *Environmental and Experimental Botany*, **60**, 504–514.

Christensen, J.H., Hewitson, B., Busuioc, A., *et al.* (2007). Regional climate projections. In: *Climate Change 2007: The Physical Science Basis. Contribution of Working Group I to the Fourth Assessment Report of the Intergovernmental Panel on Climate Change* (eds) Solomon, S., Qin, D., Manning, M., Chen, Z., Marquis, M., Averyt, K.B., Tignor, M and Miller, H.L., Cambridge University Press, Cambridge, UK and New York, USA.

Christin, P.A. Besnard, G., Samaritani, E., *et al.* (2008) Oligocene CO_2 decline promoted C_4 photosynthesis in grasses. *Current Biology*, **18**, 37–43.

Christin, P.A., Salamin, N., Savolainen, V., *et al.* (2007). C_4 photosynthesis evolved in grasses via parallel adaptive genetic changes. *Current Biology*, **17**, 1241–1247.

Christmann, A., Hoffmann, T., Teplova, I., *et al.* (2005). Generation of active pools of abscisic acid revealed by in vivo imaging of water-stressed *Arabidopsis. Plant Physiology*, **137**, 209–219.

Christmann, A., Weiler, E., Steudle, E., *et al.* (2007). A hydraulic signal in root-to-shoot signalling of water shortage. *Plant Journal*, **52**, 167–174.

Ciais, P. and Meijer, H.A.J. (1998). The $^{18}O/^{16}O$ isotope ratio of atmospheric CO_2 and its role in global carbon cycle research. In: *Stable Isotopes. Integration of Biological, Ecological and Geochemical Processes* (ed. Griffiths, H.), BIOS Scientific Publishers, Oxford, UK, pp. 409–431.

Ciais, P., Denning, A.S., Tans, P.P., *et al.* (1997a). A three-dimensional synthesis study of $\delta^{18}O$ in atmospheric CO_2: 1. Surface fluxes. *Journal of Geophysical Research Atmospheres*, **102**, 5857–5872.

Ciais, P., Friedlingstein, P., Friend, A., *et al.* (2001). Integrating global models of terrestrial primary productivity. In: *Terrestrial Global Productivity* (eds Mooney, H.A., Saugier, B. and Roy, J.), Academic Press, San Diego, CA, USA, pp. 449–478.

Ciais, P., Reichstein, M., Viovy, N., *et al.* (2005). Europe-wide reduction in primary productivity caused by the heat and drought in 2003. *Nature*, **437**, 529–533.

Ciais P., Tans, P.P., Denning, A.S., *et al.* (1997b). A three-dimensional synthesis study of $\delta^{18}O$ in atmospheric CO_2: 2. Simulation with the TM2 transport model. *Journal of Geophysical Research Atmospheres*, **102**, 5873–5883.

Ciais P., Tans, P.P., Trolier, M., *et al.* (1995a). A large northern-hemisphere terrestrial CO_2 sink indicated by the $^{13}C/^{12}C$ ratio of atmospheric CO_2. *Science*, **269**, 1098–1102.

Ciais P., Tans, P.P., White, J.W.C., *et al.* (1995b). Partitioning of ocean and land uptake of CO_2 as inferred by the $\delta^{13}C$ measurements from the NOAA climate monitoring and diagnostic laboratory global air sampling network. *Journal of Geophysical Research Atmospheres*, **100**, 5051–5070.

Cifre, J., Bota, J., Escalona, J.M., *et al.* (2005). Physiological tools for irrigation scheduling in grapevines (*Vitis vinifera* L.): an open gate to improve water-use efficiency? *Agriculture Ecosystems and Environment*, **106**, 159–170.

Cipollini, D. (2005). Interactive effects of lateral shading and jasmonic acid on morphology, physiology, seed production, and defense traits in *Arabidopsis thaliana. International Journal of Plant Science*, **166**, 955–959.

Cipollini, M.L. and Levey, D.J. (1991). Why some fruits are green when they are ripe: carbon balance in fleshy fruits. *Oecologia*, **88**, 371–377.

Cirilo, A.G. and Andrade, F.H. (1994). Sowing date and maize productivity: I. Crop growth and dry matter partitioning. *Crop Science*, **34**, 1039–1043.

Cleland, R.E. (1998). Voltammetric measurement of the plastoquinone redox state in isolated thylakoids. *Photosynthesis Research*, **58**, 183–92.

Clement, J.M.A.M., Venema, J.H. and Hasselt, P.R. (1995). Short-term exposure to atmospheric ammonia does not affect low-temperature hardening of winter wheat. *New Phytologist*, **131**, 345–351.

Clement, R.J., Burba, G.G., Grelle, A., *et al.* (2009). Improved trace gas flux estimation through IRGA sampling optimization. *Agricultural and Forest Meteorology*, **149**, 623–638.

Close, D.C., Beadle, C.L. and Hovenden, M.J. (2003). Interactive effects of nitrogen and irradiance on sustained xanthophyll cycle engagement in *Eucalyptus nitens* leaves during winter. *Oecologia*, **134**, 32–36.

Clough, B.F. (1992). Primary productivity and growth of mangrove forests. In: *Tropical Mangrove Ecosystems* (eds Robertson, A.I. and Alongi, D.M.), American Geophysical Union, Washington DC, USA, pp. 225–249.

Clough, B.F. and Sim, R.G. (1989). Changes in gas exchange characteristics and water use efficiency of mangroves in response to salinity and vapour pressure deficit. *Oecologia*, **79**, 38–44.

Cochard, H., Coll, L., Le Roux, X., *et al.* (2002). Unraveling the effects of plant hydraulics on stomatal closure during water stress in walnut. *Plant Physiology*, **128**, 282–290.

Cochard, H., Venisse, J.S., Barigah, T.S., *et al.* (2007). Putative role of aquaporins in variable hydraulic conductance of leaves in response to light. *Plant Physiology*, **143**, 122–133.

Cochrane, M.A., Alencar, A., Schulze, M. D., *et al.* (1999). Positive feedbacks in the fire dynamic of closed canopy tropical forests. *Science*, **284**, 1832–1835.

Cohen, J. and Loebenstein, G. (1975). An electron microscope study of starch lesions in cucumber cotyledons infected with tobacco mosaic virus. *Phytopathology*, **65**, 32–39.

Cohen, W.B. and Goward, S.N. (2004). Landsat's role in ecological applications of remote sensing. *BioScience*, **54**, 535–545.

Cole, D.R. and Monger, H.C. (1994). Infuence of atmospheric CO_2 on the decline of C_4 plants during the last deglaciation. *Nature*, **368**, 533–536.

Coleman, J.S., McConnaughay, K.D.M. and Bazzaz, F.A. (1993). Elevated CO_2 and plant nitrogen-use: is reduced tissue nitrogen concentration size-dependent? *Oecologia*, **93**, 195–200.

Coley, P.D. (1988). Effects of plant-growth rate and leaf lifetime on the amount and type of anti-herbivore defense. *Oecologia*, **74**, 531–536.

Coley, P.D., Bryant, J.P. and Chapin, F.S.I. (1985). Resource availability and antiherbivore defense. *Science*, **230**, 895–899.

Collatz, G.J., Ball, J.T., Grivet, C., *et al.* (1991). Physiological and environmental regulation of stomatal conductance, photosynthesis and transpiration: a model that includes a laminar boundary layer. *Agricultural and Forest Meteorology*, **54**, 107–136.

Collatz, G.J., Berry, J.A. and Clark, J.S. (1998). Effects of climate and atmospheric CO_2 partial pressure on the global distribution of C_4 grasses: present, past, and future. *Oecologia*, **114**, 441–454.

Collatz, G.J., Berry, J.A., Farquhar, G.D., *et al.* (1990). The relationship between the Rubisco reaction-mechansim and models of photosynthesis. *Plant, Cell and Environment*, **13**, 219–225.

Collatz, G.J., Ribas-Carbo, M. and Berry, J.A. (1992). Coupled photosynthesis-stomatal conductance model for leaves of C_4 plants. *Austalian Journal of Plant Physiology*, **19**, 519–538.

Collier, D.E. and Thibodeau, B.A. (1995). Changes in respiration and chemical content during autumnal senescence of *Populus tremuloides* and *Quercus rubra* leaves. *Tree Physiology*, **15**, 759–764.

Colmer, T.D. and Pedersen, O. (2008). Underwater photosynthesis and respiration in leaves of submerged wetland plants: gas films improve CO_2 and O_2 exchange. *New Phytologist*, **177**, 918–926.

Comai, L., Young, K., Till, B.J., *et al.* (2004). Efficient discovery of DNA polymorphisms in natural populations by ecotilling. *Plant Journal*, **37**, 778–786.

Comstock, J.P. (2002). Hydraulic and chemical signalling in the control of stomatal conductance and transpiration. *Journal of Experimental Botany*, **53**, 195–200.

Comstock, J.P. and Ehleringer, J.R. (1988). Contrasting photosynthetic behaviour in leaves and twigs of *Hymenoclea salsola*, a green twigged wann desert shrub. *American Journal of Botany*, **75**, 1369–1370.

Conde, L.F. and Kramer, P.J. (1975). The effect of vapour pressure deficit on diffusion resistance of *Opuntia compressa*. *Canadian Journal of Botany*, **53**, 2923–2926.

Condon, A.G., Richards, R.A., Rebetzke, G.J., *et al.* (2002). Improving intrinsic water-use efficiency and crop yield. *Crop Science*, **42**, 122–131.

Condon, A.G., Richards, R.A., Rebetzke, G.J., *et al.* (2004). Breeding for high water-use efficiency. *Journal of Experimental Botany*, **55**, 2447–2460.

Conroy, J.P., Milham, P.J., Reed, M.L., *et al.* (1990). Increases in phosphorus requirements for CO_2-enriched pine species. *Plant Physiology*, **92**, 977–982.

Conti, G.C., Vegetti, G., Bassi, M., *et al.* (1972). Some ultrastructural and cytochemical observations on Chinese cabbage leaves infected with cauliflower mosaic virus. *Virology*, **47**, 694–700.

Conway, T.J., Tans, P., Waterman, LS., *et al.* (1994). Evidence for interannual variability of the carbon cycle from the National Oceanic and Atmospheric Administration/ Climate Monitoring and Diagnostics Laboratory Global Air Sampling Network. *Journal of Geophysical Research*, **99**, 831–855.

Coops, N.C., Hilker, T., Hall, F.G. *et al.* (2010). Estimation of light-use efficiency of terrestrial ecosystems from space: a status report. *BioScience*, **60**, 788–797.

Coplen, T.B. (1995). Discontinuance of SMOW and PDB. *Nature*, **375**, 285.

Copolovici, L.O., Filella, I., Llusià, J., *et al.* (2005). The capacity for thermal protection of photosynthetic electron transport varies for different monoterpenes in *Quercus ilex*. *Plant Physiology*, **139**, 485–496.

Corcuera, L., Camarero, J.J. and Gil-Pelegrín, E. (2002). Functional groups in *Quercus* species derived from the analysis of pressure–volume curves. *Trees*, **16**, 465–472.

Cordell, S., Goldstein, G., Mueller-Dombois, D., *et al.* (1998). Physiological and morphological variation in *Metrosideros polymorpha*, a dominant Hawaiian tree species, along an altitudinal gradient: the role of phenotypic plasticity. *Oecologia*, **113**, 188–196.

Corelli-Grappadelli, L. and Magnanini, E. (1993). A whole-tree system for gas-exchange studies. *HortScience*, **28**, 41–45.

Corey, K.A. and Wheeler, R.M. (1992). Gas exchange in NASA's biomass production chamber. *BioScience*, **42**, 503–509.

Corley, R.H.V. and Lee, C.H. (1992). The physiological basis for genetic improvement of oil palm in Malaysia. *Euphytica*, **60**, 179–84.

Cornic, G. (1973). Etude de l'inhibition de la respiration par la lumière chez la moutarde blanche *Sinapis alba* L. *Physiologie Végetale*, **11**, 663–679.

Cornic, G. (2000). Drought stress inhibits photosynthesis by decreasing stomatal aperture – not by affecting ATP synthesis. *Trends in Plant Science*, **5**, 187–188.

Cornic, G. and Briantais, J.M. (1991). Partitioning of photosynthetic electron flow between CO_2 and O_2 reduction in a C_3 leaf (*Phaseolus vulgaris* L.) at different CO_2 concentrations and during drought. *Planta*, **183**, 178–184.

Cornic, G., Bukhov, N.G., Wiese, C., *et al.* (2000). Flexible coupling between light-dependent electron and vectorial proton transport in illuminated C_3 plants. Role of photosystem I-dependent proton pumping. *Planta*, **210**, 468–477.

Cornic, G., Le Gouallec, J.L., Briantais, J.M., *et al.* (1989). Effect of dehydration and high light on photosynthesis of two C_3 plants. *Phaseolus vulgaris L.* and *Elastostema repens* (hour.) Hall f.). *Planta*, **177**, 84–90.

Corp, L., McMurtrey, J., Middleton, E., *et al.* (2003). Fluorescence sensing systems: in vivo detection of biophysical variations in field corn due to nitrogen supply. *Remote Sensing of Environment*, **86**, 470–479.

Corp, L.A., McMurtrey, J.E., Chappelle, E.W., *et al.* (1997). UV band fluorescence (*in vivo*) and its implications to the remote assessment of nitrogen supply in vegetation. *Remote Sensing of the Environment*, **61**, 110–117.

Cosgrove, D.J. (2000). Expansive growth of plant cell walls. *Plant Physiology and Biochemistry*, **38**, 109–124.

Cosgrove, D.J. (2005). Growth of the plant cell wall. *Nature Reviews Molecular Cell Biology*, **6**, 850–861.

Costa, J.M., Ortuño, M.F. and Chaves, M.M. (2007). Deficit irrigation as a strategy to save water: physiology and potential application to horticulture. *Journal of Integrative Plant Biology*, **49**, 1421–1434.

Cousins, A.B., Badger, M.R. and von Caemmerer, S. (2006). Carbonic anhydrase and its influence on carbon isotope discrimination during C_4 photosynthesis. Insights from antisense RNA in *Flaveria bidentis*. *Plant Physiology*, **141**, 232–242.

Cowan, I.R. (1986). Economics of carbon fixation in higher plants. In: *On the Economy of Plant Form and Function* (ed. T.J. Givnish), Cambridge University Press, Cambridge, UK, pp. 133–170.

Cowan, I.R. and Faquhar, G.D. (1977). Stomatal function in relation to leaf metabolism and environment. *Symposia of the Society for Experimental Biology*, **31**, 471–505.

Cowling, S.A., Jones, C.D. and Cox, P.M. (2007). Consequences of the evolution of C_4 photosynthesis for surface energy and water exchange. *Journal of Geophysical Research – Biogeosciences* **112**: Article number G01020.

Cox, P.M., Betts, R.A., Jones, C.D., *et al.* (2000). Acceleration of global warming due to carbon-cycle feedbacks in a coupled climate model. *Nature*, **408**, 184–187.

Crafts-Brandner, S.J. and Salvucci, M.E. (2000). Rubisco activase constrains the photosynthetic potential of leaves at high temperature and CO_2. *Proceedings of the National Academy of Sciences of the United States of America*, **97**, 13430–13435.

Craig, H. (1953). The geochemistry of the stable carbon isotopes. *Geochimica et Cosmochimica Acta*, **3**, 53–92.

Craig, H. (1954). Carbon 13 in plants and the relationship between carbon 13 and carbon 14 variation in nature. *The Journal of Geology*, **62**, 115–149.

Craig, H. (1961). Isotopic variations in meteoric waters. *Science*, **133**, 1702–1703.

Craig, H. and Gordon, L.I. (1965). Deuterium and oxygen-18 variations in the ocean and the marine atmosphere. In: *Proceedings of a Conference on Stable Isotopes in Oceanographic Studies and Paleo-temperatures* (ed. E. Tongiorgi), Lischi and Figli, Pisa, Italy, pp. 9–130.

Craine, J.M. and Reich, P.B. (2005). Leaf-level light compensation points in shade-tolerant woody seedlings. *New Phytologist*, **166**, 710–713.

Cramer, M.D., Hawkins, H.J. and Verboom, G.A. (2009). The importance of nutritional regulation of plant water flux. *Oecologia*, **161**, 15–24.

Cramer, M.D., Kleizen, C. and Morrow, C. (2007). Does the prostrate-leaved geophyte *Brunsvigia orientalis* utilize soil-derived CO_2 for photosynthesis? *Annals of Botany*, **99**, 835–844.

Crane, J.H. and Davies, F.S. (1988). Flooding duration and seasonal effects on growth and development of young rabbiteye blueberry plants. *Journal of the American Society for Horticultural Science*, **113**, 180–184.

Crane, P.R. (1996). The fossil history of the Gnetales. *International Journal of Plant Sciences*, **157** (suppl), S50-S57.

Crane, P.R., Herendeen, P. and Friis, E.M. (2004). Fossils and plant phylogeny. *American Journal of Botany*, **91**, 1683–1699.

Crayn, D.M., Terry, R.G., Smith, J.A.C., *et al.* (2000). Molecular systematic investigations in Pitcarnioideae (Bromeliaceae) as a basis for understanding the evolution of crassaulacean acid metabolism (CAM). In: *Monocots: Systematics and Evolution* (eds Wilson, K.L. and Morrison, D.A.), CSIRO, Melbourne, Australia, pp. 569–579.

Crayn, D.M., Winter, K. and Smith, J.A.C. (2004). Multiple origins of crassulacean acid metabolism and the epiphytic habit in the neotropical family Bromeliaceae. *Proceedings of the National Academy of Sciences USA*, **101**, 3703–3708.

Crofts, A.R., Deamer, D.W. and Packer, L. (1967). Mechanisms of light-induced structural change in chloroplasts ii. The role of ion movements in volume changes. *Biochimica et Biophysica Acta*, **131**, 97–118.

Crofts, A.R., Hong, S., Ugulava, N., *et al.* (1999). Pathways for proton release during ubihydroquinone oxidation by the bc1 complex. *Proceedings of the National Academy of Sciences USA*, **96**, 10021–10026.

Croker, J.L., Witte, W.T. and Augé, R.M. (1998). Stomatal sensitivity of six temperate, deciduous tree species to non-hydraulic root-to-shoot signalling of partial soil drying. *Journal of Experimental Botany*, **49**, 761–774.

Cromer, R.N., Kriedemann, P.E., Sands, P.J., *et al.* (1993). Leaf growth and photosynthetic response to nitrogen and phosphorus in seedling trees of *Gmelina arborea*. *Australian Journal of Plant Physiology*, **20**, 83–98.

Crosatti, C., Soncini, C., Stanca, A.M., *et al.* (1995). The accumulation of a cold-regulated chloroplastic protein is light-dependent. *Planta*, **196**, 458–465.

Crous, K. and Ellsworth, D. (2004). Canopy position affects photosynthetic adjustments to long-term elevated CO_2 concentration (FACE) in aging needles in a mature *Pinus taeda* forest. *Tree Physiology*, **24**, 961–970.

Cseh, E., Fodor, F., Varga, A., *et al.* (2000). Effect of lead treatment on the distribution of essential elements in cucumber. *Journal of Plant Nutrition*, **23**, 1095–1105.

Csintalan, Z., Tuba, Z. and Lichtenthaler, HK. (1998). Changes in laser-induced chlorophyll fluorescence ratio F690/F735 in the poikilochlorophyllous desiccation tolerant plant *Xerophyta scabrida* during desiccation. *Journal of Plant Physiology*, **152**, 540–544.

Cui, M., Vogelmann, T.C. and Smith, W.K. (1991). Chlorophyll and light gradients in sun and shade leaves of *Spinacia oleracea*. *Plant Cell Environment*, **14**, 493–500.

Cui, X.H., Hao, F.S., Chen, H., *et al.* (2008). Expression of the *Vicia faba* VfPIP1 gene in *Arabidopsis thaliana* plants improves their drought resistance. *Journal of Plant Research*, **121**, 207–214.

Cullen, L.E., Adams, M.A., Anderson, M.J., *et al.* (2008). Analyses of $\delta^{13}C$ and $\delta^{18}O$ in tree rings of *Callitris columellaris* provide evidence of a change in stomatal control of photosynthesis in response to regional changes in climate. *Tree Physiology*, **28**, 1525–1533.

Cunnif, J., Osborne, C.P., Ripley, B.S., *et al.* (2008). Response of wild C_4 crop progenitors to subambient CO_2 highlights a possible role in the origin of agriculture. *Global Change Biology*, **14**, 576–587.

Cunningham, F.X. Jr. (2002). Regulation of carotenoid synthesis and accumulation in plants. *Pure and Applied Chemistry*, **74**, 1409–1417.

Curtis, P.S. (1996). A meta-analysis of leaf gas exchange and nitrogen in trees grown under elevated carbon dioxide. *Plant Cell And Environment*, **19**, 127–137.

Cushman, J.C. (2001). Crassulacean acid metabolism. A plastic photosynthetic adaptation to arid environments. *Plant Physiology*, **127**, 1439–1448.

Cushman, J.C. and Bohnert, H.J. (2002). Induction of crassulacean acid metabolism by salinity: molecular aspects.

In: *Salinity: Environmet-Plants-Molecules* (eds Läuchli, A. and Lüttge, U.), Dordrecht, Boston, London. Kluwer Academic Publishers, 361–393.

Cushman, J.C. and Borland, A.M. (2002). Induction of crassulacean acid metabolism by water limitatation. *Plant, Cell and Environment*, **25**, 295–310.

Cutler, S.R., Rodriguez, P.L., Finkelstein, R.R., *et al.* (2010). Abscisic acid: emergence of a core signaling network. *Annual Review of Plant Biology*, **61**, 651–679.

D'Antonio, C.M. (2000). Fire, plant invasions, and global changes. In: *Invasive Species in a Changing World* (ed. Mooney, H.A. and Hobbs, R.J.), Island Press, Washington, D.C., USA, pp. 65–93.

Dai, Y.J., Dickinson, R.E. and Wang, Y.P. (2004). A two-big-leaf model for canopy temperature, photosynthesis, and stomatal conductance. *Journal of Climate*, **17**, 2281–2299.

Dal Bosco, C., Busconi, M., Covoni, C., *et al.* (2003). Cor gene expression in barley mutants affected in chloroplast development and photosynthetic electron transport. *Plant Physiology*, **131**, 793–802.

Dale, J.E. (1982). The growth of leaves. International Botanical Congress, 1981. Edward Arnold Ltd., London.

Dale, V.H., Joyce, L.A., McNulty, S., *et al.* (2001). Climate change and forest disturbances. *Bioscience*, **51**, 723–734.

Daley, P.F. (1995). Chlorophyll fluorescence analysis and imaging in plant stress and disease. *Canadian Journal of Plant Pathology*, **17**, 167–173.

Daley, P.F., Raschke, K., Ball, J.T., *et al.* (1989). Topography of photosynthetic activity of leaves obtained from video images of chlorophyll fluorescence. *Plant Physiology*, **90**, 1233–1238.

Dalling, J.W. and Harms, K.E. (1999). Damage tolerance and cotyledonary resource use in the tropical tree *Gustavia superba*. *Oikos*, **85**, 257–264.

Damesin, C. and Rambal, S. (1995). Field-study of leaf photosynthetic performance by a Mediterranean deciduous oak tree (*Quercus pubescens*) during a severe summer drought. *The New Phytologist*, **131**, 159–167.

Damesin, C., Ceschia, E., Le Goff, N., *et al.* (2002). Stem and branch repiration of beech: from tree measurements to estimations at the stand level. *New Phytologist*, **153**, 159–172.

Damesin, C., Rambal, S. and Joffre, R. (1998). Co-occurrence of trees with differing leaf habit: a functional approach on Mediterranean oaks. *Acta Oecologica*, **19**, 195–204.

Damm, A., Elbers, J., Releer, A., *et al.* (2010). Remote sensing of sun induced fluorescence to improve modelling of diurnal courses of gross primary production. *Global Change Biology*, **16**, 171–186.

Damour, G., Vandame, M. and Urban, L. (2009). Long-term drought results in a reversible decline in photosynthetic capacity in mango leaves, not just a decrease in stomatal conductance. *Tree Physiology*, **29**, 675–684.

Danner, B.T. and Knapp, A.K. (2001). Growth dynamics of oak seedlings (*Quercus macrocarpa* Michx. and *Quercus muhlenbergii* Engelm.) from gallery forests: implications for forest expansion into grasslands. *Trees – Structure and Function*, **15**, 271–277.

Danon, A., Coll, N.S. and Apel, K. (2006). Cryptochrome-1-dependent execution of programmed cell death induced by singlet oxygen in *Arabidopsis thaliana*. *Proceedings of the National Academy of Sciences (USA)*, **103**, 17036–17041.

Dansgaard, W. (1954). The ^{18}O abundance in fresh water. *Geochimica et Cosmochimica Acta*, **6**, 241–260.

Dansgaard, W. (1964). Stable isotopes in precipitation. *Tellus*, **16**, 436–467.

D'Antonio, C.M., Hughes, R.F. and Vitousek, P.M. (2001). Factors influencing dynamics of two invasive C_4 grasses in seasonally dry Hawaiian woodlands. *Ecology*, **82**, 89–104.

Dardick, C. (2007). Comparative expression profiling of *Nicotiana benthamiana* leaves systemically infected with three fruit tree viruses. *Molecular Plant-Microbe Interactions*, **20**, 1004–1017.

Darley, E.F. (1966). Studies of the effect of cement-kiln dust on vegetation. *Journal of the Air Pollution Control Association*, **16**, 145–150.

Darrall, N.M. (1989). The effect of air pollutants on physiological processes in plants. *Plant, Cell and Environnment*, **12**, 1–30.

Darwin, C. (1881a). The movements of leaves. *Nature*, **23**, 603–604.

Darwin, C. (1881b). Leaves injured at night by free radiation. *Nature*, **24**, 459.

Darwin, C. (1882). The action of carbonate of ammonia on chlorophyll-bodies. *Journal of the Linnean Society of London (Botany)*, **19**, 262–284.

Darwin, F. (1916). On the relation between transpiration and stomatal aperture. *Philosophical Transactions of the Royal Society of London Series B*, **207**, 413–437.

Darwin, F. and Pertz, D.F.M. (1911) On a new method of estimating the aperture of stomata. *Proceedings of the Royal Society of London Series B*, **84**, 136–154.

Dat, J., Vandenabeele, S., Vranova, E., *et al.* (2000). Dual action of the active oxygen species during plant stress responses. *Cellular and Molecular Life Sciences: CMLS*, **57**, 779–795.

Dau, H. (1994). Molecular mechanisms and quantitative models of variable photosystem II fluorescence. *Photochemistry Photobiology*, **60**, 1–23.

Daughtry, C.S.T., Gallo, K.P., Goward, S.N., *et al.* (1992). Spectral estimates of absorbed radiation and phytomass production in corn and soybean canopies. *Remote Sensing of the Environment*, **39**, 141–52.

Daumard, F. (2010) "Contribution à la télédetection passive de la fluorescence chlorophyllienne des végétaux". PhD thesis. L.M.D, Ecole Polytechnique, Palaiseau - France. http:pastel.archives-ouvertes.fr/pastel-00563244

Daumard, F., Champagne, S., Fournier, A., *et al.* (2010). A field platform for long-term measurements of canopy fluorescence. *IEEE Trans. Geosci. Remote Sens.*, 48 (9), 3358–3368.

Daumard F., Goulas Y., Champagne S., *et al.* (2012) Canopy level chlorophyll fluorescence at 760 nm better tracks in-field sorghum growht. *IEEE Trans. Geosci. Remote Sens* (in press)

Daumard, F., Goulas, Y., Ounis, A., *et al.* (2007). Atmospheric correction of airborne passive measurements of fluorescence. 10th International Symposium on Physical Measurements and Signatures in Remote Sensing, March 12–14, 2007, Davos, Switzerland (ISPMSRS07).

Davey, P.A., Hunt, S., Hymus, G.J., *et al.* (2004). Respiratory oxygen uptake is not decreased by an instantaneous elevation of [CO_2], but is increased with long-term growth in the field at elevated [CO_2](1). *Plant Physiology*, **134**, 520–527.

Davey, P.A., Olcer, H., Zakhleniuk, O., *et al.* (2006). Can fast-growing plantation trees escape biochemical down-regulation of photosynthesis when grown throughout their complete production cycle in the open air under elevated carbon dioxide? *Plant, Cell and Environment*, **29**, 1235–1244.

Davi, H., Dufrêne, E., Granier, A., *et al.* (2005). Modelling carbon and water cycles in a beech forest. Part II.: Validation of the main processes from organ to stand scale. *Ecological Modeling*, **185**, 387–405.

Davies, B. and Sharp, B. (Ed.) (2000). Water deficits and plant growth. *Journal of Experimental Botany*, (**51**). 350 WD Special Issue.

Davies, B., Tuberosa, R., Blum, A., *et al.* (2007). Integrated approaches to sustain and improve plant production under drought stress. *Journal of Experimental Botany*, **58**, (2). Special Issue.

Davies, B.H. (1976). Carotenoids. In: *Chemistry and Biochemistry of Plant Pigments* (ed. Goodwin, T.W.) Vol 2. Academic Press, London, UK, pp. 38–165.

Davies, S.J. (1998). Photosynthesis of nine pioneer *Macaranga* species from Borneo in relation to life history. *Ecology*, **79**, 2292–2308.

Davies, W.J. and Zhang, J. (1991). Root signals and the regulation of growth and development of plants in drying soil. *Annual Review of Plant Physiology and Plant Molecular Biology*, **42**, 55–76.

Davies, W.J., Metcalfe, J., Lodge, T.A., *et al.* (1986). Plant growth substances and regulation of growth under drought. *Australian Journal of Plant Physiology*, **13**, 105–125.

Davis, S.C., Anderson-Teixeira, K.J. and DeLucia, E.H. (2009). Life-cycle analysis and the ecology of biofuels. *Trends in Plant Science*, **14**, 140–146.

Davison, P.A., Hunter, C.N. and Horton, P. (2002). Overexpression of β-carotene hydroxylase enhances stress tolerance in *Arabidopsis*. *Nature*, **418**, 203–206.

Dawson, J.O. and Gordon, J.C. (1979). Nitrogen fixation in relation to photosynthesis in *Alnus glutinosa*. *Botanical Gazette*, **140**, S70-S75.

Dawson, T.E. (1993). Water sources of plants as determined from xylem-water isotopic composition: perspectives on plant competition, distribution, and water relations. In: *Stable Isotopes and Plant Carbon-Water Relations* (eds. Ehleringer, J.R., Hall, A.E. and Farquhar, G.D.) Academic Press, San Diego, California, USA, pp. 465–496.

Dawson, T.E., Mambelli, S., Plamboeck, A.H., *et al.* (2002). Stable isotopes in plant ecology. *Annual Review of Ecology and Systematics*, **33**, 507–559.

Day, D.A., Krab, K., Lambers, H., *et al.* (1996). The cyanide-resistant oxidase: to inhibit or not to inhibit, that is the question. *Plant Physiology*, **110**, 1–2.

Day, D.A., Neuberger, M. and Douce, R. (1985). Interactions between glycine decarboxylase, the tricarboxylic acid cycle and the respiratory chain in pea leaf mitochondria. *Australian Journal of Plant Physiology*, **12**, 119–130.

Day, T.A., Ruhland, C.T., Grobe, C.W., *et al.* (1999). Growth and reproduction of Antarctic vascular plants in response to warming and UV radiation reductions in the field. *Oecologia*, **119**, 24–35.

De Chalain, T.M.B. and Berjak, B. (1979). Cell death as a functional event in the development of the leaf intercellular spaces in *Avicennia marina* (Forsskål) Vierh. *New Phytologist*, **83**, 147–155.

De Jong, G. (2005). Evolution of phenotypic plasticity: patterns of plasticity and the emergence of ecotypes. *New Phytologist*, **166**, 101–118.

De Niro, M.J. and Epstein, S. (1979). Relationship between the oxygen isotope ratio of terrestrial plant cellulose, carbon dioxide and water. *Science*, **204**, 51–53.

De Pury, D.G.G. and Farquhar, G.D. (1997). Simple scaling of photosynthesis from leaves to canopies without the errors of big-leaf models. *Plant Cell Environment*, **20**, 537–557.

De Ronde, J.A., Cress, W.A., Krüger, G.H.J., *et al.* (2004). Photosynthetic response of transgenic soybean plants, containing an *Arabidopsis* P5CR gene, during heat and drought stress. *Journal of Plant Physiology*, **161**, 1211–1224.

De Souza, A.A., Takita, M.A., Coletta-Filho, H.D., *et al.* (2007). Analysis of expressed sequence tags from *Citrus sinensis* L. Osbeck infected with *Xylella fastidiosa*. *Genetics and Molecular Biology*, **30**, 957–964.

De Souza, J., Silka, P.A. and Davis, S.D. (1986). Comparative physiology of burned and unburned *Rhus laurina* after chaparral wildfire. *Oecologia*, **71**, 63–68.

Deamer, D.W., Crofts, A.R. and Packer, L. (1967). Mechanisms of light-induced structural changes in chloroplasts. I. Light-scattering increments and ultrastructural changes mediated by proton transport. *Biochimica et Biophysica Acta*, **131**, 81–96.

Decker, H. (1969). Phytonematologie. In: *Biologie und Bekämpfung Pflanzenparasitärer Nematoden*. Deutscher Landwirtschaftsverlag, Berlin, Germany, pp. 505.

DeFries, R.S., Townshend, J.R.G. and Hansen, M.C. (1999). Continuous fields of vegetation characteristics at the global scale at 1km resolution. *Journal of Geophysical Research*, **104**, 16911–16925.

Deighton, N., Magill, W.J., Bremner, D.H., *et al.* (1997). Malondialdehyde and 4-hydroxy 2-nonenal in plant tissue cultures: LC-MS determination of 2,4-dinitrophenylhydrazone derivatives. *Free Radicals Research*, **27**, 255–265.

De Jong, G. (2005). Evolution of phenotypic plasticity: patterns of plasticity and the emergence of ecotypes. *New Phytologist*, **166**, 101–118.

DeJong, T.M. (1986). Fruit effects on photosynthesis in *Prunus perscia*. *Physiologia Plantarum*, **66**, 149–53.

DeKok, L.J., Stuiver, C.E.E. and Stulen, I. (1998). Impact of H_2S on plants. In: *Responses of Plant Metabolism to Air Pollutants and Global Change* (eds deKok, L.J. and Stulen, I.), Backhuys Publishers, Leiden, Netherlands, pp. 51–63.

Del Arco, J.M., Escudero, A. and Vega Garrido, M. (1991). Effects of site characteristics on nitrogen retranslocation from senescing leaves. *Ecology*, **72**, 701–708.

Del Pierre, N., Soudani, K., Francois, C., *et al.* (2009). Exceptional carbon uptake in European forests during the warm spring of 2007: a data–model analysis. *Global Change Biology*, **15**, 1455–1474.

Del Río, D., Stewart, A.J. and Pellegrini, N. (2005). A review of recent studies on malondialdehyde as toxic molecule and biological marker of oxidative stress. *Nutrition Metabolism and Cardiovascular Diseases*, **15**, 316–328.

Del Río, L.A., Pastori, G.M., Palma, J.M., *et al.* (1998). The activated oxygen role of peroxisomes in senescence. *Plant Physiology*, **116**, 1195–1200.

Delagrange, S., Messier, C., Lechowicz, M.J., *et al.* (2004). Physiological, morphological and allocational plasticity in understory deciduous trees: importance of plant size and light availability. *Tree Physiology*, **24**, 775–784.

Delaney, K.J. and Higley, L.G. (2006). An insect countermeasure impacts plant physiology: midrib vein cutting, defoliation and leaf photosynthesis. *Plant, Cell and Environment*, **29**, 1245–1258.

Delfine, S., Alvino, A., Villani, M.C., *et al.* (1999). Restrictions to carbon dioxide conductance and photosynthesis in spinach leaves recovering from salt stress. *Plant Physiology*, **119**, 1101–1106.

Delgado, E., Medrano, H., Keys, A.J., *et al.* (1995). Species variation in Rubisco specifity factor. *Journal of Experimental Botany*, **46**, 1775–1777.

Delgado, E., Parry, M.A.J., Vadell, J., *et al.* (1993). Photosynthesis, Ribulose-1,5-bisphosphate carboxylase and leaf characteristics of *Nicotiana tabacum* L. genotypes selected by survival at low CO_2 concentrations. *Journal of Experimental Botany*, **44**, 1–7.

Delosme, R. (1967). Étude de l'induction de fluorescence des algues et des chloroplastes au début d'une illumination intense. *Biochimica et Biophysica Acta*, **143**, 108–128.

De Lucia, E.H., Coleman, J.S., Dawson, T.E., *et al.* (2001). Plant physiological ecology: linking the organism to scales above and below. *The New Phytologist*, **149**, 9–16.

Delwiche, C.F. and Palmer, J.D. (1996). Rampant horizontal transfer and duplication of rubisco genes in eubacteria and plastids. *Molecular Biology and Evolution*, **13**, 873–882.

Demarty, M., Morvan, C. and Thellier, M. (1984). Calcium and the cell wall. *Plant Cell Environment*, **7**, 441–448.

Demicco, R.V., Lowenstein, T. K. and Hardie, L.A. (2003). Atmospheric pCO_2 since 60 Ma from records of seawater pH, calcium, and primary carbonate mineralogy. *Geology*, **31**, 793–796.

Demmig-Adams, B. (1998). Survey of thermal energy dissipation and pigment composition in sun and shade leaves. *Plant and Cell Physiology*, **39**, 474–482.

Demmig-Adams, B. and Adams, III W.W. (2002). Antioxidants in photosynthesis and human nutrition. *Science*, **298**, 2149–2153.

Demmig-Adams, B. and Adams, III W.W. (2003). Photoinhibition. In: *Encyclopedia of Applied Plant Science* (eds Thomas, B., Murphy, D. and Murray, B.), Academic Press, New York, USA, pp. 707–714.

Demmig-Adams, B. and Adams, W.W. (1996b). The role of xanthophyll cycle carotenoids in the protection of photosynthesis. *Trends Plant Science*, **1**, 21–26.

Demmig-Adams, B. and Adams, W.W. III (1996a). Chlorophyll and carotenoid composition in leaves of *Euonymus kiautschovicus* acclimated to different degrees of light stress in the field. *Australian Journal of Plant Physiology*, **23**, 649–659.

Demmig-Adams, B. and Adams, W.W. III (2006). Photoprotection in an ecological context: the remarkable complexity of thermal energy dissipation. *New Phytologist*, **172**, 11–21.

Demmig-Adams, B., Adams, III W.W., Barker, D.H., *et al.* (1996a). Using chlorophyll fluorescence to assess the fraction of absorbed light allocated to thermal dissipation of excess excitation. *Physiologia Plantarum*, **98**, 253–264.

Demmig-Adams, B., Adams, W.W. III, Winter, K., *et al.* (1989). Photochemical efficiency of photosystem II, photon yield of oxygen evolution, photosynthetic capacity, and carotenoid composition during the midday depression of net CO_2 uptake in *Arbutus unedo* frowing in Portugal. *Planta*, **177**, 377–387.

Demmig-Adams, B., Gilmore, A.M. and Adams, III W.W. (1996b). In vivo functions of carotenoids in higher plants. *FASEB Journal*, **10**, 403–412.

Den Hertog, J., Stulen, I. and Lambers, H. (1993). Assimilation, respiration and allocation of carbon in *Plantago major* as affected by atmospheric CO_2 levels. A case study. *Vegetation*, **104**, 369–378.

Deng, X. and Melis, A. (1986). Phosphorylation of the light-harvesting complex II in higher plant chloroplasts: effect on photosystem II and photosystem I absorption cross section. *Photobiochemistry and Photobiophysics*, **13**, 41–52.

Dengler, N. and Wilson, T. (1999). Leaf structure and development in C_4 plants. In: *C_4 Plant Biology* (eds Sage, R.F. and Monson, R.K.), Academic Press, San Diego, CA, USA, pp. 133–72.

Dengler, N.G. (1980). Comparative histological basis of sun and shade leaf dimorphism in *Helianthus annuus*. *Canadian Journal of Botany*, **58**, 717–730.

Dengler, N.G. and Kang, J. (2001). Vascular patterning and leaf shape. *Current Opinion in Plant Biology*, **4**, 50–56.

Dermody, O., Long, S.P. and DeLucia, E.H. (2006). How does elevated CO_2 or ozone affect the leaf-area index of soybean when applied independently? *New Phytologist*, **169**, 145–155.

Desai, A.R., Richardson, A.D., Moffat, A., *et al.* (2008). Cross-site evaluation of eddy covariance GPP and RE decomposition techniques. *Agricultural and Forest Meteorology*, **148**, 821–838.

Desikan, R., Cheung, M.K., Bright, J., *et al.* (2004). ABA, hydrogen peroxide and nitric oxide signalling in stomatal guard cells. *Journal of Experimental Botany*, **55**, 205–212.

Desikan, R., Mackerness, S.A.H., Hancock, J.T., *et al.* (2001). Regulation of the *Arabidopsis* transcriptome by oxidative stress. *Plant Physiology*, **127**, 159–172.

Dewar, R.C. (1995). Interpretation of an empirical model for stomatal conductance in terms of guard cell function. *Plant, Cell and Environment*, **18**, 365–372.

Dewar, R.C. (2002). The Ball-Berry Leuning and Tardieu-Davies stomatal models: synthesis and extension within a spatially aggregated picture of guard cell function. *Plant, Cell and Environment*, **25**, 1383–1398.

Di Cagno, R., Guidi, L., De Gara, L., *et al.* (2001). Combined cadmium and ozone treatments affect photosynthesis and ascorbate-dependent defences in sunflower. *New Phytologist*, **151**, 627–636.

Di Marco, G., Manes, F., Tricoli, D., *et al.* (1990). Fluorescence parameters measured concurrently with net photosynthesis to investigate chloroplastic CO_2 concentration in leaves of *Quercus ilex* L. *Journal of Plant Physiology*, **136**, 538–543.

Dias, M.C. and Brüggemann, W. (2010). Limitations of photosynthesis in *Phaseolus vulgaris* under drought stress: gas exchange, chlorophyll fluorescence and Calvin cycle enzymes. *Photosynthetica*, **48**, 96–102.

Díaz, M. and Granadillo, E. (2005). The significance of episodic rains for reproductive phenology and productivity of trees in semiarid regions of northwestern Venezuela. *Trees*, **19**, 336–348.

Díaz-Espejo, A., Nicolás, E. and Fernández, J.E. (2007). Seasonal evolution of diffusional limitations and photosynthetic capacity in olive under drought. *Plant, Cell and Environment*, **30**, 922–933.

Díaz-Espejo, A., Walcroft, A.S., Fernández, J.E., *et al.* (2006). Modelling photosynthesis in olive leaves under drought conditions. *Tree Physiology*, **26**, 1445–1456.

Dickmann, D.I. (1971). Photosynthesis and respiration by developing leaves of cottonwood (*Populus deltoides* Bartr.). *Botanical Gazette*, **132**, 253–259.

Diefendorf, A.F., Mueller, K.E., Wing, S.L., *et al.* (2010). Global patterns in leaf ^{13}C discrimination and implications for studies of past and future climate. *Proceedings of the National Academy of Sciences USA*, **107**, 5738–5743.

Diemer, M. and Körner, C. (1996). Lifetime leaf carbon balances of herbaceous perennial plants from low and high altitudes in the central Alps. *Functional Ecology*, **10**, 33–43.

Dietz, K.J. (1985). A possible rate limiting function of chloroplast hexosemonophosphate isomerase in starch synthesis of leaves. *Biochimica et Biophysica Acta*, **839**, 240–248.

Dietz, K.J. and Keller, F. (1997). Transient storage of photosynthates in leaves. In: *Handbook of Photosynthesis* (ed. Pessarakli, M.). Marcel Dekker, New York, USA, pp. 717–737.

Dietz, K.J., Jacob, S., Oelze, M.L., *et al.* (2006). The function of peroxiredoxins in plant organelle redox metabolism. *Journal of Experimental Botany*, **57**, 1697–1709.

Diffenbaugh, N.S., Pal, J.S., Trapp, R.J., *et al.* (2005). Fine-scale processes regulate the response of extreme events to global climate change. *Proceedings of the National Academy of Sciences*, **102**, 15774–15778.

Dinakar, C., Raghavendra, A.S. and Padmasree, K. (2010). Importance of AOX pathway in optimizing photosynthesis under high light stress: role of pyruvate and malate in activating AOX. *Physiologia Plantarum*, **139**, 13–26.

Ding, S., Lu, Q., Zhang, Y., *et al.* (2008). Enhanced sensitivity to oxidative stress in transgenic tobacco plants with decreased glutathione reductase activity leads to a decrease in ascorbate pool and ascorbate redox state. *Plant Molecular Biology*, **69**, 577–592.

Disante, K.B., Fuentes, D. and Cortina, J. (2011). Response to drought of Zn-stressed *Quercus suber* L. seedlings. *Environmental and Experimental Botany*, **70**, 96–103.

Dismukes, G.C., Klimov, V.V., Baranov, S.V., *et al.* (2001). The origin of atmospheric oxygen on Earth: the innovation of oxygenic photosynthesis. *Proceedings of the National Academy of Sciences*, **98**, 2170–2175.

Dittmar, C., Zech, W. and Elling, W. (2003). Growth variations of common beech (*Fagus sylvatica* L.) under different climatic and environmental conditions in Europe – a dendroecological study. *Forest Ecology and Management*, **173**, 63–78.

Dixon, R.K., Solomon, A.M., Brown, S., *et al.* (1994). Carbon pools and flux of global forest ecosystems. *Science*, **263**, 185–190.

Dobrowski, S.Z., Pushnik, J.C., Zarco-Tejada, P.J., *et al.* (2005). Simple reflectance indices track heat and water stress-induced changes in steady state chlorophyll fluorescence at the canopy scale. *Remote Sensing of Environment*, **97**, 403–414.

Doi, M. and Shimazaki, K.I. (2008). The stomata of the fern *Adiantum capillus-veneris* do not respond to CO_2 in the dark and open by photosynthesis in guard cells. *Plant Physiology*, **147**, 922–930.

Dolan, L. (2009). Body building on land – morphological evolution of land plants. *Current Opinion in Plant Biology*, **12**, 4–8.

Dole, M. (1935). The relative atomic weight of oxygen in water and air. *Journal of the American Chemical Society*, **57**, 2731–1935.

Dole, M., Lane, G.A., Rudd, D.P. *et al.* (1954). Isotopic composition of atmospheric oxygen and nitrogen. *Geochimica et Cosmochimica Acta*, **6**, 65–78.

Döll, P. and Siebert, S. (2002). Global modelling of irrigation water requirements. *Water Resource Research*, **38**, 1037.

Dolman, A.J., Moors, E.J. and Elbers, J.A. (2002). The carbon uptake of a mid latitude pine forest growing on sandy soil. *Agricultural and Forest Meteorology*, **111**, 157–170.

Domingo, R., Pérez-Pastor, A. and Ruiz-Sánchez, M.C. (2002). Physiological responses of apricot plants grafted on two different rootstocks to flooding conditions. *Journal of Plant Physiology*, **159**, 725–732.

Donahue, R.A., Poulson, M.E. and Edwards, G.E. (1997). A method for measuring whole plant photosynthesis in *Arabidopsis thaliana*. *Photosynthesis Research*, **52**, 263–269.

Donald, P.W. and Strobel, G.A. (1970). Adenosine diphosphate-glucose pyrophosphorylase control of starch accumulation in rust-infected wheat leaves. *Plant Physiology*, **46**, 126–135.

Donatelli, M., Hammer, G.L. and Vanderlip, R.L. (1992). Genotype and water limitation effects on phenology, growth, and transpiration efficiency in grain sorghum. *Crop Science*, **32**, 781–786.

Dongmann, G. (1974). Contribution of land photosynthesis to stationery enrichment of ^{18}O in atmosphere. *Radiation and Environmental Biophysics*, **11**, 219–225.

Donigian, J.A.S., Barnwell, J.T.O., Jackson, R.B., *et al.* (1994). Assessment of alternative management practices and policies affecting soil carbon in agroecosystems of the Central United States, U.S.-EPA, Athens, GA.

Donovan, L.A. and Ehleringer, J.R. (1991). Ecophysiological differences among juvenile and reproductive plants of several woody species. *Oecologia*, **86**, 594–597.

Donovan, L.A. and Ehleringer, J.R. (1992). Contrasting water-use patterns among size and life-history classes of a semi-arid shrub. *Functional Ecology*, **6**, 482–488.

Donovan, L.A., Dudley, S.A., Rosenthal, D.M., *et al.* (2007). Phenotypic selection on leaf water use efficiency and related ecophysiological traits for natural populations of desert sunflowers. *Oecologia*, **152**, 13–25.

Donovan, L.A., Ludwig, F., Rosenthal, D.M., *et al.* (2009). Phenotypic selection on leaf ecophysiological traits in *Helianthus. The New Phytologist*, **183**, 868–879.

Donovan, L.A., Maherali, H., Caruso, C.M. *et al.* (2011). The evolution of the worldwide leaf economics spectrum. *Trends in Plant Science*, **26**, 88–95.

Dorchin, N., Cramer, M.D. and Hoffmann, J.H. (2006). Photosynthesis and sink activity of wasp-induced galls in *Acacia pycnantha. Ecology*, **87**, 1781–1791.

Dore, S., Hymus, G.J., Johnson, D.P., *et al.* (2003). Cross-validation of open-top chamber and eddy covariance measurements of ecosystem CO_2 exchange in a Florida scrub-oak ecosystem. *Global Change Biology*, **9**, 84–95.

Dorey, S., Baillieul, F., Pierrel, M.A., *et al.* (1997). Spatial and temporal induction of cell death, defense genes, and accumulation of salicylic acid in tobacco leaves reacting hypersensitively to a fungal glycoprotein elicitor. *Molecular Plant-Microbe Interactions*, **10**, 646–655.

Dorne, A.J., Cadel, G. and Douce, R. (1986). Polar lipid composition of leaves from nine typical alpine species. *Phytochemistry*, **25**, 65–68.

Douglas, N.A. and Manos, P.A. (2007). Molecular phylogeny of Nyctaginaceae: taxonomy, biogeography, and characters associated with a radiation of xerophytic genera in North America. *American Journal of Botany*, **94**, 856–872.

Doulis, A.G., Hausladen A., Mondy B., *et al.* (1993). Antioxidant response and winter hardiness in red spruce (*Picea rubens* Sarg.). *New Phytologist*, **23**, 365–374.

Downes, R.W. (1969). Differences in transpiration rates between tropical and temperate grasses under controlled conditions. *Planta*, **88**, 261–273.

Downton, W.J.S. (1971). Adaptive and evolutionary aspects of C_4 photosynthesis. In: *Photosynthesis and Photorespiration* (eds Hatch, M.D., Osmond, C.B. and Slatyer, R.O.), Wiley Interscience, New York, USA, pp. 3–17.

Downton, W.J.S., Berry, J.A. and Seemann, J.R. (1984). Tolerance of photosynthesis to high temperature in desert plants. *Plant Physiology*, **74**, 786–90.

Downton, W.J.S., Loveys, B.R. and Grant, W.J.R. (1988a). Stomatal closure fully accounts for the inhibition of photosynthesis by abscisic acid. *The New Phytologist*, **108**, 263–266.

Downton, W.J.S., Loveys, B.R. and Grant, W.J.R. (1988b). Non-uniform stomatal closure induced by water stress causes putative non-stomatal inhibition of photosynthesis. *The New Phytologist*, **110**, 503–509.

Dowsey, A.W., English, J., Pennington, K., *et al.* (2006). Examination of 2-DE in the human proteome organisation brain proteome project pilot studies with the new RAIN gel matching technique. *Proteomics*, **6**, 5030–5047.

Doyle, J.A. (1996). Seed plant phylogeny and the relationships of the Gnetales. *International Journal of Plant Sciences*, **157 (suppl)**, S3-S39.

Drake, B. (1992). A field study of the effects of elevated CO_2 on ecosystem processes in a Chesapeake Bay wetland. *Australian Journal of Botany*, **40**, 579–595.

Drake, B., Leadley, P., Arp, W., *et al.* (1989). An open top chamber for field studies of elevated atmospheric CO_2 concentration on saltmarsh vegetation. *Functional Ecology*, 3, 363–71.

Drake, B.G. and Leadley, P.W. (1991). Canopy photosynthesis of crops and native plant communities exposed to long-term elevated CO_2. *Plant Cell Environment*, **14**, 853–860.

Drake, B.G., Gonzàlez-Meler, M.A. and Long, S.P. (1997). More efficient plants: a consequence of rising atmospheric CO_2? *Annual Review of Plant Physiology and Molecular Biology*, **48**, 607–637.

Drake, J.E., Raetz, L.M., Davis, S.C., *et al.* (2010). Hydraulic limitation not declining nitrogen availability causes the age-related photosynthetic decline in loblolly pine (*Pinus taeda* L.) *Plant, Cell and Environment*, **33**, 1756–1766.

Draper, H.H. and Hadley, M. (1990). Malondialdehyde determination as index of lipid peroxidation. *Methods in Enzymology*, **186**, 421–431.

Drennan, P.N. and Nobel, P.S. (2000). Response of CAM species to increasing atmospheric CO_2 concentrations. *Plant, Cell and Environment* **23**, 767–781.

Dreyer, E., Le Roux, X., Montpied, P., *et al.* (2001). Temperature response of leaf photosynthetic capacity in seedlings from seven temperate tree species. *Tree Physiology*, **21**, 223–232.

Driever, S.M. and Baker, N.R. (2011). The water–water cycle in leaves is not a major alternative electron sink for dissipation of excess excitation energy when CO_2 assimilation is restricted. *Plant, Cell and Environment*, **34**, 837–846.

Driscoll, S.P., Prins, A., Olmos, E., *et al.* (2006). Specification of adaxial and abaxial stomata, epidermal structure and photosynthesis to CO_2 enrichment in maize leaves. *Journal of Experimental Botany*, **57**, 381–390.

Drolet, G.G., Huemmrich, K.F., Hall, F.G., *et al.* (2005). A MODIS-derived photochemical reflectance index to detect inter-annual variations in the photosynthetic light-use efficiency of a boreal deciduous forest. *Remote Sensing of Environment*, **98**, 212–224.

Drolet, G.G., Middleton, E.M., Huemmrich, K.F., *et al.* (2008). Regional mapping of gross light-use efficiency using MODIS spectral indices. *Remote Sensing of Environment*, **112**, 3064–3078.

Drusch M., Moreno J., Goulas Y., *et al.* (2008). Candidate Earth Explorer Core Mission. FLEX - Fluorescence Explorer. Report for Assessment. Noordwijk, The Netherlands, European Space Agency. ESA SP-1313/4. www.congrex.nl/09c01/SP1313-4_FLEX.pdf

Dry, P.R., Loveys, B.R., McCarthy, M.G., *et al.* (2001). Strategic irrigation management in Australian vineyards. *Journal International de la Science de la Vigne et du Vin*, **35**, 129–139.

Du, Y.C., Kawamitsu, Y., Nose, A., *et al.* (1996). Effects of water stress on carbon exchange rate and activities of of photosynthetic enzymes in leaves of sugarcane (*Saccharum* sp.). *Australian Journal of Plant Physiology*, **23**, 719–726.

Duan, B., Li, Y., Zhang, X., *et al.* (2009). Water deficit affects mesophyll limitation of leaves more strongly in sun than in shade in two contrasting *Picea asperata* populations. *Tree Physiology*, **29**, 1551–1561.

Duarte, H.M. and Lüttge, U. (2007). Circadian rhythmicity. In: *Clusia: A Woody Neotropical Genus of Remarkable Plasticity and Diversity* (ed. Lüttge, U.), Berlin – Heidelberg – New York. Springer, 245–256.

Dubois, J.J.B., Fiscus, E.L., Booker, F.L., *et al.* (2007). Optimizing the statistical estimation of the parameters of the Farquhar–von Caemmerer–Berry model of photosynthesis. *New Phytologist*, **176**, 402–414.

Ducrey, M. (1994). Influence of shade on photosynthetic gas exchange of 7 tropical rain-forest species from guadeloupe (French-West-Indies). *Annales Des Sciences Forestieres*, **51**, 77–94.

Ducruet, J.M. (2003). Chlorophyll thermoluminescence of leaf discs: simple instruments and progress in signal interpretation open the way to new ecophysiological indicators. *Journal of Experimental Botany*, **54**, 2419–2430.

Ducruet, J.M., Gaillardon, P. and Vienot, J. (1984). Use of chlorophyll fluorescence induction kinetics to study translocation and detoxication of DCMU-type herbicides in plant leaves. *Zeitschrift für Naturforschung*, **39c**, 354–358.

Ducruet, J.M., Peeva, V. and Havaux, M. (2007). Chlorophyll thermofluorescence and thermoluminescence as complementary tools for the study of temperature stress in plants. *Photosynthesis Research*, **93**, 169–171.

Ducruet, J.M., Roman, M., Havaux, M., *et al.* (2005). Cyclic electron flow around PSI monitored by afterglow luminescence in leaves of maize inbred lines (*Zea mays* L.): correlation with chilling tolerance. *Planta*, **221**, 567–579.

Dudley, S.A. (1996a). The response to differing selection of plant physiological traits: evidence for local adaptation. *Evolution*, **50**, 103–110.

Dudley, S.A. (1996b). Differing selection on plant physiological traits in response to environmental water availability: a test of adaptive hypotheses. *Evolution*, **50**, 92–102.

Duffin, K.I. (2008). The representation of rainfall and fire intensity in fossil pollen and charcoal records from a South African savanna. *Review of Palaeobotany and Palynology*, **151**, 59–71.

Dufresne, J.L., Fairhead, L., Le Treut, H., *et al.* (2002). On the magnitude of positive feedback between future climate change and the carbon cycle. *Geophysical Research Letters*, **29**, 1405.

Dugas, W.A. (1993). Micrometeorological and chamber measurements of CO_2 flux from bare soil. *Agric Forest Meteorology*, **67**, 115–28.

Dunagan, S.C., Gilmore, M.S. and Varekamp, J.C. (2007). Effects of mercury on visible/near-infrared reflectance spectra of mustard spinach plants (*Brassica rapa* P.). *Environmental Pollution*, **148**, 301–311.

Dunbar-Co, S., Sporck, M.J. and Sack, L. (2009). Leaf trait diversification and design in seven rare taxa of the Hawaiian *Plantago* radiation. *International Journal of Plant Sciences*, **170**, 61–75.

Duniway, J.M. and Slatyer, R.O. (1971). Gas exchange studies on the transpiration and photosynthesis of tomato leaves affected by *Fusarium oxysporum* f. sp. *lycopersici*. *Phytopathology*, **61**, 1377–1381.

Duranceau, M., Ghashghaie, J., Badeck, F., *et al.* (1999). $\delta^{13}C$ of CO_2 respired in the dark in relation to $\delta^{13}C$ of leaf carbohydrates in *Phaseolus vulgaris* L. under progressive drought. *Plant, Cell and Environment*, **22**, 515–523.

Durand, L.Z. and Goldstein, G. (2001). Photosynthesis, photoinhibition, and nitrogen use efficiency in native and invasive tree ferns in Hawaii. *Oecologia*, **126**, 345–354.

Dutilleul, C., Garmier, C., Noctor, G., *et al.* (2003). Leaf mitochondria modulate whole cell redox homeostasis, set antioxidant capacity, and determine stress resistance through altered signaling and diurnal regulation. *Plant Cell*, **15**, 1212–1226.

Dutilleul, C., Lelarge, C., Prioul, J.L., *et al.* (2005). Mitochondria-driven changes in leaf NAD status exert a crucial influence on the control of nitrate assimilation and the integration of carbon and nitrogen metabolism. *Plant Physiology*, **139**, 64–78.

Dutton, R., Jiao, J., Tsujita, M.J., *et al.* (1988). Whole plant CO_2 exchange measurements for nondestructive estimation of growth. *Plant Physiology*, **86**, 355–358.

Duursma, R.A., Kolari, P., Perämäki, M., *et al.* (2009). Contributions of climate, leaf-area index and leaf physiology to variation in gross primary production of six coniferous forests across Europe: a model-based analysis. *Tree Physiology*, **29**, 621–639.

Eagleson, O.S. (1982). Ecological optimality in water limited natural soil-vegetation systems. 1. Theory and hypothesis. *Water Resources Research*, **18**, 325–340.

Earl, H.J. and Davis, R.F. (2003). Effect of drought stress on leaf and whole canopy radiation use efficiency and yield of maize. *Agronomy Journal*, **95**, 688–696.

Earl, H.J. and Ennhali, S. (2004). Estimating photosynthetic electron transport via chlorophyll fluorometry without photosystem II light saturation. *Photosynthesis Research*, **82**, 177–186.

Earl, H.J. and Tollenaar, M. (1999). Using chlorophyll fluorometry to compare photosynthetic performance of commercial maize (*Zea mays* L.) hybrids in the field. *Field Crops Res.*, **61**, 201–210.

Eastmond, P. and Rawsthorne, S. (1998). Comparison of the metabolic properties of plastids isolated from developing leaves or embryos of *Brassica napus*. *Journal of Experimental Botany*, **49**, 1105–1111.

Ebbert, V., Adams, W.W., Mattoo, A.K., *et al.* (2005). Up-regulation of a photosystem II core protein phosphatase inhibitor and sustained D1 phosphorylation in zeaxanthin-retaining, photoinhibited needles of overwintering Douglas fir. *Plant Cell Environment*, **28**, 232–240.

Eckardt, N.A., Snyder, G.W., Portis, A.R. Jr., *et al.* (1997). Growth and photosynthesis under high and low irradiance of *Arabidopsis thaliana* antisense mutants with reduced ribulose-1,5-bisphosphate carboxylase/oxygenase activase content. *Plant Physiology*, **113**, 575–586.

Edgerton, M.D. (2009). Increasing crop productivity to meet global needs for feed, food and fuel. *Plant Physiology*, **149**, 7–13.

Edmondson, D.L., Badger, M.R. and Andrews, T.J. (1990). Slow inactivation of ribulose bisphosphate carboxylase during catalysis is caused by accumulation of a slow, tight-binding inhibitor at the catalytic site. *Plant Physiology*, **93**, 1390–1397.

Edwards, D., Edwards, D.S. and Rayner, R. (1982). The cuticle of early vascular plants and its evolutionary significance. In: eds. Cutler, D.F., Alvin, K.L., Price, C.E., *The Plant Cuticle, Linnean Society Symposium Series n°10*. Academic Press, London pp. 341–361.

Edwards, D., Kerp, H. and Hass, H. (1998). Stomata in early land plants: an anatomical and ecophysiological approach. *Journal of Experimental Botany*, **49**, 255–278.

Edwards, E.J. and Still, C.J. (2008). Climate, phylogeny and the ecological distribution of C_4 grasses. *Ecology Letters*, **11**, 266–276.

Edwards, E.J., Still, C.J. and Donoghue, M.J. (2007a). The relevance of phylogeny to studies of global change. *Trends in Ecology and Evolution*, **22**, 243–249.

Edwards, G.E. and Baker, N.R. (1993). Can CO_2 assimilation in maize leaves be predicted accurately from chlorophyll fluorescence analysis? *Photosynthesis Research*, **37**, 89–102.

Edwards, G.E., Dai, Z., Cheng, S.H., *et al.* (1996). Factors affecting the induction of crassulacean acid metabolism in *Mesembryanthemum crystallinum*. In: *Crassulacean Acid Metabolism: Biochemistry, Ecophysiology and Evolution* (eds Winter, K. and Smith, J.A.C.), Springer-Verlag, Berlin, Germany, pp. 119–34.

Edwards, J., Salib, S., Thomson, F., *et al.* (2007b). The impact of *Phaeomoniella chlamydospora* infection on the grapevine's physiological response to water stress. Part 1: Zinfandel. *Phytopathologia Mediterranea*, **46**, 26–37.

Egea, G., González-Real, M.M., Baille, A. *et al.* (2011). Disentangling the contributions of ontogeny and water stress to photosynthetic limitations in almond trees. *Plant, Cell and Environment*, **34**, 962–979.

Eggers, T. and Jones, T.H. (2000). You are what you eat … or are you? *Trends in Ecology and Evolution*, **15**, 265–266.

Egli, D.B. (1998). *Seed Biology and the Yield of Grain Crops*. CAB International, Wallingford, UK and New York, USA.

Ehara, Y. and Misawa, T. (1975). Occurrence of abnormal chloroplasts in tobacco leaves infected systemically with the ordinary strain of cucumber mosaic virus. *Phytopathologische Zeitschrift*, **84**, 233–252.

Ehleringer, J. (1982). The influence of water stress and temperature of leaf pubescence in *Encelia farinosa*. *American Journal of Botany*, **69**, 670–675.

Ehleringer, J. (1983). Ecophysiology of *Amaranthus palmeri*, a Sonoran Desert summer annual. *Oecologia*, **57**, 107–112.

Ehleringer, J. (1977). Adaptive value of leaf hairs in *Encelia farinosa*. *Carnegie Institution of Washington Yearbook*, **77**, 413–418.

Ehleringer, J. and Pearcy, R.W. (1983). Variation in quantum yield for CO_2 uptake among C_3 and C_4 plants. *Plant Physiology*, **73**, 555–559.

Ehleringer, J.R. (1978). Implications of quantum yield differences on distributions of C_3 and C_4 grasses. *Oecologia*, **31**, 255–267.

Ehleringer, J.R. (1985). Annuals and perennials of warm deserts. In: *Physiological Ecology of North American Plant Communities* (eds Chabot, B.F. and Mooney, H.A.), Chapman and Hall, New York, USA, pp. 162–80.

Ehleringer, J.R. and Björkman, O. (1977). Quantum yields for CO_2 uptake in C3 and C4 plants. *Plant Physiology*, **59**, 86–90.

Ehleringer, J.R. and Björkman, O. (1978a). A comparison of photosynthetic characteristics of *Encelia* species possessing glabrous and pubescent leaves. *Plant Physiology*, **62**, 185–190.

Ehleringer, J.R. and Björkman, O. (1978b). Pubescence and leaf spectral characteristics in a desert shrub, *Encelia farinosa*. *Oecologia*, **36**, 151–162.

Ehleringer, J.R. and Cooper, T.A. (1992). On the role of orientation in reducing photoinhibitory damage in photosynthetic-twig desert shrubs. *Plant Cell and Environment*, **15**, 301–306.

Ehleringer, J.R. and Field, C.B. (1993). *Scaling Physiological processes: From Leaf to Globe*. Academic Press, San Diego, CA, USA, 1–388.

Ehleringer, J.R. and Mooney, H.A. (1978). Leaf hairs: effects on physiological activity and adaptive value to a desert shrub. *Oecologia*, **37**, 183–200.

Ehleringer, J.R., Björkman, O. and Mooney, H.A. (1976). Leaf pubescence: effects on absorptance and photosynthesis in a desert shrub. *Science*, **192**, 376–377.

Ehleringer, J.R., Cerling, T. E. and Helliker, B. R. (1997). C_4 photosynthesis, atmospheric CO_2, and climate. *Oecologia*, **112**, 285–299.

Ehleringer, J.R., Comstock, J.P. and Cooper, T.A. (1987). Leaf-twig carbon isotope ratio differences in photosynthetic-twig desert shrubs. *Oecologia*, **71**, 318–320.

Ehleringer, J.R. and Cooper, T.A. (1988). Correlations between carbon isotopes and microhabitat in desert plants. *Oecologia*, **76**, 562–566.

Ehleringer, J.R., Mooney, H.A., Gulmon, S.L., *et al.* (1980). Orientation and its consequences for *Copiapoa* (Cactaceae) in the Atacama Desert. *Oecologia*, **46**, 63–67.

Ehleringer, J.R., Mooney, H.A., Gulmon, S.L., *et al.* (1981). Parallel evolution of leaf pubescence in *Encelia* in coastal deserts of North and South America. *Oecologia*, **49**, 38–41.

Ehleringer, J.R., Roden, J. and Dawson, T.E. (2000). Assessing ecosystem-level water relations through stable isotope ratio analyses. In: *Methods in Ecosystem Science*, 181–198.

Ehleringer, J.R., Sage, R.F., Flanagan, L.B., *et al.* (1991). Climate change and the evolution of C_4 photosynthesis. *Trends in Ecology and Evolution*, **6**, 95–99.

Eichelmann, H. and Laisk, A. (2000). Cooperation of photosystems II and I in leaves as analyzed by simultaneous measurements of chlorophyll fluorescence and transmittance at 800 nm. *Plant and Cell Physiology*, **41**, 138–147.

Eichelmann, H., Oja, B., Rasulov, B., *et al.* (2004). Development of leaf photosynthetic parameters in *Betula pendula* Roth. leaves: correlations with Photosystem I density. *Plant Biology*, **6**, 307–318.

Eichelmann, H., Oja, V., Rasulov, B., *et al.* (2005). Adjustment of leaf photosynthesis to shade in a natural canopy: reallocation of nitrogen. *Plant, Cell and Environment*, **28**, 389–401.

Eilenberg, H., Hanania, U., Stein, H., *et al.* (1998). Characterization of rbcS genes in the fern *Pteris vittata* and their photoregulation. *Planta*, **206**, 204–214.

Eisenhut, M.S., Kahlon, D., Hasse, R., *et al.* (2006). The plant-like C2 glycolate cycle and the bacterial-like glycerate pathway cooperate in phosphoglycolate metabolism in cyanobacteria. *Plant Physiology*, **142**, 333–342.

Eissenstat, D., Graham, J.H., Syvertsen, J.P., *et al.* (1993). Carbon economy of sour orange in relation to mycorrhizal colonization and phosphorus status. *Annals of Botany*, **71**, 1–10.

Elagoz, V. and Manning, W. (2005). Responses of sensitive and tolerant bush beans (*Phaseolus vulgaris* L.) to ozone in open-top chambers are influenced by phenotypic differences, morphological characteristics, and the chamber environment. *Environmental Pollution*, **136**, 371–383.

Ellenberg, H. (1981). Ursachen des vorkommens und fehlens von sukkulenten in den trockengebieten der erde. *Flora*, **171**, 114–169.

Eller, B.M. and Ferrari, S. (1997). Water use efficiency of two succulents with contrasting CO_2 fixation pathways. *Plant Cell Environment*, **20**, 93–100.

Eller, B.M. and Ruess, B.R. (1986). Modulation of CAM and water balance of *Senecio medley woodii* by environmental factors and age of leaf. *Journal of Plant Physiology*, **125**, 295–309.

Eller, B.M., Ferrari, S. and Ruess, B.R. (1992). Spatial and diel variations of water relations in leaves of the CAM-plant *Senecio medley woodii*. *Botanica Helvetica*, **102**, 193–200.

Ellis, J.R. and Leech, R.M. (1985). Cell size and chloroplast size in relation to chloroplast replication in light-grown wheat leaves. *Planta*, **165**, 120–125.

Ellis, R.P., Vogel, J.C. and Fuls, A. (1980). Photosynthetic pathways and the geographical distribution of grasses in southwest Africa/Namibia. *South African Journal of Science*, **76**, 307–314.

Ellison, A.M. and Gotelli, N.J. (2009). Energetics and the evolution of carnivorous plants – Darwin's 'most wonderful plants in the world'. *Journal of Experimental Botany*, **60**, 19–42.

Ellsworth, D., Reich, P., Naumburg, E., *et al.* (2004). Photosynthesis, carboxylation and leaf nitrogen responses of 16 species to elevated pCO_2 across four free-air CO_2 enrichment experiments in forest, grassland and desert. *Global Change Biology*, **10**, 2121–2138.

Ellsworth, D.S. and Reich, P.B. (1993). Canopy structure and vertical patterns of photosynthesis and related leaf traits in a deciduous forest. *Oecologia*, **96**, 169–178.

Elmore, C.D. (1980). The paradox of no correlation between leaf photosynthetic rates and crop yields. In: *Predicting Photosynthesis for Ecosystems Models* (eds Hesketh, J.D. and Jones, J.W), CRC Press, Boca Raton, Florida, USA, pp. 155–67.

Else, M.A., Coupland, D., Dutton, L., *et al.* (2001). Decreased root hydraulic conductivity reduces leaf water potential, initiates stomatal closure and slows leaf expansion in flooded plants of castor oil (*Ricinus communis*) despite diminished delivery of ABA from roots to shoots in xylem sap. *Physiologia Plantarum*, **111**, 46–54.

Else, M.A., Davies, W.J., Whitford, P.N., *et al.* (1994). Concentrations of abscisic acid and other solutes in xylem sap from root systems of tomato and castor-oil plants are distorted by wounding and variable sap flow rates. *Journal of Experimental Botany*, **45**, 317–324.

Eltayeb, A.E., Kawano, N., Badawi, G.H., *et al.* (2007). Overexpression of monodehydroascorbate reductase in transgenic tobacco confers enhanced tolerance to ozone, salt and polyethylene glycol stresses. *Planta*, **225**, 1255–1264.

Elvira, S., Alonso, R., Castillo, F. J., *et al.* (1998). On the responses of pigments and antioxidants of *Pinus halepensis* seedlings to Mediterranean climatic factors and long-term ozone exposure. *New Phytologist*, **138**, 419–432.

Emberson, L.D., Ashmore, M.R., Cambridge, H.M., *et al.* (2000). Modelling stomatal ozone flux across Europe. *Environmental Pollution*, **109**, 403–413.

Enami, I., Kitamura, M., Tomo, T., *et al.* (1994). Is the primary cause of thermal inactivation of oxygen evolution in spinach PSII membranes release of the extrinsic 33 kDa protein or Mn? *Biochimica et Biophysica Acta*, **1186**, 52–58.

Endler, J.A. (1993). The colour of light in forests and its implications. *Ecological Monographs*, **61**, 1–27.

Engel, N., Schmidt, M., Lütz, C., *et al.* (2006). Molecular identification, heterologous expression and properties of light-insensitive plant catalases. *Plant, Cell and Environment*, **29**, 593–607.

Engelbrecht, B.M.J., Comita, L.S., Condit, R., *et al.* (2007). Drought sensitivity shapes species distribution patterns in tropical forests. *Nature*, **447**, 80-U2.

Englemann, S.C.W., Burscheidt, J., Gowik, U., *et al.* (2008). The gene for the P-subunit of glycine decarboxylase from the C_4 species *Flaveria trinervia*: analysis of transcriptional control in transgenic *Flaveria bidentis* (C_4) and *Arabidopsis* (C_3). *Plant Physiology*, **146**, 1773–1785.

Ensminger, I., Busch, F. and Huner, N.P.A. (2006). Photostasis and cold acclimation: sensing low temperature through photosynthesis. *Physiologia Planarum*, **126**, 28–44.

Ensminger, I., Schmidt, L. and Lloyd, J. (2008). Soil temperature and intermittent frost modulate the rate of recovery of photosynthesis in Scots pine under simulated spring conditions. *New Phytologist*, **177**, 428–442.

Ensminger, I., Schmidt, L., Tittmann, S., *et al.* (2005). Will photosynthetic gain of boreal evergreen conifers increase in response to a potentially longer growing season? In: *Photosynthesis: Fundamental Aspects to Global Perspectives* (eds Carpentier, R., Bruce, D. and van der Est, A.) Vol. 2, Allen Press, Lawrence, KS, USA, pp 976–978.

Ensminger, I., Sveshnikov, D., Campbell, D.A., *et al.* (2004). Intermittent low temperatures constrain spring recovery of photosynthesis in boreal Scots pine forests. *Global Change Biology*, **10**, 995–1008.

Epron, D. and Dreyer, E. (1992). Effects of severe dehydration on leaf photosynthesis in *Quercus petraea* (Matt.) Liebl.: photosystem II efficiency, photochemical and non-photochemical quenching and electrolyte leakage. *Tree Physiology*, **10**, 273–285.

Epron, D., Dreyer, E. and Bréda, N. (1992). Photosynthesis of oak trees [*Quercus petraea* (Matt) Liebl.] during drought under field conditions: diurnal course of net CO_2 assimilation and photochemical efficiency of photosystem II. *Plant, Cell and Environment*, **15**, 809–820.

Epron, D., Godard, G., Cornic, G., *et al.* (1995). Limitation of net CO_2 assimilation rate by internal resistances to CO_2 transfer in the leaves of two tree species (*Fagus sylvatica* and *Castanea sativa* Mill). *Plant, Cell and Environment*, **18**, 43–51.

Epstein, H.E., Lauenroth, W.K.,. Burke, I.C., *et al.* (1997). Productivity patterns of C_3 and C_4 functional types in the U.S. Great Plains. *Ecology*, **78**, 722–731.

Epstein, S. and Mayeda, T. (1953). Variation of ^{18}O content of waters from natural sources. *Geochimica et Cosmochimica Acta*, **4**, 213–224.

Ergova, E.A., Bukhov, N.G., Heber, U., *et al.* (2003). Effect of the pool size of stromal reductants on the alternative pathway of electron transfer to photosystem I in chloroplasts of intact leaves. *Russian Journal of Plant Physiology*, **50**, 431–440.

Esau, K. (1968). *Viruses in Plant Hosts. Form, Distribution and Pathologic Effect.* Wisconsin Press, Madison, Milwaukee.

Escalona, J.M., Flexas, J., Bota, J., *et al.* (2003). From leaf photosynthesis to grape yield: influence of soil water availability. *Vitis*, **42**, 57–64.

Eschrich, W., Fromm, J. and Essiamah, S. (1988). Mineral partitioning in the phloem during autumn senescence of beech leaves. *Trees*, **2**, 73–83.

Escoubas, J., Lomas, M., LaRoche, J., *et al.* (1995). Light intensity regulation of cab gene transcription is signaled by the redox state of the plastoquinone pool. *Proceedings of the National Academy of Sciences USA*, **92**, 10237–10241.

Esfeld, P., Siebke, K., Wacker, I., *et al.* (1995). Local defence-related shift in the carbon metabolism in chickpea leaves induced by a fungal pathogen. In: *Photosynthesis: From Light to Biosphere* (ed. P. Mathis) Vol. 5, Kluwer Academic Publisher, Dordrecht, Netherlands, pp. 663–666.

Esler, K.J. and Rundel, P.W. (1999). Comparative patterns of phenology and growth form diversity in two winter rainfall deserts: the Succulent Karoo and the Mojave Desert ecosystems. *Plant Ecology*, **142**, 97–104.

Espinoza, C., Vega, A., Medina, C., *et al.* (2007). Gene expression associated with compatible viral diseases in grapevine cultivars. *Functional and Integrative Genomics*, **7**, 95–110.

Esteban, R., Fernández-Marín, B., Becerril, J.M., *et al.* (2008). Photoprotective implications of leaf variegation in *E. dens-canis* L. and *P. officinalis* L. *Journal of Plant Physiology*, **165**, 1255–1263.

Esteban, R., Olano, J.M., Castresana, J., *et al.* (2009). Distribution and evolutionary trends of photoprotective isoprenoids (xanthophylls and tocopherols) within the plant kingdom. *Physiologia Plantarum*, **135**, 379–389.

Ethier, G.J. and Livingston, N.J. (2004). On the need to incorporate sensitivity to CO_2 transfer conductance into the Farquhar-von Caemmerer-Berry leaf photosynthesis model. *Plant, Cell and Environment*, **27**, 137–153.

Ethier, G.J., Livingston, N.J., Harrison, D.L., *et al.* (2006). Low stomatal and internal conductance to CO_2 versus Rubisco deactivation as determinants of the photosynthetic decline of ageing evergreen leaves. *Plant, Cell and Environment*, **29**, 2168–2184.

Etiope, G. and Ciccioli, P. (2009). Earth's degassing: a missing ethane and propane source. *Science*, **323**, 478.

Evain, S., Flexas, J. and Moya, I. (2004). A new instrument for passive remote sensing: 2. Measurement of leaf and canopy reflectance changes at 531 nm and their relationship with photosynthesis and chlorophyll fluorescence. *Remote Sensing of Environment*, **91**, 175–185.

Evans, J.R. (1983). Nitrogen and photosynthesis in the flag leaf of wheat (*Triticum aestivum* L.). *Plant Physiology*, **72**, 297–302.

Evans, J.R. (1986). A quantitative analysis of light distribution between the two photosystems, considering variation in both the relative amounts of the chlorophyll-protein complexes and the spectral quality of light. *Photochemistry and Photobiophysics*, **10**, 135–147.

Evans, J.R. (1989). Photosynthesis and nitrogen relationships in leaves of C3 plants. *Oecologia*, **78**, 9–19.

Evans, J.R. (1999). Leaf anatomy enables more equal access to light and CO_2 between chloroplasts. *New Phytologist*, **143**, 93–104.

Evans, J.R. (2009). Potential errors in electron transport rates calculated from chlorophyll fluorescence as revealed by a multilayer leaf model. *Plant and Cell Physiology*, **50**, 698–706.

Evans, J.R. and Poorter, H. (2001). Photosynthetic acclimation of plants to growth irradiance: the relative importance of specific leaf area and nitrogen partitioning in maximizing carbon gain. *Plant, Cell and Environment*, **24**, 755–767.

Evans, J.R. and Seemann, J.R. (1989). The allocation of protein nitrogen in the photosynthetic apparatus: costs, consequences, and control. In: *Photosynthesis. Proceedings of the C.S. French Symposium on Photosynthesis held in Stanford, California, July 17–23, 1988* (ed. Briggs, W.R.), Alan R. Liss, New York, USA. pp. 183–205.

Evans, J.R. and Terashima, I. (1987). Effects of nitrogen nutrition on electron transport components and photosynthesis in spinach. *Australian Journal of Plant Physiology*, **14**, 59–68.

Evans, J.R. and Terashima, I. (1988). Photosynthetic characteristics of spinach leaves grown with different nitrogen treatments. *Plant and Cell Physiology*, **29**, 157–165.

Evans, J.R. and Vellen, L. (1996). Wheat cultivars differ in transpiration efficiency and CO_2 diffusion inside their leaves. In: *Crop Research in Asia: Achievements and Perspectives. Proceedings of the 2nd Asian Crop Science Conference "Toward the Improvement of Food Production under Steep Population Increase and Global Environment Change" 21–23 August, 1995, the Fukui Prefectural University, Fukui, Japan* (eds Ishii, R. and Horie, T.), Crop Science Society of Japan, Asian Crop Science Association (ACSA), Fukui, Japan, pp. 326–329.

Evans, J.R. and Vogelmann, T.C. (2003). Profiles of ^{14}C fixation through spinach leaves in relation to light absorption and photosynthetic capacity. *Plant, Cell and Environment*, **26**, 547–560.

Evans, J.R. and Vogelmann, T.C. (2006). Photosynthesis within isobilateral *Eucalyptus pauciflora* leaves. *The New Phytologist*, **171**, 171–182.

Evans, J.R., Jakobsen, I. and Ögren, E. (1993). Photosynthetic light-response curves. 2. Gradients of light absorption and photosynthetic capacity. *Planta*, **189**, 191–200.

Evans, J.R., Kaldenhoff, R., Genty, B., *et al.* (2009). Resistances along the CO_2 diffusion pathway inside leaves. *Journal of Experimental Botany*, **60**, 2235–2248.

Evans, J.R. and Loreto, F. (2000). Acquisition and diffusion of CO_2 in higher plant leaves. In: *Photosynthesis: Physiology and Metabolism* (eds Leegood, R.C., Sharkey, T.D. and von Caemmerer, S.), Kluwer Academic Publishers, Dordrecht, Netherlands, pp. 321–351.

Evans, J.R., Sharkey, T.D., Berry, J.A., *et al.* (1986). Carbon isotope discrimination measured concurrently with gas exchange to investigate CO_2 diffusion in leaves of higher plants. *Australian Journal of Plant Physiology*, **13**, 281–292.

Evans, J.R., von Caemmerer, S., Setchell, B.A., *et al.* (1994). The relationship between CO_2 transfer conductance and leaf anatomy in transgenic tobacco with a reduced content of Rubisco. *Australian Journal of Plant Physiology*, **21**, 475–495.

Evans, L.T. (1975). The physiological basis of crop yield. In: *Crop Physiology* (ed. Evans, L.T.), Cambridge University Press, Cambridge, UK, pp. 327–335.

Evans, L.T. (1993). *Crop Evolution, Adaptation and Yield.* Cambridge University Press, Cambridge, UK.

Eversman, S. and Sigal, L.L. (1984). Ultrastructural effects of Peroxyacetyl Nitrate (PAN) on two lichen species. *The Bryologist*, **87**, 112–118.

Ewe, S. and Sternberg, L.D.S.L. (2005). Growth and gas exchange responses of Brazilian pepper (*Schinus terebinthifolius*) and native South Florida species to salinity. *Trees*, **19**, 119–128.

Ewers, B.E., Oren, R., Phillips, N., *et al.* (2001). Mean canopy stomatal conductance responses to water and nutrient availabilities in *Picea abies* and *Pinus taeda*. *Tree Physiology*, **21**, 841–850.

Ewers, F.W. (1982). Secondary growth in needle leaves of Pinus longaeva (bristlecone pine) and other conifers: quantitative data. *American Journal of Botany*, **69**, 1552–1559.

Eyles, A., Pinkard, E.A., O'Grady, A.P., *et al.* (2009). Role of corticular photosynthesis following defoliation in Eucalyptus globules. *Plant, Cell Environment*, **32**, 1004–1014.

Eyster, C., Brown, T., Tanner, H., *et al.* (1958). Manganese requirement with respect to growth, Hill reaction and photosynthesis. *Plant Physiology*, **338**, 235–241.

Ezcurra, E., Montaña, C. and Arizaga, S. (1991). Architecture, light interception, and distribution of *Larrea* species in the Monte Desert, Argentina. *Ecology*, **72**, 23–34.

Fader, G.M. and Koller, H.R. (1985). Seed growth rate and carbohydrate pool sizes of the soybean fruit. *Plant Physiology*, **79**, 663–666.

Fahn, A. and Cutler, D.F. (1992). *Xerophytes.* Gebruder Borntraeger, Berlin and Sttutgart.

Falge, E., Baldocchi, D., Olson, R., *et al.* (2001). Gap filling strategies for defensible annual sums of net ecosystem exchange. *Agricultural and Forest Meteorology*, **107**, 43–69.

Falge, E., Ryel, R.J., Alsheimer, M., *et al.* (1997). Effects of stand structure and physiology on forest gas exchange: a simulation study for Norway spruce. *Trees*, **11**, 436–448.

Falge, E., Tenhunen, J.D., Ryel, R., *et al.* (2000). Modelling age- and density related gas exchange of *Picea abies* canopies in the Fichtelgebirge, Germany. *Annals of Forest Science*, **57**, 229–243.

Falk, S., Maxwell, D.P., Laudenbach, D.E., *et al.* (1996). Photosynthetic adjustment to temperature. In: *Advances in Photosynthesis Vol.5: Photosynthesis and the Environment* (ed. Baker N.R.) Kluwer Academic Publishers, Dordrecht, Netherlands, pp. 367–385.

Falkowski, P.G. and Chen, Y.B. (2003). Photoacclimation of light harvesting systems in eukaryotic algae. In: *Advances in Photosynthesis and Respiration: Light Harvesting Antennas in Photosynthesis* (eds Green, B.R. and Parson, W.W.) Kluwer Academic Publishers, Dordrecht, Netherlands, pp. 423–447.

Falkowski, P.G. and Isozaki, Y. (2008). The story of O_2. *Science*, **322**, 540–542.

Falster, D.S. and Westoby, M. (2003). Leaf size and angle vary widely across species: what consequences for light interception? *New Phytologist*, **158**, 509–525.

Fan, L.-M, Zhao, Z. and Assmann, S.M. (2004). Guard cells: a dynamic signaling model. *Current Opinion in Plant Biology*, **7**, 537–546.

Fan, S.H. and Grossnickle, S.C. (1998). Comparisons of gas exchange parameters and shoot water relations of interior spruce (*Picea glauca* (Moench)Voss × *Picea engelmannii* Parry ex Engelm.) clones under repeated soil drought. *Canadian Journal of Forest Research-Revue Canadienne De Recherche Forestiere*, **28**, 820–830.

Fan, S.M., Wofsy, S.C., Bakwin, P.S., *et al.* (1990). Atmosphere-biosphere exchange of CO_2 and O3 in the central-Amazon-forest. *Journal of Geophysical Research-Atmospheres*, **95**, 16851–16864.

Fangmeier, A., Bender, J., Weigel H.J., *et al.* (2002). Effects of pollutant mixtures. In: *Air Pollution and Plant Life* (eds Bell, J.N.B. and Treshow, M.) 2nd edn. John Wiley and Sons, Chichester, USA, pp. 251–272.

FAO (2005). Report of the 22nd session of the International Poplar Commission and 42nd session of its Executive Committee, Santiago, Chile, 28 November – 2 December 2004.

Farage, P.K. (1996). The effect of ozone fumigation over one season on photosynthetic processes of *Quercus robur* seedlings. *New Phytologist*, **134**, 279–285.

Farage, P.K. and Long, S.P. (1991). The occurrence of photoinhibition in an over-wintering crop of oilseed rape (*Brassica-Napus* L.) and its correlation with changes in crop growth. *Planta*, **185**, 279–286.

Farage, P.K., Blowers, D., Long, S.P., *et al.* (2006). Low growth temperatures modify the efficiency of light use by photosystem II for CO_2 assimilation in leaves of two chilling-tolerant C4 species, *Cyperus longus* L. and *Miscanthus × giganteus*. *Plant, Cell and Environment*, **29**, 720–728.

Farage, P.K., McKee, I. and Long, S.P. (1998). Does a low nitrogen supply necessarily lead to acclimation of photosynthesis to elevated CO_2? *Plant Physiology*, **118**, 573–580.

Fares, S., Barta, C., Ederli, L., *et al.* (2006). Impact of high ozone on isoprene emission and some anatomical and physiological parameters of developing *Populus alba* leaves directly or indirectly exposed to the pollutant. *Physiologia Plantarum*, **128**, 456–465.

Farmer, A. (2002). Effects of particulates. In: *Air Pollution and Plant Life* (eds Bell, J.N.B. and Treshow, M.), 2nd edn. John Wiley and Sons, Chichester, USA, pp. 187–200.

Farmer, A.M. (1993). The effects of dust on vegetation: a review. *Environmental Pollution*, **79**, 63–75.

Farmer, A.M. (1996). Carbon uptake by roots. In: *Plant Roots: The Hidden Half* (eds Waisel, Y., Eshel, A. and Kafkaki, U.), Marcel Dekker, New York, USA, pp. 679–687.

Farnsworth, E.J. and Ellison, A.M. (2008). Prey availability directly affects physiology, growth, nutrient allocation and scaling relationships among leaf traits in 10 carnivorous plant species. *Journal of Ecology*, **96**, 213–221.

Farque, L., Sinoquet, H. and Colin, F. (2001). Canopy structure and light interception in *Quercus petraea* seedlings in relation to light regime and plant density. *Tree Physiology*, **21**, 1257–1267.

Farquhar, G.D. (1980). Carbon isotope discrimination by plants and the ratio of intercellular and atmospheric CO_2 concentrations. In: *Carbon Dioxide and Climate: Australian Research* (ed. Pearman, G.I.), Australian Academy of Science, Canberra, Australia, pp. 105–110.

Farquhar, G.D. (1983). On the nature of carbon isotope discrimination in C_4 plants. *Australian Journal of Plant Physiology*, **10**, 205–226.

Farquhar, G.D. (1989). Models of integrated photosynthesis of cells and leaves. *Philosophical Transactions of the Royal Society of London Series B-Biological Sciences*, **323**, 357–367.

Farquhar, G.D. and Cernusak, L.A. (2005). On the isotopic composition of leaf water in the non-steady state. *Functional Plant Biology*, **32**, 293–303.

Farquhar, G.D. and Gan, K.S. (2003). On the progressive enrichment of the oxygen isotopic composition of water along a leaf. *Plant, Cell and Environment*, **26**, 1579–1597.

Farquhar, G.D. and Lloyd, J. (1993). Carbon and oxygen isotope effects in the exchange of carbon dioxide between terrestrial plants and the atmosphere. In: *Stable Isotopes and Plant Carbon-water Relations* (eds Ehleringer, J.R., Hall, A.E. and Farquhar, G.D.), Academic Press, San Diego, CA, USA, pp. 47–70.

Farquhar, G.D. and Richards, R.A. (1984). Isotopic composition of plant carbon correlates with water-use efficiency of wheat genotypes. *Australian Journal of Plant Physiology*, **11**, 359–552.

Farquhar, G.D. and Sharkey, T.D. (1982). Stomatal conductance and photosynthesis. *Annual Review of Plant Physiology*, **33**, 317–345.

Farquhar, G.D. and von Caemmerer, S. (1982). Modeling of photosynthetic responses to environmental conditions. In: *Encyclopedia of Plant Physiology* (New Series) (eds Lange, O.L., Nobel, P.S., Osmond, C.B. and Ziegler, H.), Springer-Verlag, Berlin, Germany, pp. 549–587.

Farquhar, G.D. and Wong, S.C. (1984). An empirical model of stomatal conductance. *Australian Journal of Plant Physiology*, **11**, 191–210.

Farquhar, G.D., Cernusak, L.A. and Barnes, B. (2007). Heavy water fractionation during transpiration. *Plant Physiology*, **143**, 11–18.

Farquhar, G.D., Ehleringer, J.R. and Hubick, K.T. (1989). Carbon isotope discrimination and photosynthesis. *Annual Review of Plant Physiology and Plant Molecular Biology*, **40**, 503–537.

Farquhar, G.D., Firth, P.M., Wetselaar, R., *et al.* (1980c). On the gaseous exchange of ammonia between leaves and the environment: determination of the ammonia compensation point. *Plant Physiology*, **66**, 710–714.

Farquhar, G.D., Lloyd, J., Taylor, J.A., *et al.* (1993). Vegetation effects on the isotope composition of oxygen in the atmospheric CO_2. *Nature*, **363**, 439–443.

Farquhar, G.D., O'Leary, M.H. and Berry J.A. (1982). On the relationship between carbon isotope discrimination and the intercellular carbon dioxide concentration in leaves. *Australian Journal of Plant Physiology*, **9**, 121–137.

Farquhar, G.D., Schulze, E.D. and Küuppers, M. (1980b). Responses to humidity by stomata of *Nicotiana glauca* L. and *Corylus avellana* L. are consistent with the optimization of carbon-dioxide uptake with respect to water-loss. *Australian Journal of Plant Physiology*, **7**, 315–327.

Farquhar, G.D., von Caemmerer, S. and Berry J.A. (1980a). A biochemical model of photosynthetic CO_2 assimilation in leaves of C_3 species. *Planta*, **149**, 78–90.

Farquhar, G.D., von Caemmerer, S. and Berry, J.A. (2001). Models of photosynthesis. *Plant Physiology*, **125**, 42–45.

Farris, F. and Strain, B.R. (1978). The effects of water-stress on leaf $H_2^{18}O$ enrichment. *Radiation and Environmental Biophysics*, **15**, 167–202.

Fatemy, F., Trinder, P.K.E., Wingfield, J.N., *et al.* (1985). Effect of *Globodera rostochiensis* water stress and oxogeneris abscisic acid on stomatal function and water use of Cara and Pentland Dell potato plant. *Review of Nematology*, **8**, 249–255.

Favali, M.A., Pellegrini, S., and Bassi, M. (1975). Ultrastructural alterations induced by rice tungro virus in rice leaves. *Virology*, **66**, 502–507.

Febrero, A., Fernández, A., Fernández, S., *et al.* (1998). Yield, carbon isotope discrimination, canopy reflectance and cuticular conductance of barley isolines of differing glaucosness. *Journal of Experimental Botany*, **49**, 1575–1581.

Federer, C.A. and Tanner, C.B. (1966). Spectral distribution of light in the forest. *Ecology*, **47**, 555–560.

Feierabend, J., Streb, P., Schmidt, M., *et al.* (1996). Expression of catalase and its relation to light stress and stress tolerance. In: *Physical Stresses in Plants* (eds Grillo, S. and Leone, A.) Springer, Berlin, Heidelberg, Germany, pp. 223–234.

Feild, T.S. and Arens, N.C. (2007). The ecophysiology of early angiosperms. *Plant, Cell and Environment*, **30**, 291–309.

Feild, T.S., Lee, D.W. and Holbrook, M.M. (2001). Why leaves turn red in autumn. The role of anthocyanins in senescing leaves of red-osier dogwood. *Plant Physiology*, **127**, 566–574.

Felzer, B., Cronin, T., Reilly, J.M., *et al.* (2007). Impacts of ozone on trees and crops. *C. R. Geoscience*, **339**, 784–798.

Feng, L., Han, H., Liu, G., *et al.* (2007a). Overexpression of sedoheptulose-1,7-bisphosphatase enhances photosynthesis and growth under salt stress in transgenic rice plants. *Functional Plant Biology*, **34**, 822–834.

Feng, L., Wang, K., Li, Y., *et al.* (2007b). Overexpression of SBPase enhances photosynthesis against high temperature stress in transgenic rice plants. *Plant Cell Report*, **26**, 1635–1646.

Feng, Y.L., Lei, Y.B., Wang, R.F., *et al.* (2009). Evolutionary tradeoffs for nitrogen allocation to photosynthesis versus cell walls in an invasive plant. *Proceedings of the National Academy of Sciences USA*, **106**, 1853–1856.

Feng, Z., Jin, M., Zhang, F., *et al.* (2003). Effects of ground-level ozone (O_3) pollution on the yields of rice and winter wheat in the Yangtze River Delta. *Journal of Environmental Sciences-China*, **15**, 360–362.

Fenton, J.M. and Crofts, A.R. (1990). Computer aided fluorescence imaging of photosynthetic systems: application of video imaging to the study of fluorescence induction in green plants and photosynthetic bacteria. *Photosynthesis Research*, **26**, 59–66.

Fereres, E., Cruz-Romero, G., Hoffman, G.J., *et al.* (1979). Recovery of orange trees following severe water stress. *Journal of Applied Ecology*, **16**, 833–842.

Fernandes, G.W., deMattos, E.A., Franco, A.C., *et al.* (1998). Influence of the parasite *Pilostyles ingai* (Rafflesiaceae) on some physiological parameters of the host plant, *Mimosa naguirei* (Mimosaceae). *Botanica Acta*, **111**, 51.

Fernandez, A.P. and Strand, A. (2008). Retrograde signalling and plant stress: plastid signals initiate cellular stress responses. *Current Opinion in Plant Biology*, **11**, 509–513.

Ferrar, P.J. and Osmond, C.B. (1986). Nitrogen supply as a factor influencing photoinhibition and photosynthetic acclimation after transfer of shade-grown *Solanum dulcamara* to bright light. *Planta*, **168**, 563–570.

Ferrar, P.J., Slatyer, R.O. and Vranjic, J.A. (1989). Photosynthetic temperature acclimation in *Eucalyptus* species from diverse habitats, and a comparison with *Nerium oleander*. *Australian Journal of Plant Physiology*, **16**, 199–217.

Ferraro, F., Castagna, A., Soldatini, G.F., *et al.* (2003). Tomato (*Licopersicon esculentum* M.) T3238FER and T3238fer genotypes. Influence of different iron concentrations on thylakoid pigment and protein composition. *Plant Science*, **164**, 783–792.

Ferrio, J.P., Cuntz, M., Offermann, C., *et al.* (2009). Effect of water availability on leaf water isotopic enrichment in beech seedlings shows limitations of current fractionation models. *Plant, Cell and Environment*, **32**, 1285–1296.

Fetene, M., Nauke, P., Lüttge, U., *et al.* (1997). Photosyntesis and photoinhibition in a tropical alpine giant rosette plant, *Lobelia rhynchopetalum*. *New Phytologist*, **137**, 453–461.

Fey, V., Wagner, R., Bräutigam, K., *et al.* (2005a). Retrograde plastid redox signals in the expression of nuclear genes for chloroplast proteins *Arabidopsis thaliana*. *The Journal of Biological Chemistry*, **280**, 5318–5328.

Fey, V., Wagner, R., Bräutigam, K., *et al.* (2005b). Photosynthetic redox control of nuclear gene expression. *Journal Experimental Botany*, **56**, 1491–1498.

Field, C. (1983). Allocating leaf nitrogen for the maximization of carbon gain: leaf age as a control on the allocation program. *Oecologia*, **56**, 341–347.

Field, C. and Mooney, H.A. (1986). The photosynthesis – nitrogen relationship in wild plants. In: *On the Economy of Plant Form and Function. Proceedings of the Sixth Maria Moors Cabot Symposium, "Evolutionary constraints on primary productivity: adaptive patterns of energy capture in plants", Harvard Forest, August 1983* (ed. Givnish, T.J.). Cambridge University Press, Cambridge, UK, pp. 25–55.

Field, C., Merino, J. and Mooney, H.A. (1983). Compromises between water-use efficiency and nitrogen-use efficiency in 5 species of California evergreens. *Oecologia*, **60**, 384–389.

Filella, I. and Peñuelas, J. (1994). The red edge position and shape as indicators of plant chlorophyll content, biomass and hydric status. *International Journal of Remote Sensing*, **15**, 1459–1470.

Filella, I., Amaro, T., Araus, J.L., *et al.* (1996). Relationship between photosynthetic radiation-use-efficiency of barley canopies and the photochemical reflectance index (PRI). *Physiologia Plantarum*, **96**, 211–216.

Filippou, M., Fasseas, C. and Karabourniotis, G. (2007). Photosynthetic characteristics of olive tree (*Olea europaea*) bark. *Tree Physiology*, **27**, 977–984.

Finazzi, G. (2002). Redox-coupled proton pumping activity in cytochrome b6f, as evidenced by the pH dependence of electron transfer in whole cells of *Chlamydomonas reinhardtii*. *Biochemistry*, **41**, 7475–7482.

Finazzi, G., Jonson, G.N., Dall'Osto, L., *et al.* (2004). A zeaxanthin-independent nonphotochemical quenching mechanism localized in the photosystem II core complex. *Proceedings of the National Academy of Sciences USA*, **101**, 12375–12380.

Finazzi, G., Zito, F., Barbagallo, R.P., *et al.* (2001). Contrasted effects of inhibitors of cytochrome b6f complex on state transitions in *Chlamydomonas reinhardtii*: the role of Qo site occupancy in LHCII kinase activation. *Journal of Biological Chemistry*, **276**, 9770–9774.

Finn, M.W. and Tabita, F.R. (2004). Modified pathway to synthesize ribulose 1,5-bisphosphate in methanogenic archaea. *The Journal of Bacteriology*, **186**, 6360–6366.

Finnigan, J. (2006). The storage term in eddy flux calculations. *Agriculture and Forest Meteorology*, **136**, 108–113.

Finnigan, J.J., Clement, R., Malhi, Y., *et al.* (2003). A re-evaluation of long-term flux measurement techniques part I: averaging and coordinate rotation. *Boundary Layer Meteorology*, **107**, 1–48.

Finzi, A.C., Norby, R.J., Calfapietra, C., *et al.* (2007). Increases in nitrogen uptake rather than nitrogen-use efficiency support higher rates of temperate forest productivity under elevated CO_2. *Proceedings of the National Academy of Sciences of the USA*, **104**, 14014–14019.

Fischer, B.B., Krieger-Liszkay, A., Hideg, E., *et al.* (2007). Role of singlet oxygen in chloroplast to nucleus retrograde signaling in *Chlamydomonas reinhardtii. FEBS letters*, **581**, 5555–5560.

Fischer, K. and Kluge, M. (1984). Studies on carbon flow in crassulacean acid metabolism during the initial light period. *Planta*, **160**, 121–138.

Fischer, M. and Kaldenhoff, R. (2008). On the pH regulation of plant aquaporins. *Journal of Biological Chemistry*, **283**, 33889–33892.

Fiscus, E., Booker, F. and Burkey, K. (2005). Crop responses to ozone: uptake, modes of action, carbon assimilation and partitioning. *Plant, Cell and Environment*, **28**, 997–1011.

Fiscus, E., Reid, C., Miller, J., *et al.* (1997). Elevated CO_2 reduces O_3 flux and O_3-induced yield losses in soybeans: possible implications for elevated CO_2 studies. *Journal of Experimental Botany*, **48**, 307–313.

Fiscus, E.L., Booker, F.L. and Burkey, K.O. (2005). Crop responses to ozone: uptake, modes of action, carbon assimilation and partitioning. *Plant, Cell and Environment*, **28**, 997–1011.

Fish, D.A. and Earl, H.J. (2009). Water-use efficiency is negatively correlated with leaf epidermal conductance in cotton (*Gossypium* spp.). *Crop Science*, **49**, 1409–1415.

Flaig, H. and Mohr, H. (1992). Assimilation of nitrate and ammonium by the Scots pine (*Pinus sylvestris*) seedling under conditions of high nitrogen supply. *Physiologia Plantarum*, **84**, 568–576.

Flamant, P.H., Loth, C., Bruneau, D., *et al.* (2005) FACTS: Future Atmospheric Carbon dioxide Testing from Space, Final report, available from ESA on request.

Flanagan, L.B., Bain, J.F. and Ehleringer, J.R. (1991b). Stable oxygen and hydrogen isotope composition of leaf water in C_3 and C_4 plant species under field conditions. *Oecologia*, **88**, 394–400.

Flanagan, L.B., Comstock, J.P. and Ehleringer, J.R. (1991a). Comparison of modeled and observed environmental influences on the stable oxygen and hydrogen isotope composition of leaf water in *Phaseolus vulgaris* L. *Plant Physiology*, **96**, 588–596.

Fleck, I., Hogan, K. P., Llorens, L., *et al.* (1998). Photosynthesis and photoprotection in *Quercus ilex* resprouts after fire. *Tree Physiology*, **18**, 607–614.

Fleck, I., Peña-Rojas, K. and Aranda, X. (2010). Mesophyll conductance to CO_2 and leaf morphological characteristics under drought stress during *Quercus ilex* L. resprouting. *Annals of Forest Science*, **67**, 308.

Fleck, S., Niinemets, Ü., Cescatti, A., *et al.* (2003). Three-dimensional lamina architecture alters light harvesting efficiency in *Fagus*: a leaf-scale analysis. *Tree Physiology*, **23**, 577–589.

Fleischer, A., Titel, C., Ehwald, R. (1998). The boron requirement and cell wall properties of growing and stationary suspension-cultured *Chenopodium album* L. cells. *Plant Physiology*, **117**, 1401–1410.

Fleischmann, F., Koehl, J., Portz, R., *et al.* (2005). Physiological changes of *Fagus sylvatica* seedlings infected with *Phytophthora citricola* and the contribution of its elicitin "Citricolin" to pathogenesis. *Plant Pathology*, **7**, 650–658.

Fleming, A.J. (2005). The control of leaf development. Tansley review. *New Phytologist*, **166**, 9–20.

Flexas, J. and Medrano, H. (2002a). Drought-inhibition of photosynthesis in C_3 plants: stomatal and non-stomatal limitations revisited. *Annals of Botany*, **89**, 183–189.

Flexas, J. and Medrano, H. (2002b). Energy dissipation in C_3 plants under drought. *Functional Plant Biology*, **29**, 1209–1215.

Flexas, J., Badger, M., Chow, W.S., *et al.* (1999b). Analysis of the relative increase in photosynthetic O_2 uptake when photosynthesis in grapevine leaves is inhibited following low night temperatures and/or water stress. *Plant Physiology*, **121**, 675–684.

Flexas, J., Barón, M., Bota, J., *et al.* (2009). Photosynthesis limitations during water stress acclimation and recovery in the drought-adapted *Vitis* hybrid Richter-110 (*V. berlandieri* × *V. rupestris*). *Journal of Experimental Botany*, **60**, 2361–2377.

Flexas, J., Bota, J., Cifre, J., *et al.* (2004b). Understanding down-regulation of photosynthesis under water stress: future prospects and searching for physiological tools for irrigation management. *Annals of Applied Biology*, **144**, 273–283.

Flexas, J., Bota, J., Escalona, J.M., *et al.* (2002a). Effects of drought on photosynthesis in grapevines under field conditions: an evaluation of stomatal and mesophyll limitations. *Functional Plant Biology*, **29**, 461–471.

Flexas, J., Bota, J., Galmés, J., *et al.* (2006c). Keeping a positive carbon balance under adverse conditions: responses of photosynthesis and respiration to water stress. *Physiologia Plantarum*, **127**, 343–352.

Flexas, J., Bota, J., Loreto, F., *et al.* (2004a). Diffusive and metabolic limitations to photosynthesis under drought and salinity in C_3 plants. *Plant Biology*, **6**, 269–279.

Flexas, J., Briantais, J.M., Cerovic, ZG., *et al.* (2000) Steady state and maximum chlorophyll fluorescence responses to water stress in grapevine leaves: a new remote sensing system. *Remote Sensing of the Environment*, **73**, 283–297.

Flexas, J., Díaz-Espejo, A., Berry, J.A., *et al.* (2007b). Analysis of leakage in IRGA's leaf chambers of open gas exchange systems: quantification and its effects in photosynthesis parameterization. *Journal Experimental Botany*, **58**, 1533–1543.

Flexas, J., Diaz-Espejo, A., Galmés, J., *et al.* (2007a). Rapid variations of mesophyll conductance in response to changes in CO_2 concentration around leaves. *Plant, Cell and Environment*, **30**, 1284–1298.

Flexas, J., Escalona, J.M. and Medrano, H. (1999a). Water stress induces different levels of photosynthesis and electron transport rate regulations in grapevines. *Plant, Cell and Environment*, **22**, 39–48.

Flexas, J., Escalona, J.M., Evain, S., *et al.* (2002b). Steady state chlorophyll fluorescence (Fs) measurements as a tool to follow varations of net CO_2 assimilation and stomatal conductance during water-stress in C_3 plants. *Physiologia Plantarum*, **114**, 231–240.

Flexas, J., Galmés, J., Gallé, A., *et al.* (2010). Improving water use efficiency in grapevines: potential physiological targets for biotechnological improvement. *Australian Journal of Grape and Wine Research*, **16**, 106–121.

Flexas, J., Gulías, J. and Medrano, H. (2003). Leaf photosynthesis in Mediterranean vegetation. In: *Advances in Plant Physiology* Vol. V. (ed. Hemantaranjan, A.), Scientific Publishers, Jodphur, India, pp. 181–226.

Flexas, J., Gulías, J., Jonasson, S., *et al.* (2001). Seasonal patterns and control of gas-exchange in local populations of the Mediterranean evergreen shrub *Pistacia lentiscus* L. *Acta Oecologica*, **22**, 33–43.

Flexas, J., Ortuño, M.F., Ribas-Carbó, M., *et al.* (2007c). Mesophyll conductance to CO_2 in *Arabidopsis thaliana*. *New Phytologist*, **175**, 501–511.

Flexas, J., Ribas-Carbó, M., Bota, J., *et al.* (2006b). Decreased Rubisco activity during water stress is not induced by decreased relative water content but related to conditions of low stomatal conductance and chloroplast CO_2 concentration. *The New Phytologist*, **172**, 73–82.

Flexas, J., Ribas-Carbó, M., Díaz-Espejo, A., *et al.* (2008). Mesophyll conductance to CO_2: current knowledge and future prospects. *Plant Cell and Environment*, **31**, 602–621.

Flexas, J., Ribas-Carbó, M., Hanson, D.T., *et al.* (2006a). Tobacco aquaporin NtAQP1 is involved in mesophyll conductanve to CO_2 *in vivo*. *Plant Journal*, **48**, 427–439.

Flinn, A.M., Atkins, C.A. and Pate, J.S. (1977). Significance of photosynthetic and respiratory exchanges in the carbon economy of the developing pea fruit. *Plant Physiology*, **60**, 412–418.

Flood, P.J., Harbinson, J. and Aarts, M.G.M. (2011). Natural genetic variation in plant photosynthesis. *Trends in Plant Science*, **16**, 327–335.

Florez-Sarasa, I.D., Flexas, J., Rasmusson, A.G. *et al.* (2011). In vivo cytochrome and alternative pathway respiration in leaves of *Arabidopsis thaliana* plants with altered alternative oxidase under different light conditions. *Plant, Cell and Environment*, **34**, 1373–1383.

Flor-Henry, M., McCabe, T.C., de Bruxelles, G.L., *et al.* (2004). Use of a highly sensitive two-dimensional luminescence imaging system to monitor endogenous bioluminescence in plant leaves. *BMC Plant Biology*, **4**, 19.

Flors, C., Fryer, M.J., Waring, J., *et al.* (2006). Imaging the production of singlet oxygen in vivo using a new fluorescent sensor, Singlet Oxygen Sensor Green (R). *Journal of Experimental Botany*, **57**, 1725–1734.

Flowers, M.D., Fiscus, E.L., Burkey, K.O., *et al.* (2007) Photosynthesis, chlorophyll fluorescence, and yield of snap bean (*Phaseolus vulgaris* L.) genotypes differing in sensitivity to ozone. *Environmental and Experimental Botany*, **61**, 190–198.

Flügge, U.I. (1999). Phosphate translocators in plastids. *Annual Review of Plant Physiology and Plant Molecular Biology*, **50**, 27–45.

Flügge, U.I. and Heldt, H.W. (1991). Metabolite translocators of the chloroplast envelope. *Annual Review of Plant Physiology and Plant Molecular Biology*, **42**, 129–144.

Foley, J.A., Prentice, I.C., Ramankutty, N., *et al.* (1996). An integrated biosphere model of land surface processes, terrestrial carbon balance, and vegetation dynamics. *Global Biogeochemical Cycles*, **10**, 603–628.

Forest, F., Grenyer, R., Rouget, M., *et al.* (2007). Preserving the evolutionary potential of floras in biodiversity hotspots. *Nature*, **445**, 757–760.

Fork, D.C. and Murata, N. (1990). The effect of light intensity on the assay of the low temperature limit of photosynthesis using msec delayed light emission. *Photosynthesis Research*, **23**, 319–323.

Fork, D.C., Mohanty, P. and Hoshina, S. (1985). The detection of early events in heat disruption of thylakoid membranes by delayed light emission. *Physiologie Végétale*, **23**, 511–521.

Forseth, I.N. and Ehleringer, J.R. (1982). Ecophysiology of two solar tracking desert winter annuals. II. Leaf movements, water relations and microclimate. *Oecologia*, **54**, 41–49.

Forseth, I.N. and Ehleringer, J.R. (1983). Ecophysiology of two solar tracking desert winter annuals. III. Gas exchange responses to light, CO, and VPD in relation to long-term drought. *Oecologia*, **57**, 344–351.

Förstel, H. (1978). Contribution of oxygen isotope fractionation during the transpiration of plant leaves to the biogeochemical oxygen cycle. In: *Environmental Biogeochemistry and Geomicrobiology*, Vol 3 (ed. Krumbein, W.E.) Ann Arbor Science, Ann Arbor, MI, USA, pp. 811–824.

Förster, B., Mathesius, U. and Pogson, B.J. (2006). Comparative proteomics of high light stress in the model alga *Chlamydomonas reinhardtii*. *Proteomics*, **6**, 4309–4320.

Förster, B., Osmond, C.B. and Pogson, B.J. (2009). De novo synthesis and degradation of Lx and V cycle pigments during shade and sun acclimation in Avocado leaves. *Plant Physiology*, **149**, 1179–1195.

Forster, P., Ramaswamy, V., Artaxo, P., *et al.* (2007). Changes in atmospheric constituents and in eadiative forcing. In: *Climate Change 2007: The Physical Science Basis. Contribution of Working Group I to the Fourth Assessment Report of the Intergovernmental Panel on Climate Change* (eds Solomon, S., D. Qin, M. Manning, Z. Chen, M. Marquis, K.B. Averyt, M.Tignor and H.L. Miller), Cambridge University Press, Cambridge, UK and New York, USA.

Fournier, A., Daumard, F., Champagne, S., *et al.* (2012). Effects of canopy structure on sun-induced chlorophyll fluorescence. *Journal of Photogrammetry and Remote Sensing. IJPRS*, 68: 112–120.

Fowler, D., Cape, J.N., Coyle, M., *et al.* (1999a). Modelling photochemical oxidant formation, transport, deposition and exposure of terrestrial ecosystems. *Environmental Pollution*, **100**, 43–55.

Fowler, D., Cape, J.N., Coyle, M., *et al.* (1999b). The global exposure of forests to air pollutants. *Water, Air, and Soil Pollution*, **116**, 5–32.

Fowler, D., Flechard, C., Skiba, U.T.E., *et al.* (1998). The atmospheric budget of oxidized nitrogen and its role in ozone formation and deposition. *New Phytologist*, **139**, 11–23.

Fox, D.L. and Koch, P.L. (2003). Tertiary history of C_4 biomass in the Great Plains, USA. *Geology*, **31**, 809–812.

Foy, C.D., Chaney, R.L. and White, M.C. (1978). The physiology of metal toxicity in plants. *Annual Review of Plant Physiology*, **29**, 511–566.

Foyer, C. and Spencer, C. (1986). The relationship between phosphate status and photosynthesis in leaves. Effects on intracellular orthophosphate distribution, photosynthesis and assimilate partitioning. *Planta*, **167**, 369–375.

Foyer, C., Furbank, R., Harbinson, J., *et al.* (1990). The mechanisms contributing to photosynthetic control of electron transport by carbon assimilation in leaves. *Photosynthesis Research*, **25**, 83–100.

Foyer, C., Lelandais, M. and Kunert, K.J. (1994). Photooxidative stress in plants. *Physiologia Plantarum*, **92**, 696–717.

Foyer, C.H., Noctor, G. and Verrier, P. (2006a). Photosynthetic carbon-nitrogen interactions: modelling inter-pathway control and signalling. In: *Control of Primary Metabolism in Plants, Annual Plant Reviews*, Volume 22 (eds McManus, M. and Plaxton, B.), Blackwell Publishing, Oxford, UK, pp. 325–347.

Foyer, C.H. and Harbinson, J. (1994). Oxygen metabolism and the regulation of photosynthetic electron transport. In: *Causes of Photooxidative Stresses and Amelioration of Defense Systems in Plants* (eds Foyer, C.H. and Mullineaux, P.), CRC Press, Florida, USA, pp.1–42.

Foyer, C.H. and Noctor, G. (2003). Redox sensing and signalling associated with reactive oxygen in chloroplasts, peroxisomes and mitochondria. *Physiologia Plantarum*, **119**, 355–364.

Foyer, C.H. and Noctor, G. (2005). Oxidant and antioxidant signalling in plants: a re-evaluation of the concept of oxidative stress in a physiological context. *Plant, Cell and Environment*, **28**, 1056–1071.

Foyer, C.H. and Noctor, G. (2009). Redox regulation in photosynthetic organisms: signaling, acclimation and practical implications. *Antioxidants and Redox Signaling*, **11**, 861–905.

Foyer, C.H., Bloom, A., Queval, G., *et al.* (2009). Photorespiratory metabolism: genes, mutants, energetics, and redox signaling. *Annual Reviews of Plant Biology*, **60**, 455–484.

Foyer, C.H., Lelandais, M. and Harbinson, J. (1992). Control of the quantum efficiencies of PSI and PSII, electron flow and enzyme activation following dark to light transitions in pea leaves; the relationship between NADP/NADPH ratios and NADP-malate dehydrogenase activation state. *Plant Physiology*, **99**, 979–986.

Foyer, C.H., Pellny, T.K., Locato, V., *et al.* (2008). Analysis of redox relationships in the plant cell cycle: determinations of ascorbate, glutathione and poly (ADPribose) polymerase (PARP) in plant cell cultures. *Redox Mediated Signal*

Transduction. Methods in Molecular Biology Series, **476**, 193–209.

Foyer, C.H., Trebst, A. and Noctor, G. (2006b). Protective and signalling functions of ascorbate, glutathione and tocopherol in chloroplasts. In: *Advances in Photosynthesis and Respiration*, Volume 19, "Photoprotection, Photoinhibition, Gene Regulation, and Environment" (eds Demmig-Adams, B. and Adams, W.W.). Kluwer Academic Publishers. Dordrecht, Netherlands, pp. 241–268.

Fracheboud, Y., Luquez, V., Björkén, L., *et al.* (2009). The control of autumn senescence in European aspens (*Populus tremula*). *Plant Physiology*, **149**, 1982–1991.

Francey, R.J. (1985). Cape Grim isotope measurements-a preliminary assessment. *Journal of Atmospheric Chemistry*, **3**, 247–260.

Francey, R.J. and Tans, P.P. (1987). Latitudinal variation in oxygen-18 of atmospheric CO_2. *Nature*, **327**, 495–497.

Franck, F., Juneau, P. and Popovic, R. (2002). Resolution of the photosystem I and photosystem II contributions to chlorophyll fluorescence of intact leaves at room temperature. *Biochimica et Biophysica Acta*, **1556**, 239–246.

Franco, A.C., de Soyza, A.G., Virginia, R.A., *et al.* (1994). Effects of plant size and water relations on gas exchange and growth of the desert shrub *Larrea tridentata*. *Oecologia*, **97**, 171–178.

Franke, J. and Menz, G. (2007). Multi-temporal wheat disease detection by multi-spectral remote sensing. *Precision Agriculture*, **8**, 161–172.

Frankenberg C., Fisher J.B., Worden J., *et al.* (2011). New global observations of the terrestrial carbon cycle from GOSTA: Patterns of plant fluorescence with gross primary productivity. Geophys. Res. Lett. 38(17):L17706.

Franks, P.J. and Beerling, D.J. (2009). Maximum leaf conductance driven by CO_2 effects on stomatal size and density over geologic time. *Proceedings of the National Academy of Sciences USA*, **106**, 10343–10347.

Franks, P.J. and Farquhar, G.D. (2001). The effect of exogenous abscisic acid on stomatal development, stomatal mechanics, and leaf gas exchange in *Tradescantia virginiana*. *Plant Physiology*, **125**, 935–942.

Franks, P.J. and Farquhar, G.D. (2007). The mechanical diversity of stomata and its significance in gas-exchange control. *Plant Physiology*, **143**, 78–87.

Frantz, J.M., Joly, R.J. and Mitchell, C.A. (2000). Intracanopy lighting influences radiation capture, productivity, and leaf senescence in cowpea canopies. *Journal of the American Society of Horticultural Science*, **125**, 694–701.

Fravolini, A., Williams, D.G. and Thompson, T.L. (2002). Carbon isotope discrimination and bundle sheath leakiness in three C_4 subtypes grown under variable nitrogen, water and atmospheric CO_2 supply. *Journal of Experimental Botany*, **53**, 2261–2269.

Fredeen, A.L., Koch, G.W. and Field, C.B. (1995). Effects of atmospheric CO_2 enrichment on ecosystem CO_2 exchange in a nutrient and water limited grassland. *Journal of Biogeography*, **22**, 215–219.

Fredeen, A.L., Rao, I.M. and Terry, N. (1989). Influence of phosphorus nutrition on growth and carbon partition in *Glycine max*. *Plant Physiology*, **89**, 225–230.

Freitag, H. and Stichler, W. (2002). *Bienertia cycloptera* Bunge ex Boiss., Chenopodiaceae, another C_4 plant without Kranz tissues. *Plant Biology*, **4**, 121–131.

Fridlyand, L., Backhausen, J. and Scheibe, R. (1998). Flux control of the malate valve in leaf cells. *Archives of Biochemistry and Biophysics*, **349**, 290–298.

Friedli, H., Siegenthaler, U., Rauber, D., *et al.* (1987). Measurements of concentration, $^{13}C/^{12}C$ and $^{18}O/^{16}O$ ratios of tropospheric carbon dioxide over Switzerland. *Tellus*, **39B**, 80–88.

Friedman, I. (1953). Deuterium content of natural water and other substances. *Geochimica et Cosmochimica Acta*, **4**, 89–103.

Friedman, W.E. (2008). Hydatellaceae are water lilies with gymnospermous tendencies. *Nature*, **453**, 94–97.

Friend A.D. and Woodward F.I. (1990). Evolutionary and ecophysiological responses of mountain plants to the growing season environment. *Advances in Ecological Research*, **20**, 59–124.

Friend, A.D. (2001). Modelling canopy CO_2 fluxes: are 'big-leaf' simplifications justified? *Global Ecology and Biogeography*, **10**, 603–619.

Friend, A.D., Arneth, A., Kiang, N.Y., *et al.* (2007). FLUXNET and modelling the global carbon cycle. *Global Change Biology*, **13**, 610–633.

Friis, E.M., Pedersen, K.R. and Crane, P.R. (2005). When Earth started blooming: insights from the fossil record. *Current Opinion in Plant Biology*, **8**, 5–12.

Fromm, J. and Fei, H. (1998). Electrical signalling and gas exchange in maize plants of drying soil. *Plant Science*, **132**, 203–213.

Fromm, J. and Lautner, S. (2007). Electrical signals and their physiological significance in plants. *Plant, Cell and Environment*, **30**, 249–257.

Frost, D.L., Gurney, A.L., Press, M.C., *et al.* (1997). *Striga hermonthica* reduces photosynthesis in sorghum: the importance of stomatal limitations and a potential role for ABA. *Plant, Cell and Environment*, **20**, 483–492.

Frost-Christensen, H. and Floto, F. (2007). Resistance to CO_2 diffusion in cuticular membranes of amphibious plants and the implication for CO_2 acquisition. *Plant, Cell and Environment*, **30**, 12–18.

Fryer, M.J., Andrews, J.R., Oxborough, K., *et al.* (1998). Relationship between CO_2 assimilation, photosynthetic electron transport, and active O_2 metabolism in leaves of maize in the field during periods of low temperature. *Plant Physiology*, **116**, 571–580.

Fryer, M.J., Ball, L., Oxborough, K., *et al.* (2003). Control of *Ascorbate Peroxidase 2* expression by hydrogen peroxide and leaf water status during excess light stress reveals a functional organisation of *Arabidopsis* leaves. *The Plant Journal*, **33**, 691–705.

Fryer, M.J., Oxborough, K., Mullineaux, P.M., *et al.* (2002). Imaging of photo-oxidative stress responses in leaves. *Journal of Experimental Botany*, **53**, 1249–1254.

Fu, J., Momcilovic I., Clemente, T.E., *et al.* (2008). Heterologous expression of a plastid EF-Tu reduces protein thermal aggregation and enhances CO_2 fixation in wheat (*Triticum aestivum*) following heat stress. *Plant Molecular Biology*, **68**, 277–288.

Fuchs, E.E. and Livingston, N.J. (1996). Hydraulic control of stomatal conductance in Douglas fir [*Pseudotsuga menziesii* (Mirb.) Franco] and alder [*Alnus rubra* Bong.] seedlings. *Plant, Cell and Environment*, **19**, 1091–1098.

Fuchs, E.E., Livingston, N.J. and Rose, P.A. (1999). Structure-activity relationships of ABA analogs based on their effects on the gas exchange of clonal white spruce (*Picea glauca*) emblings. *Physiologia Plantarum*, **105**, 246–256.

Fuentes, D.A., Gamon, J.A., Cheng, Y., *et al.* (2006). Mapping carbon and water vapor fluxes in a chaparral ecosystem using vegetation indices derived from AVIRIS. *Remote Sensing of Environment*, **103**, 312–323.

Fuhrer, J. (2009). Ozone risk for crops and pastures in present and future climates. *Naturwissenschaften*, **96**, 173–194.

Fuhrmann, J., Johnen, T. and Heise, K.P. (1994). Compartmentation of fatty acid metabolism in zygotic rape embryo. *Journal of Plant Physiology*, **143**, 565–569.

Fujii, H., Chinnusamy, V., Rodrigues, A., *et al.* (2009). In vitro reconstitution of an abscisic acid signalling pathway. *Nature*, **462**, 660–664.

Fujita, K., Takagi, S. and Terashima, I. (2008). Leaf angle in *Chenopodium album* is determined by two processes: induction and cessation of petiole curvature. *Plant, Cell and Environment*, **31**, 1138–1146.

Fukshansky, L. (1981). Optical properties of plants. In: *Plants and the Daylight Spectrum* (ed Krumbein, W.E.), Academic Press, London, UK, pp. 21–40.

Fukuda, H. (2000). Programmed cell death of tracheary elements as a paradigm in plants. *Plant Molecular Biology*, **44**, 245–253.

Funayama, S., Hikosaka, K. and Yahara, T. (1997b). Effects of virus infection and growth irradiance on fitness components and photosynthetic properties of *Eupatorium makinoi* (Compositae). *American Journal of Botany*, **84**, 823–829.

Funayama, S., Sonoike, K. and Terashima, I. (1997a). Photosynthetic properties of leaves of *Eupatorium makinoi* infected by ageminivirus. *Photosynthesis Research*, **53**, 253–261.

Fung, I., Field, C.B., Berry, J.A., *et al.* (1997). Carbon-13 exchanges between the atmosphere and biosphere, *Global Biogeochemical Cycles*, **11**, 507–533.

Funk, J.L., Giardina, C.P., Knohl, A., *et al.* (2006). Influence of nutrient availability, stand age, and canopy structure on isoprene flux in a *Eucalyptus saligna* experimental forest. *Journal of Geophysical Research – Biogeosciences* **111**:G02012.

Furbank, R.T., Jenkins, C.L.D. and Hatch, M.D. (1989). CO_2 concentrating mechanism of C_4 photosynthesis. Permeability of isolated bundle sheath cells to inorganic carbon. *Plant Physiology*, **91**, 1364–1371.

Furley, P. (2006). Tropical savannas. *Progress in Physical Geography*, **30**, 105–121.

Furley, P.A. and Metcalfe, S.E. (2007). Dynamic changes in savanna and seasonally dry vegetation through time. *Progress in Physical Geography*, **31**, 633–642.

Gaastra, P. (1959). Photosynthesis of crop plants as influenced by light, carbon dioxide, temperature, and stomatal diffusion resistance. *Mededelingen van Landbouwhogeschool te Wageningen, Nederland*, **59**, 1–68.

Gadjev, I., Vanderauwera, S., Gechev, T.S., *et al.* (2006). Transcriptomic footprints disclose specificity of reactive oxygen species signalling in *Arabidopsis*. *Plant Physiology*, **141**, 436–445.

Gale, J. (1972). Availability of carbon dioxide for photosynthesis at high altitudes. *Ecology*, **53**, 494–497.

Gallé, A. and Feller, U. (2007). Changes of photosynthetic traits in beech saplings (*Fagus sylvatica*) under severe drought stress and during recovery. *Physiologia Plantarum*, **131**, 412–421.

Gallé, A., Florez-Sarasa, I., Thameur, A, *et al.* (2010). Effects of drought stress and subsequent rewatering on photosynthetic and respiratory pathways in *Nicotiana sylvestris* wild type and the mitochondrial complex I-deficient CMSII mutant. *Journal of Experimental Botany*, **61**, 765–775.

Gallé, A., Florez-Sarasa, I., Tomás, M., *et al.* (2009). The role of mesophyll conductance during water stress and recovery in tobacco (*Nicotiana sylvestris*): acclimation or limitation? *Journal of Experimental Botany*, **60**, 2379–2390.

Gallé, A., Haldimann, P. and Feller, U. (2007). Photosynthetic performance and water relations in young pubescent oak (*Quercus pubescens*) trees during drought stress and recovery. *New Phytologist*, **174**, 799–810.

Gallo, K., Daughtry, C.S.T. and Wiegand, C.L. (1993). Errors in measuring absorbed radiation and computing crop radiation use efficiency. *Agronomy Journal*, **85**, 1222–1228.

Gallo, K.P. and Daughtry, C.S.T. (1986). Techniques for measuring intercepted and absorbed photosynthetically active radiation in corn canopies. *Agronomy Journal*, **78**, 752–756.

Galmés J., Medrano H. and Flexas J. (2007a). Photosynthetic limitations in response to water stress and recovery in Mediterranean plants with different growth forms. *New Phytologist*, **175**, 81–93.

Galmés, J., Abadía, A., Cifre, J., *et al.* (2007b). Photoprotection processes under water stress and recovery in Mediterranean plants with different growth forms and leaf habits. *Physiologia Plantarum*, **130**, 495–510.

Galmés, J., Conesa, M.A., Ochogavía, J.M. *et al.* (2011). Physiological and morphological adaptations in relation to water use efficiency in Mediterranean accessions of *Solanum lycopersicum*. *Plant, Cell and Environment*, **34**, 245–260.

Galmés, J., Flexas, J., Keys, A.J., *et al.* (2005). Rubisco specificity factor tends to be larger in plant species from drier habitats and in species with persistent leaves. *Plant, Cell and Environment*, **28**, 571–579.

Galmés, J., Flexas, J., Ribas-Carbó, M., *et al.* (2010). Rubisco activity in Mediterranean species is regulated by chloroplastic CO_2 concentration under water stress. *Journal of Experimental Botany*, **62**, 653–665.

Galmés, J., Medrano, H. and Flexas, J. (2006). Acclimation of Rubisco specificity factor to drought in tobacco: discrepancies between *in vitro* and *in vivo* estimations. *Journal of Experimental Botany*, **57**, 3659–3667.

Galmés, J., Medrano, H. and Flexas, J. (2007c) Water relations and stomatal characteristics of Mediterranean plants with different growth forms and leaf habits: responses to water stress and recovery. *Plant and Soil*, **290**, 139–155.

Galmés, J., Medrano, H. and Flexas, J. (2007f). Photosynthesis and photoinhibition in response to drought in a pubescent (var. minor) and a glabrous (var. palaui) variety of *Digitalis minor*. *Environmental and Experimental Botany*, **60**, 105–111.

Galmés, J., Pou, A., Alsina, M.M., *et al.* (2007d). Aquaporin expression in response to different water stress intensities and recovery in Richter-110 (*Vitis* sp.): relationship with ecophysiological status. *Planta*, **226**, 671–681.

Galmés, J., Ribas-Carbó, M., Medrano, H. and Flexas, J. (2007e). Response of leaf respiration to water stress in Mediterranean species with different growth forms. *Journal of Arid Environments*, **68**, 206–222.

Galvez, D. and Pearcy, R.W. (2003). Petiole twisting in the crowns of *Psychotria limonensis*: implications for light interception and daily carbon gain. *Oecologia*, **135**, 22–29.

Gamon, J.A. and Qiu, H.L. (1999). Ecological applications of remote sensing at multiple scales. In: *Handbook of Functional Plant Ecology* (eds Pugnaire, F.I. and Valladares, F.), Marcel Dekker, New York, USA, pp. 805–846.

Gamon, J.A. and Surfus, J.S. (1999). Assessing leaf pigment content and activity whith a reflectometer. *New Phytologist*, **143**, 105–117.

Gamon, J.A., Field, C.B., Bilger, W., *et al.* (1990). Remote sensing of the xanthophyll cycle and chlorophyll fluorescence in sunflower leaves and canopies. *Oecologia*, **85**, 1–7.

Gamon, J.A., Field, C.B., Goulden, M.L., *et al.* (1995). Relationships between NDVI, canopy structure, and photosynthesis in three Californian vegetation types. *Ecological Applications*, **5**, 28–41.

Gamon, J.A., Huemmrich, K.F., Peddle, D.R., *et al.* (2004). Remote sensing in BOREAS: lessons learned. *Remote Sensing of Environment*, **89**, 139–162.

Gamon, J.A., Peñuelas, J. and Field, C.B. (1992). A narrow-waveband spectral index that traces diurnal changes in photosynthetic efficiency. *Remote Sensing of the Environment*, **41**, 35–44.

Gamon, J.A., Serrano, L. and Surfus, J.S. (1997). The photochemical reflectance index: an optical indicator of photosynthetic radiation use efficiency across species, functional types, and nutrient levels. *Oecologia*, **112**, 492–501.

Gan, K.S., Wong, S.C., Yong, J.W.H., *et al.* (2002). ^{18}O spatial patterns of vein xylem water, leaf water, and dry matter in cotton leaves. *Plant Physiology*, **130**, 1008–1021.

Gao, D., Gao, Q., Xu, H.Y., *et al.* (2009). Physiological responses to gradual drought stress in the diploid hybrid *Pinus densata* and its two parental species. *Trees*, **23**, 717–728.

Gao, Q., Zhao, P., Zeng, X., *et al.* (2002). A model of stomatal conductance to quantify the relationship between leaf transpiration, microclimate and soil water stress. *Plant, Cell and Environment*, **25**, 1373–1381.

Gao, Z., Bian, L. and Zhou, X. (2003). Measurements of turbulent transfer in the near-surface layer over a rice paddy in China. *Journal of Geophysical Research*, **108**, 4387.

Garbulsky, M.F., Peñuelas, J., Gamon, J. *et al.* (2011). The photochemical reflectance index (PRI) and the remote sensing of leaf, canopy and ecosystem radiation use efficiencies: a review and meta-analysis. *Remote Sensing of the Environment*, **115**, 281–297.

Garcia, R.L., Idso, S.B., Wall, G.W., *et al.* (1994). Changes in net photosynthesis and growth of *Pinus eldarica* seedlings in response to atmospheric CO_2 enrichment. *Plant Cell Environment*, **17**, 971–978.

García-Plazaola, J.I. and Becerril, J.M. (2001). Seasonal changes in photosynthetic pigments and antioxidants in beech (*Fagus sylvatica*) in a Mediterranean climate: implications for tree decline diagnosis. *Australian Journal of Plant Physiology*, **28**, 225–232.

García-Plazaola, J.I., Artetxe, U., Duñabeitia, M.K., *et al.* (1999). Role of photoprotective systems of holm-oak (*Quercus ilex*) in the adaptation to winter conditions. *Journal of Plant Physiology*, **155**, 625–630.

García-Plazaola, J.I., Becerril, J.M., Hernández, A., *et al.* (2004). Acclimation of antioxidant pools to the light environment in a natural forest canopy. *The New Phytologist*, **163**, 87–97.

García-Plazaola, J.I., Esteban, R., Hormaetxe, K., *et al.* (2008a). Seasonal reversibility of acclimation to irradiance in leaves of common box (*Buxus sempervirens* L.) in a deciduous forest. *Flora*, **203**, 254–260.

García-Plazaola, J.I., Esteban, R., Hormaetxe, K., *et al.* (2008b). Photoprotective responses of Mediterranean and Atlantic trees to the extreme heat-wave of summer 2003 in Southwestern Europe. *Trees: Structure and Function*, **3**, 385–392.

Garcia-Plazaola, J.I., Hernández, A. and Becerill, J.M. (2003a). Antioxidant and pigment composition during autumnal leaf senescence in woody deciduous species differing in their ecological traits. *Plant Biology*, **5**, 557–566.

García-Plazaola, J.I., Matsubara, S. and Osmond, C.B. (2007). The lutein epoxide cycle in higher plants: its relationships to other xanthophyll cycles and possible functions. *Functional Plant Biology*, **34**, 759–773.

García-Plazaola, J.I., Olano, J.M., Hernandez, J.M., *et al.* (2003b). Photoprotection in evergreen Mediterranean plants during sudden periods of intense cold weather. *Trees*, **17**, 285–291.

Gardeström, P. and Wigge, B. (1988). Influence of photorespiration on ATP/ADP ratios in the chloroplasts, mitochondria and cytosol, studies by rapid fractionation of barley protoplasts. *Plant Physiology*, **88**, 69–76.

Gardeström, P., Igamberdiev, A.U. and Raghavendra, A.S. (2002). Mitochondrial functions in the light and significance to carbon-nitrogen interactions. In: *Photosynthetic Nitrogen Assimilation and Associated Carbon and Respiratory Metabolism* (eds Fouer, C.H. and Noctor, G.), Kluwer Academic Publisher, Dordrecht, Netherlands, pp. 151–172.

Garmier, M., Priault, P., Vidal, G., *et al.* (2008). Light and oxygen are not required for harpin-induced cell death. *Journal of Biological Chemistry*, **282**, 37556–37566.

Gates, D.M. (1962) *Energy Exchange in the Biosphere.* Harper and Row, New York, USA, pp. 151.

Gates, D.M. (1970). Physical and physiological properties of plants. In: *Remote Sensing.* Nat Acad of Sci., Washington DC, USA, pp. 224–252.

Gates, D.M. (1980). *Biophysical Ecology.* Springer-Verlag, New York – Heidelberg – Berlin.

Gates, D.M. and Papian, L.E. (1971). *Atlas of Energy Budgets of Plant Leaves.* London: Academic Press.

Gauslaa, Y. (1984). Heat resistance and energy budget in different scandinavian plants. *Holarctic Ecology*, **7**, 1–78.

Gaut, B.S. and Doebley, J.F. (1997). DNA sequence evidence for the segmental allotetraploid origin of maize. *Proceedings of the National Academy of Sciences (USA)*, **94**, 68090–68094.

Gautier, H., Vavasseur, Al., Gans, P., *et al.* (1991). Relationship between respiration and photosynthesis in guard cell and mesophyll cell protoplasts of *Commelina communis* L. *Plant Physiology*, **95**, 636–641.

Geber, M.A. and Dawson, T.E. (1997). Genetic variation in stomatal and biochemical limitations to photosynthesis in the annual plant, *Polygonum arenastrum. Oecologia*, **109**, 535–546.

Gedney, N., Cox, P., Betts, R., *et al.* (2006) Detection of a direct carbon dioxide effect in continental river runoff records. *Nature,* **439**, 835–838.

Geiken, B., Masojídek, J., Rizzuto, M., *et al.* (1998). Incorporation of [35S]methionine in higher plants reveals that stimulation of the D1 reaction centre II protein turnover accompanies tolerance to heavy metal stress. *Plant, Cell and Environment*, **21**, 1265–1273.

Geissler, N., Hussin, S. and Koyro, H.W. (2009). Elevated atmospheric CO_2 concentration ameliorates effects of NaCl salinity on photosynthesis and leaf structure of *Aster tripolium* L. *Journal of Experimental Botany*, **60**, 137–151.

Gent, M.P., Ferrandino, F.J. and Elmer, W.H. (1995). Effects of *Verticillium* wilt on gas exchange of entire eggplants. *Canadian Journal of Botany*, **73**, 557–565.

Gent, M.P.N., LaMondia, J.A., Ferrandino, F.J., *et al.* (1999). The influence of compost amendement or straw mulch on the reduction of gas exchange in potato by *Verticillium dahliae* and *Pratylenchus penetrans*. *Plant Disease*, **83**, 371–376.

Gentry, A.H. and Dodson, C.H. (1987). Diversity and evolution of Neotropical Vascular epiphytes. *Annals of the Missouri Botanical Garden*, **74**, 205–233.

Genty, B. and Harbinson, J. (1996). Regulation of light utilization for photosynthetic electron transport. In: *Photosynthesis and the Environment* (ed. Baker, N.R.), Kluwer, Amsterdam, Netherlands, pp. 69–99.

Genty, B. and Meyer, S. (1995). Quantitative mapping of leaf photosynthesis using chlorophyll fluorescence imaging. *Australian Journal of Plant Physiology*, **22**, 277–284.

Genty, B., Briantais, J.M. and Baker, N.R. (1989). The relationship between the quantum yield of photosynthetic electron transport and quenching of chlorophyll fluorescence. *Biochimica et Biophysica Acta*, **990**, 87–92.

Genty, B., Meyer, S., Pile, C., *et al.* (1998). CO_2 diffusion inside leaf mesophyll of ligneous plants. In *Photosynthesis: Mechanisms and Effects* (ed. Garab, G.), Kluwer Academic Publishers, Dordrecht, Netherlands, pp. 3961–3967.

Genty, B., Wonders, J. and Baker, N. (1990). Nonphotochemical quenching of F0 in leaves is emission wavelength dependent. Consequences for quenching analysis and its interpretation. *Photosynthesis Research*, **26**, 133–139.

Gepstein, S. (2004). Leaf senescence- not just a "wear and tear" phenomenon. *Genome Biology*, **5**, 212.

Gerbaud, A. and Andre, M. (1979). Photosynthesis and photorespiration in whole plants of wheat. *Plant Physiology*, **64**, 735–738.

Gerbens-Leenes, W., Hoekstra A.Y. and van der Meer, T.H. (2009). The water footprint of bioenergy. *Proceedings of the National Academy of Science USA*, **106**, 10219–10223.

Germeraad, J.H., Hopping, C.A. and Muller, J. (1967). Palynology of tertiary sediments from tropical areas. *Review of Palaeobotany and Palynology*, **6**, 189–348.

Germino, M.J. and Smith, W.K. (2000). High resistance to low-temperature photoinhibition in two alpine, snowbank species. *Physiologia Plantarum*, **110**, 89–95.

Germino, M.J. and Smith, W.K. (2001). Relative importance of microhabitat, plant form and photosynthetic physiology to carbon gain in two alpine herbs. *Functional Ecology*, **15**, 243–251.

Gerrish, G. (1989). Comparing crown growth and phenology of juvenile, early mature and late mature *Metrosideros polymorpha* trees. *Pacific Science*, **43**, 211–222.

Gerrish, G. (1990). Relating carbon allocation patterns to tree senescence in *Metrosideros* forests. *Ecology*, **71**, 1176–1184.

Gerst, U., Schönknecht, G. and Heber, U. (1994). ATP and NADPH as the driving force of carbon reduction in leaves in relation to thylakoid energization by light. *Planta*, **193**, 421–429.

Gerwing, J.J. and Farias, D.L. (2000). Integrating liana abundance and forest stature into an estimate of total aboveground biomass for an eastern Amazonian forest. *Journal of Tropical Ecology*, **16**, 327–335.

Gessler, A., Brandes, E., Buchmann, N., *et al.* (2009b). Tracing carbon and oxygen isotope signals from newly assimilated sugars in the leaves to the tree-ring archive. *Plant, Cell and Environment*, **32**, 780–795.

Gessler, A., Schrempp, S., Matzarakis, A., *et al.* (2001). Radiation modifies the effect of water availability on the carbon isotope composition of beech (*Fagus sylvatica*). *New Phytologist*, **150**, 653–664.

Gessler, A., Tcherkez, G., Karyanto, O., *et al.* (2009a). On the metabolic origin of the carbon isotope composition of CO_2 evolved from darkened light-acclimated leaves in *Ricinus communis*. *New Phytologist*, **181**, 374–386.

Gessler, A., Tcherkez, G., Peuke, A.D., *et al.* (2008). Experimental evidence for diel variations of the carbon isotope composition in leaf, stem and phloem sap organic matter in *Ricinus communis*. *Plant, Cell and Environment*, **31**, 941–953.

Ghannoum, O. (2009). C_4 photosynthesis and water stress. *Annals of Botany*, **103**, 635–644.

Ghannoum, O., Evans, J.R., Chow, W.S., *et al.* (2005). Faster rubisco is the key to superior nitrogen-use efficiency in NADP-malic enzyme relative to NAD-malic enzyme C-4 grasses. *Plant Physiology*, **137**, 638–650.

Ghannoum O., Phillips N.G., Conroy J.P., *et al.* (2010a). Exposure to preindustrial, current and future atmospheric CO_2 and temperature differentially affects growth and photosynthesis in Eucalyptus. *Global Change Biology*, **16**, 303–319.

Ghannoum, O., Phillips, N.G., Sears, M.A., *et al.* (2010b). Photosynthetic responses of two eucalypts to industrial-age changes in atmospheric (CO_2) and temperature. *Plant, Cell and Environment*, **33**, 1671–1681.

Ghannoum, O., von Caemmerer, S. and Conroy, J.P. (2002). The effect of drought on plant water use efficiency on nine NAD-ME and nine NADP-ME Australian C_4 grasses. *Functional Plant Biology*, **29**, 1337–1348.

Ghashghaie, J. and Cornic, C. (1994). Effect of temperature on partitioning of photosynthetic electron flow between CO_2 assimilation and O_2 reduction and on the CO_2/O_2 specificity of Rubisco. *Journal of Plant Physiology*, **143**, 643–650.

Ghashghaie, J., Badeck, F.W., Lanigan, G., *et al.* (2003). Carbon isotope fractionation during dark respiration and photorespiration in C_3 plants. *Phytochemistry Reviews*, **2**, 145–161.

Ghashghaie, J., Duranceau, M., Badeck, F.W., *et al.* (2001). $\delta^{13}C$ of CO_2 respired in the dark in relation to leaf metabolites: comparisons between *Nicotiana sylvestris* and *Helinathus annuus* under drought. *Plant, Cell and Environment*, **24**, 505–515.

Gholz, H.L. (1982). Environmental limits on aboveground net primary production, leaf area, and biomass in vegetation zones of the Pacific Northwest. *Ecology*, **63**, 469–481.

Gibberd, M.R., Walker, R.R., Blackmore, D., *et al.* (2001). Transpiration efficiency and carbon-isotope discrimination of grapevines grown under well-watered conditions in either glasshouse or vineyard. *Australian Journal of Grape and Wine Research*, **7**, 110–117.

Gibert, F., Flamant, P.H., Bruneau, D., *et al.* (2006). 2-μm heterodyne differential absorption lidar measurements of at mospheric CO_2 mixing ratio in the boundary layer. *Applied Optics*, **45**, 4448–4458.

Gibert, F., Schmidt, M., Cuesta, J., *et al.* (2007). Retrieval of average CO_2 fluxes by combining in-situ CO_2 measurements and backscatter lidar information. *Journal of Geophysical Research*, **112**, D10301.

Gibson, A.C. (1981). Vegetative anatomy of *Pachycormus* (Anacardiaceae). *Linnean Society, Botany*, **83**, 273–284.

Gibson, A.C. (1983). Anatomy of photosynthetic old stems of nonsucculent dicotyledons from North American deserts. *Botanical Gazette*, **144**, 347–362.

Gibson, A.C. (1996). *Structure-Function Relations of Warm Desert Plants*. Springer-Verlag, Berlin, Germany.

Gibson, A.C. (1998). Photosynthetic organs of desert plants. *BioScience*, **48**, 911–920.

Gibson, A.C. and Nobel, P.S. (1986). *The Cactus Primer*. Harvard University Press, Cambridge, MA, USA.

Gibson, A.C., Rundel, P.W. and Sharifi, M.R. (2008). Ecology and ecophysiology of a subalpine fellfield community on Mount Pinos, southern California. *Madroño*, **55**, 41–51.

Gielen, B. and Ceulemans, R. (2001). The likely impact of rising atmospheric CO_2 on natural land and managed *Populus*: a literature review. *Environmental Pollution*, **115**, 335–358.

Gifford, R.M. (1995). Whole plant respiration and photosynthesis of wheat under increased CO_2 concentration and temperature: long-term vs. short-term distinctions for modelling. *Global Change Biology*, **1**, 385–396.

Gifford, R.M. and Evans, L.T. (1981). Photosynthesis, carbon partitioning, and yield. *Annual Review of Plant Physiology*, **32**, 485–509.

Gil, P.M., Gurovich, L., Schaffer, B., *et al.* (2008). Root to leaf electrical signalling in avocado in response to light and soil water content. *Journal of Plant Physiology*, **165**, 1070–1078.

Gilbert, M., Wagner, H., Weingart, I., *et al.* (2004). A new type of thermoluminometer: a highly sensitive tool in applied photosynthesis research and and stress physiology. *Journal of Plant Physiology*, **161**, 641–651.

Gilbert, M.E., Zwieniecki, M.A. and Holbrook, N.M. (2011). Independent variation in photosynthetic capacity and stomatal conductance leads to differences in intrinsic water use efficiency in 11 soybean genotypes before and during mild drought. *Journal of Experimental Botany*, **62**, 2875–2887.

Gillon, J. and Yakir, D. (2001). Influence of carbonic anhydrase activity in terrestrial vegetation on the ^{18}O content of atmospheric CO_2. *Science*, **291**, 2584–2587.

Gillon, J.S. and Griffiths, H. (1997). The influence of (photo) respiration on carbon isotope discrimination in plants. *Plant, Cell and Environment*, **20**, 1217–1230.

Gillon, J.S. and Yakir, D. (2000a). Internal conductance to CO_2 diffusion and $C^{18}OO$ discrimination in C_3 leaves. *Plant Physiology*, **123**, 201–213.

Gillon, J.S. and Yakir, D. (2000b). Naturally low carbonic anhydrase activity in C_4 and C_3 plants limits discrimination against (COO)-^{18}O during photosynthesis. *Plant Cell and Environment*, **23**, 903–915.

Gilmanov, T.G., Soussana, J.F., Aires, L., *et al.* (2007). Partitioning European grassland net ecosystem CO_2 exchange into gross primary productivity and ecosystem respiration using light response function analysis. *Agriculture, Ecosystems and Environment*, **121**, 93–120.

Gilmore, A.M. and Ball, M.C. (2000). Protection and storage of chlorophyll in overwintering evergreens. *Proc Natl Acad Sci* USA, **97**, 11098–11101.

Gilmore, D.W., Seymour, R.S., Halteman, W.A., *et al.* (1995). Canopy dynamics and the morphological development of *Abies balsamea*: effects of foliage age on specific leaf area and secondary vascular development. *Tree Physiology*, **15**, 47–55.

Giordano, M., Beardall, J. and Raven, J.A. (2005). CO_2 concentrating mechanisms in algae: mechanisms, environmental modulation, and evolution. *Annual Review of Plant Biology*, **56**, 99–131.

Giraud, E., Ho, L.H.M., Clifton, R., *et al.* (2008). The absence of alternative oxidase 1a in *Arabidopsis* results in acute sensitivity to combined light and drought stress. *Plant Physiology*, **147**, 595–610.

Giri, A.P., Wunsche, H., Mitra, S., *et al.* (2006). Molecular interactions between the specialist herbivore *Manduca sexta* (Lepidoptera, Sphingidae) and its natural host *Nicotiana attenuata*. VII. Changes in the plant's proteome. *Plant Physiology*, **142**, 1621–1641.

Gitelson, A., Viña, A., Verma, S., *et al.* (2006). Relationship between gross primary production and chlorophyll content in crops: implications for the synoptic monitoring of vegetation productivity. *Journal of Geophysical Research*, **111**, D08S11.

Gitelson, A.A., Buschmann, C. and Lichtenthaler H.K. (1998). Leaf chlorophyll fluorescence corrected for re-absorption by means of absorption and reflectance measurements. *Journal of Plant Physiology*, **152**, 283–296.

Givnish, T. (1979). On the adaptive significance of leaf form. In: *Topics in Plant Population Biology* (eds Solbrig, O.T., Jain, S., Johnson, G.B. and Raven, P.H.), Columbia University Press, New York, USA, pp. 375–407.

Givnish, T.J. (1984). Leaf and canopy adaptations in tropical forests. In: *Physiological Ecology of Plants of the Wet Tropics. Proceedings of an International Symposium held in Oxatepec and Los Tuxtlas, Mexico, June 29 to July 6, 1983* (eds Medina, E., Mooney, H.A., Vásquez-Yánes, C.), Dr. W. Junk Publishers, The Hague, Netherlands, pp. 51–84.

Givnish, T.J. (1988). Adaptation to sun and shade: a whole-plant perspective. *Australian Journal of Plant Physiology*, **15**, 63–92.

Gleadow, R.M., Foley, W.J. and Woodrow, I.E. (1998). Enhanced CO_2 alters the relationship between photosynthesis and defence in cyanogenic *Eucalyptus cladocalyx* F. Muell. *Plant, Cell and Environment*, **21**, 12–22.

Glover, J.R. and Lindquist, S. (1998). Hspl04, Hsp70, and Hsp40: A novel chaperone system that rescues previously aggregated proteins. *Cell*, **94**, 73–82.

Godbold, D.L. and Hüttermann, A. (1988). Inhibition of photosynthesis and transpiration in relation to mercury induced root damage in spruce seedlings. *Physiologia Plantarum*, **74**, 270–275.

Goerner, A., Reichstein, M. and Rambal, S. (2009). Tracking seasonal drought effects on ecosystem light use efficiency with satellite-based PRI in a Mediterranean forest. *Remote Sensing of the Environment*, **113**, 1101–1111.

Goffeau, A. and Bové, J.M. (1965). Virus infection and photosyntesis. I. Increased photophosphorylation by chloroplasts from Chinese cabbage infected with turnip yellow mosaic virus. *Virology*, **27**, 243–252.

Gog, L., Berenbaum, M.R., de Lucia, E.H., *et al.* (2005). Autotoxic effects of essential oils on photosynthesis in parsley, parsnip, and rough lemon. *Chemoecology*, **15**, 115–119.

Goh, C.H., Schreiber, U. and Hedrich, R. (1999). New approach of monitoring changes in chlorophyll a fluorescence of single guard cells and protoplasts in response to physiological stimuli. *Plant, Cell and Environment*, **22**, 1057–1070.

Goicoechea, N., Aguirreolea, J., Cenoz, S., *et al.* (2001). Gas exchange and flowering in verticillium-wilted pepper plants. *Journal of Phytopathology*, **149**, 281–286.

Golding, A.J. and Johnson, G.N. (2003). Down-regulation of linear and activation of cyclic electron transport during drought. *Planta*, **218**, 107–14.

Goldstein, G. and Nobel, P.S. (1994). Water relations and low temperature acclimation for cactus species varying in freezing tolerance. *Plant Physiology*, **104**, 675–681.

Goldstein, G., Sharifi, M.R., Kohorn, L.U., *et al.* (1991). Photosynthesis by inflated pods of a desert shrub, *Isomeris arborea*. *Oecologia*, **85**, 396–402.

Golem, S. and Culver, J.N. (2003). Tobacco mosaic virus induced alterations in the gene expression profile of *Arabidopsis thaliana*. *Molecular Plant-Microbe Interactions*, **16**, 681–688.

Gollan, T., Schurr, U. and Schulze, E-D. (1992). Stomatal response to drying soil in relation to changes in the xylem sap composition of *Helianthus annuus*. I. The concentration of cations, anions, amino acids in, and pH of, the xylem sap. *Plant, Cell and Environment*, **15**, 551–559.

Golluscio, R.A. and Oesterheld, M. (2007). Water use efficiency of twenty five co-existing Patagonian species growing under different soil water availability. *Oecologia*, **154**, 207–217.

Gombos, Z., Wada, H., Hideg, E., *et al.* (1994). The unsaturation of membrane lipids stabilizes photosynthesis against heat stress. *Plant Physiology*, **104**, 563–567.

Gomes, F., Oliva, M.A., Mielke, M.S., *et al.* (2008). Photosynthetic limitations in leaves of young Brazilian Green Dwarf coconut (*Cocos nucifera* L. '*nana*') palm under well-watered conditions or recovering from drought stress. *Environmental and Experimental Botany*, **62**, 195–204.

Gomes, F.P., Oliva, M.A., Mielke, M.S., *et al.* (2006). Photosynthetic irradiance-response in leaves of dwarf coconut palm (*Cocos nucifera* L. '*nana*', Arecaceae): Comparison of three models. *Scientia Horticulturae*, **109**, 101–105.

Gómez-del-Campo, M., Baeza, P., Ruiz, C., *et al.* (2007). Effect of previous water conditions on vine response to rewatering. *Vitis*, **46**, 51–55.

Gonfiantini, R., Gratziou, S. and Tongiorgi E. (1965). Oxygen isotopic composition of water in leaves. In: *Isotopes and Radiation in Soil-plant Nutrition Studies*. International Atomic Energy Agency, Vienna, 405–410.

González, A., Steffen, K.L. and Lynch, J.P. (1998). Light and excess manganese. Implications for oxidative stress in common bean. *Plant Physiology*, **118**, 493–504.

Goodman, R.N., Király, Z. and Wood, K.R. (1986). *The Biochemistry and Physiology of Plant Disease*. Columbia: University of Missouri Press.

Gordon, A.H., Lomax, J.A., Dalgarno, K., *et al.* (1985). Preparation and composition of mesophyll, epidermis and fibre cell walls from leaves of perennial ryegrass (*Lolium perenne*) and italian ryegrass (*Lolium multiflorum*). *Journal of the Science of Food and Agriculture*, **36**, 509–519.

Gordon, D.R., Welker, J.M., Menke, J.W., *et al.* (1989). Competition for soil water between annual plants and blue oak (*Quercus douglasii*) seedlings. *Oecologia*, **79**, 533–541.

Gordon, J.C. and Wheeler, C.T. (1978). Whole plant studies on photosynthesis and acetylene reduction in *Alnus glutinosa*. *New Phytologist*, **80**, 179–186.

Gorton, H.L., Herbert, S.K. and Vogelmann, T.C. (2003). Photoacoustic analysis indicates that chloroplast movement does not alter liquid-phase CO_2 diffusion in leaves of *Alocasia brisbanensis*. *Plant Physiology*, **132**, 1529–1539.

Goudarzi, S. (2006). *Scientist Reading the Leaves to Predict Violent Weather*. LiveScience, 13 March 2006.

Goudriaan, J., Van Laar, H.H., Van Keulen, H., *et al.* (1985). Photosynthesis, CO_2, and plant production. In: *Wheat Growth and Modelling. NATO ASI Series, Series A*, Vol 86 (eds Day, W. and Atkin, R.K.), Plenum Press, New York, USA, pp. 107–122.

Goulas, Y., Camenen, L., Guyot, G., *et al.* (1997). Measurements of laser-induced fluorescence decay and reflectance of plant canopies. *Remote Sensing Reviews*, **15**, 305–322.

Gould, K.S. (2004). Nature's swiss army knife: the diverse protective roles of anthocyanins in leaves. *Journal of Biomedicine and Biotechnology*, **5**, 314–320.

Gould, S.J. and Lewontin, R.C. (1979). The Spandrels of San Marco and the Panglossian paradigm: a critique of the adaptationist programme. *Proceedings of the Royal Society of London*, **205B**, 581–598.

Goulden, M.L. (1996). Carbon assimilation and water-use efficiency by neighboring Mediterranean-climate oaks that differ in water access. *Tree Physiology*, **16**, 417–424.

Goulden, M.L. and Field, C.B. (1994). Three methods for monitoring the gas exchange of individual tree canopies: ventilated-chamber, sap-flow and Penman-Monteith measurements on evergreen oaks. *Functional Ecology*, **8**, 125–135.

Goulden, M.L., Miller, S.D., da Rocha, H.R., *et al.* (2004). Diel and seasonal patterns of tropical forest CO_2 exchange. *Ecological Applications*, **14**, S42-S54.

Goulden, M.L., Munger, J.W., Fan, S.M., *et al.* (1996). Measurements of carbon sequestration by long-term eddy covariance: methods and critical evaluation of accuracy. *Global Change Biology*, **2**, 169–181.

Gounaris, K., Brain, A.P.R., Quinn, P.J., *et al.* (1984). Structural reorganization of chloroplast thylakoid membranes in response to heat stress. *Biochimica et Biophysica Acta*, **766**, 198–208.

Gout, E., Bligny, R. and Douce, R. (1992). Regulation of intracellular pH values in higher plant cells. *Journal of Biological Chemistry*, **267**, 13903–13909.

Gout, E., Boisson, A.M., Aubert, S., *et al.* (2001). Origin of the cytoplasmic pH changes during anaerobic stress in higher plant cells: carbon-13 and phosphorus-31 nuclear magnetic resonance studies. *Plant Physiology*, **125**, 912–925.

Gouveia, A.C. and Freitas, H. (2009). Modulation of leaf attributes and water use efficiency in *Quercus suber* along a rainfall gradient. *Trees*, **23**, 267–275.

Govindachary, S., Bigras, C., Harnois, J., *et al.* (2007). Changes in the mode of electron flow to photosystem I following chilling-induced photoinhibition in a C_3 plant *Cucumis sativus* L. *Photosynthesis Research*, **94**, 333–345.

Govindjee, J. (1995). Sixty three years since Kautsky: chlorophyll a fluorescence. *Australian Journal of Plant Physiology*, **22**, 131–160.

Gower, S.T. (2002). Productivity of terrestrial ecosystems. In: *Encyclopedia of Climate Change*, vol 2 (eds Mooney, H.A. and Canadell, J.), Blackwell Scientific, Oxford, UK, pp. 516–521.

Gower, S.T., Reich, P.B. and Son, Y. (1993). Canopy dynamics and aboveground production of five tree species with different leaf longevities. *Tree Physiology*, **12**, 327–345.

Gowik, U., Burscheidt, J., Akyildiz, M. *et al.* (2004). Cis-regulatory elements for mesophyll-specific gene expression in the C_4 plant *Flaveria trinervia*, the promoter of the C_4 phosphoenolpyruvate carboxylase gene. *Plant Cell*, **16**, 1077–1090.

Grabherr, G., Gottfried, M. and Paull, H. (1994). Climate effects on mountain plants. *Nature*, **369**, 448–448.

Grace J. (1990) Cuticular water loss unlikely to explain tree-line in Scotland. *Oecologia*, **84**, 64–68.

Grace, J., Berninger, F. and Nagy, L. (2002). Impacts of climate change at the treeline. *Annals of Botany*, **90**, 537–554.

Grace, J., Jose, J.S., Meir, P., *et al.* (2006). Productivity and carbon fluxes of tropical savannas. *Journal of Biogeography*, **33**, 387–400.

Grace, J., Lloyd, J., McIntyre, J., *et al.* (1995). Carbon-dioxide uptake by an undistrubed tropical rain-forest in Southwest Amazonia, 1992 to 1993. *Science*, **270**, 778–780.

Grace, J., Lloyd, J., Miranda, A.C., *et al.* (1998). Fluxes of carbon dioxide and water vapour over a C4 pasture in South Western Amazonia (Brazil). *Australian Journal of Plant Physiology*, **25**, 519–530.

Grace, J., Nichol, C., Disney, M., *et al.* (2007). Can we measure terrestrial photosynthesis from space directly, using spectral reflectance and fluorescence? *Global Change Biology*, **13**, 1484–1497.

Graham, D. and Patterson, B.D. (1982). Responses of plants to low, nonfreezing temperatures: proteins, metabolism, and acclimation. *Annual Review of Plant Physiology*, **33**, 347–372.

Graham, E.A., Mulkey, S.S., Kitajima, K., *et al.* (2003). Cloud cover limits net CO_2 uptake and growth of a rainforest tree during tropical rainy seasons. *Proceedings of the National Academy of Sciences of the United States of America*, **100**, 572–576.

Grainger, J. and Ring, J. (1962). Anomalous Fraunhofer lines profiles. *Nature*, **193**, 762–764.

Grams, T.E.E., Borland, A.M., Roberts, A., *et al.* (1997). On the mechanism of reinitiation of endogenous crassulacean acid metabolism by temperature changes. *Plant Physiology*, **113**, 1309–1317.

Grams, T.E.E., Koziolek, C., Lautner, S., *et al.* (2007). Distinct roles of electric and hydraulic signals on the reaction of leaf gas exchange upon re-irrigation in *Zea mays* L. *Plant, Cell and Environment*, **30**, 79–84.

Granier, A., Aubinet, M., Epron, D., *et al.* (2003). Deciduous forests: carbon and water fluxes, balances, ecological and ecophysiological determinants. In: *Fluxes of Carbon, Water and Energy of European Forests*, Ecological Studies 163 (ed. Valentini, R.), Springer-Verlag, Berlin Heidelberg, Germany, pp. 55–70.

Granier, A., Biron, P., Bréda, N., *et al.* (1996). Transpiration of trees and forest stands: short and long-term monitoring using sapflow methods. *Global Change Biology*, **2**, 265–274.

Granier, A., Reichstein, M., Bréda, N., *et al.* (2007) Evidence for soil water control on carbon and water dynamics in European forests during the extremely dry year: 2003. *Agricultural and Forest Meteorology*, **143**, 123–145.

Granier, C. and Tardieu, F. (1999). Leaf expansion and cell division are affected by reducing absorbed light before but not after the decline in cell division rate in the sunflower leaf. *Plant Cell Environment*, **22**, 1365–1376.

Granier, C., Turc, O. and Tardieu, F. (2000). Co-ordination of cell division and tissue expansion in sunflower, tobacco, and pea leaves: dependence or independence of both processes? *Journal of Plant Growth Regulation*, **19**, 45–54.

Grant, O.M., Tronina, L., Jones, H.G., *et al.* (2007). Exploring thermal imaging variables for the detection of stress responses in grapevine under different irrigation regimes. *Journal of Experimental Botany*, **58**, 815–825.

Grant, O.M., Tronina, Ł., Ramalho, J.C., *et al.* (2010). The impact of drought on leaf physiology of *Quercus suber* L. trees: comparison of an extreme drought event with chronic rainfall reduction. *Journal of Experimental Botany*, **61**, 4361–4371.

Grantz, D.A. and Assmann, S.M. (1991). Stomatal response to blue-light-water-use efficiency in sugarcane and soybean. *Plant, Cell and Environment*, **14**, 683–690.

Grassi, G. and Magnani, F. (2005). Stomatal, mesophyll conductance and biochemical limitations to photosynthesis as affected by drought and leaf ontogeny in ash and oak trees. *Plant, Cell and Environment*, **28**, 834–849.

Grassi, G., Ripullone, F., Borghetti, M., *et al.* (2009). Contribution of diffusional and non-diffusional limitations to midday depression of photosynthesis in *Arbutus unedo* L. *Trees*, **23**, 1149–1161.

Grassi, G., Vicinelli, E., Ponti, F., *et al.* (2005). Seasonal and interannual variability of photosynthetic capacity in relation to leaf nitrogen in a deciduous forest plantation in northern Italy. *Tree Physiology*, **25**, 349–360.

Gratani, L. and Bombelli, A. (2000). Correlation between leaf age and other leaf traits in three Mediterranean maquis shrub species: *Quercus ilex*, *Phillyrea latifolia* and *Cistus incanus*. *Environmental and Experimental Botany*, **43**, 141–153.

Gratani, L. and Ghia, E. (2002). Adaptive strategy at the leaf level of *Arbutus unedo* L. to cope with Mediterranean climate. *Flora*, **197**, 275–284.

Graves, J.D., Press, M.C. and Steward, G.R. (1989). A carbon balance model of the sorghum-*Striga hermonthica* host-parasite association. *Plant, Cell and Environment*, **12**, 101–107.

Gray, G.R. and Heath, D. (2005). A global reorganization of the metabolome in *Arabidopsis* during cold acclimation is revealed by metabolic fingerprinting. *Physiologia Plantarum*, **124**, 236–248.

Gray, G.R., Ivanov, A.G., Krol, M., *et al.* (1998). Adjustment of thylakoid plastoquinone content and electron donor pool size in response to growth temperature and growth irradiance in Winter Rye (*Secale cereale* L.). *Photosynthesis Research*, **56**, 209–221.

Gray, G.R., Chauvin, L.P., Sarhan, F., *et al.* (1997). Cold acclimation and freezing tolerance. A complex interaction of light and temperature. *Plant Physiology*, **114**, 467–474.

Gray, J.E., Holroyd, G.H., van der Lee, F.M., *et al.* (2000). The HIC signaling pathway links CO_2 perception to stomatal development. *Nature*, **408**, 713–716.

Greer, D.H. (1996). Photosynthetic development in relation to leaf expansion in kiwifruit (*Actinidia deliciosa*) vines during growth in a controlled environment. *Australian Journal of Plant Physiology*, **23**, 541–549.

Greer, D.H. and Sicard, S.M. (2009). The net carbon balance in relation to growth and biomass accumulation of grapevines (*Vitis vinifera* cv. Semillon) grown in a controlled environment. *Functional Plant Biology*, **36**, 645–653.

Greer, D.H., Seleznyova, A.N. and Green, S.R. (2004). From controlled environments to field simulations: leaf area dynamics and photosynthesis in kiwifruit vines (*Actinidia deliciosa*). *Functional Plant Biology*, **31**, 169–179.

Greger, M. and Johansson, M. (1992). Cadmium effects on leaf transpiration of sugar beet (*Beta vulgaris*). *Physiologia Plantarum*, **86**, 465–473.

Greger, M. and Ögren, E. (1991). Direct and indirect effects of Cd2+ on photosynthesis in sugar beet (*Beta vulgaris*). *Physiologia Plantarum*, **83**, 129–135.

Gregg, J.W., Jones, C.G. and Dawson, T.E. (2003). Urbanization effects on tree growth in the vicinity of New York City. *Nature*, **424**, 183–187.

Grewe, V., Dameris, M., Hein, R., *et al.* (2001). Future changes of the atmospheric composition and the impact of climate change. *Tellus B*, **53**, 103–121.

Griffin, K.L., Anderson, O.R., Gastrich, M.D., *et al.* (2001). Plant growth in elevated CO_2 alters mitochondrial number and chloroplast fine structure. *Proceedings of the National Academy of Sciences USA*, **98**, 2473–2478.

Griffin, K.L., Ross, P.D., Sims, D.A., *et al.* (1996). EcoCELLs: tools for mesocosm scale measurements of gas exchange. *Plant Cell Environment*, **19**, 1210–1221.

Griffis, T.J., Sargent, S.D., Baker, J.M., *et al.* (2008). Direct measurements of biosphere-atmosphere isotopic CO_2 exchange using the eddy covariance technique. *Journal of Geophysical Research*, **113**, D08304.

Griffith, H. and Jarvis, P. (edit) (2005). *The Carbon Balance of Forest Biomes*. SEB series vol 57. Taylor & Francis, Oxford, UK, pp. 1–720.

Griffiths, H. (1989). Carbon dioxide concentrating mechanisms and the evolution of CAM in vascular epiphytes. In: *Vascular Plants as Epiphytes: Evolution and Ecophysiology* (ed. Lüttge, U), Springer Verlag, Berlin, Germany, pp. 42–86.

Griffiths, H. (1992). Carbon isotope discrimination and the integration of carbon assimilation pathways in terrestrial CAM plants. *Plant, Cell and Environment*, **15**, 1051–1062.

Griffiths, H., Broadmeadow, M.S.J., Borland, A.M., *et al.* (1990). Short-term changes in carbon-isotope discrimination identify transitions between C_3 and C_4 carboxylation during crassulacean acid metabolism. *Planta*, **181**, 604–610.

Griffiths, H., Cousins, A.B., Badger, M., *et al.* (2007). Discrimination in the dark. Resolving the interplay between metabolic and physical constraints to phosphoenolpyruvate carboxylase activity during the crassulacean acid metabolism cycle. *Plant Physiology*, **143**, 1055–1067.

Groom, Q.J., Baker, N.R. and Long, S.P. (1991). Photoinhibition of holly (Ilex aquifolium) in the field during the winter. *Physiologia Plantarum*, **83**, 585–590.

Grünzweig, J.M. and Körner, C. (2001). Growth, water and nitrogen relations in grassland model ecosystems of the semi-arid Negev of Israel exposed to elevated CO_2. *Oecologia*, **128**, 251–262.

Grünzweig, J.M., Lin, T., Rotenberg, E., *et al.* (2003). Carbon sequestration in arid land forest. *Global Change Biology*, **9**, 791–799.

Grzesiak, M.T., Grzesiak, S. and Skoczowski, A. (2006). Changes of leaf water potential and gas exchange during and after drought in triticale and maize genotypes differing in drought tolerance. *Photosynthetica*, **44**, 561–568.

Gu, L., Baldocchi, D., Verma, S.B., *et al.* (2002). Advantages of diffuse radiation for terrestrial ecosystem productivity. *Journal of Geophysical Research*, **107**, 1–23.

Gu, L., Baldocchi, D.D., Wofsy, S.C., *et al.* (2003). Response of a deciduous forest to the Mount Pinatubo eruption: enhanced photosynthesis. *Science*, **299**, 2035–2038.

Gu, L., Hanson, P.J., Post, W.M., *et al.* (2008). The 2007 Eastern US Spring Freeze: Increased Cold Damage in a Warming World? *Bioscience*, **58**, 253–262.

Guanter, L., L. Alonso, L. Gomez-Chova, *et al.* (2010), Developments for vegetation fluorescence retrieval from spaceborne high-resolution spectrometry in the O_2A and O_2B absorption bands, *J. Geophys. Res.*, **115**, D19303, doi:10.1029/2009JD013716.

Guderian, R. (1986). Terrestrial ecosystems: particulate deposition. In: *Air Pollutants and their Effects on the Terrestrial Ecosystem. Advances in Environmental Science and Technology* (eds Legge, A.H. and Krupa, S.V.), Wiley, New York, USA, pp. 339–363.

Guehl, J.M., Domenacg A.M., Bereau, M., *et al.* (1998). Functional diversity in an Amazonian rainforest of Fench Guyana: a dual isotope approach (δ^{15}N and δ ^{13}C). *Oecologia*, **116**, 316–330.

Guenther, A., Hewitt, C.N., Erickson, D., *et al.* (1995). A global model of natural volatile compound emissions. *Journal of Geophysical Research*, **100**, 8873–8892.

Guenther, A.B., Zimmerman, P.R. and Harley, P.C. (1993). Isoprene and monoterpene emission rate variability: Model evaluations and sensitivity analysis. *Journal of Geophysical Research*, **98**, 12,609–12,617.

Guida dos Santo, M., Vasconcelos Ribeiro, R., Ferraz de Oliveira, R., *et al.* (2006). The role of inorganic phosphate on photosynthesis recovery of common bean after a mild water deficit. *Plant Science*, **170**, 659–664.

Guidi, L., Degl'Innocenti, E. and Soldatini, G.F., (2002). Assimilation of CO_2, enzyme activation and photosynthetic electron transport in bean leaves as affected by high light and ozone. *New Phytologist*, **156**, 377–388.

Guissé, B., Srivastava, A. and Strasser, R.J. (1995). The polyphasic rise of the chlorophyll-a fluorescence (O-K-J-I-P) in heat-stressed leaves. *Archives des Sciences*, **48**, 147–160.

Gulías, J., Cifre, J., Jonasson, S., *et al.* (2009). Seasonal and inter-annual variations of gas exchange in thirteen woody species along a climatic gradient in the Mediterranean island of Mallorca. *Flora*, **204**, 169–181.

Gulías, J., Flexas, J., Abadía, A., *et al.* (2002). Photosynthetic responses to water deficit in six Mediterranean sclerophyll species: possible factors explaining the declining of *Rhamnus ludovici-salvatoris*, an endemic Balearic species. *Tree Physiology*, **22**, 687–697.

Gulías, J., Flexas, J., Mus, M., *et al.* (2003). Relationship between maximum leaf photosynthesis, nitrogen content and specific leaf area in balearic endemic and non-endemic Mediterranean species. *Annals of Botany*, **92**, 215–222.

Gunasekera, D. and Berkowitz, G.A. (1992). Heterogeneous stomatal closure in response to leaf water deficits is not a universal phenomenon. *Plant Physiology*, **98**, 660–665.

Gunawardena, A.H.L.A.N. (2008). Programmed cell death and tissue remodelling in plants. *Journal of Experimental Botany*, **59**, 445–451.

Gunderson, C.A. and Wullschleger, S.D. (1994). Photosynthetic acclimation in trees to rising atmospheric CO_2: a broader perspective. *Photosynthesis Research*, **39**, 369–388.

Gunderson, C.A., Norby, R.J. and Wullschleger, S.D. (2000). Acclimation of photosynthesis and respiration to simulated climatic warming in northern and southern populations of *Acer saccharum*: laboratory and field evidence. *Tree Physiology*, **20**, 87–96.

Gunderson, C.A., Sholtis, J.D., Wullschleger, S.D., *et al.* (2002). Environmental and stomatal control of photosynthetic enhancement in the canopy of a sweetgum (*Liquidambar styraciflua* L.) plantation during 3 years of CO_2 enrichment. *Plant, Cell and Environment*, **25**, 379–393.

Gunnell, G.F., Morgan, M.E., Maas, M.C., *et al.* (1995). Comparative palaeoecology of Paleogene and Neogene mammalian faunas: trophic structure and composition. *Palaeogeography, Palaeoclimatology, Palaeoecology*, **115**, 265–286.

Günthardt-Goerg, M.S. and Vollenweider, P. (2007). Linking stress with macroscopic and microscopic leaf response in trees: new diagnostic perspectives. *Environmental Pollution*, **147**, 467–488.

Günthardt-Goerg, M.S., McQuattie, C.J., Scheidegger, C., *et al.* (1997). Ozone induced cytochemical and ultrastructural changes in leaf mesophyll cell walls. *Canadian Journal of Forest Research*, **27**, 453–463.

Guo, J. and Trotter, C.M. (2004). Estimating photosynthetic light-use efficiency using the photochemical reflectance index: variations among species. *Functional Plant Biology*, **31**, 255–265.

Guo, S.J., Zhou, H.Y., Zhang, X.S., *et al.* (2007). Overexpression of CaHSP26 in transgenic tobacco alleviates photoinhibition of PSII and PSI during chilling stress under low irradiance. *Journal of Plant Physiology*, **164**, 126–136.

Guralnick, L.J. and Jackson, M.D. (2001). The occurrence and phylogenetics of crassulacean acid metabolism in the Portulacaceae. *International Journal of Plant Science*, **162**, 257–262.

Guralnick, L.J., Ting, I.P. and Lord, E.M. (1986). Crassulacean acid metabolism in the Gesneriaceae. *American Journal of Botany*, **53**, 336–345.

Gurney, A.L., Press, M.C. and Ransom, J.K. (1995). The parasitic angiosperm *Striga hermonthica* can reduce photosynthesis of its sorghum and maize hosts in the field. *Journal of Experimental Botany*, **46**, 1817–1823.

Güsewell, S. (2004) Tansley review. N : P ratios in terrestrial plants: variation and functional significance. *The New Phytologist*, **164**, 243–266.

Gustafsson, H.G., Winter, K. and Bittrich, V. (2007). Diversity, phylogeny and classification of Clusia. In: *Clusia: A Woody Neotropical Genus of Remarkable Plasticity and Diversity*. Ecological Studies, Vol. 194 (ed. Lüttge, U.), Springer-Verlag, Berlin Heidelberg, Germany, pp. 95–116.

Gutschick, V.P. and BassiriRad, H. (2003). Tansely review: Extreme events as shaping physiology, ecology, and evolution of plants: toward a unified definition and evaluation of their consequences. *The New Phytologist*, **160**, 21–42.

Gutschick, V.P. and Simonneau, T. (2002). Modelling stomatal conductance of field-grown sunflower under varying soil water content and leaf environment: comparison of three models of stomatal response to leaf environment and coupling with an abscisic acid based model of stomatal response to soil drying. *Plant, Cell and Environment*, **25**, 1423–1434.

Gutschick, V.P. and Wiegel, F.W. (1988). Optimizing the canopy photosynthetic rate by patterns of investment in specific leaf mass. *American Naturalist*, **132**, 67–86.

Guy, C.L., Carter, J.V., Yelenosky, G., *et al.* (1984). Changes in glutathione content during cold acclimation in *Cornus sericea* and *Citrus sinensis*. *Cryobiology*, **21**, 443–453.

Guy, R.D., Berry, J.A., Fogel, M.L., *et al.* (1989). Differential fractionation of oxygen isotopes by cyanide-resistant and cyanide-sensitive respiration in plants. *Planta*, **177**, 483–491.

Guy, R.D., Fogel, M.L. and Berry, J.A. (1993). Photosynthetic fractionation of the stable isotopes of oxygen and carbon. *Plant Physiology*, **101**, 37–47.

Guy, R.D., Fogel, M.L., Berry, J.A., *et al.* (1987). Isotope fractionation during oxygen production and consumption by plants. In: *Progress in Photosynthesis Research III* (ed. Biggins, J.), Springer, Dordrecht, Netherlands, pp. 597–600.

Gwathmey, C.O., Hall, A.E. and Madore, M.A. (1992). Pod removal effects on cowpea genotypes contrasting in monocarpic senescence traits. *Crop Science*, **32**, 1003–1009.

Gyenge, J., Fernandez, M.E., Sarasola, M., *et al.* (2008). Testing a hypothesis of the relationship between productivity and water use efficiency in Patagonian forests with native and exotic species. *Forest Ecology and Management*, **255**, 3281–3287.

Haase, P., Pugnaire, F.J., Clark, S.C., *et al.* (1999a). Diurnal and seasonal changes in cladode photosynthetic rate in relation to canopy age structure in the leguminous shrub *Retama sphaerocarpa*. *Functional Ecology*, **14**, 640–649.

Haase, P., Pugnaire, P.I., Clark, S.C., *et al.* (1999b). Environmental control of canopy dynamics and photosynthetic rate in the evergreen tussock grass *Stipa tenacissima*. *Plant Ecology*, **145**, 327–339.

Habash, D., Percival, M.P. and Baker N.R. (1985). Rapid chlorophyll fluorescence technique for the study of penetration of photosynthetically active herbidies into leaf tissue. *Weed Research*, **25**, 389–395.

Habermann, G., Machado, E.C., Rodrigues, J.D. *et al.* (2003a). Gas exchange rates at different vapour pressure deficits and water relations of 'Pera' sweet orange plants with citrus variegated chlorosis (CVC). *Scientia Horticulturae*, **98**, 233–245.

Habermann, G., Machado, E.C., Rodrigues, J.D., *et al.* (2003b). CO_2 assimilation, photosynthetic light response curves, and water relations of 'Pêra' sweet orange plants infected with *Xylella fastidiosa*. *Brazilian Journal of Plant Physiology*, **15**, 79–87.

Haehnel, W. (1984). Photosynthetic electron transport in higher plants. *Annual Reviews of Plant Physiology and Plant Molelcular Biology*, **35**, 659–683.

Hafke, J.B., Hafke, Y., Smith, J.A.C., *et al.* (2003). Vacuolar malate uptake is mediated by an anion-selective inward rectifier. *The Plant Journal*, **35**, 116–128.

Haile, F.J. and Higley, L.G. (2003). Changes in soybean gas exchange after moisture stress and spider mite injury. *Environmental Entomology*, **32**, 433–440.

Hajirezaei, M., Sonnewald, U., Viola, R., *et al.* (1994). Transgenic potato plants with strongly decreased expression of PFP show no visible phenotype and only minor changes in metabolic fluxes in their tubers. *Planta*, **192**, 16–30.

Hald, S., Nandha, B., Gallois, P., *et al.* (2007). Feedback regulation of photosynthetic electron transport by NADP(H) redox poise. *Biochimica et Biophysica Acta*, **1777**, 433–440.

Haldimann, P. and Feller, U. (2004). Inhibition of photosynthesis by high temperature in oak (*Quercus pubescens* L.) leaves grown under natural conditions closely correlates with a reversible heat-dependent reduction of the activation state of ribulose-1,5-bisphosphate carboxylase/oxygenase. *Plant, Cell and Environment*, **27**, 1169–1183.

Haldimann, P., Gallé, A. and Feller, U. (2008). Impact of an exceptionally hot dry summer on photosynthetic traits in oak (*Quercus pubescens*) leaves. *Tree Physiology*, **28**, 785–795.

Haldrup, A., Jensen, P.E., Lunde, C., *et al.* (2001). Balance of power: a view of the mechanism of photosynthetic state transitions. *Trends in Plant Science*, **6**, 301–305.

Hales, S. (1727). *Vegetable Staticks, or an Account of Some Statistical Experiments on the Sap in Vegetation.* W. Innys, London.

Hall, A.E. and Schulze, E.D. (1980). Stomatal responses to environment and a possible interrelation between stomatal effects on transpiration and CO_2 assimilation. *Plant, Cell and Environment*, **3**, 467–474.

Hall, D.A., Ptacek, J.J. and Snyder, M.M. (2007). Protein microarray technology. *Mechanisms of Ageing and Development*, **128**, 161–167.

Hall, N.M., Griffiths, H., Corlett, J.A., *et al.* (2005). Relationships between water-use traits and photosynthesis in *Brassica oleracea* resolved by quantitative genetic analysis. *Plant Breeding*, **124**, 557–564.

Hallik, L., Kull, O., Niinemets, Ü., *et al.* (2009). Contrasting correlation networks between leaf structure, nitrogen and chlorophyll in herbaceous and woody canopies. *Basic and Applied Ecology*, **10**, 309–318.

Halliwell, B. and Whiteman, M. (2004). Measuring reactive species and oxidative damage in vivo and in cell culture: how should you do it and what do the results mean? *British Journal of Pharmacology*, **142**, 231–255.

Halsted, M. and Lynch, J. (1996). Phosphorus responses of C_3 and C_4 species. *Journal of Experimental Botany*, **47**, 497–505.

Ham, J., Owensby, C., Coyne, P., *et al.* (1995). Fluxes of CO_2 and water vapor from a prairie ecosystem exposed to ambient and elevated atmospheric CO_2. *Agricultural and Forest Meteorology*, **77**, 73–93.

Ham, J.M. and Knapp, A.K. (1998). Fluxes of CO_2, water vapor, and energy from a prairie ecosystem during the seasonal transition from carbon sink to carbon source. *Agricultural and Forest Meteorology*, **89**, 1–14.

Ham, J.M., Owensby, C.E. and Coyne, P.I. (1993). Technique for measuring air flow and carbon dioxide flux in large open-top chambers. *Journal of Environmental Quality*, **22**, 759–766.

Hamerlynck, E.P. and Knapp, A.K. (1994). Leaf-level responses to light and temperature in two co-occurring *Quercus* (Fagaceae) species: implications for tree distribution patterns. *Forest Ecology and Management*, **68**, 149–159.

Hamerlynck, E.P., Huxman, T.E., Loik, M.E., *et al.* (2000). Effects of extreme high temperature, drought and elevated CO_2 on photosynthesis of the Mojave Desert evergreen shrub, *Larrea tridentata*. *Plant Ecology*, **148**, 183–93.

Han, H., Li, T. and Zhou, S. (2008). Overexpression of phytoene synthase gene from *Salicornia europaea* alters response to reactive oxygen species under salt stress in transgenic *Arabidopsis*. *Biotechnology Letters*, **30**, 1501–1507.

Han, Q. (2011). Height-related decreases in mesophyll conductance, leaf photosynthesis and compensating adjustments associated with leaf nitrogen concentrations in *Pinus densiflora*. *Tree Physiology*, **31**, 976–984.

Han, T., Vogelmann, T. and Nishio, J. (1999). Profiles of photosynthetic oxygen-evolution within leaves of *Spinacia oleracea*. *The New Phytologist*, **143**, 83–92.

Hanan, N.P., Berry, J.A., Verma, S.B., *et al.* (2005). Testing a model of CO_2, water and energy exchange in Great Plains tallgrass prairie and wheat ecosystems. *Agricultural and Forest Meteorology*, **131**, 162–179.

Hanba, Y.T., Miyazawa, S.I., Kogami, H., *et al.* (2001). Effects of leaf age on internal CO_2 transfer conductance and photosynthesis in tree species having different types of shoot phenology. *Australian Journal Plant Physiology*, **28**, 1075–1084.

Hanba, Y.T., Shibasaka M., Hayashi Y., *et al.* (2004). Overexpression of the barley aquaporin HvPIP2;1 increases internal CO_2 conductance and CO_2 assimilation in the leaves of transgenic rice plants. *Plant and Cell Physiology*, **45**, 521–529.

Handa, I.T., Körner, C. and Hättenschweiler, S. (2005). A test of the tree-line carbon limitation hypothesis by in situ CO_2 enrichment and defoliation. *Ecology*, **86**, 1288–1300.

Handford, M.G. and Carr, J.P. (2007). A defect in carbohydrate metabolism ameliorates symptom severity in virus-infected *Arabidopsis thaliana*. *Journal of General Virology*, **88**, 337–341.

Hänninen H. (1989). Modelling bud dormancy release in trees from cool and temperate regions. *Acta Forestalia Fennica*, **213**, 1–47.

Hanning, I. and Heldt, H.W. (1993). On the function of mitochondrial metabolism during photosynthesis in spinach (*Spinacia oleracea* L.) leaves. *Plant Physiology*, **103**, 1147–1154.

Hansen, H.C. (1959). Der Einfluß des Lichtes auf die Bildung von Licht- und Schattenblättern der Buche, *Fagus silvatica*. *Physiologia Plantarum*, **12**, 545–550.

Hansen, J. and Moller, I.B. (1975). Percolation of starch and soluble carbohydrates from plant tissue for quantitative determination. *Biochemistry*, **68**, 87–94.

Hanson, P.J., McRoberts, R.E., Isebrands, J.G., *et al.* (1987). An optimal sampling strategy for determining CO_2 exchange rate as a function of photosynthetic photon-flux density. *Photosynthetica*, **21**, 98–101.

Harazono Y., Mano M., Zulueta R.Y., *et al.* (2003). Inter-annual carbon dioxide uptake of a wet sedge tundra ecosystem in the Arctic. *Tellus*, **55B**, 215–231.

Harbinson, J. and Foyer, C.H. (1991). Relationships between the efficiencies of photosystems I and II and stromal redox state in CO_2-free air: evidence for cyclic electron flow in Vivo. *Plant Physiology*, **97**, 41–49.

Harbinson, J. and Hedley, C.L. (1989). The kinetics of P700+ reduction in leaves: a novel in situ probe of thylakoid functioning. *Plant, Cell and Environment*, **12**, 357–369.

Harbinson, J. and Woodward, F.I. (1987). The use of light-induced absorbance changes at 820 nm to monitor the oxidation state of p-700 in leaves. *Plant, Cell and Environment*, **10**, 131–140.

Hardie, D.G. (2007). AMP-activated / SNF1 protein kinases: conserved guardians of cellular energy. *National Review of Molecular Cell Biology*, **8**, 774–785.

Hari, P., Keronen, P., Bäck, J., et al. (1999). An improvement of the method for calibrating measurements of photosynthetic CO_2 flux. *Plant, Cell and Environment*, **22**, 1297–1301.

Harley, P.C., Loreto, F., Di Marco, G., *et al.* (1992a). Theoretical considerations when estimating the mesophyll conductance to CO_2 flux by the analysis of the response of photosynthesis to CO_2. *Plant Physiology*, **98**, 1429–1436.

Harley, P.C. and Baldocchi, D.D. (1995). Scaling carbon dioxide and water vapour exchange from leaf to canopy in a deciduous forest. I. Leaf model parametrization. *Plant Cell Environment*, **18**, 1146–1156.

Harley, P.C. and Sharkey T.D. (1991). An improved model of C_3 photosynthesis at high CO_2: reversed O_2 sensitivity explained by lack of glycerate reentry into the chloroplast. *Photosynthesis Research*, **27**, 169–178.

Harley, P.C., Thomas, R.B., Reynolds, J.F., *et al.* (1992b). Modelling photosynthesis of cotton grown in elevated CO_2. *Plant, Cell and Environment*, **15**, 271–282.

Harley, P.C., Weber, J.A. and Gates, D.M. (1985). Interactive effects of light, leaf temperature, CO_2 and O_2 on photosynthesis in soybean. *Planta*, **165**, 249–263.

Harms, K.E. and Dalling, J.W. (1997). Damage and herbivory tolerance through resprouting as an advantage of large seed size in tropical trees and lianas. *Journal of Tropical Ecology*, **13**, 617–621.

Harris, S., Tapper, N., Packham, D., *et al.* (2008). The relationship between the monsoonal summer rain and dry season fire activity of northern Australia. *International Journal of Wildland Fire*, **17**, 674–684.

Harrison, M.T., Edwards, E.J., Farquhar, G.D., *et al.* (2009). Nitrogen in cell walls of sclerophyllous leaves accounts for little of the variation in photosynthetic nitrogen-use efficiency. *Plant, Cell and Environment*, **32**, 259–270.

Hartung, W. (2010). The evolution of abscisic acid (ABA) and ABA function in lower plants, fungi and lichen. *Functional Plant Biology*, **37**, 806–812.

Haselwandter, K., Hofmann, A., Holzmann, H.P., *et al.* (1983). Availability of nitrogen and phosphorus in the nival zone of the Alps. *Oecologia*, **57**, 266–269.

Hashimoto, M., Negi, J., Young, J., *et al.* (2006). Arabidopsis HT1 kinase controls stomatal movements in response to CO_2. *Nature Cell Biology*, **8**, 391–397.

Haslam, E. (1985). *Metabolites and Metabolism: A Commentary on Secondary Metabolism*. Clarendon Press, Oxford, UK.

Haslam, E. (1986). Secondary metabolism – fact and fiction. *Natural Product Reports*, **4**, 217–249.

Hassiotou, F., Evans, J.R., Ludwig, M., *et al.* (2009b). Stomatal crypts may facilitate diffusion of CO_2 to adaxial mesophyll cells in thick sclerophylls. *Plant, Cell and Environment*, **32**, 1596–1611.

Hassiotou, F., Ludwig, M., Renton, M., *et al.* (2009a). Influence of leaf dry mass per area, CO_2 and irradiance on mesophyll conductance in sclerophylls. *Journal of Experimental Botany*, **60**, 2303–2314.

Hassiotou, F., Renton, M., Ludwig, M., *et al.* (2010). Photosynthesis at an extreme end of the leaf trait spectrum: how does it relate to high leaf dry mass per area and associated structural parameters? *Journal of Experimental Botany*, **61**, 3015–3028.

Hastings, S.J., Oechel, W.C. and Sionit, N. (1989). Water relations and photosynthesis of chaparral resprouts and seedlings following fire and hand clearing. In: *The California Chaparral. Paradigms Re-examined* (ed. Keeley, S.C.), Natural History Museum of Los Angeles County, Los Angeles, USA, pp. 107–113.

Hatch, M.D. and Slack, C.R. (1966). Photosynthesis by sugarcane leaves: a new carboxylation reaction and the pathway of sugar formation. *Biochemical Journal*, **101**, 103–111.

Hatch, M.D., Agostino, A. and Jenkins, C.L.D. (1995). Measurements of the leakage of CO_2 from bundle-sheath cells of leaves during C_4 photosynthesis. *Plant Physiology*, **108**, 173–181.

Hatier, J.H.B. and Gould, K.S. (2007). Black colouration of *Ophiopogon planiscapus* "nigrescens". Leaf optics, chromaticity, and internal light gradients. *Functional Plant Biology*, **34**, 130–138.

Hättenschwiler, S. (2001). Tree seedling growth in natural deep shade: functional traits related to interspecific variation in response to elevated CO_2. *Oecologia*, **129**, 31–42.

Hattersley, P.W. (1982). $\delta^{13}C$ values of C_4 types in grasses. *Australian Journal of Plant Physiology*, **9**, 139–154.

Hattersley, P.W., Watson, L. and Osmond, C.B. (1977). In situ immunofluorescent labelling of ribulose-1,5-bisphosphate carboxylase in leaves of C_3 and C_4 plants. *Australian Journal of Plant Physiology*, **4**, 523–39.

Hauben, M., Haesendonckx, B., Standaert, E., *et al.* (2009). Energy use efficiency is characterized by an epigenetic component that can be directed through artificial selection to increase yield. *Proceedings of the National Academy of Sciences USA*, **106**, 20109–20114.

Haug, A. (1984). Molecular aspects of aluminum toxicity. *Critical Reviews in Plant Sciences*, **1**, 345–373.

Haughn, G.W. and Gilchrist, E.J. (2006). TILLING in the botanical garden: a reverse genetic technique feasible for all plant species. *Floriculture, Ornamental and Plant Biotechnology*, **1**, 476–482.

Hauglustaine, D. and Brasseur, G.P. (2001). Evolution of tropospheric ozone under anthropogenic activities and associated radiative forcing of climate. *Journal of Geophysical Research D. Atmospheres*, **106**, 32,337–32,360.

Haupt, W. and Scheuerlein, R. (1990). Chloroplast movement. *Plant, Cell and Environment*, **13**, 595–614.

Haupt-Herting, S., Klug, K. and Fock, H. (2001). A new approach to measure gross CO_2 fluxes in leaves. Gross CO_2 assimilation, photorespiration, and mitochondrial respiration in the light in tomato under drought stress. *Plant Physiology*, **126**, 388–396.

Havaux, M. (1992). Stress tolerance of photosystem II in vivo. Antagonistic effects of water, heat and photoinhibition stresses. *Plant Physiology*, **100**, 424–432.

Havaux, M. (1993). Rapid photosynthetic adaptation to heat stress triggered in potato leaves by moderately elevated temperatures. *Plant, Cell and Environment*, **16**, 461–467.

Havaux, M. (1996). Short-term responses of photosystem I to heat stress – Induction of a PS II-independent electron transport through PS I fed by stromal components. *Photosynthesis Research*, **47**, 85–97.

Havaux, M. (1998). Carotenoids as membrane stabilizers in chloroplasts. *Trends in Plant Science*, **3**, 147–151.

Havaux, M. (2003). Spontaneous and thermoinduced photon emission: new methods to detect and quantify oxidative stress in plants. *Trends Plant Science*, **8**, 409–413.

Havaux, M. and Davaud, A. (1994). Photoinhibition of photosynthesis in chilled potato leaves is not correlated with a loss of photosystem-II activity. Preferential inactivation of photosystem I. *Photosynthesis Research*, **40**, 75–92.

Havaux, M. and Lannoye, R. (1983). Temperature dependence of delayed chlorophyll fluorescence in intact leaves of higher plants. A rapid method for detecting the phase transition of thylakoid membrane lipids. *Photosynthesis Research*, **4**, 257–263.

Havaux, M. and Niyogi, K.K. (1999). The violaxanthin cycle protects plants from photooxidative damage by more than one mechanism. *Proceedings of the National Academy of Sciences USA*, **96**, 8762–8767.

Havaux, M., Dall'Osto, L. and Bassi, R. (2007). Zeaxanthin has enhanced antioxidant capacity with respect to all other xanthophylls in Arabidopsis leaves and functions independent of binding of PSII antennae. *Plant Physiology*, **145**, 1506–1520.

Havaux, M., Greppin, H. and Strasser, R.J. (1991). Functioning of photosytems I and II in pea leaves exposed to heat stress in the presence or absence of light. *Planta*, **186**, 88–98.

Havaux, M., Tardy, F., Ravenel, J., *et al.* (1996). Thylakoid membrane stability to heat stress studied by flash spectroscopic measurements of the electrochromic shift in intact potato leaves: influence of the xanthophyll content. *Plant, Cell and Environment*, **19**, 1359–1368.

Havaux, M., Triantaphylidès, C. and Genty, B. (2006). Autoluminiscence imaging: a non-invasive tool for mapping oxidative stress. *Trends in Plant Science*, **11**, 480–484.

Haveman, J. and Mathis, P. (1976). Flash-induced absorption changes of the primary donor of photosystem II at 820 nm in chloroplasts inhibited by low ph or tris-treatment. *Biochimica et Biophysica Acta (BBA) – Bioenergetics*, **440**, 346–355.

Haverkort, A.J., Rouse, D.I. and Turkensteen, L.J. (1990). The influence of *Verticillium dahliae* and drought on potato crop growth. 1. Effects on gas exchange and stomatal behaviour of individual leaves and crop canopies. *Netherlands Journal of Plant Pathology*, **96**, 273–289.

Haworth, P. and Melis, A. (1983). Phosphorylation of chloroplast membrane proteins does not increase the absorption cross-section of photosystem I. *FEBS Letters*, **160**, 277–280.

He, C., Yan, J., Shen, G., *et al.* (2005). Expression of an *Arabidopsis* vacuolar sodium/proton antiporter gene in cotton improves photosynthetic performance under salt conditions and increases fiber yield in the field. *Plant and Cell Physiology*, **46**, 1848–1854.

He, C.R., Murray, F. and Lyons, T. (2000). Monoterpene and isoprene emissions from 15 Eucalyptus species in Australia. *Atmospheric Environment*, **34**, 645–655.

He, J., Ouyang, W. and Chia, T.F. (2004). Growth and photosynthesis of virus-infected and virus-eradicated orchid plants exposed to different growth irradiances under natural tropical conditions. *Physiologia Plantarum*, **121**, 612–619.

He, J.Y., Ren, Y.F., Zhu, C., *et al.* (2008). Effect of Cd on growth, photosynthetic gas exchange, and chlorophyll fluorescence of wild and Cd-sensitive mutant rice. *Photosynthetica*, **46**(3), 466–470.

Heath, R.L. (1996). The modification of photosynthetic capacity induced by ozone exposure. In: *Photosynthesis and the Environment* (ed. Baker, N.R.), Kluwer Academic Publishers, Dordrecht, Netherlands, pp. 469–476.

Heaton, E., Voigt, T. and Long, S.P. (2004). A quantitative review comparing the yields of two candidate C_4 perennial biomass crops in relation to nitrogen, temperature and water. *Biomass and Bioenergy*, **27**, 21–30.

Heber, U. (1969). Conformational changes of chloroplasts induced by illumination of leaves in vivo. *Biochimica et Biophysica Acta*, **180**, 302–319.

Heber, U., Bilger, W., Bligny, R., *et al.* (2000). Phototolerance of lichens, mosses and higher plants in an alpine environment: analysis of photoreactions. *Planta*, **211**, 770–780.

Heber, U., Bligny, R., Streb, P., *et al.* (1996). Photorespiration is essential for the protection of the photosynthetic apparatus of C_3 plants against photoinactivation under sunlight. *Botanica Acta*, **109**, 307–315.

Heber, U., Neimanis, S., Dietz, K.J., *et al.* (1986). Assimilation power as a driving force in photosynthesis. *Biochimica et Biophysica Acta*, **852**, 144–155.

Heckathorn, S.A., Downs, C.A., Sharkey, T.D., *et al.* (1998). The small, methionine-rich chloroplast heat-shock protein protects photosystem II electron transport during heat stress. *Plant Physiology*, **116**, 439–444.

Heckathorn, S.A., Ryan, S.L., Baylis, J.A., *et al.* (2002). In vivo evidence from an *Agrostis stolonifera* selection genotype that chloroplast small heat-shock proteins can protect photosystem II during heat stress. *Functional Plant Biology*, **29**, 933–944.

Heckwolf, M., Pater, D., Hanson, D.T. *et al.* (2011). The *Arabidopsis thaliana* aquaporin AtPIP1;2 is a physiologically relevant CO_2 transport facilitator. *The Plant Journal*, 67, 795–804.

Hedberg, O. (1964). Features of afroalpine plant ecology. *Acta Phytogeographica Suecia*, **49**, 1–144.

Hedrich, R., Marten, I., Lohse, G., *et al.* (1994). Malate-sensitive anion channels enable guard cells to sense changes in the ambient CO_2. *The Plant Journal*, **6**, 741–748.

Heidorn, T. and Joern, A. (1984). Differential herbivory on C_3 and C_4 grasses by the grasshopper *Ageneotettix deorum* (Orthoptera: Acrididae). *Oecologia*, **65**, 19–25.

Heilmann, I., Mekhedov, S., King, B., *et al.* (2004). Identification of the *Arabidopsis* palmitoyl-monogalactosyldiacylglycerol D7-desaturase gene FAD5, and effects of plastidial retargeting of *Arabidopsis* desaturases on the fad5 mutant phenotype. *Plant Physiology*, **136**, 4237–4245.

Heimann, M. and Reichstein, M. (2008). Terrestrial ecosystem carbon dynamics and climate feedbacks. *Nature*, **451**, 289–292.

Heisel, F., Sowinska, M., Khalili, E., *et al.* (1997). Laser-induced fluorescence imaging for monitoring nitrogen fertilising treatements of wheat. In: *Aerosense '97* (eds Narayanan, R.M. and Kalshoven, J.E.), SPIE, Bellingham, USA, pp. 10–21.

Held, A.A., Steduto, P., Orgaz, F., *et al.* (1990). Bowen-ratio energy balance technique for estimating crop net CO_2 assimilation and comparison with a canopy chamber. *Theoretical and Applied Climatology*, **42**, 203–213.

Helliker, B.R. and Ehleringer, J.R. (2000). Establishing a grassland signature in veins: ^{18}O in the leaf water of C_3 and C_4 grasses. *Proceedings of the National Academy of Science USA*, **97**, 7894–7898.

Helliker, B.R. and Richter, S.L. (2008). Subtropical to boreal convergence of tree-leaf temperatures. *Nature*, **454**, 511–514.

Hellkvist, J., Richards, G.P. and Jarvis, P.G. (1974). Vertical gradients of water potential and tissue water relations in Sitka spruce trees measured with the pressure chamber. *J Appl Ecol*, **11**, 637–667.

Henderson, S.A., von Caemmerer, S. and Farquhar, G.D. (1992). Short-term measurements of carbon isotope discrimination in several C_4 species. *Australian Journal of Plant Physiology*, **19**, 263–285.

Hendrey, G., Lewin, K., Kolber, Z., *et al.* (1989). Control of ozone concentrations for plant effect studies. Report to the National Council of the Paper Industry for Air and Stream Improvement, BNL 43589.

Hendrickson, L., Chow, W.S. and Furbank, R.T. (2004b). Low temperature effects on grapevine photosynthesis: the role of inorganic phosphate. *Functional Plant Biology*, **31**, 789–801.

Hendrickson, L., Furbank, R.T. and Chow, W.S. (2004a). A simple alternative approach to assessing the fate of absorbed light energy using chlorophyll fluorescence. *Photosynth Research*, **82**, 73–81.

Hendrickson, L., Sharwood, R., Ludwig, M., *et al.* (2008). The effects of Rubisco activase on C4 photosynthesis and metabolism at high temperature. *Journal of Experimental Botany*, **59**, 1789–1798.

Hendrickson, L., Vlckova, A., Selstam, E., *et al.* (2006). Cold acclimation of the *Arabidopsis* dgd1 mutant results in recovery from photosystem I-limited photosynthesis. *FEBS Letters*, **580**, 4959–4968.

Heng-Moss, T., Macedo, T., Franzen, L., *et al.* (2006). Physiological responses of resistant and susceptible buffalo grasses to *Blissus occiduus* (Hemiptera: Blissidae) feeding. *Journal of Economic Entomology*, **99**, 222–228.

Henriques, F.S. (1989). Effects of copper deficiency on the photosynthetic apparatus of sugar beet (*Beta vulgaris* L.). *Journal of Plant Physiology*, **135**, 453–458.

Henriques, F.S. (2001). Loss of blade photosynthetic area and of chloroplasts' photochemical capacity account for reduced CO_2 assimilation rates in zinc-deficient sugar beet leaves. *Journal of Plant Physiology*, **158**, 915–919.

Henriques, F.S. (2003). Gas exchange, chlorophyll a fluorescence kinetics and lipid peroxidation of pecan leaves with varying manganese concentrations. *Plant Science*, **165**, 239–244.

Henry, B.K., Atkin, O.K., Day, D.A., *et al.* (1999). Calculation of the isotope discrimination factor for studying plant respiration. *Australian Journal of Plant Physiology*, **26**, 773–780.

Herbert, S.K., Fork, D.C. and Malkin, S. (1990). Photoacoustic measurements in vivo of energy storage by cyclic electron flow in algae and higher plants. *Plant Physiology*, **94**, 926–934.

Herbette, S., Le Menn, A., Rousselle, P., *et al.* (2007). Modification of photosynthetic regulation in tomato overexpressing glutathione peroxidise. *Biochimica et Biophysica Acta*, **1724**, 108–118.

Herde, O., Peña-Cortés, H., Willmitzer, L., *et al.* (1997). Stomatal responses to jasmonic acid, linolenic acid and abscisic acid in wild-type and ABA-deficient tomato plants. *Plant, Cell and Environment*, **20**, 136–141.

Hernández, J.A. and Almansa, M.S. (2002). Short-term effects of salt stress on antioxidant systems and leaf water relations of pea leaves. *Physiologia Plantarum*, **115**, 251–257.

Herold, A. (1980). Regulation of photosynthesis by sink activity – the missing link. *New Phytologist*, **86**, 131–144.

Herrera, A., Tezara, W., Marín, O., *et al.* (2008). Stomatal and non-stomatal limitations of photosynthesis in trees of a tropical seasonally flooded forest. *Physiologia Plantarum*, **134**, 41–48.

Herrick, J.D. and Thomas, R.B. (2003). Leaf senescence and late-season net photosynthesis of sun and shade leaves of overstory sweetgum (*Liquidambar styraciflua*) grown in elevated and ambient carbon dioxide concentrations. *Tree Physiology*, **23**, 109–118.

Herring, J.R. (1985). Charcoal fluxes into sediments of the North Pacific Ocean: the Cenozoic record of burning. In: *The Carbon Cycle and Atmospheric CO2: Natural Variations Archean to Present* (eds Sundquist, E.T. and Broecker, W.S.), American Geophysical Union, Washington DC, USA, pp. 419–442.

Herrmann, K.M. and Weaver, L.M. (1999). The shikimate pathway. *Annual Review of Plant Physiology and Plant Molecular Biology*, **50**, 473–503.

Herzog, B., Grams, T.E.E., Haag-Kerwer, A., *et al.* (1999a). Expression of modes of photosynthesis (C_3, CAM) in *Clusia criuva* CAMB. in a cerrado/gallery forest transect. *Plant Biology*, **1**, 357–364.

Herzog, B., Hoffmann, S., Hartung, W., *et al.* (1999b). Comparison of photosynthetic responses of the sympatric tropical C_3-species *Clusia multiflora* H.B.K. and the C_3-CAM intermediate species *Clusia minor* L. to irradiance and drought stress in a phytotron. *Plant Biology*, **1**, 460–470.

Heschel, M.S., Donohue, K., Hausmann, N., *et al.* (2002). Population differentiation and natural selection for water-use efficiency in *Impatiens capensis* (Balsaminaceae). *International Journal of Plant Sciences*, **163**, 907–912.

Hetherington, A.M. and Woodward, F.I. (2003). The role of stomata in sensing and driving environmental change. *Nature*, **424**, 901–908.

Hetherington, S.E., Smillie, R.M. and Davies, W.J. (1998). Photosynthetic activities of vegetative and fruiting tissues of tomato. *Journal of Experimental Botany*, **15**, 1173–1181.

Hew, C.S., Lee, G.L. and Wong, S.C. (1980). Occurrence of non-functional stomata in the flowers of tropical orchids. *Annals of Botany*, **46**, 195–201.

Hew, C.S., Ye, Q.S. and Pau, R.C. (1991). Relation of respiration to CO_2 fixation by *Aranda* orchid roots. *Environmental and Experimental Botany*, **31**, 327–331.

Hibberd, J.M. and Covshoff, S. (2010). The regulation of gene expression required for C_4 photosynthesis. *Annual Review of Plant Biology*, **61**, 181–207.

Hibberd, J.M. and Quick, W.P. (2002). Characteristics of C4 photosynthesis in stems and petioles of C3 flowering plants. *Nature*, **145**, 451–454.

Hibberd, J.M., Bungard, R.A., Press, M.C., *et al.* (1998a). Localization of photosynthetic metabolism in the parasitic angiosperm *Cuscuta reflexa*. *Planta*, **205**, 506–513.

Hibberd, J.M., Quick, W.P., Press, M.C., *et al.* (1998b). Can source-sink relations explain responses of tobacco to infection by the root holoparasitic angiosperm *Orobanche cernua*. *Plant, Cell and Environment*, **21**, 333–340.

Hideg, E., Kálai, T., Kós, P.B., *et al.* (2006). Singlet oxygen in plants: its significance and possible detection with double (fluorescent and spin) indicator reagents. *Photochemistry and Photobiology*, **82**, 1211–1218.

Hideg, É., Ogawa, K., Kálai, T., *et al.* (2001). Singlet oxygen imaging in *Arabidopsis thaliana* leaves under photoinhibition by excess photosynthetically active radiation. *Physiologia Plantarum*, **112**, 10–14.

Hietz, P., Wanek, W. and Dünisch, O. (2005). Long-term trends in cellulose d^{13}C and water-use efficiency of tropical Cedrela and Swietenia from Brazil. *Tree Physiology*, **25**, 745–752.

Higuchi, R.G., Dollinger, P.S., Walsh, S., *et al.* (1992). Simultaneous amplification and detection of specific DNA sequences. *Biotechnology*, **10**, 413–417.

Hikosaka, K. (1997). Modelling optimal temperature acclimation of the photosynthetic apparatus in C3 plants with respect to nitrogen use. *Annals of Botany*, **80**, 721–730.

Hikosaka, K. (2003). A model of dynamics of leaves and nitrogen in a plant canopy: an integration of canopy photosynthesis, leaf lifespan, and nitrogen use efficiency. *American Naturalist*, **162**, 149–164.

Hikosaka, K. and Shigeno, A. (2009). The role of Rubisco and cell walls in the interspecific variation in photosynthetic capacity. *Oecologia*, **160**, 443–451.

Hikosaka, K. and Terashima, I. (1996). Nitrogen partitioning among photosynthetic components and its consequence in sun and shade plants. *Functional Ecology*, **10**, 335–343.

Hikosaka, K., Hanba, Y.T., Hirose, T., *et al.* (1998). Photosynthetic nitrogen-use efficiency in leaves of woody and herbaceous species. *Functional Ecology*, **12**, 896–905.

Hikosaka, K., Ishikawa, K., Borjigidai, A., *et al.* (2006). Temperature acclimation of photosynthesis: mechanisms involved in the changes in temperature dependence of photosynthetic rate. *Journal of Experimental Botany*, **57**, 291–302.

Hikosaka, K., Murakami, A. and Hirose, T. (1999). Balancing carboxylation and regeneration of ribulose-1,5-bisphosphate in leaf photosynthesis: temperature acclimation of an evergreen tree, *Quercus myrsinaefolia*. *Plant Cell Environment*, **22**, 841–849.

Hilker, T., Hall, F.G., Coops, N.C. *et al.* (2010). Remote sensing of photosynthetic light-use efficiency across two forested biomes: spatial scaling. *Remote Sensing of the Environment*, **114**, 2863–2874.

Hill, S.A. and Bryce, J.H. (1992). Malate metabolism and LEDR in barley mesophyll protoplasts. In: *Molecular, Biochemical and Physiological Aspects of Plant Respiration*. (eds Lambers, H. and Van der Plas, L.H.W.), SPB Academic Publishing, The Hague.

Hiraki, M., van Rensen, J.J.S., Vredenberg, W.J., *et al.* (2003). Characterization of the alterations of chlorophyll a fluorescence induction curve after addition of Photosystem II inhibiting herbicides. *Photosynthesis Research*, **78**, 35–46.

Hirashima, M., Satoh, S., Tanaka, R., *et al.* (2006). Pigment shuffling in antenna systems achieved by expressing prokaryotic chlorophyllide *a* oxygenase in *Arabidopsis*. *The Journal of Biological Chemistry*, **281**, 15385–15393.

Hiroki, S., and Ichino, K. (1998). Comparison of growth habits under various light conditions between two climax species, *Castanopsis sieboldii* and *Castanopsis cuspidata*, with special reference to their shade tolerance. *Ecological Research*, **13**, 65–72.

Hirose, T. and Werger, M.J.A. (1987). Maximizing daily canopy photosynthesis with respect to the leaf nitrogen allocation pattern in the canopy. *Oecologia*, **72**, 520–526.

Hirschberg, J. (2001). Carotenoid biosynthesis in flowering plants. *Current Opinion of Plant Biology*, **4**, 210–218.

Hjelm, U. and Ögren, E. (2003). Is photosynthetic acclimation to low temperature controlled by capacities for storage and growth at low temperature? Results from comparative studies of grasses and trees. *Physioogial Plantarum*, **119**, 113–120.

Hlaváckova, V., Spundova, M., Naus, J., *et al.* (2002). Mechanical wounding caused by inoculation influences the photosynthetic response of *Nicotiana benthamiana* plants to plum pox potyvirus. *Photosynthetica*, **40**, 269–277.

Hoch, G. and Körner, C. (2003). The carbon charging of pines at the climatic treeline: a global comparison. *Oecologia*, **135**, 10–21.

Hoch, G., Popp, M. and Körner, C. (2002). Altitudinal increase of mobile carbon pools in *Pinus cembra* suggests sink limitation of growth at the Swiss treeline. *Oikos*, **98**, 361–374.

Hoch, G., Richter, A. and Körner, C. (2003). Non-structural carbon compounds in temperate forest trees. *Plant Cell Environment*, **26**, 1067–1081.

Hoch, W.A., Sinsaas, E.L. and McCown, B.H. (2003b). Resorption protection. Anthocyanins facilitate nutrient recovery in autumn by shielding leaves from potentially damaging light levels. *Plant Physiology*, **133**, 1296–1305.

Hoch, W.A., Zeldin, E.L. and McCown, B.H. (2001). Physiological significance of anthocyanins during autumnal leaf senescence. *Tree Physiology*, **21**, 1–8.

Hodges, D.M., DeLong, J.M., Forney, C.F., *et al.* (1999). Improving the thiobarbituric acid-reactive-substances assay for estimating lipid peroxidation in plant tissues containing anthocyanin and other interfering compounds. *Planta*, **207**, 604–611.

Hodges, M. (2002). Enzyme redundancy and the importance of 2-oxoglutarate in plant ammonium assimilation. *Journal of Experimental Botany*, **53**, 905–916.

Hodges, M. and Moya, I. (1986). Time-resolved chlorophyll fluorescence studies of photosynthetic membranes: resolution and characterisation of four kinetics components. *Biochimica et Biophysica Acta*, **849**, 193–202.

Hodgson, A.J.R., Beachy, N.R., Pakrasi, B.H. (1989). Selective inhibition of photosystem II spinach by Tobacco mosaic virus: an effect of the viral coat protein. *FEBS Letters*, **245**, 267–270.

Hoefnagel, H.N., Atkin, O.K. and Wiskich, J.T. (1998). Interdependence between chloroplasts and mitochondria in the light and the dark. *Biochimica et Biophysica Acta*, **1366**, 235–255.

Hoekstra, H.E. and Coyne, J.A. (2007). The locus of evolution: evo devo and the genetics of adaptation. *Evolution*, **61**, 995–1016.

Hoff, C. and Rambal, S. (2003). An examination of the interaction between clime, soil and leaf-area index in a *Quercus ilex* ecosystem. *Annals of Forest Science*, **60**, 153–161.

Hoffmann, W.A. (1999). Fire and population dynamics of woody plants in a neotropical savanna: matrix model projections. *Ecology*, **80**, 1354–1369.

Hoffmann, W.A., Franco, A.C., Moreira, M.Z., *et al.* (2005). Specific leaf area explains differences in leaf traits between congeneric savanna and forest trees. *Functional Ecology*, **19**, 932–940.

Hoffmann, W.A., Orthen, B. and Do Nascimento, P.K.V. (2003). Comparative fire ecology of tropical savanna and forest trees. *Functional Ecology*, **17**, 720–726.

Hoffmann-Benning, S., Willmitzer, L. and Fisahn, J. (1997). Analysis of growth, composition and thickness of the cell walls of transgenic tobacco plants expressing a yeast-derived invertase. *Protoplasma*, **200**, 146–153.

Högberg P., Nordgren A., Buchmann N., *et al.* (2001). Large-scale forest girdling shows that current photosynthesis drives soil respiration. *Nature*, **411**, 789–792.

Hohmann-Marriott, M.F. and Blankenship, R.E. (2011). Evolution of photosynthesis. *Annual Review of Plant Biology*, **62**, 515–548.

Holbrook, N.M., Shashidhar, V.R., James, R.A., *et al.* (2002). Stomatal control in tomato with ABA-deficient roots: response of grafted plants to soil drying. *Journal of Experimental Botany*, **53**, 1503–1514.

Holden, M. (1976). Chlorophylls. In: *Chemistry and Biochemistry of Plant Pigments* (ed. Goodwin T.W.), Academic Press, New York, USA, pp. 1–37.

Holladay, A.S., Martindale, W., Alred, R., *et al.* (1992). Changes in activities of enzymes of carbon metabolism in leaves during exposure of plants to low temperature. *Plant Physiology*, **98**, 1105–114.

Hollinger, S.E., Bernacchi, C.J. and Meyers, T. (2005). Carbon budget of mature no-till ecosystem in North Central Region of the United States. *Agricultural and Forest Meteorology*, **130**, 59–69.

Hollinger, D.Y., Goltz, S.M., Davidson, E.A., *et al.* (1999). Seasonal patterns and environmental control of carbon dioxide and water vapour exchange in an ecotonal boreal forest. *Global Change Biology*, **5**, 891–902.

Holmes, M.G. and Keiller, D.R. (2002). Effects of pubescence and waxes on the reflectance of leaves in the ultraviolet and photosynthetic wavebands: a comparison of a range of species. *Plant, Cell and Environment*, **25**, 85–93.

Holt, N.E., Zigmantas, D., Valkunas, L., *et al.* (2005). Carotenoid cation formation and the regulation of photosynthetic light harvesting. *Science*, **307**, 433–436.

Holtgrefe, S., Gohlke, J., Starmann, J., *et al.* (2007). Regulation of plant cytosolic glyceraldehyde 3-phosphate dehydrogenase isoforms by thiol modifications. *Physiologia Plantarum*, **133**, 211–218.

Holthe, P.A. and Szarek, S.R. (1985). Physiological potential for survival of propagules of crassulacean acid metabolism species. *Plant Physiology*, **79**, 219–224.

Holtum, J.A.M. and Winter, K. (1999). Degrees of crassulacean acid metabolism in tropical epiphytic and lithophytic ferns. *Australian Journal of Plant Physiology*, **26**, 749–757.

Holtum, J.A.M., Aranda, J. Virgo, A., *et al.* (2004). $\delta^{13}C$ values and crassulacean acid metabolism in Clusia species from Panama. *Trees: Structure and Function*, **18**, 658–668.

Holub, O., Seufferheld, M.J., Gohlke, C., *et al.* (2000). Fluorescence lifetime imaging (FLI) in real time: a new technique in photosynthesis research. *Photosynthetica*, **38**, 581–599.

Holzwarth, A.R., Wendler, J. and Haehnel, W. (1985). Time-resolved picosecond fluorescence spectra of the antenna chlorophylls in *Chlorella vulgaris*. Resolution of photosystem I fluorescence. *Biochimica et Biophysica Acta*, **807**, 155–167.

Homann, P.H. (1999). Reliability of photosystem II thermoluminescence measurements after sample freezing: few artifacts with photosystem II membranes but gross distortions with certain leaves. *Photosynthesis Research*, **62**, 219–229.

Hong, S.W., Lee, U. and Vierling, E. (2003). *Arabidopsis* hot mutants define multiple functions required for acclimation to high temperatures. *Plant Physiology*, **132**, 757–767.

Hoorn, C., Ohja, T. and Quade, J. (2000). Palynological evidence for vegetation development and climatic change in the sub-Himalayan Zone (Neogene, central Nepal). *Palaeogeography, Palaeoclimatology, Palaeoecology*, **163**, 133–161.

Hope, A.B., Valente, P. and Matthews, D.B. (1994). Effects of pH on the kinetics of redox reactions in and around the cytochrome bf complex in an isolated system. *Photosynthesis Research*, **42**, 111–120.

Hopkins, D.L. (1989). *Xylella fastidiosa*: xylem-limited bacterial pathogen of plants. *Annual Review of Phytopathology*, **27**, 271–290.

Hopley, P.J., Marshall, J.D., Weedon, G.P., *et al.* (2007). Orbital forcing and the spread of C4 grasses in the late Neogene: stable isotope evidence from South African speleotherms. *Journal of Human Evolution*, **53**, 620–634.

Horken, K.M. and Tabita, F.R. (1999). Closely related form I ribulose bisphosphate carboxylase/oxygenase molecules that possess different CO_2/O_2 substrate specificities. *Archives of Biochemistry and Biophysics*, **361**, 183–194.

Horler, D.N.H., Barber, J. and Barringer, A.R. (1980). Effects of heavy metals on the absorbance and reflectance spectra of plants. *International Journal of Remote Sensing*, **1**, 121–136.

Hormaetxe, K., Becerril, J.M., Fleck, I., *et al.* (2005a). Functional role of red (*retro*)-carotenoids as passive light filters in the leaves of *Buxus sempervirens* L.: increased protection of photosynthetic tissues? *Journal of Experimental Botany*, **56**, 2629–2636.

Hormaetxe, K., Esteban, R., Becerril, J.M., *et al.* (2005b), Dynamics of the α-tocopherol pool as affected by external (environmental) and internal (leaf age) factors in *Buxus sempervirens* leaves. *Physiologia Plantarum*, **125**, 333–344.

Horn, H.S. (1971). *The Adaptive Geometry of Trees*. Princeton University Press, Princeton, New Jersey.

Hörtensteiner, S. (2006). Chlorophyll degradation during senescence. *Annual Review of Plant Biology*, **57**, 55–77.

Hörtensteiner, S. and Feller, U. (2002). Nitrogen metabolism and remobilization during senescence. *Journal Experimental Botany*, **53**, 927–937.

Horton, J.L., Kolb, T. and Hart, S.C. (2001). Leaf gas exchange characteristics differ among Sonoran Desert riparian tree species. *Tree Physiology*, **21**, 233–241.

Horton, P. (2000). Prospects for crop improvement through the genetic manipulation of photosynthesis: morphological and biochemical aspects of light capture. *J. Experimental Botany*, **51**, 475–485.

Horton, P., Mathew, P.J., Perez-Bueno, M.L., *et al.* (2008). Photosynthetic acclimation: does the dynamic structure and macro-organisation of photosystem II in higher plant grana membranes regulate light harvesting states? *FEBS Journal*, **275**, 1069–1079.

Horton, P., Ruban, A.V. and Young, A.J. (1999). Regulation of the structure and function of the light harvesting complexes of photosystem II by the xanthophyll cycle. In: *Advances in Photosynthesis and Respiration: The Photochemistry of Carotenoids* (eds Frank, H.A., Young, A.J., Britton, G. and Cogdell, R.J.), Kluwer Academic Publishers, Dordrecht, Netherlands, pp. 271–291.

Horton, P., Wentworth, M. and Ruban, A. (2005). Control of the light harvesting function of chloroplast membranes:

the LHCII-aggregation model for non-photochemical quenching. *FEBS Letters*, **579**, 4201–4206.

Horváth, E.M., Peter, S.O., Joët, T., *et al.* (2000). Targeted inactivation of the plastid ndhB gene in tobacco results in an enhanced sensitivity of photosynthesis to moderate stomatal closure. *Plant Physiology*, **123**, 1337–1349.

Hosler, J.P. and Yocum, C.F. (1987). Regulation of cyclic photophosphorylation during ferredoxin-mediated electron transport: effect of DCMU and the NADPH/NADP ratio. *Plant Physiology*, **83**, 965–969.

Housman, D.C., Naumburg, E., Huxman, T.E., *et al.* (2006). Increases in desert shrub productivity under elevated carbon dioxide vary with water availability. *Ecosystems*, **9**, 374–385.

Howe, G.A. and Jander, G. (2008). Plant immunity to insect herbivores. *Annual Review of Plant Biology*, **59**, 41–66.

Howe, G.T., Aitken, S.N., Neale, D.B., *et al.* (2003). From genotype to phenotype: unraveling the complexities of cold adaptation in forest trees. *Canadian Journal of Botany*, **81**, 1247–1266.

Hsiao, T.C., Steduto, P. and Fereres, E. (2007). A systematic and quantitative approach to improve water use efficiency in agriculture. *Irrigation Sciences*, **25**, 209–231.

Hsieh, T.H., Lee, J.T., Charng, Y.Y., *et al.* (2002). Tomato plants ectopically expressing *Arabidopsis* CBF1 show enhanced resistance to water deficit stress. *Plant Physiology*, **130**, 618–626.

Hu, H., Boisson-Dernier, A., Israelsson-Nordström, M., *et al.* (2010). Carbonic anhydrases are upstream regulators of CO_2-controlled stomatal movements in guard cells. *Nature Cell Biology*, **12**, 87–93.

Hu, H., Dai, M., Yao, J., *et al.* (2006). Overexpressing a NAM, ATAF, and CUC (NAC) transcription factor enhances drought resistance and salt tolerance in rice. *Proceedings of the National Academy of Sciences of USA*, **103**, 12987–12992.

Hu, L., Wang, Z. and Huang, B. (2010). Diffusion limitations and metabolic factors associated with inhibition and recovery of photosynthesis from drought stress in a C_3 perennial grass species. *Physiologia Plantarum*, **139**, 93–106.

Hu, X.L., Jiang, M.Y., Zhang, J.H., *et al.* (2007). Calcium-calmodulin is required for abscisic acid-induced antioxidant defense and functions both upstream and downstream of H_2O_2 production in leaves of maize (*Zea mays*) plants. *New Phytologist*, **173**, 27–38.

Huang, C.Y., Bazzaz, F.A. and Vanderhoef, L.N. (1974). The inhibition of soybean metabolism by cadmium and lead. *Plant Physiology*, **54**, 122–124.

Huang, H., Ger, M., Chen, C., *et al.* (2007). Disease resistance to bacterial pathogens affected by the amount of ferredoxin-I protein in plants. *Molecular Plant Pathology*, **8**, 129–137.

Hubbard, R.M., Ryan, M.G., Stiller, V., *et al.* (2001). Stomatal conductance and photosynthesis vary linearly with plant hydraulic conductance in ponderosa pine. *Plant, Cell and Environment*, **24**, 113–121.

Hubick, K.T., Shorter, R. and Farquhar, G.D. (1988). Heritability and genotype × environment interactions of carbon isotope discrimination and transpiration efficiency in peanut (*Arachis hypogaea* L.). *Australian Journal of Plant Physiology*, **15**, 799–813.

Huc, R., Ferhi, A. and Guel, J.M. (1994). Pioneer and late stage tropical rainforest tree species (French Guyana) growing under common conditions differ in leaf gas exchange regulation, carbon isotope discrimination and leaf water potential. *Oecologia*, **99**, 297–305.

Huerta, L., Forment, J., Gadea, J., *et al.* (2008). Gene expression analysis in citrus reveals the role of gibberellins on photosynthesis and stress. *Plant, Cell and Environment*, **31**, 1620–1633.

Hughes, N.M., Vogelmann, T.C. and Smith, W.K. (2008). Optical effects of abaxial anthocyanin on absorption of red wavelengths by understory species: revisiting the back-scatter hypothesis. *Journal of Experimental Botany*, **59**, 3435–3442.

Hui, D.Q., Iqbal, J., Lehmann, K., *et al.* (2003). Molecular interactions between the specialist herbivore *Manduca sexta* (Lepidoptera, Sphingidae) and its natural host *Nicotiana attenuata*. V. Microarray analysis and further characterization of large-scale changes in herbivore-induced mRNAs. *Plant Physiology*, **131**, 1877–1893.

Hulbert, L.C. (1988). Causes of fire effects in tallgrass prairie. *Ecology*, **69**, 46–58.

Hull, R. (2002). *Matthew's Plant Virology*, 4th edn. San Diego. California: Elsevier Academic Press.

Humbeck, K., Quast, S. and Krupinska, K. (1996). Functional and molecular changes in the photosynthetic apparatus during senescence of flag leaves from field-grown barley plants. *Plant Cell Environment*, **19**, 337–344.

Hummel, I., Pantin, F., Sulice, R. *et al.* (2010). *Arabidopsis* plants acclimate to water deficit at low cost through changes of carbon usage: an integrated perspective using growth, metabolite, enzyme, and gene expression analysis. *Plant Physiology*, **154**, 357–372.

Huner, N.P.A. and Ivanov, A.G. (2006). Photoprotection of photosystem II: Reaction center quenching versus antenna quenching. In: *Advances in Photosynthesis and Respiration – Photoprotection, Photoinhibition, Gene Regulation, and Environment* (eds Demmig-Adams, B., Adams, III, W.W. and Mattoo, A.), Springer, Dordrecht, Netherlands, pp. 155–173.

Huner, N.P.A., Öquist, G. and Melis, A. (2003). Photostasis in plants, green algae and cyanobacteria: the role of light harvesting antenna complexes. In: *Advances in Photosynthesis and Respiration Light Harvesting Antennas in Photosynthesis* (eds Green, B.R. and Parson, W.W.), Kluwer Academic Publishers, Dordrecht, Netherlands, pp. 401–421.

Huner, N.P.A., Öquist, G. and Sarhan, F. (1998). Energy balance and acclimation to light and cold. *Trends in Plant Science*, **3**, 224–230.

Huner, N.P.A., Öquist, G., Hurry, V.M., *et al.* (1993). Photosynthesis, photoinhibition and low temperature acclimation in cold tolerant plants. *Photosynthesis Research*, **37**, 19–39.

Hungate, B.A., Reichstein, M., Dijkstra, P., *et al.* (2002). Evapotranspiration and soil water content in a scrub-oak woodland under carbon dioxide enrichment. *Global Change Biology*, **8**, 289–298.

Huppe, H. C. and Turpin, D. H. (1994). Integration of carbon and nitrogen metabolism in plant and algal cells. *Annual Review of Plant Physiology and Plant Molecular Biology*, **45**, 577–607.

Hura, T., Grzesiak, S., Hura, K., *et al.* (2006). Differences in the physiological state between triticale and maize plants during drought stress and followed rehydration expressed by the leaf gas exchange and spectrofluorimetric methods. *Acta Physiologiae Plantarum*, **28**, 433–443.

Hurry, V., Igamberdiev, A.U., Keergerg, O., *et al.* (2005). Respiration in photosynthetic cells: gas exchange components, interactions with photorespiration and the operation of mitochondria in the light. In: *Plant Respiration: From Cell to Ecosystem* (eds Lambers, H. and Ribas-Carbó, M.), Springer, Doordrecht, Netherlands, pp. 43–61.

Hurry, V.M., Malmberg, G., Garderstrom, P., *et al.* (1994). Effects of a short-term shift to low-temperature and of long-term cold hardening on photosynthesis and ribulose-1,5-bisphosphate carboxylase oxygenase and sucrose-phosphate synthase activity in leaves of winter rye (*Secale cereale* L). *Plant Physiology*, **106**, 983–999.

Hurry, V., Strand, A., Furbank, R., *et al.* (2000). The role of inorganic phosphate in the development of freezing tolerance and the acclimation of photosynthesis to low temperature is revealed by the *pho* mutants of *Arabidopsis thaliana*. *Plant Journal*, **24**, 383–396.

Hutchinson, B.A. and Matt, D.R. (1977). The distribution of solar radiation within a deciduous forest. *Ecological Monographs*, **47**, 185–207.

Hutchison, B.A., Matt, D.R. and McMillen, R.T. (1980). Effects of sky brightness distribution upon penetration of diffuse radiation through canopy gaps in a deciduous forest. *Agricultural Meteorology*, **22**, 137–147.

Hutchison, R.S., Groom, Q. and Ort, D.R. (2000). Differential effects of chilling-induced photooxidation on the redox regulation of photosynthetic enzymes. *Biochemistry*, **39**, 6679–6688.

Hutyra, L.R., Munger, J.W., Saleska, S.R., *et al.* (2007). Seasonal controls on the exchange of carbon and water in an Amazonian rain forest. *Journal of Geophysical Research-Biogeosciences*, **112**.

Hüve, K., Bichele, I., Rasulov, B. *et al.* (2011). When it is too hot for photosynthesis: heat-induced instability of photosynthesis in relation to respiratory burst, cell permeability changes and H_2O_2 formation. *Plant, Cell and Environment*, **34**, 113–126.

Hüve, K., Bichele, I., Tobias, M., *et al.* (2006). Heat sensitivity of photosynthetic electron transport varies during the day due to changes in sugars and osmotic potential. *Plant, Cell and Environment*, **29**, 212–228.

Hüve, K., Christ, M.M., Kleist, E., *et al.* (2007). Simultaneous growth and emission measurements demonstrate an interactive control of methanol release by leaf expansion and stomata. *Journal of Experimental Botany*, **58**, 1783–1793.

Huxman, T.E., Barron-Gafford, G., Gerst, K.L., *et al.* (2008). Photosynthetic resource-use efficiency and demographic variability in desert winter annual plants. *Ecology*, **89**, 1554–1563.

Huxman, T.E., Hamerlynck, E.P., Moore, B.D., *et al.* (1998). Photosynthetic down-regulation in *Larrea tridentata* exposed to elevated atmospheric CO_2: interaction with drought under glasshouse and field (FACE) exposure. *Plant, Cell and Environment*, **21**, 1153–1161.

Hwang, Y.H. and Morris, J.T. (1994). Whole-plant gas exchange responses of *Spartina alterniflora* (Poaceae) to a range of constant and transient salinities. *American Journal of Botany*, **81**, 659–665.

Ibrom, A., Dellwik, E., Flyvbjerg, H., *et al.* (2007). Strong low-pass filtering effects on water vapour flux measurements with closed-path eddy correlation systems. *Agricultural and Forest Meteorology*, **147**, 140–156.

ICNIRP (1996). International commission on non-ionizing radiation protection. Guidelines on limits of exposure to laser radiation of wavelengths between 180 nm and 1,000 μm. *Health Physics*, **71**, 804–819.

Ida, K., Masamoto, K., Maoka, T., *et al.* (1995). The leaves of the common box, *Buxus sempervirens* (Buxaceae), become red as the level of a red carotenoid, anhydroeschscholtzxanthin, increases. *Journal of Plant Research*, **108**, 369–376.

Idle, D.B. and Proctor, C.W. (1983). An integrating sphere leaf chamber. *Plant, Cell and Environment*, **6**, 437–439.

Igamberdiev, A.U. and Gardeström, P. (2003). Regulation of NAD- and NADP-dependent isocitrate dehydrogenases by reduction levels of pyridine nucleotides in mitochondria and cytosol of pea leaves. *Archives of Biochemistry and Biophysics*, **1606**, 117–125.

Igamberdiev, A.U. and Lea, P.J. (2006). Land plants equilibrate O_2 and CO_2 concentrations in the atmosphere. *Photosynthesis Research*, **87**, 177–194.

Igamberdiev, A.U., Mikkelsen, T.N., Ambus, P., *et al.* (2004). Photorespiration contributes to stomatal regulation and carbon isotope fractionation: a study with barley, potato and *Arabidopsis* plants deficient in glycine decarboxylase. *Photosynthesis Research*, **81**, 139–152.

Iida, S., Miyagi, A., Aoki, S., *et al.* (2009). Molecular adaptation of *rbc*L in the heterophyllous aquatic plant *Potamogeton*. *PLoS ONE*, **4**, e4633.

Inoue, Y. and Shibata, K. (1974). Comparative examination of terrestrial plant leaves in terms of light-induced absorption changes due to chloroplast rearrangements. *Plant and Cell Physiology*, **15**, 717–721.

IPCC (2007). *Fourth Assessment Report of the Intergovernmental Panel on Climate Change* (eds Solomon, S., Qin, D., Manning, M., Chen, Z., Marquis, M., Averyt, K.B., Tignor, M. and Miller, H.L.), Cambridge University Press, Cambridge, UK and New York, USA, p. 976.

Ishida, H., Makino, A. and Mae, T. (1999). Fragmentation of the large subunit of ribulose-1,5-bisphosphate carboxylase by reactive oxygen species occurs near Gly 329. *Journal of Biological Chemistry*, **274**, 5222–5226.

Ishihara, K., Nishihara, T. and Ogura, T. (1971). The relationship between environmental factors and behaviour of stomata in the rice plant. I. On the measurement of the stomatal aperture. *Proceedings of the Japanese Society of Crop Science*, **40**, 491–496.

Ish-Shalom-Gordon, N., Lin, G. and Sternberg, L.D.S.L. (1992). Isotopic fractionation during cellulose synthesis in two mangrove species: salinity effects. *Phytochemistry*, **31**, 2623–2626.

Ismail, A.M. and Hall, A.E. (1993). Inheritance of carbon isotope discrimination and water-use efficiency in cowpea. *Crop Science*, **33**, 498–503.

Israel, A.A. and Nobel, P.S. (1995). Growth temperature versus CO_2 uptake, Rubisco and PEPCase activities, and enzyme high-temperature sensitivities for a CAM plant. *Plant Physiology and Biochemistry*, **33**, 345–351.

Iturbe-Ormaetxe, I., Morán, J. F., Arrese-Igor, C., *et al.* (1995). Activated oxygen and antioxidant defences in iron-deficient pea plants. *Plant, Cell and Environment*, **18**, 421–429.

Ivanov, A.G., Morgan, R.M., Gray, R.G., *et al.* (1998). Temperature/light dependent development of selective resistance to photoinhibition of photosystem I. *FEBS Letters*, **430**, 288–292.

Ivanov, A.G., Sane, P.V., Zeinalov, Y., *et al.* (2001). Photosynthetic electron transport adjutments in overwintering Scots pine *(Pinus sylvestris* L). *Planta*, **213**, 575–585.

Ivanov, A.G., Sane, P.V., Zeinalov, Y., *et al.* (2002). Seasonal responses of photosynthetic electron transport in Scots pine *(Pinus sylvestris* L.) studies by thermoluminescence. *Planta*, **215**, 457–465.

Jack, S.B. and Long, J.N. (1992). Forest production and the organization of foliage within crowns and canopies. *Forest Ecology and Management*, **49**, 233–245.

Jackson, J.E. and Palmer, J.W. (1979). A simple model of light transmission and interception by discontinuous canopies. *Annals of Botany*, **44**, 381–383.

Jackson, M.B. (2002). Long-distance signalling from roots to shoots assessed: the flooding story. *Journal of Experimental Botany*, **53**, 175–181.

Jackson, M.B. and Hall, K.C. (1987). Early stomatal closure in water-logged pea plants is mediated by abscisic acid in the absence of foliar water deficits. *Plant, Cell and Environment*, **10**, 121–130.

Jackson, P.C., Meinzer, F.C., Bustamante, M., *et al.* (1999). Partitioning of soil water among tree species in a Brazilian cerrado ecosystem. *Tree Physiology*, **19**, 717–724.

Jacobs, B.F. (2004). Palaeobotanical studies from tropical Africa: relevance to the evolution of forest, woodland and savannah biomes. *Philosophical Transations of the Royal Society of London, Series B*, **359**, 1573–1583.

Jacoby, G.C. and D'Arrigo, R.D. (1997). Tree rings, carbon dioxide, and climatic change. *Proceedings of the National Academy of Sciences of the United States of America*, **94**, 8350–8353.

Jacquemoud, S. and Baret, F. (1990). PROSPECT: a model of leaf optical properties spectra. *Remote Sensing of Environment*, **34**, 75–91.

Jahnke, S. (2001). Atmospheric CO_2 concentration does not directly affect leaf respiration in bean or poplar. *Plant, Cell and Environment*, **24**, 1139–1151.

Jahnke, S. and Krewitt, M. (2002). Atmospheric CO_2 concentration may directly affect leaf respiration measurement in tobacco, but not respiration itself. *Plant, Cell and Environment*, **25**, 641–651.

Jahnke, S. and Pieruschka, R. (2006). Air-pressure in clamp-on leaf chambers: a neglected issue in gas exchange measurements. *Journal of Experimental Botany*, **57**, 2553–2561.

Jalink, H., Frandas, A., van der Schoor, R., *et al.* (1998). Chlorophyll fluorescence of the testa of *Brassica oleracea* seeds as an indicator of maturity and seed quality. *Scientia Agricola*, **55**, 88–93.

James, S.A. and Bell, D. (2000). Leaf orientation, light interception and stomatal conductance of *Eucalyptus globulus* ssp. *globulus* leaves., *Tree Physiology*, **20**, 815–823.

Janda, T., Szalai, G., Papp, N., *et al.* (2004). Effect of freezing on thermoluminescence in various plant species. *Photochemistry and Photobiology*, **80**, 525–530.

Janda, T., Szalai, G., Rios-Gonzalez, K., *et al.* (2003). Comparative study of frost tolerance and antioxidant activity in cereals. *Plant Science*, **164**, 301–306.

Janis, C.M., Damuth, J. and Theodor, J.M. (2000). Miocene ungulates and terrestrial primary productivity: where have all the browsers gone? *Proceedings of the National Academy of Sciences*, **97**, 7899–7904.

Jarvis, P.G. (1976). The interpretation of the variations in leaf water potential and stomatal conductance found in canopies in the field. *Philosophical Transactions of the Royal Society of London Series B – Biological Sciences*, **273**, 593–610.

Jarvis, P.G. (1995). Scaling processes and problems. *Plant, Cell and Environment*, **18**, 1079–1089.

Jarvis, P.G. and Leverenz, J.W. (1983). Productivity of temperate, deciduous and evergreen forests. In: *Physiological Plant Ecology*, vol IV (eds Lange, O.L., Nobel, P.S., Osmond, C.B. and Ziegler, H.), Springer-Verlag, Berlin, Germany, pp. 233–280.

Jarvis, P.G. and McNaughton, K.G. (1986). Stomatal control of transpiration: scaling up from leaf to region. *Advances in Ecological Research*, **15**, 1–49.

Jarvis, P.G. and Sandford, A.P. (1986). Temperate forests. In: *Photosynthesis in Contrasting Environments* (eds Baker, N.R. and Long, S.P.), Elsevier Science Publishers B.V., New York, USA, pp. 199–236.

Jarvis, P.G., James, G.B. and Landsberg, J.J. (1976). Coniferous forests. In: *Vegetation and the Atmosphere*, vol 2 (ed. Monteith, J.L.). Academic Press, New York, USA, pp. 171–240.

Jasoni, R.L, Smith, S.D. and Arnone J.A. (2005). Net ecosystem CO_2 exchange in Mojave Desert shrublands during the eighth year of exposure to elevated CO_2. *Global Change Biology*, **11**, 749–756.

Jayasekera, R. and Schleser, G.H. (1991). Seasonal changes in organic carbon content of leaves of deciduous trees. *Journal of Plant Physiology*, **138**, 507–510.

Jenkins, C.L.D., Furbank, R.T. and Hatch, M.D. (1989). Mechanisms of C_4 photosynthesis. A model describing the inorganic carbon pool in bundle sheath cells. *Plant Physiology*, **91**, 1372–1381.

Jenny, H. (1941). *Factors of Soil Formation*. McGraw Hill, New York, USA.

Jensen, P.E., Bassi, R., Boekema, E.J., *et al.* (2007). Structure, function and regulation of plant photosystem I. *Biochimica et Biophysica Acta*, **1767**, 335–352.

Jensen, P.E., Haldrup, A., Zhang, S., *et al.* (2004). The PSI-O subunit of plant photosystem I is involved in balancing the excitation pressure between the two photosystems. *Journal of Biological Chemistry*, **279**, 24212–24217.

Jensen, S.G. (1972). Metabolism and carbohydrate composition in barley yellow dwarf virus infected wheat. *Phytopathology*, **62**, 587–592.

Jeong, W.J., Park, Y.I., Suh, K., *et al.* (2002). A large population of small chloroplasts in tobacco leaf cells allows more effective chloroplast movement than a few enlarged chloroplasts. *Plant Physiology*, **129**, 112–121.

Jeschke, W.D. and Hilpert, A. (1997). Sink-stimulated photosynthesis and sink-dependent increase in nitrate uptake: nitrogen and carbon relations of the parasitic association *Cuscuta reflexa-Ricinus communis. Plant, Cell and Environment*, **20**, 47–56.

Jeschke, W.D., Baig, A. and Hilpert, A. (1997). Sink-stimulated photosynthesis, increased transpiration and increased demand-dependent stimulation of nitrate uptake: nitrogen and carbon relations in the parasitic association *Cuscuta reflexa-Coleus blumei*. *Journal of Experimental Botany*, **48**, 915–925.

Jia, G., Peng, P., Zhao, Q., *et al.*(2003). Changes in terrestrial ecosystem since 30 Ma in East Asia: stable isotope evidence from black carbon in the South China Sea. *Geology*, **31**, 1093–1096.

Jia, H., Oguchi, R., Hope, A.B., *et al.* (2008). Differential effects of severe water stress on linear and cyclic electron fluxes through photosystem I in spinach leaf discs in CO_2-enriched air. *Planta*, **228**, 803–812.

Jiang, C.D., Gao, H.Y. and Zou, Q. (2001). Enhanced thermal energy dissipation depending on xanthophyll cycle and D1 protein turnover in iron-deficient maize leaves under high irradiance. *Photosynthetica*, **39**, 269–274.

Jiang, C.Z., Rodermel, S.R. and Shibles, R.M. (1997). Regulation of photosynthesis in developing leaves of soybean chlorophyll-deficient mutants. *Photosynthesis Research*, **51**, 185–192.

Jiang, G., Tang, H., Yu, M., *et al.* (1999). Response of photosynthesis of different plant functional types to environmental changes along Northeast China transect. *Trees*, **14**, 72–82.

Jiang, H.X., Chen, L.S., Zheng, J.G., *et al.* (2008). Aluminum-induced effects on photosystem II photochemistry in *Citrus* leaves assessed by the chlorophyll a fluorescence transient. *Tree Physiology*, **28**, 1863–1871.

Jiao, D., Hang, X., Li, X., *et al.* (2002). Photosynthetic characteristics and tolerance to photo-oxidation of transgenic rice expressing C4 photosynthesis enzymes. *Photosynthesis Research*, **72**, 85–93.

Jiao, J., Goodwin, P. and Grodzinski, B. (1999). Inhibition of photosynthesis and export in geranium grown at two CO_2 levels and infected with *Xanthomonas campestris* pv. Pelargonii. *Plant, Cell and Environment*, **22**, 15–25.

Jiménez, I., Lopez, L., Alamillo, J.M., *et al.* (2006). Identification of a plum pox virus CI-interacting protein from chloroplast that has a negative effect in virus infection. *Molecular Plant-Microbe Interactions*, **19**, 350–358.

Job, C., Raijou, L., Lovigny, Y., *et al.* (2005). Patterns of protein oxidation in *Arabidopsis* seeds and during germination. *Plant Physiology*, **138**, 790–802.

Joel, G., Chapin, F.S. and Chiariello, N.R. (2001). Species-specific responses of plant communities to altered carbon and nutrient availability. *Global Change Biology*, **7**, 435–450.

Joët, T., Cournac, L., Horvath, E. M, *et al.* (2001). Increased sensitivity of photosynthesis to antimycin A induced by inactivation of the chloroplast ndhB gene. Evidence for a participation of the NADH-dehydrogenase complex to cyclic electron flow around photosystem I. *Plant Physiology*, **125**, 1919–1929.

Joët, T., Cournac, L, Peltier, G., *et al.* (2002). Cyclic electron flow around photosystem I in C-3 plants. In vivo control by the redox state of chloroplasts and involvement of the NADH-dehydrogenase complex. *Plant Physiology*, **128**, 760–769.

Johansson, E., Olsson, O. and Nystrom, T. (2004). Progression and specificity of protein oxidation in the lifecycle of *Arabidopsis thaliana*. *Journal of Biological Chemistry*, **279**, 22204–22208.

Johansson, J., Andersson M., Edner H., *et al.* (1996). Remote fluorescence measurements of vegetation spectrally resolved and by multi-colour fluorescence imaging. *Journal of Plant Physiology*, **148**, 632–637.

Johnson, D.A., Richards, R.A. and Turner, N.C. (1983). Yield, water relations, gas exchange, and surface reflectance of near-isogenic wheat lines differing in glaucousness. *Crop Science*, **24**, 1168–1173.

Johnson, F.H., Eyring, H. and Williams, R.W. (1942). The nature of enzyme inhibitions in bacterial luminescence: sulfanilamide, urethane, temperature and pressure. *Journal of Cellular and Comparative Physiology*, **20**, 247–268.

Johnson, S.C. and Brown, W.V. (1973). Grass leaf ultrastructural variations. *American Journal of Botany*, **60**, 727–735.

Joiner, J., Y. Yoshida, A. Vasilkov, *et al.* (2010), First observations of global and seasonal terrestrial chlorophyll fluorescence from space, Biogeosci. Discuss., 7(6), 8281–8318.

Joliot, P., Lavergne, J. and Beal, D. (1992). Plastoquinone compartmentation in chloroplasts. I: evidence for domains with different rates of photo-reduction. *Biochimica et Biophysica Acta* (Bioenergetics), **1101**, 1–12.

Jones, A.M.E., Thomas, V., Bennett, M.H., *et al.* (2006). Modifications to the *Arabidopsis* defense proteome occur prior to significant transcriptional change in response to inoculation with *Pseudomonas syringae*. *Plant Physiology*, **142**, 1603–1620.

Jones, H.G. (1973). Moderate-term water stresses and associated changes in some photosynthetic parameters in cotton. *The New Phytologist*, **72**, 1095–1105.

Jones, H.G. (1985). Partitioning stomatal and non-stomatal limitations to photosynthesis. *Plant, Cell and Environment*, **8**, 95–104.

Jones, H.G. (1992). *Plants and Microclimate*. Cambridge University Press, New York, USA, pp. 323.

Jones, H.G. (1998). Stomatal control of photosynthesis and transpiration. *Journal of Experimental Botany*, **49**, 387–398.

Jones, H.G. (1999). Use of thermography for quantitative studies of spatial and temporal variation of stomatal conductance over leaf surfaces. *Plant, Cell and Environment*, **22**, 1043–1055.

Jones, H.G. (2004a). Applications of thermal imaging and infrared sensing in plant physiology and ecophysiology. *Advances in Botanical Research Incorporating Advances in Plant Pathology*, **41**, 107–163.

Jones, H.G. (2004b). Irrigation scheduling: advantages and pitfalls of plant-based methods. *Journal of Experimental Botany*, **55**, 2427–2436.

Jones, H.G. and Leinonen, I. (2003). Thermal imaging for the study of plant water relations. *Journal of Agricultural Meteorology, Japan*, **59**, 205–217.

Jones, H.G. and Morison, J. (eds) (2007). Special issue: imaging stress responses in plants. *Journal of Experimental Botany*, **58**, 743–898.

Jones, H.G. and Schofield, P. (2008). Thermal and other remote sensing of plant stress. *General and Applied Plant Physiology*, **34**, 19–32.

Jones, H.G. and Slatyer, R.O. (1971). Effects of intercellular resistances on estimates of the intracellular resistance to CO_2 uptake by plant leaves. *Australian Journal of Biological Sciences*, **25**, 443–453.

Jones, H.G. and Slatyer, R.O. (1972). Estimation of the transport and carboxylation components of the intracellular limitation to leaf photosynthesis. *Plant Physiology*, **50**, 283–288.

Jones, H.G., Stoll, M., Santos, T., *et al.* (2002). Use of infrared thermography for monitoring stomatal closure in the field: application to grapevine. *Journal of Experimental Botany*, **53**, 2249–2260.

Jones, M.G.K., Outlaw, W.H. and Lowry O.H. (1977). Enzymic assay of 10–7 to 10–14 moles of sucrose in plant tissues. *Plant Physiology*, **60**, 379–383.

Jones, T.L., Tucker, D.E. and Ort, D.R. (1998). Chilling delays circadian pattern of sucrose phosphate synthase and nitrate reductase activity in tomato. *Plant Physiology*, **118**, 149–158.

Jones, T.P. (1994). 13C enriched Lower Carboniferous fossil plants from Donegal, Ireland: carbon isotope constraints on taphonomy, diagenesis and palaeoenvironment. *Review of Palaeobotany and Palynology*, **81**, 53–64.

Jongebloed, U., Szederkényi, J., Hartig, K., *et al.* (2004). Sequence of morphological and physiological events during natural ageing and senescence of a castor bean leaf: sieve tube occlusion and carbohydrate back-up precede chlorophyll degradation. *Physiologia Plantarum*, **120**, 338–346.

Jongschaap, R.E.E., Blesgraaf, R.A.R., Bogaard, T.A., *et al.* (2009). The water footprint of bioenergy from *Jatropha curcas* L. *Proceedings of the National Academy of Sciences USA*, **106**, E92.

Jordan, C.F. (1969). Derivation of leaf-area index from quality of light in the forest floor. *Ecology*, **50**, 663–666.

Jordan, D.B. and Ögren, W.L. (1981a). A sensitive assay procedure for simultaneous determination of ribulose-1,5-bisphosphate carboxylase and oxygenase activities. *Plant Physiology*, **67**, 237–245.

Jordan, D.B. and Ögren W.L. (1981b). Species variation in the specificity of ribulose bisphosphate carboxylase/oxygenase. *Nature*, **291**, 513–515.

Jordan, D.B. and Ögren W.L. (1983). Species variation in kinetic properties of ribulose 1,5-bisphosphate carboxylase oxygenase. *Archives of Biochemistry and Biophysics*, **227**, 425–433.

Jordan, D.B. and Ögren, W.L. (1984). The CO_2/O_2 specificity of ribulose 1,5-bisphosphate carboxylase/oxygenase. *Planta*, **161**, 308–313.

Jordan, G.J., Dillon, R.A. and Weston, P.H. (2005). Solar radiation as a factor in the evolution of scleromorphic leaf anatomy in Proteaceae. *American Journal of Botany*, **92**, 789–796.

June, T., Evans, J.R. and Farquhar, G.D. (2004). A simple new equation for the reversible temperature dependence of photosynthetic electron transport: a study on soybean leaf. *Functional Plant Biology*, **31**, 275–283.

Jung, M., Reichstein, M., Ciais, P. *et al.* (2010). Recent decline in the global land evapotranspiration trend due to limited moisture supply. *Nature*, **467**, 951–954.

Junge, W. and Witt, H.T. (1968). On the ion transport system in photosynthesis: investigations on a molecular level. *Zeitschrift für Naturforschung*, **23b**, 244–254.

Jurik, T.W. (1986). Temporal and spatial patterns of specific leaf weight in successional northern hardwood tree species. *American Journal of Botany*, **73**, 1083–1092.

Jurik, T.W. and Chabot, B.F. (1986). Leaf dynamics and profitability in wild strawberries. *Oecologia*, **69**, 296–304.

Jurik, T.W. and Kliebenstein, H. (2000). Canopy architecture, light extinction and self-shading of a prairie grass, *Andropogon gerardii*. *American Midland Naturalist*, **144**, 51–65.

Jurik, T.W., Weber, J.A. and Gates, D.M. (1984). Short-term effects of CO_2 on gas exchange of leaves of bigtooth aspen (*Populus grandidentata*) in the field. *Plant Physiology*, **75**, 1022–1026.

Kacperska, A. (2004). Sensor types in signal transduction pathways in plant cells responding to abiotic stressors: do they depend on stress intensity? *Physiologia Plantarum*, **122**, 159–168.

Kaimal J.C. and Finnigan, J.J. (1994). *Atmospheric Boundary Layer Flows: Their Structure and Measurement*. Oxford University Press, Oxford, UK.

Kaiser, H. (2009). The relation between stomatal aperture and gas exchange under consideration of pore geometry and diffusional resistance in the mesophyll. *Plant, Cell and Environment*, **32**, 1091–1098.

Kaiser, H. and Grams, T.E.E. (2006). Rapid hydropassive opening and subsequent active stomatal closure follow heat-induced electrical signals in *Mimosa pudica*. *Journal of Experimental Botany*, **57**, 2087–2092.

Kaiser, H. and Kappen, L. (2000). In situ observation of stomatal movements and gas exchange of *Aegopodium podagraria* L. in the understory. *Journal of Experimental Botany*, **51**, 1741–1749.

Kalapos, T., van den Boogaard, R. and Lambers, H. (1996). Effect of soil drying on growth, biomass allocation and leaf gas exchange of two annual grass species. *Plant and Soil*, **185**, 137–149.

Kalberer, S.R., Wisniewski, M. and Arora, R. (2006). Deacclimation and reacclimation of cold-hardy plants: current understanding and emerging concepts. *Plant Science*, **171**, 3–16.

Kamal-Eldin, A., Görgen, S., Pettersson, J., *et al.* (2000). Normal-phase high-performance liquid chromatography of tocopherols and tocotrienols. Comparison of different chromatographic columns. *Journal of Chromatography*, **881**, 217–227.

Kamaluddin, M. and Grace, J. (1992). Acclimation in seedlings of a tropical tree, *Bischofia javanica*, following a stepwise reduction in light. *Annals of Botany*, **69**, 557–562.

Kampfenkel, K., Van Montagu, M. and Inzé, D. (1995). Effect of iron excess on *Nicotiana plumbaginifolia* plants. Implications to oxidative stress. *Plant Physiology*, **107**, 725–735.

Kanai, R. and Edwards, G.E. (1999) The biochemistry of C4 photosynthesis. In: *C4 Plant Biology* (eds Sage, R.F. and Monson, R.K.), Academic Press, San Diego, CA, USA, pp. 49–87.

Kanai, S., Ohkura, K., Adu-Gyamfi, J.J., *et al.* (2007). Depression of sink activity precedes the inhibition of biomass production in tomato plants subjected to potassium deficiency stress. *Journal of Experimental Botany*, **58**, 2917–2928.

Kandil, F.E., Grace, M.H., Seigler, D.S., *et al.* (2004). Polyphenolics in *Rhizophora mangle* L. leaves and their changes during leaf development and senescence. *Trees*, **18**, 518–528.

Kane, H.J., Viil, J., Entsch, B., *et al.* (1994). An improved method for measuring the CO_2/O_2 specificity of ribulose-bisphosphate carboxylase-oxygenase. *Australian Journal of Plant Physiology*, **21**, 449–461.

Kanervo, E., Tasaka, Y., Murata, N., *et al.* (1997). Membrane lipid unsaturation modulates processing of the photosystem II reaction-center protein d1 at low temperatures. *Plant Physiology*, **114**, 841–849.

Kangasjärvi, J., Talvinen, M. and Karjalainen, R. (1994). Plant defence systems induced by ozone. *Plant, Cell and Environment*, **17**, 783–94.

Kaplan, A. and Reinhold, L. (1999). CO_2 concentrating mechanisms in photosynthetic microorganisms. *Annual Review of Plant Physiology and Plant Molecular Biology*, **50**, 539–570.

Kappen, L., Andresen, G. and Lösch, R. (1987). In situ observations of stomatal movements. *Journal of Experimental Botany*, **38**, 126–141.

Kapralov, M.V. and Filatov, D.A. (2006). Molecular adaptation during adaptive radiation in the Hawaiian endemic genus *Schiedea*. *PLoS ONE*, **1**, e8.

Kapralov, M.V. and Filatov, D.A. (2007). Widespread positive selection in the photosynthetic Rubisco enzyme. *BMC Evolutionary Biology*, **7**, 73.

Karabourniotis, G. and Bornman, J.F. (1999). Penetration of UV-A, UV-B and blue light through the leaf trichome layers of two xeromorphic plants, olive and oak, measured by optical fibre microprobes. *Physiologia Plantarum*, **105**, 655–661.

Karabourniotis, G., Papastergiou, N., Kabanopoulou, E., *et al.* (1994). Foliar sclereids of *Olea europaea* may function as optical fibres. *Canadian Journal of Botany*, **72**, 330–336.

Karam., F., Lahoud, R., Massad, R., *et al.* (2007). Evapotranspiration, seed yield and water use efficiency of drip irrigated sunflower under full and deficit irrigation conditions. *Agricultural Water Management*, **90**, 213–223.

Kargul, J. and Barber, J. (2008). Photosynthetic acclimation: Structural reorganisation of light harvesting antenna: role of redox-dependent phosphorylation of major and minor chlorophyll a/b binding proteins. *FEBS J*, **275**, 1056–1068.

Karnosky, D.F., Mankovska, B., Percy, K., *et al.* (1999). Effects of tropospheric O_3 on trembling aspen and interaction with CO_2: results from an O_3-gradient and a FACE experiment. *Water Air and Soil Pollution*, **116**, 311–322.

Karnosky, D.F., Zak, D.R., Pregitzer, K.S., *et al.* (2003). Tropospheric O_3 moderates responses of temperate hardwood forests to elevated CO_2: a synthesis of molecular to ecosystem results from the Aspen FACE project. *Functional Ecology*, **17**, 289–304.

Karolin, A. Moldau, H. (1976). A controlled environment chamber for recording of transpiration and CO_2 exchange of plants shoots and roots. *Fiziologia Rastenii (Sov. Plant Physiol.)*, **23**, 630–634.

Karpinski, S., Reynolds, H., Karpinska, B., *et al.* (1999). Systemic signalling and acclimation in response to excess excitation energy in Arabidopsis. *Science*, **284**, 654–657.

Kasahara, M., Kagawa, T., Oikawa, K., *et al.* (2002). Chloroplast avoidance movement reduces photodamage in plants. *Nature*, **420**, 829–832.

Kasimova, M.R., Grigiene, J., Krab, K., *et al.* (2006). The free NADH concentration is kept constant in plant mitochondria under different metabolic conditions. *Plant Cell*, **18**, 688–698.

Katahata, S., Naramoto, M., Kakubari, Y., *et al.* (2005). Photosynthetic acclimation to dynamic changes in environmental conditions associated with deciduous overstory phenology in *Daphniphyllum humile*, an evergreen understory shrub. *Tree Physiology*, **25**, 437–445.

Katahata, S.I., Naramoto, M., Kakubari, Y., *et al.* (2007). Seasonal changes in photosynthesis and nitrogen allocation in leaves of different ages in evergreen understory shrub *Daphniphyllum humile*. *Trees*, **21**, 619–629.

Kato T., Tang Y., Gu S., *et al.* (2004). Carbon dioxide exchange between the atmosphere and an alpine meadow ecosystem on the Qinghai–Tibetan Plateau, China Agricultural and Forest Meteorology, **124**, 121–134.

Kato, M.C., Hikosaka, K. and Hirose, T. (2002). Photoinactivation and recovery of photosystem II in *Chenopodium album* leaves grown at different levels of irradiance and nitrogen availability. *Functional Plant Biology*, **29**, 787–795.

Kato, T. and Tang, Y.H. (2008). Spatial variability and major controlling factors of CO_2 sink strength in Asian terrestrial ecosystems: evidence from eddy covariance data. *Global Change Biology*, **14**, 2333–2348.

Kattge, J. and Knorr W. (2007). Temperature acclimation in a biochemical model of photosynthesis: a reanalysis of data from 36 species. *Plant, Cell and Environment*, **30**, 1176–1190.

Kautsky, H. and Hirsch, A. (1931). Neue Versuche zur Kohlensäureassimilation. *Naturwissenschaften*, **19**, 964.

Kaya, C., Higgs, D., Kirnak, H., *et al.* (2003). Mycorrhizal colonisation improves fruit yield and water use efficiency in watermelon (*Citrullus lanatus* Thunb.) grown under well-watered and water-stressed conditions. *Plant and Soil*, **253**, 287–292.

Kebeish, R., Niessen, M., Thiruveedhi, K., *et al.* (2007). Chloroplastic photorespiratory bypass increases photosynthesis and biomass production in *Arabidopsis thaliana*. *Nature Biotechnology*, **25**, 593–599.

Kee, S.C., Martin, B. and Ort, D.R. (1986). The effects of chilling in the dark and in the light on photosynthesis of tomato – electron-transfer reactions. *Photosynthesis Research*, **8**, 41–51.

Keeley, J.E. (1990). Photosynthetic pathways in freshwater aquatic plants. *Trends in Ecology and Evolution*, **5**, 330–333.

Keeley, J.E. (1996). Aquatic CAM photosynthesis. In: *Crassulacean Acid Metabolism. Biochemistry, Ecophysiology and Evolution* (eds Winter, K. and Smith, J.A.C.) Ecological Studies vol. 114. Springer-Verlag, Berlin – Heidelberg – New York, pp. 281–295.

Keeley, J.E. (1998a). CAM photosynthesis in submerged aquatic plants. *Botanical Review*, **64**, 121–175.

Keeley, J.E. (1998b). C_4 photosynthetic modifications in the evolutionary transition from land to water in aquatic grasses. *Oecologia*, **116**, 85–97.

Keeley, J.E. and Busch, G. (1984). Carbon assimilation characteristics of the aquatic CAM plant, *Isoetes howellii*. *Plant Physiology*, **76**, 525–530.

Keeley, J.E. and Keeley, S.C. (1989). Crassulacean acid metabolism (CAM) in high elevation tropical cactus. *Plant, Cell and Environment*, **12**, 331–336.

Keeley, J.E. and Rundel, P.W. (2003). Evolution of CAM and C_4 carbon-concentrating mechanisms. *International Journal of Plant Sciences*, **164 (3 Suppl.)**, S55-S77.

Keeley, J.E. and Rundel, P.W. (2005). Fire and the Miocene expansion of C_4 grasslands. *Ecology Letters*, **8**, 683–690.

Keeley, J.E., Osmond, C.B. and Raven, J.A. (1984). *Stylites*, a vascular land plant without stomata absorbs CO_2 via its roots. *Nature*, **310**, 694–695.

Keeling, C.D. (1958). The concentration and isotopic abundance of atmospheric carbon dioxide in rural and marine areas. *Geochimica et Cosmochimica Acta*, **13**, 322–334.

Keeling, C.D., Chin, J.F.S. and Whorf, T.P. (1996). Increased activity of northern vegetation inferred from atmospheric CO_2 measurements. *Nature*, **382**, 146–149.

Keeling, C.D., Mook, W.G. and Tans, P.P. (1979). Recent trends in the $^{13}C/^{12}C$ ratio of atmospheric carbon dioxide. *Nature*, **277**, 121–123.

Keeling, R.F., Piper, S.C., Bollenbacher, *et al.* (2009). Atmospheric CO_2 records from sites in the SIO air sampling network. In: *Trends: A Compendium of Data on Global Change.* Carbon Dioxide Information Analysis Center, Oak Ridge National Laboratory, U.S. Department of Energy, Oak Ridge, Tenn., U.S.A.

Keenan, T., Sabate, S. and Gracia, C. (2010a). The importance of mesophyll conductance in regulating forest ecosystem productivity during drought periods. *Global Change Biology*, **16**, 1019–1034.

Keenan, T., Sabate, S. and Gracia, C. (2010b). Soil water stress and coupled photosynthesis–conductance models: bridging the gap between conflicting reports on the relative roles of stomatal, mesophyll conductance and biochemical limitations to photosynthesis. *Agricultural and Forest Meteorology*, **150**, 443–453.

Keitel, C., Adams, M.A., Holst, T., *et al.* (2003). Carbon and oxygen isotope composition of organic compounds in the phloem sap provides a short-term measure of stomatal conductance of European beech (*Fagus sylvatica* L.). *Plant, Cell and Environment*, **26**, 1157–1168.

Keitel, C., Matzarakis, A., Rennenberg, H., *et al.* (2006). Carbon isotopic composition and oxygen isotopic enrichment in phloem and total leaf organic matter of European beech (*Fagus sylvatica* L.) along a climate gradient. *Plant, Cell and Environment*, **29**, 1492–1507.

Kellogg, E.A. (1999). Phylogenetic aspects of the evolution of C4 photosynthesis. In: *C4 Plant Biology* (eds Sage, R.F. and Monson, R.K.) Academic Press, San Diego, CA, USA, pp. 411–444.

Kellogg, E.A. (2001). Evolutionary history of the grasses. *Plant Physiology*, **125**, 1198–1205.

Kellomaki, S., Wang, K.Y. and Lemettinen, M. (2000). Controlled environment chambers for investigating tree response to elevated CO_2 and temperature under boreal conditions. *Photosynthetica*, **38**, 69–81.

Kemp, P.R. and Williams, G.J., III (1980), A physiological basis for niche separation between *Agropyron smithii* (C_3) and *Bouteloua gracilis* (C_4). *Ecology*, **61**, 846–858.

Kempema, L.A., Cui, X., Holzer, F.M., *et al.* (2007). *Arabidopsis* transcriptome changes in response to phloem-feeding silverleaf whitefly nymphs. Similarities and distinctions in responses to aphids. *Plant Physiology*, **143**, 849–865.

Kenrich, P. and Crane, P.R. (1997). The origin and early evolution of plants on land. *Nature*, **389**, 33–39.

Kent, S.S. and Tomany, M.J. (1995). The differential of the ribulose 1,5-bisphosphate carboxylase/oxygenase specificity factor among higher plants and the potential for biomass enhancement. *Plant Physiology and Biochemistry*, **33**, 71–80.

Kerstiens, G. (1996). Cuticular water permeability and its physiological significance. *Journal of Experimental Botany*, **47**, 1813–1832.

Kerstiens, G. (2006). Water transport in plant cuticles: an update. *Journal of Experimental Botany*, **57**, 2493–2499.

Keskitalo, J., Bergquist, G., Gardeström, P., *et al.* (2005). A cellular timetable of autumn senescence. *Plant Physiology*, **139**, 1635–1648.

Kesselmeier, J. and Staudt, M. (1999). Biogenic volatile organic compounds (VOC): an overview on emission, physiology and ecology. *Journal of Atmospheric Chemistry*, **33**, 23–88.

Kessler, A. and Baldwin, I.T. (2002). Plant responses to insect herbivory: the emerging molecular analysis. *Annual Review of Plant Biology*, **53**, 299–328.

Kessler, M., Siorak, Y., Wunderlich, M., *et al.* (2007). Patterns of morphological leaf traits among pteridophytes along humidity and temperature gradients in the Bolivian Andes. *Functional Plant Biology*, **34**, 963–971.

Kessler, S. and Sinha, N. (2004). Shaping up: the genetic control of leaf shape. *Current Opinion in Plant Biology*, **7**, 65–72.

Kettle, A.J., Kuhn, U., von Hobe, M., *et al.* (2002). Global budget of atmospheric carbonyl sulfide: temporal and spatial variations of the dominant sources and sinks. *Journal of Geophysical Research – Atmospheres*, **107**, Article number 4658.

Khalil, A.A.M. and Grace, J. (1993). Does xylem sap ABA control the stomatal behaviour of water-stressed sycamore (*Acer pseudoplatanus* L.) seedlings? *Journal of Experimental Botany*, **44**, 1127–1134.

Khamis, S., Lamaze, T., Lemoine, Y., *et al.* (1990). Adaptation of the photosynthetic apparatus in maize leaves as a result of nitrogen limitation. Relationships between electron transport and carbon assimilation. *Plant Physiology*, **94**, 1436–1443.

Khan, M.S. (2007). Engineering photorespiration in chloroplasts: a novel strategy for increasing biomass production. *Trends in Biotechnology*, **25**, 437–40.

Khandelwal, A., Elvitigala, T., Ghosh, B., *et al.* (2008). *Arabidopsis* transcriptome reveals control circuits regulating redox homeostasis and the role of an AP2 transcription factor. *Plant Physiology*, **148**, 2050–2058.

Kikuzawa, K. (1983). Leaf survival of woody plants in deciduous broad-leaved forests. 1. Tall trees. *Canadian Journal of Botany*, **61**, 2133–2139.

Kikuzawa, K. (1991). A cost-benefit analysis of leaf habit and leaf longevity of trees and their geographical pattern. *American Naturalist*, **138**, 1250–1260.

Kilian, J., Whitehead, D., Horak, J., *et al.* (2007). The AtGenExpress global stress expression data set: protocols, evaluation and model data analysis of UV-B light, drought and cold stress responses. *The Plant Journal*, **50**, 347–363.

Killingbeck, K.T. (1996). Nutrients in senesced leaves: keys to the search for potential resorption and resorption proficiency. *Ecology*, **77**, 1716–1727.

Kiltie, R.A. (1993). New light on forest shade. *Trends in Ecology and Evolution*, **8**, 39–40.

Kim, C.S. and Jung, J. (1993). The susceptibility of mung bean chloroplasts to photoinhibition is increased by an excess supply of iron to plants: a photobiological aspect of iron toxicity in plant leaves. *Photochemistry and Photobiology*, **58**, 120–126.

Kim, J. (2007). Perception, transduction, and networks in cold signaling. *Journal of Plant Biology*, **50**, 139–147.

Kim, M.S., Lefcourt, A.M. and Chen, Y.C. (2003). Multispectral laser-induced fluorescence imaging system for large biological samples. *Applied Optics*, **42**, 3927–3934.

Kim, T.H., Böhmer, M., Hu, H., *et al.* (2010). Guard cell signal transduction network: advances in understanding abscisic acid, CO_2, and Ca^2+ signaling. *Annual Review of Plant Biology*, **61**, 561–591.

Kim, Y.X. and Steudle, E. (2009). Gating of aquaporins by light and reactive oxygen species in leaf parenchyma cells of the midrib of *Zea mays*. *Journal of Experimental Botany*, **60**, 547–556.

Kim. S.H., Sicher, R.C., Bae, H., *et al.* (2006). Canopy photosynthesis, evapotranspiration, leaf nitrogen, and transcription profiles of maize in response to CO_2 enrichment. *Global Change Biology*, **12**, 588–600.

Kimball, B. and Bernacchi, C.J. (2006). Evapotranspiration, canopy temperature, and plant water relations. In: *Managed Ecosystems and Rising CO2: Case Studies, Processes, and Perspectives* (eds Nösberger, J. and Blum, H.), Springer Verlag, Berlin, Germany, pp. 311–324.

Kimball, B.A., Pinter, P.J., Garcia, R.L., *et al.* (1995). Productivity and water use of wheat under free-air CO_2 enrichment. *Global Change Biology*, **1**, 429–442.

Kimura, K., Ishida, A., Uemura, A., *et al.* (1998). Effects of current-year and previous-year PPFDs on shoot gross morphology and leaf properties in *Fagus japonica*. *Tree Physiology*, **18**, 459–466.

King, J.S., Hanson, P.J., Bernhardt, E., *et al.* (2004). A multiyear synthesis of soil respiration responses to elevated atmospheric CO_2 from four forest FACE experiments. *Global Change Biology*, **10**, 1027–1042.

King, S.P., Badger, M.R. and Furbank, R.T. (1998). CO_2 characteristics of developing canola seeds and silique wall. *Australian Journal of Plant Physiology*, **25**, 377–386.

Kingston-Smith, A.H., Harbinson, J., Williams, J., *et al.* (1997). Effect of chilling on carbon assimilation, enzyme activation, and photosynthetic electron transport in the absence of photoinhibition in maize leaves. *Plant Physiology*, **114**, 1039–1046.

Kiniry, J.R. (1998). Biomass accumulation and radiation use efficiency of honey mesquite and eastern red cedar. *Biomass and Bioenergy*, **15**, 467–73.

Kiniry, J.R., Jones, C.A., O'Toole, J.C., *et al.* (1989). Radiation-use efficiency in biomass accumulation prior to grain-filling for five grain-crop species. *Field Crops Research*, **20**, 51–64.

Kira, T. (1975). Primary production of forests. In: *Photosynthesis and Productivity in Different Environments* (eds Winter, K. and Smith, J.A.C.), Cambridge University Press, Cambridge, UK, pp. 5–40.

Kirschbaum, M.U. (2004). Direct and indirect climate change effects on photosynthesis and transpiration. *Plant Biology*, **6**, 242–53.

Kirschbaum, M.U.F. (1987). Water-stress in *Eucalyptus pauciflora* – comparison of effects on stomatal conductance with effects on the mesophyll capacity for photosynthesis, and investigation of a possible involvement of photoinhibition. *Planta*, **171**, 466–473.

Kirschbaum, M.U.F. (1988). Recovery of photosynthesis from water stress in *Eucalyptus pauciflora* – a process in two stages. *Plant, Cell and Environment*, **11**, 685–694.

Kirschbaum, M.U.F. and Pearcy, R.W. (1988). Concurrent measurements of oxygen-and carbon-dioxide exchange during lightflecks in *Alocasia macrorrhiza* (L.) G. Don. *Planta*, **174**, 527–533.

Kirschbaum, M.U.F. and Tompkins, D. (1990). Photosynthetic responses to phosphorus nutrition in *Eucalyptus grandis* seedlings. *Australian Journal of Plant Physiology*, **17**, 527–535.

Kirschbaum, M.U.F., Keith, H., Leuning, R., *et al.* (2007) Modelling net ecosystem carbon and water exchange of a temperate *Eucalyptus delegatensis* forest using multiple constraints. *Agricultural and Forest Meteorology*, **145**, 48–68.

Kitajima, K. (1994). Relative importance of photosynthetic traits and allocation patterns as correlates of seedling shade tolerance of 13 tropical trees. *Oecologia*, **98**, 419–428.

Kitajima, K. and Hogan, K.P. (2003). Increases of chlorophyll *a*/*b* ratios during acclimation of tropical woody seedlings to nitrogen limitation and high light. *Plant, Cell and Environment*, **26**, 857–865.

Kitajima, K., Mulkey, S.S., Samaniego, M., *et al.* (2002). Decline of photosynthetic capacity with leaf age and position in two tropical pioneer tree species. *American Journal of Botany*, **89**, 1925–1932.

Kitajima, K., Mulkey, S.S., and Wright, S.J. (1997). Decline of photosynthetic capacity with leaf age in relation to leaf longevities for five tropical canopy tree species. *American Journal of Botany*, **84**, 702–708.

Kitao, M., Utsugi, H., Kuramoto, S., *et al.* (2003) Light-dependent photosynthetic characteristics indicated by chlorophyll fluorescence in five mangrove species native to Pohnpei Island, Micronesia. *Physiologia Plantarum*, **117**, 376–382.

Kitaoka, S. and Koike, T. (2004). Invasion of broad-leaf tree species into a larch plantation: seasonal light environment, photosynthesis and nitrogen allocation. *Physiologia Plantarum*, **121**, 604–611.

Kjellström, E., Bärring, L., Jacob, D., *et al.* (2007). Modelling daily temperature extremes: recent climate and future changes over Europe. *Climatic Change*, **81**, S249-S265.

Klein, T., Hemming, D., Lin, T., *et al.* (2005). Association between tree-ring and needle delta 13C and leaf gas exchange in *Pinus halepensis* under semi-arid conditions. *Oecologia*, **144**, 45–54.

Kleine, T., Kindgren, P., Benedict, C., *et al.* (2007). Genome-wide gene expression analysis reveals a critical role for CRYPTOCHROME1 in the response of *Arabidopsis* to high irradiance. *Plant Physiology* **144**, 1391–1406.

Kleinert, K. and Strecker, M.R. (2001). Climate change in response to orographic barrier uplift: paleosol and stable evidence from the late Neogene Santa Maria Basin, norwestern Argentina. *Bulletin of the Geological Society of America*, **113**, 728–742.

Kliemchen, A., Schomburg, M., Galla, H.J., *et al.* (1993). Phenotypic changes in the fluidity of the tonoplast membrane of crassulacean acid metabolism plants in response to temperature and salinity stress. *Planta*, **189**, 403–409.

Klingeman, W.E., Buntin, G.D., van Iersel, M.W., *et al.* (2000). Whole-plant gas exchange, not individual-leaf measurements, accurately assesses azalea response to insecticides. *Crop Protection*, **19**, 407–415.

Klingeman, W.E., van Iersel, M.W., Kang, J.G., *et al.* (2005). Whole-plant gas exchange measurements of mycorrhizal 'iceberg' roses exposed to cyclic drought. *Crop Protection*, **24**, 309–317.

Klingler, J.P., Batelli, G. and Zhu, J.-K. (2010). ABA receptors: the START of a new paradigm in phytohormone signalling. *Journal of Experimental Botany*, **61**, 3199–3210.

Kluge, M. and Brulfert, J. (2000). Ecophysiology of vascular plants on inselbergs. In: *Inselbergs*. Ecological Studies, Vol. 146 (eds Porembski, S. and Barthlott, W.), Springer-Verlag, Berlin – Heidelberg – New York, pp. 143–174.

Kluge, M., Kliemchen, A. and Galla, H.J. (1991). Temperature effects on crassulacean acid metabolism: EPR spectroscopic studies on the thermotropic phase behaviour of the tonoplast membranes of *Kalanchoë daigremontiana*. *Botanica Acta*, **104**, 355–360.

Klughammer, C. and Schreiber, U. (1994). An improved method, using saturating light pulses, for the determination of photosystem I quantum yield via P700+-absorbance changes at 830 nm. *Planta*, **192**, 261–268.

Knapp, A.K. (1993). Gas exchange dynamics in C_3 and C_4 grasses: consequences of differences in stomatal conductance. *Ecology*, **74**, 113–123.

Knapp, A.K. and Smith, M.D. (2001). Variation among biomes in temporal dynamics of aboveground primary production. *Science*, **291**, 481–484.

Knapp, A.K., Briggs, J.M. and Koelliker, J.K. (2001). Frequency and extent of water limitation to primary production in a mesic temperate grassland. *Ecosystems*, **4**, 19–28.

Knapp, A.K., Fay, P.A., Blair, J.M., *et al.* (2002). Rainfall variability, carbon cycling, and plant species diversity in a mesic grassland. *Science*, **298**, 202–2205.

Knight, C.A. and Ackerly, D.D. (2003a). Evolution and plasticity of photosynthetic thermal tolerance, specific leaf area and leaf size: congeneric species from desert and coastal environments.. *New Phytologist*, **160**, 337–47.

Knight, C.A. and Ackerly, D.D. (2003b). Small heat shock protein responses of a closely related pair of desert and coastal *Encelia*. *International Journal of Plant Sciences*, **164**, 53–60.

Knight, C.A. and Ackerly, D.D. (2002). An ecological and evolutionary analysis of photosynthetic thermotolerance using the temperature-dependent increase in fluorescence. *Oecologia*, **130**, 505–514.

Knoop, W. and Walker, B. (1985). Interactions of woody and herbaceous vegetation in a southern African savanna. *Journal of Ecology*, **73**, 235–253.

Knorre, A.A., Kirdyanov, A.V. and Vaganov, E.A. (2006). Climatically induced interannual variability in aboveground production in forest-tundra and northern taiga of central Siberia. *Oecologia*, **147**, 86–95.

Kobayashi, M., Sasaki, K., Enomoto, M., *et al.* (2007). Highly sensitive determination of transient generation of biophotons during hypersensitive response to cucumber mosaic virus in cowpea. *Journal of Experimental Botany*, **58**, 465–472.

Koç, M., Barutçular, C. and Genç, I. (2003). Photosynthesis and productivity of old and modern durum wheats in a Mediterranean environment. *Crop Science*, **43**, 2089–2098.

Koch, C., Noga, G., and Strittmatter, G. (1994). Photosynthetic electron transport is differentially affected during early stages of cultivar/race specific interactions between potato and *Phytophthora infestans*. Planta, **193**, 551–557.

Koch, G.W., Sillett, S.C., Jennings, G.M., *et al.* (2004). The limits to tree height. *Nature*, **428**, 851–854.

Koch, W., Lange, O.L. and Schulze, E.D. (1971). Ecophysiological investigations on wild and cultivated plants in the Negev desert. *Oecologia*, **8**, 296–309.

Kocsy, G., Galiba, G. and Brunold, C. (2001). Role of glutathione in adaptation and signalling during chilling and cold acclimation in plants. *Physiologia Plantarum*, **113**, 158–164.

Koenning, S.R. and Barker, K.R. (1995). Soybean photosynthesis and yield as influenced by *Heterodera glycines*, soil type and irrigation. *Journal of Nematology*, **27**, 51–62.

Koeppe, D.E. and Miller, R.J. (1970). Lead effects on corn mitochondrial respiration. *Science*, **167**, 1376–1378.

Kogami, H., Hanba, Y.T., Kibe, T., *et al.* (2001). CO_2 transfer conductance, leaf structure and carbon isotope composition of *Polygonum cuspidatum* leaves from low and high altitudes. *Plant, Cell and Environment*, **24**, 529–538.

Kohen, E., Santus, R. and Hirschberg, J.G. (1995). *Photobiology*. Academic Press, London.

Kohzuma, K., Cruz, J.A., Akashi, K., *et al.* (2009). The long-term responses of the photosynthetic proton circuit to drought. *Plant, Cell and Environment*, **32**, 209–219.

Koike, T. (1987). Photosynthesis and expansion in leaves of early, mid, and late successional tree species, birch, ash, and maple. *Photosynthetica*, **21**, 503–508.

Koike, T. (1990). Autumn colouring, photosynthetic performance and leaf development of deciduous broad-leaved trees in relation to forest succession. *Tree Physiology*, **7**, 21–32.

Koike, T., Kitaoka, S., Masyagina, O.V., *et al.* (2007). Nitrogen dynamics of leaves of deciduous broad-leaved tree seedlings grown in summer green forests in northern Japan. *Eurasian Journal of Forestry Research*, **10**, 115–119.

Koiwa, H., Kojima, M., Ikeda, T., et al. (1992). Fluctuations of particles on chloroplast thylakoid membranes in tomato plants infected with virulent or attenuated strain of tobacco mosaic virus. *Annals of the Phytopathological Society of Japan*, **58**, 58–64.

Koizumi, H. and Oshima, Y. (1985). Seasonal changes in photosynthesis of four understory herbs in deciduous forests. *The Botanical Magazine, Tokyo*, **98**, 1–13.

Koizumi, M., Takahashi, K., Mineuchi, K., *et al.* (1998). Light gradients and the transverse distribution of chlorophyll fluorescence in mangrove and *Camellia* leaves. *Annals of Botany London*, **81**, 527–533.

Kok, B. (1948). A critical consideration of the quantum yield of *Chlorella* photosynthesis. *Enzymologia*, **13**, 1–56.

Kokkinos, C.N., Clark, C.A.,. McGregorl, C.E., *et al.* (2006). The effect of sweet potato virus disease and its viral components on gene expression levels in sweetpotato. *Journal of American Society for Horticultural Science*, **13**, 657–666.

Kolari, P., Pumpanen, J., Rannik, U., *et al.* (2004). Carbon balance of different aged Scots pine forests in Southern Finland. *Global Change Biology*, **10**, 1106–1119.

Kolb, C.A., Kopecký, J., Riederer, R., *et al.* (2003). UV screening by phenolics in berries of grapevine (*Vitis vinifera*). *Functional Plant Biology*, **30**, 1177–1186.

Kolbe, A., Tiessen, A., Schluepmann, H., *et al.* (2005). Trehalose-6-phosphate regulates starch synthesis via post-translational redox activation of ADP-glucose pyrophosphorylase. *Proceedings of the National Academy of Sciences USA*, **102**, 11118–11123.

Kolber, Z., Klimov, D., Ananyev, G., *et al.* (2005), Measuring photosynthetic parameters at a distance: laser induced fluorescence transient (LIFT) method for remote measurements of photosynthesis in terrestrial vegetation. *Photosynthesis Research*, **84**, 121–129.

Kolber, Z. S., Prasil, O. and Falkowski, P.G. (1998). Measurements of variable chlorophyll fluorescence using fast repetition rate techniques. I. Defining methodology and experimental protocols. *Biochimica et Biophysica Acta*, **1367**, 88–106.

Kollist, T., Moldau, H., Rasulov, B., *et al.* (2007). A novel device detects a rapid ozone-induced transient stomatal closure in intact *Arabidopsis* and its absence in abi2 mutant. *Physiologia Plantarum*, **129**, 796–803.

Koltai, H. and Mckenzie Bird, D. (2000). High throughput cellular localization of specific plant mRNAs by liquid-phase in situ reverse transcription-polymerase chain reaction of tissue sections. *Plant Physiology*, **123**, 1203–1212.

Koop, H.U., Herz, S., Golds, T.J., *et al.* (2008). *Top Current Genetics*, **19**, 457–510.

Köppen, W. (1936). Das geographische System der Klimate. In: *Handbuch der Klimatologie* (eds Köppen, W. and Geiger, G.), Gebrüder Borntraeger, Berlin, Germany, pp. 1–46.

Koricheva, J., Larsson, S., Haukioja, E., *et al.* (1998). Regulation of woody plant secondary metabolism by resource availability: hypothesis testing by means of metaanalysis. *Oikos*, **83**, 212–226.

Körner, C, Asshoff, R., Bignucolo, O., *et al.* (2005). Carbon flux and growth in mature deciduous forest trees exposed to elevated CO_2. *Science*, **309**, 1360–1362.

Körner, C. (1991). Some often overlooked plant characteristics as determinants of plant growth: a reconsideration. *Functional Ecology*, **5**, 162–173.

Körner, C. (1998). A re-assessment of high elevation treeline positions and their explanation. *Oecologia*, **115**, 445–459.

Körner, C. and Pelaez Menendez-Riedl, S. (1990). The significance of developmental aspects in plant growth analysis. In: *Causes and Consequences of Variation in Growth Rate and Productivity of Higher Plants* (eds Köppen, W. and Geiger, G.). SPB Academic Publishing, The Hague, Netherlands, pp. 141–157.

Körner, C., Scheel, J.A. and Bauer, H. (1979). Maximum leaf diffusive conductance in vascular plants. *Photosynthetica*, **13**, 45–82.

Körner, C. (1999). Alpine plant life. Functional plant ecology of high mountain ecosystems. Springer Verlag, Berlin, Heidelberg.

Körner, C. (2007). The use of 'altitude' in ecological research. *Trends in Ecology and Evolution*, **22**, 569–574.

Körner, C. and Diemer, M. (1987). In situ photosynthetic responses to light, temperature and carbon dioxide in herbaceous plants from low and high altitude. *Functional Ecology*, **1**, 179–194.

Körner, C., Allison, A. and Hilscher, H. (1983). Altitudinal variations in leaf diffusive conductance and leaf anatomy in heliophytes of montane New Guinea and their interrelation with microclimate. *Flora*, **174**, 91–135.

Körner, C., Farquhar, G.D. and Wong, S.C. (1991). Carbon isotope discrimination by plants follows latitudinal and altitudinal trends. *Oecologia*, **88**, 30–40.

Kornyeyev, D., Logan, B.A., Payton, P.R., *et al.* (2003). Elevated chloroplastic glutathione reductase activities decrease chilling-induced photoinhibition by increasing rates of photochemistry, but not thermal energy dissipation, in transgenic cotton. *Functional Plant Biology*, **30**, 101–110.

Kortschak, H.P., Hartt, C.E. and Burr, G.O. (1965). Carbon dioxide fixation in sugarcane leaves. *Plant Physiology*, **40**, 209–213.

Köstner, B., Falge, E. and Tenhunen, J.D. (2002). Age-related effects on leaf area/sapwood area relationships, canopy transpiration and carbon gain of Norway spruce stands (*Picea abies*) in the Fichtelgebirge, Germany. *Tree Physiology*, **22**, 567–574.

Köstner, B., Granier, A. and Chermák, J. (1998). Sapflow measurements in forest stands: methods and uncertainties. *Annals of Forest Science*, **55**, 13–27.

Kosugi, Y., Takanashi, S., Ohkubo, S., *et al.* (2008). CO_2 exchange of a tropical rainforest at Pasoh in Peninsular Malaysia. *Agricultural and Forest Meteorology*, **148**, 439–452.

Kottapalli, K.R., Rakwal, R., Shibato, J., *et al.* (2009). Physiology and proteomics of the water-deficit stress response in three contrasting peanut genotypes. *Plant, Cell and Environment*, **32**, 380–407.

Koussevitzky, S., Nott, A., Mockler, T.C., *et al.* (2007). Signals from chloroplasts converge to regulate nuclear gene expression. *Science*, **316**, 715–719.

Kowalski, A., Loustau, D., Berbigier, P., *et al.* (2004). Paired comparison of carbon exchange between undisturbed and regenerating stands in four managed forests in Europe. *Global Change Biology*, **10**, 1707–1723.

Kozaki, A. and Takeba, G. (1996). Photorespiration protects C_3 plants from photooxidation. *Nature*, **384**, 557–560.

Kozela, C. and Regan, S. (2003). How plants make tubes. *Annals of Forest Science*, **8**, 159–164.

Kozlowski, T.T. (1984). Plant responses to flooding of soil. *BioScience*, **34**, 162–167.

Krall, J.P. and Edwards, G.E. (1992). Relationship between photosystem II activity and CO_2 fixation in leaves. *Physiologia Plantarum*, **86**, 180–187.

Kramer, D.M., Avenson, T.J. and Edwards, G.E. (2004b). Dynamic flexibility in the light reactions of photosynthesis governed by both electron and proton transfer reactions. *Trends in Plant Sciences*, **9**, 349–357.

Kramer, D.M., Cruz, J.A. and Kanazawa, A. (2003). Balancing the central roles of the thylakoid proton gradient. *Trends in Plant Sciences*, **8**, 27–32.

Kramer, D.M., Johnson, G., Kiirats, O., *et al.* (2004a). New fluorescence parameters for the determination of Qa redox state and excitation energy fluxes. *Photosynthesis Research*, **79**, 209–218.

Kramer, K., Friend, A. and Leinonen, I. (1996). Modelling comparison to evaluate the importance of phenology and spring frost damage for the effects of climate change on growth of mixed temperate-zone deciduous forests. *Climate Research*, **7**, 31–41.

Krapp, A. and Stitt, M. (1995). An evaluation of direct and indirect mechanisms for the sink-regulation of photosynthesis in spinach – changes in gas-exchange, carbohydrates, metabolites, enzyme-activities and steady state transcript levels after cold-girdling source leaves. *Planta*, **195**, 313–323.

Krause, G.H. (1988). Photoinhibition of photosynthesis. An evaluation of damaging and protective mechanisms. *Physiologia Plantarum*, **74**, 566–574.

Krause, G.H. and Weis, E. (1991). Chlorophyll fluorescence and photosynthesis: the basics. *Annual Review of Plant Physiology and Plant Molecular Biology*, **42**, 313–349.

Krause, G.H. and Winter, K. (1996). Photoinhibition of photosynthesis in plants growing in natural tropical forest gaps. A chlorophyll fluorescence study. *Botanica Acta*, **109**, 456–462.

Krause, G.H., Galle, A., Gademann, R., *et al.* (2003b). Capacity of protection against ultraviolet radiation in sun and shade leaves of tropical forest plants. *Functional Plant Biology*, **30**, 533–542.

Krause, G.H., Grube, E., Koroleva, O.Y., *et al.* (2004). Do mature shade leaves of tropical tree seedlings acclimate to high sunlight and UV radiation? *Functional Plant Biology*, **31**, 743–756.

Krause, G.H., Grube, E., Virgo, A., *et al.* (2003a). Sudden exposure to solar UV-B radiation reduces net CO_2 uptake and photosystem I efficiency in shade-acclimated tropical tree seedlings. *Plant Physiology*, **131**, 745–752.

Krause, G.H., Jahns, P., Virgo, A., *et al.* (2007). Photoprotection, photosynthesis and growth of tropical tree seedlings under near-ambient and strongly reduced solar ultraviolet-B radiation. *Journal of Plant Physiology*, **164**, 1311–1322.

Krause, G.H., Koroleva, O.Y., Dalling, J.W., *et al.* (2001). Acclimation of tropical tree seedlings to excessive light in simulated tree-fall gaps. *Plant, Cell and Environment*, **24**, 1345–1352.

Krause, G.H., Schmude, C., Garden, H., *et al.* (1999). Effects of solar ultraviolet radiation on the potential efficiency of photosystem II in leaves of tropical plants. *Plant Physiology*, **121**, 1349–1358.

Krause, G.H., Virgo, A. and Winter, K. (1995). High susceptibility to photoinhibition of young leaves of tropical forest trees. *Planta*, **197**, 583–591.

Krauss, K.W. and Allen, J.A. (2003). Influences of salinity and shade on seedling photosynthesis and growth of two mangrove species, *Rhizophora mangle* and *Bruguiera sexangula*, introduced to Hawaii. *Aquatic Botany*, **77**, 311–324.

Krauss, K.W., Lovelock, C.E., McKee, K.L., *et al.* (2008). Environmental drivers in mangrove establishment and early development: a review. *Aquatic Botany*, **89**, 105–127.

Krestov, P.V. (2003). Forest Vegetation of Easternmost Russia (Russian Far East). In: *Forest Vegetation of Northeast Asia* (eds Kolbek, J., Srutek, M. and Box, E.). Kluwer Academic Publishers, Dordrecht, Netherlands, pp. 93–180.

Kriedemann, P.E. and Anderson, J.E. (1988). Growth and photosynthetic responses to manganese and copper deficiencies in wheat (*Triticum aestivum*) and barley grass (*Hordeum glaucum* and *H. leporinum*). *Australian Journal of Plant Physiology*, **15**, 429–446.

Kriedemann, P.E. and Downton, W.J.S. (1981). Photosynthesis. In: *The Physiology and Biochemistry of Drought Resistance in Plants* (eds Paleg, L.G. and Aspinall, D.), Academic Press, Sydney, Australia, pp. 283–314.

Kriedemann, P.E., Graham, R.D. and Wiskich, J.T. (1985). Photosynthetic dysfunction and in vivo changes in chlorophyll a fluorescence from manganese-deficient wheat leaves. *Australian Journal of Agricultural Research*, **36**, 157–169.

Krieger, A., Bolte, S., Dietz, K.J., *et al.* (1998). Thermoluminescence studies on the facultative CAM plant *Mesembryanthenum crystallinum* L. *Planta*, **205**, 587–594.

Krishnan, P., Kruger, N.J. and Ratcliffe, R.G. (2005). Metabolite fingerprinting and profiling using NMR. *Journal of Experimental Botany*, **56**, 255–265.

Krömer, S. (1995). Respiration during photosynthesis. *Annual Review of Plant Physiology and Plant Molecular Biology*, **46**, 45–70.

Krömer, S. and Heldt, H.W. (1991). Respiration of pea leaf mitochondria and redox transfer between the mitochondrial and extramitochondrial compartment. *Biochim Biophys Acta*, **1057**, 42–50.

Krömer, S., Malmberg, G., and Gardeström, P. (1993). Mitochondrial contribution to photosynthetic metabolism. *Plant Physiology*, **102**, 947–955.

Kroopnick, P. and Craig, H. (1972). Atmospheric oxygen: isotopic composition and solubility fractionation. *Science*, **175**, 54–55.

Kruckeberg, A.L., Neuhaus, H.E., Feil, R., *et al.* (1989). Decreased-activity mutants of phosphoglucose isomerase in the cytosol and chloroplast of *Clarkia xantiana*. Impact on mass-action ratios and fluxes to sucrose and starch, and estimation of flux control coefficients and elasticity coefficients. *Biochemical Journal*, **261**, 457–467.

Krupa, S., Nosal, M. and Legge, A. (1998). A numerical analysis of the combined open-top chamber data from the USA and Europe on ambient ozone and negative crop responses. *Environmental Pollution*, **101**, 157–160.

Krupa, S.V. (2003). Effects of atmospheric ammonia (NH_3) on terrestrial vegetation: a review. *Environmental Pollution*, **124**, 179–221.

Krupa, Z., Öquist, G. and Huner, N.P.A. (1993). The effect of cadmium on photosynthesis of *Phaseolus vulgaris* – a fluorescence analysis. *Physiologia Plantarum*, **88**, 626–630.

Kruse, A., Fieuw, S., Heinke, D., *et al.* (1998). Antisense inhibition of cytosolic NADP-isocitrate dehydrogenase in transgenic potato plants. *Planta*, **205**, 82–91.

Kubien, D.S. and Sage, R.F. (2004). Low-temperature photosynthetic performance of a C-4 grass and a co-occurring C-3 grass native to high latitudes. *Plant Cell Environment*, **27**, 907–916.

Kubiske, M.E., Quinn, V.S., Marquardt, P.E., *et al.* (2006). Effects of elevated atmospheric CO_2 and/or O_3 on intra- and interspecific competitive ability of aspen. *Plant Biology*, **9**, 342–355.

Kucharik, C.J., Norman, J.M. and Gower, S.T. (1999). Characterization of radiation regimes in nonrandom forest canopies: theory, measurements, and a simplified modeling approach. *Tree Physiology*, **19**, 695–706.

Külheim, C., Ågren, J. and Jansson, S. (2002). Rapid regulation of light harvesting and plant fitness in the field. *Science*, **297**, 91–93.

Kull, O. and Jarvis, P.G. (1995). The role of nitrogen in a simple scheme to scale up photosynthesis from leaf to canopy. *Plant Cell Environment*, **18**, 1174–1182.

Kull, O. and Niinemets, U. (1998). Distribution of leaf photosynthetic properties in tree canopies: comparison of species with different shade tolerance. *Functional Ecology*, **12**, 472–479.

Kumagai, T., Tateishi, M., Shimizu, T., *et al.* (2008). Transpiration and canopy conductance at two slope positions in a Japanese cedar forest watershed. *Agricultural and Forest Meteorology*, **148**, 1444–1455.

Kumar, P., Kumar Tewari, R. and Sharma, P.N. (2008). Modulation of copper toxicity induced oxidative damage by excess supply of iron in maize plants. *Plant Cell Reports*, **27**, 399–409.

Kume, A., Tsuboi, N., Satomura, T., *et al.* (2000). Physiological characteristics of Japanese red pine, *Pinus densiflora* Sieb. et Zucc., in declined forests at Mt. Gokurakuji in Hiroshima Prefecture, Japan. *Trees*, **141**, 305–311.

Kump, L.R. (2008). The rise of atmospheric oxygen. *Nature*, **451**, 277–278.

Kumudini, S., Hume, D.J. and Chu, G. (2001). Genetic improvement in short season soybeans: 1. Dry matter accumulation, partitioning, and leaf area duration. *Crop Science*, **41**, 391–398.

Kumutha, D., Ezhilmathi, K., Sairam, R.K., *et al.* (2009). Waterlogging induced oxidative stress and antioxidant activity in pigeonpea genotypes. *Biologia Plantarum*, **53**, 75–84.

Kunkel, K.E., Angel, J.R., Changnon, S.A., *et al.* (2006). *The 2005 Illinois Drought*, Illinois State Water Survey, Champaign, IL.

Küpper, H., Setlík, I., Trtílek, M., *et al.* (2000). A microscope for two-dimensional measurements of in vivo chlorophyll fluorescence kinetics using pulsed measuring radiation, continuous actinic radiation, and saturating flashes. *Photosynthetica*, **38**, 553–570.

Kurepa, J., Bueno, P., Kampfenkel, K., *et al.* (1997). Effects of iron deficiency on iron superoxide dismutase expression in *Nicotiana tabacum*. *Plant Physiology and Biochemistry*, **35**, 467–474.

Kurimoto, K., Millar, A.H., Lambers, H., *et al.* (2004). Maintenance of growth rate at low temperature in rice and wheat cultivars with a high degree of respiratory homeostasis is associated with a high efficiency of respiratory ATP production. *Plant and Cell Physiology*, **45**, 1015–1022.

Kursar, T.A. and Coley, P.D. (1992a). The consequences of delayed greening during leaf development for light absorption and light use efficiency. *Plant Cell Environment*, **15**, 901–909.

Kursar, T.A. and Coley, P.D. (1992b). Delayed development of the photosynthetic apparatus in tropical rain forest species. *Functional Ecology*, **6**, 411–422.

Kursar, T.A. and Coley, P.D. (1992c). Delayed greening in tropical leaves: an antiherbivore defense? *Biotropica*, **24**, 256–262.

Kyparissis, A. and Manetas, Y. (1993). Seasonal leaf dimorphism in a semi-deciduous Mediterranean shrub: ecophysiological comparisons between winter and summer leaves. *Acta Oecologica*, **14**, 23–32.

Kyparissis, A., Drilias, P. and Manetas, Y. (2000). Seasonal fluctuations in photoprotective (xanthophyll cycle) and photoselective (chlorophylls) capacity in eight Mediterranean plant species belonging to two different growth forms. *Australian Journal of Plant Physiology*, **27**, 265–272.

Kyparissis, A., Petropoulou, Y. and Manetas, Y. (1995). Summer survival of leaves in a soft-leaved shrub *(Phlomis fruticosa* L., Labiatae) under Mediterranean field conditions: avoidance of photoinhibitory damage through decreased chlorophyll contents. *Journal of Experimental Botany*, **46**, 1825–1831.

Labrecque, M., Bellefleur, P., Simon, J.P., *et al.* (1989). Influence of light conditions on the predetermination of foliar characteristics in *Betula alleghaniensis* Britton. *Annals of Forest Science*, **46**, 497–501.

Laidler, K.J. and King, M.C. (1983). Development of transition-state theory. *The Journal of Physical Chemistry*, **87**, 2657–2664.

Laisk, A. (1983). Calculation of photosynthetic parameters considering the statistical distribution of stomatal apertures. *Journal of Experimental Botany*, **34**, 1627–1635.

Laisk, A. and Loreto, F. (1996). Determining photosynthetic parameters from leaf CO_2 exchange and chlorophyll fluorescence. Ribulose-1,5-bisphosphate carboxylase/oxygenase specificity factor, dark respiration in the light, excitation distribution between photosystems, alternative electron transport rate, and mesophyll diffusion resistance. *Plant Physiology*, **110**, 903–912.

Laisk, A. and Oja, V. (1998). *Dynamics of Leaf Photosynthesis: Rapid-Response Measurements and their Interpretations.* CSIRO Publishing, Canberra, Australia.

Laisk, A., Kull, O. and Moldau, H. (1989). Ozone concentration in leaf intercellular air spaces is close to zero. *Plant Physiology*, **90**, 1163–1167.

Laisk, A., Oja, V. and Kull, K. (1980). Statistical distribution of stomatal apertures of *Vicia faba* and *Hordeum vulgare* and *Spannungsphase* of stomatal opening. *Journal of Experimental Botany*, **31**, 49–58.

Laisk, A., Oja, V., Rasulov, B., *et al.* (2002). A computer-operated routine of gas exchange and optical measurements to diagnose photosynthetic apparatus in leaves. *Plant, Cell and Environment*, **25**, 923–943.

Laisk, A.K. (1977). Kinetics of photosynthesis and photorespiration in C_3-plants. [In Russian] *Nauka, Moscow.*

Lake, J.A. and Wade, R.N. (2009) Plant-pathogen interactions and elevated CO_2: morphological changes in favour of pathogens. *Journal of Experimental Botany*, **60**, 3123–3131.

Lal, R. (2004a) Agricultural activities and the global carbon cycle. *Nutrient Cycling in Agroecosystems*, **70**, 103–116.

Lal, R. (2004b). Soil carbon sequestration impacts on global climate change and food security. *Science*, **304**, 1623–1627.

Lal, R. (2004c). Soil carbon sequestration to mitigate climate change. *Geoderma*, **123**, 1–22.

Lal, R. and Edwards, G.E. (1996). Analysis of inhibition of photosynthesis under water stress in the C_4 species *Amaranthus cruentus* and *Zea mays*: electron transport, CO_2 fixation and carboxylation capacity. *Functional Plant Biology*, **23**, 403–412.

Laloi, C., Mestres-Ortega, D., Marco, Y., *et al.* (2004). The *Arabidopsis* cytosolic thioredoxin h5 gene induction by oxidative stress and its W-box-mediated response to pathogen elicitor. *Plant Physiology*, **134**, 1006–1016.

Lambers, H. and Poorter, H. (1992). Inherent variation in growth rate between higher plants: a search for physiological causes and ecological consequences. In: *Advances in Ecological Research* (eds Begon, M. and Fitter, A.H.), Academic Press, London, UK, pp. 187–261.

Lambers, H., Chapin III, F.S. and Pons, T.L. (2008). *Plant Physiological Ecology*, 2nd ed. Springer Science + Business Media, LLC, New York, USA, pp. 604.

Lambert, G., Monfray, P., Ardouin, B., *et al.* (1995). Year-to-year changes in atmo-spheric CO_2. *Tellus*, **47B**, 53–55.

Lamont, B.B., Groom, P.K. and Cowling, R.M. (2002). High leaf mass per area of related species assemblages may reflect low rainfall and carbon isotope discrimination rather than low phosphorus and nitrogen concentrations. *Functional Ecology*, **16**, 403–412.

Landlot, W., Bühlmann, U., Bleuler, P., *et al.* (2000). Ozone exposure-response relationships for biomass and root/shoot ratio of beech (*Fagus sylvatica*), ash (*Fraxinus excelsior*), Norway spruce (*Picea abies*) and Scots pine (*Pinus sylvestris*). *Environmental Pollution*, **109**, 473–478.

Landsberg, J.J., Prince, S.D., Jarvis, P.G., *et al.* (1996). Energy conversion and use in forests: an analysis of forest production in terms of radiation utilisation efficiency (epsilon). In: *The Use of Remote Sensing in the Modeling of Forest Productivity* (eds Gholz, H.L., Nakane, K. and Shimoda, H.), Kluwer Academic Publishers, Dordrecht, Netherlands, pp. 273–298.

Lane, G.A. and Dole, M. (1956). Fractionation of oxygen isotopes during respiration. *Science*, **123**, 574–576.

Lane, M.D., Maruyama, H. and Easterday, R.L. (1969). Phosphoenolpyruvate carboxylase from peanut cotyledons. *Methods in Enzymology*, **13**, 277–283.

Lange, O.L. (1969). CO_2-Gaswechsel von Moosen nach Wasserdamnfaufnahme aus dem Luftraum. *Planta*, **89**, 90–94.

Lange, O.L., Reichenberger, H. and Walz, H. (1997). Continuous monitoring of CO_2 exchange of lichens in the field: short-term enclosure with an automatically operating cuvette. *Lichenologist*, **29**, 259–274.

Lange, O.L., Schulze, E.D., Evenari, M., *et al.* (1974). The temperature-related photosynthetic capacity of plants under desert conditions I. Seasonal changes of the photosynthetic response to temperature. *Oecologia*, **17**, 97–110.

Lange, O.L., Schulze, E.D., Kappen, L., *et al.* (1975). CO_2 exchange pattern under natural conditions of *Caralluma negevensis*, a CAM plant of the Negev desert. *Photosynthetica*, **9**, 318–326.

Langenheim, J.H., Osmond, C.B., Brooks, A., *et al.* (1984). Photosynthetic responses to light in seedlings of selected Amazonian and Australian rainforest tree species. *Oecologia*, **63**, 215–224.

Lanigan, G., Betson, N., Griffiths, H., *et al.* (2008). Carbon isotope fractionation during photorespiration and carboxylation in *Senecio*. *Plant Physiology*, **148**, 2013–2020.

Larbi, A., Abadía, A., Abadía, J., *et al.* (2006). Down co-regulation of light absorption, photochemistry, and carboxylation in Fe-deficient plants growing in different environments. *Photosynthesis Research*, **89**, 113–126.

Larbi, A., Morales, F., Abadía, A., *et al.* (2002). Effects of Cd and Pb in sugar beet plants grown in nutrient solution: induced Fe deficiency and growth inhibition. *Functional Plant Biology*, **29**, 1453–1464.

Larcher, W., Wagner, J. and Lütz, C. (1997). The effect of heat on photosynthesis, dark respiration and cellular ultrastructure of the arctic-alpine psychrophyte *Ranunculus glacialis*. *Photosynthetica*, **34**, 219–232.

Larom, S., Salama, F., Schuster, G. *et al.* (2010). Engineering of an alternative electron transfer path in photosystem II. *Proceedings of the National Academy of Sciences USA*, **107**, 9650–9655.

Lasslop, G., Reichstein, M., Kattge, J., *et al.* (2008). Influences of observation errors in eddy flux data on inverse model parameter estimation. *Biogeosciences*, **5**, 1311–1324.

Lasslop, G., Reichstein, M., Papale, D., *et al.* (2010). Separation of net ecosystem exchange into assimilation and respiration using a light response curve approach: critical issues and global evaluation. *Biogeosciences*, **16**, 187–208.

Lassoie, J.P., Dougherty, P.M., Reich, P.B., *et al.* (1983). Ecophysiological investigations of understory eastern redcedar in Central Missouri. *Ecology*, **64**, 1355–1366.

Latham, R.E. (1992). Co-occurring tree species change rank in seedling performance with resources varied experimentally. *Ecology*, **73**, 2129–2144.

Latorre, C., Quade, J. and McIntosh, W.C. (1997). The expansion of C_4 grasses and global change in the late Miocene: stable isotope evidence from the Americas. *Earth Planetary Science Letters*, **146**, 83–96.

Latowski, D., Kruk, J. and Strzalka, K. (2005). Inhibition of zeaxanthin epoxidase activity by cadmium ions in higher plants. *Journal of Inorganic Biochemistry*, **99**, 2081–2087.

Lauer, M.J. and Boyer, J.S. (1992). Internal CO_2 measured directly in leaves. Abscisic acid and low leaf water potential cause opposing effects. *Plant Physiology*, **98**, 1310–1316.

Lauer, M.J., Pallardy, S.G., Blevins, D.G., *et al.* (1989). Whole leaf carbon exchange characteristics of phosphate deficient soybeans (*Glycine max* L.). *Plant Physiology*, **91**, 848–854.

Lauteri, M., Scartazza, A., Guido, C., *et al.* (1997). Genetic variation in photosynthetic capacity, carbon isotope discrimination and mesophyll conductance in provenances of *Castanea sativa* adapted to different environments. *Functional Ecology*, **11**, 675–683.

Lavergne, J. and Junge, W. (1992). Proton release during the redox cycle of the water oxidase. *Photosynthesis Research*, **38**, 279–296.

Law, B.E., Falge, E., Gu, L., *et al.* (2002). Environmental controls over carbon dioxide and water vapor exchange of terrestrial vegetation. *Agricultural and Forest Meteorology*, **113**, 97–120.

Lawlor, D. (2001). *Photosynthesis*. BIOS Scientific Publishers, Oxford, UK, pp. 386.

Lawlor, D.W. (1995). Photosynthesis, productivity and environment. *Journal of Experimental Botany*, **46**, 1449–1461.

Lawlor, D.W. (2002). Carbon and nitrogen assimilation in relation to yield: mechanisms are the key to understanding production systems. *Journal of Experimental Botany*, **53**, 773–787.

Lawlor, D.W. and Cornic, G. (2002). Photosynthetic carbon assimilation and associated metabolism in relation to water deficits in higher plants. *Plant, Cell and Environment*, **25**, 275–294.

Lawlor, D.W. and Tezara, W. (2009). Causes of decreased photosynthetic rate and metabolic capacity in water-deficient leaf cells: a critical evaluation of mechanisms and integration of processes. *Annals of Botany*, **103**, 561–579.

Lawson, T., Lefebvre, S., Baker, N.R., *et al.* (2008). Reductions in mesophyll and guard cell photosynthesis impact on the control of stomatal responses to light and CO_2. *Journal of Experimental Botany*, **59**, 3609–3619.

Lawson, T., Oxborough, K., Morison, J.I.L., *et al.* (2002). Responses of photosynthetic electron transport in stomatal guard cells and mesophyll cells in intact leaves to light, CO_2 and humidity. *Plant Physiology*, **128**, 52–62.

Lawson, T., Oxborough, K., Morison, J.I.L., *et al.* (2003). The responses of guard and mesophyll cell photosynthesis to CO_2, O_2, light and water stress in a range of species are similar. *Journal of Experimental Botany*, **54**, 1743–1752.

Lawyer, A.L., Cornwell, K.L., Larsen, P.O., *et al.* (1981). Effects of carbon dioxide and oxygen on the regulation of photosynthetic carbon metabolism by ammonia in spinach mesophyll cells. *Plant Physiology*, **68**, 1231–1236.

Le Roux, X., Grand, S., Dreyer, E., *et al.* (1999). Parameterization and testing of a biochemically based photosynthesis model for walnut (*Juglans regia*) trees and seedlings. *Tree Physiology*, **19**, 481–492.

Leadley, P. and Drake, B. (1993). Open top chambers for exposing plant canopies to elevated CO_2 concentration and for measuring net gas exchange. *CO2 and Biosphere*, **104**, 3–15.

Leakey, A.D.B., Ainsworth, E.A., Bernacchi, C.J., *et al.* (2009b). Elevated CO_2 effects on plant carbon, nitrogen, and water relations: six important lessons from FACE. *Journal of Experimental Botany*, **60**, 2859–2876.

Leakey, A.D.B., Bernacchi, C.J., Dohleman, F.G., *et al.* (2004). Will photosynthesis of maize (*Zea mays*) in the U.S. Corn Belt increase in future [CO_2] rich atmospheres? An analysis of diurnal courses of CO_2 uptake under Free-Air Concentration Enrichment (FACE). *Global Change Biology*, **10**, 951–962.

Leakey, A.D.B., Uribelarrea, M., Ainsworth, E.A., *et al.* (2006). Photosynthesis, productivity, and yield of maize are not affected by open-air elevation of CO_2 concentration in the absence of drought. *Plant Physiology*, **140**, 779–790.

Leakey, A.D.B., Xu, F., Gillespie, K.M., *et al.*. (2009a). The genomic basis for stimulated respiratory carbon loss to the atmosphere by plants growing under elevated [CO_2]. *Proceedings of the National Academy of Sciences*, USA, **106**, 3597–3602.

Lebaube, S., Le Goff, N., Ottorini, J.M., *et al.* (2000). Carbon balance and tree growth in a *Fagus sylvatica* stand. *Annals of Forest Science*, **57**, 49–61.

LeCain, D.R., Morgan, J.A., Mosier, A.R., *et al.* (2003). Soil and plant water relations, not photosynthetic pathway, primarily influence photosynthetic responses in a semi-arid ecosystem under elevated CO_2. *Annals of Botany*, **92**, 41–52.

LeCain, D.R., Morgan, J.A., Schuman, G.E., *et al.* (2000). Carbon exchange of grazed and ungrazed pastures of a mixed grass prairie. *Journal of Range Management*, **53**, 199–206.

LeCain, D.R., Morgan, J.A., Schuman, G.E., *et al.* (2002). Carbon exchange and species composition of grazed pastures and exclosures in the shortgrass steppe of Colorado. *Agriculture, Ecosystems and Environment*, **93**, 421–435.

Ledford, H.K. and Niyogi, K.K. (2005). Singlet oxygen and photo-oxidative stress management in plants and algae. *Plant, Cell Environment*, **28**, 1037–1045.

Lee, D.W. (1987). The spectral distribution of radiation in two neotropical rainforests. *Biotropica*, **19**, 161–166.

Lee, D.W. and Collins, T.M. (2001). Phylogenetic and ontogenetic influences on the distribution of anthocyanins and betacyanins in leaves of tropical plants. *International Journal of Plant Sciences*, **162**, 1141–1153.

Lee, D.W., Lowry, J.B. and Stone, B.C. (1979). Abaxial anthocyanin layer in leaves of tropical rainforest plants: enhancer of light capture in deep shade. *Biotropica*, **11**, 70–77.

Lee, D.W., O'Keefe, J., Holbrook, N.M., *et al.* (2003). Pigment dynamics and autumn leaf senescence in a New England deciduous forest, eastern USA. *Ecological Research*, **18**, 677–694.

Lee, H.S.J., Lüttge, U., Medina, E., *et al.* (1989). Ecophysiology of xerophytic and halophytic vegetation of a coastal alluvial plain in northern Venezuela. III. *Bromelia humilis* Jacq. a terrestrial CAM bromeliad. *New Phytologist*, **111**, 253–271.

Lee, H.Y., Hong, Y.N. and Chow, W.S. (2001a). Photoinactivation of photosystem II complexes and photoprotection by non-functional neighbours in *Capsicum annuum* L. leaves. *Planta*, **212**, 332–342.

Lee, J.H. and Schöffl, F. (1996). An Hsp70 antisense gene affects the expression of HSP70/HSC70, the regulation of HSF, and the acquisition of thermotolerance in transgenic *Arabidopsis thaliana*. Molecular and General Genetics, **252**, 11–19.

Lee, K.C., Cunningham, B.A., Paulsen, G.M., *et al.* (1976). Effects of cadmium on respiration rate and activities of several enzymes in soybean seedlings. *Physiologia Plantarum*, **36**, 4–6.

Lee, K.P., Kim, C., Landgraf, F., *et al.* (2007). EXECUTER1- and EXECUTER2- dependent transfer of stress-related signals from the plastid to the nucleus of *Arabidopsis thaliana*. *Proceedings of the National Academy of Sciences USA*, **104**, 10270–10275.

Lee, R.H., Wang, C.H., Huang, L.T., *et al.* (2001b). Leaf senescence in rice plants: cloning and characterization of senescence up-regulated genes. *Journal of Experimental Botany*, **52**, 1117–1121.

Lee, T.D., Tjoelker, M.G., Ellsworth, D.S., *et al.* (2001c). Leaf gas exchange responses of 13 prairie grassland species to elevated CO_2 and increased nitrogen supply. *New Phytologist*, **150**, 405–418.

Leegood, R.C. (1990). Enzymes of the Calvin cycle. In: *Enzymes of Primary Metabolism, Methods in Plant Biochemistry 3* (eds Lea, P.J. and Harborne, J.B.), Academic Press, London. pp. 15–37.

Leegood, R.C. (1993). Carbon metabolism. In: *Photosynthesis and Production in a Changing Environment. A Field Laboratory Manual* (eds Hall, D.O., Scurlock, J.M.O., Bolhar-Nordenkampf, J.R., Leegood, R.C. and Long, S.P.), Chapman and Hall, London, UK.

Leegood, R.C. (2002). C_4 photosynthesis: principles of CO_2 concentration and prospects for its introduction into C_3 plants. *Journal of Experimental Botany*, **53**, 581–590.

Leegood, R.C. (2008). C_4 photosynthesis: minor or major adjustments to a C3 theme? In: *Charting New Pathways to C4 in Rice* (eds Sheehy, J.E., Mitchel, P.L. and Hardy, B.) World Scientific Pub, Hackensack, USA.

Leegood, R.C. and Edwards, G.E. (1996). Carbon metabolism and photorespiration: temperature dependence in relation to other environmental factors. In: *Advances in Photosynthesis Vol.5: Photosynthesis and the Environment* (ed. Baker N.R.) Kluwer Academic Publishers, Dordrecht, Netherlands, pp. 191–221.

Lefi, E., Medrano, H. and Cifre, J. (2004). Water uptake dynamics, photosynthesis and water use efficiency in field-grown *Medicago arborea* and *Medicago citrina* under prolonged Mediterranean drought conditions. *Annals of Applied Biology*, **144**, 299–307.

Legge, A.H. and Krupa, S.V. (2002). Effects of sulphur dioxide. In: *Air Pollution and Plant Life*, 2nd edn (eds Bell, J.N.B. and Treshow, M.), John Wiley and Sons, Chichester, USA, pp. 135–162.

Legge, A.H., Jäger, H.J. and Krupa, S.V. (1998). Sulfur dioxide. In: *Recognition of Air Pollution Injury to Vegetation: A Pictorial Atlas,* 2nd ed (ed. Flagler, R.B.), Air and Waste Management Association, Pittsburgh, USA, pp. 3–42.

Lehto, K., Tikkanen, M., Hiriart, J.B., *et al.* (2003). Depletion of the photosystem II core complex in mature tobacco leaves infected by the Flavum strain of *Tobacco mosaic virus. Molecular Plant-Microbe Interactions*, **16**, 1135–1144.

Leigh, R.A. and Wyn Jones, R.G. (1984). A hypothesis relating critical potassium concentrations for growth to the distribution and functions of this ion in the plant cell. *New Phytologist*, **97**, 1–13.

Leinonen, I. and Jones, H.G. (2004). Combining thermal and visible imagery for estimating canopy temperature and identifying plant stress. *Journal of Experimental Botany*, **55**, 1423–1431.

Leinonen, I., Grant, O.M., Tagliavia, C.P.P., *et al.* (2006). Estimating stomatal conductance with thermal imagery. *Plant, Cell and Environment*, **29**, 1508–1518.

Leipner, J., Basilides, A., Stamp, P., *et al.* (2000). Hardly increased oxidative stress after exposure to low temperature in chilling-acclimated and non-acclimated maize leaves. *Plant Biology*, **2**, 243–251.

Leith, J.H. and Reynolds, J.F. (1987). The nonrectangular hyperbola as a photosynthetic light response model: geometrical interpretation and estimation of the parameter. *Photosynthetica*, **21**, 363–366.

Lemaire, S.D., Michelet, L., Zaffagnini, M., *et al.* (2007). Thioredoxins in chloroplasts. *Current Genetics*, **51**, 343–365.

Lemaître, T., Urbanczyk-Wochniak, E., Flesch, V., *et al.* (2007). NAD-dependent isocitrate dehydrogenase mutants of *Arabidopsis* suggest the enzyme is not limiting for nitrogen assimilation. *Plant Physiology*, **144**, 1546–1558.

Lenk, S. and Buschmann, C. (2006). Distribution of UV-shielding of the epidermis of sun and shade leaves of the beech (*Fagus sylvatica* L.) as monitored by multi-colour fluorescence imaging. *Journal of Plant Physiology*, **163**, 1273–1283.

Lenk, S., Chaerle, L., Pfündel, E.E., *et al.* (2007). Multispectral fluorescence and reflectance imaging at the leaf level and its possible applications. *Journal of Experimental Botany*, **58**, 807–814.

León, A.M., Palma, J.M., Corpas, F J., *et al.* (2002). Antioxidative enzymes in cultivars of pepper plants with different sensitivity to cadmium. *Plant Physiology and Biochemistry*, **40**, 813–820.

Leonardos, E.D., Savitch, L.V., Huner, N.P.A., *et al.* (2003). Daily photosynthetic and C-export patterns in winter wheat leaves during cold stress and acclimation. *Physiologia Plantarum*, **117**, 521–531.

Lerdau, M. (2007). A positive feedback with negative consequences. *Science*, **316**, 212–213.

Lerdau, M. and Slobodkin, K. (2002). Trace gas emissions and species-dependent ecosystem services. *Trends in Ecology and Evolution*, **17**, 309–312.

Leroux, X. and Mordelet, P. (1995). Leaf and canopy CO_2 assimilation in a West-African humid savanna during the early growing season. *Journal of Tropical Ecology*, **11**, 529–545.

Leshem, Y., Melamed-Book, Cagnac, O., Ronen, G., *et al.* (2006). Suppression of *Arabidopsis* vesicle-SNARE expression inhibited fusion of H_2O_2-containing vesicles with tonoplast and increased salt tolerance. *Proceedings of the National Academy of Sciences USA*, **103**, 18008–18013.

Letts, M.G., Phelan, C.A., Johnson, D.R.E., *et al.* (2008). Seasonal photosynthetic gas exchange and leaf reflectance characteristics of male and female cottonwoods in a riparian woodland. *Tree Physiology*, **28**, 1037–1048.

Leuning R. (1990). Modelling stomatal behaviour and photosynthesis of *Eucalyptus grandis. Australian Journal of Plant Physiology*, **17**, 159–175.

Leuning R. (1995). Critical-appraisal of a combined stomatal-photosynthesis model for C_3 plants. *Plant, Cell and Environment*, **18**, 339–355.

Leuning R. (2002). Temperature dependence of two parameters in a photosynthesis model. *Plant, Cell and Environment*, **25**, 1205–1210.

Leuning R. and Judd M.J. (1996). The relative merits of open- and closed-path analysers for measurements of eddy fluxes. *Global Change Biology*, **2**, 241–254.

Leuning, R., Cleugh, H.A., Zegelin, S.J., *et al.* (2005). Carbon and water fluxes over a temperate *Eucalyptus* forest and a tropical wet/dry savanna in Australia: measurements and comparison with MODIS remote sensing estimates. *Agricultural and Forest Meteorology*, **129**, 151–173

Leuning, R., Dunin, F. and Wang, Y. (1998). A two-leaf model for canopy conductance, photosynthesis and partitioning of available energy. II. Comparison with measurements. *Agricultural and Forest Meteorology*, **91**, 113–125.

Leuning, R., Kelliher, F.M., De Pury, D.G.G., *et al.* (1995). Leaf nitrogen photosynthesis, conductance and transpiration-scaling from leaves to canopies. *Plant, Cell and Environment*, **18**, 1183–1200.

Leverenz, J.W. (1988). The effects of illumination sequence, CO_2 concentration, temperature and acclimation on the convexity of the photosynthetic light response curve. *Physiologia Plantarum*, **74**, 332–341.

Leverenz, J.W. (1995). Shade shoot structure of conifers and the photosynthetic response to light at two CO_2 partial pressures. *Functional Ecology*, **9**, 413–421.

Leverenz, J.W. and Hinckley, T.M. (1990). Shoot structure, leaf-area index and productivity of evergreen conifer stands. *Tree Physiology*, **6**, 135–149.

Leverenz, J.W., Bruhn, D. and Saxe, H. (1999). Responses of two provenances of *Fagus sylvatica* seedlings to a combination of four temperature and two CO_2 treatments during their first growing season: gas exchange of leaves and roots. *New Phytologist*, **144**, 437–454.

Levi, A., Ovnat, L., Paterson, A.H., *et al.* (2009). Photosynthesis of cotton near-isogenic lines introgressed with QTLs for productivity and drought related traits. *Plant Science*, **177**, 88–96.

Levin, D.A. (1973). The role of trichomes in plant defense. *The Quarterly Review of Biology*, **48**, 3–15.

Levine, A. (1999). Oxidative stress as a regulator of environmental responses in plants. In: *Plant Responses to Environmental Stresses: From Phytohormones to Genome Reorganization* (ed. Lerner, H.R.), CRC Press, Florida, USA, pp. 248–266.

Levitt, J. (1972). *Responses of Plants to Environmental Stresses*, Academic Press, New York, USA.

Levitt, J. (1980). *Responses of Plants to Environmental Stresses*, Adademic Press, New York, USA.

Lewis, C.E., Noctor, G., Causton, D., *et al.* (2000). Regulation of assimilate partitioning in leaves. *Australian Journal of Plant Physiology*, **27**, 507–519.

Lewis, D.A. and Nobel, P.S. (1977). Thermal energy exchange model and water loss of a barrel cactus, *Ferocactus acanthodes. Plant Physiology*, **60**, 609–616.

Lewis, J.D., Griffin, K.L., Thomas, R.B., *et al.* (1994). Phosphorus supply affects the photosynthetic capacity of loblolly pine grown in elevated carbon-dioxide. *Tree Physiology*, **14**, 1229–1244.

Lewis, S.L., Lopez-Gonzalez, G., Sonké, B., *et al.* (2009). Increasing carbon storage in intact African tropical forests. *Nature*, **457**, 1003–1006.

Leyman, B., Geelen, D., Quintero, F.J., *et al.* (1999). A tobacco syntaxin with a role in hormonal control of guard cell ion channels. *Science*, **283**, 537–540.

Li, C., Liu, S. and Berninger, F. (2004). Picea seedlings show apparent acclimation to drought with increasing altitude in the eastern Himalaya. *Trees*, **18**, 277–283.

Li, C., Potuschak, T., Colón-Carmona, A., *et al.* (2005). *Arabidopsis* TCP20 links regulation of growth and cell division control pathways. *Proceedings of the National Academy of Sciences USA*, **102**, 12978–12983.

Li, S., Goodwin, S. and Pezeshki, S.R. (2007). Photosynthetic gene expression in black willow under various soil moisture regimes. *Biologia Plantarum*, **51**, 593–596.

Li, X., Gilmore, A., Caffarri, S., *et al.* (2004). Regulation of photosynthetic light harvesting involves intrathylakoid lumen pH sensing by the PsbS protein. *Journal of Biological Chemistry*, **279**, 22866–22874.

Li, X.P., Björkman, O., Shih, C., *et al.* (2000). A pigment-binding protein essential for regulation of photosynthetic light harvesting. *Nature*, **403**, 391–395.

Li, Y., Gao, Y., Xu, X., *et al.* (2009). Light-saturated photosynthetic rate in high-nitrogen rice (*Oryza sativa* L.) leaves is related to chloroplastic CO_2 concentration. *Journal of Experimental Botany*, **6**, 2351–2360.

Li, Z., Zhang, S., Hu, H., *et al.* (2008). Photosynthetic performance along a light gradient as related to leaf characteristics of a naturally occurring *Cypripedium flavum*. *Journal of Plant Research*, **121**, 559–569.

Liang, J.S. and Zhang, J.H. (1999) The relations of stomatal closure and reopening to xylem ABA concentration and leaf water potential during soil drying and rewatering. *Plant Growth Regulation*, **29**, 77–86.

Liberloo, M., Tulva, I., Raïm, O., *et al.* (2007). Photosynthetic stimulation under long-term CO_2 enrichment and fertilization is sustained across a closed *Populus* canopy profile (EUROFACE). *New Phytologist*, **173**, 537–549.

Lichtenthaler, H.K., Hak, R. and Rinderle, U. (1990). The chlorophyll fluorescence ratio F690/F730 in leaves of different chlorophyll content. *Photosynthesis Research*, **25**, 295–298.

Lichtenthaler, H.K. (1987). Chlorophylls and carotenoids: pigments of photosynthetic biomembranes. *Methods in Enzymology*, **148**, 350–382.

Lichtenthaler, H.K. (1999). The 1-deoxy D-xylulose-5-phosphate pathway of isoprenoid biosynthesis in plants. *Annual Review of Plant Physiology and Plant Molecular Biology*, **50**, 47–65.

Lichtenthaler, H.K. and Babani, F. (2000). Detection of photosynthetic activity and water stress by imaging the red chlorophyll fluorescence. *Plant Physiology and Biochemistry*, **38**, 889–895.

Lichtenthaler, H.K. and Miehé, J.A. (1997). Fluorescence imaging as a diagnostic tool for plant stress. *Trends in Plant Sciences*, **2**, 316–320.

Lichtenthaler, H.K., Buschmann C., Rinderle U., *et al.* (1986). Application of chlorophyll fluorescence in ecophysiology. *Radiation and Environmental Biophysics*, **25**, 297–308.

Lichtenthaler, H.K., Langsdorf, G., Lenk, S., *et al.* (2005). Chlorophyll fluorescence imaging of photosynthetic activity with the flash-lamp fluorescence imaging system. *Photosynthetica*, **43**, 355–369.

Lichtenthaler, H.K., Rohmer, M. and Schwender, J. (1997). Two independent biochemical pathways for isopentenyl diphosphate and isoprenoid biosynthesis in higher plants. *Physiologia Plantarum*, **101**, 643–652.

Lichtenthaler, H.K., Wenzel, O., Buschmann, C., *et al.* (1998). Plant stress detection by reflectance and fluorescence. *Annals of the New York Academy of Sciences*, **851**, 271–285.

Li-Cor Inc. (2001). Interfacing custom chambers to the LI-6400 sensor head. LI-6400 portable photosynthesis system: application note 3. Li-Cor, Inc, Lincoln, Nebraska, USA.

Li-Cor Inc. (2008). Using the 6400–17 whole plant *Arabidopsis* chamber. LI-6400 portable photosynthesis system: application note 4. Li-Cor, Inc, Lincoln, Nebraska, USA.

Lidon, F.C., Barreiro, M.G., Ramalho, J.C., *et al.* (1999). Effects of aluminum toxicity on nutrient accumulation in maize shoots: implications on photosynthesis. *Journal of Plant Nutrition*, **22**, 397–416.

Liebig, M., Scarano, F.R., Mattos de E.A., *et al.* (2001). Ecophysiological and floristic implications of sex expression in the dioecious neotropical CAM tree *Clusia hilariana* Schltdl. *Trees*, **15**, 278–288.

Lieth, H. (1975). Modelling the primary productivity of the world. In: *Primary Productivity of the Biosphere* (ed. Lieth, H. and Whittaker, R.H.). Springer-Verlag, Berlin-Heidelberg-New York, pp. 237–263.

Lieth, H. and Whittaker, R.H. (1975). Primary productivity of the biosphere. In: *Ecological Studies 14*. Springer-Verlag, Berlin, Germany, pp. 339.

Lim, P.O., Kim, H.J. and Nam, H.G. (2007). Leaf senescence. *Annual Review of Plant Biology*, **58**, 115–136.

Lim, P.O., Woo, H.R., Nam, H.G. (2003). Molecular genetics of leaf senescence in Arabidopsis. *Trends Plant Sci*, **8**, 272–278.

Limousin, J.M., Misson, L., Lavoir, A.V., *et al.* (2010). Do photosynthetic limitations of evergreen *Quercus ilex* leaves change with long-term increased drought severity? *Plant, Cell and Environment*, **33**, 863–875.

Lin, C.T. (2000). Plant blue-light receptors. *Trends in Plant Science*, **5**, 337–342.

Lin, G. and Ehleringer, J.R. (1997). Carbon isotope fractionation does not occur during dark respiration in C_3 and C_4 plants. *Plant Physiology*, **114**, 391–394.

Lin, G. and Sternberg, L.D.L. (1992a). Effect of growth form, salinity, nutrient and sulfide on photosynthesis, carbon isotope discrimination and growth of red mangrove (*Rhizophora mangle* L.). *Australian Journal of Plant Physiology*, **19**, 509–517.

Lin, G. and Sternberg, L.D.L. (1992b). Differences in morphology, carbon isotope ratios, and photosynthesis between scrub and fringe mangroves in Florida, USA. *Aquatic Botany*, **42**, 303–313.

Lin, M., Turpin, D.H. and Plaxton, W.C. (1989). Pyruvate kinase isozymes from the green alga *Selenastrum minutum.* Kinetic and regulatory properties. *Archives of Biochemistry and Biophysics*, **269**, 228–238.

Lindenthal, M., Steiner, U., Dehne, H.W., *et al.* (2005). Effect of downy mildew development on transpiration of cucumber leaves visualized by digital infrared thermography. *Phytopathology*, **95**, 233–240.

Lindner, R.C., Kirpatrick, H.C. and Weeks, T.E. (1959). Some factors affecting the susceptibility of cucumber cotyledons to infection by tobacco mosaic virus. *Phytopatology*, **49**, 78–88.

Lindquist, J.L., Arkebauer, T.J., Walters, D.T., *et al.* (2005). Maize radiation use efficiency under optimal growth conditions. *Agronomy Journal*, **97**, 72–78.

Lindquist, S. and Craig, E.A. (1988). The Heat-Shock Proteins. *Annual Review of Genetics*, **22**, 631–677.

Lindroth, A., Grelle, A. and Moren, A.S. (1998). Long-term measurements of boreal forest carbon balance reveal large temperature sensitivity. *Global Change Biology*, **4**, 443–450.

Littlejohn, R.O. and Ku, M.S.B. (1984). Characterization of early morning crassulacean acid metabolism in *Opuntia erinacea* var. *Columbiana* (Griffiths) L. Benson. *Plant Physiology*, **74**, 1050–1054.

Liu, C.C., Liu, Y.G., Guo, K., *et al.* (2010). Influence of drought on the response of six woody karst species subjected to successive cycles of drought and rewatering. *Physiologia Plantarum*, **139**, 39–54.

Liu, D., Li, T.Q., Yang, X.E., *et al.* (2008). Effect of Pb on leaf antioxidant enzyme activities and ultrastructure of the two ecotypes of *Sedum alfredii* Hance. *Russian Journal of Plant Physiology*, **55**, 68–76.

Liu, F. and Stützel, H. (2004). Biomass partitioning, specific leaf area, and water use efficiency of vegetable amaranth (*Amaranthus* spp.) in response to drought stress. *Scientia Horticulturae*, **102**, 15–27.

Liu, J., Yeo, H.C., Doniger, S.J., *et al.* (1997). Assay of aldehydes from lipid peroxidation: gas chromatography mass spectrometry compared to thiobarbituric acid. *Analytical Biochemistry*, **245**, 161–166.

Liu, L., Zhang, Y., Wang, J, *et al.* (2005). Detection solar-induced chlorophyll fluorescence from field radiance spectra based on the Fraunhofer line principle. *IEEE transaction on Geoscience and Remote Sensing*, **43**, 827–832.

Liu, L.X., Xu, S.M. and Woo, K.C. (2003). Influence of leaf angle on photosynthesis and the xanthophyll cycle in the tropical tree species *Acacia crassicarpa. Tree Physiology*, **23**, 1255–1261.

Liu, M.Z. and Osborne, C.P. (2008). Leaf cold acclimation and freezing injury in C_3 and C_4 grasses of the Mongolian plateau. *Journal of Experimental Botany*, **59**, 4161–4170.

Liu, N., Peng, C.L., Lin, Z.F., *et al.* (2006). Changes in photosystem II activity and leaf reflectance features of several subtropical woody plants under simulated SO_2 treatment. *Journal of Integrative Plant Biology*, **48**, 1274–1286.

Ljubešic, N. and Britvec, M. (2006). Tropospheric ozone-induced structural changes in leaf mesophyll cell walls in grapevine plants. *Biologia, Bratislava*, **61**, 85–90.

Lloret, F., Siscart, D. and Dalmases, C. (2004). Canopy recovery after drought dieback in holm-oak Mediterranean forests of Catalonia (NE Spain). *Global Change Biology*, **10**, 2092–2099.

Lloyd, J. (1991). Modelling stomatal responses to environment in *Macadamia integrifolia. Australian Journal of Plant Physiology*, **18**, 649–660.

Lloyd, J. and Farquhar G.D. (1994). ^{13}C discrimination during CO_2 assimilation by the terrestrial biosphere. *Oecologia*, **99**, 201–215.

Lloyd, J. and Taylor, J. A. (1994). On the temperature dependence of soil respiration. *Functional Ecology*, **8**, 315–323.

Lloyd, J., Bird, M., Vellen, L., *et al.* (2008). Contributions of woody and herbaceous vegetation to tropical savanna ecosystem productivity: a quasi-global estimate. *Tree Physiology*, **28**, 451–468.

Lloyd, J., Krujit, B., Hollinger, D.Y., *et al.*(1996). Vegetation effects on the isotopic composition of CO_2 at local and regional scales: theoretical aspects and a comparison between rain forest in Amazonia and a boreal forest in Siberia. *Australian Journal of Plant Physiology*, **23**, 371–392.

Lloyd, J., Syvertsen, J.P., Kriedemann, P.E., *et al.* (1992). Low conductances for CO_2 diffusion from stomata to the sites of carboxylation in leaves of woody species. *Plant, Cell and Environment*, **15**, 873–899.

Lloyd, J., Wong, S.C., Styles, J.M., *et al.* (1995). Measuring and modelling whole-tree gas exchange. *Australian Journal of Plant Physiology*, **22**, 987–1000.

Lobel, R., Mamane, Y., Gepstein, S., *et al.* (2001). Particulate Matter Effect on Petunia "Blue Spark" Plants. Abstract of the Eurasap workshop on air pollution and the natural environment: 25–27 April 2001 Sofia, Bulgaria.

Loescher, H.W., Law, B.E., Mahrt, L., *et al.* (2006). Uncertainties in, and interpretation of, carbon flux estimates using the eddy covariance technique *Journal of Geophysical Research*, **111**, 1–19.

Loescher, H.W., Oberbauer, S.F., Gholz, H.L., *et al.* (2003). Environmental controls on net ecosystem-level carbon exchange and productivity in a Central American tropical wet forest. *Global Change Biology*, **9**, 396–412.

Loescher, H.W., Ocheltree, T.W., Tanner, B., *et al.* (2005). Comparison of temperature and wind statistics in contrasting environments among different sonic anemometer-thermometers. *Agricultural and Forest Meteorology*, **133**, 119–139.

Logan, B.A., Demmig-Adams, B., Rosenstiel, T.N., *et al.* (1999). Effect of nitrogen limitation on foliar antioxidants in relationship to other metabolic characteristics. *Planta*, **209**, 213–220.

Lohammer, T., Larsson, S., Linder, S., *et al.* (1980). FAST-Simulation models of gaseous exchange in Scots Pine. *Ecological Bulletin*, **32**, 505–523.

Lohaus, G., Heldt, H.W. and Osmond, C.B. (2000). Infection with phloem limited *Abutilon mosaic virus* causes localized carbohydrate accumulation in leaves of *Abutilon striatum*: relationships to symptom development and effects on chlorophyll fluorescence quenching during photosynthetic induction. *Plant Biology*, **2**, 161–167.

Loik, M.E. and Holl, K.D. (1999). Photosynthetic responses to light for rainforest seedlings planted in abandoned pasture, Costa Rica. *Restoration Ecology*, **7**, 382–391.

Lomax, B.H., Woodward, F.I., Leitch, I.J., *et al.* (2009). Genome size as a predictor of guard cell length in *Arabidopsis thaliana* is independent of environmental conditions. *The New Phytologist*, **181**, 311–314.

Long, S.P. (1991). Modification of the response of photosynthetic productivity to rising temperature by atmospheric CO_2 concentrations: has its importance been underestimated. *Plant, Cell and Environment*, **14**, 729–739.

Long, S.P. (1999). Environmental responses. In: *The Biology of C_4 Photosynthesis* (eds Sage, R.F. and Monson, R.K.), Academic Press, San Diego, USA, pp. 215–249.

Long, S.P. and Bernacchi, C.J. (2003). Gas exchange measurements, what can they tell us about the underlying limitations to photosynthesis? Procedures and sources of error. *Journal of Experimental Botany*, **54**, 2393–2401.

Long, S.P. and Drake, B.G. (1992). Photosynthetic CO_2 assimilation and rising atmospheric CO_2 concentrations. In: *Crop Photosynthesis: Spatial and Temporal Determinants* (eds Baker, N.R. and Thomas, H.), Elsevier Science Publishers, Amsterdam, Netherlands, pp. 69–103.

Long, S.P. and Hallgren, J.E. (1993). Measurement of CO_2 assimilation by plants in the field and the laboratory. In: *Photosynthesis and Production in a Changing Environment. A Fi eld and Laboratory Manual* (eds Hall, D.O., Scurlock, J.M.O., Bolhar-Norden Kampf, H.R., Leegood, R.C. and Long, S.P.), pp. 129–167. Chapman and Hall, London.

Long, S.P. and Naidu, S.L. (2002). Effects of oxidants at the biochemical, cell and physiological levels, with particular reference to ozone. In: *Air Pollution and Plant Life* (eds Bell, J.N.B. and Treshow, M.) 2nd edn. John Wiley and Sons, Chichester, USA, pp. 69–88.

Long, S.P., Ainsworth, E.A., Leakey, A.D.B., *et al.* (2005). Global food insecurity. Treatment of major food crops with elevated carbon dioxide or ozone under large-scale fully open-air conditions suggests recent models may have overestimated future yields. *Philosophical Transactions of the Royal Society B*, **360**, 2011–2020.

Long, S.P., Ainsworth, E.A., Leakey, A.D.B., *et al.* (2006b). Food for thought: lower-than-expected crop yield stimulation with rising CO_2 concentration. *Science*, **312**, 1918–1921.

Long, S.P., Ainsworth, E.A., Leakey, A.D.B., *et al.* (2007). Crop models, CO_2, and climate change – Response. *Science*, **315**, 460–460.

Long, S.P., Ainsworth, E.A., Rogers, A., *et al.* (2004). Rising atmospheric carbon dioxide: plants FACE the future. *Annual Review of Plant Biology*, **55**, 591–628.

Long, S.P., Farage P.K. and García R.L. (1996). Measurement of leaf and canopy photosynthetic CO_2 exchange in the field. *Journal of Experimental Botany*, **47**, 1629–1642.

Long, S.P., Farage P.K., Bolhar-Nordenkampf H.R., *et al.* (1989). Separating the contribution of the upper and lower mesophyll to photosynthesis in *Zea mays* L. leaves. *Planta*, **177**, 207–216.

Long, S.P., Humphries, S. and Falkowski, P.G. (1994). Photoinhibition of photosynthesis in nature. *Annual Review of Plant Physiology*, **45**, 633–662.

Long, S.P., Postl, W.F. and Bolhár-Nordenkampf, H.R. (1993). Quantum yields for uptake of carbon dioxide in C_3 vascular plants of contrasting habitats and taxonomic groupings. *Planta*, **189**, 226–234.

Long, S.P., Zhu, X.I.N.G., Naidu, S.L., *et al.* (2006a). Can improvement in photosynthesis increase crop yields? *Plant, Cell and Environment*, **29**, 315–330.

Loomis, R.S. and Amthor, J.S. (1999). Yield potential, plant assimilatory capacity, and metabolic efficiencies. *Crop Science*, **39**, 1584–1596.

López, C., Soto, M., Restrepo, S., Piégu, B., *et al.* (2005). Gene expression profile in response to *Xanthomonas axonopodis* pv. *manihotis* infection in cassava using a cDNA microarray. *Plant Molecular Biology*, **57**, 393–410.

Lopez-Hoffman, L., Anten, N.P.R., Martinez-Ramos, M., *et al.* (2007). Salinity and light interactively affect neotropical mangrove seedlings at the leaf and whole plant levels. *Oecologia*, **150**, 545–556.

Lopushinsky, W. and Kaufmann, M.R. (1984). Effects of cold soil on water relations and spring growth of Douglas-fir seedlings. *Forest Science*, **30**, 628–634.

Lorenzini, G., Guidi, L., Nali, C., *et al.* (1997). Photosynthetic response of tomato plants to vascular wilt diseases. *Plant Science*, **124**, 143–152.

Loreto, F. and Fares, S. (2007). Is ozone flux inside leaves only a damage indicator? Clues from volatile isoprenoid studies. *Plant Physiology*, **143**, 1096–1100.

Loreto, F. and Sharkey, T.D. (1993). On the relationship between the isoprenene emission and photosynthetic metabolites under different environmental conditions. *Planta*, **189**, 420–424.

Loreto, F. and Velikova, V. (2001). Isoprene produced by leaves protects the photosynthetic apparatus against ozone damage, quenches ozone products, and reduces lipid peroxidation of cellular membranes. *Plant Physiology*, **127**, 1781–1787.

Loreto, F., Baker, N.R. and Ort, D.O. (2004b). Environmental constraints, chloroplast to leaf. In: *Photosynthetic Adaptation, Chloroplast to Landscape* (eds Smith, W.K., Vogelmann, T.C. and Critchley, C.), Springer, New York, USA, pp. 231–261.

Loreto, F., Centritto, M. and Chartzoulakis, K. (2003). Photosynthetic limitations in olive cultivars with different sensitivity to salt stress. *Plant, Cell and Environment*, **26**, 595–601.

Loreto, F., Delfine, S. and Di Marco, G. (1999). Estimation of photorespiratory carbon dioxide recycling during photosynthesis. *Australian Journal of Plant Physiology*, **26**, 733–736.

Loreto, F., Di Marco, G., Tricoli, D., *et al.* (1994). Measurements of mesophyll conductance, photosynthetic electron transport and alternative electron sinks of field grown wheat leaves. *Photosynthesis Research*, **41**, 397–403.

Loreto, F., Harley, P.C., di Marco, G., *et al.* (1992). Estimation of mesophyll conductance to CO_2 flux by three different methods. *Plant Physiology*, **98**, 1437–1443.

Loreto, F., Mannozzi, M., Maris, C., *et al.* (2001b). Ozone quenching properties of isoprene and its antioxidant role in leaves. *Plant Physiology*, **126**, 993–1000.

Loreto, F., Pinelli, P., Manes, F., *et al.* (2004a). Impact of ozone on monoterpene emissions and evidence for an isoprene-like antioxidant action of monoterpenes emitted by *Quercus ilex* leaves. *Tree Physiology*, **24**, 361–367.

Loreto, F., Tsonev, T. and Centritto, M. (2009). The impact of blue light on leaf mesophyll conductance. *Journal of Experimental Botany*, **60**, 2283–2290.

Loreto, F., Velikova, V. and Di Marco, G. (2001a). Respiration in the light measured by $^{12}CO_2$ emission in $^{13}CO_2$ atmosphere in maize leaves. *Australian Journal of Plant Physiology*, **28**, 1103–1108.

Loriaux, S.D., Burns, R.A., Welles, J.M., *et al.* (2006). Determination of maximal chlorophyll fluorescence using a multiphase single flash of sub-saturating intensity. Poster. August, 2006. *American Society of Plant Biologists Annual Meetings*, Boston, MA. See: www.licor.com.

Lorimer, G.H., Badger, M.R. and Andrews, T.J. (1977). D-ribulose 1,5-bisphosphate carboxylase-oxygenase. Improved methods for activation and assay of catalytic activities. *Analytical Biochemistry*, **78**, 66–75.

Lösch, R., Jensen, C.R. and Andersen, M.N. (1992). Diurnal courses and factorial dependencies of leaf conductance and transpiration of differently potassium fertilized and watered field grown barley plants. *Plant and Soil*, **140**, 205–224.

Louis, J., Cerovic, Z.G. and Moya, I. (2006). Quantitative study of fluorescent excitation and emission spectra of bean leaves. *Journal of Photochemistry and Photobiology B Biology*, **85**, 65–71.

Louis, J., Ounis, A., Ducruet, J.M., *et al.* (2005). Remote sensing of sunlight-induced chlorophyll fluorescence and reflectance on Scots pine in the boreal forest during spring recovery. *Remote Sensing of Environment*, **96**, 37–48.

Loustau, D., Ben Brahim, M., Gaudillere, J.P., *et al.* (1999). Photosynthetic responses to phosphorus nutrition in two-year-old maritime pine seedlings. *Tree Physiology*, **19**, 707–715.

Loveless, A.R. (1962). Further evidence to support a nutritional interpretation of sclerophylly. *Annals of Botany*, **26**, 551–561.

Lovelock, C.E., Clough, B.F. and Woodrow, I.E. (1992). Distribution and accumulation of ultraviolet-radiation-absorbing compounds in leaves of tropical mangroves. *Planta*, **188**, 143–154.

Lovelock, C.E., Jebb, M. and Osmond, C.B. (1994). Photoinhibition and recovery in tropical plant species: response to disturbance. *Oecologia*, **97**, 297–307.

Lovelock, C.E., Osmond, C.B. and Seppelt, R.D. (1995). Photoinhibition in the Antarctic moss *Grimmia antarctici*

Card. when exposed to cycles of freezing and thawing. *Plant, Cell and Environment*, **18**, 1395–1402.

Loveys, B.R., Schurwater, I., Pons, T.L., *et al.* (2001). The relative importance of photosynthesis and respiration in determining plant growth rate. In: *Proceedings of the 12th International Congress on Photosynthesis, Brisbane Convention and Exhibition Centre, Queensland, Australia, August 18–23, 2001*, CSIRO Publishing, Canberra, Australia, pp. S35–010.

Low, M., Haberle, K.H., Warren, C.R., *et al.* (2007). O_3 flux-related responsiveness of photosynthesis, respiration, and stomatal conductance of adult *Fagus sylvatica* to experimentally enhanced free-air O_3 exposure. *Plant Biology*, **9**, 197–206.

Löw, M., Herbinger, K., Nunn, A.J., *et al.* (2006). Extraordinary drought of 2003 overrules ozone impact on adult beech trees (*Fagus sylvatica*). *Trees: Structure and Function*, **20**, 539–548.

Lu, C., Lu, Q., Zhang, J., *et al.* (2001). Characterization of photosynthetic pigment composition, photosystem II photochemistry and thermal energy dissipation during leaf senescence of wheat plants grown in the field. *Journal of Experimental Botany*, **52**, 1805–1810.

Lu, X. and Zhuang, Q. (2010). Evaluating evapotranspiration and water-use efficiency of terrestrial ecosystems in the conterminous United States using MODIS and AmeriFlux data. *Remote Sensing of the Environment*, **114**, 1924–1939.

Lucht, W., Schaphoff, S., Erbrecht, T., *et al.* (2006). Terrestrial vegetation redistribution and carbon balance under climate change. *Carbon Balance and Management*, **6**, 1–6.

Lüdeker, W., Dahn, H.G. and Günter, K.P. (1996). Detection of fungal infection of plants by laserinduced fluorescence: an attempt to use remote sensing. *Journal of Plant Physiology*, **148**, 579–585.

Ludwig, L.J. and Canvin, D.T. (1971). The rate of photorespiration during photosynthesis and the relationship of the substrate of light respiration to the products of photosynthesis in sunflower leaves. *Plant Physiology*, **48**, 712–719.

Lunde, C., Jensen, P.E., Haldrup, A., *et al.* (2000). The PSI-H subunit of photosystem I is essential for state transitions in plant photosynthesis. *Nature*, **408**, 613–615.

Lundell, R., Saarinen, T., Åström, H., *et al.* (2008) The boreal dwarf shrub *Vaccinium vitis-idaea* retains its capacity for photosynthesis through the winter. *Botany*, **86**, 491–500.

Lundmark, M., Cavaco, A.M., Trevanion, S., *et al.* (2006). Carbon partitioning and export in transgenic *Arabidopsis thaliana* with altered capacity for sucrose synthesis grown at low temperature: a role for metabolite transporters. *Plant Cell Environment*, **29**, 1703–1714.

Lundmark, T., Bergh, J., Strand, M., *et al.* (1998). Seasonal variation of maximum photochemical efficiency in boreal Norway spruce stands. *Trees-Struct Funct*, **13**, 63–67.

Lunn, J.E., Feil, R., Hendriks, J.H.M., *et al.* (2006). Sugar-induced increases in trehalose 6-phosphate are correlated with redox activation of ADP-glucose pyrophosphorlyase and higher rates of starch synthesis in *Arabidopsis thaliana*. *Biochemical Journal*, **397**, 139–148.

Lunt, D.J., Ross, I. Hopley, P.J., *et al.* (2007). Modelling late Oligocene C_4 grasses and climate. *Palaeogeography, Palaeoclimatology, Palaeocoecology*, **251**, 239–253.

Luo, H., Oechel, W.C., Hastings, S.J., *et al.* (2007). Mature semiarid chaparral ecosystems can be a significant sink for atmospheric carbon dioxide. *Global Change Biology*, **13**, 386–396.

Luo, J., Zang, R. and Li, C. (2006). Physiological and morphological variations of *Picea asperata* populations originating from different altitudes in the mountains of southwestern China. *Forest Ecology and Management*, **221**, 285–290.

Luo, R., Wei, H., Ye, L., *et al.*. (2009). Photosynthetic metabolism of C_3 plants shows highly cooperative regulation under changing environments: a systems biological analysis. *Proceedings of the National Academy of Sciences USA*, **106**, 847–852.

Luo, S., Su, B. and Currie, W.S. (2004). Progressive nitrogen limitation of ecosystem responses to rising atmospheric carbon dioxide. *BioScience*, **54**, 731–739.

Luo, Y., Gerten, D., Le Maire, G., *et al.* (2008). Modeled interactive effects of precipitation, temperature, and $[CO_2]$ on ecosystem carbon and water dynamics in different climatic zones. *Global Change Biology*, **14**, 1986–1999.

Luo, Y., Hui, D. and Zhang, D. (2006). Elevated CO_2 stimulates net accumulations of carbon and nitrogen in land ecosystems: a meta-analysis. *Ecology*, **87**, 53–63.

Luo, Y., Su, B., Currie, W., *et al.* (2004). Progressive nitrogen limitation of ecosystem responses to rising atmospheric carbon dioxide. *Bioscience*, **54**, 731–739.

Luoma, S. (1997). Geographical pattern in photosynthetic light response of *Pinus sylvestris* in Europe. *Functional Ecology*, **11**, 273–281.

Lusk, C.H. (1996). Gradient analysis and disturbance history of temperate rain forests of the coast range summit plateau, Valdivia, Chile. *Revista Chilena de Historia Natural*, **69**, 401–411.

Lusk, C.H. and Smith, B. (1998). Life history differeces and tree species coexistence in an old-growth New Zealand rain forest. *Ecology*, **79**, 795–806.

Lusk, C.H. and Warton, D.I. (2007). Global meta-analysis shows that relationships of leaf mass per area with species shade tolerance depend on leaf habit and ontogeny. *The New Phytologist*, **176**, 764–774.

Lusk, C.H., Wright, I. and Reich, P.B. (2003). Photosynthetic differences contribute to competitive advantage of evergreen angiosperm trees over evergreen conifers in productive habitats. *New Phytologist*, **160**, 329–336.

Lüttge, U. (1986). Nocturnal water storage in plants having crassulacean acid metabolism. *Planta*, **168**, 287–289.

Lüttge, U. (1987a). Malate relations of plants and crassulacean acid metabolism: protons, carbon dioxide and water: a review. *Giornale Botanico Italiano*, **121**, 217–227.

Lüttge, U. (1987b). Carbon dioxide and water demand: crassulacean acid metabolism (CAM), a versatile ecological adaptation exemplifying the need for integration in ecophysiological work. *New Phytologist*, **106**, 593–629.

Lüttge, U. (1989). Vascular epiphytes. Setting the scene. In: *Vascular Plants as Epiphytes. Evolution and Ecophysiology.* Ecological Studies vol. 76 (ed. Lüttge, U.), Springer-Verlag, Berlin – Heidelberg – New York, pp. 1–14.

Lüttge, U. (1993). The role of crassulacean acid metabolism (CAM) in the adaptation of plants to salinity. *New Phytologist*, **125**, 59–71.

Lüttge, U. (1997). *Physiological Ecology of Tropical Plants.* Springer-Verlag, Berlin-Heidelberg, Germany.

Lüttge, U. (2000). The tonoplast functioning as the master switch for circadian regulation of crassulacean acid metabolism. *Planta*, **211**, 761–769.

Lüttge, U. (2002a). CO_2-concentrating: consequences in crassulacean acid metabolism. *Journal of Experimental Botany*, **53**, 2131–2142.

Lüttge, U. (2002b). Performance of plants with C_4-carboxylation modes of photosynthesis under salinity. In: *Salinity: Environment – Plants – Molecules* (eds Läuchli, A. and Lüttge, U.) Kluwer Academic Publishers, Dordrecht, Boston, London, pp. 113–135.

Lüttge, U. (2003). Photosynthesis: CAM plants. In: *Encyclopedia of Applied Plant Sciences* (eds Thomas, B., Murphy, D. and Murphy, B.), Academic Press, Oxford, UK, pp. 688–705.

Lüttge, U. (2004). Ecophysiology of crassulacean acid metabolism (CAM). *Annals of Botany*, **93**, 629–652.

Lüttge, U. (2005). Genotypes – phenotypes – ecotypes: relations to crassulacean acid metabolism. *Nova Acta Leopoldina* NF, **92/342**, 177–193.

Lüttge, U. (2006). Photosynthetic flexibility and ecophysiological plasticity: questions and lessons from Clusia, the only CAM tree, in the neotropics. *New Phytologist*, **171**, 7–25.

Lüttge, U. (2008a). *Physiological Ecology of Tropical Plants.* 2nd edn. Springer, Berlin – Heidelberg – New York.

Lüttge, U. (2008b). Stem CAM in arborescent succulents. *Trees*, **22**, 139–148.

Lüttge, U. and Beck, F. (1992). Endogenous rhythms and chaos in crassulacean acid metabolism. *Planta*, **188**, 28–38.

Lüttge, U. ed. (2007): *Clusia. A Woody Neotropical Genus of Remarkable Plasticity and Diversity*. Ecological Studies, Vol. 194. Springer-Verlag: Berlin-Heideiberg-New York.

Lüttge, U., Medina, E., Cram, W.J., *et al.* (1989). Ecophysiology of xerophytic and halophytic vegetation of a coastal alluvial plain in northern Venezuela. II. Cactaceae. *New Phytologist*, **111**, 245–251.

Lütz, C. (1996). Avoidance of photoinhibition and examples of photodestruction in high alpine *Eriophorum*. Journal of Plant Physiology, **148**, 120–128.

Luyssaert, S., Inglima, I., Jung, M., *et al.* (2007). CO_2 balance of boreal, temperate, and tropical forests derived from a global database. *Global Change Biology*, **13**, 2509–2537.

Luyssaert, S., Reichstein, M., Schulze, E.D., *et al.* (2009). Toward a consistency cross-check of eddy covariance flux-based and biometric estimates of ecosystem carbon balance. *Global Biogeochemical Cycles*, **23**, GB3009.

Lv, S., Zhang, K., Gao, Q., *et al.* (2008). Overexpression of an H+-PPase gene from *Thellungiella halophila* in cotton enhances salt tolerance and improves growth and photosynthetic performance. *Plant and Cell Physiology*, **49**, 1150–1164.

Lyons, J.M. (1973). Chilling injury in plants. *Annual Review of Plant Physiology*, **24**, 445–466.

Ma, S., Baldocchi, D.D., Xu, L., *et al.* (2007). Inter-annual variability in carbón dioxide exchange of and oak/grass savanna and open grassland in California. *Agricultural and Forest Metereology*, **147**, 151–171.

Maas, J.M. (1995). Conversion of tropical dry forest to pasture and agriculture. In: *Seasonal Dry Tropical Forests* (eds Bullock, S.H., Mooney, H.A. and Medina, E.), Cambridge University Press, Cambridge, UK, pp. 399–422.

Macedo, T.B., Weaver, D.K. and Peterson, R.K.D. (2007). Photosynthesis in wheat at the grain filling stage is altered by the wheat stem sawfly (Hymenoptera: Cephidae) injury and reduced water availability. *Journal of Entomological Science*, **42**, 228–238.

MacFadden, B.J. (2000). Cenozoic mammalian herbivores from the Americas: Reconstructing ancient diets and terrestrial communities. *Annual Review of Ecology and Systematics*, **31**, 33–59.

MacFadden, B.J. and Cerling, T.E. (1996). Mammalian herbivore communities, ancient feeding ecology, and carbon isotopes: A 10 million-year sequence from the Neogene of Florida. *Journal of Vertebrate Paleontology*, **16**, 103–115.

Mächler, F. and Nösberger, J. (1977). Effect of light intensity and temperature on apparent photosynthesis of altitudinal ecotypes of *Trifolium repens* L. *Oecologia*, **31**, 73–78.

Mächler, F. and Nösberger, J. (1978). The adaptation to temperature of photorespiration and of the photosynthetic carbon metabolism of altitudinal ecotypes of *Trifolium repens* L. *Oecologia*, **35**, 267–276.

Mächler, F., Nösberger, J. and Erismann, K.H. (1977). Photosynthetic $(^{14})CO_2$ fixation products in altitudinal exotypes of *Trifolium repens* L. with different temperature requirements. *Oecologia*, **31**, 79–84.

Macinnis-Ng, C., McClenahan, K. and Eamus, D. (2004). Convergence in hydraulic architecture, water relations and primary productivity amongst habitats and across seasons in Sydney. *Functional Plant Biology*, **31**, 429–439.

MacKenzie, T.D.B., Krol, M., Huner, N.P.A., *et al.* (2002). Seasonal changes in chlorophyll fluorescence quenching and the induction and capacity of the photoprotective xanthophyll cycle in *Lobaria pulmonaria*. *Canadian Journal of Botany Revue Canadienne De Botanique*, **80**, 255–261.

Macrobbie, E.A.C. (1998). Signal transduction and ion channels in guard cells. *Philosophical Transactions of the Royal Society B*, **353**, 1475–1488.

Magnani, F., Mencuccini, M., Borghetti, M., *et al.* (2007). The human footprint in the carbon cycle of temperate and boreal forests. *Nature*, **447**, 848–850.

Magnani, F. and Borghetti, M. (1995). Interpretation of seasonal changes of xylem embolism and plant hydraulic resistance in *Fagus sylvatica*. *Plant Cell and Environment*, **186**, 689–696.

Magnussen, S., Smith, V.G. and Yeatman, C.W. (1986). Foliage and canopy characteristics in relation to aboveground dry matter increment of seven jack pine provenances. *Canadian Journal of Forest Research*, **16**, 464–470.

Magrin, G., Gay García, C., Cruz Choque, D., *et al.* (2007). *Latin America. Climate Change 2007: Impacts, Adaptation and Vulnerability. Contribution of Working Group II to the Fourth Assessment Report of the Intergovernmental Panel on Climate Change* (eds Parry, O.F.C.M.L., Palutikof, J.P., van der Linden, P.J. and Hanson, C.E.), Cambridge University Press, Cambridge, UK, pp. 581–615.

Mahalingam, R., Jambunathan, N., Gunjan, S.K., *et al.* (2006). Analysis of oxidative signalling induced by ozone in *Arabidopsis thaliana*. *Plant, Cell and Environment*, **29**, 1357–1371.

Mahan, J.R. and Mauget, S.A. (2005). Antioxidant metabolism in cotton seedlings exposed to temperature stress in the field. *Crop Science*, **45**, 2337–2345.

Maherali, H., Sherrard, M.E., Clifford, M.H., *et al.* (2008). Leaf hydraulic conductivity and photosynthesis are genetically correlated in an annual grass. *The New Phytologist*, **180**, 240–247.

Mäkelä, A., Hari, P., Berninger, F., *et al.* (2004). Acclimation of photosynthetic capacity in Scots pine to the annual cycle of temperature. *Tree Physiology*, **24**, 369–376.

Makino, A. and Osmond, C.B. (1991). Effects of nitrogen nutrition on nitrogen partitioning between chloroplast and mitochondria in pea and wheat. *Plant Physiology*, **96**, 355–362.

Makino, A. and Sage, R.F. (2007). Temperature response of photosynthesis in transgenic rice transformed with 'sense' or 'antisense' *rbc*S. *Plant Cell Physiology*, **48**, 1472–1483.

Makino, A., Miyake, C. and Yokota, A. (2002). Physiological functions of the water-water cycle (Mehler reaction) and the cyclic electron flow around PSI in rice leaves. *Plant and Cell Physiology*, **43**, 1017–1026.

Makino, A., Nakano, H., Mae, T., *et al.* (2000). Photosynthesis, plant growth and N allocation in transgenic rice plants with decreased Rubisco under CO_2 enrichment. *Journal of Experimental Botany*, **51**, 383–389.

Makino, A., Sakashita, H., Hidema, J., *et al.* (1992). Distinctive responses of ribulose-1,5-bisphosphate carboxylase and carbonic anhydrase in wheat leaves to nitrogen nutrition and their possible relationship to CO_2 transfer resistance. *Plant Physiology*, **100**, 1737–1743.

Makino, A., Sato, T., Nakano, H., *et al.* (1997). Leaf photosynthesis, plant growth and nitrogen allocation in rice under different irradiances. *Planta*, **203**, 390–398.

Makino, A., Tadahiko, M. and Ohira, K. (1985). Enzymic Properties of ribulose-1,5-bisphosphate carboxylase/oxygenase purified from rice leaves. *Plant Physiology*, **79**, 57–61.

Maksymowych, R. (1973). *Analysis of Leaf Development.* Cambridge University Press, Cambridge, UK.

Malenovsky, Z., Mishra, K.B., Zemek, F., *et al.* (2009). Scientific and technical challenges in remote sensing of plant canopy reflectance and fluorescence. *Journal of Experimental Botany*, **60**, 2987–3004.

Malhi, Y., Nobre, A.D., Grace, J., *et al.* (1998). Carbon dioxide transfer over a Central Amazonian rain forest. *Journal of Geophysical Research-Atmospheres*, **103**, 31593–31612.

Malkin, S. and Cannaani, O. (1994). The use and characteristics of the photoacoustic method in the study of photosynthesis. *Annual Review of Plant Physiology and Plant Molecular Biology*, **45**, 493–526.

Malkin, S., Bilger, W. and Schreiber, U. (1994). The relationship between millisecond luminescence and fluorescence in tobacco leaves during the induction period. *Photosynthesis Research*, **39**, 57–66.

Man, R. and Lieffers, V.J. (1997). Seasonal photosynthetic responses to light and temperature in white spruce (*Picea glauca*) seedlings planted under an aspen (*Populus tremuloides*) canopy and in the open. *Tree physiology*, **17**, 437–444.

Manca, G. (2003). Analisi dei flussi di carbonio di una cronosequenza di cerro (*Quercus cerris* L.) dell'Italia centrale attraverso la tecnica della correlazione turbolenta. PhD Thesis. University of Tuscia, Viterbo, Italy.

Manetas, Y. (2004a). Photosynthesizing in the rain: beneficial effects of twig wetting on corticular photosynthesis trough changes in the periderm optical properties. *Flora*, **199**, 334–341.

Manetas, Y. (2004b). Probing corticular photosynthesis through *in vivo* chlorophyll fluorescence measurements: evidence that high internal CO_2 levels suppress electron flow and increase the risk of photoinhibition. *Physiologia Plantarum*, **120**, 509–517.

Manetas, Y. and Pfanz, H. (2005). Spatial heterogeneity of light penetration through periderm and lenticels and concomitant patchy acclimation of corticular photosynthesis. *Trees*, **19**, 409–414.

Mansfield, J.W. (2005). Biophoton distress flares signal the onset of the hypersensitive reaction. *Trends in Plant Sciences*, **10**, 307–309.

Mansfield, T.A. (1999). SO_2 pollution: a bygone problem or a continuing hazard? In: *Physiological Plant Ecology* (ed. Press, M.C.), Blackwell Science, Oxford, UK, pp. 219–240.

Mansfield, T.A. (2002). Nitrogen oxides: old problems and new challenges. In: *Air Pollution and Plant Life* (ed. Bell, J.N.B. and Treshow, M.) 2nd edn, John Wiley and Sons, Chichester, USA, pp. 119–134.

Mansfield, T.A. and Pearson, M. (1996). Disturbances in stomatal behaviour in plants exposed to air pollution. In: *Plant Response to Air Pollution* (eds Yunus, M. and Iqbal, M.), John Wiley and Sons, Chichester, USA, pp. 179–194.

Manter, D.K. and Kavanagh, K.L. (2003). Stomatal regulation in Douglas fir following a fungal-mediated chronic reduction in leaf area. *Trees*, **17**, 485–491.

Manter, D.K. and Kerrigan, J. (2004). A/C_i curve analysis across a range of woody plant species: influence of regression analysis parameters and mesophyll conductance. *Journal of Experimental Botany*, **55**, 2581–2588.

Manter, D.K., Kelsey, R.G. and Karchesy, J.J. (2007). Antimicrobial activity of extractable conifer heartwood compounds toward *Phytophthora ramorum*. *Journal of Chemical Ecology*, **33**, 2133–2147.

Manuel, N., Cornic, G., Aubert, S., *et al.* (1999). Protection against photoinhibition in the alpine plant *Geum montanum*. *Oecologia*, **119**, 149–158.

Marcelis, L.F.M. and Baan Hofman-Eijer, L.R.B. (1995). The contribution of fruit photosynthesis to the carbon requirement of cucumber fruits as affected by irradiance, temperature and ontogeny. *Physiologia Plantarum*, **93**, 476–483.

Marchi, S., Tognetti, R., Minnocci, A., *et al.* (2008). Variation in mesophyll anatomy and photosynthetic capacity during leaf development in a deciduous mesophyte fruit tree (*Prunus persica*) and an evergreen sclerophyllous Mediterranean shrub (*Olea europaea*). *Trees*, **22**, 559–571.

Marcolla, B., Pitacco, A. and Cescatti, A. (2003). Canopy architecture and turbulence structure in a coniferous forest. *Boundary Layer Meteorology*, **108**, 39–59.

Marín-Navarro, J., Manuell, A.L., Wu, J., *et al.* (2007). Chloroplast translation regulation. *Photosynthesis Research*, **94**, 359–374.

Markgraf, T. and Berry, J. (1990). Measurement of photochemical and non-photochemical quenching: correction for turnover of PS2 during steady state photosynthesis. In: *Current Research in Photosynthesis*, Vol IV (ed. Baltscheffsky, M.), Kluwer Academic Publishers, Dordrecht, Netherlands, pp. 279–282.

Marks, S. and Clay, K. (1996). Physiological responses of *Festuca arundinacea* to fungal endophyte infection. *New Phytologist*, **133**, 727–733.

Maroco, J.P., Rodrigues, M.L., Lopes, C., *et al.* (2002). Limitations to leaf photosynthesis in field-grown grapevine under drought – metabolic and modelling approaches. *Functional Plant Biology*, **29**, 451–459.

Marques da Silva, J. and Arrabaça, M.C. (2004). Photosynthesis in the water-stressed C_4 grass *Setaria sphacelata* is mainly limited by stomata with both rapidly and slowly imposed water deficits. *Physiologia Plantarum*, **121**, 409–420.

Marquez, E.J., Rada, F. and Farinas, M.R. (2006). Freezing tolerance in grasses along an altitudinal gradient in the Venezuelan Andes. *Oecologia*, **150**, 393–397.

Marschner, H. (1995). *Mineral Nutrition of Higher Plants.* Academic Press, London, UK.

Marschner, H. and Cakmak, I. (1989). High light intensity enhances chlorosis and necrosis in leaves of zinc-, potassium- and magnesium-deficient bean (*Phaseolus vulgaris*) plants. *Journal of Plant Physiology*, **134**, 308–315.

Marshall, F.M. (2002). Effects of air pollutants in developing countries. In: *Air Pollution and Plant Life*, 2nd edn (eds Bell, J.N.B. and Treshow, M.), John Wiley and Sons, Chichester, USA.

Mariscal, M.J., Orgaz, F. and Villalobos, F.J. (2000). Radiation-use efficiency and dry matter partitioning of a young olive (*Olea europea*) orchard. *Tree Physiology*, **20**, 65–72.

Martin, B. and Ruiz-Torres, N.A. (1992). Effects of water-deficit stress on photosynthesis, its components and component limitations, and on water-use efficiency in wheat (*Triticum aestivum* L). *Plant Physiology*, **100**, 733–739.

Martin, B. and Thorstenson, Y. R. (1988). Stable carbon isotope composition (δ13C), water use efficiency, and biomass productivity of *Lycopersicon esculentum*, *Lycopersicon pennellii*, and the F1 hybrid. *Plant Physiology*, **88**, 213–217.

Martin, B., Nienhuis, J., King, G., *et al.* (1989). Restriction fragment length polymorphisms associated with water-use efficiency in tomato. *Science*, **243**, 1725–1728.

Martin, C.E. (1994). Physiological ecology of the Bromeliaceae. *Botanical Review*, **60**, 1–82.

Martin, R.E., Asner, G.P. and Sack, L. (2007). Genetic variation in leaf pigment, optical and photosynthetic function among diverse phenotypes of *Metrosideros polymorpha* grown in a common garden. *Oecologia*, **151**, 287–400.

Martínez, J.P., Ledent J.F., Bajji, M., *et al.* (2003). Effect of water stress on growth, Na+ and K+ accumulation and water use efficiency in relation to osmotic adjustment in two populations of *Atriplex halimus* L. *Plant Growth Regulation*, **41**, 63–73.

Martino-Catt, S. and Ort, D.R. (1992). Low-temperature interrupts circadian regulation of transcriptional activity in chilling-sensitive plants. *Proc Natl Acad Sci USA*, **89**, 3731–3735.

Maruyama, A. and Kuwagata, T. (2008). Diurnal and seasonal variation in bulk stomatal conductance of the rice canopy and its dependence on developmental stage. *Agricultural and Forest Meteorology*, **148**, 1161–1173.

Masclaux-Daubresse, C., Purdy, S., Lemaitre, T., *et al.* (2007). Genetic variation suggests interaction between cold acclimation and metabolic regulation of leaf senescence. *Plant Physiology*, **143**, 434–446.

Masle, J., Gilmore, S.R. and Farquhar, G.D. (2005). The ERECTA gene regulates plant transpiration efficiency in *Arabidopsis*. *Nature*, **436**, 866–870.

Masle, J., Hudson, G.S. and Badger, M.R. (1993). Effects of ambient CO_2 concentration on growth and nitrogen use in tobacco (*Nicotiana tabacum*) plants transformed with an antisense gene to the small subunit of ribulose-1,5-bisphosphate carboxylase/oxygenase. *Plant Physiology*, **103**, 1075–1088.

Mason, E.A. and Marrero, T.R. (1970). The diffusion of atoms and molecules. *Advances in At Molecular Physics*, **6**, 155–232.

Masoni, A., Ercoli, L. and Mariotti, M. (1996). Spectral properties of leaves deficient in iron, sulfur, magnesium and manganese. *Agronomy Journal*, **88**, 937–943.

Massad, R.S., Tuzet, A. and Bethenod, O. (2007). The effect of temperature on C_4-type leaf photosynthesis parameters. *Plant, Cell and Environment*, **30**, 1191–1204.

Massman, W.J., Lee, X. (2002). Eddy covariance flux corrections and uncertainties in long-term studies of carbon and energy exchanges, *Agricultural and Forest Meteorology*, **113**, 121–144.

Massonnet, C., Costes, E., Rambal, S., *et al.* (2007). Stomatal regulation of photosynthesis in apple leaves: evidence for different water-use strategies between two cultivars. *Annals of Botany*, **100**: 1347–1356.

Massonnet, C., Regnard, J.L., Lauri, P.E., *et al.* (2008). Contributions of foliage distribution and leaf functions to light interception, transpiration and photosynthetic capacities in two apple cultivars at branch and tree scales. *Tree Physiology*, **28**, 665–678.

Mast, A.R. and Givnish, T.J. (2002). Historical biogeography and the origin of stomatal distributions in *Banksia* and *Dryandra* (Proteaceae) based on their cpDNA phylogeny. *American Journal of Botany*, **89**, 1311–1323.

Masterson, J. (1994). Stomatal size in fossil plants: evidence for polyploidy in majority of angiosperms. *Science*, **264**, 421–424.

Matile, P., Hörtensteiner, S. and Thomas, H. (1999). Chlorophyll degradation. *Annual Review of Plant Physiology and Plant Molecular Biology*, **50**, 67–95.

Matouš, K., Benediktyová, Z., Berger, S., *et al.* (2006). Case study of combinatorial imaging: what protocol and what chlorophyll fluorescence image to use when visualizing infection of *Arabidopsis thaliana* by *Pseudomonas syringae*. *Photosynthesis Research*, **90**, 243–253.

Matson, P.A., Parton, W.J., Power, A.G., *et al.* (2007). Agricultural intensification and ecosystem properties. *Science*, **277**, 504–509.

Matsubara, S., Gilmore, A.M. and Osmond, C.B. (2001). Diurnal and acclimatory responses of violaxanthin and lutein epoxide in the Australian mistletoe *Amyema miquelii*. *Autralian Journal of Plant Physiology*, **28**, 793–800.

Matsubara, S., Gilmore, A.M., Ball, M.C., *et al.* (2002). Sustained downregulation of photosystem II in mistletoes during winter depression of photosynthesis. *Functional Plant Biology*, **29**, 1157–1169.

Matsubara, S., Krause, G.H., Aranda, J., *et al.* (2009). Sun-shade patterns of leaf carotenoid composition in 86 species of neotropical forest plants. *Functional Plant Biology*, **36**, 20–36.

Matsubara, S., Krause, G.H., Seltmann, M., *et al.* (2008). Lutein epoxide cycle, light harvesting and photoprotection in species of the tropical tree genus *Inga*. Plant Cell and Environment, **31**, 548–561.

Matsubara, S., Morosinotto, T., Osmond, C.B., *et al.* (2007). Short- and long-term operation of the lutein-epoxide cycle in light-harvesting antenna complexes. *Plant Physiology*, **144**, 926–941.

Matsueda, H. and Inoue, H. (1996). Measurements of atmospheric CO_2 and CH_4 using a commercial airliner from 1993 to 1994. *Atmospheric Environment*, **30**, 1647–1655.

Matsumoto, K., Ohta, T. and Tanaka, T. (2005). Dependence of stomatal conductance on leaf chlorophyll concentration and meteorological variables. *Agricultural and Forest Meteorology*, **132**, 44–57.

Matsumoto, K., Ohta, T., Nakai, T., *et al.* (2008). Responses of surface conductance to forest environments in the Far East. *Agricultural and Forest Meteorology*, **148**, 1926–1940.

Matsumura, T., Tabayashi, N., Kamagata, Y., *et al.* (2002). Wheat catalase expressed in transgenic rice can improve tolerance against low temperature stress. *Physiologia Plantarum*, **116**, 317–327.

Matteucci, G. (1998). Bilancio del carbonio in una faggeta dell'Italia Centro-Meridionale: determinanti ecofisiologici, integrazione a livello di copertura e simulazione dell'impatto dei cambiamenti ambientali. PhD Thesis, University of Padua, Italy.

Matteucci, G., Valentini, R., Scarascia, Mugnosa, G., *et al.* (1995). Struttura e funzionalita' di una comunita' vegetale Mediterranea a *Quercus Ilex* L. dell'Italia Centrale. I. Bilancio del carbonio: variazioni stagionali e fattori limitanti. *Studi Trent. Sci. Nat., Acta biologica*, **69**, 1992, 127–141.

Matthews, R.E.F. and Sarkar, S. (1976). A light-induced structural change in chloroplasts of chinese cabbage cells infected with turnip yellow mosaic virus. *Journal of General Virology*, **33**, 435–446.

Matthews, S. and Donoghue, M.J. (1999). The root of angiosperm phylogeny inferred from duplicate phytochrome genes. *Science*, **286**, 947–950.

Mattoo, A.K., Pick, U., Hoffman-Falk, H., *et al.* (1981). The rapidly metabolized 32,000-dalton polypeptide of the chloroplast is the "proteinaceous shield" regulating photosystem II electron transport and mediating diuron herbicide sensitivity. *Proceedings of the National Academy of Sciences of the United States of America*, **78**, 1572–1576.

Maurel, C., Verdoucq, L., Luu, D.T., *et al.* (2008). Plant aquaporins: membrane channels with multiple integrated functions. *Annual Review of Plant Biology*, **59**, 595–624.

Maximov, N.A. (1931). The physiological significance of the xeromorphic structure of plants. *Journal of Ecology*, **19**, 273–282.

Maximov, N.A. (1929). *The Plants in Relation to Water.* Allen and Unwin, London, UK.

Maxwell, D.P., Laudenbach, D.E. and Huner, N.P.A. (1995). Redox regulation of light-harvesting complex II and cab mRNA abundance in *Dunaliella salina*. *Plant Physiology*, **109**, 787–795.

Maxwell, D.P., Wong, Y. and McIntosh, L. (1999). The alternative oxidase lowers mitochondrial reactive oxygen production in plant cells. *Proceedings of the National Academy of Sciences*, **96**, 8271–8276.

Maxwell, K. and Johnson, G.N. (2000). Chlorophyll fluorescence – a practical guide. *Journal of Experimental Botany*, **51**, 659–668.

Maxwell, K., von Caemmerer, S. and Evans, J.R. (1997). Is a low internal conductance to CO_2 diffusion a consequence of succulence in plants with crassulacean acid metabolism? *Australian Journal of Plant Physiology*, **24**, 777–786.

May, J.D. and Killingbeck, K.T. (1992). Effects of preventing nutrient resorption on plant fitness and foliar nutrient dynamics. *Ecology*, **73**, 1868–1878.

Mayr, S., Schwienbacher, F. and Bauer, H. (2003). Winter at the Alpine timberline. Why does embolism occur in Norway spruce but not in stone pine? *Plant Physiology*, **131**, 780–792.

Mazzafera, P., Kubo, R.K. and Inomotos, M.M. (2004). Carbon fixation and partitioning in coffee seedlings infested with *Pratylenchus coffeae*. *European Journal of Plant Pathology*, **110**, 861–865.

Mazzella, M.A. and Casal, J.J. (2001). Interactive signalling by phytochromes and cryptochromes generates de-etiolation homeostasis in *Arabidopsis thaliana*. *Plant Cell and Environment*, **24**, 155–161.

McAinsh, M.R., Evans, N.H., Montgomery, L.T., *et al.* (2002). Calcium signalling in stomatal responses to pollutants. *New Phytologist*, **153**, 441–447.

McCallum, C.M., Comai, L., Greene, E.A., *et al.* (2000a). Targeted screening for induced mutations. *Nature Biotechnology*, **18**, 455–457.

McCallum, C.M., Comai, L., Greene, E.A., *et al.* (2000b). Targeting induced local lesions in genomes (TILLING) for plant functional genomics. *Plant Physiology*, **123**, 439–442.

McCarron, J.K. and Knapp, A.K. (2001). C_3 woody plant expansion in a C4 grassland: are grasses and shrubs functionally distinct? *American Journal of Botany*, **88**, 1818–1823.

McCarthy, H.R., Pataki, D.E. and Jenerette, G.D. (2011). Plant water use efficiency as a metric of urban ecosystem services. *Ecological Applications*, **21**, 3115–3127.

McCarthy, I., Romero-Puertas, M.C., Palma, J.M., *et al.* (2001). Cadmium induces senescence symptoms in leaf peroxisomes of pea plants. *Plant, Cell and Environment*, **24**, 1065–1073.

McCool, M.M. (1935). Effect of light intensity on the manganese content of plants. *Contributions from Boyce Thompson Institute*, **7**, 427–437.

McCree, K.J. (1986). Measuring the whole-plant daily carbon balance. *Photosynthetica*, **20**, 82–93.

McCree, K.J. (1988). Sensitivity of sorghum grain yield to ontogenetic changes in respiration coefficients. *Crop Science*, **28**, 114–120.

McDermitt, D.K. (1990). Sources of error in the estimation of stomatal conductance and transpiration from porometer data. *HortScience*, **25**, 1538–1547.

McDonald, E.P., Agrell, J. and Lindroth, R.L. (1999). CO_2 and light effects on deciduous trees: growth, foliar chemistry, and insect performance. *Oecologia*, **119**, 389–399.

McDowell, N.G. (2011). Mechanisms linking drought, hydraulics, carbon metabolism, and vegetation mortality. *Plant Physiology*, **155**, 1051–1059.

McElrone, A.J. and Forseth, I.N. (2004). Photosynthetic responses of a temperate liana to *Xylella fastidiosa* infection and water stress. *Journal of Phytopathology*, **152**, 9–20.

McElrone, A.J., Sherald, J.L. and Forseth, I.N. (2003). Interactive effects of water stress and xylem-limited bacterial infection on the water relations of a host vine. *Journal of Experimental Botany*, **54**, 419–430.

McElwain, J.C., Beerling, D.J., Woodward, F.I. (1999). Fossil plants and global warming at the Triassic-Jurassic boundary. *Science*, **285**, 1386–1390.

McElwain, J.C. and Chaloner, W.G. (1995). Stomatal density and index of fossil plants track atmospheric carbon dioxide in the Paleozoic. *Annals of Botany*, **76**, 389–395.

McGonigle, B. and Nelson, T. (1995). C_4 isoform of NADP-malate dehydrogenase: cDNA cloning and expression in leaves of C_4, C_3, and C_3-C_4 intermediate species of Flaveria. *Plant Physiology*, **108**, 1119–1126.

McKee, I.F., Bullimore, J.F. and Long, S.P. (1997). Will elevated CO_2 concentrations protect the yield of wheat from O_3 damage? *Plant, Cell and Environment*, **20**, 77–84.

McKee, I.F., Farage, P.K. and Long, S.P. (1995). The interactive effects of elevated CO_2 and O_3 concentration on photosynthesis in spring wheat. *Photosynthesis Research*, **45**, 111–119.

McKenzie, R., Conner, B. and Bodeker, G. (1999). Increased summertime UV radiation in New Zealand in response to ozone loss. *Science*, **285**, 1709–1711.

McKinney, C.R., McCrea, J.M., Epstein, S., *et al.* (1950). Improvements in mass spectrometers for the measurement of small differences in isotope abundance ratios. *Review of Scientific Instruments*, **21**, 724–730.

McKown, A.D., Moncalvo, J.M. and Dengler, N.G. (2005). Phylogeny of *Flaveria* (Asteraceae) and inference of C-4 photosynthesis evolution. *American Journal of Botany*, **92**, 1911–1928.

McLeod, A.R. and Long, S.P. (1999). Free-air carbon dioxide enrichment (FACE) in Global Change Research. A review. *Advances in Ecological Research*, **28**, 1–15.

McMullen, C.R., Gardner, W.S. and Myers, G.A. (1978). Aberrant plastids in balley leaf tissue infected with balley stripe mosaic virus. *Phytopathology*, **68**, 317–325.

McNaughton, S.J., Milchunas, D. and Frank, D.A. (1996). How can net primary productivity be measured in grazing ecosystems? *Ecology*, **77**, 974–977.

McNevin, D.B., Badger, M.R., Whitney, S.M., *et al.* (2007). Differences in carbon isotope discrimination of three variant Rubiscos reflect differences in their catalytic mechanisms. *Journal of Biological Chemistry*, **282**, 36068–36076.

Medhurst, J., Parsby, J., Linder, S., *et al.* (2006). A whole-tree chamber system for examining tree-level physiological responses of field-grown trees to environmental variation and climate change. *Plant, Cell and Environment*, **29**, 1853–1869.

Medina, E. (1982). Temperature and humidity effects on dark CO_2 fixation by *Kalanchoë pinnata. Zeitschrift für Pflanzenphysiologie*, **107**, 251–258.

Medina, E. (1987). Ecophysiological aspects of CAM plants in the tropics. *Revista De Biologia Tropical*, **35**, 55–70.

Medina, E. and Delgado M. (1976). Photosynthesis and night CO_2 fixation in *Echeveria columbiana* Poellnitz. *Photosynthetica*, **10**, 155–163.

Medina, E. and Francisco, M. (1994). Photosynthesis and water relations of savanna tree species differing in leaf phenology. *Tree Physiology*, **14**, 1367–1381.

Medina, E., Cram, W.J., Lee, H.S.J., *et al.* (1989). Ecophysiology of xerophytic and halophytic vegetation on an alluvial plain in northern Venezuela. I. Site description and plant communities. *New Phytologist*, **111**, 233–243.

Medlyn, B.E. (2004). A MAESTRO retrospective. In: *Forests at the Land-atmosphere Interface* (eds Mencuccini, M., Grace, J.C., Moncrieff, J. and McNaughton, K.), CAB International, Wallingford, UK, pp. 105–121.

Medlyn, B.E., Badeck, F.W., De Pury, D.G.G., *et al.* (1999). Effects of elevated [CO_2] on photosynthesis in European forest species: a meta-analysis of model parameters. *Plant, Cell and Environment*, **22**, 1475–1495.

Medlyn, B.E., Barton, C.V.M., Broadmeadow, M.S.J., *et al.* (2001). Stomatal conductance of European forest species after long-term exposure to elevated [CO_2]: a synthesis of experimental data. *New Phytologist*, **149**, 247–264.

Medlyn, B.E., Berbigier, P., Clement, R., *et al.* (2005). The carbon balance of coniferous forests growing in contrasting climatic conditions: a model-based analysis. *Agricultural and Forest Meteorology*, **131**, 97–124.

Medlyn, B.E., Dreyer, E., Ellsworth, D., *et al.* (2002a). Temperature response of parameters of a biochemically based model of photosynthesis. II. A review of experimental data. *Plant, Cell and Environment*, **25**, 1167–1179.

Medlyn, B.E., Loustau, D. and Delzon, S. (2002b). Temperature response of parameters of a biochemically based model of photosynthesis. I. Seasonal changes in mature maritime pine (*Pinus pinaster* Ait). *Plant, Cell and Environment*, **25**, 1155–1165.

Medrano H., Keys, A.J., Lawlor, D.W., *et al.* (1995). Improving plant production by selection for survival at low CO_2 concentrations. *Journal of Experimental Botany*, **46**, 1389–1396.

Medrano, H., Bota, J., Abadía, A., *et al.* (2002a). Effects of drought on light-energy dissipation mechanisms in high-light-acclimated, field-grown grapevines. *Functional Plant Biology*, **29**, 1197–1207.

Medrano, H., Escalona, J.M., Bota, J., *et al.* (2002b). Regulation of photosynthesis of C_3 plants in response to progressive drought: stomatal conductance as a reference parameter. *Annals of Botany*, **89**, 895–905.

Medrano, H., Flexas, J. and Galmés, J. (2009). Variability in water use efficiency at the leaf level among Mediterranean plants with different growth forms. *Plant and Soil*, **317**, 17–29.

Medrano, H. and Primo-Millo, E. (1985). Selection of *Nicotiana tabacum* haploids of high photosynthetic effeciency. *Plant Physiology*, **79**, 505–508.

Meehl, G.A., Stocker, T.F., Collins, W.D., *et al.* (2007). Global climate projections. In: *Climate Change 2007: The Physical Science Basis. Contribution of Working Group I to the Fourth Assessment Report of the Intergovernmental Panel on Climate Change* (eds Solomon, S., Qin, D., Manning, M., Chen, Z., Marquis, M., Averyt, K.B., Tignor, M. and Miller, H.L.), Cambridge University Press, Cambridge, UK and New York, USA.

Meidner, H. (1975). Water supply, evaporation, and vapour diffusion in leaves. *Journal of Experimental Botany*, **26**, 666–672.

Meinzer, F.C. and Zhu, J. (1998). Nitrogen stress reduces the efficiency of the C_4 CO_2 concentrating system, and therefore quantum yield, in *Saccharum* (sugarcane) species. *Journal of Experimental Botany*, **49**, 1227–1234.

Meinzer, F.C.., Wisdom, C. S., Gonzalez-Coloma, A., *et al.* (1990). Effects of leaf resin on stomatal behaviour and gas exchange of *Larrea tridentata* (DC.) Cov. *Functional Ecology*, **4**, 579–84.

Meir, P., Kruijt, B., Broadmeadow, M., *et al.* (2002). Acclimation of photosynthetic capacity to irradiance in tree canopies in relation to leaf nitrogen concentration and leaf mass per unit area. *Plant Cell Enviroment*, **25**, 343–357.

Meister, M., Agostino, A. and Hatch, M.D. (1996). The roles of malate and aspartate in C_4 photosynthetic metabolism in *Flaveria bidentis* (L.). *Planta*, **199**, 262–269.

Melakeberhan, H., Ferris, H. and Dias, J.M. (1990). Physiological response of resistant and susceptible *Vitis vinifera* to *Meloidogyne incognita*. *Journal of Nematology*, **22**, 224–230.

Melillo, J.M., McGuire, A.D., Kicklighter, D.W., *et al.* (1993). Global climate change and terrestrial net primary production. *Nature*, **263**, 234–239.

Melis, A. (1999). Photosystem-II damage and repair cycle in chloroplasts: what modulates the rate of photodamage *in vivo*? *Trends in Plant Science*, **4**, 130–135.

Mellerowicz, E.J., Coleman, W.K., Riding, R.T., *et al.* (1992). Periodicity of cambial activity in *Abies balsamea*. I. Effects of temperature and photoperiod on cambial dormancy and frost hardiness. *Physiologia Plantarum*, **85**, 515–525.

Melotto, M., Underwood, W., Koczan, J., *et al.* (2006). Plant stomata function in innate immunity against bacterial invasion. *Cell*, **126**, 969–980.

Mencuccini, M., Mambelli, S. and Comstock, J. (2000). Stomatal responsiveness to leaf water status in common bean (*Phaseolus vulgaris* L.) is a function of time of day. *Plant Cell Environment*, **23**, 1109–1118.

Mendez, M., Jones, G.D. and Manetas, Y. (1999). Enhanced UV-B radiation under field conditions increases anthocyanin and reduces the risk of photoinhibition but does not affects growth in the carnivorous plant *Pinguicula vulgaris*. *New Phytologist*, **144**, 1–8.

Mendham, N.J., Rao, M.S.S. and Buzza, G.C. (1991). The apetalous flower character as a component of a high yielding ideotype. *Proceedings of the 8th International Rapeseed Congress*, Saskatoon, Canada, **2**, 596–600.

Merlot, S., Leonhardt, N., Fenzi, F. *et al.* (2007). Constitutive activation of a plasma membrane H+-ATPase prevents abscisic acid-mediated stomatal closure. *The EMBO Journal*, **26**, 3216–3226.

Meroni, M. and Colombo, R. (2006). Leaf level detection of solar induced chlorophyll fluorescence by means of a subnanometer resolution spectroradiometer. *Remote Sens. Environ.*, **103**, 438–448.

Meroni, M., Rossini, M., Guanter, L., *et al.* (2009). Remote sensing of solar- induced chlorophyll fluorescence: review of methods and applications. *Remote Sensing of the Environment*, **113**, 2037–2051.

Meroni, M., Rossini, M., Picchi, V., *et al.* (2008). Assessing steady state fluorescence and PRI from hyperspectral proximal sensing as early indicators of plant stress: the case of ozone exposure. *Sensors 2008*, **8**, 1740–1754.

Merzlyak, M.N. and Gitelson, A. (1995). Why and what for the leaves are yellow in autumn? On the interpretation of optical spectra of senescing leaves (*Acer platanoides* L.). *Journal of Plant Physiology*, **145**, 315–320.

Mesarch, M.A., Walter-Shea, E.A., Asner, G.P., *et al.* (1999). A revised measurement methology for conifer needles spectral optical properties: evaluating the influence of gaps between elements. *Remote Sensing of Environment*, **68**, 177–192.

Messinger, S., Buckley, T.N. and Mott, K.A. (2006). Evidence for involvement of photosynthetic processes in the stomatal response to CO_2. *Plant Physiology*, **140**, 771–778.

Mészáros, R., Zsely, I.G., Szinyei, D., *et al.* (2009). Sensitivity analysis of an ozone deposition model. *Atmospheric Environment*, **43**, 663–672.

Methy, M. (2000a). A two-channel hyperspectral radiometer for the assesment of photosynthetic radiation-use efficiency. *Journal of Agricole Engineering Research*, **75**, 107–110.

Méthy, M. (2000b). Stress-induced effects on *Quercus ilex* under a Mediterranean climate: contribution of chlorophyll fluorescece signatures. In: *Life and the Environment in the Mediterranean* (ed. Trabaud, L.), Wit Press, Southampton, USA, pp. 203–228.

Methy, M., Joffre, R. and Rambal, S. (1999). Remote sensing of canopy photosynthetic performances: two complementary ways for assessing the photochemical reflectance index. *Photosynthetica*, **37**, 239–247.

Meyer, A.J., Brach, T., Marty. L., *et al.* (2007). Redox-sensitive GFP in *Arabidopsis thaliana* is a quantitative biosensor for the redox potential of the cellular glutathione redox buffer. *Plant Journal*, **52**, 973–986.

Meyer, K.M., Ward, D., Moustakas, A., *et al.* (2005). Big is not better: small *Acacia mellifera* shrubs are more vital after fire. *African Journal of Ecology*, **43**, 131–136.

Meyer, M., Seibt, U. and Griffiths, H. (2008). To concentrate or ventilate? Carbon acquisition, isotope discrimination and physiological ecology of early land life forms. *Philosophical Transactions of the Royal Society B*, **363**, 2767–2778.

Meyer, S. and Genty, B. (1998). Mapping intercellular CO_2 mole fraction (C_i) in *Rosa rubiginosa* leaves fed with abscisic acid by using chlorophyll fluorescence imaging. Significance of C_i estimated from leaf gas exchange. *Plant Physiology*, **116**, 947–957.

Meyer, S., Cartelat, A., Moya, I., *et al.* (2003). UV-induced blue-green and far-red fluorescence along wheat leaves: a potential signature of leaf ageing. *Journal of Experimental Botany*, **54**, 757–769.

Meyer, S., Cerovic, Z.G., Goulas, Y., *et al.* (2006). Relationships between optically assessed polyphenols and chlorophyll concentrations, and leaf mass per area ratio in woody plants: a signature of the carbon-nitrogen balance within leaves? *Plant, Cell and Environment* **29**, 1338–1348.

Meyer, S., Saccardy Adji, K., Rizza, F., *et al.* (2001). Inhibition of photosynthesis by *Colletotrichum lindemuthianum* in bean leaves determined by chlorophyll fluorescence imaging. *Plant, Cell and Environment*, **24**, 947–955.

Miao, Z.W., Xu, M., Lathrop, R.G., *et al.* (2009). Comparison of the *A*-*Cc* curve fitting methods in determining maximum ribulose 1,5-bisphosphate carboxylase/oxygenase carboxylation rate, potential light saturated electron transport rate and leaf dark respiration. *Plant, Cell and Environment*, **32**, 109–122.

Micol, J.L. (2009). Leaf development: time to turn over a new leaf? *Current Opinion in Plant Biology*, **12**, 9–16.

Middleton, E.M., Kim, M.S., Krizek, D.T., *et al.* (2005). Evaluating UV-B effects and EDU protection in soybean leaves using fluorescence. *Photochemistry and Photobiology*, **81**, 1075–1085.

Mikkelsen, T.N. and Heide-Jųrgensen, H.S. (1996). Acceleration of leaf senescence in *Fagus sylvatica* L. by low levels of tropospheric ozone demonstrated by leaf colour, chlorophyll fluorescence and chloroplast ultrastructure. *Trees: Structure and Function*, **10**, 145–156.

Miles, C.D., Brandle, J.R., Daniel, D.J., *et al.* (1972). Inhibition of photosystem II in isolated chloroplasts by lead. *Plant Physiology*, **49**, 820–825.

Millar, A.H., Heazlewood, J.L., Kristensen, B.K., *et al.* (2005). The plant mitochondrial proteome. *Trends in Plant Science*, **10**, 36–43.

Millar, A.H., Mittova, V., Kiddle, G., *et al.* (2003). Control of ascorbate synthesis by respiration and its implications for stress responses. *Plant Physiology*, **133**, 443–447.

Miller, A., Schlagnhaufer, C., Spalding, M., *et al.* (2000). Carbohydrate regulation of leaf development: prolongation of leaf senescence in Rubisco antisense mutants of tobacco. *Photosynthesis Research*, **63**, 1–8.

Miller, A.M., van Iersel, M.W. and Armitage, A.M. (2001). Whole-plant carbon dioxide exchange responses of *Angelonia angustifolia* to temperature and irradiance. *Journal of the American Society for Horticultural Science*, **125**, 606–610.

Miller, E.W., Albers, A.E., Pralle, A., *et al.* (2005). Boronate-based fluorescent probes for imaging cellular hydrogen peroxide. *Journal of the American Chemical Society*, **127**, 16652–16659.

Miller, G., Suzuki, N., Ciftci-Yilmaz, S., *et al.* (2010). Reactive oxygen species homeostasis and signalling during drought and salinity stresses. *Plant, Cell and Environment*, **33**, 453–467.

Miller, J.B., Yakir, D., White, J.W.C., *et al.* (1999). Measurements of $^{18}O/^{16}O$ in the soil-atmosphere CO_2 flux. *Global Biogeochemical Cycles*, **13**, 761–774.

Miller, J.M., Williams, R.J. and Farquhar, G.D. (2001). Carbon isotope discrimination by a sequence of *Eucalyptus* species along a subcontinental rainfall gradient in Australia. Functional Ecology, **15**, 222–232.

Miller, P.M., Eddleman, L.E. and Miller, J.M. (1995). *Juniperus occidentalis* juvenile foliage: advantages and disadvantages for a stress-tolerant, invasive conifer. *Canadian Journal of Forest Research*, **25**, 470–479.

Miller, R.E., Watling, J.R. and Robinson, S.A. (2009). Functional transition in the floral receptacle of the sacred lotus (*Nelumbo nucifera*): from thermogenesis to photosynthesis. *Functional Plant Biology*, **36**, 471–480.

Mills, G., Ball, G., Hayes, F., *et al.* (2000). Development of a multi-factor model for predicting the critical level of ozone for white clover. *Environmental Pollution*, **109**, 533–542.

Mills, G.A. and Urey, H.C. (1940). The kinetics of isotopic exchange between carbon dioxide, bicarbonate ion, carbonate ion and water. *Journal of the American Chemical Society*, **62**, 1019–1026.

Miloslavina, Y., Nilkens, M., Jahns, P., *et al.* (2007). Reorganization of LHC II complexes in the thylakoid membrane in response to high light adaptation (NPQ). *Photosynthesis Research*, **91**, 256–256.

Minamikawa, T., Toyooka, K., Okamoto, T., *et al.* (2001). Degradation of ribulose-bisphosphate carboxylase by vacuolar enzymes of senescing French bean leaves: immunocytochemical and ultrastructural observations. *Protoplasma*, **218**, 144–153.

Miranda, T. and Ducruet, J.M. (1995). Characterization of the chlorophyll thermoluminescence afterglow in dark-adapted or far-red-illuminated plant leaves. *Plant Physiology and Biochemistry*, **33**, 689–699.

Mishra, R.K. and Singhal, G.S. (1993). Photosynthetic activity and peroxidation of thylakoid lipids during photoinhibition and high temperature treatment of isolated wheat chloroplasts. *Journal of Plant Physiology*, **141**, 286–292.

Misson, L., Baldocchi, D.D., Black, T.A., *et al.* (2007.) Partitioning forest carbon fluxes with overstory and understory eddy covariance measurements: a synthesis based on FLUXNET data. *Agricultural and Forest Meteorology*, **144**, 14–31.

Misson, L., Limousin, J.M., Rodriguez, R., *et al.* (2010). Leaf physiological responses to extreme droughts in Mediterranean *Quercus ilex* forest. *Plant, Cell and Environment*, **33**, 1898–1910.

Misson, L., Panek, J.A. and Goldstein, A.H. (2004). A comparison of three approaches to modeling leaf gas exchange in annually drought-stressed ponderosa pine forests. *Tree Physiology*, **24**, 529–541.

Mitchell, C.A. (1992). Measurement of photosynthetic gas exchange in controlled environments. *HortScience*, **27**, 764–767.

Mitchell, T.D., Carter, T.R., Jones, P.D., *et al.* (2004). A comprehensive set of high resolution grids of monthly climate for Europe and the globe: the observed record (1901–2000) and 16 scenarios (2001–2100). Working paper 55 (Tyndall Centre for Climate Change Research, July 2004); available at www.tyndall.ac.uk/publications/working_papers/wp55.pdf.

Mittler, R. (2002). Oxidative stress, antioxidants and stress tolerance. *Trends in Plant Science*, **7**, 405–410.

Mittler, R. (2006). Abiotic stress, the field environment and stress combination. *Trends in Plant Science*, **11**, 15–19.

Mittler, R. and Blumwald, E. (2010). Genetic engineering for modern agriculture: challenges and perspectives. *Annual Review of Plant Biology*, **61**, 443–462.

Mittler, R., Vanderauwera, S., Gollery, M., *et al.* (2004). Reactive oxygen gene network of plants. *Trends in Plant Science*, **9**, 490–498.

Miyaji, K., Da Silva, W.S. and De Paulo, T.A. (1997). Longeivity of leaves of a tropical tree, *Theombroma cacao*, grown under shading, in relation to position within the canopy and time of emergence. *New Phytologist*, **135**, 445–454.

Miyake, C. and Yokota, A. (2000). Determination of the rate of photoreduction of O_2 in the water-water cycle in watermelon leaves and enhancement of the rate by limitation of photosynthesis. *Plant Cell Physiology*, **41**, 335–343.

Miyashita, K., Tanakamaru, S., Maitani, T., *et al.* (2005). Recovery responses of photosynthesis, transpiration, and stomatal conductance in kidney bean following drought stress. *Environmental and Experimental Botany*, **53**, 205–214.

Miyazawa, S.I. and Terashima, I. (2001). Slow development of leaf photosynthesis in an evergreen broad-leaved tree, *Castanopsis sieboldii*: relationships between leaf anatomical characteristics and photosynthetic rate. *Plant, Cell and Environment*, **24**, 279–291.

Miyazawa, S.I., Makino, A. and Terashima, I. (2003). Changes in mesophyll anatomy and sink-source relationships during leaf development in *Quercus glauca*, an evergreen tree showing delayed leaf greening. *Plant Cell Environment*, **26**, 745–755.

Miyazawa, S.I., Satomi, S. and Terashima, I. (1998). Slow leaf development of evergreen broad-leaved tree species in Japanese warm temperate forests. *Annals of Botany*, **82**, 859–869.

Miyazawa, S.I., Yoshimura, S., Shinzaki, Y., *et al.* (2008). Deactivation of aquaporins decreases internal conductance to CO_2 diffusion in tobacco leaves grown under long term drought. *Functional Plant Biology*, **35**, 553–564.

Miyazawa, Y. and Kikuzawa, K. (2005a). Winter photosynthesis by saplings of evergreen broadleaved trees in a deciduous temperate forest. *The New Phytologist*, **165**, 857–866.

Miyazawa, Y. and Kikuzawa, K. (2005b). Physiological basis of seasonal trend in leaf photosynthesis of five evergreen broad-leaved species in a temperature deciduous forest. *Tree Physiology*, **26**, 249–256.

Miyazawa, Y. and Kikuzawa, K. (2006). Photosynthesis and physiological traits of evergreen broadleafed saplings during winter under different light environments in a temperate forest. *Canadian Journal of Botany*, **85**, 60–69.

Mochizuki, N., Tanaka, R., Tanaka, A., *et al.* (2008). The steady state level of mg-protoporphyrin IX is not a determinant of plastid-to-nucleus signalling in Arabidopsis. *Proceedings of the National Academy of Sciences USA*, **105**, 15,184–15,189.

Moffat, A.M., Papale, D., Reichstein, M., *et al.* (2007). Comprehensive comparison of gap-filling techniques for eddy covariance net carbon fluxes. *Agricultural and Forest Meteorology*, **147**, 209–232.

Moghaddam, P.R. and Wilman. D. (1998). Cell wall thickness and cell dimensions in plant parts of eight forage species. *Journal of Agricultural Science Cambridge*, **131**, 59–67.

Moghaieb, R.E.A., Tanaka, N., Saneoka, H., *et al.* (2008). Characterization of salt tolerance in ectoine-transformed tobacco plants (*Nicotiana tabaccum*): photosynthesis, osmotic adjustment, and nitrogen partitioning. *Plant, Cell and Environment*, **29**, 173–182.

Mohanty, P., Vani, B. and Prakash, J.S.S. (2002). Elevated temperature treatment induced alteration in thylakoid membrane organization and energy distribution between the two photosystems in *Pisum sativum*. Zeitschrift fur Natuforschung part C, **57**, 836–842.

Moise, N. and Moya, I. (2004a). Correlation between lifetime heterogeneity and kinetics heterogeneity during chlorophyll fluorescence induction in leaves: 1. Mono-frequency phase and modulation analysis reveals a conformational change of a PSII pigment complex during the IP thermal phase. *Biochimica et Biophysica Acta*, **1657**, 33–46.

Moise, N. and Moya, I. (2004b). Correlation between lifetime heterogeneity and kinetics heterogeneity during chlorophyll fluorescence induction in leaves: 2. Multi-frequency phase and modulation analysis evidences a loosely connected PSII pigment-protein complex. *Biochimica et Biophysica Acta*, **1657**, 47–60.

Mommer, L., Pons, T.L., Wolters-Arts, M., *et al.* (2005). Submergence-induced morphological, anatomical, and biochemical responses in a terrestrial species affect gas diffusion resistance and photosynthetic performance. *Plant Physiology*, **139**, 497–508.

Monclus, R., Dreyer, E., Delmotte, F.M., *et al.* (2005) Productivity, leaf traits and carbon isotope discrimination in 29 *Populus deltoides* × *P. nigra* clones. *New Phytologist*, **167**, 53–62.

Monclus, R., Dreyer, E., Villar, M., *et al.* (2006). Impact of drought on productivity and water use efficiency in 29 genotypes of *Populus deltoides* × *Populus nigra*. *New Phytologist*, **169**, 765–777.

Moncrieff, J.B., Malhi, Y. and Leuning, R. (1996). The propagation of errors in long-term measurements of land atmosphere fluxes of carbon and water. *Global Change Biology*, **2**, 231–240.

Monje, O. and Bugbee, B. (1998). Adaptation to high CO_2 concentration in an optimal environment; radiation capture, canopy quantum yield and carbon use efficiency. *Plant Cell and Environment*, **21**, 315–324.

Monk, C.D. (1966). An ecological significance of evergreens. *Ecology*, **47**, 504–505.

Monnet, F., Vaillant, N., Vernay, P., *et al.* (2001). Relationship between PSII activity, CO_2 fixation, and Zn, Mn and Mg contents of *Lolium perenne* under zinc stress. *Journal of Plant Physiology*, **158**, 1137–1144.

Monroy, A.F. and Dhindsa, R.S. (1995). Low-temperature signal transduction: induction of cold acclimation-specific genes of alfalfa by calcium at 25 degrees C. *Plant Cell*, **7**, 321–331.

Monsi, M. and Saeki, T. (1953). Über den lichtfaktor in den pflanzengesellschaften und seine bedeutung für die stoffproduktion. *Japanese Journal of Botany*, **14**, 22–52.

Monsi, M. and Saeki, T. (2005). On the factor light in plant communities and ist importance for matter production. *Annals of Botany*, **95**, 549–567.

Monson, R.K. (1989a). The relative contributions of reduced photorespiration, and improved water- and nitrogen-use efficiencies, to the advantages of C_3-C_4 intermediate photosynthesis in *Flaveria*. *Oecologia*, **80**, 215–221.

Monson, R.K. (1989b). On the evolutionary pathways resulting in C_4 photosynthesis and crassulacean acid metabolism. *Advances in Ecological Research*, **19**, 57–110.

Monson, R.K. (1999). The origins of C_4 genes and evolutionary pattern in the C_4 metabolic phenotype. In: *C_4 Plant Biology* (eds Sage, R.F. and Monson, R.K.), Academic Press, San Diego, CA, USA, pp. 377–410.

Monson, R.K. (2003). Gene duplication, neofunctionalization and the evolution of C_4 photosynthesis. *International Journal of Plant Sciences*, **164 (suppl)**, S43-S54.

Monson, R.K. and Rawsthorne, S. (2000). CO_2 assimilation in C_3-C_4 intermediate plants. In: *Photosynthesis: Physiology and Metabolism, Advances in Photosynthesis* (eds Leegood, R.C., Sharkey, T.D. and von Caemmerer, S.), Kluwer Academic Press, New York, USA, pp. 533–550.

Monson, R.K., Littlejohn, R.O. Jr and Williams, G.J. III (1983). Photosynthetic adaptation to temperature in four species from the Colorado shortgrass steppe: a physiological model for coexistence. *Oecologia*, **58**, 43–51.

Monson, R.K., Sparks, J.P., Rosenstiel, T.N., *et al.* (2005). Climatic influences on net ecosystem CO_2 exchange during the transition from wintertime carbon source to springtime carbon sink in a high-elevation, subalpine forest. *Oecologia*, **146**, 130–147.

Monson, R.K., Turnipseed, A.A., Sparks, J.P. *et al.* (2002). Carbon sequestration in a high-elevation, subalpine forest. *Global Change Biology*, **8**, 459–478.

Montagu, K.D. and Woo, K.C. (1999). Recovery of tree photosynthetic capacity from seasonal drought in the wet-dry tropics: the role of phyllode and canopy processes in *Acacia auriculiformis*. *Australian Journal of Plant Physiology*, **26**, 135–145.

Montalbini, P. and Lupattelli, M. (1989). Effect of localized and systemic tobacco mosaic-virus infection on some photochemical and enzymatic-activities of isolated tobacco chloroplasts. *Physiological and Molecular Plant Pathology*, **34**, 147–162.

Montanaro, G., Dichio, B. and Xiloyannis, C. (2007). Response of photosynthetic machinery of field-grown kiwifruit under Mediterranean conditions during drought and re-watering. *Photosynthetica*, **45**, 533–540.

Monteiro, J.A.F. and Prado, C.H.B.A. (2006). Apparent carboxylation efficiency and relative stomatal and mesophyll limitations of photosynthesis in an evergreen cerrado species during water stress. *Photosynthetica*, **44**, 39–45.

Monteith, J.L. (1972). Solar radiation and productivity in tropical ecosystems. *Journal of Applied Ecology*, **9**, 747–766.

Monteith, L.J. (1973). *Principles of Environmental Physics*. American Elsevier, New York, USA, pp. 243.

Monteith, J.L. (1984). Consistency and convenience in the choice of units for agricultural science. *Experimental Agriculture*, **20**, 105–117.

Monteith, J.L. and Unsworth, M.H. (1990). *Principles of Environmental Physics*. Edward Arnold, London, UK.

Montesano, M., Scheller, H.R., Wettstein, R., *et al.* (2004). Down-regulation of photosystem I by *Erwinia carotovora*-derived elicitors correlates with H_2O_2 accumulation in chloroplasts of potato. *Molecular Plant Pathology*, **5**, 115–123.

Montgomery, R., Goldstein, G. and Givnish, T.J. (2008). Photoprotection of PSII in Hawaiian lobeliads from diverse light environments. *Functional Plant Biology*, **35**, 595–605.

Montgomery, R.A. (2004a). Effects of understory foliage on patterns of light attentuation near the forest floor. *Biotropica*, **36**, 33–39.

Montgomery, R.A. (2004b). Relative importance of photosynthetic physiology and biomass allocation for tree seedling growth across a broad light gradient. *Tree Physiology*, **24**, 155–167.

Montgomery, R.A. and Givnish, T.J. (2008). Adaptive radiation of photosynthetic physiology in the Hawaiian lobeliads: dynamic photosynthetic responses. *Oecologia*, **155**, 455–467.

Monti, A., Bezzi, G. and Venturi, G. (2009). Internal conductance under different light conditions along the plant profile of Ethiopian mustard (*Brassica carinata* A. Brown). *Journal of Experimental Botany*, **60**, 2341–2350.

Monti, A., Brugnoli, E, Scartazza, A., *et al.* (2006). The effect of transient and continuous drought on yield, photosynthesis and carbon isotope discrimination in sugar beet (*Beta vulgaris* L.). *Journal of Experimental Botany*, **57**, 1253–1262.

Montpied, P., Granier, A. and Dreyer, E. (2009). Seasonal time-course of gradients of photosynthetic capacity and mesophyll conductance to CO_2 across a beech (*Fagus sylvatica* L.) canopy. *Journal of Experimental Botany*, **60**, 2407–2418.

Montzka, S.A., Calvert, P., Hall, B.D., *et al.* (2007). On the global distribution, seasonality, and budget of atmospheric carbonyl sulfide (COS) and some similarities to CO_2. *Journal of Geophysical Research – Atmospheres* 112: Article number D09302.

Mook, W.G. (1986). ^{13}C in atmospheric CO_2. *Netherlands Journal Sea Research*, **20**, 211–223.

Mook, W.G., Bommerson, J.C., Staverman, W.H. (1974). Carbon isotope fractionation between dissolved carbonate and gaseous carbon dioxide. *Earth Planet Science Letters*, **22**, 169–176.

Mook, W.G., Koopmans, M., Carter, A.F., Keeling C.D. (1983). Seasonal, latitudinal and secular variations in the abundance and isotopic ratios of atmospheric carbon dioxide. 1. Results from land stations. *Journal of Geophysical Research*, **88**, 10915–10933.

Mooney, H.A. (1972). The carbon balance of plants. *Annual Review of Ecology and Systematics*, **3**, 315–346.

Mooney, H.A. and Billings, W.D. (1961). Comparative physiological ecology of Arctic and Alpine populations of *Oxyria digyna*. *Ecological Monographs*, **31**, 1–29.

Mooney, H.A. and Ehleringer, J.R. (1978). The carbon gain benefits of solar tracking in a desert annual. *Plant, Cell and Environment*, **1**, 307–11.

Mooney, H.A., Björkman, O. and Collatz, G.J. (1978). Photosynthetic acclimation to temperature in the desert shrub *Larrea divaricata*. I. Carbon dioxide exchange characteristics of intact leaves. *Plant Physiology*, **61**, 406–410.

Mooney, H.A., Ehleringer, J.R. and Berry, J.A. (1976). High photosynthetic capacity of a winter desert annual in Death Valley. *Science*, **194**, 322–324.

Mooney, H.A., Field, C., Gulmon, S. L., *et al.* (1981). Photosynthetic capacity in relation to leaf positions in desert versus old-field annuals. *Oecologia*, **50**, 109–112.

Mooney, H.A., Pearcy, R.W. and Ehleringer J. (1987). Plant physiological ecology today. *BioScience*, **37**, 18–20.

Moore, B.D. and Seemann, J.R. (1992). Metabolism of 2'-carboxyarabinitol in leaves. *Plant Physiolology*, **99**, 1551–1555.

Moore, B.D., Cheng, S.H., Sims, D., *et al.* (1999). The biochemical and molecular basis for photosynthetic acclimation to elevated atmospheric CO_2. *Plant Cell and Environment*, **22**, 567–582.

Moore, R.C. and Purugganan, M.D. (2005). The evolutionary dynamics of plant duplicate genes. *Current Opinion in Plant Biology*, **8**, 122–128.

Morales, F., Abadía, A. and Abadía, J. (1990). Characterization of the xanthophyll cycle and other photosynthetic pigment changes induced by iron deficiency in sugar beet (*Beta vulgaris* L.). *Plant Physiology*, **94**, 607–613.

Morales, F., Abadía, A. and Abadía, J. (1991). Chlorophyll fluorescence and photon yield of oxygen evolution in iron-deficient sugar beet (*Beta vulgaris* L.) leaves. *Plant Physiology*, **97**, 886–893.

Morales, F., Abadía, A. and Abadía, J. (1998). Photosynthesis, quenching of chlorophyll fluorescence and thermal energy dissipation in iron-deficient sugar beet leaves. *Australian Journal of Plant Physiology*, **25**, 403–412.

Morales, F., Abadía, A. and Abadía, J. (2006). Photoinhibition and photoprotection under nutrient defiencies, drought and salinity. In: *Photoprotection, Photoinhibition, Gene regulation and Environment* (eds Demmig-Adams, B., Adams, W.W. III, and Mattoo, A.K.), Springer, Dordrecht, Netherlands, pp 65–85.

Morales, F., Abadía, A., Abadía, J., *et al.* (2002). Trichomes and photosynthetic pigment composition changes: responses of *Quercus ilex* subsp *ballota* (Desf.) Samp. and *Quercus coccifera* L. to Mediterranean stress conditions. *Trees, Structure and Function*, **16**, 504–510.

Morales, F., Belkhodja, R., Abadía, A., *et al.* (1994). Iron deficiency induced changes in the photosynthetic pigment composition of field-grown pear (*Pyrus communis* L.) leaves. *Plant, Cell and Environment*, **17**, 1153–1160.

Morales, F., Belkhodja, R., Abadía, A., *et al.* (2000). Photosystem II efficiency and mechanisms of energy dissipation in iron-deficient, field-grown pear trees (*Pyrus communis* L.). *Photosynthesis Research*, **63**, 9–21.

Morales, F., Belkhodja, R., Goulas, Y., *et al.* (1999). Remote and near-contact chlorophyll fluorescence during photosynthetic induction in iron-deficient sugar beet leaves. *Remote Sensing of the Environment*, **69**, 170–178.

Morales, F., Cartelat, A., Álvarez-Fernández, A., *et al.* (2005). Time-resolved spectral studies of blue-green fluorescence of artichoke (*Cynara cardunculus* L. Var. Scolymus) leaves: identification of CGA as one of the major fluorophores and age-mediated changes. *Journal of Agricultural and Food Chemistry*, **53**, 9668–9678.

Morales, F., Cerovic, Z.G. and Moya, I. (1996). Time resolved blue-green fluorescence of sugar beet (*Beta vulgaris* L.) leaves. Spectroscopic evidence for the presence of ferulic acid as the main fluorophore of the epidermis. *Biochimica et Biophysica Acta*, **1273**, 251–262.

Morales, F., Moise, N., Quílez, R., *et al.* (2001). Iron deficiency interrupts energy transfer from a disconnected part of the antenna to the rest of photosystem II. *Photosynthesis Research*, **70**, 207–220.

Mordelet, P. (1993) Influence of tree shading on carbon assimilation of grass leaves in Lamto Savanna, Côte d'Ivoire. *Acta Oecologica-International Journal of Ecology*, **14**, 119–127.

Mordelet, P. and Menaut, J.C. (1995). Influence of trees on aboveground production dynamics of grasses in a humid savanna. *Journal of Vegetation Science*, **6**, 223–228.

Morecroft, M.D. and Woodward, F.I. (1996). Experiments on the Causes of Altitudinal Differences in the Leaf Nutrient Contents, Size and $\delta^{13}C$ of *Alchemilla alpina*. *New Phytologist*, **134**, 471–479.

Morecroft, M.D., Stokes, V.J. and Morison, J.I.L. (2003). Seasonal changes in the photosynthetic capacity of canopy oak (*Quercus robur*) leaves: the impact of slow development on annual carbon uptake. *Int J Biometeorol*, **47**, 221–226.

Moreira, A.G. (2000). Effects of fire protection on savanna structure in central Brazil. *Journal of Biogeography*, **27**, 1021–1029.

Moreno, J., Gracia-Murria, M.J. and Marin-Navarro, J. (2008). Redox modulation of Rubisco conformation and activity through its cysteine residues. *Journal of Experimental Botany*, **59**, 1605–1614.

Moreno, M.T., Riera, D., Carambula, C., *et al.* (2006). Estimation of water use efficiency in three cultivars of *Dactylis glomerata L.* under different soil water contents. 21st General Meeting of the European Grasslands Federation, 2–6 April, Badajoz (Spain).

Moreno-Sotomayor, M., Weiss, A., Paparozzi, E.T., *et al.* (2002). Stability of leaf anatomy and light response curves of field grown maize as a function of age and nitrogen status. *Journal of Plant Physiology*, **159**, 819–826.

Morgan, J.A., LeCain, D.R., Mosier, A.R., *et al.* (2001). Elevated CO_2 enhances water relations and productivity and affects gas exchange in C_3 and C_4 grasses of the Colorado shortgrass steppe. *Global Change Biology*, **7**, 451–466.

Morgan, J.A., Milchunas, D.G., Lecain, D.R., *et al.* (2007). Carbon dioxide enrichment alters plant community structure and promotes shrub growth in the shortgrass steppe. *Proceedings of the National Academy of Sciences*, **104**, 14,724–14,729.

Morgan, J.A., Pataki, D.E., Korner, C., *et al.* (2004). Water relations in grassland and desert ecosystems exposed to elevated atmospheric CO_2. *Oecologia*, **140**, 11–25.

Morgan, M.E., Kingston, J.D. and Marino, B.D. (1994). Carbon isotopic evidence for the emergence of C_4 plants in the Neogene from Pakistan and Kenya. *Nature*, **367**, 162–165.

Morgan, P.B., Ainsworth, E.A. and Long, S.P. (2003). How does elevated ozone impact soybean? A meta-analysis of photosynthesis, growth and yield. *Plant, Cell and Environment*, **26**, 1317–1328.

Morgan, P.B., Bernacchi, C.J., Ort, D.R., *et al.* (2004). An in vivo analysis of the effect of season-long open-air elevation of ozone to anticipated 2050 levels on photosynthesis in soybean. *Plant Physiology*, **135**, 2348–2357.

Morgenstern, K., Black, T.A., Humpreys, E.R., *et al.* (2004). Sensitivity and uncertainty of the carbon balance of a Pacific Northwest Douglas-fir forest during an El Niño/La Niña cycle. *Agricultural and Forest Meteorology*, **123**, 201–219.

Morgenthal, K., Weckwerth, W. and Steuer, R. (2006). Metabolomic networks in plants: transitions from pattern recognition to biological interpretation. *Biosystems*, **83**, 108–117.

Moriana, A., Villalobos, F.J. and Fereres, E. (2002). Stomatal and photosynthetic responses of olive (*Olea europaea* L.) leaves to water deficits. *Plant, Cell and Environment*, **25**, 395–405.

Moriondo, M., Orlandini, S., Giuntoli, A., *et al.* (2005). The effect of downy and powdery mildew on grapevine (*Vitis vinifera* L.) leaf gas exchange. *Journal of Phytopathology*, **153**, 350–357.

Morison, J.I.L. (1985). Sensitivity of stomata and water use efficiency to high CO_2. *Plant, Cell and Environment*, **8**, 467–474.

Morison, J.I.L. and Lawson, T. (2007). Does lateral gas diffusion in leaves matter? *Plant, Cell and Environment*, **30**, 1072–1085.

Morison, J.I.L., Baker, N.R., Mullineaux, P.M., *et al.* (2008). Improving water use efficiency in crop production. *Philosophical transactions of the Royal Society*, **363**, 639–658.

Morison, J.I.L., Gallouët, E., Lawson, T., *et al.* (2005). Lateral diffusion of CO_2 in leaves is not sufficient to support photosynthesis. *Plant Physiology*, **139**, 254–266.

Morison, J.I.L., Lawson, T. and Cornic, G. (2007). Lateral CO_2 diffusion inside dicotyledonous leaves can be substantial: Quantification in different light intensities. *Plant Physiology*, **145**, 680–690.

Morita, N. (1935). The increased density of air oxygen relative to water oxygen. *Journal of the Chemical Society of Japan*, **56**, 1291.

Morley, R.J. (2000). *The Origin and Evolution of Tropical Rain Forests*. John Wiley and Sons, Chichester, UK and New York, USA.

Morley, R.J. and Richards, K. (1993). Gramineae cuticle: a key indicator of late Cenozoic climatic change in the Niger Delta. *Review of Palaeobotany and Palynology*, **77**, 119–127.

Morrison, M.J., Voldeng, H.D. and Cober, E.R. (1999). Physiological changes from 58 years of genetic improvement of short-season soybean cultivars in Canada. *Agronomy Journal*, **91**, 685–689.

Moser, W. (1973). Licht, Temperatur und Photosynthese an der Station "Hoher Nebelkogel" (3184m). In: *Ökosystemforschung* (ed. Ellenberg, H.) Springer, Berlin, Germany, pp. 203–223.

Mott, K.A. (1988). Do stomata respond to CO_2 concentrations other than intercellular? *Plant Physiology*, **86**, 200–203.

Mott, K.A. (1995). Effects of patchy stomatal closure on gas exchange measurements following abscisic acid treatment. *Plant, Cell and Environment*, **18**, 1291–1300.

Mott, K.A. and Parkhurst, D.F. (1991). Stomatal responses to humidity in air and helox. *Plant, Cell and Environment*, **14**, 509–515.

Mott, K.A., Gibson, A.C. and O'Leary, J.W. (1982). The adaptive significance of amphistomatic leaves. *Plant, Cell and Environment*, **5**, 455–460.

Mott, K.A., Sibbernsen, E.D. and Shope, J.C. (2008). The role of the mesophyll in stomatal responses to light and CO_2. *Plant, Cell Environment*, **31**, 1299–1306.

Moulin, M., McCormac, A.C., Terry, M.J., *et al.* (2008). Tetrapyrrole profiling in *Arabidopsis* seedlings reveals that retrograde plastid nuclear signalling is not due to mg-protoporphyrin IX accumulation. *Proceedings of the National Academy of Sciences* (USA), **105**, 15178–15183.

Moustakas, M. and Ouzounidou, G. (1994). Increased non-photochemical quenching in leaves of aluminum-stressed wheat plants is due to Al3+-induced elemental loss. *Plant Physiology and Biochemistry*, **32**, 527–532.

Moustakas, M., Ouzounidou, G. and Lannoye, R. (1995). Aluminum effects on photosynthesis and elemental uptake in an aluminum-tolerant and non-tolerant wheat cultivar. *Journal of Plant Nutrition*, **18**, 669–683.

Moustakas, M., Ouzounidou, G., Symeonidis, L., *et al.* (1997). Field study of the effects of excess copper on wheat photosynthesis and productivity. *Soil Science and Plant Nutrition*, **43**, 531–539.

Moya, I. (1974). Durée de vie et rendement de fluorescence de la chlorophylle in vivo. Leur relation dans differents modeles d'unites photosynthètique. *Biochimica et Biophysica Acta*, **368**, 214–227.

Moya, I. and Cerovic, Z.G. (2004). Remote sensing of chlorophyll fluorescence : instrumentation and analysis.

In: *Chlorophyll Fluorescence: The Signature of Green Plant Photosynthesis* (eds Papageorgiou, G.C. and Govindgee, J.), Kluwer, Dordrecht, Netherlands, pp. 429–445.

Moya, I., Camenen, L., Evain, S., *et al.* (2004). A new instrument for passive remote sensing: 1. Measurements of sunlight induced chlorophyll fluorescence. *Remote Sens. Environ.*, **91**, 186–197.

Moya, I., Camenen, L., Latouche, G., *et al.* (1999). An instrument for the measurement of sunlight excited plant fluorescence. In: *Photosynthesis: Mechanisms and Effects* (ed. Garab, G.), Kluwer Academic Press, Dordrecht, Netherlands, pp. 4265–4270.

Moya, I., Cartelat, A., Cerovic, Z.G., *et al.* (2003). Possible approaches to remote sensing of photosynthetic activity. IEEE International Geoscience and Remote Sensing Symposium (IGARSS 2003) 21–25 July 2003. Toulouse, France.

Moya, I., Daumard, F., Moise, N., *et al.* (2006). First airborne multiwavelength passive chlorophyll fluorescence measurements over La Mancha (Spain) fields. The 2nd International Symposium on Recent Advances in Quantitative Remote Sensing: RAQRS'II 25–29 September 2006. Torrent (Valencia)-Spain, C50 Torrent Laurent.

Moya, I., Goulas, Y., Morales, F., *et al.* (1995). Remote sensing of time-resolved chlorophyll fluorescence and back-scattering of the laser excitation by vegetation. *EARSeL Advances in Remote Sensing*, **3**, 188–197.

Moya, I., Guyot, G. and Goulas, Y. (1992). Remotely sensed blue and red fluorescence emission for monitoring vegetation. *ISPRS Journal of Photogrammetry and Remote Sensing*, **47**, 205–231.

Moya, I., Sebban, P. and Haehnel, W. (1986). Lifetime of excited states and quantum yield of chlorophyll a fluorescence in vivo. In: *Light Emission by Plants and Bacteria* (eds Govindjee, J., Amesz, J. and Fork, D.C.), Academic Press, Orlando, USA, pp. 161–190.

Muchow, R.C. and Sinclair, T.R. (1994). Nitrogen response of leaf photosynthesis and canopy radiation use efficiency in field-grown maize and sorghum. *Crop Science*, **34**, 721–727.

Muhlenbock, P., Szechynska-Hebda, M., Plaszczyca, M., *et al.* (2008). Chloroplast signalling and LESION SIMULATING DISEASE1 regulate crosstalk between light acclimation and immunity in *Arabidopsis*. *Plant Cell*, **20**, 2339–2356.

Mulkey, S.S., Wright, S.J. and Smith, A.P. (1993). Comparative phsiology and demography of three Neotropical forest shrubs: alternative shad-adaptive character syndromes. *Oecologia*, **96**, 526–536.

Mullen, J.L., Weinig, C. and Hangarter, R.P. (2006). Shade avoidance and the regulation of leaf inclination in *Arabidopsis*. *Plant Cell Environment*, **29**, 1099–1106.

Mųller, C.M. (1946). Untersuchungen über Laubmenge, Stoffverlust und Stoffproduktion des Waldes.

Müller, D.J., Dencher, N.A., Meier, T., *et al.* (2001). ATP synthase: constrained stoichiometry of the transmembrane rotor. *FEBS Letters*, **504**, 219–222.

Müller, J., Eschenröder, A. and Diepenbrock, W. (2009). Through-flow chamber CO_2/H_2O canopy gas exchange system – construction, microclimate, errors, and measurements in a barley (*Hordeum vulgare* L.) field. *Agricultural and Forest Meteorology*, **149**, 214–229.

Müller, M., Zellnig G., Tausz M., *et al.* (1997). Structural changes and physiological stress response of spruce trees to SO_2, O_3 and elevated levels of CO_2. In: *Impact of Global Change on Tree Physiology and Forest Ecosystems* (eds Mohren, G.M.J., Kramer, K. and Sabaté, S.), Kluwer Academic Publishers, Dordrecht, Netherlands, pp. 93–102.

Müller, P., Li, X.P. and Niyogi, K.K. (2001). Non-photochemical quenching. A response to excess light energy. *Plant Physiology* **125**, 1558–1566.

Mulligan, D.R. (1989). Leaf phosphorus and nitrogen concentrations and net photosynthesis in *Eucalyptus* seedlings. *Tree Physiology*, **5**, 149–157.

Mullin, L.P., Sillett, S.C., Koch, G.W., *et al.* (2009). Physiological consequences of height-related morphological variation in *Sequoia sempervirens* foliage. *Tree Physiology*, **29**, 999–1010.

Mullineaux, P. and Karpinski, S. (2002). Signal transduction in response to excess light: getting out of the chloroplast. *Current Opinion in Plant Biology*, **5**, 43–48.

Mulroy, T.W. and Rundel, P.W. (1977). Annual plants: adaptations to desert environments. *BioScience*, **27**, 109–114.

Munné-Bosch, S. (2007). Ageing in perennials. *Critical Reviews in Plant Sciences*, **26**, 123–138.

Munné-Bosch, S. (2008). Do perennials really senesce? *Trends in Plant Science*, **13**, 216–220.

Munné-Bosch, S., Alegre, L. (2000). Changes in carotenoids, tocopherols and diterpenes during drought and recovery, and the biological significance of chlorophyll loss in *Rosmarinus officinalis* plants. *Planta*, **210**, 925–931.

Munné-Bosch, S. and Alegre, L. (2002). Plant aging increases oxidative stress in chloroplasts. *Planta*, **214**, 608–615.

Munné-Bosch, S. and Alegre, L. (2004). Die and let live: leaf senescence contributes to plant survival under drought stress. *Functional Plant Biology*, **31**, 203–216.

Munné-Bosch, S. and Peñuelas, J. (2003). Photo- and antioxidative protection during summer leaf senescence in *Pistacia lentiscus* L. grown under Mediterranean field conditions. *Annals of Botany*, **92**, 385–391.

Munné-Bosch, S., Jubany Marí, T. and Alegre, L. (2001). Drought-induced senescence is characterized by a loss of antioxidant defences in chloroplasts. *Plant Cell Environment*, **24**, 1319–1327.

Munné-Bosch, S., Nogues, S. and Alegre, L. (1999). Diurnal variations of photosynthesis and dew absorption by leaves in two evergreen shrubs growing in Mediterranean field conditions. *New Phytologist*, **144**, 109–119.

Munné-Bosch, S., Peñuelas, J., Asensio, D., *et al.* (2004). Airborne ethylene may alter antioxidant protection and reduce tolerance of holm oak to heat and drought stress. *Plant Physiology*, **136**, 2937–2947.

Munns, R. (2002). Comparative physiology of salt and water stress. *Plant, Cell and Environment*, **25**, 239–250.

Munns, R. and Tester, M. (2008). Mechanisms of salinity tolerance. *Annual Review of Plant Biology*, **59**, 651–681.

Munns, R., James, R.A. and Läuchli, A. (2006). Approaches to increasing the salt tolerance of wheat and other cereals. *Journal of Experimental Botany*, **57**, 1025–1043.

Murata, N. and Los, D.A. (1997). Membrane fluidity and temperature perception. *Plant Physiology*, **115**, 875–879.

Murata, N. and Los, D.A. (2006). Histidine kinase Hik33 is an important participant in cold-signal transduction in cyanobacteria. *Physiologia Plantarum*, **126**, 17–27.

Murata, N., Takahashi, S., Nishiyama, Y., *et al.* (2007). Photoinhibition of photosystem II under environmental stress. *Biochimica et Biophysica Acta – Bioenergetics*, **1767**, 414–421.

Murchie, E.H., Pinto, M. and Horton, P. (2009). Agriculture and the new challenges for photosynthesis research. *The New Phytologist*, **181**, 532–552.

Murphy, R. and Smith, J.A.C. (1998). Determination of cell water-relation parameters using the pressure-clamp technique. *Plant, Cell and Environment*, **21**, 637–657.

Murray, M.B., Smith, R.I., Leith, I.D., *et al.* (1994). Effects of elevated CO_2, nutrition and climatic warming on bud phenology in Sitka spruce (*Picea sitchensis*) and their impact on the risk of frost damage. *Tree Physiology*, **14**, 691–706.

Murthy, R., Barron-Gafford, G.A., Dougherty, P.M., *et al.* (2005). Increased leaf area dominates carbon flux response to elevated CO_2 in stands of *Populus deltoides* (Bartr.) and underlies a switch from canopy light-limited CO_2 influx in well-watered treatments to individual leaf, stomatally limited influx under water stress. *Global Change Biology*, **11**, 716–731.

Mustilli, A.C., Merlot, S., Vavasseur, A., *et al.* (2002). *Arabidopsis* OST1 protein kinase mediates the regulation of stomatal aperture by abscisic acid and acts upstream of reactive oxygen species production. *The Plant Cell*, **14**, 3089–3099.

Myers, B.A., Duff, G.A., Eamus, D., *et al.* (1997). Seasonal variation in water relations of trees of differing leaf phenology in a wet-dry tropical savanna near Darwin, northern Australia. *Australian Journal of Botany*, **45**, 225–240.

Myers, J.A. and Kitajima, K. (2007). Carbohydrate storage enhances seedling shade and stress tolerance in a neotropical forest. *Journal of Ecology*, **95**, 383–395.

Myneni, R.B., Hoffman, S., Knyazikhin, Y., *et al.* (2002). Global products of vegetation leaf area and fraction absorbed PAR from one year of MODIS data. *Remote Sensing of Environment*, **76**, 139–155.

Nabity, P.D., Zavala, J.A. and DeLucia, E.H. (2009). Indirect suppression of photosynthesis on individual leaves by arthropod herbivory. *Annals of botany*, **103**, 655–663.

Naidu, R.A., Krishnan, M., Nayudu, M.V., *et al.* (1986). Studies on Peanut green mosaic virus infected peanut (*Arachis hypogaea* L.) leaves. III. Changes in the polypeptides of photosystem II particles. *Physiological and Molecular Plant Pathology*, **29**, 53–58.

Naidu, S.L. and DeLucia, E.H. (1997). Acclimation of shade-developed leaves on saplings exposed to late-season canopy gaps. *Tree Physiology*, **17**, 367–376.

Nainanayake, A.D. (2004). Impact of drought on coconut (*Cocos nucifera L.*): screening germplasm for photosynthetic tolerance in the field. Ph.D. thesis, University of Essex, pp. 215.

Nakano, R., Ishida, H., Makino, A., *et al.* (2006). In vivo fragmentation of the large subunit of ribulose-1,5-bisphosphate carboxylase by reactive oxygen species in an intact leaf of cucumber under chilling-light conditions. *Plant and Cell Physiology*, **47**, 270–276.

Nalborczyk, E. (1978). Dark carboxylation and its possible effect on the value of $\delta^{13}C$ in C_3 plants. *Acta Physiologiae Plantarum*, **1**, 53–58.

Nali, C., Guidi, L., Ciompi, S., *et al.* (1995). Photosynthesis of *Medicago sativa* L. plants exposed to long-term fumigation with sulphur dioxide. In: *Responses of Plants to*

Air Pollution (eds Lorenzini, G. and Soldatini, G.F.), Pacini Editore, Pisa, Italy, pp. 82–89.

Nambudiri, E.M.V., Tidwell, W.D., Smith, B.N., *et al.* (1978). A C_4 plant from the Pliocene. *Nature*, **276**, 816–817.

Nardini, A., Gortan, E., Ramani, M., *et al.* (2008). Heterogeneity of gas exchange rates over the leaf surface in tobacco: an effect of hydraulic architecture? *Plant, Cell and Environment*, **31**, 804–812.

Nasyrov, Y.S. (1978). Genetic control of photosynthesis and improving of crop productivity. *Ann. Rev. Plant Physiol.*, **29**, 215–37.

Naumann, J.C., Young, D.R. and Anderson, J.E. (2008). Leaf chlorophyll fluorescence, reflectance, and physiological response to freshwater and saltwater flooding in the evergreen shrub, *Myrica cerifera*. *Environmental and Experimental Botany*, **63**, 402–409.

Naumburg, E. and Ellsworth, D.S. (2000). Photosynthesis sunfleck utilization potential of understory saplings growing under elevated CO_2 in FACE. *Oecologia*, **122**, 163–174.

Naumburg, E., Houseman, D.C., Huxman, T.E., *et al.* (2003). Photosynthetic responses of Mojave Desert shrubs to free air CO_2 enrichment are greatest during wet years. *Global Change Biology*, **9**, 276–285.

Naumburg, E., Loik, M.E. and Smith, S.D. (2004). Photosynthetic responses of *Larrea tridentata* to seasonal temperature extremes under elevated CO_2. *New Phytologist*, **162**, 323–330.

Neals, T.F. and Incoll, L.D. (1968). The control of leaf photosynthesis rate by the level of assimilate concentration in the leaf: a review of hypotheses. *Botanical Review*, **34**, 107–125.

Nebdal, L., Soukupová, J., Whitmarsh, J., *et al.* (2000a). Postharvest imaging of chlorophyll fluorescence from lemons can be used to predict fruit quality. *Photosynthetica*, **38**, 571–579.

Nedbal, L. and Březina, V. (2002). Complex metabolic oscillations in plants forced by harmonic irradiance. *Biophysical Journal*, **83**, 2180–2189.

Nedbal, L. and Whitmarsh, J. (2004). Chlorophyll fluorescence imaging of leaves and fruits. In: *Chlorophyll a Fluorescence: A Signature of Photosynthesis*. Advances in photosynthesis and respiration, vol. 19. (eds Papageorgiou, C.G. and Govindjee, J.), Springer, Dordrecht, Netherlands, pp. 389–407.

Nedbal, L., Soukupová, J., Kaftan, D., *et al.* (2000b). Kinetic imaging of chlorophyll fluorescence using modulated light. *Photosynthesis Research*, **66**, 25–34.

Neftel, A., Friedli, H., Moor, E., *et al.* (1994). Historical CO_2 record from the Siple Station ice core. In: *Trends: A Compendium of Data on Global Change*. Carbon Dioxide Information Analysis Center, Oak Ridge National Laboratory, U.S. Department of Energy, Oak Ridge, Tenn., U.S.A.

Neill, S., Barros, R., Bright, J., *et al.* (2008). Nitric oxide, stomatal closure, and abiotic stress. *Journal of Experimental Botany*, **59**, 165–176.

Neill, S.O., Gould, K.S., Kilmartin, P.A., *et al.* (2002). Antioxidant activities of red versus green leaves in Elatostema rugosum. *Plant Cell Environment*, **25**, 539–547.

Nelson, D.E., Repetti, P.P., Adams, T.R., *et al.* (2007). Plant nuclear factor Y (NF-Y) B subunits confer drought tolerance and lead to improved corn yields on water-limited acres. *Proceedings of the National Academy of Sciences USA*, **104**, 16450–16455.

Nelson, E.A. and Sage, R.F. (2008). Functional constraints of CAM leaf anatomy: tight cell packing is associated with increased CAM function across a gradient of CAM expression. *Journal of Experimental Botany*, **59**, 1841–1850.

Nelson, E.A. and Sage, R.F. (2005). Functional leaf anatomy of plants with crassulacean acid metabolism. *Functional Plant Biology*, **32**, 409–419.

Nelson, J.A., Morgan, J.A., LeCain, D.R., *et al.* (2004). Elevated CO_2 increases soil moisture and enhances plant water relations in a long-term field study in semi-arid shortgrass steppe of Colorado. *Plant and Soil*, **259**, 169–179.

Nelson, N., Sacher, A. and Nelson, H. (2002). The significance of molecular slips in transport systems. *National Reviews of Molecular and Cellular Biology*, **3**, 876–881.

Nelson, S.V. (2005). Paleoseasonality inferred from equid teeth and intra-tooth isotopic variability. *Palaeogeography, Palaeoclimatology, Palaeoecology*, **222**, 122–144.

Nemani, R.R., Keeling, C.D., Hashimoto, H., *et al.* (2003). Climate-driven increases in global terrestrial net primary production from 1982 to 1999. *Science*, **300**, 1560–1563.

Neubauer, C. and Schreiber, U. (1987). The polyphasic rise of chlorophyll fluorescence upon onset of strong continuous illumination. I. Saturation characteristics and partial control by the photosystem II acceptor side. *Zeitschrift für Naturforschung*, **42c**, 1246–1254.

Neufeld, H.S., Meinzer, F.C., Wisdom, C.S., *et al.* (1988). Canopy architecture of *Larrea tridentata* (DC.) Cov., a desert shrub: foliage orientation and direct beam radiation interception. *Oecologia*, **75**, 54–60.

Neuner, G. and Pramsohler, M. (2006). Freezing and high temperature thresholds of photosystem 2 compared to ice nucleation, frost and heat damage in evergreen subalpine plants. *Physiologia Plantarum*, **126**, 196–204.

Neuner, G., Braun, V., Buchner, O., *et al.* (1999). Leaf rosette closure in the alpine rock species *Saxifraga paniculata* Mill.: significance for survival of drought and heat under high irradiation. *Plant Cell and Environment*, **22**, 1539–1548.

Newell, E.A., McDonald, E.P., Strain, B.R., *et al.* (1993). Photosynthetic responses of Miconia species to canopy openings in a lowland tropical rainforest. *Oecologia*, **94**, 49–56.

Newman, S.M. and Cattolico, R.A. (1990). Ribulose bisphosphate carboxylase in algae: synthesis, enzymology and evolution. *Photosynthesis Research*, **26**, 69–85.

Ngugi, M.R., Doley, D., Hunt, M.A., *et al.* (2004). Physiological responses to water stress in *Eucalyptus cloeziana* and *E. argophioia* seedlings. *Trees-Structure and Function*, **18**, 381–389.

Nichol, C.J., Huemmrich, K.F., Black, T.A., *et al.* (2000). Remote sensing of photosynthetic-light-use efficiency of boreal forest. *Agricultural and Forest Meteorology*, **101**, 131–142.

Nichol, C.J., Lloyd, J., Shibistova, O., *et al.* (2002). Remote sensing of photosynthetic-light-use efficiency of a Siberian boreal forest. *Tellus*, **54B**, 677–687.

Nichol, C.J., Rascher, U., Matsubara, S., *et al.* (2006). Assessing photosynthetic efficiency in an experimental mangrove canopy using remote sensing and chlorophyll fluorescence. *Trees*, **20**, 9–15.

Nickrent, D.L., Parkinson, C.L., Palmer, J.D., *et al.* (2000). Multigene phylogeny of land plants with special reference to Bryophytes and the earliest land plants. *Molecular Biology and Evolution*, **17**, 1885–1895.

Nicolás, E., Torrecillas, A., Dell'Amico, J., *et al.* (2005). The effect of short-term flooding on the sap flow, gas exchange and hydraulic conductivity of young apricot trees. *Trees*, **19**, 51–57.

Nicotra, A., Chazdon, R.L. and Montgomery, R.A. (2003). Sexes show contrasting patterns of leaf and crown carbon gain in a dioecious rainforest shrub. *American Journal of Botany*, **90**, 347–355.

Nicotra, A.B. and Davidson, A. (2010). Adaptive phenotypic plasticity and plant water use. *Functional Plant Biology*, **37**, 117–127.

Nicotra, A.B., Cosgrove, M.J., Cowling, A., *et al.* (2008). Leaf shape linked to photosynthetic rates and temperature optima in South African Pelargonium species. *Oecologia*, **154**, 625–635.

Niederleitner, S. and Knoppik, D. (1997). Effects of the cherry leaf spot pathogen *Blumeriella jaapii* on gas exchange before and after expression of symptoms on cherry leaves. *Physiological and Molecular Plant Pathology*, **51**, 145–153.

Niemann, G.J., van der Kerk, A., Niessen, W.M.A., *et al.* (1991). Free and cell wall-bound phenolics and other constituents from healthy and fungus-infected carnation (*Dianthus caryophyllus* L.) stems. *Physiological and Molecular Plant Pathology*, **38**, 417–432.

Nier, A.O. (1940). A mass spectrometer for routine isotope abundance measurements. *Review of Scientific Instruments*, **11**, 212–216.

Nier, A.O. and Gulbransen, E.A. (1939). Variations in the relative abundance of the carbon isotopes. *Journal of the American Chemical Society*, **61**, 697–698.

Nihlgård, B. (1972). Plant biomass, primary production and distribution of chemical elements in a beech and a planted spruce forest in South Sweden. *Oikos*, **23**, 69–81.

Niinemets, Ü. (1997a). Acclimation to low irradiance in *Picea abies*: influences of past and present light climate on foliage structure and function. *Tree Physiology*, **17**, 723–732.

Niinemets, Ü. (1997b). Distribution patterns of foliar carbon and nitrogen as affected by tree dimensions and relative light conditions in the canopy of *Picea abies*. *Trees*, **11**, 144–154.

Niinemets, Ü. (1997c). Energy requirement for foliage construction depends on tree size in young *Picea abies* trees. *Trees*, **11**, 420–431.

Niinemets, Ü. (1998). Growth of young trees of *Acer platanoides* and *Quercus robur* along a gap – understory continuum: interrelationships between allometry, biomass partitioning, nitrogen, and shade-tolerance. *International Journal of Plant Science*, **159**, 318–330.

Niinemets, Ü. (1999). Components of leaf dry mass per area – thickness and density – alter leaf photosynthetic capacity in reverse directions in woody plants. *The New Phytologist*, **144**, 35–47.

Niinemets, Ü. (2001). Global-scale climatic controls of leaf dry mass per area, density, and thickness in trees and shrubs. *Ecology*, **82**, 453–469.

Niinemets, Ü. (2002). Stomatal conductance alone does not explain the decline in foliar photosynthetic rates with increasing tree age and size in *Picea abies* and *Pinus sylvestris*. *Tree Physiology*, **22**, 515–535.

Niinemets, Ü. (2004). Adaptive adjustments to light in foliage and whole-plant characteristics depend on relative age in the perennial herb *Leontodon hispidus*. *New Phytologist*, **162**, 683–696.

Niinemets, Ü. (2006). The controversy over traits conferring shade-tolerance in trees: ontogenetic changes revisited. *Journal of Ecology*, **94**, 464–470.

Niinemets, Ü. (2007). Photosynthesis and resource distribution through plant canopies. *Plant, Cell and Environment*, **30**, 1052–1071.

Niinemets, Ü. (2009). Light interception in plant stands from leaf to canopy in different plant functional types and in species with varying shade tolerance: a review. *Ecological Research*, **25**, 693–714.

Niinemets, Ü. and Anten, N.P.R. (2009). Packing photosynthesis machinery: from leaf to canopy. In: *Photosynthesis in Silico: Understanding Complexity from Molecules to Ecosystems* (eds Laisk. A., Nedbal, L. and Govindjee, J.), Springer-Verlag, Berlin, pp. 363–399.

Niinemets, Ü. and Kull, K. (1994). Leaf weight per area and leaf size of 85 Estonian woody species in relation to shade tolerance and light availability. *Forest Ecology and Management*, **70**, 1–10.

Niinemets, Ü. and Kull, O. (1995a). Effects of light availability and tree size on the architecture of assimilative surface in the canopy of *Picea abies*: variation in shoot structure. *Tree Physiology*, **15**, 791–798.

Niinemets, Ü. and Kull, O. (1995b). Effects of light availability and tree size on the architecture of assimilative surface in the canopy of *Picea abies*: variation in needle morphology. *Tree Physiology*, **15**, 307–315.

Niinemets, Ü. and Kull, O. (1998). Stoichiometry of foliar carbon constituents varies along light gradients in temperate woody canopies: implications for foliage morphological plasticity. *Tree Physiology*, **18**, 467–479.

Niinemets, Ü. and Sack, L. (2006). Structural determinants of leaf light-harvesting capacity and photosynthetic potentials. In: *Progress in Botany*, vol. 67 (eds Esser K., Lüttge, U.E., Beyschlag, W. and Murata, J.), Springer Verlag, Berlin, pp. 385–419.

Niinemets, Ü. and Tamm, Ü. (2005). Species differences in timing of leaf fall and foliage chemistry modify nutrient resorption efficiency in deciduous temperate forest stands. *Tree Physiology*, **25**, 1001–1014.

Niinemets, Ü. and Tenhunen, J.D. (1997). A model separating leaf structural and physiological effects on carbon gain along light gradients for the shade-tolerant species *Acer saccharum*. *Plant, Cell and Environment*, **20**, 845–866.

Niinemets, Ü. and Valladares, F. (2004). Photosynthetic acclimation to simultaneous and interacting environmental stresses along natural light gradients: optimality and constraints. *Plant Biology*, **6**, 254–268.

Niinemets, Ü. and Valladares, F. (2006). Tolerance to shade, drought, and waterlogging of temperate northern hemisphere trees and shrubs. *Ecological Monographs*, **76**, 521–547.

Niinemets, Ü., Cescatti, A., Rodeghiero, M., *et al.* (2005a). Leaf internal diffusion conductance limits photosynthesis more strongly in older leaves of Mediterranean evergreen broad-leaved species. *Plant, Cell and Environment*, **28**, 1552–1566.

Niinemets, Ü., Cescatti, A., Rodeghiero, M., *et al.* (2006a). Complex adjustments of photosynthetic capacity and internal mesophyll conductance to current and previous light availabilities and leaf age in Mediterranean evergreen species *Quercus ilex*. *Plant, Cell and Environment*, **29**, 1159–1178.

Niinemets, Ü., Diaz-Espejo, A., Flexas, J., *et al.* (2009a). Importance of mesophyll-diffusion conductance in estimation of plant photosynthesis in the field. *Journal of Experimental Botany*, **60**, 2271–2282.

Niinemets, Ü., Díaz-Espejo, A., Flexas, J., *et al.* (2009b). Role of mesophyll-diffusion conductance in constraining potential photosynthetic productivity in the field. *Journal of Experimental Botany*, **60**, 2249–2270.

Niinemets, Ü., Ellsworth, D.S., Lukjanova, A., *et al.* (2001). Site fertility and the morphological and photosynthetic acclimation of *Pinus sylvestris* needles to light. *Tree Physiology*, **21**, 1231–1244.

Niinemets, Ü., Flexas, J. and Peñuelas, J. (2011). Evergreens favoured by higher responsiveness to increased CO_2. *Trends in Ecology and Evolution*, **26**, 136–142.

Niinemets, Ü., Hauff, K., Bertin, N., *et al.* (2002). Monoterpene emissions in relation to foliar photosynthetic and structural variables in Mediterranean evergreen *Quercus* species. *New Phytologist*, **153**, 243–256.

Niinemets, Ü., Kollist, H., García-Plazaola, J.I., *et al.* (2003). Do the capacity and kinetics for modification of xanthophyll cycle pool size depend on growth irradiance in temperate trees? *Plant, Cell and Environment*, **26**, 1787–1801.

Niinemets, Ü. and Kull, K. (2005). Co-limitation of plant primary productivity by nitrogen and phosphorus in a species-rich wooded meadow on calcareous soils. *Acta Oecologica*, **28**, 345–356.

Niinemets, Ü. and Kull, O. (2001). Sensitivity to photoinhibition of photosynthetic electron transport in a temperate deciduous forest canopy: photosystem II centre openness, non-radiative energy dissipation and excess irradiance under field conditions. *Tree Physiology*, **21**, 899–914.

Niinemets, Ü., Kull, O. and Tenhunen, J.D. (1998). An analysis of light effects on foliar morphology, physiology, and light interception in temperate deciduous woody species of contrasting shade tolerance. *Tree Physiology*, **18**, 681–696.

Niinemets, Ü., Kull, O. and Tenhunen, J.D. (1999a). Variability in leaf morphology and chemical composition as a function of canopy light environment in co-existing trees. *International Journal of Plant Sciences*, **160**, 837–848.

Niinemets, Ü., Kull, O. and Tenhunen, J.D. (2004a). Within canopy variation in the rate of development of photosynthetic capacity is proportional to integrated quantum flux density in temperate deciduous trees. *Plant, Cell and Environment*, **27**, 293–313.

Niinemets, Ü., Lukjanova, A., Sparrrow, A.D., *et al.* (2005c). Light-acclimation of cladode photosynthetic potentials in *Casuarina glauca*: trade-offs between physiological and structural investments. *Functional Plant Biology*, **32**, 571–582.

Niinemets, Ü., Oja, V. and Kull, O. (1999b). Shape of leaf photosynthetic electron transport versus temperature response curve is not constant along canopy light gradients in temperate deciduous trees. *Plant, Cell and Environment*, **22**, 1497–1514.

Niinemets, Ü., Portsmuth, A., Tena, D., *et al.* (2007). Do we underestimate the importance of leaf size in plant economics? Disproportionate scaling of support costs within the spectrum of leaf physiognomy. *Annals of Botany*, **100**, 283–303.

Niinemets, Ü., Sõber, A., Kull, O., *et al.* (1999d). Apparent controls on leaf conductance by soil water availability and via light-acclimation of foliage structural and physiological properties in a mixed deciduous, temperate forest. *International Journal of Plant Sciences*, **160**, 707–721.

Niinemets, Ü., Sonninen, E. and Tobias, M. (2004c). Canopy gradients in leaf intercellular CO_2 mole fractions revisited: interactions between leaf irradiance and water stress need consideration. *Plant, Cell and Environment*, **27**, 569–583.

Niinemets, Ü., Tenhunen, J.D. and Beyschlag, W. (2004b). Spatial and age-dependent modifications of photosynthetic capacity in four Mediterranean oak species. *Functional Plant Biology*, **31**, 1179–1193.

Niinemets, U., Tenhunen, J.D., Canta, N.R., *et al.* (1999c). Interactive effects of nitrogen and phosphorus on the acclimation potential of foliage photosynthetic properties of cork oak, *Quercus suber*, to elevated atmospheric CO_2 concentrations. *Global Change Biology*, **5**, 455–470.

Niinemets, Ü., Tobias, M., Cescatti, A., *et al.* (2006b). Size-dependent variation in shoot light-harvesting efficiency in shade-intolerant conifers. *International Journal of Plant Science*, **167**, 19–32.

Niinemets, Ü., Wright, I.J. and Evans, J.R. (2009c). Leaf diffusion conductance in 35 Australian sclerophylls covering a broad range of foliage structural and physiological variation. *Journal of Experimental Botany*, **60**, 2433–2449.

Niklas, K.J. (2001). Invariant scaling relationships for interspecific plant biomass production rates and body size. *Proceedings of the National Academy of Sciences USA*, **98**, 2922–2927.

Niklas, K.J. and Kutschera, U. (2009a). The evolution of the land plant lifecycle. *New Phytologist*, **185**, 27–41.

Niklas, K.J. and Kutschera, U. (2009b). The evolutionary development of plant body plans. *Functional Plant Biology*, **36**, 682–695.

Nikolopoulos, D., Liakopoulos, G., Drossopoulos, I., *et al.* (2002). The relationship between anatomy and photosynthetic performance of heterobaric leaves. *Plant Physiology*, **129**, 235–243.

Nilsen, E.T. (1992). The influence of water stress on leaf and stem photosynthesis in *Spartium junceum* L. *Plant, Cell and Environment*, **15**, 455–461.

Nilsen, E.T. and Sharifi, M.R. (1997). Carbon isotopic composition of legumes with photosynthetic stems from Mediterranean and desert habitats. *American Journal of Botany*, **84**, 1707–13.

Nilsen, E.T., Meinzer, F.C. and Rundel, P.W. (1989). Stem photosynthesis in *Psorothamnus spinosus* (smoke tree) in the Sonoran desert of California. *Oecologia*, **79**, 193–197.

Nilsen, E.T., Rundel, P.W. and Sharifi, M.R. (1996). Diurnal gas exchange characteristics of two stem photosynthesizing legumes in relation to the climate at two contrasting sites in the California desert. *Flora*, **191**, 105–16.

Nilson, S.E. and Assmann, S.M. (2007). The control of transpiration. Insights from *Arabidopsis*. *Plant Physiology*, **143**, 19–27.

Nilson, T. (1971). A theoretical analysis of the frequency of gaps in plant stands. *Agr Meteorol*, **8**, 25–38.

Nilsson, H.E. (1995). Remote sensing and image analysis in plant pathology. *Annual Review of Phytopathology*, **15**, 489–527.

Ning, L., Edwards, G.E., Strobel, G.A., *et al.* (1995). Imaging fluorometer to detect pathological and physiological change in plants. *Applied Spectroscopy*, **49**, 1381–1389.

Nippert, J.B., Fay, P.A. and Knapp, A.K. (2007). Photosynthetic traits in C_3 and C_4 grassland species in mesocosm and field environments. *Environmental and Experimental Botany*, **60**, 412–420.

Nisbet, E.G., Grassineau, N.V., Howe, C.J., *et al.* (2007). The age of Rubisco: the evolution of oxygenic photosynthesis. *Geobiology*, **5**, 311–335.

Nishida, I. and Murata, N. (1996). Chilling sensitivity in plants and cyanobacteria: the crucial contribution of membrane lipids. *Annual Review of Plant Physiology and Plant Molecular Biology*, **47**, 541–568.

Nishio, J.N. (2000). Why are higher plants green? Evolution of the higher plant photosynthetic pigment complement. *Plant, Cell and Environment*, **23**, 539–548.

Nishio, J.N. and Whitmarsh, J. (1993). Dissipation of the proton electrochemical potential in intact chloroplasts. II. The pH gradient monitored by cytochrome f reduction kinetics. *Plant Physiology*, **101**, 89–96.

Nishio, J.N., Sun, J. and Vogelmann, T.C. (1993). Carbon fixation gradients across spinach leaves do not follow internal light gradients. *Plant Cell*, **5**, 953–961.

Nitta, I. and Ohsawa, M. (1997). Leaf dynamics and shoot phenology of eleven warm-temperate evergreen broad-leaved trees near their northern limit in central Japan. *Plant Ecology*, **130**, 71–88.

Nittylä, T., Messerli, G., Trevisan, M., *et al.* (2004). A novel maltose transporter is essential for starch degradation in leaves. Science, **303**, 87–89.

Niyogi, K.K. (1999). Photoprotection revisited: genetic and molecular approaches. *Annual Review of Plant Physiology and Plant Molecular Biology*, **50**, 333–359.

Niyogi, K.K. (2000). Safety valves for photosynthesis. *Current Opinion in Plant Biology*, **3**, 455–460.

Niyogi, K.K., Li, X.P., Rosenberg, V., *et al.*(2005). Is PsbS the site of non-photochemical quenching in photosynthesis? *Journal of Experimental Botany*, **56**, 375–382.

Nobel, P.S. (1977). Internal leaf area and cellular CO_2 resistance: photosynthetic implications of variations with growth conditions and plant species. *Physiologia Plantarum*, **40**, 137–144.

Nobel, P.S. (1978). Microhabitat, water relations, and photosynthesis of a desert fern, *Notholaena parryi*. *Oecologia*, **31**, 293–309.

Nobel, P.S. (1980). Water vapor conductance and CO_2 uptake for leaves of a C_4 desert grass, *Hilaria rigida*. *Ecology*, **61**, 252–258.

Nobel, P.S. (1982). Orientation of terminal cladodes of platyopuntias. *Botanical Gazette*, **143**, 219–24.

Nobel, P.S. (1983). Nutrient levels in cacti – relation to nocturnal acid accumulation and growth. *American Journal of Botany*, **70**, 1244–1253.

Nobel, P.S. (1988). Environmental biology of Agaves and cacti. Cambridge University Press, Cambridge, UK.

Nobel, P.S. (1991). Achievable productivities of CAM plants; basis for high values compared with C_3, and C_4. *New Phytologist*, **119**, 183–205.

Nobel, P.S. (1991b). Physicochemical and environmental plant physiology. Academic Press, San Diego, USA.

Nobel, P.S. (1994). *Remarkable Agaves and Cacti.* Oxford University Press, New York, USA.

Nobel, P.S. (1996). High productivity of certain agronomic CAM species. In: Winter, K., Smith, J.A.C., eds. *Crassulacean Acid Metabolism. Biochemistry, Ecophysiology and Evolution.* Ecological studies, Vol. 145. Springer-Verlag, Berlin – Heidelberg – New York, 255–265.

Nobel, P.S. (2005). *Physicochemical and Environmental Plant Physiology*, 3rd edn. Academic Press, San Diego, USA.

Nobel, P.S. and Berry, W.L. (1985). Element responses of agaves. *American Journal of Botany* **72**, 686–694.

Nobel, P.S. and Smith, S.D. (1983). High and low temperature tolerances and their relationships to distribution of agaves. *Plant, Cell and Environment*, **6**, 711–719.

Nobel, P.S., Forseth, I.N. and Long, S.P. (1993). Canopy structure and light interception. In: *Photosynthesis and Production in a Changing Environment: A Field and Laboratory Manual* (eds Hall, D.O., Scurlock, J.M.O., Bolhar-Nordenkampf, H.R., Leegood, R.C. and Long, S.P.), Chapman and Hall, New York, USA, pp. 78–90.

Nobel, P.S., Geller, G.N., Kee, S.C., *et al.* (1986). Temperatures and thermal tolerances for cacti exposed to high temperatures near the soil surface. *Plant, Cell and Environment*, **9**, 279–287.

Nobel, P.S., Longstreth, D.J., Hartsock, T.L. (1978). Effect of water stress on the temperature optima of net CO_2 exchange for two desert species. *Physiologia Plantarum*, **44**, 97–101.

Nobel, P.S., Zaragoza, L.J. and Smith, W.R. (1975). Relation between mesophyll surface area, photosynthesis rate, and illumination level during development for leaves of *Plectranthus parviflorus* Henckel. *Plant Physiology*, **55**, 1067–1070.

Nock, C.A., Caspersen, J.P. and Thomas, S.C. (2008). Large ontogenetic declines in intra-crown leaf-area index in two temperate deciduous tree species. *Ecology*, **89**, 744–753.

Noctor, G. and Foyer, C.H. (1998a). Ascorbate and glutathione: keeping active oxygen under control. *Annual Review of Plant Physiology and Plant Molecular Biology*, **49**, 249–279.

Noctor, G. and Foyer, C.H. (1998b). A re-evaluation of the ATP:NADPH budget during C_3 photosynthesis. A contribution from nitrate assimilation and its associated respiratory activity? *Journal of Experimental Botany*, **49**, 1895–1908.

Noctor, G., Arisi, A.C.M., Jouanin, L., *et al.* (1998). Glutathione: biosynthesis, metabolism and relationship to stress tolerance explored in transformed plants. *Journal of Experimental Botany*, **49**, 623–647.

Noctor, G., De Paepe, R. and Foyer, C.H. (2007). Mitochondrial redox biology and homeostasis in plants. *Trends Plant Science*, **12**, 125–134.

Noguchi, K. and Yoshida, K. (2008). Interaction between photosynthesis and respiration in illuminated leaves. *Mitochondrion*, **8**, 87–99.

Nogueira, A., Martínez, C.A., Ferreira, L.L., *et al.* (2004). Photosynthesis and water use efficiency in twenty tropical tree species of different succession status in a Brazilian reforestation. *Photosynthetica*, **42**, 351–356.

Nogués, S., Cotxarrera, L., Alegre, L., *et al.* (2002). Limitations to photosynthesis in tomato leaves induced by *Fusarium* wilt. *New Phytologist*, **154**, 461–470.

Nogués, S., Tcherkez, G., Streb, P., *et al.* (2006). Respiratory carbon metabolites in the high mountain plant species *Ranunculus glacialis*. *Journal of Experimental Botany*, **57**, 3837–3845.

Nomura, M., Higuchi, T., Ishida, Y., *et al.* (2005). Differential expression pattern of C_4 bundle sheath expression genes in rice, a C_3 plant. *Plant and Cell Physiology*, **46**, 754–761.

Noodén, L.D. and Guiamét, J.I. (1997). Senescence mechanisms. *Physiologia Plantarum*, **101**, 746–53.

Noormets, A., McDonald, E.P., Kruger, E.L., *et al.* (2001a). The effect of elevated carbon dioxide and ozone on leaf and branch-level photosynthesis and potential plant-level carbon gain in aspen. *Trees*, **15**, 262–270.

Noormets, A., Sober, A., Pell, E.J., *et al.* (2001b). Stomatal and non-stomatal limitation to photosynthesis in two trembling aspen (*Populus tremuloides* Michx.) clones exposed to elevated CO_2 and/or O_3. *Plant, Cell and Environment*, **24**, 327–336.

Norby, R.J., Delucia, E.H., Gielen, B., *et al.* (2005). Forest response to elevated CO_2 is conserved across a broad range of productivity. *Proceedings of the National Academy of Sciences of the USA*, **102**, 18052–18056.

Norby, R.J., Sholtis, J.D., Gunderson, C.A., *et al.* (2003). Leaf dynamics of a deciduous forest canopy: no response to elevated CO_2. *Oecologia*, **136**, 574–584.

Norby, R.J., Wullschleger, S.D., Gunderson, C.A., *et al.* (1999). Tree responses to rising CO_2 in field experiments: implications for the future forest. *Plant, Cell and Environment*, **22**, 683–714.

Nordborg, M. (2000). Linkage disequilibrium, gene trees and selfing: ancestral recombination graph with partial self-fertilization. *Genetics*, **154**, 923–939.

Norman, J.M. and Jarvis, P.G. (1974). Photosynthesis in Sitka spruce (*Picea sitchensis* (Bong.) Carr.). III. Measurements of canopy structure and interception of radiation. *Journal of Applied Ecology*, **11**, 375–398.

North, P.R.J. (1996). Three-dimensional forest light interaction model using a Monte Carlo method. *IEEE Transactions on Geoscience and Remote Sensing*, **34**, 946–955.

Nowak, R.S. and Caldwell, M.M. (1984). A test of compensatory photosynthesis in the field: implications for herbivory tolerance. *Oecologia*, **61**, 311–318.

Nowak, R.S., Ellsworth, D.S. and Smith, S.D. (2004). Functional responses of plants to elevated atmospheric CO_2 – do photosynthetic and productivity data from FACE experiments support early predictions? *New Phytologist*, **162**, 253–280.

Noy Meir, I. (1973). Desert ecosystems: environment and producers. *Annual Review of Ecology and Systematics*, **4**, 25–51.

Nunes-Nesi, A., Araújo, W.L. and Fernie, A.R. (2011). Targeting mitochondrial metabolism and machinery as a means to enhance photosynthesis. *Plant Physiology*, **155**, 101–107.

Nunes-Nesi, A., Carrari, F., Gibon, Y., *et al.* (2007). Deficiency of mitochondrial fumarase activity in tomato plants impairs photosynthesis via an effect on stomatal function. *Plant Journal*, **50**, 1093–1106.

Nunes-Nesi, A., Carrari, F., Lytovchenko, A., *et al.* (2005). Enhanced photosynthetic performance and growth as a consequence of decreasing mitochondrial malate dehydrogenase activity in transgenic tomato plants. *Plant Physiology*, **137**, 611–622.

Nunes-Nesi, A., Sulpice, R., Gibon, Y. *et al.* (2008). The enigmatic contribution of mitochondrial function in photosynthesis. *Journal of Experimental Botany*, **59**, 1675–1684.

Ocheltree, T.W. and Loescher, H.W. (2007). Design of the AmeriFlux portable eddy covariance system and uncertainty analysis of carbon measurements. *Journal of Atmospheric and Oceanic Technology*, **24**, 1389–1406.

Odening, W.R., Strain, B.R. and Oechel, W.C. (1974). The effect of decreasing water potential on net CO, exchange of intact desert shrubs. *Ecology*, **55**, 1086–95.

Oechel, W.C. and Strain, B.R. (1985). Native species responses to increased atmospheric carbon dioxide concentration. In: *Direct Effects of Increasing Carbon Dioxide on Vegetation* (eds Strain, B.D. and Cure, J.D.), Department of Energy, Office of Basic Energy Sciences, Carbon Dioxide Research Division, Springfield, VA, Washington DC, USA, pp. 117–154.

Oerke, E.C. and Dehne, H.W. (2004). Safeguarding production – losses in major crops and the role of crop protection. *Crop Protection*, **23**, 275–285.

Oerke, E.C., Steiner, U., Dehne, H.W., *et al.* (2006). Thermal imaging of cucumber leaves affected by downy mildew and environmental conditions. *Journal of Experimental Botany*, **57**, 2121–2132.

Ogee, J. Cuntz, M., Peylin, P., *et al.* (2007). Non-steady state, non-uniform transpiration rate and leaf anatomy effects on the progressive stable isotope enrichment of leaf water along monocot leaves. *Plant, Cell and Environment*, **30**, 367–387.

Ogle, K. and Reynolds, J.F. (2002). Desert dogma revisited: coupling of stomatal conductance and photosynthesis in the desert shrub, *Larrea tridentata*. *Plant, Cell and Environment*, **25**, 909–21.

Ögren, E. (1988). Photoinhibition of photosynthesis in willow leaves under field conditions. *Planta*, **175**, 229–236.

Ögren, E. (1993). Convexity of the photosynthetic light-response curve in relation to intensity and direction of light during growth. *Plant Physiology*, **101**, 1013–1019.

Ögren, E. and Baker, N.R. (1985). Evaluation of a technique for the measurement of chlorophyll fluorescence from leaves exposed to continuous white light. *Plant, Cell, Environment*, **8**, 539–547.

Ögren, E. and Evans, J.R. (1993), Photosynthetic light response curves. 1. The influence of CO_2 partial pressure and leaf inversion. *Planta*, **189**, 182–190.

Ogren, W.L. (1984). Photorespiration pathways, regulation, and modification. *Annual Review of Plant Physiology*, **35**, 415–442.

Ogren, W.L. (2003). Affixing the o to Rubisco: discovering the source of photorespiratory glycolate and its regulation. *Photosynthesis Research*, **76**, 53–63.

Ogren, W.L. and Bowes, G. (1971). Ribulose diphosphate carboxylase regulates soybean photorespiration. *Nature New Biology*, **230**, 159–160.

Oguchi, R., Douwstra, P., Fujita, T. *et al.* (2011). Intra-leaf gradients of photoinhibition induced by different colour lights: implications for the dual mechanisms of photoinhibition and for the application of conventional chlorophyll fluorometers. *The New Phytologist*, **191**, 146–159.

Oguchi, R., Hikosaka, K. and Hirose, T. (2005). Leaf anatomy as a constraint for photosynthetic acclimation: differential responses in leaf anatomy to increasing growth irradiance among three deciduous trees. *Plant, Cell and Environment*, **28**, 916–927.

Oguchi, R., Hikosaka, K., Hirose, T. (2003). Does leaf photosynthetic light-acclimation need change in leaf anatomy? *Plant Cell Environment*, **26**, 505–512.

Oguchi, R., Hikosaka, K., Hiura, T., *et al.* (2006). Leaf anatomy and light acclimation in woody seedlings after gap formation in a cool-temperate forest. *Oecologia*, **149**, 571–582.

Oguchi, R., Jia, H., Barber, J., *et al.* (2008). Recovery of photoinactivated photosystem II in leaves: retardation due to restricted mobility of photosystem II in the thylakoid membrane. *Photosynthesis Research*, **98**, 621–629.

Ohki, K. (1976). Effect of zinc nutrition on photosynthesis and carbonic anhydrase activity in cotton. *Physiologia Plantarum*, **38**, 300–304.

Ohki, K. (1985). Manganese deficiency and toxicity effects on photosynthesis, chlorophyll, and transpiration in wheat. *Crop Science*, **25**, 187–191.

Ohki, K. (1986). Photosynthesis, chlorophyll, and transpiration responses in aluminum stressed wheat and sorghum. *Crop Science*, **26**, 572–575.

Ohki, K., Wilson, D.O. and Anderson, O.E. (1981). Manganese deficiency and toxicity sensitivities of soybean cultivar. *Agronomy Journal*, **72**, 713–716.

Ohsawa, M. and Nitta, I. (1997). Patterning of subtropical/warm-temperate evergreen broad-leaved forests in East Asian mountains with special reference to shoot phenology. *Tropics*, **6**, 317–334.

Ohsugi, R., Samejima, M., Chonan, N., *et al.* (1988). $\delta^{13}C$ values and the occurrence of suberized lamellae in some *Panicum* species. *Annals of Botany*, **62**, 53–59.

Ohsumi, A., Hamasaki, A., Nakagawa, H., *et al.* (2007). A model explaining genotypic and ontogenetic variation of leaf photosynthetic rate in rice (*Oryza sativa*) based on leaf nitrogen content and stomatal conductance. *Annals of Botany*, **99**, 265–273.

Ohya, T., Yoshida, S. and Kawabata, R. (2002). Biophoton emission due to drought injury in red beans: possibility of early detection of drought injury. *Japanese Journal of Applied Physics Part 1-Regular Papers Short Notes and Review Papers*, **41**, 4766–4771.

Oker-Blom, P. (1984). Penumbral effects of within-plant and between-plant shading on radiation distribution and leaf photosynthesis: a Monte Carlo simulation. *Photosynthetica*, **18**, 522–528.

O'Leary, M.H. (1984). Measurement of the isotope fractionation associated with diffusion of carbon dioxide in aqueous solution. *Journal of Physical Chemistry*, **88**, 823–825.

O'Leary, M.H. (1993). Biochemical basis of carbon isotope fractionation. In: *Stable Isotopes and Plant Carbon-water Relations* (eds Ehleringer, J.R., Hall, A.E. and Farquhar, G.D.), Academic Press, San Diego, California, USA, pp. 19–28.

Oleksyn, J., Tjoelker, M.G., Lorenc-Plucinska, G., *et al.* (1997). Needle CO_2 exchange, structure and defence traits in relation to needle age in *Pinus heldreichii* Christ – a relict of Tertiary flora. *Trees*, **12**, 82–89.

Oleksyn, J., Zhytkowiak, R., Reich, P.B., *et al.* (2000). Ontogenetic patterns of leaf CO_2 exchange, morphology and chemistry in *Betula pendula* trees. *Trees*, **14**, 271–281.

Oliveira, G. and Peñuelas, J. (2002). Comparative protective strategies of *Cistus albidus* and *Quercus ilex* facing photoinhibitory winter conditions. *Environmental Experimental Botany*, **47**, 281–289.

Ollinger, S.V., Richardson, A.D., Martin, M.E., *et al.* (2008). Canopy nitrogen, carbon assimilation, and albedo in temperate and boreal forests: functional relations and potential climate feedbacks. *Proceedings of the National Academy of Sciences USA*, **105**, 19336–19341.

Olsen, J.E. and Junttila, O. (2002). Far red end-of-day treatment restores wild-type-like plant length in hybrid aspen overexpressing phytochrome A. *Physiologia Plantarum*, **115**, 448–457.

Omami, E.N., Hammes, P.S. and Robbertse, P.J. (2006). Differences in salinity tolerance for growth and water-use efficiency in some amaranth (*Amaranthus* spp.) genotypes. *New Zealand Journal of Crop and Horticultural Sciences*, **34**, 11–22.

Omasa, K. and Takayama, K. (2003). Simultaneous measurement of stomatal conductance, non-photochemical quenching, and photochemical yield of photosystem II in intact leaves by thermal and chlorophyll fluorescence imaging. *Plant Cell Physiology*, **44**, 1290–1300.

Omasa, K., Hashimoto, Y., Kramer, P.J., *et al.* (1985). Direct observation of reversible and irreversible stomatal response of attached sunflower leaves to SO_2. *Plant Physiology*, **7**, 153–158.

Omasa, K., Shimazaki, K.I., Aiga, I., *et al.* (1987). Image analysis of chlorophyll fluorescence transients for diagnosing the photosynthetic system of attached leaves. *Plant Physiology*, **84**, 748–752.

Onoda, Y., Hikosaka, K. and Hirose, T. (2004). Allocation of nitrogen to cell walls decreases photosynthetic nitrogen-use efficiency. *Functional Ecology*, **18**, 419–425.

Onoda, Y., Hikosaka, K. and Hirose, T. (2005). Seasonal change in the balance between capacities of RuBP carboxylation and RuBP regeneration affects CO_2 response of photosynthesis in *Polygonum cuspidatum*. *Journal of Experimental Botany*, **56**, 755–763.

Oppenheimer, H.R. (1960). Adaptation to drought: xerophytism. In: *Plant-water Relationships in arid and Semi-arid Conditions*. Vol. 15. Reviews of Research, UNESCO – Arid Zone Research, pp. 105–138.

Öquist, G. and Huner N.P.A. (2003). Photosynthesis of overwintering evergreen plants. *Annual Review of Plant Physiology*, **54**, 329–355.

Öquist, G. and Huner, N.P.A. (1993). Cold-hardening induced resistance to photoinhibition in winter rye is dependent upon an increased capacity for photosynthesis. *Planta*, **189**, 150–156.

Öquist, G., Brunes, L. and Hällgren, J.E. (1982). Photosynthetic efficiency of *Betula pendula* acclimated to different quantum flux densities. *Plant Cell Environment*, **5**, 9–15.

Öquist, G., Hällgren, J.E. and Brunes, L. (1978). An apparatus for measuring photosynthetic quantum yields and quanta absorption spectra of intact plants. *Plant, Cell and Environment*, **1**, 21–27.

Oren, R., Phillips, N., Katul, G., *et al.* (1998). Scaling xylem sap flux and soil water balance and calculating variance: a method for partitioning water flux in forests. *Annals of Forest Science*, **55**, 191–216.

Oren, R., Waring, R.H., Stafford, S.G., *et al.* (1987). Twenty four years of ponderosa pine growth in relation to canopy leaf area and understory competition. *Forest Science*, **33**, 538–547.

Orians, G.H. and Milewski, A.V. (2007). Ecology of Australia: the effects of nutrient poor soils and intense fires. *Biological Reviews*, **82**, 393–423.

Ort, D.R. and Baker, N.R. (2002). A photoprotective role for O_2 as an alternative electron sink in photosynthesis? *Current Opinion in Plant Biology*, **5**, 193–198.

Ortiz-López, A, Ort, D.R. and Boyer, J.S. (1991). Photophosphorylation in attached leaves of *Helianthus annuus* at low water potentials. *Plant Physiology*, **96**, 1018–1025.

Osborne C.P., Beerling D.J., Lomax B.H., *et al.* (2004a). Biophysical constraints on the origin of leaves inferred from the fossil record. *Proceedings of the National Academy of Sciences*, **101**, 10360–10362.

Osborne, C.P. (2008). Atmosphere, ecology and evolution: what drove the Miocene expansion of C_4 grasslands? *Journal of Ecology*, **96**, 35–45.

Osborne, C.P. and Beerling, D.J. (2006). Nature's green revolution: the remarkable evolutionary rise of C_4 plants. *Philosophical Transactions of the Royal Society, B*, **361**, 173–194.

Osborne, C.P., Chaloner, W.G. and Beerling, D.J. (2004b). Falling atmospheric CO_2 – the key to megaphyll leaf origins. In: *The Evolution of Plant Physiology* (eds. Hemsley, A.R. and Poole, I.), Elsevier Academic Press, London, UK, pp. 197–215.

Osborne, C.P., Mitchell, P.L., Sheehy, J.E., *et al.* (2000). Modelling the recent historical impacts of atmospheric CO_2 climate change on Mediterranean vegetation. *Global Change Biology*, **6**, 445–458.

Osmond, B., Ananyav, G., Berry, J.A. *et al.* (2004). Changing the way we think about global change research: scaling up in experimental ecosystem science. *Global Change Biology*, **10**, 393–407.

Osmond, B., Schwartz, O. and Gunning, B. (1999). Photoinhibitory printing on leaves, visualized by chlorophyll fluorescence imaging and confocal microscopy, is due to diminished fluorescence from grana. *Australian Journal of Plant Physiology*, **26**, 717–724.

Osmond, C.B. (1979). Crassulacean acid metabolism: a curiosity in context. *Annual Reviews in Plant Physiology*, **29**, 379–414.

Osmond, C.B. (1983). Interactions between irradiance, nitrogen nutrition, and water stress in the sun-shade responses of *Solanum dulcamara*. *Oecologia*, **57**, 316–321.

Osmond, C.B. (1994). What is photoinhibition? Some insights from comparisons of shade and sun plants. In: *Photoinhibition of Photosynthesis. From Molecular Mechanisms to the Field* (eds. Baker, N.R. and Bowyer, J.R.), BIOS Scientific, Oxford, UK, pp. 1–24.

Osmond, C.B., Anderson, J.M., Ball, M.C., *et al.* (1999). Compromising efficiency: the molecular ecology of light-resource utilization in plants. In: *Physiological Plant Ecology*. The 39th Symposium of the British Ecological Society held at the University of York, 7–9 September 1998 (Press, M.C., Scholes, J.D. and Barker, M.G.), pp. 1–24, Blackwell Science, Oxford, UK.

Osmond, C.B., Badger, M., Maxwell, K., *et al.* (1997). Too many photons: photorespiration, photoinhibition and photooxidation. *Trends in Plant Science*, **2**, 119–121.

Osmond, C.B., Björkman, O. and Anderson, D.J. (1980b). Physiological processes in plant ecology. Toward a synthesis with Atriplex. Springer Verlag, Berlin – Heidelberg – New York.

Osmond, C.B., Daley, P.F., Badger, M.R., *et al.* (1998). Chlorophyll fluorescence quenching during photosynthetic induction in leaves of *Abutilon striatum* Dicks. infected with *Abutilon mosaic virus,* observed with a field-portable imaging system. *Botanica Acta*, **111**, 390–397.

Osmond, C.B., Ludlow, M.M., Davis, R., *et al.* (1979). Stomatal responses to humidity in *Opuntia inermis* in relation to control of CO_2 and H_2O exchange patterns. *Oecologia*, **41**, 65–76.

Osmond, C.B., Neales, T.F. and Stange, G. (2008). Curiosity and context revisited: crassulacean acid metabolism in the anthropocene. *Journal of Experimental Botany* **59**, 1489–1502.

Osmond, C.B., Winter, K. and Ziegler, H. (1982). Funtional significance of different pathways of CO_2 fixation in photosynthesis. In: *Encyclopedia of Plant Physiology*, Vol 12B (eds. Lange, O.L., Nobel, P.S., Osmond, C.B. and Ziegler, H.) Springer-Verlag, Berlin, Germany, pp. 479–547.

Osmond, C.B., Winter, K. and Powles, S.B. (1980a). Adaptive significance of carbon dioxide recycling during photosynthesis in water stressed plants. In: *Adaptation of Plants to Water and High Temperature Stress?* (eds Turner, N.C. and Kramer, P.J.), Wiley Interscience, New York, USA, pp. 137–154.

Osteryoung, K.W. and Nunnari, J. (2003). The division of endosymbiotic organelles. *Science*, **302**, 1698–1704.

Oswald, O., Martin, T., Dominy, P.J., *et al.* (2001). Plastid redox state and sugars: interactive regulators of nuclear-encoded photosynthetic gene expression. *Proceedings of the Natural Academy of Sciences USA*, **98**, 2047–2052.

Otegui, M.E., Nicolini, M.G., Ruiz, R.A., *et al.* (1995). Sowing date effects on grain yield components for different maize genotypes. *Agronomy Journal*, **87**, 29–33.

Ott, T., Clarke, J., Birks, K., *et al.* (1999). Regulation of the photosynthetic electron transport chain. *Planta*, **209**, 250–258.

Ottander, C., Campbell, D. and Öquist, G. (1995). Seasonal changes in photosystem II organization and pigment composition in *Pinus sylvestris*. *Planta*, **197**, 176–183.

Ottoni, T.B., Matthias, A.D., Guerra, A.F., *et al.* (1992). Comparison of three resistance methods for estimating heat flux under stable conditions. *Agricultural and Forest Meteorology*, **58**, 1–18.

Ounis, A., Cerovic, Z.G., Briantais, J.M., *et al.* (2001a). Dual-excitation FLIDAR for the estimation of epidermal UV absorption in leaves and canopies. *Remote Sensing of Environment*, **76**, 33–48.

Ounis, A., Evain, S., Flexas, J., *et al.* (2001b). Adaptation of a PAM-fluorometer for remote sensing of chlorophyll fluorescence. *Photosynthesis Research*, **68**, 113–120.

Ouzounidou, G. (1996). The use of photoacoustic spectroscopy in assessing leaf photosynthesis under copper stress: correlation of energy storage to photosystem II fluorescence parameters and redox change of P700. *Plant Science*, **113**, 229–237.

Ouzounidou, G., Ilias, I., Tranopoulou, H., *et al.* (1998). Amelioration of copper toxicity by iron on spinach physiology. *Journal of Plant Nutrition*, **21(10)**, 2089–2101.

Ouzounidou, G., Moustakas, M. and Strasser, R.J. (1997). Sites of action of copper in the photosynthetic apparatus of maize leaves: kinetic analysis of chlorophyll fluorescence, oxygen evolution, absorption changes and thermal dissipation as monitored by photoacoustic signals. *Australian Journal of Plant Physiology*, **24**, 81–90.

Ouzounidou, G., Symeonidis, L., Babalonas, D., *et al.* (1994). Comparative responses of a copper-tolerant and a copper-sensitive population of *Minuartia hirsuta* to copper toxicity. *Journal of Plant Physiology*, **144**, 109–115.

Ovaska, J., Maenpaa, P., Nurmi, A., *et al.* (1990). Distribution of chlorophyll-protein complexes during chilling in the light compared with heat-induced modifications. *Plant Physiology*, **93**, 48–54.

Ovaska, J.A., Nilsen, J., Wielgolaski, F.E., *et al.* (2005). Phenology and performance of mountain birch provenances in transplant gardens: latitudinal, altitudinal and oceanity continentality gradients. In: *Plant Ecology, Herbivory, and Human Impact in Nordic Mountain Birch Forests* (eds Wielgolaski, F.E., Karlsson, P.S., Neuvonen, S. and Thannheiser, D.), Springer Verlag, Berlin, Germany, pp. 99–115.

Overdieck, D. (1993). Effects of atmospheric CO_2 enrichment on CO_2 exchange rates of beech stands in small model ecosystems. *Water Air and Soil Pollution*, **70**, 259–277.

Overdieck, D. and Forstreuter, M. (1994). Evapotranspiration of beech stands and transpiration of beech leaves subject to atmospheric CO_2 enrichment. *Tree Physiology*, **14**, 997–1003.

Overmyer, K., Kollist, H., Tuominen, H., *et al.* (2008). Complex phenotypic profiles leading to ozone sensitivity in *Arabidopsis thaliana* mutants. *Plant, Cell and Environment*, **31**, 1237–1249.

Owen, P.C. (1957). The effect of infection with tobacco etch virus on the rates of respiration and photosynthesis of tobacco leaves. *Annals of Applied Biology*, **45**, 327–331.

Owensby, C.E., Ham, J.M., Knapp, A.K., *et al.* (1997). Water vapour fluxes and their impact under elevated CO_2 in a C_4-tallgrass prairie. *Global Change Biology*, **3**, 189–195.

Owensby, C.E., Ham, J.M., Knapp, A.K., *et al.* (1999). Biomass production and species composition change in a tallgrass prairie ecosystem after long-term exposure to elevated atmospheric CO_2. *Global Change Biology*, **5**, 497–506.

Owen-Smith, R.N. (1988). *Megaherbavores: The Influence of Very Large Body Size on Ecology*. Cambridge University Press, Cambridge, UK.

Oxborough, K. (2004). Using chlorophyll a fluorescence imaging to monitor photosynthetic performance. In: *Chlorophyll a Fluorescence: A Signature of Photosynthesis, Advances in Photosynthesis and Respiration*, vol.19 (eds Papageorgiou, C.G. and Govindjee, J.), Springer, Dordrecht, Netherlands, pp. 389–407.

Oxborough, K. and Baker, N.R. (1997). An instrument capable of imaging chlorophyll a fluorescence from intact leaves at very low irradiance and at cellular and sub-cellular levels. *Plant, Cell and Environment*, **20**, 1473–1483.

Pace, C.N., Shirley, B.A., McNutt, M., *et al.* (1996). Forces contributing to the conformational stability of proteins. *The FASEB Journal*, **10**, 75–83.

Pachepsky, L.B., Reddy, V.R. and Acock, B. (1994). Cotton canopy photosynthesis model for predicting the effect of temperature and elevated carbon dioxide concentration. *Biotronics*, **23**, 35–46.

Pagani, M., Freeman, K.H. and Arthur, M.A. (1999). Late Miocene atmospheric CO_2 concentrations and the expansion of C_4 grasses. *Science*, **285**, 876–879.

Pagani, M., Zachos, J.C., Freeman, K.H., *et al.* (2005). Marked decline in atmospheric carbon dioxide concentrations during the Paleogene. *Science*, **309**, 600–660.

Pälike, H., Norris, R.D., Herrle, J.O., *et al.* (2006). The heartbeat of the Oligocene climate system. *Science*, **314**, 1894–1898.

Palmer, J.D. (1985). Comparative organization of chloroplast genomes. *Annual Review of Genetics*, **19**, 325–354.

Palmer, J.W. (1992). Effects of varying crop load on photosynthesis, dry matter production and partitioning of Crispin/M.27 apple trees. *Tree Physiology*, **11**, 19–33.

Palmer, J.W., Giuliani, R. and Adams, H.M. (1997). Effect of crop load on fruiting and leaf photosynthesis of 'Braeburn'/M.26 apple trees. *Tree Physiology*, **17**, 741–746.

Palmer, T.N. and Räisänen, J. (2002). Quantifying the risk of extreme seasonal precipitation events in a changing climate. *Nature*, **415**, 514–517.

Palmquist, K. (2000). Carbon economy of lichen. *New Phytologist*, **148**, 11–36.

Palmqvist, K., Sundblad, L.G., Samuelsson, G., *et al.* (1986). A correlation between changes in luminescence decay kinetics and the appearance of a CO_2-accumulating mechanism in *Scenedesmus obliquus. Photosynthesis Research*, **10**, 113–123.

Palmroth, S., Berninger, F., Lloyd, J., *et al.* (1999). No water conserving behaviour is observed in Scots pine from wet to dry climates. *Oecologia*, **121**, 302–309.

Panda, S., Mishra, A.K. and Biswal, U.C. (1986). Manganese-induced modification of membrane lipid peroxidation during aging of isolated wheat chloroplasts. *Photobiochemistry and Photobiophysics*, **13**, 53–61.

Paoletti, E. (1998). UV-B and acid rain effects on beech (*Fagus sylvatica* L.) and holm oak (*Quercus ilex* L.) leaves. *Chemosphere*, **36**, 835–840.

Papadopoulos, Y.A., Gordon, R.J., McRae, K.B., *et al.* (1999). Current and elevated levels of UV-B radiation have few impacts on yields of perennial forage crops. *Global Change Biology*, **5**, 847–856.

Papageorgiou, G. (1975). Chlorophyll fluorescence: an intrinsic probe of photosynthesis. In: *Bioenergetics of Photosynthesis* (ed. Govindjee, J.), Academic Press, New York, USA, pp. 319–371.

Papageorgiou, G.C. and Govindjee, J. (eds) (2004). *Chlorophyll a Fluorescence: A Signature of Photosynthesis. Advances in Photosynthesis and Respiration*. Kluwer Academic Publishers, Dordrecht, Netherlands.

Papageorgiou, G.C., Tsimilli-Michael, M. and Stamatakis, K. (2007). The fast and slow kinetics of chlorophyll a fluorescence induction in plants, algae and cyanobacteria: a view point. *Photosynthesis Research*, **94**, 275–290.

Papale, D. and Valentini, R. (2003). A new assessment of European forests carbon exchanges by eddy fluxes and artificial neural network spacialization, *Global Change Biology*, **9**, 525–535.

Papale, D., Reichstein, M., Aubinet, M., *et al.* (2006). Towards a standardized processing of Net Ecosystem Exchange measured with eddy covariance technique: algorithms and uncertainty estimation. *Biogeosciences*, **3**, 1–13.

Parent, B., Hachez, C., Redondo, E., *et al.* (2009). Drought and abscisic acid effects on aquaporin content translate into changes in hydraulic conductivity and leaf growth rate: a trans-scale approach. *Plant Physiology*, **149**, 2000–2012.

Parida, A.K., Das, A.B. and Mitra, B. (2004). Effects of SALT on growth, ion accumulation, photosynthesis and leaf anatomy of the mangrove, *Bruguiera parviflora*, *5*, **18**, 167–174.

Park, R. and Epstein, S. (1960). Carbon isotope fractionation during photosynthesis. *Geochimica et Cosmochimica Acta*, **21**, 110–126.

Park, S.Y., Yu, J.W., Park, J.S., *et al.* (2007). The senescence-induced staygreen protein regulates chlorophyll degradation. *Plant Cell*, **19**, 1649–1664.

Parkhurst, D.F. (1994). Diffusion of CO_2, and other gases inside leaves. *New Phytologist*, **126**, 449–79.

Parkhurst, D.F. and Mott, K.A. (1990). Intercellular diffusion limits to CO_2 uptake in leaves. 1. Studies in air and helox. *Plant Physiology*, **94**, 1024–1032.

Parkhurst, D.F., Wong, S.C., Farquhar, G.D., *et al.* (1988). Gradients of intercellular CO_2 levels across the leaf mesophyll. *Plant Physiology*, **86**, 1032–1037.

Pärnik, T.R. and Keerberg, O.F. (2006). Advanced radiogasometric method for the determination of the rates of photorespiratory and respiratory decarboxylations of primary and stored photosynthates under steady state photosynthesis. *Physiologia Plantarum*, **129**, 34–44.

Pärnik, T.R., Voronin, P.Y., Ivanona, H.N., *et al.* (2002). Respiratory CO_2 fluxes in photosynthesizing leaves of C3 species varying in rates of starch synthesis. *Russian Journal of Plant Physiology*, **49**, 729–735.

Parra, M.J., Acuña, K., Corcuera, L.J., *et al.* (2009). Vertical distribution of Hymenophyllaceae species among host tree microhabitats in a temperate rainforest in Southern Chile. *Journal of Vegetation Science*, **20**, 588–595.

Parry, M.A.J., Andralojc, P.J., Parmar, S., *et al.* (1997). Regulation of Rubisco by inhibitors in the light. *Plant Cell and Environment*, **20**, 528–534.

Parry, M.A.J., Flexas, J. and Medrano, H. (2005). Prospects for crop production under drought: research priorities and futures directions. *Annals of Applied Biology*, **147**, 211–226.

Parry, M.A.J., Keys, A.J. and Gutteridge, S. (1989). Variation in the specificity factor of C-3 higher-plant Rubisco determined by the total consumption of ribulose-P_2. *Journal of Experimental Botany*, **40**, 317–320.

Parry, M.A.J., Keys, A.J., Madgwick, P.J., *et al.* (2008). Rubisco regulation: a role for inhibitors. *Journal of Experimental Botany*, **59**, 1569–1580.

Parry, M.A.J., Madgwick, P.J., Bayon, C., *et al.* (2009). Mutation discovery for crop improvement. *Journal of Experimental Botany*, **60**, 2817–2825.

Parry, M.A.J., Madgwick, P.J., Carvalho, J.F.C., *et al.* (2007). Prospects for increasing photosynthesis by overcoming the limitations of Rubisco. *Journal of Agricultural Science*, **145**, 31–43.

Parsons, W.T. and Cuthbertson, E G. (2001). *Noxious Weeds of Australia*. CSIRO Publishing, Canberra, Australia.

Passey, B.H., Cerling, T.E., Perkins, M.E., *et al.* (2002). Environmental change in the Great Plains: an isotopic record from fossil horses. *Journal of Geology*, **110**, 123–140.

Pastenes, C. and Horton, P. (1996). Effect of high temperature on photosynthesis in beans. 1. Oxygen evolution and chlorophyll fluorescence. *Plant Physiology*, **112**, 1245–1251.

Pastenes, C. and Horton, P. (1999). Resistance of photosynthesis to high temperature in two bean varieties (*Phaseolus vulgaris* L.). *Photosynthesis Research*, **62**, 197–203.

Pastenes, C., Pimentel, P. and Lillo, J. (2005). Leaf movements and photoinhibition in relation to water stress in field-grown beans. *Journal of Experimental Botany* **56**, 425–433.

Pastori, G.M. and Foyer, C.H. (2002). Common components, networks, and pathways of cross-tolerance to stress. The central role of "redox" and abscisic acid-mediated controls. *Plant Physiology*, **129**, 460–468.

Pataki, D.E., Oren, R. and Smith, W.K. (2000). Sap flux of co-occurring species in a western subalpine forest during seasonal soil drought. *Ecology*, **81**, 2557–2566.

Pate, J. and Arthur, D. (1998). $\delta^{13}C$ analysis of phloem sap carbon: novel means of evaluating seasonal water stress and interpreting carbon isotope signatures of foliage and trunk wood of *Eucalyptus globulus*. *Oecologia*, **117**, 301–311.

Patel, M., Siegel, A.J. and Berry, J.O. (2006). Untranslated regions of FbRbcS1 mRNA mediate bundle sheath cell-specific gene expression in leaves of a C_4 plant. *Journal of Biological Chemistry*, **281**, 25485–25491.

Patiño, S., Tyree, M. and Herre, E.A. (1995). Comparison of hydraulic architecture of woody plants of differing phylogeny and growth form with special reference to free-standing and hemi-epiphytic *Ficus* species from Panama. *New Phytologist*, **129**, 125–134.

Patsikka, E., Kairavuo, M., Sersen, F., *et al.* (2002). Excess copper predisposes photosystem II to photoinhibition in vivo by outcompeting iron and causing decrease in leaf chlorophyll. *Plant Physiology*, **129**, 1359–1367.

Paul, M.J. and Driscoll, S.P. (1997). Sugar repression of photosynthesis: the role of carbohydrates in signalling nitrogen deficiency through source:sink imbalance. *Plant, Cell and Environment*, **20**, 110–116.

Paul, M.J. and Foyer, C.H. (2001). Sink regulation of photosynthesis. *Journal of Experimental Botany*, **52**, 1383–1400.

Paula, S. and Pausas, J.G. (2006). Leaf traits and resprouting ability in the Mediterranean basin. *Functional Ecology*, **20**, 941–947.

Pavel, E.W. and Dejong, T.M. (1993). Seasonal CO_2 exchange patterns of developing peach (*Prunus persica*) fruits in response to temperature, light and CO_2 concentration. *Physiologia Plantarum*, **88**, 322–330.

Pavlovic, A., Masarovičová, E. and Hudák, J. (2007). Carnivorous syndrome in Asian pitcher plants of the genus *Nepenthes*. *Annals of Botany*, **100**, 527–536.

Payton, P., Allen, R.D., Trolinder, N., *et al.* (1997). Over-expression of chloroplast-targeted Mn superoxide dismutase in cotton (*Gossypium hirsutum* L., cv. Coker 312) does not alter the reduction of photosynthesis after short exposures to low temperature and high light intensity. *Photosynthesis Research*, **52**, 233–244.

Pearce, D.W., Millard, S., Bray, D.F., *et al.* (2006). Stomatal characteristics of riparian poplar species in a semi-arid environment. *Tree Physiology*, **26**, 211–218.

Pearcy, R.B. and Sims, D.A. (1994). Photosynthetic acclimation to changing light environments: scaling from leaf to the whole plant. In: *Exploitation of Environmental Heterogeneity by Plants* (eds Caldwell, M.M. and Pearcy, R.W.), Academic Press, San Diego, USA, pp. 145–174.

Pearcy, R.W. (1983). The light environment and growth of C3 and C4 tree species in the understory of a Hawaiian forest. *Oecologia*, **58**, 19–25.

Pearcy, R.W. (1988). Photosynthetic utilization of lightflecks by understory plants. *Australian Journal of Plant Physiology*, **15**, 223–238.

Pearcy, R.W. (1989). Radiation and light measurements. In: *Plant Physiological Ecology: Field Methods and Instrumentation*, Vol 457 (eds Pearcy, R.W., Ehleringer, J.R., Mooney, H.A. and Rundel, P.W.). Chapman and Hall, New York, USA, pp. 353–359.

Pearcy, R.W. (1990). Sunflecks and photosynthesis in plant canopies. *Annual Review of Plant Physiology and Plant Molecular Biology*, **41**, 421–453.

Pearcy, R.W. (2007). Responses of plants to heterogeneous light environments. In: *Functional Plant Ecology* (eds

Pugnaire, F.I. and Valladares, F.), CRC Press, New York, USA, pp. 213–258.

Pearcy, R.W. and Yang, W. (1996). A three-dimensional crown architecture model for assessment of light capture and carbon gain by understory plants. *Oecologia*, **108**, 1–12.

Pearcy, R.W., Bjorkman, O., Caldwell, M.M., *et al.* (1987). Carbon gain by plants in natural environments. *BioScience*, **37**, 21–29.

Pearcy, R.W., Chazdon, R.L., Gross, L.J., *et al.* (1994). Photosynthetic utilization of sunflecks, a temporally patchy resource on a time-scale of seconds to minutes. In: *Exploitation of Environmental Heterogeneity by Plants* (eds Caldwell, M.M. and Pearcy, R.W.), Academic Press, San Diego, USA, pp. 175–208.

Pearcy, R.W., Gross, L.J. and He, D. (1997). An improved dynamic model of photosynthesis for estimation of carbon gain in sunfleck light regimes. *Plant, Cell and Environment*, **20**, 411–424.

Pearcy, R.W., Harrison, A.T., Mooney, H.A., *et al.* (1974). Seasonal changes in net photosynthesis of *Atriplex hymenelytra* shrubs growing in Death Valley, California. *Oecologia*, **17**, 111–121.

Pearcy, R.W., Muraoka, H. and Valladares, F. (2005). Crown architecture in sun and shade environments: assessing function and trade-offs with a three-dimensional simulation model. *New Phytologist*, **166**, 791–800.

Pearcy, R.W., Valladares, F., Wright, S.J., *et al.* (2004). A functional analysis of the crown architecture of tropical forest *Psychotria* species: do species vary in light capture efficiency and consequently in carbon gain and growth? *Oecologia*, **139**, 163–177.

Pearse, I., Heath, K.D. and Cheeseman, J.M. (2005). A partial characterization of peroxidase in *Rhizophora mangle. Plant, Cell and Environment*, **28**, 612–622.

Pedrós, R., Goulas, Y., Jacquemoud, S., *et al.* (2009). FluorMODleaf: a new leaf fluorescence emission model based on the PROSPECT model. *Remote Sensing of Environment,* **114**, 155–167.

Pedrós, R., Moya, I., Goulas, Y., *et al.* (2008). Chlorophyll fluorescence emission spectrum inside a leaf. *Photochemical and Photobiological Sciences*, **7**, 498–502.

Peet, M.M. and Kramer, P.J. (1980). Effects of decreasing source-sink ratio in soybeans on photosynthesis, photo-respiration, transpiration and yield. *Plant, Cell and Environment*, **3**, 201–206.

Peguero-Pina, F.J., Gil-Pelegrín, E. and Morales, F. (2009). Photosystem II efficiency of the palisade and spongy mesophyll in *Quercus coccifera* using adaxial/abaxial illumination and excitation light sources with wavelengths varying in penetration into the leaf tissue. *Photoynthesis Research*, **99**, 49–61.

Peguero-Pina, J.J., Morales, F., Flexas, J., *et al.* (2008). Photochemistry, remotely sensed physiological reflectance index and de-epoxidation state of the xanthophylls cycle in *Quercus coccifera* under intense drought. *Oecologia*, **156**, 1–11.

Pei, Z.M., Murata, Y., Benning, G., *et al.* (2000). Calcium channels activated by hydrogen peroxide mediate abscisic acid signalling in guard cells. *Nature*. **406**, 731–734.

Peichl, M. and Arain, M.A. (2007). Allometry and partitioning of above- and belowground tree biomass in an age-sequence of white pine forests. *Forest Ecology and Management*, **253**, 68–80.

Peisker, M. and Apel, H. (2001). Inhibition by light of CO_2 evolution from dark respiration: comparison of two gas exchange methods. *Photosynthesis Research*, **70**, 291–298.

Pell, E.J., Schlagnhaufer, C.D. and Arteca, R.N. (1997). Ozone-induced oxidative stress: mechanisms of action and reaction. *Physiologia Plantarum*, **100**, 264–273.

Peltier, G. and Cournac, L. (2002). Chlororespiration. *Annual Review of Plant Physiology*, **53**, 523–550.

Peltier, G. and Thibault, P. (1985). O_2 uptake in the light in *Chlamydomonas*: evidence for persistent mitochondrial respiration. *Plant Physiology*, **79**, 225–230.

Peltzer, D. and Polle, A. (2001). Diurnal fluctuation of antioxidative systems in leaves of field-grown beech trees (*Fagus sylvatica*): responses to light and temperature. *Physiologia Plantarum*, **111**, 158–164.

Penman, H.L. (1948). Natural evaporation from open water bare soil and grass. *Proceedings of the Royal Society*, **193**, 120–145.

Penning de Vries, F.W.T., Brunsting, A.H.M. and Van Laar, H.H. (1974). Products, requirements and efficiency of biosynthesis: a quantitative approach. *Journal of Theoretical Biology*, **45**, 339–77.

Peñuelas, J. and Azcon-Bieto, J. (1992). Changes in leaf $\delta^{13}C$ of herbarium plant species during the last 3 centuries of CO_2 increase. *Plant Cell and the Environment*, **15**, 485–489.

Peñuelas, J. and Boada, M. (2003). A global change-induced biome shift in the Motseny mountains (NE Spain). *Global Change Biology*, **9**, 131–140.

Peñuelas, J. and Filella, I. (1998). Visible and near-infrared reflectance techniques for diagnosing plant physiological status. *Trends in Plant Science*, **3**, 151–156.

Peñuelas, J. and Llusià, J. (2003). BVOCs: plant defense against climatic warming? *Trends in Plant Science*, **8**, 105–109.

Peñuelas, J. and Munné-Bosch, S. (2005). Isoprenoids: an evolutionary pool for photoprotection. *Trends in Plant Science*, **10**, 166–169.

Peñuelas, J., Baret, F. and Filella, I. (1995a). Semi-empirical indices to assess carotenoids/chlorophyll a ratio from spectral reflectance. *Photosynthetica*, **31**, 221–230.

Peñuelas, J., Filella, I. and Gamon, J.A. (1995c). Assessment of photosynthetic radiation-use efficiency with spectral reflectance. *New Phytologist*, **131**, 291–296.

Peñuelas, J., Filella, I., Biel, C., *et al.* (1993a). The reflectance at the 950–970 nm region as an indicator of plant water status. *International Journal of Remote Sensing*, **14**, 1887–1905.

Peñuelas, J., Filella, I., Lloret, P., *et al.* (1995b). Reflectance assessment of mite attack on apple trees. *International Journal of Remote Sensing*, **16**, 2727–2733.

Peñuelas, J., Filella, I., Llusia, J., *et al.* (1998). Comparative field study of spring and summer leaf gas exchange and photobiology of the Mediterranean trees *Quercus ilex* and *Phillyrea latifolia*. *Journal of Experimental Botany*, **49**, 229–238.

Peñuelas, J., Gamon, J.A., Fredeen, A.L., *et al.* (1994). Reflectance indices associated with physiological changes in nitrogen- and water-limited sunflower leaves. *Remote Sensing of Environment*, **48**, 135–146.

Peñuelas, J., Gamon, J.A., Griffin, K.L., *et al.* (1993b). Assessing community type, plant biomass, pigment composition, and photosynthetic efficiency of aquatic vegetation from spectral reflectance. *Remote Sensing of the Environment*, **46**, 1–25.

Peñuelas, J., Isla, R., Filella, I., *et al.* (1997b). Visible and near-infrared reflectance assessment of salinity effects on barley. *Crop Science*, **37**, 198–202.

Peñuelas, J., Llusià, J., Piñol, J., *et al.* (1997a). Photochemical reflectance index and leaf photosynthetic radition-use-efficiency assessment in Mediterranean trees. *International Journal of Remote Sensing*, **18**, 2863–2868.

Peñuelas, J., Munné-Bosch, S., Llusià, J., *et al.* (2004). Leaf reflectance and photo- and antioxidant protection in field-grown summer-stressed *Phillyrea angustifolia*. Optical signals of oxidative stress? *New Phytologist*, **162**, 115–124.

Peñuelas, J., Prieto, P., Beier, C., *et al.* (2007). Response of plant species richness and primary productivity in shrublands along a north-south gradient in Europe to seven years of experimental warming and drought: reductions in primary productivity in the heat and drought year of 2003. *Global Change Biology*, **13**, 2563–2581.

Pepin, S. and Livingston, N.J. (1997). Rates of stomatal opening in conifer seedlings in relation to air temperature and daily carbon gain. *Plant, Cell and Environment*, **20**, 1462–1472.

Pereira, J.S. and Chaves, M.M. (1993). Plant water deficit in Mediterranean ecosystems In: *Water Deficits: Plant Responses from Cell to Community* (eds Smith, J.A.C. and Griffiths, H.), Bios Scientific Publishing, USA, pp. 237–251.

Pereira, J.S., Mateus, J.A., Aires, L.M., *et al.* (2007). Net ecosystem carbon exchange in three contrasting Mediterranean ecosystems-the effect of drought. *Biogeosciences*, **4**, 791–802.

Pérez, C., Madero, P., Pequerul, A., *et al.* (1993). Specificity of manganese in some aspects of soybean (*Glycine max* L.) physiology. In: *Optimization of Plant Nutrition* (eds. Fragoso, M.A.C. and van Beusichem, M.L.), Kluwer Academic Publishers, Dordrecht, The Netherlands, pp. 503–507.

Pérez-Bueno, M.L., Ciscato, M., vandeVen, M., *et al.* (2006). Imaging viral infection. Studies on *Nicotiana benthamiana* plants infected with the pepper mild mottle tobamovirus. *Photosynthesis Research*, **90**, 111–123.

Pérez-Bueno, M.L., Rahoutei, J., Sajnani, C., *et al.* (2004). Proteomic analysis of the oxygen-evolving complex of photosystem II under biotic stress. Studies on *Nicotiana benthamiana* infected with tobamoviruses. *Proteomics*, **4**, 418–425.

Perez-Martin, A., Flexas, J., Ribas-Carbó, M., *et al.* (2009). Interactive effects of soil water deficit and air vapour pressure deficit on mesophyll conductance to CO_2 in *Vitis vivifera* and *Olea europaea*. *Journal of Experimental Botany*, **60**, 2391–2405.

Perez-Peña, J. and Tarara, J. (2004). A portable whole canopy gas exchange system for several mature field-grown grapevines. *Vitis*, **43**, 7–14.

Pérez-Pérez, J.G., Syvertsen, J.P., Botía, P., *et al.* (2007). Leaf water relations and net gas exchange responses of salinized *Carrizo citrange* seedlings during drought stress and recovery. *Annals of Botany*, **100**, 335–345.

Pérez-Priego, O., Zarco-Tejada, P., Miller, J.R., *et al.* (2005). Detection of water stress in orchard trees with a high-resolution spectrometer through chlorophyll fluorescence *in-filling* of the O_2-a Band. *IEEE Transactions on Geoscience and Remote Sensing*, **43**, 2, December 2005.

Pérez-Torres, E., Bravo, L.A., Corcuera, L.J., *et al.* (2007). Is electron transport to oxygen an important mechanism in photoprotection? Contrasting responses from antarctic vascular plants. *Physiologia Plantarum*, **130**, 185–194.

Pérez-Torres, E., Garcia, A., Dinamarca, J., *et al.* (2004). The role of photochemical quenching and antioxidants in photoprotection of *Deschampsia antartica. Functional Plant Biology*, **31**, 731–741.

Perks, M.P., Irvine, J. and Grace, J. (2002). Canopy stomatal conductance and xylem sap abscisic acid (ABA) in mature Scots pine during a gradually imposed drought. *Tree Physiology*, **22**, 877–883.

Peschel, S., Beyer, M. and Knoche, M. (2003). Surface characteristics of sweet cherry fruit: stomata-number, distribution, functionality and surface wetting. *Scientia Horticulturae*, **97**, 265–278.

Peterhansel, C., Niessen, M. and Kebeish, R.M. (2008). Metabolic Engineering towards the enhancement of photosynthesis. *Photochemistry and Photobiology*, **84**, 1317–1323.

Peterson, B.J. and Fry B. (1987). Stable isotopes in ecosystem studies. *Annual Review of Ecology and Systematics*, **18**, 293–320.

Peterson, R.B. and Aylor, D.E. (1995). Chlorophyll fluorescence induction in leaves of *Phaseolus vulgaris* infected with bean rust (*Uromyces appendiculatus*). *Plant Physiology*, **108**, 163–171.

Peterson, R.B., Oja, V. and Laisk, A. (2001). Chlorophyll fluorescence at 680 and 730 nm and leaf photosynthesis. *Photosynthesis Research*, **70**, 185–196.

Peterson, R.K.D. (2001). Photosynthesis, yield loss and injury guilds. In: *Biotic Stress and Yield Loss* (eds Peterson, R.K.D. and Higley, L.G.), CRC Press, Boca Raton, FL, USA, pp. 83–97.

Peterson, R.K.D., Higley, L.G., Haile, F.J., *et al.* (1998). Mexican bean beetle (Coleoptera: Coccinelidae) injury affects photosynthesis of *Glycine max* and *Phaseolus vulgaris. Environmental Entomology*, **27**, 373–381.

Peterson, R.K.D., Shannon, C.L. and Lenssen, A.W. (2004). Photosynthetic responses of legume species to leaf-mass consumption injury. *Environmental Entomology*, **33**, 450–456.

Pethybridge, S.J., Haya, F., Eskerb, P., *et al.* (2008). Visual and radiometric assessments for yield losses caused by ray blight in *Pyrethrum. Crop Science*, **48**, 343–352.

Petit, A.N., Vaillant, N., Boulay, M., *et al.* (2006). Alteration of photosynthesis in grapevines affected by *Esca. Phytopathology*, **96**, 1060–1066.

Pfannschmidt, T. (2003). Chloroplast redox signals: how photosynthesis controls its own genes. *Trends in plant science*, **8**, 33–41.

Pfannschmidt, T., Brautigam, K., Wagner, R., *et al.* (2009). Potential regulation of gene expression in photosynthetic cells by redox and energy state: approaches towards better understanding. *Annals of Botany*, **103**, 599–607.

Pfanz, H., Aschan, G., Langenfeld-Heyser, R., *et al.* (2002). Ecology and ecophysiology of tree stems: corticular and wood photosynthesis. *Naturwissenschaften*, **89**, 147–162.

Pfeffer, M. and Peisker, M. (1998). CO_2 gas exchange and phosphoenolpyruvate carboxylase activity in leaves of *Zea mays* L. *Photosynthesis Research*, **58**, 281–291.

Pfündel, E. (1998). Estimating the contribution of photosystem I to total leaf chlorophyll fluorescence. *Photosynthesis Research*, **56**, 185–195.

Pfündel, E.E., Ben Ghoslen, N., Meyer, S., *et al.* (2007). Investigating UV screening in leaves by two different types of portable UV fluorimeters reveals in vivo screening by anthocyanins and carotenoids. *Photosynthesis Research*, **93**, 205–221.

Phene, C., Baker, D., Lambert, J., *et al.* (1978). SPAR – Soil-Plant-Atmosphere Research System. *Transactions of the ASAE*, **21**, 924–30.

Phillips, N.G., Buckley, T.N. and Tissue, D.T. (2008). Capacity of old trees to respond to environmental change. *Journal of Integrative Plant Biology*, **50**, 1355–1364.

Piel, C., Frank, E., Le Roux, X., *et al.* (2002). Effect of local irradiance on CO_2 transfer conductance of mesophyll in walnut. *Journal of Experimental Botany*, **53**, 2423–2430.

Pielke, Sr. R.A., Adegoke, J.O., Chase, T.N., *et al.* (2007). A new paradigm for assessing the role of agriculture in the climate system and in climate change. *Agricultural and Forest Meteorology*, **142**, 234–254.

Pierce, J., Tolbert, N.E. and Barker, R. (1980). Interaction of Rubisco with transition state analogues. *Biochemistry*, **19**, 934–962.

Pieruschka, R., Chavarría-Krauser, A., Cloos, K., *et al.* (2008). Photosynthesis can be enhanced by lateral CO_2 diffusion inside leaves over distances of several millimetres. *New Phytologist*, **178**, 335–347.

Pieruschka, R., Huber, G. and Berry, J.A. (2010) Control of transpiration by radiation. *Proceedings of the National Academy of Sciences USA*, **107**, 13372–1337.

Pieruschka, R., Schurr, U. and Jahnke, S. (2005). Lateral gas diffusion inside leaves. *Journal of Experimental Botany*, **56**, 857–864.

Pieters, A.J. and Núñez, M. (2008). Photosynthesis, water use efficiency, and $\delta^{13}C$ in two rice genotypes with contrasting response to water deficit. *Photosynthetica*, **46**, 574–580.

Pieters, A.J., Paul, M.J. and Lawlor, D.W. (2001). Low sink demand limits photosynthesis under Pi deficiency. *Journal of Experimental Botany*, **52**, 1083–1091.

Pietrini, F. and Massacci, A. (1998). Leaf anthocyanin content changes in *Zea mays* L grown at low temperature: Significance fort he relationship between the quantum yield of PSII and the apparent quantum yield of CO_2 assimilation. *Photosynthesis Research*, **58**, 213–219.

Pincebourd, S., Frak, E., Sinoquet, H., *et al.* (2006). Herbivory mitigation through increased water-use efficiency in a leaf-mining moth-apple tree relationship. *Plant Cell and Environment*, **29**, 2238–2247.

Pinder, J.E. and Jackson, P.R. (1988). Plant photosynthetic pathways and grazing by phytophagous orthopterans. *American Midland Naturalist*, **120**, 201–211.

Pinder, J.E. and Kroh, G.C. (1987). Insect herbivory and photosynthetic pathways in old-field ecosystems. *Ecology*, **68**, 254–259.

Pineda, M., Gáspár, L., Morales, F., *et al.* (2008a). Multicolour fluorescence imaging: a useful tool to visualise systemic viral infections in plants. *Photochemistry and Photobiology*, **84**, 1048–1060.

Pineda, M., Soukupová, J., Matouš, K., *et al.* (2008b). Conventional and combinatorial chlorophyll fluorescence imaging of tobamovirus-infected plants. *Photosynthetica*, **46**, 441–451.

Pinelli, P. and Loreto, F. (2003). $^{12}CO_2$ emission from different metabolic pathways measured in illuminated and darkened C_3 and C_4 leaves at low, atmospheric and elevated CO_2 concentration. *Journal of Experimental Botany*, **54**, 1761–1769.

Pinheiro, C. and Chaves, M.M. (2011). Photosynthesis and drought: can we make metabolic connections from available data? *Journal of Experimental Botany*, **62**, 869–882.

Pinheiro, C., António, C., Ortuño, M.F. *et al.* (2011). Initial water deficit effects on *Lupinus albus* photosynthetic performance, carbon metabolism, and hormonal balance: metabolic reorganization prior to early stress responses. *Journal of Experimental Botany*, **62**, 4965–4974.

Pinkard, E.A. and Mohammed, C.L. (2006). Photosynthesis of *Eucalyptus globulus* with *Mycosphaerella* leaf disease. *New Phytologist*, **170**, 119–127.

Pitzschke, A. and Hirt, H. (2009). Disentangling the complexity of mitogen-activated protein kinases and reactive oxygen species signaling. *Plant Physiology*, **149**, 606–615.

Pizon, A. (1902). *Anatomie et Physiologie Végétales*. Doin Eds, Paris.

Plascyk, J.A. (1975). The MK II Fraunhofer line discriminator (FLD-II) for airborne and orbital remote sensing of solar-stimulated luminescence. *Optical Engineering*, **14**, 339–346.

Plaxton, W.C. (1996). The organization and regulation of plant glycolysis. *Annual Review of Plant Physiology and Plant Molecular Biology*, **47**, 185–214.

Plaziat, J.C., Cavagnetto, C., Koeniguer, J.C., *et al.* (2001). History and biogeography of the mangrove ecosystem, based on a critical reassessment of the paleontological record. *Wetlands Ecology and Management*, **9**, 161–179.

Plesnicar, M., Kastori, R., Petrovic, N., *et al.* (1994). Photosynthesis and chlorophyll fluorescence in sunflower (*Helianthus annuus* L.) leaves as affected by phosphorus nutrition. *Journal of Experimental Botany*, **45**, 919–924.

Poni, S., Bernizzoni, F., Civardi, S., *et al.* (2009). Performance and water-use efficiency (single-leaf vs. whole-canopy) of well-watered and half-stressed split-root *Lambrusco* grapevines grown in Po Valley (Italy). *Agriculture, Ecosystems and Environment*, **129**, 97–106.

Poni, S., Magnani, E. and Bernizzoni, F. (2003). Degree of correlation between total light interception and whole-canopy net CO_2 exchange rate in two grapevine growth systems. *Australian Journal of Grape and Wine Research*, **9**, 2–11.

Pons, T.L. and Pearcy, R.W. (1994). Nitrogen reallocation and photosynthetic acclimation in response to partial shading in soybean plants. *Physiologia Plantarum*, **92**, 636–644.

Pons, T.L. and Welschen, R.A.M. (2002). Overestimation of respiration rates in commercially available clamp-on leaf chambers. Complications with measurement of net photosynthesis. *Plant, Cell and Environment*, **25**, 1367–1372.

Pons, T.L., Flexas, J., von Caemmerer, S., *et al.* (2009). Estimating mesophyl conductance to CO_2: methodology, potential errors and recommendations. *Journal of Experimental Botany*, **60**, 2217–2234.

Pontailler, J.Y. (1990). A cheap quantum sensor using a gallium arsenide photodiode. *Functional Ecology*, **4**, 591–596.

Ponton, S., Flanagan, L.B., Alstad, K.P., *et al.* (2006). Comparison of ecosystem water-use efficiency among Douglas-fir forest, aspen forest and grassland using eddy covariance and carbon isotope techniques. *Global Change Biology*, **12**, 294–310.

Poormohammad, Kiani, S., Grieu, P. *et al.* (2007). Genetic variability for physiological traits under drought conditions and differential expression of water stress-associated genes in sunflower (*Helianthus annuus* L.). *Theoretical and Applied Genetics*, **114**, 193–207.

Poorter, H. (1993). Interspecific variation in the growth response of plants to an elevated CO_2 concentration. *Vegetatio*, **104/105**, 77–97.

Poorter, H. and Navas, M.L. (2003). Plant growth and competition at elevated CO_2: On winners, losers and functional groups. *New Phytologist*, **157**, 175–198.

Poorter, H., Niinemets, Ü., Poorter, L., *et al.* (2009). Causes and consequences of variation in leaf mass per area (LMA): a meta-analysis. Tansley review. *New Phytologist*, **182**, 565–588.

Poorter, H., Remkes, C. and Lambers, H. (1990). Carbon and nitrogen economy of 24 wild species differing in relative growth rate. *Plant Physiology*, **94**, 621–627.

Poorter, L. and Oberbauer, S.F. (1993). Photosynthetic induction responses of 2 rain-forest tree species in relation to light environment. *Oecologia*, **96**, 193–199.

Poorter, L. and Werger, M.J.A. (1999). Light environment, sapling architecture and leaf display in six rain forest tree species. *American Journal of Botany*, **86**, 1464–1473.

Popper, Z.A. and Fry, S.C. (2003). Primary cell wall composition of bryophytes and charophytes. *Annals of Botany*, **91**, 1–12.

Popper, Z.A., Michel, G., Hervé, C. *et al.* (2011). Evolution and diversity of plant cell walls: from algae to flowering plants. *Annual Review of Plant Biology*, **62**, 567–590.

Porcar-Castell, A., Juurola, E., Berninger, F., *et al.* (2008). Seasonal acclimation of photosystem II in *Pinus sylvestris*. Studying the effect of light environment through the rate constants of sustained heat dissipation and photochemistry I. *Tree Physiology*, **28**, 1483–1491.

Porembski, S. and Barthlott, W. eds. (2000). *Inselbergs*. Ecological Studies, vol. 146, Springer-Verlag, Berlin – Heidelberg-New York.

Port, M., Tripp, J., Zielinski, D., Weber, C., *et al.* (2004). Role of Hsp17.4-CII as coregulator and cytoplasmic retention factor of tomato heat stress transcription factor HsfA2 1. *Plant Physiology*, **135**, 1457–1470.

Portis, A.R., Salvucci, M.E. and Ogren, W.L. (1986). Activation of ribulosebisphosphate carboxylase/ oxygenase at physiological CO_2 and ribulosebisphosphate concentrations by rubisco activase. *Plant Physiology*, **82**, 967–971.

Portis, Jr., A.R., Li, C., Wang, D., *et al.* (2008). Regulation of Rubisco activase and its interaction with Rubisco. *Journal of Experimental Botany*, **59**, 1597–1604.

Portsmuth, A., Niinemets, Ü., Truus, L., *et al.* (2005). Biomass allocation and growth rates in *Pinus sylvestris* are interactively modified by nitrogen and phosphorus availabilities and by tree size and age. *Canadian Journal of Forest Research*, **35**, 2346–2359.

Pospisil, P., Skotnica, J. and Naus, J. (1998). Low and high temperature dependence of minimum F-O and maximum F-M chlorophyll fluorescence in vivo. *Biochimica et Biophysica Actah*, **1363**, 95–99.

Postuka, J.W., Dropkin, V.H. and Nelson, C.J. (1986). Photosynthesis, photorespiration, and respiration of soybean after infection with root nematodes. *Photosynthetica*, **20**, 405–410.

Pou, A., Flexas, J., Alsina, M.M., Bota, J., *et al.* (2008). Adjustments of water-use efficiency by stomatal regulation during drought and recovery in the drought-adapted *Vitis* hybrid Richter-110 (*V. berlandieri* × *V. rupestris*). *Physiologia Plantarum*, **134**, 313–323.

Poulson, M.E. and Vogelmann, T.C. (1990). Epidermal focusing and effects upon photosynthetic light-harvesting in leaves of *Oxalis*. *Plant, Cell and Environment*, **13**, 803–811.

Poulson, M.E., Torres-Boeger, M.R. and Donahue, R.A. (2006). Response of photosynthesis to high light and drought for *Arabidopsis thaliana* grown under a UV-B enhanced light regime. *Photosynthesis Research*, **90**, 79–90.

Pozsar, B.I., Horvath, L., Lehoczky, J., *et al.* (1969). Effect of the grape chromemosaic and grape fanleaf yellow-mosaic virus infection on the photosynthetical carbon dioxide fixation in vine leaves. *Vitis*, **8**, 206–10.

Prändl, R., Hinderhofer, K., Eggers-Schumacher, G., *et al.* (1998). HSF3, a new heat shock factor from *Arabidopsis thaliana*, derepresses the heat shock response and confers thermotolerance when overexpressed in transgenic plants. *Molecular and General Genetics*, **258**, 269–278.

Prasad, D.D.K. and Prasad, A.R.K. (1987). Altered d-aminolevulinic acid metabolism by lead and mercury in germinating seedlings of bajra (*Pennisetum typhoideum*). *Journal of Plant Physiology*, **127**, 241–249.

Prasad, M.N.V. and Strzalka, K. (1999). Impact of heavy metals on photosynthesis. In *Heavy Metal Stress in Plants: From Molecules to Ecosystems* (eds Prasad, M.N.V. and Hagemeyer, J.), Springer, Berlin, Germany, pp. 117–138.

Prasad, T.K. (1996). Mechanisms of chilling-induced oxidative stress injury and tolerance in developing maize seedlings: changes in antioxidant system, oxidation of proteins and lipids, and protease activities. *Plant Journal*, **10**, 1017–1026.

Prendergast, H.D.V. (1989). Geographical distribution of C_4 acid decarboxylation types and associated structural variants in native Australian C_4 grasses (Poaceae). *Australian Journal of Botany*, **37**, 253–273.

Prentice, I.C., Farquhar, G.D., Fasham, M.J.R., *et al.* (2001). The carbon cycle and atmospheric carbon dioxide. In: *Climate Change 2001: The Scientific Basis* (eds Houghton, J.T., Ding, Y., Griggs, D.J., Noguer, M., van der Linden, P.J., Dai, X., Maskell, K. and Johnson, C.A.), Cambridge University Press, Camridge, UK, pp. 183–237.

Priault, P., Tcherkez, G., Cornic, G., *et al.* (2006). The lack of mitochondrial complex I in a CMSII mutant of *Nicotiana sylvestris* increases photorespiration through an increased internal resistance. *Journal of Experimental Botany*, **57**, 3195–3207.

Price, A.H., Cairns, J.E., Horton, P., *et al.* (2002). Linking drought resistance mechanisms to drought avoidance in upland rice using a QTL approach: progress and new opportunities to integrate stomatal and mesophyll responses. *Journal of Experimental Botany*, **53**, 989–1004.

Price, G.D., Von Caemmerer, S., Evans, J.R., *et al.* (1994). Specific reduction of chloroplast carbonic anhydrase activity by antisense RNA in transgenic tobacco plants has a minor effect on photosynthetic CO_2 assimilation. *Planta*, **193**, 331–340.

Price, P., Waring, G.L., Julkunen-Titto, R., *et al.* (1989). Carbon-nutrient balance hypothesis in within-species phytochemical variation of *Salix lasiolepis*. *Journal of Chemical Ecology*, **15**, 1117–1131.

Prins, A., Van Heerden, P.D.R., Olmos, E., *et al.* (2008). Cysteine proteases regulate chloroplast protein content and composition in tobacco leaves: a model for dynamic interactions with ribulose-1, 5-bisphosphate carboxylase-oxygenase (Rubisco) vesicular bodies. *Journal of Experimental Botany*, **59**, 1935–1950.

Prins, H.B.A. and de Guia, M.B. (1986). Carbon source of the water soldier (*Stratiotes aloides* L.). *Aquatic Botany*, **26**, 225–234.

Prinsley, R.T., Hunt, S., Smith, A.M., *et al.* (1986). The influence of a decrease in irradiance on photosynthetic carbon assimilation in leaves of *Spinacia oleracea* L. *Planta*, **167**, 414–420.

Prior, L.D., Eamus, D. and Duff, G.A. (1997). Seasonal and diurnal patterns of carbon assimilation, stomatal conductance and leaf water potential in *Eucalyptus tetrodonta* saplings in a wet–dry savanna in northern Australia. *Australian Journal of Botany*, **45**, 241–258.

Proctor, M.C.F. (1984). Structure and ecological adaptation. In: *The Experimental Biology of Bryophytes, Experimental Botany: An International Series of Monographs* (eds Dyer, A.F., Duckett, J.G. and Cronshaw, J.), Academic Press, London, pp. 9–37.

Proctor, M.C.F. (2000). Mosses and Alternative Adaptation to Life on Land. *New Phytologist*, **148**, 1–3.

Proctor, M.C.F. and Pence, V. (2002). Vegetative tissues: bryophytes, vascular resurrection plants and vegetative propagules. In: Black, M., Pritchard, H.V., eds. *Dessication and Survival in Plants: Drying without Dying*. Wallingford, UK: CAB International, 207–237.

Proietti, P., Famiani, F. and Tombessi, A. (1999). Gas exchange in olive fruit. *Photosynthetica*, **36**, 423–432.

Prud'homme, B., Gompel, N. and Carroll, S.B. (2007). Emerging principles of regulatory evolution. *Proceedings of the National Academy of Sciences*, **104**, 8605–8612.

Pruitt, W.O., Swann, B.D., Held, A.A., *et al.* (1987). Bowen ratio and Penman: Australia-California tests. In: *Irrigation System for the 21st Century* (eds James, L.G. and English, M.J), American Society of Civil Engineers, Portland, USA, pp.149–158.

Pryer, K.M., Schneider, H., Smith, A.R., *et al.* (2001). Horsetails and ferns are a monophyletic group and the closest living relatives to seed plants. *Nature*, **409**, 618–621.

Psaras, G.K. and Rhizopoulou, S. (1995). Mesophyll structure during leaf development in *Ballota acetabulosa*. *New Phytologist*, **131**, 303–309.

Puckeridge, D.W. (1971). Photosynthesis of wheat under field conditions III. Seasonal trends in carbon dioxide uptake of crop communities. *Australian Journal of Agriultural Research*, **22**, 1–9.

Pugnaire, F.I., Haase, P., Incoll, L.D., *et al.* (1996). Response of the tussock grass *Stipa tenacissima* to watering in a semi-arid environment. *Functional Ecology*, **10**, 265–274.

Puthiyaveetil, S., Kavanagh, T.A., Cain, P., *et al.* (2008). The ancestral symbiont sensor kinase CSK links photosynthesis with gene expression in chloroplasts. *Proceedings of the National Academy of* Sciences (USA), **105**, 10061–10066.

Pyankov, V.I., Gunin, P.D., Tsoog, S., *et al.* (2000). C4 plants in the vegetation of Mongolia: their natural occurrence and geographical distribution in relation to climate. *Oecologia*, **123**, 15–31.

Pyke, K.A. (1999). Plastid division and development. *Plant Cell*, **11**, 549–556.

Pyke, K.A., Rutherford, S.M., Robertson, E.J., *et al.* (1994). *arc6*, a fertile *Arabidopsis* mutant with only two mesophyll cell chloroplasts. *Plant Physiology*, **106**, 1169–1177.

Qian, H. and Ricklefs, R.E. (2000). Large-scale processes and the Asian bias in species diversity of temperate plants. *Nature*, **407**, 180–182.

Quade, J. and Cerling, T.E. (1995). Expansion of C_4 grasses in the late Miocene of northern Pakistan: evidence from stable isotopes in paleosols. *Palaeogeography, Palaeoclimatology, Palaecoecology*, **115**, 91–116.

Quade, J., Cater, J.M.L., Ojha, T.P., *et al.* (1995). Late Miocene environmental change in Nepal and the northern Indian subcontinent: stable isotopic evidence from paleosols. *GSA Bull.*, **107**, 1381–1397.

Quarrie, S.A. and Jones, H.G. (1977). Effect of abscisic acid and water stress on development and morphology of wheat. *Journal of Experimental Botany*, **28**, 192–203.

Queiroz-Voltan, R.B. and Paradela-Filho, O. (1999). Caracterização de estruturas anatômicas de citros infectados com *Xylella fastidiosa*. *Laranja*, **20**, 55–76.

Queitsch, C., Hong, S.W., Vierling, E., *et al.* (2000). Heat shock protein 101 plays a crucial role in thermotolerance in *Arabidopsis*. *The Plant Cell*, **12**, 479.

Queitsch, C., Sangster, T.A. and Lindquist, S. (2002). Hsp90 as a capacitor for genetic variation. *Nature*, **417**, 618–624.

Querejeta, J.I., Allen, M.F., Alguacil, M.M., *et al.* (2007). Plant isotopic composition provides insight into mechanisms underlying growth stimulation by AM fungi in a semiarid environment. *Functional Plant Biology*, **34**, 683–691.

Querejeta, J.I., Barea, J.M., Allen, M.F., *et al.* (2003). Differential response of δ13C and water use efficiency to arbuscular mycorrhizal infection in two aridland woody plant species. *Oecologia*, **135**, 510–515.

Queval, G., Hager, J., Gakière B., *et al.* (2008). Why are literature data for H_2O_2 contents so variable? A discussion of potential difficulties in quantitative assays of leaf extracts. *Journal of Experimental Botany*, **59**, 135–146.

Queval, G., Issakidis-Bourguet, E., Hoeberichts, F.A., *et al.* (2007). Conditional oxidative stress responses in the *Arabidopsis* photorespiratory mutant cat2 demonstrate that redox state is a key modulator of day length-dependent gene expression and define photoperiod as a crucial factor in the regulation of H_2O_2-induced cell death. *Plant Journal*, **52**, 640–657.

Quick, W.P. and Stitt, M. (1989). An examination of factors contributing to non-photochemical quenching of chlorophyll fluorescence in barley leaves. *Biochimica et Biophysica Acta*, **977**, 287–296.

Quinn, P.J. (1988). Effects of temperature on cell membranes. In: *Plant Membrane Stability* (ed. Society for Experimental Biology), pp. 237–258.

Quinn, P.J., Williams, W.P., Barber, J., *et al.* (1985). Environmentally induced changes in chloroplast membranes and their effects on photosynthetic function. In: *Photosynthetic Mechanisms and the Environment* (eds Barber, J. and Baker, N.J.), Elsevier Science Publishers, Dordrecht, Netherlands, pp. 1–47.

Quirino, B.F., Noh, Y.S., Himelblau, E., *et al.* (2000). Molecular aspects of leaf senescence. *Trends Plant Science*, **5**, 278–282.

Rachmilevitch, S., Cousins, A.B. and Bloom, A.J. (2004). Nitrate assimilation in plant shoots depends on photorespiration. *Proceedings of the Natural Academy of Sciences USA*, **101**, 11506–11510.

Raftoyannis, Y. and Radoglou, K. (2002). Physiological responses of beech and sessile oak in a natural mixed stand during a dry summer. *Annals of Botany*, **89**, 723–730.

Raghavendra, A.S. and Padmasree, K. (2003). Beneficial interactions of mitochondrial metabolism with photosynthetic carbon assimilation. *Trends Plant Science*, **8**, 546–553.

Rahman, A.F., Cordova, V.D., Gamon, J.A., *et al.* (2004). Potential of MODIS ocean bands for estimating CO_2 flux from terrestrial vegetation: a novel approach. *Geophysical Research Letters*, **31**, L10503.

Rahman, A.F., Gamon, J.A., Fuentes, D.A., *et al.* (2001). Modeling spatially distributed ecosystem flux of boreal forest using hyperspectral indices from AVIRIS imagery. *Journal of Geophysical Research – Atmospheres*, **106**, 33579–33591.

Rahoutei, J., Barón, M., García-Luque, I., *et al.* (1999). effect of tobamovirus infection on the thermoluminiscence characteristics of chloroplast from infected plants. *Z. Naturforsch*, **54c**, 634–639.

Rahoutei, J., García-Luque, I. and Barón, M. (2000). Inhibition of photosynthesis by viral infection: effect on PSII structure and function. *Physiologia Plantarum*, **110**, 286–292.

Raines, C.A. (2006). Transgenic approaches to manipulate the environmental responses of the C3 carbon fixation cycle. *Plant, Cell and Enviaronment*, **29**, 331–339.

Raison, J.K., Roberts, J.K.M. and Berry, J.A. (1982). Correlations between the thermal stability of chloroplast (thylakoid) membranes and the composition and fluidity of their polar lipids upon acclimation of the higher plant, *Nerium oleander*, to growth temperature. *Biochimica et Biophysica Acta*, **688**, 218–228.

Rajabi, A., Ober, E.S. and Griffiths, H. (2009). Genotypic variation for water use efficiency, carbon isotope discrimination, and potential surrogate measures in sugar beet. *Field Crops Research*, **112**, 172–181.

Rakocevic, M., Sinoquet, H., Christophe, A., *et al.* (2000). Assessing the geometric structure of a white clover

(*Trifolium repens* L.) canopy using 3-D digitising. *Annals of Botany*, **86**, 519–526.

Ramalho, J.C., Campos, P.S., Teixeira, M., *et al.* (1998). Nitrogen dependent changes in antioxidant system and in fatty acid composition of chloroplast membranes from *Coffea arabica* L. plants submitted to high irradiance. *Plant Science*, **135**, 115–124.

Rambal, S. (2001). Hierarchy and productivity of Mediterranean-type ecosystems. In: *Global Terrestrial Productivity* (eds Roy, J., Saugier, B. and Mooney, H.A.), Academic Press, San Diego, USA, pp. 315–344.

Ranieri, A., Castagna, A., Baldan, B., *et al.* (2001). Iron deficiency differently affects peroxidase isoforms in sunflower. *Journal of Experimental Botany*, **52**, 25–35.

Ranjith, S.A., Meinzer, F.S., Perry M.H., *et al.* (1995). Partitioning of carboxylase activity in nitrogen stressed sugarcane and its relationship to bundle sheath leakiness to CO_2, photosynthesis and carbon isotope discrimination. *Australian Journal of Plant Physiology*, **22**, 903–911.

Rao, D.N. (1971). A study of the air pollution problem due to coal unloading in Varanasi, India. In: *Proceedings of the Second International Clean Air Congress* (eds Englund, H.M. and Beery, W.T.), Academic Press, New York, USA, pp. 273–276.

Rao, I.M., Abadía, A. and Terry, N. (1986). Leaf phosphate status and photosynthesis in vivo: changes in light scattering and chlorophyll fluorescence during photosynthetic induction in sugar beet leaves. *Plant Science*, **44**, 133–137.

Rao, I.M. and Terry, N. (1989). Leaf phosphate status, photosynthesis and carbon partitioning in sugar beet. I. Changes in growth, gas exchange and Calvin cycle enzymes. *Plant Physiology*, **90**, 814–819.

Rao, I.M. and Terry, N. (1995). Leaf phosphate status, photosynthesis, and carbon partitioning in sugar beet. IV: Changes with time following increased supply of phosphate to low phosphate plants. *Plant Physiology*, **107**, 1313–1321.

Rascher, U. and Pieruschka, R. (2008). Spatio-temporal variations of photosynthesis: the potential of optical remote sensing to better understand and scale light use efficiency and stresses of plant ecosystems. *Precision Agriculture*, **9**, 355–366.

Rascher, U., Agati, G., Alonso, L., *et al.* (2009). CEFLES2: the remote sensing component to quantify photosynthetic efficiency from the leaf to the region by measuring sun-induced fluorescence in the oxygen absorption bands. *Biogeosciences Discussions*, **6**, 2217–2266.

Rascher, U., Hütt, M.Th., Siebke, K., *et al.* (2001). Spatiotemporal variation of metabolism in a plant circadian rhythm. The biological clock as an assembly of coupled individual oscillators. *Proceedings of the Natural Academy of Sciences USA*, **98**, 11, 801–11, 805.

Rascher, U., Liebig M. and Lüttge U. (2000). Evaluation of instant light-response curves of chlorophyll fluorescence parameters obtained with a portable chlorophyll fluorometer on site in the field. *Plant, Cell and Environment*, **23**, 1397–1405.

Raschke K. and Resemann A. (1986). Midday depression of CO_2 assimilation in leaves of *Arbutus unedo* L.: diurnal changes in photosynthetic capacity related to changes in temperature and humidity. *Planta*, **168**, 546–58.

Raschke, K. (1956). Über die physikalischen beziehungen zwischen wämeübergangszahl, strahlungsaustausch, temperatur and transpiration eines blattes. *Planta*, **48**, 200–238.

Rasmusson, A.G. and Escobar, M.A. (2007). Light and diurnal regulation of plant respiratory gene expression. *Physiologia Plantarum*, **129**, 57–67.

Rastetter, E.B., Agren, G.I. and Shaver, G.R. (1997). Responses of N-limited ecosystems to increased CO_2: A balanced-nutrition, coupled-element-cycles model. *Ecological Applications*, **7**, 444–460.

Rasulov, B., Copolovici, L., Laisk, A., *et al.* (2009). Postillumination isoprene emission: in vivo measurements of dimethylallylpyrophosphate pool size and isoprene synthase kinetics in aspen leaves. *Plant Physiology*, **149**, 1609–1618.

Rauner, J.L. (1976). Deciduous forests. In: *Vegetation and the Atmosphere, Vol 2. Case Studies* (ed. Monteith, J.L.), Academic Press, London – New York – San Francisco, pp. 241–264.

Raupach, M.R., Marland, G. Ciais, P., *et al.* (2007). Global and Regional drivers of accelerating CO_2 emissions. *Proceedings of the National Academy of Science*, **104**, 10288–10293.

Raupach, M.R., Rayner, P.J., Barrett, D.J., *et al.* (2005). Model–data synthesis in terrestrial carbon observation: methods, data requirements and data uncertainty specifications. *Global Change Biology*, **11**, 378–397.

Raven, J.A. (2002b). Evolutionary options. *Nature*, **415**, 375–376.

Raven, J.A. (2000). Land plant biochemistry. *Philosophical Transactions of the Royal Society of London B*, **355**, 833–846.

Raven, J.A. (2002a). Selection pressures on stomatal evolution. *The New Phytologist*, **153**, 371–386.

Raven, J.A. and Allen, J.F. (2003). Genomics and chloroplast evolution: what did cyanobacteria do for plants? *Genome Biology*, **4**, 209.

Raven, J.A. and Farquhar, G.D. (1990). The influence of N metabolism and organic acid synthesis on the natural abundance of isotopes of carbon in plants. *New Phytologist*, **116**, 505–529.

Raven, J.A. and Glidewell, S.M. (1981). Processes limiting photosynthetic conductance. In: *Processes Limiting Plant Productivity* (ed. Johnson, C.B.), Butterworths, London, UK, pp. 109–136.

Raven, J.A. and Spicer, R.A. (1996). The evolution of crassulacean acid metabolism. In: *Crassulacean Acid Metabolism. Biochemistry, Ecophysiology and Evolution* (eds Winter, K. and Smith, J.A.C.), Springer, New York, USA, pp. 360–385.

Raven, J.A., Cockell, C.S. and De La Roche, C.L. (2008). The evolution of inorganic carbon concentrating mechanisms in photosynthesis. *Philosophical Transactions of the Royal Society, B*, **363**, 2641–2650.

Ravenel, J., Peltier, G. and Havaux, M. (1994). The cyclic electron pathways around photosystem I in *Chlamydomonas reinhardtii* as determined in vivo by photoacoustic measurements of energy storage. *Planta*, **193**, 251–259.

Rawat, A.S. and Purohit, A.N. (1991). CO_2 and water vapour exchange in four alpine herbs at two altitudes and under varying light and temperature conditions. *Photosynthesis Research*, **28**, 99–108.

Rawsthorne, S., Hylton, C.M., Smith, A.M., *et al.* (1988). Photorespiratory metabolism and immunogold localisation of photorespiratory enzymes in leaves of C_3 and C_3-C_4 intermediate species of *Moricandia*. *Planta*, **173**, 298–308.

Rawsthorne, S., Morgan, C.L., O'Neill, C.M., *et al.* (1998). Cellular expression pattern of the glycine decarboxylase P protein in leaves of an intergenic hybrid between the C_3-C_4 intermediate species *Moricandia nitens* and the C_3 species *Brassica napus*. *Theoretical and Applied Genetics*, **96**, 922–927.

Read, J., Sanson, G.D., de Garine-Wichatitsky, M., *et al.* (2006). Sclerophylly in two contrasting tropical environments: low nutrients and low rainfall. *American Journal of Botany*, **93**, 1601–1614.

Rebetzke, G.J., Condon, A.G., Richards, R.A., *et al.* (2002). Selection for reduced carbon isotope discrimination increases aerial biomass and grain yield of rainfed bread wheat. *Crop Science*, **42**, 739–745.

Reddy, K.R., Hodges, H.F., Read, J.J., *et al.* (2001). Soil-Plant-Atmosphere-Research (SPAR) facility: a tool for plant research and modeling. *Biotronics*, **30**, 27–50.

Reich, P.B. (1993). Reconciling apparent discrepancies among studies relating life-span, structure and function of leaves in contrasting plant life forms and climates – the blind men and the elephant retold. *Functional Ecology*, **7**, 721–725.

Reich, P.B. and Amundson, R.G. (1985). Ambient levels of ozone reduce net photosynthesis in tree and crop species. *Science*, **230**, 566–570.

Reich, P.B. and Bolstad, P. (2001). Productivity of evergreen and deciduous temperature forests. In: *Terrestrial Global Productivity* (eds Mooney, H.A., Saugier, B. and Roy, J.). Academic Press, San Diego, USA, pp. 245–283.

Reich, P.B., Ellsworth, D.S. and Walters, M.B. (1998a). Leaf structure (specific leaf area) modulates photosynthesis-nitrogen relations: evidence from within and across species and functional groups. *Functional Ecology*, **12**, 948–958.

Reich, P.B., Ellsworth, D.S., Walters, M.B., *et al.* (1999). Generality of leaf trait relationships: a test across six biomes. *Ecology*, **80**, 1955–1969.

Reich, P.B., Kloeppel, B.D., Ellsworth, D.S., *et al.* (1995). Different photosynthesis-nitrogen relations in deciduous hardwood and evergreen coniferous tree species. *Oecologia*, **104**, 24–30.

Reich, P.B., Tjoelker, M.G., Machado, J.L., *et al.* (2006). Universal scaling of respiratory metabolism, size and nitrogen in plants. *Nature*, **439**, 457–461.

Reich, P.B., Uhl, C., Walters, M.B., *et al.* (2004). Leaf demography and phenology in Amazonian rain forest: A census of 40 000 leaves of 23 tree species. *Ecological Monographs*, **74**, 3–23.

Reich, P.B., Walters, M.B. and Ellsworth, D.S. (1991). Leaf age and season influence the relationships between leaf nitrogen, leaf mass per area and photosynthesis in maple and oak trees. *Plant Cell Environment*, **14**, 251–259.

Reich, P.B., Walters, M.B. and Ellsworth D.S. (1997). From tropics to tundra: global convergence in plant functioning. *Proceedings of the National Academy of Science of the USA*, **94**, 13730–13734.

Reich, P.B., Walters, M.B. and Ellsworth, D.S. (1992). Leaf life-span in relation to leaf, plant, and stand characteristics among diverse ecosystems. *Ecological Monographs*, **62**, 365–392.

Reich, P.B., Walters, M.B., Ellsworth, D.S., *et al.* (1998b). Relationships of leaf dark respiration to leaf nitrogen, specific leaf area and leaf life-span: a test across biomes and functional groups. *Oecologia*, **114**, 471–482.

Reich, P.B., Walters, M.B., Tjoelker, M.G., *et al.* (1998c). Photosynthesis and respiration rates depend on leaf and root morphology and nitrogen concentration in nine boreal tree species differing in relative growth rate. *Functional Ecology*, **12**, 395–405.

Reich, P.B., Wright, I.J., Cavender-Bares, J., *et al.* (2003). The evolution of plant functional variation: traits, spectra and strategies. *International Journal of Plant Sciences*, **164**, s143-s164.

Reich, P. B. and Schoettle, A.W. (1988). Role of phosphorus and nitrogen in photosynthetic and whole plant carbon gain and nutrient use efficiency in eastern white-pine. *Oecologia*, **77**, 25–33.

Reichle, D.E. (1981). *Dynamic Properties of Forest Ecosystems.* IBP 23, Cambridge University Press, Cambridge, UK, pp. 683.

Reichstein, M., Falge, E., Baldocchi, D., *et al.* (2005). On the separation of net ecosystem exchange into assimilation and ecosystem respiration: review and improved algorithm. *Global Change Biology*, **11**, 1424–1439.

Reichstein, M., Papale, D., Valentini, R., *et al.* (2007). Determinants of terrestrial ecosystem carbon balance inferred from European eddy covariance flux sites. *Geophysical Research Letters*, **34**, L01402.

Reichstein, M., Tenhunen, J.D., Roupsard, O., *et al.* (2002). Severe drought effects on ecosystem CO_2 and H_2O fluxes at three Mediterranean evergreen sites: revision of current hypothesis? *Global Change Biology*, **8**, 999–1017.

Reid, C.D. and Strain, B.R. (1994). Effects of CO_2 enrichment on whole-plant carbon budget of seedlings of *Fagus grandifolia* and *Acer saccharum* in low irradiance. *Oecologia*, **98**, 31–39.

Reinbothe, C., Lebedev, N. and Reinbothe, S. (1999). A protochlorophyllide light-harvesting complex involved in de-etiolation of higher plants. *Nature*, **397**, 80–84.

Reinero, A. and Beachy, N.R. (1989). Reduced photosystem II activity and accumulation of viral coat protein in chloroplast of leaves infected with tobacco mosaic virus. *Plant Physiology*, **89**, 111–116.

Reinert, F., Roberts, A., Wilson, J.M., *et al.* (1997). Gradation in nutrient composition and photosynthetic pathways across the restinga vegetation of Brazil. *Botanica Acta*, **110**, 135–142.

Reinert, F., Russo, C.A.M. and Salles, L.O. (2003). The evolution of CAM in the subfamily Pitcairnioideae (Bromeliaceae). *Biological Journal of the Linnean Society*, **80**, 261–268.

Reinikäinen, J. and Huttunen, S. (1989). The level of injury and needle ultrastructure of acid rain-irrigated pine and spruce seedlings after low temperature treatment. *New Phytologist*, **112**, 29–39.

Reiter, R., Hötberger, M., Green, R., *et al.* (2008). Photosynthesis of lichens from lichen-dominated communities in the alpine/nival belt of the Alps – II: laboratory and field measurements of CO_2 exchange and water relations. *Flora*, **203**, 34–46.

Rekika, D., Monneveux, P. and Havaux, M. (1997). The in vivo tolerance of photosynthetic membranes to high and low temperatures in cultivated and wild wheats of the *Triticum* and *Aegilops* genera. *Journal of Plant Physiology*, **150**, 734–738.

Renaut, J., Hausman, J.F. and Wisniewski, M.E. (2006). Proteomics and low-temperature studies: bridging the gap between gene expression and metabolism. *Physiologia Plantarum*, **126**, 97–109.

Rennenberg, H., Loreto, F., Polle, A., *et al.* (2006). Physiological responses of forest trees to heat and drought. *Plant Biology*, **8**, 556–571.

Renou, J.L., Gerbaud, A., Just, D., *et al.* (1990). Differing substomatal and chloroplastic concentrations in water-stressed wheat. *Planta*, **182**, 415–419.

Repka, V. (2002). Chlorophyll-deficient mutant in oak (*Quercus petraea* L.) displays an accelerated hypersensitive-like cell death and an enhanced resistance to powdery mildew disease. *Photosynthetica*, **40**, 183–193.

Repo, T., Leinonen, I., Ryyppö, A. and Finer, L. (2004). The effect of soil temperature on the bud phenology, chlorophyll fluorescence, carbohydrate content and cold hardiness of Norway spruce seedlings. *Physiologia Plantarum*, **121**, 93–100.

Resco, V., Ewers, B.E., Sun, W., *et al.* (2009). Drought-induced hydraulic limitations constrain leaf gas exchange recovery after precipitation pulses in the C_3 woody legume, *Prosopis velutina. The New Phytologist*, **181**, 672–682.

Restom, T.G. and Nepstad, D.C. (2001). Contribution of vines to the evapotranspiration of a secondary forest in eastern Amazonia. *Plant and Soil*, **236**, 155–163.

Retallack, G.J. (1997). Earliest Triassic origin of Isoetes and quillwort evolutionary radiation. *Journal of Paleontology*, **71**, 500–521.

Retallack, G.J. (2001). Cenozoic expansion of grasslands and climatic cooling. *Journal of Geology*, **109**, 407–426.

Retallack, G.J. (2005). Permian greenhouse crises. In: *The Nonmarine Permian* (eds Lucas, S.G. and Zeigler, K.E.), New Mexico Museum of Natural History and Science Bulletin n° 30, pp. 256–269.

Retuerto, R., Fernandez-Lema, B., Rodriguez-Roiloa, S., *et al.* (2004). Increased photosynthetic performance in holly trees infested by scale. *Functional Ecology*, **18**, 664–669.

Reumann, S. and Weber, A.P.M. (2006). Plant peroxisomes respire in the light: some gaps of the photorespiratory C_2

cycle have become filled, other remain. *Biochim Biophys Acta*, **1763**, 1496–1510.

Reverberi, M., Fanelli, C., Zjalic, S., *et al.* (2005). Relationship among lipoperoxides, jasmonates and indole-3-acetic acid formation in potato tubers after wounding. *Free Radical Research*, **39**, 637–647.

Reynolds, M., Foulkes, M.J., Slafer, G.A., *et al.* (2009). Raising yield potential in wheat. *Journal of Experimental Botany*, **60**, 1899–1918.

Reynolds, O. (1895). On the dynamical theory of incompressible viscous fluids and the determination of criterion. *Philosophical Transactions of Royal Society of London*, **A174**, 935–982.

Rhizopoulou, S. and Psaras, G.K. (2003). Development and Structure of drought-tolerant leaves of the Mediterranean shrub *Capparis spinosa* L. *Annals of Botany*, **92**, 377–383.

Rhoads, D.M. and Subbaiah, C.C. (2007). Mitochondrial retrograde regulation in plants. *Mitochondrion*, **7**, 177–194.

Riaño, D., Vaughan, P., Chuvieco, E., *et al.* (2005). Estimation of fuel moisture content by inversion of radiative transfer models to simulate equivalent water thickness and dry matter content: analysis at leaf and canopy level. *IEEE Transactions on Geoscience and Remote Sensing*, **43**, 819–826.

Ribas-Carbo, M., Berry, J.A., Yakir, D., *et al.* (1995). Electron partitioning between the cytochrome and alternative pathways in plant mitochondria. *Plant Physiology*, **109**, 829–837.

Ribas-Carbo, M., Giles, L., Flexas, J., *et al.* (2008). Phytochrome-driven changes in respiratory electron transport partitioning in soybean (*Glycine max.* L.) cotyledons. *Plant Biology*, **10**, 281–287.

Ribas-Carbo, M., Lennon, A.M., Robinson, S.A., *et al.* (1997). The regulation of the electron partitioning between the cytochrome and alternative pathways in soybean cotyledon and root mitochondria. *Plant Physiology*, **113**, 903–911.

Ribas-Carbo, M., Robinson, S.A. and Giles, L. (2005a). The application of the oxygen-isotope technique to assess respiratory pathway partitioning. In: *Plant Respiration: From Cell to Ecosystem* (eds Lambers, H. and Ribas-Carbo, M.) Springer, USA, pp. 31–42.

Ribas-Carbo, M., Robinson, S.A., Gonzalez-Meler, M.A., *et al.* (2000). Effects of light on respiration and oxygen isotope fractionation in soybean cotyledons. *Plant Cell Environment*, **23**, 983–989.

Ribas-Carbo, M., Taylor, N.L., Giles, L., *et al.* (2005b). Effects of water stress on respiration in soybean (*Glycine max.* L.) leaves. *Plant Physiology*, **139**, 466–473.

Ribeiro, R.V., Machado, E.C. and Oliveira, R.F. (2003a). Early photosynthetic responses of sweet orange plants infected with *Xylella fastidiosa*. *Physiological and Molecular Plant Pathology*, **62**, 167–173.

Ribeiro, R.V., Machado, E.C. and Oliveira, R.F. (2004). Growth- and leaf-temperature effects on photosynthesis of sweet orange seedlings infected with *Xylella fastidiosa*. *Plant Pathology*, **53**, 334–340.

Ribeiro, R.V., Machado, E.C., Oliveira, R.F. *et al.* (2003b). High temperature effects on the response of photosynthesis to light in sweet orange plants infected with *Xylella fastidiosa*. *Brazilian Journal of Plant Physiology*, **15**, 89–97.

Rice, S.A. and Bazzaz, F.A. (1989). Quantification of plasticity of plant traits in response to light intensity: comparing phenotypes at a common weight. *Oecologia*, **78**, 502–507.

Rich, P.R. (1988). A critical examination of the supposed variable proton stoichiometry of the chloroplast cytochrome Ibf complex. *Biochimica et Biophysica Acta*, **932**, 33–42.

Richards, A.R. (2008). Genetic opportunities to improve cereal root systems for drylands agriculture. *Plant Production Science*, **1**, 12–16.

Richards, R.A. (2000). Selectable traits to increase crop photosynthesis and yield of grain crops. *Journal Experimental Botany*, **51**, 447–58.

Richardson, A.D., Hollinger, D.Y., Burba, G.G., *et al.* (2006). A multi-site analysis of random error in tower-based measurements of carbon and energy fluxes, *Agricultural and Forest Meteorology*, **136**, 1–18.

Richter, A. and Popp, M. (1992). The physiological importance of accumulation of cyclitols in *Viscum album* L. *New Phytologist*, **121**, 431–438.

Richter, A., Wanek, W., Werner, R.A., *et al.* (2009). Preparation of starch and soluble sugars of plant material for analysis of carbon isotope composition: a comparison of methods. *Rapid Communications in Mass Spectrometry*, **23**, 2476–2488.

Ricklefs, R.E. (2004). A comprehensive framework for global patterns in biodiversity. *Ecology Letters*, **7**, 1–15.

Rintamaki, E., Keys, A.J. and Parry, M.A.J. (1988). Comparison of the specific activity of ribulose-1,5-bis-phosphate carboxylase-oxygenase from some C_3 and C_4 plants. *Physiologia Plantarum*, **74**, 326–331.

Rintamaki, E., Salonen, M., Suoranta, U.M., *et al.* (1997). Phosphorylation of light-harvesting complex II and photosystem II core proteins shows different irradiance-dependent regulation in vivo application of phosphothreonine antibodies to analysis of thylakoid phosphoproteins. *Journal of Biological Chemistry*, **272**, 30476–30482.

Rios-Estepa, R. and Lange, B.M. (2007). Experimental and mathematical approaches to modeling plant metabolic networks. *Phytochemistry*, **68**, 2351–2374.

Ripley, B.S., Gilbert, M.E., Ibrahim, D.G., *et al.* (2007). Drought constraints on C_4 photosynthesis: stomatal and metabolic limitations in C_3 and C_4 subspecies of *Alloteropsis semialata*. *Journal of Experimental Botany*, **58**, 1351–1363.

Rivero, R.M., Kojima M., Gepstein, A., *et al.* (2007). Delayed leaf senescence induces extreme drought tolerance in a flowering plant. *Proceedings of the National Academy of Sciences*, **104**, **49**, 19631–19636.

Rivero, R.M., Shulaev, V. and Blumwald, E. (2009). Cytokinin-dependent photorespiration and the protection of photosynthesis during water deficit. *Plant Physiology*, **150**, 1530–1540.

Robakowski, P. (2005). Susceptibility to low-temperature photoinhibition in three conifers differing in successional status. *Tree Physiology*, **25**, 1151–1160.

Robert, C., Bancal, M.O., Ney, B., *et al.* (2005). Wheat leaf photosynthesis loss due to leaf rust, with respect to lesion development and leaf nitrogen status. *New Phytologist*, **165**, 227–241.

Roberts, A., Borland, A.M. and Griffiths, H. (1997). Discrimination processes and shifts in carboxylation during the phases of crassulacean acid metabolism. *Plant Physiology*, **113**, 1283–1292.

Roberts, D.A. and Corbett, M.K. (1965). Reduced photosynthesis in tobacco plants infected with tobacco ringspot virus. *Phytopathology*, **55**, 370–371.

Roberts, P.L. and Wood, K.R. (1982). Effects of a severe (P6) and a mild (W) strain of cucumber mosaic virus on tobacco leaf chlorophyll, starch and cell ultrastructure. *Physiological and Molecular Plant Pathology*, **21**, 31–37.

Robinson, J.M. (1988). Does O_2 photoreduction occur within chloroplasts in vivo? *Physiologia Plantarum*, **72**, 666–680.

Robinson, S.A. and Osmond, C.B. (1994). Internal gradients of chlorophyll and carotenoid pigments in relation to photoprotection in thick leaves of plants with Crassulacean acid metabolism. *Australian Journal of Plant Physiology*, **21**, 497–506.

Robinson, S.P. and Portis, A.R., Jr. (1989). Ribulose-1, 5-bisphosphate carboxylase/oxygenase activase protein prevents the in vitro decline in activity of ribulose- 1,5-bisphosphate carboxylase/oxygenase. *Plant Physiology*, **90**, 968–971.

Rochaix, J.D. (2007). Role of thylakoid protein kinases in photosynthetic acclimation. *FEBS Letters*, **581**, 2768–2775.

Rochette, P., Desjardins, R.L., Pattey, E., *et al.* (1996). Instantaneous measurement of radiation and water use efficiencies of a maize crop. *Agronomy Journal*, **88**, 627–635.

Rockström, J., Lannerstad, M. and Falkenmark, M. (2007). Assessing the water challenge of a new green revolution in developing countries. *Proceedings of the National Academy of Sciences USA*, **104**, 6253–6260.

Rodeghiero, M., Niinemets, Ü. and Cescatti, A. (2007). Major diffusion leaks of clamp-on leaf cuvettes still unaccounted: how erroneous are the estimates of Farquhar *et al.* model parameters? *Plant Cell Environment*, **30**, 1006–1022.

Roden, J.S. and Ehleringer, J.R. (1999). Hydrogen and oxygen isotope ratio of tree-ring cellulose for riparian trees grown long term under hydroponically controlled environments. *Oecologia*, **121**, 467–477.

Roderick, M.L. (1999). Estimating the diffuse component from daily and monthly measurements of global radiation. *Agricultural and Forest Meteorology*, **95**, 169–185.

Roderick, M.L., Farquhar, G.D., Berry, S.L., *et al.* (2001). On the direct effect of clouds and atmospheric particles on the productivity and structure of vegetation. *Oecologia*, **129**, 21–30.

Rodrigues, M.L., Santos, T.P., Rodrigues, A.P., *et al.* (2008). Hydraulic and chemical signalling in the regulation of stomatal conductance and plant water use of field grapevines growing under deficit irrigation. *Functional Plant Biology*, **35** (7) 565–579.

Rodriguez, D., Ewert, F., Goudriaan, J., *et al.* (2001). Modelling the response of wheat canopy assimilation to atmospheric CO_2 concentrations. *New Phytologist*, **150**, 337–346.

Rodriguez-Moreno, L., Pineda, M., Soukupova, J., *et al.* (2008). Early detection of bean infection by *Pseudomonas syringae* in asymptomatic leaf areas using chlorophyll fluorescence imaging. *Photosynthesis Research*, **96**, 27–35.

Rodríguez-Pérez, J.R., Riaño, D., Carlisle, E., *et al.* (2007). Evaluation of hyperspectral reflectance indexes to detect grapevine water status in vineyards. *American Journal of Enology and Viticulture*, **58**, 302–317.

Rogers, A., Ellsworth, D.S. and Humphries, S.W. (2001). Possible explanation of the disparity between the *in vitro* and *in vivo* measurements of Rubisco activity: a study in loblolly pine grown in elevated pCO_2. *Journal of Experimental Botany*, **52**, 1555–1561.

Rogiers, N., Eugster, W., Furger, M., *et al.* (2005). Effect of land management on ecosystem carbon fluxes at a subalpine grassland site in the Swiss Alps. *Theoretical and Applied Climatology*, **80**, 187–203.

Rolfe, S.A. and Scholes, J.D. (2002). Extended depth-of-focus imaging of chlorophyll fluorescence from intact leaves. *Photosynthesis Research*, **72**, 107–115.

Rolletschek, H., Radchuk, R., Klukas, C., *et al.* (2005). Evidence of a key role for photosynthetic oxygen release in oil storage in developing soybean seeds. *The New Phytologist*, **167**, 777–786.

Rolletschek, H., Weber, H. and Borisjuk, L. (2003). Energy status and its control on embryogenesis of legumes. Embryo photosynthesis contributes to oxygen supply and is coupled to biosynthetic fluxes. *Plant Physiology*, **132**, 1196–1206.

Rolletschek, H., Weschke, W., Wobus, U., *et al.* (2004). Energy state and its control on seed development: starch accumulation is associated with high ATP and steep oxygen gradients within barley grains. *Journal of Experimental Botany*, **55**, 1351–1359.

Roloff, I., Scherm, H. and van Iersel, M.W. (2004). Photosynthesis of blueberry leaves as affected by *Septoria* leaf spot and abiotic leaf damage. *Plant Disease*, **88**, 397–401.

Romero-Puertas, M.C., Palma, J.M., Gómez, M., *et al.* (2002). Cadmium causes the oxidative modification of proteins in pea plants. *Plant, Cell and Environment*, **25**, 677–686.

Romero-Puertas, M.C., Rodríguez-Serrano, M., Corpas, F.J., *et al.* (2004). Cadmium-induced subcellular accumulation of O_2.- and H_2O_2 in pea leaves. *Plant, Cell and Environment*, **27**, 1122–1134.

Roscher, A., Emsley, L., Raymond, P., *et al.* (1998). Unidirectional steady state rates of central metabolism enzymes measured simultaneously in a living plant tissue. *The Journal of Biological Chemistry*, **273**, 25053–25061.

Rosema, A. and Verhoef, W., (1991). Modeling of fluorescence light-canopy interaction. Proceeding of the 5th International Colloquium-Physical Signatures in Remote Sensing Courchevel France 14–18 January 1991 (ESA SP-319 May 1991).

Rosema, A., Snel, J.F.H., Zahn, H., *et al.* (1998). The relation between laser induced chlorophyll fluorescence and photosynthesis. *Remote Sensing of the Environmentment*, **65**, 143–154.

Rosema, A., Verhoef, W., Schroote, J., *et al.* (1991). Simulating fluorescence light-canopy interaction in support of laser-induced fluorescence measurements. *Remote Sensing of Environment*, **37**, 117–130.

Rosenberg, N.J., Blad, B.L. and Verma, S.B. (1983). *Microclimate: The Biological Environment*. Wiley, New York, USA.

Rosenstiel, T.N., Ebbets, A.L., Khatri, W.C., *et al.* (2004). Induction of poplar leaf nitrate reductase: a test of extrachloroplastic control of isoprene emission rate *Plant Biology*, **6**, 12–21.

Rosenthal, S.I. and Camm, E.L. (1996). Effects of air temperature, photoperiod and leaf age on foliar senescence of western larch (*Larix occidentalis* Nutt.) in environmentally controlled chambers. *Plant Cell Environment*, **19**, 1057–1065.

Rosenthal, Y., Farnsworth, B., Rodrigo Romo, F.V., *et al.* (1999). High quality, continuous measurements of CO_2 in Biosphere 2 to assess whole mesocosm carbon cycling. *Ecological Engineering*, **13**, 249–262.

Rosenzweig, C., Tubiello, F.N., Goldberg, R., *et al.* (2002). Increased crop damage in the US from excess precipitation under climate change. *Global Environmental Change*, **12**, 197–202.

Ross, J. (1981). The radiation regime and architecture of plant stands. Dr. W. Junk Publishers, The Hague.

Ross, J. and Sulev, M. (2000). Sources of errors in measurements of PAR. *Agricultural and Forest Meteorology*, **100**, 103.

Rossa, B. and von Willert, D.J. (1999). Physiological characteristics of geophytes in a semi-arid Namaqualand, South Africa. *Plant Ecology*, **142**, 121–132.

Rossel, J.B., Wilson, P.B., Hussain, D., *et al.* (2007). Systemic and intracellular responses to photooxidative stress in *Arabidopsis*. *Plant Cell*, **19**, 4091–4110.

Rosso, D., Ivanov, A.G., Fu, A., Geisler-Lee, J., *et al.* (2006). IMMUTANS does not act as a stress-induced safety valve in the protection of the photosynthetic apparatus of *Arabidopsis* during steady state photosynthesis. *Plant Physiology*, **142**, 574–585.

Rotenberg, D., MacGuidwin, A.E., Saeed, I.A.M., *et al.* (2004). Interaction of spatially separated *Pratylenchus penetrans* and *Verticillium dahliae* on potato measured by impaired photosynthesis. *Plant Pathology*, **53**, 294–302.

Rotenberg, E. and Yakir, D. (2010). Contribution of semi-arid forests to the climate system. *Science*, **327**, 451–454.

Roth, I. (1992). *Leaf Structure: Coastal Vegetation and Mangroves of Venezuela*. Gebrüder Borntraeger, Berlin, Germany.

Roth-Nebelsick, A., Hassiotou, F. and Veneklaas, E.J. (2009). Stomatal *Crypts* have small effects on transpiration: A numerical model Analysis. *Plant Physiology*, **151**, 2018–2027.

Roth-Nebelsick, A., Uhl, D., Mosbrugger, V., *et al.* (2001). Evolution and function of leaf venation: a review. *Annals of Botany*, **87**, 553–566.

Roumet, C., Garnier, E., Suzor, H., *et al.* (2000). Short and long-term responses of whole-plant gas exchange to elevated CO_2 in four herbaceous species. *Environmental Experimental Botany*, **43**, 155–169.

Rousseaux, M.C., Scopel, A.L., Searles, P.S., *et al.* (2001). Responses to solar ultraviolet-B radiation in a shrub-dominated natural ecosystem of Tierra del Fuego (southern Argentina). *Global Change Biology*, **7**, 467–478.

Roussel, M., Dreyer, E., Montpied, P., *et al.* (2009). The diversity of ^{13}C isotope discrimination in a *Quercus robur* full-sib family is associated with differences in intrinsic water use efficiency, transpiration efficiency, and stomatal conductance. *Journal of Experimental Botany*, **60**, 2419–2431.

Rowell, D.M., Ades, P.K., Tausz, M., *et al.* (2009). Lack of genetic variation in tree ring $\delta^{13}C$ suggests a uniform, stomatally driven response to drought stress across *Pinus radiata* genotypes. *Tree Physiology*, **29**, 191–198.

Rowland, D., Dorner, J., Sorensen, R., *et al.* (2005). Tomato spotted wilt virus in peanut tissue types and physiological effects related to disease incidence and severity. *Plant Pathology*, **54**, 431–440.

Roxas, V.P., Lodhi, S.A., Garrett, D.K., *et al.* (2000). Stress tolerance in transgenic tobacco seedlings that overexpress glutathione S-transferase/glutathione peroxidase. *Plant and Cell Physiology*, **41**, 1229–1234.

Royer, D.L., Osborne, C.P. and Beerling, D.J. (2005). Contrasting seasonal patterns of carbon gain in evergreen and deciduous trees of ancient polar forests. *Paleobiology*, **31**, 141–150.

Royer, D.L., Sack, L., Wilf, P., *et al.* (2007). Fossil leaf economics quantified: calibration, Eocene case study, and implications. *Paleobiology*, **33**, 574–589.

Royer, D.L., Wing, S.L., Beerling, D.J., *et al.* (2001). Paleobotanical evidence for near present-day levels of atmospheric CO_2 during part of the Tertiary. *Science*, **292**, 2310–2313.

Royo, C. (2002). Spanish research priorities and current approaches in plant physiology and breeding of crops to improve water economies. In: *Identifying Priority Tools for Cooperation. INCO-MED Workshops* (ed. Rodríguez, R.) European Comission-CSIC, Brussels (Belgium).

Ruban, A.V., Berera, R., Ilioana, C., *et al.* (2007). Identification of photoprotective energy dissipation in higher plants. *Nature*, **450**, 575–579.

Ruban, A.V., Pascal, A.A., Robert, B., *et al.* (2002). Activation of zeaxanthin is an obligatory event in the regulation of photosynthetic light harvesting. *Journal of Biological Chemistry*, **277**, 7785–7789.

Ruban, A.V., Young, A.J. and Horton, P. (1993). Induction of nonphotochemical energy dissipation and absorbance changes in leaves. *Plant Physiology*, **102**, 741–750.

Rubio, S., Rodrigues, A., Saez, A., *et al.* (2009). Triple loss of function of protein phosphatases TYPE 2C leads to partial constitutive response to endogenous abscisic acid. *Plant Physiology*, **150**, 1345–1355.

Rudall, P.J., Sokoloff, D.D., Remizowa, M.V., *et al.* (2007). Morphology of Hydatellaceae, an anomalous aquatic family recently recognized as an early divergent angiosperm lineage. *American Journal of Botany*, **94**, 1073–1092.

Rüdiger, W., Böhm, S., Helfrich, M., *et al.* (2005). Enzymes of the last steps of chlorophyll biosynthesis: modification of the substrate structure helps to understand the topology of the active centers. *Biochemistry*, **44**, 10864–10872.

Ruelland, E., Vaulthier, M.N., Zachowski, A., *et al.* (2009). Cold signalling and cold acclimation in plants. *Advances in Botanical Research*, **49**, 35–150.

Ruess, B.R., Ferrari, S. and Eller, B.M. (1988). Water economy and photosynthesis of the CAM plant *Senecio medley woodii* during increasing drought. *Plant, Cell and Environment*, **11**, 583–589.

Rumeau, D., Peltier, G. and Cournac, L. (2007). Chlororespiration and cyclic electron flow around PSI during photosynthesis and plant stress response. *Plant Cell and Environment*, **30**, 1041–1051.

Rundel, P.W. and Gibson, A.C. (1996). *Ecological Communities and Processes in a Mojave Desert Ecosystem.* Cambridge University Press, Cambridge, UK.

Rundel, P.W. and Sharifi, M.R. (1993). Carbon isotope discrimination and resource availability in the desert shrub *Larrea tridentata*. In: *Stable Isotopes and Plant Carbon-Water Relations* (eds Ehleringer, J.R., Hall, A.E. and Farquhar, J.D.), Academic Press, San Diego, USA, pp. 173–185.

Rundel, P.W., Cowling, R.M., Esler, K.J., *et al.* (1995). Winter growth phenology and leaf orientation in *Pachypodium namaquanum* (Apocynaceae) in the Succulent Karoo of the Richtersveld, South Africa. *Oecologia*, **101**, 472–477.

Rundel, P.W., Ehleringer, J.R. and Nagy, K.A. (1988). *Stable Isotopes in Ecological Research.* Berlin: Springer-Verlag.

Rundel, P.W., Esler, K.J. and Cowling, R.M. (1999). Ecological and phylogenetic patterns of carbon isotope discrimination in the winter-rainfall flora of the Richtersveld, South Africa. *Plant Ecology*, **142**, 133–148.

Rundel, P.W., Gibson, A.C. and Sharifi, M.R. (2005). Plant functional groups in alpine fellfield habitats of the White Mountains, California. *Arctic, Antarctic, and Alpine Research*, **37**, 358–365.

Rundel, P.W., Gibson, A.C., Midgley, G.S., *et al.* (2003). Ecological and ecophysiological patterns in a pre-altiplano shrubland of the Andean Cordillera in northern Chile. *Plant Ecology*, **169**, 179–193.

Running, S.R. and Nemani, R. (1988). Relating seasonal patterns of the AVHRR vegetation index to simulated photosynthesis and transpiration of forests in different climates. *Remote Sensing of Environment*, **24**, 347–367.

Running, S.R., Nemani, R.R., Heinsch, F.A., *et al.* (2004). A continuous satellite-derived measure of global terrestrial primary production. *BioScience*, **54**, 547–560.

Running, S.W., Baldocchi, D., Turner, D., *et al.* (1999). A global terrestrial monitoring network integrating tower fluxes, flask sampling, ecosystem modeling and EOS satellite data. *Remote Sensing of Environment*, **70**, 108–127.

Runyon, J., Waring, R.H., Goward, S.N., *et al.* (1994). Environmental limits on net primary production and light-use efficiency across the Oregon transect. *Ecological Applications*, **4**, 226–237.

Russell, G., Jarvis, P.G. and Monteith, J.L. (1989). Absorption of radiation by canopies and stand growth. In: *Plant Canopies, Their Growth, Form and Function* (eds Russell, G., Marshall, B. and Jarvis, P.G.), Cambridge University Press, Cambridge, UK, pp. 21–39.

Russell, R.J. (1931). Dry climates of the Unites States: I climatic map. *University of California, Publications in Geography*, **5**, 1–41.

Russo, M. and Martelli, G.P. (1982). Ultrastructure of turnip crinckle- and Saguaro cactus virus-infected tissues. *Virology*, **118**, 109–116.

Rutherford, A.W. and Boussac, A. (2004). Water photolysis in biology. *Science*, **303**, 1782–1784.

Rutherford, A.W., Crofts, A.R. and Inoue, Y. (1982). Thermoluminescence as a probe of Photosystem II photochemistry: the origin of the flash-induced glow peaks. *Biochimica Biophysica Acta*, **689**, 457–465.

Rutherford, A.W., Govindjee, J. and Inoue, Y. (1984). Charge accumulation and photochemistry in leaves studied by thermoluminescence and delayed light emission. *Proceedings of the National Academy of Science USA*, **81**, 1107–1111.

Ruuska, S.A., Schwender, J. and Ohlrogge, J.B. (2004). The capacity of green oilseeds to utilize photosynthesis to drive biosynthetic processes. *Plant Physiology*, **136**, 2700–2709.

Ryan, M., Bond, B.J., Law, B.E., *et al.* (2000). Transpiration and whole-tree conductance in ponderosa pine trees of different heights. *Oecologia*, **124**, 553–560.

Ryan, M.G. and Waring, R.H. (1992). Maintenance respiration and stand development in a subalpine lodgepole pine forest. *Ecology*, **73**, 2100–2108.

Ryan, M.G. and Yoder, B.J. (1997). Hydraulic limits to tree height and tree growth. What keeps trees from growing beyond a certain height? *BioScience*, **47**, 235–242.

Ryan, M.G., Binkley, D. and Fownes, J.H. (1997). Age-related decline in forest productivity: pattern and process. *Advances in Ecological Research*, **27**, 213–262.

Ryan, M.G., Binkley, D., Fownes, J.H., *et al.* (2004). An experimental test of the causes of forest growth decline with stand age. *Ecological Monographs*, **74**, 393–414.

Ryel, R.J. and Beyschlag, W. (1995). Benefits associated with steep foliage orientation in two tussock grasses of the American Intermountain West. A look at water-use-efficiency and photoinhibition. *Flora*, **190**, 1–10.

Ryel, R.J., Beyschlag, W. and Caldwell, M.M. (1994). Light field heterogeneity among tussock grasses – theoretical considerations of light harvesting and seedling establishment in tussocks and uniform tiller distributions. *Oecologia*, **98**, 241–246.

Sabar, M., De Paepe, R. and Kouchkovsky, Y. (2000). Complex I impairment, respiratory compensations, and photosynthetic decrease in nuclear and mitochondrial male sterile mutants of *Nicotiana sylvestris. Plant Physiology*, **124**, 1239–1249.

Sack, L. and Holbrook, N.M. (2006). Leaf hydraulics. *Annual Review of Plant Biology*, **57**, 361–381.

Sack, L., Cowan, P.D., Jaikumar, N.J., *et al.* (2003). The 'hydrology' of leaves: coordination of structure and function in temperate woody species. *Plant, Cell and Environment*, **26**, 1343–1356.

Sack, L., Dietrich, E.M., Streeter, C.M., *et al.* (2008). Leaf palmate venation and vascular redundancy confer tolerance of hydraulic disruption. *Proceedings of the National Academy of Sciences*, **105**, 1567–1572.

Sacksteder, C.A. and Kramer, D.A. (2000). Dark-interval relaxation kinetics (DIRK) of absorbance changes as a quantitative probe of steady state electron transfer. *Photosynthesis Research*, **66**, 145–158.

Sade, N., Gebretsadik, M., Seligmann, R., *et al.* (2010). The role of tobacco aquaporin1 in improving water use efficiency, hydraulic conductivity, and yield production under salt stress. *Plant Physiology*, **152**, 245–254.

Saeed, I.A.M., MacGuidwin, A.E., Rouse, D.I., *et al.* (1999). Limitation to photosynthesis in *Pratylenchus penetrans*- and *Verticillium dahliae*-infected potato. *Crop Science*, **39**, 1340–1346.

Sáez, A., Apostolova, N., Gonzalez-Guzman, M., *et al.* (2004). Gain-of-function and loss-of-function phenotypes of the protein phosphatase 2C HAB1 reveal its role as a negative regulator of abscisic acid signalling. *The Plant Journal*, **37**, 354–369.

Sáez, A., Robert, N., Maktabi, M.H., *et al.* (2006). Enhancement of abscisic acid sensitivity and reduction of water consumption in Arabidopsis by combined inactivation of the protein phosphatases type 2C ABI1 and HAB1. *Plant Physiology*, **141**, 1389–1399.

Sagardoy, R., Morales, F., López-Millán, A.F., *et al.* (2009). Effects of zinc toxicity in sugar beet (*Beta vulgaris* L.) plants grown in hydroponics. *Plant Biology*, **11**, 339–350.

Sagardoy, R., Vázquez, S., Florez-Sarasa, I.D. *et al.* (2010). Stomatal and mesophyll conductances to CO_2 are the main limitations to photosynthesis in sugar beet (*Beta vulgaris*) plants grown with excess zinc. *The New Phytologist*, **187**, 145–158.

Sage, R.F. (1999) Why C_4 photosynthesis?. In: *C4 Plant Biology* (eds Sage, R.F. and Monson, R.K.), Academic Press, San Diego, CA, USA, pp. 3–16.

Sage, R.F. (2001). Environmental and evolutionary preconditions for the origin and diversification of the C_4 phtosynthetic syndrome. *Plant Biology*, **3**, 202–213.

Sage, R.F. (2004). The evolution of C_4 photosynthesis. *New Phytologist*, **161**, 341–370.

Sage, R.F. and Kubien, D.S. (2003). *Quo vadis* C_4? An ecophysiological perspective on global change and the future of C_4 plants. *Photosynthesis Research*, **77**, 209–225.

Sage, R.F. and Kubien, D.S. (2007). The temperature response of C_3 and C_4 photosynthesis. *Plant, Cell and Environment*, **30**, 1086–1106.

Sage, R.F. and Monson, R.K. (eds.). (1999). *C4 Plant Biology*. Academic Press, San Diego, USA.

Sage, R.F. and Pearcy, R.W. (1987a). The nitrogen use efficiency of C_3 and C_4 plants. I. Leaf nitrogen effects on the gas exchange characteristics of *Chenopodium album* L. and *Amaranthus retroflexus* L. *Plant Physiology*, **84**, 954–958.

Sage, R.F. and Pearcy, R.W. (1987b). The nitrogen use efficiency of C_3 and C_4 plants. II. Leaf nitrogen effects on the gas exchange characteristics of *Chenopodium album* L. and *Amaranthus retroflexus* L. *Plant Physiology*, **84**, 959–963.

Sage, R.F., Cen, Y.P. and Li, M. (2002). The activation state of Rubisco directly limits photosynthesis at low CO_2 and low O_2 partial pressures. *Photosynthesis Research*, **71**, 241–250.

Sage, R.F., Li, M. and Monson, R.K. (1999). The taxonomic distribution of C_4 photosynthesis. In: *C4 Plant Biology* (eds Sage, R.F. and Monson, R.K.), Academic Press, San Diego, USA, pp. 551–584.

Sage, R.F. and Sharkey, T.D. (1987).The effect of temperature on the occurrence of O_2 and CO_2 insensitive photosynthesis in field grown plants. *Plant Physiology*, **84**, 658–664.

Sage, R.F., Schäppi, B. and Körner, C. (1997). Effect of atmospheric CO_2 enrichment on Rubisco content in herbaceous species from high and low altitude. *Acta Ecologica*, **18**, 183–192.

Sage, R.F., Sharkey, T.D. and Seeman, J.R. (1989). Acclimation of photosynthesis to elevated carbon dioxide in five C_3 species. *Plant Physiology*, **89**, 590–596.

Sage, R.F., Way, D.A. and Kubien, D.S. (2008) Rubisco, Rubisco activase, and global climate change. Journal of Experimental Botany, **59**, 1581–1595.

Sage, R.F., Wedin, D.A. and Li, M. (1999). The biogeography of C4 photosynthesis, patterns and controlling factors. In: *C4 Plant Biology* (eds Sage, R.F. and Monson, R.K.), Academic Press, San Diego, USA, pp. 313–373.

Saito, T., Soga, K., Hoson, T. *et al.* (2006). The bulk elastic modulus and the reversible properties of cell walls in developing *Quercus* leaves. *Plant and Cell Physiology*, **47**, 715–725.

Saito, Y., Saito, R., Nomura, E., *et al.* (1999). Performance check of vegetation fluorescence imaging lidar through *in vivo* and remote estimation of chlorophyll concentration inside plant leaves. *Optical Review*, **6**, 155–159.

Sajnani, C., Zurita, J., Roncel, M., *et al.* (2007). Changes in photosynthetic metabolism induced by tobamovirus infection in *Nicotiana benthamiana* studied in vivo by chlorophyll thermoluminescence. *New Phytologist*, **175**, 120–130.

Sakai, A. (1960). Survival of the twigs of woody plants at –196°C. *Nature*, **185**, 393–394.

Sakai, A. and Larcher, W. (1987). Frost survival of plants. responses and adaptation to freezing stress. *Ecological Studies*, **62**.

Sakata, T. and Yokoi, Y. (2002). Analysis of the O_2 dependency in leaf-level photosynthesisof two *Reynoutria japonica* populations growing at different altitudes. *Plant, Cell and Environment*, **25**, 65–74.

Sakata, T., Nakano, T., Iino, T., *et al.* (2006). Contrastive seasonal changes in ecophysiological traits of leaves of two perennial Polygonaceae herb species differing in leaf longevity and altitudinal distribution. *Ecological Research*, **21**, 633–640.

Sakugawa, H. and Cape, J.N. (2007). Harmful effects of atmospheric nitrous acid on the physiological status of Scots pine trees. *Environmental Pollution*, **147**, 532–534.

Sala, A. and Hoch, G. (2009). Height-related growth declines in ponderosa pine are not due to carbon limitation. *Plant Cell Environment*, **32**, 22–30.

Sala, A. and Tenhunen, J.D. (1996). Simulations of canopy net photosynthesis and transpiration in *Quercus ilex* L. under the influence of seasonal drought. *Agricultural and Forest Meteorology*, **78**, 203–222.

Sala, O.E., Parton, W.J., Joyce, L.A., *et al.* (1988). Primary production of the central grassland region of the United States. *Ecology*, **69**, 40–45.

Saleska, S.R., Miller, S.D., Matross, D.M., *et al.* (2003). Carbon in Amazon forests: unexpected seasonal fluxes and disturbance-induced losses. *Science*, **302**, 1554–1557.

Salleo, S. and Lo Gullo, M.A. (1990). Sclerophylly and plant water relations in three Mediterranean *Quercus* species. *Annals of Botany*, **65**, 259–270.

Salleo, S. and Nardini, A. (2000). Sclerophylly: evolutionary advantage or mere epiphenomenon? *Plant Biosystems*, **134**, 247–259.

Salleo, S., Nardini, A. and Lo Gullo, M.A. (1997). Is sclerophylly of Mediterranean evergreens and adaptation to drought? *The New Phytologist*, **135**, 603–612.

Salvetat, R., Juneau, P. and Popovic, R. (1998). Measurement of chlorophyll fluorescence by synchronous detection in integrating sphere: a modified analytical approach for the accurate determination of photosynthesis parameters for whole plants. *Environmental Science and Technology*, **32**, 2640–2645.

Salvi, S. and Tuberosa, R. (2005). To clone or not to clone plant QTLs: present and future challenges. *Trends in Plant Science*, **10**, 297–304.

Salvucci, M.E. and Crafts-Brandner, S.J. (2004a). Relationship between the heat tolerance of photosynthesis and the thermal stability of rubisco activase in plants from contrasting thermal environments. *Plant Physiology*, **134**, 1460–1470.

Salvucci, M.E., Crafts-Brandner, S.J. (2004b). Inhibition of photosynthesis by heat stress: the activation state of Rubisco as a limiting factor in photosynthesis. *Physiologia Plantarum*, **120**, 179–186.

Salvucci, M.E., DeRidder, B.P. and Portis, Jr A.R. (2006). Effect of activase level and isoform on the thermotolerance of photosynthesis in *Arabidopsis*. *Journal of Experimental Botany*, **57**, 3793–3799.

Sampol, B., Bota, J., Riera, D., *et al.* (2003). Analysis of the virus-induced inhibition of photosynthesis in malmsey grapevines. *New Phytologist*, **160**, 403–412.

Sampson, D.A., Janssens, I.A. and Ceulemans, R. (2001). Simulated soil CO_2 efflux and net ecosystem exchange in a 70-year-old Belgian Scots pine stand using the process model SECRETS. *Annals of Forest Science*, **58**, 31–36.

Samson, G., Tremblay, N., Dudelzak, A.E., *et al.* (2000). Nutrient stress of corn plants: early detection and discrimination using a compact multiwavelength fluorescent lidar. In: *EARSeL Proceedings* (ed. Reuter, R.), EARSeL, Dresden, Germany, pp. 214–223.

Samsuddin, Z. and Impens, I. (1979). Photosynthesis and diffusion resistances to carbon dioxide in *Hevea brasiliensis* muel. agr. clones. *Oecologia*, **37**, 361–363.

Sandermann, H. Jr. (1996). Ozone and plant health. *Annual Review of Phytopathology*, **34**, 347–366.

Sandmann, G. and Boger, P. (1980). Copper-mediated lipid peroxidation processes in photosynthetic membranes. *Plant Physiology*, **66**, 797–800.

Sandquist, D.R. and Ehleringer, J.R. (2003). Population- and family level variation of brittlebush (*Encelia farinosa*, Asteraceae) pubescence: its relation to drought and implications for selection in variable environments. *American Journal of Botany*, **90**, 1481–1486.

Sandquist, D.R., Schuster, W.S.F., Donovan, L.A., *et al.* (1993). Differences in carbon isotope discrimination between seedlings and adults of southwestern desert perennial plants. *The Southwestern Naturalist*, **38**, 212–217.

Sands, P.J. (1995). Modelling canopy production. I. Optimal distribution of photosynthetic resources. *Australian Journal of Plant Physiology*, **22**, 593–601.

Sandström, J. (2000). Nutritional quality of phloem sap in relation to host plant-alternation in the bird cherry oat aphid. *Chemoecology*, **10**, 17–24.

Sane, P.V., Ivanov, A.G., Hurry, V., *et al.* (2003). Changes in the redox potential of primary and secondary electron-accepting quinones in photosystem II confer increased resistance to photoinhibition in low-temperature-acclimated *Arabidopsis*. *Plant Physiology*, **132**, 2144–2151.

Sangsing, K., Poonpipope, K., Thanisawanyangkura, S., *et al.* (2004) Respiation rate and a two-component model of growth and maintenance respiration in leaves of rubber (*Hevea brasiliensis* Muell. Arg.). *Kasetsart Journal (Natural Science)*, **38**, 320–330.

Sankaran, M., Hanan, N.P., Scholes, R.J., *et al.* (2005). Determinants of woody cover in African savannas. *Nature*, **438**, 846–849.

Santarius, K.A. (1975). Sites of heat sensitivity in chloroplasts and differential inactivation of cyclic and noncyclic photophosphorylation by heating. *Journal of Thermal Biology*, **1**, 101–107.

Santarius, K.A. and Müller, M. (1979). Investigations on heat resistance of spinach leaves. *Planta*, **146**, 529–538.

Santiago, J., Dupeux, F., Round, A., *et al.* (2009). The abscisic acid receptor PYR1 in complex with abscisic acid. *Nature*, **462**, 665–668.

Santiago, L.S. and Wright, S.J. (2007). Leaf functional traits of tropical forest plants in relation to growth form. *Functional Ecology*, **21**, 19–27.

Sanitago, L.S., Goldstein, G., Meinzer F.C., *et al.* (2004). Leaf photosynthetic traits scale with hydraulic conductivity and wood density in Panamanian forest canopy trees. *Oecologia*, **140**, 543–550.

Šantrůček, J. and Sage, R.F. (1996). Acclimation of stomatal conductance to a CO_2 enriched atmosphere and elevated temperature in *Chenopodium album*. *Australian Journal of Plant Physiology*, **23**, 467–478.

Šantrůček, J., Šimáňová, E. Karbulková, J., *et al.* (2004). A new technique for measurement of water permeability of stomatous cuticular membranes isolated from *Hedera helix* leaves. *Journal of Experimental Botany*, **55**, 1411–1422.

Sarris, D., Christodoulakis, D. and Körner C. (2007). Recent decline in precipitation and tree growth in the eastern Mediterranean. *Global Change Biology*, **13**, 1187–1200.

Sasaki, H., Samejima, M. and Ishii, R. (1996). Analysis by delta-13C measurement on mechanism of cultivar difference in leaf photosynthesis on rice (*Oryza sativa* L.). *Plant and Cell Physiology*, **37**, 1161–1166.

Sassenrath, G.F., Ort, D.R. and Portis, A.R. (1990). Impaired reductive activation of stromal bisphosphatases in tomato leaves following low-temperature exposure at high light. *Archives of Biochemistry and Biophysics*, **282**, 302–308.

Sassenrath-Cole, G.F., Pearcy, R.W. and Steinmaus, S. (1994). The role of enzyme activation state in limiting carbon assimilation under variable light conditions. *Photosynthesis Research*, **41**, 295–302.

Saugier, B., Roy, J. and Mooney, H.A. (2001). Estimation of global terrestrial productivity: converging toward a single number? In: *Terrestrial Global Productivity* (eds Mooney, H.A., Saugier, B. and Roy, J.), Academic Press, San Diego, USA, pp. 543–557.

Savé, R., Biel, C. and de Herralde, F. (2000). Leaf pubescence, water relations and chlorophyll fluorescence in two subspecies of *Lotus creticus* L. *Biologia Plantarum*, **43**, 239–244.

Saveyn, A., Steppe, K., Ubierna, N., *et al.* (2010). Woody tissue photosynthesis and bud development in young plants. *Plant, Cell and Environment*, **33**, 1949–1958.

Savitch, L.V., Allard, G., Seki, M., *et al.* (2005). The effect of over-expression of two Brassica CBF/DREB1-like transcription factors on photosynthetic capacity and freezing tolerance in *Brassica napus*. *Plant and Cell Physiology*, **46**, 1525–1539.

Savitch, L.V., Barker-Astrom, J., Ivanov, A.G., *et al.* (2001). Cold acclimation of *Arabidopsis thaliana* results in incomplete recovery of photosynthetic capacity, associated with an increased reduction of the chloroplast stroma. *Planta*, **214**, 295–303.

Savitch, L.V., Gray, G.R. and Huner, N.P.A. (1997). Feedback-limited photosynthesis and regulation of sucrose-starch accumulation during cold acclimation and low-temperature stress in a spring and winter wheat. *Planta*, **201**, 18–26.

Savitch, L.V., Harney, T. and Huner, N.P.A. (2000a). Sucrose metabolism in spring and winter wheat in response to high irradiance, cold stress and cold acclimation. *Physiologia Plantarum*, **108**, 270–278.

Savitch, L.V., Leonardos, E.D., Krol, M., *et al.* (2002). Two different strategies for light utilization in photosynthesis in relation to growth and cold acclimation. *Plant Cell Environment*, **25**, 761–771.

Savitch, L.V., Massacci, A., Gray, G.R., *et al.* (2000b). Acclimation to low temperature or high light mitigates sensitivity to photoinhibition: roles of the Calvin cycle and the Mehler reaction. *Austral Journal of Plant Physiology*, **27**, 253–264.

Sawada, S., Usuda, H. and Tsukui, T. (1992). Participation of inorganic orthophosphate in regulation of the Ribulose-1,5-bisphosphate carboxylase activity in response to changes in the photosynthetic source-sink balance. *Plant and Cell Physiology*, **33**, 943–949.

Saxe, H. (1986). Effects of NO, NO_2 and CO_2 on net photosynthesis, dark respiration and transpiration of pot plants. *New Phytologist*, **103**, 185–197.

Saxe, H. and Christensen, O.V. (1985). Effects of carbon dioxide with and without nitric oxide pollution on growth, morphogenesis and production time of pot plants. *Environmental Pollution*, **38**, 159–169.

Saxe, H., Cannell, M.G.R., Johnsen, Ã, *et al.* (2001). Tansley review 123: Tree and forest functioning in response to global warming. *New Phytologist*, **149**, 369–400.

Saxe, H., Ellsworth, D.S. and Heath, J. (1998). Tree and forest functioning in an enriched CO_2 atmosphere. *New Phytologist*, **139**, 395–436.

Sayed, O.H. (2001). Crassulacean acid metablism 1975–2000, a check list. *Photosynthetica*, **39**, 339–352.

Scafaro, A.P., von Caemmerer, S., Evans, J.R. *et al.* (2011). Temperature response of mesophyll conductance in cultivated and wild *Oryza* species with contrasting mesophyll cell wall thickness. *Plant, Cell and Environment*, **34**, 1999–2008

Scanlon, T.M. and Albertson, J.D. (2004). Canopy scale measurements of CO_2 and water vapor exchange along a precipitation gradient in southern Africa. *Global Change Biology*, **10**, 329–341.

Scarascia-Mugnozza, G., De Angelis, P., Matteucci, G., *et al.* (1996). Long term exposure to elevated CO_2 in a natural *Quercus ilex* L. community: net photosynthesis and photochemical efficiency of PSII at different levels of water stress. *Plant, Cell and Environment*, **19**, 643–654.

Scartazza, A., Lauteri, M., Guido, M.C., *et al.* (1998). Carbon isotope discrimination in leaf and stem sugars, water-use efficiency and mesophyll conductance during different developmental stages in rice subjected to drought. *Australian Journal of Plant Physiology*, **25**, 489–498.

Scartazza, A., Mata, C., Matteucci, G., *et al.* (2004). Comparisons of $\delta^{13}C$ of photosynthetic products and ecosystem respiratory CO_2 and their response to seasonal climate variability. *Oecologia*, **140**, 340–351.

Schaefer, H.M. and Wilkinson, D.M. (2004). Red leaves, insects and coevolution: a red herring? *Trends in Ecology and Evolution*, **19**, 616–618.

Schaeffer, S.M., Anderson, D.E., Burns, S.P., *et al.* (2008). Canopy structure and atmospheric flows in relation to the $\delta^{13}C$ of respired CO_2 in a subalpine coniferous forest. *Agricultural and Forest Meteorology*, **148**, 592–605.

Schäfer, K.V.R., Oren, R. and Tenhunen, J.D. (2000). The effect of tree height on crown level stomatal conductance. *Plant Cell Environment*, **23**, 365–375.

Schans, J. and Arntzen, F.K. (1991). Photosynthesis, transpiration and plant growth characters of different potato cultivars at various densities of *Globodera pallida*. *Netherland Journal of Plant Pathology*, **97**, 297–310.

Schansker, G., Tóth, S.Z. and Strasser, R. J. (2005). Methylviologen and dibromothymoquinone treatments of pea leaves reveal the role of photosystem I in the Chl a fluorescence rise OJIP. *Biochimica et Biophysica Acta*, **1706**, 250–261.

Scharte, J., Schon, H. and Weiss, E. (2005). Photosynthesis and carbohydrate metabolism in tobacco leaves during an incompatible interaction with *Phytophthora nicotianae*. *Plant, Cell and Environment*, **28**, 1421–1435.

Schäufele, R., Santrucek, J. and Schynder, H. (2011). Dynamic changes of canopy scale mesophyll conductance to CO_2 diffusion of sunflower as affected by CO_2 concentration and abscisic acid. *Plant, Cell and Environment*, **34**, 127–136.

Scheibe, R. (1990). Light/dark modulation: regulation of chloroplast metabolism in a new light. *Botanica Acta*, **103**, 327–334.

Scheibe, R. (2004). Malate valves to balance cellular energy supply. *Physiologia Plantarum*, **120**, 21–26.

Scheibe, R., Backhausen, J. E., Emmerlich, V., *et al.* (2005). Strategies to maintain redox homeostasis during photosynthesis under changing conditions. *Journal of Experimental Botany*, **56**, 1481–1489.

Scheible, W.R., Krapp, A. and Stitt, M. (2000). Reciprocal diurnal changes of PEPc expression, cytosolic pyruvate kinase, citrate synthase and NADP-isocitrate dehydrogenase expression regulate organic acid metabolism during nitrate assimilation in tobacco leaves. *Plant, Cell Environment*, **23**, 1155–1167.

Scheirs, J., De Bruyn, L. and Verhagen, R. (2001). A test of the C_3-C_4 hypothesis with two grass miners. *Ecology*, **82**, 410–421.

Scheller, H.V. and Haldrup, A. (2005). Photoinhibition of photosystem I. *Planta*, **221**, 5–8.

Schidlowski, M. (1988). A 3,800-million-year isotopic record of life from carbon in sedimentary rocks. *Nature*, **333**, 313–318.

Schidlowski, M. (1995). Isotope fractionations in the terrestrial carbon cycle: a brief over view. *Advances in Space Research*, **15**, 441–449.

Schimel, D. (2007). Carbon cycle conundrums. *Proceedings of the National Academy of Sciences of the USA*, **104**, 18353–18354.

Schimel, D.S., Braswell, B.H., McKeown, R., *et al.* (1996). Climate and nitrogen controls on the geography and timescales of terrestrial biogeochemical cycling. *Global Biogeochemical Cycles*, **10**, 677–692.

Schlegel, H., Godbold, D.L. and Hüttermann, A. (1987). Whole plant aspects of heavy metal induced changes in CO_2 uptake and water relations of spruce (*Picea abies*) seedlings. *Physiologia Plantarum*, **69**, 265–270.

Schlesinger, W.H., Belnap, J. and Marion, G. (2009). On carbon sequestration in desert ecosystems. *Global Change Biolog*, **15**, 1488–90.

Schleucher, J., Vanderveer, P., Markley, J.L., *et al.* (1999). Intramolecular deuterium distributions reveal disequilibrium of chloroplast phosphoglucose isomerase. *Plant, Cell and Environment*, **22**, 525–533.

Schmid, H.P. (2002). Footprint modeling for vegetation atmosphere exchange studies: a review and perspective. *Agricultural and Forest Meteorology*, **113**, 159–183.

Schmitgen, S., Ciais, P., Geiss, H., *et al.* (2004). Carbon dioxide uptake of a forested region in southwest France derived from airborne CO_2 and CO observations in a Lagrangian budget approach. *Journal of Geophysical Research*, **109**, D14302.

Schmitz, A., Tartachnyk, I., Kiewnick, S., *et al.* (2006). Detection of *Heterodera schachtii* infestation in sugar beet by means of laser-induced and pulse amplitude modulated chlorophyll fluorescente. *Nematology*, **8**, 273–286.

Schnablová, R., Synková, H. and Čeřovská, N. (2005). The influence of potato virus Y infection on the ultrastructure of Pssu-ipt transgenic tobacco *International Journal of Plant Sciences*, **166**, 713–721.

Schneckenburger, H. and Frenz, M. (1986). Time-resolved fluorescence of conifers exposed to environmental pollutants. *Radiation Environment Biophysics*, **25**, 289–295.

Schoefs, B. and Franck, F. (2003). Protochlorophyllide reduction: mechanisms and evolution. *Photochemistry and Photobiology*, **78**, 543–557.

Schoettle, A.W. and Smith, W.K. (1991). Interrelation between shoot characteristics and solar irradiance in the crown of *Pinus contorta* ssp. *latifolia*. *Tree Physiology*, **9**, 245–254.

Schoettle, A.W. and Smith, W.K. (1999). Interrelationships among light, photosynthesis and nitrogen in the crown of mature *Pinus contorta* ssp. *latifolia*. *Tree Physiology*, **19**, 13–22.

Scholberg, J., McNeal, B.L., Jones, J.W., *et al.* (2000). Growth and canopy characteristics of field-grown tomato. *Agronomy Journal*, **92**, 152–159.

Scholes, J.D. and Rolfe, S.A. (1996). Photosynthesis in localised regions of oat leaves infected with crown rust (*Puccinia coronata*): quantitative imaging of chlorophyll fluorescence. *Planta*, **199**, 573–582.

Scholes, J.D., Lee, P.J., Horton, P. *et al.* (1994). Invertase: understanding changes in the photosynthetic and carbohydrate metabolism of barley leaves infected with powdery mildew. *New Phytologist*, **126**, 213–222.

Scholes, R.J. and Archer, S. (1997) Tree-grass interactions in savannas. *Annual Review of Ecology and Systematics*, **28**, 517–544.

Scholes, R.J., Frost, P.G. and Tian, Y. (2004). Canopy structure in savannas along a moisture gradient on Kalahari sands. *Global Change Biology*, **10**, 292–302.

Scholze, M., Ciais, P. and Heimann, M. (2008). Modeling terrestrial ^{13}C cycling: climate, land use and fire. *Global Biogeochemical Cycles*, **22**, GB1009.

Schomburg, M. and Kluge, M. (1994). Phenotypic adaptation to elevated temperatures of tonoplast fluidity in the CAM plant *Kalanchoë daigremontiana* is caused by membrane proteins. *Botanica Acta*, **107**, 328–332.

Schöner, S. and Krause, G.H. (1990). Protective systems against active oxygen species in spinach: response to cold acclimation in excess light. *Planta*, **180**, 383–389.

Schrader, S.M., Wise, R.R., Wacholtz, W.F., *et al.* (2004). Thylakoid membrane responses to moderately high leaf temperature in pima cotton. *Plant, Cell and Environment*, **27**, 725–735.

Schrauder, M., Langebartels, C. and Sandermann, H. (1997). Changes in the biochemical status of plant cells induced by the environmental pollutant ozone. *Physiologia Plantarum*, **100**, 274–280.

Schreiber, U. and Berry, J.A. (1977). Heat-induced changes in chlorophyll fluorescence in intact leaves correlated with damage of the photosynthetic apparatus. *Planta*, **136**, 233–238.

Schreiber, U. and Neubauer, C. (1987). The polyphasic rise of chlorophyll upon onset of continuous illumination. II. Partial control by the photosystem II donor side and possible ways of interpretation. *Zeitschrift für Naturforschung*, 42c, 1255–1264.

Schreiber, U., Bilger, W. and Neubauer, C. (1994). Chlorophyll fluorescence as a noninvasive indicator for rapid assessment of in vivo photosynthesis. In: *Ecophysiology of Photosynthesis* (eds Schulze, E.D. and Caldwell, M.M.), Springer-Verlag, Berlin, Germany, pp. 49–70.

Schreiber, U., Klughammer, C. and Neubauer, C. (1989). Measuring P700 absorbance changes around 830 nm with a new type of pulse modulation system. *Zeitschrift für Naturforschung*, **43c**, 686–698.

Schreiber, U., Neubauer, C. and Schliwa, U. (1993). PAM fluorometer based on medium-frequency pulsed Xe-flash measuring light: a highly sensitive new tool in basic and applied photosynthesis research. *Photosynthesis Research*, **36**, 65–72.

Schreiber, U., Schliwa, U. and Bilger, W. (1986). Continuous recording of photochemical and non-photochemical fluorescence quenching with a new type of pulse modulation fluorometer. *Photosynthesis Research*, **10**, 51–62.

Schröder, R., Forstreuter, M. and Hilker, M. (2005). A plant notices insect egg deposition and changes its rate of photosynthesis. *Plant Physiology*, **138**, 470–477.

Schroeder, J.I. (2003). Knockout of the guard cell K+ out channel and stomatal movements. *Proceedings of the National Academy of Sciences USA*, **100**, 4976–4977.

Schroeder, J.I., Allen, G.J., Hugouvieux, V., *et al.* (2001a). Guard cell signal transduction. *Annual Review of Plant Physiology and Plant Molecular Biology*, **52**, 627–658.

Schroeder, J.I., Kwak, J.M. and Allen, G.J. (2001b). Guard cell abscisic acid signalling and engineering drought hardiness in plants. *Nature*, **410**, 327–330.

Schuerger, A.C., Capelle, G.A., Di Benedetto, J.A., *et al.* (2003). Comparison of two hyperspectral imaging and two laser-induced fluorescence instruments for the detection of zinc stress and chlorophyll concentration in bahia grass (*Paspalum notatum* Flugge.). *Remote Sensing of Environment*, **84**, 572–588.

Schulze, E.D. (1986). Carbon dioxide and water vapor exchange in response to drought in the atmosphere and in the soil. *Annual Review of Plant Physiology*, **37**, 247–274.

Schulze, E.D. and Hall, A.E. (1982). Stomatal responses to water loss and CO_2 assimilation rates in plants of contrasting environments. In: *Encyclopedia of Plant Physiology. Physiological Plant Ecology*. Vol. 12B (eds Lange, O. L., Nobel, P., Osmond, C.B. and Ziegler, H.), Springer-Verlag, Berlin, Germany, pp. 181–230.

Schulze, E.D. and Kock, W. (1971). Measurement of primary production with cuvettes. In: *Productivity of Forest Ecosystems* (ed. Duvigneaud, P.), UNESCO, Paris, France, pp. 141–157.

Schulze, E.D., Ellis, R., Schulze, W., *et al.* (1996). Diversity, metabolic types and delta ^{13}C carbon isotope ratios in the grass flora of Namibia in relation to growth form, precipitation and habitat conditions. *Oecologia*, **106**, 352–369.

Schulze, E.D., Fuchs, M. and Fuchs, M.I. (1977). Spacial distribution of photosynthetic capacity and performance in a mountain spruce forest of northern Germany. III. The significance of the evergreen habit. *Oecologia*, **30**, 239–248.

Schulze, E.D., Hall, A.E., Lange, O.L., *et al.* (1982). A portable steady state porometer for measuring the carbon dioxide and water vapour exchanges of leaves under natural conditions. *Oecologia*, **53**, 141–145.

Schulze, E.D., Kelliher, F.M., Körner, C., *et al.* (1994). Relationships among maximum stomatal conductance, ecosystem surface conductance, carbon assimilation rate, and plant nitrogen nutrition: a global scaling exercise. *Annual Review of Ecological Systems*, **25**, 629–660.

Schürmann, P. and Buchanan, B. (2008). The ferredoxin-thioredoxin system of oxygenic photosynthesis. *Antioxidants and Redox Signalling*, **10**, 1235–1273.

Schurr, U. (1997). Growth physiology: approaches to spatially and temporarily varying problem. *Progress in Botany*, **59**, 355–373.

Schurr, U. and Schulze, E.D. (1996). Effects of drought on nutrient and ABA transport in *Ricinus communis*. *Plant, Cell and Environment*, **19**, 665–674.

Schurr, U., Gollan, T. and Schulze, E.D. (1992). Stomatal response to drying soil in relation to changes in the xylem sap composition of *Helianthus annuus*. II. Stomatal sensitivity to abscisic acid imported from the xylem sap. *Plant, Cell and Environment*, **15**, 561–567.

Schurr, U., Heckenberger, U., Herdel, K., *et al.* (2000). Leaf development in *Ricinus communis* during drought stress: dynamics of growth processes, of cellular structure and of sink-source-transition. *Journal of Experimental Botany*, **51**, 1515–1529.

Schuster, W.S. and Monson, R.K. (1990). An examination of the advantages of C_3-C_4 intermediate photosynthesis in warm environments. *Plant, Cell and Environment*, **13**, 903–912.

Schuster, W.S.F., Sandquist, D.R., Phillips, S.L., *et al.* (1992). Comparisons of carbon isoptope discrimination in populations of aridland plant species differing in lifespan. *Oecologia*, **91**, 332–337.

Schwaller, M.R., Schnetzler, C.C. and Marshall, P.E. (1983). The changes in leaf reflectance of sugar maple (*Acer saccharum* Marsh) seedlings in response to heavy metal stress. *International Journal of Remote Sensing*, **4**, 93–100.

Schwender, J., Goffman, F., Ohlrogge, J.B., *et al.* (2004). Rubisco without the Calvin cycle improves the carbon efficiency of developing green seeds. *Nature*, **432**, 779–782.

Scott, L. and Vogel, J.C. (2000). Evidence for environmental conditions during the last 20,000 years in Southern Africa from C^{13} in fossil hyrax dung. *Global and Planetary Change*, **26**, 207–215.

Seel, W.E. and Press, M.C. (1996). Effects of repeated parasitism by *Rhinanthus minor* on the growth and photosynthesis of a perennial grass, *Poa alpina*. *New Phytologist*, **134**, 495–502.

Seel, W.E., Cooper, R.E. and Press, M.C. (1993). Growth, gas exchange and water use efficiency of the facultative hemiparasite *Rhinanthus minor* associated with hosts differing in foliar nitrogen concentration. *Physiologia Plantarum*, **89**, 64–70.

Seelert, H., Poetsch, A., Dencher, N.A., *et al.* (2000). Proton-powered turbine of a plant motor. *Nature*, **405**, 418–419.

Seibt, U., Rajabi, A., Griffiths, H., *et al.* (2008). Carbon isotopes and water use efficiency: sense and sensitivity. *Oecologia*, **155**, 441–454.

Seki, M., Narusaka, M., Ishida, J., *et al.* (2002). Monitoring the expression profiles of 7000 *Arabidopsis* genes under drought, cold and high-salinity stresses using a full-length cDNA microarray. *Plant Journal*, **31**, 279–292.

Sellers, P.J. (1987). Canopy reflectance, photosynthesis and transpiration: II. The role of biophysics in the linearity of their interdependence. *Remote Sensing of Environment*, **21**, 143–183.

Sellers, P.J., Berry, J.A., Collatz, G.J., *et al.* (1992). Canopy reflectance, photosynthesis and transpiration. Part III: a reanalysis using enzyme kinetics-electron transport models of leaf physiology. *Remote Sensing of Environment*, **42**, 187–216.

Sellers, P.J., Bounoua, L, Collatz, G.J., *et al.* (1996). Comparison of radiative and physiological effects of doubled atmospheric CO_2 on climate. *Science*, **271**, 1402–1406.

Sellers, P.J., Dickinson, R.E., Randall, D.A., *et al.* (1997). Modeling the exchanges of energy, water and carbon between continents and the atmosphere. *Science*, **275**, 502–509.

Sepulcre-Cantó, G., Zarco-Tejada, P.J., Jiménez-Muñoz, J.C., *et al.* (2007). Monitoring yield and fruit quality parameters in open-canopy tree crops under water stress. Implications for ASTER. *Remote Sensing of the Environment*, **107**, 455–470.

Serraj, R., Hash, C.T., Rizvi, S.M.H., *et al.* (2005). Recent advances in marker-assisted selection for drought tolerance in pearl millet. *Plant Production Science*, **8**, 334–337.

Sersen, F., Kralova, K. and Bumbalova, A. (1998). Action of mercury on the photosynthetic apparatus of spinach chloroplasts. *Photosynthetica*, **35**, 551–559.

Sestak, Z., Catsky, J. and Jarvis, P. (1971). *Plant Photosynthetic Production. Manual of Methods*. Dr. W. Junk NV, Publishers, The Hague 818.

Setterdahl, A.T., Chivers, P.T., Hirasawa, M., *et al.* (2003). Effect of pH on the oxidation-reduction properties of thioredoxins. *Biochemistry*, **42**, 14877–14884.

Seyfried, M. and Fukshansky, L. (1983). Light gradients in plant tissue. *Applied Optics*, **22**, 1402–1408.

Shaberg, P.G., DeHayes, D.H., Hawley, G.J., *et al.* (2002). Effects of chronic N fertilization on foliar membranes, cold tolerance, and carbon storage in montane red spruce. *Canadian Journal of Forest Research*, **32**, 1351–1359.

Shahbazi, M., Gilbert, M., Labouré, A.M., *et al.* (2007). Dual role of the plastid terminal oxidase in tomato. *Plant Physiology*, **145**, 691–702.

Shalitin, D. and Wolf, S. (2000). Cucumber mosaic virus infection affects sugar transport in melon plants. *Plant Physiology*, **123**, 597–604.

Shang, W. and Feierabend, J. (1998). Slow turnover of the D1 reaction center protein of photosystem II in leaves of high mountain plants. *FEBS Letters*, **425**, 97–100.

Shantz, H.J. and Piemeisel, L.N. (1927). The water requirement of plants at Akron, Colo. *Journal of Agriculture Research*, **34**, 1093–1190.

Shapiro, J.B., Griffin, K.L., Lewis, J.D., *et al.* (2004). Response of *Xanthium strumarium* leaf respiration in the light to elevated CO_2 concentration, nitrogen availability and temperature. *New Phytologist*, **162**, 377–386.

Sharifi, M.R., Gibson, A.C. and Rundel, P.W. (1997). Surface dust impacts on gas exchange in Mojave Desert shrubs. *Journal of Applied Ecology*, **34**, 837–46.

Sharifi, M.R., Nilsen, E.T. and Rundel, P.W. (1982). Biomass and net primary production of *Prosopis glandulosa* (Fabaceae) in the Sonoran Desert of California. *American Journal of Botany*, **69**, 760–767.

Sharkey, T.D. (1979). Stomatal responses to light in *Xanthium strumarium* and other species. Ph.D. Thesis. Michigan State University, East Lansing, p. 95.

Sharkey, T.D. (1985). Photosynthesis in intact leaves of C_3 plants: physics, physiology and rate limitations. *Botanical Review*, **51**, 53–105.

Sharkey, T.D. (1988). Estimating the rate of photorespiration in leaves. *Physiologia Plantarum*, **73**, 147–152.

Sharkey, T.D. and Schrader, S.M. (2006). High temperature stress. In: *Physiology and Molecular Biology of Stress Tolerance* in Plants (eds Rao, K.V.M., Raghavendra, A.S. and Reddy, K.J.), Springer, Dordrecht, Netherlands, pp. 101–130.

Sharkey, T.D. and Vassey, T.L. (1989). Low oxygen inhibition of photosynthesis is caused by inhibition of starch synthesis. *Plant Physiology*, **90**, 385–387.

Sharkey, T.D. and Yeh, S. (2001). Isoprene emission from plants. *Annual Review of Plant Physiology and Plant Molecular Biology*, **52**, 407–436.

Sharkey, T.D., Badger, M.R., von Caemmerer, S., *et al.* (2001). Increased heat sensitivity of photosynthesis in tobacco plants with reduced Rubisco activase. *Photosynthesis Research*, **67**, 147–156.

Sharkey, T.D., Bernacchi, C.J., Farquhar, G.D., *et al.* (2007). Fitting photosynthetic carbon dioxide response curves for C3 leaves. *Plant, Cell and Environment*, **30**, 1035–1040.

Sharkey, T.D., Imai, K., Farquhar, G.D., *et al.* (1982). A direct confirmation of the standard method of estimating intercellular partial pressure of CO_2. *Plant Physiology*, **69**, 657–659.

Sharkey, T.D., Vassey, T.L., Vanderveer, P.J., *et al.* (1991). Carbon metabolism enzymes and photosynthesis in transgenic tobacco (*Nicotiana tabaccum* L.) having excess phytochrome. *Planta*, **185**, 287–296.

Sharkey, T.D., Wiberley, A.E. and Donohue, A.R. (2008). Isoprene emission from plants: why and how. *Annals of Botany*, **101**, 5–18.

Sharkey, T.D., Wise, S.E., Standish, A.J., *et al.* (2004). CO_2 processing, chloroplast to leaf. In: *Photosynthetic Adaptation, Chloroplast to Landscape* (eds Smith, W.K., Vogelmann, T.C. and Critchley, C.), Springer, New York, USA, pp. 171–206.

Sharma, P.N., Tripathi, A. and Bisht, S.S. (1995). Zinc requirement for stomatal opening in cauliflower. *Plant Physiology*, **107**, 751–756.

Sharma-Natu, P. and M.C. Ghildiyal. (2005). Potential targets for improving photosynthesis and crop yield. *Curr Sci.*, **88**, 1918–1928.

Sharp, R.E. (2002). Interaction with ethylene: changing views on the role of abscisic acid in root and shoot growth responses to water stress. *Plant, Cell and Environment*, **25**, 211–222.

Sharwood, R.E., von Caemmerer, S., Maliga, P., *et al.* (2008). The catalytic properties of hybrid Rubisco comprising tobacco small and sunflower large subunits mirror the kinetically equivalent source Rubiscos and can support tobacco growth. *Plant Physiology*, **146**, 83–96.

Shatil-Cohen, A., Attia, Z. and Moshelion, M. (2011). Bundle-sheath cell regulation of xylem-mesophyll water transport via aquaporins under drought stress: a target of xylem-borne ABA? *The Plant Journal*, **67**, 72–80.

Shaw, M. and MacLachan, G.A. (1954). Chlorophyll content and carbon dioxide uptake of stomatal cells. *Nature*, **173**, 29–30.

Sheahan, J.J. (1996). Sinapate esters provide greater UV-B attenuation than flavonoids in *Arabidopsis thaliana* (Brassicaceae). *American Journal of Botany*, **83**, 679–686.

Sheard, L.B. and Zheng, N. (2009). Signal advance for abscisic acid. *Nature*, **462**, 575–576.

Sheehy, J.E., Mitchell, P.L. and Ferrer, A.B. (2004). Bi-phasic growth patterns in rice. *Annals of Botany*, **94**, 811–817.

Shen, B., Jensen, R.G. and Bohnert, H.J. (1997). Increased resistance to oxidative stress in transgenic plants by targeting mannitol biosynthesis to chloroplasts. *Plant Physiology*, **113**, 177–183.

Shepherd, T. and Griffiths, D.W. (2006). The effects of stress on plant cuticular waxes. *New Phytologist*, **171**, 469–499.

Shesták, Z. (ed) (1985). *Photosynthesis during leaf development.* Dr. W. Junk Publishers, Dordrecht – Boston – Lancaster.

Shi, Z., Liu S., Liu X., *et al.* (2006). Altitudinal variation in photosynthetic capacity, diffusional conductance and $\delta^{13}C$ of butterfly bush (*Buddleja davidii*) plants growing at high elevations. *Physiologia Plantarum*, **128**, 722–731.

Shikanai, T. (2007). Cyclic electron transport around Photosystem I: genetic approaches. *Annual Review of Plant Biology*, **58**, 199–217.

Shimazaki, K., Doi, M., Assmann, S.M., *et al.* (2007). Light regulation of stomatal movement. *Annual Reviews Plant Biology*, **58**, 219–247.

Shinozaki, K., Yamaguchi-Shinozaki, K. and Seki, M. (2003). Regulatory network of gene expression in the drought and cold stress responses. *Current Opinion in Plant Biology*, **6**, 410–417.

Shirano, Y., Shimada, H., Kanamaru, K., *et al.* (2000). Chloroplast development in *Arabidopsis thaliana* requires the nuclear-encoded transcription factor Sigma B. *FEBS Lett*, **485**, 178–182.

Shope, J.C., Peak, D. and Mott, K.A. (2008). Stomatal responses to humidity in isolated epidermis. *Plant, Cell and Environment*, **31**, 1290–1298.

Shtienberg, D. (1992). Effects of foliar diseases on gas exchange processes: a comparative study. *Phytopathology*, **82**, 760–765.

Shu, G.G., Baum, D.A. and Mets, L.J. (1999). Detection of gene expression patterns in various plant tissues using non-radioactive mRNA *in situ* hybridization. *The World Wide Web Journal of Biology* **4**, http://www.epress.com/w3jbio/vol4/shu.

Shulaev, V. and Oliver, D.J. (2006). Metabolic and proteomic markers for oxidative stress. New tools for reactive oxygen species research. *Plant Physiology*, **141**, 367–372.

Sicher, R.C., Bahr, J.T. and Jensen, R.G. (1979). Measurement of ribulose-1,5-biphosphate from spinach chloroplasts. *Plant Physiology*, **64**, 876–879.

Siddiqui, Z.A. and Mahmood, I. (1999). Effects of *Meloidogyne incognita*, *Fusarium oxysporum* f.sp. *pisi*, *Rhizobium* sp. And different soil types on growth, chlorophyll, and carotenoid pigments of pea. *Israel Journal of Plant Sciences*, **47**, 251–256.

Siebke, K. and Weis, E. (1995). Imaging of chlorophyll a fluorescence in leaves: topography of photosynthetic oscillations in leaves of *Glechoma hederacea*. *Photosynthesis Research*, **45**, 225–237.

Siegenthaler, U. and Sarmiento, J.L. (1993). Atmospheric carbon dioxide and the ocean. *Nature*, **365**, 119–125.

Sienkiewicz- Porzucek, A., Sulpice, R., Osorio, S., *et al.* (2010). Mild reductions in mitochondrial NAD-dependent isocitrate dehydrogenase activity result in altered nitrate assimilation and pigmentation but do not impact growth. *Molecular Plant*, **3**, 156–173.

Sienkiewicz-Porzucek, A., Nunes-Nesi, A., Sulpice, R., *et al.* (2008). Mild reductions in mitochondrial citrate synthase activity result in a compromised nitrate assimilation and reduced leaf pigmentation but have no effect on photosynthetic performance or growth. *Plant Physiology*, **147**, 115–127.

Sierra-Almeida, A., Cavieres, L.A. and Bravo, L.A. (2009). Freezing resistance varies within the growing season and with elevation in high-Andean species of central Chile. *New Phytologist*, **182**, 461–469.

Sigfridsson, K.G.V., Bernát, G., Mamedov, F., *et al.* (2004). Molecular interference of Cd2+ with Photosystem II. *Biochimica et Biophysica Acta*, **1659**, 19–31.

Silim, S.N., Guy, R.D., Patterson, T.B. and Livingston, N.J. (2001). Plasticity in water-use efficiency of *Picea* sitchensis, *P. glauca* and their natural hybrids. *Oecologia*, **128**, 317–325.

Silvera, K., Santiago, L.S. and Winter, K. (2005). Distribution of crassulacean acid metabolism in orchids of Panama: evidence of selection for strong and weak modes. *Functional Plant Biology*, **32**, 397–407.

Silvera, K., Santiago, L.S., Cushman, J.C., *et al.* (2009). Crassulacean acid metabolism and epiphytism linked to adaptive radiations in the Orchidaceae. *Plant Physiology*, **149**, 1838–1847.

Simioni, G., Le Roux, X., Gignoux, J., *et al.* (2004). Leaf gas exchange characteristics and water- and nitrogen-use efficiencies of dominant grass and tree species in a West African savanna. *Plant Ecology*, **173**, 233–246.

Simpson, D. (1995). Biogenic emissions in Europe. 2. Implications for ozone control strategies. *Journal of Geophysical Research*, **100**, 22891–22906.

Sims, D.A. and Pearcy, R.W. (1992). Response of leaf anatomy and photosynthetic capacity in *Alocasia macrorrhiza* (Araceae) to a transfer from low to high light. *American Journal of Botany*, **79**, 449–455.

Sims, D.A., Luo, H., Hastings, S., *et al.* (2006). Parallel adjustments in vegetation greenness and ecosystem CO_2 exchange in response to drought in a Southern California chaparral ecosystem. *Remote Sensing of the Environment*, **103**, 289–303.

Sims, D.A., Pearcy and R.W. (1991). Photosynthesis and respiration in *Alocasia macrorrhiza* following transfers to high and low light. *Oecologia*, **86**, 447–453.

Sims, D.A., Seemann, J.R. and Luo, Y. (1998a). Elevated CO_2 concentration has independent effects on expansion rates and thickness of soybean leaves across light and nitrogen gradients. *Journal of Experimental Botany*, **49**, 583–591.

Sims, D.A., Seemann, J.R. and Luo, Y. (1998b). The significance of differences in the mechanisms of photosynthetic acclimation to light, nitrogen and CO_2 for return on investment in leaves. *Functional Ecology*, **12**, 185–194.

Sinclair, T.R. and Muchow, R.C. (1999). Radiation use efficiency. *Advances in Agronomy*, **65**, 215–215.

Sinclair, T.R. and Purcell, L.C. (2005). Is a physiological perspective relevant in a "genocentric" age? *Journal of Experimental Botany*, **56**, 2777–2782.

Sinclair, T.R., Purcell, L.C. and Sneller, C.H. (2004). Crop transformation and the challenge to increase yield potential. *Trends in Plant Science*, **9**, 70–75.

Singh, A.K., Li, H. and Sherman, L.A. (2004). Microarray analysis and redox control of gene expression in the cyanobacterium *Synechocystis* sp. PCC 6803. *Physiologia Plantarum*, **120**, 27–35.

Singh, K.K., Chen, C. and Gibbs, M. (1993a). Photoregulation of fructose and glucose respiration in the intact chloroplasts of *Chlamydomonas reinhardtii* F-60 and spinach. *Plant Physiology*, **101**, 1289–1294.

Singh, KK., Chen, C. and Epstein, D.K., *et al.* (1993b). Respiration of sugars in spinach (*Spinacia oleracea*), maize (*Zea mays*), and *Chlamydomonas reinhardtii* F-60 chloroplasts with emphasis on the hexose kinases. *Plant Physiology*, **102**, 587–593.

Singh, R. (1993). Photosynthesis characteristics of fruiting structures of cultivated crops. In: *Photosynthesis Photoreactions to Plant Productivity* (eds Abrol, Y.P., Mohanty, P. and Govindjee, J.), Kluwer Academic Publishers, Dordrecht, Netherlands, pp. 390–415.

Singsaas, E.L. and Sharkey, T.D. (2000). The effects of high temperature on isoprene synthesis in oak leaves. *Plant, Cell and Environment*, **23**, 751–757.

Singsaas, E.L., Laporte, M.M., Shi, J.Z., *et al.* (1999). Kinetics of leaf temperature fluctuation affect isoprene emission from red oak (*Quercus rubra*) leaves. *Tree Physiology*, **19**, 917–924.

Singsaas, E.L., Ort, D.R. and DeLucia, E.H. (2001). Variation in measured values of photosynthetic quantum yield in ecophysiological studies. *Oecologia*, **128**, 15–23.

Singsaas, E.L., Ort, D.R. and Delucia, E.H. (2004). Elevated CO_2 effects on mesophyll conductance and its consequences for interpreting photosynthetic physiology. *Plant, Cell and Environment*, **27**, 41–50.

Sinoquet, H., Le Roux, X., Adam, B., *et al.* (2001). RATP: a model for simulating the spatial distribution of radiation absorption, transpiration and photosynthesis within canopies: application to an isolated tree crown. *Plant Cell Environment*, **24**, 395–406.

Sinoquet, H., Sonohat, G., Phattaralerphong, J., *et al.* (2005). Foliage randomness and light interception in 3D digitized trees: an analysis of 3D discretization of the canopy. *Plant Cell Environment*, **29**, 1158–1170.

Sinoquet, H., Stephan, J., Sonohat, G., *et al.* (2007). Simple equations to estimate light interception by isolated trees from canopy structure features: assessment with three-dimensional digitized apple trees. *New Phytologist*, **175**, 94–106.

Sioris, C.E. and Evans, W.F.J. (1999). Filling in of Fraunhofer and gas-asorption lines in sky spectra as caused rotational Raman scattering. *Applied Optics*, **38**, 2706–2713.

Sitch, S., Smith, B., Prentice, I.C., *et al.* (2003). Evaluation of ecosystem dynamics, plant geography and terrestrial carbon cycling in the LPJ dynamic global vegetation model. *Global Change Biology*, **9**, 161–185.

Siwko, M.E., Marrink, S.J., de Vries, A.H., *et al.* (2007). Does isoprene protect plant membranes from thermal shock? A molecular dynamics study. *Biochimica et Biophysica Acta – Biomembranes*, **1768**, 198–206.

Skillman, J.B., Strain, B.R. and Osmond, C.B. (1996). Contrasting patterns of photosynthetic acclimation and photoinhibition in two evergreen herbs from a winter deciduous forest. *Oecologia*, **107**, 446–455.

Skre, O. and Oechel, W.C. (1981). Moss functioning in different taiga ecosystems in interior Alaska. I. Seasonal, phenotypic, and drought effects on photosynthesis and response patterns. *Oecologia*, **48**, 50–59.

Slafer, G.A., Satorre, E.H. and Andrade, F.H. (1994). Increases in grain yield in bread wheat from breeding and associated physiological changes. In: *Genetic Improvement of Field Crops* (ed. Slafer, G.A.), Marcel Dekker, New York, USA, pp. 1–68.

Slaton, M.R. and Smith, W.K. (2002). Mesophyll architecture and cell exposure to intercellular air space in alpine desert, and forest species. *International Journal of Plant Sciences*, **163**, 937–948.

Slatyer, R.O. (1967). *Plant-Water Relationships*. Academic Press, London, UK.

Slatyer, R.O. (1977a). Altitudinal variation in the photosynthetic characteristics of snow gum, *Eucalyptus pauciflora* Sieb ex Spreng. III. Temperature response of material grown in contrasting thermal environments. *Australian Journal of Plant Physiology*, **4**, 301–312.

Slatyer, R.O. (1977b). Altitudinal variation in the photosynthetic characteristics of snow gum, *Eucalyptus pauciflora* Sieb. ex Spreng. VI. Comparison of field and phytotron responses to growth temperature. *Australian Journal of Plant Physiology*, **4**, 901–916.

Slatyer, R.O. and Ferrar, P.J. (1977). Altitudinal variation in the photosynthetic characteristics of snow gum, *Eucalyptus pauciflora* Sieb. ex Spreng. II. Effects of growth temperature under controlled conditions. *Australian Journal of Plant Physiology*, **4**, 289–299.

Slot, M., Wirth, C., Schumacher, J., *et al.* (2005). Regeneration patterns in boreal Scots pine glades linked to cold-induced photoihibition. *Tree Physiology*, **25**, 1139–1150.

Slovik, S. (1996). Early needle senescence and thinning of the crown structure of *Picea abies* as induced by chronic SO_2 pollution. 1. Model deduction and analysis. *Global Change Biology*, **2**, 443–458.

Smeets, K., Cuypers, A., Lambrechts, A., *et al.* (2005). Induction of oxidative stress and anti-oxidative mechanisms in *Phaseolus vulgaris* after Cd application. *Plant Physiology and Biochemistry*, **43**, 437–444.

Smillie, R.M., Hetherington, S.E. and Davies, W.J. (1999). Photosynthetic activity of the calyx, green shoulder, pericarp, and locular parenchyma of tomato fruit. *Journal of Experimental Botany*, **50**, 707–718.

Smirnoff, N. (1998). Plant resistance to environmental stress. *Current Opinion in Biotechnology*, **9**, 6.

Smith, A.M., Zeeman, S.C. and Smith, S.M. (2005). Starch degradation. *Annual Review of Plant Biology*, **56**, 73–98.

Smith, B.N. and Epstein, S. (1971). Two categories of $^{13}C/^{12}C$ ratios for higher plants. *Plant Physiology*, **47**, 380–384.

Smith, D.M. and Allen, S.J. (1996). Measurement of sap flow in plant trunks. *Journal of Experimental Botany*, **47**, 1833–1844.

Smith, J.A.C. (1989). Epiphytic bromeliads. In: *Vascular Plants as Epiphytes*. Evolution and Ecophysiology. Ecological Studies, vol. 76 (ed. Lüttge, U.), Springer-Verlag, Berlin – Heidelberg – New York, pp. 109–138.

Smith, J.A.C. and Lüttge, U. (1985). Day night changes in leaf water relations associated with the rhythm of crassulacean acid metabolism in *Kalanchoë daigremontiana*. *Planta*, **163**, 272–283.

Smith, L.H., Langdale, J.A. and Chollet, R. (1998). A functional Calvin cycle is not indispensable for the light activation of C_4 phosphoenolpyruvate carboxylase kinase and its target enzyme in the maize mutant *bundle sheath defective2-mutable1*. *Plant Physiology*, **118**, 191–197.

Smith, M.D., Wilcox, J.C., Kelly, T., *et al.* (2004). Dominance not richness determines invasibility of tallgrass prairie. *Oikos*, **106**, 253–262.

Smith, P.R. and Neales, T.F. (1977). Analysis of the effects of virus infection on the photosynthetic properties of peach leaves. *Australian Journal of Plant Physiology*, **4**, 723–732.

Smith, S.D., Hartsock, T.L. and Nobel P.S. (1983). Ecophysiology of *Yucca brevifolia*, an arborescent monocot of the Mojave Desert. *Oecologia*, **60**, 10–17.

Smith, S.D., Monson, R.K. and Anderson, J.E. (1997). *Physiological Ecology of North American Desert Plants.* Springer-Verlag, Berlin, Germany.

Smith, W.K., Knapp, A.K. and Reiners, W.A. (1989). Penumbral effects on sunlight penetration in plant communities. *Ecology*, **70**, 1603–1609.

Smith, W.K., Vogelmann, T.C., DeLucia, E.H., *et al.* (1997). Leaf form and photosynthesis. Do leaf structure and orientation interact to regulate internal light and carbon dioxide? *BioScience*, **47**, 785–793.

Snyder, A.M., Clarck, B.M. and Bungard, R.A. (2005). Light-dependent conversion of carotenoids in the parasitic angiosperm *Cuscuta reflexa* L. *Plant, Cell and Environment*, **28**, 1326–1333.

Soar, C.J., Speirs, J., Maffei, S.M., *et al.* (2004). Gradients in stomatal conductance, xylem sap ABA and bulk leaf ABA along canes of *Vitis vinifera* cv. Shiraz: molecular and physiological studies investigating their source. *Functional Plant Biology*, **31**, 659–669.

Soares, A.S., Driscoll, S.P., Olmos, E., *et al.* (2008). Adaxial/abaxial specification in the regulation of photosynthesis and stomatal opening with respect to light orientation and growth with CO_2 enrichment in the C_4 species *Paspalum dilatatum*. *New Phytologist*, **177**, 186–198.

Sobrado, M.A. (1996). Leaf photosynthesis and water loss as influenced by leaf age and seasonal drought in an evergreen tree. *Photosynthetica*, **32**, 563–568.

Sofo, A., Dichio, B. Xiloyannis C., *et al.* (2004). Effects of different irradiance levels on some antioxidant enzymes and on malondialdehyde content during rewatering in olive tree. *Plant Science*, **166**, 293–302.

Soitamo, A.J., Piippo, M., Allahverdiyeva, Y., Battchikova, N., Aro, E.M. (2008). Light has a specific role in modulating *Arabidopsis* gene expression at low temperature. *BMC Plant Biology*, **8**, 13.

Solhaug, K.A. and Haugen, J. (1998). Seasonal variation of photosynthess in bark from *Populus tremula* L. *Photosynthetica*, **35**, 411–417.

Solti, A., Gáspár, L., Mészáros, I., *et al.* (2008). Impact of iron supply on the kinetics of recovery of photosynthesis in Cd-stressed poplar (*Populus glauca*). *Annals of Botany*, **102**, 771–782.

Soltis, D.E. and Soltis, P.S. (2003). The role of phylogenetics in comparative genetics. *Plant Physiology*, **132**, 1790–1800.

Soltis, D.E., Soltis, P.S. and Zanis, M.J. (2002). Phylogeny of seed plants based on evidence from eight genes. *American Journal of Botany*, **89**, 1670–1681.

Soltis, P.S. (2005). Ancient and recent polyploidy in angiosperms. *The New Phytologist*, **166**, 5–8.

Soltis, P.S., Soltis, D.E., Wolf, P.G., *et al.* (1999). The phylogeny of land plants inferred from 18S rDNA sequences: pushing the limits of rDNA signal? *Molecular Biology and Evolution*, **16**, 1774–1784.

Sonohat, G., Sinoquet, H., Varlet-Grancher, C., *et al.* (2002). Leaf dispersion and light partitioning in three-dimensionally digitized tall fescue-white clover mixtures. *Plant Cell Environment*, **25**, 529–538.

Sonoike, K. (1996). Photoinhibition of photosystem I: its physiological significance in the chilling sensitivity of plants. *Plant and Cell Physiology*, **37**, 239–247.

Soolanayakanahally, R.Y., Guy, R.D., Silim, S.N., *et al.* (2009). Enhanced assimilation rate and water use efficiency with latitude through increased photosynthetic capacity and internal conductance in balsam poplar (*Populus balsamifera* L.). *Plant Cell and the Environment*, **32**, 1821–1832.

Soriano, M.A., Orgaz, F., Villalobos, F., *et al.* (2004). Efficiency of water use of early planting of sunflower. *European Journal of Agronomy*, **21**, 465–476.

Soukupová, J., Cséfalvay, L., Urbam, O., *et al.* (2008). Annual variation of the steady state chlorophyll fluorescence emission of evergreen plants in temperate zone. *Functional Plant Biology*, **35**, 63–76.

Soukupová, J., Smatanová, S., Nedbal, L., *et al.* (2003). Plant response to destruxins visualized by imaging of chlorophyll fluorescence. *Physiologia Plantarum*, **118**, 399–405.

Soussana, J.F., Allard, V., Pilegaard, K., *et al.* (2007). Full accounting of the greenhouse gas (CO_2, N_2O, CH_4) budget of nine European grassland sites. *Agriculture, Ecosystems and Environment*, **121**, 121–134.

Souza, C.R., Maroco, J., Santos, T., *et al.* (2005). Impact of deficit irrigation on water use efficiency and carbon isotope composition ($\delta^{13}C$) of field-grown grapevines under Mediterranean Climate. *Journal of Experimental Botany*, **56**, 2163–2172.

Souza, R.P. de, Machado, E.C., Silva, J.A.B., *et al.* (2004). Photosynthetic gas exchange, chlorophyll fluorescence and some associated metabolic changes in cowpea (*Vigna unguiculata*) during water stress and recovery. *Environmental and Experimental Botany*, **51**, 45–56.

Sowinska, M., Cunin, B., Heisel, F., *et al.* (1999). New UV-A laser-induced fluorescence imaging system for near-field remote sensing of vegetation: characteristics and performances. In: *Proceedings of the SPIE Conference on Laser Radar Technology, Orlando Florida: The International Society for Optical Engineering*, pp. 91–102.

Sowinska, M., Heisel, F., Miehe, J.A., *et al.* (1996). Remote sensing of plants by streak camera lifetime measurements of the chlorophyll a emission. *Journal of Plant Physiology*, **148**, 638–644.

Specht, R.L. and Moll, E.J. (1983). Mediterranean-type shrublands and sclerophyllous shrublands of the world: an overview. In: *Mediterranean-type Ecosystems. The Role of Nutrients* (eds Kruger, F.J., Mitchell, D.T. and Jarvis, J.U.M.), Springer-Verlag, Berlin – Heidelberg – New York – Tokyo, pp. 41–65.

Spicer, R.A. (1993). Palaeoecology, past climate systems, and C_3/C_4 photosynthesis. *Chemosphere*, **27**, 947–978.

Spikes, J.D. and Stout, M. (1955). Photochemical activity of chloroplasts isolated from sugar beet infected with virus yellows. *Science*, **12**, 375–376.

Spitters, C.J.T., Toussaint, H.A.J.M. and Goudriaan, J. (1986). Separating the diffuse and direct component of global radiation and its implications for modelling canopy photosynthesis. Part I. Components of incoming radiation. *Agricultural and Forest Meteorology*, **38**, 217–229.

Spreitzer, R.J. and Salvucci M.E. (2002). Rubisco: structure, regulatory interactions, and possibilities for a better enzyme. *Annual Review of Plant Biology*, **53**, 449–475.

Stanghellini, C. (2005). Irrigation water: use, efficiency and economics. In: *Improvement in Water Use Efficiency in Protected Crops*. Junta de Andalucía, Sevilla, Spain, pp. 23–33.

Stanhill, G. and Cohen, S. (2001). Global dimming: a review of the evidence for a widespread and significant reduction in global radiation with discussion of its probable causes and possible agricultural consequences. *Agricultural and Forest Meterorology*, **107**, 255–278.

Starr, G. and Oberbauer, S.F. (2003). Photosynthesis of Arctic evergreens under snow: implications for tundra ecosystem carbon balance. *Ecology*, **84**, 1415–1420.

Staudt, M. and Lhoutellier, L. (2007). Volatile organic compound emission from holm oak infested by gypsy moth larvae: evidence for distinct responses in damaged and undamaged leaves. *Tree Physiology*, **27**, 1433–1440.

Steduto, P. and Hsiao, T.C. (1998). Maize canopies under two soil water regimes – II. Seasonal trends of evapotranspiration, carbon dioxide assimilation and canopy conductance, and as related to leaf-area index. *Agricultural and Forest Meteorology*, **89**, 185–200.

Steele, K., Price, A.H., Shashidhar, H.E., *et al.* (2006). Marked-assisted selection to intogress rice QTLs controlling root traits into an Indian upland rice variety. *Theoretical and Applied Genetics*, **112**, 208–221.

Steele, M.R., Gitelson, A.A. and Rundquist, D. (2008). A comparison of two techniques for nondestructive measurement of chlorophyll content in grapevine leaves. *Agronomy Journal*, **100**, 779–782.

Steigmiller, S., Turina, P. and Gräber, P. (2008). The thermodynamic H+/ATP ratios of the H+-ATP synthases from chloroplasts and *Escherichia coli*. *Proceedings of the National Academy of Sciences USA*, **105**, 3745–3750.

Stenberg, P., DeLucia, E.H., Schoettle, A.W., *et al.* (1995). Photosynthetic light capture and processing from cell to canopy, In: *Resource Physiology of Conifers Acquisition, Allocation, and Utilization* (eds Smith, W.K. and Hinckley, T.M.), Academic Press, London, pp. 3–38.

Stenström, A., Jónsdóttir, I.S. and Augner, M. (2002). Genetic and Environmental Effects on Morphology in Clonal Sedges in the Eurasian Arctic. *American Journal of Botany*, **89**, 1410–1421.

Stepien, P. and Johnson, G.N. (2009). Contrasting responses of photosynthesis to salt stress in the glycophyte *Arabidopsis* and the halophyte *Thellungiella*: role of the plastid terminal oxidase as an alternative electron sink. *Plant physiology*, **149**, 1154–1165.

Sterck, F.J. and Bongers, F. (1998). Ontogenetic changes in size, allometry, and mechanical design of tropical rain forest trees. *American Journal of Botany*, **85**, 266–272.

Stern, D.B., Hanson, M.R. and Barkan, A. (2004). Genetics and genomics of chloroplast biogenesis: maize as a model system. *Trends Plant Science*, **9**, 293–301.

Sternberg, L. and De Niro, M. (1983). Biogeochemical implications of the isotopic equilibrium fractionation factor between oxygen atoms of acetone and water. *Geochimica et Cosmochimica Acta*, **47**, 2271–2274.

Sternberg, L., De Niro, M. and Savidge, R. (1986). Oxygen isotope exchange between metabolites and water during biochemical reactions leading to cellulose synthesis. *Plant Physiology*, **82**, 423–427.

Stevens, C.L.R., Schultz, D., Van Baalen, C., *et al.* (1975). Oxygen isotope fractionation during photosynthesis in a blue-green and a green alga. *Plant Physiology*, **65**, 126–129.

Stewart, G.R. and Press, M.C. (1990). The physiology and biochemistry of parasitic angiosperms. *Annual Review of Plant Physiology and Plant Molecular Biology*, **41**, 127–151.

Stewart, G.R., Turnbull, M.H., Schmidt, S., *et al.* (1995). C^{13} Natural abundance in plant communities along a rainfall gradient – a biological integrator of water availability. *Australian Journal of Plant Physiology*, **22**, 51–55.

Stewart, J.B. (1988). Modelling surface conductance of Pine forest. *Agricultural and Forest Meteorology*, **43**, 19–35.

Steyn, W.J., Wand, S.J.E., Holcroft, D.M., *et al.* (2002). Anthocyanins in vegetative tissues: a proposed unified function in photoprotection. *New Phytologist*, **155**, 349–361.

Steyn, W.J., Wand, S.J.E., Jacobs, G., *et al.* (2009). Evidence for a photoprotective function of low-temperature-induced anthocyanin accumulation in apple and pear peel. *Physiologia Plantarum*, **136**, 461–472.

Still, C.J., Berry, J.A., Collatz, G.J., *et al.* (2003). Global distribution of C_3 and C_4 vegetation: carbon cycle implications. *Global Biogeochemical Cycles*, **17**, Article number 1006.

Stiller, I., Dulai, S., Kondrák, M., *et al.* (2008). Effects of drought on water content and photosynthetic parameters in potato plants expressing the trehalose-6-phosphate synthase gene of *Saccharomyces cerevisiae. Planta*, **227**, 299–308.

Stiller, W.N., Read, J.J., Constable, G.A., *et al.* (2005). Selection for water use efficiency traits in a cotton breeding program: cultivar differences. *Crop Sciences*, **45**, 1107–113.

Stimler, K., Montzka, S.A., Berry, J.A., *et al.* (2010). Relationships between carbonyl sulfide (COS) and CO_2 during leaf gas exchange. *New Phytologist*, **186**, 869–878.

Stirling, C.M., Aguilera, C., Baker, N.R., *et al.* (1994). Changes in the photosynthetic light response curve during leaf development of field grown maize with implications for modelling canopy photosynthesis. *Photosynthesis Research*, **42**, 217–225.

Stitt, M. (1991). Rising CO_2 levels and their potential significance for carbon flow in photosynthetic cells. *Plant, Cell and Environment*, **14**, 741–762.

Stitt, M. and Fernie, A.R. (2003). From measurements of metabolites to metabolomics: an 'on the fly' perspective illustrated by recent studies of carbon-nitrogen interactions. *Current Opinion in Biotechnology*, **14**, 136–144.

Stitt, M. and Hurry, V. (2002). A plant for all seasons: alterations in photosynthetic carbon metabolism during cold acclimation in *Arabidopsis. Current Opinion in Plant Biology*, **5**, 199–206.

Stitt, M. and Krapp, A. (1999). The interaction between elevated carbon dioxide and nitrogen nutrition: the physiological and molecular background. *Plant, Cell and Environment*, **22**, 583–621.

Stitt, M., Müller, C., Matt, P., *et al.* (2002). Steps towards an integrated view of nitrogen metabolism. *Journal of Experimental Botany*, **53**, 959–970.

Stobart, A.K., Griffiths, W.T., Ameen-Bukhari, *et al.* (1985). The effect of Cd2+ on the biosynthesis of chlorophyll in leaves of barley. *Physiologia Plantarum*, **63**, 293–298.

Stokes, V.J., Morecroft, M.D. and Morison, J.I.L. (2006). Boundary layer conductance for contrasting leaf shapes in a deciduous broadleaved forest canopy. *Agriculture for Meteorology*, **139**, 40–54.

Stoll, M.P., Court, A., Smorenburg, K., *et al.* (1999). FLEX – Fluorescence Explorer. In: *Remote Sensing for Earth Sciences, Ocean and Sea Ice Applications*, SPIE, Florence, Italy, pp. 487–494.

Strain, B.R. and Bazzaz, F.A. (1983). Terrestrial plant communities. In: *CO_2 and Plants: The Response of Plants to Rising Levels of Atmospheric Carbon Dioxide* (ed. Lemon, E.R.), Westview Press, Boulder, USA, pp. 177–222.

Strand, A., Foyer, C.H., Gustafsson, P., *et al.* (2003). Altering flux through the sucrose biosynthesis pathway in transgenic *Arabidopsis thaliana* modifies photosynthetic acclimation at low temperatures and the development of freezing tolerance. *Plant Cell Environment*, **26**, 523–535.

Strand, A., Hurry, V., Gustafsson, P., *et al.* (1997). Development of *Arabidopsis thaliana* leaves at low temperatures releases the suppression of photosynthesis and photosynthetic gene expression despite the accumulation of soluble carbohydrates. *Plant Journal*, **12**, 605–614.

Strand, A., Hurry, V., Henkes, S., *et al.* (1999). Acclimation of *Arabidopsis* leaves developing at low temperatures. Increasing cytoplasmic volume accompanies increased activities of enzymes in the Calvin cycle and in the sucrose-biosynthesis pathway. *Plant Physiology*, **119**, 1387–1397.

Strand, M., Lundmark, T., Söderbergh, I., *et al.* (2002). Impacts of seasonal air and soil temperatures on photosynthesis in Scots pine trees. *Tree Physiology*, **22**, 839–847.

Strasser, R.J., Srivasta, A. and Govindjee, J. (1995). Polyphasic chlorophyll-alpha fluorescence transient in plants and cyanobacteria. *Photochemistry and Photobiology*, **61**, 32–42.

Strauss-Debenedetti, S. and Bazzaz, F.A. (1991). Plasticity and acclimation to light in tropical Moraceae of different sucessional positions. *Oecologia*, **87**, 377–387.

Streb, P. and Feierabend, J. (1999). Significance of antioxidants and electron sinks for the cold-hardening-induced resistance of winter rye leaves to photo-oxidative stress. *Plant, Cell Environment*, **22**, 1225–1237.

Streb, P., Aubert, S. and Bligny, R. (2003b). High temperature effects on light sensitivity in the two high mountain plant species *Soldanella alpina* (L) and *Ranunculus glacialis* (L). *Plant Biology*, **5**, 432–440.

Streb, P., Aubert, S., Gout, E., *et al.* (2003a). Cold- and light-induced changes of metabolite and antioxidant levels in two high mountain plant species *Soldanella alpina* and *Ranunculus glacialis* and a lowland species *Pisum sativum*. *Physiologia Plantarum*, **118**, 96–104.

Streb, P., Aubert, S., Gout, E., *et al.* (2003c). Reversibility of cold- and light-stress tolerance and accompanying changes of metabolite and antioxidant levels in the two high mountain plant species *Soldanella alpina* and *Ranunculus glacialis*. *Journal of Experimental Botany*, **54**, 405–418.

Streb, P., Feierabend, J. and Bligny, R. (1997). Resistance to photoinhibition of photosystem II and catalase and antioxidative protection in high mountain plants. *Plant Cell and Environment*, **20**, 1030–1040.

Streb, P., Josse, E.M., Gallouët, E., *et al.* (2005). Evidence for alternative electron sinks to photosynthetic carbon assimilation in the high mountain plant species *Ranunculus glacialis*. *Plant, Cell Environment*, **28**, 1123–1135.

Streb, P., Shang, W. and Feierabend, J. (1999). Resistance of cold-hardened winter rye leaves *(Secale cereale* L.) to photo-oxidative stress. *Plant, Cell Environment*, **22**, 1211–1223.

Streb, P., Shang, W., Feierabend, J., *et al.* (1998). Divergent strategies of photoprotection in high-mountain plants. *Planta*, **207**, 313–324.

Strimbeck, G.R., Kjellsen, T.D., Schaberg, P.G., *et al.* (2007). Cold in the common garden: comparative low-temperature tolerance of boreal and temperate conifer foliage. *Trees*, **21**, 557–567.

Strong, G.L., Bannister, P. and Burritt, D. (2000). Are mistletoes shade plants? CO_2 assimilation and chlorophyll fluorescence of temperate mistletoes and their hosts. *Annals of Botany*, **85**, 511–519.

Studart-Guimaraes, C., Fait, A., Nunes- Nesi, A., *et al.* (2007). Reduced expression of succinyl-coenzyme a ligase can be compensated for by up-regulation of the γ-aminobutyrate shunt in illuminated tomato leaves. *Plant Physiology*, **145**, 626–639.

Stuntz, S. and Zotz, G. (2001). Photosynthesis in vascular epiphytes: a survey of 27 species of diverse taxonomic origin. *Flora*, **196**, 132–141.

Stylinski, C.D., Gamon, J.A. and Oechel, W.C. (2002). Seasonal patterns of reflectance indices, carotenoid pigments and photosynthesis of evergreen chaparral species. *Oecologia*, **131**, 366–374.

Stylinski, C.D., Oechel, W.C., Gamon, J.A., *et al.* (2000). Effects of lifelong [CO_2] enrichment on carboxylation and light utilization of *Quercus pubescens* Willd. examined with gas exchange, biochemistry and optical techniques. *Plant, Cell and Environment*, **23**, 1353–1362.

Suárez, L., Zarco-Tejada, P.J., Berni, J.A.J., *et al.* (2009) Modeling PRI for water stress detection using radiative transfer models. *Remote Sensing of the Environment*, **113**, 730–744.

Suárez, L., Zarco-Tejada, P.J., Sepulcre-Cantó, G., *et al.* (2008). Assessing canopy PRI for water stress detection with diurnal airborne imagery. *Remote Sensing of the Environment*, **112**, 560–575.

Suess, H.E. (1955). Radiocarbon concentration in modern wood. *Science*, **122**, 415–417.

Sugano, S.S., Shimada, T., Imai, Y., *et al.* (2010). Stomagen positively regulates stomatal density in *Arabidopsis*. *Nature*, **463**, 241–244.

Sugita, M. and Sugiura, M. (1996). Regulation of gene expression in chloroplasts of higher plants. *Plant Molecular Biology*, **32**, 315–326.

Suh, H.J., Kim, C.S., Lee, J.Y., *et al.* (2002). Photodynamic effect of iron excess on photosystem II function in pea plants. *Photochemistry and Photobiology*, **75**, 513–518.

Sulpice, R., Tschoep, H., Von Korff, M., *et al.* (2007). Description and applications of a rapid and sensitive non-radioactive microplate-based assay for maximum and initial activity of D-ribulose-1,5-bisphosphate carboxylase/oxygenase. *Plant Cell Environment*, **30**, 1163–1175.

Sun, G., Ji, Q., Dilcher, D.L., *et al.* (2002). Archaefructaceae, a new basal angiosperm family. *Science*, **296**, 899–904.

Sun, J., Nishio, J.N. and Vogelmann, T.C. (1996). High-light effects on CO_2 fixation gradients across leaves. *Plant Cell Environment*, **19**, 1261–1271.

Sun, J., Nishio, J.N. and Vogelmann, T.C. (1998). Green light drives CO_2 fixation deep within leaves. *Plant Cell Physiology*, **39**, 1020–1026.

Sun, J., Okita, T.W. and Edwards, G.E. (1999). Modification of carbon partitioning, photosynthetic capacity, and O_2 sensitivity in *Arabidopsis* plants with low ADP-glucose pyrophosphorylase activity. *Plant Physiology*, **119**, 367–276.

Sun, P., Grignetti, A., Liu, S., *et al.* (2008). Associated changes in physiological parameters and spectral reflectance indices in olive (*Olea europaea* L.) leaves in response to different levels of water stress. *International Journal of Remote Sensing*, **29**, 1725–1743.

Sun, Q., Yoda, K., Suzuki, M., *et al.* (2003). Vascular tissue in the stems and roots of woody plants can conduct light. *Journal of Experimental Botany*, **54**, 1627–1635.

Sun, W.H, Verhoeven, A.S., Bugos, R.C., *et al.* (2001). Suppression of zeaxanthin formation does not reduce photosynthesis and growth of transgenic tobacco under field conditions. *Photosynthesis Research*, **67**, 41–50.

Sun, Z.J., Livingston, N.J., Guy, R.D., *et al.* (1996). Stable carbon isotopes as indicators of increased water use efficiency and productivity in white spruce (*Picea glauca* (Moench) Voss) seedlings. *Plant, Cell and Environment*, **19**, 887–894.

Sundberg, M.D. (1986). A comparison of stomatal distribution and length in succulent and non-succulent desert plants. *Phytomorphology*, **36**, 53–66.

Sundblad, L.G., Schröder, W.P. and Akerlung, H.E. (1988). S-state distribution and redox state of Q_A in barley in relation to luminescence decay kinetics. *Biochimica et Biophysica Acta*, **973**, 47–52.

Sundbom, E., Strand, M. and Hällgreen, J.E. (1982). Temperature-induced fluorescence changes. A screening method for frost tolerance of potato (*Solanum* sp.). *Plant Physiology*, **70**, 1299–1302.

Sundby, C. and Andersson, B. (1985). Temperature-induced reversible migration along the thylakoid membrane of photosystem II regulates it association with LHC-II. *FEBS Letters*, **191**, 24–28.

Suni, T., Berninger, F., Vesala, T., *et al.* (2003). Air temperature triggers the commencement of evergreen boreal forest photosynthesis in spring. *Global Change Biology*, **9**, 1410–1426.

Susiluoto, S., Perämäki, M., Nikinmaa, E., *et al.* (2007). Effects of sink removal on transpiration at the treeline: implications for growth limitation hypothesis. *Environmental and Experimental Botany*, **60**, 334–339.

Susiluoto, S., Pumpanen, J. and Berninger, F. (2008). Effects of grazing on the vegetation structure and carbon balance of Scandinavian fell vegetation. *Arctic, Antarctic, and Alpine Research*, **40**, 422–431.

Suwignyo, R.A., Nose, A., Kawamitsu, Y., e al. (1995). Effects of manipulations of source and sink on carbon exchange rate and some enzymes of sucrose metabolism in leaves of soybean [*Glycine max* (L.) Merr.]. *Plant Cell Physiology*, **36**, 1439–1446.

Suyker, A.E. and Verma, S.B. (1993). Eddy correlation measurements of CO_2 flux using a closed-path sensor: theory and field tests against an open-path sensor. *Boundary Layer Meteorology*, **64**, 391–407.

Suyker, A.E. and Verma, S.B. (2001). Year-round observations of the net ecosystem exchange of carbon dioxide in a native tallgrass prairie. *Global Change Biology*, **7**, 279–289.

Suzuki, T., Eiguchi, M., Kumamaru, T., *et al.* (2008). MNU-induced mutant pools and high performance TILLING enable finding of any gene mutation in rice. *Molecular Genetics and Genomics*, **279**, 213–223.

Sveshnikov, D., Ensminger, I., Ivanov, A.G., *et al.* (2005). Excitation energy partitioning and quenching during cold acclimation in Scots pine. *Tree Physiology*, **26**, 325–336.

Swarbrick, P.J., Schulze-Lefert, P. and Scholes, J.D. (2006). Metabolic consequences of susceptibility and resistance (race-specific and broad-spectrum) in barley leaves challenged with powdery mildew. *Plant, Cell and Environment*, **29**, 1061–1076.

Swarthout, D., Harper, E., Judd, S., *et al.* (2009). Measures of leaf-level water-use efficiency in drought stressed endophyte infected and non-infected tall fescue grasses. *Environmental and Experimental Botany*, **66**, 88–93.

Sweetlove, L.J., Lytovchenko, A., Morgan, M., *et al.* (2006). Mitochondrial uncoupling protein is required for efficient photosynthesis. *Proceedings of the National Academy of Sciences USA*, **103**, 19, 587–19, 592.

Swinbank, W.C. (1951). The measurement of vertical transfer of heat and water vapor by eddies in the lower atmosphere. *Journal of Meterology*, **8**, 135–145.

Sykes, M.T. and Prentice, I.C. (1996). Climate change, tree species distributions and forest dynamics: a case study in the mixed conifer/northern hardwoods zone of northern Europe. *Climatic Change*, **34**, 161–177.

Syvertsen, J.P., Lloyd, J., McConchie, C., *et al.* (1995). On the relationship between leaf anatomy and CO_2 diffusion through the mesophyll of hypostomatous leaves. *Plant Cell and Environment*, **18**, 149–157.

Szarek, S.R. and Ting, I.P. (1975). Physiological responses to rainfall in *Opuntia basiliaris* (Cactaceae). *American Journal of Botany*, **62**, 602–609.

Szarek, S.R., Holthe, P.A. and Ting, I.P. (1987). Minor physiological response to elevated CO_2 by the CAM plant *Agave vilmoriniana*. *Plant Physiology*, **83**, 938–940.

Sziráki, I.B., Mustarky, L.A., Faludi-Daniel, A., *et al.* (1984). Alterations in chloroplast ultraestructure and chlorophyll content in rust infected pinto beans at different stages of disease development. *Phytopathology*, **74**, 77–84.

Tadaki, Y. (1977). Forest biomass. Leaf biomass. In: *Primary Productivity of Japanese Forests. Productivity of Terrestrial Communities. Japanese Commitee for the International Biological Program* (eds Shidei, T. and Kira, T.), University of Tokyo Press, Tokyo, pp. 39–44.

Takahashi, H. and Ehara, Y. (1992). Changes in the activity and the polypeptide composition of the oxygen-evolving complex in photosystem II of tobacco leaves infected with cucumber mosaic virus strain Y. *Molecular Plant-Microbe Interactions*, **5**, 269–272.

Takahashi, N., Ling, P.P. and Frantz, J.M. (2008). Considerations for accurate whole plant photosynthesis measurement. *Environmental Control in Biology*, **46**, 91–101.

Takahashi, S. and Badger, M.R. (2011). Photoprotection in plants: a new light on photosystem II damage. *Trends in Plant Science*, **16**, 53–60.

Takahashi, S. and Murata, N. (2008). How do environmental stresses accelerate photoinhibition? *Trends in Plant Science*, **13**, 179–182.

Takahashi, S., Milward, S.E., Fan, D.Y., *et al.* (2009). How does cyclic electron flow alleviate photoinhibition in *Arabidopsis*? *Plant Physiology*, **149**, 1560–1567.

Takamiya, K.I., Tsuchiya, T. and Ohta, H. (2000). Degradation pathway(s) of chlorophyll: what has gene cloning revealed? *Trends Plant Science*, **10**, 426–431.

Takashima, T., Hikosaka, K. and Hirose, T. (2004). Photosynthesis or persistence: nitrogen allocation in leaves of evergreen and deciduous *Quercus* species. *Plant, Cell and Environment*, **27**, 1047–1054.

Takenaka, A. (1994). Effects of leaf blade narrowness and petiole length on the light capture efficiency of a shoot. *Ecologcal Research*, **9**, 109–114.

Takenaka, A., Takahashi, K. and Kohyama, T. (2001). Optimal leaf display and biomass partitioning for efficient light capture in an understory palm, *Licuala arbuscula*. *Functional Ecology*, **15**, 660–668.

Takeuchi, Y., Kubiske, M.E., Isebrands, J.G., *et al.* (2001). Photosynthesis, light and nitrogen relationships in a young deciduous forest canopy under open-air CO_2 enrichment. *Plant, Cell and Environment*, **24**, 1257–1268.

Takizawa, K., Kanazawa, A. and Kramer, D.M. (2008). Depletion of stromal Pi induces high 'energy dependent' antenna exciton quenching (qE) by decreasing proton conductivity at CFO-CF1 ATP synthase. *Plant, Cell and Environment*, **31**, 235–243.

Talbott, L.D. and Zeiger, E. (1998). The role of sucrose in guard cell osmoregulation. *Journal of Experimental Botany*, **49**, 329–337.

Tan, W. and Hogan, G.D. (1995). Limitations to net photosynthesis as affected by nitrogen status in jack pine (*Pinus banksiana* Lamb.) seedlings. *Journal of Experimental Botany*, **46**, 407–413.

Tanaka, R. and Tanaka, A. (2007). Tetrapyrrole biosynthesis in higher plants. *Annual Review of Plant Biology*, **58**, 321–346.

Taneda, H. and Tateno, M. (2005). Hydraulic conductivity, photosynthesis and leaf water balance in six evergreen woody species from fall to winter. *Tree Physiology*, **25**, 299–306.

Tang, A.Ch. and Boyer, J.S. (2007). Leaf shrinkage decreases porosity at low water potentials in sunflower. *Functional Plant Biology*, **34**, 24–30.

Tang, J.Y., Zielinski, R.E., Zangerl, A.R., *et al.* (2006). The differential effects of herbivory by first and fourth instars of *Trichoplusia ni* (Lepidoptera: Noctuidae) on photosynthesis in *Arabidopsis thaliana*. *Journal of Experimental Botany*, **57**, 27–536.

Tang, Y.H., Kachi, N., Furukawa, A., *et al.* (1999). Heterogeneity of light availability and its effects on simulated carbon gain of tree leaves in a small gap and the understory in a tropical rain forest. *Biotropica*, **31**, 268–278.

Tanner, C.B. and Sinclair, T.R. (1983). Efficient water use in crop production: research or re-search? In: *Limitations to efficient water use in crop production* (ed. Taylor, H.M.. Jordan, W.R. and Sinclair T.R.), ASA, CSSA, SSSA, Madison, Wisconsin, USA, pp. 1–25.

Tans, P.P., Berry, J.A. and Keeling, R.F. (1993). Oceanic ^{13}C data: a new window on CO_2 uptake by oceans. *Global Biogeochemical Cycles*, **7**, 353–368.

Tans, P.P., Fung, , I.Y. and Takahashi, T. (1990). Observational constraints on the global atmospheric CO_2 budget. *Science*, **247**, 1431–1438.

Tanz, S.K., Tetu, S.G., Vella, N.G.F., *et al.* (2009). Loss of the transit peptide and an increase in gene expression of an ancestral chloroplastic carbonic anhydrase were instrumental in the evolution of the cytosolic C_4 carbonic anhydrase in *Flaveria*. *Plant Physiology*, **150**, 1515–1529.

Tao, Y., Xie, Z., Chen, W., *et al.* (2003). Quantitative nature of *Arabidopsis* responses during compatible and incompatible interactions with the bacterial pathogen *Pseudomonas syringae*. *Plant Cell*, **15**, 317–330.

Tardieu, F. and Simonneau, T. (1998). Variability among species of stomatal control under fluctuating soil water status and evaporative demand: modelling isohydric and anisohydric behaviours. *Journal of Experimental Botany*, **49**, 419–432.

Tardieu, F., Granier, C. and Muller, B. (1999). Research review. Modelling leaf expansion in a fluctuating environment: are changes in specific leaf area a consequence of changes in expansion rate? *New Phytologist*, **143**, 33–44.

Tardieu, F., Katerji, N., Bethenod, O., *et al.* (1991a). Leaf stomatal conductance in the field: its relationship with measured plant water potentials, mechanical constraints and ABA concentration in the xylem sap. *Plant, Cell and Environment*, **14**, 121–126.

Tardieu, F., Katerji, N., Bethenod, O., *et al.* (1991b). Maize stomatal conductance in the field – its relationship with soil and plant water potentials, mechanical constraints and aba concentration in the xylem sap. *Plant, Cell and Environment*, **14**, 121–126.

Taschler, D. and Neuner, G. (2004). Summer frost resistance and freezing patterns measured in situ in leaves of major alpine plant growth forms in relation to their upper distribution boundary. *Plant Cell and Environment*, **27**, 737–746.

Tattini, M., Lombardini, L. and Gucci, R. (1997). The effect of NaCl stress and relief on gas exchange properties of two olive cultivars differing in tolerance to salinity. *Plant and Soil*, **197**, 87–93.

Taub, D.R. (2000). Climate and the US distribution of C_4 grass subfamilies and decarboxylation variants of C_4 photosynthesis. *American Journal of Botany*, **87**, 1211–1215.

Taub, D.R., Seemann J.R. and Coleman, J.S. (2000). Growth in elevated CO_2 protects photosynthesis against high-temperature damage. *Plant, Cell and Environment*, **23**, 649–656.

Tausz, M., Warren, C.R. and Adams, M.A. (2005a). Is the bark of shining gum (*Eucalyptus nitens*) a sun or shade leaf? *Trees*, **19**, 415–421.

Tausz, M., Warren, C.R. and Adams, M.A. (2005b). Dynamic light use and protection from excess light in upper canopy and coppice leaves of *Nothofagus cunninghamii* in an old growth, cool temperate rainforest in Victoria, Australia. *New Phytologist*, **165**, 143–155.

Taylor, A., Marin, J. and Seel, W.E. (1996). Physiology of the parasitic association between maize and witchweed (*Striga hermonthica*): is ABA involved. *Journal of Experimental Botany*, **47**, 1057–1065.

Taylor, G. and Davies, W.J. (1986). Leaf growth of *Betula* and *Acer* in simulated shadelight. *Oecologia*, **69**, 589–593.

Taylor, S.E. and Terry, N. (1986). Variation in photosynthetic electron transport capacity and its effect on the light modulation of ribulose bisphosphate carboxylase. *Photosynthesis Research*, **8**, 249–256.

Taylor, S.H., Ripley, B.S., Woodward, F.I. *et al.* (2011). Drought limitation of photosynthesis differs between C_3 and C_4 grasses in a comparative experiment. *Plant, Cell and Environment*, **34**, 65–75.

Taylor, W.C., Rosche, E., Marshall, J.S., *et al.* (1997). Diverse mechanisms regulate the expression of genes coding for C_4 enzymes. *Australian Journal of Plant Physiology*, **24**, 437–442.

Tazoe, Y., von Caemmerer, S., Badger, M.R., *et al.* (2009). Light and CO_2 do not affect the mesophyll conductance to CO_2 diffusion in wheat leaves. *Journal of Experimental Botany*, **60**, 2291–2301.

Tazoe, Y., von Caemmerer, S., Estavillo, G. *et al.* (2011). Using tunable diode laser spectroscopy to measure carbon isotope discrimination and mesophyll conductance to CO_2 diffusion dynamically at different CO_2 concentrations. *Plant, Cell and Environment*, **34**, 580–591.

Tcherkez, G. and Farquhar, G.D. (2007). On the $^{16}O/^{18}O$ isotope effect associated with photosynthetic O_2 production. *Functional Plant Biology*, **34**, 1049–1052.

Tcherkez, G. and Farquhar, G.D. (2008). On the effect of heavy water (D_2O) on carbon isotope fractionation in photosynthesis. *Functional Plant Biology*, **35**, 201–212.

Tcherkez, G. and Hodges, M. (2008). How isotopes may help to elucidate primary nitrogen metabolism and its interactions with (photo)respiration in C_3 leaves. *Journal of Experimental Botany*, **59**, 1685–1693.

Tcherkez, G., Cornic, G., Bligny, R., *et al.* (2005). In vivo respiratory metabolism of illuminated leaves. *Plant Physiology*, **138**, 1596–1606.

Tcherkez, G., Cornic, G., Bligny, R., *et al.* (2008). Respiratory metabolism of illuminated leaves depends on CO_2 and O_2 conditions. *Proceedings of the Natural Academy of Sciences USA*, **105**, 797–802.

Tcherkez, G., Farquhar, G.D., Badek, F., *et al.* (2004). Theoretical considerations about carbon isotope distribution in glucose of C_3 plants. *Functional Plant Biology*, **31**, 857–877.

Tcherkez, G.G.B., Farquhar, G.D. and Andrews, T.J. (2006). Despite slow catalysis and confused substrate specificity, all ribulose bisphosphate carboxylases may be nearly perfectly optimized. *Proceedings of the National Academy of Sciences*, **103**, 7246–7251.

Técsi, L.I., Smith, A.M., Maule, A.J., *et al.* (1996). A spatial analysis of physiological changes associated with infection of cotyledons of marrow plants with *Cucumber mosaic virus*. *Plant Physiology*, **111**, 975–985.

Tedeschi, V., Rey, A., Manca, G., *et al.* (2006). Soil respiration in a Mediterranean oak forest at different developmental stages after coppicing. *Global Change Biology*, **12**, 110–121.

Teklemariam, T.A. and Sparks, J.P. (2004). Gaseous fluxes of peroxyacetyl nitrate (PAN) into plant leaves. *Plant, Cell and Environment*, **27**, 1149–1158.

Telfer, A. (2005). Too much light? How beta-carotene protects the photosystem II reaction centre. *Photochemical and Photobiological Sciences*, **4**, 950–956.

Telfer, A., Bottin, H., Barber, J., *et al.* (1984). The effect of magnesium and phosphorylation of light-harvesting chlorophyll a/b-protein on the yield of P-700-photooxidation in pea chloroplasts. *Biochimica et Biophysica Acta*, **764**, 324–330.

Tenhunen, J.D., Pearcy, R.W. and Lange, O.L. (1987). Diurnal variations in leaf conductance and gas exchange in natural environments. In: *Stomatal Function* (eds Zeiger, E., Farquhar, G.D. and Cowan, I.R.), Stanford University Press, Stanford, USA, pp. 323–352.

Tenhunen, J.D., Falge, E., Ryel, R., *et al.* (2001). Modelling of fluxes in a spruce forest catchment of the Fichtelgebirge. In: *Ecosystem Approaches to Landscape Management in Central Europe* (eds Tenhunen, J.D., Lenz, R. and Hantschel, R.), Springer-Verlag, Berlin, Germany, pp. 417–462.

Tenhunen, J.D., Hanano, R., Abril, M., *et al.* (1994). Above- and below-ground environmental influences on leaf conductance of *Ceanothus thyrsiflorus* growing in a chaparral environment: drought response and the role of abscisic acid. *Oecologia*, **99**, 306–314.

Tenhunen, J.D., Lange, O.L., Gebel, J., *et al.* (1984). Changes in photosynthetic capacity, carboxylation efficiency, and CO_2 compensation point associated with midday stomatal closure and midday depression of net CO_2 exchange of leaves of *Quercus suber*. *Planta*, **162**, 193–203

Tenhunen, J.D., Sala Serra, A., Harley, P.C., *et al.* (1990). Factors influencing carbon fixation and water use by mediterranean sclerophyll shrubs during summer drought. *Oecologia*, **82**, 381–393.

Terahima, I. and Ono, K. (2002). Effects of $HgCl_2$ on CO_2 dependence of leaf photosynthesis: evidence indicating involvement of aquaporins in CO_2 diffussion across the plasma membrane. *Plant Cell Physiology*, **43**, 70–78.

Terashima, I. (1992). Anatomy of non-uniform leaf photosynthesis. *Photosynthesis Research*, **31**, 195–212.

Terashima, I. and Hikosaka, K. (1995). Comparative ecophysiology of leaf and canopy photosynthesis. *Plant Cell Environment*, **18**, 1111–1128.

Terashima, I. and Inoue, Y. (1985a). Palisade tissue chloroplasts and spongy tissue chloroplasts in spinach: Biochemical and ultrastructural differences. *Plant Cell Physiology*, **26**, 63–75.

Terashima, I. and Inoue, Y. (1985b). Vertical gradient in photosynthetic properties of spinach chloroplasts dependent on intra-leaf light environment. *Plant Cell Physiology*, **26**, 781–785.

Terashima, I. and Ono, K. (2002). Effects of $HgCl_2$ on CO_2 dependence of leaf photosynthesis: evidence indicating involvement of aquaporins in CO_2 diffusion across the plasma membrane. *Plant and Cell Physiology*, **43**, 70–78.

Terashima, I., Araya, T., Miyazawa, S.I., *et al.* (2005), Construction and maintenance of the optimal photosynthetic systems of the leaf, herbaceous plant and tree: an eco-developmental treatise. *Annals of Botany*, **95**, 507–519.

Terashima, I. and Evans, J.R. (1988). Effects of light and nitrogen nutrition on the organization of the photosynthetic apparatus in spinach. *Plant and Cell Physiology*, **29**, 143–155.

Terashima, I., Fujita, T., Inoue, T., *et al.* (2009). Green light drives leaf photosynthesis more efficiently than red light in strong white light: revisiting the enigmatic question of why leaves are green. *Plant Cell Physiology*, **50**, 684–697.

Terashima, I., Hanba, Y.T., Tholen, D. *et al.* (2011). Leaf functional anatomy in relation to photosynthesis. *Plant Physiology*, **155**, 108–116.

Terashima, I., Masuzawa, T., Ohba, H., *et al.* (1995). Is photosynthesis suppressed at higher elevations due to low CO_2 pressure? *Ecology*, **76**, 2663–2668.

Terashima, I., Noguchi, K., Itoh-Nemoto, T., *et al.* (1998). The cause of PSI photoinhibition at low temperatures in leaves of *Cucumis sativus*, a chilling-sensitive plant. *Physiologia Plantarum*, **103**, 295–303.

Terashima, I., Sakaguchi, S. and Hara, N. (1986). Intra-leaf and intracellular gradients in chloroplast ultrastructure of dorsiventral leaves illuminated from the adaxial or abaxial side during their development. *Plant Cell Physiology*, **27**, 1023–1031.

Terjung, F. (1998). Reabsorption of chlorophyll fluorescence and its effects on the spectral distribution and the picosecond decay of higher plant leaves. *Z Naturforsch C*, **53**, 924–926.

Terry, N. (1980). Limiting factors in photosynthesis I. Use of iron stress to control photochemical capacity in vivo. *Plant Physiology*, **65**, 114–120.

Terry, N. and Abadía, J. (1986). Function of iron in chloroplasts. *Journal of Plant Nutrition*, **9**, 609–646.

Terry, N. and Ulrich, A. (1973a). Effects of phosphorus deficiency on the photosynthesis and respiration of leaves of sugar beet. *Plant Physiology*, **51**, 43–47.

Terry, N. and Ulrich, A. (1973b). Effects of potassium deficiency on the photosynthesis and respiration of leaves of sugar beet. *Plant Physiology*, **51**, 783–786.

Terry, N. and Ulrich, A. (1974). Photosynthetic and respiratory CO_2 exchange of sugar beet as influenced by manganese deficiency. *Crop Science*, **14**, 502–504.

Terzaghi, W.B., Fork, D.C., Berry, J.A., *et al.* (1989). Low and high temperature limits to PSII. *Plant Physiology*, **91**, 1494–1500.

Teskey, R.O. (1997). Combined effects of elevated CO_2 and air temperature on carbon assimilation of *Pinus taeda* trees. *Plant, Cell and Environment*, **20**, 373–380.

Teskey, R.O., Grier, C.C. and Hinckley, T.M. (1984). Change in photosynthesis and water relations with age and season in *Abies amabilis. Canadian Journal of Forest Research*, **14**, 77–84.

Teskey, R.O., Saveyn, A., Steppe, K., *et al.* (2008). Origin, fate and significance of CO_2 in tree stems. *New Phytologist*, **177**, 17–32.

Tew, R.K. (1970). Seasonal variation in the nutrient content of aspen foliage. *Journal of Wildlife Management*, **34**, 475–478.

Tezara, W., Mitchell, V.J., Driscoll, S.D., *et al.* (1999). Water stress inhibits plant photosynthesis by decreasing coupling factor and ATP. *Nature*, **401**, 914–917.

Thiele, A., Krause, G.H. and Winter, K. (1998). In situ study of photoinhibition of photosynthesis and xanthophyll cycle activity in plants growing in natural gaps of the tropical forest. *Australian Journal of Plant Physiology*, **25**, 189–195.

Tholen, D. and Zhu, X.-G. (2011). The mechanistic basis of internal conductance: a theoretical analysis of mesophyll cell photosynthesis and CO_2 diffusion. *Plant Physiology*, **156**, 90–105.

Tholen, D., Boom, C., Noguchi, K., *et al.* (2008). The chloroplast avoidance response decreases internal conductance to CO_2 diffusion in *Arabidopsis thaliana* leaves. *Plant Cell and Environment*, **31**, 1688–1700.

Tholen, D., Pons, T.L., Voesenek, L.A.C.J., *et al.* (2007). Ethylene insensitivity results in down-regulation of Rubisco expression and photosynthetic capacity in tobacco. *Plant Physiology*, **144**, 1305–1315.

Tholl, D., Boland, W., Hansel, A., *et al.* (2006). 'Practical approaches to plant volatile analysis'. *The Plant Journal*, **45**(4), 540–560.

Thomas, D.S., Montagu, K.D. and Conroy, J.P. (2006). Leaf inorganic phosphorus as a potential indicator of phosphorus status, photosynthesis and growth of *Eucalyptus grandis* seedlings. *Forest Ecology and Management*, **223**, 267–274.

Thomas, H., Ougham, H.J., Wagstaff, C., *et al.* (2003). Defining senescence and death. *Jounral of Experimental Botany*, **385**, 1127–1132.

Thomas, R.B., Reid, C.D., Ybema, R., *et al.* (1993). Growth and maintenance components of leaf respiration of cotton grown in elevated carbon dioxide partial pressure. *Plant Cell Environment*, **16**, 539–546.

Thomas, S.C. and Winner, W.E. (2002). Photosynthetic differences between saplings and adult trees: an integration of field results by meta-analysis. *Tree Physiology*, **22**, 117–127.

Thomashow, M.F. (1999). Plant cold acclimation: Freezing tolerance genes and regulatory mechanisms. *Annual Review of Plant Physiology and Plant Molecular Biology*, **50**, 571–599.

Thompson, L.K., Blaylock, R., Sturtevant, J.M., *et al.* (1989). Molecular basis of the heat denaturation of photosystem II. *Biochemistry*, **28**, 6686–6695.

Thomson, V.P., Cunningham, S.A., Ball, M.C., *et al.* (2003). Compensation for herbivory by *Cucumis sativus* through increased photosynthetic capacity and efficiency. *Oecologia*, **134**, 167–175.

Thornley, J.M. and Johnson, I.R. (1990). *Plant and crop modelling. A mathematical approach to plant and crop physiology.* Clarendon Press, Oxford, UK.

Tieszen, L.L., Reed, B.C., Bliss, N.B., *et al.* (1997). NDVI, C_3 and C_4 production, and distributions in Great Plains grassland land cover classes. *Ecological Applications*, **7**, 59–78.

Tikhonov, A.N., Khomutov, G.B. and Ruuge, E.K. (1984). Electron transport control in chloroplasts. Effects of magnesium ions on the electron flow between two photosystems. *Photobiochemistry and Photobiology*, **8**, 261–269.

Tikkanen, M., Grieco, M. and Aro, E.-M. (2011). Novel insights into plant light-harvesting complex II phosphorylation and 'state transitions'. *Trends in Plant Science*, **16**, 126–131.

Tikkanen, M., Nurmi, M., Suorsa, M., *et al.* (2008). Phosphorylation-dependent regulation of excitation energy distribution between the two photosystems in higher plants. *Biochimica et Biophysica Acta*, **1777**, 425–432.

Tilman, D. and Wedin, D. (1991). Plant traits and resource reduction for five grasses growing on a nitrogen gradient. *Ecology*, **72**, 685–700.

Tilman, D., Cassman, K.G., Matson, P.A., *et al.*. (2002) Agricultural sustainability and intensive production practices. *Nature*, **418**, 671–677.

Timlin, D., Rahman, S.M.L., Baker, J., *et al.* (2006). Whole plant photosynthesis, development, and carbon partitioning in potato as a function of temperature. *Agronomy Journal*, **98**, 1195–1203.

Timperio, A.M., D'Amici, G.M., Barta, C., *et al.* (2007). Proteomics, pigment composition, and organization of thylakoid membranes in iron-deficient spinach leaves. *Journal of Experimental Botany*, **58**, 3695–3710.

Tinoco-Ojanguren, C. and Pearcy, R.W. (1992). Dynamic stomatal behavior and its role in carbon gain during lightflecks of a gap phase and an understory *Piper* species acclimated to high and low light. *Oecologia*, **92**, 222–228.

Tinoco-Ojanguren, C. and Pearcy, R.W. (1993a). Stomatal dynamics and its importance to carbon gain in two rainforest *Piper* species. II. Stomatal versus biochemical limitations during photosynthetic induction. *Oecologia*, **94**, 395–402.

Tipple, B.J. and Pagani, M. (2007). The early origins of terrestrial C_4 photosynthesis. *Annual Review of Earth Planetary Science*, **35**, 435–461.

Tissue, D.T., Griffin, K.L., Turnbull, M.H., *et al.* (2005). Stomatal and non-stomatal limitations to photosynthesis in four tree species in a temperate rainforest dominated by *Dacrydium cupressinum* in New Zealand. *Tree Physiology*, **25**, 447–456.

Titel, C., Woehlecke, H., Afifi, I., *et al.* (1997). Dynamics of limiting cell wall porosity in plant suspension cultures. *Planta*, **203**, 320–326.

Tjoelker, M.G., Oleksyn, J. and Reich, P.B. (1999). Acclimation of respiration to temperature and CO_2 in seedlings of boreal tree species in relation to plant size and relative growth rate. *Global Change Biology*, **5**, 679–691.

Tjoelker, M.G., Oleksyn, J. and Reich, P.B. (2001). Modelling respiration of vegetation: evidence for a general temperature dependent Q_{10}. *Global Change Biology*, **7**, 223–230.

Tocquin, P. and Périlleux, C. (2004). Design of a versatile device for measuring whole plant gas exchanges in *Arabidopsis thaliana*. *New Phytologist*, **162**, 223–229.

Toft, N. and Pearcy, R.W. (1982). Gas exchange characteristics and temperature relations of two desert annuals: a comparison of a winter-active and a summer-active species. *Oecologia*, **55**, 170–7.

Tognetti, R., Johnson, J.D. and Michelozzi, M. (1995). The response of european beech (Fagus sylvatica L) seedlings from 2 italian populations to drought and recovery. *Trees, Structure and Function*, **9**, 348–354.

Tognetti, R., Longobucco, A., Miglietta, F., *et al.* (1999). Water relations, stomatal response and transpiration of *Quercus pubescens* trees during summer in a Mediterranean carbon dioxide spring. *Tree Physiology*, **25**, 261–270.

Tognetti, R., Minnocci, A., Peñuelas, J., *et al.* (2000). Comparative field water relations of three Mediterranean shrub species co-occurring at a natural CO_2 vent. *Journal of Experimental Botany*, **51**, 1131–1146.

Tollenaar, M. and Aguilera, A. (1992). Radiation use efficiency of an old and a new maize hybrid. *Agronomy Journal*, **84**, 536–41.

Tollenaar, M.,. McCullough, D.E. and Dwyer, L.M. (1994). Physiological basis of the genetic improvement in corn. In: *Genetic Improvement of Field Crops* (ed. Slafer, G.A.), Marcel Dekker, New York, USA, pp. 183–236.

Tomescu, A.M.F. (2008). Megaphylls, microphylls and the evolution of leaf development. *Trends in Plant Science*, **14**, 5–12.

Tominaga, M., Kinoshita, T. and Shimazaki, K. (2001). Guard cell chloroplasts provide ATP required for H^+ pumping in the plasma membrane and stomatal opening. *Plant and Cell Physiology*, **42**, 795–802.

Tomlinson, J.A. and Webb, M.J.W. (1978). Ultrastructural changes in chloroplasts of lettuce infected with beet western yellows virus. *Physiological Plant Pathology*, **12**, 13–18.

Tong, H. and Hipps, L.E. (1996). The effect of turbulence on the light environment of alfalfa. *Agricultural and Forest Meteorology*, **80**, 249–261.

Törnroth-Horsefield, S., Wang, Y., Hedfalk, K., *et al.* (2006). Structural mechanism of plant aquaporin gating. *Nature*, **439**, 688–694.

Török, Z., Tsvetkova, N. M., Balogh, G., *et al.* (2003). Heat shock protein coinducers with no effect on protein denaturation specifically modulate the membrane lipid phase. *Proceedings of the National Academy of Sciences of the United States of America*, **100**, 3131–3136.

Tortell, P.D. (2000). Evolutionary and ecological perspectives on carbon acquisition in phytoplankton. *Limnology and Oceanography*, **45**, 744–750.

Tóth, S.T., Schansker, G., Garab, G., *et al.* (2007). Photosynthetic electron transport activity in heat-treated barley leaves: the role of internal alternative electron donors to photosystem II. *Biochimica et Biophysica Acta*, **1767**, 295–305.

Tóth, S.Z., Puthur, J.T., Nagy, V., *et al.* (2009). Experimental evidence for ascorbate-dependent electron transport in leaves with inactive oxygen-evolving complexes. *Plant Physiology*, **149**, 1568–1578.

Tournaire-Roux, C., Sutka, M., Javot, H., *et al.* (2003). Cytosolic pH regulates root water transport during anoxic stress through gating of aquaporins. *Nature*, **425**, 393–397.

Tournebize, R. and Sinoquet, H. (1995). Light interception and partitioning in a shrub/grass mixture. *Agricultural and Forest Meteorology*, **72**, 277–294.

Tovar-Mendez, A., Miernyk, J.A. and Randall, D.D. (2003). Regulation of pyruvate dehydrogenase complex activity in plant cells. *European Journal of Biochemistry*, **270**, 1043–1049.

Townsend, A.R., Asner, G.P. and White, J.W.C. (2002). Land use effects of atmosphere ^{13}C imply a sizable terrestrial CO_2 sink in tropical latitudes. *Geophysical Research Letters*, **29**, 1426.

Tranquilini, W. (1957). Standortsklima, Wasserbilanz un CO_2-gaswechsel junger Zirben (*Pinus cembra* L.) an de alpinen Waldgrenze. *Planta*, **49**, 612–661.

Trejo, C.L. and Davies, W.J. (1991). Drought-induced closure of *Phaseolus vulgaris* L. stomata precedes leaf water deficit and any increase in xylem ABA concentration. *Journal of Experimental Botany*, **42**, 1507–1515

Trenberth, K.E. and Jones, P.D. (2007). Observations: Surface and atmospheric climate change. In: *Climate Change 2007: The Physical Science Basis. Contribution of Working Group I to the Fourth Assessment Report of the Intergovernmental Panel on Climate Change* (eds Solomon, S., Qin, D., Manning, M., Chen, Z., Marquis, M., Averyt, K.B., Tignor, M. and Miller, H.L.), Cambridge University Press, Cambridge, UK, pp. 235–336.

Triantaphylidès, C., Krischke, M., Hoeberichts, F.A., *et al.* (2008). Singlet oxygen is the major reactive oxygen species involved in photooxidative damage to plants. *Plant Physiology*, **148**, 960–968.

Triggs, J., Kimball, B., Pinter, P., *et al.* (2004). Free-air CO_2 enrichment effects on the energy balance and evapotranspiration of sorghum. *Agricultural and Forest Meteorology*, **124**, 63–79.

Troughton, J.H. and Slatyer, R.O. (1969). Plant water status, leaf temperature, and the calculated mesophyll resistance to carbon dioxide of cotton leaves. *Australian Journal of Biological Sciences*, **22**, 815–827.

Truman, W., Torres de Zabala. M. and Grant, M. (2006). Type III effectors orchestrate a complex interplay between transcriptional networks to modify basal defence responses during pathogenesis and resistance. *The Plant Journal*, **46**, 14–33.

Trumble, J.T., Kolodny Hirsch, D.M. and Ting, I.P. (1993). Plant compensation for arthropod herbivory. *Annual Reviews of Entomology*, **38**, 93–119.

Tseng, M.J., Liu, C.W. and Yiu, J.C. (2007). Enhanced tolerance to sulfur dioxide and salt stress of transgenic Chinese cabbage plants expressing both superoxide dismutase and catalase in chloroplasts. *Plant Physiology and Biochemistry*, **45**, 822–833.

Tsialtas, J.T., Handley, L.L., Kassioumi, M.T., *et al.* (2001). Interspecific variation in potential water-use efficiency and its relation to plant species abundance in a water-limited grassland. *Functional Ecology*, **15**, 605–614.

Tsuchihira, A., Hanba, Y.T., Kato, N., *et al.* (2010). Effect of overexpression of radish plasma membrane aquaporins on water-use efficiency, photosynthesis and growth of *Eucalyptus* trees. *Tree Physiology*, **30**, 417–430.

Tsukaya, H., Okada, H. and Mohamed, M. (2004). A novel feature of structural variegation in leaves of the tropical plant *Schismatoglottis calyptrate*. *Journal of Plant Research*, **117**, 477–480.

Tsuyama, M., Shibata, M. and Kobayashi, Y. (2003). Leaf factors affecting the relationship between chlorophyll fluorescence and the rate of photosynthetic electron transport as determined from CO_2 uptake. *Journal of Plant Physiology*, **160**, 1131–1139.

Tsvetkova, N.M., Horváth, I., Török, Z., *et al.* (2002). Small heat-shock proteins regulate membrane lipid polymorphism. *Proceedings of the National Academy of Sciences of the United States of America*, **99**, 13504–13509.

Tu, J.C. and Ford, R.E. (1968). Effect of maize dwarf mosaic virus infection on respiration and photosynthesis of corn. *Phytopathology*, **58**, 282–284.

Tucker, C.J., Townshend, J.R.G. and Goff, T.E. (1985). African landcover classification using satellite data. *Science*, **227**, 369–374.

Turgeon, R. and Wolf, S. (2009). Phloem transport: Cellular pathways and molecular trafficking. *Annual Review of Plant Biology*, **60**, 207–221.

Turgut, R. and Kadioglu, A., (1998). The effect of drought, temperature and irradiation on leaf rolling in *Ctenanthe setosa*. *Biologia Plantarum*, **41**, 629–633.

Turnbull, T.L., Adams, M.A. and Warren, C.R. (2007a). Increased photosynthesis following partial defoliation of field-grown *Eucalyptus globulus* is not caused by increased leaf nitrogen. *Tree Physiology*, **27**, 1481–1492.

Turnbull, T.L., Warren, C.R. and Adams, M.A. (2007b). Novel mannose-sequestration technique reveals variation in subcellular orthophosphate pools do not explain the effects of phosphorus nutrition on photosynthesis in *Eucalyptus globulus* seedlings. *New Phytologist*, **176**, 849–861.

Turner, I.M. (1994). Sclerophylly: primarily protective? *Functional Ecology*, **8**, 669–675.

Turner, N.C. (2004). Agronomic options for improving rainfall-use efficiency of crops in dryland farming systems. *Journal of Experimental Botany*, **55**, 2413–2425.

Tyree, M.T. and Sperry, J.S. (1989). Vulnerability of xylem to cavitation and embolism. *Annual Review of Plant Physiology and Plant Molecular Biology*, **40**, 19–38

Tyree, M.T. and Wilmot, T.R. (1990). Errors in the calculation of evaporation and leaf conductance in steady state porometry: the importance of accurate measurement of leaf temperature. *Canadian Journal of Forestry Research*, **20**, 1031–1035.

Tyree, M.T. and Zimmermann, M.H. (2002). *Xylem Structure and Ascent of Sap*. Springer-Verlag, Berlin, Germany, p. 283.

Tyystjärvi, E. and Vass, I. (2004). Light emission as a probe of charge separation and recombination in the photosynthetic apparatus: relation of prompt fluorescence to delayed light emission and thermoluminescence. In: *Chlorophyll a Fluorescence: A Signature of Photosynthesis*. Advances in photosynthesis and respiration, vol 19 (eds. Papageorgiou, G.C. and Govindjee, J.), Kluwer Academic Publishers, Dordrecht, Nederlands, pp. 363–388.

Tyystjärvi, E., Koski, A., Keranen, M., *et al.* (1999). The Kautsky curve is a built-in barcode. *Biophysical Journal*, **77**, 1159–1167.

Tzvetkova-Chevolleau, T., Franck, T., Alawady, A.E., *et al.* (2007). The light stress-induced protein ELIP2 is a regulator of chlorophyll synthesis in *Arabidopsis thaliana*. *The Plant Journal*, **50**, 795–809.

U.S. DOE. (2006). Breaking the biological barriers to cellulosic ethanol: A joint research agenda, DOE/SC/EE-0095, US Department of Energy Office of Science and Office of Energy Efficiency and Renewable Energy, http://genomicsgtl.energy.gov/biofuels/.

Ubierna, N. and Marshall, J.D. (2011). Estimation of canopy average mesophyll conductance using $\delta^{13}C$ of phloem contents. *Plant, Cell and Environment*, **34**, 1521–1535.

Uedan, K. and Sugiyama, T. (1976). Purification and characterization of phosphoenolpyruvate carboxylase from maize leaves. *Plant Physiology*, **57**, 906–910.

Uehlein, N., Lovisolo, C., Siefritz, F., *et al.* (2003). The tobacco aquaporin NtAQP1 is a membrane CO_2 transporter with physiological functions. *Nature*, **425**, 734–737.

Uehlein, N., Otto, B., Hanson, D.T., *et al.* (2008). Function of *Nicotiana tabacum* aquaporins as chloroplast gas pores challenges the concept of membrane CO_2 permeability. *Plant Cell*, **20**, 648–657.

Uemura, A., Ishida, A., Nakano, T., *et al.* (2000). Acclimation of leaf characteristics of *Fagus* species to previous-year and current-year solar irradiances. *Tree Physiology*, **20**, 945–951.

Uemura, K., Anwaruzzaman, M.S. and Yokota, A. (1997). Ribulose-1,5-bisphosphate carboxylase/oxygenase from thermophilic red algae with a strong specificity for CO_2 fixation. *Biochemical and Biophysical Research Communications*, **233**, 568–571.

Uemura, K., Suzuki, Y., Shikani, T., *et al.* (1996). A rapid and sensitive method for determination of relative specificity of RuBisCO from various species by anion-exchange chromatography. *Plant and Cell Physiology*, **37**, 325–331.

Uhl, C., Clark, K., Dezzeo, N., *et al.* (1988).Vegetation dynamics in amazonian treefall gaps. *Ecology*, **69**, 751–763.

UNIS, (2000). Secretary General address to Developing Countries "South Summit", UN Information Service Press Release, 13 April 2000. (See www.unis/unvienna.org/unis/pressrels/2000/sg2543.html.)

Upchurch, Jr., G.R. (1984). Cuticular anatomy of angiosperm leaves from the Lower Cretaceous Potomac group. I. Zone I leaves. *American Journal of Botany*, **71**, 192–202.

Urban, L., Jegouzo, L., Damour, G., *et al.* (2008). Interpreting the decrease in leaf photosynthesis during flowering in mango. *Tree Physiology*, **28**, 1025–1036.

Urbanczyk-Wochniak, E., Baxter, C., Kolbe, A., *et al.* (2005). Profiling of diurnal patterns of metabolite and transcript abundance in potato (*Solanum tuberosum*) leaves. *Planta*, **221**, 891–903.

Urey, H.C. (1947). The thermodynamic properties of isotopic substances. *Journal of Chemical Society*, **1947**, 562–581.

Uribe, E.G. and Stark, B. (1982). Inhibition of photosynthetic energy conversion by cupric ion. Evidence for Cu2+-coupling factor 1 interaction. *Plant Physiology*, **69**, 1040–1045.

Ustin, S.L., Roberts, D.R., Gamon, J.A., *et al.* (2004). Using imaging spectroscopy to study ecosystem processes and properties. *BioScience*, **54**, 523–534.

Usuda, H. (2004). Evaluation of the effect of photosynthesis on biomass production with simultaneous analysis of growth and continuous monitoring of CO_2 exchange in the whole plants of radish cv Kosena under ambient and elevated CO_2. *Plant Production Science*, **7**, 386–396.

Vahisalu, T., Kollist, H., Wang, Y.F., *et al.* (2008). SLAC1 is required for plant guard cell S- type anion channel function in stomatal signalling. *Nature*, **452**, 487–491.

Vaillant, N., Monnet, F., Hitmi, A., *et al.* (2005). Comparative study of responses in four *Datura* species to a zinc stress. *Chemosphere*, **59**, 1005–1013.

Val, J., Sanz, M., Montañés, L., *et al.* (1995). Application of chlorophyll fluorescence to study iron and manganese deficiencies in peach tree. *Acta Horticulturae*, **383**, 201–209.

Valcke, R. (2003). Fluorescence imaging: the stethoscope of the plant physiologist. *Advances in Plant Physiology*, **6**, 445–462.

Valente, M.A.S., Faria, J.A.Q.A., Soares-Ramos, J.R.L., *et al.* (2009). The ER luminal binding protein (BiP) mediates an increase in drought tolerance in soybean and delays drought-induced leaf senescence in soybean and tobacco. *Journal of Experimental Botany*, **60**, 533–546.

Valentini, R. (ed.) (2003). *Fluxes of Carbon, Water and Energy of European Forests, Ecological Studies 163*, Springer-Verlag, Berlin Heidelberg, Germany, pp. 1–450.

Valentini, R., Cecchi, G., Mazzinghi, P., *et al.* (1994). Remote sensing of chlorophyll *a* fluorescence of vegetation canopies: 2. Physiological significance of fluorescence signal in response to environmental stresses. *Remote Sensing of the Environment*, **47**, 29–35.

Valentini, R., Deangelis, P., Matteucci, G., *et al.* (1996). Seasonal net carbon dioxide exchange of a beech forest with the atmosphere. *Global Change Biology*, **2**, 199–207.

Valentini, R., Epron, D., De Angelis, P., *et al.* (1995a). *In situ* estimation of net CO_2 assimilation, photosynthetic electron flow and photorespiration in Turkey oak (*Quercus cerris* L.) leaves: diurnal cycles under different levels of water supply. *Plant Cell Environment*, **18**, 631–640.

Valentini, R., Gamon, J.A. and Field, C.B. (1995b). Ecosystem gas exchange in a California grassland: seasonal patterns and implications for scaling. *Ecology*, **76**, 1940–1952.

Valentini, R., Matteucci, G., Dolman, A.J., *et al.* (2000). Respiration as the main determinant of carbon balance in European forests. *Nature*, **404**, 861–865.

Valentini, R., Scarascia Mugnozza, G.E., De Angelis, P., *et al.* (1991). An experimental test of the eddy correlation technique over a Mediterranean macchia canopy. *Plant Cell Environment*, **14**, 987–994.

Valiela, I., Bowen, J.L. and York, J.K. (2001) Mangrove forests: one of the world's threatened major tropical environments. *BioScience*, **51**, 807–815.

Valjakka, M., Luomala, E.M., Kangasjärvi, J., *et al.* (1999). Expression of photosynthesis- and senescence-related genes during leaf development and senescence in silver birch (*Betula pendula*) seedlings. *Physiologia Plantarum*, **106**, 302–310.

Valladares, F. (2003). Light heterogeneity and plants: from ecophysiology to species coexistence and biodiversity. In: *Progress in Botany*, vol 64 (eds Esser, K., Lüttge, U., Beyschlag, W. and Hellwig, F.), Springer Verlag, Heidelberg, Germany, pp. 439–471.

Valladares, F. and Niinemets, Ü. (2007). The architecture of plant crowns: from design rules to light capture and performance. In: *Handbook of Functional Plant Ecology* (eds Pugnaire, F. I. and Valladares, F.), CRC Press, Boca Raton, Florida, USA, pp. 101–149.

Valladares, F. and Niinemets, Ü. (2008). Shade tolerance, a key plant feature of complex nature and consequences. *Annual Review of Ecology, Evolution and Systematics* **39**, 237–257.

Valladares, F. and Pearcy, R.W. (1997). Interactions between water stress, sun-shade acclimation, heat tolerance and photoinhibition in the sclerophyll *Heteromeles arbutifolia*. *Plant, Cell and Environment*, **20**, 25–36.

Valladares, F. and Pearcy, R.W. (1998). The functional ecology of shoot architecture in sun and shade plants of *Heteromeles arbutifolia* M. Roem., a Californian chaparral shrub. *Oecologia*, **114**, 1–10.

Valladares, F. and Pearcy, R.W. (1999). The geometry of light interception by shoots of *Heteromeles arbutifolia*: morphological and physiological consequences for individual leaves. *Oecologia*, **121**, 171–182.

Valladares, F. and Pearcy, R.W. (2000). The role of crown architecture for light harvesting and carbon gain in extreme light environments assessed with a realistic 3-D model. *Anales Jardin Botanico de Madrid*, **58**, 3–16.

Valladares, F. and Pugnaire, F.I. (1999). Tradeoffs between irradiance capture and avoidance in semi-arid environments assessed with a crown architecture model. *Annals of Botany*, **83**, 459–469.

Valladares, F., Allen, M.T. and Pearcy, R.W. (1997). Photosynthetic response to dynamic light under field conditions in six tropical rainforest shrubs occurring along a light gradient. *Oecologia*, **111**, 505–514.

Valladares, F., Dobarro, I., Sánchez-Gómez, D., *et al.* (2005). Photoinhibition and drought in Mediterranean woody saplings: scaling effects and interactions in sun and shade phenotypes. *Journal of Experimental Botany*, **56**, 483–494.

Valladares, F., Gianoli, E. and Gómez, J.M. (2007). Ecological limits to plant phenotypic plasticity. Tansley review. *New Phytologist*, **176**, 749–763.

Valladares, F., Martinez-Ferri, E., Balaguer, L., *et al.* (2000a). Low leaf-level response to light and nutrients in Mediterranean evergreen oaks: a conservative resource-use strategy? *New Phytologist*, **148**, 79–91.

Valladares, F., Sancho, L.G. and Ascaso, C. (1996). Functional analysis of intrathalline and intracellular chlorophyll concentrations in the lichen family *Umbilicariaceae*. *Annals of Botany*, **78**, 471–477.

Valladares, F., Skillman, J. and Pearcy, R.W. (2002). Convergence in light capture efficiencies among tropical forest understory plants with contrasting crown architectures: a case of morphological compensation. *American Journal of Botany*, **89**, 1275–1284.

Valladares, F., Wright, S.J., Lasso, E., *et al.* (2000b). Plastic phenotypic response to light of 16 congeneric shrubs from a Panamanian rainforest. *Ecology*, **81**, 1925–1936.

van Assche, F.V. and Clijsters, H. (1986a). Inhibition of photosynthesis by treatment of *Phaseolus vulgaris* with toxic concentration of zinc: effects on electron transport and photophosphorylation. *Physiologia Plantarum*, **66**, 717–721.

van Assche, F.V. and Clijsters, H. (1986b). Inhibition of photosynthesis by treatment of *Phaseolus vulgaris* with toxic concentration of zinc: effect on ribulose-1,5-bisphosphate carboxylase/oxygenase. *Journal of Plant Physiology*, **125**, 355–360.

van Assche, F.V. and Clijsters, H. (1990). Effects of metals on enzyme activity in plants. *Plant, Cell and Environment*, **13**, 195–206.

van Camp, W., Capiau, K., Van Montagu, M., *et al.* (1996). Enhancement of oxidative stress tolerance in transgenic tobacco plants overproducing Fe-superoxide dismutase in chloroplasts. *Plant Physiology*, **112**, 1703–1714.

van der Tol, C., Verhoef, W., Timmermans, J., *et al.* (2009). An integrated model of soil-canopy spectral radiances, photosynthesis, fluorescence, temperature and energy balance. Biogeosciences, **6**, 3109–3129.

van der Velde, M., Green, S.R., Vanclooster, M., *et al.* (2006). Transpiration of squash under a tropical maritime climate. *Plant and Soil*, **280**, 323–337.

van Doorn, W.G. (2008). Is the onset of senescence in leaf cells of intact plants due to low or high sugar levels? *Journal of Experimental Botany*, **59**, 1963–1972.

van Gardingen, P., Grace, J. and Jeffree, C.E. (1991). Abrasive damage by wind to the needles surface of *Pinus sylvestris* L. and *Picea stichensis* (Bong.). *Plant Cell Environment*, **14**, 185–193.

van Gorsel, E., Delpierre, N., Leuning, R., *et al.* (2009). Estimating nocturnal ecosystem respiration from the vertical turbulent flux and change in storage of CO_2. *Agricultural and Forest Meteorology*, **149**, 1919–1930.

van Gorsel, E., Leuning, R., Cleugh, H.A., *et al.* (2007). Nocturnal carbon efflux: Reconciliation of eddy covariance and chamber measurements using an alternative to the u*-threshold filtering. technique. *Tellus*, **59B**, 397–403.

van Gorsel, E., Leuning, R., Cleugh, H.A., *et al.* (2008). Application of an alternative method to derive reliable estimates of nighttime respiration from eddy covariance measurements in moderately complex topography. *Agricultural and Forest Meteorology*, **148**, 1174–1180.

van Iersel, M.W. (2003). Carbon use efficiency depends on growth respiration, maintenance respiration, and relative growth rate. A case study with lettuce. *Plant Cell Environment*, **26**, 1441–1446.

van Iersel, M.W. and Lindstrom, O.M. (1999). Temperature response of whole plant CO_2 exchange rates of three magnolia (*Magnolia grandiflora* L.) cultivars. *Journal of the American Society of Horticultural Science*, **124**, 277–282.

van Kooten, O., Meurs, C. and van Loon, L.C. (1990). Photosynthetic electron transport in tobacco leaves infected with *Tobacco mosaic virus*. *Physiologia Plantarum*, **80**, 446–452.

van Oosten, J.J.M., Gerbaud, A., Huijser, C., *et al.* (1997). An *Arabidopsis* mutant showing reduced feedback inhibition of photosynthesis. *Plant Journal*, **12**, 1011–1020.

van Volkenburgh, E. (1999). Leaf expansion – an integrating plant behaviour. *Plant Cell Environment*, **22**, 1463–1473.

van Wijk, M.T., Clemmensen, K.E., Shaver, G.R., *et al.* (2004). Long-term ecosystem level experiments at Toolik Lake, Alaska, and at Abisko, Northern Sweden: Generalizations and differences in ecosystem and plant type responses to global change. *Global Change Biology*, **10**, 105–123.

van Wijk, M.T., Dekker, S.C., Bouten, W., *et al.* (2000). Modeling daily gas exchange of Douglas-fir forest: comparison of three stomatal conductance models with and without a soil water stress function. *Tree Physiology*, **20**, 115–122.

Vandenbroucke, K., Robbens, S., Vandepoele, K., *et al.* (2007). Hydrogen peroxide-induced gene expression across kingdoms: a comparative analysis. *Molecular Biology and Evolution*, **25**, 507–516.

Vandervoet H. and Mohren G.M.J. (1994). An uncertainty analysis of the process-based growth-model FORGRO. *Forest Ecology and Management*, **69**, 157–166.

Vanninen, P., Ylitalo, H., Sievänen, R., *et al.* (1996). Effects of age and site quality on the distribution of biomass in Scots pine (*Pinus sylvestris* L.). *Trees*, **10**, 231–238.

Vapaavuori, E.M., Vuorinen, A.J., Aphalo, P.J., *et al.* (1995). Relationship between net photosynthesis and nitrogen in Scots pine: seasonal variation in seedlings and shoots. *Plant and Soil*, **168–169**, 263–270.

Vareschi, V. (1980). *Vegetationsökologie der Tropen*. Stuttgart, Eugen Ulmer.

Vass, I. and Govindjee, J. (1996). Thermoluminescence of the photosynthetic apparatus. *Photosynthesis Research*, **48**, 117–126.

Vass, I., Horvath, G., Herczeg, T. and Demeter, S. (1981). Photosynthetic energy conservation investigated by thermoluminescence. activation energies and half-lives of thermoluminescence bands of chloroplasts determined by mathematical resolution of glow curve. *Biochimica et Biophysica Acta*, **634**, 140–152.

Vaughn, K.C., Ligrone, R., Owen, H.A., *et al.* (1992). The anthocerote chloroplast :a review. *The New Phytologist*, **120**, 169–190.

Vavilin, D.V. and Ducruet, J.M. (1998). The origin of 120–130°C thermoluminescence bands in chlorophyll-containing material. *Photochemistry Photobiology*, **68**, 191–198.

Vavilin, D.V., Matorin, D.N., Kafarov, R.S., *et al.* (1991). High-temperature thermoluminescence of chlorophyll in lipid peroxidation. *Biologischke Membranyii*, **8**, 89–98.

Vavrus, S.J., Walsh, J.E., Chapman, W.L., *et al.* (2006). The behavior of extreme cold air outbreaks under greenhouse warming. *International Journal of Climatology*, **26**, 1133–1147.

Vedrova, E.F., Pleshikov, F.I. and Kaplunov, V.Y. (2006). Net ecosystem production of boreal larch ecosystems on the Yenisei transect. *Mitigation And Adaptation Strategies For Global Change*, **11**, 173–190

Veenendaal, E.M., Kolle, O. and Lloyd, J. (2004) Seasonal variation in energy fluxes and carbon dioxide exchange for a broad-leaved semi-arid savanna (Mopane woodland) in Southern Africa. *Global Change Biology*, **10**, 318–328.

Veenendaal, E.M., Shushu, D.D. and Scurlock, J.M.O. (1993). Responses to shading of seedlings of savanna grasses (with different C_4 photosynthetic pathways) in Botswana. *Journal of Tropical Ecology*, **9**, 213–229.

Veeranjaneyulu, K., Charland, M., Charlebois, D.C.N., *et al.* (1991). Photosynthetic energy storage of photosystems I and II in the spectral range of photosynthetically active radiation in intact sugar maple leaves. *Photosynthesis Research*, **30**, 131–138.

Velikova, V., Tsonev, T., Barta, C., *et al.* (2009). BVOC emissions, photosynthetic characteristics and changes in chloroplast ultrastructure of *Platanus orientalis* L. exposed to elevated CO_2 and high temperature. *Environmental Pollution*, **157**, 2629–2637.

Vendramini, F., Díaz, S., Gurvich, D.E., *et al.* (2001). Leaf traits as indicators of resource-use strategy in floras with succulent species. *New Phytologist*, **154**, 147–157.

Vener, A.V., van Kan, P.J., Rich, P.R., *et al.* (1997). Plastoquinol at the quinol oxidation site of reduced cytochrome bf mediates signal transduction between light and protein phosphorylation: Thylakoid protein kinase deactivation by a single-turnover flash. *Proceedings of the National Academy of Sciences (USA)*, **94**, 1585–1590.

Venmos, S.N. and Goldwin, G.K. (1993). Stomatal and chlorophyll distribution of Cox's orange pippin apple flowers relative to other cluster parts. *Annals of Botany*, **71**, 245–250.

Verboven, P., Kerckhofs, G., Mebatsion, H.K., *et al.* (2008). Three-dimensional gas exchange pathways in pome fruit characterized by synchrotron X-ray computed tomography. *Plant Physiology*, **147**, 518–527.

Verhoef, A. (1997). The effect of temperature differences between porometer head and leaf surface on stomatal conductance measurements. *Plant, Cell and Environment*, **20**, 641–646.

Verhoeven, A.S., Adams, W.W., III and Demmig-Adams, B. (1999). The xanthophyll cycle and acclimation of *Pinus ponderosa* and *Malva neglecta* to winter stress. *Oecologia*, **118**, 277–287.

Verhoeven, A.S., Demmig-Adams, B. and Adams, W.W, III. (1997). Enhanced employment of the xanthophyll cycle and thermal energy dissipation in spinach exposed to high light and N stress. *Plant Physiology*, **113**, 817–824.

Verma, S.B., Baldocchi, D.D., Anderson, D.E., *et al.* (1986). Eddy fluxes of CO_2, water vapor and sensible heat over a deciduous forest. *Bounday Layer Meteorology*, **36**, 71–91.

Verma, S.B., Dobermann, A., Cassman, K.G., *et al.* (2005). Annual carbon dioxide exchange in irrigated and rainfed maize-based agroecosystems. *Agricultural and Forest Meteorology*, **131**, 77–96.

Verma, S.B., Sellers, P.J., Walthall, C.L., *et al.* (1993). Photosynthesis and stomatal conductance related to reflectance on the canopy scale. *Remote Sensing of the Environment*, **44**, 103–116.

Vernon, L.P. (1960). Spectrophotometric determination of chlorophylls and pheophytins in plant extracts. *Analytical Chemistry*, **32**, 1144–1150.

Verslues, P.E. and Bray, E.A. (2006). Role of abscisic acid (ABA) and *Arabidopsis thaliana* ABA-insensitive loci in low water potential-induced ABA and proline accumulation. *Journal of Experimental Botany*, **57**, 201–212.

Vervuren, P.J.A., Blom, C.W.P.M. and De Kroon, H. (2003). Extreme flooding events on the Rhine and the survival and distribution of riparian plant species. *Journal of Ecology*, **91**, 135–146.

Vicentini, A., Barber, J.C., Aloscioni, S.S., *et al.* (2008). The age of the grasses and clusters of origins of C_4 photosynthesis. *Global Change Biology*, **14**, 2963–2977.

Vieira Dos Santos, C. and Rey, P., (2006). Plant thioredoxins are key actors in the oxidative stress response. *Trends in Plant Sciences*, **11**, 329–334.

Vieira, F.C.B., He, Z.L., Wilson, P.C., *et al.* (2008). Response of representative cover crops to aluminum toxicity, phosphorus deprivation, and organic amendment. *Australian Journal of Agricultural Research*, **59**, 52–61.

Vierling, E. (1991). The roles of heat shock proteins in plants. *Annual Review of Plant Physiology and Plant Molecular Biology*, **42**, 579–620.

Vígh, L., Combos, Z., Horváth, I., *et al.* (1989). Saturation of membrane lipids by hydrogenation induces thermal stability in chloroplast inhibiting the heat-dependent stimulation of photo-system I-mediated electron transport. *Biochimica et Biophysica Acta (BBA) – Bioenergetics*, **979**, 361–364.

Villar, R. and Merino, J. (2001). Comparison of leaf construction costs in woody species with differing leaf life-spans in contrasting ecosystems. *New Phytologist*, **151**, 213–226.

Vincent, D., Ergül, A., Bohlman, M.C., *et al.* (2007). Proteomic analysis reveals differences between *Vitis vinifera* L. cv. Chardonnay and cv. Cabernet Sauvignon and their responses to water deficit and salinity. *Journal of Experimental Botany*, **58**, 1873–1892.

Vingarzan, R. (2004) A review of surface ozone background levels and trends. *Atmospheric Environment*, **38**, 3431–3442.

Vitousek, P.M. and Denslow, J.S. (1986). Nitrogen and Phosphorus Availability in Treefall Gaps of a Lowland Tropical Rain-Forest. *Journal of Ecology*, **74**, 1167–1178.

Voesenek, L.A.C.J., Colmer, T.D., Pierik, R., *et al.* (2006). How plants cope with complete submergence. *New Phytologist*, **170**, 213–226.

Voesenek, L.A.C.J., Runders, J.H.G.M., Peeters, A.J.M., *et al.* (2004). Plant hormones regulate fast shoot elongation under water: from genes to communities. *Ecology*, **85**, 16–27.

Vogel, J.C. (1980). Fractionation of the carbon isotopes during photosynthesis. In: *Sitzungsberichte der Heidelberger Akademie der wissenschaften, matematisch-naturwissenschaftliche Klasse Jahrgang 1980*, 3, Abhandlung, Springer-Verlag, Heidelberg; Berlin; New York, USA, pp. 111–135.

Vogel, J.C. and Lerman, J.C. (1969). Groningen radiocarbon dates. VIII. *Radiocarbon*, **11**, 351–390.

Vogel, J.C., Fuls, A. and Ellis, R.P. (1978). Geographical distribution of Kranz grasses in South Africa. *South African Journal of Science*, **74**, 209–215.

Vogelmann, T.C. (1989). Penetration of light into plants. *Photochemistry and Photobiology*, **50**, 895–902.

Vogelmann, T.C. (1993). Plant tissue optics. *Annual Review of Plant Physiology and Plant Molecular Biology*, **44**, 231–251.

Vogelmann, T.C. and Evans, J.R. (2002). Profiles of light absorption and chlorophyll within spinach leaves from chlorophyll fluorescence. *Plant, Cell and Environment*, **25**, 1313–1323.

Vogelmann, T.C. and Han, T. (2000). Measurement of gradients of absorbed light in spinach leaves from chlorophyll fluorescence profiles. *Plant Cell Environment*, **23**, 1303–1311.

Vogelmann, T.C. and Martin, G. (1993). The functional significance of palisade tissue: penetration of directional versus diffuse light. *Plant Cell Environment*, **16**, 65–72.

Vogelmann, T.C., Bornman, J.F. and Josserand, S. (1989). Photosynthetic light gradients and spectral regime within leaves of *Medicago sativa*. *Philosophical Transactions of the Royal Society of London B*, **323**, 411–421.

Vogelmann, T.C., Nishio, J.N. and Smith, W.K. (1996). Leaves and light capture: light propagation and gradients of carbon fixation within leaves. *Trends in Plant Science*, **1**, 65–71.

Volz, A. and Kley, D. (1988). Evaluation of the Montsouris series of ozone measurements made in the nineteenth century. *Nature*, **332**, 240–242.

von Caemmerer, S. (2000). *Biochemical Models of Leaf Photosynthesis*, Techniques in Plant Sciences No. 2. CSIRO Publishing, Collingwood, Victoria, Australia.

von Caemmerer, S. and Evans, J.R. (1991). Determination of the average partial pressure of CO_2 in chloroplasts from leaves of several C_3 plants. *Australian Journal Plant Physiology*, **18**, 287–305.

von Caemmerer, S. and Furbank, R.T. (1999). The modelling of C_4 photosynthesis. In: *The Biology of C_4 Photosynthesis* (ed. Sage, R.M.), Academic Press, New York, USA, pp. 173–211.

von Caemmerer, S. and Furbank, R.T. (2003). The C_4 pathway: an efficient CO_2 pump. *Photosynthesis Research*, 77, 191–207.

von Caemmerer, S. and Quick, P. (2000). Rubisco: physiology in vivo. In: *Advances in Photosynthesis: Physiology and Metabolism* (eds Leegood, R.C., Sharkey, T.D. and von Caemmerer, S.), Kluwer Academic Publishers, Dodrecht, Boston, London, pp. 86–107.

von Caemmerer, S., Evans, J.R., Hudson, G.S., *et al.* (1994). The kinetics of ribulose-1,5-bisphosphate carboxylase/oxygenase in-vivo inferred from measurements of photosynthesis in leaves of transgenic tobacco. *Planta*, **195**, 88–97.

von Caemmerer, S., Lawson, T., Oxborough, K., *et al.* (2004). Stomatal conductance does not correlate with photosynthetic capacity in transgenic tobacco with reduced amounts of Rubisco. *Journal of Experimental Botany*, **55**, 1157–1166.

von Caemmerer, S.V. and Farquhar, G.D. (1981). Some relationships between the biochemistry of photosynthesis and the gas-exchange of leaves. *Planta*, **153**, 376–387.

von Willert, D.J., Armbruster, N. Drees, T., *et al.* (2005). *Welwitschia mirabilis*: CAM or not CAM – what is the answer? *Functional Plant Biology*, **32**, 389–395.

Voronin, P.Y., Ivanova, L.A., Ronzhina, D.A., *et al.* (2003). Structural and functional changes in the leaves of plants from steppe communities as affected by aridization of the eurasian climate. *Russian Journal of Plant Physiology*, **50**, 604–611.

Vountas, M., Rozanov, V.V. and Burrows, J.P. (1998). Filling in of Fraunhofer and gas-asorption lines in sky spectra as caused rotational Raman scattering. *Journal of Quantitative Spectroscopy Radiative Transfer*, **60**, 943–961.

Voznesenskaya, E.V., Chuong, S.D.X., Kiirats, O., *et al.* (2005a). Evidence that C_4 species in genus *Stipagrostis*, family *Poaceae*, are NADP-malic enzyme subtype with nonclassical type of Kranz anatomy (Stipagrostoid). *Plant Science*, **168** (3), 731–739.

Voznesenskaya, E.V., Chuong, S.D.X., Koteyeva, N.K., *et al.* (2005b). Functional compartmentation of C_4 photosynthesis in the triple-layered chlorenchyma of *Aristida* (*Poaceae*). *Functional Plant Biology*, **32**, 67–77.

Voznesenskaya, E.V., Franceschi, V.R., Kiirats, O., *et al.* (2001). Kranz anatomy is not essential for terrestrial C_4 plant photosynthesis. *Nature*, **414**, 543–546.

Vrábl, D., Vašková, M., Hronková, M., *et al.* (2009). Mesophyll conductance to CO_2 transport estimated by two independent methods: effect of ambient CO_2 concentration and abscisic acid. *Journal of Experimental Botany*, **60**, 2315–2323.

Vrba, E.S., Denton, G.H., Partridge, T.C., *et al.* (eds) (1995). Paleoclimate and evolution, with emphasis on human origins. Yale University Press, New Haven, CT. 547 p.

Vu, J.C.V. and Yelenosky, G. (1991). Photosynthesis responses of citrus trees to soil flooding. *Physiologia Plantarum*, **81**, 7–14.

Vu, J.C.V., Allen, L.H., Jr., Boote, K.J., *et al.* (1997). Effects of elevated CO_2 and temperature on photosynthesis and Rubisco in rice and soybean. *Plant Cell and Environment*, **20**, 68–76.

Vyas, P., Bisht, M.S., Miyazawa, S.I., *et al.* (2007). Effects of polyploidy on photosynthetic properties and anatomy in leaves of *Phlox drummondii*. *Functional Plant Biology*, **34**, 673–682.

Wada, M., Kagawa, T. and Sato, Y. (2003). Chloroplast movement. *Annual Review of Plant Physiology and Plant Molecular Biology*, **54**, 455–468.

Wagner, J. and Larcher, W. (1981). Dependence of CO_2 gas exchange and acid metabolism of the alpine CAM plant *Sempervivum montanum* on temperature and light. *Oecologia*, **50**, 88–93.

Wainwright, C.M. (1977). Sun-tracking and related leaf movements in a desert lupine (*Lupinus arizonicus*). *American Journal of Botany*, **64**, 1032–1041.

Walcroft, A.S., Whitehead, D., Silvester, W.B., *et al.* (1997). The response of photosynthetic model parameters to temperature and nitrogen concentration in *Pinus radiata* D. Don. *Plant, Cell and Environment*, **20**, 1338–1348.

Walker, B.H. and Noy Meir, I. (1982). Aspects of the stability and resilience of savanna ecosystems. In: *Ecology of Tropical Savanna Ecosystems* (ed. Huntley, B.J. and Walker, B.H.), Springer-Verlag, New York, USA, pp. 556–590.

Walker, D.A. (1993). Polarographic measurement of oxygen. In: *Photosynthesis and Production in a Changing Environment. A Field and Laboratory Manual* (eds Hall,

D.O., Scurlock, J.M.O, Bolhar-Nordenkampf, H.R., Leegood, R.C. and Long, S.P.), Chapman and Hall, London, UK, pp. 168–180.

Walker, D.A. and Robinson, S.P. (1978). Chloroplast and cell. A contemporary view of photosynthetic carbon assimilation. *Berichte Der Deutschen Botanischen Gesellschaft*, **91**, 513–526.

Walker, D.A. and Sivak, M.N. (1985). Can phosphate limit photosynthetic carbon assimilation in vivo? *Physiologie Vegetale*, **23**, 829–841.

Walker, D.A. and Sivak, M.N. (1986). Photosynthesis and phosphate: a cellular affair? *Trends in Biochemical Sciences*, **11**, 176–179.

Wallin, G., Linder, S., Lindroth, A., *et al.* (2001). Carbon dioxide exchange in Norway spruce at the shoot, tree and ecosystem scale. *Tree Physiology*, **21**, 969–976.

Walter, A. and Schurr, U. (1999). The modular character of growth in *Nicotiana tabacum* plants under steady state nutrition. *Journal of Experimental Botany*, **50**, 1169–1177.

Walter, A. and Schurr, U. (2000). Spatial variability of leaf development, growth and function. In: *Leaf Development and Canopy Growth* (eds Marshall, B. and Roberts, J.), Sheffield Academic Press, Sheffield, UK.

Walter, A. and Schurr, U. (2005). Dynamics of leaf and root growth – endogenous control versus environmental impact. *Annals Botany*, **95**, 891–900.

Walter, A., Scharr, H., Gilmer, F., *et al.* (2007). Dynamics of seedling growth acclimation towards altered light conditions can be quantified via GROWSCREEN: a setup and procedure designed for rapid optical phenotyping of different plant species. *New Phytologist*, **174**, 447–455.

Walters, M.B., Kruger, E.L. and Reich, P.B. (1993a). Growth, biomass distribution and CO_2 exchange of northern hardwood seedlings in high and low light: relationships with successional status and shade tolerance. *Oecologia*, **94**, 7–16.

Walters, M.B., Kruger, E.L. and Reich, P.B. (1993b). Relative growth rate in relation to physiological and morphological traits for northern hardwood tree seedlings: species, light environment and ontogenetic considerations. *Oecologia*, **96**, 219–231.

Walters, R.G. and Horton, P. (1991). Resolution of components of non-photochemical chlorophyll fluorescence quenching in barley leaves. *Photosynthesis Research*, **27**, 121–133.

Wand, S.J.E., Esler, K.J. and Bowie, M.R. (2001). Seasonal photosynthetic temperature responses and changes in $\delta^{13}C$ under varying temperature regimes in leaf-succulent and drought-deciduous shrubs from the Succulent Karoo, South Africa. *South African Journal of Botany*, **67**, 235–243.

Wand, S.J.E., Esler, K.J., Rundel, P.W., *et al.* (1999). A preliminary study of the responsiveness to seasonal atmospheric and rainfall patterns of wash woodland species in the arid Richtersveld. *Plant Ecology*, **142**, 149–160.

Wanek, W., Heintel, S. and Richter, A. (2001). Preparation of starch and other carbon fractions from higher plant leaves for stable carbon isotope analysis. *Rapid Communications in Mass Spectrometry*, **15**, 1136–1140.

Wang, D. and Luthe, D. S. (2003). Heat sensitivity in a bentgrass variant. Failure to accumulate a chloroplast heat shock protein isoform implicated in heat tolerance. *Plant Physiology*, **133**, 319–327.

Wang, D., Heckathorn, S.A., Barua, D., *et al.* (2008). Effects of elevated CO_2 on the tolerance of photosynthesis to acute heat stress in C_3, C_4, and CAM species. *American Journal of Botany*, **95**, 165–176.

Wang, D.S., Naidu, S.L., Portis, A.R., *et al.* (2008). Can the cold tolerance of C-4 photosynthesis in *Miscanthusxgiganteus* relative to *Zea mays* be explained by differences in activities and thermal properties of Rubisco? *Journal of Experimental Botany*, **59**, 1779–1787.

Wang, F.Z., Wang, Q.B. and Kwon, S.Y. (2005). Enhanced drought tolerance of transgenic *O. sativa* plants expressing a pea manganese superoxide dismutase. *Journal of Plant Physiology*, **162**, 465–472.

Wang, H. and Jin, J.Y. (2005). Photosynthetic rate, chlorophyll fluorescence parameters, and lipid peroxidation of maize leaves as affected by zinc deficiency. *Photosynthetica*, **43**, 591–596.

Wang, N. and Nobel, P.S. (1996). Doubling the CO_2 concentration enhanced the activity of carbohydrate-metabolism enzymes, source carbohydrate production, photoassimilate transport, and sink strength for *Opuntia ficus-indica*. *Plant Physiology*, **110**, 893–902.

Wang, P. and Song, C.P. (2008). Guard-cell signalling for hydrogen peroxide and abscisic acid. *New Phytologist*, **178**, 703–718.

Wang, X., Manning, W., Feng, Z., *et al.* (2007). Ground-level ozone in China: Distribution and effects on crop yields. *Environmental Pollution*, **147**, 394–400.

Wang, X.F. and Yakir, D. (1995). Temporal and spatial variation in the oxygen-18 content of leaf water in different plant species. *Plant, Cell and Environment*, **18**, 1377–1385.

Wang, X.Q., Wu, W.H. and Assmann, S.M. (1998). Differential responses of adaxial and abaxial guard cell of broad bean to abscisic acid and calcium. *Plant Physiology*, **118**, 1421–1429.

Wang, Y. and Deng, T. (2005). A 25 m.y. isotopic record of paleodiet and environmental change from fossil mammals and paleosols from the NE margin of the Tibetan Plateau. *Earth and Planetary Science Letters*, **236**, 322–338.

Wang, Y. and Jacob, D.J. (1998). Anthropogenic forcing on tropospheric ozone and OH since preindustrial times. *Journal of Geophysical Research*, **103**, D23.

Wang, Y., Cerling, T.E. and MacFadden, B.J. (1994). Fossil horses and carbon isotopes – New evidence for Cenozoic dietary habit and ecosystem changes in North America. *Palaeogeography, Palaeoclimatology and Palaeoecology*, **107**, 269–279.

Wang, Y., Guignard, G., Thévenard, F., *et al.* (2005). Cuticular anatomy of *Sphenobaiera huangii* (Ginkgoales) from the Lower Jurassic of Hubei, China. *American Journal of Botany*, **92**, 709–721.

Wang, Z.Q. (1996). Recovery of vegetation from the terminal Permian mass extinction in north China. *Review of Palaeobotany and Palynology*, **91**, 121–142.

Ward, J., Baker, H. and Beale, M.H. (2007). Recent applications of NMR spectroscopy in plant metabolomics. *FEBS Journal*, **274**, 1126–1131.

Wardle, D.A., Hornberg, G., Zackrisson, O., *et al.* (2003). Long-term effects of wildfire on ecosystem properties across an island area gradient. *Science*, **300**, 972–976.

Waring, R.H. and Running, S.W. (1998). *Forest Ecosystems. Analysis at multiple scales.* Academic Press, San Diego, USA, pp. 1–370.

Warner, D.A. and Edwards, G.E. (1989). Effects of polyploidy on photosynthetic rates, photosynthetic enzymes, contents of DNA, chlorophyll, and sizes and numbers of photosynthetic cells in the C_4 dicot *Atriplex confertifolia*. *Plant Physiology*, **91**, 1143–1451.

Warner, D.A. and Edwards, G.E. (1993). Effects of polyploidy on photosynthesis. *Photosynthesis Research*, **35**, 135–147.

Warner, D.A., Ku, M.S.B. and Edwards, G.E. (1987). Photosynthesis, leaf anatomy, and cellular constituents in the polyploid C_4 grass *Panicum virgatum*. *Plant Physiology*, **84**, 461–466.

Warren, C.R. (2004). The photosynthetic limitation posed by internal conductance to CO_2 movement is increased by nutrient supply. *Journal of Experimental Botany*, **55**, 2313–2321.

Warren, C.R. (2006a). Why does photosynthesis decrease with needle age in *Pinus pinaster*? *Trees-Structure and Function*, **20**, 157–164.

Warren, C.R. (2006b). Estimating the internal conductance to CO_2 movement. *Functional Plant Biology*, **33**, 431–442.

Warren, C.R. (2007). Does growth temperature affect the temperature response of photosynthesis and internal conductance to CO_2? A test with *Eucalyptus regnans*. *Tree Physiology*, **28**, 11–19.

Warren, C.R. (2008a). Stand aside stomata, another actor deserves centre stage: the forgotten role of the internal conductance to CO_2 transfer. *Journal Experimental Botany*, **59**, 1475–1487.

Warren, C.R. (2008b). Soil water deficits decrease the internal conductance to CO_2 transfer but atmospheric water deficits do not. *Journal of Experimental Botany*, **59**, 327–334.

Warren, C.R. and Adams, M.A. (2000). Trade-offs between the persistence of foliage and productivity in two *Pinus* species. *Oecologia*, **124**, 487–494.

Warren, C.R. and Adams, M.A. (2001). Distribution of N, Rubisco and photosynthesis in *Pinus pinaster* and acclimation to light. *Plant, Cell and Environment*, **24**, 597–609.

Warren, C.R. and Adams, M.A. (2002). Phosphorus affects growth and partitioning of nitrogen to Rubisco in *Pinus pinaster*. *Tree Physiology*, **22**, 11–19.

Warren, C.R. and Adams, M.A. (2004). Evergreen trees do not maximize instantaneous photosynthesis. *Trends in Plant Science*, **9**, 270–274.

Warren, C.R. and Adams, M.A. (2006). Internal conductance does not scale with photosynthetic capacity: implications for carbon isotope discrimination and the economics of water and N use in photosynthesis. *Plant, Cell and Environment*, **29**, 192–201.

Warren, C.R. and Dreyer, E. (2006). Temperature response of photosynthesis and internal conductance to CO_2: results from two independent approaches. *Journal of Experimental Botany*, **57**, 3057–3067.

Warren, C.R., Adams, M.A. and Chen, Z.L. (2000). Is photosynthesis related to concentrations of nitrogen and Rubisco in leaves of Australian native plants? *Australian Journal of Plant Physiology*, **27**, 407–416.

Warren, C.R., Aranda, I. and Cano, F.J. (2011). Responses to water stress of gas exchange and metabolites in *Eucalyptus* and *Acacia* spp. *Plant, Cell and Environment*, **34**, 1609–1629.

Warren, C.R., Dreyer, E. and Adams, M.A. (2003b). Photosynthesis-Rubisco relationships in foliage of *Pinus sylvestris* in response to nitrogen supply and the proposed role of Rubisco and amino acids as nitrogen stores. *Trees: Structure and Function*, **17**, 359–366.

Warren, C.R., Dreyer, E., Tausz, M., *et al.* (2006). Ecotype adaptation and acclimation of leaf traits to rainfall in 29 species of 16-year-old Eucalyptus at two common gardens. *Functional Ecology*, **20**, 929–940.

Warren, C.R., Ethier, G.H., Livingston, N.J., *et al.* (2003a). Transfer conductance in second growth Douglas-fir (*Pseudotsuga menziesii* (Mirb.) Franco) canopies. *Plant, Cell and Environment*, **26**, 1215–1227.

Warren, C.R., Hovenden, M.J., Davidson, N.J., *et al.* (1998). Cold hardening reduces photoinhibition of *Eucalypts nitens* and *E. pauciflora* at frost temperatures. *Oecologia*, **113**, 350–359.

Warren, C.R., Livingston, N.J. and Turpin, D.H. (2004). Water stress decreases the transfer conductance of Douglas fir *(Pseudotsuga menziesii)* seedlings. *Tree Physiology*, **24**, 971–979.

Warren, C.R., Low, M., Matyssek, R., *et al.* (2007). Internal conductance to CO_2 transfer of adult *Fagus sylvatica*: Variation between sun and shade leaves and due to free-air ozone fumigation. *Environmental and Experimental Botany*, **59**, 130–138.

Warren, C.R., McGrath, J.F. and Adams M.A. (2001). Water availability and carbon isotope discrimination in conifers. *Oecologia*, **127**, 476–486.

Wartinger, A., Heilmeier, H., Hartung, W., *et al.* (1990). Daily and seasonal courses of leaf conductance and abscisic acid in the xylem sap of almond trees [Prunus dulcis (Miller) D.A. Webb] under desert conditions. *The New Phytologist*, **116**, 581–587.

Waters, E.R., Lee, G.J. and Vierling, E. (1996). Evolution, structure, and function of the small heat shock proteins in plants. *Journal of Experimental Botany*, **47**, 325–338.

Watkins, J.E., Rundel, P.W. and Cardelus, C.L. (2007). The influence of life form on carbon and nitrogen relationships in tropical rainforest ferns. *Oecologia*, **153**, 225–232.

Watkins, Jr., J.E., Mack, M.C., Sinclair, T.R., *et al.* (2007). Ecological and evolutionary consequences of desiccation tolerance in tropical fern gametophytes. *The New Phytologist*, **176**, 708–717.

Watkinson, J.I., Sioson, A.A., Vasquez-Robinet, C., *et al.* (2003). Photosynthetic acclimation is reflected in specific patterns of gene expression in drought-stressed loblolly pine. *Plant Physiology*, **133**, 1702–1716.

Watling, J.R. and Press, M.C. (2001). Impacts of infection by parasitic angiosperms on host photosynthesis. *Plant Biology*, **3**, 244–250.

Watling, J.R., Robinson, S.A., Woodrow, I.E., *et al.* (1997). Responses of rainforest understory plants to excess light during sunflecks. *Australian Journal of Plant Physiology*, **24**, 17–25.

Watson, D.J. (1947). Comparative physiological studies on the growth of field crops. *Annals of Botany*, **11**, 41–76.

Watson, D.J. (1952). The physiological basis of variation in yield. *Advances in Agronomy*, **4**, 101–145.

Watson, M.A. and Caspar, B.B. (1984). Morphogenetic constraints on patterns of carbon distribution in plants. *Annual Review of Ecology and Systematics*, **15**, 233–258.

Weaver, J.C., Powell, K.T., Mintzer, R.A., *et al.* (1984). The diffusive permeability of bilayer membranes the contribution of transient aqueous pores. *Bioelectrochemistry and Bioenergetics*, **12**, 405–412.

Weaver, L.M. and Amasino, R.M. (2001). Senescence is induced in individually darkened *Arabidopsis* leaves, but inhibited in whole darkened plants. *Plant Physiology*, **127**, 876–886.

Webb, E.K., Pearman, G.I. and Leuning, R. (1980). Correction of flux measurements for density effects due to heat and water vapor transfer. *Quarterly Journal of the Royal Meteorological Society*, **106**, 85–100.

Wedin, D.A. and Tilman, D. (1993). Competition among grasses along a nitrogen gradient: initial conditions and mechanisms of competition. *Ecological Monographs*, **63**, 199–229.

Wedin, D.A. and Tilman, D. (1996). Influence of nitrogen loading and species composition on the carbon balance of grasslands. *Science*, **274**, 1720–1723.

Wedler, M., Köstner, B. and Tenhunen, J.D. (1996). Understory contribution to stand total water loss at an old Norway spruce forest. In: *Verhandlungen der Gessellschaft für Ökologie*, vol 26 (eds Pfadenhauer, J., Kappen, L., Mahn, E.G., Otte, A. and Plachter, H.) Gustav Fischer Verlag, Stuttgart, Germany, pp. 69–77.

Weigel, H.J. (1985). Inhibition of photosynthetic reactions of isolated chloroplasts by cadmium. *Journal of Plant Physiology*, **119**, 179–189.

Weih, M. and Karlsson, P.S. (2001). Growth response of Mountain birch to air and soil temperature: is increasing leaf-nitrogen content an acclimation to lower air temperature? *The New Phytologist*, **150**, 147–155.

Weikert, R.M., Wedler, M., Lippert, M., *et al.* (1989). Photosynthetic performance, chloroplast pigments, and mineral content of various needle age classes of spruce *(Picea abies)* with and without the new flush: an experimental approach for analysing forest decline phenomena. *Trees*, **3**, 161–172.

Weinbaum, S.A., Muraoka, T.T. and Plant, R.E. (1994). Intracanopy variation in nitrogen cycling through leaves is influenced by irradiance and proximity to developing fruit in mature walnut trees. *Trees*, **9**, 6–11.

Weis, E. (1982). Influence of light on the heat sensitivity of the photosynthetic apparatus in isolated spinach chloroplasts. *Plant Physiology*, **70**, 1530–1534.

Weis, E. (1985a). Chlorophyll fluorescence at 77K in intact leaves: characterization of a technique to eliminate artifacts related to self-absorption. *Photosynthesis Research*, **6**, 73–86.

Weis, E. (1985b). Light- and temperature-induced changes in the distribution of excitation energy between photosystem I and photosystem II in spinach leaves. *Biochimica et Biophysica Acta*, **807**, 118–126.

Weis, E., Meng, Q., Siebke, K., *et al.* (1998). Topography of leaf carbon metabolism as analyzed by chlorophyll a-fluorescence. In: *Photosynthesis: Mechanisms and Effects*, vol 5 (ed. Garab, G.), Kluwer Academic Publishers, Dordrecht, Netherlands, pp. 4259–4264.

Weise, S.E., Schrader, S.M., Kleinbeck, K.R., *et al.* (2006). Carbon balance and circadian regulation of hydrolytic and phosphorolytic breakdown of transitory starch. *Plant Physiology*, **141**, 879–886.

Weiss, I., Mizrahi, Y. and Raveh, E. (2009). Chamber response time: a neglected issue in gas exchange measurements. *Photosynthetica*, **47**, 121–124.

Welling, A. and Palva, E.T. (2008). Involvement of CBF transcription factors in winter hardiness in birch. *Plant Physiology*, **147**, 1199–1211.

Wells, R., Schulze, L.L., Ashley, D.A., *et al.* (1982). Cultivar differences in canopy apparent photosynthesis and their relationship to seed yield in soybeans. *Crop Science*, **22**, 886–890.

Welp, L.R., Randerson, T.R. and Liu, H.P. (2007). The sensitivity of carbon fluxes to spring warming and summer drought depends on plant functional type in boreal forest ecosystems. *Agricultural and Forest Meteorology*, **147**, 172–185.

Welsch, R., Beyer, P., Hugueney, P., *et al.* (2000). Regulation and activation of phytoene synthase, a key enzyme in carotenoid biosynthesis, during photomorphogenesis. *Planta*, **211**, 846–854.

Welter, S.C. (1989). Arthropod impact on plant gas exchange. In: *Insect-Plant Interactions* (ed. Bernays, E.A.), CRC Press, Boca Raton, FL, USA, pp. 135–151.

Wen, X.F., Sun, X.M., Zhang, S.C., *et al.* (2008). Continuous measurements of water vapour D/H and $^{18}O/^{16}O$ isotope ratios in the atmosphere. *Journal of Hydrology*, **349**, 489–500.

Weng, X.Y., Zheng, C.J., Xu, H.X., *et al.* (2007). Characteristics of photosynthesis and functions of the water-water cycle in rice (Oryza sativa) leaves in response to potassium deficiency. *Physiologia Plantarum*, **131**, 614–621.

Werk, K.S., Ehleringer, J., Forseth, I.N., *et al.* (1983). Photosynthetic characteristics of Sonoran Desert winter annuals. *Oecologia*, **59**, 101–105.

Werner, C. and Máguas, C. (2010). Carbon isotope discrimination as a tracer of functional traits in a Mediterranean macchia plant community. *Functional Plant Biology*, **37**, 467–477.

Werner, C., Correia, O. and Beyschlag, W. (1999). Two different strategies of Mediterranean macchia plants to avoid photoinhibitory damage by excessive radiation levels during summer drought. *Acta Oecologica*, **20**, 15–23.

Werner, C., Correia, O. and Beyschlag, W. (2002). Characteristic patterns of chronic and dynamic photoinhibition of different functional groups in a Mediterranean ecosystem. *Functional Plant Biology*, **29**, 999–1011.

Werner, C., Ryel, R.J., Correia, O., *et al.* (2001a). Structural and functional variability within the canopy and its relevance for carbon gain and stress avoidance. *Acta Oecologica*, **22**, 129–138.

Werner, C., Ryel, R.J., Correia, O., *et al.* (2001b). Effects of photoinhibition on whole-plant carbon gain assessed with a photosynthesis model. *Plant, Cell and Environment*, **24**, 27–40.

West, J.B., Espeleta, J.F. and Donovan, L.A. (2003). Root longevity and phenology differences between two co-occurring savanna bunchgrasses with different leaf habits. *Functional Ecology*, **17**, 20–28.

West, J.D., Peak, D., Peterson, J.Q., *et al.* (2005). Dynamics of stomatal patches for a single surface of *Xanthium strumarium* L. leaves observed with fluorescence and thermal images. *Plant, Cell and Environment*, **28**, 633–641.

Westbeek, M.H.M, Pons, T.L., Cambridge, M.L., *et al.* (1999). Analysis of differences in photosynthetic nitrogen use eficiency of alpine and lowland Poa species. *Oecologia*, **19**, 120–126.

Westhoff, P. and Gowik, U. (2004). Evolution of C_4 phosphoenolpyruvate carboxylase. Genes and proteins: a case study with the genus *Flaveria*. *Annals of Botany*, **93**, 13–23.

Weston, D.J., Bauerle, W.L., Swire-Clark, G.A., *et al.* (2007). Characterization of Rubisco activase from thermally contrasting genotypes pf *Acer rubrum* (*Aceraceae*). *American Journal of Botany*, **94**, 926–934.

Wheeler, R.M. (1992). Gas-exchange measurements using a large, closed plant growth chamber. *HortScience*, **27**, 777–780.

Wheeler, R.M., Corey, K.A., Sager, J.C., *et al.* (1993). Gas exchange characteristics of wheat stands grown in a closed, controlled environment. *Crop Science*, **33**, 161–168.

Whiley, A.W., Schaffer, B. and Lara, S.P. (1992). Carbon dioxide exchange of developing avocado (*Persea americana* Mill.) fruit. *Tree Physiology*, **11**, 85–94.

White, J.G. and Zasoski, R.H. (1999). Mapping soil micronutrients. *Field Crop Research*, **60**, 11–26.

White, T.A., Campbell, B.D., Kemp, P.D., *et al.* (2000). Sensitivity of three grassland communities to simulated extreme temperature and rainfall events. *Global Change Biology*, **6**, 671–684.

Whitehead, D. (1998). Regulation of stomatal conductance and transpiration in forest canopies. *Tree Physiology*, **18**, 633–644.

Whitehead, D. and Beadle, C.L. (2004). Physiological regulation of productivity and water use in Eucalyptus: a review. *Forest Ecology and Management*, **193**, 113–140.

Whitehead, D., Barbour, M.M., Griffin, K.L. *et al.* (2011). Effects of leaf age and tree size on stomatal and mesophyll limitations to photosynthesis in mountain beech (*Nothofagus solandrii* var. *cliffortiodes*). *Tree Physiology*, **31**, 985–996.

Whitehead, D., Livingston, N.J., Kelliher, P.M., *et al.* (1996). Response of transpiration and photosynthesis to a transient change in illuminated foliage area for a *Pinus radiata* D. Don tree. *Plant, Cell and Environment*, **19**, 949–957.

Whitelam, G. (1995). Plant photomorphogenesis – a green light for cryptochrome research. *Current Biology*, **5**, 1351–1353.

Whitham, S., Quan, S., Chang, H.S., *et al.* (2003). Diverse RNA viruses elicit the expression of common sets of genes in susceptible *Arabidopsis thaliana* plants. *Plant Journal*, **32**, 271–283.

Whiting, M.D. and Lang, G.A. (2001). Canopy architecture and cuvette flow patterns influence whole-canopy net CO_2 exchange and temperature in sweet cherry. *HortScience*, **36**, 691–698.

Whitmore, T.C. (1990). *An introduction to tropical rain forests.* Oxford University Press, Oxford, UK.

Whitney, S.M. and Andrews, T.J. (2001). The gene for the ribulose-1,5-bisphosphate carboxylase/oxygenase (Rubisco) small subunit relocated to the plastid genome of tobacco directs the synthesis of small subunits that assemble into Rubisco. *Plant Cell*, **13**, 193–205.

Whitney, S.M. and Sharwood, R.E. (2008). Construction of a tobacco master line to improve Rubisco engineering in chloroplasts. *Journal of Experimental Botany*, **59**, 1909–1921.

Wi, S.J., Kim, W.T. and Park, K.Y. (2006). Overexpression of carnation S-adenosylmethionine decarboxylase gene generates a broad-spectrum tolerance to abiotic stresses in transgenic tobacco plants. *Plant Cell Report*, **25**, 1111–1121.

Wickman, F.E. (1952). Variations in the relative abundance of the carbon isotopes in plants. *Geochimica et Cosmochimica Acta*, **2**, 243–252.

Wild, M., Gilgen, H., Roesch, A., *et al.* (2005). From dimming to brightening: decadal changes in solar radiation at Earth's surface. *Science*, **308**, 847–850.

Wildi, B. and Lütz, C. (1996). Antioxidant composition of selected high alpine plant species from different altitudes. *Plant, Cell and Environment*, **19**, 138–146.

Wilhelmová, N., Procházková, D., Sindelárová, M., *et al.* (2005). Photosynthesis in leaves of *Nicotiana tabacum* L. infected with tobacco mosaic virus. *Photosynthetica*, **43**, 597–602.

Wilkins, M.B. (1992). Circadian rhythms: their origin and control. *New Phytologist*, **121**, 347–375.

Wilkinson, S. and Davies, W.J. (1997). Xylem sap pH increase: a drought signal received at the apoplastic face of the guard cells that involves the suppression of saturable abscisic acid uptake by the epidermal symplast. *Plant Physiology*, **113**, 559–573.

Wilkinson, S. and Davies, W.J. (2002). ABA-based chemical signalling: the co-ordination of responses to stress in plants. *Plant, Cell and Environment*, **25**, 195–210.

Wilkinson, S. and Davies, W.J. (2008). Manipulation of the apoplastic pH of intact plants mimics stomatal and growth responses to water availability and microclimatic variation. *Journal of Experimental Botany*, **59**, 619–631.

Wilkinson, S. and Davies, W.J. (2009). Ozone suppresses soil drying- and abscisic acid (ABA)- induced stomatal closure via an ethylene-dependent mechanism. *Plant, Cell and Environment*, **32**, 949–959.

Wilkinson, S. and Davies, W.J. (2010). Drought, ozone, ABA and ethylene: new insights from cell to plant to community. *Plant, Cell and Environment*, **33**, 510–525.

Wilkinson, S., Clephan, A.L. and Davies, W.J. (2001). Rapid low temperature-induced stomatal closure occurs in cold-tolerant *Commelina communis* leaves but not in cold-sensitive tobacco leaves, via a mechanism that involves apoplastic calcium but not abscisic acid. *Plant Physiology*, **126**(4), 1566–1578.

Williams, E.L., Hovenden, M.J. and Close, D.C. (2003). Strategies of light energy utilisation, dissipation and attenuation in six co-occurring alpine heath species in Tasmania. *Functional Plant Biology*, **30**, 1205–1218.

Williams, G.J. III (1974). Photosynthetic adaptation to temperature in C_3 and C_4 grasses: possible ecological role in shortgrass prarie. *Plant Physiology*, **54**, 709–711.

Williams, K., Field, C.B. and Mooney, H.A. (1989). Relationships among leaf construction costs, leaf longevity, and light environment in rainforest species of the genus *Piper. American Naturalist*, **133**, 198–211.

Williams, M., Rastetter, E.B., Fernandes, D.N., *et al.* (1996). Modelling the soil-plant-atmosphere continuum in a *Quercus-Acer* stand at Harvard Forest: the regulation of stomatal conductance by light, nitrogen and soil/plant hydraulic properties. *Plant, Cell and Environment*, **19**, 911–927.

Williams, M., Woodward, F.I., Baldocchi, D.D., *et al.* (2004). CO_2 capture from the leaf to the landscape. In: *Photosynthetic Adaptation: Chloroplast to the Landscape* (eds Smith, W.K., Vogelmann, T.C. and Critchley, C.), Springer, New York, USA, pp. 133–168.

Williams, T.G., Flanagan, L.B. and Coleman, J.R. (1996). Photosynthetic gas exchange and discrimination against $^{13}CO_2$ and $C^{18}O^{16}O$ in tobacco plants modified by an antisense construct to have low chloroplastic carbonic anhydrase. *Plant Physiology*, **112**, 319–326.

Williams, W.P., Brain, A.P.R. and Dominy, P.J. (1992). Induction of non-bilayer lipid phase separation in chloroplast thylakoid membranes by compatible co-solutes and its relation to the thermal stability of photosystem II. *Biochimica et Biophysica Acta*, **1099**, 137–144.

Willmer, C. and Fricker, M. (1996). *Stomata*, 2nd edn. London: Chapman and Hall.

Willms, J.R., Salon, C. and Layzell, D.B. (1999). Evidence for light-stimulated fatty acid synthesis in soybean fruit. *Plant Physiology*, **120**, 1117–1128.

Wilson, J.A., OgunKanmi, A.B. and Mansfield, T.A. (1978). Effects of external potassium supply on stomatal closure induced by abscisic acid. *Plant, Cell and Environment*, **1**, 199–201.

Wilson, K., Goldstein, A., Falge, E., *et al.* (2002). Energy balance closure at FLUXNET sites. *Agricultural and Forest Meteorology*, **113**, 223–243.

Wilson, K.B., Baldocchi, D.D. and Hanson P.J. (2001). Leaf age affects the seasonal pattern of photosynthetic capacity and net ecosystem exchange of carbon in a deciduous forest. *Plant, Cell and Environment*, **24**, 571–583.

Wilson, K.B., Baldocchi, D.D. and Hanson, P.J. (2000a). Quantifying stomatal and non-stomatal limitations to carbon assimilation resulting from leaf aging and drought in mature deciduous tree species. *Tree Physiology*, **20**, 787–797.

Wilson, K.B., Baldocchi, D.D. and Hanson, P.J. (2000b). Spatial and seasonal variability of photosynthetic parameters and their relationship to leaf nitrogen in a deciduous forest. *Tree Physiology*, **20**, 565–578.

Wilson, K.E., Krol, M. and Huner, N.P.A. (2003). Temperature-induced greening of *Chlorella vulgaris.* The role of the cellular energy balance and zeaxanthin-dependent nonphotochemical quenching. *Planta*, **217**, 616–627.

Winder, T.L. and Nishio, J. (1995). Early iron deficiency stress response in leaves of sugar beet. *Plant Physiology*, **108**, 1487–1494.

Wingler, A., Marés, M. and Pourtau, N. (2004). Spatial and metabolic regulation of photosynthetic parameters during leaf senescence. *New Phytologist*, **161**, 781–789.

Wingler, A., Masclaux-Daubresse, C. and Fischer, A.M. (2009). Sugars, senescence, and ageing in plants and heterotrophic organisms. *Journal of Experimental Botany*, **60**, 1063–1066.

Winkel, A. and Borum, J. (2009). Use of sediment CO_2 by submersed rooted plants. *Annals of Botany*, **103**, 1015–1023.

Winkel, T., Méthy, M. and Thénot, F. (2002). Radiation use efficiency, chlorophyll fluorescence, and reflectance indices associated with ontogenic changes in water-limited *Chenopodium quinoa* leaves. *Photosynthetica*, **40**, 227–232.

Winner, W.E., Lefohn, A.S., Cotter, I.S., *et al.* (1989). Plant responses to elevational gradients of O_3 exposures in Virginia. *Proceedings of the National Academy of Sciences of the USA*, **86**, 8828–8832.

Winner, W.E., Mooney, H.A., Williams, K., *et al.* (1985). Measuring and assessing SO_2 effects on photosynthesis and plant growth. In: *Sulfur Dioxide and Vegetation-Physiology, Ecology and Policy Issues* (eds Winner, W.E., Mooney, H.A. and Goldstein, R.A.), Stanford University Press, Stanford, USA, pp. 118–132.

Winsemius, H.C., Savenije, H.H.G. and Bastiaanssen, W.G.M. (2008). Constraining model parameters on remotely sensed evaporation: justification for distribution in ungauged basins? *Hydrology and Earth System Sciences*, **12**, 1403–1413.

Winslow, J.C., Hunt, E.R. and Piper, S.C. (2003). The influence of seasonal water availability on global C_3 versus

C_4 grassland biomass and its implications for climate change research. *Ecological Modelling*, **163**, 153–173.

Winter, K. (1973). Zum Problem der Ausbildung des Crassulaceen-Säurestoffwechsels bei *Mesembryanthemum crystallinum* unter NaCl-Einfluss. *Planta*, **109**, 135–145.

Winter, K. and Smith, J.A.C. (ed). (1996). *Crassulacean Acid Metabolism: Biochemistry, Ecophysiology, and Evolution.* Springer, Berlin, Germany.

Winter, K., Aranda, J. and Holtum, J.A.M. (2005). Carbon isotope composition and water-use efficiency in plants with crassulacean acid metabolism. *Functional Plant Biology*, **32**, 381–388.

Winter, K., Garcia, M. and Holtum, J.A.M. (2009). Canopy CO_2 exchange of two neotropical tree species exhibiting constitutive and facultative CAM photosynthesis, *Clusia rosea* and *Clusia cylindrica*. *Journal of Experimental Botany*, **60**, 3167–3177.

Winter, K., Osmond, C.B. and Pate, J.S. (1981). Coping with salinity. In: *The Biology of Australian Plants* (eds Pate, J.S. and McComb, A.J.), University of Western Australia Press, Netlands, Australia, pp. 88–113.

Winter, K., Richter, A., Engelbrecht, B., *et al.* (1997). Effect of elevated CO_2 on growth and crassulacean acid metabolism activity of *Kalanchoë pinnata* under tropical conditions. *Planta*, **201**, 389–396.

Winter, K., Wallace, B.J., Stocker, G.C., *et al.* (1983). Crassulacean acid metabolism in Australian vascular epiphytes and some related species. *Oecologia*, **57**, 129–141.

Wintermans, J. and De Mots, A. (1965). Spectrophotometric of chlorophyll a and b and their pheophytins in ethanol. *Biochemistry and Biophysics Acta*, **109**, 448–453.

Wise, R.R. (1995). Chilling-enhanced photooxidation – the production, action and study of reactive oxygen species produced during chilling in the light. *Photosynthesis Research*, **45**, 79–97.

Wise, R.R. and Ort, D.R. (1989). Photophosphorylation after chilling in the light: Effects on membrane energization and coupling factor activity. *Plant Physiology*, **90**, 657–664.

Wise, R.R., Olson, A.J., Schrader, S.M., *et al.* (2004). Electron transport is the functional limitation of photosynthesis in field-grown Pima cotton plants at high temperature. *Plant, Cell and Environment*, **27**, 717–724.

Wisniewski, M., Bassett, C. and Gusta, L.V. (2003). An overview of cold hardiness in woody plants: Seeing the forest through the trees. *Hortscience*, **38**, 952–959.

Wittig, V.E., Ainsworth, E.A. and Long, S.P. (2007). To what extent do current and projected increases in surface ozone affect photosynthesis and stomatal conductance of trees? A meta-analytic review of the last 3 decades of experiments. *Plant, Cell and Environment*, **30**, 1150–1162.

Wittmann, C. and Pfanz, H. (2007). Temperature dependency of bark photosynthesis in beech (*Fagus sylvatica* L.) and birch (*Betula pendula* Roth.) trees. *Journal of Experimental Botany*, **58**, 4293–4306.

Wittmann, C., Pfanz, H., Loreto, F., *et al.* (2006). Stem CO_2 release under illumination: corticular photosynthesis, photorespiration or inhibition of mitochondrial respiration? *Plant, Cell and Environment*, **29**, 1149–1158.

Wohlfahrt, G., Anderson-Dunn, M., Bahn, M., *et al.* (2008b). Biotic, abiotic and management controls on the net ecosystem CO_2 exchange of European mountain grassland ecosystems. *Ecosystems*, **11**, 1338–1351.

Wohlfahrt, G., Bahn, M., Tappeiner, U., *et al.* (2000). A model of whole plant gas exchange for herbaceous species from mountain grassland sites differing in land use. *Ecological Modelling*, **125**, 173–201.

Wohlfahrt, G., Fenstermaker, L.F. and Arnone, J.A. (2008a). Large annual net ecosystem CO_2 uptake of a Mojave Desert ecosystem. *Global Change Biology*, **14**, 1475–1487.

Wolfe, K.H., Morden, C.W. and Palmer, J.D. (1992). Function and evolution of a minimal plastid genome from a nonphotosynthetic parasitic plant. *Proc Natl Acad Sci USA*, **89**, 10648–10652.

Wollman, F.A. (2001). State transitions reveal the dynamics and flexibility of the photosynthetic apparatus. *EMBO Journal*, **20**, 3623–3630.

Wong, C.E., Li, Y., Labbe, A., *et al.* (2006). Transcriptional profiling implicates novel interactions between abiotic stress and hormonal responses in *Thellungiella*, a close relative of Arabidopsis. *Plant Physiology*, **140**, 1437–1450.

Wong, S.C. and Osmond, C.B. (1991). Elevated atmospheric partial pressure of CO_2 and plant growth. III. Interactions between *Triticum aestivum* (C_3) and *Echinochloa frumentacea* (C_4) during growth in mixed culture under different CO_2, N nutrition and irradiance treatments with emphasis on belowground responses estimated using the delta ^{13}C value of root biomass. *Australian Journal of Plant Physiology*, **18**, 137–152.

Wong, S.C., Cowan, I.R. and Farquhar, G.D. (1979). Stomatal conductance correlates with photosynthetic capacity. *Nature*, **282**, 424–426.

Wong, S.C., Cowan, I.R. and Farquhar, G.D. (1985). Leaf conductance relation to rate of CO_2. 2. Effects of short-term exposures to different photon-flux densities. *Plant Physiology*, **78**, 826–829.

Wong, S.C., Cowan, I.R. and Farquhar, G.D. (1985). Leaf conductance in relation to the rate of CO_2 assimilation. I. Influence of nitrogen nutrition, phosphorus nutrition, photon-flux density, and ambient partial pressure of CO_2 during ontogeny. *Plant Physiology*, **78**, 821–825.

Wongwises, S., Pornsee, A. and Siroratsakul, E. (1999). Gas-wall shear stress distribution in horizontal stratified two-phase flow. *International Communications in Heat and Mass Transfer*, **26**, 849–860.

Woo, N.S., Badger, M.R., Pogson, B.J. (2008). A rapid, non-invasive procedure for quantitative assessment of drought survival using chlorophyll fluorescence. *Plant Methods*, **4**, 27.

Woodall, G.S., Dodd, I.C. and Stewart, G.R. (1998). Contrasting leaf development within the genus *Syzygium*. *Journal of Experimental Botany*, **49**, 79–87.

Woodroffe, C.D. (1992). *Mangrove sediments and geomorphology. Tropical Mangrove Ecosystems* (eds Robertson, A.I. and Alongi, D.M.), American Geophysical Union, Washington, DC, USA, pp. 7–41.

Woodrow, L., Jiao, J., Tsujita, M.J., *et al.* (1989). Whole plant and leaf steady state gas exchange during ethylene exposure in *Xanthium strumarium* L. *Plant Physiology*, **90**, 85–90.

Woodruff, D.R., Meinzer, F.C., Lachenbruch, B., *et al.* (2009). Coordination of leaf structure and gas exchange along a height gradient in a tall conifer. *Tree Physiology*, **29**, 261–272.

Woods, H.A. and Smith, J.N. (2010). Universal model for water costs of gas exchange by animals and plants. *Proceedings of the National Academy of Sciences USA*, **107**, 8469–8474.

Woodward, F.I. (1979). The differential temperature responses of the growth of certain plant species from different altitudes. I. Growth analysis of *Phleum alpinum* L., *P. bertolonii* D.C., *Sesleria albicans* Kit. and *Dactylis glomerata* L. *New Phytologist*, **82**, 385–395.

Woodward, F.I. (1987a). Stomatal numbers are sensitive to increases in CO_2 from pre-industrial levels. *Nature*, **327**, 617–618.

Woodward, F.I. (1987b). *Climate and Plant Distribution*. Cambridge University Press, Cambridge, UK.

Woodward, F.I. (1993). Plant responses to pasts concentrations of CO_2. *Vegetatio*, **104**, 105–155.

Woodward, F.I. and Kelly, C.K. (1995). The influence of CO_2 concentration on stomatal density. *The New Phytologist*, **131**, 311–327.

Woodward, F.I., Körner, C. and Crabtree, R.C. (1986). The dynamics of leaf extension in plants with diverse altitudinal ranges. I. Field observations on temperature responses at one altitude. *Oecologia*, **70**, 222–226.

Wosfy, S.C., Goulden, M.L., Munger, J.W., *et al.* (1993). Net exchange of CO_2 in a mid-latitude forest. *Science*, **260**, 1314–1317.

WRI (2005). World Resources Institute: Freshwater resources 2005. www.wri.org/.

Wright, D.P., Baldwin, B.C., Shephard, M.C., *et al.* (1995). Source–sink relationships in wheat leaves infected with powdery mildew. I. Alterations in carbohydrate metabolism. *Physiological and Molecular Plant Pathology*, **47**, 237–253.

Wright, I.J., and Cannon, K. (2001). Relationships between leaf lifespan and structural defences in a low nutrient, sclerophyll flora. *Functional Ecology*, **15**, 351–359.

Wright, I.J., Reich, P.B. and Westoby, M. (2003). Least-cost input mixtures of water and nitrogen for photosynthesis. *The American Naturalist*, **161**, 98–111.

Wright, I.J., Reich, P.B., Cornelissen, J.H.C., *et al.* (2005). Modulation of leaf economic traits and trait relationships by climate. *Global Ecology and Biogeography*, **14**, 411–421.

Wright, I.J., Reich, P.B., Westoby, M., *et al.* (2004). The world-wide leaf economics spectrum. *Nature*, **428**, 821–827.

Wright, S.I., Ness, R.W., Foxe, J.P., *et al.* (2008). Genomic consequences of outcrossing and selfing in plants. *International Journal of Plant Sciences*, **169**, 105–118.

Wu, X., Luo, Y., Weng, E., *et al.* (2009). Conditional inversion to estimate parameters from eddy flux observations. *Journal of Plant Ecology*, **2**, 55–68.

Wullschleger, S.D. (1993). Biochemical limitations to carbon assimilation in C3 plants-a retrospective analysis of the A/Ci curves from 109 species. *Journal of Experimental Botany*, **44**, 907–920.

Wullschleger, S.D. and Norby, R.J. (2001). Sap velocity and canopy transpiration in a sweetgum stand exposed to free-air CO_2 enrichment (FACE). *New Phytologist*, **25**, 489–498.

Wullschleger, S.D., Tschaplinski, T.J. and Norby, R.J. (2002). Plant water relations at elevated CO_2 – implications for water-limited environments. *Plant, Cell and Environment*, **25**, 319–331.

Wünsche, J.N. and Palmer, J.W. (1997). Portable through-flow cuvette system for measuring whole-canopy gas exchange of apple trees in the field. *HortScience*, **32**, 653–658.

Wyrich, R., Dressen, U., Brockman, S., *et al.* (1998). The molecular basis of C_4 photosynthesis in sorghum: isolation, characterization and RFLP mapping of mesophyll- and bundle-sheath-specific cDNAs as obtained by differential screening. *Plant Molecular Biology*, **37**, 319–335.

Xie, J., Li, Y., Zhai, C., *et al.* (2008). CO_2 absorption by alkaline soils and its implication to the global carbon cycle. *Environmental Geology*, **56**, 953–61.

Xin, Z. and Browse, J. (2000). Cold comfort farm: the acclimation of plants to freezing temperatures. *Plant Cell Environment*, **23**, 893–902.

Xiong, F.S., Mueller, E.C. and Day, T.A. (2000). Photosynthetic and respiratory acclimation and growth response of Antarctic vascular plants to contrasting temperature regimes. *American Journal of Botany*, **87**, 700–710.

Xiong, F.S., Ruhland, C.T. and Day, T.A. (1999). Photosynthetic temperature response of the Antarctic vascular plants *Colobanthus quitensis* and *Deschampsia antarctica*. *Physiologia Plantarum*, **106**, 276–286.

Xiong, J. and Bauer, C.E. (2002). Complex evolution of photosynthesis. *Annual Review of Plant Biology*, **53**, 503–521.

Xu, C., Fan, J., Riekhof, W., *et al.* (2003). A permease-like protein involved in ER to thylakoid lipid transfer in *Arabidopsis*. *EMBO Journal*, **22**, 2370–2379.

Xu, H.L., Gauthier, L., Desjardins, Y., *et al.* (1997). Photosynthesis in leaves, stem and petioles of greenhouse-grown tomato plants. *Photosynthetica*, **33**, 113–123.

Xu, H.X., Weng, X.Y. and Yang, Y. (2007). Effect of phosphorus deficiency on the photosynthetic characteristics of rice plants. *Russian Journal of Plant Physiology*, **54**, 741–748.

Xu, L. and Baldocchi, D.D. (2004). Seasonal variation in carbon dioxide exchange over a Mediterranean annual grassland in California. *Agricultural and Forest Meteorology*, **123**, 79–96.

Xu, L.K. and Baldocchi, D.D. (2003). Seasonal trends in photosynthetic parameters and stomatal conductance of blue oak (*Quercus douglasii*) under prolonged summer drought and high temperature. *Tree physiology*, **23**, 865–877.

Xu, L.K., Matista, A.A. and Hsiao, T.C. (1999). A technique for measuring CO_2 and *water vapour* profiles within and above plant canopies over short periods. *Agricultural and Forest Meteorology*, **94**, 1–12.

Xu, S. (2002). QTL analysis in plants. In: *Quantitative Trait Loci: Methods and Protocols* (eds Camp, N. and Cox, A.), Humana Press, Totowa, USA, pp. 283–310.

Xu, Z. and Zhou, G. (2008). Responses of leaf stomatal density to water status and its relationship with photosynthesis in a grass. *Journal of Experimental Botany*, **59**, 3317–3325.

Xu, Z., Zhou, G. and Shimizu, H. (2009). Are plant growth and photosynthesis limited by pre-drought following rewatering in grass? *Journal of Experimental Botany*, **60**, 3737–3749.

Yabuta, Y., Mieda, T., Rapolu, M., *et al.* (2007). Light regulation of ascorbate biosynthesis is dependent on the photosynthetic electron transport chain but independent of sugars in *Arabidopsis*. *Journal of Experimental Botany*, **58**, 2661–2671.

Yakir, D. (1998). Oxygen-18 of leaf water: a crossroad for plant-associated isotopic signals. In: *Stable Isotopes. Integration of Biological, Ecological and Geochemical Processes* (ed. Griffiths, H.), BIOS Scientific Publishers, Oxford, UK, pp. 147–168.

Yakir, D. (2003). The stable isotopic composition of atmospheric CO_2. *Treatise on Geochemistry*, **4**, 175–212.

Yakir, D. and Wang, X.F. (1996). Fluxes of $CO_2$and water between terrestrial vegetation and the atmosphere estimated from isotope measurements. *Nature*, **380**, 515–517.

Yakir, D., De Niro, M.J. and Rundel, P.W. (1989). Isotopic inhomogeneity of leaf water:evidence and implications for the use of isotopic signal transduced by plants. *Geochimica et Cosmochimica Acta*, **53**, 2769–2773.

Yamaguchi-Shinozaki, K. and Shinozaki, K. (2006). Responses and tolerance to dehydration and cold stresses. *Annual Review of Plant Biology*, **57**, 781–803.

Yamamoto, H.Y., Kamite, L. and Wang, Y.Y. (1972). An ascorbate-induced absorbance change in chloroplasts from violaxanthin de-epoxidation. *Plant Physiology*, **49**, 224–228.

Yamane, Y., Kashino, Y., Koike, H., *et al.* (1997). Increases in the fluorescence F0 level and reversible inhibition of photosystem II reaction center by high-temperature treatments in higher plants. *Photosynthesis Research*, **52**, 57–64.

Yamane, Y., Kashino, Y., Koike, H., *et al.* (1998). Effects of high temperatures on the photosynthetic systems in spinach: Oxygen-evolving activities, fluorescence characteristics and the denaturation process. *Photosynthesis Research*, **57**, 51–59.

Yamasaki, H. (1997). A function of colour. *Trends in Plant Science*, **2**, 7–9.

Yamashita, N., Ishida, A., Kushima, H., *et al.* (2000). Acclimation to sudden increase in light favouring an invasive over native trees in subtropical islands, Japan. *Oecologia*, **125**, 412–419.

Yamashita, N., Koike, N. and Ishida, A. (2002). Leaf ontogenetic dependence of light acclimation in invasive and native subtropical trees of different successional status. *Plant Cell Environment*, **25**, 1341–1356.

Yamori, W., Nagai, T. and Makino, A. (2011). The rate-limiting step for CO_2 assimilation at different temperatures is influenced by the leaf nitrogen content in several C_3 crop species. *Plant, Cell and Environment*, **34**, 764–777.

Yamori, W., Noguchi, K., Hanba, Y.T., *et al.* (2006). Effects of internal conductance on the temperature dependence of the photosynthetic rate in spinach leaves from contrasting growth temperatures. *Plant and Cell Physiology*, **47**, 1069–1080.

Yamori, W., Noguchi, K.O. and Terashima, I. (2005). Temperature acclimation of photosynthesis in spinach leaves: analyses of photosynthetic components and temperature dependencies of photosynthetic partial reactions. *Plant, Cell and Environment*, **28**, 536–547.

Yan, K., Chen, N., Qu, Y.Y., *et al.* (2008). Overexpression of sweet pepper *glycerol-3-phosphate acyltransferase* gene enhanced thermotolerance of photosynthetic apparatus in transgenic tobacco. *Journal of Integrative Plant Biology*, **50**, 613–621.

Yan, Y., Stolz, S., Chetelat, A., *et al.* (2007). A downstream mediator in the growth repression limb of the jasmonate pathway. *The Plant Cell*, **19**, 2470–2483.

Yang, C., Guo, R., Jie, F., *et al.* (2007). Spatial analysis of *Arabidopsis thaliana* gene expression in response to *Turnip mosaic virus infection. Molecular Plant-Microbe Interactions*, **20**, 358–370.

Yang, C.W., Xu, H.H., Wang, L.L., *et al.* (2009). Comparative effects of salt-stress and alkali-stress on the growth, photosynthesis, solute accumulation, and ion balance of barley plants. *Photosynthetica*, **47**, 79–86.

Yang, X., Liang, Z., Wen, X., *et al.* (2008). Genetic engineering of the biosynthesis of glycinebetaine leads to increased tolerance of photosynthesis to salt stress in transgenic tobacco plants. *Plant Molecular Biology*, **66**, 73–86.

Yang, X., Wen, X., Gong, H., *et al.* (2007). Genetic engineering of the biosynthesis of glycinebetaine enhances thermotolerance of photosystem II in tobacco plants. *Planta*, **225**, 719–733.

Yang, X.J., Guignard, G., Thévenard, F., *et al.* (2009). Leaf cuticle ultrastructure of *Pseudofrenelpsis dalatzenis* (Chow et Tsao) Cao ex Zhou (*Cheirolepidaceae*) from the Lower Cretaceous Dalazi formation of Jilin, China. *Review of Paleobotany and Palynology*, **153**, 8–18.

Yano, S. and Terashima, I. (2001). Separate localization of light signal perception for sun or shade type chloroplast and palisade tissue differentiation in *Chenopodium album. Plant and Cell Physiology*, **42**, 1303–1310.

Yano, S. and Terashima, I. (2004). Developmental processes of sun and shade leaves in *Chenopodium album* L. *Plant Cell Environment*, **27**, 781–793.

Yasumura, Y., Hikosaka, K. and Hirose, T. (2007). Nitrogen resorption and protein degradation during leaf senescence in *Chenopodium album* grown in different light and nitrogen conditions. *Functional Plant Biology*, **34**, 409–417.

Ye, Z.P. and Yu, Q. (2008). A coupled model of stomatal conductance and photosynthesis for winter wheat. *Photosynthetica*, **46**, 637–640.

Yemm, E.W. and Willis, A.J. (1954). Chlorophyll and photosynthesis in stomatal guard cells. *Nature*, **173**, 726.

Yensen, N.P., Fontes, M.R., Glenn, E.P., *et al.* (1981). New salt tolerant crops for the Sonoran Desert. *Desert Plants*, **3**, 111–118.

Yeoh, H.H., Badger, M.R. and Watson, L. (1980). Variations in Km(CO_2) of ribulose-1,5-bisphosphate carboxylase among grasses. *Plant Physiology*, **66**, 1110–1112.

Yin, X. and Struik, P.C. (2009). Theoretical reconsiderations when estimating the mesophyll conductance to CO_2 diffusion in leaves of C_3 plants by analysis of combined gas exchange and chlorophyll fluorescence measurements. *Plant, Cell and Environment*, **32**, 1513–1524.

Yin, X., Goudriaan, J., Lantinga, E.A., *et al.* (2003). A flexible sigmoid function of determinate growth. *Annals of Botany*, **91**, 361–371.

Yin, X., Harbinson, J. and Struik, P.C. (2006). Mathematical review of literature to assess alternative electron transports and interphotosystem excitation partitioning of steady state C_3 photosynthesis under limiting light. *Plant, Cell and Environment*, **29**, 1771–1782.

Yin, X., Sun, Z., Struik, P.C. *et al.* (2011). Evaluating a new method to estimate the rate of leaf respiration in the light by analysis of combined gas exchange and chlorophyll fluorescence measurements. *Journal of Experimental Botany*, **62**, 3489–3499.

Yin, X.Y., Struik, P.C., Romero, P., *et al.* (2009). Using combined measurements of gas exchange and chlorophyll fluorescence to estimate parameters of a biochemical C_3 photosynthesis model: a critical appraisal and a new integrated approach applied to leaves in a wheat (*Triticum aestivum*) canopy. *Plant, Cell and Environment*, **32**, 448–464.

Ying, J., Lee, E.A. and Tollenaar, M. (2000). Response of maize leaf photosynthesis to low temperature during the grain-filling period. *Field Crops Research*, **68**, 87–96.

Yiotis, C., Manetas, Y. and Psaras, K.G. (2006). Leaf and green stem anatomy of the drought deciduous Mediterranean shrub *Calicotome villosa* (Poiret) Link. (Leguminosae). *Flora*, **201**, 102–107.

Yoder, B.J., Ryan, M.G., Waring, R. H, *et al.* (1994). Evidence of reduced photosynthetic rates in old trees. *Forest Science*, **40**, 513–527.

Yokota, A. and Calvin, D.T. (1985). Ribulose bisphosphate carboxylase/oxygenase content determined with [^{14}C] carboxypentitol bisphosphate in plants and algae. *Plant Physiology*, **77**, 735–739.

Yokthongwattana, K. and Melis, A. (2006). Photoinhibition and recovery in oxygenic photosynthesis: mechanism of photosystem II damage and repair cycle. In: *Photoprotection, Photoinhibtion, Gene Regulation, and Environment* (eds Demmig-Adams, B., Adams, WW. and Mattoo, A.K.), Springer, Dordrecht, Germany, pp. 175–191.

Yoo, S.D., Greer, D.H., Laing, W.A., *et al.* (2003). Changes in photosynthetic efficiency and carotenoid composition in leaves of white clover at different developmental stages. *Plant Physiology and Biochemestry*, **41**, 887–893.

Yoshida, K., Terashima, I.. and Noguchi, K. (2007). Up-regulation of mitochondrial alternative oxidase concomitant with chloroplast over-reduction by excess light. *Plant and Cell Physiology*, **48**, 606–614.

Yoshida, K., Watanabe, C., Kato, Y., *et al.* (2008). Influence of chloroplastic photo-oxidative stress on mitochondrial alternative oxidase capacity and respiratory properties: a case study with *Arabidopsis yellow variegated 2. Plant and Cell Physiology*, **49**, 592–603.

Yoshida, S. (2003). Molecular regulation of leaf senescence. *Current Opinion in Plant Biology*, **6**, 79–84.

Yoshimoto, M., Oue, H. and Kobayashi, K. (2005). Energy balance and water use efficiency of rice canopies under free-air CO_2 enrichment. *Agricultural and Forest Meteorology*, **133**, 226–246.

Yoshinaga, K. and Kugita, M. (2004). Evolution of early land plants : insights from chloroplast genomic sequences and RNA editing. *Endocytobiosis Cell Research*, **15**, 256–271.

Yruela, I., Alfonso, M., Ortiz de Zarate, I., *et al.* (1993). Precise location of the Cu(II)-inhibitory binding site in higher plant and bacterial photosynthetic reaction centers as probed by light-induced absorption changes. *Journal of Biological Chemistry*, **268**, 1684–1689.

Ytterberg, A.J., Peltier, J.B. and van Wijk, K.J. (2006). Protein profiling of plastoglobules in chloroplasts and chromoplasts. A surprising site for differential accumulation of metabolic enzymes. *Plant Physiology*, **140**, 984–997.

Yu, M., Xie, Y. and Zhang, X. (2005). Quantification of intrinsic water use efficiency along a moisture gradient in northeastern China. *Journal of Environmental Quality*, **34**, 1311–1318.

Yu, Q., Osborne, L. and Rengel, Z. (1998). Micronutrient deficiency changes activities of superoxide dismutase and ascorbate peroxidase in tobacco plants. *Journal of Plant Nutrition*, **21**, 1427–1437.

Yukawa, J. and Tsuda, K. (1986). Leaf longevity of *Quercus glauc Thunb.*, with reference to the influence of gall formation by *Contarinia* sp. (Diptera: *Cecidomyiidae*) on the early mortality of fresh leaves. *Memoirs of the Faculty of Agriculture, Kagoshima University*, 22, 73–77.

Zabel, B., Hawes, P., Stuart, H., *et al.* (1999). Construction and engineering of a created environment: overview of the Biosphere 2 closed system. *Ecological Engineering*, **13**, 43–63.

Zachos, J., Pagani, M., Sloan, L., *et al.* (2001). Trends, rhythms, and aberrations in global climate 65 Ma to present. *Science*, **292**, 686–693.

Zaitlin, M. and Jagendorf, A.T. (1960). Photosynthetic phosphorylation and Hill reaction activities of chloroplasts isolated from plants infected with tobacco mosaic virus. *Virology*, **12**, 477–486.

Zaluar, H.L.T. and Scarano, F.R. (2000). Facilitação em restingas de moitas: um sécolo de buscas por espécies focais. In: *Ecologia de Restingas e Lagoas Costeiras* (eds Esteves, F.A. and Lacereda, L.D.), NUPEM-UFRJ, Rio de Janeiro, Brazil, pp. 3–23.

Zamolodchikov, D.G., Karelin, D.V., Ivaschenko, A. I., *et al.* (2003). CO_2 flux measurements in Russian Far East tundra using eddy covariance and closed chamber techniques. *Tellus*, **55B**, 879–892.

Zandonadi, R.S., Pinto, F.A.C., Sena, D.G., *et al.* (2005). Identification of lesser cornstalk borer-attacked maize plants using infrared images. *Biosystems engineering*, **91**, 433–439.

Zangerl, A.R., Berenbaum, M.R., DeLucia, E.H., *et al.* (2003). Fathers, fruits and photosynthesis: pollen donor effects on fruit photosynthesis in wild parsnip. *Ecology Letters*, **6**, 966–970.

Zangerl, A.R., Hamilton, J.G., Miller, T.J., *et al.* (2002). Impact of folivory on photosynthesis is greater than the sum of its holes. *Proceeding of the National Academy of Science USA*, **99**, 1088–1091.

Zankel, K. L. (1973). Rapid fluorescence changes observed in chloroplasts: their relationship to the O_2 evolving system. *Biochimica et Biophysica Acta*, **325**, 138–148.

Zaragoza-Castells, J., Sánchez-Gómez, D., Hartley, I.P., *et al.* (2008). Climate-dependent variations in leaf respiration in a dry land, low productivity Mediterranean forest: the importance of acclimation in both high-light and shaded habitats. *Functional Ecology*, **22**, 172–184.

Zaragoza-Castells, J., Sánchez-Gómez, D., Valladares, F., *et al.* (2007). Does growth irradiance affect temperature dependence and thermal acclimation of leaf respiration? Insights from a Mediterranean tree with long-lived leaves. *Plant, Cell and Environment*, **30**, 820–833.

Zarco-Tejada, P.J. Miller, J.R. Mohammed, G.H., *et al.* (2000b). Chlorophyll fluorescence effects on vegetation apparent relectance. II. Laboratory and airborne canopy level measurements with hyperspectral data. *Remote Sensing of Environment*, **74**, 596–608.

Zarco-Tejada, P.J. Miller, J.R. Mohammed, G.H., *et al.* (2000a). Chlorophyll fluorescence effects on vegetation apparent relectance. I. Leaf-level measurements and model simulation. *Remote Sensing of Environment*, **74**, 582–595.

Zarco-Tejada, P.J., Berjón, A., López-Lozano, R., *et al.* (2005a). Assessing vineyard condition with hyperspectral indices: Leaf and canopy reflectance simulation in a row-structured discontinuous canopy. *Remote Sensing of the Environment*, **99**, 271–287.

Zarco-Tejada, P.J., Miller, J.R., Mohammed, G.H., *et al.* (2001). Scaling-up and model inversion methods with narrow-band optical indices for chlorophyll content estimation in closed forest canopies with hyperspectral data. *IEEE Transactions on Geoscience and Remote Sensing*, **39**, 1491–1507.

Zarco-Tejada, P.J., Pushnik, J.C., Dobrowski, S., *et al.* (2003b). Steady state chlorophyll a fluorescence detection from canopy derivative reflectance and double-peak red-edge effects. *Remote Sensing of Environment*, **84**, 283–294.

Zarco-Tejada, P.J., Rueda, C.A. and Ustin, S.L. (2003a). Water content estimation in vegetation with MODIS reflectance data and model inversion methods. *Remote Sensing of the Environment*, **85**, 109–124.

Zarco-Tejada, P.J., Ustin, S.L. and Whiting, M.L. (2005b). Temporal and spatial relationships between within-field yield variability in cotton and high-spatial hyperspectral remote sensing imagery. *Agronomy Journal*, **97**, 641–653.

Zarco-Tejada, P.J., Berni, J.A.J., Suárez, L., *et al.* (2009). Imaging chlorophyll fluorescence with an airborne narrow-bannd multispectral camera for vegetation detection. *Remote Sensing of the Environment*, **113**, 1262–1275.

Zarco-Tejada, P.J., Miller J.R., Pedros R., *et al.* (2006). FluorMODgui V3.0 : A graphic user interface for the spectral simulation of leaf and canopy chlorophyll fluorescence. *Computers and Geosciences*, **32**, **5**, 577–591.

Zarter, C.R., Demming-Adams, B., Ebbert, V., *et al.* (2006). Photosynthetic capacity and light harvesting efficiency during the winter-to-spring transition in subalpine conifers. *New Phytologist*, **172**, 283–292.

Zavaleta, E.S. (2001). Influences of climate and atmospheric changes on plant diversity and ecosystem function in a California grassland. Ph.D. thesis, Stanford University, Stanford, California.

Zawada, D.G. (2003). Image processing of underwater multispectral imagery IEEE. *Journal of Oceanic Engineering*, **28**, 583–594.

Zeeman, S.C., Smith, S.M. and Smith, A.M. (2004). The breakdown of starch in leaves. *New Phytologist*, **163**, 247–261.

Zeiger, E., Talbott, L.D., Frechilla, S., *et al.* (2002). The guard cell chloroplast: a perspective for the twenty first century. *The New Phytologist*, **153**, 415–424.

Zeliou, K., Manetas, Y. and Petropoulou, Y. (2009). Transient winter leaf reddening in *Cistus creticus* characterizes weak (stress-sensitive) individuals, yet anthocyanins cannot alleviate the adverse effects on photosynthesis. *Journal of Experimental Botany*, **60**, 3031–3042.

Zelitch, I. (1982). The close relationship between net photosynthesis and crop yield. *BioScience*, **32**, 792–802.

Zellnig, G., Perktold, A. and Zechmann, B. (2010). Fine structural quantification of drought-stressed *Picea abies* (L.) organelles based on 3D reconstructions. *Protoplasma*, **243**, 129–136.

Zeuthen, J., Mikkelsen, J.N., Paluda-Müller, G., *et al.* (1997) .Effects of increased UV-B radiation and elevated levels of tropospheric ozone on physiological processes in European beech (*Fagus sylvatica*). *Physiologia Plantarum*, **100**, 281–290.

Zha, T.S., Xing, Z.S., Wang, K.Y., *et al.* (2007). Total and component carbon fluxes of a Scots pine ecosystem from chamber measurements and eddy covariance. *Annals of Botany*, **99**, 345–353.

Zhang, B.Z., Kang, S.Z., Li, F.S., *et al.* (2008a). Comparison of three evapotranspiration models to Bowen ratio-energy balance method for a vineyard in an and desert region of northwest China. *Agricultural and Forest Meteorology*, **148**, 1629–1640.

Zhang, J., Griffis, T.J. and Baker, J.M. (2006). Using continuous stable isotope measurements to partition net ecosystem CO_2 exchange. *Plant, Cell and Environment*, **29**, 483–496.

Zhang, J.L., Meng, L.Z. and Cao, K.F. (2009d). Sustained diurnal photosynthetic depression in uppermost-canopy leaves of four dipterocarp species in the rainy and dry seasons: does photorespiration play a role in photoprotection? *Tree Physiology*, **29**, 217–228.

Zhang, L.M., Lohmann, C., Prändl, R., *et al.* (2003). Heat stress-dependent DNA binding of *Arabidopsis* heat shock transcription factor HSF1 to heat shock gene promoters in *Arabidopsis* suspension culture cells in vivo. *Biological Chemistry*, **384**, 959–963.

Zhang, N., Zhao, Y.S. and Yu, G.R. (2009b). Simulated annual carbon fluxes of grassland ecosystems in extremely arid conditions. *Ecological Research*, **24**, 185–206.

Zhang, R. and Sharkey, T.D. (2009). Photosynthetic electron transport and proton flux under moderate heat stress. *Photosynthesis Research*, **100**, 29–43.

Zhang, R., Cruz, J.A., Kramer, D.M., *et al.* (2009c). Moderate heat stress reduces the pH component of the transthylakoid proton motive force in light-adapted intact tobacco leaves. *Plant, Cell and Environment*, **32**, 1538–1547.

Zhang, S.B., Hu, H. and Li, Z.R. (2008b). Variation of photosynthetic capacity with leaf age in an alpine orchid, *Cypripedium flavum*. *Acta Physiologiae Plantarum*, **30**, 381–388.

Zhang, X., Wollenweber, B., Jiang, D., *et al.* (2008c). Water deficits and heat shock effects on photosynthesis of a transgenic *Arabidopsis thaliana* constitutively expressing ABP9, a bZIP transcription factor. *Journal of Experimental Botany*, **59**, 839–848.

Zhang, Y., Primavesi, L.F., Jhurreea, D., *et al.* (2009a). Inhibition of Snf1-related protein kinase (SnRK1) activity and regulation of metabolic pathways by trehalose 6-phosphate. *Plant Physiology*, **149**, 1860–1871.

Zhao, M. and Running, S.W. (2010). Drought-induced reduction in global terrestrial net primary production from 2000 through 2009. *Science*, **329**, 940–943.

Zhao, S. and Blumwald, E. (1998). Changes in oxidation-reduction state and antioxidant enzymes in the roots of jack pine seedlings during cold acclimation. *Physiologia Plantarum*, **104**, 134–142.

Zheng, C., Jiang, D., Liu, F., *et al.* (2009). Effects of salt and waterlogging stresses and their combination on leaf photosynthesis, chloroplast ATP synthesis, and antioxidant capacity in wheat. *Plant Science*, **176**, 575–582.

Zhou, J., Wang, X., Jiao, Y., *et al.* (2007b). Global genome expression analysis of rice in response to drought and high-salinity stresses in shoot, flag leaf, and panicle. *Plant Molecular Biology*, **63**, 591–608.

Zhou, X.L., Peng, C.H., Dang, Q.L., *et al.* (2008). Simulating carbon exchange in Canadian Boreal forests I. Model structure, validation, and sensitivity analysis. *Ecological Modelling*, **219**, 287–299.

Zhou, Y., Lam, H.M. and Zhang, J. (2007a). Inhibition of photosynthesis and energy dissipation induced by water and high light stresses in rice. *Journal of Experimental Botany*, **58**, 1207–1217.

Zhu, J., Goldstein G. and Bartholomew, D.P. (1999). Gas exchange and carbon isotope composition of *Ananas comosus* in response to elevated CO_2 and temperature. *Plant, Cell and Environment*, **22**, 999–1007.

Zhu, J., Talbott, L.D., Jin, X., *et al.* (1998). The stomatal response to CO_2 is linked to changes in guard cell zeaxanthin. *Plant, Cell and Environment*, **21**, 813–820.

Zhu, X.G., de Sturler, E. and Long, S.P. (2007). Optimizing the distribution of resources between enzymes of carbon metabolism can dramatically increase photosynthetic rate: A numerical simulation using an evolutionary algorithm. *Plant Physiology*, **145**, 513–526.

Zhu, X.G., Long, S.P. and Ort, D.R. (2008). What is the maximum efficiency with which photosynthesis can convert solar energy into biomass? *Current opinion in biotechnology*, **19**, 153–159.

Zhu, X.G., Long, S.P. and Ort, D.R. (2010). Improving photosynthetic efficiency for greater yield. *Annual Review of Plant Biology*, **61**, 235–261.

Zhu, X.G., Portis, A.R. and Long, S.P. (2004). Would transformation of C_3 crop plants with foreign Rubisco increase productivity? A computational analysis extrapolating from kinetic properties to canopy photosynthesis. *Plant, Cell and Environment*, **27**, 155–165.

Zielinski, R.E., Werneke, J.M. and Jenkins, M.E. (1989). Coordinate expression of rubisco activase and rubisco during barley leaf cell development. *Plant Physiology*, **90**, 516–521.

Ziska, L.H. and Bunce, J.A. (1994). Direct and indirect inhibition of single leaf respiration by elevated CO_2 concentrations: interaction with temperature. *Physiologia Plantarum*, **90**, 130–138.

Ziska, L.H. and Bunce, J.A. (1995). Growth and photosynthetic response of three soybean cultivars to simultaneous increases in growth temperature and CO_2. *Physiologia Plantarum*, **94**, 575–584.

Ziska, L.H. and Bunce, J.A. (1998). The influence of increasing growth temperatue and CO_2 concentration on the ratio of respiration to photosynthesis in soybean seedlings. *Global Change Biology*, **4**, 637–643.

Zotz, G. (2007). Johansson revisited: the spatial structure of epiphyte assemblages. *Journal of Vegetation Science*, **18**, 123–130.

Zotz, G. and Winter, K. (1993a). Short-term regulation of Crassulacean acid metabolism activity in a tropical hemiepiphyte, *Clusia uvitana*. *Plant Physiology*, **102**, 835–841.

Zotz, G. and Winter, K. (1993b). Short-term photosynthesis measurements predict leaf carbon balance in tropical rain-forest canopy plants. *Planta*, **191**, 409–412.

Zotz, G. and Winter, K. (1994a). A one-year study on carbon, water, and nutrient relationships in a tropical C_3-CAM hemiepiphyte, *Clusia uvitana*. *New Phytologist*, **127**, 45–60.

Zotz, G. and Winter, K. (1994b). Annual carbon balance and nitogen use efficiency in tropical C_3 and CAM epiphytes. *New Phytologist*, **126**, 481–492.

Zotz, G., Patiño, S. and Tyree, M.T. (1997). CO_2 gas exchange and the occurrence of CAM in tropical woody hemiepiphytes. *Flora*, **192**, 143–150.

Zotz, G., Vollrath, B. and Schmidt, G. (2003). Carbon relations of fruits of epiphytic orchids. *Flora*, **198**, 98–105.

Zou, J., Rodriguez-Sas, S., Aldea, M., *et al.* (2005). Expression profiling soybean response to *Pseudomonas syringae* reveals new defense-related genes and rapid HR-Specific downregulation of photosynthesis. *Molecular Plant-Microbe Interactions*, **18**, 1161–1174.

Index